Not to

LANGE'S HANDBOOK OF CHEMISTRY

LANGE'S HANDBOOK OF CHEMISTRY

Dr. James G. Speight

CD&W Inc., Laramie, Wyoming

Seventeenth Edition

New York   Chicago   San Francisco   Athens   London
Madrid   Mexico City   Milan   New Delhi
Singapore   Sydney   Toronto

Library of Congress Catalog Card Number 84-643191
ISSN 0748-4585

McGraw-Hill Education books are available at special quantity discounts to use as premiums and sales promotions or for use in corporate training programs. To contact a representative, please visit the Contact Us page at www.mhprofessional.com.

Lange's Handbook of Chemistry, Seventeenth Edition

1 2 3 4 5 6 7 8 9   DOC   21 20 19 18 17 16

ISBN 978-1-25-958609-5
MHID 1-25-958609-X

This book is printed on acid-free paper.

Sponsoring Editor
Amanda Quinn

Editing Supervisor
Stephen M. Smith

Production Supervisor
Lynn M. Messina

Acquisitions Coordinator
Lauren Rogers

Project Manager
Sonam Arora,
Cenveo® Publisher Services

Copy Editors
Monika Joshi and Raghu Narayan,
Cenveo Publisher Services

Proofreader
Manish Tiwari, Cenveo Publisher Services

Indexer
Vikas Makkar, Cenveo Publisher Services

Art Director, Cover
Jeff Weeks

Composition
Cenveo Publisher Services

ABOUT THE EDITOR

Dr. James G. Speight, CChem, FRSC, FCIC, FACS, earned his B.Sc. and Ph.D. degrees (in chemistry) from the University of Manchester, England. He also holds a D.Sc. (in geological sciences) from VINIGRI, St. Petersburg, Russia, and a Ph.D. (in petroleum engineering) from Dubna International University, Moscow, Russia.

Dr. Speight has more than 45 years of experience in areas associated with (1) the properties, recovery, and refining of reservoir fluids, conventional petroleum, heavy oil, and tar sands bitumen; (2) the properties and refining of natural gas and gaseous fuels; and (3) the properties and refining of biomass, biofuels, and biogas, and the generation of bioenergy. His work has also focused on safety issues, environmental effects, and remediation associated with the production and use of fuels and biofuels. Dr. Speight is the author of more than 65 books on petroleum science, petroleum engineering, biomass and biofuels, and environmental sciences.

He was elected to the Russian Academy of Sciences in 1996 and awarded the Gold Medal of Honor that same year for outstanding contributions to the field of petroleum sciences. Dr. Speight has also received the Scientists without Borders Medal of Honor from the Russian Academy of Sciences. In 2001, the Academy awarded him the Einstein Medal for outstanding contributions and service in the field of geological sciences.

CONTENTS

For the detailed contents of any section, consult the first page of that section. See also the alphabetical index in the back of this Handbook.

PREFACE TO THE SEVENTEENTH EDITION

This new edition continues the tradition of previous editions by being a one-volume source of factual information for professional chemists, technicians, and students. The aim of *Lange's Handbook of Chemistry* is to provide sufficient data to satisfy the general needs of those working in the field of chemistry without their having to consult a multitude of scattered and diverse reference sources.

The book is divided into three main sections on inorganic chemistry, organic chemistry, and naturally occurring chemicals and chemical sources.

Section 1, Inorganic Chemistry, contains information relevant to the properties and behavior of elements and compounds. The data for each element and compound include (where available) name, naturally occurring isotopes, structural formula, formula weight, density, refractive index, melting point, and solubility in water.

Section 2, Organic Chemistry, contains the descriptive properties of approximately 5000 organic compounds. Entries are listed alphabetically to the extent possible, and the data for each compound include (where available) name, structural formula, formula weight, density, refractive index, melting point, boiling point, flash point, and solubility in water and various common organic solvents. Alternative names, as well as trivial names of long-standing usage, are listed as well.

Section 3, Naturally Occurring Chemicals and Chemical Sources, is new, and offers the reader details of the behavior and properties of the various fossil fuels (coal, crude oil, natural gas, tar sands, and oil shale) as well as details of the behavior and properties of biomass, biofuels, and minerals.

Two additional sections filled with useful information are available online at www.mhprofessional .com/Langes. The first, Spectroscopy, includes ultraviolet-visible spectroscopy, fluorescence, infrared and Raman spectroscopy, and X-ray spectrometry. Detection limits are listed for the elements when using flame emission, flame atomic absorption, electrothermal atomic absorption, argon induction-coupled plasma, and flame atomic fluorescence. Nuclear magnetic resonance embraces tables for the nuclear properties of the elements, proton chemical shifts and coupling constants, and similar material for carbon-13, boron-11, nitrogen-15, fluorine-19, silicon-29, and phosphorus-31. The second section, General Information and Conversion Tables, contains the general information and conversion tables required by the chemist.

Working professionals will find this *Handbook* appropriate for their needs. It is oriented toward scientists, engineers, or technologists who are employed by consultants, public works agencies, industry, regulatory agencies, universities, or equipment manufacturers, as well as planners, corporate managers, architects, elected officials, lawyers, students, or others seeking insight into the properties and behavior of chemicals.

It is hoped that users of this *Handbook* will continue to offer suggestions of material that might be included in, or even excluded from, future editions and call attention to errors. Such communications should be directed to the editor either directly or through the publisher, McGraw-Hill Education.

Dr. James G. Speight
Laramie, Wyoming
JamesSp8@aol.com

PREFACE TO THE SIXTEENTH EDITION

This Sixteenth Edition of *Lange's Handbook of Chemistry* takes on a new format under a new editor. Nevertheless, the Handbook remains the one-volume source of factual information for chemists and chemical engineers, both professionals and students. The aim of the Handbook remains to provide sufficient data to satisfy the general needs of the user without recourse to other reference sources. The many tables of numerical data that have been compiled, as well as additional tables, will provide the user with a valuable time-saver.

The new format involves division of the Handbook into four major sections, instead of the 11 sections that were part of previous editions. Section 1, Inorganic Chemistry, contains a group of tables relating to the physical properties of the elements (including recently discovered elements) and several thousand compounds. Likewise, Sec. 2, Organic Chemistry, contains a group of tables relating to the physical properties of the elements and several thousand compounds. Following these two sections, Sec. 3, Spectroscopy, presents the user with the fundamentals of the various spectroscopic techniques. This section also contains tables that are relevant to the spectroscopic properties of elements, inorganic compounds, and organic compounds. Section 4, General Information and Conversion Tables, contains all of the general information and conversion tables that were previously found in different sections of the Handbook.

In Secs. 1 and 2, the data for each compound include (where available) name, structural formula, formula weight, density, refractive index, melting point, boiling point, flash point, dielectric constant, dipole moment, solubility (if known) in water and relevant organic solvents, thermal conductivity, and electrical conductivity. The presentation of alternative names, as well as trivial names of long-standing use, has been retained. Section 2 also contains expanded information relating to the names and properties of condensed polynuclear aromatic compounds.

Enthalpies and Gibbs Energies of Formation, Entropies, and Heat Capacities of Organic and Inorganic Compounds, and Heats of Melting, Vaporization, and Sublimation and Specific Heat at Various Temperatures, are also presented in Secs. 1 and 2 for organic and inorganic compounds, as well as information on the critical properties (critical temperature, critical pressure, and critical volume).

As in the previous edition, Sec. 3, Spectroscopy, retains subsections on infrared spectroscopy, Raman spectroscopy, fluorescence spectroscopy, mass spectrometry, and X-ray spectrometry. The section on practical laboratory information (now Sec. 4) has been retained as it offers valuable information and procedures for laboratory methods.

As stated in the prefaces of earlier editions, every effort has been made to select the most useful and reliable information and to record it with accuracy. It is hoped that users of this Handbook will continue to offer suggestions of material that might be included in, or even excluded from, future editions and call attention to errors. These communications should be directed to the editor through the publisher, McGraw-Hill.

Dr. James G. Speight
Laramie, Wyoming

PREFACE TO THE
FIRST EDITION

This book is the result of a number of years' experience in the compiling and editing of data useful to chemists. In it an effort has been made to select material to meet the needs of chemists who cannot command the unlimited time available to the research specialist, or who lack the facilities of a large technical library which so often is not conveniently located at many manufacturing centers. If the information contained herein serves this purpose, the compiler will feel that he has accomplished a worthy task. Even the worker with the facilities of a comprehensive library may find this volume of value as a time-saver because of the many tables of numerical data which have been especially computed for this purpose.

Every effort has been made to select the most reliable information and to record it with accuracy. Many years of occupation with this type of work bring a realization of the opportunities for the occurrence of errors, and while every endeavor has been made to prevent them, yet it would be remarkable if the attempts towards this end had always been successful. In this connection it is desired to express appreciation to those who in the past have called attention to errors, and it will be appreciated if this be done again with the present compilation for the publishers have given their assurance that no expense will be spared in making the necessary changes in subsequent printings.

It has been aimed to produce a compilation complete within the limits set by the economy of available space. One difficulty always at hand to the compiler of such a book is that he must decide what data are to be excluded in order to keep the volume from becoming unwieldy because of its size. He can hardly be expected to have an expert's knowledge of all branches of the science nor the intuition necessary to decide in all cases which particular value to record, especially when many differing values are given in the literature for the same constant. If the expert in a particular field will judge the usefulness of this book by the data which it supplies to him from fields other than his specialty and not by the lack of highly specialized information in which only he and his co-workers are interested (and with which he is familiar and for which he would never have occasion to consult this compilation), then an estimate of its value to him will be apparent. However, if such specialists will call attention to missing data with which they are familiar and which they believe others less specialized will also need, then works of this type can be improved in succeeding editions.

Many of the gaps in this volume are caused by the lack of such information in the literature. It is hoped that to one of the most important classes of workers in chemistry, namely the teachers, the book will be of value not only as an aid in answering the most varied questions with which they are confronted by interested students, but also as an inspiration through what it suggests by the gaps and inconsistencies, challenging as they do the incentive to engage in the creative and experimental work necessary to supply the missing information.

While the principal value of the book is for the professional chemist or student of chemistry, it should also be of value to many people not especially educated as chemists. Workers in the natural sciences—physicists, mineralogists, biologists, pharmacists, engineers, patent attorneys, and librarians—are often called upon to solve problems dealing with the properties of chemical products or materials of construction. For such needs this compilation supplies helpful information and will serve not only as an economical substitute for the costly accumulation of a large library of monographs on specialized subjects, but also as a means of conserving the time required to search for information so widely scattered throughout the literature. For this reason especial care has been taken in compiling a comprehensive index and in furnishing cross references with many of the tables.

It is hoped that this book will be of the same usefulness to the worker in science as is the dictionary to the worker in literature, and that its resting place will be on the desk rather than on the bookshelf.

N. A. Lange
Cleveland, Ohio
May 2, 1934

LANGE'S
HANDBOOK OF
CHEMISTRY

SECTION 1
INORGANIC CHEMISTRY

1.1 *NOMENCLATURE OF INORGANIC COMPOUNDS*

The following synopsis of rules for naming inorganic compounds and the examples given in explanation are not intended to cover all the possible cases.

Generally, there are two types of inorganic compounds that can be formed: ionic compounds and molecular compounds.

Compounds consisting of a metal and nonmetal are commonly known as ionic compounds, where the compound name has an ending of *-ide*. Cations have positive charges while anions have negative charges. The net charge of any ionic compound must be zero, which also means it must be electrically neutral. For example, one Na^+ is paired with one Cl^-, and one Ca^{2+} is paired with two Br^-. The rules of nomenclature state that (1) the cation (metal) is always named first with its name unchanged, and (2) the anion (nonmetal) is written after the cation, modified to end in *-ide*.

The transition metals may form more than one ion, thus it is needs to be specified which particular ion we are talking about. This is indicated by assigning a Roman numeral after the metal, which denotes the charge and the oxidation state of the transition metal ion. For example, iron can form two common ions: Fe^{2+} and Fe^{3+}. To distinguish between the two, Fe^{2+} is named iron (II) and Fe^{3+} is named iron (III).

However, some of the charges on transition metals have specific Latin names. Just like the other nomenclature rules, the ion of the transition metal that has the lower charge has the Latin name ending with *-ous* and the one with the higher charge has a Latin name ending with *-ic*.

Several exceptions apply to the Roman numeral assignment: aluminum, zinc, and silver. Although they belong to the transition metal category, these metals do not have Roman numerals written after their names because these metals only exist in one ion. Instead of using Roman numerals, the different ions can also be presented in plain words. The metal is changed to end in *-ous* or *-ic*.

Although HF can be named hydrogen fluoride, it is given a different name for emphasis that it is an acid—a substance that dissociates into hydrogen ions (H^+) and anions in water. A quick way to identify acids is to see if there is an H (denoting hydrogen) in front of the molecular formula of the compound. To name acids, the prefix *hydro-* is placed in front of the nonmetal modified to end with *-ic*. The state of acids is aqueous (aq) because acids are found in water. Some common binary acids include:

HF (g) (hydrogen fluoride) $\rightarrow$ HF (aq) (hydrofluoric acid)
HBr (g) (hydrogen bromide) $\rightarrow$ HBr (aq) (hydrobromic acid)
HCl (g) (hydrogen chloride) $\rightarrow$ HCl (aq) (hydrochloric acid)
H_2S (g) (hydrogen sulfide) $\rightarrow$ H_2S (aq) (hydrosulfuric acid)

Polyatomic ions (meaning two or more atoms) are joined together by covalent bonds. Although there may be an element with positive charge like H^+, it is not joined with another element with an ionic bond. This occurs because if the atoms formed an ionic bond, then it would have already become a compound, thus not needing to gain or lose any electrons. Polyatomic anions have negative charges while polyatomic cations have positive charges. To correctly specify how many oxygen atoms are in the ion, prefixes and suffixes are used.

Cations and Anions

+1 Charge	+2 Charge	−1 Charge	−2 Charge	−3 Charge	−4 Charge
Hydrogen: H^+	Beryllium: Be^{2+}	Hydride: H^-	Oxide: O^{2-}	Nitride: N^{3-}	Carbide: C^{4-}
Lithium: Li^+	Magnesium: Mg^{2+}	Fluoride: F^-	Sulfide: S^{2-}	Phosphide: P^{3-}	
Sodium: Na^+	Calcium: Ca^{2+}	Chloride: Cl^-			
Potassium: K^+	Strontium: Sr^{2+}	Bromide: Br^-			
Rubidium: Rb^+	Barium: Ba^{2+}	Iodide: I^-			
Cesium: Cs^+					

Transition Metals and Metal Cations

+1 Charge	+2 Charge	+3 Charge	+4 Charge
Copper(I): Cu^+	Copper(II): Cu^{2+}	Aluminum: Al^{3+}	Lead(IV): Pb^{4+}
Silver: Ag^+	Iron(II): Fe^{2+}	Iron(III): Fe^{3+}	Tin(IV): Sn^{4+}
	Cobalt(II): Co^{2+}	Cobalt(III): Co^{3+}	
	Tin(II): Sn^{2+}		
	Lead(II): Pb^{2+}		
	Nickel: Ni^{2+}		
	Zinc: Zn^{2+}		

Common Polyatomic Ions

Cation or Anion	Formula
Ammonium ion	NH_4^+
Hydronium ion	H_3O^+
Acetate ion	$C_2H_3O_2^-$
Arsenate ion	AsO_4^{3-}
Carbonate ion	CO_3^{2-}
Hypochlorite ion	ClO^-
Chlorite ion	ClO_2^-
Chlorate ion	ClO_3^-
Perchlorate ion	ClO_4^-
Chromate ion	CrO_4^{2-}
Dichromate ion	$Cr_2O_7^{2-}$
Cyanide ion	CN^-
Hydroxide ion	OH^-
Nitrite ion	NO_2^-
Nitrate ion	NO_3^-
Oxalate ion	$C_2O_4^{2-}$
Permanganate ion	MnO_4^-
Phosphate ion	PO_4^{3-}
Sulfite ion	SO_3^{2-}
Sulfate ion	SO_4^{2-}
Thiocyanate ion	SCN^-
Thiosulfate ion	$S_2O_3^{2-}$

1.1.1 Writing Formulas

1.1.1.1 Mass Number, Atomic Number, Number of Atoms, and Ionic Charge. The mass number, atomic number, number of atoms, and ionic charge of an element are indicated by means of four indices placed around the symbol:

$$\text{mass number} \atop \text{atomic number} \quad \textbf{SYMBOL} \quad {\text{ionic charge} \atop \text{number of atoms}} \qquad {}^{15}_{7}N_2^{3-}$$

Ionic charge should be indicated by an Arabic superscript numeral preceding the plus or minus sign: Mg^{2+}, PO_4^{3-}

1.1.1.2 Placement of Atoms in a Formula. The electropositive constituent (cation) is placed first in a formula. If the compound contains more than one electropositive or more than one electronegative constituent, the sequence within each class should be in alphabetical order of their symbols. The alphabetical order may be different in formulas and names; for example, $NaNH_4HPO_4$, ammonium sodium hydrogen phosphate.

Acids are treated as hydrogen salts. Hydrogen is cited last among the cations.

When there are several types of ligands, anionic ligands are cited before the neutral ligands.

1.1.1.3 Binary Compounds between Nonmetals. For binary compounds between nonmetals, that constituent should be placed first which appears earlier in the sequence:

Rn, Xe, Kr, Ar, Ne, He, B, Si, C, Sb, As, P, N, H, Te, Se, S, At, I, Br, Cl, O, F

Examples: $AsCl_3$, SbH_3, H_3Te, BrF_3, OF_2, and N_4S_4.

1.1.1.4 Chain Compounds. For chain compounds containing three or more elements, the sequence should be in accordance with the order in which the atoms are actually bound in the molecule or ion.

Examples: SCN^- (thiocyanate), HSCN (hydrogen thiocyanate or thiocyanic acid), HNCO (hydrogen isocyanate), HONC (hydrogen fulminate), and HPH_2O_2 (hydrogen phosphinate).

1.1.1.5 Use of Centered Period. A centered period is used to denote water of hydration, other solvates, and addition compounds; for example, $CuSO_4 \cdot 5H_2O$, copper(II) sulfate 5-water (or pentahydrate).

1.1.1.6 Free Radicals. In the formula of a polyatomic radical an unpaired electron(s) is (are) indicated by a dot placed as a right superscript to the parentheses (or square bracket for coordination compounds). In radical ions the dot precedes the charge. In structural formulas, the dot may be placed to indicate the location of the unpaired electron(s).

Examples: $(HO)^{\cdot}$ $(O_2)^{2-}$ $(\dot{N}H_3^+)$

1.1.1.7 Enclosing Marks. Where it is necessary in an inorganic formula, enclosing marks (parentheses, braces, and brackets) are nested within square brackets as follows:

$$[\,(\,)\,], \quad [\,\{\,(\,)\,\}\,], \quad [\,\{\,[\,(\,)\,]\,\}\,], \quad [\,\{\,\{\,[\,(\,)\,]\,\}\,\}\,]$$

1.1.1.8 Molecular Formula. For compounds consisting of discrete molecules, a formula in accordance with the correct molecular weight of the compound should be used.

Examples: S_2Cl_2, S_8, N_2O_4, and $H_4P_2O_6$; not SCl, S, NO_2, and H_2PO_3.

1.1.1.9 Structural Formula and Prefixes. In the structural formula the sequence and spatial arrangement of the atoms in a molecule are indicated.

Examples: $NaO(O{=}C)H$ (sodium formate), Cl—S—S—Cl (disulfur dichloride).

Structural prefixes should be italicized and connected with the chemical formula by a hyphen: *cis-, trans-, anti-, syn-, cyclo-, catena-, o-* or *ortho-, m-* or *meta-, p-* or *para-, sec-* (secondary), *tert-* (tertiary), *v-* (vicinal), *meso-, as-* for asymmetrical, and *s-* for symmetrical.

The sign of optical rotation is placed in parentheses, (+) for dextrorotary, (–) for levorotary, and (±) for racemic, and placed before the formula. The wavelength (in nanometers is indicated by a right subscript; unless indicated otherwise, it refers to the sodium D-line.

The italicized symbols *d*- (for deuterium) and *t*- (for tritium) are placed after the formula and connected to it by a hyphen. The number of deuterium or tritium atoms is indicated by a subscript to the symbol.

Examples: *cis*-[PtCl$_2$(NH$_3$)$_2$]   methan-d_3-ol

di-*tert*-butyl sulfate $(+)_{589}$ [Co(en)$_3$]Cl$_2$

methan-ol-*d*

1.1.2 Naming Compounds

1.1.2.1 Names and Symbols for Elements.   Names and symbols for the elements are given in Table 1.3. Wolfram is preferred to tungsten but the latter is used in the United States. In forming a complete name of a compound, the name of the electropositive constituent is left unmodified except when it is necessary to indicate the valency (see oxidation number and charge number, formerly the Stock and Ewens-Bassett systems). The order of citation follows the alphabetic listing of the names of the cations followed by the alphabetical listing of the anions and ligands. The alphabetical citation is maintained regardless of the number of each ligand.

Example: K[AuS(S$_2$)] is potassium (disulfido)thioaurate (1–).

1.1.2.2 Electronegative Constituents.   The name of a monatomic electronegative constituent is obtained from the element name with its ending (-en, -ese, -ic, -ine, -ium, -ogen, -on, -orus, -um, -ur, -y, or -ygen) replaced by -ide. The elements bismuth, cobalt, nickel, zinc, and the noble gases are used unchanged with the ending -ide. Homopolyatomic ligands will carry the appropriate prefix. A few Latin names are used with affixes: cupr- (copper), aur- (gold), ferr- (iron), plumb- (lead), argent- (silver), and stann- (tin).

For binary compounds, the name of the element standing later in the sequence in Sec. 1.1.1.3 is modified to end in -ide. Elements other than those in the sequence of Sec. 1.1.1.3 are taken in the reverse order of the following sequence, and the name of the element occurring last is modified to end in -ide; e.g., calcium stannide.

ELEMENT SEQUENCE

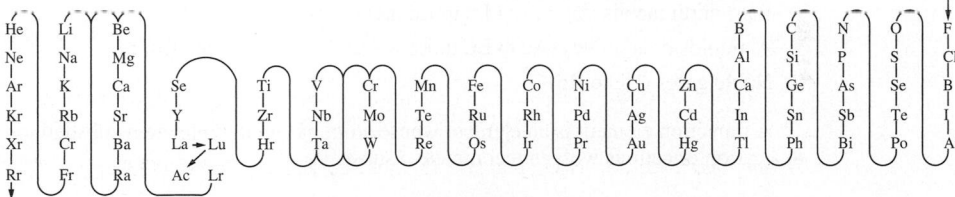

1.1.2.3 Stoichiometric Proportions.   The stoichiometric proportions of the constituents in a formula may be denoted by Greek numerical prefixes: mono-, di-, tri-, tetra-, penta-, hexa-, hepta-, octa-, nona- (Latin), deca-, undeca- (Latin), dodeca-, …, icosa- (20), henicosa- (21), …, triconta- (30), tetraconta- (40), …, hecta- (100), and so on, preceding without a hyphen the names of the elements to which they refer. The prefix mono can usually be omitted; occasionally hemi- (1/2) and sesqui- (3/2) are used. No elisions are made when using numerical prefixes except in the case of icosa- when the letter "i" is elided in docosa- and tricosa-. Beyond 10, prefixes may be replaced by Arabic numerals.

When it is required to indicate the number of entire groups of atoms, the multiplicative numerals bis-, tris-, tetrakis-, pentakis-, and so on, are used (i.e., -kis is added starting from tetra-). The entity to which they refer is placed in parentheses.

Examples: $Ca[PF_6]_2$, calcium bis(hexafluorophosphate); and $(C_{10}H_{21})_3PO_4$, tris(decyl) phosphate instead of tridecyl which is $(C_{13}H_{27}-)$.

Composite numeral prefixes are built up by citing units first, then tens, then hundreds, and so on. For example, 43 is written tritetraconta- (or tritetracontakis-).

In indexing it may be convenient to italicize a numerical prefix at the beginning of the name and connect it to the rest of the name with a hyphen; e.g., *di*-nitrogen pentaoxide (indexed under the letter "n").

1.1.2.4 Oxidation and Charge Numbers.

The *oxidation number* (Stock system) of an element is indicated by a Roman numeral placed in parentheses immediately following the name of the element. For zero, the cipher 0 is used. When used in conjunction with symbols, the Roman numeral may be placed above and to the right. The *charge number* of an ion (Ewens-Bassett system) rather than the oxidation state is indicated by an Arabic numeral followed by the sign of the charge cited and is placed in parentheses immediately following the name of the ion.

Examples: P_2O_5, diphosphorus pentaoxide or phosphorus(V) oxide; Hg_2^{2+}. mercury(I) ion or dimercury (2+) ion; $K_2[Fe(CN)_6]$, potassium hexacyanoferrate(II) or potassium hexacyanoferrate(4−); $Pb_2^{II}Pb^{IV}O_4$, dilead(II) lead(IV) oxide or trilead tetraoxide.

Where it is not feasible to define an oxidation state for each individual member of a group, the overall oxidation level of the group is defined by a formal ionic charge to avoid the use of fractional oxidation states; for example, O_2^-.

1.1.2.5 Collective Names.

Collective names include:

Halogens (F, Cl, Br, I, At)

Chalcogens (O, S, Se, Te, Po)

Alkali metals (Li, Na, K, Rb, Cs, Fr)

Alkaline-earth metals (Ca, Sr, Ba, Ra)

Lanthanoids or lanthanides (La to Lu)

Rare-earth metals (Sc, Y, and La to Lu inclusive)

Actinoids or actinides (Ac to Lr, those whose $5f$ shell is being filled)

Noble gases (He to Rn)

A transition element is an element whose atom has an incomplete *d* subshell, or which gives rise to a cation or cations with an incomplete *d* subshell.

1.1.2.6 Isotopically Labeled Compounds.

The hydrogen isotopes are given special names: ^1H (protium), ^2H or D (deuterium), and ^3H or T (tritium). The superscript designation is preferred because D and T disturb the alphabetical ordering in formulas.

Other isotopes are designated by mass numbers: ^{10}B (boron-10).

Isotopically labeled compounds may be described by inserting the italic symbol of the isotope in brackets into the name of the compound; for example, $H^{36}Cl$ is hydrogen chloride[^{36}Cl] or hydrogen chloride-36, and $^2H^{38}Cl$ is hydrogen [^2H] chloride[^{38}Cl] or hydrogen-2 chloride-38.

1.1.2.7 Allotropes.

Systematic names for gaseous and liquid modifications of elements are sometimes needed. Allotropic modifications of an element bear the name of the atom together with the descriptor to specify the modification. The following are a few common examples:

Symbol	Trivial name	Systematic name
H	Atomic hydrogen	Monohydrogen
O_2	(Common oxygen)	Dioxygen
O_3	Ozone	Trioxygen
P_4	White phosphorus	Tetraphosphorus
S_8	α-Sulfur, β-Sulfur	Octasulfur
S_n	μ-Sulfur (plastic sulfur)	Polysulfur

Trivial (customary) names are used for the amorphous modification of an element.

1.1.2.8 Heteroatomic and Other Anions.
A few heteroatomic anions have names ending in -ide. These are

—OH, hydroxide ion (not hydroxyl)

—CN, cyanide ion

—NH_2^- hydrogen difluoride ion

—NH_2, amide ion

—NH—, imide ion

—NH—NH_2, hydrazide ion

—NHOH, hydroxylamide ion

—HS^-, hydrogen sulfide ion

Added to these anions are

—triiodide ion

—N_3, axide ion

—O_3, ozonide ion

—O—O—, peroxide ion

—S—S—, disulfide ion

1.1.2.9 Binary Compounds of Hydrogen.
Binary compounds of hydrogen with the more electropositive elements are designated hydrides (NaH, sodium hydride).

Volatile hydrides, except those of Periodic Group VII and of oxygen and nitrogen, are named by citing the root name of the element (penultimate consonant and Latin affixes, Sec. 1.1.2.2) followed by the suffix -ane. Exceptions are water, ammonia, hydrazine, phosphine, arsine, stibine, and bismuthine.

Examples: B_2H_6, diborane; $B_{10}H_{14}$, decaborane (14); $B_{10}H_{16}$, decaborane (16); P_2H_4, diphosphane; Sn_2H_6, distannane; H_2Se_2, diselane; H_2Te_2, ditellane; H_2S_5, pentasulfane; and pbH_4, plumbane.

1.1.2.10 Neutral Radicals.
Certain neutral radicals have special names ending in -yl:

HO	hydroxyl	PO	phosphoryl
CO	carbonyl	SO	sulfinyl (thionyl)
ClO	chlorosyl[1]	SO_2	sulfonyl (sulfuryl)
ClO_2	chloryl[1]	S_2O_5	disulfuryl
ClO_3	perchloryl[1]	SeO	seleninyl
CrO_2	chromyl	SeO_2	selenoyl
NO	nitrosyl	UO_2	uranyl
NO_2	nitryl (nitroyl)	NpO_2	neptunyl[2]

Radicals analogous to the above containing other chalcogens in place of oxygen are named by adding the prefixes thio-, seleno-, and so on; for example, PS, thiophosphoryl; CS, thiocarbonyl.

---

[1]Similarly for the other halogens.
[2]Similarly for the other actinide elements.

1.1.3 Cations

1.1.3.1 Monatomic Cations. Monatomic cations are named as the corresponding element; for example, Fe^{2+}, iron(II) ion; Fe^{3+}, iron(III) ion.

This principle also applies to polyatomic cations corresponding to radicals with special names ending in -yl (Sec. 1.1.2.10); for example, PO^+, phosphoryl cation; NO^+, nitrosyl cation; NO_2^{2+}, nitryl cation; O_2^{2+} oxygenyl cation.

Use of the oxidation number and charge number extends the range for radicals; for example, UO_2^{2+} uranyl(VI) or uranyl(2+) cation; UO_2^+, uranyl(V) or uranyl(1+) cation.

1.1.3.2 Polyatomic Cations. Polyatomic cations derived by addition of more protons than required to give a neutral unit to polyatomic anions are named by adding the ending -onium to the root of the name of the anion element; for example, PH_4^+ phosphonium ion; H_2I^+, iodonium ion; H_3O^+, oxonium ion; $CH_3OH_2^+$ methyl oxonium ion.

Exception: The name ammonium is retained for the NH_4^+ ion; similarly for substituted ammonium ions; for example, NF_4^+, tetrafluoroammonium ion.

Substituted ammonium ions derived from nitrogen bases with names ending in -amine receive names formed by changing -amine into -ammonium. When known by a name not ending in -amine, the cation name is formed by adding the ending -ium to the name of the base (eliding the final vowel); e.g., anilinium, hydrazinium, imidazolium, acetonium, dioxanium.

Exceptions are the names uronium and thiouronium derived from urea and thiourea, respectively.

1.1.3.3 Multiple Ions from One Base. Where more than one ion is derived from one base, the ionic charges are indicated in their names: $N_2H_5^+$, hydrazinium(1+) ion; $N_2H_6^{2+}$, hydrazinium(2+) ion.

1.1.4 Anions

See Secs. 1.1.2.2 and 1.1.2.8 for naming monatomic and certain polyatomic anions. When an organic group occurs in an inorganic compound, organic nomenclature (*q.v.*) is followed to name the organic part.

1.1.4.1 Protonated Anions. Ions such as HSO_4^- are recommended to be named hydrogensulfate with the two words written as one following the usual practice for polyatomic anions.

1.1.4.2 Other Polyatomic Anions. Names for other polyatomic anions consist of the root name of the central atom with the ending -ate and followed by the valence of the central atom expressed by its oxidation number. Atoms and groups attached to the central atom are treated as ligands in a complex.

Examples: $[Sb(OH)_6^-]$, hexahydroxoantimonate(V); $[Fe(CN_6]^{3-}$, hexacyanoferrate(III); $[Co(NO_2)_6]^{3-}$, hexanitritocobaltate(III); $[TiO(C_2O_4)_2(H_2O)_2]^{2-}$, oxobisoxalatodiaquatitanate (IV); $[PCl_6]^-$, hexachlorophosphate(V).

Exceptions to the use of the root name of the central atom are antimonate, bismuthate, carbonate, cobaltate, nickelate (or niccolate), nitrate, phosphate, tungstate (or wolframate), and zincate.

1.1.4.3 Anions of Oxygen. Oxygen is treated in the same manner as other ligands with the number of -oxo groups indicated by a suffix; for example, SO_3^{2-}, trioxosulfate.

The ending -ite, formerly used to denote a lower state of oxidation, may be retained in trivial names in these cases (note Sec. 1.1.5.3 also):

AsO_3^{3-}	arsenite	NOO_2^-	peroxonitrite
BrO^-	hypobromite	PO_3^{3-}	phosphite[3]
ClO^-	hypochlorite	SO_3^{2-}	sulfite
ClO_2^-	chlorite	$S_2O_5^{2-}$	disulfite
IO^-	hypoiodite	$S_2O_4^{2-}$	dithionite
NO_2^-	nitrite	$S_2O_2^{2-}$	thiosulfite
$N_2O_2^{2-}$	hyponitrite	SeO_3^{2-}	selenite

However, compounds known to be double oxides in the solid state are named as such; for example, Cr_2CuO_4 (actually $Cr_2O_3 \cdot CuO$) is chromium(III) copper(II) oxide (and not copper chromite).

1.1.4.4   Isopolyanions.   Isopolyanions are named by indicating with numerical prefixes the number of atoms of the characteristic element. It is not necessary to give the number of oxygen atoms when the charge of the anion or the number of cations is indicated.

Examples: $Ca_3Mo_7O_{24}$, tricalcium 24-oxoheptamolybdate, may be shortened to tricalcium heptamolybdate; the anion, $Mo_7O_{24}^{6-}$, is heptamolybdate(6–); $S_2O_7^{2-}$, disulfate(2–); $P_2O_7^{4-}$, diphosphate(V)(4–).

When the characteristic element is partially or wholly present in a lower oxidation state than corresponds to its Periodic Group number, oxidation numbers are used; for example, $[O_2HP—O—PO_3H]^{2-}$, dihydrogendiphosphate(III, V)(2–).

A bridging group should be indicated by adding the Greek letter μ immediately before its name and separating this from the rest of the complex by a hyphen. The atom or atoms of the characteristic element to which the bridging atom is bonded, is indicated by numbers.

Examples: $[O_3P—S—PO_2—O—PO_3]^{5-}$, 1, 2-$\mu$-thiotriphosphate(5–)

$[S_3P—O—PS_2—O—PS_3]^{5-}$, di-$\mu$-oxo-octathiotriphosphate(5–)

1.1.5   Acids

1.1.5.1   Acids and -ide Anions.   Acids giving rise to the -ide anions (Sec. 1.1.2.2) should be named as hydrogen ... -ide; for example, HCl, hydrogen chloride; HN_3, hydrogen azide.

Names such as hydrobromic acid refer to an aqueous solution, and percentages such as 48% HBr denote the weight/volume of hydrogen bromide in the solution.

1.1.5.2   Acids and -ate Anions.   Acids giving rise to anions bearing names ending in -ate are treated as in Sec. 1.1.5.1; for example, H_2GeO_4, hydrogen germanate; $H_4[Fe(CN)_6]$, hydrogen hexacyanoferrate(II).

1.1.5.3   Trivial Names.   Acids given in Table 1.1 retain their trivial names due to long-established usage. Anions may be formed from these trivial names by changing -ous acid to -ite, and -ic acid to -ate. The prefix hypo- is used to denote a lower oxidation state and the prefix per- designates a higher oxidation state. The prefixes ortho- and meta- distinguish acids of differing water content; for example, H_4SiO_4 is orthosilicic acid and H_2SiO_3 is metasilicic acid. The anions would be named silicate (4–) and silicate(2–), respectively.

---

[3]Named for esters formed from the hypothetical acid $P(OH)_3$.

1.1.5.4   Peroxo- Group.   When used in conjunction with the trivial names of acids, the prefix peroxo- indicates substitution of —O—by—O—O—.

1.1.5.5   Replacement of Oxygen by Other Chalcogens.   Acids derived from oxoacids by replacement of oxygen by sulfur are called thioacids, and the number of replacements are given by prefixes di-, tri-, and so on. The affixes seleno- and telluro- are used analogously.

Examples: HOO—C=S, thiocarbonic acid; HSS—C=S, trithiocarbonic acid.

1.1.5.6   Ligands Other than Oxygen and Sulfur.   See Sec. 1.1.7, Coordination Compounds, for acids containing ligands other than oxygen and sulfur (selenium and tellurium).

1.1.5.7   Differences between Organic and Inorganic Nomenclature.   Organic nomenclature is largely built upon the scheme of substitution, that is, the replacement of hydrogen atoms by other atoms or groups. Although rare in inorganic nomenclature: NH_2Cl is called chloramine and $NHCl_2$ dichloroamine. Other substitutive names are fluorosulfonic acid and chlorosulfonic acid derived from HSO_3H. These and the names aminosulfonic acid (sulfamic acid), iminodisulfonic acid, and nitrilotrisulfonic acid should be replaced by the following based on the concept that these names are formed by adding hydroxyl, amide, imide, and so on, groups together with oxygen atoms to a sulfur atom:

HSO_3F	fluorosulfuric acid	$NH(SO_3H)_2$	imidobis(sulfuric) acid
HSO_3Cl	chlorosulfuric acid	$N(SO_3H)_3$	nitridotris(sulfuric) acid
NH_2SO_3H	amidosulfuric acid		

1.1.6   Salts and Functional Derivatives of Acids

1.1.6.1   Acid Halogenides.   For acid halogenides the name is formed from the corresponding acid radical if this has a special name (Sec. 1.1.2.10); for example, NOCl, nitrosyl chloride. In other cases these compounds are named as halogenide oxides with the ligands listed alphabetically; for example, BiClO, bismuth chloride oxide; VCl_2O, vanadium(IV) dichloride oxide.

1.1.6.2   Anhydrides.   Anhydrides of inorganic acids are named as oxides; for example, N_2O_5, dinitrogen pentaoxide.

1.1.6.3   Esters.   Esters of inorganic acids are named as the salts; for example, $(CH_3)_2SO_4$, dimethyl sulfate. However, if it is desired to specify the constitution of the compound, the nomenclature for coordination compounds should be used.

1.1.6.4   Amides.   Names for amides are derived from the names of the acid radicals (or from the names of acids by replacing acid by amide); for example, $SO_2(NH_2)_2$, sulfonyl diamide (or sulfuric diamide); NH_2SO_3H, sulfamidic acid (or amidosulfuric acid).

1.1.6.5   Salts.   Salts containing acid hydrogen are named by adding the word hydrogen before the name of the anion (however, see Sec. 1.1.4.1), for example, KH_2PO_4, potassium dihydrogen phosphate; $NaHCO_3$, sodium hydrogen carbonate (not bicarbonate); $NaHPHO_3$, sodium hydrogen phosphonate (only one acid hydrogen remaining).

Salts containing O^{2-} and HO^- anions are named oxide and hydroxide, respectively. Anions are cited in alphabetical order which may be different in formulas and names.

Examples: FeO(OH), iron(III) hydroxide oxide; $VO(SO_4)$, vanadium(IV) oxide sulfate.

1.1.6.6   Multiplicative Prefixes.   The multiplicative prefixes bis, tris, etc., are used with certain anions for indicating stoichiometric proportions when di, tri, etc., have been preempted to designate condensed anions; for example, $AlK(SO_4)_2 \cdot 12H_2O$, aluminum potassium bis(sulfate) 12-water (recall that disulfate refers to the anion $S_2O_7^{2-}$).

TABLE 1.1 Trivial Names for Acids

H_3ASO_4	arsenic acid	$H_4P_2O_7$	diphosphoric acid (or pyro-phosphoric acid)
H_3ASO_3	arsenious acid		
H_3BO_3	orthoboric acid (or boric acid)	$H_4P_2O_8$	peroxodiphosphoric acid
HBO_2	metaboric acid	$(HO)_2OP$	diphosphoric(IV) acid or hypophosphoric acid
$HBrO_3$	bromic acid		
$HBrO_2$	bromous acid	$(HO)_2OP$	
$HBrO$	hypobromous acid	$(HO)_2P$—O	diphosphoric(III,V) acid
H_2CO_3	carbonic acid		
$HOCN$	cyanic acid	$(HO)_2P$—O	
$HNCO$	isocyanic acid	H_2PHO_3	phosphonic acid
$HONC$	fulminic acid	$H_2P_2H_2O_5$	diphosphonic acid
$HClO_4$	perchloric acid	HPH_2O_2	phosphinic acid (formerly hypophosphorous acid)
$HClO_3$	chloric acid		
$HClO_2$	chlorous acid	$HReO_4$	perrhenic acid
$HClO$	hypochlorous acid	H_2ReO_4	rhenic acid
H_2CrO_4	chromic acid	H_2SO_4	sulfuric acid
$H_2Cr_2O_7$	dichromic acid	$H_2S_2O_7$	disulfuric acid
H_5IO_6	orthoperiodic acid	H_2SO_5	peroxomonosulfuric acid
HIO_4	periodic acid	$H_2S_2O_3$	thiosulfuric acid
HIO_3	iodic acid	$H_2S_2S_6$	dithionic acid
HIO	hypoiodous acid	H_2SO_3	sulfurous acid
$HMnO_4$	permanganic acid	$H_2S_2O_5$	disulfurous acid
H_2MnO_4	manganic acid	$H_2S_2O_2$	thiosulfurous acid
HNO_4	peroxonitric acid	$H_2S_2O_4$	dithionous acid
HNO_3	nitric acid	$H_2S_xO_6$	polythionic acid
HNO_2	nitrous acid	$(x = 3, 4, ...)$	(tri-, tetra-, . . .)
H_2NO_2	nitroxylic acid	H_2SO_2	sulfoxylic acid
$H_2N_2O_2$	hyponitrous acid	$HSb(OH)_6$	hexahydrooxoantimonic acid
$HOONO$	peroxonitrous acid	H_2SeO_4	selenic acid
H_3PO_4	orthophosphoric acid (or phosphoric acid)	H_2SeO_3	selenious acid
		H_4SiO_4	orthosilicic acid
HPO_3	metaphosphoric acid	H_2SiO_3	metasilicic acid
H_3PO_5	peroxomonophosphoric acid	$HTcO_4$	pertechnetic acid
		H_2TcO_4	technetic acid
		H_6TeO_6	orthotelluric acid

1.1.6.7 Crystal Structure. The structure type of crystals may be added in parentheses and in italics after the name; the latter should be in accordance with the structure. When the typename is also the mineral name of the substance itself, italics are not used.

Examples: $MgTiO_3$, magnesium titanium trioxide (*ilmenite* type); $FeTiO_3$, iron(II) titanium trioxide (ilmenite).

1.1.7 Coordination Compounds

1.1.7.1 Naming a Coordination Compound. To name a coordination compound, the names of the ligands are attached directly in front of the name of the central atom. The ligands are listed in alphabetical order regardless of the number of each and with the name of a ligand treated as a unit. Thus "diammine" is listed under "a" and "dimethylamine" under "d." The oxidation number of the central atom is stated last by either the oxidation number or charge number.

1.1.7.2 Anionic Ligands. Whether inorganic or organic, the names for anionic ligands end in -o (eliding the final -e, if present, in the anion name). Enclosing marks are required for inorganic anionic ligands containing numerical prefixes, and for thio, seleno, and telluro analogs of oxo anions containing more than one atom.

If the coordination entity is negatively charged, the cations paired with the complex anion (with -ate ending) are listed first. If the entity is positively charged, the anions paired with the complex cation are listed immediately afterward.

The following anions do not follow the nomenclature rules:

F^-	fluoro	HO_2^-	hydrogen peroxo
Cl^-	chloro	S^{2-}	thio (only for single sulfur)
Br^-	bromo	S_2^{2-}	disulfido
I^-	iodo	HS^-	mercapto
O^{2-}	oxo	CN^-	cyano
H^-	hydrido (or hydro)	CH_3O^-	methoxo or methanolato
OH^-	hydroxo	CH_3S^-	methylthio or methanethiolato
O_2^{2-}	peroxo		

1.1.7.3   Neutral and Cationic Ligands.

Neutral and cationic ligands are used without change in name and are set off with enclosing marks. Water and ammonia, as neutral ligands, are called "aqua" and "ammine," respectively. The groups NO and CO, when linked directly to a metal atom, are called nitrosyl and carbonyl, respectively.

1.1.7.4   Attachment Points of Ligands.

The different points of attachment of a ligand are denoted by adding italicized symbol(s) for the atom or atoms through which the attachment occurs at the end of the name of the ligand; e.g., glycine-*N* or glycinato-*O, N*. If the same element is involved in different possible coordination sites, the position in the chain or ring to which the element is attached is indicated by numerical superscripts: e.g., tartrato(3–)-O^1, O^2, or tartrato(4–)-O^2, O^3 or tartrato(2–) O^1, O^4

1.1.7.5   Abbreviations for Ligand Names.

Except for certain hydrocarbon radicals, for ligand (L) and metal (M), and a few with H, all abbreviations are in lowercase letters and do not involve hyphens. In formulas, the ligand abbreviation is set off with parentheses. Some common abbreviations are

Ac	acetyl	en	ethylenediamine
acac	acetylacetonato	Him	imidazole
Hacac	acetylacetone	H_2ida	iminodiacetic acid
Hba	benzoylacetone	Me	methyl
Bzl	benzyl	H_3nta	nitrilotriacetic acid
Hbg	biguanide	nbd	norbornadiene
bpy	2, 2'-bipyridine	ox	oxalato(2–) from parent H_2ox
Bu	Butyl	phen	1, 10-phenanthroline
Cy	cyclohexyl	Ph	phenyl
D_2dea	diethanolamine	pip	piperidine
dien	diethylenetriamine	Pr	propyl
dmf	dimethylformamide	pn	propylenediamine
H_2dmg	dimethylglyoxime	Hpz	pyrazole
dmg	dimethylglyoximato(2–)	py	pyridine
Hdmg	dimethylglyoximato(1–)	thf	tetrahydrofuran
dmso	dimethylsulfoxide	tu	thiourea
Et	ethyl	H_3tea	triethanolamine
H_4edta	ethylenediaminetetraacetic acid	tren	2, 2', 2''-triaminotriethylamine
Hedta, edta	coordinated ions derived from H_4edta	trien	triethylenetetraamine
		tn	trimethylenediamine
Hea	ethanolamine	ur	urea

Examples: Li[B(NH$_2$)$_4$], lithium tetraamidoborate(1–) or lithium tetraamidoborate(III); [Co(NH$_3$)$_5$Cl] Cl$_3$, pentaamminechlorocobalt(III) chloride or pentaamminechlorocobalt(2+) chloride; K$_3$[Fe(CN)$_5$CO], potassium carbonylpentacyanoferrate(II) or potassium carbonylpentacyanoferrate(3–); [Mn{C$_6$H$_4$(O) (COO)}$_2$(H$_2$O)$_4$]$^-$, tetraaquabis[salicylato(2–)]manganate(III) ion; [Ni(C$_4$H$_7$N$_2$O$_2$)$_2$] or [Ni(dmg)] which can be named bis-(2, 3-butanedione dioximate)nickel(II) or bis[dimethylglyoximato(2–)] nickel(II).

1.1.8 Addition Compounds

The names of addition compounds are formed by connecting the names of individual compounds by a dash (—) and indicating the numbers of molecules in the name by Arabic numerals separated by the solidus (diagonal slash). All molecules are cited in order of increasing number; those having the same number are cited in alphabetic order. However, boron compounds and water are always cited last and in that order.

Examples: 3CdSO$_4$ · 8H$_2$O, cadmium sulfate—water (3/8); Al$_2$(SO$_4$)$_3$ · K$_2$SO$_4$ · 24H$_2$O, aluminum sulfate—potassium sulfate—water (1/1/24); AlCl$_3$ · 4C$_2$H$_5$OH, aluminum chloride—ethanol (1/4).

1.1.9 Synonyms and Mineral Names

TABLE 1.2 Synonyms and Mineral Names

Acanthite, *see* Silver sulfide	Borax, *see* Sodium tetraborate 10-water
Alabandite, *see* Manganese sulfide	Braunite, *see* Manganese(III) oxide
Alamosite, *see* Lead(II) silicate(2–)	Brimstone, *see* Sulfur
Altaite, *see* Lead telluride	Bromellite, *see* Beryllium oxide
Alumina, *see* Aluminum oxide	Bromosulfonic acid, *see* Hydrogen bromosulfate
Alundum, *see* Aluminum oxide	Bromyrite, *see* Silver bromide
Alunogenite, *see* Aluminum sulfate 18-water	Brookite, *see* Titanium(IV) oxide
Amphibole, *see* Magnesium silicate(2–)	Brucite, *see* Magnesium hydroxide
Andalusite, *see* Aluminum silicon oxide (1/1)	Bunsenite, *see* Nickel oxide
Anglesite, *see* Lead sulfate	Cacodylate, *see* Sodium dimethylarsonate 3-water
Anhydrite, *see* Calcium sulfate	Caesium, *see* under Cesium
Anhydrone, *see* Magnesium perchlorate	Calamine, *see* Zinc carbonate
Aragonite, *see* Calcium carbonate	Calcia, *see* Calcium oxide
Arcanite, *see* Potassium sulfate	Calcite, *see* Calcium carbonate
Argentite, *see* Silver sulfide	Calomel, *see* Mercury(I) chloride
Argol, *see* Potassium hydrogen tartrate	Caro's acid, *see* Hydrogen peroxosulfate
Arkansite, *see* Titanium(IV) oxide	Cassiopeium, *see* Lutetium
Arsenolite, *see* Arsenic(III) oxide dimer	Cassiterite, *see* Tin(IV) oxide
Arsine, *see* Arsenic hydride	Caustic potash, *see* Potassium hydroxide
Auric and aurous, *see* under Gold	Caustic soda, *see* Sodium hydroxide
Azoimide, *see* Hydrogen azide	Celestite, *see* Strontium sulfate
Azurite, *see* Copper(II) carbonate—dihydroxide (2/1)	Cementite, *see* tri-Iron carbide
	Cerargyrite, *see* Silver chloride
Baddeleyite, *see* Zirconium(IV) oxide	Cerussite, *see* Lead carbonate
Baking soda, *see* Sodium hydrogen carbonate	Chalcanthite, *see* Copper(II) sulfate 5-water
Barite (barytes), *see* Barium sulfate	Chalcocite, *see* Copper(I) sulfide
Bieberite, *see* Cobalt sulfate 7-water	Chalk, *see* Calcium carbonate
Bismuthine, *see* Bismuth hydride	Chile nitre, *see* Sodium nitrate
Bismuthinite, *see* Bismuth sulfide	Chile saltpeter, *see* Sodium nitrate
Bleaching powder, *see* Calcium hydrochlorite	Chloromagnesite, *see* Magnesium chloride
Bleaching solution, *see* Sodium hydrochlorite	Chlorosulfonic acid, *see* Hydrogen chlorosulfate
Blue copperas, *see* Copper(II) sulfate 7-water	Cinnabar, *see* Mercury(II) sulfide
Boracic acid, *see* Hydrogen borate	Claudetite, *see* Arsenic(III) oxide dimer

(Continued)

TABLE 1.2 Synonyms and Mineral Names (*Continued*)

Clausthalite, *see* Lead selenide
Clinoenstatite, *see* Magnesium silicate(2−)
Columbium, *see* under Niobium
Corrosive sublimate, *see* Mercury(II) chloride
Corundum, *see* Aluminum oxide
Cotunite, *see* Lead chloride
Covellite, *see* Copper(H) sulfide
Cream of tartar, *see* Potassium hydrogen tartrate
Crocoite, *see* Lead chromate(VI)(2−)
Cryolite, *see* Sodium hexafluoroaluminate
Cryptohalite, *see* Ammonium hexafluorosilicate
Cupric and cuprous, *see* under Copper
Cuprite, *see* Copper(I) oxide

Dakin's solution, *see* Sodium hypochlorite
Dehydrite, *see* Magnesium perchlorate
Dental gas, *see* Nitrogen(I) oxide
Diamond, *see* Carbon
Dichlorodisulfane, *see* di-Sulfur dichloride
Diuretic salt, *see* Potassium acetate
Dolomite, *see* Calcium magnesium carbonate (1/1)
Dry ice, *see* Carbon dioxide (solid)

Enstatite, *see* Magnesium silicate(2−)
Epsom salts, *see* Magnesium sulfate 7-water
Epsomite, *see* Magnesium sulfate 7-water
Eriochalcite, *see* Copper(II) chloride

Fayalite, *see* Iron(II) silicate(4−)
Ferric and ferrous, *see* under Iron
Fluorine oxide, *see* Oxygen difluoride
Fluoristan, *see* Tin(II) fluoride
Fluorite, *see* Calcium fluoride
Fluorosulfonic acid, *see* Hydrogen fluorosulfate
Fluorspar, *see* Calcium fluoride
Forsterite, *see* Magnesium silicate(4−)
Freezing salt, *see* Sodium chloride
Fulminating mercury, *see* Mercury fulminate

Galena, *see* Lead sulfite
Glauber's salt, *see* Sodium sulfate 10-water
Goethite, *see* Iron(II) hydroxide oxide
Goslarite, *see* Zinc sulfate 7-water
Graham's salt, *see* Sodium phosphate(1−)
Graphite, *see* Carbon
Greenockite, *see* Cadmium sulfide
Gruenerite, *see* Iron(II) silicate(2−)
Guanajuatite, *see* Bismuth selenide
Gypsum, *see* Calcium sulfate 2-water

Halite, *see* Sodium chloride
Hausmannite, *see* Manganese(II,IV) oxide
Heavy hydrogen, *see* Hydrogen[^2H] or name followed
    by -*d*
Heavy water, *see* Hydrogen [^2H] oxide
Heazlewoodite, *see* tri-Nickel disulfide
Hematite, *see* Iron(III) oxide
Hermannite, *see* Manganese silicate
Hessite, *see* Silver telluride

Hieratite, *see* Potassium hexafluorosilicate
Hydroazoic acid, *see* Hydrogen azide
Hydrophilite, *see* Calcium chloride
Hydrosulfite, *see* Sodium dithionate(III)
Hypo (photographic), *see* Sodium thiosulfate
    5-water
Hypophosphite, *see* under Phosphinate

Ice, *see* Hydrogen oxide (solid)
Iceland spar, *see* Calcium carbonate
Iodyrite, *see* Silver iodide

Jeweler's borax, *see* Sodium tetraborate 10-water
Jeweler's rouge, *see* Iron(III) oxide

Kalinite, *see* Aluminum potassium bis(sulfate)
Kemite, *see* Sodium tetraborate
Kyanite, *see* Aluminum silicon oxide (1/1)

Laughing gas, *see* Nitrogen(I) oxide
Lautarite, *see* Calcium iodate
Lawrencite, *see* Iron(II) chloride
Lechatelierite, *see* Silicon dioxide
Lime, *see* Calcium oxide
Litharge, *see* Lead(II) oxide
Lithium aluminum hydride, *see* Lithium
    tetrahydri-doaluminate
Lodestone, *see* Iron(II,III) oxide
Lunar caustic, *see* Silver nitrate
Lye, *see* Sodium hydroxide

Magnesia, *see* Magnesium oxide
Magnesite, *see* Magnesium carbonate
Magnetite, *see* Iron(II,III) oxide
Malachite, *see* Copper carbonate dihydroxide
Manganosite, *see* Manganese(II) oxide
Marcasite, *see* Iron disulfide
Marshite, *see* Copper(I) iodide
Mascagnite, *see* Ammonium sulfate
Massicotite, *see* Lead oxide
Mercuric and mercurous, *see* under Mercury
Metacinnabar, *see* Mercury(II) sulfide
Millerite, *see* Nickel sulfide
Mirabilite, *see* Sodium sulfate
Mohr's salt, *see* Ammonium iron(II) sulfate 6-water
Moissanite, *see* Silicon carbide
Molybdenite, *see* Molybdenum disulfide
Molybdite, *see* Molybdenum(VI) oxide
Molysite, *see* Iron(III) chloride
Montroydite, *see* Mercury(II) oxide
Morenosite, *see* Nickel sulfate 7-water
Mosaic gold, *see* Tin disulfide
Muriatic acid, *see* Hydrogen chloride, aqueous
    solutions

Nantokite, *see* Copper(I) chloride
Natron, *see* Sodium carbonate
Naumannite, *see* Silver selenide
Neutral verdigris, *see* Copper(II) acetate
Nitre (niter), *see* Potassium nitrate

TABLE 1.2   Synonyms and Mineral Names (*Continued*)

Nitric oxide, *see* Nitrogen(II) oxide
Nitrobarite, *see* Barium nitrate
Nitromagnesite, *see* Magnesium nitrate 6-water
Nitroprusside, *see* Sodium pentacyanonitrosylfer-
  rate(II) 2-water

Oldhamite, *see* Calcium sulfide
Opal, *see* Silicon dioxide
Orpiment, *see* Arsenic trisulfide
Oxygen powder, *see* Sodium peroxide

Paris green, *see* Copper acetate arsenate(III) (1/3)
Pawellite, *see* Calcium molybdate(VI)(2−)
Pearl ash, *see* Potassium carbonate
Perborax, *see* Sodium peroxoborate
Periclase, *see* Magnesium oxide
Persulfate, *see* Peroxodisulfate
Phosgene, *see* Carbonyl chloride
Phosphine, *see* Hydrogen phosphide
Pickling acid, *see* Hydrogen sulfate
Pitchblende, *see* Uranium(IV) oxide
Plaster of Paris, *see* Calcium sulfate hemihydrate
Plattnerite, *see* Lead(IV) oxide
Polianite, *see* Manganese(IV) oxide
Polishing powder, *see* Silicon dioxide
Potash, *see* Potassium carbonate
Potassium acid phthalate, *see* Potassium hydrogen
  phthalate
Prussic acid, *see* Hydrogen cyanide
Pyrite, *see* Iron disulfide
Pyrochroite, *see* Manganese(H) hydroxide
Pyrohytpophosphite, *see* diphosphate(IV)
Pyrolusite, *see* Manganese(IV) oxide
Pyrophanite, *see* Manganese titanate(IV)(2−)
Pyrophosphate, *see* Diphosphate(V)
Pyrosulfuric acid, *see* Hydrogen disulfate

Quartz, *see* Silicon dioxide
Quicksilver, *see* Mercury

Realgar, *see* di-Arsenic disulfide
Red lead, *see* Lead(II,IV) oxide
Rhodochrosite, *see* Manganese carbonate
Rhodonite, *see* Manganese silicate(1−)
Rochelle salt, *see* Potassium sodium tartrate 4-water
Rock crystal, *see* Silicon dioxide
Rutile, *see* Titanium(IV) oxide

Sal soda, *see* Sodium carbonate 10-water
Saltpeter, *see* Potassium nitrate
Seacehite, *see* Manganese chloride
Scheelite, *see* Calcium tungstate(VI)(2−)
Sellaite, *see* Magnesium fluoride
Senarmontite, *see* Antimony(III) oxide
Siderite, *see* Iron(II) carbonate
Siderotil, *see* Iron(II) sulfate 5-water
Silica, *see* Silicon dioxide
Silicotungstic acid, *see* Silicon oxide—tungsten
  oxide—water (1/12/26)
Sillimanite, *see* Aluminum silicon oxide (1/1)

Smithsonite, *see* Zinc carbonate
Soda ash, *see* Sodium carbonate
Spelter, *see* Zinc metal
Sphalerite, *see* Zinc sulfide
Spherocobaltite, *see* Cobalt(II) carbonate
Spinel, *see* Magnesium aluminate(2−)
Stannic and stannous, *see* under Tin
Stibine, *see* Antimony hydride
Stibnite, *see* Antimony(III) sulfide
Stolzite, *see* Lead tungstate(VI)(2−)
Strengite, *see* Iron(III) phosphate
Strontianite, *see* Strontium carbonate
Sugar of lead, *see* Lead acetate
Sulfamate, *see* Amidosulfate
Sulphate, *see* Sulfate
Sulfurated lime, *see* Calcium sulfide
Sulfuretted hydrogen, *see* Hydrogen sulfide
Sulphur, *see* Sulfur
Sulfuryl, *see* Sulfonyl
Sycoporite, *see* Cobalt sulfide
Sylvite, *see* Potassium chloride
Szmikite, *see* Manganese(II) sulfate hydrate

Tarapacaite, *see* Potassium chromate(VI)
Tellurite, *see* Tellurium dioxide
Tenorite, *see* Copper(II) oxide
Tephroite, *see* Manganese silicate(1−)
Thenardite, *see* Sodium sulfate
Thionyl, *see* Sulfinyl
Thorianite, *see* Thorium dioxide
Topaz, *see* Aluminum hexafluorosilicate
Tridymite, *see* Silicon dioxide
Troilite, *see* Iron(II) sulfide
Trona, *see* Sodium carbonate—hydrogen carbonate
  dihydrate
Tschermigite, *see* Aluminum ammonium bis(sulfate)
Tungstenite, *see* Tungsten disulfide
Tungstite, *see* Hydrogen tungstate

Uraninite, *see* Uranium(IV) oxide

Valentinite, *see* Antimony (III) oxide
Verdigris, *see* Copper acetate hydrate
Vermillion, *see* Mercury(II) sulfide
Villiaumite, *see* Sodium fluoride
Vitamin B_3, *see* Calcium (+)pantothenate

Washing soda, *see* Sodium carbonate 10-water
Whitlockite, *see* Calcium phosphate
Willemite, *see* Zinc silicate(4−)
Wolfram, *see* Tungsten
Wuestite, *see* Iron(II) oxide
Wulfenite, *see* Lead molybdate(VI)(2−)
Wurtzite, *see* Zinc sulfide

Zincite, *see* Zinc oxide
Zincosite, *see* Zinc sulfate
Zincspar, *see* Zinc carbonate
Zirconia, *see* Zirconium oxide

1.1.10   Classification of Inorganic Substances

Simple substances. Molecules consist of one-type atoms (atoms of one element). In chemical reactions, molecules cannot be decomposed with formation of other substances.

Complex substances (*or chemical compounds*). Molecules consist of different types of atoms (atoms of different chemical elements). In chemical reactions, molecules are decomposed with the formation of several other substances.

Simple	Metals
	Nonmetals
Complex	Oxides
	Bases
	Acids
	Salts

A sharp transition border between metals and nonmetals does not exist, since they are simple substances showing dual properties.

Bases.   Complex substances in which atoms of metals bond with one or several hydroxyl groups [according to electrolytic dissociation theory, bases are complex substances which under the dissociating in water solution form metal cations (NH_4^+) and hydroxide anions (OH^-)].

Classification.   Soluble in water (alkalis) and insoluble. Amphoteric bases also show properties of weak acids.

Preparation

1. Reactions of active metals (alkaline and alkaline earth metals) with water

$$2Na + 2H_2O \rightarrow 2NaOH + H_2$$
$$Ca + 2H_2O \rightarrow Ca(OH)_2 + H_2$$

2. Interaction oxides of active metals with water

$$BaO + H_2O \rightarrow Ba(OH)_2$$

3. Electrolysis water solutions of salts

$$2NaCl + 2H_2O \rightarrow 2NaOH + 2HCl$$

Chemical Properties

Alkalis	Insoluble bases
Action to indicators Litmus: blue Methyl orange: yellow Phenolphthalein: crimson	
Interaction with acid oxides $2KOH + CO_2 \rightarrow K_2CO_3 + H_2O$ $KOH + CO_2 \rightarrow KHCO_3$	
Interaction with acids (reaction of neutralization) $NaOH + HNO_3 \rightarrow NaNO_3 + H_2O$	$Cu(OH)_2 + 2HCl \rightarrow CuCl_2 + 2H_2O$
Reaction of exchange with salts $Ba(OH)_2 + K_2SO_4 \rightarrow 2KOH + BaSO_4^-$ $3KOH + Fe(NO_3)_3 \rightarrow Fe(OH)_3^- + 3KNO_3$	
Thermal decomposition —	$Cu(OH)_2 \rightarrow CuO + H_2O$

Oxides. Complex substances consisting of two elements, one of which is oxygen.
Classification

Nonsalts forming (CO, N_2O, and NO)

Salts forming	Basic: A metal oxide in which metals display low oxidation number +1, +2 Na_2O; MgO; and CuO
	Amphoteric: (for metals with oxidation number +3, +4) As hydrates it corresponds with amphoteric hydroxide ZnO; Al_2O_3; Cr_2O_3; and SnO_2
	Acid: An oxide of nonmetals and metals with oxidation number from +5 to +7 SO_2; SO_3; P_2O_5; Mn_2O_7; and CrO_3

Basic, amphoteric-corresponding bases; amphoteric, acid-corresponding acids.

Preparation

1. Interaction of simple and complex substances with oxygen

   $2Mg + O_2 \rightarrow 2MgO$

   $4P + 5O_2 \rightarrow 2P_2O_5$

   $S + O_2 \rightarrow SO_2$

   $2CO + O_2 \rightarrow 2CO_2$

   $2CuS + 3O_2 \rightarrow 2CuO + 2SO_2$

   $CH_4 + 2O_2 \rightarrow CO_2 + 2H_2O$

$$4NH_3 + 5O_2 \xrightarrow{\text{cat.}} 4NO + 6H_2O$$

2. Decomposition some substances containing oxygen (bases, acids, and salts) under the heating

$$Cu(OH)_2 \xrightarrow{t°} CuO + H_2O$$

$$(CuOH)_2CO_3 \xrightarrow{·\ t°} 2CuO + CO_2 + H_2O$$

$$2Pb(NO_3)_2 \xrightarrow{t°} ® \ 2PbO + 4NO_2 + O_2$$

$$2HMnO_4 \xrightarrow[t°]{H_2SO_4 \text{ (conc.)}} Mn_2O_7 + H_2O$$

Chemical Properties

Basic oxides	Acid oxides
Interaction with water	
Base formed:	Acid formed:
$Na_2O + H_2O \rightarrow 2NaOH$	$SO_3 + H_2O \rightarrow H_2SO_4$
$CaO + H_2O \rightarrow Ca(OH)_2$	$P_2O_5 + 3H_2O \rightarrow 2H_3PO_4$
Interaction with acid or base	
On reactions with acid, salt, and water are formed	On reactions with base, salt, and water are formed
$\overset{t°C}{MgO + H_2SO_4 \rightarrow MgSO_4 + H_2O}$	$CO_2 + Ba(OH)_2 \rightarrow BaCO_3 + H_2O$
$\overset{t°C}{CuO + 2HCl \rightarrow CuCl_2 + H_2O}$	$SO_2 + 2NaOH \rightarrow Na_2SO_3 + H_2O$
Amphoteric oxides interact	
with acids as basic:	with bases as acid:
$ZnO + H_2SO_4 \rightarrow ZnSO_4 + H_2O$	$ZnO + 2NaOH \rightarrow Na_2ZnO_2 + H_2O$
	$(ZnO + 2NaOH + H_2O \rightarrow Na_2[Zn(OH)_4])$

Interaction of basic and acid oxide with each other leads to salt formation
$Na_2O + CO_2 \rightarrow Na_2CO_3$

Reduction up to simple substances
$3CuO + 2NH_3 \rightarrow 3Cu + N_2 + 3H_2O$
$P_2O_5 + 5C \rightarrow 2P + 5CO$

Acids.   Complex substances consisting of hydrogen atoms and acid radicals (according to electrolytic dissociation theory, acids—electrolytes, which under the dissociating form only H^+ in the capacity of cations).

Classification

1. On composition: oxygenless and oxoacids.

2. On hydrogen atoms number, which are capable of being substituted on metal: mono-, di-, tribasic.

Oxygenless		Salt name
HCl—hydrogen chloride (hydrochloric)	Monobasic	Chloride
HBr—hydrogen bromide	Monobasic	Bromide
HI—hydrogeniodide	Monobasic	Iodide
HF—hydrogen fluorine (hydrofluoric)	Monobasic	Fluoride
H_2S—hydrogensulphide	Bibasic	Sulphide
Containing oxygen		
HNO_3—nitric	Monobasic	Nitrate
H_2SO_3—sulphurous	Bibasic	Sulphite
H_2SO_4—sulphuric	Bibasic	Sulphate
H_2CO_3—carbonic	Bibasic	Carbonate
H_2SiO_3—silicon	Bibasic	Silicate
H_3PO_4—orthophosphoric	Tribasic	Orthophosphate

Preparation

1. Interaction of acid oxides with water (for oxoacids)

$$SO_3 + H_2O \rightarrow H_2SO_4$$
$$P_2O_5 + 3H_2O \rightarrow 2H_3PO_4$$

2. Interaction of hydrogen with nonmetals and following dissolution product in water (for oxygenless acids)

$$H_2 + Cl_2 \rightarrow 2HCl$$
$$H_2 + S \rightarrow H_2S$$

3. Reactions of exchange between salt and acid

$$Ba(NO_3)_2 + H_2SO_4 \rightarrow BaSO_4^- + 2HNO_3$$

including displacing weak, flying, or slightly soluble acid from its salts by means of stronger acids.

$$Na_2SiO_3 + 2HCl \rightarrow H_2SiO_3 + 2NaCl$$
$$\overset{t^\circ}{2NaCl \text{ (hard)} + H_2SO_4 \text{ (conc.)} \rightarrow Na_2SO_4 + 2HCl^-}$$

Chemical Properties

1. Action to indicators

Litmus—red

Methylorange—pink

2. Interaction with bases (reaction of neutralization)

$$H_2SO_4 + 2KOH \rightarrow K_2SO_4 + 2H_2O$$
$$2HNO_3 + Ca(OH)_2 \rightarrow Ca(NO_3)_2 + 2H_2O$$

3. Interaction with basic oxides

$$\overset{t^\circ C}{CuO + 2HNO_3 \rightarrow Cu(NO_3)_2 + H_2O}$$

4. Interaction with metals

$$Zn + 2HCl \rightarrow ZnCl_2 + H_2$$
$$2Al + 6HCl \rightarrow 2AlCl_3 + 3H_2$$

(metals standing in the electrochemical series before hydrogen, acid-oxidizers).

5. Interaction with salts (reactions of exchange) at which stands out gas or formed residual

$$H_2SO_4 + BaCl_2 \rightarrow BaSO_4^- + 2HCl$$
$$2HCl + K_2CO_3 \rightarrow 2KCl + H_2O + CO_2$$

Salts. Complex substances which consist of atoms of metal and acid residuals. This is the most numerous class of inorganic compounds.

Classification

Medium salts. In the time of dissociation, medium salts give only metal cations (or NH_4^+) and anions of acid radical. Products of full substitution hydrogen atoms of acids to atoms of metals:

$$Na_2SO_4 \rightarrow 2Na^+ + SO_4^{2-}$$
$$CaCl_2 \rightarrow Ca^{2+} + 2Cl^-$$

Acid salts. In the time of dissociation, acid salts give only metal cations (or NH_4^+), hydrogen anions, and anions of acid radical. Products of full substitution hydrogen atoms of multibasic acid to atoms of metal:

$$NaHCO_3 \rightarrow Na^+ + HCO_3^- \rightarrow Na^+ + H^+ + CO_3^{2-}$$

Basic salts. In the time of dissociation, basic salts give only metal cations, hydroxyl anions, and anions of acid radical. Products of incomplete substitution OH groups, corresponding bases to acid radicals:

$$Zn(OH)Cl \rightarrow [Zn(OH)]^+ + Cl^- \rightarrow Zn^{2+} + OH^- + Cl^-$$

Double salts. In the time of dissociation, double salts give two cations and one anion:

$$KAl(SO_4)_2 \rightarrow K^+ + Al^{3+} + 2SO_4^{2-}$$

Mixed salts. Formed by means of one cation and two anions:

$$CaOCl_2 \rightarrow Ca^{2+} + Cl^- + OCl^-$$

Complex salts. Contain complex cations and anions:

$$[Ag(NH_3)_2]Br \rightarrow [Ag(NH_3)_2]^+ + Br^-$$

$$Na[Ag(CN)_2] \rightarrow Na^+ + [Ag(CN)_2]^-$$

Medium Salts

Preparation.   Most of ways of getting the salts is based on the interaction of substances with opposite properties:

1. Metal with nonmetal:

$$2Na + Cl_2 \rightarrow 2NaCl$$

2. Metal with acid:

$$Zn + 2HCl \rightarrow ZnCl_2 + H_2$$

3. Metal with solution of salt of less active metal:

$$Fe + CuSO_4 \rightarrow FeSO_4 + Cu$$

4. Basic oxide with the acid oxide:

$$MgO + CO_2 \rightarrow MgCO_3$$

5. Basic oxide with acid:

$$CuO + H_2SO_4 \xrightarrow{t^\circ} CuSO_4 + H_2O$$

6. Bases with acid oxide:

$$Ba(OH)_2 + CO_2 \rightarrow BaCO_3^- + H_2O$$

7. Bases with acid:

$$Ca(OH)_2 + 2HCl \rightarrow CaCl_2 + 2H_2O$$

8. Salts with the acid:

$$MgCO_3 + 2HCl \rightarrow MgCl_2 + H_2O + CO_2$$

$$BaCl_2 + H_2SO_4 \rightarrow BaSO_4^- + 2HCl$$

9. Bases solution with salt solution:

$$Ba(OH)_2 + Na_2SO_4 \rightarrow 2NaOH + BaSO_4^-$$

10. Solutions of two salts:

$$3CaCl_2 + 2Na_3PO_4 \rightarrow Ca_3(PO_4)_2 + 6NaCl$$

Chemical Properties

1. Thermal decomposition

$$CaCO_3 \rightarrow CaO + CO_2$$

$$2Cu(NO_3)_2 \rightarrow 2CuO + 4NO_2 + O_2$$

$$NH_4Cl \rightarrow NH_3 + HCl$$

2. Hydrolysis

$$Al_2S_3 + 6H_2O \rightarrow 2Al(OH)_3^- + 3H_2S$$

$$FeCl_3 + H_2O \rightarrow Fe(OH)Cl_2 + HCl$$

$$Na_2S + H_2O \rightarrow NaHS + NaOH$$

3. Exchange reactions with acids, bases, and other salts

$$AgNO_3 + HCl \rightarrow AgCl^- + HNO_3$$

$$Fe(NO_3)_3 + 3NaOH \rightarrow Fe(OH)_3 + 3NaNO_3$$

$$CaCl_2 + Na_2SiO_3 \rightarrow CaSiO_3 + 2NaCl$$

4. Oxidation–reduction reactions, stipulated by properties of cation or anion

$$2KMnO_4 + 16HCl \rightarrow 2MnCl_2 + 2KCl + 5Cl_2 + 8H_2O$$

Acid Salts
   Preparation
1. Interaction of acid with the deficit of basis

$$KOH + H_2SO_4 \rightarrow KHSO_4 + H_2O$$

2. Interaction of bases with plenty acid oxides

$$Ca(OH)_2 + 2CO_2 \rightarrow Ca(HCO_3)_2$$

3. Interaction of medium salts with acid

$$Ca_3(PO_4)_2 + 4H_3PO_4 \rightarrow 3Ca(H_2PO_4)_2$$

   Chemical Properties
1. Thermal decomposition with medium salts formation

$$Ca(HCO_3)_2 \rightarrow CaCO_3 + CO_2 + H_2O$$

2. Interaction with the alkali. Reception of medium salts

$$Ba(HCO_3)_2 + Ba(OH)_2 \rightarrow 2BaCO_3^- + 2H_2O$$

Basic Salts
Preparation

1. Hydrolysis of salts, formed by weak base and strong acid

$$ZnCl_2 + H_2O \rightarrow [Zn(OH)]Cl + HCl$$

2. Addition (by drops) a small quantities of alkalis to solutions of medium salts of metals

$$AlCl_3 + 2NaOH \rightarrow [Al(OH)_2]Cl + 2NaCl$$

3. Interaction of weak acids salts with medium salts

$$2MgCl_2 + 2Na_2CO_3 + H_2O \rightarrow [Mg(OH)]_2CO_3^- + CO_2 + 4NaCl$$

Chemical Properties

1. Thermal decomposition

$$[Cu(OH)]_2CO_3 \rightarrow 2CuO + CO_2 + H_2O$$
malachite

2. Interaction with the acid: formation of medium salts

$$Sn(OH)Cl + HCl \rightarrow SnCl_2 + H_2O$$

Complex Salts
Structure

$K_4[Fe(CN)_6]$

$K_4[Fe(CN)_6]$	External sphere
$K_4[Fe(CN)_6]$	Internal sphere
$K_4[Fe(CN)_6]$	Central atom
$K_4[Fe(CN)_6]$	Coordinate relation
$K_4[Fe(CN)_6]$	Ligand

Central, complex forming, atoms usually serve ions of metals of greater periods (Co, Ni, Pt, Hg, Ag, Cu). Typical ligands are OH^-, CN^-, NH_3, CO, and H_2O; they connected with the central atom by donor-acceptor bound.

Preparation.   Reactions of salts with ligands

$$AgCl + 2NH_3 \rightarrow [Ag(NH_3)_2]Cl$$

$$FeCl_3 + 6KCN \rightarrow K_3[Fe(CN)_6] + 3KCl$$

Chemical Properties

1. Destruction of complexes at the expense of forming the slightly soluble compounds

$$2[Cu(NH_3)_2]Cl + K_2S \rightarrow CuS^- + 2KCl + 4NH_3$$

2. The exchange of ligands between external and internal spheres

$$K_2[CoCl_4] + 6H_2O \rightarrow [Co(H_2O)_6]Cl_2 + 2KCl$$

Genetic Relationship between Different Classes of Compounds

1. Metal; nonmetal—salt
2. Basic oxide; acid oxide—salt
3. Basic; acid—salt
4. Metal—basic oxide
5. Nonmetal—acid oxide
6. Basic oxide—bases
7. Acid oxide—acid

Examples:

1. $\underset{\text{metal}}{Hg} + \underset{\text{nonmetal}}{S} \rightarrow \underset{\text{salt}}{HgS}$ $2Al + 3I_2 \rightarrow 2AlI_3$

2. $\underset{\substack{\text{basic}\\\text{oxide}}}{Li_2O} + \underset{\substack{\text{acid}\\\text{oxide}}}{CO_2} \rightarrow \underset{\text{salt}}{Li_2CO_3}$ $CaO + SiO_2 \rightarrow CaSiO_3$

3. $\underset{\text{bases}}{Cu(OH)_2} + \underset{\text{acid}}{2HCl} \rightarrow \underset{\text{salt}}{CuCl_2} + 2H_2O$ $\underset{\text{salt}}{FeCl_3} + \underset{\text{acid}}{3HNO_3} \rightarrow \underset{\text{salt}}{Fe(NO_3)_3} + \underset{\text{acid}}{3HCl}$

4. $\underset{\text{metal}}{2Ca} + O_2 \rightarrow \underset{\substack{\text{basic}\\\text{oxide}}}{2CaO}$ $4Li + O_2 \rightarrow 2Li_2O$

5. $\underset{\text{nonmetal}}{S} + O_2 \rightarrow \underset{\substack{\text{acid}\\\text{oxide}}}{SO_2}$ $4As + 5O_2 \rightarrow 2As_2O_5$

6. $\underset{\substack{\text{basic}\\\text{oxide}}}{BaO} + H_2O \rightarrow \underset{\text{bases}}{Ba(OH)_2}$ $Li_2O + H_2O \rightarrow 2LiOH$

7. $\underset{\substack{\text{acid}\\\text{oxide}}}{P_2O_5} + 3H_2O \rightarrow \underset{\text{acid}}{2H_3PO_4}$ $SO_3 + H_2O \rightarrow H_2SO_4$

1.2   PHYSICAL PROPERTIES OF INORGANIC COMPOUNDS

Names follow the IUPAC Nomenclature. Solvates are listed under the entry for the anhydrous salt. Acids are entered under hydrogen and acid salts are entered as a subentry under hydrogen.

Formula weights are based upon the International Atomic Weights and are computed to the nearest hundredth when justified. The actual significant figures are given in the atomic weights of the individual elements. Each element that has neither a stable isotope nor a characteristic natural isotopic composition is represented in this table by one of that element's commonly known radioisotopes identified by mass number and relative atomic mass.

1.2.1 Density

Density is the mass of a substance contained in a unit volume. In the SI system of units, the ratio of the density of a substance to the density of water at 15°C is known as the *specific gravity* (*relative density*). Various units of density, such as kg/m^3, lb-mass/ft^3, and g/cm^3, are commonly used. In addition, molar densities or the density divided by the molecular weight is often specified.

Density values are given at room temperature unless otherwise indicated by the superscript figure; for example, 2.487^{15} indicates a density of 2.487 g/cm^3 for the substance at 15°C. A superscript 20 over a subscript 4 indicates a density at 20°C relative to that of water at 4°C. For gases the values are given as grams per liter (g/L).

1.2.2 Melting Point (Freezing Temperature)

The *melting point* of a solid is the temperature at which the vapor pressure of the solid and the liquid are the same and the pressure totals one atmosphere and the solid and liquid phases are in equilibrium. For a pure substance, the *melting point* is equal to the *freezing point*. Thus, the *freezing point* is the temperature at which a liquid becomes a solid at normal atmospheric pressure.

The *triple point* of a material occurs when the vapor, liquid, and solid phases are all in equilibrium. This is the point on a *phase diagram* where the solid–vapor, solid–liquid, and liquid–vapor equilibrium lines all meet. A *phase diagram* is a diagram that shows the state of a substance at different temperatures and pressures.

Melting point is recorded in a certain case as 250 d and in some other cases as d 250, the distinction being made in this manner to indicate that the former is a melting point with decomposition at 250°C while in the latter decomposition only occurs at 250°C and higher temperatures. Where a value such as –$6H_2O$, 150 is given it indicates a loss of 6 moles of water per formula weight of the compound at a temperature of 150°C. For hydrates the temperature stated represents the compound melting in its water of hydration.

1.2.3 Boiling Point

The normal boiling point (boiling temperature) of a substance is the temperature at which the vapor pressure of the substance is equal to atmospheric pressure.

At the boiling point, a substance changes its state from liquid to gas. A stricter definition of boiling point is the temperature at which the liquid and vapor (gas) phases of a substance can exist in equilibrium. When heat is applied to a liquid, the temperature of the liquid rises until the *vapor pressure* of the liquid equals the pressure of the surrounding atmosphere (gases). At this point there is no further rise in temperature, and the additional heat energy supplied is absorbed as *latent heat* of vaporization to transform the liquid into gas. This transformation occurs not only at the surface of the liquid (as in the case of *evaporation*) but also throughout the volume of the liquid, where bubbles of gas are formed. The boiling point of a liquid is lowered if the pressure of the surrounding atmosphere (gases) is decreased. On the other hand, if the pressure of the surrounding atmosphere (gases) is increased, the boiling point is raised. For this reason, it is customary when the boiling point of a substance is given to include the pressure at which it is observed, if that pressure is other than standard, i.e., 760 mm of mercury or 1 atmosphere (STP, Standard Temperature and Pressure). The boiling point of a solution is usually higher than that of the pure solvent; this boiling-point elevation is one of the colligative properties common to all solutions.

Boiling point is given at atmospheric pressure (760 mm of mercury or 101 325 Pa) unless otherwise indicated; thus 82^{15mm} indicates that the boiling point is 82°C when the pressure is 15 mm of mercury. Also, subl 550 indicates that the compound sublimes at 550°C. Occasionally decomposition products are mentioned.

1.2.4 Refractive Index

The refractive index n is the ratio of the velocity of light in a particular substance to the velocity of light in vacuum. Values reported refer to the ratio of the velocity in air to that in the substance saturated with air. Usually the yellow sodium doublet lines are used; they have a weighted mean of 589.26 nm and are symbolized by D. When only a single refractive index is available, approximate values over a small temperature range may be calculated using a mean value of 0.000 45 per degree for dn/dt, and remembering that n_D decreases with an increase in temperature. If a transition point lies within the temperature range, extrapolation is not reliable.

The *specific refraction* r_D is given by the Lorentz and Lorenz equation,

$$r_D = \frac{n_D^2 - 1}{n_D^2 + 2} \cdot \frac{1}{\rho}$$

where ρ is the density at the same temperature as the refractive index, and is independent of temperature and pressure. The molar refraction is equal to the specific refraction multiplied by the molecular weight. It is a more or less additive property of the groups or elements comprising the compound. An extensive discussion will be found in Bauer, Fajans, and Lewin, in *Physical Methods of Organic Chemistry*, 3d ed., A. Weissberger (ed.), vol. 1, part II, chap. 28, Wiley-Interscience, New York, 1960.

The empirical Eykman equation

$$\frac{n_D^2 - 1}{n_D + 0.4} \cdot \frac{1}{\rho} = \text{constant}$$

offers a more accurate means for checking the accuracy of experimental densities and refractive indices, and for calculating one from the other, than does the Lorentz and Lorenz equation.

The refractive index of moist air can be calculated from the expression

$$(n-1) \times 10^6 = \frac{103.49}{T} p_1 + \frac{177.4}{T} p_2 + \frac{86.26}{T} \left(1 + \frac{5748}{T}\right) p_3$$

where p_1 is the partial pressure of dry air (in mmHg), p_2 is the partial pressure of carbon dioxide (in mmHg), p_3 is the partial pressure of water vapor (in mmHg), and T is the temperature (in kelvins).

Example: 1-Propynyl acetate has $n_D = 1.4187$ and density = 0.9982 at 20°C; the molecular weight is 98.102. From the Lorentz and Lorenz equation,

$$r_D = \frac{(1.4187)^2 + 1}{(1.4187)^2 + 2} \cdot \frac{1}{0.9982} = 0.2528$$

The molar refraction is

$$Mr_D = (98.102)(0.2528) = 24.80$$

From the atomic and group refractions, the molar refraction is computed as follows:

6 H	6.600
5 C	12.090
1 C≡C	2.398
1 O(ether)	1.643
1 O(carbonyl)	2.211
$Mr_D =$	24.942

TABLE 1.3   Physical Constants of Inorganic Compounds

Abbreviations Used in the Table

a, acid	ca., approximately	fctetr, face-centered
abs, absolute	chl, chloroform	tetragonal
abs alc, anhydrous ethanol	conc, concentrated	FP, flash point
acet, acetone	cub, cubic	fum, fuming
alk, alkali (aq NaOH or KOH)	d, decomposes	fus, fusion, fuses
anhyd, anhydrous	dil, dilute	g, gas, gram
aq, aqueous	disprop, disproportionates	glyc, glycerol
aq reg, aqua regia	EtOAc, ethyl acetate	h, hot
atm, atmosphere	eth, diethyl ether	hex, hexagonal
BuOH, butanol	EtOH, 95% ethanol	HOAc, acetic acid
bz, benzene	expl, explodes	i, insoluble
c, solid state	fcc, face-centered cubic	ign, ignites

L, liter	soln, solution
lq, liquid	solv, solvent (s)
MeOH, methanol	subl, sublimes
min, mineral	sulf, sulfides
mL, milliliter	tart, tartrate
org, organic	THF, tetrahydrofuran
oxid, oxidizing	v, very
PE, petroleum ether	vac, vacuum
pyr, pyridine	viol, violently
s, soluble	volat, volatilizes
satd, saturated	<, less than
sl, slightly	>, greater than

Name	Formula	Formula weight	Density	Melting point, °C	Boiling point, °C	Solubility in 100 parts solvent
Actinium-227	Ac	227.0278	10.07	1050(50)	ca. 3200	d aq; s acids
bromide	$AcBr_3$	466.74	5.85	subl 800		s aq
Aluminum	Al	26.981539	2.70	660.323	2518	s HCl, H_2SO_4, alk
acetylacetonate	$Al(C_5H_7O_2)_3$	324.31	1.27	190–193	315	i aq; v s alc; s bz, eth
ammonium bis(sulfate) 12-water	$AlNH_4(SO_4)_2 \cdot 12H_2O$	453.33	1.65	anhyd >280		14.3 g/100 mL aq; s glyc; i alc
antimonide	AlSb	148.74	4.26	1060		v sl s aq, alc, eth
arsenide	AlAs	101.90	3.76	1740		i aq
bis(acetylsalicylate)	$Al(OOCC_6H_4OCOCH_3)_2OH$	402.30				
borate (2/1)	$2Al_2O_3 \cdot B_2O_3$	273.54		ca. 1050		d (viol) aq; s alc, acet, bz, CS_2
bromide	$AlBr_3$	266.69	3.205^{18}_{0}	97.5	subl 253	FP 27; v s org solv
butoxide, sec-	$Al(C_4H_9O)_3$	246.33	0.967		$200\text{–}206^{30mm}$	v s org solv
butoxide, tert-	$Al(C_4H_9O)_3$	246.33	1.025^{20}_{0}		subl 180	d aq; fire hazard
carbide (4/3)	Al_4C_3	143.96	2.360	2100	$d >2200^{400mm}$	v s aq; s alc
chlorate	$Al(ClO_3)_3$	277.35				
chloride	$AlCl_3$	133.34	2.440^{25}	192.6	subl 181.1	g/100 mL: 70 aq (viol), 100^{12} abs alc; s CCl_4, eth; sl s bz
ethoxide	$Al(C_2H_5O)_3$	162.16	1.142^{20}_{0}	140	205^{14mm}	s hot aq d; v sl s alc, eth
fluoride	AlF_3	83.98	2.882^{25}_{4}	1090	subl 1272	0.56 aq; i a, alk, alc, acet
hydroxide	$Al(OH)_3$	78.01	2.42	to Al_2O_3, 300		i aq; s acids, alkalis
iodide	AlI_3	407.69	3.98^{17}	191.0	382	d aq; s alc, eth, CS_2
isopropoxide	$Al(C_3H_7O)_3$	204.25	1.0346^{20}_{0}	118.5	135^{10mm}	d aq; s alc, bz, chl, PE
methoxide	$Al(CH_3O)_3$	72.07		0	130	

				mp	bp	
nitrate 9-water	$Al(NO_3)_3 \cdot 9H_2O$	375.13	1.72	73	d 135	g/100 mL: 64 aq, 100 alc; s acet
nitride	AlN	40.99	3.05	d 2517	2980	d aq, acid, alkali
oxide (alpha-)	Al_2O_3	101.96	3.97	2054(6)		i aq; v sl s a, alk
perchlorate 6-water	$Al(ClO_4)_3 \cdot 6H_2O$	433.43	2.020	120.8	anhyd 178	133 g/100 mL[20] aq
phenoxide	$Al(C_6H_5O)_3$	306.27	1.23	d 265		d aq; s alc, chl, eth
phosphate	$AlPO_4$	121.95	2.56	>1460		i aq; sl s a
phosphide	AlP	57.96	2.85^{15}_{4}	2550		d aq
phosphinate (hypophosphite)	$Al(H_2PO_2)_3$	221.94		d to PH_3, 220		i aq; s HCl, warm alkali
potassium bis(sulfate) 12-water	$AlK(SO_4)_2 \cdot 12H_2O$	474.39	1.757^{20}	$-9H_2O$, 92	anhyd, 200	11.4 g/100 mL aq; v s glyc; i alc
propoxide	$Al(C_3H_7O)_3$	204.25	1.0578^{20}_{20}	106	248^{14mm}	d aq; s alc
selenide	Al_2Se_3	290.84	3.4371^{20}_{4}	947		d aq; acid
silicon oxide (1/1)	$Al_2O_3 \cdot SiO_2$	162.05	3.247			i aq; d HF; s fused alkali
sodium bis(sulfate) 12-water	$AlNa(SO_4)_2 \cdot 12H_2O$	458.28	1.675^{20}	61		110 g/100 mL[15] aq; i alc
stearate	$Al(C_{18}H_{35}O_2)_3$	877.41	1.070	117–120		i aq, alc; s bz, alk
sulfate	$Al_2(SO_4)_3$	342.15	1.61	770 d		36.4 g/100 mL[20] aq; sl s alc
sulfate 18-water	$Al_2(SO_4)_3 \cdot 18H_2O$	666.46	1.69^{17}	d 86.5		87 g/100 mL[0] aq; i alc
sulfide	Al_2S_3	150.16	2.20^{13}	1097	subl 1500	hyd aq; s acid
tetrahydridoborate	$Al(BH_4)_3$	71.53		−64.5	44.5	d aq; ign air; expl in O_2, 20
Americium	Am	243	12	1176	2011	s a
Ammonia	NH_3	17.03	lq: 0.6818 at bp; g: $0.6175^{15.7.2atm}$	−77.75	−33.35	g/100 mL: 34 aq; 13.2 alc; s eth, organic solvents
Ammonium acetate	$NH_4C_2H_3O_2$	77.08	1.17^{20}	114	d	g/100 mL: 148[4] aq, 7.9[15] MeOH; s alc
amidosulfate	$NH_4SO_3NH_2$	114.13	1.260	131	d 160	v s aq; sl s alc
benzoate	$NH_4C_6H_5O_2$	139.15		198	subl 160	g/100 mL: 20[15] aq, 2.8 alc; s glyc
bromide	NH_4Br	97.94	2.429	452 (subl under pressure)	d 397 vacuo	76 g/100 mL[20] aq; v s acet, alc, eth
calcium arsenate 6-water	$NH_4CaAsO_4 \cdot 6H_2O$	305.13	1.905^{15}	d 140		0.02 aq; s NH_4Cl
carbamate	NH_4COONH_2	78.07		subl 60		v s aq; sl s alc; i eth
carbonate 1-water	$(NH_4)_2CO_3 \cdot H_2O$	114.10		volatilizes 60		v s aq; i alc
chloride	NH_4Cl	53.49	1.5274^{25}	237.8	520	g/100 mL: 26[15] aq, 0.6[19] abs alc; i acet, eth
chromate(VI)	$(NH_4)_2CrO_4$	152.07	1.91^{12}	d 185		34 g/100 mL[20] aq; sl s MeOH
chromium(III) bissulfate 12-water	$NH_4Cr(SO_4)_2 \cdot 12H_2O$	478.34	1.72	94 d		7.2 g/100 mL[0] aq
copper(II) tetrachloride 2-water	$(NH_4)_2CuCl_4 \cdot 2H_2O$	277.46	1.993	anhyd, 110	d >120	40.3 g/100 mL[20] aq; s alc

(Continued)

TABLE 1.3 Physical Constants of Inorganic Compounds (*Continued*)

Name	Formula	Formula weight	Density	Melting point, °C	Boiling point, °C	Solubility in 100 parts solvent
cyanide	NH_4CN	44.06	1.10	d 36		v s aq, alc
dichromate(VI)	$(NH_4)_2Cr_2O_7$	252.07	2.155	d 180 to Cr_2O_3		35.6 g/100 mL[20] aq; s alc; flammable
dihydrogen arsenate	$NH_4H_2AsO_4$	158.97	2.311	d 300		v s aq
dihydrogen phosphate	$NH_4H_2PO_4$	115.03	1.803[19]	d 190		37 g/100 mL[20] aq; sl s alc; i acet
disulfatocobatate(II) 6-water	$(NH_4)_2[Co(SO_4)_2] \cdot 6H_2O$	395.23	1.902			18 g/100 mL[20] aq; v sl s alc
disulfatoferrate(II) 6-water	$(NH_4)_2[Fe(SO_4)_2] \cdot 6H_2O$	392.14	1.864	d 100		36.4 g/100 mL[20] aq; i alc
disulfatoferrate(III) 12-water	$NH_4[Fe(SO_4)_2] \cdot 12H_2O$	482.19	1.71	39–41	d 230	124 g/100 mL aq
disulfatonickelate(II) 6-water	$(NH_4)_2 [Ni(SO_4)_2] \cdot 6H_2O$	395.00	1.923			8.95 g/100 mL[20] aq
dithiocarbamate	$NH_4S(C{=}S)NH_2$	110.20	1.451[20]4	99 d		v s aq; s alc; sl s eth
diuranate(VI)	$(NH_4)_2U_2O_7$	624.22				v sl s aq, alk; s acids
fluoride	NH_4F	37.04	1.009[25]	d to NH_3 + HF		100 g/100 mL[20] aq; s alc
formate	NH_4OOCH	63.06	1.27	116	d 180	143 g/100 mL[20] aq; s alc, eth
heptamolybdate(VI)(6–) 4-water	$(NH_4)_2Mo_7O_{24} \cdot 4H_2O$	1235.86	2.498	anhyd 90	d 190	43 g/100 mL aq; s acids; i alc
hexachloropalladate(IV)	$(NH_4)_2[PdCl_6]$	355.20	2.418	d		sl s aq
hexachloroplatinate(IV)	$(NH_4)_2[PtCl_6]$	443.87	3.065	d 380		0.5 aq
hexadecanoate	$NH_4OOC(CH_2)_{14}CH_3$	273.45		21–22		s aq; sl s bz; i alc, acet
hexafluoroaluminate(3–)	$(NH_4)_3[AlF_6]$	195.09	1.78	d >100		v s aq
hexafluorogallate	$(NH_4)_3GaF_6$	237.83	2.10	d 200		
hexafluorogermanate	$(NH_4)_2GeF_6$	222.68	2.564	380	subl	s aq; i eth
hexafluorophosphate	$NH_4[PF_6]$	163.00	2.180[18]4	d 68		74.8 g/100 mL[20] aq; s alc, acet
hexafluorosilicate	$(NH_4)_2[SiF_6]$	178.15	2.011	d		18.6 g/100 mL[20] aq; i alc, acet
hexanitratocerate(IV)	$(NH_4)_2[Ce(NO_3)_6]$	548.22				135 g/100 mL[20] aq; s alc, HNO
hydrogen carbonate	NH_4HCO_3	79.06	1.586	107 (rapid heating)		g/100 mL: 17.4[20] aq, 10 glyc
hydrogen citrate	$(NH_4)_2HC_6H_5O_7$	226.19	1.48		240 d	100 g/100 mL aq; sl s alc
hydrogen difluoride	NH_4HF_2	57.04	1.51	124.6		v s aq; sl s alc
hydrogen oxalate hydrate	$NH_4HC_2O_4 \cdot H_2O$	125.08	1.556	anhyd, 170		s aq, alc; i bz, eth
hydrogen phosphate	$(NH_4)_2HPO_4$	132.06	1.619	d 155		69 g/100 mL[20] aq; i alc, acet
hydrogen sulfate	NH_4HSO_4	115.11	1.78	146.9	d 350	100 g/100 mL aq; i alc, acet

Name	Formula	Formula wt	Density	mp	bp	Solubility
hydrogen sulfide	NH_4HS	51.11	1.17	d 25 to NH_3 + H_2S		128 g/100 mL[0] aq; s glyc; i alc, acet
hydrogen sulfite	NH_4HSO_3	99.11	2.03	subl 150 in N_2		267 g/100 mL[10] aq
hydrogen (±)tartrate	$NH_4HC_4H_4O_6$	167.12	1.68	d 200		2.2[15] aq; i alc
hydroxide	NH_4OH	35.05		−77		49% dissolved NH_3
hypophosphite	$NH_4H_2PO_2$	83.03		d		v s aq; sl s alc; i acet
iodate	NH_4IO_3	192.94	3.309	d 150		2.6[15] aq
iodide	NH_4I	144.94	2.514[25]	subl 551		167 g/100 mL[20] aq; v s alc, acet
lactate	$NH_4C_3H_5O_3$	107.11	1.2[15]	92		v s aq, alc, glyc; i acet, eth
magnesium arsenate 6-water	$NH_4MgAsO_4 \cdot 6H_2O$	289.36	1.923	d	220 vacuo	0.038[20] aq
molybdate(VI)(2−)	$(NH_4)_2MoO_4$	196.04	2.276[25]	d		s acids
nitrate	NH_4NO_3	80.04	1.725[25]	169.6	d 210	g/100 mL: 192[20] aq; 3.8[20] alc; 17[20] MeOH; s acet
octadecanoate	$NH_4OOC(CH_2)_{16}CH_3$	301.50		21–22		sl s aq; s alc; i acet
octanoate	$NH_4OOC(CH_2)_6CH_3$	161.24		d on standing		v s aq, alc, acet; sl s eth
oxalate hydrate	$(NH_4)_2C_2O_4 \cdot H_2O$	142.11	1.50	d 70		5.1[20] aq; s alc
oxodioxalatotitanate(IV)	$(NH_4)_2TiO(C_2O_4)_2$	276.02				v s aq
perchlorate	NH_4ClO_4	117.49	1.95	d 240		g/100 mL[25]: 21.9 aq, 1.49 EtOH, 0.014 BuOH, 0.029 EtOAc
permanganate	NH_4MnO_4	136.97	2.208[10]	explodes, 110		0.8[15] aq
peroxodisulfate	$(NH_4)_2S_2O_8$	228.20	1.982	d 120	expl 180	58 g/100 mL[0] aq
phosphinate	$NH_4PH_2O_2$	83.04	1.634	200	d 240	g/100 mL: 100 aq, 5 alc; i acet
phosphomolybdate hydrate	$(NH_4)_3PO_4 \cdot 12MoO_3 \cdot H_2O$	1894.36		d		sl s aq
picrate	$NH_4C_6H_2N_3O_7$	246.14	1.719	d	expl 423	1.1[20] aq; sl s alc
selenate(VI)	$(NH_4)_2SeO_4$	179.04	2.193[4]	d		117 g/100 mL[7] aq; s HOAC; i alc
stearate	$NH_4C_{18}H_{35}O_2$	301.51	0.89	22		sl s aq, bz; s alc; i acet
sulfamate	$NH_4NH_2SO_3$	114.13		131	d 160	v s aq; sl s alc
sulfate	$(NH_4)_2SO_4$	132.14	1.769[20]	d >280		43.5 g/100 mL[20] aq; i alc, acet
sulfide	$(NH_4)_2S$	68.14		d ≈ 0		v s aq; s alc, alk
sulfite hydrate	$(NH_4)_2SO_3 \cdot H_2O$	134.16	1.41	d 60		75 g/100 mL[20] aq; i alc, acet
(±)tartrate	$(NH_4)_2C_4H_4O_6$	184.15	1.601	d		58 g/100 mL[15] aq; sl s alc
tetraborate 4-water	$(NH_4)_2B_4O_7 \cdot 4H_2O$	263.44				s aq; i alc

(*Continued*)

31

TABLE 1.3 Physical Constants of Inorganic Compounds (*Continued*)

Name	Formula	Formula weight	Density	Melting point, °C	Boiling point, °C	Solubility in 100 parts solvent
tetrachloroaluminate	NH$_4$[AlCl$_4$]	186.83	2.170	304		s aq, eth
tetrachloropalladate(II)	(NH$_4$)$_2$[PdCl$_4$]	284.29	2.936	d		v s aq; i abs alc
tetrachloroplatinate(II)	(NH$_4$)$_2$[PtCl$_4$]	372.97	2.936	140 d		s aq; i alc
tetrachlorozincate	(NH$_4$)$_2$[ZnCl$_4$]	243.28	1.879	150 d	subl 341	v s aq
tetrafluoroborate	NH$_4$[BF$_4$]	104.84	1.871	subl		25 g/100 mL^{16} aq
thiocyanate	NH$_4$SCN	76.12	1.305	149.6	d 170	128 g/100 mL^0 aq; v s alc; s acet
thiosulfate	(NH$_4$)$_2$S$_2$O$_3$	148.21	1.679	d 150		2.15^{15} aq; i alc, eth
vanadate(V)(1−)	NH$_4$VO$_3$	116.98	2.326	d 200		0.48^{20} aq
Antimony	Sb	121.760(1)	6.697^{25}	630.7	1587	s hot conc H$_2$SO$_4$, aqua regia
arsenide	SbAs	196.68	6.0	≈680		s acet, bz, chl
(III) bromide	SbBr$_3$	361.47	4.35	96.6	280	10 g/100 mL^{20} aq; s alc, bz, chl
(III) chloride	SbCl$_3$	228.12	3.14$^{20}_4$	73.4	220.3	d aq; s HCl, chl, CCl$_4$
(V) chloride	SbCl$_5$	299.02	2.336$^{20}_4$	3.5	79^{22mm}	444 g/100 mL^{20} aq
(III) fluoride	SbF$_3$	178.75	4.379$^{20}_{20}$	292	376	d viol aq; s HOAc; forms solids with alc, bz, CS$_2$, eth
(V) fluoride	SbF$_3$	216.75	2.99^{23}	8.3	141	20 mL/100 mL^{20} aq; s CS$_2$, eth
hydride (stibine)	SbH$_3$	124.78	5.475 g/L	−91.5	−18.4	g/100 g^{25}: 1.16 bz, 1.24 tol, 0.16 chl
(III) iodide	SbI$_3$	502.47	4.92	168	401	v sl s aq; s HCl, KOH
(III) oxide (valentinite)	Sb$_2$O$_3$	291.52	5.7	655	1425	v sl s aq; sl s warm KOH, eth
(V) oxide	Sb$_2$O$_5$	323.52	3.78	−O$_2$, >300		v sl s aq; s conc HCl
(III) selenide	Sb$_2$Se$_3$	480.40	5.81	612		sl s aq
(III) sulfate	SB$_2$(SO$_4$)$_3$	531.71	3.62	d		0.002^{20} aq (d); s H$_2$SO$_4$
(III) sulfide	Sb$_2$S$_3$	339.72	4.56	546		i aq; s HCl (d), NaOH
(V) sulfide	Sb$_2$S$_5$	403.85	4.120	75 d		i aq; s HNO$_3$
(III) telluride	Sb$_2$Te$_3$	626.32	6.52	620		i aq
triethyl	Sb(C$_2$H$_5$)$_3$	209.0	1.324^{14}	−29	159.5	sl s aq
trimethyl	Sb(CH$_3$)$_3$	166.9	1.523^{15}		80.6	
Argon	Ar	39.948(1)	1.7824 g/L^0	−189.38	−185.87	3.36 mL/100 mL^{20} aq
Arsenic	As	74.92159(2)	5.727^{25}	817	subl 615	i aq; s HNO$_3$
(III) bromide	AsBr$_3$	314.63	3.3972^{25}	31.1	220.0	hyd aq; s HCl, CS$_2$, PE
(III) chloride	AsCl$_3$	181.28	2.1497$^{25}_4$	−16.2	130.2	misc chl, CCl$_4$, eth; s HCl
(di-) disulfide	As$_2$S$_2$	213.97	3.254^{19}	320	565	s alkali; v sl s bz
(III) fluoride	AsF$_3$	131.92	2.73^{15}	−5.95	57.8	s alc, bz, eth, HF
(V) fluoride	AsF$_5$	169.91	7.46 g/L	−79.8	−52.8	hyd aq; s alc, bz, eth

Name	Formula	MW	Density	mp	bp	Solubility
(III) hydride (arsine)	AsH_3	77.95	3.420 g/L	−116.9	−62.5	28 mL/100 mL^{20} aq; s bz, chl
(III) iodide	AsI_3	455.63	4.73	140.9	424	s bz, tol; sl s aq, alc, eth
(III) oxide (arsenolite)	As_2O_3	197.84	3.86	274	460	1.8^{20} aq; s alc
(III) oxide (claudetite)	As_2O_3	197.84	3.74	313	460	sl s aq; s dil acid, alk
(V) oxide	As_2O_5	229.84	4.32	315	d 800	66 g/100 mL^{20} aq; s alc
(III) selenide	As_2Se_3	386.72	4.75	260		s alkali, HNO_3
(III) sulfide	As_2S_3	246.04	3.460	310	707	i aq; s alk, slowly s hot HCl
(V) sulfide	As_2S_5	310.17		subl 500		0.0003 aq; s alkali, HNO_3
(III) telluride	As_2Te_3	532.64	6.50	621		
Astatine	At	210		302		
Barium	Ba	137.33	3.51^{20}	726.9	1845	d aq to Ba(OH)
acetate hydrate	$Ba(C_2H_3O_2)_2 \cdot H_2O$	273.43	2.19	anhyd 110	d 150	58.8 g/100 mL^0 aq; 0.014 alc
benzenesulfonate	$Ba(O_3SC_6H_5)_2$	451.70				s aq; sl s alc
bromate hydrate	$Ba(BrO_3)_2 \cdot H_2O$	411.14	3.99^{18}	d 260		0.96^{30} aq; s acet; i alc
bromide	$BaBr_2$	297.14	4.781	856	1835	92 g/100 mL^0 aq; s MeOH, acet
carbonate	$BaCO_3$	197.34	4.2865	d 1300 to BaO + CO_2		0.0024 aq; s acids
chlorate hydrate	$Ba(ClO_3)_2 \cdot H_2O$	322.24	3.179	anhyd 120	−O_2, 250	34 g/100 mL^{20} aq; sl s alc, acet
chloride	$BaCl_2$	208.24	3.856^{24}	962	1560	36 g/100 mL^{20} aq; s MeOH; i acet, EtAc
chloride dihydrate	$BaCl_2 \cdot 2H_2O$	244.26	3.097	anhyd113		31.7 g/100 mL^0 aq
chromate(VI)	$BaCrO_4$	253.33	4.498^{20}	d		0.001^{20} aq; s mineral acids
cyanide	$Ba(CN)_2$	189.36				80 g/100 mL^{14} aq; s alc
fluoride	BaF_2	175.32	4.89	1368	2260	0.161^{20} aq; s acids
hexafluorosilicate	$Ba[SiF_6]$	279.40	4.29^{21}	d 300		0.0235^{25} aq; s NH_4Cl soln; i alc
hydrogen phosphate	$BaHPO_4$	233.31	4.165^{15}	d 410		0.01 aq; s HCl, HNO_3
hydroxide 8-water	$Ba(OH)_2 \cdot 8H_2O$	315.48	2.18^{16}	78		3.9^{20} aq
iodate	$Ba(IO_3)_2$	487.13	5.23^{20}	d 476		0.033^{20} aq; s HCl
iodide	BaI_2	391.14	5.15	711	2027	169 g/100 mL^{20} aq; s alc, acet
manganate(VI)(2−)	$BaMnO_4$	256.26	4.85			disprop to $Ba(MnO_4)_2$ + MnO_2
molybdate	$BaMoO_4$	297.27	4.975	1450		0.0058^{25} aq
niobate	$Ba(NbO_3)_2$	419.14	5.44	1455		i aq
nitrate	$Ba(NO_3)_2$	261.34	3.24^{23}	592	d	5.0 aq; v sl s alc, acet
nitrite hydrate	$Ba(NO_2)_2 \cdot H_2O$	247.35	3.173^{30}	d 115		54.8 g/100 mL^{20} aq; i alc

(Continued)

TABLE 1.3 Physical Constants of Inorganic Compounds (*Continued*)

Name	Formula	Formula weight	Density	Melting point, °C	Boiling point, °C	Solubility in 100 parts solvent
oxalate	BaC_2O_4	225.35	2.658	400 d		i aq; s acids, EtOH
oxide	BaO	153.33	5.72	1973	3088	3.5^{20} aq; s acids, EtOH
perchlorate	$Ba(ClO_4)_2$	336.23	3.20	505		$g/100\ mL^{25}$: 129 aq, 78 EtOH, 42 BuOH, 81 EtOAc; i eth
perchlorate 3-water	$Ba(ClO_4)_2 \cdot 3H_2O$	390.27	2.74	d 400		$198\ g/100\ mL^{25}$ aq; s MeOH; sl s acet
permanganate	$Ba(MnO_4)_2$	375.20	3.77	d 200		v s aq
peroxide	BaO_2	169.33	4.96	450 d	$-O_2$, 800	1.5^{0} aq
selenide	$BaSe$	216.29	5.02	1780		d aq
stearate	$Ba(C_{18}H_{35}O_2)_2$	704.28	1.145	160		i aq
sulfate	$BaSO_4$	233.39	4.50^{15}	1580	d >1600	0.00285 aq
sulfide	BaS	169.39	4.25^{15}	2230		7.9^{20} aq; dec in acids
sulfite	$BaSO_3$	217.39	4.44	d		0.02^{0} aq; i alc
tetracyanoplatinate(II)-4-water	$Ba[Pt(CN)_4] \cdot 4H_2O$	508.54	2.076			2.86 aq; i alc
thiocyanate 2-water	$Ba(SCN)_2 \cdot 2H_2O$	289.53	2.286^{18}	d 160		$170\ g/100\ mL^{20}$ aq; s alc, acet
thiosulfate hydrate	$BaS_2O_3 \cdot H_2O$	267.47	3.5^{18}	d 220		0.21^{20} aq; i alc, acet, eth, CS
titanate(IV)(2−)	$BaTiO_3$	233.19	6.02	1625		i aq
vanadate	$Ba_3(VO_4)_2$	641.86	5.14	707		
zirconate	$BaZrO_3$	276.55	5.52	2500		i aq, alk; sl s acids
Berkelium (α form)	Bk	247	14.78	1050		
(β form)	Bk	247	13.25	986		
Beryllium	Be	9.012	1.8477^{20}	1287	2467	i aq; s acid, alk
bromide	$BeBr_2$	168.82	3.465^{25}	508	521	v s aq; s alc; 18.6 pyr
carbide	Be_2C	30.04	1.90^{15}	d >2127		d aq; s acids, alkali giving CH_4
chloride	$BeCl_2$	79.92	1.899^{25}	415 (alpha)	482.3	$42\ g/100\ mL$ aq; s alc, eth, pyr, CS_2
fluoride	BeF_2	47.01	1.986	555	subl 1036^{mm}	v s aq (slowly)
hydride	BeH_2	11.03	0.65	$-H_2$, 220		d aq (slowly), acids (rapidly)
hydroxide	$Be(OH)_2$	43.03	1.909	93		s hot conc acids and alkali (viol)
iodide	BeI_2	262.82	4.32	480	487	hyd aq violently; s alc, eth, CS_2
nitrate 3-water	$Be(NO_3)_2 \cdot 3H_2O$	187.07	1.557	60.5	d 125	$166\ g/100\ mL^{20}$ aq
nitride	Be_3N_2	55.05	2.71	2200		d hot aq, alkali
oxide	BeO	25.01	3.025	2578 (alpha)	3787	s conc H_2SO_4
selenate 4-water	$BeSeO_4 \cdot 4H_2O$	224.03	2.03	anhyd 300	d 560	$49\ g/100\ mL^{25}$ aq

	Formula	Mol wt	Density	MP	BP	Solubility
silicate	Be_2SiO_4	110.11	3.0	1560	d 580	i aq
sulfate 4-water	$BeSO_4 \cdot 4H_2O$	177.14	1.713	anhyd 270		39 g/100 mL^{20} aq; i alc
sulfide	BeS	41.08	2.36	d		i aq; s HNO_3
Bismuth	Bi	208.9804	9.78	271.5	1564	i aq; s hot H_2SO_4
(III) bromide	$BiBr_3$	448.69	5.72	218	453	d aq; s dil acids, acet
bromide oxide	BiBrO	304.88	8.082^{15}	d		i aq; s acids
(III) chloride	$BiCl_3$	315.34	4.75	233.5	447	d aq; s HCl, alc, eth, acet
chloride oxide	BiClO	260.43	7.72^{15}	d		i aq; s HCl
(III) fluoride	BiF_3	265.98	8.32	727	900	i aq; s HF
(V) fluoride	BiF_5	303.97	5.55^{25}	154.4	subl 550	d (viol) aq giving O_3 + BiF_3
hydride	BiH_3	212.00	9.303 g/L	−67	16.8	very unstable liquid
(III) hydroxide	$Bi(OH)_3$	260.00	4.962^{15}	−water, 100		d aq; s HCl
(III) iodide	BiI_3	589.69	5.778_4^{20}	408.6	subl 439	i aq; s HCl, alc
iodide oxide	BiIO	351.88	7.922	d red heat		i aq; s HCl
(III) nitrate 5-water	$Bi(NO_3)_3 \cdot 5H_2O$	485.07	2.83	anhyd 80		d aq; s HNO_3, acet, glyc
(III) oxide	Bi_2O_3	465.96	8.76	817	1890	i aq; s HCl, HNO_3
(V) oxide	Bi_2O_5	497.96	5.10	d 150		i aq; s KOH
(III) phosphate	$BiPO_4$	303.95	6.323^{15}	d		s conc HCl, HNO_3
(III) selenide	Bi_2Se_3	654.84	7.70^{20}	710 d		i aq; d aq reg
(III) sulfate	$Bi_2(SO_4)_3$	706.14	5.08	d 405	d	d aq, alc; s HCl
(III) sulfide	Bi_2S_3	514.16	6.78	850		i aq, EtAe; s HNO_3, HCl
(III) telluride	Bi_2Te_3	800.76	7.74	588.5		i aq; s alc
Boranes						
diborane(6)	B_2H_6	27.67	1.214 g/L	−165.5	−92.5	FP −68; s NH_4OH, conc H_2SO_4
tetraborane(10)	B_4H_{10}	53.32	2.340 g/L	−120	18	sl s aq; s bz
pentaborane(9)	B_5H_9	63.13	0.60	−46.81	60.0	hyd aq
pentaborane(11)	B_5H_{11}	65.14	0.745	−123	63	d aq
hexaborane(10)	B_6H_{10}	74.95	0.67	−62.3	108 d	d hot aq
decaborane(14)	$B_{10}H_{14}$	122.22	0.948	99.5	213	sl s aq; s bz, CS_2, eth
Borazine	$B_3H_6N_3$	80.50	lq: 0.81^{bp}	−58	55	sl s aq (d)
Boric acids, *see under* Hydrogen						
Boron	B	10.811	2.34	2076	3864	i aq
carbide	B_4C	55.25	2.510^{25}	2350	>3500	s fused alkalis
tribromide	BBr_3	250.52	2.6	−46.0	91.3	d aq, alc
trichloride	BCl_3	117.17	5.141 g/L	−107	12.7	d aq, alc

(Continued)

TABLE 1.3 Physical Constants of Inorganic Compounds (*Continued*)

Name	Formula	Formula weight	Density	Melting point, °C	Boiling point, °C	Solubility in 100 parts solvent
trifluoride	BF_3	67.81	3.077 g/L^{STP}	−127.1	−100.4	332 g/100 mL^0 aq; s bz, chl, CCl$_4$
trifluoride 1-diethyl ether	$BF_3 \cdot O(C_2H_5)_2$	141.94	1.125	−60.4	125.7	d aq
trifluoride 1-methanol	$BF_3 \cdot HOCH_3$	131.89	1.203		59^{4mm}	
nitride	BN	24.82	2.18	2967		sl s hot acids
oxide	B_2O_3	69.62	2.55	450.0	2065	3.3 aq (slowly); s alc, glyc
Bromine	Br_2	159.808	3.1023$^{25}_{4}$	−7.25	58.8	3.4 g/100 mL^{20} aq; v s alc, chl, eth
pentafluoride	BF_5	174.90	2.460	−60.5	40.76	explodes with water; s HF
trifluoride	BF_3	136.90	2.803^{25}	8.77	125.74	d viol aq; d alk; smokes in air
Cadmium	Cd	112.411	8.65^{25}	321	765	i aq, alk; s HNO$_3$, hot HCl
acetate	$Cd(C_2H_3O_2)_2$	230.50	2.341	255	d	v s aq; s alc
bromide	$CdBr_2$	272.22	5.192	566	963	99 g/100 mL^{20} aq; s acet; sl s eth
carbonate	$CdCO_3$	172.42	4.258^4	d 500		s acids, NH$_4$OH
chloride	$CdCl_2$	183.32	4.05^{25}	568	960	120 g/100 mL^{25} aq
cyanide	$Cd(CN)_2$	164.44	2.226	d 200		1.71 g/100 mL^{15} aq; sl s alc
fluoride	CdF_2	150.41	6.33	1110	1748	4.3 g/100 mL^{25} aq
hydroxide	$Cd(OH)_2$	146.43	4.79	−H$_2$O, 130	CaO, 200	0.00026^{20} aq; s acids
iodide	CdI_2	366.22	5.670	388	742	84.7 g/100 mL^{20} aq; s alc, acet, eth
nitrate 4-water	$Cd(NO_3)_2 \cdot 4H_2O$	308.48	2.455	59.4		167 g/100 mL^{25} aq; s alc, acet
oxide	CdO	128.41	8.15 cubic	1540		i aq; s acids
phosphide	Cd_3P_2	399.18	5.96	700		s dil acid
selenide	$CdSe$	191.37	5.81^{15}	1350		i aq; d acids
sulfate-water (3/8)	$3CdSO_4 \cdot 8H_2O$	769.56	3.08	monohydrate, 80		94.4 g/100 mL^{25} aq; i alc, EtAc
sulfide	CdS	144.48	4.83	1750		0.13^{18} aq; s acids
telluride	$CdTe$	240.01	6.20$^{5}_{4}$	1041		i aq; d HNO$_3$
tungstate(VI)	$CdWO_4$	360.25	8.0			i aq; dil acids; s alkali CN's
Calcium	Ca	40.078(4)	1.55	842	1484	d aq; s acids
acetate	$Ca(C_2H_3O_2)_2$	158.17	1.50	d >160		37.4 g/100 mL^0 aq; i alc, bz, acet
arsenate	$Ca_3(AsO_4)_2$	398.07	3.620			0.013^{25} aq
bromide	$CaBr_2$	199.89	3.38	742	1815	143 g/100 mL^{20} aq; v s alc, acet
carbide	CaC_2	64.10	2.222	2300		reacts with aq giving C$_2$H$_2$
carbonate (aragonite)	$CaCO_3$	100.09	2.83	d 825 to CaO		s dil acids
carbonate (calcite)	$CaCO_3$	100.09	2.711	d 825 to CaO		0.0013 g/100 mL^{20}; s acids
chlorate 2-water	$Ca(ClO_3)_2 \cdot 2H_2O$	243.01	2.711	anhyd 100		167 g/100 mL^{20} aq; s alc

(Continued)

chloride	$CaCl_2$	110.98	2.16_4^{25}	775	ca. 1940	42 g/100 mL^{20} aq; s alc, acet
chloride 6-water	$CaCl_2 \cdot 6H_2O$	219.07	1.71	anhyd 200		74.5 g/100 mL^{20} aq; v s alc
chlorite	$Ca(ClO_2)_2$	174.99	2.71^{25}	100		167 g/100 mL aq; s alc
chromate(VI) 2-water	$CaCrO_4 \cdot 2H_2O$	192.10	2.50	anhyd 200		si s aq; s dil acids
citrate 4-water	$Ca_3(C_6H_5O_7)_2 \cdot 4H_2O$	570.51		anhyd 120		0.10 aq; i alc
cyanamide	$CaCN_2$	80.10	2.29	ca. 1340	subl	no known solv without dec
cyanide	$Ca(CN)_2$	92.11		s >350		s aq
dichromate(VI)	$CaCr_2O_7$	256.10	2.370_4^{30}	d >100		v s aq; i eth; d alc
dihydrogen phosphate hydrate	$Ca(H_2PO_4)_2 \cdot H_2O$	252.07	2.220_4^{18}	anhyd 100	d 200	1.8^{30} aq
diphosphate (pyrophosphate)	$Ca_2P_2O_7$	254.10	3.09	1353		i aq; s HCl, HNO₃
fluoride	CaF_2	78.08	3.180	1418	2533	0.0015^{20} aq; s conc mineral acids
formate	$Ca(CHO_2)_2$	130.11	2.015	300 d		16.6 g/100 mL^{20} aq; i alc
(+)gluconate	$Ca[OOC(CHOH)_4CH_2OH]_2$	430.38				3.72^{20} aq
glycerophosphate	$Ca[C_3H_5(OH)_2PO_4]$	210.16		d >170		1.66^{20} aq; i alc
hexafluorosilicate	$Ca[SiF_6]$	182.17	2.662			i aq, acet
hydride	CaH_2	42.09	1.70	1000		d aq, alc
hydroxide	$Ca(OH)_2$	74.09	2.343	$-H_2O$, 580		0.17^{10} aq; s acids
hypochlorite	$Ca(OCl)_2$	142.99	2.35	100 d		d aq evolving Cl₂; i alc
iodate	$Ca(IO_3)_2$	389.88	4.519_4^{15}	d >540	1755	0.10^0 aq; i alc
iodide	CaI_2	293.89	3.956	783	anhyd 120	68 g/100 mL^{20} aq; v s alc, acet; i eth
lactate 5-water	$Ca(C_3H_5O_3)_2 \cdot 5H_2O$	308.30		$-3H_2O$, 100		5.4^{15} aq; v sl s alc
magnesium carbonate	$Ca[Mg(CO_3)_2]$	184.41	2.872	d 730		0.032^{18} aq; s HCl
molybdate(VI)(2−)	$CaMoO_4$	200.02	4.35			s conc mineral acids
nitrate	$Ca(NO_3)_2$	164.09	2.504	561		152 g/100 mL^{30} aq
nitride	Ca_3N_2	148.25	2.67	1195		d aq; s dilute acids (d)
nitrite 4-water	$Ca(NO_2)_2 \cdot 4H_2O$	204.15	1.674	d		84.5 g/100 mL^{18} aq; sl s alc
oleate	$Ca(C_{18}H_{33}O_2)_2$	603.01		83–84	d >400	0.04 aq; s chl, bz; v sl s alc, eth
oxalate hydrate	$CaC_2O_4 \cdot H_2O$	146.11	2.2	anhyd 200		0.0006 aq; s acids
oxide	CaO	56.08	3.34	2900	3500	0.13^{25} aq; s acids
palmitate	$Ca(C_{16}H_{31}O_2)_2$	550.93		d >155		0.003 aq; sl s bz, chl, HOAc
(+)panthothenate (vitamin B₃)	$Ca[O_2CH_2CH_2NHOCH(OH)C(CH_3)_2CH_2OH]_2$	476.55		d 195–196		36 g/100 mL aq; sl s alc, acet
perchlorate	$Ca(ClO_4)_2$	238.98	2.65	d 270		g/100 mL^{25}: 112 aq, 89.5 EtOH, 68 BuOH, 57 EtOAc, 43 acet

TABLE 1.3 Physical Constants of Inorganic Compounds (*Continued*)

Name	Formula	Formula weight	Density	Melting point, °C	Boiling point, °C	Solubility in 100 parts solvent
permanganate 5-water	$Ca(MnO_4)_2 \cdot 5H_2O$	368.03	2.4	d		338 g/100 mL aq; si s aq; s acids
peroxide	CaO_2	72.08	2.92	explodes 275		sl s aq, alc
phenoxide	$Ca(OC_6H_5)_2$	226.28	d in air			0.03^{25} aq; s HCl, HNO$_3$; i alc
phosphate	$Ca_3(PO_4)_2$	310.18	3.14	1670		d aq; s acids; i alc, eth
phosphide	Ca_3P_2	182.18	2.51	ca. 1600		15.4 g/100 mL aq; sl s glyc
phosphinate	$Ca(PH_2O_2)_2$	170.06		d >300		s aq; sl s alc; i acet, bz
propanoate	$Ca(OOCC_3H_5)_2$	186.22				2.8^{15} aq; 0.015^{16} EtOH
salicylate 2-water	$Ca(C_7H_5O_3)_2 \cdot 2H_2O$	350.34		anhyd 200	d 240	9.2 g/100 mL^{25} aq
selenate 2-water	$CaSeO_4 \cdot 2H_2O$	219.07	2.75	anhyd 200	d 698	i aq
selenide	CaSe	119.04	3.82			
silicate	Ca_2SiO_4	172.24	3.27	2130		0.004^{15} aq; s hot pyr; i acet, chl
stearate	$Ca(C_{18}H_{35}O_2)_2$	607.04		179–180		1.28^{20} aq; s acids; i alc
succinate 3-water	$CaC_4H_6O_4 \cdot 3H_2O$	212.22				0.20 aq; s acids
sulfate	$CaSO_4$	136.14	2.960	1460		0.3^{20} aq; s acids, glyc
sulfate hemihydrate	$CaSO_4 \cdot 0.5H_2O$	145.15		anhyd 163		0.26^{20} aq; s acid, glyc
sulfate 2-water	$CaSO_4 \cdot 2H_2O$	172.17	2.32	-1.5 H$_2$O 128	anhyd 163	0.02 (d) aq; d acids
sulfide	CaS	72.14	2.59	2525		0.004 aq; s acids d; sl s alc
sulfite 2-water	$CaSO_3 \cdot 2H_2O$	156.17		anhyd 100		0.0045^{25} aq; s acids; sl s alc
(±)tartrate 4-water	$CaC_4H_4O_6 \cdot 4H_2O$	260.21		anhyd 200		
telluride	CaTe	167.68	4.873			s dil acids
tetraborate	CaB_4O_7	195.36				
tetrahydridoaluminate	$Ca[AlH_4]_2$	102.10		ign moist air		d viol aq, alc; i bz, eth
thiocyanate 3-water	$Ca(SCN)_2 \cdot 3H_2O$	210.29		d > 160		150 g/100 mL aq; v s alc
thioglycollate 3-water	$Ca(—OOCCH_2S—) \cdot 3H_2O$	184.24		$-H_2O$, >95	d >220	s aq; v sl s alc, chl; i bz, eth
thiosulfate 6-water	$CaS_2O_3 \cdot 6H_2O$	260.30	1.872	d >45		92 g/100 mL^{25} aq; i alc
titanate	$CaTiO_3$	135.84	3.98	1980		
tungstate(VI)(2−)	$CaWO_4$	287.93	6.062^{20}			0.0032 aq; d hot acids
Californium-252	Cf	252.1		900		
chloride	$CfCl_3$	358.5	5.88			i aq, alc
Carbon (diamond)	C	12.011	3.513	$3500^{63.5atm}$	3930	i aq, alc
(graphite)	C		2.267	subl 3915–4020		
dioxide	CO_2	44.01	c: 1.56^{-79}; g: 1.975 g/L^0	-78.44 subl		88 mL/100 mL^{20} aq
diselenide	CSe_2	169.93	2.6626^{25}_4	-45.5	125.1	i aq; s acet, eth; misc CCl$_4$; d alc
disulfide	CS_2	76.14	1.2555	-111.6	46.56	FP -30; 0.29^{20} aq; s alc, eth

Name	Formula	Molar mass	Density	mp	bp	Solubility
hydride (methane)	CH_4	16.04		−182.48	−161.49	s bz
monoxide	CO	28.01	0.415^{-164}; lq: 0.814^{-195}	−205.05	−191.49	2.3 mL/100 mL^{20} aq; 16 mL/100 ml alc; s HOAc, EtAc
suboxide	C_3O_2	68.03	g: 1.250 g/L^0; 1.114_4^0; 2.985 g/L	−111.3	6.8	d aq to malonic acid; sl s CS_2
tetrabromide	CBr_4	331.65	3.42	90.1	190	i aq; s alc, chl, eth
tetrachloride	CCl_4	153.82	1.589_{25}^{25}	−22.9	76.7	0.05 mL/100 mL aq; s alc, chl, eth
tetrafluoride	CF_4	88.00	1.96^{-184}	−183.6	−127.8	sl s aq
tetraiodide	CI_4	519.63	4.34_4^{20}	171	subl 130	slowly hyd aq; s bz, chl, eth
Carbonyl bromide	$COBr_2$	187.82	2.5		64.5	hyd aq
chloride	$COCl_2$	98.92	4.340 g/L	−127.9	8.2	hyd aq; s bz, HOAc
fluoride	COF_2	66.01	lq: 1.139; g: 2.896 g/L	−114.0	−83.1	hyd aq
sulfide	COS	60.07	2.636 g/L	−138.81	−50.23	54 mL/100 mL^{20} aq; s alc, CS_2
Cerium	Ce	140.11	6.773	795	3440	i aq; s acids
(III) bromide	$CeBr_3$	379.83	5.18	733	1460	s aq, alc
(III) chloride	$CeCl_3$	246.47	3.97_{25}	817	1730	s aq, alc
(III) fluoride	CeF_3	197.11	6.157	1430	2327	i but slowly hyd aq; s H_2SO_4
(IV) fluoride	CeF_4	216.11	4.77	d >550		i aq
(III) iodide	CeI_3	520.83		766	1400	s aq
(III) nitrate 3-water	$Ce(NO_3)_3\ 3H_2O$	380.17		anhyd 150	d 200	234 g/100 mL^{20} aq
(IV) oxide	CeO_2	172.11	7.65	2400		i aq; s acids
(III) sulfate	$Ce_2(SO_4)_3$	568.42	3.912	d 1000		9.72 g/100 mL^{21} aq
(IV) sulfate	$Ce(SO_4)_2$	332.24	3.91	d 195		hyd aq; s dil H_2SO_4
Cesium	Cs	132.9054	1.8875^{15}	28.44	668.2	d aq; s acids
bromide	$CsBr$	212.81	4.44	636	≈1300	107 g/100 mL^{18} aq; s alc; i acet
carbonate	Cs_2CO_3	325.82	4.24	792		v s aq; 11 g/100 mL^{20} alc; s eth
chloride	$CsCl$	168.36	3.99	646	1300	g/100 mL: 187^{20} aq; 34^{25} MeOH; v s alc
fluoride	CsF	151.90	4.115	703	1231	322 g/100 mL^{18} aq
hydroxide	$CsOH$	149.91	3.68	272	990	386 g/100 mL^{15} aq; s alc
iodate	$CsIO_3$	307.81	4.934^{20}	565		2.6^{25} aq
iodide	CsI	259.81	4.510	621	≈1280	76.5 g/100 mL^{20} aq; s EtOH; i acet
nitrate	$CsNO_3$	194.91	3.66	414	d 849	23 g/100 mL^{20} aq; s acet; v sl s alc

(Continued)

TABLE 1.3 Physical Constants of Inorganic Compounds (*Continued*)

Name	Formula	Formula weight	Density	Melting point, °C	Boiling point, °C	Solubility in 100 parts solvent
oxide	Cs_2O	281.81	4.65	490		v s aq
perchlorate	$CsClO_4$	232.36	3.327	250		g/100 mL[25]: 1.96, 0.0086 EtOH, 0.118 acet, 0.0048 BuOH; i EtOAc, eth
selenate	Cs_2SeO_4	408.77	4.453	1005		244 g/100 mL[12] aq
sulfate	Cs_2SO_4	361.87	4.243			179 g/100 mL[20] aq; i alc, acet, pyr
Chlorine	Cl_2	70.905	g: 2.98^{20} g/L lq: 1.5649^{-35}	−101.5	−34.04	199 mL/100 mL[25] aq
dioxide	ClO_2	67.45	2.960 g/L	−59.6	10.9	11.2 g/100 mL[10] aq
fluoride	ClF	54.45	4.057 g/L	−155.6	−100.1	d viol aq; organics burst into flame
heptoxide	Cl_2O_7	182.90	1.805^{25}	−91.5	82	hyd aq slowly; explodes on concussion or on contact with flame or I_2
monoxide	Cl_2O	86.90	3.813 g/L	−120.6	2.2	v s aq (forms HClO); s CCl_4
pentafluoride	ClF_5	130.44	5.724 g/L	−103	−13.1	
trifluoride	ClF_3	92.45	g: 4.057 g/L lq: 1.825^{bp}_{20}	−76.3	11.75	hyd viol aq; organic matter and glass wool burst into flame
trioxide (dimer)	$(ClO_3)_2$	166.90	1.92^{20}	3.5	≈200	reacts with aq
Chromium	Cr	51.996	7.15	1907	2679	s dil HCl
(II) acetate	$Cr(C_2H_3O_2)_2$	170.09	1.79			sl s aq, alc; s a; i eth
(III) acetate	$Cr(C_2H_3O_2)_3$	229.13				s aq
(II) bromide	$CrBr_2$	211.80	4.236	842		s aq, alc
(III) bromide	$CrBr_3$	291.71	4.68			s hot aq; v s alc
(II) chloride	$CrCl_2$	122.90	2.88^{25}	814	subl 1300	v s aq
(III) chloride	$CrCl_3$	158.35	2.87	1152	d > 1300	s aq, alc (slow); i acet
(II) fluoride	CrF_2	89.99	3.79	894		sl s aq; s hot HCl
(III) fluoride	CrF_3	108.99	3.8	1400		aq, alc; s HF, HCl
(III) formate 6-water	$Cr(CHO_2)_3 \cdot 6H_2O$	295.15		d >300		s aq
hexacarbonyl	$Cr(CO)_6$	220.06	1.77	d 130	explodes 210	i aq, alc; s eth, chl
(III) hydroxide	$Cr(OH)_3$	101.02		d		i aq; s acids
(III) nitrate 9-water	$Cr(NO_3)_3 \cdot 9H_2O$	400.15	1.80	66	d >100	208 g/100 mL[15] aq; s alc
(III) oxide	Cr_2O_3	151.99	5.21	2330	≈3000	i aq, alc; sl s acids, alkalis
(IV) oxide	CrO_2	84.00	4.89	197	$-O_2$, 250	i aq; s HNO_3
(VI) oxide	CrO_3	99.99	2.70^{25}	198	d 250	61.7 g/100 mL aq; may ign organics
(III) phosphate	$CrPO_4$	146.97	4.6	>1800		i aq, acids, aq reg

Name	Formula	Mol. wt.	Density	mp, °C	bp, °C	Solubility
potassium bissulfate 12-water	$CrK(SO_4)_2 \cdot 12H_2O$	499.41	1.826^{25}	89	anhyd 400	22 g/100 mL^{25} aq; i alc
(II) sulfate 7-water	$CrSO_4 \cdot 7H_2O$	274.17	1.7	d 100		22.9 g/100 mL^{0} aq; sl s alc
(III) sulfate 18-water	$Cr_2(SO_4)_3 \cdot 18H_2O$	716.45				220 g/100 mL^{20} aq
Chromyl chloride	CrO_2Cl_2	154.90	1.9145^{25}_4	−96.5	117	d aq; s bz, chl, eth, CCl$_4$
fluoride	CrO_2F_2	121.99		31.6^{885mm}	subi 29.6	
Cobalt	Co	58.9332	8.90	1494	2927	i aq; s dil HNO$_3$
(II) acetate 4-water	$Co(C_2H_3O_2)_2 \cdot 4H_2O$	249.08	1.705^{19}	anhyd 140		s aq; 2.1 g/100 mL^{15} MeOH
(III) acetate	$Co(C_2H_3O_2)_3$	236.07		d >100		s aq, HOAc, alc
(II) bromide	$CoBr_2$	218.74	4.909^{25}_4	678 (in N$_2$)		112 g/100 mL^{20} aq; s alc, acet
(II) carbonate	$CoCO_3$	118.94	4.13	d		0.18^{15} aq; s hot acids
(II) chloride	$CoCl_2$	129.84	3.367^{25}_4	735	1049	53 g/100 mL^{20} aq; s alc, acet, eth, glyc, pyr
(II) chloride 6-water	$CoCl_2 \cdot 6H_2O$	237.93	1.924	anhyd 110		97 g/100 mL^{20} aq
(II) chromate	$CoCrO_4$	174.93	≈4.0	d		i aq; s acids
(II) cyanide	$Co(CN)_2$	110.97	1.872^{25}_4	d 300		0.0042^{18} aq; s KCN
(II) fluoride	CoF_2	96.93	4.46	1127	≈1400	1.36^{20} aq; s warm mineral acids
(III) fluoride	CoF_3	115.93	3.88	926		d aq
(II) formate 2-water	$Co(CHO_2)_2 \cdot 2H_2O$	185.00	2.129^{22}_4	anhyd 140	d 175	5.03 g/100 mL^{30} aq; i alc
(II) hydroxide	$Co(OH)_2$	92.95	3.37	168 (vacuo)		0.00018 aq; v s acids
(III) hydroxide	$Co(OH)_3$	109.96	4.46	−H$_2$O, 100	d	0.00032 aq; s acids
(II) iodide (alpha, black)	CoI_2	312.74	5.584^{25}_4	515 (vacuo)	570 (vacuo)	203 aq
(II) nitrate 6-water	$Co(NO_3)_2 \cdot 6H_2O$	291.03	1.88	55	d >74	155 g/100 mL^{30} aq; v s alc
(II) oxalate	CoC_2O_4	146.95	3.021	d 250		0.002^{18} aq
(II) oxide	CoO	74.93	6.44	−s1935		i aq; s acids, alkalis
(II,III) oxide	Co_3O_4	240.80	6.07	d >900		i aq; s acids, alkalis
(II) phosphate 8-water	$Co_3(PO_4)_2 \cdot 8H_2O$	510.87	2.769	anhyd 200		v sl s aq; s mineral acids
(II) sulfate 7-water	$CoSO_4 \cdot 7H_2O$	281.10	2.03	anhyd 420	d 1140	65 g/100 mL^{20} aq; sl s alc
(II) sulfide	CoS	91.00	5.45^{18}	1180		i aq; s acids
(II) thiocyanate 3-water	$Co(SCN)_2 \cdot 3H_2O$	229.14		anhyd 105		7.8^{18} aq; s alc, eth
Copper	Cu	63.546	8.96^{20}	1084.62	2561.5	i; s HNO$_3$, hot H$_2$SO$_4$
(II) acetate 1-water	$Cu(C_2H_3O_2)_2 \cdot H_2O$	199.65	1.882	115	d 240	8 g/100 mL aq; 0.48 MeOH; sl s eth
acetate meta-arsenate (1/3)	$Cu(C_2H_3O_2)_2 \cdot 3Cu(AsO_2)_2$	1013.80				unstable in acids, bases; s NH$_4$OH
(II) borate(1−)	$Cu(BO_2)_2$	149.17	3.859			s a; i aq
(I) bromide	CuBr	143.45	4.98	497	1345	v sl s aq; s HCl, HBr, NH$_4$OH
(II) bromide	$CuBr_2$	223.35	4.71	498	900	126 g/100 mL aq; s alc, acet, pyr; i

(Continued)

TABLE 1.3 Physical Constants of Inorganic Compounds (*Continued*)

Name	Formula	Formula weight	Density	Melting point, °C	Boiling point, °C	Solubility in 100 parts solvent
(II) carbonate hydroxide (1/1) (malachite)	$CuCO_3 \cdot Cu(OH)_2$	221.12	4.0	d 200		i aq; s acids
(II) chlorate 6-water	$Cu(ClO_3)_2 \cdot 6H_2O$	338.54		65	d 100	242 g/100 mL^{18} aq; v s alc; s acet
(I) chloride	$CuCl$	99.00	4.14	430	≈1400	0.024 aq; s conc HCl, conc NH$_4$OH
(II) chloride	$CuCl_2$	134.45	3.386	300 d		73 g/100 mL^{20} aq; s alc, acet
(II) chloride 2-water	$CuCl_2 \cdot 2H_2O$	170.48	2.51	anhyd 200	d >300	76.4 g/100 mL^{25} aq; v s alc; s acet
(I) chromium(III) oxide (1/1)	$Cr_2O_3 \cdot Cu_2O$	295.07	5.24^{20}	d >900		i aq; s HNO$_3$
(II) citrate 2.5-water	$Cu_2C_6H_4O_7 \cdot 2.5H_2O$	360.22		anhyd 100		0.17 aq; s acids
(I) cyanide	$CuCN$	89.56	2.92	473 (in N_2)	d	i aq; s NH$_4$OH, KCN; d hot dil HCl
(II) fluoride	CuF_2	101.54	4.23	836	1676	4.75 g/100 mL^{20} aq; s acids
(II) formate	$Cu(CHO_2)_2$	153.58	1.831			12.5 aq
(II) hexafluorosilicate 4-water	$Cu[SiF_6] \cdot 4H_2O$	277.60	2.56	d		124 g/100 mL^{20} aq
(II) hydroxide	$Cu(OH)_2$	97.56	3.368	d 160		i aq; s acids
(I) iodide	CuI	190.45	5.67	606	≈1290	i aq; s KCN, NH$_4$OH, KI
(II) nitrate 3-water	$Cu(NO_3)_2 \cdot 3H_2O$	241.60	2.32	114.5	170 d	138 g/100 mL^{0} aq; v s alc
(II) oleate	$Cu(OOCC_{17}H_{33})_2$	626.46				i aq; sl s alc; s eth
(II) oxalate hemihydrate	$CuC_2O_4 \cdot 0.5H_2O$	160.57	6.0^{25}	anhydr >200	d 310	0.002 aq; s NH$_4$OH
(I) oxide	Cu_2O	143.09	6.315^{14}	1235	$-O_2$, 1800	i aq; s HCl
(II) oxide	CuO	79.54	2.225^{23}	1450		i aq, alc; s acids, KCN
(II) perchlorate	$Cu(ClO_4)_2$	262.45		d >130		146 g/100 mL^{30} aq; s eth, EtAc; i bz
(II) phosphate 3-water	$Cu_3(PO_4)_2 \cdot 3H_2O$	434.63		dehyd in air		i aq; s acids
(II) salicylate 4-water	$Cu(C_7H_5O_3)_2 \cdot 4H_2O$	409.83				v s aq; s alc
(II) selenate 5-water	$CuSeO_4 \cdot 5H_2O$	296.58	2.559	anhyd 265		25 g/100 mL^{20} aq; v sl s acet
(I) selenide	Cu_2Se	206.05	6.84^{21}	1113		d HCl
(II) selenide	$CuSe$	142.51	6.0	d 550	d ca. 480	s acids
(II) stearate	$Cu(OOCC_{17}H_{35})_2$	630.50	3.603	≈250		i aq, alc, eth; s hot bz, pyr
(II) sulfate	$CuSO_4$	159.61	2.284^{16}	d >560		14.3 g/100 mL^{0} aq; i alc
(II) sulfate 5-water	$CuSO_4 \cdot 5H_2O$	249.69	5.6^{20}	anhyd 200		32 g/100 mL^{20} aq; s MeOH, glyc
(I) sulfide	Cu_2S	159.16	4.76	1130		i aq; d HNO$_3$; s KCN
(II) sulfide	CuS	95.61				i aq; s hot HNO$_3$, KCN

(*Continued*)

(I) sulfite hydrate	$Cu_2SO_3 \cdot H_2O$	225.16	3.83^{15}			sl s aq; s HCl
(II) tartrate 3-water	$CuC_4H_4O_6 \cdot 3H_2O$	265.66		d		0.42^{20} aq; s acids, alkalis
(I) thiocyanate	CuSCN	121.62	2.85	1084		0.0044 aq; s NH_4OH, eth, alkali SCN
(II) tungstate(VI)(2−)	$CuWO_4 \cdot 2H_2O$	347.41				0.1^{15} aq; d acids; s NH_4OH
Curium-244	Cm	244.063	13.51	1340	≈3110	s acids
Cyanogen	NC—CN	52.03	2.335 g/L	−27.84	−21.15	mL/100 mL: 450^{20} aq, 230 alc;
azide	NC—N₃	68.04				s acetonitrile; pure azide detonates upon shock. Handle only in solvents.
bromide	NCBr	105.92	2.005	52	61.5	v s aq, alc, eth
chloride	NCCl	61.47	2.697 g/L	−6.5	13.8	s aq, alc, eth
fluoride	NCF	45.02	1.975 g/L	−82	−46	
Deuterium	D_2 or $^2H_2$	4.03	0.169^{mp} lq	−252.89	−249.49	sl s aq
oxide	D_2O	20.03	1.1056^{20}	3.82	101.43	misc aq
Dysprosium	Dy	162.50	8.540^{25}	1412	2567	s acids
bromide	$DyBr_3$	402.21	4.78	880	1480	s aq
chloride	$DyCl_3$	268.86	3.67	680	1530	s aq
fluoride	DyF_3	219.50	7.465	1154	2230	i aq
oxide	Dy_2O_3	373.00	7.81^{27}	2408		s aq
Einsteinium	Es	252.083	8.84	860		
Erbium	Er	167.26	9.066	1529	2868	s acid
chloride	$ErCl_3$	273.62	4.1	776	1500	s aq; sl s alc
oxide	Er_2O_3	382.52	8.640	2418		0.0005^{25} aq; s acids
sulfate 8-water	$Er_2(SO_4)_3 \cdot 8H_2O$	766.83	3.205	anhyd 110	d 630	16.0 g/100 mL ^{20} aq
Europium	Eu	151.965	5.244	822	1527	s acids
(III) chloride	$EuCl_3$	258.32	4.89	623 d		s aq
(III) oxide	Eu_2O_3	351.93	7.42	2350		i aq; s acids
(III) sulfate 8-water	$Eu_2(SO_4)_3 \cdot 8H_2O$	736.24	$-8H_2O$, 375	1527		2.56^{20} aq
Fermium-257	Fm	257.0951				
Fluorine	F_2	38.00	1.513^{bp} lq; 1.667 g/L	−219.61	−188.13	d aq viol; ignites organics and silicates
nitrate	$FONO_2$	81.00	1.507^{bp} lq	−175	−45.9	hyd aq; s acet; ignites alc, eth; liquid explodes on slight concussion
perchlorate	$FOClO_3$	118.45	5.20 g/L	−167.3	−15.9	explodes on slightest provocation
Francium-223	Fr	223.02				

TABLE 1.3 Physical Constants of Inorganic Compounds (*Continued*)

Name	Formula	Formula weight	Density	Melting point, °C	Boiling point, °C	Solubility in 100 parts solvent
Gadolinium	Gd	157.25	7.90	1312	3273	s acids
chloride	$GdCl_3$	263.61	4.52^0	~609	1580	s aq
fluoride	GdF_3	214.25	7.047	1231	2277	i aq
nitrate 6-water	$Gd(NO_3)_3 \cdot 6H_2O$	451.36	2.332	91		s aq, alc
oxide	Gd_2O_3	362.50	7.407^{15}	2340		s acids
sulfate 8-water	$Gd_2(SO_4)_3 \cdot 8H_2O$	746.81	3.010^{18}	anhyd 400	d 500	4.08 aq
Gallium	Ga	69.723	$5.904^{29.6}$ (c) $6.095^{29.8}$ (lq)	29.7646	2203	s conc HCl, halogens, alkalis
antimonide	GaSb	191.48	5.614	712		s HCl
arsenide	GaAs	144.65	5.318_4^{25}	1238		s HCl
chloride	$GaCl_3$	176.08	2.47	77.9	201.2	d aq; s bz, CCl_4, CS_2
fluoride	GaF_3	126.72	4.47	>1000	subl 950	0.004^{25} aq; s HF
nitrate	$Ga(NO_3)_3$	255.74		d 110	$\rightarrow Ga_2O_3$, 200	v s aq
phosphide	GaP	100.70		1465		
selenide	GaSe	148.68	5.03^{25}	960	d	
triethyl	$Ga(C_2H_5)_3$	146.90	1.058^{30}	−82.3	142.8	
trimethyl	$Ga(CH_3)_3$	114.84	1.151^{15}	−15.7	55.8	
Germanium	Ge	72.61	5.323	937.3	2830	i aq; s hot H_2SO_4
(IV) bromide	$GeBr_4$	392.23	3.132	26.1	186.4	hyd aq; s bz, eth
(IV) chloride	$GeCl_4$	214.42	1.879	−49.5	86.5	hyd aq; s bz, eth; sl s dil HCl
(IV) fluoride	GeF_4	148.60	6.521 g/L	−15	d >1000	hyd aq; s dil HCl
hydride (germane)	GeH_4	76.64	3.363 g/L	−164.8	−88.1	sl s hot HCl
(IV) oxide	GeO_2	104.61	4.25	1115	1200	0.43^{20} aq; s acids, alkalis
sulfide	GeS_2	136.74	3.01	530		
Gold	Au	196.967	19.3	1064.18	2856	s aq reg, KCN, hot H_2SO_4
(I) chloride	AuCl	232.42	7.57	289		s HCl, HBr, KCN
(III) chloride	$AuCl_3$	303.33	4.7	d >160	subl 180	68 g/100 mL^{20} aq; s EtOH
(I) cyanide	AuCN	222.99	7.14_4^{20}	d		s aq reg, KCN, NH_4OH
(III) cyanide 3-water	$Au(CN)_3 \cdot 3H_2O$	329.07		d 50		v s aq; sl s alc
diantimonide	$AuSb_2$	440.47		460		
(III) fluoride	AuF_3	253.96	6.75	subl 300		
(III) oxide	Au_2O_3	441.93		d 150	d 500	s HCl, KCN

(Continued)

(I) sodium thiosulfate 2-water	AuNa$_3$(S$_2$O$_3$)$_2$·2H$_2$O	526.24		anhyd 160		50 g/100 mL aq; i alc
stannide	AuSn	315.66	3.09	418		i aq; s Na$_2$S
(III) sulfide	Au$_2$S$_3$	490.13	8.754	d 197		s HF
Hafnium	Hf	178.49	13.31	2227	4450	
chloride	HfCl$_4$	320.30		432	subl 317	hyd aq; s acet, MeOH
oxide	HfO$_2$	210.49	9.68^{20}	2774		i aq
Helium	He	4.00260	0.176 g/L; 0.1249 (lq)	$-272.15^{25\,atm}$	-268.935	0.861 mL/100 mL^{20} aq
Holmium	Ho	164.9304	8.79	1474	2720	s acids; oxidizes in moist air
bromide	HoBr$_3$	404.64	4.86	914	1470	s aq
chloride	HoCl$_3$	271.29	3.7	718	1510	s aq
Hydrazine	H$_2$N—NH$_2$	32.05	1.0036_4^{25}	2.0	113.5	FP 52; misc aq, alc
hydrate	H$_2$N—NH$_2$·H$_2$O	50.06	1.030	-51.7 & -65	118–119	misc aq, alc; i chl, eth
Hydrazinium(1+) chloride	H$_2$N—NH$_3$Cl	68.51	1.5	89	d 240	v s aq; i org solv
(2+) chloride	ClH$_3$N—NH$_3$Cl	104.97	1.423	198	d 200	v s aq; sl s alc
(1+) iodide	H$_2$N—NH$_3$I	159.96	1.939^{15}	125	d 145	s aq
(+1) perchlorate	H$_2$N—NH$_3$ClO$_4$	132.51	1.378	137	d	d aq; s alc
(2+) sulfate	(H$_3$NNH$_3$)SO$_4$	130.13		254		3.4^{20} aq; i alc
(1+) tartrate	(H$_2$N—NH$_3$)$_2$C$_4$H$_4$O$_6$	182.13		183		6.0 g/100 mL^0 aq
Hydrogen	H$_2$	2.0159; 0.07099^{bp} (lq)	0.088 g/L	-259.35	-252.88	1.9 mL aq
amidosulfate (sulfamate)	H$_2$NSO$_3$H	97.09	2.126	205	d	14.7 g/100 mL aq; sl s alc, acet
azide	HN$_3$	43.03	1.126^0	-80	37	v s aq; (very explosive)
borate(1−) (cubic)	HBO$_2$	43.83	2.486	236	d 357	v sl s aq
borate(3−) (ortho)	H$_3$BO$_3$	61.83	1.435^{15}	171.0		5.56 g/100 mL^{30} aq
bromide	HBr	80.91	3.388 g/L^{20}	-86.87	-66.71	193 g/100 mL^{25} aq; misc alc
bromide (constant boiling)	48% HBr + H$_2$O		1.49		126	v s aq
bromide-d	^{2}HBr	81.91	3.39 g/L^{20}	-11	-66.5	v s aq
bromosulfate	HOSO$_2$Br	240.90		-87.46	d	hyd aq
chlorate (40% solution)	HClO$_3$	84.46	1.282_4^{20}	-6 to -8		72 g/100 mL^{20} aq
chloride	HCl	36.46	1.526 g/L^{20}	-114.18	-85.05	v s aq
chloride (constant boiling)	20.24% HCl + H$_2$O		1.097		110	v s aq
chloride-d	^{2}HCl	37.47	1.49 g/L^{25}	-114.64	-84.72	
chlorosulfate	HSO$_3$Cl	116.52	1.753	-80	152	hyd viol → HCl + H$_2$SO$_4$
cyanate	HOCN	43.03	1.140_4^{-20}	-86	23.5	s aq d; s bz, eth

TABLE 1.3 Physical Constants of Inorganic Compounds (*Continued*)

Name	Formula	Formula weight	Density	Melting point, °C	Boiling point, °C	Solubility in 100 parts solvent
cyanide	HCN	27.03	0.687	−13.4	25.6	misc aq
deuteride	$^1H^2H$ or HD	3.02		−256.56	−251.03	aq
diphosphate(IV)	$(HO)_2OP{-}PO(OH)_2$	162.01	70	d 100	d	709 g/100 mL[23] aq
diphosphate(V)	$H_4P_2O_7$	177.98		61		v s aq, alc; 2.54 g/100 g[5] bz
fluoride	HF	20.01	0.922 g/L[0]	−83.57	19.52	v s aq
fluoride (constant boiling)	35.35% HF + H_2O				120	s aq
fluoride-*d*	^2HF	21.02		−83.6	18.65	v s aq
fluoroborate	$H[BF_4]$	87.81		d 130		
fluorophosphate	H_2PO_3F	99.99	1.818	−80		s aq
fluorosulfate	$HOSO_2F$	100.07	1.726[25]	−87.3	165.5	60–70% aq solution
hexafluorosilicate 2-water	$H_2[SiF_6] \cdot 2H_2O$	180.11	1.463	19		269 g/100 mL[20] aq; s alc; i eth
iodate	HIO_3	175.91	4.629[0]	110→H_5IO_6	220→I_2O_5	234 g/100 mL[10] aq; misc alc
iodide	HI	127.91	5.37 g/L[20]	−50.8	−35.1	v s aq
iodide (constant boiling)	57% HI + H_2O		1.70		127	v s aq
iodide-*d*	HI	128.91	3.124[15]	−51.87	−35.7	v s aq
molybdate hydrate	$H_2MoO_4 \cdot H_2O$	179.97		−H O, 70		0.133[18] aq; s alk
nitrate	HNO_3	63.02	1.5492[0] lq	−41.59	83	v s
nitrate (constant boiling)	69% HNO_3 + H_2O		1.41[20]		120.5	misc aq
oxide (water)	H_2O	18.02	1.000	0.00	100.00	misc aq
oxide-d_2	D_2O or ^2H_2O	20.03	1.1044[25]	3.81	101.42	misc aq
perchlorate 2-water	$HClO_4 \cdot 2H_2O$	136.49	1.67[20]	−17.8	203	v s aq (commercial 72% acid)
periodate(1−) (meta)	HIO_4	191.91		subl 110	d 138	440 g/100 mL[25] aq
periodate(5−)	H_5IO_6	227.94		122	d 130–140	misc aq; s alc
peroxide	H_2O_2	34.01	1.463[0]	−0.43	152	misc aq; s alc, eth
peroxodisulfate	$HO_3S{-}O{-}OSO_3H$	194.14		d 60		v s aq
phosphate(V)(1−) (meta)	HPO_3	79.98	2.2–2.5	subl	red heat	slowly s aq → H_3PO_4; s alc
phosphate(V)(3−) (ortho)	H_3PO_4	98.00	1.868[25]	42.35	d 213	v s aq
commercial 85% acid			1.685	anhyd 150	$H_4P_2O_7$, 200	→ HPO_3, >300
phosphate(V)(3−)-d_3	$^2H_3PO_4$	101.03	1.908[25]	46.0		v s aq
phosphide, *see* Phosphine						
phosphinate	HPH_2O_2	66.0	1.493[19]	26.5	d 50	s aq
phosphonate (phosphorous acid)	H_2PHO_3	82.00	1.651[25]	≈73	d >180	v s aq, alc

Name	Formula		Density	mp	bp	Solubility
selenate	H_2SeO_4	144.98	2.9508^{15}	58	260	vs aq (viol)
selenide	H_2Se	80.98	2.12^{-bp}	−65.73	−41.4	9.5 mL/100 mL^{20} aq; s CS_2
sulfate	H_2SO_4	98.08	1.8318^{20}	10.38	335.5	misc aq
sulfate-d_2	$^2H_2SO_4$ or D_2SO_4	100.09	1.8620	14.35		misc aq
sulfide	H_2S	34.08	1.5392 g/L^0	−85.49	−60.33	0.334 mL^{25} aq
tellurate(IV)	H_2TeO_3	177.63	3.0	d to TeO_2		0.0007 aq; s acid, alkali
tellurate(VI) (monoclinic)	H_6TeO_6	229.66	3.068	−2H_2O, 120	320 → TeO	30 g/100 mL^{18} aq
telluride	H_2Te	129.62	5.687 g/L	−49	−2	s aq d
trithiocarbonate	$(HS)_2CS$	110.21	1.483^{34}	−26.9	57.8	d aq, alc
tungstate(VI)(2−)	H_2WO_4	249.86	5.5	anhyd 100		i aq; s HF, alkalis
Hydroxylamine	$HONH_2$	33.03	1.204^{40}	33	58^{22mm}	v s aq, MeOH; sl s bz, eth
Hydroxylammonium chloride	$HONH_3Cl$	69.49	1.680^{20}	150.5	d	g/100 mL: 83^{17} aq, 12.5^{20} MeOH, 5.1^{20} EtOH; s glyc
sulfate	$(HONH_3)_2SO_4$	164.14		170		69 g/100 mL^{20} aq
Indium	In	114.82	7.31	156.60	2072	s acids
antimonide	InSb	236.58	5.77	525		i aq
arsenide	InAs	189.74	5.67	942		
chloride	$InCl_3$	221.18	4.0	583	subl 500	212 g/100 mL^{25} aq
fluoride	InF_3	171.82	4.39	1170		0.040^{25} aq; s dilute acids
oxide	In_2O_3	277.63	7.179			s hot mineral acids
phosphide	InP	145.79	4.81	1062		v s acids
telluride	In_2Te_3	612.44	5.75	667		
trimethyl	$In(CH_3)_3$	159.93	1.568	88.4	135.8	d aq; s acet, bz
Iodine	I_2	253.809	4.63^{25}	113.60	185.24	g/100 mL^{25}: 0.029 aq, 14.1 bz, 16.5 CS_2, 21.4 EtOH, 25.2 eth, 2.6 CCl_4; s chl, HOAc
heptafluoride	IF_7	259.89	lq: 2.8^6	6.45	4.77 subl	s aq (d), s NaOH
monobromide	IBr	206.81	4.416	40	116 d	s aq, alc, eth, CS_2
monochloride	ICl	162.36	3.10^{29}	27.2 α-form	97 d	d aq; s alc, eth, HOAc
pentafluoride	IF_5	221.90	3.19^{25}	9.43	100.5	d aq viol
pentoxide	I_2O_5	333.81	4.98	d 275		187 g/100 mL^{13} aq
trichloride	ICl_3	233.26	3.202^{-4}	~33	64 subl	d aq; s alc, bz, HCl
Iridium	Ir	192.217	22.65^{20}	2447	~2550	s K_2SO_4 fusion, KOH + KNO_3 fusion
hexafluoride	IrF_6	306.21	4.82	44.4	53.6	d aq
(III) oxide	Ir_2O_3	432.43		d ~1000 to Ir + O_2		s boiling HCl
(IV) oxide	IrO_2	224.22	11.7	d 1100		0.0002^{20} aq; s HCl
trichloride	$IrCl_3$	298.58	5.30	d 763		i acids, alkalis

(Continued)

TABLE 1.3 Physical Constants of Inorganic Compounds (*Continued*)

Name	Formula	Formula weight	Density	Melting point, °C	Boiling point, °C	Solubility in 100 parts solvent
Iron	Fe	55.845	7.86	1535	2861	i aq; s acids
(III) arsenate 2-water	$FeAsO_4 \cdot 2H_2O$	2?0.79	3.18	1020		v sl s aq; s acids
(II) bromide	$FeBr_2$	126.75	3.16	677	1023	117 g/100 mL²⁰ aq; v s alc
(III) bromide	$FeBr_3$	295.67	4.5	d		s aq, alc, eth, HOAc
(*tri-*) carbide	Fe_3C	179.55	7.694	1227		s acids
(II) carbonate	$FeCO_3$	115.85	3.9	d		0.072¹⁸ aq; s acids
(II) chloride	$FeCl_2$	126.75	3.16	677	1024	62.5 g/100 mL²⁰ aq; v s alc, acet
(III) chloride	$FeCl_3$	162.20	2.898	304	≈316	74 g/100 mL⁰ aq; s alc, acet, eth
disulfide (pyrite)	FeS_2	119.98	5.02	d 602		s acids d
(II) fluoride	FeF_2	93.84	4.09	1100		sl s aq; s dil HF; i alc, bz, eth
(III) fluoride	FeF_3	112.84	3.87	subl 1000	1837	0.091²⁵ aq; s HF
(III) hexacyanoferrate(II)	$Fe_4[Fe(CN)_6]_3$	859.23	1.80	250 d		i aq; s HCl
(II) hydroxide	$Fe(OH)_2$	89.86	3.4			0.006 aq; s acids
(III) hydroxide oxide	$FeO(OH)$	88.85	4.26	anhyd 136		i aq, alc; s HCl
(II) iodide	FeI_2	309.65	5.315	587	1093	s aq
(III) nitrate 9-water	$Fe(NO_3)_3 \cdot 9H_2O$	404.00	1.684	47	d 100	138 g/100 mL²⁰ aq
(*di-*) nitride	Fe_2N	125.70	6.35	d 200		sHCl
(II) oxalate 2-water	$FeC_2O_4 \cdot 2H_2O$	179.89	2.28	d 150		0.044¹⁸ aq; s mineral acids
(II) oxide	FeO	71.84	6.0	1377	d 3414	i aq; s acids
(II,III) oxide	Fe_3O_4	231.53	5.17	1597		i aq; s acids
(III) oxide	Fe_2O_3	159.69	5.25	1565		i aq; s HCl
pentacarbonyl	$Fe(CO)_5$	195.90	1.49	−20.0	103.9	FP −20; i aq; s alc, bz, eth
(II) phosphate 8-water	$Fe_3(PO_4)_2 \cdot 8H_2O$	501.60	2.58			i aq; s acids
phosphide	Fe_2P	142.66	6.85	1370		s hot mineral acids
(II) selenide	$FeSe$	134.81	6.78	d		sHCl
(II) silicate(2−)	$FeSiO_3$	131.93	3.5	1140		
(II) silicate(4−)	Fe_2SiO_4	203.77	4.30	1220		d HCl
(II) sulfate 7-water	$FeSO_4 \cdot 7H_2O$	278.01	1.89	anhyd 300	d 671	48 g/100 mL²⁰ aq
(III) sulfate	$Fe_2(SO_4)_3$	399.88	3.097¹⁸	d 1178		slowly s aq (hyd); sl s alc
(II) sulfide	FeS	87.92	4.7	1190	d	0.0006¹⁸ aq; s acid
(III) thiocyanate	$Fe(SCN)_3$	230.09				v s aq
Krypton	Kr	83.80	3.7493 g/L	−157.36	−153.22	5.94 mL/100 mL²⁰ aq
difluoride	KrF_2	121.80	3.24	subl−60		s anhyd HF
Lanthanum	La	138.9055	6.162	920	3464	i aq; s HCl
chloride	$LaCl_3$	245.26	3.84	852	1812	v s aq
chloride 7-water	$LaCl_3 \cdot 7H_2O$	371.37		anhyd 852 (in HCl atm)		v s aq; s alc

fluoride	LaF$_3$	195.90	5.9	1493	2327	181 g/100 mL^{20} aq; v s alc
nitrate 6-water	La(NO$_3$)$_3$ · 6H$_2$O	433.01		40	d 126	s acids
oxide	La$_2$O$_3$	325.81	6.51	2305	4200	2.33 g/100 mL^{20} aq; i alc
sulfate	La$_2$(SO$_4$)$_3$	566.00	3.60	d white heat		2.92 g/100 mL^{20} aq; i alc
sulfate 9-water	La$_2$(SO$_4$)$_3$ · 9H$_2$O	728.14	2.821	anhyd 400		
Lawrencium	Lr	262		1627		
Lead	Pb	207.2	11.34$^{20}_4$ (fcc)	327.43	1749	s hot conc HNO$_3$, HCl, H$_2$SO$_4$
(II) acetate 3-water	Pb(C$_2$H$_3$O$_2$)$_2$ · 3H$_2$O	427.3	2.55	75	d >200	g/100 mL: 63^{15} aq, 3.3 alc
(IV) acetate	Pb(C$_2$H$_3$O$_2$)$_4$	443.4	2.228	≈75–180		s hot HOAc, bz, chl, conc HX acids
(II) azide	Pb(N$_3$)$_2$	291.2	4.7	expl 350 or when shocked		0.023^{18} aq; v s HOAc
(II) borate(1−) hydrate	Pb(BO$_2$)$_2$ · H$_2$O	310.8	5.598 anhyd	anhyd 160	mp 500	s acids
(II) bromide	PbBr$_2$	367.0	6.69	371	912	0.450^0 aq; s acids; i alc
(II) carbonate	PbCO$_3$	267.2	6.61	d 340 → PbO		i aq; s acids, alkalis
(II) chlorate	Pb(ClO$_3$)$_2$	374.1	3.89	d 230		140 g/100 mL^{18} aq; v s alc
(II) chloride	PbCl$_2$	278.1	5.98	501	950	0.99^{20} aq
(II) chloride fluoride	PbClF	261.7	7.05			
(II) chromate(VI)(2−)	PbCrO$_4$	323.2	6.12	844	d	i aq; s dil HNO$_3$, alkalis
(II) fluoride	PbF$_2$	245.2	8.445	830	1297	0.064^{20} aq
(IV) fluoride	PbF$_4$	283.2	6.7	≈600		hyd aq
(II) formate	Pb(CHO$_2$)$_2$	297.2	4.63	d 190		1.6 g/100 mL^{20} aq
(II) hydrogen arsenate	PbHAsO$_4$	347.1	5.94	d 280 to Pb$_2$As$_2$O$_7$		s HNO$_3$, alkalis
(II) hydroxide	Pb(OH)$_2$	241.2	7.59	d 145		0.016^{20} aq; s acids, alkalis
(II) iodide	PbI$_2$	461.0	6.16	410	872	0.063^{20} aq; s KI, Na$_2$S$_2$O$_3$, alkalis
(II) molybdate(VI)(2−)	PbMoO	367.1	6.7	1065		s acids, alkalis
(II) nitrate	Pb(NO$_3$)$_2$	331.2	4.53	470		g/100 mL: 56^{20} aq, 1.3 MeOH
(II) oleate	Pb(C$_{18}$H$_{33}$O$_2$)$_2$	770.1				s alc, bz, eth
(II) oxalate	PbC$_2$O$_4$	295.2	5.28	d 300		s acids, alkalis
(II) oxide (litharge)	PbO	223.2	9.35 (red)	886	1472 d	0.0017^{20} aq; s HNO$_3$
(IV) oxide	PbO$_2$	239.2	9.64	d 290, Pb$_3$O$_4$		s HCl, dil HNO$_3$ + H$_2$O$_2$$_3$, H$_2C_2O_4$
(II,IV) oxide (red lead)	Pb$_3$O$_4$	685.6	8.92	d 595 → PbO	d 595, PbO	s HNO$_3$, hot HCl
(II) phosphate	Pb$_3$(PO$_4$)$_2$	811.5	7.0	1014		s HNO$_3$, alkalis
(II) selenide	PbSe	286.2	8.15	1078		s HNO$_3$
(II) silicate(2−)	PbSiO$_3$	283.3	6.5	764		s acids
(II) silicate(4−)	Pb$_2$SiO$_4$	506.5	7.60	743		
(II) stearate	Pb(C$_{18}$H$_{35}$O$_2$)$_2$	774.2	1.4	≈125		0.05^{35} aq; s hot alc
(II) sulfate	PbSO$_4$	303.3	6.29	1170		0.00425 aq; s NaOH
(II) sulfide	PbS	239.3	7.60	1118	1300 subl	0.0006^{18} aq; s HNO$_3$, hot dil HCl

(Continued)

TABLE 1.3 Physical Constants of Inorganic Compounds (*Continued*)

Name	Formula	Formula weight	Density	Melting point, °C	Boiling point, °C	Solubility in 100 parts solvent
(II) telluride	PbTe	334.8	8.16	924	≈200	i acids and alkalis
tetraethyl	Pb(C₂H₅)₄	323.45	1.653	−137	110	i aq; s bz, hydrocarbons
tetramethyl	Pb(CH₃)₄	267.35	1.995	−30.2		s hydrocarbons
(II) thiocyanate	Pb(SCN)₂	323.4	3.82	d 190		0.44[18] aq, s HNO₃, NaOH
Lithium	Li	6.941	0.534²⁰	180.54	1341	d aq to LiOH
acetate 2-water	LiC₂H₃O₂ · 2H₂O	102.02	1.3	58	d	63 g/100 mL²⁰ aq; v s alc
aluminate(1−)	LiAlO₂	65.92	2.554	1700		
amide	LiNH₂	22.96	1.178	380	d 450 vacuo	d aq (→ LiOH + NH₃); i bz, eth
benzoate	LiC₆H₅O₂	128.06		>300		g/100 mL: 33 aq; 7.7 alc
borate(1−)	LiBO₂	49.75	2.18	849	1719	2.7 g/100 mL²⁰ aq; i alc
borohydride	Li[BH₄]	21.78	0.66	268	d 380	s aq, eth, THF, aliphatic amines
bromate	LiBrO₃	134.85	3.62			179 g/100 mL²⁰ aq
bromide	LiBr	86.84	3.464	552	1289	164 g/100 mL aq; s alc, eth
carbonate	Li₂CO₃	73.89	2.11	720	d 1300	1.3 g/100 mL²⁰ aq; i alc; s acids
chloride	LiCl	42.39	2.07	613	1360	77 g/100 mL²⁰ aq; s alc, acet
chromate(VI)(2−) 2-water	Li₂CrO₄ · 2H₂O	165.91	2.15	anhyd 75		142 g/100 mL¹⁸ aq; s EtOH
citrate 4-water	Li₃C₆H₅O₇ · 4H₂O	281.98		anhyd 105		61 g/100 mL¹⁵ aq; sl s alc
fluoride	LiF	25.94	2.640	848	1681	0.13²⁵ s acids
hexafluoroaluminate(3−)	Li₃[AlF₆]	161.79		1012		
hydride	LiH	7.95	0.76–0.77	680	d 950	no solvent known; flammable
hydride-d	Li²H or LiD	8.96	0.881	686		
hydroxide	LiOH	23.95	1.45	471.2	1626	12.4 g/100 mL²⁰ aq; sl s alc
iodate	LiIO₃	181.84	4.502	450		66 g/100 mL aq; in alc
iodide	LiI	133.84	4.061	469	1174	165 g/100 mL²⁰ aq & alc; v s acet
nitrate	LiNO₃	68.95	2.38	−255		50 g/100 mL²⁰ aq; s alc
nitride	Li₃N	34.83	1.27	813		d aq
oxide	Li₂O	29.88	2.013	1570	2563	forms LiOH in aq
perchlorate	LiClO₄	106.39	2.43	236	d ∼ 400 LiCl + O₂	47.4 g/100 mL²⁵ aq; v s organic solv
peroxide	Li₂O₂	45.88	2.31	d >195 to Li₂O		d dil HCl
silicate(2−)	Li₂SiO₃	89.97	2.52²⁵₄	1201		34.5 g/100 mL²⁰ aq; i alc
sulfate	Li₂SO₄	109.95	2.22	859		sl s aq
tetraborate(2−)	Li₂B₄O₇	169.12		917		
tetrahydridoaluminate	Li[AlH₄]	37.95	0.917	d 137		d aq, alc; g/100 mL: 30 eth, 13 THF; flammable
tetrahydridoborate	LiBH₄	21.79	0.666	268	d 380	s aq pH >7; s eth, THF
Lutetium	Lu	174.967	9.841	1663	3402	s acids
chloride	LuCl₃	281.33	3.98	892	subl >750	s aq
sulfate 8-water	Lu₂(SO₄)₃ · 8H₂O	782.25				42.3 g/100 mL²⁰ aq

Name	Formula	Mol. wt.	Density	m.p.	b.p.	Solubility
Magnesium	Mg	24.305		651	1100	i aq; s dilute acids
acetate	$Mg(C_2H_3O_2)_2$	142.00	1.738^{20}	323 d		53.4 g/100 mL^{20} aq; v s alc
aluminate(2−)	$MgAl_2O_4$	142.25	1.42	2135		v sl s HCl
amide	$Mg(NH_2)_2$	56.35	3.6	ign in air		d viol water giving NH_3
borate(1−) 8-water	$Mg(BO_2)_2 \cdot 8H_2O$	254.04	1.39^{25}_4			si s aq; s acids
bromide	$MgBr_2$	184.11	2.30	711 d	1158	101 g/100 mL^{20} aq
carbonate	$MgCO_3$	84.31	3.722	990		0.01 aq; s acids
chloride	$MgCl_2$	95.21	3.05	714	1412	54.6 g/100 mL^{20} aq
fluoride	MgF_2	62.30	2.33	1263	2270	0.013^{25} aq; s HNO_3
(*di*-) germanide	Mg_2Ge	121.22	3.148	1115		
hexafluorosilicate 6-water	$Mg[SiF_6] \cdot 6H_2O$	274.47	3.09	− SiF_4, 120		51 g/100 mL^{20} aq; i alc
hydride	MgH_2	26.32	1.788	d 200 vacuo	ign in air	d aq and alc violently
hydrogen phosphate 3-water	$MgHPO_4 \cdot 3H_2O$	174.33	1.45	anhyd 205	d 550	sl s aq; s acids
hydroxide	$Mg(OH)_2$	58.32	2.13^{15}	350 d		0.00125 aq; s acids
iodide	MgI_2	278.12	2.36	634	0	140 g/100 mL^{20} aq; s alc
lactate 3-water	$MgC_6H_{10}O_6 \cdot 3H_2O$	256.51	4.43			4 g/100 mL aq; sl s alc
mandelate	$MgC_{16}H_{14}O_6$	326.59				0.004^{100} aq; i alc
nitrate 6-water	$Mg(NO_3)_2 \cdot 6H_2O$	256.41	1.464	95	d 129	120 g/100 mL^{20} aq; v s alc
nitride	Mg_3N_2	100.93	2.712	d 270		s aq; s acids
oleate	$Mg(C_{18}H_{33}O_2)_2$	587.22				sl s alc, eth, PE
oxide	MgO	40.30	3.65–3.75	2800	3600	i aq, alc; s acids
perchlorate	$Mg(ClO_4)_2$	223.21	2.21	d >251		g/100 mL^{25}. 73 aq, 18 EtOH, 44.6 BuOH, 54 EtOAc, 32 acet
permanganate	$Mg(MnO_4)_2$	262.19	≈3.0	d 100		v s aq
peroxide	MgO_2	56.30				s acids
peroxoborate 7-water	$Mg(BO_3)_2 \cdot 7H_2O$	268.09		anhyd ~400		sl s aq d; s dilute acids
phosphate 5-water	$Mg_3(PO_4)_2 \cdot 5H_2O$	352.96	1.64^{15}	d 1557		0.02 aq; s acids
silicate(2−)	$MgSiO_3$	100.39	3.192^{25}_4	1898		i aq; v sl s HF
silicate(4−)	Mg_2SiO_4	140.69	3.21	1100		i aq; d hot HCl
(*di*-) silicide	Mg_2Si	76.70	2.0	1100		d aq, HCl
(*di*-) stamide	Mg_2Sn	167.32	3.60	778		s aq, HCl
sulfate 7-water	$MgSO_4 \cdot 7H_2O$	246.47	1.67	anhyd 250		27.2 g/100 mL aq; sl s alc
sulfate 6-water	$MgSO_4 \cdot 6H_2O$	212.46	1.725	anhyd 200		0.66^{25} aq
tungstate(VI)(2−)	$MgWO_4$	272.14	6.89			i aq; d acids
Manganese	Mn	54.9380	7.21^{20}	1244 fctetr	mp: 2227	d aq; s acids
acetate 4-water	$Mn(C_2H_3O_2)_2 \cdot 4H_2O$	245.09	1.589	80	2095	38 g/100 mL^{50} aq; v s alc

(*Continued*)

TABLE 1.3 Physical Constants of Inorganic Compounds (*Continued*)

Name	Formula	Formula weight	Density	Melting point, °C	Boiling point, °C	Solubility in 100 parts solvent
bromide	$MnBr_2$	214.75	4.39	698	1027	147 g/100 mL^{20} aq; s alc
(*tri-*) carbide	Mn_3C	176.83	6.89	1520		d aq; s acid
carbonate	$MnCO_3$	114.95	3.125	d >200		0.0065^{25} aq; s acids
chloride	$MnCl_2$	125.84	2.977	650	1210	74 g/100 mL^{20} aq; s alc, pyr; i eth
chloride 4-water	$MnCl_2 \cdot 4H_2O$	187.91	2.01	97.5	anhyd 198	143 g/100 mL aq; s alc; i eth
decacarbonyl	$Mn_2(CO)_{10}$	389.98	1.75	d 110		i aq; s organic solvents
diphosphate	$Mn_2P_2O_7$	283.82	3.707	1196		i aq; s acid
(II) fluoride	MnF_2	92.93	3.98	930	1820	0.66^{40} aq; s HF, conc HCl
(III) fluoride	MnF_3	111.93	3.54	d >600		hyd aq; s acid
hydroxide	$Mn(OH)_2$	88.95	3.258	d		0.002^{18} aq; s acids
iodide	MnI_2	308.75	5.04	638	1017	s aq
nitrate 6-water	$Mn(NO_3)_2 \cdot 6H_2O$	287.04	1.8	25.8		v s aq, alc
(II) oxide	MnO	70.94	5.37	1840		i aq; s acids
(III) oxide	Mn_2O_3	157.87	4.89	877 d		i aq; s HCl giving off Cl$_2$
(IV) oxide	MnO_2	86.94	5.08	$-O_2$, 530		s HCl; i HNO$_3$, cold H$_2$SO$_4$
(II,IV) oxide	Mn_3O_4	228.81	4.84	1567		i aq; s HCl
(VII) oxide	Mn_2O_7	221.87	2.396	ca. −20	ca. 25	explodes 85; v s aq
phosphinate hydrate	$Mn(PH_2O_2)_2 \cdot H_2O$	202.93		d to PH$_3$		15 g/100 mL aq; i alc
silicate, meta-	$MnSiO_3$	131.02	3.48	1290		i aq, HCl
sulfate	$MnSO_4$	151.00	3.25	700	d 850	52 g/100 mL aq; i alc
sulfate hydrate	$MnSO_4 \cdot H_2O$	169.02	2.95	anhyd 400–450		70 g/100 mL^{20} aq
sulfate 7-water	$MnSO_4 \cdot 7H_2O$	277.11	2.09	anhyd 280		115 g/100 mL^{20} aq
sulfide	MnS	87.00	3.99	1610		0.0006^{18} aq; s acids
titanate(IV)(2−)	Mn_2TiO_4	150.84	4.54	1360		
Mercury	Hg	200.59	13.534	−38.83	356.7	i aq; s HNO$_3$, hot conc H$_2$SO$_4$
(II) acetate	$Hg(C_2H_3O_2)_2$	318.68	3.28	178–180 d		g/100 mL: 40^{10} aq, 7.5^{15} MeOH
(II) benzoate	$Hg(C_7H_5O_2)_2$	424.83		165		v s NaCl soln; sl s alc
(I) bromide	Hg_2Br_2	560.99	7.307	subl 393 d		i aq, alc, eth; d hot HCl
(II) bromide	$HgBr_2$	360.40	6.05	237	322 subl	g/100 mL: 0.56^{20} aq; 20^{25} alc; v s HCl, HBr
(I) chloride	Hg_2Cl_2	472.09	7.16	subl 382	d without melting	s aqua regia; i aq, alc, eth
(II) chloride	$HgCl_2$	271.50	5.4	277	304	g/100 mL^{20}: 7.15 aq, 26 alc, 4 eth 8.3 glyc. 0.5 bz; s HOAc, EtAc
(II) cyanide	$Hg(CN)_2$	252.63	4.00	d 320		g/100 mL^{20}: 9.3 aq, 25 MeOH, 8 EtOH
(I) fluoride	Hg_2F_2	439.18	8.73	>570 d		hydrolyses in water

(II) fluoride	HgF₂	238.59	8.95	d 645	d >650	hyd aq; s HF
(II) fulminate	Hg(ONC)₂	284.62	4.42	explodes		si s aq; s alc; dangerously flammable
(I) iodide	Hg₂I₂	654.99	7.70	290 d	subl 140	i aq, alc, eth; s KI
(II) iodide	HgI₂	454.40	6.28	259	350 subl	g/100 mL: 0.006^{25} aq, 0.8 alc, 0.8 eth, 1.7 acet
(I) nitrate 2-water	Hg₂(NO₃)₂ · 2H₂O	561.22	4.79	70 d		hyd aq; s HNO₃
(II) nitrate	Hg(NO₃)₂	324.60	4.3	79	d	v s aq; s acet
(I) oxide	Hg₂O	417.18	9.8	d 100		i aq; s HNO₃
(II) oxide	HgO	216.59	11.14	d 500		0.005^{25} aq; s dil HCl, HNO, I⁻, CN⁻
(I) sulfate	Hg₂SO₄	497.24	7.56	d		0.06^{25} aq; s HNO₃
(II) sulfate	HgSO₄	296.65	6.47	d		d aq; s acid
(II) sulfide (cinnabar)	HgS	232.66	8.17	subl 583	→ blk HgO, 386	i aq; s aqua regia
(II) thiocyanate	Hg(SCN)₂	316.76	3.71	d 165		0.063^{25} aq; s HCl
Molybdenum	Mo	95.94	10.28	2622	4825	s hot H₂SO₄, HNO₃, fused KNO₃
(III) bromide	MoBr₃	335.65	4.89	subl 977		d alkalis
(IV) chloride	MoCl₄	237.75		317	407	s conc acids
(V) chloride	MoCl₅	273.19	2.928	194	268	s conc acids, dry eth, dry alc
(VI) fluoride	MoF₆	209.93	2.54	17.6	35.0	hyd aq; s alkalis; 31 g/100 g HF
hexacarbonyl	Mo(CO)₆	264.00	1.96	150 d	subl	s bz
(IV) oxide	MoO₂	127.94	6.47	d ≈1100		i aq
(VI) oxide	MoO₃	143.94	4.696^{26}_4	801	1155	0.05^{28} aq; s conc mineral acids, alk
(III) sulfide	Mo₂S₃	288.07	5.91^{15}	1807	d 1867	d hot HNO₃
(IV) sulfide	MoS₂	160.07	5.06^{15}	2375	subl 450	s aqua regia
Neodymium	Nd	144.24	7.01	1024	3074	s hot aq, acids
chloride	NdCl₃	250.60	4.134	760	1600	98 g/100 mL²⁰ aq; s alc
oxide	Nd₂O₃	336.48	7.28	1900		s dilute acids
sulfate 8-water	Nd₂(SO₄)₃ · 8H₂O	720.79	2.85	d 700–800		8.87 g/100 mL²⁰ aq
Neon	Ne	20.180	0.8999 g/L⁰	−248.67	−246.05	1.05 mL²⁰ aq
Neptunium	Np	237.0482	20.2	644	>3900	s HCl
(IV) oxide	NpO₂	269	11.1	2547		
Nickel	Ni	58.69	8.908^{20}	1453	2884	i aq; s HNO₃
acetate 4-water	Ni(C₂H₃O₂)₂ · 4H₂O	248.86	1.744	d		16 g/100 mL aq; s alc

(Continued)

TABLE 1.3 Physical Constants of Inorganic Compounds (*Continued*)

Name	Formula	Formula weight	Density	Melting point, °C	Boiling point, °C	Solubility in 100 parts solvent
acetylacetonate	$Ni(C_5H_7O_2)_2$	256.91	1.455^{17}	230	235^{11atm}	s aq, alc, bz, chl; i eth
bromide	$NiBr_2$	218.50	5.098	963	subl	100 g/100 mL^{20} aq
carbonate hydroxide (1/2)	$NiCO_3 \cdot 2Ni(OH)_2$	304.12	2.6			s dilute acids
carbonyl	$Ni(CO)_4$	170.73	1.31	−19.3	43 (expl 60)	s EtOH, bz, acet
chloride	$NiCl$	129.60	3.51	1009	subl 973	61 g/100 mL^{20} aq
chloride 6-water	$NiCl_2 \cdot 6H_2O$	237.69		anhyd 400		100 g/100 mL^{20} aq; s alc
cyanide 4-water	$Ni(CN)_2 \cdot 4H_2O$	182.79		subl 250		$0.006^{1.8}$ aq; s KCN, NH$_4$OH
dimethylglyoxime	$Ni(HC_2H_6N_2O_2)_2$	288.92				i aq; s abs alc, dilute acids
(tri-) disulfide	Ni_3S_2	240.21	5.87	790	d 2967	s HNO$_3$
fluoride	NiF_2	96.69	4.72	1450	1740	4 g/100 mL^{20} aq; i alc, eth
formate 2-water	$Ni(CHO_2)_2 \cdot 2H_2O$	184.78	2.154^{20}	anhyd 130	d 180–200	s aq; i alc
nitrate 6-water	$Ni(NO_3)_2 \cdot 6H_2O$	290.81	2.05	56.7	136.7	150 g/100 mL^{20} aq
(II) oxide	NiO	74.71	7.45	2000		s acids
(III) oxide	Ni_2O_3	165.42	4.83	−O$_2$, 600		s hot HCl, HNO$_3$, H$_2$SO$_4$
sulfate	$NiSO_4$	154.78	3.68	−SO$_3$, 840		29 g/100 mL^0 aq
sulfate 6-water	$NiSO_4 \cdot 6H_2O$	262.86	2.07	anhyd 280		40 g/100 mL^{20} aq
sulfide	NiS	90.77	5.3–5.6	976		s HNO$_3$, KHS
tetracarbonyl	$Ni(CO)_4$	170.74	1.3185^{17}	−19.3	42.3	explodes 63; FP −4; s organic solvents
Niobium	Nb	92.9064	8.57^{20}	2468	4860	s fused alkali hydroxides
(V) chloride	$NbCl_5$	270.20	2.75	206	247.0	s HCl, CCl$_4$
(V) fluoride	NbF_5	187.91	2.696^{80}_{4}	80.0	234.9	hyd aq, alc; sl s CS$_2$, CCl$_4$
(V) oxide	Nb_2O_5	265.82	4.55	1512		s HF, hot H$_2$SO$_4$
Nitrogen	N_2	28.0341	1.165 g/L^{20}	−210.01	−195.79	mL/100 mL: 1.6^{20} aq, 0.112 alc
	$^{15}N_2$	30.01	1.25 g/L^{20}	−209.952	−195.73	
(I) oxide	N_2O	44.02	1.843 g/L^{20}	−90.81	−88.46	130^0 mL aq; s alc, eth
(II) oxide	NO	30.01	1.249 g/L^{20}	−163.64	−151.76	4.6 mL/100 mL^{20} aq
(III) oxide	N_2O_3	76.02	1.447 g/L^2	−100.7	2	s eth
(IV) oxide dimer	N_2O_4	92.02	1.448^{20}_{4}	−9.3	21.15 d	s conc HNO$_3$, conc H$_2$SO$_4$, chl
(V) oxide	N_2O_5	108.01	2.05	30	47.0	v s chl; s CCl$_4$
selenide	N_4Se_4	371.87	4.2	explosive		si s bz, CS$_2$
sulfide	N_4S_4	184.28	2.24^{18}	180	185	s organic solvents
trichloride	NCl_3	120.37	1.653^{20}	−27	71	i aq; s bz, CS$_2$, CCl$_4$
trifluoride	NF_3	70.01	2.96 g/L^{-5}	−208.5	−129.06	hyd aq; s fuming H$_2$SO$_4$
Nitrosyl chloride	$NOCl$	65.47	1.592^{-5}	−61.5	−5.5	hyd aq
fluoride	NOF	49.01	2.788 g/L^{20}	−132.5	−59.9	d aq; s H$_2$SO$_4$
hydrogen sulfate	$NOHSO_4$	127.08		d 73.5		d aq
tetrafluoroborate	$NO[BF_4]$	116.83	2.185^{25}_{4}	subl $250^{0.01mm}$		d aq

Name	Formula		Density	mp	bp	Solubility
Nitryl chloride	NO_2Cl	81.46	2.81 g/L[100]	−145	−14.3	d aq
fluoride	NO_2F	65.00	2.7 g/L[20]	−166.0	−72.4	d aq
Osmium	Os	190.2	22.61[20]	3045	5225	s molten alkali or oxidizing fluxes
hexafluoride	OsF_6	304.2		32.1	45.9	hyd aq
tetrachloride	$OsCl_4$	332.0	4.38[4]	subl 450		slow hyd aq
tetraoxide	OsO_4	254.20	4.91	40.6	130.0	g/100 mL: 7.24[25] aq; 375[25] CCl_4; s bz, eth, alc
Oxygen	O_2	31.9988	1.331 g/L[20]	−218.4	−182.96	mL/100 mL[20]. 3.13 aq, 14.3 alc
difluoride	OF_2	54.00	2.26 g/L[20]	−223.8	−145.3	6.8 mL/100 mL[0] aq
(di-) difluoride	O_2F_2	70.00	1.45[bp] (lq)	−154	d − 100	
Ozone	O_3	48.00	1.998 g/L[20]	−192.5	−111.9	49.4 mL/100 mL[0] aq
Palladium	Pd	106.42	12.023[20]	1555	3167	s hot HNO_3, H_2SO_4
acetate	$Pd(C_2H_3O_2)_2$	224.49		205 d		i aq; s acet, chl, eth
chloride	$PdCl_2$	177.30	4.0[18]	680	d >680	s alc, acet, HCl
nitrate	$Pd(NO_3)_2$	230.42		879 d		s dil HNO_3
oxide	PdO	122.40	8.70[20]	d		s 48% HBr; sl s aqua regia
Perchloryl fluoride	ClO_3F	102.46	0.637 g/L	−147.74	−46.67	
Phosphorus (white)	P_4 molecules	123.8950	1.823[25]	44.15	280.3	g/100 mL: 2.86 bz, 2.50 chl, 1.25 CS_2; 0.025 abs alc, 1.0 eth
(red)	P_4 molecules	123.8950	2.34	597	subl 416	i aq; ignites in air, 260
hydride, *see* Phosphine						
pentabromide	PBr_5	430.56	3.46[20]	106 d	166 d	d aq; s CCl_4, CS_2
pentachloride	PCl_5	208.27	2.119[20]	subl 100	subl 100	hyd aq; s CCl_4, CS_2
pentafluoride	PF_5	125.98	5.805 g/L	−93.8	−84.6	hyd aq
pentoxide (dimer)	P_4O_{10}	283.88	2.30	340	subl 360	d aq; s H_2SO_4
pentasulfide	P_2S_5	222.29	2.09	288	514	hyd aq; s alkali; 0.222[17] CS_2
tribromide	PBr_3	270.73	2.85[15]	−41.5	173.2	d aq, alc; s acet, CS_2
trichloride	PCl_3	137.35	1.575[20]	−93.6	76.1	d aq, alc; s'bz, chl
trifluoride	PF_3	87.98	3.907 g/L	−151.30	−101.38	hyd aq
trioxide (dimer)	P_4O_6	219.90	2.136[4]	23.8	173 (N_2 atm)	hyd aq; s bz, CS_2
(tetra-) triselenide	P_4Se_3	360.80	1.31	245−246	360−400	flammable in air; s bz, acet, chl, CS_2
(tetra-) trisulfide	P_4S_3	220.09	2.03[17]	167	407	100 g/100 mL[17] CS_2; s tolune

(*Continued*)

TABLE 1.3 Physical Constants of Inorganic Compounds (*Continued*)

Name	Formula	Formula weight	Density	Melting point, °C	Boiling point, °C	Solubility in 100 parts solvent
Phosphine	PH_3	34.00	1.529 g/L	−133.81	−87.78	mL/100 mL^{17}: 1025 CS_2, 726 bz, 319 HO Ac, 26 aq; s alc, eth
Phosphonium iodide	PH_4I	161.91	2.86	18.5	subl 62.5	d aq
Phosphoryl chloride difluoride	$POClF_2$	120.43	1.656^0	−96.4	3.1	
dichloride fluoride	$POCl_2F$	136.89	1.5497^{20}	−80.1	52.90	s bz, CS_2, eth
tribromide	$POBr_3$	286.72	2.822	56	191.7 d	d aq, alc
trichloride	$POCl_3$	153.35	1.645^{25}	1.25	105	s aqua regia, fused alkali
Platinum	Pt	195.08	21.09^{20}	1769	3824	i aq, alc; s HCl, NH_4OH
(II) chloride	$PtCl_2$	266.00	5.87	d 581		143 g/100 mL^{25} aq
(IV) chloride	$PtCl_4$	336.90	4.303^{25}	d 370		
(VI) fluoride	PtF_6	309.08	3.826 (lq)	61.3	69.14	i aq; s HCl
(II) oxide	PtO	211.09	14.9^{15}	d 550		i aqua regia
(IV) oxide	PtO_2	227.09	10.2	450		s HCl, HNO_3
(IV) sulfide	PtS_2	259.22	7.66	d 225		i aq; s acids
Plutonium	Pu	239.052	19.816^{20}_4	639.5	3230	s aq
(III) bromide	$PuBr_3$	478.79	6.69	681	d >1300	i aq; s acids
(III) chloride	$PuCl_3$	345.42	5.70	760	1767	hyd aq
(III) fluoride	PuF_3	296.06	9.32	1425	d 2000	i aq
(IV) fluoride	PuF_4	315.05	7.00	1037 d		
(VI) fluoride	PuF_6	353.05	4.86	51.59	62.16	
(II) hydride	PuH_2	241.08	10.40	ca. 727		
(III) hydride	PuH_3	242.08	9.61	ca. 327		
(II) oxide	PuO	255.05	13.9	1900		
(III) oxide	Pu_2O_3	526.12	10.2	2085 (in He)		
(IV) oxide	PuO_2	271.05	11.46	2390 (in He)	d 2800	
(III) sulfide	Pu_2S_3	574.30	9.95	1727		
Polonium	Po	208.9824	9.196 alpha 9.398 beta	254	962	sl s aq; s acids
(IV) chloride	$PoCl_4$	350.79		300 (in Cl_2)	390 (in Cl_2)	sl hyd aq; v s HCl; s alc, acet
(IV) oxide	PoO	240.98	d 550			v s dilute HCl

Name	Formula					Solubility
Potassium	K	39.0983	0.89	63.38	759	d aq to KOH; s acids
acetate	$KC_2H_3O_2$	98.14	1.57	292		g/100 mL: 200 aq, 34 alc
arsenate	K_3AsO_4	256.21	2.8	1310		19 g/100 mL aq; slowly s glyc; s alc
borate(1−)	KBO_2	81.91		947	1401	71 g/100 mL[30] aq
bromate	$KBrO_3$	167.00	3.27	≈350	d 370	6.9 g/100 mL[20] aq
bromide	KBr	119.00	2.75	734	1435	g/100 mL: 65[20] aq, 22 glyc, 0.4 alc
carbonate	K_2CO_3	138.21	2.29	901	d to K_2O	90 g/100 mL[20] aq; i alc
chlorate	$KClO_3$	122.55	2.32	368	d >400	g/100 mL: 7.3[20] aq, 2 glyc
chloride	KCl	74.55	1.988	771	1437	g/100 mL: 34[20] aq, 7 glyc, 0.4 alc
chromate(VI)	K_2CrO_4	194.19	2.732	975		64 g/100 mL[20] aq; i alc
citrate hydrate	$K_3C_6H_5O_7 \cdot H_2O$	324.42	1.98	anhyd 180		g/100 mL: 154 aq, 40 glyc
cyanate	$KOCN$	81.11	2.05	d ≈700		s aq; sl s alc
cyanide	KCN	65.12	1.55	634	1625	g/100 mL: 50 aq, 50 glyc, 4 MeOH
dichromate(VI)	$K_2Cr_2O_7$	294.19	2.676^{25}	398	d 500	11.7 g/100 mL[20] aq
dicyanoargentate(I)	$K[Ag(CN)_2]$	199.01	2.36			25 g/100 mL[30] aq
dihydrogen arsenate	KH_2AsO_4	180.03	2.867	288		g/100 mL: 19[6] aq, 63 glyc; i alc
dihydrogen phosphate	KH_2PO_4	136.09	2.338	d 400 (KPO_3)		22.6 g/100 mL[20] aq; i alc
dioxide	KO_2	71.10	2.14	509		v s aq with decomposition
diphosphate(V) 3-water	$K_4P_2O_7 \cdot 3H_2O$	384.38	2.33	anhyd 300	d	s aq; i alc
disulfate(IV)	$K_2S_2O_5$	222.32				s aq; flammable if ground
disulfate(VI) (pyrosulfate)	$K_2S_2O_7$	254.32	2.28	≈325	mp: 1090	s aq
ethyldithiocarbonate	$KOCSSC_2H_5$	160.30	1.558	d 200		v s aq
fluoride	KF	58.10	2.48	859.9	1505	95 g/100 mL[20] aq
formate	$KCHO$	84.12	1.91	167.5	d >mp	250 g/100 mL aq
gluconate	$KC_6H_{11}O_7$	234.25		d 180		v s aq; i alc, bz, chl
heptaiodobis-muthate(III)(4−)	$K_4[BiI_7]$	1253.82				d aq; s alkali iodide solutions
hexachloroplatmate(IV)	$K_2[PtCl_6]$	485.99	3.50	d 250		0.48^{20} aq
hexacyanoferrate(II) 3-water	$K_4[Fe(CN)_6] \cdot 3H_2O$	422.39	1.85	anhyd 100	d	28 g/100 mL[20] aq
hexacyanoferrate(III)	$K_3[Fe(CN)_6]$	329.25	1.89	d		40 g/100 mL[20] aq (slow); sl s alc
hexafluorosilicate	$K_2[SiF_6]$	220.27	2.27	d		sl s aq; i alc
hexafluorozirconate	$K_2[ZrF_6]$	283.41	3.58			2.7 g/100 mL[20] aq; i alc
hexanitritocobaltate(III) 1.5-water	$K_3[Co(NO_2)_6] \cdot 1.5H_2O$	479.30		d 200		0.089^{18} aq; s HOAc; v sl s alc
hydride	KH	40.11	1.43	417 d		d aq

(Continued)

TABLE 1.3 Physical Constants of Inorganic Compounds (*Continued*)

Name	Formula	Formula weight	Density	Melting point, °C	Boiling point, °C	Solubility in 100 parts solvent
hydrogen carbonate	$KHCO_3$	100.11	2.17	d >100		34 g/100 mL[20] aq; i alc
hydrogen difluoride	KHF_2	78.10	2.37	238.80	d 477	39 g/100 mL[20] aq; s alc
hydrogen phosphate	K_2HPO_4	174.18		d to $K_2P_2O_7$	d to $K_2P_2O_7$	150 g/100 mL aq
hydrogen phthalate	$KHC_8H_4O_4$	204.22	1.636	d		8.3 g/100 mL aq; sl s alc
hydrogen sulfate	$KHSO_4$	136.17	2.24	197	d to $K_2S_2O_7$	48 g/100 mL[20] aq
hydrogen sulfide	KHS	72.17	1.70	≈455		s aq, alc
hydrogen tartrate	$KHC_4H_4O_6$	188.18	1.956			0.5[20] aq; s acids; v sl s alc
hydroxide	KOH	56.11	2.044	406	1323	g/100 mL: 112[20] aq, 33 alc, 40 glyc
iodate	KIO_3	214.00	3.89	560 d		8.1 g/100 mL[20] aq; i alc
iodide	KI	166.00	3.12	681	1345	g/100 mL: 144[20] aq, 4.5 alc, 50 glyc aq; stable in KOH
manganate(VI)	K_2MnO_4	197.13		190 d		s aq
molybdate(VI)	K_2MoO_4	238.14	2.3	919	d 1400	160 g/100 mL aq
nitrate	KNO_3	101.10	2.11	333	d 400	g/100 mL: 32[20] aq, 0.16 alc, s glyc
nitrite	KNO_2	85.10	1.915	441	d 350	306 g/100 mL[20] aq; sl s alc
oxalate hydrate	$K_2C_2O_4 \cdot H_2O$	184.23	2.13	anhyd 160	d to K_2CO_3	36 g/100 mL[20] aq
oxide	K_2O	94.20	2.35	350 d		d aq to KOH, s alc
oxobisoxalatodiaquatitanate(IV)	$K_2[TiO(C_2O_4)_2(H_2O)_2]$	354.18				v s aq
perchlorate	$KClO_4$	138.55	2.52	d 400		2.04[25] aq; 0.0036[25] BuOH; 0.0013 EtOAc
periodate	KIO_4	230.010	3.618	582		0.42[20] aq, sl s KOH
permanganate	$KMnO_4$	158.03	2.7	d 240 → O_2		6.34 g/100 mL[20] aq; d HCl
peroxide	K_2O_2	110.20		490		d aq
peroxodicarbonate hydrate	$K_2C_2O_6 \cdot H_2O$	216.24		d 100		6.5 g/100 mL aq; d hot aq
peroxodisulfate	$K_2S_2O_8$	270.32	2.48			2.5 g/100 mL[20] aq; i alc
perrhenate	$KReO_4$	289.30	4.38	555	1370	0.99[20] aq
phenolsulfonate hydrate	$KC_6H_4(OH)SO_3 \cdot H_2O$	240.28	1.87			s aq, alc
phosphate	K_3PO_4	212.27	2.564_4^{17}	1340		50.8 g/100 mL[20] aq; i alc
selenocyanate	$KSeCN$	144.08		d 100		s aq
silicate(2−)	K_2SiO_3	154.29		976		s aq
sodium hexanitritocobaltate(III) hydrate	$K_2Na[Co(NO_2)_6] \cdot H_2O$	454.18	1.633	d 135		0.07 aq
sodium tartrate 4-water	$KNaC_4H_4O_6 \cdot 4H_2O$	282.23	1.790	70–80	anhyd 130–140	54 g/100 mL[15] aq
sorbate	$KC_6H_7O_2$	150.22	1.363_{20}^{25}	d >270		g/100 mL: 58.2[20] aq, 6.5 alc
stannate(IV) 3-water	$K_2SnO_3 \cdot 3H_2O$	298.94	3.197	anhyd 140		100 g/100 mL[20] aq; i alc

Name	Formula	Molar mass	Density	mp	bp	Solubility
stearate	$KOOCC_{17}H_{35}$	322.57				readily soluble hot aq or alc
sulfate	K_2SO_4	174.26	2.66	1069	1670	g/100 mL: 11[20] aq, 1.3 glyc, i alc
sulfide	K_2S	110.26	1.74	948		
sulfite 2-water	$K_2SO_3 \cdot 2H_2O$	194.29		d		28.6 g/100 mL[20] aq
tartrate hemihydrate	$K_2C_4H_4O_6 \cdot 0.5H_2O$	235.28	1.98	anhyd 155		138 g/100 mL[20] aq
tellurate(IV)	K_2TeO_3	253.79				s aq
tetrachloroaurate(III)	$K[AuCl_4]$	377.88	3.75	d 357		61.8 g/100 mL[20] aq
tetrafluoroborate	$K[BF_4]$	125.90	2.505[20]	530		0.45[20] aq
tetrahydridoborate	$K[BH_4]$	53.94	1.11	d 497		g/100 mL: 21[25] aq, 3.5[20] MeOH
tetraiodocadmate 2-water	$K_4[CdI_4] \cdot 2H_2O$	698.21	3.359[4]			g/100 mL: 137[15] aq, 71[15] alc, 4 eth
tetraiodomercurate(II)	$K_2[HgI_4]$	786.48				v s aq; s alc, acet, eth
thiocyanate	KSCN	97.18	1.89	173		g/100 mL: 217[20] aq, 200 acet, 8 alc
thiosulfate	$K_2S_2O_3$	190.33		d 400		155 g/100 mL[20] aq; i alc
trihydrogen bisoxalate 2-water	$KH_3(C_2O_4)_2 \cdot 2H_2O$	254.20	1.836	d		1.8 aq
trisoxalatoantimonate(III)	$K_3[Sb(C_2O_4)_3]$	503.12				a aq
trithiocarbonate	K_2CS_3	186.41				v s aq
uranyl(VI) acetate hydrate	$K(UO_2)(C_2H_3O_2)_2 \cdot H_2O$	504.28	3.296[15]	d		s aq
Praseodymium	Pr	140.9077	6.475 α-form	anhyd 275	3520	s hot water and acids
chloride	$PrCl_3$	247.27	4.0	935	1710	104 g/100 mL[13] aq; s alc
(III) oxide	Pr_2O_3	329.81	7.07	769–782 oxidizes to Pr_6O_{11}		i aq; s acids
(IV)	PrO_2	172.91	6.82	tr 350 to Pr_6O_{11}		
Promethium-147	Pm	146.915	7.22	1080	3000 est	
bromide	$PmBr_3$	386.7	5.38	727	1667	s aq
chloride	$PmCl_3$	153.4		737	1670	s aq
Protactinium	Pa	231.0359	15.37	1568(8)	4227	
(IV) chloride	$PaCl_4$	372.85	4.72	subl 400		i aq; s HCl
(V) chloride	$PaCl_5$	408.31	3.74	301		hyd aq; s THF, CH_3CN
Radium	Ra	226.03	5.5	700.1	420	d aq; s acids
bromide	$RaBr_2$	385.88	5.79	728	1737	s aq
chloride	$RaCl_2$	296.93	4.91	1000	subl 900	s aq
Radon	Rn	222.0	9.73 g/L	−71	−62	23 mL/100 mL[20] aq; s org solv
Rhenium	Re	186.207	21.02	3180	5678	s HNO_3

(Continued)

TABLE 1.3 Physical Constants of Inorganic Compounds (*Continued*)

Name	Formula	Formula weight	Density	Melting point, °C	Boiling point, °C	Solubility in 100 parts solvent
chloride trioxide	ReClO$_3$	269.66	5.38	4.5	128	hyd in water to HReO$_4$; s CCl$_4$
(IV) fluoride	ReF$_4$	262.20		124.5	795	hyd aq
(VI) fluoride	ReF$_6$	300.20	3.58	18.5	33.8	52.5 g/100 mL anhyd HF; s HNO$_3$
(VII) fluoride	ReF$_7$	319.20	3.65	48.3	73.7	hyd aq
(VI) oxide	ReO$_3$	234.20	6.9–7.4	disprop 400	750	s HNO$_3$
(VII) oxide	Re$_2$O$_7$	484.41	6.1	300.3	360.3	v s aq, org solv
(VII) sulfide	Re$_2$S$_7$	596.88	4.866	d 460		i aq; s HNO$_3$
(VII) tetrachloride oxide	ReCl$_4$O	344.02	3.309	29.3	225	hyd aq; s CCl$_4$
Rhodium	Rh	102.9055	12.41^{20}	1963	3727	s fused KHSO$_4$
(III) chloride	RhCl$_3$	209.26	5.38	d 450		i aq; s KOH, KCN
(III) fluoride	RhF$_3$	159.90	5.4	subl 600		i acids, alkalis
(III) oxide	Rh$_2$O$_3$	253.81	8.20	d 1100		i aq reg, KOH
tetracarbonyldi-μ-chlorodichloride	Rh$_2$(CO)$_4$Cl$_2$	388.76		124–125		s org solv except hydrocarbons
Rubidium	Rb	85.4678	1.532	39.31	691	d aq to RbOH
acetate	RbC$_2$H$_3$O$_2$	144.52		246		86 g/100 mL^{45} aq
bromide	RbBr	165.37	3.35	682	1346	108 g/100 mL^{20} aq
carbonate	Rb$_2$CO$_3$	230.95		837	d 900	g/100 mL: 450^{20} aq, 0.74$_{19}$ alc
chlorate	RbClO$_3$	168.94	3.184	342		5.4 g/100 mL^{20} aq
chloride	RbCl	120.92	2.76	715	1390	g/100 mL: 91^{20} aq, 1.1 MeOH
dihydrogen phosphate	RbH$_2$PO$_4$	182.47		840		s aq
fluoride	RbF	104.47	3.2	833	1410	131 g/100 mL^{18} aq
hexachloroplatinate(IV)	Rb$_2$[PtCl$_6$]	578.75	3.94	d		0.028^{20} aq
hydroxide	RbOH	102.47	3.20	301		180 g/100 mL^{18} aq; s alc
iodide	RbI	212.37	3.55	642	1304	163 g/100 mL^{25} aq; s alc
nitrate	RbNO$_3$	147.47	3.11	305		19.5 g/100 mL^{20} aq
oxide	Rb$_2$O	186.93	4.0	400 d		s aq → RbOH
sulfate	Rb$_2$SO$_4$	267.00	3.5	1050		48 g/100 mL^{20} aq
Ruthenium	Ru	101.07	12.45$^{20}_4$	2334	4150	s fused alkali, oxidizing fluxes
(III) chloride (hexagonal)	RuCl$_3$	207.43	3.11	d > 500		i aq; s HCl, alc
(V) fluoride	RuF$_5$	196.06	3.90	86.5	227	d aq
(IV) oxide	RuO$_2$	133.07	6.97	d		i aq; s fused alkali
Samarium	Sm	150.36	7.52	1074	1794	s acids
(II) chloride	SmCl$_2$	221.27	3.687	855	2030	s aq dec; i alc
(III) chloride	SmCl$_3$	256.72	4.46	682	d	93.4 g/100 mL^{20} aq
(III) fluoride	SmF$_3$	207.36	6.643	1306	2427	i aq; s H$_2$SO$_4$
(III) oxide	Sm$_2$O$_3$	348.72	8.347	2335		s acids
(III) sulfate 8-water	Sm$_2$(SO$_4$)$_3$ · 8H$_2$O	733.03	2.93	anhyd 450		2.7 g/100 mL^{20} aq

Name	Formula	Mol. wt.	Density	m.p.	b.p.	Solubility
Scandium	Sc	44.956	2.985 hex	1541	2836	d aq
chloride	$ScCl_3$	151.31	2.39	967	967	v s aq; i alc
oxide	Sc_2O_3	137.91	3.864	2485		s hot or conc acids
sulfate 5-water	$Sc_2(SO_4)_3 \cdot 5H_2O$	468.17	2.519	anhyd 250	d 550	54.6 g/100 mL^{25} aq
Selenium (hexagonal)	Se	78.96	4.81_4^{20}	217	685	s eth, KOH, KCN; i aq, alc
(IV) bromide	$SeBr_4$	398.58	4.029	123		d aq; s HBr, chl, CS_2
(IV) chloride	$SeCl_4$	220.77	2.6	305	subl 196	d aq
(di-) dibromide	Se_2Br_2	317.73	3.604_4^{15}		225 d	d aq; s chl, CS_2
dibromide oxide	$SeBr_2O$	254.77	3.38^{50}	41.6	217 d	d aq
(di-) dichloride	Se_2Cl_2	228.83	2.774_4^{25}	−85	127 dec	d aq; s bz, chl, CS_2
dichloride oxide	$SeCl_2O$	165.867	2.44	8.5	177.2	d aq; misc bz, chl, CCl_4, CS_2
difluoride oxide	SeF_2O	132.96	2.8	15	125	d aq
(IV) fluoride	SeF_4	154.95	2.75	−10	106	reacts aq viol; misc alc, eth; s chl
(VI) fluoride	SeF_6	192.95	8.467 g/L	−34.6		s CS_2; 1.2 g/100 mL^{20} bz
(di-) hexasulfide	Se_2S_6	350.32	2.44	121.5		w/w %: 38^{14} aq, 10^{12} MeOH, 4.35 acet, 6.7^{14} EtOH, 1.1^{12} HOAc; s H_2SO_4
(IV) oxide	SeO_2	110.96	3.95	340	subl 315	i aq; 0.04 g/100 mL^{20} bz; s CS_2
(tetra-) tetrasulfide	Se_4S_4	444.10	3.20	113 d		d aq slowly; i alc, bz, chl, eth
Silane	SiH_4	32.12	1.409 g/L	−185	−111.9	d aq
chloro-	SiH_3Cl	66.56	2.921 g/L	−118	−30.4	d aq
dichloro-	SiH_2Cl_2	101.01	4.432 g/L	−122	8.3	d aq
iodo-	SiH_3I	158.01	2.035	−57	45.5	d aq
trichloro-	$SiHCl_3$	135.45	1.331	−128	33	d aq; s bz, chl
Silicon	Si	28.0855	2.33	1412	3265	s HF + HNO_3, fused alkali oxides
carbide (beta)	SiC	40.10	3.16	2830		s fused alkali oxides
dioxide (α quartz)	SiO_2	60.08	2.648	573 tr β quartz	2950	i aq; s HF
dioxide–tungsten trioxide–water (silicotungstic acid)	$SiO_2 \cdot 12WO_3 \cdot 26H_2O$	3310.66				v s aq, alc
disulfide	SiS_2	92.22	2.04	1090		s d aq, alc; i bz

(Continued)

TABLE 1.3 Physical Constants of Inorganic Compounds (*Continued*)

Name	Formula	Formula weight	Density	Melting point, °C	Boiling point, °C	Solubility in 100 parts solvent
tetrabromide	$SiBr_4$	347.70	2.81	5.2	154	hyd aq viol
tetrachloride	$SiCl_4$	169.90	1.5	-68.8	57.6	hyd aq; s bz, CCl_4, eth
tetrafluoride	SiF_4	104.08	4.567 g/L	-90.3	-86	hyd aq; s HF
tetraiodide	SiI_4	535.70	4.1	120.5	287.3	d aq; 2.2 g/100 mL[27] CS_2
(*tri-*) tetranitride	Si_3N_4	140.28	3.17	1878		i aq; s HF
Silver	Ag	107.8682	10.49	961.78	2164	s HNO_3
acetate	$AgC_2H_3O_2$	166.91	3.259	d		1.04^{20} aq; s dil HNO_3
antimonide	Ag_3Sb	445.35		559		
azide	AgN_3	149.89	4.9	exp ~-252		i aq; s KCN, HNO_3 (explosive)
bromide	AgBr	187.77	6.473	432	1500	i aq; s KCN
carbonate	Ag_2CO_3	275.75	6.077	218		0.003^{20} aq; s KCN, HNO_3, NH_4OH
chlorate	$AgClO_3$	191.32	4.430^{20}_4	231	d 270	10 g/100 mL[15] aq
chloride	AgCl	143.32	5.56	455	1547	i aq; 7.7 g/100 mL NH_4OH, KCN, $Na_2S_2O_3$
chromate(VI)	Ag_2CrO_4	331.73	5.625^{25}			0.002^{20} aq; s HNO_3, NH_4OH
cyanide	AgCN	133.89	3.95	320 d		i aq; s KCN
fluoride	AgF	126.87	5.852	435	≈1150	182 g/100 mL[20] aq; s HF, CH_3CN
(II) fluoride	AgF_2	145.87	4.57	690	d 700	hyd viol aq
iodate	$AgIO_3$	282.77	5.525^{20}	>200	d	0.053^{25} aq; 40 g/100 mL 10% NH_4OH
iodide (alpha)	AgI	234.77	5.683^{30}	558	1505	i aq; s KCN, KI, $(NH_4)_2CO_3$
nitrate	$AgNO_3$	169.87	4.352^{19}	212	d 440	g/100 mL: 216^{20} aq, 3.3 alc, 0.4 acet
nitrite	$AgNO_2$	153.87	4.453	d >140		0.33^{25} aq; d dilute acids
oxalate	$Ag_2C_2O_4$	303.76	5.03^4	explodes 140		0.004^{20} aq; s HNO_3, NH_4OH
oxide	Ag_2O	231.73	7.22^{25}_4	d 200 (d light)		0.002^{25} aq; s dil HNO_3, NH_4OH
(II) oxide	AgO	123.87	7.483^{25}_4	d >100		i aq; d alk and acids
perchlorate	$AgClO_4$	207.32	2.806^{25}	d 486		557 g/100 mL[20] aq; s bz, glyc, pyr
permanganate	$AgMnO_4$	226.80	4.49	d by light		0.9 aq; d alc
phosphate	Ag_3PO_4	418.62	6.37	849		0.006 aq; v s dil HNO_3, KCN, $(NH_4)_2CO_3$
selenate(IV)	Ag_2SeO_3	342.69	5.93	530	d > 530	sl s aq; s HNO_3
sulfate	Ag_2SO_4	311.80	5.45	660	d 1085	0.80^{20} aq (slow); s HNO_3, NH_4OH, H_2SO_4
sulfide (agentite)	Ag_2S	247.80	7.234^{20}_4	845	d	i aq; s HNO_3, alk CN's
Sodium	Na	22.98977	0.968^{20}	97.82	881.4	d aq to NaOH
acetate	$NaC_2H_3O_2$	82.03	1.528	324		75 g/100 mL[20] aq
acetate 3-water	$NaC_2H_3O_2 \cdot 3H_2O$	136.08	1.45	anhyd 120	d >120	g/100 mL: 125^{20} aq, 5.1 alc
aluminate(1−)	$NaAlO_2$	81.97	4.63	1650		v s aq; i alc

(Continued)

aluminum sulfate 12-water	$NaAl(SO_4)_2\cdot 12H_2O$	458.28	1.61	−60		110 g/100 mL^{15} aq; i alc
amide	$NaNH_2$	39.01	1.39	210	subl 400	d >500, reacts aq viol
ammonium phosphate 4-water	$NaNH_4HPO_4\cdot 4H_2O$	209.07	1.54	≈80	anhyd >280	14.3 g/100 mL aq
arsenate(III)(1−)	$NaAsO_2$	129.91	1.87	d 218		v s aq; sl s alc
ascorbate	$NaC_6H_7O_6$	198.11				62 g/100 mL^{20} aq
azide	NaN_3	65.01	1.846^{20}	d to Na + N$_2$		41 g/100 mL^{20} aq; 0.3 alc
benzoate	$NaO_2C_6H_5$	144.11		d		g/100 mL: 63^{25} aq; 1.3 alc
bismuthate(V)(1−)	$NaBiO_3$	279.96				i cold aq; dec by hot aq & acids
bismuthide	Na_3Bi	277.95	3.34	766		d aq
bromate	$NaBrO_3$	150.89	3.200$^{20}_{4}$	381 d		40 g/100 mL^{20} aq; i alc
bromide	$NaBr$	102.89	2.533^{20}	755	1390	g/100 mL: 90^{20} aq, 6 alc; 16 MeOH
carbonate	Na_2CO_3	105.99	2.25	858.1	d	29 g/100 mL^{20} aq; s glyc; i alc
carbonate hydrate	$Na_2CO_3\cdot H_2O$	124.00		anhyd 100		g/100 mL: 33 aq, 14 glyc; i alc
carbonate 10-water	$Na_2CO_3\cdot 10H_2O$	286.14	1.46	34 d		50 g/100 mL aq; s glyc
carbonate–hydrogen carbonate 2-water (trona)	$Na_2CO_3\cdot NaHCO_3\cdot 2H_2O$	226.02	2.112			13 g/100 mL^{0} aq
chlorate(V)	$NaClO_3$	106.44	2.5	248	d >300 → O$_2$	g/100 mL: 96^{20} aq, 0.77 alc, 25 glyc
chloride	$NaCl$	58.44	2.17	800.8	1465	g/100 mL: 36^{20} aq, 10 glyc
chlorite	$NaClO_2$	90.44		d 180–200		34 g/100 mL^{17} aq
chromate(VI)	Na_2CrO_4	161.97	2.72	792		84 g/100 mL^{20} aq
citrate 2-water	$Na_3C_6H_5O_7\cdot 2H_2O$	294.10		anhyd 150		77 g/100 mL^{25} aq; i alc
cyanate	$NaOCN$	65.01	1.89	550		s aq d; 0.22^{0} alc
cyanide	$NaCN$	49.01	1.6	563		58.7 g/100 mL^{20} aq
cyanohydridoborate	$Na[BH_3CN]$	62.84	1.12	>240 d		g/100 mL: 212 aq, 37.2 THF; v s
dichromate 2-water	$Na_2Cr_2O_7\cdot 2H_2O$	298.00	2.348$^{25}_{4}$	anhyd 100; mp 356	d 400	73.1 g/100 mL^{20} aq
diethyldithiocarbamate	$NaS_2CN(C_2H_5)_2\cdot 3H_2O$	225.31		anhyd 94–96		s aq, alc
dihydrogen arsenate(V) hydrate	$NaH_2AsO_4\cdot H_2O$	181.94	2.53	anhyd 130	d 200	s aq
dihydrogen diphosphate(V)	$Na_2H_2P_2O_7$	221.94	1.9	d 220	d NaPO$_3$, 200	4.5 g/100 mL^{0} aq
dihydrogen phosphate(V) dihydrate	$NaH_2PO_4\cdot 2H_2O$	156.01	1.91	anhyd 100		71 g/100 mL^{0} aq; i alc

TABLE 1.3 Physical Constants of Inorganic Compounds (*Continued*)

Name	Formula	Formula weight	Density	Melting point, °C	Boiling point, °C	Solubility in 100 parts solvent
dimethylarsonate 3-water (cacodylate)	$NaO_2As(CH_3)_2$	214.03		anhyd 120		g/100 mL: 200 aq, 40 alc
dioxide	NaO_2	54.99	2.53	552		2.26^0 aq
diphosphate(V)	$Na_4P_2O_7$	265.90		988		13.4 g/100 mL^{20} aq; i alc
dithionate(V) 2-water	$Na_2S_2O_6 \cdot 2H_2O$	242.14	2.19	anhyd 110	d 267 to $Na_2SO_4 + SO_2$	
dithionate(III)	$Na_2S_2O_4$	174.11		d		22 g/100 mL^{20} aq; sl s alc
diuranate(VI)	$Na_2U_2O_7$	634.03				i aq; s acids
dodecylbenzenesulfonate	$NaO_3SC_6H_4C_{12}H_{25}$	348.49				10 g/100 mL aq
dodecylsulfate	$NaO_3SOC_{12}H_{25}$	288.38		>300		d aq; s abs alc
ethoxide	$NaOC_2H_5$	68.06				103 g/100 mL aq
ethylenebis(imino-diacetate) (EDTA)	$(NaOOCCH_2)_2NC_2H_4\text{-}N(CH_2COONa)_2$	380.20				
ethylsulfate	$NaO_3SOC_2H_5$	148.12				140 g/100 mL aq; s alc
fluoride	NaF	41.99	2.78	996	1704	4 g/100 mL^{15} aq; i alc
formate	$NaHCO_2$	68.01	1.92	253	d >253	81 g/100 mL^{20} aq; s glyc; sl s alc
gluconate	$NaC_6H_{10}O_7$	218.14				59 g/100 mL^{25} aq; sl s alc; i eth
glycerophosphate	$Na_2C_3H_5(OH)_2PO_4$	216.04		d > 130		67 g/100 mL aq; i alc
hexachloroplatinate(IV) 6-water	$Na_2[PtCl_6] \cdot 6H_2O$	561.88	2.50	$-6H_2O$, 110		v s aq; s alc
hexacyanoferrate(II) 10-water	$Na_4[Fe(CN)_6] \cdot 10H_2O$	484.06	1.46	anhyd 82	d 435	28 g/100 mL^{20} aq
hexacyanoferrate(III) 1-water	$Na_3[Fe(CN)_6] \cdot H_2O$	298.93				18.9 g/100 mL^0 aq
hexafluoroaluminate	$Na_3[AlF_6]$	209.94	2.97	1009		s aq
hexanitritocobaltate(III)	$Na_3[Co(NO_2)_6]$	403.98	1.39	425 d		v s aq; sl s alc
hydride	NaH	24.00	1.87	anhyd 130		ign spontaneously moisture; d alc viol
hydrogen arsenate(V) 7-water	$Na_2HAsO_4 \cdot 7H_2O$	312.01			d 150	61 g/100 mL^{15} aq; s glyc; sl s alc
hydrogen carbonate	$NaHCO_3$	84.01	2.20	to Na_2CO_3	270	8 g/100 mL^{20} aq; i alc
hydrogen difluoride	$NaHF_2$	62.00	2.08	d > 160		3.7 g/100 mL^{20} aq
hydrogen phosphate 7-water	$Na_2HPO_4 \cdot 7H_2O$	268.07	1.7	d		25 g/100 mL^{40} aq; v sl s alc
hydrogen sulfate	$NaHSO_4$	120.06	2.435	315	d	50 g/100 mL^{20} aq; d alc
hydrogen sulfide	$NaHS$	56.06	1.79	350		s aq, alc, eth
hydrogen sulfite	$NaHSO_3$	104.06	1.48	d		g/100 mL: 29 aq, 1.4 alc

(Continued)

hydroxide	NaOH	40.00	2.130	323	1388	g/100 mL: 108^{20} aq, 14 abs alc, 24 MeOH; s glyc
hydroxymethanesulfinate dihydrate	$Na[HOCH_2SO_2] \cdot 2H_2O$	154.12		63–64	d >64	v s aq; i abs alc, bz, eth
hypochlorite 5-water	$NaClO \cdot 5H_2O$	164.52	1.6	18	d by CO$_2$ from air	29 g/100 mL^0 aq
iodate	$NaIO_3$	197.89	4.28	d		8.1 g/100 mL^{20} aq
iodide	NaI	149.89	3.67	660	1304	g/100 mL: 200^{20} aq, 100 glyc, 50 alc; s acet
lactate	$NaOOCCHOHCH_3$	112.06		d		misc aq, alc
methoxide	$NaOCH_3$	54.02		>300		d aq; s alc
molybdate(VI) 2-water	$Na_2MoO_4 \cdot 2H_2O$	241.95	≈3.5	anhyd 100	mp 687	65 g/100 mL^{20} aq
nitrate	$NaNO_3$	85.00	2.26	307	d ≈500	g/100 mL: 88^{20} aq, 0.8 alc
nitrite	$NaNO_2$	69.00	2.17	271	d > 320	67 g/100 mL^{20} aq
oxalate	$Na_2C_2O_4$	134.00	2.34	d ≈250		3.4 g/100 mL^{20} aq; i alc
oxide	Na_2O	61.98	2.27	dull red heat	d > 400	d aq to NaOH violently
pentacyanonitrosylferrate(III) 2-water (nitroprusside)	$Na_2[Fe(CN)_5NO] \cdot 2H_2O$	297.65	1.72			40 g/100 mL^{16} aq
perchlorate	$NaClO_4$	122.44	2.52	480 d		g/100 mL^{25}: 114 aq, 1.5 BuOH, 8.4 EtOAc
periodate	$NaIO_4$	213.89	3.865	d ≈300		10.3 g/100 mL^{20} aq
peroxide	Na_2O_2	77.98	2.805	675	d	v s aq (dec)
peroxoborate 4-water	$NaBO_3 \cdot 4H_2O$	153.88		d >60		2.5 g/100 mL aq
peroxodisulfate(VI)	$Na_2S_2O_8$	238.11		d		55 g/100 mL aq; d by alc
perrhenate	$NaReO_4$	273.19	5.24	300		33 g/100 mL^{20} aq
phosphate	Na_3PO_4	163.94	2.537	1340		12.1 g/100 mL^{20} aq
phosphate 12-water	$Na_3PO_4 \cdot 12H_2O$	380.12	1.62	73.4	$-11H_2O$, 100	28.3 g/100 mL^{20} aq; i alc
phosphinate hydrate	$NaPH_2O_2 \cdot H_2O$	105.99		anhyd 200	d to PH$_3$	100 g/100 mL^{20} aq; s glyc, alc
propanoate	$NaOOCC_2H_5$	96.06				g/100 mL^{25}: 100 aq, 4.1 alc
salicylate	$NaOOCC_6H_4OH$	160.10				g/100 mL: 110^{20} aq, 11 alc, 25 glyc
selenate(VI)	Na_2SeO_4	188.94	3.098			27 g/100 mL^{20} aq
silicate(2−) meta-	Na_2SiO_3	122.06	2.614	1089	anhyd 100	s aq; hyd by hot aq; i alc
silicate(2−) 5-water	$Na_2SiO_3 \cdot 5H_2O$	212.14	1.749	72.2		v s aq
silicate(4−)	Na_4SiO_4	184.04		1018		s aq

TABLE 1.3 Physical Constants of Inorganic Compounds (*Continued*)

Name	Formula	Formula weight	Density	Melting point, °C	Boiling point, °C	Solubility in 100 parts solvent
stannate(IV) 3-water	$Na_2SnO_3 \cdot 3H_2O$	266.71		d 140 (slow)		59 g/100 mL^{20} aq; i alc
stearate	$NaOOCC_{17}H_{35}$	306.47		d		sl s aq
sulfate	Na_2SO_4	142.04	2.7	8800		28 g/100 mL^{20} aq
sulfate 10-water	$Na_2SO_4 \cdot 10H_2O$	322.20	1.46	32.4	anhyd 100	67 g/100 mL^{25} aq; s glyc; i alc
sulfide	Na_2S	78.05	1.856	1172 vacuo		18.6 g/100 mL^{20} aq; sl s alc
sulfide 9-water	$Na_2S \cdot 9H_2O$	240.18	1.43	d ≈50		200 g/100 mL^{20} aq; sl s alc
sulfite	Na_2SO_3	126.04	2.63	d		31 g/100 mL^{20} aq; s glyc; i alc
tartrate dihydrate	$Na_2C_4H_4O_6 \cdot 2H_2O$	230.08	1.82	anhyd ~120		29 g/100 mL^6 aq; i alc
tetraborate	$Na_2B_4O_7$	201.22	2.4	742.5		2.6^{20} aq
tetraborate 10-water (borax)	$Na_2B_4O_7 \cdot 10H_2O$	381.37	1.73	75 d	anhyd 320	g/100 mL: 6.3 aq, 100 glyc
tetrachloroaluminate	$Na[AlCl_4]$	191.78	2.01	151		s aq
tetrachloroaurate	$Na[AuCl_4] \cdot 2H_2O$	397.80		d >100		166 g/100 mL^{27} aq; s alc, chl
tetrafluoroborate	$Na[BF_4]$	109.82	2.47	384	d	108 g/100 mL^{27} aq
tetrahydridoborate	$Na[BH_4]$	37.83	1.074	497	d 315	18^{25} DMF; 16.4^{20} MeOH (reacts)
thiocyanate	$NaSCN$	81.07		287		134 g/100 mL^{20} aq
thiosulfate	$Na_2S_2O_3$	158.11	2.345			s aq; i alc
thiosulfate 5-water	$Na_2S_2O_3 \cdot 5H_2O$	248.19	1.69	anhyd 100	d >100	70 g/100 mL^{20} aq (dec slowly)
trimetaphosphate 6-water	$(NaPO_3)_3 \cdot 6H_2O$	414.04	1.786	53	anhyd 100	22 g/100 mL aq; i alc
tungstate(VI) dihydrate	$Na_2WO_4 \cdot 2H_2O$	329.85	3.25	anhyd 100	mp: 695.6	88 g/100 mL^0 aq; i alc
vanadate(V)	$NaVO_3$	121.93				s hot aq
Strontium	Sr	87.62	2.64	757	1366	d to Sr(OH)$_2$ in water
bromide	$SrBr_2$	247.43	4.216	657	2045	100 g/100 mL^{20} aq
carbonate	$SrCO_3$	147.63	3.5	d 1100 to SrO + CO$_2$		i aq; s acids
chlorate	$Sr(ClO_3)_2$	254.52	3.152	120 d → O$_2$		167 g/100 mL^{20} aq
chloride	$SrCl_2$	158.53	3.052	874	1250	52.9 g/100 mL^{20} aq
chromate(VI)	$SrCrO_4$	203.61	3.89	d		0.12^{20} aq; s HCl
fluoride	SrF_2	125.62	4.24	1477	2460	0.011^{20} aq; s hot HCl
hydrogen phosphate	$SrHPO_4$	183.60	3.544			i aq; s acids
hydroxide	$Sr(OH)_2$	121.64	3.625	535	−H$_2$O, 744	0.8^{20} aq
iodate	$Sr(IO_3)_2$	437.43	5.045^{15}			0.03^{15} aq
iodide	SrI_2	341.43	4.42	402	1773 d	178 g/100 mL^{20} aq; s alc
lactate 3-water	$Sr(OOCCHOHCH_3)_2 \cdot 3H_2O$	319.81		anhyd 150		33 g/100 mL aq
nitrate	$Sr(NO_3)_2$	211.63	2.99	570	645	69.5 g/100 mL^{20} aq; sl s alc, acet
oxide	SrO	103.62	4.7	2430		0.69^{20} aq
perchlorate	$Sr(ClO_4)_2$	286.52	3.00^{25}			g/100 mL^{25}: 157 aq, 71 BuOH, 77 EtOAc, 90 acet

Name	Formula	mol wt	density	mp	bp	solubility
peroxide	SrO$_2$	119.62	4.78	215 d		0.018^{20} aq; d hot aq
sulfate	SrSO$_4$	183.68	3.96	1607		0.013^{20} aq; sl s acid
sulfide	SrS	119.69	3.70	2227		sl s aq; s acid (dec)
Sulfinyl bromide (Thionyl)	SOBr$_2$	207.87	2.688$_4^{20}$	−52	140	hyd aq (slow); misc bz, chl, CCl$_4$
chloride	SOCl$_2$	118.97	1.638	−104.5	76	hyd aq; misc bz, chl, CCl$_4$
fluoride	SOF$_2$	86.06	3.776 g/L	−129.5	−43.8	hyd aq; s bz, chl, eth
Sulfonyl chloride (Sulfuryl)	SO$_2$Cl$_2$	134.97	1.6674$_4^{20}$	−54.1	69.3	hyd aq; misc bz, eth, HOAc
diamide	SO$_2$(NH$_2$)$_2$	96.11	1.807	93	d 250	s aq, hot EtOH, acet
fluoride	SO$_2$F$_2$	102.06	4.478 g/L	−135.8	−55.38	mL gas/100 mL; 4 aq, 24 alc, 136 CCl$_4$, 210 toluene
Sulfur (gamma)	S	32.066	1.92	106.8	444.72	23 g/100 mL^0 CS$_2$; s alc, bz
(alpha) orthorhombic	S$_8$	256.53	2.08^{20}	tr 94.5 to beta form	444.6	i aq; s organic solvents
(beta) monoclinic tr slowly to rhombic	S$_8$	256.53	1.96	115.21	444.6	23 g/100 mL^0 CS ; s alc, bz
(di-) decafluoride	S$_2$F$_{10}$	254.11	2.08	−52.7	30	d fusion with KOH
(di-) dichloride	ClSSCl	135.04	1.688	−77	137	hyd aq; s alc, bz, eth, CS$_2$, CCl$_4$
dichloride	SCl$_2$	102.97	1.622	−122	59.5	hyd aq
dioxide	SO$_2$	64.07	2.811 g/L	−75.47	−10	mL/100 mL: 3937^{20} aq, 25 alc, 32 MeOH; s chl, eth
hexafluoride	SF$_6$	146.06	6.409 g/L	−50.8	subl −63.8	si s aq; s alc, KOH
tetrafluoride	SF$_4$	108.06	4.742 g/L	−121.0	−38	d aq viol; v s bz
trioxide (alpha)	SO$_3$	80.06		62.3	vp 73mm at 25	stable modification
(beta)	SO$_3$	80.06		32.5	vp 344mm at 25	
(gamma)	SO$_3$	80.06	1.92	16.8	44.8	v s aq (slow)
Sulfuryl, *see* Sulfonyl						
Tantalum	Ta	180.9479	16.69	2996	5429	s HF, fused alkali (slowly)
(V) bromide	TaBr$_5$	580.47	4.99	265	349	hyd aq; s abs alc, eth
carbide	TaC	192.96	14.3	3880	4780	si s HF
(di-) carbide	Ta$_2$C	373.91	15.1	3327		
(V) chloride	TaCl$_5$	358.21	3.68	216	239.3	hyd aq; s abs alc
diboride	TaB$_2$	202.57	11.2	3140		
(V) fluoride	TaF$_5$	275.94	4.74^{20}	96.8	229.5	s aq, eth, conc HNO$_3$
(V) iodide	TaI	815.47	5.80	496	543	hyd aq; s eth
nitride	TaN	194.95	13.7	3090		sl s aq reg; reacts alkalis
(V) oxide	Ta$_2$O$_5$	441.89	8.2	1785		s HF; d fused KHSO$_4$ or KOH
Technetium-98	Tc	97.9072	11	2157	4265	s HNO$_3$, aq reg, conc H$_2$SO$_4$
(VI) fluoride	TcF$_6$	212.91	3.0	37.4	55.3	sHCl
(IV) oxide	TcO$_2$	130.91	6.9	subl 1000		s acid, alkali
(VII) oxide	Tc$_2$O$_7$	309.81		119.5	310.6	s aq

(Continued)

TABLE 1.3 Physical Constants of Inorganic Compounds (*Continued*)

Name	Formula	Formula weight	Density	Melting point, °C	Boiling point, °C	Solubility in 100 parts solvent
Tellurium	Te	127.60	6.24	449.8	989.9	s HNO_3, KOH, conc H_2SO_4
(IV) bromide	$TeBr_4$	447.22	4.3	380	≈20 d	s HBr, eth, HOAc
(II) chloride	$TeCl_2$	198.51	6.9	208	328	disprop with eth, diox; s acid
(IV) chloride	$TeCl_4$	269.41	3.0	225	380	hyd aq; s HCl, abs alc, bz
(IV) fluoride	TeF_4	203.59		129	d > 195	d aq
(VI) fluoride	TeF_6	241.59	10.601 g/L	−37.68	subl − 38.9	hyd aq, KOH
(IV) iodide	TeI_4	635.22	5.05	280	1245	hyd aq; s HI, alkali; sl s acet
(IV) oxide	TeO_2	159.60	5.9	733		s HCl, HF, NaOH
Terbium	Tb	158.9254	8.23	1356	3230	s acids
chloride	$TbCl_3$	265.28	4.35	588	1550	v s aq
nitrate 6-water	$Tb(NO_3)_3 \cdot 6H_2O$	453.03		89.3		s aq
Thallium	Tl	204.383	11.85	303.5	1457	i aq; s HNO_3
(I) bromide	TlBr	284.29	7.5	460	820	0.05^{20} aq; s alc
(I) carbonate	Tl_2CO_3	468.78	7.11	272		4.1 g/100 mL^{20} aq; i alc
(I) chloride	TlCl	239.84	7.00	430	720	0.33^{20} aq; i alc
(I) cyanide	TlCN	230.40	6.523	d		16.8 g/100 mL^{28} aq; s alc, acid
(I) ethoxide	$TlOC_2H_5$	249.44	3.49	−3	d 130	s eth; sl s alc; d aq
(I) fluoride	TlF	223.38	8.36	326	826	78.6%^{15} aq
(III) fluoride	TlF_3	261.38	8.65	550 d		d aq
(I) iodide (rhombic)	TlI	331.29	7.1	442	823	i aq, alc; s KI
(I) nitrate	$TlNO_3$	266.39	5.55	206	d 450	9.55 g/100 mL^{20} aq; i alc
(I) oxide	Tl_2O	424.77	9.52	579	1080	v s aq; s acid, alc
(III) oxide (hexagonal)	Tl_2O_3	456.77	10.2	834	$-O_2$, 875	i aq; d by HCl, H_2SO_4
(I) selenate(VI)	Tl_2SeO_4	551.73	6.875	>400		2.8 g/100 mL^{20} aq; i alc, eth
(I) selenide	Tl_2Se	487.73	9.05	340		i aq, acid
(I) sulfate	Tl_2SO_4	504.83	6.77	632	d	4.87 g/100 mL^{20} aq
(I) sulfide	Tl_2S	440.83	8.39	448	1367	0.02^{20} aq; s mineral acids
Thiocarbonyl chloride	S=CCl	114.98	1.509^{15}	ca. −2	73.5	d aq; s eth
Thiocyanogen	$(SCN)_2$	116.16				d aq; s alc, CS_2, eth
Thionyl, *see* Sulfinyl						
Thiophosphoryl tribromide	$PSBr_3$	302.78	2.85^{17}	38.0	209 d	s aq, eth, CS_2
trichloride (alpha)	$PSCl_3$	169.41	1.635	−40.8	125	hyd aq; s bz, chl, CS_2
trifluoride	PSF_3	120.03		−148.8	−52.2	
Thiosulfinyl difluoride	$S=SF_2$	102.13		−165	−10.6	hyd aq
Thorium	Th	232.038	11.7	1750	4788	s acids
chloride	$ThCl_4$	373.85	4.59	770	921	s aq, alc
fluoride	ThF_4	308.03	6.1	1110	1680	s acids
iodide	ThI_4	739.66	6.00	570	837	hyd aq
nitrate	$Th(NO_3)_4$	400.06		d 630, ThO_2		191 g/100 mL^{20} aq; v s alc

Name	Formula		Density	mp	bp	Solubility
oxide 9-water	ThO$_2$·9H$_2$O	264.04	10.0	3390	4400	s hot H$_2$SO$_4$
sulfate 9-water	Th(SO$_4$)$_2$·9H$_2$O	586.30	2.77	anhyd 400		1.57 g/100 mL^{25} aq
Thullium	Tm	168.9342	9.32	1545	1950	s acids
chloride	TmCl$_3$	275.29		824	1490	s aq, alc
fluoride	TmF$_3$	225.93	7.971	1158	2230	s H$_2$SO$_4$
Tin (white)	Sn	118.710	7.265	231.928	2602	s conc HCl, hot H$_2$SO$_4$
(II) acetate	Sn(C$_2$H$_3$O$_2$)$_2$	236.80	2.31	182.5	240	d aq; s dilute HCl
(II) bromide	SnBr$_2$	278.52	5.12	215	639	85 g/100 mL^0 aq; s alc, eth
(IV) bromide	SnBr$_4$	438.33	3.34	31	205	v a (hyd) aq; s acet, alc
(II) chloride	SnCl$_2$	189.61	3.90	246.9	623	84 g/100 mL^0 aq; s acet, alc, eth
(IV) chloride	SnCl$_4$	260.52	2.234	−3.3	114.1	s aq (hyd), alc, acet, bz, eth
(II) fluoride	SnF$_2$	156.71	4.57	213	850	30% aq
(IV) fluoride	SnF$_4$	194.70	4.78		subl 705	hyd aq
hexafluorozirconate	Sn[ZrF$_6$]	323.92	4.21			s aq
(II) iodide	SnI$_2$	372.52	5.285	320	714	0.98^{20} aq (d); s bz, chl, alk Cl$^-$ or I$^-$
(IV) iodide	SnI$_4$	626.33	4.46	143	364	hyd aq; s alc, bz, chl, eth, CCl$_4$, CS$_2$
(II) oxalate	SnC$_2$O$_4$	206.73	3.56	280 d		s dilute HCl
(II) oxide	SnO	134.71	6.45	to SnO$_2$, 300		s acids, conc KOH
(IV) oxide	SnO$_2$	150.71	6.95	1630		s hot conc KOH (slow)
(II) selenide	SnSe	197.67	6.179	861		s aqua regia, alkali sulfides
(II) sulfate	SnSO$_4$	214.77	4.15	to SnO$_2$, 378		18.9 g/100 mL^{20} aq; s dilute H$_2$SO$_4$
(II) sulfide	SnS	150.78	5.08	880	1210	s conc HCl, hot conc H$_2$SO$_4$
(IV) sulfide	SnS$_2$	182.84	4.5	d 600		s aq reg, alkali hydroxides & sulfides
(II) telluride	SnTe	246.31	6.5	790		i aq
Titanium (hexagonal)	Ti	47.867	4.506	1668	3287	s hot acid, HF
(III) bromide	TiBr$_3$	287.58	4.24		subl 794	
(IV) bromide	TiBr$_4$	367.48	3.37	39	230	hyd aq; 187 g/100 mL abs alc
(II) chloride	TiCl$_2$	118.77	3.13	1035	1500	d aq; s alc
(III) chloride	TiCl$_3$	154.23	2.64	425 d		s aq (heat evolved), alc
(IV) chloride	TiCl$_4$	189.68	1.73	−25	136.4	s cold aq, alc
dihydride	TiH$_2$	49.88	3.752	d 450		
(IV) fluoride	TiF$_4$	123.86	2.798	>400	subl 285.5	s aq (slow hyd); s alc, pyr
(IV) iodide	TiI$_4$	555.49	4.3	150	377	s dry nonpolar solvents
(IV) isopropoxide	Ti[OCH(CH$_3$)$_2$]$_4$	284.22	0.9711^{20}	−20	220	d aq; s bz, chl, eth
(II) oxide	TiO	63.87	4.95	1750	3660	s H$_2$SO$_4$

(Continued)

TABLE 1.3 Physical Constants of Inorganic Compounds (*Continued*)

Name	Formula	Formula weight	Density	Melting point, °C	Boiling point, °C	Solubility in 100 parts solvent
(III) oxide	Ti_2O_3	143.73	4.486	1842		s H_2SO_4, hot HF
(IV) oxide (rutile)	TiO_2	79.87	4.23	1843		s HF, hot conc H_2SO_4
oxide sulfate	$TiOSO_4$	159.94				d aq
(III) sulfate	$Ti_2(SO_4)_3$	383.93				s dilute HCl, dilute H_2SO_4
Tungsten	W	183.84	19.25	3387	5900	s HNO_3 + HF, fusion NaOH + $NaNO_3$
(V) bromide	WBr_5	583.36		286	333	hyd aq; s chl, eth
(VI) bromide	WBr_6	663.26	6.9	309	subl 327	hyd aq; s eth CS_2
(V) chloride	WCl_5	361.10	3.875	242	286	hyd aq
(VI) chloride	WCl_6	396.56	3.52	279	347	hyd aq; s CS_2, CCl_4
dichloride dioxide	WCl_2O_2	286.74	4.67	265	d 369	hyd aq; s HCl
(VI) fluoride	WF_6	297.83	3.441	2.3	17.5	hyd aq; s anhyd HF
(IV) oxide	WO_2	215.84	10.8	1550	d 1724	s acids, KOH
(VI) oxide	WO_3	231.84	7.16	1472	1837	i aq; s hot alkali
(IV) sulfide	WS_2	247.97	7.6	d 1250		s HNO_3 + HF
tetrachloride oxide	WCl_4O	341.65	11.92	211	227	hyd aq
tetrafluoride oxide	WF_4O	275.83	5.07	106	186	hyd aq
Uranium	U	238.0289	19.1	1135	4131	s acid
(IV) bromide	UBr_4	557.65	5.55	519	777	v s aq
(III) chloride	UCl_3	344.39	5.51	837	1657	v s aq
(IV) chloride	UCl_4	379.84	4.725	590	790	v s aq (d); s polar org solvents
(V) chloride	UCl_5	415.29		287	527	d aq; s CS_2
(VI) chloride	UCl_6	450.75	3.6	177	392	hyd aq; s chl
(IV) fluoride	UF_4	314.02	6.70	1036	1417	s conc acids (d); alk (d)
(VI) fluoride	UF_6	352.02	5.09	64.0	subl 56.5	hyd aq; s chl, CCl_4
(III) hydride	UH_3	241.05	11.1			i aq
(IV) iodide	UI_4	745.65	5.6	506	757	s aq
(IV) oxide (pitchblende)	UO_2	270.03	10.97	2827		s conc HNO_3
(VI) oxide	UO_3	286.03	7.29	d 1300		i aq; s HCl, HNO_3
octaoxide [(V,VI) oxide]	U_3O_8	842.08	8.38	d 1300 to UO_2		s HNO_3
peroxide 2-water	$UO_4 \cdot 2H_2O$	338.06		d 90–195 to U_2O_7 (slow)	d >200 to UO_2	d by HCl
Uranyl(VI) acetate 2-water	$UO_2(C_2H_3O_2)_2 \cdot 2H_2O$	422.13	2.893	anhyd 110	d 275	7.7 g/100 mL[15] aq; sl s alc
chloride	UO_2Cl_2	340.93	5.43	577		320 g/100 mL[18] aq; s acet, alc
fluoride	UO_2F_2	308.03	6.37	d 300		v s aq

nitrate 6-water	$UO_2(NO_3)_2 \cdot 6H_2O$	502.13	2.807	60	d 118	155 g/100 mL^{20} aq; v s alc, eth
sulfate 3-water	$UO_2SO_4 \cdot 3H_2O$	420.14	3.28	d 100		g/100 mL: 21 aq, 4 alc
Vanadium	V	50.9415	6.11^{19}	1917	3421	s HF, HNO$_3$, hot H$_2$SO$_4$, aq reg
(IV) chloride	VCl_4	192.75	1.82	−25.7	148	hyd aq; s nonpolar solvents
dichloride oxide	VCl_2O	137.86	2.88	disprop 384		hyd (slow) aq; s abs alc, HOAc
(III) fluoride	VF_3	107.94	3.363	≈1400	subl 800	i almost all organic solvents
(IV) fluoride	VF_4	126.94	3.15	subl 120 (vac) & disprop		s aq, acet, HOAc
(V) fluoride	VF_5	145.93	2.50	19.5	48	hyd aq; v s anhyd HF, acet, alc
(II) oxide	VO	66.94	5.76	1790		s HCl
(III) oxide	V_2O_3	149.88	4.87	1940		sl s acids
(IV) oxide	VO_2	82.94	4.34	1967		s acids, alkalis
(V) oxide	V_2O_5	181.88	3.35	670	d 1800	0.07 aq; s conc acids, alkalis
(IV) oxide sulfate	$VOSO_4$	163.00				s aq
(III) sulfate	$V_2(SO_4)_3$	390.07	4.72	410 (vac)		s (slow) aq, HNO$_3$
(III) sulfide	V_2S_3	198.08		d 600		s hot acids, alkali sulfides
Xenon	Xe	131.29	5.761 g/L	−111.8	−108.04	10.8 mL/100 mL^{20} aq
difluoride	XeF	169.29	4.32	129.0	subl 114.3	2.5 g/100 mL^0 aq
hexafluoride	XeF_6	245.28	3.56	49.5	75.6	hyd aq
tetrafluoride	XeF_4	207.28	4.04	117.1	subl 115.7	hyd aq; s F$_3$CCOOH
trioxide	XeO_3	179.29	4.55	explodes 25		s aq giving xenic acid
Yterbium	Yb	173.04	6.90	819	1196	s acids
(II) chloride	$YbCl_2$	243.95	5.27	721	1930	s aq
(III) chloride 6-water	$YbCl_3 \cdot 6H_2O$	387.49	2.57	anhyd 180	mp: 865	v s aq
(III) fluoride	YbF_3	230.04	8.17	1157	2230	v s H$_2$SO$_4$
(III) nitrate 4-water	$Yb(NO_3)_3 \cdot 4H_2O$	431.12				s aq
(III) oxide	Yb_2O_3	394.08	9.18	2435		s dilute acids
(III) sulfate 8-water	$Yb_2(SO_4)_3 \cdot 8H_2O$	778.39	3.3			34.8 g/100 mL^{20} aq
Yttrium	Y	88.9059	4.472	1522	3345	s hot water (d)
chloride	YCl_3	195.26	2.61	721	1510	79 g/100 mL^{20} aq; s alc
fluoride	YF_3	145.90	4.0	1152	2230	s conc acids (d)
nitrate 6-water	$Y(NO_3)_3 \cdot 6H_2O$	383.01	2.68	−3H$_2$O, 100		171 g/100 mL^{20} aq
oxide	Y_2O_3	225.81	5.03	2440	4300	s acids
sulfate 8-water	$Y_2(SO_4)_3 \cdot 8H_2O$	610.12	2.56	anhyd 400	d >1000	9.6 g/100 mL^{20} aq
Zinc	Zn	65.39	7.14	419.527	907	i aq; s acids, alkalis (slow)
acetate dihydrate	$Zn(C_2H_3O_2)_2 \cdot 2H_2O$	219.51	1.735	237 d		g/100 mL: 41.6^{20} aq, 3.3 alc
arsenate(III)(1−)	$Zn(AsO_2)_2$	279.23				s acids

(Continued)

TABLE 1.3 Physical Constants of Inorganic Compounds (*Continued*)

Name	Formula	Formula weight	Density	Melting point, °C	Boiling point, °C	Solubility in 100 parts solvent
arsenate(V)(3−) 8-water	$Zn_3(AsO_4)_2 \cdot 8H_2O$	618.13	3.33			s acids and alkalis
bromide	$ZnBr_2$	225.20	4.5	394	697	g/100 mL: 471[25] aq, 200 alc; s KOH, eth
carbonate	$ZnCO_3$	125.40	4.4	$-CO_2$, 300		0.02[25] aq; s acids, KOH, NH_4 salts
chloride	$ZnCl_2$	136.29	2.907	290	732	g/100 ml: 395[20] aq, 77 alc, 50 glyc; v s acet
chromate(VI)	$ZnCrO_4$	181.39	3.40			s acids
cyanide	$Zn(CN)_2$	117.43	1.852	d 800		0.058[18] aq; s acids, KCN, KOH
fluoride	ZnF_2	103.39	4.9	872	1500	s HNO_3, HCl, NH_4OH
hexafluorosilicate 6-water	$Zn[SiF_6] \cdot 6H_2O$	315.56	2.104	d 100		v s aq
iodate	$Zn(IO_3)_2$	415.20	5.063	d		0.87[20] aq; s HNO_3, KOH
iodide	ZnI_2	319.20	4.74	446	625 d	g/100 mL: 332[20] aq, 50 glyc; v s alc
nitrate 6-water	$Zn(NO_3)_2 \cdot 6H_2O$	297.49	2.067	$-6H_2O$, 131		146 g/100 mL[0] aq; v s alc
oxide	ZnO	81.39	5.60	1975		i aq; s acids, KOH, NH_4OH
peroxide	ZnO_2	97.39	1.57	d >150	explodes 212	d (slow) aq; s dilute acids (d)
1,4-phenolsulfonate 8-water	$Zn[C_6H_4(OH)S\,O_3]_2 \cdot 8H_2O$	555.84		anhyd 120		g/100 mL: 63 aq, 56 alc
phosphate(V)	$Zn_3(PO_4)_2$	386.11	3.998	900		s acids, NH_4OH
phosphide	Zn_3P_2	258.12	4.55	420	1100	d aq, HCl (viol); s bz, CS_2
propionate	$Zn(C_3H_5O_2)_2$	211.53				32%[15] aq; 2.8%[15] alc
selenide	$ZnSe$	144.35	5.65	>1100		d dilute HNO_3
silicate(2−)	Zn_2SiO_4	222.86	4.10	1512		i aq or dilute acids
stearate	$Zn(C_{18}H_{35}O_2)_2$	632.34	1.095	130		d dil acids; s bz; i aq, alc, eth
sulfate	$ZnSO_4$	161.45	3.8	680 d		53.8%[20] aq
sulfate 7-water	$ZnSO_4 \cdot 7H_2O$	287.56	1.97	anhyd 280	d >500	g/100 mL: 167 aq, 40 glyc; i alc
sulfide (wirzite)	ZnS	97.46	4.09	1722		i aq; s dilute mineral acids
telluride	$ZnTe$	192.99	6.34	1239		d (slow) aq or dilute HCl
thiocyanate	$Zn(SCN)_2$	181.56				0.14 aq; s alc
Zirconium	Zr	91.224	6.52	1852	3577	s aq reg, HF, hot H_3PO_4, fusion with KOH + KNO_3
(IV) bromide	$ZrBr_4$	410.84	3.98	450	subl 357	sl s conc H_2SO_4
carbide	ZrC	103.23	6.73	3532	5100	
(II) chloride	$ZrCl_2$	162.13	3.6	727	1292	d aq

(IV) chloride	$ZrCl_4$	233.03	2.80	437 (25 atm)	subl 334	hyd aq to $ZrCl_2O$; s alc, eth
diboride	ZrB_2	112.85	6.17	3245	d 4193	v s aq, alc
dichloride oxide 8-water	$ZrCl_2O \cdot 8H_2O$	322.25	1.91	anhyd 210	d 410	i aq
dihydride	ZrH_2	93.24	5.61			1.32 g/100 mL[20] aq
(IV) fluoride	ZrF_4	167.22	4.436	932[tp]	subl 912	s mineral acids
(IV) hydroxide	$Zr(OH)_4$	159.25	3.25	to ZrO_2, 500		s aq (d), eth
(IV) iodide	ZrI_4	598.84		499 (sealed tube)	subl 432.5	
(IV) nitrate 5-water	$Zr(NO_3)_4 \cdot 5H_2O$	429.32		d 100		v s aq; s alc
(IV) oxide	ZrO_2	123.22	5.68	2678	4300	s hot H_2SO_4, HF (slow)
(IV) silicate(4−)	$ZrSiO_4$	183.31	4.56	d 1540 to ZrO_2 + SiO_2		unaffected by aqueous reagents
sulfate 4-water	$Zr(SO_4)_2 \cdot 4H_2O$	355.41	2.80	anhyd 380		52.5 g/100 g aqueous solution

TABLE 1.4   Color, Crystal Symmetry, and Refractive Index of Inorganic Compounds

Abbreviations Used in the Table

Color				Crystal symmetry	
B	brown	R	red	C	cubic
BE	blue	SL	silver	H	hexagonal
BK	black	V	violet	M	monoclinic
CL	colorless	W	white	R	rhombic
G	gray	Y	yellow	RH	rhombohedral
GN	green			T	tetragonal
O	orange			TG	trigonal
P	purple			TR	triclinic

Compound	Formula	Molecular weight	Color	Crystal symmetry	Refractive index n_D
Actinium					
Bromide	$AcBr_3$	466.7	W	H	
Chloride	$AcCl_3$	333.4	W	H	
Fluoride	AcF_3	284.0	W	H	
Oxide	Ac_2O_3	502.0	W	H	
Aluminum					
Bromide	$AlBr_3$	266.7	CL	R	
Carbide	Al_4C_3	143.9	Y	H	2.70
Chloride	ACl_3	133.3	W	H	1.56
Fluoride	AlF_3	84.0	CL	TR	1.38
Hydroxide	$Al(OH)_3$	78.0	W	M	
Iodide	AlI_3	407.7	W		
Nitrate	$Al(NO_3)_3 \cdot 9H_2O$	375.1	CL	R	1.54
Nitride	AlN	41.0	W	H	
Oxide	Al_2O_3	102.0	CL	H	1.68
Phosphate	$AlPO_4$	122.0	W	R	1.56
Silicate	Al_2SiO_5	162.0	W	R	1.66
Sulfate	$Al_2(SO_4)_3$	342.2	W	R	1.47
Sulfide	Al_2S_3	150.2	Y	H	
Americium					
Oxide IV	AmO_2	275.1	B	C	
Ammonium					
Bromide	NH_4Br	98.0	W	C	1.711
Carbonate	$(NH_4)_2CO_3 \cdot H_2O$	114.1	W	C	
Chlorate	NH_4ClO_3	101.5	W	M	
Chloride	NH_4Cl	53.5	W	C	1.642
Chromate	$(NH_4)_2CrO_4$	152.1	Y	M	
Fluoride	NH_4F	37.0	W	H	1.315
Iodate	NH_4IO_3	192.9	W	R	
Iodide	NH_4I	144.9	W	C	1.703
Nitrate	NH_4NO_3	80.0	W	R	1.413
Nitrite	NH_4NO_2	64.0	Y		
Oxalate	$(NH_4)_2C_2O_4 \cdot H_2O$	142.1	CL	R	1.44–1.59
Perchlorate	NH_4ClO_4	117.5	W	R	1.49
Hydrogen Phosphate	$(NH_4)_2HPO_4$	132.1	W	M	1.53
Dihydrogen Phosphate	$NH_4H_2PO_4$	115.0	W	T	1.48–1.53
Sulfate	$(NH_4)_2SO_4$	132.1	W	R	1.53
Hydrogen sulfide	NH_4HS	51.1	W	R	1.74
Thiocyanate	NH_4SCN	76.1	CL	M	1.61–1

TABLE 1.4   Color, Crystal Symmetry, and Refractive Index of Inorganic Compounds (*Continued*)

Compound	Formula	Molecular weight	Color	Crystal symmetry	Refractive index n_D
Antimony					
Bromide III	$SbBr_3$	361.5	CL	R	1.74
Chloride III	$SbCl_3$	228.1	CL	R	1.74
Chloride V	$SbCl_5$	299.0	W	LIQ	1.601[1]
Fluoride III	SbF_3	178.8	CL	R	
Fluoride V	SbF_5	216.7	CL	LIQ	
Hydride III	SbH_3	124.8	CL	GAS	
Iodide III	SbI_3	502.5	RD	H	
Iodide V	SbI_5	756.3	B		
Oxide III	Sb_2O_3	291.5	CL	R	2.35
Oxide V	Sb_2O_5	323.5	Y	C	
Oxychloride III	SbOCl	173.2	W	M	
Sulfate III	$Sb_2(SO_4)_3$	531.7	W		
Sulfide III	Sb_2S_3	339.7	BK	R	4.064
Sulfide V	Sb_2S_5	403.8	Y		
Arsenic					
Acid, ortho	$H_3AsO_4 \cdot \frac{1}{2}H_2O$	151.0	CL		
Bromide III	$AsBr_3$	314.7	CL	R	
Chloride III	$AsCl_3$	181.3	CL	LIQ	1.598
Chloride V	$AsCl_5$	252.2	CL		
Fluoride III	AsF_3	131.9	CL	LIQ	
Fluoride V	AsF_5	169.9	CL	GAS	
Hydride III	AsH_3	77.9	CL	GAS	
Iodide III	AsI_3	455.6	R	H	
Iodide V	AsI_5	709.5	B	M	
Oxide III	As_2O_3	197.2	CL	C	
Oxide V	As_2O_5	229.9	W		
Sulfide II	As_2S_2	214.0	R	M	2.46–2.52
Sulfide III	As_2S_3	246.0	Y	M	2.4–2.6
Sulfide V	As_2S_5	310.2	Y	M	
Barium					
Bromate	$Ba(BrO_3)_2 \cdot H_2O$	411.2	CL	M	
Bromide	$BaBr_2$	297.2	CL	R	1.75
Carbide	BaC_2	161.4	G	T	
Carbonate	$BaCO_3$	197.4	W	R	1.676
Chlorate	$Ba(ClO_3)_2 \cdot H_2O$	322.3	CL	M	1.56–1
Chloride	$BaCl_2$	208.3	CL	M	1.736
Chromate	$BaCrO_4$	253.3	Y	R	
Fluoride	BaF_2	175.3	CL	C	1.474
Hydride	BaH_2	139.4	G		
Hydroxide	$Ba(OH)_2 \cdot 8H_2O$	315.5	CL	M	1.502
Iodide	BaI_2	391.2	CL	M	
Nitrate	$Ba(NO_3)_2$	261.4	CL	C	1.572
Oxalate	BaC_2O_4	225.4	W		
Oxide	BaO	153.3	CL	C	1.98
Perchlorate	$Ba(ClO_4)_2$	336.2	CL	H	
Sulfate	$BaSO_4$	233.4	W	R	1.636
Sulfide	BaS	169.4	CL	C	2.155
Titanate	$BaTiO_3$	233.3		T/H	2.40

(*Continued*)

TABLE 1.4   Color, Crystal Symmetry, and Refractive Index of Inorganic Compounds (*Continued*)

Compound	Formula	Molecular weight	Color	Crystal symmetry	Refractive index n_D
Beryllium					
Bromide	$BeBr_2$	168.8	W	OR	
Carbide	Be_2C	30.0	Y	H	
Chloride	$BeCl_2$	79.9	W	OR	
Fluoride	BeF_2	47.0	CL	T	
Hydroxide	$Be(OH)_2$	43.0	W	R	
Iodide	BeI_2	262.8	CL	RH	
Nitrate	$Be(NO_3)_2 \cdot 3H_2O$	187.1	W		
Nitride	Be_3N_2	55.1	CL	C	
Oxide	BeO	25.0	W	H	1.72
Sulfate	$BeSO_4$	105.1	CL	T	
Sulfate	$BeSO_4 \cdot 4H_2O$	177.1	CL	T	1.44–1.47
Bismuth					
Bromide III	$BiBr_3$	448.7	Y		
Chloride III	$BiCl_3$	315.4	W		
Fluoride III	BiF_3	266.0	G	C	1.74
Hydroxide III	$Bi(OH)_3$	260.0	W		
Iodide III	BiI_3	589.7	RD	H	
Nitrate III	$Bi(NO_3)_3 \cdot 5H_2O$	485.1	CL	TR	
Nitrate, Basic III	$BiO(NO_3) \cdot H_2O$	305.0	W	H	
Oxide III	Bi_2O_3	466.0	Y	R	1.91
Oxide IV	$Bi_2O_4 \cdot 2H_2O$	518.0	B		
Oxide V	Bi_2O_5	498.0	B		
Oxychloride III	$BiOCl$	260.5	W	T	2.15
Phosphate III	$BiPO_4$	304.0	W	M	
Sulfate III	$Bi_2(SO_4)_3$	706.1	W		
Sulfide III	Bi_2S_3	514.2	B	R	1.34–1.46
Boron					
Arsenate	$BAsO_4$	149.7	W	T	1.68
Boric Acid	H_3BO_3	61.8	W	TR	
Bromide	BBr_3	250.5	CL	LIQ	1.5312^{16}
Carbide	B_4C	55.3	BK	RH	
Chloride	BCl_3	117.2	CL	LIQ	
Diborane	B_2H_6	27.7	CL	GAS	
Fluoride	BF_3	67.8	CL	GAS	
Iodide	BI_3	391.6	W		
Nitride	BN	24.8	W	H	
Oxide	B_2O_3	69.6	W	C	
Sulfide	B_2S_3	117.8	W		
Bromine					
Chloride I	$BrCl$	115.4	R	GAS	
Fluoride I	BrF	98.9	B	GAS	
Fluoride III	BrF_3	136.9	CL	LIQ	1.4536^{25}
Fluoride V	BrF_5	174.9	CL	LIQ	1.3529^{25}
Hydride I	$H\,Br$	80.9	CL	GAS	1.325^{10}
Cadmium					
Bromide	$CdBr_2$	272.2	W	H	
Carbonate	$CdCO_3$	172.4	W	TG	
Chloride	$CdCl_2$	228.4	W	H	

TABLE 1.4   Color, Crystal Symmetry, and Refractive Index of Inorganic Compounds (*Continued*)

Compound	Formula	Molecular weight	Color	Crystal symmetry	Refractive index n_D
Cadmium (*Continued*)					
Fluoride	CdF_2	150.4	W	C	1.56
Hydroxide	$Cd(OH)_2$	146.4	W	TR	
Iodide	CdI_2	366.2	B	H	
Nitrate	$Cd(NO_3)_2 \cdot 4H_2O$	308.5	W		
Oxide	CdO	128.4	B	C	
Sulfate	$CdSO_4$	208.5	W	R	
Sulfate	$3CdSO_4 \cdot 8H_2O$	769.6	CL	M	1.565
Sulfide	CdS	144.5	Y	H	2.51
Calcium					
Bromate	$CaBrO_3 \cdot H_2O$	313.9		M	
Bromide	$CaBr_2 \cdot 6H_2O$	308.0	CL	H	
Carbide	CaC_2	64.1	CL	T	1.75
Carbonate	$CaCO_3$	100.1	CL	R	1.681
Chloride	$CaCl_2$	111.0	CL	C	1.52
Chloride	$CaCl_2 \cdot 6H_2O$	219.1	C	T	1.417
Chromate	$CaCrO_4 \cdot 2H_2O$	192.1	Y	M	
Fluoride	CaF_2	78.1	CL	C	1.434
Hydride	CaH_2	42.1	W	R	
Hydroxide	$Ca(OH)_2$	74.1	CL	H	1.574
Iodide	CaI_2	293.9	W	H	
Nitrate	$Ca(NO_3)_2$	164.1	CL	C	
Nitrate	$Ca(NO_3)_2 \cdot 4H_2O$	236.2	CL	M	1.498
Nitride	Ca_3N_2	148.3	B	H	
Oxalate	CaC_2O_4	128.1	CL	C	
Oxide	CaO	56.1	CL	C	1.838
Perchlorate	$Ca(ClO_4)_2$	239.0	CL		
Peroxide	CaO_2	72.1	W	T	
Sulfate	$CaSO_4$	136.1	CL	M	1.576
Sulfate	$CaSO_4 \cdot 2H_2O$	172.2	CL	M	1.5226
Sulfide	CaS	72.1	CL	C	2.137
Carbon					
Dioxide	CO_2	44.0	CL	GAS	
Disulfide	CS_2	76.1	CL	LIQ	1.6290
Monoxide	CO	28.0	CL	GAS	
Oxybromide	$COBr_2$	187.8	CL	LIQ	
Oxychloride	$COCl_2$ (Phosgene)	98.9	CL	GAS	
Oxysulfide	COS	60.1	CL	GAS	
Cerium					
Bromide III	$CeBr_3$	380.0		H	
Chloride III	$CeCl_3$	246.5	CL	H	
Fluoride III	CeF_3	197.1	W	H	
Iodate IV	$Ce(IO_3)_4$	839.7	Y		
Iodide III	CeI_3	520.8	Y	R	
Molybdate III	$Ce_2(MoO_4)_3$	760.0	Y	T	2.01
Nitrate III	$Ce(NO_3)_3 \cdot 6H_2O$	434.2	CL		
Oxide III	Ce_2O_3	328.2	GN	H	
Oxide IV	CeO_2	172.1	W	C	
Sulfate III	$Ce_2(SO_4)_3$	568.4	CL	M/R	
Sulfide	Ce_2S_3	376.4	Y	C	

(*Continued*)

TABLE 1.4 Color, Crystal Symmetry, and Refractive Index of Inorganic Compounds (*Continued*)

Compound	Formula	Molecular weight	Color	Crystal symmetry	Refractive index n_D
Cesium					
Bromide	CsBr	212.8	CL	C	1.642
Carbonate	Cs$_2$CO$_3$	325.8	CL		
Chloride	CsCl	168.4	CL	C	1.534
Fluoride	CsF	151.9	CL	C	1.481
Hydroxide	CsOH	149.9	W		
Iodide	CsI	259.8		C	1.661; 1.669
Iodide III	CsI$_3$	513.7	BK	R	
Nitrate	CsNO$_3$	194.9	W	H	1.55
Oxide	Cs$_2$O	281.8	R		
Perchlorate	CsClO$_4$	232.4	CL	R	1.479
Periodate	CsIO$_4$	323.8	W	R	
Peroxide	Cs$_2$O$_2$	297.8	Y	R	
Sulfate	Cs$_2$SO$_4$	361.9	CL	R	1.564
Superoxide	CsO$_2$	164.9	Y		
Trioxide	Cs$_2$O$_3$	313.8	B	C	
Chlorine					
Dioxide	ClO$_2$	67.5	Y	GAS	
Fluoride	ClF	54.5	CL	GAS	
Trifluoride	ClF$_3$	92.5	CL	GAs	
Monoxide	Cl$_2$O	86.9	B	GAS	
Hydrochloric Acid	HCl	36.5	CL	GAS	1.254[10]
Perchloric Acid	HClO$_4$	100.5	CL	LIQ	
Chromium					
Bromide II	CrBr$_2$	211.8	W	M	
Carbide III	Cr$_3$C$_2$	180.0	G	R	
Chloride II	CrCl$_2$	122.9	W	R	
Chloride III	CrCl$_3$	158.4	V	R	
Fluoride II	CrF$_2$	90.0	GN	M	
Fluoride III	CrF$_3$	109.0	GN	R	
Iodide II	CrI$_2$	305.8	B	M	
Nitrate III	Cr(NO$_3$)$_3$	238.0	GN		
Nitrate III	CrN	66.0		C	
Oxide II	CrO	68.0	BK	H	
Oxide III	Cr$_2$O$_3$	152.0	GN	H	2.551
Oxide IV	CrO$_2$	84.0	B		
Oxide VI	CrO$_3$	100.0	RD	R	
Phosphate III	CrPO$_4$ · 6H$_2$O	255.1	V	TR	
Sulfate III	Cr$_2$(SO$_4$) · 18H$_2$O	716.5	V	C	1.564
Sulfide II	CrS	84.1	BK	M	
Sulfide III	Cr$_2$S$_3$	200.2	B	TG	
Cobalt					
Bromide II	CoBr$_2$	218.8	GN	H	
Chlorate II	Co(ClO$_3$)$_2$ · 6H$_2$O	333.9	R	C	1.55
Chloride II	CoCl$_2$	129.8	BE	H	
Fluoride II	CoF$_2$	96.9	R	M	
Fluoride III	CoF$_3$	115.9	B	H	
Hydroxide II	Co(OH)$_2$	92.9	R	R	
Iodate II	Co(IO$_3$)$_2$	408.7	V		

TABLE 1.4 Color, Crystal Symmetry, and Refractive Index of Inorganic Compounds (*Continued*)

Compound	Formula	Molecular weight	Color	Crystal symmetry	Refractive index n_D
Cobalt (*Continued*)					
Iodide II	CoI_2	312.7	BK	H	
Nitrate II	$Co(NO_3)_2 \cdot 6H_2O$	291.0	R	M	
Oxide II	CoO	74.9	GN	C	
Oxide III	Co_2O_3	165.9	B	R	
Oxide II–III	Co_3O_4	240.8	BK	C	
Perchlorate II	$Co(ClO_4)_2$	257.8	R		1.50
Sulfate II	$CoSO_4$	155.0	BE	C	
Sulfate II	$CoSO_4 \cdot 7H_2O$	281.1	R	M	1.48
Sulfide II	CoS	91.0	R	H	
Sulfide III	Co_2S_3	214.1	BK		
Copper					
Bromide I	$CuBr$	143.5	W	C	
Bromide II	$CuBr_2$	223.4	BK	M	
Carbonate, Basic II	$2CuCO_3 \cdot Cu(OH)_2$	344.7	BE	M	1.731
Chloride I	$CuCl$	99.0	W	C	
Chloride II	$CuCl_2$	134.5	Y	M	
Chloride II	$CuCl_2 \cdot 2H_2O$	170.5	Y	R	
Fluoride II	$CuF_2 \cdot 2H_2O$	137.6	W	M	
Hydroxide I	$CuOH$	80.6	Y		
Hydroxide II	$Cu(OH)_2$	97.6	BE		
Iodide I	CuI	190.5	W	C	2.346
Nitrate II	$Cu(NO_3)_2 \cdot 3H_2O$	241.6	BE		
Oxide I	Cu_2O	143.1	R	C	2.705
Oxide II	CuO	79.5	BK	TR	2.63
Sulfate II	$CuSO_4$	159.6	W	R	
Sulfate II	$CuSO_4 \cdot 5H_2O$	249.7	BE	TR	1.52
Sulfide I	Cu_2S	159.1	BK	C	
Sulfide II	CuS	95.6	BK	H	
Thiocyanate I	$CuSCN$	121.6	W		
Curium					
Bromide III	$CmBr_3$	488		R	
Chloride III	$CmCl_3$	353	W	H	
Fluoride III	CmF_3	304	W	H	
Fluoride IV	CmF_4	323	B	M	
Iodide III	CmI_3	628	W	H	
Dysprosium					
Bromide	$DyBr_3$	402.3	CL	R	
Chloride	$DyCl_3$	268.9	Y	M	
Fluoride	DyF_3	219.5	CL	H	
Iodide	DyI_3	543.2	GN	H	
Nitrate	$Dy(NO_3)_3 \cdot 5H_2O$	438.6	Y	TR	
Oxide	Dy_2O_3	373.0	W	C	
Sulfate	$Dy_2(SO_4)_3 \cdot 8H_2O$	757.3	Y	M	
Erbium					
Bromide	$ErBr_3$	407.1	V	R	
Chloride	$ErCl_3$	273.6	V	M	
Fluoride	ErF_3	224.3	RD	R	

(*Continued*)

TABLE 1.4   Color, Crystal Symmetry, and Refractive Index of Inorganic Compounds (*Continued*)

Compound	Formula	Molecular weight	Color	Crystal symmetry	Refractive index n_D
Erbium (*Continued*)					
Iodide	ErI_3	548.0	V	H	
Oxide	Er_2O_3	382.6	R	C	
Sulfate	$Er_2(SO_4)_3$	622.7	W		
Sulfide	Er_2S_3	263.5	R	M	
Europium					
Bromide II	$EuBr_2$	311.8		R	
Bromide III	$EuBr_3$	391.7	G	R	
Chloride II	$EuCl_2$	222.9	W	R	
Chloride III	$EuCl_3$	258.3	Y	H	
Fluoride II	EuF_2	190.0	Y	C	
Fluoride III	EuF_3	209.0	W	R	
Iodide II	EuI_2	405.8	GN	M	
Iodide III	EuI_3	532.7			
Oxide III	Eu_2O_3	351.9	R	C	
Sulfate III	$Eu_2(SO_4)_3 \cdot 8H_2O$	736.2	R	M	
Fluorine					
Dioxide	F_2O_2	70.0	B	GAS	
Hydride	HF	20.0	CL	GAS	
Oxide	F_2O	54.0	CL	GAS	
Cadolinium					
Bromide	$GdBr_3$	397.0	W	H	
Chloride	$GdCl_3$	263.6	W	H	
Fluoride	GdF_3	214.3	W	R	
Iodide	GdI_3	538.0	Y	H	
Nitrate	$Gd(NO_3)_3 \cdot 6H_2O$	451.4		T	
Oxide	Gd_2O_3	362.5	W	C	
Sulfate	$Gd_2(SO_4)_3$	602.7	CL		
Sulfide	Gd_2S_3	410.7	Y	C	
Gallium					
Arsenide III	GaAs	144.6	G	C	
Bromide III	$GaBr_3$	309.5	CL		
Chloride II	Ga_2Cl_4	281.3	W		
Chloride III	$GaCl_3$	176.0	CL	TR	
Fluoride III	GaF_3	126.7	W	RH	
Iodide III	GaI_3	450.4	Y		
Oxide I	Ga_2O	155.4	G		
Oxide III	Ga_2O_3	187.4	G	M (β)	1.95
Sulfide I	Ga_2S	171.5	G		
Sulfide II	Ga_2S_3	235.6	Y	H	
Germanium					
Bromide IV	$GeBr_4$	392.2	G		1.627
Chloride IV	$GeCl_4$	214.4	CL	LIQ	1.464
Fluoride IV	GeF_4	148.6	CL	GAS	
Hydride IV	GeH_4 (Germane)	76.6	CL	GAS	1.00089
Iodide IV	GeI_4	580.2	R	C	
Oxide II	GeO	88.6	G		1.607

TABLE 1.4 Color, Crystal Symmetry, and Refractive Index of Inorganic Compounds (*Continued*)

Compound	Formula	Molecular weight	Color	Crystal symmetry	Refractive index n_D
Germanium (*Continued*)					
Oxide IV	GeO_2	104.6	CL	H	
Sulfide II	GeS	104.7	Y	R	
Sulfide IV	GeS_2	136.7	W	R	
Gold					
Bromide I	AuBr	276.9	G		
Bromide III	$AuBr_3$	436.7	B		
Chloride I	AuCl	232.4	Y	R	
Chloride III	$AuCl_3$	303.3	R		
Hydroxide III	$Au(OH)_3$	248.0	B		
Iodide	AuI	323.9	Y	TR	
Iodide III	AuI_3	577.7	G		
Sulfate III	$Au_2(SO_4)_3 \cdot H_2O$	490.5	B		
Sulfide I	Au_2S	426.0	B		
Sulfide III	Au_2S_3	490.1	B		
Hafnium					
Bromide	$HfBr_4$	498.1	W		
Carbide	HfC	190.5		C	
Chloride	$HfCl_4$	320.3	W		
Fluoride	HfF_4	254.5	CL	M	1.56
Iodide	HfI_4	686.1			
Nitride	HfN	192.5	Y	C	
Oxide	HfO_2	210.5	W	T	
Sulfide	HfS_2	242.6		H	
Holmium					
Bromide	$HoBr_3$	404.7	Y	R	
Chloride	$HoCl_3$	271.3	Y	M	
Fluoride	HoF_3	221.9	B	H	
Iodide	HoI_3	545.6	Y		
Oxide	Ho_2O_3	377.9		C	
Hydrogen					
Bromide	HBr	80.9	CL	GAS	2.77^{-67}
Chloride	HCl	36.5	CL	GAS	
Fluoride	HF	20.0	CL	GAS	
Iodide	HI	127.9	CL	GAS	1.466
Oxide	H_2O	18.0	CL	LIQ	1.3333
Oxide-Deutero	$2H_2O$	20.0	CL	LIQ	1.3284
Peroxide	H_2O_2	34.0	CL	LIQ	1.414^{22}
Selenide	H_2Se	81.0	CL	GAS	
Sulfide	H_2S	34.1	CL	GAS	1.374
Telluride	H_2Te	129.9	CL	GAS	
Indium					
Bromide I	InBr	194.7	B		
Bromide III	$InBr_3$	354.5	CL		
Chloride I	InCl	150.3	R	C	
Chloride III	$InCl_3$	221.2	CL	M	
Fluoride III	InF_3	171.8	CL	H	

(*Continued*)

TABLE 1.4 Color, Crystal Symmetry, and Refractive Index of Inorganic Compounds (*Continued*)

Compound	Formula	Molecular weight	Color	Crystal symmetry	Refractive index n_D
Indium (*Continued*)					
Iodide I	InI	241.7	B		
Iodide III	InI$_3$	495.5	Y	M	
Oxide III	In$_2$O$_3$	277.6	Y	C	
Sulfate III	In$_2$(SO$_4$)$_3$	517.8	W	M	
Sulfide III	In$_2$S$_3$	325.8	R (β)	C	
Iodine					
Bromide I	IBr	206.8	BK	OR	
Chloride I, α	ICl	162.4	R	C	
Chloride I, β	ICl	162.4	R	LIQ	
Chloride III	ICl$_3$	233.3	Y	R	
Fluoride V	IF$_5$	221.9	CL	LIQ	
Fluoride VII	IF$_7$	259.9	CL	GAS	
Oxide IV	I$_2$O$_4$	317.8	Y		
Oxide V	I$_2$O$_5$	333.8	CL		
Iodic Acid	HIO$_3$	175.9	W	R	
Hydrogen Iodide	HI	127.9	CL	GAS	1.466
Iridium					
Bromide II	IrBr$_3$ · 4H$_2$O	504.0	GN		
Bromide IV	IrBr$_4$	511.8	BK		
Chloride III	IrCl$_3$	298.6	GN	H	
Chloride IV	IrCl$_4$	334.0	R	C	
Fluoride VI	IrF$_6$	306.2	Y	T	
Iodide III	IrI$_3$	572.9	GN		
Iodide IV	IrI$_4$	699.8	BK		
Oxide IV	IrO$_2$	224.2	BK		
Sulfide IV	IrS$_2$	256.3	BK		
Iron					
Arsenide	FeAs	130.8	W	R	
Arsenide, di–	FeAs$_2$	205.7	G	R	
Bromide II	FeBr$_2$	215.7	GN	H	
Bromide III	FeBr$_3$ · 6H$_2$O	403.7	R		
Carbide	Fe$_3$C	179.6	G	C	
Carbonate II	FeCO$_3$	115.9	G		
Chloride II	FeCl$_2$	126.8	G	H	
Chloride III	FeCl$_3$	162.2	GN	H	
Fluoride III	FeF$_3$	112.9	W	R	
Hydroxide II	Fe(OH)$_2$	89.9	GN	H	
Hydroxide III	Fe(OH)$_3$	106.9	B		
Iodide II	FeI$_2$	309.7	BK	H	
Nitrate II	Fe(NO$_3$)$_2$ · 6H$_2$O	288.0	GN	R	
Nitrate III	Fe(NO$_3$)$_3$ · 9H$_2$O	404.0	CL	M	
Nitride	Fe$_2$N	125.7	G		
Oxide II	FeO	71.9	BK	C	2.32
Oxide III	Fe$_2$O$_3$	159.7	B	TG	3.04
Oxide II–III	Fe$_3$O$_4$	231.6	BK	C	2.42
Phosphate III	FePO$_4$ · 2H$_2$O	186.9	W	M	1.35
Phosphide	Fe$_2$P	142.7	G	H	
Sulfate II	FeSO$_4$ · 7H$_2$O	278.0	GN	M	1.48

TABLE 1.4 Color, Crystal Symmetry, and Refractive Index of Inorganic Compounds (*Continued*)

Compound	Formula	Molecular weight	Color	Crystal symmetry	Refractive index n_D
Iron (*Continued*)					
Sulfate III	$Fe_2(SO_4)_3$	399.9	Y	R	1.81
Sulfate II, Ammonium	$(NH_4)_2 Fe(SO_4) \cdot 6H_2O$	392.2	GN	M	1.49
Sulfide II	FeS	87.9	BK	H	
Sulfide III	Fe_2S_3	207.9	BK	H	
Sulfide, di	FeS_2	120.0	Y	C	
Lanthanum					
Bromate	$La(BrO_3)_3 \cdot 9H_2O$	684.8		H	
Bromide	$LaBr_3$	378.6	W	H	
Chloride	$LaCl_3$	245.3	W	H	
Fluoride	LaF_3	195.9	W	H	
Iodide	LaI_3	519.6	G	R	
Molybdate	$La_2(MoO_4)_3$	757.6		T	
Oxide	La_2O_3	325.8	W	R	
Sulfate	$La_2(SO_4)_3$	566.0	W		
Sulfide	La_2S_3	374.0	Y	H	
Lead					
Acetate II	$Pb(C_2H_3O_2)_2$	325.3	W		
Acetate IV	$Pb(C_2H_3O_2)_4$	443.4	CL	M	
Arsenate II	$Pb_3(AsO_4)_2$	899.4	W		
Bromide II	$PbBr_2$	367.0	W	R	
Carbonate II	$PbCO_3$	267.2	CL	R	1.80–2.08
Chloride II	$PbCl_2$	278.1	W	R	2.22
Chloride IV	$PbCl_4$	349.0	Y	LIQ	
Chromate II	$PbCrO_4$	323.2	Y	M	2.33
Fluoride II	PbF_2	245.2	CL	R	
Hydroxide II	$Pb(OH)_2$	241.2	W	H	
Iodate II	$Pb(IO_3)_2$	557.0	W		
Iodide II	PbI_2	461.0	Y	H	
Molybdate II	$PbMoO_4$	367.2	CL	T	2.30
Nitrate II	$Pb(NO_3)_2$	331.2	CL	C	1.782
Oxide II	PbO	223.2	R	T	
Oxide IV	PbO_2	239.2	B	T	
Oxide II–IV	Pb_3O_4	685.6	R	T	
Phosphate, III	$Pb_3(PO_4)_2$	811.6	W	H	1.95
Sulfate II	$PbSO_4$	303.3	W	R	1.85
Sulfide II	PbS	239.3	BK	C	3.911
Tungstate II	$PbWO_4$	455.1	CL	M	
Lithium					
Aluminum Hydride	$LiAlH_4$	37.9	W		
Bromide	LiBr	86.9	W	C	1.784
Carbonate	Li_2CO_3	73.9	W	M	1.43; 1.5
Chloride	LiCl	42.4	W	C	1.662
Fluoride	LiF	25.9	W	C	1.391
Hydride	LiH	8.0	CL	C	
Hydroxide	LiOH	24.0	W	T	1.46
Iodide	LiI	133.9	W	C	1.955
Nitrate	$LiNO_3$	68.9	W	TG	1.435;1.439
Oxide	Li_2O	29.9	W	C	1.644

(*Continued*)

TABLE 1.4   Color, Crystal Symmetry, and Refractive Index of Inorganic Compounds (*Continued*)

Compound	Formula	Molecular weight	Color	Crystal symmetry	Refractive index n_D
Lithium (*Continued*)					
Peroxide	Li_2O_2	45.9		H	
Perchlorate	$LiClO_4$	160.4	W	H	
Phosphate	Li_3PO_4	115.8	CL	R	
Sulfate,	Li_2SO_4	109.9	CL	M	1.465
Sulfide	Li_2S	45.9	W	C	
Lutetium					
Bromide	$LuBr_3$	414.7	W	TG	
Chloride	$LuCl_3$	281.3	W	M	
Fluoride	LuF_3	232.0	W	R	
Iodide	LuI_3	555.7	B	H	
Oxide	Lu_2O_3	397.9		C	
Magnesium					
Aluminate	$MgO \cdot Al_2O_3$	142.3	CL	C	1.723
Bromide	$MgBr_2$	184.1	W	H	
Carbonate	$MgCO_3$	84.3	W	TG	1.51; 1.70
Chloride	$MgCl_2$	95.2	W	H	1.59; 1.67
Fluoride	MgF_2	62.3	CL	T	1.38
Hydroxide	$Mg(OH)_2$	58.3	CL	H	1.57
Iodide	MgI_2	278.2	W	H	
Nitrate	$Mg(NO_3)_2 \cdot 6H_2O$	256.4	CL	M	
Oxide	MgO	40.3	CL	C	1.736
Silicide	Mg_2Si	76.7	BE	C	
Silicate, m	$MgSiO_3$	100.4	W	M	1.66
Silicate, o	Mg_2SiO_4	140.7	W	R	1.65
Sulfate	$MgSO_4$	120.4	CL	R	
Sulfide	MgS	56.4	R	C	2.271
Manganese					
Bromide II	$MnBr_2$	214.8	W	H	
Carbonate II	$MnCO_3$	114.9	W	R	1.817
Chloride II	$MnCl_2$	125.9	W	H	
Fluoride II	MnF_2	92.9	R	T	
Iodide II	MnI_2	308.8	W	H	
Oxide II	MnO	70.9	GN	C	2.16
Oxide III	Mn_2O_3	157.9	BK	C	
Oxide IV	MnO_2	86.9	BK	R	
Oxide II–IV	Mn_3O_4	228.8	BK	R	
Potassium Permanganate	$KMnO_4$	158.0	P	R	1.59
Silicide	$MnSi$	83.0		C	
Sulfate II	$MnSO_4$	151.0	R		
Sulfide II	MnS	87.0	GN	C	
Mercury					
Bromide I	Hg_2Br_2	561.1	W	T	
Bromide II	$HgBr_2$	360.4	CL	R	
Chloride I	Hg_2Cl_2	472.1	W	T	1.97; 2.66
Chloride II	$HgCl_2$	271.5	CL	R	1.72; 1.97
Cyanide II	$Hg(CN)_2$	252.7	CL	T	1.645
Fluoride I	Hg_2F_2	439.2	Y	C	

TABLE 1.4 Color, Crystal Symmetry, and Refractive Index of Inorganic Compounds (*Continued*)

Compound	Formula	Molecular weight	Color	Crystal symmetry	Refractive index n_D
Mercury (*Continued*)					
Fluoride II	HgF_2	238.6	CL	C	
Iodide I	Hg_2I_2	655.0	Y	T	
Iodide II	HgI_2	454.4	R/Y	T/R	2.45; 2.7
Nitrate I	$Hg_2(NO_3)_2 \cdot 2H_2O$	561.2	CL	M	
Nitrate II	$Hg(NO_3)_2 \cdot {}^1\!/_2H_2O$	333.6	W		
Oxide I	Hg_2O	417.2	BK		
Oxide II	HgO	216.6	Y/R	R	2.37; 2.6
Sulfate I	Hg_2SO_4	497.3	CL	M	
Sulfate II	$HgSO_4$	296.7	CL	R	
Sulfide III	HgS	232.7	R	H	2.85; 3.2
Molybdenum					
Carbide II	Mo_2C	203.9	W	H	
Carbide IV	MoC	108.0	G	H	
Chloride II	$MoCl_2$	166.9	Y		
Chloride III	$MoCl_3$	202.3	R		
Chloride V	$MoCl_5$	273.2	BK	M	
Fluoride VI	MoF_6	202.9	Cl		
Iodide II	MoI_2	349.8	B		
Molybdic Acid	$H_2MoO_4 \cdot 4H_2O$	180.0	Y	M	
Oxide IV	MoO_2	127.9	G	T	
Oxide VI	MoO_3	143.9	CL	R	
Silicide IV	$MoSi_2$	152.1	G	T	
Sulfide IV	MoS_2	160.1	BK	H	4.7
Neodymium					
Bromide	$NdBr_3$	384.0	V	R	
Chloride	$NdCl_3$	250.6	V	H	
Fluoride	NdF_3	201.2	V	H	
Iodide	NdI_3	524.9	G	R	
Oxide	Nd_2O_3	336.5	BE	H	
Sulfide	Nd_2S_3	384.7	GN		
Neptunium					
Bromide II	$NpBr_3$	476.7	GN	R	
Chloride III	$NpCl_3$	343.4	GN	H	
Chloride IV	$NpCl_4$	378.8	BN	T	
Fluoride III	NpF_3	294.0	P	H	
Fluoride VI	NpF_6	351.0	O	R	
Iodide III	NpI_3	617.7	B	R	
Oxide IV	NpO_2	269.0	GN	C	
Nickel					
Arsenide	$NiAs$	133.6	W	H	
Bromide II	$NiBr_2$	218.5	Y		
Carbonyl	$Ni(CO)_4$	170.7	CL	LIQ	1.458[10]
Chloride II	$NiCl_2$	129.6	Y	H	
Fluoride II	NiF_2	96.7	Y	T	
Hydroxide II	$Ni(OH)_2$	92.7	GN		
Iodide II	NiI_2	312.5	BK	H	
Nitrate II	$Ni(NO_3)_2 \cdot 6H_2O$	290.8	GN	M	
Oxide II	NiO	74.7	G	C	2.37

(*Continued*)

TABLE 1.4 Color, Crystal Symmetry, and Refractive Index of Inorganic Compounds (*Continued*)

Compound	Formula	Molecular weight	Color	Crystal symmetry	Refractive index n_D
Nickel (*Continued*)					
Phosphide	Ni_2P	148.4	G		
Sulfate II	$NiSO_4$	154.8	Y	C	
Sulfide II	NiS	90.8	BK	TR	
Niobium					
Bromide	$NbBr_5$	492.5	R	R	
Carbide	NbC	104.9	BK	C	
Chloride	$NbCl_5$	270.2	W	M	
Fluoride	NbF_5	187.9	CL	M	
Iodide	NbI_5	727.4	BRASS	M	
Oxide	Nb_2O_5	265.8	W	R	
Nitrogen					
Ammonia	NH_3	17.0	CL	GAS	1.325
Hydrazine	N_2H_4	32.0	CL	LIQ	1.4707
Hydrazoic Acid	NH_3	43.0	CL	LIQ	
Hydroxylamine	NH_2OH	33.0	W	R	1.440[23.5]
Nitric Acid	HNO_3	63.0	CL	LIQ	1.397[16]
Chloride	NCl_3	120.4	Y	LIQ	
Fluoride	NF_3	71.0	CL	GAS	
Iodide	NI_3	394.7	BK		
Oxide I (nitrous-)	N_2O	44.0	CL	GAS	
Oxide II (nitric-)	NO	30.0	CL	GAS	1.193[16]
Oxide III (tri-)	N_2O_3	76.0	B	GAS	
Oxide IV (per-)	NO_2	46.0	B	GAS	
Oxide V (penta-)	N_2O_5	108.0	W	R	
Sulfide II	N_4S_4	184.3	O	M	2.046
Nitrosyl Chloride	$NOCl$	65.5	O	GAS	
Nitrosyl Fluoride	NOF	49.0	CL	GAS	
Nitryl Chloride	NO_2Cl	81.5	CL	GAS	
Osmium					
Chloride IV	$OsCl_4$	332.0	R		
Fluoride V	OsF_5	285.2	G	M	
Fluoride VI	OsF_6	304.2	GN	C	
Fluoride VIII	OsF_8	342.2	Y		
Iodide IV	OsI_4	697.8	BK		
Oxide IV	OsO_2	222.2	BK	T	
Oxide VIII	OsO_4	254.1	CL	M	
Sulfide IV	OsS_2	254.3	BK	C	
Oxygen					
Fluoride	OF_2	54.0	B	GAS	
Ozone	O_3	48.0	CL	GAS	
Palladium					
Bromide II	$PdBr_2$	266.6	B		
Chloride II	$PdCl_2$	177.3	R	C	
Fluoride II	PdF_2	144.4	B	T	
Iodide II	PdI_2	360.2	BK		
Oxide II	PdO	122.4	G	T	
Sulfide II	PdS	138.5	BK	T	

TABLE 1.4 Color, Crystal Symmetry, and Refractive Index of Inorganic Compounds (*Continued*)

Compound	Formula	Molecular weight	Color	Crystal symmetry	Refractive index n_D
Phosphorus					
Hypophosphorous Acid	H_3PO_2	66.0	CL		
Phosphoric Acid	H_3PO_4	98.0	CL	R	
Phosphorous Acid	H_3PO_3	82.0	CL		
Bromide III	PBr_3	270.7	CL	LIQ	1.6945[19]
Bromide V	PBr_5	430.5	Y	R	
Chloride III	PCl_3	137.3	CL	LIQ	
Chloride V	PCl_5	208.3	W	T	
Fluoride III	PF_3	88.0	CL	GAS	
Fluoride V	PF_5	126.0	CL	GAS	
Hydride (Phosphine)	PH_3	34.0	CL	GAS	
Iodide III	PI_3	411.7	R	H	
Oxide III	P_4O_6	219.9	W	M	
Oxide IV	PO_2	63.0	CL	R	
Oxide V	P_2O_5	142.0	W	H	
Oxybromide V	$POBr_3$	286.7	CL		
Oxychloride	$POCl_3$	153.4	CL	LIQ	
Oxyfluoride	POF_3	104.0	CL	GAS	
Sulfide	P_4S_7	348.4	Y		
Sulfide V	P_2S_5	222.3	Y		
Thiobromide V	$PSBr_3$	302.8	Y	C	
Thiochloride V	$PSCl_3$	169.4	CL	LIQ	1.635[25]
Platinum					
Bromide II	$PtBr_2$	354.9	B	C	
Bromide IV	$PtBr_4$	514.8	B		
Chloride II	$PtCl_2$	260.0	GN	H	
Chloride IV	$PtCl_4$	336.9	B		
Fluoride IV	PtF_4	271.2	R		
Fluoride VI	PtF_6	309.1	R		
Hydroxide II	$Pt(OH)_2$	229.1	BK		
Hydroxide IV	$Pt(OH)_4$	263.1	B		
Iodide II	PtI_2	448.9	BK		
Oxide II	PtO	211.1	G	T	
Oxide IV	PtO_2	227.1	BK		
Sulfate IV	$Pt(SO_4)_2 \cdot 4H_2O$	459.4	Y		
Sulfide II	PtS	227.2	BK	T	
Sulfide III	Pt_2S_3	486.6	G		
Sulfide IV	PtS_2	259.2	G		
Plutonium					
Bromide III	$PuBr_3$	481.7	GN	R	
Carbide IV	PuC	256.0	SL	C	
Chloride III	$PuCl_3$	346.4	GN	H	
Fluoride III	PuF_3	299.0	P	H	
Fluoride IV	PuF_4	318.0	B	M	
Fluoride VI	PuF_6	356.0	B	R	
Iodide III	PuI_3	622.7	GN	R	
Nitride III	PuN	256.0	BK	C	
Oxide IV	PuO_2	274.0	GN	C	2.4

(*Continued*)

TABLE 1.4 Color, Crystal Symmetry, and Refractive Index of Inorganic Compounds (*Continued*)

Compound	Formula	Molecular weight	Color	Crystal symmetry	Refractive index n_D
Polonium (*Continued*)					
Bromide IV	$PoBr_4$	529.7	R	C	
Chloride II	$PoCl_2$	281.0	R	R	
Chloride IV	$PoCl_4$	351.9	Y	M	
Oxide IV	PoO_2	242.0	R/Y	T/C	
Potassium					
Bromate	$KBrO_3$	167.0	CL	TR	
Bromide	KBr	119.0	CL	C	1.559
Carbonate	K_2CO_3	138.2	CL	M	1.426; 1.431
Chlorate	$KClO_3$	122.6	CL	M	1.409; 1.423
Chloride	KCl	74.6	CL	C	1.490
Cyanide	KCN	65.1	CL	C	1.410
Dichromate	$K_2Cr_2O_7$	294.2	O	M/TR	1.738 TR
Ferrocyanide	$K_4[Fe(CN)_6] \cdot 3H_2O$	422.4	Y	M/T	1.577
Fluoride	KF	58.1	CL	C	1.35
Hydroxide	KOH	56.1	W	C/R	
Iodate	KIO_3	214.0	CL	M	
Iodide	KI	166.0	W	C	1.677
Nitrate	KNO_3	101.1	CL	R/TR	1.335; 1.?
Oxide	K_2O	94.2	CL	C	
Perchlorate	$KClO_4$	138.6	CL	R	1.47
Periodate	KIO_4	230.0	CL	T	1.63
Permanganate	$KMnO_4$	158.0	P	R	1.59
Peroxide	K_2O_2	110.2	Y	R	
Phosphate, o	K_3PO_4	212.3	CL	TR	
Sulfate	K_2SO_4	174.3	CL	R/H	1.495
Sulfide	K_2S	110.3	B	C	
Superoxide	KO_2	71.1	Y	T	
Thiocyanate	KSCN	97.2	CL	R	
Praseodymium					
Bromide	$PrBr_3$	380.6	GN	H	
Chloride	$PrCl_3$	247.3	GN	H	
Fluoride	PrF_3	197.9	GN	H	
Iodide	PrI_3	521.6	G	R	
Oxide	Pr_2O_3	329.8	Y	H	
Sulfate	$Pr_2(SO_4)_3 \cdot 8H_2O$	714.1	GN	M	1.55
Sulfide	Pr_2S_3	378.0	B		
Protactinium					
Bromide IV	$PaBr_4$	470.9	R	T	
Chloride IV	$PaCl_4$	372.9	GN	T	
Fluoride IV	PaF_4	307.1	B	M	
Iodide III	PaI_3	611.8	BK	R	
Oxide IV	PaO_2	263.1	BK	C	
Radium					
Bromide	$RaBr_2$	385.8	Y	M	
Chloride	$RaCl_2$	296.1	Y	M	
Sulfate	$RaSO_4$	322.1	CL	R	

TABLE 1.4   Color, Crystal Symmetry, and Refractive Index of Inorganic Compounds (*Continued*)

Compound	Formula	Molecular weight	Color	Crystal symmetry	Refractive index n_D
Rhenium					
Bromide III	$ReBr_3$	425.9	B		
Chloride III	$ReCl_3$	292.6	R		
Chloride V	$ReCl_5$	363.5	B		
Fluoride IV	ReF_4	262.5	GN	T	
Flouride VI	ReF_6	300.2	Y	LIQ	
Flouride VII	ReF_7	319.2	O	C	
Oxide IV	ReO_2	218.2	BK	M	
Oxide VI	ReO_3	234.2	R	C	
Oxide VII	Re_2O_7	484.4	Y	H	
Oxybromide VII	ReO_3Br	314.1	W		
Oxychloride VII	ReO_3Cl	269.7	CL	LIQ	
Sulfide IV	ReS_2	250.4	BK	H	
Sulfide VII	Re_2S_7	596.9	BK	T	
Rhodium					
Chloride III	$RhCl_3$	209.3	R		
Fluoride III	RhF_3	159.9	R	R	
Hydroxide III	$Rh(OH)_3$	155.9	Y		
Oxide III	Rh_2O_3	253.8	G		
Oxide IV	RhO_2	134.9	B		
Sulfide III	Rh_2S_3	302.0	BK		
Rubidium					
Bromate	$RbBrO_3$	213.4	CL	C	
Bromide	$RbBr$	165.4	CL	C	1.5530
Carbonate	Rb_2CO_3	231.0	CL		
Chloride	$RbCl$	120.9	CL	C	1.493
Fluoride	RbF	104.5	CL	C	1.398
Hydroxide	$RbOH$	102.5	W	R	
Iodide	RbI	212.4	CL	C	1.6474
Nitrate	$RbNO_3$	147.5	CL		1.52
Oxide	Rb_2O	187.0	Y	C	
Perchlorate	$RbClO_4$	189.4		C/R	1.4701
Peroxide	Rb_2O_2	202.9	Y	C	
Sulfate	Rb_2SO_4	267.0	CL	R	1.513
Sulfide	Rb_2S	203.0	Y		
Superoxide	RbO_2	117.5	Y	T	
Ruthenium					
Chloride III	$RuCl_3$	207.4	R	TR/H	
Fluoride V	RuF_5	196.1	GN	M	
Oxide IV	RuO_2	133.1	BE	T	
Oxide VIII	RuO_4	165.1	Y	R	
Sulfide IV	RuS_2	165.2	BK	C	
Samarium					
Bromate III	$Sm(BrO_3)_3 \cdot 9H_2O$	696.2	Y	H	
Bromide II	$SmBr_2$	310.2	B		
Bromide III	$SmBr_3$	390.1	Y	R	
Chloride II	$SmCl_2$	221.3	B	R	

(*Continued*)

TABLE 1.4 Color, Crystal Symmetry, and Refractive Index of Inorganic Compounds (*Continued*)

Compound	Formula	Molecular weight	Color	Crystal symmetry	Refractive index n_D
Samarium (*Continued*)					
Chloride III	$SmCl_3$	256.7	Y	H	
Fluoride II	SmF_2	188.4	Y	C	
Fluoride III	SmF_3	207.4	W	R	
Iodide II	SmI_2	404.2	Y	M	
Iodide III	SmI_3	531.1	Y	H	
Nitrate III	$Sm(NO_3)_3 \cdot 6H_2O$	444.5	Y	TR	
Oxide III	Sm_2O_3	348.7	Y	M	
Sulfate III	$Sm_2(SO_4)_3 \cdot 8H_2O$	733.0	Y	M	1.55
Sulfide III	Sm_2S_3	396.9	Y	C	
Scandium					
Bromide	$ScBr_3$	284.7	W		
Chloride	$ScCl_3$	151.3	CL	RH	
Fluoride	ScF_3	102.0		RH	
Iodide	ScI_3	425.7	W	H	
Nitrate	$Sc(NO_3)_3$	231.0	CL		
Oxide	Sc_2O_3	137.9	W	C	
Sulfate	$Sc_2(SO_4)_3$	378.1	CL		
Selenium					
Bromide I	Se_2Br_2	317.7	R	LIQ	
Bromide IV	$SeBr_4$	398.6	B		
Chloride I	Se_2Cl_2	228.8	B	LIQ	
Chloride IV	$SeCl_4$	220.8	CL	C	1.807
Fluoride IV	SeF_4	154.9	CL	LIQ	
Fluoride VI	SeF_6	192.9	CL	GAS	1.895
Hydride II	H_2Se	81.0	CL	GAS	
Oxide IV	SeO_2	111.0	CL	T	>1.76
Oxide VI	SeO_3	127.0	W	T	
Oxybromide	$SeOBr_2$	254.8	O	LIQ	
Oxychloride	$SeOCl_2$	165.9	Y	LIQ	1.651
Oxyfluoride	$SeOF_2$	133.0	CL	LIQ	
Selenic Acid	H_2SeO_4	145.0	W	R	
Selenous Acid	H_2SeO_3	129.0	CL	H	
Silicon					
Bromide	$SiBr_4$	347.7	CL	LIQ	1.5797^1
Carbide	SiC	40.1	BK	C/H	2.67
Chloride	$SiCl_4$	169.9	CL	LIQ	
Fluoride	SiF_4	104.1	CL	GAS	
Hydride (silane)	SiH_4	32.1	CL	GAS	
Hydride (disilane)	Si_2H_6	62.2	CL	GAS	
Hydride (trisilane)	Si_3H_8	92.3	CL	LIQ	
Iodide	SiI_4	535.7	CL	C	
Nitride	Si_3N_4	140.3	G	H	
Oxide II	SiO	44.1	W	C	
Oxide IV (amorph)	SiO_2	60.1	CL		1.4588
Oxychloride	Si_2OCl_6	284.9	CL	LIQ	
Sulfide	SiS_2	92.2	W	R	

TABLE 1.4   Color, Crystal Symmetry, and Refractive Index of Inorganic Compounds (*Continued*)

Compound	Formula	Molecular weight	Color	Crystal symmetry	Refractive index n_D
Silver					
Bromate	$AgBrO_3$	235.8	CL	T	1.874,1.904
Bromide	$AgBr$	187.8	Y	C	2.253
Carbonate	Ag_2CO_3	257.8	Y		
Chlorate	$AgClO_3$	191.3	W	T	
Chloride	$AgCl$	143.3	W	C	2.071
Cyanide	$AgCN$	133.9	W	H	1.685,1.9
Fluoride	AgF	126.9	Y	C	
Iodate	$AgIO_3$	282.8	CL	R	
Iodide	AgI	234.8	Y	H/C	2.21
Nitrate	$AgNO_3$	169.9	CL	R	1.74
Nitrite	$AgNO_2$	153.9	Y	R	
Oxide	Ag_2O	231.8	B	C	
Perchlorate	$AgClO_4$	207.4	W	C	
Phosphate, o	Ag_3PO_4	418.6	Y	C	
Sulfate	Ag_3SO_4	311.8	W	R	
Sulfide	Ag_2S	247.8	BK	C/R	
Telluride	Ag_2Te	343.4	G	M	
Thiocyanate	$AgSCN$	166.0	CI		
Sodium					
Bicarbonate	$NaHCO_3$	84.0	W	M	1.500
Bromate	$NaBrO_3$	150.9	CL	C	1.594
Bromide	$NaBr$	102.9	Cl	C	1.6412
Carbonate	Na_2CO_3	106.0	W		1.535
Chlorate	$NaClO_3$	106.4	CL	C	1.513
Chloride	$NaCl$	58.4	CL	C	1.544
Cyanide	$NaCN$	49.0	CL	C	1.452
Fluoride	NaF	42.0	CL	C	1.336
Hydride	NaH	24.0	SL	C	1.470
Hydroxide	$NaOH$	40.0	W	R/C	1.358
Iodate	$NaIO_3$	197.9	W	R	
Iodide	NaI	149.9	CL	C	1.775
Nitrate	$NaNO_3$	85.0	CL	TR	1.34;1
Nitrite	$NaNO_2$	69.0	Y	R	
Oxide	Na_2O	62.0	G	C	
Perchlorate	$NaClO_4$	122.4	W	C/R	1.46
Periodate	$NaIO_4$	213.9	CL	T	
Peroxide	Na_2O_2	78.0	Y	H	
Phosphate, o	Na_3PO_4	163.9	W		
Silicate, m	Na_2SiO_3	122.1	CL	M	1.52
Sulfate	Na_2SO_4	142.1	CL	R	1.48
Sulfide	Na_2S	78.1	W	C	
Sulfite	Na_2SO_3	126.1	W	H	1.5
Thiosulfate	$Na_2S_2O_3$	158.1	CL	M	
Strontium					
Bromide	$SrBr_2$	247.5	W	R	1.575
Carbonate	$SrCO_3$	147.6	CL	R	1.521
Chloride	$SrCl_2$	158.5	CL	C	1.650
Fluoride	SrF_2	125.6	CL	C	1.442
Hydride	SrH_2	89.6	W	R	

(*Continued*)

TABLE 1.4 Color, Crystal Symmetry, and Refractive Index of Inorganic Compounds (*Continued*)

Compound	Formula	Molecular weight	Color	Crystal symmetry	Refractive index n_D
Strontium (*Continued*)					
Hydroxide	$Sr(OH)_2$	121.7	W		
Iodate	$Sr(IO_3)_2$	437.4		TR	
Iodide	SrI_2	341.4	CL	—	
Nitrate	$Sr(NO_3)_2$	211.7	CL	C	1.567
Oxide	SrO	103.6	W	C	1.870
Peroxide	SrO_2	119.6	CL	T	
Sulfate	$SrSO_4$	183.7	CL	R	1.62
Sulfide	SrS	119.7	CL	C	2.107
Sulfur					
Bromide I	S_2Br_2	224.0	R	LIQ	1.736
Chloride I	S_2Cl_2	135.0	Y	LIQ	1.666[14]
Chloride II	SCl_2	103.0	R	LIQ	1.557
Chloride IV	SCl_4	173.9	R	LIQ	
Fluoride I	S_2F_2	102.1	CL	GAS	
Fluoride VI	SF_6	146.0	CL	GAS	
Hydride	H_2S	34.1	CL	GAS	1.374
Oxide IV	SO_2	64.1	CL	GAS	
Oxide VI	SO_3	80.1	CL	LIQ	
Pyrosulfuric Acid	$H_2S_2O_7$	178.1	CL	LIQ	
Sulfuric Acid	H_2SO_4	98.1	CL	LIQ	1.429[23]
Sulfuryl Chloride	SO_2Cl_2	135.0	CL	LIQ	1.444[12]
Thionyl Bromide	$SOBr_2$	207.9	Y	LIQ	
Thionyl Chloride	$SOCl_2$	119.0	CL	LIQ	1.527[10]
Tantalum					
Bromide	$TaBr_5$	580.5	Y	R	
Carbide	TaC	193.0	BK	C	
Chloride	$TaCl_5$	358.2	Y	M	
Fluoride	TaF_5	275.9	CL	M	
Iodide	TaI_5	815.4	BK	R	
Nitride	TaN	194.9	BK	H	
Oxide	Ta_2O_5	441.9	CL	R	
Sulfide	Ta_2S_4	490.1	BK	H	
Tellurium					
Bromide II	$TeBr_2$	287.4	GN		
Bromide V	$TeBr_4$	447.3	Y		
Chloride II	$TeCl_2$	198.5	GN		
Chloride IV	$TeCl_4$	269.4	W	M	
Fluoride VI	TeF_6	241.6	CL	GAS	
Hydride	H_2Te	129.6	CL	GAS	
Iodide IV	TeI_4	635.2	BK	R	
Oxide IV	TeO_2	159.6	W	T/R	2.00–2.35
Oxide VI	TeO_3	175.6	Y		
Telluric Acid, o	H_2TeO_6	229.7	W	C	
Terbium					
Bromide	$TbBr_3$	398.6	W		
Chloride	$TbCl_3$	265.3	W		

TABLE 1.4 Color, Crystal Symmetry, and Refractive Index of Inorganic Compounds (*Continued*)

Compound	Formula	Molecular weight	Color	Crystal symmetry	Refractive index n_D
Terbium (*Continued*)					
Fluoride	TBF_3	215.9	W	R	
Iodide	TbI_3	539.6		H	
Nitrate	$Tb(NO_3)_3 \cdot 6H_2O$	453.0	CL	M	
Oxide	Tb_2O_3	365.8	W	C	
Thalliun					
Bromide I	$TlBr$	284.3	W	C	2.4–2.8
Carbonate I	Tl_2CO_3	468.8	CL	M	
Chloride I	$TlCl$	239.8	W	C	2.247
Chloride III	$TlCl_3$	310.8	W	H	
Fluoride	TlF	223.4	CL	R	
Hydroxide I	$TlOH$	221.4	Y	R	
Iodide I	TlI	331.3	Y/R	R/C	2.78
Nitrate I	$TlNO_3$	266.4	W	C/TR	
Oxide I	Tl_2O	424.7	BK	RH	
Oxide III	Tl_2O_3	456.7	CL	C	
Sulfate I	Tl_2SO_4	504.8	CL	R	1.87
Sulfide I	Tl_2S	440.8	BK	T	
Thorium					
Bromide	$ThBr_4$	551.7	W	T	
Carbide	ThC_2	256.1	Y	T	
Chloride	$ThCl_4$	373.9	W	T	
Fluoride	ThF_4	308.0	W	M	
Iodide	ThI_4	739.7	Y	M	
Oxide	ThO_2	264.0	W	C	
Sulfate	$Th(SO_4)_2$	424.2	W	M	
Sulfide	ThS_2	296.2	BK	R	
Thulium					
Bromide	$TmBr_3$	408.7	W	H	
Chloride	$TmCl_3$	275.2	Y	M	
Fluoride	TmF_3	225.9	W	R	
Iodide	TmI_3	549.6	Y	H	
Oxide	Tm_2O_3	385.9	Y	C	
Tin					
Bromide II	$SnBr_2$	278.5	Y	R	
Bromide IV	$SnBr_4$	438.4	CL	R	
Chloride II	$SnCl_2$	189.6	W	R	
Chloride IV	$SnCl_4$	260.5	CL	LIQ	1.512
Fluoride II	SnF_2	156.7	W	M	
Fluoride IV	SnF_4	194.7	W	M	
Hydride	SnH_4	122.7		GAS	
Iodide II	SnI_2	372.5	R	R	
Iodide IV	SnI_4	626.3	R	C	2.106
Oxide II	SnO	143.7	BK	T	
Oxide IV	SnO_2	150.7	W	T	1.996
Sulfide II	SnS	150.8	BK	R	
Sulfide IV	SnS_2	182.8	Y	H	

(*Continued*)

TABLE 1.4   Color, Crystal Symmetry, and Refractive Index of Inorganic Compounds (*Continued*)

Compound	Formula	Molecular weight	Color	Crystal symmetry	Refractive index n_D
Titanium					
Bromide IV	TiBr$_4$	367.6	O	M	
Carbide IV	TiC	59.9	G	C	
Chloride II	TiCl$_2$	118.8	BK	H	
Chloride III	TiCl$_3$	154.3	V	H	
Chloride IV	TiCl$_4$	189.7	Y	LIQ	1.61
Fluoride IV	TiF$_4$	123.9	W		
Iodide IV	TiI$_4$	555.5	B	C	
Nitride	TiN	61.9	Y	C	
Oxide II	TiO	63.9	BK	C	
Oxide IV	TiO$_2$	79.9	BK	T	2.55
Sulfide IV	TiS$_2$	112.0	Y	H	
Tungsten					
Bromide V	WBr$_5$	583.4	B		
Carbide II	W$_2$C	379.7	G	H	
Carbide IV	WC	195.9	G	C	
Chloride V	WCl$_5$	361.1	GN		
Chloride VI	WCl$_6$	396.6	BE	C	
Fluoride VI	WF$_6$	297.8	CL	GAS	
Oxide IV	WO$_2$	215.9	B	T	
Oxide VI	WO$_3$	231.9	Y	M	
Sulfide IV	WS$_2$	248.0	BK	H	
Tungstic Acid	H$_2$WO$_4$	250.0	Y	R	2.24
Uranium					
Bromide III	UBr$_3$	477.8	R	H	
Bromide IV	UBr$_4$	557.7	B	M	
Carbide	UC	250.0	BK	C	
Carbide	UC$_2$	262.0	BK	T	
Chloride III	UCl$_3$	344.4	R	H	
Chloride IV	UCl$_4$	379.9	GN	T	
Fluoride IV	UF$_4$	314.1	GN	M	
Fluoride VI	UF$_6$	352.1	Y	R	1.38
Nitride	UN	252.0	B	C	
Oxide IV	UO$_2$	270.1	BK	C	
Oxide VI	UO$_3$	286.1	R	H	
Oxide IV–VI	U$_3$O$_8$	842.2	BK	R	
Uranyl Acetate	UO$_2$(C$_2$H$_3$O$_2$)$_2 \cdot$ 6H$_2$O	422.1	Y	R	
Uranyl Nitrate	UO$_2$(NO$_3$)$_2 \cdot$ 6H$_2$O	502.1	Y	R	1.49
Vanadium					
Carbide IV	VC	62.9	BK	C	
Chloride IV	VCl$_4$	192.7	R	LIQ	1
Fluoride III	VF$_3$	107.9	GN	R	
Fluoride V	VF$_5$	145.9	CL	R	
Iodide II	VI$_2$	304.7	V	H	
Oxide III	V$_2$O$_3$	149.9	BK	RH	
Oxide IV	VO$_2$	82.9	BE	T	
Oxide V	V$_2$O$_5$	181.9	R	R	
Oxychloride V	VOCl$_3$	173.3	Y	LIQ	
Sulfide II	VS	83.0	BK	H	

TABLE 1.4 Color, Crystal Symmetry, and Refractive Index of Inorganic Compounds (*Continued*)

Compound	Formula	Molecular weight	Color	Crystal symmetry	Refractive index n_D
Xenon					
Fluoride II	XeF_2	169.3	CL	T	
Fluoride IV	XeF_4	207.3	CL	M	
Fluoride VI	XeF_6	245.3	CL	M	
Oxide VI	XeO_3	179.3	CL	R	1.79
Yttebium					
Bromide III	$YbBr_3$	412.8	CL		
Chloride II	$YbCl_2$	244.0	GN	R	
Chloride III	$YbCl_3$	279.3	W	M	
Fluoride III	YbF_3	230.0	W	R	
Iodide II	YbI_2	426.9	BK	H	
Iodide III	YbI_3	553.8	Y	H	
Oxide III	Yb_2O_3	394.1	CL	C	
Sulfate III	$Yb_2(SO_4)_3$	634.3	CL		
Yttrium					
Bromide	YBr_3	328.6	W		
Chloride	YCl_3	195.3	W	M	
Fluoride	YF_3	145.9	W		
Iodide	YI_3	469.6	W	H	
Oxide	Y_2O_3	225.8	W	C	
Sulfate	$Y_2(SO_4)_3$	466.0	W		
Zinc					
Acetate	$Zn(C_2H_3O_2)_2$	183.5	CL	M	
Bromide	$ZnBr_2$	225.2	CL	R	1.5452
Calbonate	$ZnCO_3$	125.4	CL	TR	1.168
Chloride	$ZnCl_2$	136.3	W	H	1.687
Fluoride	ZnF_2	103.4	CL	M	
Hydroxide	$Zn(OH)_2$	99.4	CL	R	
Iodide	ZnI_2	319.2	CL	C	
Nitrate	$Zn(NO_3)_2 \cdot 6H_2O$	297.5	CL	T	
Oxide	ZnO	81.4	W	H	2.01
Sulfate	$ZnSO_4$	161.4	CL	R	1.669
Sulfide	ZnS	97.5	CL	C/H	2.36
Zirconium					
Bromide	$ZrBr_4$	410.9	W		
Carbide	ZrC	103.2	G	C	
Chloride	$ZrCl_4$	233.1	W	C	
Fluoride	ZrF_4	167.2	W	M	1.59
Iodide	ZrI_4	598.8	W		
Nitride	ZrN	105.2	B		
Oxide	ZrO_2	123.2	W	M	

TABLE 1.5 Refractive Index of Minerals

Mineral name	Refractive index	Mineral name	Refractive index
Actinolite	1.618–1.641	Crocoite	2.31–2.66
Adularia moonstone	1.525	Cuprite	2.85
Adventurine feldspar	1.532–1.542		
Adventurine quartz	1.544–1.533	Danburite	1.633
Agalmatoite	1.55	Demantoid garnet	1.88
Agate	1.544–1.553	Diamond	2.417–2.419
Albite feldspar	1.525–1.536	Diopsite	1.68–1.71
Albite moonstone	1.535	Dolomite	1.503–1.682
Alexandrite	1.745–1.759	Dumortierite	1.686–1.723
Almandine garnet	1.76–1.83		
Almandite garnet	1.79	Ekanite	1.60
Amazonite feldspar	1.525	Elaeolite	1.532–1.549
Amber	1.540	Emerald	1.576–1.582
Amblygonite	1.611–1.637	Enstatite	1.663–1.673
Amethyst	1.544–1.553	Epidote	1.733–1.768
Anatase	2.49–2.55	Euclase	1.652–1.672
Andalusite	1.634–1.643		
Andradite garnet	1.82–1.89	Fibrolite	1.659–1.680
Anhydrite	1.571–1.614	Fluorite	1.434
Apatite	1.632–1.648		
Apophyllite	1.536	Gaylussite	1.517
Aquamarine	1.577–1.583	Glass	1.44–1.90
Aragonite	1.530–1.685	Grossular garnet	1.738–1.745
Augelite	1.574–1.588		
Axinite	1.675–1.685	Hambergite	1.559–1.631
Azurite	1.73–1.838	Hauynite	1.502
		Hematite	2.94–3.22
Barite	1.636–1.648	Hemimorphite	1.614–1.636
Barytocalcite	1.684	Hessonite garnet	1.745
Benitoite	1.757–1.8	Hiddenite	1.655–1.68
Beryl	1.577–1.60	Howlite	1.586–1.609
Beryllonite	1.553–1.562	Hypersthene	1.67–1.73
Brazilianite	1.603–1.623		
Brownite	1.567–1.576	Idocrase	1.713–1.72
		Iolite	1.548
Calcite	1.486–1.658	Ivory	1.54
Cancrinite	1.491–1.524		
Cassiterite	1.997–2.093	Jadeite	1.66–1.68
Celestite	1.622–1.631	Jasper	1.54
Cerussite	1.804–2.078	Jet	1.66
Ceylanite	1.77–1.80		
Chalcedony	1.53–1.539	Kornerupine	1.665–1.682
Chalybite	1.63–1.87	Kunzite	1.655–1.68
Chromite	2.1	Kyanite	1.715–1.732
Chrysoberyl	1.745		
Chrysocolla	1.50	Labradorite feldspar	1.565
Chrysoprase	1.534	Lapis gem	1.50
Citrine	1.55	Lazulite	1.615–1.645
Clinozoisite	1.724–1.734	Leucite	1.5085
Colemanite	1.586–1.614		
Coral	1.486–1.658	Magnesite	1.515–1.717
Cordierite	1.541	Malachite	1.655–1.909
Corundum	1.766–1.774	Meerschaum	1.53.... none

TABLE 1.5   Refractive Index of Minerals (*Continued*)

Mineral name	Refractive index	Mineral name	Refractive index
Microcline feldspar	1.525	Serpentine	1.555
Moldavite	1.50	Shell	1.53–1.686
Moss agate	1.54–1.55	Sillimanite	1.658–1.678
		Sinhalite	1.699–1.707
Natrolite	1.48–1.493	Smaragdite	1.608–1.63
Nephrite	1.60–1.63	Smithsonite	1.621–1.849
Nephrite jade	1.600–1.627	Sodalite	1.483
		Spessartite garnet	1.81
Obsidian	1.48–1.51	Spinel	1.712–1.736
Oligoclase feldspar	1.539–1.547	Sphalerite	2.368–2.371
Olivine	1.672	Sphene	1.885–2.05
Onyx	1.486–1.658	Spodumene	1.65–1.68
Opal	1.45	Staurolite	1.739–1.762
Orthoclase feldspar	1.525	Steatite	1.539–1.589
		Stichtite	1.52–1.55
Painite	1.787–1.816	Sulfur	1.96–2.248
Pearl	1.52–1.69		
Periclase	1.74	Taaffeite	1.72
Peridot	1.654–1.69	Tantalite	2.24–2.41
Peristerite	1.525–1.536	Tanzanite	1.691–1.70
Petalite	1.502–1.52	Thomsonite	1.531
Phenakite	1.65–1.67	Tiger eye	1.544–1.553
Phosgenite	2.117–2.145	Topaz (white)	1.638
Prase	1.54–1.533	Topaz (blue)	1.611
Prasiolite	1.54–1.553	Topaz (pink, yellow)	1.621
Prehnite	1.61–1.64	Tourmaline	1.616–1.652
Proustite	2.79–3.088	Tremolite	1.60–1.62
Purpurite	1.84–1.92	Tugtupite	1.496–1.50
Pyrite	1.81	Turquoise	1.61–1.65
Pyrope	1.74	Turquoise gem	1.61
Quartz	1.55	Ulexite	1.49–1.52
		Uvarovite	1.87
Rhodizite	1.69		
Rhodochrisite	1.60–1.82	Variscite	1.55–1.59
Rhodolite garnet	1.76	Vivianite	1.580–1.627
Rhodonite	1.73–1.74		
Rock crystal	1.544–1.553	Wardite	1.59–1.599
Ruby	1.76–1.77	Willemite	1.69–1.72
Rutile	2.61–2.90	Witherite	1.532–1.68
		Wulfenite	2.300–2.40
Sanidine	1.522		
Sapphire	1.76–1.77	Zincite	2.01–1.03
Scapolite	1.54–1.56	Zircon	1.801–2.01
Scapolite (yellow)	1.555	Zirconia (cubic)	2.17
Scheelite	1.92–1.934	Zoisite	1.695

TABLE 1.6 Properties of Molten Salts

Material	Melting point T_m (K)	Boiling point (K)	Density at melting point (g · cm^{-3})	Critical temperature (K)	Volume change on melting $\Delta V_f/\Delta V_s$ 100	Surface tension at melting point (dynes · cm^{-1})	Viscosity at melting point (centipoise)	Sound velocity at melting point (m · cm^{-1})	Cryoscopic constant (K/mole · kg)
LiF	1121	1954	1.83	4140	29.4	252		2546	2.77
NaF	1268	1977	1.96	4270	27.4	185		2080	16.6
KF	1131	1775	1.91	3460	17.2	141		1827	21.8
RbF	1048	1681	—	3280	—	167			38.4
LiCl	883	1655	1.60	3080	26.2	137	1.73	2038	13.7
NaCl	1073	1738	1.55	3400	25.0	116	1.43	1743	20.0
KCl	1043	1680	1.50	3200	17.3	99	1.38	1595	25.4
LiBr	823	1583	2.53	3020	24.3	—		1470	27.6
NaBr	1020	1665	2.36	3200	22.4	100		1325	34.0
KBr	1007	1656	2.133	3170	16.6	90		1256	55.9
NaNO$_2$	544	$d > 593$	1.81		—	120			
KNO$_2$	692	$d623$	—		—	109			
LiNO$_3$	527	—	1.78		21.4	116	5.46	1853	5.93
NaNO$_3$	583	$d653$	1.90		10.7	116	2.89	1808	15.4
KNO$_3$	610	$d > 613$	1.87		3.32	110	2.93	1754	30.8
RbNO$_3$	589	—	2.48		−0.23	109			89.0
AgNO$_3$	483	$d > 485$	3.97			148	4.25	1607	25.9
TlNO$_3$	480	706	4.90			94			58
Li$_2$SO$_4$	1132	—	2.00			225			142
Na$_2$SO$_4$	1157	—	2.07			192			66.3
K$_2$SO$_4$	1347	—	1.88			144			68.7
ZnCl$_2$	548	1005	2.39			53		1002	
HgCl$_2$	550	577	4.37			—			
PbCl$_2$	771	1227	3.77			137	4.25	4952	39.3
Na$_2$WO$_4$	969	—	3.85			202			
Na$_3$AlF$_6$	1273	—	1.84			135			
KCNS	450	—	1.60			101			12.7

Notes: [a]5893 Å; [b]5890 Å.

Material	Heat capacity, C_p (cal./K · mole)	Heat of fusion at melting point (kcal · mole⁻¹)	Entropy of fusion at melting point (entropy units)	Equivalent conductance at 1.1 T_m [(ohm)⁻¹ cm² (equiv)⁻¹]	Decomposition potential of melt (volts)	Measurement temperature for decomposition potential (K)	Molar refractivity at 5461 Å (cm³ · mole⁻¹)	Refractive index at 5461 Å	Measurement temperature for refractive index (K)
LiF	15.50	6.47	5.77	151	2.20	1273	2.89	1.32	1223
NaF	16.40	8.03	6.33	120	2.76	1273	3.41	1.25	1273
KF	16.00	6.75	5.97	148	2.54	1273	5.43	1.28	1173
RbF		6.15	5.76						
LiCl	15.0	4.76	5.39	178.5	3.30	1073	8.32	1.501	883
NaCl	16.0	6.69	6.23	152.3	3.25	1073	9.65	1.320	1173
KCl	16.0	6.34	6.08	122.4	3.37	1073	11.75	1.329	1173
LiBr		4.22	5.13	181	2.95	1073	11.81	1.60	843
NaBr		6.24	6.12	149	2.83	1073	13.19	1.486	1173
KBr		6.10	6.06	108	2.97	1073	15.40	1.436	1173
NaNO₂				58			9.63[a]	1.416[a]	573
KNO₂				~87			11.67	1.356[a]	873
LiNO₃	26.6	5.961	11.66	44			10.74	1.467	573
NaNO₃	37.0	3.696	6.1	58			11.54	1.431	573
KNO₃	29.5	2.413	4.58	46			13.57	1.426	573
RbNO₃		1.105	1.91	35			15.31[b]	1.431[b]	573
AgNO₃	30.6	2.886		38			16.20[a]	1.660[a]	573
TlNO₃		2.264		27			21.38	1.688[b]	573
Li₂SO₄		1.975		123			14.87	1.452	1173
Na₂SO₄		5.67		90			16.53	1.395	1173
K₂SO₄	47.8	9.06		157			20.93	1.388	1173
ZnCl₂	24.1	2.45		~0.08	1.43	973	18.2	1.588	593
HgCl₂	25.0	4.15		0.00096	0.86	973	22.9	1.661	563
PbCl₂		4.40		52.3	1.12	973	26.1	2.024	873
Na₂WO₄				46			24.58	1.542	1173
Na₃AlF₆		27.64					17.2	1.290	1273
KCNS		3.07		17.3			19.65	1.537	573

TABLE 1.7 Triple Points of Various Materials

Substance	Triplet point, K	Pressure, mmHg
Ammonia	195.46	45.58
Argon	83.78	516
Boron tribromide	226.67	
Bromine	280.4	44.1
Carbon dioxide	216.65	
Cyclopropane	145.59	
Deuterium oxide	276.97	
1-Hexene	133.39	
Hydrogen, normal	13.95	54
Hydrogen, para	13.81	
Hydrogen bromide	186.1	~232
Hydrogen chloride	158.8	
Iodine heptafluoride	279.6	
Krypton	115.95	548
Methane	90.67	87.60
Methane-d_1	90.40	84.52
Methane-d_2	90.14	81.80
Methane-d_3	89.94	80.12
Methane-d_4	89.79	79.13
Molybdenum oxide tetrafluoride	370.3	
Molybdenum pentafluoride	340	
Neon	24.55	324
Neptunium hexafluoride	328.25	758.0
Niobium pentabromide	540.6	
Niobium pentachloride	476.5	
Nitrogen	63.15	94
1-Octene	171.45	
Oxygen	54.34	
Phosphorus, white	863	32 760
Plutonium hexafluoride	324.74	533.0
Propene	103.95	
Radon	202	~500
Rhenium dioxide trifluoride	363	
Rhenium heptafluoride	321.4	
Rhenium oxide pentafluoride	313.9	
Rhenium pentafluoride	321	
Succinonitrile (NIST standard)	331.23	
Sulfur dioxide	197.68	1.256
Tantalum pentabromide	553	
Tantalum pentachloride	489.0	
Tungsten oxide tetrafluoride	377.8	
Uranium hexafluoride	337.20	1 139.6
Water	273.16	
Xenon	161.37	612

TABLE 1.8 Density of Mercury and Water

The density of mercury and pure air-free water under a pressure of 101, 325 Pa(1 atm) is given in units of grams per cubic centimeter (g · cm^{-3}). For mercury, the values are based on the density at 20°C being 13.545 884 g · cm^{-3}. Water attains its maximum density of 0.999 973 g · cm^{-3} at 3.98°C. For water, the temperature (t_m, °C) of maximum density at different pressures (p) in atmospheres is given by

$$t_m = 3.98 - 0.0225(p - 1)$$

Density of water	Temp., °C	Density of mercury	Density of water	Temp., °C	Density of mercury
	−20	13.644 59	0.987 12	52	13.467 68
	−18	13.639 62	0.986 18	54	13.462 82
	−16	13.634 66	0.985 21	56	13.457 96
	−14	13.629 70	0.984 22	58	13.453 09
	−12	13.624 75	0.983 20	60	13.448 23
	−10	13.619 79	0.982 16	62	13.443 37
	−8	13.614 85	0.981 09	64	13.438 52
	−6	13.609 90	0.980 01	66	13.433 67
	−4	13.604 96	0.978 90	68	13.428 82
	−2	13.600 02	0.977 77	70	13.423 97
0.999 84	0	13.595 08	0.976 61	72	13.419 13
0.999 94	2	13.590 15	0.975 44	74	13.414 28
0.999 97	4	13.585 22	0.974 24	76	13.409 43
0.999 94	6	13.580 29	0.973 03	78	13.404 60
0.999 85	8	13.575 36	0.971 79	80	13.399 77
0.999 70	10	13.570 44	0.970 53	82	13.394 92
0.999 50	12	13.565 52	0.969 26	84	13.390 09
0.999 24	14	13.560 60	0.967 96	86	13.385 26
0.998 94	16	13.555 70	0.966 65	88	13.380 42
0.998 60	18	13.550 79	0.965 31	90	13.375 60
0.998 20	20	13.545 88	0.963 96	92	13.370 77
0.997 77	22	13.540 97	0.962 59	94	13.365 94
0.997 30	24	13.536 06	0.961 20	96	13.361 12
0.996 78	26	13.531 17	0.959 79	98	13.356 30
0.996 23	28	13.526 26	0.958 36	100	13.351 48
0.995 65	30	13.521 37		120	13.303 4
0.995 03	32	13.516 47		140	13.255 4
0.994 37	34	13.511 58		160	13.207 6
0.993 69	36	13.506 70		180	13.159 8
0.992 97	38	13.501 82		200	13.112 0
0.992 22	40	13.496 93		220	13.064 5
0.991 44	42	13.492 07		240	13.016 9
0.990 63	44	13.487 18		260	12.969 2
0.989 79	46	13.482 29		280	12.921 5
0.988 93	48	13.477 42		300	12.873 7
0.988 04	50	13.472 56			

TABLE 1.9 Specific Gravity of Air at Various Temperatures

The table below gives the weight in grams $\cdot$ 10^4 of 1 mL of air at 760 mm of mercury pressure and at the temperature indicated. Density in grams per milliliter is the same as the specific gravity referred to water at 4°C as unity. To convert to density referred to air at 70°F as unity, divide the values below by 12.00.

t°C	Sp.Gr. $\times 10^4$	t°C	Sp.Gr. $\times 10^4$	t°C	Sp.Gr. $\times 10^4$	t°C	Sp.Gr. $\times 10^4$
−25	14.240	15	12.255	60	10.596	140	8.541
−24	14.182	16	12.213	62	10.532	142	8.500
−23	14.125	17	12.170	64	10.470	144	8.459
−22	14.069	18	12.129	66	10.408	146	8.419
−21	14.013	19	12.087	68	10.347	148	8.379
−20	13.957	20	12.046	70	10.286	150	8.339
−19	13.902	21	12.004	72	10.227	155	8.242
−18	13.847	22	11.964	74	10.168	160	8.147
−17	13.793	23	11.923	76	10.109	165	8.054
−16	13.739	24	11.883	78	10.052	170	7.963
−15	13.685	25	11.843	80	9.995	175	7.874
−14	13.632	26	11.803	82	9.938	180	7.787
−13	13.580	27	11.764	84	9.882	185	7.702
−12	13.527	28	11.725	86	9.828	190	7.619
−11	13.476	29	11.686	88	9.773	195	7.537
−10	13.424	30	11.647	90	9.719	200	7.457
−9	13.373	31	11.609	92	9.666	205	7.379
−8	13.322	32	11.570	94	9.613	210	7.303
−7	13.272	33	11.533	96	9.561	215	7.228
−6	13.222	34	11.495	98	9.509	220	7.155
−5	13.173	35	11.458	100	9.458	230	7.013
−4	13.124	36	11.420	102	9.408	240	6.881
−3	13.075	37	11.383	104	9.358	250	6.753
−2	13.026	38	11.347	106	9.308	260	6.624
−1	12.978	39	11.310	108	9.259	270	6.504
0	12.931	40	11.274	110	9.211	280	6.389
+1	12.883	41	11.238	112	9.163	290	6.277
2	12.836	42	11.202	114	9.116	300	6.166
3	12.790	43	11.167	116	9.069	310	6.062
4	12.743	44	11.132	118	9.022	320	5.942
5	12.697	45	11.097	120	8.976	330	5.847
6	12.652	46	11.062	122	8.931	340	5.755
7	12.606	47	11.027	124	8.886	350	5.664
8	12.561	48	10.993	126	8.841	360	5.578
9	12.517	49	10.958	128	8.797	370	5.493
10	12.472	50	10.924	130	8.753	380	5.407
11	12.428	52	10.857	132	8.710	400	5.248
12	12.385	54	10.791	134	8.667	420	5.101
13	12.341	56	10.725	136	8.625	440	4.952
14	12.298	58	10.660	138	8.583	460	4.812

TABLE 1.10 Boiling Points of Water

psi	Boiling point, °F	psi	Boiling point, °F	psi	Boiling point, °F
0.5	79.6	44	273.1	150	358.5
1	101.7	46	275.8	175	371.8
2	126.0	48	278.5	200	381.9
3	141.4	50	281.0	225	391.9
4	125.9	52	283.5	250	401.0
5	162.2	54	285.9	275	409.5
6	170.0	56	288.3	300	417.4
7	176.8	58	290.5	325	424.8
8	182.8	60	292.7	350	431.8
9	188.3	62	294.9	375	438.4
10	193.2	64	297.0	400	444.7
11	197.7	66	299.0	425	450.7
12	201.9	68	301.0	450	456.4
13	205.9	70	303.0	475	461.9
14	209.6	72	304.9	500	467.1
14.69	212.0	74	306.7	525	472.2
15	213.0	76	308.5	550	477.1
16	216.3	78	310.3	575	481.8
17	219.4	80	312.1	600	486.3
18	222.4	82	313.8	625	490.7
19	225.2	84	315.5	650	495.0
20	228.0	86	317.1	675	499.2
22	233.0	88	318.7	700	503.2
24	237.8	90	320.3	725	507.2
26	242.3	92	321.9	750	511.0
28	246.4	94	323.4	775	514.7
30	250.3	96	324.9	800	518.4
32	254.1	98	326.4	825	521.9
34	257.6	100	327.9	850	525.4
36	261.0	105	331.4	875	528.8
38	264.2	110	334.8	900	532.1
40	267.3	115	338.1	950	538.6
42	270.2	120	341.3	1000	544.8

TABLE 1.11 Boiling Points of Water

A. Barometric Pressures at Various Temperatures

Temp. °C	0.0°	0.2°	0.4°	0.6°	0.8°
	mm of Hg	mm of Hg	mm of Hg	mm of Hg	mm of Hg
80	355.40	358.28	361.19	364.11	367.06
81	370.03	373.01	376.02	379.05	382.09
82	385.16	388.25	391.36	394.49	397.64
83	400.81	404.00	407.22	410.45	413.71
84	416.99	420.29	423.61	426.95	430.32
85	433.71	437.12	440.55	444.01	447.49
86	450.99	454.51	458.06	461.63	465.22
87	468.84	472.48	476.14	479.83	483.54
88	487.28	491.04	494.82	498.63	502.46
89	506.32	510.20	514.11	518.04	521.99
90	525.97	529.98	534.01	538.07	542.15
91	546.26	550.40	554.56	558.75	562.96
92	567.20	571.47	575.76	580.08	584.43
93	588.80	593.20	597.63	602.09	606.57
94	611.08	615.62	620.19	624.79	629.41
95	634.06	638.74	643.45	648.19	652.96
96	657.75	662.58	667.43	672.32	677.23
97	682.18	687.15	692.15	697.19	702.25
98	707.35	712.47	717.63	722.81	728.03
99	733.28	738.56	743.87	749.22	754.59
100	760.00	765.44	770.91	776.42	781.95

B. Boiling Points of Water at Various Pressures

Pressure, atm	Boiling point, °C	Pressure, atm	Boiling point, °C	Pressure, atm	Boiling point, °C	Pressure, atm	Boiling point, °C
0.5	80.9	7	164.2	14	194.1	21	213.9
1	100.0	8	169.6	15	197.4	22	216.2
2	119.6	9	174.5	16	200.4	23	218.5
3	132.9	10	179.0	17	203.4	24	220.8
4	142.9	11	183.2	18	206.1	25	222.9
5	151.1	12	187.1	19	208.8	26	225.0
6	158.1	13	190.7	20	211.4	27	227.0

TABLE 1.12 Refractive Index, Viscosity, Dielectric Constant, and Surface Tension of Water at Various Temperatures

Temp., °C	Refractive index, n_D	Viscosity mN · s · m^{-2}	Dielectric constant, ε	Surface tension mN · s · m^{-2}
0	1.333 95	1.793	87.90	75.83
5	1.333 88	1.521	85.84	75.09
10	1.333 69	1.307	83.96	74.36
15	1.333 39	1.135	82.00	73.62
20	1.333 00	1.002	80.20	72.88
25	1.332 50	0.890 3	78.35	72.14
30	1.331 94	0.797 7	76.60	71.40
35	1.331 31	0.719 0	74.83	70.66
40	1.330 61	0.653 2	73.17	69.92
50	1.329 04	0.547 0	69.58	68.45
60	1.327 25	0.466 5	66.73	66.97
70	1.325 11	0.404 0	63.73	65.49
80		0.354 4	60.86	64.01
90		0.314 5	58.12	62.54
100		0.281 8	55.51	61.07

TABLE 1.13 Compressibility of Water

In the table below are given the relative volumes of water at various temperatures and pressures. The volume at 0°C and one normal atmosphere (760 mm of Hg) is taken as unity.

P, atm	−10°C	0°C	10°C	20°C	40°C	60°C	80°C
1	1.0017	1.0000	1.0001	1.0016	1.0076	1.0168	1.0287
500	0.9788	0.9767	0.9778	0.9804	0.9867	0.9967	1.0071
1000	0.9581	0.9566	0.9591	0.9619	0.9689	0.9780	0.9884
1500	0.9399	0.9394	0.9424	0.9456	0.9529	0.9617	0.9717
2000	0.9223	0.9241	0.9277	0.9312	0.9386	0.9472	0.9568
2500	0.9083	0.9112	0.9147	0.9183	0.9257	0.9343	0.9437
3000	0.8962	0.8993	0.9028	0.9065	0.9139	0.9225	0.9315
3500	0.8852	0.8884	0.8919	0.8956	0.9030	0.9115	0.9203
4000	0.8751	0.8783	0.8818	0.8855	0.8931	0.9012	0.9097
4500	0.8658	0.8692	0.8725	0.8762	0.8838	0.8919	0.9001
5000	0.8573	0.8606	0.8639	0.8675	0.8752	0.8832	0.8913
6000	......	0.8452	0.8481	0.8517	0.8595	0.8674	0.8752
7000	......	......	0.8340	0.8374	0.8456	0.8534	0.8610
8000	......	......	......	0.8244	0.8330	0.8408	0.8483
9000	......	......	......	0.8128	0.8219	0.8297	0.8371
10000	......	......	......	0.8027	0.8119	0.8196	0.8268
11000	......	......	......	......	0.8023	0.8101	0.8172
12000	......	......	......	......	0.7931	0.8009	0.8080

TABLE 1.14   Flammability Limits of Inorganic Compounds in Air

| Compound | Limits of flammability | |
	Lower volume, %	Upper volume, %
Ammonia	15.50	27.00
Carbon monoxide	12.50	74.20
Carbonyl sulfide	11.90	28.50
Cyanogen	6.60	42.60
Hydrocyanic acid	5.60	40.00
Hydrogen	4.00	74.20
Hydrogen sulfide	4.30	45.50

TABLE 1.15   Cyanide Compounds (Inorganic)

Cyanide compounds are a group of compounds based on a structure formed when carbon and nitrogen are combined. When the cyanide function (—C≡N) combines with chemicals from the metal groups, it forms simple salts. Calcium cyanide, potassium cyanide, and sodium cyanide all will liberate hydrogen cyanide (which is also soluble in water and smells like bitter almonds) in acidic water.

Cyanide salts are mainly used in electroplating, metallurgy, the production of organic chemicals (such as acrylonitrile, methyl methacrylate, and adiponitrile), photographic development, the extraction of gold and silver from ores, tanning leather, and the making of plastics and fibers. They are also used to manufacture fumigation chemicals, insecticides, and rodenticides.

The cyanide function is found in combination with other chemicals in the environment. The more common cyanide derivatives are hydrogen cyanide (CASR# 74-90-8), sodium cyanide (CASR# 143-33-9), potassium cyanide (CASR# 151-50-8), and calcium cyanide (CASR# 592-01-8).

Hydrogen cyanide exists as colorless or pale blue liquid or gas with a bitter almond odor detectable at 1 to 5 ppm. Calcium cyanide, potassium cyanide, and sodium cyanide are all examples of simple cyanide salts. They are all white solids, are soluble in water, and have the odor of bitter almonds.

Melting point	Hydrogen cyanide: −13.4°C
	Sodium cyanide: 563.7°C
	Potassium cyanide: 634.5°C
Boiling point	Hydrogen cyanide: 25.6°C
	Sodium cyanide: 1496°C
	Potassium cyanide: 1625°C
Density	Hydrogen cyanide: 0.699 g/mL (liquid)
	Sodium cyanide: 1.6 g/cm^3
	Potassium cyanide: 1.52 g/cm^3

1.3   THE ELEMENTS

The chemical elements are the fundamental materials of which all matter is composed. From the modern viewpoint a substance that cannot be broken down or reduced further is, by definition, an element.

The periodic table displays all the chemical elements organized by atomic number and electron—the elements are presented by increasing atomic number. The trends in the periodic table can help to predict the properties of various elements and the relations between properties. As a result, the periodic table is a useful framework for analyzing chemical behavior and as such is widely used in chemistry and in other sciences.

1.3.1 Groups

A group, or family, is a vertical column in the periodic table. Elements in the same group show patterns in atomic radius, ionization energy, and electronegativity. From top to bottom in a group, the atomic radii of the elements increase: Since there are more filled energy levels, valence electrons are found farther from the nucleus. From top to bottom, each successive element has a lower ionization energy because it is easier to remove an electron since the atoms are less tightly bound. Similarly, from top to bottom, elements decrease in electronegativity due to an increasing distance between the valence electrons and the nucleus. However, there are exceptions to these trends—in group 11, for example, the electronegativity increases down the group.

1.3.2 Periods

A period is a horizontal row in the periodic table. Although groups generally show stronger trends, in some cases horizontal trends can be more significant. For example, in the "f" block, the lanthanides and actinides are two horizontal series of elements that follow significant trends.

Elements in the same period show trends in atomic radius, ionization energy, electron affinity, and electronegativity. Moving left to right across a period, from the alkali metals to the noble gases, atomic radius usually decreases. This is because each successive element has an additional proton and electron, which causes the electrons to be drawn closer to the nucleus. This decrease in atomic radius also causes ionization energy to increase from left to right across a period; the more tightly bound an element is, the more energy is required to remove an electron. Electronegativity increases in the same manner as ionization energy because of the pull exerted on the electrons by the nucleus. Electron affinity also shows a slight trend across a period: Metals (the left side of a period) generally have a lower electron affinity than nonmetals (the right side of a period), with the exception of the noble gases.

1.3.3 Blocks

Because of the importance of the outermost electron shell, the different regions of the periodic table are sometimes referred to as blocks, named according to the subshell in which the "last" electron resides. The "s" block includes the first two groups (alkali metals and alkaline earth metals) as well as hydrogen and helium. The "p" block includes the last six groups, which are groups 13 to 18 in the IUPAC system (3A to 8A in the US group numbering system), and contains, among others, all of the metalloids. The "d" block includes groups 3 to 12 in IUPAC (or 3B to 2B in the US group numbering) and contains all of the transition metals. The "f" block, usually offset below the rest of the periodic table, includes the lanthanide elements (the lanthanide series) and the actinide elements (the actinide series).

1.3.4 Periodicity

The primary determinant of the chemical properties of an element lies in the electron configuration particularly that of the valence shell electrons. For example, all atoms with four valence electrons occupying p orbitals will share some similar characteristics. The type of orbital in which the outermost electrons of an atom reside determines the "block" to which that element belongs. The number of valence shell electrons determines the family, or group, to which the element belongs. The total number of electron shells an atom has determines the period to which it belongs.

The first ionization energy is the energy it takes to remove one electron from an atom, and the second ionization energy is the energy it takes to remove a second electron from that atom. Future ionization energies follow this same pattern. Higher (e.g., third) ionization energies tend to be larger than lower (e.g., second) ionization energies.

The electronic configuration for an element's ground state is a shorthand representation giving the number of electrons (superscript) found in each of the allowed sublevels (s, p, d, f) above a noble gas core (indicated by brackets). In addition, values for the thermal conductivity, the electrical resistance, and the coefficient of linear thermal expansion are included.

Hund's Rule states that for a set of equal-energy orbitals, each orbital is occupied by one electron before any orbital has two. Therefore, the first electrons to occupy orbitals within a sublevel have parallel spins.

TABLE 1.16   Subdivision of Main Energy Levels

Main energy level	1	2		3			4			
Number of sublevels (n)	1	2		3			4			
Number of orbitals (n^2)	1	4		9			16			
Kind and no. of orbitals	s	s	p	s	p	d	s	p	d	f
per sublevel	1	1	3	1	3	5	1	3	5	7
Maximum no. of electrons per sublevel	2	2	6	2	6	10	2	6	10	14
Maximum no. of electrons per main level ($2n^2$)	2	8		18			32			

Hydrogen (1)   Symbol, H. A colorless, odorless gas at room temperature. The most common isotope has atomic weight 1.00794. The lightest and most abundant element in the universe.

• Electrons in first energy level: 1

Helium (2)   Symbol, He. A colorless, odorless gas at room temperature. The most common isotope has atomic weight 4.0026. The second lightest and second most abundant element in the universe.

• Electrons in first energy level: 2

Lithium (3)   Symbol, Li. Classified as an alkali metal. In pure form it is silver-colored. The lightest elemental metal. The most common isotope has atomic weight 6.941.

• Electrons in first energy level: 2
• Electrons in second energy level: 1

Beryllium (4)   Symbol, Be. Classified as an alkaline earth. In pure form it has a grayish color similar to that of steel. Has a relatively high melting point. The most common isotope has atomic weight 9.01218.

• Electrons in first energy level: 2
• Electrons in second energy level: 2

Boron (5)   Symbol, B. Classified as a metalloid. The most common isotope has atomic weight 10.82. Can exist as a powder or as a black, hard metalloid. Boron is not found free in nature.

• Electrons in first energy level: 2
• Electrons in second energy level: 3

TABLE 1.17   Chemical Symbols, Atomic Numbers, and Electron Arrangements of the Elements*

Element name	Chemical symbol	Atomic number
Actinium	Ac	89
Aluminum	Al	13
Americium	Am	95
Antimony	Sb	51
Argon	Ar	18
Arsenic	As	33
Astatine	At	85
Barium	Ba	56
Berkelium	Bk	97
Beryllium	Be	4
Bismuth	Bi	83
Bohrium	Bh	107
Boron	B	5
Bromine	Br	35
Cadmium	Cd	48
Calcium	Ca	20
Californium	Cf	98
Carbon	C	6
Cerium	Ce	58
Cesium	Cs	55
Chlorine	Cl	17
Chromium	Cr	24
Cobalt	Co	27
Copernicium	Cn	112
Copper	Cu	29
Curium	Cm	96
Darmstadtium	Ds	110
Dubnium	Db	105
Dysprosium	Dy	66
Einsteinium	Es	99
Erbium	Er	68
Europium	Eu	63
Fermium	Fm	100
Flerovium	Fl	114
Fluorine	F	9
Francium	Fr	87
Gadolinium	Gd	64
Gallium	Ga	31
Germanium	Ge	32
Gold	Au	79
Hafnium	Hf	72
Hassium	Hs	108
Helium	He	2
Holmium	Ho	67
Hydrogen	H	1
Indium	In	49
Iodine	I	53
Iridium	Ir	77
Iron	Fe	26
Krypton	Kr	36
Lanthanum	La	57
Lawrencium	Lr or Lw	103
Lead	Pb	82
Lithium	Li	3

(Continued)

TABLE 1.17 Chemical Symbols, Atomic Numbers, and Electron Arrangements of the Elements* (*Continued*)

Element name	Chemical symbol	Atomic number
Livermorium	Lv	116
Lutetium	Lu	71
Magnesium	Mg	12
Manganese	Mn	25
Meitnerium	Mt	109
Mendelevium	Md	101
Mercury	Hg	80
Molybdenum	Mo	42
Neodymium	Nd	60
Neon	Ne	10
Neptunium	Np	93
Nickel	Ni	28
Niobium	Nb	41
Nitrogen	N	7
Nobelium	No	102
Osmium	Os	76
Oxygen	O	8
Palladium	Pd	46
Phosphorus	P	15
Platinum	Pt	78
Plutonium	Pu	94
Polonium	Po	84
Potassium	K	19
Praseodymium	Pr	59
Promethium	Pm	61
Protactinium	Pa	91
Radium	Ra	88
Radon	Rn	86
Rhenium	Re	75
Rhodium	Rh	45
Roentgenium	Rg	111
Rubidium	Rb	37
Ruthenium	Ru	44
Rutherfordium	Rf	104
Samarium	Sm	62
Scandium	Sc	21
Seaborgium	Sg	106
Selenium	Se	34
Silicon	Si	14
Silver	Ag	47
Sodium	Na	11
Strontium	Sr	38
Sulfur	S	16
Tantalum	Ta	73
Technetium	Tc	43
Tellurium	Te	52
Terbium	Tb	65
Thallium	Tl	81
Thorium	Th	90
Thulium	Tm	69
Tin	Sn	50
Titanium	Ti	22
Tungsten	W	74
Ununoctium	Uuo	118
Uranium	U	92

TABLE 1.17 Chemical Symbols, Atomic Numbers, and Electron Arrangements of the Elements* (*Continued*)

Element name	Chemical symbol	Atomic number
Vanadium	V	23
Xenon	Xe	54
Ytterbium	Yb	70
Yttrium	Y	39
Zinc	Zn	30
Zirconium	Zr	40

*New names have been proposed, but not yet been accepted, for elements 113, 115, 117, and 118: 113, nihonium; 115, moscovium; 117, tennessine; 118, oganesson.

Carbon (6) Symbol, C. A nonmetallic element that is a solid at room temperature. Has a characteristic hexagonal crystal structure. Known as the basis of life on Earth. The most common isotope has atomic weight 12.011. Exists in three well-known forms: *graphite* (a black powder) which is common, *diamond* (a clear solid) which is rare, and *amorphous*.

Another form of carbon is graphite. Used in electrochemical cells, air-cleaning filters, thermocouples, and noninductive electrical resistors. Also used in medicine to absorb poisons and toxins in the stomach and intestines. Abundant in mineral rocks such as

- Electrons in first energy level: 2
- Electrons in second energy level: 4

Nitrogen (7) Symbol, N. A nonmetallic element that is a colorless, odorless gas at room temperature. The most common isotope has atomic weight 14.007. The most abundant component of the earth's atmosphere (approximately 78 percent at the surface). Reacts to some extent with certain combinations of other elements.

- Electrons in first energy level: 2
- Electrons in second energy level: 5

Oxygen (8) Symbol, O. A nonmetallic element that is a colorless, odorless gas at room temperature. The most common isotope has atomic weight 15.999. The second most abundant component of the earth's atmosphere (approximately 21 percent at the surface).

Combines readily with many other elements, particularly metals. One of the oxides of iron, for example, is known as common rust. Normally, two atoms of oxygen combine to form a molecule (O_2). In this form, oxygen is essential for the sustenance of many forms of life on Earth. When three oxygen atoms form a molecule (O_3), the element is called *ozone*. This form of the element is beneficial in the upper atmosphere because it reduces the amount of ultraviolet radiation reaching the earth's surface. Ozone is, ironically, also known as an irritant and pollutant in the surface air over heavily populated areas.

- Electrons in first energy level: 2
- Electrons in second energy level: 6

Fluorine (9) Symbol, F. The most common isotope has atomic weight 18.998. A gaseous element of the halogen family. Has a characteristic greenish or yellowish color. Reacts readily with many other elements.

- Electrons in first energy level: 2
- Electrons in second energy level: 7

Neon (10)   Symbol, Ne. The most common isotope has atomic weight 20.179. A noble gas present in trace amounts in the atmosphere.

- Electrons in first energy level: 2
- Electrons in second energy level: 8

Sodium (11)   Symbol, Na. The most common isotope has atomic weight 22.9898. An element of the alkali-metal group. A solid at room temperature.

- Electrons in first energy level: 2
- Electrons in second energy level: 8
- Electrons in third energy level: 1

Magnesium (12)   Symbol, Mg. The most common isotope has atomic weight 24.305. A member of the alkaline earth group. At room temperature it is a whitish metal.

- Electrons in first energy level: 2
- Electrons in second energy level: 8
- Electrons in third energy level: 2

Aluminum (13)   Symbol, Al. The most common isotope has atomic weight 26.98. A metallic element and a good electrical conductor. Has many of the same characteristics as magnesium, except it reacts less easily with oxygen in the atmosphere.

- Electrons in first energy level: 2
- Electrons in second energy level: 8
- Electrons in third energy level: 3

Silicon (14)   Symbol, Si. The most common isotope has atomic weight 28.086. A metalloid abundant in the earth's crust. Especially common in rocks such as granite, and in many types of sand.

- Electrons in first energy level: 2
- Electrons in second energy level: 8
- Electrons in third energy level: 4

Phosphorus (15)   Symbol, P. The most common isotope has atomic weight 30.974. A nonmetallic element of the nitrogen family. Found in certain types of rock.

- Electrons in first energy level: 2
- Electrons in second energy level: 8
- Electrons in third energy level: 5

Sulfur (16)   Symbol, S. Also spelled *sulphur*. The most common isotope has atomic weight 32.06. A nonmetallic element. Reacts with some other elements.

- Electrons in first energy level: 2
- Electrons in second energy level: 8
- Electrons in third energy level: 6

Chlorine (17)   Symbol, Cl. The most common isotope has atomic weight 35.453. A gas at room temperature and a member of the halogen family. Reacts readily with various other elements.

- Electrons in first energy level: 2
- Electrons in second energy level: 8
- Electrons in third energy level: 7

Argon (18)   Symbol, A or Ar. The most common isotope has atomic weight 39.94. A gas at room temperature; classified as a noble gas. Present in small amounts in the atmosphere.

- Electrons in first energy level: 2
- Electrons in second energy level: 8
- Electrons in third energy level: 8

Potassium (19)   Symbol, K. The most common isotope has atomic weight 39.098. A member of the alkali metal group.

- Electrons in first energy level: 2
- Electrons in second energy level: 8
- Electrons in third energy level: 8
- Electrons in fourth energy level: 1

Calcium (20)   Symbol, Ca. The most common isotope has atomic weight 40.08. A metallic element of the alkaline-earth group. Calcium carbonate, or calcite, is abundant in the earth's crust, especially in limestone.

- Electrons in first energy level: 2
- Electrons in second energy level: 8
- Electrons in third energy level: 8
- Electrons in fourth energy level: 2

Scandium (21)   Symbol, Sc. The most common isotope has atomic weight 44.956. In the pure form it is a soft metal. Classified as a transition metal.

- Electrons in first energy level: 2
- Electrons in second energy level: 8
- Electrons in third energy level: 9
- Electrons in fourth energy level: 2

Titanium (22)   Symbol, Ti. The most common isotope has atomic weight 47.88. Classified as a transition metal.

- Electrons in first energy level: 2
- Electrons in second energy level: 8
- Electrons in third energy level: 10
- Electrons in fourth energy level: 2

Vanadium (23)   Symbol, V. The most common isotope has atomic weight 50.94. Classified as a transition metal. In its pure form it is whitish in color.

- Electrons in first energy level: 2
- Electrons in second energy level: 8
- Electrons in third energy level: 11
- Electrons in fourth energy level: 2

Chromium (24) Symbol, Cr. The most common isotope has atomic weight 51.996. Classified as a transition metal. In its pure form it is grayish in color.

- Electrons in first energy level: 2
- Electrons in second energy level: 8
- Electrons in third energy level: 13
- Electrons in fourth energy level: 1

Manganese (25) Symbol, Mn. The most common isotope has atomic weight 54.938. Classified as a transition metal. In its pure form it is grayish in color.

- Electrons in first energy level: 2
- Electrons in second energy level: 8
- Electrons in third energy level: 13
- Electrons in fourth energy level: 2

Iron (26) Symbol, Fe. The most common isotope has atomic weight 55.847. In its pure form it is a dull gray metal.

- Electrons in first energy level: 2
- Electrons in second energy level: 8
- Electrons in third energy level: 14
- Electrons in fourth energy level: 2

Cobalt (27) Symbol, Co. The most common isotope has atomic weight 58.94. Classified as a transition metal. In the pure form it is silvery in color.

- Electrons in first energy level: 2
- Electrons in second energy level: 8
- Electrons in third energy level: 15
- Electrons in fourth energy level: 2

Nickel (28) Symbol, Ni. The most common isotope has atomic weight 58.69. Classified as a transition metal. In its pure form it is light gray to white.

- Electrons in first energy level: 2
- Electrons in second energy level: 8
- Electrons in third energy level: 16
- Electrons in fourth energy level: 2

Copper (29) Symbol, Cu. The most common isotope has atomic weight 63.546. Classified as a transition metal. In its pure form it has a characteristic red or wine color.

- Electrons in first energy level: 2
- Electrons in second energy level: 8
- Electrons in third energy level: 18
- Electrons in fourth energy level: 1

Zinc (30) Symbol, Zn. The most common isotope has atomic weight 65.39. Classified as a transition metal. In pure form, it is a dull blue-gray color.

- Electrons in first energy level: 2
- Electrons in second energy level: 8

- Electrons in third energy level: 18
- Electrons in fourth energy level: 2

Gallium (31)   Symbol, Ga. The most common isotope has atomic weight 69.72. A semiconducting metal. In pure form it is light gray to white.

- Electrons in first energy level: 2
- Electrons in second energy level: 8
- Electrons in third energy level: 18
- Electrons in fourth energy level: 3

Germanium (32)   Symbol, Ge. The most common isotope has atomic weight 72.59. A semiconducting metalloid.

- Electrons in first energy level: 2
- Electrons in second energy level: 8
- Electrons in third energy level: 18
- Electrons in fourth energy level: 4

Arsenic (33)   Symbol, As. The most common isotope has atomic weight 74.91. A metalloid used as a dopant in the manufacture of semiconductors. In its pure form it is gray in color.

- Electrons in first energy level: 2
- Electrons in second energy level: 8
- Electrons in third energy level: 18
- Electrons in fourth energy level: 5

Selenium (34)   Symbol, Se. The most common isotope has atomic weight 78.96. Classified as a nonmetal. In its pure form it is gray in color.

- Electrons in first energy level: 2
- Electrons in second energy level: 8
- Electrons in third energy level: 18
- Electrons in fourth energy level: 6

Bromine (35)   Symbol, Br. The most common isotope has atomic weight 79.90. A nonmetallic element of the halogen family. A reddish-brown liquid at room temperature. Has a characteristic unpleasant odor. Reacts readily with various other elements.

- Electrons in first energy level: 2
- Electrons in second energy level: 8
- Electrons in third energy level: 18
- Electrons in fourth energy level: 7

Krypton (36)   Symbol, Kr. The most common isotope has atomic weight 83.80. Classified as a noble gas. Colorless and odorless. Present in trace amounts in the earth's atmosphere. Some common isotopes of this element are radioactive.

- Electrons in first energy level: 2
- Electrons in second energy level: 8
- Electrons in third energy level: 18
- Electrons in fourth energy level: 8

Rubidium (37)   Symbol, Rb. The most common isotope has atomic weight 85.468. Classified as an alkali metal. In its pure form it is silver-colored. Reacts easily with oxygen and chlorine.

- Electrons in first energy level: 2
- Electrons in second energy level: 8
- Electrons in third energy level: 18
- Electrons in fourth energy level: 8
- Electrons in fifth energy level: 1

Strontium (38)   Symbol, Sr. The most common isotope has atomic weight 87.62. A metallic element of the alkaline-earth group. In pure form it is gold-colored.

- Electrons in first energy level: 2
- Electrons in second energy level: 8
- Electrons in third energy level: 18
- Electrons in fourth energy level: 8
- Electrons in fifth energy level: 2

Yttrium (39)   Symbol, Y. The most common isotope has atomic weight 88.906. Classified as a transition metal. In its pure form it is silver-colored.

- Electrons in first energy level: 2
- Electrons in second energy level: 8
- Electrons in third energy level: 18
- Electrons in fourth energy level: 9
- Electrons in fifth energy level: 2

Zirconium (40)   Symbol, Zr. The most common isotope has atomic weight 91.22. Classified as a transition metal. In its pure form it is grayish in color.

- Electrons in first energy level: 2
- Electrons in second energy level: 8
- Electrons in third energy level: 18
- Electrons in fourth energy level: 10
- Electrons in fifth energy level: 2

Niobium (41)   Symbol, Nb. The most common isotope has atomic weight 92.91. Classified as a transition metal. This element is sometimes called *columbium*. In pure form it is shiny, and is light gray to white in color.

- Electrons in first energy level: 2
- Electrons in second energy level: 8
- Electrons in third energy level: 18
- Electrons in fourth energy level: 12
- Electrons in fifth energy level: 1

Molybdenum (42)   Symbol, Mo. The most common isotope has atomic weight 95.94. Classified as a transition metal. In its pure form, it is hard and silver-white.

Used as a catalyst, as a component of hard alloys for the aeronautical and aerospace industries, and in steel-hardening processes. It is known for high thermal conductivity, low thermal-expansion coefficient, high melting point, and resistance to corrosion. Most molybdenum compounds are relatively nontoxic.

- Electrons in first energy level: 2
- Electrons in second energy level: 8
- Electrons in third energy level: 18
- Electrons in fourth energy level: 13
- Electrons in fifth energy level: 1

Technetium (43)   Symbol, Tc. Formerly called *masurium*. The most common isotope has atomic weight 98. Classified as a transition metal. In its pure form, it is grayish in color. This element is not found in nature; it occurs when the uranium atom is split by nuclear fission. It also occurs when molybdenum is bombarded by high-speed deuterium nuclei (particles consisting of one proton and one neutron). This element is radioactive.

- Electrons in first energy level: 2
- Electrons in second energy level: 8
- Electrons in third energy level: 18
- Electrons in fourth energy level: 14
- Electrons in fifth energy level: 1

Ruthenium (44)   Symbol, Ru. The most common isotope has atomic weight 101.07. A rare element, classified as a transition metal. In pure form it is silver-colored.

- Electrons in first energy level: 2
- Electrons in second energy level: 8
- Electrons in third energy level: 18
- Electrons in fourth energy level: 15
- Electrons in fifth energy level: 1

Rhodium (45)   Symbol, Rh. The most common isotope has atomic weight 102.906. Classified as a transition metal. In its pure form it is silver-colored. Occurs in nature along with platinum and nickel.

- Electrons in first energy level: 2
- Electrons in second energy level: 8
- Electrons in third energy level: 18
- Electrons in fourth energy level: 16
- Electrons in fifth energy level: 1

Palladium (46)   Symbol, Pd. The most common isotope has atomic weight 106.42. Classified as a transition metal. In its pure form it is light gray to white. In nature, palladium is found with copper ore.

- Electrons in first energy level: 2
- Electrons in second energy level: 8
- Electrons in third energy level: 18
- Electrons in fourth energy level: 18
- Electrons in fifth energy level: 0

Silver (47)   Symbol, Ag. The most common isotope has atomic weight 107.87. Classified as a transition metal. In its pure form it is a bright, shiny, and silverish-white colored metal.

- Electrons in first energy level: 2
- Electrons in second energy level: 8
- Electrons in third energy level: 18

- Electrons in fourth energy level: 18
- Electrons in fifth energy level: 1

Cadmium (48)   Symbol, Cd. The most common isotope has atomic weight 112.41. Classified as a transition metal. In its pure form it is silver-colored.

- Electrons in first energy level: 2
- Electrons in second energy level: 8
- Electrons in third energy level: 18
- Electrons in fourth energy level: 18
- Electrons in fifth energy level: 2

Indium (49)   Symbol, In. The most common isotope has atomic weight 114.82. A metallic element used as a dopant in semiconductor processing. In pure form it is silver-colored. In nature, it is often found along with zinc.

- Electrons in first energy level: 2
- Electrons in second energy level: 8
- Electrons in third energy level: 18
- Electrons in fourth energy level: 18
- Electrons in fifth energy level: 3

Tin (50)   Symbol, Sn. The most common isotope has atomic weight 118.71. In pure form it is a white or grayish metal. It changes color (from white to gray) when it is cooled through a certain temperature range. It is ductile and malleable.

- Electrons in first energy level: 2
- Electrons in second energy level: 8
- Electrons in third energy level: 18
- Electrons in fourth energy level: 18
- Electrons in fifth energy level: 4

Antimony (51)   Symbol, Sb. The most common isotope has atomic weight 121.76. Classified as a metalloid. In pure form, it is blue-white or blue-gray in color. Has a characteristic flakiness and brittleness.

- Electrons in first energy level: 2
- Electrons in second energy level: 8
- Electrons in third energy level: 18
- Electrons in fourth energy level: 18
- Electrons in fifth energy level: 5

Tellurium (52)   Symbol, Te. The most common isotope has atomic weight 127.60. A rare metalloid element related to selenium. In pure form, it is silverish-white and has high luster. In nature it is found along with other metals such as copper. It has a characteristic brittleness.

- Electrons in first energy level: 2
- Electrons in second energy level: 8
- Electrons in third energy level: 18
- Electrons in fourth energy level: 18
- Electrons in fifth energy level: 6

Iodine (53)   Symbol, I. The most common isotope has atomic weight 126.905. A member of the halogen family. In pure form it has a black or purple-black color.

- Electrons in first energy level: 2
- Electrons in second energy level: 8
- Electrons in third energy level: 18
- Electrons in fourth energy level: 18
- Electrons in fifth energy level: 7

Xenon (54)   Symbol, Xe. The most common isotope has atomic weight 131.29. Classified as a noble gas. Colorless and odorless; present in trace amounts in the earth's atmosphere.

- Electrons in first energy level: 2
- Electrons in second energy level: 8
- Electrons in third energy level: 18
- Electrons in fourth energy level: 18
- Electrons in fifth energy level: 8

Cesium (55)   Symbol, Cs. Also spelled *caesium* (in Britain). The most common isotope has atomic weight 132.91. Classified as an alkali metal. In pure form, it is silver-white in color, is ductile, and is malleable.

- Electrons in first energy level: 2
- Electrons in second energy level: 8
- Electrons in third energy level: 18
- Electrons in fourth energy level: 18
- Electrons in fifth energy level: 8
- Electrons in sixth energy level: 1

Barium (56)   Symbol, Ba. The most common isotope has atomic weight 137.36. Classified as an alkaline earth. In pure form it is silver-white in color, and is relatively soft; it is sometimes mistaken for lead.

- Electrons in first energy level: 2
- Electrons in second energy level: 8
- Electrons in third energy level: 18
- Electrons in fourth energy level: 18
- Electrons in fifth energy level: 8
- Electrons in sixth energy level: 2

Lanthanum (57)   Symbol, La. The most common isotope has atomic weight 138.906. Classified as a rare earth. In pure form it is white in color, malleable, and soft.

- Electrons in first energy level: 2
- Electrons in second energy level: 8
- Electrons in third energy level: 18
- Electrons in fourth energy level: 18
- Electrons in fifth energy level: 9
- Electrons in sixth energy level: 2

Cerium (58)   Symbol, Ce. The most common isotope has atomic weight 140.13. Classified as a rare earth. In pure form it is light silvery-gray. It reacts readily with various other elements and is malleable and ductile.

- Electrons in first energy level: 2
- Electrons in second energy level: 8
- Electrons in third energy level: 18
- Electrons in fourth energy level: 20
- Electrons in fifth energy level: 8
- Electrons in sixth energy level: 2

Praseodymium (59)   Symbol, Pr. The most common isotope has atomic weight 140.908. Classified as a rare earth. In pure form it is silver-gray, soft, malleable, and ductile.

- Electrons in first energy level: 2
- Electrons in second energy level: 8
- Electrons in third energy level: 18
- Electrons in fourth energy level: 21
- Electrons in fifth energy level: 8
- Electrons in sixth energy level: 2

Neodymium (60)   Symbol, Nd. The most common isotope has atomic weight 144.24. Classified as a rare earth. In pure form it is shiny and is silvery in color.

- Electrons in first energy level: 2
- Electrons in second energy level: 8
- Electrons in third energy level: 18
- Electrons in fourth energy level: 22
- Electrons in fifth energy level: 8
- Electrons in sixth energy level: 2

Promethium (61)   Symbol, Pm. Formerly called *illinium*. The most common isotope has atomic weight 145. Classified as a rare earth. In pure form it is gray in color, and is highly radioactive.

- Electrons in first energy level: 2
- Electrons in second energy level: 8
- Electrons in third energy level: 18
- Electrons in fourth energy level: 23
- Electrons in fifth energy level: 8
- Electrons in sixth energy level: 2

Samarium (62)   Symbol, Sm. The most common isotope has atomic weight 150.36. Classified as a rare earth. In pure form it is silvery-white in color with high luster.

- Electrons in first energy level: 2
- Electrons in second energy level: 8
- Electrons in third energy level: 18
- Electrons in fourth energy level: 24

- Electrons in fifth energy level: 8
- Electrons in sixth energy level: 2

Europium (63)   Symbol, Eu. The most common isotope has atomic weight 151.96. Classified as a rare earth. In pure form it is silver-gray in color, and has ductility similar to that of lead.

- Electrons in first energy level: 2
- Electrons in second energy level: 8
- Electrons in third energy level: 18
- Electrons in fourth energy level: 25
- Electrons in fifth energy level: 8
- Electrons in sixth energy level: 2

Gadolinium (64)   Symbol, Gd. The most common isotope has atomic weight 157.25. Classified as a rare earth. In pure form it is silver in color, is ductile, and is malleable.

- Electrons in first energy level: 2
- Electrons in second energy level: 8
- Electrons in third energy level: 18
- Electrons in fourth energy level: 25
- Electrons in fifth energy level: 9
- Electrons in sixth energy level: 2

Terbium (65)   Symbol, Tb. The most common isotope has atomic weight 158.93. Classified as a rare earth. In pure form it is silver-gray, soft, malleable, and ductile.

- Electrons in first energy level: 2
- Electrons in second energy level: 8
- Electrons in third encrgy level: 18
- Electrons in fourth energy level: 27
- Electrons in fifth energy level: 8
- Electrons in sixth energy level: 2

Dysprosium (66)   Symbol, Dy. The most common isotope has atomic weight 162.5. Classified as a rare earth. In pure form it has a bright, shiny silver color. It is soft and malleable, but it has a relatively high melting point.

- Electrons in first energy level: 2
- Electrons in second energy level: 8
- Electrons in third energy level: 18
- Electrons in fourth energy level: 28
- Electrons in fifth energy level: 8
- Electrons in sixth energy level: 2

Holmium (67)   Symbol, Ho. The most common isotope has atomic weight 164.93. Classified as a rare earth. In pure form it is silver in color. It is soft and malleable.

- Electrons in first energy level: 2
- Electrons in second energy level: 8

- Electrons in third energy level: 18
- Electrons in fourth energy level: 29
- Electrons in fifth energy level: 8
- Electrons in sixth energy level: 2

Erbium (68)   Symbol, Er. The most common isotope has atomic weight 167.26. Classified as a rare earth. In pure form it is silverish, soft, malleable, and ductile.

- Electrons in first energy level: 2
- Electrons in second energy level: 8
- Electrons in third energy level: 18
- Electrons in fourth energy level: 30
- Electrons in fifth energy level: 8
- Electrons in sixth energy level: 2

Thulium (69)   Symbol, Tm. The most common isotope has atomic weight 168.93. Classified as a rare earth. In pure form this element is grayish in color, soft, malleable, and ductile.

- Electrons in first energy level: 2
- Electrons in second energy level: 8
- Electrons in third energy level: 18
- Electrons in fourth energy level: 31
- Electrons in fifth energy level: 8
- Electrons in sixth energy level: 2

Ytterbium (70)   Symbol, Yb. The most common isotope has atomic weight 173.04. Classified as a rare earth. In pure form it is silver-white in color, soft, malleable, and ductile.

- Electrons in first energy level: 2
- Electrons in second energy level: 8
- Electrons in third energy level: 18
- Electrons in fourth energy level: 32
- Electrons in fifth energy level: 8
- Electrons in sixth energy level: 2

Lutetium (71)   Symbol, Lu. The most common isotope has atomic weight 174.967. Classified as a rare earth. In its pure form, it is silver-white and radioactive, with a half-life on the order of thousands of millions of years.

- Electrons in first energy level: 2
- Electrons in second energy level: 8
- Electrons in third energy level: 18
- Electrons in fourth energy level: 32
- Electrons in fifth energy level: 9
- Electrons in sixth energy level: 2

Hafnium (72)   Symbol, Hf. The most common isotope has atomic weight 178.49. Classified as a transition metal. In pure form, it is silver-colored, shiny, and ductile.

- Electrons in first energy level: 2
- Electrons in second energy level: 8
- Electrons in third energy level: 18
- Electrons in fourth energy level: 32
- Electrons in fifth energy level: 10
- Electrons in sixth energy level: 2

Tantalum (73)   Symbol, Ta. The most common isotope has atomic weight 180.95. Classified as a transition metal; an element of the vanadium family. In pure form it is grayish-silver in color, ductile, and hard, with a high melting point.

- Electrons in first energy level: 2
- Electrons in second energy level: 8
- Electrons in third energy level: 18
- Electrons in fourth energy level: 32
- Electrons in fifth energy level: 11
- Electrons in sixth energy level: 2

Tungsten (74)   Symbol, W. Also known as *wolfram.* The most common isotope has atomic weight 183.85. Classified as a transition metal. In pure form it is silver-colored. It has an extremely high melting point.

- Electrons in first energy level: 2
- Electrons in second energy level: 8
- Electrons in third energy level: 18
- Electrons in fourth energy level: 32
- Electrons in fifth energy level: 12
- Electrons in sixth energy level: 2

Rhenium (75)   Symbol, Re. The most common isotope has atomic weight 186.207. Classified as a transition metal. In pure form it is silver-white, has high density, and has a high melting point.

- Electrons in first energy level: 2
- Electrons in second energy level: 8
- Electrons in third energy level: 18
- Electrons in fourth energy level: 32
- Electrons in fifth energy level: 13
- Electrons in sixth energy level: 2

Osmium (76)   Symbol, Os. The most common isotope has atomic weight 190.2. A transition metal of the platinum group. In pure form it is bluish-silver in color, dense, hard, and brittle.

- Electrons in first energy level: 2
- Electrons in second energy level: 8
- Electrons in third energy level: 18
- Electrons in fourth energy level: 32

- Electrons in fifth energy level: 14
- Electrons in sixth energy level: 2

Iridium (77)   Symbol, Ir. The most common isotope has atomic weight 192.22. A transition metal of the platinum group. In pure form it is yellowish-white in color with high luster; it is hard, brittle, and has high density.

- Electrons in first energy level: 2
- Electrons in second energy level: 8
- Electrons in third energy level: 18
- Electrons in fourth energy level: 32
- Electrons in fifth energy level: 15
- Electrons in sixth energy level: 2

Platinum (78)   Symbol, Pt. The most common isotope has atomic weight 195.08. Classified as a transition metal. In pure form it has a brilliant, shiny, white luster. It is malleable and ductile.

- Electrons in first energy level: 2
- Electrons in second energy level: 8
- Electrons in third energy level: 18
- Electrons in fourth energy level: 32
- Electrons in fifth energy level: 17
- Electrons in sixth energy level: 1

Gold (79)   Symbol, Au. The most common isotope has atomic weight 196.967. A transition metal. In pure form it is shiny, yellowish, ductile, malleable, and comparatively soft.

- Electrons in first energy level: 2
- Electrons in second energy level: 8
- Electrons in third energy level: 18
- Electrons in fourth energy level: 32
- Electrons in fifth energy level: 18
- Electrons in sixth energy level: 1

Mercury (80)   Symbol, Hg. The most common isotope has atomic weight 200.59. Classified as a transition metal. In pure form it is silver-colored and liquid at room temperature.

- Electrons in first energy level: 2
- Electrons in second energy level: 8
- Electrons in third energy level: 18
- Electrons in fourth energy level: 32
- Electrons in fifth energy level: 18
- Electrons in sixth energy level: 2

Thallium (81)   Symbol, Tl. The most common isotope has atomic weight 204.38. A metallic element. In pure form it is bluish-gray or dull gray, soft, malleable, and ductile.

- Electrons in first energy level: 2
- Electrons in second energy level: 8

- Electrons in third energy level: 18
- Electrons in fourth energy level: 32
- Electrons in fifth energy level: 18
- Electrons in sixth energy level: 3

Lead (82)   Symbol, Pb. The most common isotope has atomic weight 207.2. A metallic element. In pure form it is dull gray or blue-gray, soft, and malleable; relatively low melting temperature.

- Electrons in first energy level: 2
- Electrons in second energy level: 8
- Electrons in third energy level: 18
- Electrons in fourth energy level: 32
- Electrons in fifth energy level: 18
- Electrons in sixth energy level: 4

Bismuth (83)   Symbol, Bi. The most common isotope has atomic weight 208.98. A metallic element. In pure form it is pinkish-white and brittle.

- Electrons in first energy level: 2
- Electrons in second energy level: 8
- Electrons in third energy level: 18
- Electrons in fourth energy level: 32
- Electrons in fifth energy level: 18
- Electrons in sixth energy level: 5

Polonium (84)   Symbol, Po. The most common isotope has atomic weight 209. Classified as a metalloid. It is produced from the decay of radium and is sometimes called radium-F. Polonium is radioactive; it emits primarily alpha particles.

- Electrons in first energy level: 2
- Electrons in second energy level: 8
- Electrons in third energy level: 18
- Electrons in fourth energy level: 32
- Electrons in fifth energy level: 18
- Electrons in sixth energy level: 6

Astatine (85)   Symbol, At. The most common isotope has atomic weight 210. Formerly called *alabamine*. Classified as a halogen. The element is radioactive.

- Electrons in first energy level: 2
- Electrons in second energy level: 8
- Electrons in third energy level: 18
- Electrons in fourth energy level: 32
- Electrons in fifth energy level: 18
- Electrons in sixth energy level: 7

Radon (86)   Symbol, Rn. The most common isotope has atomic weight 222. Classified as a noble gas. It is radioactive, emitting primarily alpha particles, and has a short half-life. Radon is a colorless gas that results from the disintegration of radium.

- Electrons in first energy level: 2
- Electrons in second energy level: 8
- Electrons in third energy level: 18
- Electrons in fourth energy level: 32
- Electrons in fifth energy level: 18
- Electrons in sixth energy level: 8

Francium (87)   Symbol, Fr. The most common isotope has atomic weight 223. Classified as an alkali metal. This element is radioactive, and all isotopes decay rapidly. Produced as a result of the radioactive disintegration of actinium.

- Electrons in first energy level: 2
- Electrons in second energy level: 8
- Electrons in third energy level: 18
- Electrons in fourth energy level: 32
- Electrons in fifth energy level: 18
- Electrons in sixth energy level: 8
- Electrons in seventh energy level: 1

Radium (88)   Symbol, Ra. The most common isotope has atomic weight 226. Classified as an alkaline earth. In pure form it is silver-gray, but darkens quickly when exposed to air. This element is radioactive, emitting alpha particles, beta particles, and gamma rays. It has a moderately long half-life.

- Electrons in first energy level: 2
- Electrons in second energy level: 8
- Electrons in third energy level: 18
- Electrons in fourth energy level: 32
- Electrons in fifth energy level: 18
- Electrons in sixth energy level: 8
- Electrons in seventh energy level: 2

Actinium (89)   Symbol, Ac. The most common isotope has atomic weight 227. Classified as a rare earth. In pure form it is silver-gray in color. This element is radioactive, emitting beta particles. The most common isotope has a half-life of 21.6 years.

- Electrons in first energy level: 2
- Electrons in second energy level: 8
- Electrons in third energy level: 18
- Electrons in fourth energy level: 32
- Electrons in fifth energy level: 18
- Electrons in sixth energy level: 9
- Electrons in seventh energy level: 2

Thorium (90)   Symbol, Th. The most common isotope has atomic weight 232.038. Classified as a rare earth and a member of the actinide series. In pure form it is silver-colored, soft, ductile, and malleable.

- Electrons in first energy level: 2
- Electrons in second energy level: 8

- Electrons in third energy level: 18
- Electrons in fourth energy level: 32
- Electrons in fifth energy level: 18
- Electrons in sixth energy level: 10
- Electrons in seventh energy level: 2

Protactinium (91)   Symbol, Pa. Formerly called *protoactinium*. The most common isotope has atomic weight 231.036. Classified as a rare earth. In pure form it is silver-colored.

- Electrons in first energy level: 2
- Electrons in second energy level: 8
- Electrons in third energy level: 18
- Electrons in fourth energy level: 32
- Electrons in fifth energy level: 20
- Electrons in sixth energy level: 9
- Electrons in seventh energy level: 2

Uranium (92)   Symbol, U. The most common isotope has atomic weight 238.029. Classified as a rare earth. In pure form it is silver-colored, malleable, and ductile.

- Electrons in first energy level: 2
- Electrons in second energy level: 8
- Electrons in third energy level: 18
- Electrons in fourth energy level: 32
- Electrons in fifth energy level: 21
- Electrons in sixth energy level: 9
- Electrons in seventh energy level: 2

Neptunium (93)   Symbol, Np. The most common isotope has atomic weight 237. Classified as a rare earth. In pure form it is silver-colored, and reacts with various other elements to form compounds.

- Electrons in first energy level: 2
- Electrons in second energy level: 8
- Electrons in third energy level: 18
- Electrons in fourth energy level: 32
- Electrons in fifth energy level: 23
- Electrons in sixth energy level: 8
- Electrons in seventh energy level: 2

Plutonium (94)   Symbol, Pu. The most common isotope has atomic weight 244. Classified as a rare earth. In pure form it is silver-colored; when it is exposed to air, a yellow oxide layer forms. Plutonium reacts with various other elements to form compounds.

- Electrons in first energy level: 2
- Electrons in second energy level: 8
- Electrons in third energy level: 18
- Electrons in fourth energy level: 32
- Electrons in fifth energy level: 24

- Electrons in sixth energy level: 8
- Electrons in seventh energy level: 2

Americium (95)   Symbol, Am. The most common isotope has atomic weight 243. Classified as a rare earth. In pure form it is silver-white and malleable.

- Electrons in first energy level: 2
- Electrons in second energy level: 8
- Electrons in third energy level: 18
- Electrons in fourth energy level: 32
- Electrons in fifth energy level: 25
- Electrons in sixth energy level: 8
- Electrons in seventh energy level: 2

Curium (96)   Symbol, Cm. The most common isotope has atomic weight 247. Classified as a rare earth. In pure form it is silvery in color, and it reacts readily with various other elements. This element, like most transuranic elements, is dangerously radioactive.

- Electrons in first energy level: 2
- Electrons in second energy level: 8
- Electrons in third energy level: 18
- Electrons in fourth energy level: 32
- Electrons in fifth energy level: 25
- Electrons in sixth energy level: 9
- Electrons in seventh energy level: 2

Berkelium (97)   Symbol, Bk. The most common isotope has atomic weight 247. Classified as a rare earth. It is radioactive with a short half-life. Berkelium is a human-made element and is not known to occur in nature.

- Electrons in first energy level: 2
- Electrons in second energy level: 8
- Electrons in third energy level: 18
- Electrons in fourth energy level: 32
- Electrons in fifth energy level: 26
- Electrons in sixth energy level: 9
- Electrons in seventh energy level: 2

Californium (98)   Symbol, Cf. The most common isotope has atomic weight 251. Classified as a rare earth. It is radioactive, emitting neutrons in large quantities. It is human-made element, not known to occur in nature.

- Electrons in first energy level: 2
- Electrons in second energy level: 8
- Electrons in third energy level: 18
- Electrons in fourth energy level: 32

- Electrons in fifth energy level: 28
- Electrons in sixth energy level: 8
- Electrons in seventh energy level: 2

Einsteinium (99)   Symbol, E or Es. The most common isotope has atomic weight 252. Classified as a rare earth. It is radioactive with a short half-life. Einsteinium is a human-made element and is not known to occur in nature.

- Electrons in first energy level: 2
- Electrons in second energy level: 8
- Electrons in third energy level: 18
- Electrons in fourth energy level: 32
- Electrons in fifth energy level: 29
- Electrons in sixth energy level: 8
- Electrons in seventh energy level: 2

Fermium (100)   Symbol, Fm. The most common isotope has atomic weight 257. Classified as a rare earth. It has a short half-life, is human-made, and is not known to occur in nature.

- Electrons in first energy level: 2
- Electrons in second energy level: 8
- Electrons in third energy level: 18
- Electrons in fourth energy level: 32
- Electrons in fifth energy level: 30
- Electrons in sixth energy level: 8
- Electrons in seventh energy level: 2

Mendelevium (101)   Symbol, Md or Mv. The most common isotope has atomic weight 258. Classified as a rare earth. It has a short half-life, is human-made, and is not known to occur in nature.

- Electrons in first energy level: 2
- Electrons in second energy level: 8
- Electrons in third energy level: 18
- Electrons in fourth energy level: 32
- Electrons in fifth energy level: 31
- Electrons in sixth energy level: 8
- Electrons in seventh energy level: 2

Nobelium (102)   Symbol, No. The most common isotope has atomic weight 259. Classified as a rare earth. It has a short half-life (seconds or minutes, depending on the isotope), is human-made, and is not known to occur in nature.

- Electrons in first energy level: 2
- Electrons in second energy level: 8
- Electrons in third energy level: 18
- Electrons in fourth energy level: 32
- Electrons in fifth energy level: 32

- Electrons in sixth energy level: 8
- Electrons in seventh energy level: 2

Lawrencium (103)   Symbol, Lr or Lw. The most common isotope has atomic weight 262. Classified as a rare earth. It has a half-life less than one minute, is human-made, and is not known to occur in nature.

- Electrons in first energy level: 2
- Electrons in second energy level: 8
- Electrons in third energy level: 18
- Electrons in fourth energy level: 32
- Electrons in fifth energy level: 32
- Electrons in sixth energy level: 9
- Electrons in seventh energy level: 2

Rutherfordium (104)   Symbol, Rf. Also called *unnilquadium* (Unq) and *Kurchatovium* (Ku). The most common isotope has atomic weight 261. Classified as a transition metal. It has a half-life on the order of a few seconds to a few tenths of a second (depending on the isotope), is human-made, and is not known to occur in nature.

- Electrons in first energy level: 2
- Electrons in second energy level: 8
- Electrons in third energy level: 18
- Electrons in fourth energy level: 32
- Electrons in fifth energy level: 32
- Electrons in sixth energy level: 10
- Electrons in seventh energy level: 2

Dubnium (105)   Symbol, Db. Also called *unnilpentium* (Unp) and *Hahnium* (Ha). The most common isotope has atomic weight 262. Classified as a transition metal. It has a half-life on the order of a few seconds to a few tenths of a second (depending on the isotope), is human-made, and is not known to occur in nature.

- Electrons in first energy level: 2
- Electrons in second energy level: 8
- Electrons in third energy level: 18
- Electrons in fourth energy level: 32
- Electrons in fifth energy level: 32
- Electrons in sixth energy level: 11
- Electrons in seventh energy level: 2

Seaborgium (106)   Symbol, Sg. Also called *unnilhexium* (Unh). The most common isotope has atomic weight 263. Classified as a transition metal. It has a half-life on the order of one second or less, is human-made, and is not known to occur in nature.

- Electrons in first energy level: 2
- Electrons in second energy level: 8
- Electrons in third energy level: 18

- Electrons in fourth energy level: 32
- Electrons in fifth energy level: 32
- Electrons in sixth energy level: 12
- Electrons in seventh energy level: 2

Bohrium (107)   Symbol, Bh. Also called *unnilseptium* (Uns). The most common isotope has atomic weight 262. Classified as a transition metal. It is human-made and is not known to occur in nature.

- Electrons in first energy level: 2
- Electrons in second energy level: 8
- Electrons in third energy level: 18
- Electrons in fourth energy level: 32
- Electrons in fifth energy level: 32
- Electrons in sixth energy level: 13
- Electrons in seventh cnergy level: 2

Hassium (108)   Symbol, Hs. Also called *unniloctium* (Uno). The most common isotope has atomic weight 265. Classified as a transition metal. It is human-made and not known to occur in nature.

- Electrons in first energy level: 2
- Electrons in second energy level: 8
- Electrons in third energy level: 18
- Electrons in fourth energy level: 32
- Electrons in fifth energy level: 32
- Electrons in sixth energy level: 14
- Electrons in seventh energy level: 2

Meitnerium (109)   Symbol, Mt. Also called *unnilenium* (Une). The most common isotope has atomic weight 266. Classified as a transition metal. It is human-made and not known to occur in nature.

- Electrons in first energy level: 2
- Electrons in second energy level: 8
- Electrons in third energy level: 18
- Electrons in fourth energy level: 32
- Electrons in fifth energy level: 32
- Electrons in sixth energy level: 15
- Electrons in seventh energy level: 2

Darmstadtium (110)   Symbol, Ds. The most common isotope has atomic weight 281. Classified as a transition metal. It is human-made and not known to occur in nature.

- Electrons in first energy level: 2
- Electrons in second energy level: 8
- Electrons in third energy level: 18
- Electrons in fourth energy level: 32
- Electrons in fifth energy level: 32
- Electrons in sixth energy level: 17
- Electrons in seventh energy level: 1

Roentgenium (111)   Symbol, Rg. The most common isotope has atomic weight 280. Classified as a transition metal. It is human-made and not known to occur in nature.

- Electrons in first energy level: 2
- Electrons in second energy level: 8
- Electrons in third energy level: 18
- Electrons in fourth energy level: 32
- Electrons in fifth energy level: 32
- Electrons in sixth energy level: 18
- Electrons in seventh energy level: 1

Copernicium (112)   Symbol, Cn. The most common isotope has atomic weight 285. Classified as a transition metal. It is human-made and not known to occur in nature.

- Electrons in first energy level: 2
- Electrons in second energy level: 8
- Electrons in third energy level: 18
- Electrons in fourth energy level: 32
- Electrons in fifth energy level: 32
- Electrons in sixth energy level: 18
- Electrons in seventh energy level: 2

Ununtrium (113)   Symbol, Uut. The most common isotope has atomic weight 284. The synthesis of or appearance of such an atom is believed possible because of the observation of flerovium (Fl, element 114) in the laboratory.

Flerovium (114)   Symbol, Fl. The most common isotope has atomic weight 289. First reported in January 1999. It is human-made and not known to occur in nature.

Ununpentium (115)   Symbol, Uup. The most common isotope has atomic weight 288. The synthesis or appearance of such an atom is believed possible because of the observation of livermorium (Lv, element 116) in the laboratory.

Livermorium (116)   Symbol, Lv. The most common isotope has atomic weight 293. First reported in January 1999. It is a decomposition product of ununoctium, and it in turn decomposes into flerovium. It is not known to occur in nature.

Ununseptium (117)   Symbol, Uus. The most common isotope has atomic weight 294. The synthesis or appearance of such an atom is believed possible because of the observation of ununoctium (Uuo, element 118) in the laboratory.

Ununoctium (118)   Symbol, Uuo. The most common isotope has atomic weight 294. It is the result of the fusion of krypton and lead and decomposes into livermorium. It is not known to occur in nature.

TABLE 1.18 Atomic Numbers, Periods, and Groups of the Elements (The Periodic Table)

Group / Period	1	2	3	4	5	6	7	8	9	10	11	12	13	14	15	16	17	18
1	1 H																1 H	2 He
2	3 Li	4 Be											5 B	6 C	7 N	8 O	9 F	10 Ne
3	11 Na	12 Mg											13 Al	14 Si	15 P	16 S	17 Cl	18 Ar
4	19 K	20 Ca	21 Sc	22 Ti	23 V	24 Cr	25 Mn	26 Fe	27 Co	28 Ni	29 Cu	30 Zn	31 Ga	32 Ge	33 As	34 Se	35 Br	36 Kr
5	37 Rb	38 Sr	39 Y	40 Zr	41 Nb	42 Mo	43 Tc	44 Ru	45 Rh	46 Pd	47 Ag	48 Cd	49 In	50 Sn	51 Sb	52 Te	53 I	54 Xe
6	55 Cs	56 Ba	71 Lu *	72 Hf	73 Ta	74 W	75 Re	76 Os	77 Ir	78 Pt	79 Au	80 Hg	81 Tl	82 Pb	83 Bi	84 Po	85 At	86 Rn
7	87 Fr	88 Ra	103 Lr **	104 Unq	105 Db	106 Sg	107 Bh	108 Hs	109 Mt	110 Ds	111 Rg	112 Cn	113 Uut	114 Fl	115 Uup	116 Lv	117 Uus	118 Uuo

*Lanthanides	57 La	58 Ce	59 Pr	60 Nd	61 Pm	62 Sm	63 Eu	64 Gd	65 Tb	66 Dy	67 Ho	68 Er	69 Tm	70 Yb
†Actinides	89 Ac	90 Th	91 Pa	92 U	93 Np	94 Pu	95 Am	96 Cm	97 Bk	98 Cf	99 Es	100 Fm	101 Md	102 No

TABLE 1.19 Atomic Weights of the Elements*

Name	Atomic number	Symbol	Atomic weight
Actinium	89	Ac	[227]
Aluminum	13	Al	26.981538
Americium	95	Am	[243]
Antimony	51	Sb	121.76
Argon	18	Ar	39.948
Arsenic	33	As	74.9216
Astatine	85	At	[210]
Barium	56	Ba	137.327
Berkelium	97	Bk	[247]
Beryllium	4	Be	9.012182
Bismuth	83	Bi	8.98038
Bohrium	107	Bh	[264]
Boron	5	B	10.811
Bromine	35	Br	79.904
Cadmium	48	Cd	112.411
Caesium	55	Cs	132.90545
Calcium	20	Ca	40.078
Californium	98	Cf	[251]
Carbon	6	C	12.0107
Cerium	58	Ce	140.116
Chlorine	17	Cl	35.4527
Chromium	24	Cr	51.9961
Cobalt	27	Co	8.9332
Copernicium	112	Cn	[285]
Copper	29	Cu	63.546
Curium	96	Cm	[247]
Darmstadtium	110	Ds	[281]
Dubnium	105	Db	[262]
Dysprosium	66	Dy	162.5
Einsteinium	99	Es	[252]
Erbium	68	Er	167.26
Europium	63	Eu	151.964
Fermium	100	Fm	[257]
Fluorine	9	F	18.9984032
Francium	87	Fr	[223]
Gadolinium	64	Gd	157.25
Gallium	31	Ga	69.723
Germanium	32	Ge	72.61
Gold	79	Au	196.96655
Hafnium	72	Hf	178.49
Hassium	108	Hs	[265]
Helium	2	He	4.002602
Holmium	67	Ho	164.93032
Hydrogen	1	H	1.00794
Indium	49	In	114.818
Iodine	53	I	126.90447
Iridium	77	Ir	192.217
Iron	26	Fe	55.845
Krypton	36	Kr	83.8
Lanthanum	57	La	138.9055
Lawrencium	103	Lr	[262]
Lead	82	Pb	207.2
Lithium	3	Li	6.941
Lutetium	71	Lu	174.967
Magnesium	12	Mg	24.305
Manganese	25	Mn	54.938049
Meitnerium	109	Mt	[268]

TABLE 1.19   Atomic Weights of the Elements* (*Continued*)

Name	Atomic number	Symbol	Atomic weight
Mendelevium	101	Md	[258]
Mercury	80	Hg	200.59
Molybdenum	42	Mo	95.94
Neodymium	60	Nd	144.24
Neon	10	Ne	20.1797
Neptunium	93	Np	[237]
Nickel	28	Ni	58.6934
Niobium	41	Nb	92.90638
Nitrogen	7	N	14.00674
Nobelium	102	No	[259]
Osmium	76	Os	190.23
Oxygen	8	O	15.9994
Palladium	46	Pd	106.42
Phosphorus	15	P	30.973761
Platinum	78	Pt	195.078
Plutonium	94	Pu	[244]
Polonium	84	Po	[209]
Potassium	19	K	39.0983
Praseodymium	59	Pr	140.90765
Promethium	61	Pm	[145]
Protactinium	91	Pa	231.03588
Radium	88	Ra	[226]
Radon	86	Rn	[222]
Rhenium	75	Re	186.207
Rhodium	45	Rh	102.9055
Roentgenium	111	Rg	[280]
Rubidium	37	Rb	85.4678
Ruthenium	44	Ru	101.07
Rutherfordium	104	Rf	[261]
Samarium	62	Sm	150.36
Scandium	21	Sc	44.95591
Seaborgium	106	Sg	[263]
Selenium	34	Se	78.96
Silicon	14	Si	28.0855
Silver	47	Ag	107.8682
Sodium	11	Na	22.98977
Strontium	38	Sr	87.62
Sulfur	16	S	32.066(6)
Tantalum	73	Ta	180.9479
Technetium	43	Tc	[98]
Tellurium	52	Te	127.6
Terbium	65	Tb	158.92534
Thallium	81	Tl	204.3833
Thorium	90	Th	232.0381
Thulium	69	Tm	168.93421
Tin	50	Sn	118.71
Titanium	22	Ti	47.867
Tungsten	74	W	183.84
Uranium	92	U	238.0289
Vanadium	23	V	50.9415
Xenon	54	Xe	131.29
Ytterbium	70	Yb	173.04
Yttrium	39	Y	88.90585
Zinc	30	Zn	65.39
Zirconium	40	Zr	91.224

*New names have been proposed, but not yet been accepted, for elements 113, 115, 117, and 118: 113, nihonium; 115, moscovium; 117, tennessine; 118, oganesson.

TABLE 1.20 Physical Properties of the Elements*

The relative atomic masses in the following table are based on the $^{12}C = 12$ scale; a value in brackets denotes the mass number of the most stable isotope. The data are based on the most recent values adopted by IUPAC, with a maximum of six significant figures.

ρ denotes density, $\theta_{C,m}$ denotes melting temperature, $\theta_{C,b}$ denotes boiling temperature, and c_p denotes specific heat capacity. subl. denotes sublimes

Element	Symbol	Atomic number	Relative atomic mass	ρ/g cm^{-3}	$\theta_{C,m}$/°C	$\theta_{C,b}$/°C	c_p/J kg^{-1} K^{-1}	Oxidation states
Actinium	Ac	89	227.028	10.1	1050	3200	900	3
Aluminium	Al	13	26.9815	2.70	660	2470	140	3
Americium	Am	95	(243)	11.7	(1200)	(2600)	209	3, 4, 5, 6
Antimony	Sb	51	121.75	6.62	630	1380	519	3, 5
Argon	Ar	18	39.948	1.40 (87 K)	−189	−186	326	
Arsenic (α, grey)	As	33	74.9216	5.72	(302)	613 subl.	(140)	3, 5
Astatine	At	85	(210)			(380)		
Barium	Ba	56	137.33	3.51	714	1640	192	2
Berkelium	Bk	97	(247)					3, 4
Beryllium	Be	4	9.01218	1.85	1280	2477	1.82×10^3	2
Bismuth	Bi	83	208.980	9.80	271	1560	121	3, 5
Boron	B	5	10.81	2.34	2300	3930	1.03×10^3	3
Bromine	Br	35	79.904	3.12	−7.2	58.8	448	1, 3, 4, 5, 6
Cadmium	Cd	48	112.41	8.64	321	765	230	2
Caesium	Cs	55	132.905	1.90	28.7	690	234	1
Calcium	Ca	20	40.08	1.54	850	1487	653	2
Californium	Cf	98	(251)					3
Carbon	C	6	12.011	2.25 (graphite)	3730 subl.	4830	711 (graphite)	2, 4
				3.51 (diamond)			519 (diamond)	
Cerium	Ce	58	140.12	6.78	795	3470	184	3, 4
Chlorine	Cl	17	35.453	1.56 (238 K)	−101	−34.7	477	1, 3, 4, 5, 6, 7
Chromium	Cr	24	51.996	7.19	1890	2482	448	2, 3, 6
Cobalt	Co	27	58.9332	8.90	1492	2900	435	2, 3
Copper	Cu	29	63.546	8.92	1083	2595	385	1,2
Curium	Cm	96	(247)					3
Dysprosium	Dy	66	162.50	8.56	1410	2600	172	3
Einsteinium	Es	99	(252)					3
Erbium	Er	68	167.26	9.16	1500	2900	167	3
Europium	Eu	63	151.96	5.24	826	1440	138	2, 3

Fermium	Fm	100	(257)					3
Fluorine	F	9	18.9984	1.11 (85 K)	−220	−188	824	1
Francium	Fr	87	(223)		(27)	(680)	(140)	1
Gadolinium	Gd	64	157.25	7.95	1310	3000	234	3
Gallium	Ga	31	69.72	5.91	29.8	2400	381	3
Germanium	Ge	32	72.59	5.35	937	2830	322	4
Gold	Au	79	196.967	19.3	1063	2970	130	1, 3
Hafnium	Hf	72	178.49	13.3	2220	5400	146	4
Helium	He	2	4.00260	0.147 (4 K)	−270	−269	5.19×10^3	
Holmium	Ho	67	164.930	8.80	1460	2600	163	3
Hydrogen	H	1	1.0079	0.070 (20 K)	−259	−252	1.43×10^4	1
Indium	In	49	114.82	7.30	157	2000	238	1, 3
Iodine	I	53	126.905	4.93	114	184	218	1, 3, 5, 7
Iridium	Ir	77	192.22	22.5	2440	5300	134	2, 3, 4, 6
Iron	Fe	26	55.847	7.86	1535	3000	448	2, 3, 6
Krypton	Kr	36	83.80	2.16 (121 K)	−157	−152	247	2
Lanthanum	La	57	138.906	6.19	920	3470	201	3
Lawrencium	Lr	103	(260)					3
Lead	Pb	82	207.2	11.3	327	1744	130	2, 4
Lithium	Li	3	6.941	0.53	180	1330	3.39×10^3	1
Lutetium	Lu	71	174.967	9.84	1650	3330	155	3
Magnesium	Mg	12	24.305	1.74	650	1110	1.03×10^3	2
Manganese	Mn	25	54.9380	7.20	1240	2100	477	2, 3, 4, 6, 7
Mendelevium	Md	101	(258)					3
Mercury	Hg	80	200.59	13.6	−38.9	357	138	1, 2
Molybdenum	Mo	42	95.94	10.2	2610	5560	251	2, 3, 4, 5, 6
Neodymium	Nd	60	144.24	7.00	1020	3030	188	3
Neon	Ne	10	20.179	1.20 (27 K)	−249	−246	1.03×10^3	
Neptunium	Np	93	237.048	20.4	640			3, 4, 5, 6
Nickel	Ni	28	58.69	8.90	1453	2730	439	2, 3
Niobium	Nb	41	92.9064	8.57	2470	3300	264	3, 5

(Continued)

TABLE 1.20 Physical Properties of the Elements* (*Continued*)

Element	Symbol	Atomic number	Relative atomic mass	ρ/g cm^{-3}	$\theta_{C,m}$/°C	$\theta_{C,b}$/°C	c_p/J kg^{-1} K^{-1}	Oxidation states
Nitrogen	N	7	14.0067	0.808 (77 K)	−210	−196	1.04×10^3	1, 2, 3, 4, 5
Nobelium	No	102	(259)					2
Osmium	Os	76	190.2	22.5	3000	5000	130	2, 3, 4, 6, 8
Oxygen	O	8	15.9994	1.15 (90 K)	−218	−183	916	2
Palladium	Pd	46	106.42	12.0	1550	3980	243	2, 4
Phosphorus	P	15	30.9738	1.82 (white) 2.34 (red)	44.2 (white) 590 (red)	280 (white)	757 (white) 670 (red)	3, 5
Platinum	Pt	78	195.08	21.4	1769	4530	134	2, 4, 6
Plutonium	Pu	94	(244)	19.8	640	3240		3, 4, 5, 6
Polonium	Po	84	(209)	9.4	254	960	126	2, 4
Potassium	K	19	39.0983	0.86	63.7	774	753	1
Praseodymium	Pr	59	140.908	6.78	935	3130	192	3, 4
Promethium	Pm	61	(145)		1030	2730	184	3
Protoactinium	Pa	91	231.036	15.4	1230		121	4, 5
Radium	Ra	88	226.025	5.0	700	1140	121	2
Radon	Rn	86	(222)	4.4 (211 K)	−71	−61.8	92	
Rhenium	Re	75	186.207	20.5	3180	5630	138	2, 4, 5, 6, 7
Rhodium	Rh	45	102.906	12.4	1970	4500	243	2, 3, 4
Rubidium	Rb	37	85.4678	1.53	38.9	688	360	1
Ruthenium	Ru	44	101.07	12.3	2500	4900	238	3, 4, 5, 6, 8
Samarium	Sm	62	150.36	7.54	1070	1900	197	2, 3
Scandium	Sc	21	44.9559	2.99	1540	2730	556	3
Selenium	Se	34	78.96	4.81	217	685	322	2, 4, 6
Silicon	Si	14	28.0855	2.33	1410	2360	711	4
Silver	Ag	47	107.868	10.5	961	2210	234	1
Sodium	Na	11	22.9898	0.97	97.8	890	1.23×103	1
Strontium	Sr	38	87.62	2.62	768	1380	284	2
Sulphur (α, rhombic)	S	16	32.06	2.07 (α) 1.96 (β)	113 (α) 119 (β)	445	732	2, 4, 6
Tantalum	Ta	73	180.948	16.6	3000	5420	138	5
Technetium	Tc	43	(98)	11.5	2200	3500	243	7
Tellurium	Te	52	127.60	6.25	450	990	201	2, 4, 6

Terbium	Tb	65	158.925	8.27	1360	2800	184	3, 4
Thallium	Tl	81	204.383	11.8	304	1460	130	1, 3
Thorium	Th	90	232.038	11.7	1750	3850	113	3, 4
Thulium	Tm	69	168.934	9.33	1540	1730	159	2, 3
Tin (white)	Sn	50	118.71	7.28 (white)	232	2270	218	2, 4
				5.75 (grey)				
Titanium	Ti	22	47.88	4.54	1675	3260	523	2, 3, 4
Tungsten	W	74	183.85	19.4	3410	5930	134	2, 4, 5, 6
Uranium	U	92	238.029	19.1	1130	3820	117	3, 4, 5, 6
Vanadium	V	23	50.9415	5.96	1900	3000	481	2, 3, 4, 5
Xenon	Xe	54	131.29	3.52 (165 K)	−112	−108	159	2, 4, 6, 8
Ytterbium	Yb	70	173.04	6.98	824	1430	146	2, 3
Yttrium	Y	39	88.9059	4.34	1500	2930	297	3
Zinc	Zn	30	65.39	7.14	420	907	385	2
Zirconium	Zr	40	91.224	6.49	1850	3580	276	2, 3, 4

*New names have been proposed, but not yet been accepted, for elements 113, 115, 117, and 118: 113, nihonium; 115, moscovium; 117, tennessine; 118, oganesson.

139

TABLE 1.21 Conductivity and Resistivity of the Elements*

Name	Symbol	Atomic number	Electronic configuration	Thermal conductivity, $W \cdot (m \cdot K)^{-1}$ at 25°C	Electrical resistivity, $\mu\Omega \cdot cm$ at 20°C	Coefficient of linear thermal expansion (25°C), $m \cdot m^{-1}$ ($\times 10^6$)
Actinium	Ac	89	[Rn] $6d^2\,7s$	12		23.1
Aluminum	Al	13	[Ne] $3s^2\,3p$	237	2.6548	
Americium	Am	95	[Rn] $5f^7\,7s^2$	10		11.0
Antimony	Sb	51	[Kr] $4d^{10}\,5s^2\,5p^3$	24.4	41.7	
Argon	Ar	18	[Ne] $3s^2\,3p^6$	0.017 72		
Arsenic	As	33	[Ar] $3d^{10}\,4s^2\,4p^3$	50.2	33.3	
Astatine	At	85	[Xe] $4f^{14}\,5d^{10}\,6s^2\,6p^5$	1.7		
Barium	Ba	56	[Xe] $6s^2$	18.4	33.2	20.6
Berkelium	Bk	97	[Rn] $5f^8\,6d\,7s^2$	10		
Beryllium	Be	4	[He] $2s^2$	200	3.56	11.3
Bismuth	Bi	83	[Xe] $4f^{14}\,5d^{10}\,6s^2\,6p^3$	7.97	129	13.4
Boron	B	5	[He] $2s^2\,2p$	27.4	1.5×10^{12}	5–7
Bromine	Br	35	[Ar] $3d^{10}\,4s^2\,4p^5$	0.122	7.8×10^{18}	
Cadmium	Cd	48	[Kr] $4d^{10}\,5s^2$	96.6	7.27 (22°C)	30.8
Calcium	Ca	20	[Ar] $4s^2$	201	3.36	22.3
Californium	Cf	98	[Rn] $5f^{10}\,7s^2$			
Carbon	C	6	[He] $2s^2\,2p^2$			
(amorphous)				1.59		
(diamond)				900–2320	0.8	
(graphite)				119–165	1375	
Cerium	Ce	58	[Xe] $4f\,5d\,6s^2$	11.3	82.8 (β, hex)	6.3
Cesium	Cs	55	[Xe] $6s$	35.9	20.5	
Chlorine	Cl	17	[Ne] $3s^2\,3p^5$	0.0089	$>10^9$	
Chromium	Cr	24	[Ar] $3d^5\,4s$	93.9	12.5	4.9
Cobalt	Co	27	[Ar] $3d^7\,4s^2$	100	6.24	13.0
Copper	Cu	29	[Ar] $3d^{10}\,4s$	401	1.678	16.5
Curium	Cm	96	[Rn] $5f^7\,6d\,1s^2$			
Dysprosium	Dy	66	[Xe] $4f^{10}\,6s^2$	10.7	92.6	9.9
Einsteinium	Es	99	[Rn] $5f^{11}\,1s^2$			
Erbium	Er	68	[Xe] $4f^{14}\,6s^2$	14.5	86.0	12.2
Europium	Eu	63	[Xe] $4f^7\,6s^2$	13.9	90.0	35.0

Element	Symbol	Z	Configuration			
Fermium	Fm	100	$[\mathrm{Rn}]\,5f^{12}\,1s^2$		0.0277	9.4 (100°C)
Fluorine	F	9	$[\mathrm{He}]\,2s^2\,2p^5$			
Francium	Fr	87	$[\mathrm{Rn}]\,7s$			
Gadolinium	Gd	64	$[\mathrm{Xe}]\,4f^7\,5d\,6s^2$	131	10.5	
Gallium	Ga	31	$[\mathrm{Ar}]\,3d^{10}\,4s2\,4p$	25.795 (30°C)	29.4(lq) 40.6(c)	120
Germanium	Ge	32	$[\mathrm{Ar}]\,3d^{10}\,4s^2\,4p^2$	53 000	60.2	6.0
Gold (aurum)	Au	79	$[\mathrm{Xe}]\,4f^{14}\,5d^{10}\,6s$	2.214	318	14.2
Hafnium	Hf	72	$[\mathrm{Xe}]\,4f^{14}\,5d^2\,6s^2$	33.1	23.0	5.9
Helium	He	2	$1s^2$		0.1513	
Holmium	Ho	67	$[\mathrm{Xe}]\,4f^{11}\,6s^2$	81.4	16.2	11.2
Hydrogen	H	1	$1s$		0.1805	
Indium	In	49	$[\mathrm{Kr}]\,4d^{10}\,5s^2\,5p$	8.37	81.8	32.1
Iodine	I	53	$[\mathrm{Kr}]\,4d^{10}\,5s^2\,5p^5$	449	449	
Iodine	I	53		1.3×10^{15} (0°C)		
Iridium	Ir	77	$[\mathrm{Xe}]\,4f^{14}\,5d^7\,6s^2$	4.71	147	6.4
Iron	Fe	26	$[\mathrm{Ar}]\,3d^6\,4s^2$	9.61	80.4	11.8
Krypton	Kr	36	$[\mathrm{Ar}]\,3d^{10}\,4s^2\,4p^6$		9.43	
Lanthanum	La	57	$[\mathrm{Xe}]\,5d\,6s^2$	61.5	13.4	12.1
Lawrencium	Lr	103	$[\mathrm{Rn}]\,4f^{14}\,6d\,7s^2$			
Lead	Pb	82	$[\mathrm{Xe}]\,4f^{14}\,5d^{10}\,6s^2\,6p^2$	20.8	35.3	28.9
Lithium	Li	3	$1s^2 2s$	9.28	84.8	46
Lutetium	Lu	71	$[\mathrm{Xe}]\,4f^{14}\,5d\,6s^2$	58.2	16.4	9.9
Magnesium	Mg	12	$[\mathrm{Ne}]\,3s^2$	4.39	156	24.8
Manganese	Mn	25	$[\mathrm{Ar}]\,3d^5\,4s^2$	144	7.81	21.7
Mendelevium	Md	101	$[\mathrm{Rn}]\,5f^{13}\,7s^2$			
Mercury	Hg	80	$[\mathrm{Xe}]\,4f^{14}\,5d^{10}\,6s^2$	95.8(lq); 21(c)	8.30	
Molybdenum	Mo	42	$[\mathrm{Kr}]\,4d^5\,5s$	5.34	138	4.8
Neodymium	Nd	60	$[\mathrm{Xe}]\,4f^4\,6s^2$	64.3	16.5	9.6
Neon	Ne	10	$1s^2\,2s^2\,2p^6$		0.0491	
Neptunium	Np	93	$[\mathrm{Rn}]\,5f^4\,6d\,7s^2$	122.0 (22°C)	6.3	13.4
Nickel	Ni	28	$[\mathrm{Ar}]\,3d^8\,4s^2$	6.93	90.9	7.3
Niobium	Nb	41	$[\mathrm{Kr}]\,4d^4\,5s$	15.2 (0°C)	53.7	
Nitrogen	N	7	$1s^2\,2s^2\,2p^3$		0.025 83	
Nobelium	No	102	$[\mathrm{Rn}]\,5f^{14}\,7s^2$			
Osmium	Os	76	$[\mathrm{Xe}]\,4f^{14}\,5d^6\,6s^2$	8.12 (0°C)	87.6	5.1
Oxygen	O	8	$1s^2\,2s^2\,2p^4$		0.026 58 (g)	
Oxygen	O	8			0.149 (lq)	
Palladium	Pd	46	$[\mathrm{Kr}]\,4d^{10}$	10.54	71.8	11.8

(Continued)

TABLE 1.21 Conductivity and Resistivity of the Elements* (*Continued*)

Name	Symbol	Atomic number	Electronic configuration	Thermal conductivity, W · (m · K)$^{-1}$ at 25°C	Electrical resistivity, $\mu\Omega$ · cm at 20°C	Coefficient of linear thermal expansion (25°C), m · m^{-1} ($\times 10^6$)
Phosphorus	P	15	[Ne] $3s^2\,3p^3$	0.236 17	10	8.8
Platinum	Pt	78	[Xe] $4f^{14}\,5d^9\,6s$	71.6	10.6	46.7
Plutonium	Pu	94	[Rn] $5f^6\,7s^2$	6.74	146.0 (0°C)	
Polonium	Po	84	[Xe] $4f^{14}\,5d^{10}\,6s^2\,6p^4$	0.2	40.0 (0°C) alpha	
Potassium	K	19	[Ar] $4s$	102.5	7.2	6.7
Praseodymium	Pr	59	[Xe] $4f^3\,6s^2$	12.5	70.0	est [11.]
Promethium	Pm	61	[Xe] $4f^5\,6s^2$	17.9	64.0 (25°C)	
Protactinium	Pa	91	[Rn] $5f^2\,6d\,7s^2$	47	19.1 (22°C)	
Radium	Ra	88	[Rn] $7s^2$	18.6	100	
Radon	Rn	86	[Xe] $4f^{14}\,5d^{10}\,6s^2\,6p^6$	0.003 61		
Rhenium	Re	75	[Xe] $5f^{14}\,5d^5\,6s^2$	48.0	19.3	6.2
Rhodium	Rh	45	[Kr] $4d^8\,5s$	150	4.33 (0°C)	8.2
Rubidium	Rb	37	[Kr] $5s$	58.2	12.8	
Ruthenium	Ru	44	[Kr] $4d^7\,5s$	117	7.1 (0°C)	6.4
Samarium	Sm	62	[Xe] $4f^6\,6s^2$	13.3	94.0	12.7
Scandium	Sc	21	[Ar] $3d\,4s^2$	15.8	56.2	10.2
Selenium	Se	34	[Ar] $3d^{10}\,4s^2\,4p^4$	0.519	1.2 (0°C)	37
Silicon	Si	14	[Ne] $3s^2\,3p^2$	149	10^5	
Silver	Ag	47	[Kr] $4d^{10}\,5s$	429	1.587	18.9
Sodium	Na	11	[Ne] $3s$	142	4.77	71
Strontium	Sr	38	[Kr] $5s^2$	35.4	13.2	22.5
Sulfur	S	16	[Ne] $3s^2\,3p^4$	0.205	2×10^{23}	

Tantalum	Ta	73	[Xe] $4f^{14}\,5d^3\,6s^2$	57.5	13.5	6.3
Technetium	Tc	43	[Kr] $4d^6\,5s^2$	50.6	22.6 (100°C)	
Tellurium	Te	52	[Kr] $4d^{10}\,5s^2\,5p^4$	1.97–3.38	$(5.8–33) \times 10^3$	
Terbium	Tb	65	[Xe] $4f^9\,6s^2$	11.1	115	10.3
Thallium	Tl	78	[Xe] $4f^{14}\,5d^{10}\,6s^2\,6p$	46.1	18	29.9
Thorium	Th	90	[Rn] $6d^2\,7s^2$	54.0	15.4 (22°C)	11.1
Thullium	Tm	69	[Xe] $4f^{13}\,6s^2$	16.9	67.6	13.3
Tin (stannum)	Sn	50	[Kr] $4d^{10}\,5s^2\,5p^2$	66.8	11.5 (0°C)	22.0
Titanium	Ti	22	[Ar] $3d^2\,4s^2$	21.9	42.0	8.6
Tungsten (wolframium)	W	74	[Xe] $4f^{14}\,5d^4\,6s^2$	173	5.28	4.5
Uranium	U	92	[Rn] $5f^3\,6d\,7s^2$	27.5	28.0 (0°C)	13.9
Vanadium	V	23	[Ar] $3d^3\,4s^2$	30.7	19.7	8.4
Xenon	Xe	54	[Kr] $4d^{10}\,5s^2\,5p^6$	0.005 65		
Ytterbium	Yb	70	[Xe] $4f^{14}\,6s^2$	38.5	25	26.3
Yttrium	Y	39	[Kr] $4d\,5s^2$	17.2	59.6	10.6
Zinc	Zn	30	[Ar] $3d^{10}\,4s^2$	116	5.9	30.2
Zirconium	Zr	40	[Kr] $4d^2\,5s^2$	22.6	42.1	5.7

*The element in square brackets indicates the base electronic configuration.
New names have been proposed, but not yet been accepted, for elements 113, 115, 117, and 118: 113, nihonium; 115, moscovium; 117, tennessine; 118, oganesson.

TABLE 1.22   Work Functions of the Elements

The work function ϕ is the energy necessary to just remove an electron from the metal surface in thermoelectric or photoelectric emission. Values are dependent upon the experimental technique (vacua of 10^{-9} or 10^{-10} torr, clean surfaces, and surface conditions including the crystal face identification).

Element	ϕ, eV	Element	ϕ, eV	Element	ϕ, eV
Ag	4.64	Hg	4.50	Ru	4.80
Al	4.19	In	4.08	Sb	4.56
As	(3.75)	Ir	5.6	Sc	3.5
Au	5.32	K	2.30	Se	5.9
B	(4.75)	La	3.40	Si	4.85
Ba	2.35	Li	3.10	Sm	2.95
Be	5.08	Mg	3.66	Sn	4.35
Bi	4.36	Mn	3.90	Sr	2.76
C	(5.0)	Mo	4.30	Ta	4.22
Ca	2.71	Na	2.70	Tb	3.0
Cd	4.12	Nb	4.20	Te	4.70
Ce	2.80	Nd	3.1	Th	3.71
Co	4.70	Ni	5.15	Ti	4.10
Cr	4.40	Os	4.83	Tl	4.02
Cs	1.90	Pb	4.18	U	3.70
Cu	4.70	Pd	5.00	V	4.44
Eu	2.50	Po	4.6	W	4.55
Fe	4.65	Pr	2.7	Y	3.1
Ga	4.25	Pt	5.40	Zn	4.30
Ge	5.0	Rb	2.20	Zr	4.00
Gd	3.1	Re	4.95		
Hf	3.65	Rh	4.98		

TABLE 1.23   Relative Abundances of Naturally Occurring Isotopes

Element	Mass number	Percent	Element	Mass number	Percent
Aluminum	27	100	Cadmium	106	1.25(4)
Antimony	121	57.21(5)		108	0.89(2)
	123	42.79(5)		110	12.49(12)
Argon	36	0.337(3)		111	12.80(8)
	38	0.063(1)		112	24.13(14)
	40	99.600(3)		113	12.22(8)
Arsenic	75	100		114	28.7(3)
Barium	130	0.106(2)		116	7.49(9)
	132	0.101(2)	Calcium	40	96.941(18)
	134	2.42(3)		42	0.647(9)
	135	6.59(2)		43	0.135(6)
	136	7.85(4)		44	2.088(12)
	137	11.23(4)		46	0.004(3)
	138	71.70(7)		48	0.187(4)
Beryllium	9	100	Carbon	12	98.89(1)
Bismuth	209	100		13	1.11(1)
Boron	10	19.9(2)	Cerium	136	0.19(1)
	11	80.1(2)		138	0.25(1)
Bromine	79	50.69(7)		140	88.43(10)
	81	49.31(7)		142	11.13(10)

TABLE 1.23   Relative Abundances of Naturally Occurring Isotopes (*Continued*)

Element	Mass number	Percent	Element	Mass number	Percent
Cesium	133	100	Iodine	127	100
Chlorine	35	75.77(7)	Iridium	191	37.27(9)
	37	24.23(7)		193	62.73(9)
Chromium	50	4.345(13)	Iron	54	5.85(4)
	52	83.79(2)		56	91.75(4)
	53	9.50(2)		57	2.12(1)
	54	2.365(7)		58	0.26(1)
Cobalt	59	100	Krypton	78	0.35(2)
Copper	63	69.17(3)		80	2.25(2)
	65	30.83(3)		82	11.6(1)
Dysprosium	156	0.06(1)		83	11.5(1)
	158	0.10(1)		84	57.0(3)
	160	2.34(6)		86	17.3(2)
	161	18.9(2)	Lanthanum	138	0.0902(2)
	162	25.5(2)		139	99.9098(2)
	163	24.9(2)	Lead	204	1.4(1)
	164	28.2(2)		206	24.1(1)
Erbium	162	0.14(1)		207	22.1(1)
	164	1.61(2)		208	52.4(1)
	166	33.6(2)	Lithium	6	7.5(2)
	167	22.95(15)		7	92.5(2)
	168	26.8(2)	Lutetium	175	97.41(2)
	170	14.9(2)		176	2.59(2)
Europium	151	47.8(5)	Magnesium	24	78.99(3)
	153	52.2(5)		25	10.00(1)
Fluorine	19	100		26	11.01(2)
Gadolinium	152	0.20(1)	Manganese	55	100
	154	2.18(3)	Mercury	196	0.15(1)
	155	14.80(5)		198	9.97(8)
	156	20.47(4)		199	16.87(10)
	157	15.65(3)		200	23.10(16)
	158	24.84(12)		201	13.18(8)
	160	21.86(4)		202	29.86(20)
Gallium	69	60.108(9)		204	6.87(4)
	71	39.892(9)	Molybdenum	92	14.84(4)
Germanium	70	21.23(4)		94	9.25(3)
	72	27.66(3)		95	15.92(5)
	73	7.73(1)		96	16.68(5)
	74	35.94(2)		97	9.55(3)
	76	7.44(2)		98	24.13(7)
Gold	197	100		100	9.63(3)
Hafnium	174	0.162(3)	Neodymium	142	27.13(12)
	176	5.206(5)		143	12.18(6)
	177	18.606(13)		144	23.80(12)
	178	27.297(4)		145	8.30(6)
	179	13.629(6)		146	17.19(9)
	180	35.100(7)		148	5.76(3)
Helium	4	100		150	5.64(3)
Holmium	165	100	Neon	20	90.48(3)
Hydrogen	1	99.985(1)		21	0.27(1)
	2	0.015(1)		22	9.25(3)
Indium	113	4.29(2)	Nickel	58	68.077(9)
	115	95.71(2)		60	26.223(8)

(*Continued*)

TABLE 1.23 Relative Abundances of Naturally Occurring Isotopes (*Continued*)

Element	Mass number	Percent	Element	Mass number	Percent
	61	1.140(1)		154	22.7(2)
	62	3.634(2)	Scandium	45	100
	64	0.926(1)	Selenium	74	0.89(2)
Niobium	93	100		76	9.36(11)
Nitrogen	14	99.634(9)		77	6.63(6)
	15	0.366(9)		78	23.78(9)
Osmium	184	0.020(3)		80	49.61(10)
	186	1.58(2)		82	8.73(6)
	187	1.6(4)	Silicon	28	92.23(2)
	188	13.3(1)		29	4.67(2)
	189	16.1(1)		30	3.10(1)
	190	26.4(2)	Silver	107	51.839(7)
	192	41.0(3)		109	48.161(7)
Oxygen	16	99.76(1)	Sodium	23	100
	17	0.04	Strontium	84	0.56(1)
	18	0.20(1)		86	9.86(1)
Palladium	102	1.02(1)		87	7.00(1)
	104	11.14(8)		88	82.58(1)
	105	22.33(8)	Sulfur	32	95.02(9)
	106	27.33(3)		33	0.75(4)
	108	26.46(9)		34	4.21(8)
	110	11.72(9)		36	0.02(1)
Phosphorus	31	100	Tantalum	180	0.012(2)
Platinum	190	0.01(1)		181	99.988(2)
	192	0.79(6)	Tellurium	120	0.096(2)
	194	32.9(6)		122	2.603(4)
	195	33.8(6)		123	0.908(2)
	196	25.3(6)		124	4.816(6)
	198	7.2(2)		125	7.139(6)
Potassium	39	93.258(4)		126	18.952(11)
	40	0.0117(1)		128	31.687(11)
	41	6.730(3)		130	33.799(10)
Praseodymium	141	100	Terbium	159	100
Protoactinium	230	100	Thallium	203	29.52(1)
Rhenium	185	37.40(2)		205	70.48(1)
	187	62.60(2)	Thorium	228	100
Rhodium	103	100	Thullium	169	100
Rubidium	85	72.17(2)	Tin	112	0.97(1)
	87	27.83(2)		114	0.65(1)
Ruthenium	96	5.52(6)		115	0.34(1)
	98	1.88(6)		116	14.53(11)
	99	12.7(1)		117	7.68(7)
	100	12.6(1)		118	24.23(11)
	101	17.0(1)		119	8.59(4)
	102	31.6(2)		120	32.59(10)
	104	18.7(2)		122	4.63(3)
Samarium	144	3.1(1)		124	5.79(5)
	147	15.0(2)	Titanium	46	8.25(3)
	148	11.3(1)		47	7.44(2)
	149	13.8(1)		48	73.72(3)
	150	7.4(1)		49	5.41(2)
	152	26.7(2)		50	5.4(1)

TABLE 1.23 Relative Abundances of Naturally Occurring Isotopes (*Continued*)

Element	Mass number	Percent	Element	Mass number	Percent
Tungsten	180	0.12(1)		170	3.05(6)
	182	26.50(3)		171	14.3(2)
	183	14.31(1)		172	21.9(3)
	184	30.64(1)		173	16.12(2)
	186	28.43(4)		174	31.8(4)
Uranium	234	0.0055(5)		176	12.7(2)
	235	0.720(1)	Yttrium	89	100
	238	99.275(2)	Zinc	64	48.6(3)
Vanadium	50	0.250(2)		66	27.9(2)
	51	99.750(2)		67	4.1(1)
Xenon	124	0.10(1)		68	18.8(4)
	126	0.09(1)		70	0.6(1)
	128	1.91(3)	Zirconium	90	51.45(3)
	129	26.4(6)		91	11.22(4)
	130	4.1(1)		92	17.15(2)
	131	21.2(4)		94	17.38(4)
	132	26.9(5)		96	2.80(2)
	134	10.4(2)			
	136	8.9(1)			
Ytterbium	168	0.13(1)			

TABLE 1.24 Radioactivity of the Elements (Neptunium Series)

Element	Symbol	Radiation	Half-life
Plutonium ↓	^{241}Pu	β	13.2 years
Americium ↓	^{241}Am	α	462 years
Neptunium ↓	^{237}Np	α	2.20×10^6 years
Protactinium ↓	^{233}Pa	β	27.4 days
Uranium ↓	^{233}U	α	1.62×10^5 years
Thorium ↓	^{229}Th	α	7.34×10^3 years
Radium ↓	^{225}Ra	β	14.8 days
Actinium ↓	^{225}Ac	α	10.0 days
Francium ↓	^{221}Fr	α	4.8 min
Astatine ↓	^{217}At	α	1.8×10^{-2} sec

(*Continued*)

TABLE 1.24 Radioactivity of the Elements (Neptunium Series) (*Continued*)

Element	Symbol	Radiation	Half-life
Bismuth 98% | 2%	^{213}Bi	β and α	47 min
Polonium	^{213}Po	α	4.2×10^{-6} sec
Thallium	^{209}Tl	β	2.2 min
Lead	^{209}Pb	β	3.32 hr
Bismuth (End Product)	^{209}Bi	Stable	—

TABLE 1.25 Radioactivity of the Elements (Thorium Series)

Radioelement	Corresponding element	Symbol	Radiation	Half-life
Thorium	Thorium	^{232}Th	α	1.39×10^{10} years
Mesothorium I	Radium	^{228}Ra	β	6.7 years
Mesothorium II	Actinium	^{228}Ac	β	6.13 hr
Radiothorium	Thorium	^{228}Th	α	1.91 years
Thorium X	Radium	^{224}Ra	α	3.64 days
Th Emanation	Radon	^{220}Rn	α	52 sec
Thorium A	Polonium	^{216}Po	α	0.16 sec
Thorium B	Lead	^{212}Pb	β	10.6 hr
Thorium C 66.3% | 33.7%	Bismuth	^{212}Bi	β and α	60.5 min
Thorium C′	Polonium	^{212}Po	α	3×10^{-7} sec
Thorium C″	Thallium	^{208}Tl	β	3.1 min
Thorium D (End Product)	Lead	^{208}Pb	Stable	—

TABLE 1.26   Radioactivity of the Elements (Actinium Series)

Radioelement	Corresponding element	Symbol	Radiation	Half-life
Actinouranium ↓	Uranium	^{235}U	α	7.13×10^8 years
Uranium Y ↓	Thorium	^{231}Th	β	25.6 hr
Protactinium ↓	Protactinium	^{231}Pa	α	3.43×10^4 years
Actinium 98.8% | 1.2% ↓	Actinium	^{227}Ac	β and α	21.8 years
Radioactinium	Thorium	^{227}Th	α	18.4 days
Actinium K	Francium	^{223}Fr	β	21 min
Actinium X ↓	Radium	^{223}Ra	α	11.7 days
Ac Emanation ↓	Radon	^{219}Rn	α	3.92 sec
Actinium A ~100% | ~5 × 10⁻⁴% ↓	Polonium	^{215}Po	α and β	1.83×10^{-3} s
Actinium B	Lead	^{211}Pb	β	36.1 min
Astatine-215	Astatine	^{215}At	α	~10^{-4} sec
Actinium C 99.7% | 0.3%	Bismuth	^{211}Bi	α and β	2.16 min
Actinium C′	Polonium	^{211}Po	α	0.52 sec
Actinium C″	Thallium	^{207}Tl	β	4.8 min
Actinium D (End Product)	Lead	^{207}Pb	Stable	—

TABLE 1.27   Radioactivity of the Elements (Uranium Series)

Radioelement	Corresponding element	Symbol	Radiation	Half-life
Uranium I ↓	Uranium	^{238}U	α	4.51×10^9 years
Uranium X_1 ↓	Thorium	^{234}Th	β	24.1 days
Uranium X_2^* ↓	Protactinium	^{234}Pa	β	1.18 min
Uranium II ↓	Uranium	^{234}U	α	2.48×10^5 years
Ionium ↓	Thorium	^{230}Th	α	8.0×10^4 years
Radium ↓	Radium	^{226}Ra	α	1.62×10^3 years

(Continued)

TABLE 1.27   Radioactivity of the Elements (Uranium Series) (*Continued*)

Radioelement	Corresponding element	Symbol	Radiation	Half-life
Ra Emanation ↓	Radon	^{222}Rn	α	3.82 days
Radium A 99.98% | 0.02%	Polonium	^{218}Po	α and β	3.05 min
Radium B	Lead	^{214}Pb	β	26.8 min
Astatine-218	Astatine	^{218}At	α	2 sec
Radium C 99.96% | 0.04%	Bismuth	^{214}Bi	β and α	19.7 min
Radium C′	Polonium	^{214}Po	α	1.6×10^{-4} sec
Radium C″	Thallium	^{210}Tl	β	1.32 min
Radium D ↓	Lead	^{210}Pb	β	19.4 years
Radium E ~100% | 2×10^{-4}%	Bismuth	^{210}Bi	β and α	5.0 days
Radium F	Polonium	^{210}Po	α	138.4 days
Thallium-206	Thallium	^{206}Tl	β	4.20 min
Radium G (End Product)	Lead	^{206}Pb	Stable	—

*Uranium X_2 is an excited state of ^{234}Pa and undergoes isomeric transition to a small extent to form uranium Z (^{234}Pa in its ground state); the latter has a half-life of 6.7 h, emitting beta radiation and forming uranium II (^{234}U).

Please note that elements after Element No. 118 are hypothetical only.

TABLE 1.28   Electronic Configuration of the Elements*

1. H—Hydrogen: $1s^1$
2. He—Helium: $1s^2$
3. Li—Lithium: [He] $2s^1$
4. Be—Beryllium: [He] $2s^2$
5. B—Boron: [He] $2s^2 2p^1$
6. C—Carbon: [He] $2s^2 2p^2$
7. N—Nitrogen: [He] $2s^2 2p^3$
8. O—Oxygen: [He] $2s^2 2p^4$
9. F—Fluorine: [He] $2s^2 2p^5$
10. Ne—Neon: [He] $2s^2 2p^6$
11. Na—Sodium: [Ne] $3s^1$
12. Mg—Magnesium: [Ne] $3s^2$
13. Al—Aluminum: [Ne] $3s^2 3p^1$

TABLE 1.28 Electronic Configuration of the Elements* (*Continued*)

14. Si—Silicon: [Ne] $3s^2 3p^2$
15. P—Phosphorus: [Ne] $3s^2 3p^3$
16. S—Sulfur: [Ne] $3s^2 3p^4$
17. Cl—Chlorine: [Ne] $3s^2 3p^5$
18. Ar—Argon: [Ne] $3s^2 3p^6$
19. K—Potassium: [Ar] $4s^1$
20. Ca—Calcium: [Ar] $4s^2$
21. Sc—Scandium: [Ar] $3d^1 4s^2$
22. Ti—Titanium: [Ar] $3d^2 4s^2$
23. V—Vanadium: [Ar] $3d^3 4s^2$
24. Cr—Chromium: [Ar] $3d^5 4s^1$
25. Mn—Manganese: [Ar] $3d^5 4s^2$
26. Fe—Iron: [Ar] $3d^6 4s^2$
27. Co—Cobalt: [Ar] $3d^7 4s^2$
28. Ni—Nickel: [Ar] $3d^8\, 4s^2$
29. Cu—Copper: [Ar] $3d^{10} 4s^1$
30. Zn—Zinc: [Ar] $3d^{10} 4s^2$
31. Ga—Gallium: [Ar] $3d^{10} 4s^2 4p^1$
32. Ge—Germanium: [Ar] $3d^{10} 4s^2 4p^2$
33. As—Arsenic: [Ar] $3d^{10} 4s^2 4p^3$
34. Se—Selenium: [Ar] $3d^{10} 4s^2 4p^4$
35. Br—Bromine: [Ar] $3d^{10} 4s^2 4p^5$
36. Kr—Krypton: [Ar] $3d^{10} 4s^2 4p^6$
37. Rb—Rubidium: [Kr] $5s^1$
38. Sr—Strontium: [Kr] $5s^2$
39. Y—Yttrium: [Kr] $4d^1 5s^2$
40. Zr—Zirconium: [Kr] $4d^2 5s^2$
41. Nb—Niobium: [Kr] $4d^4 5s^1$
42. Mo—Molybdenum: [Kr] $4d^5 5s^1$
43. Tc—Technetium: [Kr] $4d^5 5s^2$
44. Ru—Ruthenium: [Kr] $4d^7 5s^1$
45. Rh—Rhodium: [Kr] $4d^8 5s^1$
46. Pd—Palladium: [Kr] $4d^{10}$
47. Ag—Silver: [Kr] $4d^{10} 5s^1$
48. Cd—Cadmium: [Kr] $4d^{10} 5s^2$
49. In—Indium: [Kr] $4d^{10} 5s^2 5p^1$
50. Sn—Tin: [Kr] $4d^{10} 5s^2 5p^2$
51. Sb—Antimony: [Kr] $4d^{10} 5s^2 5p^3$
52. Te—Tellurium: [Kr] $4d^{10} 5s^2 5p^4$
53. I—Iodine: [Kr] $4d^{10} 5s^2 5p^5$
54. Xe—Xenon: [Kr] $4d^{10} 5s^2 5p^6$
55. Cs—Caesium: [Xe] $6s^1$
56. Ba—Barium: [Xe] $6s^2$
57. La—Lanthanum: [Xe] $5d^1 6s^2$
58. Ce—Cerium: [Xe] $4f^1 5d^1 6s^2$
59. Pr—Praseodymium: [Xe] $4f^3 6s^2$
60. Nd—Neodymium: [Xe] $4f^4 6s^2$
61. Pm—Promethium: [Xe] $4f^5 6s^2$
62. Sm—Samarium: [Xe] $4f^6 6s^2$

(*Continued*)

TABLE 1.28 Electronic Configuration of the Elements* (*Continued*)

63. Eu—Europium: [Xe] $4f^76s^2$
64. Gd—Gadolinium: [Xe] $4f^75d^1 6s^2$
65. Tb—Terbium: [Xe] $4f^96s^2$
66. Dy—Dysprosium: [Xe] $4f^{10}6s^2$
67. Ho—Holmium: [Xe] $4f^{11}6s^2$
68. Er—Erbium: [Xe] $4f^{12}6s^2$
69. Tm—Thulium: [Xe] $4f^{13}6s^2$
70. Yb—Ytterbium: [Xe] $4f^{14}6s^2$
71. Lu—Lutetium: [Xe] $4f^{14}5d^16s^2$
72. Hf—Hafnium: [Xe] $4f^{14}5d^26s^2$
73. Ta—Tantalum: [Xe] $4f^{14}5d^36s^2$
74. W—Tungsten: [Xe] $4f^{14}5d^46s^2$
75. Re—Rhenium: [Xe] $4f^{14}5d^56s^2$
76. Os—Osmium: [Xe] $4f^{14}5d^66s^2$
77. Ir—Iridium: [Xe] $4f^{14}5d^76s^2$
78. Pt—Platinum: [Xe] $4f^{14}5d^96s^1$
79. Au—Gold: [Xe] $4f^{14}5d^{10}6s^1$
80. Hg—Mercury: [Xe] $4f^{14}5d^{10}6s^2$
81. Tl—Thallium: [Xe] $4f^{14}5d^{10}6s^26p^1$
82. Pb—Lead: [Xe] $4f^{14}5d^{10}6s^26p^2$
83. Bi—Bismuth: [Xe] $4f^{14}5d^{10}6s^26p^3$
84. Po—Polonium: [Xe] $4f^{14}5d^{10}6s^26p^4$
85. At—Astatine: [Xe] $4f^{14}5d^{10}6s^26p^5$
86. Rn—Radon: [Xe] $4f^{14}5d^{10}6s6p^6$
87. Fr—Francium: [Rn] $7s^1$
88. Ra—Radium: [Rn] $7s^2$
89. Ac—Actinium: [Rn] $6d^17s^2$
90. Th—Thorium: [Rn] $6d^27s^2$
91. Pa—Protactinium: [Rn] $5f^26d^17s^2$
92. U—Uranium: [Rn] $5f^36d^17s^2$
93. Np—Neptunium: [Rn] $5f^46d^17s^2$
94. Pu—Plutonium: [Rn] $5f^67s^2$
95. Am—Americium: [Rn] $5f^7 7s^2$
96. Cm—Curium: [Rn] $5f^76d^17s^2$
97. Bk—Berkelium: [Rn] $5f^97s^2$
98. Cf—Californium: [Rn] $5f^{10}7s^2$
99. Es—Einsteinium: [Rn] $5f^{11}7s^2$
100. Fm—Fermium: [Rn] $5f^{12}7s^2$
101. Md—Mendelevium: [Rn] $5f^{13}7s^2$
102. No—Nobelium: [Rn] $5f^{14}7s^2$
103. Lr—Lawrencium: [Rn] $5f^{14}7s^27p^1$
104. Rf—Rutherfordium: [Rn] $5f^{14}6d^27s^2$
105. Db—Dubnium: [Rn] $5f^{14}6d^37s^2$
106. Sg—Seaborgium: [Rn] $5f^{14}6d^47s^2$
107. Bh—Bohrium: [Rn] $5f^{14}6d^57s^2$
108. Hs—Hassium: [Rn] $5f^{14}6d^67s^2$
109. Mt—Meitnerium: [Rn] $5f^{14}6d^77s^2$
110. Ds—Darmstadtium: [Rn] $5f^{14}6d^87s2$
111. Rg—Roentgenium: [Rn] $5f^{14}6d^97s^2$
112. Cn—Copernicium: [Rn] $5f^{14}6d^{10}7s^2$
113. Uut—Ununtrium: [Rn] $5f^{14}6d^{10}7s^27p^1$

TABLE 1.28 Electronic Configuration of the Elements* (*Continued*)

114. Fl—Flerovium: [Rn] $5f^{14}6d^{10}7s^27p^2$
115. Uup—Ununpentium: [Rn] $5f^{14}6d^{10}7s^27p^3$
116. Lv—Livermorium: [Rn] $5f^{14}6d^{10}7s^27p^4$
117. Uus—Ununseptium: [Rn] $5f^{14}6d^{10}7s^27p^5$
118. Uuo—Ununoctium: [Rn] $5f^{14}6d^{10}7s^27p^6$
119. Uue—Ununennium: [Uuo] $8s^1$
120. Ubn—Unbinilium: [Uuo] $8s^2$
121. Ubu—Unbiunium: [Uuo] $8s^28p^1$
122. Ubb—Unbibium: [Uuo] $7d^18s^28p^1$
123. Ubt—Unbitrium: [Uuo] $6f^17d^18s^28p^1$
124. Ubq—Unbiquadium: [Uuo] $6f^38s^28p^1$
125. Ubp—Unbipentium: [Uuo] $5g\,6f^38s^28p^1$
126. Ubh—Unbihexium: [Uuo] $5g^26f^27d^18s^28p^1$
127. Ubs—Unbiseptium: [Uuo] $5g^36f^28s^28p^2$
128. Ubo—Unbioctium: [Uuo] $5g^46f^28s^28p^2$
129. Ube—Unbiennium: [Uuo] $5g^56f^28s^28p^2$
130. Utn—Untrinilium: [Uuo] $5g^66f^28s^28p^2$
131. Utu—Untriunium: [Uuo] $5g^76f^28s^28p^2$
132. Utb—Untribium: [Uuo] $5g^86f^28s^2\,p^2$
133. Utt—Untritrium: [Uuo] $5g^86f^38s^28p^2$
134. Utq—Untriquadium: [Uuo] $5g^86f^48s^28p^2$
135. Utp—Untripentium: [Uuo] $5g^96f^48s^28p^2$
136. Uth—Untrihexium: [Uuo] $5g^{10}6f^48s^28p^2$
137. Uts—Untriseptium: [Uuo] $5g^{11}6f^37d^18s^28p^2$
138. Uto—Untrioctium: [Uuo] $5g^{12}6f^37d^18s^28p^2$
139. Ute—Untriennium: [Uuo] $5g^{13}6f^27d^28s^28p^2$
140. Uqn—Unquadnilium: [Uuo] $5g^{14}6f^37d^18s^28p^2$
141. Uqu—Unquadunium: [Uuo] $5g^{15}6f^27d^28s^28p^2$
142. Uqb—Unquadbium: [Uuo] $5g^{16}6f^27d^28s^28p^2$
143. Uqt—Unquadtrium: [Uuo] $5g^{17}6f^27d^28s^28p^2$
144. Uqq—Unquadquadium: [Uuo] $5g^{18}6f^17d^38s^28p^2$
145. Uqp—Unquadpentium: [Uuo] $5g^{18}6f^37d^28s^28p^2$
146. Uqh—Unquadhexium: [Uuo] $5g^{18}6f^47d^28s^28p^2$
147. Uqs—Unquadseptium: [Uuo] $5g^{18}6f^57d^28s^28p^2$
148. Uqo—Unquadoctium: [Uuo] $5g^{18}6f^67d^28s^28p^2$
149. Uqe—Unquadennium: [Uuo] $5g^{18}6f^67d^38s^28p^2$
150. Upn—Unpentnilium: [Uuo] $5g^{18}6f^67d^48s^28p^2$
151. Upu—Unpentunium: [Uuo] $5g^{18}6f^87d^38s^28p^2$
152. Upb—Unpentbium: [Uuo] $5g^{18}6f^97d^38s^28p^2$
153. Upt—Unpenttrium: [Uuo] $5g^{18}6f^{11}7d^28s^28p^2$
154. Upq—Unpentquadium: [Uuo] $5g^{18}6f^{12}7d^28s^28p^2$
155. Upp—Unpentpentium: [Uuo] $5g^{18}6f^{13}7d^28s^28p^2$
156. Uph—Unpenthexium: [Uuo] $5g^{18}6f^{14}7d^28s^28p^2$
157. Ups—Unpentseptium: [Uuo] $5g^{18}6f^{14}7d^38s^28p^2$
158. Upo—Unpentoctium: [Uuo] $5g^{18}6f^{14}7d^48s^28p^2$
159. Upe—Unpentennium: [Uuo] $5g^{18}6f^{14}7d^48s^28p^29s^1$

(*Continued*)

TABLE 1.28 Electronic Configuration of the Elements* (*Continued*)

160.	Uhn—Unhexnilium:	[Uuo] $5g^{18}6f^{14}7d^58s^28p^29s^1$
161.	Uhu—Unhexunium:	[Uuo] $5g^{18}6f^{14}7d^68s^28p^29s^1$
162.	Uhb—Unhexbium:	[Uuo] $5g^{18}6f^{14}7d^88s^28p^2$
163.	Uht—Unhextrium:	[Uuo] $5g^{18}6f^{14}7d^98s^28p^2$
164.	Uhq—Unhexquadium:	[Uuo] $5g^{18}6f^{14}7d^{10}8s^28p^2$
165.	Uhp—Unhexpentium:	[Uuo] $5g^{18}6f^{14}7d^{10}8s^28p^29s^1$
166.	Uhh—Unhexhexium:	[Uuo] $5g^{18}6f^{14}7d^{10}8s^28p^29s^2$
167.	Uhs—Unhexseptium:	[Uuo] $5g^{18}\ 6f^{14}7d^{10}8s^28p^29s^29p^1$
168.	Uho—Unhexoctium:	[Uuo] $5g^{18}\ 6f^{14}7d^{10}8s^28p^29s^29p^2$
169.	Uhe—Unhexennium:	[Uuo] $5g^{18}6f^{14}7d^{10}8s^28p^39s^29p^2$
170.	Usn—Unseptnilium:	[Uuo] $5g^{18}6f^{14}7d^{10}8s^28p^49s^29p^2$
171.	Usu—Unseptunium:	[Uuo] $5g^{18}6f^{14}7d^{10}8s^28p^59s^29p^2$
172.	Usb—Unseptbium:	[Uuo] $5g^{18}6f^{14}7d^{10}8s^28p^69s^29p^2$

*The element in square brackets indicates the base electronic configuration.

New names have been proposed, but not yet been accepted, for elements 113, 115, 117, and 118: 113, nihonium; 115, moscovium; 117, tennessine; 118, oganesson.

1.4 IONIZATION ENERGY

The ionization energy or ionization potential is the energy necessary to remove an electron from the neutral atom. It is at a minimum for the alkali metals, which have a single electron outside a closed shell. The ionization energy generally increases across a row on the periodic maximum for the noble gases, which have closed shells. The ionization energy is one of the primary energy considerations used in quantifying chemical bonds.

TABLE 1.29 Ionization Energy of the Elements

The minimum amount of energy required to remove the least strongly bound electron from a gaseous atom (or ion) is called the ionization energy and is expressed in $MJ \cdot mol^{-1}$.

At. no.	Element	Spectrum (in $MJ \cdot mol^{-1}$)					
		I	II	III	IV	V	VI
1	H	1.312					
2	He	2.372	5.251				
3	Li	0.520	7.298	11.815			
4	Be	0.899	1.757	14.849	21.007		
5	B	0.801	2.427	3.660	25.027	32.828	
6	C	1.086	2.353	4.620	6.223	37.832	47.191
7	N	1.402	2.856	4.578	7.475	9.445	53.268
8	O	1.314	3.388	5.300	7.469	10.989	13.326
9	F	1.681	3.374	6.147	8.408	11.022	15.164

TABLE 1.29   Ionization Energy of the Elements (*Continued*)

At. no.	Element	Spectrum (in MJ · mol⁻¹)					
		I	II	III	IV	V	VI
10	Ne	2.081	3.952	6.122	9.370	12.177	15.238
11	Na	0.496	4.562	6.912	9.543	13.353	16.610
12	Mg	0.738	1.451	7.733	10.540	13.629	17.994
13	Al	0.578	1.817	2.745	11.577	14.831	18.377
14	Si	0.786	1.577	3.231	4.355	16.091	19.784
15	P	1.012	1.903	2.912	4.956	6.274	21.268
16	S	1.000	2.251	3.361	4.564	7.004	8.495
17	Cl	1.251	2.297	3.822	5.158	6.54	9.362
18	Ar	1.521	2.666	3.931	5.771	7.238	8.787
19	K	0.419	3.051	4.411	5.877	7.976	9.649
20	Ca	0.590	1.145	4.912	6.474	8.144	10.496
21	Sc	0.631	1.235	2.389	7.089	8.844	10.719
22	Ti	0.658	1.310	2.652	4.175	9.573	11.516
23	V	0.650	1.414	2.828	4.507	6.299	12.362
24	Cr	0.653	1.592	2.987	4.743	6.70	8.738
25	Mn	0.717	1.509	3.248	4.94	6.99	9.22
26	Fe	0.759	1.561	2.957	5.63	7.24	9.56
27	Co	0.758	1.646	3.232	4.95	7.67	9.84
28	Ni	0.737	1.753	3.393	5.30	7.34	10.4
29	Cu	0.745	1.958	3.555	5.536	7.70	9.9
30	Zn	0.906	1.733	3.833	5.73	7.95	10.4
31	Ga	0.579	1.979	2.963	6.2		
32	Ge	0.762	1.537	3.302	4.410	9.022	
33	As	0.947	1.798	2.735	4.837	6.043	12.31
34	Sc	0.941	2.045	2.974	4.143	6.99	7.883
35	Br	1.140	2.10	3.47	4.56	5.76	8.55
36	Kr	1.351	2.350	3.565	5.07	6.24	7.57
37	Rb	0.403	2.632	3.9	5.08	6.85	8.14
38	Sr	0.549	1.064	4.138	5.5	6.91	8.76
39	Y	0.616	1.181	1.980	5.96	7.43	8.97
40	Zr	0.660	1.267	2.218	3.313	7.75	
41	Nb	0.664	1.382	2.416	3.695	4.877	9.847
42	Mo	0.685	1.558	2.621	4.477	5.91	6.641
43	Tc	0.702	1.472	2.850			
44	Ru	0.711	1.617	2.747			
45	Rh	0.720	1.744	2.997			
46	Pd	0.805	1.875	3.177			
47	Ag	0.731	2.073	3.361			
48	Cd	0.868	1.631	3.616			
49	In	0.558	1.821	2.704	5.2		
50	Sn	0.709	1.412	2.943	3.930	6.974	
51	Sb	0.834	1.595	2.44	4.26	5.4	10.4
52	Te	0.869	1.795	2.698	3.610	5.668	6.82
53	I	1.008	1.846	3.2			
54	Xe	1.170	2.046	3.099			
55	Cs	0.376	2.234				
56	Ba	0.503	0.965				
57	La	0.538	1.067	1.850	4.820	5.94	
58	Ce	0.528	1.047	1.949	3.547	6.325	7.487
59	Pr	0.523	1.018	2.086	3.761	5.551	
60	Nd	0.530	1.035	2.13	3.90		

(*Continued*)

TABLE 1.29 Ionization Energy of the Elements (*Continued*)

At. no.	Element	Spectrum (in MJ · mol⁻¹)					
		I	II	III	IV	V	VI
61	Pm	0.535	1.052	2.15	3.97		
62	Sm	0.543	1.068	2.26	3.99		
63	Eu	0.547	1.085	2.40	4.12		
64	Gd	0.592	1.167	1.99	4.26		
65	Tb	0.564	1.112	2.114	3.839		
66	Dy	0.572	1.126	2.20	3.99		
67	Ho	0.581	1.139	2.204	4.10		
68	Er	0.589	1.151	2.194	4.13		
69	Tm	0.596	1.163	2.285	4.13		
70	Yb	0.603	1.174	2.417	4.203		
71	Lu	0.524	1.34	2.022	4.366		
72	Hf	0.68	1.44	2.25	3.216		
73	Ta	0.761					
74	W	0.770					
75	Re	0.760					
76	Os	0.84					
77	Ir	0.88					
78	Pt	0.87	1.791				
79	Au	0.890	1.98				
80	Hg	1.007	1.810	3.30			
81	Tl	0.589	1.971	2.878			
82	Pb	0.716	1.450	3.081	4.083	6.64	
83	Bi	0.703	1.610	2.466	4.371	5.40	8.52
84	Po	0.812					
85	At						
86	Rn	1.037					
87	Fr						
88	Ra	0.509	0.979				
89	Ac	0.67	1.17				
90	Th	0.587	1.11	1.93	2.78		
91	Pa	0.568					
92	U	0.598					
93	Np	0.605					
94	Pu	0.585					
95	Am	0.578					
96	Cm	0.581					
97	Bk	0.601					
98	Cf	0.608					
99	Es	0.619					
100	Fm	0.627					
101	Md	0.635					
102	No	0.642					

TABLE 1.30 Ionization Energy of Molecular and Radical Species

Species	Ionization energy		$\Delta_f H$ (ion) in kJ · mol^{-1}
	In MJ · mol^{-1}	In electron volts	
Aluminum tribromide	1.00	10.4	593
Aluminum trichloride	1.159	12.01	573
Aluminum trifluoride	1.394	14.45	282
Aluminum triiodide	0.88	9.1	673
Amidogen (NH$_2$)	1.075(1)	11.14(1)	1264
Ammonia	0.980(1)	10.16(1)	934
Antimony trichloride	0.97(1)	10.1(1)	661
Arsenic trichloride	1.018(3)	10.55(3)	754
Arsenic trifluoride	1.239(5)	12.84(5)	452
Arsine	0.954	9.89	1021
Barium oxide	0.667(6)	6.91(6)	543
Bismuth trichloride	1.00	10.4	736
Borane (BH$_3$)	1.19(1)	12.3(1)	1287
Boron dioxide (BO$_2$)	1.30(3)	13.5(3)	1001
Boron oxide (B$_2$O$_3$)	1.303(14)	13.50(15)	460
Boron tribromide	1.014(2)	10.51(2)	809
Boron trichloride	1.119(2)	11.60(2)	718
Boron trifluoride	1.501(3)	15.56(3)	365
Boron triiodide	0.893(3)	9.25(3)	964
Bromine (Br$_2$)	1.0146(5)	10.515(5)	1046
Bromine chloride (BrCl)	1.062	11.01	1079
Bromine fluoride (BrF)	1.136(1)	11.77(1)	1077
Bromine pentafluoride	1.271(1)	13.17(1)	840
Bromosilane (BrSiH$_3$)	1.02	10.6	943
Calcium oxide	0.67	6.9	691
Cesium chloride	0.756(5)	7.84(5)	510
Cesium fluoride	1.221(1)	12.65(1)	1170
Cesium fluoride	0.849(10)	8.80(10)	489
Chlorine (Cl$_2$)	1.1424(5)	11.840(5)	1108
Chlorine difluoride	1.232(5)	12.77(5)	1128
Chlorine dioxide	1.000(2)	10.36(2)	1096
Chlorine oxide	1.057	10.95	1159
Chlorine trifluoride	1.221(5)	12.65(5)	1057
Chlorosilane (ClSiH$_3$)	1.10	11.4	899
Chromyl chloride (CrO$_2$Cl$_2$)	1.12	11.6	580
Diborane (B$_2$H$_6$)	1.098(3)	11.38(3)	1134
Dichlorosilane (Cl$_2$SiH$_2$)	1.10	11.4	765
Difluoramine (HNF$_2$)	1.112(8)	11.53(8)	1046
Difluoroamidogen (NF$_2$)	1.122(1)	11.628(1)	1155
Difluorosilane (F$_2$SiH$_2$)	1.18	12.2	386
Dioxygen fluoride	1.22(2)	12.6(2)	1228
Disilane	0.94	9.7	1015
Disulfur oxide	1.017(4)	10.54(4)	967
Fluorine (F$_2$)	1.5146(3)	15.697(3)	1515
Fluorosilane (FSiH$_3$)	1.13	11.7	752
Gallium bromide	1.003	10.40	711
Gallium chloride	1.112	11.52	648
Gallium triiodide	0.907	9.40	765
Gallium(I) fluoride	0.93(5)	9.6(5)	700

(Continued)

TABLE 1.30 Ionization Energy of Molecular and Radical Species (*Continued*)

Species	Ionization energy		$\Delta_f H$ (ion) in kJ · mol^{-1}
	In MJ · mol^{-1}	In electron volts	
Germane (GeH$_4$)	1.093	11.33	1185
Germanium oxide (GeO)	1.085(1)	11.25(1)	1044
Germanium sulfide (GeS)	0.963(2)	9.98(2)	1055
Germanium tetrachloride	1.1270(5)	11.68(5)	629
Germanium tetrafluoride	1.50	15.5	307
Germanium tetraiodide	0.909	9.42	850
Hafnium bromide	1.05	10.9	366
Hafnium chloride	1.13	11.7	246
Hexaborane (B$_6$H$_{10}$)	0.87	9.0	965
Hydrazine	7.82(14)	8.10(15)	877
Hydrazoic acid (HN$_3$)	1.0344(24)	10.720(25)	1328
Hydrogen (H$_2$)	1.488413(5)	15.42589(5)	1488
Hydrogen bromide	1.125(3)	11.66(3)	1087
Hydrogen chloride	1.2299	12.747	1137
Hydrogen fluoride	1.5481(3)	16.044(3)	1276
Hydrogen iodide	1.0004(1)	10.368(1)	1028
Hydrogen peroxide	1.017	10.54	881
Hydrogen selenide	0.9535(1)	9.882(1)	983
Hydrogen sulfide	1.0085(8)	10.453(8)	988
Hydroperoxy (HOO)	1.095(1)	11.35(1)	1106
Hydroxyl (OH)	1.254	13.00	1293
Hydroxylamine (NH$_2$OH)	0.947	10.00	923
Hypochlorous acid (HOCl)	1.073(1)	11.12(1)	993
Hypofluorous acid (HOF)	1.226(1)	12.71(1)	1130
Imidogen (NH)	1.302(1)	13.49(1)	1678
Iodine (I$_2$)	0.90694(12)	9.3995(12)	969
Iodine bromide	0.9446(4)	9.790(4)	986
Iodine chloride	0.9734(10)	10.088(10)	991
Iodine fluoride	1.025	10.62	930
Iodine pentafluoride	1.2488(5)	12.943(5)	408
Lead oxide (PbO)	0.976(10)	9.08(10)	939
Lead(II) chloride	0.96	10.0	789
Lead(II) fluoride	1.11	11.5	679
Lead(II) sulfide	0.825	8.5(5)	954
Lithium bromide	0.84	8.7	685
Lithium chloride	0.923	9.57	727
Lithium hydride	0.74	7.7	882
Lithium iodide	0.72	7.5	633
Lithium oxide	0.815	8.45(20)	895
Magnesium fluoride	1.29	13.4	569
Magnesium oxide	0.93	9.7	992
Mercapto (SH)	1.001	10.37	1140
Mercury(II) bromide	1.019(3)	10.560(3)	935
Mercury(II) chloride	1.0988(3)	11.380(3)	952
Mercury(II) iodide	0.91748(22)	9.5088(22)	900
Molybdenum hexafluoride	1.40(1)	14.5(1)	−159
Molybdenum(V) chloride	0.84	8.7	392
Niobium(V) chloride	1.058	10.97	656
Nitric acid	1.153(1)	11.95(1)	1019

TABLE 1.30 Ionization Energy of Molecular and Radical Species (*Continued*)

Species	Ionization energy		$\Delta_f H$ (ion) in kJ · mol^{-1}
	In MJ · mol^{-1}	In electron volts	
Nitric oxide	0.893900(6)	9.26436(6)	985
Nitrogen (N$_2$)	1.59336	15.5808	1503
Nitrogen dioxide	0.941(1)	9.75(1)	974
Nitrogen pentoxide	1.15	11.9	1161
Nitrogen tetroxide	1.04(2)	10.8(2)	1050
Nitrogen trichloride	0.9765(10)	10.12(10)	1244
Nitrogen trifluoride	1.254(2)	13.00(2)	1125
Nitrosyl bromide	0.981(3)	10.17(3)	1065
Nitrosyl chloride (NOCl)	1.049(1)	10.87(1)	1099
Nitrosyl fluoride (NOF)	1.219(3)	12.63(3)	1152
Nitrous acid (HONO)	1.09	11.3	977
Nitrous oxide (N$_2$O)	1.2433	12.886	1325
Nitryl chloride (NO$_2$Cl)	1.142	11.84	1155
Nitryl fluoride (NO$_2$F)	1.263	13.09	1154
Osmium tetroxide	1.1895	12.320	850
Oxygen (O$_2$)	1.1647(1)	12.071(1)	1165
Oxygen dichloride	1.056	10.94	1135
Oxygen difluoride (OF$_2$)	1.265(1)	13.11(1)	1290
Oxygen fluoride	1.232	12.77	1341
Ozone (O$_3$)	1.199	12.43	1342
Pentaborane (B$_5$H$_9$)	0.955(4)	9.90(4)	1028
Perchloryl fluoride (ClO$_3$F)	1.2490(5)	12.945(5)	1224
Phosphine (PH$_3$)	0.9522(2)	9.869(2)	958
Phosphorus (P$_2$)	1.016	10.53	1160
Phosphorus nitride	1.143	11.85	1248
Phosphorus pentachloride	1.03	10.7	656
Phosphorus pentafluoride	1.46	15.1	−137
Phosphorus sulfur trichloride (PSCl$_3$)	0.956	9.91	668
Phosphorus tribromide	0.94	9.7	798
Phosphorus trichloride	0.956	9.91	668
Phosphorus trifluoride	1.104	11.44	146
Phosphoryl chloride (POCl$_3$)	1.096(2)	11.36(2)	540
Phosphoryl trifluoride (POF$_3$)	1.231(1)	12.76(1)	−24
Potassium bromide	0.757(10)	7.85(10)	578
Potassium chloride	0.77(4)	8.0(4)	557
Potassium iodide	0.696(29)	7.21(30)	570
Rhenium(VII) oxide	1.23(2)	12.7(2)	125
Rubidium bromide	0.766(3)	7.94(3)	583
Rubidium chloride	0.820(3)	8.50(3)	590
Ruthenium tetroxide	1.172(3)	12.15(3)	988
Silane	1.124	11.65	1158
Silicon oxide (SiO)	1.103	11.43	1002
Silicon tetrachloride	1.136(1)	11.79(1)	527
Silicon tetrafluoride	1.51	15.7	−100
Silver chloride	0.973	10.08	1065
Silver fluoride	1.06(3)	11.0(3)	1071
Sodium bromide	0.802(10)	8.31(10)	660
Sodium chloride	0.861(6)	8.92(6)	681
Sodium iodide	0.737(2)	7.64(2)	659
Stibine (SbH$_3$)	0.920(3)	9.54(3)	1067

(Continued)

TABLE 1.30   Ionization Energy of Molecular and Radical Species (*Continued*)

Species	Ionization energy		$\Delta_f H$ (ion) in kJ · mol^{-1}
	In MJ · mol^{-1}	In electron volts	
Strontium oxide	0.675(14)	7.00(15)	662
Sulfur (S$_2$)	0.9027(2)	9.356(2)	1031
Sulfur chloride pentafluoride	1.1921(5)	12.335(5)	144
Sulfur dichloride	0.912(3)	9.45(3)	895
Sulfur difluoride	0.973	10.08	676
Sulfur dioxide	1.189(2)	12.32(2)	892
Sulfur hexafluoride	1.479(3)	15.33(3)	259
Sulfur oxide (SO)	0.996(2)	10.32(2)	1001
Sulfur pentafluoride	1.01(1)	10.5(1)	97
Sulfur trioxide	1.235(4)	12.80(4)	839
Sulfuryl chloride (SO$_2$Cl$_2$)	1.163	12.05	807
Sulfuryl fluoride (SO$_2$F$_2$)	1.110	11.5	679
Tantalum(V) chloride	1.069	11.08	348
Tetraborane (B$_4$H$_{10}$)	1.038(4)	10.76(4)	1105
Tetrafluorohydrazine (gauche)	1.152(3)	11.94(3)	1119
Thallium(I) bromide	0.882(2)	9.14(2)	844
Thallium(I) chloride	0.936(3)	9.70(3)	869
Thallium(I) fluoride	1.015	10.52	835
Thionitrosyl fluoride (NSF)	1.111(4)	11.51(4)	1090
Thionyl chloride	1.058	10.96	844
Thionyl fluoride	1.182	12.25	688
Thiophosphoryl trifluoride (PSF$_3$)	1.066(4)	11.05(4)	58
Thorium(IV) oxide	0.847(14)	8.70(15)	342
Tin(II) bromide	0.87	9.0	830
Tin(II) chloride	0.965	10.0	760
Tin(II) fluoride	1.07	11.1	586
Tin(II) oxide	0.926(2)	9.60(2)	944
Tin(II) sulfide	0.85	8.8	966
Tin(IV) bromide	1.02	10.6	709
Tin(IV) chloride	1.146(5)	11.88(5)	673
Tin(IV) hydride	1.037	10.75	1200
Titanium(IV) bromide	0.99	10.3	375
Titanium(IV) chloride	1.124(14)	11.65(15)	363
Titanium(IV) oxide	0.920(10)	9.54(10)	623
tran-Difluorodiazine	1.24	12.8	1315
Trifluoramine oxide (NOF$_3$)	1.279(1)	13.26(1)	1116
Trifluorosilane (F$_3$SiH)	1.35	14.0	150
Trisilane	0.89	9.2	1009
Tungsten(VI) chloride	0.92	9.5	348
Uranium hexafluoride	1.350(10)	14.00(10)	−796
Uranium(IV) oxide	5.2(1)	5.4(1)	57
Uranium(VI) oxide	1.01(5)	10.5(5)	214
Vanadium(IV) chloride	0.89	9.2	210
Vanadium(V) oxychloride (VOCl$_3$)	1.120	11.61	425
Water	1.2170(10)	12.612(10)	975
Xenon difluoride	1.192(1)	12.35(1)	1083
Xenon tetrafluoride	1.221(10)	12.65(10)	1016
Zirconium bromide	1.03	10.7	388
Zirconium chloride	1.08	11.2	392

Source: Sharon, G., et al., *J. Phys. Chem. Ref. Data*, **17**:Suppl. No 1 (1988).

1.5   ELECTRONEGATIVITY

Electronegativity χ is the relative attraction of an atom for the valence electrons in a covalent bond. It is proportional to the effective nuclear charge and inversely proportional to the covalent radius:

$$\chi = \frac{0.31(n+1\pm c)}{r} + 0.50$$

where n is the number of valence electrons, c is any formal valence charge on the atom and the sign before it corresponds to the sign of this charge, and r is the covalent radius. Originally the element fluorine, whose atoms have the greatest attraction for electrons, was given an arbitrary electronegativity of 4.0. A revision of Pauling's values based on newer data assigns -3.90 to fluorine. Values in Table 1.31 refer to the common oxidation states of the elements.

TABLE 1.31   Electronegativity Values of the Elements

H
2.20

Li	Be												B	C	N	O	F
0.98	1.57												2.04	2.55	3.04	3.44	3.90
Na	Mg												Al	Si	P	S	Cl
0.93	1.31												1.61	1.90	2.19	2.58	3.16

K	Ca	Sc	Ti	V	Cr	Mn	Fe	Co	Ni	Cu	Zn	Ga	Ge	As	Se	Br
0.82	1.00	1.36	1.54	1.63	1.66	1.55	1.83	1.88	1.91	1.90	1.65	1.81	2.01	2.18	2.55	2.96
Rb	Sr	Y	Zr	Nb	Mo	Tc	Ru	Rh	Pd	Ag	Cd	In	Sn	Sb	Te	I
0.82	0.95	1.22	1.33	1.6	2.16	2.10	2.2	2.28	2.20	1.93	1.69	1.78	1.96	2.05	2.1	2.66
Cs	Ba	La	Hf	Ta	W	Re	Os	Ir	Pt	Au	Hg	Tl	Pb	Bi	Po	At
0.79	0.89	1.10	1.3	1.5	1.7	1.9	2.2	2.2	2.2	2.4	1.9	1.8	1.8	1.9	2.0	2.2
Fr	Ra	Ac														
0.7	0.9	1.1														

	Ce	Pr	Nd		Sm		Gd		Dy	Ho	Er	Tm		Lu
Lanthanides	1.12	1.13	1.14		1.17		1.20		1.22	1.23	1.24	1.25		1.0

	Th	Pa	U	Np	Pu	Am	Cm	Bk	Cf	Es	Fm	Md	No
Actinides	1.3	1.5	1.7	1.3	1.3	1.3	1.3	1.3	1.3	1.3	1.3	1.3	1.3

The greater the difference is electronegativity, the greater is the ionic character of the bond. The amount of ionic character I is given by:

$$I = 0.46\,|\,\chi_A - \chi_B\,| + 0.035(\chi_A - \chi_B)^2$$

The bond is fully covalent when $(\chi_A - \chi_B) < 0.5$ (and $I < 6\%$).

1.6 ELECTRON AFFINITY

TABLE 1.32 Electron Affinities of Elements, Molecules, and Radicals

Electron affinity of an atom (molecule or radical) is defined as the energy difference between the lowest (ground) state of the neutral and the lowest state of the corresponding negative ion in the gas phase.

$$A(g) + e^- = A^-(g)$$

Data are limited to those negative ions which, by virtue of their positive electron affinity, are stable. Uncertainty in the final data figures is given in parentheses. Calculated values are enclosed in brackets.

A. Atoms

Atom	Electron affinity, in eV	in kJ · mol^{-1}
Aluminum	0.441(10)	42.5(10)
Antimony	1.046(5)	100.9(5)
Arsenic	0.81(3)	78.(3)
Astatine	[2.8(3)]	[270.(30)]
Barium	[0.15]	[14.]
Bismuth	0.946(10)	91.3(10)
Boron	0.277(10)	26.7(10)
Bromine	3.363590(3)	324.5367(3)
Calcium	0.0185(25)	1.78(24)
Carbon	1.2629(3)	121.85(3)
Cesium	0.471626(25)	45.5048(24)
Chlorine	3.61269	348.570
Chromium	0.666(12)	64.3(12)
Cobalt	0.662(3)	63.9(3)
Copper	1.235(5)	119.2(5)
Fluorine	3.401190(4)	328.1638(4)
Francium	[0.46]	[44]
Gallium	0.30(15)	29.(15)
Germanium	1.233(3)	119.0(3)
Gold	2.30863(3)	222.748(3)
Hafnium	[≈0.]	[≈0.]
Hydrogen	0.75195(19)	72.552(18)
Hydrogen-d_1 deuterium	0.75459(7)	72.807(7)
Indium	0.3(2)	29.(2)
Iodine	3.05904(1)	295.151(1)
Iridium	1.565(8)	151.0(8)
Iron	0.151(3)	14.6(3)
Lanthanum	[0.5(3)]	[48.(30)]
Lead	0.364(8)	35.1(8)
Lithium	0.6180(5)	59.63(5)
Molybdenum	0.748(2)	72.2(2)
Nickel	1.156(10)	111.5(10)
Niobium	0.893(25)	86.2(24)
Osmium	[0.2(1)]	[19.(10)]
Oxygen	1.4611103(7)	140.97523(7)
Palladium	0.562(5)	54.2(5)
Phosphorus	0.7465(3)	72.03(3)
Platinum	2.128(2)	205.3(2)
Polonium	[1.9(3)]	[183.(30)]

TABLE 1.32 Electron Affinities of Elements, Molecules, and Radicals (*Continued*)

A. Atoms

Atom	Electron affinity,	
	in eV	in kJ · mol⁻¹
Potassium	0.50147(10)	48.384(10)
Rhenium	[0.15(15)]	[14.(14)]
Rubidium	0.48592(2)	46.884(2)
Ruthenium	[1.05(15)]	[101.(14)]
Scandium	0.188(20)	18.1(19)
Selenium	2.020670(25)	194.9643(24)
Silver	1.302(7)	125.6(7)
Sodium	0.547926(25)	52.86666(24)
Strontium	0.048(6)	4.6(6)
Sulfur	2.077104(1)	200.4094(1)
Tantalum	0.322(12)	31.1(12)
Technetium	[0.55(20)]	[53.(19)]
Tellurium	1.9708(3)	190.15(3)
Thallium	0.2(2)	19.(19)
Tin	1.112(4)	107.3(4)
Titanium	0.079(14)	7.6(14)
Tungsten	0.815(2)	78.6(2)
Vanadium	0.525(12)	50.7(12)
Yttrium	0.307(12)	29.6(12)
Zirconium	0.426(14)	41.1(14)

B. Molecules

Molecule	Electron affinity,	
	in eV	in kJ · mol⁻¹
BF_3	2.65	256
BH_3	0.038(15)	3.7(15)
1,4-Benzoquinone	1.91(10)	184.(10)
Br_2	2.55(10)	246.(10)
$CBrF_3$	0.91(20)	89.(19)
CF_3I	1.57(20)	151.(19)
COS	0.46(20)	44.(19)
CS_2	0.895(20)	86.3(19)
C_6F_6 hexafluorobenzene	0.52(10)	50.(10)
$1,2-C_6H_4(NO_3)_2$ (also 1,3-)	1.65(10)	159.(10)
$1,4-C_6H_4(NO_3)_2$	2.00(10)	193.(10)
C_6H_5Br bromobenzene	1.15(11)	111.(11)
C_6H_5Cl chlorobenzene	0.82(11)	79.(11)
C_6H_5I iodobenzene	1.41(11)	136.(11)
$C_6H_5NO_2$ nitrobenzene	1.01(10)	97.(10)
$1,4-C_6H_4(CN)NO_2$	1.72(10)	166.(10)
Cl_2	2.38(10)	229.(10)
CoH_2	1.450(14)	139.9(13)
CsCl	0.455(10)	43.9(10)
CuO	1.777(6)	171.5(6)
F_2	3.08(10)	297.(10)
FeO	1.493(5)	144.1(5)
I_2	2.55(5)	246.(5)

(*Continued*)

TABLE 1.32   Electron Affinities of Elements, Molecules, and Radicals (*Continued*)

B. Molecules (*Continued*)

Molecule	Electron affinity,	
	in eV	in kJ · mol⁻¹
IBr	2.55(10)	246.(10)
IrF_6	6.5(4)	627.(40)
KBr	0.642(10)	61.9(10)
KCl	0.582(10)	56.1(10)
KI	0.728(10)	70.2(10)
LiCl	0.593(10)	54.3(10)
LiH	0.342(12)	33.0(12)
MoO_3	2.9(2)	280.(20)
NO	0.026(5)	2.5(5)
NO_2	2.273(5)	219.3(5)
N_2O	0.22(10)	21.(10)
NaBr	0.788(10)	76.0(10)
NaCl	0.727(10)	70.1(10)
NaI	0.865(10)	83.5(10)
NaK	0.465(30)	44.9(30)
O_2	0.451(7)	43.5(7)
O_3	2.103(3)	202.9(9)
OsF_6	6.0(3)	579.(29)
PBr_3	1.59(15)	153.(14)
PCl_3	0.82(10)	79.(10)
PF_5	0.75(15)	72.(14)
$POCl_3$	1.41(2)	136.(2)
PbO	0.722(6)	69.7(6)
PtF_6	7.0(4)	675.(40)
RbCl	0.544(10)	52.5(10)
RuF_6	7.5(3)	724.(28)
SF_4	1.5(2)	145.(19)
SF_6	1.05(10)	101.(10)
SO_2	1.107(8)	106.8(8)
SeF_6	2.9(2)	280.(19)
SeO	1.456(20)	140.5(19)
SeO_2	1.823(50)	175.9(48)
TeF_6	3.34(17)	322.(16)
TeO	1.695(22)	163.5(21)
UF_6	5.1(2)	492.(19)
V_4O_{10}	4.2(6)	405.(60)
WO_3	3.9(2)	376.(19)

C. Radicals

Radical	Electron affinity	
	in eV	in kJ · mol⁻¹
AsH_2	1.27(3)	123.(3)
CCl_2	1.591(10)	153.5(10)
CF_2	0.165(10)	15.9(10)
CH	1.238(8)	119.4(8)
CHBr	1.454(5)	140.3(5)
CHCl	1.210(5)	117.5(5)
CHF	0.542(5)	52.3(5)

TABLE 1.32 Electron Affinities of Elements, Molecules, and Radicals (*Continued*)

C. Radical

Radical	Electron affinity,	
	in eV	in kJ · mol^{-1}
CHI	1.42(17)	137.(17)
CHO$_2$	3.498(5)	337.5(5)
CH$_2$	0.652(6)	62.9(6)
CH$_2$S	0.465(23)	44.9(22)
CH$_2$=SiH	2.010(10)	193.9(10)
CH$_3$	0.08(3)	7.7(3)
CH$_3$CH$_2$O ethoxide	1.726(33)	166.5(32)
CH$_3$O	1.570(22)	151.5(21)
CH$_3$S	1.861(4)	179.6(4)
CH$_3$SCH$_2$	0.868(51)	83.7(49)
CH$_3$Si	0.852(10)	82.2(10)
CH$_3$SiH$_2$	1.19(4)	115.(4)
C$_2$F$_2$ difluorovinylidene	2.255(6)	217.6(6)
C$_2$H$_2$ vinylidene	0.490(6)	47.3(6)
CH$_2$=CH vinyl	0.667(24)	64.3(23)
C$_2$H$_3$O acetaldehyde enolate	1.82476(12)	176.062(12)
CH$_3$CH$_2$S	1.953(6)	188.4(6)
HC≡C—CH$_2$	0.893(25)	86.2(24)
CH$_3$CHCN	1.247(12)	120.3(12)
C$_2$H$_8$O ethoxide	1.726(33)	166.5(31)
C$_2$H$_5$S ethyl sulfide	1.953(6)	188.4(6)
C$_3$H$_3$ propargyl radical	0.893(25)	86.2(24)
CH$_3$CH—CN	1.247(12)	120.3(12)
C$_3$H$_5$ allyl	0.362(19)	34.9(18)
C$_3$H$_5$O acetone enolate	1.758(19)	169.2(18)
propionaldehyde enolate	1.621(6)	156.4(6)
C$_3$H$_5$O$_2$ methyl acetate enolate	1.80(6)	174.(6)
C$_3$H$_7$O propoxide	1.789(33)	172.6(31)
isopropyl oxide	1.839(29)	177.4(28)
C$_3$H$_7$S propyl sulfide	2.00(2)	193.(2)
isopropyl sulfide	2.02(2)	195.(2)
C$_4$H$_5$O cyclobutanone enolate	1.801(8)	173.8(8)
C$_4$H$_7$O butyraldehyde enolate	1.67(5)	161.(5)
C$_4$H$_9$O *tert*-butoxyl	1.912(54)	184.5(52)
C$_4$H$_9$S butyl sulfide	2.03(2)	196.(2)
tert-butyl sulfide	2.07(2)	200.(2)
C$_5$H$_5$ cyclopentadienyl	1.804(7)	174.1(7)
C$_5$H$_7$ pentadienyl	0.91(3)	88.(3)
C$_5$H$_7$O cyclopentanone enolate	1.598(7)	154.2(7)
C$_5$H$_9$O 3-pentanone enolate	1.69(5)	163.(5)
C$_5$H$_{11}$S pentyl sulfide	2.09(2)	202.(2)
C$_6$H$_5$ phenyl	1.096(6)	105.7(6)
C$_6$H$_5$NH anilide	1.70(3)	164.(3)
C$_6$H$_8$O phenoxyl	2.253(6)	217.4(6)
C$_6$H$_5$S thiophenoxide	≤2.47(6)	≤238.(6)
C$_6$H$_5$CH$_2$ benzyl	0.912(6)	88.0(6)
C$_6$H$_5$CH$_2$O benzyl oxide	2.14(2)	206.(2)
C$_6$H$_9$O cyclohexanone enolate	1.526(10)	147.2(10)
H$_2$C=CH—CH=CH—CH=CH—CH$_2$ heptatrienyl	1.27(3)	122.(3)
CN	3.862(4)	372.6(4)

(*Continued*)

TABLE 1.32   Electron Affinities of Elements, Molecules, and Radicals (*Continued*)

C. Radical

Radical	Electron affinity	
	in eV	in kJ · mol^{-1}
CNCH$_2$ cyanomethyl	1.543(14)	148.9(14)
CO$_3$	2.69(14)	259.(14)
CS	0.205(21)	19.8(20)
ClO	2.275(6)	219.5(6)
HCO	0.313(5)	30.2(5)
HNO	0.338(15)	32.6(14)
HO$_2$	1.078(17)	104.0(6)
FO	2.272(6)	219.2(6)
N$_3$	2.70(12)	260.(12)
NCO	3.609(5)	348.2(5)
NCS	3.537(5)	341.3(5)
NH	0.370(4)	35.7(4)
NO$_3$	3.937(14)	379.9(14)
NS	1.194(11)	115.2(11)
O$_2$Aryl	0.52(2)	50.(2)
OClO	2.140(8)	206.5(8)
OH	1.82767(2)	176.343(2)
OIO	2.577(8)	248.6(8)
PH	1.028(10)	99.2(10)
PH$_2$	1.27(1)	123.(1)
PO	1.092(10)	105.4(10)
PO$_2$	3.42(1)	330.(1)
SF	2.285(6)	220.5(6)
SH	2.314344(4)	223.300(4)
SO	1.125(5)	108.5(5)
SeH	2.21252(3)	213.475(3)
SiF$_3$	≤ 2.95(10)	285.(10)
SiH	1.277(9)	123.2(9)
SiH$_2$	1.124(20)	108.4(19)
SiH$_3$	1.406(14)	106.7(14)

Source: H. Hotop and W. C. Lineberger, *J. Phys. Chem. Reference Data* **14**:731 (1985).

1.7   BOND LENGTHS AND STRENGTHS

Distances between centers of bonded atoms are called *bond lengths*, or *bond distances*. Bond lengths vary depending on many factors, but in general, they are very consistent. Of course the bond orders affect bond length, but bond lengths of the same order for the same pair of atoms in various molecules are very consistent.

The *bond order* is the number of electron pairs shared between two atoms in the formation of the bond. Bond order for C=C and O=O is 2. The amount of energy required to break a bond is called *bond dissociation energy* or simply *bond energy*. Since bond lengths are consistent, bond energies of similar bonds are also consistent.

Bonds between the same type of atom are *covalent bonds*, and bonds between atoms when their electronegativity differs slightly are also predominant covalent in character. Theoretically, even ionic bonds have some covalent character. Thus, the boundary between ionic and covalent bonds is not a clear line of demarcation.

For covalent bonds, bond energies and bond lengths depend on many factors: electron affinities, sizes of atoms involved in the bond, differences in their electronegativity, and the overall structure of the molecule. There is a general trend in that *the shorter the bond length, the higher the bond energy* but there is no formula to show this relationship, because of the widespread variation in bond character.

1.7.1 Atom Radius

The *atom radius* of an element is the shortest distance between like atoms. It is the distance of the centers of the atoms from one another in metallic crystals and for these materials the atom radius is often called the metal radius. Except for the lanthanides (CN = 6), CN = 12 for the elements.

1.7.2 Ionic Radii

One of the major factors in determining the structures of the substances that can be thought of as made up of cations and anions packed together is ionic size. It is obvious from the nature of wave functions that no ion has a precisely defined radius. However, with the insight afforded by electron density maps and with a large base of data, new efforts to establish tables of ionic radii have been made.

Effective ionic radii are based on the assumption that the ionic radius of O^{2-} (CN 6) is 140 pm and that of F^- (CN 6) is 133 pm. Also taken into consideration is the coordination number (CN) and electronic spin state (HS and LS, high spin and low spin) of first-row transition metal ions. These radii are empirical and include effects of covalence in specific metal-oxygen or metal-fluorine bonds. Older "crystal ionic radii" were based on the radius of F^- (CN 6) equal to 119 pm; these radii are 14–18 percent larger than the effective ionic radii.

1.7.3 Covalent Radii

Covalent radii are the distance between two kinds of atoms connected by a covalent bond of a given type (single, double, etc.).

TABLE 1.33   Atom Radii and Effective Ionic Radii of Elements

Element	Atom radius, pm	Ion charge	Effective ionic radii, pm Coordinator number 4	6	8	12
Actinium	187.8	3 +		111		
Aluminum	143.1	3+	39	53.5		
Americium	173	2+			126	
		3 +		97.5	109	
		4+		89	95	
		5 +		86		
		6+		80		
Antimony	145	3–		245		
		1 +		89		
		3 +	76	76		
		5+		60		

(Continued)

TABLE 1.33 Atom Radii and Effective Ionic Radii of Elements (*Continued*)

Element	Atom radius, pm	Ion charge	Effective ionic radii, pm			
			Coordinator number			
			4	6	8	12
Arsenic	124.8	3−		222		
		3+		58		
		5+	33.5	46		
Astatine		1−		227		
		5+		57		
		7+		62		
Barium	217.3	2+		136	142	160
Berkelium		2+		118		
		3+		98		
		4+		87	93	
Beryllium	111.3	1−	195			
		2+	27	45		
Bismuth	154.7	3−		213		
		3+		103	111	
		5+		76		
Boron	86	1+	35			
		3+	11	27		
Bromine		1−		196		
		3+	59			
		5+	31*	47		
		7+		25		
Cadmium	148.9	2+	78	95	110	131
Calcium	197	2+		100	112	135
Californium	186(2)	2+		117		
		3+		95		
		4+		82.1		
Carbon		4−	260			
		4+	15	16		
Cerium	181.8	3+		102	114.3	134
		4+		87	97	114
Cesium	265	1+		167	174	188
Chlorine		1−		181		
		5+	34			
		7+	8	27		
Chromium	128	1+	81			
		2+		73 LS		
				80 HS		
		3+		61.5		
		4+	41	55		
		5+	34.5	49	57	
		6+	26	44		
Cobalt	125	2+	38	65 LS	90	
				74.5 HS		
		3+		54.5 LS		
				61 HS		
		4+	40	53 HS		
Copper	128	1+	60	77		
		2+	57	73		
		3+		54 LS		

*CN = 3.

TABLE 1.33 Atom Radii and Effective Ionic Radii of Elements (*Continued*)

Element	Atom radius, pm	Ion charge	Effective ionic radii, pm			
			Coordinator number			
			4	6	8	12
Curium	174	3+		97		
		4+		85	95	
Dysprosium	178.1	2+		107	119	
		3 +		91.2	102.7	
Einsteinium	186(2)	3+		98		
Erbium	176.1	3+		89.0	100.4	
Europium	208.4	2+		117	125	135
		3+		94.7	106.6	
Fluorine	71.7	1−	131	133		
		7+		8		
Francium	270	1+		180		
Gadolinium	180.4	3+		93.8	105.3	
Gallium	135	2+		120		
		3+	47	62.0		
Germanium	128	2+		73		
		4+	39.0	53.0		
Gold	144	1+		137		
		3+	68	85		
Hafnium	159	4+	58	71	83	
Holmium	176.2	3+		90.1	101.5*	112
Hydrogen		1−		154		
Indium	167	1+		140		
		3+	62	80.0	92	
Iodine		1−		220		
		5+		95		
		7+	42	53		
Iridium	135.5	3+		68		
		4+		62.5		
		5+		57		
Iron	126	2+		61 LS		
			63 HS	78 HS	92 HS	
		3+		55 LS		
			49 HS	64.5 HS	78 HS	
		4+		58.5		
		6+	25			
Lanthanum	183	3+		103.2	116.0	136
Lead	175	2+	98	119	129	149
		4+		78	94	
Lithium	152	1+	59	76		
Lutetium	173.8	3+		86.1	97.7	
Magnesium	160	2+	57	72.0	89	
Manganese	127	2+	66 HS	67 LS	96	
				83 HS		
		3+		58 LS		
				64.5 HS		
		4+	39	53		
		5+	33			
		6+	25.5			
		7+	25	46		

*CN = 10

(*Continued*)

TABLE 1.33 Atom Radii and Effective Ionic Radii of Elements (*Continued*)

Element	Atom radius, pm	Ion charge	Effective ionic radii, pm Coordinator number 4	6	8	12
Mercury	151	1+	111*	119		
		2+	96	102	114	
Molybdenum	139	3+		69		
		4+		65.0		
		5+	46	61		
		6+	41	59	73†	
Neodymium	181.4	2+			129	
		3+		98.3	110.9	127
Neptunium	155	2+		110		
		3+		101		
		4+		87	98	
		5+		75		
		6+		72		
		7+		71		
Nickel	124	2+	55	69.0		
		3+		56 LS		
				60 HS		
		4+		48 LS		
Niobium	146	3+		72		
		4+		68	79	
		5+	48	64	74	
Nitrogen		3−	146			
		1+	25			
		3+		16		
		5+		13		
Nobelium		2+		110		
Osmium	135	4+		63.0		
		5+		57.5		
		6+		54.5		
		7+		52.5		
		8+	39			
Oxygen		2−	138	140	142	
Palladium	137	2+	64	86		
		3+		76		
		4+		61.5		
Phosphorus	108	3−		212		
		3+		44		
		5+	17	38		
Platinum	138.5	2+		80		
		4+		62.5		
		5+		57		
Plutonium	159	3+		100		
		4+		86	96	
		5+		74		
		6+		71		

*CN = 3.
†CN = 7.

TABLE 1.33   Atom Radii and Effective Ionic Radii of Elements (*Continued*)

Element	Atom radius, pm	Ion charge	Effective ionic radii, pm Coordinator number 4	6	8	12
Polonium	164	2−		(230)		
		4+		94	108	
		6+		67		
Potassium	232	1+	137	138	151	164
Praseodymium	182.4	3+		99	112.6	
		4+		85	96	
Promethium	183.4	3+		97	109.3	
Protoactinium	163	3+		104		
		4+		90	101	
		5+		78	91	
Radium	(220)	2+			148	170
Rhenium	137	4+		63		
		5+		58		
		6+		55		
		7+	38	53		
Rhodium	134	3+		66.5		
		4+		60		
		5+		55		
Rubidium	248	1+		152	161	172
Ruthenium	134	3+		68		
		4+		62.0		
		5+		56.5		
		7+	38			
		8+	36			
Samarium	180.4	2+			127	
		3+		95.8	107.9	124
Scandium	162	3+		74.5	87.0	
Selenium	116	2−		198		
		4+		50		
		6+		42		
Silicon	118	4+	26	40.0		
Silver	144	1+	100	115	130	
		2+	79	94		
		3+	67	75		
Sodium	186	1+	99	102	118	139
Strontium	215	2+		118	126	144
Sulfur	106	2−		184		
		4+		37		
		6+	12	29		
Tantalum	146	3+		72		
		4+		68		
		5+		64	74	
Technetium	136	4+		64.5		
		5+		60		
		7+	37	56		
Tellurium	142	2−		221		
		4+	66	97		
		6+	43	56		

(*Continued*)

TABLE 1.33   Atom Radii and Effective Ionic Radii of Elements (*Continued*)

| Element | Atom radius, pm | Ion charge | Effective ionic radii, pm | | | |
| | | | Coordinator number | | | |
			4	6	8	12
Terbium	177.3	3+		92.3	104.0	
		4+		76	88	
Thallium	170	1+		150	159	170
		3+	75	88.5	98	
Thorium	179	4+		94	105	121
Thullium	175.9	2+		103		
		3+		88.0	99.4	105*
Tin	151	2+		118		
		4+	55	69.0	81	
Titanium	147	2+		86		
		3+		67.0		
		4+	42	60.5	74	
Tungsten	139	4+		66		
		5+		62		
		6+	42	60		
Uranium	156	3+		102.5		
		4+		89	100	117
		5+		76		
		6+	52	73	86	
Vanadium	134	2+		79		
		3+		64.0		
		4+		58	72	
		5+	35.5	54		
Xenon		8+	40	48		
Ytterbium	193.3	2+		102	114	
		3+		86.8	98.5	104*
Yttrium	180	3+		90.0	101.9	108*
Zinc	134	2+	60	74.0	90	
Zirconium	160	4+	59	72	84	89*

*CN = 11.

TABLE 1.34 Approximate Effective Ionic Radii in Aqueous Solutions at 25°C

å (in Å)	Inorganic ions	å (in Å)	Organic ions
2.5	Rb^-, Cs^+, NH_4^+, Tl^+, Ag^+	3.5	$HCOO^-$, H_2Cit^-, $CH_3NH_3^+$, $(CH_3)_2NH_2^+$
3	K^+, Cl^-, Br^-, I^-, CN^-, NO_2^-, NO_3^-	4	$H_3N^+CH_2COOH$, $(CH_3)_3NH^+$, $C_2H_5NH_3^+$
3.5	OH^-, F^-, SCN^-, OCN^-, HS^-, ClO_3^-, ClO_4^-, BrO_3^-, IO_4^-, MnO_4^-	4.5	CH_3COO^-, $ClCH_2COO^-$, $(CH_3)_4N^+$, $(C_2H_5)_2NH_2^+$, $H_2NCH_2COO^-$, oxalate^{2-}, $HCit^{2-}$
4	Na^-, $CdCl^+$, Hg_2^{2+}, ClO_2^-, IO_3^-, HCC_3^-, $H_2PO_4^-$, HSO_3^-, $H_2AsO_4^-$, SO_4^{2-}, $S_2O_3^{2-}$, $S_2O_8^{2-}$, SeO_4^{2-}, CrO_4^{2-}, HPO_4^{2-}, $S_2O_6^{2-}$, PO_4^{3-}, $Fe(CN)_6^{3-}$, $Cr(NH_3)_6^{3+}$, $Co(NH_3)_6^{3+}$, $Co(NH_3)_5H_2O^{3+}$	5	Cl_2CHCOO^-, Cl_3COO^-, $(C_2H_5)_3NH^+$, $C_3H_7NH_3^+$, Cit^{3-}, succinate^{2-}, malonate^{2-}, tartrate^{2-}
4.5	Pb^{2+}, CO_3^{2-}, SO_3^{2-}, MoO_4^{2-}, $Co(NH_3)_5Cl^{2+}$, $Fe(CN)_5NO^{2-}$	6	benzoate$^-$, hydroxybenzoate$^-$, chlorobenzoate$^-$, phenylacetate$^-$, vinylacetate$^-$, $(CH_3)_2C{=}CHCOO^-$, $(C_2H_5)_4N^+$, $(C_3H_7)_2NH_2^+$, phthalate^{2-}, glutarate^{2-}, adipate^{2-}
5	Sr^{2+}, Ba^{2+}, Ra^{2+}, Cd^{2+}, Hg^{2+}, S^{2-}, $S_2O_4^{2-}$, WO_4^{2-}, $Fe(CN)_6^{4-}$	7	trinitrophenolate$^-$, $(C_3H_7)_3NH^+$, methoxybenzoate$^-$, pimelate^{2-}, suberate^{2-}, Congo red anion^{2-}
6	Li^+, Ca^{2+}, Cu^{2+}, Zn^{2+}, Sn^{2+}, Mn^{2+}, Fe^{2+}, Ni^{2+}, Co^{2+}, $Co(en)_3^{3+}$, $Co(S_2O_3)(CN)_5^{4-}$	8	$(C_6H_5)_2CHCOO^-$, $(C_3H_7)_4N^+$
8	Mg^{2+}, Be^{2+}		
9	H^+, Al^{3+}, Fe^{3+}, Cr^{3+}, Sc^{3+}, Y^{3+}, La^{3+}, In^{3+}, Ce^{3+}, Pr^{3+}, Nd^{3+}, Sm^{3+}, $Co(SO_3)_2(CN)_4^{5-}$		
11	Th^{4+}, Zr^{4+}, Ce^{4+}, Sn^{4+}		

TABLE 1.35 Covalent Radii for Atoms

Element	Single-bond radius, pm*	Double-bond radius, pm	Triple-bond radius, pm
Aluminum	126		
Antimony	141	131	
Arsenic	121	111	
Beryllium	106		
Boron	88		
Bromine	114	104	
Cadmium	148		
Carbon	77.2	66.7	60.3
Chlorine	99	89	
Copper	135		
Fluorine	64	54	
Gallium	126		
Germanium	122	112	
Hydrogen	30		
Indium	144		
Iodine	133	123	
Magnesium	140		
Mercury	148		
Nitrogen	70	60	55
Oxygen	66	55	
Phosphorus	110	100	93
Silicon	117	107	100
Selenium	117	107	
Silver	152		
Sulfur	104	94	87
Tellurium	137	127	
Tin	140	130	
Zinc	131		

*Single-bond radii are for a tetrahedral (CN = 4) structure.

TABLE 1.36 Octahedral Covalent Radii for CN = 6

Atom	Octahedral covalent radius, pm
Cobalt(II)	132
Cobalt(III)	122
Gold(IV)	140
Iridium(III)	132
Iron(II)	123
Iron(IV)	120
Nickel (II)	139
Nickel(III)	130
Nickel(IV)	121
Osmium(II)	133
Palladium(IV)	131
Platinum(IV)	131
Rhodium(III)	132
Ruthenium(II)	133

TABLE 1.37   Bond Lengths between Elements

Elements	Bond type	Bond length, pm	Elements	Bond type	Bond length, pm
	Boron			Oxygen	
B-B	B_2H_6	177(1)	O-H	H_2O	95.8
B-Br	BBr_3	187(2)		ROH	97(1)
B-Cl	BCl_3	172(1)		OH^+	102.89
B-F	BF_3, R_2BF	129(1)		HOOH	96.0(5)
B-H	Boranes	121(2)		D_2O (^2H_2O)	95.75
	Bridge	139(2)		OD	96.99
B-N	Borazoles	142(1)	O-O	HO—OH	148(1)
B-O	$B(OH)_3, (RO)_3B$	136(5)		O_2^+	122.7
				O_2^-	126(2)
	Hydrogen			O_3^{2-}	149(2)
				O_3	127.8(5)
H-Al	AlH	164.6	O-Al	AlO	161.8
H-As	AsH_3	151.9	O-As	As_2O_6 bridges	179
H-Be	BeH	134.3	O-Ba	BaO	190.0
H-Br	HBr	140.8	O-Cl	ClO_2	148.4
H-Ca	CaH	200.2		OCl_2	168
H-Cl	HCl	127.4	O-Mg	MgO	174.9
H-F	HF	91.7	O-Os	OsO_4	166
H-Ge	GeH_4	153	O-Pb	PbO	193.4
H-I	HI	160.9		Phosphorus	
H-K	KH	224.4			
H-Li	LiH	159.5	P-Br	PBr_3	223(1)
H-Mg	MgH	173.1	P-Cl	PCl_3	200(2)
H-Na	NaH	188.7	P-F	$PFCl_2$	155(3)
H-Sb	H_3Sb	170.7	P-H	PH_3, PH_4^+	142.4(5)
H-Se	H_2Se	146.0	P-I	PI_3	252(1)
H-Sn	SnH_4	170.1	P-N	Single bond	149.1
D-Br	DBr (^2HBr)	141.44	P-O	Single bond	144.7
D-Cl	DCl	127.46		p^3 bonding	167
D-I	DI	161.65		sp^3 bonding	154(4)
T-Br	TBr (^3HBr)	141.44	P-S	p^3 bonding	212(5)
T-Cl	TCl	127.40		sp^3 bonding	208(2)
				In rings	220(3)
	Nitrogen		P-C	Single bond	156.2
				p^3 bonding	187(2)
N-Cl	NO_2Cl	179(2)		Silicon	
N-F	NF_3	136(2)			
N-H	NH_4^+	103.4(3)	Si-Br	$SiBr_4, R_3SiBr$	216(1)
	NH_3, RNH_2	101.2	Si-Cl	$SiCl_4, R_3SiCl$	201.9(5)
	H_2NNH_2	103.8	Si-F	SiF_4, R_3SiF	156.1(3)
	R—CO—NH_2	99(3)		SiF_6	158
	HN=C=S	101.3(3)	Si-H	SiH_4	148.0(5)
N-D	ND (N^2H)	104.1		R_3SiH	147.6(5)
N-N	HN_3	102(1)	Si-I	SiI_4	234
	R_2NNH_2	145.1(5)		R_3SiI	246(2)
	N_2O	112.6(2)	Si-O	R_3SiOR	153.3(5)
	N_2^+	111.6	Si-Si	H_3SiSiH_3	230(2)
N-O	NO_2Cl	124(1)		Sulfur	
	RO—NO_2	136(2)			
	NO_2	118.8(5)	S-Br	$SOBr_2$	227(2)
N=O	N_2O	118.6(2)	S-Cl	S_2Cl_2	158.5(5)
	RNO_2	122(I)	S-F	SOF_2	158.5(5)
	NO^+	106.19	S-H	H_2S	133.3
N-Si	SiN	157.2		RSH	132.9(5)
				D_2S	134.5
			S-O	SO_2	143.21
				$SOCl_2$	145(2)
			S-S	RSSR	205(1)

TABLE 1.38 Bond Dissociation Energies

The bond dissociation energy (enthalpy change) for a bond A—B which is broken through the reaction

$$AB \rightarrow A + B$$

is defined as the standard-state enthalpy change for the reaction at a specified temperature, here at 298 K. That is,

$$\Delta Hf_{298} = \Delta Hf_{298}(A) + \Delta Hf_{298}(B) - \Delta Hf_{298}(AB)$$

All values refer to the gaseous state and are given at 298 K. Values of 0 K are obtained by subtracting $RT from the value at 298 K.

To convert the tabulated values to kcal/mol, divide by 4.184.

Bond	ΔHf_{298}, kJ/mol	Bond	ΔHf_{298}, kJ/mol
	Aluminum		**Antimony** (*continued*)
Al—Al	186(9)	Sb—O	372(84)
Al—As	180	Sb—P	357
Al—Au	326(6)	Sb—S	379
Al—Br	439(8)	Sb—Te	277.4(38)
Al—C	255		
Al—Cl	494(13)		**Arsenic**
AlCl—Cl	402(8)		
AlCl₂—Cl	372(8)	As—As	382(11)
AlO—Cl	515(84)	As—Cl	448
Al—Cu	216(10)	As—Ga	209.6(12)
Al—D	291	As—H	272(12)
Al—F	664(6)	As—N	582(126)
AlF—F	546(42)	As—O	481(8)
AlF₂—F	544(46)	As—P	534(13)
AlO—F	761(42)	As—S	(478)
Al—H	285(6)	As—Se	96
Al—I	368(4)	As—Tl	198(15)
Al—Li	176(15)		
Al—N	297(96)		**Astatine**
Al—O	512(4)		
AlCl—O	540(41)	At—At	(115.9)
AlF—O	582		
Al—P	213(13)		**Barium**
Al—Pd	259(12)		
Al—S	374(8)	Ba—Br	370(8)
Al—Se	334(10)	Ba—Cl	444(13)
Al—Si	251(3)	Ba—F	487(7)
Al—Te	268(10)	Ba—I	>431(4)
Al—U	326(29)	Ba—O	563(42)
		Ba—OH	477(42)
	Antimony	Ba—S	400(19)
Sb—Sb	299(6)		
Sb—Br	314(59)		**Béryllium**
Sb—Cl	360(50)		
Sb—F	439(96)	Be—Be	59
Sb—N	301(50)	Be—Br	381(84)
		Be—Cl	388(9)

TABLE 1.38 Bond Dissociation Energies (*Continued*)

Bond	ΔHf_{298}, kJ/mol	Bond	ΔHf_{298}, kJ/mol
Beryllium (*continued*)		Bromine	
BeCl—Cl	540(63)	Br—Br	193.870(4)
Be—F	577(42)	Br—C	280(21)
Be—H	226(21)	Br—CH₃	284(8)
Be—O	448(21)	Br—CH₂Br	255(13)
Be—S	372(59)	Br—CHBr₂	259(17)
		Br—CBr₃	209(13)
Bismuth		Br—CCl₃	218(13)
		Br—CF₃	285(13)
Bi—Bi	197(4)	Br—CF₂CF₃	287.4(63)
Bi—Br	267(4)	Br—CF₂CF₂CF₃	278.2(63)
Bi—Cl	305(8)	Br—CHF₂	289
Bi—D	284	Br—Cl	218.84(4)
Bi—F	259(29)	Br—CN	381
Bi—Ga	159(17)	Br—CO—C₆H₅	268
Bi—H	279	Br—F	233.8(2)
Bi—O	343(6)	Br—N	276(21)
Bi—P	280(13)	Br—NF₂	222
Bi—Pb	142(15)	Br—NO	120.1(63)
Bi—S	316(5)	Br—O	235.1(4)
Bi—Sb	251(4)		
Bi—Se	280(6)	Cadmium	
Bi—Te	232(11)		
Bi—Tl	121(13)	Cd—Cd	11.3(8)
		Cd—Br	159(96)
Boron		Cd—Cl	206.7(34)
		Cd—F	305(21)
B—B	297(21)	Cd—H	69.0(4)
H₃B—BH₃	146	Cd—I	138(21)
OB—BO	506(84)	Cd—In	138
B—Br	435(21)	Cd—O	142(42)
B—C	448(29)	Cd—S	196
B—Cl	536(29)	Cd—Se	310
BO—Cl	460(42)		
B—D	341(6)	Calcium	
D—F	766(13)		
BF—F	523(63)	Ca—Ca	14.98(46)
BF₂—F	557(84)	Ca—Br	321(23)
B—H	330(4)	Ca—Cl	398(13)
B—I	384(21)	Ca—F	527(21)
B—N	389(21)	Ca—H	167.8
B—O	806(5)	Ca—I	285(63)
BCl—O	715(41)	Ca—O	464(84)
B—P	347(17)	Ca—S	314(19)
B—S	581(9)		
B—Se	462(15)	Cerium	
B—Si	289(29)		
B—Te	354(20)	Ce—Ce	243(21)
		Ce—F	582(42)
		Ce—N	519(21)
		Ce—O	795(13)
		Ce—S	573(13)
		Ce—Se	495(15)
		Ce—Te	389(42)

(*Continued*)

TABLE 1.38   Bond Dissociation Energies (*Continued*)

Bond	ΔHf_{298}, kJ/mol	Bond	ΔHf_{298}, kJ/mol
Cesium		Chromium (*continued*)	
Cs—Cs	41.75(93)	Cr—Cu	155(21)
Cs—Br	397.5(42)	Cr—F	437(20)
Cs—Cl	439(21)	Cr—Ge	170(29)
Cs—F	514(8)	Cr—H	280(50)
Cs—H	178.1(38)	Cr—I	287(24)
Cs—I	339(4)	Cr—N	378(19)
Cs—O	297(25)	Cr—O	427(29)
Cs—OH	385(13)	OCr—O	531(63)
Chlorine		O_2Cr—O	477(84)
		Cr—S	339(21)
Cl—Cl	242.580(16)	Cobalt	
Cl—C	338(42)		
Cl—CH$_3$	339(21)	Co—Co	167(25)
Cl—CH$_3^+$	213	Co—Br	331(42)
Cl—C(CH$_3$)$_3$	328.4	Co—Cl	398(8)
Cl—CH$_2$Cl	310(13)	Co—Cu	162(17)
Cl—CCl$_3$	293(21)	Co—F	435(63)
Cl—CF$_3$	360(33)	Co—Ge	239(25)
Cl—CCl$_2$F	305(8)	Co—I	235(81)
Cl—CClF$_2$	318(8)	Co—O	368(21)
Cl—CF$_2$CF$_2$	346.0(71)	Co—S	343(21)
Cl—CH=CH$_2$	351	Copper	
Cl—CN	439		
Cl—COCl	328	Cu—Cu	202(4)
Cl—COCH$_3$	349.4	Cu—Br	331(25)
Cl—COC$_6$H$_5$	310(13)	Cu—Cl	383(21)
Cl—Cl$^+$	393	Cu—F	431(13)
Cl—ClO	143.3(42)	Cu—Ga	216(15)
O$_3$Cl—ClO$_4$	243	Cu—Ge	209(21)
Cl—F	250.54(8)	Cu—H	280(8)
O$_3$Cl—F	255	Cu—I	197(21)
Cl—N	389(50)	Cu—Ni	206(17)
Cl—NCl	280	Cu—O	343(63)
Cl—NCl$_2$	381	Cu—S	285(17)
Cl—NF$_2$	*ca.* 134	Cu—Se	293(38)
Cl—NH$_2$	251(25)	Cu—Sn	177(17)
Cl—NO	159(6)	Cu—Te	176(38)
Cl—NO$_2$	142(4)	Curium	
Cl—O	272(4)		
OCl—O	243(13)	Cm—O	736
O$_2$Cl—O	201(4)	Dysprosium	
Cl—P	289(42)		
Cl—SiCl$_3$	464	Dy—F	527(21)
Chromium		Dy—O	611(42)
		Dy—Se	322(42)
Cr—Cr	155(21)	Dy—Te	234(42)
Cr—Br	328(24)		
Cr—Cl	366(24)		

TABLE 1.38 Bond Dissociation Energies (*Continued*)

Bond	ΔHf_{298}, kJ/mol	Bond	ΔHf_{298}, kJ/mol
Erbium		**Gallium** (*continued*)	
Er—F	565(17)	Ga—O	285(63)
Er—O	611(13)	Ga—P	230(13)
Er—S	418(42)	Ga—Sb	209(13)
Er—Se	326(42)	Ga—Te	251(25)
Er—Te	239(42)	**Germanium**	
Europium		Ge—Ge	274(21)
Eu—Eu	33.5(165)	Ge—Br	255(29)
Eu—Cl	*ca.* 326	Ge—Cl	431.8(4)
Eu—F	528(18)	Ge—F	485(21)
Eu—O	557(13)	Ge—H	321.3(8)
Eu—S	364(15)	Ge—O	662(13)
Eu—Se	301(15)	Ge—S	551.0(25)
Eu—Te	243(15)	Ge—Se	490(21)
		Ge—Si	301(21)
Fluorine		Ge—Te	402(8)
F—F	156.9(96)	**Gold**	
F—F$^+$	>251		
F—CH$_3$	452(21)	Au—Au	221.3(21)
F—C(CH$_3$)$_3$	439	Au—B	368(11)
F—C$_6$H$_5$	485	Au—Be	285(8)
F—CCl$_3$	444(21)	Au—Bi	293(84)
F—CCl$_2$F	460(25)	Au—Cl	343(10)
F—CClF$_2$	490(25)	Au—Co	215(13)
F—CF$_3$	523(17)	Au—Cr	215(6)
F—COCH$_3$	498	Au—Cu	232(9)
F—FO	272(13)	Au—Fe	187(17)
F—FO$_2$	81.0	Au—Ga	294(15)
F—N	301(42)	Au—Ge	277(15)
F—NF	318(25)	Au—H	314(10)
F—NF$_2$	243(8)	Au—La	80(5)
F—NO	235.6(42)	Au—Li	68.0(16)
F—NO$_2$	197(25)	Au—Mg	243(42)
		Au—Mn	185(13)
Gadolinium		Au—Ni	274(21)
		Au—Pb	130(42)
Gd—F	590(27)	Au—Pd	143(21)
Gd—O	716(17)	Au—Rh	231(29)
Gd—S	525(15)	Au—S	418(25)
Gd—Se	431(15)	Au—Si	312(12)
		Au—Sn	244(17)
Gallium		Au—Te	247(67)
Ga—Ga	138(21)	Au—U	318(29)
Ga—Br	444(17)	**Hafnium**	
(CH$_3$)$_3$Ga—CH$_3$	253		
Ga—Cl	481(13)	Hf—C	548(63)
Ga—F	577(15)	Hf—N	534(29)
Ga—H	<274	Hf—O	791(8)
Ga—I	339(10)		

(*Continued*)

TABLE 1.38 Bond Dissociation Energies (*Continued*)

Bond	ΔHf_{298}, kJ/mol	Bond	ΔHf_{298}, kJ/mol
Hydrogen		Hydrogen (*continued*)	
H—H	436.002(4)	H—CHCl$_2$	414.2
H—^{2}H or H—D	439.446(4)	H—CCl$_3$	377(8)
^{2}H—^{2}H or D—D	443.546(4)	H—CBr$_3$	377(8)
H—Br	365.7(21)	H—CCl$_2$CHCl$_2$	393(8)
H—C	337.2(8)	H—CH$_2$F	423(8)
H—CH	452(33)	H—CHF$_2$	423(8)
H—CH$_2$	473(4)	H—CF$_3$	444(13)
H—CH$_3$	431(8)	H—CF$_2$Cl	435(4)
^{2}H—C^{2}H$_3$ or D—CD$_3$	442.75(25)	H—CH$_2$CF$_3$	446(45)
H—C≡CH	523(4)	H—CF$_2$CH$_3$	416(4)
H—CH=CH$_2$	427	H—CF$_2$CF$_3$	431(63)
H—CH$_2$CH$_3$	410(4)	H—CH$_2$I	431(8)
H—CH$_2$C≡CH	392.9(50)	H—CHI$_2$	431(8)
H—CH$_2$CH=CH$_2$	356	H—CN	540(25)
H—cyclopropyl	423(13)	H—CH$_2$CN	*ca.* 389
H—CH$_2$CH$_2$CH$_3$	410(8)	H—CH(CH$_3$)CN	377(8)
H CH(CH$_3$)$_2$	395.4	H—C(CH$_3$)$_2$CN	364(8)
H—cyclobutyl	397(13)	H—CH$_2$NH$_2$	397(8)
H—CH$_2$CH(CH$_3$)$_2$	360	H—CH$_2$Si(CH$_3$)$_3$	414(4)
H—CH(CH$_3$)CH$_2$CH$_3$	397(4)	H—CH$_2$COCH$_3$	393(75)
H—C(CH$_3$)$_3$	381	H—Cl	431.8(4)
		H—CO	126(8)
H—(cyclopentadienyl)	339(4)	H—CHO	364(4)
		H—COOH	377
H—CH(CH=CH$_2$)$_2$	335(4)	H—COCH$_3$	364(4)
		H—COCH$_2$CH$_3$	364(4)
H—(cyclopentenyl)	343(4)	H—(tetrahydrofuranyl)	385
H—C(CH$_2$)(CH$_3$)(CH$_3$)(CH$_3$)	414(4)	H—COC$_6$H$_5$	364(4)
		H—COCF$_3$	381(8)
		H—F	568.6(13)
		H—I	298.7(8)
		H—N	314(17)
		H—NH	377(8)
H—C(CH$_3$)$_2$CH=CH$_2$	331	H—NH$_2$	435(8)
H—cyclopentyl	395(42)	H—NHCH$_3$	431(8)
H—CH$_2$C(CH$_3$)$_3$	418(4)	H—N(CH$_3$)$_2$	397(8)
H—C$_6$H$_5$	431	H—NHC$_6$H$_5$	335(13)
H—CH$_2$C$_6$H$_5$	356(4)	H—N(CH$_3$)C$_6$H$_5$	310(13)
H—C(C$_6$H$_5$)$_3$	314	HNF$_2$	318(13)
		H—N$_3$	356
H—(cyclohexadienyl)	310	H—NO	<205
		H—O	428.0(21)
H—cyclohexyl	399.6(42)	H—OH	498.7(8)
H—cycloheptyl	387.0(42)	H—OCH$_3$	436.8(42)
H—norbornyl	406(13)	H—OCH$_2$CH$_3$	436.0
H—CH$_2$Br	410(25)	H—OC(CH$_3$)$_3$	439(4)
H—CHBr$_2$	435	H—OC$_6$H$_5$	368(25)
H—CH$_2$Cl	423	H—ONO	327.6(25)

TABLE 1.38   Bond Dissociation Energies (*Continued*)

Bond	ΔHf_{298}, kJ/mol	Bond	ΔHf_{298}, kJ/mol
Hydrogen (*continued*)		**Iridium**	
H—ONO$_2$	423.4(25)	Ir—O	352(21)
H—OOH	374(8)	Ir—Si	463(21)
H—OOCCH$_3$	469(17)		
H—OOCCH$_2$CH$_3$	460(17)	**Iron**	
H—OOCC$_3$H$_7$	431(17)		
H—P	343(29)	Fe—Fe	100(21)
H—S	344(12)	Fe—Br	247(96)
H—SH	381(4)	Fe—Cl	*ca.* 352
H—SCH$_3$	*ca.* 368	Fe—O	409(13)
H—Se	305(2)	Fe—S	339(21)
H—Si	298.49(46)	Fe—Si	297(25)
H—SiH$_3$	393(13)		
H—Si(CH$_3$)$_3$	377(13)	**Krypton**	
H—Te	268(2)		
		Kr—Kr	5.4(8)
Indium		Kr—F	54
In—In	100(8)	**Lanthanum**	
In—Br	418(21)		
In—Cl	439(8)	La—La	247(21)
In—F	506(15)	La—C	506(63)
In—O	360(21)	La—F	598(42)
In—P	197.9(85)	La—N	519(42)
In—S	289(17)	La—O	799(13)
In—Sb	152(11)	La—S	577(25)
In—Se	247(17)	**Lead**	
In—Te	218(17)		
		Pb—Pb	339(25)
Iodine		Pb—Br	247(38)
		Pb(CH$_3$)$_3$—CH$_3$	207(42)
I—I	152.549(8)	Pb—Cl	301(29)
I—Br	179.1(4)	Pb—F	356(8)
I—CH$_3$	232(13)	Pb—H	176(21)
I—C$_2$H$_5$	223.8	Pb—I	197(38)
I—CH(CH$_3$)$_2$	222	Pb—O	378(4)
I—C(CH$_3$)$_3$	207.1	Pb—S	346.0(17)
I—CH$_2$CF$_3$	234(4)	Pb—Se	303(4)
I—CF$_2$CH$_3$	216(4)	Pb—Te	251(13)
I—C$_3$F$_7$	209(4)		
I—CH=CHCH$_3$	172	**Lithium**	
I—C$_6$H$_5$	268(4)		
I—C$_6$F$_5$	276	Li—Li	106(4)
I—Cl	213.3(4)	Li—Br	423(21)
I—COCH$_3$	219.7	Li—Cl	469(13)
I—CN	305(4)	Li—F	577(21)
I—F	280(4)	Li—H	247
I—N	159(17)	Li—I	352(13)
I—NO	71(4)	Li—Na	88
I—NO$_2$	75(4)	Li—O	341(6)
I—O	184(21)	Li—OH	427(21)

(Continued)

TABLE 1.38 Bond Dissociation Energies (*Continued*)

Bond	ΔHf_{298}, kJ/mol	Bond	ΔHf_{298}, kJ/mol
Lutetium		**Molybdenum**	
Lu—Lu	142(34)	Mo—I	372
Lu—F	569(42)	Mo—O	607(34)
Lu—O	695(13)	MoO—O	678(84)
Lu—S	507(15)	MoO$_2$—O	565(84)
Lu—Te	326(17)		
Magnesium		**Neodymium**	
Mg—Mg	8.522(4)	Nd—F	545(13)
Mg—Br	297(63)	Nd—O	703(34)
Mg—Cl	318(13)	Nd—S	474(15)
Mg—F	462(21)	Nd—Se	385(17)
MgF—F	569(42)	Nd—Te	305(17)
Mg—H	197(50)	**Neon**	
Mg—I	*ca.* 285		
Mg—O	394(35)	Ne—Ne	3.93
Mg—OH	238(21)	**Neptunium**	
Mg—S	310(75)	Np—O	720(29)
Manganese		**Nickel**	
Mn—Mn	42(29)		
Mn—Br	314(10)	Ni—Ni	261.9(25)
Mn—Cl	361(10)	Ni—Br	360(13)
Mn—F	423(15)	Ni—Cl	372(21)
Mn—I	283(10)	Ni—F	435
Mn—Cu	159(17)	Ni—H	289(13)
Mn—O	402(34)	Ni—I	293(21)
Mn—S	301(17)	Ni—O	391.6(38)
Mn—Se	201(13)	Ni—S	360(21)
		Ni—Si	318(17)
Mercury		**Niobium**	
Hg—Hg	17.2(21)		
Hg—Br	72.8(42)	Nb—O	753(13)
CH$_3$—HgCH$_3$	240.6	**Nitrogen**	
C$_2$H$_5$—HgC$_2$H$_5$	182.8(42)		
C$_3$H$_7$—HgC$_3$H$_7$	197.1	N—N	945.33(59)
Isopropyl—Hgisopropyl	170.3	N—Br	276(21)
C$_6$H$_5$—HgC$_6$H$_5$	285	ON—Br	28.7(15)
Hg—Cl	100(8)	N—Cl	389(50)
Hg—F	130(38)	ON—Cl	159(6)
Hg—H	39.8	O$_2$N—Cl	142(4)
Hg—I	38	N—F	301(42)
Hg—K	8.24(21)	FN—F	318(21)
Hg—Na	>6.7	F$_2$F—N	243(8)
Hg—S	213	ON—F	236(4)
Hg—Se	(167)	O$_2$N—F	188(21)
Hg—Te	(142)		

TABLE 1.38   Bond Dissociation Energies (*Continued*)

Bond	ΔHf_{298}, kJ/mol	Bond	ΔHf_{298}, kJ/mol
Nitrogen (*continued*)		Oxygen (*continued*)	
N—I	159(17)	C_2H_5O—OC_2H_5	159
F_2N—NF_2	88(4)	C_3H_7O—OC_3H_7	155
H_2N—NH_2	297(8)	Palladium	
H_2N—$NHCH_3$	271		
H_2N—$N(CH_3)_2$	264		
H_2N—NHC_6H_5	213	Pd—O	234(29)
HN—N_2	38		
ON—N⁻	480.7(42)	Phosphorus	
ON—NO_2	39.8(8)		
O_2N—NO_2	57.3(21)	P—P	490(11)
HN=NH	456(42)	P—Br	266.5
N≡N	946	P—C	513(8)
N—O	630.57(13)	P—Cl	289(42)
HN=O	481	P—F	439(96)
NN—O	167	P—H	343(29)
ON—O	305	P—N	617(21)
N—P	617(21)	P—O	596.6
N—S	464(21)	Br_3P=O	498(21)
		Cl_3P=O	510(21)
Osmium		F_3P=O	544(21)
		P—S	346.0(17)
O_3Os—O	301(21)	P=S	347
		P—Se	363(10)
		P—Te	298(10)
Oxygen		Platinum	
O—O	498.34(20)		
O—Br	235.1(4)	Pt—B	478(17)
HO—CH_3	377(13)	Pt—H	352(38)
HO—CH=CH_2	364	Pt—O	347(34)
HO—CH_2CH=CH_2	456	Pt—P	417(17)
HO—C_6H_5	431	Pt—Si	501(18)
HO—$CH_2C_6H_5$	322		
HO—CHO	402(13)	Potassium	
HO—$COCH_3$	452(21)		
HO—COC_2H_5	180	K—K	57.3(42)
O—Cl	272(4)	K—Br	383(8)
HO—Cl	251(13)	K—Cl	427(8)
O—F	222(17)	K—F	497.5(25)
O—FO	467	K—H	183(15)
FO—OF	261(84)	K—I	331(13)
O—I	184(21)	K—Na	63.6(29)
HO—I	234(13)	K—O	239(34)
O—N	630.57(13)	K—OH	343(8)
HO—NCH_3	209		
HO—$OC(CH_3)_3$	192(8)	Praseodymium	
HO—OH	213.8(21)		
O—OH	268(4)	Pr—F	582(46)
CF_3O—OCF_3	192	Pr—O	753(17)
CH_3O—OCH_3	157.3(8)	Pr—S	492.5(46)

(*Continued*)

TABLE 1.38 Bond Dissociation Energies (*Continued*)

Bond	ΔHf_{298}, kJ/mol	Bond	ΔHf_{298}, kJ/mol
Praseodymium (*continued*)		Scandium	
Pr—Se	446(23)	Sc—Sc	163(21)
Pr—Te	326(42)	Sc—Br	444(63)
		Sc—C	393(63)
Promethium		Sc—Cl	318
		Sc—F	589(13)
Pm—F	540(42)	Sc—N	469(84)
Pm—O	674(63)	Sc—O	674(13)
Pm—S	423(63)	Sc—S	478(13)
Pm—Se	339(63)	Sc—Se	385(17)
Pm—Te	255(63)	Sc—Te	289(17)
Radium		Selenium	
Ra—Cl	343(75)	Se—Se	332.6(4)
		Se—Br	297(84)
Rhodium		Se—C	582(96)
		Se—Cl	322
Rh—Rh	285(21)	Se—F	339(42)
Rh—B	476(21)	Se—H	305(2)
Rh—C	583.7(63)	Se—N	381(63)
Rh—O	377(63)	Se—O	423(13)
Rh—Si	395(18)	Se—P	364(10)
Rh—Ti	391(15)	Se—S	381(21)
		Se—Si	531(25)
Rubidium		Se—Te	268(8)
Rb—Rb	45.6(21)	Silicon	
Rb—Br	389(13)		
Rb—Cl	448(21)	Si—Si	327(10)
Rb—F	494(21)	Si—Br	343(50)
Rb—H	167(21)	Si—C	435(21)
Rb—I	335(13)	Si—Cl	456(42)
Rb—O	255(84)	Si—F	540(13)
Rb—OH	351(8)	Si—H	298.49(46)
		Si—I	339(84)
Ruthenium		Si—N	439(38)
		Si—O	798(8)
Ru—O	481(63)	Si—S	619(13)
O_3Ru—O	439	Si—Se	531(25)
Ru—Si	397(21)	H_3Si—SiH$_3$	339(17)
Ru—Th	592(42)	(CH$_3$)$_3$Si—Si(CH$_3$)$_3$	339
		(Aryl)$_3$Si—Si(aryl)$_3$	368(31)
Samarium		Si—Te	506(38)
Sm—Cl	423(13)	Silver	
Sm—F	531(18)		
Sm—O	619(13)	Ag—Ag	163(8)
Sm—S	389	Ag—Au	203(9)
Sm—Se	331(15)	Ag—Bi	193(42)
Sm—Te	272(15)		

TABLE 1.38   Bond Dissociation Energies (*Continued*)

Bond	ΔHf_{298}, kJ/mol	Bond	ΔHf_{298}, kJ/mol
Silver (*continued*)		Tantalum	
Ag—Br	293(29)	Ta—N	611(84)
Ag—Cl	341.4	Ta—O	805(13)
Ag—Cu	176(8)	Tellurium	
Ag—F	354(16)		
Ag—Ga	180(15)	Te—B	354(20)
Ag—Ge	175(21)	Te—H	268(2)
Ag—H	226(8)	Te—I	193(42)
Ag—I	234(29)	Te—O	391(8)
Ag—In	176(17)	Te—P	298(10)
Ag—O	213(84)	Te—S	339(21)
Ag—Sn	136(21)	Te—Se	268(8)
Ag—Te	293(96)		
Sodium		Terbium	
Na—Na	77.0	Tb—F	561(42)
Na—Br	370(13)	Tb—O	707(13)
Na—Cl	410(8)	Tb—S	515(42)
Na—F	481(8)	Tb—Te	339(42)
Na—H	201(21)	Thallium	
Na—I	301(8)		
Na—K	63.6(29)	Tl—Tl	63
Na—O	257(17)	Tl—Br	333.9(17)
Na—OH	381(13)	Tl—Cl	372.8(21)
Na—Rb	59(4)	Tl—F	445(19)
Strontium		Tl—H	188(8)
		Tl—I	272(8)
Sr—Br	332(19)	Thorium	
Sr—Cl	406(13)		
Sr—F	542(7)	Th—Th	289
Sr—H	163(8)	Th—C	484(25)
Sr—I	263(42)	Th—N	577.4(21)
Sr—O	454(15)	Th—O	854(13)
Sr—OH	381(42)	Th—P	377
Sr—S	314(21)	Thullium	
Sulfur		Tm—F	569(42)
S—S	429(6)	Tm—O	557(13)
S—Cl	255	Tm—S	368(42)
S—F	343(5)	Tm—Se	276(42)
O_2S—F	71	Tm—Te	276(42)
S—N	464(21)	Tin	
S—O	521.70(13)		
OS—O	551.4(84)	Sn—Sn	195(17)
O_2S—O	348.1(42)	Sn—Br	339(4)
HS—SH	272(21)		

(*Continued*)

TABLE 1.38 Bond Dissociation Energies (*Continued*)

Bond	ΔHf_{298}, kJ/mol	Bond	ΔHf_{298}, kJ/mol
Tin (*continued*)		Vanadium (*continued*)	
BrSn—Br	326	V—Cl	477(63)
Br$_3$Sn—Br	272	V—F	590(63)
(C$_2$H$_5$)$_3$Sn—C$_2$H$_5$	*ca.* 238	V—N	477(8)
Sn—Cl	406(13)	V—O	644(21)
Sn—F	467(13)	V—S	490(16)
Sn—H	267(17)	V—Se	347(21)
Sn—I	234(42)		
Sn—O	548(21)	Xenon	
Sn—S	464(3)		
Sn—Se	401.3(59)	Xe—Xe	6.53(30)
Sn—Te	319.2(8)	Xe—F	13.0(4)
		Xe—O	36.4
Titanium			
		Ytterbium	
Ti—Ti	141(21)		
Ti—Br	439	Yb—Cl	322
Ti—C	435(25)	Yb—F	521(10)
Ti—Cl	494	Yb—H	159(38)
Ti—F	569(34)	Yb—O	397.9(63)
Ti—H	*ca.* 159	Yb—S	167
Ti—I	310(42)		
Ti—N	464	Yttrium	
Ti—O	662(16)		
Ti—S	426(8)	Y—Y	159(21)
Ti—Se	381(42)	Y—Br	485(84)
Ti—Te	289(17)	Y—C	418(63)
		Y—Cl	527(42)
Tungsten		Y—F	605(21)
		Y—N	481(63)
W—Cl	423(42)	Y—O	715.1(30)
W—F	548(63)	Y—S	528(11)
W—O	653(25)	Y—Se	435(13)
OW—O	632(84)	Y—Te	339(13)
O$_2$W—O	598(42)		
W—P	305(4)	Zinc	
Uranium		Zn—Zn	29
		Zn—Br	142(29)
U—O	761(17)	C$_2$H$_5$C—C$_2$H$_5$	*ca.* 201
OU—O	678(59)	Zn—Cl	229(20)
O$_2$U—O	644(88)	Zn—F	368(63)
U—S	523(10)	Zn—H	85.8(21)
		Zn—I	138(29)
Vanadium		Zn—O	284.1
		Zn—S	205(13)
V—V	242(21)	Zn—Se	136(13)
V—Br	439(42)	Zn—Te	205
V—C	469(63)		

TABLE 1.38   Bond Dissociation Energies (*Continued*)

Bond	$\Delta Hf_{298},$ kJ/mol	Bond	$\Delta Hf_{298},$ kJ/mol
	Zirconium		Zirconium (*continued*)
Zr—C	561(25)	Zr—O	760(8)
Zr—F	623(63)	Zr—S	575(17)
Zr—N	565(25)		

1.8   DIPOLE MOMENTS

The dipole moment is the mathematical product of the distance between the centers of charge of two atoms multiplied by the magnitude of that charge. Thus, the dipole moment (μ) of a compound or molecule is:

$$\mu = Q \times r$$

where Q is the magnitude of the electrical charge(s) that are separated by the distance r; the unit of measurement is the Debye (D)

All bonds between equal atoms are given zero values. Because of their symmetry, methane and ethane molecules are nonpolar. The principle of bond moments thus requires that the CH_3 group moment equal one H—C moment. Hence the substitution of any aliphatic H by CH_3 does not alter the dipole moment, and all saturated hydrocarbons have zero moments as long as the tetrahedral angles are maintained.

TABLE 1.39   Bond Dipole Moments

Bond	Moment, D*	Bond	Moment, D*
H—C		C—N, aliphatic	0.45
Aliphatic	0.3	C=N	1.4
Aromatic	0.0	C≡N (nitrile)	3.6
C—C	0.0	NC (isonitrile)	3.0
C=C	0.0	N—H	1.31
C—O		N—O	0.3
Ether, aliphatic	0.74	N=O	2.0
Alcohol, aliphatic	0.7	N (lone pair on sp^3 N)	1.0
C=O		C—P, aliphatic	0.8
Aliphatic	2.4	P—O	(0.3)
Aromatic	2.65	P=O	2.7
O—H	1.51	P—S	0.5
C—S	0.9	P=S	2.9
C=S	2.0	B—C, aliphatic	0.7
S—H	0.65	B—O	0.25
S—O	(0.2)	Se—C	0.7
S=O		Si—C	1.2
Aliphatic	2.8	Si—H	1.0
Aromatic	3.3	Si—N	1.55

*To convert Debye units D into coulomb-meters, multiply by 3.33564×10^{-30}.

TABLE 1.40 Group Dipole Moments

Bond	Moment, D*	Bond	Moment, D*
H—Sb	−0.08	Br—F	1.3
H—As	−0.10	Cl—F	0.88
H—P	0.36	Li—C	1.4
H—I	0.38	K—Cl	10.6
H—Br	0.78	K—F	7.3
H—Cl	1.08	Cs—Cl	10.5
H—F	1.94	Cs—F	7.9
C—Te	0.6		
N—F	0.17	Dative (coordination) bonds	
P—I	0.3		
P—Br	0.36	N → B	2.6
P—Cl	0.81	O → B	3.6
As—I	0.78	S → B	3.8
As—Br	1.27	P → B	4.4
As—Cl	1.64	N → O	4.3
As—F	2.03	P → O	2.9
Sb—I	0.8	S → O	3.0
Sb—Br	1.9	As → O	4.2
Sb—Cl	2.6	Se → O	3.1
S—Cl	0.7	Te → O	2.3
Cl—O	0.7	P → S	3.1
I—Br	1.2	P → Se	3.2
I—Cl	1	Sb → S	4.5
Br—Cl	0.57		

*To convert Debye units D into coulomb-meters, multiply by 3.33564×10^{-30}.

The group moment always includes the C—X bond. When the group is attached to an aromatic system, the moment contains the contributions through resonance of those polar structures postulated as arising through charge shifts around the ring.

1.8.1 Dielectric Constant

The *dielectric constant* (also referred to as the *relative permittivity*, K) is the ratio of the permittivity of the material to the permittivity of free space and is the property of a material that determines the relative speed with which an electrical signal will travel in that material.

$$K = \epsilon_T / \epsilon_0$$

Signal speed is roughly inversely proportional to the square root of the dielectric constant. A low dielectric constant will result in a high signal propagation speed and a high dielectric constant will result in a much slower signal propagation speed.

The *dielectric loss factor* is the tangent of the loss angle and the *loss tangent* ($\tan \Delta$) is defined by the relationship:

$$\tan \Delta = 2\sigma / \varepsilon \upsilon$$

σ is the electrical conductivity, ε is the dielectric constant, and υ is the frequency. The loss tangent is roughly wavelength independent.

TABLE 1.41   Dipole Moments and Dielectric Constants

Substance	Dielectric constant, ε	Dipole moment, D	Substance	Dielectric constant, ε	Dipole moment, D
Air	1.000 536 4		$GeClH_3$		2.13
$AlBr_3$	3.38^{100}	5.2	H_2(g)	1.000 253 8	0
Ar			t		
(g)	1.000 517 2		(lq)	$1.279^{13.5\,K}$,	
(lq)	1.538^{-191},			$1.228^{20.4\,K}$	
	1.325^{-132}	0	HBr(g)	$1.003\ 13^0$	0.827
$AsBr_3$	8.83^{35}	1.61	(lq)	8.23^{-86}, 3.82^{25}	
$AsCl_3$	12.6^{20}	1.59	He (g)	1.000 0565 0	0
AsH_3 (arsine)	2.40^{-72}, 2.05^{20}	0.20	(lq) (II)	$1.055^{2.055\,K}$	
BBr_3	2.58^0	0	(III)		
BCl_3		0	(IV)		
BF_3		0	HCl (g)	1.0046^0	1.109
B_2H_6 (diborane)	$1.872^{-92.5}$	0	(lq)	14.3^{-114},	
B_4H_{10}		0.486		4.60^{28}	
B_5H_9	21.1^{25}	2.13	HClO		1.3
B_6H_{10}		2.50	HCN	114.9^{20}	2.98
$B_3H_6N_3$		0	HCNO (isocyanate)		1.6
Br_2 (g)	1.0128^{20}		HCNS		1.7
(lq)	3.1484^{25}	0	HF	83.6^0	1.826
BrF_3	106.8^{25}	1.1	HFO		2.23
BrF_5	$7.91^{24.5}$	1.51	HI (g)	$1.002\ 34^0$	0.448
Cl_2 (g)		0	(lq)	3.87^{-53}, 2.90^{22}	
(lq)	2.147^{-65},		HN_3 (azide)		1.70
	1.91^{14}		H_2O (see Table 1.12)		
ClF_3	4.394^{20}, 4.29^{25}	0.554	H_2O_2	84.2^0, 74.6^{17}	1.573
ClF_5	4.28^{-80}		HNO_3		2.17
ClO_3F	2.194^{-123}	0.023	H_2S (g)	1.0040^0	0.97
CO (g)	$1.000\ 70^0$	0.112	(lq)	5.93^{10}	
(lq)			H_2Se		0.24
CO_2 (g)	1.000 922	0	HSO_3Cl	60^{60}	
(lq)	$1.60^{\circ}C^{50\,atm}$,		HSO_3F	ca. 120^{25}	
	1.449^{23}		H_2SO_4	100^{25}	
$COCl_2$	4.34^{22}	1.17	H_2Te		<0.2
COF_2		0.95	Hg		0
COS	4.47^{-88}	0.712	I_2	11.1^{118}	0
COSe	3.47^{10}	0.73	IBr		0.726
CS		1.98	IF		1.95
CS_2 (g)	1.0029^0	0	IF_5	37.13^{20}	2.18
(lq)	2.632^{20}		IF_7	1.97^{23}	
CrO_2Cl_2	2.6^{20}	0.47	IOF_5	1.75^{25}	
D_2 (deuterium)	1.290^{-255},		Kr (g)		<0.05
	1.277^{-253}		(lq)	$1.644^{-153.4}$	
DH	$1.269^{16.78\,K}$		Mn_2O_7	3.28^{20}	
D_2O	79.75^{20},	1.87	Ne (g)	$1.000\ 063\ 9^{20}$	0
	78.25^{25}		(lq)	$1.1907^{-247.1}$	
F_2	1.491^{-220},		N_2 (g)	$1.000\ 548\ 0^{20}$	0
	1.54^{-202}		(lq)	1.468^{-210},	
$GaCl_3$		0.85		1.454^{-203}	
$GeBr_4$			NH_3 (g)	1.0072^0	1.471
$GeBr_4$	2.955^{26}		(lq)	$22.4^{-33.5}$	
$GeCl_4$	2.463^0, 2.430^{25}	0		16.61^{20}	

(Continued)

TABLE 1.41 Dipole Moments and Dielectric Constants (*Continued*)

Substance	Dielectric constant, ε	Dipole moment, D	Substance	Dielectric constant, ε	Dipole moment, D
N_2H_4 (hydrazine)	52.9^{20}, 51.7^{25}	1.75	S_2Cl_2 dimer	4.79^{15}	1.0
$Ni(CO)_4$			S_2F_2		
NO		0.159	FSSF isomer		1.45
N_2O (g)	$1.001\ 13^0$	0.161	$S=SF_2$ isomer		1.03
(lq)	1.52^{15}		SF_4		0.632
NO_2		0.316	SF_6	1.81^{-50}	0
N_2O_4	2.56^{25}, 2.44^{20}	0.5	S_2F_{10}	2.020^{20}	0
N_2O_3		2.122	SO_2 (g)	1.0093^0	1.63
NOBr	13.4^{15}	1.8	(lq)	16.3^{25}	
NOCl	18.2^{12}	1.9	SO_3	3.11^{18}	0
NO_2Cl		0.53	$SOBr_2$	9.06^{20}	9.11
NOF		1.73	$SOCl_2$	9.25^{20}, 8.675^{25}	1.45
NO_2F		0.47	SOF_2		1.63
NO_3	31.13^{-70}		SO_2Cl_2	9.15^{20}	1.81
O_2 (g)	$1.000\ 494\ 7^{20}$	0	SO_2F_2		1.12
(lq)	$1.568^{-218.7}$, 1.507^{-193}		$SbCl_3$	33.2^{75}	3.93
			$SbCl_5$	3.22^{20}	0
O_3	4.75^{-183}	0.534	SbF_5		
OF_2		0.297	SbH_3		0.12
O_2F_2 (FOOF)		1.44	Se (lq)	$5.44^{237.5}$	
OsO_4		0	SeF_4		1.78
P (lq)	4.096^{34}		SeF_6		0
PBr_3	3.9^{20}	0.56	$SeOCl_2$	46.2^{20}	2.64
PCl_3	3.43^{25}, 3.50^{17}	0.78	SeO_2		2.62
PCl_5	2.85^{160}, 2.7^{165}	0.9	$SiCl_4$	2.248^0	0
PCl_2F_3	2.813^{-45}		SiF_4		0
PCl_3F_2	2.375^{-5}		SiH_4		0
PCl_4F	$2.65^{0.5}$		$SiHCl_3$		0.86
PF_3		1.03	SiH_3Cl		1.31
PF_5		0	$SnBr_4$	3.169^{30}	0
PH_3	2.9^{15}	0.574	$SnCl_4$	3.014^0, 2.89^{20}	0
PI_3	4.12^{65}	0	TeF_6		0
PO_3			$TiCl_4$	2.843^{14}, 2.80^{20}	0
$POCl_3$	13.7^{25}	2.54	UF_6 (g)	$1.002\ 92^{67}$	0
POF_3		1.868	(lq)	2.18^{65}	
$PSCl_3$	5.8^{22}	1.42	VCl_4	3.05^{25}	0
PSF_3		0.64	$VOBr_3$	3.6^{25}	
$PbCl_4$	2.78^{20}		$VOCl_3$	3.4^{25}	0.3
ReO_2Cl_3			Xe (g)	$1.001\ 23$	0
ReO_3Cl			(lq, II)	$1.880^{-111.9}$	
S	3.499^{134}		XeF_6	4.10^{125}	
SCl_2	2.915^{25}	0.36			

1.9 MOLECULAR GEOMETRY

Molecular geometry is the specific three-dimensional arrangement of atoms and the positions of the atomic nuclei in a molecule.

Various instrumental techniques such as x-ray crystallography and other experimental techniques can be used to derive information about the locations of atoms in a molecule.

Thus, molecular geometry is associated with the specific orientation of bonding atoms. A careful analysis of electron distribution in various orbitals will usually result in correct determination of the molecular geometry.

TABLE 1.42   Spatial Orientation of Common Hybrid Bonds

On the assumption that the pairs of electrons in the valency shell of a bonded atom in a molecule are arranged in a definite way which depends on the number of electron pairs (coordination number), the geometrical arrangement or shape of molecules may be predicted. A multiple bond is regarded as equivalent to a single bond as far as molecular shape is concerned.

Coordination number	Orbitals hybridized	Geometrical arrangement	Minimum radius ratio
2	sp dp	Linear	
	P^2 ds d^2	Bent (angular)	
3	sp^2 ds^2	Trigonal planar	0.155
	P^3 d^2p	Trigonal pyramidal	
	sp^2d p^2d^2	Square planar	
4	sp^3 d^3s	Tetrahedral	0.225
	d^4	Tetragonal pyramidal	
5	sp^3d d^3sp	Trigonal bipyramidal	0.155
6	d^2sp^3	Octahedral	0.414
	d^4sp^3	Trigonal prism	
7		One atom above the face of an octahedron, which is distorted chiefly by separating the atoms at the corners of this face.	0.592
8	d^4sp^3	Square antiprism (dodecahedral)	0.645
		Cube	0.732
9		Formed by adding atoms beyond each of the vertical faces of a right triangular prism.	0.732
12		Cube-octahedron	1.000

TABLE 1.43 Crystal Lattice Types

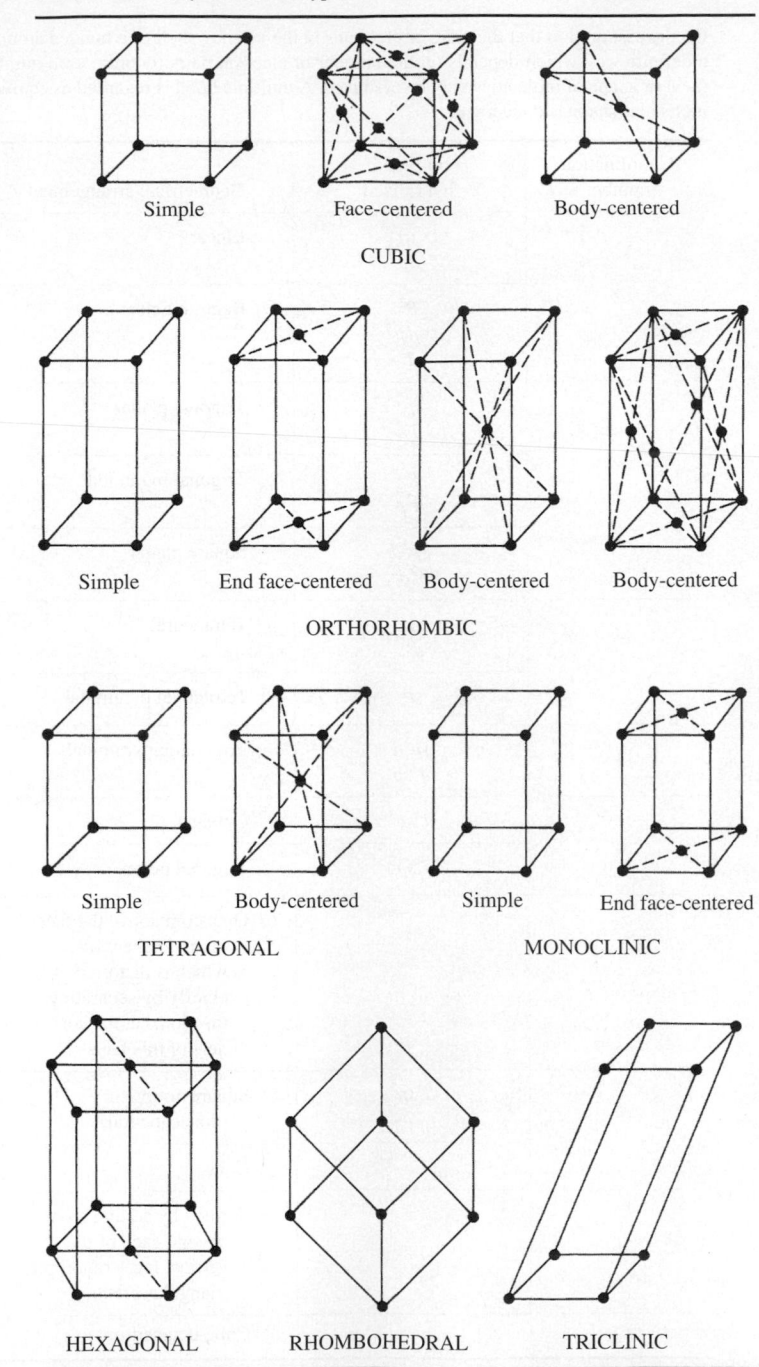

Simple Face-centered Body-centered

CUBIC

Simple End face-centered Body-centered Body-centered

ORTHORHOMBIC

Simple Body-centered Simple End face-centered

TETRAGONAL MONOCLINIC

HEXAGONAL RHOMBOHEDRAL TRICLINIC

TABLE 1.44   Crystal Structure

Unit cells of the different lattice types in each system are illustrated in Table 1.43

System	Characteristics	Essential symmetry	Axes in unit cell	Angles in unit cell
Cubic	Three axes equal and mutually perpendicular	Four threefold axes	$a = b = c$	$\alpha = \beta = \gamma = 90°$
Tetragonal	Two equal axes and one unequal axis mutually perpendicular	One fourfold axis	$a = b \neq c$	$\alpha = \beta = \gamma = 90°$
Orthorhombic (or rhombic)	Three unequal axes mutually perpendicular	Three mutually perpendicular two-fold axes, or two planes intersecting in a twofold axis	$a \neq b \neq c$	$\alpha = \beta = \gamma = 90°$
Hexagonal or trigonal	Three equal axes inclined at 120° with a fourth axis un equal and perpendicular to the other three	One sixfold axis or one threefold axis	$a = b \neq c$ $a = b = c$	$\alpha = \beta = 90°;$ $\gamma = 120°$ $\alpha = \beta = \gamma \neq 90°$
Monoclinic	Two axes at an oblique angle with a third perpendicular to the other two	One twofold axis or one plane	$a \neq b \neq c$	$\alpha = \beta = 90°;$ $\gamma \neq 90°$
Triclinic	Three unequal axes intersecting obliquely	No planes or axes of symmetry	$a \neq b \neq c$	$\alpha \neq \beta \neq \gamma \neq 90°$
Rhombohedral	Two equal axes making equal angle with each other			

1.10   NUCLIDES

The nuclide is the nucleus of a particular isotope.

TABLE 1.45   Table of Nuclides

Explanation of Column Headings

Nuclide. Each nuclide is identified by element name and the mass number A, equal to the sum of the numbers of protons Z and neutrons N in the nucleus. The m following the mass number (for example, ^{69m}Zn) indicates a metastable isotope. An asterisk preceding the mass number indicates that the radionuclide occurs in nature.

Half-life. The following abbreviations for time units are employed: y = years, d = days, h = hours, min = minutes, s = seconds, ms = milliseconds, and ns = nanoseconds.

Natural abundance. The natural abundances listed are on an "atom percent" basis for the stable nuclides present in naturally occurring elements in the earth's crust.

Thermal neutron absorption cross section. Simply designated "cross section," it represents the ease with which a given nuclide can absorb a thermal neutron (energy less than or equal to 0.025 eV) and become a different nuclide. The cross section is given here in units of barns (1 barn = 10^{-24} cm^2). If the mode of reaction is other than (n, γ), it is so indicated.

Major radiations. In the last column are the principal modes of disintegration and energies of the radiations in million electronvolts (MeV). Symbols used to represent the various modes of decay are:

α, alpha particle emission   K, electron capture
β^-, beta particle, negatron   IT, isomeric transition
β^+, positron   x, X-rays of indicated element (e.g., O-x,
γ, gamma radiation   oxygen X-rays, and the type, K or L)

For β^- and β^+, values of E$_{max}$ are listed. Radiation types and energies of minor importance are omitted unless useful for identification purposes.

(Continued)

TABLE 1.45 Table of Nuclides (*Continued*)

Element	A	Half-life	Natural abundance, %	Cross section, barns	Radiation (MeV)
Hydrogen	1		99.985(1)	0.332(2)	
	2		0.015(1)	0.000 52(1)	
	3	12.32 y			β^- (0.0186)
Beryllium	7	53.28 d			K, γ(0.478)
	9		100	0.008(1)	
	10	1.52×10^6 y			β^-(0.555)
Boron	10		19.9(2)	3837(10)(n, α)	
	11		80.1(6)	0.005(3)	
Carbon	11	20.3 min			β^+(0.961)
	12		98.89(1)	0.0035(1)	
	14	5715 y			β^- (0.156)
Nitrogen	13	9.965 min			β^+(1.190)
	14		99.634(9)	1.8(1)(n, p)	
Oxygen	15	122.2 s			β^+(2.754)
	19	26.9 s			β^-(4.82); γ(0.197, 1.357)
Fluorine	18	1.8295 h			β^+(0.635); K, O-x
	19		100	0.0095(7)	β^+(2.754)
	20	11.00 s			β^-(5.40); γ(1-63)
Sodium	22	2.605 y		2800.(300)(n, p)	β^+(0.545, 1.83); K, Ne-x, γ(1.275)
	23		100	0.53	
	24	14.659 h			β^-(1.39); γ(2.75, 1.37)
Magnesium	24		78.89(3)	0.053(6)	
	25		10.00(1)	0.17(5)	
	27	9.45 min		0.07(2)	β^-(1.75, 1.59); γ(0.844, 1.014)
	28	20.90 h			β^-(0.459); γ(1.342, 0.942, 0.401, 0.031)
Aluminum	26	7.1×10^5 y			β^+(1.16); K, Mg-x; γ(1.809)
	27		100	0.230(2)	
	28	2.25 min			β^-(2.865); γ(1-778)
Silicon	28		92.23(2)	0.17(1)	
	29		4.67(2)	0.12(1)	
	30		3.10(1)	0.107(4)	
	31	2.62 h		0.073(6)	β^-(1.471); γ(1.266)
	32	1.6×10^2 y			β^-(0.213)
Phosphorus	30	2.50 min			β^+(3.245)
	31		100	0.16(2)	
	32	14.28 d			β^-(1.710)
	33	25.3 d			β^-(0.249)
Sulfur	32		95.02(9)	0.55(2)	
	34		4.21(8)	0.29(6)	
	35	87.51 d			β^-(0.167)
	37	5.05 min			β^-(4.75, 1.64); γ(3.103, 0.908)
	38	2.84 h			β^-(1.00, 3.0); γ(1.942, 0.196)

TABLE 1.45 Table of Nuclides (*Continued*)

Element	A	Half-life	Natural abundance, %	Cross section, barns	Radiation (MeV)
Chlorine	35		75.77(5)	43.7(4)	
	36	3.01×10^5 y		46.(2)	β^-(0.709); K, S-x
	37		24.23(5)	0.4	
	38	37.24 min			β^-(4.91, 1.11, 2.77); γ(2.168, 1.642)
	39	55.6 min			β^-(1.91, 2.18, 3.45); γ(1.267, 0.250, 1.52)
Argon	37	35.0 d			K, Cl-x
	39	268 y			β^-(0.565)
	40		99.600(3)	0.64(3)	
	41	1.82 h		0.5(1)	β^-(1.20, 2.49); γ(1.29)
	42	33 y			β^-(0.60)
Potassium	39		93.258(4)	2.1(2)	
	*40	1.26×10^9 y	0.0117(1)	30.(8)	β^-(1.312); K, Ar-x; γ(1.461)
	41		6.730(4)	1.46(3)	
	42	12.360 h			β^-(3.523, 1.97); γ(1.525)
	43	22.3 h			β^-(0.825, 0.45, 1.24, 1.814); 7(0.618, 0.373, 0.39, 0.221)
Calcium	40		96.941(18)	0.41(3)	
	42	1.02×10^5 y		≈4	
	43		0.135(6)	6.(1)	
	44		2.086(12)	0.8(2)	
	45	162.7 d		≈15	β^-(0.257)
	47	4.536 d			β(1.98, 0.684); γ(1.297)
	49	8.72 min			β^-(1.95,0.89); γ(3.084, 4.07)
Scandium	42m	61.6 s			β^+(2.82); 7(0.438, 1.227, 1.524)
	43	3.89 h			β^+(1.22)
	44m	2.442 d			I'I, Sc-x; γ(0.271)
	44	3.927 h			β^+(1.47); K, γ(1.16)
	45		100	27	
	46m	19.5 s			γ(0.142)
	46	83.81 d		8.(1)	β^-(0.357); γ(1.12, 0.889); Ti-x
	47	3.341 d			β^-(0.439, 0.60); γ(0.159)
	48	1.821 d			β^-(0.65); γ(1.31, 1.04, 0.984)
Titanium	44	47.3 y			K, γ(0.68, 0.078)
	45	3.08 h			β^+(1.044); K, Sc-x
	48		73.72(3)	7.9(9)	
	49		5.41(2)	1.9(5)	
	50		5.18(2)	0.179(3)	
	51	5.76 min			β^-(2.14, 1.50); γ(0.320, 0.928)
Vanadium	48	16.0 d			β^+(0.698); γ(0.511, 0.945, 0.983, 1.312, 2.24)

(*Continued*)

TABLE 1.45   Table of Nuclides (*Continued*)

Element	A	Half-life	Natural abundance, %	Cross section, barns	Radiation (MeV)
Vanadium	49	330 d			K, Ti-x
(*cont.*)	50	>1.4×10^{17} y	0.250(2)	40.(20)	
	51		99.750(2)	4.9(1)	
	52	3.75 min			β^-(2.47); γ(1.434)
Chromium	48	21.6 h			K, V-x; γ(0.116, 0.305)
	50		4.345(13)	15.(1)	
	51	27.70 d			K, V-x; γ(0.320)
	52		83.79(2)	0.8(1)	
	53		9.50(2)	18.(2)	
Manganese	51	46.2 min			β^+(2.2); γ(0.749, 1.15)
	52	5.60 d			β^+(0.575); γ(0.511, 0.744, 1.434)
	53	3.7×10^6 y		70.(10)	
	54	312.2 d		<10	γ(0.834)
	55		100	13.3(1)	
	56	2.5785 h			β^-(1.028, 1.03, 0.718); γ(0.847, 1.81, 2.11)
Iron	52	8.275 h			β^+(0.804); K, Mn-x; γ(0.169)
	54		5.85(4)	2.7(5)	
	55	2.73 y		13.(2)	K, Mn-x
	56		91.75(4)	2.6(2)	
	57		2.12(1)	2.5(5)	
	59	44.51 d		13.(3)	β^-(0.273, 0.475); γ(1.10 1.29)
Cobalt	55	17.53 h			β^+(1.04, 1.50); K, Fe-x; γ(0.932, 0.480, 1.41)
	56	77.3 d			β^+(1.46); K, Fe-x; γ(0.847, 1.04, 1.24, 1.77, 2.60, 3.26, 2.02)
	57	271.77 d			K, Fe-x; γ(0.136, 0.122)
	58*m*	9.1 h		$1.4(1) \times 10^5$	γ(0.025)
	58	70.88 d		$1.9(2) \times 10^3$	K, β^+(0.474); Fe-x; γ(0.811)
	59		100	19	
	60*m*	10.47 min		58.(8)	β^-(1.55)
	60	5.271 y		2.0(2)	β^-(0.318); γ(1.173, 1.332)
	61	1.650 h			β^-(1.22); 7(0.842–0.909)
Nickel	56	6.08 d			K, Co-x; γ(0.158, 0.270, 0.480, 0.75, 0.812, 1.56)
	57	35.6 h			K, β^+(0.849, 0.712); Co-x, γ(1.378, 0.0127, 1.76)
	58		68.077(9)	4.6(4)	
	60		26.22(1)	2.9(3)	
	63	100 y		24.(3)	β^-(0.067)

TABLE 1.45   Table of Nuclides (*Continued*)

Element	A	Half-life	Natural abundance, %	Cross section, barns	Radiation (MeV)
Nickel (*cont.*)	64		0.926(1)	1.8(1)	
	65	2.517 h		22.(2)	β^-(2.14, 0.65, 1.020); γ(1.48, 0.366, 1.116)
	66	2.275 d			β^-(0.23)
Copper	61	3.408 h			β^+(1.220); K, Ni-x; γ(0.283, 0.656)
	63		69.17(3)	4.5(2)	
	64	12.701 h		≈270	β^-(0.578); β^+(0.65); Ni-x; γ(1.346)
	65		30.83(3)	2.17(3)	
	66	5.07 min		$1.4(1) \times 10^2$	β^-(2.74); γ(1.039)
	67	2.580 d			β^-(0.395, 0.484, 0.577); γ(0.185, 0.092)
Zinc	62	9.26 h			K, β^+(0.66); Cu-x; γ(0.041, 0.597)
	64		48.6(3)	0.46	
	65	243.8 d		66.(8)	K, β^+(0.325), Cu-x; γ(1.116)
	66		27.9(2)	1.0(2)	
	67		4.1(1)	6.9(1)	
	68		18.8(4)	0.87	
	69*m*	13.76 h			IT, Zn-x, γ(0.439)
	69	56 min			β^-(0.905)
	71*m*	3.97 h			β^-(1.45); γ(0.386, 0.487, 0.620)
	72	46.5 h			β^-(0.30, 0.25); γ(0.145, 0.191)
Gallium	66	9.5 h			β^+(1.84, 4.153); γ(1.039, 2.752)
	67	3.260 d			K, Zn-x; γ(0.093, 0.184, 0.300)
	68	1.130 h			β^+(1.83); K, Zn-x; γ(1.077)
	69		60.108(9)	1.68(7)	
	70	21.1 min			β^-(1.65); γ(0.175, 1.042)
	71		39.892(9)	4.7(2)	
	72	14.10 h			β^-(0.64, 1.51, 2.52, 3.15); γ(0.63, 2.20, 2.50)
	73	3.120 d			β^-(1.59); γ(0.053, 0.297)
Germanium	66	2.66 h			K, β^+(1.02); Ga-x; γ(0.044, 0.382)
	68	270.8 d			Ga, K-x
	69	1.63 d			β^+(0.70, 1.22); γ(1,107, 0.574)
	71	11.2 d			Ga-x
	72		27.66(3)	0.9(2)	
	73		7.73(1)	15.(1)	
	74		35.94(2)	0.3	
	75	1.380 h			β^-(1.19); γ(0.265, 0.419)

(*Continued*)

TABLE 1.45   Table of Nuclides (*Continued*)

Element	A	Half-life	Natural abundance, %	Cross section, barns	Radiation (MeV)
Germanium	77	11.30 h			β^-(0.71, 1.38, 2.19);
(*cont.*)					γ(0.211,0.215, 0.264)
	78	1.45 h			β^-(0.95); γ(0.277, 0.294)
Arsenic	71	2.70 d			K, β^+(0.81); Ge-x;
					γ(0.175, 1.096)
	72	1.083 d			β^+(3.339, 2.498, 1.884);
					K, Ge-x; γ(0.834,
					1.051)
	73	80.30 d			K, γ(0.0534, 0.0133)
	74	17.78 d			β^+(0.94); β^-(0.71, 1.35);
					γ(0.596, 0.635)
	75		100	4.0(4)	
	76	1.096 d			β^-(2.97, 2.41, 1.79);
					γ(0.559, 0.657)
	77	38.8 h			β^-(0.683); γ(0.239,
					0.250, 0.521)
	78	91 min			β^-(4.21); γ(0.614, 0.70,
					1.31)
Selenium	72	8.40 d			K, As-x; γ(0.046)
	73	7.1 h			β^+(1.32); γ(0.361,0.067)
	74		0.89(2)	50.(4)	
	75	119.78 d			K, γ(0.265, 0.136); As-x
	77*m*	17.5 s			γ(0.162)
	77		7.63(6)	42.(4)	
	80		49.61(10)	0.5	
	81	18.5 min			β^-(1.58); γ(0.276, 0.290,
					0.828)
Bromine	75	1.62 h			β^+(3.03); γ(0.287)
	76	16.2 h		224.(42)	β^+(1.9, 3.68); K, Se-x;
					γ(0.559, 1.86)
	77	2.376 d			γ(0.239, 0.521)
	79		50.69(7)	10.8	
	80*m*	4.42 h			IT, Br-x; γ(0.037, 0.049)
	80	17.66 min			β^-(1.997, 1.38); K,
					β^+(0.85), Se-x;
					γ(0.617)
	81		49.31(7)	2.6	
	82	1.4708 d			β^-(0.444); γ(0.554,
					0.619, 0.776)
Krypton	76	14.8 h			K, γ(0.252)
	77	1.24 h			β^+(1.875, 1.700, 1.550);
					K, Br-x; γ(0.130,
					0.147)
	79	1.455 d			β^+(1.626); γ(0.261,
					0.398, 0.606)
	81*m*	13 s			IT, Kr-x; γ(0.190)
	81	2.10×10^5 y			K, Br-x; γ(0.276)
	83		11.5(1)	183.(30)	
	84		57.0(3)	0.10	
	85*m*	4.48 h			β^-(0.83); γ(0.151, 0.305)

TABLE 1.45 Table of Nuclides (*Continued*)

Element	A	Half-life	Natural abundance, %	Cross section, barns	Radiation (MeV)
Krypton	85	10.72 y			$\beta^-(0.67)$; $\gamma(0.517)$
(*cont.*)	87	1.27 h			$\beta^-(3.49, 0.389, 1.38)$; $\gamma(0.403, 2.55)$
	88	2.84 h			$\beta^-(2.91)$; $\gamma(0.196, 2.392)$
Rubidium	84	32.9 d			$\beta^-(0.894)$; $\beta^+(2.681)$; $\gamma(0.882)$
	85		72.17(2)	0.5	
	86	18.65 d		<20	$\beta^-(1.775)$; $\gamma(1.08)$
	87	4.88×10^{10} y	27.83(2)	0.10(1)	$\beta^-(0.283)$
	88	17.7 min		1.2(3)	$\beta^-(5.31)$; $\gamma(1.836, 0.898)$
	89	15.4 min			$\beta^-(1.26, 2.2, 4.49)$; $\gamma(1.032, 1.248, 2.196)$
Strontium	82	25.36 d			K, Rb-x
	85*m*	1.126 h			K, Rb-x, Sr-x; $\gamma(0.150, 0.231)$
	85	64.84 d			K, Rb-x; $\gamma(0.514)$
	87*m*	2.795 h			IT, $\gamma(0.388)$
	88		82.58(1)	0.0058(4)	
	89	50.52 d		0.42(4)	$\beta^-(1.497)$; $\gamma(0.909)$
	90	29.1 y		0.0097(7)	$\beta^-(0.546)$
	91	9.5 h			$\beta^-(1.09, 1.36, 2.66)$; $\gamma(0.556, 0.750, 1.024)$
	92	2.71 h			$\beta^-(0.55, 1.5)$; $\gamma(1.383)$
Yttrium	85*m*	4.86 h			$\beta^+(2.24)$; K, Sr-x; $\gamma(0.767, 0.232, 2.124)$
	85	2.68 h			$\beta^+(1.58, 1.15)$; K, Sr-x; $\gamma(0.504, 0.232)$
	86	14.74 h			$\beta^+(5.24)$; 7(0.307, 0.628, 1.077, 1.153, 1.921)
	87*m*	12.9 h			Y-x; $\gamma(0.381)$
	88	106.6 d			$\beta^-(0.76)$; $\gamma(0.898, 1.836, 2.734, 3.219)$
	90	2.67 d		<7	$\beta^-(2.28)$; $\gamma(2.186)$
	91*m*	49.71 min			Y-x; IT; $\gamma(0.556)$
	91	58.5 d		1.4(3)	$\beta^-(1.545)$; $\gamma(1-21)$
	92	3.54 h			$\beta^-(3.64)$; $\gamma(0.448, 0.561, 0.934, 1.405)$
	93	10.2 h			$\beta^-(2.88)$; $\gamma(0.267, 0.947, 1.918)$
Zirconium	86	16.5 h			K, Y-x; $\gamma(0.243, 0.612)$
	87	1.73 h			$\beta^+(2.260)$; K, Y-x; $\gamma(0.381, 1.228)$
	88	83.4 d			K, Y-x; $\gamma(0.393)$
	89	3.27 d			K, $\beta^+(0.897)$; Y-x; $\gamma(0.909)$
	91		11.22(4)	1.2(3)	
	93	1.5×10^6 y			$\beta^-(0.091)$
	95	64.02 d			$\beta^-(0.366, 0.400)$; $\gamma(0.724, 0.757)$
	97	16.90 h			$\beta^-(1.91)$; $\gamma(0.743)$

(*Continued*)

TABLE 1.45   Table of Nuclides (*Continued*)

Element	A	Half-life	Natural abundance, %	Cross section, barns	Radiation (MeV)
Niobium	89	2.03 h			β^+(3.320); γ(1.627)
	90	14.60 h			β^+(1.50); K, Zr-x; γ(0.141, 1.129, 2.186, 2.319)
	91*m*	62 d			IT, Nb-x; γ(0.1045, 1.205)
	91	700 y			Mo-x
	92*m*	10.15 d			K, γ(0.913, 0.934, 1.848)
	93*m*	16.1 y			Nb-x
	93		100	1.1	
	94*m*	6.26 min			γ(0.871)
	94	2.4×10^4 y			β^-(0.473); γ(0.703, 0.871)
	95*m*	3.61 d			γ(0.204, 0.236)
	95	35.0 d		<7	β^-(0.160); γ(0.765)
	96	23.4 h			β^-(0.748, 0.500); γ(0.778, 1.091)
	97*m*	58.1 s			IT; γ(0.766)
	97	1.23 h			β^-(1.267); γ(0.481, 0.658)
Molybdenum	90	5.67 h			K, β^+(1.085); Nb-x; γ(0.122, 0.257)
	93*m*	6.85 h			IT, Mo-x; γ(0.264, 0.685, 1.477)
	95		15.92(5)	13.4(5)	
	97		9.55(3)	2.5(3)	
	98		24.13(7)	0.14(1)	
	99	2.75 d			β^-(1.357); Tc-x; γ(0.181, 0.366, 0.739)
	101	14.6 min			β^-(2.23, 0.7); γ(0.192, 0.591)
Technetium	93	2.73 h			β^+(0.81); γ(1.363, 1.477, 1.520)
	94	4.88 h			β^+(4.256); γ(0.449, 0.703, 0.850, 0.871)
	95*m*	61 d			β^+(0.71); γ(0.204, 0.582, 0.835)
	95	20.0 h			K, Mo-x; γ(0.766, 1.074)
	96	4.3 d			K, Mo-x; γ(0.778, 0.813, 0.850, 1.122)
	97*m*	90 d			K, Tc-x; γ(0.0965)
	97	2.6×10^6 y			K, Mo-x
	98	4.2×10^6 y			β^-(0.40); γ(0.652, 0.745)
	99*m*	6.012 h			IT, Tc-x; γ(0.141, 0.143)
	99	2.13×10^5 y		20	β^-(0.292)
Ruthenium	95	1.64 h			β^+(1.20, 0.91); γ(0.290, 0.336, 0.627)
	97	2.88 d			K, Tc-x; γ(0.216, 0.324, 0.461)
	100		12.6(1)	5.8(6)	

TABLE 1.45   Table of Nuclides (*Continued*)

Element	A	Half-life	Natural abundance, %	Cross section, barns	Radiation (MeV)
Ruthenium	101		17.0(1)	5.(1)	
(*cont.*)	102		31.6(2)	1.2(1)	
	103	39.27 d		<20	β^-(0.12, 0.223); γ(0.295, 0.4444, 0.497, 0.557, 0.610)
	105	4.44 h			β^-(1.187, 0.11, 1.134); γ(0.149, 0.263, 0.317, 0.469, 0.676, 0.724)
	106	1.020 y			β^-(0.0394)
Rhodium	99*m*	4.7 h			β^+(0.74); γ(0.277, 0.341, 0.618, 1.261)
	99	16 d			β^+(0.54, 0.68); 7(0.089, 0.353, 0.528)
	100	20.8 h			β^+(2.62, 2.07); γ(0.446, 0.540, 0.588, 0.823, 1.553, 2.376)
	101*m*	4.35 d			K, IT, Ru-x, Rh-x; γ(0.127, 0.307, 0.545)
	101	3.3 y			K, Ru-x; γ(0.127, 0.198, 0.325)
	102*m*	207 d			β^-(1.15); β^+(1.29, 0.82); γ(0.469, 0.475, 0.557, 0.628, 1.103)
	102	2.9 y			K, Ru-x; γ(0.475, 0.631, 0.697, 0.767, 1.047, 1.103)
	103*m*	56.12 min			IT, Rh-x, γ(0.0.040)
	103		100	145	
	104*m*	4.36 min		800.(100)	γ(0.051, 0.097, 0.556)
	104	42.3 s		40.(30)	β^-(2.44), γ(0.358, 0.556, 1.237)
	105*m*	40 s			IT, Rh-x; γ(0.130)
	105	35.4 h		$1.1(3) \times 10^4$	β^-(0.567, 0.247); γ(0.280, 0.306, 0.319)
	106*m*	2.18 h			β^-(0.92); γ(0.222, 0.451, 0.512, 0.616, 0.717, 0.784, 1.046, 1.528)
	106	29.80 s			β^-(3.54, 3.0, 2.4); γ(0.512, 0.622)
Palladium	100	3.63 d			K, Rh-x; γ(0.0748, 0.0840, 0.0327)
	101	8.47 h			K, Rh-x; β^+(0.776); γ(0.296, 0.590)
	103	16.99 d			K, Rh-x; γ(0.357, 0.497)
	105		22.33(8)	22.(2)	
	107	6.5×10^6 y		1.8(2)	β^-(0.03)
	108		26.46(9)	8.7	
	109	13.5 h			β^-(1.028); Ag-x; γ(0.088, 0.311, 0.636)

(*Continued*)

TABLE 1.45 Table of Nuclides (*Continued*)

Element	A	Half-life	Natural abundance, %	Cross section, barns	Radiation (MeV)
Palladium (*cont.*)	111*m*	5.5 h			β^-(0.35, 0.77); γ(0.070, 0.172, 0.391)
	111	23.4 min			β^-(2.2); γ(0.060, 0.245, 0.580, 0.650, 1.389, 1.459)
	112	21.4 h			β^-(0.28); γ(0.018)
Silver	103	1.10 h			β^+(1.7, 1.3); γ(0.119, 0.148)
	104	69 min			β^+(0.99); γ(0.556, 0.926, 0.942)
	105	41.29 d			K, Pd-x; γ(0.064, 0.280, 0.344, 0.443)
	106*m*	8.4 d			K, Pd-x; γ(0.451, 0.512, 0.717, 1.046)
	107*m*	44.2 s			K, Ag-x; γ(0.093)
	107		51.839(7)	35	
	108*m*	130 y			γ(0.434, 0.614, 0.723)
	108	2.42 min			β^-(1.65); β^+(0.90);
	109		48.161(7)	91	γ(0.434, 0.619, 0.633)
	110*m*	249.8 d		82.(11)	β^-(0.087, 0.530); IT, γ(0.658, 0.764, 0.885, 0.937, 1.384)
	111*m*	1.08 min			K, Ag-x; γ(0.060, 0.245)
	111	7.47 d		3.(2)	β^-(1.04); γ(0.245, 0.342)
	112	3.13 h			β^-(3.94, 3.4); γ(0.607, 0.617, 1.39)
Cadmium	107	6.52 h			β^+(0.302); K, Ag-x; γ(0.093, 0.829)
	109	462 d			K, Ag-x; γ(0.088)
	111*m*	48.5 min			K, Cd-x; γ(0.151, 0.245)
	111		12.80(8)	24.(3)	
	113*m*	14.1 y			β^-(0.59); γ(0.264)
	113	9×10^{15} y	12.22(6)	20 060.(40)	
	115*m*	44.6 d			β^-(1.62); γ(0.934, 1.29, 0.485)
	115	2.228 d			β^-(1.11, 0.593); In-x; γ(0.231, 0.260, 0.336, 0.492, 0.528)
	117*m*	3.4 h			β^-(0.72); γ(0.159, 0.553); In-x
	117	2.49 h			β^-(0.67, 2.2); γ(0.221, 0.273, 0.345, 1.303)
Indium	109	4.2 h			K, Cd-x; β^+(0.79); γ(0.203, 0.623)
	110*m*	4.9 h			γ(0.658, 0.885, 0.937)
	110	1.15 h			β^+(2.22); K, Cd-x; γ(0.658)
	111	2.805 d			K, Cd-x; γ(0.171, 0.245)

TABLE 1.45   Table of Nuclides (*Continued*)

Element	A	Half-life	Natural abundance, %	Cross section, barns	Radiation (MeV)
Indium	113*m*	1.658 h			IT, In-x; γ(0.392)
(*cont.*)	114*m*	49.51 d			IT, K, In-x; γ(0.190)
	114	1.1983 min			β^-(1.99); K, Cd-x, β^+(0.40); γ(0.558, 0.573, 1.30)
	115*m*	4.486 h			β^-(0.83); K, In-x; γ(0.336, 0.497)
	*115	4.4×10^{14} y	95.71(2)	205	β^-(0.495)
	116*m*	54.1 min			β^-(1.00); γ(0.138, 0.417, 1.09, 1.293)
	117*m*	1.94 h			β^-(1.77); γ(0.159, 0.315, 0.553)
	117	44 min			β^-(0.74); γ(0.159, 0.397, 0.553)
Tin	110	4.1 h			K, In-x; γ(0.283)
	113	115.1 d		≈9	K, In-x; γ(0.392, 0.255)
	116		14.53(11)	1.1(1)	
	117*m*	13.60 d			K, Sn-x; γ(0.159)
	119*m*	293 d			K, Se-x; γ(0.239)
	119		8.59(4)	2.(1)	
	121*m*	≈55 y			β^-(0.354); K, In-x; γ(0.0372)
	121	1.128 d			β^-(0.383)
	123	129.2 d			β^-(1.42); γ(0.160, 1.030, 1.089)
	125	9.63 d			β^-(2.35); γ(1.067)
	127	2.10 h			β^-(2.42, 3.2); γ(0.823, 1.096)
Antimony	115	32.1 min			β^+(1.51); γ(0.499)
	116*m*	1.00 h			β^+(1.16); γ(0.407, 0.543, 0.973, 1.293)
	117	2.80 h			β^+(0.57); γ(0.159)
	118*m*	5.00 h			γ(0.254, 1.051, 1.280)
	118	3.6 min			β^+(2.65); γ(1.230)
	119	38.1 h			γ(0.0239)
	120	15.89 min			β^+(1.72); γ(0.704, 1.171)
	121		57.21(5)	6	
	122	2.72 d			β^-(1.414); β^+(1.980); γ(0.564, 0.693, 1.141, 1.257)
	123		42.7(9)	3.3	
	124	60.20 d			β^-(0.61, 2.301); γ(0.603, 0.646, 1.69, 0.723)
	126	12.4 d			β^-(1.9); γ(0.279, 0.415, 0.666, 0.695, 0.720)
	127	3.84 d			β^-(0.89, 1.10, 1.50); γ(0.252, 0.291, 0.412, 0.437, 0.686, 0.784)
	128	9.1 h			β^-(2.3); γ(0.215, 0.314, 0.527, 0.743, 0.754)

(*Continued*)

TABLE 1.45 Table of Nuclides (*Continued*)

Element	A	Half-life	Natural abundance, %	Cross section, barns	Radiation (MeV)
Antimony (*cont.*)	129	4.40 h			β^-(0.65); γ(0.181, 0.359, 0.460, 0.545, 0.813, 0.915, 1.030)
Tellurium	116	2.49 h			γ(0.0937)
	117	1.03 h			β^+(1.78); γ(0.920, 1.716, 2.300)
	119m	4.69 d			γ(0.154, 0.271, 1.213)
	119	16.0 h			β^+(0.627; γ(0.644, 0.700)
	121m	≈154 d			γ(0.212)
	121	16.8 d			γ(0.508, 0.573)
	123m	119.7 d			γ(0.159)
	125		7.139(6)	1.6(2)	
	127m	109 d			β^-(0.77); γ(0.088)
	127	9.35 h			β^-(0.696); γ(0.360)
	129m	33.6 d			β^-(1.60); γ(0.460, 0.696)
	129	1.160 h			β^-(1.453, 0.989); I-x, γ(0.460, 0.487)
	131m	1.35 d			β^-(0.42); IT, Te-x, I-x; γ(0.150, 0.774, 0.794, 0.852)
	131	25.0 min			β^-(2.14, 1.69, 1.35); I-x; γ(0.150, 0.453, 0.493)
	132	25.0 min			β^-(0.215); γ(0.050, 0.112, 0.228)
Iodine	121	2.12 h			β^+(1-2); γ(2.12)
	122	3.6 min			β^+(3.1); γ(0.564)
	123	13.2 h			K, Te-x; γ(0.159)
	124	4.18 d			β^+(1.54, 2.14, 0.75); γ(0.603, 0.723, 1.691)
	125	59.4 d		$9.(1) \times 10^2$	K, Te-x; γ(0.035)
	126	13.0 d			β^+(1.13); β^-(0.87, 1.25); γ(0.389, 0.662)
	127		100	6.15(10)	
	128	24.99 min		22.(4)	β^-(2.13); γ(0.443, 0.527)
	129	1.7×10^7 y			β^-(0.15); γ(0.040)
	130	12.36 h		18.(3)	β^+(1.13); β^-(0.87, 1.25); γ(0.389, 0.662)
	131	8.040 d		≈0.7	β^-(0.606); γ(0.284, 0.364, 0.637)
	132	208 h			β^-(0.80, 1.03, 1.2, 1.6, 2.16); γ(0.098, 0.506, 0.523, 0.630, 0.651, 0.667, 0.723, 0.955)
	133	20.8 h			β^-(1.24); γ(0.511, 0.530, 0.875)
	135	6.57 h			β^-(0.9, 1.3); γ(0.418, 0.527, 1.132, 1.260)
Xenon	123	2.00 h			β^+(1.51); γ(0.149, 0.178)
	125	17.1 h			γ(0.188, 0.243)

TABLE 1.45 Table of Nuclides (*Continued*)

Element	A	Half-life	Natural abundance, %	Cross section, barns	Radiation (MeV)
Xenon	127*m*	1.15 min			γ(0.127, 0.173)
(*cont.*)	127	36.4 d			γ(0.172, 0.203, 0.375)
	129*m*	8.89 d			γ(0.040, 0.197)
	129		26.4(6)	22.(5)	
	131*m*	11.9 d			γ(0.164)
	131		21.2(4)	90.(10)	
	133*m*	2.19 d			γ(0.233)
	133	5.243 d		190.(90)	β^-(0.346); Cs-x; γ(0.081)
	135*m*	15.3 min			γ(0.527)
	135	9.1 h			β^-(0.91); γ(0.250, 0.608)
Cesium	126	1.64 min			β^+(3.4, 3.7); γ(0.0389, 0.491, 0.925)
	127	6.2 h			β^+(0.65, 1.06); γ(0.125, 0.412)
	127	3.62 min			β^+(2.44, 2.88); γ(0.443)
	129	1.336 d			γ(0.372, 0.412)
	132	6.48 d			γ(0.465, 0.630, 0.668)
	133		100	28	
	134*m*	2.91 h			IT, K, Cs-x; γ(0.127)
	134	2.065 y		140.(10)	β^-(0.658, 0.089); γ(0.563, 0.569, 0.605, 0.796)
	135	2.3×10^6 y		8.9(5)	β^-(0.205)
	136	13.16 d			β^-(0.341); γ(0.341, 0.819, 1.048)
	137	30.2 y			β^-(0.514); K, Ba-x; γ(0.662)
Barium	126	1.65 h			γ(0.218, 0.234, 0.258)
	128	2.43 d			γ(0.273); K, Cs-x
	129*m*	2.17 h			γ(0.177, 0.182, 0.202, 1.459)
	129	2.2 h			β^+(1.42); γ(0.129, 0.214, 0.221)
	131	11.7 d			γ(0.124, 0.216, 0.496)
	133*m*	1.621 d			γ(0.276)
	133	10.53 y		4.(1)	γ(0.081, 0.356)
	135*m*	1.196 d			IT, Ba-x; γ(0.268)
	135		6.59(2)	5.8	
	137		11.23(4)	5.(1)	
	137*m*	2.552 min			IT, K, Ba-x; γ(0.662)
	138		71.70(7)	0.41(2)	
	139	1.396 h		5.1	β^-(2.27, 2.14); K, La-x; γ(0.166, 1.254, 1.421)
	140	12.75 d			β^-(0.48, 1.02); γ(0.163, 0.305, 0.537)
	142	10.7 min			β^-(1.0, 1.1); γ(0.231, 0.255, 0.309, 1.204)

(*Continued*)

TABLE 1.45 Table of Nuclides (*Continued*)

Element	A	Half-life	Natural abundance, %	Cross section, barns	Radiation (MeV)
Lanthanum	131	59 min			β^+(1.42, 1.94); γ(0.526, 0.109, 0.366)
	132	4.8 h			β^+(2.6, 3.2, 3.7); γ(0.465, 0.567)
	133	3.91 h			β^+(1.2); γ(0.279, 0.290, 0.302)
	134	6.5 min			β^+(2.67); γ(0.605)
	135	19.5 h			γ(0.481)
	136	8.87 min			β^+(1.8); γ(0.816)
	*138	1.06×10^{11} y		57.(6)	
	139		99.9098(2)	9.2(2)	
	140	1.68 d		2.7(3)	β^-(1.670, 1.35)
	141	3.90 h			β^-(2.43)
	142	1.54 h			β^-(2.11, 2.98, 4.52)
Cerium	132	3.5 h			γ(0.154, 0.182)
	133	5.4 h			β^+(1.3); γ(0.058, 0.131, 0.472, 0.510)
	135	17.7 h			β^+(0.8); γ(0.266, 0.300, 0.607)
	137m	1.43 d			IT K, Ce-x; γ(0.169, 0.254)
	137	9.0 h			γ(0.447)
	139	137.6 d			γ(0.166)
	140		88.43(10)	0.58(4)	
	141	32.50 d			β^-(0.436, 0.581); K, Pr-x; γ(0.145)
	142		11.13(10)	0.97(3)	
	143	1.38 d		6.1(7)	β^-(1.404, 1.110); K, Pr-x; γ(0.293)
	144	284.6 d		1.0(1)	β^-(0.318, 0.185); K, Pr-x; γ(0.080, 0.134)
Praseodymium	136	13.1 min			β^+(2.98); γ(0.540, 0.552)
	137	1.28 h			β^+(1.68); γ(0.434, 0.514, 0.837)
	138m	2.1 h			β^+(1.65); γ(0.304, 0.789, 1.038)
	139	4.41 h			β^+(1.09); γ(0.255, 1.347, 1.631)
	141		100	11.5	
	142	19.12 h		20.(3)	β^-(2.164); γ(1.576)
	143	13.57 d		90.(10)	β^-(0.933); γ(0.742)
	145	5.98 h			β^-(1.80); γ(0.073, 0.676, 0.748)
Neodymium	139m	5.5 h			β^+(1.17); γ(0.114, 0.738)
	141	2.49 h			β^+(0.802)
	142		27.13(2)	19.(1)	
	143		12.18(6)	220.(10)	
	*144	2.1×10^{15} y	23.8(1)	3.6(3)	
	145		8.3(6)	47.(6)	

TABLE 1.45   Table of Nuclides (*Continued*)

Element	A	Half-life	Natural abundance, %	Cross section, barns	Radiation (MeV)
Neodymium	146		17.19(9)	1.5(2)	
(*cont.*)	147	10.98 d		440.(150)	β^-(0.805); γ(0.091, 0.531)
	149	1.73 h			β^-(1.03, 1.13); γ(0.211, 0.114)
Promethium	143	265 d			K, Nd-x; γ(0.742)
	144	360 d			K, Nd-x; γ(0.618, 0.696)
	146	5.53 y		$8.4(2) \times 10^3$	K, β^-(0.795); Nd-x; γ(0.453, 0.75)
	147	2.6234 y		180	β^-(0.224); γ(0.122, 0.197)
	148*m*	41.29 d		$106.(8) \times 10^2$	β^-(0.69, 0.50, 0.40); IT, Pm-x, Sm-x; γ(0.550, 0.630, 0.726)
	148	5.37 d		≈ 1000	β^-(1.02, 2.47); γ(0.550, 0.915, 1.465)
	149	2.212 d		$14.(2) \times 10^2$	β^-(1.072, 0.78); γ(0.286, 0.591, 0.859)
	150	2.68 h			β^-(1.6, 2.3, 1.8); γ(0.334, 1.166, 0.132)
	151	1.183 d		≈ 150	β^-(0.84); γ(0.168, 0.275, 0.340)
Samarium	142	1.208 h			β^+(1.0); K, Pr-x
	144		3.1(1)	1.6(1)	
	145	340 d		280.(20)	γ(0.061, 0.492); K, Pm-x
	146	1.03×10^8 y			α(2.50)
	*147	1.06×10^{11} y	15.0(2)	56.(4)	α(2.23)
	148	7×10^{15} y	11.3(1)	2.4(6)	α(1.96)
	149	10^{16} y	13.8(1)	$401.(6) \times 10^2$	
	150		7.4(1)	102.(5)	
	151	90 y			β^-(0.076)
	152		26.7(2)	206.(15)	
	153	1.929 d		420.(180)	β^-(0.64, 0.69); γ(0.103)
	154		22.7(2)	7.5(3)	
	155	22.2 min			β^-(1.52); γ(0.104)
	156	9.4 h			β^-(0.43, 0.71); γ(0.166, 0.204)
Europium	148	54.5 d			β^+(0.92); γ(0.550, 0.630)
	149	93.1 d			K, Sm-x; γ(0.277, 0.328)
	150*m*	12.8 h			β^-(1.013); γ(0.334, 0.407)
	150	36 y			γ(0.334, 0.439, 0.584)
	151		47.8(5)	9000	
	152*m*	9.30 h			β^-(1.85); γ(0.122, 0.841, 0.963)
	152	13.48 y		$11.(2) \times 10^3$	K, β^-(1.47, 0.690); K, Gd-x, K, Sm-x; γ(0.122, 0.344, 1.408)
	153		52.2(5)	320.(20)	

(*Continued*)

TABLE 1.45 Table of Nuclides (*Continued*)

Element	A	Half-life	Natural abundance, %	Cross section, barns	Radiation (MeV)
Europium (*cont.*)	154	8.59 y		$1.5(3) \times 10^3$	β^-(0.27, 0.58, 0.843, 1.87); γ(0.123, 0.723, 1.274)
	155	4.76 y		$3.9(2) \times 10^3$	β^-(0.15); γ(0.087, 0.105)
	156	15.2 d			β^-(0.30, 0.49, 1.2, 2.45); γ(0.089, 0.646, 0.723, 0.812)
	157	15.13 h			β^-(1.30); γ(0.064, 0.371, 0.411)
	158	45.9 min			β^-(2.5); γ(0.898, 0.944, 0.977)
Gadolinium	146	48.3 d			β^+(0.35); γ(0.115, 0.155)
	147	1.588 d			β^+(0.93); γ(0.229, 0.370, 0.396, 0.929)
	151	124 d			α(2.73); γ(0.154, 0.243)
	153	241.6 d			γ(0.94, 0.103)
	155		14.80(5)	$61.(1) \times 10^3$	
	157		15.65(3)	$2.54(3) \times 10^5$	
	158		24.84(12)	2.3(5)	
	159	18.56 h			β^-(0.971); Tb-x; γ(0.363)
	160		21.86(4)	1.5(7)	
Terbium	158	180 y			γ(0.944, 0.962)
	159		100	23.2(5)	
	160	72.3 d		$5.7(11) \times 10^2$	β^-(0.57, 0.86); γ(0.299, 0.879, 0.966)
Dysprosium	159	144 d		$8.(2) \times 10^3$	K, Tb-x; γ(0.326)
	161		18.9(2)	600.(150)	
	162		25.5(2)	170.(20)	
	163		24.9(2)	120.(10)	
	164		28.2(2)	2000	
	165	2.33 h		$3.5(3) \times 10^3$	β^-(1.29); Ho-x; γ(0.095)
	165m	1.26 min			γ(0.108, 0.515)
Holmium	156	56 min			γ(0.138, 0.267)
	159	33.0 min			γ(0.121, 0.132, 0.253, 0.310)
	167	3.1 h			β^-(0.31, 0.62, 0.96); γ(0.238, 0.321, 0.347)
	165		100	61	
	166m	1.2×10^3 y		$9.14(65) \times 10^3$	Er-x; γ(0.810, 0.712, 0.184)
	166	1.117 d			β^-(1.855, 1.776); γ(1.379)
Erbium	166		33.6(2)	20	
	167		22.95(15)	$7.(2) \times 10^2$	
	168		26.8(2)	2.0(6)	
	169	9.40 d			β^-(0.35)
	170		14.9(2)	6.2(2)	

TABLE 1.45   Table of Nuclides (*Continued*)

Element	A	Half-life	Natural abundance, %	Cross section, barns	Radiation (MeV)
Erbium (*cont.*)	171	7.52 h		370.(40)	$\beta^-(1.49)$; Tm-x; $\gamma(0.112,$ 0.296, 0.308)
	172	2.05 d			$\beta^-(0.28, 0.36)$; $\gamma(0.407,$ 0.610)
Thullium	166	7.70 h			$\gamma(0.184, 0.779, 1.273,$ 2.052)
	169		100	106	
	170	128.6 d		100.(20)	$\beta^-(0.968, 0.884)$
	171	1.92 y		≈160	$\beta^-(0.096)$; $\gamma(0.067)$
	172	2.65 d			$\beta^-(1.79, 1.86)$; $\gamma(1.387,$ 1.466, 1.530, 1.609)
	173	8.2 h			$\beta^-(0.80, 0.86)$; $\gamma(0.399,$ 0.461)
Ytterbium	165	9.9 min			$\beta^+(1.58)$; $\gamma(1.090)$
	166	2.363 d			$\gamma(0.184, 0.779, 1.273,$ 2.052)
	169	32.03 d		$3.6(3) \times 10^3$	$\gamma(0.110, 0.177, 0.198)$
	171		14.3(2)	50.(10)	
	173		16.12(21)	16.(2)	
	174		31.8(4)	120	
	175	4.19 d			$\beta^-(0.466)$; Lu-x; $\gamma(0.396)$
	176		12.7(2)	3.1(2)	
	177	1.9 h			$\beta^-(1.40)$; K, Lu-x; $\gamma(0.150)$
	178	1.23 h			$\beta^-(0.25)$; $\gamma(0.141, 0.325,$ 0.352, 0.381, 0.613)
Lutetium	164	3.14 min			$\beta^+(1.6, 3.8)$; $\gamma(0.124,$ 0.262, 0.740, 0.864, 0.880)
	165	16.7 min			$\beta^+(2.06)$; $\gamma(0.121, 0.132,$ 0.174, 0.204)
	175		97.41(2)	24	
	176m	3.66 h			$\beta^-(1.229, 1.317)$; Hf-x; $\gamma(0.0884)$
	176	3.8×10^{16} y		2100	$\gamma(0.202, 0.307)$
	177	6.75 d		$10.(3) \times 10^2$	$\beta^-(0.497)$, Hf-x; $\gamma(0.113,$ 0.208)
Hafnium	178		27.297(4)	85	
	179		13.629(6)	46	
	†179m₁	18.7 s			$\gamma(0.161, 0.214)$
	†179m₂	25.1 d			$\gamma(0.123, 0.146, 0.363,$ 0.454)
	180		35.100(7)	13.(1)	
	180m	5.519 h			IT, Hf-x; $\gamma(0.215, 0.332,$ 0.443)
	181	42.4 d		30.(25)	$\beta^-(0.408)$; Ta-x; $\gamma(0.133, 0.346, 0.482)$

†Two different metastable states possessing the same mass number but different half-lives.

(*Continued*)

TABLE 1.45   Table of Nuclides (*Continued*)

Element	A	Half-life	Natural abundance, %	Cross section, barns	Radiation (MeV)
Hafnium (*cont.*)	183	1.07 h			β^-(1.18, 1.54); γ(0.459, 0.784)
	184	4.1 h			β^-(0.74, 0.85, 1.10); γ(0.139, 0.345)
Tantalum	181		99.988(2)	20	
	182*m*	16.5 min			γ(0.147, 0.172, 0.184)
	182	114.43 d		$8.2(6) \times 10^3$	β^-(0.25, 0.44, 0.52); γ(0.068, 1.121)
	183	5.1 d			β^-(0.62); γ(0.108, 0.246, 0.304)
	184	8.7 h			β^-(1.17); γ(0.253, 0.414)
Tungsten	182		26.50(3)	20.(1)	
	183		14.31(1)	10.5(3)	
	184		30.64(1)	2	
	185	74.8 d		$\approx$3.3	β^-(0.433); γ(0.125)
	186		28.43(4)	37.(2)	
	187	23.9 h		70.(10)	β^-(1.315, 0.624; K, Re-x; γ(0.072, 0.480, 0.686)
	188	69.4 d			β^-(0.349); γ(0.227, 0.291)
Rhenium	182*m*	12.7 h			β^+(0.55, 1.74); γ(1.121, 1.221)
	184	38 d			γ(0.790, 0.903)
	185		37.40(2)	110	
	186	3.718 d			β^-(1.07, 0.933); K, W-x, Os-x; γ(0.123, 0.137, 0.632, 0.768)
	*187	4.2×10^{10}	62.60(2)	74	
	188	16.94 h			β^-(2.12, 1.96); Os-x; γ(0.155)
	189	24 h			β^-(1.01); γ(0.147, 0.22, 0.245)
Osmium	186	2×10^{15} y	1.58(2)	$\approx$80	
	188		13.3(1)	$\approx$5	
	190*m*	9.9 min			IT, Os-x; γ(0.187,0.361, 0.503, 0.616)
	190		26.4(2)	13	
	191	15.4 d		$3.8(6) \times 10^2$	β^-(0.143); Os-x; γ(0.129)
	192		41.0(3)	3.(1)	
	193	30.5 h			β^-(1.04); Ir-x; γ(0.139, 0.460)
	196	34.9 min			β^-(0.84); γ(0.126, 0.408)
Iridium	184	3.0 h			β^+(2.3, 2.9); γ(0.120, 0.264, 0.390)
	185	14 h			γ(0.254, 1.829)
	186	15.7 h			γ(0.137, 0.296, 0.435)
	188	1.72 d			γ(0.155, 0.478, 0.633, 2.215)

TABLE 1.45 Table of Nuclides (*Continued*)

Element	A	Half-life	Natural abundance, %	Cross section, barns	Radiation (MeV)
Iridium (*cont.*)	189	13.2 d			K, Os-x; γ(0.245)
	190	11.8 d			γ(0.187, 0.407, 0.519, 0.558, 0.605)
	191		37.27(9)	920	
	192	73.83 d			β^-(0.672); K, Pt-x; γ(0.316, 0.468)
	193		62.73(9)	116	
	194	19.3 h		$1.5(3) \times 10^3$	β^-(2.25); γ(0.294, 0.328, 0.645)
	195m	3.9 h			β^-(0.41, 0.97); γ(0.320, 0.365, 0.433, 0.685)
Platinum	187	2.35 h			γ(0.105,0.110, 0.201, 0.285, 0.709)
	188	10.2 d			γ(0.188, 0.195)
	189	10.89 h			K, Ir-x; γ(0.094, 0.608, 0.721)
	194		32.9(6)	1.2	
	195m	4.02 d			IT, Pt-x; γ(0.099)
	195		33.8(6)	28.(1)	
	196		25.3(6)	55	
	197m	1.573 h			IT, Pt-x; γ(0.053, 0.346)
	197	18.3 h			β^-(0.719); K, Au-x; γ(0.191, 0.269)
	199m	14.1 s			γ(0.392)
	199	30.8 min		$\approx$16	β^-(0.90, 1.14); γ(0.186, 0.317, 0.494, 0.549)
	200	12.5 h			γ(0.136, 0.227, 0.244)
Gold	197		100	98.7(1)	
	197m	7.8 s			IT, K, Au-x; γ(0.130, 0.279)
	198	2.694 d		$26.5(15) \times 10^3$	β^-(0.961); K, Hg-x; γ(0.412)
	199	3.139 d			β^-(0.292, 0.250); K, Hg-x; γ(0.158, 0.208)
	200m	18.7 h			β^-(0.56); γ(0.111,0.368, 0.498, 0.597, 0.760)
	200	48.4 min			β^-(2.2); γ(0.368, 1.225)
Mercury	196		0.15(1)	3150	
	197m	23.8 h			IT, K, Hg-x; γ(0.134)
	197	2.6725 d			K, Au-x; γ(0.077)
	199m	42.6 min			γ(0.158)
	199		16.87(10)	$2.1(2) \times 10^3$	
	200		23.10(16)	<60	
	202		29.86(20)	4.9(5)	
	203	46.61 d			β^-(0.213); γ(0.279)
Thallium	201	3.040 d			K, Hg-x; γ(0.135,0.167)
	202	12.23 d			K, Hg-x; γ(0.440)
	203		29.52(1)	11.(1)	
	204	3.78 y		22.(2)	β^-(0.763); K, Hg-x

(Continued)

TABLE 1.45 Table of Nuclides (*Continued*)

Element	A	Half-life	Natural abundance, %	Cross section, barns	Radiation (MeV)
Thallium	205		70.48(1)	0.11(2)	
(*cont.*)	*206	4.20 min			β^-(1.53); K, Pb-x; γ(0.803)
	*207	4.77 min			β^-(1.43); γ(0.897)
	208	3.053 min			β^-(1.796, 1.28, 1.52); γ(0.277, 0.511,0.583, 0.614)
	209	2.16 min			β^-(1.8); γ(1.567, 0.465)
	210	1.30 min			β^-(1.9, 1.3); γ(0.298, 0.798)
Lead	201	9.33 h			γ(0.331, 0.361)
	203	2.1615 d			γ(0.279)
	204*m*	1.120 h			IT, Pb-x; γ(0.375, 0.899, 0.912)
	207		22.1(1)	0.70(1)	
	209	3.253 h			β^-(0.645)
	*210	22.6 y			α(3.72)
	*211	36.1 min			β^-(1.36); γ(0.405, 0.427, 0.832)
	*212	10.64 h			β^-(0.569, 0.28); Bi-x; γ(0.239)
	*214	26.9 min			β^-(0.67, 0.73); γ(0.24, 0.30, 0.352)
Bismuth	205	15.31 d			γ(0.703, 1.764)
	206	6.243 d			γ(0.516, 0.803, 0.881)
	209		100	0.034	
	*210	5.013 d			β^-(1.16); γ(0.266, 0.352)
	212	1.0092 h			β^-(2.25); γ(0.288, 0.727, 0.786, 1.621); Tl-x; α(6.05, 6.09)
	*214	19.7 min			β^-(3.26); γ(0.609, 1.120, 1.764)
Polonium	204	3.53 h			γ(0.270, 0.884, 1.016)
	205	1.7 h			γ(0.837, 0.850, 0.872, 1.001)
	206	8.8 d			α(5.233); γ(0.286, 0.312, 0.807)
	208	2.898 y			α(5.116)
	209	102 y			α(4.88); IT, K, Bi-x; γ(0.260, 0.896)
	210	138.38 d			α(5.304); γ(0.803)
	212	298 ns			α(8.784)
	214	0.1637 ms			α(7.686)
	216	145 ms			α(6.778)
	218	3.04 min			α(5.18)
Astatine	207	1.81 h			α(5.76); γ(0.168, 0.588, 0.814)
	208	1.63 h			α(5.641); K, Po-x, γ(0.177,0.660, 0.685, 0.845, 1.028)

TABLE 1.45 Table of Nuclides (*Continued*)

Element	A	Half-life	Natural abundance, %	Cross section, barns	Radiation (MeV)
Astatine (*cont.*)	209	5.41 h			α(5.65), K, Po-x; γ(0.545, 0.782, 0.790)
	210	8.1 h			K, Po-x; γ(0.245, 0.528, 1.181, 1.437, 1.483)
	211	7.214 h			α(5.87); K, Po-x; γ(0.669, 0.742)
Radon	210	2.4 h			α(6.039); γ(0.196, 0.458, 0.571, 0.649)
	211	14.68 h			α(5.784, 5.851); γ(0.169, 0.250, 0.370, 0.674, 0.678, 1.363)
	212	24 min			α(6.260)
	220	55.6 s			α(6.288)
	222	2.8235 d		0.74(5)	α(5.49); γ(0.510)
Francium	212	20 min			α(6.41, 6.26); γ(1.186, 1.275)
	220	27.4 s			
	221	4.8 min			α(6.686, 0.641, 6.582); γ(0.106, 0.154, 0.162)
	222	14.3 min			α(6.341); 7(0.218, 0.409)
	223	22.0 min			β^-(0.178) β^-(0.117)
Radium	*224	3.66 d		12.0(5)	α(5.685, 5.45); K, Rn-x; γ(0.241, 0.409, 0.650)
	*226	1599 y		$\approx$13	α(4.78, 4.60); K, Rn-x; γ(0.186, 0.262)
	*228	5.76 y		36.(5)	γ(0.0135)
Actinium	*227	21.77 y		$8.8(7) \times 10^2$	β^-(0.045); α(4.95, 4.94); K, Th-x; γ(0.084, 0.160, 0.270)
	*228	6.15 h			β^-(2.18, 1.85, 1.11); K, Th-x; γ(0.339, 0.911, 0.969)
Thorium	226	30.6 min		$1.2(2) \times 10^2$	α(6.337, 6.228); γ(0.206, 0.242)
	228	1.913 y		23.4(5)	α(5.42, 5.34, 5.18); K, Ra-x
	*230	7.54×10^4 y		7.37(4)	
	231	1.063 d		$1.5(1) \times 10^3$	α(4.68, 4.62); K, Ra-x; γ(0.068)
	*232	1.405×10^{10} y		1.8(5)	β^-(0.305, 0.218, 0.138)
	233	22.3 min			α(4.01, 3.95); γ(0.059)
	*234	24.10 d			β^-(1.245); γ(0.459) β^-(0.198, 0.102); K, Pa-x
Protactinium	230	17.4 d		$1.5(3) \times 10^3$	β^-(0.51); γ(0.444, 0.455, 0.899, 0.952)
	*231	3.25×10^4 y		$2.0(1) \times 10^2$	α(5.06, 5.03, 5.01, 4.95, 4.73); K, Ac-x; γ(0.260, 0.284, 0.300, 0.330)

(Continued)

TABLE 1.45 Table of Nuclides (*Continued*)

Element	A	Half-life	Natural abundance, %	Cross section, barns	Radiation (MeV)
Protactinium (*cont.*)	232	1.31 d		$4.6(10) \times 10^2$	$\beta^-(1.34)$; $\gamma(0.109, 0.150, 0.894, 0.969)$
	233	27.0 d			$\beta^-(0.256, 0.15, 0.568)$; K,L U-x; $\gamma(0.300, 0.312, 0.341)$
	234*m*	1.17 min			$\beta^-(2.29)$; IT, K, U-x
	235	24.4 min			$\beta^-(1.4)$
Uranium	230	20.8 d			$\alpha(5.89, 5.82)$
	232	68.9 y		73.(2)	$\alpha(5.320, 5.263)$
	233	1.592×10^5 y		47.(2)	$\alpha(4.825, 4.783)$; L, Th-x; $\gamma(0.029, 0.042, 0.055, 0.097, 0.119, 0.146, 0.164, 0.22, 0.291, 0.32)$
	*234	2.454×10^5 y	0.0055(5)	96.(2)	$\alpha(4.776, 4.723)$; L, Th-x; $\gamma(0.121)$
	*235	7.037×10^8 y	0.720(1)	95.(5)	$\alpha(4.40, 4.37, 4.22)$; K,L Th-x; $\gamma(0.14, 0.16, 0.186, 0.20)$
	237	6.75 d		$\approx$100	
	*238	4.46×10^9 y	99.2745(15)	2.7(1)	$\alpha(4.196, 4.147)$
	239	23.47 min		22.(2)	$\beta^-(1.21, 1.29)$
Neptunium	236	1.55×10^5 y			$\beta^-(0.49)$, $\gamma(0.104, 0.160)$
	237	2.14×10^6 y		180	$\alpha(4.79, 4.77)$; K,L Pa-x
	238	2.117 d		51	$\beta^-(1.2)$; $\gamma(0.984, 1.029)$
	239	2.355 d		$5.1(2) \times 10^2$	$\beta^-(0.438, 0.341)$; $\gamma(0.228, 0.278)$
Plutonium	237	45.7 d			K,L Np-x
	238	87.74 y			$\alpha(5.50, 5.46)$; K, U-x; $\gamma(0.0435)$
	239	2.411×10^4 y		$2.7(1) \times 10^2$	$\alpha(5.16, 5.14, 5.11)$; K, U-x; $\gamma(0.375, 0.414, 0.129)$
	240	6.537×10^3 y		$2.9(1) \times 10^2$	$\alpha(5.168, 5.124)$; L, U-x
	242	3.763×10^5 y		19.(1)	$\alpha(4.90, 4.86)$; $\gamma(0.045, 0.103)$
	244	8.2×10^7 y		1.7(1)	$\alpha(4.59, 4.55)$; L, U-x
	246	10.85 d			$\beta^-(0.150, 0.35)$; $\gamma(0.224)$
Americium	241	432.2 y		600	$\alpha(5.49, 5.44)$; $\gamma(0.12, 0.14)$
	243	7370 y		80	$\alpha(5.277, 5.234)$; $\gamma(0.075)$
Curium	242	162.8 d		$\approx$20	$\alpha(6.113, 6.069)$; L, Pu-x
	243	28.5 y		$1.3(1) \times 10^2$	$\alpha(5.786, 5.742)$
	244	18.11 y		15.(1)	$\alpha(5.805, 5.753)$; $\gamma(0.099, 1.526)$

TABLE 1.45   Table of Nuclides (*Continued*)

Element	A	Half-life	Natural abundance, %	Cross section, barns	Radiation (MeV)
Berkelium	247	1.4×10^3 y			α(5.532, 5.678, 5.712)
	249	320 d		7.(1) $\times 10^2$	α(5.42); β^-(0.125)
	250	3.217 h			β^-(0.74); γ(0.989, 1.032)
Californium	251	900 y		2.9(2) $\times 10^2$	α(5.677, 5.851, 6.014)
	252	2.645 y		20.(2)	α(6.118, 6.076); L, Cm-x; γ(0.043, 0.100)
Einsteinium	253	20.47 d		186	α(6.64); γ(0.389)
	254	275.7 d		28.(3)	α(6.43)
	255	40 d		$\approx$55	β^-(0.29); α(6.26)
Fermium	255	20.1 h		26.(3)	α(7.023)
	257	100.5 d			α(6.519); L, Cf-x; γ(0.179, 0.241)
Mendelevium	258	51.5 d			α(6.718, 6.763); γ(0.368)
	260	32 d			
Nobelium	255	3.1 min			α(8.12, 7.93); γ(0.187)
	259	58 min			α(7.52, 7.55)
Lawrencium	260	3 min			
	261	40 min			
	262	3.6 h			

1.11 VAPOR PRESSURE

Vapor pressure is the pressure exerted by a pure component at equilibrium, at any temperature, when both liquid and vapor phases exist and thus extends from a minimum at the triple point temperature to a maximum at the critical temperature (the critical pressure), and is the most important of the basic thermodynamic properties affecting liquids and vapors.

Except at very high total pressures (above about 10 MPa), there is no effect of total pressure on vapor pressure. If such an effect is present, a correction can be applied. The pressure exerted above a solid–vapor mixture may also be called vapor pressure but is normally only available as experimental data for common compounds that sublime.

1.11.1 Vapor Pressure Equations

Numerous mathematical formulas relating the temperature and pressure of the gas phase in equilibrium with the condensed phase have been proposed. The Antoine equation (Eq. 1) gives good correlation with experimental values. Equation 2 is simpler and is often suitable over restricted temperature ranges. In these equations, and the derived differential coefficients for use in the Haggenmacher and Clausius–Clapeyron equations, the p term is the vapor pressure of the compound, the t term is the temperature in degrees Celsius, and the T term is the absolute temperature in kelvins ($t°C + 273.15$).

Eq.	Vapor-pressure equation	dp/dT	$-[d(\ln p)/d(1/T)]$
1	$\log p = A - \dfrac{B}{t+C}$	$\dfrac{2.303\,pB}{(t+C)^2}$	$\dfrac{2.303\,BT^2}{(t+C)^2}$
2	$\log p = A - \dfrac{B}{T}$	$\dfrac{2.303\,pB}{T^2}$	$2.303B$
3	$\log p = A - \dfrac{B}{T} - C\,\log T$	$p\left(\dfrac{2.303\,B}{T^2} - \dfrac{C}{T}\right)$	$2.303B - CT$

TABLE 1.46 Vapor Pressures of Selected Elements at Different Temperatures

Element	Atomic number	Atomic symbol	Boiling point, °C	Vapor pressure temperature, °C								
				E-08	E-07	E-06	E-05	E-04	E-03	E-02	E-01	1
Aluminum	13	Al	2467	685	742	812	887	972	1082	1217	1367	1557
Antimony	52	Sb	1750	279	309	345	383	425	475	533	612	757
Arsenic	33	As	613	104	127	150	174	204	237	277	317	372
Barium	56	Ba	1140	272	310	354	402	462	527	610	711	852
Beryllium	4	Be	2970	707	762	832	907	997	1097	1227	1377	1557
Bismuth	83	Bi	1560	347	367	409	459	517	587	672	777	897
Boron	5	B	2550	1282	1367	1467	1582	1707	1867	2027	2247	2507
Cadmium	48	Cd	765	74	95	119	146	177	217	265	320	392
Calcium	20	Ca	1484	282	317	357	405	459	522	597	689	802
Carbon	6	C	4827	1657	1757	1867	1987	2137	2287	2457	2657	2897
Cobalt	27	Co	2870	922	992	1067	1157	1257	1382	1517	1687	1907
Chromium	24	Cr	2672	837	902	977	1062	1157	1267	1397	1552	1737
Copper	29	Cu	2567	722	787	852	937	1027	1132	1257	1417	1617
Dysprosium	66	Dy	2562	625	682	747	817	897	997	1117	1262	1437
Erbium	68	Er	2510	649	708	777	852	947	1052	1177	1332	1527
Europium	63	Eu	1597	283	319	361	409	466	532	611	708	827
Gallium	31	Ga	2403	619	677	742	817	907	1007	1132	1282	1472
Germanium	32	Ge	2830	812	877	947	1037	1137	1257	1397	1557	1777
Gold	79	Au	2807	807	877	947	1032	1132	1252	1397	1567	1767
Indium	77	In	2000	488	539	597	664	742	837	947	1082	1247
Iron	26	Fe	2750	892	957	1032	1127	1227	1342	1477	1647	1857
Lanthanum	57	La	3469	1022	1102	1192	1297	1422	1562	1727	1927	2177
Lead	82	Pb	1740	342	383	429	485	547	625	715	832	977
Lithium	49	Li	1347	235	268	306	350	404	467	537	627	747
Magnesium	12	Mg	1107	185	214	246	282	327	377	439	509	605
Manganese	25	Mn	1962	505	554	611	675	747	837	937	1082	1217
Mercury	80	Hg	357	−72	−59	−44	−27	7	16	46	80	125
Molybdenum	42	Mo	4612	1592	1702	1822	1957	2117	2307	2527	2787	3117
Nickel	28	Ni	2732	927	997	1072	1157	1262	1382	1527	1697	1907
Niobium	41	Nb	4927	1762	1867	1987	2127	2277	2447	2657	2897	3177
Palladium	46	Pd	2927	842	912	992	1082	1192	1317	1462	1647	1877
Phosphorus	15	P	2804	54	69	88	108	129	157	185	222	261

(Continued)

TABLE 1.46 Vapor Pressures of Selected Elements at Different Temperatures (*Continued*)

Element	Atomic number	Atomic symbol	Boiling point, °C	Vapor pressure temperature, °C								
				E-08	E-07	E-06	E-05	E-04	E-03	E-02	E-01	1
Platinum	78	Pt	3827	1292	1382	1492	1612	1747	1907	2097	2317	2587
Potassium	19	K	774	21	42	65	91	123	161	208	267	345
Praseodymium	59	Pr	3127	797	867	947	1042	1147	1277	1427	1617	1847
Rhenium	75	Re	5627	1947	2077	2217	2387	2587	2807	3067	3407	3807
Rhodium	45	Rh	3727	1277	1372	1472	1582	1707	1857	2037	2247	2507
Scandium	21	Sc	2832	772	837	917	1007	1107	1232	1377	1567	1797
Selenium	34	Se	685	63	83	107	133	164	199	243	297	363
Silicon	14	Si	4827	992	1067	1147	1237	1337	1472	1632	1817	2057
Silver	47	Ag	2212	574	626	685	752	832	922	1027	1162	1322
Sodium	11	Na	553	74	97	123	155	193	235	289	357	441
Strontium	38	Sr	1384	241	273	309	353	394	465	537	627	732
Sulfur	16	S	45	−10	3	17	37	55	80	109	147	189
Tantalum	73	Ta	5425	1957	2097	2237	2407	2587	2807	3057	3357	3707
Tellurium	52	Te	990	155	181	209	242	280	323	374	433	518
Thallium	81	Tl	1457	283	319	359	407	463	530	609	706	827
Tin	50	Sn	2270	682	747	807	897	997	1107	1247	1412	1612
Titanium	22	Ti	3287	1062	1137	1227	1327	1442	1577	1737	1937	2177
Tungsten	74	W	5660	2117	2247	2407	2567	2757	2977	3227	3537	3917
Ytterbium	70	Yb	1466	247	279	317	365	417	482	557	647	787
Yttrium	39	Y	3337	957	1032	1117	1217	1332	1467	1632	1832	2082
Zinc	30	Zn	907	123	147	177	209	247	292	344	408	487

TABLE 1.47 Vapor Pressures of Inorganic Compounds up to 1 Atmosphere

Compound name	Formula	Pressure, mm Hg — Temperature, °C										Melting point, °C
		1	5	10	20	40	60	100	200	400	760	
Aluminum	Al	1284	1421	1487	1555	1635	1684	1749	1844	1947	2056	660
borohydride	Al(BH$_4$)$_3$		−52.2	−42.9	−32.5	−20.9	−13.4	−3.9	+11.2	28.1	45.9	−64
bromide	AlBr$_3$	81.3	103.8	118.0	134.0	150.6	161.7	176.1	199.8	227.0	256.3	97
chloride	Al$_2$Cl$_6$	100.0	116.4	123.8	131.8	139.9	145.4	152.0	161.8	171.6	180.2	192.4
fluoride	AlF$_3$	1238	1298	1324	1350	1378	1398	1422	1457	1496	1537	1040
iodide	AlI$_3$	178.0	207.7	225.8	244.2	265.0	277.8	294.5	322.0	354.0	385.5	
oxide	Al$_2$O$_3$	2148	2306	2385	2465	2549	2599	2665	2766	2874	2977	2050
Ammonia	NH$_3$	−109.1	−97.5	−91.9	−85.8	−79.2	−74.3	−68.4	−57.0	−45.4	−33.6	−77.7
heavy	ND$_3$						−74.0	−67.4	−57.0	−45.4	−33.4	−74.0
Ammonium bromide	NH$_4$Br	198.3	234.5	252.0	270.6	290.0	303.8	320.0	345.3	370.8	396.0	
carbamate	N$_2$H$_6$CO$_2$	−26.1	−10.4	−2.9	+5.3	14.0	19.6	26.7	37.2	48.0	58.3	
chloride	NH$_4$Cl	160.4	193.8	209.8	226.1	245.0	256.2	271.5	293.2	316.5	337.8	520
cyanide	NH$_4$CN	−50.6	−35.7	−28.6	−20.9	−12.6	−7.4	−0.5	+9.6	20.5	31.7	36
hydrogen sulfide	NH$_4$HS	−51.1	−36.0	−28.7	−20.8	−12.3	−7.0	0.0	+10.5	21.8	33.3	
iodide	NH$_4$I	210.9	247.0	263.5	282.8	302.8	316.0	331.8	355.8	381.0	404.9	
Antimony	Sb	886	984	1033	1084	1141	1176	1223	1288	1364	1440	630.5
tribromide	SbBr$_3$	93.9	126.0	142.7	158.3	177.4	188.1	203.5	225.7	250.2	275.0	96.6
trichloride	SbCl$_3$	49.2	71.4	85.2	100.6	117.8	128.3	143.3	165.9	192.2	219.0	73.4
pentachloride	SbCl$_5$	22.7	48.6	61.8	75.8	91.0	101.0	114.1				2.8
triiodide	SbI$_3$	163.6	203.8	223.5	244.8	267.8	282.5	303.5	333.8	368.5	401.0	167
trioxide	Sb$_4$O$_6$	574	626	666	729	812	873	957	1085	1242	1425	656
Argon	A	−218.2	−213.9	−210.9	−207.9	−204.9	−202.9	−200.5	−195.6	−190.6	−185.6	−189.2
Arsenic	As	372	416	437	459	483	498	518	548	579	610	814
Arsenic tribromide	AsBr$_3$	41.8	70.6	85.2	101.3	118.7	130.0	145.2	167.7	193.6	220.0	
trichloride	AsCl$_3$	−11.4	+11.7	+23.5	36.0	50.0	58.7	70.9	89.2	109.7	130.4	−18
trifluoride	AsF$_3$					−2.5	+4.2	13.2	26.7	41.4	56.3	−5.9
pentafluoride	AsF$_5$	−117.9	−108.0	−103.1	−98.0	−92.4	−88.5	−84.3	−75.5	−64.0	−52.8	−79.8
trioxide	As$_2$O$_3$	212.5	242.6	259.7	279.2	299.2	310.3	332.5	370.0	412.2	457.2	312.8
Arsine	AsH$_3$	−142.6	−130.8	−124.7	−117.7	−110.2	−104.8	−98.0	−87.2	−75.2	−62.1	−116.3
Barium	Ba		984	1049	1120	1195	1240	1301	1403	1518	1638	850

(Continued)

219

TABLE 1.47 Vapor Pressures of Inorganic Compounds up to 1 Atmosphere (*Continued*)

Compound name	Formula	Pressure, mm Hg — Temperature, °C										Melting point, °C
		1	5	10	20	40	60	100	200	400	760	
Beryllium borohydride	Be(BH₄)₂	+1.0	19.8	28.1	36.8	46.2	51.7	58.6	69.0	79.7	90.0	123
bromide	BeBr₂	289	325	342	361	379	390	405	427	451	474	490
chloride	BeCl₂	291	328	346	365	384	395	411	435	461	487	405
iodide	BeI₂	283	322	341	361	382	394	411	435	461	487	488
Bismuth	Bi	1021	1099	1136	1177	1217	1240	1271	1319	1370	1420	271
tribromide	BiBr₃		261	282	305	327	340	360	392	425	461	218
trichloride	BiCl₃		242	264	287	311	324	343	372	405	441	230
Diborane hydrobromide	B₂H₅Br	−93.3	−75.3	−66.3	−56.4	−45.4	−38.2	−29.0	−15.4	0.0	+16.3	−104.2
Borine carbonyl	BH₃CO	−139.2	−127.3	−121.1	−114.1	−106.6	−101.9	−95.3	−85.5	−74.8	−64.0	−137.0
triamine	B₃N₃H₆	−63.0	−45.0	−35.3	−25.0	−13.2	−5.8	+4.0	18.5	34.3	50.6	−58.2
Boron hydrides dihydrodecaborane	B₁₀H₁₄	60.0	80.8	90.2	100.0	117.4	127.8	142.3	163.8			99.6
dihydrodiborane	B₂H₆	−159.7	−149.5	−144.3	−138.5	−131.6	−127.2	−120.9	−111.2	−99.6	−86.5	−169
dihydropentaborane	B₅H₉	−50.2	−40.4	−30.7	−20.0	−8.0	−0.4	+9.6	24.6	40.8	58.1	−47.0
tetrahydropentaborane	B₅H₁₁	−41.4	−29.9	−19.9	−9.2	+2.7	10.2	20.1	34.8	51.2	67.0	−119.9
tetrahydrotetraborane	B₄H₁₀	−90.9	−73.1	−64.3	−54.8	−44.3	−37.4	−28.1	−14.0	+0.8	16.1	
Boron tribromide	BBr₃		−20.4	−10.1	+1.5	14.0	22.1	33.5	50.3	70.0	91.7	−45
trichloride	BCl₃	−91.5	−75.2	−66.9	−57.9	−47.8	−41.2	−32.4	−18.9	−3.6	+12.7	−107
trifluoride	BF₃	−154.6	−145.4	−141.3	−136.4	−131.0	−127.6	−123.0	−115.9	−108.3	−100.7	−126.8
Bromine	Br₂	−48.7	−32.8	−25.0	−16.8	−8.0	−0.6	+9.3	24.3	41.0	58.2	−7.3
pentafluoride	BrF₅	−69.3	−51.0	−41.9	−32.0	−21.0	−14.0	−4.5	+9.9	25.7	40.4	−61.4
Cadmium	Cd	394	455	484	516	553	578	611	658	711	765	320.9
chloride	CdCl₂		618	656	695	736	762	797	847	908	967	568
fluoride	CdF₂	1112	1231	1286	1344	1400	1436	1486	1561	1651	1751	520
iodide	CdI₂	416	481	512	546	584	608	640	688	742	796	385
oxide	CdO	1000	1100	1149	1200	1257	1295	1341	1409	1484	1559	
Calcium	Ca		926	983	1046	1111	1152	1207	1288	1388	1487	851
Carbon (graphite)	C	3586	3828	3946	4069	4196	4273	4373	4516	4660	4827	
dioxide	CO₂	−134.3	−124.4	−119.5	−114.4	−108.6	−104.8	−100.2	−93.0	−85.7	−78.2	−57.5
disulfide	CS₂	−73.8	−54.3	−44.7	−34.3	−22.5	−15.3	−5.1	+10.4	28.0	46.5	−110.8
monoxide	CO	−222.0	−217.2	−215.0	−212.8	−210.0	−208.1	−205.7	−201.3	−196.3	−191.3	−205.0

Compound	Formula											
oxyselenide	COSe	−117.1	−102.3	−95.0	−86.3	−76.4	−70.2	−61.7	−49.8	−35.6	−21.9	−138.8
oxysulfide	COS	−132.4	−119.8	−113.3	−106.0	−98.3	−93.0	−85.9	−75.0	−62.7	−49.9	−75.2
selenosulfide	CSeS	−47.3	−26.5	−16.0	−4.4	+8.6	17.0	28.3	45.7	65.2	85.6	+0.4
subsulfide	C₃S₂	14.0	41.2	54.9	69.3	85.6	96.0	109.9	130.8	139.7	189.5	90.1
tetrabromide	CBr₄			96.3	106.3	119.7	130.8	139.7	163.5	189.5		
tetrachloride	CCl₄	−50.0	−30.0	−19.6	−8.2	+4.3	12.3	23.0	38.3	57.8	76.7	−22.6
tetrafluoride	CF₄	−184.6	−174.1	−169.3	−164.3	−158.8	−155.4	−150.7	−143.6	−135.5	−127.7	−183.7
Cesium	Cs	279	341	375	409	449	474	509	561	624	690	28.5
bromide	CsBr	748	838	887	938	993	1026	1072	1140	1221	1300	636
chloride	CsCl	744	837	884	934	989	1023	1069	1139	1217	1300	646
fluoride	CsF	712	798	844	893	947	980	1025	1092	1170	1251	683
iodide	CsI	738	828	873	923	976	1009	1055	1124	1200	1280	621
Chlorine	Cl₂	−118.0	−106.7	−101.6	−93.3	−84.5	−79.0	−71.7	−60.2	−47.3	−33.8	−100.7
fluoride	ClF	−106.7	−143.4	−139.0	−134.3	−128.8	−125.3	−120.8	−114.4	−107.0	−100.5	−145
trifluoride	ClF₃		−80.4	−71.8	−62.3	−51.3	−44.1	−34.7	−20.7	−4.9	+11.5	−83
monoxide	Cl₂O	−98.5	−81.6	−73.1	−64.3	−54.3	−48.0	−39.4	−26.5	−12.5	+2.2	−116
dioxide	ClO₂		−59.0	−51.2	−42.8	−37.2	−29.4	−17.8	−4.0	+11.1		−59
heptoxide	Cl₂O₇	−45.3	−23.8	−13.2	−2.1	+10.2	+18.3	29.1	44.6	62.2	78.8	−91
Chlorosulfonic acid	HSO₃Cl	32.0	53.5	64.0	75.3	87.6	95.2	105.3	120.0	136.1	151.0	−80
Chromium	Cr	1616	1768	1845	1928	2013	2067	2139	2243	2361	2482	1615
carbonyl	Cr(CO)₆	36.0	58.0	68.3	79.5	91.2	98.3	108.0	121.8	137.2	151.0	
oxychloride	CrO₂Cl₂	−18.4	+3.2	13.8	25.7	38.5	46.7	58.0	75.2	95.2	117.1	
Cobalt chloride	CoCl₂					770	801	843	904	974	1050	735
nitrosyl tricarbonyl	Co(CO)₃NO			−1.3	+11.0	18.5	29.0	44.4	62.0	80.0		−11
Columbium fluoride	CbF₃		86.3	103.0	121.5	133.2	148.5	172.2	198.0	225.0		75.5
Copper	Cu	1628	1795	1879	1970	2067	2127	2207	2325	2465	2595	1083
Cuprous bromide	Cu₂Br₂	572	666	718	777	844	887	951	1052	1189	1355	504
chloride	Cu₂Cl₂	546	645	702	766	838	886	960	1077	1249	1490	422
iodide	Cu₂I₂		610	656	716	786	836	907	1018	1158	1336	605
Cyanogen	C₂N₂	−95.8	−83.2	−76.8	−70.1	−62.7	−57.9	−51.8	−42.6	−33.0	−21.0	−34.4
bromide	CNBr	−35.7	−18.3	−10.0	−1.0	+8.6	14.7	22.6	33.8	46.0	61.5	58
chloride	CNCl	−76.7	−61.4	−53.8	−46.1	−37.5	−32.1	−24.9	−14.1	−2.3	+13.1	−6.5
fluoride	CNF	−134.4	−123.8	−118.5	−112.8	−106.4	−102.3	−97.0	−89.2	−80.5	−72.6	
Deuterium cyanide	DCN	−68.9	−54.0	−46.7	−38.8	−30.1	−24.7	−17.5	−5.4	+10.0	26.2	−12
Fluorine	F₂	−223.0	−216.9	−214.1	−211.0	−207.7	−205.6	−202.7	−198.3	−193.2	−187.9	−223
oxide	F₂O	−196.1	−186.6	−182.3	−177.8	−173.0	−170.0	−165.8	−159.0	−151.9	−144.6	−223.9

(Continued)

TABLE 1.47 Vapor Pressures of Inorganic Compounds up to 1 Atmosphere (*Continued*)

Compound name	Formula	Pressure, mm Hg										Melting point, °C
		1	5	10	20	40	60	100	200	400	760	
		Temperature, °C										
Germanium bromide	$GeBr_4$		43.3	56.8	71.8	88.1	98.8	113.2	135.4	161.6	189.0	26.1
chloride	$GeCl_4$	−45.0	−24.9	−15.0	−4.1	+8.0	16.2	27.5	44.4	63.8	84.0	−49.5
hydride	GeH_4	−163.0	−151.0	−145.3	−139.2	−131.6	−126.7	−120.3	−111.2	−100.2	−88.9	−165
Trichlorogermane	$GeHCl_3$	−41.3	−22.3	−13.0	−3.0	+8.8	16.2	26.5	41.6	58.3	75.0	−71.1
Tetramethylgermane	$Ge(CH_3)_4$	−73.2	−54.6	−45.2	−35.0	−23.4	−16.2	−6.3	+8.8	26.0	44.0	−88
Digermane	Ge_2H_6	−88.7	−69.8	−60.1	−49.9	−38.2	−30.7	−20.3	−4.7	+3.3	31.5	−109
Trigermane	Ge_3H_6	−36.9	−12.8	−0.9	+11.8	26.3	35.5	47.9	67.0	88.6	110.8	−105.6
Gold	Au	1869	2059	2154	2256	2363	2431	2521	2657	2807	2966	1063
Helium	He	−271.7	−271.5	−271.3	−271.1	−270.7	−270.6	−270.3	−269.8	−269.3	−268.6	
para-Hydrogen	H_2	−263.3	−261.9	−261.3	−260.4	−259.6	−258.9	−257.9	−256.3	−254.5	−252.5	−259.1
Hydrogen bromide	HBr	−138.8	−127.4	−121.8	−115.4	−108.3	−103.8	−97.7	−88.1	−78.0	−66.5	−87.0
chloride	HCl	−150.8	−140.7	−135.6	−130.0	−123.8	−119.6	−114.0	−105.2	−95.3	−84.8	−114.3
cyanide	HCN	−71.0	−55.3	−47.7	−39.7	−30.9	−25.1	−17.8	−5.3	+10.2	25.9	−13.2
fluoride	H_2F_2		−74.7	−65.8	−56.0	−45.0	−37.9	−28.2	−13.2	+2.5	19.7	−83.7
iodide	HI	−123.3	−109.6	−102.3	−94.5	−85.6	−79.8	−72.1	−60.3	−48.3	−35.1	−50.9
oxide(water)	H_2O	−17.3	+1.2	11.2	22.1	34.0	41.5	51.6	66.5	83.0	100.0	0.0
sulfide	H_2S	−134.3	−122.4	−116.3	−109.7	−102.3	−97.9	−91.6	−82.3	−71.8	−60.4	−85.5
disulfide	$HSSH$	−43.2	−24.4	−15.2	−5.1	+6.0	12.8	22.0	35.3	49.6	64.0	−89.7
selenide	H_2Se	−115.3	−103.4	−97.9	−91.8	−84.7	−80.2	−74.2	−65.2	−53.6	−41.1	−64
telluride	H_2Te	−96.4	−82.4	−75.4	−67.8	−59.1	−53.7	−45.7	−32.4	−17.2	−2.0	−49.0
Iodine	I_2	38.7	62.2	73.2	84.7	97.5	105.4	116.5	137.3	159.8	183.0	112.9
heptafluoride	IF	−87.0	−70.7	−63.0	−54.5	−45.3	−39.4	−31.9	−20.7	−8.3	+4.0	5.5
Iron	Fe	1787	1957	2039	2128	2224	2283	2360	2475	2605	2735	1535
pentacarbonyl	$Fe(CO)_5$		−6.5	+4.6	16.7	30.3	39.1	50.3	68.0	86.1	105.0	−21
Ferric chloride	Fe_2Cl_6	194.0	221.8	235.5	246.0	256.8	263.7	272.5	285.0	298.0	319.0	304
Ferrous chloride	$FeCl_2$			700	737	779	805	842	897	961	1026	
Krypton	Kr	−199.3	−191.3	−187.2	−182.9	−178.4	−175.7	−171.8	−165.9	−159.0	−152.0	−156.7
Lead	Pb	973	1099	1162	1234	1309	1358	1421	1519	1630	1744	327.5
bromide	$PbBr_2$	513	578	610	646	686	711	745	796	856	914	373
chloride	$PbCl_2$	547	615	648	684	725	750	784	833	893	954	501
fluoride	PbF_2		861	904	950	1003	1036	1080	1144	1219	1293	855

iodide	PbI₂	479	540	571	605	644	668	701	750	807	872	402
oxide	PbO	943	1039	1085	1134	1189	1222	1265	1330	1402	1472	890
sulfide	PbS	852	928	975	1005	1048	1074	1108	1160	1221	1281	1114
Lithium	Li	723	838	881	940	1003	1042	1097	1178	1273	1372	186
bromide	LiBr	748	840	888	939	994	1028	1076	1147	1126	1310	547
chloride	LiCl	783	880	932	987	1045	1081	1129	1203	1290	1382	614
fluoride	LiF	1047	1156	1211	1270	1333	1372	1425	1503	1591	1681	870
iodide	LiI	723	802	841	883	927	955	993	1049	1110	1171	446
Magnesium	Mg	621	702	743	789	838	868	909	967	1034	1107	651
chloride	MgCl₂	778	877	930	968	1050	1088	1142	1223	1316	1418	712
Manganese	Mn	1292	1434	1505	1583	1666	1720	1792	1900	2029	2151	1260
chloride	MnCl₂		736	778	825	879	913	960	1028	1108	1190	650
Mercury	Hg	126.2	164.8	184.0	204.6	228.8	242.0	261.7	290.7	323.0	357.0	−38.9
Mercuric bromide	HgBr₂	136.5	165.3	179.8	194.3	211.5	221.0	237.8	262.7	290.0	319.0	237
chloride	HgCl₂	135.2	166.0	180.2	195.8	212.5	222.2	237.0	256.5	275.5	304.0	277
iodide	HgI₂	157.5	189.2	204.5	220.0	238.2	249.0	261.8	291.0	324.2	354.0	259
Molybdenum	Mo	3102	3393	3535	3690	3859	3964	4109	4322	4553	4804	2622
hexafluoride	MoF₆	−65.5	−49.0	−40.8	−32.0	−22.1	−16.2	−8.0	+4.1	17.2	36.0	17
oxide	MoO₃	734	785	814	851	892	917	955	1014	1082	1151	795
Neon	Ne	−257.3	−255.5	−254.6	−253.7	−252.6	−251.9	−251.0	−249.7	−248.1	−246.0	−248.7
Nickel	Ni	1810	1979	2057	2143	2234	2289	2364	2473	2603	2732	1452
carbonyl	Ni(CO)₄					−23.0	−15.9	−6.0	+8.8	25.8	42.5	−25
chloride	NiCl₂	671	731	759	789	821	840	866	904	945	987	1001
Nitrogen	N₂	−226.1	−221.3	−219.1	−216.8	−214.0	−212.3	−209.7	−205.6	−200.9	−195.8	−210.0
Nitric oxide	NO	−184.5	−180.6	−178.2	−175.3	−171.7	−168.9	−166.0	−162.3	−156.8	−151.7	−161
Nitrogen dioxide	NO₂	−55.6	−42.7	−36.7	−30.4	−23.9	−19.9	−14.7	−5.0	+8.0	21.0	−9.3
Nitrogen pentoxide	N₂O₅	−36.8	−23.0	−16.7	−10.0	−2.9	+1.8	7.4	15.6	24.4	32.4	30
Nitrous oxide	N₂O	−143.4	−133.4	−128.7	−124.0	−118.3	−114.9	−110.3	−103.6	−96.2	−85.5	−90.9
Nitrosyl chloride	NOCl					−60.2	−54.2	−46.3	−34.0	−20.3	−6.4	−64.5
Nitrosyl fluoride	NOF	−132.0	−120.3	−114.3	−107.8	−100.3	−95.7	−88.8	−79.2	−68.2	−56.0	−134
Osmium tetroxide (yellow)	OsO₄	3.2	22.0	31.3	41.0	51.7	59.4	71.5	89.5	109.3	130.0	56
(white)	OsO₄	−5.6	+15.6	26.0	37.4	50.5	59.4	71.5	89.5	109.3	130.0	42
Oxygen	O₂	−219.1	−213.4	−210.6	−207.5	−204.1	−201.9	−198.8	−194.0	−188.8	−183.1	−218.7
Ozone	O₃	−180.4	−168.6	−163.2	−157.2	−150.7	−146.7	−141.0	−132.6	−122.5	−111.1	−251
Phosgene	COCl₂	−92.9	−77.0	−69.3	−60.3	−50.3	−44.0	−35.6	−22.3	−7.6	+8.3	−104
Phosphorus (yellow)	P	76.6	111.2	128.0	146.2	166.7	179.8	197.3	222.7	251.0	280.0	44.1
(violet)	P	237	271	287	306	323	334	349	370	391	417	590
tribromide	PBr₃	7.8	34.4	47.8	62.4	79.0	89.8	103.6	125.2	149.7	175.3	−40

(Continued)

TABLE 1.47 Vapor Pressures of Inorganic Compounds up to 1 Atmosphere (*Continued*)

Compound name	Formula	Pressure, mm Hg										Melting point, °C
		Temperature, °C										
		1	5	10	20	40	60	100	200	400	760	
trichloride	PCl_3	−51.6	−31.5	−21.3	−10.2	+2.3	10.2	21.0	37.6	56.9	74.2	−111.8
pentachloride	PCl_5	55.5	74.0	83.2	92.5	102.5	108.3	117.0	131.3	147.2	162.0	
Phosphine	PH_3					−129.4	−125.0	−118.8	−109.4	−98.3	−87.5	−132.5
Phosphonium bromide	PH_4Br	−43.7	−28.5	−21.2	−13.3	−5.0	+0.3	7.4	17.6	28.0	38.3	
chloride	PH_4Cl	−91.0	−79.6	−74.0	−68.0	−61.5	−57.3	−52.0	−44.0	−35.4	−27.0	−28.5
iodide	PH_4I	−25.2	−9.0	−1.1	+7.3	16.1	21.9	29.3	39.9	51.6	62.3	
Phosphorus trioxide	P_4O_6		39.7	53.0	67.8	84.0	94.2	108.3	129.0	150.3	173.1	22.5
pentoxide	P_4O_{10}	384	424	442	462	481	493	510	532	556	591	569
oxychloride	$POCl_3$			2.0	13.6	27.3	35.8	47.4	65.0	84.3	105.1	2
thiobromide	$PSBr_3$	50.0	72.4	83.6	95.5	108.0	116.0	126.3	141.8	157.8	175.0	38
thiochloride	$PSCl_3$	−18.3	+4.6	16.1	29.0	42.7	51.8	63.8	82.0	102.3	124.0	−36.2
Platinum	Pt	2730	3007	3146	3302	3469	3574	3714	3923	4169	4407	1755
Potassium	K	341	408	443	483	524	550	586	643	708	774	62.3
bromide	KBr	795	892	940	994	1050	1087	1137	1212	1297	1383	730
chloride	KCl	821	919	968	1020	1078	1115	1164	1239	1322	1407	790
fluoride	KF	885	988	1039	1096	1156	1193	1245	1323	1411	1502	880
hydroxide	KOH	719	814	863	918	976	1013	1064	1142	1233	1327	380
iodide	KI	745	840	887	938	995	1030	1080	1152	1238	1324	723
Radon	Rn	−144.2	−132.4	−126.3	−119.2	−111.3	−106.2	−99.0	−87.7	−75.0	−61.8	−71
Rhenium heptoxide	Re_2O_7	212.5	237.5	248.0	261.0	272.0	280.0	289.0	307.0	336.0	362.4	296
Rubidium	Rb	297	358	389	422	459	482	514	563	620	679	38.5
bromide	RbBr	781	876	923	975	1031	1066	1114	1186	1267	1352	682
chloride	RbCl	792	887	937	990	1047	1084	1133	1207	1294	1381	715
fluoride	RbF	921	982	1016	1052	1096	1123	1168	1239	1322	1408	760
iodide	RbI	748	839	884	935	991	1026	1072	1141	1223	1304	642
Selenium	Se	356	413	442	473	506	527	554	594	637	680	217
dioxide	SeO_2	157.0	187.7	202.5	217.5	234.1	244.6	258.0	277.0	297.7	317.0	340
hexafluoride	SeF_6	−118.6	−105.2	−98.9	−92.3	−84.7	−80.0	−73.9	−64.8	−55.2	−45.8	−34.7
oxychloride	$SeOCl_2$	34.8	59.8	71.9	84.2	98.0	106.5	118.0	134.6	151.7	168.0	8.5
tetrachloride	$SeCl_4$	74.0	96.3	107.4	118.1	130.1	137.8	147.5	161.0	176.4	191.5	

Silicon	Si	1724	1835	1888	1942	2000	2036	2083	2151	2220	2287	1420
dioxide	SiO_2			1732	1798	1867	1911	1969	2053	2141	2227	1710
tetrachloride	$SiCl_4$	-63.4	-44.1	-34.4	-24.0	-12.1	-4.8	+5.4	21.0	38.4	56.8	-68.8
tetrafluoride	SiF_4	-144.0	-134.8	-130.4	-125.9	-120.8	-117.5	-113.3	-170.2	-100.7	-94.8	-90
Trichlorofluorosilane	$SiFCl_3$	-92.6	-76.4	-68.3	-59.0	-48.8	-42.2	-33.2	-19.3	-4.0	+12.2	-120.8
Iodosilane	SiH_3I		-53.0	-47.7	-33.4	-21.8	-14.3	-4.4	+10.7	27.9	45.4	-57.0
Diiodosilane	SiH_2I_2		3.8	18.0	34.1	52.6	64.0	79.4	101.8	125.5	149.5	-1.0
Disiloxan	$(SiH_3)_2O$	-112.5	-95.8	-88.2	-79.8	-70.4	-64.2	-55.9	-43.5	-29.3	-15.4	-144.2
Trisilane	Si_3H_8	-58.9	-49.7	-40.0	-29.0	-16.9	-9.0	+1.6	17.8	35.5	53.1	-117.2
Trisilazane	$(SiH_3)_3N$	-68.7	-49.9	-40.4	-30.0	-18.5	-11.0	-1.1	+14.0	31.0	48.7	-105.7
Tetrasilane	Si_4H_{10}	-27.7	-6.2	+4.3	15.8	28.4	36.6	47.4	63.6	81.7	100.0	-93.6
Octachlorotrisilane	Si_3Cl_8	46.3	74.7	89.3	104.2	121.5	132.0	146.0	166.2	189.5	211.4	
Hexachlorodisiloxane	$(SiCl_3)_2O$	-5.0	17.8	29.4	41.5	55.2	63.8	75.4	92.5	113.6	135.6	-33.2
Hexachlorodisilane	Si_2Cl_6	+4.0	27.4	38.8	51.5	65.3	73.9	85.4	102.2	120.6	139.0	-1.2
Tribromosilane	$SiHBr_3$	-30.5	-8.0	+3.4	16.0	30.0	39.2	51.6	70.2	90.2	111.8	-73.5
Trichlorosilane	$SiHCl_3$	-80.7	-62.6	-53.4	-43.8	-32.9	-25.8	-16.4	-1.8	+14.5	31.8	-126.6
Trifluorosilane	$SiHF_3$	-152.0	-142.7	-138.2	-132.9	-127.3	-123.7	-118.7	-111.3	-102.8	-95.0	-131.4
Dibromosilane	SiH_2Br_2	-60.9	-40.0	-29.4	-18.0	-5.2	+3.2	14.1	31.6	50.7	70.5	-70.2
Difluorosilane	SiH_2F_2	-146.7	-136.0	-130.4	-124.3	-117.6	-113.3	-107.3	-98.3	-87.6	-77.8	
Monobromosilane	SiH_3Br	-117.8	-85.7	-77.3	-68.3	-57.8	-51.1	-42.3	-28.6	-13.3	+2.4	-93.9
Monochlorosilane	SiH_3Cl	-153.0	-104.3	-97.7	-90.1	-81.8	-76.0	-68.5	-57.0	-44.5	-30.4	
Monofluorosilane	SiH_3F	-46.1	-145.5	-141.2	-136.3	-130.8	-127.2	-122.4	-115.2	-106.8	-98.0	
Tribromofluorosilane	$SiFBr_3$	-124.7	-25.4	-15.1	-3.7	+9.2	17.4	28.6	45.7	64.6	83.8	-82.5
Dichlorodifluorosilane	SiF_2Cl_2		-110.5	-102.9	-94.5	-85.0	-78.6	-70.3	-58.0	-45.0	-31.8	-139.7
Trifluorobromosilane	SiF_3Br	-144.0	-133.0	-127.0	-120.5	-112.8	-108.2	-101.7	-91.7	-81.0	-41.7	-70.5
Trifluorochlorosilane	SiF_3Cl			-63.1	-57.0	-50.6	-46.7	-41.7	-34.2	-26.4	-70.0	-142
Hexafluorodisilane	Si_2F_6	-81.0	-68.8	-59.0	-48.8	-37.0	-29.0	-19.5	-3.2	+15.4	-18.9	-18.6
Dichlorofluorobromosilane	$SiFCl_2Br$	-86.5	-68.4	-35.6	-24.5	-12.0	-4.7	+6.3	23.0	43.0	35.4	-112.3
Dibromochlorofluorosilane	$SiFClBr_2$	-65.2	-45.5								59.5	-99.3
Silane	SiH_4	-179.3	-168.6	-163.0	-156.9	-150.3	-146.3	-140.5	-131.6	-122.0	-111.5	-185
Disilane	Si_2H_6	-114.8	-99.3	-91.4	-82.7	-72.8	-66.4	-57.5	-44.6	-29.0	-14.3	-132.6
Silver	Ag	1357	1500	1575	1658	1743	1795	1865	1971	2090	2212	960.5
chloride	$AgCl$	912	1019	1074	1134	1200	1242	1297	1379	1467	1564	455
iodide	AgI	820	927	983	1045	1111	1152	1210	1297	1400	1506	552
Sodium	Na	439	511	549	589	633	662	701	758	823	892	97.5
bromide	$NaBr$	806	903	952	1005	1063	1099	1148	1220	1304	1392	755
chloride	$NaCl$	865	967	1017	1072	1131	1169	1220	1296	1379	1465	800

(Continued)

TABLE 1.47 Vapor Pressures of Inorganic Compounds up to 1 Atmosphere (*Continued*)

Pressure, mm Hg — Temperature, °C

Compound name	Formula	1	5	10	20	40	60	100	200	400	760	Melting point, °C
cyanide	$NaCN$	817	928	983	1046	1115	1156	1214	1302	1401	1497	564
fluoride	NaF	1077	1186	1240	1300	1363	1403	1455	1531	1617	1704	992
hydroxide	$NaOH$	739	843	897	953	1017	1057	1111	1192	1286	1378	318
iodide	NaI	767	857	903	952	1005	1039	1083	1150	1225	1304	651
Strontium	Sr		847	898	953	1018	1057	1111	1192	1285	1384	800
Strontium oxide	SrO	2068	2198	2262	2333	2410						2430
Sulfur	S	183.8	223.0	243.8	264.7	288.3	305.5	327.2	359.7	399.7	444.6	112.8
monochloride	S_2Cl_2	−7.4	+15.7	27.5	40.0	54.1	63.2	75.3	93.5	115.4	138.0	−80
hexafluoride	SF_6	−132.7	−120.6	−114.7	−108.4	−101.5	−96.8	−90.9	−82.3	−72.6	−63.5	−50.2
Sulfuryl chloride	SO_2Cl_2		−35.1	−24.8	−13.4	−1.0	+7.2	17.8	33.7	51.3	69.2	−54.1
Sulfur dioxide	SO_2	−95.5	−83.0	−76.8	−69.7	−60.5	−54.6	−46.9	−35.4	−23.0	−10.0	−73.2
trioxide (α)	SO_3	−39.0	−23.7	−16.5	−9.1	−1.0	+4.0	10.5	20.5	32.6	44.8	16.8
trioxide (β)	SO_3	−34.0	−19.2	−12.3	−4.9	+3.2	8.0	14.3	23.7	32.6	44.8	32.3
trioxide (γ)	SO_3	−15.3	−2.0	+4.3	11.1	17.9	21.4	28.0	35.8	44.0	51.6	62.1
Tellurium	Te	520	605	650	697	753	789	838	910	997	1087	452
chloride	$TeCl_4$			233	253	273	287	304	330	360	392	224
fluoride	TeF_6	−111.3	−98.8	−92.4	−83.0	−78.4	−73.8	−67.9	−57.3	−48.2	−38.6	−37.8
Thallium	Tl	825	931	983	1040	1103	1143	1196	1274	1364	1457	303.5
Thallous bromide	$TlBr$		490	522	559	598	621	653	703	759	819	460
chloride	$TlCl$		487	517	550	589	612	645	694	748	807	430
iodide	TlI	440	502	531	567	607	631	663	712	763	823	440
Thionyl bromide	$SOBr_2$	−6.7	+18.4	31.0	44.1	58.8	68.3	80.6	99.0	119.2	139.5	−52.2
Thionyl chloride	$SOCl_2$	−52.9	−32.4	−21.9	−10.5	+2.2	10.4	21.4	37.9	56.5	75.4	−104.5
Tin	Sn	1492	1634	1703	1777	1855	1903	1968	2063	2169	2270	231.9
Stannic bromide	$SnBr_4$		58.3	72.7	88.1	105.5	116.2	131.0	152.8	177.7	204.7	31.0
Stannous chloride	$SnCl_2$	316	366	391	420	450	467	493	533	577	623	246.8
Stannic chloride	$SnCl_4$	−22.7	−1.0	+10.0	22.0	35.2	43.5	54.7	72.0	92.1	113.0	−30.2
Stannic iodide	SnI_4		156.0	175.8	196.2	218.8	234.2	254.2	283.5	315.5	348.0	144.5
hydride	SnH_4	−140.0	−125.8	−118.5	−111.2	−102.3	−96.6	−89.2	−78.0	−65.2	−52.3	−149.9

Tin tetramethyl	Sn(CH$_3$)$_4$	−51.3	−31.0	−20.6	−9.3	+3.5	11.7	22.8	39.8	58.5	78.0	
trimethyl-ethyl	Sn(CH$_3$)$_3$ · C$_2$H$_5$	−30.0	−7.6	+3.8	16.1	30.0	38.4	50.0	67.3	87.6	108.8	
trimethyl-propyl	Sn(CH$_3$)$_3$ · C$_3$H$_7$	−12.0	+10.7	21.8	34.0	48.5	57.5	69.8	88.0	109.6	131.7	
Titanium chloride	TiCl$_4$	−13.9	+9.4	21.3	34.2	48.4	58.0	71.0	90.5	112.7	136.0	−30
Tungsten	W	3990	4337	4507	4690	4886	5007	5168	5403	5666	5927	3370
Tungsten hexafluoride	WF$_6$	−71.4	−56.5	−49.2	−41.5	−33.0	−27.5	−20.3	−10.0	+1.2	17.3	−0.5
Uranium hexafluoride	UF$_6$	−38.8	−22.0	−13.8	−5.2	+4.4	10.4	18.2	30.0	42.7	55.7	69.2
Vanadyl trichloride	VOCl$_3$	−23.2	+0.2	12.2	26.6	40.0	49.8	62.5	82.0	103.5	127.2	
Xenon	Xe	−168.5	−158.2	−152.8	−147.1	−141.2	−137.7	−132.8	−125.4	−117.1	−108.0	−111.6
Zinc	Zn	487	558	593	632	673	700	736	788	844	907	419.4
chloride	ZnCl$_2$	423	481	508	536	566	584	610	648	689	732	365
fluoride	ZnF$_2$	970	1055	1086	1129	1175	1207	1254	1329	1417	1497	872
diethyl	Zn(C$_2$H$_5$)$_2$	−22.4	0.0	+11.7	24.2	38.0	47.2	59.1	77.0	97.3	118.0	−28
Zirconium bromide	ZrBr$_4$	207	237	250	266	281	289	301	318	337	357	450
chloride	ZrCl$_4$	190	217	230	243	259	268	279	295	312	331	437
iodide	ZrI$_4$	264	297	311	329	344	355	369	389	409	431	499

TABLE 1.48   Vapor Pressures of Various Inorganic Compounds

Substance	State	Eq.	Range,°C	A	B	C
Aluminum						
AlCl$_3$		2	70–190	16.24	6 006	
Al$_2$O$_3$		2	1840–2000	14.22	28 200	
Ammonium						
NH$_3$	c*	1		9.963 82	1 617.907	272.55
	liq	1		7.360 50	926.132	240.17
NH$_4$Br	subl c	1		9.220 0	3 947	227.0
NH$_4$Cl	subl c	1		9.355 7	3 703.7	232.0
NH$_4$I	subl c	1		9.147 0	3 858	226.0
NH$_4$N$_3$	c	1		10.433 4	2 821.0	240.0
Antimony						
Sb	c	2	1070–1325	9.051	9 871	
SbBr$_3$		2	235–324	8.005	2 873	
SbCl$_3$		2	170–253	8.090	2 582.3	
SbI$_3$		2	330–445	7.831	3 350.55	
Sb$_2$Se$_3$	subl c	2		8.790 6	6 432.3	
Argon						
Ar	c	1		7.505 81	399.085	272.63
	liq	1		6.616 51	304.227	267.32
Arsenic						
As		2	440–815	10.800	6 947	
		2	800–860	6.692	2 460	
AsCl$_3$		2	50–100	7.953	2 042.7	
As$_2$O$_3$		2	100–310	12.127	5 815.81	
		2	315–490	6.513	2 722.2	
Barium						
Ba		2	930–1130	15.765	18 280	
BaH$_2$ [97% pure]		2	500–1000	6.86	4 000	
Bismuth						
Bi		2	1210–1420	8.876	10 446	
BiCl$_3$		2	91–213	2.681	685.519	
Boron						
BBr$_3$		2	−40 to 90	7.655	1 740.3	
BCl$_3$		1		6.188 11	756.89	214.0
B(CH$_3$)$_3$		2	−118 to −20	7.459 5	1 157.99	
B$_2$H$_6$	liq	1		6.366 38	521.490	241.98
B$_5$H$_{11}$	liq	2	−43 to 8.4	7.901	1 690.3	
Bromine						
Br$_2$	c	1		9.7209	2 041.3	260.1
	liq	1		6.877 80	1 119.68	221.38
BrF$_3$	liq	1		7.729 74	1 673.95	219.48
BrF$_5$	liq	1		7.273 68	1 219.28	236.40
BrO$_2$F	liq	1		7.436 51	1 195.8	260.1
Cadmium						
Cd		2	150–321	8.564	5 693	
		2	500–840	7.897	5 218	
CdI$_2$		2	385–450	9.269	6 383	
Calcium						
Ca		2	500–700	9.697	10 185	
		2	960–1100	16.240	19 325	

*Crystalline solid.

TABLE 1.48   Vapor Pressures of Various Inorganic Compounds (*Continued*)

Substance	State	Eq.	Range,°C	A	B	C
Carbon						
C [as C(g)]	liq	1		11.042 8	37 736	302.2
[as C$_2$(g)]	liq	1		12.583 2	43 281	318.3
[all species]	liq	1		9.381 3	27 240	264.0
Carbon						
CNBr	subl c	1		9.488 9	2 041.8	251.70
CNF		1	−76 to −47	6.778 9	697.61	224.95
CO	c I	1		7.414 8	342.50	269.0
	liq	1		6.694 22	291.743	267.99
CO$_2$	c	1		9.810 66	1 347.786	273.00
C$_3$O$_2$	liq	1	−71 to 7	7.188 99	1 100.94	249.15
COCl$_2$	liq	1		6.971 33	998.770	236.68
COF$_2$		1	−109 to −84	6.885 5	576.70	228.58
COS		1	−111 to −49	6.907 23	804.48	250.0
CS$_2$		1	3−80	6.942 79	1 169.11	241.59
CSe$_2$		1	0−50	6.776 73	1 353.20	219.95
CSeS		1	−16 to 84	6.699 6	1 161.97	219.59
Cesium						
Cs		2	200−350	6.949	3 833.7	
CsBr		2	978−1305	7.990	8 022.53	
CsCl		2	986−1295	8.340	8 523.94	
CsF		2	1033−1255	7.703	7 359.21	
CsH		2	245−378	11.79	5 900	
		2	340−440	9.25	4 410	
CsI		2	1052−1280	9.124	9 699.11	
Chlorine						
Cl$_2$	c	1		9.705 12	1 444.19	267.13
	liq	1		6.937 90	861.34	246.33
ClF	liq	1		6.989	682.1	256
ClF$_3$	liq	1		7.366 85	1 096.28	232.63
ClF$_5$		1		6.269 33	653.06	206.6
ClO$_2$	liq	1		6.036 11	590.09	176.15
Cl$_2$O	liq	1		7.132 68	1 021.56	238.16
ClOClO$_3$	liq	1		7.538 67	1 404.18	257.00
Cl$_2$O$_7$	liq	1		6.869 29	1 214.00	220.79
ClO$_2$F	liq	1		6.677 15	809.78	218.96
ClO$_3$F	liq	1		6.895 19	791.73	243.88
Copper						
CuBr		2	997−1351	5.460	4 173.2	
CuCl		2	878−1369	5.454	4 215.0	
CuI		2	991−1154	5.570	4 215.0	
Fluorine						
F$_2$	liq	1		6.765 88	304.35	266.54
FNO$_3$	liq	1		6.658 6	769.5	248.0
Germanium						
GeCl$_4$		2	10.4−86	7.340	2 010.9	
Helium						
^{3}He	liq	1	−271.13 to −270.86	4.272 7	5.594	273.840
	liq	1	−271.13 to −269.92	5.100 0	11.062	274.950
^{4}He		1	−271.4 to −270.1	4.558 7	8.1548	273.710
		1	−271.4 to −268.9	5.320 75	14.6515	274.950
		1	−271.4 to −268.1	6.004 60	24.0668	276.650

(*Continued*)

TABLE 1.48 Vapor Pressures of Various Inorganic Compounds (*Continued*)

Substance	State	Eq.	Range,°C	A	B	C
Hydrogen						
$^1H_2$ normal, 25% para	c	1		6.043 86	66.507	274.630
	liq	1		5.824 38	67.5078	275.700
equilibrium	c	1		6.042 07	65.961	274.60
	liq	1		5.814 64	66.7945	275.650
$^1H^2H$ (DH)	c	1		6.960 08	99.968	276.590
	liq	1		6.016 12	77.1349	275.620
$^2H_2$ (D_2) normal, 66.7% ortho	c	1		7.726 05	135.461	278.550
	liq	1		6.128 25	83.5251	275.216
$^2H_2$ equilibrium, 97.8% ortho	c	1		7.751 10	135.58	278.50
	liq	1		6.044 68	79.5888	274.680
$^3H_2$ (T_2) normal, 25% para	c	1		6.184 03	76.7445	271.850
	liq	1		6.089 21	81.8971	273.650
^1HBr	c	1		7.667 61	878.57	253.2
	liq	1		6.287 53	540.82	225.44
^2HBr (DBr)	c	1		7.500 93	820.68	247.3
	liq	1		6.162 38	505.68	220.6
^1HCl	c	1		8.134 73	941.57	268.06
	liq	1		7.170 00	745.80	258.88
^2HCl (DCl)	c	1		7.850 47	843.32	258.32
	liq	1		6.935 96	668.20	249.50
HCN	liq	1	−16 to 46	7.528 2	1329.5	260.4
^1HF	liq	1		7.680 98	1475.60	287.88
^2HF (DF)	liq	1		7.217 04	1268.37	273.87
^1HI	c	1		7.315 6	894.32	239.6
	liq	1		5.608 9	416.04	188.1
^2HI (DI)	c	1		7.314 9	889.52	238.8
	liq	1		5.601 8	413.98	187.8
HN_3	liq	1		6.857	1 066	232
HNO_3	liq	1		7.511 9	1 406	221.0
^1H_2O			[See Tables 5.4 and 5.6]			
^2H_2O (D_2O)			[See Table 5.7]			
$H_2^{18}O$		1	0–60	8.133 2	1 762.39	235.660
		1	60–120	7.972 08	1 668.84	227.700
H_2O_2	liq	1		7.969 17	1 886.76	220.6
HPO_2F	liq	1		6.735 3	1 342.9	232.0
H_2S	c	1		7.614 18	885.319	250.25
	liq	1		6.993 92	768.130	249.09
H_2S_2	liq	1		6.974	1 232	225
H_2S_3	liq	1		6.807	1 488	209
H_2S_4	liq	1		6.945	1 772	196
H_2S_5	liq	1		7.320	2 104	189
HSO_3Cl	liq	1		7.049	1 480	201
HSO_3F	liq	1		7.399 5	1 521	174.0
H_2Se	c	1		7.635 4	927.6	240.0
	liq	1		6.966 0	787.67	235.0
H_2Te	liq	1		7.000	935	229
Iodine						
I_2	c	1		9.810 9	2 901.0	256.00
	liq	1		7.018 1	1 610.9	205.0
ICl	liq	1		7.702 1	1 517.9	217.0
IF_5	c	1		10.964	2 538	245
	liq	1		7.464 8	1 460	216.0
IF_7	c	1		7.998	1 340	256

TABLE 1.48   Vapor Pressures of Various Inorganic Compounds (*Continued*)

Substance	State	Eq.	Range,°C	A	B	C
Iridium						
IrF$_6$	c	2	0.4–44	8.618	1 868	
	liq	2	44–54	7.952	1 657	
Iron						
FeCl$_2$	liq	2	708–834	9.794	7 455	
	liq	2	700–930	8.33	7 061	
FeCl$_3$	c	2	160–304	15.11	7 142	
FeI$_2$		2	517–577	13.183	10 778	
		2	601–686	9.674	7 716	
Krypton						
Kr	c	1		7.539 55	539.48	269.8
	liq	1		6.630 70	416.38	264.45
Lead						
Pb		2	525–1325	7.827	9 845.4	
PbBr$_2$		2	735–918	8.064	6 163.1	
PbCl$_2$		2	500–950	8.961	7 411.4	
PbF$_2$		2	1078–1289	8.391	8 623.2	
Lithium						
LiBr		2	1010–1265	8.068	7 975.5	
LiCl		2	1045–1325	7.939	8 142.7	
LiF		2	1398–1666	8.753	11 407	
LiH		2	500–650	11.227	9 600	
		2	700–800	9.926	8 204	
LiI		2	940–1140	8.011	7 500	
Magnesium						
Mg		2	900–1070	12.993	13 579.8	
MgH$_2$		2	337–415	9.78	3 857	
Mercury						
Hg				[See Table 5.3]		
HgBr$_2$		2	130–270	10.094	4 168.0	
HgCl$_2$		2	130–270	10.094	4 118.34	
		2	275–309	8.409	3 187.1	
Hg$_2$Cl$_2$		1		8.521 51	3 110.96	168.0
HgI$_2$		2	266–360	8.115	3 278.5	
Neon						
Ne	c	1		7.065 16	110.61	272.00
	liq	1		6.084 44	78.380	270.550
Neptunium						
NpF$_6$	liq	3	55.1–76.8	0.010 23	1 191.1	−2.582 5
Nickel						
Ni(CO)$_4$		2	2–40	7.780	1 556.5	
Niobium						
NbBr$_5$	liq	2		8.92	3 850	
NbCl$_5$	liq	2	210–254	8.37	2 827	
NbF$_5$	liq	2		8.439	2 824	
Nitrogen						
N$_2$ natural	c	1		7.345 12	322.222	269.980
	liq	1		6.494 57	255.680	266.550
^{15}N$_2$	c	1		7.363 96	323.17	269.88
	liq	1		6.494 14	255.535	266.451
NCl$_3$		1		6.956	1 190	221
NF$_3$	liq	1		6.779 66	501.913	257.79
NH$_3$				[See Table 1.49]		

(*Continued*)

TABLE 1.48   Vapor Pressures of Various Inorganic Compounds (*Continued*)

Substance	State	Eq.	Range, °C	A	B	C
Nitrogen (*cont.*)						
N_2H_4	liq	1		7.801 9	1 679.07	227.7
NO natural	c	1		9.628 26	758.736	266.00
	liq	1		8.743 00	682.938	268.27
N_2O	c	1		9.437 00	1 174.020	268.22
	liq	1		7.003 94	654.260	247.16
N_2O_4 equilibrium	c	1		10.736 31	2 075.53	252.80
mixture	liq	1		8.917 12	1 798.54	276.80
N_2O_5	c	1		11.644 5	2 510	253.0
NOCl	c	1		8.540 8	1 397.3	261.0
	liq	1		7.361 54	1 094.73	249.70
N_2O_3		2	−25 to 0	10.30	2 057.9	
NOF	liq	1		6.443 5	556.13	216.0
NO_2Cl	liq	1		5.372 3	395.40	174.0
NO_2F	liq	1		6.833 4	654.55	238.0
Osmium						
OsF_5		2	75−180	9.75	3 429	
OsF_6		2	34−48	7.470	1 473	
OsF_8		2	38−47	7.650	1 525	
OsO_4		2	−38 to 40	10.710 0	2 951.00	
OsO_3F_2		2	59−105	7.994	1 911	
Oxygen						
O_2	liq	1		6.691 44	319.013	266.697
O_3	liq	1		6.837	552.5	251.0
OF_2	liq	1		7.236 19	545.05	269.91
O_2F_2	liq	1		6.779 02	756.39	250.16
O_3F_2		2	79−114	6.134 3	675.57	
Palladium						
$PdCl_2$		2	680−857	6.32	5 032	
Phosphorus						
P red, V	subl c	1		11.060	5 323	220
white	subl c	1		6.936 9	1 907.6	190.0
P_4 black, o-rh		1		12.405	6 671	247
PBr_3	liq	1	−40 to 173	6.915 5	1 590.5	221.0
PBr_5	liq	1	to 104	6.948	1 320	214
$PBrF_2$	liq	1	−133 to − 16	6.904 2	885.12	236.0
PBr_2F	liq	1	−115 to 78	6.858 0	1 210.3	226.0
PCl_3	liq	1	−92 to 76	6.826 7	1 196	227.0
PCl_5	c	1	to 160	10.206 8	2 903.1	237.0
	liq	1		7.033	1 490	200.0
$PClF_2$	liq	1	−165 to −47	6.639 6	780.88	255.0
PCl_2F	liq	1	−144 to 14	6.796 56	982.332	237.00
$P(OCN)_3$	liq	2	−2 to 169	8.745 5	2 595	
PF_3	liq	1	−152 to −101	6.860 4	620.22	257.0
PF_5	liq	1	−93.8 to −84.5	6.914 4	647.21	245.0
PH_3	c	1		7.482 35	794.496	265.20
	liq	1		6.715 59	645.512	256.066
P_2H_4	liq	1		6.862 8	1 137	227.0
P_4O_6	liq	1	24−175	6.716 37	1 412.8	193.0
P_4O_{10}	c III	1		9.707 0	3 822	201.0
	c I	1		10.843 2	6 424	213
	liq	1		6.935 2	3 069	152
$POBr_3$	liq	1	51−192	7.007 8	1 609.2	198.0
$POBrCl_2$	liq	1	31−165	6.924	1 411	213
POBrClF	liq	1		6.914	1 214	222

TABLE 1.48 Vapor Pressures of Various Inorganic Compounds (*Continued*)

Substance	State	Eq.	Range,°C	A	B	C
Phosphorus (*continued*)						
$POBrF_2$	liq	1	−85 to 32	7.101 9	1 118.9	233.0
$POBr_2F$	liq	1	−117 to 110	6.721 2	1 328.9	236.0
$POCl_3$	liq	1	1.2–105	6.865 8	1 297.2	220.0
$POClF_2$	liq	1	−96 to 3	6.926 6	946.96	231.0
$POCl_2F$	liq	1	−80 to 53	7.084 65	1 201.86	233.00
POF_3	c	1		10.930 5	1 783	261.0
	liq	1		7.115 5	810.1	231.0
$PO(OCN)_3$		2	5–193	9.168 2	2 931	
$PO(SCN)_3$		2	14–300	8.533 0	3 240	
P_4S_{10}		2		9.17	4 940	
$PSBr_3$	c	2		10.105	3 196.2	
	liq	2		8.338 3	2 641.9	
$PS(OCN)_3$		2		10.032	3 492	
Platinum						
Pt		2	1425–1765	7.786	25 384	
PtF_6	liq	1	61.3–81.7	89.15	5 686	27.49
Polonium						
Po	liq	1		7.041 4	5 017.6	241.0
$PoCl_4$	liq	1		7.554	2 360	115
Potassium						
K		2	260–760	7.183	4 434.33	
KBr		2	1095–1375	7.936	8 555.3	
KCl		2	1116–1418	8.130	8 863.4	
KF		2	1278–1500	9.000	10 838	
KOH		2	1170–1327	7.330	7 103.3	
KI		2	1063–1333	7.949	8 132.2	
Protactinium	liq	2		17.27	7 377	
Radon						
Rn	c	1		7.495 5	884.41	255.0
	liq	1		6.701 5	718.25	250.0
Rhenium						
ReF_5	c	2		9.024	3 037	
ReF_6	c	3	−3.45 to 18.5	9.123 0	1 765.4	0.1790
	liq	3	18.5–48	18.208 1	1 956.7	3.599
ReF_7	c	3	−14.5 to 48.3	13.043 2	2 205.8	1.470 3
	liq	3	48.3–74.6	−21.583 5	244.28	−9.908 3
ReO_2	c	2	650–785	11.65	14 437	
	liq	2	480–660	5.345	4 742	
ReO_3	c	2	325–420	15.16	10 882	
	liq	2	300–480	7.745	4 966	
Re_2O_7	liq	2	230–360	8.98	3 868	
$ReOF_4$	liq	2	108–172	10.09	3 206	
$ReOF_5$	liq	2	41–73	7.727	1 679	
ReS_2	c	2	500–700	3.214	4 976	
Re_2S_7	c	2	260–410	8.86	4 800	
Rubidium						
Rb		2	250–370	6.976	3 969.5	
RbCl		2	1142–1395	9.111	10 373	
RbF		2	1142–1400	8.570	9 568.4	
Ruthenium						
$RuOF_4$		2	120–160	8.60	2 616	
Selenium						
Se	liq	1		7.631 6	4 213.0	202.0
$SeCl_4$	c	1		10.250 9	3 068.8	225.0

(*Continued*)

TABLE 1.48 Vapor Pressures of Various Inorganic Compounds (*Continued*)

Substance	State	Eq.	Range,°C	A	B	C
Selenium (*Continued*)						
SeF$_4$	liq	1		7.888 7	1 603.0	215.0
SeF$_6$	c	1		8.385 4	1 121.4	250.0
SeO$_2$		1		6.577 81	1 879.81	179.0
SeOCl$_2$	liq	1		6.257 3	970.87	112.0
SeOF$_2$	liq	1		7.420	1 380	178
Silicon						
SiCl$_4$	liq	1	0–53	6.857 26	1 138.92	228.88
SiH$_4$		2	−160 to −112	6.881	645.9	
Si$_2$H$_6$		2	−115 to −14.6	7.258	1 133.4	
Si$_3$H$_8$		2	−70 to 52	7.676	1 559.1	
Silver						
AgCl		2	1255–1442	8.179	9 688.7	
Sodium						
Na		2	180–883	7.553	5 395.4	
NaCl		2	976–1155	8.329 7	9 417.07	
NaCl		2	1156–1430	8.548	9 704.3	
NaCN		2	800–1360	7.472	8 122.81	
NaF		2	1562–1701	8.640	11 396.6	
NaI		2	1063–1307	8.371	8 623.2	
NaOH		2	1010–1402	7.030	6 894	
Strontium						
Sr		2	940–1140	16.056	18 802.8	
Sulfur						
S equilibrium	liq	1		6.843 59	2 500.12	186.30
S$_2$Br$_2$	liq	1		7.177	1 660	185
SCl$_2$	liq	1		8.454	1 594	227
S$_2$Cl$_2$	liq	1		6.783 6	1 341	206.0
S$_2$F$_2$	liq	1		6.684	628	256
SF$_4$	liq	1		6.839 5	823.4	248.0
SF$_6$	c	1		8.416 0	1 096.5	262.0
S$_2$F$_{10}$	liq	1		7.067 6	1 100.6	234.0
SO$_2$	c	1		9.754 3	1 553.8	225.0
	liq	1		7.282 28	999.900	237.190
SO$_3$ "icelike"	c III	1		10.565 7	2 273.8	255.0
"woollike"	c II	1		11.590 1	2 665.6	264.0
	c I	1		14.255 9	3 692.1	273.0
	liq	1		9.050 85	1 735.31	236.50
SOBr$_2$	liq	1		7.056	1 445	206
SOCl$_2$	liq	1		7.287 45	1 446.7	252.7
SOClF	liq	1		7.173 1	1 100.1	244.00
SOF$_2$	liq	1		6.959 06	775.48	234.00
SOF$_4$	liq	1		7.071 8	840.3	249.0
S$_2$O$_2$F$_{10}$	liq	1		6.874	1 110	229
S$_2$O$_5$Cl$_2$	liq	1		7.019	1 460	202
S$_2$O$_5$ClF	liq	1		7.015 6	1 257.4	204.0
S$_2$O$_5$F$_2$	liq	1		6.881	1 120	229
S$_2$O$_5$F$_4$	liq	1		6.885	1 140	227
SO$_2$BrF	liq	1		7.142 8	1 155	231.0
SO$_2$Cl$_2$	liq	1		7.001 7	1 209	224.0
SO$_2$ClF	liq	1		6.521 5	793.73	210.70
SO$_2$F$_2$	liq	1		6.907 0	784.3	250
Tantalum						
TaBr$_5$	liq	2		8.11	3 260	
TaCl$_5$	liq	2	220–240	8.68	2 970	

TABLE 1.48   Vapor Pressures of Various Inorganic Compounds (*Continued*)

Substance	State	Eq.	Range,°C	A	B	C
Tantalum (*Continued*)						
TaF$_5$	liq	2		8.524	2 834	
TaI$_5$	liq	2		7.67	3 950	
Technetium						
TcF$_6$	liq	3	37.4–51.7	24.808 7	2 405	5.803 6
TcO$_3$F	liq	2	18.3–51.8	8.417	2 065	
Tc$_2$O$_7$	c	2		18.279	7 205	
	liq	2		8.999	3 571	
Tellurium						
Te	liq	1		7.301 0	5 370.6	221
TeCl$_4$	liq	1		7.558 6	2 355	115
TeF$_6$	liq	1		6.748 8	807.0	247.0
Te$_2$F$_{10}$	liq	1		6.901 8	1 150	227.0
TeO$_2$		2	450–733	12.328 4	13 222	
Thallium						
Tl		2	950–1200	6.1240	6 268	
TlF		2	282–298	12.52	5 484	
Thorium						
ThF$_4$	liq	2		10.821	15 270	
ThH$_2$		2	up to 883	9.50	7 650	
Tin						
SnCl$_4$		2	−52 to −38	9.824	2 441.23	
SnH$_4$		2	−148 to −49	7.400	999.68	
Titanium						
TiCl$_2$	subl c	2		9.30	8 500	
TiCl$_3$	subl c	2	455–550	10.401	8 296	
TiCl$_4$	liq	2	−23 to 136	7.683	1 964	
TiI$_4$	liq	2	160–360	7.577	3 054	
Tungsten						
W		2	2230–2770	9.920	46 850	
Uranium						
UF$_6$	liq	1	64–116	6.994 64	1 126.288	221.963
	liq	1	116–230	7.690 69	1 683.165	302.148
UH$_3$ dissociation		2	200–430	9.39	4 590	
U^{2}H$_3$ (UD$_3$)		2		9.43	4 500	
U^{3}H$_3$ (UT$_3$)		2		9.46	4 471	
Vanadium						
VBr$_2$	c	2	541–716	9.08	10 460	
	subl c	2	800–905	5.9	9 830	
VBr$_3$		2	314–427	11.12	7 470	
VCl$_2$	subl c	2	910–1100	5.725	9 721	
VCl$_3$		2	352–567	11.20	9 777	
VCl$_4$	liq	2	30–153	7.62	2 020	
VF$_3$	subl c	2	650–920	12.357	15 603	
VF$_5$	subl c	2	−20 to 19.5	8.168	2 608	
	liq	2	19.5–45.5	7.549	2 423	
VI$_2$	subl c	2	850–1016	2.56	5 600	
VOCl$_3$	liq	2	15.4–125	7.69	1 920	
Xenon						
Xe	c	1		7.484 5	714.896	264.0
	liq	1		6.642 89	566.282	258.660
XeF$_2$	subl c	1		10.019 47	2 683.96	261.68
XeF$_4$	subl c	1		10.913 87	3 095.06	269.56
Zinc						
Zn	c	2	250–419	9.200	6 946.6	

TABLE 1.49 Vapor Pressure of Mercury

Temp., °C	mm of Hg	Temp., °C	mm of Hg	Temp., °C	mm of Hg
0	0.000 185	92	0.1769	184	10.116
2	0.000 228	94	0.1976	186	10.839
4	0.000 276	96	0.2202	188	11.607
6	0.000 335	98	0.2453	190	12.423
8	0.000 406	100	0.2729	192	13.287
10	0.000 490	102	0.3032	194	14.203
12	0.000 588	104	0.3366	196	15.173
14	0.000 706	106	0.3731	198	16.200
16	0.000 846	108	0.4132	200	17.287
18	0.001 009	110	0.4572	202	18.437
20	0.001 201	112	0.5052	204	19.652
22	0.001 426	114	0.5576	206	20.936
24	0.001 691	116	0.6150	208	22.292
26	0.002 000	118	0.6776	210	23.723
28	0.002 359	120	0.7457	212	25.233
30	0.002 777	122	0.8198	214	26.826
32	0.003 261	124	0.9004	216	28.504
34	0.003 823	126	0.9882	218	30.271
36	0.004 471	128	1.084	220	32.133
38	0.005 219	130	1.186	222	34.092
40	0.006 079	132	1.298	224	36.153
42	0.007 067	134	1.419	226	38.318
44	0.008 200	136	1.551	228	40.595
46	0.009 497	138	1.692	230	42.989
48	0.010 98	140	1.845	232	45.503
50	0.012 67	142	2.010	234	48.141
52	0.014 59	144	2.188	236	50.909
54	0.016 77	146	2.379	238	53.812
56	0.019 25	148	2.585	240	56.855
58	0.022 06	150	2.807	242	60.044
60	0.025 24	152	3.046	244	63.384
62	0.028 83	154	3.303	246	66.882
64	0.032 87	156	3.578	248	70.543
66	0.037 40	158	3.873	250	74.375
68	0.042 51	160	4.189	252	78.381
70	0.048 25	162	4.528	254	82.568
72	0.054 69	164	4.890	256	86.944
74	0.061 89	166	5.277	258	91.518
76	0.069 93	168	5.689	260	96.296
78	0.078 89	170	6.128	262	101.28
80	0.088 80	172	6.596	264	106.48
82	0.100 0	174	7.095	266	111.91
84	0.112 4	176	7.626	268	117.57
86	0.126 1	178	8.193	270	123.47
88	0.1413	180	8.796	272	129.62
90	0.1582	182	9.436	274	136.02

TABLE 1.49 Vapor Pressure of Mercury (*Continued*)

Temp., °C	mm of Hg	Temp., °C	mm of Hg	Temp., °C	mm of Hg
276	142.69	332	478.13	388	1299.1
278	149.64	334	497.12	390	1341.9
280	156.87	336	516.74	392	1386.1
282	164.39	338	537.00	394	1431.3
284	172.21	340	557.90	396	1477.7
286	180.34	342	579.45	398	1525.2
288	188.79	344	601.69	400	1574.1
290	197.57	346	624.64	430	2464
292	206.70	348	648.30	460	3715
294	216.17	350	672.69	490	5420
296	226.00	352	697.83	520	7691
298	236.21	354	723.73	550	10650
300	246.80	356	750.43	600	22.87 atm
302	257.78	358	777.92	650	35.49 atm
304	269.17	360	806.23	700	52.51 atm
306	280.98	362	835.38	750	74.86 atm
308	293.21	364	865.36	800	103.31 atm
310	305.89	366	896.23	850	138.42 atm
312	319.02	368	928.02	900*	180.92 atm
314	332.62	370	960.66	950	226.58 atm
316	346.70	372	994.34	1000	290.5 atm
318	361.26	374	1028.9	1050	358.1 atm
320	376.33	376	1064.4	1100	437.3 atm
322	391.92	378	1100.9	1150	521.3 atm
324	408.04	380	1138.4	1200	616.8 atm
326	424.71	382	1177.0	1250	721.4 atm
328	441.94	384	1216.6	1300	835.9 atm
330	459.74	386	1257.3		

*Critical point.

TABLE 1.50   Vapor Pressure of Ice in Millimeters of Mercury

For temperatures from −99 to 0°C.

The values in the table are for ice in contact with its own vapor. Where the ice is in contact with air at a temperature $t°C$, this correction must be added: Correction $= 20p/(100)(t + 273)$.

t, °C	p, mm Hg	t, °C	p, mm Hg	t, °C	p, mm Hg
−99	0.000 012	−51	0.026 1	−16.5	1.080
−98	0.000 015	−50	0.029 6	−16.0	1.132
−97	0.000 018	−49	0.033 4	−15.5	1.186
−96	0.000 022	−48	0.037 8	−15.0	0.241
−95	0.000 027	−47	0.042 6	−14.5	1.300
−94	0.000 033	−46	0.048 1	−14.0	1.361
−93	0.000 040	−45	0.054 1	−13.5	1.424
−92	0.000 048	−44	0.060 9	−13.0	1.490
−91	0.000 058	−43	0.068 4	−12.5	1.559
−90	0.000 070	−42	0.076 8	−12.0	1.632
−89	0.000 084	−41	0.086 2	−11.5	1.707
−88	0.000 10	−40	0.096 6	−11.0	1.785
−87	0.000 12	−39	0.108 1	−10.5	1.866
−86	0.000 14	−38	0.120 9	−10.0	1.950
−85	0.000 17	−37	0.135 1	−9.8	1.985
−84	0.000 20	−36	0.150 7	−9.6	2.021
−83	0.000 24	−35	0.168 1	−9.4	2.057
−82	0.000 29	−34	0.187 3	−9.2	2.093
−81	0.000 34	−33	0.208 4	−9.0	2.131
−80	0.000 40	−32	0.231 8	−8.8	2.168
−79	0.000 47	−31	0.257 5	−8.6	2.207
−78	0.000 56	−30.0	0.285 9	−8.4	2.246
−77	0.000 66	−29.5	0.301	−8.2	2.285
−76	0.000 77	−29.0	0.317	−8.0	2.326
−75	0.000 90	−28.5	0.334	−7.8	2.367
−74	0.001 05	−28.0	0.351	−7.6	2.408
−73	0.001 23	−27.5	0.370	−7.4	2.450
−72	0.001 43	−27.0	0.389	−7.2	2.493
−71	0.001 67	−26.5	0.409	−7.0	2.537
−70	0.001 94	−26.0	0.430	−6.8	2.581
−69	0.002 25	−25.5	0.453	−6.6	2.626
−68	0.002 61	−25.0	0.476	−6.4	2.672
−67	0.003 02	−24.5	0.500	−6.2	2.718
−66	0.003 49	−24.0	0.526	−6.0	2.765
−65	0.004 03	−23.5	0.552	−5.8	2.813
−64	0.004 64	−23.0	0.580	−5.6	2.862
−63	0.005 34	−22.5	0.609	−5.4	2.912
−62	0.006 14	−22.0	0.640	−5.2	2.962
−61	0.007 03	−21.5	0.672	−5.0	3.013
−60	0.008 08	−21.0	0.705	−4.8	3.065
−59	0.009 25	−20.5	0.740	−4.6	3.117
−58	0.010 6	−20.0	0.776	−4.4	3.171
−57	0.012 1	−19.5	0.814	−4.2	3.225
−56	0.013 8	−19.0	0.854	−4.0	3.280
−55	0.015 7	−18.5	0.895	−3.8	3.336
−54	0.017 8	−18.0	0.939	−3.6	3.393
−53	0.020 3	−17.5	0.984	−3.4	3.451
−52	0.023 0	−17.0	1.031	−3.2	3.509

TABLE 1.50 Vapor Pressure of Ice in Millimeters of Mercury (*Continued*)

t, °C	p, mm Hg	t, °C	p, mm Hg	t, °C	p, mm Hg
−3.0	3.568	−1.8	3.946	−0.8	4.287
−2.8	3.360	−1.6	4.012	−0.6	4.359
−2.6	3.691	−1.4	4.079	−0.4	4.431
−2.4	3.753	−1.2	4.147	−0.2	4.504
−2.2	3.816	−1.0	4.217	0.0	4.579
−2.0	3.880				

TABLE 1.51 Vapor Pressure of Liquid Ammonia, NH_3

t, °C	p in atm	t, °C	p in atm	t, °C	p in atm
−78	0.0582	−6	3.3677	66	29.784
−76	0.0683	−4	3.6405	68	31.211
−74	0.0797	−2	3.9303	70	32.687
−72	0.0929	0	4.2380	72	34.227
−70	0.1078	+2	4.5640	74	35.813
−68	0.1246	4	4.9090	76	37.453
−66	0.1437	6	5.2750	78	39.149
−64	0.1651	8	5.6610	80	40.902
−62	0.1891	10	6.0685	82	42.712
−60	0.2161	12	6.4985	84	44.582
−58	0.2461	14	6.9520	86	46.511
−56	0.2796	16	7.4290	88	48.503
−54	0.3167	18	7.9310	90	50.558
−52	0.3578	20	8.4585	92	52.677
−50	0.4034	22	9.0125	94	54.860
−48	0.4536	24	9.5940	96	57.111
−46	0.5087	26	10.2040	98	59.429
−44	0.5693	28	10.8430	100	61.816
−42	0.6357	30	11.512	102	64.274
−40	0.7083	32	12.212	104	66.804
−38	0.7875	34	12.943	106	69.406
−36	0.8738	36	13.708	108	72.084
−34	0.9676	38	14.507	110	74.837
−32	1.0695	40	15.339	112	77.668
−30	1.1799	42	16.209	114	80.578
−28	1.2992	44	17.113	116	83.570
−26	1.4281	46	18.056	118	86.644
−24	1.5671	48	19.038	120	89.802
−22	1.7166	50	20.059	122	93.045
−20	1.8774	52	21.121	124	96.376
−18	2.0499	54	22.224	126	99.796
−16	2.2349	56	23.372	128	103.309
−14	2.4328	58	24.562	130	106.913
−12	2.6443	60	25.797	132	110.613
−10	2.8703	62	27.079	132.3	111.3(c.p.)
−8	3.1112	64	28.407		

TABLE 1.52   Vapor Pressure of Water

For temperatures from −10 to 120°C.

The values in the table are for water in contact with its own vapor. Where the water is in contact with air at a temperature t in degrees. Celsius, the following correction must be added: Correction (when $t \leq 40°C$) = $p(0.775 − 0.000\ 313t)/100$; correction (when $t > 50°C$) = $p(0.0652 − 0.000\ 087\ 5t)/100$.

t, °C	p, mm Hg	t, °C	p, mm Hg	t, °C	p, mm Hg	t, °C	p, mm Hg
−10.0	2.149	13.0	11.231	23.4	21.583	32.6	36.891
−9.5	2.236	13.5	11.604	23.6	21.845	32.8	37.308
−9.0	2.326	14.0	11.987	23.8	22.110	33.0	37.729
−8.5	2.418	14.5	12.382	24.0	22.387	33.2	38.155
−8.0	2.514	15.0	12.788	24.2	22.648	33.4	38.584
−7.5	2.613	15.2	12.953	24.4	22.922	33.6	39.018
−7.0	2.715	15.4	13.121	24.6	23.198	33.8	39.457
−6.5	2.822	15.6	13.290	24.8	23.476	34.0	39.898
−6.0	2.931	15.8	13.461	25.0	23.756	34.2	40.344
−5.5	3.046	16.0	13.634	25.2	24.039	34.4	40.796
−5.0	3.163	16.2	13.809	25.4	24.326	34.6	41.251
−4.5	3.284	16.4	13.987	25.6	24.617	34.8	41.710
−4.0	3.410	16.6	14.166	25.8	24.912	35.0	42.175
−3.5	3.540	16.8	13.347	26.0	25.209	35.2	42.644
−3.0	3.673	17.0	14.530	26.2	25.509	35.4	43.117
−2.5	3.813	17.2	14.715	26.4	25.812	35.6	43.595
−2.0	3.956	17.4	14.903	26.6	26.117	35.8	44.078
−1.5	4.105	17.6	15.092	26.8	26.426	36.0	44.563
−1.0	4.258	17.8	15.284	27.0	26.739	36.2	45.054
−0.5	4.416	18.0	15.477	27.2	27.055	36.4	45.549
0.0	4.579	18.2	15.673	27.4	27.374	36.6	46.050
0.5	4.750	18.4	15.871	27.6	27.696	36.8	46.556
1.0	4.926	18.6	16.071	27.8	28.021	37.0	47.067
1.5	5.107	18.8	16.272	28.0	28.349	37.2	47.582
2.0	5.294	19.0	16.477	28.2	28.680	37.4	48.102
2.5	5.486	19.2	16.685	28.4	29.015	37.6	48.627
3.0	5.685	19.4	16.894	28.6	29.354	37.8	49.157
3.5	5.889	19.6	17.105	28.8	29.697	38.0	49.692
4.0	6.101	19.8	17.319	29.0	30.043	38.2	50.231
4.5	6.318	20.0	17.535	29.2	30.392	38.4	50.774
5.0	6.543	20.2	17.753	29.4	30.745	38.6	51.323
5.5	6.775	20.4	17.974	29.6	31.102	38.8	51.879
6.0	7.013	20.6	18.197	29.8	31.461	39.0	52.442
6.5	7.259	20.8	18.422	30.0	31.824	39.2	53.009
7.0	7.513	21.0	18.650	30.2	32.191	39.4	54.580
7.5	7.775	21.2	18.880	30.4	32.561	39.6	54.156
8.0	8.045	21.4	19.113	30.6	32.934	39.8	54.737
8.5	8.323	21.6	19.349	30.8	33.312	40.0	55.324
9.0	8.609	21.8	19.587	31.0	33.695	40.5	56.81
9.5	8.905	22.0	19.827	31.2	34.082	41.0	58.34
10.0	9.209	22.2	20.070	31.4	34.471	41.5	59.90
10.5	9.521	22.4	20.316	31.6	34.864	42.0	61.50
11.0	9.844	22.6	20.565	31.8	35.261	42.5	63.13
11.5	10.176	22.8	20.815	32.0	35.663	43.0	64.80
12.0	10.518	23.0	21.068	32.2	36.068	43.5	66.51
12.5	10.870	23.2	21.324	32.4	36.477	44.0	68.26

TABLE 1.52 Vapor Pressure of Water (*Continued*)

t, °C	p, mm Hg	t, °C	p, mm Hg	t, °C	p, mm Hg	t, °C	p, mm Hg
44.5	70.05	63.0	171.38	81.5	377.3	97.0	682.07
45.0	71.88	63.5	175.35	82.0	384.9	97.2	687.04
45.5	73.74	64.0	179.31	82.5	392.8	97.4	692.05
46.0	75.65	64.5	183.43	83.0	400.6	97.6	697.10
46.5	77.61	65.0	187.54	83.5	408.7	97.8	702.17
47.0	79.60	65.5	191.82	84.0	416.8	98.0	707.27
47.5	81.64	66.0	196.09	84.5	425.2	98.2	712.40
48.0	83.71	66.5	200.53	85.0	433.6	98.4	717.56
48.5	85.85	67.0	204.96	85.5	442.3	98.6	722.75
49.0	88.02	67.5	209.57	86.0	450.9	98.8	727.98
49.5	90.24	68.0	214.17	86.5	459.8	99.0	733.24
50.0	92.51	68.5	218.95	87.0	468.7	99.2	738.53
50.5	94.86	69.0	223.73	87.5	477.9	99.4	743.85
51.0	97.20	69.5	228.72	88.0	487.1	99.6	749.20
51.5	99.65	70.0	233.7	88.5	496.6	99.8	754.58
52.0	102.09	70.5	238.8	89.0	506.1	100.0	760.00
52.5	104.65	71.0	243.9	89.5	515.9	101.0	787.57
53.0	107.20	71.5	249.3	90.0	525.76	102.0	815.86
53.5	109.86	72.0	254.6	90.5	535.83	103.0	845.12
54.0	112.51	72.5	260.2	91.0	546.05	104.0	875.06
54.5	115.28	73.0	265.7	91.5	556.44	105.0	906.07
55.0	118.04	73.5	271.5	92.0	566.99	106.0	937.92
55.5	120.92	74.0	277.2	92.5	577.71	107.0	970.60
56.0	123.80	74.5	283.2	93.0	588.60	108.0	1004.42
56.5	126.81	75.0	289.1	93.5	599.66	109.0	1038.92
57.0	129.82	75.5	295.3	94.0	610.90	110.0	1074.56
57.5	132.95	76.0	301.4	94.5	622.31	111.0	1111.20
58.0	136.08	76.5	307.7	95.0	633.90	112.0	1148.74
58.5	139.34	77.0	314.1	95.2	638.59	113.0	1187.42
59.0	142.60	77.5	320.7	95.4	643.30	114.0	1227.25
59.5	145.99	78.0	327.3	95.6	648.05	115.0	1267.98
60.0	149.38	78.5	334.2	95.8	652.82	116.0	1309.94
60.5	152.91	79.0	341.0	96.0	657.62	117.0	1352.95
61.0	156.43	79.5	348.1	96.2	662.45	118.0	1397.18
61.5	160.10	80.0	355.1	96.4	667.31	119.0	1442.63
62.0	163.77	80.5	362.4	96.6	672.20	120.0	1489.14
62.5	167.58	81.0	369.7	96.8	677.12		

TABLE 1.53 Vapor Pressure of Deuterium Oxide

t, °C	p, mm Hg	t, °C	p, mm Hg	t, °C	p, mm Hg
0	3.65	20	15.2	80	331.6
1	3.93	30	28.0	90	495.5
2	4.29	40	49.3	100	722.2
3	4.65	50	83.6	101.43	760.0
3.8	5.05	60	136.6		
10	7.79	70	216.1		

1.12   *VISCOSITY AND SURFACE TENSION*

Viscosity is the shear stress per unit area at any point in a confined fluid divided by the velocity gradient in the direction perpendicular to the direction of flow. If this ratio is constant with time at a given temperature and pressure for any species, the fluid is called a Newtonian fluid.

The *absolute viscosity* (μ) is the sheer stress at a point divided by the velocity gradient at that point. The most common unit is the poise (1 kg/m sec) and the SI unit is the Pa.sec (1 kg/m sec). As many common fluids have viscosities in the hundredths of a poise the centipoise (cp) is often used. One centipoise is then equal to one mPa sec.

The *kinematic viscosity* (ν) is ratio of the absolute viscosity to density at the same temperature and pressure. The most common unit corresponding to the poise is the stoke (1 cm^2/sec) and the SI unit is m^2/sec.

The molecules in a gas-liquid interface are in tension and tend to contract to a minimum surface area. This tension may be quantified by the surface tension (σ), which is the force in the plane of the surface per unit length.

TABLE 1.54   Viscosity and Surface Tension of Inorganic Substances

For the majority of compounds the dependence of the surface tension γ on the temperature can be given as:

$$\gamma = a - bt$$

where a and b are constants and t is the temperature in degrees Celsius. The values of the dipole moment are for the gas phase.

Substance	Viscosity, mN · s · m^{-2}	Surface tension, mN · m^{-1}	
		a	b
Air	0.0182[20], 0.0231[127]		
AlBr$_3$			
Ar			
(g)	0.0233[20], 0.0288[127]		
(lq)		34.28	0.2493
AsBr$_3$		54.41	0.1043
AsCl$_3$		41.67	0.097 81
AsH$_3$ (arsine)			
BBr$_3$		31.90	0.1280
BCl$_3$			
BF$_3$	0.0171[27], 0.0217[127]	−2.92	0.2030
B$_2$H$_6$ (diborane)		−3.13	0.1783
B$_4$H$_{10}$			
B$_5$H$_9$			
B$_6$H$_{10}$			
B$_3$H$_6$N$_3$			
Br$_2$ (g)			
(lq)	1.252[0], 1.03[16], 0.744[25]	45.5	0.1820
BrF$_3$	2.22[20]	38.30	0.0999
BrF$_5$	0.62[24]	25.24	0.1098

TABLE 1.54 Viscosity and Surface Tension of Inorganic Substances (*Continued*)

Substance	Viscosity, $mN \cdot s \cdot m^{-2}$	Surface tension, $mN \cdot m^{-1}$	
		a	*b*
Cl_2 (g)	0.0132^{20}		
(lq)		19.87	0.1897
ClF_3	0.48^{12}	26.9	0.1660
ClF_5			
ClO_3F		12.24	0.1576
CO (g)	0.0175^{20}, 0.0221^{127}		
(lq)		−30.20	0.2073
CO_2 (g)	0.0147^{20}, 0.0197^{127}		
(lq)	0.071^{20}	6.14^{-10}	2.67^{10}
$COCl_2$		22.59	0.1456
COF_2			
COS		12.12	0.1779
COSe			
CS			
CS_2 (g)			
(lq)	0.429^{0}, 0.375^{20}, 0.352^{25}	35.29	0.1484
CrO_2Cl_2			
D_2 (deuterium)	0.0126^{27}, 0.0154^{127}		
DH		6.537	0.1883
D_2O	0.0111^{25} (g), 1.098^{25} (lq)	71.72^{20}	68.38^{40}
F_2		−16.10	0.1646
$GaCl_3$		35.0	0.1000
$GeBr_4$		35.51^{30}	33.70^{50}
$GeBr_4$		35.51^{30}	33.70^{50}
$GeCl_4$		22.44^{30}	
$GeClH_3$			
H_2 (g)	0.0088^{20}, 0.109^{127}		
t			
(lq)		2.80^{-258}	2.12^{-254}
HBr (g)			
(lq)	0.83^{-67}	13.10	0.2079
He (g)	0.0196^{27}, 0.0244^{27}		
(lq) (II)		$0.351^{0.50 K}$	$0.317^{2.00 K}$
(III)		$0.151^{3.61 K}$	$0.131^{1.13 K}$
(IV)		$0.372^{0.50 K}$	$0.354^{1.40 K}$
HCl (g)	0.0146^{27}, 0.0197^{127}		
(lq)	0.51^{-95}		

(*Continued*)

TABLE 1.54 Viscosity and Surface Tension of Inorganic Substances (*Continued*)

Substance	Viscosity, $mN \cdot s \cdot m^{-2}$	Surface tension, $mN \cdot m^{-1}$	
		a	*b*
HClO			
HCN	0.235^0, 0.206^{18}, 0.183^{25}	19.45^{10}	18.33^{20}
HCNO (iso-cyanate)			
HCNS			
HF	0.256^0	10.41	0.078 67
HFO			
HI (g)			
(lq)			
HN$_3$ (azide)			
H$_2$O (see Table 5.19)			
H$_2$O$_2$	1.25^{20}	78.97	0.1549
HNO$_3$			
H$_2$S (g)			
(lq)	0.412^0	48.95	0.1758
H$_2$Se		22.32	0.1482
HSO$_3$Cl	2.43^{20}		
HSO$_3$F	1.56^{25}		
H$_2$SO$_4$	24.54^{25}		
H$_2$Te		29.03	0.2619
Hg	1.552^{20}, 1.526^{25}, 1.402^{50}	490.6	0.2049
I$_2$	1.98^{116}		
IBr			
IF			
IF$_5$		33.16	0.1318
IF$_7$			
IOF$_5$			
Kr (g)	0.0250^{20}, 0.0331^{127}		
(lq)		40.576 (in K)	0.2890 (in K)
Mn$_2$O$_7$			
Ne (g)	0.0303^{20}, 0.0389^{127}		
(lq)			
N$_2$(g)	0.0176^{20}, 0.0222^{127}		
(lq)		26.42 (in K)	0.2265 (in K)
NH$_3$ (g)			
(lq)	$0.254^{-33.5}$	37.91^{-50}	35.38^{-40}
N$_2$H$_4$ (hydrazine)	0.97^{20}, 0.876^{25}, 0.628^{50}	72.41	0.2407
Ni(CO)$_4$		18.11	0.1117
NO	0.0192^{27}, 0.0238^{127}	−67.48	0.5853

TABLE 1.54   Viscosity and Surface Tension of Inorganic Substances (*Continued*)

Substance	Viscosity, $mN \cdot s \cdot m^{-2}$	Surface tension, $mN \cdot m^{-1}$	
		a	b
N_2O (g)	$0.0146^{20},$ 0.0194^{127}		
(lq)		5.09	0.2032
NO_2	$0.532^0, 0.402^{25}$		
N_2O_4			
N_2O_3			
NOBr			
NOCl		29.49	0.1493
NO_2Cl			
NOF		14.00	0.1165
NO_2F		8.26	0.1854
NO_3			
O_2 (g)	$0.0204^{20},$ 0.0261^{127}		
(lq)		−33.72	0.2561
O_3		38.1^{-183}	
OF_2			
O_2F_2 (FOOF)			
OsO_4			
P (lq)			
PBr_3		45.34	0.1283
PCl_3	$0.662^0, 0.529^{25},$ 0.439^{50}	31.14	0.1266
PCl_5			
PCl_2F_3			
PCl_3F_2			
PCl_4F			
PF_3			
PF_5			
PH_3			
PI_3		61.66	0.067 71
PO_3		40.44	0.1158
$POCl_3$	1.065^{25}	35.22	0.1275
POF_3			
$PSCl_3$		37.00	0.1272
PSF_3			
$PbCl_4$			
ReO_2Cl_3		57.00	0.2485
ReO_3Cl		54.05	0.1979
S			
SCl_2			
S_2Cl_2 dimer		46.23	0.1464
S_2F_2			
FSSF isomer			
$S = SF_2$ isomer			
SF_4		12.87	0.1734
SF_6	$0.0153^{27},$ 0.0198^{127}	5.66	0.1190

(*Continued*)

TABLE 1.54   Viscosity and Surface Tension of Inorganic Substances (*Continued*)

Substance	Viscosity, mN · s · m^{-2}	Surface tension, mN · m^{-1}	
		a	b
S_2F_{10}			
SO_2 (g)	0.0129[27], 0.0175[127]		
(lq)		26.58	0.1948
SO_3			
$SOBr_2$		46.28	0.0750
$SOCl_2$		36.10	0.1416
SOF_2			
SO_2Cl_2		32.10	0.1328
SO_2F_2			
$SbCl_3$		47.87	0.1238
$SbCl_5$			
SbF_5		49.07	0.1937
SbH_3			
Se (lq)			
SeF_4		38.61	0.1274
SeF_6			
$SeOCl_2$			
SeO_2			
$SiCl_4$	99.4[25], 96.2[50]	20.78	0.099 62
SiF_4			
SiH_4			
$SiHCl_3$	0.415[0], 0.326[25]	20.43	0.1076
SiH_3Cl			
$SnBr_4$			
$SnCl_4$		29.92	0.1134
TeF_6			
$TiCl_4$		33.54[20]	31.06[40]
UF_6 (g)			
(lq)		25.5	0.1240
VCl_4			
$VOBr_3$			
$VOCl_3$		36.36[20]	33.60[40]
Xe (g)	0.0228[20], 0.030[127]		
(lq, II)		0.345[1.00 K]	0.317[2.00 K]
XeF_6			

1.13 THERMAL CONDUCTIVITY

The thermal conductivity is a measure of the effectiveness of a material as a thermal insulator. The energy transfer rate through a body is proportional to the temperature gradient across the body and the cross sectional area of the body. In the limit of infinitesimal thickness and temperature difference, the fundamental law of heat conduction is:

$$Q = \lambda A dT/dx$$

where Q is the heat flow, A is the cross-sectional area, dT/dx is the temperature/thickness gradient, and λ is the thermal conductivity.

A substance with a large thermal conductivity value is a good conductor of heat; one with a small thermal conductivity value is a poor heat conductor i.e. a good insulator.

TABLE 1.55 Thermal Conductivity of the Elements

Element number	Element symbol	Thermal conductivity (W/m)/K 27°C, 81°F	Element number	Element symbol	Thermal conductivity (W/m)/K 27°C, 81°F
1	H	0.1815	2	He	0.152
3	Li	84.7	4	Be	200
5	B	27	6	C	155
7	N	0.02598	8	O	0.02674
9	F	0.0279	10	Ne	0.0493
11	Na	141	12	Mg	156
13	Al	237	14	Si	148
15	P	0.235	16	S	0.269
17	Cl	0.0089	18	Ar	0.0177
19	K	102.5	20	Ca	200
21	Sc	15.8	22	Ti	21.9
23	V	30.7	24	Cr	93.7
25	Mn	7.82	26	Fe	80.2
27	Co	100	28	Ni	90.7
29	Cu	401	30	Zn	116
31	Ga	40.6	32	Ge	59.9
33	As	50	34	Se	2.04
35	Br	0.122	36	Kr	0.00949
37	Rb	58.2	38	Sr	35.3
39	Y	17.2	40	Zr	22.7
41	Nb	53.7	42	Mo	138
43	Tc	50.6	44	Ru	117
45	Rh	150	46	Pd	71.8
47	Ag	429	48	Cd	96.8
49	In	81.6	50	Sn	66.6
51	Sb	24.3	52	Te	2.35
53	I	0.449	54	Xe	0.00569
55	Cs	35.9	56	Ba	18.4
57	La	13.5	58	Ce	11.4
59	Pr	12.5	60	Nd	16.5
61	Pm	17.9	62	Sm	13.3
63	Eu	13.9	64	Gd	10.6
65	Tb	11.1	66	Dy	10.7
67	Ho	16.2	68	Er	14.3

TABLE 1.56 Thermal Conductivity of Various Solids

All values of thermal conductivity, k, are in millijoules $cm^{-1} \cdot s^{-1} \cdot K^{-1}$. To convert to $mW \cdot m^{-1} \cdot K^{-1}m$, divide values by 10. For values in millicalories, divide by 4.184.

Substance	t, °C	k
Asphalt	20	7.447
Basalt	20	21.76
Bauxite	600	5.56
Boiler scale	66	13.1
Brick, common	20	6.3
Blotting paper	20	0.628
Cardboard	20	2.1
Cement, Portland	90	2.97
Chalk	20	9.2
Chemical elements, *see* Table 4.1		
Coal	0	1.69
Concrete	20	9.2
Cork, sp. grav. = 0.2	30	0.54
Cork meal	100	0.556
Cotton, sp. grav. = 0.081	0	0.569
Diatomaceous earth	20	0.54
Ebonite	0	1.58
Eiderdown	20	0.046
Feathers (with air)	9	0.238
Feldspar	20	23.4
Felt (dark gray)	40	0.623
Fire brick	20	4.6
Flannel	60	0.148
Flint	20	10.0
Glass, crown	12.5	6.82
flint	12.5	5.98
Jena	22	9.50
quartz	0	13.89
	100	19.12
soda	20	7.1
	100	7.5
Granite	20	34.2
Graphite, sp. grav. = 1.58	50	441.4
Graphite powder, sp. grav. = 0.7	40	11.92
Gypsum	0	13.0
Horse hair, sp. grav. = 0.172	20	0.510
Ice		23.8
Leather, cowhide	84	1.76
Linen	20	0.879
Magnesia brick	20	11.3
	1130	30.1
Marble, white		32.6
Mica	41	3.60
Naphthalene	0	3.77
Paper	20	1.3
Paraffin	0	2.88
Plaster of Paris	20	2.93
Porcelain	95	10.38
Quartz, parallel to axis	0	136.0
	100	90.0

TABLE 1.56   Thermal Conductivity of Various Solids (*Continued*)

Substance	t, °C	k
Quartz, perpendicular to axis	0	72.43
	100	55.77
Plastics, see Section 10		
Roofing paper	0	1.90
Rubber, natural and synthetic, *see* Section 10		
Sand, dry	20	3.89
Sandstone, sp. grav. = 2.259	40	18.37
Silk, sp. grav. = 0.101	0	0.510
Slate	20	19.66
Soil, dry	20	1.38
Wax, bees	20	0.866
Wood, maple, parallel to face	20	4.25
perpendicular to face	50	1.82
Wood, oak, parallel to face	15	3.49
perpendicular to face	15	2.09
Wood, pine, parallel to face	20	3.49
perpendicular to face	15	1.51

1.14 CRITICAL PROPERTIES

All substances have chemical and physical properties that can be used as a means of identification of the substance. There are two basic types of properties that are associated with matter: (1) chemical properties, which are properties that change the chemical nature of matter, and (2) physical properties, which are properties that do not change the chemical nature of matter. The more properties can be used for identification of a substance, the better the understanding of the nature of that substance, which can assist in understanding how the substance will behave under various conditions. The critical properties are part of the property assessment and behavior of any substance.

Critical temperature (T_c), critical pressure (P_c), and critical volume (V_c) represent three widely used pure component constants. These critical constants are very important properties in chemical engineering field because almost all other thermo chemical properties are predictable from boiling point and critical constants with using corresponding state theory. Therefore, precise prediction of critical constants is very necessary.

1.14.1 Critical Temperature

The critical temperature of a compound is the temperature above which a liquid phase cannot be formed, no matter what the pressure on the system. The critical temperature is important in determining the phase boundaries of any compound and is a required input parameter for most phase equilibrium thermal property or volumetric property calculations using analytic equations of state or the theorem of corresponding states. Critical temperatures are predicted by various empirical methods according to the type of compound or mixture being considered.

Another somewhat simpler method for estimating the critical temperature of pure compounds requires the normal boiling point, the relative density, and the compound family.

$$\log T_c = A + B \log_{10} (\text{relative density}) + C \log T_b$$

where T_c and T_b are the critical and normal boiling temperatures, respectively, expressed in kelvins. The relative density of the liquid at 15°C is 0.1 MPa. The regression constants A, B, and C are available by family.

For pure inorganic compounds, the method only requires the normal boiling point as input.

$$T_c = 1.64 T_b$$

1.14.2   Critical Pressure

The critical pressure of a compound is the vapor pressure of that compound at the critical temperature. Below the critical temperature, any compound above its vapor pressure will be a liquid.

1.14.3   Critical Volume

The critical volume of a compound is the volume occupied by a specified mass of a compound at its critical temperature and critical pressure.

1.14.4   Critical Compressibility Factor

The critical compressibility factor of a compound is calculated from the experimental or predicted values of the critical properties.

$$Z_c = (P_c V_c)/(RT_c)$$

Critical compressibility factors are used as characterization parameters in corresponding states methods to predict volumetric and thermal properties. The factor varies from approximately 0.23 for water to 0.26–0.28 for most hydrocarbons to above 0.30 for light gases.

TABLE 1.57   Critical Properties

Substance	T_c, °C	P_c, atm	P_c, MPa	V_c, cm$^3 \cdot$ mol^{-1}	p_c, g $\cdot$ cm^{-3}
Air	−140.6	37.2	3.77	92.7	0.313
Aluminum tribromide	490	28.5	2.89	310	0.860
Aluminum trichloride	356	26	2.63	261	0.510
Ammonia	132.4	111.3	11.28	72.5	0.235
Antimony tribromide	631.4	56	5.67		
Antimony trichloride	521			270	0.84
Argon	−122.3	48.1	4.87	74.6	0.536
Arsenic	1400				
Arsenic trichloride	318	58.4	5.91	252	0.720
Arsine	99.9	63.3	6.41	133	0.588
Arsine-d_3	98.9				
Bismuth tribromide	946			301	1.49
Bismuth trichloride	906	118	11.96	261	1.21
Boron pentafluoride	205				
Boron tribromide	308	48.1	4.87	272	0.921
Boron trichloride	178.8	38.2	3.87	266	0.441
Boron trifluoride	−12.3	49.2	4.98	124	0.549
Bromine	315	102	10.3	135	1.184
Benzaldehyde	422	45.9	4.65	324	0.327
Benzene	288.90	48.31	4.895	255	0.306
Benzoic acid	479	41.55	4.21	341	0.358
Benzonitrile	426.3	41.55	4.21	339	0.304
Benzyl alcohol	422	42.4	4.3	334	0.324
Biphenyl	516	38.0	3.85	502	0.307
Carbon dioxide	31.1	72.8	7.38	94.0	0.468
Carbon disulfide	279	78.0	7.90	173	0.41
Carbon monoxide	−140.2	34.5	3.50	93.1	0.301

TABLE 1.57 Critical Properties (*Continued*)

Substance	T_c, °C	P_c, atm	P_c, MPa	V_c, cm³·mol⁻¹	p_c, g·cm⁻³
Carbonyl chloride	182	56	5.67	190	0.52
Carbonyl sulfide	102	58	5.88	140	0.44
Cesium	1806			300	0.44
Chlorine	143.8	76.1	7.71	124	0.573
Chlorine pentafluroide	142.6	51.9	5.26	230.9	0.565
Chlorine trifluoride	153.5				
Deuterium (equilibrium)	−234.8	16.28	1.650	60.4	0.0668
Deuterium (normal)	−234.7	16.43	1.665	60.3	0.0669
Deuterium bromide	88.8				
Deuterium chloride	50.3				
Deuterium hydride (DH)	−237.3	14.64	1.483	62.8	0.0481
Deuterium iodide	148.6				
Deuterium oxide	370.9	213.8	21.66	55.6	0.360
Diborane	166	39.5	4.00		
Dihydrogen disulfide	299	58.3	5.91		
Dihydrogen heptasulfide	742	33	3.34		
Dihydrogen hexasulfide	707	36	3.65		
Dihydrogen octasulfide	767	32	3.24		
Dihydrogen pentasulfide	657	38.4	3.89		
Dihydrogen tetrasulfide	582	43.1	4.37		
Dihydrogen trisulfide	465	50.6	5.13		
Flurorine	−129.0	51.47	5.215	66.2	0.574
Germanium tetrachloride	276.9	38	3.85	330	0.650
Hafnium tetrabromide	473			415	1.20
Hafnium tetrachloride	450	57.0	5.86	304	1.05
Hafnium tetraiodide	643			528	1.30
Helium (equilibrium)	−267.96	2.261	0.2289		0.06930
Helium-3	−269.85	1.13	0.1182	72.5	0.0414
Helium-4	−267.96	2.24	0.227	57.3	0.0698
Hydrazine	380	14.5	1.47	96.1	0.333
Hydrogen (equilibrium)	−240.17	12.77	1.294	65.4	0.0308
Hydrogen (normal)	−239.91	12.8	1.297	65.0	0.0310
Hydrogen bromide	89.8	84.4	8.55	100.0	0.809
Hydrogen chloride	51.40	82.0	8.31	81.0	0.45
Hydrogen cyanide	183.5	53.2	5.39	139	0.195
Hydrogen deuteride	−237.25	14.64	1.483	62.8	0.048
Hydrogen fluoride	188	64	6.5	69	0.29
Hydrogen iodide	150.7	82.0	8.31	131	0.976
Hydrogen selenide	137	88	8.9		
Hydrogen sulfide	100.4	88.2	8.94	98.5	0.31
Iodine	546	115	11.7	155	0.164
Krypton	−63.75	54.3	5.50	91.2	0.9085
Mercury	1477	1587	160.8		
Mercury(II) bromide	789				
Mercury(II) chloride	700				
Mercury(II) iodide	799				
Neon	−228.71	27.2	2.77	41.7	0.4835
Niobium pentabromide	737			469	1.05
Niobium pentachloride	534			400	0.68
Niobium pentafluoride	464	62	6.28	155	1.21

(*Continued*)

TABLE 1.57   Critical Properties (*Continued*)

Substance	T_c, °C	P_c, atm	P_c, MPa	V_c, cm$^3 \cdot$ mol^{-1}	p_c, g $\cdot$ cm^{-3}
Nitric oxide	−92.9	64.6	6.55	58	0.52
Nitrogen-14	146.94	33.5	3.39	89.5	0.313
Nitrogen-15	146.8	33.5	3.39	90.4	0.332
Nitrogen chloride difluoride	64.3	50.8	5.15		
Nitrogen dioxide (equilibrium)	158.2	100	10.1	170	0.557
Nitrogen trideuteride (ND$_3$)	132.4				
Nitrogen trifluoride	−39.3	44.7	4.53		
Nitrous oxide	36.434	71.596	7.2545	97.4	0.4525
Nitrosyl chloride	167	90	9.12	139	0.471
Nitryl fluoride	76.3				
Osmium tetroxide	132	170	17.2		
Oxygen	−118.56	49.77	5.043	73.4	0.436
Oxygen difluoride	−58.0	48.9	4.95	97.7	0.553
Ozone	−12.10	53.8	5.45	88.9	0.540
Phosgene	182	56	5.67	190	0.52
Phosphine	51.3	64.5	6.54		
Phosphine-d_3	50.4				
Phosphonium chloride	49.1	72.7	7.37		
Phosphorus	721				
Phosphorus bromide difluoride	113				
Phosphorus chloride difluoride	89.2	44.6	4.52		
Phosphorus dibromide fluoride	254				
Phosphorus dichloride fluoride	189.9	49.3	5.00		
Phosphorus pentachloride	372				
Phosphorus trichloride	290			260	0.528
Phosphorus trifluoride	−1.9	42.7	4.33		
Phosphoryl chloride difluoride	150.7	43.4	4.40		
Phosphoryl trichloride	329				
Phosphoryl trifluoride	73.4	41.8	4.24		
Radon	104	62	6.28	139	1.6
Rhenium(VII) oxide	669			334	
Rhenium(VI) oxide tetrachloride	508			161	0.95
Rubidium	1832			250	0.34
Selenium	1493				
Silane	−3.5	47.8	4.84		
Silicon chloride trifluoride	34.5	34.2	3.47		
Silicon tetrabromide	390				
Silicon tetrachloride	234	37	3.75	326	0.521
Silicon tetrafluoride	−14.0	36.7	3.72		
Silicon trichloride fluoride	165.4	35.3	3.57		
Sulfur	1041	116	11.7		
Sulfur dioxide	157.7	77.8	7.88	122	0.5240
Sulfur hexafluoride	45.6	37.1	3.76	198	0.734
Sulfur tetrafluoride	91.7				
Sulfur trioxide	217.9	81	8.2	130	0.633
Tantalum pentabromide	701			461	1.26
Tantalum pentachloride	494			400	0.89
Tin(IV) chloride	318.7	37.0	3.75	351	0.742
Titanium tetrachloride	365	46	4.66	340	0.558
Tungsten(VI) oxide tetrachloride	509			338	1.01
Uranium hexafluoride	232.7	45.5	4.61	250	1.41
Water	374.2	217.6	22.04	56.0	0.325

TABLE 1.57 Critical Properties (*Continued*)

Substance	T_c, °C	P_c, atm	P_c, MPa	V_c, cm$^3 \cdot$ mol^{-1}	p_c, g $\cdot$ cm^{-3}
Xenon	16.583	57.64	5.84	118	1.105
Zirconium tetrabromide	532			415	0.99
Zirconium tetrachloride	505	56.9	5.77	319	0.730
Zirconium tetraiodide	687			528	1.13

1.15 THERMODYNAMIC FUNCTIONS (CHANGE OF STATE)

All substances can exist in one of three forms (also called *states* or *phases*) that basically depend on the temperature of the substance. These states or phases are (1) solid, (2) liquid, and (3) gas.

The solid-to-liquid transition is a melting process, and the heat required is the heat of melting. The liquid-to-solid transition is the reverse process, and the heat liberated is the heat of freezing. The solid-to-gas transition is a sublimation process, and the heat required is the heat of sublimation. The liquid-to-gas transition is a vaporization process, and the heat required is the heat of vaporization (heat of boiling). Both the gas-to-solid and the gas-to-liquid processes are condensation processes and have an associated heat of condensation.

Each change of state is accompanied by a change in the energy of the system. Wherever the change involves the disruption of intermolecular forces, energy must be supplied. The disruption of intermolecular forces accompanies the state going toward a less ordered state. As the strengths of the intermolecular forces increase, greater amounts of energy are required to overcome them during a change in state. The melting process for a solid is also referred to as fusion, and the enthalpy-change associated with melting a solid is often called the heat of fusion (ΔH_{fus}). The heat needed for the vaporization of a liquid is called the heat of vaporization (ΔH_{vap}).

The specific heat is the amount of heat per unit mass required to raise the temperature by one degree Celsius. The relationship between heat and temperature change is usually expressed in the form shown below where c is the specific heat. The relationship does not apply if a phase change is encountered, because the heat added or removed during a phase change does not change the temperature.

$$Q = cm\Delta T$$

That is, heat added is equal to the specific heat multiplied by the mass (weight) multiplied by the temperature difference ($\Delta T = t_{final} - t_{initial}$)

TABLE 1.58 Enthalpies and Gibbs Energies of Formation, Entropies, and Heat Capacities of the Elements and Inorganic Compounds

Substance	Physical state	$\Delta_f H°$ kJ $\cdot$ mol^{-1}	$\Delta_f G°$ kJ $\cdot$ mol^{-1}	$S°$ J $\cdot$ deg$^{-1} \cdot$ mol^{-1}	$C°_p$ J $\cdot$ deg$^{-1} \cdot$ mol^{-1}
Ac Actinium	c	0	0	56.5	27.2
Al Aluminum	c	0	0	28.30(10)	24.4
	g	330.0(40)	289.4	164.554(4)	21.4
Al^{3+} std. state	aq	−538.4(15)	−485.3	−325.(10)	
Al$_6$BeO$_{10}$	c	−5624	−5317	175.6	265.19
Al(BH$_4$)$_3$	lq	−16.3	145.0	289.1	194.6
AlBr$_3$	c	−527.2	−488.5	180.2	100.58
std. state	aq	−895	−799	−74.5	
Al$_4$C$_3$	c	−216	−203	89	
Al(CH$_3$)$_3$	lq	136.4	−10.0	209.4	155.6
Al(OAc)$_3$	c	−1892.4			
AlCl$_3$	c	−704.2	−628.8	109.29	91.13
std. state	aq	−1033	−878	−152.3	
AlCl$_3 \cdot$ 6H$_2$O	c	−2692	−2269	377	
AlF$_3$	c	−1510.4(13)	−1431.1	66.5(5)	75.13
std. state	aq	−1531.0	−1322	−363.2	

(*Continued*)

TABLE 1.58 Enthalpies and Gibbs Energies of Formation, Entropies, and Heat Capacities of the Elements and Inorganic Compounds (*Continued*)

Substance	Physical state	$\Delta_f H°$ kJ · mol⁻¹	$\Delta_f G°$ kJ · mol⁻¹	$S°$ J · deg⁻¹ · mol⁻¹	$C°_p$ J · deg⁻¹ · mol⁻¹
$AlF_3 \cdot H_2O$	c	−2297	−2052	209	
AlH_3	c	−46.0		30.0	40.2
AlI_3	c	−313.8	−300.8	159.0	98.7
std. state	aq	−699	−640	12.1	
$AlK(SO_4)_2 \cdot 12H_2O$	c	−6061.8	−5141.7	687.4	651.0
AlN	c	−318.1	−287.0	20.14	30.10
$Al(NO_3)_3$ std. state	aq	−1155	−820	117.6	
$Al(NO_3)_3 \cdot 6H_2O$	c	−2850.5	−2203.9	467.8	433.0
$Al(NO_3)_3 \cdot 9H_2O$	c	−3757.1	−2929.6	569	
AlO_2^- std. state	aq	−930.9	−830.9	−36.8	
Al_2O_3 corundum	c	−1675.7(13)	−1582.3	50.92(10)	79.15
$Al(OH)_3$	c	−1284	−1306	71	93.1
$Al(OH)_4^-$ std. state	aq	−1502.5	−1305.3	102.9	
AlP	c	−166.5			
$AlPO_4$ berlinite	c	−1733.8	−1618.0	90.79	93.18
Al_2S_3	c	−724.0	−640	116.85	105.06
Al_2Se_3	c	−565			
Al_2SiO_5 andalusite	c	−2592.0	−2444.8	93.2	122.76
$Al_2(SO_4)_3$	c	−3435	−3507	239.3	259.4
std. state	aq	−3790	−3205	−583.3	
Al_2Te_3	c	−326			
Americium					
Am	c	0	0	62.7	
Am^{3+}	aq	−682.8	−671.5	−159.0	
Am^{4+}	aq	−511.7	−461.1	−372	
Am_2O_3	c	−1757	−1678	154.7	
AmO_2	c	−1005.0	950.2	96.2	
Ammonium					
NH_3	g	−45.94(35)	−16.4	192.776(5)	35.65
undissoc; std. state	aq	−80.29	−26.57	111.3	
ND_3	g	−58.6	−26.0	203.9	38.23
NH_4^+ std. state	aq	−133.26(25)	−79.37	111.17(40)	79.9
NH_4OH undissoc; std. state	aq	−361.2	−254.0	165.5	
ionized; std. state	aq	−362.50	−236.65	102.5	−68.6
NH_4OAc	c	−616.14			
std. state	aq	−618.52	−448.78	200.0	73.6
$NH_4Al(SO_4)_2$	c	−2352.2	−2038.4	216.3	226.44
std. state	aq	−2481	−2054	−168.2	
NH_4AsO_2 std. state	aq	−561.54	−429.41	154.8	
$NH_4H_2AsO_3$ std. state	c	−847.30	−666.60	223.8	
$NH_4H_2AsO_4$	c	−1059.8	−833.0	172.05	151.17
std. state	aq	−1042.07	−832.66	230.5	
$(NH_4)_2HAsO_4$ std. state	aq	−1171.1	−873.20	225.1	
$(NH_4)_3AsO_4$ std. state	aq	−1286.7	−886.63	177.4	
NH_4Br	c	−271.8	−175.2	113.0	96.0
std. state	aq	−254.05	−183.34	194.97	−61.9
NH_4BrO_3	aq	−199.58	−60.84	275.10	
NH_4 carbamate	c	−657.60	−448.07	133.5	
NH_4Cl	c	−314.5	−202.9	94.6	84.1
std. state	aq	−299.66	−210.62	169.9	−56.5
NH_4ClO_3 std. state	aq	−236.48	−87.40	275.7	

TABLE 1.58   Enthalpies and Gibbs Energies of Formation, Entropies, and Heat Capacities of the Elements and Inorganic Compounds (*Continued*)

Substance	Physical state	$\Delta_f H°$ kJ · mol^{-1}	$\Delta_f G°$ kJ · mol^{-1}	$S°$ J · deg^{-1} · mol^{-1}	$C°_p$ J · deg^{-1} · mol^{-1}
NH_4ClO_4	c	−295.3	−88.8	186.2	128.1
std. state	aq	−261.84	−87.99	295.4	
NH_4CN	c	0.4			134.0
std. state	aq	18.0	92.9	207.5	
NH_4CNO cyanate	aq	−278.7	−177.0	220.1	
std. state					
$(NH_4)_2CO_3$ std. state	aq	−942.15	−686.64	169.9	
$(NH_4)_2C_2O_4$ oxalate	c	−1123.0			226.0
$(NH_4)_2CrO_4$	c	−1167.3			
std. state	aq	−1144.3	−886.59	277.0	
$(NH_4)_2Cr_2O_7$	aq	−1755.2	−1459.5	488.7	
NH_4 dithiocarbonate	c	−126.8			
NH_4F	c	−463.96	−348.78	71.97	65.27
std. state	aq	−465.14	−358.19	99.6	−26.8
NH_4 formate std. state	aq	−558.06	−430.5	205.0	−7.9
NH_4HCO_3	c	−849.4	−665.9	120.9	
	aq	−824.5	−666.1	204.6	
NH_4I	c	−201.4	−112.5	117.0	81.8
std. state	aq	−187.69	−130.96	224.7	−62.3
NH_4IO_3	c	−385.8			
std. state	aq	−354.0	−207.5	231.8	
NH_4N_3 azide	c	115.5	274.1	112.6	
	aq	142.7	268.6	221.3	
NH_4NO_2	aq	−237.2	−111.6	236.4	−17.6
NH_4NO_3	c	−365.56	−184.01	151.08	139.3
std. state	aq	−339.87	−190.71	259.8	−6.7
$NH_4H_2PO_4$	c	−1145.07	−1210.56	151.96	142.26
std. state	aq	−1428.79	−1209.76	203.8	
$(NH_4)_2HPO_4$	c	−1556.91		188.0	
std. state	aq	−1557.16	−1248.00	193.3	
$NH_4H_3P_2O_7$	aq	−2409.1	−2102.6	326.0	
NH_4HS	c	−156.9	−50.6	97.5	
	aq	−150.2	−67.2	176.1	
NH_4HSO_3	aq	−758.7	−607.0	253.1	
NH_4HSO_4	c	−1026.96			
std. state	aq	−1019.85	−835.38	245.2	−3.8
$(NH_4)_3PO_4$	c	−1671.9			
std. state	aq	−1674.9	−1256.9	117	
$(NH_4)_4P_2O_7$ std. state	aq	−2801.2	−2236.8	335	
$(NH_4)_2PtCl_6$	c	−803.3			237.7
NH_4ReO_4	c	−945.6	−774.9	232.6	
$(NH_4)_2S$	aq	−231.8	−72.8	212.1	
NH_4SCN	aq	−56.1	13.4	257.7	39.7
NH_4HSeO_4 std. state	aq	−714.2	−531.6	262.8	
$(NH_4)_2SeO_4$	aq	−864.0	−599.8	280.7	
$(NH_4)_2SiF_6$	c	−2681.69	−2365.3	280.24	228.11
$(NH_4)_2SO_3$	aq	−900.4	−645.0	197.5	
$(NH_4)_2SO_4$	c	−1180.9	−901.70	220.1	187.49
std. state	aq	−1174.28	−903.37	246.9	−133.1
$(NH_4)_2S_2O_8$	c	−1648.08			
std. state	aq	−1610.0	−1273.6	471.1	
NH_4VO_3	c	−1053.1	−888.3	140.6	129.33

(*Continued*)

TABLE 1.58   Enthalpies and Gibbs Energies of Formation, Entropies, and Heat Capacities of the Elements and Inorganic Compounds (*Continued*)

Substance	Physical state	$\Delta_f H°$ kJ · mol^{-1}	$\Delta_f G°$ kJ · mol^{-1}	$S°$ J · deg^{-1} · mol^{-1}	$C°_p$ J · deg^{-1} · mol^{-1}
Antimony					
Sb	c	0	0	45.7	25.2
	g	262.3	222.1	180.3	20.8
SbBr$_3$	c	−259.4	−239.3	207.1	
	g	−194.6	−223.9	372.9	80.2
SbCl$_3$	c	−382.0	−323.7	184.1	107.9
SbCl$_5$	lq	−440.16	−350.2	301	
SbF$_3$	c	−915.5			
SbH$_3$	g	145.11	147.74	232.8	41.05
SbI$_3$	c	−100.4		215.5	97.57
Sb$_2$O$_3$	c	−708.8		123.01	101.25
Sb$_2$O$_5$	c	−971.9	−829.2	125.1	117.61
Sb$_2$S$_3$	c	−174.9		182.0	117.74
Sb$_2$Te$_3$	c	−56.5	−55.2	234	
Argon					
Ar	g	0	0	154.846(3)	20.79
Arsenic					
As gray	c	0	0	35.1	24.64
AsBr$_3$	g	−130.0	−159.0	363.9	79.16
AsCl$_3$	lq	−305.0	−259.4	216.3	133.5
	g	−261.5	−248.9	327.06	75.73
AsF$_3$	lq	−821.3	−774.2	181.2	126.2
	g	−785.8	−770.8	289.1	65.6
AsH$_3$	g	66.44	68.91	222.8	38.07
AsI$_3$	c	−58.2	−59.4	213.05	105.77
AsO$_2^-$	aq	−429.0	−350.0	40.6	
AsO$_4^{3-}$	aq	−888.1	−648.4	−162.8	
As$_2$O$_5$	c	−924.87	−782.3	105.4	116.5
As$_4$O$_6$ octahedral	c	−1313.94	−1152.52	214.2	191.29
As$_2$S$_3$	c	−169.0	−168.6	163.6	116.3
Astatine					
At	c	0	0	121.3	
Barium					
Ba	c	0	0	62.48	28.10
Ba^{2+} std. state	aq	−537.64	−560.74	9.6	
Ba(OAc)$_2$ acetate	c	−1484.5			
std. state	aq	−1509.67	−1299.55	182.8	
BaBr$_2$	c	−757.3	−736.8	146.0	77.0
std. state	aq	−780.73	−768.68	174.5	
BaBr$_2$ · 2H$_2$O	c	−1366.1	−1230.5	226	
Ba(BrO$_3$)$_2$	c	−752.66	−577.4	243	
BaC$_2$O$_4$ oxalate	c	−1368.6			
BaCl$_2$	c	−855.0	−806.7	123.67	75.14
BaCl$_2$ · 2H$_2$O	c	−1456.9	−1293.2	202.9	161.96
Ba(ClO$_3$)$_2$	c	−762.7			
Ba(ClO$_3$)$_2$ · H$_2$O	c	−1691.6	−1270.7	393	
BaCO$_3$ witherite	c	−1213.0	−1134.4	112.1	86.0
BaCrO$_4$	c	−1446.0	−1345.3	158.6	
BaF$_2$	c	−1207.1	−1156.8	96.4	71.20
std. state	aq	−1202.90	−1118.38	−17.0	
Ba(HCO$_3$)$_2$ std. state	aq	−1921.63	−1734.4	192.1	

TABLE 1.58   Enthalpies and Gibbs Energies of Formation, Entropies, and Heat Capacities of the Elements and Inorganic Compounds (*Continued*)

Substance	Physical state	$\Delta_f H°$ kJ · mol⁻¹	$\Delta_f G°$ kJ · mol⁻¹	$S°$ J · deg⁻¹ · mol⁻¹	$C_p°$ J · deg⁻¹ · mol⁻¹
Ba(H₂PO₂)₂	c	−1762.3			
BaI₂	c	−602.1	−601.4	165.1	77.49
std. state	aq	−648.02	−663.92	232.2	
Ba(IO₃)₂	c	−1027.2	−864.8	249.4	187.4
std. state	aq	−980.3	−816.7	246.4	
BaMnO₄	c	−1548	−1439.7	138	140.6
BaMoO₄	c	−1507.5	−1439.7	144.3	114.7
Ba(NO₂)₂	c	−768.2			
Ba(NO₃)₂	c	−988.0	−792.6	213.8	151.38
std. state	aq	−952.36	−783.41	302.5	
BaO	c	−548.0	−520.4	72.07	47.28
BaO₂	c	−634.3			
Ba(OH)₂	c	−944.7	−859.5	107	101.6
Ba(OH)₂ · H₂O	c	−3342.2	−2793.2	427	
BaS	c	−460.0	−456.0	78.2	49.37
BaSe	c	−372			
BaSeO₃	c	−1040.6	−968.2	167	
BaSiF₆	c	−1952.2	−2794.1	163	
BaSO₃	c	−1179.5			
BaSO₄	c	−1473.19	−1362.2	132.2	101.75
BaTiO₃	c	−1659.8	−1572.4	108.0	102.47
Beryllium					
Be	c	0	0	9.50(8)	16.38
	g	324.(5)		136.275(3)	
Be²⁺ std. state	aq	−382.8	−379.7	−129.7	
BeAl₂O₄ chrysoberyl	c	−2301.0	−2178.5	66.29	105.38
BeBr₂	c	−353.5	−337	108.0	69.4
Be₂C	c	91	−88	16.3	43.2
BeCl₂ β form	c	−490.4	−445.6	75.81	62.43
BeCO₃	c	1025.0		52.0	65.0
BeF₂ α form	c	−1026.8	−979.4	53.35	51.82
BeI₂	c	−192.5	−187	121.0	71.1
Be₃N₂ cubic	c	−588.3	−532.9	34.13	64.36
BeO α form	c	−609.4(25)	−580.1	13.77(4)	25.56
BeO₂²⁻	aq	−790.8	−640.1	−159.0	
3BeO · B₂O₃	c	−3105	−2939	100	139.7
Be(OH)₂ β form	c	−902.5	−815.0	45.5	62.1
BeS	c	−234.3	−233.0	34.0	34.0
BeSeO₄	c	−1205.2	−1093.8	77.9	85.7
std. state	aq	−982.0	−820.9	−75.7	
Be₂SiO₄	c	−2117	−2003	64.19	95.6
BeSO₄	c	−1200.8	−1089.4	77.97	85.70
std. state	aq	−1290.0	−1124.3	−109.6	
BeSO₄ · H₂O	c	−2423.75	−2080.66	232.97	216.61
BeWO₄	c	−1513	−1405	88.4	97.3
Bismuth					
Bi	c	0	0	56.7	25.5
	g	207.1	168.2	187.0	20.8
BiBr₃	c	264	234	226	109
BiCl₃	c	−379.1	−315.1	177.0	105.0
BiH₃	g	277.8			

(*Continued*)

TABLE 1.58 Enthalpies and Gibbs Energies of Formation, Entropies, and Heat Capacities of the Elements and Inorganic Compounds (*Continued*)

Substance	Physical state	$\Delta_f H°$ kJ · mol^{-1}	$\Delta_f G°$ kJ · mol^{-1}	$S°$ J · deg^{-1} · mol^{-1}	$C_p°$ J · deg^{-1} · mol^{-1}
BiI_3	c	−100.4	−175.3		
Bi_2O_3	c	−574.0	−493.7	151.5	113.5
BiOCl	c	−366.9	−322.2	120.5	
Bi_2S_3	c	−143.1	−140.6	200.4	122.2
$Bi_2(SO_4)_3$	c	−2544.3			
Bi_2Te_3	c	−78.24		260.91	152.21
Boron					
B	c	0	0	5.90(8)	11.1
	g	565.(5)		153.436(15)	
BBr_3	lq	−239.7	−238.5	229.7	128.03
B_4C	c	−62.7	−62.1	27.18	53.76
BCl_3	g	−403.8	−388.7	290.1	62.7
BF_3	g	−1136.0(8)	−1119.4	254.42(20)	50.45
BF_4^- std. state	aq	−1574.9	−1487.0	179.9	
BH_3	g	100.0	111	187.9	36.22
BH_4^- std. state	aq	48.16	114.27	110.5	
B_2H_6 diborane(6)	g	35.6	86.7	232.1	56.9
B_5H_9 pentaborane(9)	lq	42.7	171.8	184.2	151.13
$B_{10}H_{14}$ decaborane(14)	c	−29.83	212.9	234.9	221.2
BN	c	−254.4	−228.4	14.80	19.72
$B_3N_3H_6$ borazine	lq	−541.0	−392.7	199.6	
	g	−510	−389	288.61	96.94
BO_2^- std. state	aq	−772.37	−678.94	−37.24	
B_2O_3	c	−1273.5(14)	−1194.3	53.97(30)	62.8
$B(OH)_4^-$ std. state	aq	−1344.03	−1153.32	102.5	
$B_3O_3H_3$ boroxin	c	−1262	−11.56	167	98.3
B_2S_3	c	−240.6		100.0	111.7
Bromine					
Br atomic	g	111.87(12)	82.4	175.018(4)	20.8
Br$^-$ std. state	aq	−121.41(15)	−103.97	82.55(20)	−141.8
Br_2	lq	0	0	152.21(30)	75.67
	g	30.91(11)		245.468(5)	
Br_3^- std. state	aq	−130.42	−107.07	215.5	
BrCl	g	14.6	−0.96	239.91	34.98
BrF	g	−93.8	−109.2	229.0	32.97
BrF_3	lq	−300.8	−240.5	178.2	124.6
	g	−255.6	229.4	292.5	66.6
BrF_5	lq	−458.6	−351.9	225.1	
	g	−428.9	−351.6	323.2	99.6
BrO$^-$ std. state	aq	−94.1	−33.5	42.0	
BrO_3^- std. state	aq	−67.07	18.6	161.71	
BrO_4^-	aq	13.0	118.1	199.6	
Cadmium					
Cd	c	0	0	51.80(15)	25.9
	g	111.80(20)		167.749(4)	20.8
Cd^{2+}	aq	−75.92(60)		−72.8(15)	
$CdBr_2$	c	−316.18	−296.31	137.2	76.7
std. state	aq	−318.99	−285.52	91.6	
$CdCl_2$	c	−391.6	−343.9	115.3	74.7
std. state	aq	−410.20	−340.12	39.8	
$CdCl_2 \cdot 5/2H_2O$	c	−1131.94	−944.08	227.2	

TABLE 1.58 Enthalpies and Gibbs Energies of Formation, Entropies, and Heat Capacities of the Elements and Inorganic Compounds (*Continued*)

Substance	Physical state	$\Delta_f H°$ kJ · mol⁻¹	$\Delta_f G°$ kJ · mol⁻¹	$S°$ J · deg⁻¹ · mol⁻¹	$C°_p$ J · deg⁻¹ · mol⁻¹
$Cd(CN)_2$	c	162.3			
std. state	aq	225.5	267.4	115.1	
$CdCO_3$	c	−750.6	−669.4	92.5	
$Cd(OAc)_2$ std. state	aq	−1047.9	−816.4	100	
CdF_2	c	−700.4	−647.7	77.4	
std. state	aq	−741.15	−635.21	−100.8	
CdI_2	c	−203.3	−201.4	161.1	80.0
std. state	aq	−186.3	−180.8	149.4	
CdI_4^- std. state	aq	−341.8	−315.9	326	
$Cd(NH_3)_4^{2+}$ std. state	aq	−450.2	−226.4	336.4	
$Cd(NO_3)_2$	c	−456.3			
std. state	aq	−490.6	−300.2	219.7	
CdO	c	−258.35(40)	−228.7	54.8(15)	43.4
$Cd(OH)_2$	c	−560.7	−473.6	96.0	
CdS	c	−161.9	−156.5	64.9	55.5
$CdSO_4$	c	−933.4	−822.7	123.0	99.6
std. state	aq	−985.2	−822.2	−53.1	
$CdSO_4 · 8/3H_2O$	c	−1729.30(80)	−1465.3	229.65(40)	213.3
$CdSeO_4$	c	−633.0	−531.8	164.4	
std. state	aq	−674.9	−518.8	−19.3	
$CdTe$	c	−92.5	−92.0	100.0	
Calcium					
Ca	c	0	0	41.59(40)	25.9
	g	177.8(8)		154.887(4)	
Ca^{2+} std. state	aq	−543.0(10)	−553.54	−56.2(10)	
$Ca(OAc)_2$	c	−1479.5			
std. state	aq	−1514.73	−1292.35	120.1	
$Ca_3(AsO_4)_2$	c	−3298.7	−3063.1	226	
$Ca(BO_2)_2$	c	−2030.9	−1924.1	104.85	103.98
CaB_4O_7	c	−3360.3	−3167.1	134.7	157.9
$CaBr_2$	c	−682.8	−663.6	130.0	75.04
std. state	aq	−785.9	−761.5	111.7	
CaC_2	c	−59.8	64.9	69.96	62.72
$CaCl_2$	c	−795.4	−748.8	108.4	72.9
std. state	aq	−877.13	−816.05	59.8	
$CaCl_2 · 2H_2O$	c	−1402.9			738
$CaCN_2$ cyanamide	c	−350.6			
$Ca(CN)_2$	c	−184.5			
$CaCO_3$ calcite	c	−1207.6	−1129.1	91.7	83.5
aragonite	c	−1207.8	−1128.2	88.0	82.3
	aq	−1220.0	−1081.4	−110.0	
CaC_2O_4	c	−1360.6			
$CaC_2O_4 · H_2O$	c	−1674.9	−1514.0	156.5	152.8
$CaCrO_4$	c	−1379.1	−1277.4	134	
CaF_2	c	−1228.0	−1175.6	68.6	67.0
	aq	−1208.1	−1111.2	−80.8	
$Ca(formate)_2$	c	1386.6			
CaH_2	c	−181.5	−142.5	41.4	41.0
$CaHPO_4 · 2H_2O$	c	−2403.58	−2154.75	189.45	197.07
$Ca(H_2PO_2)_2$ hypophosphite	c	−1752.7			
$Ca(H_2PO_4)_2$ std. state	aq	−3135.41	−2814.33	127.6	

(Continued)

TABLE 1.58 Enthalpies and Gibbs Energies of Formation, Entropies, and Heat Capacities of the Elements and Inorganic Compounds (*Continued*)

Substance	Physical state	$\Delta_f H°$ kJ · mol^{-1}	$\Delta_f G°$ kJ · mol^{-1}	$S°$ J · deg^{-1} · mol^{-1}	$C°_p$ J · deg^{-1} · mol^{-1}
Ca(H$_2$PO$_4$)$_2$. H$_2$O	c	−3409.67	−3058.42	259.8	258.82
CaI$_2$	c	−533.5	−528.9	142.0	77.16
std. state	aq	−653.2	−656.7	169.5	
Ca(IO$_3$)$_2$	c	−1002.5	−839.3	230	
Ca[Mg(CO$_3$)$_2$] dolomite	c	−2326.3	−2163.6	155.18	157.53
CaMoO$_4$	c	−1541.4	−1434.7	122.6	114.3
Ca$_3$N$_2$	c	−439.3		105.0	113.0
Ca(NO$_2$)$_2$	c	−741.4			
Ca(NO$_3$)$_2$	c	−938.2	−742.8	193.3	149.37
std. state	aq	−957.55	−776.22	239.7	
CaO	c	−634.92(90)	−603.3	38.1(4)	42.0
Ca(OH)$_2$	c	−985.2	−897.5	83.4	87.5
Ca$_3$P$_2$	c	−506			
Ca$_3$(PO$_4$)$_2$	c	−4120.8	−3884.8	236.0	227.8
Ca$_2$P$_2$O$_7$	c	−3338.8	−3132.1	189.24	187.8
Ca$_{10}$(PO$_4$)$_6$F$_2$ fluoroapatite	c	−13,744	−12,983	775.7	751.9
CaS	c	−482.4	−477.4	56.5	47.4
CaSe	c	−368.2	−363.2	67	
CaSiO$_3$	c	−1634.9	−1549.7	81.92	85.27
Ca$_2$SiO$_4$	c	−2307.5	−2192.8	127.7	128.8
3CaO · SiO$_2$	c	−2929.2	−2784.0	168.6	171.9
CaSO$_3$ · 2H$_2$O	c	−1752.7	−1555.2	184	178.7
CaSO$_4$	c	−1425.2	−1309.1	108.4	99.0
	aq	−1451.1	−1298.1	−33.1	
CaSO$_4$. ½ H$_2$O	c	−1576.7	−1436.8	130.5	119.4
CaSO$_4$ · 2H$_2$O	c	−2022.6	−1797.5	194.1	186.0
Ca(VO$_3$)$_2$	c	−2329.3	−2169.7	179.1	166.8
CaWO$_4$	c	−1645.15	−1538.50	126.40	114.14
Carbon					
C graphite	c	0	0	5.74(10)	8.517
	g	716.68(45)		158.100(3)	
diamond	c	1.897	2.900	2.377	6.116
CN$^-$	aq	150.6	172.4	94.1	
(CN)$_2$ cyanogen	g	306.7	297.2	241.9	56.9
CNBr	g	186.2	165.3	248.36	46.9
CNCl	g	137.95	131.02	236.2	45.0
CNF	g			224.7	41.8
CNI	c	166.2	185.0	96.2	
	g	225.5	196.6	256.8	48.3
CNN$_3$ cyanogen azide	c	387.4			
OCN$^-$	aq	−146.0	−97.4	106.7	
CO	g	−110.53(17)	−137.16	197.660(4)	29.14
CO$_2$	g	−393.51(13)	394.39	213.785(10)	37.13
undissoc; std. state	aq	−413.26(20)	−386.0	119.36(60)	
CO$_3^{2-}$	aq	−675.23(25)	−527.9	−50.0(10)	
C$_3$O$_2$ suboxide	g	−93.7	−109.8	276.4	67.0
COBr$_2$	g	−96.2	−110.9	309.1	61.8
COCl$_2$ phosgene	g	−219.1	−204.9	283.50	57.70
COClF	g			276.7	52.4
COF$_2$	g	−639.8	−623.33	258.89	46.8

TABLE 1.58 Enthalpies and Gibbs Energies of Formation, Entropies, and Heat Capacities of the Elements and Inorganic Compounds (*Continued*)

Substance	Physical state	$\Delta_f H°$ kJ · mol^{-1}	$\Delta_f G°$ kJ · mol^{-1}	$S°$ J · deg^{-1} · mol^{-1}	$C°_p$ J · deg^{-1} · mol^{-1}
COS carbonyl sulfide	g	−142.0	−166.9	231.56	41.50
CS$_2$	lq	89.0			74.6
	g	117.7	67.1	237.8	45.4
CTe$_2$	lq	164.8			
Cerium					
Ce γ, fcc	c	0	0	72.0	26.9
Ce^{3+} std. state	aq	−696.2	−672.0	−205.0	
Ce^{4+} std. state	aq	−537.2	−503.8	−301.0	
CeCl$_3$	c	−1060.5	−984.8	151.0	87.4
std. state	aq	−1197.5	−1065.7	−38.0	
CeF$_3$	c	−1635.9	−1556	115.1	99.3
CeI$_3$	c	−669.3	−674	209	
Ce(NO$_3$)$_3$	c	−1225.9			
CeO$_2$	c	−1088.7	−1024.7	62.30	61.63
Ce$_2$O$_3$	c	−1796.2	−1706.2	150.6	114.6
CeS	c	−459.4	−451.5	78.2	50.0
Ce$_2$(SO$_4$)$_3$	c	−3954.3			
std. state	aq	−4176.9	−3652.6	−318	
Ce$_2$(SO$_4$)$_3$ · 8H$_2$O	c	−5522.9	−5607.4		
Cesium					
Cs	c	0	0	85.23(40)	32.20
	lq	2.087	0.025	92.1	32.4
	g	76.5(10)		175.601(3)	
Cs$^+$ std. state	aq	−258.00(50)	−292.0	132.1(5)	−10.5
Cs acetate	aq	−744.3	−661.3	219.7	
CsBO$_2$	c	−972.0	−915.0	104.4	80.6
CsBr	c	−405.8	−391.4	113.05	52.93
std. state	aq	−379.8	−396.0	215.5	
CsCl	c	−442.8	414.4	101.18	52.44
std. state	aq	−425.4	−423.3	189.4	−146.9
CsClO$_4$	c	−443.1	−314.3	175.1	108.3
Cs$_2$CO$_3$	c	−1139.7	−1054.4	204.5	123.9
std. state	aq	−1193.7	−1111.9	209.2	
CsF	c	−553.5	−525.5	92.8	51.1
std. state	aq	−590.9	−570.8	119.2	
Cs formate	aq	−683.8	−643.0	226.0	
CsHCO$_3$	c	−966.1			
CsHF	c	−923.8	−858.9	135.2	87.3
CsHSO$_4$	c	−1158.1			
	aq	−1145.6	−1047.9	264.8	
CsI	c	−346.6	−340.6	123.1	52.8
std. state	aq	−313.5	−343.6	244.4	−152.7
CsIO$_3$	c	−525.9	−433.9		167
CsNO$_3$	c	−506.0	−406.6	155.2	
std. state	aq	−465.6	−403.3	279.5	−99.2
Cs$_2$O	c	−345.8	−308.2	146.9	76.0
CsOH	c	−417.2	370.7	98.7	67.9
std. state	aq	−488.3	−449.3	122.3	
Cs$_2$PtCl$_6$ std. state	aq	−1184.9	−1066.9	485.8	
Cs$_2$S	aq	−483.7	−498.3	251.0	
Cs$_2$Se	aq		454.8		

(*Continued*)

TABLE 1.58 Enthalpies and Gibbs Energies of Formation, Entropies, and Heat Capacities of the Elements and Inorganic Compounds (*Continued*)

Substance	Physical state	$\Delta_f H°$ kJ · mol^{-1}	$\Delta_f G°$ kJ · mol^{-1}	$S°$ J · deg^{-1} · mol^{-1}	$C°_p$ J · deg^{-1} · mol^{-1}
Cs$_2$SO$_4$	c	−1443.0	−1323.6	211.9	134.9
std. state	aq	−1425.8	−1328.6	286.2	
Chlorine					
Cl atomic	g	121.301(8)		165.190(4)	
Cl$^-$ std. state	aq	−167.08(10)	−131.3	56.60(20)	−136.4
Cl$_2$	g	0	0	233.08(10)	33.95
ClF	g	−50.3	−51.84	217.9	32.08
ClF$_3$	g	−163.2	−123.0	281.6	63.85
ClF$_5$	g	−239	−147	310.74	97.17
ClO	g	101.8	98.1	226.6	31.5
ClO$^-$ std. state	aq	−107.1	−36.8	41.8	
ClO$_2$	g	102.5	120.5	256.8	42.00
ClO$_2^-$ std. state	aq	−66.5	17.2	101.3	
ClO$_3^-$ std. state	aq	−104.0	−8.0	162.3	
ClO$_3$F perchloryl fluoride	g	−23.8	48.2	279.0	64.9
ClO$_4^-$ std. state	aq	−128.10(40)	−8.62	184.0(15)	
Cl$_2$O	g	80.3	97.9	266.2	45.4
Cl$_2$O$_7$	lq	238.1			
	g	1138			
Chromium					
Cr	c	0	0	23.8	23.43
Cr^{2+} std. state	aq	−143.5			
CrBr$_2$	c	−302.1			
CrCl$_2$	c	−395.4	−356.0	115.3	71.2
CrCl$_3$	c	−556.5	−486.1	123.0	91.8
Cr(CO)$_6$ hexacarbonyl	c	−1077.8		293.01	226.23
CrF$_2$	c	−778.0			
CrF$_3$	c	−1159	−1088	93.9	78.7
Cr$_2$FeO$_4$	c	−1444.7	−1343.8	146.0	133.6
CrI$_2$	c	−156.9			
CrI$_3$	c	−205.0			
CrN	c	−117	−93	38	52.7
CrO$_2$	c	−598.0			
Cr$_2$O$_3$	c	−1140	−1058.1	81.2	118.7
Cr$_3$O$_4$	c	−1131.0			
CrO$_2$Cl$_2$	g	−538.1	−501.6	329.8	84.5
CrO$_4^{2-}$ std. state	aq	−881.15	−727.85	50.21	
HCrO$_4^-$ std. state	aq	−878.22	−764.84	184.1	
Cr$_2$O$_7^{2-}$ std. state	aq	−1490.3	−1301.2	261.9	
Cr$_2$(SO$_4$)$_3$	c	−609.6		269.9	302.6
Cobalt					
Co	c	0	0	30.0	24.8
Co^{2+} std. state	aq	−58.2	−54.4	−113	
Co^{3+} std. state	aq	92	134	−305	
CoBr$_2$	c	−220.9			79.5
std. state	aq	−301.3	−262.3	50	
CoCl$_2$	c	−312.5	−269.8	109.2	78.49
std. state	aq	−392.5	−316.7	0	
CoCO$_3$	c	−713.0			
CoF$_2$	c	−692	−647	82.4	68.9
CoF$_3$	c	−790	−719	95	92

TABLE 1.58 Enthalpies and Gibbs Energies of Formation, Entropies, and Heat Capacities of the Elements and Inorganic Compounds (*Continued*)

Substance	Physical state	$\Delta_f H°$ kJ · mol^{-1}	$\Delta_f G°$ kJ · mol^{-1}	$S°$ J · deg^{-1} · mol^{-1}	$C°_p$ J · deg^{-1} · mol^{-1}
CoI$_2$	c	−88.7			
	aq	−168.6	−157.7	109.0	
Co(NH$_3$)$_6^{2+}$ std. state	aq	−584.9	−157.3	146	
Co(NH$_3$)$_6^{3+}$ std. state	aq		−189.5		
Co(NO$_3$)$_2$	c	−420.5			
std. state	aq	−472.8	−277.0	180	
CoO	c	−237.7	−214.0	53.0	55.3
Co$_3$O$_4$	c	−891	−774	102.5	123.4
Co(OH)$_2$	c	−539.7	−454.4	79.0	
CoS	c	−82.8			
Co$_2$S$_3$	c	−147.3			
CoSO$_4$	c	−888.3	−782.4	118.0	103
std. state	aq	−967.3	−799.1	92.0	
CoSO$_4$ · 7H$_2$O	c	−2979.93	−2473.83	406.06	390.49
Copper					
Cu	c	0	0	33.15(8)	24.44
	g	337.4(12)		166.398(4)	
Cu$^+$ std. state	aq	71.67	50.00	40.6	
Cu^{2+} std. state	aq	64.9(10)	65.52	−98.(4)	
Cu(OAc)$_2$ acetate	c	−893.3			
std. state	aq	−907.25	−673.29	73.6	
Cu$_3$(AsO$_4$)$_2$ std. state	aq	−1581.97	−1100.48	−804.2	
CuBr	c	−104.6	−100.8	96.2	54.7
CuBr$_2$	c	−141.84			
CuCl	c	−137.2	−119.9	86.2	48.5
CuCl$_2$	c	−220.1	−175.7	108.09	71.88
Cu(ClO$_4$)$_2$ std. state	aq	−193.89	48.28	264.4	
CuCN	c	95.0	108.4	90.00	61.04
CuCNS std. state	aq	138.11	142.67	184.93	
Cu(CNS)$_2$ std. state	aq	217.65	250.87	189.1	
CuF	c	−280	−260	64.9	51.9
CuF$_2$	c	−542.7	−492	77.45	65.55
Cu(formate)$_2$	aq	−786.34	−636.4	84	
CuI	c	67.8	−69.5	96.7	54.1
Cu(NH$_3$)$_4^{2+}$ std. state	aq	−348.5	−111.3	273.6	
Cu(NO$_3$)$_2$	c	−302.9			
std. state	aq	−349.95	−157.15	193.3	
CuO	c	−157.3	−129.7	42.6	42.2
Cu$_2$O	c	−168.6	−149.0	93.1	63.6
Cu(OH)$_2$	c	−450	−373	108.4	95.19
CuS	c	−53.1	−53.7	66.5	47.8
Cu$_2$S	c	−79.5	−86.2	120.9	76.3
CuSe	c	−39.5			
Cu$_2$Se	c	−59.4		157.3	88.70
CuSO$_4$	c	−771.4(12)	−662.2	109.2(4)	98.87
std. state	aq	−844.50	−679.11	−79.5	
CuSO$_4$ · 5H$_2$O	c	−2279.65	−1880.04	300.4	280
CuWO$_4$	c	−1105.0			
Dysprosium					
Dy	c	0	0	75.6	27.7
Dy^{3+} std. state	aq	−699.0	−665.0	−231.0	21.0

(*Continued*)

TABLE 1.58 Enthalpies and Gibbs Energies of Formation, Entropies, and Heat Capacities of the Elements and Inorganic Compounds (*Continued*)

Substance	Physical state	$\Delta_f H°$ kJ · mol^{-1}	$\Delta_f G°$ kJ · mol^{-1}	$S°$ J · deg^{-1} · mol^{-1}	$C°_p$ J · deg^{-1} · mol^{-1}
DyCl$_3$	c	−1000			100.0
	aq	−1197.0	−1059.0	−61.9	−389.0
DyF$_3$	c	−1711.0			
Dy$_2$O$_3$	c	−1863.1	−1771.5	149.8	116.27
Erbium					
Er	c	0	0	73.18	28.12
Er^{3+} std. state	aq	−705.4	−669.1	−244.3	21.0
ErCl$_3$	c	−998.7			100.0
	aq	−1207.1	−1062.7	−75.3	−389.0
Er$_2$O$_3$	c	−1897.9	−1808.7	155.6	108.49
Europium					
Eu	c	0	0	77.78	27.66
Eu^{2+} std. state	aq	−527.0	540.2	−8.0	
Eu^{3+}	aq	−605.0	−574.0	−222.0	8.0
EuCl$_2$	aq	−862.0			
EuCl$_3$	c	−936.0	−856	144.1	
	aq	−1106.2	−967.7	−54.0	−402.0
EuF$_3$	c	−1571			
Eu$_2$O$_3$ monoclinic	c	−1651.4	−1556.9	146	122.2
Eu$_3$O$_4$	c	−2272.0	−2142.0	205.0	
Eu(OH)$_3$	c	−1332	−1195	119.9	
Fluorine					
F atomic	g	79.38(30)	62.3	158.751(4)	22.7
F$^-$	aq	−335.35(65)	−278.8	−13.8(8)	−106.7
F$_2$	g	0	0	202.791(5)	31.30
FNO$_3$	g	10.5	73.7	292.9	65.22
FO	g	109.0	105.0	216.8	30.5
F$_2$O	g	24.7	41.9	247.4	43.3
F$_2$O$_2$	g	18.0			
Francium					
Fr	c	0	0	95.40	31.80
FrCl	c	−439		113.0	53.56
Fr$_2$O	c	−338	299.2	156.9	
Gadolinium					
Gd	c	0	0	68.07	37.03
Gd^{3+} std. state	aq	−686.0	−661.0	−205.9	
GdCl$_3$	c	−1008.0	−933	151.4	88.0
std. state	aq	−1188.0	−1059.0	−36.8	−410.0
GdF$_3$	lq	−1297			
Gd$_2$O$_3$ monoclinic	c	−1819.6	−1730	150.6	106.7
Gallium					
Ga	c	0	0	40.8	26.06
	lq	5.6			
	g	272.0	233.7	169.0	25.3
Ga^{3+}	aq	−211.7	−159.0	−331.0	
GaAs	c	−71.0	−67.8	64.2	46.2
GaBr$_3$	c	−386.6	−359.8	180.0	
GaCl$_3$	c	−524.7	−454.8	142.0	
GaF$_3$	c	−1163.0	−1085.3	84	
GaI$_3$	c	−238.9		205.0	100
Ga$_2$O$_3$ rhombic	c	−1089.1	−998.3	84.98	92.1

TABLE 1.58 Enthalpies and Gibbs Energies of Formation, Entropies, and Heat Capacities of the Elements and Inorganic Compounds (*Continued*)

Substance	Physical state	$\Delta_f H°$ kJ · mol⁻¹	$\Delta_f G°$ kJ · mol⁻¹	$S°$ J · deg⁻¹ · mol⁻¹	$C°_p$ J · deg⁻¹ · mol⁻¹
Ga(OH)₃	c	−964.4	−831.3	100.0	
GaSb	c	−41.8	−38.9	76.07	48.53
Germanium					
Ge	c	0	0	31.09(15)	23.3
	g	372.0(30)	331.2	167.904(5)	30.7
GeBr₄	lq	−347.7	−331.4	280.8	
	g	−300.0	−318.0	396.2	101.8
GeCl₄	lq	−531.8	−462.8	245.6	
	g	−495.8	−457.3	347.7	96.1
GeF₄	g	−1190.20(50)	−1150.0	301.9(10)	81.84
GeH₄	g	90.8	113.4	217.02	45.02
GeI₄	c	−141.8	−144.4	271.1	
	g	−56.9	−106.3	428.9	104.1
GeO₂ tetragonal	c	−580.0(10)	−521.4	39.71(15)	52.1
GeP	c	−21.0	−17.0	63.0	
GeS	c	−69.0	−71.6	71	
Gold					
Au	c	0	0	47.4	25.36
AuBr	c	−14.0			
AuBr₃	c	−53.3			
AuCl	c	−34.7		92.9	48.74
AuCl₃	c	−117.6		148.1	94.81
AuCl₄⁻ std. state	aq	−322.2	−237.32	266.9	
Au(CH)₂⁻ std. state	aq	242.3	285.8	172	
AuF₃	c	−363.6		114.2	91.29
AuSb₂	c	−19.46		119.2	77.40
AuSn	c	−30.5		93.7	49.41
Hafnium					
Hf hexagonal	c	0	0	43.56	25.69
HfC	c	−230.1		41.21	34.43
HfCl₄	c	−990.4	−901.3	190.8	120.46
HfF₄ monoclinic	c	−1930.5	−1830.5	113	
HfO₂	c	−1144.7	−1088.2	59.3	60.25
Helium					
He	g	0	0	126.153(2)	20.786
Holmium					
Ho	c	0	0	75.3	27.15
Ho³⁺ std. state	aq	−705.0	−673.7	226.8	17.0
HoCl₃	c	−1005.4		88	
std. state	aq	−1206.7	−1067.3	−57.7	−393.0
HoF₃	c	−1707.0			
Ho₂O₃	c	−1880.7	−1791.2	158.2	115.0
Hydrogen					
H atomic	g	217.998(6)	203.3	114.717(2)	20.8
H⁺ std. state	aq	0	0	0	0
H₂	g	0	0	130.680(3)	28.84
H²H	g	0.321	−1.463	143.80	29.20
²H₂ (D₂) deuterium	g	0	0	144.96	29.19
HAsO₂ undissoc; std. state	aq	−456.5	−402.71	125.9	

(*Continued*)

TABLE 1.58 Enthalpies and Gibbs Energies of Formation, Entropies, and Heat Capacities of the Elements and Inorganic Compounds (*Continued*)

Substance	Physical state	$\Delta_f H°$ kJ · mol^{-1}	$\Delta_f G°$ kJ · mol^{-1}	$S°$ J · deg^{-1} · mol^{-1}	$C°_p$ J · deg^{-1} · mol^{-1}
$H_2AsO_3^-$ undissoc; std. state	aq	−714.79	−587.22	110.5	
H_3AsO_3 undissoc; std. state	aq	−742.2	−639.90	195.0	
$HAsO_4^{2-}$ undissoc; std. state	aq	−906.34	−714.70	−1.7	
$H_2AsO_4^-$ undissoc; std. state	aq	−909.56	−753.29	117	
H_3AsO_3	c	−906.30			
undissoc; std. state	aq	−902.5	−766.1	184	
HBO_2	c	−794.3	−723.4	38	54.4
H_3BO_3	c	−1094.8(8)	−968.9	89.95(60)	86.1
undissoc	aq	−1072.8(8)		162.4(6)	
HBr	g	−36.29(16)	−53.4	198.700(4)	29.1
std. state	aq	−121.55	−103.97	82.4	−141.8
HBrO undissoc; std. state	aq	−113.0	−82.4	142	
$HBrO_3$ std. state	aq	−67.07	18.54	161.71	
HCl	g	−92.31(10)	−95.30	186.902(5)	29.12
std. state	aq	−167.15	−131.25	56.5	−136.4
^{2}HCl deuterium chloride	g	−93.35	−95.94	192.63	29.17
HClO	g	−78.7	−66.1	236.7	37.15
undissoc; std. state	aq	−120.9	−79.9	142	
$HClO_2$ undissoc; std. state	aq	−51.9	5.9	188.3	
$HClO_3$ std. state	aq	−103.97	−8.03	162.3	
$HClO_4$	lq	−40.58			
std. state	aq	−129.33	−8.62	182.0	
$HClO_4H_2O$	c	−302.21			
$HClO_4 \cdot 2H_2O$	lq	−677.98			
HCN	lq	108.87	124.93	112.84	70.63
	g	135.1	124.7	201.81	35.86
ionized; std. state	aq	150.6	172.4	94.1	
undissoc; std. state	aq	107.11	119.66	124.7	
HCNO ionized; std. state	aq	−146.0	−97.5	106.7	
undissoc; std. state	aq	−154.39	−117.2	144.8	
HCNS ionized; std. state	aq	76.44	92.68	144.4	−40.2
HCOO$^-$ formate	aq	−425.6	−351.0	92.0	−87.9
CH_3COO^- acetate	aq	−486.0	−369.3	86.6	−6.3
HCO_3^- std. state	aq	−689.93(20)	−586.85	98.4(5)	
H_2CO_3 std. state	aq	−699.65	−623.16	187.4	
$HC_2O_4^-$	aq	−818.4	−698.3	149.4	
$H_2C_2O_4$	c	−821.7	−723.7	109.8	91.0
$C_2O_4^{2-}$	aq	−825.1	−673.9	45.6	
H_2CS_3 trithiocarbonic acid	lq	25.1	27.82	233.0	149.8
HF	g	−273.30(70)	−275.4	173.779(3)	29.14
	lq	−299.78	75.40	51.67	
undissoc; std. state	aq	−320.08	−296.86	88.7	
F$^-$	aq	−332.63	−278.8	−13.8	−106.7
^{2}HF	g	−275.5	−277.27	179.70	29.14
HF_2^- std. state	aq	−649.94	−578.15	92.5	
H_2F_2 dimer	g	−572.66	−544.51	238	44.89
$H_2Fe(CN)_6^{2-}$ std. state	aq	455.6	658.44	218	

TABLE 1.58 Enthalpies and Gibbs Energies of Formation, Entropies, and Heat Capacities of the Elements and Inorganic Compounds (*Continued*)

Substance	Physical state	$\Delta_f H°$ kJ · mol⁻¹	$\Delta_f G°$ kJ · mol⁻¹	$S°$ J · deg⁻¹ · mol⁻¹	$C_p°$ J · deg⁻¹ · mol⁻¹
HFO	g	98	−86	226.8	35.93
HI	g	26.50(10)	1.7	206.590(4)	29.16
std. state	aq	−55.19	−51.59	111.3	−142.3
HIO undissoc; std. state	aq	−138.1	−99.2	95.4	
HIO₃	c	−230.1			
H₂MoO₄	c	−1046.0			
HN	g	351.5	345.6	181.2	29.2
HN₃	lq	264.0	327.2	140.6	
	g	294.1	328.1	239.0	43.7
H₂N	g	184.9	194.6	195.0	33.9
²H₂N₂ *cis*-diazine	g	207	241	224.09	39.02
HNCO isocyanic acid	g	−116.73	−107.36	238.11	44.85
HNCS isothiocyanic acid	g	127.61	112.88	248.03	46.40
HNO₂	g	−79.5	−46.0	254.1	45.5
HNO₃	lq	−174.1	−80.7	155.60	109.9
	g	−133.9	−73.54	266.9	54.1
std. state	aq	−207.36	−111.34	146.4	−86.6
H₂N₂O₂ hyponitrous acid	aq	−57.3	36.0	218	
HO hydroxyl	g	39.0	34.2	183.64	30.00
HO⁻	aq	−230.015	−157.28	−10.90	−148.5
HO₂	g	10.5	22.6	229.0	34.9
HO₂⁻ std. state	aq	−160.33	67.4	23.9	
H₂O	c	−292.72			37.11
	lq	−285.830(40)	−237.14	69.95(3)	75.35
	g	−241.826(40)	−228.61	188.835(10)	33.60
¹H²HO	g	−245.37	−233.18	199.51	33.79
²H₂O deuterium oxide	g	−249.20	−234.54	198.33	34.25
H₂O₂ hydrogen peroxide	lq	−187.78	−120.42	109.6	89.1
	g	−136.3	−105.6	232.7	43.14
undissoc; std. state	aq	−191.17	−134.10	143.9	
HOCN undissoc; std. state	aq	−154.39	−117.2	144.8	
OCN⁻ cyanate std. state	aq	−146.02	−97.5	106.7	
HPO₃	c	−948.51			
HPO₄²⁻ std. state	aq	−1299.0(15)	−1089.26	−33.5(15)	
H₂PO₄⁻ std. state	aq	−1302.6(15)	−1130.39	92.5(15)	
HPH₂O₂ hypophosphorous acid	c	−604.6			
H₃PO₃	c	−964.4			
H₃PO₄	c	−1284.4	−1124.3	110.5	106.1
	lq	−1271.7	−1123.6	150.8	145.06
ionized; std. state	aq	−1277.4	−1018.8	222	
undissoc; std. state	aq	−1288.34	−1142.65	158.2	
HP₂O₇³⁻	aq	−2274.8	−1972.2	46.0	
H₂P₂O₇²⁻	aq	−2278.6	−2010.2	163.0	
H₄P₂O₇	c	−2241.0			
undissoc; std. state	aq	−2268.6	−2032.2	268	
HReO₄	c	−762.3	−656.4	158.2	
HS	g	142.7	113.3	195.7	32.3
HS⁻ std. state	aq	−16.3(15)	12.05	67.(5)	
H₂S	g	−20.6(5)	−33.4	205.81(5)	34.19
undissoc; std. state	aq	−38.6(15)	−27.87	126.(5)	

TABLE 1.58 Enthalpies and Gibbs Energies of Formation, Entropies, and Heat Capacities of the Elements and Inorganic Compounds (*Continued*)

Substance	Physical state	$\Delta_f H°$ kJ · mol⁻¹	$\Delta_f G°$ kJ · mol⁻¹	$S°$ J · deg⁻¹ · mol⁻¹	$C°_p$ J · deg⁻¹ · mol⁻¹
²H₂S	g	−23.9	−35.3	215.3	35.76
H₂S₂	g	15.5			51.5
HSbO₂ undissoc; std. state	aq	−487.9	−407.5	46.6	
HSCN undissoc; std. state	aq	76.4	97.7	144.3	−40.2
SCN⁻ std. state	aq	76.44	92.68	144.5	−40.2
HSe⁻ std. state	aq	15.9	43.9	79.0	
H₂Se	g	29.7	15.9	219.0	34.7
HSeO₃⁻ std. state	aq	−514.55	−411.54	135.1	
H₂SeO₃	c	−524.46			
undissoc; std. state	aq	−507.48	−426.22	207.9	
HSeO₄⁻ std. state	aq	−581.6	−452.3	149.4	
H₂SeO₄	c	−530.1			
H₂SiO₃	c	−1188.67	−1092.4	134.0	
undissoc; std. state	aq	−1182.8	−1079.5	109	
H₄SiO₄	c	−1481.1	−1333.0	192	
undissoc; std. state	aq	−1468.6	−1316.7	180	
HSO₃⁻ std. state	aq	−626.22	−527.8	139.8	
HSO₄⁻	aq	−886.9(10)	−755.9	131.7(30)	−84.0
HSO₃Cl	lq	−601.2			
HSO₃F	lq	−795.0			
	g	−753	−691	297	75.24
H₂SO₃ undissoc; std. state	aq	−608.81	−537.90	232.2	
H₂SO₄	lq	−814.0	−689.9	156.90	138.9
std. state	aq	−909.27	−744.63	20.1	293
H₂SO₄ · H₂O	lq	−1127.6	−950.3	211.5	214.3
H₂SO₄ · 2H₂O	lq	−1427.1	−1199.6	276.4	261.5
H₂SO₄ · 3H₂O	lq	−1720.4	−1443.9	345.4	319.1
H₂SO₄ · 4H₂O	lq	−2011.2	−1685.8	414.5	386.4
H₂S₂O₇	c	−1273.6			
H₂Te	g	99.6		228.9	35.56
H₂WO₄	c	−1131.8	−1003.9	145	113
Indium					
In	c	0	0	57.8	26.7
In³⁺	aq	−105.0	−98.0	−151.0	
InAs	c	−58.6	−53.6	75.7	47.78
InBr₃	c	−428.9			
InCl₃	c	−537.2			
InF	g	−203.4			
InH	g	215.5	190.3	207.53	29.58
InI	c	−116.3	−120.5	130.0	
InI₃	c	−238.0			
InOH²⁺	aq	−370.3	−313.0	−88.0	
In(OH)₂⁺	aq	−619.0	−525.0	25.0	
In₂O₃	c	−925.27	−830.73	104.2	92
InP	c	−88.7	−77.0	59.8	45.44
InS	c	−138.1	−131.8	67	
In₂S₃	c	−427	−412.5	163.6	118.0
In₂Se₃	c	−343			
InSb	c	−30.5	−25.5	86.2	49.5
Iodine					
I atomic	g	106.76(4)	70.2	180.787(4)	20.8

TABLE 1.58 Enthalpies and Gibbs Energies of Formation, Entropies, and Heat Capacities of the Elements and Inorganic Compounds (*Continued*)

Substance	Physical state	$\Delta_f H°$ kJ · mol^{-1}	$\Delta_f G°$ kJ · mol^{-1}	$S°$ J · deg^{-1} · mol^{-1}	$C°_p$ J · deg^{-1} · mol^{-1}
I$^-$ std. state	aq	−56.78(5)	−51.59	106.45(30)	−142.3
I$_2$	c	0	0	116.14(30)	54.44
	g	62.42(8)	19.37	260.687(5)	36.86
std. state	aq	22.6	16.40	137.2	
I$_3^-$ std. state	aq	−51.5	−51.5	239.3	
IBr	c	−10.5			
	g	40.8	3.7	258.8	36.4
ICl	c	−35.4	−14.05	97.93	55.23
	lq	−23.93	−13.6	135.1	
	g	17.8	−5.5	247.6	35.6
ICl$_3$	c	−89.5	−22.34	167.4	
IF	g	−95.7	−118.5	236.3	33.4
IF$_5$	lq	−864.8			
	g	−822.5	−751.5	327.7	99.2
IF$_7$	g	−961.1	−835.8	347.7	134.5
IO	g	175.1	149.8	245.5	32.9
IO$^-$ std. state	aq	−107.5	−38.5	−5.4	
IO$_3^-$ std. state	aq	−221.3	−128.0	118.4	
IO$_4^-$ std. state	aq	−151.5	−58.6	222	
I$_2$O$_5$	c	−158.07			
Iridium					
Ir	c	0	0	35.48	25.06
IrCl$_3$	c	−245.6	180	113	
IrF$_6$	c	−579.65	−461.66	247.7	
IrO$_2$	c	−274.1		57.3	57.32
IrS$_2$	c	−138.0			
Iron					
Fe alpha	c	0	0	27.32	25.09
Fe^{2+} std. state	aq	−89.1	−78.87	−137.7	
Fe^{3+} std. state	aq	−48.5	−4.7	−315.9	
FeBr$_2$	c	−249.8	−238.1	140.7	80.2
std. state	aq	−332.2	−286.81	27.2	
FeBr$_3$	c	−286.2			
	aq	−413.4	−316.7	−68.6	
Fe$_3$C α-cementite	c	25.1	20.1	104.6	105.9
FeCl$_2$	c	−341.8	−302.3	118.0	76.7
	aq	−423.4	−341.3	−24.7	
FeCl$_3$	c	−399.4	−333.9	142.34	96.65
std. state	aq	−550.2	−398.3	−146.4	
Fe(CN)$_6^{3-}$ std. state	aq	561.9	729.3	270.3	
Fe(CN)$_6^{4-}$ std. state	aq	455.6	694.9	95.0	
FeCNS^{2+} std. state	aq	23.4	71.1	−130	
FeCO$_3$	c	−740.6	−666.7	92.9	82.1
Fe(CO)$_5$	lq	−774.0	−705.3	338.1	240.6
FeCr$_2$O$_4$	c	−1446.0	−1343.9	146.2	133.8
FeF$_2$	c	−711.3	−668.6	86.99	68.12
std. state	aq	−754.4	−636.5	−165.3	
FeF$_3$	c	−1042	−972	98	91.0
	aq	−1046.4	−840.9	−357.3	
FeI$_2$	c	−113.0	−111.7	167.4	83.7
std. state	aq	−199.6	−182.1	84.9	

(*Continued*)

TABLE 1.58 Enthalpies and Gibbs Energies of Formation, Entropies, and Heat Capacities of the Elements and Inorganic Compounds (*Continued*)

Substance	Physical state	$\Delta_f H°$ kJ · mol^{-1}	$\Delta_f G°$ kJ · mol^{-1}	$S°$ J · deg^{-1} · mol^{-1}	$C°_p$ J · deg^{-1} · mol^{-1}
FeI$_3$	aq	−214.2	−159.4	18.0	
FeMoO$_4$	c	−1075.0	−975.0	129.3	118.5
Fe$_2$N	c	−3.8		101.3	70.0
Fe(NO$_3$)$_3$ std. state	aq	−670.7	−338.5	123.4	
FeO	c	−272.0	−251.4	60.75	49.91
Fe$_2$O$_3$ hematite	c	−824.2	−742.2	87.40	103.9
Fe$_3$O$_4$ magnetite	c	−1118.4	−1015.4	145.27	143.4
FeOH$^+$ std. state	aq	−324.7	−277.4	−29	
Fe(OH)$^{2+}$ std. state	aq	−290.8	−229.4	−142	
Fe(OH)$_2$	c	−574.0	−490.0	87.9	97.1
Fe(OH)$_3$	c	−833	−705	104.6	101.7
FeS	c	−100.0	−100.4	60.32	50.52
FeS$_2$ marcasite	c	−167.4	−156.1	53.87	62.39
FeS$_2$ pyrite	c	−178.2	−166.9	52.92	62.12
FeSiO$_3$	c	−1155		87.5	89.4
Fe$_2$SiO$_4$	c	−1479.9	−1379.0	145.18	132.9
FeSO$_4$	c	−928.4	−820.8	107.5	100.6
std. state	aq	−998.3	−823.4	−117.6	
Fe$_2$(SO$_4$)$_3$	c	−2583.0	−2262.7	307.5	264.8
std. state	aq	−2825.0	−2243.0	−571.5	
FeTiO$_3$	c	−1246.4		105.9	99.5
FeWO$_4$	c	−1155.0	−1054.0	131.8	114.4
Krypton					
Kr	g	0	0	164.085(3)	20.786
Lanthanum					
La	c	0	0	56.9	27.11
La^{3+}	aq	−707.1	683.7	−217.6	−13.0
LaCl$_3$	c	−1072.2		144.4	108.8
std. state	aq	−1208.8	−1077.4	−50.0	−423.0
LaCl$_3$ · 7H$_2$O	c	−3178.6	−2713.3	462.8	431.0
LaI$_3$	c	−668.9			
La(NO$_3$)$_3$	c	−1254.4			
std. state	aq	−1329.3			
La$_2$O$_3$	c	−1793.7	−1705.8	127.32	108.78
La$_2$(SO$_4$)$_3$	c	−3941.3		280	
La$_2$Te$_3$	c	−724	−714.6	231.63	132.13
Lead					
Pb	c	0	0	64.80(30)	26.84
	g	195.2(8)	162.2	175.375(5)	20.8
Pb^{2+}	aq	0.92(25)	−24.4	18.5(10)	
Pb(OAc)$_2$	c	−964.4			
Pb(BO$_2$)$_2$	c	−1556	−1450	131	107.1
PbB$_4$O$_7$	c	−2858	−2667	167	168
PbBr$_2$	c	−278.7	−261.9	161.5	80.1
	aq	−244.8	−232.3	175.3	
Pb(CH$_3$)$_4$	lq	97.9			
Pb(C$_2$H$_5$)$_4$	lq	52.7		464.6	307.4
PbCl$_2$	c	−359.4	−314.1	136	77.1
	aq	−336.0	−286.9	123.4	
PbCl$_4$	lq	−329.3			
PbClF	c	−534.7	−488.3	121.8	

TABLE 1.58   Enthalpies and Gibbs Energies of Formation, Entropies, and Heat Capacities of the Elements and Inorganic Compounds (*Continued*)

Substance	Physical state	$\Delta_f H°$ kJ · mol^{-1}	$\Delta_f G°$ kJ · mol^{-1}	$S°$ J · deg^{-1} · mol^{-1}	$C°_p$ J · deg^{-1} · mol^{-1}
$PbCO_3$	c	−699.2	−625.5	131.0	87.40
PbC_2O_4	c	−851.4	−750.2	146.0	105.4
$PbCrO_4$	c	−930.9			
PbF_2	c	−664	−617.1	110.5	72.3
	aq	−666.9	−582.0	−17.2	
PbF_4	c	−941.8			
PbI_2	c	−175.5	−173.58	174.9	77.4
	aq	−112.1	−127.6	233.0	
$PbMoO_4$	c	−1051.9	−951.4	166.1	119.70
$Pb(N_3)_2$ monoclinic	c	478.2	624.7	148.1	
$Pb(NO_3)_2$	c	−451.9			
	aq	−416.3	−246.9	303.3	
PbO litharge	c	−219.0	−188.9	66.5	45.8
PbO_2	c	−277.4	−217.3	68.60	64.6
Pb_3O_4	c	−718.4	−601.2	211.3	146.9
$Pb_3(PO_4)_2$	c	−2595.3	−2432.6	353.1	256.3
PbS	c	−100.4	−98.7	91.3	49.4
PbSe	c	−102.9	−101.7	102.5	50.2
$PbSeO_4$	c	−609.2	505.0	167.8	
$PbSiO_3$	c	−1145.7	−1062.1	109.6	90.04
$PbSiO_4$	c	−2023.8	−1909.6	84.01	98.66
Pb_2SiO_4	c	−1363.1	−1252.6	186.6	137.2
$PbSO_3$	c	−669.9			
$PbSO_4$	c	−919.97(40)	−813.0	148.50(60)	103.2
$PbSO_4 · PbO$	c	−1182.0		225.06	150.16
PbTe	c	−70.7	−69.5	110.0	50.5
Lithium					
Li	c	0	0	29.12(20)	24.8
	g	159.3(10)		138.782(10)	
Li^+ std. state	aq	−278.47(8)	−293.30	12.24(15)	68.6
Li_3AlF_6 cryolite	c	−3317	−3152	238.5	215.7
$LiAlH_4$	c	−116.3	−44.7	78.7	83.2
$LiAlO_2$	c	−1188.7	−1126.3	53.3	67.78
$LiBeF_3$	c	−1651.8	−1576.3	89.2	91.8
$LiBH_4$	c	−190.8	−125.0	75.9	82.6
$LiBH_4 ·$ tetrahydrofuran	c	−415.5	−220.5	289	
Li_2BeF_4	c	−2274	−2171	130.6	135.3
$LiBO_2$	c	−1032.2	−976.1	51.5	59.8
$Li_2B_4O_7$	c	−3362	−3170	156	183.0
LiBr	c	−351.2	−342.00	74.27	48.91
std. state	aq	−400.03	−397.27	95.81	−73.2
$LiBrO_3$	c	−346.98			
std. state	aq	−345.56	−274.89	174.9	
LiCl	c	−408.6	−384.4	59.3	48.03
	aq	−445.6	−424.6	69.9	−67.8
$LiClO_4$	c	−381.0	−254	126	105
std. state	aq	−407.81	−302.1	195.4	−7.5
Li_2CO_3	c	−1215.9	−1132.12	90.4	99.1
	aq	−1234.1	−1114.6	−29.7	
LiF	c	−616.0	−587.7	35.66	41.6
std. state	aq	−611.12	−571.9	−0.4	−38.1

TABLE 1.58 Enthalpies and Gibbs Energies of Formation, Entropies, and Heat Capacities of the Elements and Inorganic Compounds (*Continued*)

Substance	Physical state	$\Delta_f H°$ kJ · mol^{-1}	$\Delta_f G°$ kJ · mol^{-1}	$S°$ J · deg^{-1} · mol^{-1}	$C°_p$ J · deg^{-1} · mol^{-1}
LiH	c	−90.5	−68.45	20.04	27.96
LiI	c	−210.4	−270.3	86.8	51.0
std. state	aq	−333.67	−344.8	124.7	−73.6
LiIO$_3$	c	−503.38			
std. state	aq	−499.82	−421.33	131.4	−55.2
Li$_3$N	c	−164.6	−128.6	62.59	75.27
LiNO$_2$	c	−372.4	−302.0	96.0	
LiNO$_3$	c	−483.1	−381.1	90.0	
std. state	aq	−485.9	−404.5	160.2	−18.0
Li$_2$O	c	−597.9	−561.2	37.6	
Li$_2$O$_2$	c	−634.3	−578.9	56.5	70.6
LiOH	c	−484.9	−439	42.82	49.7
std. state	aq	−508.40	−451.9	7.1	
Li$_3$PO$_4$	c	−2095.8			
Li$_2$SiO$_3$	c	−1648.1	−1557.2	79.8	99.1
Li$_2$Si$_2$O$_5$	c	−2561	−2417	125.5	138.1
Li$_2$SO$_4$	c	−1436.4	−1321.7	115.1	117.6
std. state	aq	−1466.2	−1331.2	7.3	−155.6
Li$_2$TiO$_3$	c	−1670.7	−1579.8	91.8	109.9
Lutetium					
Lu	c	0	0	50.96	26.86
Lu^{3+}	aq	−665.0	−628.0	−264.0	25.0
LuCl$_3$	c	−945.6			
std. state	aq	−1167.0	−1021.0	−96.0	−385.0
LuI$_3$	c	−548.0			
Lu$_2$O$_3$	c	−1878.2	−1789.1	109.96	101.75
Magnesium					
Mg	c	0	0	32.67(10)	24.87
	g	147.1(8)		148.648(3)	
Mg^{2+} std. state	aq	−467.0(6)	−454.8	−137.(4)	
MgAl$_2$O$_4$	c	−2299	−2177	89.0	116.20
MgBr$_2$	c	−524.3	−503.8	117.2	73.16
std. state	aq	−709.94	−662.8	26.8	
MgBr$_2$ · 6H$_2$O	c	−2410.0	−2056.0	397	
MgCl$_2$	c	−641.3	−591.8	89.63	71.38
std. state	aq	−801.15	−717.1	−25.1	
MgCl$_2$ · 6H$_2$O	c	−2499.0	−2115.0	315.1	
Mg(ClO$_4$)$_2$	c	−568.90			
std. state	aq	−725.51	−472.0	225.4	
Mg(ClO$_4$)$_2$ · 6H$_2$O	c	−2445.6	−1863.1	520.1	
MgCO$_3$	c	−1095.8	−1012.1	65.7	75.51
MgC$_2$O$_4$	c	−1269.0			
std. state	aq	−1292.0	−1128.8	−92.5	
MgF$_2$	c	−1124.2(12)	1071.1	57.2(5)	61.5
Mg$_2$Ge	c	−108.8	−105.9	86.48	69.54
MgH$_2$	c	−75.3	−35.9	31.1	35.4
MgI$_2$	c	−364.0	−358.2	129.7	74.8
std. state	aq	−577.22	−558.1	84.5	
Mg$_3$N$_2$	c	−461.1	−400.9	87.9	104.5
MgNH$_4$PO$_4$ · 6H$_2$O	c	−3681.9			
Mg(NO$_3$)$_2$	c	−790.65	−589.5	164.0	141.9
std. state	aq	−881.6	−677.4	154.8	

TABLE 1.58   Enthalpies and Gibbs Energies of Formation, Entropies, and Heat Capacities of the Elements and Inorganic Compounds (*Continued*)

Substance	Physical state	$\Delta_f H°$ kJ · mol^{-1}	$\Delta_f G°$ kJ · mol^{-1}	$S°$ J · deg^{-1} · mol^{-1}	$C°_p$ J · deg^{-1} · mol^{-1}
Mg(NO$_3$)$_2$ · 6H$_2$O	c	−2613.3	−2080.7	452	
MgO microcrystal	c	−601.6(3)	−569.3	26.95(15)	37.2
Mg(OH)$_2$	c	−924.7	−833.7	63.24	77.25
std. state	aq	−926.8	−769.4	−149.0	
Mg$_3$(PO$_4$)$_2$	c	−3780.7	−3538.8	189.20	213.47
MgS	c	−346.0	−341.8	50.3	45.6
MgSeO$_4$	c	−968.51			
std. state	aq	−1066.1	−896.2	−84.1	
Mg$_2$Si	c	−77.8	−77.1	81.6	67.9
MgSiO$_3$ clinoenstatite	c	−1548.9	−1462.0	67.8	81.9
Mg$_2$SiO$_4$ forsterite	c	−2174.0	−2055.1	95.1	118.5
Mg$_3$Si$_4$O$_{10}$(OH)$_2$ talc	c	−5922.5	−5543.0	260.7	321.8
MgSO$_3$ · 3H$_2$O	c	−1931.8			
MgSO$_3$ · 6H$_2$O	c	−2817.5			
MgSO$_4$	c	−1284.9	−1170.6	91.6	96.5
std. state	aq	−1376.1	−1199.5	−118.01	
MgSO$_4$ · H$_2$O kieserite	c	−1602.1	−1428.8	126.4	
MgSO$_4$ · 7H$_2$O epsomite	c	−3388.71	−2871.9	372	
MgTiO$_3$	c	−1497.6	−1420.1	111.08	91.88
Mg$_2$TiO$_4$	c	−2164.0	−2048	115.0	129
MgTi$_2$O$_5$	c	−2509	−2369	135.6	146.9
Mg$_2$V$_2$O$_7$ triclinic	c	−2835.9	−2645.29	200.4	203.47
MgWO$_4$	c	−1516	−1404	101.2	109.1
Manganese					
Mn	c	0	0	32.01	26.30
Mn^{2+} std. state	aq	−220.75	−228.1	−73.6	50
MnBr$_2$	c	−384.2	−372	138.1	75.31
std. state	aq	−464.0	−409.2		
Mn$_3$C	c	−4.6	5.4	98.7	93.51
MnCl$_2$	c	−481.3	−440.5	118.20	72.9
std. state	aq	−555.05	−490.8	38.9	−222
MnCO$_3$	c	−894.1	−816.7	85.8	81.5
Mn$_2$(CO)$_{10}$	c	1677.4			
MnF$_2$	c	−795.0	−749	92.26	67.99
MnI$_2$	c	−242.7		150.6	75.35
	aq	−331.0			
Mn(NO$_3$)$_2$	c	−576.26			
std. state	aq	−635.6	−451.0	218.0	−121.0
MnO	c	−385.2	−362.9	59.8	45.4
MnO$_2$	c	−520.1	−465.2	53.1	54.1
Mn$_2$O$_3$	c	−959.0	−881.2	110.5	107.7
MnO$_4^-$	aq	−541.4	−447.3	191.2	−82.0
MnO$_4^{2-}$	aq	−653.0	−500.8	59	
Mn$_3$O$_4$	c	−1387.8	−1283.2	155.6	139.7
Mn$_3$(PO$_4$)$_2$	c	−3116.7			
MnS	c	−214.2	−218.4	78.2	50.0
MnSe	c	−106.7	−111.7	90.8	51.0
MnSiO$_3$	c	−1320.9	−1240.6	89.1	86.4
MnSiO$_4$	c	−1730.5	−1632.1	163.2	129.9
MnSO$_4$	c	−1065.3	−957.42	112.1	100.4
std. state	aq	−1130.1	−972.8	−53.6	−243
MnTiO$_3$	c	−1355.6		105.9	99.8

(*Continued*)

TABLE 1.58 Enthalpies and Gibbs Energies of Formation, Entropies, and Heat Capacities of the Elements and Inorganic Compounds (*Continued*)

Substance	Physical state	$\Delta_f H°$ kJ · mol⁻¹	$\Delta_f G°$ kJ · mol⁻¹	$S°$ J · deg⁻¹ · mol⁻¹	$C°_p$ J · deg⁻¹ · mol⁻¹
Mercury					
Hg	lq	0	0	75.90(12)	28.00
	g	61.38(4)	31.8	174.971(5)	20.8
Hg^{2+}	aq	170.21(20)		−36.19(80)	
Hg^+	aq	166.87(50)		65.74(80)	
$HgBr_2$	c	−170.7	−153.1	172.0	75.3
Hg_2Br_2	c	−206.9	−181.1	218.0	104.6
$Hg(CH_3)_2$	lq	59.8	140.2	209	
$Hg(C_2H_5)_2$	lq	30.1			
$HgCl_2$	c	−224.3	−178.6	146.0	73.9
Hg_2Cl_2	c	−265.37(40)	−210.7	191.6(8)	102.0
$Hg(CN)_2$	c	263.6			
Hg_2CO_3	c	−553.5	−468.1	180.0	
HgC_2O_4	c	−678.2			
HgF_2	c	−405	−362	134.3	74.86
Hg_2F_2	c	−485	−469	161	100.4
HgI_2	c	−105.4	−101.7	180.0	77.75
Hg_2I_2	c	−121.3	−111.1	233.5	105.9
$Hg_2(N_3)_2$	c	594.1	746.4	205	
HgO	c	−90.79(12)	−58.49	70.25(30)	44.06
HgS	c	−58.2	−50.6	82.4	48.4
$HgSO_4$	c	−707.5	−594		
Hg_2SO_4	c	−743.09(40)	−625.8	200.70(20)	131.96
HgTe	c	−42.0			
Molybdenum					
Mo	c	0	0	28.71	24.13
$MoBr_3$	c	−284	−259	175	105.4
$MoCl_4$	c	−477	−402	224	128
$MoCl_5$	c	−527	−423	238	155.6
$MoCl_6$	c	−523	−391	255	175
$Mo(CO)_6$	c	−982.8	−877.8	325.9	242.3
MoF_6	lq	−1585.66	−1473.17	259.69	169.8
MoO_2	c	−588.9	−533.0	46.3	56.0
MoO_3	c	−745.2	−668.1	77.8	75.0
MoO_4^{2-} std. state	aq	−997.9	−836.4	27.2	
MoS_2	c	−235.1	−225.9	62.57	63.56
Mo_2S_3	c	−270.3	−278.6	181.2	109.3
Neodymium					
Nd	c	0	0	71.6	27.5
Nd^{3+} std. state	aq	−696.2	−671.5	−206.7	−21
$NdCl_3$	c	−1041.0			113
std. state	aq	−1197.9	−1065.7	−37.7	−431
NdF_3	c	−1657.0			
$Nd(NO_3)_3$	c	−1230.9			
Nd_2O_3	c	−1807.9	−1720.9	158.6	111.3
Neon					
Ne	g	0	0	146.328(3)	20.786
Neptunium					
Np	c	0	0		29.46
NpF_6	c	−1937			
N_pO_2	c	−1029	−979	80.3	66.1

TABLE 1.58 Enthalpies and Gibbs Energies of Formation, Entropies, and Heat Capacities of the Elements and Inorganic Compounds (*Continued*)

Substance	Physical state	$\Delta_f H°$ kJ · mol^{-1}	$\Delta_f G°$ kJ · mol^{-1}	$S°$ J · deg^{-1} · mol^{-1}	$C_p°$ J · deg^{-1} · mol^{-1}
Nickel					
Ni	c	0	0	29.87	26.1
Ni^{2+} std. state	aq	−54.0	−45.6	−128.9	
Ni(OAc)$_2$ std. state	aq	−1025.9	−784.5	44.4	
NiBr$_2$	c	−212.1			
	aq	−297.1	−253.6	36.0	
NiCl$_2$	c	−305.3	−259.0	97.7	71.66
std. state	aq	−388.3	−307.9	−15.1	
Ni(CN)$_4^{2-}$ std. state	aq	367.8	472.0	218	
Ni(CO)$_4$	lq	−633.0	−588.2	313	404.6
	g	−602.9	−587.2	410.6	145.2
NiC$_2$O$_4$	c	−856.9			
NiF$_2$	c	−651.5	−604.2	73.6	64.1
	aq	−719.2	−603.3	−156.5	
NiI$_2$	c	−78.8			
	aq	−164.4	−149.0	93.7	
Ni(NO$_3$)$_2$	c	−415.1			
std. state	aq	−468.6	−268.6	164.0	
NiO	c	−240.6	−211.7	38.00	44.31
Ni$_2$O$_3$	c	−489.5			
NiOH$^+$	aq	−287.9	−227.6	−71.0	
Ni(OH)$_2$	c	−529.7	−447.3	88.0	
NiS	c	−82.0	−79.5	53.0	47.1
Ni$_3$S$_2$	c	−216.0	−210	133.9	117.7
NiS$_2$	c	−131.4	−124.7	72	70.6
NiSO$_4$	c	−872.9	−759.8	92.0	138.0
std. state	aq	−963.2	−790.3	−108.8	327.9
NiSO$_4$ · 7H$_2$0	c	−2976.3	−2462.2	378.94	364.59
NiWO$_4$	c	−1128.4		118.0	136.0
Niobium					
Nb	c	0	0	36.4	24.67
NbBr$_5$	c	−556	−508	258.8	147.9
NbC	c	−138.9	−136.8	34.98	36.23
NbCl$_5$	c	−797.5	−683.3	210.5	148.1
NbF$_5$	c	−1813.8	−1699.0	160.3	134.7
NbI$_5$	c	−268.6		343	155.6
NbN	c	−236.4	−205.9	34.5	39.0
NbO	c	−405.8	−392.6	48.1	41.3
NbO$_2$	c	−796.2	−740.5	54.5	57.45
Nb$_2$O$_5$	c	−1899.5	−1765.8	137.3	132.0
NbOCl$_3$	c	−879.5	−782	159	120.0
Nitrogen					
N atomic	g	472.68(40)		153.301(3)	
N$_2$	g	0	0	191.609(4)	29.124
N$_3^-$	aq	275.1	348.2	107.9	
NCl$_3$	lq	230.0			
NF$_2$	g	43.1	57.8	249.9	41.0
NF$_3$	g	−132.1	−90.6	260.8	53.37
H$_2$NOH	c	−114.2			
N$_2$F$_2$ *cis*	g	69.5	109	259.8	49.96
trans	g	82.0	120.5	262.6	53.47

(*Continued*)

TABLE 1.58 Enthalpies and Gibbs Energies of Formation, Entropies, and Heat Capacities of the Elements and Inorganic Compounds (*Continued*)

Substance	Physical state	$\Delta_f H°$ kJ · mol⁻¹	$\Delta_f G°$ kJ · mol⁻¹	$S°$ J · deg⁻¹ · mol⁻¹	$C°_p$ J · deg⁻¹ · mol⁻¹
N_2F_4	g	−8.4	79.9	301.2	79.2
N_2H_4 hydrazine	lq	50.6	149.3	121.2	98.84
$N_2^2H_4$ hydrazine-d_4	g	81.6	150.9	248.86	55.52
$N_2H^+_5$ std. state	aq	−7.5	82.4	151	70.3
N_2H_5Br	c	−155.6			
std. state	aq	−128.9	−21.8	233.1	−71.6
N_2H_5Cl	c	−197.1			
std. state	aq	−174.9	−49.0	207.1	−66.1
$N_2H_5Cl · HCl$	c	−367.4			
N_2H_5OH	lq	−242.7			
undissoc; std. state	aq	−251.50	−109.2	207.9	73.2
$N_2H_5NO_3$	c	−251.58			
std. state	aq	−215.10	−28.91	297	
$(N_2H_5)_2SO_4$	c	−959.0			
std. state	aq	−924.7	−579.9	322	−151
NO	g	91.29	87.60	210.76	29.85
NOBr	g	82.23	82.42	273.7	45.48
NOCl	g	51.71	66.10	261.68	44.7
NOF	g	−66.5	−51.0	248.02	41.3
NOF_3	g	−163	−96	278.40	67.86
NO_2	g	33.1	51.3	240.1	37.2
NO^-_2	aq	−104.6	−32.2	123.0	−97.5
NO_2Cl	g	12.6	54.4	272.19	53.19
NO_2F	g	−109	−66	260.2	49.8
NO_3	g	69.41	114.35	252.5	46.9
NO^-_3	aq	−206.85(40)	−111.3	146.70(40)	−86.6
N_2O	g	81.6	103.7	220.0	38.62
N_2O_2	g	170.37	202.88	287.52	63.51
$N_2O^{2-}_2$ hyponitrite	aq	−17.2	138.9	27.6	
N_2O_3	g	86.6	142.4	314.7	72.72
N_2O_4	lq	−19.5	97.5	209.20	142.71
	g	11.1	99.8	304.38	79.2
N_2O_5	g	11.3	117.1	355.7	95.30
NSF	g			259.8	44.1
Osmium					
Os	c	0	0	32.6	24.7
$OsCl_3$	c	−190.4	−121	130	
$OsCl_4$	c	−254.8	−159	155	
OsF_6	g			358.1	120.8
OsO_4	c	−394.1	−305.0	143.9	
	g	−337.2	−292.8	293.8	74.1
Oxygen					
O atomic	g	249.18(10)	231.7	161.059(3)	21.9
O_2	g	0	0	205.152(5)	29.4
O_3	g		142.7	163.2	238.92
OF_2	g	24.5	41.8	247.5	57.11
O_2F_2	g	18.0	61.42	268.11	54.06
OH^-	aq	−230.015(40)	−157.28	−10.90(20)	−148.5
Palladium					
Pd	c	0	0	37.61	25.94
Pd^{2+} std. state	aq	149.0	176.6	−184.0	

TABLE 1.58   Enthalpies and Gibbs Energies of Formation, Entropies, and Heat Capacities of the Elements and Inorganic Compounds (*Continued*)

Substance	Physical state	$\Delta_f H°$ kJ · mol^{-1}	$\Delta_f G°$ kJ · mol^{-1}	$S°$ J · deg^{-1} · mol^{-1}	$C_p°$ J · deg^{-1} · mol^{-1}
PdBr$_2$	c	−104.2			
PdBr$_4^{2-}$ std. state	aq	−384.9	−318.0	247	
PdCl$_2$	c	−171.5	−125.1	105	
PdCl$_4^{2-}$ std. state	aq	−550.2	−416.7	167	
Pd$_2$H	c	−19.7	−5.0	91.6	
PdO	c	−85.4		56.1	31.5
PdS	c	−75	−67	46	
PdS$_2$	c	−81.2	−74.5	80	
Phosphorus					
P white	c	0	0	41.09(25)	23.83
	g	316.5(10)	280.1	163.1199(3)	20.8
red, V	c	−17.46	−12.46	22.85	21.19
P$_2$	g	144.0(20)		218.123(4)	
P$_4$	g	58.9(3)	24.4	280.01(50)	67.16
PBr$_3$	lq	−184.5	−175.5	240.2	
	g	−139.3	−162.8	348.15	76.02
PBr$_5$	c	−269.9			
PCl$_3$	lq	−319.7	−272.4	217.2	
	g	−227.1	−267.8	311.8	71.8
PCl$_5$	c	−443.5			
	g	−374.9	−305.0	364.6	112.8
PF$_3$	g	−958	−937	273.1	58.69
PF$_5$	g	−1594.4	−1520.7	300.8	84.8
PH$_3$	g	5.4	13.4	210.24	37.10
std. state	aq	−9.50	25.31	120.1	
PH$_4$Br	c	−127.6	−47.7	110.0	
PH$_4$Cl	c	−145.2			
PH$_4$I	c	−69.9	0.8	123.0	109.6
PH$_4$OH undissoc; std. state	aq	−295.35	−211.88	190.0	
PI$_3$	c	−45.6			
PO$_2$	g	−279.9	−281.6	252.1	39.5
PO$_3^-$	aq	−977.0			
PO$_4^{3-}$ std. state	aq	−1277.4	−1018.8	−220.5	
P$_2$O$_7^{4-}$ std. state	aq	−2271.1	−1919.2	−117.0	
(P$_2$O$_3$)$_2$ dimer	c	−1640.1			
P$_4$O$_{10}$	c	−3009.9	−2723.3	228.78	211.71
POBr$_3$	c	−458.6			
	g	−389.11	−390.91	−359.84	89.87
POCl$_3$	lq	−597.1	−520.9	222.46	138.82
	g	−558.5	−512.9	325.5	84.94
POClF$_2$	g	−970.7	−924.1	301.68	68.83
POCl$_2$F	g	−765.7	−721.6	320.38	79.32
POF$_3$	g	−1254.0	−1206	285.4	68.82
PSCl$_3$	g	−363.2	−347.7	337.23	89.83
PSF$_3$	g	−1009	−985	298.1	74.55
P$_4$S$_3$	c	−155	−159	201	146
Platinum					
Pt	c	0		41.63	25.87
PtBr$_2$	c	−82.0			
PtBr$_3$	c	−120.9			
PtBr$_4$	c	−156.5			

TABLE 1.58   Enthalpies and Gibbs Energies of Formation, Entropies, and Heat Capacities of the Elements and Inorganic Compounds (*Continued*)

Substance	Physical state	$\Delta_f H°$ kJ · mol⁻¹	$\Delta_f G°$ kJ · mol⁻¹	$S°$ J · deg⁻¹ · mol⁻¹	$C°_p$ J · deg⁻¹ · mol⁻¹
$PtCl_2$	c	−123.4		117	
$PtCl_3$	c	−182.0	−134	151	
$PtCl_4$	c	−3218			
$PtCl_4^{2-}$	c	−231.8	−172	176	
$PtCl_4^{2-}$ std. state	aq	−499.2	−361.5	155	
$PtCl_6^{2-}$ std. state	aq	−668.2	−482.8	220.1	
PtF_6	g			348.3	122.8
PtI_4	c	−72.8			
PtS	c	−81.6	−76.2	55.06	43.39
PtS_2	c	−108.8	−99.6	74.68	65.90
Plutonium					
Pu	c	0	0	51.5	35.5
Pu^{3+}	aq	−579.9	−587.9	−163	
Pu^{4+}	aq	−579.9	−1490		
$PuBr_3$	c	−831.8	−804.6	192.88	107.86
$PuCl_3$	c	−961.5	−892.7	159.00	102.84
$PuCl_4$	c	−1381			
PuF_3	c	−1552	−1478.8	112.97	96.82
PuF_4	c	−1732	−1644.7	161.9	120.8
PuF_6	c	25.48	27.2	222.59	167.36
PuH_2	c	−139.3	−101.7	59.8	39.0
PuH_3	c	−138	−82.4	64.9	43.2
PuI_3	c	−648.5	−643.9	214.2	111.8
PuO	c	−565	−538.9	70.7	51.3
PuO_2	c	−1058.1	−1005.8	82.4	68.6
Pu_2O_3 beta	c	−1715.4	−1632.3	152.3	131.0
$Pu(SO_4)_2$	c	−2200.8	−1969.5	163.18	181.96
PuS	c	−439.3	−436.7	78.24	53.97
Pu_2S_3	c	−989.5	−985.5	192.46	129.66
Polonium					
Po	c	0	0	62.8	26.4
PoO_2	c	−251	−197	71	61.5
Potassium					
K	c	0	0	64.68(20)	29.60
	lq	2.284	0.264	71.46	32.72
	g	89.0(8)		160.341(3)	
K^+ std. state	aq	−252.14(8)	−283.26	101.20(20)	21.8
KOAc acetate	c	−723.0			
	aq	−738.39	−652.66	189.1	15.5
$KAg(CN)_2$	aq	18.0	22.2	297	
$KAgCl_2$	aq	−497.4	−498.7	333.9	
K_2AgI_3	aq	−686.6	−720.5	458.1	
$KAlCl_4$	c	97	−1094	197	156.4
K_3AlCl_6	c	−2092.0	−1938	377	248.9
K_3AlF_6	c	−3358.1		284.5	221.1
$KAl(SO_4)_2$	c	−2470.2	−2240.1	204.47	192.92
K_3AsO_4 std. state	aq	−1645.27	−1498.29	144.8	
KBF_4	c	−1887	−1785	133.9	114.48
std. state	aq	−1827.2	−1770.3	285	
KBH_4	c	−227.4	−160.2	106.31	96.57
std. state	aq	−204.22	−168.99	212.97	

TABLE 1.58 Enthalpies and Gibbs Energies of Formation, Entropies, and Heat Capacities of the Elements and Inorganic Compounds (*Continued*)

Substance	Physical state	$\Delta_f H°$ kJ · mol^{-1}	$\Delta_f G°$ kJ · mol^{-1}	$S°$ J · deg^{-1} · mol^{-1}	$C_p°$ J · deg^{-1} · mol^{-1}
KBO_2	c	−981.6	−923.4	79.98	66.7
std. state	aq	−1024.75	−962.19	65.3	
$K_2B_4O_7$	c	−3334.2	−3136.8	208	170.5
KBr	c	−393.8	−380.7	95.9	52.3
std. state	aq	−373.92	−387.23	184.9	−120.1
$KBrO_3$	c	−360.2	−271.2	149.2	105.2
	aq	−319.45	−264.72	264.22	
$KBrO_4$	c	−287.86	−174.47	170.01	120.2
KCl	c	−436.5	−408.5	82.55	51.29
std. state	aq	−419.53	−414.51	159.0	−114.6
KClO std. state	aq	−359.4	−320.1	146	
$KClO_2$ std. state	aq	−318.8	−266.1	203.8	
$KClO_3$	c	−397.73	−296.31	143.1	100.3
std. state	aq	−356.35	−291.29	264.9	
$KClO_4$	c	−432.8	−303.1	151.0	112.41
std. state	aq	−381.71	−291.88	284.5	
KCN	c	−113.1	−101.9	128.52	66.3
std. state	aq	−101.7	−110.9	196.7	
K_2CO_3	c	−1151.0	−1063.5	155.5	114.44
std. state	aq	−1181.90	−1094.41	148.1	
$K_2C_2O_4$	c	−1346.0			
	aq	−1329.72			
K_2CrO_4	c	−1403.7	−1295.8	200.12	145.98
std. state	aq	−1385.91	−1294.36	255.2	
$K_2Cr_2O_7$	c	−2061.5	−1882.0	291.2	219.2
$K_2CuCl_4 · 2H_2O$	c	−1707.1	−1492.9	355.43	253.22
KF	c	−567.2	−537.8	66.5	48.98
std. state	aq	−585.01	−562.08	88.7	−84.9
$K_3Fe(CN)_6$	c	−249.8	−129.7	426.06	
std. state	aq	−139.4	−120.5	577.8	
$K_4Fe(CN)_6$	c	−594.1	−453.1	418.8	322.2
std. state	aq	−554.0	−438.11	505.0	
K formate	c	−679.73			
std. state	aq	−677.93	−634.3	192	−66.1
K glycinate	aq	−722.16	−598.23	221.8	
KH	c	−57.72	−53.01	50.21	37.91
K_2HAsO_4 std. state	aq	−1411.10	−1281.22	203.3	
KH_2AsO_4	c	−1180.7	−1036.0	155.02	126.73
std. state	aq	−1161.94	−1036.54	218	
$KHCrO_4$ std. state	aq	−1130.5	−1048.1	286.6	
$KHCO_3$	c	−963.2	−863.6	115.5	
std. state	aq	−944.33	−870.10	193.7	
KHC_2O_4 std. state	aq	−1070.7	−981.7	251.9	
KHF_2	c	−927.7	−859.7	104.3	76.94
	aq	−902.32	−861.40	195.0	
$KHgBr_3$	c	−550.20			
std. state	aq	−545.6	−542.7	360	
K_2HgBr_4	c	−963.6			
std. state	aq	−935.5	−937.6	515	
$KHgCl_3$	c	−671.1			
std. state	aq	−641.0	−592.5	314	

(*Continued*)

TABLE 1.58 Enthalpies and Gibbs Energies of Formation, Entropies, and Heat Capacities of the Elements and Inorganic Compounds (*Continued*)

Substance	Physical state	$\Delta_f H°$ kJ · mol^{-1}	$\Delta_f G°$ kJ · mol^{-1}	$S°$ J · deg^{-1} · mol^{-1}	$C°_p$ J · deg^{-1} · mol^{-1}
K$_2$Hg(CN)$_4$	c	−32.2			
std. state	aq	21.8	51.9	510	
K$_2$HgI$_4$	c	−775.0			
std. state	aq	−739.7	−778.2	565	
KH$_2$PO$_4$	c	−1568.33	−1415.95	134.85	116.57
std. state	aq	−1548.67	−1622.85	192.9	
K$_2$HPO$_4$ std. state	aq	−1796.90	−1655.78	171.5	
K$_2$H$_2$P$_2$O$_7$	c	−2815.8			
	aq	−2783.2	−2576.9	368	
K$_3$HP$_2$O$_7$	aq	−3032.1	−2822.1	351	
KHS	c	−265.10			75.3
std. state	aq	−269.9	−271.21	165.3	
KHSO$_3$	aq	−878.60	−811.07	242.3	
KHSO$_4$	c	−1160.6	−1131.4	138.1	
std. state	aq	−1139.72	−1039.26	234.3	−63.0
KI	c	−327.9	−324.9	106.3	52.9
	aq	−307.57	−334.85	213.8	−120.5
KIO$_3$	c	−510.43	−418.4	151.46	106.48
	aq	−473.6	−411.3	220.9	
KIO$_4$	c	−467.23	−361.41	175.7	
	aq	−403.8	−341.8	322	
KMnO$_4$	c	−837.2	−737.6	171.71	117.6
K$_2$MoO$_4$	c	−1498.71			
std. state	aq	−1502.5	−1402.9	232.2	
KNH$_2$ amide	c	−128.9			
KNO$_2$	c	−369.82	−306.60	152.09	107.40
std. state	aq	−356.9	−315.5	225.5	
KNO$_3$	c	−494.63	−394.93	133.05	96.4
std. state	aq	−459.74	−394.59	249.0	−64.9
K$_2$Ni(CN)$_4$ std. state	aq	−136.8	−94.6	423	
K$_2$O	c	−361.5	−322.1	94.1	83.7
KO$_2$	c	−284.9	−239.4	122.5	77.53
K$_2$O$_2$	c	−494.1	−425.1	102.0	110
KOCN cyanate	c	−418.65			
std. state	aq	−398.3	−380.7	209.2	
KOH	c	−424.7	−378.7	78.9	64.9
std. state	aq	−482.37	−440.53	91.6	−126.8
K$_2$PdBr$_4$	c	−938.1			
std. state	aq	−889.5	−884.5	452	
K$_3$PO$_4$	c	−1950.2			
std. state	aq	−2034.7	−1868.6	87.9	
K$_4$P$_2$O$_7$	aq	−3280.7	−3052.2	293	
K$_2$PtBr$_4$	c	−915.0			
std. state	aq	−872.8	−828.4	326.4	
K$_2$PtBr$_6$	c	−1021.3			
std. state	aq	−975.3	−898.7	368	
K$_2$PtCl$_4$	c	−1054.4			180.2
std. state	aq	−1003.7	−928.0	360	
K$_2$PtCl$_6$	c	−1229.3	−1078.6	333.9	205.60
std. state	aq	−1171.8	−1049.4	424.7	
K$_2$ReCl$_6$	c	−1310.4	−1172.8	371.71	214.68
std. state	aq	−1266.92	−1156.0	460	

TABLE 1.58 Enthalpies and Gibbs Energies of Formation, Entropies, and Heat Capacities of the Elements and Inorganic Compounds (*Continued*)

Substance	Physical state	$\Delta_f H°$ kJ · mol^{-1}	$\Delta_f G°$ kJ · mol^{-1}	$S°$ J · deg^{-1} · mol^{-1}	$C_p°$ J · deg^{-1} · mol^{-1}
KReO$_4$	c	−1097.0	−994.5	167.82	122.55
std. state	aq	−1039.7	−977.8	303.8	8.4
K$_2$S	c	−380.7	−364.0	105.0	74.7
std. state	aq	−471.5	−480.7	190.4	
K$_2$S$_2$	c	−432.2			
	aq	−474.5	−487.0	233.5	
KSCN	c	−200.16	−178.32	124.26	88.53
std. state	aq	−175.94	−190.58	246.9	−18.4
K$_2$SeO$_3$	c	−979.5			
std. state	aq	−1013.8	−936.4	218.0	
K$_2$SeO$_4$	c	−1110.02	−1002.9	222	
std. state	aq	−1103.7	−1007.9	259.0	
K$_2$SiF$_6$	c	−2956.0	−2798.7	225.9	
std. state	aq	−2893.7	−2766.0	327.2	
K$_2$SiO$_3$	c	−1548.1	−1455.7	146.1	118.4
K$_2$SnBr$_6$	c	−1218.0	−1160.2	443.1	246.0
K$_2$SnCl$_6$	c	−1477.0	−1333.0	366.5	246.0
K$_2$SO$_3$	c	−1125.5			
std. state	aq	−1140.1	−1053.1	176	
K$_2$SO$_4$	c	−1437.8	−1321.4	175.6	131.5
	aq	−1414.0	−1311.1	225.1	−251.0
K$_2$SO$_6$	c	−1437.7	−1319.6	175.5	131.3
std. state	aq	−1414.02	−1311.14	225.1	−251
K$_2$S$_2$O$_3$	c	−1173.6			
std. state	aq	−1156.9	−1089.1	272	
K$_2$S$_2$O$_4$	aq	−1258.1	−1166.9	297	
K$_2$S$_2$O$_7$	c	−1986.6	−1791.6	255	
K$_2$S$_2$O$_8$	c	−1916.10	−1697.41	278.7	213.2
std. state	aq	−1849.3	−1681.6	449.4	
K$_2$S$_4$O$_6$	c	−1780.7	−1613.43	309.66	230.79
std. state	aq	−1728.8	−1607.1	462.3	−24.3
KSO$_3$F	c	−1159.0			
K$_2$UO$_4$	c	1921.3			
KVO$_4$	c	−1154.8			
std. state	aq	−1140.6	−1066.9	155	
K$_2$Zn(CN)$_4$	c	−100.0			
std. state	aq	−162.3	−119.7	431	
Praseodymium					
Pr	c	0	0	73.2	27.20
Pr^{3+} std. state	aq	−704.6	−679.1	−209.0	−29.0
Pr(OAc)$_3$ std. state	aq	−2147.52	−1805.56	164.9	
PrCl$_3$	c	−1056.9			100.0
std. state	aq	−1206.3	−1072.8	−42.0	−439.0
Pr(NO$_3$)$_3$	c	−1229.3			
Pr$_2$O$_3$	c	−1809.6			117.40
Promethium					
PmCl$_3$	c	−1054.0			
Protactinium					
Pa	c	0	0	51.8	
Pa^{4+}	aq	−619.2			
PaBr$_4$	c	−824.0	−787.9	234.0	
PaBr$_5$	c	−862	−820	289	

(*Continued*)

TABLE 1.58 Enthalpies and Gibbs Energies of Formation, Entropies, and Heat Capacities of the Elements and Inorganic Compounds (*Continued*)

Substance	Physical state	$\Delta_f H°$ kJ · mol^{-1}	$\Delta_f G°$ kJ · mol^{-1}	$S°$ J · deg^{-1} · mol^{-1}	$C°_p$ J · deg^{-1} · mol^{-1}
PaCl$_4$	c	−1043.1	−953.0	192.0	
PaCl$_5$	c	−1144.7	−1034.3	238.0	
Radium					
Ra	c	0	0	71	
Ra^{2+}	aq	−527.6	−561.5	54.0	
RaCl$_2$ std. state	aq	−861.9	−823.8	167.0	
Ra(NO$_3$)$_2$	c	−992	−796.2	222	
std. state	aq	−942.2	−784.1	347.0	
RaSO$_4$	c	−1471.1	−1365.7	138	
std. state	aq	−1436.8	−1306.2	75.0	
Radon					
Rn	g	0	0	176.235	20.79
Rhenium					
Re	c	0	0	36.9	25.5
	g	769.9	724.6	188.9	20.8
Re$^-$ std. state	aq	46.0	10.1	230.0	
ReBr$_3$	c	−167.0			
ReCl$_3$	c	−264	−188	123.9	92.4
ReCl$_6^{2-}$ std. state	aq	−761	−590	251	
ReO$_2$	c	−423	−368	172	
ReO$_3$	c	−605.0	−531	257.3	
Re$_2$O$_7$	c	−1240.1	−1066.1	207.1	166.1
	g	−1100.0	−994.0	452.0	
Rhodium					
Rh	c	0	0	31.51	24.98
RhCl$_3$	c	−299.2			
Rh$_2$O$_3$	c	−343.0		110.9	104.0
Rubidium					
Rb	c	0	0	76.78(30)	31.06
	g	80.9(8)	53.1	170.094(3)	20.8
Rb$^+$ std. state	aq	−251.12(10)	−283.97	121.75(25)	
Rb acetate	aq	−737.2	−653.3	207.9	
RbBO$_2$	c	−971.0	−913.0	94.3	74.1
RbBr	c	−394.59	−381.79	109.96	52.84
std. state	aq	−372.71	−387.94	203.93	
RbBrO$_3$	c	−367.27	−278.11	161.1	
Rb$_2$CO$_3$	c	−1136.0	−1051.0	181.33	117.61
std. state	aq	−1179.5	−1095.8	186.2	
RbCl	c	−435.35	−407.81	95.90	52.41
std. state	aq	−418.32	−415.22	178.0	
RbClO$_3$	c	−402.9	−300.4	151.9	103.2
std. state	aq	−355.14	−291.9	283.68	
RbClO$_4$	c	−437.19	−306.9	161.1	
std. state	aq	−380.49	−292.59	303.3	
RbF	c	−557.7		75.3	50.5
std. state	aq	−583.79	−562.79	107.53	
Rb formate	aq	−676.7	−635.1	213.0	
RbHCO$_3$	c	−963.2	−863.6	121.3	
std. state	aq	−943.16	−870.82	212.71	
RbHF$_2$	c	−922.6	−855.6	120.08	79.37
std. state	aq	−901.11	−862.11	213.8	

TABLE 1.58 Enthalpies and Gibbs Energies of Formation, Entropies, and Heat Capacities of the Elements and Inorganic Compounds (*Continued*)

Substance	Physical state	$\Delta_f H°$ kJ $\cdot$ mol^{-1}	$\Delta_f G°$ kJ $\cdot$ mol^{-1}	$S°$ J $\cdot$ deg^{-1} $\cdot$ mol^{-1}	$C°_p$ J $\cdot$ deg^{-1} $\cdot$ mol^{-1}
$RbHSO_4$	c	−1159.0			
std. state	aq	−1138.51	−1039.98	253.1	
RbI	c	−333.8	−328.9	118.4	53.18
std. state	aq	−306.35	−335.56	232.6	
$RbNO_2$	c	−367.4	−306.2	172.0	
$RbNO_3$	c	−495.05	−395.85	147.3	102.1
std. state	aq	−458.52	−395.30	267.8	
Rb_2O	c	−339			
Rb_2O_2	c	−472.0			
RbOH	c	−418.19			
std. state	aq	−481.16	−441.24	110.75	
Rb_2PtCl_6	c	−1245.6	−1109.6	406	
std. state	aq	−1170.7	−1056.6	464	
$RbReO_4$	c	−1102.9	−996.2	167	
std. state	aq	−1038.5	−978.6	322.6	
Rb_2S	aq	−469.4	−482.0	228.4	
Rb_2SeO_4	c	−1114.2			
std. state	aq	−1101.7	−1009.2	297.1	
Rb_2SO_4	c	−1435.61	−1316.96	197.44	134.06
std. state	aq	−1411.60	−1312.56	263.2	
Ruthenium					
Ru	c	0	0	28.53	24.1
$RuBr_3$	c	−138.0			
$RuCl_3$	c	−205.0			
RuI_3	c	−65.7			
RuO_2	c	−305.0			
RuO_4	c	−239.3	−152.3	146.4	
	lq	−228.5	−152.3	183.3	
Samarium					
Sm	c	0	0	69.58	29.54
Sm^{3+} std. state	aq	−691.6	−666.5	−211.7	−21
$SmCl_2$	c	−815.5			
$SmCl_3$	c	−1025.9			
std. state	aq	−1193.3	−1060.2	−42.7	−431
SmF_3	c	−1778.0			
$SmF_3 \cdot \frac{1}{2}H_2O$	c	−1825.1			
SmI_3	c	−620.1			
$Sm(IO_3)_3$	c	−1381			
$Sm(NO_3)_2$	c	−1212.1			
Sm_2O_3	c	−1823.0	−1734.7	151.0	114.5
$Sm_2(SO_4)_3$	c	−3899.1			
Scandium					
Sc	c	0	0	34.64	25.52
Sc^{3+} std. state	aq	−614.2	−586.6	−255.0	
$ScBr_3$	c	−743.1			
$ScCl_3$	c	−925.1		121.3	93.64
ScF_3	c	−1629.2	−1555.6	92	
$ScOH^{2+}$	aq	−861.5	−801.2	−134.0	
Sc_2O_3	c	−1908.8	−1819.41	76.99	94.2

(*Continued*)

TABLE 1.58   Enthalpies and Gibbs Energies of Formation, Entropies, and Heat Capacities of the Elements and Inorganic Compounds (*Continued*)

Substance	Physical state	$\Delta_f H°$ kJ · mol⁻¹	$\Delta_f G°$ kJ · mol⁻¹	$S°$ J · deg⁻¹ · mol⁻¹	$C_p°$ J · deg⁻¹ · mol⁻¹
Selenium					
Se	c	0	0	41.97	24.98
	g	227.1	187.0	174.8	22.1
SeBr₂	g	−21.0			
SeCl₄	c	−188.3			
SeF₆	g	−1117.0	−1017.0	313.8	110.5
SeO	g	53.4	26.8	234.0	31.3
SeO₂	c	−225.4			
SeO₃	c	−166.9			
SeO₃²⁻ std. state	aq	−509.2	−369.9	13	
SeO₄²⁻	aq	−599.2	−441.4	54.0	
Silicon					
Si	c	0	0	18.81(8)	20.00
	g	450.(8)		167.981(4)	
SiBr₄	lq	−457.3	−433.9	277.5	146.4
	g	−415.5	−431.8	377.9	97.1
SiBrCl₃	g			350.1	90.9
SiC alpha	c	−62.8	−60.2	16.49	26.76
beta	c	−65.3	−62.8	16.61	26.9
SiCl₄	lq	−686.93	−620.0	239.7	145.3
	g	−657.0	−617.0	330.7	90.26
SiClBr₃	g			377.1	95.3
SiClF₃	g	−1318	−1280	309	79.4
SiF₄	g	−1615.0(8)	−1572.7	282.76(50)	73.62
SiH₄	g	34.3	56.8	204.65	42.83
SiHBr₃	g	−317.6	−328.5	348.6	80.8
SiHCl₃	lq	−539.3	−482.5	227.6	
	g	−513.0	−482.0	313.7	75.8
SiHF₃	g			271.9	60.5
SiH₂Cl₂	g	−320.5	−295.0	285.7	60.5
SiH₃Cl	g	−142	−119	250.8	51.10
SiH₃F	g	−377	−353	238.4	47.20
Si₂H₆	g	80.3	127.2	272.7	80.79
SiI₄	c	−189.5	−191.6	258.1	108.0
	lq	−174.60	−187.49	294.30	159.79
Si₃N₄	c	−743.5	−642.1	101.3	99.5
SiO	g	−99.6	−126.4	211.6	29.9
SiO₂ quartz	c	−910.7(10)	−856.4	41.46(20)	44.4
high cristobalite	c	−905.5	−853.6	50.05	26.58
SiOF₂	g	−967	−951	271.3	53.69
SiS₂	c	−213.4	−212.6	80.3	77.5
Silver					
Ag	c	0	0	42.55(20)	25.4
	g	284.9(8)		172.997(4)	
Ag⁺ std. state	aq	105.79(8)	77.12	73.45(40)	21.8
Ag²⁺ in 4*M* HClO₄	aq	268.6	269.0	−88	
AgAt	c	−45.2		133.1	55.7
AgBr	c	−100.37	−96.90	107.11	52.38
AgBrO₃	c	−10.5	71.3	151.9	
AgCl	c	−127.01(5)	−109.8	96.25(20)	50.79
AgClO₂	c	8.79	75.7	134.56	87.32

TABLE 1.58 Enthalpies and Gibbs Energies of Formation, Entropies, and Heat Capacities of the Elements and Inorganic Compounds (*Continued*)

Substance	Physical state	$\Delta_f H°$ kJ · mol^{-1}	$\Delta_f G°$ kJ · mol^{-1}	$S°$ J · deg^{-1} · mol^{-1}	$C°_p$ J · deg^{-1} · mol^{-1}
AgClO$_3$	c	−30.3	64.5	142.0	
AgClO$_4$	c	−31.13		162.3	
std. state	aq	−23.77	68.49	254.8	
AgCN	c	146.0	156.9	107.19	66.73
Ag(CN)$_2^-$ std. state	aq	270.3	305.4	192	
Ag$_2$CrO$_4$	c	−731.74	−641.83	217.6	142.26
Ag$_2$CO$_3$	c	−505.9	−436.8	167.4	112.26
Ag$_2$C$_2$O$_4$	c	−673.2	−584.1	209	
AgF	c	−204.6		83.7	51.92
AgF$_2$	c	−360.0			
AgI	c	−61.84	−66.19	115.5	56.82
AgIO$_3$	c	−171.1	−93.7	149.4	102.93
AgN$_3$	c	308.8	376.1	104.2	
Ag(NH$_3$)$_2^+$ std. state	aq	−111.29	−17.24	245.2	
AgNO$_3$	c	−124.4	−33.47	140.92	93.05
std. state	aq	−101.80	−34.23	219.2	−64.9
AgO	c	−12.15	13.83	58.5	44.0
Ag$_2$O	c	−31.1	−11.21	121.3	65.86
Ag$_2$O$_3$	c	33.9	121.4	100.0	
Ag$_2$S argentite	c	−32.59	−40.67	143.9	76.53
Ag$_3$Sb	c	−23.0		171.5	101.7
AgSCN	c	87.9	101.38	131.0	63
Ag$_2$Se	c	−38	−44.4	150.71	81.76
Ag$_2$SO$_4$	c	−715.9	−618.4	200.4	131.4
std. state	aq	−698.10	−590.36	165.7	−251
Ag$_2$Te	c	−37.2	−43.1	154.8	87.5
Sodium					
Na	c	0	0	51.30(20)	28.15
	g	107.5(7)		153.718(3)	
Na$^+$ std. state	aq	−240.34(6)	−261.88	58.45(15)	46.4
NaAg(CN)$_2$ std. state	aq	30.12	43.5	251	
NaOAc	c	−708.81	−607.27	123.0	79.9
std. state	aq	−726.13	−631.28	145.6	40.2
NaAlCl$_4$	c	−1142.0	−996.4	188.3	154.98
Na$_3$AlCl$_6$	c	−1979.0	−1829	347.0	244.1
NaAlF$_4$	g	−1869.0	−1827.5	345.7	105.9
Na$_3$AlF$_6$	c	−3361.2	−3136.7	239.5	215.89
NaAlH$_4$	c	−115.5			
NaAlO$_2$	c	−1137.3	−1069.2	70.40	73.64
NaAl(SO$_4$)$_2$ std. state	aq	−2590	−2238	−222.6	
NaAlSiO$_4$	c	−2092.8	−1978.2	124.3	
NaAsO$_2$	c	−660.53			
std. state	aq	−669.15	−611.91	99.6	
Na$_3$AsO$_4$	c	−1540			
std. state	aq	−1608.50	−1434.19	14.2	
NaAu(CN)$_2$	aq	2.1	23.9	230	
NaBF$_4$	c	−1844.7	−1750.1	145.31	120.3
std. state	aq	−1812.1	−1748.9	243	
NaBH$_4$	c	−188.6	−123.9	101.3	86.8
std. state	aq	−199.60	−147.61	169.5	

(Continued)

TABLE 1.58 Enthalpies and Gibbs Energies of Formation, Entropies, and Heat Capacities of the Elements and Inorganic Compounds (*Continued*)

Substance	Physical state	$\Delta_f H°$ kJ · mol⁻¹	$\Delta_f G°$ kJ · mol⁻¹	$S°$ J · deg⁻¹ · mol⁻¹	$C°_p$ J · deg⁻¹ · mol⁻¹
NaBO₂	c	−977.0	−920.7	73.54	65.94
std. state	aq	−1012.49	−940.81	21.8	
NaBO₃ · 4H₂O	c	−2114.2			
Na₂B₄O₇	c	−3291.1	−3096.0	189.0	186.8
std. state	aq	−3271.1	−3076.9	192.9	
Na₂B₄O₇ · 10H₂O	c	−6298.6	−5516.6	586	614.5
NaBr	c	−361.08	−349.00	86.82	51.38
std. state	aq	−361.66	−365.85	141.4	−95.4
NaBr₃ std. state	aq	−370.54	−368.95	274.5	
NaBrO std. state	aq	−384.3	−295.4	100	
NaBrO₃	c	−334.09	−242.6	128.9	
std. state	aq	−307.19	−243.34	220.9	
NaBrO₄ std. state	aq	−227.19	−143.93	−258.57	
Na₂[Cd(CN)₄]	aq	−52.3	−16.3	439	
NaCl	c	−411.2	−384.1	72.1	50.51
std. state	aq	−407.27	−393.17	115.5	−90.0
NaClO std. state	aq	−347.3	−298.7	100	
NaClO₂	c	−307.02		115.9	
std. state	aq	−306.7	−244.8	160.3	
NaClO₃	c	−365.77	−262.34	123.4	
std. state	aq	−344.09	−269.91	221.3	
NaClO₄	c	−383.3	−254.9	142.3	111.3
std. state	aq	−369.45	−270.50	241.0	
NaCN	c	−87.5	−76.4	115.6	70.4
std. state	aq	−89.5	−89.5	153.1	
Na₃[Co(NO₂)₆]	c	−1423.0			
Na₂CO₃	c	−1130.7	−1044.4	135.0	112.3
	aq	−1157.4	−1051.6	61.6	
Na₂CO₃ · H₂O	c	−1431.26	−1285.41	168.11	145.60
Na₂CO₃ · 10H₂O	c	−4081.32	−3428.20	564.0	550.32
Na₂C₂O₄	c	−1318.0			142
std. state	aq	−1305.4	−1197.9	163.6	
Na₂CrO₄	c	−1342.2	−1235.0	176.61	142.13
std. state	aq	−1361.39	−1251.64	168.2	
Na₂Cr₂O₇	c	−1978.6			
std. state	aq	−1970.7	−1825.1	379.9	
Na ethoxide	c	−413.80			
NaF	c	−576.6	−546.3	51.11	46.85
std. state	aq	−572.75	−540.70	45.2	−60.3
Na₃[Fe(CN)₆] std. state	aq	−158.6	−56.5	447.3	
Na₄[Fe(CN)₆] std. state	aq	−505.0	−352.63	231.0	
Na formate	c	−666.5	−600.00	103.76	82.68
std. state	aq	−666.67	−613.0	151	−41.4
NaH	c	−56.34	−33.55	40.02	36.39
Na₂HAsO₄ std. state	aq	−1386.58	−1238.51	116.3	
NaH₂AsO₄ std. state	aq	−1149.68	−1015.16	176	
NaHCO₃	c	−950.81	−851.0	101.7	87.61
std. state	aq	−932.11	−848.72	150.2	
NaHCrO₄ std. state	aq	−1118.4	−1026.8	243.1	
NaHF₂	c	−920.27	−852.20	90.92	75.02
std. state	aq	−890.06	−840.02	151.5	

TABLE 1.58 Enthalpies and Gibbs Energies of Formation, Entropies, and Heat Capacities of the Elements and Inorganic Compounds (*Continued*)

Substance	Physical state	$\Delta_f H°$ kJ · mol^{-1}	$\Delta_f G°$ kJ · mol^{-1}	$S°$ J · deg^{-1} · mol^{-1}	$C_p°$ J · deg^{-1} · mol^{-1}
Na$_2$H$_2$[Fe(CN)$_6$]	aq	−24.7	134.64	335	
NaH$_2$PO$_4$	c	−1536.8	−1386.2	127.49	116.86
std. state	aq	−1536.4	−1392.27	149.4	
Na$_2$HPO$_4$	c	−1748.1	−1608.3	150.50	135.31
std. state	aq	−1772.38	−1613.06	84.5	
Na$_2$H$_2$P$_2$O$_7$	c	−2764.8	−2522.5	220.20	198.15
NaHS	c	−237.23			
std. state	aq	−257.73	−249.83	121.8	
NaHSeO$_3$	c	−759.23			
std. state	aq	−754.67	−673.41	194.1	
NaHSeO$_4$	c	−821.40			
std. state	aq	−821.74	−714.2	208.4	
NaHSO$_4$	c	−1125.5	−992.9	113.0	
std. state	aq	−1127.46	−1017.88	190.8	−38
NaI	c	−287.9	−286.1	98.50	52.1
std. state	aq	−295.31	−313.47	170.3	−95.8
NaI$_3$	aq	−291.6	−313.4	298.3	
NaIO$_3$	c	−481.79		135.1	92.1
std. state	aq	−461.50	−389.95	177.4	
NaIO$_4$	c	−429.28	−323.09	163.0	
std. state	aq	−391.62	−320.49	280	
Na methoxide	c	−367.8	−294.80	110.58	69.45
std. state	aq	−433.59	−332.46	17.6	
NaMnO$_4$ std. state	aq	−781.6	−709.2	250.2	
Na$_2$MnO$_4$	c	−1156.0			
std. state	aq	−1134	−1024.7	176	
Na$_2$MoO$_4$	c	−1468.12	−1354.30	159.70	141.71
std. state	aq	−1478.2	−1360.2	145.2	
Na$_2$Mo$_2$O$_7$	c	−2245.05	−2058.19	250.6	217.15
NaN$_3$	c	21.71	93.76	96.86	76.61
std. state	aq	35.02	86.2	166.9	
NaNH$_2$	c	−123.9	−64.0	76.90	66.15
NaNbO$_3$	c	−1315.9	1233.0	117	
std. state	aq	−1265.7	−1194.1	155	
NaNO$_2$	c	−358.65	−284.60	103.8	
std. state	aq	−344.8	−294.1	182.0	−51.0
NaNO$_3$	c	−467.85	−367.06	116.52	92.88
std. state	aq	−447.48	−373.21	205.4	−40.2
Na$_2$[Ni(CN)$_4$]	aq	−112.6	−51.9	335	
NaO$_2$	c	−260.2	−218.4	115.9	72.14
Na$_2$O	c	−414.2	−375.5	75.04	69.10
Na$_2$O$_2$	c	−510.9	−449.6	94.8	89.3
NaOCN cyanate	c	−405.39	−358.2	96.7	86.6
std. state	aq	−386.2	−359.4	165.7	
NaOH	c	−425.6	−379.4	64.4	59.5
std. state	aq	−469.15	−419.20	48.1	−102.1
Na$_3$PO$_4$	c	−1917.40	−1788.87	173.80	153.47
std. state	aq	−1997.9	−1804.6	−46	
Na$_4$P$_2$O$_7$	c	−3188	−2969.4	270.29	241.12
std. state	aq	−3231.7	−2966.9	117	

(*Continued*)

TABLE 1.58 Enthalpies and Gibbs Energies of Formation, Entropies, and Heat Capacities of the Elements and Inorganic Compounds (*Continued*)

Substance	Physical state	$\Delta_f H°$ kJ · mol^{-1}	$\Delta_f G°$ kJ · mol^{-1}	$S°$ J · deg^{-1} · mol^{-1}	$C°_p$ J · deg^{-1} · mol^{-1}
NaReO$_4$	c	−1057.09	−953.74	151.5	133.89
std. state	aq	−1027.6	−956.5	260.2	
Na$_2$S	c	−364.8	−349.8	83.7	82.8
std. state	aq	−443.3	−438.1	103.3	
Na$_2$S$_2$	c	−397.0	−392	151	
std. state	aq	−450.2	−444.3	146.4	
NaSCN	c	−170.50			
std. state	aq	−163.68	−169.20	203.84	6.3
Na$_2$Se	c	−341.4			
Na$_2$SeO$_3$	c	−958.6			
std. state	aq	−989.5	−893.7	130	
Na$_2$SeO$_4$	c	−1069.0			
Na$_2$SiF$_6$	c	−2909.6	−2754.2	207.1	187.1
Na$_2$SiO$_3$	c	−1554.9	−1462.8	113.8	111.9
Na$_2$Si$_2$O$_5$	c	−2470.1	−2324.1	164.1	157.0
NaSnBr$_3$	aq	−615.1	−608.8	310	
NaSnCl$_3$	aq	−727.2	−692.0	318	
Na$_2$SO$_3$	c	−1100.8	−1012.5	145.94	120.25
std. state	aq	−1115.87	−1010.44	87.9	
Na$_2$SO$_4$	c	−1387.1	−1270.2	149.6	128.2
std. state	aq	−1389.51	−1268.40	138.1	−201
Na$_2$SO$_4$ · 10H$_2$O	c	−4327.26	−3647.40	592.0	
Na$_2$S$_2$O$_3$	c	−1123.0	−1028.0	155	
std. state	aq	−1132.40	−1046.0	184.1	
Na$_2$S$_2$O$_3$ · 5H$_2$O	c	−2607.93	−2230.1		
Na$_2$S$_2$O$_4$ dithionate	c	−1232.2			
std. state	aq	−1233.9	−1124.2	209.2	
Na$_2$S$_2$O$_7$	c	−1925.1	−1722.1	202.1	
Na$_2$S$_2$O$_8$	aq	−1825.1	−1638.9	362.3	
Na$_2$Te	c	−349.4			
Na$_2$TeO$_4$	c	−1270.7			
Na$_2$TiO$_3$	c	−1591.2	−1496.2	121.67	125.65
Na$_2$UO$_4$ beta	c	−1893.3	−1777.78	166.02	146.65
Na$_3$UO$_4$	c	−2025.1	−1901.2	198.20	173.01
NaVO$_3$	c	−1145.79	−1064.12	113.68	97.57
std. state	aq	−1128.4	−1045.6	109	
Na$_3$VO$_4$	c	−1757.87	−1637.83	190.0	164.85
Na$_2$V$_2$O$_7$	c	−2918.84	−2712.52	318.4	269.74
Na$_2$WO$_4$	c	−1544.7	−1429.8	160.3	139.8
Na$_2$[Zn(CN)$_4$]	aq	−138.1	−77.0	343	
Strontium					
Sr	c	0	0	55.0	26.79
Sr^{2+} std. state	aq	−545.8	−559.44	−32.6	
Sr(OAc)$_2$	c	−1487.4			
Sr$_3$(AsO$_4$)$_2$	c	−3317.1	−3080.3	255	
SrBr$_2$	c	−717.6	−697.1	135.1	75.3
	aq	−788.89	−767.39	132.2	
SrCl$_2$	c	−828.9	−781.1	114.9	75.59
std. state	aq	−880.10	−821.95	80.3	
Sr(ClO$_4$)$_2$	c	−762.69			
std. state	aq	−804.46	−576.68	331.4	

TABLE 1.58 Enthalpies and Gibbs Energies of Formation, Entropies, and Heat Capacities of the Elements and Inorganic Compounds (*Continued*)

Substance	Physical state	$\Delta_f H°$ kJ · mol^{-1}	$\Delta_f G°$ kJ · mol^{-1}	$S°$ J · deg^{-1} · mol^{-1}	$C_p°$ J · deg^{-1} · mol^{-1}
SrCO$_3$	c	−1220.1	−1140.1	97.1	81.42
	aq	−1222.9	−1087.3	−89.5	
SrC$_2$O$_4$	c	−1370.7			
SrF$_2$	c	−1216.3	−1164	82.1	70.0
Sr fonnate	c	−1393.3			
SrHPO$_4$	c	−1821.7	−1688.7	121	
Sr(H$_2$PO$_4$)$_2$	c	−3134.7			
SrI$_2$	c	−558.1	−557.7	159.1	77.95
std. state	aq	−656.18	−662.62	190.0	
Sr(IO$_3$)$_2$	c	−1019.2	−855.2	234	
SrMoO$_4$	c	−1561.1		128.9	117.07
Sr(NO$_2$)$_2$	c	−762.3			
Sr(NO$_3$)$_2$	c	−978.22	−780.0	194.56	149.87
std. state	aq	−960.52	−782.12	260.2	
SrO	c	−592.0	−561.9	54.4	45.0
SrO$_2$	c	−654.4		54	79.45
Sr(OH)$_2$	c	−959	−881	97	74.9
Sr$_3$(PO$_4$)$_2$	c	−4122.9			
SrS	c	−472.4	−467	68.2	48.7
SrSe	c	−385.8			
SrSeO$_3$	c	−1047.7			
SrSeO$_4$	c	−1142.7			
SrSiO$_3$	c	−1633.9	−1549.8	96.7	88.53
Sr$_2$SiO$_4$	c	−2304.6	−2191.2	153.1	134.26
SrSO$_3$	c	−1177.0			
SrSO$_4$	c	−1453.1	−1341.0	117.0	107.78
	aq	−1455.1	−1304.0	−12.6	
Sr$_2$TiO$_4$	c	−2287.4	−2178.6	159.0	143.68
Sulfur					
S rhombic	c	0	0	32.054(50)	22.60
monoclinic	c	0.360	−0.070	33.03	23.23
	g	277.17(15)		167.829(6)	
S$^{2-}{}_2$	aq	33.1	85.8	−14.6	
S$_2$	g	128.60(30)		228.167(10)	
S$_8$	g	101.25	49.16	430.20	156.06
S$_2$Br$_2$	lq	−13.0			
SCl$_2$	lq	−50.0	−28.5	184	91.0
SClF$_5$	lq	−1065.7			
S$_2$Cl$_2$	lq	−59.4	−39	224	124.3
SCN$^-$	aq	76.4	92.7	144.3	−40.2
SF$_4$	g	−763.2	−722.0	299.6	77.60
SF$_6$	g	−1220.5	−1116.5	291.5	96.96
S$_2$F$_{10}$	g	−2064	−1861	397	176.7
So	g	6.3	−19.9	222.0	30.2
SO$_2$	g	−296.81(20)	−300.13	248.223(50)	39.88
SO$_3$	g	−395.7	−371.02	256.77	50.66
SOCl$_2$	g	−212.50	−198.3	309.8	66.5
SOF$_2$	g	−544	−502	278.7	56.81
SO$_2$Cl$_2$	g	−364.0	−320.0	311.9	77.01
SO$_2$ClF	g	−556	−513	303	71.6
SO$_2$F$_2$	g	−759	−712	284.0	66.0

(*Continued*)

TABLE 1.58 Enthalpies and Gibbs Energies of Formation, Entropies, and Heat Capacities of the Elements and Inorganic Compounds (*Continued*)

Substance	Physical state	$\Delta_f H°$ kJ · mol^{-1}	$\Delta_f G°$ kJ · mol^{-1}	$S°$ J · deg^{-1} · mol^{-1}	$C°_p$ J · deg^{-1} · mol^{-1}
SO_3^{2-}	aq	−635.5	−486.5	−29.0	
SO_4^{2-}	aq	−909.34(40)	−744.5	18.50(40)	−293.0
$S_2O_3^{2-}$	aq	−652.3	−522.5	67.0	
$S_2O_4^{2-}$	aq	−753.5	−600.3	92.0	
$S_2O_8^{2-}$	aq	−1344.7	−1114.9	244.3	
Tantalum					
Ta	c	0	0	41.47	25.40
TaB_2	c	−209.2		44.4	48.12
$TaBr_5$	c	−598.3		305.4	155.73
TaC	c	−144.1	−142.7	42.37	36.79
Ta_2C	c	−197.5		83.7	60.96
$TaCl_5$	c	−859.0	−746	222	148
TaF_5	c	−1903.6		195.0	130.46
Ta_2H	c	−32.6	−69.0	79.1	90.8
TaI_5	c	−490		343	155.6
TaN	c	−251		50.6	42.1
TaO_2	g	−201	−209	280	44.0
Ta_2O_5	c	−2046	−1911.0	143.1	135.0
$TaOCl_3$	g	−780.7		361.5	98.53
Technetium					
Tc	c	0	0	33.47	24.27
Tc_2O_7	c	−1113			
Tellurium					
Te	c	0	0	49.70	25.70
$TeBr_4$	c	−190.4			
$TeCl_4$	c	−326.4		209	138.5
TeF_6	g	−1318.0		335.77	116.90
TeO_2	c	−322.6	−270.3	79.5	63.89
$Te(OH)^+_3$	aq	−322.6	−496.1	111.7	
Terbium					
Tb	c	0	0	73.22	28.91
Tb^{3+} std. state	aq	−682.8	−651.9	−226.0	17.0
$TbCl_3$	c	−997.1			
std. state	aq	−1184.1	−1045.6	−59.0	−393.0
TbO_2	c	−971.5			
Tb_2O_3	c	−1865.2			115.9
$Tb_2(SO_4)_3$ std. state	aq	−4131.7	−3597.4		
Thallium					
Tl	c	0	0	64.18	26.32
Tl^+ std. state	aq	5.36	−32.38	125.5	
Tl^{3+} std. state	aq	196.6	214.6	−192.0	
TlBr	c	−173.2	−167.36	120.5	50.50
std. state	aq	−116.19	−136.36	207.9	
$TlBr_3$	aq	−168.2	−97.1	54.0	
$TlBrO_3$	c	−136.4	−53.14	168.6	
std. state	aq	−78.2	−30.5	288.7	
TlCl	c	−204.10	−184.93	111.30	50.92
std. state	aq	−161.80	−163.64	182.00	
$TlCl_3$	c	−315.1			
std. state	aq	−305.0	−179.1	−23.0	
$TlClO_3$	aq	−93.7	−35.6	287.9	

TABLE 1.58 Enthalpies and Gibbs Energies of Formation, Entropies, and Heat Capacities of the Elements and Inorganic Compounds (*Continued*)

Substance	Physical state	$\Delta_f H°$ kJ · mol^{-1}	$\Delta_f G°$ kJ · mol^{-1}	$S°$ J · deg^{-1} · mol^{-1}	$C°_p$ J · deg^{-1} · mol^{-1}
Tl_2CO_3	c	−700	−614.6	155.2	
TlF	c	−324.6		83.3	54.77
std. state	aq	−327.27	−311.21	111.7	
TlI	c	−123.9	−125.39	127.6	52.51
std. state	aq	−49.83	−83.97	236.8	
$TlNO_3$	c	−243.93	−152.46	160.7	99.50
	aq	−202.0	−143.7	272.0	
Tl_2O	c	−178.7	−147.3	126	
TlOH	c	−238.9	−195.8	88	
std. state	aq	−224.64	−189.66	114.6	
Tl_2S	c	−97.1	−93.7	151.0	
Tl_2Se	c	−59.0	−59.0	172.0	
Tl_2SO_4	c	−931.8	−830.48	230.5	
std. state	aq	−898.56	−809.40	271.1	
Thorium					
Th	c	0	0	51.8(5)	27.32
	g	602.(6)		190.17(5)	
Th^{4+} std. state	aq	−769.0	−705.1	−422.6	
$ThBr_4$	c	−965.3	−927.2	230	
$ThC_{1.94}$	c	−146	−147.7	68.49	56.69
$ThCl_4$	c	−1186.2	−1094.1	190.4	120.3
ThF_3	g	−1166.1	−1160.6	339.2	73.3
ThF_4	c	−2097.8	−2003.4	142.05	110.7
undissoc; std. state	aq	−2115.0	−1947.2	−105	
ThH_2	c	−139.8	−100.0	50.71	36.69
ThI_4	c	−664.8	−655.2	255	
ThN	c	−391.2	−363.6	56.07	45.2
Th_3N_4	c	−1315.0	−1212.9	201	155.90
$Th(NO_3)_4$	c	−1441.4			
ThO_2	c	−1226.4(35)	−1169.20	65.23(20)	61.76
$ThOCl_2$	c	−1232.2	−1156.0	123.4	91.25
$ThOF_2$	c	−1665.2	−1589.5	105	
$Th(OH)^{3+}$	aq	−1030.1	−920.5	−343.0	
$Th(OH)_2^{2+}$	aq	−1282.4	−1140.9	−218.0	
Th_3P_4	c	−1140.2	−1112.9	221.8	
ThS_2	c	−626.3	−620.1	96.2	
Th_2S_3	c	−1083.7	−1077.0	180	
$Th(SO_4)_2$	c	−2542.6	−2310.4	159.0	173.47
Thullium					
Tm	c	0	0	74.01	27.03
Tm^{3+} std. state	aq	−697.9	−661.9	−243.0	25.0
$TmCl_3$	c	−986.6			
std. state	aq	−1199.1	−1055.6	−75.0	−385.0
Tm_2O_3	c	−1888.7	−1794.5	139.8	116.7
Tin					
Sn white	c	0	0	51.08(8)	26.99
	aq	301.2(15)		168.492(4)	
gray	c	−2.09	0.13	44.14	25.77
Sn^{2+} in aqueous HCl	aq	−8.9(10)	−27.2	−16.7(40)	
Sn^{4+} in aqueous HCl	aq	30.5	2.5	−117	
$SnBr_2$	c	−243.5			

(*Continued*)

TABLE 1.58 Enthalpies and Gibbs Energies of Formation, Entropies, and Heat Capacities of the Elements and Inorganic Compounds (*Continued*)

Substance	Physical state	$\Delta_f H°$ kJ · mol^{-1}	$\Delta_f G°$ kJ · mol^{-1}	$S°$ J · deg^{-1} · mol^{-1}	$C°_p$ J · deg^{-1} · mol^{-1}
SnBr$_4$	c	−377.4	−350.2	264.4	136.44
	g	−314.6	−331.4	411.9	103.4
SnCl$_2$	c	−325.1		130	79.33
std. state	aq	−329.7	−299.6	172	
SnCl$_4$	lq	−511.3	−440.2	258.6	165.3
	g	−471.5	−432.2	365.8	98.3
SnH$_4$	g	162.8	188.3	227.7	48.95
SnI$_2$	c	−143.5			
SnI$_4$	g			446.1	105.4
SnO tetragonal	c	−280.71(20)	−251.9	57.17(30)	44.31
SnO$_2$ tetragonal	c	−577.63(20)	−515.8	49.04(10)	52.59
Sn(OH)$^+$	aq	−286.2	−254.8	50.0	
Sn(OH)$_2$	c	−561.1	−491.6	155.0	
SnS	c	−100	−98.3	77.0	49.25
SnS$_2$	c	−167.4		87.4	70.12
Titanium					
Ti	c	0	0	30.72(10)	25.0
	g	473.(3)		180.298(10)	
TiB	c	−160	−160	35	29.7
TiB$_2$	c	−280	−275	28.5	44.3
TiBr$_2$	c	−402	−383	108	78.7
TiBr$_3$	c	−548.5	−523.8	176.6	101.7
TiBr$_4$	c	−616.7	−589.5	243.5	131.5
TiC	c	−184	−180	24.2	33.81
TiCl$_2$	c	−513.8	−464.4	87.4	69.8
TiCl$_3$	c	−720.9	−653.5	139.7	97.2
TiCl$_4$	lq	−804.2	−737.2	252.3	145.2
	g	−763.2(30)	−726.3	353.2(40)	95.4
TiF$_3$	c	−1435	−1362	88	92
TiF$_4$	c	−1649	−1559	133.96	114.27
TiH$_2$	c	−144	−105.1	29.71	30.09
TiI$_4$	c	−375	−371.5	249.4	125.6
TiN	c	−265.8	−243.8	52.73	37.08
TiO	c	−519.7	−495.0	50.0	39.9
TiO$_2$	c	−944.0(8)	−888.8	50.62(30)	55.0
Ti$_2$O$_3$	c	−1520.9	−1434.2	78.8	97.4
Ti$_3$O$_5$	c	−2459.4	−2317.4	129.3	154.8
Tungsten					
W	c	0	0	32.6	24.3
WBr$_5$	c	−312	−270	272	155
WBr$_6$	c	−348.5	−290.8	314	181.4
W(CO)$_6$	c	−953.5		331.8	242.5
WCl$_4$	c	−443	−360	198.3	129.7
WCl$_5$	c	−515	−402	217.6	155.6
WCl$_6$	c	−602.5	−456	238.5	175.4
WF$_6$	lq	−1747.7	−1631.4	251.5	
	g	−1721.7	−1631.4	341.1	119.0
WO$_2$	c	−589.9	−533.86	50.5	56.1
WO$_3$	c	−842.9	−764.1	75.9	73.8
WO$_4^{2-}$	aq	−1075.7			
WOCl$_4$	c	−671	−549	173	146

TABLE 1.58 Enthalpies and Gibbs Energies of Formation, Entropies, and Heat Capacities of the Elements and Inorganic Compounds (*Continued*)

Substance	Physical state	$\Delta_f H°$ kJ · mol^{-1}	$\Delta_f G°$ kJ · mol^{-1}	$S°$ J · deg^{-1} · mol^{-1}	$C°_p$ J · deg^{-1} · mol^{-1}
WOF$_4$	c	−1407	−1298	176.0	133.6
WO$_2$Cl$_2$	c	−780	−703	200.8	104.4
Uranium					
U	c	0	0	50.20(20)	27.66
	g	533.(8)		199.79(10)	
U^{3+}	aq	−489.1	−476.2	−188.0	
U^{4+}	aq	−591.2	−531.9	−410.0	
UB$_2$	c	−161.6	−159.4	55.52	55.77
UBr$_3$	c	−699.2	−673.6	192	108.8
UBr$_4$	c	−802.5	−767.8	238.0	128.0
UBr$_5$	c	−810.9	−769.9	293	160.7
UC	c	−98.3	−99.2	59.20	50.12
UCl$_3$	c	−866.5	−799.1	159.0	102.5
UCl$_4$	c	−1019.2	−930.1	197.1	122.0
	aq	−1259.8	−1056.8	−184.0	
UCl$_5$	c	−1058	−950	242.7	144.6
UCl$_6$	c	−1092	−962	285.8	175.7
UF$_3$	c	−1502.1	−1433.4	123.43	95.10
UF$_4$	c	−1921.2	−1823.3	151.67	116.02
UF$_5$	c	−2075.3	−1958.6	199.6	132.3
UF$_6$	c	−2197.0	−2068.6	227.6	166.8
UH$_3$	c	−127.2	−72.8	63.68	49.29
UI$_3$	c	−460.7	−459.8	222	112.1
UI$_4$	c	−512.1	−506.7	264	134.3
UN	c	−290.8	−265.7	62.43	47.57
UO$_2$	c	−1085.0(10)	−1031.8	77.03(20)	63.60
UO$_2^{2+}$ std. state	aq	−1019.0(15)	−953.5	−98.2(30)	
UO$_3$ gamma	c	−1223.8(12)	−1145.7	96.11(40)	81.67
U$_3$O$_7$	c	−3427.1	−3242.9	250.5	215.5
U$_3$O$_8$	c	−3574.8(25)	−3369.8	282.55(50)	238.36
U$_4$O$_9$	c	−4510.4	−4275.1	334.1	293.3
UOBr$_2$	c	−973.6	−929.7	158.00	98.00
UOCl$_2$	c	−1066.9	−996.2	138.32	95.06
UOF$_2$	c	−1499.1	−1428.8	119.2	
UO$_2$(OAc)$_2$	c	−1963.55			
UO$_2$Br$_2$	c	−1137.6	−1066.5	169.5	
UO$_2$Cl$_2$	c	−1243.9	−1146.4	150.5	107.86
std. state	aq	−1353.9	−1215.9	15.5	
UO$_2$CO$_3$	c	−1691.2	−1562.7	138	
std. state	aq	−1696.6	−1481.6	−154.4	
UO$_2$C$_2$O$_4$	c	−1796.94			
UO$_2$F$_2$	c	−1653.5	−1557.4	135.56	103.22
std. state	aq	−1684.0	−1551.3	−125.1	
UO$_2$(NO$_3$)$_2$	c	−1349.3	−1105.0	243	
std. state	aq	−1434.3	−1176.1	195.4	
UO$_2$(OH)$_2$ std. state	aq	−1479.5	−1267.8	−118.8	
UO$_2$SO$_4$	c	−1845.1	−1683.6	154.8	145.2
std. state	aq	−1928.8	−1698.3	−77.4	
US$_2$	c	−527	−526.4	110.42	74.64
US$_3$	c	−549.4	−547.3	138.49	95.60

(*Continued*)

TABLE 1.58 Enthalpies and Gibbs Energies of Formation, Entropies, and Heat Capacities of the Elements and Inorganic Compounds (*Continued*)

Substance	Physical state	$\Delta_f H°$ kJ · mol⁻¹	$\Delta_f G°$ kJ · mol⁻¹	$S°$ J · deg⁻¹ · mol⁻¹	$C°_p$ J · deg⁻¹ · mol⁻¹
Vanadium					
V	c	0	0	28.94	24.90
VBr_4	g	−336.8			
VCl_2	c	−452	−406	97.1	72.22
VCl_3	c	−580.7	−511.3	131.0	93.18
VCl_4	lq	−569.4	−503.8	255.0	161.7
VF_5	lq	−1480.3	−1373.2	175.7	
	g	−1433.9	−1369.8	320.9	98.58
VN	c	−217.15	−191.08	37.28	38.00
VO	c	−431.8	−404.2	39.0	45.5
VO_2	c	−717.6		51.5	62.59
VO^+_2 std. state	aq	−649.8	−587.0	−42.3	
VO^{2+}_2 std. state	aq	−486.6	−446.4	−133.9	
VO^-_3 std. state	aq	−888.3	−783.7	50.2	
V_2O_3	c	−1218.8	−1139.3	98.3	103.2
V_2O_4	c	−1427	−1318.4	103	115.4
V_2O_5	c	−1550	−1419.3	130	130.6
V_3O_5	c	−1933	−1803	163	
$VOCl_3$	lq	−734.7	−668.6	244.4	150.62
	g	−695.6	−659.3	344.4	89.9
$VOSO_4$	c	−1309.2	−1169.9	108.8	
Xenon					
Xe	g	0	0	169.685(3)	20.786
XeF_2	c	−164.0			
XeF_4	c	−261.5	−123.0		
XeF_6	c	−360			
	g	−297			
XeO_3	c	402			
$XeOF_4$	lq	146			
Ytterbium					
Yb	c	0	0	59.87	26.74
Yb^{2+} std. state	aq		−527.0		
Yb^{3+} std. state	aq	−674.5	−643.9	238.0	25.0
$Yb(OAc)_3$ undissoc; std. state	aq	−2105.0	−1772.84	183.3	
$YbCl_2$	c	−799.6			
$YbCl_3$	c	−959.8			
std. state	aq	−1176.1	−1037.6	−71.0	−385.0
$Yb(NO_3)_3$ std. state	aq	−1296.6			
Yb_2O_3	c	−1814.6	−1726.7	133.1	115.35
Yttrium					
Y	c	0	0	44.4	26.51
Y^{3+} std. state	aq	−723.4	−693.7	−251.0	
YCl_3	c	−1000		136.8	75.0
YF_3	c	−1718.8	−1644.7	100	
Y_2O_3	c	−1905.31	−1816.65	99.08	102.51
$Y(OH)_3$	c	−1435	−1291	99.2	
Zinc					
Zn	c	0	0	41.63(15)	25.40
	g	130.40(40)		160.990(4)	
Zn^{2+} std. state	aq	−153.39(20)	−147.1	−109.8(5)	46.0

TABLE 1.58 Enthalpies and Gibbs Energies of Formation, Entropies, and Heat Capacities of the Elements and Inorganic Compounds (*Continued*)

Substance	Physical state	$\Delta_f H°$ kJ · mol^{-1}	$\Delta_f G°$ kJ · mol^{-1}	$S°$ J · deg^{-1} · mol^{-1}	$C°_p$ J · deg^{-1} · mol^{-1}
$ZnBr_2$	c	−328.65	−312.13	138.5	65.7
std. state	aq	−396.98	−354.97	52.72	−238.0
$ZnCl_2$	c	−415.05	−369.45	111.46	71.34
std. state	aq	−488.19	−409.53	0.84	−226.0
$Zn(CN)^{2-}_4$ std. state	aq	342.3	446.9	226	
$ZnCO_3$	c	−812.78	−731.57	82.4	79.71
ZnF_2	c	−764.4	−713.3	73.68	65.7
std. state	aq	−819.14	−704.67	−139.8	−167.0
ZnI_2	c	−208.03	−208.95	161.1	65.69
	aq	−264.3	−250.2	110.5	−238.0
$Zn(NO_3)_2$	c	−483.7			
	aq	−568.6	−369.6	180.7	−126.0
ZnO	c	−350.46(27)	−320.52	43.65(40)	40.25
$Zn(OH)_2$	c	−641.91	−553.59	81.2	
std. state	aq	−613.88	−461.62	−133.5	−251
ZnS sphalerite	c	−205.98	−201.29	57.7	46.02
wurtzite	c	−192.6			
$ZnSe$	c	−163	−163	84.0	
$ZnSO_4$	c	−982.84	−871.5	110.5	99.2
	aq	−1063.2	−891.6	−92.0	−247.0
Zn_2SiO_4	c	−1636.7	−1523.2	131.42	123.3
Zirconium					
Zr	c	0	0	39.0	25.40
ZrB	c	−322	−318.2	35.94	48.24
$ZrBr_2$	c	−405	−382	116	86.7
$ZrBr_4$	c	−760.7	−725.3	224	124.8
ZrC	c	197	−193	33.32	37.90
$ZrCl_2$	c	−502.0	−386	110	72.6
$ZrCl_3$	c	−714	−646	146	96
$ZrCl_4$	c	−981	−890	181.4	119.8
ZrF_2	c	−962	−913	75	66
ZrF_4	c	−1911.3	−1810.0	104.7	103.6
ZnH_2	c	−169.0	−128.8	35.0	31.0
ZrI_2	c	−259	−258	150.2	94.1
ZrI_3	c	−397.5	−394.9	204.6	103.8
ZrI_4	c	−488	−485.4	260	127.8
ZrN	c	−365	−336.7	38.86	40.44
ZrO_2	c	−1100.6	−1042.8	50.36	56.19
$ZrSiO_4$	c	−2033.4	−1919.1	84.1	98.7
$ZrSO_4$	c	−2217.1			172.0

TABLE 1.59 Heats of Fusion, Vaporization, and Sublimation and Specific Heat at Various Temperatures of the Elements and Inorganic Compounds

Abbreviation Used in the Table

Hm, enthalpy of melting (at the melting point) in kJ · mol^{-1}
Hv, enthalpy of vaporization (at the boiling point) in kJ · mol^{-1}
Hs, enthalpy of sublimation (or vaporization at 298 K) in kJ · mol^{-1}
C_p, specific heat (at temperature specified on the Kelvin scale) for the physical state in existence (or specified: c, lq, g) at that temperature in J · K^{-1} (mol^{-1})
Ht, enthalpy of transition (at temperature specified, superscript, measured in degrees Celsius) in kJ · mol^{-1}

Substance	ΔHm	ΔHv	ΔHs	C_p 400 K	600 K	800 K	1000 K
Aluminum							
Al	10.71	294.0	326.4	25.8	27.9	30.6	34.9(lq)
Al(BH$_4$)$_3$		30					
Al$_6$BeO$_{10}$	402			324.3	380.6	407.8	425.2
AlBr$_3$	11.25	23.5		125.0	125.0	125.0	125.0
Al$_4$C$_3$				138.5	159.2	169.7	176.1
AlCl$_3$	35.4		116	100.1	117.7	135.2	152.8
AlF$_3$, AHt = 0.56^{455}	98			86.3	97.3	98.5	100.8
AlI$_3$	15.9	32.2	112	108.5	121.3		
AlN				36.7	43.5	46.8	48.5
Al$_2$O$_3$ corundum	111.4			96.1	112.5	120.1	124.8
AlOCl				64.3	72.6	76.9	79.3
Al$_2$SiO$_5$ andalusite				149.6	174.5	186.1	194.0
Kyanite				148.3	176.2	188.3	196.2
Sillimanite				147.5	173.0	185.0	193.5
Al$_6$Si$_2$O$_{13}$ mullite				390.7	459.8	494.1	513.4
Al$_2$S$_3$	55			115.0	124.1	129.7	134.0
Al$_2$TiO$_5$				162.0	182.8	192.9	200.0
Americium							
Am	14.39						
Ammonium							
NH$_3$	5.66	23.35	19.86	38.7	45.3	51.1	56.2
ND$_3$ ammonia-d_3				42.9	51.5	58.6	64.3
NH$_4$Br, ΔHt = 3.22^{138}							
NH$_4$Cl, ΔHt = 1.046$^{-30.6}$				103			
ΔHt = 3.950$^{184-6}$							
NH$_4$ClO$_4$				148.7			
NH$_4$I, ΔHt = 2.93^{-13}	20.9		168.5^{525}	89.0	103.3	117.7	
NH$_4$NO$_3$	6.40						
Antimony							
Sb	19.87	193.43		25.9	27.7	29.5	31.4
SbBr$_3$	14.6	59		125.5(lq)	81.6(g)	82.2	82.5
SbCl$_3$	12.7	45.2		123.4(lq)	81.6(g)	82.2	82.5
SbCl$_5$	10.0	48.4					
SbH$_3$		21.3					
SbI$_3$	22.8	68.6		106.6(lq)	143.5(lq)	82.2(g)	82.5(g)
Sb$_2$O$_3$, ΔHt = 7.1^{573}	54.4	74.6		108.5	122.8	137.1	150.6
Sb$_2$S$_3$				123.3	134.4	145.4	
Argon							
Ar	1.12	6.43		20.8	20.8	20.8	20.8

TABLE 1.59 Heats of Fusion, Vaporization, and Sublimation and Specific Heat at Various Temperatures of the Elements and Inorganic Compounds (*Continued*)

Substance	ΔHm	ΔHv	ΔHs	C_p 400 K	600 K	800 K	1000 K
Arsenic							
As	24.44			25.6	27.5	29.3	
$AsBr_3$	11.7	41.8					
$AsCl_3$	10.1	35.0		133.5(lq)	88.3(g)	88.3	
AsF_3	10.4	29.7					
AsF_5		20.8					
AsH_3		16.7		45.4	53.2	58.8	63.9
AsI_3		59.3					
As_2O_3	18.4			116.4			
Barium							
Ba	7.12	140.3		33.2	33.9(c)		39. 1(lq)
$BaBr_2$	32.2			79.2	83.5	87.9	92.2
$BaCl_2, \Delta Ht = 16.9^{925}$	15.85	246.4		77.3	80.4	84.3	89.5
$BaCO_3, \Delta Ht = 18.8^{806}$	40			99.0	113.0	124.2	134.6
$BaF_2, \Delta Ht = 2.67^{1207}$	17.8	285.4	405.1	75.9	80.3	84.9	94.6
BaH_2	25						
BaI_2	26.5	43.9	302.5	79.5	83.5	87.5(c)	113.0(lq)
$BaMoO_4$				129.5	143.5	152.2	159.3
BaO	46	330.6	424.3	49.9	53.2	55.4	57.1
$Ba(OH)_2$	16			112.6	122.7(c)	141.0(lq)	
BaS	63						
$BaSO_4$	40			119.4	131.6	135.9	137.9
$BaTiO_3, \Delta Ht = 0.067^{75}$				111.5	121.8	126.1	128.7
Beryllium							
Be	7.895	297	291	20.0	23.3	25.5	27.3
$BeAl_2O_4$, chrysoberyl	170.0			130.3	155.0	166.8	174.2
$BeBr_2$	18	100.0	515	70.6	77.6(c)	113.0(lq)	113.0
Be_2C	75.3			47.6	51.9	64.7	73.2
$BeCl_2, \Delta Ht = 6.8^{403}$	8.66	105	136.0	68.7	75.8(c)	121.4(lq)	121.4
$BeF_2, \Delta Ht = 0.92^{227}$	4.77	199.4		62.5	67.5	74.1(c)	85.6(lq)
BeI_2	18	70.5	125	76.9	84.2		
Be_3N_2	129.3			84.4	106.5	117.6	123.6
$BeO, \Delta Ht = 6.7^{2100}$	86			33.8	42.4	46.7	49.3
BeS				120.8	149.2	166.0	174.1
Be_2SiO_4				103.9	126.8	149.8	174.4
$BeSO_4, \Delta Ht = 1.113^{590}$ $\Delta Ht = 19.55^{635}$	6			103.9	126.8	149.8	174.4
$BeWO_4$				113.0	131.3	142.9	153.0
Bismuth							
Bi	11.30	151		27.0(c)	31.8(lq)	31.8	31.8
$BiBr_3$	21.7	75.4					
$BiCl_3$	10.9	72.6					
BiI_3		20.9					
$Bi_2O_3, \Delta Ht = 116.7^{717}$	28.5			116.9	123.6	130.3	137.0
Bi_2S_3				131.1	136.2	141.3	146.4
Bi_2Te_3	120.5			164.3	179.7	192.3	
Boron							
B	50.2	480	552	15.7	20.8	23.4	25.0
BBr_3		30.5		72.6(g)	77.6	79.8	81.1
B_4C	105			76.4	98.4	107.7	114.3

(*Continued*)

TABLE 1.59   Heats of Fusion, Vaporization, and Sublimation and Specific Heat at Various Temperatures of the Elements and Inorganic Compounds (*Continued*)

Substance	ΔHm	ΔHv	ΔHs	C_p 400 K	600 K	800 K	1000 K
BCl_3	2.10	23.8	23.1	68.4(g)	75.0	78.2	79.8
BF_3	4.20	19.3	57.5	67.1	72.6	75.8	
$F_2B\text{-}BF_2$		28					
BH_3				38.9	45.4	52.3	58.4
B_2H_6	4.44	14.3		74.3	101.3	121.7	136.4
B_4H_9	6.13	28.4		130.2(g)	187.6	227.4	254.4
B_4H_{10}		27.1					
B_5H_{11}		31.8					
$B_{10}H_{14}$	32.5	48.5	76.7	250.0(lq)	351.6(g)	417.2	460.4
BI_3		40.5					
BN	81		728	26.3	35.2	40.5	44.3
$B_3N_3H_6$ borazine		32.1		126.9	169.4	197.2	216.6
B_2O_3	24.56	390.4		77.9	98.1(c)	129.7(lq)	129.7
$B_3O_3H_3$ boroxin			44.8	120.1	162.8	194.6	214.2
Bromine							
Br_2	10.57	29.96	30.9	36.7(g)	37.3	37.6	37.8
$BrCl$	10.4	34.7					
BrF		25.1					
BrF_3	12.05	47.6		72.6	78.0	80.1	81.2
BrF_5	5.67	30.6		113.0	123.2	127.3	129.3
Cadmium							
Cd	6.19	99.9		27.1(c)	29.7(lq)	29.7	29.7
$CdBr_2$	20.9	115					
$CdCl_2$	48.58	124.3		79.8	86.3	92.7	104.6
CdF_2	22.6	214					
CdI_2	15.3	115					
$Cd(NO_3)_2 \cdot 4H_2O$	32.6						
CdO			225.1	43.8	45.6	47.3	49.1
CdS			209.6	55.5	56.2	57.0	57.7
$CdSO_4$				108.3	123.8	139.2	154.7
Calcium							
Ca, $\Delta Ht = 0.93^4$	8.54	154.7		26.9	30.0	33.8	39.7
$Ca(BO_2)_2$	74.1			125.0	144.9	157.2	176.2
CaB_4O_7	113.4			202.0	243.0	267.7	287.8
$CaBr_2$	29.1	200	298.3	78.0	80.5	83.5	88.6
CaC_2 carbide	32						
$CaCl_2$	28.05	235		75.6	78.2	80.9	85.8
$CaCN_2$ cyanamide	0.432						
$CaCO_3$	36						
CaF_2, $\Delta Ht = 4.8^{1151}$	29.3	308.9	441	73.9	78.5	83.9	90.1
CaH_2	6.7						
CaI_2	41.8	179.4	243	79.2	83.1	87.1	91.0
$Ca[Mg(CO_3)_2]$ dolomite				143.3	163.3	176.8	188.3
$CaMoO_4$				131.3	144.9	153.5	150.6
Ca_3N_2				122.2	140.8	159.2	
$Ca(NO_3)_2$	21.4			173.7	210.5	243.4	
CaO	79.5			46.6	50.5	52.4	53.7
$Ca(OH)_2$, $\Delta Hdec = 99.2$				98.4	107.4		
$Ca_3(PO_4)_2$, $\Delta Ht = 15.5^{1100}$				255.1	295.6	331.3	365.7
CaS	70			49.2	51.5	53.0	54.1

TABLE 1.59   Heats of Fusion, Vaporization, and Sublimation and Specific Heat at Various Temperatures of the Elements and Inorganic Compounds (*Continued*)

Substance	ΔHm	ΔHv	ΔHs	C_p 400 K	600 K	800 K	1000 K
$CaSiO_3$, $\Delta Ht = 7.1^{1190}$	56.1			100.4	113.0	119.2	123.8
$Ca2SiO_4$, $\Delta Ht = 4.44^{675}$				146.4	162.8	179.2	184.0
$\Delta Ht = 3.26^{1420}$							
$3CaO \cdot SiO_2$				196.4	218.4	230.8	240.4
$CaSO_4$	28.0			109.7	129.5	149.2	169.0
$CaSO_4 \cdot \frac{1}{2}H_2O$				147.4	167.2	186.9	206.7
$CaSO_4 \cdot 2H_2O$				260.7	280.3	300.0	319.8
$CaTiO_3$, $\Delta Ht = 2.30^{1257}$				112.3	123.1	127.7	130.4
$Ca(VO_2)_2$				182.9	206.7	230.5	254.4
$CaWO_4$				127.6	140.2	147.3	152.8
Carbon							
C graphite	117			12.0	16.6	19.7	21.7
$(CN)_2$ cyanogen	8.1	23.3	19.7	61.9(g)	68.2	72.9	76.4
CNBr			45.4	50.19(g)	53.7	56.2	58.1
CNCl	11.4			48.7	52.8	55.7	57.7
CNI			59.4	50.8	53.7	55.8	57.4
CO, $\Delta Ht = 0.632^{-211.6}$	0.837	6.04		29.3	30.4	31.9	33.2
CO_2	9.02	15.8	25.2	41.3	47.3	51.4	54.3
C_2O_3	5.40	$26.9^{4.35}$		75.0	85.5	92.7	97.7
$COCl_2$	5.74	24.4		63.9	71.1	75.0	77.4
CoF_2		16.1		54.8	64.9	70.8	74.4
COS	7.73	18.6		45.9	51.3	54.7	57.0
CS_2	4.40	26.7	27.5	49.7	54.6	57.4	59.3
Cerium							
Ce, $\Delta Ht = 3.01^{730}$	5.46	398	419	30.6	30.8	32.1	33.8
$CeCl_3$	54.4	170.1	326				
CeI_3	51.9						
CeO_2				66.9	69.0	71.1	73.2
Cesium							
Cs	2.09	63.9	76.6	31.5	31.0	30.9(lq)	20.8(g)
CsBr	23.6	151		52.9	55.0	57.2(c)	77.4(lq)
CsCl, $\Delta Ht = 3.77^{470}$	15.9	115.1		54.7	59.1	63.7(c)	77.4(lq)
CsF	21.7	115.5		53.8	57.4	60.9(c)	74.1(lq)
CsI	23.9	150.2		51.9	57.8(c)	65.5(lq)	67.8
$CsIO_3$	13.0						
CsOH, $\Delta Ht = 1.30^{137}$	4.56	120		74.4(c)	81.6(lq)	81.6	81.6
$\Delta Ht = 6.1^{220}$							
Cs_2SO_4, $\Delta Ht = 4.3^{667}$	35.7		76.5	112.1	132.2	163.2	194.2
Chlorine							
Cl_2	6.406	20.41	17.65	35.3	36.6	37.1	37.4
ClF		24		33.8	35.6	36.5	37.0
ClF_3	7.61	27.5		70.6(g)	76.8	79.4	80.7
ClF_5		22.9		110.0	121.6	126.3	128.6
ClO				33.2	35.3	36.3	36.9
ClO_2		30		46.1	51.4	54.2	55.8
ClO_3F	3.83	19.33		75.9	89.2	96.1	100.0
Cl_2O		25.9		51.4	54.7	56.2	56.9
Cl_2O_7		34.69					
Chromium							
Cr, $\Delta Ht = 0.0008^{38.5}$	21.0	339.5	397	25.2	27.7	29.4	31.9

TABLE 1.59   Heats of Fusion, Vaporization, and Sublimation and Specific Heat at Various Temperatures of the Elements and Inorganic Compounds (*Continued*)

Substance	ΔHm	ΔHv	ΔHs	C_p 400 K	600 K	800 K	1000 K
$CrCl_2$	32.2	196.7		72.6	77.0	81.5	85.9
$CrCl_3$			237.7	93.1	99.0	104.9	110.7
$Cr(CO)_6$			72.0	233.9			
$CrN, \Delta Hdec = 112$			49.1	50.4	51.7	53.0	
CrO_2Cl_2		35.1					
CrO_2F_2	23.4	34.3					
CrO_3	15.77			63.9	72.5	76.7	78.8
Cr_2O_3	129.7			112.7	120.5	124.3	127.0
$Cr_2(SO_4)_3$				316.9	345.2	373.5	401.8
Cobalt							
$Co, \Delta Ht = 0.452^{427}$	16.2	377	424	26.5	29.7	32.4	37.0
$CoCl_2$	45	146	219	81.7	84.6	86.8	88.2
CoF_2	59	202	315	75.7	80.8	82.9	84.2
CoF_3				97	100	102	104
CoO				52.9	54.3	54.8	56.0
Co_3O_4				143	163	185	210
$CoSO_4, \Delta Ht = 2.1^{691}$				119	141	152	158
Copper							
Cu	13.26	300.4	337.7	25.3	26.5	27.4	28.7
$CuBr, \Delta Ht = 5.86^{380}$ $\Delta Ht = 2.9^{465}$	9.6			56.5	59.8(c)	66.9(lq)	66.9
$CuCl$	10.2	54	241.8	56.9	61.5(c)	66.9(lq)	66.9
$CuCl_2, \Delta Ht = 0.700^{402}$ $\Delta Ht = 15.001^{598}$	20.4			76.3	80.2(c)	82.4(lq)	100.0
$CuCN$		12			66.7	73.1	78.0
CuF			268	55.5	59.6		
CuF_2	55	156	261	72.4	81.9	87.0	90.4
CuI	10.9			55.4	57.8	60.2	66.9
CuO	11.8			46.8	50.8	53.2	55.0
Cu_2O	64.8			67.6	73.3	77.6	81.5
CuS				48.8	51.0	53.2	55.4
$Cu_2S, \Delta Ht = 3.85^{103}$ $\Delta Ht = 0.84^{350}$	10.9			97.3	97.3	85.0	85.0
$Cu_2Se, \Delta Ht = 4.85^{110}$				90.9	91.7	92.5	93.4
$CuSO_4$				114.9	136.3	147.7	153.8
Dysprosium							
Dy	11.06	280	290.4				
Erbium							
Er	19.90	280	317.2				
Europium							
Eu	9.21	176	178				
Fluorine							
$F_2, \Delta Ht = 0.728^{227.6}$	0.510	6.62		33.0	35.2	36.3	37.1
FNO_3				75.1	87.8	94.8	98.9
Gadolinium							
Gd	10.05	301.3		36.6	35.5	34.5	33.5
Gd_2O_3				113.4	120.1	124.4	127.9
Gallium							
Ga	5.59	254		27.1(lq)	26.7	26.6	26.6
$GaBr_3$	12.1	38.9					

TABLE 1.59 Heats of Fusion, Vaporization, and Sublimation and Specific Heat at Various Temperatures of the Elements and Inorganic Compounds (*Continued*)

Substance	ΔHm	ΔHv	ΔHs	C_p 400 K	600 K	800 K	1000 K
$GaCl_3$	11.13	23.9					
GaI_3	12.9	56.5					
Ga_2O_3	100			91.4	112.5	133.5	
GaSb	25.1						
Germanium							
Ge, $\Delta Ht = 37.03^{938.3}$	36.94	334		24.3	25.4	26.2	26.9
$GeBr_4$		41.4					
$GeCl_4$		27.9		100.7	104.6	106.1	106.8
GeH_4		14.1					
Ge_2H_6		25.1					
Ge_3H_8		32.2					
GeO_2	43.9			61.39	69.1	72.4	75.0
Gold							
Au	12.55	324		25.8	26.8	27.8	28.8
AuSn	25.6			54.1	63.3(c)	60.6(lq)	
Hafnium							
Hf, $\Delta Ht = 5.9^{1750}$	27.2	571	618.4	26.7	28.6	30.3	31.9
$HfCl_4$	75		99.6	125.4	105.8	106.7	107.1
HfO_2, $\Delta Ht = 10.5^{1700}$	104.6			67.7	73.9	77.3	79.9
Helium							
He	0.0138	0.0829		20.79	20.79	20.79	20.79
Holmium							
Ho	16.8	71		280	317		
Hydrogen							
H_2	0.117	0.904		29.2	29.3	29.6	30.2
$^1H^2H$				29.2	29.4	29.9	30.7
$^2H_2$				29.2	29.6	30.5	31.6
HBO_2	14.3		242.1	61.5(c)			
H_3BO_3	22.3						
HBr	2.406	17.61	12.7	29.2	29.8	31.1	32.3
HCl, $\Delta Ht = 1.188^{-174.77}$	1.992	16.14	9.1	19.2	29.2	29.6	31.6
^2HCl				29.4	30.6	32.1	33.5
HClO				40.0	44.0	46.6	48.5
HCN	8.406	25.22		39.4	44.2	47.9	51.0
HF	4.58			29.1	29.2	29.5	30.2
^2HF				29.2	29.5	30.5	31.6
H_2F_2 dimer				49.7	56.5	61.0	64.4
HFO				38.6	42.8	45.7	47.9
HI	2.87	19.77	17.4	29.3	30.3	31.8	33.1
HNCO isocyanic acid				50.6	58.3	63.5	67.5
HNCS isothiocyanic acid				53.2	61.0	65.9	69.3
HNO_2 *cis*				51.4	59.9	65.4	69.2
trans				52.1	60.3	65.6	69.3
HNO_3	10.47	39.46	39.1	63.1	76.8	85.0	90.4
HN_3		30.5					
H_2O	6.009	40.66	44.0	34.3(g)	36.4	38.8	41.4
$^1H^2HO$				34.8	37.5	40.4	43.3
^2H_2O				35.6	38.8	42.2	45.4
H_2O_2	12.50		51.63	48.5	55.7	59.8	66.7
$^2H_2O_2$	12.68		52.4				

(*Continued*)

TABLE 1.59 Heats of Fusion, Vaporization, and Sublimation and Specific Heat at Various Temperatures of the Elements and Inorganic Compounds (*Continued*)

Substance	ΔHm	ΔHv	ΔHs	C_p 400 K	600 K	800 K	1000 K
HPH_2O_2	9.67						
H_3PO_3	12.84						
H_3PO_4	13.4			175.7	236.0	296.2	365.5
H_2S, $\Delta Ht = 1.531^{-169.61}$	23.8	18.67	14.1	38.9	42.5	45.8	
H_2S_2		33.8					
H_2Se		19.7					
HSO_3F				87.5	102.6	111.0	116.3
H_2SO_4	10.71	50.2		158.2	197.0(lq)	125.9(g)	132.7
$H_2SO_4 \cdot H_2O$	19.46			228.5			
$H_2SO_4 \cdot 2H_2O$	18.24			294.6			
$H_2SO_4 \cdot 3H_2O$	24.0			347.8			
$H_2SO_4 \cdot 4H_2O$	30.64			410.3			
H_2Te		19.2					
Indium							
In	3.28	231.8	243.1	28.5(c)	30.1(lq)	30.1	30.1
InBr	15	92					
$InBr_3$	26						
InCl	21.3						
$InCl_3$	27						
InF_3	64						
InI	17.3	90.8					
InI_3	18.5						
In_2O_3	105						
InSb	25.5						
Iodine							
I_2	150.66	41.6	62.4	79.6(lq)	37.6(g)	37.9	38.1
ICl	11.60		52.9	98.3(lq)	90.0	81.6	73.2
IF				35.1	36.6	37.3	37.7
IF_5		41.3		476.1(g)	516.7	533.0	541.4
IF_7				152.0(g)	167.6	173.9	177.0
Iridium							
It	41.12	231.8	243.1	28.5(c)	30.1(lq)	30.1	30.1
IrF_6	8.40	36					
IrO_2				63.8	76.5	89.2	102.0
Iron							
Fe, $\Delta Ht = 0.90^{911}$	13.81	340	415.5	27.4	32.1	38.0	54.4
$\Delta Ht = 0.837^{1392}$							
$FeBr_2$	50.2						
$FeBr_3$, $\Delta Ht = 0.418^{377}$	50.2		207.5	83.0	87.0	91.4	95.9
Fe_2C, $\Delta Ht = 0.75^{190}$	51.5			115.7	114.7	117.2	119.8
$FeCl_2$	43.01	26.3		79.7	83.1	85.5	101.2
$FeCl_3$	43.1	43.76		106.7(c)	133.9(lq)	82.3(g)	81.5
$FeCO_3$				93.5	115.9	138.3	
$Fe(CO)_5$	13.23	33.72		189.0	209.8	223.1	232.2
$FeCr_2O_4$				152.0	167.7	175.9	182.2
FeF_2	51.9	224.4	316	72.0	77.1	80.3	82.1
FeF_3			274	96.4	96.8	99.3	101.8
FeI_2, $\Delta Ht = 0.8^{377}$	45	104.6	192	83.9	84.4	110.9	113.0(lq)
Fe_3N				72.6	77.7	82.8	87.9
FeO	24.06			51.8	54.9	57.3	59.4

TABLE 1.59   Heats of Fusion, Vaporization, and Sublimation and Specific Heat at Various Temperatures of the Elements and Inorganic Compounds (*Continued*)

Substance	ΔHm	ΔHv	ΔHs	C_p 400 K	600 K	800 K	1000 K
Fe_2O_3, $\Delta Ht = 0.67^{677}$				120.1	141.2	158.2	150.6
Fe_3O_4	138.1			171.1	212.5	252.9	
$Fe(OH)_2$			243.5	102.1	111.3	118.9	123.4
$Fe(OH)_3$				118.0	140.6	154.8	164.9
FeS, $\Delta Ht = 0.40^{138}$	31.5			89.2	62.0	58.6	59.0
$\Delta Ht = 0.095^{325}$							
FeS_2 marcasite				69.2	74.6	78.7	82.8
pyrite				68.9	74.3	78.3	82.5
$FeSiO_3$				100.8	114.3	124.5	133.9
Fe_2SiO_4	92			150.9	168.5	179.7	189.1
$FeSO_4$				116.7	138.0	149.4	
$Fe_2(SO_4)_3$				307.0	363.3	393.3	409.2
$FeTiO_3$ ilminite	90.8	111.4	122.0	128.1	132.8		
Krypton							
Kr	1.37	9.08					
Lanthanum							
La, $\Delta Ht = 2.85^{868}$	6.20	402.1		28.5	29.8	31.2	32.5
$LaCl_3$	43.1	192.1		105.8	110.1	114.3	118.7
La_2O_3				117.3	124.7	128.9	132.3
Lead							
Pb	4.77	179.5	195.2	27.7	29.4	30.0	29.4
$Pb(BO_2)_2$				129.7	162.3		
PbB_4O_7				207	265	305	330
$PbBr_2$	16.44	133	173	81.3	88.8	112. l(lq)	112.1
$Pb(CH_3)_4$	10.86						
$Pb(C_2H_5)_4$	8.80						
$PbCl_2$	21.9	127	185.3	80.1	85.9	111.5(lq)	111.5
$PbCO_3$				99.7	123.6	147.6	
PbF_2, $\Delta Ht = 1.46^{310}$	14.7	157		76.1	82.5	89.1	95.6
PbI_2	23.4	104	172	78.9	83.7(c)	108.6(lq)	108.6
$PbMoO_4$				135.3	148.9	159.0	168.2
PbO, $\Delta Ht = 0.17^{488}$	25.5	207		50.4	55.4	55.0	57.8
PbO_2				67.6			
Pb_3O_4				173.1	190.8	199.2	
PbS	18.8	230		50.5	52.4	54.3	56.2
$PbSiO_3$	26.0			101.5	113.5	125.6	138.4
Pb_2SiO_4	51.0			152.0	173.3	184.2	189.1
$PbSO_4$, $\Delta Ht = 17.2^{866}$	40.2			108.7	128.6	152.4	177.3
$PbSO_4 \cdot PbO$				157.3	182.5	211.7	242.0
Lithium							
Li	3.00	147.1	159.3	27.6(c)	29.5(lq)	28.9	28.8
Li_2AlF_6, $\Delta Ht = 9.5^{562}$	110.5			236.4	262.8	290.8	318.6
$LiAlO_2$	87			81.5	92.7	98.2	102.0
$LiBH_4$				91.0			
$LiBeF_3$	27.2			104.6	129.7(c)	159.0(lq)	159.0
Li_2BeF_4	44.0			150.5	180.2(c)	232.1(lq)	232.1
$LiBO_2$	33.8	265		81.1	85.1	96.9	108.3
$Li_2B_4O_7$	121			197.6	241.1	274.4	300.2
LiBr	17.6	107.1		51.3	56.1	64.5(c)	65.3(lq)
LiCl	19.9			51.0	55.6	65.8	

(*Continued*)

TABLE 1.59 Heats of Fusion, Vaporization, and Sublimation and Specific Heat at Various Temperatures of the Elements and Inorganic Compounds (*Continued*)

Substance	ΔHm	ΔHv	ΔHs	C_p 400 K	600 K	800 K	1000 K
LiClO$_4$	29			130.0(c)	161.0(lq)	161	161
Li$_2$CO$_3$, $\Delta Ht = 0.561^{350}$	41			112.2	149.4	159.0	
$\Delta Ht = 2.238^{410}$							
LiF	27.09	146.8	276.1	46.5	51.6	55.7	59.6
LiH	22.6		231.3	34.8	46.4	57.3	
LiI	14.6						
LiIO$_3$, $\Delta Ht = 2.22^{260}$							
Li$_3$N				87.1	106.4	124.4	141.0
LiNO$_3$	24.9						
Li$_2$O	58.6			64.0	73.8	80.6	86.2
Li$_2$O$_2$				82.7(c)	80.2(g)	81.4	82.1
LiOH	20.88	187.9	250.6	58.0	68.2(c)	87.1(lq)	87.1
Li$_2$SiO$_3$	28.0			118.8	134.3	144.4	152.3
Li$_2$Si$_2$O$_5$, $\Delta Ht = 0.941^{936}$	53.8			174.9	205.7	222.6	235.4
Li$_2$SO$_4$, $\Delta Ht = 28.5^{575}$	7.50			139.2	168.5	196.1	223.4
Li$_2$TiO$_3$, $\Delta Ht = 11.51^{1212}$	110.7			127.4	141.5	149.0	153.9
Lutetium							
Lu	(22)	414					
Magnesium							
Mg	8.48	128	147	26.1	28.2	30.5	
MgAl$_2$O$_4$	192			138.0	157.9	169.5	178.7
MgBr$_2$	39.3	149	222	77.3	81.4	84.5	
MgCl$_2$	43.1	156.2	249.2	75.7	79.9	82.5	
MgCO$_3$	59			89.9	109.0	122.3	131.8
MgF$_2$	58.5	274.1	399.5	68.5	75.3	78.6	80.5
MgH$_2$	14						
MgI$_2$	26		206	78.4	83.0	96.3(c)	100.4(lq)
Mg$_3$N$_2$, $\Delta Ht = 0.46^{550}$			107.6	113.8	119.9	123.8	
$\Delta Ht = 0.92^{788}$							
Mg(NO$_3$)$_2$				168.5	225.5		
MgO	77			42.6	47.4	49.7	51.2
Mg(OH)$_2$				91.7			
Mg$_3$(PO$_4$)$_2$	121			240.2	282.2	320.6	351.5
MgS	63						
Mg$_2$Si	85.8			73.8	79.8	83.9	87.4
MgSiO$_3$, $\Delta Ht = 0.67^{630}$	71			94.2	107.0	115.8	120.3
$\Delta Ht = 1.63^{985}$							
Mg$_2$SiO$_4$				137.6	156.4	167.1	174.6
MgSO$_4$	14.6			110.0	127.6	140.5	151.7
MgTiO$_3$				105.2	118.5	125.4	129.9
Mg$_2$TiO$_4$				146	164	175	184
MgWO$_4$				123.4	137.0	146.1	154.8
Manganese							
Mn, $\Delta Ht = 2.23^{727}$	12.9	221		28.5	31.9	34.9	37.5
$\Delta Ht = 2.12^{1101}$							
$\Delta Ht = 1.88^{1137}$							
MnBr$_2$	33	113		77.8	82.8	87.7	
Mn$_3$C, $\Delta Ht = 14.94^{1037}$				104.4	115.0	121.7	127.4
MnCl$_2$	30.7	149.0		77.2	81.8	85.1	96.2(lq)
Mn$_2$(CO)$_{10}$			62.8				

TABLE 1.59 Heats of Fusion, Vaporization, and Sublimation and Specific Heat at Various Temperatures of the Elements and Inorganic Compounds (*Continued*)

Substance	ΔHm	ΔHv	ΔHs	C_p 400 K	600 K	800 K	1000 K
MnF_2	23.0			70.6	75.7	80.7	85.9
MnI_2	42			78.1	83.6	89.0	108.8
MnO	54.4			47.5	50.3	52.4	54.2
MnO_2				63.4	71.1	75.1	
Mn_2O_3				109.0	120.8	129.4	137.2
Mn_3O_4, $\Delta Ht = 20.79$[1172]				157.3	169.5	179.7	189.3
MnS	26.4			50.7	52.2	53.7	55.2
$MnSiO_3$	66.9			100.9	113.1	119.5	124.2
$MnSO_4$				119.0	136.7	147.7	
$MnTiO_3$				111.7	121.2	125.7	128.8
Mercury							
Hg	2.29	59.1	61.4	27.4	27.1(lq)	20.8(g)	20.8
$HgBr_2$	17.9	58.9		78.3	102.1(lq)	102.1	102.1
Hg_2Br_2				109.6	115.6		
$HgCl_2$	19.41	58.9		77.0(c)	102.9(lq)		
Hg_2Cl_2				106.0	112.1		
HgF_2	23.0	92		77.0	81.2	85.4(c)	102.9(lq)
Hg_2F_2				104.7	111.7	116.9	
HgI_2, $\Delta Ht = 2.52$[129]	18.9	59.2		82.0(c)	84.1(lq)	62.2(g)	62.2
Hg_2I_2	27.8			110.4(c)	136.4(lq)		
HgO				48.3	54.1		
HgS, $\Delta Ht = 4.2$[386]				48.0	51.0	54.1	
Molybdenum							
Mo	37.48	617	664	25.1	26.5	27.4	28.4
$MoBr_3$				106.9	109.8	112.7	
$MoCl_4$	17	61.5		135.0(c)	146.4(lq)		
$MoCl_5$	18.8	62.8		167.4(c)	175.7(lq)	175.7	175.7
$Mo(CO)_6$		72.5	69.9				
MoF_6, $\Delta Ht = 8.17^{-9.65}$	4.33	27.2	28.0	133.1	145.3	150.4	153.0
MoO_2				63.5	71.2	76.5	81.4
MoO_3	48	138		83.1	91.8	100.0	109.0
MoS_2				68.9	73.6	76.2	78.2
Mo_2S_3	130			117.5	127.4	135.2	142.3
Neodymium							
Nd, $\Delta Ht = 2.98$[862]	7.14	289		28.2	32.1	36.9	42.0
Nd_2O_3				120.3	130.0	137.7	144.4
Neon							
Ne	0.335	1.71					
Neptunium							
Np, $\Delta Ht = 8.37$[280]	3.20	336		34.8			
Nickel							
Ni	17.48	377.5		28.5	30.0	31.0	32.2
$NiCl_2$	71.2		231.0	76.3	79.9	80.9	
$Ni(CO)_4$	13.8	29.3		160.4(g)	173.2	182.1	188.6
NiF_2				76.4	78.5	82.6	
NiO				52.2	51.8	53.6	55.2
NiS, $\Delta Ht = 6.4$[379]	30.1			12.1	13.2	13.7	15.1
Ni_3S_2, $\Delta Ht = 56.2$[556]	19.7			127.1	139.9	150.7	188.6
NiS_2	65.7			72.8	70.0	81.0	85.2
$NiSO_4$				142.6	150.8	159.2	167.4
$NiWO_4$				138.9	144.6	150.3	155.9

(*Continued*)

TABLE 1.59   Heats of Fusion, Vaporization, and Sublimation and Specific Heat at Various Temperatures of the Elements and Inorganic Compounds (*Continued*)

Substance	ΔHm	ΔHv	ΔHs	C_p 400 K	600 K	800 K	1000 K
Niobium							
Nb	30	689.9	726	25.4	26.3	27.2	28.0
NbBr$_5$	24.0	50.2	112.5	147.9(c)	147.9(lq)		
NbCl$_5$	38.3	52.7		170.7(c)	127.9(g)	129.8	130.7
NbF$_5$	12.2	52.3		43.5(lq)			
NbI$_5$	37.7	58.6		182.0(c)			
NbN, $\Delta Ht = 4.2^{1370}$	46.0			45.4	49.9	51.6	53.2
NbO	85	618		44.0	47.2	49.5	51.5
NbO$_2$, $\Delta Ht = 3.42^{817}$	92		598.0	63.5	71.7	70.5	87.5
Nb$_2$O$_5$	104.3			145.0	160.7	170.0	175.5
Nitrogen							
N$_2$, $\Delta Ht = 0.230^{-237.53}$	0.720	5.577		29.2	30.1	31.4	32.7
NF$_3$		11.6		61.9	71.4	76.0	78.4
N$_2$F$_2$ *cis*	15.4	91.6		58.2	68.3	73.6	76.6
trans	14.2	87.9		60.2	68.9	73.8	76.7
N$_2$F$_4$		13.3					
NH$_3$ (*see* Ammonium)							
N$_2$H$_4$	12.66	41.8	44.7	61.7(g)	77.6	88.2	96.4
NO	2.30	13.83		29.9	31.2	32.8	34.0
NOCl		25.8		47.1	50.7	53.2	54.9
NOF		19.3		44.6	48.9	51.7	53.5
NOF$_3$				78.7	90.9	97.0	100.5
NO$_2$				40.5	46.4	50.4	53.0
NO$_2$Cl		25.7		59.6	68.1	73.1	76.1
NO$_2$F		18.0		57.0	66.4	71.9	75.3
NO$_3$				55.9	67.4	73.3	76.5
N$_2$O	6.54	16.53		42.7	48.4	52.2	54.9
N$_2$O$_4$	14.65	38.12		88.5	104.0	113.4	119.2
N$_2$O$_5$			62.3	110.9	128.4	137.0	141.4
NSF		22.2					
Osmium							
Os	57.85	738		25.1	25.9	26.7	27.4
OsF$_6$		28.62					
OsO$_4$	9.8	39.54					
Oxygen							
O$_2$, $\Delta Ht = 0.092^{249.49}$ $\Delta Ht = 0.745^{-229.38}$	0.444	6.820	8.204	30.11	32.09	33.74	34.88
O$_3$		10.84		43.74	49.86	53.15	55.02
OF$_2$		11.09		64.3	72.4	76.4	78.6
O$_2$F$_2$		19.1					
Palladium							
Pd	16.74	362		26.5	27.7	28.8	30.0
PdCl$_2$	40.1						
PdO				37.6	49.5	61.3	
Phosphorus							
P		0.66	12.4	14.2			
P$_4$, $\Delta Ht = 0.521^{-77.8}$	0.659	56.5	58.9	73.3(g)	78.4	80.4	81.4
PBr$_3$		38.8		78.9	81.2	82.0	82.4
PClF$_2$		17.6					

TABLE 1.59   Heats of Fusion, Vaporization, and Sublimation and Specific Heat at Various Temperatures of the Elements and Inorganic Compounds (*Continued*)

Substance	ΔHm	ΔHv	ΔHs	C_p 400 K	600 K	800 K	1000 K
$PClF_3$		17.6					
PCl_2F		24.9					
PCl_3	7.10	30.5	32.1	76.0(g)	79.7	81.2	81.9
PCl_5			64.9	120.1(g)	126.8	129.5	130.7
PF_3		16.5		66.3(g)	74.0	77.6	79.5
PF_5		17.2		99.2(g)	114.7	121.9	125.6
PH_3	1.130	14.60		41.8	50.9	58.5	64.3
P_2H_4		28.8					
PI_3		43.9					
P_4O_6	14.06	43.43		172.1	200.8	213.5	220.0
P_4O_{10}	27.2		106.0	260.3	336.0(c)		
$POBr_3$	38						
$POCl_3$	13.1	34.3	38.6	92.0(g)	99.1	102.5	108.5
$POClF_2$		25.4		79.3	91.6	97.7	101.1
$POCl_2F$		30.96		87.7	96.6	100.9	103.2
POF_3	15.06	23.22	21.1	79.1	91.2	97.4	100.9
$PSCl_3$				96.5	102.4	104.8	105.9
PSF_3		19.58		84.5	95.3	100.3	102.9
P_4S_3	9.2	59.8		184.1	184. l(lq)	155.0(g)	155.0
Platinum							
Pt	22.17	469	545	26.4	27.5	28.5	29.6
PtS				51.4	53.8	56.2	58.6
PtS_2				69.9	75.9	81.9	87.9
Plutonium							
Pu, $\Delta Ht = 13.4^{122}$	2.82	333.5		39.5	46.9	40.6	40.6
$\Delta Ht = 2.9^{206}$							
$\Delta Ht = 3.3^{319}$							
$\Delta Ht = 66.9^{480}$							
$PuBr_3$	55.2	236.4	292.5				
$PuCl_3$	63.6	241.0	304.6				
PuF_3	59.8		374.9				
PuF_4	65.3		299.6				
PuF_6	17.6	29.9	48.5				
PuI_3	50.2						
PuO_2		559.8					
Polonium							
Po		102.91					
Potassium							
K	2.321	76.90	88.8	31.5(lq)	30.1	29.8	30.7
$KAlCl_4$				165.5	183.2	196.6	202.1
K_3AlCl_6				259.2	279.5	295.8	
K_3AlF_6				244.5	269.4	286.8	302.0
KBF_4, $\Delta Ht = 14.06^{283}$	17.7			130.8	142.1	150.9	167.2
KBH_4				100.9	106.0	118.4	
KBO_2	31	238.9		76.7	89.8	98.5	
$K_2B_4O_7$	104			206.3	250.5	271.1	283.3
KBr	25.5	149.2		53.8	56.4	60.4	68.0
KCl	26.53	124.3		53.0	55.9	59.2	64.0
$KClO_4$, $\Delta Ht = 13.77^{299.6}$				138.5	165.3		
KCN, $\Delta Ht = 1.167^{-104.9}$	14.6	157.1		66.3	66.4	66.5(c)	66.5(lq)

(*Continued*)

TABLE 1.59   Heats of Fusion, Vaporization, and Sublimation and Specific Heat at Various Temperatures of the Elements and Inorganic Compounds (*Continued*)

Substance	ΔHm	ΔHv	ΔHs	C_p 400 K	600 K	800 K	1000 K
K_2CO_3	27.6			128.1	150.7	170.0	189.0
K_2CrO_4	29.0						
$K_2Cr_2O_7$	36.7						
KF	27.2	141.8	231.8	51.0	54.3	57.4	61.2
KH				44.1	51.9		
KHF_2, $\Delta Ht = 11.22^{196.7}$	6.62			86.1(c)	104.6(lq)		
KI	24.0	190.9	202.4	53.9	57.3	62.6(c)	72.4(lq)
KNO_3, $\Delta Ht = 5.10^{128}$	10.1			108.4	120.5		
K_2O, $\Delta Ht = 6.20^{372}$				79.1	100.0	100.0	100.0
KO_2, $\Delta Ht = 0.302^{-79.7}$				83.9	90.2		
$\quad \Delta Ht = 0.157^{-42.3}$							
K_2O_2				107	121		
KOH, $\Delta Ht = 6.4^{243}$	8.60	142.7	192	72.5	79.0(c)	83.0(lq)	83.0
KPO_3	8.8						
K_3PO_4	37.2						
$K_2P_2O_7$	58.6						
$KReO_4$	85.4						
K_2S	16.15	77.3	82.5	87.7			
K_2SiO_3	50			135.6	157.7	170.7	179.1
K_2SO_4, $\Delta Ht = 8.45^{584}$	34.39			147.6	172.5	199.6	226.1
K_2WO_4	19.5						
K_2ZrCl_6	23.0						
Praseodymium							
Pr	6.89	331	356				
Promethium							
Pm	7.13	289	328				
Protactinium							
Pa	12.34	481					
$PaCl_3$	92.9	61.3					
Radium							
Ra	8.5	113					
Radon							
Rn	3.247	18.10					
Rhenium							
Re	60.43	704	779	26.0	26.9	28.0	29.1
ReF_5		58.1					
ReF_6	4.6	28.7					
ReF_7	7.5	38.3					
ReO_2			274.6				
ReO_3	21.8		208.4				
Re_2O_7	64.2	74.1					
$ReOCl_4$		45.6					
$ReOF_4$	13.5	61.0					
$ReOF_5$		32.0	37.4				
Rhodium							
Rh	26.59	494	556	26.0	28.0	30.0	32.0
Rh_2O_3				109.9	121.4	133.0	144.5
Rubidium							
Rb	2.19	75.77		31.7	30.9	30.7	
RbBr	15.5	154.8		52.8	54.9	57.1(c)	66.9(lq)

TABLE 1.59   Heats of Fusion, Vaporization, and Sublimation and Specific Heat at Various Temperatures of the Elements and Inorganic Compounds (*Continued*)

Substance	ΔHm	ΔHv	ΔHs	C_p 400 K	600 K	800 K	1000 K
RbCl	18.4	165.7		52.3	54.3	56.4(c)	64.0(lq)
RbClO$_4$, $\Delta Ht = 12.59^{284}$							
RbF	17.3	177.8		51.9	57.9	64.9	72.3
RbI	12.5	150.6			55.1	57.3(c)	66.9(lq)
RbNO$_3$	5.61						
RbOH	6.78						
Ruthenium							
Run, $\Delta Ht = 0.13^{1035}$	38.59	591.6		24.5	25.7	27.0	28.2
$\quad \Delta Ht = 0.96^{1500}$							
Samarium							
Sm, $\Delta Ht = 3.11^{917}$	8.62	165	207	33.3	39.1	44.3	49.3
Sm$_2$O$_3$, $\Delta Ht = 1.05^{922}$				125.2	135.3	141.4	146.3
Scandium							
Sc	14.1	332.7	376				
ScCl$_3$				96.7	102.7	108.7	114.6
Sc$_2$O$_3$				106.4	111.1	115.8	120.5
Selenium							
Se, $\Delta Ht = 0.75^{150}$	6.69	95.48		28.1(c)	35.2(lq)	35.1	
SeF$_4$		47.2					
SeF$_6$	8.4		26.8	127.9	141.3	147.1	150.7
SeO$_2$		94.5					
SeOCl$_2$	4.23	42.7					
Silicon							
Si	50.21	359	450	22.3	24.5	25.7	26.5
SiBr$_4$		37.9		146.4(lq)	104.9(g)	106.2	106.2
SiC beta				34.1	41.8	45.9	48.4
SiCl$_4$	7.60	28.7	29.7	96.9(g)	102.6	104.8	106.0
SiClF$_3$		18.7		88.3	97.5	101.7	103.8
SiCl$_2$F$_2$		21.2					
SiF$_4$			25.7	83.1	94.1	99.4	102.3
SiH$_4$	0.67	12.1		51.5	65.9	76.7	84.5
Si$_2$H$_6$		21.2					
Si$_3$H$_8$		28.5					
SiH$_3$Br		24.4					
SiH$_2$Br$_2$		31					
SiHBr$_3$		34.8					
SiH$_3$Cl		21		60.7	74.0	83.1	89.4
SiH$_2$Cl$_2$		25.2	24.2	71.5	82.9	90.0	94.6
SiHCl$_3$		26.6	25.7	83.7	92.5	97.2	100.2
SiH$_3$F		18.8		57.2	71.8	81.7	88.3
SiH$_2$F$_2$		16.3					
SiHF$_3$		16.2					
SiI$_4$	19.7	56.9	79	164.0(lq)	106.0(g)	106.9	107.3
Si$_3$N$_4$				110.7	129.7	145.8	158.2
SiO$_2$ cristobalite	8.51						
SiO$_2$ quartz	7.7		600	53.5	64.4	76.2	68.94
$\quad \Delta Ht = 0.73^{574}$							
$\quad \Delta Ht = 2.0^{806}$							
SiOF$_2$				61.3	70.4	75.0	77.6
SiS$_2$	20.9			78.6	81.7	83.4	85.4

(*Continued*)

TABLE 1.59 Heats of Fusion, Vaporization, and Sublimation and Specific Heat at Various Temperatures of the Elements and Inorganic Compounds (*Continued*)

Substance	ΔHm	ΔHv	ΔHs	C_p 400 K	600 K	800 K	1000 K
Silver							
Ag	11.95	258		25.7	26.8	28.4	30.0
AgBr	9.12	198		59.0	71.8(c)	62.3(lq)	62.3
AgCl	13.2	199		56.9	54.4	54.4	54.4
Ag_2CO_3					122.6		
AgF	16.7	179.1		54.1(c)	58.4		
AgI, $\Delta Ht = 6.15^{147}$	9.41	143.9		64.7	56.5	56.5	58.6(lq)
$AgNO_3$, $\Delta Ht = 2.5^{160}$	11.5			112.5	128.0		
Ag_2O				73.0			
Ag_2S, $\Delta Ht = 5.86^{176}$ $\Delta Ht = 5.86^{586}$	14.1			86.6	90.5	90.5	90.5
Sodium							
Na	2.60	97.42	107.5	31.5(lq)	29.3	29.9	29.0
$NaAlCl_4$				164.8(c)			
Na_3AlCl_6				254.4	273.0		
Na_3AlF_6, $\Delta Ht = 8.37^{565}$ $\Delta Ht = 0.42^{880}$	107.28			234.6	261.8	196.8	282.8
$NaAlO_2$, $\Delta Ht = 1.297^{467}$				83.4	94.3	98.7	102.3
$NaBH_4$, $\Delta Ht = 0.999^{-83.3}$				94.6	108.6		
$NaBO_2$	36.2	239.7	322.2	75.4	88.6	97.2	103.2
$Na_2B_4O_7$	76.9			221.7	268.6	444.9(lq)	
NaBr	26.11	160.7	217.5	53.5	56.1	58.6	61.1
$NaBrO_3$	28.11						
NaCl	28.16			52.3	55.5	59.3	72.5
$NaClO_3$	22.1						
$NaClO_4$, $\Delta Ht = 13.98^{308}$				136.0(c)			
NaCN	8.79	148.1	172.8	68.7	68.8	69.0	
Na_2CO_3, $\Delta Ht = 0.690^{450}$	29.64			125.1	163.3	153.3	179.8
NaF	33.35	176.1	284.9	49.6	52.7	55.7	59.5
NaH				42.5	50.7		
NaI	23.60			53.8	56.2	58.5(c)	64.9(lq)
$NaIO_s$, $\Delta Ht = 35.1^{422}$							
$NaNO_3$	15						
NaO_2, $\Delta Ht = 1.464^{-76.7}$ $\Delta Ht = 1.548^{-49.9}$				76.3	84.5	92.6	
Na_2O, $\Delta Ht = 1.76^{750.1}$ $\Delta Ht = 11.92^{970.1}$	47.7			75.8	85.7	91.3	94.9
Na_2O_2, $\Delta Ht = 5.73^{512}$				97.7	108.4	113.6	
NaOH, $\Delta Ht = 72^{299.6}$	6.60	175.3	228.2	64.9(c)	86.1(lq)	84.9	83.7
Na_2S	19.3			20.1	20.9	21.5	22.0
Na_2S_2				104.3	115.4(c)	124.7(lq)	124.7
Na_2SiO_3	51.8			127.8	147.1	159.7	169.4
$Na_2Si_2O_5$, $\Delta Ht = 0.42^{678}$	35.6			183.4	217.6	235.2	292.9
Na_2SO_4, $\Delta Ht = 10.91^{241}$	23.6			145.1	175.3	187.3	200.3
Na_2TiO_3	70.3						
Na_2WO_4, $\Delta Ht = 30.85^{587.7}$ $\Delta Ht = 4.113^{588.9}$	23.80			155.3	178.2	198.7	
Strontium							
Sr, $\Delta Ht = 0.84^{547}$	7.43	136.9	164.0	27.8	29.8	31.9	34.1
$SrBr_2$, $\Delta Ht = 12.2^{645}$	10.1	194.1	310	79.0	82.7	87.6(c)	116.4(lq)

TABLE 1.59 Heats of Fusion, Vaporization, and Sublimation and Specific Heat at Various Temperatures of the Elements and Inorganic Compounds (*Continued*)

Substance	ΔHm	ΔHv	ΔHs	C_p 400 K	600 K	800 K	1000 K
$SrCl_2$, $\Delta Ht = 6.0^{727}$	17.5	248.1	356	78.9	83.7	90.8	105.8
$SrCO_3$, $\Delta Ht = 19.7^{924}$	40			95.1	107.1	116.1	124.0
SrF_2, $\Delta Ht = 0.04^{1148}$	28.5	320	451.0	74.7	79.8	81.0	85.8
$\quad \Delta Ht = 0.04^{1211}$							
SrI_2	19.67	189.7	286.6	80.7	86.3	91.8(c)	110.0(lq)
SrH_2	23						
$SrMoO_4$				131.5	145.4	154.0	161.2
SrO	81			48.5	52.0	54.3	56.1
SrO_2				81.3	85.0		
$Sr(OH)_2$	23			88.5	115.0(c)	157.8(lq)	157.8
SrS	63			50.2	53.2	54.9	56.2
$SrSO_4$	36			113.5	124.6	135.7	146.9
Sulfur							
S monoclinic	1.727	45	62.2	23.2	23.3(lq)	21.8(g)	21.5
$\quad \Delta Ht = 0.400^{95.2}$							
S_8				167.1	177.9	186.7	193.6
SCl_2		32.4		53.6	56.0	56.9	57.4
S_2Cl_2		36.0		124.3(lq)	80.8(g)	82.6	83.5
SF_4		26.4		87.5	97.3	101.7	103.8
SF_6	5.02	17.1	9.0	116.4	136.1	144.8	149.3
S_2F_{10}				211.4	246.4	261.8	269.2
SO_2	7.40	24.94	22.92	43.43	48.9	52.3	54.3
SO_3	8.60	40.7	43.14	57.7	67.3	72.8	76.0
$SOCl_2$		31.7	31	71.3	76.4	78.9	80.3
SOF_2		21.8		64.3	72.4	76.4	78.6
SO_2Cl_2		31.38	30.1	85.2	94.5	99.4	102.1
SO_2ClF				81.1	92.1	97.9	101.1
SO_2F_2		20.0		76.5	89.3	96.1	99.9
Tantalum							
Ta	36.57	732.8	778	25.8	26.8	27.5	27.9
TaB_2	83.7			57.6	66.6	72.2	83.3
$TaBr_5$	45.6	62.3		168.2			
TaC	105			41.7	46.5	49.1	51.1
Ta_2C				66.7	72.4	76.2	79.5
$TaCl_5$	41.6	54.8	94.1	148(c)	129(g)	131	132
TaF_5	18.8	56.9		182.0(lq)			
TaI_5	41.8	64.9		164.6	182.0(c)	120.0(g)	120.6
TaN	67			45.4	51.9	58.5	65.0
TaO_2				47.7	52.3	54.6	55.7
Ta_2O_5	120			147.5	164.4	175.2	182.8
Technetium							
Tc	33.29	585.2		25.1	26.8	28.5	30.1
TcF_6	4.72	31.1					
TcO_3F	22.5	39.5					
Tellurium							
Te	17.49	114.1		28.0	32.3(c)	37.7(lq)	37.7
$TeCl_4$	18.8	77		138.9(c)	222.6(lq)	108.8(g)	108.8
TeF_4		34.3					
TeF_6				28.2	132.2	143.8	148.7	151.7
Te_2F_{10}		39.5					

(*Continued*)

TABLE 1.59 Heats of Fusion, Vaporization, and Sublimation and Specific Heat at Various Temperatures of the Elements and Inorganic Compounds (*Continued*)

Substance	ΔHm	ΔHv	ΔHs	C_p 400 K	600 K	800 K	1000 K
TeH_2		23.9					
TeO_2	29.1			67.9	72.5	76.1	79.2
Terbium							
Tb	10.15	293	389				
Thallium							
Tl, $\Delta Ht = 0.38^{234}$	4.14	165	181	27.5(c)	30.1(lq)	30.1	30.1
TlBr	16.4	99.6		53.5	59.5(c)	75.5(lq)	67.8
TlCl	15.56	102.2		53.6	55.2(c)	59.4(lq)	59.4
Tl_2CO_3	18.4						
TlF	13.87	115.9			66.8(lq)	67.3	
TlI	14.73	104.7		53.9	60.6(c)	72.0(lq)	72.0
$TlNO_3$	9.56						
Tl_2O	30.3						
Tl_2O_3	53						
Tl_2S	12	154					
Tl_2SO_4	23.0						
Thorium							
Th, $\Delta Ht = 2.73^{1360}$	13.81	514		28.4	30.5	32.7	34.4
$ThBr_4$	66.9						
$ThCl_4$, $\Delta Ht = 5.0^{406}$	40.2	146.4		126.7	132.7	136.4	139.6
ThF_4	44.0	258					
ThI_4	61.4	56.9					
Th_3N_4				169.5	196.5	222.7	
ThO_2	1218.0			67.4	72.4	75.3	77.7
$ThOCl_2$				97.0	102.5	105.9	108.6
$Th(SO_4)_2$				197.0	243.2	289.4	
Thullium							
Tm	16.84	247	232.2				
Tin							
Sn white, $\Delta Ht = 2.09^{13}$	7.03	296.1		28.9	28.9(c)	28.7(lq)	28.7
$SnBr_2$	7.2	102					
$SnBr_4$	11.9	43.5		158.0(lq)	106.8(g)	107.3	107.5
$SnCl_2$	12.8	86.8		83.3(c)	92.1(lq)	92.1	92.1
$SnCl_4$	9.20	34.9					
SnH_4		19.1					
SnI_2		105					
SnO				45.8	48.7	51.7	54.6
SnO_2, $\Delta Ht = 1.88^{410}$				64.4	73.9	78.5	81.8
$\Delta Ht = 1.26^{540}$							
SnS, $\Delta Ht = 0.67^{602}$				50.5	55.5	61.3	
SnS_2				71.9	75.4	79.0	82.5
Titanium							
Ti, $\Delta Ht = 4.2^{893}$	14.15	425	469	26.9	28.6	29.5	32.1
TiB				40.3	48.6	50.9	51.9
TiB_2	100.4			54.9	66.2	72.1	76.9
$TiBr_2$			206.2	79.9	82.1	84.4	86.7
$TiBr_3$			138.8	105.8	125.5	147.3	156.7
$TiBr_4$	12.9	44.4		151.9(lq)	106.1(g)	106.9	107.3
TiC	71			40.7	47.7	49.9	51.2
$TiCl_2$		232	212	73.4	78.4	82.2	85.9

TABLE 1.59 Heats of Fusion, Vaporization, and Sublimation and Specific Heat at Various Temperatures of the Elements and Inorganic Compounds (*Continued*)

Substance	ΔHm	ΔHv	ΔHs	C_p 400 K	600 K	800 K	1000 K
$TiCl_3$		124	166.3	98.6	102.0	104.4	106.7
$TiCl_4$	9.97	36.2		146.2(lq)	104.4(g)	106.0	106.7
TiF_3			222	93	98	103	109
TiF_4			97.9	126.7(c)	100.2(g)	103.3	104.9
TiH_2				39.3	53.8	63.1	68.5
TiI_2			217	87.0	88.4	89.9	91.3
TiI_3				117.5	119.0	120.4(c)	20.6(g)
TiI_4, $\Delta Ht = 9.9^{106}$	19.8	58.4		148.1(c)	156.6(lq)	25.7(g)	27.8
TiN	66.9			43.8	48.7	50.6	52.1
TiO, $\Delta Ht = 4.2^{992}$	41.8			45.0	50.8	55.2	59.1
TiO_2 rutile	58.0		673	63.6	70.9	73.9	75.3
Ti_2O_3, $\Delta Ht = 1.138^{197}$	105			117.5	136.4	143.0	146.4
Tungsten							
W	52.31	806.7	851	24.9	25.9	26.7	27.6
WBr_5	17.1	81.5		166.(c)	182.(lq)	132.2(g)	132.5
WBr_6				192.5(c)	156.3(g)	157.0	157.4
WCl_4				135.3	146.2(c)	106.7(g)	107.2
WCl_5	20.5	68.1	100	167.4(c)	129.5(g)	131.0	131.8
WCl_6, $\Delta Ht = 4.1^{177}$	6.60	52.7	79.2	192.5(c)	200.8(lq)	155.8(g)	156.6
$W(CO)_6$			72.0				
WF_6, $\Delta Ht = 2.067^{-8.5}$	4.10	27.05	26.65	132.4(g)	145.0	150.3	153.0
WO_2			666.3	63.4	71.3	75.5	78.2
WO_3, $\Delta Ht = 1.49^{777}$	73.4	76.6	550.2	82.2	93.1	98.2	101.7
$WOCl_4$	45	67.8		157.(c)	123.2(g)	127.0	129.1
WOF_4	5.0	56		107.8	119.8	125.0	127.8
WO_2Cl_2				115.1	135.6(c)		
Uranium							
U, $\Delta Ht = 2.93^{672}$	9.14	417.1	525	29.0	34.8	41.6	41.8
$\Delta Ht = 4.791^{772}$							
UBr_3	43.9						
UBr_4	55.2	119.2		131.4	140.1(c)	163.2(lq)	163.2
UC				64.6	58.3	60.3	62.2
UCl_3	46.4	193.0		102.8	107.7	113.6	119.9
UCl_4	44.8	141.4		126.1	134.4	142.0	162.5
UCl_5	35.6	75.3		150.9	159.8(c)	186.7(lq)	134.5(g)
UCl_6	20.9	50.2		182.8	214.0	158.8	168.0
UF_3				99.0	104.9	111.0	117.2
UF_4	42.7	221.8		119.1	125.0	130.9	136.8
UF_5	33.5			136.4	143.1(c)	166.6(lq)	
UF_6	19.19	28.90	48.20	140.5(g)	148.7	152.2	154.4
UH_3				50.9	57.4	66.1	
UI_4	70.7	130.6		140.6	149.5(c)	165.7(lq)	165.7
UN				52.2	56.3	58.3	59.8
UO_2				72.7	79.8	83.2	85.5
UO_3				88.9	95.3	99.0	
U_3O_8				266.0	290.7	304.2	
$UOCl_2$				101.9	109.6	115.1	
UO_2Cl_2				118.1	126.2	130.0	
UO_2F_2				113.9	122.5	126.7	129.5

(*Continued*)

TABLE 1.59 Heats of Fusion, Vaporization, and Sublimation and Specific Heat at Various Temperatures of the Elements and Inorganic Compounds (*Continued*)

Substance	ΔHm	ΔHv	ΔHs	C_p 400 K	600 K	800 K	1000 K
Vanadium							
V	21.5	459	516	26.2	27.5	28.7	30.1
VCl_4	2.30	41.4	42.5	161.7(lq)	100.1(g)	102.6	104.7
VF_5	50.0	44.5					
VN, $\Delta Hdec = 227.6^{2346}$			741	43.3	48.2	51.2	53.7
VO	63			49.6	53.5	57.1	60.5
VO_2, $\Delta Ht = 4.21^{72}$	56.9			67.2	74.3	77.8	80.2
V_2O_3, $\Delta Ht = 1.623^{-104.3}$	117.2			117.5	127.3	132.6	138.0
V_2O_4, $\Delta Ht = 9.0^{67}$	112.1			135.3	148.4	155.5	160.7
V_2O_5	64.5	263.6		151.0	168.3	177.3	183.7
$VOCl_3$		36.8					
Xenon							
Xe	1.81	12.64		20.79(g)	20.79	20.79	20.79
Ytterbium							
Yb	7.66	159					
Yttrium							
Y, $\Delta Ht = 4.97^{1485}$	11.42	365	425	27.3	28.5	29.9	31.5
Y_2O_3, $\Delta Ht = 1.30^{1057}$	105			113.3	121.3	124.7	126.9
Zinc							
Zn	7.32	123.6		26.3	28.6(c)	31.4(lq)	31.4
$ZnBr_2$	16.7	118		70.1(c)	78.8(lq)	113.8	61.5(g)
$ZnCl_2$	10.25	126		69.9(c)	100.8(lq)	100.8	100.8
ZnF_2		190.1		66.9	69.1	71.4	73.7
ZnO, $\Delta Ht = 13.4^{1020}$	52.3			49.4	52.4	54.1	55.5
Zn_2SiO_4				129.4	141.4	153.4	165.4
$ZnSO_4$, $\Delta Ht = 20.3^{740}$				116.0	137.4	139.7	142.0
Zirconium							
Zr, $\Delta Ht = 4.02^{862}$	21.00	573	610.0	25.9	27.3	29.0	31.1
ZrB_2	104.6			57.5	65.8	69.7	72.1
$ZrBr_2$	63	131.5	230	87.9	90.2	92.5	94.8
$ZrBr_4$				129.3	133.3(c)	107.2(g)	107.6
ZrC	79.5			43.6	49.4	52.3	53.4
$ZrCl_2$	27	45.0		76.0	80.0	83.1	85.9
$ZrCl_3$			190	101	106	109	112
$ZrCl_4$	50		110.5	125.4	131.1(c)	106.5(g)	107.1
ZrF_2	33	289	404	70	76	81	84
ZrF_4	64.2		237.7	113.5	124.0	129.4	134.1
ZrI_2	25.1	113		95.0	96.6	106.1	123.6
ZrI_3			176	105.9	106.7	107.1(c)	82.9(g)
ZrI_4			126.4	131.0	134.6(c)	107.6(g)	107.6
ZrN	67.4			44.8	48.7	50.9	52.7
ZrO_2, $\Delta Ht = 5.02^{1205}$	87.0	624		63.9	70.2	73.5	75.7
$ZrSiO_4$				114.6	133.7	142.7	147.3

1.16 ACTIVITY COEFFICIENTS

The activity coefficient is the ratio of the chemical activity of any substance to its molar concentration. The measured concentration of a substance may not be an accurate indicator of its chemical effectiveness, as represented by the equation for a particular reaction, in which case an activity coefficient is arbitrarily established and used instead of the concentration...

Although it is not possible to measure an individual ionic activity coefficient, f_i, it may be estimated from the following equation of the Debye-Hückel theory:

$$-\log f_i = \frac{A z_i^2 \sqrt{I}}{1 + B \mathring{a} \sqrt{I}}$$

where I is the ionic strength of the medium, and $\mathring{a}$ is the ion-size parameter—the effective ionic radius (Table 1.34). The values of A and B vary with the temperature and dielectric constant of the solvent; values from 0 to 100C for aqueous medium ($\mathring{a}$ in angstrom units) are listed in Table 1.61. Corresponding values of A and B for unit weight of solvent (when employing molality) can be obtained by multiplying the corresponding values for unit volume (molarity units) by the square root of the density of water at the appropriate temperature.

The ionic strength can be estimated from the summation of the product molarity times ionic charge squared for all the ionic species present in the solution, i.e., $I = 0.5\,(c_1 z_1^2 + c_2 z_2^2 + \cdots + c_i z^2 i)$.

Values for the activity coefficients of ions in water at 25°C are given in Table 8.1 in terms of their effective ionic radii.

At moderate ionic strengths a considerable improvement is effected by subtracting a term bI from the Debye-Hückel expression; b is an adjustable parameter which is 0.2 for water at 25°C. Table 1.60 gives the values of the ionic activity coefficients (for zi from 1 to 6) with $\mathring{a}$ taken to be 4.6Å.

In general, the mean ionic activity coefficient is given by

$$f_\pm = {}^{(x+y)}\sqrt{f_+^x f_-^y}$$

where f_+, f_- are the individual ionic activity coefficients, and x, y are the charge numbers (z_+, z_-) of the respective ions. In binary electrolyte solution,

$$f_\pm = \sqrt{f_+ f_-}$$

In ternary electrolytes, e.g., $BaCl_2$ or K_2SO_4,

$$f_\pm = \sqrt[3]{f_+ f_-^2} \quad \text{or} \quad f_\pm = \sqrt[3]{f_+^2 f_-}$$

In quaternary electrolytes, e.g., $LaCl_3$ or $K_3[Fe(CN)_6]$,

$$f_\pm = \sqrt[4]{f_+ f_-^3} \quad \text{or} \quad f_\pm = \sqrt[4]{f_+^3 f_-}$$

TABLE 1.60 Individual Activity Coefficients of Ions in Water at 25°C

Effective ionic radii å (in Å)	f_i at ionic strength of				
	0.001	0.005	0.01	0.05	0.1
Univalent Ions					
9	0.967	0.933	0.914	0.86	0.83
8	0.966	0.931	0.912	0.85	0.82
7	0.965	0.930	0.909	0.845	0.81
6	0.965	0.929	0.907	0.835	0.80
5	0.964	0.928	0.904	0.83	0.79
4	0.964	0.928	0.902	0.82	0.775
3.5	0.964	0.926	0.900	0.81	0.76
3	0.964	0.925	0.899	0.805	0.755
2.5	0.964	0.924	0.898	0.80	0.75
Divalent Ions					
8	0.872	0.755	0.69	0.52	0.45
7	0.872	0.755	0.685	0.50	0.425
6	0.870	0.749	0.675	0.485	0.405
5	0.868	0.744	0.67	0.465	0.38
4.5	0.868	0.741	0.663	0.45	0.36
4	0.867	0.740	0.660	0.445	0.355
Trivalent Ions					
6	0.731	0.52	0.415	0.195	0.13
5	0.728	0.51	0.405	0.18	0.115
4	0.725	0.505	0.395	0.16	0.095
Tetravalent Ions					
11	0.588	0.35	0.255	0.10	0.065
5	0.57	0.31	0.20	0.048	0.021
Pentavalent Ions					
9	0.43	0.18	0.105	0.020	0.009

TABLE 1.61 Constants of the Debye-Hückel Equation from 0 to 100°C

$$-\log f_i = \frac{Az_i^2 \sqrt{I}}{I + Bå \sqrt{I}}$$

Temp, °C	Unit volume of solvent		Temp, °C	Unit volume of solvent	
	A	B		A	B
0	0.4918	0.3248	55	0.5432	0.3358
5	0.4952	0.3256	60	0.5494	0.3371
10	0.4989	0.3264	65	0.5558	0.3384
15	0.5028	0.3273	70	0.5625	0.3397
20	0.5070	0.3282	75	0.5695	0.3411
25	0.5115	0.3291	80	0.5767	0.3426
30	0.5161	0.3301	85	0.5842	0.3440
35	0.5211	0.3312	90	0.5920	0.3456
40	0.5262	0.3323	95	0.6001	0.3471
45	0.5317	0.3334	100	0.6086	0.3488
50	0.5373	0.3346			

The values for unit weight of solvent (molality scale) can be obtained by multiplying the corresponding values for unit volume by the square root of the density of water at the appropriate temperature.

TABLE 1.62   Individual Ionic Activity Coefficients at Higher Ionic Strengths at 25°C

The values were calculated from the modified Debye-Hückel equation utilizing the modifications proposed by Robinson and by Guggenheim and Bates:

$$-\frac{\log f_i}{z_i^2} = \frac{0.511I}{1+1.5I} - 0.2I$$

where I is the ionic strength and $\mathring{a}$ is assumed to be 4.6 Å.

I	$-\dfrac{\log_{10} f_i}{z_i^2}$	f_i for $z_i =$					
		1	2	3	4	5	6
0.05	0.0756	0.840	0.498	0.209	0.0617	0.0129	0.00190
0.1	0.0896	0.814	0.438	0.156	0.0369	0.00576	0.000595
0.2	0.0968	0.800	0.410	0.138	0.0283	0.00380	0.000328
0.3	0.0936	0.806	0.422	0.144	0.0318	0.00457	0.000427
0.4	0.0858	0.821	0.454	0.169	0.0424	0.00716	0.000815
0.5	0.0753	0.841	0.500	0.210	0.0624	0.0131	0.00195
0.6	0.0631	0.865	0.559	0.270_5	0.0978	0.0265	0.00535
0.7	0.0496	0.892	0.633	0.358	0.161	0.0575_5	0.0164
0.8	0.0352	0.922	0.723	0.482	0.273	0.132	0.0541
0.9	0.0201	0.955	0.831	0.659	0.477	0.314	0.189
1.0	0.0044	0.900	0.960	0.913	0.850	0.776	0.694

1.17   BUFFER SOLUTIONS

A buffer solution is a solution that resists changes in pH when small quantities of an acid or an alkali are added.

An acidic buffer solution is a solution that has a pH less than 7. Acidic buffer solutions are commonly made from a weak acid and one of its salts. A common example is a mixture of ethanoic acid and sodium ethanoate in solution. In this case, if the solution contained equal molar concentrations of both the acid and the salt, the pH would be 4.76. The pH of the buffer solution can be changed by changing the ratio of acid to salt, or by choosing a different acid and one of its salts.

An alkaline buffer solution has a pH greater than 7. Alkaline buffer solutions are commonly made from a weak base and one of its salts. An example is a mixture of ammonia solution and ammonium chloride solution. If these were mixed in equal molar proportions, the solution would have a pH of 9.25.

To prepare the standard pH buffer solutions recommended by the National Bureau of Standards (U.S.), the indicated weights of the pure materials should be dissolved in water of specific conductivity not greater than 5 micromhos. The tartrate, phthalate, and phosphates can be dried for 2 h at 100°C before use. Potassium tetroxalate and calcium hydroxide need not be dried. Fresh-looking crystals of borax should be used. Before use, excess solid potassium hydrogen tartrate and calcium hydroxide must be removed. Buffer solutions pH 6 or above should be stored in plastic containers and should be protected from carbon dioxide with soda-lime traps. The solutions should be replaced within 2 to 3 weeks, or sooner if formation of mold is noticed. A crystal of thymol may be added as a preservative.

1.17.1   Standards for pH Measurement of Blood and Biological Media

Blood is a well-buffered medium. In addition to the NBS phosphate standard of $0.025M$ (pH$_s$ = 6.480 at 38°C), another reference solution containing the same salts, but in the molal ratio 1:4, has an ionic strength of 0.13. It is prepared by dissolving 1.360 g of KH_2PO_4 and 5.677 g of Na_2HPO_4 (air weights) in carbon dioxide-free water to make 1 liter of solution. The pH_s is 7.416 ± 0.004 at 37.5 and 38°C.

The compositions and pH_s values of *tris*(hydroxymethyl)aminomethane, covering the pH range 7.0 to 8.9, are listed in Table 1.65.

When there are two or more acid groups per molecule, or a mixture is composed of several overlapping acids, the useful range is larger. Universal buffer solutions consist of a mixture of acid groups which overlap such that successive pK_a values differ by 2 pH units or less. The Prideaux-Ward mixture comprises phosphate, phenyl acetate, and borate plus HCl and covers the range from 2 to 12 pH units. The McIlvaine buffer is a mixture of citric acid and Na_2HPO_4 that covers the range from pH 2.2 to 8.0. The Britton-Robinson system consists of acetic acid, phosphoric acid, and boric acid plus NaOH and covers the range from pH 4.0 to 11.5. A mixture composed of Na_2CO_3, NaH_2PO_4, citric acid, and 2-amino-2-methyl-1,3-propanediol covers the range from pH 2.2 to 11.0.

General directions for the preparation of buffer solutions of varying pH but fixed ionic strength are given by Bates.[5] Preparation of McIlvaine buffered solutions at ionic strengths of 0.5 and 1.0 and Britton-Robinson solutions of constant ionic strength have been described by Elving et al.[6] and Frugoni,[7] respectively.

[5]Bates, *Determination of pH: Theory and Practice*, Wiley, New York, 1964, pp. 121–122.
[6]Elving, Markowitz, and Rosenthal, *Anal. Chem.*, **28:**1179 (1956).
[7]Frugoni, *Gazz. Chim. Ital.*, **87:**L403 (1957).

TABLE 1.63   National Bureau of Standards (U.S.) Reference pH Buffer Solutions

Temperature, °C	Secondary standard 0.05M K tetraoxalate	KH tartrate (saturated at 25°C)	0.05M KH₂ citrate	0.05M KH phthalate	0.025M KH₂PO₄, 0.025M Na₂HPO₄	0.0087M KH₂PO₄, 0.0302M Na₂HPO₄	0.01M Na₂B₄O₇	0.025M NaHCO₃, 0.025M Na₂CO₃	Secondary standard Ca(OH)₂ (saturated at 25°C)
0	1.666		3.860	4.003	6.984	7.534	9.464	10.317	13.423
5	1.668		3.840	3.999	6.951	7.500	9.395	10.245	13.207
10	1.638		3.820	3.997	6.923	7.472	9.332	10.179	13.003
15	1.642		3.802	3.998	6.900	7.448	9.276	10.118	12.810
20	1.644		3.788	4.002	6.881	7.429	9.225	10.062	12.627
25	1.646	3.557	3.776	4.005	6.865	7.413	9.180	10.012	12.454
30	1.648	3.552	3.766	4.011	6.853	7.400	9.139	9.966	12.289
35		3.549	3.759	4.018	6.844	7.389	9.102	9.925	12.133
38	1.649	3.548	3.756	4.030	6.840	7.384	9.088	9.910	12.043
40	1.650	3.547	3.753	4.035	6.838	7.380	9.068	9.889	11.984
45		3.547		4.047	6.834	7.373	9.038		11.841
50	1.653	3.549	3.749	4.050	6.833	7.367	9.011	9.828	11.705
55		3.554		4.075	6.834		8.985		11.574
60	1.660	3.560		4.081	6.836		8.962		11.449
70	1.671	3.580		4.116	6.845		8.921		
80	1.689	3.609		4.164	6.859		8.885		
90	1.72	3.650		4.205	6.877		8.850		
95	1.73	3.674		4.227	6.886		8.833		
Dilution value ΔpH½	+0.186	+0.049	0.024	+0.052	+0.080	+0.070	+0.01	0.079	−0.28

Source: R. G. Bates, *J. Res. Natl. Bur. Stand. (U.S.),* **66A:**179 1962) and B. R. Staples and R. G. Bates, *J. Res. Natl. Bur. Stand. (U.S.),* **73A:**37 (1969).

Note: The uncertainty is ±0.003 in pH in the range 0–50°C, rising to ±0.02 above 70°C.

TABLE 1.64 Compositions of Standard pH Buffer Solutions [National Bureau of Standards (U.S.)]

Standard	Weight, g
$KH_3(C_2O_4)_2 \cdot 2H_2O$, 0.05$M$	12.61
Potassium hydrogen tartrate, about 0.034M	Saturated at 25°C
Potassium hydrogen phthalate, 0.05M	10.12
Phosphate:	
$\quad KH_2PO_4$, 0.025M	3.39
$\quad Na_2HPO_4$, 0.025M	3.53
Phosphate:	
$\quad KH_2PO_4$, 0.008665M	1.179
$\quad Na_2HPO_4$, 0.03032M	4.30
$Na_2B_4O_7 \cdot 10H_2O$, 0.01M	3.80
Carbonate:	
$\quad NaHCO_3$, 0.025M	2.10
$\quad Na_2CO_3$, 0.025M	2.65
$Ca(OH)_2$, about 0.0203M	Saturated at 25°C

TABLE 1.65 Composition and pH Values of Buffer Solutions 8.107

Values based on the conventional activity pH scale as defined by the National Bureau of Standards (U.S.) and pertain to a temperature of 25°C [Ref: Bower and Bates, *J. Research Natl. Bur. Standards (U.S.)*, **55**:197 (1955) and Bates and Bower, *Anal. Chem.*, **28**:1322 (1956)]. Buffer value is denoted by column headed β.

25 ml 0.2M KCl + x ml 0.2M HCl, Diluted to 100 ml			50 ml 0.1M KH Phthalate + x ml 0.1M HCl, Diluted to 100 ml			50 ml 0.1M KH Phthalate + x ml 0.1M NaOH, Diluted to 100 ml		
pH	X	β	pH	X	β	pH	X	β
1.00	67.0	0.31	2.20	49.5		4.20	3.0	0.017
1.20	42.5	0.34	2.40	42.2	0.036	4.40	6.6	0.020
1.40	26.6	0.19	2.60	35.4	0.033	4.60	11.1	0.025
1.60	16.2	0.077	2.80	28.9	0.032	4.80	16.5	0.029
1.80	10.2	0.049	3.00	22.3	0.030	5.00	22.6	0.031
2.00	6.5	0.030	3.20	15.7	0.026	5.20	28.8	0.030
2.20	3.9	0.022	3.40	10.4	0.023	5.40	34.1	0.025
			3.60	6.3	0.018	5.60	38.8	0.020
			3.80	2.9	0.015	5.80	42.3	0.015

TABLE 1.65 Composition and pH Values of Buffer Solutions 8.107 (*Continued*)

50 ml 0.1M KH$_2$PO$_4$ + x ml 0.1M NaOH, Diluted to 100 ml			50 ml 0.1M Tris(hydroxy-methyl)aminomethane + x ml 0.1M HCl, Diluted to 100 ml ΔpH/$\Delta t \approx -0.028$ $I = 0.00$lx			50 ml of a Mixture 0.1M with Respect to Both KCl and H$_3$BO$_3$ + x ml 0.1M NaOH, Diluted to 100 ml		
pH	x	β	pH	x	β	pH	x	β
5.80	3.6		7.00	46.6		8.00	3.9	
6.00	5.6	0.010	7.20	44.7	0.012	8.20	6.0	0.011
6.20	8.1	0.015	7.40	42.0	0.015	8.40	8.6	0.015
6.40	11.6	0.021	7.60	38.5	0.018	8.60	11.8	0.018
6.60	16.4	0.027	7.80	34.5	0.023	8.80	15.8	0.022
6.80	22.4	0.033	8.00	29.2	0.029	9.00	20.8	0.027
7.00	29.1	0.031	8.20	22.9	0.031	9.20	26.4	0.029
7.20	34.7	0.025	8.40	17.2	0.026	9.40	32.1	0.027
7.40	39.1	0.020	8.60	12.4	0.022	9.60	36.9	0.022
7.60	42.4	0.013	8.80	8.5	0.016	9.80	40.6	0.016
7.80	44.5	0.009	9.00	5.7		10.00	43.7	0.014
8.00	46.1					10.20	46.2	

50 ml 0.025M Borax + x ml 0.1M HCl, Diluted to 100 ml ΔpH/$\Delta t \approx -0.008$ $I = 0.025$			50 ml 0.025M Borax + x ml 0.1M NaOH, Diluted to 100 ml ΔpH/$\Delta t \approx -0.008$ $I = 0.001(25 + x)$			50 ml 0.05M NaHCO$_3$ + x ml 0.1M NaOH, Diluted to 100 ml ΔpH/$\Delta t \approx -0.009$ $I = 0.001(25 + 2x)$		
pH	x	β	pH	x	β	pH	x	β
8.00	20.5		9.20	0.9		9.60	5.0	
8.20	19.7	0.010	9.40	3.6	0.026	9.80	6.2	0.014
8.40	16.6	0.012	9.60	11.1	0.022	10.00	10.7	0.016
8.60	13.5	0.018	9.80	15.0	0.018	10.20	13.8	0.015
8.80	9.4	0.023	10.00	18.3	0.014	10.40	16.5	0.013

50 ml 0.025M Borax + x ml 0.1M HCl, Diluted to 100 ml ΔpH/$\Delta t \approx -0.008$ $I = 0.025$			50 ml 0.025M Borax + x ml 0.1M NaOH, Diluted to 100 ml ΔpH/$\Delta t \approx -0.008$ $I = 0.001(25 + x)$			50 ml 0.05M NaHCO$_3$ + x ml 0.1M NaOH, Diluted to 100 ml ΔpH/$\Delta t \approx -0.009$ $I = 0.001(25 + 2x)$		
pH	x	β	pH	x	β	pH	x	β
9.00	4.6	0.026	10.20	20.5	0.009	10.60	19.1	0.012
9.10	2.0		10.40	22.1	0.007	10.80	21.2	0.009
			10.60	23.3	0.005	11.00	22.7	

(*Continued*)

TABLE 1.65   Composition and pH Values of Buffer Solutions 8.107 (*Continued*)

	50 ml 0.05*M* Na$_2$HPO$_4$ + *x* ml 0.1*M* NaOH, Diluted to 100 ml ΔpH/$\Delta t \simeq -0.025$ $I = 0.001(77 + 2x)$			25 ml 0.2*M* KCl + *x* ml 0.2*M* NaOH, Diluted to 100 ml ΔpH/$\Delta t \simeq -0.033$ $I = 0.001(50 + 2x)$	
pH	*x*	β	pH	*x*	β
11.00	4.1	0.009	12.00	6.0	0.028
11.20	6.3	0.012	12.20	10.2	0.048
11.40	9.1	0.017	12.40	16.2	0.076
11.60	13.5	0.026	12.60	25.6	0.12
11.80	19.4	0.034	12.80	41.2	0.21
11.90	23.0	0.037	13.00	66.0	0.30

The phosphate-succinate system gives the values of pH$_s$

Molality KH$_2$PO$_4$ = Molality Na$_2$HC$_6$H$_5$O$_7$	pH$_s$	Δ(pH$_s$/Δt)
0.005	6.251	$-0.000\,86$ deg^{-1}
0.010	6.197	$-0.000\,71$
0.015	6.162	
0.020	6.131	
0.025	6.109	-0.004

TABLE 1.66   Standard Reference Values pH for the Measurement of Acidity in 50 Weight Percent Methanol-Water

Temperature, °C	0.02*M* HOAc, 0.02*M* NaOAc, 0.02*M* NaCl	0.02*M* NaHSuc, 0.02*M* NaCl	0.02*M* KH$_2$PO$_4$, 0.02*M* Na$_2$HPO$_4$, 0.02*M* NaCl
10	5.560	5.806	7.937
15	5.549	5.786	7.916
20	5.543	5.770	7.898
25	5.540	5.757	7.884
30	5.540	5.748	7.872
35	5.543	5.743	7.863
40	5.550	5.741	7.858

OAc = acetate   Suc = succinate

Reference: R. G. Bates, *Anal Chem.*, **40**(6):35A (1968).

TABLE 1.67   pH Values for Buffer Solutions in Alcohol-Water Solvents at 25°C

Liquid-junction potential not included.

Solvent composition (weight percent alcohol)	0.01M $H_2C_2O_4$, 0.01M $NH_4HC_2O_4$	0.01M H_2Suc, 0.01M LiHSuc	0.01M HSal, 0.01M NaSal
	Methanol-Water Solvents		
0	2.15	4.12	
10	2.19	4.30	
20	2.25	4.48	
30	2.30	4.67	
40	2.38	4.87	
50	2.47	5.07	
60	2.58	5.30	
70	2.76	5.57	
80	3.13	6.01	
90	3.73	6.73	
92	3.90	6.92	
94	4.10	7.13	
96	4.39	7.43	
98	4.84	7.89	
99	5.20	8.23	
100	5.79	8.75	7.53
	Ethanol-Water Solvents		
0	2.15	4.12	
30	2.32	4.70	
50	2.51	5.07	
71.9	2.98	5.71	
100			8.32

Suc = succinate   Sal = salicylate

1.17.2   Buffer Solutions Other Than Standards

The range of the buffering effect of a single weak acid group is approximately one pH unit on either side of the pK_a. The ranges of some useful buffer systems are collected in Table 1.68. After all the components have been brought together, the pH of the resulting solution should be determined at the temperature to be employed with reference to standard reference solutions. Buffer components should be compatible with other components in the system under study; this is particularly significant for buffers employed in biological studies. Check tables of formation constants to ascertain whether metal-binding character exists.

TABLE 1.68 pH Values of Biological and Other Buffers for Control Purposes

Materials	Acronym	PK_a	pH range
p-Toluenesulfonate and p-toluenesulfonic acid		1.7	1.1–3.3
Glycine and HCl		2.35	1.0–3.7
Citrate and HCl		3.13	1.3–4.7
Formate and HCl		3.71	2.8–4.6
Succinate and borax		4.21, 5.64	3.0–5.8
Phenyl acetate and HCl		4.31	3.5–5.0
Acetate and acetic acid		4.76	3.7–5.6
Succinate and succinic acid		4.21, 5.64	4.8–6.3
2-(N-Morpholino)ethanesulfonic acid	MES	6.1	5.5–6.7
Bis(2-hydroxyethyl)iminotris(hydroxymethyl)methane	BIS-TRIS	6.5	5.8–7.2
KH_2PO_4 and borax		2.2, 7.2; 9	5.8–9.2
N-(2-Acetamido)-2-iminodiacetic acid	ADA	6.6	6.0–7.2
2-[(2-Amino-2-oxoethyl)amino]ethanesulfonic acid	ACES	6.8	6.1–7.5
Piperazine-N,N′-bis(2-ethanesulfonic acid)	PIPES	6.8	6.1–7.5
3-(N-Morpholino)-2-hydroxypropanesulfonic acid	MOPSO	6.9	6.2–1.6
l,3-Bis[tris(hydroxymethyl)methylamino]propane	BIS-TRIS PROPANE	6.8, 9.0	6.3–9.5
KH_2PO_4 and Na_2HPO_4		7.2	6.1–7.5
N,N-Bis(2-hydroxyethyl)-2-aminoethanesulfonic acid	BES	7.1	6.4–7.8
3-(N-Morpholino)propanesulfonic acid	MOPS	7.2	6.5–7.9
N-(2-Hydroxyethyl)piperazine-N′-(2-ethanesulfonic acid)	HEPES	7.5	6.8–8.2
N-Tris(hydroxymethyl)methyl-2-aminoethanesulfonic acid	TES	7.5	6.8–8.2
3-[N,N-Bis(2-hydroxyethyl)amino]-2-hydroxypropanesulfonic acid	DIPSO	7.6	7.0–8.2
3-[N-tris(hydroxymethyl)methylamino]-2-hydroxypropanesulfonic acid	TAPSO	7.6	7.0–8.2
5,5-Diethylbarbiturate (veronal) and HCl		8.0	7.0–8.5
Tris(hydroxymethyl)aminoethane	TRIZMA	8.1	7.0–9.1
N-(2-hydroxyethyl)piperazine-N′-(2-hydroxypropanesulfonic acid)	HEPPSO	7.8	7.1–8.5
Piperazine-N,N′-bis(2-hydroxypropanesulfonic acid)	POPSO	7.8	7.2–8.5
Triethanolamine	TEA	7.8	6.9–8.5
N-Tris(hydroxymethyl)methylglycine	TRICINE	8.1	7.4–8.8
Borax and HCl			7.6–8.9
N,N-Bis(2-hydroxyethyl)glycine	BICINE	8.3	7.6–9.0
N-Tris(hydroxymethyl)methyl-3-aminopropanesulfonic acid	TAPS	8.4	7.7–9.1
3-[(1,1-Dimethyl-2-hydroxyethyl)-2-hydroxypropanesulfonic acid	AMPSO	9.0	8.3–9.7
Ammonia (aqueous) and NH_4Cl		9.2	8.3–9.2
2-(N-Cyclohexylamino)-2-hydroxy-1-propanesulfonic acid	CHES	9.3	8.6–10.0
Glycine and NaOH		9.7	8.2–10.1
Ethanolamine (2-aminoethanol) and HCl		9.5	8.6–10.4
3-(Cyclohexylamino)-2-hydroxy-1-propanesulfonic acid	CAPSO	9.6	8.9–10.3
2-Amino-2-methyl-1-propanol	AMP	9.7	9.0–10.5
Carbonate and hydrogen carbonate		10.3	9.2–11.0
Borax and NaOH			9.4–11.1
3-(Cyclohexylamino)-1-propanesulfonic acid	CAPS	10.4	9.7–11.1
Na_2HPO_4 and NaOH		11.9	11.0–12.0

TABLE 1.68 pH Values of Biological and Other Buffers for Control Purposes (*Continued*)

x mL of 0.2M Sodium Acetate (27.199 g NaOAc · 3H$_2$O per liter) plus y mL of 0.2M Acetic Acid			x mL of 0.1M KH$_2$PO$_4$ (13.617 g · L^{-1}) plus y mL of 0.05M Borax Solution (19.404 g Na$_2$B$_4$O$_7$ · 10H$_2$O per Liter)					
pH	NaOAc, mL	Acetic Acid, mL	pH	KH$_2$PO$_4$, mL	Borax, mL	pH	KH$_2$PO$_4$, mL	Borax, mL
3.60	7.5	92.5	5.80	92.1	7.9	7.60	51.7	48.3
3.80	12.0	88.0	6.00	87.7	12.3	7.80	49.2	50.8
4.00	18.0	82.0	6.200	83.0	17.0	8.00	46.5	53.5
4.20	26.5	73.5	6.40	77.8	22.2	8.20	43.0	57.0
4.40	37.0	63.0	6.60	72.2	27.8	8.40	38.7	61.3
4.60	49.0	51.0	6.80	66.7	33.3	8.60	34.0	66.0
4.80	60.0	40.0	7.00	62.3	37.7	8.80	27.6	72.4
5.00	70.5	29.5	7.20	58.1	41.9	9.00	17.5	82.5
5.20	79.0	21.0	7.40	55.0	45.0	9.20	5.0	95.0
5.40	85.5	14.5						
5.60	90.5	9.5						

x mL of Veronal (20.6 g Na Diethylbarbiturate per Liter) plus y mL of 0.1M HCl			x mL of 0.2M Aqueous NH$_3$ Solution plus y mL of 0.2M NH$_4$Cl (10.699 g · L^{-1})			x mL of 0.1M Citrate (21.0 g Citric Acid Monohydrate + 200 mL 1M NaOH per Liter) plus y mL of 0.1M NaOH		
pH	Veronal, mL	HCl, mL	pH	Aq NH$_3$, mL	NH$_4$Cl, mL	pH	Citrate, mL	NaOH mL
7.00	53.6	46.4	8.00	5.5	94.5	5.10	90.0	10.0
7.20	55.4	44.6	8.20	8.5	91.5	5.30	80.0	20.0
7.40	58.1	41.9	8.40	12.5	87.5	5.50	71.0	29.0
7.60	61.5	38.5	8.60	18.5	81.5	5.70	67.0	33.0
7.80	66.2	33.8	8.80	26.0	74.0	5.90	62.0	38.0
8.00	71.6	28.4	9.00	36.0	64.0			
8.20	76.9	23.1	9.25	50.0	50.0			
8.40	82.3	17.7	9.40	58.5	41.5			
8.60	87.1	12.9	9.60	69.0	31.0			
8.80	90.8	9.2	9.80	78.0	22.0			
9.00	93.6	6.4	10.00	85.0	15.0			

x mL of 0.2M NaOH Added to 100 mL of Stock Solution (0.04M Acetic Acid, 0.04M H$_3$PO$_4$, and 0.04M Boric Acid)

pH	NaOH, mL	pH	NaOH, mL	pH	NaOH, mL	pH	NaOH, mL
1.81	0.0	4.10	25.0	6.80	50.0	9.62	75.0
1.89	2.5	4.35	27.5	7.00	52.5	9.91	77.5
1.98	5.0	4.56	30.0	7.24	55.0	10.38	80.0
2.09	7.5	4.78	32.5	7.54	57.5	10.88	82.5
2.21	10.0	5.02	35.0	7.96	60.0	11.20	85.0
2.36	12.5	5.33	37.5	8.36	62.5	11.40	87.5
2.56	15.0	5.72	40.0	8.69	65.0	11.58	90.0
2.87	17.5	6.09	42.5	8.95	67.5	11.70	92.5
3.29	20.0	6.37	45.0	9.15	70.0	11.82	95.0
3.78	22.5	6.59	47.5	9.37	72.5	11.92	97.5

(*Continued*)

TABLE 1.68   pH Values of Biological and Other Buffers for Control Purposes (*Continued*)

pH	HCl, mL	Glycine, mL	pH	HCl, mL	Citrate, mL	pH	Succinic Acid, mL	Borax, mL
	x mL of 0.1M HCl plus y mL of 0.1M Glycine (7.505 g Glycine + 5.85 g NaCl per Liter)			x mL of 0.1M HCl plus y mL of 0.1M Citrate (21.008 g Citric Acid Monohydrate + 200 ml 1M NaOH per Liter)			x mL of 0.05M Succinic Acid (5.90 g·L⁻¹) plus y mL of Borax Solution (19.404 g Na₂B₄O₇·10H₂O per Liter)	
1.20	84.0	16.0	3.50	52.8	47.2	3.60	90.5	9.5
1.40	71.0	29.0	3.60	51.3	48.7	3.80	86.3	13.7
1.60	61.8	38.2	3.80	48.6	51.4	4.00	82.2	17.8
1.80	55.2	44.8	4.00	43.8	56.2	4.20	77.8	22.2
2.00	49.1	50.9	4.20	38.6	61.4	4.40	73.8	26.2
2.20	42.7	57.3	4.40	34.6	65.4	4.60	70.0	30.0
2.40	36.5	63.5	4.60	24.3	75.7	4.80	66.5	33.5
2.60	30.3	69.7	4.80	11.0	89.0	5.00	63.2	36.8
2.80	24.0	76.0				5.20	60.5	39.5
3.00	17.8	82.2				5.40	57.9	42.1
3.30	10.8	89.2				5.60	55.7	44.3
3.60	6.0	94.0				5.80	54.0	46.0

x mL of 0.2M Na₂HPO₄·2H₂O (35.599 g·L⁻¹) plus y mL of 0.1M Citric Acid (19.213 g·L⁻¹)

pH	Na₂HPO₄, mL	Citric Acid, mL	pH	Na₂HPO₄, mL	Citric Acid, mL	pH	Na₂HPO₄, mL	Citric Acid, mL
2.20	2.00	98.00	4.20	41.40	58.60	6.20	66.10	33.90
2.40	6.20	93.80	4.40	44.10	55.90	6.40	69.25	30.75
2.60	10.90	89.10	4.60	46.75	53.25	6.60	72.75	27.25
2.80	15.85	84.15	4.80	49.30	50.70	6.80	77.25	22.75
3.00	20.55	79.45	5.00	51.50	48.50	7.00	82.35	17.65
3.20	24.70	75.30	5.20	53.60	46.40	7.20	86.95	13.05
3.40	28.50	71.50	5.40	55.75	44.25	7.40	90.85	9.15
3.60	32.20	67.80	5.60	58.00	42.00	7.60	93.65	6.35
3.80	35.50	64.50	5.80	60.45	39.55	7.80	95.75	4.25
4.00	38.55	61.45	6.00	63.15	36.85	8.00	97.25	2.75

1.18   SOLUBILITY AND EQUILIBRIUM CONSTANT

The equilibrium constant is the value of the reaction quotient for a system at equilibrium. The reaction quotient is the ratio of molar concentrations of the reactants to those of the products, each concentration being raised to the power equal to the coefficient in the equation.

For the hypothetical chemical reaction

$$A + B \leftrightarrow C + D$$

the equilibrium constant, K, is:

$$K = [C][D]/[A][B]$$

The notation [A] signifies the molar concentration of species A. An alternative expression for the equilibrium constant can involve the use of partial pressures.

The equilibrium constant can be determined by allowing a reaction to reach equilibrium, measuring the concentrations of the various solution-phase or gas-phase reactants and products, and substituting these values into the relevant equation.

TABLE 1.69   Solubility of Gases in Water

The column (or line entry) headed "α" gives the volume of gas (in milliliters) measured at standard conditions (0°C and 760 mm or 101.325 kN · m^{-2}) dissolved in 1 mL of water at the temperature stated (in degrees Celsius) and when the pressure of the gas without that of the water vapor is 760 mm. The line entry "A" indicates the same quantity except that the gas itself is at the uniform pressure of 760 mm when in equilibrium with water.

The column headed "l" gives the volume of the gas (in milliliters) dissolved in 1 mL of water when the pressure of the gas plus that of the water vapor is 760 mm.

The column headed "q" gives the weight of gas (in grams) dissolved in 100 g of water when the pressure of the gas plus that of the water vapor is 760 mm.

Temp., °C	Acetylene		Air*		Ammonia		Bromine	
	α	q	$\alpha(\times 10^3)$	% oxygen in air	α	q	α	q
0	1.73	0.200	29.18	34.91	1130	89.5	60.5	42.9
1	1.68	0.194	28.42	34.87	—	—	—	—
2	1.63	0.188	27.69	34.82	—	—	54.1	38.3
3	1.58	0.182	26.99	34.78	—	—	—	—
4	1.53	0.176	26.32	34.74	1047	79.6	48.3	34.2
5	1.49	0.171	25.68	34.69	—	—	—	—
6	1.45	0.167	25.06	34.65	—	—	43.3	30.6
7	1.41	0.162	24.47	34.60	—	—	—	—
8	1.37	0.157	23.90	34.56	947	72.0	38.9	27.5
9	1.34	0.154	23.36	34.52	—	—	—	—
10	1.31	0.150	22.84	34.47	870	68.4	35.1	24.8
11	1.27	0.146	22.34	34.43	—	—	—	—
12	1.24	0.142	21.87	34.38	857	65.1	31.5	22.2
13	1.21	0.138	21.41	34.34	837	63.6	—	—
14	1.18	0.135	20.97	34.30	—	—	28.4	20.0
15	1.15	0.131	20.55	34.25	770	—	—	—
16	1.13	0.129	20.14	34.21	775	58.7	25.7	18.0
17	1.10	0.125	19.75	34.17	—	—	—	—
18	1.08	0.123	19.38	34.12	—	—	23.4	16.4
19	1.05	0.119	19.02	34.08	—	—	—	—
20	1.03	0.117	18.68	34.03	680	52.9	21.3	14.9
21	1.01	0.115	18.34	33.99	—	—	—	—
22	0.99	0.112	18.01	33.95	—	—	19.4	13.5
23	0.97	0.110	17.69	33.90	—	—	—	—
24	0.95	0.107	17.38	33.86	639	48.2	17.7	12.3
25	0.93	0.105	17.08	33.82	—	—	—	—
26	0.91	0.102	16.79	33.77	—	—	16.3	11.3
27	0.89	0.100	16.50	33.73	—	—	—	—
28	0.87	0.098	16.21	33.68	586	44.0	15.0	10.3
29	0.85	0.095	15.92	33.64	—	—	—	—
30	0.84	0.094	15.64	33.60	530	41.0	13.8	9.5
35	—	—	—	—	—	—	—	—
40	—	—	14.18	—	400	31.6	9.4	6.3
45	—	—	—	—	—	—	—	—
50	—	—	12.97	—	290	23.5	6.5	4.1
60	—	—	12.16	—	200	16.8	4.9	2.9
70	—	—	—	—	—	11.1	3.8	1.9
80	—	—	11.26	—	—	6.5	3.0	1.2
90	—	—	—	—	—	3.0	—	—
100	—	—	11.05	—	—	0.0	—	—

*Free from NH_3 and CO_2; total pressure of air + water vapor is 760 mm.

TABLE 1.69 Solubility of Gases in Water

Temp., °C	Carbon dioxide		Carbon monoxide		Chlorine		Ethane		Ethylene		Hydrogen	
	α	q	α	q	l	q	α	q	α	q	α	q
0	1.713	0.334 6	0.035 37	0.004 397	—	—	0.098 74	0.013 17	0.226	0.028 1	0.021 48	0.000 192 2
1	1.646	0.321 3	0.034 55	0.004 293	—	—	0.094 76	0.012 63	0.219	0.027 2	0.021 26	0.000 190 1
2	1.584	0.309 1	0.033 75	0.004 191	—	—	0.090 93	0.012 12	0.211	0.026 2	0.021 05	0.000 188 1
3	1.527	0.297 8	0.032 97	0.004 092	—	—	0.087 25	0.011 62	0.204	0.025 3	0.020 84	0.000 186 2
4	1.473	0.287 1	0.032 22	0.003 996	—	—	0.083 72	0.011 14	0.197	0.024 4	0.020 64	0.000 184 3
5	1.424	0.277 4	0.031 49	0.003 903	—	—	0.080 33	0.010 69	0.191	0.023 7	0.020 44	0.000 182 4
6	1.377	0.268 1	0.030 78	0.003 813	—	—	0.077 09	0.010 25	0.184	0.022 8	0.020 25	0.000 180 6
7	1.331	0.258 9	0.030 09	0.003 725	—	—	0.074 00	0.009 83	0.178	0.022 0	0.020 07	0.000 178 9
8	1.282	0.249 2	0.029 42	0.003 640	—	—	0.071 06	0.009 43	0.173	0.021 4	0.019 89	0.000 177 2
9	1.237	0.240 3	0.028 78	0.003 559	—	—	0.068 26	0.009 06	0.167	0.020 7	0.019 72	0.000 175 6
10	1.194	0.231 8	0.028 16	0.003 479	3.148	0.997 2	0.065 61	0.008 70	0.162	0.020 0	0.019 55	0.000 174 0
11	1.154	0.223 9	0.027 57	0.003 405	3.047	0.965 4	0.063 28	0.008 38	0.157	0.019 4	0.019 40	0.000 172 5
12	1.117	0.216 5	0.027 01	0.003 332	2.950	0.934 6	0.061 06	0.008 08	0.152	0.018 8	0.019 25	0.000 171 0
13	1.083	0.209 8	0.026 46	0.003 261	2.856	0.905 0	0.058 94	0.007 80	0.148	0.018 3	0.019 11	0.000 169 6
14	1.050	0.203 2	0.025 93	0.003 194	2.767	0.876 8	0.056 94	0.007 53	0.143	0.017 6	0.018 97	0.000 168 2
15	1.019	0.197 0	0.025 43	0.003 130	2.680	0.849 5	0.055 04	0.007 27	0.139	0.017 1	0.018 83	0.000 166 8
16	0.985	0.190 3	0.024 94	0.003 066	2.597	0.823 2	0.053 26	0.007 03	0.136	0.016 7	0.018 69	0.000 165 4
17	0.956	0.184 5	0.024 48	0.003 007	2.517	0.797 9	0.051 59	0.006 80	0.132	0.016 2	0.018 56	0.000 164 1
18	0.928	0.178 9	0.024 02	0.002 947	2.440	0.773 8	0.050 03	0.006 59	0.129	0.015 8	0.018 44	0.000 162 8
19	0.902	0.173 7	0.023 60	0.002 891	2.368	0.751 0	0.048 58	0.006 39	0.125	0.015 3	0.018 31	0.000 161 6
20	0.878	0.168 8	0.023 19	0.002 838	2.299	0.729 3	0.047 24	0.006 20	0.122	0.014 9	0.018 19	0.000 160 3
21	0.854	0.164 0	0.022 81	0.002 789	2.238	0.710 0	0.045 89	0.006 02	0.119	0.014 6	0.018 05	0.000 158 8
22	0.829	0.159 0	0.022 44	0.002 739	2.180	0.691 8	0.044 59	0.005 84	0.116	0.014 2	0.017 92	0.000 157 5
23	0.804	0.154 0	0.022 08	0.002 691	2.123	0.673 9	0.043 35	0.005 67	0.114	0.013 9	0.017 79	0.000 156 1
24	0.781	0.149 3	0.021 74	0.002 646	2.070	0.657 2	0.042 17	0.005 51	0.111	0.013 5	0.017 66	0.000 154 8
25	0.759	0.144 9	0.021 42	0.002 603	2.019	0.641 3	0.041 04	0.005 35	0.108	0.013 1	0.017 54	0.000 153 5
26	0.738	0.140 6	0.021 10	0.002 560	1.970	0.625 9	0.039 97	0.005 20	0.106	0.012 9	0.017 42	0.000 152 2
27	0.718	0.136 6	0.020 80	0.002 519	1.923	0.611 2	0.038 95	0.005 06	0.104	0.012 6	0.017 31	0.000 150 9
28	0.699	0.132 7	0.020 51	0.002 479	1.880	0.597 5	0.037 99	0.004 93	0.102	0.012 3	0.017 20	0.000 149 6
29	0.682	0.129 2	0.020 24	0.002 442	1.839	0.584 7	0.037 09	0.004 80	0.100	0.012 1	0.017 09	0.000 148 4
30	0.665	0.125 7	0.019 98	0.002 405	1.799	0.572 3	0.036 24	0.004 68	0.098	0.011 8	0.016 99	0.000 147 4

35	0.592	0.110 5	0.018 77	0.002 231	1.602	0.510 4	0.032 30	0.004 12	—	—	0.016 66	0.000 142 5
40	0.530	0.097 3	0.017 75	0.002 075	1.438	0.459 0	0.029 15	0.003 66	—	—	0.016 44	0.000 138 4
45	0.479	0.086 0	0.016 90	0.001 933	1.322	0.422 8	0.026 60	0.003 27	—	—	0.016 24	0.000 134 1
50	0.436	0.076 1	0.016 15	0.001 797	1.225	0.392 5	0.024 59	0.002 94	—	—	0.016 08	0.000 128 7
60	0.359	0.057 6	0.014 88	0.001 522	1.023	0.329 5	0.021 77	0.002 39	—	—	0.016 00	0.000 117 8
70	—	—	0.014 40	0.001 276	0.862	0.279 3	0.019 48	0.001 85	—	—	0.016 0	0.000 102
80	—	—	0.014 30	0.000 980	0.683	0.222 7	0.018 26	0.001 34	—	—	0.016 0	0.000 079
90	—	—	0.014 2	0.000 57	0.39	0.127	0.017 6	0.000 8	—	—	0.016 0	0.000 046
100	—	—	0.014 1	0.000 00	0.00	0.000	0.017 2	0.000 0	—	—	0.016 0	0.000 000

0	4.670	0.706 6	0.055 63	0.003 959	0.073 81	0.009 833	0.023 54	0.002 942	0.048 89	0.006 945	79.789	22.83
1	4.522	0.683 9	0.054 01	0.003 842	0.071 84	0.009 564	0.022 97	0.002 869	0.047 58	0.006 756	77.210	22.09
2	4.379	0.661 9	0.052 44	0.003 728	0.069 93	0.009 305	0.022 41	0.002 798	0.046 33	0.006 574	74.691	21.37
3	4.241	0.640 7	0.050 93	0.003 619	0.068 09	0.009 057	0.021 87	0.002 730	0.045 12	0.006 400	72.230	20.66
4	4.107	0.620 1	0.049 46	0.003 513	0.066 32	0.008 816	0.021 35	0.002 663	0.043 97	0.006 232	69.828	19.98
5	3.977	0.600 1	0.048 05	0.003 410	0.064 61	0.008 584	0.020 86	0.002 600	0.042 87	0.006 072	67.485	19.31
6	3.852	0.580 9	0.046 69	0.003 312	0.062 98	0.008 361	0.020 37	0.002 537	0.041 80	0.005 918	65.200	18.65
7	3.732	0.562 4	0.045 39	0.003 217	0.061 40	0.008 147	0.019 90	0.002 477	0.040 80	0.005 773	62.973	18.02
8	3.616	0.544 6	0.044 13	0.003 127	0.059 90	0.007 943	0.019 45	0.002 419	0.039 83	0.005 632	60.805	17.40
9	3.505	0.527 6	0.042 92	0.003 039	0.058 46	0.007 747	0.019 02	0.002 365	0.038 91	0.005 498	58.697	16.80
10	3.399	0.511 2	0.041 77	0.002 955	0.057 09	0.007 560	0.018 61	0.002 312	0.038 02	0.005 368	56.647	16.21
11	3.300	0.496 0	0.040 72	0.002 879	0.055 87	0.007 393	0.018 23	0.002 263	0.037 18	0.005 246	54.655	15.64
12	3.206	0.481 4	0.039 70	0.002 805	0.054 70	0.007 233	0.017 86	0.002 216	0.036 37	0.005 128	52.723	15.09
13	3.115	0.467 4	0.038 72	0.002 733	0.053 57	0.007 078	0.017 50	0.002 170	0.035 59	0.005 014	50.849	14.56
14	3.028	0.454 0	0.037 79	0.002 665	0.052 50	0.006 930	0.017 17	0.002 126	0.034 86	0.004 906	49.033	14.04
15	2.945	0.441 1	0.036 90	0.002 599	0.051 47	0.006 788	0.016 85	0.002 085	0.034 15	0.004 802	47.276	13.54
16	2.865	0.428 7	0.036 06	0.002 538	0.050 49	0.006 652	0.016 54	0.002 045	0.033 48	0.004 703	45.578	13.05
17	2.789	0.416 9	0.035 25	0.002 478	0.049 56	0.006 524	0.016 25	0.002 006	0.032 83	0.004 606	43.939	12.59
18	2.717	0.405 6	0.034 48	0.002 422	0.048 68	0.006 400	0.015 97	0.001 970	0.032 20	0.004 514	42.360	12.14
19	2.647	0.394 8	0.033 76	0.002 369	0.047 85	0.006 283	0.015 70	0.001 935	0.031 61	0.004 426	40.838	11.70
20	2.582	0.384 5	0.033 08	0.002 319	0.047 06	0.006 173	0.015 45	0.001 901	0.031 02	0.004 339	39.374	11.28
21	2.517	0.374 5	0.032 43	0.002 270	0.046 25	0.006 059	0.015 22	0.001 869	0.030 44	0.004 252	37.970	10.88
22	2.456	0.364 8	0.031 80	0.002 222	0.045 45	0.005 947	0.014 98	0.001 838	0.029 88	0.004 169	36.617	10.50
23	2.396	0.355 4	0.031 19	0.002 177	0.044 69	0.005 838	0.014 75	0.001 809	0.029 34	0.004 087	35.302	10.12
24	2.338	0.346 3	0.030 61	0.002 133	0.043 95	0.005 733	0.014 54	0.001 780	0.028 81	0.004 007	34.026	9.76
25	2.282	0.337 5	0.030 06	0.002 091	0.043 23	0.005 630	0.014 34	0.001 751	0.028 31	0.003 931	32.786	9.41
26	2.229	0.329 0	0.029 52	0.002 05C	0.042 54	0.005 530	0.014 13	0.001 724	0.027 83	0.003 857	31.584	9.06

(Continued)

TABLE 1.69 Solubility of Gases in Water (*Continued*)

Temp., °C	Carbon dioxide		Carbon monoxide		Chlorine		Ethane		Ethylene		Hydrogen	
	α	q	α	q	l	q	α	q	α	q	α	q
26	2.229	0.329 0	0.029 52	0.002 050	0.042 54	0.005 530	0.014 13	0.001 724	0.027 83	0.003 857	31.584	9.06
27	2.177	0.320 8	0.029 01	0.002 011	0.041 88	0.005 435	0.013 94	0.001 698	0.027 36	0.003 787	30.422	8.73
28	2.128	0.313 0	0.028 52	0.001 974	0.041 24	0.005 342	0.013 76	0.001 672	0.026 91	0.003 718	29.314	8.42
29	2.081	0.305 5	0.028 06	0.001 938	0.040 63	0.005 252	0.013 58	0.001 647	0.026 49	0.003 651	28.210	8.10
30	2.037	0.298 3	0.027 62	0.001 904	0.040 04	0.005 165	0.013 42	0.001 624	0.026 08	0.003 588	27.161	7.80
35	1.831	0.264 8	0.025 46	0.001 733	0.037 34	0.004 757	0.012 56	0.001 501	0.024 40	0.003 315	22.489	6.47
40	1.660	0.236 1	0.023 69	0.001 586	0.035 07	0.004 394	0.011 84	0.001 391	0.023 06	0.003 082	18.766	5.41
45	1.516	0.211 0	0.022 38	0.001 466	0.033 11	0.004 059	0.011 30	0.001 300	0.021 87	0.002 858	—	—
50	1.392	0.188 3	0.021 34	0.001 359	0.031 52	0.003 758	0.010 88	0.001 216	0.020 90	0.002 657	—	—
60	1.190	0.148 0	0.019 54	0.001 144	0.029 54	0.003 237	0.010 23	0.001 052	0.019 46	0.002 274	—	—
70	1.022	0.110 1	0.018 25	0.000 926	0.028 10	0.002 668	0.009 77	0.000 851	0.018 33	0.001 856	—	—
80	0.917	0.076 5	0.017 70	0.000 695	0.027 00	0.001 984	0.009 58	0.000 660	0.017 61	0.001 381	—	—
90	0.84	0.041	0.017 35	0.000 40	0.026 5	0.001 13	0.009 5	0.000 38	0.017 2	0.000 79	—	—
100	0.81	0.000	0.017 0	0.000 00	0.026 3	0.000 00	0.009 5	0.000 00	0.017 0	0.000 00	—	—

*Atmospheric nitrogen containing 98.815% N_2 by volume + 1.185% inert gases.

TABLE 1.69 Solubility of Gases in Water

Substance		0°	10°	20°	30°	40°	60°	80°
Argon	α	0.0528	0.0413	0.0337	0.0288	0.0251	0.0209	0.0184
Helium	A	0.0098	0.00911	0.0086	0.00839	0.00841	0.00902	$0.009\ 42^{70°}$
Hydrogen bromide	l	612	582		$533^{25°}$		$469^{50°}$	$406^{75°}$
Hydrogen chloride	α	512	475	442	412	385	339	
Krypton	α	0.1105	0.0810	0.0626	0.0511	0.0433	0.0357	
Neon	A		$0.011\ 7^{9°}$	0.0106	0.0100	$0.009\ 48^{42°}$		$0.009\ 84^{73°}$
Nitrous oxide	A		0.88	0.63				
Ozone	$g \cdot L^{-1}$	0.0394	$0.029\ 9^{12°}$	$0.021\ 0^{19°}$	$0.0139^{27°}$	0.0042	0	
Radon	α	0.510	0.326	0.222	0.162	0.126	0.085	
Xenon	α	0.242	0.174	0.123	0.098	0.082		

TABLE 1.70 Solubility of Inorganic Compounds and Metal Salts of Organic Acids in Water at Various Temperatures

Solubilities are expressed as the number of grams of substance of stated molecular formula which when dissolved in 100 g of water make a saturated solution at the temperature stated (°C).

Substance	Formula	0°	10°	20°	30°	40°	60°	80°	90°	100°
Aluminum chloride	$AlCl_3$	43.9	44.9	45.8	46.6	47.3	48.1	48.6		49.0
fluoride	AlF_3	0.56	0.56	0.67	0.78	0.91	1.1	1.32		1.72
nitrate	$Al(NO_3)_3$	60.0	66.7	73.9	81.8	88.7	106	132	153	160
perchlorate	$Al(ClO_4)_3$	122	128	133						182
sulfate	$Al_2(SO_4)_3$	31.2	33.5	36.4	40.4	45.8	59.2	73.0	80.8	89.0
thallium(I) sulfate	$Al_2Tl_2(SO_4)_4$	3.15	4.60	6.39	9.37	14.39	35.35			
Ammonium aluminum sulfate	$NH_4Al(SO_4)_2$	2.10	5.00	7.74	10.9	14.9	26.7			
azide	NH_4N_3	16.0		25.3		37.1				
bromide	NH_4Br	60.5	68.1	76.4	83.2	91.2	108	125	135	145
chloride	NH_4Cl	29.4	33.2	37.2	41.4	45.8	55.3	65.6	71.2	77.3
chloroiridate(IV)	$(NH_4)_2IrCl_6$	0.56	0.71	0.95	1.20	1.56	2.45	4.38		
chloroplatinate(IV)	$(NH_4)_2PtCl_6$	0.289	0.374	0.499	0.637	0.815	1.44	2.16	2.61	3.36
chromate	$(NH_4)_2CrO_4$	25.0	29.2	34.0	39.3	45.3	59.0	76.1		
chromium(III) sulfate	$(NH_4)Cr(SO_4)_2$	3.95	9.5	13.0	18.8	32.6				
cobalt(II) sulfate	$(NH_4)_2Co(SO_4)_2$	6.0			17.0	22.0	33.5	49.0	58.0	75.1
dichromate	$(NH_4)_2Cr_2O_7$	18.2	25.5	35.6	46.5	58.5	86.0	115	122	156
dihydrogen arsenate	$NH_4H_2AsO_4$	33.7		48.7		63.8	83.0	107		
dihydrogen phosphate	$NH_4H_2PO_4$	22.7	29.5	37.4	46.4	56.7	82.5	118		173
dithionate	$(NH_4)_2S_2O_6$	133	151	166	179	204	311	533		
formate	NH_4CHO_2	102		143				109	170	354
hydrogen carbonate	NH_4HCO_3	11.9	16.1	21.7	28.4	36.6	59.2			
hydrogen phosphate	$(NH_4)_2HPO_4$	42.9	62.9	68.9	75.1	81.8	97.2			
hydrogen tartrate	$NH_4C_4H_5O_6$	1.00	1.88	2.70						
iodide	NH_4I	155	163	172	182	191	209	229		250
iron(II) sulfate	$(NH_4)_2Fe(SO_4)_2$	12.5	17.2	26.4	33	46				
Ammonium magnesium sulfate	$(NH_4)_2Mg(SO_4)_2$	11.8	14.6	18.0	21.7	25.8	35.1	48.3		65.7
nickel sulfate	$(NH_4)_2Ni(SO_4)_2$	1.00	4.00	6.50	9.20	12.0	17.0			
nitrate	NH_4NO_3	118	150	192	242	297	421	580	740	871
oxalate	$(NH_4)_2C_2O_4$	2.2	3.21	4.45	6.09	8.18	14.0	22.4	27.9	34.7
perchlorate	NH_4ClO_4	12.0	16.4	21.7	27.7	34.6	49.9	68.9		
selenite	$(NH_4)_2SeO_3$	96	105	115	126	143	192			

	Formula									
sulfate	$(NH_4)_2SO_4$	70.6	73.0	75.4	78.0	81	88	95		103
sulfite	$(NH_4)_2SO_3$	47.9	54.0	60.8	68.8	78.4	104	144		153
tartrate	$(NH_4)_2C_4H_4O_6$	45.0	55.0	63.0	70.5	76.5	86.9			
thioantimonate(V)	$(NH_4)_3SbS_4$	71.2					91.2	120	150	
thiocyanate	NH_4SCN	120	144	170	208	234	346			
vanadate	NH_4VO_3			0.48	0.84	1.32	2.42			
zinc sulfate	$(NH_4)_2Zn(SO_4)_2$	7.0	9.5	12.5	16.0	20.0	30.0	46.6	58.0	72.4
Antimony(III) chloride	$SbCl_3$	602		910	1087	1368				
fluoride	SbF_3	385		444	562					
Arsenic hydride (760 mm), cc	AsH_3	42	30	28	[completely miscible at 72°]					
oxide (pent-)	As_2O_5	59.5	62.1	65.8	69.8	71.2	73.0	75.1		76.7
oxide (tri-)	As_2O_3	1.20	1.49	1.82	2.31	2.93	4.31	6.11		8.2
Barium acetate	$Ba(C_2H_3O_2)_2 \cdot 3H_2O$	58.8	62	72	75	78.5	75.0	74.0		74.8
azide	$Ba(N_3)_2$	12.5	16.1	$17.4^{17°}$						
bromate	$Ba(BrO_3)_2 \cdot H_2O$	0.29	0.44	0.65	0.95	1.31	2.27	3.52	4.26	5.39
bromide	$BaBr_2 \cdot 2H_2O$	98	101	104	109	114	123	135		149
n-butyrate	$Ba(C_4H_7O_2)_2$	37.0	36.1	35.4	34.9	35.2	37.2	41.7	45.5	$48.1^{95°}$
caproate	$Ba(C_6H_{11}O_2)_2 \cdot 3.5H_2O$	11.71	8.38	6.89	5.87	5.79	8.39	14.71	19.28	
chlorate	$Ba(ClO_3)_2 \cdot H_2O$	20.3	26.9	33.9	41.6	49.7	66.7	84.8		105
chloride	$BaCl_2 \cdot 2H_2O$	31.2	33.5	35.8	38.1	40.8	46.2	52.5	55.8	59.4
chlorite	$Ba(ClO_2)_2$	43.9	44.6	45.4		47.9	53.8	66.6		80.8
fluoride	BaF_2	0.159	0.159	0.160	0.162					
formate	$Ba(CHO_2)_2$	26.2	28.0	29.9	31.9	34.0	38.6	44.2	47.6	51.3
hydroxide	$Ba(OH)_2$	1.67	2.48	3.89	5.59	8.22	20.94	101.4		
iodate	$Ba(IO_3)_2$			0.035	0.046	0.057				
iodide	$BaI_2 \cdot 2H_2O$	182	201	223	250	264	291			301
nitrate	$Ba(NO_3)_2$	4.95	6.67	9.02	11.48	14.1	20.4	27.2		34.4
nitrite	$Ba(NO_2)_2 \cdot H_2O$	50.3	60	72.8		102	151	222	261	325
perchlorate	$Ba(ClO_4)_2 \cdot 3H_2O$	239		336	416	495	575			653
propionate	$Ba(C_3H_5O_2)_2 \cdot H_2O$	57.2	56.8	56.8	57.5	59.0	62.0	67.8	73.0	82.7
*iso*succinate	$BaC_4H_4O_4$	0.421	0.432	0.418	0.393	0.366	0.306	0.237		
sulfamate	$Ba(SO_3NH_2)_2$	18.3	22.3	26.8	32.5	38.5	49.6	61.5	67.34	73.5
sulfide	BaS	2.88	4.89	7.86	10.38	14.89	27.69	49.91		60.29
tartrate	$Ba(C_2H_2O_3)_2$	0.021	0.024	0.028	0.032	0.035	0.044	0.053		
Beryllium nitrate	$Be(NO_3)_2$	97	102	108	113	125	178			
sulfate	$BeSO_4$	37.0	37.6	39.1	41.4	45.8	53.1	67.2		82.8
Boric acid	H_3BO_3	2.67	3.73	5.04	6.72	8.72	14.81	23.62	30.38	40.25
Cadmium bromide	$CdBr_2$	56.3	75.4	98.8	129	152	153	156		160

(Continued)

TABLE 1.70 Solubility of Inorganic Compounds and Metal Salts of Organic Acids in Water at Various Temperatures (*Continued*)

Substance	Formula	0°	10°	20°	30°	40°	60°	80°	90°	100°
chlorate	Cd(ClO₃)₂	299	308	322	348	376	455			147
chloride	CdCl₂ · 2.5H₂O	90	100	113	132	135	136	140		
	CdCl₂ · H₂O		135	135	135					
formate	Cd(CHO₂)₂	8.3	11.1	14.4	18.6	25.3	59.5	80.5	85.2	94.6
iodide	CdI₂	78.7		84.7	87.9	92.1	100	111		125
nitrate	Cd(NO₃)₂	122	136	150	167	194	310	713		
perchlorate	Cd(ClO₄)₂ · 6H₂O		180	188	195	203	221	243		272
selenate	CdSeO₄	72.5	68.4	64.0	58.9	55.0	44.2	32.5	27.2	22.0
sulfate	CdSO₄	75.4	76.0	76.6		78.5	81.8	66.7	63.1	60.8
Calcium acetate	Ca(OAc)₂ · 2H₂O	37.4	36.0	34.7	33.8	33.2	32.7	33.5	31.1	29.7
benzoate	Ca(OBz)₂ · 3H₂O	2.32	2.45	2.72	3.02	3.42	4.71	6.87	8.55	8.70
bromide	CaBr₂ · 6H₂O	125	132	143	$185^{34°}$	213	278	295		$312^{105°}$
butyrate	Ca(C₄H₇O₂)₂	20.31	19.15	18.20	17.25	16.40	15.15	14.95		15.85
cacodylate	Ca(C₂H₆AsO₂)₂ · 9H₂O	48	52	59	71					
chloride	CaCl₂ · 6H₂O	59.5	64.7	74.5	100	128	137	147	154	159
chromate	CaCrO₄ · 2H₂O	4.5		2.25	1.83	1.49	0.83			
(mn)	CaCrO₄ · 2H₂O	17.3		16.6	16.1					
formate	Ca(CHO₂)₂	16.15		16.60		17.05	17.50	17.95		18.40
gluconate	Ca(C₆H₁₁O₇)₂ · H₂O			3.72		5.29		12.11	36.80	$57.2^{96°}$
hydrogen carbonate	Ca(HCO₃)₂	16.15		16.60		17.05	17.50	17.95		18.40
hydroxide	Ca(OH)₂	0.189	0.182	0.173	0.160	0.141	0.121		0.086	0.076
Calcium iodide	Ca(IO₃)₂ · 6H₂O	0.090		0.24	0.38	0.52	0.65	0.66	0.67	
iodide	CaI₂	64.6	66.0	67.6	69.0	70.8	74	78		81
lactate	Ca(C₃H₅O₃)₂ · 5H₂O	3.1		$5.4^{15°}$	7.9					
levulinate	Ca(C₁₀H₁₄O₆) · 2H₂O	38.1		$45.1^{16°}$	55.0	$70.3^{45°}$	$88.7^{55°}$			
malonate	Ca(C₃H₂O₄)	0.29	0.33	0.36	0.40	0.42	0.46	0.48		
nitrate	Ca(NO₃)₂ · 4H₂O	102	115	129	152	191		358		363
nitrite	Ca(NO₂)₂ · 4H₂O	63.9		$84.5^{18°}$	104		134	151	166	178
propionate	Ca(C₃H₅O₂)₂ · H₂O	42.80		39.85			38.25	39.85	42.15	48.44
selenate	CaSeO₄ · 2H₂O	9.73	9.77	9.22	8.79	7.14				
succinate	Ca(C₃H₂O₂)₂ · 3H₂O	1.127	1.22	1.28		1.18	0.89	0.68	0.67	0.66
sulfamate	Ca(SO₃NH₂)₂	56.5	62.8	72.3	84.5	100.1	150.0	215.2	$242^{95°}$	
sulfate	CaSO₄ · ½H₂O			0.32	$0.29^{25°}$	$0.26^{35°}$	$0.21^{45°}$	$0.145^{65°}$	$0.12^{75°}$	
	CaSO₄ · 2H₂O	0.223	0.244	$0.255^{18°}$	0.264	0.265	$0.244^{65°}$	$0.234^{75°}$		
tartrate	CaC₄H₄O₆ · 4H₂O	0.026	0.029	0.034	0.046	0.063	0.091	0.130		0.071
uranyl carbonate	Ca₂UO₂(CO₃)₃ · 10H₂O	0.1		0.4^{23}		0.8	1.555			0.205

(Continued)

Name	Formula									
valerate	Ca(C$_5$H$_9$O$_2$)$_2$	9.82	9.25	8.80	8.40	8.05	7.78	7.95	8.20	8.78
isovalerate	Ca(C$_5$H$_9$O$_2$)$_2$ · 3H$_2$O	26.05	22.70	21.80	21.68	22.00	18.38	16.88	16.65	16.55
Carbon disulfide	CS$_2$	0.204	0.194	0.179	0.155	0.111				
oxide sulfide (STP) mL/100 mL	COS	133.3	83.6	56.1	40.3					
tetrafluoride (STP) mL/100 g	CF$_4$	0.595	0.595	0.490	0.415	0.356				
Cerium(III) ammonium nitrate	Ce(NH$_4$)$_2$(NO$_3$)$_5$		242	276	318	376	681			
(IV) ammonium nitrate	Ce(NH$_4$)$_2$(NO$_3$)$_6$			135	150	169	213			
(III) ammonium sulfate	Ce(NH$_4$)(SO$_4$)$_2$			5.53	4.49	3.48	2.02	1.33		
(III) selenate	Ce$_2$(SeO$_3$)$_3$	39.5	37.2	35.2	33.2	32.6	13.7	4.6	2.1	
(III) sulfate	Ce$_2$(SO$_4$)$_3$ · 9H$_2$O	21.4		9.84	7.24	5.63	3.87			
	Ce$_2$(SO$_4$)$_3$ · 8H$_2$O	18.8		9.43	7.10	5.70	4.04			
Cesium aluminum sulfate	Cs$_2$Al$_2$(SO$_4$)$_4$	0.21	0.30	0.40	0.61	0.85	2.00	5.40	10.5	22.7
bromate	CsBrO$_3$			3.66$^{25°}$	4.53	5.30$^{35°}$				
chlorate	CsClO$_3$	2.46	3.8	6.2	9.5	13.8	26.2	45.0	58.0	79.0
chloride	CsCl	151	175	187	197	208	230	250	260	271
chloroaurate(III)	CsAuCl$_4$		0.5	0.8	1.7	3.3	8.9	19.5	27.7	37.9
chloroplatinate(IV)	Cs$_2$PtCl$_6$	0.0047	0.0064	0.0087	0.0119	0.0158	0.0290	0.0525	0.0675	0.0914
formate	CsCHO$_2$	335	381	450	533	694				
iodide	CsI	44.1	58.5	76.5	96	124$^{45°}$	150	190	205	
nitrate	CsNO$_3$	9.33	14.9	23.0	33.9	47.2	83.8	134	163	197
perchlorate	CsClO$_4$	0.8	1.0	1.6	2.6	4.0	7.3	14.4	20.5	30.0
sulfate	Cs$_2$SO$_4$	167	173	179	184	190	200	210	215	220
Chlorine dioxide	ClO$_2$	2.76	6.00	8.70$^{15°}$						
Chromium(III) nitrate	Cr(NO$_3$)$_3$	108$^{5°}$	124$^{15°}$	130$^{25°}$	152$^{35°}$					
(VI) oxide	CrO$_3$	164.8		167.2		172.5	183.9	191.6		206.8
(III) perchlorate	Cr(ClO$_4$)$_3$	104	123	130		163				
Cobalt(II) bromide	CoBr$_2$	91.9		112	128		227	241		257
chlorate	Co(ClO$_3$)$_2$	135	162	180	195	214	316			
chloride	CoCl$_2$	43.5	47.7	52.9	59.7	69.5	93.8	97.6	101	106
iodate	Co(IO$_3$)$_2$			1.02	0.90	0.88	0.82	0.73		0.70
nitrate	Co(NO$_3$)$_2$	84.0	89.6	97.4	111	125	174	204	300	
nitrite	Co(NO$_2$)$_2$	0.076	0.24	0.40	0.61	0.85				
sulfate	CoSO$_4$	25.5	30.5	36.1	42.0	48.8	55.0	53.8	45.3	38.9
	CoSO$_4$ · 7H$_2$O	44.8	56.3	65.4	73.0	88.1	101			

TABLE 1.70   Solubility of Inorganic Compounds and Metal Salts of Organic Acids in Water at Various Temperatures (*Continued*)

Substance	Formula	0°	10°	20°	30°	40°	60°	80°	90°	100°
Copper(II) ammonium chloride	$CuCl_2 \cdot 2NH_4Cl$	28.2	$32.0^{12°}$	35.0	38.3	43.8	56.6	76.5	76.5	107
ammonium sulfate	$CuSO_4 \cdot (NH_4)_2SO_4$	11.5	15.1	19.4	24.4	30.5	46.3	69.7	86.1	
bromide	$CuBr_2$	107	116	126	128	$131^{50°}$				120
chloride	$CuCl_2$	68.6	70.9	73.0	77.3	87.6	96.5	104	108	120
fluorosilicate	$CuSiF_6$	73.5	76.5	81.6	$84.1^{25°}$	$91.2^{50°}$		$932^{75°}$		
nitrate	$Cu(NO_3)_2$	83.5	100	125	156	163	182	208	222	247
potassium sulfate	$CuSO_4K_2SO_4$	5.1	7.2	10.0	13.6	18.2				
selenate	$CuSeO_4$	12.04	14.53	17.51	21.04	25.22	36.50	53.68		
sulfate	$CuSO_4 \cdot 5H_2O$	23.1	27.5	32.0	37.8	44.6	61.8	83.8		114
tartrate	$CuC_4H_4O_6 \cdot 3H_2O$		$0.020^{15°}$	0.042	0.089	0.142	0.197	0.144		
Gadolinium bromate	$Gd(BrO_3)_3 \cdot 9H_2O$	50.2	70.1	95.6	126	166				
sulfate	$Gd_2(SO_4)_3$	3.98	3.30	2.60	2.32	0.61				
Germanium(IV) oxide	GeO_2		0.49	0.43	0.50	4.52				
Holmium sulfate	$Ho_2(SO_4)_3 \cdot 8H_2O$			8.18	$6.71^{25°}$					
Hydrazinium (1+) nitrate	$N_2H_5NO_3$		175	266	402	607	2127			130.0
(2+) sulfate	$N_2H_6SO_4$			2.87	3.89	4.15	9.08	14.39		
(1+) sulfate	$(N_2H_5)_2SO_4$				221	300	554			
Hydrogen bromide	HBr	221.2	210.3	$204.0^{15°}$		$171.5^{50°}$		$150.5^{75°}$		
chloride	HCl	82.3	77.2	72.1	67.3	63.3	56.1			
selenide, mL at STP	H_2Se	386	351	289						
Iodine	I_2	0.014	0.020	0.029	0.039	0.052	0.100	0.225	0.315	0.445
Iridium(IV) ammonium chloride	$(NH_4)_2IrCl_6$	0.556	0.706	0.77	1.21	1.57	2.46	4.38	dec	
sodium chloride	Na_2IrCl_6		$34.46^{15°}$		56.17	96.00	191.2	279.3		
Iron(II) ammonium sulfate	$FeSO_4 \cdot (NH_4)_2SO_4 \cdot 6H_2O$	17.23	31.0	36.47	45.0					
(II) bromide	$FeBr_2$	101	109	117	124	133	144	168	176	184
(II) chloride	$FeCl_2$	49.7	59.0	62.5	66.7	70.0	78.3	88.7	92.3	94.9
(III) chloride	$FeCl_3 \cdot 6H_2O$	74.4	74.4	91.8	106.8		$83.7^{50°}$	$88.1^{75°}$		$100.1^{106°}$
(III) fluorosilicate	$FeSiF_6 \cdot 6H_2O$	72.1			$77.0^{25°}$		266			
(II) nitrate	$Fe(NO_3)_2 \cdot 6H_2O$	113	134	137.7		175.0				
(III) nitrate	$Fe(NO_3)_3 \cdot 9H_2O$	112.0			422	478	772			
(III) perchlorate	$Fe(ClO_4)_3$	289		368						
(II) sulfate	$FeSO_4 \cdot 7H_2O$	28.8	40.0	48.0	60.0	73.3	100.7	79.9	68.3	57.8
Lanthanum bromate	$La(BrO_3)_3$	98	120	149	200	168	247			
nitrate	$La(NO_3)_3$	100		136						

(Continued)

Name	Formula									
selenate	$La_2(SeO_3)_3$	50.5	45	45	45	45	18.5	5.4	2.2	0.68
Lead(II) sulfate	$La_2(SO_4)_3$	3.00	2.72	2.33	1.90	1.67	1.26	0.91	0.79	
Lead(II) acetate	$Pb(C_2H_3O_2)_2$	19.8	29.5	44.3	69.8	116				
bromide	$PbBr_2$	0.45	0.63	0.86	1.12	1.50	2.29	3.23	3.86	4.55
chloride	$PbCl_2$	0.67	0.82	1.00	1.20	1.42	1.94	2.54	2.88	3.20
flcorosilicate	$PbSiF_6$	190		222			403	428		463
Germanium(IV) oxide	GeO_2		0.49	0.43	0.50	0.61				
Holmium sulfate	$Ho_2(SO_4)_3 \cdot 8H_2O$		14.39	8.18	$6.71^{25°}$	4.52				
Hydrazinium (1+) nitrate	$N_2H_5NO_3$		175	266	402	607	2127			
(2+) sulfate	$N_2H_6SO_4$			2.87	3.89	4.15	9.08			
(1+) sulfate	$(N_2H_5)_2SO_4$					300	554			
Hydrogen bromide	HBr	221.2	210.3	$204.0^{15°}$		$171.5^{50°}$		$150.5^{75°}$		130.0
chloride	HCl	82.3	77.2	72.1	67.3	63.3	56.1			
selenide, mL at STP	H_2Se	386	351	289	221					
Iodine	I_2	0.014	0.020	0.029	0.039	0.052	0.100	0.225	0.315	0.445
Iridium(IV) ammonium chloride	$(NH_4)_2IrCl_6$	0.556	0.706	0.77	1.21	1.57	2.46	4.38		
sodium chloride	Na_2IrCl_6		$34.46^{15°}$		56.17	96.00	191.2	279.3		
Iron(II) ammonium sulfate	$FeSO_4 \cdot (NH_4)_2SO_4 \cdot 6H_2O$	17.23	31.0	36.47	45.0					
(II) bromide	$FeBr_2$	101	109	117	124	133	144	168	176	184
(II) chloride	$FeCl_2$	49.7	59.0	62.5	66.7	70.0	78.3	88.7	92.3	94.9
(III) chloride	$FeCl_3 \cdot 6H_2O$	74.4		91.8	106.8					
(II) fluoro-silicate	$FeSiF_6 \cdot 6H_2O$	72.1	74.4		$77.0^{25°}$	$83.7^{50°}$		$88.1^{75°}$		100.1^{106}
(II) nitrate	$Fe(NO_3)_2 \cdot 6H_2O$	113	134	137.7		175.0				
(III) nitrate	$Fe(NO_3)_3 \cdot 9H_2O$	112.0		368	422	478	772			
(III) perchlorate	$Fe(ClO_4)_3$	289								
(II) sulfate	$FeSO_4 \cdot 7H_2O$	28.8	40.0	48.0	60.0	73.3	100.7	79.9	68.3	57.8
Lanthanum bromate	$La(BrO_3)_3$	98	120	149						
nitrate	$La(NO_3)_3$	100		136	200	168	247			
selenate	$La_2(SeO_3)_3$	50.5	45	45	45	45	18.5	5.4	2.2	0.68
sulfate	$La_2(SO_4)_3$	3.00	2.72	2.33	1.90	1.67	1.26	0.91	0.79	
Lead(II) acetate	$Pb(C_2H_3O_2)_2$	19.8	29.5	44.3	69.8	116				
bromide	$PbBr_2$	0.45	0.63	0.86	1.12	1.50	2.29	3.23	3.86	4.55
chloride	$PbCl_2$	0.67	0.82	1.00	1.20	1.42	1.94	2.54	2.88	3.20
fluorosilicate	$PbSiF_6$	190		222			403	428		463

TABLE 1.70 Solubility of Inorganic Compounds and Metal Salts of Organic Acids in Water at Various Temperatures (*Continued*)

Substance	Formula	0°	10°	20°	30°	40°	60°	80°	90°	100°
iodide	PbI₂	0.044	0.056	0.069	0.090	0.124	0.193	0.294		0.42
nitrate	Pb(NO₃)₂	37.5	46.2	54.3	63.4	72.1	91.6	111		133
Lithium acetate	LiC₂H₃O₂	31.2	35.1	40.8	50.6	68.6				
ammonium sulfate	LiNH₄SO₄		55.2		55.9	56.1	56.5			100
azide	LiN₃	61.3	64.2	67.2	71.2	75.4	86.6			
benzoate	LiC₇H₅O₂	38.9	41.6	44.7	53.8					
borate (meta-)	LiBO₂	0.90	1.3	2.7	5.7	10.9				
bromate	LiBrO₃	154	166	179	198	221	269	308	329	355
bromide	LiBr	143	147	160	183	211	223	245		266
carbonate	Li₂CO₃	1.54	1.43	1.33	1.26	1.17	1.01	0.85		0.72
chlorate	LiClO₃	241	283	372	488	604	777			
chloride	LiCl	69.2	74.5	83.5	86.2	89.8	98.4	112	121	128
chloroaurate(III)	LiAuCl₄	105	113	136	167	206	324	599		
cyanoplatinate(II)	Li₂Pt(CN)₄			141	153	160	178	216	239	
formate	LiCHO₂	32.3	35.7	39.3	44.1	49.5	64.7	92.7	116	138
hydrogen phosphite	Li₂HPO₃	9.97			7.61	7.11	6.03			4.43
hydroxide	LiOH	11.91	12.11	12.35	12.70	13.22	14.63	16.56		19.12
iodide	LiI	151	157	165	171	179	202	435	440	481
molybdate	Li₂MoO₄	82.6		79.5	79.4	78.0				73.9
nitrate	LiNO₃	53.4	60.8	70.1	138	152	175	233	272	324
nitrite	LiNO₂	70.9	82.5	96.8	114	133	177	128	151	
perchlorate	LiClO₄	42.7	49.0	56.1	63.6	72.3	92.3			
phosphate (meta-)	LiPO₃	0.101		0.058²⁵°		0.048				
selenite	Li₂SeO₃	25.0	23.3	21.5	19.6	17.9	14.7	11.9	11.1	9.9
sulfate	Li₂SO₄	36.1	35.5	34.8	34.2	33.7	32.6	31.4	30.9	
tartrate (d-)	Li₂C₄H₄O₆	42.0	31.8	27.1	26.6	27.2	29.5			
thiocyanate	LiSCN			114	131	153				
vanadate	Li₃VO₄	2.50		4.82	6.28	4.38	2.67			
Magnesium acetate	Mg(C₂H₃O₂)₂	56.7	59.7	53.4	68.6	75.7	118			125
bromide	MgBr₂	98	99	101	104	106	112		268	
chlorate	Mg(ClO₃)₂	114	123	135	155	178	242			
chloride	MgCl₂	52.9	53.6	54.6	55.8	57.5	61.0	66.1	69.5	73.3
fluorosilicate	MgSiF₆	26.3		30.8		34.9	44.4			
formate	Mg(CHO₂)₂	14.0	14.2	14.4	14.9	15.9	17.9	20.5	22.2	23.9
iodate	Mg(IO₃)₂		7.2	8.6	10.0	11.7	15.2	15.5	15.6	
iodide	MgI₂	120		140		173		186		

Magnesium nitrate	Mg(NO$_3$)$_2$	52.1	66.0	69.5	73.6	78.9	78.9	91.6	106	
selenate	MgSeO$_4$	20.0	30.4	38.3	44.3	48.6	55.8	55.8	52.9	
sulfate	MgSO$_4$	22.0	28.2	33.7	38.9	44.5	54.6			50.4
sulfite	MgSO$_3$	0.339	0.446	0.573	0.751	0.959	0.779	0.642	0.622	
tartrate	MgC$_4$H$_4$O$_6$	0.54	0.78	1.06		1.02				
Manganese bromide	MnBr$_2$	127	136	147	157	169	197	225	226	228
chloride	MnCl$_2$	63.4	68.1	73.9	80.8	88.5	109	113	114	115
fluoride	MnF$_2$			1.06		0.67	0.44			0.48
nitrate	Mn(NO$_3$)$_2$	102	118	139	206					
oxalate	MnC$_2$O$_4$	0.020	0.024	0.028	0.033					
sulfate	MnSO$_4$	52.9	59.7	62.9	62.9	60.0	53.6	45.6	40.9	35.3
Mercury(II) bromide	HgBr$_2$	0.30	0.40	0.56	0.66	0.91	1.68	2.77		4.9
(II) chloride	HgCl$_2$	3.63	4.82	6.57	8.34	10.2	16.3	30.0		61.3
(I) perchlorate	Hg$_2$(ClO$_4$)$_2$	282	325	367	407	455	499	541		580
Molybdenum trioxide	MoO$_3$			0.134	0.285	0.454	1.08	1.74		
Neodymium bromate	Nd(BrO$_3$)$_3$	43.9	59.2	75.6	95.2	116				
chloride	NdCl$_3$		96.7	98.0	99.6	102	105			
nitrate	Nd(NO$_3$)$_3$	127	133	142	145	159	211			
selenate	Nd$_2$(SeO$_3$)$_3$	46.2	44.6	41.8	39.9	39.9	43.9		3.3	
sulfate	Nd$_2$(SO$_4$)$_3$	13.0	9.7	7.1	5.3	4.1	2.8		1.2	
Nickel bromide	NiBr$_2$	113	122	131	138	144	153	154	155	155
chlorate	Ni(ClO$_3$)$_2$	111	120	133	155	181	221	308		
chloride	NiCl$_2$	53.4	56.3	60.8	70.6	73.2	81.2	86.6	86.6	87.6
fluoride	NiF$_2$		2.55	2.56					2.59	
iodate	Ni(IO$_3$)$_2$	0.74		1.09	1.15		1.06		1.00	
	Ni(IO$_3$)$_2$·4H$_2$O				1.43		1.06		1.00	
iodide	NiI$_2$	124	135	148	161	174	184	187	188	
nitrate	Ni(NO$_3$)$_2$	79.2		94.2	105	119	158	187	188	
perchlorate	Ni(ClO$_4$)$_2$	105	107	110	113	117				
Nickel sulfate	NiSO$_4$·6H$_2$O	(pale blue)(green)		40.1	43.6	47.6				
	NiSO$_4$·7H$_2$O	26.2	32.4	44.4	46.6	49.2	55.6	64.5	70.1	76.7
Osmium tetroxide	OsO$_4$	5.26	5.75	6.43	43.4	50.4				
Oxalic acid	H$_2$C$_2$O$_4$	3.54	6.08	9.52	14.23	21.52	44.32	84.5	120	
Potassium acetate	KC$_2$H$_3$O$_2$	216	233	256	283	324	350	381	398	
aluminum sulfate	KAl(SO$_4$)$_2$	3.00	3.99	5.90	8.39	11.7	24.8	71.0	109	
azide	KN$_3$	41.4	46.2	50.8	55.8	61.0				106
benzoate	KC$_7$H$_5$O$_2$		65.8	70.7	76.7	82.1				

(Continued)

TABLE 1.70 Solubility of Inorganic Compounds and Metal Salts of Organic Acids in Water at Various Temperatures (*Continued*)

Substance	Formula	0°	10°	20°	30°	40°	60°	80°	90°	100°
bromate	$KBrO_3$	3.09	4.72	6.91	9.64	13.1	22.7	34.1		49.9
bromide	KBr	53.6	59.5	65.3	70.7	75.4	85.5	94.9	99.2	104
cadmium bromide	$KCdBr_3$	116	133	150	170	191	233	276	298	325
cadmium chloride	$KCdCl_3$	26.6	32.3	38.9	45.6	53.1	67.5	83.5		101
carbonate	K_2CO_3	105	108	111	114	117	127	140	148	156
chlorate	$KClO_3$	3.3	5.2	7.3	10.1	13.9	23.8	37.6	46.0	56.3
chloride	KCl	28.0	31.2	34.2	37.2	40.1	45.8	51.3	53.9	56.3
chloroaurate(III)	$KAuCl_4$		38.3	61.8	94.9	145	405			
chloroplatinate(IV)	K_2PtCl_6	0.48	0.60	0.78	1.00	1.36	2.45	3.71		5.03
chromate	K_2CrO_4	56.3	60.0	63.7	66.7	67.8	70.1		74.5	
citrate	$K_3C_6H_5O_7$		153	172	194					
cobalt(II) sulfate	$K_2Co(SO_4)_2$	8.5	11.7	15.5	19.3	23.3	32.5	47.7		
copper(II) sulfate	$K_2Cu(SO_4)_2$	5.1	7.2	10.0	13.6	18.2				
cyanoplatinate(II)	$K_2Pt(CN)_4$	11.6	19.8	33.9	52.0	78.3	139	177	194	
dichromate	$K_2Cr_2O_7$	4.7	7.0	12.3	18.1	26.3	45.6	73.0		
dihydrogen phosphate	KH_2PO_4	14.8	18.3	22.6	28.0	33.5	50.2	70.4	83.5	
dithionate	$K_2S_2O_6$	2.6	4.2	6.6	9.3					
ferricyanide	$K_3Fe(CN)_6$	30.2	38	46	53	59.3	70			91
ferrocyanide	$K_4Fe(CN)_6$	14.3	21.1	28.2	35.1	41.4	54.8	66.9	71.5	74.2
fluoride	KF	44.7	53.5	94.9	108	138	142	150		
fluorogermanate(IV)	K_2GeF_6	0.25	0.36	0.50	0.66	0.96				
fluorosilicate	K_2SiF_6	0.077	0.102	0.151	0.202	0.253				
fluorotitanate(IV)	K_2TiF_6	0.55	0.91	1.28						
formate	$KCHO_2$		313	337	361	398	471	580	658	
hydrogen carbonate	$KHCO_3$	22.5	27.4	33.7	39.9	47.5	65.6	114		
Potassium hydrogen fluoride	KHF_2	24.5	30.1	39.2	46.8	56.5	78.8			
hydrogen selenite	$KH_3(SeO_3)_2$	115	162	215	300	408	900			
hydrogen sulfate	$KHSO_4$	36.2		48.6	54.3	61.0	76.4	96.1		122
hydrogen tartrate	$KC_4H_5O_6$	0.231	0.358	0.523	0.762					
hydroxide	KOH	95.7	103	112	126	134	154			178
iodate	KIO_3	4.60	6.27	8.08	10.3	12.6	18.3	24.8		32.3
iodide	KI	128	136	144	153	162	176	192	198	206
iron(II) sulfate	$K_2Fe(SO_4)_2$	19.6	24.5	32.1	39.1	44.9	57.2			
magnesium sulfate	$K_2Mg(SO_4)_2$	14.0	19.5	25.0	30.4	36.6	50.2	63.4		

Substance	Formula									
nickel sulfate	$K_2Ni(SO_4)_2$	3.37	4.50	5.94	7.72	9.85	15.4	23.0	27.8	33.4
nitrate	KNO_3	13.9	21.2	31.6	45.3	61.3	106	167	203	245
nitrite	KNO_2	279	292	306	320	329	348	376	390	410
oxalate	$K_2C_2O_4$	25.5	31.9	36.4	39.9	43.8	53.2	63.6	69.2	75.3
perchlorate	$KClO_4$	0.76	1.06	1.68	2.56	3.73	7.3	13.4	17.7	22.3
periodate	KIO_4	0.17	0.28	0.42	0.65	1.0	2.1	4.4	5.9	
permanganate	$KMnO_4$	2.83	4.31	6.34	9.03	12.6	22.1			
peroxodisulfate	$K_2S_2O_8$	1.65	2.67	4.70	7.75	11.0				
perrhenate	$KReO_4$	0.34	0.63	0.99	1.47	2.2	4.58	8.7		
phosphate	K_3PO_4		81.5	92.3	108	133				
salicylate	$KC_7H_5O_3$	21.2	32.4	47.1	61.3	78.6	116	156		
selenate	K_2SeO_4	107	109	111	113	115	119	121		122
selenite	K_2SeO_3	169	186	203	217	217	220			217
sulfate	K_2SO_4	7.4	9.3	11.1	13.0	14.8	18.2	21.4	22.9	24.1
sulfite	K_2SO_3	106		106	107	107	108			112
tellurate	K_2TeO_4	8.8		27.5	50.4					
thioantimonate(V)	K_3SbS_4	306	320	224	302	315	372	381		
thiocyanate	$KSCN$	177	198	155	255	289	238	492	571	675
thiosulfate	$K_2S_2O_3$	96	73.0	91.8	175	205		293	312	
zinc sulfate	$K_2Zn(SO_4)_2 \cdot 6H_2O$	13.0	18.9	25.9	35.0	44.9	72.1			
Praseodymium bromate	$Pr(BrO_3)_3$	55.9		112	114	144				
nitrate	$Pr(NO_3)_3$				162	178				
selenate	$Pr_2(SeO_3)_3$	36.2			32.4	31.2	30.4	5.43	3.6	
sulfate	$Pr_2(SO_4)_3$	19.8	15.6	12.6	9.89	2.56	5.04	3.5	1.1	0.91
Rubidium aluminum sulfate	$Rb_2Al_2(SO_4)_4$	0.72	1.05	1.50	2.20	3.25	7.40	21.6		
bromate	$RbBrO_3$				3.6	5.1				
bromide	$RbBr$	90	99	108	119	132	158			
chlorate	$RbClO_3$	2.1	3.4	5.4	8.0	11.6	22	38	49	63
chloride	$RbCl$	77	84	91	98	104	115	127	133	143
chloroaurate(III)	$RbAuCl_4$		4.8	9.9	15.5	21.5	36.2	54.6	65.8	79.2
chloroplatinate(IV)	Rb_2PtCl_6	0.014	0.020	0.028	0.040	0.056	0.090	0.182	0.247	0.33
chromate	Rb_2CrO_4	62.0	67.5	73.6	78.9	85.6	95.7			
cobalt sulfate	$Rb_2Co(SO_4)_2$	5.10	7.47	10.8	14.5	18.2	30.2	44.9	55.0	70.1
dichromate (mn)	$Rb_2Cr_2O_7$			5.9	10.0	15.2	32.3			
(tric)				5.8	9.5	14.8	32.4			
formate	$RbCHO_2$		443	554	614	694	900			
iron(III) sulfate	$RbFe(SO_4)_2 \cdot 12H_2O$		8.0	20	35	52				
nitrate	$RbNO_3$	19.5	33.0	52.9	81.2	117	200	310	374	452

(Continued)

TABLE 1.70 Solubility of Inorganic Compounds and Metal Salts of Organic Acids in Water at Various Temperatures (*Continued*)

Substance	Formula	0°	10°	20°	30°	40°	60°	80°	90°	100°
perchlorate	RbClO$_4$	1.09	1.19	1.55	2.20	3.26	6.27	11.0	15.5	22.0
salicylate	RbC$_7$H$_5$O$_3$		187	212	238	268	324			
sulfate	Rb$_2$SO$_4$	37.5	42.6	48.1	53.6	58.5	67.5	75.1	78.6	81.8
Samarium bromate	Sm(BrO$_3$)$_3$	34.2	47.6	62.5	79.0	98.5				
chloride	SmCl$_3$		92.4	93.4	94.6	96.9				
Selenic acid	H$_2$SeO$_4$	426		567	1328					
Selenious acid	H$_2$SeO$_3$	90.1	122.2	166.7	235.6	344.4	383.1	383.1	385.4	
Selenium dioxide	SeO$_2$		222	257	291	335	440			
Silver acetate	AgC$_2$H$_3$O$_2$	0.73	0.89	1.05	1.23	1.43	1.93	2.59		
bromate	AgBrO$_3$		0.11	0.16	0.23	0.32	0.57	0.94	1.33	
chlorate	AgClO$_3$		10.4	15.3	20.9	26.8				
fluoride	AgF	85.9	120	172	190	203				
nitrate	AgNO$_3$	122	167	216	265	311	440	585	652	733
nitrite	AgNO$_2$	0.16	0.22	0.34	0.51	0.73	1.39			
perchlorate	AgClO$_4$	455	484	525	594	635				793
sulfamate	AgNH$_2$SO$_3$	2.30	4.82	7.53	10.3	15.3	28.5			
sulfate	Ag$_2$SO$_4$	0.57	0.70	0.80	0.89	0.98	1.15	1.30	1.36	1.41
Sodium acetate	NaC$_2$H$_3$O$_2$	36.2	40.8	46.4	54.6	65.6	139	153	161	170
aluminum sulfate	Na$_2$Al$_2$(SO$_4$)$_4$	37.4	39.3	39.7	41.7	43.8				
azide	NaN$_3$	38.9	39.9	40.8						
benzoate	NaC$_7$H$_5$O$_2$	62.6	62.8	62.8	62.9	63.1	64.5	68.6	70.6	73.3
borate (penta-)	Na$_2$B$_{10}$O$_{16}$	6.4	8.6	12.0	16.4	22.0	37.9	63.4	83.5	108
borate (tetra-)	Na$_2$B$_4$O$_7$	1.11	1.60	2.56	3.86	6.67	19.0	31.4	41.0	52.5
bromate	NaBrO$_3$	24.2	30.3	36.4	42.6	48.8	62.6	75.7		90.8
bromide	NaBr	80.2	85.2	90.8	98.4	107	118	120	121	121
carbonate	Na$_2$CO$_3$	7.00	12.5	21.5	39.7	49.0	46.0	43.9	43.9	
chlorate	NaClO$_3$	79.6	87.6	95.9	105	115	137	167	184	204
chloride	NaCl	35.7	35.8	35.9	36.1	36.4	37.1	38.0	38.5	39.2
chloroaurate(III)	NaAuCl$_4$		139	151	178	227	900			
chloroiridate(IV)	Na$_2$IrCl$_6$		31.6	39.3	56.2	96.1	192	279		
chromate	Na$_2$CrO$_4$	31.7	50.1	84.0	88.0	96.0	115	125		126
cyanide	NaCN	40.8	48.1	58.7	71.2					
dichromate	Na$_2$Cr$_2$O$_7$	163	172	183	198	215	269	376	405	415
diethyl barbiturate	NaC$_8$H$_{11}$N$_2$O$_3$		12.7	21.5	24.7				48.0	
dihydrogen phosphate (ortho-)	NaH$_2$PO$_4$	56.5	69.8	86.9	107	133	172	211	234	

dihydrogen phosphate (pyro-)	Na$_2$H$_2$P$_2$O$_7$	4.47	6.95	12.0	17.1	18.4	36.1	49.3	56.3	64.7
dithionate	Na$_2$S$_2$O$_6$	6.3	11.1	15.1	19.6	24.7				
dodecanesulfonate	NaC$_{12}$H$_{25}$SO$_3$			0.13	0.25	6.54	22.7			27.0[98°]
EDTA (Y)*	Na$_2$H$_2$Y·2H$_2$O	10.6	11.1	12.8	14.2	17.0	22.2	24.3		
ferrocyanide	Na$_4$Fe(CN)$_6$	11.2	14.8	18.8	23.8	29.9	43.7	62.1		
fluoride	NaF	3.66	4.06	4.22	4.40	4.68	4.89			5.08
fluoroberyllate	Na$_2$BeF$_4$	1.33	1.44	1.92	2.24	2.25	2.62	2.73		
fluorogermanate	Na$_2$GeF	1.52	1.68	2.83	3.36					
fluorosilicate	Na$_2$SiF$_6$	4.35	5.7	7.2	8.6	10.3	14.3	18.7	21.5	24.5
formate	NaCHO$_2$	43.9	62.5	81.2	102	108	122	138	147	160
germanate	Na$_2$GeO$_3$	14.4	18.8	23.8	28.7	37.2	65.0	116	138	
hydrogen arsenate	Na$_2$HAsO$_4$	5.9	13.0	33.9	49.3	69.5	144	186	188	198
hydrogen carbonate	NaHCO$_3$	7.0	8.1	9.6	11.1	12.7	16.0			
hydrogen phosphate	Na$_2$HPO$_4$	1.68	3.53	7.83	22.0	55.3	82.8	92.3	102	104
hydrogen phosphite	Na$_2$HPO$_3$	418	424	429	566					
hydrogen succinate	NaC$_4$H$_5$O$_4$	17.5	25.3	34.8	47.7	61.6	74.5	90.1		
hydroxide	NaOH		98	109	119	129	174			
hydroxostannate(IV)	Na$_2$Sn(OH)$_6$	29.4	38.9	42.7						
hypochlorite	NaClO	46.0	53.4							
iodate	NaIO$_3$	2.48	4.59	8.08	10.7	13.3	19.8	26.6	29.5	33.0
iodide	NaI	159	167	178	191	205	257	295		302
molybdate	Na$_2$MoO$_4$	44.1	64.7	65.3	66.9	68.6	71.8			
nitrate	NaNO$_3$	73.0	80.8	87.6	94.9	102	122	148		180
nitrite	NaNO$_2$	71.2	75.1	80.8	87.6	94.9	111	133	148	160
oxalate	Na$_2$C$_2$O$_4$	2.69	3.05	3.41	3.81	4.18	4.93	5.71		6.50
perchlorate	NaClO$_4$	167	183	201	222	245	288	306		329
periodate	NaIO$_4$	1.83	5.6	10.3	19.9					
phosphate	Na$_3$PO$_4$	4.5	8.2	12.1	16.3	20.2	30.4			
potassium tartrate	NaKC$_4$H$_4$O$_6$	31.9	46.6	67.8					68.1	77.0
salicylate	NaC$_7$H$_5$O$_3$	44.7	95.3	111	117	130	144			
selenate	Na$_2$SeO$_4$	13.3	25.2	26.9	77.0	81.8	78.6	74.8	73.0	72.7
selenite	Na$_2$SeO$_3$	78.6	81.2	86.2	94.2	96.5	91.6	86.6	84.5	82.5
sulfate	Na$_2$SO$_4$	4.9	9.1	19.5	40.8	48.8	45.3	43.7	42.7	42.5
sulfate·7H$_2$O	Na$_2$SO$_4$·7H$_2$O	19.5	30.0	44.1						
sulfide	Na$_2$S	9.6	12.1	15.7	20.5	26.6	39.1	55.0	65.3	
sulfite	Na$_2$SO$_3$	14.4	19.5	26.3	35.5	37.2	32.6	29.4	27.9	
thioantimonate(V)	Na$_3$SbS$_4$	13.4	20.0	27.9	37.2	53.8			88.3	
thiocyanate	NaSCN	111	134	164	176	192	210	218		

(*Continued*)

343

TABLE 1.70 Solubility of Inorganic Compounds and Metal Salts of Organic Acids in Water at Various Temperatures (*Continued*)

Substance	Formula	0°	10°	20°	30°	40°	60°	80°	90°	100°
thiosulfate	$Na_2S_2O_3 \cdot 5H_2O$	50.2	59.7	70.1	83.2	104		90.8		97.2
tungstate	Na_2WO_4	71.5		73.0		77.6		40.8		
vanadate	$NaVO_3$			19.3	22.5	26.3	33.0			
Strontium acetate	$Sr(C_2H_3O_2)_2$	37.0	42.9	41.1	39.5	38.3	36.8	36.1	36.2	36.4
bromide	$SrBr_2$	85.2	93.4	102	112	123	150	182		223
chloride	$SrCl_2$	43.5	47.7	52.9	58.7	65.3	81.8	90.5		101
chromate	$SrCrO_4$		0.085	0.090				0.058		
Strontium fluoride	SrF_2	0.0113		0.0117	0.0119					
formate	$Sr(CHO_2)_2$	9.1	10.6	12.7	15.2	17.8	25.0	31.9	32.9	34.4
hydroxide	$Sr(OH)_2$	0.91	1.25	1.77	2.64	3.95	8.42	20.2	44.5	91.2
iodide	SrI_2	165		178		192	218	270	365	383
nitrate	$Sr(NO_3)_2$	39.5	52.9	69.5	88.7	89.4	93.4	96.9	98.4	
nitrite	$Sr(NO_2)_2$			65	72	79	97	130	134	
oxide	SrO				1.03	1.05	3.40	9.15	13.13	12.15
sulfate	$SrSO_4$	0.0113	0.0129	0.0132	0.0138	0.0141	0.0131	0.0116	0.0115	
Sulfamic acid	H_2NSO_3H	14.7	18.6	21.3	26.1	29.5	37.1	47.1		
Telluric acid	H_2TeO_4	16.2	33.8	41.6	50.0	57.2	77.5	106		155
Terbium bromate	$Tb(BrO_3)_3 \cdot 9H_2O$	66.4	89.7	117	152	198				
Thallium(I) azide	TlN_3	0.171	0.236	0.364						
bromide	$TlBr$	0.022	0.032	0.048	0.068	0.097	0.177			
carbonate	Tl_2CO_3			5.3		12.7[50°]	12.2	36.6		27.2
chlorate	$TlClO_3$	2.00		3.92						57.3
chloride	$TlCl$	0.21	0.25	0.33	0.42	0.52	0.80	1.20		1.80
hydroxide	$TlOH$	25.4	29.6	35.0	40.4	49.4	73.3	106	126	150
iodide	TlI	0.002		0.006		0.015	0.035	0.070		0.120
nitrate	$TlNO_3$	3.90	6.22	9.55	14.3	21.0	46.1	110	200	414
nitrite	$TlNO_2$	17.9	28.9	40.3	53.2	83.6	216	1150	750	
perchlorate	$TlClO_4$	6.00	8.04	13.1	19.7	28.3	50.8	81.5		
picrate	$TlOC_6H_2(NO_2)_3$	0.135		0.40	0.57	0.83	1.73			
selenate	Tl_2SeO_4		2.17	2.80				8.50		10.8
sulfate	Tl_2SO_4	2.73	3.70	4.87	6.16	7.53	11.0	14.6	16.5	18.4
Thorium nitrate	$Th(NO_3)_4 \cdot 4H_2O$	186	187	191						
sulfate	$Th(SO_4)_2 \cdot 4H_2O$	0.74	0.99	1.38	1.99	4.04				
	$Th(SO_4)_2 \cdot 9H_2O$					3.00	1.63			
Tin(II) iodide	SnI_2			0.99	1.17	1.42	2.11	3.04	3.58	4.20
Uranium(IV) sulfate	$U(SO_4)_2 \cdot 4H_2O$				10.1	9.0	7.7			
	$U(SO_4)_2 \cdot 8H_2O$			11.9	17.9	29.2	55.8			

Uranyl nitrate	UO$_2$(NO$_3$)$_2$	98	107	122	141	167	317	388	426	474
oxalate	UO$_2$C$_2$O$_4$		0.45	0.50	0.61	0.80	1.22	1.94		3.16
Ytterbium sulfate	Yb$_2$(SO$_4$)$_3$	44.2	37.5		22.2	17.2	10.4	6.4	5.8	4.7
Yttrium bromide	YBr$_3$	63.9	78.1	75.1	79.6	87.3	101	116	123	
chloride	YCl$_3$	77.3		78.8		30.8				
nitrate	Y(NO$_3$)$_3$	93.1	106	123	143	163	200			
sulfate	Y$_2$(SO$_4$)$_3$	8.05	7.67	7.30	6.78	6.09	4.44	2.89	2.2	
Zinc bromide	ZnBr$_2$	389		446	528	591	618	645		672
chlorate	Zn(ClO$_3$)$_2$	145	152	200	209	223				
chloride	ZnCl$_2$	342	363	395	437	452	488	541		614
formate	Zn(CHO$_2$)$_2$	3.70	4.30	5.20	6.10	7.40	11.8	21.2	28.8	38.0
iodide	ZnI$_2$	430		432		445	467	490		510
nitrate	Zn(NO$_3$)$_2$	98			138	211				
sulfate (rh)	ZnSO$_4$	41.6	47.2	53.8	61.3	70.5	75.4	71.1		60.5
sulfate (mn)			54.4	60.0	65.5					
tartrate	ZnC$_4$H$_4$O$_6$			0.022	0.041	0.060	0.104	0.059		

*Properly called dihydrogen ethylenediaminetetraacetate (Na$_2$H$_2$ EDTA · 2H$_2$O).

TABLE 1.71   Dissociation Constants of Inorganic Acids

The *dissociation constant* of an acid K_a may conveniently be expressed in terms of the pK_a value where p$K_a = -\log_{10}(K_a/\text{mol dm}^{-3})$. The values given in the following table are for aqueous solutions at 298 K: the pK_1, pK_2, and pK_3 values refer to the first, second, and third ionizations respectively.

Name	Formula	pK_a
Aluminium ion (hydrated)	$[Al(H_2O)_6]^{3+}$	4.9 (pK_1)
Ammonium ion	NH_4^+	9.25
Arsenic(III) acid	H_3AsO_3	9.22 (pK_1)
Arsenic(V) acid	H_3AsO_4	2.30 (pK_1)
Boric acid	H_3BO_3	9.24 (pK_1)
Bromic(1) acid	HOBr	8.70
Carbonic acid	H_2CO_3	$\left\{\begin{array}{l} 6.38^a\ (\text{p}K_1) \\ 10.32\ (\text{p}K_2) \end{array}\right.$
Chloric(I) acid	HOCl	7.43
Chloric(III) acid	$HClO_2$	2.0
Chromium(III) ion (hydrated)	$[Cr(H_2O)_6]^{3+}$	3.9 (pK_1)
Hydrazinium ion	$N_2H_5^+$	7.93
Hydrocyanic acid	HCN	9.40
Hydrofluoric acid	HF	3.25
Hydrogen peroxide	H_2O_2	11.62 (pK_1)
Hydrogen sulphide	H_2S	$\left\{\begin{array}{l} 7.05\ (\text{p}K_1) \\ 12.92\ (\text{p}K_2) \end{array}\right.$
Hydroxyammonium ion	NH_3OH^+	5.82
Iodic(I) acid	HOI	10.52
Iodic(V) acid	HIO_3	0.8
Iron(III) ion (hydrated)	$[Fe(H_2O)_6]^{3+}$	2.22 (pK_1)
Lead(II) ion (hydrated)	$[Pb(H_2O)_n]^{2+}$	7.8 (pK_1)
Nitrous acid	HNO_2	3.34
Phosphinic acid	H_3PO_2	2.0
Phosphoric(V) acid	H_3PO_4	$\left\{\begin{array}{l} 2.15\ (\text{p}K_1) \\ 7.21\ (\text{p}K_2) \\ 12.36\ (\text{p}K_3) \end{array}\right.$
Phosphonic acid	H_3PO_3	$\left\{\begin{array}{l} 2.00\ (\text{p}K_1) \\ 6.58\ (\text{p}K_2) \end{array}\right.$
Silicic acid	H_2SiO_3	$\left\{\begin{array}{l} 9.9\ (\text{p}K_1) \\ 11.9\ (\text{p}K_2) \end{array}\right.$
Sulphuric acid	H_2SO_4	1.92 (pK_2)
Sulphurous acid	H_2SO_3	$\left\{\begin{array}{l} 1.92\ (\text{p}K_1) \\ 7.21\ (\text{p}K_2) \end{array}\right.$

[a]Some of the unionized acid exists as dissolved CO_2 molecules rather than H_2CO_3: pK_1 for the molecular species H_2CO_3 is approximately 3.7.

TABLE 1.72   Ionic Product Constant of Water

This table gives values of pKw on a modal scale, where Kw is the ionic activity product constant of water. Values are from W. L. Marshall and E. U. Franck, *J. Phys. Chem. Ref. Data*, **10**:295 (1981).

Temp., °C	pKw	Temp., °C	pKw	Temp., °C	pKw
0	14.938	45	13.405	95	12.345
5	14.727	50	13.275	100	12.264
10	14.528	55	13.152	125	11.911
15	14.340	60	13.034	150	11.637
18	14.233	65	12.921	175	11.431
20	14.163	70	12.814	200	11.288
25	13.995	75	12.711	225	11.207
30	13.836	80	12.613	250	11.192
35	13.685	85	12.520	275	11.251
40	13.542	90	12.431	300	11.406

TABLE 1.73   Solubility Product Constants

The data refer to various temperatures between 18 and 25°C, and were complied from values cited by Bjerrum, Schwarzenbach, and Sillen, *Stability Constants of Metal Complexes*, Part II, Chemical Society, London, 1958, and values taken from publications of the IUPAC Solubility Data Project: *Solubility Data Series*, international Union of Pure and Applied Chemistry, Pergamon Press, Oxford, 1979–1992; H. L. Clever, and F. J. Johnston, *J. Phys Chem. Ref. Data*, **9**:751 (1980); Y. Marcus, *Ibid.* **9**:1307 (1980); H. L. Clever, S. A. Johnson, and M. E. Derrick, *Ibid.* **14**:631 (1985), and **21**:941 (1992).

In the table, "L" is the abbreviation of the organic ligand.

Compound	Formula	pK_{sp}	K_{sp}
Actinium			
hydroxide	$Ac(OH)_3$	15	1×10^{-15}
Aluminum			
arsonate	$AlAsO_4$	15.80	1.6×10^{-16}
cupferrate	AlL_3	18.64	2.3×10^{-19}
hydroxide	$Al(OH)_3$	32.89	1.3×10^{-33}
phosphate	$AlPO_4$	20.01	9.84×10^{-21}
8-quinolinolate	$AlL3$	29.00	1.00×10^{-29}
selenide	Al_2Se_3	24.4	4×10^{-25}
sulfide	Al_2S_3	6.7	2×10^{-7}
Americium			
(III) hydroxide	$Am(OH)_3$	19.57	2.7×10^{-20}
(IV) hydroxide	$Am(OH)_4$	56	1×10^{-56}
Ammonium			
uranyl arsenate	$NH_4UO_2AsO_4$	23.77	1.7×10^{-24}
Arsenic			
(III) sulfide	As_2S_3	21.68	2.1×10^{-22}

(Continued)

TABLE 1.73 Solubility Product Constants (*Continued*)

Compound	Formula	pK_{sp}	K_{sp}
Barium			
arsenate	$Ba_3(AsO_4)_2$	50.11	8.0×10^{-51}
bromate	$Ba(BrO_3)_2$	5.50	2.43×10^{-4}
carbonate	$BaCO_3$	8.59	2.58×10^{-9}
chromate	$BaCrO_4$	9.93	1.17×10^{-10}
ferricyanide 6-hydrate	$Ba_2[Fe(CN)_6] \cdot 6H_2O$	7.49	3.2×10^{-8}
fluoride	BaF_2	6.74	1.84×10^{-7}
hexafluorosilicate	$BaSiF_6$	6	1×10^{-6}
hydrogen phosphate	$BaHPO_4$	6.49	3.2×10^{-7}
hydroxide 8-hydrate	$Ba(OH)_2 \cdot 8H_2O$	3.59	2.55×10^{-4}
iodate hydrate	$Ba(IO_3)_2 \cdot H_2O$	8.40	4.01×10^{-9}
molybdate	$BaMoO_4$	7.45	3.54×10^{-8}
niobate	$Ba(NbO_3)_2$	16.50	3.2×10^{-17}
nitrate	$Ba(NO_3)_2$	2.33	4.64×10^{-3}
oxalate	BaC_2O_4	6.79	1.6×10^{-7}
oxalate hydrate	$BaC_2O_4 \cdot H_2O$	7.64	2.3×10^{-8}
permanganate	$Ba(MnO_4)_2$	9.61	2.5×10^{-10}
perrhenate	$Ba(ReO_4)_2$	1.28	5.2×10^{-2}
phosphate	$Ba_3(PO_4)_2$	22.47	3.4×10^{-23}
pyrophosphate	$Ba_2P_2O_7$	10.50	3.2×10^{-11}
8-quinolinolate	BaL_2	8.30	5.0×10^{-9}
selenate	$BaSeO_4$	7.47	3.40×10^{-8}
sulfate	$BaSO_4$	9.97	1.08×10^{-10}
sulfite	$BaSO_3$	9.30	5.0×10^{-10}
thiosulfate	BaS_2O_3	4.79	1.6×10^{-5}
Beryllium			
carbonate 4-hydrate	$BeCO_3 \cdot 4H_2O$	3	1×10^{-3}
hydroxide (amorphous)	$Be(OH)_2$	21.16	6.92×10^{-22}
molybdate	$BeMoO_4$	1.49	3.2×10^{-2}
niobate	$Be(NbO_3)_2$	15.92	1.2×10^{-16}
Bismuth			
arsenate	$BiAsO_4$	9.35	4.43×10^{-10}
cupferrate	BiL_3	27.22	6.0×10^{-28}
hydroxide	$Bi(OH)_3$	30.4	6.0×10^{-31}
iodide	BiI_3	18.11	7.71×10^{-19}
oxide bromide	$BiOBr$	6.52	3.0×10^{-7}
oxide chloride	$BiOCl$	30.75	1.8×10^{-31}
oxide hydroxide	$BiO(OH)$	9.4	4×10^{-10}
oxide nitrate	$BiO(NO_3)$	2.55	2.82×10^{-3}
oxide nitrite	$BiO(NO_2)$	6.31	4.9×10^{-7}
oxide thiocyanate	$BiO(SCN)$	6.80	1.6×10^{-7}
phosphate	$BiPO_4$	22.89	1.3×10^{-23}
sulfide	Bi_2S_3	97	1×10^{-97}
Cadmium			
anthranilate	CdL_2	8.27	5.4×10^{-9}
arsenate	$Cd_3(AsO_4)_2$	32.66	2.2×10^{-33}
benzoate 2-hydrate	$CdL_2 \cdot 2H_2O$	2.7	2×10^{-3}
borate, *meta*	$Cd(BO_2)_2$	8.64	2.3×10^{-9}
carbonate	$CdCO_3$	12.0	1.0×10^{-12}
cyanide	$Cd(CN)_2$	8.0	1.0×10^{-8}
ferrocyanide	$Cd_2[Fe(CN)_6]$	16.49	3.2×10^{-17}
fluoride	CdF_2	2.19	6.44×10^{-3}

TABLE 1.73 Solubility Product Constants (*Continued*)

Compound	Formula	pK_{sp}	K_{sp}
hydroxide	$Cd(OH)_2$ fresh	14.14	7.2×10^{-15}
iodate	$Cd(IO_3)_2$	7.60	2.5×10^{-8}
oxalate 3-water	$CdC_2O_4 \cdot 3H_2O$	7.85	1.42×10^{-8}
phosphate	$Cd_3(PO_4)_2$	32.60	2.53×10^{-33}
quinaldate	CdL_2	12.30	5.0×10^{-13}
sulfide	CdS	26.10	8.0×10^{-27}
tungstate	$CdWO_4$	5.7	2×10^{-6}
Calcium			
acetate 3-water	$Ca(OAc)_2 \cdot 3H_2O$	2.4	4×10^{-3}
arsenate	$Ca_3(AsO_4)_2$	18.17	6.8×10^{-19}
benzoate 3-water	$CaL_2 \cdot 3H_2O$	2.4	4×10^{-3}
carbonate	$CaCO_3$	8.54	2.8×10^{-9}
carbonate (calcite)	$CaCO_3$	8.47	3.36×10^{-9}
carbonate (aragonite)	$CaCO_3$	8.22	6.0×10^{-9}
carbonatomagnesium	$Ca[Mg(CO_3)_2]$ dolomite	11	1×10^{-11}
chromate	$CaCrO_4$	3.15	7.1×10^{-4}
fluoride	CaF_2	8.28	5.3×10^{-9}
hexafluorosilicate	$Ca[SiF_6]$	3.09	8.1×10^{-4}
hydrogen phosphate	$CaHPO_4$	7.0	1.0×10^{-7}
hydroxide	$Ca(OH)_2$	5.26	5.5×10^{-6}
iodate 6-water	$Ca(IO_3)_2 \cdot 6H_2O$	6.15	7.10×10^{-7}
molybdate	$CaMoO_4$	7.84	1.46×10^{-8}
niobate	$Ca(NbO_3)_2$	17.06	8.7×10^{-18}
oxalate hydrate	$CaC_2O_4 \cdot H_2O$	8.63	2.32×10^{-9}
phosphate	$Ca_3(PO_4)_2$	28.68	2.07×10^{-29}
8-quinolinolate	CaL_2	11.12	7.6×10^{-12}
selenate	$CaSeO_4$	3.09	8.1×10^{-4}
selenite	$CaSeO_3$	5.53	8.0×10^{-6}
silicate, *meta*	$CaSiO_3$	7.60	2.5×10^{-8}
sulfate	$CaSO_4$	4.31	4.93×10^{-5}
sulfate dihydrate	$CaSO_4 \cdot 2H_2O$	4.50	3.14×10^{-5}
sulfite	$CaSO_3$	7.17	6.8×10^{-8}
sulfite 0.5-water	$CaSO_3 \cdot 0.5H_2O$	6.51	3.1×10^{-7}
tartrate dihydrate	$CaL \cdot 2H_2O$	6.11	7.7×10^{-7}
tungstate	$CaWO_4$	8.06	8.7×10^{-9}
Cerium			
(III) fluoride	CeF_3	15.1	8×10^{-16}
(III) hydroxide	$Ce(OH)_3$	19.80	1.6×10^{-20}
(IV) hydroxide	$Ce(OH)_4$	47.7	2×10^{-48}
(III) iodate	$Ce(IO_3)_3$	9.50	3.2×10^{-10}
(IV) iodate	$Ce(IO_3)_4$	16.3	5×10^{-17}
(III) oxalate 9-water	$Ce_2(C_2O_4)_3 \cdot 9H_2O$	25.50	3.2×10^{-26}
(III) phosphate	$CePO_4$	23	1×10^{-23}
(III) selenite	$Ce_2(SeO_3)_3$	24.43	3.7×10^{-25}
(III) sulfide	Ce_2S_3	10.22	6.0×10^{-11}
(III) tartrate	Ce_2L_3	19.0	1.0×10^{-19}
Cesium			
bromate	$CsBrO_3$	1.7	5×10^{-2}
chlorate	$CsClO_3$	1.4	4×10^{-2}
cobaltihexanitrite	$Cs_3[Co(NO_2)_6]$	15.24	5.7×10^{-16}
hexachloroplatinate (IV)	$Cs_2[PtCl_6]$	7.50	3.2×10^{-8}
hexafluoroplatinate(IV)	$Cs_2[PtF_6]$	5.62	2.4×10^{-6}
hexafluorosilicate	$Cs_2[SiF_6]$	4.90	1.3×10^{-5}

(*Continued*)

TABLE 1.73   Solubility Product Constants (*Continued*)

Compound	Formula	pK_{sp}	K_{sp}
perchlorate	$CsClO_4$	2.40	3.95×10^{-3}
periodate	$CsIO_4$	5.29	5.16×10^{-6}
permanganate	$CsMnO_4$	4.08	8.2×10^{-5}
perrhanate	$CsReO_4$	3.40	4.0×10^{-4}
tetrafluoroborate	$CS[BF_4]$	4.7	5×10^{-5}
Chromium(II)			
hydroxide	$Cr(OH)_2$	15.7	2×10^{-16}
Chromium(III)			
arsenate	$CrAsO_4$	20.11	7.7×10^{-21}
fluoride	CrF_3	10.18	6.6×10^{-11}
hydroxide	$Cr(OH)_3$	30.20	6.3×10^{-31}
phosphate 4-water	$CrPO_4 \cdot 4H_2O$ green	22.62	2.4×10^{-23}
	violet	17.00	1.0×10^{-17}
Cobalt			
anthranilate	CoL_2	9.68	2.1×10^{-10}
arsenate	$Co_3(AsO_4)_2$	28.17	6.80×10^{-29}
carbonate	$CoCO_3$	12.84	1.4×10^{-13}
ferrocyanide	$Co_2[Fe(CN)_6]$	14.74	1.8×10^{-15}
hydrogen phosphate	$CoHPO_4$	6.7	2×10^{-7}
(II) hydroxide	$Co(OH)_2$ fresh	14.23	5.92×10^{-15}
(III) hydroxide	$Co(OH)_3$	43.80	1.6×10^{-44}
iodate	$Co(IO_3)_2$	4.0	1.0×10^{-4}
phosphate	$Co_3(PO_4)_2$	34.69	2.05×10^{-35}
selenite	$CoSeO_3$	6.80	1.6×10^{-7}
quinaldate	CoL_2	10.80	1.6×10^{-11}
8-quinolinolate	CoL_2	24.80	1.6×10^{-25}
sulfide	α-CoS	20.40	4.0×10^{-21}
	β-CoS	24.70	2.0×10^{-25}
Copper(I)			
azide	CuN_3	8.31	4.9×10^{-9}
bromide	$CuBr$	8.20	6.27×10^{-9}
chloride	$CuCl$	6.76	1.72×10^{-7}
cyanide	$CuCN$	19.46	3.47×10^{-20}
hydroxide	$CuOH$	14	1×10^{-14}
iodide	CuI	11.90	1.27×10^{-12}
sulfide	Cu_2S	47.60	2.5×10^{-48}
tetraphenylborate	CuL	8.0	1.0×10^{-8}
thiocyanate	$CuSCN$	12.75	1.77×10^{-13}
Copper(II)			
anthranilate	CuL_2	13.22	6.0×10^{-14}
arsenate	$Cu_3(AsO_4)_2$	35.10	7.95×10^{-36}
azide	$Cu(N_3)_2$	9.20	6.3×10^{-10}
carbonate	$CuCO_3$	9.86	1.4×10^{-10}
chromate	$CuCrO_4$	5.44	3.6×10^{-6}
dithiooxamide	CuL	15.12	7.67×10^{-16}
ferrocyanide	$Cu_2[Fe(CN)_6]$	15.89	1.3×10^{-16}
hydroxide	$Cu(OH)_2$	19.66	2.2×10^{-20}
iodate	$Cu(IO_3)_2$	7.16	6.94×10^{-8}
oxalate	CuC_2O_4	9.35	4.43×10^{-10}
phosphate	$Cu_3(PO_4)_2$	36.85	1.40×10^{-37}
pyrophosphate	$Cu_2P_2O_7$	15.08	8.3×10^{-16}
quinaldate	CuL_2	16.80	1.6×10^{-17}
8-quinolinolate	CuL_2	29.70	2.0×10^{-30}

TABLE 1.73 Solubility Product Constants (*Continued*)

Compound	Formula	pK_{sp}	K_{sp}
selenite	$CuSeO_3$	7.68	2.1×10^{-8}
sulfide	CuS	35.20	6.3×10^{-36}
Dysprosium			
chromate 10-water	$Dy_2(CrO_4)_3 \cdot 10H_2O$	8	1×10^{-8}
hydroxide	$Dy(OH)_3$	21.85	1.4×10^{-22}
Erbium			
hydroxide	$Er(OH)_3$	23.39	4.1×10^{-24}
Europium			
hydroxide	$Eu(OH)_3$	23.03	9.38×10^{-24}
Gadolinium			
hydrogen carbonate	$Gd(HCO_3)_3$	1.7	2×10^{-2}
hydroxide	$Gd(OH)_3$	22.74	1.8×10^{-23}
Gallium			
ferrocyanide	$Ga_4[Fe(CN)_6]_3$	33.82	1.5×10^{-34}
hydroxide	$Ga(OH)_3$	35.14	7.28×10^{-36}
8-quinolinolate	GaL_3	40.80	1.6×10^{-41}
Germanium			
oxide	GeO_2	57.0	1.0×10^{-57}
Gold(I)			
chloride	$AuCl$	12.70	2.0×10^{-13}
iodide	AuI	22.80	1.6×10^{-23}
Gold(III)			
chloride	$AuCl_3$	24.50	3.2×10^{-25}
hydroxide	$Au(OH)_3$	45.26	5.5×10^{-46}
iodide	AuI_3	46	1×10^{-46}
oxalate	$Au_2(C_2O_4)_3$	10	1×10^{-10}
Hafnium			
hydroxide	$Hf(OH)_3$	25.40	4.0×10^{-26}
Holmium			
hydroxide	$Ho(OH)_3$	22.3	5.0×10^{-23}
Indium			
ferrocyanide	$In_4[Fe(CN)_6]_3$	43.72	1.9×10^{-44}
hydroxide	$In(OH)_3$	33.2	6.3×10^{-34}
quinolinolate	InL_3	31.34	4.6×10^{-32}
selenite	$In_2(SeO_3)_3$	32.60	4.0×10^{-33}
sulfide	In_2S_3	73.24	5.7×10^{-74}
Iron(II)			
carbonate	$FeCO_3$	10.50	3.13×10^{-11}
fluoride	FeF_2	5.63	2.36×10^{-6}
hydroxide	$Fe(OH)_2$	16.31	4.87×10^{-17}
oxalate dihydrate	$FeC_2O_4 \cdot 2H_2O$	6.50	3.2×10^{-7}
sulfide	FeS	17.20	6.3×10^{-18}
Iron(III)			
arsenate	$FeAsO_4$	20.24	5.7×10^{-21}
ferrocyanide	$Fe_4[Fe(CN)_6]_3$	40.52	3.3×10^{-41}
hydroxide	$Fe(OH)_3$	38.55	2.79×10^{-39}
phosphate dihydrate	$FePO_4 \cdot 2H_2O$	15.00	9.91×10^{-16}
quinaldate	FeL_3	16.89	1.3×10^{-17}
selenite	$Fe_2(SeO_3)_3$	30.70	2.0×10^{-31}
Lanthanum			
bromate 9-water	$La(BrO_3)_3 \cdot 9H_2O$	2.50	3.2×10^{-3}
fluoride	LaF_3	16.2	7×10^{-17}

(*Continued*)

TABLE 1.73   Solubility Product Constants (*Continued*)

Compound	Formula	pK_{sp}	K_{sp}
hydroxide	La(OH)$_3$	18.70	2.0×10^{-19}
iodate	La(IO$_3$)$_3$	11.12	7.50×10^{-12}
molybdate	La$_2$(MoO$_4$)$_3$	20.4	4×10^{-21}
oxalate 9-water	La$_2$(C$_2$O$_4$)$_3$	26.60	2.5×10^{-27}
phosphate	LaPO$_4$	22.43	3.7×10^{-23}
sulfide	La$_2$S$_3$	12.70	2.0×10^{-13}
tungstate trihydrate	La$_2$(WO$_4$)$_3 \cdot$ 3H$_2$O	3.90	1.3×10^{-4}
Lead			
acetate	Pb(OAc)$_2$	2.75	1.8×10^{-3}
anthranilate	PbL$_2$	9.81	1.6×10^{-10}
arsenate	Pb$_3$(AsO$_4$)$_3$	35.39	4.0×10^{-36}
azide	Pb(N$_3$)$_2$	8.59	2.5×10^{-9}
borate, *meta*	Pb(BO$_2$)$_3$	10.78	1.6×10^{-11}
bromate	Pb(BrO$_3$)$_2$	1.70	2.0×10^{-2}
bromide	PbBr$_2$	6.82	6.60×10^{-6}
carbonate	PbCO$_3$	13.13	7.4×10^{-14}
chloride	PbCl$_2$	4.77	1.70×10^{-5}
chloride fluoride	PbClF	8.62	2.4×10^{-9}
chlorite	Pb(ClO$_2$)$_2$	8.4	4×10^{-9}
chromate	PbCrO$_4$	12.55	2.8×10^{-13}
ferrocyanide	Pb$_2$[Fe(CN)$_6$]	14.46	3.5×10^{-15}
fluoride	PbF$_2$	7.48	3.3×10^{-8}
fluoride iodide	PbFI	8.07	8.5×10^{-9}
hydrogen phosphate	PbHPO$_4$	9.90	1.3×10^{-10}
hydrogen phosphite	PbHPO$_3$	6.24	5.8×10^{-7}
hydroxide	Pb(OH)$_2$	14.84	1.43×10^{-15}
hydroxide bromide	PbOHBr	14.70	2.0×10^{-15}
hydroxide chloride	PbOHCl	13.7	2×10^{-14}
hydroxide nitrate	PbOHNO$_3$	3.55	2.8×10^{-4}
iodate	Pb(IO$_3$)$_2$	12.43	3.69×10^{-13}
iodide	PbI$_2$	8.01	9.8×10^{-9}
molybdate	PbMoO$_4$	13.00	1.0×10^{-13}
niobate	Pb(NbO$_3$)$_2$	16.62	2.4×10^{-17}
oxalate	PbC$_2$O$_4$	9.32	4.8×10^{-10}
phosphate	Pb$_3$(PO$_4$)$_2$	42.10	8.0×10^{-43}
quinaldate	PbL$_2$	10.60	2.5×10^{-11}
selenate	PbSeO$_4$	6.84	1.37×10^{-7}
selenite	PbSeO$_3$	11.50	3.2×10^{-12}
sulfate	PbSO$_4$	7.60	2.53×10^{-8}
sulfide	PbS	27.10	8.0×10^{-28}
thiocyanate	Pb(SCN)$_2$	4.70	2.0×10^{-5}
thiosulfate	PbS$_2$O$_3$	6.40	4.0×10^{-7}
tungstate	PbWO$_4$	6.35	4.5×10^{-7}
Lead(IV)			
hydroxide	Pb(OH)$_4$	65.50	3.2×10^{-66}
Lithium			
carbonate	Li$_2$CO$_3$	1.60	2.5×10^{-2}
fluoride	LiF	2.74	1.84×10^{-3}
phosphate	Li$_3$PO$_4$	10.63	2.37×10^{-11}
uranylarsenate	LiUO$_2$AsO$_4$	18.82	1.5×10^{-19}
Lutetium			
hydroxide	Lu(OH)$_3$	23.72	1.9×10^{-24}

TABLE 1.73 Solubility Product Constants (*Continued*)

Compound	Formula	pK_{sp}	K_{sp}
Magnesium			
ammonium phosphate	$MgNH_4PO_4$	12.60	2.5×10^{-13}
arsenate	$Mg_3(AsO_4)_2$	19.68	2.1×10^{-20}
carbonate	$MgCO_3$	5.17	6.82×10^{-6}
carbonate trihydrate	$MgCO_3 \cdot 3H_2O$	5.62	2.38×10^{-6}
fluoride	MgF_2	10.29	5.16×10^{-11}
hydroxide	$Mg(OH)_2$	11.25	5.61×10^{-12}
iodate 4-water	$Mg(IO_3)_2 \cdot 4H_2O$	2.50	3.2×10^{-3}
niobate	$Mg(NbO_3)_2$	16.64	2.3×10^{-17}
oxalate dihydrate	$MgC_2O_4 \cdot 2H_2O$	5.32	4.83×10^{-6}
phosphate	$Mg_3(PO_4)_2$	23.98	1.04×10^{-24}
8-quinolinolate	MgL_2	15.40	4.0×10^{-16}
selenite	$MgSeO_3$	4.89	1.3×10^{-5}
sulfite	$MgSO_3$	2.50	3.2×10^{-3}
Manganese			
anthranilate	MnL_2	6.75	1.8×10^{-3}
arsenate	$Mn_3(AsO_4)_2$	28.72	1.9×10^{-29}
carbonate	$MnCO_3$	10.63	2.34×10^{-11}
ferrocyanide	$Mn_2[Fe(CN)_6]$	12.10	8.0×10^{-13}
iodate	$Mn(IO_3)_2$	6.36	4.37×10^{-7}
hydroxide	$Mn(OH)_2$	12.72	1.9×10^{-13}
oxalate dihydrate	$MnC_2O_4 \cdot 2H_2O$	6.77	1.70×10^{-7}
8-quinolinolate	MnL_2	21.70	2.0×10^{-22}
selenite	$MnSeO_3$	6.90	1.3×10^{-7}
sulfide	MnS amorphous	9.60	2.5×10^{-10}
	MnS crystalline	12.60	2.5×10^{-13}
Mercury(I)			
azide	$Hg_2(N_3)_2$	9.15	7.1×10^{-10}
bromide	Hg_2Br_2	22.19	6.40×10^{-23}
carbonate	Hg_2CO_3	16.44	3.6×10^{-17}
chloride	Hg_2Cl_2	17.84	1.43×10^{-18}
cyanide	$Hg_2(CN)_2$	39.3	5×10^{-40}
chromate	Hg_2CrO_4	8.70	2.0×10^{-9}
ferricyanide	$(Hg_2)_3[Fe(CN)_6]_2$	20.07	8.5×10^{-21}
fluoride	Hg_2F_2	5.51	3.10×10^{-6}
hydrogen phosphate	Hg_2HPO_4	12.40	4.0×10^{-13}
hydroxide	$Hg_2(OH)_2$	23.70	2.0×10^{-24}
iodate	$Hg_2(IO_3)_2$	13.71	2.0×10^{-14}
iodide	Hg_2I_2	28.72	5.2×10^{-29}
oxalate	$Hg_2C_2O_4$	12.76	1.75×10^{-13}
quinaldate	Hg_2L_2	17.90	1.3×10^{-18}
selenite	Hg_2SeO_3	14.20	8.4×10^{-15}
sulfate	Hg_2SO_4	6.19	6.5×10^{-7}
sulfite	Hg_2SO_3	27.0	1.0×10^{-27}
sulfide	Hg_2S	47.0	1.0×10^{-47}
thiocyanate	$Hg_2(SCN)_2$	19.49	3.2×10^{-20}
tungstate	Hg_2WO_4	16.96	1.1×10^{-17}
Mercury(II)			
bromide	$HgBr_2$	19.21	6.2×10^{-20}
hydroxide	$Hg(OH)_2$	25.52	3.2×10^{-26}
iodate	$Hg(IO_3)_2$	12.49	3.2×10^{-13}
iodide	HgI_2	28.54	2.9×10^{-29}
1,10-phenanthroline	HgL_2	24.70	2.0×10^{-25}

(*Continued*)

TABLE 1.73 Solubility Product Constants (*Continued*)

Compound	Formula	pK_{sp}	K_{sp}
quinaldate	HgL_2	16.80	1.6×10^{-17}
selenite	$HgSeO_3$	13.82	1.5×10^{-14}
sulfide	HgS red	52.4	4×10^{-53}
	HgS black	51.80	1.6×10^{-52}
Neodymium			
carbonate	$Nd_2(CO_3)_3$	32.97	1.08×10^{-33}
hydroxide	$Nd(OH)_3$	21.49	3.2×10^{-22}
Neptunyl(VI)			
hydroxide	$NpO_2(OH)_2$	21.60	2.5×10^{-22}
Nickel			
ammine perrhenate	$[Ni(NH_3)_6][ReO_4]_2$	3.29	5.1×10^{-4}
anthranilate	NiL_2	9.09	8.1×10^{-10}
arsenate	$Ni_3(AsO_4)_2$	25.51	3.1×10^{-26}
carbonate	$NiCO_3$	6.85	1.42×10^{-7}
ferrocyanide	$Ni_2[Fe(CN)_6]$	14.89	1.3×10^{-15}
hydrazine sulfate	$[Ni(N_2H_4)_3]SO_4$	13.15	7.1×10^{-15}
hydroxide	$Ni(OH)_2$ fresh	15.26	5.48×10^{-16}
iodate	$Ni(IO_3)_2$	4.33	4.71×10^{-5}
oxalate	NiC_2O_4	9.4	4×10^{-10}
phosphate	$Ni_3(PO_4)_2$	31.32	4.74×10^{-32}
pyrophosphate	$Ni_2P_2O_7$	12.77	1.7×10^{-13}
quinaldate	NiL_2	10.1	8×10^{-11}
8-quinolinolate	NiL_2	26.1	8×10^{-27}
selenite	$NiSeO_3$	5.0	1.0×10^{-5}
α-sulfide	α-NiS	18.50	3.2×10^{-19}
β-sulfide	β-NiS	24.0	1.0×10^{-24}
γ-sulfide	γ-NiS	25.70	2.0×10^{-26}
Palladium			
(II) hydroxide	$Pd(OH)_2$	31.0	1.0×10^{-31}
(IV) hydroxide	$Pd(OH)_4$	70.20	6.3×10^{-71}
quinaldate	PdL_2	12.90	1.3×10^{-13}
thiocyanate	$Pd(SCN)_2$	22.36	4.39×10^{-23}
Platinum			
(IV) bromide	$PtBr_4$	40.50	3.2×10^{-41}
(II) hydroxide	$Pt(OH)_2$	35	1×10^{-35}
Plutonium			
(III) fluoride	PuF_3	15.60	2.5×10^{-16}
(IV) fluoride	PuF_4	19.20	6.3×10^{-20}
(IV) hydrogen phosphate	$Pu(HPO_4)_2 \cdot xH_2O$	27.7	2×10^{-28}
(III) hydroxide	$Pu(OH)_3$	19.70	2.0×10^{-20}
(IV) hydroxide	$Pu(OH)_4$	55	1×10^{-55}
(IV) iodate	$Pu(IO_3)_4$	12.3	5×10^{-13}
(VI) carbonate	PuO_2CO_3	12.77	1.7×10^{-13}
(V) hydroxide	$PuO_2(OH)$	9.3	5×10^{-10}
(VI) hydroxide	$PuO_2(OH)_2$	24.7	2×10^{-25}
Polonium			
sulfide	PoS	28.26	5.6×10^{-29}
Potassium			
hexabromoplatinate	$K_2[PtBr_6]$	4.20	6.3×10^{-5}
hexachloropalladinate	$K_2[PdCl_6]$	5.22	6.0×10^{-6}
hexachloroplatinate	$K_2[PtCl_6]$	5.13	7.48×10^{-6}
hexafluoroplatinate	$K_2[PtF_6]$	4.54	2.9×10^{-5}

TABLE 1.73 Solubility Product Constants (*Continued*)

Compound	Formula	pK_{sp}	K_{sp}
hexafluorosilicate	$K_2[SiF_6]$	6.06	8.7×10^{-7}
hexafluorozirconate	$K_2[ZrF_6]$	3.3	5×10^{-4}
iodate	KIO_4	3.43	3.74×10^{-4}
perchlorate	$KClO_4$	1.98	1.05×10^{-2}
sodium cobaltinitrite hydrate	$K_2Na[Co(NO_2)_6] \cdot H_2O$	10.66	2.2×10^{-11}
tetraphenylborate	$K[B(C_6H_5)_4]$	7.66	2.2×10^{-8}
uranyl arsenate	$K[UO_2AsO_4]$	22.60	2.5×10^{-23}
uranyl carbonate	$K_4[UO_2(CO_3)_3]$	4.20	6.3×10^{-5}
Praseodymium			
hydroxide	$Pt(OH)_3$	23.45	3.39×10^{-24}
Promethium			
hydroxide	$Pm(OH)_3$	21	1×10^{-21}
Radium			
iodate	$Ra(IO_3)_2$	8.94	1.16×10^{-9}
sulfate	$RaSO_4$	10.44	3.66×10^{-11}
Rhodium			
hydroxide	$Rh(OH)_3$	23	1×10^{-23}
Rubidium			
cobaltinitrite	$Rb_3[Co(NO_2)_6]$	14.83	1.5×10^{-15}
hexachloroplatinate	$Rb_2[PtCl_6]$	7.20	6.3×10^{-8}
hexafluoroplatinate	$Rb_2[PtF_6]$	6.12	7.7×10^{-7}
hexafluorosilicate	$Rb_2[SiF_6]$	6.30	5.0×10^{-7}
perchlorate	$RbClO_4$	2.52	3.0×10^{-3}
periodate	$RbIO_4$	3.26	5.5×10^{-4}
Ruthenium			
hydroxide	$Ru(OH)_3$	36	1×10^{-36}
Samarium			
hydroxide	$Sm(OH)_3$	22.08	8.3×10^{-23}
Scandium			
fluoride	ScF_3	23.24	5.81×10^{-24}
hydroxide	$Sc(OH)_3$	30.65	2.22×10^{-31}
Silver			
acetate	$AgOAc$	2.71	1.94×10^{-3}
arsenate	Ag_3AsO_4	21.99	1.03×10^{-22}
azide	AgN_3	8.54	2.8×10^{-9}
bromate	$AgBrO_3$	4.27	5.38×10^{-5}
bromide	$AgBr$	12.27	5.35×10^{-13}
carbonate	Ag_2CO_3	11.07	8.46×10^{-12}
chloride	$AgCl$	9.75	1.77×10^{-10}
chlorite	$AgClO_2$	3.70	2.0×10^{-4}
chromate	Ag_2CrO_4	11.95	1.12×10^{-12}
cobaltinitrite	$Ag_3[Co(NO_2)_6]$	20.07	8.5×10^{-21}
cyanamide	Ag_2CN_2	10.14	7.2×10^{-11}
cyanate	$AgOCN$	6.64	2.3×10^{-7}
cyanide	$AgCN$	16.22	5.97×10^{-17}
dichromate	$Ag_2Cr_2O_7$	6.70	2.0×10^{-7}
dicyanimide	$AgN(CN)_2$	8.85	1.4×10^{-9}
ferrocyanide	$Ag_4[Fe(CN)_6]$	40.81	1.6×10^{-41}
hydroxide	$AgOH$	7.71	2.0×10^{-8}
hyponitrite	$Ag_2N_2O_2$	18.89	1.3×10^{-19}
iodate	$AgIO_3$	7.50	3.17×10^{-8}

(*Continued*)

TABLE 1.73   Solubility Product Constants (*Continued*)

Compound	Formula	pK_{sp}	K_{sp}
iodide	AgI	16.07	8.52×10^{-17}
molybdate	Ag_2MoO_4	11.55	2.8×10^{-12}
nitrite	$AgNO_2$	3.22	6.0×10^{-4}
oxalate	$Ag_2C_2O_4$	11.27	5.40×10^{-12}
phosphate	Ag_3PO_4	16.05	8.89×10^{-17}
quinaldate	AgL	16.89	1.3×10^{-17}
perrhenate	$AgReO_4$	4.10	8.0×10^{-5}
selenate	Ag_2SeO_4	7.25	5.7×10^{-8}
selenite	Ag_2SeO_3	15.00	1.0×10^{-15}
selenocyanate	AgSeCN	15.40	4.0×10^{-16}
sulfate	Ag_2SO_4	4.92	1.20×10^{-5}
sulfite	Ag_2SO_3	13.82	1.50×10^{-14}
sulfide	Ag_2S	49.20	6.3×10^{-50}
thiocyanate	AgSCN	11.99	1.03×10^{-12}
vanadate	$AgVO_3$	6.3	5×10^{-7}
tungstate	Ag_2WO_4	11.26	5.5×10^{-12}
Sodium			
ammonium cobaltinitrite	$Na(NH_4)_2[Co(NO_2)_6]$	10.66	2.2×10^{-11}
antimonate	$Na[Sb(OH)_6]$	7.4	4×10^{-8}
hexafluoroaluminate	$Na_2[AlF_6]$	9.39	4.0×10^{-10}
uranyl arsenate	$NaUO_2AsO_4$	21.87	1.3×10^{-22}
Strontium			
arsenate	$Sr_3(AsO_4)_2$	18.37	4.29×10^{-19}
carbonate	$SrCO_3$	9.25	5.60×10^{-10}
chromate	$SrCrO_4$	4.65	2.2×10^{-5}
fluoride	SrF_2	8.36	4.33×10^{-9}
iodate	$Sr(IO_3)_2$	6.94	1.14×10^{-7}
iodate hydrate	$Sr(IO_3)_2 \cdot H_2O$	6.42	3.77×10^{-7}
molybdate	$SrMoO_4$	6.7	2×10^{-7}
niobate	$Sr(NbO_3)_2$	17.38	4.2×10^{-18}
oxalate hydrate	$SrC_2O_4 \cdot H_2O$	6.80	1.6×10^{-7}
phosphate	$Sr_3(PO_4)_2$	27.39	4.0×10^{-28}
8-quinolinolate	SrL_2	9.3	5×10^{-10}
selenate	$SrSeO_4$	3.09	8.1×10^{-4}
selenite	$SrSeO_3$	5.74	1.8×10^{-6}
sulfate	$SrSO_4$	6.46	3.44×10^{-7}
sulfite	$SrSO_3$	7.4	4×10^{-8}
tungstate	$SrWO_4$	9.77	1.7×10^{-10}
Terbium			
hydroxide	$Tb(OH)_3$	21.70	2.0×10^{-22}
Tellurium			
hydroxide	$Te(OH)_4$	53.52	3.0×10^{-54}
Thallium(I)			
azide	TlN_3	3.66	2.2×10^{-4}
bromate	$TlBrO_3$	4.96	1.10×10^{-5}
bromide	TlBr	5.43	3.71×10^{-6}
chloride	TlCl	3.73	1.86×10^{-4}
chromate	Tl_2CrO_4	12.06	8.67×10^{-13}
ferrocyanide dihydrate	$Tl_4[Fe(CN)_6] \cdot 2H_2O$	9.3	5×10^{-10}
hexachloroplatinate	$Tl_2[PtCl_6]$	11.40	4.0×10^{-12}
iodate	$TlIO_3$	5.51	3.12×10^{-6}
iodide	TlI	7.26	5.54×10^{-8}

TABLE 1.73 Solubility Product Constants (*Continued*)

Compound	Formula	pK_{sp}	K_{sp}
oxalate	$Tl_2C_2O_4$	3.7	2×10^{-4}
selenate	Tl_2SeO_4	4.00	1.0×10^{-4}
selenite	Tl_2SeO_3	38.7	2×10^{-39}
sulfide	Tl_2S	20.30	5.0×10^{-21}
thiocyanate	$TlSCN$	3.80	1.57×10^{-4}
Thallium(III)			
hydroxide	$Tl(OH)_3$	43.77	1.68×10^{-44}
8-quinolinolate	TlL_3	32.40	4.0×10^{-33}
Thorium			
hydrogen phosphate	$Th(HPO_4)_2$	20	1×10^{-20}
hydroxide	$Th(OH)_4$	44.40	4.0×10^{-45}
iodate	$Th(IO_3)_4$	14.60	2.5×10^{-15}
oxalate	$Th(C_2O_4)_2$	22	1×10^{-22}
phosphate	$Th_3(PO_4)_4$	78.60	2.5×10^{-79}
Thullium			
hydroxide	$Tm(OH)_3$	23.48	3.3×10^{-24}
Tin			
(II) hydroxide	$Sn(OH)_2$	27.26	5.45×10^{-28}
(IV) hydroxide	$Sn(OH)_4$	56	1×10^{-56}
(II) sulfide	SnS	25.00	1.0×10^{-25}
Titanium			
(III) hydroxide	$Ti(OH)_3$	40	1×10^{-40}
(IV) oxide hydroxide	$TiO(OH)_2$	29	1×10^{-29}
Uranium(IV)			
fluoride 2.5-water	$UF_4 \cdot 2.5H_2O$	21.24	5.7×10^{-22}
Uranyl(VI)(2+)			
carbonate	UO_2CO_3	11.73	1.8×10^{-12}
ferrocyanide	$UO_2[Fe(CN)_6]$	13.15	7.1×10^{-14}
hydrogen arsenate	UO_2HAsO_4	10.50	3.2×10^{-11}
hydrogen phosphate	UO_2HPO_4	10.67	2.1×10^{-11}
hydroxide	$UO_2(OH)_2$	21.95	1.1×10^{-22}
iodate hydrate	$UO_2(IO_3)_2 \cdot H_2O$	7.50	3.2×10^{-8}
oxalate trihydrate	$UO_2C_2O_4 \cdot 3H_2O$	3.7	2×10^{-4}
phosphate	$(UO_2)_3(PO_4)_2$	46.7	2×10^{-47}
sulfite	UO_2SO_3	8.58	2.6×10^{-9}
thiocyanate	$(UO_2)(SCN)_2$	3.4	4×10^{-4}
Vanadium			
(IV) hydroxide	$VO(OH)_2$	22.13	5.9×10^{-23}
(III) phosphate	$(VO_2)_3PO_4$	24.1	8×10^{-25}
Ytterbium			
hydroxide	$Yt(OH)_3$	23.60	2.5×10^{-24}
Yttrium			
carbonate	$Y_2(CO_3)_3$	2.99	1.03×10^{-3}
fluoride	YF_3	20.06	8.62×10^{-21}
hydroxide	$Y(OH)_3$	22.00	1.00×10^{-22}
iodate	$Y(IO_3)_3$	9.95	1.12×10^{-10}
oxalate	$Y_2(C_2O_4)_3$	28.28	5.3×10^{-29}
Zinc			
anthranilate	ZnL_2	9.23	5.9×10^{-10}
arsenate	$Zn_3(AsO_4)_2$	27.55	2.8×10^{-28}
borate hydrate	$Zn(BO_2)_2 \cdot H_2O$	10.18	6.6×10^{-11}
carbonate	$ZnCO_3$	9.94	1.46×10^{-10}
ferrocyanide	$Zn_2[Fe(CN)_6]$	15.40	4.0×10^{-15}

(*Continued*)

TABLE 1.73 Solubility Product Constants (*Continued*)

Compound	Formula	pK_{sp}	K_{sp}
fluoride	ZnF_2	1.52	3.04×10^{-2}
hydroxide	$Zn(OH)_2$	16.5	3×10^{-17}
iodate dihydrate	$Zn(IO_3)_2 \cdot 2H_2O$	5.37	4.1×10^{-6}
oxalate dihydrate	$ZnC_2O_4 \cdot 2H_2O$	8.86	1.38×10^{-9}
phosphate	$Zn_3(PO_4)_2$	32.04	9.0×10^{-33}
quinaldate	ZnL_2	13.80	1.6×10^{-14}
8-quinolinolate	ZnL_2	24.30	5.0×10^{-25}
selenide	$ZnSe$	25.44	3.6×10^{-26}
selenite hydrate	$ZnSeO_3 \cdot H_2O$	6.80	1.57×10^{-7}
sulfide	α-ZnS	23.80	1.6×10^{-24}
	β-ZnS	21.60	2.5×10^{-22}
Zirconium			
oxide hydroxide	$ZrO(OH)_2$	48.20	6.3×10^{-49}
phosphate	$Zr_3(PO_4)_4$	132	1×10^{-132}

TABLE 1.74 Stability Constants of Complex Ions

The stability constant of a complex ion is a measure of its stability with respect to dissociation into its constituent species at a given temperature, e.g., the formation of the tetra-amminecopper(II) ion may be represented by the equation

$$Cu^{2+} + 4NH_3 = [Cu(NH_3)_4]^{2+}$$

and the stability constant is given by

$$K_{stab} = \frac{[Cu(NH_3)_4^{2+}]}{[Cu^{2+}][NH_3]^4}$$

The higher the stability constant the more stable the complex ion. v denotes the stoichiometric number of a molecule, atom or ion, and is positive for a product and negative for a reactant.

Equilibrium	$\dfrac{K_{stab}}{(mol \cdot dm^{-3})^{\Sigma v}}$	$\log_{10}\left\{\dfrac{K_{stab}}{(mol \cdot dm^{-3})^{\Sigma v}}\right\}$
$Ag^+ + 2CN^- = [Ag(CH)_2]^-$	1.0×10^{21}	21.0
$Ag^+ + NH_3 = [Ag(NH_3)]^+$	2.5×10^3	3.4
$[Ag(NH_3)]^+ + NH_3 = [Ag(NH_3)_2]^+$	6.3×10^3	3.8
$Ag^+ + 2NH_3 = [Ag(NH_3)_2]^+$	1.7×10^7	7.2
$Ag^+ + 2S_2O_3^{2-} = [Ag(S_2O_3)_2]^{3-}$	1.0×10^{13}	13.0
$Al^{3+} + 6F^- = [AlF_6]^{3-}$	6×10^{19}	19.8
$Al(OH)_3 + OH^- = [Al(OH)_4]^-$	40	1.6
$Cd^{2+} + 4CN^- = [Cd(CN)_4]^{2-}$	7.1×10^{16}	16.9
$Cd^{2+} + 4I^- = [CdI_4]^{2-}$	2×10^6	6.3
$Cd^{2+} + 4NH_3 = [Cd(NH_3)_4]^{2+}$	4.0×10^6	6.6
$Co^{2+} + 6NH_3 = [Co(NH_3)_6]^{2+}$	7.7×10^4	4.9
$Co^{3+} + 6NH_3 = [Co(NH_3)_6]^{3+}$	4.5×10^{33}	33.7
$Cr(OH)_3 + OH^- = [Cr(OH)_4]^-$	1×10^{-2}	-2
$Cu^+ + 4CN^- = [Cu(CN)_4]^{3-}$	2.0×10^{27}	27.3
$Cu^{2+} + 4Cl^- = [CuCl_4]^{2-}$	4.0×10^5	5.6
$Cu^+ + 2NH_3 = [Cu(NH_3)_2]^+$	1×10^{11}	11

TABLE 1.74   Stability Constants of Complex Ions (*Continued*)

Equilibrium	$\dfrac{K_{stab}}{(\text{mol} \cdot \text{dm}^{-3})^{\Sigma v}}$	$\log_{10}\left\{\dfrac{K_{stab}}{(\text{mol} \cdot \text{dm}^{-3})^{\Sigma v}}\right\}$
$Cu^{2+} + NH_3 = [Cu(NH_3)]^{2+}$	$2{\cdot}0 \times 10^4$ (K_1)	4·3
$[Cu(NH_3)]^{2+} + NH_3 = [Cu(NH_3)_2]^{2+}$	$4{\cdot}2 \times 10^3$ (K_2)	3·6
$[Cu(NH_3)_2]^{2+} + NH_3 = [Cu(NH_3)_3]^{2+}$	$1{\cdot}0 \times 10^3$ (K_3)	3·0
$[Cu(NH_3)_3]^{2+} + NH_3 = [Cu(NH_3)_4]^{2+}$	$1{\cdot}7 \times 10^2$ (K_4)	2·2
$Cu^{2+} + 4NH_3 = [Cu(NH_3)_4]^{2+}$	$1{\cdot}4 \times 10^{13}$	13·1
	$(K = K_1 K_2 K_3 K_4)$	
$Fe^{2+} + 6CN^- = [Fe(CN)_6]^{4-}$	*ca.* 10^{24}	*ca.* 24
$Fe^{3+} + 6CN^- = [Fe(CN)_6]^{3-}$	*ca.* 10^{31}	*ca.* 31
$Fe^{3+} + 4Cl^- = [FeCl_4]^-$	8×10^{-2}	−1·1
$Fe^{3+} + SCN^- = [Fe(SCN)]^{2+}$	$1{\cdot}4 \times 10^2$	2·1
$[Fe(SCN)]^{2+} + SCN^- = [Fe(SCN)_2]^+$	16	1·2
$[Fe(SCN)_2]^+ + SCN^- = Fe(SCN)_3$	1	0
$Hg^{2+} + 4CN^- = [Hg(CN)_4]^{2-}$	$2{\cdot}5 \times 10^{41}$	41·4
$Hg^{2+} + 4Cl^- = [HgCl_4]^{2-}$	$1{\cdot}7 \times 10^{16}$	16·2
$Hg^{2+} + 4I^- = [HgI_4]^{2-}$	$2{\cdot}0 \times 10^{30}$	30·3
$I^- + I_2 = I_3^-$	$7{\cdot}1 \times 10^2$	2·9
$Ni^{2+} + 6NH_3 = [Ni(NH_3)_6]^{2+}$	$4{\cdot}8 \times 10^7$	7·7
$Pb(OH)_2 + OH^- = [Pb(OH)_3]^-$	50	1·7
$Sn(OH)_4 + 2OH^- = [Sn(OH)_6]^{2-}$	5×10^3	3·7
$Zn^{2+} + 4CN^- = [Zn(CN)_4]^{2-}$	5×10^{16}	16·7
$Zn^{2+} + 4NH_3 = [Zn(NH_3)_4]^{2+}$	$3{\cdot}8 \times 10^9$	9·6
$Zn(OH)_2 + 2OH^- = [Zn(OH)_4]^{2-}$	10	1·0

TABLE 1.75   Saturated Solutions

The following table provides the data for making saturated solutions of the substances listed at the temperature designated. Data are provided for making saturated solutions by weight (g of substance per 100 g of saturated solution) and by volume (g of substance per 100 ml of saturated solution and the ml of water required to make such a solution).

To make one *fluid ounce* of a saturated solution: multiply the grams of substance per 100 ml of saturated solution by 4.55 to obtain the number of grains required, by 0.01039 to obtain the number of avoirdupois ounces, by 0.00947 to obtain the number of apothecaries (Troy) ounces; also multiply the ml of water by 16.23 to obtain the number of minims, or divide by 100 to obtain the number of fluid ounces.

To make one *fluid dram*: multiply the grams of substance per 100 ml of saturated solution by 0.5682 to obtain the number of grains required; also multiply the ml of water by 0.60 to obtain the number of minims required.

Substance	Formula	Temp, °C	g/100 g satd soln	g/100 ml satd soln	ml water/ 100 ml satd soln	Specific gravity
acetanilide	$C_6H_5NHCOCH_3$	25	0.54	0.54	99.2	0.997
p-acetophenetidin	$C_6H_4(OC_2H_5)NHCH_3CO$	25	0.0766	0.0766	99.92	1.00
p-acetotoluide	$CH_3CONHC_6H_4CH_3$	25	0.12	0.12	99.7	0.9979
alanine	$CH_3CH(NH_2)COOH$	25	14.1	14.7	89.5	1.042
aluminum ammonium sulfate	$Al_2(SO_4)_3(NH_4)_2SO_4 \cdot 24H_2O$	25	12.4	13	92	1.05
aluminum chloride hydrated	$AlCl_3 \cdot 6H_2O$	25	55.5	75	60	1.35
aluminum fluoride	$Al_2F_6 \cdot 5H_2O$	20	0.499	0.5015	100.0	1.0051
aluminum potassium sulfate	$AlK(SO_4)_2$	25	6.62	7.02	99.1	1.061
aluminum sulfate	$Al_2(SO_4)_3 \cdot 18H_2O$	25	48.8	63	66	1.29

(*Continued*)

TABLE 1.75 Saturated Solutions (*Continued*)

Substance	Formula	Temp, °C	g/100 g satd soln	g/100 ml satd soln	ml water/ 100 ml satd soln	Specific gravity
o-aminobenzoic acid	$C_6H_4NH_2COOH$	25	0.52	0.519	99.4	0.999
DL-α-amino-*n*-butyric acid	$CH_3CH_2CH(NH_2)COOH$	25	17.8	18.6	86.2	1.046
DL-α-aminoisobutyric acid	$(CH_3)_2C(NH_2)COOH$	25	13.3	13.7	89.5	1.031
ammonium arsenate	$NH_4H_2AsO_4$	20	32.7	40.2	83.0	1.228
ammonium benzoate	$NH_4C_7H_5O_2$	25	18.6	19.4	84.7	1.040
ammonium bromide	NH_4Br	15	41.7	53.8	75.2	1.290
ammonium carbonate		25	20	22	88	1.10
ammonium chloride	NH_4Cl	15	26.3	28.3	79.3	1.075
ammonium citrate, dibasic	$(NH_4)_2HC_6H_5O_7$	25	48.7	60.5	61.5	1.22
ammonium dichromate	$(NH_4)_2Cr_2O_7$	25	27.9	33	85	1.18
ammonium iodide	NH_4I	25	64.5	106.2	58.3	1.646
ammonium molybdate	$(NH_4)_6Mo_7O_{24} \cdot 4H_2O$	25	30.6	39	88	1.27
ammonium nitrate	NH_4NO_3	25	68.3	90.2	41.8	1.320
ammonium oxalate	$(NH_4)_2C_2O_4 \cdot H_2O$	25	4.95	5.06	97.0	1.019
ammonium perchlorate	NH_4ClO_4	25	21.1	23.7	88.7	1.123
ammonium periodate	NH_4IO_4	16	2.63	2.68	99.2	1.018
ammonium persulfate	$(NH_4)_2S_2O_8$	25	42.7	53	71	1.24
ammonium phosphate, dibasic	$(NH_4)_2 \cdot HPO_4$	14.5	56.2	75.5	58.8	1.343
ammonium phosphate, monobasic	$NH_4H_2PO_4$	25	28.4	33	83	1.16
ammonium salicylate	$NH_4C_7H_5O_3$	25	50.8	58.2	56.4	1.145
ammonium silicofluoride	$(NH_4)_2SiF_6$	17.5	15.7	17.2	92.3	1.095
ammonium sulfate	$(NH_4)_2SO_4$	20	42.6	53.1	71.7	1.248
ammonium sulfite	$(NH_4)_2SO_3.H_2O$	25	39.3	47.3	73.2	1.204
ammonium thiocyanate	NH_4CNS	25	62.2	71	43	1.14
amyl alcohol	$C_5H_{11}OH$	25	2.61	2.60	96.9	0.995
aniline	$C_6H_5NH_2$	22	3.61	3.61	96.2	0.998
aniline hydrochloride	$C_6H_5NH_2 \cdot HCl$	25	49	54	56	1.10
aniline sulfate	$(C_6H_5NH_2)_2 \cdot H_2SO_4$	25	5.88	6	96	1.02
L-asparagine	$NH_2COCH_2CH(NH_2)COOH$	25	2.44	2.46	98.2	1.007
barium bromide	$BaBr_2$	20	51	87.2	83.8	1.710
barium chlorate	$Ba(ClO_3)_2$	25	28.5	36.8	92.6	1.294
barium chloride	$BaCl_2$	20	26.3	33.4	93.8	1.27
barium iodide	$BaI_2 \cdot 7\frac{1}{2}H_2O$	25	68.8	157.0	71.1	2.277
barium nitrate	$Ba(NO_3)_2$	25	9.4	10.2	97.9	1.080
barium nitrite	$Ba(NO_2)_2$	17	40	59.6	89.4	1.490
barium perchlorate	$Ba(ClO_4)_2$	25	75.3	145.8	47.8	1.936
benzamide	$C_6H_5CONH_2$	25	1.33	1.33	98.6	0.999
benzoic acid	$C_7H_6O_2$	25	0.367	0.367	99.63	1.00
beryllium sulfate	$BeSO_4 \cdot 4H_2O$	25	28.7	37.3	93.0	1.301
boric acid	H_3BO_3	25	4.99	5.1	97	1.02
n-butyl alcohol	$CH_3(CH_2)_2CH_2OH$	25	79.7	67.3	17.1	0.845
cadmium bromide	$CdBr_2 \cdot 4H_2O$	25	52.9	94.0	83.9	1.775
cadmium chlorate	$Cd(ClO_3)_2 \cdot 12H_2O$	18	76.4	174.5	54.0	2.284
cadmium chloride	$CdCl_2 \cdot 2\frac{1}{2}H_2O$	25	54.7	97.2	80.8	1.778
cadmium iodide	CdI_2	20	45.9	73.0	86.3	1.590
cadmium sulfate	$3(CdSO_4) \cdot 8H_2O$	25	43.4	70.3	91.8	1.619
calcium bromide	$CaBr_2$	20	58.8	107.2	75.0	1.82

TABLE 1.75 Saturated Solutions (*Continued*)

Substance	Formula	Temp, °C	g/100 g satd soln	g/100 ml satd soln	ml water/ 100 ml satd soln	Specific gravity
calcium chlorate	$Ca(ClO_3)_2 \cdot 2H_2O$	18	64.0	110.7	62.3	1.729
calcium chloride	$CaCl_2 \cdot 6H_2O$	25	46.1	67.8	79.2	1.47
calcium chromate	$CaCrO_4 \cdot 2H_2O$	18	14.3	16.4	98.7	1.149
calcium ferrocyanide	$Ca_2Fe(CN)_6$	25	36.5	49.6	86.2	1.357
calcium iodide	CaI_2	20	67.6	143.8	69.0	2.125
calcium lactate	$Ca(C_3H_5O_3)_2 \cdot 5H_2O$	25	4.95	5	96	1.01
calcium nitrite	$Ca(NO_2)_2 \cdot 4H_2O$	18	45.8	65.7	77.8	1.427
calcium sulfate	$CaSO_4 \cdot 2H_2O$	25	0.208	0.208	99.70	0.999
camphoric acid	$C_8H_{14}(COOH)_2$	25	0.754	0.754	99.246	1.00
carbon disulfide	CS_2	22	0.173	0.173	99.63	0.998
cerium nitrate	$Ce(NO_3)_3 \cdot 6H_2O$	25	63.7	119.9	68.2	1.880
cesium bromide	$CsBr$	21.4	53.1	89.8	79.5	1.693
cesium chloride	$CsCl$	25	65.7	126.3	65.9	1.923
cesium iodide	CsI	22.8	48.0	74.1	80.5	1.545
cesium nitrate	$CsNO_3$	25	21.9	26.1	92.9	1.187
cesium perchlorate	$CsClO_4$	25	2.01	2.03	99.0	1.010
cesium periodate	$CsIO_4$	15	2.10	2.13	99.5	1.017
cesium sulfate	Cs_2SO_4	25	64.5	129.8	71.7	2.013
chloral hydrate	$CCl_3CHO \cdot H_2O$	25	79.4	120	31	1.51
chloroform	$CHCl_3$	29.4	0.703	0.705	99.57	1.0028
chromic oxide	CrO_3	18	62.5	106.3	64.0	1.703
chromium potassium sulfate	$Cr_2K_2(SO_4)_4 \cdot 24H_2O$	25	19.6	22	90	1.12
citric acid	$(CH_2)_2COH(COOH)_3 \cdot H_2O$	25	67.5	88.6	42.7	1.311
cobalt chlorate	$Co(ClO_3)_2$	18	64.2	119.3	66.5	1.857
cobalt nitrate	$Co(NO_3)_2$	18	49.7	78.2	79.1	1.572
cobalt perchlorate	$Co(ClO_4)_2$	26	71.8	113.5	44.7	1.581
cupric ammonium chloride	$CuCl_2 \cdot 2NH_4Cl \cdot 2H_2O$	25	30.3	35.5	82	1.17
cupric ammonium sulfate	$CuSO_4 \cdot (NH_4)_2SO_4$	19	15.3	17.3	96.0	1.131
cupric bromide	$CuBr_2$	25	55.8	102.5	81.2	1.84
cupric chlorate	$Cu(ClO_3)_2$	18	62.2	105.2	64.1	1.692
cupric chloride	$CuCl_2 \cdot 2H_2O$	25	53.3	80	70	1.50
cupric nitrate	$Cu(NO_3)_2 \cdot 6H_2O$	20	56.0	94.5	74.3	1.688
cupric selenate	$CuSeO_4$	21.2	14.7	17.2	99.4	1.165
cupric sulfate	$CuSO_4 \cdot 5H_2O$	25	18.5	22.3	98.7	1.211
dextrose	$C_6H_{12}O_6 \cdot H_2O$	25	49.5	59	60	1.19
ether	$(C_2H_5)_2O$	22	5.45	5.34	93.0	0.985
ethyl acetate	$CH_3COOC_2H_5$	25	7.47	7.44	92.1	0.996
ferric ammonium citrate		25	67.7	97	46	1.43
ferric ammonium oxalate	$Fe(NH_4)_3(C_2O_4)_3 \cdot 3H_2O$	25	51.5	65	61	1.26
ferric ammonium sulfate	$FeSO_4 \cdot (NH_4)_2SO_4$	16.5	19.1	22.4	94.3	1.165
ferric chloride	$FeCl_3$	25	73.1	131.1	48.3	1.793
ferric nitrate	$Fe(NO_3)_3$	25	46.8	70.2	79.8	1.50
ferric perchlorate	$Fe(ClO_4)_3 \cdot 10H_2O$	25	79.9	132.1	33.2	1.656
ferrous sulfate	$FeSO_4 \cdot 7H_2O$	25	42.1	52.8	72.7	1.255
gallic acid	$C_6H_2(OH)_3COOH \cdot H_2O$	25	1.15	1.15	99.05	1.002
D-glutamic acid	$C_5H_9O_4N$	25	0.86	0.86	99.15	1.0002
glycine	NH_2CH_2COOH	25	20.0	21.7	86.8	1.083
hydroquinone	$C_6H_4(OH)_2$	20	6.7	6.78	94.4	1.012
m-hydroxybenzoic acid	$C_6H_4OHCOOH$	25	0.975	0.975	99.03	1.000

(*Continued*)

TABLE 1.75 Saturated Solutions (*Continued*)

Substance	Formula	Temp, °C	g/100 g satd soln	g/100 ml satd soln	ml water/ 100 ml satd soln	Specific gravity
lactose	$C_{12}H_{22}O_{11} \cdot H_2O$	25	15.9	17	90	1.07
lead acetate	$Pb(C_2H_3O_2)_2$	25	36.5	49.0	85.1	1.340
lead bromide	$PbBr_2$	25	0.97	0.98	99.6	1.006
lead chlorate	$Pb(ClO_3)_2$	18	60.2	117.0	77.3	1.944
lead chloride	$PbCl_2$	25	1.07	1.08	99.6	1.007
lead iodide	PbI_2	25	0.08	0.08	99.7	0.998
lead nitrate	$Pb(NO_3)_2$	25	37.1	53.6	91.0	1.445
DL-leucine	$C_6H_{13}O_2N$	25	0.976	0.975	98.9	0.999
L-leucine	$C_6H_{13}O_2N$	25	2.24	2.24	97.85	1.0012
lithium benzoate	$LiC_7H_5O_2$	25	27.7	30.4	79.6	1.100
lithium bromate	$LiBrO_3$	18	60.4	110.5	72.5	1.830
lithium carbonate	Li_2CO_3	15	1.36	1.38	100.0	1.014
lithium chloride	$LiCl \cdot H_2O$	25	45.9	59.5	70.2	1.296
lithium citrate	$Li_3C_6H_5O_7$	25	31.8	38.6	82.8	1.213
lithium dichromate	$Li_2Cr_2O_7 \cdot H_2O$	18	52.6	82.9	74.8	1.574
lithium fluoride	LiF	18	0.27	0.27	99.9	1.002
lithium formate	$LiCHO_2$	18	27.9	31.8	80.4	1.140
lithium iodate	$LiIO_3$	18	44.6	69.9	86.8	1.566
lithium nitrate	$LiNO_3$	19	48.9	64.5	67.5	1.318
lithium perchlorate	$LiClO_4 \cdot 3H_2O$	25	37.5	47.6	79.5	1.269
lithium salicylate	$LiC_7H_5O_3$	25	52.7	63.6	57.1	1.206
lithium sulfate	$Li_2SO_4 \cdot H_2O$	25	27.2	33	88.5	1.21
magnesium bromide	$MgBr_2 \cdot 6H_2O$	18	50.1	83.1	82.8	1.655
magnesium chlorate	$Mg(ClO_3)_2$	18	56.3	90.0	69.7	1.594
magnesium chloride	$MgCl_2 \cdot 6H_2O$	25	62.5	79	47.5	1.26
magnesium chromate	$MgCr_2O_4 \cdot 7H_2O$	18	42.0	59.7	82.5	1.422
magnesium dichromate	$MgCrO_7 \cdot 5H_2O$	25	81.0	138.8	32.6	1.712
magnesium iodate	$Mg(IO_3)_2 \cdot 4H_2O$	18	6.44	6.95	100.8	1.078
magnesium iodide	$MgI_2.8H_2O$	18	59.7	114.0	77.1	1.909
magnesium molybdate	$MgMoO_4$	25	15.9	18.4	97.4	1.159
magnesium nitrate	$Mg(NO_3)_2 \cdot 6H_2O$	25	42.1	58.6	80.5	1.388
magnesium perchlorate	$Mg(ClO_4)_2 \cdot 6H_2O$	25	49.9	73.6	73.9	1.472
magnesium selenate	$MgSeO_4$	20	35.3	50.8	93.0	1.440
magnesium sulfate	$MgSO_4 \cdot 7H_2O$	25	55.3	72	58.5	1.30
manganese chloride	$MnCl_2$	25	43.6	63.2	82.0	1.449
manganese nitrate	$Mn(NO_3)_2 \cdot 6H_2O$	18	57.3	93.2	69.2	1.624
manganese silicofluoride	$MnSiF_6$	17.5	37.7	54.5	90.1	1.446
manganese sulfate	$MnSO_4$	25	39.4	59.1	90.8	1.499
mercuric acetate	$Hg(C_2H_3O_2)_2$	25	30.2	38	88	1.26
mercuric bromide	$HgBr_2$	25	0.609	0.610	99.6	1.0023
mercury bichloride	$HgCl_2$	25	6.6	6.96	98.5	1.054
methylene blue	$C_{16}H_{18}N_3ClS \cdot 3H_2O$	25	4.25	4.3	97	1.01
methyl salicylate	$C_6H_4OHCOOCH_3$	25	0.12	0.12	99.88	1.00
monochloracetic acid	$CH_2ClCOOH$	25	78.8	105	28	1.33
β-naphthalenesulfonic acid	$C_{10}H_7SO_3H$	30	56.9	67.9	51.4	1.193
nickel ammonium sulfate	$NiSO_4(NH_4)_2SO_4 \cdot 6H_2O$	25	9.0	9.5	96	1.05
nickel chlorate	$Ni(ClO_3)_2$	18	56.7	94.2	72.0	1.658
nickel chlorate	$Ni(ClO_3)_2 \cdot 6H_2O$	18	64.5	107.2	59.1	1.661
nickel nitrate	$Ni(NO_3)_2 \cdot 6H_2O$	25	77	122	36	1.58

TABLE 1.75 Saturated Solutions (*Continued*)

Substance	Formula	Temp, °C	g/100 g satd soln	g/100 ml satd soln	ml water/ 100 ml satd soln	Specific gravity
nickel perchlorate	$Ni(ClO_4)_2$	26	70.8	112.2	46.4	1.584
nickel perchlorate	$Ni(ClO_4)_2 \cdot 9H_2O$	18	52.4	82.7	75.1	1.576
nickel sulfate	$NiSO_4 \cdot 6H_2O$	25	47.3	64	71	1.35
DL-norleucine	$C_6H_{13}NO_2$	25	1.13	1.13	98.97	0.999
oxalic acid	$H_2C_2O_4 \cdot 2H_2O$	25	9.81	10.3	94.2	1.044
phenol	C_6H_5OH	20	6.1	6.14	94.5	1.0057
β-phenylalanine	$C_6H_5CH_2CH(NH_2)COOH$	25	2.88	2.89	97.5	1.0035
m-phenylenediamine	$C_6H_8N_2$	20	23.1	23.8	79.3	1.032
p-phenylenediamine	$C_6H_8N_2$	20	3.69	3.70	96.67	1.0038
phenyl salicylate	$C_6H_4OHCOOC_6H_5$	25	0.015	0.015	99.84	0.999
phenyl thiourea	$CS(NH_2)NHC_6H_5$	25	0.24	0.24	99.6	0.998
phosphomolybdic acid	$20MoO_3 \cdot 2H_3PO_4 \cdot 48H_2O$	25	74.3	135	46	1.81
phosphotungstic acid	Approx. $20WO_3 \cdot 2H_3PO_4 \cdot 25H_2O$	25	71.4	160	64	2.24
potassium acetate	$KC_2H_3O_2$	25	68.7	97.1	44.3	1.413
potassium antimony tartrate	$KSbOC_4H_4O_6$	25	7.64	8.02	96.9	1.049
potassium bicarbonate	$KHCO_3$	25	26.6	31.6	87.5	1.188
potassium bitartrate	$KC_4H_5O_6$	25	0.65	0.65	99.3	0.999
potassium bromate	$KBrO_3$	25	7.53	7.89	97.5	1.054
potassium bromide	KBr	25	40.6	56.0	82.0	1.380
potassium carbonate	$K_2CO_3 \cdot 1\frac{1}{2}H_2O$	25	52.9	82.2	73.5	1.559
potassium chlorate	$KClO_3$	25	8.0	8.41	96.6	1.051
potassium chloride	KCl	25	26.5	31.2	86.8	1.178
potassium chromate	K_2CrO_4	25	39.4	54.1	83.7	1.381
potassium citrate	$K_3C_6H_5O_7$	25	60.91	92.1	59.2	1.514
potassium dichromate	$K_2Cr_2O_7$	25	13.0	14.2	95.0	1.092
potassium ferricyanide	$K_3Fe(CN)_6$	22	32.1	38.1	80.8	1.187
potassium ferrocyanide	$K_4Fe(CN)_6$	25	24.0	28.2	89.2	1.173
potassium fluoride	$KF \cdot 2H_2O$	18	48.0	72.0	78.0	1.500
potassium formate	$KCHO_2$	18	76.8	120.6	36.4	1.571
potassium hydroxide	KOH	15	51.7	79.2	74.2	1.536
potassium iodate	KIO_3	25	8.40	8.99	98.0	1.071
potassium iodide	KI	25	59.8	103.2	69.1	1.721
potassium meta-antimonate	$KSbO_3$	18	2.73	2.81	99.7	1.025
potassium nitrate	KNO_3	25	28.0	33.4	86.0	1.193
potassium nitrite	KNO_2	20	74.3	121.5	42.3	1.649
potassium oxalate	$K_2C_2O_4 \cdot H_2O$	25	28.3	34	86	1.20
potassium perchlorate	$KClO_4$	25	2.68	2.72	99.0	1.014
potassium periodate	KIO_4	13	0.658	0.661	99.83	1.005
potassium permanganate	$KMnO_4$	25	7.10	7.43	97.3	1.046
potassium sodium tartrate	$KNaC_4H_4O_6 \cdot 4H_2O$	25	39.71	51.9	78.8	1.308
potassium stannate	K_2SnO_3	15.5	42.7	69.2	92.9	1.620
potassium sulfate	K_2SO_4	25	10.83	11.8	96.9	1.086
quinine salicylate	$C_{20}H_{24}N_2O_2 \cdot C_6H_4(OH)COOH.2H_2O$	25	0.065	0.065	99.84	0.999
resorcinol	$C_6H_4(OH)_2$	25	58.8	67.2	47.2	1.142
rubidium bromate	$RbBrO_3$	16	2.15	2.18	99.4	1.016
rubidium bromide	$RbBr$	25	52.7	85.6	76.9	1.625
rubidium chloride	$RbCl$	25	48.6	72.8	77.1	1.050
rubidium iodate	$RbIO_3$	15.6	2.72	2.78	99.5	1.022
rubidium iodide	RbI	24.3	63.6	117.7	67.3	1.850
rubidium nitrate	$RbNO_3$	25	40.1	55.0	82.4	1.375

(*Continued*)

TABLE 1.75   Saturated Solutions (*Continued*)

Substance	Formula	Temp, °C	g/100 g satd soln	g/100 ml satd soln	ml water/ 100 ml satd soln	Specific gravity
rubidium perchlorate	$RbClO_4$	25	1.88	1.90	99.3	1.012
rubidium periodate	$RbIO_4$	16	0.645	0.648	99.85	1.0052
rubidium sulfate	Rb_2SO_4	25	33.8	45.6	89.7	1.354
silicotungstic acid	$H_4SiW_{12}O_{40}$	18	90.6	258	26.8	2.843
silver acetate	$Ag(C_2H_3O_2)$	25	1.10	1.11	99.40	1.0047
silver bromate	$AgBrO_3$	25	0.204	0.2037	99.65	0.9985
silver fluoride	$AgF \cdot 2H_2O$	15.8	64.5	168.4	92.7	2.61
silver nitrate	$AgNO_3$	25	71.5	164	65.5	2.29
silver perchlorate	$AgClO_4 \cdot H_2O$	25	84.5	237.1	43.5	2.806
sodium acetate	$NaC_2H_3O_2$	25	33.6	40.5	80.0	1.205
sodium ammonium sulfate	$NaNH_4SO_4$	15	25.2	29.6	87.9	1.174
sodium arsenate	$Na_3AsO_4 \cdot 12H_2O$	17	21.1	23.5	88.0	1.119
sodium benzenesulfonate	$NaC_6H_5SO_3$	25	16.4	17.6	90.1	1.076
sodium benzoate	$NaC_7H_5O_2$	25	36.0	41.5	73.9	1.152
sodium bicarbonate	$NaHCO_3$	15	8.28	8.80	97.6	1.061
sodium bisulfate	$NaHSO_4 \cdot H_2O$	25	59	87	60	1.47
sodium bromide	$NaBr \cdot 2H_2O$	25	48.6	75.0	79.4	1.542
sodium carbonate	$Na_2CO_3 \cdot 10H_2O$	25	22.6	28.1	96.5	1.242
sodium chlorate	$NaClO_3$	25	51.7	74.3	69.6	1.440
sodium chloride	$NaCl$	25	26.5	31.7	88.1	1.198
sodium chromate	Na_2CrO_4	18	40.1	57.4	85.7	1.430
sodium citrate	$Na_3C_6H_5O_7 \cdot 5H_2O$	25	48.1	61.2	66.0	1.272
sodium dichromate	$Na_2Cr_2O_7$	18	63.9	111.4	63.0	1.743
sodium ferrocyanide	$Na_4Fe(CN)_6$	25	17.1	19.4	93.9	1.131
sodium fluoride	NaF	25	3.98	4.14	99.7	1.038
sodium formate	$NaCHO_2$	18	44.7	58.9	73.0	1.316
sodium hydroxide	$NaOH$	25	50.8	77	74	1.51
sodium hypophosphite	NaH_2PO_2	16	52.1	72.4	66.6	1.386
sodium iodate	$NaIO_3 \cdot H_2O$	25	8.57	9.21	98.5	1.075
sodium iodide	NaI	25	64.8	124.3	67.7	1.919
sodium molybdate	Na_2MoO_4	18	39.4	56.6	87.0	1.435
sodium nitrate	$NaNO_3$	25	47.9	66.7	72.5	1.391
sodium nitrite	$NaNO_2$	20	45.8	62.3	73.8	1.359
sodium oxalate	$Na_2(CO_2)_2$	25	3.48	3.58	99.1	1.025
sodium paratungstate	$(Na_2O)_3(WO_3)_7 \cdot 16H_2O$	0	26.7	35.2	96.5	1.316
sodium perchlorate	$NaClO_4$	25	67.8	114.1	54.1	1.683
sodium periodate	$NaIO_4 \cdot 3H_2O$	25	12.6	13.9	96.2	1.103
sodium phenolsulfonate	$C_6H_4(OH)SO_3Na$	25	16.1	17.4	90.5	1.079
sodium phosphate dibasic	Na_2HPO_4	17	4.2	4.4	99.9	1.043
sodium phosphate tribasic	Na_3PO_4	14	9.5	10.5	99.8	1.103
sodium pyrophosphate	$Na_2H_2P_2O_7 \cdot 6H_2O$	25	13.0	14.4	95.8	1.104
sodium salicylate	$NaC_7H_5O_3$	25	53.6	67.0	58.0	1.248
sodium selenate	Na_2SeO_4	18	29.0	38.1	93.4	1.313
sodium silicofluoride	$NaSiF_6$	20	0.773	0.737	99.76	1.0054
sodium sulfate	Na_2SO_4	25	21.8	26.4	94.5	1.208
sodium sulfate	$Na_2SO_4 \cdot 10H_2O$	25	27.7	33.3	87.0	1.207
sodium sulfide	$Na_2S \cdot 9H_2O$	25	52.3	63	57	1.20
sodium sulfite, anhydrous	Na_2SO_3	25	23	28.5	95.5	1.24
sodium thiocyanate	$NaCNS$	25	62.9	87	51	1.38

TABLE 1.75   Saturated Solutions (*Continued*)

Substance	Formula	Temp, °C	g/100 g satd soln	g/100 ml satd soln	ml water/ 100 ml satd soln	Specific gravity
sodium thiosulfate	$Na_2S_2O_3 \cdot 5H_2O$	25	66.8	93	46	1.39
sodium tungstate	$Na_2WO_4 \cdot 10H_2O$	18	42.0	66.1	91.3	1.573
stannous chloride	$SnCl_2$	15	72.9	133.1	49.5	1.827
strontium chlorate	$Sr(ClO_3)_2$	18	63.6	117.0	67.0	1.839
strontium chloride	$SrCl_2 \cdot 6H_2O$	15	33.4	45.5	90.7	1.36
strontium iodide	$SrI_2 \cdot 6H_2O$	20	64.0	137.8	77.5	2.15
strontium nitrate	$Sr(NO_3)_2$	25	44.2	65.3	82.5	1.477
strontium nitrite	$Sr(NO_2)_2$	19	39.3	56.8	87.8	1.445
strontium perchlorate	$Sr(ClO_4)_2$	25	75.6	158.5	50.8	2.084
strontium salicylate	$Sr(C_7H_5O_3)_2$	25	4.58	4.68	97.5	1.019
succinic acid	$(CH_2)_2(COOH)_2$	25	7.67	7.82	94.5	1.021
succinimide	$(CH_2CO)_2NH \cdot H_2O$	25	30.6	32.7	74.2	1.067
sucrose	$C_{12}H_{22}O_{11}$	25	67.89	90.9	43.0	1.340
tartaric acid	$C_2H_2(OH)_2(COOH)_2$	15	58.5	76.9	54.7	1.31
tetraethyl ammonium iodide	$N(C_2H_5)_4I$	25	32.9	36.2	74.0	1.102
tetramethyl ammonium iodide	$N(CH_3)_4I$	25	5.51	5.60	96.1	1.016
thallium chloride	$TlCl$	25	0.40	0.40	99.6	1.0005
thallium nitrate	$TlNO_3$	25	10.4	11.4	98.0	1.093
thallium nitrite	$TlNO_2$	25	32.1	43.7	92.5	1.360
thallium perchlorate	$TlClO_4$	25	13.5	15.2	97.1	1.122
thallium sulfate	Tl_2SO_4	25	5.48	5.74	99.0	1.047
trichloroacetic acid	CCl_3COOH	25	92.3	149.6	12.41	1.615
uranyl chloride	UO_2Cl_2	18	76.2	208.5	65.2	2.736
uranyl nitrate	$UO_2(NO_3)_2 \cdot 6H_2O$	25	68.9	120	54.5	1.74
urea	$(NH_2)_2CO$	25	53.8	62	53.5	1.15
urea phosphate	$CO(NH_2)_2 \cdot H_3PO_4$	24.5	52.4	66.1	60.1	1.26
urethan	$NH_2CO_2C_2H_5$	25	82.8	88.8	18.5	1.073
D-valine	$(CH_3)_2CHCH(NH_2)COOH$	25	8.14	8.26	93.3	1.015
DL-valine	$(CH_3)_2CHCH(NH_2)COOH$	25	6.61	6.68	94.5	1.012
zinc acetate	$Zn(C_2H_3O_2)_2$	25	25.7	30.0	86.5	1.165
zinc benzenesulfonate	$Zn(C_6H_5SO_3)_2$	25	29.5	34.9	83.4	1.182
zinc chlorate	$Zn(ClO_3)_2$	18	65.0	124.4	67.0	1.914
zinc chloride	$ZnCl_2$	25	67.5	128	61	1.89
zinc iodide	ZnI_2	18	81.2	221.3	51.2	2.725
zinc phenolsulfonate	$(C_6H_5OSO_3)_2Zn \cdot 8H_2O$	25	39.8	47.3	71.5	1.185
zinc selenate	$ZnSeO_4$	22	37.8	58.9	97.0	1.559
zinc silicofluoride	$ZnSiF_6 \cdot 6H_2O$	20	32.9	47.2	96.3	1.434
zinc sulfate	$ZnSO_4 \cdot 7H_2O$	25	36.7	54.6	94.7	1.492
zinc valerate	$Zn(C_5H_9O_2)_2$	25	1.27	1.27	98.8	1.001

1.19 *PROTON TRANSFER REACTIONS*

A proton transfer reaction is a reaction in which the main feature is the intermolecular or intramolecular transfer of a proton from one binding site to another.

In the detailed description of proton transfer reactions, especially of rapid proton transfers between electronegative atoms, it should always be specified whether the term is used to refer to the overall process, including the more-or-less *encounter-controlled* formation of a hydrogen bonded complex and the separation of the products or, alternatively, the proton transfer event (including solvent rearrangement) by itself.

For the general proton transfer reaction:

$$HB = H^+ + B$$

the acidic dissociation constant is formulated as follows:

$$K_a = \frac{[H^+][B]}{[HB]}$$

The most common charge types for the acid HB and its conjugate base B are

$$CH_3COOH = H^+ + CH_3COO-\text{(acetic acid, acetate ion)}$$
$$HSO_4^- = H^+ + SO_4^{2-} \text{ (hydrogen sulfate ion, sulfate ion)}$$
$$NH_4^+ = H^+ + NH_3 \text{ (ammonium ion, ammonia)}$$

Acids which have more than one acidic hydrogen ionize in steps, as shown for phosphoric acid:

$H_3PO_4 = H^+ + H_2PO_4^-$	$pK_1 = 2.148$	$K_1 = 7.11 \times 10^{-3}$
$H_2PO_4^- = H^+ + HPO_4^{2-}$	$pK_2 = 7.198$	$K_2 = 6.34 \times 10^{-8}$
$HPO_4^{2-} = H^+ + PO_4^{3-}$	$pK_3 = 11.90$	$K_3 = 1.26 \times 10^{-12}$

If the basic dissociation constant K_b for the equilibrium such as

$$NH_3 + H_2O = NH_4 + OH$$

is required, pK_b may be calculated from the relationship

$$pK_b = pK_w - pK_a$$

I$_a$ general, for an organic acid, a useful estimate of its pK_a value can sometimes be obtained by making a comparison with recognizably similar compounds for which pK_a values are known: (1) alkyl chains, alicyclic rings, or saturated carbocyclic rings fused to aromatic or heterocyclic rings can be replaced by methyl or ethyl groups; (2) acid-strengthening inductive and mesomeric effects of a nitro group attached to an aromatic ring are very similar to those of a nitrogen atom located at the same position in a heteroaromatic ring (e.g., 3-hydroxypyridine and 3-nitrophenol).

1.19.1 Calculation of the Approximate pH Value of Solutions

Strong acid:	pH = −log [acid]
Strong base:	pH = 14.00 + log [base]
Weak acid:	pH = 1/2pK_a − 1/2 log [acid]
Weak base:	pH = 14.00 − 1/2pK_b + 1/2 log [base]

Salt formed by a weak acid and a strong base:

$$pH = 7.00 + 1/2pK_a + 1/2 \log[\text{salt}]$$

Acid salts of a dibasic acid:

$$pH = 1/2pK_1 + 1/2pK_2 - 1/2 \log[\text{salt}] + 1/2 \log(K_1 + [\text{salt}])$$

Buffer solution consisting of a mixture of a weak acid and its salt:

$$pH = pK_a + \log\left(\frac{[\text{salt}] + [H_3O^+] - [OH^-]}{[\text{acid}] + [H_3O^+] - [OH^-]}\right)$$

1.19.2 Calculation of Concentrations of Species Present at a Given pH

$$\alpha_0 = \frac{[H^+]^n}{[H^+]^n + K_1[H^+]^{n-1} + K_1K_2[H^+]^{n-2} + \cdots + K_1K_2\cdots K_n} = \frac{[H_nA]}{C_{\text{acid}}}$$

$$\alpha_1 = \frac{K_1[H^+]^{n-1}}{[H^+]^n + K_1[H^+]^{n-1} + K_1K_2[H^+]^{n-2} + \cdots + K_1K_2\cdots K_n} = \frac{[H_{n-1}A^-]}{C_{\text{acid}}}$$

$$\alpha_2 = \frac{K_1K_2[H^+]^{n-2}}{[H^+]^n + K_1[H^+]^{n-1} + K_1K_2[H^+]^{n-2} + \cdots + K_1K_2\cdots K_n} = \frac{[H_{n-2}A^{2-}]}{C_{\text{acid}}}$$

$$\vdots$$

$$\alpha_n = \frac{K_1K_2\cdots K_n}{[H^+]^n + K_1[H^+]^{n-1} + K_1K_2[H^+]^{n-2} + \cdots + K_1K_2\cdots K_n} = \frac{[A^{n-}]}{C_{\text{acid}}}$$

TABLE 1.76 Proton Transfer Reactions of Inorganic Materials in Water at 25°C

Substance	Formula or remarks	pK_1	pK_2
Aluminic acid	H_3AlO_3	11.2	
Aluminum ion (aquo)	Al^{3+} (aquo)	4.98(4)	
Americium(III) ion	Am^{3+} (aquo) $\mu = 0.1$	5.92	
Ammonium ion	NH_4^+	9.246(2)	
Ammonium-d_3	ND_3H^+	9.757	
Antimonic acid	$HSb(OH)_6 = Sb(OH)_6^- + H^+$ $\mu = 0.5$	2.55	
Antimony(III) ion	$SbO^+ + H_2O = Sb(OH)_3 + H^+$ $\mu = 1.0$	1.42	
Barium ion	pK_b of $Ba(OH)^+$ $\mu = 0.1$	0.64	
Berkelium(III) ion	pK for hydrolysis of Bk^{3+} $\mu = 0.1$	5.66	
Beryllium(II) ion	Be^{2+} (aquo) $= BeOH^+ + H^+$ $\mu = 1.0$	6.5	
Bismuth(III) ion	$Bi^{3+} = BiOH^{2+} + H^+$ $\mu = 3.0$	1.58	
Boric acid, tetra-	$H_2B_4O_7$	4	9
Bromine	$Br_2 + H_2O = HBrO + H^+ + Br^-$	7.92	
Cadmium ion	Cd^{2+} (aquo) hydrolysis	9.2(1)	
Calcium ion	Ca^{2+} (aquo) hydrolysis	12.67(3)	
Californium(III) ion	Cf^{3+} (aquo) hydrolysis $\mu = 0.1$	5.62	
Carbon dioxide	CO_2 (aquo)	6.352(1)	10.329
	CO_2 in D_2O	6.77	10.93
Cerium(III) ion	Ce^{3+} (aquo) hydrolysis	ca. 9.3	
Cerium(IV) ion	Hydrolysis to $Ce(OH)^{3+}$ and $Ce(OH)_2^{2+}$	−1.15	0.82
Chromium(III) ion	Cr^{3+} (aquo) hydrolysis	3.95	
Cobalt(II) ion	Co^{2+} (aquo) hydrolysis	8.9	
Cobalt(III) ion	Co^{3+} (aquo) hydrolysis $m = 1$	1.75	
Copper(II) ion	Cu^{2+} (aquo) hydrolysis	7.34	
Curium(III) ion	Cm^{3+} (aquo) hydrolysis $m = 0.1$	6.00(5)	
Deuterium oxide	D_2O (molal scale)	14.956(1)	
Dysprosium(III) ion	Dy^{3+} (aquo) hydrolysis	8.10	
Erbium(III) ion	Er^{3+} (aquo) hydrolysis $\mu = 3$	9.0	
Europium(III) ion	Eu^{3+} (aquo) hydrolysis	8.03	
Fermium(III) ion	Fm^{3+} hydrolysis $\mu = 0.1$	3.8	
Gadolinium(III) ion	Gd^{3+} hydrolysis	8.27	
Gallium(III) ion	Ga^{3+} (successive values for hydrolysis)	2.92	3.77
		pK_3 4.75	
Gold(III) hydroxide	H_3AuO_3	<11.7	13.36
Hafnium(IV) ion	Hf^{4+} hydrolysis $\mu = 1$	−0.12	0.23
Hexaminotriphosphazene	$N_3P_3(NH_2)_6$	<3.2	7.68(3)
Holmium(III) ion	Ho^{3+} hydrolysis $\mu = 0.3$	8.04	
Hydrazinium(2+) ion	$^+H_3N—NH_3^+$	0.27	7.94(3)
Hydrogen amidodisulfonate	$HNSO(OH)_2$	pK_3 8.50	
Hydrogen amidophosphate	$H_2NPO(OH)_2$ (26°C)	2.739	8.102
Hydrogen arsenate	H_3AsO_2	2.223	6.760
Hydrogen-d_3 arsenate	D_3AsO_4	2.596	
Hydrogen arsenite	$HAsO_2$	9.28(10)	
Hydrogen azide	HN_3	4.62	
Hydrogen-d azide	DN_3 (in D_2O)	5.115	
Hydrogen borate (3−)	H_3BO_3	9.236	
Hydrogen bromate	$HBrO_3$ (in formamide)	1.02	
Hydrogen bromide	HBr	−8.72(15)	
Hydrogen chlorate	$HClO_2$ (theoretical prediction)	−2.7	
Hydrogen chloride	HCl	−6.2(1)	
Hydrogen-d chloride	DCl (in dimethylformamide)	3.58	
Hydrogen chlorite	$HClO_2$	1.94	
Hydrogen chromate	H_2CrO_4	0.74	6.488
Hydrogen cyanate	$HOCN$	3.46	
Hydrogen cyanide	HCN	9.21	
Hydrogen-d cyanide	DCN (in D_2O) $\mu = 0.11$	8.97	
Hydrogen diamidophosphate	$(NH_2)PO(OH)$ (30°C)	1.279(+1)	4.889
Hydrogen diamidothiophosphate	$(NH_2)PO(SH)$ (20°C)	2.0(+1)	4.3
Hydrogen diimidotriphosphate	$(HO)_2PO(NH)PO(OH)(NH)PO(OH)_2$ $\mu = 0.1$	~1	~2
		pK_3 3.03	pK_4 6.61
		pK_s 9.84	
Hydrogen diphosphate	$H_4P_2O_7$	0.91	2.10
		pK_3 6.70	pK_4 9.35

TABLE 1.76 Proton Transfer Reactions of Inorganic Materials in Water at 25°C (*Continued*)

Substance	Formula or remarks	pK_1	pK_2
Hydrogen disulfate	$H_2S_2O_7$ (theoretical prediction)	−12	−8
Hydrogen dithionate	$H_2S_2O_6$	−3.4	−0.2
Hydrogen dithionite	$H_2S_2O_4$	0.35	2.45
Hydrogen fluoride	H_2F_2	3.20(4)	
Hydrogen germanate	H_2GeO_4	9.01	12.30
Hydrogen hexafluorosilicate	H_2SiF_6		1.92
Hydrogen hydrosulfite	$H_2S_2O_4$	0.35	2.50
Hydrogen hypobromite	HBrO	8.55	
Hydrogen hypochlorite	HClO	7.537	
Hydrogen hypoiodite	HIO	10.5(5)	
Hydrogen hyponitrite	$H_2N_2O_2$	7.21	11.45(10)
Hydrogen iodate	HIO_3	0.804	
Hydrogen-*d* iodate	DIO_3 (in D_2O)	1.15	
Hydrogen iodide	HI	−8.56	
Hydrogen manganate(VI)	H_2MnO_4 (35°C) $\mu = 0.1$		10.15
Hydrogen nitrate	HNO_3	−1.37(7)	
Hydrogen nitrite	HNO_2	3.14(1)	
Hydrogen perchlorate	$HClO_4$	−1.6	
Hydrogen periodate	HIO_4	1.64	
Hydrogen peroxide	H_2O_2	11.64(2)	
Hydrogen peroxophosphate	H_3PO_5 $\mu = 0.2$	1.1 pK_3 12.8	5.5
Hydrogen peroxosulfate	H_2SO_5	1.0	9.86
Hydrogen perrhenate	$HReO_4$	−1.25	
Hydrogen pertechnetate	$HTcO_4$	0.3	
Hydrogen perthiocarbonate	H_2CS_4	3.54	7.24
Hydrogen perxenate	H_4XeO_6	pK_3 10.5	
Hydrogen phosphate(3−)	H_3PO_4	2.148(20) pK_3 12.32(6)	7.198(10)
Hydrogen-d_2 phosphate	D_2PO_4 (in D_2O)	7.780	
Hydrogen phosphinate	H_2PHO_2	1.23	
Hydrogen phosphonate	H_2PHO_3	1.43	6.68(14)
Hydrogen selenate	H_2SeO_4		1.66
Hydrogen selenide	H_2Se $\mu = 0.03$	3.89	11.0
Hydrogen selenite	H_2SeO_3	2.62	8.30(15)
Hydrogen silicate(4−)	H_4SiO_4	9.60(10)	11.8(1)
Hydrogen sulfamate	H_2NSO_3H	0.99	
Hydrogen sulfate	H_2SO_4		1.99(1)
Hydrogen sulfide	H_2S	6.97	12.90
Hydrogen sulfite	$SO_2 + H_2O = HSO_3^- = H^+$	1.89	7.205
Hydrogen tellurate	H_6TeO_6	7.65(5)	11.00(5)
Hydrogen telluride	H_2Te (18°C)	2.64	11-12
Hydrogen tellurite	H_2TeO_3 (20°C)	6.27	8.43
Hydrogen tetrafluoroborate	HBF_4	0.5	
Hydrogen tetracyanonickelate	$H_2Ni(CN)_4$	4.69	6.59
Hydrogen tetraperoxochromate	H_3CrO_8 (30°C) $\mu = 3$	7.16	
Hydrogen tetrapolyphosphate	$H_4P_4O_{13}$ $\mu = 0.034$	1.99 pK_3 6.62	2.64 pK_4 8.2
Hydrogen tetrathiophosphate	H_3PS_4	1.5 pK_3 6.6	3.5
Hydrogen thiocyanate	HSCN $\mu = 3$	−1.8	
Hydrogen thiophosphate	H_3PO_3S	1.788 pK_3 10.08	5.427
Hydrogen thiosulfate	$H_2S_2O_3$	0.6	1.74
Hydrogen tripolyphosphate	$H_3P_3O_9$	~1 pK_3 2.00(10) pK_4 5.83(7) pK_5 8.51(6)	1.7
Hydrogen triselenocarbonate	H_2CSe_3	1.16	7.70
Hydrogen trithiocarbonate	H_2CS_3 (20°C)	2.68	8.18
Hydrogen tungstate	H_2WO_4	2.20	3.70
Hydrogen vanadate(−1)	HVO_3	3.80	

(Continued)

TABLE 1.76   Proton Transfer Reactions of Inorganic Materials in Water at 25°C (*Continued*)

Substance	Formula or remarks	pK_1	pK_2
Hydrogen vanadate(3−)	H_3VO_4	3.78	7.78(4)
Hydroxylamine-*N,N*-disulfonic acid	$HON(SO_3H)_2$ $\mu = 1.6$	pK_3 11.85	
Hydroxylamine *O*-sulfonate	$^+H_3NOSO_3^-$ $\mu = 1$	1.48	
Imidodiphosphoric acid	$(HO)_2PO(NH)PO(OH)_2$ $\mu = 0.2$	~2	2.85
		pK_3 7.08	pK_4 9.72
Indium(III) ion	In^{3+} hydrolysis	3.54	4.28
Iridium(III) ion	Ir^{3+} hydrolysis $\mu = 1$	4.37	5.20
Iron(II) ion	Fe^{2+} hydrolysis $\mu = 1$	6.8	
Iron(III) ion	Fe^{3+} hydrolysis	2.19	
Lanthanum(III) ion	La^{3+} hydrolysis	9.06	
Lead(II) ion	Pb^{2+} hydrolysis $\mu = 0.3$	7.8	
Lead(IV) ion	Pb^{4+} hydrolysis	1.8	3.2
Lithium(I) ion	Li^+	13.8	
Lutetium(III) ion	Lu^{3+} hydrolysis	7.94	
Magnesium(II) ion	Mg^{2+} hydrolysis	11.41	
Manganese(II) ion	Mn^{2+} hydrolysis	10.59	
Manganese(III) ion	Mn^{3+} hydrolysis	0.4	
Mercury(I) ion	Hg_2^{2+} hydrolysis $\mu = 0.5$	5.0	
Mercury(II) ion	Hg^{2+} hydrolysis $\mu = 0.5$	3.70	2.65
Neodymium(III) ion	Nd^{3+} hydrolysis $\mu = 3$	9.0(5)	
Neptunium(III) ion	Np^{3+} hydrolysis $\mu = 0.3$	7.43	
Neptunium(IV) ion	Np^{4+} hydrolysis $\mu = 2$	2.30	
Neptunium(V) ion	NpO_2^+ hydrolysis	8.90(2)	
Nickel(II) ion	Ni^{2+} hydrolysis	9.86	
Osmium tetroxide	OsO_4 hydrolysis $\mu = 1$	12.1	
Palladium(II) ion	Pd^{2+} (stepwise pK_b values)	13.0	12.8
Pentacyanoaquoferrate(II) ion	$Fe(CN)_5(H_2O)^{3-}$ $\mu = 0.1$	2.63	
Plutonium(III) ion	Pu^{3+} hydrolysis $\mu = 0.07$	7.2(2)	
Plutonium(IV) ion	Pu^{4+} hydrolysis $\mu = 2$	1.26	
Plutonium(V) ion	PuO_2^+ hydrolysis $\mu = 0.003$	9.7	
Plutonium(VI) ion	PuO_2^{2+} hydrolysis	3.33	4.05
Polonium(IV) ion	Po^{4+} hydrolysis	0.48	2.74
		pK_3 5.58	
Praseodymium(III) ion	Pr^{3+} hydrolysis $\mu = 0.3$	8.55	
Protoactinium(IV) ion	Pa^{4+} hydrolysis $\mu = 3$	0.14	0.38
Protoactinium(V) ion	Pa^{5+} hydrolysis $\mu = 3$	1.05	
Scandium(III) ion	Sc^{3+} hydrolysis $\mu = 0.05$	4.58(3)	
Silver(I) ion	Ag^+ hydrolysis	>11.1	
Sodium ion	Na^+ (aquo)	14.67(10)	
Strontium ion	Sr^{2+} (aquo)	13.18	
Terbium(III) ion	Tb^{3+} hydrolysis $\mu = 0.3$	8.16	
Thallium(I) ion	Tl^+	13.36(15)	
Thallium(III) ion	Tl^{3+} hydrolysis $\mu = 3$	1.14	
Thorium(IV) ion	Th^{4+} hydrolysis $\mu = 0.5$	3.89	4.20
Tin(II) ion	Sn^{2+} hydrolysis $\mu = 3$	3.81(10)	
Titanium(III)	Ti^{3+} hydrolysis $\mu = 3$	2.55	
Titanium(IV)	$TiO^{2+} + H_2O = TiO(OH)^+ + H^+$	1.3	
Tritium oxide	pK_w for $T_2O = T^+ + OH^-$	15.21	
Uranium(IV) ion	U^{4+} hydrolysis	0.68	
Uranyl(VI) ion	UO_2^{2+} $\mu = 0.035$	5.82	
Vanadium(II) ion	V^{2+} hydrolysis	6.85	
Vanadium(III) ion	V^{3+} hydrolysis	2.92	3.5
Vanadyl(IV) ion	VO^{2+} hydrolysis	6.86(10)	
Vanadyl(V) ion	$VO_2^+(20°C)$ $\mu = 0.1$	1.83	
Xenon trioxide	$XeO_3 + H_2O = HXeO_4^- + H^+$	10.5	
Ytterbium(III) ion	Yb^{3+} hydrolysis	7.99(6)	
Yttrium(III) ion	Y^{3+} hydrolysis $\mu = 0.3$	8.34	
Zinc ion	Zn^{2+} hydrolysis	8.96	
Zirconium(IV) ion	Zr^{4+} hydrolysis $\mu = 1$	−0.32	0.06
		pK_3 0.35	

Source: J. J. Christensen, L. D. Hansen, and R. M. Izatt, *Handbook of Proton Ionization Heats and Related Thermodynamic Quantities,* Wiley-Interscience, New York, 1976; D. D. Perrin, *Ionisation Constants of Inorganic Acids and Bases in Aqueous Solution,* 2d ed., Pergamon Press, 1982.

1.20 FORMATION CONSTANTS

The formation constant of a metal complex is the equilibrium constant for the formation of a complex ion from its components in solution.

Each value listed is the logarithm of the overall formation constant for the cumulative binding of a ligand L to the central metal cation M, viz.:

	Comulative formation constant	Stepwise stability constants
$M + L = ML$	K_1	k_1
$M + 2L = ML_2$	K_2	$k_1 k_2$
....		
$M + nL = ML_n$	K_n	$k_1 k_2 \cdots k_n$

As an example, the entries in Table 1.77 for the zinc ammine complexes represent these equilibria:

$$Zn^{2+} + NH_3 = Zn(NH_3)^{2+} \qquad K_1 = \frac{[Zn(NH_3)^{2+}]}{[Zn^{2+}][NH_3]}$$

$$Zn^{2+} + 2NH_3 = Zn(NH_3)_2^{2+} \qquad K_2 = \frac{[Zn(NH_3)_2^{2+}]}{[Zn^{2+}][NH_3]^2}$$

$$Zn^{2+} + 3NH_3 = Zn(NH_3)_3^{2+} \qquad K_3 = \frac{[Zn(NH_3)_3^{2+}]}{[Zn^{2+}][NH_3]^3}$$

$$Zn^{2+} + 4NH_3 = Zn(NH_3)_4^{2+} \qquad K_4 = \frac{[Zn(NH_3)_4^{2+}]}{[Zn^{2+}][NH_3]^4}$$

If the stepwise stability or formation constants of the reactions are desired, for the first step $\log K_1 = \log k_1 = 2.37$. For the second and succeeding steps the equilibria and corresponding constants are as follows:

$$Zn(NH_3)^{2+} + NH_3 = Zn(NH_3)_2^{2+} \qquad \log k_2 = \log k_2 - \log k_1 = 2.44$$

$$Zn(NH_3)_2^{2+} + NH_3 = Zn(NH_3)_3^{2+} \qquad \log k_3 = \log k_2 - \log k_1 = 3.50$$

$$Zn(NH_3)_3^{2+} + NH_3 = Zn(NH_3)_4^{2+} \qquad \log k_4 = \log k_4 - \log k_3 = 2.15$$

The reverse of the association or formation reactions would represent the dissociation or instability constant for the systems, i.e., $-\log K_f = \log K_{instab}$.

The data in the tables generally refer to temperatures of about 20 to 25°C. Most of the values in Table 1.77 refer to zero ionic strength, but those in Table 1.78 often refer to a finite ionic strength.

TABLE 1.77 Cumulative Formation Constants for Metal Complexes with Inorganic Ligands

	$\log K_1$	$\log K_2$	$\log K_3$	$\log K_4$	$\log K_5$	$\log K_6$
Ammonia						
Cadmium	2.65	4.75	6.19	7.12	6.80	5.14
Cobalt(II)	2.11	3.74	4.79	5.55	5.73	5.11
Cobalt(III)	6.7	14.0	20.1	25.7	30.8	35.2
Copper(I)	5.93	10.86				
Copper(II)	4.31	7.98	11.02	13.32	12.86	
Iron(II)	1.4	2.2				
Manganese(II)	0.8	1.3				
Mercury(II)	8.8	17.5	18.5	19.28		
Nickel	2.80	5.04	6.77	7.96	8.71	8.74
Platinum(II)						35.3
Silver(I)	3.24	7.05				
Zinc	2.37	4.81	7.31	9.46		
Bromide						
Astatine	2.51 [AtBr]					
Bismuth(III)	4.30	5.55	5.89	7.82		9.70
Bromine	1.24 [Br_3^-]					
Cadmium	1.75	2.34	3.32	3.70		
Cerium(III)	0.42					
Copper(I)		5.89				
Copper(II)	0.30					
Gold(I)		12.46				
Indium	1.30	1.88	2.48			
Iodine	2.64 [IBr]					
Iron(III)	−0.30	−0.50				
Lead	1.2	1.9		1.1		
Mercury(II)	9.05	17.32	19.74	21.00		
Palladium(II)				13.1		
Platinum(II)				20.5		
Rhodium(III)		14.3	16.3	17.6	18.4	17.2
Scandium	2.08	3.08				
Silver(I)	4.38	7.33	8.00	8.73		
Thallium(I)	0.93					
Thallium(III)	9.7	16.6	21.2	23.9	29.2	31.6
Tin(II)	1.11	1.81	1.46			
Uranium(IV)	0.18					
Yttrium	1.32					
Chloride						
Americium(III)	1.17					
Antimony(III)	2.26	3.49	4.18	4.72		
Bismuth(III)	2.44	4.7	5.0	5.6		
Cadmium	1.95	2.50	2.60	2.80		
Cerium(III)	0.48					
Copper(I)		5.5	5.7			
Copper(II)	0.1	−0.6				
Curium(III)	1.17					
Gold(III)		9.8				
Indium	1.42	2.23	3.23			
Iron(II)	0.36					
Iron(III)	1.48	2.13	1.99	0.01		
Lead	1.62	2.44	1.70	1.60		
Manganese(II)	0.96					
Mercury(II)	6.74	13.22	14.07	15.07		

TABLE 1.77 Cumulative Formation Constants for Metal Complexes with Inorganic Ligands (*Continued*)

	$\log K_1$	$\log K_2$	$\log K_3$	$\log K_4$	$\log K_5$	$\log K_6$
Palladium(II)	6.1	10.7	13.1	15.7		
Platinum(II)		11.5	14.5	16.0		
Plutonium(III)	1.17					
Silver(I)	3.04	5.04		5.30		
Thallium(I)	0.52					
Thallium(III)	8.14	13.60	15.78	18.00		
Thorium	1.38	0.38				
Tin(II)	1.51	2.24	2.03	1.48		
Tin(IV)						4
Uranium(IV)	0.8					
Uranium(VI)	0.22					
Zinc	0.43	0.61	0.53	0.20		
Zirconium	0.9	1.3	1.5	1.2		
Cyanide						
Cadmium	5.48	10.60	15.23	18.78		
Copper(I)		24.0	28.59	30.30		
Gold(I)		38.3				
Iron(II)						35
Iron(III)						42
Mercury(II)				41.4		
Nickel				31.3		
Silver(I)		21.1	21.7	20.6		
Zinc				16.7		
Fluoride						
Aluminum	6.10	11.15	15.00	17.75	19.37	19.84
Beryllium	5.1	8.8	12.6			
Cerium(III)	3.20					
Chromium(III)	4.41	7.81	10.29			
Gadolinium	3.46					
Gallium	5.08					
Indium	3.70	6.25	8.60	9.70		
Iron(III)	5.28	9.30	12.06			
Lanthanum	2.77					
Magnesium	1.30					
Manganese(II)	5.48					
Plutonium(III)	6.77					
Scandium						17.3
Thallium(I)	0.1					
Thallium(III) [TlO$^+$]	6.44					
Thorium	7.65	13.46	17.97			
Titanium(IV) [TiO^{2+}]	5.4	9.8	13.7	18.0		
Uranium(VI)	4.59	7.93	10.47	11.84		
Yttrium	4.81	8.54	12.14			
Zirconium	8.80	16.12	21.94			
Hydroxide						
Aluminum	9.27			33.03		
Antimony(III)		24.3	36.7	38.3		
Arsenic [as AsO$^+$]	14.33	18.73	20.60	21.20		
Beryllium	9.7	14.0	15.2			
Bismuth(III)	12.7	15.8		35.2		
Cadmium	4.17	8.33	9.02	8.62		
Cerium(III)	14.6					
Cerium(IV)	13.28	26.46				

(*Continued*)

TABLE 1.77 Cumulative Formation Constants for Metal Complexes with Inorganic Ligands (*Continued*)

	log K_1	log K_2	log K_3	log K_4	log K_5	log K_6
Chromium(III)	10.1	17.8		29.9		
Copper(II)	7.0	13.68	17.00	18.5		
Dysprosium	5.2					
Erbium(III)	5.4					
Gadolinium	4.6					
Gallium	11.0	21.7		34.3	38.0	40.3
Indium	9.9	19.8		28.7		
Iodine	9.49	11.24				
Iron(II)	5.56	9.77	9.67	8.58		
Iron(III)	11.87	21.17	29.67			
Lanthanum	3.3					
Lead(II)	7.82	10.85	14.58			61.0
Lutetium	6.6					
Magnesium	2.58					
Manganese(II)	3.90		8.3			
Neodymium	5.5					
Nickel	4.97	8.55	11.33			
Praseodymium	4.30					
Plutonium(III)	7.0					
Plutonium(IV)	12.39					
Plutonium [as PuO_2^{2+}]	8.3	16.6	20.9			
Samarium(III)	4.8					
Scandium	8.9					
Tellurium(IV)			41.6	53.0	64.8	72.0
Thallium(III)	12.86	25.37				
Titanium(III)	12.71					
Uranium(IV)	13.3				41.2	
Uranium(VI) [as UO_2^{2+}]	9.5	22.80		32.4		
Vanadium(III)	11.1	21.6				
Vanadium(IV) [as VO^{2+}]	8.6		[25.8 for $V_2O_4(OH)^-$]			
Vanadium(V) [as VO^{3+}]		25.2		46.2	58.5	
Yttrium	5.0					
Zinc	4.40	11.30	14.14	17.66		
Zirconium	14.3	28.3	41.9	55.3		
Iodide						
Bismuth	3.63			14.95	16.80	18.80
Cadmium	2.10	3.43	4.49	5.41		
Copper(I)		8.85				
Indium	1.00	2.26				
Iodine	2.89	5.79				
Iron(III)	1.88					
Lead	2.00	3.15	3.92	4.47		
Mercury(II)	12.87	23.82	27.60	29.83		
Silver	6.58	11.74	13.68			
Thallium(I)	0.72	0.90	1.08			
Thallium(III)	11.41	20.88	27.60	31.82		
Iodate						
Barium	1.05					
Calcium	0.89					
Magnesium	0.72					
Strontium	1.00					
Thorium	2.88	4.79	7.15			

TABLE 1.77 Cumulative Formation Constants for Metal Complexes with Inorganic Ligands (*Continued*)

	$\log K_1$	$\log K_2$	$\log K_3$	$\log K_4$	$\log K_5$	$\log K_6$
Nitrate						
Barium	0.92					
Beryllium	1.62					
Bismuth(III)	1.26					
Cadmium	0.40					
Calcium	0.28					
Cerium(III)	1.04	2.55				
Curium(III)	0.57					
Hafnium	0.92	2.43	4.32	6.40	8.48	10.29
Iron(III)	1.0					
Lanthanum	0.26	0.69	1.27			
Lead	1.18					
Mercury(II)	0.35					
Neodymium	0.52	1.18				
Neptunium(IV)	0.38					
Plutonium(III)	0.77	1.93	3.09			
Plutonium(IV)	0.54					
Strontium	0.82					
Thallium(I)	0.33					
Thallium(III)	0.92					
Thorium	0.78	1.89	2.89	3.63		
Uranium(IV)	0.20	0.37				
Uranium(VI)	0.34	0.45				
Ytterbium	0.45	1.30	2.42			
Zirconium [as ZrO^{2+}]		1.91		3.54		
Pyrophosphate						
Barium	4.6					
Calcium	4.6					
Cadmium	5.6					
Copper(II)	6.7	9.0				
Lead		5.3				
Magnesium	5.7					
Nickel	5.8	7.4				
Strontium	4.7					
Yttrium		9.7				
Zirconium		6.5				
Sulfate						
Cerium(III)	3.40					
Erbium	3.58					
Gadolinium	3.66					
Holmium	3.58					
Indium	1.78	1.88	2.36			
Iron(III)	2.03	2.98				
Lanthanum	3.64					
Neodymium	3.64					
Nickel	2.4					
Plutonium(IV)	3.66					
Praseodymium	3.62					
Samarium	3.66					
Thorium	3.32	5.50				
Uranium(IV)	3.24	5.42				
Uranium(VI)	1.70	2.45	3.30			

(*Continued*)

TABLE 1.77   Cumulative Formation Constants for Metal Complexes with Inorganic Ligands (*Continued*)

	$\log K_1$	$\log K_2$	$\log K_3$	$\log K_4$	$\log K_5$	$\log K_6$
Yttrium	3.47					
Ytterbium	3.58					
Zirconium	3.79	6.64	7.77			
Sulfite						
Copper(I)	7.5	8.5	9.2			
Mercury(II)		22.66				
Silver	5.30	7.35				
Thiocyanate						
Bismuth	1.15	2.26	3.41	4.23		
Cadmium	1.39	1.98	2.58	3.6		
Chromium(III)	1.87	2.98				
Cobalt(II)	−0.04	−0.70	0	3.00		
Copper(I)	12.11	5.18				
Gold(I)		23		42		
Indium	2.58	3.00	4.63			
Iron(III)	2.95	3.36				
Mercury(II)		17.47		21.23		
Nickel	1.18	1.64	1.81			
Ruthenium(III)	1.78					
Silver		7.57	9.08	10.08		
Thallium(I)	0.80					
Uranium(IV)	1.49	2.11				
Uranium(VI)	0.76	0.74	1.18			
Vanadium(III)	2.0					
Vanadium(IV)	0.92					
Zinc	1.62					
Thiosulfate						
Cadmium	3.92	6.44				
Copper(I)	10.27	12.22	13.84			
Iron(III)	2.10					
Lead		5.13	6.35			
Mercury(II)		29.44	31.90	33.24		
Silver	8.82	13.46				

TABLE 1.78 Cumulative Formation Constants for Metal Complexes with Organic Ligands

Temperature is 25°C and ionic strengths are approaching zero unless indicated otherwise: (*a*) At 20°C, (*b*) at 30°C, (*c*) 0.1 *M* uni-univalent salt, (*d*) 1.0 *M* uni-univalent salt, (*e*) 2.0 *M* uni-univalent salt present.

	log K_1	log K_2	log K_3	log K_4
Acetate				
Ag(I)	0.73	0.64		
Ba(II)	0.41			
Ca(II)	0.6			
Cd(II)	1.5	2.3	2.4	
Ce(III)	1.68	2.69	3.13	3.18
Co(II)	1.5	1.9		
Cr(III)	1.80	4.72		
Cu(II) *a*	2.16	3.20		
Fe(II) *c*	3.2	6.1	8.3	
Fe(III) *a,d*	3.2			
In(III)	3.50	5.95	7.90	9.08
Hg(II)		8.43		
La(III) *a,e*	1.56	2.48	2.98	2.95
Mg(II)	0.8			
Mn(II)	9.84	2.06		
Ni(II)	1.12	1.81		
Pb(II)	2.52	4.0	6.4	8.5
Rare earths *a,e*	1.6–1.9	2.8–3.0	3.3–3.7	
Sr(II)	0.44			
Tl(III)				15.4
UO$_2$(II) *a,e*	2.38	4.36	6.34	
Y(III) *a,e*	1.53	2.65	3.38	
Zn(II)	1.5			
Acetylacetone				
Al(III) *b*	8.6	15.5		
Be(II)	7.8	14.5		
Cd(II)	3.84	6.66		
Ce(III)	5.30	9.27	12.65	
Cr(II)	5.9	11.7		
Co(II)	5.40	9.54		
Cu(II)	8.27	16.34		
Dy(III) *b*	6.03	10.70	14.04	
Er(III) *b*	5.99	10.67	14.09	
Eu(III) *b*	5.87	10.35	13.64	
Fe(II)	5.07	8.67		
Fe(III)	11.4	22.1	26.7	
Ga(III)	9.5	17.9	23.6	
Gd(III) *b*	5.90	10.38	13.79	
Hf(IV)	8.7	15.4	21.8	28.1
Ho(III)	6.05	10.73	14.13	
In(III)	8.0	15.1		
La(III) *b*	5.1	8.90	11.90	
Lu(III) *b*	6.23	11.00	13.63	
Mg(II)	3.65	6.27		
Mn(II)	4.24	7.35		
Mn(III)			3.86	
Nd(III)	5.6	9.9	13.1	
Ni(II) *a*	6.06	10.77	13.09	

(*Continued*)

TABLE 1.78   Cumulative Formation Constants for Metal Complexes with Organic Ligands (*Continued*)

	$\log K_1$	$\log K_2$	$\log K_3$	$\log K_4$
Pd(II) *b*	16.2	27.1		
Pr(III) *b*	5.4	9.5	12.5	
Pu(IV) *c*	10.5	19.7	28.1	34.1
Sc(III) *b*	8.0	15.2		
Sm(III) *b*	5.9	10.4		
Tb(III) *b*	6.02	10.63	14.04	
Th(IV)	8.8	16.2	22.5	26.7
Tm(IV) *b*	6.09	10.85	14.33	
U(IV) *a,c*	8.6	17.0	23.4	29.5
UO$_2$(II) *b*	7.74	14.19		
VO(II)	8.68	15.79		
V(II)	5.4	10.2	14.7	
Y(III) *b*	6.4	11.1	13.9	
Yb(III) *b*	6.18	11.04	13.64	
Zn(II) *b*	4.98	8.81		
Zr(IV)	8.4	16.0	23.2	30.1
Alizarin red				
Cr(VI)	4.7			
Cu(II)	4.1			
Hf(IV)		10.4		
Mo(VI)		9.6		
Pb(II)	6.0			
Th(IV)		8.24		
UO$_2$(II)	4.22			
V(V)		8.6		
W(VI)		7.8		
Arsenazo				
Hf(IV)	10.07			
Zr(IV)	12.95			
Aurintricarboxylic acid				
Be(II)	4.54			
Cu(II)	4.1	8.81		
Fe(III)	4.68			
Th(IV)	5.04			
UO$_2$(II)	4.77			
Benzoylacetone (75% dioxane)				
Ba(II)		9.4		
Be(II)	12.59	24.01		
Cd(II)	7.79	14.36		
Ce(III)	10.09	19.42	27.04	
Co(II)	9.42	17.83		
Cu(II)	12.05	23.01		
La(III)	6.33	11.66	16.78	
Mg(II)	7.69	14.09		
Mn(II)	8.66	15.78		
Ni(II)	9.58	18.00		
Pb(II)	8.84	16.35		
Pr(III)	7.02	13.62	18.74	
UO$_2$(II)	12.15	23.27		
Y(III)	8.24	14.98	20.57	
Zn(II)	9.62	17.90		

TABLE 1.78 Cumulative Formation Constants for Metal Complexes with Organic Ligands (*Continued*)

	$\log K_1$	$\log K_2$	$\log K_3$	$\log K_4$
Calmagite				
Ca	6.05			
Mg	8.05			

	Complex of HL^{2-} Anion		Complex of L^{3-} Anion		Complex of H_2L^-
	$\log K_1$	$\log K_2$	$\log K_1$	$\log K_2$	$\log K_3$
Citric acid					
Ag	7.1				
Al	7.0		20.0		
Ba	2.98				
Be	4.52				
Ca	4.68				
Cd	3.98		11.3		
Ce(III)		6.18		9.65	3.2
Co(II)	4.8		12.5		
Cu(II)	4.35		14.2		
Eu(III)		6.46		9.80	
Fe(II)	3.08		15.5		
Fe(III)	12.5		25.0		
La		6.97		9.45	6.22
Mg	3.29				
Mn(II)	3.67				
Nd(III)		6.32		9.70	
Ni	5.11		14.3		
Pb	6.50				
Pr					3.4
Ra	2.36				
Sr	2.8				
Tl(I)	1.04				
UO_2	8.5	10.8			
Y					3.6
Yb				8	
Zn	4.71		11.4		

	$\log K_1$	$\log K_2$	$\log K_3$	
1,2-Diaminocyclohexane-*N, N, N′, N′*-tetraacetic acid				
Al *c*	17.63			
Ba *c*	8.64			
Ca *c*	12.3			
Cd *c*	19.88			
Ce(III) *c*	16.76			
Co(II) *c*	19.57			
Cu(II) *c*	21.95			
Dy(III) *c*	19.69			
Er(III) *c*	20.20			
Eu(III) *c*	18.77			
Fe(III) *c*	27.48			
Ga *c*	22.91			

(*Continued*)

TABLE 1.78 Cumulative Formation Constants for Metal Complexes with Organic Ligands (*Continued*)

	log K_1	log K_2	log K_3	log K_4
Gd *c*	18.80			
Hg(II) *c*	24.4			
Ho *c*	19.89			
La *c*	16.35			
Lu *c*	21.51			
Mg *c*	10.41			
Mn(II) *c*	17.43			
Nd *c*	17.69			
Ni *c*	19.4			
Pb *c*	20.33			
Pr *c*	17.23			
Sm(III) *c*	18.63			
Sr *c*	8.92			
Tb *c*	19.30			
Tm *c*	20.46			
VO(II) *c*	19.40			
Y *c*	19.41			
Yb *c*	20.80			
Zn *c*	18.6			
Dibenzoylmethane (75% dioxane)				
Ba	6.10	11.50		
Be	13.62	26.03		
Ca	7.17	13.55		
Cd	8.67	16.63		
Ce(III)	10.99	21.53	30.38	
Co(II)	10.35	20.05		
Cu(II)	12.98	24.98		
Cs	3.42			
Fe(II)	11.15	21.50		
K	3.67			
Li	5.95			
Mg	8.54	16.21		
Mn(II)	9.32	17.79		
Na	4.18			
Ni	10.83	20.72		
Pb	9.75	18.79		
Rb	3.52			
Sr	6.40	12.10		
Zn	10.23	19.65		

	log K_1	log K_2	log K_3	log K_f[MHL]
4,5-Dihydroxybenzene-1,3-disulfonic acid (Tiron)				
Al	19.02	31.10	33.5	
Ba	4.10			14.6
Ca	5.80			14.8
Cd *d*	7.69	13.29		
Ce(III)		3.75		
Co(II) *d*	8.19	14.41		15.7
Cu(II) *d*	12.76	23.73		18.1

TABLE 1.78   Cumulative Formation Constants for Metal Complexes with Organic Ligands (*Continued*)

	$\log K_1$	$\log K_2$	$\log K_3$	$\log K_f$[MHL]
Fe(III) *a,c*	20.7	35.9	46.9	22.6
La	12.9			18.6 [La(OH)L]
Mg *a,c*	6.86			14.6
Mn(II) *c*	8.6			
Ni *a,c*	8.56	14.90		15.6
Pb *d*	11.95	18.28		
Sr *c*	4.55			
UO$_2$(II) *c*	15.90			
VO(II)	15.88			
Zn *d*	9.00	16.91		15.9

	$\log K_1$	$\log K_2$	$\log K_f$[M$_2$L$_3$]
2,3-Dimercaptopropan-1-of (BAL)			
Fe(II)	15.8		
Fe(III)	30.6 [Fe(OH)L]		28
Mn(II)	5.23	10.43	
Ni		22.78	
Zn	13.48	23.3	40.6

	$\log K_1$	$\log K_2$	$\log K_3$	$\log K_4$
Dimethylglyoxime (50% dioxane)				
Cd	5.7	10.7		
Co(II)	9.80	18.94		
Cu(II)	12.00	33.44		
Fe(II)		7.25		
La	6.6	12.5		
Ni	11.16			
Pb	7.3			
Zn	7.7	13.9		
2,2′-Dipyridyl				
Ag	3.65	7.15		
Cd	4.26	7.81	10.47	
Co(II)	5.73	11.57	17.59	
Cr(II)	4.5	10.5	14.0	
Cu(I)		14.2		
Cu(II)	8.0	13.60	17.08	
Fe(II)	4.36	8.0	17.45	
Hg(II)	9.64	16.74	19.54	
Mg	0.5			
Mn(II) *d*	4.06	7.84	11.47	
Ni	6.80	13.26	18.46	
Pb	3.0			
Ti(III)			25.28	
V(II)	4.9	9.6	13.1	
Zn	5.30	9.83	13.63	
Eriochrome Black T				
Ca	5.4			
Mg	7.0			
Zn	13.5	20.6		

(*Continued*)

TABLE 1.78   Cumulative Formation Constants for Metal Complexes with Organic Ligands (*Continued*)

	log K_1	log K_2	log K_3	log K_4
Ethanolamine				
Ag	3.29	6.92		
Cu(II)		6.68		16.48
Hg(II)	8.51	17.32		
Ethylenediamine				
Ag	4.70	7.70		
Cd *a*	5.47	10.09	12.09	
Co(II)	5.91	10.64	13.94	
Co(III)	18.7	34.9	48.69	
Cr(II)	5.15	9.19		
Cu(I)		10.8		
Cu(II)	10.67	20.00	21.0	
Fe(II)	4.34	7.65	9.70	
Hg(II)	14.3	23.3		
Mg	0.37			
Mn(II)	2.73	4.79	5.67	
Ni	7.52	13.84	18.33	
Pd(II)		26.90		
V(II)	4.6	7.5	8.8	
Zn	5.77	10.83	14.11	
Ethylenediamine-*N, N, N′, N′*-tetraacetic acid				
Ag	7.32			
Al	16.11			
Am(III)	18.18			
Ba	7.78			
Be	9.3			
Bi	22.8			
Ca	11.0			
Cd	16.4			
Ce(III)	16.80			
Cf(III)	19.09			
Cm(III)	18.45			
Co(II)	16.31			
Co(III)	36			
Cr(II)	13.6			
Cr(III)	23			
Cu(II)	18.7			
Dy	18.0			
Er	18.15			
Eu(III)	17.99			
Fe(II)	14.33			
Fe(III)	24.23			
Ga	20.25			
Gd	17.2			
Hg(II)	21.80			
Ho	18.1			
In	24.95			
La	16.34			
Li	2.79			
Lu	19.83			
Mg	8.64			
Mn(II)	13.8			
Mo(V)	6.36			

TABLE 1.78 Cumulative Formation Constants for Metal Complexes with Organic Ligands (*Continued*)

	$\log K_1$	$\log K_2$	$\log K_3$	$\log K_4$
Na	1.66			
Nd	16.6			
Ni	18.56			
Pb	18.3			
Pd(II)	18.5			
Pm(III)	17.45			
Pr	16.55			
Pu(III)	18.12			
Pu(IV)	17.66			
Pu(VI)	17.66			
Ra	7.4			
Sc	23.1			
Sm	16.43			
Sn(II)	22.1			
Sr	8.80			
Tb	17.6			
Th	23.2			
Ti(III)	21.3			
TiO(II)	17.3			
Tl(III)	22.5			
Tm	19.49			
U(IV)	17.50			
V(II)	12.70			
V(III)	25.9			
VO(II)	18.0			
V(V)	18.05			
Y	18.32			
Yb	18.70			
Zn	16.4			
Zr	19.40			
Glycine				
Ag	3.41	6.89		
Ba	0.77			
Be		4.95		
Ca	1,38			
Cd	4.74	8.60		
Co(II)	5.23	9.25	10.76	
Cu(II)	8.60	15.54	16.27	
Dy		12.2		
Er		12.7		
Fe(II) *a*	4.3	7.8		
Fe(III) *a,d*	10.0			
Gd		11.9		
Hg(II)	10.3	19.2		
La		11.2		
Mg	3.44	6.46		
Mn(II)	3.6	6.6		
Ni	6.18	11.14	— 15	
Pb	5.47	8.92		
Pd(II)	9.12	17.55		
Pr		11.5		
Sm		11.7		

(*Continued*)

TABLE 1.78   Cumulative Formation Constants for Metal Complexes with Organic Ligands (*Continued*)

	$\log K_1$	$\log K_2$	$\log K_3$	$\log K_4$
Sr	0.91			
Y		12.5		
Yb		13.0		
Zn	5.52	9.96		
N'-(2-Hydroxyethyl)ethylenediamine-N, N, N'-triacetic acid				
Ba *c*	5.54			
Ca *c*	8.43			
Cd *c*	13.0			
Ce(III) *c*	14.11			
Co(II) *c*	14.4			
Cu(II) *c*	17.40			
Dy *c*	15.30			
Er *c*	15.42			
Eu(III) *c*	15.35			
Fe(II) *c*	11.6			
Fe(III) *c*	19.8			
Gd *c*	15.22			
Hg(II) *c*	20.1			
Ho *c*	15.32			
La *c*	13.46			
Lu *c*	15.88			
Mg *c*	5.78			
Mn(II) *c*	10.7			
Nd *c*	14.86			
Ni *c*	17.0			
Pb *c*	15.5			
Pr *c*	14.61			
Sm *c*	15.28			
Sr *c*	6.92			
Tb *c*	15.32			
Th *c*	18.5			
Tm *c*	15.59			
Y *c*	14.65			
Yb *c*	15.88			
Zn *c*	14.5			
8-Hydroxy-2-methylquinoline (50% dioxane)				
Cd	9.00	9.00	16.60	
Ce(III)	7.71			
Co(II)	9.63	18.50		
Cu(II)	12.48	24.00		
Fe(II)	8.75	17.10		
Mg	5.24	9.64		
Mn(II)	7.44	13.99		
Ni	9.41	17.76		
Pb	10.30	18.50		
UO$_2$(II)	9.4	17		
Zn	9.82	18.72		
8-Hydroxyquinoline-5-sulfonic acid				
Ba	2.31			
Ca	3.52			
Cd	7.70	14.20		
Ce(III)	6.05	11.05	14.95	

TABLE 1.78   Cumulative Formation Constants for Metal Complexes with Organic Ligands (*Continued*)

	$\log K_1$	$\log K_2$	$\log K_3$	$\log K_4$
Co(II)	8.11	15.05	20.41	
Cu(II)	11.92	21.87		
Er	7.16	13.34	18.56	
Fe(II)	8.4	15.7	21.75	
Fe(III)	11.6	22.8	35.65	
Gd	6.64	12.37	17.27	
La	5.63	10.13	13.83	
Mg	4.79	8.19		
Mn(II)	5.67	10.72		
Nd	6.3	11.6	16.0	
Ni	9.57	18.27	22.9	
Pb	8.53	16.13		
Pr	6.17	11.37	15.67	
Sm	6.58	12.28	17.04	
Sr	2.75			
Th	9.56	18.29	25.92	32.04
UO$_2$(II)	8.52	15.67		
Zn	8.65	16.15		
Lactic acid				
Ba	0.64			
Ca	1.42			
Cd	1.70			
Ce(III) *a,c*	2.76	4.73	5.96	
Co(II)	1.90			
Cu(II)	3.02	4.85		
Er	2.77	5.11	6.70	
Eu(III)	2.53	4.60	5.88	
Fe(III)	7.1			
Gd	2.53	4.63	5.91	
Ho	2.71	4.97	6.55	
La *a,c*	2.60	4.34	5.64	
Li	0.20			
Mg	1.37			
Mn(II)	1.43			
Nd	2.47	4.37	5.60	
Ni	2.22			
Pb	2.40	3.80		
Pr *a,c*	2.85	4.90	6.10	
Rare earths *a,c*	2.8–3.0	4.9–5.4	6.1–7.8	
Sm	2.56	4.58	5.90	
Sr	0.98			
Tb	2.61	4.73	6.01	
Y	2.53	4.70	6.12	
Yb	2.85	5.27	7.96	
Zn	2.20	3.75		
Nitrilotriacetic acid				
Al	>10			
Ba *a*	5.88			
Ca	7.60	11.61		
Cd *c*	9.80	15.2		
Ce(III) *c*	10.83	18.67		

(*Continued*)

TABLE 1.78 Cumulative Formation Constants for Metal Complexes with Organic Ligands (*Continued*)

	$\log K_1$	$\log K_2$	$\log K_3$	$\log K_4$
Co(II) *c*	10.38	14.5		
Cr(III)	>10			
Cu(II) *c*	13.10			
Dy *c*	11.74	21.15		
Er *c*	12.03	21.29		
Eu(III) *c*	11.52	20.70		
Fe(II) *c*	8.84			
Fe(III) *c*	15.87	24.32		
Gd *c*	11.54	20.80		
Hg(II)	12.7			
Ho *c*	11.90	21.25		
In	15			
La *c*	10.36	17.60		
Li *a*	3.28			
Lu *c*	12.49	21.91		
Mg *c*	5.36	10.2		
Mn(II)	8.60	11.1		
Na	2.15			
Nd *c*	11.26	19.73		
Ni	11.26	16.0		
Pb *a,c*	11.8			
Pr *c*	11.07	19.25		
Sm(III) *c*	11.53	20.53		
Sr	6.73			
Tb *c*	11.59	20.97		
Tl(I)	3.44			
Th *c*	12.4			
Tm *c*	12.22	21.45		
Y *c*	11.48	20.43		
Yb *c*	12.40	21.69		
Zn *c*	10.45	13.45		
Zr *c*	20.8			
1-Nitroso-2-naphthol (75% dioxane)				
Ag	7.74			
Cd	6.18	11.38		
Co(II)	10.67	22.81		
Cu(II)	12.52	23.37		
Mg	6.2	10.60		
Nd	9.5	17.7	25.6	
Ni	10.75	21.29	28.09	
Pb	9.73	17.31		
Pr	9.04	17.06	23.85	
Th *c*	8.50	16.13	24.03	30.29
Y	9.02	17.74	25.04	
Zn	9.32	17.02		
Zr	3.6			
Oxalate				
Ag	2.41			
Al	7.26	13.0	16.3	
Am(III)		9.8		$[Am(HL)_4^-$ 11.0]
Ba	2.31			

TABLE 1.78　Cumulative Formation Constants for Metal Complexes with Organic Ligands (*Continued*)

	$\log K_1$	$\log K_2$	$\log K_3$	$\log K_4$
Be	4.90			
Ca	3.0			
Cd	3.52	5.77		
Ce(III)	6.52	10.5	11.3	
Co(II)	4.79	6.7	9.7	
Co(III)			~20	
Cu(II)	6.16	8.5		
Er	4.82	8.21	10.03	
Fe(II)	2.9	4.52	5.22	
Fe(III)	9.4	16.2	20.2	
Gd	7.04			
Hg(II)		6.98		
Mg	3.43	4.38		
Mn(II)	3.97	5.80		
Mn(III)　*e*	9.98	16.57	19.42	
Mo(III)	3.38			
Mo(VI)				[$MoO_3(L)_2^-$ 13.0]
Nd	7.21	11.5	>14	
Ni	5.3	7.64	~8.5	
NpO_2(II)	3.30	7.07		
Pb		6.54		
Pu(III)	9.31	18.70	28	
Pu(IV)	8.74	16.91	23.39	27.50
PuO_2(II)		11.4		
Sr	2.54			
Th				24.48
TiO(II)	2.67			
Tl(I)	2.03			
UO_2(II)		10.57		
VO(II)		9.80		
V(II)	~2.7			
Y	6.52	10.10	11.47	
Yb	7.30	11.7	>14	
Zn	4.89	7.60	8.15	
Zr	9.80	17.14	20.86	21.15
1,10-Phenanthroline				
Ag	5.02	12.07		
Ca	0.7			
Cd	5.93	10.53	14.31	
Co(II)	7.25	13.95	19.90	
Cu(II)	9.08	15.76	20.94	
Fe(II)	5.85	11.45	21.3	
Fe(III)	6.5	11.4	23.5	
Hg(II)		19.65	23.35	
Mg	1.2			
Mn(II)	3.88	7.04	10.11	
Ni	8.80	17.10	24.80	
Pb	4.65	7.5	9	
VO(II)	5.47	9.69		
Zn	6.55	12.35	17.55	

(*Continued*)

TABLE 1.78 Cumulative Formation Constants for Metal Complexes with Organic Ligands (*Continued*)

	log K_1	log K_2	log K_3	log K_4
Phthalic acid				
Ba	2.33			
Ca	2.43			
Cd	2.5			
Co(II)	1.81	4.51		
Cu(II)	3.46	4.83		
La		7.74		
Ni	2.14			
Pb *d*	3.4			
UO$_2$(II)	4.38			
Zn	2.2			
Piperidine				
Ag	3.30	6.48		
Hg(II)	8.70	17.44		
Pt(II)			log K_5 5.7	log K_6 8.2
Propylene-1,2-diamine				
Cd *b,c*		9.97	12.12	
Co(II) *d*	5.42	11.47	14.72	
Cu(II) *c*	6.41	20.06		
Hg(II) *c*	10.78	23.53	23.25	
Ni *d*	7.43	13.62	17.89	
Zn *b,c*	5.89	10.87	12.57	
Pyridine				
Ag	1.97	4.35		
Cd	1.40	1.95	2.27	2.50
Co(II)	1.14	1.54		
Cu(I)		3.34	4.51	5.44
				log K_6 6.89
Cu(II)	2.59	4.33	5.93	6.54
			log K_5 7.00	log K_6 10.2
Fe(II)	0.71			
Hg(II)	5.1	10.0	10.4	
Mn(II)	1.92	2.77	3.37	3.50
VO(II)	−1.70			
Zn	1.41	1.11	1.61	1.93
Pyridine-2,6-dicarboxylic acid				
Ba *a,d*	3.46			
Ca *a,d*	4.6	7.2		
Cd *a,d*	5.7	10.0		
Ce(III) *a,d*	8.34	14.42	18.80	
Co(II) *a,d*	7.0	12.5		
Cu(II) *a,d*	9.14	16.52		
Dy *a,d*	8.69	16.19	22.14	
Er *a,d*	8.77	16.39	22.14	
Eu(III) *a,d*	8.84	15.98	21.00	
Fe(II) *a,d*	5.71	10.36		
Fe(III) *a,d*	10.91	17.13		
Gd *a,d*	8.74	16.06	21.83	
Ho *a,d*	8.72	16.23	22.08	
La *a,d*	7.98	13.79	18.06	
Lu *a,d*	9.03	16.80	21.48	

TABLE 1.78   Cumulative Formation Constants for Metal Complexes with Organic Ligands (*Continued*)

	$\log K_1$	$\log K_2$	$\log K_3$	$\log K_4$
Hg(II) *a,d*	20.28			
Mg *a,d*	2.7			
Mn(II) *a,d*	5.01	8.49		
Nd *a,d*	8.78	15.60	20.66	
Ni *a,d*	6.95	13.50		
Pb *a,d*	8.70	10.60		
Pr *a,d*	8.63	15.10	19.94	
Sm *a,d*	8.86	15.88	21.23	
Sr *a,d*	3.89			
Tb *a,d*	8.68	16.11	22.03	
Tm *a,d*	8.83	16.54	22.04	
Y *a,d*	8.46	15.73	21.34	
Yb *a,d*	8.85	16.61	21.83	
Zn *a,d*	6.35	11.88		
1-(2-Pyridylazo)-2-naphthol (PAN)				
Co(II)	>12			
Cu(II)	16			
Mn(II)	8.5	16.4		
Ni	12.7	25.3		
Tl(III)	2.29			
Zn	11.2	21.7		

	$\log K_f$ [ML]	$\log K_f$ [MHL]	$\log K_f$ [M(HL)$_2$]
4-(2-Pyridylazo)resorcinal (PAR)			
Co(II)		>12	
Cu(II)	10.3		
Mn(II)		9.7	18.9
Ni		13.2	26.0
Sc	4.8		
Tl(III)	4.23		
Zn		12.4	23.5

	$\log K_f$ [ML]	$\log K_f$ [M$_2$L]	$\log K_f$ [MHL]
Pyrocatechol-3,5-disulfonate (Pyrocatechol Violet)			
Al	19.13	4.95	
Bi	27.07	5.25	
Cd	8.13		5.86
Co(II)	9.01		6.53
Cu(II)	16.47		11.18
Ga	22.18	4.65	
In	18.10	4.81	
Mg	4.42	4.6	3.66
Mn(II)	7.13		5.36
Ni	9.35	4.38	6.85
Pb	13.25		10.19
Th	23.36	4.42	
Zn	10.41	6.21	7.21
Zr	27.40	4.18	

(*Continued*)

TABLE 1.78 Cumulative Formation Constants for Metal Complexes with Organic Ligands (*Continued*)

	$\log K_1$	$\log K_2$	$\log K_3$	$\log K_4$
8-Quinolinol				
Ba	2.07			
Be	3.36			
Ca (75% dioxane)	7.3	13.2		
Cd	7.2	13.4		
Ce(III) (50% dioxane)	9.15	17.13		
Co(II)	9.1	17.2		
Cu(II)	12.2	23.4		
Fe(II)	8.58	16.93	22.23	
Fe(III)	12.3	23.6	33.9	
La	5.85	16.95		
Mg (50% dioxane)	6.38	11.81		
Mn(II) (50% dioxane)	8.28	15.45		
Ni (50% dioxane)	11.44	21.38		
Pb (50% dioxane)	10.61	18.70		
Sm	6.84		19.50	
Sr	2.89	6.08		
Th	10.45	20.40	29.85	38.80
UO$_2$(II) (50% dioxane)	11.25	20.89		
V(II)	12.8	23.6		
VO(II)	10.97	20.19		
Y	8.15	14.90	20.25	
Zn (50% dioxane)	9.96	18.86		

	$\log K_f\,[\text{MHL}^+]$	$\log K_f\,[\text{M(HL)}_2]$
Salicylaldoxime		
Ba	0.53	3.72
Be	<7	
Ca	0.92	3.72
Cd	<4.4	
Co(II)		8.13
Cu(II)		8.13
Mg	0.64	4.10
Ni		3.77
Sr		3.77
Zn	<5.2	

	$\log K_1$	$\log K_2$	$\log K_3$	$\log K_4$
Salicylic acid				
Al	14.11			
Be	17.4			
Cd	5.55			
Ce(III)	2.66			
Co(II)	6.72	11.42		
Cr(II)	8.4	15.3		
Cu(II)	10.60	18.45		
Fe(II)	6.55	11.25		
Fe(III) *a,c*	16.48	28.12	36.80	
La	2.64			

TABLE 1.78 Cumulative Formation Constants for Metal Complexes with Organic Ligands (*Continued*)

	$\log K_1$	$\log K_2$	$\log K_3$	$\log K_4$
Mg (75% dioxane)	4.7			
Mn(II)	5.90	9.80		
Nd	2.70			
Ni	6.95	11.75		
Pr	2.68			
Th	4.25	7.60	10.05	11.60
TiO(II)	6.09			
UO_2(II)	13.4			
V(II)	6.3			
Zn	6.85			
Succinic acid				
Ba	2.08			
Be	3.08			
Ca	2.0			
Cd	2.2			
Co(II)	2.22			
Cu(II)	3.33			
Fe(III)	7.49			
Hg(II)		7.28		
La	3.96			
Mg	1.20			
Mn(II)	2.26			
Nd	8.1			
Ni	2.36			
Pb	2.8			
Ra	1.0			
Sr	1.06			
Zn	1.6			
5-Sulfosalicylic acid				
Al *c*	13.20	22.83	28.89	
Be *c*	11.71	20.81		
Cd *c*	16.68	29.08		
Co(II) *c*	6.13	9.82		
Cr(II) *c*	7.1	12.9		
Cr(III) *c*	9.56			
Cu(II) *c*	9.52	16.45		
Fe(II) *c*	5.90			
Fe(III) *c*	14.64	25.18	32.12	
La *c*	9.11			
Mn(II) *c*	5.24	8.24		
NbO(III) *c*	4.0	7.7		
Ni *c*	6.42	10.24		
UO_2(II) *c*	11.14	19.20		
Zn *c*	6.05	10.65		
Tartaric acid				
Ba		1.62		
Bi			8.30	
Ca	2.98	9.01		
Cd	2.8			
Co(II)	2.1			
Cu(II)	3.2	5.11	4.78	6.51
				$\log K_f$ 19.14 [$Cu(OH)_2L^{2-}$]

(*Continued*)

TABLE 1.78   Cumulative Formation Constants for Metal Complexes with Organic Ligands (*Continued*)

	$\log K_1$	$\log K_2$	$\log K_3$	$\log K_4$
Eu(III)	4.98	8.11		
Fe(III)	7.49			
La	3.06			
Mg		1.36		
Nd	9.0			
Pb	3.78		4.7	$\log K_f$ 14.1 [Pb(OH)$_2$L^{2-}]
Ra	1.24			
Sr	1.60			
Zn	2.68	8.32		
Thioglycolic acid				
Ce(III) *a,c*	1.99	3.03		
Co(II)	5.84	12.15		
Fe(II)		10.92		
Hg(II)		43.82		
La *a,c*	1.98	2.98		
Mn(II)	4.38	7.56		
Pb	8.5			
Ni	6.98	13.53		
Rare earths *a,c*	1.9–2.1	3.0–3.3		
Y *a,c*	1.91	3.19		
Zn	7.86	15.04		
Thiourea				
Ag	7.4	13.1		
Bi				$\log K_6$ 11.9
Cd	0.6	1.6	2.6	4.6
Cu(I)			13	15.4
Hg(II)		22.1	24.7	26.8
Pb	1.4	3.1	4.7	8.3
Ru(III)	1.21		0.72	
Thoron				
Th		10.15		
Triethanolamine				
Ag	2.30	3.64		
Co(II)	1.73			
Cu(II)	4.30			
Hg(II)	6.90	13.08		
Ni	2.7			
Zn	2.00			
Triethylenetetramine (Trien)				
Ag	7.7			
Cd	10.75	13.9		
Co(II)	11.0			
Cu(II)	20.4			
Fe(II)	7.8			
Fe(III)	21.9			
Hg(II)	25.26			
Mn(II)	4.9			
Ni	14.0			
Pb	10.4			
Zn	11.9			

TABLE 1.78   Cumulative Formation Constants for Metal Complexes with Organic Ligands (*Continued*)

	$\log K_1$	$\log K_2$	$\log K_3$	$\log K_4$
1,1,1-Trifluoro-3-2′-Thenoylacetone (TTA)				
Ba		10.6		
Cu(II)	6.55	13.0		
Fe(III)	6.9			
Ni	10.0			
Pr	9.53			
Pu(III)	9.53			
Pu(IV)	8.0			
Th	8.1			
U(IV)	7.2			
Zr	3.03 [as ZrL^{3+}]			
Xylenol orange				
Bi	5.52			
Fe(III)	5.70			
Hf	6.50			
Tl(III)	4.90			
Zn	6.15			
Zr	7.60			
Zincon				
Zn	13.1			

1.21 ELECTRODE POTENTIALS

The electrode potential is the difference between the charge on an electrode and the charge in the solution.

The electrode potential is denoted as the electromotive force (EMF) and the electromotive force of any electrolytic cell is the sum of the potentials produced at two electrodes.

1.21.1 Electromotive Force

The *electromotive force* (emf, ε; measured in volts) is the voltage developed by any source of electrical energy such as a battery or dynamo and is the electrical potential for a source in a circuit. The electromotive force converts chemical, mechanical, and other forms of energy into electrical energy and the device that supplies electrical energy is the *seat* of the electromotive force. In electromagnetic induction, the electromotive force around a closed loop is the electromagnetic work that would be done on a charge if it travels once around that loop. For a time-varying magnetic flux linking a loop, the electronic potential scalar field is not defined due to circulating electric vector field, but nevertheless an electromotive force does work that can be measured as a virtual electric potential around that loop.

A source of the electromotive force acts to move positive charge from a point of low potential through its interior to a point of high potential. By chemical, mechanical, or other means, the source of an electromotive force performs work (dW) on that charge to move it to the high-potential terminal. The electromotive force (ε) of the source is the work (dW) done per charge dq:

$$\varepsilon = dW/dq$$

TABLE 1.79 Potentials of the Elements and Their Compounds at 25°C

Standard potentials are tabulated except when a solution composition is stated; the latter are formal potentials and the concentrations are in mol/liter.

Half-reaction	Standard or formal potential	Solution composition
Actinium		
$Ac^{3+} + 3e^- = Ac$	−2.13	
Aluminum		
$Al^{3+} + 3e^- = Al$	−1.676	
$AlF_6^3 + 3e^- = Al + 6F^-$	−2.07	
$Al(OH)_4 + 3e^- = Al + 4OH^-$	−2.310	
Americium		
$AmO_2^{2+} + 4H^+ + 2e^- = Am^{4+} + 2H_2O$	1.20	
$AmO_2^{2+} + e^- = AmO_2^+$	1.59	
$AmO_2^+ + 4H^+ + e^- = Am^{4+} + 2H_2O$	0.82	
$AmO_2^+ + 4H^+ + 2e^- = Am^{3+} + 2H_2O$	1.72	
$Am^{4+} + e^- = Am^{3+}$	2.62	
$Am^{4+} + 4e^- = Am$	−0.90	
$Am^{3+} + 3e^- = Am$	−2.07	

TABLE 1.79 Potentials of the Elements and Their Compounds at 25°C (*Continued*)

Half-reaction	Standard or formal potential	Solution composition
Antimony		
$Sb(OH)_4^- + 2e^- = SbO_2^- + 2OH^- + 2H_2O$	−0.465	1 NaOH
$SbO_2^- + 2H_2O + 3e^- = Sb + 4OH^-$	0.639	1 NaOH
$Sb + 3H_2O + 3e^- = SbH_3 + 3OH^-$	−1.338	1 NaOH
$Sb_2O_5 + 6H^+ + 4e^- = 2SbO^+ + 3H_2O$	0.605	
$Sb_2O_5 + 4H^+ + 4e^- = Sb_2O_3 + 2H_2O$	0.699	
$Sb_2O_5 + 2H^+ + 2e^- = Sb_2O_4 + H_2O$	1.055	
$Sb_2O_4 + 2H^+ + 2e^- = Sb_2O_3 + H_2$	0.342	
$SbO^+ + 2H^+ + 3e^- = Sb + H_2O$	0.204	
$Sb + 3H^+ + 3e^- = SbH_3$	−0.510	
Arsenic		
$H_3AsO_4 + 2H^+ + 2e^- = HAsO_2 + 2H_2O$	0.560	
$HAsO_2 + 3H^+ + 3e^- = As + 2H_2O$	0.240	
$As + 3H^+ + 3e^- = AsH_3$	−0.225	
$AsO_4^{3-} + 2H^+ + 2e^- = AsO_2^- + 4OH^-$	−0.67	
$AsO_2^- + 2H_2O + 3e^- = As + 4OH^-$	−0.68	
$As + 3H_2O + 3e^- = AsH_3 + 3OH^-$	−1.37	
Astatine		
$HAtO_3 + 4H^+ + 4e^- = HAtO + 2H_2$	*ca.* 1.4	
$2HAtO + 2H^+ + 2e^- = At_2 + 2H_2O$	*ca.* 0.7	
$At_2 + 2e^- = 2At^-$	0.20	
Barium		
$BaO_2 + 4H^+ + 2e^- = Ba^{2+} + 2H_2O$	2.365	
$Ba^{2+} + 2e^- = Ba$	−2.92	
Berkelium		
$Bk^{4+} + 4e^- = Bk$	−1.05	
$Bk^{4+} + e^- = Bk^{3+}$	1.67	
$Bk^{3+} + 3e^- = Bk$	−2.01	
Beryllium		
$Be^{2+} + 2e^- = Be$	−1.99	
Bismuth		
$Bi_2O_4 \text{ (bismuthate)} + 4H^+ + 2e^- = 2BiO^+ + 2H_2O$	1.59	
$Bi^{3+} + 3e^- = Bi$	0.317	
$Bi + 3H^+ + 3e^- = BiH_3$	−0.97	
$BiCl_4^- + 3e^- = Bi + 4Cl^-$	0.199	
$BiBr_4^- + 3e^- = Bi + 4Br^-$	0.168	
$BiOCl + 2H^+ + 3e^- = Bi + H_2O + Cl^-$	0.170	

(*Continued*)

TABLE 1.79 Potentials of the Elements and Their Compounds at 25°C (*Continued*)

Half-reaction	Standard or formal potential	Solution composition
Boron		
$B(OH)_3 + 3H^+ + 3e^- = B + 3H_2O$	-0.890	
$BO_2^- + 6H_2O + 8e^- = BH_4^- + 8OH^-$	-1.241	
$B(OH)_4^- + 3e^- = B + 4OH^-$	-1.811	
Bromine		
$BrO_4^- + 2H^+ + 2e^- = BrO_3^- + H_2O$	1.853	
$BrO_3^- + 6H^+ + 6e^- = Br^- + 3H_2O$	1.478	
$BrO_3^- + 5H^+ + 4e^- = HBrO + 2H_2O$	1.444	
$2BrO_3^- + 12H^+ + 10e^- = Br_2 + 6H_2O$	1.5	
$2HBrO + 2H^+ + 2e^- = Br_2 + 2H_2O$	1.604	
$HBrO + H^+ + 2e^- = Br^- + H_2O$	1.341	
$BrO^- + H_2O + 2e^- = Br^- + 2OH^-$	0.76	1 NaOH
$Br_3^- + 2e^- = 3Br^-$	1.050	
$Br_2(aq) + 2e^- = 2Br^-$	1.087	
Cadmium		
$Cd^{2+} + 2e^- = Cd$	-0.403	
$Cd^{2+} + Hg + 2e^- = Cd(Hg)$	-0.352	
$CdCl_4^{2-} + 2e^- = Cd + 4Cl^-$	-0.453	
$Cd(CN)_4^{2-} + 2e^- = Cd + 4CN^-$	-0.943	
$Cd(NH_3)_4^{2+} + 2e^- = Cd + 4NH_3$	-0.622	
$Cd(OH)_4^{2-} + 2e^- = Cd + 4OH^-$	-0.670	
Calcium		
$CaO_2 + 4H^+ + 2e^- = Ca^{2+} + H_2O$	2.224	
$Ca^{2+} + 2e^- = Ca$	-2.84	
$Ca + 2H^+ + 2e^- = CaH_2$	0.776	
Californium		
$Cf^{3+} + 3e^- = Cf$	-1.93	
$Cf^{3+} + e^- = Cf^{2+}$	-1.6	
$Cf^{2+} + 2e^- = Cf$	-2.1	
Carbon		
$CO_2 + 2H^+ + 2e^- = CO + H_2O$	-0.106	
$CO_2 + 2H^+ + 2e^- = HCOOH$	-0.20	
$2CO_2 + 2H^+ + 2e^- = H_2C_2O_4$	-0.481	
$C_2O_4^{2-} + 2H^+ + 2e^- = 2HCOO^-$	0.145	
$HCOOH + 2H^+ + 2e^- = HCHO + H_2O$	0.034	
$C_2N_2 + 2H^+ + 2e^- = 2HCN$	0.373	
$HCNO + 2H^+ + 2e^- = CO + H_2O$	0.330	
$HCHO + 2H^+ + 2e^- = CH_3OH$	0.2323	
$CNO^- + H_2O + 2e^- = CN^- + 2OH^-$	-0.97	

TABLE 1.79   Potentials of the Elements and Their Compounds at 25°C (*Continued*)

Half-reaction	Standard or formal potential	Solution composition
Cerium		
$Ce(IV) + e^- = Ce(III)$	1.70	1 $HClO_4$
	1.61	1 HNO_3
	1.44	0.5 H_2SO_4
	1.28	1 HCl
$Ce^{3+} + 3e^- = Ce$	−2.34	
Cesium		
$Cs^+ + e^- = Cs$	−2.923	
$Cs^+ + Hg + e^- = Cs(Hg)$	−1.78	
Chlorine		
$ClO_4^- + 2H^+ + 2e^- = ClO_3^- + H_2O$	1.201	
$2ClO_4^- + 16H^+ + 14e^- = Cl_2 + 8H_2O$	1.392	
$ClO_4^- + 8H^+ + 8e^- = Cl^- + 4H_2O$	1.388	
$ClO_3^- + 2H^+ + e^- = ClO_2(g) + H_2O$	1.175	
$ClO_3^- + 3H^+ + 2e^- = HClO_2 + H_2O$	1.181	
$2ClO_3^- + 12H^+ + 10e^- = Cl_2 + 6H_2O$	1.468	
$ClO_3^- + 6H^+ + 6e^- = Cl^- + 3H_2O$	1.45	
$ClO_2(g) + H^+ + e^- = HClO_2$	1.188	
$HClO_2 + 2H^+ + 2e^- = HClO + H_2O$	1.64	
$HClO_2 + 3H^+ + 4e^- = Cl^- + 2H_2O$	1.584	
$2HClO_2 + 6H^+ + 6e^- = Cl_2(g) + 4H_2O$	1.659	
$2ClO^- + 2H_2O + 2e^- = Cl_2(g) + 4OH^-$	0.421	1 $NaOH$
$ClO^- + H_2O + 2e^- = Cl^- + 2OH^-$	0.890	1 $NaOH$
$Cl_3^- + 2e^- = 3Cl^-$	1.415	
$Cl_2(aq) + 2e^- = 2Cl^-$	1.396	
Chromium		
$Cr_2O_7^{2-} + 14H^+ + 6e^- = 2Cr^{3+} + 7H_2O$	1.36	
	1.15	0.1 H_2SO_4
	1.03	1 $HClO_4$
$CrO_7^{2-} + 4H_2O + 3e^- = Cr(OH)_4^- + 4OH^-$	−0.13	1 $NaOH$
$Cr^{3+} + e^- = Cr^{2+}$	−0.424	
$Cr^{3+} + 3e^- = Cr$	−0.74	
$Cr^{2+} + 2e^- = Cr$	0.90	
Cobalt		
$CoO_2 + 4H^+ + e^- = Co^{3+} + 2H_2O$	1.416	
$Co(H_2O)_6^{3+} + e^- = Co(H_2O)_6^{2+}$	1.92	
$Co(NH_3)_6^{3+} + e^- = Co(NH_3)_6^{2+}$	0.058	7 NH_3
$Co(OH)_3 + e^- = Co(OH)_2 + OH^-$	0.17	
$Co(en)_3^{3+} + e^- = Co(en)_3^{2+}$ [en = ethylenediamine]	−0.2	0.1 en
$Co(CN)_6^{3+} + e^- = Co(CN)_5^{2-} + CN^-$	−0.8	0.8 KOH
$Co^{2+} + 2e^- = Co$	−0.277	
$Co(NH_3)_6^{2+} + 2e^- = Co + 6NH_3$	−0.422	
$[Co(CO)_4]_2 + 2e^- = 2Co(CO)_4^-$	−0.40	

(*Continued*)

TABLE 1.79 Potentials of the Elements and Their Compounds at 25°C (*Continued*)

Half-reaction	Standard or formal potential	Solution composition
Copper		
$Cu^{2+} + 2e^- = Cu$	0.340	
$Cu^{2+} + e^- = Cu^+$	0.159	
$Cu^+ + e^- = Cu$	0.520	
$Cu^{2+} + Cl^- + e^- = CuCl$	0.559	
$Cu^{2+} + 2Br^- + e^- = CuBr_2^-$	0.52	1 KBr
$Cu^{2+} + I^- + e^- + CuI$	0.86	
$Cu^{2+} + 2CN^- + e^- = Cu(CN)_2^-$	1.12	
$Cu(NH_3)_4^{2+} + e^- = Cu(NH_3)_2^+ + 2NH_3$	0.10	1 NH$_3$
$Cu(en)_2^{2+} + e^- = Cu(en)^+ + en$	−0.35	
$Cu(CN)_2^- + e^- = Cu + 2CN^-$	−0.44	
$CuCl_3^{2-} + e^- = Cu + 3Cl^-$	0.178	1 HCl
$Cu(NH_3)_2^+ + e^- = Cu + 2NH_3$	−0.100	
Curium		
$Cm^{4+} + e^- = Cm^{3+}$	3.2	1 HClO$_4$
$Cm^{3+} + 3e^- = Cm$	−2.06	
Dysprosium		
$Dy^{3+} + 3e^- = Dy$	−2.29	
$Dy^{3+} + e^- = Dy^{2+}$	−2.5	
$Dy^{2+} + 2e^- = Dy$	−2.2	
Einsteinium		
$Es^{3+} + 3e^- = Es$	−2.0	
$Es^{3+} + e^- = Es^{2+}$	−1.5	
$Es^{2+} + 2e^- = Es$	−2.2	
Erbium		
$Er^{3+} + 3e^- = Er$	−2.32	
Europium		
$Eu^{3+} + 3e^- = Eu$	−1.99	
$Eu^{3+} + e^- = Eu^{2+}$	−0.35	
$Eu^{2+} + 2e^- = Eu$	−2.80	
Fermium		
$Fm^{3+} + 3e^- = Fm$	−1.96	
$Fm^{3+} + e^- = Fm^{2+}$	−1.15	
$Fm^{2+} + 2e^- = Fm$	−2.37	
Fluorine		
$F_2 + 2H^+ + 2e^- = 2HF$	3.053	
$F_2 + H^+ + 2e^- = HF_2^-$	2.979	
$F_2 + 2e^- = 2F^-$	2.87	
$OF_2 + 3H^+ + 4e^- = HF_2^- + H_2O$	2.209	
Francium		
$Fr^+ + e^- = Fr$	*ca.* −2.9	
Gadolinium		
$Gd^{3+} + 3e^- = Gd$	−2.28	
Gallium		
$Ga^{3+} + 3e^- = Ga$	−0.529	
$Ga^{3+} + e^- = Ga^{2+}$	−0.65	
$Ga^{2+} + 2e^- = Ga$	−0.45	

TABLE 1.79 Potentials of the Elements and Their Compounds at 25°C (*Continued*)

Half-reaction	Standard or formal potential	Solution composition
Germanium		
$GeO_2(tetr) + 2H^+ + 2e^- = GeO(yellow) + H_2O$	−0.255	
$GeO_2(tetr) + 4H^+ + 2e^- = Ge^{2+} + 2H_2O$	−0.210	
$GeO_2(hex) + 4H^+ + 2e^- = Ge^{2+} + 2H_2O$	−0.132	
$H_2GeO_3 + 4H^+ + 4e^- = Ge + 3H_2O$	0.012	
$Ge^{4+} + 2e^- = Ge^{2+}$	0.0	
$Ge^{2+} + 2e^- = Ge$	0.247	
$GeO + 2H^+ + 2e^- = Ge + H_2O$	−0.255	
$Ge + 4H^+ + 4e^- = GeH_4$	−0.29	
Gold		
$Au^{3+} + 3e^- = Au$	1.52	
$Au^{3+} + 2e^- = Au^+$	1.36	
$Au^+ + e^- = Au$	1.83	
$AuCl_4^- + 2e^- = AuCl_2^- + 2Cl^-$	0.926	
$AuBr_4^- + 2e^- = AuBr_2^- + 2Br-$	0.802	
$Au(SCN)_4^- + 2e^- = Au(SCN)_2^- + 2SCN^-$	0.623	
$AuBr_4^- + 3e^- = Au + 4Br^-$	0.854	
$AuCl_4^- + 3e^- = Au + 4Cl^-$	1.002	
$Au(SCN)_4^- + 3e^- = Au + 4SCN^-$	0.662	
$Au(OH)_3 + 3H^+ + 3e^- = Au + 3H_2O$	1.45	
$AuBr_2^- + e^- = Au + 2Br^-$	0.960	
$AuCl_2^- + e^- = Au + 2Cl^-$	1.15	
$AuI_2^- + e^- = Au + 2I^-$	0.576	
$Au(CN)_2^- + e^- = Au + 2CN^-$	−0.596	
$Au(SCN)_2 + e^- = Au + 2SCN^-$	0.69	
Hafnium		
$Hf^{4+} + 4e^- = Hf$	−1.70	
$HfO_2 + 4H^+ + 4e^- = Hf + 2H_2O$	−1.57	
Holmium		
$Ho^{31} + 3e^- = Ho$	−2.23	
Hydrogen		
$2H^+ + 2e^- = H_2$	0.0000	
$2D^+ + 2e^- = D_2$	0.029	
$2H_2O + 2e^- = H_2 + 2OH^-$	−0.828	
Indium		
$In^{3+} + 3e^- = In$	−0.338	
$In^{3+} + 2e^- = In^+$	−0.444	
$In^+ + e^- = In$	−0.126	
Iodine		
$H_5IO_6 + H^+ + 2e^- = IO_3^- + 3H_2O$	1.603	
$IO_3^- + 5H^+ + 4e^- = HIO + 2H_2O$	1.14	
$HIO_3 + 5H^+ + 2Cl^- + 4e^- = ICl_2 + 3H_2O$	1.214	
$2IO_3 + 12H^+ + 10e^- = I_2(c) + 3H_2O$	1.195	
$IO_3^- + 3H_2O + 6e^- = I^- + 6OH^-$	0.257	
$2IBr_2^- + 2e^- = I_2Br- + 3Br-$	0.821	
$2IBr_2^- + 2e^- = I_2(c) + 4Br^-$	0.874	
$2IBr + 2e^- = I_2Br^- + Br^-$	0.973	
$2IBr + 2e^- = I_2 + 2Br^-$	1.02	
$2ICl + 2e^- = I_2(c) + 2Cl^-$	1.20	

(*Continued*)

TABLE 1.79   Potentials of the Elements and Their Compounds at 25°C (*Continued*)

Half-reaction	Standard or formal potential	Solution composition
$2ICl_2 + 2e^- = I_2(c) + 4Cl^-$	1.07	
$2ICN + 2H^+ + 2e^- = I_2(c) + 2HCN$	0.695	
$2ICN + 2H^+ + 2e^- = I_2(aq) + 2HCN$	0.609	
$2HIO + 2H^+ + 2e^- = I_2 + 2H_2O$	1.45	
$HIO + H^+ + 2e^- = I^- + H_2O$	0.985	
$I_3^- + 2e^- = 3I^-$	0.536	
$I_2(aq) + 2e^- = 2I^-$	0.621	
$I_2(c) + 2e^- = 2I^-$	0.5355	
Iridium		
$IrBr_6^{2-} + e^- = IrBr_6^{3-}$	0.805	
$IrCl_6^{2-} + e- = IrCl_6^{3-}$	0.867	
$IrI_6^{2-} + e^- = IrI_6^{3-}$	0.49	
$IrO_2 + 4H^+ + e^- = Ir^{3+} + 2H_2O$	0.223	
$IrO_2 + 4H^+ + 4e^- = Ir + 2H_2O$	0.935	1 H_2SO_4
$Ir^{3+} + 3e^- = Ir$	1.156	
$IrCl_6^{2-} + 4e^- = Ir + 6Cl^-$	0.835	
$IrCl_6^{3-} + 3e^- = Ir + 6Cl^-$	0.77	
Iron		
$FeO_4^{2-} + 8H^+ + 3e^- = Fe^{3+} + 4H_2O$	2.2	
$FeO_4^{2-} + 2H_2O + 3e^- = FeO_2 + 4OH^-$	0.55	10 NaOH
$Fe^{3+} + e^- = Fe^{2+}$	0.771	
	0.70	1 HCl
	0.67	0.5 H_2SO_4
	0.44	0.3 H_3PO_4
$Fe(CN)_6^{3-} + e^- = Fe(CN)_6^{4-}$	0.361	
	0.71	1 HCl
$Fe(EDTA)^- + e^- = Fe(EDTA)^{2-}$	0.12	0.1 EDTA, pH 4–6
$Fe(OH)_4^- + e^- = Fe(OH)_4^{2-}$	−0.73	1 NaOH
$Fe^{2+} + 2e^- = Fe$	−0.44	
$[Fe(CO)_4]_3 + 6e^- = 3Fe(CO)_4^{2-}$	−0.70	
Lanthanum		
$La^{3+} + 3e^- = La$	−2.38	
Lawrencium		
$Lr^{3+} + 3e^- = Lr$	−2.0	
Lead		
$Pb^{4+} + 2e^- = Pb^{2+}$	1.65	
$PbO_2(alpha) + SO_4^{2-} + 4H^+ + 2e^- = PbSO_4 + 2H_2O$	1.690	
$PbO_2 + 4H^+ + 2e^- = Pb^{2+} + 2H_2O$	1.46	
$PbO_2 + 2H^+ + 2e^- = PbO + H_2O$	0.28	
$PbO_2^{2-} + H_2O + 2e^- = HPbO_2^- + 3OH^-$	0.3	2 NaOH
$Pb^{2+} + 2e^- = Pb$	−0.126	
$HPbO_2^- + H_2O + 2e^- = Pb + 3OH^-$	−0.54	
$PbHPO_4 + 2e^- = Pb + HPO_4^{2-}$	−0.465	
$PbSO_4 + 2e^- = Pb + SO_4^{2-}$	−0.356	
$PbF_2 + 2e^- = Pb + 2F^-$	−0.344	
$PbCl_2 + 2e^- = Pb + 2Cl^-$	−0.268	
$PbBr_2 + 2e^- = Pb + 2Br^-$	−0.280	
$PbI_2 + 2e^- = Pb + 2I^-$	−0.365	
$Pb + 2H^+ + 2e^- = PbH_2$	−1.507	

TABLE 1.79 Potentials of the Elements and Their Compounds at 25°C (*Continued*)

Half-reaction	Standard or formal potential	Solution composition
Lithium		
$Li^+ + e^- = Li$	-3.040	
$Li^+ + Hg + e^- = Li(Hg)$	-2.00	
Lutetium		
$Lu^{3+} + 3e^- = Lu$	-2.30	
Magnesium		
$Mg^{2+} + 2e^- = Mg$	-2.356	
$Mg(OH)_2 + 2e^- = Mg + 2OH^-$	-2.687	
Manganese		
$MnO_4^- + e^- = MnO_4^{2-}$	0.56	
$MnO_4^- + 4H^+ + 3e^- = MnO_2(beta) + 2H_2O$	1.70	
$MnO_4^- + 2H_2O + 3e^- = MnO_2 + 4OH^-$	0.60	
$MnO_4^- + 8H^+ + 5e^- = Mn^{2+} + 4H_2O$	1.51	
$MnO_4^{2-} + e^- = MnO_4^{3-}$	0.27	
$MnO_4^{2-} + 2H_2O + 2e^- = MnO_2 + 4OH^-$	0.62	
$MnO_4^{3-} + 2H_2O + e^- = MnO_2 + 4OH^-$	0.96	
$MnO_2 + 4H^+ + e^- = Mn^{3+} + 2H_2O$	0.95	
$MnO_2(beta) + 4H^+ + 2e^- = Mn^{2+} + 2H_2O$	1.23	
$Mn^{3+} + e^- = Mn^{2+}$	1.5	
$Mn(H_2P_2O_7)_3^{3-} + 2H^+ + e^- = Mn(H_2P_2O_7)_2^{2-} + H_4P_2O_7$	1.15	$0.4\ H_2P_2O_7^{2-}$
$Mn(CN)_6^{3-} + e^- = Mn(CN)_6^{4-}$	-0.24	$1.5\ NaCN$
$Mn^{2+} + 2e^- = Mn$	-1.17	
Mendelevium		
$Md^{3+} + 3e^- = Md$	-1.7	
$Md^{3+} + e^- = Md^{2+}$	-0.15	
$Md^{2+} + 2e^- = Md$	-2.4	
Mercury		
$2Hg^{2+} + 2e^- = Hg_2^{2+}$	0.911	
$2HgCl_2 + 2e^- = Hg_2Cl_2 + 2Cl^-$	0.63	
$Hg^{2+} + 2e^- = Hg(lq)$	0.8535	
$HgO(c,red) + 2H^+ + 2e^- = Hg + H_2O$	0.926	
$Hg_2^{2+} + 2e^- = 2Hg$	0.7960	
$Hg_2F_2 + 2e^- = 2Hg + 2F^-$	0.656	
$Hg_2Cl_2 + 2e^- = 2Hg + 2Cl^-$	0.2682	
$Hg_2Br_2 + 2e^- = 2Hg + 2Br^-$	0.1392	
$Hg_2I_2 + 2e^- = 2Hg + 2I^-$	-0.0405	
$Hg_2SO_4 + 2e^- = 2Hg + SO_4^{2-}$	0.614	
Molybdenum		
$MoO_4^{2-} + 4H_2O + 6e^- = Mo + 8OH^-$	-0.913	
$H_2MoO_4 + 6H^+ + 6e^- = Mo + 4H_2O$	0.114	
$H_2MoO_4 + 2H^+ + 2e^- = MoO_2 + 2H_2O$	0.646	
$MoO_2 + 4H^+ + 4e^- = Mo + 2H_2O$	-0.152	
$H_2MoO_4 + 6H^+ + 3e^- = Mo^{3+} + 4H_2O$	0.428	
$Mo(CN)_8^{3-} + e^- = Mo(CN)_8^{4-}$	0.725	
$Mo^{3+} + 3e^- = Mo$	-0.2	
Neodynium		
$Nd^{3+} + 3e^- = Nd$	-2.32	
$Nd^{3+} + e^- = Nd^{2+}$	-2.6	
$Nd^{2+} + 2e^- = Nd$	-2.2	

(*Continued*)

TABLE 1.79 Potentials of the Elements and Their Compounds at 25°C (*Continued*)

Half-reaction	Standard or formal potential	Solution composition
Neptunium		
$NpO_3^+ + 2H^+ + e^- = NpO_2^{2+} + H_2O$	2.04	
$NpO_2^{2+} + e^- = NpO_2^+$	1.34	
$NpO_2^{2+} + 4H^+ + 2e^- = Np^{4+} + 2H_2O$	0.95	
$Np^{4+} + e^- = Np^{3+}$	0.18	
$Np^{4+} + 4e^- = Np$	−1.30	
$Np^{3+} + 3e^- = Np$	−1.79	
Nickel		
$NiO_4^{2-} + 4H^+ + 2e^- = NiO_2 + 2H_2O$	1.8	
$NiO_2 + 4H^+ + 2e^- = Ni^{2+} + 2H_2O$	1.593	
$NiO_2 + 2H_2O + 2e^- = Ni(OH)_2 + 2OH^-$	0.490	
$Ni(CN)_4^{2-} + e^- = Ni(CN)_3^{2-} + CN^-$	−0.401	
$Ni^{2+} + 2e^- = Ni$	−0.257	
$Ni(OH)_2 + 2e^- = Ni + 2OH^-$	−0.72	
$Ni(NH_3)_6^{2+} + 2e^- = Ni + 6NH_3$	−0.49	
Niobium		
$Nb_2O_5 + 10H^+ + 4e^- = 2Nb^{3+} + 5H_2O$	−0.1	
$Nb_2O_5 + 10H^+ + 10e^- = 2Nb + 5H_2O$	−0.65	
$Nb^{3+} + 3e^- = Nb$	−1.1	
Nitrogen		
$2NO_3^- + 4H^+ + 2e^- = N_2O_4 + 2H_2O$	0.803	
$NO_3^- + 3H^+ + 2e^- = HNO_2 + H_2O$	0.94	
$N_2O_4 + 2H^+ + 2e^- = 2HNO_2$	1.07	
$HNO_2 + H^+ + e^- = NO + H_2O$	0.996	
$2HNO_2 + 4H^+ + 4e^- = N_2O(g) + 3H_2O$	1.297	
$2HNO_2 + 4H^+ + 4e^- = H_2N_2O_2 + 2H_2O$	0.86	
$2NO + 2H^+ + 2e^- = H_2N_2O_2$	0.71	
$2NO + 2H^+ + 2e^- = N_2O + H_2O$	1.59	
$H_2N_2O_2 + 6H^+ + 4e^- = 2HONH_3^+$	0.496	
$N_2O + 2H^+ + 2e^- = N_2 + H_2O$	1.77	
$N_2O + 6H^+ + H_2O + 4e^- = 2HONH_3^+$	−0.05	
$N_2 + 2H_2O + 4H^+ + 2e^- = 2HONH_3^+$	−1.87	
$N_2 + 5H^+ + 4e^- = N_2H_5^+$	−0.23	
$HONH_3^+ + 2H^+ + 2e^- = NH_4^+ + H_2O$	1.35	
$2HONH_3 + H^+ + 2e^- = N_2H_5^+ + 2H_2O$	1.41	
$N_2H_5^+ + 3H^+ + 2e^- = 2NH_4^+$	1.275	
$3N_2 + 2H^+ + 2e^- = 2HN_3$	−3.40	
Nobelium		
$No^{3+} + 3e^- = No$	−1.2	
$No^{3+} + e^- = No^{2+}$	1.4	
$No^{2+} + 2e^- = No$	−2.5	
Osmium		
$OsO_4(aq) + 4H^+ + 4e^- = OsO_2 \cdot 2H_2O + 2H_2O$	0.964	
$OsO_4(c, yellow) + 8H^+ + 8e^- = Os + 4H_2O$	0.85	
$OsO_2 + 4H^+ + 4e^- = Os + 2H_2O$	0.687	
$OsCl_6^{2-} + e^- = OsCl_6^{3-}$	0.45	
$OsBr_6^{2-} + e^- = OsBr_6^{3-}$	0.35	
Oxygen		
$O_3 + 2H^+ + 2e^- = O_2 + H_2O$	2.075	
$O_3 + H_2O + 2e^- = O_2 + 2OH^-$	1.240	1 NaOH

TABLE 1.79   Potentials of the Elements and Their Compounds at 25°C (*Continued*)

Half-reaction	Standard or formal potential	Solution composition
$O_2 + 4H^+ + 4e^- = 2H_2O$	1.229	
$O_2 + 2H^+ + 2e^- = H_2O$	0.695	
$O_2 + H_2O + 2e^- = HO_2^- + OH^-$	−0.076	
$H_2O_2 + 2H^+ + 2e^- = 2H_2O$	1.763	
$HO_2 + H_2O + 2e^- = 3OH^-$	0.867	1 NaOH
$O_2 + 2H_2O + 4e^- = 4OH^-$	0.401	
Palladium		
$PdO_3 + 2H^+ + 2e^- = PdO_2 + H_2O$	2.030	
$PdCl_6^{2-} + 2e^- = PdCl_4^{2-} + 2Cl^-$	1.470	
$PdBr_6^{2-} + 2e^- = PdBr_4^{2-} + 2Br^-$	0.99	
$PdI_6^{2-} + 2e^- = PdI_4^{2-} + 2I^-$	0.48	
$Pd^{2+} + 2e^- = Pd$	0.915	
$PdCl_4^{2-} + 2e^- = Pd + 4Cl^-$	0.62	1 HCl
$PdBr_4^{2-} + 2e^- = Pd + 4Br^-$	0.49	
$Pd(NH_3)_4^{2+} + 2e^- = Pd + 4NH_3$	0.0	1 NH$_3$
$Pd(CN)_4^{2-} + 2e^- = Pd + 4CN^-$	−1.35	1 KCN
Phosphorus		
$H_3PO_4 + 2H^+ + 2e^- = H_3PO_3 + H_2O$	−0.276	
$2H_3PO_4 + 2H^+ + 2e^- = H_4P_2O_6 + 2H_2O$	−0.933	
$H_4P_2O_6 + 2H^+ + 2e^- = 2H_3PO_3$	0.380	
$H_3PO_3 + 2H^+ + 2e^- = HPH_2O_2 + H_2O$	−0.499	
$HPH_2O_2 + H^+ + e^- = P + 2H_2O$	−0.365	
$H_3PO_3 + 3H^+ + 3e^- = P + 3H_2O$	−0.502	
$2P(white) + 4H^+ + 4e^- = P_2H_4$	−0.100	
$P_2H_4 + 2H^+ + 2e^- = 2PH_3$	−0.006	
$P(white) + 3H^+ + 3e^- = PH_3$	−0.063	
Platinum		
$PtO_3 + 2H^+ + 2e^- = PtO_2 + H_2O$	2.0	
$PtO_2 + 2H^+ + 2e^- = PtO + H_2O$	1.045	
$PtCl_6^{2-} + 2e^- = PtCl_4^{2-} + 2Cl^-$	0.726	
$PtBr_6^{2-} + 2e^- = PtBr_4^{2-} + 2Br^-$	0.613	1 KBr
$PtI_6^{2-} + 2e^- = PtI_4^{2-} + 2I^-$	0.321	1 KI
$Pt^{2+} + 2e^- = Pt$	1.188	
$PtCl_4^{2-} + 2e^- = Pt + 4Cl^-$	0.758	
$PtBr_4^{2-} + 2e^- = Pt + 4Br^-$	0.698	
Plutonium		
$PuO_2^{2+} + e^- = PuO_2^+$	1.02	
$PuO_2^{2+} + 4H^+ + 2e^- = Pu^{4+} + 2H_2O$	1.04	
$Pu^{4+} + e^- = Pu^{3+}$	1.01	
	0.80	1 H$_3$PO$_4$
	0.50	1 HF
$Pu^{4+} + 4e^- = Pu$	−1.25	
$Pu^{3+} + 3e^- = Pu$	−2.00	
Polonium		
$PoO_2 + 4H^+ + 2e^- = Po^{2+} + 2H_2O$	1.1	
$Po^{4+} + 4e^- = Po$	0.73	
$Po^{2+} + 2e^- = Po$	0.37	
$Po + 2H^+ + 2e^- = H_2Po$	*ca.* −1.0	

(*Continued*)

TABLE 1.79 Potentials of the Elements and Their Compounds at 25°C (*Continued*)

Half-reaction	Standard or formal potential	Solution composition
Potassium		
$K^+ + e^- = K$	-2.924	
$K^+ + Hg + e^- = K(Hg)$	*ca.* -1.9	
Praseodymium		
$Pr^{4+} + e^- = Pr^{3+}$	3.2	
$Pr^{3+} + e^- = Pr$	-2.35	
Promethium		
$Pm^{3+} + 3e^- = Pm$	-2.42	
Protoactinium		
$PaOOH^{2+} + 3H^+ + e^- = Pa^{4+} + 2H_2O$	-0.10	
$PaOOH^{2+} + 3H^+ + 5e^- = Pa + 2H_2O$	-1.19	
$Pa^{4+} + 4e^- = Pa$	-1.46	
Radium		
$Ra^{2+} + 2e^- = Ra$	-2.916	
Rhenium		
$ReO_4^- + 2H^+ + e^- = ReO_3 + H_2O$	0.768	
$ReO_4^- + 4H^+ + 3e^- = ReO_2 + 2H_2O$	0.51	
$ReO_4^- + 2H_2O + 3e^- = ReO_2 + 4OH^-$	-0.594	
$ReO_4^- + 6Cl^- + 8H^+ + 3e^- = ReCl_6^{2-} + 4H_2O$	0.12	
$2ReO_4^- + 10H^+ + 8e^- = Re_2O_3 + 5H_2O$	-0.808	
$ReO_3 + 2H^+ + 2e^- = ReO_2 + H_2O$	0.63	
$ReO_2 + 4H^+ + 4e^- = Re + 2H_2O$	0.22	
$ReCl_6^{2} + 4e^- = Re + 6Cl^-$	0.51	
$Re + e^- = Re^-$	-0.10	
Rhodium		
$RhO_2 + 4H^+ + e^- = Rh^{3+} + 2H_2O$	1.881	
$Rh^{3+} + 3e^- = Rh$	0.76	
$RhCl_6^{3-} + 3e^- = Rh + 6Cl^-$	0.5	
Rubidium		
$Rb^+ + e^- = Rb$	-2.924	
$Rb^+ + Hg + e^- = Rb(Hg)$	-1.81	
Ruthenium		
$RuO_4 + e^- = RuO_4^-$	0.89	
$RuO_4 + 4H^+ + 4e^- = RuO_2 + 2H_2O$	1.4	
$RuO_4 + 8H^+ + 8e^- = Ru + 4H_2O$	1.04	
$RuO_4^- + e^- = RuO_4^{2-}$	0.593	
$RuO_4^{2-} + 4H^+ + 2e^- = RuO_2 + 2H_2O$	2.0	
$RuO_2 + 4H^+ + 4e^- = Ru + 2H_2O$	0.68	
$Ru(H_2O)_6^{3+} + e^- = Ru(H_2O)_6^{2+}$	0.249	
$Ru(NH_3)_6^{3+} + e^- = Ru(NH_3)_6^{2+}$	0.10	
$Ru(CN)_6^{3-} + e^- = Ru(CN)_6^{4-}$	0.86	
$Ru^{3+} + e^- = Ru^{2+}$	0.249	
Samarium		
$Sm^{3+} + 3e^- = Sm$	-2.30	
$Sm^{3+} + e^- = Sm^{2+}$	-1.55	
$Sm^{2+} + 2e^- = Sm$	-2.67	
Scandium		
$Sc^{3+} + 3e^- = Sc$	-2.03	

TABLE 1.79 Potentials of the Elements and Their Compounds at 25°C (*Continued*)

Half-reaction	Standard or formal potential	Solution composition
Selenium		
$SeO_4^{2-} + 4H^+ + 2e^- = H_2SeO_3 + H_2O$	1.151	
$H_2SeO_3 + 4H^+ + 4e^- = Se + 3H_2O$	0.74	
$Se(c) + 2H^+ + 2e^- = H_2Se(aq)$	−0.115	
$Se + H^+ + 2e^- = HSe^-$	−0.227	
$Se + 2e^- = Se^{2-}$	−0.670	1 NaOH
Silicon		
$SiO_2(quartz) + 4H^+ + 4e^- = Si + 2H_2O$	−0.909	
$SiO_2 + 2H^+ + 2e^- = SiO + H_2O$	−0.967	
$SiO_2 + 8H^+ + 8e^- = SiH_4 + 2H_2O$	−0.516	
$SiF_6^{2-} + 4e^- = Si + 6F^-$	−1.37	
$SiO + 2H^+ + 2e^- = Si + H_2O$	−0.808	
$Si + 4H^+ + 4e^- = SiH_4(g)$	−0.143	
Silver		
$AgO^+ + 2H^+ + e^- = Ag^{2+} + H_2O$	1.360	
$Ag_2O_3 + 2H^+ + 2e^- = 2AgO + H_2O$	1.569	
$Ag_2O_3 + H_2O + 2e^- = 2AgO + 2OH^-$	0.739	1 NaOH
$Ag_2O_3 + 6H^+ + 4e^- = 2Ag^+ + 3H_2O$	1.670	
$Ag^{2+} + e^- = Ag^+$	1.980	
$AgO + 2H^+ + e^- = Ag^+ + H_2O$	1.772	
$Ag^+ + e^- = Ag$	0.7991	
$Ag_2SO_4 + 2e^- = 2Ag + SO_4^{2-}$	0.653	
$Ag_2C_2O_4 + 2e^- = 2Ag + C_2O_4^{2-}$	0.47	
$Ag_2CrO_4 + 2e^- = 2Ag + CrO_4^{2-}$	0.447	
$Ag(NH_3)_2^+ + e^- = Ag + 2NH_3$	0.373	
$AgCl + e^- = Ag + Cl^-$	0.2223	
$AgBr + e^- = Ag + Br^-$	0.071	
$AgCN + e^- = Ag + CN^-$	−0.017	
$AgI + e^- = Ag + I^-$	−0.152	
$Ag(CN) + e^- = Ag + 2CN^-$	−0.31	
$AgSCN + e^- = Ag + SCN^-$	0.09	
$Ag_2S + 2e^- = 2Ag + S^{2-}$	−0.71	
Sodium		
$Na^+ + e^- = Na$	−2.713	
$Na^+ + Hg + e^- = Na(Hg)$	−1.84	
Strontium		
$SrO_2 + 4H^+ + 2e^- = Sr^{2+}$	2.33	
$Sr^{2+} + 2e^- = Sr$	−2.89	
Sulfur		
$S_2O_8^{2-} + 2e^- = 2SO_4^{2-}$	1.96	
$S_2O_8^{2-} + 2H^+ + 2e^- = 2HSO_4^-$	2.08	
$2SO_4^{2-} + 4H^+ + 2e^- = S_2O_6^{2-} + 2H_2O$	−0.25	
$SO_4^{2-} + 4H^+ + 2e^- = SO_2(aq) + H_2O$	0.158	
$SO_4^{2-} + H_2O + 2e^- = SO_3^{2-} + 2OH^-$	−0.936	
$S_2O_6^{2-} + 4H^+ + 2e^- = 2H_2SO_3$	0.569	
$S_2O_6^{2-} + 2e^- = 2SO_3^{2-}$	0.037	
$2HSO_3^- + 2H^+ + 2e^- = S_2O_4^{2-} + 2H_2O$	0.099	
$2SO_3^{2-} + 2H_2O + 2e^- = S_2O_4^{2-} + 4OH^-$	−1.13	
$4H_2SO_3 + 4H^+ + 6e^- = S_4O_6^{2-} + 6H_2O$	0.507	

(*Continued*)

TABLE 1.79   Potentials of the Elements and Their Compounds at 25°C (*Continued*)

Half-reaction	Standard or formal potential	Solution composition
$4HSO_3^- + 8H^+ + 6e^- = S_4O_6^{2-} + 6H_2O$	0.577	
$2SO_2(aq) + 2H^+ + 4e^- = S_2O_3^{2-} + H_2O$	0.400	
$2SO_3^{2-} + 3H_2O + 4e^- = S_2O_3^{2-} + 6OH^-$	−0.576	1 NaOH
$SO_3^{2-} + 3H_2O + 4e^- = S + 6OH^-$	−0.59	1 NaOH
$S_4O_6^{2-} + 2e^- = 2S_2O_3^{2-}$	0.080	
$S_2O_3^{2-} + 6H^+ + 4e^- = 2S + 3H_2O$	0.5	
$SF_4(g) + 4e^- = S + 4F^-$	0.97	
$S_2Cl_2(g) + 2e^- = 2S + 2Cl^-$	1.19	
$S + H^+ + 2e^- = HS^-$	0.287	
$S + 2H^+ + 2e^- = H_2S(aq)$	0.144	
$S + 2H^+ + 2e^- = H_2S(g)$	0.174	
$S + 2e^- = S^{2-}$	−0.407	
Tantalum		
$Ta_2O_5 + 10H^+ + 10e^- = 2Ta + 5H_2O$	−0.81	
$TaF_7^{2-} + 5e^- = Ta + 7F_7^{2-}$	−0.45	
Technetium		
$TcO_4^- + 4H^+ + 3e^- = TcO_2 + 2H_2O$	0.738	
$TcO_4^- + 2H^+ + e^- = TcO_3 + H_2O$	0.700	
$TcO_4^- + e^- = TcO_4^{2-}$	0.569	
$TcO_4^- + 8H^+ + 7e^- = Tc + 4H_2O$	0.472	
$TcO_4^{2-} + 4H^+ + 2e^- = TcO_2 + 2H_2O$	1.39	
$TcO_2 + 4H^+ + 4e^- = Tc + 2H_2O$	0.272	
$Tc + e^- = Tc^-$	*ca.* −0.5	
Tellurium		
$H_2TeO_4 + 6H^+ + 2e^- = Te^{4+} + 4H_2O$	0.929	
$H_2TeO_4 + 2H^+ + 2e^- = TeO_2(c) + 2H_2O$	1.02	
$TeO_4^{2-} + 2H^+ + 2e^- = TeO_3^{2-} + H_2O$	0.897	
$TeOOH^+ + 3H^+ + 4e^- = Te + 2H_2O$	0.559	
$H_2TeO_3 + 4H^+ + 4e^- = Te + 3H_2O$	0.589	
$TeO_3^{2-} + 6H^+ + 4e^- = Te + 3H_2O$	0.827	
$TeO_3^{2-} + 3H_2O + 4e^- = Te + 6OH^-$	−0.415	
$TeO_2(c) + 4H^+ + 4e^- = Te + 2H_2O$	0.521	
$Te + 2H^+ + 2e^- = H_2Te(aq)$	−0.740	
$Te + H^+ + 2e^- = HTe^-$	−0.817	
$Te^{2-} + 2H^+ + 2e^- = 2HTe^-$	−0.794	
Terbium		
$Tb^{3+} + 3e^- = Tb$	−2.31	
Thallium		
$Tl^{3+} + 2e^- = Tl^+$	1.25	1 HClO_4
	0.77	1 HCl
$Tl^{3+} + 3e^- = Tl$	0.72	
$Tl^+ + e^- = Tl$	−0.336	
$TlCl + e^- = Tl + Cl^-$	−0.557	
$TlBr + e^- = Tl + Br^-$	−0.658	
$TlI + e^- = Tl + I^-$	−0.752	
Thorium		
$Th^{4+} + 4e^- = Th$	−1.83	

TABLE 1.79 Potentials of the Elements and Their Compounds at 25°C (*Continued*)

Half-reaction	Standard or formal potential	Solution composition
Thullium		
$Tm^{3+} + 3e^- = Tm$	-2.32	
Tin		
$Sn^{4+} + 2e^- = Sn^{2+}$	0.154	
$SnCl_6^{2-} + 2e^- = SnCl_4^{2-} + 2Cl^-$	0.14	
$SnO_3^{2-} + 6H^+ + 2e^- = Sn^{2+} + 3H_2O$	0.849	
$SnF_6^{2-} + 4e^- = Sn + 6F^-$	-0.200	
$Sn^{2+} + 2e^- = Sn$	-0.1375	
$SnCl_4^{2-} + 2e^- = Sn + 4Cl^-$	-0.19	1 HCl
$HSnO_2 + H_2O + 2e^- = Sn + 3OH^-$	-0.91	
$Sn + 4H^+ + 4e = SnH_4$	-1.07	
Titanium		
$TiO^{2+} + 2H^+ + e^- = Ti^{3+} + H_2O$	-0.10	
$TiO^{2+} + 2H^+ + 4e^- = Ti + H_2O$	-0.86	
$Ti^{3+} + e^- = Ti^{2+}$	-0.37	
$Ti^{3+} + 3e^- = Ti$	-1.21	
$Ti^{2+} + 2e^- = Ti$	-1.63	
Tungsten		
$2WO_3 + 2H^+ + 2e^- = W_2O_5 + H_2O$	-0.029	
$WO_3 + 6H^+ + 6e^- = W + 3H_2O$	-0.090	
$WO_4^{2-} + 4H_2O + 6e^- = W + 8OH^-$	-1.074	
$WO_4^{2-} + 2H_2O + 2e^- = WO_2 + 4OH^-$	-1.259	
$W_2O_5 + 2H^+ + 2e^- = 2WO_2 + H_2O$	-0.031	
$W(CN)_8^{3-} + e^- = W(CN)_8^{4-}$	0.457	
$WO_2 + 4H^+ + 4e^- = W + 2H_2O$	-0.119	
$WO_2 + 2H_2O + 4e^- = W + 4OH^-$	-0.982	
Uranium		
$UO_2^{2+} + e^- = UO_2^+$	0.16	
$UO_2^{2+} + 4H^+ + 2e^- = U^{4+} + 2H_2O$	0.27	
$UO_2^{2+} + 4H^+ + e^- = U^{4+} + 2H_2O$	0.38	
$U^{4+} + e^- = U^{3+}$	-0.52	
$U^{4+} + 4e^- = U$	-1.38	
$U^{3+} + 3e^- = U$	-1.66	
Vanadium		
$VO_2^+ + 2H^+ + e^- = VO^{2+} + H_2O$	1.000	
$VO_2^+ + 4H^+ + 2e^- = V^{3+} + 2H_2O$	0.668	
$VO_2^+ + 4H^+ + 3e^- = V^{2+} + 2H_2O$	0.361	
$VO_2^+ + 4H^+ + 5e^- = V + 4H_2O$	-0.236	
$VO^{2+} + 2H^+ + e^- = V^{3+} + H_2O$	0.337	
$V^{3+} + e^- = V^{2+}$	-0.255	
$V^{2+} + 2e^- = V$	-1.13	
Xenon		
$H_4XeO_6 + 2H^+ + 2e^- = XeO_3 + 3H_2O$	2.42	
$HXeO_6^{3-} + 2H_2O + e^- = HXeO_4 + 4OH^-$	0.9	
$XeO_3 + 6H^+ + 2F^- + 4e^- = XeF_2 + 3H_2O$	1.6	
$XeO_3 + 6H^+ + 6e^- = Xe(g) + 3H_2O$	2.10	
$XeF_2 + e^- = XeF + F^-$	0.9	
$XeF_2 + 2H^+ + 2e^- = Xe(g) + 2HF$	2.64	
$XeF + e^- = Xe(g) + F^-$	3.4	

(*Continued*)

TABLE 1.79   Potentials of the Elements and Their Compounds at 25°C (*Continued*)

Half-reaction	Standard or formal potential	Solution composition
Ytterbium		
$Yb^{3+} + e^- = Yb^{2+}$	-1.05	
$Yb^{2+} + 2e^- = Yb$	-2.8	
$Yb^{3+} + 3e^- = Yb$	-2.22	
Yttrium		
$Y^{3+} + 3e^- = Y$	-2.37	
Zinc		
$Zn^{2+} + 2e^- = Zn$	-0.7626	
$Zn(NH_3)_4^{2+} + 2e^- = Zn + 4NH_3$	-1.04	
$Zn(CN)_4^{2+} + 2e^- = Zn + 4CN^-$	-1.34	
$Zn(tartrate)_4^{6-} + 2e^- = Zn + 4(tartrate)^{2-}$	-1.15	
$Zn(OH)_4^{2-} + 2e^- = Zn + 4OH^-$	-1.285	
Zirconium		
$Zr^{4+} + 4e^- = Zr$	-1.55	
$ZrO_2 + 4H^+ + 4e^- = Zr + 2H_2O$	-1.45	

Source: A. J. Bard, R. Parsons, and J. Jordan (eds.), *Standard Potentials in Aqueous Solution* (prepared under the auspices of the International Union of Pure and Applied Chemistry), Marcel Dekker, New York, 1985; G. Charlot et al., *Selected Constants: Oxidation-Reduction Potentials of Inorganic Substances in Aqueous Solution*, Butterworths, London, 1971.

TABLE 1.80   Potentials of Selected Half-Reactions at 25°C

A summary of oxidation-reduction half-reactions arranged in order of decreasing oxidation strength and useful for selecting reagent systems.

Half-reaction	$E°$, volts
$F_2(g) + 2H^+ + 2e^- = 2HF$	3.053
$O_3 + H_2O + 2e^- = O_2 + 2OH^-$	1.246
$O_3 + 2H^+ + 2e^- = O_2 + H_2O$	2.075
$Ag^{2+} + e^- = Ag^+$	1.980
$S_2O_8^{2-} + 2e^- = 2SO_4^{2-}$	1.96
$HN_3 + 3H^+ + 2e^- = NH_4^+ + N_2$	1.96
$H_2O_2 + 2H^+ + 2e^- = 2H_2O$	1.763
$Ce^{4+} + e^- = Ce^{3+}$	1.72
$MnO_4^- + 4H^+ + 3e^- = MnO_2(c) + 2H_2O$	1.70
$2HClO + 2H^+ + 2e^- = Cl_2 + H_2O$	1.630
$2HBrO + 2H^+ + 2e^- = Br_2 + H_2O$	1.604
$H_5IO_6 + H^+ + 2e^- = IO_3^- + 3H_2O$	1.603
$NiO_2 + 4H^+ + 2e^- = Ni^{2+} + 2H_2O$	1.593
$Bi_2O_4(bismuthate) + 4H^+ + 2e^- = 2BiO^+ + 2H_2O$	1.59
$MnO_4^- + 8H^+ + 5e^- = Mn^{2+} + 4H_2O$	1.51
$2BrO_3^- + 12H^+ + 10e^- = Br_2 + 6H_2O$	1.478
$PbO_2 + 4H^+ + 2e^- = Pb^{2+} + 2H_2O$	1.468
$Cr_2O_7^{2-} + 14H^+ + 6e^- = 2Cr^{3+} + 7H_2O$	1.36
$Cl_2 + 2e^- = 2Cl^-$	1.3583
$2HNO_2 + 4H^+ + 4e^- = N_2O + 3H_2O$	1.297
$N_2H_5^+ + 3H^+ + 2e^- = 2NH_4^+$	1.275
$MnO_2 + 4H^+ + 2e^- = Mn^{2+} + 2H_2O$	1.23
$O_2 + 4H^+ + 4e^- = 2H_2O$	1.229
$ClO_4^- + 2H^+ + 2e^- = ClO_3^- + H_2O$	1.201

TABLE 1.80 Potentials of Selected Half-Reactions at 25°C (*Continued*)

Half-reaction	$E°$, volts
$2IO_3^- + 12H^+ + 10e^- = I_2 + 3H_2O$	1.195
$N_2O_4 + 2H^+ + 2e^- = 2HNO_3$	1.07
$2ICl_2^- + 2e^- = 4Cl^- + I_2$	1.07
$Br_2(lq) + 2e^- = 2Br^-$	1.065
$N_2O_4 + 4H^+ + 4e^- = 2NO + 2H_2O$	1.039
$HNO_2 + H^+ + e^- = NO + H_2O$	0.996
$NO_3^- + 4H^+ + 3e^- = NO + 2H_2O$	0.957
$NO_3^- + 3H^+ + 2e^- = HNO_2 + H_2O$	0.94
$2Hg^{2+} + 2e^- = Hg_2^{2+}$	0.911
$Cu^{2+} + I^- + e^- = CuI$	0.861
$OsO_4(c) + 8H^+ + 8e^- = Os + 4H_2O$	0.84
$Ag^+ + e^- = Ag$	0.7991
$Hg_2^{2+} + 2e^- = 2Hg$	0.7960
$Fe^{3+} + e^- = Fe^{2+}$	0.771
$H_2SeO_3 + 4H^+ + 4e^- = Se + 3H_2O$	0.739
$HN_3 + 11H^+ + 8e^- = 2NH_4^+$	0.695
$O_2 + 2H^+ + 2e^- = H_2O_2$	0.695
$Ag_2SO_4 + 2e^- = 2Ag + SO_4^{2-}$	0.654
$Cu^{2+} + Br^- + e^- = CuBr(c)$	0.654
$Au(SCN)_4^- + 3e^- = Au + 4SCN^-$	0.636
$2HgCl_2 + 2e^- = Hg_2Cl_2(c) + 2Cl^-$	0.63
$Sb_2O_5 + 6H^+ + 4e^- = 2SbO^+ + 3H_2O$	0.605
$H_3AsO_4 + 2H^+ + 2e^- = HAsO_2 + 2H_2O$	0.560
$TeOOH^+ + 3H^+ + 4e^- = Te + 2H_2O$	0.559
$Cu^{2+} + Cl^- + e^- = CuCl(c)$	0.559
$I_3^- + 2e^- = 3I^-$	0.536
$I_2 + 2e^- = 2I^-$	0.536
$Cu^+ + e^- = Cu$	0.53
$4H_2SO_3 + 4H^+ + 6e^- = S_4O_6^{2-} + 6H_2O$	0.507
$Ag_2CrO_4 + 2e^- = 2Ag + CrO_4^{2-}$	0.449
$2H_2SO_3 + 2H^+ + 4e^- = S_2O_3^{2-} + 3H_2O$	0.400
$UO_2^+ + 4H^+ + e^- = U^{4+} + 2H_2O$	0.38
$Fe(CN)_6^{3-} + e^- = Fe(CN)_6^{4-}$	0.361
$Cu^{2+} + 2e^- = Cu$	0.340
$VO^{2+} + 2H^+ + e^- = V^{3+} + H_2O$	0.337
$BiO^+ + 2H^+ + 3e^- = Bi + H_2O$	0.32
$UO_2^{2+} + 4H^+ + 2e^- = U^{4+} + 2H_2O$	0.27
$Hg_2Cl_2(c) + 2e^- = 2Hg + 2Cl^-$	0.2676
$AgCl + e^- = Ag + Cl^-$	0.2223
$SbO^+ + 2H^+ + 3e^- = Sb + H_2O$	0.212
$CuCl_3^{2-} + e^- = Cu + 3Cl^-$	0.178
$SO_4^{2-} + 4H^+ + 2e^- = H_2SO_3 + H_2O$	0.158
$Sn^{4+} + 2e^- = Sn^{2+}$	0.15
$S + 2H^+ + 2e^- = H_2S$	0.144
$Hg_2Br_2(c) + 2e^- = 2Hg + 2Br^-$	0.1392
$CuCl + e^- = Cu + Cl^-$	0.121
$TiO^{2+} + 2H^+ + e^- = Ti^{3+} + H_2O$	0.100
$S_4O_6^{2-} + 2e^- = 2S_2O_3^{2-}$	0.08
$AgBr + e^- = Ag + Br^-$	0.0711
$HCOOH + 2H^+ + 2e^- = HCHO + H_2O$	0.056
$CuBr + e^- = Cu + Br^-$	0.033
$2H^+ + 2e^- = H_2$	0.0000

(*Continued*)

TABLE 1.80 Potentials of Selected Half-Reactions at 25°C (*Continued*)

Half-reaction	$E°$, volts
$Hg_2I_2 + 2e^- = 2Hg + 2I^-$	−0.0405
$Pb^{2+} + 2e^- = Pb$	−0.125
$Sn^{2+} + 2e^- = Sn$	−0.136
$AgI + e^- = Ag + I^-$	−0.1522
$N_2 + 5H^+ + 4e^- = N_2H_5^+$	−0.225
$V^3 + e^- = V^{2+}$	−0.255
$Ni^{2+} + 2e^- = Ni$	−0.257
$Co^{2+} + 2e^- = Co$	−0.277
$Ag(CN)_2^- + e^- = Ag + 2CN^-$	−0.31
$PbSO_4 + 2e^- = Pb + SO_4^{2-}$	−0.3505
$Cd^{2+} + 2e^- = Cd$	−0.4025
$Cr^{3+} + e^- = Cr^{2+}$	−0.424
$Fe^{2+} + 2e^- = Fe$	−0.44
$H_3PO_3 + 2H^+ + 2e^- = HPH_2O_2 + H_2O$	−0.499
$2CO_2 + 2H^+ + 2e^- = H_2C_2O_4$	−0.49
$U^{4+} + e^- = U^{3+}$	−0.52
$Zn^{2+} + 2e^- = Zn$	−0.7626
$Mn^{2+} + 2e^- = Mn$	−1.18
$Al^{3+} + 3e^- = Al$	−1.67
$Mg^{2+} + 2e^- = Mg$	−2.356
$Na^+ + e^- = Na$	−2.714
$K^+ + e^- = K$	−2.925
$Li^+ + e^- = Li$	−3.045
$3N_2 + 2H^+ + 2e^- = 2HN_3$	−3.10

TABLE 1.81 Overpotentials for Common Electrode Reactions at 25°C

The overpotential is defined as the difference between the actual potential of an electrode at a given current density and the reversible electrode potential for the reaction.

Electrode	Current Density, A/cm^2					
	0.001	0.01	0.1	0.5	1.0	5.0
	Overpotential, volts					
Liberation of H$_2$ from 1M H$_2$SO$_4$						
Ag	0.097	0.13	0.3		0.48	0.69
Al	0.3	0.83	1.00		1.29	
Au	0.017		0.1		0.24	0.33
Bi	0.39	0.4			0.78	0.98
Cd		1.13	1.22		1.25	
Co		0.2				
Cr		0.4				
Cu			0.35		0.48	0.55
Fe		0.56	0.82		1.29	
Graphite	0.002		0.32		0.60	0.73
Hg	0.8	0.93	1.03		1.07	
Ir	0.0026	0.2				
Ni	0.14	0.3			0.56	0.71
Pb	0.40	0.4			0.52	1.06
Pd	0	0.04				
Pt (smooth)	0.0000	0.16	0.29		0.68	
Pt (platinized)	0.0000	0.030	0.041		0.048	0.051
Sb		0.4				
Sn		0.5	1.2			
Ta		0.39	0.4			
Zn	0.48	0.75	1.06		1.23	
Liberation of O$_2$ from 1M KOH						
Ag	0.58	0.73	0.98		1.13	
Au	0.67	0.96	1.24		1.63	
Cu	0.42	0.58	0.66		0.79	
Graphite	0.53	0.90	1.09		1.24	
Ni	0.35	0.52	0.73		0.85	
Pt (smooth)	0.72	0.85	1.28		1.49	
Pt (platinized)	0.40	0.52	0.64		0.77	
Liberation of Cl$_2$ from saturated NaCl solution						
Graphite			0.25	0.42	0.53	
Platinized Pt	0.006		0.026	0.05		
Smooth Pt	0.008	0.03	0.054	0.161	0.236	
Liberation of Br$_2$ from saturated NaBr solution						
Graphite		0.002	0.027	0.16	0.33	
Platinized Pt		0.002	0.012	0.069	0.21	
Smooth Pt		0.002	0.006*	0.26	0.38[†]	
Liberation of I$_2$ from saturated NaI solution						
Graphite	0.002	0.014	0.097			
Platinized Pt		0.006	0.032		0.196	
Smooth Pt		0.003	0.03	0.12	0.22	

*At 0.23 A/cm^2.† At 0.72 A/cm^2.

The overpotential required for the evolution of O$_2$ from dilute solutions of HClO$_4$, HNO$_3$, H$_3$PO$_4$ or H$_2$SO$_4$ onto smooth platinum electrodes is approximately 0.5 V.

TABLE 1.82 Half-Wave Potentials of Inorganic Materials

All values are in volts vs. The saturated calomel electrode.

Element	$E_{1/2}$, volts	Solvent system
Aluminum		
3+	−0.5	0.2M acetate, pH 4.5–4.7, plus 0.07% azo dye Pontochrome Violet SW; reduction wave of complexed dye is 0.2 V more negative than that of the free dye.
Antimony		
3+ to 0	−0.15	1M HCl
	−0.31(1)	1M HNO$_3$ (or 0.5M H$_2$SO$_4$)
	−0.8	0.5M tartrate, pH 4.5
	−1.0; −1.2	0.5M tartrate, pH 9 (waves not distinct)
	−1.26	1M NaOH; also anodic wave (3+ to 5+) at −0.45
	−1.32	0.5M tartrate plus 0.1M NaOH
5+	0.0; −0.257	6M HCl. First wave (5+ to 3+) starts at the oxidation potential of Hg; second wave is 3+ to 0.
5+ to 0	−0.35	1M HCl plus 4M KBr
Arsenic		
3+ to 5+	−0.26	0.5M KOH (anodic wave); only suitable wave
3+	−0.8; −1.0	0.1M HCl; ill-defined waves
	−0.7; −1.0	0.5M H$_2$SO$_4$ (or 1M HNO$_3$)
Barium		
2+ to 0	−1.94	0.1M (C$_2$H$_5$)$_4$NI
Bismuth		
3+ to 0	−0.025(15)	1M HNO$_3$ (or 0.5M H$_2$SO$_4$)
	−0.09	1M HCl
	−0.29	0.5M tartrate, pH 4.5
	−0.7	0.5M tartrate (pH 9), wave not well-developed
	−1.0	0.5M tartrate plus 0.1M NaOH, poor wave
Bromine		
5+ to 1−	−1.75	0.1M alkali chlorides (or 0.1M NaOH)
	0.13	0.05M H$_2$SO$_4$
0 to 1−	0.0	Wave (anodic) starts at zero; Hg$_2$Br$_2$ forms
Br$^-$	0.1	Oxidation of Hg to form mercury(I) bromide
Cadmium		
2+ to 0	−0.60	0.1M KCl, or 0.5M H$_2$SO$_4$, or 1M HNO$_3$
	−0.64	0.5M tartrate at pH 4.5 or 9
	−0.81	1M NH$_4$Cl plus 1M NH$_3$
Calcium		
2+ to 0	−2.22	0.1M (C$_2$H$_5$)$_4$NCl
	−2.13	0.1M (C$_2$H$_5$)$_4$NCl in 80% ethanol
Cerium		
3+ to 0	−1.97	0.02M alkali sulfate
Cesium		
1+ to 0	−2.05	0.1M (C$_2$H$_5$)$_4$NOH in 50% ethanol
Chlorine		
Cl$^-$	0.25	Oxidation of Hg to form Hg$_2$Cl$_2$
Chromium		
6+ to 3+	−0.85	CrO$_4^{2-}$ to CrO$_4^-$ in 0.1 to 1M NaOH
3+ to 0	−0.35; −1.70	1M NH$_4$Cl—NH$_3$ buffer (pH 8–9); 3+ to 2+ to 0
3+ to 2+	−0.95	0.1M pyridine–0.1M pyridinium chloride

TABLE 1.82   Half-Wave Potentials of Inorganic Materials (*Continued*)

Element	$E_{1/2}$, volts	Solvent system
2+ to 0	−1.54	1*M* KCl
2+ to 3+	−0.40	1*M* KCl (anodic wave)
Cobalt		
3+ to 0	−0.5; −1.3	1*M* NH$_4$Cl plus 1*M* NH$_3$; 3+ to 2+ to 0
2+ to 0	−1.07	0.1*M* pyridine plus pyridinium chloride
	−1.03	Neutral 1*M* potassium thiocyanate
	−1.4	Co(H$_2$O)$_6^{2+}$ in noncomplexing systems
3+ to 2+	0.0	1*M* sodium oxalate in acetate buffer (pH 5); diffusion current measured between 0 and −0.1 V
Copper		
2+ to 0	0.04	0.1*M* KNO$_3$, 0.1*M* NH$_4$ClO$_4$, or 1*M* Na$_2$SO$_4$
	−0.085	0.1*M* Na$_4$P$_2$O$_7$ plus 0.2*M* Na acetate, pH 4.5
	−0.09	0.5*M* Na tartrate, pH 4.5
	−0.20	0.1*M* potassium oxalate, pH 5.7 to 10
	−0.22	0.5*M* potassium citrate, pH 7.5
	−0.4	0.5*M* Na tartrate plus 0.1*M* NaOH (pH 12)
	−0.568	0.1*M* KNO$_3$ plus 1*M* ethylenediamine
2+	0.04; −0.22	1*M* KCl; consecutive waves: 2+ to 1+ to 0
	−0.02; −0.39	0.1*M* KSCN; consecutive waves: 2+ to 1+ to 0
	0.05; −0.25	0.1*M* pyridine plus 0.1*M* pyridinium chloride; consecutive waves: 2+ to 1+ to 0
	−0.24; −0.50	1*M* NH$_4$Cl plus 1*M* NH$_3$; consecutive waves
Gallium		
3+ to 0	−1.1	Not more than 0.001*M* HCl or wave masked by hydrogen wave which immediately follows
Germanium		
2+ to 0	−0.45	6*M* HCl; prior reduction with HPH$_2$O$_2$ to 2+
Gold		
3+ to 1+	0	1*M* KCN; wave starts at 0 V
1+ to 0	−1.4	Au(CN)$_2^-$ wave best for analytical purposes
Indium		
3+ to 0	−0.60	1*M* KCl
		In Na acetate, pH 3.9 to 4.2
Iodine		
IO$_4^-$	0.36	First wave at pH 0 (shifts to −0.08 at pH 12); second wave corresponds to iodate reduction
IO$_3^-$	−0.075	0.2*M* KNO$_3$ (shifts −0.13 V/pH unit increase)
	−0.305	0.1*M* hydrogen phthalate, pH 3.2
	−0.500	0.1*M* acetate plus 0.1*M* KCl, pH 4.9
	−0.650	0.1 M citrate, pH 5.95
	−1.050	0.2*M* phosphate, pH 7.10
	−1.20	0.05*M* borax + 0.1*M* KCl, pH 9.2; or NaOH plus 0.1*M* KCl, pH 13.0
0 to 1−	0.0	Wave starts from zero in acid media; Hg$_2$I$_2$ formed
1−	−0.1	Oxidation of Hg to form Hg$_2$I$_2$
Iron		
3+	−0.44; −1.52	1*M* (NH$_4$)$_2$CO$_3$; two waves; 3+ to 2+ to 0
	−0.17; −1.50	0.5*M* Na tartrate, pH 5.8; two waves; 3+ to 2+ to 0
	−0.9; −1.5	0.1 to 5*M* KOH plus 8% mannitol; 3+ to 2+ to 0

(*Continued*)

TABLE 1.82   Half-Wave Potentials of Inorganic Materials (*Continued*)

Element	$E_{1/2}$, volts	Solvent system
3+ to 2+	−0.13	0.1M EDTA plus 2M Na acetate, pH 6–7
	−0.27	0.2M Na oxalate, pH 7.9 or less
	−0.28	0.5M Na citrate, pH 6.5
	−1.46(2)	1M NH$_4$ClO$_4$
	−1.36	0.1M KHF$_2$, pH 4 or less
2+ to 3+	−0.28	0.5M Na citrate, pH 6.5
	−0.27	0.2M Na oxalate, pH 7.9 or less
	−0.17	0.5M Na tartrate, pH 5.8
	−1.36	0.1M KHF$_2$, pH 4 or less
Lead		
2+ to 0	−0.405	1M HNO$_3$
	−0.435	1M KCl (or HCl)
	−0.49(1)	0.5M Na tartrate, pH 4.5 or 9
	−0.72	1M KCN
	−0.75	1M KOH or 0.5M Na tartrate plus 0.1M NaOH
Lithium		
1+ to 0	−2.31	0.1M (C$_2$H$_5$)$_4$ NOH in 50% ethanol
Magnesium		
2+ to 0	−2.2	0.1M (C$_2$H$_5$)$_4$NCl (poorly defined wave)
Manganese		
2+ to 0	−1.65	1M NH$_4$Cl plus 1M NH$_3$
	−1.55	1M KCNS
	−1.33	1.5M KCN
Molybdenum		
6+	−0.26; −0.63	0.3M HCl, two waves: 6+ to 5+ to 3+
Nickel		
2+ to 0	−0.70	1M KSCN
	−0.78	1M KCl plus 0.5M pyridine
	−1.09	1M NH$_4$Cl plus 1M NH$_3$
	−1.1	Ni(H$_2$O)$_6^{2+}$ in NH$_4$ClO$_4$ or KNO$_3$
	−1.36	Ni(CN)$_4^{2-}$ in 1M KCN (alkaline media)
Niobium		
5+ to 3+	−0.80(4)	1M HNO$_3$
Nitrogen		
Nitrate	−1.45	0.011M LaCl$_3$ (reduced to hydroxylamine)
HNO$_2$	−0.77	0.1M HCl
C$_2$N$_2$	−1.2; −1.55	0.1M Na acetate, two waves
Oxamic acid	−1.55	0.1M Na acetate
Cyanide	−0.45	0.1M NaOH; anodic wave starts at −0.45
Thiocyanate	0.18	Anodic wave; neutral or weakly alkaline medium
Osmium		
OsO$_4$	0.0; −0.41; −1.16	Sat'd Ca(OH)$_2$. Three waves: first starts at 0; second wave is OsO$_4^{2-}$ to Os(V); and third wave is Os(V) to Os(III)
Oxygen		
O$_2$	−0.05; −0.9	Buffer solutions of pH 1 to 10. Two waves: O$_2$ to H$_2$O$_2$, and H$_2$O$_2$ to H$_2$O. Second wave extends from −0.5 to −1.3
H$_2$O$_2$	−0.9	Very extended wave (see above); sharper in presence of Aerosol OT

TABLE 1.82 Half-Wave Potentials of Inorganic Materials (*Continued*)

Element	$E_{1/2}$, volts	Solvent system
Palladium		
2+ to 0	−0.31	1M pyridine plus 1M KCl
	−0.64	0.1M ethylenediamine plus 1M KCl
	−0.72	1M NH$_4$Cl plus 1M NH$_3$
Potassium		
1+ to 0	−2.10	0.1M (C$_2$H$_5$)$_4$NOH in 50% ethanol
Rhenium		
7+ to 4+	−0.44	2M HCl or (better) 4M HClO$_4$
4+ to 3+	−0.51	ReCl$_6^{2-}$ ion in 1M HCl
Rhodium		
3+ to 2+	−0.41	1M pyridine plus 1M KCl
Rubidium		
1+ to 0	−1.99	0.1M (C$_2$H$_5$)$_4$NOH in 50% ethanol
Scandium		
3+ to 0	−1.80	0.1M LiCl, KCl, or BaCl$_2$
Selenium		
4+ to 2−	−1.44	1M NH$_4$Cl plus NH$_3$, pH 8.0
	−1.54	Same system adjusted to pH 9.5
2−	−0.49	Anodic wave at pH 0 due to HgSe
	−0.94	Anodic wave at pH 12 (0.01M NaOH)
Silver		
1+ to 0		Wave starts at oxidation potential of Hg
1+ to 0	−0.3	0.0014M KAg(CN)$_2$ without excess cyanide
Sodium		
1+ to 0	−2.07	0.1M (C$_2$H$_5$)$_4$NOH in 50% ethanol
Strontium		
2+ to 0	−2.11	0.1M (C$_2$H$_5$)$_4$NI, water or 80% ethanol
Sulfur		
SO$_2$	−0.38	1M HNO$_3$ (or other strong acid); 4+ to 2+
S$_2$O$_4^{2-}$	−0.43	0.5M (NH$_4$)$_2$HPO$_4$ plus 1M NH$_3$ (anodic wave)
S$_2$O$_3^{2-}$	−0.15	1M strong acid; anodic mercury wave
0 to 2−	−0.50	90% methanol, 9.5% pyridine, 0.5% HCl (pH 6)
HS$^-$	−0.76	0.1M NaOH (anodic mercury wave)
Tellurium		
4+ to 0	−0.4	Citrate buffer, pH 1.6 (second of two waves)
	−0.63	Ammoniacal buffer, pH 9.4
4+ to 2−	−1.22	0.1M NaOH
2− to 0	−0.72	1M HCl (true anodic reversible wave)
	−0.08	1M NaOH (same as above; intermediate values at pH 1 to 13)
Thallium		
3+ to 0	−0.48	1M KCl, KNO$_3$, K$_2$SO$_4$, KOH, or NH$_3$
Tin		
4+ to 2+	−0.25; −0.52	4M NH$_4$Cl + 1M HCl; two waves: 4+ to 2+ to 0
2+ to 0	−0.59	0.5M tartrate, pH 4.3
	−1.22	1M NaOH (stannite ion to tin)
2+ to 4+	−0.28	0.5M Na tartrate, pH 4.3 (anodic wave)
	−0.73	1M NaOH (stannite ion to stannate ion)

(*Continued*)

TABLE 1.82   Half-Wave Potentials of Inorganic Materials (*Continued*)

Element	$E_{1/2}$, volts	Solvent system
Titanium		
4+ to 3+	−0.173	0.1M K$_2$C$_2$O$_4$ plus 1M H$_2$SO$_4$
	−1.22	0.4M tartrate, pH 6.5
Tungsten		
6+	0.0; −0.64	6M HCl; two waves: first wave starts at zero and is W(VI) to W(V), the second wave is W(V) to W(III)
Uranium		
6+	−0.180; −0.92	UO$_2^{2+}$ to UO$_2^+$, then U^{3+} in 0.02M HCL
Vanadium		
5+ to 4+ to 2+	−0.97; −1.26	1M NH$_4$Cl plus 1M NH$_3$ and 0.08M Na$_2$SO$_3$
4+ to 2+	−0.98	0.05M H$_2$SO$_4$
3+ to 2+	−0.55	0.5M H$_2$SO$_4$
4+ to 5+	−0.32	1M NH$_4$Cl, 1M NH$_3$, and 0.08M Na$_2$SO$_3$
4+ to 5+	0.76	0.05M H$_2$SO$_4$; anodic wave starting from zero
2+ to 3+	−0.55	0.5M H$_2$SO$_4$; anodic wave
Zinc		
2+ to 0	−0.995	0.1M KCl
	−1.01	0.1M KSCN
	−1.15	0.5M tartrate, pH 9
	−1.23	0.5M tartrate, pH 4.5
	−1.33	1M NH$_4$Cl plus 1M NH$_3$
	−1.53	1M NaOH

TABLE 1.83   Standard Electrode Potentials for Aqueous Solutions

Acidic solutions ([H$^+$] = 1.0 mol kg^{-1})	
Half-reaction	$E°(V)$
Li$^+$ + e^- ⇌ Li	−3.045
K$^+$ + e^- ⇌ K	−2.925
Na$^+$ + $e-$ ⇌ Na	−2.714
La^{3+} + 3e^- ⇌ La	−2.37
Mg^{2+} + 2e^- ⇌ Mg	−2.356
$\frac{1}{2}$H$_2$ + e^- ⇌ H$^-$	−2.25
Be^{2+} + 2e^- ⇌ Be	−1.97
Zr^{4+} + 4e^- ⇌ Zr	−1.70
Al^{3+} + 3e^- ⇌ Al	−1.67
Ti^{3+} + 3e^- ⇌ Ti	−1.21
Mn^{2+} + 2e^- ⇌ Mn	−1.18
V^{2+} + 2e^- ⇌ V	−1.13
SiO$_2$(glass) + 4H$^+$ + 4e^- ⇌ Si + 2H$_2$O	−0.888
Zn^{2+} + 2e^- ⇌ Zn	−0.763
U^{4+} + e^- ⇌ U^{3+}	−0.52
Fe^{2+} + 2e^- ⇌ Fe	−0.44
Cr^{3+} + e^- ⇌ Cr^{2+}	−0.424
Cd^{2+} + 2e^- ⇌ Cd	−0.403
PbSO$_4$ + 2e^- ⇌ Pb + SO$_4^{2-}$	−0.351
Eu^{3+} + e^- ⇌ Eu^{2+}	−0.35

TABLE 1.83 Standard Electrode Potentials for Aqueous Solutions (*Continued*)

Acidic solutions ($[H^+] = 1.0$ mol kg^{-1})	
Half-reaction	$E°(V)$
$Co^{2+} + 2e^- \rightleftharpoons Co$	-0.277
$H_3PO_4 + 2H^+ + 2e^- \rightleftharpoons H_3PO_3 + H_2O$	-0.276
$Ni^{2+} + 2e^- \rightleftharpoons Ni$	-0.257
$V^{3+} + e^- \rightleftharpoons V^{2+}$	-0.255
$2SO_4^{2-} + 4H^+ + 2e^- \rightleftharpoons S_2O_6^{2-} + 2H_2O$	-0.253
$N_2 + 5H^+ + 4e^- \rightleftharpoons N_2H_5^+$	-0.23
$CO_2 + 2H^+ + 2e^- \rightleftharpoons HCOOH$	-0.16
$AgI + e^- \rightleftharpoons Ag + I^-$	-0.152
$Sn^{2+} + 2e^- \rightleftharpoons Sn$	-0.136
$Pb^{2+} + 2e^- \rightleftharpoons Pb$	-0.125
$2H^+ + 2e^- \rightleftharpoons H_2$	0.000
$HCOOH + 2H^+ + 2e^- \rightleftharpoons HCHO + H_2O$	$+0.056$
$AgBr + e^- \rightleftharpoons Ag + Br^-$	$+0.071$
$TiO^{2+} + 2H^+ \rightleftharpoons + e^- Ti^{3+} + H_2O$	$+0.100$
$S + 2H^+ + 2e^- \rightleftharpoons H_2S$	$+0.144$
$Sn^{4+} + 2e^- \rightleftharpoons Sn^{2+}$	$+0.15$
$SO_4^{2-} + 4H^+ + 2e^- \rightleftharpoons H_2SO_3 + H_2O$	$+0.158$
$Cu^{2+} + e^- \rightleftharpoons Cu^+$	$+0.159$
$AgCl + e^- \rightleftharpoons Ag + Cl^-$	$+0.222$
$HCHO + 2H^+ + 2e^- \rightleftharpoons CH_3OH$	$+0.232$
$UO_2^{2+} + 4H^+ + 2e^- \rightleftharpoons U^{4+} + 2H_2O$	$+0.27$
$VO^{2+} + 2H^+ + e^- \rightleftharpoons V^{3+} + H_2O$	$+0.337$
$Cu^{2+} + 2e^- \rightleftharpoons Cu$	$+0.340$
$Fe(CN)_6^{3-} + e^- \rightleftharpoons Fe(CN)_6^{4-}$	$+0.361$
$2H_2SO_3 + 2H^+ + 4e^- \rightleftharpoons S_2O_3^{2-} + 3H_2O$	$+0.400$
$H_2SO_3 + 4H^+ + 4e^- \rightleftharpoons S + 3H_2O$	$+0.500$
$4H_2SO_3 + 4H^+ + 6e^- \rightleftharpoons S_4O_6^{2-} + 6H_2O$	$+0.507$
$Cu^+ + e^- \rightleftharpoons Cu$	$+0.520$
$I_2 + 2e^- \rightleftharpoons 2I^-$	$+0.5355$
$I_3^- + 2e^- \rightleftharpoons 3I^-$	$+0.536$
$MnO_4^- + e^- \rightleftharpoons MnO_4^{2-}$	$+0.56$
$S_2O_6^{2-} + 4H^+ + 2e^- \rightleftharpoons 2H_2SO_3$	$+0.569$
$CH_3OH + 2H^+ + 2e^- \rightleftharpoons CH_4 + H_2O$	$+0.59$
$HN_3 + 11H^+ + 8e^- \rightleftharpoons 3NH_4^+$	$+0.695$
$O_2 + 2H^+ + 2e^- \rightleftharpoons H_2O_2$	$+0.695$
$Rh^{3+} + 3e^- \rightleftharpoons Rh$	$+0.76$
$(NCS)_2 + 2e^- \rightleftharpoons 2NCS^-$	$+0.77$
$Fe^{3+} + e^- \rightleftharpoons Fe^{2+}$	$+0.771$
$Hg_2^{2+} + 2e^- \rightleftharpoons 2Hg$	$+0.796$
$Ag^+ + e^- \rightleftharpoons Ag$	$+0.799$
$2NO_3^- + 4H^+ + 2e^- \rightleftharpoons N_2O_4 + 2H_2O$	$+0.803$
$Hg^{2+} + 2e^- \rightleftharpoons Hg$	$+0.911$
$NO_3^- + 3H^+ + 2e^- \rightleftharpoons HNO_2 + H_2O$	$+0.94$
$NO_3^- + 4H^+ + 3e^- \rightleftharpoons NO + 2H_2O$	$+0.957$
$NHO_2 + H^+ + e^- \rightleftharpoons NO + H_2O$	$+0.996$
$N_2O_4 + 4H^+ + 4e^- \rightleftharpoons 2NO + 2H_2O$	$+1.039$
$Br_2 + 2e^- \rightleftharpoons 2Br^-$	$+1.065$
$N_2O_4 + 2H^+ + 2e^- \rightleftharpoons 2HNO_2$	$+1.07$
$H_2O_2 + H^+ + e^- \rightleftharpoons \cdot OH + H_2O$	$+1.14$
$ClO_4^- + 2H^+ + 2e^- \rightleftharpoons ClO_3^- + H_2O$	$+1.201$
$O_2 + 4H^+ + 4e^- \rightleftharpoons 2H_2O$	$+1.229$
$MnO_2 + 4H^+ + 2e^- \rightleftharpoons Mn^{2+} + 2H_2O$	$+1.23$

(*Continued*)

TABLE 1.83 Standard Electrode Potentials for Aqueous Solutions (*Continued*)

Acidic solutions ([H$^+$] = 1.0 mol kg^{-1})	
Half-reaction	$E°(V)$
$N_2H_5^+ + 3H^+ + 2e^- \rightleftharpoons 2NH_4^+$	+1.275
$Cl_2 + 2e^- \rightleftharpoons 2Cl^-$	+1.358
$Cr_2O_7^{2-} + 14H^+ + 6e^- \rightleftharpoons 2Cr^{3+} + 7H_2O$	+1.36
$PbO_2 + 4H^+ + 2e^- \rightleftharpoons Pb^{2+} + 2H_2O$	+1.468
$2BrO_3^- + 12H^+ + 10e^- \rightleftharpoons Br_2 + 6H_2O$	+1.478
$Mn^{3+} + e^- \rightleftharpoons Mn^{2+}$	+1.51
$Au^{3+} + 3e^- \rightleftharpoons Au$	+1.52
$NiO_2 + 4H^+ + 2e^- \rightleftharpoons Ni^{2+} + 2H_2O$	+1.593
$2HBrO + 2H^+ + 2e^- \rightleftharpoons Br_2 + 2H_2O$	+1.604
$2HClO + 2H^+ + 2e^- \rightleftharpoons Cl_2 + 2H_2O$	+1.630
$PbO_2 + SO_4^{2-} + 4H^+ + 2e^- \rightleftharpoons PbSO_4 + 2H_2O$	+1.698
$MNO_4^- + 4H^+ + 3e^- \rightleftharpoons MnO_2 + 2H_2O$	+1.70
$Ce^{4+} + e^- \rightleftharpoons Ce^{3+}$	+1.72
$H_2O_2 + 2H^+ + 2e^- \rightleftharpoons 2H_2O$	+1.763
$Au^+ + e^- \rightleftharpoons Au$	+1.83
$Co^{3+} + e^- \rightleftharpoons Co^{2+}$	+1.92
$HN_3 + 3H^+ + 2e^- \rightleftharpoons NH_4^+ + N_2$	+1.96
$S_2O_8^{2-} + 2e^- \rightleftharpoons 2SO_4^{2-}$	+1.96
$O_3 + 2H^+ + 2e^- \rightleftharpoons O_2 + H_2O$	+2.075
$OH + H^+ + e^- \rightleftharpoons H_2O$	+2.38
$F_2 + 2H^+ + 2e^- \rightleftharpoons 2HF$	+3.053

Basic solutions ([OH$^-$] = 1.0 mol kg^{-1})	
Half-reaction	$E°(V)$
$Ca(OH)_2 + 2e^- \rightleftharpoons Ca + 2OH^-$	−3.026
$Mg(OH)_2 + 2e^- \rightleftharpoons Mg + 2OH^-$	−2.687
$Al(OH)_4^- + 3e^- \rightleftharpoons Al + 4OH^-$	−2.310
$SiO_3^{2-} + 3H_2O + 4e^- \rightleftharpoons Si + 6OH^-$	−1.7
$Mn(OH)_2 + 2e^- \rightleftharpoons Mn + 2OH^-$	−1.56
$2TiO_2 + H_2O + 2e^- \rightleftharpoons Ti_2O_3 + 2OH^-$	−1.38
$Cr(OH)_3 + 3e^- \rightleftharpoons Cr + 3OH^-$	−1.33
$Zn(OH)_4^{2-} + 2e^- \rightleftharpoons Zn + 4OH^-$	−1.285
$Zn(NH_3)_4^{2+} + 2e^- \rightleftharpoons Zn + 4NH_3$	−1.04
$MnO_2 + 2H_2O + 4e^- \rightleftharpoons Mn + 4OH^-$	−0.980
$Cd(CN)_4^{2-} + 2e^- \rightleftharpoons Cd + 4CN^-$	−0.943
$SO_4^{2-} + H_2O + 2e^- \rightleftharpoons SO_3^{2-} + 2OH^-$	−0.94
$2H_2O + 2e^- \rightleftharpoons H_2 + 2OH^-$	−0.828
$HFeO_2^- + H_2O + 2e^- \rightleftharpoons Fe + 3OH^-$	−0.8
$Co(OH)_2 + 2e^- \rightleftharpoons Co + 2OH^-$	−0.733
$CrO_4^{2-} + 4H_2O + 3e^- \rightleftharpoons Cr(OH)_4^- + 4OH^-$	−0.72
$Ni(OH)_2 + 2e^- \rightleftharpoons Ni + 2OH^-$	−0.72
$FeO_2^- + H_2O + e^- \rightleftharpoons HFeO_2^- + OH^-$	−0.69
$2SO_3^{2-} + 3H_2O + 4e^- \rightleftharpoons S_2O_3^{2-} + 6OH^-$	−0.58
$Ni(NH_3)_6^{2+} + 2e^- \rightleftharpoons Ni + 6NH_3$	−0.476
$S + 2e^- \rightleftharpoons S^{2-}$	−0.45
$O_2 + e^- \rightleftharpoons O_2^-$	−0.33
$CuO + H_2O + 2e^- \rightleftharpoons Cu + 2OH^-$	−0.29
$Mn_2O_3 + 3H_2O + 2e^- \rightleftharpoons 2Mn(OH)_2 + 2OH^-$	−0.25
$2CuO + H_2O + 2e^- \rightleftharpoons Cu_2O + 2OH^-$	−0.22
$O_2 + H_2O + 2e^- \rightleftharpoons HO_2^- + OH^-$	−0.065
$MnO_2 + 2H_2O + 2e^- \rightleftharpoons Mn(OH)_2 + 2OH^-$	−0.05

TABLE 1.83 Standard Electrode Potentials for Aqueous Solutions (*Continued*)

Basic solutions ($[OH^-] = 1.0$ mol kg^{-1})	
Half-reaction	$E°(V)$
$NO_3^- + H_2O + 2e^- \rightleftharpoons NO_2^- + 2OH^-$	+0.01
$Co(NH_3)_6^{3+} + e^- \rightleftharpoons Co(NH_3)_6^{2+}$	+0.058
HgO (red form) $+ H_2O + 2e^- \rightleftharpoons Hg + 2OH^-$	+0.098
$N_2H_4 + 2H_2O + 2e^- \rightleftharpoons 2NH_3 + 2OH^-$	+0.1
$Co(OH)_3 + e^- \rightleftharpoons Co(OH)_2 + OH^-$	+0.17
$HO_2^- + H_2O + e^- \rightleftharpoons {}^-OH + 2OH^-$	+0.184
$O_2^- + H_2O + e^- \rightleftharpoons HO_2^- + OH^-$	+0.20
$ClO_3^- + H_2O + 2e^- \rightleftharpoons ClO_2^- + 2OH^-$	+0.295
$Ag_2O + H_2O + 2e^- \rightleftharpoons 2Ag + 2OH^-$	+0.342
$Ag(NH_3)_2^+ + e^- \rightleftharpoons Ag + 2NH_3$	+0.373
$ClO_4^- + H_2O + 2e^- \rightleftharpoons ClO_3^- + 2OH^-$	+0.374
$O_2 + 2H_2O + e^- \rightleftharpoons 4OH^-$	+0.401
$NiO_2 + 2H_2O + 2e^- \rightleftharpoons Ni(OH)_2 + 2OH$	+0.490
$FeO_4^{2-} + 2H_2O + 3e^- \rightleftharpoons FeO_2^- + 4OH^-$	+0.55
$BrO_3^- + 3H_2O + 6e^- \rightleftharpoons Br^- + 6OH^-$	+0.584
$MnO_4^{2-} + 2H_2O + 2e^- \rightleftharpoons MnO_2 + 4OH^-$	+0.62
$ClO_2^- + H_2O + 2e^- \rightleftharpoons ClO^- + 2OH^-$	+0.681
$BrO^- + H_2O + 2e^- \rightleftharpoons Br^- + 2OH^-$	+0.766
$HO_2^- + H_2O + 2e^- \rightleftharpoons 3OH^-$	+0.867
$ClO^- + H_2O + 2e^- \rightleftharpoons Cl^- + 2OH^-$	+0.890
$ClO_2 + e^- \rightleftharpoons ClO_2^-$	+1.041
$O_3 + H_2O + 2e^- \rightleftharpoons O_2 + 2OH^-$	+1.246
$OH + e^- \rightleftharpoons OH^-$	+1.985

TABLE 1.84 Potentials of Reference Electrodes in Volts as a Function of Temperature

Liquid-junction potential included.

Temp., °C	0.1*M* KCl Calomel[*]	1.0*M* KCl Calomel*	3.5*M* KCl Calomel[*]	Satd. KCl Calomel[*]	1.0*M* KCl Ag/AgCl[†]	1.0*M* KBr Ag/AgBr[‡]	1.0*M* KI Ag/AgI[§]
0	0.3367	0.2883		0.25918	0.23655	0.08128	
5					0.23413	0.07961	−0.14637
10	0.3362	0.2868	0.2556	0.25387	0.23142	0.07773	−0.14719
15	0.3361			0.2511	0.22857	0.07572	−0.14822
20	0.3358	0.2844	0.2520	0.24775	0.22557	0.07349	−0.14942
25	0.3356	0.2830	0.2501	0.24453	0.22234	0.07106	−0.15081
30	0.3354	0.2815	0.2481	0.24118	0.21904	0.06856	−0.15244
35	0.3351			0.2376	0.21565	0.06585	−0.15405
38	0.3350		0.2448	0.2355			−0.15590
40	0.3345	0.2782	0.2439	0.23449	0.21208	0.06310	
45					0.20835	0.06012	−0.15788
50	0.3315	0.2745		0.22737	0.20449	0.05704	−0.15998
55					0.20056		−0.16219
60	0.3248	0.2702		0.2235	0.19649		
70					0.18782		
80				0.2083	0.1787		
90					0.1695	0.0251	

[*]Bates et al., *J. Research Natl. Bur. Standards*, **45**, 418 (1950).
[†]Bates and Bower, *J. Research Natl. Bur. Standards*, **53**, 283 (1954).
[‡]Hetzer, Robinson and Bates, *J. Phys. Chem.*, **66**, 1423 (1962).
[§]Hetzer, Robinson and Bates, *J. Phys. Chem.*, **68**, 1929 (1964).

TABLE 1.85   Potentials of Reference Electrodes (in Volts) at 25°C for Water-Organic Solvent Mixtures

Solvent, wt %	Methanol, Ag/AgCl	Ethanol, Ag/AgCl	2-Propanol, Ag/AgCl	Acetone, Ag/AgCl	Dioxane, Ag/AgCl	Ethylene glycol, Ag/AgCl	Methanol, calomel	Dioxane, calomel
5			0.2180	0.2190		0.2190		
10	0.2153	0.2146	0.2138	0.2156		0.2160		
20	0.2090	0.2075	0.2063	0.2079	0.2031	0.2101	0.255	0.2501
30		0.2003				0.2036		
40	0.1968	0.1945		0.1859		0.1972	0.243	
45					0.1635			0.2104
50		0.1859		0.158				
60	0.1818	0.173				0.1807		
70		0.158			0.0659		0.216	0.1126
80	0.1492	0.136						
82					−0.0614			−0.0014
90	0.1135	0.096		−0.034				
94.2	0.0841							
98		0.0215						
99							0.103	
100	−0.0099	−0.0081		−0.53				

1.22 CONDUCTANCE

1.22.1 Electrolytes

An electrolyte is a substance that produces an electrically conducting solution when dissolved in a polar solvent, such as water. The dissolved electrolyte separates into cations (atoms or molecules with a net negative charge) and anions (atoms or molecules with a net positive charge), which disperse uniformly through the solvent. Electrically, such a solution is neutral, but if an electrical potential (voltage) is applied to such a solution, the cations of the solution will be drawn to the electrode that has an abundance of electrons, whereas the anions will be drawn to the electrode that has a deficit of electrons. The movement of anions and cations in opposite directions within the solution amounts to a *current*. This includes most soluble salts, acids, and bases. Some gases, such as hydrogen chloride (HCl), under conditions of high temperature or low pressure can also function as electrolytes. Electrolyte solutions can also result from the dissolution of some biological polymers (such as deoxyribonucleic acid, e.g., DNA, and polypeptides) as well as synthetic polymers (such as polystyrene sulfonate), termed *polyelectrolytes*, which contain charged functional groups. A substance that dissociates into ions in solution acquires the capacity to conduct electricity. Sodium, potassium, calcium, chloride, and phosphate are examples of electrolytes. In medicine, electrolyte replacement is needed when a patient has prolonged vomiting or loss of bodily fluids and as a response to strenuous athletic activity.

Electrolyte solutions are normally formed when a salt is placed into a solvent such as water and the individual components dissociate due to the thermodynamic interactions between solvent and solute molecules (solvation). For example, when table salt (sodium chloride, NaCl) is placed in water, the salt (a solid) dissolves into its component ions (Na^+ and Cl^-), according to the dissociation reaction:

$$NaCl_{(s)} \rightarrow Na^+_{(aq)} + Cl^-_{(aq)}$$

It is also possible for substances to react with water, producing ions. Molten salts can also be electrolytes, because, for example, when sodium chloride is molten, the liquid conducts electricity. In particular, ionic liquids, which are molten salts with melting points below 100°C (212°F), are a type of highly conductive nonaqueous electrolytes and thus have found more and more applications in fuel cells and batteries.

Conductivity. The standard unit of conductance is electrolytic conductivity (formerly called specific conductance) κ, which is defined as the reciprocal of the resistance $[\Omega^{-1}]$ of a 1-m cube of liquid at a specified temperature $[\Omega^{-1} \cdot m^{-1}]$. See Table 1.88 and the definition of the cell constant.

In accurate work at low concentrations it is necessary to subtract the conductivity of the pure solvent (Table 2.69) from that of the solution to obtain the conductivity due to the electrolyte.

Resistivity (Specific Resistance)

$$\rho = \frac{1}{k} \quad [\Omega \cdot m]$$

Conductance of an Electrolyte Solution

$$\frac{1}{R} = k\frac{S}{d} \quad [\Omega^{-1}]$$

where S is the surface area of the electrode, or the mean cross-sectional area of the solution $[m^2]$, and d is the mean distance between the electrodes [m].

Equivalent Conductivity

$$\Lambda = \frac{k}{C} \quad [\Omega^{-1} \cdot m^2 \cdot equiv^{-1}]$$

In the older literature, C is the concentration in equivalents per liter. The volume of the solution in cubic centimeters per equivalent is equal to $1000/C$, and $\Lambda = 1000\ \kappa/C$, the units employed in Table 8.32 $[\Omega^{-1} \cdot cm^2 \cdot equiv^{-1}]$. The formula unit used in expressing the concentration must be specified; for example, NaCl, ½K$_2$SO$_4$, 1/3LaCl$_3$.

The equivalent conductivity of an electrolyte is the sum of contributions of the individual ions. At infinite dilution: $\Lambda° = \lambda_c° + \lambda_a°$, where $\lambda_c°$ and $\lambda_a°$ are the ionic conductances of cations and anions, respectively, at infinite dilution (Table 1.89).

Ionic Mobility and Ionic Equivalent Conductivity

$$\lambda_c = Fu_c \quad \text{and} \quad \lambda_a = Fu_a \quad [\Omega^{-1} \cdot m^2 \cdot equiv^{-1}]$$

where F is the Faraday constant, and u_c, u_a are the ionic mobilities $[m^2 \cdot s^{-1} \cdot V^{-1}]$.

$$\Lambda = \alpha F(u_c + u_a) = \alpha(\lambda_c + \lambda_a)$$

where α is the degree of electrolytic dissociation, $\Lambda/\Lambda°$. The electric mobility u of a species is the magnitude of the velocity in an electric field $[m \cdot s^{-1}]$ divided by the magnitude of the strength of the electric field $E[V \cdot m^{-1}]$.

Ostwald Dilution Law

$$K_d = \frac{\alpha^2 C}{1-\alpha}$$

where K_d is the dissociation constant of the weak electrolyte. In general for an electrolyte which yields n ions:

$$K_d = \frac{C^{(n-1)}\Lambda^n}{\Lambda°^{(n-1)}(\Lambda° - \Lambda)}$$

Transference Numbers or Hittorf Transport Numbers

$$T_c = \frac{\lambda_c}{\lambda_c + \lambda_a} \quad T_a = \frac{\lambda_a}{\lambda_c + \lambda_a} \quad T_c + T_a = 1$$

$$\frac{T_c}{T_a} = \frac{u_c}{u_a} = \frac{\lambda_c}{\lambda_a}$$

$$\lambda_c = T_c\Lambda \quad \lambda_a = T_a\Lambda$$

TABLE 1.86 Properties of Liquid Semi-Conductors

Material	Melting point (K)	Density at K* (g cm⁻³) Solid	Liquid	Electrical conductivity at K* ($\Omega^{-1} \cdot cm^{-1}$) Solid	Liquid	Atomization energy (kcal/mole)	Heat of fusion (kcal/mole)	Entropy of fusion (e.u.)	Thermoelectric power (μV per K) Solid	Liquid	Activation energy for viscous flow (kcal per mole)	Entropy of viscous flow (e.u.)	Viscosity of liquid at K* (centipoises)
Si	1693	2.30	2.53	580	12000	204	12.1	7.1	−90	0	8.63	2.1	0.348
Ge	1210	5.26	5.51	1250	14000	178	8.35	6.9	−160	−60	2.74	2.85	0.135
AlSb	1353	4.18	4.72	160	9900	160	14.2	5.2	−60	0	10	2.2	0.250
GaSb	985	5.60	6.06	280	10600	134	12.0	6.1	−120	−20	2.7	5.0	0.368
InSb	809	5.76	6.48	2900	10000	121	11.6	7.2			2.0	9.4	0.363
GaAs	1511	5.16	5.71	300	7900	146	23.2	7.7	—	—	6.5	7.8	0.320
InAs	1215	5.5	5.89	3600	6800	130	12.6	5.2	—	—	6.2	3	0.174
ZnTe	1512					109					9.0	6.5	0.868
CdTe	1365					99					5.75	7.7	0.435
CuI	875	5.36	4.84				2.6		550	490			0.432
Ga₂Te₃	1063	5.35	5.086						−290	−85	11	7	0.546
In₂Te₃	940	5.77	5.54						−50	30	13	7.5	0.323
Mg₂Si	1375	1.84	2.27	1120	9800		20.4	5.0			13.9	2.3	0.299
Mg₂Ge	1388		3.20	1140	8400						9.5	3.8	0.311
Mg₂Sn	1051	3.45	3.52	2040	10600		11.4	3.6			9.5	5.8	0.520
Mg₂Pb	823	5.00	5.20	3530	8600		9.3	3.8			9.6	6.2	0.560
GeTe	998	5.97	5.57	2400	2600		11.3	5.7	130	21	4.70	5.8	0.375
SnTe	1063	6.15	5.85	1440	1800		8.0	3.7	140	28	4.90	6.1	0.348
PbTe	1190	7.69	7.45	420	1520		7.5	3.1	−60	−10	6.85	6.0	0.243
PbSe	1361	7.57	7.10	300	450		8.5	3.1	−120	−60	6.85	7.5	0.240
PbS	1392	7.07	6.45	250	220		8.7	3.1	−220	−220	9.80	7.5	0.319
Bi₂Se₃	979	7.27	6.97	450	900				−90	−35	9.7	5.05	0.540
Bi₂Te₃	858	7.5	7.26	1250	2580		28.35	6.6	−45	−3	2.7	7.80	0.198
Sb₂Te₃	895	6.29	6.09	900	1850		23.65	5.3	90	11	6.1	7.35	0.513
Se (hex)	493	4.69	3.975				1.5	3			3.94	6.7	6.63
Te	725	6.1	5.775				4.17	5.7			1.18	6.7	0.357

*At melting point.

TABLE 1.87 Limiting Equivalent Ionic Conductances in Aqueous Solutions

In 10^{-4} $m^2 \cdot S \cdot equiv^{-1}$ or mho $\cdot cm^2 \cdot equiv^{-1}$.

Ion	Temperature, °C		
	0	18	25
Inorganic cations			
Ag^+	33	54.5	61.9
Al^{3+}	29		61
Ba^{2+}	33.6	54.3	63.9
Be^{2+}			45
Ca^{2+}	30.8	51	59.5
Cd^{2+}	28	45.1	54
Ce^{3+}			70
Co^{2+}	28	45	53
$Co(NH_3)_6^{3+}$			100
$Co(ethylenediamine)_3^{3+}$			74.7
Cr^{3+}			67
Cs^+	44	68	77.3
Cu^{2+}	28	45.3	56.6
D^+ (deuterium)		213.7	
Dy^{3+}			65.7
Er^{3+}			66.0
Eu^{3+}			67.9
Fe^{2+}	28	45.3	53.5
Fe^{3+}			69
Gd^{3+}			67.4
H^+	224.1	315.8	350.1
Hg_2^{2+}			68.7
Hg^{2+}			63.6
Ho^{3+}			66.3
K^+	40.3	64.6	73.5
La^{3+}	35.0	59.2	69.6
Li^+	19.1	33.4	38.69
Mg^{2+}	28.5	46	53.06
Mn^{2+}	27	44.5	53.5
NH_4^+	40.3	64	73.7
$N_2H_5^+$ (hydrazinium 1 +)			59
Na^+	25.85	43.5	50.11
Nd^{3+}			69.6
Ni^{2+}	28	45	50
Pb^{2+}	37.5	60.5	71
Pr^{3+}			69.6
Ra^{2+}	33	56.6	66.8
Rb^+	43.5	67.5	77.8
Sc^{3+}			64.7
Sm^{3+}			68.5
Sr^{2+}	31	51	59.46
Tl^+	43.3	66	74.9
Tm^{3+}			65.5
UO_2^{2+}			32
Y^{3+}			62
Yb^{3+}			65.2
Zn^{2+}	28	45.0	52.8

(Continued)

TABLE 1.87   Limiting Equivalent Ionic Conductances in Aqueous Solutions (*Continued*)

Ion	Temperature, °C		
	0	18	25
Inorganic anions			
$Au(CN)_2^-$			50
$Au(CN)_4^-$			36
$B(C_6H_5)_4^-$			21
Br^-	43.1	67.6	78.1
Br_3^-			43
BrO_3^-	31.0	49.0	55.7
Cl^-	41.4	65.5	76.31
ClO_2^-			52
ClO_3^-	36	55.0	64.6
ClO_4^-	37.3	59.1	67.3
CN^-			78
CO_3^{2+}	36	60.5	69.3
$Co(CN)_6^{3-}$			98.9
CrO_4^{2-}	42	72	85
F^-		46.6	55.4
$Fe(CN)_6^{4-}$			110.4
$Fe(CN)_6^{3-}$			100.9
$H_2AsO_4^-$			34
HCO_3^-			44.5
HF_2^-			75
HPO_4^{2-}			33
$H_2PO_4^-$		28	33
HS^-	40	57	65
HSO_3^-	27		50
HSO_4^-			50
$H_2SbO_4^-$			31
I^-	42.0	66.5	76.9
IO_3^-	21.0	33.9	40.5
IO_4^-		49	54.5
MnO_4^-	36	53	61.3
MoO_4^{2-}			74.5
N_3^-			69.5
$N(CN)_2^-$			54.5
NO_2^-	44	59	71.8
NO_3^-	40.2	61.7	71.42
NH_2SO_3 (sulfamate)			48.6
OCN^- (cyanate)		54.8	64.6
OH^-	117.8	175.8	198
PF_6^-			56.9
PO_3F^{2-}			63.3
PO_4^{3-}			69.0
$P_2O_7^{4-}$			96
$P_3O_9^{3-}$			83.6
$P_3O_{10}^{5-}$			109
ReO_4^-		46.5	54.9
SCN^- (thiocyanate)	41.7	56.6	66.5
$SeCN^-$			64.7
SeO_4^{2-}		65	75.7
SO_3^{2-}			79.9

TABLE 1.87 Limiting Equivalent Ionic Conductances in Aqueous Solutions (*Continued*)

Ion	Temperature, °C		
	0	18	25
SO_4^{2-}	41	68.3	80.0
$S_2O_3^{2-}$			85.0
$S_2O_4^{2-}$	34		66.5
$S_2O_6^{2-}$			93
$S_2O_8^{2-}$			86
WO_4^{2-}	35	59	69.4
Organic cations			
Decylpyridinium$^+$			29.5
Diethylammonium$^+$			42.0
Dimethylammonium$^+$			51.5
Dipropylammonium$^+$			30.1
Dodecylammonium$^+$			23.8
Ethylammonium$^+$			47.2
Ethyltrimethylammonium$^+$			40.5
Isobutylammonium$^+$			38.0
Methylammonium$^+$			58.3
Piperidinium$^+$			37.2
Propylammonium$^+$			40.8
Tetrabutylammonium$^+$			19.5
Tetraethylammonium$^+$			32.6
Tetramethylammonium$^+$			44.9
Tetrapropy lammonium$^+$			23.5
Triethylsulfonium$^+$			36.1
Trimethylammonium$^+$			47.2
Trimethylsulfonium$^+$			51.4
Tripropylammonium$^+$			26.1
Organic anions			
Acetate$^-$	20	34	41
Benzoate$^-$			32.4
Bromoacetate$^-$			39.2
Bromobenzoate$^-$			30
Butanoate$^-$			32.6
Chloroacetate$^-$			42.2
m-Chlorobenzoate$^-$			31
o-Chlorobenzoate$^-$			30.5
Citrate(3−)			70.2
Crotonate$^-$			33.2
Cyanoacetate$^-$			43.4
Cyclohexanecarboxylate$^-$			28.7
Cyclopropane-1,3-dicarboxylate^{2-}			53.4
Decylsulfonate$^-$			26
Dichloroacetate$^-$			38.3
Diethylbarbiturate(2−)			26.3
Dihydrogencitrate$^-$			30
Dimethylmalonate(2−)			49.4
3,5-Dinitrobenzoate$^-$			28.3
Dodecylsulfonate$^-$			24
Ethylmalonate$^-$			49.3
Ethylsulfonate$^-$			39.6

TABLE 1.88 Standard Solutions for Calibrating Conductivity Vessels

The values of conductivity κ are corrected for the conductivity of the water used. The cell constant θ of a conductivity cell can be obtained from the equation

$$\theta = \frac{KRR_{\text{solv}}}{R_{\text{solv}} - R}$$

where R is the resistance measured when the cell is filled with a solution of the composition stated in the table below, and R_{solv} is the resistance when the cell is filled with solvent at the same temperature.

Grams KCl per kilogram solution (in vacuo)	Conductivity in $\text{ohm}^{-1} \cdot \text{cm}^{-1}$ at		
	0°C	18°C	25°C
71.135 2	0.065 14$_4$	0.097 79$_0$	0.111 28$_7$
7.419 13	0.007 134$_4$	0.011 161$_2$	0.012 849$_7$
0.745 263*	0.000 773 2$_6$	0.001 219 9$_2$	0.001 408 0$_8$

*Virtually 0.0100 M.

From the data of Jones and Bradshaw, *J. Am. Chem. Soc.*, **55,** 1780 (1933). The original data have been converted from $(\text{int. ohm})^{-1}\text{cm}^{-1}$.

TABLE 1.89 Equivalent Conductivities of Electrolytes in Aqueous Solutions at 18°C

The unit of Λ in the table is $\Omega^{-1} \cdot cm^{-2} \cdot equiv^{-1}$. The entities to which the equivalent relates are given in the first column.

Electrolyte	Concentration, N										
	0.001	0.005	0.01	0.05	0.1	0.5	1.0	2.0	3.0	4.0	5.0
Acetic acid	41	20.0	14.3	6.48	4.60	2.01	1.32		0.54		0.29
AgNO$_3$	113.2	110.0	107.8	99.5	94.3	77.8	67.8	56.0	48.2	42.1	37.2
½Ag$_2$SO$_4$	116.3	108.4	102.9								
1/3 AlBr$_3$ (25°)	132	124	119	103	97						
1/3 AlCl$_3$	121.1	105.0	93.8			65.0	56.2	44.2	34.7	27.2	
1/3 AlI$_3$ (25°)	131	124	119	108							
1/3 Al(NO$_3$)$_3$ (25°)	123	115	110	94	88						
1/6 Al$_2$(SO$_4$)$_3$ (25°)	107.2	76.8	60.6								
½Ba(OAc)$_2$	85.0	80.4	77.1	65.7	60.2	43.8	34.3				
½Ba(BrO$_3$)$_2$ (25°)	113.6	106.8	102.7								
½BaCl$_2$	115.6	112.3	106.7	96.0	90.8	77.3	70.1	60.3	52.3		
½Ba(NO$_3$)$_2$	111.7	105.3	101.0	86.8	78.9	56.6	48.4		29.8	23.4	
½Ba(OH)$_2$	216	213	207	191	180						
Butyric acid						1.66	0.98	0.46	0.26	0.18	0.11
½Ca(OAc)$_2$	79.6	75.0	71.9	60.3	54.0	36.3	26.3				
½CaCl$_2$	112.0	106.7	103.4	93.3	88.2	74.9	67.5	58.3	49.7	42.4	35.6
½Ca(NO$_3$)$_2$	108.5	103.0	99.5	88.4	82.5	65.7	55.9	43.5	35.5	26.0	21.5
½Ca(OH)$_2$		233	226								
½CaSO$_4$	104.3	86.3	77.4								
½CdBr$_2$		86.5	76.3	53.2	44.6	25.3	18.3	12.5	9.1	6.8	5.3
½CdCl$_2$		91	83	59	50	30.8	22.4	14.4	9.9	7.1	5.4
½CdI$_2$		76.7	65.6	40.1	31.0	18.3	15.4	12.3	9.7	8.0	
½Cd(NO$_3$)$_2$		100	96	86.4	80.8	63.9	54.5	41.0	31.4	23.7	17.6
½CdSO$_4$	97.7	79.7	70.3	49.6	42.2	28.7	23.6	17.7	14.0	11.0	8.35
1/3 CeCl$_3$ (25°)	137.4		122.1		99.0						
1/6 Ce$_2$(C$_2$O$_4$)$_3$ (25°)	85.5	54	45.8	29							
Chloroacetic acid (25°)	88.4				42.9	20.2	13.6	8.1	5.6	4.2	3.3
Citric acid		54	42.5	22.0	16.1	7.3	5.4				
½COCl$_2$						51.5	45.3	40.3	35.4		
1/3 CrCl$_3$		99.3	95.6	82.3	75.0	68.6	56.8	44.8	35.2	30.5	26.4

(Continued)

TABLE 1.89 Equivalent Conductivities of Electrolytes in Aqueous Solutions at 18°C (*Continued*)

The unit of Λ in the table is $\Omega^{-1} \cdot cm^{-2} \cdot equiv^{-1}$. The entities to which the equivalent relates are given in the first column.

Electrolyte	Concentration, N										
	0.001	0.005	0.01	0.05	0.1	0.5	1.0	2.0	3.0	4.0	5.0
$\frac{1}{2}CrO_3(H_2CrO_4)$ (25°)	201	195	193	191	186						
CsCl	130.7	127.5	125.2		113.5	104.3	100.3	95.7	85.1		
$\frac{1}{2}Cu(OAc)_2$ (25°)	55.7	50.6	47.2	34.9	28.4						
$\frac{1}{2}CuCl_2$	107.9	97.1	93.7	83.7	78.2	67.5	56.8	41.2	31.5	24.5	19.1
$\frac{1}{2}Cu(NO_3)_2$ (15°)								45.4	35.3	27.8	21.4
$\frac{1}{2}CuSO_4$	98.5	81.0	71.7	53.6	43.8	30.5	25.6	19.7	16.5	16.3	9.6
Dichloroacetic acid (25°)					207.5	119	82	44.6	26.5		
$\frac{1}{2}FeCl_2$ (25°)	131	125	120	103	93	66.5	52.9	37.6	28.1	20.5	15.9
$\frac{1}{3}FeCl_3$	82	75	70	54	44.5	30.8	25.8	19.5	15.37		
$\frac{1}{2}FeSO_4$	125.6						5.18	3.68	2.93	2.39	1.92
Formic acid	13.5										
H_3AsO_4 (1 M) (25°)	308.2	230.0	187.0	103.4	80.4						
H_3BO_3											
HBr	401	387	373		356	306	282	243	214	179	
$HBrO_3$ (25°)				272	156						
HCl	377	373	370	360	351	327	301	247	215		152.2
$HClO_3$					343	317	292		207		
$HClO_4$ (25°)	413	406	402	392	386	358					
HF		90	60	35.9	31.3	27.0	25.7		24.2		24.0
HI				357	347	322	297	255	215	179	
HIO_3	343.3	332.8	323.9		253	175	141	106	87	71	
HNO_3	375	371	368		350	324	310		220		156
H_3PO_4 (1M)	318	279	255				66		53.1		51.3
HSCN (25°)	399	394	390	377	370						
$\frac{1}{2}H_2SO_4$	361	330	308	253	225	205	198		166.8		135.0
$\frac{1}{2}HgCl_2$				1.85	1.23						
$\frac{1}{3}$ InBr$_3$	98.3	95.7	94.0	87.7	53.9	37.0	28.7	19.8	14.4	10.1	
KOAc					83.8	71.6	63.4	50.0	40.7	31.4	24.5
KBr	129.4	126.4	124.4	117.8	114.2	105.4	102.5	98.0	93.3	87.9	
$KBrO_3$	109.9	106.9	104.7	97.3	93.0						
$\frac{1}{3}$ K$_3$ citrate		109.9	103	87.8	80.8						
KCl	127.3	124.4	122.4	115.8	112.0	102.4	98.3	92.0	88.9		
$KClO_3$	116.9	113.6	111.6	103.7	99.2	85.3					
$KClO_4$ (25°)	137.9	134.2	131.5	121.6	115.2						

KCN (15°)	133.0	121.6	115.5	100.7	94.1	104.2	99.7	65.0	55.6	49.2	42.9
$\frac{1}{2}K_2CO_3$	122.4	116.7	112.5	100.8	94.9	77.8	70.7				
$\frac{1}{2}K_2C_2O_4$					100.5	80.4	73.7				
$\frac{1}{2}K_2CrO_4$					98.2	86.4	79.5	72.0	59.9		
$\frac{1}{2}K_2Cr_2O_7$						85.4					
KF	108.9	106.2	104.3	97.7	94.0	82.6	76.0	63.4	56.5	51.7	46.5
$\frac{1}{3}K_3[Fe(CN)_6]$	163.1	150.7	134.8	107.7	97.9						
$\frac{1}{4}K_4[Fe(CN)_6]$	167.2	146.1									
$KHCO_3$ (25°)	115.3	112.2	110.1			86.5	78.9				
KH phthalate	119.3	103.7	99.9	89.3	83.8						
KHS						92.5	91.7	86.4	80.7		69.3
$KHSO_4$						21.0	18.4	15.2			
KH_2PO_4 (1 M) (25°)	107.1	100.8	98.0	90.7	85.6	60.0[18]	45.8[18]				
KI	128.2	125.3	123.4	117.3	114.0	106.2	103.6	101.3	96.4	89.0	81.2
KIO_3	96.0	93.2	91.2	84.1	79.7						
KIO_4 (25°)	124.9	121.2	118.5	106.7	98.1						
$KMnO_4$ (25°)	133.3		126.5		113						
KNO_3	123.6	120.5	118.2	109.9	104.8	89.2	80.5	69.4	61.3		
KOH	234	230	228	219	213	197	184				
$KReO_4$ (25°)	125.1	121.3	118.5	106.4	97.4						
$\frac{1}{2}K_2S$							135.6	119.7	140.6		105.8
KSCN	118.6	115.8	113.9	107.7	104.3	95.7	91.6	86.8	108.3	97.2	86.1
$\frac{1}{2}K_2SO_4$	126.9	120.3	115.8	101.9	94.9	78.5	71.6		74.6		
$\frac{1}{2}LaCl_3$ (25°)	137.0	127.5	121.8	106.2	99.1	65.4	54.0	39.1			
$\frac{1}{3}La(NO_3)_3$				86.1	72.1				28.5	19.9	
$\frac{1}{6}La_2(SO_4)_3$				25.7	25.5						
Lactic acid	108.9	53.5	39	18.1	13.2						
LiOAc					51.3	37.7	28.9	18.2	11.9	7.2	
LiBr				87.9	84.4	73.9	67.2	57.7			
LiCl	96.5	93.9	92.1	86.1	82.4	70.7	63.4	53.1	45.3	44.2	
$LiClO_4$ (25°)	103.4	100.6	98.6	92.2	88.6						33.3

(Continued)

TABLE 1.89 Equivalent Conductivities of Electrolytes in Aqueous Solutions at 18°C (*Continued*)

The unit of Λ in the table is $\Omega^{-1} \cdot cm^{-2} \cdot equiv^{-1}$. The entities to which the equivalent relates are given in the first column.

Electrolyte	\multicolumn{11}{c}{Concentration, N}										
	0.001	0.005	0.01	0.05	0.1	0.5	1.0	2.0	3.0	4.0	5.0
½Li₂CO₃				64.2	59.1	75.3	69.2	61.0			
LiI	65.3	62.9	61.2	55.3	51.5	39.0	31.2	21.4	14.6		
LiIO₃	92.9	90.3	88.6								
LiNO₃				82.7	79.2	68.0	60.8	50.3	34.9	27.3	
LiOH						149.0	134.5	113.5	95.7		
½Li₂SO₄	96.4		86.9	74.7	68.2	50.5	41.3	30.7	23.3	18.1	13.9
½MgCl₂	106.4	101.3	98.1	88.5	83.4	69.6	61.5	52.3	43.3	35.0	28.0
½Mg(NO₃)₂	102.6	97.7	94.7	85.3	80.5	67.0	59.0	47.0	39.8		
½MgSO₄	99.8	84.5	76.2	56.9	49.7	35.4	28.9	23.0	17.3	12.9	9.3
½MnCl₂					86.0	68.5	61.0	48.5	38.8	30.2	23.0
½MnSO₄						27.6	24.4	18.3	14.0	10.5	7.3
NH₃(aq)	28.0	13.2	9.6	4.6	3.3	1.35	0.89		0.36		0.20
NH₄OAc		92.9	91.4	84.9		60.5	54.7	42.9	34.0	26.5	
NH₄Cl	127.3	124.3	122.1	115.2	110.7	101.4	97.0	92.1	88.2	85.0	80.7
NH₄F					90.1	74.5	65.7	55.3	47.9	42.2	
NH₄I	124.5		118.0	118.0	115.0	106.0	103.1	100.0		91.4	84.5
NH₄NO₃				110.0	106.6	94.5	88.8	85.1	79.2	71.9	47.6
NH₄SCN					104.3	94.0	89.9	84.7		74.0	
½(NH₄)₂SO₄		120.0	116.5		89.0	79.5	73.0	65.0		55.2	10.5
NaOAc	75.2	72.4	70.2	64.2	61.1	49.4	41.2	29.8	21.5	15.3	
NaBr				99.1	96.0	84.6	78.1	69.1		53.0	
NaBrO₃						61.8	54.5	44.1			
Na *n*-butyrate (25°)	80.3	77.6	75.8	69.3	65.3						
NaCl	106.5	103.8	102.0	95.7	92.0	80.9	74.3	64.8	56.5	49.4	
NaClO₄	114.9^{25}	111.7^{25}	109.6^{25}	102.4^{25}	98.4^{25}	71.7	65.0	55.1	46.0	38.8	42.7
½Na₂CO₃	112	102.5	96.2	80.3	72.9	54.5	45.5	34.5	27.2		
½Na₂CrO₄					82.5	66.4	57.7	46.6	38.3	31.1	
½Na₂Cr₂O₇ (25°)		103		98.3	94.9						
NaF	87.8	85.2	83.5	77.0	73.1	60.0	51.9				
¼Na₄[Fe(CN)₆] (25°)		129.6	120.0	97.0	88.2						
Na formate	88.6					61.4	53.7	43.1	34.8	28.2	
NaHCO₃ (25°)	93.5	90.5	88.4	80.6	76.0						
1/3Na₂HPO₄	58.4		54.0		44.0	33.5	28.0				
NaH₂PO₄	67.9	65.8	64.4	57.8	54.1						
¼Na₂H₂P₂O₇	41.1	39.4	38.2	34.6	32.5	25.4					
NaI	124.2	121.2	119.2	112.8	108.8	97.5	89.9	78.6	69.9	62.2	

NaIO₃	75.2	72.6	70.9	64.4	60.5						
½Na₂MoO₄	120.8	113	110								
NaN₃ (25°)	117.1	113.8	110.5	101.3	95.7						
NaNO₂ (25°)	102.9	100.1	98.2	91.4	87.2	74.1	68.0	63.1	53.6		39.7
NaNO₃	208	203	200	190	183	172	75.9	54.5	46.0	39.0	
NaOH	78.6	75.7	73.7	66.3	61.8	160	65.9		108.0		69.0
Na picrate (25°)	125	122	119	91							
1/3Na₃PO₄											
Na propionate (25°)	83.5	80.9	79.1								
½Na₂S	144	139	136	124	116	117.0	104.3	85.0	71.0	59.0	47.2
NaSCN	106.7	100.8	96.8	83.9	78.4	74.3	68.9	59.8	50.9	43.7	
½Na₂SiO₃	120	81.5	74.8	64.3	60.4	88	72	51	38	27	19
½Na₂SO₄	116.1	109.2	104.8	92.2	85.8	59.7	50.8	40.0	33.5		
(mono) Na tartrate											
½Na₂WO₄ (25°)	96.3	79.5	70.8	51.0	43.8	30.4	25.1	19.3	15.1		
½NiSO₄	180.7	158.2	132.9	116.9	116.9	75.9	59.4	31.0			
½Oxalic acid	116.1	108.6	103.5	86.3	77.3	53.2	42.0				
½Pb(NO₃)₂					1.57	1.00	0.54				
Propionic acid	130.3	127.4	125.3	117.8	113.9	101.9	97.1	92.7		0.20	
RbCl	220.6	216.8	204.8	192.0	170.0	148.3	87.2				
RbOH	114.5	108.9	105.4	94.4	90.2	75.7	66.9	58.7	47.9	32.7	
½SnCl₄	108.3	102.7	99.0	87.3	80.9	62.7	68.5	58.7	49.9	42.2	16.4
½SrCl₂						52.1	38.0		29.3	29.3	1.83
½Sr(NO₃)₂						7.03		4.58	3.32	2.48	
Tartaric acid (15°)	128.2	123.7	120.2	97.4	92.6	61.0	54.0	44.3	36.3	29.8	
¼ThCl₄	113.3	108.2	105.4	107.9	101.2	78.8	71.5	62.7			
TlCl	124.7	121.1	118.4								
TlF	127.4	118.4	112.3	92.7	83.1	71.5					
TlNO₃	26.10	12.31	9.17	5.43	4.74	273	207	127	79	44	19
½Tl₂SO₄	106.5	63.2	49.2	27.6	22.2	3.75	3.22				2.7
Trichloroacetic acid (25°)	129	122	118	109		14.4	11.6				
½UO₂F₂ (25°)	83	77	73	58	49						
½UO₂SO₄ (25°)	107	101	98	87	82	65	55	44			
1/3YCl₃ (25°)	120	114	111	100		32.3	55	39.6	29.6	23.2	18.5
½Zn(OAc)₂ (25°)	98.4	82.1	73.2	53.0	45.6	26.6	20.0	15.9	12.0	9.0	

TABLE 1.90   Conductivity of Very Pure Water at Various Temperatures and the Equivalent Conductances of Hydrogen and Hydroxyl Ions

Temp., °C	Conductivity, $\mu S \cdot cm^{-1}$	Resistivity, $M\Omega \cdot cm$	Equivalent conductance, $cm^2 \cdot ohm^{-1} \cdot equivalent^{-1}$	
			λ^0, H$^+$	λ^0, OH$^-$
0	0.01161	86.14	224.1	117.8
5	0.016 61	60.21	250.0	133.6
10	0.023 15	43.21	275.6	149.6
15	0.031 53	31.71	300.9	165.9
18	0.037 54	26.64	315.8	491.6
20	0.042 05	23.78	325.7	182.5
25	0.055 08	18.15	350.1	199.2
30	0.070 96	14.09	374.0	216.1
35	0.090 05	11.10	397.4	233.0
40	0.1127	8.88	420.0	267.2
45	0.139 3	7.18	442.0	267.2
50	0.170 2	5.88	463.3	284.3
55	0.205 5	4.86	483.8	301.4
60	0.245 7	4.06	503.4	318.5
65	0.291 2	3.43	522.0	335.4
70	0.341 6	2.93	539.7	352.2
75	0.397 8	2.51	556.4	368.8
80	0.459 3	2.18	572.0	385.2
85	0.525 8	1.90	586.4	401.4
90	0.597 7	1.67	599.6	417.3
95	0.675 3	1.48	611.6	432.8
100	0.756 9	1.32	622.2	448.1
150	1.84	0.543		
200	2.99	0.334	824	701
250	3.31	0.302		
300	2.42	0.413	894	821

Source: Data from T. S Light and S.L. Licht. *Anal Chem.*, **59**: 2327–2330(1987).

1.23   THERMAL PROPERTIES

TABLE 1.91   Eutectic Mixtures

The *eutectic temperature* $\theta_{C,E}$ is the lowest temperature at which both the solid components of a mixture are in equilibrium with the liquid phase. $\theta_{C,m}$ denotes melting temperature.

Component 1	$\theta_{c,m}/°C$	Component 2	$\theta_{C,m}/°C$	$\theta_{C,E}/°C$	Composition of eutectic mixture (per cent by mass)	
Sn	232	Pb	327	183	Sn, 63·0	Pb, 37·0
Sn	232	Zn	420	198	Sn, 91·0	Zn, 9·0
Sn	232	Ag	961	221	Sn, 96·5	Ag, 3·5
Sn	232	Cu	1083	227	Sn, 99·2	Cu, 0·8
Sn	232	Bi	271	140	Sn, 42·0	Bi, 58·0
Sb	630	Pb	327	246	Sb, 12·0	Pb, 88·0
Bi	271	Pb	327	124	Bi, 55·5	Pb, 44·5
Bi	271	Cd	321	146	Bi, 60·0	Cd, 40·0
Cd	321	Zn	420	270	Cd, 83·0	Zn, 17·0

TABLE 1.92 Transition Temperatures

$\theta_{C,t}$ denotes transition temperature

Substance	System	$\theta_{C,t}/°C$
sulphur	Rhombic (α) $\rightleftharpoons$ Monoclinic (β)	95.6
Tin	Grey (α) White (β)	
Iron	α (body-centered cubic) $\rightleftharpoons$ γ (face-centered cubic)	906
	γ (body-centered cubic) $\rightleftharpoons$ δ (face-centered cubic)	1401
Sodium sulphate	Na_2So_4 $10H_2O$ $\rightleftharpoons$ Na_2SO_4 + $10H_2O$	32.4
Mercury(II) iodide	Tetragonal (red) $\rightleftharpoons$ Orthorhombic (yellow)	126
Ammonium chloride	α (CsCl structure) $\rightleftharpoons$ β (NaCl structure)	184
Caesium chloride	CsCl structure $\rightleftharpoons$ NaCl structure	445
Copper(I) mercury(II) Iodide	Tetragonal (red) $\rightleftharpoons$ Cubic (dark brown)	69

SECTION 2
ORGANIC CHEMISTRY

2.1 NOMENCLATURE OF ORGANIC COMPOUNDS

The following synopsis of rules for naming organic compounds and the examples given in explanation are not intended to cover all the possible cases.

2.1.1 Nonfunctional Compounds

Alkanes.   The saturated open-chain (acyclic) hydrocarbons (C_nH_{2n+2}) have names ending in -ane. The first four members have the trivial names *methane* (CH_4), *ethane* (CH_3CH_3 or C_2H_6), *propane* (C_3H_8), and *butane* (C_4H_{10}). For the remainder of the alkanes, the first portion of the name is derived from the Greek prefix that cites the number of carbons in the alkane followed by -ane with elision of the terminal -a from the prefix.

TABLE 2.1   Straight-Chain Alkanes

n*	Name	n*	Name	n*	Name	n*	Name
1	Methane	11	Undecane‡	21	Henicosane	60	Hexacontane
2	Ethane	12	Dodecane	22	Docosane	70	Heptacontane
3	Propane	13	Tridecane	23	Tricosane	80	Octacontane
4	Butane	14	Tetradecane			90	Nonacontane
5	Pentane	15	Pentadecane	30	Triacontane	100	Hectane
6	Hexane	16	Hexadecane	31	Hentriacontane	110	Decahectane
7	Heptane	17	Heptadecane	32	Dotriacontane	120	Icosahectane
8	Octane	18	Octadecane			121	Henicosahectane
9	Nonane†	19	Nonadecane	40	Tetracontane		
10	Decane	20	Icosane§	50	Pentacontane		

*n = total number of carbon atoms.
†Formerly called enneane.
‡Formerly called hendecane.
§Formerly called eicosane.

For branching compounds, the parent structure is the longest continuous chain present in the compound. Consider the compound to have been derived from this structure by replacement of hydrogen by various alkyl groups. Arabic number prefixes indicate the carbon to which the alkyl group is attached. Start numbering at whichever end of the parent structure that results in the lowest-numbered locants. The arabic prefixes are listed in numerical sequence, separated from each other by commas and from the remainder of the name by a hyphen.

If the same alkyl group occurs more than once as a side chain, this is indicated by the prefixes di-, tri-, tetra-, etc. Side chains are cited in alphabetical order (before insertion of any multiplying prefix). The name of a complex radical (side chain) is considered to begin with the first letter of its complete name. Where names of complex radicals are composed of identical words, priority for citation is given to that radical which contains the lowest-numbered locant at the first cited point of difference in the radical. If two or more side chains are in equivalent positions, the one to be assigned the lowest-numbered locant is that cited first in the name. The complete expression for the side chain may be enclosed in parentheses for clarity or the carbon atoms in side chains may be indicated by primed locants.

If hydrocarbon chains of equal length are competing for selection as the parent, the choice goes in descending order to (1) the chain that has the greatest number of side chains, (2) the chain whose side chains have the lowest-numbered locants, (3) the chain having the greatest number of carbon atoms in the smaller side chains, or (4) the chain having the least-branched side chains.

These trivial names may be used for the unsubstituted hydrocarbon only:

Isobutane	$(CH_3)_2CHCH_3$	Neopentane	$(CH_3)_4C$
Isopentane	$(CH_3)_2CHCH_2CH_3$	Isohexane	$(CH_3)_2CHCH_2CH_2CH_3$

Univalent radicals derived from saturated unbranched alkanes by removal of hydrogen from a terminal carbon atom are named by adding -yl in place of -ane to the stem name. Thus the alkane *ethane* becomes the radical *ethyl*. These exceptions are permitted for unsubstituted radicals only:

Isopropyl	$(CH_3)_2CH—$	Isopentyl	$(CH_3)_2CHCH_2CH_2—$
Isobutyl	$(CH_3)_2CHCH_2—$	Neopentyl	$(CH_3)_3CCH_2—$
sec-Butyl	$CH_3CH_2CH(CH_3)—$	*tert*-Pentyl	$CH_3CH_2C(CH_3)_2—$
tert-Butyl	$(CH_3)_3C—$	Isohexyl	$(CH_3)_2CHCH_2CH_2CH_2—$

Note the usage of the prefixes iso-, neo-, *sec*-, and *tert*-, and note when italics are employed. Italicized prefixes are never involved in alphabetization, except among themselves; thus *sec*-butyl would precede isobutyl, isohexyl would precede isopropyl, and *sec*-butyl would precede *tert*-butyl.

Examples of alkane nomenclature are

$$\overset{4}{CH_3}-\overset{3}{CH_2}-\overset{2}{CH}-\overset{1}{CH_3} \qquad \text{2-Methylbutane (or the trivial name, isopentane)}$$
$$\underset{CH_3}{|}$$

$$\overset{5}{CH_3}-\overset{4}{CH_2}-\overset{3}{CH}-CH_3 \qquad \text{3-Methylpentane (not 2-ethylbutane)}$$
$$\underset{\overset{2}{CH_2}-\overset{1}{CH_3}}{|}$$

$$\overset{8}{CH_3}-\overset{7}{CH_2}-\overset{6}{CH_2}-\overset{5}{CH}-\overset{4}{CH_2}-\overset{3}{CH_2}-\overset{2}{C}-\overset{1}{CH_3}$$

5-Ethyl-2,2-dimethyloctane (note cited order)

$$\overset{8}{CH_3}-\overset{7}{CH_2}-\overset{6}{CH}-\overset{5}{CH_2}-\overset{4}{CH_2}-\overset{3}{CH}-\overset{2}{CH_2}-\overset{1}{CH_3}$$

3-Ethyl-6-methyloctane (note locants reversed)

$$\overset{8}{CH_3}-\overset{7}{CH_2}-\overset{6}{CH_2}-\overset{5}{CH_2}-\overset{4}{C}-\overset{3}{CH_2}-\overset{2}{CH}-\overset{1}{CH_3}$$

4,4-Bis (1,1-dimethylethyl)-2-methyloctane
4,4-Bis-1′,1′-dimethylethyl-2-methyloctane
4,4-Bis (*tert*-butyl)-2-methyloctane

Bivalent radicals derived from saturated unbranched alkanes by removal of two hydrogen atoms are named as follows: (1) If both free bonds are on the same carbon atom, the ending -ane of the hydrocarbon is replaced with -ylidene. However, for the first member of the alkanes it is methylene rather than methylidene. Isopropylidene, *sec*-Butylidene, and neopentylidene may be used for the unsubstituted group only. (2) If the two free bonds are on different carbon atoms, the straight-chain group terminating in these two carbon atoms is named by citing the number of methylene groups comprising the chain. Other carbon groups are named as substituents. Ethylene is used rather than dimethylene for the first member of the series, and propylene is retained for CH_3—CH—CH_2— (but trimethylene is —CH_2—CH_2—CH_2—).

Trivalent groups derived by the removal of three hydrogen atoms from the same carbon are named by replacing the ending -ane of the parent hydrocarbon with -ylidyne.

Alkenes and Alkynes.   Each name of the corresponding saturated hydrocarbon is converted to the corresponding alkene by changing the ending -ane to -ene. For alkynes the ending is -yne. With more than one double (or triple) bond, the endings are -adiene, -atriene, etc. (or -adiyne, -atriyne, etc.). The position of the double (or triple) bond in the parent chain is indicated by a locant obtained by numbering from the end of the chain nearest the double (or triple) bond; thus CH_3CH_2CH=CH_2 is 1-butene and CH_3C≡CCH_3 is 2-butyne.

For multiple unsaturated bonds, the chain is so numbered as to give the lowest possible locants to the unsaturated bonds. When there is a choice in numbering, the double bonds are given the lowest locants, and the alkene is cited before the alkyne where both occur in the name. Examples:

$CH_3CH_2CH_2CH_2CH$=CH—CH=CH_2	1,3-Octadiene
CH_2=CHC≡CCH=CH_2	1,5-Hexadiene-3-yne
CH_3CH=$CHCH_2C$≡CH	4-Hexen-1-yne
CH≡CCH_2CH=CH_2	1-Penten-4-yne

Unsaturated branched acyclic hydrocarbons are named as derivatives of the chain that contains the maximum number of double and/or triple bonds. When a choice exists, priority goes in sequence to (1) the chain with the greatest number of carbon atoms and (2) the chain containing the maximum number of double bonds.

These nonsystematic names are retained.

Ethylene	CH_2=CH_2
Allene	CH_2=C=CH_2
Acetylene	HC≡CH

An example of nomenclature for alkenes and alkynes is

$$\overset{6}{HC}≡\overset{5}{C}-\overset{4}{C}=\overset{3}{C}-\overset{2}{CH}=\overset{1}{CH_2}$$

with CH_2—CH_2—CH_3 on carbon 3 and CH=CH_2 on carbon 4   4-Propyl-3-vinyl-1,3-hexadien-5-yne

Univalent radicals have the endings -enyl, -ynyl, -dienyl, -diynyl, etc. When necessary, the positions of the double and triple bonds are indicated by locants, with the carbon atom with the free valence numbered as 1. Examples:

CH_2=CH—CH_2—	2-Propenyl
CH_3—C≡C—	1-Propynyl
CH_3—C≡C—CH_2CH=CH_2—	1-Hexen-4-ynyl

These names are retained:

Vinyl (for ethenyl) $CH_2{=}CH{-}$

Allyl (for 2-propenyl) $CH_2{=}CH{-}CH_2{-}$

Isopropenyl (for 1-methylvinyl but for unsubstituted radical only) $CH_2{=}C(CH_3){-}$

Should there be a choice for the fundamental straight chain of a radical, that chain is selected which contains (1) the maximum number of double and triple bonds, (2) the largest number of carbon atoms, and (3) the largest number of double bonds. These are in descending priority.

Bivalent radicals derived from unbranched alkenes, alkadienes, and alkynes by removing a hydrogen atom from each of the terminal carbon atoms are named by replacing the endings -ene, -diene, and -yne by -enylene, -dienylene, and -ynylene, respectively. Positions of double and triple bonds are indicated by numbers when necessary. The name *vinylene* instead of ethenylene is retained for $-CH{=}CH-$.

Monocyclic Aliphatic Hydrocarbons.   Monocyclic aliphatic hydrocarbons (with no side chains) are named by prefixing cyclo- to the name of the corresponding open-chain hydrocarbon having the same number of carbon atoms as the ring. Radicals are formed as with the alkanes, alkenes, and alkynes. Examples:

Cyclohexane   Cyclohexyl- (for the radical)

Cyclohexene   1-Cyclohexenyl- (for the radical with the free valence at carbon 1)

1,3-Cyclohexandiene   Cyclohexadienyl- (the unsaturated carbons are given numbers as low as possible, numbering from the carbon atom with the free valence given the number 1)

For convenience, aliphatic rings are often represented by simple geometric figures: a triangle for cyclopropane, a square for cyclobutane, a pentagon for cyclopentane, a hexagon (as illustrated) for cyclohexane, etc. It is understood that two hydrogen atoms are located at each corner of the figure unless some other group is indicated for one or both.

Monocyclic Aromatic Compounds.   Except for six retained names, all monocyclic substituted aromatic hydrocarbons are named systematically as derivatives of benzene. Moreover, if the substituent introduced into a compound with a retained trivial name is identical with one already present in that compound, the compound is named as a derivative of benzene. These names are retained:

Cumene

Cymene (all three forms; *para*-shown)

Mesitylene

Styrene   Toluene   Xylene (all three
                              forms; *meta*-shown)

The position of substituents is indicated by numbers, with the lowest locant possible given to substituents. When a name is based on a recognized trivial name, priority for lowest-numbered locants is given to substituents implied by the trivial name. When only two substituents are present on a benzene ring, their position may be indicated by *o-* (*ortho-*), *m-* (*meta-*), and *p-* (*para-*) (and alphabetized in the order given) used in place of 1,2-, 1,3-, and 1,4-, respectively.

Radicals derived from monocyclic substituted aromatic hydrocarbons and having the free valence at a ring atom (numbered 1) are named phenyl (for benzene as parent, since benzyl is used for the radical $C_6H_5CH_2$—), cumenyl, mesityl, tolyl, and xylyl. All other radicals are named as substituted phenyl radicals. For radicals having a single free valence in the side chain, these trivial names are retained:

Benzyl	$C_6H_5CH_2$—	Phenethyl	$C_6H_5CH_2CH_2$—
Benzhydryl (alternative to		Styryl	$C_6H_5CH=CH$—
diphenylmethyl)	$(C_6H_5)_2CH$—	Trityl	$(C_6H_5)_3C$—
Cinnamyl	$C_6H_5CH=CH—CH_2$—		

Otherwise, radicals having the free valence(s) in the side chain are named in accordance with the rules for alkanes, alkenes, or alkynes.

The name *phenylene* (*o-*, *m-*, or *p-*) is retained for the radical —C_6H_4—. Bivalent radicals formed from substituted benzene derivatives and having the free valences at ring atoms are named as substituted phenylene radicals, with the carbon atoms having the free valences being numbered 1,2-, 1,3-, or 1,4-, as appropriate.

Radicals having three or more free valences are named by adding the suffixes -triyl, -tetrayl, etc. to the systematic name of the corresponding hydrocarbon.

Fused Polycyclic Hydrocarbons.   The names of polycyclic hydrocarbons containing the maximum number of conjugated double bonds end in -ene. Here the ending does not denote one double bond. Names of hydrocarbons containing five or more fixed benzene rings in a linear arrangement are formed from a numerical prefix followed by -acene.

Numbering of each ring system is fixed but it follows a systematic pattern. The individual rings of each system is oriented so that the greatest number of rings are (1) in a horizontal row and (2) the maximum number of rings is above and to the right (upper-right quadrant) of the horizontal row. When two orientations meet these requirements, the one is chosen that has the fewest rings in the lower-left quadrant. Numbering proceeds in a clockwise direction, commencing with the carbon atom not engaged in ring fusion that lies in the most counterclockwise position of the uppermost ring (upper-right quadrant); omit atoms common to two or more rings. Atoms common to two or more rings are designated by adding lowercase roman letters to the number of the position immediately preceding. Interior atoms follow the highest number, taking a clockwise sequence wherever there is a choice. Anthracene and phenanthrene are two exceptions to the rule on numbering. Two examples of numbering follow:

When a ring system with the maximum number of conjugated double bonds can exist in two or more forms differing only in the position of an "extra" hydrogen atom, the name can be made specific by indicating the position of the extra hydrogen(s). The compound name is modified with a locant followed by an italic capital H for each of these hydrogen atoms. Carbon atoms that carry an indicated hydrogen atom are numbered as low as possible. For example, 1H-indene is illustrated in Table 2.2; 2H-indene would be

Names of polycyclic hydrocarbons with less than the maximum number of noncumulative double bonds are formed from a prefix dihydro-, tetrahydro-, etc., followed by the name of the corresponding unreduced hydrocarbon. The prefix perhydro- signifies full hydrogenation. For example, 1,2-dihydronaphthalene is

Examples of retained names and their structures are as follows:

Indan   Acenaphthene   Aceanthrene

Acephenanthrene

Polycyclic compounds in which two rings have two atoms in common or in which one ring contains two atoms in common with each of two or more rings of a contiguous series of rings and which contain at least two rings of five or more members with the maximum number of noncumulative double bonds and which have no accepted trivial name are named by prefixing to the name of the parent ring or ring system designations of the other components. The parent name should contain as many rings as possible (provided it has a trivial name). Furthermore, the attached component(s) should be as simple as possible. For example, one writes dibenzophenanthrene and not naphthophenanthrene because the attached component benzo- is simpler than napththo-. Prefixes designating attached components are formed by changing the ending -ene into -eno-; for example, indeno- from indene. Multiple prefixes are arranged in alphabetical order. Several abbreviated prefixes are recognized; the parent is given in parentheses:

Acenaphtho-	(acenaphthylene)	Naphtho-	(naphthalene)
Anthra-	(anthracene)	Perylo-	(perylene)
Benzo-	(benzene)	Phenanthro-	(phenanthrene)

TABLE 2.2   Fused Polycyclic Hydrocarbons

Listed in order of increasing priority for selection as parent compound.

1. Pentalene	9. Acenaphthylene
2. Indene	10. Fluorene
3. Naphthalene	11. Phenalene
4. Azulene	12. Phenanthrene*
5. Heptalene	13. Anthracene*
6. Biphenylene	14. Fluoranthene
7. *asym*-Indacene	15. Acephenanthrylene
8. *sym*-Indacene	16. Aceanthrylene

*Asterisk after a compound denotes exception to systematic numbering.

TABLE 2.2 Fused Polycyclic Hydrocarbons (*Continued*)

17. Triphenylene 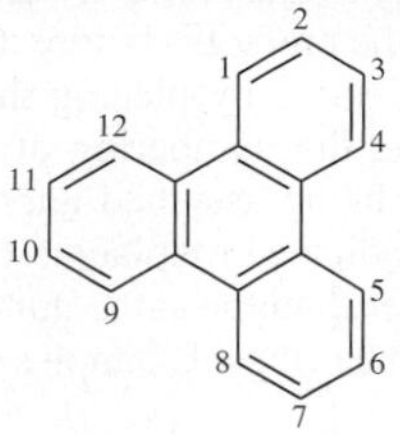	19. Chrysene 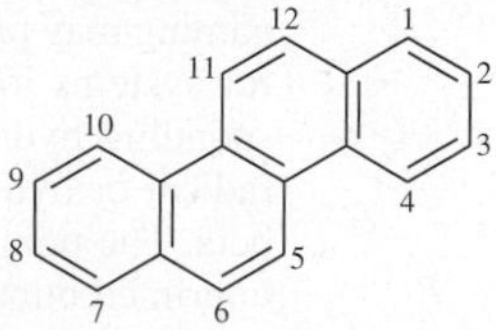
18. Pyrene 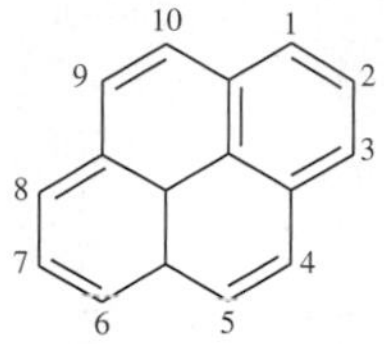	20. Naphthacene 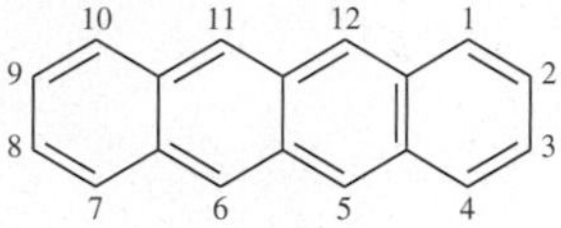

For monocyclic prefixes other than benzo-, the following names are recognized, each to represent the form with the maximum number of noncumulative double bonds: cyclopenta-, cyclohepta-, cycloocta-, etc.

Isomers are distinguished by lettering the peripheral sides of the parent beginning with *a* for the side 1,2, and so on, lettering every side around the periphery. If necessary for clarity, the numbers of the attached position (1,2, for example) of the substituent ring are also denoted. The prefixes are cited in alphabetical order. The numbers and letters are enclosed in square brackets and placed immediately after the designation of the attached component. Examples are

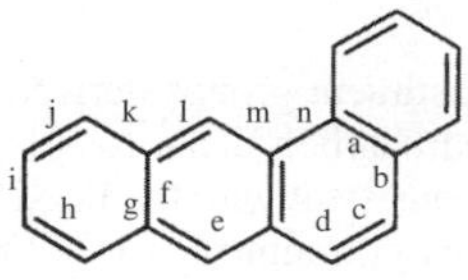

Benz[α]anthracene

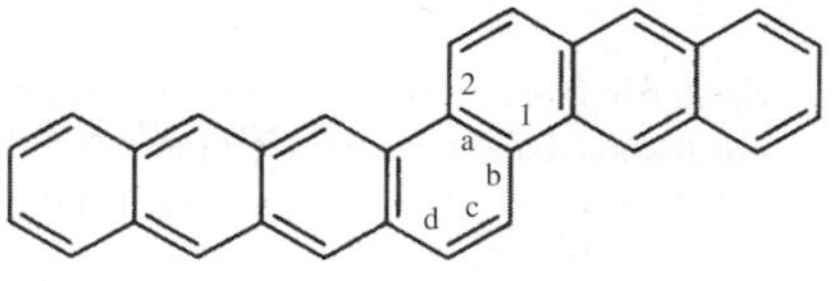

Anthra[2,1-α]naphthacene

Bridged Hydrocarbons. Saturated alicyclic hydrocarbon systems consisting of two rings that have two or more atoms in common take the name of the open-chain hydrocarbon containing the same total number of carbon atoms and are preceded by the prefix bicyclo-. The system is numbered commencing with one of the bridgeheads, numbering proceeding by the longest possible path to the second bridgehead. Numbering is then continued from this atom by the longer remaining unnumbered path back to the first bridgehead and is completed by the shortest path from the atom next to the first bridgehead. When a choice in numbering exists, unsaturation is given the lowest numbers. The number of carbon atoms in each of the bridges connecting the bridgeheads is indicated in brackets in descending order. Examples are

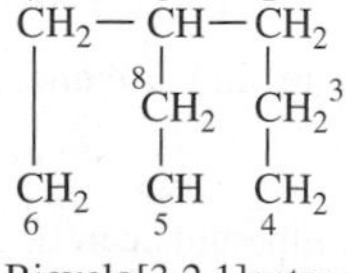

Bicyclo[3.2.1]octane

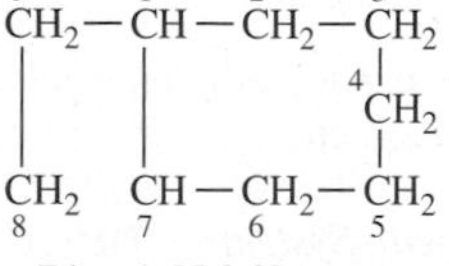

Bicyclo[5.2.0]nonane

Hydrocarbon Ring Assemblies.   Assemblies are two or more cyclic systems, either single rings or fused systems, that are joined directly to each other by double or single bonds. For identical systems naming may proceed (1) by placing the prefix bi- before the name of the corresponding radical or (2), for systems joined through a single bond, by placing the prefix bi- before the name of the corresponding hydrocarbon. In each case, the numbering of the assembly is that of the corresponding radical or hydrocarbon, one system being assigned unprimed numbers and the other primed numbers. The points of attachment are indicated by placing the appropriate locants before the name; an unprimed number is considered lower than the same number primed. The name *biphenyl* is used for the assembly consisting of two benzene rings. Examples are

1,1′-Bicyclopropyl or 1,1′-bicyclopropane   2-Ethyl-2′-propylbiphenyl

For nonidentical ring systems, one ring system is selected as the parent and the other systems are considered as substituents and are arranged in alphabetical order. The parent ring system is assigned unprimed numbers. The parent is chosen by considering the following characteristics in turn until a decision is reached: (1) the system containing the larger number of rings, (2) the system containing the larger ring, (3) the system in the lowest state hydrogenation, and (4) the highest-order number of ring systems. Examples are given, with the deciding priority given in parentheses preceding the name:

(1) 2-Phenylnaphthalene

(2) and (4) 2-(2′-Naphthyl)azulene

(3) Cyclohexylbenzene

Radicals from Ring Systems.   Univalent substituent groups derived from polycyclic hydrocarbons are named by changing the final *e* of the hydrocarbon name to -yl. The carbon atoms having free valences are given locants as low as possible consistent with the fixed numbering of the hydrocarbon. Exceptions are naphthyl (instead of naphthalenyl), anthryl (for anthracenyl), and phenanthryl (for phenanthrenyl). However, these abbreviated forms are used only for the simple ring systems. Substituting groups derived from fused derivatives of these ring systems are named systematically.

Cyclic Hydrocarbons with Side Chains.   Hydrocarbons composed of cyclic and aliphatic chains are named in a manner that is the simplest permissible or the most appropriate for the chemical intent. Hydrocarbons containing several chains attached to one cyclic nucleus are generally named as derivatives of the cyclic compound, and compounds containing several side chains and/or cyclic radicals attached to one chain are named as derivatives of the acyclic compound. Examples are

2-Ethyl-1-methylnaphthalene   Diphenylmethane

1,5-Diphenylpentane   2,3-Dimethyl-1-phenyl-1-hexene

Recognized trivial names for composite radicals are used if they lead to simplifications in naming. Examples are

1-Benzylnaphthalene   1,2,4-Tris(3-*p*-tolylpropyl)benzene

Fulvene, for methylenecyclopentadiene, and stilbene, for 1,2-diphenylethylene, are trivial names that are retained.

Heterocyclic Systems.   Heterocyclic compounds can be named by relating them to the corresponding carbocyclic ring systems by using replacement nomenclature. Heteroatoms are denoted by prefixes ending in *a*. If two or more replacement prefixes are required in a single name, they are cited in the order of their listing in the table. The lowest possible numbers consistent with the numbering of

TABLE 2.3 Heterocyclic Systems

Heterocyclic atoms are listed in decreasing order of priority.

Element	Valence	Prefix	Element	Valence	Prefix
Oxygen	2	Oxa-	Antimony	3	Stiba-*
Sulfur	2	Thia-	Bismuth	3	Bisma-
Selenium	2	Selena-	Silicon	4	Sila-
Tellurium	2	Tellura-	Germanium	4	Germa-
Nitrogen	3	Aza-	Tin	4	Stanna-
Phosphorus	3	Phospha-*	Lead	4	Plumba-
Arsenic	3	Arsa-*	Boron	3	Bora-
			Mercury	2	Mercura-

*When immediately followed by -in or -ine, phospha- should be replaced by phosphor-, arsa- by arsen-, and stiba- by antimon-. The saturated six-membered rings corresponding to phosphorin and arsenin are named *phosphorinane* and *arsenane*. A further exception is the replacement of borin by borinane.

the corresponding carbocyclic system are assigned to the heteroatoms and then to carbon atoms bearing double or triple bonds. Locants are cited immediately preceding the prefixes or suffixes to which they refer. Multiplicity of the same heteroatom is indicated by the appropriate prefix in the series: di-, tri-, tetra-, penta-, hexa-, etc.

If the corresponding carbocyclic system is partially or completely hydrogenated, the additional hydrogen is cited using the appropriate *H*- or hydro- prefixes. A trivial name along with the state of hydrogenation may be used. In the specialist nomenclature for heterocyclic systems, the prefix or prefixes (Table 2.3) are combined with the appropriate stem from Table 2.4, ending in an *a* where necessary. Examples of acceptable usage, including (1) replacement and (2) specialist nomenclature, are

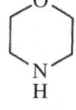

(1) 1-Oxa-4-azacyclo-
hexane

(2) 1,4-Oxazoline
Morpholine

(1) 1,3-Diazacyclo-
hex-5-ene

(2) 1,2,3,4-Tetra-
hydro-1,3 diazine

(1) Thiacyclopropane

(2) Thiirane
Ethylene sulfide

TABLE 2.4 Suffixes for Heterocyclic Systems

Number of ring members	Rings containing nitrogen		Rings containing nitrogen	
	Unsaturation*	Saturation	Unsaturation*	Saturation
3	-irine	-iridine	-irene	-irane
4	-ete	-etidine	-ete	-etane
5	-ole	-olidine	-ole	-olane
6	-ine†	‡	-in	-ane§
7	-epine	‡	-epin	-epane
8	-ocine	‡	-ocin	-ocane
9	-onine	‡	-onin	-onane
10	-ecine	‡	-ecin	-ecane

*Unsaturation corresponding to the maximum number of noncumulative double bonds. Heteroatoms have the normal valences.
†For phosphorus, arsenic, antimony, and boron, there are special provisions (Table 2.3).
‡Expressed by prefixing perhydro- to the name of the corresponding unsaturated compound.
§Not applicable to silicon, germanium, tin, and lead; perhydro- is prefixed to the name of the corresponding unsaturated compound.

TABLE 2.5   Trivial Names of Heterocyclic Systems Suitable for Use in Fusion Names

Listed in order of increasing priority as senior ring system.

Structure	Parent name	Radical name	Structure	Parent name	Radical name
	Thiophene	Thienyl		2H-Pyrrole	2H-Pyrrolyl
	Thianthrene	Thianthrenyl		Pyrrole	Pyrrolyl
	Furan	Furyl		Imidazole	Imidazolyl
	Pyran (2H-shown)	Pyranyl		Pyrazole	Pyrazolyl
	Isobenzofuran	Isobenzo-furanyl		Isothiazole	Isothiazolyl
				Isoxazole	Isoxazolyl
	Chromene (2H-shown)	Chromenyl		Pyridine	Pyridyl
				Pyrazine	Pyrazinyl
	Xanthene*	Xanthenyl		Pyrimidine	Pyrimidinyl
	Phenoxathiin	Phenoxa-thiinyl		Pyridazine	Pyridazinyl

*Asterisk after a compound denotes exception to systematic numbering.

TABLE 2.5 Trivial Names of Heterocyclic Systems Suitable for Use in Fusion Names (*Continued*)

Structure	Parent name	Radical name	Structure	Parent name	Radical name
	Indolizine	Indolizinyl		Phthalazine	Phthalazinyl
	Isoindole	Isoindolyl		Naphthyri-dine (1,8-shown)	Naphthyri-dinyl
	3*H*-Indole	3*H*-Indolyl		Quinoxaline	Quinoxalinyl
	Indole	Indolyl		Quinazoline	Quinazolinyl
	1*H*-Indazole	1*H*-Indazolyl		Cinnoline	Cinnolinyl
	Purine*	Purinyl		Pteridine	Pteridinyl
	4*H*-Quin-olizine	4*H*-Quin-olizinly		4α*H*-Carbazole*	4α*H*-Carbazolyl
	Isoquinoline	Isoquinolyl		Carbazole*	Carbazolyl
	Quinolone	Quinolyl			

*Asterisk after a compound denotes exception to systematic numbering.

(*Continued*)

TABLE 2.5 Trivial Names of Heterocyclic Systems Suitable for Use in Fusion Names (*Continued*)

Structure	Parent name	Radical name	Structure	Parent name	Radical name
	β-Carboline	β-Carbolinyl		Phenazine	Phenazinyl
	Phenanthri-dine	Phenanthri-dinyl		Phenarsazine	Phenarsazinyl
	Acridine*	Acridinyl		Phenothiazine	Phenothiazinyl
	Perimidine	Perimidinyl		Furazan	Furazanyl
	Phenanthroline (1,10-shown)	Phenanthrolinyl		Phenoxazine	Phenoxazinyl

*Asterisk after a compound denotes exception to systematic numbering.

Radicals derived from heterocyclic compounds by removal of hydrogen from a ring are named by adding -yl to the names of the parent compounds (with elision of the final *e,* if present). These exceptions are retained:

Furyl (from furan)	Furfuryl (for 2-furylmethyl)
Pyridyl (from pyridine)	Furfurylidene (for 2-furylmethylene)
Piperidyl (from piperidine)	Thienyl (from thiophene)
Quinolyl (from quinoline)	Thenylidyne (for thienylmethylidyne)
Isoquinolyl	Furfurylidyne (for 2-furylmethylidyne)
Thenylidene (for thienylmethylene)	Thenyl (for thienylmethyl)

Also, piperidino- and morpholino- are preferred to 1-piperidyl- and 4-morpholinyl-, respectively.

TABLE 2.6   Trivial Names for Heterocyclic Systems That Are Not Recommended for Use in Fusion Names

Listed in order of increasing priority.

Structure	Parent name	Radical name	Structure	Parent name	Radical name
	Isochroman	Isochromanyl		Pyrazoline (3-shown*)	Pyrazolinyl
	Chroman	Chromanyl		Piperidine	Piperidyl[†]
	Pyrrolidine	Pyrrolidinyl		Piperazine	Piperazinyl
	Pyrroline (2-shown*)	Pyrrolinyl		Indoline	Indolinyl
	Imidazolidine	Imidazolidinyl		Isoindoline	Isoindolinyl
	Imidazoline (2-shown*)	Imidazolinyl		Quinuclidine	Quinuclidinyl
	Pyrazolidine	Pyrazolidinyl		Morpholine	Morpholinyl[‡]

*Denotes position of double bond.
[†]For 1-piperidyl, use piperidino.
[‡]For 4-morpholinyl, use morpholino.

If there is a choice among heterocyclic systems, the parent compound is decided in the following order of preference:

1. A nitrogen-containing component

2. A component containing a heteroatom, in the absence of nitrogen, as high as possible (Table 2.3)

3. A component containing the greater number of rings

4. A component containing the largest possible individual ring

5. A component containing the greatest number of heteroatoms of any kind

6. A component containing the greatest variety of heteroatoms

7. A component containing the greatest number of heteroatoms first listed in Table 2.3

If there is a choice between components of the same size containing the same number and kind of heteroatoms, choose as the base component that one with the lower numbers for the heteroatoms before fusion. When a fusion position is occupied by a heteroatom, the names of the component rings to be fused are selected to contain the heteroatom.

2.1.2 Functional Compounds

There are several types of nomenclature systems that are recognized. Which type to use is sometimes obvious from the nature of the compound. Substitutive nomenclature, in general, is preferred because of its broad applicability, but radicofunctional, additive, and replacement nomenclature systems are convenient in certain situations.

Substitutive Nomenclature.   The first step is to determine the kind of characteristic (functional) group for use as the principal group of the parent compound. A characteristic group is a recognized combination of atoms that confers characteristic chemical properties on the molecule in which it occurs. Carbon-to-carbon unsaturation and heteroatoms in rings are considered nonfunctional for nomenclature purposes.

Substitution means the replacement of one or more hydrogen atoms in a given compound by some other kind of atom or group of atoms, functional or nonfunctional. In substitutive nomenclature, each substituent is cited as either a prefix or a suffix to the name of the parent (or substituting radical) to which it is attached; the latter is denoted the parent compound (or parent group if a radical).

When oxygen is replaced by sulfur, selenium, or tellurium, the priority for these elements is in the descending order listed. The higher valence states of each element are listed before considering the successive lower valence states. Derivative groups have priority for citation as principal group after the respective parents of their general class.

Systematic names formed by applying the principles of substitutive nomenclature are single words except for compounds named as acids. First, select the parent compound, and thus the suffix, from the characteristic group (Table 2.7). All remaining functional groups are handled as prefixes that precede, in alphabetical order, the parent name. Two examples are:

Structure I   Structure II

Structure I contains an ester group and an ether group. Since the ester group has higher priority, the name is ethyl 2-methoxy-6-methyl-3-cyclohexene-1-carboxylate. Structure II contains a carbonyl group, a hydroxy group, and a bromo group. The latter is never a suffix. Between the other two, the carbonyl group has higher priority, the parent has -one as suffix, and the name is 4-bromo-1-hydroxy-2-butanone.

Selection of the principal alicyclic chain or ring system is governed by these selection rules:

1. For purely alicyclic compounds, the selection process proceeds successively until a decision is reached: (a) the maximum number of substituents corresponding to the characteristic group

(Table 2.7) (b) the maximum number of double and triple bonds considered together, (c) the maximum length of the chain, and (d) the maximum number of double bonds.

2. If the characteristic group occurs only in a chain that carries a cyclic substituent, the compound is named as an aliphatic compound into which the cyclic component is substituted; a radical prefix is used to denote the cyclic component. This chain need not be the longest chain.

3. If the characteristic group occurs in more than one carbon chain and the chains are not directly attached to one another, then the chain chosen as parent should carry the largest number of the characteristic group. If necessary, the selection is continued as in rule 1.

4. If the characteristic group occurs only in one cyclic system, that system is chosen as the parent.

5. If the characteristic group occurs in more than one cyclic system, that system is chosen as parent which (a) carries the largest number of the principal group or, failing to reach a decision, (b) is the senior ring system.

6. If the characteristic group occurs both in a chain and in a cyclic system, the parent is that portion in which the principal group occurs in largest number. If the numbers are the same, that portion is chosen which is considered to be the most important or is the senior ring system.

TABLE 2.7 Characteristic Groups for Substitutive Nomenclature

Listed in order of decreasing priority for citation as principal group or parent name.

Class	Formula*	Prefix	Suffix
1. Cations:		-onio-	-onium
	H_4N^+	Ammonio-	-ammonium
	H_3O^+	Oxonio-	-oxonium
	H_3S^+	Sulfonio-	-sulfonium
	H_3Se^+	Selenonio-	-selenonium
	H_2Cl^+	Chloronio-	-chloronium
	H_2Br^+	Bromonio-	-bromonium
	H_2I^+	Iodonio-	-iodonium
2. Acids:			
Carboxylic	—COOH	Carboxy-	-carboxylic acid
	—(C)OOH		-oic acid
	—C(=O)OOH		-peroxy⋯carboxylic acid
	—(C=O)OOH		-peroxy⋯oic acid
Sulfonic	—SO$_3$H	Sulfo-	-sulfonic acid
Sulfinic	—SO$_2$H	Sulfino-	-sulfinic acid
Sulfenic	—SOH	Sulfeno-	-sulfenic acid
Salts	—COOM		Metal⋯carboxylate
	—(C)OOM		Metal⋯oate
	—SO$_3$M		Metal⋯sulfonate
	—SO$_2$M		Metal⋯sulfinate
	—SOM		Metal⋯sulfenate
3. Derivatives of acids:			
Anhydrides	—C(=O)OC(=O)—		-carboxylic anhydride
	—(C=O)O(C=O)—		-oic anhydride
Esters	—COOR	R-oxycarbonyl-	R⋯carboxylate
	—C(OOR)		R⋯oate
Acid halides	—CO—halogen	Haloformyl	-carbonyl halide
Amides	—CO—NH$_2$	Carbamoyl-	-carboxamide
	(C)O—NH$_2$		-amide

(Continued)

TABLE 2.7 Characteristic Groups for Substitutive Nomenclature (*Continued*)

Class	Formula*	Prefix	Suffix
Hydrazides	—CO—NHNH$_2$	Carbonyl- hydrazino-	-carbohydrazide
	—(CO)—NHNH$_2$		-ohydrazide
Imides	—CO—NH—CO—	R-imido-	-carboximide
Amidines	—C(=NH)—NH$_2$	Amidino-	-carboxamidine
	—(C=NH)—NH$_2$		-amidine
4. Nitrile (cyanide)	—CN	Cyano-	-carbonitrile
	— (C)N		-nitrile
5. Aldehydes	—CHO	Formyl-	-carbaldehyde
	—(C=O)H	Oxo-	-al
	(then their analogs and derivatives)		
6. Ketones	⩾(C=O)	Oxo-	-one
	(then their analogs and derivatives)		
7. Alcohols (and phenols)	—OH	Hydroxy-	-ol
Thiols	—SH	Mercapto-	-thiol
8. Hydroperoxides	—O—OH	Hydroperoxy-	
9. Amines	—NH$_2$	Amino-	-amine
Imines	⩾NH	Imino-	-imine
Hydrazines	—NHNH$_2$	Hydrazino-	-hydrazine
10. Ethers	—OR	R-oxy-	
Sulfides	—SR	R-thio-	
11. Peroxides	—O—OR	R-dioxy-	

*Carbon atoms enclosed in parentheses are included in the name of the parent compound and not in the suffix or prefix.

TABLE 2.8 Characteristic Groups Cited Only as Prefixes in Substitutive Nomenclature

Characteristic group	Prefix	Characteristic Group	Prefix
—Br	Bromo-	—IX$_2$	X may be halogen or a radical; dihalogenoiodo- or diacetoxyiodo-, e.g., —ICl$_2$ is dichloroido-
—Cl	Chloro-		
—ClO	Chlorosyl-		
—ClO$_2$	Chloryl-	⩾N$_2$	Diazo-
—ClO$_3$	Perchloryl-	—N$_3$	Azido-
—F	Fluoro-	—NO	Nitroso-
—I	Iodo-	—NO$_2$	Nitro-
—IO	Iodosyl-	⩾N(=O)OH	*aci*-Nitro-
—IO$_2$	Iodyl*	—OR	R-oxy-
—I(OH)$_2$	Dihydroxyiodo-	—SR	R-thio-
		—SeR (—TeR)	R-seleno- (R-telluro-)

*Formerly iodoxy.

7. When a substituent is itself substituted, all the subsidiary substituents are named as prefixes and the entire assembly is regarded as a parent radical.

8. The seniority of ring systems is ascertained by applying the following rules successively until a decision is reached: (a) all heterocycles are senior to all carbocycles, (b) for heterocycles, the preference follows the decision process described under Heterocyclic Systems (p. 1.11) (c) the largest number of rings, (d) the largest individual ring at the first point of difference, (e) the largest number of atoms in common among rings, (f) the lowest letters in the expression for ring functions, (g) the lowest numbers at the first point of difference in the expression for ring junctions, (h) the lowest state of hydrogenation, (i) the lowest-numbered locant for indicated hydrogen, (j) the lowest-numbered locant for point of attachment (if a radical), (k) the lowest-numbered locant for an attached group expressed as a suffix, (l) the maximum number of substituents cited as prefixes, (m) the lowest-numbered locant for substituents named as prefixes, hydro prefixes, -ene, and -yne, all considered together in one series in ascending numerical order independent of their nature, and (n) the lowest-numbered locant for the substituent named as prefix which is cited first in the name.

Numbering of Compounds.   If the rules for aliphatic chains and ring systems leave a choice, the starting point and direction of numbering of a compound are chosen so as to give lowest-numbered locants to these structural factors, if present, considered successively in the order listed below until a decision is reached. Characteristic groups take precedence over multiple bonds.

1. Indicated hydrogen, whether cited in the name or omitted as being conventional

2. Characteristic groups named as suffix following ranking order (Table 2.7)

3. Multiple bonds in acyclic compounds; in bicycloalkanes, tricycloalkanes, and polycycloalkanes, double bonds having priority over triple bonds; and in heterocyclic systems whose names end in -etine, -oline, or -olene

4. The lowest-numbered locant for substituents named as prefixes, hydro prefixes, -ene, and -yne, all considered together in one series in ascending numerical order

5. The lowest locant for that substituent named as prefix which is cited first in the name

For cyclic radicals, indicated hydrogen and thereafter the point of attachment (free valency) have priority for the lowest available number.

Prefixes and Affixes.   Prefixes are arranged alphabetically and placed before the parent name; multiplying affixes, if necessary, are inserted and *do not* alter the alphabetical order already attained. The parent name includes any syllables denoting a change of ring number or relating to the structure of a carbon chain. Nondetachable parts of parent names include

1. Forming rings: cyclo-, bicyclo-, spiro-

2. Fusing two or more rings: benzo-, naphtho-, imidazo-

3. Substituting one ring or chain member atom for another: oxa-, aza-, thia-

4. Changing positions of ring or chain members: iso-, *sec-*, *tert-*, neo-

5. Showing indicated hydrogen

6. Forming bridges: ethano-, epoxy-

7. Hydro-

Prefixes that represent complete terminal characteristic groups are preferred to those representing only a portion of a given group. For example, for the prefix $—C(=O)CH_3$, the name (formylmethyl) is preferred to (oxoethyl).

The multiplying affixes di-, tri-, tetra-, penta-, hexa-, hepta-, octa-, nona-, deca-, undeca-, and so on are used to indicate a set of *identical* unsubstituted radicals or parent compounds. The forms bis-, tris-, tetrakis-, pentakis-, and so on are used to indicate a set of identical radicals or parent compounds *each substituted in the same way*. The affixes bi-, ter-, quater-, quinque-, sexi-, septi-, octi-, novi-, deci-, and so on are used to indicate the number of identical rings joined together by a single or double bond.

Although multiplying affixes may be omitted for very common compounds when no ambiguity is caused thereby, such affixes are generally included throughout this handbook in alphabetical listings. An example would be ethyl ether for diethyl ether.

Conjunctive Nomenclature.   Conjunctive nomenclature may be applied when a principal group is attached to an acyclic component that is directly attached by a carbon-carbon bond to a cyclic component. The name of the cyclic component is attached directly in front of the name of the acyclic component carrying the principal group. This nomenclature is not used when an unsaturated side chain is named systematically. When necessary, the position of the side chain is indicated by a locant placed before the name of the cyclic component. For substituents on the acyclic chain, carbon atoms of the side chain are indicated by Greek letters proceeding from the principal group to the cyclic component. The terminal carbon atom of acids, aldehydes, and nitriles is omitted when allocating Greek positional letters. Conjunctive nomenclature is not used when the side chain carries more than one of the principal group, except in the case of malonic and succinic acids.

The side chain is considered to extend only from the principal group to the cyclic component. Any other chain members are named as substituents, with appropriate prefixes placed before the name of the cyclic component.

When a cyclic component carries more than one identical side chain, the name of the cyclic component is followed by di-, tri-, etc., and then by the name of the acyclic component, and it is preceded by the locants for the side chains. Examples are

H$_3$C—⟨4   1⟩—CH$_2$—CH$_2$OH   4-Methyl-1-cyclohexaneethanol

α-Ethyl-β,β-dimenthylcyclohexaneethanol

When side chains of two or more different kinds are attached to a cyclic component, only the senior side chain is named by the conjunctive method. The remaining side chains are named as prefixes. Likewise, when there is a choice of cyclic component, the senior is chosen. Benzene derivatives may be named by the conjunctive method only when two or more identical side chains are present. Trivial names for oxo carboxylic acids may be used for the acyclic component. If the cyclic and acyclic components are joined by a double bond, the locants of this bond are placed as superscripts to a Greek capital delta that is inserted between the two names. The locant for the cyclic component precedes that for the acyclic component, e.g., indene-$\Delta^{1,\alpha}$-acetic acid.

Radicofunctional Nomenclature.   The procedures of radicofunctional nomenclature are identical with those of substitutive nomenclature except that suffixes are never used. Instead, the functional class name (Table 2.9) of the compound is expressed as one word and the remainder of the molecule as another that precedes the class name. When the functional class name refers to a characteristic group that is bivalent, the two radicals attached to it are each named, and when different, they are written as separate words arranged in alphabetical order. When a compound contains more than one kind of group, that kind is cited as the functional group or class name that occurs higher in the table, all others being expressed as prefixes.

Radicofunctional nomenclature finds some use in naming ethers, sulfides, sulfoxides, sulfones, selenium analogs of the preceding three sulfur compounds, and azides.

TABLE 2.9   Radicofunctional Nomenclature

Groups are listed in order of decreasing priority.

Group	Functional class names
X in acid derivatives	Name of X (in priority order: fluoride, chloride, bromide, iodide, cyanide, azide; then the sulfur and selenium analogs)
—CN, —NC	Cyanide, isocyanide
$>$CO	Ketone; then S and Se analogs
—OH	Alcohol; then S and Se analogs
—O—OH	Hydroperoxide
$>$O	Ether or oxide
$>$S, $>$SO, $>$SO$_2$	Sulfide, sulfoxide, sulfone
$>$Se, $>$SeO, $>$SeO$_2$	Selenide, selenoxide, selenone
—F, —Cl, —Br, —I	Fluoride, chloride, bromide, iodide
—N$_3$	Azide

TABLE 2.10   Functional Groups and Their Properties

Group	Polarity	Volatility/M.P./B.P.	Solubility in Water	Acid-Base Behavior	Examples
Alkanes C_nH_{2n+2}	Nonpolar	Volatile, smaller molecules are gases Boiling point increases with size (higher mass) and increased dispersion forces	No	None	Methane Propane Butane Octane Wax
Alkenes C_nH_{2n} Alkynes C_nH_{2n-2}	Nonpolar	Have similar b.p. to alkanes with same number of carbons but a few degrees lower since they have fewer electrons = less dispersion forces	No	None	
Halo-alkanes RX	Slightly polar to nonpolar	CH_3Cl, CH_3Br, and CH_3CH_2Cl are gases Typically, liquids at room temperature Boiling point decreases with the number of substituents	Slightly	None	Polyvinyl chloride Chlorofluorocarbons Teflon
Alcohols ROH	Polar bond capable of H-bonds	Higher boiling points than alkanes due to hydrogen bonding Volatility and boiling point increase with chain length	Decreases with chain length	Both acidic and basic (amphoteric)	Methanol Ethanol Isopropyl alcohol
Ethers ROR	No hydrogen bonds	• Volatile • Lower boiling points than alcohols			Diethyl ether
Amines RNH_2, R_2NH, R_3N	Polar bonds, hydrogen bonding	Low volatility High boiling points due to hydrogen bonding	Yes	Basic due to lone pair of electrons on N atom	Ammonia Amino acids

(Continued)

TABLE 2.10 Functional Groups and Their Properties (*Continued*)

Group	Polarity	Volatility/M.P./B.P.	Solubility in Water	Acid-Base Behavior	Examples
Aldehydes RCHO	Hydrogen bonds between carbonyl group (C=O) and water	Boiling point is higher than that of similarly sized alkanes but lower the corresponding alcohols Boiling point increases with size of carbon chain	Yes, decreases with length of carbon chain	Carbonyl can act as a base	Formaldehyde
Ketones RCOR	Hydrogen bonds between carbonyl group (C=O) and water	See aldehydes	Decreases with chain length	Carbonyl can act as a base	Acetone Methyl ethyl ketone (MEK)
Carboxylic acids RCOOH	Capable of hydrogen bonds	Low volatility Higher boiling points due to hydrogen bonding	Decreases with chain length	Acid	Acetic acid
Esters RCOOR	Will form hydrogen bonds with water	More volatile than carboxylic acids Liquids at room temperature for lower molecular weight compounds	Slight; decreases with chain size	Variable depending upon location of function	Polyester Fatty acids
Amides RCONH$_2$	Polar, amine can hydrogen bond	High melting point due to hydrogen bonding Liquids or solids at room temperature	Yes	Amine part of amide is basic	Polypeptides Proteins

Replacement Nomenclature. Replacement nomenclature is intended for use only when other nomenclature systems are difficult to apply in the naming of chains containing heteroatoms. When no group is present that can be named as a principal group, the longest chain of carbon and heteroatoms terminating with carbon is chosen and named as though the entire chain were that of an acyclic hydrocarbon. The heteroatoms within this chain are identified by means of prefixes aza-, oxa-, thia-, etc. Locants indicate the positions of the heteroatoms in the chain. Lowest-numbered locants are assigned to the principal group when such is present. Otherwise, lowest-numbered locants are assigned to the heteroatoms considered together and, if there is a choice, to the heteroatoms cited earliest in Table 2.3. An example is

$$\overset{13}{HO-CH_2}-O-\overset{12}{CH_2}-\overset{11}{CH_2}-\overset{10}{CH_2}-\overset{9}{O}-\overset{8}{CH_2}-\overset{7}{CH_2}-\overset{6}{\underset{H}{N}}-\overset{5}{CH_2}-\overset{4}{CH_2}-\overset{3}{\underset{H}{N}}-\overset{2}{CH_2}-\overset{1}{COOH}$$

13-Hydroxy-9,12-dioxa-3,6-diazatridecanoic acid

2.1.3 Specific Functional Groups

Acetals and Acylals. Acetals, which contain the group >C(OR)$_2$, where R may be different, are named (1) as dialkoxy compounds or (2) by the name of the corresponding aldehyde or ketone followed by the name of the hydrocarbon radical(s) followed by the word *acetal*. For example, CH$_3$—CH(OCH$_3$)$_2$ is named either (1) 1,1-dimethoxyethane or (2) acetaldehyde dimethyl acetal.

A cyclic acetal in which the two acetal oxygen atoms form part of a ring may be named (1) as a heterocyclic compound or (2) by use of the prefix methylenedioxy for the group —O—CH_2—O— as a substituent in the remainder of the molecule. For example,

(1) 1,3-Benzo[*d*]dioxole-5-carboxylic acid

(2) 3,4-Methylenedioxybenzoic acid

Acylals, $R^1R^2C(OCOR^3)_2$, are named as acid esters;

Butylidene acetate propionate

α-Hydroxy ketones, formerly called acyloins, had been named by changing the ending -ic acid or -oic acid of the corresponding acid to -oin. They are preferably named by substitutive nomenclature; thus

$$CH_3—CH(OH)—CO—CH_3 \quad \text{3-Hydroxy-2-butanone (formerly acetoin)}$$

Acid Anhydrides. Symmetrical anhydrides of monocarboxylic acids, when unsubstituted, are named by replacing the word *acid* by *anhydride*. Anhydrides of substituted monocarboxylic acids, if symmetrically substituted, are named by prefixing bis- to the name of the acid and replacing the word *acid* by *anhydride*. Mixed anhydrides are named by giving in alphabetical order the first part of the names of the two acids followed by the word *anhydride*, e.g., acetic propionic anhydride or acetic propanoic anhydride. Cyclic anhydrides of polycarboxylic acids, although possessing a heterocyclic structure, are preferably named as acid anhydrides. For example,

1,8;4,5-Naphthalenetetracarboxylic dianhydride (note the use of a semicolon to distinguish the pairs of locants)

Acyl Halides. Acyl halides, in which the hydroxyl portion of a carboxyl group is replaced by a halogen, are named by placing the name of the corresponding halide after that of the acyl radical. When another group is present that has priority for citation as principal group or when the acyl halide is attached to a side chain, the prefix haloformyl- is used as, for example, in fluoroformyl-.

Alcohols and Phenols. The hydroxyl group is indicated by a suffix -ol when it is the principal group attached to the parent compound and by the prefix hydroxy- when another group with higher priority for citation is present or when the hydroxy group is present in a side chain. When confusion may arise in employing the suffix -ol, the hydroxy group is indicated as a prefix; this terminology is also used when the hydroxyl group is attached to a heterocycle, as, for example, in the name 3-hydroxythiophene to avoid confusion with thiophenol (C_6H_5SH). Designations such as isopropanol, *sec*-butanol, and *tert*-butanol are incorrect because no hydrocarbon exists to which the suffix can be added. Many trivial names are retained. (Table 2.11).

The radicals (RO—) are named by adding -oxy as a suffix to the name of the R radical, e.g., pentyloxy for $CH_3CH_2CH_2CH_2CH_2O$—. These contractions are exceptions: methoxy (CH_3O—), ethoxy (C_2H_5O—), propoxy (C_3H_7O—), butoxy (C_4H_9O—), and phenoxy (C_6H_5O—). For unsubstituted radicals only, one may use isopropoxy [$(CH_3)_2CH$—O—], isobutoxy [$(CH_3)_2CH_2CH$—O—], *sec*-butoxy [$CH_3CH_2CH(CH_3)$—O—], and *tert*-botoxy [$(CH_3)_3C$—O—].

TABLE 2.11 Alcohols and Phenols

Ally alcohol	$CH_2=CHCH_2OH$
tert-Butyl alcohol	$(CH_3)_3COH$
Benzyl alcohol	$C_6H_5CH_2OH$
Phenethyl alcohol	$C_6H_5CH_2CH_2OH$
Ethylene glycol	$HOCH_2CH_2OH$
1,2-Propylene glycol	$CH_3CHOHCH_2OH$
Glycerol	$HOCH_2CHOHCH_2OH$
Pentaerythritol	$C(CH_2OH)_4$
Pinacol	$(CH_3)_2COHCOH(CH_3)_2$
Phenol	C_6H_5OH

Xylitol

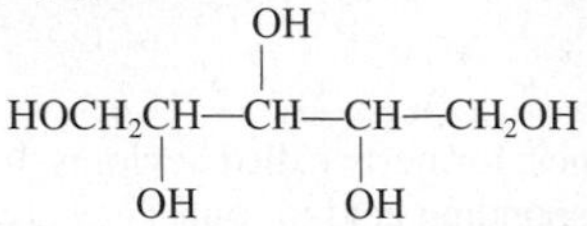

$$\underset{\underset{OH}{|}}{HOCH_2CH}-\underset{\underset{OH}{|}}{CH}-\underset{\underset{OH}{|}}{CH}-CH_2OH$$

Geraniol

$$(CH_3)_2C=CHCH_2CH_2\underset{\underset{CH_3}{|}}{C}=CHCH_2OH$$

Phytol

$$\underset{\underset{CH_3}{|}}{CH_2CH_2CHCH_2CH_2CH_2CH(CH_3)_2}$$
$$CH_2\underset{\underset{CH_3}{|}}{CHCH_2CH_2CH_2}\underset{\underset{CH_3}{|}}{C}=CHCH_2OH$$

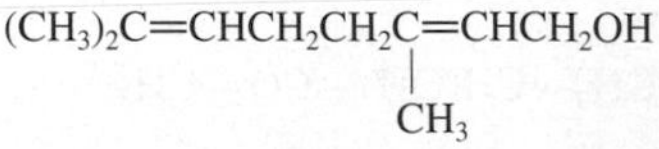

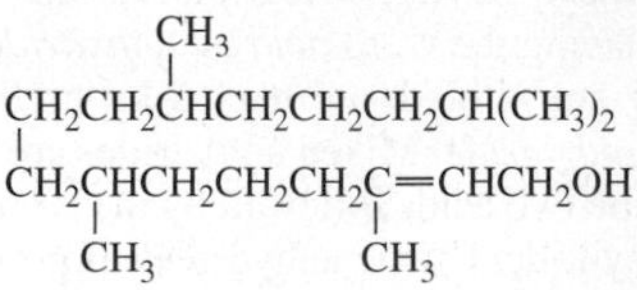

Menthol   Borneol

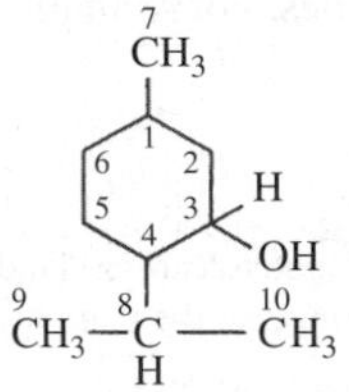

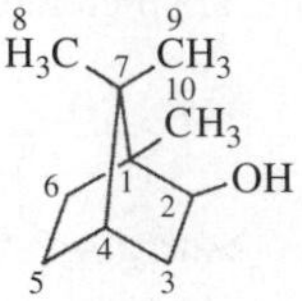

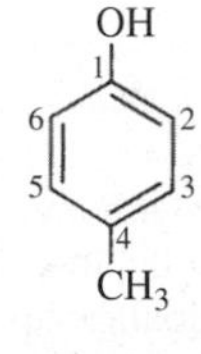

Cresol (1,4-isomer shown) 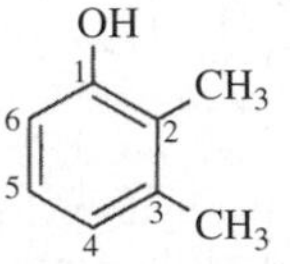   Xylenol (2,3-isomer shown) 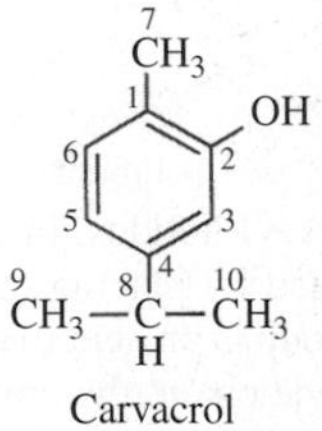   Carvacrol 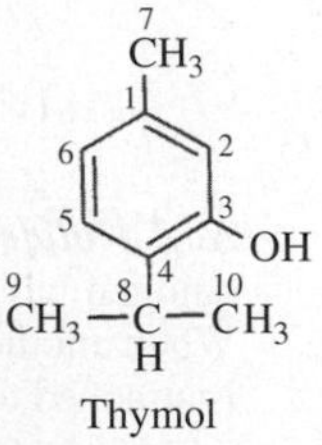   Thymol

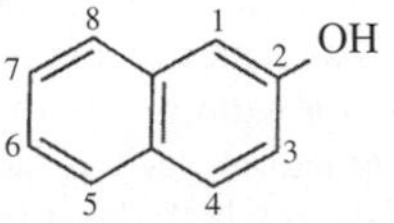

Naphthol (2-isomer shown)
2-Hydroxynaphthalene

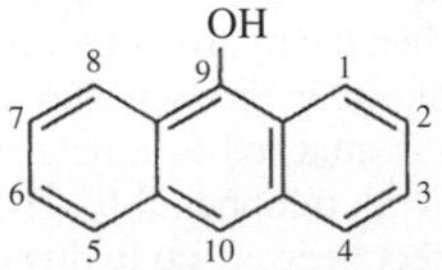

Anthrol (9-isomer shown)
9-Hydroxyanthracene

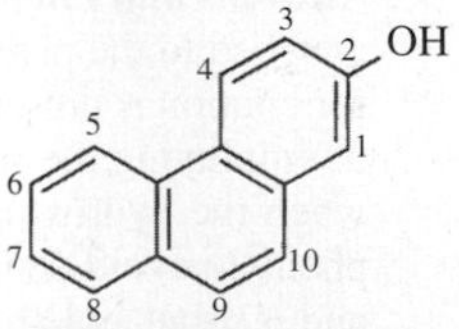

Phenanthrol (2-isomer shown)
2-Hydroxyphenanthrene

TABLE 2.11   Alcohols and Phenols (*Continued*)

Pyrocatechol	Resorcinol	Hydroquinone	Pyrogallol
1,2-Dihydroxybenzene	1, 3-Dihydroxybenzene	1,4-Dihydroxybenzene	1,2,3-Trihydroxybenzene

Phloroglucinol	Picric acid	Styphnic acid
1,3,5-Trihydroxybenzene	2,4,6-Trinitrophenol	1,3-Dihydroxy-2,4,6-trinitroben-zene

Bivalent radicals of the form O—Y—O are named by adding -dioxy to the name of the bivalent radicals except when forming part of a ring system. Examples are —O—CH_2—O— (methylenedioxy), —O—CO—O— (carbonyldioxy), and —O—SO_2—O— (sulfonyldioxy). Anions derived from alcohols or phenols are named by changing the final -ol to -olae.

Salts composed of an anion, RO—, and a cation, usually a metal, can be named by citing first the cation and then the RO anion (with its ending changed to -yl oxide), e.g., sodium benzyl oxide for $C_6H_5CH_2ONa$. However, when the radical has an abbreviated name, such as methoxy, the ending -oxy is changed to -oxide. For example, CH_3ONa is named sodium methoxide (not sodium methylate).

Aldehydes.   When the group —C(=O)H, usually written —CHO, is attached to carbon at one (or both) end(s) of a linear acyclic chain the name is formed by adding the suffix -al (or -dial) to the name of the hydrocarbon containing the same number of carbon atoms. Examples are butanal for $CH_3CH_2CH_2CHO$ and propanedial for, $OHCCH_2CHO$.

Naming an acyclic polyaldehyde can be handled in two ways. First, when more than two aldehyde groups are attached to an unbranched chain, the proper affix is added to -carbaldehyde, which becomes the suffix to the name of the longest chain carrying the maximum number of aldehyde groups. The name and numbering of the main chain do not include the carbon atoms of the aldehyde groups. Second, the name is formed by adding the prefix formyl- to the name of the -dial that incorporates the principal chain. Any other chains carrying aldehyde groups are named by the use of formylalkyl- prefixes. Examples are

(1) 1,2,5-Pentanetricarbaldehyde
(2) 3-Formylheptanedial

(1) 4-(2-Formylethyl)-3-(formylmethyl)-1,2,7-heptanetricarbaldehyde
(2) 3-Formyl-5-(2-formylethyl)-4-(formylmethyl)nonanedial

When the aldehyde group is directly attached to a carbon atom of a ring system, the suffix-carbaldehyde is added to the name of the ring system, e.g., 2-naphthalenecarbaldehyde. When the aldehyde group is separated from the ring by a chain of carbon atoms, the compound is named (1) as a derivative of the acyclic system or (2) by conjunctive nomenclature, for example, (1) (2-naphthyl) propionaldehyde or (2) 2-naphthalenepropionaldehyde.

An aldehyde group is denoted by the prefix formyl- when it is attached to a nitrogen atom in a ring system or when a group having priority for citation as principal group is present and part of a cyclic system.

When the corresponding monobasic acid has a trivial name, the name of the aldehyde may be formed by changing the ending -ic acid or -oic acid to -aldehyde. Examples are

Formaldehyde	Acrylaldehyde (not acrolein)
Acetaldehyde	Benzaldehyde
Propionaldehyde	Cinnamaldehyde
Butyraldehyde	2-Furaldehyde (not furfural)

The same is true for polybasic acids, with the proviso that all the carboxyl groups must be changed to aldehyde; then it is not necessary to introduce affixes. Examples are

Glyceraldehyde	Succinaldehyde
Glycolaldehyde	Phthalaldehyde (*o-*, *m-*, *p-*)
Malonaldehyde	

These trivial names may be retained: citral (3,7-dimethyl-2,6-octadienal), vanillin (4-hydroxy-3-methoxybenzaldehyde), and piperonal (3,4-methylenedioxybenzaldehyde).

Amides. For primary amides the suffix -amide is added to the systematic name of the parent acid. For example, CH_3—CO—NH_2 is acetamide. Oxamide is retained for H_2N—CO—CO—NH_2. The name -carboxylic acid is replaced by -carboxamide.

For amino acids having trivial names ending in -ine, the suffix -amide is added after the name of the acid (with elision of *e* for monomides). For example, H_2N—CH_2—CO—NH_2 is glycinamide.

In naming the radical R—CO—NH—, either (1) the -yl ending of RCO— is changed to -amido or (2) the radicals are named as acylamino radicals. For example,

CH_3—CO—NH—⟨ ⟩—COOH   (1) 4-Acetamidobenzoic acid
(2) 4-Acetylaminobenzoic acid

The latter nomenclature is always used for amino acids with trivial names.

N-substituted primary amides are named either (1) by citing the substitutents as *N* prefixes or (2) by naming the acyl group as an *N* substituent of the parent compound. For example,

⟨ ⟩—CO—NH—CH_3 (1) *N*-Methylbenzamide
(2) Benzoylaminomethane

Amines. Amines are preferably named by adding the suffix -amine (and any multiplying affix) to the name of the parent radical. Examples are

$CH_3CH_2CH_2CH_2CH_2NH_2$ Pentylamine
$H_2NCH_2CH_2CH_2CH_2CH_2NH_2$ 1,5-Pentyldiamine or pentamethylenediamine

Locants of substituents of symmetrically substituted derivatives of symmetrical amines are distinguished by primes or else the names of the complete substituted radicals are enclosed in parentheses. Unsymmetrically substituted derivatives are named similarly or as *N*-substituted products of a primary amine (after choosing the most senior of the radicals to be the parent amine). For example,

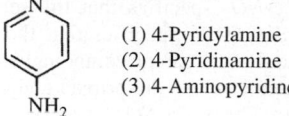

(1) 1,3′-Difluorodipropylamine
(2) 1-Fluoro-*N*-(3-fluoropropyl)propylamine
(3) (1-Fluoropropyl)(3-fluoropropyl)amine

Complex cyclic compounds may be named by adding the suffix -amine or the prefix amino- (or aminoalkyl-) to the name of the parent compound. Thus three names are permissible for

(1) 4-Pyridylamine
(2) 4-Pyridinamine
(3) 4-Aminopyridine

Complex linear polyamines are best designated by replacement nomenclature. These trivial names are retained: aniline, benzidene, phenetidine, toluidine, and xylidine.

The bivalent radical —NH— linked to two identical radicals can be denoted by the prefix imino-, as well as when it forms a bridge between two carbon ring atoms. A trivalent nitrogen atom linked to three identical radicals is denoted by the prefix nitrilo-. Thus ethylenediaminetetraacetic acid (an allowed exception) should be named ethylenedinitrilotetraacetic acid.

Ammonium Compounds. Salts and hydroxides containing quadricovalent nitrogen are named as a substituted ammonium salt or hydroxide. The names of the substituting radicals precede the word *ammonium*, and then the name of the anion is added as a separate word. For example, $(CH_3)_4N^+I^-$ is tetramethylammonium iodide.

When the compound can be considered as derived from a base whose name does not end in -amine, its quaternary nature is denoted by adding ium to the name of that base (with elision of *e*), substituent groups are cited as prefixes, and the name of the anion is added separately at the end. Examples are

$C_6H_5NH_3^+HSO_4^-$ Anilinium hydrogen sulfate

$[(C_6H_5NH_3)^+]_2PtCl_6^{2-}$ Dianilinium hexachloroplatinate

The names *choline* and *betaine* are retained for unsubstituted compounds.

In complex cases, the prefixes amino- and imino- may be changed to ammonio- and iminio- and are followed by the name of the molecule representing the most complex group attached to this nitrogen atom and are preceded by the names of the other radicals attached to this nitrogen. Finally the name of the anion is added separately. For example, the name might be 1-trimethylammonio-acridine chloride or 1-acridinyltrimethylammonium chloride.·

When the preceding rules lead to inconvenient names, then (1) the unaltered name of the base may be used followed by the name of the anion or (2) for salts of hydrohalogen acids only the unaltered name of the base is used followed by the name of the hydrohalide. An example of the latter would be 2-ethyl-*p*-phenylenediamine monohydrochloride.

Azo Compounds. When the azo group (—N=N—) connects radicals derived from identical unsubstituted molecules, the name is formed by adding the prefix azo- to the name of the parent unsubstituted molecules. Substituents are denoted by prefixes and suffixes. The azo group has priority for lowest-numbered locant. Examples are azobenzene for C_6H_5—N=N—C_6H_5, azobenzene-4-sulfonic acid for C_6H_5—N=N—$C_6H_5SO_3H$, and 2′,4-dichloroazobenzene-4′-sulfonic acid for ClC_6H_4—N=N—$C_6H_3ClSO_3H$.

When the parent molecules connected by the azo group are different, azo is placed between the complete names of the parent molecules, substituted or unsubstituted. Locants are placed between the affix azo and the names of the molecules to which each refers. Preference is given to the more complex parent molecule for citation as the first component, e.g., 2-aminonaphthalene-1-azo-(4'-chloro-2'-methylbenzene).

In an alternative method, the senior component is regarded as substituted by RN=N-, this group R being named as a radical. Thus 2-(7-phenylazo-2-naphthylazo)anthracene is the name by this alternative method for the compound named anthracene-2-azo-2'-naphthalene-7'-azobenzene.

Azoxy Compounds.   Where the position of the azoxy oxygen atom is unknown or immaterial, the compound is named in accordance with azo rules, with the affix azo replaced by azoxy. When the position of the azoxy oxygen atom in an unsymmetrical compound is designated, a prefix *NNO-* or *ONN-* is used. When both the groups attached to the azoxy radical are cited in the name of the compound, the prefix *NNO-* specifies that the second of these two groups is attached directly to —N(O)—; the prefix *ONN-* specifies that the first of these two groups is attached directly to —N(O)—. When only one parent compound is cited in the name, the prefixed *ONN-* and *NNO-* specify that the group carrying the-primed and unprimed substituents is connected, respectively, to the —N(O)— group. The prefix *NON-* signifies that the position of the oxygen atom is unknown; the azoxy group is then written as —N_2O—. For-example,

2,2',4-Trichloro-*NNO*-azoxybenzene

Boron Compounds.   Molecular hydrides of boron are called boranes. They are named by using a multiplying affix to designate the number of boron atoms and adding an Arabic numeral within parentheses as a suffix to denote the number of hydrogen atoms present. Examples are pentaborane(9) for B_5H_9 and pentaborane(11) for B_5H_{11}.

Organic ring systems are named by replacement nomenclature. Three- to ten-membered monocyclic ring systems containing uncharged boron atoms may be named by the specialist nomenclature for heterocyclic systems. Organic derivatives are named as outlined for substitutive nomenclature.

Carboxylic Acids.   Carboxylic acids may be named in several ways. First, —COOH groups replacing CH_3— at the end of the main chain of an acyclic hydrocarbon are denoted by adding -oic acid to the name of the hydrocarbon. Second, when the —COOH group is the principal group, the suffix -carboxylic acid can be added to the name of the parent chain whose name and chain numbering *does not include* the carbon atom of the —COOH group. The former nomenclature is preferred unless use of the ending -carboxylic acid leads to citation of a larger number of carboxyl groups as suffix. Third, carboxyl groups are designated by the prefix carboxy- when attached to a group named as a substituent or when another group is present that has higher priority for citation as principal group. In all cases, the principal chain should be linked to as many carboxyl groups as possible even though it might not be the longest chain present. Examples are

$CH_3CH_2CH_2CH_2CH_2CH_2COOH$ (1) Heptanoic acid
                                                            (2) 1-Hexanecarboxylic acid

$C_6H_{11}COOH$ (2) Cyclohexanecarboxylic acid

(3) 2-(Carboxymethyl)-1,4-hexanedicarboxylic acid

TABLE 2.12 Names of Some Carboxylic Acids

Systematic name	Trivial name	Systematic name	Trivial name
Methanoic	Formic	trans-Methylbutenedioic	Mesaconic*
Ethanoic	Acetic		
Propanoic	Propionic	1,2,2-Trimethyl-1,3-cyclopen-	Camphoric
Butanoic	Butyric	tanedicarboxylic acid	
2-Methylpropanoic	Isobutyric*		
Pentanoic	Valeric	Benzenecarboxylic	Benzoic
3-Methylbutanoic	Isovaleric*	1,2-Benzenedicarboxylic	Phthalic
2,2-Dimethylpropanoic	Pivalic*	1,3-Benzenedicarboxylic	Isophthalic
Hexanoic	(Caproic)	1,4-Benzenedicarboxylic	Terephthalic
Heptanoic	(Enanthic)	Naphthalenecarboxylic	Naphthoic
Octanoic	(Caprylic)	Methylbenzenecarboxylic	Toluic
Decanoic	(Capric)	2-Phenylpropanoic	Hydratropic
Dodecanoic	Laurie*	2-Phenylpropenoic	Atropic
Tetradecanoic	Myristic*	trans-3-Phenylpropenoic	Cinnamic
Hexadecanoic	Palmitic*	Furancarboxylic	Furoic
Octadecanoic	Stearic*	Thiophenecarboxylic	Thenoic
		3-Pyridinecarboxylic	Nicotinic
Ethanedioic	Oxalic	4-Pyridinecarboxylic	Isonicotinic
Propanedioic	Malonic		
Butanedioic	Succinic	Hydroxyethanoic	Glycolic
Pentanedioic	Glutaric	2-Hydroxypropanoic	Lactic
Hexanedioic	Adipic	2,3-Dihydroxypropanoic	Glyceric
Heptanedioic	Pimelic*	Hydroxypropanedioic	Tartronic
Octanedioic	Suberic*	Hydroxybutanedioic	Malic
Nonanedioic	Azelaic*	2,3-Dihydroxybutanedioic	Tartaric
Decanedioic	Sebacic*	3-Hydroxy-2-phenylpropanoic	Tropic
Propenoic	Acrylic	2-Hydroxy-2,2-diphenyl-	Benzilic
Propynoic	Propiolic	ethanoic	
2-Methylpropenoic	Methacrylic	2-Hydroxybenzoic	Salicylic
trans-2-Butenoic	Crotonic	Methoxybenzoic	Anisic
cis-2-Butenoic	Isocrotonic	4-Hydroxy-3-methoxybenzoic	Vanillic
cis-9-Octadecenoic	Oleic		
trans-9-Octadecenoic	Elaidic	3,4-Dimethoxybenzoic	Veratric
cis-Butenedioic	Maleic	3,4-Methylenedioxybenzoic	Piperonylic
trans-Butenedioic	Fumaric	3,4-Dihydroxybenzoic	Protocatechuic
cis-Methylbutenedioic	Citraconic*	3,4,5-Trihydroxybenzoic	Gallic

*Systematic names should be used in derivatives formed by substitution on a carbon atom.
Note: The names in parentheses have been discontinued.

Removal of the OH from the —COOH group to form the acyl radical results in changing the ending -oic acid to -oyl or the ending -carboxylic acid to -carbonyl. Thus the radical $CH_3CH_2CH_2CH_2CO—$ is named either pentanoyl or butanecarbonyl. When the hydroxyl has not been removed from all carboxyl groups present in an acid, the remaining carboxyl groups are denoted by the prefix carboxy-. For example, $HOOCCH_2CH_2CH_2CH_2CH_2CO—$ is named 6-carboxyhexanoyl.

Many trivial names exist for acids (Table 2.12). Generally, radicals are formed by replacing -ic acid by -oyl.[1] When a trivial name is given to an acyclic monoacid or diacid, the numeral 1 is

---

[1]Exceptions: formyl, acetyl, propionyl, butyryl, isobutyryl, valeryl, isovaleryl, oxalyl, malonyl, succinyl, glutaryl, furoyl, and thenoyl.

always given as locant to the carbon atom of a carboxyl group in the acid or to the carbon atom with a free valence in the radical RCO—.

Ethers (R¹—O—R²).   In substitutive nomenclature, one of the possible radicals, R—O—, is stated as the prefix to the parent compound that is senior from among R^1 or R^2. Examples are methoxy-ethane for $CH_3OCH_2CH_3$ and butoxyethanol for $C_4H_9OCH_2CH_2OH$.

When another principal group has precedence and oxygen is linking two identical parent compounds, the prefix oxy- may be used, as with 2,2'-oxydiethanol for $HOCH_2CH_2OCH_2CH_2OH$.

Compounds of the type RO—Y—OR, where the two parent compounds are identical and contain a group having priority over ethers for citation as suffix, are named as assemblies of identical units. For example, $HOOC—CH_2—O—CH_2CH_2—O—CH_2—COOH$ is named 2,2'-(ethylenedioxy) diacetic acid.

Linear polyethers derived from three or more molecules of aliphatic dihydroxy compounds, particularly when the chain length exceeds ten units, are most conveniently named by open-chain replacement nomenclature. For example, $CH_3CH_2—O—CH_2CH_2—O—CH_2CH_3$ could be 3,6-dioxaoctane or (2-ethoxy) ethoxyethane.

An oxygen atom directly attached to two carbon atoms already forming part of a ring system or to two carbon atoms of a chain may be indicated by the prefix epoxy-. For example, $CH_2—CH—CH_2Cl$ is named 1-chloro-2,3-epoxypropane. $\overset{\diagdown}{O}\diagup$

Symmetrical linear polyethers may be named (1) in terms of the central oxygen atom when there is an odd number of ether oxygen atoms or (2) in terms of the central hydrocarbon group when there is an even number of ether oxygen atoms. For example, $C_2H_5—O—C_4H_8—O—C_4H_8—O—C_2H_5$ is bis-(4-ethoxybutyl)ether, and 3,6-dioxaoctane (earlier example) could be named 1,2-bis(ethoxy)ethane.

Partial ethers of polyhydroxy compounds may be named (1) by substitutive nomenclature or (2) by stating the name of the polyhydroxy compound followed by the name of the etherifying radical(s) followed by the word *ether*. For example,

$$
\begin{array}{l}
CH_2O—C_4H_9 \\
| \\
HCOH \\
| \\
CH_2OH
\end{array}
\qquad
\begin{array}{l}
\text{(1) 3-Butoxy-1,2-propanediol} \\
\text{(2) Glycerol 1-butyl ether; also, 1-}O\text{-butylglycerol}
\end{array}
$$

Cyclic ethers are named either as heterocyclic compounds or by specialist rules of heterocyclic nomenclature. Radicofunctional names are formed by citing the names of the radicals R^1 and R^2 followed by the word *ether*. Thus methoxyethane becomes ethyl methyl ether and ethoxyethane becomes diethyl ether.

Halogen Derivatives.   Using substitutive nomenclature, names are formed by adding prefixes listed in Table 2.8 to the name of the parent compound. The prefix perhalo- implies the replacement of all hydrogen atoms by the particular halogen atoms.

Cations of the type $R^1R^2X^+$ are given names derived from the halonium ion, H_2X^+, by substitution, e.g., diethyliodonium chloride for $(C_2H_5)_2I^+Cl^-$.

Retained are these trivial names; bromoform $(CHBr_3)$, chloroform $(CHCl_3)$, fluoroform (CHF_3), iodoform (CHI_3), phosgene $(COCl_2)$, thiophosgene $(CSCl_2)$, and dichlorocarbene radical $(=CCl_2)$. Inorganic nomenclature leads to such names as carbonyl and thiocarbonyl halides $(COX_2$ and $CSX_2)$ and carbon tetrahalides (CX_4).

Hydroxylamines and Oximes.   For RNH—OH compounds, prefix the name of the radical R to hydroxylamine. If another substituent has priority as principal group, attach the prefix hydroxyamino- to the parent name. For example, C_6H_5NHOH would be named N-phenylhydroxylamine, but HOC_6H_4NHOH would be (hydroxyamino)phenol, with the point of attachment indicated by a locant preceding the parentheses.

Compounds of the type $R^1NH\text{—}OR^2$ are named (1) as alkoxyamino derivatives of compound R^1H, (2) as *N,O*-substituted hydroxylamines. (3) as alkoxyamines (even if R^1 is hydrogen), or (4) by the prefix aminooxy- when another substituent has priority for parent name. Examples of each type are

1. 2-(Methoxyamino)-8-naphthalenecarboxylic acid for $CH_3ONH\text{—}C_{10}H_6COOH$
2. *O*-Phenylhydroxylamine for $H_2N\text{—}O\text{—}C_6H_5$ or *N*-phenylhydroxylamine for $C_6H_5NH\text{—}OH$
3. Phenoxyamine for $H_2N\text{—}O\text{—}C_6H_5$ (not preferred to *O*-phenylhydroxylamine)
4. Ethyl (aminooxy)acetate for $H_2N\text{—}O\text{—}CH_2CO\text{—}OC_2H_5$

Acyl derivatives, $RCO\text{—}NH\text{—}OH$ and $H_2N\text{—}O\text{—}CO\text{—}R$, are named as *N*-hydroxy derivatives of amides and as *O*-acylhydroxylamines, respectively. The former may also be named as hydroxamic acids. Examples are *N*-hydroxyacetamide for $CH_3CO\text{—}NH\text{—}OH$ and *O*-acetylhydroxylamine for $H_2N\text{—}O\text{—}CO\text{—}CH_3$. Further substituents are denoted by prefixes with *O*- and/or *N*-locants. For example, $C_6H_5NH\text{—}O\text{—}C_2H_5$ would be *O*-ethyl-*N*-phenylhydroxylamine or *N*-ethoxyaniline.

For oximes, the word *oxime* is placed after the name of the aldehyde or ketone. If the carbonyl group is not the principal group, use the prefix hydroxyimino-. Compounds with the group $>N\text{—}OR$ are named by a prefix alkyloxyimino- oxime *O*-ethers or as *O*-substituted oximes. Compounds with the group $>C=N(O)R$ are named by adding *N*-oxide after the name of the alkylideneamine compound. For amine oxides, add the word *oxide* after the name of the base, with locants. For example, $C_5H_5N\text{—}O$ is named pyridine *N*-oxide or pyridine 1-oxide.

Imines. The group $>C=NH$ is named either by the suffix -imine or by citing the name of the bivalent radical $R^1R^2C<$ as a prefix to amine. For example, $CH_3CH_2CH_2CH=NH$ could be named 1-butanimine or butylideneamine. When the nitrogen is substituted, as in $CH_2=N\text{—}CH_2CH_3$, the name is *N*-(methylidene)ethylamine.

Quinones are exceptions. When one or more atoms of quinonoid oxygen have been replaced by $>NH$ or $>NR$, they are named by using the name of the quinone followed by the word *imine* (and preceded by proper affixes). Substituents on the nitrogen atom are named as prefixes. Examples are

p-Benzoquinone monoimine

p-Benzoquinone diimine

Ketenes. Derivatives of the compound ketene, $CH_2=C=O$, are named by substitutive nomenclature. For example, $C_4H_9CH=C=O$ is butyl ketene. An acyl derivative, such as $CH_3CH_2\text{—}CO\text{—}CH_2CH=C=O$, may be named as a polyketone, 1-hexene-1,4-dione. Bisketene is used for two to avoid ambiguity with diketene (dimeric ketene).

Ketones. Acyclic ketones are named (1) by adding the suffix -one to the name of the hydrocarbon forming the principal chain or (2) by citing the names of the radicals R^1 and R^2 followed by the word *ketone*. In addition to the preceding nomenclature, acyclic monoacyl derivatives of cyclic compounds may be named (3) by prefixing the name of the acyl group to the name of the cyclic compound. For example, the three possible names of

(1) 1-(2-Furyl)-1-propanone
(2) Ethyl 2-furyl ketone
(3) 2-Propionylfuran

When the cyclic component is benzene or naphthalene, the -ic acid or -oic acid of the acid corresponding to the acyl group is changed to -ophenone or -onaphthone, respectively. For example, $C_6H_5\text{—}CO\text{—}CH_2CH_2CH_3$ can be named either butyrophenone (or butanophenone) or phenyl propyl ketone.

Radicofunctional nomenclature can be used when a carbonyl group is attached directly to carbon atoms in two ring systems and no other substituent is present having priority for citation.

When the methylene group in polycarbocyclic and heterocyclic ketones is replaced by a keto group, the change may be denoted by attaching the suffix -one to the name of the ring system. However, when $\geq$CH in an unsaturated or aromatic system is replaced by a keto group, two alternative names become possible. First, the maximum number of noncumulative double bonds is added after introduction of the carbonyl group(s), and any hydrogen that remains to be added is denoted as indicated hydrogen with the carbonyl group having priority over the indicated hydrogen for lower-numbered locant. Second, the prefix oxo- is used, with the hydrogenation indicated by hydro prefixes; hydrogenation is considered to have occurred before the introduction of the carbonyl group. For example,

(1) 1-(2H)-Naphthalenone
(2) 1-Oxo-1,2-dihydronaphthalene

When another group having higher priority for citation as principal group is also present, the ketonic oxygen may be expressed by the prefix oxo-, or one can use the name of the carbonyl-containing radical, as, for example, acyl radicals and oxo-substituted radicals. Examples are

4-(4'-Oxohexyl)-1-benzoic acid

1,2,4-Triacetylbenzene

Diketones and tetraketones derived from aromatic compounds by conversion of two or four $\geq$CH groups into keto groups, with any necessary rearrangement of double bonds to a quinonoid structure, are named by adding the suffix -quinone and any necessary affixes.

Polyketones in which two or more contiguous carbonyl groups have rings attached at each end may be named (1) by the radicofunctional method or (2) by substitutive nomenclature. For example,

(1) 2-Naphthyl 2-pyridyl diketone
(2) 1-(2-Naphthyl)-2-(2-pyridyl) ethanedione

Some trivial names are retained: acetone (2-propanone), biacetyl (2,3-butanedione), propiophenone (C_6H_5—CO—CH_2CH_3), chalcone (C_6H_5—CH=CH—CO—C_6H_5), and deoxybenzoin (C_6H_5—CH_2—CO—C_6H_5).

These contracted names of heterocyclic nitrogen compounds are retained as alternatives for systematic names, sometimes with indicated hydrogen. In addition, names of oxo derivatives of fully saturated nitrogen heterocycles that systematically end in -idinone are often contracted to end in -idone when no ambiguity might result. For example,

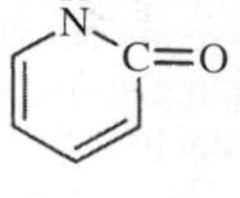

2-Pyridone
2(1*H*)-Pyridone

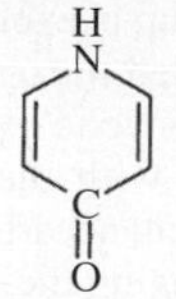

4-Pyridone
4(1*H*)-Pyridone

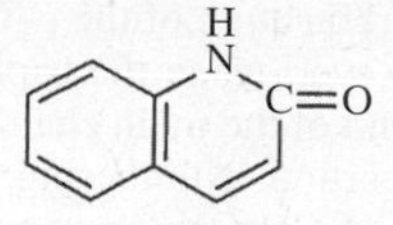

2-Quinolone
2(1*H*)-Quinolone

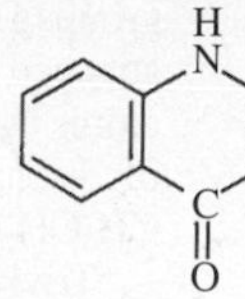

4-Quinolone
4(1*H*)-Quinolone

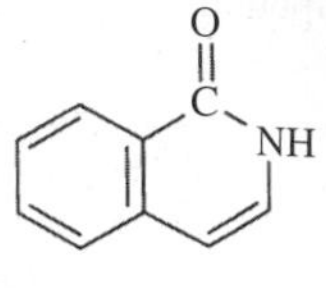

1-Isoquinolone
1(2*H*)-Isoquinolone

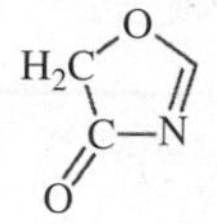

4-Oxazolone
4(5*H*)-Oxazolone

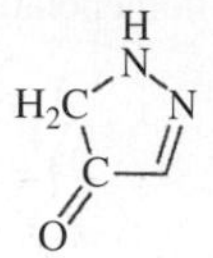

4-Pyrazolone
4(5*H*)-Pyrazolone

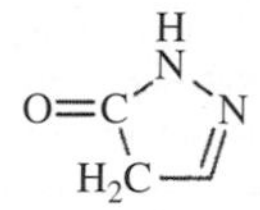

5-Pyrazolone
5(4*H*)-Pyrazolone

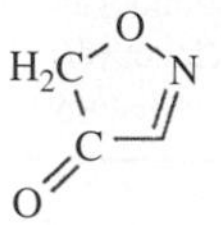

4-Isoxazoline
4(5*H*)-Isoxazolone

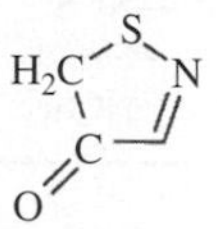

4-Thiazolone
4(5*H*)-Thiazolone

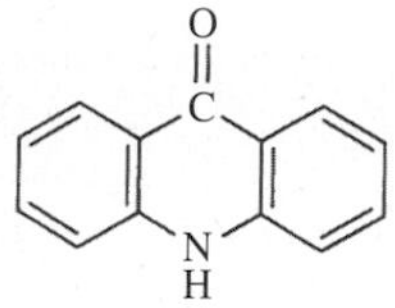

9-Acridone
9(10*H*)-Acridone

Lactones, Lactides, Lactams, and Lactims.   When the hydroxy acid from which water may be considered to have been eliminated has a trivial name, the lactone is designated by substituting -olactone for -ic acid. Locants for a carbonyl group are numbered as low as possible, even before that of a hydroxyl group.

Lactones formed from aliphatic acids are named by adding -olide to the name of the nonhydroxy-lated hydrocarbon with the same number of carbon atoms. The suffix -olide signifies the change of $>CH\cdots CH_3$ into $>C\cdots C=O$.

Structures in which one or more (but not all) rings of an aggregate are lactone rings are named by placing -carbolactone (denoting the —O—CO— bridge) after the names of the structures that remain when each bridge is replaced by two hydrogen atoms. The locant for —CO— is cited before that for the ester oxygen atom. An additional carbon atom is incorporated into this structure as compared to the -olide.

These trivial names are permitted: γ-butyrolactone, γ-valerolactone, and δ-valerolactone. Names based on heterocycles may be used for all lactones. Thus, γ-butyrolactone is also tetrahydro-2-furanone or dihydro-2(3*H*)-furanone.

Lactides, intermolecular cyclic esters, are named as heterocycles. *Lactams* and *lactims*, containing a —CO—NH— and —C(OH)=N— group, respectively, are named as heterocycles, but they may also be named with -lactam or -lactim in place of -olide. For example,

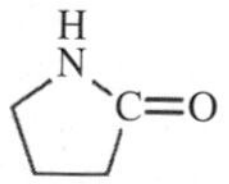

(1) 2-Pyrrolidinone
(2) 4-Butanelactam

Nitriles and Related Compounds.   For acids whose systematic names end in -carboxylic acid, nitriles are named by adding the suffix -carbonitrile when the —CN group replaces the —COOH group. The carbon atom of the —CN group is excluded from the numbering of a chain to which it is attached. However, when the triple-bonded nitrogen atom is considered to replace three hydrogen atoms at the end of the main chain of an acyclic hydrocarbon, the suffix -nitrile is added to the name of the hydrocarbon. Numbering begins with the carbon attached to the nitrogen. For example, $CH_3CH_2CH_2CH_2CH_2CN$ is named (1) pentanecarbonitrile or (2) hexanenitrile.

Trivial acid names are formed by changing the endings -oic acid or -ic acid to -onitrile. For example, CH_3CN is acetonitrile. When the —CN group is not the highest priority group, the —CN group is denoted by the prefix cyano-.

In order of decreasing priority for citation of a functional class name, and the prefix for substitutive nomenclature, are the following related compounds:

Functional group	Prefix	Radicofunctional ending
—NC	Isocyano-	Isocyanide
—OCN	Cyanato-	Cyanate
—NCO	Isocyanato-	Isocyanate
—ONC	—	Fulminate
—SCN	Thiocyanato-	Thiocyanate
—NCS	Isothiocyanato-	Isothiocyanate
—SeCN	Selenocyanato-	Selenocyanate
—NCSe	Isoselenocyanato-	Isoselenocyanate

Peroxides.   Compounds of the type R—O—OH are named (1) by placing the name of the radical R before the word *hydroperoxide* or (2) by use of the prefix hydroperoxy- when another parent name has higher priority. For example, C_2H_5OOH is ethyl hydroperoxide.

Compounds of the type R^1O—OR^2 are named (1) by placing the names of the radicals in alphabetical order before the word *peroxide* when the group —O—O— links two chains, two rings, or a ring and a chain, (2) by use of the affix dioxy to denote the bivalent group —O—O— for naming assemblies of identical units or to form part of a prefix, or (3) by use of the prefix epidioxy- when the peroxide group forms a bridge between two carbon atoms, a ring, or a ring system. Examples are methyl propyl peroxide for CH_3—O—O—C_3H_7 and 2,2'-dioxydiacetic acid for HOOC—CH_2—O—O—CH_2—COOH.

Phosphorus Compounds.   Acyclic phosphorus compounds containing only one phosphorus atom, as well as compounds in which only a single phosphorus atom is in each of several functional groups, are named as derivatives of the parent structures (Table 2.13). Often these are purely hypothetical parent structures. When hydrogen attached to phosphorus is replaced by a hydrocarbon group, the derivative is named by substitution nomenclature. When hydrogen of an —OH group is replaced, the derivative is named by radicofunctional nomenclature. For example, $C_2H_5PH_2$ is ethylphosphine; $(C_2H_5)_2PH$, diethylphosphine; $CH_3P(OH)_2$, dihydroxy-methyl-phosphine or methylphosphonous acid; C_2H_5—$PO(Cl)(OH)$, ethylchlorophosphonic acid or ethylphosphonochloridic acid or hydrogen-chlorodioxoethylphosphate(V); $CH_3CH(PH_2)COOH$, 2-phosphinopropionic acid; $HP(CH_2COOH)_2$, phosphinediyldiacetic acid; $(CH_3)HP(O)OH$, methylphosphinic acid or hydrogen-hydridomethyldioxophosphate(V); $(CH_3O)_3PO$, trimethyl phosphate; and $(CH_3O)_3P$, trimethyl phosphite.

Salts and Esters of Acids.   Neutral salts of acids are named by citing the cation(s) and then the anion, whose ending is changed from -oic to -oate or from -ic to -ate. When different acidic residues are present in one structure, prefixes are formed by changing the anion ending -ate to -ato- or -ide to -ido. The prefix carboxylato- denotes the ionic group —COO⁻. The phrase (metal) salt of (the acid) is permissible when the carboxyl groups are not all named as affixes.

Acid salts include the word *hydrogen* (with affixes, if appropriate) inserted between the name of the cation and the name of the anion (or word *salt*).

TABLE 2.13   Phosphorus-Containing Compounds

Formula	Parent name	Substitutive prefix		Radicofunctional ending
H_3P	Phosphine	H_2P— Phosphino-		Phosphide
H_5P	Phosphorane	H_4P— Phosphoranyl-		
		$H_3P{<}$ Phosphoroanediyl-		
		$H_2P{\leqslant}$ Phosphoranetriyl-		
H_3PO	Phosphine oxide			
H_3PS	Phosphine sulfide			
H_3PNH	Phosphine imide			
$P(OH)_3$	Phosphorous acid			Phosphite
$HP(OH)_2$	Phosphonous acid			Phosphonite
H_2POH	Phosphinous acid			Phosphinite
$P(O)(OH)_3$	Phosphoric acid	$P(O){\leqslant}$	Phosphoryl-	Phosphate(V)
$HP(O)(OH)_2$	Phosphonic acid	$HP(O){<}$	Phosphonoyl-	Phosphonate
		—$P(O)OH_2$	Phosphono-	
$H_2P(O)OH$	Phosphinic acid	$H_2P(O)$—	Phosphinoyl-	Phosphinate
		${>}P(O)OH$	Phosphinoco-	
			Phosphinato-	

Esters are named similarly, with the name of the alkyl or aryl radical replacing the name of the cation. Acid esters of acids and their salts are named as neutral esters, but the components are cited in-the order: cation, alkyl or aryl radical, hydrogen, and anion. Locants are added if necessary. For example,

$$\begin{array}{l} CH_2-CO-OC_2H_5 \\ | \\ HOC-COO^- \qquad K^+ \quad H^+ \qquad \text{Potassium 1-ethyl hydrogen citrate} \\ | \\ CH_2-COO^- \end{array}$$

Ester groups in R^1—CO—OR^2 compounds are named (1) by the prefix alkoxycarbonyl- or aryloxycarbonyl- for —CO—OR^2 when the radical R^1 contains a substituent with priority for citation as principal group or (2) by the prefix acyloxy- for R^1—CO—O— when the radical R^2 contains a substituent with priority for citation as principal group. Examples are

$$\text{(naphthalene)} \begin{array}{l} CH_2CH_2CH_2CO-OCH_3 \\ CO-OCH_3 \end{array} \qquad \text{Methyl 3-methoxycarbonyl-2-napthalenebutyrate}$$

$$[CH_3O-CO-CH_2CH_2\overset{+}{N}(CH_3)_3]Cl^- \qquad \text{[(2-Methoxycarbonyl)ethyl]trimethylammonium chloride}$$

$$C_6H_5-CO-OCH_2CH_2COOH \qquad \text{3-Benzoyloxypropionic acid}$$

The trivial name *acetoxy* is retained for the CH_3—CO—O— group. Compounds of the type $R^2C(OR^2)_3$ are named as R^2 esters of the hypothetical ortho acids. For example, $CH_3C(OCH_3)_3$ is trimethyl orthoacetate.

Silicon Compounds. SiH_4 is called silane; its acyclic homologs are called disilane, trisilane, and so on, according to the number of silicon atoms present. The chain is numbered from one end to the other so as to give the lowest-numbered locant in radicals to the free valence or to substitutents on a chain. The abbreviated form silyl is used for the radical SiH_3—. Numbering and citation of side chains proceed according to the principles set forth for hydrocarbon chains. Cyclic nonaromatic structures are designated by the prefix cyclo-.

When a chain or ring system is composed entirely of alternating silicon and oxygen atoms, the parent name *siloxane* is used with a multiplying affix to denote the number of silicon atoms present. The parent name *silazane* implies alternating silicon and nitrogen atoms; multiplying affixes denote the number of silicon atoms present.

The prefix sila- designates replacement of carbon by silicon in replacement nomenclature. Prefix names for radicals are formed analogously to those for the corresponding carbon-containing compounds. Thus silyl is used for SiH_3—, silyene for —SiH_2—, silylidyne for —$SiH<$, as well as trily, tetrayl, and so on for free valences(s) on ring structures.

Sulfur Compounds   *Bivalent Sulfur.* The prefix thio, placed before an affix that denotes the oxygen-containing group or an oxygen atom, implies the replacement of that oxygen by sulfur. Thus the suffix -thiol denotes —SH, -thione denotes —(C)=S and implies the presence of an =S at a nonterminal carbon atom, -thioic acid denotes $[(C)=S]OH \rightleftharpoons [(C)=O]SH$ (i.e., the *O*-substituted acid and the *S*-substituted acid, respectively), -dithioic acid denotes [—C(S)SH], and -thial denotes —(C)HS (or -carbothialdehyde denotes —CHS). When -carboxylic acid has been used for acids, the sulfur analog is named -carbothioic acid or -carbodithioic acid.

Prefixes for the groups HS— and RS— are mercapto- and alkylthio-, respectively; this latter name may require parentheses for distinction from the use of thio- for replacement of oxygen in a trivially named acid. Examples of this problem are 4-C_2H_5—C_6H_4—CSOH named *p*-ethyl(thio) benzoic acid and 4-C_2H_5—S—C_6H_4—COOH named *p*-(ethylthio)benzoic acid. When —SH is not the principal group, the prefix mercapto- is placed before the name of the parent compound to denote an unsubstituted —SH group.

The prefix thioxo- is used for naming =S in a thioketone. Sulfur analogs of acetals are named as alkylthio- or arylthio-. For example, $CH_3CH(SCH_3)OCH_3$ is 1-methoxy-1-(methylthio)ethane. Prefix forms for -carbothioic acids are hydroxy(thiocarbonyl)- when referring to the *O*-substituted acid and mercapto(carbonyl)- for the *S*-substituted acid.

Salts are formed as with oxygen-containing compounds. For example, C_2H_5—S—Na is named either sodium ethanethiolate or sodium ethyl sulfide. If mercapto- has been used as a prefix, the salt is named by use of the prefix sulfido- for —S^-.

Compounds of the type R^1—S—R^2 are named alkylthio- (or arylthio-) as a prefix to the name of R^1 or R^2, whichever is the senior.

Sulfonium Compounds.   Sulfonium compounds of the type $R^1R^2R^3S^+X^-$ are named by citing in alphabetical order the radical names followed by -sulfonium and the name of the anion. For heterocyclic compounds, -ium is added to the name of the ring system. Replacement of $\geq$CH by sulfonium sulfur is denoted by the prefix thionia-, and the name of the anion is added at the end.

Organosulfur Halides.   When sulfur is directly linked only to an organic radical and to a halogen atom, the radical name is attached to the word *sulfur* and the name(s) and number of the halide(s) are stated as a separate word. Alternatively, the name can be formed from R—SOH, a sulfenic acid whose radical prefix is sulfenyl-. For example, CH_3CH_2—S—Br would be named either ethylsulfur monobromide or ethanesulfenyl bromide. When another principal group is present, a composite prefix is formed from the number and substitutive name(s) of the halogen atoms in front of the syllable thio. For example, BrS—COOH is (bromothio)formic acid.

Sulfoxides.   Sulfoxides, R^1—SO—R^2, are named by placing the names of the radicals in alphabetical order before the word *sulfoxide.* Alternatively, the less senior radical is named followed by sulfinyl- and concluded by the name of the senior group. For example, CH_3CH_2—SO—$CH_2CH_2CH_3$ is named either ethyl propyl sulfoxide or 1-(ethylsulfinyl)propane.

When an $\geq$SO group is incorporated in a ring, the compound is named an oxide.

Sulfones.   Sulfones, R^1—SO_2—R^2, are named in an analogous manner to sulfoxides, using the word *sulfone* in place of *sulfoxide*. In prefixes, the less senior radical is followed by -sulfonyl-. When the $>SO_2$ group is incorporated in a ring, the compound is named as a dioxide.

Sulfur Acids.   Organic oxy acids of sulfur, that is, —SO_3H, —SO_2H, and —SOH, are named sulfonic acid, sulfinic acid, and sulfenic acid, respectively. In subordinate use, the respective prefixes are sulfo-, sulfino-, and sulfeno-. The grouping —SO_2—O—SO_2— or —SO—O—SO is named sulfonic or sulfinic anhydride, respectively.

Inorganic nomenclature is employed in naming sulfur acids and their derivatives in which sulfur is linked only through oxygen to the organic radical. For example, $(C_2H_5O)_2SO_2$ is diethyl sulfate and C_2H_5O—SO_2—OH is ethyl hydrogen sulfate. Prefixes *O-* and *S-* are used where necessary to denote attachment to oxygen and to sulfur, respectively, in sulfur replacement compounds. For example, CH_3—S—SO_2—ONa is sodium *S*-methyl thiosulfate.

When sulfur is linked only through nitrogen, or through nitrogen and oxygen, to the organic radical, naming is as follows: (1) *N*-substituted amides are designated as *N*-substituted derivatives of the sulfur amides and (2) compounds of the type R—NH—SO_3H may be named as *N*-substituted sulfamic acids or by the prefix sulfoamino- to denote the group HO_3S—NH—. The groups —N=SO and —N=SO_2 are named sulfinylamines and sulfonylamines, respectively.

Sultones and Sultams.   Compounds containing the group —SO_2—O— as part of the ring are called -sultone. The —SO_2— group has priority over the —O— group for lowest-numbered locant.

Similarly, the —SO_2—N= group as part of a ring is named by adding -sultam to the name of the hydrocarbon with the same number of carbon atoms. The —SO_2— has priority over —N= for lowest-numbered locant.

2.1.4 Stereochemistry

Concepts in stereochemistry, that is, chemistry in three-dimensional space, are in the process of rapid expansion. This section will deal with only the main principles. The compounds discussed will be those that have identical molecular formulas but differ in the arrangement of their atoms in space. *Stereoisomers* is the name applied to these compounds.

Stereoisomers can be grouped into three categories: (1) Conformational isomers differ from each other only in the way their atoms are oriented in space, but can be converted into one another by rotation about sigma bonds. (2) Geometric isomers are compounds in which rotation about a double bond is restricted. (3) Configurational isomers differ from one another only in configuration about a chiral center, axis, or plane. In subsequent structural representations, a broken line denotes a bond projecting behind the plane of the paper and a wedge denotes a bond projecting in front of the plane of the paper. A line of normal thickness denotes a bond lying essentially in the plane of the paper.

Conformational Isomers.   A molecule in a conformation into which its atoms return spontaneously after small displacements is termed a *conformer*. Different arrangements of atoms that can be converted into one another by rotation about single bonds are called *conformational isomers* (see Fig. 2.1). A pair of conformational isomers can be but do not have to be mirror images of each other. When they are not mirror images, they are called *diastereomers*.

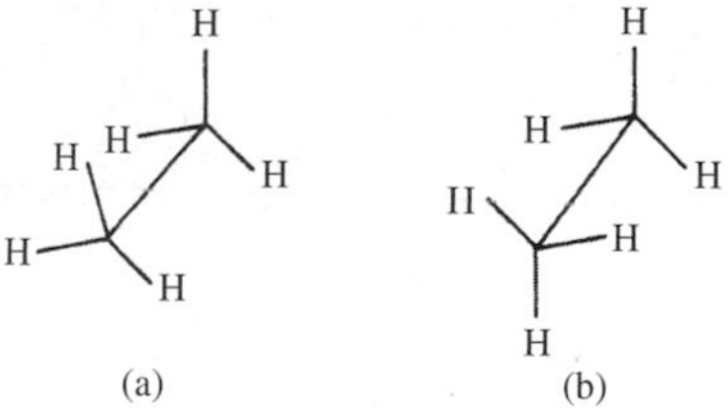

Acyclic Compounds.   Different conformations of acyclic compounds are best viewed by construction of ball-and-stick molecules or by use of Newman projections (see Fig. 2.2). Both types of representations are shown for ethane. Atoms or groups that are attached at opposite ends of a single bond should be viewed along the bond axis. If two atoms or groups attached at opposite ends of the bond

FIGURE 2.1 Conformations of ethane. (*a*) Eclipsed; (*b*) staggered.

appear one directly behind the other, these atoms or groups are described as eclipsed. That portion of the molecule is described as being in the eclipsed conformation. If not eclipsed, the atoms or groups and the conformation may be described as staggered. Newman projections show these conformations clearly.

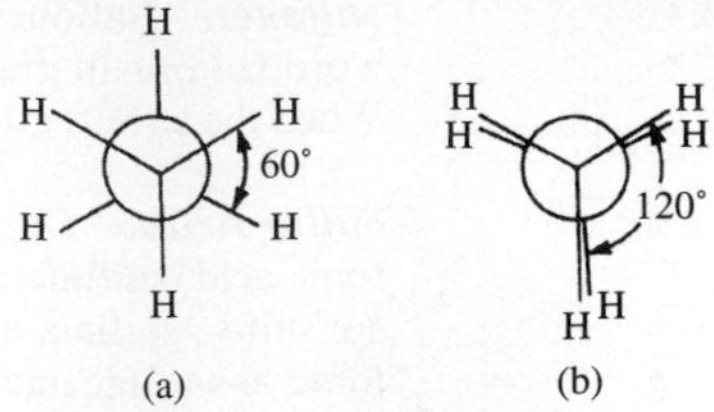

Certain physical properties show that rotation about the single bond is not quite free. For ethane there is an energy barrier of about 3 kcal · mol^{-1} (12 kJ · mol^{-1}). The potential energy of the molecule is at a minimum for the staggered conformation, increases with rotation, and reaches a maximum at the eclipsed conformation. The energy required to rotate the atoms or groups about the carbon-carbon bond is called *torsional energy*. Torsional strain is the cause of the relative instability of the eclipsed conformation or any intermediate skew conformations.

FIGURE 2.2 Newman projections for ethane. (*a*) Staggered; (*b*) eclipsed.

In butane, with a methyl group replacing one hydrogen on each carbon of ethane, there are several different staggered conformations (see Fig. 2.3). There is the *anti*-conformation in which the methyl groups are as far apart as they can be (dihedral angle of 180°). There are two *gauche* conformations in which the methyl groups are only 60° apart; these are two nonsuperimposable mirror images of each other. The *anti*-conformation is more stable than the *gauche* by about 0.9 kcal · mol^{-1} (4 kJ · mol^{-1}). Both are free of torsional strain. However, in a *gauche* conformation the methyl groups are closer together than the sum of their van der Waals' radii. Under these conditions van der Waals' forces are repulsive and raise the energy of conformation. This strain can affect not only the relative stabilities of various staggered conformations but also the heights of the energy barriers between them. The energy maximum (estimated at 4.8 to 6.1 kcal · mol^{-1} or 20 to 25 kJ · mol^{-1}) is reached when two methyl groups swing past each other (the eclipsed conformation) rather than past hydrogen atoms.

Cyclic Compounds. Although cyclic aliphatic compounds are often drawn as if they were planar geometric figures (a triangle for cyclopropane, a square for cyclobutane, and so on), their structures are not that simple. Cyclopropane does possess the maximum angle strain if one considers the difference between a tetrahedral angle (109.5°) and the 60° angle of the cyclopropane structure. Nevertheless the cyclopropane structure is thermally quite stable. The highest electron density of the carbon-carbon bonds does not lie along the lines connecting the carbon-carbon bonds does not lie

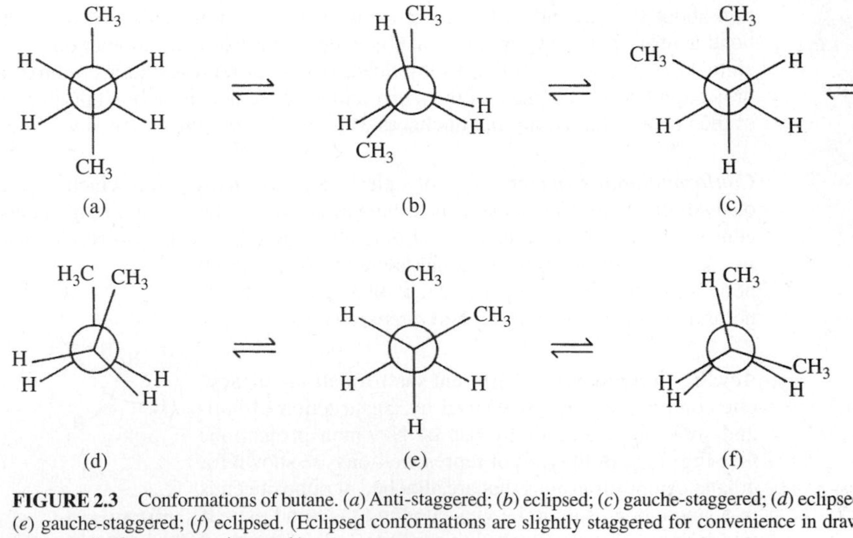

FIGURE 2.3 Conformations of butane. (*a*) Anti-staggered; (*b*) eclipsed; (*c*) gauche-staggered; (*d*) eclipsed; (*e*) gauche-staggered; (*f*) eclipsed. (Eclipsed conformations are slightly staggered for convenience in drawing; actually they are superimposed.)

along the lines connecting the carbon atoms. Bonding electrons lie principally outside the triangular internuclear lines and result in what is known as *bent bonds* (see Fig. 2.4).

Cyclobutane has less angle strain than cyclopropane (only 19.5°). It is also believed to have some bent-bond character associated with the carbon-carbon bonds. The molecule exists in a nonplanar conformation in order to minimize hydrogen-hydrogen eclipsing strain.

Cyclopentane is nonplanar, with a structure that resembles an envelope (see Fig. 2.5). Four of the carbon atoms are in one plane, and the fifth is out of that plane. The molecule is in continual motion so that the out-of-plane carbon moves rapidly around the ring.

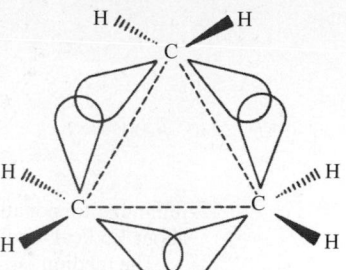

FIGURE 2.4 The bent bonds ("tear drops") of cyclopropane.

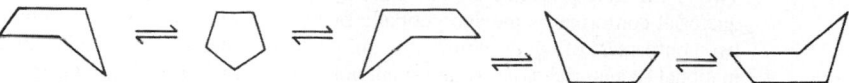

FIGURE 2.5 The conformations of cyclopentane.

The 12 hydrogen atoms of cyclohexane do not occupy equivalent positions. In the chair conformation six hydrogen atoms are perpendicular to the average plane of the molecule and six are directed outward from the ring, slightly above or below the molecular plane (see Fig. 2.6). Bonds which are perpendicular to the molecular plane are known as *axial bonds*, and those which extend outward from the ring are known as *equatorial bonds*. The three axial bonds directed upward originate from alternate carbon atoms and are parallel with each other; a similar situation exists for the three axial bonds directed downward. Each equatorial bond is drawn so as to be parallel with the ring carbon-carbon bond once removed from the point of attachment to that equatorial bond. At room temperature, cyclohexane is interconverting rapidly between two chair conformations. As one chair form converts to the other, all the equatorial hydrogen atoms become axial and all the axial hydrogens become equatorial. The interconversion is so rapid that all hydrogen atoms on cyclohexane can be considered equivalent. Interconversion is believed to take place by movement of one side of the chair structure to produce the twist boat, and then movement of the other side of the twist boat to give the other chair form. The chair conformation is the most favored structure for cyclohexane. No angle strain is encountered since all bond angles remain tetrahedral. Torsional strain is minimal because all groups are staggered.

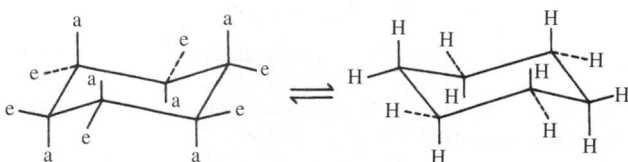

FIGURE 2.6 The two chair conformations of cyclohexane; *a* = axial hydrogen atom and *e* = equatorial hydrogen atom.

In the boat conformation of cyclohexane (see Fig. 2.7) eclipsing torsional strain is significant, although no angle strain is encountered. Nonbonded interaction between the two hydrogen atoms across the ring from each other (the "flagpole" hydrogens) is unfavorable. The boat conformation is about 6.5 kcal · mol^{-1} (27 kJ · mol^{-1}) higher in energy than the chair form at 25°C.

A modified boat conformation of cyclohexane, known as the twist boat (see Fig. 2.8), or skew boat, has been suggested to

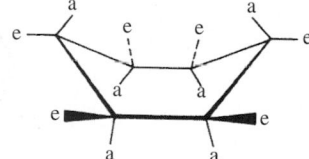

FIGURE 2.7 The boat conformation of cyclohexane. *a* = axial hydrogen atom and *e* = equatorial hydrogen atom.

FIGURE 2.8   Twist-boat conformation of cyclohexane.

minimize torsional and nonbounded interactions. This particular conformation is estimated to be about 1.5 kcal · mol⁻¹ · (6 kJ · mol⁻¹) lower in energy than the boat form at room temperature.

The medium-size rings (7 to 12 ring atoms) are relatively free of angle strain and can easily take a variety of spatial arrangements. They are not large enough to avoid all nonbonded interactions between atoms.

Disubstituted cyclohexanes can exist as *cis-trans* isomers as well as axial-equatorial conformers. Two isomers are predicted for 1,4-dimethylcyclohexane (see Fig. 2.9). For the *trans* isomer the diequatorial conformer is the energetically favorable form. Only one *cis* isomer is observed, since the two conformers of the *cis* compound are identical. Interconversion takes place between the conformational (equatorial-axial isomers) but not configurational (*cis-trans*) isomers.

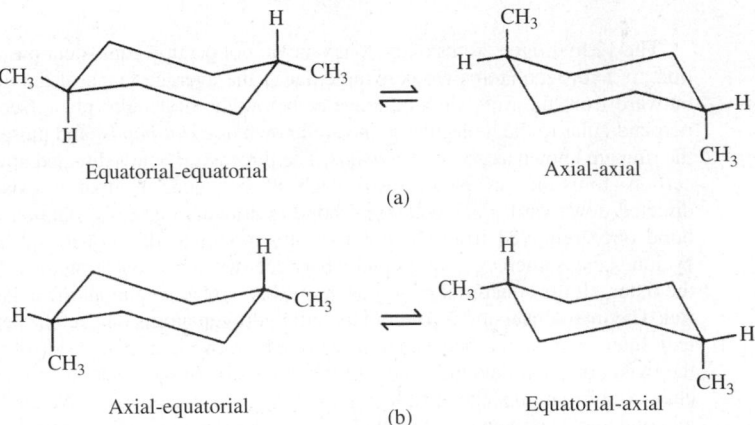

FIGURE 2.9   Two isomers of 1,4-dimethylcyclohexane. (*a*) *Trans* isomer; (*b*) *cis* isomer.

The bicyclic compound decahydronaphthalene, or bicyclo[4.4.0]decane, has two fused six-membered rings. It exists in *cis* and *trans* forms (see Fig. 2.10), as determined by the configurations at the bridgehead carbon atoms. Both *cis*- and *trans*-decahydronaphthalene can be constructed with two chair conformations.

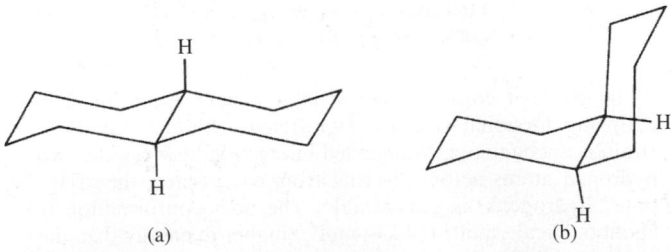

FIGURE 2.10   Two isomers of decahydronaphthalene, or bicyclo[4.4.0]decane. (*a*) *Trans* isomer; (*b*) *cis* isomer.

Geometrical Isomerism.   Rotation about a carbon-carbon double bond is restricted because of interaction between the *p* orbitals which make up to pi bond. Isomerism due to such restricted rotation about a bond is known as *geometric isomerism*. Parallel overlap of the *p* orbitals of each carbon atom of the double bond forms the molecular orbital of the pi bond. The relatively large barrier to rotation about the pi bond is estimated to be nearly 63 kcal · mol^{-1} (263 kJ · mol^{-1}).

When two different substituents are attached to each carbon atom of the double bond, *cis-trans* isomers can exist. In the case of *cis*-2-butene (see Fig. 2.11a), both methyl groups are on the same side of the double bond. The other isomer has the methyl groups on opposite sides and is designated as *trans*-2-butene (see Fig. 2.11b). Their physical properties are quite different. Geometric isomerism can also exist in ring systems; examples were cited in the previous discussion on conformational isomers.

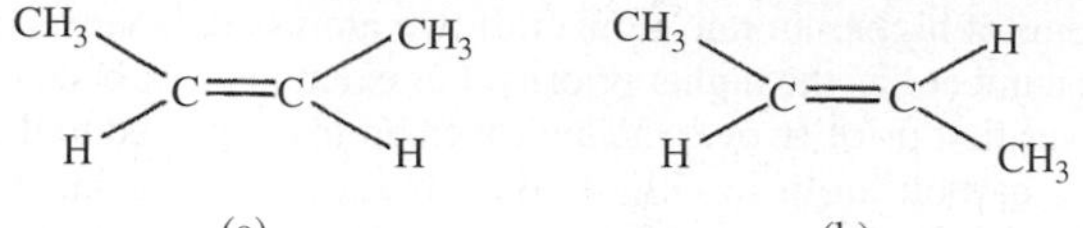

FIGURE 2.11 Two isomers of 2-butene. (*a*) *Cis* isomer, bp 3.8°C, mp −138.9°C, dipole moment 0.33 D; (*b*) *trans* isomer, bp 0.88°C, mp −105.6°C, dipole moment 0 D.

For compounds containing only double-bonded atoms, the reference plane contains the double bonded atoms and is perpendicular to the plane containing these atoms and those directly attached to them. It is customary to draw the formulas so that the reference plane is perpendicular to that of the paper. For cyclic compounds the reference plane is that in which the ring skeleton lies or to which it approximates. Cyclic structures are commonly drawn with the ring atoms in the plane of the paper.

Sequence Rules for Geometric Isomers and Chiral Compounds.   Although *cis* and *trans* designations have been used for many years, this approach becomes useless in complex systems. To eliminate confusion when each carbon of a double bond or a chiral center is connected to different groups, the Cahn, Ingold, and Prelog system for designating configuration about a double bond or a chiral center has been adopted by IUPAC. Groups on each carbon atom of the double bond are assigned a first (1) or second (2) priority. Priority is then compared at one carbon relative to the other. When both first priority groups are on the *same side* of the double bond, the configuration is designated as *Z* (from the German *zusammen*, "together"), which was formerly *cis*. If the first priority groups are on *opposite sides* of the double bond, the designation is *E* (from the German *entgegen*, "in opposition to"), which was formerly *trans*. (See Fig. 2.12).

FIGURE 2.12 Configurations designated by priority groups. (*a*) *Z* (*cis*); (*b*) *E* (*trans*).

When a molecule contains more than one double bond, each *E* or *Z* prefix has associated with it the lower-numbered locant of the double bond concerned. Thus (see also the rules that follow)

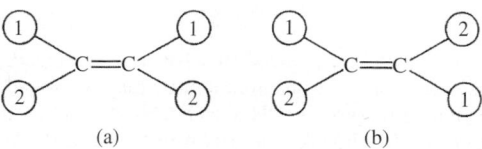

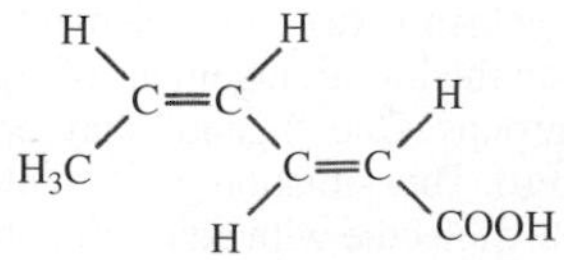

(2*E*,4*Z*)-2,4-Hexadienoic acid

When the sequence rules permit alternatives, preference for lower-numbered locants and for inclusion in the principal chain is allotted as follows in the order stated: *Z* over *E* groups and *cis* over *trans* cyclic groups. If a choice is still not attained, then the lower-numbered locant for such a preferred group at the first point of difference is the determining factor. For example,

(2Z,5E)-2,5-Heptadienedioic acid

Rule 1.   Priority is assigned to atoms on the basis of atomic number. Higher priority is assigned to atoms of higher atomic number. If two atoms are isotopes of the same element, the atom of higher mass number has the higher priority. For example, in 2-butene, the carbon atom of each methyl group receives first priority over the hydrogen atom connected to the same carbon atom. Around the asymmetric carbon atom in chloroiodomethanesulfonic acid, the priority sequence is I, Cl, S, H. In 1-bromo-1-deuteroethane, the priority sequence is Cl, C, D, H.

Rule 2.   When atoms attached directly to a double-bonded carbon have the same priority, the second atoms are considered and so on, if necessary, working outward once again from the double bond or chiral center. For example, in 1-chloro-2-methylbutene, in CH_3 the second atoms are H, H, H and in CH_2CH_3 they are C, H, H. Since carbon has a higher atomic number than hydrogen, the ethyl group has the next highest priority after the chlorine atom.

(Z)-1-Chloro-2-methylbutene   (E)-1-Chloro-2-methylbutene

Rule 3.   When groups under consideration have double or triple bonds, the multiple-bonded atom is replaced conceptually by two or three single bonds to that same kind of atom.

Thus, $=$ A is considered to be equivalent to two A's, $<^A_{}$ or and $\equiv$ A equals $\Longleftarrow A$. However, a real $<^A_A$ has priority over $=$ A; likewise a real $\Longleftarrow A$ has priority over $\equiv$ A. Actually, both atoms of a multiple bond are duplicated, or triplicated, so that C $=$ O is treated as $\overset{C-O}{\underset{O\quad C}{|\quad|}}$, that is $\overset{C-O}{\underset{(O)}{|}}$ and $\overset{C-O}{\underset{(C)}{|}}$, and C $\equiv$ N is treated as $\overset{C}{_{(N)}}\diagdown^{N}_{(N)}$ $\overset{C}{_{(C)}}\diagdown^{N}_{(C)}$. A phenyl carbon becomes $-C\overset{CH}{\underset{CH}{\diagup\diagdown}}C$.

Only the double-bonded atoms themselves are duplicated, not the atoms or groups attached to them. The duplicated atoms (or phantom atoms) may be considered as carrying atomic number zero. For example, among the groups OH, CHO, CH_2OH, and H, the OH group has the highest priority, and the C(O, O, H) of CHO takes priority over the C(O, H, H) of CH_2OH.

Chirality and Optical Activity.   A compound is chiral (the term *dissymmetric* was formerly used) if it is not superimposable on its mirror image. A chiral compound does not have a plane of symmetry. Each chiral compound possesses one (or more) of three types of chiral element, namely, a chiral center, a chiral axis, or a chiral plane.

Chiral Center.   The chiral center, which is the chiral element most commonly met, is exemplified by an asymmetric carbon with a tetrahedral arrangement of ligands about the carbon. The ligands comprise four different atoms or groups. One "ligand" may be a lone pair of electrons; another, a phantom atom of atomic number zero. This situation is encountered in sulfoxides or with a nitrogen atom. Lactic acid is an example of a molecule with an asymmetric (chiral) carbon. (See Fig. 2.13.)

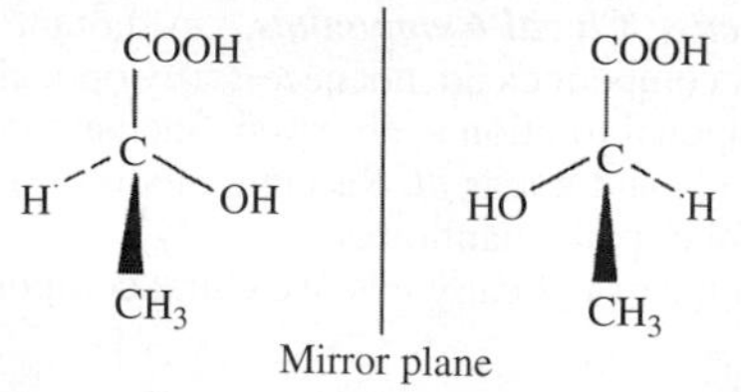

COOH COOH

H—C—OH HO—C—H

CH₃ CH₃

Mirror plane

FIGURE 2.13 Asymmetric (chiral) carbon in the lactic acid molecule.

A simpler representation of molecules containing asymmetric carbon atoms is the Fischer projection, which is shown here for the same lactic acid configurations. A Fischer projection involves

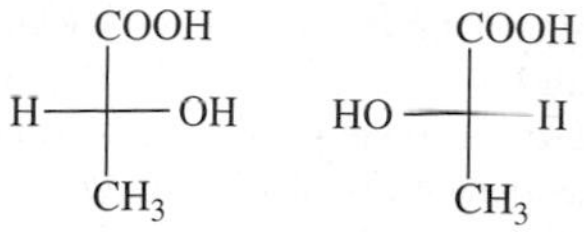

COOH COOH

H——OH HO——II

CH₃ CH₃

drawing a cross and attaching to the four ends the four groups that are attached to the asymmetric carbon atom. The asymmetric carbon atom is understood to be located where the lines cross. The horizontal lines are understood to represent bonds coming toward the viewer out of the plane of the paper. The vertical lines represent bonds going away from the viewer behind the plane of the paper as if the vertical line were the side of a circle. The principal chain is depicted in the vertical direction; the lowest-numbered (locant) chain member is placed at the top position. These formulas may be moved sideways or rotated through 180° in the plane of the paper, but they may not be removed from the plane of the paper (i.e., rotated through 90°). In the latter orientation it is essential to use thickened lines (for bonds coming toward the viewer) and dashed lines (for bonds receding from the viewer) to avoid confusion.

Enantiomers. Two nonsuperimposable structures that are mirror images of each other are known as *enantiomers*. Enantiomers are related to each other in the same way that a right hand is related to a left hand. Except for the direction in which they rotate the plane of polarized light, enantiomers are identical in all physical properties. Enantiomers have identical chemical properties except in their reactivity toward optically active reagents.

Enantiomers rotate the plane of polarized light in opposite directions but with equal magnitude. If the light is rotated in a clockwise direction, the sample is said to be dextrorotatory and is designed as (+). When a sample rotates the plane of polarized light in a counterclockwise direction, it is said to be levorotatory and is designed as (−). Use of the designations *d* and *l* is discouraged.

Specific Rotation. Optical rotation is caused by individual molecules of the optically active compound. The amount of rotation depends upon how many molecules the light beam encounters in passing through the tube. When allowances are made for the length of the tube that contains the sample and the sample concentration, it is found that the amount of rotation, as well as its direction, is a characteristic of each individual optically active compound.

Specific rotation is the number of degrees of rotation observed if a 1-dm tube is used and the compound being examined is present to the extent of 1 g per 100 mL. The density for a pure liquid replaces the solution concentration.

$$\text{Specific rotation} = [\alpha] = \frac{\text{observed rotation (degrees)}}{\text{length (dm)} \times (\text{g/100 ml})}$$

The temperature of the measurement is indicated by a superscript and the wavelength of the light employed by a subscript written after the bracket; for example, $[\alpha]_{590}^{20}$ implies that the measurement was made at 20°C using 590-nm radiation.

Optically Inactive Chiral Compounds.   Although chirality is a necessary prerequisite for optical activity, chiral compounds are not necessarily optically active. With an equal mixture of two enantiomers, no net optical rotation is observed. Such a mixture of enantiomers is said to be *racemic* and is designated as ($\pm$) and not as *dl*. Racemic mixtures usually have melting points higher than the melting point of either pure enantiomer.

A second type of optically inactive chiral compounds, *meso* compounds, will be discussed in the next section.

Multiple Chiral Centers.   The number of stereoisomers increases rapidly with an increase in the number of chiral centers in a molecule. A molecule possessing two chiral atoms should have four optical isomers, that is, four structures consisting of two pairs of enantiomers. However, if a compound has two chiral centers but both centers have the same four substituents attached, the total number of isomers is three rather than four. One isomer of such a compound is not chiral because it is identical with its mirror image; it has an internal mirror plane. This is an example of a diastereomer. The achiral structure is denoted as a *meso* compound. Diastereomers have different physical and chemical properties from the optically active enantiomers. Recognition of a plane of symmetry is usually the easiest way to detect a *meso* compound. The stereoisomers of tartaric acid are examples of compounds with multiple chiral centers (see Fig. 2.14), and one of its isomers is a *meso* compound.

Mirror plane

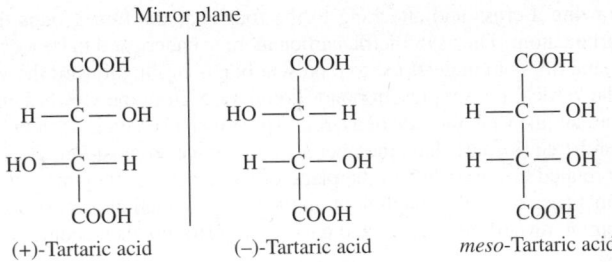

| (+)-Tartaric acid | (−)-Tartaric acid | *meso*-Tartaric acid |

FIGURE 2.14   Isomers of tartaric acid.

When the asymmetric carbon atoms in a chiral compound are part of a ring, the isomerism is more complex than in acyclic compounds. A cyclic compound which has two different asymmetric carbons with different sets of substituent groups attached has a total of $2^2 = 4$ optical isomers: an enantiometric pair of *cis* isomers and an enantiometric pair of *trans* isomers. However, when the two asymmetric centers have the same set of substituent groups attached, the *cis* isomer is a *meso* compound and only the *trans* isomer is chiral. (See Fig. 2.15.)

Torsional Asymmetry.   Rotation about single bonds of most acyclic compounds is relatively free at ordinary temperatures. There are, however, some examples of compounds in which nonbonded interactions between large substitutent groups inhibit free rotation about a sigma bond. In some cases these compounds can be separated into pairs of enantiomers.

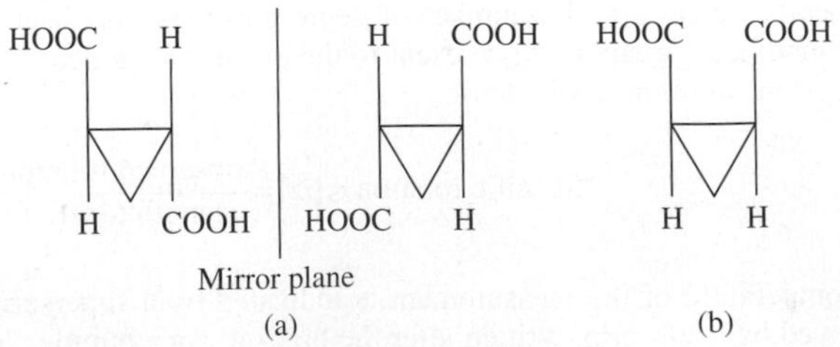

FIGURE 2.15   Isomers of cyclopropane-1,2-dicarboxylic acid. (*a*) *Trans* isomer; (*b*) *meso* isomer.

A *chiral axis* is present in chiral biaryl derivatives. When bulky groups are located at the *ortho* positions of each aromatic ring in biphenyl, free rotation about the single bond connecting the two rings is inhibited because of torsional strain associated with twisting rotation about the central single bond. Interconversion of enantiomers is prevented (see Fig. 2.16).

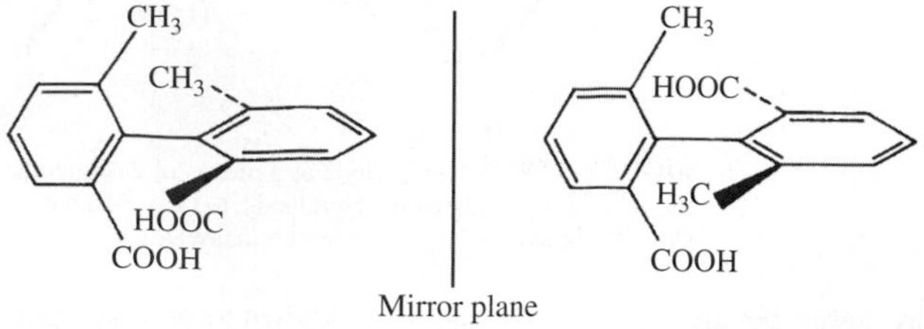

FIGURE 2.16   Isomers of biphenyl compounds with bulky groups attached at the *ortho* positions.

For compounds possessing a chiral axis, the structure can be regarded as an elongated tetrahedron to be viewed along the axis. In deciding upon the absolute configuration it does not matter from which end it is viewed; the nearer pair of ligands receives the first two positions in the order of precedence (see Fig. 2.17).

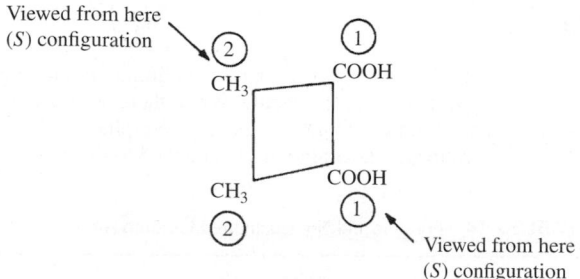

FIGURE 2.17   Example of a chiral axis.

A *chiral plane* is exemplified by the plane containing the benzene ring and the bromine and oxygen atoms in the chiral compound (see Fig. 2.18). Rotation of the benzene ring around the oxygen-to-ring single bonds is inhibited when *x* is small (although no critical size can be reasonably established).

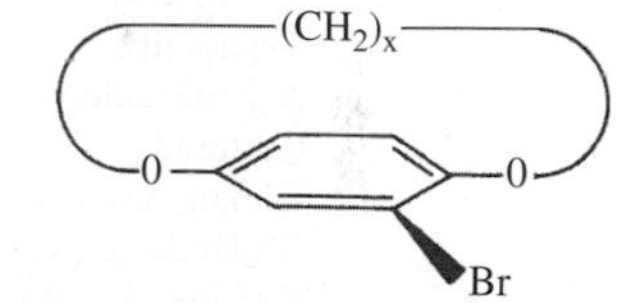

FIGURE 2.18   Example of a chiral plane.

Absolute Configuration.   The terms absolute stereochemistry and absolute configuration are used to describe the three-dimensional arrangement of substituents around a chiral element. A general system for designating absolute configuration is based upon the priority system and sequence rules. Each group attached to a chiral center is assigned a number, with number one the highest-priority group. For example, the groups attached to the chiral center of 2-butanol (see Fig. 2.19) are assigned these priorities: 1 for OH, 2 for CH_2CH_3, 3 for CH_3, and 4 for H. The molecule is then viewed from the side opposite the group of lowest priority (the hydrogen atom), and the arrangement of the remaining groups is noted. If, in proceeding from the group of highest priority to the group of second priority and thence to the third, the eye travels in a clockwise direction, the configuration is specified *R* (from the Latin *rectus*, "right"); if the eye travels in a counterclockwise direction, the configuration is specified *S* (from the Latin *sinister*, "left"). The complete name includes both configuration and direction of optical rotation, as for example, (*S*)-(+)-2-butanol.

The relative configurations around the chiral centers of many compounds have been established. One optically active compound is converted to another by a sequence of chemical reactions which

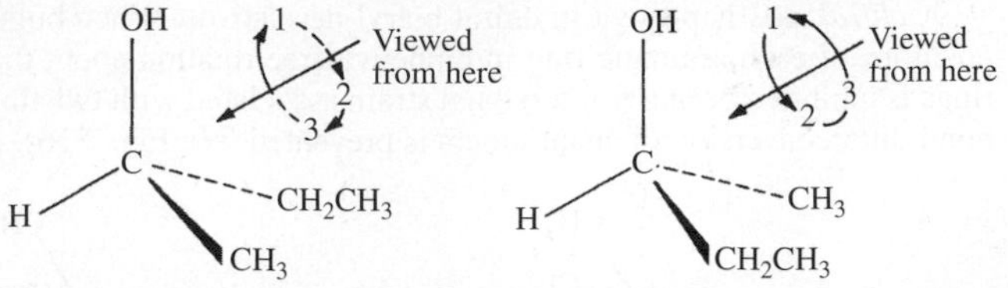

FIGURE 2.19 Viewing angle as a means of designating the absolute configuration of compounds with a chiral axis. (*a*) (*R*)-2-Butanol (sequence clockwise); (*b*) (*S*)-2-butanol (sequence counterclockwise).

are stereospecific; that is, each reaction is known to proceed spatially in a specific way. The configuration of one chiral compound can then be related to the configuration of the next in sequence. In order to establish absolute configuration, one must carry out sufficient stereospecific reactions to relate a new compound to another of known absolute configuration. Historically the configuration of D-(+)-2,3-dihydroxypropanal has served as the standard to which all configuration has been compared. The absolute configuration assigned to this compound has been confirmed by an X-ray crystallographic technique.

2.1.5   Amino Acids

An *amino acid* is an organic compound containing an amine group ($-NH_2$) and a carboxylic acid group ($-CO_2H$) in the same molecule. While there are many forms of amino acids, all of the important amino acids found in living organisms are alpha-amino acids. Alpha amino acids have the carboxylic acid group and the amino group attached to the same carbon atom.

TABLE 2.14   Formula and Nomenclature of Amino Acids

Name	Abbr.	Linear structural formula
Alanine	ala	$CH_3-CH(NH_2)-COOH$
Arginine	arg	$HN=C(NH_2)-NH-(CH_2)_3-CH(NH_2)-COOH$
Asparagine	asn	$H_2N-CO-CH_2-CH(NH_2)-COOH$
Aspartic acid	asp	$HOOC-CH_2-CH(NH_2)-COOH$
Cysteine	cys	$HS-CH_2-CH(NH_2)-COOH$
Glutamine	gln	$H_2N-CO-(CH_2)_2-CH(NH_2)-COOH$
Glutamic acid	glu	$HOOC-(CH_2)_2-CH(NH_2)-COOH$
Glycine	gly	NH_2-CH_2-COOH
Histidine	his	$NH-CH=N-CH=C-CH_2CH(NH_2)-COOH$
Isoleucine	ile	$CH_3-CH_2-CH(CH_3)-CH(NH_2)-COOH$
Leucine	leu	$(CH_3)_2-CH-CH_2-CH(NH_2)-COOH$
Lysine	lys	$H_2N-(CH_2)_4-CH(NH_2)-COOH$
Methionine	met	$CH_3-S-(CH_2)_2-CH(NH_2)-COOH$
Phenylalanine	phe	$C_6H_5-CH_2-CH(NH_2)-COOH$
Proline	pro	$NH-(CH_2)_3-CH-COOH$
Serine	ser	$HO-CH_2-CH(NH_2)-COOH$
Threonine	thr	$CH_3-CH(OH)-CH(NH_2)-COOH$
Tryptophan	trp	$C_6H_4-NH-CH=C-CH_2-CH(NH_2)-COOH$
Tyrosine	tyr	$HO-p-C_6H_4-CH_2-CH(NH_2)-COOH$
Valine	val	$(CH_3)_2-CH-CH(NH_2)-COOH$

TABLE 2.15   Acid-Base Properties of Amino Acids

Amino acid	pK_{a1}*	pK_{a2}*	pI
Glycine	2.34	9.60	5.97
Alanine	2.34	9.69	6.00
Valine	2.32	9.62	5.96
Leucine	2.36	9.60	5.98
Isoleucine	2.36	9.60	6.02
Methionine	2.28	9.21	5.74
Proline	1.99	10.60	6.30
Phenylalanine	1.83	9.13	5.48
Tryptophan	2.83	9.39	5.89
Asparagine	2.02	8.80	5.41
Glutamine	2.17	9.13	5.65
Serine	2.21	9.15	5.68
Threonine	2.09	9.10	5.60
Tyrosine	2.20	9.11	5.66

*In all cases pK_{a1} corresponds to ionization of the carboxyl group; pK_{a2} corresponds to deprotonation of the ammonium ion.

The simplest amino acid is glycine (H_2NCH_2COOH) and contains no asymmetric carbon atoms (tetrahedral carbon atoms with four different groups attached). All of the other amino acids contain an asymmetric carbon atom and are therefore optically active. Under physiological aqueous conditions a proton transfer from the acid to the base occurs, forming a dipolar ion or zwitterion, because the carboxylic acid is a much stronger acid than is the ammonium ion. The actual structure of glycine in solution, for example, is $^+H_3NCH_2COO^-$ at pH 7 rather than H_2NCH_2COOH. At very low pH the acid group can be protonated and at very high pH the ammonium group can be deprotonated, but the forms of amino acids relevant to living organisms are the zwitterions.

TABLE 2.16   Acid-Base Properties of Amino Acids with Ionizable Side Chains

Amino acid	pK_{a1}*	pK_{a2}	pK_a of side chain	pI
Aspartic acid	1.88	9.60	3.65	2.77
Glutamic acid	2.19	9.67	4.25	3.22
Lysine	2.18	8.95	10.53	9.74
Arginine	2.17	9.04	12.48	10.76
Histidine	1.82	9.17	6.00	7.59

*In all cases pK_{a1} corresponds to ionization of the carboxyl group of $\underset{\underset{+}{NH_3}}{RCHCO_2H}$, and pK_{a2} to ionization of the ammonium ion.

2.1.6   Carbohydrates

Carbohydrates consist of the elements carbon, hydrogen, and oxygen. In their basic form, carbohydrates are simple sugars or *monosaccharides*. These simple sugars can combine with each other to form more complex carbohydrates. The combination of two simple sugars is a *disaccharide*. Carbohydrates consisting of two to ten simple sugars are called *oligosaccharides*, and those with a larger number are called *polysaccharides*.

Sugars.   Sugars are white crystalline carbohydrates that are soluble in water and generally have a sweet taste. Monosaccharides are simple sugars

The classification system of monosaccharides is based on the number of carbons in the sugar:

Number of carbon atoms	Category name	Examples
4	Tetrose	Erythrose, Threose
5	Pentose	Arabinose, Ribose, Ribulose, Xylose, Xylulose, Lyxose
6	Hexose	Allose, Altrose, Fructose, Galactose, Glucose, Gulose, Idose, Mannose, Sorbose, Talose
7	Heptose	Sedoheptulose

Many saccharide structures differ only in the orientation of the hydroxyl groups (—OH). This slight structural difference makes a big difference in the biochemical properties, organoleptic properties (e.g., taste), and in the physical properties such as melting point and Specific Rotation (how polarized light is distorted). A chain-form monosaccharide that has a carbonyl group (C=O) on an end carbon forming an aldehyde group (—CHO) is classified as an *aldose*. When the carbonyl group is on an inner atom forming a ketone, it is classified as a *ketose*.

Tetroses

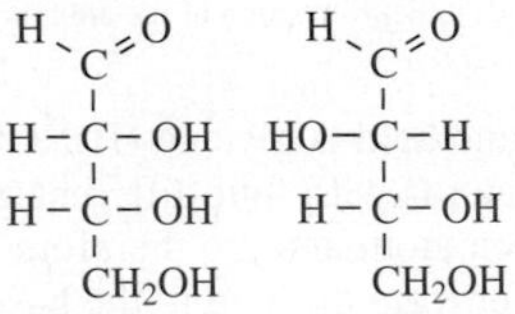

D-Erythrose   D-Threose

Pentoses. The ribose structure is a component of deoxyribonucleic acid (DNA) and ribonucleic acids (RNA).

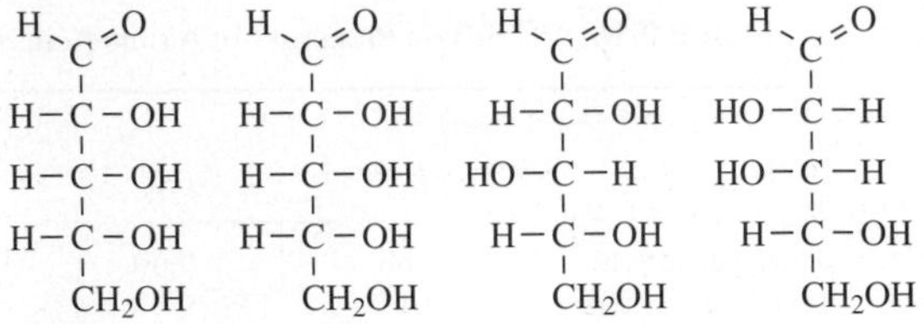

D-Ribose   D-Arabinose   D-Xylose   D-Lyxose

Hexoses. Hexoses, such as the ones illustrated here, have the molecular formula $C_6H_{12}O_6$.

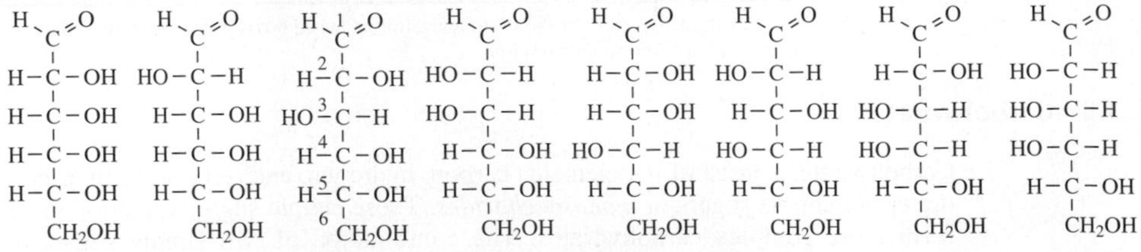

D-Allose   D-Altrose   D-Glucose   D-Mannose   D-Gulose   D-Idose   D-Galactose   D-Talose

Structures that have opposite configurations of a hydroxyl group at only one position, such as glucose and mannose, are called *epimers*.

Glucose, also called dextrose, is the most widely distributed sugar in the plant and animal king-doms and it is the sugar present in blood as "blood sugar". The chain form of glucose is a polyhydric aldehyde, meaning that it has multiple hydroxyl groups and an aldehyde group. Fructose, also called levulose, is shown here in the chain and ring forms.

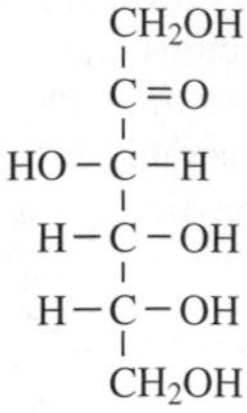

| D-Fructose (a ketose) | Fructose | Galactose | Mannose |

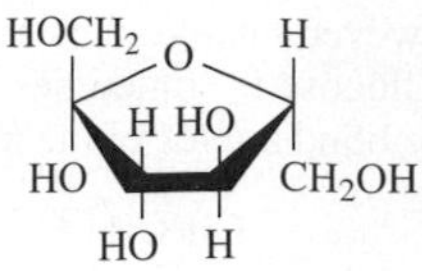

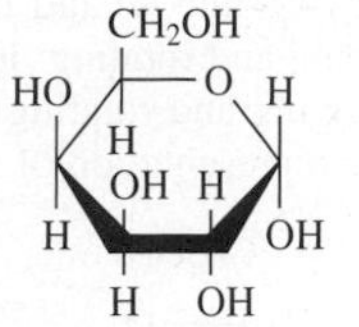

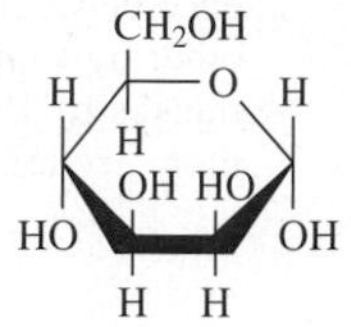

Heptoses. Sedoheptulose has the same structure as fructose, but it has one extra carbon.

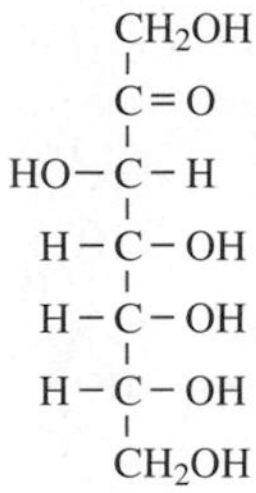

D-Sedoheptulose

Chain and Ring Structure. Many simple sugars can exist in a chain form or a ring form, as illustrated by the hexoses above. The ring form is favored in aqueous solutions, and the mechanism of ring formation is similar for most sugars. The glucose ring form is created when the oxygen on carbon number 5 links with the carbon comprising the carbonyl group (carbon number 1) and transfers its hydrogen to the carbonyl oxygen to create a hydroxyl group. The rearrangement produces *alpha*-glucose when the hydroxyl group is on the opposite side of the —CH₂OH group, or *beta*-glucose when the hydroxyl group is on the same side as the —CH₂OH group. Isomers that differ only in their configuration about their carbonyl carbon atom are called *anomers*.

The symbol 'd' (or 'D') is used to indicate that the shows that a sugar is *dextrorotary*, i.e., it rotates polarized light to the right, but can also denote a specific configuration. On the other hand, the symbol '1' (or 'L') indicates that the sugar is *laevorotatory*, i.e., it rotates polarized light to the left. Again the symbol may be used to indicate a specific configuration.

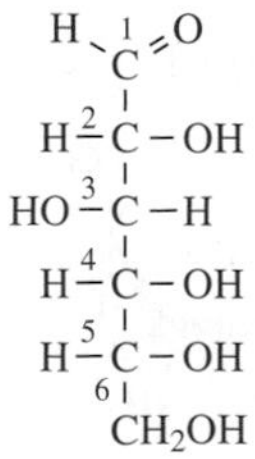

| d-Glucose (an aldose) | α-d-Glucose | β-d-Glucose |

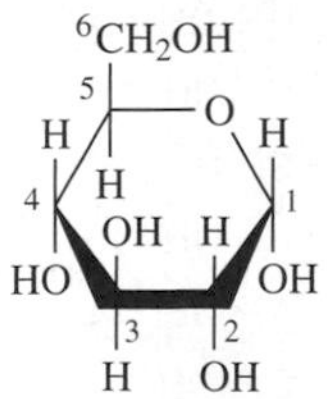

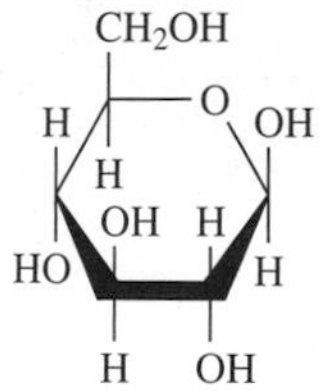

Stereochemistry.   Saccharides with identical functional groups but with different spatial configurations have different chemical and biological properties. Stereochemistry is the study of the arrangement of atoms in three-dimensional space. Stereoisomers are compounds in which the atoms are linked in the same order but differ in their spatial arrangement. Compounds that are mirror images of each other but are not identical are called *enantiomers*. The following structures illustrate the difference between β-D-glucose and β-L-glucose. Identical molecules can be made to correspond to each other by flipping and rotating. However, enantiomers cannot be made to correspond to their mirror images by flipping and rotating. Glucose is sometimes illustrated as a "chair form" because it is a more accurate representation of the bond angles of the molecule.

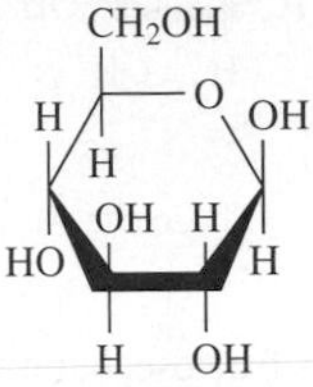

β-D-Glucose 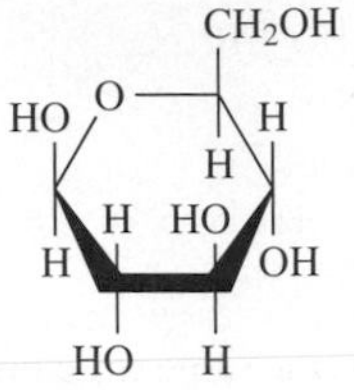   β-L-Glucose 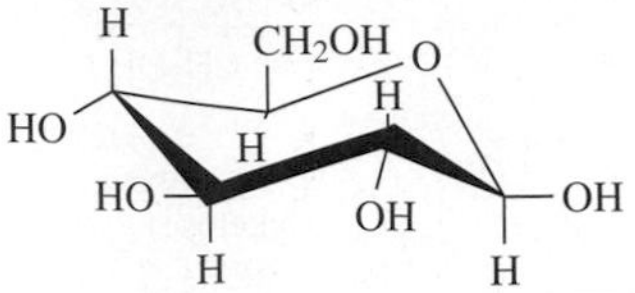   β-D-Glucose (chair form)

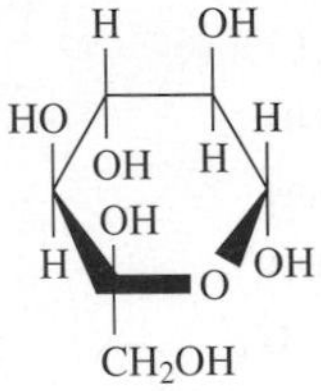

β-D-Glucose 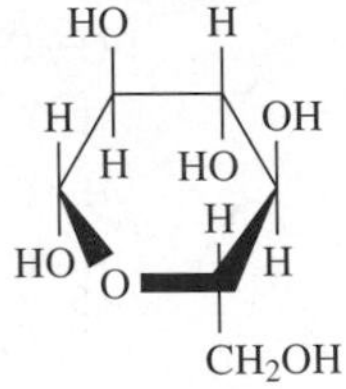   β-L-Glucose

Sugar Alcohols, Amino Sugars, and Uronic Acids.   Sugars may be modified by natural or laboratory processes into compounds that retain the basic configuration of saccharides, but have different functional groups. *Sugar alcohols*, also known as polyols, polyhydric alcohols, or polyalcohols, are the hydrogenated forms of the aldoses or ketoses. For example, glucitol, also known as sorbitol, has the same linear structure as the chain form of glucose, but the aldehyde (—CHO) group is replaced with a —CH$_2$OH group. Other common sugar alcohols include the monosaccharides erythritol and xylitol and the disaccharides lactitol and maltitol. Sugar alcohols have about half the calories of sugars and are frequently used in low-calorie or "sugar-free" products.

Amino sugars or aminosaccharides replace a hydroxyl group with an amino (—NH$_2$) group. Glucosamine is an amino sugar used to treat cartilage damage and reduce the pain and progression of arthritis.

Uronic acids have a carboxyl group (—COOH) on carbon number six.

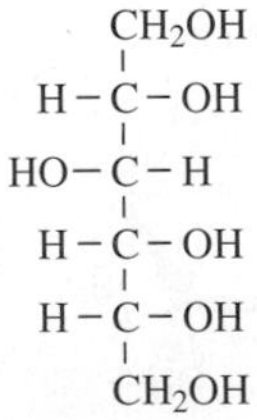

Glucitol   or Glucosamine   Glucuronic acid
Sorbitol   (an amino sugar)   (a uronic acid)
(a sugar alcohol)

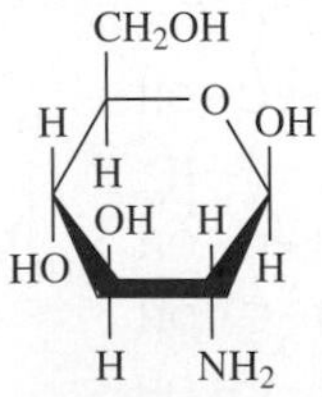

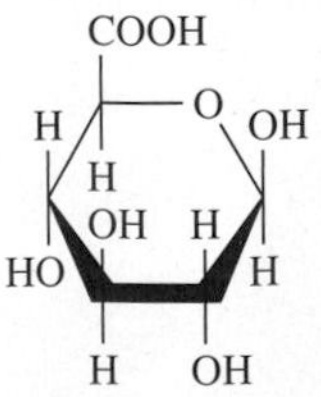

Disaccharides. Disaccharides consist of two simple sugars and the common disaccharides are sucrose, lactose, and maltose.

Disaccharide	Description	Component monosaccharides
Sucrose	common table sugar	Glucose + fructose
Lactose	main sugar in milk	galactose + glucose
Maltose	product of starch hydrolysis	glucose + glucose

Sucrose   Lactose   Maltose

Lactose has a molecular structure consisting of galactose and glucose. It is of interest because it is associated with lactose intolerance, which is the intestinal distress caused by a deficiency of lactase, an intestinal enzyme needed to absorb and digest lactose in milk. Undigested lactose ferments in the colon and causes abdominal pain, bloating, gas, and diarrhea. Yogurt does not cause these problems because lactose is consumed by the bacteria that transform milk into yogurt.

Maltose consists of two α-D-glucose molecules with the alpha bond at carbon 1 of one molecule attached to the oxygen at carbon 4 of the second molecule. This is called a $1\alpha\rightarrow4$ linkage.

Cellobiose is a disaccharide consisting of two β-D-glucose molecules that have a $1\beta\rightarrow4$ linkage. Cellobiose has no taste, whereas maltose is about one-third as sweet as sucrose.

Polysaccharides. Polysaccharides are polymers of simple sugars but, unlike sugars, polysaccharides are insoluble in water.

Starch. Starch is the major form of stored carbohydrate in plants. Starch is composed of a mixture of two substances: *amylose*, an essentially linear polysaccharide, and *amylopectin*, a highly branched polysaccharide. Both forms of starch are polymers of α-D-glucose. Natural starch contains 10–20% amylose and 80–90% amylopectin.

Amylose molecules consist typically of 200 to 20,000 glucose units that form a helix as a result of the bond angles between the glucose units.

Amylose

Amylopectin differs from amylose in being highly branched. Short side chain of about 30 glucose units are attached approximately every twenty to thirty glucose units along the chain. Amylopectin molecules may contain up to two million glucose units.

Amylopectin

Starches are transformed into many commercial products by hydrolysis with acids or enzymes. The resulting products are assigned a Dextrose Equivalent (DE) value that is related to the degree of hydrolysis. A DE value of 100 corresponds to completely hydrolyzed starch, which is pure glucose (dextrose). Maltodextrins are not sweet and have DE values less than 20. Syrups, such as corn syrup, have DE values from 20 to 95. "High fructose corn syrup," commonly used to sweeten soft drinks, is made by enzymatically isomerizing a portion of the glucose into fructose, which is about twice as sweet as glucose.

Glycogen. Glucose is stored as glycogen in animal tissues by the process of glycogenesis. When glucose cannot be stored as glycogen or used immediately for energy, it is converted to fat. Glycogen is a polymer of α-D-glucose identical to amylopectin, but the branches in glycogen tend to be shorter (about 13 glucose units) and more frequent. The glucose chains are organized globularly, like the branches of a tree, surrounding a pair of molecules of glycogenin, a protein with a molecular weight of 38,000 that acts as a primer at the core of the structure. Glycogen is easily converted back to glucose to provide energy.

Cellulose. Cellulose is a polymer of β-D-glucose, which in contrast to starch, is oriented with —CH_2OH groups alternating above and below the plane of the cellulose molecule thus producing long, unbranched chains. The absence of side chains allows cellulose molecules to lie close together and form rigid structures. Cellulose is the major structural material of plants. Wood is largely cellulose, and cotton is almost pure cellulose. Cellulose can be hydrolyzed to its constituent glucose units by microorganisms that inhabit the digestive tract of termites and ruminants. Cellulose may be modified in the laboratory by treating it with nitric acid (HNO_3) to replace all the hydroxyl groups with nitrate groups (—ONO_2) to produce cellulose nitrate that is an explosive component of smokeless powder.

Cellulose

2.1.7 Miscellaneous Compounds

TABLE 2.17   Representative Terpenes

Monoterpenes

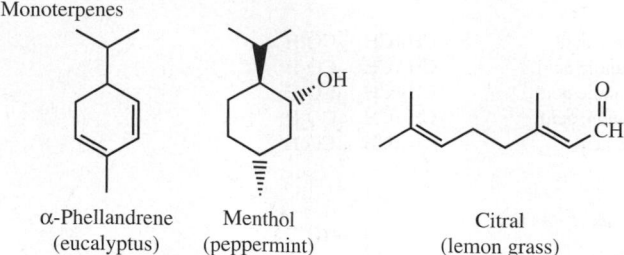

α-Phellandrene
(eucalyptus)

Menthol
(peppermint)

Citral
(lemon grass)

Sesquiterpenes

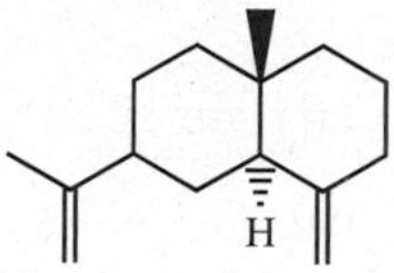

α-Selinene
(celery)

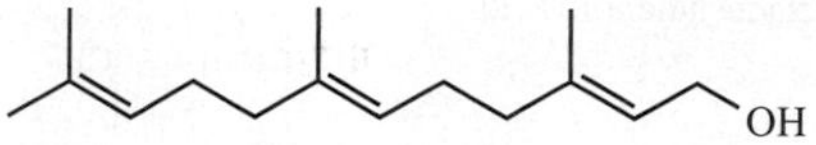

Farnesol
(ambrette)

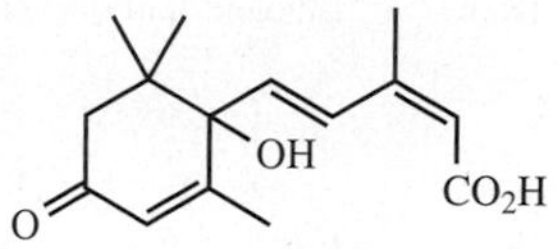

Abscisic acid
(a plant hormone)

Diterpenes

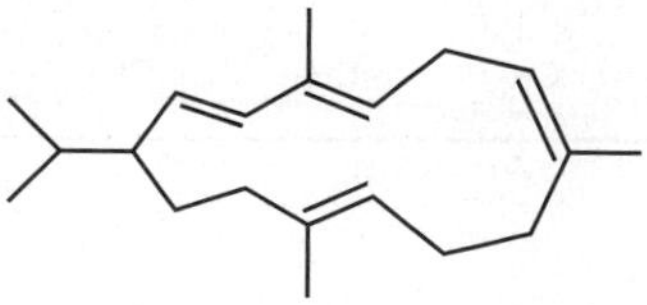

Cembrene
(pine)

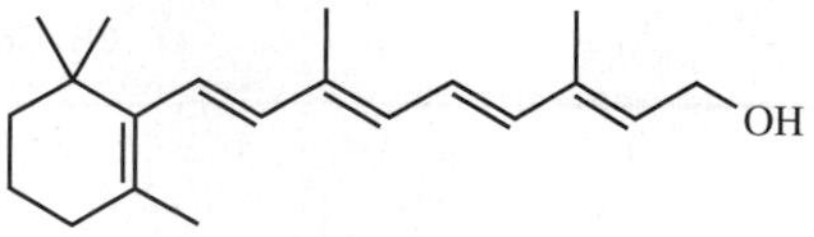

Vitamin A
(present in mammalian tissue and fish oil;
important substance in the chemistry of vision)

Triterpenes

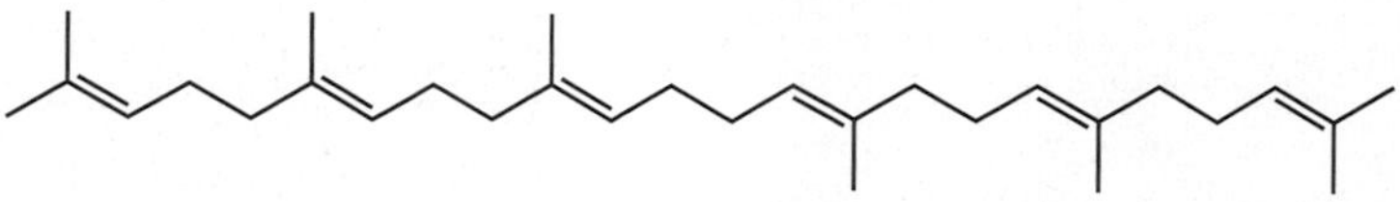

Squalene
(shark liver oil)

Tetraterpenes

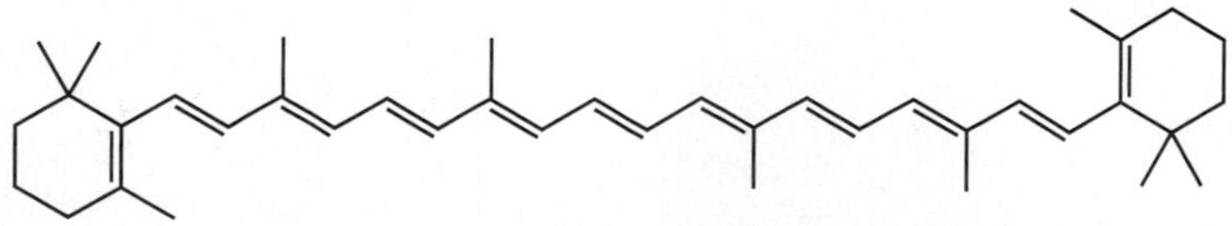

β-Carotene
(present in carrots and other vegetables;
enzymes in the body cleave β-carotene to vitamin A)

TABLE 2.18 Representative Fatty Acids

Number of Carbons	Common name	Systematic name	Structural formula	Melting point °C
Saturated fatty acids				
12	Lauric acid	Dodecanoic acid	$CH_3(CH_2)_{10}CO_2H$	44
14	Myristic acid	Tetradecanoic acid	$CH_3(CH_2)_{12}CO_2H$	58
16	Palmitic acid	Hexadecanoic acid	$CH_3(CH_2)_{14}CO_2H$	63
18	Stearic acid	Octadecanoic acid	$CH_3(CH_2)_{16}CO_2H$	69
20	Arachidic acid	Icosanoic acid	$CH_3(CH_2)_{18}CO_2H$	75
Unsaturated fatty acids				
18	Oleic acid	*cis*-9-Octadecenoic acid		4
18	Linoleic acid	*cis,cis*-9, 12-Octadecadienoic acid		−12
18	Linoleimic acid	*cis,cis,cis*-9, 12, 15-Octadecatrienoic acid		—
20	Arachidonic acid	*cis,cis,cis,cis*-5, 8, 11, 14-Icosatetraenoic acid		−49

TABLE 2.19 Pyrimidines and Purines That Occur in DNA and RNA

Name	Structure	Occurrence
Pyrimidines		
Cytosine		DNA and RNA
Thymine		DNA
Uracil		RNA
Purines		
Adenine		DNA and RNA
Guanine		DNA and RNA

TABLE 2.20   Names and Formulas of Organic Radicals

For more comprehensive lists, see the various lists of radicals given in the subject indexes of the annual and decennial indexes of Chemical Abstracts.

Name	Formula	Name	Formula
Acenaphthenyl	$C_{12}H_9$—	Azido	N_3—
Acenaphthenylene	—$C_{12}H_8$—	Azino	$=N$—$N=$
Acenaphthenylidene	$C_{12}H_8=$	Azo	—$N=N$—
Acetamido	CH_3—CO—NH—	Azoxy	—$N(O)$—N—
Acetimidoyl	$CH_3C(=NH)$—	Azulenyl	$C_{10}H_7$—
Acetoacetyl	CH_3—CO—CH_2—CO—	Benzamido	C_6H_5—CO—NH—
Acetohydrazonoyl	CH_3—$C(=NNH_2)$—	Benzeneazo	C_6H_5—$N=N$—
Acetohydroximoyl	CH_3—$C(=NOH)$—	Benzeneazoxy	C_6H_5—N_2O—
Acetonyl	CH_3—CO—CH_2—	1,2-Benzenedicarbonyl, *see* Phthaloyl	
Acetonylidene	CH_3—CO—$CH=$	1,3-Benzenedicarbonyl (*or* isophthaloyl)	—CO—C_6H_4—CO— (*m*-)
Acetoxy	CH_3—CO—O—	1,4-Benzenedicarbonyl (*or* terephthaloyl)	—CO—C_6H_4—CO— (*p*-)
Acetyl (*not ethanoyl*)	CH_3—CO—	Benzenesulfinyl	C_6H_5—SO—
Acetylamino	CH_3—CO—NH—	Benzenesulfonamido	C_6H_5—SO_2—NH—
Acetylhydrazino	CH_3—CO—NH—NH—	Benzenesulfonyl	C_6H_5—SO_2—
Acetylimino	CH_3—CO—$N=$	Benzenesulfonylamino	C_6H_5—SO_2—NH—
Acridinyl (*from acridine*)	$NC_{13}H_8$—	Benzenetriyl	C_6H_3—
Acroyloyl (*or propenoyl*)	$CH_2=CH$—CO—	Benzhydryl (*or diphenyl-methyl*)	$(C_6H_5)_2CH$—
Adipoyl (*or hexanedioyl*)	—CO—$[CH_2]_4$—CO—	Benzidino	p-H_2N—C_6H_4—C_6H_4—NH—
Alanyl	CH_3—$CH(NH_2)$—CO—		
β-Alanyl	H_2N—CH_2—CH_2—CO—	Benziloyl (*or 2-hydroxy-2,2-diphenylethanoyl*)	$(C_6H_5)_2C(OH)$—CO—
Allyl (*or 2-propenyl*)	$CH_2=CH$—CH_2—	Benzimidazolyl	$N_2C_7H_5$—
Allylidene	$CH_2=CH$—$CH=$	Benzimidoyl	C_6H_5—$C(=NH)$—
Allyloxy	$CH_2=CH$—CH_2—O—	Benzofuranyl	OC_8H_5—
Amidino	H_2N—$C(=NH)$—	Benzopyranyl	OC_9H_7—
Amino	H_2N—	Benzoquinonyl (1,2- or 1,4-)	$(O=)_2C_6H_3$—
Aminomethyleneamino	H_2N—$CH=N$—	Benzo[*b*]thienyl	SC_8H_5—
Aminooxy	H_2N—O—	Benzoyl	C_6H_5—CO—
Ammonio	^+H_3N—	Benzoylamino	C_6H_5—CO—NH—
Amyl, *see* Pentyl		Benzoylhydrazino	C_6H_5—CO—NH—NH—
Anilino	C_6H_5—NH—	Benzoylimino	C_6H_5—CO—$N=$
Anisidino (*o-, m-, or p-*)	CH_3O—C_6H_4—NH—	Benzoyloxy	C_6H_5—CO—O—
Anisoyl (*o-, m-, or p-; or methoxyben-zoyl*)	CH_3O—C_6H_4—CO—	Benzyl	C_6H_5—CH_2—
		Benzylidene	C_6H_5—$CH=$
Anthraniloyl	o-NH_2—C_6H_4—CO—	Benzylidyne	C_6H_5—$C\equiv$
Anthryl (*from anthracene*)	$C_{14}H_9$—	Benzyloxy	C_6H_5—CH_2—O—
Anthrylene	—$C_{14}H_8$—	Benzyloxycarbonyl	C_6H_5—CH_2—O—CO—
Arginyl	H_2N—$C(=NH)$—NH—$[CH_2]_3 CH(NH)$—CO—	Benzylthio	C_6H_5—CH_2—S—
		Biphenylenyl	$C_{12}H_7$—
Asparaginyl	H_2N—CO—CH_2—$CH(NH_2)$—CO—	Biphenylyl	C_6H_5—C_6H_4—
		Bornenyl	$C_{10}H_{15}$—
Aspartoyl	—CO—CH_2—$CH(NH_2)$—CO—	Bornyl (*not camphyl or bornylyl*)	$C_{10}H_{17}$—
α-Aspartyl	HO_2C—$CH_2CH(NH_2)$—		
Atropoyl (*or 2-phenylpro-penoyl*)	C_6H_5—$C(=CH_2)$—CO—	Bromo	Br—
Azelaoyl, *see* Nonane-dioyl		Bromoformyl	Br—CO—

TABLE 2.20 Names and Formulas of Organic Radicals (*Continued*)

Name	Formula	Name	Formula
Bromonio	$^+$HBr—	Cinnamoyl (*or 3-phenyl-propenoyl*)	C_6H_5—CH=CH—CO
Butadienyl (1,3- shown)	CH_2=CH—CH=CH—	Cinnamyl	C_6H_5—CH=CH—CH_2—
Butanedioyl, *see* Succinyl		Cinnamylidene	C_6H_5—CH=CH—CH=
Butanediylidene	=CH—CH_2—CH_2—CH=	Citraconoyl (*unsubstituted only*)	HC—CO— ‖ CH_3—C—CO—
Butanediylidyne	≡C—CH_2—CH_2—C≡	Crotonoyl	CH_3—CH=CH—CO— (*trans*)
Butanoyl, *see* Butyryl			
cis-Butenedioyl, *see* Maleoyl		Crotyl, *see* 2-Butenyl	
trans-Butenedioyl, *see* Fumaroyl Butenoyl, *see* Crotonoyl and Isocrotonoyl		Cumenyl (*o-, m-,* or *p-*)	$(CH_3)_2$CH—C_6H_4—
		Cyanato	NCO—
		Cyano	NC—
		Cyclobutyl	C_4H_7—
1-Butenyl	CH_3—CH_2—CH=CH—	Cycloheptyl	C_7H_{13}—
2-Butenyl (*not crotyl*)	CH_3—CH=CH—CH_2—		CH=CH—CH— CH=CH—CH_2
2-Butenylene	—CH_2—CH=CH—CH_2—		
Butenylidene (*2- shown*)	CH_3CH=CH—CH=		CH—CH_2—C< ‖ CH—CH=CH
Butenylidyne (*2- shown*)	CH_3—CH=CH—C≡	Cyclohexadienylidene (*2,4- shown*)	
Butoxy	CH_3—$[CH_2]_3$—O—	Cyclohexanecarbonyl	C_6H_{11}—CO—
sec-Butoxy (*unsubstituted only*)	C_2H_5—CH(CH_3)—O—	Cyclohexanecarbothioyl	C_6H_{11}—CS—
tert-Butoxy (*unsubstituted only*)	$(CH_3)_3$C—O—	Cyclohexanecarboxamido	C_6H_{11}—CO—NH—
Butyl	CH_3—$[CH_2]_3$— or C_4H_9—	Cyclohexanecarboximidoyl	C_6H_{11}—C(=NH)—
		Cyclohexenyl	C_6H_9—
		2-Cyclohexenylidene	CH=CH—C< H_2C—CH_2—CH_2
sec-Butyl (*unsubstituted only*)	C_2H_5—CH(CH_3)—		
		Cyclohexyl	C_6H_{11}—
tert-Butyl (*unsubstituted only*)	$(CH_3)_3$C—	Cyclohexylcarbonyl	C_6H_{11}—CO—
		Cyclohexylene	—C_6H_{10}—
Butylidene	CH_3—CH_2—CH_2—CH=	Cyclohexylidene	CH_2—CH_2—C< CH_2—CH_2—CH_2
sec-Butylidene (*unsubstituted only*)	C_2H_5C(CH_3)=		
Butylidyne	CH_3—$[CH_2]_2$—C≡	Cyclohexylthiocarbonyl	C_6H_{11}—CS—
Butyryl (*or butanoyl*)	CH_3—CH_2—CH_2—CO—	Cyclopentadienyl	C_5H_5—
Camphoroyl	$C_{10}H_{14}O_2$—	Cyclopentadienylidene	CH=CH—CH=CH—C=
Carbamoyl	H_2N—CO—	Cyclopenta[*a*]phenanthryl	$C_{17}H_{17}$—
Carbazolyl	$NC_{12}H_8$—	1,2-Cyclopentenophenanthryl	$C_{17}H_{11}$—
Carbazoyl	H_2N—NH—CO—		
Carbonimidoyl	—C(=NH)—	Cyclopentenyl	C_5H_7—
Carbonohydrazido (*preferred to carbohydrazido or carbazido*)	H_2N—NH—CO—NH—NH—	Cyclopentyl	C_5H_9—
		Cyclopentylene	—C_5H_8—
		Cyclopropyl	C_3H_5—
Carbonyl	—CO— or =C(O)	Cysteinyl	HS—CH_2—CH(NH_2)—CO—
Carbonyldioxy	—O—CO—O—		
Carboxy	HO_2C—	Cystyl	—CO—CH(NH_2)—CH_2—S—S—CH_2—CH(NH_2)—CO—
Carboxylato	—O_2C—		
Chloro	Cl—		
Chlorocarbonyl, *see* Chloroformyl		Decanedioyl	—CO—$[CH_2]_8$—CO—
		Decanoyl	CH_3—$[CH_2]_8$—CO—
Chloroformyl	Cl—C(O)—	Decyl	CH_3—$[CH_2]_9$—
Chlorosyl	OCl—	Diacetoxyiodo	$(CH_3$—CO—O$)_2$I—
Chlorothio	ClS—	Diacetylamino	$(CH_3$—CO$)_2$N—
Chloryl	O_2Cl—	Diaminomethyleneamino	$(NH_2)_2$C=N—

(*Continued*)

TABLE 2.20 Names and Formulas of Organic Radicals (*Continued*)

Name	Formula	Name	Formula
Diazo	$=N_2$	Fluorenyl	$C_{13}H_9-$
Diazoamino	$-N=N-NH-$	Fluoro	$F-$
Dibenzoylamino	$(C_6H_5-CO)_2N-$	Fluoroformyl	$F-CO-$
Dichloroiodo	Cl_2I-	Formamido	$OCH-NH-$
Diethylamino	$(C_2H_5)_2N-$	Formimidoyl	$CH(=NH)-$
3,4-Dihydroxybenzoyl, *see* Protocatechuoyl		Formyl (*not methanoyl*)	$OCH-$ or $-C(O)H$
2,3-Dihydroxybutanedioyl, *see* Tartaroyl		Formylamino	$H-CO-NH-$
		Formylimino	$H-CO-N=$
Dihydroxyiodo	$(HO)_2I-$	Formyloxy	$H-CO-O-$
2,3-Dihydroxypropanoyl, *see* Glyceroyl		Fumaroyl (*or trans-butene-dioyl*)	$-CO-CH=CH-CO-$ (*trans*)
3,4-Dimethoxybenzoyl, *see* Veratroyl		Furancarbonyl, *see* Furoyl	
3,4-Dimethoxyphenethyl	$3,4-$ $(CH_3O)_2C_6H_3CH_2CH-$	Furfurylidene (2- *only*; preferred to 2-furylmethyl)	
3,4-Dimethoxyphenyl-lacetyl	$3,4-$ $(CH_3O)_2C_6H_3CH_2CO-$	Furfurylidene (2- *only*)	
Dimethylamino	$(CH_3)_2N-$	Furoyl (3- *shown; preferred to furancarbonyl*)	
Dimethylbenzoyl	$(CH_3)_2C_6H_3-CO-$		
Dioxy	$-O-O-$	Furyl	OC_4H_3-
Diphenylamino	$(C_6H_5)_2N-$	3-Furylmethyl	
Diphenylmethylene	$(C_6H_5)_2C=$		
Dithio	$-S-S-$		
Diethiocarboxy	$HSSC-$	Galloyl (*or 3,4,5-trihydroxybenzoyl*)	$3,4,5-(HO)_3C_6H_2-CO-$
Dithiosulfo	HOS_2-		
Dodecanoyl	$CH_3[CH_2]_{10}-CO-$	Geranyl (*from geraniol*)	$C_{10}H_{17}-$
Dodecyl	$CH_3[CH_2]_{11}-$	Glutaminyl	$H_2N-CO-CH_2-CH_2-$ $CH(NH_2)-CO-$
Elaidoyl (*or trans-9-octa-decenoyl*)	$CH_3[CH_2]_7CH=CH-$ $[CH_2]_7-CO-$	Glutamoyl	$-CO-CH_2-CH_2-$ $CH(NH_2)-CO-$
Epidioxy (as a bridge)	$-O-O-$	α-Glutamyl	$HOOC[CH_2]_2CH(NH_2)-$ $CO-$
Epidiseleno (as bridge)	$-Se-Se-$		
Epidithio (as a bridge)	$-S-S-$	γ-Glutamyl	$HOOC-CH(NH_2)-$ $[CH_2]_2-CO-$
Epimino (as a bridge)	$-NH-$		
Episeleno (as a bridge)	$-Se-$	Glutaryl (*or pentanedioyl*)	$-CO-[CH_2]_3-CO-$
Epithio (as a bridge)	$-S-$	Glyceroyl (*or 2,3- dihydroxypropanoyl*)	$HO-CH_2-CH(OH)-$ $CO-$
Epoxy (as a bridge)	$-O-$		
Ethanesulfonamide	$C_2H_5-SO_2-NH-$	Glycoloyl (*or hydroxy-ethanoyl*)	H_2N-CH_2-CO-
Ethanoyl, *see* Acetyl			
Ethenyl, *see* Vinyl			
Ethoxalyl	$C_2H_5-OOC-CO-$	Glycyl	H_2N-CH_2-CO-
Ethoxy	C_2H_5-O-	Glycylamino	$H_2N-CH_2-CO-NH-$
Ethoxycarbonyl	$C_2H_5-O-CO-$	Glyoxyloyl	$OHC-CO-$
Ethyl	C_2H_5- or CH_3-CH_2-	Guanidino	$H_2N-C(=NH)-NH-$
Ethylamino	C_2H_5-NH-	Guanyl, *see* Amidino	
Ethylene	$-CH_2-CH_2-$	Heptanamido	$CH_3-[CH_2]_5-CO-NH-$
Ethylenedioxy	$-O-CH_2-CH_2-O-$	Heptanedioyl	$-CO-[CH_2]_5-CO-$
Ethylidene	$CH_3-CH=$	Heptanoyl	$CH_3-[CH_2]_5-CO-$
Ethylidyne	$CH_3-C\equiv$	Heptyl	$CH_3-[CH_2]_5-CH_2-$
Ethylsulfonylamino	$C_2H_5-SO_2-NH-$	Hexadecanoyl	$CH_3-[CH_2]_{14}-CO-$
Ethylthio	C_2H_5-S-	Hexadecyl	$CH_3-[CH_2]_{14}-CH_2-$
Ethynyl	$HC\equiv C-$	Hexamethylene	$-[CH_2]_6-$
Ethynylene	$-C\equiv C-$	Hexanamido	$CH_3-[CH_2]_4-CO-NH-$
Fluoranthenyl	$C_{16}H_9-$	Hexanedioyl (*or adipoyl*)	$-CO-[CH_2]_4-CO-$

TABLE 2.20 Names and Formulas of Organic Radicals (*Continued*)

Name	Formula	Name	Formula
Hexanimidoyl	CH_3—$[CH_2]_4$—$C(=NH)$—	Iodonio	^+HI—
Hexanoyl	CH_3—$[CH_2]_4$—CO—	Iodosyl	OI—
Hexanoylamino	CH_3—$[CH_2]_4$—CO—NH—	Iodyl	O_2I—
Hexyl	CH_3—$[CH_2]_4$—CH_2—	Isobutoxy (*unsubstituted only*)	$(CH_3)_2CH$—CH_2—O—
Hexylidene	CH_3—$[CH_2]_4$—$CH\equiv$	Isobutyl (*unsubstituted only*)	$(CH_3)_2CH$—CH_2—
Hexyloxy	$CH_3[CH_2]_5$—O—	Isobutylidene (*unsubstituted only*)	$(CH_3)_2CH$—$CH=$
Hippuroyl	C_6H_5—CO—NH—CH_2—CO—	Isobutylidyne (*unsubstituted only*)	$(CH_3)_2CH$—$C\equiv$
Histidyl	$N_2C_3H_3$—CH_2—$CH(NH_2)$—CO—	Isobutyryl (*unsubstituted only; or 2-methylpropanoyl*)	$(CH_3)_2CH$—CO—
Homocysteinyl	HS—CH_2—CH_2—$CH(NH_2)$—CO—		
Homoseryl	HO—CH_2—CH_2—$CH(NH_2)$—CO—	Isocarbonohydrazido	H_2N—$N=C(OH)$—NH—NH—
Hydantoyl	H_2N—CO—NH—CH_2—CO—	Isocrotonoyl	CH_3—$CH=CH$—CO— (*cis*)
Hydratropoyl (*or 2-phenylpropanoyl*)	C_6H_5—$CH(CH_3)$—CO—	Isocyanato	OCN—
Hydrazi	—NH—NH— (to single atom)	Isocyano	CN—
Hydrazino	H_2N—NH—	Isohexyl (*unsubstituted only*)	$(CH_3)_2CH$—$[CH_2]_3$—
Hydrazo	—NH—NH— (to different atoms)	Isoleucyl	C_2H_5—$CH(CH_3)$—$CH(NH_2)$—CO
Hydrazono	H_2N—$N=$	Isonicotinoyl (*or 4-pyridinecarbonyl*)	NC_5H_4—CO— (4-)
Hydroperoxy	HO—O—		
Hydroseleno	HSe—	Isopentyl (*unsubstituted only*)	$(CH_3)_2CH$—CH_2—CH_2—
Hydroxy	HO—		
Hydroxyamino	HO—NH—	Isophthaloyl (*or 1,3-benzenedicarbonyl*)	—CO—C_6H_4—CO— (*m-*)
o-Hydroxybenzoyl (*or salicyloyl*)	*o*-HO—C_6H_4—CO—	Isopropenyl (*unsubstituted only; or 1-methylvinyl*)	$CH_2=C(CH_3)$—
m-Hydroxybenzoyl	*m*-HO—C_6H_4—CO—	Isopropoxy (*unsubstituted only*)	$(CH_3)_2CH$—O—
p-Hydroxybenzoyl	*p*-HO—C_6H_4—CO—		
Hydroxybutanedioyl, *see* Maloyl		Isopropyl (*unsubstituted only*)	$(CH_3)_2CH$—
2-Hydroxy-2,2-diphenyl ethanoyl, *see* Benziloyl		*p*-Isopropylbenzoyl	*p*-$(CH_3)_2CH$—C_6H_4—CO—
Hydroxyethanoyl, *see* Glycoloyl		Isopropylbenzyl	$(CH_3)_2CH$—C_6H_4—CH_2—
Hydroxyimino	HO—$N=$	Isopropylidene	$(CH_3)_2C=$
4-Hydroxy-3-methoxybenzoyl (*or vanilloyl*)	4-HO,3-CH_3O—C_6H_3—CO—	Isoselenocyanato	$SeCN$—
		Isosemicarbazido	H_2N—NH—$C(OH)=N$—
3-Hydroxy-2-phenylpropanoyl (*or tropoyl*)	C_6H_5—$CH(CH_2OH)$—CO—	Isothiocyanato	SCN—
Hydroxypropanedioyl (*or tartronoyl*)	—CO—$CH(OH)$—CO—	Isothioureido	$HN=C(SH)$—NH—, H_2N—$C(SH)=N$—
2-Hydroxypropanoyl (*or lactoyl*)	CH_3—$CH(OH)$—CO—	Isoureido	$HN=C(OH)$—NH—, H_2N—$C(OH)=N$—
Icosyl	CH_3—$[CH_2]_{18}$—CH_2—	Isovaleryl (*unsubstituted only; or 3-methylbutanoyl*)	$(CH_3)_2CH$—CH_2—CO—
Imino	—NH—, $HN=$		
Iminomethylamino	$HN=CH$—NH—	Lactoyl	CH_3—$CH(OH)$—CO—
Iodo	I—	Lauroyl (*unsubstituted only*)	CH_3—$[CH_2]_{10}$—CO—
Iodoformyl	I—CO—		

(*Continued*)

TABLE 2.20   Names and Formulas of Organic Radicals (*Continued*)

Name	Formula	Name	Formula
Leucyl	$(CH_3)_2CH—CH_2—$ $CH(NH_2)—CO—$	5-Methylhexyl	$(CH_3)_2CH—[CH_2]_4—$
		Methylidyne	$HC\equiv$
Lysyl	$H_2N—[CH_2]_4—$ $CH(NH_2)—CO—$	Methylsulfinimidoyl	$CH_3—S(=NH)—$
		Methylsulfinohydrazonoyl	$CH_3—S(=NNH_2)—$
Maleoyl	$—CO—CH=CH—CO—$	Methylsulfinohydroxi- moyl	$CH_3—S(=N—OH)—$
Malonyl	$—CO—CH_2—CO—$		
Maloyl	$—CO—CH(OH)—CH_2—$ $CO—$	Methylsulfinyl	$CH_3—SO—$
		Methylsulfinylamino	$CH_3—SO—NH—$
Mercapto-	$HS—$	Methylsulfonohydrazo- noyl	$CH_3—S(O)(NNH_2)—$
Mesaconoyl (*unsubstituted only*)	$—CO—CH$ $\parallel$ $CH_3—C—CO—$	Methylsulfonimidoyl	$CH_3—S(O)(=NH)—$
		Methylsulfonohydroxa- moyl	$CH_3—S(O)(N—OH)—$
Mesityl	$2,4,6-(CH_3)_3C_6H_2—$	Methylsulfonyl	$CH_3—SO_2—$
Mesoxalo	$HOOC—CO—CO—$	Methylthio	$CH_3S—$
Mesoxalyl	$—CO—CO—CO—$	(Methylthio)sulfonyl	$CH_3S—SO_2—$
Mesyl	$CH_3—SO_2—$	1-Methylvinyl, *see* Isopro- penyl	
Methacryloyl (*or 2-methyl- propenoyl*)	$CH_2=C(CH_3)—CO—$	Morpholino (*4- only*)	$O{<}^{CH_2—CH_2}_{CH_2—CH_2}{>}N—$
Methaneazo	$CH_3—N=N—$		
Methaneazoxy	$CH_3—N_2O—$		
Methanesulfinamido	$CH_3—SO—NH—$	Morpholinyl (*3- shown*)	$O{<}^{CH_2—CH_2}_{CH_2—CH}{>}NH$
Methanesulfinyl	$CH_3—SO—$		
Methanesulfonamido	$CH_3—SO_2—NH—$	Myristoyl (*unsubstituted only*)	$CH_3—[CH_2]_{12}—CO—$
Methanesulfonyl, *see* Mesyl			
Methanoyl, *see* Formyl		Naphthalenazo	$C_{10}H_7—N=N—$
Methionyl	$CH_3—S—CH_2—CH_2—$ $CH(NH_2)—CO—$	Naphthalenecarbonyl, *see* Naphthoyl	
Methoxalyl	$CH_3OOC—CO—$	Naphthoyl	$C_{10}H_7—CO—$
Methoxy	$CH_3O—$	Naphthoyloxy	$C_{10}H_7—CO—O—$
Methoxybenzoyl (*o-, m-, or p-*)	$CH_3O—C_6H_4—CO—$	Naphthyl	$C_{10}H_7—$
Methoxycarbonyl	$CH_3O—CO—$	Naphthylazo	$C_{10}H_7—N=N—$
Methoxyimino	$CH_3O—N=$	Naphthylene	$—C_{10}H_6—$
Methoxyphenyl	$CH_3O—C_6H_4—$	Naphthylenebisazo	$—N=N—C_{10}H_6—$ $N=N—$
Methoxysulfinyl	$CH_3O—SO—$		
Methoxysulfonyl	$CH_3O—SO_2—$	Naphthyloxy	$C_{10}H_7—O—$
Methoxy(thiosulfonyl)	$CH_3O—S_2O—$	Neopentyl (*unsubstituted only*)	$(CH_3)_3C—CH_2—$
Methyl	$CH_3—$	Nicotinoyl	$NC_5H_4—CO—$ (3-)
Methylallyl	$CH_2=C(CH_3)—CH_2—$	Nitrilo	$N\equiv$
Methylamino	$CH_3—NH—$	Nitro	$O_2N—$
Methylazo	$CH_3—N=N—$	aci-Nitro	$HO—(O=)N=$
Methylazoxy	$CH_3—N_2O—$	Nitroso	$ON—$
α-Methylbenzyl	$C_6H_5—CH(CH_3)—$	Nonanedioyl	$—CO—[CH_2]_7—CO—$
Methylbenzyl	$CH_3—C_6H_4—CH_2—$	Nonanoyl	$CH_3—[CH_2]_7—CO—$
3-Methylbutanoyl	$(CH_3)_2CH—CH_2—CO—$	Nonyl	$CH_3—[CH_2]_7—CH_2—$
cis-Methylbutenedioyl	$HC—CO—$ $\parallel$ $CH_3—C—CO—$	Norbomyl	$C_7H_{11}—$
		Norbomylyl, *see* Norbornyl	
trans-Methylbutenedioyl	$—CO—CH$ $\parallel$ $CH_3—C—CO—$	Norcamphyl, *see* Norbornyl	
		Norleucyl	$CH_3—[CH_2]_3—CH(NH_2)—$ $CO—$
Methyldithio	$CH_3—S—S—$		
Methylene	$—CH_2—, H_2C=$	Norvalyl	$CH_3—CH_2—CH_2—$ $CH(NH_2)—CO—$
Methylenedioxy	$—O—CH_2—O—$		
3,4-Methylenedioxy benzoyl	$3,4-CH_2O_2: C_6H_3—CO—$	Octadecanoyl	$CH_3[CH_2]_{16}—CO—$

TABLE 2.20   Names and Formulas of Organic Radicals (*Continued*)

Name	Formula	Name	Formula
cis-9-Octadecenoyl	H[CH$_2$]$_8$—CH=CH—[CH$_2$]$_7$—CO—	Phenylsulfamoyl	C$_6$H$_5$—NH—SO$_2$
		Phenylsulfinyl	C$_6$H$_5$—SO—
Octadecyl	CH$_3$—[CH$_2$]$_{16}$—CH$_2$—	Phenylsulfonyl	C$_6$H$_5$—SO$_2$—
Octanedioyl	—CO—[CH$_2$]$_6$—CO—	Phenylsulfonylamino	C$_6$H$_5$—SO$_2$—NH—
Octanoyl	CH$_3$—[CH$_2$]$_6$—CO—	Phenylthio	C$_6$H$_5$—S—
Octyl	CH$_3$—[CH$_2$]$_6$—CH$_2$—	3-Phenylureido	C$_6$H$_5$—NH—CO—NH—
Oleoyl	H[CH$_2$]$_8$—CH=CH—[CH$_2$]$_7$—CO—	Phthalamoyl	H$_2$N—CO—C$_6$H$_4$—CO— (*o*-)
Ornithyl	H$_2$N—[CH$_2$]$_3$—CH(NH$_2$)—CO—	Phthalidyl	C$_6$H$_4$—CO—O—CH—
Oxalacetyl	—CO—CH$_2$—CO—CO—	Phthalimido	CO—C$_6$H$_4$—CO—N—
Oxalaceto	HOOC—CO—CH$_2$—CO—	Phthaloyl	—CO—C$_6$H$_4$—CO— (*o*-)
		Picryl	2,4,6-(NO$_2$)$_3$C$_6$H$_2$—
Oxalo	HOOC—CO—	Pimeloyl (*unsubstituted only*)	—CO—[CH$_2$]$_5$—CO—
Oxalyl	—CO—CO—		
Oxamoyl	H$_2$N—CO—CO—	Piperidino (*1- only*)	C$_5$H$_{10}$N—
Oxido	$^-$O— (ion)	Piperidyl (*2-, 3-, 4-*)	NC$_5$H$_{10}$—
Oxo	O=	Piperonyl	3,4-CH$_2$O$_2$: C$_6$H$_3$—CH$_2$—
Oxonio	$^+$H$_2$O—	Pivaloyl (*unsubstituted only*)	(CH$_3$)$_3$C—CO—
Oxy	—O—	Polythio	—S$_4$—
Palmitoyl (*unsubstituted only*)	CH$_3$—[CH$_2$]$_{14}$—CO—	Propanedioyl, *see* Malonyl	
Pentafluorothio	F$_5$S—	Propanoyl, *see* Propionyl	
Pentamethylene	—CH$_2$—CH$_2$—CH$_2$—CH$_2$—CH$_2$—	Propargyl, *see* 2-Propynyl	
		Propenoyl, *see* Acryloyl	
Pentanedioyl, *see* Glu-taryl		1-Propenyl	CH$_3$—CH=CH—
Pentanoyl, *see* Valeryl		2-Propenyl, *see* Allyl	
Pentenyl (*2- shown*)	CH$_3$—CH$_2$—CH=CH—CH$_2$—	Propenylene	—CH$_2$—CH=CH—
		Propioloyl	CH≡C—CO—
Pentyl	CH$_3$—CH$_2$—CH$_2$—CH$_2$—CH$_2$—	Propionamido	CH$_3$—CH$_2$—CO—NH—
		Propionyl	CH$_3$—CH$_2$—CO—
Pentyloxy	CH$_3$—[CH$_2$]$_4$ O —	Propionylamino	CH$_3$—CH$_2$—CO—NH—
Perchloryl	O$_3$Cl—	Propionyloxy	CH$_3$—CH$_2$— CO O—
Phenacyl	C$_6$H$_5$—CO—CH$_2$—	Propoxy	CH$_3$—CH$_2$—CH$_2$—O
Phenacylidene	C$_6$H$_5$—CO—CH=	Propyl	CH$_3$—CH$_2$—CH$_2$—
Phenanthryl	C$_{14}$H$_9$—	Propylene	—CH(CH$_3$)—CH$_2$—
Phenethyl	C$_6$H$_5$—CH$_2$—CH$_2$—	Propylidene	CH$_3$—CH$_2$—CH=
Phenetidino (*o-, m-, or p-*)	C$_2$H$_5$O—C$_6$H$_4$—NH—	Propylidyne	CH$_3$—CH$_2$—C≡
Phenoxy	C$_6$H$_5$—O—	Propynoyl, *see* Propiolyl	
Phenyl	C$_6$H$_5$—	1-Propynyl	CH$_3$—C≡C—
Phenylacetyl	C$_6$H$_5$—CH$_2$—CO—	2-Propynyl	HC≡C—CH$_2$—
Phenylazo	C$_6$H$_5$—N=N—	Protocatechuoyl	3,4-(HO)$_2$C$_6$H$_3$—CO—
Phenylazoxy	C$_6$H$_5$—N$_2$O—	3-Pyridinecarbonyl	NC$_5$H$_4$—CO— (3-)
Phenylcarbamoyl	C$_6$H$_5$—NH—CO	4-Pyridinecarbonyl	NC$_5$H$_4$—CO— (4-)
Phenylene	—C$_6$H$_4$—	Pyridinio	$^+$NC$_5$H$_5$— (ion)
Phenylenebisazo	—N=N—C$_6$H$_4$—N=N—	Pyridyl	NC$_5$H$_4$—
		2-Pyridylcarbonyl	NC$_5$H$_4$—CO— (2-)
Phenylimino	C$_6$H$_5$—N=	Pyridyloxy	NC$_5$H$_4$—O—
2-Phenylpropanoyl	C$_6$H$_5$—CH(CH$_3$)—CO—	Pyruvoyl	CH$_3$—CO—CO—
3-Phenylpropenoyl, *see* Cinnamoyl		Salicyl	*o*-HO—C$_6$H$_4$—CH$_2$—
		Salicylidene	*o*-HO—C$_6$H$_4$—CH=
3-Phenylpropyl	C$_6$H$_5$—CH$_2$—CH$_2$—CH$_2$—	Salicyloyl	*o*-HO—C$_6$H$_4$—CO—
		Sarcosyl	CH$_3$—NH—CH$_2$—CO—

(*Continued*)

TABLE 2.20 Names and Formulas of Organic Radicals (*Continued*)

Name	Formula	Name	Formula
Sebacoyl (*unsubstituted only*)	$-CO-[CH_2]_8-CO-$	(Terthiophen)yl	$SC_4H_3-SC_4H_2-SC_4H_2-$
Seleneno	$HOSe-$	Tetradecanoyl	$CH_3-[CH_2]_{12}-CO-$
Selenino	HO_2Se-	Tetradecyl	$CH_3-[CH_2]_{12}-CH_2-$
Seleninyl	$OSe=$	Tetramethylene	$-CH_2-CH_2-CH_2-$
Seleno	$-Se-$		CH_2-
Selenocyanato	$NC-Se-$	Thenoyl (2- *shown*)	$CH=C$
Selenoformyl	$HSeC-$		
Selenonio	$^+H_2Se-$ (ion)		$CH=CH$
Selenono	HO_3Se-	Thenyl	$SC_4H_3-CH_2-$
Selenonyl	O_2Se-	Thienyl	SC_4H_3-
Selenoureido	$H_2N-CSe-NH-$	Thio	$-S-$
Selenoxo	$(C)=Se$	Thioacetyl	CH_3-CS-
Semicarbazido	$H_2N-CO-NH-NH-$	Thiobenzoyl	C_6H_5-CS-
Semicarbazono	$H_2N-CO-NH-N=$	Thiocarbamoyl	H_2N-CS-
Seryl	$HO-CH_2-CH(NH_2)-$	Thiocarbazono	$HN=N-CS-NH-$
	$CO-$		$NH-$
Stearoyl (*unsubstituted only*)	$CH_3-[CH_2]_{16}-CO-$	Thiocarbodiazono	$HN=N-CS-N=N-$
		Thiocarbonohydrazido	$H_2N-NH-CS-NH-$
Styryl	$C_6H_5-CH=CH-$		$NH-$
Suberoyl (*unsubstituted only*)	$-CO-[CH_2]_6-CO-$	Thiocarbonyl	$-CS-, SC=$
		Thiocarboxy	$HSOC-, HS-CO-$
		Thiocyanato	$NCS-$
Succinamoyl	$H_2N-CO-CH_2-CH_2-$	Thioformyl	$SHC-, HCS-$
	$CO-$	Thiophenecarbonyl, *see* Thenoyl	
Succinimido	CH_2-C O $N-$ CH_2-C O	Thiosemicarbazido	$H_2N-CS-NH-NH-$
		Thiosulfino	HOS_2-
		Thiosulfo	HO_2S_2-
Succinimidoyl	$-C(=NH)-CH_2-$	Thioreido	$H_2N-CS-NH-$
	$CH_2C(=NH)-$	Thioxo	$S=$
Succinyl	$-CO-CH_2-CH_2-CO-$	Threonyl	$CH_3-CH(OH)-$
Sulfamoyl	H_2N-SO_2-		$CH(NH_2)-CO-$
Sulfanilamido	$p\text{-}H_2N-C_6H_4-SO_2-$	Toluenesulfonyl (*o-, m-*)	$CH_3-C_6H_4-SO_2-$
	$NH-$	Toluidino (*o-, m-, or p-*)	$CH_3-C_6H_4-NH-$
Sulfanilyl	$p\text{-}H_2N-C_6H_4-SO_2-$	Toluoyl (*om- or p-*)	$CH_3-C_6H_4-CO-$
Sulfenamoyl	H_2N-S-	Tolyl (*om- or p-*)	$CH_3-C_6H_4-$
Sulfeno	$HO-S-$	Tolylsulfonyl	$CH_3-C_6H_4-SO_2-$
Sulfido	$^-S-$ (ion)	Tosyl (*p- only*)	$p\text{-}CH_3\text{-}C_6H_4-SO_2-$
Sulfinamoyl	H_2N-SO-	Triazano	$H_2N-NH-NH-$
Sulfino	HO_2S-	Triazeno	$H_2N-N=N-$
Sulfinyl	$-SO-$	Trichlorothio	Cl_3S-
Sulfo	$HO-SO_2-$	Tridecanoyl	$CH_3-[CH_2]_{11}-CO-$
Sulfoamino	HO_2S-NH-	Tridecyl	$CH_3-[CH_2]_{12}-$
Sulfonato	$^-O_3S-$ (ion)	Trifluorothio	F_3S-
Sulfonio	$^+H_2S-$ (ion)	3,4,5-Trihydroxybenzoyl	$3,4,5\text{-}(HO)_3C_6H_2-CO-$
Sulfonyl	$-SO_2-$	Trimethylammonio	$(CH_3)_3N^+-$(ion)
Sulfonyldioxy	$-O-SO_2-O-$	Trimethylanilino (*all isomers*)	$(CH_3)_3C_6H_2-NH-$
Tartaroyl	$-CO-CH(OH)-$		
	$CH(OH)-CO-$	Trimethylene	$-CH_2-CH_2-CH_2-$
Tartronoyl	$-CO-CH(OH)-CO-$	Trimethylenedioxy	$-O-CH_2-CH_2-$
Tauryl	$H_2N-CH_2-CH_2-SO_2-$		CH_2-O-
Telluro	Te replacing O	Triphenylmethyl	$(C_6H_5)_3C-$
Terephthaloyl	$-CO-C_6H_4-CO-$ (*p-*)	Trithio	$-S_3-$
Terphenylyl	$C_6H_5-C_6H_4-C_6H_4-$	Trithiosulfo	$HS-S_3-$

TABLE 2.20   Names and Formulas of Organic Radicals (*Continued*)

Name	Formula	Name	Formula
Trityl	$(C_6H_5)_3C$—	Vanilloyl	$3,4$-$CH_3O(HO)C_6H_3$—
Tropoyl	C_6H_5—$CH(CH_2OH)$—		CO—
	CO—	Vanillyl	$3,4$-$CH_3O(HO)C_6H_3$—
Tyrosyl	p-HO—C_6H_4—CH_2—		CH_2—
	$CH(NH_2)$—CO—	Veratroyl	$3,4$-$(CH_3O)_2C_6H_3$—
Undecanoyl	CH_3—$[CH_2]_9$—CO—		CO—
Undecyl	CH_3—$[CH_2]_9$—CH_2—	Veratryl	$3,4$-$(CH_3O)_2C_6H_2$—
Ureido	H_2N—CO—NH—		CH_2—
Ureylene	—NH—CO—NH—	Vinyl	CH_2=CH—
Valery	CH_3—$[CH_2]_3$—CO—	Vinylene	—CH=CH—
Valyl	$(CH_3)_2CH$—$CH(NH_2)$—	Xylidino (*all isomers*)	$(CH_3)_2C_6H_3$—NH—
	CO —	Xylyl (*all isomers*)	$(CH_3)_2C_6H_3$—

2.2   *PHYSICAL PROPERTIES OF ORGANIC COMPOUNDS*

Names of the compounds (Table 2.21) are arranged alphabetically. Usually substitutive nomenclature is employed; exceptions generally involve ethers, sulfides, sulfones, and sulfoxides. Each compound is given a number within its letter classification; thus compound c209 is 3-chlorophenol.

Formula Weights are based on the International Atomic Weights of 1993 and are computed to the nearest hundredth when justified. The actual significant figures are given in the atomic weights of the individual elements; see Table 3.2.

Density values are given at room temperature unless otherwise indicated by the superscript figure; thus 0.9711^{112} indicates a density of 0.9711 for the substance at 112°C. A density of 0.899^{16}_{14} indicates a density of 0.899 for the substance at 16°C relative to water at 4°C.

Refractive Index, unless otherwise specified, is given for the sodium line at 589.6 nm. The temperature at which the measurement was made is indicated by the superscript figure; otherwise it is assumed to be room temperature.

Melting Point is recorded in certain cases as 250 d and in some other cases as d 250, the distinction being made in this manner to indicate that the former is a melting point with decomposition at 250°C, while the latter decomposition occurs only at 250°C and higher temperatures. Where a value such as —$2H_2O$, 120 is given, it indicates a loss of 2 moles of water per formula weight of the compound at a temperature of 120°C.

Boiling Point is given at atmospheric pressure (760 mm Hg) unless otherwise indicated; thus 82^{15mm} indicates that the boiling point is 82°C when the pressure is 15 mm Hg. Also, subl 550 indicates that the compound sublimes at 550°C.

Flash Point is given in degrees Celsius, usually using a closed cup. When the method is known, the acronym appears in parentheses after the value: closed cup (CC), Cleveland closed cup (CCC), open cup (OC), Tag closed cup (TCC), and Tag open cup (TOC). Because values will vary with the specific procedure employed, and many times the method was not stated, the values listed for the flash point should be considered only as indicative.

Solubility is given in parts by weight (of the formula weight) per 100 parts by weight of the solvent and at room temperature. Other temperatures are indicated by the superscript. Another way in which solubility is explicitly stated is in weight (in grams) per 100 mL of the solvent. In the case of gases, the solubility is often expressed as 5 mL^{10}, which indicates that at 10°C, 5 mL of the gas is soluble in 100 g (or 100 mL, if explicitly stated) of the solvent.

TABLE 2.21 Physical Constants of Organic Compounds

Abbreviations Used in the Table

abs, absolute	*DMF*, dimethylformamide	*misc*, miscible; soluble in all proportions
acet, acetone	*E, trans* (German "entgegen")	*NaOH*, aqueous sodium hydroxide
alc, alcohol (ethanol usually)	*EtOAc*, ethyl acetate	*o*, ortho configuration
alk, alkali (aqueous NaOH or KOH)	*eth*, diethyl ether	*org*, organic
anhyd, anhydrous	*EtOH*, ethanol, 95%	*p*, para configuration
aq, aqueous, water	*expl*, explodes	*PE*, petroleum ether
as, asymmetrical	*glyc*, glycerol	*pyr*, pyridine
atm, atmosphere	*h*, hot	*s*, soluble
BuOH, 1-butanol	*HOAc*, acetic acid	*sec*, secondary
bz, benzene	*hyd*, hydrolysis	*sl*, slight, slightly
c, cold	*hygr*, hygroscopic	*soln*, solution
chl, chloroform	*i*, insoluble	*solv*, solvent
conc, concentrated	*ign*, ignites	*subl*, sublimes
d, decomposes or decomposed	*i-PrOH*, isopropyl alcohol,	*s*, symmetrical
D, dextrorotatory	2-propanol	*sym*, symmetrical
deliq, deliquescent	*L*, levorotatory	*tert*, tertiary
dil, dilute	*m*, meta configuration	*v*, very
diox, 1,4-dioxane	*Me*, methyl	*v sl s*, very slightly soluble
DL, inactive (50% D and 50% L)	*MeOH* methanol	*v s*, very soluble

Abbreviations in right column (continued):
vac, vacuo or vacuum
vols, volumes
Z, cis (German "zusamman")
>, greater than
<, less than
~, approximately
±, inactive [50% (+) and 50% (−)]
α, alpha (first) position
β, beta (second) position
γ, gamma (third) position
δ, delta (fourth) position
ω, omega position (farthest from parent functional group)

No.	Name	Formula	Formula weight	Beilstein reference	Density, g/mL	Refractive index	Melting point, °C	Boiling point, °C	Flash point, °C	Solubility in 100 parts solvent
a1	(−)-Abietic acid		302.46	9², 424			172–175			i aq; s acet, alc, bz, chl, CS_2, eth, dil alk
a2	Acenaphthene		154.21	5, 586	1.189	1.6048⁹⁵	93.4	279		i aq; 3.2 alc; 20 bz; 10 chl; 1.8 MeOH; 3.2 g in 100 mL HOAc
a3	Acenaphthylene		152.20	5, 625	0.899₄¹⁶		88–91	280		i aq; v s alc, eth
a4	Acetaldehyde	CH_3CHO	44.05	1, 594	0.788₆¹⁶	1.3316²⁰	−123	21	−38(CC)	misc aq, alc, eth
a5	Acetaldoxime	$CH_3CH{=}NOH$	59.07	1, 608	0.966	1.415²⁰	46.5(α) 12(β)	114.5	40	v s aq, alc, eth
a6	Acetamide	CH_3CONH_2	59.07	2², 177	0.999⁷⁸	1.4158¹¹⁰	81	222		70 aq; 50 alc; 16 pyr; s chl, glyc, hot bz

No.	Name	Formula	M.W.	Beil.	Density	n_D	M.P., °C	B.P., °C	Flash pt	Solubility
a7	Acetamidine HCl	CH$_3$C(=NH)NH$_2$·HCl	94.54	2, 185			164–166			v s aq; s alc; i acet, eth
a8	N-(2-Acetamido)-2-aminoethanesulfonic acid	H$_2$N(CO)CH$_2$NHCH$_2$CH$_2$SO$_2$H	182.20				>220 dec			
a9	4-Acetamidobenzaldehyde	CH$_3$CONHC$_6$H$_4$CHO	163.18	14, 38			156–158			s aq, bz; sl s alc
a10	4-Acetamidobenzenesulfonyl chloride	CH$_3$CONHC$_6$H$_4$SO$_3$Cl	233.67	14, 702			148 dec			d aq; v s alc, bz, eth, acet
a11	2-Acetamidobenzoic acid	CH$_3$CONHC$_6$H$_4$CO$_2$H	179.18	14, 337			185–187			sl s aq; v s alc, bz, eth, acet
a12	4-Acetamidobenzoic acid	CH$_3$CONHC$_6$H$_4$CO$_2$H	179.18	14, 432			262 dec			i aq; s alc; sl s eth
a13	2-Acetamidofluorene		223.28	12, 1331			192–196			i aq; s alc, glycols
a14	N-(2-Acetamido)iminodiacetic acid	H$_2$NCOCH$_2$N(CH$_2$CO$_2$H)$_2$	190.16				219 d			
a15	2-Acetamidophenol	CH$_3$CONHC$_6$H$_4$OH	151.17	13, 370			207–210			
a16	3-Acetamidophenol	CH$_3$CONHC$_6$H$_4$OH	151.17	13, 415			146–149			
a17	4-Acetamidophenol	CH$_3$CONHC$_6$H$_4$OH	151.17	13, 460			170–172			
a18	Acetanilide	CH$_3$CONHC$_6$H$_5$	135.17	12, 237	1.2934^{21} 1.2194^{15}		114	304–305	173	s alc, acet
a19	Acetic acid	CH$_3$CO$_2$H	60.05	2, 96	1.0492^{20}	1.3718^{20}	16.7	118	39 (CC)	0.56 aq^{25}; 25 acet; 29 alc; 2 bz; 27 chl; 5 eth
a20	Acetic acid-d	CH$_3$CO$_2$D	61.06	2[3], 202	1.059	1.2715^{20}		115.5	40	misc aq, alc, eth, CCl$_4$
a21	Acetic-d$_3$ acid-d	CD$_3$CO$_2$D	64.08	2[3], 203	1.137	1.3687^{20}		114.4	40	misc aq, alc, eth, CCl$_4$
a22	Acetic anhydride	(CH$_3$CO)$_2$O	102.09	2, 166	1.0804^{15}	1.3904^{20}	−73	139	54 (CC)	s chi, eth; slowly s aq forming HOAc, alc forming EtOAc
a23	Acetic anhydride-d$_6$	(CD$_3$CO)$_2$O	108.14			1.3875^{20}		65^{65mm}	54	see acetic anhydride
a24	Acetoacetanilide	CH$_3$COCH$_2$CONHC$_6$H$_5$	177.20	12, 518	1.260^{20}		85	dec	185	s alc, hot bz, acids, alkalis, chl, eth
a25	Acetoacetic acid	CH$_3$COCH$_2$COOH	102.09	3, 630			36–37	d viol 100		misc aq, alc
a26	Acetone	CH$_3$COCH$_3$	58.08	1, 635	0.7908^{20}	1.3591^{20}	−94	56	−20	misc aq, alc, chl, DMF
a27	Acetone-d$_6$	CD$_3$COCD$_3$	64.13		0.872	1.3554^{20}	−93.8	55.5	−17	see acetone

(*Continued*)

TABLE 2.21 Physical Constants of Organic Compounds (*Continued*)

No.	Name	Formula	Formula weight	Beilstein reference	Density, g/mL	Refractive index	Melting point, °C	Boiling point, °C	Flash point, °C	Solubility in 100 parts solvent
a28	Acetone oxime	$(CH_3)_2C{=}NOH$	73.10	1, 649	0.9111^{62}		60	135		v s aq, alc, eth
a29	Acetonitrile	CH_3CN	41.05	2, 183	0.7875^{15}_4	1.3460^{15}	−44	81.6	6	misc aq, acet, alc, chl, eth, EtOAc
a30	Acetonitrile-d_3	CD_3CN	44.08	2^4, 428	0.844	1.3406^{20}		80.7	5	misc aq, alc, chl
a31	Acetophenone	$C_6H_5COCH_3$	120.15	7, 271	1.0262^{20}_4	1.5372^{20}	20	202	77	0.55 aq; s alc, chl, eth, glyc
a32	Acetophenone-methyl-d_3	$C_6H_5COCD_3$	123.18	7^4, 626	1.055	1.5325^{20}		201–202	82	
a33	4-Acetylbenzenesulfonic acid, sodium salt	$CH_3COC_6H_5SO_3^-Na^+$	222.20	11^2, 186			>300			
a34	Acetylbiphenyl	$C_6H_5C_6H_4COCH_3$	196.25	7^2, 337	1.663^{16}	1.4486^{20}	116–118	325–327	>110	i aq; v s alc, acet dec viol by aq or alc; misc bz, chi, eth
a35	Acetyl bromide	CH_3COBr	122.95	2, 174			−96	76	>110	20% v/v aq dec viol aq or alc; misc bz, chl, eth, HOAc, PE
a36	2-Acetylbutyrolactone		128.13	17^3, 5837	1.1846^{20}_4	1.4585^{20}		107^{5mm}	>110	
a37	Acetyl chloride	CH_3COCl	78.50	2, 173	1.104^{20}_4	1.3896^{20}	−113	51	4 (CC)	v s aq (dec by hot aq or alkalis); s alc; i eth
a38	Acetylcholine bromide	$(CH_3)_3N(Br)CH_2CH_2-O_2CCH_3$	226.11	4^1, 428			144–146			v s aq, alc; dec by hot aq or alkalis; i eth
a39	Acetylcholine chloride	$(CH_3)_3N(Cl)CH_2CH_2-O_2CCH_3$	181.66	4, 281			150–152			1 vol in 1 vol aq in 6 vol HOAc or alc; s bz, eth; acet dissolves 25 vol[15] but 300 vols at 12 atm
a40	2-Acetylcyclopentan-one		126.16	7, 558	1.043	1.4905^{20}		75^{8mm}	72	
a41	Acetylene	$HC{\equiv}CH$	26.04	1, 228	0.90(g)		−85(subl)		−18	v s aq, alc, eth
a42	Acetylenedicarboxylic acid	$HO_2CC{\equiv}CCO_2H$	114.06	2, 801			180 d			5 aq(dec); sl s acet, alc, bz, eth
a43	Acetyl fluoride	CH_3OF	62.04	2, 172	1.002^{15}_4		<−60	20.8	71	
a44	2-Acetylfuran		110.11	17, 286	1.098	1.5065^{20}	29–30	67^{10mm}		

502

(Continued)

No.	Name	Formula	Formula weight	Beilstein reference	Density	Refractive index	mp, °C	bp, °C	Flash point, °C	Solubility
a45	N-Acetyl-(−)-glutamic acid	$HO_2CCH_2CH_2CH(NHCOCH_3)CO_2CH_3$	189.17	4^2, 908			200–201			2.7 aq^{15}; s alc; sl s acet, chl, HOAc; i bz, eth
a46	N-Acetylglycine	$CH_3CONHCH_2CO_2H$	117.10	4, 354			206–208			dec aq, alc; s bz, eth
a47	1-Acetylimidazole		110.12	2, 174			103–105			v s aq, alc, chl; i eth; dec by alkalis, eth
a48	Acetyl iodide	CH_3COI	169.95	Merck: 12, 6003	2.0674^{20}_4	1.5491^{20}		108		
a49	Acetyl-2-methylcholine chloride	$CH_3CO_2CH(CH_3)CH_2\text{-}N(Br)(CH_3)_3$	195.69				172–173			
a50	2-Acetylphenothiazine		241.31				180–185			
a51	2-Acetylphenylaceto-nitrile	$C_6H_5CH(CN)COCH_3$	159.19	10, 699				92–94		
a52	1-Acetyl-4-pipidone		141.17		1.146	1.5026^{20}		218	>110	v s alc, eth
a53	2-Acetylpyridine	$(C_5H_4N)COCH_3$	121.14	21, 279	1.080	1.5203^{20}		188–189	73	v s acids, alc, eth; s aq
a54	3-Acetylpyridine	$(C_5H_4N)COCH_3$	121.14	21, 279	1.102	1.5336^{20}		220	150	v s alc, eth
a55	4-Acetylpyridine	$(C_5H_4N)COCH_3$	121.14	21, 279	1.095	1.5350^{20}		212	>110	
a56	Acetylsalicylic acid	$HO_2C_6H_4\text{-}2\text{-}O_2CCH_3$	180.16	10, 67	1.35		135			0.33 aq^{25}; 29 acet; 20 alc; 5.9 chl; 5 eth; s bz
a57	2-Acetylthiophene	$(C_4H_3S)COCH_3$	126.18	17, 287	1.1684^{22}_4	1.5564^{20}	10–11	214		sl s aq; misc alc, eth
a58	1-Acetyl-2-thiourea	$CH_3C(O)NHC(S)NH_2$	118.16	3, 191			167			s hot aq, alc; sl s eth
a59	N-Acetyl-(±)-tryptophan		246.27	22^2, 469			206			s aq, alc; v s eth
a60	Acridine		179.22	20, 459	1.0054^{20}_4		106–110 subl 100	346		s alc, eth, CS$_2$, PE; sl s hot aq
a61	Acrylamide	$H_2C=CHCONH_2$	71.08	2, 400	1.1222^{30}_4		84.5	192.6		at 30°, g/100 mL: 215 aq, 155 MeOH, 86 EtOH, 63 acet, 12.6 EtOAc, 2.7 chl, 0.3 bz
a62	Acrylic acid	$H_2C=CHCO_2H$	72.06	2, 397	1.0511^{20}_4	1.4224^{20}	12–14	141	50	misc aq, alc, bz, eth, chl, acet
a63	Acrylonitrile	$H_2C=CHCN$	53.06	2, 400	0.8060^{20}_4	1.3911^{20}	−83.5	77.3	0	7.3 aq; misc org solv
a63a	Acryloyl chloride	$H_2C=CHCOCl$	90.51	2, 400	1.114	1.4350^{20}		72–76	15	d aq; v s chl
a64	1-Adamantanamine		151.25	Merck: 12, 389			160–190			sl s aq
a65	Adamantane		136.24	Merck: 12, 149	1.09	1.568	270 (sealed tube)	205 subl		s acet

TABLE 2.21 Physical Constants of Organic Compounds (*Continued*)

No.	Name	Formula	Formula weight	Beilstein reference	Density, g/mL	Refractive index	Melting point, °C	Boiling point, °C	Flash point, °C	Solubility in 100 parts solvent
a66	Adenine		135.13	26, 420			360 dec	subl 220		0.005 aq; sl s alc; i chl, eth
a67	(–)-Adenosine		267.24	31, 27			235	subl >200		s aq; i alc
a68	(±)-α-Alanine	CH₃CH(NH₂)CO₂H	89.09	4, 387	1.424		264–269 (depends on heating rate)			16.7 aq^{25}, 0.009 alc^{25}; i eth
a69	(–)-α-Alanine	CH₃CH(NH₂)CO₂H	89.09	4, 381	1.401		dec 297			16.7 aq^{25}; 0.2 alc^{25}; i eth
a70	β-Alanine	H₂NCH₂CH₂CO₂H	89.09	4, 401	1.437^{-5}		197 dec			v s aq; sl s alc; i eth
a71	Allantoin		158.12	25, 474			238			0.45 aq; 0.2 alc; i eth
a72	Allene	H₂C=C=CH₂	40.06	1, 248	1.787	1.4168	–136	–34		
a73	Alloxan monohydrate		160.09	24, 500			anhyd: 256 dec			s aq, alc, acet, HOAc; sl s chl, EtOAc, PE
a74	Allyl acetate	H₂C=CHCH₂OCOCH₃	100.12	2, 136	0.9774^{20}	1.4040^{20}		104	22	i aq; misc alc, eth
a75	Allyl alcohol	H₂C=CHCH₂OH	58.08	2, 436	0.8540^{40}_{2}	1.4134^{20}	–129	97	21	misc aq, alc, chl, eth
a76	Allylamine	H₂C=CHCH₂NH₂	57.10	4, 205	0.7612^{40}_{2}	1.4185^{20}	–88.2	53–55	–29	misc aq, alc, chl, eth
a77	N-Allylaniline	C₆H₅NHCH₂CH=CH₂	133.19	12, 170	0.982^{25}	1.5630^{20}		220	89	i aq; s alc, eth
a78	Allylbenzene	C₆H₅CH₂CH=CH₂	118.18	5, 484	0.892^{20}	1.5122^{20}		157	33	i aq; s alc, eth
a79	Allyl bromide	H₂C=CHCH₂Br	120.98	1, 201	1.398^{20}_{4}	1.4654^{20}	–119	70	–2	sl s aq; misc org solv
a80	Allyl butanoate	CH₃CH₂CH₂COOCH₂CH=CH₂	128.17	2, 272	0.902	1.4142^{20}		44^{15mm}		
a81	Allyl chloride	H₂C=CHCH₂Cl	76.53	1, 198	0.9384^{20}	1.4154^{20}	–134.5	44–46	–31 (CC)	sl s aq; misc alc, chl, eth, PE
a82	Allyl chloroformate	H₂C=CHCH₂OOCCl	120.54	3, 12	1.136	1.4223		110	31	
a83	Allylcyclohexylamine	(C₆H₁₁)NHCH₂CH=CH₂	139.24		0.962	1.4664^{20}		66^{12mm}	53	
a84	4-Allyl-1,2-dimethoxybenzene	H₂C=CHCH₂C₆H₃(OCH₃)₂	178.23	6, 963	1.036	1.5344^{20}	–4	255		
a85	N-Allyl-N,N-dimethylamine	H₂C=CHCH₂N(CH₃)₂	85.0			1.4010^{20}		64		
a86	Allyl ethyl ether	H₂C=CHCH₂OCH₂CH₃	86.13	1, 438	0.7655^{20}	1.3881^{20}		68	–20	i aq; misc alc, eth
a87	Allyl iodide	H₂C=CHCH₂I	167.98	1, 202	1.825^{20}	1.5540^{21}	–99	103		i aq; misc alc, eth
a88	Allyl isothiocyanate	H₂C=CHCH₂NCS	99.16	4, 214	1.013^{25}	1.5248^{25}	–80	152	46	0.2 aq; misc org solv
a89	Allyl methacrylate	H₂C=C(CH₃)COOCH₂CH=CH₂	126.16	2^3, 1290	0.938	1.4360		61^{43mm}	33	
a90	Allyl methyl sulfide	H₂C=CHCH₂SCH₃	88.17	1, 440	0.803	1.4714^{20}		91–93	18	

No.	Name	Formula	Mol wt	Beilstein	Density	n_D	mp	bp	Flash P.	Solubility
a91	1-Alloxy-2,3-epoxy-propane	H₂C—CHCH₂OCH₂-CH=CH₂ (epoxide O)	114.14		0.962	1.4332^{20}		154	57	s alc, eth
a92	3-Alloxy-1,2-propane-diol	$H_2C=CHCH_2$-$CH_2CH(OH)CH_2OH$	132.16	1, 513	1.068	1.4620^{20}		142^{28mm}	>110	i aq; s alc, misc eth
a93	Allyloxytrimethyl-silane	$H_2C=CHCH_2OSi(CH_3)_3$	130.26		0.7830	1.4075^{25}		102	0	s alc; misc eth
a94	2-Allylphenol	$H_2C=CHCH_2C_6H_4OH$	134.18	6, 572	1.0335^{15}_4	1.5450^{20}	10	220	88	3.3 aq; s alc; i bz; v sl s eth
a95	Allyl phenyl ether	$H_2C=CHCH_2OC_6H_5$	134.18	6, 144	0.9834^{15}_4	1.5200^{20}		192	62	
a96	Allyl propyl ether	$H_2C=CHCH_2OC_3H_7$	100.16	1, 438	0.7674^{20}	1.3990^{20}		90–92	−5	
a97	1-Allyl-2-thiourea	$H_2C=CHCH_2NHC(S)NH_2$	116.19	4, 211	1.219^{20}_{20}		70–72			
a98	Allyltrichlorosilane	$H_2C=CHCH_2SiCl_3$	175.52	4^3, 1909	1.2011^{20}	1.4550^{20}		117.5	31	
a99	Allyltriethoxysilane	$H_2C=CHCH_2Si(OC_2H_5)_3$	204.34	4^3, 1909	0.9030^{20}	1.4062^{20}		176^{740mm}	21	
a100	Allyl trifluoroacetate	$CF_3COOCH_2CH=CH_2$	154.09	2^4, 464	1.183	1.3350^{20}		66–67	−1	
a101	Allyltrimethylsilane	$H_2C=CHCH_2Si(CH_3)_3$	114.27		0.7193^{20}_4	1.4080^{20}		84–88	7	
a102	Allylurea	$H_2C=CHCH_2NHCONH_2$	100.12	4, 209			85			v s aq, alc; i chl, CS₂ eth, toluene
a103	Aminoacetonitrile	H_2NCH_2CN	56.07	4, 344				58^{15mm} d		s acids, alc
a104	Aminoacetonitrile hydrogen sulfate	$H_2NCH_2CN \cdot H_2SO_4$	154.14	4, 344			121	d 165		v s aq; sl s alc; i eth
a105	2'-Aminoaceto-phenone	$H_2NC_6H_4COCH_3$	135.17	14, 41				70^{3mm}	>110	v sl s aq; s alc, eth
a106	3'-Aminoaceto-phenone	$H_2NC_6H_4COCH_3$	135.17	14, 45			99	290		
a107	4'Aminoacetophenone	$H_2NC_6H_4COCH_3$	135.17	14, 46			106	293–295		s hot aq, alc, eth, HOAC
a108	1-Aminoanthra-quinone		223.23	14, 177			ca. 250	subl		i aq; v s alc, bz, chl, eth, HOAc, HCl
a109	2-Aminoanthra-quinone		223.23	14, 191			295 d	subl		i aq, eth; s alc, bz
a110	4-Aminoantipyrine		203.25	24, 273			109			s aq, alc, bz; sl s eth
a111	p-Aminoazobenzene	$C_6H_5N=NC_6H_4NH_2$	197.24				128			sl a aq; v s alc, bz, chl, eth
a112	2-Aminobenzamide	$H_2NC_6H_4CONH_2$	136.15	14, 320			110	>360		v s hot aq, alc; i bz; sl s eth
a113	4-Aminobenzene-arsonic acid	$H_2NC_6H_4AsO(OH)_2$	217.06	16, 878			232	300 sl d		s hot aq; alk CO₃, conc'd mineral acids; i acet, bz, chl, eth

(Continued)

TABLE 2.21 Physical Constants of Organic Compounds (*Continued*)

No.	Name	Formula	Formula weight	Beilstein reference	Density, g/mL	Refractive index	Melting point, °C	Boiling point, °C	Flash point, °C	Solubility in 100 parts solvent
a114	5-Aminobenzene-1,3-dicarboxylic acid	$H_2NC_6H_3(COOH)_2$	181.15	14[1], 636			>300			
a115	2-Aminobenzenesulfonic acid	$H_2NC_6H_4SO_3H$	173.19	14, 681			ca. d 325			1.5 aq[15]; v sl s alc, eth
a116	3-Aminobenzenesulfonic acid	$H_2NC_6H_4SO_3H$	173.19	14, 688	1.69		>300			2 aq[15]; sl s alc, MeOH
a117	4-Aminobenzenesulfonic acid	$H_2NC_6H_4SO_3H$	173.19	14, 695			d 288			1 aq[20], sl s hot MeOH; i alc, bz, eth
a118	2-Aminobenzoic acid	$H_2NC_6H_4COOH$	137.14	14, 310			144–146	subl		v s hot aq, alc, eth
a119	3-Aminobenzoic acid	$H_2NC_6H_4COOH$	137.14	14, 383	1.511[4]		172–174			v s hot aq, alc; s eth
a120	4-Aminobenzoic acid	$H_2NC_6H_4COOH$	137.14	14, 418	1.374		187			0.59 aq; 12 alc; 2 eth; s EtOAc, HOAc
a121	2-Aminobenzonitrile	$H_2NC_6H_4CN$	118.14	14, 322			49	268	>110	s alc, eth
a122	3-Aminobenzonitrile	$H_2NC_6H_4CN$	118.14	14, 391			53	288–290	>110	s hot aq; v s alc, eth
a123	4-Aminobenzonitrile	$H_2BC_6H_4CN$	118.14	14, 425			85	dec		v s hot aq, alc, eth
a124	2-Aminobenzophenone	$H_2NC_6H_4COC_6H_5$	197.24	14, 76			108	223–226		sl s alc; s alc, eth
a125	2-Aminobenzothiazole		150.20	27, 182			132	dec		v s conc'd acids, alc, chl, eth
a126	2-Aminobenzotrifluoride	$H_2NC_6H_4CF_3$	161.13	12[12], 453	1.290[25]	1.4785[25]	34	175	55	i aq; s alc, bz, chi
a127	3-Aminobenzotrifluoride	$H_2NC_6H_4CF_3$	161.13	12, 870	1.290	1.4800[20]	6	187	85	sl s aq; s alc
a128	4-Aminobenzotrifluoride	$H_2NC_6H_4CF_3$	161.13	12[3], 2151	1.283[27]	1.4815[25]	38	83[12mm]	86	s alc, bz, chl, eth, HOAc; v s acet
a129	N-(4-Aminobenzoyl)-glycine	$H_2NC_6H_4CONHCH_2COOH$	194.19	14[2], 258			198–199			i aq; s alc, bz, chi
a130	2-Aminobiphenyl	$H_2NC_6H_4C_6H_5$	169.23	12, 1317			50–53	299	>110	sl s aq; s alc
a131	4-Aminobiphenyl	$H_2NC_6H_4C_6H_5$	169.23	12, 1318			52–54	191[15mm]	>110	s hot aq, alc, eth
a132	2-Amino-5-bromo-benzoic acid	$Br(NH_2)C_6H_3COOH$	216.03	14, 370			218–219			21 aq[25], 0.18 hot alc; i eth
a133	(±)-2-Aminobutanoic acid	$CH_3CH_2CH(NH_2)COOH$	103.12	4, 408			304 d	subl >300		125 aq; i alc, eth
a133a	3-Aminobutanoic acid	$H_3CCH_2CH(NH_2)COOH$	103.12	4, 412			193–194			v s aq; i org solv
a134	4-Aminobutanoic acid	$H_2NCH_2CH_2CH_2COOH$	103.12	4, 413			195 d			misc aq; s alc
a135	2-Amino-1-butanol	$CH_3CH_2CH(NH_2)CH_2OH$	89.14	4, 291	0.9442[20]	1.4521[20]	−2	176–178	74 (OC)	

No.	Name	Formula	Mol. wt.	Ref.	Density	n_D	m.p.	b.p.	Solubility	Solubility
a136	3-(4-Aminobutyl)-piperidine	$(HNC_5H_9)(CH_2)_4NH_2$	156.27	22^3, 3788	0.910		39–42	148^{10mm}	>110	
a137	4-Amino-6-chloro-1,3-benzenedisulfonamide	$H_2NC_6H_2(Cl)(SO_2NH_2)_2$	285.73	14^4, 2810			257–261			
a138	2-Amino-4-chlorobenzoic acid	$H_2N(Cl)C_6H_3COOH$	171.58	14, 365			231–233			
a139	5-Amino-2-chlorobenzoic acid	$H_2N (Cl)C_6H_3COOH$	171.58	14, 412			188 d			
a140	2-Amino-4′-chlorobenzophenone	$H_2NC_6H_4COC_6H_4Cl$	231.68	14^1, 389			104			
a141	2-Amino-5-chlorobenzophenone	$H_2N(Cl)C_6H_3COC_6H_5$	231.68	14, 79			98–100			
a142	2-Amino-5-chlorobenzotrifluoride	$H_2N(Cl)C_6H_3CF_3$	195.57	12^3, 1921	1.386	1.5069^{20}	36–38	67^{3mm}	none	
a143	5-Amino-2-chlorobenzotrifluoride	$H_2N(Cl)C_6H_3CF_3$	195.57						>110	
a144	2-(3-Amino-4-chlorobenzoyl)benzoic acid	$H_2N(Cl)C_6H_3COC_6H_4COOH$	275.69	14, 661			171–173			i aq; s alc, acet, bz, chl, HOAc
a145	4-Amino-4′-chlorobiphenyl	$H_2NC_6H_4-C_6H_4Cl$	203.67				128–134			
a146	4-Amino-5-chloro-2-methoxybenzoic acid	$H_2NC_6H_2(Cl)(OCH_3)COOH$	201.61				206 d			
a147	2-Amino-4-chlorophenol	$H_2N(Cl)C_6H_3OH$	143.57	13, 383			139–143			
a148	2-Amino-5-chloropyridine	$H_2N(Cl)(C_5H_3N)$	129.56	22^2, 332			135–138	128^{11mm}		
a149	3-Aminocrotononitrile	$CH_3C(NH_2)=CHCN$	82.11	3, 660						
a150	1-[(2-Aminoethyl)-amino]-2-propanol	$CH_3CH(OH)CH_2NHCH_2CH_2NH_2$	118.18	25^1, 698	0.9837^{25}_4	1.4788^{25}		112^{10mm}		
a151	5-Amino-2,3-dihydro-1,4-phthalazinedione		177.16				319–320			
a152	2-Amino-4,6-dihydroxypyrimidine		127.10	24, 468			>300			
a153	4-Amino-2,6-dihydroxypyrimidine		127.10	24, 469			>300			
a154	2-Amino-3,3-dimethylbutane	$(CH_3)_3CCH(NH_2)CH_3$	101.19	4, 193	0.755	1.4130^{20}	−20	102–103	1	

(Continued)

TABLE 2.21 Physical Constants of Organic Compounds (*Continued*)

No.	Name	Formula	Formula weight	Beilstein reference	Density, g/mL	Refractive index	Melting point, °C	Boiling point, °C	Flash point, °C	Solubility in 100 parts solvent
a155	2-Amino-4,6-dimethyl-pyridine	$(CH_3)_2(NH_2)(C_5H_2N)$	122.17	22, 435			63–64	235		156 aq; 18.9 alc
a156	4-Amino-2,6-dimethyl-pyrimidine		123.16	24^2, 45			184–186			
a157	6-Amino-1,3-dimethyl-uracil		155.16	24, 471			295 d			
a158	5-Amino-2,6-dioxo-1,2,3,6-tetrahydro-4-pyrimidinecarbox-ylic acid		171.11	25, 264			>300			
a159	α-Aminodiphenyl-methane	$(C_6H_5)_2CHNH_2$	183.25	12, 1323	1.0635^{22}	1.5950^{20}	34	304	>110	sl s aq; s acids
a160	2-Aminoethanesulfonic acid	$H_2NCH_2CH_2SO_3H$	125.15	4, 528			d ca. 300			5.45 aq[12]; 0.004 alc[17]
a161	2-Aminoethanethiol	$HSCH_2CH_2NH_2$	77.14	4, 286			97–99	110 d		v s aq; s alc
a162	1-Aminoethanol	$CH_2CH(OH)NH_2$	61.08	4, 274			97			s aq; sl s eth
a163	2-Aminoethanol	$H_2NCH_2CH_2OH$	61.08	4^3, 642	1.0117^{25}_{4}	1.4539^{20}	10.3	171	93	misc aq; org solv
a164	2-(2-Aminoethoxy)-ethanol	$H_2NCH_2CH_2OCH_2CH_2OH$	105.14		1.048			218–224		
a165	2-(2-Aminoethyl-amino)ethanol	$H_2NCH_2CH_2NHCH_2CH_2OH$	104.15	4, 286	1.030	1.4861^{20}		240^{753mm}	>110	v s aq, alc; sl s eth
a166	1-[(2-Aminoethyl)-amino]-2-propanol	$CH_3CH(OH)CH_2NHCH_2CH_2NH_2$	118.18	Merck: 12, 458	0.9837^{25}_{4}	1.4738^{25}		112^{10mm}		s acids
a167	3-(2-Aminoethyl-amino)propyltri-methoxysilane	$H_2NCH_2CH_2NHCH_2CH_2Si(OCH_3)_3$	222.1		1.01^{25}_{4}	1.4418^{25}		140^{15mm}	150	
a168	3-Amino-9-ethylcarba-zole		210.28	22^1, 642			98–100			
a169	2-Aminoethyl hydro-gen sulfate	$H_2NCH_2CH_2OSO_3H$	141.15	4, 276			277 d			i aq, bz, chl, eth; s alc, acet, HCl
a170	3-(2-Aminoethyl)-indole		160.22	22^1, 636			118	$137^{0.15mm}$		
a171	S-2-Aminoethyliso-thiouronium bro-mide HBr		281.01	Merck: 12, 176			194–195			
a172	N-(2-Aminoethyl)-morpholine		130.19	27^3, 370	0.992	1.4755^{20}	25.6	205	175	s aq, alc, bz, acet, acids
a173	4-(2-Aminoethyl)-phenol	$HOC_6H_4CH_2CH_2NH_2$	137.18	13, 625			164–165	166^{2mm}		1 aq[15]; 10 boiling alc; s HCl

No.	Name	Formula	Formula weight	Beilstein reference	Density	n_D	Melting point	Boiling point	Flash point	Solubility
a174	N-(2-Aminoethyl)-piperazine		129.21		0.985^{20}_{20}	1.4983^{20}	−26	218–222	93 (OC)	
a175	N-(2-Aminoethyl)-1,3-propanediamine	$H_2NCH_2CH_2CH_2NHCH_2CH_2NH_2$	117.20		0.928	1.4815^{20}		152^{10mm}	96	
a176	2-Amino-2-ethyl-1,3-propanediol	$HOCH_2C(NH_2)(C_2H_5)CH_2OH$	119.16	4,3,850	1.099^{20}	1.490^{20}	38		>110	misc aq; s alc
a177	2-(2-Aminoethyl)-pyridine	$H_2NCH_2CH_2(C_5H_4N)$	122.17	22, 434	1.021	1.5360^{20}		93^{12mm}	100	
a178	4-(2-Aminoethyl)-pyridine	$H_2NCH_2CH_2(C_5H_4N)$	122.17		1.012	1.5403^{20}		104^{9mm}		
a179	2-Amino-5-fluorobenzotrifluoride	$H_2N(F)C_6H_3CF_3$	179.12	12^3, 1991	1.3781	1.4608^{20}		81^{20mm}	70	
a180	Aminoguanidine hydrogen carbonate	$H_2NNHC(=NH)NH_2 \cdot H_2CO_3$	136.11	3, 117			172 d			i aq; d hot aq
a181	N-Aminohexamethyleneimine	$(C_6H_{12}N)NH_2$	114.19		0.984	1.4850^{20}		165	56	
a182	(±)-2-Aminohexanoic acid	$CH_3(CH_2)_3CH(NH_2)COOH$	131.17	4, 433	1.172		301			1.15 aq^{25}, 0.42 alc^{25}; s acids
a183	6-Aminohexanoic acid	$H_2N(CH_2)_5COOH$	131.17	4, 434			204–206			v s aq; i alc, s acids
a184	6-Amino-1-hexanol	$H_2N(CH_2)_5CH_2OH$	117.19	4^2, 748			56–58	135^{30mm}		
a185	(−)-2-Amino-3-hydroxybutanoic acid	$CH_3CH(OH)CH(NH_2)COOH$	119.12	4, 514			d 255			v s aq; i alc, chl, eth
a186	(±)-4-Amino-3-hydroxybutanoic acid	$H_2NCH_2CH(OH)CH_2COOH$	119.12	4^2, 938			218 d			s aq; sl s alc, chl, eth, EtOAc
a187	4-Amino-6-hydroxy-2-mercaptopyrimidine hydrate		161.18	24, 476			>300			
a188	2-Amino-4-hydroxy-6-methylpyrimidine		125.13	24, 343			>300			
a189	4-Amino-3-hydroxy-1-naphthalenesulfonic acid		239.25	14, 846			295 d			i aq, alc, bz, eth
a190	4-Amino-5-hydroxy-1-naphthalenesulfonic acid		239.25	14, 835						sl s aq; i alc, eth
a191	5-Amino-6-hydroxy-2-naphthalenesulfonic acid		239.25				>300			
a192	6-Amino-7-hydroxy-2-naphthalenesulfonic acid		239.25	14, 849						sl s hot aq; i eth

(Continued)

TABLE 2.21 Physical Constants of Organic Compounds (*Continued*)

No.	Name	Formula	Formula weight	Beilstein reference	Density, g/mL	Refractive index	Melting point, °C	Boiling point, °C	Flash point, °C	Solubility in 100 parts solvent
a193	2-Amino-3-hydroxy-pyridine	$H_2N(HO)(C_5H_3N)$	110.12	12², 408			172–174			
a194	4-Amino-2-hydroxy-pyrimidine		111.10	24, 314			>300			0.77 aq; sl s alc
a195	1-Aminoindane		133.19	12, 1191	1.038_4^{15}	1.5613^{20}	1.5	97^{8mm}	94	sl s aq
a196	5-Aminoindane		133.19	12¹, 511			36	249^{745mm}	>110	sl s aq
a197	5-Aminoindazole		133.15	25², 308			175–178			
a198	6-Aminoindazole		133.15	25, 317			206 d			
a199	2-Amino-5-iodobenzoic acid	$H_2N(I)C_6H_3COOH$	263.03	14, 373			221 d			sl s aq, PE; s alc
a200	(±)-2-Amino-4-mercaptobutanoic acid	$HSCH_2CH_2CH(NH_2)COOH$	135.19	4³, 1647			232–233			
a201	Aminomethanesulfonic acid	$H_2NCH_2SO_3H$	111.12	1, 583			185 d			v s aq
a202	3-Amino-4-methoxy-benzoic acid	$CH_3O(NH_2)C_6H_3COOH$	167.16	14¹, 657			210			
a203	2-Amino-1-methoxy-propane	$CH_3OCH_2CH(CH_3)NH_2$	84.14	4⁴, 1615	0.845	1.4065^{20}		93	8	
a204	5-Amino-2-methoxy-pyridine	$CH_3O(NH_2)(C_5H_3N)$	124.14	22², 408		1.5745^{20}	31	90^{1mm}	>110	
a205	4'-Amino-N-methyl-acetanilide	$CH_3ON(CH_3)C_6H_4NH_2$	164.21	13¹, 30			90–92			
a206	4-Amino-3-methyl-benzenesulfonic acid	$H_2NC_6H_3(CH_3)SO_3H$	187.22	14, 726			>300			
a207	2-Amino-5-methyl-benzoic acid	$H_2N(CH_3)C_6H_3COOH$	151.17	14, 481			175 d			sl s aq; s alc, eth
a208	3-Amino-4-methyl-benzoic acid	$H_2N(CH_3)C_6H_3COOH$	151.17	14, 487			167–169			s aq
a209	2-Amino-3-methyl-1-butanol	$(CH_3)_2CHCH(NH_2)CH_2OH$	103.17	4³, 805	0.906	1.4543^{20}	35–36	80^{8mm}	90	
a210	2-(Aminomethyl)-1-ethylpyrrolidine		128.22		0.887	1.4665^{20}		60^{16mm}	60	
a211	2-Amino-3-methyl-1-pentanol	$CH_2CH_2CH(CH_3)CH(NH_2)CH_2OH$	117.19			1.4589^{20}	30	97^{14mm}	100	
a212	2-Amino-4-methyl-1-pentanol	$CH_3CH(CH_3)CH_2CH(NH_2)CH_2OH$	117.19	4, 298	0.917	1.4496^{20}		200	90	
a213	4-Amino-3-methyl-phenol	$H_2N(CH_3)C_6H_3OH$	123.16				179			

(Continued)

No.	Name	Formula	Mol. wt.	Beilstein ref.	Density	n_D	m.p., °C	b.p., °C	Flash p., °C	Solubility
a214	4-(Aminomethyl)-piperidine		114.19			1.4900^{20}	25	200	78	250 aq^{20}; s alc
a215	2-Amino-2-methyl-1,3-propanediol	HOCH$_2$C(CH$_3$)(NH$_2$)CH$_2$OH	105.14	4^3, 783			108–110	151^{10mm}	67	
a216	2-Amino-2-methyl-1-propanol	(CH$_3$)$_2$C(NH$_2$)CH$_2$OH	89.14	4, 414	0.934^{20}_{20}	1.4480^{20}	25	165		misc aq; s alc, org solv
a217	2-Amino-2-methyl-propionic acid	(CH$_3$)$_2$C(NH$_2$)COOH	103.12				335 (sealed tube)	280 subl		v s aq
a218	2-(Aminomethyl)-pyridine	H$_2$NCH$_2$(C$_5$H$_4$N)	108.14	22^3, 4181	1.049	1.5440^{20}		85^{12mm}	90	
a219	3-(Aminomethyl)-pyridine	H$_2$NCH$_2$(C$_5$H$_4$N)	108.14	22^2, 342	1.062	1.5510^{20}	–21	74^{1mm}	100	
a220	4-(Aminomethyl)-pyridine	H$_2$NCH$_2$(C$_5$H$_4$N)	108.14	22^2, 342	1.065	1.5515^{20}	–8	230	108	
a221	2-Amino-3-methyl-pyridine	H$_2$N(CH$_3$)(C$_5$H$_3$N)	108.14	22^2, 342	1.073	1.5823^{20}	32–34	222	111	
a222	2-Amino-4-methyl-pyridine	H$_2$N(CH$_3$)(C$_4$H$_3$N)	108.14	22^2, 342			98–100	230		v s aq, alc, DMF
a223	2-Amino-6-methyl-pyridine	H$_2$N(CH$_3$)(C$_4$H$_3$N)	108.14	22^1, 633			42–45	209	103	v s aq
a224	2-Amino-4-methyl-pyrimidine		109.13	24, 84			160	subl		s hot aq; s alc
a225	2-Amino-4-methyl-thiazole		114.17	27, 159			44–46	232	>110	v s aq, alc, eth
a226	2-Aminomethyl-3,5,5-trimethylcyclo-hexanol		171.29		0.969	1.4904^{20}	43–48	265	>110	
a227	N-Aminomorpholine		102.14	27, 8	1.059	1.4772^{20}		168	58	
a228	1-Aminonaphthalene	(C$_{10}$H$_7$)NH$_2$	143.18	12, 1212	1.13		48–50	301	157	0.17 aq; v s alc, eth
a229	2-Aminonaphthalene	(C$_{10}$H$_7$)NH$_2$	143.18	12, 1212			111–113 dec	306		s hot aq, alc, eth
a230	2-Amino-1-naphtha-lenesulfonic acid	H$_2$N(C$_{10}$H$_6$)SO$_3$H	223.25	14, 736						0.031 aq; sl s hot aq; s dil alkali
a231	5-Amino-2-naphtha-lenesulfonic acid	H$_2$N(C$_{10}$H$_6$)SO$_3$H	223.25	14, 758			180			sl s aq; s hot aq
a232	8-Amino-2-naphthol	H$_2$NC$_{10}$H$_6$OH	159.19	13, 685			207			
a233	2-Amino-4-nitro-benzoic acid	H$_2$N(NO$_2$)C$_6$H$_3$COOH	182.14	14, 374			270 d			i aq; v s alc, eth
a234	2-Amino-5-nitro-benzonitrile	H$_2$N(NO$_2$)C$_6$H$_3$CN	163.14	14^2, 234			200–207			
a235	5-Amino-5-nitro-benzophenone	C$_6$H$_5$COC$_6$H$_4$(NH$_2$)NO$_2$	242.23	14, 79			166–168			

TABLE 2.21 Physical Constants of Organic Compounds (*Continued*)

No.	Name	Formula	Formula weight	Beilstein reference	Density, g/mL	Refractive index	Melting point, °C	Boiling point, °C	Flash point, °C	Solubility in 100 parts solvent
a236	2-Amino-6-nitrobenzothiazole		195.20	27[2], 232			247–249			
a237	4-Amino-3-nitrobenzotrifluoride	$H_2N(NO_2)C_6H_3CF_3$	206.12				105–106			
a238	2-Amino-4-nitrophenol	$O_2NO(NH_2)C_6H_3OH$	154.13	13[2], 192			143–145			
a239	2-Amino-5-nitrophenol	$O_2N(NH_2)C_6H_3OH$	154.13	13, 390			202 d			
a240	4-Amino-2-nitrophenol	$O_2N(NH_2)C_6H_3OH$	154.13	13, 520			125–127			
a241	D-(−)-*threo*-2-Amino-1-(4-nitrophenyl)-1,3-propanediol	$HOCH_2C(NH_2)C(OH)\text{-}C_6H_4NO_2$	212.21				163–165			
a242	2-Amino-5-(4-nitrophenylsulfonyl-thiazole)		285.30				222–226			
a243	2-Amino-5-nitropyridine	$H_2N(C_5H_3N)NO_2$	139.11	22[1], 631			186–188			sl s aq, bz, eth
a244	2-Amino-5-nitrothiazole		145.14	Merck: 12, 477			d 202			s sl s aq; 0.7 alc; 0.4 ether; s dil acids
a245	*exo*-2-Aminonorbornane	$H(CH_2)_3\,CH(NH_2)CH_3$	111.19	12[3], 160	0.938	1.4807^{20}		$49^{10\text{mm}}$	35	s aq, alc, eth, PE
a246	2-Aminopentane	$C_2H_5CH(NH_2)C_2H_5$	87.17	4, 177	0.739^{20}	1.4047^{20}		91–92		misc aq, alc, eth
a247	3-Aminopentane	$H(CH_2)_3CH(NH_2)COOH$	87.17	4, 179	0.749^{20}_4	1.4055^{20}		91	1	5.5 aq[18], v sl s alc, chl, eth, PE
a248	DL-2-Aminopentanoic acid		117.15	4, 416			303	320 subl		
a249	5-Aminopentanoic acid	$H_2N(CH_2)_4COOH$	117.15	4, 418			158–161			v s aq; sl s alc; i eth
a250	5-Amino-1-pentanol	$H_2N(CH_2)_5OH$	103.17	4[1], 441	0.949	1.4615^{20}	35–37	$122^{16\text{mm}}$	65	
a251	2-Aminophenethyl alcohol	$H_2NC_6H_4CH_2CH_2OH$	137.18	13[3], 1679	1.045	1.5849^{20}		$148^{4\text{mm}}$	>112	
a252	2-Aminophenol	$H_2NC_6H_4OH$	109.13	13, 354			170–174	$164^{11\text{mm}}$		2 aq; 4.3 alc; v s eth
a253	3-Aminophenol	$H_2NC_6H_4OH$	109.13	13, 401			122–123			2.5 aq; v s hot aq, alc, eth
a254	4-Aminophenol	$H_2NC_6H_4OH$	109.13	13, 427			190	$150^{3\text{mm}}$		0.65 aq; 4.5 alc; 9.3 EtMeKetone[58], s eth
a255	4'-Aminophenylacetonitrile	$H_2NC_6H_4CH_2CN$	132.17				45–48	312		sl s hot aq; s alc
a256	1-(3-Aminophenyl)ethanol	$H_2NC_6H_4CH(CH_3)OH$	137.18	13[3], 1654			68–71		>110	
a257	2-Amino-1-phenylethanol	$H_2NCH_2CH(C_6H_5)OH$	137.18	13[2], 361			56–58	$160^{17\text{mm}}$		v s aq; s alc

(Continued)

No.	Name	Formula	M.W.	Beilstein ref.	Density	n_D	m.p., °C	b.p., °C	Flash pt	Solubility
a258	1S,2S-(+)-2-Amino-1-phenyl-1,3-propane-diol	$C_6H_5CH(OH)CH(NH_2)CH_2OH$	167.21	13^4, 2968			109–113			
a259	L-2-Amino-3-phenyl-1-propanol	$C_6H_5CH_2(NH_2)CH_2OH$	151.21	13^3, 1757			92–94			
a260	3-Amino-1-phenyl-2-pyrazolin-5-one		175.19				210 d			
a261	N-Aminopiperidine		100.17	20, 89	0.928	1.4750^{20}		146^{730mm}	36	v s aq, alc; i eth
a262	3-Amino-1,2-propane-diol	$H_2NCH_2CH(OH)CH_2OH$	91.11	4, 301	1.175	1.4920^{20}		265^{739mm}	>110	v s aq, alc, eth
a263	DL-1-Amino-2-propanol	$CH_3CH(OH)CH_2NH_2$	75.11	4, 289	0.973	1.4483^{20}	–2	160	76	
a264	DL-2-Amino-1-propanol	$CH_3CH(NH_2)CH_2OH$	75.11	4^1, 432	0.943	1.4495^{20}		173–176	83	
a265	S-(+)-2-Amino-1-propanol	$CH_3CH(NH_2)CH_2OH$	75.11	4^3, 735	0.965	1.4498^{20}		176	62	v s aq, alc, eth
a266	3-Amino-1-propanol	$H_2NCH_2CH_2CH_2OH$	75.11	4, 288	0.982	1.4610^{20}	10–12	188	79 (TOC)	s aq, alc
a267	2-Amino-1-propene-1,1,3-tricarbonitrile	$NCC(CN){=}C(NH_2)CH_2CN$	132.13	Merck: 11, 495			171–173			s aq
a268	3-Aminopropyl-(diethoxy)methylsilane	$H_2N(CH_2)_3Si(CH_3)(OCH_2CH_3)_2$	191.4		0.916_4^{20}	1.427^{20}		88^{8mm}		
a269	1-(3-Aminopropyl)-imidazole		125.18	23^3, 577	1.049	1.5190^{20}			>110	
a270	N-(3-Aminopropyl)-iminodiethanol	$H_2N(CH_2)_3N(CH_2CH_2OH)_2$	162.23		0.1071	1.4980^{20}		170^{2mm}	137	
a271	N-(3-Aminopropyl)-morpholine		144.22		0.9872_{20}^{20}	1.4761^{20}	–15	224	98	misc aq, alc, bz
a272	N-(3-Aminopropyl)-2-pyrolidinone		142.20		1.014	1.500^{20}		123^{1mm}	>110	
a273	3-Aminopropyltriethoxysilane	$H_2N(CH_2)_3Si(OC_2H_5)_3$	221.37		0.9506_4^{20}	1.4225^{20}		217	104	
a274	3-Aminopropyltrimethoxysilane	$H_2N(CH_2)_3Si(OCH_3)_3$	179.29		1.01_4^{25}	1.420^{25}		80^{8mm}	83	
a275	2-Aminopyridine	$(C_5H_4N)NH_2$	94.12	22, 428			58.1	210.6	92	s aq, alc, bz, eth
a276	3-Aminopyridine	$(C_5H_4N)NH_2$	94.12	22, 431			64	250–252		s aq, alc, bz, eth
a277	4-Aminopyridine	$(C_5H_4N)NH_2$	94.12	22, 433			160–162	273		s aq, alc; sl s bz, eth
a278	2-Aminopyrimidine		95.11	24, 80			125–127	subl		v s aq
a279	4-Aminoquinaldine		158.20	22, 453			167–169	333		sl s aq; v s alc, eth, acet; s hot bz

TABLE 2.21 Physical Constants of Organic Compounds (*Continued*)

No.	Name	Formula	Formula weight	Beilstein reference	Density, g/mL	Refractive index	Melting point, °C	Boiling point, °C	Flash point, °C	Solubility in 100 parts solvent
a280	4-Aminosalicylic acid	$H_2NC_6H_3(OH)CO_2H$	153.14	14, 579			150–151			0.2 aq; 4.8 alc; s dil acids, alk; sl s eth
a281	5-Aminosalicylic acid	$H_2NC_6H_3(OH)CO_2H$	153.14	14, 579			280 d			
a282	2-Aminoterephthalic acid	$H_2NC_6H_3(CO_2H)_2$	181.15	14, 558			324 d			
a283	5-Amino-1,2,3,4-tetrazole hydrate		103.08	26, 403			204 d			
a284	2-Amino-1,3,4-thiadiazole		101.13	27, 624			190–192			sl s aq, alc, eth; s hot aq, HCl
a285	2-Aminothiazole		100.14	27, 155			93			s HCl
a286	2-Amino-2-thiazoline	$H_2NC_6H_4SH$	100.14	27, 136			79–82			
a287	2-Aminothiophenol		125.19	13, 397	1.170	1.6420^{20}	19–21	$72^{0.1mm}$	79	i aq^{12}; v s hot aq
a288	2-Aminotoluene-5-sulfonic acid	$H_2NC_6H_3(CH_3)SO_3H$	187.22	14, 726			>300			
a289	3-Amino-1,2,4-triazole		84.08	26, 137			150–153			s aq, alc, chl
a290	5-Amino-1,3,3-tri-methylcyclohexane-methylamine	$H_2N(C_6H_7)(CH_3)_3CH_2NH_2$	170.30		0.922	1.4880^{20}	10	247	>110	
a291	5-Amino-2,2,4-tri-methylcyclopentane-methylamine		156.27		0.901	1.4733^{20}		221	97	
a292	11-Aminoundecanoic acid	$H_2N(CH_2)_{10}CO_2H$	201.31				190–192			
a293	Aniline	$C_6H_5NH_2$	93.12	12, 59	1.027^{20}_{20}	1.5863^{20}	-6	184–186	70	3.5 aq^{25}; s acids; misc most org solv
a294	Aniline hydrochloride	$C_6H_5NH_2HCl$	129.59	Merck: 12, 696	1.222		198	245	193 (CC)	100 aq; v s alc
a295	2-Anilinoethanol	$C_6H_5NHCH_2CH_2OH$	137.18	12, 182	1.085	1.5793^{20}		152^{10mm}	153	sl s aq; v s alc, chl, eth
a296	3-Anilinopropionitrile	$C_6H_5NHCH_2CH_2CN$	146.19				52–53		>110	
a297	Anthracene		178.23	5, 657	1.25^{27}		215–218	339–342	121 (CC)	1.5 alc; 1.6 bz; 1.2 chl; 3.1 CS_2; 0.5 eth; i aq
a298	9,10-Anthraquinone		208.20	7, 781	1.43^{20}_{4}		286	377	185 (CC)	0.44 alc^{25}, 0.6 chl^{20}, 0.2 bz^{20}; 0.11 eth^{25}
a299	Antipyrine		188.23	24, 27	1.088^{113}_{4}		111–114	319		100 aq; 77 alc; 100 chl; 2.3 eth
a300	L-(+)-Arabinose		150.13	31, 32			157–160			100 aq; 0.4 alc

No.	Name	Formula	Formula wt	Beilstein ref	Density	n_D	mp, °C	bp, °C	Flash pt	Solubility
a301	L-(+)-Arginine	$H_2NC(=NH)NH(CH_2)_3$-$CH(NH_2)CO_2H$	174.20	4, 420			d 240			15 aq^{21}; sl s alc
a302	L-(+)-Ascorbic acid		176.12	18^3, 3038	1.65^{25}		190–192			33 aq; 3.3 alc; 1 glyc; i bz, chl, eth, PE
a303	L-(+)-Asparagine	$H_2NCOCH_2CH(NH_2)CO_2H$	132.12	4, 476			235			3.5 aq^{28}; s alkalis, ac-ids; i alc, bz, eth
a304	L-(+)-Aspartic acid	$HO_2CCH_2CH(NH_2)CO_2H$	133.10	4, 472	$1.661^{12.5}$		270–272			0.45 aq; s alkalis, acids; i alc, eth
a305	Atropine		289.38	21, 27			114–116	subl 110 high vac		0.22 aq; 50 alc; 4 eth; 100 chl; 3.9 glyc; s bz, dil acids
a306	Aurintricarboxylic acid, triammonium salt		473.44	10^2, 775			225 d			v s aq
a307	2-Azacyclooctanone		127.19	21, 242			35–38	148^{10mm}	>110	
a308	2-Azacyclotridecanone		197.32				150–153	high vac		
a309	Azidotrimethylsilane	$(CH_3)_3SiN_3$	115.21		0.868	1.4140^{20}	–95	95–96	23	
a310	Azidotriphenylsilane	$(C_6H_5)_3SiN_3$	301.4				83–84	$100^{0.01mm}$		
a311	1-Aziridineethanol	$(C_2H_4N)CH_2CH_2OH$	87.12		1.088	1.4560^{20}		168	67	
a312	Azobenzene	$C_6H_5N=NC_6H_5$	182.23	16, 8	1.203_4^{20}		67–68	293		4.2 alc^{20}; s eth, HOAc
a313	2,2'-Azobis(2-methyl-propionitrile)	$(CH_3)_2C(CN)N=N$-$C(CN)(CH_3)_2$	164.21	4, 563			107 d			2 EtOH^{20}; 5 MeOH^{20}; can explode in acetone
a314	Azodicarbonamide	$H_2NCON=NCONH_2$	116.08	3, 123			225 d			i aq; alc; s hot aq
a315	4,4'-Azoxydianisole	$H_3OC_6H_4N=N(\rightarrow O)C_6H_4$-$OCH_3$	258.28	16, 637			120			
a316	Azulene		128.17	5^2, 432			99–100	242		i aq; s org solvents
b1	Barbituric acid		128.09	24, 467	1.22		252 d			s hot aq, dil acids
b2	Basic fuchsin		337.86	13, 765			250 d			0.3 aq; s alc, acids
b3	Benzaldehyde	C_6H_5CHO	106.12	7, 174	1.050_4^{15}	1.5456^{20}	–26	179	63	0.3 aq; misc alc, eth
b4	Benzamide	$C_6H_5CONH_2$	121.13	9, 195	1.341^4		129–130	288–290		1.3 aq; 17 alc; 30 pyr
b5	Benzanihde	$C_6H_5CONHC_6H_5$	197.24	12, 262	1.315		163	117^{10mm}		i aq; 1.7 alc; sl s eth
b6	1,2-Benzanthracene		228.29	5, 718			155–157	437.6		sl s hot aq; s org solv
b7	2,3-Benzanthracene		228.29	5^2, 628	1.35		357 (Cu block)	subl		sl s most org solv
b8	Benzene	C_6H_6	78.11	5, 179	0.8787_4^{15}	1.5011^{20}	5.5	80.0	–11 (CC)	0.17 aq; misc most org solv
b9	Benzene-1,3,5-d_3	C_6D_3	81.14	5^3, 518	0.908	1.4990^{20}	5.5	80	–11 (CC)	similar to ordinary benzene
b10	Benzene-$^{13}C_6$	$^{13}C_6H_6$	84.07		0.949	1.5010^{20}	5.5	80	–11 (CC)	similar to ordinary benzene

(Continued)

TABLE 2.21 Physical Constants of Organic Compounds (*Continued*)

No.	Name	Formula	Formula weight	Beilstein reference	Density, g/mL	Refractive index	Melting point, °C	Boiling point, °C	Flash point, °C	Solubility in 100 parts solvent
b11	Benzene-d_6	C_6D_6	84.16	5[3], 519	0.950	1.4986^{20}	6.8	79.1	−11 (CC)	similar to ordinary benzene
b12	Benzenearsonic acid	$C_6H_5AsO(OH)_2$	202.03	16, 868	1.760^{25}		162			2.5 aq; 2 alc; i chl
b13	Benzeneboronic acid	$C_6H_5B(OH)_2$	121.94	16, 920			216			2.5 aq; 1.8 bz; 30 eth; 178 MeOH
b14	1,4-Benzenedicarb-aldehyde	$C_6H_4(CHO)_2$	134.13	7, 675			113	248		i aq; 6 bz; 17 acet; 2 eth; 14 diox; 46 MeOH
b15	1,2-Benzenedicarbonyl dichloride	$C_6H_4(COCl)_2$	203.02	9, 834	1.409^{20}		15–16	280–282		d aq, alc; s eth
b16	1,4-Benzenedicarbonyl dichloride	$C_6H_4(COCl)_2$	203.02	9, 844			81	266	180	37 bz; 9 CCl_4
b17	1,3-Benzenedicarbox-ylic acid	$C_6H_4(COOH)_2$	166.13	9, 832			345–348	subl		0.012 aq; v s alc, HOAc; i bz, PE
b18	1,4-Benzenedicarbox-ylic acid	$C_6H_4(COOH)_2$	166.13	9, 841			subl 402			sl s alc; s alkalis; v sl s aq, chl, eth
b19	1,4-Benzenedimethanol	$C_6H_4(CH_2OH)_2$	138.17	6, 919	1.100^{117}		117–119	143^{1mm}	188	v s aq, alc
b20	Benzenehexacar-boxylic acid	$C_6(COOH)_6$	342.17	9, 1008			286 d			
b21	Benzenesulfinic acid	$C_6H_5S(=O)OH$	142.16	11, 2			85	100 d		sl s aq; s alc, bz, eth
b22	Benzenesulfonamide	$C_6H_5SO_2NH_2$	157.19	11, 39			150–152			i aq; sl s alc; s eth
b23	Benzenesulfonic acid	$C_6H_5SO_2OH$	158.18	11, 26			50–51			v s aq, alc; sl s bz; i CS_2, eth
b24	Benzenesulfonyl chloride	$C_6H_5SO_2Cl$	176.62	11, 34	1.3842^{15}_{15}	1.5518^{20}	14.5	120^{10mm}	>110	i aq; s alc, eth
b25	Benzenesulfonyl fluoride	$C_6H_5SO_2F$	160.17	11[2], 23	1.3286^{20}_4	1.4920^{20}	−5	207–208	87	s alc, eth
b26	Benzenesulfonyl hy-drazide	$C_6H_5SO_2NHNH_2$	172.21	11, 52			d 104			flammable solid
b27	1,2,4,5-Benzenetetra-carboxylic acid	$C_6H_2(COOH)_4$	254.15	9, 997			276			1.5 aq; v s alc
b28	1,2,4,5-Benzenetetra-carboxyl dianhy-dride		218.12	19, 196			283–286	397–400		
b29	1,2,3-Behzenetricarb-oxylic acid dihyrate	$C_6H_3(COOH)_3 \cdot 2H_2O$	246.18	9, 976			192 d			sl s aq; v s eth

No.	Name	Formula	Beilstein Ref.	M.W.	Density	n_D	M.P.	B.P.	Flash Point	Solubility
b30	1,2,4-Benzenetricarboxylic acid	$C_6H_3(COOH)_3$	9, 997	210.14			231 d			2.1 aq; 25.3 alc; 7.9 acet; v s eth
b31	1,3,5-Benzenetricarboxylic acid	$C_6H_3(COOH)_3$	9, 978	210.14			>330			sl s aq; v s alc; s eth
b32	1,2,4-Benzenetricarboxylic anhydride		18, 468	192.13			161–163	245^{14mm}		50 acet; 22 EtOAc; 15 DMF
b33	1,3,5-Benzenetricarboxylic trichloride	$C_6H_3(COCl)_3$		265.48			35–36	180^{16mm}	>110	
b34	1,2,4-Benzenetriol	$C_6H_3(OH)_3$	6, 1087	126.11			141			v s aq, alc, eth, EtOAc
b35	Benzil	C_6H_5CO—COC_6H_5	7, 747	210.23	1.23^{15}_4		95	346–348		i aq; s alc, bz, chl, EtOAc, eth
b36	Benzil dioxime	$C_6H_5C(=NOH)$—$C(=NOH)C_6H_5$	7^3, 3816	240.25			(α) 240 (β) 214			i aq, HOAc, eth; sl s alc; s NaOH
b37	Benzilic acid	$(C_6H_5)_2C(OH)COOH$	10, 342	228.24			150			sl s aq; v s alc, eth hot
b38	Benzil monohydrazone	$C_6H_5C(=NNH_2)COC_6H_5$	7^1, 394	224.26			150–152			aq
b39	Benzimidazole		23, 131	118.13			170.5	>360		sl s aq, eth; v s alc
b40	7,8-Benzo-1,3-diazaspiro[4,5]decane-2,4-dione		Merck: 12, 9372	216.23			268			s alc, HOAc
b41	1,4-Benzodioxan		17, 54	136.15	1.142	1.5490^{20}		103^{6mm}	87	i aq; misc alc, bz, eth, PE
b42	2,3-Benzofuran			118.13	1.072	1.5660^{20}	<−18	173–175	56	
b43	Benzofurazan-1-oxide		27^1, 740	136.11			69–71			
b44	Benzoic acid	C_6H_5COOH	9, 92	122.12	1.321		122.4	249	121 (CC)	0.29 aq^{25}, 43 alc; 10 bz; 22 chl; 33 eth; 33 acet; 30 CS_2
b45	Benzoic anhydride	$(C_6H_5CO)_2O$	9, 164	226.22	1.1989^{15}_4		42	360	110	i aq; s alc, acet, chl bz, HOAc, EtOAc
b46	DL-Benzoin	$C_6H_5COCH(OH)C_6H_5$	8, 167	212.25	1.3100^{20}_4		137	194^{12mm}		s hot alc, acet; 20 pyr; sl s eth
b47	Benzoin ethyl ether	$C_6H_5CH(C_2H_5)COC_6H_5$	8, 174	240.30	1.1016^{17}_4	1.5727^{17}	62	195^{20mm}		s alc, bz, eth
b48	Benzoin isobutyl ether	$C_6H_5CH[OCH_2CH(CH_3)_2]COC_6H_5$	8, 175	268.36	0.985	1.5485^{20}		$133^{0.5mm}$	85	s alc, bz, eth
b49	Benzoin methyl ether	$C_6H_5CH(OCH_3)COC_6H_5$	8, 174	226.28	1.1278^{14}_4		48	189^{15mm}	>110	v s alc, bz, eth
b50	α-Benzoinoxime	$C_6H_5CH(OH)C(=NOH)C_6H_5$	8, 175	227.26			152–156			sl s aq; s alc, NH_4OH
b51	Benzonitrile	C_6H_5CN	9, 275	103.12	1.010	1.5289^{20}	−12.7	191	71	0.2 aq; misc org solv
b52	1,2-Benzophenanthrene		5, 718	202.26	1.274^{20}_4		258	448		i aq; s alc, eth

(Continued)

517

TABLE 2.21 Physical Constants of Organic Compounds (*Continued*)

No.	Name	Formula	Formula weight	Beilstein reference	Density, g/mL	Refractive index	Melting point, °C	Boiling point, °C	Flash point, °C	Solubility in 100 parts solvent
b53	Benzophenone	$C_6H_5COC_6H_5$	182.22	7, 411	1.1108^{18}_4	1.5975^{45}	48	305	>110	13.3 alc; 17 eth; s chl
b54	Benzophenone hydrazone	$C_6H_5C(=NNH_2)C_6H_5$	196.25	7, 417			95–98	230^{55mm}		
b55	1-Benzopyran-4(4H)-one		146.15	17, 327			55–60			s bz; sl s alc
b56	1,2-Benzol[a]pyrene		252.32	Merck: 12, 1134			179	312^{10mm}		s bz
b57	4,5-Benzo[e]pyrene		252.32	Merck: 12, 1105			179			
b58	1,4-Benzoquinone	$C_6H_4(=O)_2$	108.10	7, 609	1.318^{20}_4		116			sl s aq; s alc, hot bz, eth, hot PE; alkalis with dec
b59	Benzothiazole		135.19	Merck: 12, 1139	1.2460^{20}_4	1.6379^{20}	2	131^{34mm}	>110	sl s aq; v s alc, CS₂
b60	Benzo[b]thiophene		134.20	17, 59	1.1937^{40}	1.6302^{40}	32	221	>110	s alc, bz, chl, eth
b61	1,2,3-Benzotriazole		119.13	26, 38	1.238	1.6420^{20}	98.5	204 may explode		sl s aq; s alc, bz, chl, DMF
b62	Benzoxazole		119.12	27, 42		1.5594	30	182	58	sl s aq
b63	1-Benzoylacetone	$C_6H_5COCH_2COCH_3$	162.19	7, 680	1.090^{60}_{60}		60	260 sl d		sl s aq; v s alc, eth
b64	2-Benzoylbenzoic acid	$C_6H_5COC_6H_4COOH$	226.23	10, 747			129	265		sl s aq; v s alc, eth
b65	Benzoyl bromide	C_6H_5COBr	185.03	9, 195	1.5467^{20}	1.5883^{20}	−24	219	90	d aq, alc; misc eth
b66	Benzoyl chloride	C_6H_5COCl	140.57	9, 182	1.211^{20}_4	1.5537^{20}	−1.0	197.2	88(CC)	d aq, alc; misc bz, eth CS₂
b67	Benzoyl cyanide	C_6H_5COCN	131.13	10, 659	1.106		32	206		i aq
b68	Benzoyl fluoride	C_6H_5COF	124.11	9, 181	1.140	1.4960^{20}	−28	161	48	d hot aq; v s alc, eth
b69	Benzoylformic acid	$C_6H_5COCOOH$	150.13	10, 654			67–69			0.4 aq; 0.1 chl; 0.25 eth; sl s alc; i bz, PE
b70	N-Benzoylglycine	$C_6H_5CONHCH_2COOH$	179.18	9, 225			179			
b71	Benzoylhydrazine	$C_6H_5CONHNH_2$	136.15	9, 319			117			2.5 CS₂; s bz, chl, eth
b71a	Benzoyl peroxide	$(C_6H_5CO)_2O_2$	242.23	9, 179			103–106	explodes		sl s aq; s alc
b72	3-Benzoylpropanoic acid	$C_6H_5COCH_2CH_2COOH$	178.19	10, 696			117–119			
b73	2-Benzoylpyridine	$C_6H_5CO(C_5H_4N)$	183.21	21, 330			44	317	150	s alc, bz, eth
b74	3-Benzoylpyridine	$C_6H_5CO(C_5H_4N)$	183.21	21, 331			40	397	150	s alc, bz, eth
b75	4-Benzoylpyridine	$C_6H_5CO(C_5H_4N)$	183.21	21, 331			71	315	150	s alc, bz, eth
b76	Benzyl acetate	$CH_3CO_2CH_2C_6H_5$	150.18	6, 435	1.050^{25}_4	1.4998^{25}	−51.5	213.5	102 (CC)	i aq; misc alc, eth
b77	Benzyl acetoacetate	$CH_3COCH_2CO_2CH_2C_6H_5$	192.21	6, 438	1.112	1.5121^{20}		159^{10mm}	>110	

(Continued)

No.	Name	Formula	Mol. wt.	Beilstein reference	Density	n_D	m.p., °C	b.p., °C	Flash p., °C	Solubility
b77a	Benzylacetone	$C_6H_5CH_2CH_2COCH_3$	148.21	7, 314	0.989	1.5122^{20}		235	98	0.08 aq; misc alc, chl, eth
b78	Benzyl alcohol	$C_6H_5CH_2OH$	108.14	6, 428	1.0453^{20}_{4}	1.5403^{20}	−15.2	205	93 (CC)	misc aq, alc, eth
b79	Benzylamine	$C_6H_5CH_2NH_2$	107.16	12, 1013	0.983^{19}_{4}	1.5401^{20}	10	185	60	
b80	N-Benzylaminoethanol	$C_6H_5CH_2NHCH_2CH_2OH$	151.21	12, 1040	1.065	1.5435^{20}		156^{12mm}	>110	misc alc, chl, eth
b81	3-(Benzylamino)propanonitrile	$C_6H_5CH_2NHCH_2CH_2CN$	160.22		1.024	1.5308^{20}			>110	
b82	N-Benzylbenzamide	$C_6H_5CONHCH_2C_6H_5$	211.26	9, 121			106			
b83	Benzyl benzoate	$C_6H_5CO_2CH_2C_6H_5$	212.25	9^2, 471	1.118^{25}_{4}	1.5681^{21}	21	323	148	sl s aq; s alc, bz, chl, eth
b84	2-Benzylbenzoic acid	$C_6H_5CH_2C_6H_4COOH$	212.24	5, 306			110–113			
b85	Benzyl bromide	$C_6H_5CH_2Br$	171.04	6^1, 220	1.4380^{22}_{0}	1.5752^{20}	−3.9	199	86	slowly dec aq
b86	Benzyl 2-bromoacetate	$BrCH_2CO_2CH_2C_6H_5$	229.08	6, 548	1.446	1.5440^{20}		170^{22mm}	>110	
b87	Benzyl-tert-butanol	$C_6H_5CH_2CH_2C(CH_3)_2OH$	164.25	9^2, 594		1.5090^{20}	31–33	144^{85mm}	>110	
b88	Benzyl butyl 1,2-phthalate	$C_6H_5CH_2O_2C_6H_4CO_2C_4H_9$	312.37		1.119^{25}_{25}	1.5400^{20}			199	
b89	Benzyl carbamate	$C_6H_5CH_2OCONH_2$	151.17	6, 437			87–89	220 d	67	v s alc; sl s eth
b90	Benzyl chloride	$C_6H_5CH_2Cl$	126.59	5, 292	1.100^{20}_{20}	1.5381^{20}	−43 to −49	179		misc alc, chl, eth
b91	Benzyl chloroformate	$C_6H_5CH_2OCOCl$	170.60	6, 437	1.195	1.5190^{20}		103^{20mm}	91	dec aq; s eth
b92	Benzyl chlorothiolformate	$C_6H_5CH_2SCOCl$	186.5		1.237^{30}_{4}	1.5711^{30}		$80^{0.13mm}$	118	
b93	Benzyl cinnamate	$C_6H_5CH=CHCO_2CH_2C_6H_5$	238.29	9, 584			39	200^{5mm}	>110	s alc, eth; i aq, glyc
b94	S-Benzyl-L-cysteine	$C_6H_5CH_2SCH_2CH(NH_2)COOH$	211.28	6, 465			214 d			
b95	Benzyl N,N-dimethyldithiocarbamate	$(CH_3)_2NCS_2CH_2C_6H_5$	211.35				41		>110	
b96	Benzyldimethylstearylammonium chloride hydrate	$C_6H_5CH_2N[(CH_2)_{17}CH_3](CH_3)_2Cl·H_2O$	442.18	12^3, 2212			67–69			
b96a	N-Benzyl-N-ethylaniline	$C_6H_5N(CH_2C_6H_5)C_2H_5$	211.31	12, 1026; Merck: 12, 1168	1.029	1.5950^{20}		164^{6mm}	>110	misc alc, eth; i aq
b97	Benzyl ethyl ether	$C_6H_5CH_2OC_2H_5$	136.20	12, 1043	0.9478^{20}	1.4955^{20}		186		
b98	N-Benzylformamide	$C_6H_5CH_2NHCHO$	135.17	Merck: 12, 1169			61			
b99	Benzyl formate	$C_6H_5CH_2O_2CH$	136.15		1.0814^{20}_{4}			203		i aq; s alc
b100	Benzyl 4-hydroxybenzoate	$HOC_6H_4CO_2CH_2C_6H_5$	228.25	10^3, 311			110–112			
b101	O-Benzylhydroxylamine hydrochloride	$C_6H_5CH_2ONH_2·HCl$	159.62	6, 440				238 subl	>110	

TABLE 2.21 Physical Constants of Organic Compounds (*Continued*)

No.	Name	Formula	Formula weight	Beilstein reference	Density, g/mL	Refractive index	Melting point, °C	Boiling point, °C	Flash point, °C	Solubility in 100 parts solvent
b102	Benzylidineaniline	C$_6$H$_5$N=CHC$_6$H$_5$	181.24	12, 195	1.045^{50}_{4}		56	300	>110	s alc, chl, CS$_2$
b103	Benzylidenemalono-nitrile	C$_6$H$_5$CH=C(CN)$_2$	154.17	9, 895			83–85			
b104	N-Benzylidenemethyl-amine	C$_6$H$_5$CH=NCH$_3$	119.17	7, 213	0.967	1.5520^{20}		80^{18mm}	>112	
b105	3-Benzylidene-phthalide		124.21	17, 376			99–102			
b106	Benzyl mercaptan	C$_6$H$_5$CH$_2$SH	222.24	6, 453	1.058^{20}	1.5751^{20}		206^{30mm}	>110	
b107	Benzyl methacrylate	H$_2$C=C(CH$_3$)CO$_2$CH$_2$C$_6$H$_5$	176.22	6^3, 1481	1.040	1.5120^{20}		98^{4mm}	77	
b108	N-Benzylmethylamine	C$_6$H$_5$CH$_2$NHCH$_3$	138.23	12, 1019	0.939	1.5230^{20}		184–189	77	
b109	3-(N-Benzyl-N-methyl-amino)-1,2-propane-diol	C$_6$H$_5$CH$_2$N(CH$_3$)CH$_2$-CH(OH)CH$_2$OH	195.26		1.084	1.5341^{20}		206^{30mm}	>110	
b110	Benzyl methyl sulfide	C$_6$H$_5$CH$_2$SCH$_3$	138.23	6, 453	1.015	1.5620^{20}		195–198	73	
b111	1-Benzyl-3-methyl-2-thiourea	C$_6$H$_5$CH$_2$NHC(=S)NHCH$_3$	180.27	12, 1052			74–76		>110	
b112	Benzyl nicotinate	(C$_5$H$_4$N)CO$_2$CH$_2$C$_6$H$_5$	213.24	22^3, 366	1.165	1.5700^{20}	21–23	189^{12mm}	>110	
b113	4-Benzyloxybenz-aldehyde	C$_6$H$_5$CH$_2$OC$_6$H$_4$CHO	212.25	8, 73			73–74			
b114	4-Benzyloxybenzyl alcohol	C$_6$H$_5$CH$_2$OC$_6$H$_4$CH$_2$OH	214.26				86–87			0.4 aq
b115	2-Benzyloxyethanol	C$_6$H$_5$CH$_2$OCH$_2$CH$_2$OH	152.20	6^2, 413	1.07^{20}_{20}	1.5210^{20}		265	129	
b116	4-Benzyloxy-3-meth-oxybenzaldehyde	C$_6$H$_5$CH$_2$OC$_6$H$_4$(OCH$_3$)CHO	242.29				63–65			
b117	4-(Benzyloxymethyl)-2,2-dimethyl-1,3-dioxolane		222.28	19^2, 73	1.051	1.4940^{20}		$91^{0.1mm}$	>110	
b118	Benzyl phenyl sulfide	C$_6$H$_5$CH$_2$SC$_6$H$_5$	200.30	6, 454	1.014	1.5467^{20}	41–44	197^{27mm}	>110	i aq; sl s alc; s eth
b119	1-Benzylpiperazine		176.26	20, 296	0.997	1.5379^{20}	6–7	279	>110	s aq, alc, eth
b120	4-Benzylpiperidine		175.28		1.021	1.5399^{20}		134^{7mm}	>110	
b121	1-Benzyl-4-piperidone		189.26		1.054	1.5790^{20}	8–10	276	125	
b122	2-Benzylpyridine	C$_6$H$_5$CH$_2$(C$_5$H$_4$N)	169.23	20, 425	1.061^{20}_{0}	1.5818^{20}		287	115	i aq; v s alc, eth
b123	4-Benzylpyridine	C$_6$H$_5$CH$_2$(C$_5$H$_4$N)	169.23	20, 426	1.095	1.5525^{20}			>110	s alc; v s eth
b124	1-Benzyl-2-pyrrolidi-none		175.23							
b125	Benzyl salicylate	HOC$_6$H$_4$CO$_2$CH$_2$C$_6$H$_5$	228.25	Merck: 12, 1181	1.175^{20}			208^{25mm}		sl s aq; misc alc, eth
b126	Benzyl thiocyanate	C$_6$H$_5$CH$_2$SCN	149.22	6, 460			43	235	>110	i aq; s alc; v s eth

No.	Name	Formula	Mol. wt.	Beilstein ref.	Density	n_D	M.p., °C	B.p., °C	Flash p., °C	Solubility
b127	Benzyltributyl-ammonium chloride	$C_6H_5CH_2N(C_4H_9)_3^+Cl^-$	312.94				164 d		93	
b128	BenzyltriChlorosilane	$C_6H_5CH_2SiCl_3$	225.28	16, 912	1.288_4^{20}	1.5250^{20}		142^{100mm} 175^{70mm}		
b129	Benzyltriethoxysilane	$C_6H_5CH_2Si(OC_2H_5)_3$	254.40	12, 1021	0.986_4^{20}					160 aq; 55 MeOH; 8.7 EtOH
b130	Benzyltriethylammo-nium Chloride	$C_6H_5CH_2N(C_2H_5)_3^+Cl^-$	227.78	12, 1021			185 d			
b131	Benzyltrimethylam-monium CHloride	$C_6H_5CH_2N(CH_3)_3^+Cl^-$	185.70				239 d	190	none	i aq; s eth
b132	Benzyltrimethylsilane	$C_6H_5CH_2Si(CH_3)_3$	164.32	$16^1, 526$	0.8933^{20}	1.4941^{20}		190	57	
b133	Betaine	$(CH_3)_3N^+CH_2COO^-$	117.15	4, 347			dec 310			
b134	Bicyclo[2.2.1]hepta-2,5-diene		92.14		0.909^{20}	1.4707^{20}	-20	89	-11	i aq; s PE
b135	Bicyclo[2.2.1]-2-heptene		94.16				44–46	96	-15	s eth
b136	Bicyclo[2.2.1]-2-heptene-2-carbaldehyde		122.16		1.108	1.4883^{20}		70^{12mm}	51	
b137	Biguanide	$H_2NC(=NH)NH-C(=NH)NH_2$	101.11	3, 93			130	dec 142		s aq, alc; i bz, chl, eth
b138	Biphenyl	$C_6H_5·C_6H_5$	154.20	5, 578	0.991_4^{75}	1.588^{77}	69–71	256 subl	113 (CC)	i aq; s alc, eth
b139	4-Biphenylcarboxylic acid	$C_6H_5—C_6H_4COOH$	198.22	9, 671			226			v s alc, eth; s bz; i aq
b140	4,4′-Biphenyldiamine	$H_2NC_6H_4—C_6H_4NH_2$	184.24	13, 214			120	$ca.$ 400		s alc; 2 eth; 20 hot alc
b141	2,2′-Biphenyldi-carboxylic acid	$HOOCC_6H_4—C_6H_4COOH$	242.23	9, 922			228–229			0.06 aq; s org solvents
b142	4-Biphenylsulfonic acid	$C_6H_5—C_6H_4SO_3H$	234.26				138			
b143	2-Biphenylyl glycidyl ether		226.28				30–32	$120^{0.1mm}$		
b144	2,2-Bis[4-(allyloxy)-phenyl]-propane	$H_2C=CHCH_2OC_6H_4-C(CH_3)_2C_6H_4—OCH_2CH=CH_2$	308.42		1.022	1.5636^{20}			>110	
b145	N,N′-Bis(3-amino-propyl)ethylenedi-amine	$H_2N(CH_2)_3NHCH_2-CH_2NH(CH_2)_3NH_2$	174.29		0.952	1.4910^{20}		160^{5mm}	>110	
b146	N,N′-Bis(3-amino-propyl)piperazine		200.33	$23^2, 12$	0.973	1.5015^{20}	15	152^{2mm}	162	
b147	N,N′-Bis(3-amino-propyl)-1,3-propane-diamine	$H_2N(CH_2)_3NHCH_2CH_2CH_2-NH(CH_2)_3NH_2$	188.32	$4^4, 1278$	0.920	1.4915^{20}		103^{1mm}		
b148	Bis(2-bromoethyl) ether	$BrCH_2CH_2OCH_2CH_2Br$	231.92					107^{20mm}		

(Continued)

TABLE 2.21 Physical Constants of Organic Compounds (*Continued*)

No.	Name	Formula	Formula weight	Beilstein reference	Density, g/mL	Refractive index	Melting point, °C	Boiling point, °C	Flash point, °C	Solubility in 100 parts solvent
b149	1,3-Bis(bromoethyl)-tetramethyldisiloxane	[BrCH$_2$Si(CH$_3$)$_2$]$_2$O	320.17		1.3918$^{20}_4$	1.4719^{20}				
b150	2,2-Bis(bromomethyl)-1,3-propanediol	HOCH$_2$CH(CH$_2$Br)$_2$CH$_2$OH	261.95	1^1, 251			114			
b151	Bis(2-butoxyethyl)-ether	(C$_4$H$_9$OCH$_2$CH$_2$)$_2$O	218.34		0.8853$^{20}_{20}$	1.4240^{20}		256	118	0.3 aq; misc alc, esters, eth, CCl$_4$ ketones
b152	Bis[2-(2-butoxyethoxy)-ethyl] adipate	[-CH$_2$CH$_2$CO$_2$(CH$_2$CH$_2$O)$_2$-(CH$_2$)$_3$CH$_3$]$_2$	434.58	2^3, 1718	1.010	1.4480^{20}			110	
b153	2,5-Bis(5-tert-butyl-2-2'-benzoxazolyl)-thiophene		430.57				201			
b154	Bis(sec-butyl) disulfide	[CH$_3$CH$_2$CH(CH$_3$)]$_2$S$_2$	178.36	1^3, 1549	0.957	1.4920^{20}		164^{739mm}	112	
b155	Bis(tert-butyl) disulfide	(CH$_3$)$_3$CSSC(CH$_3$)$_3$	178.36	1, 379	0.909	1.4930^{20}		204	79	
b156	1,1-Bis(tert-butylperoxy)cyclohexane	C$_6$H$_{10}$[OOC(CH$_3$)$_3$]$_2$	260.38		0.970	1.4570^{20}		54^{15mm}	90	
b157	2,5-Bis(tert-butylperoxy)-2,5-dimethyl-hexane	[(CH$_3$)$_3$COOC(CH$_3$)$_2$CH$_2$]$_2$	290.45		0.877	1.4230^{20}	-11	57^{7mm}	41	
b158	2,5-Bis(tert-butylperoxy)-2,5-dimethyl-3-hexyne	(CH$_3$)$_3$COOC(CH$_3$)$_2$C≡C-C(CH$_3$)$_2$OOC(CH$_3$)$_3$	286.41	1^4, 2701	0.881	1.4320^{20}		67^{2mm}	85	
b159	Bis[1-(tert-butylperoxy)-1-methylethyl]-benzene	C$_6$H$_4$[C(CH$_3$)$_2$OOC(CH$_3$)$_3$]$_2$	338.49				44-48			flammable solid oxidizer
b160	1,1-Bis(tert-butylperoxy)-3,3,5-tri-methyl-cyclohexane	[(CH$_3$)$_3$COO]$_2$C$_6$H$_7$(CH$_3$)$_3$	302.46		0.906	1.4410^{20}			87	
b161	1,2-Bis(2-Chloro-ethoxy)ethane	(ClCH$_2$CH$_2$OCH$_2$)$_2$	187.07	1^3, 2079	1.197$^{20}_4$	1.4610^{20}		235	121	
b162	Bis(2-Chloroethoxy)-methylsilane	H(CH$_3$)Si(OCH$_2$CH$_2$Cl)$_2$	203.1		1.1643$^{20}_4$	1.4431^{20}		97^{18mm}		
b163	Bis(2-Chloroethyl) ether	ClCH$_2$CH$_2$OCH$_2$CH$_2$Cl	143.01	1^2, 335	1.2220$^{20}_{20}$	1.4575^{20}	-50 to -52	178.5	55	s most org solvents
b164	Bis(2-Chloroethyl)-N-methylamine	CH$_3$N(CH$_2$CH$_2$Cl)$_2$	156.07		1.118$^{25}_4$		-60	75^{10mm}		v sl s aq; misc most org solvents

No.	Name	Formula	Mol. wt.	Beilstein ref.	Density	n_D	M.p., °C	B.p., °C	Flash p., °C	Solubility
b165	Bis(chloromethyl)dimethylsilane	$(CH_3)_2Si(CH_2Cl)_2$	157.12	4^3, 1845	1.975_4^{20}	1.4600^{20}		160	46	
b165a	Bis(chloromethyl) ether	$ClCH_2OCH_2Cl$	114.96	Merck: 12, 3119	1.315_4^{20}	1.4346	−41.5	106		dec aq
b166	Bis(2-chloro-1-methyl)ethyl ether	$ClCH_2CH(CH_3)OCH(CH_3)CH_2Cl$	171.07		1.1122_4^{20}			187.3	85	
b167	1,3-Bis(chloromethyl)tetramethyldisiloxane	$[ClCH_2Si(CH_3)_2]_2O$	231.3	4^3, 1864	1.050	1.4405^{20}		205	73	
b168	Bis(4-chlorophenoxy)acetic acid	$(ClC_6H_4O)_2CHCOOH$	313.14				140–142			
b169	2,2-Bis(4-chlorophenyl)-1,1-dichloroethane	$(ClC_6H_4)_2CHCHCl_2$	320.05	5^3, 1830			110			similar to b168
b170	1,1-Bis(4'-chlorophenyl)ethanol	$(ClC_6H_4)_2C(OH)CH_3$	267.16	6^3, 3396			69			s org solvents
b171	Bis(4-chlorophenyl)sulfone	$ClC_6H_4SO_2C_6H_4Cl$	287.16	6, 327			145–148	250^{10mm}		
b172	Bis(4-chlorophenyl)sulfoxide	$ClC_6H_4S(O)C_6H_4Cl$	271.17	6^1, 149			141–144			
b173	1,1-Bis(4-chlorophenyl)-2,2,2-trichloroethane	$(ClC_6H_4)_2CHCCl_3$	354.49	5^3, 1833			109–111			58 acet; 78 bz; 45 chl; v s pyr, 1,4-dioxane
b174	1,2-Bis(dichloromethylsilyl)ethane	$[-CH_2Si(CH_3)Cl_2]_2$	256.11	4^4, 192	1.263	1.4760^{20}	33–35	210	90	
b175	1,3-Bis(dichloromethyl)tetramethyldisiloxane	$[ClCH(CH_3)_2Si]_2O$	300.16		1.2213_4^{20}	1.4660^{20}		149^{40mm}		
b176	N,N-Bis(2,2-diethoxyethyl)methylamine	$[(C_2H_5O)_2CHCH_2]_2NCH_3$	263.38	4, 311	0.945	1.4259^{20}		222^{244mm}	60	
b177	4,4'-Bis(diethylamino)benzophenone	$[(C_2H_5)_2NC_6H_4]_2C{=}O$	324.47	14, 98			95			
b178	4,4'-Bis(dimethylamino)benzophenone	$[(CH_3)_2NC_6H_4]_2C{=}O$	268.35	14, 89			172	>360 d		s alc, warm bz; v sl s eth; i aq
b179	Bis(dimethylamino)dimethylsilane	$[(CH_3)_2N]Si(CH_3)_2$	146.31	4^4, 4143	0.810^{22}	1.4170^{20}	−98	128–129	−7	
b180	1,3-Bis(dimethylamino)-2-propanol	$[(CH_3)_2NCH_2]_2CHOH$	146.23	4, 290	0.897	1.4422^{20}			>110	

(Continued)

TABLE 2.21 Physical Constants of Organic Compounds (*Continued*)

No.	Name	Formula	Formula weight	Beilstein reference	Density, g/mL	Refractive index	Melting point, °C	Boiling point, °C	Flash point, °C	Solubility in 100 parts solvent
b181	2,4-Bis(α,α-dimethyl-benzyl)phenol	[C$_6$H$_5$C(CH$_3$)$_2$]$_2$C$_6$H$_3$OH	330.47	6^4, 5076			63–65	206^{15mm}		
b182	1,1-Bis(3,4-dimethyl-phenyl)ethane	[(CH$_3$)$_2$C$_6$H$_3$]$_2$CHCH$_3$	238.38	5^3, 1908	0.982	1.5640^{20}		174^{5mm}	>110	s alc, eth; sl s bz, acet; i aq
b183	Bis(dimethylthio-carbamyl) disulfide	[(CH$_3$)$_2$NC(=S)S-]$_2$	240.43	4, 76	1.29		155–156			
b184	Bis(3,4-epoxycyclo-hexylmethyl) adi-pate		366.46		1.149	1.4930			>110	
b185	1,4-Bis(2,3-epoxy-pro-poxy)butane	[H$_2$C—CHCH$_2$OCH$_2$CH$_2$-]$_2$ O	202.25		1.049	1.4530^{20}		160^{11mm}	>110	
b186	Bis(2-ethoxyethyl) ether	(C$_2$H$_5$OCH$_2$CH$_2$)$_2$O	162.23	1^2, 519	0.907$^{20}_4$	1.4110^{20}	−45	188	82	v s aq, alc, org sol-vents
b187	Bis(2-ethylhexyl) adipate	[-CH$_2$CH$_2$CO$_2$CH(C$_2$H$_5$)-(CH$_2$)$_3$CH$_3$]$_2$	370.58	2^3, 1715	0.990	1.4425^{20}		167^{1mm}	>110	
b188	Bis(2-ethylhexyl)-amine	[CH$_3$(CH$_2$)$_3$CH(C$_2$H$_5$)-(CH$_2$)$_3$CH$_3$]$_2$	241.46	4^3, 388	0.805	1.4425^{20}		123^{5mm}	>110	
b189	Bis(2-ethylhexyl) chlorendate		613.28		1.240	1.500^{20}		233$^{0.3mm}$	>110	
b190	Bis(2-ethylhexyl) decanedioate	CH$_3$(CH$_2$)$_3$CH(C$_2$H$_5$)CH$_2$-OOC(CH$_2$)$_8$COOCH$_2$-CH(C$_2$H$_5$)(CH$_2$)$_3$CH$_3$	426.66		0.9119$^{25}_{25}$	1.4496^{25}				
b191	Bis(2-ethylhexyl) hydrogen phosphate	[CH$_3$(CH$_2$)$_3$CH(C$_2$H$_5$)-CH$_2$O]$_2$P(O)OH	322.43	1^4, 1786	0.965	1.4430^{20}	−60	209^{10mm}	>110	
b192	Bis(2-ethylhexyl) hydrogen phosphite	[CH$_3$(CH$_2$)$_3$CH(C$_2$H$_5$)-CH$_2$O]$_2$POH	306.43		0.916	1.4420^{20}			>110	
b193	Bis(2-ethylhexyl) o-phthalate	[CH$_3$(CH$_2$)$_3$CH(C$_2$H$_5$)-CH$_2$OOC]$_2$C$_6$H$_4$	390.56	Merck: 12, 1291	0.9843^{20}	1.4859^{20}	−50 to −55	384	218	0.01aq
b194	Bis(2-ethylhexyl) 1,4-phthalate	[CH$_3$(CH$_2$)$_3$CH(C$_2$H$_5$)-CH$_2$OOC]$_2$C$_6$H$_4$	390.56	9^4, 3306	0.980	1.4900^{20}	30–34	400	>110	
b195	Bis(4-fluorophenyl)-methane	(FC$_6$H$_4$)$_2$CH$_2$	204.22	5^3, 1789	1.145	1.5362^{20}	29–30	260^{742mm}	>110	
b196	Bis(hexamethylene)-tri amine	[H$_2$N(CH$_2$)$_6$]$_2$NH	215.39				33–36	165^{4mm}	>110	
b197	1,4-Bis(2-hydroxy-ethoxy)-2-butyne	HOCH$_2$CH$_2$OCH$_2$C≡CCH$_2$-OCH$_2$CH$_2$OH	174.20		1.144	1.4850^{20}			>110	

No.	Name	Formula	Formula weight	Beilstein reference	Density	n_D	Melting point, °C	Boiling point, °C	Flash point, °C	Solubility
b198	Bis(2-hydroxyethyl) ether	$HOCH_2CH_2OCH_2CH_2OH$	106.12	1, 468	1.1184^{20}_{20}	1.4460^{20}	−10.4	246	118	misc aq, alc, acet, eth
b199	N,N-Bis(2-hydroxyethyl)glycine	$(HOCH_2CH_2)_2NCH_2COOH$	163.17	Merck: 12, 1248			193–195			17.9 aq°
b200	2,6-Bis(hydroxymethyl)-p-cresol	$CH_3C_6H_2(CH_2OH)_2OH$	168.19	6, 1127			128–130			
b201	2,2-Bis(hydroxymethyl)propanoic acid	$(HOCH_2)_2C(CH_3)COOH$	134.13	3, 401			181–185			s aq, MeOH; sl s acet; i bz
b202	4,8-Bis(hydroxymethyl)tricyclo[5.2.1.0²·⁶]decane		196.29	6^4, 5538		1.5280^{20}			110	
b203	4,4-Bis(4-hydroxyphenyl)pentanoic acid	$CH_3C(C_6H_4OH)_2CH_2CH_2COOH$	286.33	Merck: 12, 3370			171–172 higher melting form			s hot aq, acet, alc, HOAc, MeEtKe
b204	Bis(2-hydroxypropyl) ether	$HO(CH_2)_3O(CH_2)_3OH$	134.18	1^2, 537	1.0252^{20}_{20}	1.4410^{20}		231.8	137	misc aq, alc
b205	1,3-Bis(isocyanatomethyl)benzene	$C_6H_4(CH_2NCO)_2$	188.19	13^3, 334	1.202	1.5910^{20}	−7	130^{2mm}	>110	
b206	1,3-Bis(isocyanatomethyl)cyclohexane	$C_6H_{10}(CH_2NCO)_2$	194.24		1.101	1.4850^{20}			>110	
b207	1,3-Bis(1-isocyanato-1-methylethyl)benzene	$C_6H_4[C(CH_3)_2NCO]_2$	244.30	1^3, 2107	1.060	1.5110^{20}		$106^{0.9mm}$	153	
b208	Bis(2-mercaptoethyl) ether	$(HSCH_2CH_2)_2O$	138.25		1.114		−80	217	98	
b209	Bis(2-mercaptoethyl) sulfide	$(HSCH_2CH_2)_2S$	154.32		1.183	1.5961^{20}		136^{10mm}	90	
b210	1,4-Bis(methanesulfonyloxy)butane	$(CH_3SO_2OCH_2CH_2)_2$	246.30				115–117			sl hyd aq; 0.1 alc; 1.4 acet
b211	1,2-Bis(methoxyethoxy)ethane	$(CH_3OCH_2CH_2OCH_2)_2$	178.23		0.990^{20}_{4}	1.4224^{20}	−45	216	110	misc aq
b212	Bis[2-(2-methoxyethoxy)ethyl] ether	$(CH_3OCH_2CH_2OCH_2CH_2)_2O$	228.28	4^3, 691	1.0087_{4}	1.4330^{20}	−27	275	140	s aq
b213	Bis(2-methoxyethyl)amine	$(CH_3OCH_2CH_2)_2NH$	133.19		0.902	1.4190^{20}		172	58	
b214	Bis(2-methoxyethyl) ether	$(CH_3OCH_2CH_2)_2O$	134.18	1^2, 520	0.9440^{25}	1.4043^{25}	−64 to −68	162	67	misc aq
b214a	2,2-Bis(4-methoxyphenyl)-1,1,1-trichloroethane	$(CH_3OC_6H_4)_2CHCCl_3$	345.66	6, 1007			86–88			v sl s aq; s alc

(Continued)

TABLE 2.21 Physical Constants of Organic Compounds (*Continued*)

No.	Name	Formula	Formula weight	Beilstein reference	Density, g/mL	Refractive index	Melting point, °C	Boiling point, °C	Flash point, °C	Solubility in 100 parts solvent
b215	Bis(2-methylallyl) carbonate	$[H_2C{=}C(CH_3)CH_2O]_2C{=}O$	170.21		0.943	1.4370^{20}		202	72	i aq; s alc; v s eth
b216	Bis(3-nitrophenyl) disulfide	$O_2NC_6H_4SSC_6H_4NO_2$	308.33	6, 339			83			
b217	Bis(octadecyl)pentaerythritol diphosphite	$[C_{18}H_{37}OP(OCH_2)_2]_2$	721.01		0.925	1.457	40		261	
b218	1,4-Bis(5-phenyloxazol-2-yl)benzene		364.40				244			
b219	N,N′-Bis(salicylidene)-1,4-butanediamine	$HOC_6H_4CH{=}N(CH_2)_4{-}N{=}CHC_6H_4OH$	296.37	8^3, 163			88–90			
b220	N,N′-Bis(salicylidene)-ethylenediamine	$({-}CH_2N{=}CHC_6H_4OH)_2$	268.32	8, 48			128			
b221	N,N′-Bis(salicylidene)-1,6-hexanediamine	$HOC_6H_4CH{=}N(CH_2)_6{-}N{=}CHC_6H_4OH$	324.44	8^3, 165			69			
b222	Bis(p-tolyl) disulfide	$CH_3C_6H_4SSC_6H_4CH_3$	246.39	6, 425			43–46			i aq; s alc; v s eth
b223	Bis(p-tolyl) sulfoxide	$CH_3C_6H_4S({\to}O)C_6H_4CH_3$	230.33	6, 419			94–96			v s alc, bz, chl, eth
b224	Bis(tributyltin) oxide	$(C_4H_9)_3SnOSn(C_4H_9)_3$	596.08		1.170	1.4860^{20}		180^{2mm}	>110	i aq; s alc; 26 acet; 38 bz
b225	1,4-Bis(trichloromethyl)benzene	$Cl_3CC_6H_4CCl_3$	312.84	5, 385			108–110			
b226	Bis(2,4,5-trichlorophenyl) disulfide	$Cl_3C_6H_2SSC_6H_2Cl_3$	425.01				140–144			
b227	1,2-Bis(trichlorosilyl)ethane	$Cl_3SiCH_2CH_2SiCl_3$	296.94	4^4, 4266	1.4834^{20}_4	1.4750^{20}	24.5	202	65	
b228	3,5-Bis(trifluoromethyl)aniline	$(F_3C)_2C_6H_3NH_2$	229.13		1.467	1.4340^{20}		85^{15mm}	83	
b229	1,3-Bis(trifluoromethyl)benzene	$(F_3C)_2C_6H_4$	214.11	5^3, 834	1.3790^{25}	1.3916^{25}		116	26	
b230	N,O-Bis(trimethylsilyl)acetamide	$CH_3{-}C{=}N{-}Si(CH_3)_3$ $\;\;\;\;\;O{-}Si(CH_3)_3$	203.43		0.8324^{20}_4	1.4170^{20}		75^{35mm}	11	
b231	Bis(trimethylsilyl)acetylene	$(CH_3)_3SiC{\equiv}CSi(CH_3)_3$	170.41		0.7704^{20}_4	1.4270^{20}		137	2	
b232	Bis(trimethylsilyl)formamide	$HC{=}NSi(CH_3)_3$ $\;\;\;\;OSi(CH_3)_3$	189.41		0.885	1.4381^{20}		55^{13mm}	28	
b233	N,O-Bis(trimethylsilyl)hydroxylamine	$(CH_3)_3SiONHSi(CH_3)_3$	177.40		0.830	1.4112^{20}		80^{100mm}	28	

(Continued)

No.	Name	Formula	Mol. wt.	Beilstein	Density	n_D	M.P.	B.P.	Flash pt.	Solubility
b234	1,2-Bis(trimethylsilyl)oxyethane	$(CH_3)_3SiOCH_2CH_2OSi(CH_3)_3$	206.43		0.842	1.4034^{20}		166	46	
b235	N,O-Bis(trimethylsilyl)trifluoroacetamide	$F_3C[=NSi(CH_3)_3]OSi(CH_3)_3$	257.40		0.969	1.3839^{20}	−10	50^{14mm}	23	
b236	1,3-Bis(trimethylsilyl)urea	$(CH_3)_3SiNHCONHSi(CH_3)_3$	204.42				232 dec			
b237	1,3-Bis[tris(hydroxymethyl)methylamino]propane	$CH_2[CH_2NHC(CH_2OH)_3]_2$	282.34	4^3, 859			170			s aq
b238	Biuret	$H_2NC(=O)NHC(=O)NH_2$	103.08	3, 70	1.467_4^{-5}	1.4600^{20}	anhyd 110; 100 dec	dec 190	40	v s alc; 2 aq[25]
b239	Borane-tert-butylamine	$(CH_3)_3CNH_2 \cdot BH_3$	86.97							
b240	Borane-N,N-diethylaniline	$C_6H_5N(C_2H_5)_2 \cdot BH_3$	163.07		0.822		−30		21	
b241	Borane-N,N-diisopropylethylamine	$[(CH_3)_2CH]_2C_2H_5 \cdot BH_3$	143.08				15–17			
b242	Borane-dimethylamine	$(CH_3)_2NH \cdot BH_3$	58.92				36		43	
b243	Borane-dimethyl sulfide	$(CH_3)_2S \cdot BH_3$	75.97		0.801				18	
b244	Borane-pyridine	$C_5H_4N \cdot BH_3$	92.93	6, 72	0.920	1.5320^{20}	10–11	210^{779mm}	21	i aq; 176 alc; s eth
b245	(1S-endo)-(−)-Borneol		154.25	6, 82	1.011_4^{20}		204	224	65	v sl s aq; s alc, eth
b246	(−)-1-Bornyl acetate		196.29		0.982	1.4626	27		84	sl s aq; v s eth
b247	N-Bromoacetamide	$CH_3CON(Br)H$	137.96	2, 181			102–105			
b248	p-Bromoacetanilide	$BrC_6H_4NHCOCH_3$	214.06	12, 642	1.717	1.4804^{50}	168	208	>110	s alc, bz, Chl, EtOAc
b249	Bromoacetic acid	$BrCH_2COOH$	138.95	2, 213	1.934_4^{50}	1.4800^{20}	50	62^{24mm}	>110	v s aq, alc
b250	Bromoacetonitrile	$BrCH_2CN$	119.95	2, 216	1.722			135^{18mm}	>110	
b251	2-Bromoacetophenone	$C_6H_5COCH_2Br$	199.05	7, 283	1.647_4^{20}		50	255	>110	v s alc, bz, chl, eth
b253	p-Bromoacetophenone	$BrC_6H_4COCH_3$	199.05	7, 283	1.647		54	150	>110	s alc, bz, CS_2, HO Ac PE
b254	Bromoacetyl bromide	$BrCH_2COBr$	201.86	2, 215	2.317_{22}^{22}	1.5480^{20}		128	none	dec aq, alc
b255	Bromoacetyl chloride	$BrCH_2COCl$	157.40	2, 215	1.908	1.4960^{20}		229	none	dec aq, alc
b256	2-Bromoaniline	$BrC_6H_4NH_2$	172.03	12, 631	1.578_4^{20}	1.6223^{20}	31	251	>110	i aq; s alc, eth
b257	3-Bromoaniline	$BrC_6H_4NH_2$	172.03	12, 633	1.580_4^{20}	1.6250^{20}	16.8		>110	sl s aq; s alc, eth
b258	4-Bromoaniline	$BrC_6H_4NH_2$	172.03	12, 636	1.4970_4^{100}		66.3	223		i aq; v s alc, eth
b259	2-Bromoanisole	$BrC_6H_4OCH_3$	187.04	6, 197	1.502	1.5740^{20}	2	223	96	i aq; v s alc, eth
b260	4-Bromoanisole	$BrC_6H_4OCH_3$	187.04	6, 199	1.494	1.5640^{20}	9–10	230	94	
b261	3-Bromobenzaldehyde	BrC_6H_4CHO	185.03	7, 238	1.587	1.5935^{20}		156	96	
b262	Bromobenzene	C_6H_5Br	157.01	5, 206	1.4952_4^{24}	1.5602^{20}	−30.6		51	i aq; v s alc, eth; 0.045 aq[30], 10.4 alc[25], 71.6 eth[25]; misc bz, chl, PE

TABLE 2.21 Physical Constants of Organic Compounds (*Continued*)

No.	Name	Formula	Formula weight	Beilstein reference	Density, g/mL	Refractive index	Melting point, °C	Boiling point, °C	Flash point, °C	Solubility in 100 parts solvent
b263	Bromobenzene-d_5	C_6D_5Br	162.06		1.539	1.5585^{20}		53^{23mm} 153^{15mm}	51	i aq; s alc (dec); v s eth
b264	4-Bromobenzene-sulfonyl chloride	$BrC_6H_4SO_2Cl$	255.52	11, 57			74.5			
b265	2-Bromobenzoic acid	BrC_6H_4COOH	201.02	9, 347			148–150			
b266	4-Bromobenzoic acid	BrC_6H_4COOH	201.02	9, 351	1.929^{25}_4		251–253		350	0.18 aq^{25}; s alc, eth
b267	4-Bromobenzo-phenone	$BrC_6H_4COC_6H_5$	261.12	7, 422				82		i alc; sl s bz, eth
b268	2-Bromobenzotri-fluoride	$BrC_6H_4CF_3$	225.01		1.652^{20}	1.4820^{20}		168	51	s aq, alc, bz, eth, CS_2, HOAc
b269	3-Bromobenzotri-fluoride	$BrC_6H_4CF_3$	225.01		1.613	1.4730^{20}		152	43	
b270	3-Bromobenzoyl chloride	BrC_6H_4COCl	219.47	9, 350	1.662	1.5965^{20}		$75^{0.5mm}$	107	
b271	4-Bromobenzyl bromide	$BrC_6H_4CH_2Br$	249.94	5, 308		1.6193^{20}	61	124^{12mm}	>110	sl s aq; v s alc, acet, eth. A war gas.
b272	α-Bromobenzyl cyanide	$C_6H_5CH(Br)CN$	196.05		1.539^{29}_4	1.5696^{20}	29	242 dec	>110	i aq; s alc, bz, eth
b273	4-Bromobiphenyl	$BrC_6H_4C_6H_5$	233.11	5, 580	0.9327^{25}		90–92	310		i aq; s alc, bz, eth
b274	1-Bromobutane	$CH_3CH_2CH_2CH_2Br$	137.02	1, 119	1.2686^{25}_4	1.4374^{25}	–112.4	101.6	18	<0.1 aq; v s alc, eth
b275	2-Bromobutane	$CH_3CH_2CHBrCH_3$	137.02	1, 119	1.2585^{20}	1.4360^{20}	–112.7	91.4	21	
b276	1-Bromo-2-butene	$CH_3CH{=}CHCH_2Br$	135.01	1, 205	1.312	1.4765^{20}		99	11	Mixture of *cis, trans*
b277	2-Bromo-2-butene	$CH_3CH{=}C(Br)CH_3$	135.01	1, 205	1.328	1.4590^{20}		90^{740mm}	1	
b278	4-Bromo-1-butene	$BrCH_2CH_2CH{=}CH_2$	135.01	1^1, 84	1.3230^{20}_4	1.4608^{20}		100	9	i aq; s alc, eth
b279	4-Bromobutyl acetate	$CH_3CO_2(CH_2)_4Br$	195.06	2^3, 39	1.348	1.4600^{20}		93^{12mm}	109	
b280	1-Bromo-4-*tert*-butyl-benzene	$(CH_3)3CC_6H_4Br$	213.12	5, 416	1.229	1.5330^{20}	15–16	81^{8mm}	97	
b281	4-Bromobutyl phenyl ether	$C_6H_5O(CH_2)_4Br$	229.12	6^2, 82			41–43	156^{18mm}	>110	
b282	2-Bromobutyric acid	$CH_3CH_2CH(Br)COOH$	167.00	2, 281	1.5669^{20}_{20}	1.4720^{20}	–4	103^{10mm}	>110	6.7 aq; s alc, eth
b283	α-Bromo-γ-butyro-lactone		164.99		1.990^{20}	1.5080^{20}		138^{6mm}	>110	
b284	[1R-*endo*]-(+)-3-Bromocamphor		231.14	7, 120	1.449		75–78	244		15 alc; 200 Chl; 62 eth; s olive oil
b285	1-Bromocarbonyl-1-methylethyl acetate	$CH_3CO_2C(CH_3)_2COBr$	209.05		1.431	1.4570^{20}		77^{12mm}	110	
b286	2-Bromo-4'-chloro-acetophenone	$ClC_6H_4COCH_2Br$	233.50							

No.	Name	Formula	Ref.	FW	Density	n_D	mp	bp	fp	Solubility
b287	2-Bromochlorobenzene	BrC_6H_4Cl	5, 209	191.46	1.6384^{25}_4	1.5789^{25}		204	79	i aq; v s bz
b288	3-Bromochlorobenzene	BrC_6H_4Cl	5, 209	191.46	1.6302^{20}_4	1.5770^{20}	−21	196	80	i aq; v s alc, bz, eth
b296	4-Bromochlorobenzene	BrC_6H_4Cl	5, 209	191.46	1.576^{71}_4	1.5531^{70}	66	196		0.1 aq; misc MeOH, eth
b297	3-Bromo-4-chlorobenzotrifluoride	$Br(Cl)C_6H_3CF_3$	5^3, 715	259.46	1.726	1.4990^{20}	−22	190	94	
b298	1-Bromo-4-chlorobutane	$ClCH_2CH_2CH_2CH_2Br$	5^3, 294	171.47	1.488	1.4875^{20}		82^{30mm}	60	i aq; s alc, chl, eth
b299	4′-Bromo-4-chlorobutyrophenone	$BrC_6H_4CO(CH_2)_3Cl$		261.55			36–38		>110	
b300	4-Bromo-6-chloro-o-cresol	$Br(Cl)C_6H_2(OH)CH_3$	6, 360	221.49			45–47		>110	
b301	Bromochlorodifluoromethane	$Br(Cl)CF_2$		165.36	6.579 g/L		−160	−3.7		
b302	3-Bromo-1-chloro-5,5-dimethyl-hydantoin			241.48			160–164			
b303	1-Bromo-2-chloroethane	$ClCH_2CH_2Br$	1, 89	143.41	1.7392^{20}_4	1.4917^{20}	−18.4	106.6		0.7 aq; misc org solv
b303a	Bromochlorofluoromethane	$Br(Cl)CHF$		149.37	1.9771^{0}	1.4144^{55}	−115	36		
b304	7-Bromo-5-chloro-8-hydroxyquinoline		21^1, 222	258.51			177–179			
b305	Bromochloromethane	$ClCH_2Br$	1, 67	129.38	1.923^{25}_4	1.480^{25}	−88	68		0.9 aq; misc MeOH, eth
b306	1-Bromo-3-chloro-2-methylpropane	$ClCH_2CH(CH_3)CH_2Br$	1^3, 324	171.47	1.467	1.4809^{20}		154	>110	
b307	1-Bromo-3-chloropropane	$ClCH_2CH_2CH_2Br$	1, 109	157.44	1.492	1.4851^{20}	<−50	143.5		0.1 aq; misc org solv
b308	2-Bromo-2-chloro-1,1,1-trifluoroethane	$BrCH(Cl)CF_3$	1^4 156	197.39	1.8636^{25}	1.3691^{20}		50.2	none	
b309	2-Bromocinnamaldehyde	$C_6H_4CH=C(Br)CHO$	7, 358	211.06			66–68			
b310	Bromocycloheptane	$Br(C_7H_{13})$	5, 29	177.09	1.2887^{22}_2	1.5052^{20}		72^{10mm}	68	i aq; v s chl, eth
b311	Bromocyclohexane	$Br(C_6H_{11})$	5, 24	163.06	1.3264^{15}_4	1.4956^{15}		165.8	62	0.1 aq; 10 MeOH; 71 eth
b312	3-Bromocyclohexene		5^2, 40	161.04	1.3890^{20}_4	1.5292^{20}		65^{15mm}	54	
b313	Bromocyclopentane	$Br(C_5H_9)$	5, 19	149.04	1.3900^{20}_4	1.4881^{20}		137–139	35	
b314	Bromocyclopropane	$Br(C_3H_5)$		120.98	1.510	1.4605^{29}		69	−6	

(Continued)

TABLE 2.21 Physical Constants of Organic Compounds (*Continued*)

No.	Name	Formula	Formula weight	Beilstein reference	Density, g/mL	Refractive index	Melting point, °C	Boiling point, °C	Flash point, °C	Solubility in 100 parts solvent
b315	1-Bromodecane	$CH_3(CH_2)_9Br$	221.18	1^2, 130	1.0658^{20}	1.4560^{20}	−30	238–240	94	i aq; v s Chl, eth
b316	Bromodichloro-methane	$BrCHCl_2$	163.83	1, 67	1.980^{20}	1.4967^{20}	−55	87	none	sl s aq; misc org solv
b317	2-Bromo-1,1-di-ethoxyethane	$BrCH_2CH(OC_2H_5)_2$	197.08	1, 625	1.310	1.4385^{20}		67^{18mm}	51	s hot alc
b318	4-Bromo-1,2-di-methoxybenzene	$BrC_6H_3(OCH_3)_2$	217.07	6, 784	1.702	1.5743^{20}	256	109		
b319	2-Bromo-1,1-di-methoxyethane	$BrCH_2CH(OCH_3)_2$	169.02	1, 624	1.430	1.4450^{20}		150	53	
b320	1-Bromo-2,2-di-methoxypropane	$CH_3C(OCH_3)_2CH_2Br$	185.05		1.355	1.4475^{20}		87^{80mm}	40	
b321	4-Bromo-2,6-di-methylphenol	$BrC_6H_2(CH_3)_2OH$	201.07	6, 485			79–81			
b322	3-Bromo-2,2-dimethyl-1-propanol	$BrCH_2C(CH_3)_2CH_2OH$	167.05	1^1, 201	1.358	1.4794^{20}		184–187	75	v s hot alc, hot acet
b323	2-Bromo-4,6-dinitro-aniline	$BrC_6H_2(NO_2)_2NH_2$	262.03	12, 761			154	subl		
b324	1-Bromo-2,4-dinitro-benzene	$BrC_6H_3(NO_2)_2$	247.01				71–73			
b325	4-Bromodiphenyl ether	$BrC_6H_4OC_6H_5$	249.11	6^1, 105	1.423	1.6070^{20}	18	305	>110	
b326	1-Bromodiphenyl-methane	$C_6H_5CH(Br)C_6H_5$	247.14	5, 592			40–42	184^{20mm}	>110	
b327	1-Bromododecane	$CH_3(CH_2)_{11}Br$	249.24	1^2, 133	1.038	1.4580^{20}	−11	135^{6mm}	>110	0.1 aq; s alc, eth
b328	1-Bromo-2,3-epoxy-propane	$H_2C{-}CHCH_2Br$ (epoxide, O)	136.98	17, 9	1.601^{20}	1.4820^{20}	−40	134–136	56	i aq; sl s alc; s eth
b329	Bromoethane	CH_3CH_2Br	108.97	1, 88	1.4612^{20}_{4}	1.4242^{20}	−119	38.2	−23	0.91 aq^{20}; misc alc, chl, eth
b330	2-Bromoethanesulfonic acid, sodium salt	$BrCH_2CH_2SO_2^-\ Na^+$	211.02	4, 7			283 dec			
b331	2-Bromoethanol	$BrCH_2CH_2OH$	124.98	1, 338	1.7629^{20}_{4}	1.4936^{20}		57^{20mm}	>110	misc aq; s org solv except PE
b332	2-Bromoethyl acetate	$CH_3CO_2CH_2CH_2Br$	167.01	2^1, 57	1.514^{20}_{4}	1.4547^{20}	−13.8	159	71	v s aq; misc alc, eth
b333	2-Bromoethylamine HBr	$BrCH_2CH_2NH_2 \cdot HBr$	204.90	4, 134			172–174			v s aq, alc
b334	(1-Bromoethyl)-benzene	$C_6H_5CH(CH_3)Br$	185.07	5, 355	1.356	1.5600^{20}		94^{16mm}	81	
b334a	(2-Bromoethyl)-benzene	$C_6H_5CH_2CH_2Br$	185.07	5, 355	1.355	1.5560^{20}		221	89	

No.	Name	Formula	Mol. wt.	Beilstein ref.	Density	n_D	m.p., °C	b.p., °C	Flash pt, °C	Solubility
b335	1-Bromo-2-ethylbenzene	BrC$_6$H$_4$CH$_2$CH$_3$	185.07	5, 355	1.338	1.5490^{20}		194^{16mm}	71	i aq; misc alc, eth
b336	Bromoethylene	H$_2$C=CHBr	106.95	1, 188	1.493^{20}	1.4380^{20}	−139	15.8	none	sl s aq; misc alc, eth
b337	2-Bromoethyl ethyl ether	BrCH$_2$CH$_2$OCH$_2$CH$_3$	153.02	1, 338	1.3572^{20}_{4}	1.4450^{20}		150	21	i aq; s alc, eth
b338	2-Bromoethyl phenyl ether	BrCH$_2$CH$_2$OC$_6$H$_5$	201.07	6, 142			34	144^{40mm}	65	
b339	N-(2-Bromoethyl)phthalimide		254.09	21, 461			81–84			s hot aq; s eth
b340	1-Bromo-2-fluorobenzene	BrC$_6$H$_4$F	175.01		1.601	1.5337^{20}		156	43	
b341	1-Bromo-3-fluorobenzene	BrC$_6$H$_4$F	175.01		1.567	1.5257^{20}		150	38	
b342	1-Bromo-4-fluorobenzene	BrC$_6$H$_4$F	175.01	5, 209	1.593^{15}	1.5310^{15}	−17.4	152	60	
b343	1-Bromoheptane	H(CH$_2$)$_7$Br	179.11	1, 155	1.1384^{20}_{4}	1.4505^{20}	−58	180	60	i aq; v s alc, eth
b344	2-Bromoheptane	H(CH$_2$)$_5$CH(Br)CH$_3$	179.11	1, 155	1.142	1.4470^{20}		66^{21mm}	47	i aq; misc org solv
b345	1-Bromohexadecane	H(CH$_2$)$_{16}$Br	305.35	1^2, 138	0.9991	1.4618^{20}	17.8	336	177	i aq; misc alc, eth
b346	1-Bromohexane	H(CH$_2$)$_6$Br	165.08	1, 144	1.1763^{20}_{4}	1.4475^{20}	−85	154–158	57	s alc, eth
b347	dl-2-Bromohexanoic acid	CH$_3$(CH$_2$)$_3$CH(Br)COOH	195.06	2, 325	1.370	1.4720^{20}		138^{18mm}	>110	
b348	5-Bromoisatin		226.03	21, 453			251–253			i aq; 50 MeOH; misc org solvents
b350	(2-Bromoisopropyl)benzene	C$_6$H$_5$CH(CH$_3$)CH$_2$Br	199.10	5^1, 191	1.316	1.5480^{20}	−20	108^{18mm}	91	
b351	2-Bromo-4-isopropyl-1-methylbenzene	CH$_3$(Br)C$_6$H$_3$CH(CH$_3$)$_2$	213.0		1.253^{25}_{25}	1.535^{25}		120		
b352	Bromomaleic anhydride		176.96	17, 435	1.905	1.5400^{20}		215	>110	
b353	2-Bromomesitylene	1,3,5-(CH$_3$)$_3$C$_6$H$_2$Br	199.10	5, 408	1.301	1.5520^{20}	2	255	96	
b354	Bromomethane	CH$_3$Br	94.94	1, 67	1.732^{0}	1.4234^{10}	−94	3.56	none	
b355	4-Bromomandelic acid	BrC$_6$H$_4$CH(OH)COOH	231.05	10, 210			117–118			0.1 aq; s alc, Chl, eth
b356	5-Bromo-2-methoxybenzaldehyde	BrC$_6$H$_3$(OCH$_3$)CHO	215.05	8, 55			116–119			sl s aq
b357	2-Bromo-1-methoxybenzene	BrC$_6$H$_4$OCH$_3$	187.04	6, 197	1.5018^{20}	1.5737^{20}	2	223	96	i aq; v s alc, eth
b358	3-Bromo-1-methoxybenzene	BrC$_6$H$_4$OCH$_3$	187.04	6, 198	1.477	1.5635^{20}		211	93	i aq; s alc, eth
b359	4-Bromo-1-methoxybenzene	BrC$_6$H$_4$OCH$_3$	187.04	6, 199	1.4564^{20}_{4}	1.5630^{20}	10	223	94	
b360	4-Bromo-2-methylaniline	CH$_3$(Br)C$_6$H$_3$NH$_2$	186.06	12, 838			57–59	240	>110	sl s aq; v s alc, eth
b361	1-Bromo-3-methylbenzyl alcohol	BrC$_6$H$_4$CH(CH$_3$)OH	201.07	6^2, 447	1.460		36–38	121^{7mm}	63	sl s aq; v s alc

(Continued)

TABLE 2.21 Physical Constants of Organic Compounds (*Continued*)

No.	Name	Formula	Formula weight	Beilstein reference	Density, g/mL	Refractive index	Melting point, °C	Boiling point, °C	Flash point, °C	Solubility in 100 parts solvent
b362	1-Bromo-3-methyl-butane	$(CH_3)_2CHCH_2CH_2Br$	151.05	1, 136	1.210^{15}	1.4409^{20}	−112	119.7	32	0.02 aq; misc alc, eth
b363	2-Bromo-2-methyl-butane	$C_2H_5C(CH_3)_2Br$	151.05	1, 136	1.182	1.4423^{20}		107^{735mm}	5	
b364	2-Bromo-3-methyl-butanoic acid	$(CH_3)_2CHCH(Br)COOH$	181.04	2, 317			44	126^{20mm}	107	sl s aq; s alc, eth
b365	4-Bromo-2-methyl-2-butene	$BrCH_2C{=}C(CH_3)_2$	149.04	1^2, 189	1.293	1.4898^{20}		60^{60mm}	32	
b366	(Bromomethyl)Chloro-dimethylsilane	$BrCH_2Si(CH_3)_2Cl$	187.5	4^4, 4024	1.375	1.4650^{20}		130^{740mm}	41	
b367	(Bromomethyl)cyclo-hexane	$(C_6H_{11})CH_2Br$	177.09	5^2, 18	1.269	1.4907^{20}		77^{26mm}	57	
b368	2-Bromomethyl-1,3-dioxalane		167.01	19^2, 8	1.613	1.4817^{20}		82^{27mm}	62	
b369	Bromomethyl methyl ether	$BrCH_2OCH_3$	124.97	1, 582	1.531	1.4550^{20}		87	26	
b370	1-Bromo-2-methyl-naphthalcne	$Br(C_{10}H_6)CH_3$	221.10	5, 568	1.418	1.6486^{20}		296	>110	0.06 aq; misc alc, eth
b371	1-Bromo-2-methyl-propane	$(CH_3)_2CHCH_2Br$	137.03	1, 126	1.2641^{20}	1.4362^{20}	−119	91.5	18	i aq; misc org solv
b372	2-Bromo-2-methyl-propane	$(CH_3)_3CBr$	137.03	1, 127	1.2125^{25}_4	1.425^{25}	−16.2	73.1	18	sl s aq; s alc, eth; dec by hot aq
b373	2-Bromo-2-methyl-propanoic acid	$BrC(CH_3)_2COOH$	167.01	2, 295	1.52		48–49	200	>110	
b374	2-Bromo-2-methyl-propionyl bromide	$(CH_3)_2C(Br)COBr$	229.91	2, 297	1.860	1.5064^{24}		164	110	
b375	2-Bromo-2-methyl-propiophenone	$C_6H_5CO(CH_3)_2Br$	227.11	7, 316	1.350	1.5561^{20}		148^{30mm}	>112	
b376	1-Bromonaphthalcne	$(C_{10}H_7)Br$	207.07	5, 547	1.4834^{20}_4	1.6580^{20}	−1.8	281	>110	misc alc, bz, Chl, eth
b377	1-Bromo-1-naphthol	$BrC_{10}H_6OH$	233.07	6, 650			78	130 dec		i aq; s alc, bz, eth
b378	1-Bromo-2-naphthol	$BrC_{10}H_6OH$	223.07	6, 650	1.6245^{80}_4		78–81			
b379	1-Bromo-2-nitroben-zene	$BrC_6H_4NO_2$	202.01	5^1, 247			43	261	110	v s alc; s bz, eth
b380	5-Bromo-2-nitrobenzo-trifluoride	$O_2N(Br)C_6H_3CF_3$	270.02	5^3, 755	1.7992^{25}	1.5180^{25}	33–35	100^{5mm}	>110	s aq, alc, EtOAc; sl s bz, acet, Chl, eth
b381	2-Bromo-2-nitro-1,3-propanediol	$(HOCH_2)_2C(Br)NO_2$	199.99	1, 476			120–122			
b382	1-Bromononane	$H(CH_2)_9Br$	207.16	1^1, 63	1.084	1.4540^{20}		201	90	i aq; s Chl, eth

No.	Name	Formula	Formula wt.	Beilstein ref.	Density	n_D	mp, °C	bp, °C	Flash pt, °C	Solubility
b383	*exo*-2-Bromo-norbornane		175.07		1.363	1.5148^{20}		82^{29mm}	60	i aq; s alc, eth
b384	1-Bromooctadecane	$H(CH_2)_{18}Br$	333.41	1[1], 69	0.976		23	216^{12mm}	>110	i aq; misc alc, eth
b385	1-Bromooctane	$H(CH_2)_8Br$	193.13	1, 160	1.108^{24}	1.4518^{25}	−55	201	78	
b386	Bromopentafluoro-benzene	BrC_6F_5	246.97		1.947^{20}	1.4490^{20}	−31	137	87	i aq; s alc; misc eth
b387	1-Bromopentane	$H(CH_2)_5Br$	151.05	1, 131	1.2237^{15}	1.4444^{20}	−88	129.6	31	i aq; s alc, eth
b388	2-Bromopentane	$CH_3CH_2CH_2CH(Br)CH_3$	151.05	1, 131	1.2039^{20}	1.4403^{20}		117	20	
b389	3-Bromopentane	$C_2H_5CH(Br)C_2H_5$	151.05	1[1], 43	1.216	1.4445^{20}		119	18	
b390	5-Bromopentyl acetate	$CH_3CO_2(CH_2)_5Br$	209.09	2[3], 249	1.255	1.4620^{20}		110^{15mm}	>110	
b391	9-Bromophenanthrene		257.14	5, 671	1.409^{4}		54–58	190^{2mm}	>110	i aq; s alc, eth
b392	2-Bromophenol	BrC_6H_4OH	173.01	6, 197	1.492	1.5892^{20}	6	194	42	s aq; misc Chl, eth
b393	3-Bromophenol	BrC_6H_4OH	173.01	6, 198			32	236	>110	
b394	4-Bromophenol	BrC_6H_4OH	173.01	6, 198	1.5875^{80}		64	238	106	14 aq; v s alc, Chl
b395	1-(4-Bromophenoxy)-1-ethoxy ethane	$CH_3CH(OC_6H_4Br)OC_2H_5$	245.12		1.348	1.5229^{20}		125^{8mm}		
b396	4-Bromophenylacetic acid	$BrC_6H_4CH_2COOH$	215.05	9, 451			119		>110	sl s aq; v s alc, eth
b397	4-Bromophenylaceto-nitrile	$BrC_6H_4CH_2CN$	196.05	9, 451			47–49			i aq; sl s alc; v s bz
b398	4-Bromophenyl phenyl ether	$BrC_6H_4OC_6H_5$	249.11	6[1], 105	1.423	1.6070^{20}	18	305	>110	
b399	1-Bromo-3-phenyl-propane	$BrC_6H_4CH_2CH_2CH_2Br$	199.10	5, 391	1.310	1.5450^{20}		220	101	
b400	1-Bromopropane	$CH_3CH_2CH_2Br$	122.99	1, 108	1.3597^{15}	1.4370^{15}	−110.1	71.0	19	0.23 aq^{30}; misc alc
b401	2-Bromopropane	$CH_3CH(Br)CH_3$	123.99	1, 108	1.3222^{15}	1.4285^{15}	−89.0	59.5		0.3 aq^{18}; misc alc, bz, chl, eth
b402	3-Bromo-1-propanol	$BrCH_2CH_2CH_2OH$	139.00	1, 356	1.5374^{20}	1.4858^{20}		62^{5mm}	93	s aq; misc alc, eth
b403	1-Bromo-2-propanone	CH_3OCH_2Br	136.98	Merck: 12, 1422	1.634^{23}	1.4697^{15}	−36.5	137		v sl s aq; s alc; s acet
b404	1-Bromo-1-propene	$CH_3CH{=}CHBr$	120.98	1, 200	1.4133^{20}	1.4538^{20}	−116	70	−6	i aq
b405	2-Bromo-2-propene	$CH_2C(Br){=}CH_2$	120.98	1, 200	1.362^{4}	1.4425^{20}	−125	47–49	4	
b406	2-Bromopropionic acid	$CH_3CH(Br)COOH$	152.98	2, 254	1.7000^{20}	1.4750^{20}	25.7	203	100	v s aq, alc, bz, chl, eth
b407	3-Bromopropionic acid	$BrCH_2CH_2COOH$	152.98	2, 256	1.480		62.5		65	s aq, alc, bz, chl, eth
b408	3-Bromopropionitrile	$BrCH_2CH_2CN$	133.98	2[2], 231	1.6152^{20}	1.4800^{20}		78^{10mm}	98	v s alc, eth
b409	2-Bromopropionyl bromide	$CH_3CH(Br)COBr$	215.88	2, 256	2.061	1.5182^{20}		50^{10mm}	>110	
b410	2-Bromopropionyl chloride	$CH_3CH(Br)COCl$	171.43	2, 256	1.700^{11}	1.4800^{20}		133	51	d aq; s chl, eth
b411	3-Bromopropionyl chloride	$CH_3CH(Br)COCl$	171.43	2[2], 231	1.701	1.4968^{20}		57^{17mm}	79	

(Continued)

TABLE 2.21 Physical Constants of Organic Compounds (*Continued*)

No.	Name	Formula	Formula weight	Beilstein reference	Density, g/mL	Refractive index	Melting point, °C	Boiling point, °C	Flash point, °C	Solubility in 100 parts solvent
b412	2-Bromopropiophenone	$C_6H_5COCH(Br)CH_3$	213.08	7, 302	1.430_4^{20}	1.5715^{20}		250	>110	s alc, bz, eth, acet
b413	3-Bromopropyl phenyl ether	$C_6H_5OCH_2CH_2CH_2Br$	215.10	6, 142	1.365	1.5464^{20}	10–11	134^{14mm}	96	
b414	3-Bromopropyltrichlorosilane	$Br(CH_2)_3SiCl_3$	256.44		1.605	1.4900^{20}		202–204	76	
b415	3-Bromopropyne	$BrCH_2C{\equiv}CH$	118.97	1, 248	1.335	1.4905^{20}		88–90	18	i aq; s org solv
b416	2-Bromopyridine	$Br(C_5H_4N)$	158.00	20, 233	1.657^{18}	1.5720^{20}		194	54	s aq; v s alc, eth
b417	3-Bromopyridine	$Br(C_5H_4N)$	158.00	20, 233	1.645^0	1.5695^{20}	142–143	173	51	s HOAc
b418	3-Bromoquinoline		208.06	20, 363	1.533	1.6640^{20}	15	276	>110	0.3 aq^{80}, 85 alc^{25}; 70 eth^{25}
b419	5-Bromosalicylic acid	$Br(HO)C_6H_3COOH$	217.02	10, 107			166			
b420	β-Bromostyrene	$C_6H_5CH{=}CHBr$	183.05	5, 477	1.422_4^{20}	1.6066^{20}	7	112^{20mm}	79	i aq; misc alc, eth
b421	(±)-Bromosuccinic acid	$HOOCCH_2CH(Br)COOH$	196.99	2, 621	2.073		161			18 aq; s alc, acet, eth
b422	N-Bromosuccinimide		177.99	21, 380	2.098		173 sl dec			1.5 aq^{25}, 14.4 acet^{25}, 3.1 HOAc^{25}
b423	1-Bromotetradecane	$H(CH_2)_{14}Br$	277.30	1^2, 136	1.0124_4^{25}	1.4600^{20}	6	178^{20mm}	>110	s alc; v s chl; misc bz, acet
b424	3-Bromotetrahydro-2H-pyran		179.06	17^3, 75	1.366	1.4830^{20}		61^{17mm}	57	
b425	3-Bromothioanisole	$BrC_6H_4SCH_3$	203.11	6, 330			38–40		>110	v s acet, eth
b426	2-Bromothiophene	$Br(C_5H_3S)$	163.04	17, 33	1.684_4^{20}	1.5860^{20}		151	60	
b427	3-Bromothiophene	$Br(C_5H_3S)$	163.04	17, 33	1.740	1.5910^{20}		150	56	
b428	4-Bromothiophenol	BrC_6H_4SH	189.08	6, 330			76	239		0.1 aq; misc alc, bz, chl, eth
b429	2-Bromotoluene	$BrC_6H_4CH_3$	171.04	5, 304	1.422_{25}^{25}	1.552^{25}	–26	181	78	s alc, bz, eth
b430	3-Bromotoluene	$BrC_6H_4CH_3$	171.04	5, 305	1.4099^{20}	1.5517^{20}	–39.8	183.7	60	s alc, bz, eth
b431	4-Bromotoluene	$BrC_6H_4CH_3$	171.04	5, 305	1.3959^{35}	1.5490	28.5	184.5	85	misc org solv
b432	Bromotrichloromethane	$BrCCl_3$	198.28	1, 67	1.9972^{25}	1.5063^{20}	–6	104–105		
b433	1-Bromotridecane	$H(CH_2)_{13}Br$	263.27	1^2, 134	1.0262_4^{24}	1.4592^{20}	7	150^{10mm}	>110	v s chl
b434	Bromotrifluoromethane	$BrCF_3$	148.91	1^3, 83	6.087 g/L		–168 to –172	–57.8		v s chl
b435	5-Bromo-1,2,4-trimethylbenzene	$BrC_6H_2(CH_3)_3$	199.10	5, 403			73	235		i aq; s alc
b436	2-Bromo-1,3,5-trimethylbenzene	$BrC_6H_2(CH_3)_3$	199.10	5, 408	1.301	1.5511^{20}	2	225	96	i aq; s bz; v s eth

No.	Name	Formula	M.W.	Beil. ref.	Density	n_D	m.p.	b.p.	Flash pt.	Solubility
b437	Bromotrimethyl-germane	(CH$_3$)$_3$GeBr	197.60		1.544^{18}	1.4705^{20}	-25	113.7	37	i aq; v s alc
b438	Bromotrimethylsilane	(CH$_3$)$_3$SiBr	153.10		1.160	1.4140^{20}		79	32	
b439	Bromotriphenyl-ethylene	(C$_6$H$_5$)$_2$C=C(Br)C$_6$H$_5$	335.22	5, 722			115–117			s alc, eth
b440	Bromotriphenyl-methane	(C$_6$H$_5$)$_3$CBr	323.24	5, 704			152–154	230^{15mm}		s alc, eth
b441	1-Bromoundecane	CH$_3$(CH$_2$)$_{10}$Br	235.22	1^2, 132	1.954	1.4563^{20}	-9	138^{18mm}	>110	v s chl, hot ether
b442	11-Bromoundecanoic acid	Br(CH$_2$)$_{10}$COOH	265.20	2^2, 315			51	174^{2mm}	>110	v s chl, hot ether
b443	α-Bromo-1,2-xylene	BrCH$_2$C$_6$H$_3$CH$_3$	185.07	5, 365	1.381^{23}	1.581^{20}	21	224	82	s alc, eth
b444	α-Bromo-1,3-xylene	BrCH$_2$C$_6$H$_3$CH$_3$	185.07	5, 374	1.370^{23}	1.5560^{20}		185^{340mm}	82	s alc, eth
b445	2-Bromo-1,4-xylene	BrCH$_2$C$_6$H$_3$CH$_3$	185.07	5, 385	1.340	1.5505^{20}	9–10	199–201	79	v s alc, eth
b446	4-Bromo-1,2-xylene	BrCH$_2$C$_6$H$_3$CH$_3$	185.07	5, 365	1.370^{15}	1.5560^{20}		215	80	v s alc, eth
b447	Brucine		394.45	27^2, 797			178			77 alc; 1 bz; 20 chl; 4 EtOAc
b448	1,2-Butadiene	CH$_3$CH=C=CH$_2$	54.09	1, 249	0.676^{10}	1.4205^1	-136.2	10.9		misc alc, eth
b449	1,3-Butadiene	H$_2$C=CHCH=CH$_2$	54.09	1, 249	2.211 g/L	1.4293^{-25}	-108.9	-4.4	-76	misc alc, eth
b450	Butadiene sulfone		118.15	17^3, 144			66		>110	
b451	1,3-Butadienyl acetate	CH$_3$CO$_2$CH=CHCH=CH$_2$	112.13	2^3, 295	0.945	1.4690^{20}	-36	60^{40mm}	33	v s eth; s acet, bz
b452	1,3-Butadiyne	HC≡CC≡CH	50.06	1^3, 1056	0.7364^0	1.4189^5		10.3		
b453	2-Butanamine	CH$_3$CH$_2$CH(NH$_2$)CH$_3$	73.14	4, 160	0.7308^{15}	1.3963^{15}	-104.5	66	-19	misc aq, alc
b454	Butane	CH$_3$CH$_2$CH$_2$CH$_3$	58.12	1, 118	0.6011^0	1.3562^{-13}	-138.3	-0.50	-60	1 vol aq dissolves 0.15 vol and 1 vol alc 18 vols at 17° and 770 mm; 1 vol ether or CHCl$_3$ dissolves 25 or 30 vols, resp.
b455	1,4-Butanediamine	H$_2$NCH$_2$CH$_2$CH$_2$CH$_2$NH$_2$	88.15	4, 264	0.877^{25}_4	1.4569^{20}	28	158–160	51	s aq
b456	Butanedinitrile	NCCH$_2$CH$_2$CN	80.09	2, 615	0.9867^{60}_4	1.4173^{60}	54.5	266	132	11.5 aq; s acet, chl, 1,4-dioxane; sl s bz
b457	1,2-Butanediol	CH$_3$CH(OH)CH$_2$OH	90.12	1, 477	1.0060^{18}_0	1.4380^{20}		207.5	93	s aq, alc, acet
b457a	1,3-Butanediol	CH$_3$CH(OH)CH$_2$CH$_2$OH	90.12	1, 477	1.0053^{20}_4	1.441^{20}	<-50	207.5	121	s aq, alc, acet; 9 eth
b457b	1,4-Butanediol	HOCH$_2$CH$_2$CH$_2$CH$_2$OH	90.12	1, 478	1.016^{25}_4	1.4452^{20}	20	235	121	misc aq, alc, acet; 0.3 bz; 3.1 eth; 0.9 PE
b458	meso-2,3-Butanediol	CH$_3$CH(OH)CH(OH)CH$_3$	90.12	1, 479	0.9939^{25}_4	1.4324^{35}	25	182	85	misc aq, alc
b459	1,4-Butanediol di-methanesulfonate	CH$_3$SO$_2$O(CH$_2$)$_4$OSO$_2$CH$_3$	246.30	4^4, 19			114–117			2.4 acet^{25}; 0.1 alc^{25}
b460	1,3-Butanediol di-acetate	CH$_3$CO$_2$CH$_2$CH$_2$CH(CH$_3$)-O$_2$CCH$_3$	174.20	2, 143	1.028	1.4199^{20}		99^{8mm}	85	
b461	1,4-Butanediol di-acrylate	(H$_2$C=CHCO$_2$CH$_2$CH$_2$-)$_2$	198.22	2^4, 170	1.051	1.4560^{20}		$83^{0.3mm}$	>110	

(Continued)

TABLE 2.21 Physical Constants of Organic Compounds (*Continued*)

No.	Name	Formula	Formula weight	Beilstein reference	Density, g/mL	Refractive index	Melting point, °C	Boiling point, °C	Flash point, °C	Solubility in 100 parts solvent
b462	1,3-Butanediol di–methacrylate	$H_2C{=}C(CH_3)CO_2CH_2CH_2$-$CH(CH_3)O_2CC(CH_3){=}CH_2$	226.28		1.010	1.4520^{20}		290	>110	
b463	1,4-Butanediol di–methacrylate	$[H_2C{=}C(CH_3)CO_2CH_2CH_2]_2$	226.28	2^4, 1534	1.010	1.4560^{20}		134^{4mm}	>110	
b464	1,4-Butanediol divinyl ether	$(—CH_2CH_2OCH{=}CH_2)_2$	142.20	1^4, 2518	0.898	1.444^{20}	−8	64^{10mm}	62	
b465	1,4-Butanediol vinyl ether	$H_2C{=}CHO(CH_2)_4OH$	116.16	1^4, 2518	0.939	1.4440^{20}		95^{20}	85	
b466	2,3-Butanedione	$CH_3C({=}O)C({=}O)CH_3$	86.09	1, 769	0.990^{15}	1.3951^{20}		86	7	25 aq; misc alc, eth
b467	2,3-Butanedione mon-oxide	$CH_3C({=}NOH)C({=}O)CH_3$	101.11	1, 772			75–78	186		
b468	1,4-Butanedithiol	$HSCH_2CH_2CH_2CH_2SH$	122.25	1, 479	1.042	1.5290^{20}	−112	106^{30mm}	70	i aq; v s alc
b468a	Butanenitrile	$CH_3CH_2CH_2CN$	69.11	2^2, 252	0.7936	1.4440^{20}		117.6	24	3.3 aq; misc alc, eth
b469	1,2,3,4-Butanetetra-carboxylic acid	$[—CH(COOH)CH_2COOH]_2$	234.16	2, 863			196			v s alc
b470	1-Butanethiol	$CH_3CH_2CH_2CH_2SH$	90.19	1, 370	0.8367_4^{25}	1.4430^{25}	−116	98.5	2	0.06 aq; v s alc, eth
b471	2-Butanethiol	$CH_3CH_2CH(SH)CH_3$	90.19	1, 373	0.8246_4^{25}	1.4338^{25}	−165	85.0	21	sl s aq; v s alc, eth
b472	1,2,4-Butanetriol	$HOCH_2CH_2CH(OH)CH_2OH$	106.12	1, 519	1.190^{20}	1.4748^{20}		191^{18mm}	167	v s aq, alc
b473	1-Butanol	$CH_3CH_2CH_2CH_2OH$	74.12	1, 367	0.8097^{20}	1.3993^{20}	−89.5	117.7	37	7.4 aq; misc alc, eth
b474	2-Butanol	$CH_3CH_2CH(OH)CH_3$	74.12	1, 371	0.8069^{20}	1.3972^{20}	−114.7	99.5	24	12.5 aq; misc alc, eth
b475	2-Butanone	$CH_3CH_2COCH_3$	72.11	1, 666	0.8054^{20}	1.3788^{20}	−86.7	79.6	−9	24 aq; misc alc, bz, eth
b476	2-Butanone oxime	$CH_3CH_2C({=}NOH)CH_3$	87.12	1, 668	0.924	1.4420^{20}		60^{15mm}	60	
b477	1-Butene	$CH_3CH_2CH{=}CH_2$	56.11	1, 203	0.6255^{mp}	1.3962^{20}	−185.3	−6.5	−80	i aq; v s alc, eth
b478	cis-2-Butene	$CH_3CH{=}CHCH_3$	56.11	1^3, 728	0.6213	1.3931^{-25}	−139.3	3.7	−73	i aq; v s alc, eth
b479	trans-2-Butene	$CH_3CH{=}CHCH_3$	56.11	1, 205	0.6041	1.3848^{-25}	−105.8	0.9	−73	i aq; v s alc, eth
b480	cis-2-Butene-1,4-diol	$HOCH_2CH{=}CHCH_2OH$	88.11	1^2, 567	1.0700_4^{20}	1.4780^{20}	2	234	128	s aq; v s alc
b481	trans-2-Butene-1,4-diol	$HOCH_2CH{=}CHCH_2OH$	88.11	1^3, 2252	1.070_4^{20}	1.4755^{20}	25	132		v s aq, alc
b482	3-Butenenitrile	$H_2C{=}CHCH_2CN$	67.09	2, 408	0.8341_4^{20}	1.4060^{20}	−87	119	21	sl s aq; misc alc, eth
b483	cis-2-Butenoic acid	$CH_3CH{=}CHCOOH$	86.09	2, 412	1.0267^{20}	1.4483^{14}	14–15	168–169		v s aq; s alc
b484	trans-2-Butenoic acid	$CH_3CH{=}CHCOOH$	86.09	2, 408	0.9604_4^{80}	1.4248^{77}	72	185	87	55 aq; 52 EtOH; 53 acet; 37 toluene
b485	3-Butenoic acid	$H_2C{=}CHCH_2COOH$	86.09	2, 407	1.0091^{20}	1.4249^{20}	−39	163	65	s aq; misc alc, eth
b486	cis-2-Buten-1-ol	$CH_3CH{=}CHCH_2OH$	72.11	1, 442	0.8662^{20}	1.4342^{20}	−89.4	123.6	56	16.6 aq; misc alc
b487	trans-2-Buten-1-ol	$CH_3CH{=}CHCH_2OH$	72.11	1, 442	0.8524_4^{20}	1.4289^{20}	<−30	121.2	56	16.6 aq; misc alc
b488	3-Buten-2-one	$H_2C{=}CHCOCH_3$	70.09	1, 728	0.8636^{20}	1.4086^{20}		81.4	−6	v s aq, alc, acet, eth
b489	1-Buten-3-yne	$HC{\equiv}CCH{=}CH_2$	52.07	1^3, 1032	0.7095_4	1.4161		5.1		
b490	4-Butoxyaniline	$CH_3(CH_2)_3OC_6H_4NH_2$	165.24	13^2, 226	0.992	1.5543^{20}		149^{13mm}	>110	

No.	Name	Formula	Mol wt	Beilstein ref	Density	n_D	mp, °C	bp, °C	Flash pt, °C	Solubility
b491	4-Butoxybenzoic acid	$CH_3(CH_2)_3OC_6H_4COOH$	194.23	10[2], 93	1.100	1.4900^{20}	150	219^{5mm}	>110	5 aq; s most org solv
b492	Butoxycarbonylmethyl butyl phthalate	$2-[CH_3(CH_2)_3]O_2CCH_2O_2C]-C_6H_4CO_2(CH_2)_3CH_3$	336.39	9[3], 4187						
b493	2-Butoxyethanol	$CH_3(CH_2)_3OCH_2CH_2OH$	118.18	1[2], 519	0.9012^{20}	1.4198^{20}	−75	168	69	
b494	1-tert-Butoxy-2-ethoxyethane	$(CH_3)_3COCH_2CH_2OC_2H_5$	146.23	1[3], 2085	0.834	1.4015^{20}		148	33	
b495	2-(2-Butoxyethoxy)-Ethanol	$HOCH_2CH_2OCH_2CH_2OC_4H_9$	162.23	1[2], 521	0.9536^{20}_{20}	1.4306^{20}	−68.1	230.4	100	misc aq, alc, bz, acet, CCl$_4$, PE
b496	2-(2-Butoxyethoxy)-ethyl acetate	$CH_3CO_2(CH_2O_2CH_2CH_2CH_2CH_2CH_3$	204.27	2[3], 308	0.978	1.4260^{20}		245	>110	
b497	2-Butoxyethyl acetate	$CH_3CO_2CH_2CH_2O(CH_2)_3CH_3$	160.22	2[3], 307	0.942	1.4136^{20}		192	76	
b498	2-tert-Butoxy-2-methoxyethane	$(CH_3)_3CO_2CH_2CH_2OCH_3$	132.20	1[3], 2084	0.840	1.3985^{20}		132	25	
b499	1-tert-Butoxy-2-propanol	$(CH_3)_3COCH_2CH(OH)CH_3$	132.10	1[3], 2148	0.874	1.4130^{20}		143–145	44	
b500	3-Butoxypropylamine	$CH_3(CH_2)_3O(CH_2)_3NH_2$	131.22	4[3], 739	0.853	1.4260^{20}	−77/−78	170	63	0.43 aq; misc alc, eth; s most org solvents
b501	Butyl acetate	$C_4H_9O_2CCH_3$	116.16	2, 130	0.8813^{4}	1.3941^{20}		126	22	0.62 aq; s alc, eth
b502	dl-sec-Butyl acetate	$CH_3CO_2CH(CH_3)C_2H_5$	116.16	2[2], 131	0.8748^{20}	1.3888^{20}	−99	112	31	i aq; misc alc, eth
b503	tert-Butyl acetate	$(CH_3)_3CO_2CCH_3$	116.16	2, 131	0.8665^{20}	1.3870^{20}		95.1	16	
b504	tert-Butylacetic acid	$(CH_3)_3CCH_2COOH$	116.16	2, 337	0.912	1.4115^{20}	6–7	190		
b505	tert-Butyl acetoacetate	$(CH_3)_3COC(=O)CH_2-C(=O)CH_3$	158.20	1[4], 3482	0.954	1.4180^{20}			60	
b506	2-Butylacrolein	$CH_3(CH_2)_3C(=CH_2)CHO$	112.17	4[4], 664	0.843	1.4348^{20}		139	33	
b507	N-tert-Butylacrylamide	$H_2C=CHCONHC(CH_3)_3$	127.19	2[2], 388			128–129			
b507a	Butyl acrylate	$H_2C=CHCO_2(CH_2)_3CH_3$	128.17	2[3], 1228	0.894	1.4180^{20}	−64	145	39	0.14 aq[20]
b508	tert-Butyl acrylate	$H_2C=CHCO_2C(CH_3)_3$	128.17	4, 156	0.875	1.4108^{20}		63^{60mm}	17	
b509	Butylamine	$CH_3CH_2CH_2CH_2NH_2$	73.14	4, 160	0.7327^{25}	1.3992^{25}	−50/−49	77	−12	misc aq, alc, eth
b510	(±)-sec-Butylamine	$C_2H_5CH(NH_2)CH_3$	73.14	4, 173	0.7244^{20}	1.3928^{20}	−104	63	−9	misc aq, alc
b511	tert-Butylamine	$(CH_3)_3CNH_2$	73.14	14[2], 249	0.6951^{4}	1.3788^{20}	−66	44	−9	misc aq, alc
b512	Butyl-4-aminobenzoate	$H_2NC_6H_4CO_2(CH_2)_3CH_3$	193.25				57–59	174^{8mm}		v sl s aq; s dil acids, alc, chl, eth
b513	2-(tert-Butylamino)-ethanol	$(CH_3)_3CNHCH_2CH_2OH$	117.19		0.914	1.4420^{20}	42–45	92^{25mm}	68	
b514	2-(tert-Butylamino)-ethyl methacrylate	$H_2C=C(CH_3)CO_2CH_2-CH_2NC(CH_3)_3$	185.27	4[4], 1509				82^{10mm}	71	
b515	3-(tert-Butylamino)-1,2-propanediol	$(CH_3)_3CNHCH_2CH(OH)-CH_2OH$	147.22				70	92^{1mm}		
b516	2-Butylaniline	$CH_3(CH_2)_3C_6H_4NH_2$	149.24	12[2], 633	0.953	1.5380^{20}		123^{12mm}	108	
b517	2-sec-Butylaniline	$C_2H_5CH(CH_3)C_6H_4NH_2$	149.24	12[3], 2721	0.957	1.5410^{20}		122^{16mm}	>110	
b518	4-Butylaniline	$CH_3(CH_2)_3C_6H_4NH_2$	149.24	12[1], 503	0.945	1.5350^{20}		120^{15mm}	101	
b519	4-sec-Butylaniline	$C_2H_5CH(CH_3)C_6H_4NH_2$	149.24	12[2], 635	0.977	1.5370^{20}		245^{727mm}	107	

(Continued)

TABLE 2.21 Physical Constants of Organic Compounds (*Continued*)

No.	Name	Formula	Formula weight	Beilstein reference	Density, g/mL	Refractive index	Melting point, °C	Boiling point, °C	Flash point, °C	Solubility in 100 parts solvent
b520	2-*tert*-Butylanthraquinone		264.32				98–100			
b521	Butylbenzene	$CH_3CH_2CH_2CH_2C_6H_5$	134.22	5, 413	0.8604^{20}_4	1.4898^{20}	−88	183	71	misc alc, bz, eth
b522	*sec*-Butylbenzene	$C_2H_5CH(CH_3)C_6H_5$	134.22	5, 414	0.8608^{20}_4	1.4890^{20}	−82.7	173	52	misc alc, bz, eth
b523	*tert*-Butylbenzene	$(CH_3)_3CC_6H_5$	134.22	5, 415	0.8669^{20}_4	1.4923^{20}	−58.1	168.5	60	misc alc, bz, eth
b524	Butyl benzoate	$C_6H_5CO_2C_4H_9$	178.23	9, 112	1.0000^{20}	1.496	−22	250	106	i aq; s alc, eth
b525	2-Butylbenzofuran		174.25		0.987	1.5330^{20}			101	
b526	4-*tert*-Butylbenzoic acid	$(CH_3)_3CC_6H_4COOH$	178.23	9, 560	1.142^{20}_4		166.3			i aq; v s alc, bz
b527	4-*tert*-Butylbenzoyl chloride	$(CH_3)_3CC_6H_4COCl$	196.68		1.007	1.5364^{20}		135^{20mm}	87	
b528	*N*-(*tert*-Butyl)benzyl-amine	$C_6H_5CH_2NHC(CH_3)_3$	163.27	12, 1022	0.881	1.4968^{20}		80^{5mm}	80	
b529	*tert*-Butyl bromo-acetate	$BrCH_2CO_2C(CH_3)_3$	195.06	2^1, 96	1.321	1.4450^{20}		50^{10mm}	49	
b530	Butyl 2-butoxy-2-hydroxyacetate	$CH_3(CH_2)_3OCH(OH)CO_2(CH_2)_3CH_3$	204.27	3^4, 1497	0.996	1.4291^{20}		90^{40mm}	74	
b531	Butyl butyrate	$CH_3CH_2CH_2CCO_2C_4H_9$	144.22	2, 271	0.8692^{20}_4	1.4064^{20}	−91.5	166	49	i aq; misc alc, eth
b532	Butyl carbamate	$H_2NCO_2(CH_2)_3CH_3$	117.15				53–55		108	
b533	*tert*-Butyl carbazate	$H_2NNHCO_2C(CH_3)_3$	132.16				39–42	$65^{0.03mm}$	91	
b534	4-*tert*-Butylcatechol	$(CH_3)_3C_6H_3$-1,2-$(OH)_2$	166.22		1.049^{60}_{25}		52–55	285	151	0.2 aq;[80] 240 eth;[25] s alc; v s acet
b535	*tert*-Butyl chloro-acetate	$ClCH_2CO_2C(CH_3)_3$	150.61	2^3, 444	1.053	1.4230^{20}		49^{11mm}	46	
b536	4-*tert*-Butyl-1-chloro-benzene	$(CH_3)_3C_6H_4Cl$	158.67	5, 416	1.006	1.5108^{20}	23–25	217		
b537	*tert*-Butylchlorodi-phenylsilane	$(CH_3)_3CSi(C_6H_5)_2Cl$	274.87		1.057	1.5675^{20}		$90^{0.02mm}$	>110	d aq. alc; misc eth
b538	Butyl chloroformate	$ClCO_2C_4H_9$	136.58	3^2, 11	1.074^{25}	1.4114^{20}		142	25	
b539	Butyl cyanoacetate	$NCCH_2CO_2C_4H_9$	141.17	2^1, 255	0.993	1.4254^{20}		115^{15mm}	87	
b540	*tert*-Butyl cyanoacetate	$NCCH_2CO_2C(CH_3)_3$	141.17		0.972	1.4200^{20}		108	91	
b541	Butylcyclohexane	$(C_6H_{11})_6C_4H_9$	140.27	5^1, 20	0.818	1.4400^{20}		178–180	41	
b542	*tert*-Butylcyclohexane	$(C_6H_{11})C(CH_3)_3$	140.27	5^1, 20	0.831	1.4470^{20}	−78	167	42	

(Continued)

No.	Name	Formula	Formula weight	Beilstein reference	Density	n_D	mp, °C	bp, °C	Flash point, °C	Solubility
b543	2-*tert*-Butylcyclohexanol	$(CH_3)_3C(C_6H_{10})OH$	145.27	6^3, 126	0.902		43–46		79	i aq
b544	4-*tert*-Butylcyclohexanol	$(CH_3)_3C(C_6H_{10})OH$	156.27	6^1, 18			62–70	115^{15mm}	105	i aq
b545	2-*tert*-Butylcyclohexanone	$(CH_3)_3C(C_6H_9)(=O)$	154.25	7^3, 143	0.896	1.4565^{20}		63^{4mm}	72	
b546	4-*tert*-Butylcyclohexanone	$(CH_3)_3C(C_6H_9)(=O)$	154.25	7^1, 29	0.994^{25}_{25}		47–50	116^{20mm}	96	i aq
b547	Butyl decyl *o*-phthalate	$C_4H_9O_2C_6H_4CO_2C_{10}H_{21}$	362.51	6^3, 2094	0.902				202	
b548	4-*sec*-Butyl-2,6-di-*tert*-Butylphenol	$C_2H_5CH(CH_3)C_6H_2(OH)$ $[C(CH_3)_3]$	262.44				25	142^{10mm}	>110	
b549	N-Butyldiethanolamine	$C_4H_9N(CH_2CH_2OH)_2$	161.25	4, 285	0.986^{20}_{20}	1.4625^{20}	–70	276	126	
b550	Butyl 3,4-dihydro-2,2-dimethyl-4-oxo-2*H*-pyran-6-carboxylate		226.27		1.054^{25}_{25}	1.4767^{20}		256–270	>110	
b551	*tert*-Butyldimethylchlorosilane	$(CH_3)_3CSi(CH_3)_2Cl$	150.73	4^4, 4076			89	124–126	22	
b552	6-*tert*-Butyl-2,4-dimethylphenol	$(CH_3)_3CC_6H_2(CH_3)_2OH$	178.28	6^3, 2020		1.5178^{20}	23	249	111	
b553	N-Butylethanolamine	$HOCH_2CH_2NHC_4H_9$	117.19		0.89^{20}	1.444^{20}	–3.5	192	77	i aq; misc alc, eth
b554	Butyl ethyl ether	$C_4H_9OC_2H_5$	102.18	1, 369	0.7495^{20}_4	1.3818^{20}	–124	92	4	
b555	2-Butyl-2-ethyl-1,5-pentanediamine	$H_2N(CH_2)_3C[(CH_2)_3CH_3]$ $(C_2H_5)CH_2NH_2$	186.34		0.876	1.4700^{20}		269^{750mm}	>110	
b556	2-Butyl-2-ethyl-1,3-propanediol	$HOCH_2C(C_2H_5)(C_4H_9)$ CH_2OH	160.25	1^3, 2228	0.931^{50}_{20}	1.4587^{25}	41–44	178^{50mm}	>110	0.8 aq
b557	Butyl ethyl sulfide	$C_4H_9SC_2H_5$	118.24	1^3, 1522	0.8376^{20}_4	1.4491^{20}	–95.1	144.2	95	s chl
b558	N-*tert*-Butylformamide	$HCONHC(CH_3)_3$	101.15	4^3, 324	0.903	1.4330^{20}	16	202	18	
b559	Butyl formate	$HCO_2C_4H_9$	102.13	2, 21	0.892	1.3889^{20}	–91.5	106		
b560	Butyl glycidyl ether	$\overset{O}{H_2C{-}CH}CH_2OC_4H_9$	130.19							
b561	*tert*-Butyl glycidyl ether	$\overset{O}{H_2C{-}CH}CH_2OC_4H_9(CH_3)_3$		17^3, 988	0.917	1.4166^{20}			43	
b562	*tert*-Butylhydrazine HCl	$(CH_3)_3CNHNH_2 \cdot HCl$	124.61	4^3, 1734			194			
b563	*tert*-Butyl hydroperoxide	$(CH_3)_3C{-}O{-}OH$	90.12	1^3, 1579	0.896^{20}_4	1.4007^{20}	–8	34^{17mm}	37	s aq, alc, chl, eth
b564	1-Butylimidazole		124.19	23^2, 36	0.945	1.4800^{20}		116^{12mm}	>110	
b565	Butyl isocyanate	C_4H_9NCO	99.13		0.880	1.4061^{20}		115	17	

TABLE 2.21 Physical Constants of Organic Compounds (*Continued*)

No.	Name	Formula	Formula weight	Beilstein reference	Density, g/mL	Refractive index	Melting point, °C	Boiling point, °C	Flash point, °C	Solubility in 100 parts solvent
b566	*tert*-Butyl isocyanate	(CH$_3$)$_3$CNCO	99.13	4^3, 175	0.868	1.3865^{20}		86	−4	
b567	Butyl lactate	CH$_3$CH(OH)CO$_2$C$_4$H$_9$	148.19	3^2, 207	0.984	1.4210^{20}	−28	185–187	69	
b568	Butyl levulinate	CH$_3$COCH$_2$CH$_2$CO$_2$C$_4$H$_9$	172.22		0.974	1.4270^{20}		$108^{5.5mm}$	91	
b569	Butyl 3-mercapto-propionate	HSCH$_2$CH$_2$CO$_2$C$_4$H$_9$	162.25		0.795	1.4100^{20}		101^{12mm}	93	
b570	Butyl methacrylate	H$_2$C=C(CH$_3$)CO$_2$C$_4$H$_9$	142.19	2^3, 1286	0.889_{15}^{25}	1.4230^{25}		170	50	i aq; misc alc, eth
b571	*sec*-Butyl-2-methyl-2-butenoate	CH$_3$CH=C(CH$_3$)CO$_2$-CH(CH$_3$)C$_2$H$_5$	156.23		0.889	1.4350^{20}		85^{27mm}	66	
b572	*tert*-Butyl methyl ether	(CH$_3$)$_3$C—O—CH$_3$	88.15	1, 381	0.7404_4^{20}	1.3689^{20}	−109	52	−28	4.8 aq; v s alc, eth; unstable acid solns
b573	2-*tert*-Butyl-4-methyl-phenol	(CH$_3$)$_3$CC$_6$H$_3$(CH$_3$)OH	164.25		0.9247_4^{75}	1.4969^{75}	51.7	237	100	i aq; s org solv
b574	2-*tert*-Butyl-5-methyl-phenol	(CH$_3$)$_3$CC$_6$H$_3$(CH$_3$)OH	164.25	6^2, 507	0.964	1.5192^{20}		118^{12mm}	105	
b575	2-*tert*-Butyl-6-methyl-phenol	(CH$_3$)$_3$CC$_6$H$_3$(CH$_3$)OH	164.25			1.5190^{20}	30–32	230	107	
b576	*tert*-Butyl-1-methy-2-propynyl ether	(CH$_3$)$_3$COCH(CH$_3$)C≡CH	126.20		0.795	1.4100^{20}		41^{25mm}	10	
b577	*tert*-Butyl methyl sulfide	(CH$_3$)$_3$CSCH$_3$	104.21	1^3, 1591	0.826_4^{20}	1.441^{20}	−97.8	102	−3	v s alc
b578	Butyl nitrite	C$_4$H$_9$ONO	103.12	1, 369	0.9114_4^{0}	1.3768		78	−13	misc alc, eth
b579	*tert*-Butyl nitrite	(CH$_3$)$_3$CONO	103.12	1, 382	0.8671_4^{20}	1.3687^{20}		63	−13	sl s aq; v s alc, chl, eth, CS$_2$
b580	Butyl 4-nitrobenzoate	O$_2$NC$_6$H$_4$CO$_2$C$_4$H$_9$	223.23	9^2, 259	0.8551_{14}	1.4422^{25}	35–39	160^{8mm}	>110	s alc; v s acet
b581	Butyl octadecanoate	CH$_3$(CH$_2$)$_{16}$CO$_2$C$_4$H$_9$	340.60	2^2, 352	0.8704_4^{15}	1.4480^{25}	26.3	343	160	s eth
b581a	Butyl *cis*-9-octa-decenoate	CH$_3$(CH$_2$)$_8$CH=CH(CH$_2$)$_7$-CO$_2$C$_4$H$_7$	338.57				−26		180	
b582	Butyl 4-oxopentanoate	CH$_3$C(=O)CH$_2$CH$_2$CO$_2$C$_4$H$_9$	172.22	20^3, 2872	0.9735_4^{20}	1.4270^{20}		107^{6mm}	91	s alc, acet, eth
b583	4-(1-Butylpentyl)-pyridine	C$_4$H$_9$CH(CH$_2$)$_3$CH$_3$ | C$_5$H$_4$N	205.35		0.887	1.4877^{20}		267	>110	
b584	*tert*-Butyl peroxo-benzoate	C$_6$H$_5$C(=O)O—O—C(CH$_3$)$_3$	194.23		1.021	1.4990^{20}		$76^{0.2mm}$	93	
b585	2-*sec*-Butylphenol	CH$_3$CH$_2$CH(CH$_3$)C$_6$H$_4$OH	150.22	6^2, 489	0.982	1.5222^{20}	12	228	112	i aq; s alc; v s eth
b586	2-*tert*-Butylphenol	(CH$_3$)$_3$CC$_6$H$_4$OH	150.22		0.9783_4^{20}	1.5228^{20}	−7	221–224	>110	

No.	Name	Formula	Mol wt	Beilstein	Density	n_D	mp	bp	fp	Solubility
b587	4-sec-Butylphenol	$CH_3CH_2CH(CH_3)C_6H_4OH$	150.22	6, 522	0.969^{20}_4	1.5150	62	136^{25mm}	115	s hot aq, alc, eth
b588	4-tert-Butylphenol	$(CH_3)_3CC_6H_4OH$	150.22	6, 524	0.908^{114}_4	1.4787^{114}	98	237	>110	i aq; s alc, eth
b589	tert-Butyl 4-phenoxyphenol ketone	$C_6H_5OC_6H_4C(=O)C(CH_3)_3$	254.33	8^3, 491			52–54	175^{3mm}		
b590	tert-Butyl phenyl carbonate	$C_6H_5OC(=O)OC(CH_3)_3$	194.23		1.047	1.4805^{20}		$79^{0.8mm}$	101	
b591	Butyl phenyl ether	$CH_3CH_2CH_2CH_2OC_6H_5$	150.22	6, 143	0.935^{20}_4	1.4970^{20}	–19	210.3		<0.1 aq; 79 alc; 153 EtOAc; 158 toluene
b592	4-tert-Butylphenyl salicylate	$HOC_6H_4CO_2C_6H_4C(CH_3)_3$	270.31				62–64		82 (OC)	v s alc, eth; v sl s aq
b593	Butyl propionate	$CH_3CH_2CO_2C_4H_9$	130.19	2, 241	0.8818^{15}	1.3982^{25}		146.8	38	
b594	tert-Butyl propionate	$CH_3CH_2CO_2C(CH_3)_3$	130.19	2^3, 528	0.865	1.3930^{20}	–89	118	20	
b595	4-tert-Butyl pyridine	$(CH_3)_3C(C_5H_4N)$	135.21	20, 252	0.915	1.4952^{20}		197	63	
b596	tert-Butyl 1-pyrrole-carboxylate	$(C_4H_4N)CO_2C(CH_3)_3$	167.21		1.000	1.4685^{20}		92^{20mm}	75	
b597	1-Butylpyrrolidine	$(C_4H_8N)C_4H_9$	127.23	20^2, 4	0.814	1.4440^{20}		157	36	
b598	4-tert-Butylstyrene	$(CH_3)_3CC_6H_4CH=CH_2$	160.26	5^3, 1254	0.875	1.5260^{20}	–37	92^{9mm}	80	
b599	1-Butyl-3-sulfanilyl-urea	$4\text{-}(H_2N)C_6H_4SO_2NH\text{-}CONHC_4H_9$	271.34	14^4, 2667			143–145			
b600	Butyltin trichloride	$C_4H_9SnCl_3$	282.17	4^4, 4346	1.693	1.5229^{20}		93^{10mm}	81	
b601	Butyltin tris(2-ethyl-hexanoate)	$[CH_3(CH_2)_3CH(C_2H_5)CO_2]_3\text{-}SnC_4H_9$	605.43		1.105	1.4650^{20}			>110	
b602	4-tert-Butyltoluene	$(CH_3)_3CC_6H_4CH_3$	148.25	5, 439	0.8612^{20}	1.4918^{20}	–52	190	68	
b603	Butyltrichlorosilane	$C_4H_9SiCl_3$	191.56	4^1, 582	1.160	1.4370^{20}		149	45	
b604	tert-Butyltrichloro-silane	$(CH_3)_3CSiCl_3$	191.56	4^3, 1905			97–100	132–134	40	
b605	Butyl trifluoroacetate	$CF_3CO_2C_4H_9$	170.1		1.0268^{22}	1.353^{22}		100.1		
b606	Butyltrimethoxysilane	$C_4H_9Si(OCH_3)_3$	178.3		0.9312^{20}_4	1.3979^{20}		164–165		
b607	tert-Butyl trimethyl-silyl peroxide	$(CH_3)_3C—O—O—Si(CH_3)_3$	162.3		0.8219^{20}_4	1.3935^{20}	dec 135	41^{41mm}		
b608	Butylurea	$C_4H_9NHCONH_2$	116.16	4^1, 371			96–98			s aq, alc, eth
b609	Butyl vinyl ether	$C_4H_9OCH=CH_2$	100.16	5, 447	0.7792^{20}	1.4007^{20}	–92	94.2	–9	0.3 aq
b610	5-tert-Butyl-m-xylene	$(CH_3)_3CC_6H_3(CH_3)_2$	162.28		0.867	1.4946^{20}		205–206	72	
b610a	1-Butyne	$CH_3CH_2C≡CH$	54.09		2.211 g/L		–126	8.1		
b610b	2-Butyne	$CH_3C≡C—CH_3$	54.09		0.688		–32	27		
b611	2-Butyne-1,4-diol	$HOCH_2C≡CCH_2OH$	86.09	1^1, 261		1.450^{25}	56–58	238	152	374 aq; 83 als; 0.04 bz; 2.6 eth; 70 acet
b612	Butyraldehyde	$CH_3CH_2CH_2CHO$	72.11	1, 662	0.8016^{20}_4	1.3843^{20}	–96/–99	74.8	–22	7.1 aq; misc alc, acet, eth, EtOAc

(Continued)

TABLE 2.21 Physical Constants of Organic Compounds (*Continued*)

No.	Name	Formula	Formula weight	Beilstein reference	Density, g/mL	Refractive index	Melting point, °C	Boiling point, °C	Flash point, °C	Solubility in 100 parts solvent
b613	Butyramide	CH$_3$CH$_2$CH$_2$CONH$_2$	87.12	2, 275			116	216	72	16 aq; s alc
b614	Butyric acid	CH$_3$CH$_2$CH$_2$COOH	88.11	2, 264	0.9582^{20}_4	1.3991^{20}	−5.3/−5.7	163.5	54	misc aq, alc, eth
b615	Butyric anhydride	[CH$_3$CH$_2$CH$_2$C(=O)]$_2$O	158.20	2, 274	0.9668^{20}_4	1.4070^{20}	−75/−66	199.5		s aq (dec); alc (dec); eth
b616	β–Butyrolactone		86.09	17[1], 130	1.056	1.4109^{20}	−43.5	204	60	misc aq; s alc, acet, bz, eth
b617	γ–Butyrolactone		86.09	17, 234	1.124^{25}_4	1.4348^{25}	−43.5	204	98	3.3 aq; misc alc, eth
b618	Butyronitrile	CH$_3$CH$_2$CH$_2$CN	69.11	2[2], 252	0.7954^{15}_4	1.4440^{20}	−112	117.6	24	s aq (dec), alc (dec); misc eth
b619	Butyrophenone	C$_6$H$_5$C(=O)C$_3$H$_7$	148.21	7, 313	1.021	1.5195^{20}	11–13	230	88	
b620	Butyryl chloride	CH$_3$CH$_2$CH$_2$COCl	106.55	2, 274	1.0263^{21}_4	1.412^{20}	−89	102	21	
c1	Caffeine		194.19	26, 461	1.23^{18}_4		238	subl 178		2.1 aq; 1.5 alc; 18 chl; 0.19 eth; 1 bz; 2 acet
c2	(±)-Camphene		136.24	5, 156	0.8422^{54}	1.4551^{54}	51–52	159	36	i aq; s alc, chl, eth
c3	(1R)-(+)-Camphor		152.24	7, 101	0.992^{25}_4	1.5462	179	207	66	100 alc; 100 eth; 200 chl; 250 acet
c4	(1R,3S)-Camphoric acid		200.23	9, 745	1.186^{20}_4		186–188			at 25°C: 0.8 aq, 100 alc, 250 acet, 200 eth, 200 HOAc; s chl
c5	(±)-10-Camphorsulfonic acid		232.30	11, 314			194 dec			deliq moist air; sl s HOAc, EtOAc; i eth
c6	Carbazole		167.21	20, 433	1.10^{18}_4		245	355		16 pyr; 11 acet; 3 eth; 0.8 bz; sl s HOAc, PE
c7	4-Carbethoxy-2-methyl-3-cyclohexen-1-one		182.22	10, 631	1.078	1.4880^{20}		268–272	>110	
c8	Carbobenzyloxyglycine	C$_6$H$_5$CH$_2$OC(=O)NH-CH$_2$COOH	209.20				122			
c9	Carbohydrazide	H$_2$NNHC(=O)NHNH$_2$	90.08	3, 121			157–158			v s aq; i alc, bz, eth; forms salts with acids

No.	Name	Formula	Mol. wt.	Beil. ref.	Density	n_D	mp/°C	bp/°C	Flash pt	Solubility
c10	Carbon disulfide	CS_2	76.14	3, 197	1.26322^{20}_{4}	1.6270^{20}	−111.6	46.5	−30	0.3 aq; misc bz, chl, eth, CCl$_4$
c11	Carbon monoxide	CO	28.01	Merck: 12, 1861	1.145 g/L		−205	−191.5		2.3 aq; 16 alc; s chl, EtOAc, HOAc
c12	Carbon oxide sulfide	COS	60.07		2.456 g/L		−138.8	−50		
c13	Carbon tetrabromide	CBr_4	331.65	1, 68	3.42		90	190	none	0.05 aq; misc alc, bz chl, eth, CS$_2$, PE
c14	Carbon tetrachloride	CCl_4	153.82	1, 64	1.589^{25}_{25}	1.4607^{20}	−23	76.7	none	
c15	Carbon tetrafluoride	CF_4	88.01	1, 59	1.89^{-183} Liq		−183.6	−127.8		
c16	Carbon tetraiodide	CI_4	519.63	1, 74			171			s bz, chl; dec hot alc
c17	4-Carboxybenzene-sulfonamide	$HOOCC_6H_4SO_2NH_2$	201.20	11, 390	4.32^{20}_{4}		dec 280			v s alc; s alkalis; i aq, bz, eth
c18	(4-Carboxybutyl)tri-phenylphosphonium bromide	$HOOC(CH_2)_4(C_6H_5)_3Br$	443.33				205–207			
c19	1-(Carboxymethyl)-pyridinium chloride		173.60				189 dec			
c20	R-(—)-Carvone		150.22	7, 157	0.9652^{20}_{4}	1.4989^{20}	<15	230	88	i aq; misc alc
c21	Catechol	$C_6H_4\text{-}1,2\text{-}(OH)_2$	110.11		1.344		104–106	245	137	43 aq; v s alkalis, pyr; s alc, bz, chl, eth
c22	Catecholborane		119.92		1.125	1.5070^{20}	12	50^{50mm}	2	v s bz, chl, CS$_2$, eth; sl s alc
c23	Chalcone	$C_6H_5CH=CHCOC_6H_5$	208.26	7, 478	1.0712^{62}_{4}		55–57	208^{25mm}	>110	s aq, alc, eth
c23a	Chloroacetaldehyde	$ClCH_2CHO$	78.50	1, 610			−16	85–86		10 aq; 10 alc; sl s eth s alc
c24	2-Chloroacetamide	$ClCH_2CONH_2$	93.51	2, 199			119	225 dec		v s alc, CS$_2$
c25	2'-Chloroacetanilide	$ClC_6H_4NHCOCH_3$	169.61	12, 559			88–90			i aq; v s alc, eth, CS$_2$
c26	3'-Chloroacetanilide	$ClC_6H_4NHCOCH_3$	169.61	12, 604			79–81			v s alc, bz, eth
c26a	4'-Chloroacetanilide	$ClC_6H_4NHCOCH_3$	169.61	12, 611			179			v s chl, eth; sl s bz; dec by aq, alc
c27	Chloroacetic acid	$ClCH_2COOH$	94.50	2, 194	1.3854^{20}	1.4297^{65}	61	189	126	
c28	Chloroacetic anhydride	$[ClCH_2C(=O)]_2O$	170.98	2, 199	1.580 (c); 1.5494^{20}_{4}		46	203		
c29	4'-Chloroacetoacet-anilide	$CH_3COCH_2CH_2CO\text{-}NHC_6H_4Cl$	211.65				134	dec	160 (CC)	
c30	Chloroacetonitrile	$ClCH_2CN$	75.50	2, 201	1.193	1.4225^{20}		126	47	
c31	2-Chloroacetophenone	$C_6H_5COCH_2Cl$	154.60	7, 282	1.324^{15}		54–56	245		i aq; v s alc, bz, eth
c32	o-Chloroacetophenone	$ClC_6H_4COCH_3$	154.60	7^1, 151	1.188	1.5438^{20}		228^{738mm}	88	sl s aq; s eth

TABLE 2.21 Physical Constants of Organic Compounds (*Continued*)

No.	Name	Formula	Formula weight	Beilstein reference	Density, g/mL	Refractive index	Melting point, °C	Boiling point, °C	Flash point, °C	Solubility in 100 parts solvent
c33	*p*-Chloroacetophenone	$ClC_6H_4COCH_3$	154.60	7, 281	1.1922_4^{20}	1.555^{20}	20–21	237	90	i aq; misc alc, eth
c34	Chloroacetyl chloride	$ClCH_2COCl$	112.94	2, 199	1.420_4^{20}	1.4541^{20}	−21.8	106	none	dec by aq, MeOH
c36	2-Chloroacrylonitrile	$H_2C{=}C(Cl)CN$	87.51	12, 988	1.096	1.4290^{20}	−65	89	6	
c37	2-Chloro-4-amino-toluene	$ClC_6H_3(CH_3)NH_2$	141.60		1.1671	1.5840^{20}	24–25	238	100	
c38	2-Chloroaniline	$ClC_6H_4NH_2$	127.57	12, 597	1.2125_4^{20}	1.5895^{20}	−14	208.8	97	0.88 aq; s acids, most common org solvents
c39	3-Chloroaniline	$ClC_6H_4NH_2$	127.57	12, 602	1.2150_4^{22}	1.5931^{20}	−10.4	230.5	123	i aq; s most common org solvents
c40	4-Chloroaniline	$ClC_6H_4NH_2$	127.57	12, 607	1.169_4^{77}	1.5546^{85}	72.5	232		s hot aq; v s alc, acet, eth, CS_2
c41	1-Chloroanthra-quinone		242.66	7, 787			160	sublimes		sl s alc; s hot bz; misc eth
c42	2-Chloroanthra-quinone		242.66	7, 787			211	sublimes		sl s alc, bz; i eth
c43	2-Chlorobenzaldehyde	ClC_6H_4CHO	140.57	7, 233	1.2483_4^{20}	1.5658	11	215	87	sl s aq; s alc, bz, eth
c44	3-Chlorobenzaldehyde	ClC_6H_4CHO	140.57	7, 234	1.241	1.5545^{20}	18	214	88	s aq; v s alc, bz, eth
c45	4-Chlorobenzaldehyde	ClC_6H_4CHO	140.57	7, 235	1.196_4^{61}	1.552^{61}	47	214	87	s aq; v s alc, bz, eth
c46	2-Chlorobenzamide	$ClC_6H_4CONH_2$	155.58	9, 336			142–144			
c47	Chlorobenzene	C_6H_5Cl	112.56	5, 199	1.1063^{20}	1.5248^{20}	−45.3	131.7	28	0.049 aq^{30}; v s alc, bz, chl, eth
c48	4-Chlorobenzene-sulfonamide	$ClC_6H_4SO_2NH_2$	191.64	11, 55			146			s hot aq, hot alc, hot eth
c49	4-Chlorobenzene-sulfonic acid	$ClC_6H_4SO_3H$	192.62	11, 54				149^{22mm}	107	dec aq, alc; v s bz, eth
c50	4-Chlorobenzene-sulfonyl chloride	$ClC_6H_4SO_2Cl$	211.07	11, 55			55	141^{15mm}	107	
c51	2-Chlorobenzoic acid	ClC_6H_4COOH	156.57	9, 334	1.544_4^{20}		140			0.11 aq; v s alc, eth
c52	3-Chlorobenzoic acid	ClC_6H_4COOH	156.57	9, 337	1.496_4^{25}		158			0.04 aq; v s alc, eth
c53	4-Chlorobenzoic acid	ClC_6H_4COOH	156.57	9, 340			241–243			0.02 aq; v s alc, eth
c54	2-Chlorobenzonitrile	ClC_6H_4CN	137.57	9, 336			46	232	108	s alc, eth
c55	4-Chlorobenzonitrile	ClC_6H_4CN	137.57	9, 341			93	223		s alc, eth
c56	2-Chlorobenzo-Phenone	$ClC_6H_4COC_6H_5$	216.67	7, 419			44–47	300	>110	s alc, bz, chl, eth

c57	4-Chlorobenzo-phenone	$ClC_6H_4COC_6H_5$	216.67	7, 419			77	196^{17mm}		s alc, acet, bz, eth
c58	2-Chlorobenzotri-chloride	$ClC_6H_4CCl_3$	229.92	5, 302	1.508	1.5817^{20}	29	264	98	
c59	4-Chlorobenzotri-chloride	$ClC_6H_4CCl_3$	229.92	5, 303	1.495	1.5722^{20}		245	>110	
c60	2-Chlorobenzotri-fluoride	$ClC_6H_4CF_3$	180.56	5^3, 692	1.3540^{25}	1.4513^{25}	-6.4	152	58	
c61	3-Chlorobenzotri-fluoride	$ClC_6H_4CF_3$	180.56	5^3, 692	1.3311^{25}	1.4438^{25}	-56.7	137.7	38	
c62	4-Chlorobenzotri-fluoride	$ClC_6H_4CF_3$	180.56		1.353^{20}	1.4463	-36	138.7	47	
c63	2-(4-Chlorobenzoyl)-benzoic acid	$ClC_6H_4COC_6H_4COOH$	260.68	10, 750			150			s alc, bz, eth
c64	2-Chlorobenzoyl chloride	ClC_6H_4COCl	175.01	9, 336	1.382	1.5718^{20}	-3	238	>110	dec by aq & alc
c65	4-Chlorobenzoyl chloride	ClC_6H_4COCl	175.01	9, 341	1.377	1.5780^{20}	14	222	105	dec by aq & alc
c66	4-Chlorobenzyl alcohol	$ClC_6H_4CH_2OH$	142.59	6, 444	1.173	1.5630^{20}	72	234	88	v s alc, eth
c67	2-Chlorobenzylamine	$ClC_6H_4CH_2NH_2$	141.60	12, 1073	1.164	1.5586^{20}		104^{11mm}	90	
c68	4-Chlorobenzylamine	$ClC_6H_4CH_2NH_2$	141.60	12, 1074	1.274	1.5591^{20}	-17	215	82	
c69	2-Chlorobenzyl chloride	$ClC_6H_4CH_2Cl$	161.03	5, 297				214		
c70	4-Chlorobenzyl chloride	$ClC_6H_4CH_2Cl$	161.03	5, 308		1.5540^{20}	30	222	97	s alc, v s eth
c71	2-Chlorobenzyl cyanide	$ClC_6H_4CH_2CN$	151.60	9, 448			24	242	>110	
c72	4-Chlorobenzyl cyanide	$ClC_6H_4CH_2CN$	151.60	9, 448	1.202	1.5893^{20}	30.3	267	>110	
c73	4-Chlorobenzyl mercaptan	$ClC_6H_4CH_2SH$	158.65	6, 466			20		76	
c74	1-Chloro-1,3-butadiene	$H_2C=CHCH=CHCl$	88.54	1^3, 949	0.9601^{20}_4	1.4712^{20}		68	-20	v s chl
c74a	2-Chloro-1,3-butadiene	$H_2C=C(Cl)=CH_2$	88.54		0.952			59		
c75	1-Chlorobutane	$CH_3CH_2CH_2CH_2Cl$	92.57	1, 118	0.8864^{20}	1.4021^{20}	-123.1	78.4	-9	0.11 aq; misc alc, eth
c76	2-Chlorobutane	$CH_3CH_2CH(Cl)CH_3$	92.57	1, 119	0.8732^{20}_4	1.3971^{20}	-131.3	68.2	-15	0.1 aq; misc alc, eth
c77	4-Chloro-1-butanol	$ClCH_2CH_2CH_2CH_2OH$	108.56	1^2, 398	1.0883^{20}_4	1.4518^{20}		89^{20mm}	32	s alc, eth
c78	3-Chloro-2-butanone	$CH_3CH(Cl)C(=O)CH_3$	106.55	1, 669	1.055	1.4172^{20}		117	21	v s alc, eth
c79	cis-1-Chloro-2-butene	$CH_3CH=CHCH_2Cl$	90.55	1^2, 176	0.9426^{20}	1.4390^{20}		84.1	-15	s alc, acet

(Continued)

TABLE 2.21 Physical Constants of Organic Compounds (*Continued*)

No.	Name	Formula	Formula weight	Beilstein reference	Density, g/mL	Refractive index	Melting point, °C	Boiling point, °C	Flash point, °C	Solubility in 100 parts solvent
c80	*tarns*-1-Chloro-2-butene	$CH_3CH=CHCH_2Cl$	90.55	1[2], 176	0.929	1.4390^{20}		85	−5	s alc, acet
c81	3-Chloro-1-butene	$CH_3CH(Cl)CH=CH_2$	90.55	1[2], 174	0.9001_4	1.4155^{20}		65	−20	v s acet
c82	4-Chlorobutyl acetate	$CH_3CO_2CH_2CH_2CH_2CH_2Cl$	150.61	2[2], 141	1.072	1.4338^{20}		92^{22mm}	64	
c83	3-Chloro-1-butyne	$CH_3CH(Cl)C\!\equiv\!CH$	88.54	1[4], 970	0.961	1.4280^{20}		68–70	1	
c84	3-Chlorobutyric acid	$CH_3CH(Cl)CH_2COOH$	122.55	2, 277	1.186_4^{20}	1.4421^{20}	16.3	109^{17mm}	>110	s alc, eth
c85	4-Chlorobutyric acid	$ClCH_2CH_2CH_2COOH$	122.55	2, 278	1.2236_4^{20}	1.4521^{20}	12–16	196^{22mm}	>110	sl s aq; v s eth
c86	4-Chlorobutyronitrile	$ClCH_2CH_2CH_2CN$	103.55	2, 278	1.158	1.4413^{20}		197	85	s alc, eth
c87	4-Chlorobutyryl chloride	$ClCH_2CH_2CH_2COCl$	141.00	2, 278	1.258	1.4609^{20}		174	72	dec by aq, alc; s eth
c88	Chloro(chloromethyl)-dimethylsilane	$ClCH_2Si(CH_3)_2Cl$	143.09		1.086	1.4373^{20}		114^{752mm}	21	
c89	3-Chloro-2-chloro-methy-1-propene	$H_2C=C(CH_2Cl)_2$	125.00	1[2], 181	1.080	1.4753^{20}	−14	138	36	
c90	*trans*-2-Chloro-cinnamic acid	$ClC_6H_4CH=CHCO_2H$	182.61	9, 594			208–210			
c91	Chlorocyclohexane	ClC_6H_{11}	118.61	5, 21	1.000_4^{20}	1.4620^{20}	−44	142	28	i aq; s alc, eth
c92	1-Chloro-3-cyclo-hexylpropane	$C_6H_{11}(CH_2)_3Cl$	160.69	5[2], 23	0.997	1.4662^{20}		79^{5mm}	78	
c93	Chlorocyclopentane	C_5H_9Cl	104.58	5, 19	1.0051^{20}	1.4512^{20}		114	15	i aq
c94	1-Chlorodecane	$CH_3(CH_2)_9Cl$	176.73	1, 168	0.868	1.4362^{20}	−34	223	83	i aq
c95	Chlorodicyclohexyl-borane	$(C_6H_{11})_2BCl$	212.57	16[4], 1637	0.970			101^{1mm}		
c96	2-Chloro-1,1-diethoxy-ethane	$ClCH_2CH(OC_2H_5)_2$	152.62	1, 611	1.018	1.4157^{20}		157	29	
c97	3-Chloro-1,1-diethoxy-propane	$ClCH_2CH_2CH(OC_2H_5)_2$	166.65	1, 632	0.995	1.4240^{20}		84^{25mm}	36	
c98	Chlorodifluoroacetic acid	$F_2C(Cl)COOH$	130.48	2, 201	1.540	1.3559^{20}	24–26	122	none	
c99	1-Chloro-2,4-difluoro-benzene	$ClC_6H_3F_2$	148.54	5[4], 653	1.353	1.4750^{20}		127	32	
c100	1-Chloro-1,1-difluoro-ethane	$CH_3C(Cl)F_2$	100.50	1[3], 138	4.108 g/L		−131	−10		
c100a	1-Chloro-2,2-difluoro-ethylene	$ClCH=CF_2$	98.48		4.025 g/L		−138.5	−18.5		0.19 aq

No.	Name	Formula	Formula wt.	Beilstein ref.	Density	n_D	m.p., °C	b.p., °C	Flash P., °C	Solubility
c101	Chlorodifluoromethane	$HCClF_2$	86.47	1^3, 41	1.4909^{-69}		-157	-40.8		0.30 aq
c102	1-Chloro-2,4-dihydroxybenzene	$ClC_6H_3(OH)_2$	144.56	6^2, 818			107	147^{18mm}		v s aq, alc, chl, eth
c103	2-Chloro-1,4-dihydroxybenzene	$ClC_6H_3(OH)_2$	144.56	6, 849			101–102	263		v s aq; i alc, s eth
c104	2-Chloro-1,4-dimethoxybenzene	$ClC_6H_3(OCH_3)_2$	172.61	6^3, 4432	1.211	1.5467^{20}		234	110	
c105	2-Chloro-1,1-dimethoxyethane	$ClCH_2CH(OCH_3)_2$	124.57		1.094^{20}_{20}	1.4148^{20}		130	28	
c107	2-Chloro-4,6-dimethylaniline	$ClC_6H_2(CH_3)_2NH_2$	155.63	12, 1125	1.110		38–40	246	>110	
c108	4-Chloro-3,5-dimethylphenol	$ClC_6H_2(CH_3)_2OH$	156.61	6^2, 463			115.5			0.03 aq; 100 alc; s bz, eth, alkalis
c109	1-Chloro-2,2-dimethylpropane	$(CH_3)_3CCH_2Cl$	106.59	1, 141	0.866^{20}_4	1.4042^{20}	-20	84.4	32	
c110	3-Chloro-2,2-dimethyl-1-propanol	$ClCH_2C(CH_3)_2CH_2OH$	122.60			1.4504^{20}	34–36	87^{35mm}	71	
c111	Chlorodimethylsilane	$(CH_3)_2Si(Cl)H$	94.62	4^4, 4080	0.852^{20}_4	1.3827^{20}	-111	36	-28	
c112	Chlorodimethylvinylsilane	$(CH_3)_2Si(Cl)CH=CH_2$	120.7		0.884^{25}_4	1.414^{25}		82.5	-5	
c113	6-Chloro-2,4-dinitroaniline	$ClC_6H_2(NO_2)_2NH_2$	217.57	12^1, 367			159			
cl 14	1-Chloro-2,4-dinitrobenzene	$ClC_6H_3(NO_2)_2$	202.55	5, 263	1.4982^{25}_4	1.5857^{60}	52–54	315	186	sl s alc; s hot alc, bz, eth
c115	2-Chloro-3,5-dinitrobenzoic acid	$ClC_6H_2(NO_2)_2COOH$	246.56	9, 415			198	241 explodes		0.3 aq
c116	Chlorodiphenylmethane	$C_6H_5CH(Cl)C_6H_5$	202.68	5^2, 500	1.140^{20}_4	1.5951^{20}	17	140^{3mm}	>110	
c117	Chlorodiphenylmethylsilane	$(C_6H_5)_2Si(Cl)CH_3$	232.8	16^2, 606	1.1277^{20}_4	1.5742^{20}		295	>110	
c118	Chlorodiphenylphosphine	$(C_6H_5)_2PCl$	220.64	16, 763	1.229	1.6338^{20}		320	>110	
c119	1-Chlorododecane	$CH_3(CH_2)_{11}Cl$	204.79		0.8673^{20}_4	1.4426	-9	116	93	v s alc; s bz
c120	1-Chloro-2,3-epoxypropane	$\underset{O}{H_2C{-}CHCH_2Cl}$	92.53	17, 6	1.1812^{20}_4	1.4358^{20}	-57.2	116.1	31	5.9 aq; misc alc, chl
c121	Chloroethane	CH_3CH_2Cl	64.52	1, 82	0.9214^0_4	1.3742^{10}	-139	12.3	-50	0.45 aq^0, 48 alc; misc eth
c122	2-Chloroethanol	$ClCH_2CH_2OH$	80.52	1, 337	1.2019^{20}	1.4422^{20}	-57.5	128.6	60	misc aq, alc

(Continued)

TABLE 2.21 Physical Constants of Organic Compounds (*Continued*)

No.	Name	Formula	Formula weight	Beilstein reference	Density, g/mL	Refractive index	Melting point, °C	Boiling point, °C	Flash point, °C	Solubility in 100 parts solvent
c123	2-(2-Chloroethoxy)-ethanol	$ClCH_2CH_2OCH_2CH_2OH$	124.57	1, 467	1.180	1.4529^{20}		81^{5mm}	90	
c124	2-[2-(2-Chloroethoxy)-ethoxy]ethanol	$ClCH_2CH_2OCH_2CH_2O\text{-}CH_2CH_2OH$	168.62	1, 468	1.160	1.4580^{20}		120^{5mm}	107	
c125	2-Chloroethoxytri-methylsilane	$ClCH_2CH_2OSi(CH_3)_3$	152.70	4^3, 1856	0.944	1.4140^{20}		134	30	
c126	2-Chloroethylamine hydrochloride	$ClCH_2CH_2NH_2 \cdot HCl$	115.99	4, 133			146			i aq; misc alc, eth
c127	1-Chloro-2-ethyl-benzene	$ClC_6H_4C_2H_5$	140.61	5, 354	1.055$^{25}_{25}$	1.5300^{20}	−81	179.2	66	s alc, bz, eth
c128	(2-Chloroethyl)-benzene	$C_6H_5CH_2CH_2Cl$	140.61		1.069			84^{16mm}	66	sl s aq; s alc
c129	Chloroethylene	$H_2C{=}CHCl$	62.50	1, 186	0.97^{-14}		−154	−13.4	−78	
c130	N-(2-Chloroethyl)-N-ethylamine	$C_6H_5N(C_2H_5)CH_2CH_2Cl$	183.68	12^3, 263	1.075	1.5584^{20}		164^{42mm}	>110	
c131	2-Chloroethyl ethyl ether	$ClCH_2CH_2OCH_2CH_3$	108.57	1, 337	0.989	1.4120^{20}		107	15	
c132	2-Chloroethyl methyl ether	$ClCH_2CH_2OCH_3$	94.54	1, 337	1.035	1.4090^{20}		90	15	
c133	N-(2-Chloroethyl)-morpholine HCl		186.08				186			
c133a	2-Chloroethyl phenyl ether	$C_6H_5OCH_2CH_2Cl$	156.61	6^3, 675	1.129	1.5340^{20}		98^{15mm}	100	
c134	N-(2-Chloroethyl)-piperidine HCl		184.11	20, 17			236			
c135	2-Chloroethyl p-toluenesulfonate	$CH_3C_6H_4SO_3CH_2CH_2Cl$	234.70	11^2, 45	1.294	1.5290^{20}		153$^{0.3mm}$	>110	
c136	2-Chloroethyl vinyl ether	$H_2C{=}CHOCH_2CH_2Cl$	106.55	1^2, 473	1.0525$^{15}_{15}$	1.4370^{20}	−69.7	110	16	0.6 aq
c137	1-Chloro-2-fluoro-benzene	ClC_6H_4F	130.55	5^1, 110	1.244	1.5010^{20}	−42.4	138.5	31	s alc, eth
c138	1-Chloro-3-fluoro-benzene	ClC_6H_4F	130.55		1.219	1.4944^{20}		126	20	s alc, eth
c139	2-Chloro-6-fluoro-benzyl chloride	$Cl(F)C_6H_3CH_2Cl$	179.02		1.401	1.5372^{20}			93	

(Continued)

No.	Name	Formula	Mol. wt.	Beil. ref.	Density	n_D	mp/°C	bp/°C	Flash pt	Solubility
c140	4-Chloro-4′-fluoro-butyrophenone	$FC_6H_4C(=O)CH_2CH_2CH_2Cl$	200.64		1.220	1.5255^{20}	41.5	127^{17mm}	>110	0.50 aq^{25}; misc alc, bz, eth, PE, CCl$_4$
c141	3-Chloro-4-fluoro-nitrobenzene	$Cl(F)C_6H_3NO_2$	175.55	5^1, 130	1.6028^{17}	1.5674^{17}	23	88^{4mm}	75	
c142	2-Chloro-4-fluoro-phenol	$Cl(F)C_6H_3OH$	146.55	6^4, 880	1.344	1.5300		156	46	misc alc, eth
c143	2-Chloro-6-fluoro-toluene	$Cl(F)C_6H_3CH_3$	144.58		1.191	1.5026^{20}		156		i aq
c144	4-Chloro-2-fluoro-toluene	$Cl(F)C_6H_3CH_3$	144.58	5^4, 813	1.186	1.4998^{20}		158^{743mm}	51	sl s aq; v s alc, eth
c145	Chloroform	$CHCl_3$	119.39	1, 61	1.4832^{20}	1.4459^{20}	−63.6	61.1		
c146	Chloroform-d	$CDCl_3$	120.39	1^3, 63	1.500	1.4445^{20}	−64	60.9		see under chloroform
c147	1-Chloroheptane	$CH_3(CH_2)_6Cl$	134.65	1, 154	0.881^{16}_{0}	1.4250^{20}	−69	159–161	41	
c148	1-Chlorohexadecane	$CH_3(CH_2)_{15}Cl$	260.89	1, 172	0.865	1.4490^{20}		149^{1mm}	>110	
c149	1-Chlorohexane	$CH_3(CH_2)_5Cl$	120.62	1, 143	0.8780^{20}_{4}	1.4195^{20}	−94	134	26	
c150	6-Chloro-1-hexanol	$Cl(CH_2)_6OH$	136.62		1.204	1.4560^{20}		110^{14mm}	98	
c151	4-Chloro-4′-hydroxy-benzophenone	$ClC_6H_4C(=O)C_6H_4OH$	232.67	8^2, 187			175–178	257^{13mm}		
c152	5-Chloro-8-hydroxy-7-iodoquinoline		305.50	21, 98			172			i alc, eth; 0.8 chl; 0.6 HOAc
c153	5-Chloro-8-hydroxy-quinoline		179.61	21, 95			130			sl s aq HCl
c154	1-Chloro-4-iodo-benzene	ClC_6H_4I	238.46	5, 221	1.1864^{57}		53–54	227	108	s alc
c155	1-Chloro-3-iodo-propane	$Cl(CH_2)_3I$	204.44	1, 114	1.904	1.5463^{20}		170–172	>110	
c156	1-Chloro-3-mercapto-2-propanol	$HSCH_2CH(OH)CH_2Cl$	126.61	1^3, 2156	1.277	1.5276^{20}		$57^{1.3mm}$	97	
c157	Chloromethane	CH_3Cl	50.49	1, 59	2.064 g/L	1.3712^{-24}	−97.7	−24.2	<0	0.48 aq^{25} s alc; misc chl, eth, HOAc
c158	3-Chloro-4-methoxy-aniline	$ClC_6H_3(OCH_3)NH_2$	157.60	13, 511			50–55		110	
c159	5-Chloro-2-methoxy-aniline	$ClC_6H_3(OCH_3)NH_2$	157.60	13, 383			83–85			
c160	1-Chloro-2-methoxy-benzene	$ClC_6H_4OCH_3$	142.59	6, 184	1.123	1.5445^{20}		196		i aq; s alc, eth
c161	5-Chloro-2-methoxy-benzoic acid	$ClC_6H_3(OCH_3)COOH$	186.59	10, 103			98–100		76	

TABLE 2.21 Physical Constants of Organic Compounds (*Continued*)

No.	Name	Formula	Formula weight	Beilstein reference	Density, g/mL	Refractive index	Melting point, °C	Boiling point, °C	Flash point, °C	Solubility in 100 parts solvent
c162	2-Chloro-6-methoxy-pyridine	$CH_3O(Cl)(C_5H_3N)$	143.57		1.207	1.5263^{20}		186		
c163	2-Chloro-6-methyl-aniline	$CH_3O(Cl)C_6H_3NH_2$	141.60	12^1, 388	1.152	1.5761^{20}	2	215	98	s alc
c164	3-Chloro-2-methyl-aniline	$CH_3O(Cl)C_6H_3NH_2$	141.60	12, 836	1.185	1.5874^{20}	2	117^{10mm}	>110	
c165	3-Chloro-4-methyl-aniline	$CH_3O(Cl)C_6H_3NH_2$	141.60	12, 988		1.5830^{20}	25	238	100	
c166	4-Chloro-2-methyl-aniline	$CH_3O(Cl)C_6H_3NH_2$	141.60	12, 835		1.5848^{20}	27	241	99	s hot alc
c167	5-Chloro-2-methyl-aniline	$CH_3O(Cl)C_6H_3NH_2$	141.60	12, 835		1.5840^{20}	22	237	160	
c168	3-(Chloromethyl)-benzoyl chloride	$ClCH_2C_6H_4COCl$	189.04	9^2, 325	1.330	1.5748^{20}		150^{20mm}	>110	
c169	DL-4-Chloro-2-(α-methylben-zyl)phenol	$C_6H_5CH(CH_3)C_6H_3(Cl)OH$	232.71	6^4, 4710	1.238	1.5994^{20}		155^{2mm}	>110	
c169a	1-Chloro-3-methyl-butane	$ClCH_2CH_2CH(CH_3)CH_3$	106.60	1, 134	0.8750^{20}	1.4084^{20}	−104	99	<21	sl s aq; misc alc, eth
c170	2-Chloro-2-methyl-butane	$CH_3CH_2CCl(CH_3)_2$	106.59		0.8650^{20}_4	1.4052^{20}	−73.7	85	−9	i aq; s alc, eth
c171	Chloromethyldichloro-methylsilane	$ClCH_2Si(Cl)_2CH_3$	163.5	4^3, 1888	1.286	1.4494^{20}		121	110	
c172	Chloromethyl ethyl ether	$ClCH_2OCH_2CH_3$	94.54	1^2, 645	1.044^{20}_4	1.4040^{20}		79–83	19	s alc; v eth
c172a	3-(Chloromethyl)-heptane	$CH_3CH_2CH_2CH_2CH(CH_2Cl)$-$CH_2CH_3$	148.68		0.8769^{20}	1.4319^{20}		172	60	
c173	Chloromethyl methyl ether	$ClCH_2OCH_3$	80.51	1, 580	1.0703^{20}_4	1.3961^{20}	−103.5	57–59	15	dec by aq; s acet, CS_2
c174	Chloromethyl methyl sulfide	$ClCH_2SCH_3$	95.48		1.153	1.4963^{20}		105	17	
c175	1-(Chloromethyl)-naphthalene	$C_{10}H_7$-CH_2Cl	176.65	5, 566	1.180	1.6380^{20}	32	169^{25mm}	>110	
c176	4-Chloro-2-methyl-phenol	$CH_3(Cl)C_6H_3OH$	142.59	6, 359			45–48	220–225	>110	sl s aq

No.	Name	Formula	Mol. wt.	Beilstein	Density	n_D	mp	bp	Flash pt	Solubility
c177	4-Chloro-3-methyl-phenol	$CH_3(Cl)C_6H_3OH$	142.59	6, 381			65–68	235		i aq; s alc, bz, chl, eth, acet
c178	1-Chloro-2-methyl-2-phenylpropane	$C_6H_5(CH_3)_2CH_2Cl$	168.67	5^2, 320	1.047	1.5240^{20}		96^{10mm}	92	0.09 aq; misc alc, eth
c179	1-Chloro-2-methyl-propane	$(CH_3)_2CHCH_2Cl$	92.57	1, 124	0.8829^{15}	1.4010^{15}	−130.3	68.9	<21	sl s aq; misc alc, eth
c180	2-Chloro-2-methyl-propane	$(CH_3)_3CCl$	92.57	1, 125	0.8420^{20}	1.3856^{20}	−26	50.8	<0	misc alc, eth
c181	1-Chloro-2-methyl-propene	$(CH_3)_2C{=}CHCl$	90.55	1, 209	0.9186^{20}_{4}	1.4225^{20}	−1	68.1	−1	misc alc, eth
c182	3-Chloro-2-methyl-propene	$ClCH_2C(CH_3){=}CH_2$	90.55	1, 209	0.9210^{15}_{4}	1.4272^{20}	−80	72	−12	
c183	Chloromethyltri-methylsilane	$ClCH_2Si(CH_3)_3$	122.67	4^3, 1844	0.8861^{20}_{4}	1.4180^{20}		99	−2	
c184	6-(Chloromethyl)uracil		160.56	23^1, 328			257 dec		121	s alc, bz, PE
c185	1-Chloronaphthalene	$C_{10}H_7Cl$	162.62	5, 541	1.1938^{20}_{4}	1.6326^{20}	−2.3	259		s alc, bz, chl, eth
c186	2-Chloronaphthalene	$C_{10}H_7Cl$	162.62	7^3, 995	1.1377^{71}	1.6079^{71}	60	256		
c187	4'-Chloro-3-nitro-acetophenone	$ClC_6H_3(NO_2)C({=}O)CH_3$	199.60				101			sl s aq; v s alc, eth
c188	2-Chloro-4-nitroaniline	$ClC_6H_3(NO_2)NH_2$	172.57	12, 733			107–109			
c189	2-Chloro-5-nitroaniline	$ClC_6H_3(NO_2)NH_2$	172.57	12, 732			119–121			v s alc, eth
c190	4-Chloro-2-nitroaniline	$ClC_6H_3(NO_2)NH_2$	172.57	12, 729			117–119			v s alc; s eth
c191	4-Chloro-3-nitroaniline	$ClC_6H_3(NO_2)NH_2$	172.57	12, 731			99–101			s alc, bz, eth
c192	1-Chloro-2-nitrobenzene	$ClC_6H_4NO_2$	157.56	5, 241	1.348		33	246	123	sl s alc; v s chl, eth
c193	1-Chloro-3-nitrobenzene	$ClC_6H_4NO_2$	157.56	5, 243	1.534^{20}_{4}		44	236	103	sl s alc; v s eth, CS_2
c194	1-Chloro-4-nitrobenzene	$ClC_6H_4NO_2$	157.56	5, 243	1.520		83–84	242	>110	s hot aq, hot bz
c195	2-Chloro-4-nitrobenzoic acid	$ClC_6H_3(NO_2)COOH$	201.57	9, 404			139–141			
c196	2-Chloro-5-nitrobenzoic acid	$ClC_6H_3(NO_2)COOH$	201.57	9, 403	1.608^{18}		166–168			sl s aq; s alc, bz, eth
c197	4-Chloro-3-nitrobenzoic acid	$ClC_6H_3(NO_2)COOH$	201.57	9, 402	1.645^{18}		180–183			
c198	4-Chloro-3-nitro-benzophenone	$ClC_6H_3(NO_2)C({=}O)C_6H_5$	261.66	7^1, 230			104–106	235^{13mm}		sl s alc; s hot aq

(Continued)

TABLE 2.21 Physical Constants of Organic Compounds (*Continued*)

No.	Name	Formula	Formula weight	Beilstein reference	Density, g/mL	Refractive index	Melting point, °C	Boiling point, °C	Flash point, °C	Solubility in 100 parts solvent
c199	2-Chloro-5-nitro-benzotrifluoride	$ClC_6H_3(NO_2)CF_3$	225.55		1.527	1.5083^{20}		231	98	
c200	4-Chloro-3-nitro-benzotrifluoride	$ClC_6H_3(NO_2)CF_3$	225.55		1.511	1.4893^{20}	-2.5	222	101	
c201	4-Chloro-2-nitrophenol	$ClC_6H_3(NO_2)OH$	173.56	6, 238			85-87	238	125	i aq
c202	2-Chloro-6-nitrotoluene	$ClC_6H_3(NO_2)CH_3$	171.58	5, 327		1.5377^{70}	36	240^{718mm}	>110	i aq
c203	4-Chloro-2-nitrotoluene	$ClC_6H_3(NO_2)CH_3$	171.58	5, 327			39	$158^{1.5mm}$		
c203a	1-Chlorooctadecane	$CH_3(CH_2)_{17}Cl$	288.95	1^3, 566	0.849	1.4516^{20}		182	>110	0.02 aq; misc alc, eth
c204	1-Chlorooctane	$CH_3(CH_2)_7Cl$	148.68	1, 159	0.875	1.4298^{20}	-58	107-108	70	
c204a	1-Chloropentane	$CH_3(CH_2)_4Cl$	106.60	1, 130	0.8820^{20}	1.4115^{20}	-99	52^{18mm}	13	
c205	3-Chloro-2,4-pentane-dione	$CH_3COCH(Cl)COCH_3$	134.56	1,785	1.129	1.4830^{20}			12	
c206	5-Chloro-2-pentanone	$ClCH_2CH_2CH_2COCH_3$	120.58	1^2, 738	1.0571^{18}	1.4390^{20}		72^{20mm}	35	
c207	3-Chloroperoxy-benzoic acid	$ClC_6H_5C(O)OOH$	172.57	9^4, 972			69-71			s acet, eth
c208	2-Chlorophenol	ClC_6H_4OH	128.56	6, 183	1.2573^{23}	1.5565^{20}	9.8	175	63	sl s aq; v s alc, eth, caustic alkali
c209	3-Chlorophenol	ClC_6H_4OH	128.56	6, 185	1.245^{45}	1.5565^{40}	33	214	>110	sl s aq; s alc, eth
c210	4-Chlorophenol	ClC_6H_4OH	128.56	6, 186	1.2238^{78}	1.5479^{40}	43	220	115	sl s aq; v s alc, chl, eth, CHCl₃, glyc
c211	4-Chlorophenoxyacetic acid	$ClC_6H_4OCH_2COOH$	186.59	6, 187			157-159			s aq; MeOH
c212	2-(4-Chlorophenoxy)-2-methylpropanoic acid	$ClC_6H_4OC(CH_3)_2COOH$	214.65	Merck: 12, 2437			118-119			
c213	(±)-2-(4-Chlorophen-oxy)propanoic acid	$ClC_6H_4OCH(CH_3)COOH$	200.62	6^3, 695			117			
c214	4-Chlorophenylacetic acid	$ClC_6H_4CH_2COOH$	170.60	9, 448			108			v s aq, alc, eth; s bz
c215	(4-Chlorophenyl)-acetonitrile	$ClC_6H_4CH_2CN$	151.60	9, 448			30.5	265-267	>110	
c216	2-Chloro-1,4-phenyl-enediamine sulfate	$H_2NC_6H_3(Cl)NH_2 \cdot H_2SO_4$	240.67	13, 117			251-253			s aq
c217	4-Chloro-1,2-phenyl-enediamine	$ClC_6H_3(NH_2)_2$	142.59	13, 25			70-73			s mineral acids

(Continued)

No.	Name	Formula	MW	Ref.	Density	n_D	mp	bp	Flash point	Solubility
c218	1-(4-Chlorophenyl)-ethanol	$ClC_6H_4CH(CH_3)OH$	156.61	6[1], 236	1.171	1.5410^{20}		119^{10mm}	>110	at 20°C: 74 acet; 44 bz; 5 CCl₄; 65 diox; 21 i-PrOH
c219	3-Chlorophenyl isocyanate	ClC_6H_4NCO	153.57	12, 606	1.260	1.5576^{20}	-4.4	114^{43mm}	86	
c220	4-Chlorophenyl isocyanate	ClC_6H_4NCO	153.57	12, 616	1.200	1.5618^{20}	29-31	204	>110	
c221	4-Chlorophenyl phenyl sulfone	$ClC_6H_4SO_2C_6H_5$	252.72	6[1], 149			94			
c222	1-Chloro-3-phenyl-propane	$C_6H_5(CH_2)_3Cl$	154.64	5, 391	1.080	1.5207^{20}		219	87	
c223	4-Chlorophenyl sulfone	$(ClC_6H_5)_2SO_2$	287.17	6, 327			145-148	250^{10mm}		
c224	3-Chlorophthalide		168.58	17[1], 162			58	150^{10mm}		
c225	1-Chloropropane	$CH_3CH_2CH_2Cl$	78.54	1, 104	0.8899^{20}	1.3886^{20}	-122.8	46-47	-31	0.27 aq; misc alc, eth
c226	2-Chloropropane	$CH_3CHClCH_3$	78.54	1, 105	0.8563^{20}	1.3777^{20}	-117	35-36	-35	0.2 aq^{20}; misc alc, bz, chl, eth
c227	3-Chloro-1,2-propane-diol	$ClCH_2CH(OH)CH_2OH$	110.54	1, 473	1.3218^{20}_{4}	1.4805^{20}		213	>110	s aq, alc, eth
c228	2-Chloropropanoic acid	$CH_3CH(Cl)COOH$	108.52	2, 248	1.182	1.4345^{20}		170-190	101	misc aq, alc, eth
c229	3-Chloropropanoic acid	$ClCH_2CH_2COOH$	108.52	2, 249			41	200^{765mm}	>110	v s aq, alc, chl; s eth
c230	1-Chloro-2-propanol	$CH_3CH(OH)CH_2Cl$	94.54	1, 363	1.115^{20}	1.4375^{4}		126-127	51	misc aq; s alc
c231	3-Chloro-1-propanol	$ClCH_2CH_2CH_2OH$	94.54	1, 356	1.1309^{20}_{4}	1.4450^{20}		160-162	73	
c232	Chloro-2-propanone	$ClCH_2COCH_3$	92.53	1, 653	1.135^{15}	1.4320^{20}	-44.5	119.7	27	10 aq; misc alc, chl, eth
c233	3-Chloropropano-nitrile	$ClCH_2CH_2CN$	89.53	2, 250	1.1443^{18}	1.4341^{20}	-51	95^{50mm} d > 130	75	
c234	3'-Chloropropano-phenone	$ClC_6H_4C(=O)CH_2CH_3$	168.62	7[3], 1028			45-47	124^{14mm}	>110	
c235	2-Chloropropanyl chloride	$CH_3CH(Cl)COCl$	126.97	2, 248	1.308	1.4400^{20}		109-111	31	dec aq, alc
c236	3-Chloropropanyl chloride	$ClCH_2CH_2COCl$	126.97	2, 250	1.3307^{13}	1.4570^{20}		143-145	61	i aq; d hot aq, hot alc; s alc; v s eth
c236a	3-Chloro-1-propene	$ClCH_2CH=CH_2$	76.53	1, 198	0.9384_{4}	1.4154^{20}	-134.5	45	-32	
c237	3-Chloropropylacetate	$CH_3CO_2(CH_2)_3Cl$	130.02	4, 148				148-150		
c238	3-Chloropropyl thiolactate	$CH_3C(=O)SCH_2CH_2Cl$	152.64	2[3], 493	1.159	1.4946^{20}		84^{10mm}	77	0.36 aq; misc alc, PE

TABLE 2.21 Physical Constants of Organic Compounds (*Continued*)

No.	Name	Formula	Formula weight	Beilstein reference	Density, g/mL	Refractive index	Melting point, °C	Boiling point, °C	Flash point, °C	Solubility in 100 parts solvent
c239	(3-Chloropropyl)triethoxysilane	Cl(CH$_2$)$_3$Si(OC$_2$H$_5$)3	240.81		1.009^{20}_{4}	1.420^{20}		102^{10mm}		misc alc, bz, eth, EtOAc
c240	(3-Chloropropyl)trimethoxysilane	Cl(CH$_2$)$_3$Si(OCH$_3$)3	198.72	1,248	1.077^{25}_{4}	1.4183^{25}		195^{750mm}	78	
c241	3-Chloropropyne	ClCH$_2$C≡CH	74.51		1.0306^{25}_{4}	1.4560^{20}	−78	57	−13	sl s aq; s alc, eth
c242	2-Chloropyridine	Cl(C$_5$H$_4$N)	113.55	20, 230	1.205^{15}	1.5320^{20}		166^{714mm}	65	s aq; s alc, eth
c243	3-Chloropyridine	Cl(C$_5$H$_4$N)	113.55	20, 230	1.194	1.5300^{20}		148	65	
c244	4-Chlororesorcinol	ClC$_6$H$_3$-1,3(OH)$_2$	144.56	6^2, 818			106–108	147^{18mm}		
c245	4-Chlorosalicylic acid	ClC$_6$H$_3$(2-OH)COOH	172.57	10, 101			210–212			
c246	5-Chlorosalicylic acid	ClC$_6$H$_3$(2-OH)COOH	172.57	10, 102			172			
c247	N-Chlorosuccinimide	ClC$_6$H$_3$(2-OH)COOH	133.53	21, 380	1.65		150–151			1.4 aq; 0.67 alc; 2 bz; sl s chl, CCl$_4$, eth
c248	Chlorosulfonic acid	ClHO$_3$S	116.52	Merck: 12, 2218	1.7534^{20}_{4}	1.437^{14}	−80	152^{755mm}	none	s pyr, dichloroethane; aq dec with violence
c249	Chlorosulfonyl isocyanate	ClSO$_2$NCO	141.53		1.626	1.4470^{20}	−44	107	none	
c250	1-Chlorotetradecane	CH$_3$(CH$_2$)$_{13}$Cl	232.84	1^2, 135	0.859	1.4460^{20}	−72	142^{4mm}	>110	i aq; misc alc, eth
c251	2-Chlorothiophene	Cl(C$_5$H$_5$S)	118.59	17, 32	1.286	1.5483^{20}		127–129	22	misc alc, eth
c252	4-Chlorothiophenol	ClC$_6$H$_4$SH	144.62	6, 326			49–52	205–207	>110	
c253	8-Chlorotheophylline		214.61	26, 473			dec 290			s alkali
c254	Chlorotitanium triisopropoxide	[(CH$_3$)$_2$CHO]$_3$TiCl	260.62		1.091				22	
c255	2-Chlorotoluene	ClC$_6$H$_4$CH$_3$	126.59	5, 290	1.0826^{20}_{4}	1.5268^{20}	−35.6	159.0	47	sl s aq; v s alc, bz, chl, eth
c256	3-Chlorotoluene	ClC$_6$H$_4$CH$_3$	126.59	5, 291	1.0760^{19}_{4}	1.5218^{20}	−47.8	161.8	50	s alc, bz, chl; misc eth
c257	4-Chlorotoluene	ClC$_6$H$_4$CH$_3$	126.59	5, 292	1.0697^{20}_{4}	1.5150^{20}	7.5	162.4	49	sl s aq; s alc, bz, eth
c258	N-Chloro-p-toluene sulfonamide, sodium salt	CH$_3$C$_6$H$_4$SO$_2$NCl$^-$ Na$^+$	227.67				167 dec			s aq; i bz, chl, eth
c259	4-(4-Chloro-o-tolyloxy)butyric acid	ClC$_6$H$_3$(CH$_3$)O(CH$_2$)$_3$COOH	228.68	4^3, 1912			99–100			
c260	Chlorotriethylgermane	(C$_2$H$_5$)$_3$GeCl	195.23	4, 624	1.175	1.4590^{20}		142–144	>110	
c261	Chlorotriethylsilane	(C$_2$H$_5$)$_3$SiCl	150.73	1^3, 138	0.898	1.4300^{20}			29	
c262	Chloro-2,2,2-trifluoroethane	CF$_3$CH$_2$Cl	118.5		1.389^{0}	1.3090^{0}	−105	6.9		

c263	Chlorotrifluoroethylene	$CF_2=CFCl$	116.47	1^3, 646	1.315		-158.2	-28		
c264	Chlorotrifluoromethane	$ClCF_3$	104.46	1^3, 42	4.270 g/L		-181	-81		
c265	Chlorotrimethylgermane	$(CH_3)_3GeCl$	153.16		1.2382^{22}	1.4283^{20}	-13	102	1	
c266	Chlorotrimethylsilane	$(CH_3)_3SiCl$	108.64	4^3, 1857	0.8580^{20}_{4}	1.3870^{20}	-40	57	-27	
c267	Chlorotriphenylmethane	$(C_6H_5)_3CCl$	278.78	5, 700			110–112	235^{20mm}		v s bz, chl, eth
c268	Chlorotriphenyltin	$(C_6H_5)_3SnCl$	385.46	12, 914			108 dec	240^{14mm}		
c268a	Chloro-tris(dimethylamino)silane	$[(CH_3)_2N]_3SiCl$	195.8		0.975^{20}_{4}	1.442^{20}		63^{12mm}		
c269	α-Chloro-*o*-xylene	$CH_3C_6H_4CH_2Cl$	140.61	5, 364	1.063	1.5391^{20}		96^{25mm}	73	i aq; misc alc, eth
c270	α-Chloro-*m*-xylene	$CH_3C_6H_4CH_2Cl$	140.61	5, 373	1.064^{20}	1.5350^{20}		195–196	75	i aq; misc alc, eth
c271	α-Chloro-*p*-xylene	$CH_3C_6H_4CH_2Cl$	140.61	5, 384		1.5330^{20}	4.5	200	75	misc alc, bz, eth, acet
c272	2-Chloro-*p*-xylene	$ClC_6H_3(CH_3)_2$	140.61	5, 384	1.049	1.5240^{20}	2	186	57	
c273	4-Chloro-*p*-xylene	$ClC_6H_3(CH_3)_2$	140.61	5, 363	1.047	1.5280^{20}		221–223	66	misc alc, bz, eth, acet
c274	Cholesterol		386.66	6^3, 2607	1.052^{19}_{19}		148.5	$203^{0.5mm}$		1.3 alc; 35 eth; 22 chl; s bz, PE
c275	Cholic acid		408.58	10^3, 2162			198			(15°): 0.03 aq; 3.1 alc; 2.8 acet; 15.2 HOAc; 0.5 chl; 0.036 bz
c276	Cinchonine		294.40	23^2, 369			ca. 260			1.6 alc; 0.9 chl; 0.2 eth
c277	1,8-Cineole		154.25	17, 23	0.921^{25}_{25}	1.4572^{20}	1	176.4	48	misc alc, chl, eth
c278	*trans*-Cinnamaldehyde	$C_6H_5CH=CHCHO$	132.16	7, 348	1.050^{25}_{25}	1.6219^{20}	-7.5	136^{20mm}	71	0.014 aq; misc alc, chl, eth
c279	*trans*-Cinnamic acid	$C_6H_5CH=CHCOOH$	148.16	9, 573	1.2475^{4}_{4}		133	300	>110	0.05 aq; 16 alc; 8 chl
c280	*trans*-Cinnamoyl chloride	$C_6H_5CH=CHCOCl$	166.61	9^2, 390	1.1617^{25}_{25}	1.614^{43}	35–36	258	>110	s hot alc, CCl_4
c281	Cinnamyl acetate	$CH_3CO_2CH_2CH=CHC_6H_5$	176.22	6^2, 527	1.0571	1.5421^{20}		265	>110	s aq; v s common organic solvents
c282	Cinnamyl alcohol	$C_6H_5CH=CHCH_2OH$	134.18	6, 570	1.0397^{35}	1.5758^{33}	33	250.0		
c283	Cinnamyl chloride	$C_6H_5CH=CHCH_2Cl$	152.62	5, 482	1.096	1.5840^{20}	-19	108^{12mm}	79	
c284	Citraconic acid	$CH_3C(COOH)=CHCOOH$	130.10	2, 768	1.62		92 dec			
c285	Citraconic anhydride		112.08	17, 440	1.247	1.4712^{20}	8	214	101	v s aq, alc, eth; sl s chl; i bz, PE
c286	Citral (geranial plus neral, *cis* and *trans* forms, resp.)	$(CH_3)_2C=CHCH_2CH_2C(CH_3)=CHCHO$	152.24		0.888	1.4876^{20}		229	101	

(Continued)

TABLE 2.21 Physical Constants of Organic Compounds (*Continued*)

No.	Name	Formula	Formula weight	Beilstein reference	Density, g/mL	Refractive index	Melting point, °C	Boiling point, °C	Flash point, °C	Solubility in 100 parts solvent
c287	Citral dimethyl acetal	$(CH_3)_2C=CHCH_2CH_2$-$C(CH_3)=CH(OCH_3)_2$	198.31	1[4], 3570	0.890	1.4540[20]		106[10mm]	92	i aq; s alkali
c288	Citrazinic acid		155.11	22, 254			carbonizes without melting >300			
c289	Citric acid	$HOOCCH_2C(OH)(COOH)$-CH_2COOH	192.12	3, 556	1.665		154			59 aq
c290	β-Citronellol	$(CH_3)_2C=CHCH_2CH_2$-$CH(CH_3)CH_2CH_2OH$	156.27	1[1], 232	0.8570[20][4]	1.4560[20]		222	98	v sl s aq; misc alc, eth
c299	Cocaine		303.35	22[2], 150		1.5022[98]	98	187[0.1mm]		0.17 aq; 15 alc; 140 chl; 28 eth; s acet; EtOAc, CS$_2$
c300	Coumarin		146.15	17, 328	0.9354[20][4]		68–70	298		0.25 aq; v s alc, chl, eth; s alkali
c301	Creatine	$HOOCCH_2N(CH_3)$-$C(=NH)NH_2$	131.14	4, 363			dec 303			1.3 aq; 0.11 alc; i eth
c302	Creatinine		113.12	24, 245			255 dec			8 aq; sl s alc; i eth
c303	o-Cresol	$CH_3C_6H_4OH$	108.14	6, 349	1.0273[41]	1.5361[41]	30	191	81	3.1 aq[40]; misc alc, chl, eth; s alkali
c304	m-Cresol	$CH_3C_6H_4OH$	108.14	6, 373	1.034[20][4]	1.5438[20]	12	202.2	86	2.5 aq[40]; misc alc, chl, eth; s alkali
c305	p-Cresol	$CH_3C_6H_4OH$	108.14	6, 389	1.0179[41]	1.5312[41]	34.8	201.9	86	2.3 aq[40]; misc alc, chl, eth; s alkali
c306	trans-Crotonaldehyde	$CH_3CH=CHCHO$	70.09	1, 728	0.8516[20]	1.4373[20]	−76	102–104	13	18.1 aq[20]
c307	Crotonic acid	$CH_3CH=CHCOOH$	86.19	2, 408	0.964[8][4]	1.4228[80]	71.6	185	87	54.6 aq[20], 52.5 EtOH[25], 53 acet; 37.5 toluene
c308	Crotonic anhydride	$(CH_3CH=CHO)_2O$	154.17	2, 411	1.040	1.4740[20]		248	110	
c309	Crotononitrile	$CH_3CH=CHCN$	67.09	2, 412	1.4190[20]	1.4190[20]		121	20	
c310	Crotonyl chloride	$CH_3CH=CHCOCl$	104.54	2, 411	1.091	1.4600[20]		120–123	35	
c311	Crotyl alcohol	$CH_3CH=CHCH_2OH$	72.11	1, 442	0.845	1.4270[20]		122	37	17 aq; misc alc
c312	Crotyl chloride	$CH_3CH=CHCH_2Cl$	90.55	1[2], 176	0.929	1.4360[20]		85	−5	
c313	12-Crown-4		176.21		1.089	1.4630[20]		70[0.5mm]	>110	specific for Li$^+$
c314	18-Crown-6		264.32				42–45		>110	

No.	Name	Formula	Mol. wt.	Beilstein ref.	Density	n_D	m.p., °C	b.p., °C	Flash p., °C	Solubility
c315	Crystal Violet		407.99	13, 756						
c316	Cumene hydro-peroxide	$C_6H_5C(CH_3)_2OH$	152.20	6^3, 1814	1.030	1.5210^{20}	215 dec	101^{8mm}	56	v s aq, alc
c316a	Cumylphenol	$C_6H_5C(CH_3)_2C_6H_4OH$	212.29				74–76	335		78 aq; 29 BuOH; 42 EtOAc; s alc, eth
c317	Cupferron	$C_6H_5N(NO)O^-\,NH_4^+$	155.16	16^1, 395			163–164			25 aq; 3.1 alc
c318	Cyanamide	H_2NCN	42.04	3^2, 63	1.282_4^{20}		46	83^{380mm}	>110	s aq, alc, eth; sl s bz
c319	2-Cyanoacetamide	$NCCH_2CONH_2$	84.08	2, 589			119.5	108^{15mm}	215	v s aq; s alc; i eth
c320	Cyanoacetic acid	$NCCH_2COOH$	85.06	2, 583			66	dec		
c321	Cyanoacetohydrazide	$NCCH_2C(=O)NHNH_2$	99.09	Merck: 11, 2688			115		107	
c322	Cyanoacetylurea	$NCCH_2C(=O)NH\text{-}C(=O)NH_2$	127.10	3, 66			214 dec			misc aq, alc; sl s eth
c323	2-Cyanoethanol	$NCCH_2CH_2OH$	71.08	3^2, 213	1.0588^0			108^{11mm}		v s aq, alc, eth
c324	2-Cyanoethyl acrylate	$H_2C=CHCO_2CH_2CH_2CN$	125.13	3^3, 543	1.052	1.4470^{20}		108^{12mm}		
c325	Cyanogen bromide	$BrCN$	105.93	3, 39	2.015_4^{20}		52	61–62	103	
c326	1-Cyano-3-methyliso-thiourea, sodium salt	$CH_2NH(=NCN)S^-\,Na^+$	137.14	4, 71			290 dec		5	
c327	1-Cyanonaphthalene	$C_{10}H_7CN$	153.18	9, 649	1.1113_{25}^{25}	1.6298^{18}	38	299		i aq; v s alc, eth
c328	2-Cyanopyridine	$NC(C_5H_4N)$	104.11	22, 36	1.081	1.5288^{20}	26–28	215	89	s aq; v s alc, bz, eth
c329	3-Cyanopyridine	$NC(C_5H_4N)$	104.11	22, 41			50–52	201	84	v s aq, alc, bz, eth
c330	4-Cyanopyridine	$NC(C_5H_4N)$	104.11	22, 46			78–80			s aq, alc, bz, eth
c331	Cyanotrimethylsilane	$(CH_3)_3SiCN$	99.21	4^4, 3893	0.783_0^{20}	1.3924^{20}	11–12	118–119	1	0.5 aq; s hot alc, pyr; i acet, bz, chl, eth
c332	Cyanuric acid		129.08	26, 239	1.768^0		>360; dec to HOCN			
c333	Cyclobutane	C_4H_8	56.10	5, 17	0.7038^0	1.3752^0	−91	13	83	i aq; v s alc, acet
c334	Cyclobutanecarboxylic acid	$(C_4H_7)COOH$	100.12	9, 5	1.047	1.4433^{20}	−20 to −7.5	195		
c335	Cyclodecane	$C_{10}H_{20}$	140.27		0.871	1.4707^{20}		201	65	
c336	Cyclododecanol	$C_{12}H_{23}OH$	184.32	7^2, 48	0.906^{62}		77	85^{1mm}		
c337	Cyclododecanone	$C_{12}H_{22}(=O)$	182.31	5^1, 1115	0.8925_4^{20}	1.5070^{20}	59–61	231	87	
c338	trans, trans, cis-1,5,9-cyclododecatriene		162.28				−18	232–245		
c339	Cyclododecene		166.31		0.863	1.4822^{20}	28–30	124^{7mm}	93	
c340	Cyclododecylamine	$(C_{12}H_{23})NH_2$	183.34						121	
c341	Cycloheptane	C_7H_{14}	98.18	5, 29	0.811_{20}^{20}	1.4455^{20}	−8.0	118	6	v s alc, eth
c342	Cycloheptanol	$C_7H_{13}OH$	114.19	6, 10	0.948_4^{20}	1.4760^{20}	2	185	71	sl s aq; v s alc, eth
c343	Cycloheptanone	$C_7H_{12}(=O)$	112.17	7, 13	0.9490_4^{20}	1.4611^{20}		179–181	55	i aq; v s alc; s eth

(Continued)

TABLE 2.21 Physical Constants of Organic Compounds (*Continued*)

No.	Name	Formula	Formula weight	Beilstein reference	Density, g/mL	Refractive index	Melting point,°C	Boiling point,°C	Flash point,°C	Solubility in 100 parts solvent
c344	1,3,5-Cyclohepta-triene		92.13	5, 280	0.888	1.5211^{20}	−75.3	115.5	26	s alc, eth; v s bz, chl
c345	Cycloheptene	C_7H_{12}	96.17	5, 65	0.824	1.4585^{20}		114.7	−6	s alc, eth
c346	8-Cyclohexadecene-1-one		236.40	7^3, 521		1.4890^{20}		$195^{19\text{mm}}$	>110	
c347	Cyclohexane	C_6H_{12}	84.16	5, 20	0.7786_{4}^{20}	1.4262^{20}	6.6	80.7	−20	0.01 aq; misc acet, alc, bz, CCl_4, eth
c348	Cyclohexane-d_{12}	C_6D_{12}	92.26	5^3, 36	0.893	1.4210^{20}		78	−18	
c349	1,3-Cyclohexanebis-(methylamine)	$C_{10}H_{10}(NHCH_3)_2$	142.25		0.945	1.4930^{20}			106	
c350	1,3-Cyclohexane-carbonitrile	$C_6H_{11}CN$	109.17	9, 9	0.919	1.4505^{20}		$75^{16\text{mm}}$	65	
c351	Cyclohexanecarbonyl chloride	$C_6H_{11}COCl$	146.62	9, 9	1.096	1.4700^{20}		184	66	
c352	Cyclohexanecarbox-aldehyde	$C_6H_{11}CHO$	112.17	7, 19	0.926	1.4500^{20}		163	40	
c353	Cyclohexanecarboxylic acid	$C_6H_{11}COOH$	128.17	9, 7	1.0480_{4}^{15}	1.4530^{20}	29	232.5	>110	0.21 aq; s alc, bz, eth
c354	*trans*-1,2-Cyclo-hexanediamine	$C_6H_{10}(NH_2)_2$	114.19	13^3, 8	0.951	1.4884^{20}	14–15	$92^{18\text{mm}}$	68	
c355	1,3-Cyclohexanedi-carboxylic acid	$C_6H_{10}(COOH)_2$	172.18	9, 732			132–141			
c356	*cis*-1,2-Cyclohexanedi-carboxylic anhydride		154.17	17, 452			32–34	$158^{17\text{mm}}$	>110	
c357	1,4-Cyclohexanedi-methanol	$C_6H_{10}(CH_2OH)_2$	144.21		0.978_{4}^{100}	1.4893^{20}	43	283	161	misc aq, alc; 2.5 eth
c358	1,4-Cyclohexane-divinyl ether	$C_6H_{10}(OCH=CH_2)_2$	196.29		0.919	1.4720^{20}		$126^{14\text{mm}}$	>110	
c359	1,4-Cyclohexanediol	$C_6H_{10}(OH)_2$	116.16	6, 741			98–100		65	
c360	1,3-Cyclohexanedione	$C_6H_8(=O)_2$	112.13	7, 554	1.0861^{91}	1.4576^{102}	103–105	$150^{20\text{mm}}$		s aq, alc, acet, chl
c361	1,2-Cyclohexanedione dioxime	$C_6H_8(=NOH)_2$	142.16	7^2, 526			185–188			s aq
c362	Cyclohexanemethyl-amine	$C_6H_{11}CH_2NH_2$	113.20	12, 12	0.870	1.4630^{20}		145–147	43	
c363	Cyclohexanepropionic acid	$C_6H_{11}CH_2CH_2COOH$	156.23	9, 82	0.912	1.4636^{20}	14–17	275.8	>110	
c364	Cyclohexanethiol	$C_6H_{11}SH$	116.23	6, 8	0.950	1.4921^{20}		158–160	43	
c365	Cyclohexanol	$C_6H_{11}OH$	100.16	6, 5	0.9416^{30}	1.4629^{30}	25.4	161	68	3.8 aq^{25}; misc alc, bz

No.	Name	Formula	Mol. wt.	Beil. ref.	Density	n_D	m.p.	b.p.	f.p.	Solubility
c366	Cyclohexanone	$C_6H_{10}(=O)$	98.15	7, 8	0.94784^{20}_4	1.4510^{20}	−31	155.7	44	15 aq[10]; s alc, eth
c367	Cyclohexanone oxime	$C_6H_{10}(=NOH)$	113.16	7, 10			89–91	206–210	−12	s aq, eth; sl s alc
c368	Cyclohexene	C_6H_{10}	82.15	5, 63	0.80944^{20}_4	1.4464^{20}	−103.5	83.0		0.02 aq; misc alc, bz, acet, eth
c369	3-Cyclohexene-1-methanol	$C_6H_9CH_2OH$	112.17	6^3, 215	0.961	1.4853^{20}		85^{18mm}	76	v s alc
c370	Cyclohexene oxide	$C_6H_8(=O)$	98.15	17, 21	0.970	1.4520^{20}		130	27	
c371	2-Cyclohexene-1-one		96.13	7^2, 55	0.993	1.4885^{20}	−53	168	56	
c372	4-(3-Cyclohexene-1-yl)pyridine		159.23	20^3, 3239	1.021	1.5480^{20}		141^{20mm}	>110	
c373	Cyclohexyl acetate	$CH_3CO_2C_6H_{11}$	142.20	6, 7	0.966	1.4395^{20}		173	57	sl s aq; s org solv
c374	Cyclohexylacetic acid	$C_6H_{11}CH_2COOH$	142.20	9^2, 9	1.007	1.4630^{20}	31–33	242–244	>110	
c375	Cyclohexylamine	$C_6H_{11}NH_2$	99.18	12, 5	0.8671^{20}	1.4593^{20}	−18	134	31	misc aq, alc, chl, eth
c376	Cyclohexylbenzene	$C_6H_{11}C_6H_5$	160.26	5, 503	0.9502^{20}_4	1.5258^{20}	7	240	98	i aq; v s alc, eth
c377	Cyclohexyldimethoxymethylsilane	$C_6H_{11}Si(OCH_3)_2CH_3$	188.35		0.940	1.4390^{20}		201.2	73	
c378	2-Cyclohexylethanol	$C_6H_{11}CH_2CH_2OH$	128.22	6, 17	0.919	1.4647^{20}		207^{745mm}	86	s alc, eth
c379	Cyclohexylethyl acetate	$CH_3CO_2CH_2CH_2C_6H_{11}$	170.25		0.949	1.4461		98^{15mm}	81	
c380	N-Cyclohexylformamide	$C_6H_{11}NHCHO$	127.18	12^2, 11			38–40	113^{10mm}	>110	
c381	Cyclohexyl isocyanate	$C_6H_{11}NCO$	125.17	12^2, 12	0.980	1.4551^{20}		168–170	48	
c382	Cyclohexyl isothiocyanate	$C_6H_{11}NCS$	141.24	12^2, 12	0.996	1.5350^{20}		219	95	
c383	Cyclohexyl methacrylate	$H_2C=C(CH_3)CO_2C_5H_{11}$	168.24	6^3, 25	0.964	1.4580^{20}		70^{4mm}	82	
c384	Cyclohexylmethanol	$C_6H_{11}CH_2OH$	114.19	6, 14	0.9215^{25}_4	1.4640^{25}		181	71	s alc, eth
c385	3-Cyclohexyl-1-propanol	$C_6H_{11}CH_2CH_2CH_2OH$	142.24	6^1, 15	1.007	1.4975^{20}		218	101	
c386	N-Cyclohexyl-2-pyrrolidinone		167.25	21^3, 3149	1.026	1.495	12	284	>110	
c387	cis,cis-1,3-Cyclooctadiene		108.18	5^4, 401	0.869	1.4928^{20}	−53 to −51	55^{34mm}	24	
c388	1,5-Cyclooctadiene		108.18	5, 116	0.8818^{25}_4	1.4905^{25}	−69	149–150	31	s CCl_4
c389	Cyclooctane	C_8H_{16}	112.22	5, 35	0.834	1.4574^{20}	14.8	151.1	30	
c390	trans-1,2-Cyclooctanediol	$C_8H_{14}(OH)_2$	144.21	6^3, 4094	1.080	1.4980^{20}	32	$94^{0.5mm}$	>110	

(Continued)

TABLE 2.21 Physical Constants of Organic Compounds (*Continued*)

No.	Name	Formula	Formula weight	Beilstein reference	Density, g/mL	Refractive index	Melting point, °C	Boiling point, °C	Flash point, °C	Solubility in 100 parts solvent
c391	Cyclooctanol	$C_8H_{15}OH$	128.22	6^2, 25	0.9740^{20}	1.4850^{20}	14–15	108^{22mm}	86	misc alc, bz, CCl₄, eth; s aniline, HOAc, CS₂
c392	Cyclooctanone	$C_8H_{14}(=O)$	126.20	7, 21	0.9584^{20}_{4}	1.6494^{20}	41–43	195–197	72	i aq; misc alc, eth
c393	cis-Cyclooctene	C_8H_{14}	110.20	5^1, 35	0.846	1.4698^{20}	−16	145–146	25	sl s aq; s MeOH
c394	Cyclooctylamine	$C_8H_{15}NH_2$	127.23		0.928	1.4804^{20}	−48	190	62	
c395	Cyclopentadiene	C_5H_6	66.10	Merck: 12, 2807	0.8021^{24}	1.4463^{16}	−85	41–42		
c396	Cyclopentane	C_5H_{10}	70.13	5, 19	0.7460^{20}_{4}	1.4068^{20}	−94	49.3	−37	sl s aq; s alc
c397	Cyclopentane-carboxylic acid	C_5H_9COOH	114.14	9, 6	1.053^{20}_{4}	1.4540^{20}	4	216	93	sl s aq; misc alc, eth
c398	Cyclopentanol	C_5H_9OH	86.13	6, 5	0.9488^{20}_{4}	1.4521^{20}	−19	140	51	s aq, alc, bz, chl, eth
c399	Cyclopentanone	$C_5H_8(=O)$	84.12	7, 5	0.9509^{18}_{4}	1.4366^{20}	−51	130.6	26	
c400	Cyclopentanone oxime	$C_5H_8(=NOH)$	99.13	7, 7			53–55	196	92	
c401	Cyclopentene	C_5H_8	68.11	5, 61	0.7720^{20}	1.4228^{20}	−135.1	44.2	−29	
c402	2-Cyclopentene-1-acetic acid	$C_5H_7CH_2COOH$	126.16	9, 42	1.047	1.4675^{20}	19	$94^{2.5mm}$	>110	
c403	N-(1-Cyclopenten-1-yl)morpholine		153.23		0.957	1.5105^{20}		106^{12mm}	60	
c404	Cyclopentylamine	$C_5H_9NH_2$	85.15	12, 4	0.863	1.4482^{20}		106–108	17	
c405	3-Cyclopentyl-propanoic acid	$C_5H_9CH_2CH_2COOH$	142.20		0.996	1.4570^{20}		130^{12mm}	46	
c406	Cyclopropane	C_3H_6	42.08	5, 15	0.720^{-79}_{4}		−127	−32.8		37 mL/100 mL aq^{15}; v s alc, eth
c407	Cyclopropanecarbo-nitrile	C_3H_5CN	67.09	9, 4	0.911^{16}	1.4207^{20}		135	32	s eth
c408	Cyclopropanecarbonyl chloride	C_3H_5COCl	104.54	9, 4	1.152	1.4522^{20}		119	23	
c409	Cyclopropane-carboxylic acid	C_3H_5COOH	86.09	9, 4	1.088	1.4380^{20}	17–19	182–184	71	sl s hot aq; s alc, eth
c410	Cyclopropyl methyl ketone	$C_3H_5COCH_3$	84.12	7, 7	0.8993^{20}_{4}	1.4240^{20}		114	21	s aq, alc, eth
c411	L-Cysteine	$HSCH_2CH(NH_2)COOH$	121.16	4, 506			220 dec			v s aq, alc; i bz, eth
c412	L-Cystine	$HOOCCH(NH_2)SSCH_2-CH(NH_2)COOH$	240.30	4, 507			dec 240			0.01 aq; s acid, alkali; i alc
d1	1,9-Decalene	$H_2C=CH(CH_2)_6CH=CH_2$	138.25	1^1, 123	0.750	1.4320^{20}		169	41	
d2	cis-Decahydro-naphthalene	$C_{10}H_{18}$	138.25	5, 92	0.8963^{20}_{4}	1.4810^{20}	−43	195.8	58 (CC)	v s alc, chl, eth; misc most ketones, esters

No.	Name	Formula	Mol. wt.	Beil. ref.	Density	n_D	M.p., °C	B.p., °C	Flash pt	Solubility
d3	*trans*-Decahydro-naphthalene	$C_{10}H_{18}$	138.25	5², 56	0.8700^{20}_4	1.4690^{20}	−30.4	187.3	54	see under *cis*
d4	Decahydro-2-naphthol	$C_{10}H_{17}OH$	154.25	6, 67	0.996	1.500^{20}		109^{14mm}	>110	i aq
d5	Decamethylcyclo-pentasiloxane	[—Si(CH₃)₂O—]₅	370.78	4⁴, 4128	0.9593^{20}_4	1.3982^{20}	−38	101^{20mm}	72	
d6	Decamethyltetra-siloxane	(CH₃)₃SiO[Si(CH₃)₂O]₂-Si(CH₃)₃	310.69	4³, 1879	0.8536^{20}_4	1.3895^{20}	−68	194	62	sl s alc; s bz, PE
d7	Decanal	H(CH₂)₉CHO	156.27	1, 711	0.830^{15}	1.4280^{20}	−5	208–209	85	i aq; s alc, eth
d8	Decane	CH₃(CH₂)₈CH₃	142.29	1, 168	0.7301^{20}_4	1.4110^{20}	−29.7	174.1	46	0.07 aq
d9	1,10-Decanediamine	H₂N(CH₂)₁₀NH₂	172.32	4, 273			62–63	140^{12mm}		
d10	Decanedioic acid	HOOC(CH₂)₈COOH	202.25	2, 718	1.207^{20}_4	1.422^{134}	134.5	232^{10mm}	>110	0.1 aq[20], eth[17]; v s alc, esters, ketones
d11	1,2-Decanediol	CH₃(CH₂)₇CH(OH)CH₂OH	174.28	1, 494			48–50	255	>110	sl s aq, eth; v s alc
d12	1,10-Decanediol	HO(CH₂)₁₀OH	174.28	1², 560			74	170^{8mm}	>110	dec aq, alc
d13	Decanedioyl dichloride	ClC(=O)(CH₂)₈COCl	239.14	2, 719	1.1212^{20}_4	1.4678^{20}		220^{75mm}	>110	misc alc, chl, eth
d13a	Decanenitrile	CH₃(CH₂)₈CN	153.27	2, 356	0.8295^{15}_4	1.4295^{20}	−15	235–237	>110	
d14	1-Decanethiol	CH₃(CH₂)₉SH	174.35	1², 459	0.841	1.4565^{20}	−26	114^{13mm}	98	
d15	Decanoic acid	CH₃(CH₂)₈COOH	172.27	2², 309	0.8752^{50}_4	1.4288^{40}	32	270	>110	0.015 aq; s alc, bz, chl, CS_2
d16	1-Decanol	CH₃(CH₂)₉OH	158.29	1, 425	0.8297^{20}_4	1.4359^{20}	6.9	232	82	i aq; s alc, eth
d17	δ-Decanolactone		170.25	17⁵, 91	0.954	1.4580^{20}		$120^{0.02mm}$	>110	
d18	2-Decanone	CH₃(CH₂)₇COCH₃	156.27	1, 711	0.825	1.4250^{20}	3.5	211	71	i aq; misc alc, eth
d19	3-Decanone	CH₃(CH₂)₆COC₂H₅	156.27	1¹, 367	0.825	1.4241^{20}	−3.8	205	25	dec aq, alc; s eth
d20	4-Decanone	CH₃(CH₂)₅C(=O)(CH₂)₂CH₃	156.27	1, 711	0.824^{20}_0	1.4237^{20}		207	71	i aq; misc alc, eth
d21	Decanoyl chloride	CH₃(CH₂)₈COCl	190.71	2, 356	0.919	1.4410^{20}	−34.5	96^{5mm}	106	i aq; misc alc, eth
d22	1-Decene	H(CH₂)₈CH=CH₂	140.27	1³, 858	0.7408^{20}_4	1.4210^{20}	−66	170.6	47	
d23	Decylamine	H(CH₂)₁₀NH₂	157.30	4, 199	0.787	1.4360^{20}	12–14	216–218	85	sl s aq; misc alc, bz, eth, acet
d24	Dehydroabietylamine		285.48	12⁴, 3005		1.5460^{20}			>110	at 25°: 22 acet; 18 bz; 5 eth; 3 EtOH; 5 MeOH
d25	Dehydroacetic acid		168.15	17, 559			111–113	270		
d26	Deoxybenzoin	C₆H₅CH₂COC₆H₅	196.25	7², 368	1.201^{0}_4		55–56	320	110	i aq; v s alc, eth
d27	Diacetoxydimethyl-silane	(CH₃)₂Si(OOCCH₃)₂	176.3		1.054^{20}_4	1.4030^{20}		164–166		
d28	*trans*-1,1-Diacetoxy-2-butene	(CH₃CO₂)₂CHCH=CHCH₃	172.18	2, 154	1.057	1.4290^{20}		106^{20mm}	87	
d29	1,1-Diacetoxy-2-propene	(CH₃CO₂)₂CHCH=CH₂	158.16	2, 154	1.078	1.4190^{20}		184	78	

(Continued)

561

TABLE 2.21 Physical Constants of Organic Compounds (*Continued*)

No.	Name	Formula	Formula weight	Beilstein reference	Density, g/mL	Refractive index	Melting point, °C	Boiling point, °C	Flash point, °C	Solubility in 100 parts solvent
d30	Diallylamine	$(H_2C=CHCH_2)_2NH$	97.16	4, 208	0.787	1.4405^{20}	−88	112	15	i aq; misc alc, eth
d31	Diallyl ether	$(H_2C=CHCH_2)_2O$	98.15	1, 438	0.805^{18}_0	1.4160^{20}	−47	94–95	−6 (OC)	
d32	Diallyl maleate	$H_2C=CHCH_2O_2CCH=CH-CO_2CH_2CH=CH_2$	196.20	2^3, 1926	1.073	1.4702^{20}		116^{4mm}	>110	
d33	Diallyl 1,2-phthalate	$C_6H_4(CO_2CH_2CH=CH_2)_2$	246.27	9^3, 4120	1.121	1.5187^{20}		167^{5mm}	>110	sl s aq; misc alc, eth
d34	Diallyl sulfide	$(H_2C=CHCH_2)_2S$	114.21	1, 440	0.8877^{27}_4	1.4889^{20}	−85	138	46	
d35	(+)-N,N-Diallyl-tartardiamide	$[-CH(OH)CONHCH_2-CH=CH_2]_2$	228.25	4, 218			186–188			
d36	1,2-Diaminoanthra-quinone		238.25	14^1, 459			289–291			sl s alc, eth
d37	1,4-Diaminoanthra-quinone		238.25	14, 197			265–269			sl s aq, alc; v s bz
d38	1,5-Diaminoanthra-quinone		238.25	14, 203			308 dec			sl s hot aq, pyr
d39	2,6-Diaminoanthra-quinone		238.25	14, 215			>325			sl s aq; s alc, eth
d40	3,5-Diaminobenzoic acid	$(H_2N)_2C_6H_3COOH$	152.15	14, 453			228	−H₂O, 110		
d41	1,4-Diaminobutane	$H_2N(CH_2)_4NH_2$	88.15	4, 264	0.877	1.4569^{20}	27.3	158–160	51	s aq
d42	4,4'-Diaminodiphenyl-amine sulfate	$H_2NC_6H_4NHC_6H_4NH_2 \cdot H_2SO_4$	297.33	13, 110			300			
d43	trans-1,2-Diamino-cyclohexane	$C_6H_{10}(NH_2)_2$	114.19	13^3, 8	0.951	1.2886^{20}	14–15	81^{15mm}	68	
d44	trans-1,4-Diamino-cyclohexane	$C_6H_{10}(NH_2)_2$	114.19	13^1, 3			69–72	197	71	
d45	trans-1,2-Diamino-cyclohexane-N,N,N',N'-tetra-acetic acid hydrate	$C_6H_{10}[N(CH_2COOH)_2]_2 \cdot H_2O$	364.36	13^3, 10			213–216			v s aq
d46	4,4'-Diaminodiphenyl-methane	$H_2NC_6H_4CH_2C_6H_4NH_2$	198.27	13, 238			91–92	398	221	sl s aq; v s alc, bz, eth
d47	3,3'-Diaminodiphenyl sulfone	$H_2NC_6H_4SO_2C_6H_4NH_2$	248.30	13, 426			170–173			i aq; s alc, bz
d48	4,4'-Diaminodiphenyl sulfone	$H_2NC_6H_4SO_2C_6H_4NH_2$	248.30	13, 536			175–176			i aq; s alc, acet, dil HCl

(Continued)

No.	Name	Formula	Mol. wt.	Beilstein ref.	Density	n	m.p., °C	b.p., °C	Flash pt	Solubility
d49	2,4-Diamino-6-hydroxypyrimidine		126.12	24, 469			285 dec			s aq
d50	Diaminomaleonitrile	$NCC(NH_2){=}C(NH_2)CN$	108.10	4^2, 949			178–179			
d51	1,8-Diamino-p-men-thane		170.30	13, 4	0.914	1.4805^{20}	−45	125^{10mm}	93	
d52	3,3'-Diamino-N-methyldipropylam(ine)	$CH_3N[(CH_2)_3NH_2]_2$	145.25	4^4, 1279	0.901	1.4725^{20}		112^{6mm}	102	
d53	2,4-Diamino-6-phenyl-1,3,5-triazine		187.21	26^1, 69	1.40^{25}_4		227–228			0.06 aq; s alc, eth, dil HCl; sl s DMF
d54	1,2-Diaminopropane	$CH_3CH(NH_2)CH_2NH_2$	74.13	4, 257	0.878	1.4460^{20}		119–120	33	v s aq
d55	1,3-Diaminopropane	$H_2N(CH_2)_3NH_2$	74.13	4, 261	0.888	1.4570^{20}	−12	140	48	v s aq
d56	1,3-Diamino-2-propanol	$H_2NCH_2CH(OH)CH_2NH_2$	90.13	4, 290			40–45	235	>110	
d58	2,6-Diaminopyridine	$(H_2N)_2C_5H_3N$	109.13	22^1, 647			120–122	283–285		s aq, alc
d59	2,4-Diaminotoluene	$(H_2N)_2C_6H_3CH_3$	122.17	13, 124			97–99	156^{18mm}		
d60	3,4-Diaminotoluene	$(H_2N)_2C_6H_3CH_3$	122.17	13, 148			91–93	174		
d61	1,4-Diazabicyclo[2.2.2]octane		112.18	23^3, 484			158–160		62	45 aq; 77 EtOH; 51 bz; 13 acet; 26 MeEtKe
d62	1,8-Diazabicyclo[5.4.0]undec-7-ene		152.24		1.018	1.5219^{20}		$80^{0.6mm}$	>110	VERY EXPLOSIVE; s eth, dioxane
d63	Diazomethane	$CH_2{=}N{=}N$	42.04	23, 25			−145	−23		
d64	1-Diazo-2-naphthol-4-sulfonic acid		272.22	16, 595			160 dec			
d65	1,2,5,6-Dibenzanthracene		278.33	5^1, 369			266 subl	524		s bz, PE; sl s alc, eth
d66	Dibenzofuran		168.20	17, 70	1.0886^{99}_4	1.6079^{99}	81–83	285		s alc, bz, eth; i aq
d67	Dibenzothiophene		184.26	17, 72			97–100	332–333		s aq; v s alc, bz
d68	Dibenzoylmethane	$C_6H_5COCH_2COC_6H_5$	224.26	7, 769			78–79	220^{18mm}		4.4 alc; s eth, aq NaOH
d69	Dibenzoyl peroxide	$C_6H_5C({=}O)OOC({=}O)C_6H_5$	242.23	9, 179			103–106	may explode when heated		sl s aq, alc; s bz, chl, eth
d70	(−)-Dibenzoyl-L-tartaric acid hydrate	$[C_6H_5COOCH(COOH){-}]_2 \cdot H_2O$	376.34	9, 170			90–92			
d71	Dibenzylamine	$C_6H_5CH_2NHCH_2C_6H_5$	197.28	12, 1035	1.026	1.5731^{20}	−26	300	143	i aq; s alc, eth
d72	Dibenzyldisulfide	$C_6H_5CH_2SSCH_2C_6H_5$	246.39	6, 465			69	d > 270		s hot alc, bz, eth
d73	Dibenzyl ether	$C_6H_5CH_2OCH_2C_6H_5$	198.27	6, 434	1.0014^{20}_4	1.5168^{20}	2	298	135 (CC)	misc alc, acet, chl, eth

TABLE 2.21 Physical Constants of Organic Compounds (*Continued*)

No.	Name	Formula	Formula weight	Beilstein reference	Density, g/mL	Refractive index	Melting point, °C	Boiling point, °C	Flash point, °C	Solubility in 100 parts solvent
d74	N,N-Dibenzyl-ethylenediamine	(C₆H₅CH₂NHCH₂)₂	240.35	12, 1067	1.0244_4^{20}	1.5624^{20}	26	195^{4mm}	>110	v s alc, bz, chl, eth
d75	Dibenzyl malonate	CH₂[CO₂CH₂C₆H₅]₂	284.31	6, 436	1.137	1.5447^{20}		$188^{0.2mm}$	>110	
d76	Dibromoacetic acid	Br₂CHCOOH	217.86	2, 218			39–41	130^{16mm}	>110	
d77	Dibromoacetonitrile	Br₂CHCN	198.86	2, 219	2.296	1.5393^{20}		69^{24mm}		v s warm alc; s eth
d78	2,4′-Dibromoaceto-phenone	BrC₆H₄C(=O)CH₂Br	277.96	7, 285			108–110			s eth
d79	1,4-Dibromobenzene	C₆H₄Br₂	235.92	5, 211	0.9641^{100}	1.5743^{100}	87.3	220	>110	1.4 alc; v s eth; s bz
d80	4,4′-Dibromobiphenyl	BrC₆H₄C₆H₄Br	312.00	5, 580			167–170	355–360		s bz; sl s hot alc
d81	1,2-Dibromobutane	CH₃CH₂CH(Br)CH₂Br	215.93	1, 120	1.789	1.5141^{20}		60^{20mm}	>110	s chl, eth
d82	1,3-Dibromobutane	CH₃CH(Br)CH₂CH₂Br	215.93	1, 120	1.800^{20}	1.5085^{20}		175	110	s chl
d83	1,4-Dibromobutane	BrCH₂CH₂CH₂CH₂Br	215.93	1, 120	1.8080_4^{20}	1.5186^{20}	−20	198	>110	
d84	meso-2,3-Dibromo-butane	CH₃CH(Br)CH(Br)CH₃	215.93	1, 121	1.767	1.5100^{20}		74^{47mm}		
d85	2,3-Dibromo-1,4-butanediol	HOCH₂CH(Br)CH(Br)CH₂OH	247.93	1^3, 2176			88–90	$150^{1.5mm}$	>110	
d86	1,4-Dibromo-2,3-butanediol	BrCH₂C(=O)C(=O)CH₂Br	243.89	1, 774			117–119			
d87	trans-2,3-Dibromo-2-butene-1,4-diol	HOCH₂C(Br)=C(Br)CH₂OH	245.91	1^1, 260			112–114			
d88	Dibromochloro-methane	HCClBr₂	208.29	1, 67	2.451	1.5465^{20}	−22	120^{748mm}	none	misc alc, bz, eth
d89	trans-1,2-Dibromo-cyclohexane	C₆H₁₀Br₂	241.96	5, 24	1.784	1.5515^{20}		146^{10mm}	>110	
d90	1,2-Dibromo-2-chloro-1,1,2-trifluoroethane	FCCl(Br)C(Br)F₂	276.5		2.2478^{20}	1.4275^{20}		93–94	none	
d91	1,10-Dibromodecane	Br(CH₂)₁₀Br	300.09	1^1, 64	1.335^{30}	1.4912^{20}	27	160^{15mm}	>110	sl s alc; s eth
d92	1,2-Dibromo-1,1-difluoroethane	CH₂BrC(Br)F₂	223.87	1, 92	2.2238^{20}	1.4456^{20}	−61.3	92.4	none	i aq
d93	Dibromodifluoro-methane	Br₂CF₂	209.81	1^1, 16	2.2884^{15}	1.4016^{20}	−110	25	none	0.1 aq; misc alc, bz, chl, eth
d94	1,2-Dibromo-3,3-di-methylbutane	(CH₃)₃CCH(Br)CH₂Br	243.98	1, 151	1.610	1.5053^{20}		73^{3mm}	83	
d95	1,3-Dibromo-5,5-di-methylhydantoin		185.93				197 dec			
d96	1,1-Dibromoethane	CH₃CHBr₂	187.86	1, 90	2.0554_4^{20}	1.5379^{20}		113	none	i aq; v s alc, eth

No.	Name	Formula	Mol. wt.	Ref.	Density	n	m.p.	b.p.	Flash p.	Solubility
d97	1,2-Dibromoethane	BrCH$_2$CH$_2$Br	187.86	1, 90	2.1802_4^{20}	1.5387^{20}	10.0	131.7	none	0.43 aq; misc alc, eth
d98	(1,2-Dibromoethyl)-benzene	C$_6$H$_5$CH(Br)CH$_2$Br	263.97	5, 356			70–74	140^{15mm}		
d99	cis-1,2-Dibromo-ethylene	BrCH=CHBr	185.86	1, 190	2.21^{17}	1.5431^{18}	−53	112.5	none	s alc, bz, chl, eth
d100	trans-1,2-Dibromo-ethylene	BrCH=CHBr	185.86	1, 190	2.246	1.5505^{18}	−6.5	108	none	
d101	1,2-Dibromoethyltri-chlorosilane	BrCH$_2$CH(Br)SiCl$_3$	321.3		2.0046_4^{20}	1.537^{20}		90^{11mm}		
d102	4'5'-Dibromo-fluorescein		490.12	19, 228			270–273		101	s hot alc, HOAc
d103	1,4-Dibromo-2-fluoro-benzene	Br$_2$C$_6$H$_3$F	253.91	5^4, 684			33–36	216	92	
d104	2,4-Dibromo-1-fluoro-benzene	Br$_2$C$_6$H$_3$F	253.91		2.047^{20}	1.5840^{20}		105^{22mm}		
d104a	Dibromofluoro-methane	Br$_2$CHF	191.83				−78	65		
d105	1,2-Dibromohexa-fluoropropane	CF$_3$CF(Br)C(Br)F$_2$	309.84	1^4, 218	2.169	1.3605^{20}	−95	72^{734mm}	none	
d106	1,6-Dibromohexane	Br(CH$_2$)$_6$Br	243.98	1, 145	1.586_4^{18}	1.5066^{20}		243	>110	misc eth
d107	2,5-Dibromo-3,4-hexanedione	CH$_3$CHBrC(=O)C(=O)-CH(Br)CH$_3$	271.95	1^3, 3132	1.766	1.5120^{20}		103^{10mm}	>110	
d108	5,7-Dibromo-8-hydroxyquinoline		302.96	21, 97			200–201	subl		s alc, bz; v s eth
d109	2,4-Dibromomesitylene	1,3,5-(CH$_3$)$_3$-C$_6$HBr$_2$	278.00	5, 408	2.4956_4^{20}	1.5419^{20}	61–63	278–279	none	1.15 aq; misc alc, bz, acet, chl, eth
d110	Dibromomethane	CH$_2$Br$_2$	173.85	1, 67			−52.7	96–97		
d111	2,6-Dibromo-4-methyl-phenol	Br$_2$C$_6$H$_2$(CH$_3$)OH	265.94	6, 406			49–50		>110	
d112	5,7-Dibromo-2-methyl-8-quinolinol		316.99	21^3, 1240			126–130			
d113	1,6-Dibromo-2-naphthol	Br$_2$C$_{10}$H$_5$OH	301.98	6, 652			105–107			
d114	2,6-Dibromo-4-nitro-aniline	Br$_2$C$_6$H$_2$(NO$_2$)NH$_2$	295.93	12, 743			206–208			sl s aq; s HOAc
d115	2,5-Dibromonitro-benzene	Br$_2$C$_6$H$_3$NO$_2$	280.91	5, 250	2.374		82–84			
d116	1,8-Dibromooctane	Br(CH$_2$)$_8$Br	272.03	1, 160	1.477	1.4981^{20}	15–16	272	>110	s bz, hot alc
d117	1,4-Dibromopentane	CH$_3$CH(Br)CH$_2$CH$_2$CH$_2$Br	229.95	1, 131	1.687	1.5085^{20}	−34	99^{25mm}	>110	
d118	1,5-Dibromopentane	Br(CH$_2$)$_3$Br	229.95	1, 131	1.6879^{15}	1.5092^{20}	−34	110^{15mm}	>110	
d119	2,4-Dibromophenol	Br$_2$C$_6$H$_3$OH	251.92	6, 202			40–42	154^{11mm}	>110	

(Continued)

TABLE 2.21 Physical Constants of Organic Compounds (*Continued*)

No.	Name	Formula	Formula weight	Beilstein reference	Density, g/mL	Refractive index	Melting point, °C	Boiling point, °C	Flash point, °C	Solubility in 100 parts solvent
d120	1,2-Dibromopropane	$CH_3CH(Br)CH_2Br$	201.90	1, 109	1.933^{20}	1.5203^{20}	−55.5	142	none	0.2 aq; misc alc, bz, chl, eth
d121	1,3-Dibromopropane	$BrCH_2CH_2CH_2Br$	201.90	1, 110	1.9712^{25}_{4}	1.5233^{20}	−36	166.8	54	0.17 aq; s alc, eth
d122	1,3-Dibromo-2-propanol	$BrCH_2CH(OH)CH_2Br$	217.90	1, 365	2.136	1.5514^{20}		83^{7mm}	46	
d123	2,3-Dibromo-1-propanol	$BrCH_2CH(Br)CH_2OH$	217.90	1, 357	2.120^{20}_{4}	1.5599^{20}		97^{10mm}	>110	sl s aq; misc alc, bz, acet, eth
d124	2,3-Dibromopropene	$BrCH_2C(Br){=}CH_2$	199.88	1, 201	1.9336^{20}_{4}	1.5470^{20}	64–66	140–143	81	s aq, alc, bz
d125	2,3-Dibromopropionic acid	$BrCH_2CH(Br)COOH$	231.88	2, 258				160^{20mm}		
d126	2,3-Dibromopropionitrile	$BrCH_2CH(Br)CN$	212.88	2, 259	2.140	1.5450^{20}		173		v s aq, alc
d127	2,6-Dibromopyridine	BrC_5H_3N	236.91	20^2, 153			118–119	255	none	
d128	*meso*-2,3-Dibromosuccinic acid	$HOOCCH(Br)CH(Br)COOH$	275.89	2, 625			275 subl			
d129	1,2-Dibromotetrachloroethane	$BrCCl_2CCl_2Br$	325.65	1, 93	2.713		222 dec		none	
d130	1,2-Dibromotetrafluoroethane	$BrCF_2CF_2Br$	259.83		2.149^{25}	1.367^{25}	−110.5	47	none	i aq; v s alc, eth
d131	2,5-Dibromothiophene	$Br_2C_4H_2S$	241.94	17, 33	2.147^{23}_{23}	1.6289^{20}	−6	211	99	i aq; misc alc, eth
d132	α,α-Dibromotoluene	$C_6H_5CHBr_2$	249.94	5, 308	1.510^{15}	1.6147^{20}		156^{23mm}	>110	
d133	1,2-Dibromo-1,1,2-trifluoroethane	$HC(Br)FC(Br)F_2$	241.8	1, 92	2.274^{27}	1.4191^{24}		76.5	none	
d134	α,α'-Dibromo-*o*-xylene	$C_6H_4(CH_2Br)_2$	263.97	5, 366	1.960	1.4740^{20}	92–94			sl s alc, chl, eth
d135	α,α'-Dibromo-*p*-xylene	$C_6H_4(CH_2Br)_2$	263.97	5, 386	1.012^{0}		72–74	261		v s alc, chl; s eth
d136	Dibutoxydibutyltin	$[CH_3(CH_2)_3O]_2Sn[(CH_2)_3CH_3]_2$	379.15		1.110		−69.1	$138^{0.05mm}$	40	0.2 aq; misc alc, acet
d137	1,2-Dibutoxyethane	$C_4H_9OCH_2CH_2OC_4H_9$	174.28		0.8374^{20}_{20}	1.4131^{20}		203.6	85	
d138	Dibutyl adipate	$[-CH_2CH_2CO_2(CH_2)_3CH_3]_2$	258.36	2^2, 575	0.962	1.4360^{20}		305	>110	0.47 aq; s alc, acet, eth EtOAc, PE
d139	Dibutylamine	$(C_4H_9)_2NH$	129.25	4, 157	0.7670^{20}	1.4177^{20}	−62	159.6	47	
d140	Di-sec-butylamine	$[C_2H_5CH(CH_3)]_2NH$	129.25	4, 162	0.753	1.4100^{20}		135	20	
d141	N,N-Dibutylaminoethanol	$(C_4H_9)_2NCH_2CH_2OH$	173.29	4^3, 682	0.860^{20}_{20}	1.444^{20}	<−70	229–230	91	
d142	N,N-Dibutylaniline	$C_6H_5N(C_4H_9)_2$	205.34	12^3, 95	0.904^{20}	1.5297^{20}		267–275	>110	i aq, MeOH; s acet, bz, EtOH, EtOAc, eth
d143	Dibutyl decanedioate	$C_4H_9O_2C(CH_2)_8CO_2C_4H_9$	214.45	2, 719	0.9366^{20}	1.4415^{20}	−10	344–345	178	0.004 aq

No.	Name	Formula	Mol. wt.	Beilstein ref.	Density	n_D	mp, °C	bp, °C	Flash pt, °C	Solubility
d144	Di-tert-butyl dicarbonate	$(CH_3)_3COC(=O)OC(CH_3)_3$	218.25		0.950	1.4103^{20}	23	$56^{0.5mm}$	37	
d145	2,5-Di-tert-butyl-1,4-dihydroxybenzene	$[(CH_3)_3C]_2C_6H_2(OH)_2$	222.33				217–219			
d146	Dibutyl disulfide	$C_4H_9SSC_4H_9$	178.36	1^2, 400	0.9383^{20}_{4}	1.4920^{20}	−71	231.2	93	i aq; misc alc, eth
d147	Di-tert-butyl disulfide	$(CH_3)_3CSSC(CH_3)_3$	178.36		0.935	1.4920		229–233	93	
d148	Dibutyl ether	$C_4H_9OC_4H_9$	130.22	1, 369	0.7689^{20}_{4}	1.3992^{20}	−95	140	25	0.03 aq; misc alc, eth
d149	2,6-Di-tert-butyl-4-(dimethylaminomethyl)phenol	$(CH_3)_2NCH_2C_6H_2[C(CH_3)_3]_2OH$	263.43	13^4, 2014			93–94	172^{30mm}		
d150	N,N-Dibutylethylenediamine	$[CH_3(CH_2)_3]_2NCH_2CH_2NH_2$	172.32	4^4, 1182	0.823	1.4430^{20}		117^{24mm}	87	
d151	N,N-Dibutylformamide	$HC(=O)N(C_4H_9)_2$	157.26		0.864	1.4429^{20}		120^{15mm}	100	
d152	Dibutyl hexanedioate	$[-CH_2CH_2CO_2(CH_2)_3CH_3]_2$	258.36	2^2, 575	0.962	1.4358^{20}		305	>110	
d153	2,5-Di-tert-butylhydroquinone	$[(CH_3)_3C]_2C_6H_2\text{-}1,4\text{-}(OH)_2$	222.33	6^3, 4741			217–219			
d154	Dibutyl maleate	$C_4H_9O_2CCH=CHCO_2C_4H_9$	228.29	2^3, 1925	0.9950^{20}	1.4454^{20}	<−80	281	141	0.05 aq
d155	Di-tert-butyl malonate	$CH_2O_2CC(CH_3)_3\ CO_2C(CH_3)_3$	216.27	2^3, 1621		1.4184^{20}	−6.0	93^{10mm}	88	
d156	2,6-Di-tert-butyl-4-methylphenol	$[(CH_3)_3C]_2C_6H_2(CH_3)OH$	220.36	6^3, 2073	1.0484	1.4859^{75}	70	265	127	s alc, bz, acet, PE
d157	Dibutyl octanedioate	$[-(CH_2)_3CO_2(CH_2)_3CH_3]_2$	286.41	2^3, 1767	0.948	1.4390^{20}		$176^{4.5mm}$	>110	
d158	Dibutyl oxalate	$C_4H_9O_2CCO_2C_4H_9$	202.25	2, 540	0.986^{20}_{20}	1.4232^{20}	−30.0	239–240	108	misc alc, ketones, PE
d159	Di-tert-butyl peroxide	$(CH_3)_3CO-OC(CH_3)_3$	146.23	1^3, 1580	0.794^{20}	1.3890^{20}	−40	110	1	misc acet, octane
d160	2,4-Di-tert-butylphenol	$[(CH_3)_3C]_2C_6H_3OH$	206.33				56.5	263.5	115	s hot alc; i alk
d161	2,6-Di-sec-butylphenol	$[CH_3CH_2CH(CH_3)]_2C_6H_3OH$	206.23		0.918	1.5100^{20}	−42	255–260	127	
d162	2,6-Di-tert-butylphenol	$[(CH_3)_3C]_2C_6H_3OH$	206.23	6^3, 2061			35–38	253	118	s hot alc; i alk
d163	3,5-Di-tert-butylphenol	$[(CH_3)_3C]_2C_6H_3OH$	206.23				87–89			
d164	Dibutyl phosphite	$(C_4H_9O)_2P(O)H$	194.21	1^1, 187	0.995	1.4239^{20}		119^{11mm}	121	
d165	Dibutyl 1,2-phthalate	$C_6H_4\text{-}1,2\text{-}[CO_2C_4H_9]_2$	278.35	9^2, 586	1.0465^{20}_{4}	1.4911^{20}	−35	340	157	0.01 aq; v s alc, bz, acet, eth
d166	N,N-Dibutyl-1,3-propanediamine	$C_4H_9NH(CH_2)_3NHC_4H_9$	186.34		0.827	1.4463^{20}		205	103	
d167	Dibutyl suberate	$CH_3(CH_2)_3O_2C(CH_2)_6CO_2(CH_2)_3CH_3$	286.41	2^3, 1767	0.948	1.4390^{20}		$175^{4.5mm}$	>110	
d168	Dibutyl succinate	$[C_4H_9O_2CCH_2]_2$	230.30	2^2, 551	0.9768^{20}_{4}	1.4299^{20}		274.5		i aq; s alc, eth
d169	Dibutyl sulfate	$C_4H_9OSO_2OC_4H_9$	210.29		1.059^{25}_{4}	1.4213^{20}	−29.0	132^{11mm}		
d170	Dibutyl sulfide	$C_4H_9SC_4H_9$	146.30	1, 370	0.8386^{20}	1.4530^{20}	−80	185	76	i aq; v s alc, eth

(Continued)

TABLE 2.21 Physical Constants of Organic Compounds (*Continued*)

No.	Name	Formula	Formula weight	Beilstein reference	Density, g/mL	Refractive index	Melting point, °C	Boiling point, °C	Flash point, °C	Solubility in 100 parts solvent
d171	Di-*tert*-butyl sulfide	(CH$_3$)$_3$CSC(CH$_3$)$_3$	146.30	1[2], 397	0.815	1.4506^{20}		151	48	
d172	Dibutyl sulfite	(C$_4$H$_9$O)$_2$S(=O)	194.29	1, 371	0.9944^{22}_{4}	1.4310^{20}		108^{15}mm		i aq; s alc, eth
d173	Dibutyl sulfone	(C$_4$H$_9$)$_2$SO$_2$	178.29	3, 518			46	295	143	
d174	Dibutyl L-tartrate	[-CH(OH)CO$_2$(CH$_2$)$_3$CH$_3$]$_2$	262.31		1.091	1.4465^{20}	22	175^{5}mm	>110	i aq; s alc; sl s eth
d175	*N,N*-Dibutyl-2-thiourea	C$_4$H$_9$NC(=S)NHC$_4$H$_9$	188.34				63–65		>110	
d176	Dibutyltin diacetate	(CH$_3$CO$_2$)$_2$Sn(C$_4$H$_9$)$_2$	351.01		1.320	1.4700^{20}		145^{10}mm	>110	
d177	Dibutyltin dichloride	(C$_4$H$_9$)$_2$SnCl$_2$	303.83				39–41	135^{10}mm	>110	
d178	Dibutyltin dilaurate	[CH$_3$(CH$_2$)$_{10}$CO$_2$]$_2$Sn(C$_4$H$_9$)$_2$	631.56	Merck: 12, 3089	1.066	1.4683^{20}	22–24		>110	s PE, bz, acet, eth, org esters
d179	Dibutyltin maleate		346.98				135–140			
d180	Dibutyltin oxide	(C$_4$H$_9$)$_2$SnO	248.92	4[1], 588			>300			
d181	Dicaprolactone 2-(acryloxy)ethyl ester	HO(CH$_2$)$_5$CO$_2$(CH$_2$)$_5$CO$_2$-CH$_2$CH$_2$O$_2$CCH=CH$_2$	344.41		1.100	1.4660^{20}			>110	
d182	Dichloroacetic acid	Cl$_2$CHCOOH	128.94	2, 202	1.5634^{20}	1.4462^{20}	9–11	193–194	>110	misc aq, alc, eth
d183	1,1-Dichloroacetone	CH$_3$C(=O)CHCl$_2$	126.97	1, 654	1.305^{18}_{15}	1.4455^{20}		120	24	s sl aq; s alc, eth
d184	1,3-Dichloroacetone	ClCH$_2$C(=O)CH$_2$Cl	126.97	1, 655	1.383	1.5635^{20}	39–41	173	89	i aq
d185	2',4'-Dichloroacetophenone	Cl$_2$C$_6$H$_3$C(=O)CH$_3$	189.04	7, 282			33–34	145^{15}mm	>110	i aq
d186	Dichloroacetyl chloride	Cl$_2$CHC(=O)Cl	147.39	2, 204	1.5315^{16}_{4}	1.4603^{20}		107–108	none	dec aq; alc; misc eth
d187	2,3-Dichloroaniline	Cl$_2$C$_6$H$_3$NH$_2$	162.02	12, 621		1.5969^{20}	23–24	252	>110	s alc; v s eth
d188	2,4-Dichloroaniline	Cl$_2$C$_6$H$_3$NH$_2$	162.02	12, 621	1.567^{20}		59–62	245		sl s aq; s alc, eth
d189	2,5-Dichloroaniline	Cl$_2$C$_6$H$_3$NH$_2$	162.02	12, 625			49–51	251	>110	s alc, bz, eth
d190	2,6-Dichloroaniline	Cl$_2$C$_6$H$_3$NH$_2$	162.02	12, 625			38–41		>110	s alc, eth; sl s bz
d191	3,4-Dichloroaniline	Cl$_2$C$_6$H$_3$NH$_2$	162.02	12, 626			70–72	272		i aq; s alc, eth
d192	3,5-Dichloroaniline	Cl$_2$C$_6$H$_3$NH$_2$	162.02	12, 626			51–53	259^{741}mm	>110	sl s alc, bz, acet
d193	1,5-Dichloroanthraquinone		277.11	7, 787			245–247			
d194	2,3-Dichlorobenzaldehyde	Cl$_2$C$_6$H$_3$CHO	175.01	7[3], 878			64–67			
d195	2,4-Dichlorobenzaldehyde	Cl$_2$C$_6$H$_3$CHO	175.01	7, 236			69–73	233		i aq; s alc
d196	2,4-Dichlorobenzamide	Cl$_2$C$_6$H$_3$CONH$_2$	190.03	9[3], 1376			191–194			

No.	Name	Formula	Mol. wt.	Beilstein ref.	Density	n_D	mp/°C	bp/°C	Fl. pt./°C	Solubility
d197	2,6-Dichlorobenzamide	Cl$_2$C$_6$H$_3$CONH$_2$	190.03	9^1, 149			196–199		66	misc alc, bz, eth
d198	1,2-Dichlorobenzene	C$_6$H$_4$Cl$_2$	147.00	5, 201	1.3059^{20}_4	1.5510^{20}	−17.0	180.4	72	0.01 aq; s alc, eth
d199	1,3-Dichlorobenzene	C$_6$H$_4$Cl$_2$	147.00	5, 202	1.2884^{20}_4	1.5460^{20}	−24.8	173.1	66	s alc, bz, chl, eth
d200	1,4-Dichlorobenzene	C$_6$H$_4$Cl$_2$	147.00	5, 203	1.2417^6	1.5285^{20}	53	174.1	>110	d hot alc, hot aq
d201	2,5-Dichlorobenzenesulfonyl chloride	Cl$_2$C$_6$H$_3$SO$_2$Cl	245.51	11^1, 15			36–37			
d202	2,4-Dichlorobenzoic acid	Cl$_2$C$_6$H$_3$COOH	191.01	9, 342			157–160			s hot aq, alc, bz, chl
d203	2,5-Dichlorobenzoic acid	Cl$_2$C$_6$H$_3$COOH	191.01	9, 342			154–157	301		sl s aq; s alc, eth
d204	3,4-Dichlorobenzoic acid	Cl$_2$C$_6$H$_3$COOH	191.01	9, 343			207–209			s hot aq, eth; v s alc
d205	4,4′-Dichlorobenzophenone	(ClC$_6$H$_4$)$_2$C=O	251.11	7, 420			144–146	353		s hot alc, v s chl, eth
d206	2,4-Dichlorobenzotrifluoride	Cl$_2$C$_6$H$_3$CF$_3$	215.00	5^3, 698	1.484	1.4810^{20}		117–118	72	
d207	3,4-Dichlorobenzotrifluoride	Cl$_2$C$_6$H$_3$CF$_3$	215.00	5^3, 698	1.478	1.4750^{20}	−12	173–174	65	
d208	2,4-Dichlorobenzoyl chloride	Cl$_2$C$_6$H$_3$C(=O)Cl	209.46	9, 342	1.494	1.5297^{20}	16–18	150^{34mm}	137	dec aq, alc
d209	3,4-Dichlorobenzoyl chloride	Cl$_2$C$_6$H$_3$C(=O)Cl	209.46	9, 344			30–33	242	142	dec aq, alc
d210	1,4-Dichlorobutane	ClCH$_2$CH$_2$CH$_2$CH$_2$Cl	127.01	1, 119	1.1598^{20}_4	1.4566^{20}	−38	161–163	40	i aq; s chl
d211	cis-1,4-Dichloro-2-butene	ClCH$_2$CH=CHCH$_2$Cl	125.00	1^3, 743	1.188^{25}_4	1.4887^{25}	−48	152	55	i aq; s org solvents
d212	3,4-Dichloro-1-butene	ClCH$_2$CH(Cl)CH=CH$_2$	125.00	1^3, 725	1.150	1.4658^{20}	−61	123	28	
d213	1,4-Dichloro-2-butyne	ClCH$_2$C≡CCH$_2$Cl	122.98	1^3, 927	1.258^{20}_4	1.5048^{20}		165–168	160	
d214	Dichloro(2-chloroethyl)methylsilane	ClCH$_2$CH$_2$SiCl$_2$(CH$_3$)	177.53	4^3, 1892	1.261	1.4580^{20}		157^{744mm}	32	
d215	Dichloro(3-chloropropyl)methylsilane	Cl(CH$_2$)$_3$Si(CH$_3$)Cl$_2$	191.56	4^4, 4170	1.227	1.4620^{20}		80^{18mm}	59	
d216	1,10-Dichlorodecane	Cl(CH$_2$)$_{10}$Cl	211.18	1^3, 522	0.999	1.4605^{20}	15.6	168^{28mm}	>110	
d217	1,1-Dichloro-2,2-diethoxyethane	Cl$_2$CHCH(OC$_2$H$_5$)$_2$	187.07	1, 614	1.138	1.4360^{20}		183–184	60	
d218	Dichlorodifluoromethane	Cl$_2$CF$_2$	120.91	1, 61	1.486^{-30}		−158	−29.8		
d219	1,1-Dichloro-3,3-dimethylbutane	(CH$_3$)$_3$CCH$_2$CHCl$_2$	155.07	1^3, 409	1.027	1.4388^{20}	−56	148	36	0.01 aq; 9 bz; 5.5 chl; 6 diox; s alc, eth
d220	1,3-Dichloro-3,5-dimethylhydantoin		197.02	24^2, 158			134–136			
d221	Dichlorodiphenylmethane	(C$_6$H$_5$)$_2$CCl$_2$	237.13	5, 590	1.235	1.6040^{20}		305	>110	

(Continued)

TABLE 2.21 Physical Constants of Organic Compounds (*Continued*)

No.	Name	Formula	Formula weight	Beilstein reference	Density, g/mL	Refractive index	Melting point, °C	Boiling point, °C	Flash point, °C	Solubility in 100 parts solvent
d222	Dichlorodimethylsilane	(CH$_3$)$_2$SiCl$_2$	129.06		1.064^{20}_{4}	1.4038^{20}	−16	70	−16	
d223	Dichlorodiphenylsilane	(C$_6$H$_5$)$_2$SiCl$_2$	253.20	16, 910	1.222^{20}		308–309	157	dec aq, alc	
d224	1,12-Dichlorododecane	Cl(CH$_2$)$_{12}$Cl	239.23	1^1, 67			28–30	172^{10mm}	>110	
d225	1,1-Dichloroethane	CH$_3$CHCl$_2$	98.96	1, 83	1.1757^{20}_{4}	1.4164^{20}	−97	57.3	−17	0.51 aq; misc alc
d226	1,2-Dichloroethane	ClCH$_2$CH$_2$Cl	98.96	1, 84	1.235^{20}_{4}	1.4448^{20}	−35.7	83.5	13	0.8 aq; misc alc, chl, eth
d227	1,1-Dichloroethylene	H$_2$C=CCl$_2$	96.94	1, 186	1.2129^{20}_{4}	1.4247^{20}	−122.6	31.6	−28	0.01 aq; s alc, bz, chl, eth
d228	*cis*-1,2-Dichloroethylene	ClCH=CHCl	96.94	1, 188	1.2838^{20}_{4}	1.4490^{20}	−80.1	60	2	0.7 aq; s alc, eth
d229	*trans*-1,2-Dichloroethylene	ClCH=CHCl	96.94	1, 188	1.2565^{20}	1.4452^{20}	−49.8	48.7	2	0.6 aq; s alc, eth
d230	2,2'-Dichloroethyl ether	ClCH$_2$CH$_2$OCH$_2$CH$_2$Cl	143.01	1^2, 335	1.2220^{20}_{20}	1.457^{20}		178.5	55	1.1 aq; s alc, bz, eth
d231	2,2-Dichloroethyl methyl ether	Cl$_2$CHCH$_2$OCH$_3$	128.99		1.226	1.4375^{20}			33	
d232	Dichloroethylmethylsilane	(C$_2$H$_5$)Si(CH$_3$)Cl$_2$	143.09		1.063	1.4190^{20}		100	43	
d233	Dichlorofluoromethane	FCHCl$_2$	102.92	1, 61	1.405^9	1.3724^9	−135	8.9		
d234	1,6-Dichlorohexane	Cl(CH$_2$)$_6$Cl	155.07	1, 144	1.068	1.4568^{20}		87^{15mm}	73	69 HOAc; 108 diox; s alc, eth; i aq s chl
d235	Dichloromethane	CH$_2$Cl$_2$	84.93	1, 60	1.3265^{20}	1.4246^{20}	−95	40	none	1.3 aq; misc alc, eth
d236	Dichloromethane-*d$_2$*	CD$_2$Cl$_2$	86.95	1^4, 39	1.3621	1.4218^{20}		40	none	
d237	α,α-Dichloromethyl methyl ether	Cl$_2$CHOCH$_3$	114.96	4^4, 4182	1.271	1.4300^{20}		85	42	
d238	Dichloro(methyl)octylsilane	CH$_3$(CH$_2$)$_7$Si(CH$_3$)Cl$_2$	227.25		0.973	1.4440^{20}		94^{6mm}	98	
d239	Dichloro(methyl)phenylsilane	C$_6$H$_5$Si(CH$_3$)Cl$_2$	191.13		1.176	1.5190^{20}		205	82	
d240	Dichloro(methyl)silane	HSi(CH$_3$)Cl$_2$	115.04	4^1, 581	1.105	1.398^{20}	−93	41	−32	
d241	Dichloro(methyl)vinylsilane	H$_2$C=CHSi(CH$_3$)Cl$_2$	141.07		1.0874^{20}_{4}	1.4300^{20}		92	4	
d242	2,4-Dichloro-1-naphthol	Cl$_2$C$_{10}$H$_5$OH	213.06	6, 612			108			

d243	2,3-Dichloro-1,4-naphthoquinone		227.05	7, 729			190–192			sl s alc, bz, eth
d244	2,6-Dichloro-4-nitroaniline	$Cl_2C_6H_2(NO_2)NH_2$	207.02	12, 735			190–192		123	s PE
d245	2,3-Dichloronitrobenzene	$Cl_2C_6H_3NO_2$	192.00	5, 245	1.721^{14}		61–62	257–258	>110	s hot alc; misc eth
d246	2,4-Dichloronitrobenzene	$Cl_2C_6H_3NO_2$	192.00	5, 245	1.439^{80}		29–32	258	>110	
d247	2,5-Dichloronitrobenzene	$Cl_2C_6H_3NO_2$	192.00	5, 245			54–57	266–269	>110	
d248	3,4-Dichloronitrobenzene	$Cl_2C_6H_3NO_2$	192.00	5, 246	1.456^{75}		41–44	256	123	
d249	2,4-Dichloro-6-nitrophenol	$Cl_2C_6H_2(NO_2)OH$	208.00	6, 241			118–120	222		
d250	1,7-Dichlorooctamethyltetrasiloxane	$[Cl(CH_3)_2SiOSi(CH_3)_2]_2$	351.53	4^3, 1884	1.0111^{14}	1.403^{20}	−62			
d251	1,5-Dichloropentane	$Cl(CH_2)_5Cl$	141.04	1, 131	1.10584^{20}	1.4553^{20}	−72	$66^{10\text{mm}}$	26	i aq; s alc, eth
d252	2,3-Dichlorophenol	$Cl_2C_6H_3OH$	163.00	6^1, 102			58–60	206		s alc, eth
d253	2,4-Dichlorophenol	$Cl_2C_6H_3OH$	163.00	6, 189			42–43	210	113	v s alc, bz, chl, eth
d254	2,5-Dichlorophenol	$Cl_2C_6H_3OH$	163.00	6, 189			56–58	211		v s alc, bz, eth
d255	2,6-Dichlorophenol	$Cl_2C_6H_3OH$	163.00	6, 190			65–68	218–220		v s alc, eth
d256	2,4-Dichlorophenoxyacetic acid	$Cl_2C_6H_3OCH_2COOH$	221.04				136–140	$160^{0.4\text{mm}}$		s alc, bz, chl, eth
d257	4-(2,4-Dichlorophenoxy)butanoic acid	$Cl_2C_6H_3O(CH_2)_3CO_2H$	249.10	6^3, 708			117–119			46 ppm aq^{25}; s acet, alc, eth; sl s bz
d258	2-(2,4-Dichlorophenoxy)propanoic acid	$Cl_2C_6H_3OCH(CH_3)CO_2H$	235.07	6, 189			110–112			350 ppm aq^{20}; v s org-solvents
d259	3,4-Dichlorophenyl isocyanate	$Cl_2C_6H_3NCO$	188.01	12^3, 1405			42–44	$120^{18\text{mm}}$	>110	
d260	Dichlorophenylphosphine	$C_6H_5PCl_2$	178.99	16, 763	1.319	1.5980^{20}	−51	222	>112	s aq; v s eth
d261	4,5-Dichloro-o-phthalic acid	$Cl_2C_6H_2(CO_2H_2)_2$	235.02	9^1, 366			201–203			
d262	1,2-Dichloropropane	$CH_3CH(Cl)CH_2Cl$	112.99	1, 105	1.1558^{20}	1.4390^{20}	−100	96	4	0.26 aq; misc alc, bz, chl, eth
d263	1,3-Dichloropropane	$ClCH_2CH_2CH_2Cl$	112.99	1, 105	1.1878^{20}	1.4487^{20}	−99.5	120–122	32	v s alc, eth
d264	1,3-Dichloro-2-propanol	$ClCH_2CH(OH)CH_2Cl$	128.99	1, 364	1.198	1.4835^{20}	−4	174.3	85	9.1 aq; misc alc, eth

(Continued)

TABLE 2.21 Physical Constants of Organic Compounds (*Continued*)

No.	Name	Formula	Formula weight	Beilstein reference	Density, g/mL	Refractive index	Melting point, °C	Boiling point, °C	Flash point, °C	Solubility in 100 parts solvent
d265	1,3-Dichloropropene	$ClCH_2CH=CHCl$	110.97	1, 199	1.217_4^{20}	1.470^{20}		97–112	25	i aq; s chl, eth
d266	2,3-Dichloro-1-propene	$ClCH_2C(Cl)=CH_2$	110.97	1, 199	1.204_4^{25}	1.4611^{20}		94	10	misc alc; s eth
d267	3,6-Dichloropyridazine		148.98				66–69			
d268	2,6-Dichloropyridine	$Cl_2C_5H_3N$	147.99	20, 231			86–88			
d269	3,5-Dichloropyridine	$Cl_2C_5H_3N$	147.99	20, 231			65–67			
d270	4,7-Dichloroquinoline		198.05	20^3, 3384			84–86			
d270a	Dichlorosilane	Cl_2SiH_2	101.01			1.3092^0	−122	8.3		
d270b	1,1-Dichlorotetrafluoroethane	F_3CCFCl_2	170.92		1.455^{25} satd pressure		−57	4		
d271	1,2-Dichloro-1,1,2,2-tetrafluoroethane	$ClCF_2CF_2Cl$	170.93	1^3, 152	1.470^{20} satd pressure	1.3092^{20}	−94	3.6		s alc, eth
d272	2,5-Dichlorothiophene	$Cl_2(C_4H_2S)$	153.03	17, 33	1.442	1.5621^{20}	−40.5	162	59	i aq; misc alc, eth
d273	α,α-Dichlorotoluene	$C_6H_5CHCl_2$	161.03	5, 297	1.254	1.5500^{20}	−16/−17	205	92	v s alc, eth
d274	2,4-Dichlorotoluene	$Cl_2C_6H_3CH_3$	161.03	5, 295	1.2460^{20}	1.5511^{20}	−13	200.5	79	i aq
d275	2,6-Dichlorotoluene	$Cl_2C_6H_3CH_3$	161.03	5, 296	1.254	1.5507^{20}		196–203	82	i aq; s chl
d276	3,4-Dichlorotoluene	$Cl_2C_6H_3CH_3$	161.03	5, 296	1.251_{25}^{25}	1.5472^{20}	−15	209	85	i aq
d277	α,α-Dichloro-o-xylene	$C_6H_4(CH_2Cl)_2$	175.06	5, 364			55–57	239–241	107	
d278	α,α-Dichloro-p-xylene	$C_6H_4(CH_2Cl)_2$	175.06	5, 384			99–101	254		22.5 acet; 20 bz; 4.5 CCl₄; 11 eth; 18 EtOAc
d279	2,5-Dichloro-p-xylene	$Cl_2C_6H_2(CH_3)_2$	175.06	5, 384			71	222		27 acet; 44 bz; 39 eth; 32 EtOAc; 5 MeOH
d280	Dicumyl peroxide	$[C_6H_5C(CH_3)_2]_2O_2$	270.37				39–41		>110	
d281	Dicyandiamide	$H_2NC(=NH)NHCN$	84.08	3, 91	1.400_4^{25}		208–211			2.3 aq; 1.3 alc; i bz
d282	1,2-Dicyanobenzene	$C_6H_4(CN)_2$	128.13	9, 815			139–141			v s bz, alc; s hot eth
d283	1,3-Dicyanobenzene	$C_6H_4(CN)_2$	128.13	9, 836			158–160			s alc, bz, chl, eth
d284	1,4-Dicyanobutane	$NC(CH_2)_4CN$	108.14	2, 653	0.951	1.4380^{20}	1–3	295	93	
d285	1,6-Dicyanohexane	$NC(CH_2)_6CN$	136.20	2, 694	0.954	1.4436^{20}	−3.5	185^{15mm}	>110	
d286	2,4-Dicyano-3-methyl-glutaramide	$CH_3CH[CH(CN)CONH_2]_2$	194.19	2^2, 704			159–160			

No.	Name	Formula	Mol wt	Beilstein ref.	Density	n_D	mp, °C	bp, °C	Flash P, °C	Solubility
d287	1,5-Dicyanopentane	$NC(CH_2)_5CN$	122.17	2, 671	0.951	1.4410^{20}	3–4	176^{14mm}	>110	7 MeOH; misc bz, acet, eth
d288	Dicyclohexyl	$C_6H_{11}C_6H_{11}$	166.31	5, 108	0.864	1.4782^{20}		227	92	misc alc, bz, chl, eth
d289	Dicyclohexylamine	$(C_6H_{11})_2NH$	181.32	12, 6	0.910	1.4842^{20}	−2	255.8	96	
d290	N,N'-Dicyclohexyl-carbodiimide	$C_6H_{11}N{=}C{=}NC_6H_{11}$	206.33	Merck: 12, 3146			35–36	124^{6mm}	110	
d291	Dicyclohexyl o-phthalate	$C_6H_4\text{-}1,2\text{-}(CO_2C_6H_{11})_2$	330.43	9, 799			64–66			
d292	Dicyclopentadiene		132.21	5, 495	0.9304^{25}	1.5050^{25}	−1	170	26	s alc, eth
d293	Dicyclopentenyl methacrylate		218.30	6^3, 1942	1.050	1.5080^{30}		137^{13mm}	>110	
d294	Dicyclopropyl ketone	$(C_3H_5)_2C{=}O$	110.16	3^3, 556	0.977	1.4670^{20}		160–162	39	
d295	Didodecyl 3,3'-thiodi-propionate	$S[CH_2CH_2CO_2(CH_2)_{11}CH_3]_2$	514.86		0.915		40–42		>110	
d296	Dieldrin		380.92	17^3, 526			176–177			i aq; s common org solvents except PE
d297	Diethanolamine	$HOCH_2CH_2NHCH_2CH_2OH$	105.14	4, 283	1.0881^{30}	1.4747^{30}	28.0	269	172	96 aq; 4 bz; 0.8 eth; misc MeOH, acet
d298	2,2-Diethoxyacet-ophenone	$C_6H_5C({=}O)CH(OC_2H_5)_2$	208.26	7^1, 361	1.034	1.4995^{20}		134^{10mm}	>110	
d299	4,4-Diethoxybutyl-amine	$H_2N(CH_2)_3CH(OC_2H_5)_2$	161.25	4, 319	0.933	1.4275^{20}		196	62	
d300	2,2-Diethoxy-N,N-di-methylethylamine	$(C_2H_5O)_2CHCH_2N(CH_3)_2$	161.25	4, 308	0.883	1.4129^{20}		170	45	
d301	Diethoxydimethyl-silane	$(C_2H_5O)_2Si(CH_3)_2$	148.28		0.840^{20}	1.3811^{20}	−87	114	11	
d302	Diethoxydiphenyl-silane	$(C_2H_5O)_2Si(C_6H_5)_2$	272.42	16^2, 608	1.0329^{20}	1.5269^{20}		139^{2mm}	>110	
d303	1,1-Diethoxyethane	$CH_3CH(OC_2H_5)_2$	118.18	1, 603	0.8254^{20}	1.3819^{20}	−100	102.2	−21	5 aq; misc alc, eth
d304	1,2-Diethoxyethane	$C_2H_5OCH_2CH_2OC_2H_5$	118.18	1, 468	0.842	1.3922^{20}	−74	121.4	27	21 aq
d305	2,2-Diethoxyethanol	$(C_2H_5O)_2CHCH_2OH$	134.18	1, 818	0.888^{24}	1.4160^{20}		167	67	s alc, eth
d306	2,2-Diethoxyethyl-amine	$(C_2H_5O)_2CHCH_2NH_2$	133.19	4, 308	0.916	1.4170		162–163	45	
d307	Diethoxymethane	$(C_2H_5O)_2CH_2$	104.15		0.839	1.3732^{20}		87–88	−5	

(Continued)

TABLE 2.21 Physical Constants of Organic Compounds (*Continued*)

No.	Name	Formula	Formula weight	Beilstein reference	Density, g/mL	Refractive index	Melting point, °C	Boiling point, °C	Flash point, °C	Solubility in 100 parts solvent
d308	3-(Diethoxymethyl-silyl)propylamine	$CH_3Si(OC_2H_5)_2(CH_2)_3NH_2$	191.35	4^4, 4201	0.916	1.4260^{20}		88^{8mm}	75	
d309	2,5-Diethoxynitro-benzene	$(C_2H_5O)_2C_6H_5NO_2$	211.22	6, 857			48–51	169^{13mm}	>110	v s alc, eth
d310	Diethoxymethylvinyl-silane	$(C_2H_5O)_2Si(CH_3)CH=CH_2$	160.29	4^4, 4183	0.8584^{20}	1.400^{20}		133–134	17	
d311	1,1-Diethoxypropane	$CH_3CH_2CH(OC_2H_5)_2$	132.20	1, 630	0.8232^{20}	1.3884^{20}		122.8	7	
d312	3,3-Diethoxy-1-propene	$(C_2H_5O)_2CHCH=CH_2$	130.19	1, 727	0.854	1.4000^{20}		125	4	
d313	2,2-Diethoxytri-ethylamine	$(C_2H_5O)_2CHCH_2N(C_2H_5)_2$	189.30	4, 309	0.850	1.4189^{20}		194–195	65	
d314	N,N-Diethylacetamide	$CH_3C(=O)N(C_2H_5)_2$	115.18	4, 110	0.925	1.4401^{20}		182–186	70	
d315	Diethyl 1,3-acetone-dicarboxylate	$C_2H_5OOCCH_2C(=O)CH_2\text{-}CO_2C_2H_5$	202.21	3, 791	1.113	1.4385^{20}		250	86	
d316	Diethyl 2-acetyl-glutarate	$C_2H_5O_2CCH_2CH_2CH\text{-}[C(=O)CH_3]CO_2C_2H_5$	230.26	3, 809	1.071	1.4386^{20}		154^{11mm}	>110	
d317	Diethyl acetylsuccinate	$C_2H_5O_2CCH_2CH[C(=O)\text{-}CH_3]CO_2C_2H_5$	216.23	3, 801	1.081	1.4346^{20}		183^{50mm}	>110	
d318	Diethyl adipate	$C_2H_5O_2C(CH_2)_4CO_2C_2H_5$	202.25	2, 652	1.009	1.4270^{20}	−18	251	110	
d319	Diethyl allylmalonate	$C_2H_5O_2CCH(CH_2CH=CH_2)\text{-}CO_2C_2H_5$	200.23	2, 776	1.015	1.4304^{20}		222–223	71	
d320	Diethylaluminum chloride	$(C_2H_5)_2AlCl$	120.56	4^3, 1972	0.961		−50	126^{50mm}	−18	
d321	Diethylaluminum ethoxide	$(C_2H_5)_2AlOC_2H_5$	130.17	4^3, 1972	0.850		2.5–4.5	109^{10mm}	−18	
d322	Diethylaluminum iodide	$(C_2H_5)_2AlI$	212.01	4^2, 1024	1.609			120^{4mm}	−18	
d323	Diethylamine	$(C_2H_5)_2NH$	73.14	4, 95	0.7074^{20}_{4}	1.3864^{10}	−50.0	55.5	−23	misc aq, alc
d324	Diethylamine HCl	$(C_2H_5)_2NH \cdot HCl$	109.60	4, 95	1.048^{21}_{4}		227–230	320–330	53	s aq, alc, chl; i eth
d325	2-(Diethylamino)-acetonitrile	$(C_2H_5)_2NCH_2CN$	112.18	4, 350	0.866	1.4260^{20}		170		
d326	4-(Diethylamino)-benzaldehyde	$(C_2H_5)_2NC_6H_4CHO$	177.25	14^2, 25			39–41	174^{7mm}	>110	
d327	2-Diethylaminoethanol	$(C_2H_5)_2NCH_2CH_2OH$	117.19	4, 282	0.8800^{25}	1.4389^{20}	−70	163	48	s aq, alc, bz, eth

No.	Name	Formula	Mol. wt.	Beilstein reference	Density	n_D	Melting point	Boiling point	Flash point	Solubility
d328	2-Diethylaminoethyl chloride HCl	$ClCH_2CH_2N(C_2H_5)_2 \cdot HCl$	172.10	4^2, 618			108–210	80^{10mm}	76	
d329	2-(Diethylamino)ethyl methacrylate	$H_2C{=}C(CH_3)CO_2CH_2CH_2{-}N(C_2H_5)_2$	185.27	4^3, 676	0.922	1.4440^{20}		170^{15mm}	107	s aq, alc, eth
d330	3-(Diethylamino)phenol	$(C_2H_5)_2NC_6H_4OH$	165.24	13, 408			65–69	233–235	33	s aq, alc, chl, eth
d331	3-Diethylamino-1,2-propanediol	$(C_2H_5)_2NCH_2CH(OH){-}CH_2OH$	147.22	4, 302	0.9730^{20}_{20}	1.4602^{20}		59^{13mm}	65	s alc
d332	1-Diethylamino-2-propanol	$(C_2H_5)_2NCH_2CH(OH)CH_3$	131.22	4^2, 737	0.889	1.4255^{20}	13.5	83^{15mm}	58	
d333	3-Diethylamino-1-propanol	$(C_2H_5)_2NCH_2CH_2CH_2OH$	131.22	4, 288	0.884	1.4435		159		
d334	3-Diethylaminopropylamine	$(C_2H_5)_2NCH_2CH_2CH_2NH_2$	130.24		0.826	1.4416^{20}				
d335	N,N-Diethylaniline	$C_6H_5N(C_2H_5)_2$	149.24	12, 164	0.9302^{25}_{4}	1.5394^{25}	−38	216	97	1 aq; sl s alc, eth
d336	2,6-Diethylaniline	$(C_2H_5)_2C_6H_3NH_2$	149.24		0.906	1.5452^{20}	3	243	123	
d337	Diethyl azelate	$C_2H_5O_2C(CH_2)_7CO_2C_2H_5$	244.33	2, 709	0.973	1.4350^{20}	−16	172^{18mm}	>110	s alc, eth
d338	Diethyl azodicarboxylate	$C_2H_5O_2CN{=}NCO_2C_2H_5$	174.16	3, 123	1.106	1.4280^{20}		106^{13mm}	>110	s alc, eth
d339	5,5-Diethylbarbituric acid		184.19	24^2, 279	1.220		188–192			0.7 aq; 7 alc; 1.3 chl; 3.2 eth; s acet, HOAc
d340	Diethyl benzalmalonate	$C_6H_5CH{=}C(CO_2C_2H_5)_2$	248.28	9, 892	1.107	1.5365^{20}		215^{30mm}	>110	
d340a	1,2-Diethylbenzene	$C_6H_4(C_2H_5)_2$	134.22	5, 426	0.880	1.5020^{20}	−31	184	49	
d341	1,3-Diethylbenzene	$C_6H_4(C_2H_5)_2$	134.22	5, 426	0.8640^{20}_{4}	1.4950^{20}	−83.9	181.1	50	s alc, eth
d342	1,4-Diethylbenzene	$C_6H_4(C_2H_5)_2$	134.22	5, 426	0.8620^{20}_{4}	1.4940^{20}	−42.8	183.8	56	s alc, eth
d343	Diethyl benzylmalonate	$C_6H_5CH_2CH(CO_2C_2H_5)_2$	250.29	9, 869	1.064	1.4868^{20}		162^{10mm}	>110	
d344	Diethyl benzophosphonate	$C_6H_5CH_2P(O)(OC_2H_5)_2$	228.23	12, 164	1.095	1.4970^{20}		108^{1mm}	>110	
d345	Diethyl bis(hydroxymethyl)malonate	$(HOCH_2)_2C(CO_2C_2H_5)_2$	220.22				49–51		>110	
d346	Diethyl bromomalonate	$BrCH(CO_2C_2H_5)_2$	239.07	2, 594	1.4022^{25}_{4}	1.4550^{20}	−54	235 dec	>110	i aq; misc alc, eth
d347	Diethyl butylmalonate	$C_4H_9CH(CO_2C_2H_5)_2$	216.28	2^1, 282	0.983	1.4220		235–240	93	v s alc, eth
d348	Diethylcarbamoyl chloride	$(C_2H_5)_2N(O)Cl$	135.59	4, 120	1.070	1.4515^{20}	−32	187–190	75	d hot aq, hot alc
d349	Diethyl carbonate	$(C_2H_5O)_2C{=}O$	118.13	3, 5	0.9764^{20}_{4}	1.3843^{20}	−43.0	126	25	69 aq; misc alc, bz, eth, esters
d350	Diethyl chloro-phosphate	$(C_2H_5O)_2P(O)Cl$	172.55	1, 332	1.194	1.4165^{20}		60^{2mm}	61	

(Continued)

TABLE 2.21 Physical Constants of Organic Compounds (*Continued*)

No.	Name	Formula	Formula weight	Beilstein reference	Density, g/mL	Refractive index	Melting point, °C	Boiling point, °C	Flash point, °C	Solubility in 100 parts solvent
d351	Diethyl chlorothio-phosphate	$(C_2H_5O)_2P(S)Cl$	188.61	1^3, 1332	1.200	1.4715^{20}		45^{3mm}	>110	
d352	Diethylcyano-phosphate	$(C_2H_5O)_2P(O)CN$	163.11	12, 6	1.075	1.4012^{20}		105^{19mm}	80	
d353	N,N-Diethylcyclo-hexylamine	$C_6H_{11}N(C_2H_5)_2$	155.29		0.850	1.4562^{20}		194–195	57	
d354	Diethyl diethyl-malonate	$(C_2H_5)_2C(CO_2C_2H_5)_2$	216.28	2, 686	0.990	1.4230^{20}		228–230	94	
d355	1,3-Diethyl-1,3-diphenylurea	$[C_6H_5N(C_2H_5)]_2C{=}O$	268.36	12, 422			73–75			
d356	Diethyl disulfide	$C_2H_5SSC_2H_5$	122.25	1, 347	0.998_4^{20}	1.5063^{20}	−101.5	154.0	40	sl s aq; misc alc, eth
d357	Diethyldithiocarbamic acid, sodium salt	$(C_2H_5)_2NC({=}S)S^-{\cdot}Na^+{\cdot}3H_2O$	225.31	4^2, 613			95–99			
d358	Diethyl dithio-phosphate	$(C_2H_5O)_2P(S)SH$	186.23	1, 333	1.111	1.5120^{20}		60^{1mm}	82	
d359	N,N-Diethyldodecan-amide	$CH_3(CH_2)_{10}C({=}O)N(C_2H_5)_2$	255.45		0.847	1.4545^{20}		166^{2mm}	>110	
d360	Diethyl dodecane-dioate	$C_2H_5O_2C(CH_2)_{10}CO_2C_2H_5$	186.41	2^2, 616	0.951	1.4402^{20}	15	193^{14mm}	>110	misc aq, alc, bz, eth
d361	Diethylene glycol	$(HOCH_2CH_2)_2O$	106.12	1, 468	1.1197^{15}	1.4460^{20}	−10	246	124	
d362	Diethylenetriamine	$(H_2NCH_2CH_2)_2NH$	103.17	4, 255	0.9542_4^{20}	1.4826^{20}	−35/−39	207	98	
d363	Diethylenetriamine-pentaacetic acid	$[(HO_2CCH_2)_2NCH_2CH_2]_2N$-$(CH_2CO_2H)N(CH_2CO_2H)_2$	393.35	4^4, 2454			219–220			
d364	N,N-Diethylethanol-amine	$HOCH_2CH_2N(C_2H_5)_2$	117.19	4, 282	0.884	1.4410^{20}		161	48	6 aq; misc alc, bz, chl
d365	Diethyl ether	$C_2H_5OC_2H_5$	74.12	1, 314	0.7134_4^{20}	1.3527^{20}	−116.3	34.6	−45	
d366	Diethyl ethoxymethyl-enemalonate	$(C_2H_5O_2C)_2C{=}CHOC_2H_5$	216.23	3, 469	1.070	1.4620^{20}		279–281	155	
d367	N,N-Diethylethylene-diamine	$(C_2H_5)_2NCH_2CH_2NH_2$	116.21	4, 251	0.827	1.4360^{20}		145–147	30	
d368	Diethyl ethylmalonate	$C_2H_5CH(CO_2C_2H_5)_2$	188.22	2, 644	1.004^{20}	1.4158^{20}		77^{5mm}	88	sl s aq; v s alc, eth
d369	N,N-Diethylformamide	$(C_2H_5)_2NCHO$	101.15	4, 109	0.908	1.4340^{20}		176–177	60	misc aq; v s alc, eth
d370	Diethyl fumarate	$C_2H_5O_2CCH{=}CHCO_2C_2H_5$	172.18	2, 742	1.052^{20}	1.4406^{20}	1–2	218–219	91	
d371	Diethyl glutarate	$C_2H_5O_2CCH_2CH_2CH_2$-$CO_2C_2H_5$	188.22	2, 633	1.022	1.4240^{20}	−23.8	237	96	0.9 aq; v s alc; s eth
d372	2,4-Diethyl-2,6-heptadienal	$H_2C{=}CHCH_2CH(C_2H_5)$-$CH{=}C(C_2H_5)CHO$	166.27		0.862	1.4676^{20}		91^{12mm}	86	
d373	Diethyl heptanedioate	$C_2H_5O_2C(CH_2)_5CO_2C_2H_5$	216.28	2, 671	0.9945^{20}	1.4280^{20}	−24	192^{100mm}	>110	i aq; s alc, eth

No.	Name	Formula	Mol. wt.	Beilstein/Merck ref.	Density	n_D	m.p., °C	b.p., °C	Flash point, °C	Solubility
d374	Di-(2-ethylhexyl)-o-phthalate	$C_6H_4[CO_2CH_2CH(C_2H_5)C_4H_9]_2$	390.56	10, 1248	0.981^{25}_{25}	1.4853^{20}	−50	384	207	hyd aq; s alc, eth
d375	Diethyl hydrogen phosphonate	$(C_2H_5O)_2P(O)H$	138.10	1, 330	1.079^{20}_{4}	1.4076^{20}		51^{2mm}	90	
d376	N,N-Diethylhydroxyl-amine	$(C_2H_5)_2NOH$	89.14	4, 536	1.867	1.4195^{20}	−25	125–130	45	
d377	Diethyl maleate	$C_2H_5O_2CCH=CHCO_2C_2H_5$	172.18	2, 751	1.0687^{20}	1.4400^{20}	−8.8	225.3	93	1.4 aq; s alc, eth
d378	Diethyl malonate	$C_2H_5O_2CCH_2CO_2C_2H_5$	160.17	2, 573	1.0550	1.4136^{20}	−49.9	199.3	93	2.7 aq; misc alc, eth
d379	Diethylmalonic acid	$HO_2CC(C_2H_5)_2CO_2H$	160.17	2, 686			127	170–180		v s aq, alc, eth
d380	N,N-Diethylmethyl-amine	$(C_2H_5)_2NCH_3$	87.17	4, 99	0.720	1.3887^{20}		63–65	−23	
d381	Diethyl methyl-malonate	$C_2H_5O_2CCH(CH_3)CO_2C_2H_5$	174.20	2, 629	1.018^{20}_{4}	1.4130^{20}		198	76	
d382	Diethyl 2-methyl-2'-oxosuccinate	$C_2H_5O_2CCH(CH_3)C(=O)CO_2C_2H_5$	202.21	3, 794	1.073	1.4313^{20}		138^{23mm}	>110	
d383	N,N-Diethyl-4-nitroso-aniline	$C_6H_4(NO)N(C_2H_5)_2$	178.24	12, 684			82–84			
d384	Diethyl octanedioate	$C_2H_5O_2C(CH_2)_6CO_2C_2H_5$	230.30	2, 693	0.9822^{24}_{4}	1.4323^{20}	5.9	282	>112	i aq; s alc, eth
d385	Diethyl oxalate	$C_2H_5O_2CCO_2C_2H_5$	146.14	2, 535	1.0785^{24}_{4}	1.4102^{20}	−40.6	185.4	76	3.6 aq (gradual dec); misc alc, eth
d386	Diethyl oxydiformate	$[C_2H_5OC(=O)]_2O$	162.14	Merck: 12, 8182	1.12^{20}_{4}	1.3980^{20}		93^{18mm}	69	50 alc; s esters, ketones; s aq
d386a	3,3-diethylpentane	$C(C_2H_5)_4$	128.26		0.7536^{20}	1.4206^{20}	−33	146	−33	
d387	N1,N1-Diethyl-1,4-pentanediamine	$CH_3CH(NH_2)(CH_2)_3N(C_2H_5)_2$	158.29	Merck: 12, 6819	0.817	1.4429^{20}		200	68	s aq; alc, eth
d388	N1,N1-Diethyl-M-phenylenediamine	$(C_2H_5)_2NC_6H_4NH_2$	164.25	13, 75	0.988	1.5710^{20}		116^{5mm}	>110	i aq; s alc, eth
d389	Diethyl phenyl-malonate	$C_6H_5CH(CO_2C_2H_5)_2$	236.27	9, 854	1.0950^{20}_{4}	1.4913^{20}	16	170^{14mm}	>110	i aq; s alc
d390	Diethyl phosphite	$(C_2H_5)_2P(O)H$	138.10	1, 330	1.079^{20}_{4}	1.4079^{20}		51^{2mm}	90	hyd aq; s alc, eth
d391	Diethyl o-phthalate	$C_6H_4(CO_2C_2H_5)_2$	222.24	9, 798	1.232^{14}	1.5049^{14}	−40	295	160	i aq; misc alc, eth
d392	N,N-Diethyl-1,3-propanediamine	$(C_2H_5)_2NCH_2CH_2CH_2NH_2$	130.24		0.826	1.4416^{20}		159	58	
d393	2,2-Diethyl-1,3-propanediol	$(C_2H_5)_2C(CH_2OH)_2$	132.20		1.052^{20}	1.4574^{25}	61.3	125^{10mm}		25 aq; v s alc, eth

(Continued)

TABLE 2.21 Physical Constants of Organic Compounds (*Continued*)

No.	Name	Formula	Formula weight	Beilstein reference	Density, g/mL	Refractive index	Melting point, °C	Boiling point, °C	Flash point, °C	Solubility in 100 parts solvent
d394	Diethyl propyl-malonate	$C_2H_5O_2CCH(C_3H_7)CO_2C_2H_5$	202.25	2, 657	0.987	1.4185^{20}	1–2	221–222	91	0.14 aq; misc alc, eth
d395	Diethyl sebacate	$C_2H_5O_2C(CH_2)_8CO_2C_2H_5$	258.36	2, 717	0.963	1.4360^{20}	–21	312	>110	i aq; misc alc, eth
d396	Diethyl succinate	$C_2H_5O_2C(CH_2)_2CO_2C_2H_5$	174.20	2, 609	1.040^{20}_4	1.4200^{20}	–25	217.7	100	i aq; misc alc, eth
d397	Diethyl sulfate	$(C_2H_5O)_2SO_2$	154.18	1, 327	1.172^{25}_4	1.4004^{20}		208	78	i aq; misc alc, eth
d398	Diethyl sulfide	$(C_2H_5)_2S$	90.19	1, 344	0.8367^{20}_4	1.4430^{20}	–103.9	92.1	–9	i aq; misc alc, eth
d399	Diethyl sulfite	$(C_2H_5O)_2SO$	138.19	1, 325	1.883	1.450^{20}		158	53	s aq(dec), alc
d400	(+)-Diethyl-L-tartrate	$[-CH(OH)CO_2C_2H_5]_2$	206.19	3, 512	1.205^{20}_4	1.4460^{20}	17	280	93	sl s aq; misc alc, eth
d401	(–)-Diethyl-D-tartrate	$[-CH(OH)CO_2C_2H_5]_2$	206.19	3^1, 181	1.205	1.4460^{20}		162^{19mm}	93	sl s aq; misc alc, eth
d402	N,N-Diethyl-m-toluamide	$CH_3C_6H_4C(=O)N(C_2H_5)_2$	191.27	9^2, 325	0.996^{20}_4	1.5212^{20}		111^{1mm}	>110	i aq; v s alc, bz, eth
d403	N,N-Diethyl-m-toluidine	$CH_3C_6H_4CN(C_2H_5)_2$	163.26	12, 857	0.922	1.5360^{20}		231–232	100	
d404	N,N-Diethyl-1,1,1-tri-methylsilylamine	$(C_2H_5)_2NSi(CH_3)_3$	145.32	4^3, 1861	0.767	1.4110^{20}		125–126	10	
d405	Diethylzinc	$(C_2H_5)_2Zn$	123.49	6, 672	1.2065^{20}_4	1.4983^{20}	–28	118	–23	sl s alc; v s eth
d406	1,2-Difluorobenzene	$C_6H_4F_2$	114.09	5^2, 147	1.158	1.4430^{20}	–34	92	2	i aq; s alc, eth
d406a	1,4-Difluorobenzene	$C_6H_4F_2$	114.09	5, 199	1.1701^{20}	1.4410^{20}	–13	89	2	0.32 aq
d407	1,1-Difluoroethane	CH_3CHF_2	66.05	1^3, 130	0.909^{21}	1.3011^{-72}	–117	–24.7		
d408	1,1-Difluoroethylene	$CH_2=CF_2$	64.04	1, 186			–144	–86		FLAMMABLE GAS
d409	Difluoromethane	CH_2F_2	52.02	1, 59	2.126 g/L		–136	–51.6		
d410	2,4-Difluoronitro-benzene	$F_2C_6H_3NO_2$	159.09	5^1, 129	1.451	1.5110^{20}	9–10	203–204	90	sl s alc; v s eth
d411	1,1-Difluorotetra-chloroethane	ClF_2CCCl_3	203.83	1, 86	1.649	1.413	41	91	none	i aq; s alc, eth
d412	1,2-Difluorotetra-chloroethane	FCl_2CCCl_2F	203.83	1^3, 365	1.6444^{25}_4	1.413^{25}	23.8	203.8		i aq; s alc, eth
d413	Dihexylamine	$(C_6H_{13})_2NH$	185.36	4^1, 384	0.795	1.4320^{20}		192–195	95	s alc, eth
d414	Dihexyl ether	$(C_6H_{13})_2O$	186.34	1^3, 1656	0.7936^{20}_4	1.4204^{20}		226.2	77	i aq; s ethers
d415	9,10-Dihydro-anthracene		180.25	5, 641	0.880		108–110	312		i aq; s alc, bz, eth
d416	(+)-Dihydrocarvone		152.24	7^3, 337	0.929^{19}	1.4718^{20}		221–222	81	
d417	Dihydrocoumarin		148.16	17, 315	1.169^{18}	1.5563^{20}	25	272	>110	sl s alc, eth; s chl
d418	2,5-Dihydro-2,5-dimethoxyfurfuryl-amine		159.19	18^3, 7426	1.102	1.4600^{20}		96^{12mm}	96	
d419	2,3-Dihydro-2,2-dimethyl-7-benzo-furanol		164.21	17^5, 4, 47	1.101	1.5410^{20}			110	

No.	Name	Formula	Formula weight	Beilstein reference	Density	Refractive index	Melting point, °C	Boiling point, °C	Flash point, °C	Solubility
d420	3,4-Dihydro-2-ethoxy-2H-pyran		128.17		0.957	1.4394^{20}		42^{16mm}	24	s aq, alc
d421	2,3-Dihydrofuran		70.09	17^3, 141	0.927	1.4239^{20}		54–55	–24	s warm alc, HOAc, pyr; i bz, eth
d422	3,4-Dihydro-2-methoxy-2H-pyran		114.14			1.4425^{20}			16	0.005 alc; 0.2 eth; s chl
d423	3,4-Dihydro-1 (2H)-naphthalenone		146.19	7, 370	1.099	1.5685^{20}	5–6	116^{6mm}	>110	v s aq, alc, chl, eth
d424	3,4-Dihydro-2H-pyran		84.12		0.922^{19}	1.4410^{20}	–70	86	–15	43 aq; s alc, bz, chl, eth; v s pyr, alkalis
d425	2′,4′-Dihydroacetophenone	$(HO)_2C_6H_3C(=O)CH_3$	152.15	8, 266	1.180		145–147			110 aq; 110 alc; v s eth, glyc; sl s chl
d426	1,8-Dihydroxyanthraquinone		240.21	8, 458			193–197	subl		7 aq; v s alc, eth; s hot aq, alc, eth
d427	2,4-Dihydroxybenzaldehyde	$(HO)_2C_6H_3CHO$	138.12	8, 241			135–136	226^{22mm}		0.5 aq; s alc, eth
d428	1,2-Dihydroxybenzene	$C_6H_4(OH)_2$	110.11	6, 759	1.344^4		104–106	245.5	137	2 aq; s alc, eth
d429	1,3-Dihydroxybenzene	$C_6H_4(OH)_2$	110.11	6^2, 802	1.272^{15}		109–110	276	171	sl s aq; s alc, eth
d430	1,4-Dihydroxybenzene	$C_6H_4(OH)_2$	110.11	6, 836	1.332^{15}		170–171	285–287		v s alc, eth, HOAc
d431	2,4-Dihydroxybenzoic acid	$(HO)_2C_6H_3CO_2H$	154.12	10, 377			213 rapid heating			
d432	2,5-Dihydroxybenzoic acid	$(HO)_2C_6H_3CO_2H$	154.12	10, 384			199–200			
d433	3,4-Dihydroxybenzoic acid	$(HO)_2C_6H_3CO_2H$	154.12	10, 389	1.54		200–202			
d434	3,5-Dihydroxybenzoic acid	$(HO)_2C_6H_3CO_2H$	154.12	10, 404			236 dec			
d435	2,4-Dihydroxybenzophenone	$(HO)_2C_6H_3C(=O)C_6H_5$	214.22	8, 312			144–145			
d436	2,2′-Dihydroxybiphenyl	$HOC_6H_4C_6H_4OH$	186.21	6, 989			110	315		s alc, bz, eth; sl s aq
d437	4,6-Dihydroxy-2-mercaptopyrimidine		144.15	24, 476			236			
d438	1,2-Dihydroxy-4-methylbenzene	$(HO)_2C_6H_3CH_3$	124.14	6, 878	1.129^{74}	1.5425^{74}	67–69	251		v s aq, alc, eth
d439	1,5-Dihydroxynaphthalene	$C_{10}H_6(OH)_2$	160.17	6, 980			259 dec			sl s aq; s alc; v s eth

(Continued)

579

TABLE 2.21 Physical Constants of Organic Compounds (*Continued*)

No.	Name	Formula	Formula weight	Beilstein reference	Density, g/mL	Refractive index	Melting point, °C	Boiling point, °C	Flash point, °C	Solubility in 100 parts solvent
d440	1,6-Dihydroxynaphthalene	$C_{10}H_6(OH)_2$	160.17	6, 981			138–140			v s alc, eth
d441	2,3-Dihydroxynaphthalene	$C_{10}H_6(OH)_2$	160.17	6, 982			162–164			v s alc, eth
d442	2,7-Dihydroxynaphthalene	$C_{10}H_6(OH)_2$	160.17	6, 985			187 dec			sl s aq; v s alc, eth
d443	1,4-Dihydroxy-2-naphthoic acid	$(HO)_2C_{10}H_5CO_2H$	204.19	10, 442			220 dec			
d444	3,5-Dihydroxy-2-naphthoic acid	$(HO)_2C_{10}H_5CO_2H$	204.19	10, 444			277 dec			
d445	1,3-Dihydroxy-2-propanone	$HOCH_2C(=O)CH_2OH$	90.08	1, 846			65–71			v s aq, alc, acet, eth
d446	7-(2,3-Dihydroxypropyl)theophylline		254.25				158			33 aq; 2 alc; 1 chl
d447	3,6-Dihydroxypyridazine		112.09	24, 312			306–308			sl s ahot alc; s hot aq
d448	2,3-Dihydroxypyridine	$(HO)_2C_5H_3N$	111.10	21^2, 107			245 dec			sl s alc; v s eth
d449	1,4-Diiodobenzene	$C_6H_4I_2$	329.91	5, 227	2.350	1.6212^{20}	131–133	285	none	sl s aq; s alc, eth
d450	1,4-Diiodobutane	$I(CH_2)_4I$	309.92	1, 123	2.132^{10}		6	152^{26mm}		
d451	1,2-Diiodoethane	ICH_2CH_2I	281.86	1, 99			81–84	200	>110	0.12 aq; misc alc, bz, eth, PE
d452	Diiodomethane	CH_2I_2	267.84	1, 71	3.325^{20}	1.7425^{20}	6	181	>110	
d453	1,5-Diiodopentane	$I(CH_2)_5I$	323.94	1, 133	2.177	1.6002^{20}		102^{3mm}	>110	i aq; s chl, eth
d454	1,3-Diiodopropane	$I(CH_2)_3I$	295.88	1, 115	2.5755^{20}_{4}	1.6423^{20}	−13	222	>110	
d455	Diisobutylaluminum chloride	$[(CH_3)_2CHCH_2]_2AlCl$	176.67	4^4, 4403	0.905	1.4506^{20}	−40	152^{10mm}	−18	
d456	Diisobutylaluminum hydride	$[(CH_3)_2CHCH_2]_2AlH$	142.22	4^4, 4400	0.798			118^{1mm}	−18	
d457	Diisobutylamine	$[(CH_3)_2CHCH_2]_2NH$	129.25	4, 166	0.740	1.4081^{20}	−77	137–139	29	s alc, acet, eth, chl
d458	Diisobutyl ether	$[(CH_3)_2CHCH_2]_2O$	130.22		0.761^{15}			122–124	8	i aq; misc alc, eth
d459	Diisobutyl hexanedioate	$[(CH_3)_2CHCH_2O_2CCH_2CH_2]_2$	258.36		0.950^{25}_{25}				160	
d460	Diisobutyl o-phthalate	$C_6H_4[CO_2CH_2CH(CH_3)_2]_2$	278.35	9^2, 587	1.038^{25}_{25}	1.4900^{20}			174	
d461	1,6-Diisocyanatohexane	$OCN(CH_2)_6NCO$	168.20	4^2, 711	1.040	1.4525^{20}		255	140	
d462	Diisodecyl phenyl phosphite	$(C_{10}H_{21}O)_2P(O)C_6H_5$	438.64		0.940	1.4800^{20}		17^{65mm}		

No.	Name	Formula	M.W.	Ref.	d	n	m.p.	b.p.	fp	Solubility
d463	Diisoheptyl o-phthalate	$C_6H_4(CO_2C_7H_{15})_2$			0.990	1.4860^{20}			>110	
d464	Diisononyl o-phthalate	$C_6H_4(CO_2C_9H_{19})_2$			0.972	1.4850^{20}			>110	
d465	Diisooctyl nonanedioate	$C_8H_{17}O_2C(CH_2)_7CO_2C_8H_{17}$	412.66		0.905	1.4510^{10}		210^{2mm}	>110	
d466	Diisooctyl o-phthalate	$C_6H_4(CO_2C_8H_{17})_2$	390.56		0.983	1.4860^{20}			>110	
d466a	Diisopentyl ether	$[(CH_3)_2CHCH_2CH_2]_2O$	158.28		0.7777^{20}	1.4085^{20}		172.5	>110	
d467	1,3-Diisopropenyl-benzene	$C_6H_4[C(CH_3){=}CH_2]_2$	158.25		0.925	1.5571^{20}		231	91	
d468	Diisopropylamine	$[(CH_3)_2CH]_2NH$	101.19	4, 154	0.7153^{20}	1.3924^{20}	-61	83.5	-1	11 aq; s alc
d469	2-(Diisopropylamino)-ethanol	$[(CH_3)_2CH]_2NCH_2CH_2OH$	145.25	4^1, 430	0.826	1.4417^{20}		187–192	57	
d470	3-Diisopropylamino-1,2-propanediol	$[(CH_3)_2CH]_2NCH_2CH(OH)CH_2OH$	175.27		0.962	1.4583^{20}		131^{10mm}	>110	
d471	2,6-Diisopropylaniline	$[(CH_3)_2CH]_2C_6H_3NH_2$	177.29	12, 168	0.940	1.5332^{20}	-45	257	123	misc alc, bz, eth, acet
d472	Diisopropyl azodicarboxylate	$(CH_3)_2CHO_2CNCO_2CH(CH_3)_2$	202.21		1.027	1.4200^{20}		$75^{0.25mm}$	106	
d473	1,3-Diisopropyl-benzene	$C_6H_4[CH(CH_3)_2]_2$	162.28	5, 447	0.856^{20}_4	1.4890^{20}	-63	203	76	misc alc, bz, acet, eth
d474	1,4-Diisopropyl-benzene	$C_6H_4[CH(CH_3)_2]_2$	162.28	5^2, 339	0.857^{20}_4	1.4889^{20}	-17	204	76	
d475	Diisopropylcyanamide	$[(CH_3)_2CH]_2NCN$	126.20	4^3, 279	0.839	1.4270^{20}		93^{25mm}	78	1.2 aq; misc alc, bz, chl, eth
d476	Diisopropyl ether	$[(CH_3)_2CH]_2O$	102.17	1, 362	0.7258^{20}_4	1.3679^{20}	-86.9	68.4	-28	
d477	N,N-Diisopropyl-ethylamine	$[(CH_3)_2CH]_2NC_2H_5$	129.25	4^4, 511	0.742	1.4133^{20}	<-50	127	10	
d478	Diisopropyl malonate	$(CH_3)_2CHO_2CCH_2CO_2CH(CH_3)_2$	188.22	2^3, 1620	0.991	1.4120^{20}		95^{12mm}	88	
d479	2,6-Diisopropylphenol	$[(CH_3)_2CH]_2C_6H_3OH$	178.28	6^1, 272	0.962	1.5140^{20}	18	256	110	
d480	Diisopropyl phosphite	$[(CH_3)_2CHO]_2P(O)H$	166.16	1, 363	0.997	1.4070^{20}		$72-75^{20}$	>110	
d481	(+)-Diisopropyl L-tartrate	$[-CH(OH)CO_2CH(CH_3)_2]_2$	234.25	3, 517	1.114	1.4387^{20}		152^{12mm}	109	
d482	1,3-Diisopropyl-2-thiourea	$(CH_3)_2CHNHCSNHCH(CH_3)_2$	160.28	4, 155			143–145			
d483	Diketene		84.07	17^3, 4297	1.090	1.4330^{20}		127	34	v s aq, alc, chl, eth
d484	threo-1,4-Dimercapto-2,3-butanediol	$HSCH_2CH(OH)CH(OH)CH_2SH$	154.25				42.43			
d485	2,3-Dimercapto-1-propanol	$HSCH_2CH(SH)CH_2OH$	124.22		1.2385^{25}_4	1.5270^{25}		120^{15mm}	>110	8 aq(dec); s alc, eth

TABLE 2.21 Physical Constants of Organic Compounds (*Continued*)

No.	Name	Formula	Formula weight	Beilstein reference	Density, g/mL	Refractive index	Melting point, °C	Boiling point, °C	Flash point, °C	Solubility in 100 parts solvent
d486	2,5-Dimercapto-1,3,4-thiadiazole		150.24	27, 677			162 dec			
d487	3'4'-Dimethoxyacetophenone	$(CH_3O)_2C_6H_3COCH_3$	180.20	8^2, 298			49–51	286–288	>110	sl s aq, alc, eth
d488	2,4-Dimethoxyaniline	$(CH_3O)_2C_6H_3NH_2$	153.18	13, 784	1.075		34–37	270	>110	s alc, bz, eth
d489	2,5-Dimethoxyaniline	$(CH_3O)_2C_6H_3NH_2$	153.18	13, 788			80–82	270		s aq, alc
d490	3,4-Dimethoxyaniline	$(CH_3O)_2C_6H_3NH_2$	153.18	13, 780			88	176^{22mm}		s hot eth
d491	2,5-Dimethoxybenzaldehyde	$(CH_3O)_2C_6H_3CHO$	166.18	8, 245			49–52	146^{10mm}	>110	v s alc, eth
d492	3,4-Dimethoxybenzaldehyde	$(CH_3O)_2C_6H_3CHO$	166.18	8, 255			42–43	281	>110	sl s aq; s alc, eth
d493	1,2-Dimethoxybenzene	$C_6H_4(OCH_3)_2$	138.17	6, 771	1.0819^{25}	1.5232^{25}	22.5	206.3	87	s alc, bz, eth
d494	1,3-Dimethoxybenzene	$C_6H_4(OCH_3)_2$	138.17	6, 813	1.055	1.5240	−55	217^{mm}	87	s alc; v s bz, eth
d495	1,4-Dimethoxybenzene	$C_6H_4(OCH_3)_2$	138.17	6, 843	1.036^{65}_{8}		55–60	213		0.05 aq; v s alc, eth
d496	3,4-Dimethoxybenzoic acid	$(CH_3O)_2C_6H_3CO_2H$	182.18	10^1, 188			180–181			
d497	3,5-Dimethoxybenzoic acid	$(CH_3O)_2C_6H_3CO_2H$	182.18	10, 405			182–184			
d498	2,6-Dimethoxybenzoyl chloride	$(CH_3O)_2C_6H_3COCl$	200.62	10^3, 1402			64–66			
d499	3,4-Dimethoxybenzyl alcohol	$(CH_3O)_2C_6H_3CH_2OH$	168.19	5, 1113	1.157	1.5520^{20}		297^{732mm}	>110	
d500	2,2-Dimethoxycyclohexanol	$(CH_3O)_2C_6H_9OH$	160.22		1.072	1.4620^{20}		90^{9mm}	40	
d501	2,5-Dimethoxy-2,5-dihydrofuran		130.14		1.073	1.4339^{20}		160–162	47	
d502	Dimethoxydimethylsilane	$(CH_3O)_2Si(CH_3)_2$	120.23		0.880	1.3690^{20}		81.4	10	s aq, alc, chl, eth
d503	Dimethoxydiphenylsilane	$(C_6H_5)_2Si(OCH_3)_2$	244.4		1.0771^{20}_{4}	1.5447^{20}		161^{15mm}		misc aq, alc; s PE
d504	1,1-Dimethoxyethane	$CH_2CH(OCH_3)_2$	90.12	1, 603	0.8502^{20}	1.3668^{20}	−113	64.5	−17	
d505	1,2-Dimethoxyethane	$CH_3OCH_2CH_2OCH_3$	90.12	1, 467	0.8620^{20}	1.3796^{20}	−68	85.2	1	
d506	(2,2'-Dimethoxy)-ethyiamine	$H_2NCH_2CH(OCH_3)_2$	105.14	4^2, 758	0.965	1.4170^{20}		135^{95mm}	53	
d507	Dimethoxymethane	$CH_2(OCH_3)_2$	76.10	1, 574	0.8601^{20}	1.3514^{20}	−104.8	42.3	−32	32 aq
d508	1,1-Dimethoxy-2-methylaminoethane	$C \cdot H_3NHCH_2CH(OCH_3)_2$	119.16	4^2, 759	0.928	1.4115^{20}		140	29	
d509	Dimethoxymethylvinylsilane	$CH_3Si(OCH_3)_2CH{=}CH_2$	132.24	6, 789	0.884	1.3950^{20}		106	3	v s alc, eth; s chl
d510	Dimethoxymethylphenylsilane	$(CH_3O)_2Si(CH_3)C_6H_5$	182.3		0.993^4	1.469^{20}		199–200		s alc, alk; v s eth
d511	1,2-Dimethoxy-4-nitrobenzene	$(CH_3O)_2C_6H_3NO_2$	183.16	6, 1081	1.1888^{133}		95–98	230^{17mm}		
d512	2,6-Dimethoxyphenol	$(CH_3O)_2C_6H_3OH$	154.17				53–56	261	>110	
d513	3,4-Dimethoxyphenyl-acetic acid	$(CH_3O)_2C_6H_3CO_2H$	196.20	10, 409			96–98			s aq; v s alc, eth

No.	Name	Formula	M.W.	Beil. Ref.	Density	n_D	m.p.	b.p.	Flash P.	Solubility
d514	3,4-Dimethoxyphenyl-acetonitrile	$(CH_3O)_2C_6H_3CN$	177.20	10^1, 198			62–63	178^{10}mm	83	
d515	2,2-Dimethoxy-2-phenylacetophenone	$C_6H_5C(O)C(OCH_3)_2C_6H_5$	256.30				67–70		>110	
d516	1,1-Dimethoxy-2-phenylethane	$C_6H_5CH_2CH(OCH_3)_2$	166.22	7, 293	1.004	1.4950^{20}		221	0	
d517	2-(3,4-Dimethoxyphenyl)ethylamine	$(CH_3O)_2C_6H_3CH_2CH_2NH_2$	181.24	13, 800	1.074	1.5464^{20}		188^{15}mm	−11	
d518	1,2-Dimethoxypropane	$CH_3CH(OCH_3)CH_2OCH_3$	104.15	1^4, 2471	0.855	1.3835^{20}		96	37	
d519	2,2-Dimethoxypropane	$(CH_3)_2C(OCH_3)_2$	104.15	1, 648	0.847	1.3780^{20}		83	−2	
d520	1,1-Dimethoxy-2-propanone	$CH_3C(O)CH(OCH_3)_2$	118.13	1^1, 395	0.976	1.3978^{20}		143–147		
d521	3,3-Dimethoxy-1-propene	$(CH_3O)_2CHCH=CH_2$	102.13	1^1, 378	0.862	1.3954^{20}		89–90		
d522	1,2-Dimethoxy-4-propenylbenzene	$CH_3CH=CHC_6H_5(OCH_3)_2$	178.23	6, 956	1.055	1.5680^{20}		262–264	>110	
d523	3,3-Dimethoxypropionitrile	$(CH_3O)_2CHCH_2CN$	115.13	3^4, 521	1.026	1.4130^{20}		92^{30}mm	86	
d524	2,6-Dimethoxypyridine	$(CH_3O)_2C_5H_3N$	139.15		1.053	1.5129^{20}		178–180	61	
d525	2,5-Dimethoxytetrahydrofuran	$(CH_3O)_2C_4H_6O$	132.16		1.020	1.4180^{20}		145–147	35	
d526	N,N-Dimethylacetamide	$CH_3C(O)N(CH_3)_2$	87.12	4, 59	0.9366^{25}	1.4376^{20}	−20	165.5	70	misc aq, alc, bz, eth
d527	2′,6′-Dimethylacetanilide	$CH_3C(O)NHC_6H_3(CH_3)_2$	163.22	12, 1109			182–184			
d528	Dimethyl 1,3-acetone-dicarboxylate	$[CH_3O_2CCH_2]_2C=O$	174.15	3, 790	1.185	1.4434^{20}		150^{25}mm	>110	
d529	Dimethyl acetylenedi-carboxylate	$CH_3O_2CC\!\equiv\!CCO_2CH_3$	142.11	2, 803	1.156	1.4470^{20}		98^{19}mm	86	
d530	Dimethyl acetyl-succinate	$CH_3O_2CC_2CH(COCH_3)CO_2CH_3$	188.18	3^4, 1825	1.160		33	134^{12}mm	>110	
d531	N,N-Dimethylacrylamide	$H_2C=CHC(O)N(CH_3)_2$	99.13	4^3, 130	0.962	1.4730^{20}		81^{20}mm	71	
d532	3,3-Dimethylacrylic acid	$(CH_3)_2C=CHCO_2H$	100.12	2, 432	0.996		69	195		
d533	Dimethylaluminum chloride	$(CH_3)_2AlCl$	92.51	4^3, 1971			−21	126–127	−18	
d534	Dimethylamine	$(CH_3)_2NH$	45.08	4, 39	0.680_4^{0}	1.350^{17}	−92.2	6.9	20	v s aq; s alc, eth
d535	Dimethylaminoacetonitrile	$(CH_3)_2NCH_2CN$	84.12	4, 346	0.863	1.4101^{20}		138		
d536	4-(Dimethylamino)benzaldehyde	$(CH_3)_2NC_6H_4CHO$	149.19	14, 31			74	176^{17}mm	36	s alc, chl, eth, HOAc
d537	3-Dimethylamino-benzoic acid	$(CH_3)_2NC_6H_4CO_2H$	165.19	14, 392			148–152			
d538	4-Dimethylamino-benzoic acid	$(CH_3)_2NC_6H_4CO_2H$	165.19	14, 426			241 dec			
d539	2-(Dimethylamino)ethanol	$(CH_3)_2NCH_2CH_2OH$	89.14	4, 276	0.8876^{20}	1.4294^{20}		135	40	s alc; sl s eth
d540	2-[2-(Dimethylamino)ethoxy]ethanol	$(CH_3)_2NCH_2CH_2OCH_2CH_2OH$	133.19	4^2, 719	0.954	1.4420^{20}		95^{15}mm	92	misc aq, alc, eth
d541	2-(Dimethylamino)ethyl acrylate	$H_2C=CHCO_2CH_2CH_2N(CH_3)_2$	143.19	4^3, 649	0.943	1.4280^{20}		64^{12}mm	58	

(Continued)

TABLE 2.21 Physical Constants of Organic Compounds (*Continued*)

No.	Name	Formula	Formula weight	Beilstein reference	Density, g/mL	Refractive index	Melting point, °C	Boiling point, °C	Flash point, °C	Solubility in 100 parts solvent
d542	2-(Dimethylamino)-ethyl benzoate	$C_6H_5CO_2CH_2CH_2N(CH_3)_2$	193.26		1.014	1.5077^{20}		159^{20mm}	>110	
d543	2-(Dimethylamino)-ethyl methacrylate	$H_2C{=}C(CH_3)CO_2CH_2CH_2{-}N(CH_3)_2$	157.22	4^3, 649	0.933	1.4400^{20}		182–192	70	
d544	3-Dimethylamino-phenol	$(CH_3)_2NC_6H_4OH$	137.18	13, 405	1.5895^{25}	1.4609^{20}	82–84	265–268	105	v s alc, bz, eth, acet
d545	3-Dimethylamino-1,2-propanediol	$(CH_3)_2NCH_2CH(OH)CH_2OH$	119.16	4, 302	1.004			216–217		s aq, alc, chl, eth
d546	1-Dimethylamino-2-propanol	$CH_3CH(OH)CH_2N(CH_3)_2$	103.17	4^1, 433	0.837	1.4193^{20}		121–127	35	
d547	3-Dimethylamino-1-propanol	$(CH_3)_2NCH_2CH_2OH$	103.17		0.872	1.4360^{20}		163–164	36	
d548	3-(Dimethylamino)-propionitrile	$(CH_3)_2NCH_2CH_2CN$	98.15	4^3, 1265	0.870	1.4258^{20}	−43	171^{750mm}	62	
d549	3-Dimethylamino-propylamine	$(CH_3)_2N(CH_2)_3NH_2$	102.18	4^3, 554	0.812	1.4350		133	15	
d550	N-[3-(Dimethylamino)-propyl]methacrylamide	$H_2C{=}C(CH_3)CONH(CH_2)_3{-}N(CH_3)_2$	170.26		0.940	1.4790^{20}		134^{2mm}	>110	v s aq, alc, bz, chl
d551	4-(Dimethylamino)-pyridine	$(CH_3)_2N(C_5H_4N)$	122.17	22^2, 341			112–114			
d552	Dimethyl 2-amino-1,4-phthalate	$H_2NC_6H_3(CO_2CH_3)_2$	209.20	14, 559			127–130			
d553	N,N-Dimethylaniline	$C_6H_5N(CH_3)_2$	121.18	12, 141	0.9559^{20}_4	1.5584^{20}	2.5	194.2	63	v s alc, chl, eth
d554	2,3-Dimethylaniline	$(CH_3)_2C_6H_3NH_2$	121.18	12, 1101	0.9933^{20}	1.5685^{20}	<−15	221–222	97	sl s aq; s alc, eth
d555	2,4-Dimethylaniline	$(CH_3)_2C_6H_3NH_2$	121.18	12, 1111	0.9723^{20}	1.55686^{20}	−14.3	214	90	s alc, bz, eth
d556	2,5-Dimethylaniline	$(CH_3)_2C_6H_3NH_2$	121.18	12, 1135	0.9790^{21}	1.5592^{20}	15.5	214	93	sl s aq; s alc, eth
d557	2,6-Dimethylaniline	$(CH_3)_2C_6H_3NH_2$	121.18	12, 1107	0.9842^{20}	1.5601^{20}	11.2	215	96	sl s aq; s alc, eth
d558	3,4-Dimethylaniline	$(CH_3)_2C_6H_3NH_2$	121.18	12, 1103	1.076^{18}		51	228	98	sl s aq; s alc
d559	3,5-Dimethylaniline	$(CH_3)_2C_6H_3NH_2$	121.18	12, 1131	0.9706^{20}	1.5578^{20}	9.8	220.5	93	sl s aq; s alc
d560	Dimethylarsinic acid	$(CH_3)_2As(O)OH$	138.00	4, 610			195–196			v s alc; 200 aq; i eth
d561	1,3-Dimethylbarbituric acid		156.14	24, 471			124–126	133^{15mm}	>110	
d562	N,N-Dimethylbenzamide	$C_6H_5CON(CH_3)_2$	149.19	9, 201			43–45	subl		s alc, bz
d563	3,4-Dimethylbenzoic acid	$(CH_3)_2C_6H_3CO_2H$	150.18	9^2, 353			165–167	223^{730mm}	92	
d564	2,5-Dimethylbenzonitrile	$(CH_3)_2C_6H_3CN$	131.18	9, 535	0.957	1.5284^{20}	13–14		54	
d565	N,N-Dimethylbenzylamine	$C_6H_5CH_2N(CH_3)_2$	135.21	12, 1019	0.900	1.5011^{20}	−75	183		
d566	2,3-Dimethyl-1,3-butadiene	$H_2C{=}C(CH_3)C(CH_3){=}CH_2$	82.15	1^3, 991	0.72224^{25}	1.4362^{25}	−76.0	69.2	−22	
d567	2,2-Dimethylbutane	$CH_3CH_2C(CH_3)_3$	86.18	1, 150	0.6492^{20}	1.3688^{20}	−99.9	49.7	−48	

No.	Name	Formula	M.W.	Ref.	Density	n_D	M.P.	B.P.	Flash	Solubility
d568	2,3-Dimethylbutane	$(CH_3)_2CHCH(CH_3)_2$	86.18	1, 151	0.6616^{20}	1.3750^{20}	−128.5	58.0	−29	v s hot aq, alc, eth
d569	2,3-Dimethyl-2,3-butanediol	$(CH_3)_2C(OH)C(OH)(CH_3)_2$	86.18	1, 487	0.8236^{20}_{4}	1.4176^{20}	41.1	174.4	77	s aq; misc alc, eth
d570	2,3-Dimethyl-2-butanol	$(CH_3)_2CHC(CH_3)_2OH$	102.18	1, 413	0.824^{20}	1.4176^{20}	−14	118	29	
d570a	3,3-Dimethyl-1-butanol	$(CH_3)_3CCH_2CH_2OH$	102.18	1^3, 1677	0.8185^{20}_{4}	1.4151^{20}	−60	143	47	
d571	3,3-Dimethyl-2-butanol	$(CH_3)_3CCH(OH)CH_3$	102.18	1, 412			5.6	120	28	s alc; misc eth
d572	3,3-Dimethyl-2-butanone	$(CH_3)_3CCOCH_3$	100.16	1, 694	0.7250^{25}_{25}	1.3939^{25}	−52.5	106	23	2.5 aq; s alc, eth
d572a	2,3-Dimethyl-1-butene	$(CH_3)_2C=C(CH_3)=CH_2$	84.16	1^3, 816	0.680	1.3890^{20}	−157	55.6	−18	
d573	2,3-Dimethyl-2-butene	$(CH_3)_2C=C(CH_3)_2$	84.16	1, 218	0.7081^{24}	1.4124^{20}	−75	73	−16	s alc, eth
d574	3,3-Dimethyl-1-butene	$(CH_3)_3CCH=CH_2$	84.16	1, 217	0.6531^{24}	1.3762^{20}	−115	41	−28	
d575	N,N-Dimethylbutylamine	$CH_3(CH_2)_3N(CH_3)_2$	101.19	4^1, 371	0.721	1.3980^{20}		93^{750mm}	−3	
d576	2,2-Dimethylbutyric acid	$C_2H_5C(CH_3)_2CO_2H$	116.16	2, 335	0.928	1.4154^{20}	6–7	96^{5mm}	79	s alc, eth
d577	3,3-Dimethylbutyric acid	$(CH_3)_3CCH_2CO_2H$	116.16	2, 337	0.9124^{24}_{4}	1.4100^{20}	−4.5	190	88	dec aq; s PE
d578	Dimethylcadmium	$(CH_3)_2Cd$	142.48		1.9846^{17}_{4}	1.5488		105.5	>150 explodes	
d579	Dimethylcarbamyl chloride	$(CH_3)_2NCOCl$	107.54	4, 73	1.168	1.4540^{20}	−33	168	68	i aq; misc alc, eth
d580	Dimethyl carbonate	$(CH_3O)_2C=O$	90.08	3, 4	1.0651^{17}_{4}	1.3682^{20}	0.5	90–91	18	s alc, bz
d581	Dimethyl chloromalonate	$ClCH(CO_2CH_3)_2$	166.56	2, 592	1.305	1.4370^{20}		106^{19mm}	106	
d582	Dimethyl chlorothiophosphate	$(CH_3O)_2P(S)Cl$	160.56	1^1, 143	1.322	1.4819^{20}		67^{16mm}	105	
d583	Dimethylcyanamide	$(CH_3)_2NCN$	70.09	4, 74	0.867	1.4100^{20}		161–163	58	
d584	Dimethyl N-cyanothioiminocarbonate	$(CH_3S)_2C=NCN$	146.23	3, 220			46–50		110	
d584a	1,1-Dimethylcyclohexane	$(CH_3)_2C_6H_{10}$	112.22	5, 35	0.777	1.4280^{20}	−33	120	7	
d585	cis-1,2-Dimethylcyclohexane	$(CH_3)_2C_6H_{10}$	112.22	5, 36	0.7963^{20}	1.4335^{20}	−49.9	129.7	16	i aq; s alc, bz
d586	trans-1,2-Dimethylcyclohexane	$(CH_3)_2C_6H_{10}$	112.22	5, 36	0.7760^{20}	1.4273^{20}	−90	123.4	11	i aq; s alc, bz
d587	cis-1,3-Dimethylcyclohexane	$(CH_3)_2C_6H_{10}$	112.22	5, 36	0.784	1.4230^{20}	−76	120	5	
d587a	trans-1,3-Dimethylcyclohexane	$(CH_3)_2C_6H_{10}$	112.22	5^2, 21	0.780	1.4305^{20}	−90	124.5	7	
d588	cis-1,4-Dimethylcyclohexane	$(CH_3)_2C_6H_{10}$	112.22	5^2, 22	0.783	1.4297^{20}	−88	125	6	
d589	5,5-Dimethyl-1,3-cyclohexanedione		140.18	7, 559			dec 149			0.4 aq; s alc, bz
d590	2,3-Dimethylcyclohexanol	$(CH_3)_2C_6H_9OH$	128.22	6, 18	0.934	1.4653^{20}	11–12	186	65	
d591	3,5-Dimethylcyclohexanol	$(CH_3)_2C_6H_9OH$	128.22	7, 23	0.892	1.4552		175	73	
d592	2,6-Dimethylcyclohexanone	$(CH_3)_2C_6H_8(=O)$	126.20		0.925	1.4460^{20}		159	51	
d593	N,N-Dimethylcyclohexylamine	$C_6H_{11}N(CH_3)_2$	127.23		0.849	1.4535^{20}		160	42	i aq; s alc, eth
d594	2,3-Dimethylcyclohexylamine	$(CH_3)_2C_6H_9NH_2$	127.23	5	0.835	1.4595^{20}			51	

(Continued)

TABLE 2.21 Physical Constants of Organic Compounds (*Continued*)

No.	Name	Formula	Formula weight	Beilstein reference	Density, g/mL	Refractive index	Melting point,°C	Boiling point,°C	Flash point,°C	Solubility in 100 parts solvent
d595	1,5-Dimethyl-1,5-cyclooctadiene		136.24		0.867	1.4896^{20}		74^{16mm}	55	
d596	Dimethyl 1,1-cyclo-propanedicarboxylate	$C_3H_4(CO_2CH_3)_2$	158.16	9^1, 314	1.147	1.4410^{20}		196–198	95	
d597	Dimethyl decanedioate	$CH_3O_2C(CH_2)_8CO_2CH_3$	230.30	2, 719	0.983^{30}_{20}	1.4335^{28}	23	144^{5mm}	145	i aq; s alc, eth
d598	2,2-Dimethyl-1,3-dioxane-4,6-dione		144.13				94–96			s aq, acet
d599	2,2-Dimethyl-1,3-dioxolane-4-methanol		132.16	19, 65	1.063	1.4340^{20}		188–189	80	misc aq, alc, bz, esters, eth, PE, acetals
d600	Dimethyl disulfide	CH_3SSCH_3	94.20	1, 291	1.0625^{20}	1.5289^{20}	–84.7	109.8	24	i aq; misc alc, eth
d601	Dimethyldithio-carbamic acid, Zn salt	$[(CH_3)_2NCS_2]_2Zn$	305.80	4^3, 149	1.66		250–252			< 0.2 alc, eth; < 0.5 acet, bz; 0.5 naphtha
d602	N,N-Dimethyldodecylamine	$CH_3(CH_2)_{11}N(CH_3)_2$	213.41	4^3, 409	0.775	1.4375^{20}	–20	112^{3mm}	>110	35 aq(5 atm); 15 bz; 11.8 acet
d603	Dimethyl ether	$(CH_3)_2O$	46.07	1, 281	0.661^{20}		–141.5	–24.9	–41	
d604	N,N-Dimethylethylamine	$C_2H_5N(CH_3)_2$	73.14	4, 94	0.675	1.3720^{20}	–140	36–38	–36	sl s alc, eth
d605	N,N-Dimethylethylenediamine	$C_2H_5NCH_2CH_2NH_2$	88.15	4^2, 690	0.803	1.4260^{20}		106	23	i aq; misc alc, eth
d606	N,N-Dimethylformamide	$(CH_3)_2NCHO$	73.10	4, 58	0.9445^{25}_{4}	1.4305^{20}	–60.4	153.0	57	misc aq, alc, bz, eth
d607	N,N-Dimethylformamide dimethyl acetal	$(CH_3)_2NCH(OCH_3)_2$	119.16		0.897	1.3972^{20}		103^{720mm}	7	
d608	Dimethyl fumarate	$CH_3O_2CCH=CHCO_2CH_3$	144.13	2, 741	1.045^{106}	1.4664^{20}	105	193		sl s alc, eth
d609	2,5-Dimethylfuran	$(CH_3)_2(C_4H_2O)$	96.13	17, 41	0.9000^{20}_{4}	1.4414^{20}	–62	93	–1	i aq; misc alc, eth
d610	Dimethylglyoxime	$CH_3C(=NOH)-C(=NOH)CH_3$	116.12	1, 772			240			s alc, acet, eth, pyr
d611	2,4-Dimethyl-1,6-heptadienal	$H_2C=CHCH_2CH(CH_3)-CH=C(CH_3)CHO$	138.21		0.870	1.4664^{20}		47^{2mm}	64	
d612	2,4-Dimethyl-2,6-heptadien-1-ol	$H_2C=CHCH_2CH(CH_3)-CH=C(CH_3)CH_2OH$	140.23		1.351	1.4640^{20}		86^{10mm}	78	
d613	2,6-Dimethyl-2,5-heptadien-4-one	$(CH_3)_2C=CHC(=O)-CH=(CH_3)_2$	138.21	1, 751	0.885^{20}_{4}	1.4968^{21}	28	198–199	79	sl s aq; s alc, eth
d613a	2,2-Dimethylheptane	$(CH_3)_3C(CH_2)_4CH_3$	128.26	2^1, 281	0.7105^{20}	1.4016^{20}	–113	132.7	>110	s alc
d614	Dimethyl heptanedioate	$CH_3O_2C(CH_2)_5CO_2CH_3$	188.22		1.0625^{20}_{4}	1.4314^{20}	–21	122^{11mm}	66	
d615	2,6-Dimethyl-4-heptanol	$(CH_3)_2CHCH_2CH(OH)-CH_2CH(CH_3)_2$	144.26	1, 425	0.809	1.4236^{20}		178		
d616	2,6-Dimethyl-4-heptanone	$[(CH_3)_2CHCH_2]_2C=O$	142.24	1, 710	0.806^{20}_{20}	1.4114^{20}	–41.5	169.4	49	0.06 aq; misc alc, bz, chl, eth
d616a	2,4-Dimethylhexane	$C_2H_5CH(CH_3)CH_2CH(CH_3)_2$	114.23	1, 162	0.6962^{25}	1.3929^{25}		109.5	10	
d617	Dimethyl hexanedioate	$CH_3O_2C(CH_2)_4CO_2CH_3$	174.20	1, 652	1.0600^{20}_{4}	1.4285^{20}	8	112^{10mm}	107	i aq; s alc, eth

(Continued)

No.	Name	Formula	Mol. wt.	Beilstein ref.	Density	n_D	m.p., °C	b.p., °C	Flash P., °C	Solubility
d618	2,5-Dimethyl-2,5-hexanediol	$[(CH_3)_2C(OH)CH_2]_2$	146.23	1, 492	0.767	1.4209^{20}	86–90	214–215	126	v s aq, alc, bz, chl, eth, acet
d619	1,5-Dimethylhexylamine	$(CH_3)_2CH(CH_2)_3CH(NH_2)CH_3$	129.25	Merck: 11, 6678				154–156	48	
d620	2,5-Dimethyl-3-hexyne-2,5-diol	$(CH_3)_2C(OH)C{\equiv}C{-}C(OH)(CH_3)_2$	142.20	1, 501			94–95	205–206		
d621	3,5-Dimethyl-1-hexyn-3-ol	$(CH_3)_2CHCH_2C(CH_3)(OH)C{\equiv}CH$	126.20	1^2, 507	0.859	1.4335^{20}		151	44	
d622	5,5-Dimethylhydantoin		128.13	24, 289			176–178			s aq(hyd); misc alc, acet, eth
d623	1,1-Dimethylhydrazine	$(CH_3)_2NNH_2$	60.10	4, 547	0.7911^{22}_{4}	1.4075^{20}	−58	63.9	1	misc aq, alc, eth, PE
d624	1,2-Dimethylhydrazine	$CH_3NHNHCH_3$	60.10	4, 547	0.8274^{20}_{4}	1.4209^{20}		81	flamma-ble	misc aq, alc, eth, PE
d625	Dimethyl hydrogen phosphonate	$(CH_3O)_2P(O)H$	110.05	1, 285	1.200^{20}_{4}	1.4009^{20}		170–171	29	s aq(hyd); misc alc, acet, eth
d626	1,2-Dimethylimidazole		96.13	23, 66	1.084		29–30	204	92	
d627	1,3-Dimethyl-2-imidazolidinone		114.15		1.044	1.4720^{20}		108^{17mm}	80	
d628	N,N-Dimethylisopropylamine	$(CH_3)_2CHN(CH_3)_2$	87.17	4^2, 630	0.715	1.3905^{20}	−19	66	−9	8.7 aq
d629	Dimethyl maleate	$CH_3O_2CCH{=}CHCO_2CH_3$	144.13	2, 751	1.1606^{20}	1.4422^{20}		202	113	sl s aq; misc alc, eth
d630	Dimethyl malonate	$CH_3O_2CCH_2CO_2CH_3$	132.12	2, 572	1.154^{20}_{4}	1.4135^{20}	−62	180–181	90	i aq; s alc, eth
d631	Dimethylmercury	$(CH_3)_2Hg$	230.66	4, 678	3.1874^{20}	1.5452^{20}	−43	92–94	5	
d632	3,4-Dimethyl-1-methoxybenzene	$(CH_3)_2C_6H_3OCH_3$	136.19	6, 481	0.9744^{14}_{4}	1.5198^{14}		200	65	i aq; s alc, eth
d633	3,5-Dimethyl-1-methoxybenzene	$(CH_3)_2C_6H_3OCH_3$	136.19	6, 493	0.9627^{15}_{4}	1.5107^{15}		193	76	i aq; s alc, bz, eth
d634	Dimethyl methyl-malonate	$CH_3CH(CO_2CH_3)_2$	146.14	2, 628	1.098	1.4140^{20}		176–177	68	
d635	Dimethyl methyl-phosphonate	$(CH_3O)_2P(O)CH_3$	124.08	4^1, 572	1.145	1.4130^{20}		181	83	
d636	Dimethyl methyl-succinate	$CH_3O_2CCH_2CH(CH_3)CO_2CH_3$	160.17	2^3, 1696	1.076	1.4200^{20}		196		
d637	2,6-Dimethylmorpholine		115.18	5^1, 267	0.9346^{26}_{4}	1.4470^{20}	−85	147	48	misc aq, alc, bz
d637a	1,2-Dimethylnaphthalene	$C_{10}H_6(CH_3)_2$	156.23	5, 367	1.0179^{20}	1.6166^{20}	0.8	266.5	>110	
d638	1,2-Dimethyl-3-nitro-benzene	$(CH_3)_2C_6H_3NO_2$	151.17	5, 367	1.129	1.5434^{20}	7–9	245	107	i aq; s alc
d639	1,2-Dimethyl-4-nitro-benzene	$(CH_3)_2C_6H_3NO_2$	151.17	5, 368	1.139		29–31	143^{20mm}	>110	i aq; s alc
d640	1,3-Dimethyl-2-nitro-benzene	$(CH_3)_2C_6H_3NO_2$	151.17	5, 378	1.112	1.5220^{20}	14–16	225^{744mm}	87	i aq; s alc, eth
d641	1,3-Dimethyl-4-nitro-benzene	$(CH_3)_2C_6H_3NO_2$	151.17	5, 378	1.117	1.5497^{20}	2	237–239	107	s alc, bz, chl, eth
d642	N,N-Dimethyl-4-nitrosoaniline	$(CH_3)_2NC_6H_4NO$	150.18	12, 677			86	flamma-ble solid		i aq; s alc, eth
d643	Dimethyl 2-nitro-1,4-phthalate	$O_2NC_6H_3{-}1,4{-}(CO_2CH_3)_2$	239.18	9, 826			72–75			
d644	cis-3,7-Dimethyl-2,6-octadienal		152.24		0.8886^{20}_{4}	1.4898^{20}		229	101	misc alc, eth, glyc
d645	trans-3,7-Dimethyl-2,6-octadienal		152.24		0.8869^{20}_{4}	1.4869^{20}		229	101	misc alc, eth, glyc
d646	3,7-Dimethyl-1-octanol	$(CH_3)_2CH(CH_2)_3CH(CH_3)CH_2CH_2OH$	158.29	1, 426	0.840	1.4355^{20}		96^{9mm}	95	
d647	3,7-Dimethyl-3-octanol	$(CH_3)_2CH(CH_2)_3C(OH)(CH_3)C_2H_5$	158.29	1, 426	0.826	1.4336^{20}		73^{6mm}	76	

TABLE 2.21 Physical Constants of Organic Compounds (*Continued*)

No.	Name	Formula	Formula weight	Beilstein reference	Density, g/mL	Refractive index	Melting point, °C	Boiling point, °C	Flash point, °C	Solubility in 100 parts solvent
d648	2,6-Dimethyl-2,4,6-octatriene	CH$_3$CH=C(CH$_3$)CH=CH-CH=C(CH$_3$)$_2$	136.24	1^3, 1050	0.811	1.5429^{20}		75^{14mm}	68	
d649	N,N-Dimethyloctylamine	CH$_3$(CH$_2$)$_7$N(CH$_3$)$_2$	157.30	4^1, 386	0.765	1.4243^{20}	−57	195	65	
d650	3,6-Dimethyl-4-octyne-3,6-diol	C$_2$H$_5$C(CH$_3$)(OH)C≡C-C(CH$_3$)(OH)C$_2$H$_5$	170.35	1^1, 263			53–55	214^{680mm}	>110	
d651	Dimethyl octanedioate	CH$_3$O$_2$C(CH$_2$)$_6$CO$_2$CH$_3$	202.25	2, 693	1.0210^{20}_{4}	1.4325^{20}	−4.8	268	75	i aq; s alc
d652	Dimethyl oxalate	CH$_3$O$_2$CCO$_2$CH$_3$	118.09	2, 534	1.148^{54}	1.379^{80}	50–54	163.5	−9	6 aq; s alc, eth
d653	3,3-Dimethyloxetane		86.13	17^2, 21	0.835	1.3990		81		
d654	2,3-Dimethylpentane	C$_2$H$_5$CH(CH$_3$)CH(CH$_3$)$_2$	100.21	1^2, 120	0.6951_{4}	1.3920^{20}		89.8	<−1	i aq; s alc, eth
d655	2,4-Dimethylpentane	(CH$_3$)$_2$CHCH$_2$CH(CH$_3$)$_2$	100.21	1, 158	0.6727^{20}	1.3815^{20}	−120	80.4	−12	
d656	Dimethyl pentanedioate	CH$_3$O$_2$C(CH$_2$)$_3$CO$_2$CH$_3$	160.17	2, 633	1.0876^{20}	1.4244^{20}	−42.5	214	102	v s alc, eth
d657	2,4-Dimethyl-3-pentanol	(CH$_3$)$_2$CHCH(OH)CH(CH$_3$)$_2$	116.20	1, 417	0.829^{20}_{4}	1.4254^{20}		140	37	sl s aq; s alc, eth
d658	2,4-Dimethyl-3-pentanone	(CH$_3$)$_2$CHC(O)CH(CH$_3$)$_2$	114.19	1, 703	0.8062^{20}_{4}	1.3986^{20}	−69	125	15	v s alc, bz, chl, eth
d659	2,3-Dimethylphenol	(CH$_3$)$_2$C$_6$H$_3$OH	122.17	6, 480		1.5420^{20}	72.8	217		v s alc, bz, chl, eth
d660	2,4-Dimethylphenol	(CH$_3$)$_2$C$_6$H$_3$OH	122.17	6, 486	1.0276^{14}_{4}	1.5420^{14}	24.5	211	>110	v s alc, bz, chl, eth
d661	2,5-Dimethylphenol	(CH$_3$)$_2$C$_6$H$_3$OH	122.17	6, 494	0.965^{80}		74.5	211.5		v s alc, bz, chl, eth
d662	2,6-Dimethylphenol	(CH$_3$)$_2$C$_6$H$_3$OH	122.17	6, 485			45.7	201	73	v s alc, bz, chl, eth
d663	3,4-Dimethylphenol	(CH$_3$)$_2$C$_6$H$_3$OH	122.17	6, 480	0.9830^{20}		60.8	227		v s alc, bz, chl, eth
d664	3,5-Dimethylphenol	(CH$_3$)$_2$C$_6$H$_3$OH	122.17	6, 492	0.9680^{20}		64	222		v s alc, bz, ehl, eth
d665	N,N-Dimethyl-1,4-phenylenediamine	(CH$_3$)$_2$NC$_6$H$_4$NH$_2$	136.20	13, 72			36	262	90	v s aq; s alc
d666	4,4'-Dimethyl-2-phenyl-2-oxazoline		175.23	27^4, 1114	1.025	1.5322^{20}	20–24	124^{20mm}	102	
d667	2,2-Dimethyl-3-phenyl-1-propanol	C$_6$H$_5$CH$_2$C(CH$_3$)$_2$CH$_2$OH	164.25				35	126^{15mm}	109	
d668	Dimethyl 1,2-phthalate	C$_6$H$_4$(CO$_2$CH$_3$)$_2$	194.19	9, 797	1.1905^{20}	1.5138^{20}	5.5	283.7	146	0.4 aq; misc alc, chl, eth; i PE
d669	Dimethyl 1,3-phthalate	C$_6$H$_4$(CO$_2$CH$_3$)$_2$	194.19	9, 834	1.194^{20}_{4}	1.5168^{20}	67–68	282		i aq
d670	Dimethyl 1,4-phthalate	C$_6$H$_4$(CO$_2$CH$_3$)$_2$	194.19	9, 843			140–142	288		0.3 hot aq; s hot alc; s eth
d671	1,4-Dimethylpiperazine		114.19	23, 7	0.844	1.4463^{20}		132^{750mm}	18	
d672	cis-2,6-Dimethylpiperidine		113.20	20, 108	0.840	1.4394^{20}		127	11	

No.	Name	Formula	Mol wt	Merck	Density	n_D	mp, °C	bp, °C	Flash pt, °C	Solubility
d673	2,2-Dimethylpropane	$(CH_3)_4C$	72.15	12, 6545	0.613^0	1.3476^6	−16.6	9.5	−65	
d674	2,2-Dimethyl-1,3-propanediamine	$H_2NCH_2C(CH_3)_2CH_2NH_2$	102.18	4^3, 595	0.851	1.4566^{20}	31	154	47	180 aq; 12 bz; 60 acet; v s alc, eth
d675	2,2-Dimethyl-1,3-propanediol	$(CH_3)_2C(CH_2OH)_2$	104.15	1, 483	1.11^{25}		127–128	208–210	107	
d676	2,2-Dimethyl-1-propanol	$(CH_3)_3CCH_2OH$	88.15	1, 406	0.812^{20}_4		52.5	113.1	36	
d677	2,2-Dimethylpropionaldehyde	$(CH_3)_3CCHO$	86.13		0.793	1.3794^{20}	6	74^{730mm}	<1	3.6 aq; misc alc, eth
d678	N,N-Dimethylpropionamide	$C_2H_5C(O)N(CH_3)_2$	101.15	4^3, 126	0.920	1.4400^{20}	−45	175	62	
d679	2,2-Dimethylpropionic acid	$(CH_3)_3CCO_2H$	102.13	2, 319	0.905^{50}	1.3931^{37}	35.5	163.8	63	2.5 aq; v s alc, eth
d680	2,2-Dimethylpropionic anhydride	$[(CH_3)_3CC(O)]_2O$	186.25	2, 320	0.918	1.4092^{20}		193	57	
d681	2,2-Dimethylpropionyl chloride	$(CH_3)_3CC(O)Cl$	120.58	2, 320	0.979	1.4120^{20}		105–106	<1	dec aq, alc; v s eth
d682	1,1-Dimethylpropylamine	$CH_3CH_2C(CH_3)_2NH_2$	87.17	4, 179	0.731^{25}_4	1.3996^{20}	−105	77	65	misc aq, alc, eth
d683	1,1-Dimethyl-2-propynylamine	$HC{\equiv}CC(CH_3)_2NH_2$	83.13		0.790	1.4235^{20}		79–80	2	
d684	3,5-Dimethylpyrazole	$(CH_3)_2(C_5H_3N)$	96.13	23, 74	0.945		108	218		s aq; v s bz, eth
d685	2,3-Dimethylpyridine	$(CH_3)_2(C_5H_3N)$	107.16	20, 243	0.9309^{20}	1.5080	−15	163	50	
d686	2,4-Dimethylpyridine	$(CH_3)_2(C_5H_3N)$	107.16	20, 244	0.9226^{20}	1.5010^{20}	<−64	158.3	37	17 aq; v s alc, bz, eth
d687	2,6-Dimethylpyridine	$(CH_3)_2(C_5H_3N)$	107.16	20, 244	0.954^{25}_4	1.4956^{20}	−6.0	144	33	43 aq^{45}, s alc, eth
d688	3,4-Dimethylpyridine	$(CH_3)_2(C_5H_3N)$	107.16	20, 246	0.939^{25}_4	1.5100^{25}	−12	164	53	sl s aq; s alc, eth
d689	3,5-Dimethylpyridine	$(CH_3)_2(C_5H_3N)$	107.16	20, 246	1.250	1.5033^{25}	−9	170	53	s aq, alc, eth
d690	Dimethyl pyrocarbonate	$O(CO_2CH_3)_2$	134.09	3^4, 17	1.1198^{20}	1.3933^{20}		46^{5mm}	80	
d691	Dimethyl succinate	$CH_3O_2CCH_2CH_2CO_2CH_3$	146.14	2, 609	1.337	1.4190^{20}	19	196.4	85	0.83 aq; 2.9 alc
d692	Dimethylsulfamoyl chloride	$(CH_3)_2NSO_2Cl$	143.59	4, 84	1.3322^{20}_4	1.4518^{20}		114^{75mm}	94	
d693	Dimethyl sulfate	$(CH_3O)_2SO_2$	126.13	1, 283	0.8483^{20}	1.3874^{20}	−31.8	188 dec	83	2.8 aq(hyd); s acet, bz, dioxane, eth
d694	Dimethyl sulfide	$(CH_3)_2S$	62.13	1, 288	1.294	1.4438^{20}	−98.3	37.3	−36	2 aq; s alc, eth
d695	Dimethyl sulfite	$(CH_3O)_2SO$	110.13	1, 282		1.4083^{20}		126–127	30	
d696	Dimethyl sulfone	$(CH_3)_2SO_2$	94.13	1, 289			109	238	143	v s aq, alc, acet
d697	Dimethyl sulfoxide	$(CH_3)_2SO$	78.13	1, 289	1.101^{20}_4	1.4170^{20}	18.5	189.0	95	s alc, acet, bz, chl

(Continued)

TABLE 2.21 Physical Constants of Organic Compounds (*Continued*)

No.	Name	Formula	Formula weight	Beilstein reference	Density, g/mL	Refractive index	Melting point, °C	Boiling point, °C	Flash point, °C	Solubility in 100 parts solvent
d698	Dimethyl-d$_6$ sulfoxide	(CD$_3$)$_2$SO	84.18	1^4, 1279	1.190	1.4758^{20}		55^{5mm}	95	s aq; 200 alc[15]; v s bz
d699	(+)-Dimethyl L-tartrate	CH$_3$O$_2$CH(OH)CH(OH)CO$_2$CH$_3$	178.14	3, 510	1.3284^{20}		48–50	163^{23mm}	>110	
d700	Dimethyltelluride	(CH$_3$)$_2$Te	157.68	1, 291	0.833	1.4041	–10	91–92	26	dec aq; v s alc; i eth
d701	2,5-Dimethyltetrahydrofuran	(CH$_3$)$_2$(C$_4$H$_2$O)	100.16	17, 14				90–92		
d702	1,3-Dimethyl-3,4,5,6-tetrahydro-2(1*H*)-pyrimidinone		128.18	24^3, 32	1.060	1.4880^{20}		146^{44mm}	>110	
d703	Dimethyl 3,3′-dithiopropionate	(CH$_3$O$_2$CCH$_2$CH$_2$)$_2$S	206.26		1.198	1.4740^{20}		148^{18mm}	>110	
d704	N,N-Dimethylthioformamide	(CH$_3$)$_2$NC(S)H	89.16	4, 70	1.047	1.5757^{20}		58^{1mm}	99	v s aq, alc, acet
d705	N,N′-Dimethylthiourea	(CH$_3$NH)$_2$C=S	104.18	4, 70			60–62			
d706	N,N-Dimethyl-p-toluidine	CH$_3$C$_6$H$_4$N(CH$_3$)$_2$	135.21	12, 902	0.937	1.5458^{20}		211	83	
d707	N,N-Dimethyl trimethylsilylamine	(CH$_3$)$_3$SiN(CH$_3$)$_2$	117.27		0.732	1.3970^{20}		84	–19	
d708	1,3-Dimethylurea	(CH$_3$NH)$_2$C=O	88.11	4, 65			101–104	268–270	–1	v s aq, alc; i eth
d709	Dimethylzine	(CH$_3$)$_2$Zn	95.45	Merck: 12, 3312	0.724		–40	46		misc bz, PE; s eth
d710	2,4-Dinitroaniline	(O$_2$N)$_2$C$_6$H$_3$NH$_2$	183.12	12, 747	1.615^{14}		176–178			i aq; 0.75 alc
d711	1,3-Dinitrobenzene	C$_6$H$_4$(NO$_2$)$_2$	168.11	5, 258	1.368		89–90	297		0.05 aq; 2.7 alc; v s bz, chl, EtOAc
d712	2,4-Dinitrobenzenesulfenyl chloride	(O$_2$N)$_2$C$_6$H$_3$SCl	234.62	6^2, 316			96			s bz, HOAc; dec alc
d713	3,5-Dinitrobenzoic acid	(O$_2$N)$_2$C$_6$H$_3$CO$_2$H	212.12	9, 413			205–207			1.9 hot aq; v s alc; sl s bz, eth
d714	3,5-Dinitrobenzoyl chloride	(O$_2$N)$_2$C$_6$H$_3$COCl	230.56	9, 414			69–71	196^{11mm}		dec aq, alc; s eth
d715	2,6-Dinitro-p-cresol	(O$_2$N)$_2$C$_6$H$_2$(OH)CH$_3$	198.13	6, 414			77–79			v s alc, acet, eth, alk
d716	4,6-Dinitro-o-cresol	(O$_2$N)$_2$C$_6$H$_2$(OH)CH$_3$	198.13	6, 368			83–87			
d717	2,4-Dinitrodiphenylamine	(O$_2$N)$_2$C$_6$H$_3$NHC$_6$H$_5$	259.22	12, 751			159–161			
d718	2,4-Dinitro-1-fluorobenzene	FC$_6$H$_3$(NO$_2$)$_2$	186.10	5, 262	1.482	1.5690^{20}	27–30	178^{25mm}	>110	s bz, eth, glyc
d719	1,5-Dinitronaphthalene	C$_{10}$H$_6$(NO$_2$)$_2$	218.17	5, 558			216–217	subl		s bz; v s eth; sl s alc

(Continued)

d720	2,4-Dinitrophenol	$(O_2N)_2C_6H_3OH$	184.11	6, 251	1.683		106–108			s alc, bz; 16 EtOAc; 36 acet; 5 chl; 20 pyr
d721	2,4-Dinitrophenyl-hydrazine	$(O_2N)_2C_6H_3NHNH_2$	198.14	15, 489			ca. 200	300 sl d		sl s aq, alc; s acid
d722	3,5-Dinitrosalicylic acid	$(O_2N)_2C_6H_2(OH)CO_2H$	228.12	10, 122	1.321^{71}	1.442	169–172			s alc; v s alc, eth
d723	2,4-Dinitrotoluene	$CH_3C_6H_3(NO_2)_2$	182.14	5, 339	1.2833^{111}	1.479	67–70			1.2 alc; 9 eth
d724	2,6-Dinitrotoluene	$CH_3C_6H_3(NO_2)_2$	182.14	5, 341	0.9172^{25}		64–66			s alc
d725	Dinonyl hexanedioate	$C_9H_{19}O_2C(CH_2)_4CO_2C_9H_{19}$	398.63						218	
d726	Dioctadecyl phosphite	$(C_{18}H_{37}O)_2P(O)H$	586.97				57–59			
d727	Dioctadecyl 3,3'-thiopropionate	$S[CH_2CH_2CO_2(CH_2)_{17}CH_3]_2$	683.18				65–67			
d728	Dioctylamine	$(C_8H_{17})_2NH$	241.46	4, 196	0.799	1.4432^{20}	14–16	298	>110	i aq; v s alc, eth
d729	Dioctyl ether	$(C_8H_{17})_2O$	242.45	1, 419	0.806	1.4318^{20}	−7.6	287	>110	
d730	Dioctyl sulfide	$(C_8H_{17})_2S$	258.51	1, 419	0.842	1.4610^{20}		180^{10mm}	>110	
d731	4,9-Dioxa-1,12-dodecanediamine	$H_2N(CH_2)_3O(CH_2)_4O(CH_2)_3NH_2$	204.32	12, 247	0.962	1.4609^{20}		136^{4mm}	>110	
d732	1,3-Dioxane		88.11	19, 2	1.032	1.4180^{20}	−45	106	15	misc aq, alc, bz, chl, eth, PE
d733	1,4-Dioxane		88.11	19, 3	1.0329^{20}_4	1.4224^{20}	11.8	101.2	12	misc aq; s alc, eth
d734	1,3-Dioxolane		74.08	19^2, 2	1.060^{20}_4	1.4000^{20}	−95	78	2	
d735	Dipentaerythritol	$(HOCH_2)_3CCH_2OCH_2C(CH_2OH)_3$	254.28				215–218			
d736	Dipentene		136.24	5, 137	0.8402^{21}_4	1.4739^{20}	−95.5	178	45	i aq; misc alc
d737	Dipentylamine	$(C_5H_{11})_2NH$	157.29	4^1, 378	0.777	1.4272		195–202	52	v s alc, eth
d738	Dipentyl ether	$(C_5H_{11})_2O$	158.29	1^1, 193	0.7833^{20}	1.4120^{20}	−69.4	190	57	misc alc, eth; s acet
d739	N,N-Diphenylacetamide	$CH_3CON(C_6H_5)_2$	211.26	12, 247			103	$130^{0.02mm}$		sl s aq; s alc, eth
d740	Diphenylacetic acid	$(C_6H_5)_2CHCO_2H$	212.25	9, 673	1.258^{15}		148	195^{5mm}		s hot aq, alc, chl, eth
d741	Diphenylacetonitrile	$(C_6H_5)_2CHCN$	193.25	9, 674			71–73	181^{12mm}		
d742	Diphenylacetylene	$C_6H_5C{\equiv}CC_6H_5$	178.23	5, 656	0.990		62.5	300		v s eth, hot alc
d743	Diphenylamine	$(C_6H_5)_2NH$	169.23	12, 174	1.160		53	302	152	45 alc; v s bz, eth
d744	cis, trans-1,4-Diphenyl-1,3-butadiene	$C_6H_5CH{=}CHCH{=}CHC_6H_5$	206.29	5, 676	0.9974^{22}_4	1.0653^{22}	149.7	350^{720mm}		s alc; sl s eth
d745	Diphenylcarbamoyl chloride	$(C_6H_5)_2NC(O)Cl$	231.68				82–84			

TABLE 2.21 Physical Constants of Organic Compounds (*Continued*)

No.	Name	Formula	Formula weight	Beilstein reference	Density, g/mL	Refractive index	Melting point, °C	Boiling point, °C	Flash point, °C	Solubility in 100 parts solvent
d746	1,5-Diphenylcarbohydrazide	$(C_6H_5NHNH)_2C{=}O$	242.28	15, 292			168–171			s hot alc, acet, HOAc
d747	Diphenyl carbonate	$(C_6H_5O)_2C{=}O$	214.22	6, 158			80–81	301–302	>110	s hot alc, bz, eth
d748	Diphenyl chlorophosphate	$(C_6H_5O)_2P(O)Cl$	268.64	6, 179	1.296	1.5500^{20}		$316^{272\text{mm}}$		s hot alc
d749	Diphenyl diselenide	$C_6H_5SeSeC_6H_5$	312.13	6, 346	1.557^{80}		61–63			s alc, bz, eth; i aq
d750	Diphenyl disulfide	$C_6H_5SSC_6H_5$	218.34	6, 323	1.353^{20}		58–60	310		0.8 bz; 3 eth; 16 pyr; 11 acet; i aq
d751	Diphenylenimine		167.21	20, 433	1.104^{18}		246	355		s alc; v s chl, eth
d752	1,2-Diphenylethane	$C_6H_5CH_2CH_2C_6H_5$	182.27	5, 598	0.995^{20}	1.5338	52.5	284		s alc, bz, eth, HOAc
d753	Diphenyl ether	$C_6H_5OC_6H_5$	170.21	6, 146	1.0661^{30}	1.5763^{30}	26.9	258	112	s eth; v s chl
d754	N,N′-Diphenylformamidine	$C_6H_5N{=}CHNHC_6H_5$	196.25	12, 236			138–141			
d755	1,3-Diphenylguanidine	$C_6H_5NHC({=}NH)NHC_6H_5$	211.27	12, 369	1.13		148–150	dec 170		s alc, hot bz, chl
d756	5,5-Diphenylhydantoin		252.27	24, 410			294–297			i aq; 1.7 alc; 3.3 acet
d757	1,2-Diphenylhydrazine	$C_6H_5NHNHC_6H_5$	184.24	15, 123	1.158^{16}		123–126			v s alc; sl s bz
d758	Diphenylmercury	$(C_6H_5)_2Hg$	354.81	16, 946	2.318^4		128–129	dec >306		s chl; sl s hot alc
d759	Diphenylmethane	$C_6H_5CH_2C_6H_5$	168.24	5^2, 498	1.006	1.5768^{20}	25	265	>110	v s alc, bz, chl, eth
d760	Diphenylmethanol	$C_6H_5CH(OH)C_6H_5$	184.24	6, 678			66.7	298		0.05 aq; v s alc, chl, eth
d761	1,1-Diphenylmethylamine	$C_6H_5CH(NH_2)C_6H_5$	183.25	12, 1323	1.0635^{22}	1.5956^{99}	34	295	>112	sl s aq
d762	2,5-Diphenyloxazole		221.26	27, 78			72–74	360	176	
d763	Diphenyl phosphite	$(C_6H_5O)_2P(O)H$	234.19	6^1, 94	1.223	1.5575^{20}	12	$219^{26\text{mm}}$		
d764	Diphenylphosphoryl azide	$(C_6H_5O)_2P(O)N_3$	275.20		1.277	1.5518^{20}		$157^{0.17\text{mm}}$	>110	
d765	Diphenyl o-phthalate	$C_6H_4(CO_2C_6H_5)_2$	318.33	9, 801			74–76			
d766	2,2-Diphenyl-1-picryl-hydrazyl		394.32	16^2, 363			127 dec			
d767	1,3-Diphenyl-2-propanone	$C_6H_5CH_2C({=}O)CH_2C_6H_5$	210.28	7, 445	1.2		32–34	330		i aq; v s alc, eth
d768	2,2-Diphenylpropionic acid	$CH_3C(C_6H_5)_2CO_2H$	226.28	9^2, 474			175–177	300		s alc; v s bz, eth
d769	Diphenylsilanediol	$(C_6H_5)_2Si(OH)_2$	216.31	16, 909			140 dec			
d770	Diphenyl sulfide	$(C_6H_5)_2S$	186.28	6, 299	1.118^{15}_{15}	1.6327^{20}	−40	296	53	misc bz, eth, CS_2
d771	Diphenyl sulfone	$(C_6H_5)_2SO_2$	218.27	6, 300			128–129	379	>110	i aq; s hot alc, bz
d772	Diphenyl sulfoxide	$(C_6H_5)_2SO$	202.28	6, 300			69–71	$207^{13\text{mm}}$		
d773	Diphenylthiocarbazone	$C_6H_5N{=}NC(S)NHNHC_6H_5$	256.33	16, 26			168 dec			i aq; v s chl, CCl_4

No.	Name	Formula	Mol. wt.	Beilstein/Merck ref.	Density	n_D	m.p.	b.p.	Flash pt.	Solubility
d774	1,3-Diphenyl-2-thiourea	$C_6H_5NHC(S)NHC_6H_5$	228.32	12, 394	1.32		154	260 dec		i aq; v s alc, eth
d775	1,3-Diphenylurea	$C_6H_5NHC(O)NHC_6H_5$	212.35	12, 352	1.239		238			0.015 aq; s eth, HOAc
d776	Dipiperidinomethane		182.31	4, 138	0.915	1.4820^{20}		123^{15mm}	91	
d777	Dipropylamine	$(C_3H_7)_2NH$	101.19		0.7375^{20}_{4}	1.4043^{20}	−63	109.2	17	
d778	3-Dipropylamino-1,2-propanediol	$(C_3H_7)_2NCH_2CH(OH)\text{-}CH_2OH$	175.27	4^3, 841	0.949	1.4554^{20}		143^{9mm}	>110	4 aq; v s alc, eth, PE
d779	Dipropylene glycol	$HO(CH_2)_3O(CH_2)_3OH$	134.18	1^2, 537	1.023	1.4410^{20}		229	137	
d780	Dipropylene glycol butyl ether	$CH_3(OH)CH_2OCH_2\text{-}CH(OC_4H_9)CH_3$	190.29	1^4, 2474	0.9172^{25}_{25}	1.425^{25}			96	
d781	Dipropylene glycol tert-butyl ether	$(CH_3)_3CO\,(CH_2)_3O(CH_2)_3OH$	190.29		0.900	1.4240^{20}		220–222	87	
d782	Dipropylene glycol dibenzoate	$[C_6H_5CO_2(CH_2)_3]_2O$	342.40	9^2, 108	1.120	1.5280^{20}		232^{5mm}	>110	
d783	Dipropylene glycol isopropyl ether	$CH_3CH(OH)CH_2OCH\bar{-}CH[OCH(CH_3)_2]CH_3$	176.2		0.878^{25}_{25}	1.421^{25}		80.1	90	0.4 aq
d784	Dipropylene glycol methyl ether	$CH_3CH(OH)CH_2OCH\bar{-}CH(OCH_3)CH_3$	148.2		0.951^{20}_{20}	1.419^{20}	−117	188.3	74	i aq; s alc, eth v s PE
d785	Dipropylene glycol acetate	$CH_3CO_2(CH_2)_3O(CH_2)_3\text{-}OCH_3$	190.24		0.970	1.4180^{20}		200	85	
d786	Dipropyl ether	$(C_3H_7)_2O$	102.18	1, 354	0.7466^{20}_{4}	1.3803^{20}	−126.2	89.6	21	0.5 aq; v s alc, bz, chl, eth, PE
d787	Dipropyl hexanedioate	$C_3H_7O_2C(CH_2)_4CO_2C_3H_7$	230.30	2^2, 574	0.9790^{20}_{4}	1.4314^{20}	−20	144^{10mm}		s alc, bz, CS_2
d788	Dipropyl sulfate	$(C_3H_7O)_2SO_2$	182.24	1, 354	1.106^{20}_{4}		dec 140	120^{20mm}	126	
d789	Dipropyl sulfone	$(C_3H_7)_2SO_2$	150.24	1, 359	1.0284^{50}		28–30	270		
d790	2,2'-Dipyridyl		156.19	23, 199			70–73	273		
d791	Disilane	H_3SiSiH_3	62.22	Merck: 12, 3419	0.686^{-25}_{4}		−132	−14.3	ignites in air	
d792	1,3-Dithiane		120.24				53–55		90	
d793	4,4'-Dithiobutyric acid	$HO_2C(CH_2)_3SS(CH_2)_3CO_2H$	238.32	3, 312			110			
d794	3,3'-Dithiodipropionic acid	$HO_2C(CH_2)_2SS(CH_2)_2CO_2H$	210.27				157–159			
d795	Dithiooxamide	$H_2NC(S)C(S)NH_2$	120.20	2, 565			245			sl s aq; s alc; i eth
d796	2,2'-Dithiosalicylic acid	$S_2(C_6H_4CO_2H)_2$	306.36	10, 129			287–290			
d797	1,3-Di-o-tolylguanidine	$(CH_3C_6H_4NH)_2C{=}NH$	239.32	12, 803	1.10^{20}_{4}		176–178			s hot alc, eth
d798	Divinyl ether	$H_2C{=}CHOCH{=}CH_2$	70.09	Merck: 12, 10133	0.7734^{20}_{4}	1.3989^{20}	−101	28.3	<−30	0.53 aq; misc alc, eth

(Continued)

TABLE 2.21 Physical Constants of Organic Compounds (*Continued*)

No.	Name	Formula	Formula weight	Beilstein reference	Density, g/mL	Refractive index	Melting point, °C	Boiling point, °C	Flash point, °C	Solubility in 100 parts solvent
d799	1,3-Divinyltetramethyldisiloxane	$[CH_2{=}CHSi(CH_3)_2]_2O$	186.39	4^4, 4080	0.8114_4^{20}	1.4110^{20}	−9	139	24	
d800	3,9-Divinyl-2,4,8,10-tetraoxaspiro[5.5]-undecane		212.25	19^3, 5679	1.251		43–46	110^{2mm}	110	
d801	Docosane	$CH_3(CH_2)_{20}CH_3$	310.61	1, 174	0.7782^{45}	1.4358^{45}	43–45	369	>110	i aq; sl s alc; v s eth
d802	1-Docosanol	$CH_3(CH_2)_{21}OH$	326.61	1, 431			65–72	180^{22mm}		sl s eth; s alc, chl
d803	Dodecane	$CH_3(CH_2)_{10}CH_3$	170.34	1, 171	0.7490_4^{20}	1.4216^{20}	−10	216.2	74	
d804	1,12-Dodecanediamine	$H_2N(CH_2)_{12}NH_2$	200.37	4, 273			71	245^{10mm}	155	
d805	Dodecanedioic acid	$HO_2C(CH_2)_{10}CO_2H$	230.30	1^3, 2237			128–130			
d806	1,2-Dodecanediol	$CH_3(CH_2)_9CH(OH)CH_2OH$	202.34	1^2, 562	0.845^{20}	1.4587^{20}	58–60	189^{12mm}	87	
d807	1,12-Dodecanediol	$HOCH_2(CH_2)_{10}CH_2OH$	202.34		0.869^{14}	1.4183^{82}	81–84	266–283	>110	
d808	1-Dodecanethiol	$CH_3(CH_2)_{11}SH$	202.40						>110	i aq; s alc, eth
d809	Dodecanoic acid	$CH_3(CH_2)_{10}CO_2H$	200.32	2, 359			43			i aq; 100 alc; v s bz, eth; 40 PrOH
d810	1-Dodecanol	$CH_3(CH_2)_{11}OH$	186.34	1, 428	0.8308_4^{25}	1.4413^{25}	24	259	>110	i aq; s alc, eth
d811	δ-Dodecanolactone		198.31	17^5, 100	0.942	1.4602^{20}	−12	126^{1mm}	>110	
d812	Dodecanoyl peroxide	$[CH_3(CH_2)_{10}CO]_2O_2$	398.63	2^3, 893			55–57		79	
d813	1-Dodecene	$CH_3(CH_2)_9CH{=}CH_2$	168.32	1, 225	0.7584_4^{20}	1.4294^{20}	−35.2	213.4		s alc, eth, PE
d814	2-Dodecen-1-ylsuccinic anhydride		266.38				41–43	180^{5mm}	177	
d815	Dodecyl acetate	$CH_3CO_2(CH_2)_{11}CH_3$	228.38	2, 136	0.865	1.4318^{20}		150^{15mm}	>110	
d816	Dodecyl acrylate	$H_2C{=}CHCO_2(CH_2)_{11}CH_3$	240.39	2^3, 1230	0.884	1.4450^{20}			>110	
d817	Dodecyl aldehyde	$CH_3(CH_2)_{10}CHO$	184.32	1, 714	0.835	1.4344^{20}		185^{100mm}	101	
d818	Dodecylamine	$CH_3(CH_2)_{11}NH_2$	185.36	4, 200	0.808		30–32	247–249	>110	misc alc, bz, chl, eth
d819	Dodecyl methacrylate	$H_2C{=}C(CH_3)CO_2(CH_2)_{11}CH_3$	254.42	2^3, 1290	0.868	1.4460^{20}	−7	142^{4mm}	>110	
d820	Dodecyl sulfate, sodium salt	$CH_3(CH_2)_{11}SO_3Na^+$	288.38	1^3, 1786			204–207		>110	10 aq
d821	Dodecyltrichlorosilane	$CH_3(CH_2)_{11}SiCl_3$	303.8	4^3, 1907	1.020	1.458^{20}		294	>110	
d822	Dodecyl vinyl ether	$CH_3(CH_2)_{11}OCH{=}CH_2$	212.38		0.817	1.4382^{20}		117–120	>110	sl s alc, bz, eth
d823	Dotriacontane	$CH_3(CH_2)_{30}CH_3$	450.88	1, 177	0.8124_4^{70}	1.4364^{70}	68–70	467	>110	
d824	Dulcitol		182.17	1, 544	1.47^{20}		188–191	280^{1mm}		3.3 aq; sl s alc
e1	Eicosane	$CH_3(CH_2)_{18}CH_3$	282.56	1, 174	0.7823 (s)		37	343	>110	
e2	1R,2S-(−)-Ephedrine	$CH_3NHCH(CH_3)CH(OH)C_6H_5$	165.24	13, 636	1.124		39	255	85	s aq; alc, chl, eth
e3	1,2-Epoxybutane	$H_2C{-}CHCH_2CH_3$ (epoxide O)	72.11	17^2, 17	0.8297^{20}	1.3850^{20}	−150	63	−22	6 aq; misc alc, bz, chl, eth

No.	Name	Formula	M	Beilstein ref.	Density	n_D	mp/°C	bp/°C	Flash pt/°C	Solubility
e4	1,2-Epoxy-5,9-Cyclo-dodecadiene		178.28		0.980	1.5045^{20}		83^{1mm}	>110	
e5	1,2-Epoxycyclo-dodecane		182.31		0.939	1.4773^{20}			>110	
e6	1,2-Epoxycyclopentane		84.12	17, 21	0.964	1.4336^{20}		102	10	
e7	1,2-Epoxydecane	H_2C—$CHCH_2(CH_2)_6CH_3$	156.27	17, 18	0.840	1.4290^{20}		94^{15mm}	78	
e8	1,2-Epoxydodecane	H_2C—$CHCH_2(CH_2)_8CH_3$	184.32	17[3], 136	0.844	1.4355^{20}		125^{15mm}	105	
e9	1,2-Epoxyethylbenzene	H_2C—CHC_6H_5	120.15	17, 49	1.0523^{16}_{4}	1.5338^{20}	-37	194	79	i aq; s alc, eth
e10	1,2-Epoxyhexadecane	H_2C—$CHCH_2(CH_2)_{12}CH_3$	240.43	17, 20	0.846	1.4452^{20}	21–22	180^{12mm}	93	
e11	1,2-Epoxyhexane	H_2C—$CHCH_2CH_2CH_2CH_3$	100.16	17[4], 86	0.831	1.4056^{20}		118–120	15	
e12	1,2-Epoxy-5-hexene	H_2C—$CHCH_2CH_2CH$=CH_2	98.15	17[3], 163	0.870	1.4252^{20}		121	15	
e13	1,2-Epoxyoctadecane	H_2C—$CHCH_2(CH_2)_{14}CH_3$	268.49	17[3], 140			33–35	$137^{0.5mm}$	>110	
e14	1,2-Epoxy-3-phenoxy-propane	H_2C—$CHCH_2OC_6H_5$	150.18	17, 105	1.109	1.530^{20}	3.5	245	>110	41 aq; misc alc, eth
e15	1,2-Epoxypropane	H_2C—$CHCH_3$	58.08	17, 6	0.8592	1.3660^{20}	-112	35	-37	misc aq
e16	2,3-Epoxy-1-propanol	H_2C—$CHCH_2OH$	74.08	17, 104	1.1143^{25}_{4}	1.4315^{20}		$66^{2.5mm}$	81	misc aq
e17	2,3-Epoxypropyl-methacrylate	H_2C—$CHCH_2O_2C(CH_3)$—CH=CH_2	142.16		1.042	1.4494^{20}		189	76	
e18	1,2-Epoxy-3,3,3-tri-chloropropane	H_2C—$CHCCl_3$	161.42	17[2], 14	1.495	1.4778^{20}		151^{745mm}	66	
e19	*meso*-Erythritol	$HOCH_2[CH(OH)]_2CH_2OH$	122.12	1, 525			120–123	329–331		4.7 mL aq; 46 mL alc[4]
e20	Ethane	CH_3CH_3	30.07	1, 80	1.356^{0} g/L		-182.8	-88	-135	misc aq, alc; i bz
e21	1,2-Ethanediamine	$H_2NCH_2CH_2NH_2$	60.10	4, 230	0.8977^{20}_{4}	1.4568^{20}	11	117.3	33	

(Continued)

TABLE 2.21 Physical Constants of Organic Compounds (*Continued*)

No.	Name	Formula	Formula weight	Beilstein reference	Density, g/mL	Refractive index	Melting point, °C	Boiling point, °C	Flash point, °C	Solubility in 100 parts solvent
e21a	1,2-Ethanediol	$HOCH_2CH_2OH$	62.07	1, 465	1.1135^{20}_4	1.4318^{20}	−12.6	197.3	110	misc aq, alc, glyc, pyr
e22	1,2-Ethanediol diacetate	$CH_3CO_2CH_2CH_2O_2CCH_3$	146.14	2, 142	1.1043^{20}	1.4150^{20}	−31	190.2	82	misc alc, eth
e23	1,2-Ethanediol dimethacrylate	$[H_2C{=}C(CH_3)CO_2CH_2]_2$	198.22	2^3, 1292	1.051	1.4549^{20}		100^{5mm}	>110	
e24	1,2-Ethanedithiol	$HSCH_2CH_2SH$	94.20	1, 471	1.123^{24}	1.5580^{20}		146	50	v s alc, alk
e25	Ethanesulfonic acid	$C_2H_5SO_3H$	110.13	4, 5	1.350	1.4340^{20}	−17	$123^{0.01mm}$	>110	dec aq, alc; v s eth
e26	Ethanesulfonyl chloride	$CH_3CH_2SO_2Cl$	128.57	4, 6	1.357^{22}	1.4330^{20}		177	83	
e26a	Ethanethiol	CH_3CH_2SH	62.13	1, 340	0.8315^{25}	1.420^{25}	−147.9	35.0	−17	0.7 aq; s alc, eth
e27	Ethanol	CH_3CH_2OH	46.07	1, 292	0.7894^{20}_4	1.3611^{20}	−114	78.3	13	misc aq, alc, chl, eth
e28	Ethanol-*d*	CH_3CH_2OD	47.08	1^3, 1287	0.801	1.3595^{20}		78.8	12	misc aq, alc, eth
e29	Ethanolamine	$H_2NCH_2CH_2OH$	61.08	Merck: 12, 3712	1.0180^{20}	1.4539^{20}	10.5	170.8	86	misc aq, alc, acet
e30	Ethoxyacetic acid	$CH_3CH_2OCH_2CO_2H$	104.11	3, 233	1.1021^{20}_4	1.4190^{20}		97^{11mm}	97	s aq, alc, eth
e31	3-Ethoxyacrylonitrile	$CH_3OCH{=}CHCN$	97.12	3^3, 681	0.944	1.4545^{20}		91^{19mm}	81	i aq; s alc
e32	4-Ethoxyaniline	$CH_3CH_2OC_6H_5NH_2$	137.18	13, 436	1.0652^{16}	1.5609^{20}	4	250	115	misc alc, eth
e33	2-Ethoxybenzaldehyde	$CH_3CH_2OC_6H_4CHO$	150.18	8, 43	1.074	1.5422	20	136^{24mm}	107	v s alc, bz, eth
e34	4-Ethoxybenzaldehyde	$CH_3CH_2OC_6H_4CHO$	150.18	8, 73	1.080^{25}_4	1.5584^{20}	13–14	255	>110	sl s aq; s alc, eth
e35	2-Ethoxybenzamide	$CH_3CH_2OC_6H_4CONH_2$	165.19	10, 93			132–134			v s alc, eth
e36	Ethoxybenzene	$CH_3CH_2OC_6H_5$	122.17	6, 140	0.967^{20}_4	1.5074^{20}	−29.5	169.8	63	sl s aq; s alc, eth
e37	2-Ethoxybenzoic acid	$CH_3CH_2OC_6H_4CO_2H$	166.18	10, 64	1.105	1.5400^{20}	19.4	174^{15mm}	>110	sl s aq
e38	4-Ethoxybenzoic acid	$CH_3CH_2OC_6H_4CO_2H$	166.18	10, 156			197–199			sl s hot aq
e39	Ethoxycarbonyl isothiocyanate	$CH_3CH_2OC({=}O)NCS$	131.15	3^3, 279	1.112	1.5000^{20}		56^{18mm}	50	
e40	2-Ethoxyethanol	$CH_3CH_2OCH_2CH_2OH$	90.12	1, 467	0.9295^{20}	1.4075^{20}	−70	134.8	43	misc aq, alc, acet, eth
e41	2-(2-Ethoxyethoxy)ethanol	$C_2H_5OCH_2CH_2OCH_2CH_2OH$	134.18	1, 520	0.9841^{25}_4	1.4254^{25}	−76	196	96	misc aq, alc, bz, chl, acet, pyr
e41a	2-(2-Ethoxyethoxy)ethanol acetate	$CH_3CO_2CH_2CH_2OCH_2CH_2{-}OCH_2CH_3$	176.21		1.0096^{20}	1.4213^{20}	−25	218.5	110	
e42	2-Ethoxyethyl acetate	$CH_3CO_2CH_2CH_2OCH_2CH_3$	132.16	2^2, 155	0.9749^{20}_4	1.4023^{20}	−61.7	156.3	57	29 aq; misc alc, eth
e43	2-Ethoxyethyl acrylate	$H_2C{=}CHCO_2CH_2CH_2OC_2H_5$	144.17	2^3, 1232	0.982	1.4270^{20}		78^{23mm}	65	
e44	2-Ethoxyethylamine	$CH_3CH_2OCH_2CH_2NH_2$	89.14	4^2, 718	0.8512^{20}	1.4101^{20}		107	21	misc aq, alc, eth
e45	2-Ethoxyethyl methacrylate	$H_2C{=}C(CH_3)CO_2CH_2CH_2OC_2H_5$	158.20	2^3, 1291	0.964	1.4285^{20}		93^{35mm}	71	
e46	3-Ethoxy-4-hydroxybenzaldehyde	$C_2H_5OC_6H_3(OH)CHO$	166.18	8, 256			76–78			s eth, glycols; 50 alc

No.	Name	Formula	Formula Weight	Beilstein Reference	Density	n_D	mp (°C)	bp (°C)	Flash Point (°C)	Solubility
e47	3-Ethoxy-4-methoxybenzaldehyde	$C_2H_5OC_6H_3(OCH_3)CHO$	180.2	8, 256			51–53		>110	s alc, bz, chl, eth
e48	1-Ethoxy-2-methoxybenzene	$C_2H_5OC_6H_4OCH_3$	152.19	6, 771	1.044	1.5240^{20}		217–218	90	
e49	Ethoxymethylenemalononitrile	$CH_3CH_2OCH=C(CN)_2$	122.13	3^1, 162			64–66	160^{12mm}		i aq; v s alc, eth
e50	1-Ethoxynaphthalene	$C_{10}H_7OCH_2CH_3$	172.23	6, 606	1.060^{20}_{4}	1.6040^{20}	5.5	280	>110	
e51	2-Ethoxyphenol	$C_2H_5OC_6H_4OH$	138.17	6, 771	1.090	1.5288^{20}	29	217	91	
e52	trans-2-Ethoxy-5-(1-Propenyl)Phenol	$C_2H_5OC_6H_3(CH=CHCH_3)OH$	178.23	6^2, 918			86–88			
e53	3-Ethoxypropionitrile	$C_2H_5OCH_2CH_2CN$	99.14	3, 298	0.911	1.4065^{20}		171–172	63	
e54	3-Ethoxypropylamine	$C_2H_5OCH_2CH_2CH_2NH_2$	103.17	4^3, 739	0.861	1.4178^{20}		136–138	32	
e55	3-Ethoxysalicylaldehyde	$C_2H_5OC_6H_3(OH)CHO$	166.18	8^2, 267			66–68	264		
e56	Ethoxytrimethylsilane	$(CH_3)_3SiOC_2H_5$	118.3	4^3, 1856	0.7573^{20}_{4}	1.3742^{20}	−84	75–76	−18	9.7 aq; misc alc, acet, chl, eth
e57	Ethyl Acetate	$CH_3CO_2C_2H_5$	88.11	2, 125	0.9006^{20}_{4}	1.3724^{20}	−84	77	−4	2.9 aq; misc alc, chl
e58	Ethyl Acetoacetate	$CH_3COCH_2CO_2C_2H_5$	130.15	3, 632	1.0213^{25}_{4}	1.4174^{20}	−45	180.8	57	1.5 aq; s alc, eth
e59	p-Ethylacetophenone	$C_2H_5C_6H_4COCH_3$	148.21	7^4, 1101	0.993	1.5293^{20}	−20.6	114^{11mm}	90	
e60	Ethyl acrylate	$H_2C=CHCO_2C_2H_5$	100.12	2, 399	0.9234^{20}	1.4060^{20}	−71	99	10	
e61	Ethylaluminum dichloride	$C_2H_5AlCl_2$	126.95	4^3, 1973	1.207^{50}		32	113^{50mm}	−18	
e62	Ethylaluminum sesquichloride	$C_2H_5AlCl_2 \cdot ClAl(C_2H_5)_2$	247.51		1.092		−50	204	−18	
e63	Ethylamine	$C_2H_5NH_2$	45.09	4, 87	0.689^{15}_{15}	1.3663^{20}	−81	16.6	<−18	misc aq, alc, eth
e64	Ethyl 2-aminobenzoate	$H_2NC_6H_4CO_2C_2H_5$	165.19	14, 319	1.088^{15}	1.5640^{20}	13–15	266–268	>110	i aq; s alc, eth
e65	Ethyl 4-aminobenzoate	$H_2NC_6H_4CO_2C_2H_5$	165.19	14, 422			88–90	310		0.04 aq; 20 alc; 50 chl, 25 eth; s dil acid
e66	Ethyl 3-aminocrotonate	$CH_3C(NH_2)=CHCO_2C_2H_5$	129.16	3, 654	1.021^{20}_{4}		33–35	210–215	97	i aq; s alc, bz, eth
e67	2-(Ethylamino)ethanol	$CH_3CH_2NHCH_2CH_2OH$	89.14	4, 282	0.914^{20}_{4}	1.4402^{20}	−90	170	71	v s aq, alc, eth
e68	N-Ethylaniline	$C_6H_5NHC_2H_5$	121.18	12, 159	0.958^{25}	1.5559^{20}	−63.5	203	85	i aq; misc alc, eth
e69	2-Ethylaniline	$CH_3CH_2C_6H_4NH_2$	121.18	12^2, 584	0.983	1.5590^{20}	−44	210	91	sl s aq; v s alc, eth
e70	4-Ethylaniline	$CH_3CH_2C_6H_4NH_2$	121.18	12, 1090	0.975	1.5542^{20}	−5	216	85	sl s aq; v s alc, eth
e71	2-Ethylanthraquinone		236.27	7^1, 425			108–111			
e72	4-Ethylbenzaldehyde	$C_2H_5C_6H_4CHO$	134.18	7, 307	0.979	1.5390^{20}		221	92	
e73	Ethylbenzene-d_{10}	$C_6D_5CD_2CD_3$	116.25		0.949	1.4920^{20}		134.6	31	

(Continued)

TABLE 2.21 Physical Constants of Organic Compounds (*Continued*)

No.	Name	Formula	Formula weight	Beilstein reference	Density, g/mL	Refractive index	Melting point, °C	Boiling point, °C	Flash point, °C	Solubility in 100 parts solvent
e74	Ethylbenzene	$C_6H_5CH_2CH_3$	106.17	5[2], 274	0.8670_4^{20}	1.4959^{20}	−95.0	136.2	22	0.01 aq; misc alc, bz, chl, eth
e75	4-Ethylbenzene-sulfonic acid	$C_2H_5C_6H_4SO_3H$	186.23	11, 120	1.229	1.5331			>110	
e76	Ethyl benzoate	$C_6H_5CO_2C_2H_5$	150.18	9, 110	1.051^{15}	1.5000^{20}	−34.7	212.4	84	0.05 aq; misc alc, chl, bz, eth, PE
e77	Ethyl benzoylacetate	$C_6H_5C(=O)CH_2CO_2C_2H_5$	192.21	10, 674	1.110	1.5338^{20}		265–270	63	i aq; misc alc, eth
e78	Ethyl 3-benzoylacrylate	$C_6H_5C(=O)CH=CHCO_2C_2H_5$	204.23	10[2], 501	1.112	1.5435^{20}		185^{25mm}	>110	
e79	Ethyl 2-benzylaceto-acetate	$CH_3C(=O)CH(CH_2C_6H_5)CO_2C_2H_5$	220.27	10, 710	1.036	1.4996^{20}		276	>110	
e80	N-Ethylbenzylamine	$C_6H_5CH_2NHC_2H_5$	135.21	12, 1020	0.909	1.5117^{20}		194	66	i aq; misc alc, eth
e81	Ethyl (2-benzyl)-benzoylacetate	$C_6H_5C(=O)CH(CH_2C_6H_5)CO_2C_2H_5$	282.34	10, 764	1.110	1.5567^{20}		270^{80mm}	>110	
e82	Ethyl N-benzyl-N-cyclopropylcarbamate	$C_6H_5CH_2N(C_3H_5)CO_2C_2H_5$	219.28		0.997	1.5104^{20}			>110	
e83	Ethyl bromoacetate	$BrCH_2CO_2CH_2CH_3$	167.01	2, 214	1.5062_{20}^{20}	1.4510^{20}	<−20	159	47	i aq; misc alc, eth
e84	Ethyl 4-bromobenzoate	$BrC_6H_4CO_2C_2H_5$	229.08	9, 352	1.403	1.5440^{20}		131^{14mm}	>110	
e85	Ethyl 2-bromobutyrate	$CH_3CH_2CH(Br)CO_2C_2H_5$	195.06	2[2], 255	1.329_{20}^{20}	1.4470^{20}		177 dec	58	i aq; misc alc, eth
e86	Ethyl 4-bromobutyrate	$BrCH_2CH_2CH_2CO_2C_2H_5$	195.06	2, 283	1.363	1.4559^{20}		82^{10mm}	90	
e87	Ethyl 2-bromoheptanoate	$CH_3(CH_2)_3CH(Br)CO_2C_2H_5$	237.14	2, 341	1.211	1.4524^{20}		109^{10mm}	104	
e88	Ethyl 6-bromo-hexanoate	$Br(CH_2)_5CO_2C_2H_5$	223.12	2[3], 737	1.254	1.4590^{20}		130^{16mm}	>110	
e89	Ethyl 2-bromoiso-butyrate	$(CH_3)_2C(Br)CO_2C_2H_5$	195.06	2, 296	1.329_4^{20}	1.4446^{20}		67^{11mm}	60	i aq; misc alc, eth
e90	Ethyl 2-bromo-octanoate	$CH_3(CH_2)_5CH(Br)CO_2C_2H_5$	251.17	2, 349	1.167	1.4520^{20}			106	
e91	Ethyl 3-bromo-2-oxo-propionate	$BrCH_2C(=O)CO_2C_2H_5$	195.02	3[2], 409	1.554	1.4695^{20}		100^{10mm}	98	
e92	Ethyl 2-bromo-pentanoate	$CH_3(CH_2)_2CH(Br)CO_2C_2H_5$	209.09	2, 302	1.116	1.4486^{20}		190–192	77	i aq; misc alc, eth
e93	Ethyl 2-bromo-propionate	$CH_3CH(Br)CO_2C_2H_5$	181.03	2, 255	1.394	1.4460^{20}		156–160	51	i aq; misc alc, eth
e94	Ethyl 3-bromo-propionate	$BrCH_2CH_2CO_2C_2H_5$	181.03	2, 256	1.4123_4^{18}	1.4569^{18}		136^{50mm}	79	i aq; misc alc, eth

No.	Name	Formula	M	Beilstein ref.	Density	n_D	mp, °C	bp, °C	Flash pt, °C	Solubility
e95	2-Ethyl-1-butanol	$(C_2H_5)_2CHCH_2OH$	102.18	1, 412	0.8330^{20}	1.4224^{20}	<−15	146	58	0.63 aq
e95a	2-Ethyl-1-butene	$(C_2H_5)_2C{=}CH_2$	84.16	1^2, 95	0.689	1.3960^{20}	−131	65	−26	
e96	2-Ethylbutyl acetate	$CH_3CO_2CH_2CH(C_2H_5)_2$	144.21	2^3, 257	0.876	1.4100^{20}		160^{740mm}	52	
e97	N-Ethylbutylamine	$CH_3(CH_2)_3NHC_2H_5$	101.19	4, 157	0.740_4	1.4050^{20}		108	18	
e98	2-Ethylbutyraldehyde	$(C_2H_5)_2CHCHO$	100.16	1, 693	0.8162^{20}	1.4018^{20}	−89	116.7	21	0.31 aq
e99	Ethyl butyrate	$CH_3CH_2CH_2CO_2C_2H_5$	116.16	2, 270	0.8794_4^{20}	1.3998^{20}	−98	121	24	0.49 aq; misc alc, eth
e100	2-Ethylbutyric acid	$(C_2H_5)_2CHCO_2H$	116.16	2, 333	0.9225_4^{20}	1.4133^{20}	−14	194	87	
e101	Ethyl butyrylacetate	$CH_3(CH_2)_2C({=}O)CH_2CO_2C_2H_5$	158.20	3, 684	1.001	1.4270^{20}		104^{22mm}	78	
e102	Ethyl carbamate	$H_2NCO_2C_2H_5$	89.09	3, 22	1.056		49–50	182–184	92	200 aq; 125 alc; 111 chl; 67 eth
e103	Ethyl carbazate	$H_2NNHCO_2C_2H_5$	104.11	3, 98			44–47	110^{22mm}	86	
e104	N-Ethylcarbazole		195.27	20, 436			68–70			
e105	Ethyl chloroacetate	$ClCH_2CO_2C_2H_5$	122.55	2, 197	1.1498_4^{20}	1.4227^{20}	−21	144	65	i aq; misc alc, eth
e106	Ethyl 2-chloro-acetoacetate	$CH_3C({=}O)CH(Cl)CO_2C_2H_5$	164.59	3, 662	1.190	1.4430^{20}		107^{14mm}	50	i aq; s alc, eth
e107	Ethyl 4-chloro-acetoacetate	$ClCH_2C({=}O)CH_2CO_2C_2H_5$	164.59	3, 663	1.218_4^{17}	1.4520^{20}		115^{14mm}	96	i aq; misc alc, eth
e108	Ethyl 4-chlorobutyrate	$ClCH_2CH_2CH_2CO_2C_2H_5$	150.61	2, 278	1.0754_4^{20}	1.4306^{20}		186	51	s alc, acet, eth
e109	Ethyl chloroformate	$ClCO_2C_2H_5$	108.52	3, 10	1.1403_4^{20}	1.3941^{20}	−81	93	13	misc alc, bz, chl, eth
e110	Ethyl 2-chloro-propionate	$CH_3CH(Cl)CO_2C_2H_5$	136.58	2, 248	1.087_4^{20}	1.4185^{20}		146–149	38	
e111	Ethyl 3-chloro-propionate	$ClCH_2CH_2CO_2C_2H_5$	136.58	2, 250	1.1086_4^{20}	1.4249^{20}		162–163	54	misc alc, eth
e112	Ethyl chrysanthemumate		196.29	9^2, 45	0.906	1.4600^{20}		112^{10mm}	84	
e113	Ethyl *trans*-cinnamate	$C_6H_5CH{=}CHCO_2C_2H_5$	176.22	9^2, 385	1.0495_4^{20}	1.5598^{20}	10	271	>110	misc alc, eth; i aq
e114	Ethyl crotonate	$CH_3CH{=}CHCO_2C_2H_5$	114.14	2, 411	0.9175_4^{20}	1.4240^{20}		138	28	i aq; s alc, eth
e115	Ethyl cyanoacetate	$NCCH_2CO_2C_2H_5$	113.12	2, 585	1.0564_4^{25}	1.4176^{20}	−22	206	110	i aq; misc alc, eth
e116	Ethyl 2-cyano-3,3-diphenylacrylate	$(C_6H_5)_2C{=}C(CN)CO_2C_2H_5$	277.33	9^3, 4601			97–99	$174^{0.2mm}$		
e117	Ethylcyclohexane	$C_6H_{11}CH_2CH_3$	112.22	5, 35	0.7879_4^{20}	1.4330^{20}	−111	131.8	35	
e118	4-Ethylcyclohexanol	$CH_3CH_2C_6H_{10}OH$	128.22	6^2, 26	0.889	1.4625^{20}		84^{10mm}	77	
e118a	Ethylcyclopentane	$C_2H_5(C_5H_9)$	98.19	5^2, 19	0.763	1.4190^{20}	−138	103	15	
e119	Ethyl cyclopropane-carboxylate	$C_3H_5CO_2CH_2CH_3$	114.14	9, 4	0.960	1.4197^{20}		129–133	18	
e120	Ethyl decanoate	$CH_3(CH_2)_8CO_2C_2H_5$	200.32	2, 356	0.862^{20}	1.4248^{20}		245	102	misc alc, chl, eth
e121	Ethyl diazoacetate	$N_2CH_2CO_2C_2H_5$	114.10	3^1, 211	1.0852_4^{18}	1.4588^{18}	−22	141^{710mm}	26	misc alc, bz, eth

(Continued)

TABLE 2.21 Physical Constants of Organic Compounds (*Continued*)

No.	Name	Formula	Formula weight	Beilstein reference	Density, g/mL	Refractive index	Melting point, °C	Boiling point, °C	Flash point, °C	Solubility in 100 parts solvent
e122	Ethyl 2,3-dibromo-propionate	$BrCH_2CH(Br)CO_2C_2H_5$	259.94	2, 259	1.788^{16}_4	1.4986^{20}		214	91	s alc, eth
e123	Ethyl dichloro-phosphate	$CH_3CH_2OP(O)Cl_2$	162.94	1, 332	1.373	1.4338^{20}		65^{10mm}	>110	
e124	Ethyl dichlorothio-phosphate	$CH_3CH_2OP(S)Cl_2$	179.01	1, 353	1.353	1.5040^{20}		68^{10mm}	>110	
e125	N-Ethyldiethanolamine	$CH_3CH_2N(CH_2CH_2OH)_2$	133.19	4, 284	1.014	1.4665^{20}	−50	246–252	123	
e126	Ethyl 3,3-dimethyl-acrylate	$(CH_3)_2C{=}CHCO_2C_2H_5$	128.17	2, 433	0.9247^{20}_4	1.4350^{20}		155	33	
e127	Ethyl 4-dimethyl-aminobenzoate	$(CH_3)_2NC_6H_4CO_2C_2H_5$	193.25	14[1], 571			64–66			s alc, eth
e128	Ethyl 2,2-dimethyl-propionate	$(CH3)_3CCO_2C_2H_5$	130.19	2, 320	0.8584^{18}_4	1.3922^{18}		118.2	16	
e129	Ethyl 3,5-dinitro-benzoate	$(O_2N)_2C_6H_3CO_2C_2H_5$	240.17	9, 414			94–95			
e130	5-Ethyl-1,3-dioxane-5-methanol		146.19	19[5], 382	1.090	1.4630^{20}		105^{5mm}	>110	
e131	Ethylene	$H_2C{=}CH_2$	28.05	1, 180	1.147 g/L		−169.4	104		11 mL aq[25]; 200 alc[25]; v s eth; s acet, bz
e132	Ethylene carbonate		88.06	19, 100	1.3214^{39}	1.4199^{40}	36.4	248	143	misc aq
e133	Ethylenediamine	$H_2NCH_2CH_2NH_2$	60.10	4, 230	0.879^{20}	1.4566^{20}	11	117	40	
e134	Ethylenediamine-*N*,*N*,*N′*,*N′*-tetra-acetic acid	$(HO_2CCH_2)_2NCH_2CH_2{-}N(CH_2CO_2H)_2$	292.24	4[3], 1187			250 dec			0.05 aq
e135	Ethylene glycol	$HOCH_2CH_2OH$	62.07	1, 465	1.113	1.4310^{20}		196–198	>110	
e136	Ethylene glycol bis-(mercaptoacetate)	$(HSCH_2CO_2CH_2{-})_2$	210.27		1.313	1.5211^{20}		139^{2mm}	>110	
e137	Ethylene glycol diacetate	$CH_3CO_2CH_2CH_2O_2CCH_3$	146.14	2, 142	1.1043^{20}	1.4159^{20}	−31	190	88	
e138	Ethylene glycol diethyl ether	$C_2H_5OCH_2CH_2OC_2H_5$	118.18	1, 468	0.8484^{20}	1.3860^{20}	−74	119	35	
e139	Ethylene glycol diglycidyl ether	$(H_2C{-}CHCH_2OCH_2{-})_2$ O	174.20	1, 468	0.842	1.3923^{20}	−74	121	20	
e140	Ethylene glycol dimethacrylate	$[H_2C{=}C(CH_3)CO_2CH_2{-}]_2$	198.22	2[3], 1292	1.051	1.4549^{20}		100^{5mm}	>110	

(Continued)

No.	Name	Formula	Mol. wt.	Beilstein ref.	Density	n_D	m.p., °C	b.p., °C	Flash p., °C	Solubility
e141	Ethylene glycol dimethyl ether	$CH_3OCH_2CH_2OCH_3$	90.12	1, 467	0.8691^{20}	1.3796^{20}	−58	85	−2	misc aq; s alc
e142	Ethylene glycol divinyl ether	$H_2C{=}CHOCH_2CH_2OCH{=}CH_2$	114.14	1^3, 2807	0.914	1.4350^{20}		125–127	27	
e143	Ethylene glycol methyl ether acrylate	$H_2C{=}CHCO_2CH_2CH_2OCH_3$	130.14	2^3, 1232	1.012	1.4270^{20}		56^{12mm}	60	
e144	Ethylene glycol methyl ether methacrylate	$H_2C{=}C(CH_3)CO_2CH_2CH_2OCH_3$	144.17	2^3, 1291	0.993	1.4310^{20}		65^{12mm}	60	
e145	Ethylene glycol phenyl ether acrylate	$H_2C{=}CHCO_2CH_2CH_2OC_6H_5$	192.21	6^3, 572	1.104	1.5180^{20}		$84^{0.2mm}$	>110	
e146	Ethyleneimine	$H_2C{-}CH_2$ ring closed by NH	43.07		0.8321^{25}_{4}	1.4123^{25}	−78	56	−11	
e147	Ethylene oxide	$H_2C{-}CH_2$ ring closed by O	44.05	17, 4	0.891^{0}_{4}	1.3597^{7}	−111	10.6	−18	misc aq; s alc, eth
e148	Ethylene sulfide	$H_2C{-}CH_2$ ring closed by S	60.12	17^2, 12	1.010	1.4935^{20}		55–56	10	sl s alc, eth
e149	Ethyl 2-ethoxy-2-hydroxyacetate	$HOCH(OC_2H_5)CO_2C_2H_5$	148.16	3, 601	1.079	1.4200^{20}		137	49	
e150	Ethyl (ethoxymethylene)cyanoacetate	$C_2H_5OCH{=}C(CN)CO_2C_2H_5$	169.18	3, 470			51–53	190^{30mm}	>110	
e151	Ethyl 3-ethoxypropionate	$C_2H_5OCH_2CH_2CO_2C_2H_5$	146.19	3, 298	0.949	1.4050^{20}		166	52	
e152	Ethyl 4-{[(ethylphenylamino)methylene]amino}benzoate	$C_6H_5N(C_2H_5)CH{=}N{-}C_6H_4{-}CO_2C_2H_5$	296.37				62–65	215^{2mm}		
e153	Ethyl fluoroacetate	$FCH_2CO_2C_2H_5$	106.10	2, 193	1.0926^{21}	1.3755^{20}		119	30	s aq
e154	Ethyl formate	$HCO_2C_2H_5$	74.08	2, 19	0.917^{20}_{20}	1.3590^{20}	−80	54	−20	10 aq; misc alc, eth
e155	Ethyl 2-furoate		140.14	18, 275	1.117^{20}		35–37	196	70	i aq; s alc, eth
e156	Ethyl heptanoate	$CH_3(CH_2)_5CO_2C_2H_5$	158.24	2^2, 295	0.8685^{20}_{4}	1.4144^{15}	−66	189	66	s alc, eth
e157	Ethyl hexadecanoate	$CH_3(CH_2)_{14}CO_2C_2H_5$	284.48	2^2, 336	0.8577^{25}_{4}	1.4347^{34}	22	191^{10mm}		s alc, eth
e158	2-Ethylhexanaldehyde	$CH_3(CH_2)_3CH(C_2H_5)CHO$	128.22	1, 707	0.822	1.4155		$55^{13.5mm}$	42	s alc, eth
e158a	3-Ethylhexane	$(C_2H_5)CHCH_2CH_2CH_3$	114.23	1^4, 431	0.7136^{20}_{20}	1.4018^{20}		118.6		
e159	2-Ethyl–1,3-hexanediol	$CH_3(CH_2)_2CH(OH){-}CH(C_2H_5)CH_2OH$	146.23	Merck: 12, 3790	0.9325^{22}_{4}	1.4530^{22}	−40	244	127	0.6% (w/w) aq; s alc, propylene glycol

TABLE 2.21 Physical Constants of Organic Compounds (*Continued*)

No.	Name	Formula	Formula weight	Beilstein reference	Density, g/mL	Refractive index	Melting point, °C	Boiling point, °C	Flash point, °C	Solubility in 100 parts solvent
e160	Ethyl hexanoate	$CH_3(CH_2)_4CO_2C_2H_5$	144.21	2, 323	0.8714^{20}	1.4075^{20}	−67	166–168	49	i aq; misc alc, eth
e161	2-Ethylhexanoic acid	$CH_3(CH_2)_3CH(C_2H_5)CO_2H$	144.21	2, 349	0.9077	1.4241^{20}	−118.4	228	127	0.25 aq
e162	2-Ethyl-1-hexanol	$CH_3(CH_2)_3CH(C_2H_5)CH_2OH$	130.23	Merck: 12, 3854	0.8319^{25}	1.4300^{20}	−70	184.6	73	0.07 aq; s alc, bz, chl
e163	2-Ethylhexanoyl chloride	$CH_3(CH_2)_3CH(C_2H_5)COCl$	162.66	2^2, 304	0.939	1.4335^{20}		68^{11mm}	69	0.03 aq; misc alc, oils, org liquids
e164	2-Ethylhexyl acetate	$CH_3(CH_2)_3CH(C_2H_5)-CH_2O_2CCH_3$	172.27	Merck: 12, 6860	0.8718	1.4204^{20}	−80	199	71	
e165	2-Ethylhexyl acrylate	$H_2C=CCO_2CH(C_2H_5)(CH_2)_3CH_3$	184.28	2^3, 1229	0.885	1.4358		214–219	79	i aq; s alc, acet, eth
e166	2-Ethylhexylamine	$CH_3(CH_2)_3CH(C_2H_5)CH_2NH_2$	129.31	4^3, 388	0.789	1.4300^{20}	−76	169	60	
e167	2-Ethylhexyl chloroformate	$CH_3(CH_2)_3CH(C_2H_5)CH_2O_2CCl$	192.69	3^4, 28	0.981	1.4312^{20}		107^{30mm}	81	
e168	2-Ethylhexyl cyanoacetate	$NCCH_2CO_2CH_2CH(C_2H_5)-(CH_2)_3CH_3$	197.28		0.975	1.4380^{20}		150^{11mm}	>110	
e169	2-Ethylhexyl 2-cyano-3,3-diphenylacrylate	$(C_6H_5)_2C=C(CN)CO_2CH_2-CH(C_2H_5)(CH_2)_3CH_3$	361.49		1.051	1.5670^{20}	−10	$218^{1.5mm}$	>110	
e170	2-Ethylhexyl 4-(di-methylamino)-benzoate	$(CH_3)_2NC_6H_4CO_2CH_2-CH(C_2H_5)(CH_2)_3CH_3$	277.41		0.995	1.5420^{20}		325	>110	
e171	2-Ethylhexyl glycidyl ether	$CH_3(CH_2)_3CH(C_2H_5)CH_2-OCH_2CH-CH_2$ (O)	186.30		0.891	1.4340^{20}		$61^{0.3mm}$	96	
e172	2-Ethylhexyl methacrylate	$H_2C=C(CH_3)CO_2CH_2-CH(C_2H_5)(CH_2)_3CH_3$	198.31	2^3, 1289	0.885	1.4381^{20}		120^{18mm}	92	
e173	2-Ethylhexyl nitrate	$CH_3(CH_2)_3CH(C_2H_5)CH_2ONO_2$	175.23		0.963	1.4320^{20}			75	explodes when heated
e174	2-Ethylhexyl salicylate	$2-(HOC_6H_4CO_2CH_2-CH(C_2H_5)(CH_2)_3CH_3$	250.34	10^3, 124	1.014	1.5020^{20}		190^{21mm}	>110	
e175	2-Ethylhexyl vinyl ether	$CH_3(CH_2)_3CH(C_2H_5)-CH_2OCH=CH_2$	156.26		0.8102	1.4273^{20}	−85	177–178	52	0.01 aq
e176	Ethyl hydrocinnamate	$C_6H_5CH_2CH_2CO_2C_2H_5$	178.23	9, 511	1.010	1.4940^{20}		247–248	107	
e177	Ethyl hydrogen hexanedioate	$HO_2C(CH_2)_4CO_2C_2H_5$	174.20	2^1, 277		1.4387^{20}	28–29	180^{18mm}	>110	
e178	Ethyl 4-hydroxy-benzoate	$HOC_6H_4CO_2C_2H_5$	166.18	10, 159			116–118	297–298		0.07 aq; v s alc, eth

No.	Name	Formula	Mol. wt.	Beilstein ref.	Density	n_D	m.p. (°C)	b.p. (°C)	Flash pt. (°C)	Solubility
e179	Ethyl 3-hydroxy-butyrate	$CH_3CH(OH)CH_2CO_2C_2H_5$	132.16	3, 309	1.017^{20}_4	1.4205^{20}		170	64	s aq, alc
e180	Ethyl 2-hydroxyethyl sulfide	$HOCH_2CH_2SCH_2CH_3$	106.19	1², 525	1.020	1.4869^{20}		180–184	>110	s eth
e181	Ethyl 6-hydroxy-hexanoate	$HO(CH_2)_5CO_2C_2H_5$	160.22	3³, 628	0.985	1.4370^{20}		128^{12mm}	>110	
e182	Ethyl 2-hydroxyiso-butyrate	$(CH_3)_2C(OH)CO_2C_2H_5$	132.16	3, 315	0.965	1.4078^{20}		150	44	dec by hot aq
e183	2-Ethyl-2-(hydroxy-methyl)-1,3-propanediol	$C_2H_5C(CH_2OH)_3$	134.18	1³, 2349			60–62	161^{2mm}		
e184	2-Ethyl-2-(hydroxy-methyl)-1,3-propanediol-acrylate	$(H_2C{=}CHCO_2CH_2)_3CC_2H_5$	296.32		1.100	1.4736^{20}		157	>110	
e185	2-Ethyl-2-(hydroxy-methyl)-1,3-propanedioltri-methacrylate	$[H_2C{=}C(CH_3)CO_2CH_2]_3CC_2H_5$	338.40		1.060	1.4724^{20}			>110	
e186	N-Ethyl-3-hydroxy-piperidine		129.20	Merck: 12, 3890	0.970	1.4754^{20}		95^{15mm}	47	
e187	2,2'-Ethylidenebis-(4,6-di-tert-butyl-phenol)	$CH_3CH\{C_6H_2[C(CH_3)_3]_2OH\}_2$	438.70				162–164			
e188	2,2'-Ethylidenebis-(4,6-di-tert-butyl-phenyl) fluoro-phosphite		486.66				201–203			
e189	4,4'-Ethylidenebis-phenol	$CH_3CH_2CH(C_6H_4OH)_2$	214.26	6, 1006			123–127			
e190	5-Ethylidene-2-riorborene		120.20		0.893	1.4895			38	
e191	2-Ethylimidazole		96.13	23, 78			86	268	13	misc alc, eth; sl s aq
e192	Ethyl isobutyrate	$(CH_3)_2CHCO_2C_2H_5$	116.16	2, 291	0.870^{20}	1.3903^{20}	−88	110	32	i aq; misc alc, eth
e193	Ethyl isothiocyanate	CH_3CH_2NCS	87.14	4, 123	1.003^{18}_4	1.5142^{18}	−6	130–132	46	misc aq, alc, eth, esters, PE
e194	Ethyl (−)-lactate	$CH_3CH(OH)CO_2C_2H_5$	118.13	3, 264	1.0328^{20}	1.4124^{20}	−26	154–155		
e195	Ethyl (±)-mandelate	$C_6H_5CH(OH)CO_2C_2H_5$	180.21	10, 202	1.115	1.5120^{20}	33–34	253–255	>110	s alc, eth
e196	Ethyl 2-mercapto-acetate	$HSCH_2CO_2C_2H_5$	120.17	3, 255	1.0964	1.4571^{20}		54^{12mm}	47	
e197	Ethyl 3-mercapto-propionate	$HSCH_2CH_2CO_2C_2H_5$	134.20	3³, 555	1.039	1.4570^{20}		76^{10mm}	72	

(Continued)

TABLE 2.21 Physical Constants of Organic Compounds (*Continued*)

No.	Name	Formula	Formula weight	Beilstein reference	Density, g/mL	Refractive index	Melting point, °C	Boiling point, °C	Flash point, °C	Solubility in 100 parts solvent
e198	Ethylmercuric chloride	CH$_3$CH$_2$HgCl	165.13		3.5		192	sublimes		0.78 eth; 2.6 chl
e199	Ethyl methacrylate	H$_2$C=C(CH$_3$)CO$_2$C$_2$H$_5$	114.14	2, 423	0.917	1.4116^{25}		118	15	i aq; s alc, eth
e200	Ethyl 4-methoxy-phenylacetate	CH$_3$OC$_6$H$_4$CO$_2$C$_2$H$_5$	194.23	10^1, 83	1.097	1.5075^{20}		138^{7mm}	46	
e201	Ethyl 2-methyleaceto-acetate	CH$_3$C(=O)CH(CH$_3$)CO$_2$C$_2$H$_5$	144.17	3, 679	1.019	1.4280^{20}		187	62	i aq; s alc, eth
e202	N-Ethyl-2-methyl-allylamine	H$_2$C=C(CH$_3$)CH$_2$NHC$_2$H$_5$	99.18	4^4, 1104	0.753	1.4221^{20}		105	7	
e203	N-Ethyl-N-methylaniline	C$_6$H$_5$N(CH$_3$)C$_2$H$_5$	135.21	12, 162	0.947	1.5470^{20}		203–205	74	i aq; misc alc, eth
e204	Ethyl 2-methylbenzoate	CH$_3$C$_6$H$_4$CO$_2$C$_2$H$_5$	164.21	9, 463	1.032	1.5070^{20}		221^{731mm}	91	
e205	Ethyl 3-methylbenzoate	CH$_3$C$_6$H$_4$CO$_2$C$_2$H$_5$	164.21	9, 476	1.030	1.5054^{20}		110^{20mm}	101	
e206	Ethyl 4-methylbenzoate	CH$_3$C$_6$H$_4$CO$_2$C$_2$H$_5$	164.21	9, 484	1.025	1.5085^{20}		235	99	
e207	Ethyl 2-methylbutyrate	CH$_3$CH$_2$CH(CH$_3$)CO$_2$C$_2$H$_5$	130.19	2, 305	0.869	1.3969^{20}		133	26	
e208	Ethyl 3-methylbutyrate	(C$_2$H$_5$)$_2$CHCH(CH$_3$)$_2$	130.19	2^2, 275	0.8656^{20}	1.3962^{20}	−99	135	26	0.2 aq; misc alc, bz
e209	2-Ethyl-2-methyl-1,3-dioxolane		116.16	19^2, 11	0.929	1.4090^{20}		116–117	12	
e210	Ethyl methyl ether	C$_2$H$_5$OCH$_3$	60.10	1, 314	2.456 g/L		−113	7.4		s aq; misc alc, eth
e210a	3-Ethyl-4-methylhexane	(C$_2$H$_5$)$_2$CHCH(CH$_3$)C$_2$H$_5$	128.26		0.7420^{20}	1.4134^{20}		140	24	
e211	2-Ethyl-4-methylimidazole		110.16	23^2, 72	0.975	1.5000^{20}	47–54	292–295	137	
e212	Ethyl 4-methyl-5-imidazolecarboxylate		154.17	25^1, 534			204–206			
e213	4-Ethyl-2-methyl-2-(3-methylbutyl)-oxazolidine		185.3		0.877	1.4420^{20}		194	82	
e214	3-Ethyl-2-methylpentane	(C$_2$H$_5$)$_2$CHCH(CH$_3$)$_2$	114.24	1^3, 489	0.7193$^{20}_{4}$	1.4040^{20}	−115.0	115.7	<21	i aq; sl s alc; s eth
e215	3-Ethyl-3-methylpentane	(C$_2$H$_5$)$_3$CCH$_3$	114.24		0.7274^{20}	1.4078^{20}	−90.9	118.3		i aq; s eth
e216	Ethyl 1-methyl-2-piperidinecarboxylate		171.24	22^1, 485	0.975	1.4519^{20}		96^{11mm}	73	

e217	Ethyl 1-methyl-3-piperidinecarboxylate		171.24	2², 59	0.954	1.4510^{20}		89^{11mm}	68	s alc, bz, eth, acid
e218	Ethyl 3-methyl-1-piperidine propionate		199.30	1, 487	0.945	1.4530^{20}		112^{13mm}	99	i aq; misc alc, eth
e219	2-Ethyl-2-methyl-1,3-propanediol	HOCH2C(C2H5)(CH3)CH2OH	118.18				41–44	226	>110	
e220	5-Ethyl-2-methyl-pyridine	C2H5(CH3)C5H3N	121.18	20, 248	0.919	1.4970^{20}		178	66	misc aq, alc, eth
e221	Ethyl methyl sulfide	CH3CH2SCH3	76.15	1, 343	0.842	1.4392^{20}	−106	66.7	−15	i aq; misc alc, eth
e222	Ethyl (methylthio)-acetate	CH3SCH2CO2C2H5	134.20		1.043	1.4587^{20}		72^{25mm}	59	misc alc, eth
e223	N-Ethylmorpholine		115.18	27¹, 203	0.905	1.4410^{20}	−63	139	27	v s alc, eth
e224	Ethyl nitrate	CH3CH2ONO2	91.13	1, 329	1.100^{25}	1.3849^{22}	−94.6	87.7	10 (CC)	
e225	Ethyl nitrite	CH3CH2ONO	75.07	1, 329	1.90^{15}			17	−35	
e226	4-Ethylnitrobenzene	C2H5C6H4NO2	151.17	5, 358	1.118	1.5445^{20}	−32	245–246	>110	i aq; misc alc, eth
e227	Ethyl 4-nitrobenzoate	O2NC6H4CO2C2H5	195.17	9, 390			55–59			
e228	Ethyl nonanoate	CH3(CH2)7CO2C2H5	186.30	2, 353	0.866	1.4219^{20}	−37	227	94	misc DMF, oils
e229	Ethyl cis, cis-9,12-octa-decadienoic acid	H(CH2)5CH=CHCH2-CH=CH(CH2)7CO2C2H5	308.51	2², 461	0.8846	1.4675^{20}		193^{6mm}	>110	
e230	Ethyl cis-9-octa-decenoate	CH3(CH2)7CH=CH(CH2)7CO2C2H5	310.53	2, 467	0.869	1.4500^{20}	−32	216^{15mm}	>110	i aq; misc alc, eth
e231	Ethyl octanoate	CH3(CH2)6CO2C2H5	172.27	2, 348	0.878	1.4166	−43	208	75	i aq; misc alc, eth
e232	Ethyl oxalyl chloride	CH3CH2OC(=O)C(=O)Cl	136.53	2, 541	1.2223	1.4164^{20}		135	41	d aq, alc; s bz, eth
e233	Ethyl oxamate	CH3CH2OC(=O)C(=O)NH2	117.10	2, 544			114–116			s aq, eth; i bz
e234	2-Ethyl-2-oxazoline		99.13	10, 597	0.982	1.4370^{20}	−62	128	29	
e235	Ethyl 2-oxocyclo-pentanecarboxylate	(O=)(C5H7)CO2C2H5	156.18		1.054	1.4485^{20}		102^{11mm}	77	
e236	Ethyl 4-oxopentanoate	CH3C(=O)CH2CH2CO2C2H5	144.17	3, 675	1.012	1.4222^{20}		205–206		v s aq; misc alc
e237	Ethyl 2-oxopropionate	CH3C(=O)CO2C2H5	116.12	3, 616	1.060^{16}	1.408^{16}		144	45	sl s aq; misc alc, eth
e238	3-Ethylpentane	(C2H5)3CH	100.20	1³, 441	0.6982^{20}	1.3934^{20}	−118.6	93.5		i aq; s alc, eth
e239	Ethyl pentanoate	CH3(CH2)3CO2C2H5	130.19	2, 301	0.877^{4}	1.3732^{20}	−91.3	145.5		0.2 aq; misc alc, eth
e240	2-Ethylphenol	C2H5C6H4OH	122.17	5, 470	1.037	1.5372^{20}	−18	204	78	
e241	3-Ethylphenol	C2H5C6H4OH	122.17	6, 471	1.001	1.5330^{20}	−4	110^{15mm}	94	
e242	4-Ethylphenol	C2H5C6H4OH	122.17	6, 472	1.011	1.5239	45	218	100	i aq; misc alc, eth
e243	Ethyl phenylacetate	C6H5CH2CO2C2H5	164.20	9, 434	1.031	1.4980^{20}		229	77	i aq; misc alc, eth

(Continued)

TABLE 2.21 Physical Constants of Organic Compounds (*Continued*)

No.	Name	Formula	Formula weight	Beilstein reference	Density, g/mL	Refractive index	Melting point,°C	Boiling point,°C	Flash point,°C	Solubility in 100 parts solvent
e244	Ethyl 3-phenyl-glycidate		192.21		1.102	1.5180[20]		96[0.5mm]	>110	
e245	1-Ethylpiperazine		114.19	23[2], 5	0.899	1.4690[20]		157	43	
e246	Ethyl N-piperazino-carboxylate		158.20	23[2], 9	1.080	1.4765[20]		273	>110	
e247	1-Ethylpiperidine		113.20	20, 17	0.834	1.4440[20]		131	18	s aq
e248	2-Ethylpiperidine		113.20	20, 104	0.858	1.4510[20]		143	31	
e249	Ethyl 3-piperidine-carboxylate		157.21		1.012	1.4601[20]		104[7mm]	90	
e250	Ethyl 4-piperidine-carboxylate		157.21		1.010	1.4591[20]		204	80	s aq, alc, bz, eth
e251	Ethyl N-piperidine-propionate		185.27	20, 62	0.927	1.4545[20]		217–219	87	
e252	Ethyl 1-propenyle ether	$CH_3CH=CHOC_2H_5$	86.13	1, 435	0.778	1.3980[20]	−73.9	67–76	−18	1.7 aq; misc alc, eth
e253	Ethyl propionate	$CH_3CH_2CO_2C_2H_5$	102.13	2, 240	0.8917[20]	1.3839[20]	−73.9	99	12	sl s aq; misc alc, eth
e254	Ethyl propyl ether	$CH_3CH_2OCH_2CH_2CH_3$	88.15	1, 354	0.739	1.3695[20]	−79	62–63	32	s alc
e255	Ethyl propyl sulfide	$CH_3CH_2SCH_2CH_2CH_3$	104.21	1[3], 1432	0.8270	1.4462[20]	−117.0	118.5	29	sl s aq; s alc, eth
e256	2-Ethylpyridine	$CH_3CH_2(C_5H_4N)$	107.16	20, 241	0.937	1.4964[20]		149	48	v s alc, eth; sl s aq
e257	3-Ethylpyridine	$CH_3CH_2(C_5H_4N)$	107.16	20, 242	0.954	1.5015[20]		162–165	47	sl s aq; s alc, eth
e258	4-Ethylpyridine	$CH_3CH_2(C_5H_4N)$	107.16	20, 243	0.942	1.5009[20]		168		misc aq, alc, eth
e259	Ethyl 2-pyridine-carboxylate		151.17	22, 35	1.1194	1.5088[20]	2	240–241	107	
e260	1-Ethyl-2-pyrrolidinone		113.16		0.992	1.4652[20]		97[20mm]	76	misc alc, eth; sl s aq
e261	Ethyl salicylate	$C_6H_4(OH)CO_2C_2H_5$	166.18	10, 73	1.131	1.5219[20]	2–3	232–234	107	
e262	Ethyl sorbate	$CH_3CH=CHCH=CHCO_2C_2H_5$	140.18	2, 484	0.956	1.4942[20]		195.5	69	
e262a	2-Ethyltoluene	$CH_3C_6H_4C_2H_5$	120.19	5[1], 192	0.865	1.5040[20]	−81	165	39	
e262b	3-Ethyltoluene	$CH_3C_6H_4C_2H_5$	120.19	5, 398	0.865	1.4960[20]	−95	161	38	
e262c	4-Ethyltoluene	$CH_3C_6H_4C_2H_5$	120.19	5, 397	0.861	1.4950[20]	−62	162	36	
e263	Ethyl 4-toluene-sulfonate	$CH_3C_6H_4SO_2OC_2H_5$	200.26	11, 99	1.1166[45]	1.5110[20]	33	173[15mm]	157	i aq; s alc, eth
e264	N-Ethyl-m-toluidine	$CH_3C_6H_4NHC_2H_5$	135.21	12, 857	0.957	1.5451[20]		221	89	
e265	N-Ethyl-o-toluidine	$CH_3C_6H_4NHC_2H_5$	135.21		0.938	1.5470[20]		218	88	
e266	6-Ethyl-o-toluidine	$C_2H_5C_6H_3(CH_3)NH_2$	135.21		0.968	1.5525[20]	−33	231	89	
e267	2-(N-Ethyl-m-toluidino)ethanol	$CH_3C_6H_4N(C_2H_5)CH_2CH_2OH$	179.26		1.019	1.5540[20]		151[1mm]	>110	

No.	Name	Mol. formula	Mol. wt.	Beilstein ref.	Density	n_D	mp/°C	bp/°C	Flash pt/°C	Solubility
e268	Ethyl trichloroacetate	$Cl_3CCO_2C_2H_5$	191.44	2, 209	1.38334^{20}	1.4447^{20}		168	65	i aq; s alc, eth
e269	Ethyltrichlorosilane	$C_2H_5SiCl_3$	163.51	4, 630	1.238	1.4252^{20}	−106	99	13	
e270	Ethyltriethoxysilane	$C_2H_5Si(OC_2H_5)_3$	192.33	4^4, 4223	0.895	1.3920^{20}		158–166	38	
e271	Ethyltriphenyl-phosphonium iodide	$C_2H_5P(C_6H_5)_3I$	418.26	16, 760			169–171			
e272	Ethyl undecanoate	$CH_3(CH_2)_{10}CO_2C_2H_5$	214.35	2, 358	0.859	1.4280^{20}		105^{4mm}	>110	i aq; s org solvents
e273	Ethyl 10-undecenoate	$H_2C=CH(CH_2)_8CO_2C_2H_5$	212.34	2, 459	0.879	1.4390^{20}		258–259	>110	
e274	Ethylurea	$CH_3CH_2NHC(=O)NH_2$	88.11	4, 115	1.213^{18}		93–96	85^{20mm}	75	v s aq; 80 alc; i eth
e275	N-Ethylurethane	$CH_3CH_2NHCO_2C_2H_5$	117.15	4, 114	0.9814^{20}	1.4211^{20}		35		63 aq
e276	Ethyl vinyl ether	$CH_3CH_2OCH=CH_2$	72.11	1, 433	0.7589^{20}	1.3767^{20}	−116	228	<−45	0.9 aq; s alc, eth
e277	N-Ethyl-2,3-xylidine	$(CH_3)_2C_6H_3NHC_2H_5$	149.24	12, 1101	0.917	1.5468^{20}	31–33	180	71	
e278	1-Ethynyl-1-cyclohexanol	$HOC_6H_{10}C≡CH$	124.18	6^2, 100	0.967				62	
e279	Eugenol	$4-(H_2C=CHCH_2)C_6H_3-2-(OCH_3)OH$	164.20	6, 961	1.066	1.5410^{20}	−12/−10	254	>110	2.4 aq; misc alc, bz, acet, ketones, PE
f1	Fluoranthene		202.26	5, 685	1.2524^{0}		108	384		sl s alc; s bz, eth
f2	Fluorene		166.22	5, 625	1.2034^{0}		115	295		v s HOAc; s bz, eth
f3	Fluorenone		180.21	7, 465	1.1300^{99}	1.6369^{99}	82–85	342		s alc, bz; v s eth
f4	Fluorescein		332.31	19, 222			320			s hot alc, hot HOAc
f5	Fluoroacetic acid	FCH_2CO_2H	78.04	2, 193			33	165		sl s aq, alc
f6	4-Fluoroacetophenone	$FC_6H_4COCH_3$	138.14	12^1, 296	1.138	1.5110^{20}		196	71	
f7	2-Fluoroaniline	$FC_6H_4NH_2$	111.12	12, 597	1.151	1.5420^{20}	−29	183	60	sl s aq; s alc, eth
f8	4-Fluoroaniline	$FC_6H_4NH_2$	111.12	7^1, 132	1.1725	1.5395^{20}	−2	187	73	
f9	2-Fluorobenzaldehyde	FC_6H_4CHO	124.11	7^1, 132	1.178	1.5220^{20}	−44.5	91^{46mm}	55	
f10	4-Fluorobenzaldehyde	FC_6H_4CHO	124.11	5, 198	1.157	1.5200^{20}	−10	181	56	
f11	Fluorobenzene	C_6H_5F	96.11	9, 333	1.0240^{20}	1.4657^{20}	−42.2	84.7	−15	0.15 aq; misc alc
f12	2-Fluorobenzoic acid	$FC_6H_4CO_2H$	140.11	9, 333	1.4604^{25}		123–125			sl s aq; s alc, eth
f13	4-Fluorobenzoic acid	$FC_6H_4CO_2H$	140.11	9^1, 136	1.4794^{25}		184–187			0.1 aq; s alc, eth
f14	2-Fluorobenzoyl chloride	FC_6H_4COCl	158.56	9^1, 137	1.328	1.5365^{20}	4	92^{15mm}	82	
f15	4-Fluorobenzoyl chloride	FC_6H_4COCl	158.56		1.342	1.5296^{20}	9	82^{20mm}	82	
f16	4-Fluorobenzyl chloride	$FC_6H_4CH_2Cl$	144.58		1.207	1.5130^{20}		82^{26mm}	60	
f17	Fluoroethane	CH_3CH_2F	48.06	1, 82	1.195 g/L		−143.2	−37.7		198 mL aq; v s alc, eth
f18	Fluoromethane	CH_3F	34.04	1, 59			−141.8	−78.4		166 mL aq; v s alc, eth
f19	3-Fluoro-1-methoxy-benzene	$FC_6H_4OCH_3$	126.13		1.104	1.4880^{20}		158^{743mm}	43	
f20	4-Fluoro-1-methoxy-benzene	$FC_6H_4OCH_3$	126.13	6^1, 98	1.114	1.4877^{20}	−45	157	43	s eth

(Continued)

TABLE 2.21   Physical Constants of Organic Compounds (*Continued*)

No.	Name	Formula	Formula weight	Beilstein reference	Density, g/mL	Refractive index	Melting point, °C	Boiling point, °C	Flash point, °C	Solubility in 100 parts solvent
f21	2-Fluoro-2-methyl-propane	$(CH_3)_3CF$	76.11	1^4, 286			−77	12	−12	
f22	4-Fluoro-3-nitroaniline	$FC_6H_3(NO_2)NH_2$	156.12	12, 729	1.3300^{20}_{4}	1.5312^{20}	96–98	205	91	i aq; s alc, eth
f23	1-Fluoro-4-nitro-benzene	$FC_6H_4NO_2$	141.10	5, 241			21		83	
f24	4-Fluoro-3-nitro-toluene	$CH_3C_6H_3(NO_2)F$	155.13		1.262	1.5240^{20}	28–30	241	>110	
f25	4-Fluorophenol	FC_6H_4OH	112.10	6, 183	1.128	1.4680^{20}	46–48	185	68	
f26	2-Fluoropyridine	$F(C_5H_4N)$	97.09	20^1, 80	1.0014^{17}	1.4716^{17}	−62	126	28	v s alc, eth
f27	2-Fluorotoluene	$FC_6H_4CH_3$	110.13	5, 290	0.9974^{20}	1.4691^{20}	−87	115	12	s alc, eth
f28	3-Fluorotoluene	$FC_6H_4CH_3$	110.13	5, 290	0.9975^{20}	1.4698^{20}	−56	115	9	s alc, eth
f29	4-Fluorotoluene	$FC_6H_4CH_3$	110.13	5, 290				117	17	s alc, eth
f30	Fluorotrichloro-methane	$FCCl_3$	137.37	1, 64	1.494	1.3821^{20}	−110	24	none	
f31	Formaldehyde	$H_2C{=}O$	30.03	1, 558	0.8154^{-20}	0.8153^{-20}	−92	−19.5	56	122 aq; s alc, eth
f32	Formamide	$HC(=O)NH_2$	45.04	2, 26	1.1334^{20}_{4}	1.4475^{20}	2.6	220	154	misc aq, alc, acet
f33	Formamidine acetate	$HC(=NH)NH_2 \cdot HO_2CCH_3$	104.11				158 dec			
f34	Formamidinesulfinic acid	$H_2NC(=NH)S(O)OH$	108.12	3^1, 36			126 dec			
f35	Formanilide	C_6H_5NHCHO	121.14	12, 230	1.144		47	271	>110	2.5 aq
f36	Formic acid	HCO_2H	46.03	2, 8	1.220^{10}	1.3704^{20}	8.3	100.8	68	misc aq, alc, eth
f37	2-Formylbenzoic acid	$HO_2CC_6H_4CHO$	150.13	10, 666	1.404		96–98		>110	s aq; v s alc, eth
f38	Formylhydrazine	$HC(=O)NHNH_2$	60.06	2, 93			54–56	236–237	>110	v s alc, chl, eth; s bz
f39	4-Formylmorpholine		115.13	27^3, 274	1.145	1.4848^{20}		222	91	
f40	N-Formylpiperidine		113.16	20, 45	1.019	1.4780^{20}		122 dec		
f41	D-(−)-Fructose		180.16	31, 321	1.63553^{20}					v s aq; 6.7 alc; s pyr
f42	Fumaric acid	$HO_2CCH=CHCO_2H$	116.07	2, 737			287	subl 300		0.6 aq; 9 alc; 0.7 eth
f43	Fumaroyl dichloride	$ClC(=O)CH=CHC(=O)Cl$	152.96	2, 743	1.408^{20}	1.4988^{20}		161–164	73	dec aq, alc
f44	2-Furaldehyde		96.09	17^2, 305	1.1598^{20}_{4}	1.5262^{20}	−36.5	161.8	60	8 aq; misc alc, eth
f45	Furan		68.07	17, 27	0.9514^{20}	1.4214^{20}	−85.6	31.4	−35	1 aq; misc alc, eth
f46	2-Furanacrylic acid		138.12	18, 300			142–144	286		0.2 aq; 1.1 bz; s alc, eth, HO Ac
f47	2,5-Furandimethanol		128.13	17^1, 90	1.132	1.5304^{20}	74–76	155	45	
f48	2-Furanmethanethiol		114.17	17^2, 116						
f49	Furfuryl acetate		140.14	17^2, 115	1.1175^{20}_{4}	1.4618^{20}		175–177	65	i aq; s alc, eth

No.	Name	Formula	Beil./Merck Ref.	Density	n_D	mp, °C	bp, °C	Flash P., °C	Solubility
f50	Furfuryl alcohol		17, 112	1.1295^{20}_4	1.4868^{20}	−31	171	75	misc aq(dec); v s alc, eth
f51	Furfurylamine		18, 584	1.0995^{20}_4	1.4900^{20}	−70	145–146	46	misc aq; s alc, eth
f52	Furfuryl methacrylate		17[3], 1248	1.078	1.4820^{20}		82^{5mm}	90	v s alc, eth; sl s bz
f53	α-Furildioxime		19, 166			166–168			
f54	2-Furoic acid		18, 272			133–134	230–232	85	4 aq; s alc; v s eth
f55	2-Furoyl chloride		18, 276	1.324	1.5310^{20}	−2	170		dec aq, alc; s eth
g1	D-(+)-Galactose		31, 295			167			200 aq; s pyr; sl s alc
g2	Geraniol	(CH₃)₂C=CHCH₂CH₂-C(CH₃)=CHCH₂OH	1, 457	0.8894^{20}_4	1.4766^{20}		230	76	i aq; misc alc, eth
g3	Geranyl acetate	(CH₃)₂C=CHCH₂CH₂-C(CH₃)=CHCH₂O₂CCH₃	2, 140	0.9174^{15}_{15}	1.4628^{15}		138^{25mm}	104	v s alc; misc eth
g4	Gerard reagent P	[(C₅H₅N)CH₂C(=O)NHNH₂]⁺ Cl⁻	Merck: 12, 4436			dec 200			less soluble in polar solvents than T
g5	Gerard reagent T	[(CH₃)₃NCH₂C(=O)NHNH₂]⁺ Cl⁻	Merck: 12, 4436			192			v s aq, HOAc, glyc, ethylene glycol
g6	D-Gluconic acid		3, 542			131			v s aq; sl s alc; i eth
g7	δ-Gluconolactone		18[1], 405			153			50 aq; 1 alc; i eth
g8	α-D-(+)-Glucose		31, 83	1.5620^{18}_4		153–156			91 aq; 0.83 MeOH; s pyr
g9	α-D-Glucose penta-acetate		31, 119			109–111			0.15 aq; 1.3 alc; 3 eth
g11	D-Glucurono-3,6-lactone		Merck: 11, 4362			176–178	subl 200		27 aq; 2.8 MeOH
g12	(S)-(+)-Glutamic acid	HO₂CCH₂CH₂CH(NH₂)CO₂H	4, 488			d 247			0.8 aq; i alc, eth
g13	(S)-(+)-Glutamine	H₂NC(=O)CH₂CH₂-CH(NH₂)CO₂H	4, 491			185 dec			5 aq; 0.0035 MeOH; i bz, chl, eth, acet
g14	Glutaric acid	HO₂CCH₂CH₂CH₂CO₂H	2, 631	1.429^{20}_4	1.4188^{106}	98	303		43 aq^{20}; v s alc, eth; s bz, chl; sl s PE
g15	Glutaric anhydride		17, 411			55–57	150^{10mm}	>110	s aq, alc
g16	Glutaric dialdehyde	OCHCH₂CH₂CH₂CHO	1, 776		1.4338^{25}		187–189	none	s aq, alc, chl; i eth
g17	Glutaronitrile	NCCH₂CH₂CH₂CN	2, 635	0.9888^{23}	1.4345^{20}	−29	286	>110	
g18	Glutaryl dichloride	ClC(=O)(CH₂)₃C(=O)Cl	2, 634	1.324	1.4720^{20}		216–218	106	dec aq, alc; s eth
g19	Glycerol	HOCH₂CH(OH)CH₂OH	1, 502	1.2613^{20}_4	1.4746^{20}	18	290	199	misc aq, alc; 0.2 eth
g20	Glyceryl tris(butyrate)	(CH₃CH₂CH₂CO₂)₂CH-CH₂O₂CCH₂CH₂CH₃	2, 273	1.032^{20}_4	1.4359^{20}	−75	287–288	173	i aq; v s alc, eth
g21	Glyceryl tris-(dodecanoate)	[CH₃(CH₂)₁₀CO₂CH₂]₂CH-O₂C(CH₂)₁₀CH₃	2, 362	0.894^{60}_4	1.4404^{60}	46			v s bz, eth; sl s alc
g22	Glyceryl tris(nitrate)	O₂NOCH₂CH(ONO₂)CH₂ONO₂	1, 516	1.594^{20}_4	1.4786^{12}	13.3	160^{5mm}	explodes 270	0.18 aq; 54 alc; misc eth

(Continued)

609

TABLE 2.21 Physical Constants of Organic Compounds (*Continued*)

No.	Name	Formula	Formula weight	Beilstein reference	Density, g/mL	Refractive index	Melting point, °C	Boiling point, °C	Flash point, °C	Solubility in 100 parts solvent
g23	Glyceryl tris(oleate)	[CH$_3$(CH$_2$)$_7$CH=CH(CH$_2$)$_7$-CO$_2$CH$_2$]$_2$CHO$_2$C(CH$_2$)$_7$-CH=CH(CH$_2$)$_7$CH$_3$	885.46	4, 468	0.9154^{15}	1.4621^{40}	−4/−5	235^{15mm}		s chl, eth, CCl$_4$
g24	Glyceryl tris(palmitate)	[CH$_3$(CH$_2$)$_{14}$CO$_2$CH$_2$]$_2$CH-O$_2$C(CH$_2$)$_{14}$CH$_3$	807.35	2, 373	0.8663^{80}	1.4381^{80}	65–66	310–320		v s bz, chl, eth
g25	Glyceryl tris-(tridecanoate)	[CH$_3$(CH$_2$)$_{11}$CO$_2$CH$_2$]$_2$CH-O$_2$C(CH$_2$)$_{11}$CH$_3$	723.18	2, 367	0.8854^{65}	1.4428^{60}	57			v s alc, bz, chl
g26	Glycine	H$_2$NCH$_2$CO$_2$H	75.07	4, 333	1.1607		dec 240			25 aq; 0.6 pyr; i eth
g27	N-Glycylglycine	H$_2$NCH$_2$C(=O)NHCH$_2$CO$_2$H	132.12	4, 371			260 dec			s hot aq; sl s alc
g28	Glyoxal	HC(=O)CHO	58.04	1, 759	1.14	1.3826^{20}	15	50.4		viol rxn aq; s anhyd solvents; mixtures with air may explode
g29	Glyoxylic acid	HC(=O)CO$_2$H	74.04	3, 594			98			v s aq; sl s alc, eth
g30	Guanidine	H$_2$NC(=NH)NH$_2$	59.07	3, 82			ca. 50			v s aq, alc
g31	Guanine		151.13	26, 449			>300	dec 160		s alk soln, dil acids; sl s alc, eth
h1	Heptadecane	CH$_3$(CH$_2$)$_{15}$CH$_3$	140.41	1, 173	0.7767^{22}	1.4360^{25}	22.0	302.2	148	s eth; sl s alc
h1a	1-Heptadecanol	CH$_3$(CH$_2$)$_{16}$OH	256.48	1^1, 220			53.8	333	>110	
h2	Heptafluorobutyric acid	CF$_3$CF$_2$CF$_2$CO$_2$H	214.04		1.625	<1.300^{20}		120	none	
h3	Heptaldehyde	CH$_3$(CH$_2$)$_5$CHO	114.19	1^2, 750	0.8216$^{15}_4$	1.4285^{20}	−43	153	35	misc alc, eth; sl s aq
h4	2,2,4,6,8,8-Hepta-methylnonane	(CH$_3$)$_3$CCH$_2$C(CH$_3$)$_2$CH$_2$-CH(CH$_3$)CH$_2$C(CH$_3$)$_3$	226.45		0.793	1.4391^{20}		240	95	
h5	1,1,3,5,5,5-Hepta-methyltrisiloxane	[(CH$_3$)$_3$SiO]$_2$SiHCH$_3$	222.51	4^3, 1874	0.819	1.3820^{20}		142	27	
h6	Heptane	CH$_3$(CH$_2$)$_5$CH$_3$	100.21	1, 154	0.68384^{20}	1.3877^{20}	−90.6	98.4	−4 (CC)	s alc, chl, eth
h7	Heptanedioic acid	HO$_2$C(CH$_2$)$_5$CO$_2$H	160.17	2, 670	1.329^{15}		105.8	212^{10mm}	>110	5 aq; v s alc, eth
h8	1-Heptanethiol	CH$_3$(CH$_2$)$_6$SH	132.27	1, 415			−43.2	176.9	46	i aq
h9	Heptanoic acid	CH$_3$(CH$_2$)$_5$CO$_2$H	130.19	2, 338	0.9181^{20}	1.4221^{20}	−8	222	>110	0.25 aq; s alc, eth
h10	Heptanoic anhydride	[CH$_3$(CH$_2$)$_5$CO]$_2$O	242.36	2, 340	0.923	1.4332^{20}	−12.4	268	>110	i aq; s alc, eth
h11	1-Heptanol	CH$_3$(CH$_2$)$_6$OH	116.20	1, 414	0.8219$^{20}_4$	1.4242^{20}	−34	176.4	73	misc alc, eth
h12	2-Heptanol	CH$_3$(CH$_2$)$_4$CH(OH)CH$_3$	116.20	1, 415	0.8167^{20}	1.4210^{10}		159	71	0.35 aq; s alc, bz, eth
h13	3-Heptanol	CH$_3$(CH$_2$)$_3$CH(OH)CH$_2$CH$_3$	116.20	1^1, 205	0.8227^{20}	1.4214^{20}	−70	157	60	sl s aq
h14	2-Heptanone	HC(CH$_2$)$_4$C(=O)CH$_3$	114.19	1, 699	0.8197$^{15}_4$	1.4116^{15}	−35	151	39	s alc, eth
h15	3-Heptanone	CH$_3$(CH$_2$)$_3$C(=O)CH$_2$CH$_3$	114.19	1, 699	0.8197^{20}	1.4055^{20}	−39	147	46	0.43 aq; s alc, eth
h16	4-Heptanone	CH$_3$(CH$_2$)$_2$C(=O)(CH$_2$)$_2$CH$_3$	114.19	1, 699	0.817	1.4068^{20}	−32.1	143.7	48 (CC)	0.53 aq; misc alc, eth

No.	Name	Formula	Mol. wt.	Beilstein ref.	Density	n_D	m.p. (°C)	b.p. (°C)	Flash pt. (°C)	Solubility
h17	Heptanoyl chloride	CH₃(CH₂)₅C(=O)Cl	148.63	2, 340	0.960	1.4300^{20}		173	58	dec aq, alc; s eth
h18	1-Heptene	CH₃(CH₂)₄CH=CH₂	98.90	1, 219	0.6970^{20}	1.3999^{20}	−120	93.6	−8	0.1 aq; s alc, eth
h18a	cis-2-Heptene	CH₃(CH₂)₃CH=CHCH₃	98.19	1^3, 825	0.708^{20}	1.406^{20}		98.4	−6	
h18b	trans-2-Heptene	CH₃(CH₂)₃CH=CHCH₃	98.19	1, 219	0.7012^{20}	1.4045^{20}	−109.5	98	−1	
h19	1-Heptylamine	CH₃(CH₂)₆NH₂	115.22	4, 193	0.777	1.4243^{20}	−23	154–56	35	s alc, acet, eth, PE
h20	1-Heptyne	CH₃(CH₂)₄C≡CH	96.17	1, 256	0.733	1.4075^{20}	−81	99–100	−2	
h21	Hexachloroacetone	Cl₃CC(=O)CCl₃	264.75	1, 657	1.743	1.5112^{20}	−30	66^{6mm}	none	sl s aq; s acet
h22	Hexachlorobenzene	C₆Cl₆	284.78	5, 205	2.044^{24}		232	325	242	s bz, chl, eth
h23	Hexachloro-1,3-butadiene	Cl₂C=CClCCl=CCl₂	260.76	1, 250	1.655	1.5550^{20}	−21	215	none	s alc, eth
h24	1,2,3,4,5,6-Hexachlorocyclohexane, γ-isomer	C₆H₆Cl₆	290.83	5^1, 8	1.87^{20}		113–115		none	s bz, chl
h25	Hexachlorocyclo-1,3-pentadiene		272.77		1.7014^{25}	1.5644^{20}	−10	239	none	none
h27	Hexachloroethane	Cl₃CCCl₃	236.74	1, 87	2.091		187	sublimes	none	s alc, bz, chl, eth
h28	1,4,5,6,7,7-Hexachloro-5-norbornene-2,3-dicarboxylic anhydride		370.83	9^3, 4049			239–242		none	
h29	Hexachlorophene	CH[C₆H(Cl)₂OH]₂	406.91	6^3, 5407			163–165		none	
h30	Hexachloropropene	Cl₃CC(Cl)=CCl₂	248.75	1, 200	1.765	1.5480^{20}		210	none	misc eth
h31	Hexadecane	CH₃(CH₂)₁₄CH₃	226.45	1, 172	0.7733^{20}	1.4345^{20}	18.2	286.8	135	
h32	1,2-Hexadecanediol	CH₃(CH₂)₁₃CH(OH)CH₂OH	258.45	1^3, 2244	0.840		72–74	184^{7mm}		
h33	1-Hexadecanethiol	CH₃(CH₂)₁₅SH	258.51	1, 430		1.4720^{20}	18–20		101	sl s alc, s eth
h34	Hexadecanoic acid	CH₃(CH₂)₁₄CO₂H	256.43	2, 370	0.8562^{62}	1.4273^{80}	62	351	135	s hot: chl, eth
h35	1-Hexadecanol	CH₃(CH₂)₁₅OH	242.45	1, 429	0.8116^{60}	1.4355^{60}	49.3	334	132	s alc, chl, eth
h36	1-Hexadecene	CH₃(CH₂)₁₃CH=CH₂	224.43	1, 226	0.7834^{20}	1.4401	4.1	284		s alc, eth, PE
h37	1-Hexadecylamine	CH₃(CH₂)₁₅NH₂	241.46	4, 202			45–48	330	140	v s alc, eth; s bz, chl
h38	2,4-Hexadienal	CH₃CH=CHCH=CHCHO	96.13	1^2, 809	0.871	1.5386^{20}		76^{30mm}	67	s alc, eth
h39	1,5-Hexadiene	H₂C=CHCH₂CH₂CH=CH₂	82.15	1, 253	0.6923^{24}	1.4042^{20}	−140.7	59.5	−27	0.2 aq; 13 alc; 9 acet; 2.3 bz; 11 diox; 1 CCl₄
h40	2,4-Hexadienoic acid	CH₃CH=CHCH=CHCO₂H	112.13	2, 483			134.5	119^{10mm}	127	
h41	Hexafluorobenzene	C₆F₆	186.05	5^3, 523	1.6182^{20}	1.3781^{20}	5.1	80.3	10	sl s alc, eth
h42	Hexafluoroethane	F₃CCF₃	138.01	1^3, 132	1.590^{-78}		−100.7	−78.3		
h43	1,1,1,3,3,3-Hexafluoro-2-propanol	(CF₃)₂CHOH	168.04		1.596^{25}	1.2750^{20}	−3	58.2	none	s aq, bz, CCl₄
h44	Hexafluoropropene	CF₃CF=CF₂	150.02	1^3, 697			−153	−28		

TABLE 2.21 Physical Constants of Organic Compounds (*Continued*)

No.	Name	Formula	Formula weight	Beilstein reference	Density, g/mL	Refractive index	Melting point, °C	Boiling point, °C	Flash point, °C	Solubility in 100 parts solvent
h45	Hexamethylcyclotrisiloxane	[-Si(CH$_3$)$_2$O-]$_3$	222.48	4^3, 1884			64–66	133–135	35	
h46	1,1,1,3,3,3-Hexamethyldisilazane	(CH$_3$)$_3$SiNHSi(CH$_3$)$_3$	161.40	4^3, 1861	0.774^{20}_4	1.4071^{20}		126	8	
h47	Hexamethyldisiloxane	(CH$_3$)$_3$SiOSi(CH$_3$)$_3$	162.38	4^3, 1859	0.764^{20}_4	1.3775^{20}	−67	101	−2	
h48	Hexamethyleneimine		99.18	20, 94	0.880	1.4631^{20}		138^{749mm}	18	
h49	Hexamethylenetetramine		140.19	1, 583	1.331^{-5}		280 subl		250	67 aq; 8 alc; 10 chl
h50	Hexamethylphosphoramide	[(CH$_3$)$_2$N]$_3$P(=O)	179.20		1.027^{20}	1.4588^{20}	7	232^{740mm}	105	misc aq
h51	Hexanaldehyde	CH$_3$(CH$_2$)$_4$CHO	100.16	1^2, 745	0.8335^{20}_4	1.4035^{20}	−56	131	32	v s alc, eth; sl s aq
h52	Hexane	CH$_3$(CH$_2$)$_4$CH$_3$	86.18	1, 142	0.6594^{20}_4	1.3749^{20}	−95.4	68.7	−22	misc alc, chl, eth
h53	1,6-Hexanediamine	H$_2$N(CH$_2$)$_6$NH$_2$	116.21	4, 269			42	205	81	v s aq; sl s alc, bz
h54	1,6-Hexanedioic acid	HO$_2$C(CH$_2$)$_4$CO$_2$H	146.14	2, 649	1.360^{25}	1.4425^{20}	152–154	337.5	196	1.4 aq; v s alc; s acet
h55	DL-1,2-Hexanediol	CH$_3$(CH$_2$)$_3$CH(OH)CH$_2$OH	118.18	1^1, 251	0.951	1.4579^{25}	42.8	223–224	>110	v s aq, alc
h56	1,6-Hexanediol	HO(CH$_2$)$_6$OH	118.18	1, 484	0.958			208	101	s aq, alc, eth
h57	2,5-Hexanediol	CH$_3$CH(OH)CH$_2$CH$_2$CH(OH)CH$_3$	118.18	1, 485	0.9617^{45}_{16}	1.4465^{20}	−50	220.8	101	
h58	1,6-Hexanediol diacrylate	[H$_2$C=CHCO$_2$(CH$_2$)$_3$-]$_2$	226.28		1.010	1.4562^{20}			>110	
h59	1,6-Hexanediol dimethacrylate	[H$_2$C=C(CH$_3$)CO$_2$(CH$_2$)$_3$-]$_2$	254.33		0.995	1.4580^{20}		>350	>110	
h60	2,5-Hexanedione	CH$_3$C(=O)CH$_2$CH$_2$C(=O)CH$_3$	114.14	1, 788	0.9732_4	1.4260^{20}	−9	188	78	misc aq, alc, eth
h61	Hexanenitrile	CH$_3$(CH$_2$)$_4$CN	97.16	2, 324	0.8052^{20}	1.4069^{20}	−80.3	163.6	43	i aq; s alc, eth
h62	1-Hexanethiol	CH$_3$(CH$_2$)$_5$SH	118.24	1^3, 1659	0.8424^{20}_4	1.4496^{20}	−80.5	152.7	20	i aq; v s alc, eth
h63	1,2,6-Hexanetriol	HOCH$_2$CH(OH)(CH$_2$)$_3$CH$_2$OH	134.17	1^4, 2784	1.1063^{20}	1.58^{20}	−32.8	178^{5mm}	191	misc alc, acet; i bz
h64	Hexanoic acid	CH$_3$(CH$_2$)$_4$CO$_2$H	116.16	2, 321	0.9265^{20}_4	1.4168^{20}	−3	205	102	1.1 aq; v s alc, eth
h65	Hexanoic anhydride	[CH$_3$(CH$_2$)$_4$C(=O)]$_2$O	214.31	2, 324	0.926	1.4280^{20}	−41	246–248	>110	s alc
h66	1-Hexanol	CH$_3$(CH$_2$)$_5$OH	102.18	1, 407	0.8136^{20}	1.4182^{20}	−44.6	157.5	63	8 aq; misc bz, eth; s alc
h67	2-Hexanol	CH$_3$(CH$_2$)$_3$CH(OH)CH$_3$	102.18	1, 408	0.8108^{25}	1.4128^{25}	−47	139.9	41	sl s aq; s alc, eth
h68	3-Hexanol	CH$_3$CH$_2$CH$_2$CH(OH)CH$_2$CH$_3$	102.18	1, 408	0.8193^{20}_4	1.4160^{20}		135	41	
h69	6-Hexanolactone		114.14	17^2, 290	1.030	1.4630^{20}	−18	215	109	v s alc, eth
h70	2-Hexanone	CH$_3$(CH$_2$)$_3$C(=O)CH$_3$	100.16	1, 689	0.8113^{20}	1.4007^{20}	−55.5	127.6	25	v s alc, eth
h71	3-Hexanone	CH$_3$CH$_2$CH$_2$C(=O)CH$_2$CH$_3$	100.16	1, 690	0.815	1.4002^{20}	−87	123	35	
h72	Hexanoyl chloride	CH$_3$(CH$_2$)$_4$C(=O)Cl	134.61	2, 324	0.9754^{20}_4	1.4263^{20}		153	50	dec aq, alc; s eth
h73	1-Hexene	CH$_3$(CH$_2$)$_3$CH=CH$_2$	84.16	1, 215	0.6732^{20}	1.3879^{20}	−139.8	63.5	−9	0.005 aq
h74	*trans*-2-Hexenoic acid	CH$_3$(CH$_2$)$_2$CH=CHCO$_2$H	114.14	2^4, 1563	0.965	1.4885^{20}	33–35	217	>110	

No.	Name	Formula	Mol. wt.	Beilstein	Density	n_D	mp (°C)	bp (°C)	fp (°C)	Solubility
h75	*trans*-3-Hexenoic acid	CH$_3$CH$_2$CH=CHCH$_2$CO$_2$H	114.14	2, 435	0.963	1.4398^{20}	11–12	119^{22mm}	>110	0.13 aq; v s alc, eth
h76	*trans*-2-Hexen-1-ol	CH$_3$CH$_2$CH$_2$CH=CHCH$_2$OH	100.16	1^2, 486	0.849	1.4343^{20}		158–160	54	
h77	5-Hexen-2-one	H$_2$C=CHCH$_2$CH$_2$C(=O)CH$_3$	98.15	1, 734	0.847	1.4197^{20}		128–129	23	
h78	*trans*-2-Hexenyl acetate	CH$_3$C(=O)OCH$_2$CH=CHCH$_2$CH$_2$CH$_3$	142.20	2^2, 151	0.898	1.4275^{20}		166	58	
h79	Hexyl acetate	CH$_3$(CH$_2$)$_5$O$_2$CCH$_3$	144.21	2, 132	0.860$^{20}_{20}$	1.4090^{20}	−81	171	45	sl s aq; v s alc, eth
h80	Hexyl acrylate	H$_2$C=CHCO$_2$(CH$_2$)$_5$CH$_3$	156.23	2^3, 1228	0.888	1.4280^{20}		90^{24mm}	68	i aq; s alc, eth
h81	Hexylamine	CH$_3$(CH$_2$)$_5$NH$_2$	101.19	4, 188	0.763$^{25}_4$	1.4180^{20}	−23	133	8	s aq; misc alc, eth
h82	1-Hexyne	CH$_3$(CH$_2$)$_3$C≡CH	82.14	1, 253	0.7152$^{20}_4$	1.3989^{20}	−131.9	71.3	−21	41 aq; v s l s alc
h83	L-Histidine		155.16	25, 513			282 dec			s alc; alk; sl s eth
h84	Hydantoin		100.08	24, 242			221–223			misc aq, alc
h85	Hydrazine	H$_2$NNH$_2$	32.05	Merck: 12, 4809	1.00363$^{25}_4$	1.4700^{20}	1.4	113.5	52	
h86	1,4-Hydroquinone	C$_6$H$_4$-1,4-(OH)$_2$	110.11	6, 836	1.332^{15}		172	286		7 aq; v s alc, eth; sl s bz
h87	Hydroxyacetaldehyde	HOCH$_2$CHO	60.05	1, 817	1.366^{100}		93–94	110^{12mm}		v s aq, alc; sl s eth
h88	Hydroxyacetic acid	HOCH$_2$CO$_2$H	76.05	3, 228			80	100		s aq, alc, acet, eth
h89	1'-Hydroxy-2'-acetonaphthone	C$_{10}$H$_6$(OH)C(=O)CH$_3$	186.21	8, 149			98–100	325 sl d		i aq; v s bz; s HOAc
h90	Hydroxyacetone	HOCH$_2$C(=O)CH$_3$	74.08	1^1, 84	1.082	1.4315^{20}	−17	146	56	misc aq, alc, eth
h91	2'-Hydroxyacetophenone	HOC$_6$H$_4$C(=O)CH$_3$	136.15	8, 85	1.131$^{21}_4$	1.5584^{20}	4–6	213^{717mm}	>110	misc alc, eth; sl s aq
h92	3'-Hydroxyacetophenone	HOC$_6$H$_4$C(=O)CH$_3$	136.15	8, 86	1.100^{100}	1.535^{100}	87–89	296		s aq; v s alc, bz, eth
h93	4'-Hydroxyacetophenone	HOC$_6$H$_4$C(=O)CH$_3$	136.15	8, 87	1.109^{100}		109–111	148^{3mm}		v s alc, eth; sl s aq
h94	2-Hydroxybenzaldehyde	C$_6$H$_4$(OH)CHO	122.12	8, 31	1.1674^{20}	1.5740^{20}	−7	196.7	78	1.7 aq[86]; s alc, eth
h95	3-Hydroxybenzaldehyde	C$_6$H$_4$(OH)CHO	122.12	8, 58			103–105	191^{50mm}		s alc, bz, eth; sl s aq
h96	4-Hydroxybenzaldehyde	C$_6$H$_4$(OH)CHO	122.12	8, 64	1.129$^{130}_4$		117–119			1 aq; 70 acet; 4 bz[65]; v s alc, eth
h97	2-Hydroxybenzaldehyde oxime	C$_6$H$_4$(OH)CH=NOH	137.14	8, 49			57	dec		v s alc, bz, eth, acids
h98	2-Hydroxybenzamide	C$_6$H$_4$(OH)C(=O)NH$_2$	137.14	10, 87			140	dec 270		0.2 aq; s alc, chl, eth
h99	2-Hydroxybenzoic acid	C$_6$H$_4$(OH)CO$_2$H	138.12	10, 43	1.443$^{20}_4$		157–159	211^{20mm}		0.2 aq; 37 alc; 33 eth; 33 acet; 2 chl; 0.7 bz

(Continued)

TABLE 2.21 Physical Constants of Organic Compounds (*Continued*)

No.	Name	Formula	Formula weight	Beilstein reference	Density, g/mL	Refractive index	Melting point, °C	Boiling point, °C	Flash point, °C	Solubility in 100 parts solvent
h100	3-Hydroxybenzoic acid	$C_6H_4(OH)CO_2H$	138.12	10, 134	1.473		201–203			0.8 aq; 10 eth
h101	4-Hydroxybenzoic acid	$C_6H_4(OH)CO_2H$	138.12	10, 149	1.468[4]		215–217			0.2 aq; v s alc; 23 eth
h102	4-Hydroxybenzoic hydrazide	$HOC_6H_4C(=O)NHNH_2$	152.15	10, 174			266 dec			
h103	4-Hydroxybenzophenone	$HOC_6H_4C(=O)C_6H_5$	198.22	8[2], 184			132–135			v s alc, eth; sl s aq
h104	1-Hydroxybenzotriazole		135.13	26, 41			155–158			
h105	6-Hydroxy-1,3-benzoxathiol-2-one		168.17	19[4], 2508			158–160			
h106	2-Hydroxybenzylalcohol	$HOC_6H_4CH_2OH$	124.13	6, 891	1.161[25]		83–85	subl 100		6.6 aq; v s alc, chl, eth; s bz
h107	1-Hydroxy-2-butanone	$CH_3CH_2C(=O)CH_2OH$	88.11	1, 826	1.026	1.4282[20]		78[60mm]	60	misc aq, alc; sl s eth
h108	3-Hydroxy-2-butanone	$CH_3C(=O)CH(OH)CH_3$	88.11	1, 827	0.9972[17,4]	1.4171[20]	15	148	50	s alc, eth; sl s aq
h109	4-Hydroxycinnamic acid	$HOC_6H_4CH=CHCO_2H$	164.16	10, 297			210–213			v s alc, chl, alk, HOAc
h111	7-Hydroxycoumarin		162.14	18, 27			226–228			
h112	1-Hydroxy-1-cyclohexanecarbonitrile	$C_6H_{10}(OH)CN$	125.17	10, 5	1.031	1.4576[20]	29		60	v s alc, eth; i bz, chl
h113	2-Hydroxy-3,5-diiodobenzoic acid	$I_2C_6H_2(OH)CO_2H$	389.91	10, 113			232–235			
h114	4-Hydroxy-3,5-dinitrobenzoic acid	$HOC_6H_2(NO_2)CO_2H$	228.12	1, 183			245 dec			
h115	3-Hydroxydiphenyl-amine	$HOC_6H_4NHC_6H_5$	185.23	13, 410			80–82	340		s organic solvents, alk
h116	(2-Hydroxydiphenyl)-methane	$HOC_6H_4CH_2C_6H_5$	184.24	6, 675		1.5994[20]	54	312	>110	
h117	(4-Hydroxydiphenyl)-methane	$HOC_6H_4CH_2C_6H_5$	184.24	6, 675			84	322		s hot aq, org solvents, HOAc, alkalis
h118	2-(2-Hydroxyethoxy)-phenol	$HOCH_2CH_2C_6H_4OH$	154.17	6[2], 782			99–100	128[0.7mm]		
h119	N-(2-Hydroxyethyl)-acetamide	$HOCH_2CH_2NHC(=O)CH_3$	103.12	4[1], 430	1.1233[20,20]	1.4575[20]	63–65	155[5mm]	176	misc aq; sl s bz
h120	2-Hydroxyethyl acetate	$CH_3CO_2CH_2CH_2OH$	104.11	2, 141	1.108[15]	1.4201[20]		188	88	
h121	2-Hydroxyethyl acrylate	$H_2C=CHCO_2CH_2CH_2OH$	116.12	2[4], 1469	1.011	1.4500[20]		92[12mm]	98	misc aq, alc, chl, eth

No.	Name	Formula	Mol. wt.	Beilstein Ref.	Density	n_D	M.p., °C	B.p., °C	Flash pt., °C	Solubility
h122	3-(1-Hydroxyethyl)-aniline	$CH_3CH(OH)C_6H_4NH_2$	137.18	13^3, 1654			66–69		>110	
h123	2-Hydroxyethyl disulfide	$HOCH_2CH_2SSCH_2CH_2OH$	154.25	1, 471	1.261	1.5655^{20}	25–27	$158^{3.5mm}$	73	misc aq; s alc
h124	N-(2-Hydroxyethyl)-ethylenediamine-N,N',N'-triacetic acid	$HO_2CCH_2N(CH_2CH_2OH)CH_2CH_2N(CH_2CO_2H)_2$	278.26				212 dec			
h125	2-Hydroxyethyl-hydrazine	$HOCH_2CH_2NHNH_2$	76.10	4^1, 562	1.123	1.4961^{20}	−70	220	97	
h126	2-Hydroxyethyl methacrylate	$HOCH_2CH_2O_2CC(CH_3)=CH_2$	130.14		1.073	1.4520^{20}		$67^{3.5mm}$	99	
h127	N-(2-Hydroxyethyl)-morpholine		131.18	27, 7	1.083	1.4760^{20}		227		misc aq
h128	N-(2-Hydroxyethyl)-phthalimide		191.19	21, 469			126–128	246		
h129	1-(2-Hydroxyethyl)-piperazine		130.19	23^2, 6	1.061	1.5065^{20}			>110	
h130	N-(2-Hydroxyethyl)-piperazine-N'-ethane-sulfonic acid		238.31	Merck: 12, 4687			234 dec			sat'd aq: 2.25M^0
h131	N-(2-Hydroxyethyl)-piperidine		129.20	20, 25	1.0059_4^{15}	1.4804^{20}		199–202	68	
h132	N-(2-Hydroxyethyl)-pyridine	$HOCH_2CH_2NC_5H_4$	123.16	21, 50	1.093	1.5368^{20}		116^{9mm}	92	v s aq, alc, chl
h133	N-(2-Hydroxyethyl)-pyrrolidine	$HOCH_2CH_2NC_4H_8$	115.8	20^2, 5	0.985	1.4713^{20}		81^{13mm}	56	
h134	N-(2-Hydroxyethyl)-2-pyrrolidinone		129.16	21^4, 3142	1.143	1.4960^{20}		142^{2mm}	>110	
h135	2-Hydroxyethyl salicylate	$(HO)C_6H_4CO_2CH_2CH_2OH$	182.18	10, 81	1.224	1.5480^{20}		166^{13mm}	>110	
h136	(2-Hydroxyethyl)tri-phenylphosphonium bromide	$HOCH_2CH_2P(C_6H_5)_3Br$	387.26	16, 761			217–219			
h137	8-Hydroxy-7-iodo-5-quinolinesulfonic acid		351.12	22, 408			269–270 dec			

(*Continued*)

TABLE 2.21 Physical Constants of Organic Compounds (*Continued*)

No.	Name	Formula	Formula weight	Beilstein reference	Density, g/mL	Refractive index	Melting point, °C	Boiling point, °C	Flash point, °C	Solubility in 100 parts solvent
h138	2-Hydroxyisobutyric acid	$(CH_3)_2C(OH)CO_2H$	104.11	3, 313			82	$84^{1.5mm}$		v s aq, alc, eth
h138a	2-Hydroxyisobutyronitrile	$(CH_3)_2C(OH)CN$	85.11	3, 316	0.932	1.3990^{20}	−19	82^{23mm}	63	
h139	Hydroxylamine HCl	$H_2NOH \cdot HCl$	69.49		1.670		159 dec			v s aq NH$_3$, alkalis; sl s alc, acet
h140	4-Hydroxy-2-mercapto-6-methylpyrimidine		142.18	24, 351			330 dec			0.1 aq; 1.7 alc; 1.7 acet; v s alkalis
h141	4-Hydroxy-2-mercapto-6-propylpyrimidine		170.23				219–221			1 aq; s alc, chl, pyr
h142	4-Hydroxy-3-methoxybenzaldehyde	$CH_3OC_6H_3(OH)CHO$	152.15	8, 247	1.056		80–81	285		0.12 aq; v s alc
h143	4-Hydroxy-3-methoxybenzoic acid	$CH_3OC_6H_3(OH)CO_2H$	168.15	10, 392			210–213			v s alc, chl, eth
h144	2-Hydroxy-4-methoxybenzophenone	$CH_3OC_6H_3(OH)C(=O)C_6H_5$	228.25	8, 312			63–66	160^{5mm}		
h145	4-Hydroxy-3-methoxybenzyl alcohol	$CH_3OC_6H_3(OH)CH_2OH$	154.17	6, 1113			113–115			
h146	N-(Hydroxymethyl)-acrylamide	$H_2C=CHC(=O)NHCH_2OH$	101.11	2^4, 1472	1.074	1.430^{20}			none	
h147	4-Hydroxy-3-methyl-2-butanone	$HOCH_2CH(CH_3)C(=O)CH_3$	102.13	1^1, 422	0.993	1.4340^{20}		92^{15mm}	81	
h148	7-Hydroxy-4-methylcoumarin		176.17	18, 31			190–192			s alc, HOAc; sl s eth
h149	N-(Hydroxymethyl)-nicotinamide	$(C_5H_4N)C(=O)NHCH_2OH$	152.15	10, 4750			152–154			
h150	4-Hydroxy-4-methyl-2-pentanone	$(CH_3)_2C(OH)CH_2C(=O)CH_3$	116.16	Merck: 12, 3008	0.9306^{25}_{4}	1.4235^{20}	−44	167.91	58	misc aq
h151	N-(Hydroxymethyl)-phthalimide		177.16	21, 475			147–149	200		sl s aq, alc, bz
h152	4-Hydroxy-N-methyl-piperidine		115.18	21^1, 188		1.4775^{20}	29–31			
h153	2-Hydroxy-2-methyl-propionitrile	$(CH_3)_2C(OH)CN$	85.10	3, 316	0.9267^{25}_{4}	1.3992^{20}	−19	95	63	
h154	2-Hydroxy-2-methyl-propiophenone	$C_6H_5C(=O)C(CH_3)_2OH$	164.20	8^1, 553	1.077	1.5330^{20}		103^{4mm}	>110	
h155	5-Hydroxy-2-methyl-pyridine	$HO(C_5H_2N)CH_3$	109.13	21^3, 480			168–170			s aq, alc, chl, eth

No.	Name	Formula	Mol. wt.	Beilstein ref.	Density	n_D	m.p., °C	b.p., °C	Flash pt	Solubility
h156	3-Hydroxy-2-methyl-4-pyrone		126.11				161–162			1.2 aq; v s hot aq; s alc, alk; sl s bz, eth
h157	2-Hydroxy-1-naphth-aldehyde	$C_{10}H_6(OH)CHO$	172.18	8, 143			82–85	192^{27mm}		v s alc, bz, eth, alk
h158	1-Hydroxy-2-naphthoic acid	$C_{10}H_6(OH)CO_2H$	188.18	10, 331			191–192			
h159	2-Hydroxy-1-naphthoic acid	$C_{10}H_6(OH)CO_2H$	188.18	10, 328			167 dec			
h160	3-Hydroxy-2-naphthoic acid	$C_{10}H_6(OH)CO_2H$	188.18	10, 333			222–223			v s alc, eth; s bz, chl
h161	2-Hydroxy-1,4-naphthoquinone		174.16	8, 300			dec > 191			s HOAc
h162	4-Hydroxy-3-nitro-benzenearsonic acid	$HOC_6H_3(NO_2)AsO(OH)_2$	263.04	16[1], 456			>300			v s alc, acet, HOAc, alk; sl s aq; i eth
h163	4-Hydroxy-3-nitro-benzoic acid	$HOC_6H_3(NO_2)CO_2H$	183.12	10, 181			184–185			
h164	5-Hydroxy-2-pentanone	$CH_3C(=O)CH_2CH_2CH_2OH$	102.13	1, 831	1.0074^{20}_4	1.4372^{20}		144^{100mm}	93	misc aq; s alc, eth
h165	4-Hydroxyphenylacetic acid	$HOC_6H_4CH_2CO_2H$	152.15	10, 190			149–151			v s alc, eth; sl s aq
h166	4-(4-Hydroxyphenyl)-2-butanone	$HOC_6H_4CH_2CH_2C(=O)CH_3$	164.20	8[2], 117			82–83			
h167	4-Hydroxyphenyl-glycine	$HOC_6H_4CH(NH_2)CO_2H$	167.16	14[1], 659			240 dec			sl s aq, alc, bz, acet s alk, acid;
h168	N-(4-Hydroxyphenyl)-glycine	$HOC_6H_4NHCH_2CO_2H$	167.16	13, 488			244 dec			v sl s aq, alc, acet, bz, eth
h169	2′-Hydroxy-3-phenyl-propiophenone	$HOC_6H_4C(=O)CH_2CH_2C_6H_5$	226.28	8[2], 202		1.5968^{20}	36–37		>110	
h170	1-(3-Hydroxyphenyl)-urea	$HOC_6H_4NHC(=O)NH_2$	152.15	13, 417			182–184			
h171	N-Hydroxyphthalimide		163.13	21, 500			233 dec			
h172	2-Hydroxypropionitrile	$CH_3CH(OH)CN$	71.08	3[2], 209	0.9834^{25}_4	1.4027^{25}	-40	103^{50mm}	76	misc aq, alc; s eth
h173	3-Hydroxypropionitrile	$HOCH_2CH_2CN$	71.08	3, 298	1.0404^{25}_4	1.4248^{20}	-46	221	129	misc aq, alc, acet; 2,3 eth; i bz, PE
h174	2′-Hydroxypropio-phenone	$HOC_6H_4C(=O)CH_2CH_3$	150.18	8, 102	1.094	1.5480^{20}		115^{15mm}	>110	v s alc, eth; sl s aq
h175	4′-Hydroxypropio-phenone	$HOC_6H_4C(=O)CH_2CH_3$	150.18	8, 102			148			v s alc, eth; sl s aq

(Continued)

TABLE 2.21 Physical Constants of Organic Compounds (*Continued*)

No.	Name	Formula	Formula weight	Beilstein reference	Density, g/mL	Refractive index	Melting point, °C	Boiling point, °C	Flash point, °C	Solubility in 100 parts solvent
h176	1-(2-Hydroxy-1-propoxy)-2-propanol	$CH_3CH(OH)CH_2OCH_2CH(OH)CH_3$	134.18		1.0252^{20}_{20}	1.4440^{20}		231.8	138	misc aq, alc
h177	Hydroxypropyl acrylate	$H_2C{=}CHCO_2(CH_2)_3OH$	130.14	2^4, 1469	1.044	1.4450^{20}		77^{5mm}	89	
h178	Hydroxypropyl methacrylate	$H_2C{=}C(CH_3)CO_2(CH_2)_3OH$	144.17	2^4, 1532	1.066	1.4470^{20}		$57^{0.5mm}$	96	
h179	2-Hydroxypyridine	HOC_5H_4N	95.10	21, 43			105–107	280–281		aq, alc, bz, sl s eth
h180	3-Hydroxypyridine	HOC_5H_4N	95.10			126–129		151^{3mm}		v s aq, alc; sl s eth
h181	4-Hydroxypyridine	HOC_5H_4N	95.18					230^{12mm}		v s aq; i alc, bz, eth
h182	2-Hydroxypyridine-5-carboxylic acid	$HO(C_5H_3N)CO_2H$	139.11	22, 215			>300			sl s aq, alc, eth
h183	3-Hydroxypyridine-N-oxide	$(HO)C_5H_4N{=}O$	111.10				190–192			
h184	8-Hydroxyquinoline		145.16	21, 91			72–74	267^{742mm}		v s alc, acet, bz, chl
h185	8-Hydroxyquinoline-5-sulfonic acid		225.22	22, 407			>300			v s aq; sl s alc, eth
h186	DL-Hydroxysuccinic acid	$HO_2CCH(OH)CH_2CO_2H$	134.09	3, 435			131–133			56 aq; 45 EtOH; 18 acet; 0.8 eth; 23 diox
h187	(−)-Hydroxysuccinic acid	$HO_2CCH(OH)CH_2CO_2H$	134.09	3, 419			100			36 aq; 87 EtOH; 61 acet; 2.7 eth; 75 diox
h188	N-Hydroxysuccinimide		115.09	21, 380			95–98			v s aq
i1	Icosane	$CH_3(CH_2)_{18}CH_3$	282.56	1, 174	0.7777^{37}	1.4346^{40}	36.4	343.8	>112	
i2	1-Icosene	$CH_3(CH_2)_{17}CH{=}CH_2$	280.54	1^3, 881			28.7	342.4		
i3	1H-Imidazole		68.08	23, 45			90–91	257	145	v s aq, alc, chl, eth
i4	2-Imidazolidinethione		102.16	24, 4			203–204			2 aq; s alc, pyr; i bz, acet, chl, eth
i5	2-Imidazolidone		86.09	24, 16			133–135			v s aq, hot alc
i6	3,3′-Iminobis(N,N-di-methyl)propylamine	$HN[(CH_2)_3N(CH_3)_2]_2$	187.33	4^3, 565	0.841	1.4490^{20}	−78	131^{20mm}	98	
i7	Iminodiacetic acid	$HO_2CCH_2NHCH_2CO_2H$	133.10	4, 365			243 dec			2 aq; v sl s bz, eth
i8	Iminodiacetonitrile	$NCCH_2NHCH_2CN$	95.11	4, 367			77			s aq, alc; sl s eth
i9	Iminodibenzyl		195.27				105–108	178	50	s alc, chl, eth; i aq
i10	Indane		118.18	Merck: 12, 4966	0.9639^{20}_{4}	1.5383^{20}	−51.4			

618

No.	Name	Formula	Mol. wt.	Beil. ref.	Density	n_D	m.p. °C	b.p. °C	Sol.	Solubility
i11	5-Indanol		134.18	6, 575	1.1090^{45}_{4}	1.561^{45}	51–53	255	>110	v s alc, eth; sl s aq
i12	1-Indanone		132.16	7, 360			40–42	243–245	111	s alc, eth; sl s aq
i13	1,2,3-Indantrione hydrate		178.14	Merck: 12, 6645			dec 241			v s aq; s alc
i14	Indene		116.16	5, 515	0.9968^{20}_{4}	1.5762^{20}	−1.8	181.6	58	misc alc, bz, chl, eth
i15	Indole		117.15	20, 304	1.0643	1.609^{60}	52.54	253–254	>110	s hot aq, bz, eth
i16	Indole-3-acetic acid		175.19	22, 66			168–170			v s alc; s acet, eth
i17	Indole-2,3-dione		147.13	21, 432			203.5 dec			s hot aq, hot alc, alk
i18	Indoline		119.17	20, 257	1.063	1.5906^{20}		221	92	sl s aq
i19	Inositol		180.16	6^2, 1157	1.752		225			14 aq; sl s alc; i eth
i20	Iodoacetamide	ICH_2CONH_2	184.96	2, 223			93–96			s hot aq
i21	Iodoacetic acid	ICH_2CO_2H	185.95	2, 222			79–82			s aq, alc; v sl s eth
i22	3-Iodoaniline	$IC_6H_4NH_2$	219.03	12, 670	1.821	1.6820^{20}	25	146^{15mm}	>110	i aq; s alc, eth
i23	Iodobenzene	C_6H_5I	204.01	5, 215	1.8308^{20}	1.6200^{20}	−31	188	74	misc alc, chl, eth
i24	Iodobenzene diacetate	$C_6H_5I(O_2CCH_3)_2$	322.10	5, 218			163–165			
i25	2-Iodobenzoic acid	$IC_6H_4CO_2H$	248.02	9, 363	2.249^{25}_{4}		162–164			s alc, eth; sl s aq
i26	1-Iodobutane	$HC_3CH_2CH_2CH_2I$	184.02	1, 123	1.6154^{20}	1.4999^{20}	−103.5	130–131	33	i aq; s alc, eth
i27	2-Iodobutane	$CH_3CH_2CH(I)CH_3$	184.02	1, 123	1.5920^{20}	1.4991^{20}	−104.0	120	23	i aq; s alc, eth
i28	Iodocyclohexane	$C_6H_{11}I$	210.06	5^2, 13	1.626^{15}_{15}	1.5472^{20}		180		i aq; s eth
i29	1-Iododecane	$CH_3(CH_2)_9I$	268.19	1, 168	1.257^{20}_{4}	1.4850^{20}	−3	132^{15mm}	>110	i aq; s alc, eth
i30	2-Iodododecane	$CH_3(CH_2)_{11}I$	296.24	1^1, 67	1.201	1.4844		160^{15mm}	>110	i aq; s alc, eth
i31	Iodoethane	CH_3CH_2I	155.97	1, 96	1.9358^{20}	1.5130^{20}	−111	72.4	none	0.4 aq; misc alc, bz, chl, eth
i32	2-Iodoethanol	ICH_2CH_2OH	171.97	1, 339	2.2197^{20}_{4}	1.5694^{20}		75^{5mm}	65	s aq; v s alc, eth
i33	Iodoform	CHI_3	393.73	1, 73	4.008		120–123		none	1.4 alc; 10 chl; 13 eth; v s bz, acet
i34	1-Iodoheptane	$CH_3(CH_2)_6I$	226.10	1, 155	1.3734^{20}	1.4900^{20}	−48	204	78	i aq; s alc, eth
i35	1-Iodohexadecane	$CH_3(CH_2)_{15}I$	352.35	1, 172	1.121	1.4806^{20}	23	207^{10mm}	>110	i aq; misc alc, eth
i36	1-Iodohexane	$CH_3(CH_2)_5I$	212.08	1, 146	1.4374^{20}	1.4920^{20}		179–180	61	i aq
i37	1-Iodomethane	CH_3I	141.94	1, 69	2.2789^{4}_{4}	1.5308^{20}	−66.5	42.5	none	1.4 aq; misc alc, eth
i38	1-Iodo-2-methyl-propane	$(CH_3)_2CHCH_2I$	184.02	1, 128	1.6035^{20}	1.4960^{20}	−93.5	121	12	i aq; misc alc, eth
i39	2-Iodo-2-methyl-propane	$(CH_3)_3Cl$	184.02	1^3, 326	1.5710^{0}	1.4918^{20}	−38	100	7	dec aq; misc alc, eth
i40	1-Iodo-3-nitrobenzene	$IC_6H_4NO_2$	249.01	5, 253	1.9477^{50}_{4}		36–38	280	71	i aq; s alc, eth
i41	1-Iodo-4-nitrobenzene	$IC_6H_4NO_2$	249.01	5, 252			175–177	289^{772mm}	>110	
i42	1-Iodononane	$CH_3(CH_2)_8I$	254.18	1, 166	1.288	1.4870^{20}		108^{8mm}	85	
i43	1-Iodooctadecane	$CH_3(CH_2)_{17}I$	380.40	1, 173			33–35	197^{2mm}	>110	
i44	1-Iodooctane	$CH_3(CH_2)_7I$	240.13	1, 160	1.330^{20}_{4}	1.4889^{20}	−46	226	95	s alc, eth
i47	1-Iodopentane	$CH_3(CH_2)_4I$	198.06	1, 133	1.512^{20}_{4}	1.4954^{20}	−85	155	51	sl s aq; s alc, eth

(Continued)

TABLE 2.21 Physical Constants of Organic Compounds (*Continued*)

No.	Name	Formula	Formula weight	Beilstein reference	Density, g/mL	Refractive index	Melting point, °C	Boiling point, °C	Flash point, °C	Solubility in 100 parts solvent
i48	1-Iodopropane	$CH_3CH_2CH_2I$	169.99	1, 113	1.7489^{20}	1.5058^{20}	−101	102	44	0.1 aq; misc alc, eth
i49	2-Iodopropane	$(CH_3)_2CHI$	169.99	1, 114	1.7042^{20}	1.4992^{20}	−90	89.5	42	0.14 aq; misc alc, eth
i50	3-Iodo-1-propene	$ICH_2CH{=}CH_2$	167.97	1, 202	1.845^{22}_{4}	1.5540^{21}	−99	103	18	misc alc, chl, eth
i51	5-Iodosalicylic acid	$IC_6H_3(OH)CO_2H$	264.02	10, 112			189–191			v s alc; i bz, chl
i52	2-Iodothiophene		210.04	17, 34	1.902	1.6520^{20}	−40	73^{15mm}	71	v s eth
i53	2-Iodotoluene	$IC_6H_4CH_3$	218.04	5, 310	1.713	1.6079^{20}		211	90	i aq; s alc, eth
i54	3-Iodotoluene	$IC_6H_4CH_3$	218.04	5, 311	1.698	1.6040^{20}		82^{10mm}	82	i aq; misc alc, eth
i55	4-Iodotoluene	$IC_6H_4CH_3$	218.04	5, 312			34–36	211	90	i aq; misc alc, eth
i56	Iodotrimethylsilane	$(CH_3)_3SiI$	200.10	1^1, 66	1.406^{4}	1.4710^{20}		106	−31	s alc, bz, chl, eth
i57	1-Iodoundecane	$CH_3(CH_2)_{10}I$	282.21		1.220	1.4849^{20}		130^{5mm}	>110	s alc, bz, chl, eth
i58	α-Ionone		192.30	7, 168	0.932^{20}	1.4980^{20}		124^{11mm}	104	sl s aq, hot alc, acet
i59	β-Ionone		192.30	7, 167	0.946^{17}	1.521^{17}		128^{12mm}	>110	s aq, alc, acet, pyr
i60	Isatoic anhydride		163.13	27, 264			233 dec			v s alc, chl, eth
i61	D-(−)-Isoascorbic acid		176.12	6^2, 80			169 dec			0.7 aq; v s alc
i62	DL-Isoborneol		154.25				214 subl			
i63	Isobutyl acetate	$(CH_3)_2CHCH_2O_2CCH_3$	116.16	2, 131	0.8712^{20}	1.3902^{20}	−99	116.5	18	misc aq, alc, acet, eth
i64	Isobutyl acetoacetate	$CH_3COCH_2CO_2CH_2CH(CH_3)_2$	158.20		0.980	1.4240^{20}		100^{22mm}	78	misc alc, eth
i65	Isobutyl acrylate	$H_2C{=}CHCO_2CH_2CH(CH_3)_2$	128.19	2^3, 1227	0.890	1.4140		132	32	misc bz, chl, eth
i66	Isobutylamine	$(CH_3)_2CHCH_2NH_2$	73.14	4, 163	0.724^{20}	1.3972^{20}	−86.6	68	−9	1 aq; misc alc, eth
i67	Isobutylbenzene	$C_6H_5CH_2CH(CH_3)_2$	134.22	5, 414	0.8532^{20}	1.4866^{20}	−51.5	172.8	55	0.5 aq; misc alc
i68	Isobutyl chloroformate	$ClCO_2CH_2CH(CH_3)_2$	136.58	3, 12	1.053	1.4070^{20}		128.8	27	misc bz, chl, eth
i69	Isobutyl formate	$HCO_2CH_2CH(CH_3)_2$	102.13	2, 21	0.8776^{20}	1.3855^{20}	−95.5	98.4	10	1 aq; misc alc, eth
i70	Isobutyl isobutyrate	$(CH_3)_2CHCH_2O_2CCH(CH_3)_2$	144.22	2, 291	0.8542^{20}	1.3999^{20}		148.5	38	0.5 aq; misc alc
i71	Isobutyl methacrylate	$H_2C{=}C(CH_3)CO_2CH_2CH(CH_3)_2$	142.19	2^3, 1287	0.882^{25}_{15}	1.4170^{25}	−80.7	155	41	misc alc, eth
i72	Isobutyl nitrate	$(CH_3)_2CHCH_2ONO_2$	119.12	1, 377	1.015^{24}	1.4028^{20}		123	21	i aq; misc alc, eth
i73	Isobutyl nitrite	$(CH_3)_2CHCH_2ONO$	103.12	1, 377	0.870^{22}_{4}	1.3715^{22}		67	−21	misc alc; sl s aq (dec)
i74	Isobutyl propionate	$C_2H_5CO_2CH_2CH(CH_3)_2$	130.19	2, 241	0.888^{0}	1.3974^{20}	−71	137	26	i aq; misc alc
i75	Isobutyl stearate	$CH_3(CH_2)_{16}CO_2CH_2CH(CH_3)_2$	340.57				ca. 20	190–191	60	
i76	Isobutyltriethoxy-silane	$(CH_3)_2CHCH_2Si(OC_2H_5)_3$	220.39		0.880	1.400^{20}		137	39	
i77	Isobutyltrimethoxy-silane	$(CH_3)_2CHCH_2Si(OCH_3)_3$	178.30		0.930	1.3960^{20}				
i78	Isobutyl vinyl ether	$(CH_3)_2CHCH_2OCH{=}CH_2$	100.16	1^3, 1862	0.7702^{20}	1.3950^{20}	−112	83.4	−13	0.2 aq
i79	Isobutyraldehyde	$(CH_3)_2CHCHO$	72.11	1, 671	0.7988^{20}	1.3723^{20}	−65.9	64.5	−18 (CC)	11 aq; misc alc, bz, acet, chl, eth
i80	Isobutyramide	$(CH_3)_2CHCONH_2$	87.12	2, 293	1.013		127–129	216–220		

No.	Name	Formula	Formula wt.	Beilstein ref.	Density	n_D	mp (°C)	bp (°C)	Flash pt.	Solubility
i81	Isobutyric acid	$(CH_3)_2CHCO_2H$	88.11	2, 288	0.9681^{20}	1.3925^{20}	−46	154	56	17 aq; misc alc, chl, eth
i82	Isobutyric anhydride	$[(CH_3)_2CHCO]_2O$	158.20	2, 292	0.954	1.4062^{20}	−56	182	59	v s alc, eth; sl s aq
i83	Isobutyronitrile	$(CH_3)_2CHCN$	69.11	2, 294	0.7704^{20}	1.3720^{20}	−71.5	104	8	
i84	Isobutyrophenone	$C_6H_5COCH(CH_3)_2$	148.21	7, 316	0.988^{20}	1.5172		217	84	
i85	Isobutyryl chloride	$(CH_3)_2CHCOCl$	106.55	2, 293	1.017	1.4073^{20}	−90	91–93	1	dec aq, dec alc; s eth
i86	Isodecyl acrylate	$H_2C{=}CHCO_2C_{10}H_{21}$	212.34		0.875	1.4420^{20}		121^{10mm}	106	
i87	Isodecyl methacrylate	$H_2C{=}C(CH_3)CO_2C_{10}H_{21}$	226.36		0.878	1.4430^{20}		126^{10mm}	>110	
i88	L-Isoleucine	$C_2H_5C(CH_3)CH(NH_2)CO_2H$	131.18	4, 454			288 dec	subl 168		4 aq; sl s hot alc
i89	Isooctyl acrylate	$H_2C{=}CHCO_2C_8H_{17}$	184.25		0.880	1.4370^{20}		125^{20mm}	80	
i90	Isooctyl diphenyl phosphite	$(C_6H_5O)_2POC_8H_{17}$	346.41		1.045	1.5220^{20}		188		
i91	Isopentyl acetate	$CH_3CO_2CH_2CH_2CH(CH_3)_2$	130.19	2, 132	0.876^{15}	1.4007^{20}	−78.5	142	25	0.25 aq; misc alc, eth
i92	Isopentyl nitrite	$ONOCH_2CH_2CH(CH_3)_2$	117.15	1, 402	0.872	1.3860^{20}		99	10	misc alc, eth; sl s aq
i93	Isophorone		138.21	7, 65	0.955^{20}	1.4759^{20}	−8.1	215.2	84	1.2 aq
i94	Isophorone diisocyanate		222.29		1.049	1.4841^{20}		159^{15mm}	>110	
i95	Isopropenyl acetate	$CH_3CO_2C(CH_3){=}CH_2$	100.12	2^2, 278	0.909	1.4005^{20}		94	18	
i96	3-Isopropenyl-α,α-dimethylbenzyl isocyanate	$H_2C{=}C(CH_3)C_6H_4C(CH_3)_2NCO$	201.27		1.108	1.5300^{20}		268–271	>110	
i97	2-Isopropoxyethanol	$(CH_3)_2CHOCH_2CH_2OH$	104.15	1^2, 519	0.903	1.4104^{20}		44^{13mm}	45	3 aq; misc alc, eth
i98	3-Isopropoxypropylamine	$(CH_3)_2CHO(CH_2)_3NH_2$	117.19	4^3, 739	0.845	1.4195^{20}		79^{85mm}	39	misc aq, alc, eth
i99	Isopropyl acetate	$(CH_3)_2CHO_2CCH_3$	102.13	2, 130	0.8718^{20}	1.3770^{20}	−73	89	2	3 aq; misc alc, eth
i100	Isopropylamine	$(CH_3)_2CHNH_2$	59.11	4, 152	0.686^{25}	1.3711^{25}	−95	31.7	−37	misc aq, alc, eth
i101	2-Isopropylaniline	$(CH_3)_2CHC_6H_4NH_2$	135.2	12, 1147	0.955	1.5477^{20}		222	95	
i102	4-Isopropylbenzaldehyde	$(CH_3)_2CHC_6H_4CHO$	148.21	7, 318	0.977	1.5298^{20}		236	93	
i103	Isopropylbenzene	$(CH_3)_2CHC_6H_5$	120.20	5, 393	0.864^{20}	1.4915^{20}	−96	152–154	36	s alc, bz, eth
i104	4-Isopropylbenzyl alcohol	$(CH_3)_2CHC_6H_4CH_2OH$	150.22	6, 543	0.982^{15}	1.5206^{20}	28	248.4	>110	misc alc, eth; i aq
i105	N-Isopropylbenzylamine	$C_6H_5CH_2NHCH(CH_3)_2$	149.24		0.892	1.5025^{20}		200	87	
i106	Isopropyl butyrate	$CH_3CH_2CH_2CO_2CH(CH_3)_2$	130.19	2, 271	0.859	1.3932^{20}		131	30	
i107	Isopropyl chloroacetate	$ClCH_2CO_2CH(CH_3)_2$	136.58	2, 198	1.096	1.4190^{20}		149–150	70	v s alc, eth
i108	Isopropylcyclohexane	$C_6H_{11}CH(CH_3)_2$	126.24	5, 41	0.8023^{20}	1.4399^{20}	−90	155	35	

(Continued)

TABLE 2.21 Physical Constants of Organic Compounds (*Continued*)

No.	Name	Formula	Formula weight	Beilstein reference	Density, g/mL	Refractive index	Melting point, °C	Boiling point, °C	Flash point, °C	Solubility in 100 parts solvent
i109	Isopropyl hexadecanoate	$CH_3(CH_2)_{14}CO_2CH(CH_3)_2$	298.51	2[2], 336	0.862	1.4385[20]			>110	
i110	4,4'-Isopropylidene-bis(2,6-dibromo-phenoxy)ethanol	$(CH_3)_2C[C_6H_2(Br)_2OCH_2CH_2OH]_2$	632.01				107			
i111	4,4'-Isopropylidene-bis(diisodecyl phenyl phosphite)	$[(C_{10}H_{21}O)_2POC_6H_4]_2C(CH_3)_2$	917.34		0.964	1.4980[20]		336	>110	
i112	4,4'-Isopropylidene-dicyclohexanol	$(CH_3)_2C(C_6H_{10}OH)_2$	240.39	6[2], 761				234[14mm]	>110	
i113	4,4'-Isopropylidene-diphenol	$(CH_3)_2C(C_6H_4OH)_2$	228.29	6, 1011			137–140	220[4mm]		
i114	2-Isopropylimidazole	$(CH_3)_2CHCNO$	110.16	23, 83			129–131	256–260		
i115	Isopropyl isocyanate		85.11	4, 155	0.866	1.3825[20]		74–75	−2	
i116	Isopropyl S-(−)-lactate	$(CH_3)_2CHO_2CCH(OH)CH_3$	132.16	3, 282	0.998[20]	1.4082[25]		166–168	57	s aq, alc, eth
i117	2-Isopropyl-6-methylaniline	$(CH_3)_2CHC_6H_3(CH_3)NH_2$	149.24		0.957	1.5440[20]			41	
i118	2-Isopropyl-1-methylbenzene	$(CH_3)_2CHC_6H_4CH_3$	134.21	5, 419	0.8766[20]	1.5006[20]	−71.5	178.2		misc alc, eth
i119	3-Isopropyl-1-methyl-benzene	$(CH_3)_2CHC_6H_4CH_3$	134.21	5, 419	0.8610[20]	1.4930[20]	−63.8	175.1		misc alc, eth
i120	4-Isopropyl-1-methyl-benzene	$(CH_3)_2CHC_6H_4CH_3$	134.21	5, 420	0.8573[20]	1.4909[20]	−68.9	177.1	47	misc alc, eth
i121	2-Isopropyl-5-methyl-phenol	$(CH_3)_2CHC_6H_3(CH_3)OH$	150.22	6, 532	0.9254[80]		51.5	232.5		i aq; v s alc, chl, eth
i122	4-Isopropyl-3-methyl-phenol	$(CH_3)_2CHC_6H_3(CH_3)OH$	150.22	6[2], 491			111–114			
i123	5-Isopropyl-3-methyl-phenol	$(CH_3)_2CHC_6H_3(CH_3)OH$	150.22	6, 526			51		>110	
i124	Isopropyl nitrate	$(CH_3)_2CHONO_2$	105.09	1, 363	1.036[19]	1.391[20]		102	12	
i125	Isopropyl nitrite	$(CH_3)_2CHONO$	89.09	Merck: 12, 5235	0.8444[25]	1.3520[20]		39[752mm]		
i126	1-Isopropyl-4-nitro-benzene	$(CH_3)_2CHC_6H_4NO_2$	165.19	5[2], 308	1.090	1.5380[20]		107[11mm]	>110	misc alc, eth
i127	2-Isopropylphenol	$(CH_3)_2CHC_6H_4OH$	136.19	6, 504	1.012[20]	1.5259[20]	15–16	212–213	88	misc alc, eth

(Continued)

No.	Name	Formula	Mol. wt.	Beilstein	Density	n_D	m.p.	b.p.	Solubility no.	Solubility
i128	3-Isopropylphenol	(CH$_3$)$_2$CHC$_6$H$_4$OH	136.19	6, 505	0.994	1.5250^{20}	25	228	104	316 alc; 350 eth
i129	4-Isopropylphenol	(CH$_3$)$_2$CHC$_6$H$_4$OH	136.19	6, 505	0.990^{20}	1.4980^{20}	59–61	212	66	s caster oil, cottonseed oil, acet, EtOAc, EtOH, toluene, mineral oil
i130	4-Isopropylpyridine	(CH$_3$)$_2$CH(C$_5$H$_4$N)	121.18	20, 248	0.938	1.4350^{20}	ca. 3	173	>110	v sl s aq
i131	Isopropyl tetradecanoate	(CH$_3$)$_2$CHO$_2$C(CH$_2$)$_{12}$CH$_3$	270.46	2^3, 923	0.850			193^{20mm}		sl s aq; s acid
i132	Isopulegol		154.25	6, 65	0.912	1.4725^{20}		91^{12mm}	78	s acet, eth; dec aq
i133	Isoquinoline		129.16	20, 380	1.0910^{30}	1.6208^{30}	26.5	243.5	107	s aq, alc; i chl, PE
k1	Ketene	H$_2$C=C=O	42.04	1, 724			−151	−49.8		v s aq, alc, eth
k2	8-Ketotricyclo[5.2.1.0$^{2.6}$]decane		150.22	7^2, 133	1.063	1.5020^{20}		132^{30mm}	101	20 aq; v sl s alc
L1	DL-Lactic acid	CH$_3$CH(OH)CO$_2$H	90.08	3, 268	1.249^{15}		16.8	122^{14mm}	>110	
L2	L-(+)-Lactic acid	CH$_3$CH(OH)CO$_2$H	90.08	3, 261	1.2060^{25}_{4}	1.4270^{20}	53	119^{12mm}	>110	
L3	α-Lactose		342.32	31, 408	1.525^{20}		202	subl 293		
L4	β-Lactose		342.32	31, 408			202	subl 145		45 aq; i alc, eth
L5	DL-Leucine	(CH$_3$)$_2$CHCH$_2$CH(NH$_2$)CO$_2$H	131.18	4, 447			dec 332			1 aq; 0.13 alc; i eth
L6	L-Leucine	(CH$_3$)$_2$CHCH$_2$CH(NH$_2$)CO$_2$H	131.18	4, 437	1.293^{18}		293 dec			2.4 aq^{25}, 0.07 alc; 1 HOAc; i eth
L7	R-(+)-Limonene		136.24	5, 133	0.8411^{20}_{4}	1.4730	−96.5	178	49	misc alc, eth
L8	S-(−)-Limonene		136.24	5, 136	0.8412^{20}_{4}	1.4746^{20}	−96.5	178	48	misc alc, eth
L9	(+)-Limonene oxide		152.24	17, 44	0.929	1.4661^{20}		114^{50mm}	65	misc alc, eth
L10	Linalool		154.25	1, 462	0.865^{15}	1.4615^{20}		197^{720mm}	76	misc alc, eth
L11	Linalyl acetate		196.29	2, 141	0.8954	1.4460^{20}		220	90	
L12	S-(+)-Lysine	H$_2$N(CH$_2$)$_4$CH(NH$_2$)CO$_2$H	146.19	4, 435			212 dec			v s aq; sl s alc; i eth
m1	Maleic acid	HO$_2$CCH=CHCO$_2$H	116.07	2, 748	1.590		130.5			70 aq; 70 alc; s acet, HOAc; sl s eth
m2	Maleic anhydride		98.06	17, 432	1.48		52.8	202	103	s aq (to acid), alc (to ester); 227 acet; 53 chl; 50 bz; 112 EtOAc
m3	Malonic acid	HO$_2$CCH$_2$CO$_2$H	104.06	2, 566	1.63		135–137			154 aq; 42 alc; 8 eth; 14 pyr
m4	Malonodiamide	H$_2$NCOCH$_2$CONH$_2$	102.09	2, 582			172–175			9 aq; i alc, eth
m5	Malononitrile	NCCH$_2$CN	66.06	2, 589	1.1910^{20}_{4}	1.4146^{34}	32–34	220	112	13 aq; 40 alc; 20 eth
m6	Malonyl dichloride	ClCOCH$_2$COCl	140.95	2^1, 252	1.4486^{19}_{4}	1.4620^{20}		551^{9mm}	47	dec hot aq; s eth
m7	D-(+)-Maltose hydrate		360.32	31, 386	1.540^{17}		119–121	dec 130		v s aq; sl s alc; i eth
m8	DL-Mandelic acid	C$_6$H$_5$CH(OH)CO$_2$H	152.15	10, 192	1.300^{20}_{4}		120–122			16 aq; 100 alc; s eth
m9	Mandelonitrile	C$_6$H$_5$CH(OH)CN	133.15	10, 193	1.117	1.5315^{20}	−10	170	97	v s alc, cho, eth; i aq
m10	Mannitol		182.17	1, 534	1.52^{20}		166–168	$290^{3.5mm}$		18 aq; 1.2 alc; i eth

TABLE 2.21 Physical Constants of Organic Compounds (*Continued*)

No.	Name	Formula	Formula weight	Beilstein reference	Density, g/mL	Refractive index	Melting point, °C	Boiling point, °C	Flash point, °C	Solubility in 100 parts solvent
m11	D-(+)-Mannose		180.16	31, 284	1.54^{20}		128–130			250 aq; 28 pyr; 0.8 alc
m12	(−)-Menthol		156.27	6, 28	0.890^{15}_{15}	1.458^{25}	41–43	212	93	v s alc, chl, eth, PE
m13	(−)-Menthone		154.25	7, 38	0.895^{24}_{24}	1.4510^{20}	−6	207	72	misc alc, eth; sl s aq
m14	S-(+)-Menthyl acetate		198.31	6, 32	1.4480^{20}			229–230	77	
m15	Menthyl anthranilate		275.40	14^3, 885	1.040	1.5420^{20}	−16.5	179^{3mm}	>110	misc aq, alc, bz, eth
m16	Mercaptoacetic acid	$HSCH_2CO_2H$	92.12	3, 245	1.325	1.5030^{20}		96^{5mm}	>110	sl s aq; s alc
m17	2-Mercaptobenzimidazole		150.20	24, 119			301–305			v s alc, HOAc
m18	2-Mercaptobenzoic acid	$HSC_6H_4CO_2H$	154.19	10, 125			165–168			2 alc; 1 eth; 10 acet; 1 bz; s alk; i aq
m19	2-Mercaptobenzothiazole		167.25	27, 185	1.42^{20}_4		180–181	dec		misc aq, alc, bz, eth
m20	2-Mercaptoethanol	$HSCH_2CH_2OH$	78.13	1, 470	1.1143^{20}_4	1.5006^{20}		156.9	73	misc alc; v s acet
m21	3-Mercapto-1,2-propanediol	$HSCH_2CH(OH)CH_2OH$	108.16	1, 519	1.295^{14}_{14}	1.5243^{20}		118^{5mm}	>110	
m22	2-Mercaptopropionic acid	$CH_3CH(SH)CO_2H$	106.14	3, 289	1.220^{15}_4	1.4809^{20}	10–14	102^{16mm}	87	misc aq, alc, eth, acet
m23	3-Mercaptopropionic acid	$HSCH_2CH_2CO_2H$	106.14	3, 299	1.218	1.4911^{20}	17–19	111^{15mm}	93	
m24	(3-Mercaptopropyl)-trimethoxysilane	$HS(CH_2)_3Si(OCH_3)_3$	196.34		1.039^{20}_4	1.4440^{20}		198	48	
m25	Mercaptosuccinic acid	$HO_2CCH_2CH(SH)CO_2H$	150.15	3, 439			5–7			50 aq; 50 alc; s eth
m26	2-Mercaptothiazoline		119.21	27, 140			105–107			
m27	Methacrylaldehyde	$H_2C=C(CH_3)CHO$	70.09	1, 731	0.847	1.4160^{20}	−81	69	−15	6 aq; misc alc, eth
m28	Methacrylamide	$H_2C=C(CH_3)CONH_2$	85.11	2^2, 399			109–111			s alc; sl s eth
m29	Methacrylic acid	$H_2C=C(CH_3)CO_2H$	86.09	2, 421	1.0153^{20}_4	1.4314^{20}	16	163	77	9 aq; misc alc, eth
m30	Methacrylic anhydride	$[H_2C=C(CH_3)CO]_2O$	154.17	2^3, 1293	1.035	1.4530^{20}		87^{13mm}	84	
m30a	Methacrylonitrile	$H_2C=C(CH_3)CN$	67.91	2, 423	0.8001^{20}_4	1.4007^{20}	−35.8	90.3	1.1	2.6 aq; misc acet, bz
m31	Methacryloyl chloride	$H_2C=C(CH_3)COCl$	104.54	2^2, 394	1.070	1.4420^{20}		95–96	2	
m32	Methallylidene diacetate	$(CH_3CO_2)_2CHC(CH_3)=CH_2$	172.18	2^4, 292	1.039	1.4245^{20}	−15	191	83	
m33	Methane	CH_4	16.04	1, 56	0.7168 g/L / 0.4240^{bp}		−182.5	−161.5		3.3 mL aq; 47 mL alc
m34	Methanesulfonic acid	CH_3SO_3H	96.10	4, 4	1.4812^{18}_4	1.4303^{20}	20	167^{10mm}	>110	1.5 bz; misc aq
m35	Methanesulfonic anhydride	$(CH_3SO_2)_2O$	174.19	4, 5			71	138^{10mm}		v s aq (dec)
m36	Methanesulfonyl chloride	CH_3SO_2Cl	114.55	4, 5	1.4805^{18}_4	1.4518^{20}	−32	161	>110	s alc, eth

No.	Name	Formula								Solubility
m37	Methanethiol	CH_3SH	48.11	1, 288	1.966 g/L	1.3284^{20}	-123	6.0	11	2.3 aq; v s alc, eth
m38	Methanol	CH_3OH	32.04	1, 273	0.7913^{20}_{4}		-97.7	64.7	11	misc aq, alc, bz, chl, eth
m39	Methanol-d	CH_3OD	33.05	1^3, 1186	0.8127^{20}_{4}	1.3270^{20}	-110	65.5	11	misc aq, alc, eth
m40	Methanol-d_4	CD_3OD_1	36.07	1^3, 1187	0.888	1.3256^{20}		65.4	11	misc aq, alc, eth
m41	Methanol-^{13}C	$^{13}CH_3OH$	33.03	1^3, 1187	0.815	1.3290^{20}	-97.8	64	12	
m42	DL-Methionine	$CH_3SCH_2CH_2CH(NH_2)CO_2H$	149.21	4^2, 938	1.340		281 dec			3 aq; i eth; v sl s alc
m43	Methoxyacetic acid	$CH_3OCH_2CO_2H$	90.08	3, 232	1.174	1.4158^{20}		202–204	>110	misc aq, alc, eth
m44	2′-Methoxyacetophenone	$CH_3OC_6H_4COCH_3$	150.18	8, 85	1.090^{20}_{4}	1.5393^{20}		131^{18mm}	108	
m45	3′-Methoxyacetophenone	$CH_3OC_6H_4COCH_3$	150.18	8, 86	1.094	1.5410^{20}		239–241	>110	s aq
m46	4′-Methoxyacetophenone	$CH_3OC_6H_4COCH_3$	150.18	8, 87	1.082^{41}_{4}	1.5335	36–38	154^{26mm}	>110	v s alc, eth
m47	3-Methoxyacrylonitrile	$CH_3OCH{=}CHCN$	83.09		0.990	1.4550^{20}			76	
m48	2-Methoxyaniline	$CH_3OC_6H_4NH_2$	123.16	13, 358	1.098^{15}	1.5730^{20}	5–6	225	98	i aq; misc alc, eth
m49	3-Methoxyaniline	$CH_3OC_6H_4NH_2$	123.16	13, 404	1.096	1.5794^{20}	-10	251	>110	s alc, acid; sl s aq
m50	4-Methoxyaniline	$CH_3OC_6H_4NH_2$	123.16	13, 435	1.087		57–60	240–243		v s alc; sl s aq
m51	2-Methoxybenzaldehyde	$CH_3OC_6H_4CHO$	136.15	8, 43	1.127	1.560^{20}	37–39	238	117	sl s alc, bz; i eth
m52	3-Methoxybenzaldehyde	$CH_3OC_6H_4CHO$	136.15	8, 59	1.119	1.5533^{20}		143^{50mm}	>110	
m53	4-Methoxybenzaldehyde	$CH_3OC_6H_4CHO$	136.15	8, 67	1.119	1.5713^{20}	-1	248	108	misc alc
m54	4-Methoxybenzamide	$CH_3OC_6H_4CONH_2$	151.17	10^2, 100			164–167	295	108	s aq; v s alc; sl s eth
m55	Methoxybenzene	$CH_3OC_6H_5$	108.14	6, 138	0.9942^{20}	1.5170^{20}	-37.5	153.8	51	1 aq; misc alc, eth
m56	4-Methoxybenzene-sulfonyl chloride	$CH_3OC_6H_4SO_2Cl$	206.65	11, 243			40–43		>110	dec aq; s alc, eth
m57	2-Methoxybenzoic acid	$CH_3OC_6H_4CO_2H$	152.15	10, 64	1.180		100	200		0.5 aq; v s alc, eth
m58	3-Methoxybenzoic acid	$CH_3OC_6H_4CO_2H$	152.15	10, 137			104	172^{10mm}		s hot aq, alc, eth
m59	4-Methoxybenzoic acid	$CH_3OC_6H_4CO_2H$	152.15	10, 154	1.385^{4}		185	275–280		0.04 aq; v s alc, chl
m60	4-Methoxybenzoyl chloride	$CH_3OC_6H_4COCl$	170.60	10, 163		1.5810^{20}	22	145^{14mm}	87	i aq (dec); s alc (dec); s acet, bz

(Continued)

TABLE 2.21 Physical Constants of Organic Compounds (*Continued*)

No.	Name	Formula	Formula weight	Beilstein reference	Density, g/mL	Refractive index	Melting point, °C	Boiling point, °C	Flash point, °C	Solubility in 100 parts solvent
m61	4-Methoxybenzyl alcohol	$CH_3OC_6H_4CH_2OH$	138.17	6, 897	1.109^{25}_4	1.5442^{20}	23–25	259	>110	i aq; s alc, eth
m62	4-Methoxybenzylamine	$CH_3OC_6H_4CH_2NH_2$	137.18	13, 606	1.050^{15}	1.5462^{20}		236–237	>110	v s aq, alc, eth
m63	2-Methoxybiphenyl	$CH_3OC_6H_4C_6H_5$	184.24	6, 672	1.023	1.6105^{20}	30–33	274	>110	
m64	3-Methoxy-1-butanol	$CH_3OCH(CH_3)CH_2CH_2OH$	104.15		0.9229^{20}_{20}	1.4145^{20}	−85	161.1	46	misc aq
m65	4-Methoxy-3-buten-2-one	$CH_3OCH=CHCOCH_3$	100.12		0.982	1.4680^{20}		200	63	
m66	2-Methoxycinnamaldehyde	$CH_3OC_6H_4CH=CHCHO$	162.19				44–48	$130^{0.6mm}$	>110	
m67	1-Methoxy-1,4-cyclo-hexadiene		110.16	6^3, 367	0.940	1.4819^{20}		148–150	36	
m68	2-Methoxydibenzofuran		198.22	17^3, 1590			42–45		>110	
m69	7-Methoxy-3,7-dimethyloctanal	$(CH_3)_2C(OCH_3)(CH_2)_3-CH(CH_3)CH_2CHO$	186.30		0.877	1.4374^{20}		$60^{0.45mm}$	98	
m70	2-Methoxy-1,3-dioxolane		104.11	19^4, 617	1.092	1.4091^{20}		129–130	31	misc aq
m71	2-Methoxyethanol	$CH_3OCH_2CH_2OH$	76.10	1, 467	0.9646^{20}	1.4021^{20}	−85.1	124	39	
m72	2-(2-Methoxyethoxy)-acetic acid	$CH_3OCH_2CH_2OCH_2CO_2H$	134.13	3^3, 374	1.180	1.4380^{20}		245–250	>110	
m73	2-(2-Methoxyethoxy)-ethanol	$CH_3OCH_2CH_2OCH_2CH_2OH$	120.15		1.035^{20}_4	1.4264^{20}	−50	194	96	misc aq, alc, eth, ketones
m74	2-Methoxyethoxy-methyl chloride	$CH_3OCH_2CH_2OCH_2Cl$	124.57		1.091	1.4270^{20}		50^{13mm}	>110	
m75	2-Methoxyethyl acetate	$CH_3CO_2CH_2CH_2OCH_3$	118.13	2, 141	1.0049^{20}	1.4002^{20}	−70	144	49	misc aq
m76	2-Methoxyethyl acetoacetate	$CH_3COCH_2CO_2CH_2CH_2OCH_3$	160.17		1.090	1.4339^{20}		120^{20mm}	103	
m77	2-Methoxyethylamine	$CH_3OCH_2CH_2NH_2$	75.11	4^2, 718	0.864	1.4054^{20}		95	9	v s aq, alc
m78	2-Methoxyethyl cyanoacetate	$CH_3OCH_2CH_2O_2CCH_2CN$	143.14	2^4, 1891	1.127	1.4340^{20}		100^{1mm}	>110	
m79	1-Methoxy-2-indanol		164.20	6, 970	1.128	1.5482^{20}	52–54	146^{11mm}	>110	
m80	2-Methoxy-5-methyl-aniline	$CH_3OC_6H_3(CH_3)NH_2$	137.18	13^2, 388	1.065	1.5647^{20}	13–14	235	>110	s aq; v s alc, bz, eth
m81	4-Methoxy-2-methyl-aniline	$CH_3OC_6H_3(CH_3)NH_2$	137.18	13^2, 330				248–249	>110	s alc
m82	3-Methoxy-3-methyl-1-butanol	$CH_3OC(CH_3)_2CH_2CH_2OH$	118.18	1^3, 2198	0.926	1.4280^{20}		173–175	71	

No.	Name	Formula								Solubility
m83	2-Methoxy-1-methyl-ethyl cyanoacetate	$NCCH_2CO_2CH(CH_3)CH_2OCH_3$	157.17		1.030	1.4310^{20}		105^{2mm}	62	
m84	2-Methoxy-4-methylphenol	$CH_3OC_6H_3(CH_3)OH$	138.17	6, 878	1.092	1.5372^{20}	5	222	99	s bz, eth, CS$_2$
m85	5-Methoxy-2-methyl-4-nitroaniline	$CH_3OC_6H_3(CH_3)(NO_2)NH_2$	182.18	13^3, 1575			168–170			
m86	1-Methoxy-2-methyl-propylene oxide	$(CH_3)_2C\!\!-\!\!CH(OCH_3)$ epoxide O	102.13	17^3, 1035	0.904	1.3929^{20}		94	6	
m87	1-Methoxynaphthalene	$C_{10}H_7OCH_3$	158.20	6, 606	1.090	1.6220^{20}		135^{12mm}	>110	s alc, hot bz, HOAc
m88	2-Methoxynaphthalene	$C_{10}H_7OCH_3$	158.20	6, 640			73–75	274		sl s aq; s alc, eth
m89	2-Methoxy-4-nitroaniline	$CH_3OC_6H_3(NO_2)NH_2$	168.15	13, 390			140–142			0.17 aq; s alc, eth
m90	2-Methoxy-5-nitroaniline	$CH_3OC_6H_3(NO_2)NH_2$	168.15	13, 389			117–119			
m91	4-Methoxy-2-nitroaniline	$CH_3OC_6H_3(NO_2)NH_2$	168.15	13, 521			123–126			
m92	2-Methoxynitrobenzene	$CH_3OC_6H_4NO_2$	153.14	6, 217	1.2527^{20}_4	1.5161^{20}	10.5	277	>110	
m93	4-Methoxy-3-nitrobenzoic acid	$CH_3OC_6H_3(NO_2)CC_2H$	197.15	10, 181			192–194			
m94	2-Methoxy-5-nitropyridine	$CH_3O(C_5H_3N)NO_2$	154.13	21^3, 33			108–109			
m95	4-Methoxy-2-nitrotoluene	$CH_3OC_6H_3(NO_2)CH_3$	167.16	6, 411	1.207	1.5525^{20}	17	267	>110	
m96	4-Methoxyphenethylamine	$CH_3OC_6H_3CH_2CH_2NH_2$	151.21	13, 626	1.033	1.5379^{20}		140^{20mm}	>110	1.5 aq; misc alc, eth
m97	2-Methoxyphenol	$CH_3OC_6H_4OH$	124.14	6, 768	1.112(lg)	1.5429	28	205	82	misc alc, eth; sl s aq
m98	3-Methoxyphenol	$CH_3OC_6H_4OH$	124.14	6, 813	1.131	1.5510^{20}	<−17.5	115^{5mm}	>110	
m99	4-Methoxyphenol	$CH_3OC_6H_4OH$	124.14	6, 843			55–57	243	>110	v s bz; s alk
m100	3-(4-Methoxy-phenoxy)-1,2-propanediol	$CH_3OC_6H_4OCH_2CH(OH)CH_2OH$	198.22	6^3, 4411			76–80			
m101	4-Methoxyphenylacetic acid	$CH_3OC_6H_4CH_2CO_2H$	166.18	10, 190			86–88	140^{3mm}		1 aq; v s alc; s eth
m102	2-Methoxyphenyl-acetone	$CH_3OC_6H_4CH_2OCH_3$	164.20	8^3, 397	1.054	1.5250^{20}		130^{10mm}	>110	s alc, eth
m103	2-(Methoxyphenyl)-acetonitrile	$CH_3OC_6H_4CH_2CN$	147.18	10, 188			65–68	143^{15mm}		s hot bz

(Continued)

TABLE 2.21 Physical Constants of Organic Compounds (*Continued*)

No.	Name	Formula	Formula weight	Beilstein reference	Density, g/mL	Refractive index	Melting point, °C	Boiling point, °C	Flash point, °C	Solubility in 100 parts solvent
m104	4-(Methoxyphenyl)-acetonitrile	$CH_3OC_6H_4CH_2CN$	147.18	10, 191	1.085	1.5300^{20}		286–287	>110	misc aq, acet, bz, eth
m105	1-Methoxy-2-propanol	$CH_3OCH_2CH(OH)CH_3$	90.12	1^2, 536	0.9193^{20}_{20}	1.4021^{21}	−97	120.1	33	misc chl, eth; 50 alc; s bz, EtOAc
m106	2-Methoxypropene	$CH_3C(OCH_3)=CH_2$	72.11	1, 435	0.735	1.3820^{20}		34–36	−29	
m107	*trans*-1-Methoxy-4-(1-propenyl)benzene	$CH_3OC_6H_4CH=CHCH_3$	148.21	6, 566	0.9883^{20}_{4}	1.5615^{20}	21.4	237	90	misc alc, eth; sl s aq
m108	2-Methoxy-4-propenyl-phenol	$CH_3OC_6H_3(OH)CH=CHCH_3$	164.20	6, 955	1.087^{20}_{4}	1.5748^{20}	−10	266	>112	misc alc, chl, eth; s HOAc, alk; i aq
m109	2-Methoxy-4-(2-propenyl)phenol	$CH_3OC_6H_3(OH)CH_2CH=CH_2$	164.20	6, 961	1.0664^{20}_{4}	1.5408^{20}	−9.2	255	>112	
m110	3-Methoxypropionitrile	$CH_3OCH_2CH_2CN$	85.11	3^1, 113	0.937	1.4030^{20}		165	61	
m111	4-Methoxypropio-phenone	$CH_3OC_6H_4COCH_2CH_3$	164.20	8, 103	1.071	1.5465^{20}	27–29	274	61	misc aq
m112	3-Methoxypropylamine	$CH_3O(CH_2)_3NH_2$	89.14	4^3, 739	0.874	1.4175^{20}		$118^{733\text{mm}}$	22	
m113	2-Methoxypyridine	$CH_3O(C_5H_4N)$	109.13	21, 44	1.038	1.5029^{29}		142	32	
m114	6-Methoxy-1,2,3,4-tetrahydronaphthalene		162.23	6^2, 537	1.033	1.5402^{20}		$90^{1\text{mm}}$	>110	
m115	6-Methoxy-1-tetralone		176.22	9^2, 889	0.9851^{15}	1.5161^{20}	77–79	$171^{11\text{mm}}$	51	i aq; v s alc, eth
m116	2-Methoxytoluene	$CH_3OC_6H_4CH_3$	122.17	6, 352	0.9697^{25}	1.5131^{20}		170–172	54	s alc, bz, eth; i aq
m117	3-Methoxytoluene	$CH_3OC_6H_4CH_3$	122.17	6, 376	0.9692^{25}	1.5112^{20}		175–176	53	s alc, eth; i aq
m118	4-Methoxytoluene	$CH_3OC_6H_4CH_3$	122.17	6, 392	0.7560^{20}_{4}	1.3678^{20}		174	−30	
m119	Methoxytrimethylsilane	$CH_3OSi(CH_3)_3$	104.23	4^3, 1856		1.4253^{35}		57–58		
m120	*N*-Methylacetamide	$CH_3CONHCH_3$	73.10	4, 58	0.9460^{35}		30.6	206	108	s aq
m121	4′-Methyl lacetanilide	$CH_3OCONHC_6H_4CH_3$	149.19	12, 920			150	307		
m122	Methyl acetate	$CH_3CO_2CH_3$	74.08	2, 224	0.9342^{20}_{4}	1.3619^{20}	−98	57	−10 (CC)	24 aq; misc alc, eth
m123	Methyl acetoacetate	$CH_3COCH_2CO_2CH_3$	116.12	3, 632	1.0757^{20}	1.4186^{20}	27.5	171.7	77	50 aq; misc alc
m124	4′-Methylaceto-phenone	$CH_3C_6H_4COCH_3$	134.18	7, 307	1.0051	1.5328^{20}	22–24	226	92	i aq; v s alc, eth
m125	Methyl 4-acetoxy-benzoate	$CH_3CO_2C_6H_4CO_2CH_3$	194.19	10, 159			82–84			
m126	Methyl acrylate	$H_2C=CHCO_2CH_3$	86.09	2, 399	0.9541^{20}_{4}	1.4040^{20}	−76.5	80.2	−3 (CC)	6 aq; s alc, eth
m127	Methylamine	CH_3NH_2	31.06	4, 32	0.699^{-11}_{4}		−93.5	−6.3	0	959 mL aq; 10.5 bz
m128	1-(Methylamino)-anthraquinone		237.26	14, 179			170–172			

No.	Name	Formula								Solubility
m129	Methyl 2-aminobenzoate	$H_2NC_6H_4CO_2CH_3$	151.17	14, 317	1.1684^{19}_4	1.5820^{20}	24	256	104	sl s aq; v s alc, eth
m130	Methyl 3-aminocrotonate	$CH_3C(NH_2)=CHCO_2CH_3$	115.13	3, 632			81–83			misc aq, alc, eth
m131	2-(Methylamino)-ethanol	$CH_3NHCH_2CH_2OH$	75.11	4, 276	0.937^{20}	1.4387^{20}		159	72	4 aq; sl s alc; i eth
m132	4-Methylaminophenol sulfate	$(CH_3NC_6H_4OH)_2 \cdot H_2SO_4$	344.39	13, 441			260 dec			
m133	Methyl 2-(aminosulfonyl)benzoate	$H_2HSO_2C_6H_4CO_2CH_3$	215.23	11, 377			126–128			
m134	N-Methylaniline	$C_6H_5NHCH_3$	107.16	12, 135	0.989^{20}_4	1.5684^{20}	−57	196	78	sl s aq; s alc, eth
m135	N-Methylanilinium trifluoroacetate	$C_6H_5NHCH_3 \cdot HO_2CCF_3$	221.18				65–66			
m136	2-Methyl-anthraquinone		222.24	7, 809			170–173			v s bz; s alc, eth
m137	Methylarsonic acid	$CH_3AsO(OH)_2$	139.96	4, 613			161			v s aq; s alc
m138	4-Methylbenzaldehyde	$CH_3C_6H_4CHO$	120.15	7, 297	1.0194^{17}_4	1.5447^{20}		205	80	misc alc, eth; sl s aq
m139	Methyl benzene-sulfonate	$C_6H_5SO_2OCH_3$	172.20	11^2, 20	1.2889^0_4	1.5151^{20}	−4	154^{20mm}		v s alc, chl, eth
m140	2-Methylbenzimidazole		132.17	23, 145			176–177			s alk, hot aq; sl s alc
m141	Methyl benzoate	$C_6H_5CO_2CH_3$	136.15	9, 109	1.0933^{15}_4	1.5205^{15}	−15	199.5	83	0.2 aq; misc alc, eth
m142	2-Methylbenzoic acid	$CH_3C_6H_4CO_2H$	136.15	9, 462	1.062		103.7	258–259		sl s aq; v s alc
m143	3-Methylbenzoic acid	$CH_3C_6H_4CO_2H$	136.15	9, 475	1.054		111–113	263		0.09 aq; v s alc
m144	4-Methylbenzoic acid	$CH_3C_6H_4CO_2H$	136.15	9, 483			180	274–275		v s alc, eth
m145	4-Methylbenzo-phenone	$CH_3C_6H_4COC_6H_5$	196.25	7, 440			57	326		v s bz, eth
m146	2-Methylbenzothiazole		149.22	27, 46	1.173	1.6170^{20}	12–14	238	102	s alc, HOAc; i aq
m147	2-Methylbenzoxazole		133.15	27, 46	1.121	1.5497^{20}	8–10	178	75	sl s aq; v s alc
m148	α-Methylbenzyl acetate	$CH_3CO_2CH(CH_3)C_6H_5$	164.20	6, 476	1.028	1.4945^{20}	20	95^{12mm}	91	0.09 aq; v s alc
m149	α-Methylbenzyl alcohol	$C_6H_5CH(CH_3)OH$	122.17	6, 475	1.0191^{13}_4	1.5265^{20}	20	204^{745mm}	85	v s alc; s bz, chl
m150	2-Methylbenzyl alcohol	$CH_3C_6H_4CH_2OH$	122.17	6, 484		1.5408^{20}	33–36	110^{14mm}	104	5 aq; 5 alc; s eth
m151	(±)-α-Methylbenzyl-amine	$C_6H_5CH(CH_3)NH_2$	121.18	12, 1094	0.940	1.5260^{20}		185	79	4.2 aq; misc alc, eth
m152	4-Methylbenzylamine	$CH_3C_6H_4CH_2NH_2$	121.18	12, 1141	0.952	1.5340^{20}	12–13	195	75	

(Continued)

TABLE 2.21 Physical Constants of Organic Compounds (*Continued*)

No.	Name	Formula	Formula weight	Beilstein reference	Density, g/mL	Refractive index	Melting point, °C	Boiling point, °C	Flash point, °C	Solubility in 100 parts solvent
m153	Methylbis(trimethyl-silyloxy)vinyl ether	$CH_3Si[OSi(CH_3)_2]CH=CH_2$	148.55	4^4, 4184	0.864	1.3970^{20}		$48^{8.8mm}$	51	
m154	Methyl bromoacetate	$BrCH_2CO_2CH_3$	152.98	2, 213	1.616	1.4586^{20}		52^{15mm}	62	s alc
m155	(±)-Methyl 2-bromobutyrate	$CH_3CH_2CH(Br)CO_2CH_3$	181.04	2, 282	1.573	1.452^{20}		138^{50mm}	68	s alc
m156	Methyl 2-bromopropionate	$CH_3CH(Br)CO_2CH_3$	167.01	2, 253	1.497	1.5420^{20}		51^{19mm}	51	s alc
m157	2-Methyl-1,3-butadiene	$H_2C=C(CH_3)CH=CH_2$	68.12	1, 252	0.6814^{20}	1.4216^{20}	-146.0	34.1	-53	misc alc, eth
m158	2-Methylbutane	$CH_3CH_2CH(CH_3)_2$	72.15	1, 134	0.6197^{20}	1.3537^{20}	-159.9	27.8	-56	0.005 aq; misc alc
m159	2-Methyl-1-butenethiol	$CH_3CH_2CH(CH_3)CH_2SH$	104.22	1^2, 421	0.848	1.4465^{20}		117	19	s alc, eth; i aq
m160	2-Methyl-2-butanethiol	$CH_3CH_2C(CH_3)_2SH$	104.22	1^1, 196	0.842	1.4385^{20}	-103.9	99.1	-1	s alc, eth; i aq
m161	2-Methyl-1-butanol	$CH_3CH_2CH(CH_3)CH_2OH$	88.15	1, 388	0.8160^{20}	1.4100^{20}	< -70	128	43	3 aq; misc alc, eth
m162	2-Methyl-2-butanol	$CH_3CH_2C(CH_3)_2OH$	88.15	1, 388	0.8096^{20}	1.4050^{20}	-9.0	102.0	21	11 aq; misc alc, bz, chl, eth
m163	3-Methyl-1-butanol	$(CH_3)_2CHCH_2CH_2OH$	88.15	1, 392	0.8129^{15}	1.4085^{15}	-117	131	45	2 aq; misc alc, bz, chl, eth, PE, HOAc
m164	3-Methyl-2-butanol	$(CH_3)_2CHCH(OH)CH_3$	88.15	1, 391	0.8179^{20}	1.4091^{20}	-92	112.9	38	2.8 aq; misc alc, eth
m165	3-Methyl-2-butanone	$(CH_3)_2CHCOCH_3$	86.13	1, 682	0.8024^{20}	1.3880^{20}	-92	94.3	6	misc alc, eth
m165a	2-Methyl-1-butene	$C_2H_5C(CH_3)=CH_2$	70.14	1, 211	0.650	1.3780^{20}	-137.6	31	< -34	
m166	2-Methyl-2-butene	$CH_3CH=C(CH_3)_2$	70.14	1, 211	0.6620^{20}	1.3878^{20}	-133.8	38.6	-45	misc alc, eth; i aq
m167	3-Methyl-1-butene	$(CH_3)_2CHCH=CH_2$	70.14	1, 213	0.6272^{20}	1.3638^{20}	-168	20	-56	misc alc, eth
m168	*cis*-2-Methyl-2-butenoic acid	$CH_3CH=C(CH_3)CO_2H$	100.12	2, 428	0.9834^{47}	1.4437^{47}	45	185		s alc, eth; v s hot aq
m169	*trans*-2-Methyl-2-butenoic acid	$CH_3CH=C(CH_3)CO_2H$	100.12	2, 430	0.969	1.4342^{81}	64	198		s alc, eth; v s hot aq
m170	3-Methyl-2-butenoic acid	$(CH_3)_2C=CHCO_2H$	100.12	2, 432	1.006^{24}	1.4170^{20}	69	194–195	13	s aq, alc, eth
m171	2-Methyl-3-buten-2-ol	$(CH_3)_2C(OH)CH=CH_2$	86.13	1, 444	0.824	1.4440^{20}	2.6	98–99	43	
m172	3-Methyl-2-buten-1-ol	$(CH_3)_2C=CHCH_2OH$	86.13	1, 444	0.848	1.4337^{20}		140	36	
m173	3-Methyl-3-buten-1-ol	$H_2C=C(CH_3)CH_2CH_2OH$	86.13	1^1, 126	0.853	1.4140^{20}			-6	
m174	2-Methyl-1-buten-3-yne	$H_2C=C(CH_3)C≡CH$	66.10		0.695		-113	32		
m175	N-Methylbutylamine	$CH_3CH_2CH_2CH_2NCH_3$	87.17	4, 157	0.736	1.3995^{20}	-75	91	1	
m176	1-Methylbutylamine	$CH_3CH_2CH_2CH(CH_3)NH_2$	87.17	4, 177	0.7384^{24}	1.4029^{20}		91	35	misc aq, alc, eth
m177	3-Methylbutyl 3-methylbutyrate	$(CH_3)_2CHCH_2CH_2O_2CCH_2CH(CH_3)_2$	172.27	2, 312	0.8541^{25}	1.4100^{25}		190.4	84	misc alc, eth

(Continued)

No.	Name	Formula	Mol. wt.	Beilstein ref.	Density	n	m.p., °C	b.p., °C	Flash, °C	Solubility
m178	3-Methyl-1-butyne	(CH3)2CHC≡CH	68.12	1, 251	0.666_4^{20}	1.3740^{20}	−89.8	26.4	25	misc alc, eth
m179	2-Methyl-3-butyne-2-ol	(CH3)2C(OH)C≡CH	84.12	1^1, 235	0.8672^{20}	1.4209^{20}	2.6	104	4	misc aq, acet, bz
m180	2-Methylbutyraldehyde	CH3CH2CH(CH3)CHO	86.13	1^1, 352	0.804	1.3919^{20}	−51	90–92	19	misc alc, eth; sl s aq
m181	3-Methylbutyraldehyde	(CH3)2CHCH2CHO	86.13	1, 684	0.785^{20}	1.3882^{20}	−85.8	92–93	11	1.4 aq; misc alc, eth
m182	Methyl butyrate	CH3CH2CH2CO2CH3	102.13	2, 270	0.898^{20}	1.3860^{20}		103	73	4 aq; s alc, chl, eth
m183	2-Methylbutyric acid	CH3CH2CH(CH3)CO2H	102.13	2, 305		1.4055^{20}		176.5	70	misc alc, eth
m184	3-Methylbutyric acid	(CH3)2CHCH2CO2H	102.13	2, 309	0.9308^{20}	1.4033^{20}	−29.3	176.5	18	dec aq, alc; s eth
m185	3-Methylbutyronitrile	(CH3)2CHCH2CN	83.13	2^2, 278	0.7925^{19}	1.3927^{20}	−101	129	51	220 aq; 73 alc; s eth
m186	3-Methylbutyryl chloride	(CH3)2CHCH2COCl	120.58	2, 315	0.9854	1.4161^{20}		115–117	71	i aq; misc alc, eth
m187	Methyl carbamate	H2NCO2CH3	75.07	3, 21	1.136_4^{56}		56–58	177		s alc
m188	Methyl chloroacetate	ClCH2CO2CH3	108.52	2, 197	1.2382^{20}	1.4220^{20}	−32	130	102	
m189	Methyl 2-chloroacetoacetate	CH3COCH(Cl)CO2CH3	150.56	3^2, 426	1.236	1.4465^{20}	−32.7	137	104	v s eth; s alc, acet
m190	Methyl 4-chloroacetoacetate	ClCH2COCH2CO2CH3	150.56		1.305	1.4564^{20}		85^{4mm}	106	misc alc, bz, chl, eth
m191	Methyl 3-chlorobenzoate	ClC6H4CO2CH3	170.60	9, 338	1.227	1.4923^{20}	21	101^{12mm}	59	
m192	Methyl 4-chlorobenzoate	ClC6H4CO2CH3	170.60	9, 340	1.382^{14}		42–44			
m193	Methyl 4-chlorobutyrate	ClCH2CH2CH2CO2CH3	136.58	2, 278	1.1268^{14}	1.4321^{20}		175–176	17	
m194	Methyl chloroformate	ClCO2CH3	94.50	3, 9	1.2234	1.3865^{20}		70–72	73	
m195	Methyl 3-(chloroformyl)propionate	CH3O2CCH2CH2COCl	150.56	2^2, 553	1.223	1.4402^{20}		65^{3mm}	38	
m196	Methyl 2-chloropropionate	CH3CH(Cl)CO2CH3	122.55	2, 248	1.075	1.4193^{20}		132–133	79	s alc
m197	2-Methylcinnamaldehyde	C6H5CH=C(CH3)CHO	146.19	7, 369	1.0407_4^{17}	1.6045^{20}		149^{27mm}	>110	
m198	Methyl trans-cinnamate	C6H5CH=CHCO2CH3	162.19	9, 581			36–38	262		
m199	6-Methylcoumarin		160.17	17, 337			75–76	303^{725mm}		
m200	Methyl crotonate	CH3CH=CHCO2CH3	100.12	2, 410	0.9444_4^{20}	1.4242^{20}		121	4	v s alc, eth; i aq
m201	Methyl cyanoacetate	NCCH2CO2CH3	99.09	2, 584	1.1225^{25}	1.4166^{25}	−22.5	201	>110	misc alc, eth
m202	Methylcyclohexane	C6H11CH3	98.19	5, 29	0.7694^{20}	1.4221^{20}	−126.6	100.9	−4	i aq; s alc, eth
m203	Methyl cyclohexanecarboxylate	C6H11CO2CH3	142.20	9, 8	0.9954_4^{16}	1.4430^{20}		183	60	
m204	4-Methyl-1,2-cyclohexanedicarboxylic anhydride		168.19		1.162	1.4774^{20}			>110	

TABLE 2.21 Physical Constants of Organic Compounds (*Continued*)

No.	Name	Formula	Formula weight	Beilstein reference	Density, g/mL	Refractive index	Melting point, °C	Boiling point, °C	Flash point, °C	Solubility in 100 parts solvent
m205	1-Methylcyclohexanol	$CH_3C_6H_{10}OH$	114.19	6, 11	0.9251^{25}	1.4587^{25}	25	155	67	i aq; b bz, chl
m206	cis-2-Methylcyclohexanol	$CH_3C_6H_{10}OH$	114.19	6^2, 17	0.9360^{20}_4	1.4640^{30}	7	165	58	misc alc, eth
m207	trans-2-Methylcyclohexanol	$CH_3C_6H_{10}OH$	114.19	6, 11	0.9247^{20}_4	1.4616^{20}	−2	167.5	65	misc alc; s eth
m208	cis-3-Methylcyclohexanol	$CH_3C_6H_{10}OH$	114.19	6, 12	0.9155^{20}	1.4572^{20}	−6	168	62	misc alc, eth
m209	trans-3-Methylcyclohexanol	$CH_3C_6H_{10}OH$	114.19	6, 12	0.9214^{20}	1.4580^{20}	−0.5	167	62	misc alc, eth
m210	cis-4-Methylcyclohexanol	$CH_3C_6H_{10}OH$	114.19	6, 14	0.9170^{20}	1.4614^{20}	−9.2	173	70	misc alc, eth
m211	trans-4-Methylcyclohexanol	$CH_3C_6H_{10}OH$	114.19	6, 14	0.9118^{21}_4	1.4559^{20}		174	70	misc alc; s eth
m212	2-Methylcyclohexanone	$CH_3C_6H_9(=O)$	112.17	7, 14	0.925^{20}_4	1.4478^{20}		162	46 (CC)	i aq; s alc, eth
m213	3-Methylcyclohexanone	$CH_3C_6H_9(=O)$	112.17	7, 15	0.9155^{20}	1.4460^{20}		169	51	i aq; s alc, eth
m214	4-Methylcyclohexanone	$CH_3C_6H_9(=O)$	112.17	7, 18	0.916^{20}_4	1.4455^{20}		171	40	i aq; s alc, eth
m215	1-Methyl-1-cyclohexene		96.17	5, 66	0.809^{20}_4	1.4502^{20}	−121	111	−3	i aq; s alc, eth
m216	4-Methyl-1-cyclohexene		96.17	5, 67	0.799	1.4412^{20}	−115.5	102	−1	i aq; s alc, eth
m217	6-Methyl-3-cyclohexene-1-methanol		126.20		0.954	1.4830^{20}				
m218	N-Methylcyclohexylamine	$C_6H_{11}NHCH_3$	113.20	12, 6	0.868	1.4560^{20}		149	??	
m219	3-Methylcyclohexylamine	$CH_3C_6H_{10}NH_2$	113.20	12, 10	0.855	1.4525^{20}		150^{730mm}	22	
m220	4-Methylcyclohexylamine	$CH_3C_6H_{10}NH_2$	113.20	12, 12	0.955	1.4531^{20}		151–154	26	
m221	Methylcyclopentadiene dimer		160.26	5^4, 1435	0.941	1.4976^{20}	−51	200	26	
m222	Methylcyclopentane	$C_5H_9CH_3$	84.16	5, 27	0.7487^{20}	1.4097^{20}	−142.4	71.8	−23	
m223	3-Methyl-1,2-cyclopentanedione		112.13	7^3, 310			105–107			0.013 aq
m224	2-Methylcyclopentanone		98.15	7^2, 13	0.9200^{20}_4	1.4347^{20}	−76	139	26	s aq; v s alc, eth

No.	Name	Formula	Formula weight	Beilstein reference	Density	n_D	m.p., °C	b.p., °C	Flash point, °C	Solubility
m225	Methyl cyclopropanecarboxylate	$C_3H_5CO_2CH_3$	100.12	9^1, 3	0.985	1.4181^{20}		119	17	i aq; misc alc, eth
m226	Methyl decanoate	$CH_3(CH_2)_8CO_2CH_3$	186.30	2, 356	0.873	1.4255^{20}	−18	223	94	i aq; s alc
m227	Methyl dichloroacetate	$Cl_2CHCO_2CH_3$	142.97	2, 203	1.3808^{19}	1.4421^{20}	−52	143	80	
m228	Methyl 2,2-dichloro-1-methylcyclopropanecarboxylate		183.03		1.245	1.4639^{20}		74^{8mm}	74	
m229	Methyl 2,3-dichloropropionate	$ClCH_2CH(Cl)CO_2CH_3$	157.00	2^1, 111	1.3282_4	1.4447^{20}		92^{50mm}	42	s alc
m230	N-Methyldiethanolamine	$CH_3N(CH_2CH_2OH)_2$	119.16	4, 284	1.0377^{20}	1.4685^{20}		248	126	misc aq, alc
m231	Methyl 3,4-dimethoxybenzoate	$(CH_3O)_2C_6H_3CO_2CH_3$	196.20	10, 396			59–62	283	>110	
m232	Methyl 3,5-dimethoxybenzoate	$(CH_3O)_2C_6H_3CO_2CH_3$	196.20	10, 405			43	298		
m233	Methyl 3-(dimethylamino)propionate	$(CH_3)_2NCH_2CH_2CO_2CH_3$	131.18	4, 403	0.917	1.4184^{20}		154	51	
m234	Methyl 2,5-dimethyl-3-furoate		154.17	18, 398	1.037	1.4750^{20}		198	80	
m235	Methyl 2,2-dimethylpropionate	$(CH_3)_3CCO_2CH_3$	116.16	2^1, 139	0.873	1.3880^{20}		101–103	−1	misc alc, eth; sl s aq
m236	N-Methyldioctylamine	$(C_8H_{17})_2NCH_3$	255.49	4^3, 381	1.066	1.4424^{20}	−30.1	165^{15mm}	>110	
m237	4-Methyl-1,3-dioxane		102.13	19^4, 49	0.976	1.4150^{20}	−45	114	22	
m238	N-Methyldiphenylamine	$(C_6H_5)_2NCH_3$	183.26	12, 180	1.048_4^{20}	1.6193^{20}	−7.6	135^{6mm}		i aq; s alc, eth
m239	Methyl diphenylglycolate	$(C_6H_5)_2C(OH)CO_2CH_3$	242.27	10, 344			74–76	187^{13mm}		
m240	3-Methyl-1,1-diphenylurea	$(C_6H_5)_2NCONHCH_3$	226.28	12^2, 852			172–174			
m241	Methyleneaminoacetonitrile	$CH_2{=}NCH_2CN$	68.08	Merck: 11, 5976			129			s hot aq, alc; sl s bz
m242	N,N′-Methylenebisacrylamide	$H_2C{=}CHC({=}O)NHCH_2NHC({=}O)CH{=}CH_2$	154.17				>300			
m243	2,2′-Methylenebis(4-chlorophenol)	$CH_2[C_6H_3(Cl)OH]_2$	269.13	6^3, 5408			168–172			100 EtOH; 100 eth; s PE
m244	4,4′-Methylenebis(2,6-di-tert-butylphenol)	$CH_2\{C_6H_2[C(CH_3)_3]_2OH\}_2$	424.67	6^4, 6811			156–158	289^{40mm}		

(Continued)

TABLE 2.21 Physical Constants of Organic Compounds (*Continued*)

No.	Name	Formula	Formula weight	Beilstein reference	Density, g/mL	Refractive index	Melting point, °C	Boiling point, °C	Flash point, °C	Solubility in 100 parts solvent
m245	4,4'-Methylenebis-(N,N-dimethyl-aniline)	$CH_2[C_6H_4N(CH_3)_2]_2$	254.38	13, 239			88–89			
m246	1,1'-Methylenebis(3-methylpiperidine)	$CH_2[CH_3C_5H_9N]_2$	210.37		0.887	1.4734^{20}		160^{50mm}	>110	
m247	4,4'-Methylenebis-(phenylisocyanate)	$CH_2(C_6H_4NCO)_2$	250.26	13^3, 461	1.180		42–44	200^{5mm}	>110	
m248	Methylene blue		373.90	27, 393			190 dec			4 aq; 1.3 alc; s chl
m249	4,4'-Methylenedianiline	$CH_2(C_6H_4NH_2)_2$	198.26	13, 238			89–91	399	221	v s alc, bz, eth; sl s aq
m250	3,4-Methylenedioxy-benzaldehyde		150.13	19, 115			37	264	>110	0.2 aq; v s alc, eth
m251	1,2-Methylenedioxy-benzene		122.12	19, 20	1.064	1.5398		173	55	
m252	3,4-Methylenedioxy-6-propylbenzyldi-ethyleneglycol butyl ether		338.45	19^3, 779	1.059	1.498		180^{1mm}	171	misc alc, bz, geons
m253	Methylenesuccinic acid	$H_2C{=}C(CO_2H)CH_2CO_2H$	130.10	2, 760	1.573		167			8.2 aq; 20 alc; v sl s bz, chl, eth, PE
m254	N-Methylethylene-diamine	$CH_3NHCH_2CH_2NH_2$	74.13	4^1, 415	0.841	1.4395^{20}		114–116	42	
m255	N-Methylformamide	$HC({=}O)NHCH_3$	59.07	4, 58	0.9988^{25}	1.4300^{25}	−4	199.5	98	misc aq
m256	N-Methylformanilide	$C_6H_5N(CH_3)CHO$	135.17	12, 234	1.095	1.5610^{20}	8–13	244	126	30 aq; misc alc
m257	Methyl formate	HCO_2CH_3	60.05	2, 18	0.9815^{15}	1.3465^{15}	−99	31.7	−19	s aq; v s alc; misc eth
m258	5-Methylfurfuraldehyde		110.11	17, 289	1.1072^{18}	1.5263^{20}		187	72	0.3 aq
m259	2-Methylfuran		82.10	17, 36	0.915^{20}	1.4332^{20}	−88	63–66	−22	s alc, eth; sl s aq
m259a	Methyl 2-furoate		126.11	18, 274	1.179^{20}	1.4879^{20}		181	73	
m260	Methylgermanium tribromide	CH_3GeBr_3	327.35		2.6337^{20}	1.5770^{20}		168		
m261	N-Methylglucamine		195.22	Merck: 12, 6154			128–129			100 aq^{25}; 1.2 alc^{70}
m262	Methyl-α-D-gluco-pyranoside		194.18	31, 179	1.46^{30}		168	$200^{0.2mm}$		63 aq; 1.6 alc; i eth
m263	(±)-2-Methylglutaro-nitrile	$NCCH_2CH_2CH(CH_3)CN$	108.14	2, 656	0.950	1.4340^{20}	−45	269–271	126	
m264	N-Methylglycine	$CH_3NHCH_2CO_2H$	89.09	4, 345			208 dec			42 aq; sl s alc
m265	Methyl glycolate	$HOCH_2CO_2CH_3$	90.08	3, 236	1.168^{18}	1.4170^{20}	74	151	67	s aq; misc alc, eth

	Name	Formula								
m266	Methyl heptanoate	$CH_3(CH_2)_5CO_2CH_3$	144.22	2, 339	0.8815^{20}_{4}	1.4115^{20}	−55.8	173.5	52	s alc, eth; sl s aq
m267	5-Methyl-2-heptanol	$(CH_3)_2CH(CH_2)_3CH(OH)CH_3$	130.23	1, 421	0.803	1.4240^{20}		172	67	
m268	5-Methyl-3-heptanone	$C_2H_5CH(CH_3)CH_2COC_2H_5$	128.22	1^1, 363	0.823	1.4142^{20}		157–162	43	misc alc, eth
m269	6-Methyl-5-hepten-2-one	$(CH_3)_2C{=}CHCH_2CH_2COCH_3$	126.20	1^3, 3010	0.855^{16}_{4}	1.4392^{20}	−67	73^{18mm}	50	
m269a	Methyl hexadecanoate	$CH_3(CH_2)_{14}CO_2CH_3$	270.46	2, 372	0.852	1.4512^{20}	32–34	196^{15mm}	>110	s alc, chl, eth
m270	Methyl hexanoate	$CH_3(CH_2)_4CO_2CH_3$	130.19	2, 323	0.9038^{0}_{4}	1.4038^{23}	−71	151	45	v s alc, eth
m271	5-Methyl-2-hexanone	$(CH_3)_2CHCH_2CH_2COCH_3$	114.19	1^2, 756	0.888^{20}_{4}	1.4062^{20}	−73.9	144	3641	0.5 aq; misc alc, eth
m272	1-Methylhexylamine	$CH_3(CH_2)_4CH(NH_2)CH_3$	115.22	4, 194	0.7665^{18}	1.4175^{20}		144	54	sl s aq; s alc, eth
m273	1-Methylhydantoin		114.10	24, 244			157	subl		s aq, alc; 3 eth
m274	Methylhydrazine	CH_3NHNH_2	46.07	4^2, 957	0.866	1.4225^{20}	−52.4	87.5	21	misc aq, alc; s PE
m275	Methyl hydrazino-carboxylate	$H_2NNHCO_2CH_3$	90.08	3^1, 46			70–73	108^{12mm}		
m276	Methyl hydrogen glutarate	$HO_2CCH_2CH_2CH_2CO_2CH_3$	146.14	2^2, 565	1.169	1.4381^{20}	8–9	151^{10mm}	>110	
m277	Methyl hydrogen hexanedioate	$HO_2C(CH_2)_4CO_2H$	160.17	2, 652	1.081	1.4401^{20}	56–59	162^{10mm}	>110	s alc
m278	Methyl hydrogen succinate	$HO_2CCH_2CH_2CO_2H$	132.12	2, 608				151^{20mm}		v s aq, alc, eth
m279	Methyl hydroperoxide	CH_3OOH	48.04	1^2, 270	1.997^{15}_{4}	1.3642^{15}		38^{65mm}		misc aq, alc, eth; s bz
m280	Methylhydroquinone	$CH_3C_6H_3{-}1,4{-}(OH)_2$	124.14	6, 874			128–130	270 dec		v s alc, eth, acet; 0.25 aq
m281	Methyl 4-hydroxy-benzoate	$HOC_6H_4CO_2CH_3$	152.15	10, 158			126–128	270 dec		v s alc, eth; s hot aq
m282	Methyl 2-hydroxy-isobutyrate	$(CH_3)_2C(OH)CO_2CH_3$	118.13	3^2, 223	1.023	1.4112^{20}	57–60	127	42	v s aq, alc
m283	Methyl 4-hydroxy-phenylacetate	$HOC_6H_4CH_2CO_2CH_3$	166.18	10, 191				163^{5mm}		
m284	2-Methylimidazole		82.11	23, 46	1.030	1.4960^{20}	−60	198	92	misc aq
m285	2-Methylimidazole		82.11	23, 65			142–143	268		
m286	4-Methylimidazole		82.11	23, 69			53–56	263	>110	
m287	2-Methyl-1-H-indole		131.18	20, 311	1.07^{20}_{4}		58–60	273		v s alc, eth; s hot aq
m288	2-Methylindoline		133.19	20, 279	1.023	1.5681^{20}		229	93	
m289	N-Methylisatoic anhydride		177.16	27, 265			165 dec			
m290	Methyl isobutyrate	$(CH_3)_2CHCO_2CH_3$	102.13	2, 290	0.891^{20}	1.3840^{20}	−84.7	92.5	3	misc alc, eth; sl s aq
m291	Methyl isocyanate	CH_3NCO	57.05	4, 77	0.967	1.3695^{20}	−45	39	−6	s aq
m292	Methyl isodehydr-acetate		182.18	18, 410			68–70	167^{14mm}		

(Continued)

TABLE 2.21 Physical Constants of Organic Compounds (*Continued*)

No.	Name	Formula	Formula weight	Beilstein reference	Density, g/mL	Refractive index	Melting point, °C	Boiling point, °C	Flash point, °C	Solubility in 100 parts solvent
m293	N-Methylisopropyl-amine	$(CH_3)_2CHNHCH_3$	73.14	4[1], 153	0.702	1.3840^{20}	50–53	118	–31	v s alc, eth; sl s aq
m294	Methyl isothiocyanate	CH_3NCS	73.12	4, 77	1.069	1.5258^{37}	35	122	32	s aq (dec), alc, eth
m295	5-Methylisoxazole		83.09	27, 16	1.018	1.4386^{20}			30	s aq, alc, bz, chl
m296	Methyl lactate	$CH_3CH(OH)CO_2CH_3$	104.10	3, 280	1.088^{20}_{4}	1.4131^{20}		144–145	49	s alc, eth
m297	Methyl mandelate	$C_6H_5CH(OH)CO_2CH_3$	166.18	10, 202	1.1756^{20}		54–56	135^{12mm}	>110	
m298	Methyl mercapto-acetate	$HSCH_2CO_2CH_3$	106.14		1.187	1.4657^{20}		43^{10mm}	30	s alc, eth
m299	Methyl 3-mercapto-propionate	$HSCH_2CH_2CO_2CH_3$	120.17	3[2], 214	1.085	1.4660^{20}		55^{14mm}	60	
m300	Methyl methacrylate	$H_2C{=}C(CH_3)CO_2CH_3$	100.12	2[2], 398	0.9433^{20}	1.4140^{20}	–48	100	10	1,6 aq; s ketones, esters, CCl₄
m301	Methyl methane-sulfonate	$CH_3SO_2OCH_3$	110.13	4, 4	1.2943^{20}_{4}	1.4138^{20}		202–203	104	20 aq; 100 DMF
m302	Methyl methoxyacetate	$CH_3OCH_2CO_2CH_3$	104.11	3, 236	1.0511^{20}_{4}	1.3964^{20}		130	35	v s alc, eth; sl s aq
m303	Methyl 4-methoxy-acetoacetate	$CH_3OCH_2COCH_2CO_2CH_3$	146.14	3[4], 1939	1.129	1.4316^{20}		$89^{8.5mm}$	89	
m304	Methyl 2-methoxy-benzoate	$CH_3OC_6H_4CO_2CH_3$	166.18	10, 71	1.157	1.5335^{20}		248	>110	
m305	Methyl 4-methoxy-benzoate	$CH_3OC_6H_4CO_2CH_3$	166.18	10, 159			51	245	>110	
m306	Methyl 4-methoxy-phenylacetate	$CH_3OC_6H_4CH_2CO_2CH_3$	180.20	10, 191	1.135	1.5165^{20}		158^{19mm}	36	
m307	Methyl 4-methoxy-propionate	$CH_3OCH_2CH_2CO_2CH_3$	118.13	3, 297	1.009	1.4020		142–143	47	
m308	1-Methyl-4-(methyl-amino)piperidine		128.22		0.882	1.4672^{20}			55	
m309	Methyl 2-methyl-benzoate	$CH_3C_6H_4CO_2CH_3$	150.18	9, 463	1.073	1.5190^{20}		207–208	82	
m310	Methyl 3-methyl-benzoate	$CH_3C_6H_4CO_2CH_3$	150.18	9, 475	1.063	1.5160^{20}		113^{27mm}	95	
m311	Methyl 4-methyl-benzoate	$CH_3C_6H_4CO_2CH_3$	150.18	9, 484			33–36	104^{15mm}	90	
m312	Methyl 2-methyl-butyrate	$C_2H_5CH(CH_3)CO_2CH_3$	116.16	2, 304	0.885	1.3931^{20}		115	32	sl s aq; misc alc, eth
m313	2-Methyl-6-methylene-2-octanol	$C_2H_5C({=}CH_2)(CH_2)_3C(CH_3)_2OH$	156.27		0.784	1.4431^{20}		84^{10mm}	76	

m314	Methyl 2-methyl-3-furancarboxylate		140.14		1.116	1.4730^{20}		75^{20mm}	63	
m315	Methyl 5-methylthio-methyl sulfoxide	$CH_3S(=O)CH_2SCH_3$	124.22		1.191	1.5487^{20}		$95^{2.5mm}$	>110	s aq, alc, eth
m316	Methyl 3-(methyl-thio)propionate	$CH_3SCH_2CH_2CO_2CH_3$	134.20		1.077	1.4650^{20}		75^{13mm}	72	v s alc, eth
m317	4-Methylmorpholine		101.15	27, 6	0.920	1.4349^{20}	−66	116	23	v s alc, eth
m318	1-Methylnaphthalene	$C_{10}H_7CH_3$	142.20	5, 566	1.0202^{20}	1.6170^{20}	−30.4	245	82	
m319	2-Methylnaphthalene	$C_{10}H_7CH_3$	142.20	5, 567	1.029^{20}_{4}	1.6026^{40}	34.4	241	97	
m320	Methyl 1-naphthalene-acetate	$C_{10}H_7CH_2CO_2CH_3$	200.24	9^3, 3206	1.142	1.5961^{20}		162^{5mm}	>110	
m321	2-Methyl-1,4-naphtho-quinone		172.18	7^2, 656			105–107			1.4 alc; 10 bz; s chl
m322	Methyl 1-naphthyl ketone	$C_{10}H_7COCH_3$	170.21	7, 401	1.1336^{0}_{4}	1.6284^{20}	11	302	>110	s alc, eth; i aq
m323	Methyl 2-naphthyl ketone	$C_{10}H_7COCH_3$	170.21	7, 402			53–55	301	>110	sl s alc; s CS_2
m324	Methyl nitrate	CH_3ONO_2	77.04	1, 284	1.2075^{20}	1.3748^{20}	−83	64 expl		sl s aq; s alc, eth
m325	Methyl nitrite	CH_3ONO	61.04	1, 284	0.991(lq)			−17.3		s alc, eth
m326	N-Methyl-4-nitro-aniline	$O_2NC_6H_4NHCH_3$	152.15	12, 714			152–154			
m327	2-Methyl-3-nitro-aniline	$CH_3C_6H_3(NO_2)NH_2$	152.15	12, 848			88–90	305		
m328	2-Methyl-4-nitro-aniline	$CH_3C_6H_3(NO_2)NH_2$	152.15	12, 846	1.1586^{140}_{4}		131–133			v s alc; s bz
m329	2-Methyl-5-nitro-aniline	$CH_3C_6H_3(NO_2)NH_2$	152.15	12, 844			104–107			s alc, acet, eth
m330	4-Methyl-2-nitro-aniline	$CH_3C_6H_3(NO_2)NH_2$	152.15	12, 1000			115–116			v s alc; s eth
m331	Methyl 2-nitro-benzoate	$O_2NC_6H_4CO_2CH_3$	181.15	9, 372	1.280	1.5340^{20}	−13	$106^{0.1mm}$	>110	s alc, eth
m332	Methyl 3-nitro-benzoate	$O_2NC_6H_4CO_2CH_3$	181.15	9, 378			78–80	279		
m333	Methyl 4-nitro-benzoate	$O_2NC_6H_4CO_2CH_3$	181.15	9, 390			94–96			

(Continued)

TABLE 2.21 Physical Constants of Organic Compounds (*Continued*)

No.	Name	Formula	Formula weight	Beilstein reference	Density, g/mL	Refractive index	Melting point, °C	Boiling point, °C	Flash point, °C	Solubility in 100 parts solvent
m334	2-Methyl-3-nitro-benzoic acid	$CH_3C_6H_3(NO_2)CO_2H$	181.15	9, 471			182–184			
m335	3-Methyl-4-nitro-benzoic acid	$CH_3C_6H_3(NO_2)CO_2H$	181.15	9, 481			216–218			
m336	4-Methyl-3-nitro-benzoic acid	$CH_3C_6H_3(NO_2)CO_2H$	181.15	9, 502			187–190			
m337	5-Methyl-2-nitro-benzoic acid	$CH_3C_6H_3(NO_2)CO_2H$	181.15	9, 482			134–136			
m338	2-Methyl-5-nitro-imidazole		127.10	23[1], 23			252–254			
m339	3-Methyl-4-nitro-phenol	$CH_3C_6H_3(NO_2)OH$	153.14	6, 386			127–129			
m340	4-Methyl-2-nitro-phenol	$CH_3C_6H_3(NO_2)OH$	153.14	6, 412	1.240^{40}_4	1.574^{40}	32–35	125^{22mm}	108	v s alc, eth
m341	2-Methyl-2-nitro-1-propanol	$O_2NC(CH_3)_2CH_2OH$	119.12	1, 378			86–89	95^{10mm}		350 aq
m342	2-Methyl-2-nitropropyl methacrylate	$H_2C=C(CH_3)CO_2CH_2-C(CH_3)_2NO_2$	187.20	2[3], 1288	1.087	1.4500^{20}		102^{4mm}	>110	
m343	N-Methyl-N-nitroso-4-toluenesulfonamide	$CH_3C_6H_4SO_2N(CH_3)NO$	214.24	11[1], 29			62			
m344	Methyl 2-nonynoate	$CH_3(CH_2)_5C\equiv CCO_2CH_3$	168.24	2, 490	0.915	1.4484^{20}		121^{20mm}	100	
m345	Methyl-5-norbornene-2,3-dicarboxylic anhydride		178.19	17[2], 461	1.232	1.5060^{20}			>110	
m346	Methyl octadecanoate	$CH_3(CH_2)_{16}CO_2CH_3$	298.51	2, 379	0.839^{20}	1.4521^{20}	38	215^{15mm}	>110	s alc, eth
m347	Methyl cis-9-octa-decenoate	$CH_3(CH_2)_7CH=CH-(CH_2)_7CO_2CH_3$	296.50	2, 467			−19.9	168^{2mm}	>110	misc abs alc, eth
m348	7-Methyl-1,6-octadiene	$(CH_3)_2C=CH(CH_2)_3CH=CH_2$	124.23	1[4], 1049	0.753	1.4360^{20}		143–144	26	v s alc, eth; i aq
m349	Methyl octanoate	$CH_3(CH_2)_6CO_2CH_3$	158.24	2, 348	0.8775^{20}_4	1.4160^{25}	−40	192.9	72	
m350	Methyl 2-octynoate	$CH_3(CH_2)_4C\equiv CCO_2CH_3$	154.21	2, 487	0.920	1.4460^{20}		217–220	88	
m351	3-Methyl-2-oxazolidinone		101.11		1.170	1.4541^{20}	15	90^{1mm}	>110	
m352	2-Methyl-2-oxazoline		85.11	27, 13	1.005	1.4340^{20}		110	20	
m353	3-Methyl-3-oxetane-methanol		102.13	17[3], 1128	1.024	1.4460^{20}		80^{40mm}	98	
m354	Methyl 2-oxocyclo-pentanecarboxylate	$(O=)C_5H_7-CO_2CH_3$	142.16	10, 597	1.145	1.4560^{20}		105^{19mm}	>110	

No.	Name	Formula	Mol. wt.	Spectra	Density	n_D	M.p.	B.p.	Flash	Solubility
m355	Methyl 2-oxo-propionate	$CH_3C(=O)CO_2CH_3$	102.09	3, 616	1.130	1.4065^{20}		134–137	39	misc alc, eth; sl s aq
m356	*trans*-2-Methyl-1,3-pentadiene	$CH_3CH=CHC(CH_3)=CH_2$	82.15	1, 255	0.718	1.4469^{20}		75–76	−12	
m357	2-Methylpentane	$CH_3CH_2CH_2CH(CH_3)_2$	86.18	1, 148	0.6532^{20}	1.3725^{20}	−154	60.3	<−29	
m358	3-Methylpentane	$(CH_3CH_2)_2CHCH_3$	86.18	1, 149	0.6643^{20}	1.3765^{20}	−163	63	<−7	
m359	2-Methyl-1,5-pentane-diamine	$H_2N(CH_2)_3CH(CH_3)CH_2NH_2$	116.21	4, 270	0.860	1.4590^{20}	80			
m360	2-Methyl-2,4-pentanediol	$(CH_3)_2C(OH)CH_2CH(OH)CH_3$	118.18	1, 486	0.9216^{20}_{4}	1.4270^{20}	−50	198	102	misc aq
m361	4-Methylpentanenitrile	$(CH_3)_2CHCH_2CH_2CN$	97.16	2, 329	0.8035^{20}_{4}	1.4061^{20}	−51.1	156.5	45	s alc; misc eth
m362	Methyl pentanoate	$CH_3(CH_2)_3CO_2CH_3$	116.16	2, 301	0.875	1.3962^{20}		128	22	sl s aq; misc alc, eth
m363	2-Methylpentanoic acid	$CH_3CH_2CH_2CH(CH_3)CO_2H$	116.16	2[2], 288	0.9242^{20}	1.4135^{20}	−85	196.4	107	1.3 aq
m364	2-Methyl-1-pentanol	$CH_3CH_2CH_2CH(CH_3)CH_2OH$	102.18	1, 409	0.8262^{20}	1.4180^{20}	−23.6	148	54	s alc, eth
m365	3-Methyl-3-pentanol	$(CH_3CH_2)_2C(CH_3)OH$	102.18	1, 411	0.8281^{20}	1.4186^{20}		123	46	misc alc, eth; sl s aq
m366	4-Methyl-2-pentanol	$(CH_3)_2CHCH_2CH(OH)CH_3$	102.18	1, 410	0.8080^{20}	1.4112^{20}	−90	132	41	1.6 aq
m367	4-Methyl-2-pentanone	$(CH_3)_2CHCH_2COCH_3$	100.16	1, 691	0.7978^{20}	1.3958^{20}	−84	116.5	18	1.7 aq; misc alc, bz, eth
m368	2-Methyl-2-pentenal	$CH_3CH_2CH=C(CH_3)CHO$	98.15	1[4], 3471	0.861	1.4503^{20}		138	31	s alc
m369	4-Methyl-2-pentenoic acid	$(CH_3)_2CHCH=CHCO_2H$	114.14	2[2], 406	0.9529	1.4489	35	115^{20mm}	46	i aq; v s alc
m370	4-Methyl-3-penten-2-one	$(CH_3)_2C=CHCOCH_3$	98.15	1, 736	0.8653^{25}	1.4440^{20}	−59	129.5	31	3.1 aq
m370a	4-Methyl-2-pentyl	$(CH_3)_2CHCH_2CH(CH_3)O_2CCH_3$	144.21		0.8805^{25}	1.3980^{20}		147.5	45	
m371	1-Methylpentylamine	$CH_3(CH_2)_3CH(NH_2)CH_3$	101.19	4, 190	0.767^{20}_{4}		−19	116–118	13	s aq, alc, PE
m372	3-Methyl-1-pentyn-3-ol	$CH_3CH_2C(CH_3)(OH)C≡CH$	98.15	1[2], 506	0.8688^{20}_{4}	1.4318^{20}	−30.6	122	26	13 aq, misc bz, acet PE, EtOAc; s eth
m373	4-Methylphenetole	$CH_3C_6H_4OCH_2CH_3$	136.19	6, 393	0.945	1.5044^{20}		189–191	70	
m374	N-(4-Methylphenyl)-acetamide	$CH_3C_6H_4NHCOCH_3$	149.19	12, 920	1.212^{15}		150–153	307		s alc, EtOAc, HOAc
m375	Methyl phenylacetate	$C_6H_5CH_2CO_2CH_3$	150.18	9, 434	1.044	1.5075^{20}		218	90	i aq; misc alc, eth
m376	2-Methyl-1-phenyl-2-propanol	$C_6H_5CH_2C(CH_3)_2OH$	150.22	6, 523	0.974	1.5140^{20}	25–26	96^{18mm}	81	
m377	1-Methyl-3-phenyl-propylamine	$C_6H_5CH_2CH_2CH(CH_3)NH_2$	149.24	12, 1165	0.922	1.5123^{20}		222	97	

(Continued)

TABLE 2.21 Physical Constants of Organic Compounds (*Continued*)

No.	Name	Formula	Formula weight	Beilstein reference	Density, g/mL	Refractive index	Melting point, °C	Boiling point, °C	Flash point, °C	Solubility in 100 parts solvent
m378	3-Methyl-1-phenyl-2-pyrazolin-5-one		174.20	24, 20			129–130	287[265mm]		
m379	Methyl phenyl sulfide	$C_6H_5SCH_3$	124.21	6, 297	1.058	1.5882^{20}	-15	188	57	i aq; s alc
m380	N-Methyl-N-phenyl-urethane	$C_6H_5N(CH_3)CO_2CH_2CH_3$	179.22	12, 417	1.074	1.5149^{20}		243–244	>110	
m381	N-Methylpiperazine		100.17	23, 17	0.903	1.4655^{20}		138	42	v s aq, alc, eth
m382	2-Methylpiperazine		100.17				65–67	155.6	65	78 aq, 37 acet, 32 bz
m383	N-Methylpiperidine	$C_5H_{10}NCH_3$	99.19	20, 19	0.816	1.4378^{20}		106–107	3	v s aq; misc alc, eth
m384	2-Methylpiperidine	$CH_3C_5H_9N$	99.19	20, 95	0.844	1.4459^{20}		119	8	v s aq; misc alc, eth
m385	3-Methylpiperidine	$CH_3C_5H_9N$	99.19	20, 100	0.845	1.4470^{20}	-5	126	17	v s aq
m386	4-Methylpiperidine	$CH_3C_5H_9N$	99.19	20, 101	0.838	1.4458^{20}		124	7	v s aq
m387	1-Methyl-3-piperdine-methanol		129.20	21^2, 8	1.013	1.4772^{20}		140–145	94	
m388	1-Methyl-4-piperidone		113.16	21^2, 215	0.920	1.4614^{20}			60	9 aq; misc alc, bz, chl, eth
m389	2-Methylpropan-aldehyde	$(CH_3)_2CHCHO$	72.11	1, 671	0.7891^{20}	1.3727^{20}	-65	64.1	-40	13 mL aq; 1320 mL alc; 2890 mL eth
m390	2-Methylpropane	$(CH_3)_3CH$	58.12	1, 124		1.3810^{-25}	-138	-11.7	-87	
m391	N-Methyl-1,3-propane-diamine	$H_2NCH_2CH_2CH_2NHCH_3$	88.15	4^1, 419	0.844	1.4468^{20}		139–141	35	
m392	2-Methyl-1,2-propane-diamine	$(CH_3)_2C(NH_2)CH_2NH_2$	88.15	4, 266	0.841	1.4410^{20}			23	
m393	2-Methyl-1,3-propane-diol	$HOCH_2CH(CH_3)CH_2OH$	90.12	1, 480	1.015	1.4450^{20}	-91	125^{20mm}	>110	
m394	1-Methyl-1-propane-thiol	$CH_3CH_2CH(SH)CH_3$	90.19	1, 373	0.8246^{25}_4	1.4338^{25}	-165	84–85	21	sl s aq; v s alc, eth
m395	2-Methyl-1-propane-thiol	$(CH_3)_2CHCH_2SH$	90.19	1, 378	0.8357^{20}	1.4396^{20}	-79	88.5	-9	v s alc, eth
m396	2-Methyl-2-propane-thiol	$(CH_3)_3CSH$	90.19	1, 383	0.7943^{25}_4	1.4198^{25}	1.1	64.1	-4	i aq
m397	2-Methyl-1-propanol	$(CH_3)_2CHCH_2OH$	74.12	1, 373	0.8016^{20}	1.3958^{20}	-108	108	28	10 aq; misc alc, eth
m398	2-Methyl-2-propanol	$(CH_3)_3COH$	74.12	1, 379	0.7888^{20}	1.3877^{20}	25.8	82.4	11	misc aq, alc, eth
m399	2-Methylpropene	$(CH_3)_2C{=}CH_2$	56.11	1, 207	0.6266^{mp}		-140	-6.9		v s alc, eth
m400	2-Methyl-2-propen-1-ol	$H_2C{=}C(CH_3)CH_2OH$	72.11	1, 443	0.857	1.4260^{20}		113–115	33	
m401	Methyl propionate	$CH_3CH_2CO_2CH_3$	85.11	2, 239	0.915^{20}_4	1.3770^{20}	-88	79.7	6	6 aq; misc alc, eth

No.	Name	Formula	Mol. wt.	Beilstein ref.	Density	n_D	mp (°C)	bp (°C)	Flash	Solubility
m402	Methyl propionyl-acetate	$C_2H_5COCH_2CO_2CH_3$	130.15	3^3, 1212	1.037	1.4220^{20}		74^{5mm}	71	
m403	4'-Methylpropio-phenone	$CH_3C_6H_4COCH_2CH_3$	148.21	7, 317	0.993	1.5280^{20}	7.2	238–239	96	sl s aq; misc alc, eth
m404	Methyl propyl ether	$CH_3CH_2CH_2OCH_3$	74.12	1, 354	0.738^{20}			39.1		
m405	2-Methyl-2-propyl-1,3-propanediol	$CH_3CH_2CH_2C(CH_3)(CH_2OH)_2$	132.20	1^1, 254			58–60	232	>110	
m406	Methyl propyl sulfide	$CH_3SCH_2CH_2CH_3$	90.18	1^3, 1432	0.8424^{20}	1.4442^{20}	−113.0	95.5		s aq
m407	Methyl 2-propynyl ether	$CH_3OCH_2C{\equiv}CH$	70.09	1, 4541	0.830	1.3961^{20}		62	−18	
m408	2-Methylpyrazine		94.12	23, 94	1.030	1.5042^{20}	−29	135	50	v s aq, alc, eth
m409	2-Methylpyridine	$CH_3C_5H_4N$	93.13	20, 234	0.9443^{20}	1.4957^{20}	−66.7	129	39	misc aq; s alc, eth
m410	3-Methylpyridine	$CH_3C_5H_4N$	93.13	20, 239	0.9566^{20}	1.5040^{20}	−18.3	144	36	misc aq, alc, eth
m411	4-Methylpyridine	$CH_3C_5H_4N$	93.13	20, 240	0.9548^{20}	1.5037^{20}	3.8	145	57	misc aq, alc, eth
m412	Methyl 3-pyridine-carboxylate	$(C_5H_4N)CO_2CH_3$	137.14	22, 39			39	209		s aq, alc, bz
m413	Methyl 4-pyridine-carboxylate	$(C_5H_4N)CO_2CH_3$	137.14	22, 46	1.001	1.5122^{20}	8.5	207–209	82	
m414	1-Methyl-2-pyridone		109.13	21, 268	1.112	1.5690^{20}	30–32	250^{740mm}	>110	
m415	Methyl 3-pyridyl-carbamate		152.15	22^3, 4076			121–123			
m416	2-[3-(6-Methyl-2-pyridyl)propoxy]-ethanol		195.26		1.052	1.5150^{20}			>110	
m417	N-Methylpyrrole		81.12	20, 163	0.914	1.4875^{20}	−57	112–113	15	i aq; misc alc, eth
m418	N-Methylpyrrolidine		85.15	20, 4	0.819^{20}_{4}	1.4247^{20}		80–81	−21	misc aq, eth
m419	N-Methyl-2-pyrrolidinone		99.13	21, 237	1.0279^{25}	1.4680^{25}	−24.4	202	96	misc aq, alc, bz, eth
m420	2-Methylquinoline		143.19	20, 387	1.058	1.6108^{20}	−2	248	79	i aq; s chl, eth
m421	4-Methylquinoline		143.19	20, 395	1.0826^{20}_{4}	1.6200^{20}	9–10	263	>110	misc alc, bz, eth
m422	6-Methylquinoline		143.19	20, 397	1.063	1.6140^{20}		259	>110	
m423	2-Methylquinozaline		144.18	23^1, 44	1.118	1.6156^{20}	180	245–247	107	misc aq
m424	Methyl salicylate	$HOC_6H_4CO_2CH_3$	152.15	10, 70	1.1831^{20}	1.5360^{20}	−8	223	96	0.7 aq; misc alc, HOAc; s chl, eth
m425	α-Methylstyrene	$C_6H_5C(CH_3){=}CH_2$	118.18	5, 484	0.909	1.5375^{20}	−24	165.5	45	
m426	4-Methylstyrene	$CH_3C_6H_4CH{=}CH_2$	118.18	5, 485	0.897	1.5412^{20}		170–175	45	
m427	mono-Methyl succinate	$HO_2CCH_2CH_2CO_2CH_3$	132.12	2, 608			56–59	151^{20mm}		
m428	Methyl tetradecanoate	$CH_3(CH_2)_{12}CO_2CH_3$	242.40	2^2, 326	0.855	1.4362^{20}	18.4	323	>110	misc alc, bz, eth

(Continued)

TABLE 2.21 Physical Constants of Organic Compounds (*Continued*)

No.	Name	Formula	Formula weight	Beilstein reference	Density, g/mL	Refractive index	Melting point, °C	Boiling point, °C	Flash point, °C	Solubility in 100 parts solvent
m429	2-Methyltetrahydrofuran		86.13	17, 12	0.8552^{20}	1.4056^{20}		78	−11	
m430	3-Methyltetrahydropyran		100.16	17^3, 77	0.863	1.4204^{20}		109^{733mm}	6	
m431	3-Methyltetrahydrothiophene-1,1-dioxide		134.20		1.191	1.4772^{20}		276	>110	
m432	4-Methylthiazole		99.16	27.16	1.090	1.5257^{20}		134	32	
m433	4-Methyl-5-thiazole-ethanol		143.21	27^3, 1754	1.196	1.5508^{20}		135^{mm}	>110	
m434	2-Methyl-2-thiazoline		101.17	27, 13	1.067	1.5200^{20}	−101	145	37	
m435	(Methylthio)acetonitrile	CH_3SCH_2CN	87.14		1.039	1.4826^{20}		63^{15mm}	67	
m436	3-(Methylthio)aniline	$CH_3SC_6H_4NH_2$	139.22	13^1, 141	1.130	1.6423^{20}		165^{16mm}	>110	
m437	4-(Methylthio)benzaldehyde	$CH_3SC_6H_4CHO$	152.22	8^1, 533	1.144	1.6452^{20}		90^{1mm}	>110	
m438	2-(Methylthio)benzothiazole		181.28	27, 109			43–46		>110	
m439	3-(Methylthio)-2-butanone	$CH_3CH(SCH_3)COCH_3$	118.20	1^4, 3993	0.975	1.4710^{20}		50–54^{20mm}	44	
m440	Methyl thiocyanate	CH_3SCN	73.12	3, 175	1.068^{20}	1.4680^{20}	−5	133	38	i aq; misc alc, eth
m441	2-Methylthiophene		98.17	17, 37	1.0193^{20}	1.5199^{20}	−63	113	7	
m442	3-Methylthiophene		98.17	17, 38	1.0218^{20}	1.5180^{20}	−69	115.4	11	i aq; misc alc, eth
m443	5-Methyl-2-thiophenecarboxaldehyde		126.18	17^1, 151	1.170	1.5860^{20}		114^{25mm}	82	
m444	N-Methyl-2-thiourea	$CH_3NHC(=S)NH_2$	90.15	4, 70			119–121			
m445	N-Methyl-o-toluamide	$CH_3C_6H_4CONHCH_3$	149.19	9, 465	1.158^{15}		69–71			v s aq, alc
m446	N-Methyl-p-toluenesulfonamide	$CH_3C_6H_4SO_2NHCH_3$	185.25	11, 105			76–79			
m447	Methyl p-toluenesulfonate	$CH_3C_6H_4SO_2OCH_3$	186.23	11, 99	1.234		27.5	145^{5mm}	>110	
m448	Methyltriacetoxysilane	$CH_3Si(O_2CCH_3)_3$	220.26	4^3, 1896	1.175^{20}_4	1.408^{20}	40–45	88^{3mm}	85	
m449	Methyl trichloroacetate	$Cl_3CCO_2CH_3$	177.42	2, 208	1.488	1.4558^{20}		153	72	
m450	Methyltrichlorosilane	CH_3SiCl_3	149.48	4^3, 1896	1.273	1.4110^{20}		66	−15	
m451	Methyltriethoxysilane	$CH_3Si(OCH_3)_3$	178.30	4, 629	0.895	1.3840^{20}		141–143	23	
m452	Methyl trifluoroacetate	$F_3CCO_2CH_3$	128.05	2^3, 427	1.273	1.2907^{20}		43	−7	

No.	Name	Formula	Formula wt.	Beilstein ref.	Density	n_D	mp (°C)	bp (°C)	Flash pt. (°C)	Solubility
m453	Methyl trifluoromethanesulfonate	$F_3CSO_2OCH_3$	164.10	3⁴, 34	1.450	1.3244^{20}		94–99	38	
m454	Methyl 3,4,5-trihydroxybenzoate	$(HO)_3C_6H_2CO_2CH_3$	184.15	10, 483			201–203			
m455	Methyltrimethoxysilane	$CH_3Si(OCH_3)_3$	136.22	4⁴, 4203	0.955	1.3703^{20}		102	11	
m456	Methyl trimethylacetate	$(CH_3)_3CCO_2CH_3$	116.16	2, 320	0.873	1.3900^{20}		101	6	
m457	N-Methyl-N-(trimethylsilyl)trifluoroacetamide	$F_3CC(=O)N(CH_3)Si(CH_3)_3$	199.25		1.075	1.3802^{20}		132	25	
m458	(Methyl)triphenylphosphonium bromide	$[CH_3P(C_6H_5)_3]^+Br^-$	357.24	16, 760			230–234			
m459	2-Methylundecanal	$CH_3(CH_2)_8CH(CH_3)CHO$	184.32	4, 64	0.830^{15}_{4}	1.4321^{20}		171	93	s alc, eth
m460	Methyl urea	$CH_3NHCONH_2$	74.08	4³, 442	1.204		101–102			v s aq, alc; i eth
m461	N-Methyl-N-vinylacetamide	$CH_3CON(CH_3)CH=CH_2$	99.13		0.959	1.4829^{20}		70^{25mm}	58	
m462	Methyl vinyl ether	$CH_3OCH=CH_2$	58.08	1³, 1857	0.7511^{20}_{4}	1.3947	−123	5.5	−56	0.8 aq; v s alc
m463	Morpholine		87.12	27, 5	1.0005^{20}	1.4548^{20}	−4.9	128	35	misc aq, alc, bz, eth
m464	4-Morpholinepropionitrile		140.19	27³, 337	1.037	1.4715^{20}	21	121^{2mm}		
m465	N-Morpholino-1-cyclohexene		167.25		0.995	1.5128^{20}		120^{10mm}	68	
m466	3-(N-Morpholino)-1,2-propanediol		161.20		1.157		37–38	191^{30mm}	>110	
m467	Myrcene	$(CH_3)_2C=CHCH_2CH_2-C(=CH_2)CH=CH_2$	136.24	1, 264	0.8013^{20}_{4}	1.4709^{20}		167	39	s alc, chl, eth, HOAc
n1	1-Naphthaldehyde	$C_{10}H_7CHO$	156.18	7, 400	1.150^{20}_{4}	1.6520^{20}	1–2	161^{15mm}	>110	s alc, eth
n2	Naphthalene	$C_{10}H_8$	128.17	5, 531	1.162^{20}	1.5821^{100}	80	217.7	79	0.3 aq; 7 alc; 33 bz; 50 chl
n3	1-Naphthalenecarboxylic acid	$C_{10}H_7CO_2H$	172.18	9, 647			160–162	300		sl s aq; v s hot alc, eth
n4	1,5-Naphthalenediamine	$C_{10}H_6(NH_2)_2$	158.20	13, 203			185–187			s hot aq, hot alc
n5	1,8-Naphthalenediamine	$C_{10}H_6(NH_2)_2$	158.20	13, 204	1.1265^{99}_{4}	1.6828^{99}	66.5	205^{12mm}		sl s aq; s alc, eth

(Continued)

TABLE 2.21 Physical Constants of Organic Compounds (*Continued*)

No.	Name	Formula	Formula weight	Beilstein reference	Density, g/mL	Refractive index	Melting point, °C	Boiling point, °C	Flash point, °C	Solubility in 100 parts solvent
n6	1-Naphthalenesulfonic acid	$C_{10}H_7SO_3H$	208.24	11, 155			90 de-hydrates			v s aq, alc, sl s eth
n7	2-Naphthalenesulfonic acid	$C_{10}H_7SO_3H$	208.24	11, 171			124 de-hydrates			v s aq, alc
n8	1,8-Naphthalic anhydride		198.18	17, 521			268			sl s HOAc
n9	1-Naphthol	$C_{10}H_7OH$	144.17	6, 596	1.0954^{99}	1.6206^{99}	96	288		v s alc, bz, chl, eth
n10	2-Naphthol	$C_{10}H_7OH$	144.17	6, 627	1.217^4		123	285	161	0.1 aq; 125 alc; 6 chl; 77 eth; s alk
n11	1,4-Naphthoquinone		158.16	7, 724	1.422		126			s bz, chl, eth, hot alc
n12	(2-Naphthoxy)acetic acid	$C_{10}H_7OCH_2CO_2H$	202.21	6, 645			155–157			i aq; s bz, CS_2
n13	2-(1-Naphthyl)-acetamide	$C_{10}H_7CH_2ONH_2$	185.23	9, 666			182			
n14	1-Naphthyl acetate	$C_{10}H_7O_2CCH_3$	186.21	6, 608			43–46	dec	>110	s alc, eth
n15	1-Naphthylacetic acid	$C_{10}H_7CH_2CO_2H$	186.21	9, 666			135	194^{18mm}		3.3 alc; v s chl, eth
n16	1-Naphthylacetonitrile	$C_{10}H_7CH_2CN$	167.21	9, 667	1.123^{25}	1.6192^{20}	33–35	301	>110	s alc
n17	1-Naphthylamine	$C_{10}H_7NH_2$	143.18	12, 1212	1.177	1.6703	50	267	157	0.2 aq; v s alc, eth
n18	1-Naphthyl isocyanate	$C_{10}H_7NCO$	169.19	12, 1244		1.6344^{20}	4	123^{17mm}	>110	
n19	Nicotine		162.24	23, 117	1.0097^{20}	1.5882^{20}	−79		101	misc aq; v s alc, eth, PE
n20	Nitrilotriacetic acid	$N(CH_2CO_2H)_3$	191.14	4, 369			242 dec			0.1 aq; s hot alc
n21	3′-Nitroacetophenone	$O_2NC_6H_4COCH_3$	165.15	7, 288			76–78	202		s alc, eth
n22	4′-Nitroacetophenone	$O_2NC_6H_4COCH_3$	165.15	7, 288			78–80	202		s alc
n23	2-Nitroaniline	$O_2NC_6H_4NH_2$	138.13	12, 687	1.442^{15}		71	284		s hot aq, alc, chl
n24	3-Nitroaniline	$O_2NC_6H_4NH_2$	138.13	12, 698	1.43		114	306		0.1 aq; 5 alc; 6 eth
n25	4-Nitroaniline	$O_2NC_6H_4NH_2$	138.13	12, 711	1.437^{14}		147	332	165	4 alc; 3.3 eth; s bz
n26	3-Nitrobenzaldehyde	$O_2NC_6H_4CHO$	151.12	7, 250	1.2792^{20}		58	164^{23mm}		s alc, chl, eth
n27	4-Nitrobenzaldehyde	$O_2NC_6H_4CHO$	151.12	7, 256	1.496		106–107			s alc, bz, HOAc
n28	2-Nitrobenzamide	$O_2NC_6H_4CONH_2$	166.12	9, 373	1.462^{32}		174–178	317		s hot aq, hot alc, eth
n29	3-Nitrobenzamide	$O_2NC_6H_4CONH_2$	166.12	9, 381			140–143			
n30	Nitrobenzene	$C_6H_5NO_2$	123.11	5, 233	1.205^{15}	1.5546^{15}	5.8	210.8	88	v s alc, bz, eth
n31	3-Nitrobenzene-1,2-dicarboxylic acid	$O_2NC_6H_3(CO_2H)_2$	211.13	9, 823			216 dec			2 aq; v s hot alc
n32	5-Nitrobenzene-1,3-dicarboxylic acid	$O_2NC_6H_3(CO_2H)_2$	211.13	9, 840			260			0.15 aq; v s alc, eth

No.	Name	Formula	Beilstein reference	Formula weight	Density	n_D	mp, °C	bp, °C	Flash point, °C	Solubility
n33	2-Nitrobenzenesulfonyl chloride	$O_2NC_6H_4SO_2Cl$	11, 67	221.62			65–67			s eth; d hot aq, alc
n34	5-Nitrobenzimidazole		23, 135	163.14			207–209			s alc, acid
n35	2-Nitrobenzoic acid	$O_2NC_6H_4CO_2H$	9, 370	167.12	1.58		146–148			0.7 aq; 33 alc; 22 eth
n36	3-Nitrobenzoic acid	$O_2NC_6H_4CO_2H$	9, 376	167.12	1.494		140–142			0.3 aq; 33 alc; 40 acet
n37	4-Nitrobenzoic acid	$O_2NC_6H_4CO_2H$	9, 389	167.12	1.58		242.8			9 alc; 2 eth; 5 acet
n38	4-Nitrobenzonitrile	$O_2NC_6H_4CN$	9, 397	148.12			146–149			s HOAc; sl s aq, alc
n39	3-Nitrobenzoyl chloride	$O_2NC_6H_4COCl$	9, 381	185.57			32–35	275–278	>110	dec aq, alc; v s eth
n40	4-Nitrobenzoyl chloride	$O_2NC_6H_4COCl$	9, 394	185.57			75	$205^{105\text{mm}}$	>110	dec aq, alc; s eth
n41	2-Nitrobenzyl alcohol	$O_2NC_6H_4CH_2OH$	6, 447	153.14			70–72	270		s aq, alc, eth
n42	3-Nitrobenzyl alcohol	$O_2NC_6H_4CH_2OH$	6, 449	153.14			30–32	$180^{3\text{mm}}$		v s alc, eth; sl s aq
n43	4-Nitrobenzyl alcohol	$O_2NC_6H_4CH_2OH$	6, 450	153.14			92–94	$185^{12\text{mm}}$		2 alc; v s eth
n44	4-Nitrobenzyl bromide	$O_2NC_6H_4CH_2Br$	5, 334	216.04			98–100			8 alc; s eth
n45	4-Nitrobenzyl chloride	$O_2NC_6H_4CH_2Cl$	5, 329	171.58			70–73			s alc, acet, CCl_4
n46	2-Nitrobiphenyl	$O_2NC_6H_4C_6H_5$	5, 582	199.21	1.44^{25}_{4}	1.613^{25}	36.7	325	179	sl s alc; s chl, eth
n47	4-Nitrobiphenyl	$O_2NC_6H_4C_6H_5$	5, 583	199.21			112–114	340		sl s aq; misc alc, eth
n48	1-Nitrobutane	$CH_3CH_2CH_2CH_2NO_2$	1, 123	103.18	0.9752^{20}_{4}	1.4112	–81.3	152.8	47	1 alc
n49	3-Nitro-2-butanol	$CH_3CH(NO_2)CH(OH)CH_3$	1, 373	119.12	1.1296^{25}_{4}	1.4414^{20}		$92^{10\text{mm}}$	91	
n50	3-Nitrocinnamic acid	$O_2NC_6H_4CH=CHCO_2H$	Merck: 12, 6692	193.16			200–201			
n51	2-Nitrodiphenylamine	$O_2NC_6H_4NHC_6H_5$	12, 690	214.22			76			i aq; s alc
n52	Nitroethane	$CH_3CH_2NO_2$	1, 99	75.07	1.0528^{20}_{20}	1.3920^{20}	–90	114	28	4.5 aq; misc alc, eth; s alk, chl
n53	5-Nitro-2-furaldehyde semicarbazone		17^{3}, 4467	198.14			242–244			s alk, chl, alk; 0.2 alc
n54	1-nitroguanidine	$O_2NNHC(=NH)NH_2$	3, 126	104.07			dec >225			0.4 aq; sl s MeOH
n55	5-Nitro-1*H*-indazole		23, 129	163.14			207–209			s alc, bz, eth, acet
n56	Nitromethane	CH_3NO_2	1, 74	61.04	1.1322^{24}_{4}	1.3795^{25}	–28.4	101.2	35	11 aq; s alc, eth
n57	1-Nitronaphthalene	$C_{10}H_7NO_2$	5, 553	173.17	1.223		59–60	304	90	s alc; v s chl, eth
n58	3-Nitro-2-pentanol	$CH_3CH_2CH(NO_2)CH(OH)CH_3$	1, 385	133.15	1.0818^{25}_{4}	1.4430^{20}		$100^{10\text{mm}}$		
n59	2-Nitrophenol	$O_2NC_6H_4OH$	6, 213	139.11	1.495		45	216		s alc, bz, eth, alk
n60	4-Nitrophenol	$O_2NC_6H_4OH$	6, 226	139.11	1.2704^{120}_{4}		113–114	279		s aq; v s alc, chl, eth
n61	4-Nitrophenyl acetate	$O_2NC_6H_4O_2CCH_3$	6, 233	181.15			77–79			s aq; v s alc, bz, eth
n62	2-Nitrophenylacetic acid	$O_2NC_6H_4CH_2CO_2H$	9, 454	181.15			139–142			s hot aq, alc
n63	4-Nitrophenylacetic acid	$O_2NC_6H_4CH_2CO_2H$	9, 455	181.15			153–155			s alc, bz, eth; sl s aq

(Continued)

TABLE 2.21   Physical Constants of Organic Compounds (*Continued*)

No.	Name	Formula	Formula weight	Beilstein reference	Density, g/mL	Refractive index	Melting point, °C	Boiling point, °C	Flash point, °C	Solubility in 100 parts solvent
n64	4-Nitrophenylaceto-nitrile	$O_2NC_6H_4CH_2CN$	162.15	9, 456			115–117			s alc, eth; i aq
n65	2-Nitro-1,4-phenylene-diamine	$O_2NC_6H_3(NH_2)_2$	153.14	13, 120			137–140			
n66	4-Nitro-1,2-pheny lene-diamine	$O_2NC_6H_3(NH_2)_2$	153.14	13, 29			199–201			sl s aq; s HCl
n67	4-Nitrophenyl-hydrazine	$O_2NC_6H_4NHNH_2$	153.14	15, 468			156 dec			s alc, chl, eth, hot bz
n68	2-Nitrophenyl phenyl ether	$O_2NC_6H_4OC_6H_5$	215.21	6^2, 222	1.2539^{20}	1.575^{20}	<−20	184^{8mm}		s alc, eth
n69	4-Nitrophenyl phenyl ether	$O_2NC_6H_4OC_6H_5$	215.21	6, 232			53–56	320	>110	s bz, eth
n70	3-Nitro-1,2-phthalic acid	$O_2NC_6H_3(CO_2H)_2$	211.13	9, 823			213–216 dec			
n71	4-Nitro-1,2-phthalic acid	$O_2NC_6H_3(CO_2H)_2$	211.13	9, 828			170–172			
n72	3-Nitrophthalic anhydride		193.11	17, 486			163–165			sl s aq, bz
n73	1-Nitropropane	$CH_3CH_2CH_2NO_2$	89.09	1, 115	1.0009^{20}	1.4016^{20}	−108	131.1	36	1.4 aq; misc org solv
n74	2-Nitropropane	$(CH_3)_2CHNO_2$	89.09	1, 116	0.9821^{20}	1.3949^{20}	−91.3	120.3	24	1.7 aq; misc org solv
n75	2-Nitro-1-propanol	$CH_3CH(NO_2)CH_2OH$	105.09	1, 358	1.1841_4^{25}	1.4379^{20}		99^{10mm}	100	s aq, alc, eth
n76	4-Nitropyridine-*N*-oxide	$O_2NC_5H_4N(\rightarrow O)$	140.10	20^3, 2528			159–162			
n77	Nitrosobenzene	C_6H_5NO	107.11	6, 230	1.0048^{20}_4	1.4368^{20}	67–69	59^{18mm}		v s aq, alc, eth
n78	N-Nitrosodimethyl-amine	$(CH_3)_2NNO$	74.08	8, 84				151	61	
n79	4-Nitrosodiphenyl-amine	$C_6H_5NC_6H_4NO$	198.22	Merck: 12, 6737			144–145			v s alc, bz, chl, eth
n80	1-Nitroso-2-naphthol	$C_{10}H_6(NO)OH$	173.16	7, 712			109–110			3 alc; s bz, eth, alk; 0.1 aq
n81	1-Nitroso-2-naphthol-3,6-disulfonic acid disodium salt hydrate		377.26	11^2, 190			>300			2.5 aq; sl s alc

(Continued)

No.	Name	Formula	Beilstein ref.	Formula wt	Density	n_D	mp, °C	bp, °C	Flash pt, °C	Solubility
n82	4-Nitrosophenol	HOC_6H_4NO	7, 622	123.11			126	dec 144		s aq; v s alc, eth; explodes on contact with conc acid, alk, or fire
n83	2-Nitrotoluene	$CH_3C_6H_4NO_2$	5, 318	137.14	1.1622^{19}_4	1.5472^{20}	−10	222	106	s alc, bz
n84	3-Nitrotoluene	$CH_3C_6H_4NO_2$	5, 321	137.14	1.1581^{20}_4	1.5459^{20}	15.5	231.9	101	misc alc, eth; s bz
n85	4-Nitrotoluene	$CH_3C_6H_4NO_2$	5, 323	137.14	1.392		52	238	106	s alc, bz, chl, eth
n86	2-Nitro-α,α,α-trifluorotoluene	$CF_3C_6H_4NO_2$	5^2, 251	191.11			31–32	105^{20mm}	95	v s alc, bz
n87	3-Nitro-α,α,α-trifluorotoluene	$CF_3C_6H_4NO_2$	5, 327	191.11	1.436^{16}_4	1.4715^{20}	−2.4	200–205	87	s alc, eth
n88	5-Nitrouracil		24, 320	157.09			>300			
n89	Nonadecane	$CH_3(CH_2)_{17}CH_3$	1, 174	268.51	0.7776^{32}_4	1.4335^{38}	32	330	168	s eth; sl s alc
n90	Nonane	$CH_3(CH_2)_7CH_3$	1, 165	128.26	0.7176^{20}_4	1.4054^{20}	−53.5	150.8	31	s abs alc, eth
n91	1,9-Nonanediamine	$H_2N(CH_2)_9NH_2$	4, 272	158.29	0.929		37–38	258	>110	
n92	Nonanedinitrile	$NC(CH_2)_7CN$	2, 709	150.23		1.4460^{20}		176^{11mm}	>110	v s alc, bz, eth
n93	1,9-Nonanedioic acid	$HO_2C(CH_2)_7CO_2H$	2, 707	188.22	1.029^{20}_4		106.5	286^{100mm}		0.24 aq; v s alc; 3 eth
n94	1,9-Nonanediol	$HO(CH_2)_9OH$	1, 493	160.26			47–49	177^{15mm}	>110	
n95	Nonanenitrile	$CH_3(CH_2)_7CN$	2, 354	139.24	0.851^{15}_4	1.4260^{20}	−34.2	224.0	81	s alc, eth
n96	Nonanoic acid	$CH_3(CH_2)_7CO_2H$	2, 352	158.24	0.906^{20}_4	1.4330^{20}	12.5	254.5	100	s alc, chl, eth
n97	γ-Nonanoic lactone		17, 245	156.23	0.976	1.4475^{20}		122^{6mm}	>110	
n98	1-Nonanol	$CH_3(CH_2)_8OH$	1, 423	144.26	0.8279^{20}_4	1.4338^{20}	−5.5	215	75	0.6 aq; misc alc, eth
n99	2-Nonanone	$CH_3(CH_2)_6COCH_3$	1, 709	142.24	0.832	1.4210^{20}	−21	192^{743mm}	64	
n100	3-Nonanone	$CH_3(CH_2)_5COCH_2CH_3$	1, 709	142.24	0.821	1.4204^{20}		187–188	67	misc alc, eth
n101	5-Nonanone	$(CH_3CH_2CH_2CH_2)_2CO$	1, 710	142.24	0.806^{20}_4	1.4190^{20}	−50	186–187	60	
n102	Nonanoyl chloride	$CH_3(CH_2)_7COCl$	2, 353	176.69	0.946^{15}_4	1.4377^{20}	−60.5	215.4	95	dec aq, alc; s eth
n103	3-Nonen-2-one	$CH_3(CH_2)_4CH{=}CHCOCH_3$	1^3, 3017	140.23	0.848	1.4484^{20}		85^{12mm}	81	
n104	Nonyl aldehyde	$CH_3(CH_2)_7CHO$	1, 708	142.24	0.827^{19}	1.4240^{20}		185	63	sl s aq; s alc, eth
n105	Nonylamine	$CH_3(CH_2)_8NH_2$	4, 198	143.27	0.782	1.4330^{20}		201	62	
n106	Nopol		6^3, 396	166.26	0.973	1.4930^{20}		230–240	98	s alc
n107	Norbomane		5^1, 45	96.17			82–84			
n108	2-Norbomanone		7, 57	110.16	1.048		94–96	168–172	33	
n109	exo-2-Norbomyl formate		6^3, 219	140.18		1.4622^{20}		67^{16mm}	53	
n110	(+)-Norephedrine	$C_6H_5CH(OH)CH(CH_3)NH_2$	13^2, 371	151.21			51–54			v s eth; 10 PE; s abs alc
o1	cis,cis-9,12-Octadecadienoic acid	$CH_3(CH_2)_4CH{=}CHCH_2{-}CH{=}CH(CH_2)_7CO_2H$	2, 496	280.44	0.9025^{20}_4	1.4699^{20}	−5	230^{16mm}	>110	

TABLE 2.21 Physical Constants of Organic Compounds (*Continued*)

No.	Name	Formula	Formula weight	Beilstein reference	Density, g/mL	Refractive index	Melting point, °C	Boiling point, °C	Flash point, °C	Solubility in 100 parts solvent
o2	Octadecanamide	$CH_3(CH_2)_{16}CONH_2$	283.50	2, 383			102–104	251^{12mm}	165	s hot alc, hot eth
o3	Octadecane	$CH_3(CH_2)_{16}CH_3$	254.50	1, 173	0.7776^{28}_{4}	1.4367^{28}	28.2	316.3	185	s acet, eth; sl s alc
o4	1-Octadecanethiol	$CH_3(CH_2)_{17}SH$	286.57	1^3, 1838		1.4648	31–35	360		s eth; sl s alc
o5	Octadecanoic acid	$CH_3(CH_2)_{16}CO_2H$	284.48	2, 377	0.847^{70}	1.4299^{80}	69	383		4.9 alc; 20 bz; 50 chl; 3.9 acet; 16.6 CCl_4; s toluene, pentyl acetate
o6	1-Octadecanol	$CH_3(CH_2)_{17}OH$	270.50	1, 431	0.8123^{58}_{4}	1.4388^{20}	59.6	203^{10mm}		s alc, eth
o7	9,12,15-Octadecatrienoic acid	$CH_3CH_2CH{=}CH_2)_3CH_2-(CH_2)_6CO_2H$	278.44	2, 499	0.914^{18}_{4}	1.4800^{20}		230^{17mm}	>110	s alc, bz, eth
o8	1-Octadecene	$CH_3(CH_2)_{15}CH{=}CH_2$	252.49	1, 226	0.791^{18}_{4}	1.4439^{20}	17.7	314.9	148	s hot acet
o9	9-Octadecen-1-amine	$CH_3(CH_2)_7CH{=}CH(CH_2)_8NH_2$	267.50	4, 196	0.813	1.4596^{20}		360	154	s alc, bz, chl, eth
o10	*cis*-9-Octadecenoic acid	$CH_3(CH_2)_7CH{=}CH(CH_2)_7CO_2H$	282.47	2, 463	0.8936^{20}_{4}	1.4581^{20}	13.4		189	
o11	*trans*-9-Octadecenoic acid	$CH_3(CH_2)_7CH{=}CH(CH_2)_7CO_2H$	282.47	2^2, 441	0.851^{79}	1.4308^{99}	44–45	288^{100mm}		s bz, chl, eth
o12	*cis*-9-Octadecen-1-ol	$CH_3(CH_2)_7CH{=}CH(CH_2)_8OH$	268.49	1, 453	0.8504^{20}	1.4610^{20}	13–19	195^{8mm}	>110	s alc, eth; i aq
o13	9-Octadecenoyl chloride	$CH_3(CH_2)_7CH{=}CH-(CH_2)_7COCl$	300.92	2, 469	0.912	1.4630^{20}		180^{3mm}	>110	
o14	Octadecyl acrylate	$H_2C{=}CHCO_2(CH_2)_{17}CH_3$	324.55	2^4, 1468	0.800		32–34	232^{32mm}	>110	s alc, bz, eth
o15	Octadecylamine	$CH_3(CH_2)_{17}NH_2$	269.52	4, 196	0.777^{27}	1.4501^{20}	55–57	173^{5mm}	>110	
o16	Octadecyl isocyanate	$CH_3(CH_2)_{17}NCO$	299.51	4^3, 439	0.847	1.4602^{20}	15–16	223^{10mm}	148	
o17	Octadecyltrichlorosilane	$CH_3(CH_2)_{17}SiCl_3$	387.94		0.984				89	
o18	Octadecyl vinyl ether	$CH_3(CH_2)_{17}OCH{=}CH_2$	296.54		0.821^{30}_{4}	1.4440^{30}	28	187^{5mm}	177	
o19	1,7-Octadiene	$H_2C{=}CH(CH_2)_4CH{=}CH_2$	110.20		0.746	1.4220^{20}		114–121	9	
o20	1H,1H,5H-Octafluoro-1-pentanol	$HCF_2CF_2CF_2CF_2CH_2OH$	232.07	1^4, 1648	1.6647^{20}	1.3178^{20}		140–141	75	
o21	Octamethylcyclotetrasiloxane	$[-(CH_3)_2SiO-]_4$	296.62	4^3, 1885	0.956	1.3958^{20}	17–18	176	60	
o22	Octamethyltrisiloxane	$[(CH_3)_3SiO]_2Si(CH_3)_2$	236.54	4^3, 1879	0.8200^{20}	1.3848^{20}	ca. −80	153	29	s bz, PE; sl s alc
o23	Octane	$CH_3(CH_2)_6CH_3$	114.23	1, 159	0.7028^{24}_{4}	1.3974^{20}	−56.8	125.7	22	s eth; sl s alc

(Continued)

	Name	Formula	M.W.	Ref.	d	n	mp	bp	fp	Solubility
o24	1,8-Octanediamine	H2N(CH2)8NH2	144.26	4, 271			50–52	225–226	165	0.16 aq; 0.6 eth; s alc
o25	1,8-Octanedioic acid	HO2C(CH2)6CO2H	174.20	2, 691			140–144	230^{15mm}	>110	
o26	1,2-Octanediol	CH3(CH2)5CH(OH)CH2OH	146.23	1^3, 2217			36–38	132^{10mm}		v s alc; sl s aq, eth
o27	1,8-Octanediol	HO(CH2)8OH	146.23	1, 490			59–61	172^{20mm}		s eth; sl s alc
o28	Octanenitrile	CH3(CH2)6CN	125.22	2, 349	0.8135^{20}	1.4202^{20}	−45.6	198	73	s alc
o29	1-Octanethiol	CH3(CH2)7SH	146.30	1^3, 1710	0.843	1.4525^{20}	−49.2	199.0	68	
o30	Octanoic acid	CH3(CH2)6CO2H	144.21	2, 347	0.9088^{20}_{4}	1.4279^{20}	16.6	239	>110	0.07 aq; v s alc, chl, eth, PE
o31	γ-Octanoic lactone		142.20	17, 244	0.981	1.4440^{20}		234	>110	0.06 aq; misc alc, chl, eth
o32	1-Octanol	CH3(CH2)7OH	130.23	1, 418	0.8258^{20}_{4}	1.4290^{20}	−15.5	195	81	0.1 aq; misc. alc, eth
o33	(±)-2-Octanol	CH3(CH2)5CH(OH)CH3	130.23	1, 419	0.8193^{20}_{4}	1.4202^{20}	−31.6	175	71	
o34	3-Octanol	CH3(CH2)4CH(OH)CH2CH3	130.23	1^1, 208	0.819	1.4260^{20}		174–176	65	
o35	4-Octanol	CH3(CH2)3CH(OH)CH2CH2CH3	130.23		0.8192^{20}	1.425^{20}		176.6	71	
o36	2-Octanone	CH3(CH2)5COCH3	128.22	1, 704	0.819^{20}_{4}	1.4150^{20}	−16	173	52	i aq; misc alc, eth
o37	3-Octanone	CH3(CH2)4COCH2CH3	128.22	1, 706	0.8220^{20}_{4}	1.4150^{20}		167–168	46	i aq; misc alc, eth
o38	4-Octanone	CH3(CH2)3COCH2CH2CH3	128.22	1, 706	0.809	1.4139^{20}		164	45	
o39	Octanoyl chloride	CH3(CH2)6COCl	162.66	2, 348	0.955	1.4350^{20}	<−70	195	80	dec aq, alc; s eth
o40	1-Octene	CH3(CH2)5CH=CH2	112.22	1, 221	0.7149^{20}_{4}	1.4087^{20}	−102	121	21	i aq; misc alc, eth
o41	2-Octen-1-ylsuccinic anhydride		210.27		1.000	1.4694^{20}	8–12	168^{10mm}	>110	
o42	Octyl acetate	CH3CO2(CH2)7CH3	172.27	2, 134	0.868	1.4185^{20}		211	88	sl s aq; misc alc
o43	Octyl aldehyde	CH3(CH2)6CHO	128.22	1, 704	0.821^{20}_{4}	1.4183^{20}	12–15	171	51	sl s aq; misc alc
o44	Octylamine	CH3(CH2)7NH2	129.25	4, 196	0.782	1.4290^{20}	−5/−1	175–177	62	i aq; s alc, eth
o45	Octyl cyanoacetate	NCCH2CO2(CH2)7CH3	197.28	10^3, 2079	0.934	1.4490^{20}		$95^{0.11mm}$	>110	
o46	Octyl gallate	3,4,5-(HO)3C6H2CO2(CH2)7CH3	282.34		0.920	1.4650^{20}	101–104	172^{15mm}		
o47	1-Octyl-2-pyrrolidine		197.32	4^3, 1907	1.070^{20}	1.4473^{20}	−25	226^{730mm}	>110	i aq; s alc, eth
o48	Octyltrichlorosilane	CH3(CH2)7SiCl3	247.67	1, 258	0.7457^{20}	1.4159^{20}	−79.3	126.2	96	v s aq, alc; sl s eth
o49	1-Octyne	CH3(CH2)5C≡CH	110.19	1^3, 1996	0.864	1.4410^{20}		83^{19mm}	17	
o50	1-Octyn-3-ol	CH3(CH2)4CH(OH)C≡CH	126.20						63	
o51	L-(+)-Ornithine	H2N(CH2)3CH(NH2)CO2H	132.16	4, 420			140			
o52	Oxalic acid	HO2CCO2H	90.04	2, 502	1.90^{17}_{4}		190 dec			14 aq^{20}; 40 alc; 1.3 eth
o53	Oxalic acid dihydrate	HO2CCO2H · 2H2O	126.07	2, 502	1.653^{19}_{4}		−2H2O, 102			14 aq; 40 alc; 1 eth

TABLE 2.21 Physical Constants of Organic Compounds (*Continued*)

No.	Name	Formula	Beilstein reference	Formula weight	Density, g/mL	Refractive index	Melting point, °C	Boiling point, °C	Flash point, °C	Solubility in 100 parts solvent
o54	Oxalyl bromide	$BrC(=O)C(=O)Br$	2^1, 236	215.84		1.5220^{20}	−19	$103^{720\,mm}$	none	s eth; viol dec aq, alc
o55	Oxalyl chloride	$ClC(=O)C(=O)Cl$	2, 542	126.93	1.455	1.4290^{20}	−10	64	none	s hot aq; sl s alc, eth
o56	Oxalyl dihydrazide	$H_2NNHC(=O)C(=O)NHNH_2$	2, 559	118.10			240 dec			s alk; sl s aq; i eth
o57	Oxamic hydrazide	$H_2NC(=O)C(=O)NHNH_2$	2, 559	103.08	1.667^{20}_4		218 dec			sl s hot aq, alc
o58	Oxamide	$H_2NC(=O)C(=O)NH_2$	27, 135	88.07			dec 350	$220^{48\,mm}$		
o59	2-Oxazolidone			87.08			86–89	$82^{16\,mm}$	81	
o60	2-Oxobutyric acid	$CH_3CH_2C(=O)CO_2H$	3, 629	102.09	1.200^{17}	1.3972^{20}	32–34	270		v s aq, alc; v sl s eth
o61	2-Oxohexamethylene-imine		21^2, 216	113.16	1.02^{75}		69.2		125	84 aq; v s alc, eth, chlorinated HC's
o62	5-Oxohexanonitrile	$CH_3CO(CH_2)_3CN$	3^3, 1234	111.14	0.975	1.4328^{20}		240	107	v s aq, alc, bz, eth
o63	4-Oxopentanoic acid	$CH_3COCH_2CH_2CO_2H$	3, 671	116.12	1.1447^{25}_4	1.4396^{20}	33–35	246	137	s aq, alc
o64	2-Oxopropionaldehyde	$CH_3C(=O)CHO$	1, 762	72.06	1.0455^{24}_4	1.4209^{20}		72	none	misc aq, alc, eth
o65	2-Oxopropionic acid	$CH_3C(=O)CO_2H$	3, 608	88.06	1.267^{15}	1.4315^{20}	11.8	165 dec	82	misc aq, alc, eth
o66	2-Oxo-1-pyrrolidine-propionitrile			138.17	1.120	1.4880^{20}		$140^{0.3\,mm}$	>110	
o66a	2,2'-Oxybis [2-methyl]-propane	$(CH_3)_3COC(CH_3)_3$		130.23	0.7658	1.3949^{20}		107		dec acids
o67	2,2'-Oxydiacetic acid	$HO_2CCH_2OCH_2CO_2H$	3, 234	134.09			142–145	dec		v s aq, alc; sl s eth
o68	4,4'-Oxy dianiline	$H_2NC_6H_4OC_6H_4NH_2$	13, 441	200.24			190–192		218	
o69	3,3'-Oxydipropionitrile	$NCCH_2CH_2OCH_2CH_2CN$		124.14	1.043	1.4405^{20}		$112^{0.5\,mm}$	>110	
p1	Paraformaldehyde	$(CH_2O)_x$	1, 566				165 dec		71	s(slow) aq; s alk; i alc, eth
p2	Paraldehyde	$[-HC(CH_3)O-]_3$	19, 385	132.16	0.9984^{15}	1.4049^{20}	12.6	124		11 aq; misc alc, chl
p3	Parathion	$(C_2H_5O)_2P(=S)C_6H_4NO_2$		291.27	1.26^{25}_4	1.5370^{25}	6	375		v s alc, bz, eth
p4	Pentabromophenol	C_6Br_5OH	6, 206	488.62			223–226		none	sl s alc, eth
p5	Pentachloroacetone	$Cl_2CHC(=O)CCl_3$	1, 690	230.31	1.656	1.4967^{20}	21 (anhyd)	192	none	i aq; v s acet
p6	Pentachlorobenzene	C_6HCl_5	5, 205	250.34	1.8342^{16}		82–85	275–277	none	v s bz, chl, eth
p7	Pentachloroethane	Cl_2CHCCl_3	1, 87	202.30	1.6712^{24}_4	1.5030^{20}	−29.0	160		0.05 aq; misc alc, eth
p8	Pentachloronitro-benzene	$C_6Cl_5(NO_2)$	5, 247	295.34	1.718^{25}_4		140–143			s bz, chl
p9	Pentachlorophenol	C_6Cl_5OH	6, 194	266.34	1.978^{22}_4		190–191	310		v s alc; s bz; 148 eth
p10	Pentachloropyridine	C_5Cl_5N	20, 232	251.33			124–126			
p11	Pentadecane	$CH_3(CH_2)_{13}CH_3$	1, 172	212.42	0.7684^{20}_4	1.4319^{20}	9.9	270	132	v s alc, eth
p12	Pentadecanenitrile	$CH_3(CH_2)_{13}CN$	2^1, 163	223.40	0.825	1.4420^{20}	20–23	322	>110	
p13	8-Pentadecanone	$[CH_3(CH_2)_7]_2C=O$	1, 717	226.40			41–43	178	>110	s alc

p14	3-Pentadecylphenol	$CH_3(CH_2)_{14}C_6H_4OH$	304.52					195^{1mm}	>110	
p15	1,2-Pentadiene	$CH_3CH_2CH=C=CH_2$	68.12	1, 251	0.6926^{20}_4	1.4209^{20}	−137.3	44.9	−28	
p16	cis-1,3-Pentadiene	$CH_3CH=CHCH=CH_2$	68.12	1, 251	0.6910^{10}	1.4363^{20}	−140.8	44.1	−28	
p17	trans-1,3-Pentadiene	$CH_3CH=CHCH=CH_2$	68.12	1, 251	0.6760^{22}_4	1.4301^{20}	−87.5	42.0	−28	
p18	1,4-Pentadiene	$H_2C=CHCH_2CH=CH_2$	68.12	1, 251	0.66084^{22}	1.3888^{20}	−148.3	26.0	4	6 aq; v sl s alc; i eth
p19	Pentaerythritol	$C(CH_2OH)_4$	136.15	1, 528	1.38^{25}_4	1.548	260		>110	
p20	Pentaerythritol diacrylate monostrearate	$CH_3(CH_2)_{16}CO_2CH_2$-$C(CH_2O_2CCH=CH_2)_2$-CH_2OH	510.72		1.018		29–31			
p21	Pentaerythritol triacrylate	$(H_2C=CHCO_2CH_2)_3CCH_2OH$	298.30		1.180	1.4864^{20}			>110	
p22	Pentaerythrityl tetranitrate	$C(CH_2ONO_2)_4$	316.15	1^2, 602	1.1773^{20}_4	1.5096^{20}	140	explodes on shock		s acet; sl s eth, alc
p23	Pentaethylenehex-amine	$H_2N(CH_2CH_2NH)_4CH_2CH_2NH_2$	232.38	4^4, 1245	0.950				>110	
p24	Pentamethylbenzene	$C_6H(CH_3)_5$	148.25	5, 443	0.917^{20}_4	1.527^{20}	54.4	231	91	v s alc, bz
p25	1,2,3,4,5-Pentamethyl-cyclopentadiene		136.24		0.870	1.4733^{20}		58^{13mm}	44	
p26	N,N,N',N',N''-Penta-methyldiethylene-triamine	$[(CH_3)_2NCH_2CH_2]_2NCH_3$	173.30	4^4, 1245	0.830	1.4420^{20}	−20	198	53	
p27	1,5-Pentamethylene-tetrazole		138.17	26^2, 213			59–61	194^{12mm}		
p28	Pentanal	$CH_3CH_2CH_2CH_2CHO$	86.13	1, 676	0.8095^{20}_4	1.3942^{20}	−92	103	12	1.4 aq; misc alc, eth
p29	Pentane	$CH_3CH_2CH_2CH_2CH_3$	72.15	1, 130	0.6262^{20}_{20}	1.3575^{20}	−129.7	36.0	−49	misc alc, eth
p30	1,5-Pentanediamine	$H_2N(CH_2)_5NH_2$	102.18	4, 266	0.8734^{25}	1.4591^{20}	−129.7	178–180	62	s aq, alc; sl s eth
p31	1,2-Pentanediol	$CH_3CH_2CH_2CH(OH)CH_2OH$	104.15	1^2, 548	0.971	1.4397^{20}		206	104	
p32	1,5-Pentanediol	$HO(CH_2)_5OH$	104.15	1, 481	0.9941^{20}	1.4494^{20}	−18	239	129	s aq, alc; sl s eth
p33	2,3-Pentanedione	$CH_3CH_2C(=O)C(=O)CH_3$	100.11	1, 776	0.957	1.4068^{20}	−52	110–112	19	17 aq; misc alc, eth
p34	2,4-Pentanedione	$CH_3COCH_2COCH_3$	100.11	1, 777	0.9721^{25}	1.4510^{20}	−23.1	138	34	i aq; s alc, eth
p35	Pentanenitrile	$CH_3CH_2CH_2CH_2CN$	83.13	2, 301	0.80355^4	1.3991^{15}	−92	141.3	40	4 aq
p36	1-Pentanesulfonic acid, sodium salt	$CH_3(CH_2)_4SO_3^-\ Na^+$	174.19	4^3, 23			>300			

(Continued)

TABLE 2.21 Physical Constants of Organic Compounds (*Continued*)

No.	Name	Formula	Formula weight	Beilstein reference	Density, g/mL	Refractive index	Melting point, °C	Boiling point, °C	Flash point, °C	Solubility in 100 parts solvent
p37	1-Pentanethiol	$CH_3(CH_2)_4SH$	104.22	1, 384	0.840	1.4460^{20}	−75.7	126.6	18	i aq; misc alc, eth
p38	Pentanoic acid	$CH_3(CH_2)_3CO_2H$	102.13	2, 299	0.9390^{20}_4	1.4080^{20}	−33.7	186	96	2.4 aq; v s alc, eth
p39	1-Pentanol	$CH_3(CH_2)_4OH$	88.15	1, 383	0.8146^{20}_4	1.4100^{20}	−79	137.5	33	2.7 aq^{22}; misc alc, eth
p40	2-Pentanol	$CH_3CH_2CH_2CH(OH)CH_3$	88.15	1, 384	0.8098^{20}_4	1.4054^{20}	−73	119.3	34	16.6 aq^{20}; misc alc, eth
p41	3-Pentanol	$CH_3CH_2CH(OH)CH_2CH_3$	88.15	1, 385	0.8150^{20}_4	1.4077^{25}	−69	116	41	5.5 aq^{20}; s alc, eth
p42	2-Pentanone	$CH_3CH_2CH_2COCH_3$	86.13	1, 676	0.8095^{20}	1.3900^{20}	−76.8	102	7	misc acet, bz, eth, PE
p43	3-Pentanone	$CH_3CH_2COCH_2CH_3$	86.13	1, 679	0.8143^{20}	1.3920^{20}	−39.0	102.0	13	3.4 aq
p44	Pentanophenone	$C_6H_5CO(CH_2)_3CH_3$	162.23	7, 327	0.988	1.5143^{20}		107^{5mm}	102	s alc, eth
p45	Pentanoyl chloride	$CH_3(CH_2)_3COCl$	120.58	2, 301	1.016	1.4216^{20}		125–127	32	
p46	1,4,7,10,13-Pentaoxacyclopentadecane	$[-CH_2CH_2O-]_5$	220.27		1.109	1.4650^{20}		$135^{0.2mm}$	>110	
p47	2,5,8,11,14-Pentaoxapentadecane	$CH_3(OCH_2CH_2)_4OCH_3$	222.28	1^3, 2107	1.0087^{20}_4	1.4330^{20}	−27	275–276	140	s aq; misc hydrocarbon solvents
p48	1-Pentene	$CH_3CH_2CH_2CH{=}CH_2$	70.14	1, 210	0.6429^{20}	1.3714^{20}	−165	30.1	−18	misc alc, bz, eth
p49	cis-2-Pentene	$CH_3CH_2CH{=}CHCH_3$	70.14	1, 210	0.6503^{20}_4	1.3813^{20}	−151	37.0	−20	misc alc, eth
p50	trans-2-Pentene	$CH_3CH_2CH{=}CHCH_3$	70.14	1, 210	0.6482^{20}_4	1.3792^{20}	−140	36.3	−45	misc alc, eth
p51	cis-2-Pentenenitrile	$CH_3CH_2CH{=}CHCN$	81.12	2^2, 400	0.820	1.4269^{20}		128	23	
p52	trans-3-Pentenenitrile	$CH_3CH{=}CHCH_2CN$	81.12	2, 427	0.837	1.4221^{20}		144–147	40	
p53	Pentyl acetate	$CH_3(CH_2)_4O_2CCH_3$	130.19	2, 131	0.8753^{20}	1.4020^{20}	−70.8	149.2	16	0.17 aq; misc alc, eth
p54	Pentylamine	$CH_3(CH_2)_4NH_2$	87.16	4, 175	0.7544^{20}_4	1.448^{20}	−55	104	−1	v s aq; misc eth; s alc
p55	Pentylbenzene	$CH_3(CH_2)_4C_6H_5$	148.25	5, 434	0.8594^{20}_4	1.4885^{20}	−78.3	202.2	65	s alc, misc bz, eth
p56	2-Pentylcinnamaldehyde	$C_6H_5CH{=}C[(CH_2)_4CH_3]CHO$	202.30	7^2, 310	0.970	1.5571^{20}		290	>110	
p57	4-tert-Pentylphenol	$CH_3CH_2C(CH_3)_2C_6H_4OH$	164.25	6, 548	0.962^{20}_4	1.3852^{20}	93	262		s alc, eth
p58	1-Pentyne	$CH_3CH_2CH_2C{\equiv}CH$	68.11	1, 250	0.6901^{20}_4	1.3010^{20}	−106	40.2	−34	v s alc; misc eth
p59	Perfluoro-1-octanesulfonyl fluoride	$CF_3(CF_2)_7SO_2F$	502.12	2^4, 996	1.824			154–155	none	
p60	Peroxyacetic acid	$CH_3C({=}O)CO_2H$	76.05	2, 169	1.226^{15}_4	1.3876^{20}	−0.2	110	41	v s aq, alc, eth
p61	Petroleum ether	Principally pentanes and hexanes		Merck: 12, 7329	0.640	1.3630^{20}		35–60	−49	misc bz, alc, chl, eth, CCl$_4$; s glacial HOAc
p62	Phenanthrene		178.23	5, 667	1.063		100	340		1.6 alc; 50 bz; 30 eth
p63	1,10-Phenanthroline		180.21	23, 227			114–117			0.3 aq; 1.4 bz; s alc, acet
p64	Phenethylisobutyrate	$(CH_3)_2CHCO_2CH_2CH_2C_6H_5$	192.26	6^2, 451	0.988	1.4880^{20}		250	108	
p65	Phenol	C_6H_5OH	94.11	6, 110	1.0576^{41}_4	1.5418^{41}	41	182	79	6.7 aq; 8.2 bz; v s alc, chl, eth, alk

No.	Name	Formula	M.W.	Beil. ref.	Density	n	m.p.	b.p.	Flash	Solubility
p66	Phenolphthalein		318.33	18, 143	1.299		261–263			8.2 alc; 1 eth
p67	Phenothiazine		199.28	27, 63			185.1	371		v s bz; s eth; sl s alc
p68	Phenoxyacetic acid	$C_6H_5OCH_2CO_2H$	152.15	6, 161			98–100	285 sl dec		1.3 aq; v s alc, bz, HOAc, CS_2, eth
p69	Phenoxyacetyl chloride	$C_6H_5OCH_2COCl$	170.60	6, 162	1.235	1.5340^{20}		225–226	108	dec aq, alc; s eth
p70	4-Phenoxyaniline	$C_6H_5OC_6H_4NH_2$	185.23	13, 438			84	189^{14mm}		s hot aq; v s alc, eth
P71	2-Phenoxybutyric acid	$CH_3CH_2CH(OC_6H_5)CO_2H$	180.20	6, 163			79–83	258		sl s aq
p72	2-Phenoxyethanol	$C_6H_5OCH_2CH_2OH$	138.17	6, 146	1.102^{22}_{4}	1.5370^{20}	14	245.2	>110	s aq; v s alc, eth
p73	1-Phenoxy-2-propanol	$C_6H_5OCH_2CH(OH)CH_3$	152.19	6', 85	1.063^{25}_{4}	1.523^{20}	13–18	240	135	
p74	2-Phenoxypropionic acid	$CH_3CH(OC_6H_5)CO_2H$	166.18	6, 163			116–119	265		s alc; sl s aq
p75	3-Phenoxypropyl bromide	$C_6H_5O(CH_2)_3Br$	215.10	6, 142	1.365	1.5460^{20}		134^{14mm}	96	
p76	3-Phenoxytoluene	$C_6H_5OC_6H_4CH_3$	184.24	6, 377	1.051	1.5727^{20}	33–34	271–273	>110	sl s aq; s alc, eth
p77	Phenylacetaldehyde	$C_6H_5CH_2CHO$	120.15	7, 292	1.027^{25}_{25}	1.5290^{20}		195	86	
p78	Phenylacetaldehyde dimethyl acetal	$C_6H_5CH_2CH(OCH_3)_2$	166.22	7, 293	1.004	1.4930^{20}		221	83	
p79	Phenylacetaldehyde ethylene acetal		164.21	19^4, 220	1.100	1.5220^{20}		120^{12mm}	107	
p80	Phenyl acetate	$C_6H_5O_2CCH_3$	136.15	6, 152	1.073	1.5030^{20}		196	76	misc alc, eth, chl
p81	Phenylacetic acid	$C_6H_5CH_2CO_2H$	136.15	9, 431	1.091^{17}		76.5	265.5		s hot aq, alc, eth
p82	Phenylacetonitrile	$C_6H_5CH_2CN$	117.15	9, 441	1.0214	1.5233^{20}	–23.8	233.5	101	i aq; misc alc, eth
p83	Phenylacetyl chloride	$C_6H_5CH_2COCl$	154.60	9, 436	1.169	1.5325^{20}		95^{12mm}	102	dec aq, alc
p84	Phenylacetylene	$C_6H_5C{\equiv}CH$	102.14	5, 511	0.9300	1.5470^{20}	–44.9	142.4	31	misc alc, eth
p85	Phenylacetylurea	$C_6H_5CH_2CONHCONH_2$	178.19	Merck: 12, 7343			212–216			sl s alc, bz, chl, eth
p86	(±)-3-Phenylalanine	$C_6H_5CH_2CH(NH_2)CO_2H$	165.19	14, 495			271–273			1.4 aq
p87	Phenyl 4-amino-salicylate	$H_2NC_6H_3-2-(OH)CO_2C_6H_5$	229.24	Merck: 12, 7426			153			0.7 mg aq
p88	4-Phenylazoaniline	$C_6H_5N{=}NC_6H_4NH_2$	197.24	16^1, 310	1.235		123–126	>360		v s alc, bz, chl, eth
p89	Phenylazoformic acid 2-phenylhydrazide	$C_6H_5N{=}NCONHNHC_6H_5$	240.27	16, 24			156–159 dec			
p90	4-Phenylazophenol	$C_6H_5N{=}NC_6H_4OH$	198.23	16, 96			150–152	230^{20mm}		v s alc, eth
p91	2-Phenylbenzimidazole		194.24	23, 230			293–296			s abs alc; sl s bz, chl
p92	Phenyl benzoate	$C_6H_5CO_2C_6H_5$	198.22	9, 116	1.235		69–72	298–299		v s hot alc; sl s eth
p93	N-Phenylbenzylamine	$C_6H_5CH_2NHC_6H_5$	183.25	12, 1023	1.061		35–38	306–307	>110	s alc, chl, eth
p94	trans-4-Phenyl-3-buten-2-one	$C_6H_5CH{=}CHCOCH_3$	146.19	7, 364	1.0097^{45}_{4}	1.5836^{45}	41.5	260–262	65	v s alc, bz, chl, eth

(Continued)

TABLE 2.21 Physical Constants of Organic Compounds (*Continued*)

No.	Name	Formula	Formula weight	Beilstein reference	Density, g/mL	Refractive index	Melting point, °C	Boiling point, °C	Flash point, °C	Solubility in 100 parts solvent
p95	2-Phenyl-3-butyn-2-ol	$CH_3C(OH)(C_6H_5)C\equiv CH$	146.19	6^2, 559			47–49	217–218	96	0.8 aq; s alc, bz, acet
p96	3-Phenylbutyraldehyde	$CH_3CH(C_6H_5)CH_2CHO$	148.21	7^1, 168	0.997	1.5179^{20}		94^{16mm}	96	s bz, eth
p97	2-Phenylbutyric acid	$CH_3CH_2CH(C_6H_5)CO_2H$	164.20	9^2, 356	1.055	1.5160^{20}	42–44	270–2	>110	
p98	2-Phenylbutyronitrile	$CH_3CH_2CH(C_6H_5)CN$	145.21	9, 541	0.974	1.5086^{20}		114^{15mm}	105	
p99	Phenyl chloroformate	$C_6H_5O_2CCl$	156.57	6, 159	1.248	1.5107^{20}		71^{9mm}	75	
p100	Phenyl dichlorophosphate	$C_6H_5OP(O)Cl_2$	210.98	6, 179	1.412	1.5230^{20}		241–243	>110	
p101	N-Phenyldiethanolamine	$C_6H_5N(CH_2CH_2OH)_2$	181.24	12, 183	1.120^{60}_{20}		56–80	350 sl dec		5 aq; v s alc; 29 eth; 25 bz
p102	4-Phenyl-1,3-dioxane		164.21	19^1, 616	1.111	1.5300^{20}		250–251	>110	
p103	2-Phenyl-1,3-dioxolane		150.18		1.106	1.5260^{20}		$80^{0.3mm}$	98	v s alc, chl, eth; sl s aq
p104	1,2-Phenylenediamine	$C_6H_4\text{-}1,2\text{-}(NH_2)_2$	108.14	13, 6			103	257		s aq, alc, acet, chl
p105	1,3-Phenylenediamine	$C_6H_4\text{-}1,3\text{-}(NH_2)_2$	108.14	13, 33	1.139^{15}		63.5	285		1 aq; s alc, chl, eth
p106	1,4-Phenylenediamine	$C_6H_4\text{-}1,4\text{-}(NH_2)_2$	108.14	13, 61			146	267	156	
p107	1,4-Phenylene diisocyanate	$C_6H_4\text{-}1,4\text{-}(NCO)_2$	160.13	13, 105			97–98	260	>110	
p108	1-Phenyl-1,2-ethanediol	$C_6H_5CH(OH)CH_2OH$	138.17	6, 907			66–68	272–274		v s aq, alc, bz, eth, chl, HOAc
p109	1-Phenylethanol	$CH_3CH(OH)(C_6H_5)$	122.17	6, 475	1.0130^{20}	1.5270^{20}	20	204	85	2.3 aq
p110	2-Phenylethanol	$C_6H_5CH_2CH_2OH$	122.17	6, 478	1.023^{25}_{25}	1.5317^{20}	−27	221	102	2 aq; misc alc, eth
p111	2-Phenylethyl acetate	$CH_3CO_2CH_2CH_2C_6H_5$	164.20	9, 510	0.984	1.4985^{20}		238–239	101	2 aq; misc alc, eth
p112	2-Phenylethylamine	$C_6H_5CH_2CH_2NH_2$	121.18	12, 1096	0.9640^{25}_{4}	1.5290^{25}	<0	197.5	90	80 aq^{15}, s alc; i eth
p113	1-Phenylethyl propionate	$C_2H_5CO_2CH(CH_3)C_6H_5$	178.23	5^3, 1680	1.007	1.4895^{20}		92^{5mm}	94	
p114	(±)-2-Phenylglycine	$C_6H_5CH(NH_2)CO_2H$	151.17	14, 460			subl 255			s org solvents, alk
p115	1-Phenylheptane	$C_6H_5(CH_2)_6CH_3$	176.30	5, 451	0.860	1.4850^{20}		233	95	
p116	1-Phenylhexane	$C_6H_5(CH_2)_5CH_3$	162.28	5^2, 337	0.861	1.4860^{20}	−61	226	83	misc eth
p117	Phenylhydrazine	$C_6H_5NHNH_2$	108.14	15^2, 44	1.0978^{20}_{4}	1.6080^{20}	19.5	243	88	misc alc, bz, chl, eth
p118	Phenyl 1-hydroxy-2-naphthoate	$HOC_{10}H_6CO_2C_6H_5$	264.28	10, 332			94–96			
p119	Phenyl 3-hydroxy-2-naphthoate	$C_{10}H_6(OH)CO_2C_6H_5$	264.28	10, 335			129–132	261^{160mm}		
p120	2-Phenylimidazole		144.18	23, 182			144–147			
p121	2-Phenyl-2-imidazoline		146.19	23, 154			94–99			
p122	2-Phenyl-1,3-indandione		222.28	7, 808			148–150			

(Continued)

No.	Name	Formula	Form. wt.	Beilstein ref.	Density	n_D	mp (°C)	bp (°C)	Fp (°C)	Solubility
p123	2-Phenylindole		193.25	20, 467		1.5350^{20}	188–190	$250^{10\text{mm}}$	55	dec aq, alc; s eth
p124	Phenyl isocyanate	C_6H_5NCO	119.12	12, 437	1.0956^{20}_4		−30	162–163	87	i aq; s alc, eth
p125	Phenyl isothiocyanate	C_6H_5NCS	135.19	12, 453	1.1288^{25}_4	1.6497^{20}	−21	221		s alc, chl, eth
p126	N-Phenylmaleimide		173.17	21, 400			85–87	$163^{12\text{mm}}$		
p127	Phenylmalonic acid	$C_6H_5CH(CO_2H)_2$	180.16				153 dec			
p128	Phenylmercury(II) acetate	$C_6H_5HgO_2CCH_3$	336.74	Merck: 12, 7453			150–152			0.17 aq; s alc, bz, acet
p129	Phenylmercury(II) chloride	C_6H_5HgCl	313.15	Merck: 12, 7454			250–252			s bz, eth, pyr
p130	Phenylmercury(II) hydroxide	C_6H_5HgOH	294.70	16, 952			190 dec			
p131	N-Phenylmorpholine	$C_{10}H_7NHC_6H_5$	163.22	27, 6	1.058^{270}		51–54	268	>110	1.0 aq; v s hot alc
p132	N-Phenyl-1-naphthylamine		219.29	12, 1224			60–62	$226^{15\text{mm}}$		s alc, bz, chl, eth
p133	N-Phenyl-2-naphthylamine	$C_{10}H_7NHC_6H_5$	219.29	12, 1275			107–109	395		
p134	2-Phenyl-2-oxazoline		147.18	27, 47	1.118	1.5670^{20}	12	$75^{0.3\text{mm}}$		
p135	2-Phenylphenol	$C_6H_5C_6H_4OH$	170.21	6^2, 623	1.213		57–59	282	123	s alc, chl, eth, alk
p136	4-Phenylphenol	$C_6H_5C_6H_4OH$	170.21	6, 674			165–167	321	165	s alc, chl, eth, alk
p137	N-Phenyl-1,4-phenylenediamine	$C_6H_5NHC_6H_4NH_2$	184.24	13, 76			73–75			
p138	Phenylphosphinic acid	$C_6H_5PH(O)OH$	142.09	16, 791			85–87			
p139	Phenylphosphonic acid	$C_6H_5P(O)(OH)_2$	158.09	16, 803			163–166			
p140	Phenylphosphonic dichloride	$C_6H_5P(O)Cl_2$	194.99	16, 804	1.375	1.5600^{20}	3	258	>110	
p141	N-Phenylpiperazine		162.24	23^3, 49	1.06214^{20}_4	1.5875^{20}	3–4	286	>110	i aq; misc alc
p142	1-Phenyl lpiperidine		161.25	20, 22	1.001	1.5620^{20}	44–45	257–258	106	
p143	2-Phenyl-1,2-propanediol	$CH_3C(C_6H_5)(OH)CH_2OH$	152.19	6, 930				$162^{26\text{mm}}$	>110	
p144	3-Phenyl-1-propanethiol	$C_6H_5CH_2CH_2CH_2SH$	152.26	6^1, 253	1.010	1.5494^{20}		$109^{10\text{mm}}$	90	
p145	1-Phenyl-1-propanol	$C_6H_5CH(OH)CH_2CH_3$	136.19	6, 502	0.99152^{25}_4	1.5200^{20}		219	90	misc alc, bz
p146	3-Phenyl-1-propanol	$C_6H_5CH_2CH_2CH_2OH$	136.19	6, 503	1.008	1.5257^{20}	−18	235	109	s aq; misc alc, eth
p147	1-Phenyl-2-propanone	$C_6H_5CH_2COCH_3$	134.18	7^2, 233	1.01157^{20}_4	1.5160^{20}	27	$100^{13\text{mm}}$	84	v s alc, eth; misc bz
p148	2-Phenylpropionaldehyde	$CH_3CH(C_6H_5)CHO$	134.18	7, 305	1.009^{20}_4	1.5175^{20}		202–205	76	i aq; s alc

TABLE 2.21 Physical Constants of Organic Compounds (*Continued*)

No.	Name	Formula	Formula weight	Beilstein reference	Density, g/mL	Refractive index	Melting point, °C	Boiling point, °C	Flash point, °C	Solubility in 100 parts solvent
p149	3-Phenylpropion-aldehyde	$C_6H_5CH_2CH_2CHO$	134.18	7, 304	1.019	1.5230^{20}		98^{12mm}	95	0.6 aq; s bz, alc, chl, eth, HOAc, PE
p150	3-Phenylpropionic acid	$C_6H_5CH_2CH_2CO_2H$	150.18	9, 508	1.047_4^{100}		47–49	280	>110	10 hot aq; s hot alc, alk, acid
p151	1-Phenyl-3-pyrazolidinone		162.19	24, 2			121–123			s alc, eth
p152	2-Phenylpyridine	$C_6H_5\text{-}C_5H_4N$	155.20	20, 424	1.086	1.6332^{20}		268–270	>110	0.8 alc; 1 eth; 0.3 chl
p153	2-Phenyl-4-quinoline-carboxylic acid		249.27	22, 103			214–215			
p154	Phenyl salicylate	$C_6H_5(OH)CO_2C_6H_5$	214.22	10, 76	1.25		44–46	173^{12mm}	>110	17 alc; 66 bz; s acet, chl, eth; 0.015 aq
p155	Phenylsuccinic acid	$HO_2CCH_2CH(C_6H_5)CO_2H$	194.19	9, 865			167–169	$-H_2O,$ >168		s hot aq, alc, eth
p156	(Phenylthio)acetic acid	$C_6H_5SCH_2CO_2H$	168.21	6, 313			64–66			0.25 aq; s alc, alk
p157	S-Phenyl thio-isobutyrate	$(CH_3)_2CHC(=O)SC_6H_5$	152.22	6^4, 1524	1.056	1.5460^{20}		129^{10mm}	>110	
p158	1-Phenyl-2-thiourea	$C_6H_5NHC(S)NH_2$	152.22	12, 388	1.3		154			
p159	Phenyltrichlorosilane	$C_6H_5SiCl_3$	211.55	16, 911	1.329^{20}	1.5230^{20}		201	91	
p160	Phenyltriethoxysilane	$C_6H_5Si(OC_2H_5)_3$	240.38	16, 911	0.996	1.4604^{20}		113^{10mm}	42	
p161	Phenyltrimethoxy-silane	$C_6H_5Si(OCH_3)_3$	198.30	16^4, 1556	1.062	1.4680^{20}		233	99	
p162	Phenyltrimethyl-ammonium bromide	$[C_6H_5N(CH_3)_3]^+\ Br^-$	216.13	12, 158			215 dec			v s aq; s hot alc
p163	Phenyltrimethyl-ammonium chloride	$[C_6H_5N(CH_3)_3]^+\ Cl^-$	171.67	12, 158			237 subl			s aq; v s alc; sl s eth
p164	Phenyltrimethyl-ammonium iodide	$[C_6H_5N(CH_3)_3]^+\ I^-$	263.12	12, 159			227 subl			s aq, alc; sl s acet
p165	Phenyltrimethyl-ammonium tribro-mide	$[C_6H_5N(CH_3)_3]^+\ Br_3^-$	375.95	12, 159			114–116			
p166	Phenyltrimethylsilane	$C_6H_5Si(CH_3)_3$	150.30	16^1, 525	0.873	1.4907^{20}		168–170	44	
p167	Phenylurea	$C_6H_5NHCONH_2$	136.15	12, 346	1.302		145–147	238		s hot aq, hot alc, eth

No.	Name	Formula	MW	Ref.	Density	n_D	mp	bp	Flash	Solubility
p168	1,2-Phthalic acid	$C_6H_4\text{-}1,2\text{-}(CO_2H)_2$	166.13	9, 791	1.593^{20}_4		230 rapid heating	295		0.6 aq; 10 alc; 0.5 eth; v sl s chl
p169	Phthalic anhydride		148.12	17, 469	1.53		131–134	290	151	0.6 aq(dec); s alc
p170	Phthalide		134.13	17, 310	1.1641^{99}		72–74			s alc
p171	Phthalimide		147.13	21, 458			234–236			v s alk; v sl s bz, PE
p172	1,2-Phthaloyl dichloride	$C_6H_5\text{-}1,2\text{-}(COCl)_2$	203.02	9, 805	1.409^{20}	1.5684^{20}	15–16	280–282	>110	dec by aq, alc; s eth
p173	Phthalylsulfathioazole		403.44	Merck: 12, 7533			272–277			s alk; sl s alc; i chl
p174	Picric acid	$2,4,6\text{-}(O_2N)_3C_6H_2OH$	229.11	6, 265	1.763^{20}_4		122–123	explodes >300		1.3 aq; 8.2 alc; 10 bz; 2.9 chl; 1.6 eth
p175	(+)-α-Pinene		136.24	5, 146	0.8591^{20}_4	1.4650^{20}	−62	156	35	misc alc, eth
p176	(−)-β-Pinene		136.24	5, 154	0.8590^{20}	1.4780^{20}	−61	166	38	
p177	α-Pinene oxide		152.24	5, 152	0.964	1.4690^{20}		103^{50mm}	65	
p178	Piperazine		86.14	23, 4		1.446^{113}	108–110	145–146	109	v s aq; 50 alc; i eth
p179	1,4-Piperazinebis-(ethanesulfonic acid)		302.37	Merck: 12, 7633			>300			
p180	Piperidine		85.15	20, 6	0.8622^{24}	1.4525^{20}	−13	106	4	misc aq; s alc, bz, chl
p181	1-Piperidinecarbonitrile		110.16	20, 56	0.951	1.4705^{20}		102^{10mm}	97	
p182	N-Piperidineethanol		129.20	20, 25	0.8732^{25}_{25}	1.4804^{20}		199–202	68	misc aq; s alc
p183	2-Piperidineethanol		129.20	21, 2	1.010^{17}		38–40	234	102	v s aq, alc, eth
p184	1-Piperidinepropionic acid		157.21	20^{3}, 1049			105–110	$108^{0.5mm}$		
p185	Piperidinepropionitrile		138.21		0.933	1.4695^{20}		111^{16mm}	102	
p186	2-(2-Piperidineethyl)-pyridine		190.29		0.985	1.5260^{20}		150^{17mm}	>110	
p187	L-Proline		115.13	22, 2			228 dec			
p188	Propane	$CH_3CH_2CH_3$	44.10	1, 103	0.584^{-42}	1.340^{-42}	−188	−42.1	−104	volumes per 100 vols solvent: 6.5 aq; 790 alc; 926 eth; 1300 chl; 1450 bz
p189	1,2-Propanediamine	$CH_3CH(NH_2)CH_2NH_2$	74.13	4, 257	0.878^{15}	1.4460^{20}		119–120	33	misc aq, bz; s alc, eth
p190	1,3-Propanediamine	$H_2NCH_2CH_2CH_2NH_2$	74.13	4, 261	0.8844^{25}	1.4575^{20}	−12	140	48	misc alc, eth; s aq
p191	1,2-Propanediol	$CH_3CH(OH)CH_2OH$	76.10	1, 472	1.0364^{44}	1.4331^{20}	−60	188	107	misc aq, acet, chl; s alc, eth
p192	1,3-Propanediol	$HOCH_2CH_2CH_2OH$	76.10	1, 475	1.0538^{20}	1.4396^{20}	−27	214	79	misc aq, alc

(Continued)

TABLE 2.21 Physical Constants of Organic Compounds (*Continued*)

No.	Name	Formula	Formula weight	Beilstein reference	Density, g/mL	Refractive index	Melting point, °C	Boiling point, °C	Flash point, °C	Solubility in 100 parts solvent
p193	1,3-Propanediol bis-(4-aminobenzoate)	$CH_2(CH_2CO_2CC_6H_4NH_2)_2$	314.34	14^3, 1034	1.140		124–127			
p194	1,2-Propanediol dibenzoate	$C_6H_5CO_2CH_2CH(CH_3)-O_2CC_6H_5$	284.31	9, 129	1.160	1.5450^{20}	–3	232^{12mm}	>110	
p195	1,3-Propanedithiol	$HSCH_2CH_2CH_2SH$	108.23	1, 476	1.0772^{20}_4	1.5405^{20}	–79	172.9	58	misc alc, bz, eth, chl
p196	1-Propanesulfonyl chloride	$CH_3CH_2CH_2SO_2Cl$	142.60	4, 8	1.2864^{15}	1.4542^{20}		66^{8mm}	80	dec hot aq, hot alc
p197	1,3-Propane sultone		122.14	19^3, 4	1.392		31–33	180^{30mm}	>110	s alc, eth
p198	1-Propanethiol	$CH_3CH_2CH_2SH$	76.16	1, 359	0.8363_4	1.4380^{20}	–113	67–68	–20	misc alc, eth; sl s aq
p199	2-Propanethiol	$CH_3(SH)CH_3$	76.16	1, 367	0.8092^{25}	1.4255^{20}	–131	52.6	–34	misc alc, eth, bz, chl, eth
p200	1,2,3-Propanetriol tris(acetate)	$H_3CCO_2CH(CH_2O_2CCH_3)_3$	218.21	2, 147	1.1580^{20}	1.4302^{20}	–78	259	138	7.2 aq; misc alc, bz, chl, eth
p201	1-Propanol	$CH_3CH_2CH_2OH$	60.10	1, 350	0.8037^{20}	1.3840^{20}	–127	97.2	23	misc aq, alc, eth
p202	2-Propanol	$(CH_3)_2CHOH$	60.10	1, 360	0.7855^{20}	1.3772^{20}	–89.5	82.4	12	misc aq, alc, chl, eth
p203	2-Propenal	$H_2C=CHCHO$	56.07	1, 725	0.841^{20}	1.4017^{20}	–88	52.6	–18	21 aq; s alc, eth
p204	Propene	$H_2C=CHCH_3$	42.08	1, 196	0.610^{-48}_4	1.3567^{-40}	–185.2	–47.7	–108	vols in 100 vols solvent: 45 aq; 1200 alc; 500 acet
p205	2-Propene-1-thiol	$H_2C=CHCH_2SH$	74.15	1, 440	0.9254^{23}	1.4765^{20}		67–68	21	misc alc, eth
p206	trans-1,2,3-Propene-tricarboxylic acid		174.11	2, 849			190 dec			50 aq^{25}, 50 88% alc^{12}; sl s eth
p207	1-Propen-2-yl acetate	$H_2C=C(O_2CCH_3)CH_3$	100.12		0.909	1.4000^{20}		97	18	misc alc, eth
p208	4-(1-Propenyloxy-methyl)-1,3-dioxo-lan-2-one		158.16		1.100	1.4610^{20}		251–252	>110	v s aq, alc, chl, eth
p209	2-Propenylphenol	$CH_3CH=CHC_6H_4OH$	134.18	6^1, 279	1.044	1.5780^{20}		230–231	90	
p210	β-Propiolactone		72.06	17^1, 130	1.1460^{20}_4	1.4131^{20}	–33.4	162	70	37 aq(hyd); misc alc (reacts); bz, eth, acet
p211	Propionaldehyde	CH_3CH_2CHO	58.08	1, 629	0.8071^{20}_4	1.3636^{20}	–81	48	–30	30 aq; misc alc, eth
p212	Propionamide	$CH_3CH_2CONH_2$	73.10	2, 243	0.9597^{94}	1.4160^{110}	79	222.2		v s aq, alc, chl, eth
p213	Propionic acid	$CH_3CH_2CO_2H$	74.09	2, 234	0.9934^{20}	1.3809^{20}	–20.5	141.1	52	misc aq; s alc, chl, eth
p214	Propionic anhydride	$[CH_3CH_2C(=O)]_2O$	130.14	2, 242	1.0110^{20}	1.4037^{20}	–45	170	63	dec aq; s alc, chl, eth
p215	Propionitrile	CH_3CH_2CN	55.08	2, 245	0.7818^{20}	1.3658^{20}	–92.8	97.2	2	10 aq; misc alc, eth
p216	Propionyl chloride	CH_3CH_2COCl	92.53	2, 243	1.065^{20}_4	1.4051^{20}	–94	80	11	dec by aq, alc
p217	Propiophenone	$C_6H_5COCH_2CH_3$	134.18	7^2, 231	1.0105^{20}	1.5258^{20}	21	218.0	87	misc bz, eth, abs alc
p218	2-Propoxyethanol	$CH_3CH_2CH_2OCH_2CH_2OH$	104.15	1, 468	0.913	1.4130^{20}	–75	150–153	48	
p219	2-(2-Propoxyethyl)-pyridine	$C_5H_4NCH_2CH_2OCH_2CH_2CH_3$	165.24		0.954	1.4880^{20}			95	

(Continued)

No.	Name	Formula	Formula wt	Beilstein ref.	Density	n_D	mp, °C	bp, °C	Flash P, °C	Solubility
p220	1-Propoxy-2-propanol	$CH_3CH_2CH_2OCH_2CH(OH)CH_3$	118.18	1^2, 536	0.885	1.4110^{20}		140–160	48	
p221	Propoxytrimethylsilane	$CH_3CH_2CH_2OSi(CH_3)_3$	132.28	4^4, 3994	0.7684^{20}_{4}	1.3840^{20}		100^{735mm}	−2	
p222	Propyl acetate	$CH_3CH_2CH_2O_2CCH_3$	102.13	2, 129	0.8878^{20}	1.3844^{20}		101.6	13	2.3 aq; misc alc, eth
p223	Propylamine	$CH_3CH_2CH_2NH_2$	59.11	4, 136	0.7173^{20}	1.3872^{20}	−93	42.2	−37	misc aq, alc, eth
p224	2-(Propylamino)-ethanol	$C_3H_7NHCH_2CH_2OH$	103.17	4, 282	0.900	1.4415^{20}	−83	182^{746mm}	78	
p225	Propylbenzene	$CH_3CH_2CH_2C_6H_5$	120.20	5, 390	0.8621^{20}_{4}	1.4912^{20}	−99.2	159.2	47	s alc, eth
p226	Propyl benzoate	$C_6H_5CO_2CH_2CH_2CH_3$	164.20	9, 112	1.032^{20}	1.5010^{20}	−51.6	230	98	i aq; s alc, eth
p227	Propyl butyrate	$CH_3CH_2CH_2CO_2CH_2CH_2CH_3$	130.19	2, 271	0.879^{15}	1.4000^{20}	−95	143	38	sl s aq; misc alc, eth
p228	Propyl chloroformate	$ClCO_2CH_2CH_2CH_3$	122.55	3, 11	1.090	1.4034^{20}		105–106	28	misc bz, chl, eth
p229	Propylcyclohexane	$CH_3CH_2CH_2C_6H_{11}$	126.24	5^2, 23	0.7929^{20}_{4}	1.4370^{20}	−94.9	156.7	35	s bz, eth
p230	Propylene carbonate		102.09	19^3, 1564	1.2041^{20}_{4}	1.4210^{20}	−48.8	242	135	v s aq, alc, bz, eth
p231	Propyleneimine	$CH_3CH{-}CH_2$ (NH)	57.09	20, 3	0.8017^{25}	1.4084^{25}		66.0	−15	misc aq, alc, PE
p232	1,2-Propylene oxide	$CH_3CH{-}CH_2$ (O)	58.08	17, 6	0.8287^{20}	1.3660^{20}	−112	34	−35 (CC)	41 aq; misc alc, eth
p233	Propylene sulfide	$CH_3CH{-}CH_2$ (S)	74.15	17^2, 15	0.946	1.4760^{20}		72–75	10	
p234	Propyl formate	$CH_3CH_2CH_2O_2CH$	88.10	2, 21	0.9058^{20}	1.3779^{20}	−92.9	80.9	−3	2 aq; misc alc, eth
p235	Propyl 4-hydroxy-benzoate	$HOC_6H_4CO_2CH_2CH_2CH_3$	180.20	10, 160			95–98			0.05 aq; v s alc, eth
p236	Propyl isocyanate	$CH_3CH_2CH_2NCO$	85.11	4^1, 366	0.908	1.3940^{20}		83–84	0	s aq, alc, eth
p237	Propyl lactate	$CH_3CH(OH)CO_2CH_2CH_2CH_3$	132.16	3, 265	0.996^{20}	1.4167^{25}		86^{40mm}		s alc, eth
p238	Propyl nitrate	$CH_3CH_2CH_2ONO_2$	105.09	1, 355	1.0538_{4}	1.3976^{20}	−100	110.1	23 (may explode on heating)	
p239	2-Propylpentanoic acid	$(CH_3CH_2CH_2)_2CHCO_2H$	144.21	2, 350	0.921	1.4250^{20}		220	111	
p240	2-Propylphenol	$CH_3CH_2CH_2C_6H_4OH$	136.19	6, 499	1.015^{20}	1.5279^{20}		224–226	93	s alc, eth
p241	Propylphosphonic dichloride	$CH_3CH_2CH_2P(O)Cl_2$	160.97	4, 596	1.290	1.4643^{20}		90^{50mm}	>110	
p242	Propyltrichlorosilane	$CH_3CH_2CH_2SiCl_3$	177.53	4, 630	1.1851^{20}_{4}	1.429^{20}		123–124	2	s alc, eth
p243	1-Propyl-4-piperidone		141.22		0.936	1.4600^{20}		56^{1mm}	75	
p244	Propyl propionate	$CH_3CH_2CO_2CH_2CH_2CH_3$	116.16	2, 240	0.883^{20}	1.3935^{20}	−76	122.5	19	0.5 aq; 103 alc; 83 eth

TABLE 2.21 Physical Constants of Organic Compounds (*Continued*)

No.	Name	Formula	Formula weight	Beilstein reference	Density, g/mL	Refractive index	Melting point, °C	Boiling point, °C	Flash point, °C	Solubility in 100 parts solvent
p245	Propyl 3,4,5-trihydroxybenzoate	$(HO)_3C_6H_2CO_2CH_2CH_2CH_3$	212.20	Merck: 12, 8044			150			0.35 aq; 1 alc; 83 eth
p246	Propyne	$CH_3C{\equiv}CH$	40.06	1, 246	0.691^{-20}_{4}	1.3725^{20}	−102.8	−23.2		v s alc; 3000 mL eth
p247	2-Propynyl benzenesulfonate	$C_6H_5SO_3CH_2C{\equiv}CH$	196.23	11[3], 37	1.243	1.5250^{20}	−30	142^{2mm}	100	
p248	2-Propynoic acid	$HC{\equiv}CCO_2H$	70.05	2, 477	1.138^{20}_{4}	1.4320^{20}	9	102^{200mm}	58	s aq, alc, eth
p249	2-Propyn-1-ol	$HC{\equiv}CCH_2OH$	56.06	1, 454	0.9478^{20}	1.4320^{20}	−51.8	114	36	misc aq, alc, bz, chl
p250	(+)-Pulegone		152.24	7, 87	0.9346^{15}_{4}	1.4870^{20}		224	85	misc alc, chl, eth
p251	Pyrazine		80.09	23, 91	1.031^{61}	1.4953^{61}	55	115	55	v s aq, alc, eth
p252	Pyrazinecarbonitrile		105.10	25[3], 777	1.174	1.5340^{20}		87^{6mm}	96	sl s hot aq; 0.008 abs alc; i bz, chl, eth
p253	Pyrazinecarboxylic acid		124.10	25, 125			225 dec			s aq, alc, bz, eth
p254	Pyrazole		68.08	23, 39		1.4203	68	187		s org solvents
p255	Pyrene		202.26	5, 693	1.271^{23}		151	404		misc aq, bz; v s alc, eth
p256	Pyridazine		80.09	23, 89	1.10355^{25}	1.5230^{23}	−8	208	85	misc aq, alc, eth
p257	Pyridine	C_5H_5N	79.10	20, 181	0.98275^{25}_{4}	1.5067^{25}	−41.6	115.2	20	
p258	Pyridine-d_5	C_5D_5N	84.14	20[3], 2305	1.050	1.5092^{20}		114.4	20	
p259	2-Pyridinealdoxime	$(C_5H_4N)\text{-}2\text{-}CH{=}NOH$	122.13	21[1], 288			110–112			
p260	4-Pyridinealdoxime	$(C_5H_4N)\text{-}4\text{-}CH{=}NOH$	122.13	21[1], 288			130–133			
p261	2-Pyridinecarboxaldehyde	$(C_5H_4N)\text{-}2\text{-}CHO$	107.11	21[1], 287	1.126	1.5370^{20}		181	54	
p262	3-Pyridinecarboxaldehyde	$(C_5H_4N)\text{-}3\text{-}CHO$	107.11	21[1], 288	1.135	1.5493^{20}		97^{15mm}	60	
p263	4-Pyridinecarboxaldehyde	$(C_5H_4N)\text{-}4\text{-}CHO$	107.11	21, 287	1.122	1.5440^{20}		78^{12mm}	54	s aq, eth
p264	3-Pyridinecarboxamide	$(C_5H_4N)\text{-}3\text{-}CONH_2$	122.13	22, 40	1.400	1.466	130–133			100 aq; 66 alc
p265	2-Pyridinecarboxylic acid	$(C_5H_4N)\text{-}2\text{-}CO_2H$	123.11	22, 33			134–136	sublimes		s aq, alc, bz; v s HOAc
p266	3-Pyridinecarboxylic acid	$(C_5H_4N)\text{-}3\text{-}CO_2H$	123.11	22, 38	1.473		236.6	sublimes		1.4 aq; s alk; v s hot aq, hot alc
p267	4-Pyridinecarboxylic acid	$(C_5H_4N)\text{-}4\text{-}CO_2H$	123.11	22, 45			319	260^{15mm}		0.52 aq; i alc, bz, eth
p268	2,3-Pyridinedicarboxylic acid	$(C_5H_4N)\text{-}2,3\text{-}(CO_2H)_2$	167.12	22, 150			188–190 dec			0.56 aq; s alk
p269	2,5-Pyridinedicarboxylic acid	$(C_5H_4N)\text{-}2,5\text{-}(CO_2H)_2$	167.12	22, 153			256 dec			s hot acid
p270	2,6-Pyridinedicarboxylic acid	$(C_5H_4N)\text{-}2,6\text{-}(CO_2H)_2$	167.12	22, 154			248–250 dec			sl s aq; v sl s alc

No.	Name	Formula	Mol. wt.	Ref.	Density	n	m.p.	b.p.	Flash	Solubility
p271	Pyridine-N-oxide	C_5H_5NO	95.10	20[2], 131			61–65	270		
p272	Pyridinium p-toluene-sulfonate	$C_5H_5NH^+$ $^-O_3SC_6H_4CH_3$	251.31	20[2], 129			117–119			
p273	2-Pyridylcarbinol	(C_5H_4N)-2-CH_2OH	109.13	21[1], 203	1.131	1.5420^{20}		113^{16mm}	>110	v s aq, alc, eth
p274	3-Pyridylcarbinol	(C_5H_4N)-3-CH_2OH	109.13	21, 50	1.124	1.5445^{20}		154^{28mm}	>110	v s aq, eth
p275	3-(3-Pyridyl)-1-propanol	(C_5H_4N)-3-$CH_2CH_2CH_2OH$	137.18	21[3], 549	1.063	1.5300^{20}		133^{3mm}	>110	
p276	3-(4-Pyridyl)-1-propanol	(C_5H_4N)-4-$CH_2CH_2CH_2OH$	137.18	21[4], 550	1.061	1.5040^{20}	35–39	289	>110	
p277	Pyrimidine		80.09	23, 89	1.016		22	124	31	misc aq; s alc, eth
p278	2,4(1H,3H)-Pyrimidinedione		112.09	24, 312			335			0.3 aq; s alk
p279	Pyrrole		67.09	20, 159	0.9691^{20}_{4}	1.5085^{20}	−23.4	130	39	4.5 aq; v s alc, eth
p280	Pyrrolidine		71.12	20, 4	0.8586^{20}	1.4431^{20}	−58	86.5	3	misc aq; s alc, chl, eth
p281	1-Pyrrolidinebutyronitrile		138.21		0.926	1.4605^{20}		115^{18mm}	99	
p282	1-Pyrrolidinecarbo-dithioic acid, ammonium salt		164.29				153–155			
p283	1-Pyrrolidinecarbo-nitrile		96.13		0.954	1.4690^{20}		$77^{1.8mm}$	107	
p284	1-Pyrrolidino-1-cyclohexene		151.25		0.940	1.5225^{20}		115^{15mm}	39	
p285	2-Pyrrolidinone		85.11	21, 236	1.116^{25}_{4}	1.4806^{25}	25	251	129	misc aq, alc, bz, chl, eth, EtOAc
p286	3-(N-Pyrrolidino)-1,2-propanediol		145.20	20[1], 4	1.401^{20}_{4}		46–48	158^{30mm}	>110	
q1	Quinhydrone		218.20	7, 617	1.401^{20}_{4}		171–173			s hot aq, alc, eth
q2	Quinine		324.44	23, 511		1.625	173–175			125 alc; 1.2 bz; 83 chl
q3	Quinoline		129.16	20, 339	1.095^{20}_{4}	1.6273^{20}	−15	237	101	0.6 aq; misc alc, eth
q4	Quinoxaline		130.15	23, 176	1.334^{48}_{4}	1.6231^{48}	29–32	220–223	98	v s aq, alc, bz, eth
q5	2-Quinoxalinol		146.15	24, 147			271–272			14 aq; 10 MeOH
r1	D-Raffinose penta-hydrate		594.52	31, 462			80–82	dec 118		
r2	Resorcinol	C_6H_4-1,3-$(OH)_2$	110.11	6, 796	1.272	1.5030^{20}	110–112	280	>110	111 aq; 111 alc; v s eth
r3	Resorcinol 1,3-diacetate	C_6H_4-1,3-$(O_2CCH_3)_2$	194.19	6, 816	1.178			146^{12mm}		

(Continued)

TABLE 2.21 Physical Constants of Organic Compounds (*Continued*)

No.	Name	Formula	Formula weight	Beilstein reference	Density, g/mL	Refractive index	Melting point, °C	Boiling point, °C	Flash point, °C	Solubility in 100 parts solvent
r4	Resorcinol monoacetate	$CH_3CO_2C_6H_4$-3-(OH)	152.15	6, 816	1.223	1.5370^{20}		ca 283	>110	i aq; misc alc, bz, chl, acet; s alk OH's
r5	Resorcinol monobenzoate	$C_6H_5CO_2C_6H_4$-3-(OH)	214.20					133–135		
r6	Rhodamine B		479.02	19, 345			210–211 dec			v s aq, alc
r7	Rhodanine		133.19	27, 242	0.868		167–170 may explode on rapid heating			v s hot aq, alc, eth
r8	Riboflavin		376.37	Merck: 12, 8367			dec 278–282			v s alk(dec); i acet, bz, eth; sl s pentyl acetate, cyclohexanol
r9	D-Ribose		150.13	1, 859			88–92			s aq; sl s alc
s1	Saccharin		183.19	27, 168	0.828		228–230			0.34 aq; 3 alc; 8 acet
s2	Safrole		162.19	19, 39	1.095^{20}	1.5370^{20}	11.2	232–234	97	v s alc; misc chl, eth
s3	Semicarbazide hydrochloride	$H_2NNHCONH_2 \cdot HCl$	111.53	3, 98			175–177 dec			v s aq, alc; i eth
s4	L-Serine	$HOCH_2CH(NH_2)CO_2H$	105.09	4, 505			222 dec			s aq; v sl s alc, eth
s5	D-Sorbitol		182.17	1, 533	1.472^{-5}		98–100 if hydrated; 111 anhyd			83 aq; s hot alc, acet
s6	L-Sorbose		180.16	1, 927	1.65^{15}	1.4530^{15}	163–165			55 aq; v sl s alc
s7	Squalane	$[(CH_3)_2CH(CH_3)_2CH(CH_3)-(CH_2)_3CH(CH_3)CH_2CH_2]_2$	422.83	1, 72	0.8115^{15}		−38	350	218	s bz, chl, eth, PE
s8	Squalene	$CH_3[C(CH_3)=CHCH_2CH_2]_5-C(CH_3)=C(CH_3)_2$	470.73	1^1, 130	0.8584^{20}_{4}	1.4965^{20}	−75	285^{25mm}	200	v s eth, acet, PE
s9	*trans*-Stilbene	$C_6H_5CH=CHC_6H_5$	180.25	5, 630	0.970		122–124	307		v s bz, eth
s10	(−)-Strychnine		334.42	27^2, 723	1.36^{20}_{4}		284–286	270^{5mm}		0.66 alc; 20 chl; 0.55 bz; 0.15 mg aq
s11	Styrene	$C_6H_5CH=CH_2$	104.15	5, 474	0.9060^{20}	1.5463^{20}	−31	145	31	s alc, acet, eth, CS_2
s12	Styrene oxide	$H_2C\!-\!CHC_6H_5$ (O)	120.15	17, 49	1.054	1.5338^{20}	−37	194	79	
s13	Succinamic acid	$H_2NCOCH_2CH_2CO_2H$	117.10	2, 614			153–156			s aq; sl s alc; i eth

No.	Name	Formula	Form. wt.	Beilstein/Merck ref.	Density	n_D	mp, °C	bp, °C	Solubility
s14	Succinamide	$H_2NCOCH_2CH_2CONH_2$	116.12	2, 614			265 dec		0.45 aq; i alc, eth
s15	Succinic acid	$HO_2CCH_2CH_2CO_2H$	118.09	2, 601	1.552		188	235 dec	7.7 aq; 5.4 alc; 2.8 acet; 0.88 eth; i bz
s16	Succinic anhydride		100.07	17, 407			119.6	261	s alc, chl; v sl s eth
s17	Succinimide		99.09	21, 369	1.41		123–125	285–290	33 aq; 4 alc; i eth
s18	Succinonitrile	$NCCH_2CH_2CN$	80.09	2, 615	0.9864^{60}	1.4173^{60}	54.5	266	see b456
s19	Succinyl chloride	$ClCOCH_2CH_2COCl$	154.98	2, 613	1.395_4^{15}	1.473^{15}	16–17	190; 132	dec by aq, alc; s bz
s20	Sucrose		342.30	31, 424	1.587_4^{25}		185–187	76	200 aq; 0.59 alc
s21	Sulfadiazine		250.28	Merck: 12, 9071			252–256		sl s aq, alc, acet; v s dil mineral acids, alk
s22	Sulfamethazine		278.34	Merck: 12, 9083			198–201		0.15 aq; s alk
s23	Sulfamic acid	HSO_3NH_2	97.09	Merck: 12, 9090	2.15		205 dec		15 aq; sl s alc, acet; s bases
s24	Sulfanilamide	$H_2NC_6H_4SO_2NH_2$	172.21	14, 698			164–166		0.76 aq; 2.7 alc; 20 acet; s acid, alk
s25	Sulfanilic acid	$4-(H_2N)-C_6H_4SO_3H$	173.19	14, 695			d 288		1.45 aq; sl s hot MeOH
s26	Sulfoacetic acid	$HCO_2CH_2SO_3H$	140.11	4, 21			84–86		s aq, alc; i eth, chl
s27	2-Sulfobenzoic acid cyclic anhydride		184.17	19, 110				245 dec / 186^{18mm}	s bz, chl, eth; i aq
s28	4,4′-Sulfonylbis(2,6-dibromophenol)	$[2,6-(Br)_2—C_6H_2OH]_2SO_2$	565.88	6, 865			303–306		
s29	4,4′-Sulfonylbis(methylbenzoate)	$(CH_3O_2CC_6H_4)_2SO_2$	334.35	10^2, 109			195–196		
s30	4,4′-Sulfonyldiphenol	$(HOC_6H_4)_2SO_2$	250.27	6, 861	1.3663^{15}		245–247		s alc, eth, acet; i aq
s31	5-Sulfosalicylic acid	$HO_3SC_6H_3(OH)CO_2H$	254.21	11, 411			120 anhyd		v s aq, alc; s eth
t1	D-(−)-Tartaric acid	$HO_2CCH(OH)CH(OH)CO_2H$	150.09	3, 520	1.75984^{20}, 1.7598_4^{20}		172–174		139 aq^{20}, 59 MeOH; 33 EtOH; s glyc; 0.4 eth
t2	L-(+)-Tartaric acid		150.09	3, 481	1.7598_4^{20}		168–170		139 aq^{20}, 59 MeOH; 33 EtOH; s glyc; 0.4 eth
t3	meso-Tartaric acid monohydrate	$HO_2CCH(OH)CH(OH)-CO_2H \cdot H_2O$	168.11	3, 528	1.666_4^{20}, 1.737 also		140; also 159–160		125 aq^{20}

(Continued)

TABLE 2.21 Physical Constants of Organic Compounds (*Continued*)

No.	Name	Formula	Formula weight	Beilstein reference	Density, g/mL	Refractive index	Melting point, °C	Boiling point, °C	Flash point, °C	Solubility in 100 parts solvent
t4	DL-Tartaric acid monohydrate	HO$_2$CCH(OH)CH(OH)-CO$_2$H · H$_2$O	168.11	3, 522	1.697^{20}_{4}		210–212			20.6 aq^{20}; 5 alc^{25}; 1 eth
t5	Tartrazine		534.37	25, 252						v s aq
t6	Terephthaldicarboxaldehyde	C$_6$H$_4$-1,4-(CHO)$_2$	134.13	7, 675			115–116	245–248		
t7	m-Terphenyl	C$_6$H$_5$—C$_6$H$_4$—C$_6$H$_5$	230.31	5, 695	1.195		87	363		misc alc, eth
t8	o-Terphenyl	C$_6$H$_5$—C$_6$H$_4$—C$_6$H$_5$	230.31	5^2, 611	1.16		56.2	332	>110	
t9	p-Terphenyl	C$_6$H$_5$—C$_6$H$_4$—C$_6$H$_5$	230.31	5, 695	1.213		210	376	>110	
t10	α-Terpinene		136.24	5, 126	0.8375^{20}_{4}	1.4775^{20}		174	46	misc alc, eth
t11	γ-Terpinene		136.24	5, 128	0.853^{15}	1.4754^{16}		183	51	
t12	Terpinen-4-ol		154.25	6, 55	0.9338^{20}	1.4820^{20}	36.4	90^{6mm}	79	v s alc, eth
t13	α-Terpineol		154.25	6, 57	0.9337^{20}	1.4813^{20}	40.5	220	90	
t14	1,2,4,5-Tetrabromobenzene	C$_6$H$_2$Br$_4$	393.72	5, 214			180–182			
t15	3,4,5,6-Tetrabromocresol	CH$_3$C$_6$Br$_4$(OH)	423.75	6, 362			205–208			s alc, eth, alk
t16	1,1,2,2,-Tetrabromoethane	Br$_2$CHCHBr$_2$	345.67	1, 94	2.9655^{20}	1.6358^{20}	0	243.5	none	misc alc, chl, eth, HOAc
t17	Tetrabromophthalic anhydride		463.72	17, 485			274–276			sl s bz; i aq, alc
t18	α,α,α',α'-Tetrabromo-o-xylene	C$_6$H$_4$-1,2-(CHBr$_2$)$_2$	421.77	5, 367			114–116			v s chl
t19	α,α,α',α'-Tetrabromo-m-xylene	C$_6$H$_4$-1,3-(CHBr$_2$)$_2$	421.77	5, 375			105–108			
t20	α,α,α',α'-Tetrabromo-p-xylene	C$_6$H$_4$-1,4-(CHBr$_2$)$_2$	421.77	5, 386			254–256			
t21	Tetrabutylammonium bromide	(C$_4$H$_9$)$_4$N$^+$ Br$^-$	322.38	4^2, 634			102–104			
t22	Tetrabutylammonium chloride	(C$_4$H$_9$)$_4$N$^+$ Cl$^-$	277.92	4^3, 292			73–75			
t23	Tetrabutylammonium hydrogen sulfate	(C$_4$H$_9$)$_4$N$^+$ HSO$_4^-$	339.54				171–173			
t24	Tetrabutylammonium iodide	(C$_4$H$_9$)$_4$N$^+$ I$^-$	369.38	4, 157			145–147			sl s aq; s alc, eth
t25	Tetrabuty lammonium tetrafluoroborate	(C$_4$H$_9$)$_4$N$^+$ BF$_4^-$	329.28	4^3, 293			160–162			
t26	Tetrabuty lammonium tribromide	(C$_4$H$_9$)$_4$N$^+$ Br$_3^-$	482.20	4^4, 557			74–76			

(Continued)

No.	Name	Formula	Mol. wt.	Beil. ref.	Density	n_D	mp (°C)	bp (°C)	Flash point	Solubility
t27	N,N,N'-Tetrabutyl-1,6-hexanediamine	$\{-(CH_2)_3N[(CH_2)_3CH_3]_2\}_2$	340.64		0.820	1.4510^{20}		83^{2mm}	57	v s acet, chl
t28	Tetrabutyl orthosilicate	$Si[O(CH_2)_3CH_3]_4$	320.55	1^2, 398	0.8992^{20}_{4}	1.4131^{20}	100–103	275	78	
t29	Tetrabutyl phosphonium bromide	$[CH_3(CH_2)_3]_4PBr$	339.35							
t30	Tetrabutyltin	$(C_4H_9)_4Sn$	347.15	1, 656	1.057	1.4742^{20}	–97	145^{10mm}	107	
t31	1,1,3,3,-Tetrachloro-acetone	$Cl_2CHC(=O)CHCl_2$	195.86		1.624^{15}_{4}	1.497^{18}		182^{745mm}	none	v s eth; sl s alc
t32	1,2,3,4-Tetrachloro-benzene	$C_6H_2Cl_4$	215.89	5, 204			46–47	254	>110	
t33	1,2,4,5-Tetrachloro-benzene	$C_6H_2Cl_4$	215.89	5, 205	1.858^{22}		139–142	240–246	>110	s bz, chl, eth
t34	Tetrachloro-1,2-benzoquinone	$C_6Cl_4\text{-}1,2\text{-}(=O)_2$	245.88	7, 602			127–129			
t35	Tetrachloro-1,4-benzoquinone	$C_6Cl_4\text{-}1,4\text{-}(=O)_2$	245.88	7, 636			290 dec			s eth; sl s chl; i aq
t36	Tetrachloro-1,2-difluoroethane	$Cl_2CFCFCl_2$	203.83		1.6447^{25}	1.4130^{25}	26.0	92.8		0.012 aq
t36a	1,1,1,2-Tetrachloro-ethane	$ClCH_2CCl_3$	167.85	1, 86	1.5406^{20}	1.4821^{20}	–70.2	130.5	47	
t37	1,1,2,2-Tetrachloro-ethane	$Cl_2CHCHCl_2$	167.85	1, 86	1.5866^{25}_{4}	1.4910^{25}	–44	147	62	0.3 aq; misc alc, chl, eth, PE
t38	Tetrachloroethylene	$Cl_2C=CCl_2$	165.83	1, 187	1.6230^{20}_{4}	1.5057^{20}	–22	121	45	misc alc, chl, eth
t39	2,3,5,6-Tetrachloro-nitrobenzene	$HC_6Cl_4NO_2$	260.89	5, 247	1.744^{25}_{4}		98–101	304		s alc, bz, chl
t40	Tetrachlorophthalic anhydride		285.90	17, 484			254–258	371		dec hot aq; sl s eth
t41	Tetracosane	$CH_3(CH_2)_{22}CH_3$	338.66	1, 175	0.7786^{51}	1.4283^{70}	51	391	>110	9.4 chl; s eth
t42	Tetradecafluorohexane	$CF_3(CF_2)_4CF_3$	338.05	1^3, 388	1.669	1.2515^{20}	–4	58–60	none	
t43	Tetradecane	$CH_3(CH_2)_{12}CH_3$	198.40	1, 171	0.7627^{20}_{20}	1.4290^{20}	5.5	253.6	99	v s alc, eth
t44	Tetradecanoic acid	$CH_3(CH_2)_{12}CO_2H$	228.38	2, 365	0.8525^{70}_{4}	1.4273^{70}	54	250^{100mm}	>110	v s bz, chl, eth; s alc
t45	1-Tetradecanol	$CH_3(CH_2)_{13}OH$	214.39	1, 428	0.8151^{50}	1.4358^{50}	39.5	289	>110	s eth; sl s alc
t46	Tetradecanoyl chloride	$CH_3(CH_2)_{12}COCl$	246.82	2, 368	0.908	1.4490^{20}	–1	168^{15mm}		dec aq, alc; s eth
t47	1-Tetradecene	$CH_3(CH_2)_{11}CH=CH_2$	196.38	1, 226	0.7751^{15}_{4}	1.4360^{20}	–12.9	251.2	115	v s alc, eth
t48	Tetraethoxysilane	$(CH_3CH_2O)_4Si$	208.33	1, 334	0.9342^{20}	1.383^{20}	–77	168	46	dec aq; s alc
t49	Tetraethylammonium bromide	$(CH_3CH_2)_4N^+ Br^-$	210.16	4, 104	1.397^{20}_{4}		285 dec			v s aq, alc, acet, chl

TABLE 2.21 Physical Constants of Organic Compounds (*Continued*)

No.	Name	Formula	Formula weight	Beilstein reference	Density, g/mL	Refractive index	Melting point, °C	Boiling point, °C	Flash point, °C	Solubility in 100 parts solvent
t50	Tetraethylammonium chloride	$(CH_3CH_2)_4N^+\ Cl^-$	165.71	4, 104	1.0801^{21}_4					141 aq; s alc; 8.2 chl
t51	Tetra(ethylene glycol)	$(HOCH_2CH_2OCH_2CH_2)_2O$	194.23	1, 468	1.1252^{20}_{20}	1.4577^{20}	−6	328	182	misc aq, alc, bz, eth
t52	Tetra(ethylene glycol) diacrylate	$(H_2C{=}CHCO_2CH_2CH_2O{-}CH_2CH_2)_2O$	302.33		1.110	1.4650^{20}			>110	
t53	Tetra(ethylene glycol) diethyl ether	$C_2H_5(OCH_2CH_2)_4OC_2H_5$	250.34	1^3, 2107	0.970	1.4324^{20}		159^{11mm}	>110	s aq
t54	Tetra(ethylene glycol) dimethacrylate	$[H_2C{=}C(CH_3)CO_2CH_2CH_2{-}OCH_2CH_2]_2O$	330.37	2^4, 1531	1.080	1.4630^{20}		220	>110	
t55	Tetra(ethylene glycol) dimethyl ether	$CH_3(OCH_2CH_2)_4OCH_3$	222.28	1^3, 2107	1.0087^{20}_4	1.4330^{20}	−30	275–276	140	s aq
t56	Tetraethylenepentamine	$H_2NCH_2CH_2NHCH_2CH_2)_2NH$	189.31	4^3, 543	0.999^{20}_{20}	1.5055^{20}	−40	340	185	misc aq, alc, eth
t57	N,N,N',N'-Tetraethyl-ethylenediamine	$(C_2H_5)_2NCH_2CH_2N(C_2H_5)_2$	172.32	4, 251	0.808	1.4343^{20}		189–192	58	
t58	Tetraethylgermanium	$(C_2H_5)_4Ge$	188.84	4, 631	0.998	1.4420^{20}	−90	165.5	35	s alc, eth; i aq
t59	Tetraethyllead	$(C_2H_5)_4Pb$	323.45	4, 639	1.653^{34}	1.5190^{20}	−136	85^{15mm}	72	s bz; misc eth
t60	Tetraethylsilane	$(C_2H_5)_4Si$	144.34	4, 625	0.7658^{20}	1.4268^{20}	−82	154.7	26	i aq
t61	N,N,N',N'-Tetraethyl-sulfamide	$(C_2H_5)_2NSO_2N(C_2H_5)_2$	208.33	4, 129	1.030	1.4480^{20}		249–251	>110	
t62	Tetraethylthiuram disulfide	$[(C_2H_5)_2NC({=}S)S^-]_2$	296.54	4, 122	1.30		71–72			3.8 alc; 7.1 eth; s bz, acet, chl; 0.02 aq
t63	Tetraethyltin	$(C_2H_5)_4Sn$	234.94	4, 632	1.199^{20}	1.4730^{20}	−112	181	53	i aq; s eth
t64	1,1,1,2-Tetrafluoroethane	FCH_2CF_3	102.03	1^4, 123			−26.5			
t65	Tetrafluoroethylene	$F_2C{=}CF_2$	100.02	1^3, 638	1.151^{-40}		−142.5	−76		i aq
t66	2,2,3,3-Tetrafluoro-1-propanol	$HCF_2CF_2CH_2OH$	132.06	1^4, 1438	1.4853^4	1.3197^{20}	−15	109–110	43	
t67	1,2,3,6-Tetrahydro-benzaldehyde	C_6H_9CHO	110.16	7^1, 48	0.940	1.4745^{20}		163–164	57	
t68	1,2,3,4-Tetrahydro-carbazole		171.24	20, 416			118–120	325–330		
t69	Tetrahydrofuran		72.11	17, 10	0.8892^{20}	1.4052^{20}	−108.5	65	−14	misc aq, alc, eth, PE
t70	2,5-Tetrahydrofuran-dimethanol		132.16		1.1542^{25}_4	1.4766^{25}	<−50	265		misc aq, alc, bz, chl; s eth
t71	Tetrahydro-2-furan-methanol		102.13	17^2, 106	1.0524^{20}	1.4520^{20}	<−80	178	75	misc aq, alc, bz, chl, eth, acet
t72	Tetrahydro-2-furan-methylamine		101.15	18^2, 415	0.980	1.4560^{20}		154^{744mm}	45	

No.	Name	Formula	Mol wt	Beil. ref.	Density	n_D	mp, °C	bp, °C	Flash pt, °F	Solubility
t73	Tetrahydrofurfuryl acetate		144.17	17^{2}, 107	1.061	1.4370^{20}		196	84	
t74	Tetrahydrofurfuryl acrylate		156.18	17^{3}, 1104	1.064	1.4600^{20}		87^{9mm}	>110	
t75	Tetrahydrofurfuryl chloride		120.58	17^{3}, 61	1.110	1.4550^{20}		150–151	47	
t76	Tetrahydrofurfuryl methacrylate		170.21	17^{3}, 1105	1.044	1.4580^{20}		$52^{0.4mm}$	90	
t77	2(3)-(Tetrahydrofuryloxy)tetrahydropyran		186.25		1.030	1.4610^{20}			97	
t78	1,2,3,4-Tetrahydroisoquinoline		133.19	20, 275	1.064	1.5668^{20}	−30	232–233	98	
t79	Tetrahydrolinalool	$(CH_3)_2CHCH_2CH_2CH_2$ $C(CH_3)(OH)CH_2CH_3$	158.29	1, 426	0.826	1.4340^{20}	76	73^{6mm}	76	misc alc, bz, chl, eth, acet, PE
t80	1,2,3,4-Tetrahydronaphthalene	$C_{10}H_{12}$	132.21	5, 491	0.9702^{20}_{4}	1.5414^{20}	−35.8	207.6	77	
t81	cis-1,2,3,6-Tetrahydrophthalic anhydride		152.15	17, 462			97–103		157	
t82	cis-1,2,3,6-Tetrahydrophthalimide		151.17				129–133			
t83	Tetrahydropyran		86.14	17, 12	0.8814^{20}_{4}	1.4200^{20}	−45	88	−155	misc aq, alc, eth
t84	Tetrahydropyran-2-methanol		116.16		1.0254^{20}	1.4580^{20}	−70	187	93	misc aq, alc, bz, eth
t85	3,4,5,6-Tetrahydropyrimidinethiol		116.19	24, 5			210–212			
t86	1,2,3,4-Tetrahydroquinoline		133.19	20, 262	1.061	1.5940^{20}	15–16	249	100	s aq; misc alc, eth
t87	Tetrahydrothiophene		88.17	17^{1}, 5	0.9987²⁰	1.5040^{20}	−96	121	12	misc alc, eth; i aq
t88	2,2′,4,4′-Tetrahydroxybenzophenone	$[(HO)_2C_6H_3]_2C{=}O$	246.22	8, 496			200–203			
t89	Tetrakis(dimethylamino)ethylene	$[(CH_3)_2N]_2C{=}C[N(CH_3)_2]_2$	200.23	4^{4}, 167	0.861	1.4800^{20}		$59^{0.9mm}$	53	
t90	N,N,N′,N′-Tetrakis(2-hydroxypropyl)ethylenediamine	$[CH_3CH(OH)CH_2]_2NCH_2$ $CH_2N[CH_2CH(OH)CH_3]_2$	292.40	4^{4}, 1685	1.013	1.4812^{20}		$181^{0.8mm}$	>110	
t91	1,1,8,8-Tetramethoxyoctane	$(CH_3O)_2CH(CH_2)_6CH(OCH_3)_2$	234.34		0.949	1.4300^{20}		130^{5mm}	52	
t92	1,1,3,3-Tetramethoxypropane	$[(CH_3O)_2CH]_2CH_2$	164.20		0.997	1.4081^{20}		183	54	

(Continued)

TABLE 2.21 Physical Constants of Organic Compounds (*Continued*)

No.	Name	Formula	Formula weight	Beilstein reference	Density, g/mL	Refractive index	Melting point, °C	Boiling point, °C	Flash point, °C	Solubility in 100 parts solvent
t93	Tetramethyl-ammonium bromide	$(CH_3)_4N^+ Br^-$	154.06	4, 51	1.56		>300			55 aq
t94	Tetramethyl-ammonium chloride	$(CH_3)_4N^+ Cl^-$	109.60	4, 51	1.169_4^{20}		>300			s aq, hot alc
t95	Tetramethyl-ammonium iodide	$(CH_3)_4N^+ I^-$	201.06	4, 51	1.829		>300			sl s aq; v s abs alc
t96	N,N-3,5-Tetramethyl-aniline	$(CH_3)_2C_6H_3N(CH_3)_2$	149.24	12, 1131	0.913	1.5443^{20}		226–228	90	
t97	1,2,3,4-Tetramethyl-benzene	$C_6H_2\text{-}1,2,3,4\text{-}(CH_3)_4$	134.22	5, 430	0.905_4^{20}	1.5187^{20}	−6.2	205.0	68	misc alc, eth
t98	1,2,3,5-Tetramethyl-benzene	$C_6H_2\text{-}1,2,3,5\text{-}(CH_3)_4$	134.22	5, 430	0.8906_4^{20}	1.5134^{20}	−23.7	198.0	63	s alc; v s eth
t99	1,2,4,5-Tetramethyl-benzene	$C_6H_2\text{-}1,2,4,5\text{-}(CH_3)_4$	134.22	5, 431	0.8384_4^{81}		79.3	196.8	73	v s alc, bz, eth
t100	2,2,3,3-Tetramethyl-butane	$(CH_3)_3CC(CH_3)_3$	114.23	1, 165	0.8242^{20}		−100.7	106.5	4	
t101	N,N,N',N'-Tetra-methyl-1,3-butane-diamine	$(CH_2)_2NCH(CH_3)CH_2\text{-}CH_2N(CH_3)_2$	144.26	4^3, 570	0.787	1.4318^{20}		165	40	
t102	N,N,N',N'-Tetra-methyl-1,4-butane-diamine	$(CH_3)_2N(CH_2)_4N(CH_3)_2$	144.26	4, 265	0.786^{20}	1.4280^{20}		169	46	s aq, alc, eth
t103	1,1,3,3-Tetramethyl-butylamine	$(CH_3)_3CCH_2C(CH_3)_2NH_2$	129.25	4, 198	0.805	1.4240^{20}		137–143	32	s alc, eth, PE; i aq
t104	1,3,5,7-Tetramethyl-cyclotetrasiloxane	$[\text{-}SiH(CH_3)O\text{-}]_4$	240.51	4^4, 4099	0.9912_4^{20}	1.3870^{20}	−69	134–135	−12	
t105	N,N,N',N'-Tetra-methyldiamino-methane	$(CH_3)_2NCH_2N(CH_3)_2$	102.18	4, 54	0.749	1.4005^{20}		85		
t106	1,1,3,3-Tetramethyl-disiloxane	$[(CH_3)_2CH]_2O$	134.33	4^4, 3991	0.7574_4^{20}	1.3700^{20}		70–71	−10	
t107	Tetramethylene sulfone		120.17	17^1, 5	1.2606_4^{30}	1.4820^{30}	27.6	285	177	misc aq, acet, toluene; s octanes, olifines, naphthenes
t108	N,N,N',N'-Tetra-methylethylene-diamine	$(CH_3)_2NCH_2CH_2N(CH_3)_2$	116.21	4, 250	0.770	1.4179^{20}	−55	120–122	10	

No.	Name	Formula	Mol. wt.	Beilstein Reference	Density	n_D	mp	bp	Flash pt	Solubility
t109	Tetramethylgermanium	$(CH_3)_4Ge$	132.73	4², 1008	0.978	1.3890^{20}	−88	43.4	−37	misc alc, eth
t110	1,1,3,3-Tetramethylguanidine	$[(CH_3)_2N]_2C{=}NH$	115.18	4¹, 335	0.918	1.4692^{20}		163	60	
t111	N,N,N',N'-Tetramethyl-1,6-hexanediamine	$[(CH_3)_2N(CH_2)_3{-}]_2$	172.32	4¹, 423	0.806	1.4359^{20}		209–210	73	
t112	Tetramethyl lead	$(CH_3)_4Pb$	267.33	4, 639	1.995_4^{20}	1.4005^{20}	−27.5	110	38	
t113	N,N,N',N'-Tetramethylmethanediamine	$(CH_3)_2NCH_2N(CH_3)_2$	102.18	4, 54	0.749			85	−12	
t114	2,6,10,14-Tetramethylpentadecane	$[(CH_3)_2CH(CH_3)_3{-}CH(CH_3)CH_2]_2CH_2$	268.53	Merck: 12, 7932	0.7827_4^{20}	1.4385^{20}	−100	296	>110	s bz, chl, eth, PE
t115	2,2,6,6-Tetramethylpiperidinyl-1-oxy (free radical)		156.25				36–40		67	
t116	N,N,N',N'-Tetramethyl-1,3-propanediamine	$(CH_3)_2N(CH_2)_3N(CH_3)_2$	130.24	4, 262	0.779	1.4234^{20}		145–146	31	
t117	Tetramethylpyrazine		136.20	23, 99			84–86	190		
t118	Tetramethylsilane	$(CH_3)_4Si$	88.23	4, 625	0.6411_4^{20}	1.3580^{20}	−99.5	26.5	−27	v s alc, eth
t119	1,1,3,3-Tetramethyl-2-thiourea	$(CH_3)_2NC({=}S)N(CH_3)_2$	132.23	4¹, 336			75–77	245		0.002 alc, 0.002 eth; 0.012 acet; 0.025 bz; s chl
t120	Tetramethylthiuram disulfide	$[(CH_3)_2NCS]_2$	240.43	4, 76	1.29		155–156			
t121	Tetramethyltin	$(CH_3)_4Sn$	178.83	4, 631	1.3149_4^{25}	1.5201	−54	74–75	−12	misc aq, common org solvents
t122	1,1,3,3-Tetramethylurea	$(CH_3)_2NC({=}O)N(CH_3)_2$	116.16	4, 74	0.9687_4^{20}	1.4493^{25}	−0.6	176–177	77	
t123	Tetranitromethane	$C(NO_2)_4$	196.03	1, 80	1.6229_4^{25}	1.4358^{25}	13.8	126	>110	
t124	1,4,7,10-Tetraoxacyclododecane (12-Crown-4)		176.21		1.089	1.4630^{20}	16	$70^{0.5mm}$	>110	v s alc, eth, alk
t125	2,4,8,10-Tetraoxaspiro[5.5]undecane		160.17	19, 436			52–55	$83^{1.5mm}$	108	
t126	Tetraphenylboron sodium	$(C_6H_5)_4B^- Na^+$	342.23	Merck: 12, 8839			>300			v s aq, acet; s chl

(Continued)

TABLE 2.21 Physical Constants of Organic Compounds (*Continued*)

No.	Name	Formula	Formula weight	Beilstein reference	Density, g/mL	Refractive index	Melting point, °C	Boiling point, °C	Flash point, °C	Solubility in 100 parts solvent
t127	1,1,4,4-Tetraphenyl-1,3-butadiene	$(C_6H_5)_2C$=CHCH=$C(C_6H_5)_2$	358.49	5, 750			207–209			
t128	Tetraphenyltin	$(C_6H_5)_4Sn$	427.11	1, 355	1.490^0		224–227	>420	110	
t129	Tetrapropoxysilane	$(C_3H_7O_4)Si$	264.4	4[1], 364	0.916_4^2	1.401^{20}	270 dec	94^{5mm}	95	s aq
t130	Tetrapropylammonium bromide	$(CH_3CH_2CH_2)_4N^+$ Br^-	266.27							
t131	1*H*-Tetrazole		70.06	26, 346			157–158			s aq, alc, acet
t132	2-Thenoyltrifluoroacetone		222.18				40–44	98^{8mm}		
t133	Theobromine		180.17	26, 457			357	sublimes 290–295		100 aq; 0.045 alc; s alk; i bz, chl, eth
t134	Theophylline		180.17	26, 455			274–275			0.83 aq; 1.25 alc; 0.9 chl; s hot aq, alk, dil acids
t135	Thiamine HCl		337.27	Merck: 12, 9430			dec 260			100 aq; 1 alc; 5.5 glyc
t136	Thiazole		85.13	27, 15	1.200	1.5390^{20}	202	117–118	22	s alc, eth; sl s aq
t137	N^2-(2-Thiazolyl)-sulfanilamide		255.32	27[3], 4623						0.06 aq; 0.52 alc; s acet, dil mineral acids, alkalis
t138	Thioacetamide	CH_3C(=S)NH_2	75.13	2, 232	1.174		112–114			16 aq; 16 alc; sl s eth
t139	Thiobenzoic acid	C_6H_5C(=O)SH	138.19	9, 419		1.6050^{20}	15–18	122^{30mm}	>110	misc eth; v s alc; i aq
t140	4,4'-Thiobis(2-*tert*-butyl-6-methylphenol)		358.54	6[4], 6043			163–165	316^{40mm}	240	
t141	Thiocarbaniilide	C_6H_5NHC(=S)NHC_6H_5	228.32	12, 394	1.32^{24}		152–155			v s alc, eth
t142	*p*-Thiocresol	$HSC_6H_4CH_3$	124.21	6, 416			42–44	195	68	s alc, eth; i aq
t143	2,2'-Thiodiacetic acid	$(HO_2CCH_2)_2S$	150.15	3, 253			128–131			s aq, alc
t144	2,2'-Thiodiethanol	$(HOCH_2CH_2)_2S$	122.19	1, 470	1.1824_4^{20}	1.5203^{20}	–10.2	282	160	misc aq, alc; sl s eth
t145	4,4'-Thiodiphenol	$(HOC_6H_4)_2S$	218.27	6, 860			154–156			3.7 aq; v s hot aq, alc, acet
t146	3,3'-Thiodipropionic acid	$(HO_2CCH_2CH_2)_2S$	178.21				131–134			
t147	Thiolacetic acid	CH_3C(=O)SH	76.12	2, 230	1.065	1.4630	<–17	88–91	11	s aq; v s alc
t148	*N*-Thionylaniline	C_6H_5N=SO	139.18	12, 578	1.236	1.6270^{20}		200	84	
t149	Thionyl bromide	$SOBr_2$	207.88	Merck: 12, 9484	2.683	1.6750^{20}	–52	138		misc bz, chl, CCl_4; hyd by aq

(Continued)

No.	Name	Formula	Mol. wt.	Merck	Density	n_D	mp (°C)	bp (°C)	Flash pt	Solubility
t150	Thionyl chloride	$SOCl_2$	118.97	Merck: 12, 9485	1.635	1.517^{20}	−101	76	none	misc bz, chl, CCl4; hyd by aq
t151	Thiophene	C_4H_4S	84.14	17, 29	1.0573^{25}_{4}	1.5257^{25}	−39.4	84	−1	misc alc, eth; i aq
t152	2-Thiopheneacetic acid	$(C_4H_3S)CH_2CO_2H$	142.18	18, 293			63–67	150^{22mm}		
t153	2-Thiophenecarbonyl chloride	$(C_4H_3S)COCl$	146.60	18, 290	1.371	1.5900^{20}		206–208	90	s eth
t154	2-Thiophenecarboxaldehyde	$(C_4H_3S)CHO$	112.15	17, 285	1.200	1.5900^{20}		198	77	s aq, chl; v s alc, eth
t155	2-Thiophenecarboxylic acid	$(C_4H_3S)CO_2H$	128.15	18, 289			127–130	260		v s alc; misc bz, eth
t156	Thiophenol	C_6H_5SH	110.18	6, 294	1.073	1.5880^{20}	−14.9	169	50	s bz, chl, CCl4, CS2
t157	Thiophenoxyacetic acid	$C_6H_5SCH_2CO_2H$	168.21	6, 313			64–66			
t158	Thiophosphoryl chloride	$PSCl_3$	169.40		1.668	1.5550^{20}	−36 (β) −40 (α)	125	none	
t159	Thiopropionic acid	$CH_3CH_2C(=O)SH$	90.14	2, 264	1.014	1.4640^{20}		108–110	11	s aq, alc
t160	3-Thiosemicarbazide	$H_2NC(=S)NHNH_2$	91.14	3, 195			182–184			9 aq; s alc; sl s eth
t161	Thiourea	$H_2NC(=S)NH_2$	76.12	3, 180	1.405		176–178			v s aq; s alc; hot HO Ac
t162	Thioxanthen-9-one		212.27	17, 357			212–213	373^{715mm}		v s bz, chl, hot HOAc
t162a	Thymol		150.22	6, 532	0.9699^{25}_{4}	1.5227^{20}	51.5	233	102	0.1 aq; 100 alc; 140 eth; s HOAc, alk OH
t163	Titanium(IV) ethoxide	$Ti(OC_2H_5)_4$	228.15	1, 335	1.088	1.5043^{20}		152^{10mm}	28	s bz, chl, eth
t164	Titanium(IV) isopropoxide	$Ti[OCH(CH_3)_2]_4$	284.26	1^2, 382	0.963	1.4660^{20}	18–20	220	22	
t165	Titanium(IV) propoxide	$Ti(OCH_2CH_2CH_3)_4$	284.26	1^3, 1423	1.033	1.4986^{20}		170^{3mm}	42	misc alc, chl, eth, acet, HOAc; 0.067 aq
t166	Toluene	$C_6H_5CH_3$	92.14	5, 280	0.8660^{20}_{4}	1.4960^{20}	−94.9	110.6	4	s hot aq, alc, eth
t167	2,4-Toluenediamine	$CH_3C_6H_3$-2,4-$(NH_2)_2$	122.17	13, 124			99	292		v s aq, alc, eth
t168	2,5-Toluenediamine	$CH_3C_6H_3$-2,5-$(NH_2)_2$	122.17	13, 144			64	273–274		s aq, alc
t169	2,6-Toluenediamine	$CH_3C_6H_3$-2,6-$(NH_2)_2$	122.17	13, 148			104–106			v s aq
t170	3,4-Toluenediamine	$CH_3C_6H_3$-3,4-$(NH_2)_2$	122.17	13, 148			91–93	156^{18mm}		
t171	Toluene-2,4-diisocyanate	$CH_3C_6H_3$-2,4-$(NCO)_2$	174.16	13, 138	1.2244^{20}_{4}	1.5689^{20}	20–21	251	132	dec aq, alc; misc acet, bz, eth
t172	p-Toluenesulfinic acid	$CH_3C_6H_4SO_2H$	156.21	11, 9			85			v s alc, eth; sl s aq
t173	o-Toluenesulfonamide	$CH_3C_6H_4SO_2NH_2$	171.22	11, 86			156–158			
t174	p-Toluenesulfonamide	$CH_3C_6H_4SO_2NH_2$	171.22	11, 104			138–140			0.2 aq; 3.6 alc

TABLE 2.21 Physical Constants of Organic Compounds (*Continued*)

No.	Name	Formula	Formula weight	Beilstein reference	Density, g/mL	Refractive index	Melting point, °C	Boiling point, °C	Flash point, °C	Solubility in 100 parts solvent
t175	p-Toluenesulfonyl-hydrazide	$CH_3C_6H_4SO_2NHNH_2$	186.23	11², 66			110 dec			
t176	p-Toluenesulfonic acid	$CH_3C_6H_4SO_3H$	172.20	11, 97			107 anhyd	140^{20mm}		67 aq; s alc, eth
t177	p-Toluenesulfonyl chloride	$CH_3C_6H_4SO_2Cl$	190.65	11, 103			67–69	134^{10mm}		v s alc, bz, eth; i aq
t178	p-Toluenesulfonyl fluoride	$CH_3C_6H_4SO_2F$	174.19	11², 54		1.4355^{20}	41–42	112^{16mm}	105	
t179	p-Toluenesulfonyl isocyanate	$CH_3C_6H_4SO_2NCO$	197.21					144^{10mm}	>110	
t180	m-Toluidine	$CH_3C_6H_4NH_2$	107.16	12, 853	0.989^{20}_{4}	1.5680^{20}	−31	203	85 (CC)	misc alc, eth
t181	o-Toluidine	$CH_3C_6H_4NH_2$	107.16	12, 772	0.998^{20}_{4}	1.5720^{20}	−16.3	200	85	1.7 aq; s alc, eth
t182	p-Toluidine	$CH_3C_6H_4NH_2$	107.16	12, 880	0.9619^{20}	1.5532^{59}	43.8	200	87	7.4 aq; v s alc, eth
t183	m-Tolunitrile	$CH_3C_6H_4CN$	117.15	9, 477	0.976^{15}	1.5256^{20}	−23	210	86	0.09 aq; v s alc, eth
t184	o-Tolunitrile	$CH_3C_6H_4CN$	117.15	9, 466	0.989	1.5279^{20}	−13	205	84	i aq; misc alc, eth
t185	p-Tolunitrile	$CH_3C_6H_4CN$	117.15	9, 489	0.9785^{30}_{4}		29.5	217	85	i aq; v s alc, eth
t186	2-(p-Toluoyl) benzoic acid	$CH_3C_6H_4COC_6H_4CO_2H$	240.26	10, 759			137–139			v s alc, bz, eth, acet
t187	m-Toluoyl chloride	$CH_3C_6H_4COCl$	154.60	9,477	1.173	1.5485^{20}		86^{5mm}	76	
t188	o-Toluoyl chloride	$CH_3C_6H_4COCl$	154.60	9, 464	1.185	1.5549^{20}		90^{12mm}	76	
t189	p-Toluoyl chloride	$CH_3C_6H_4COCl$	154.60	9, 484	1.169	1.5530^{20}	−2	225–227	82	
t190	p-Tolyl acetate	$CH_3CO_2C_6H_4CH_3$	150.18	6, 397	1.048	1.5010^{20}		210–211	90	
t191	1-(o-Tolyl)biguanide	$CH_3C_6H_4NHC(=NH)NH-C(=NH)NH_2$	191.24	12³, 1873			143–145		>110	
t192	m-Tolyl isocyanate	$CH_3C_6H_4NCO$	133.15	12, 864	1.033	1.5305^{20}		76^{12mm}	65	s alc, eth; i aq
t193	1,2,4-Triacetoxy-benzene	$C_6H_3(O_2CCH_3)_3$	252.22	6, 1089			98–100			
t194	Triacetoxyvinylsilane	$(CH_3CO_2)_3SiCH=CH_2$	232.26		1.167	1.4220^{20}		128^{25mm}	76	
t195	Triallylamine	$(H_2C=CHCH_2)_3N$	137.23	4, 208	0.790	1.4510^{20}		152^{4mm}	30	
t196	Triallyl-1,3,5-triazine-2,4,6(1H,3H,5H)-trione		249.27		1.159	1.5129^{20}	150–151		>110	
t197	1H-1,2,4-Triazole		69.07	26, 13			119–121	260	65	s aq, alc
t198	Tribenzylamine	$(C_6H_5CH_2)_3N$	287.41	12, 1038	0.991^{95}_{4}	1.5850^{20}	91–94		65	s hot alc, eth
t199	Tribromoacetaldehyde	Br_3CCHO	280.76	1, 626	2.665			174		s aq, alc, chl, eth
t200	Tribromoacetic acid	Br_3CCO_2H	296.76	2, 220			130–133	245		s aq, alc, eth
t201	2,4,6-Tribromoaniline	$Br_3C_6H_2NH_2$	329.83	12, 663	2.35		120–122	300		s hot alc, chl, eth
t202	2,2,2-Tribromoethanol	Br_3CCH_2OH	282.77	1², 338			73–79	93^{10mm}		
t203	1,1,2-Tribromoethylene	$BrCH=CBr_2$	264.74	1, 191	1.708^{21}	1.6247^{25}		162.5		2 aq; s alc, bz, eth

t204	Tribromomethane	$CHBr_3$	252.77	1, 68	2.9000^{15}	1.6005^{15}	8.1	149.6	83	0.3 aq; misc eth, MeOH
t205	2,4,6-Tribromophenol	$Br_3C_6H_2OH$	330.82	6, 203	2.55		87–89	290^{746mm}	93	s alc, chl, eth; i aq
t206	1,2,3-Tribromopropane	$BrCH_2CH(Br)CH_2Br$	280.78	1, 112	2.390	1.584^{18}	16.5	220	93	s alc, eth
t207	Tributoxyborane	$(C_4H_9O)_3B$	230.16	1^2, 398	0.8567^{20}	1.4092^{20}	<–70	234	93	hyd aq
t208	Tributylamine	$(C_4H_9)_3N$	185.36	4, 157	0.7784	1.4280^{20}	–70	216	86	v s alc, eth; s acet
t209	Tributylborane	$(C_4H_9)_3B$	182.16	4^2, 1022	0.747			109^{20mm}	–36	i aq; s most org solv
t210	2,4,6-Tri-*tert*-butyl-phenol	$[(CH_3)_3C]_3C_6H_2OH$	262.44		0.864^{27}_4		129–132	277		
t211	Tributyl phosphate	$(C_4H_9O)_3P(O)$	266.32	1^2, 397	0.9727^{25}	1.4226^{25}	–79	289	146	0.04 aq; misc org solv
t212	Tributyl phosphite	$(C_4H_9O)_3P$	250.32	1^1, 187	0.925^{20}_4	1.4326^{20}		125^{7mm}	91	misc alc, bz, eth, PE
t213	Tributyltin chloride	$(C_4H_9)_3SnCl$	325.49	4^3, 1926	1.200	1.4905^{20}		173^{25mm}	>110	
t214	Tributyltin ethoxide	$(C_4H_9)_3SnOC_2H_5$	335.10		1.098	1.4672^{20}		$92^{0.1mm}$	40	
t215	Tributyltin hydride	$(C_4H_9)_3SnH$	291.05	4^4, 4312	1.082	1.4730^{20}		$80^{0.4mm}$	40	
t216	Tributyltin methoxide	$(C_4H_9)_3SnOCH_3$	321.07	4^4, 4331	1.115	1.4720^{20}		$97^{0.06mm}$	98	
t217	Trichloroacetamide	Cl_3CCONH_2	162.40	2, 211			141–143	238–240	>110	dec aq, alc; s eth
t218	Trichloroacetaldehyde	Cl_3CCHO	147.40	Merck: 12, 9755	1.510^{20}_4	1.4557^{20}	–57.5	97.8		
t219	Trichloroacetic acid	Cl_3CCO_2H	163.39	2, 206	1.629^{61}_4	1.6200^{20}	57.5	196.5	>110	120 aq; v s alc, eth
t220	Trichloroacetic anhydride	$(Cl_3CCO)_2O$	308.75	2, 210	1.690	1.4838^{20}		141^{60mm}	none	
t221	1,1,3-Trichloroacetone	$ClCH_2COCHCl_2$	161.42	1, 655	1.508	1.4892^{20}	13–15	172	79	s alc
t222	Trichloroacetonitrile	Cl_3CCN	144.39	2, 212	1.4440^{25}	1.4409^{20}	–42	86	none	s alc, eth
t223	2,2′,4′-Trichloro-acetophenone	$Cl_2C_6H_3COCH_2Cl$	223.49	7, 283			52–55	135^{4mm}	>110	
t224	Trichloroacetyl chloride	Cl_3CCOCl	181.83	2, 210	1.629	1.4689^{20}	–146	118		
t225	2,4,5-Trichloroaniline	$Cl_3C_6H_2NH_2$	196.46	12, 627		1.5776^{20}	93–95	270	126	s acet, bz, eth
t226	2,4,6-Trichloroaniline	$Cl_3C_6H_2NH_2$	196.46	12, 627		1.5707^{20}	73–75	262	110	s alc, eth
t227	1,2,3-Trichlorobenzene	$C_6H_3Cl_3$	181.45	5, 203	1.69		53–55	218–220	126	v s bz, CS_2; sl s alc
t228	1,2,4-Trichlorobenzene	$C_6H_3Cl_3$	181.45	5, 204	1.454^{20}	1.5707^{20}	17	213–214	110	misc bz, eth, PE
t229	1,3,5-Trichlorobenzene	$C_6H_3Cl_3$	181.45	5, 204	1.66	1.5662^{19}	63.5	208	107	v s bz, eth, PE
t230	Trichloro-3-chloro-propylsilane	$Cl(CH_2)_3SiCl_3$	211.98		1.350	1.4666^{20}		181–183		
t231	1,1,1-Trichloroethane	CH_3CCl_3	133.41	1, 85	1.3390^{20}	1.4379^{20}	–30.4	74	–1	s acet, bz, eth
t232	1,1,2-Trichloroethane	$ClCH_2CHCl_2$	133.41	1, 85	1.4397^{20}	1.4714^{20}	–37	114	32	misc alc, eth
t233	2,2,2-Trichloroethanol	Cl_3CCH_2OH	149.40	1, 338	1.557	1.4900^{20}	18	151–153		8 aq; misc alc, eth
t234	2,2,2-Trichloroethyl chloroformate	$ClCO_2CH_2CCl_3$	211.86		1.539	1.4703^{20}		171–172		

(Continued)

673

TABLE 2.21 Physical Constants of Organic Compounds (*Continued*)

No.	Name	Formula	Formula weight	Beilstein reference	Density, g/mL	Refractive index	Melting point, °C	Boiling point, °C	Flash point, °C	Solubility in 100 parts solvent
t235	Trichloroethylene	ClCH=CCl$_2$	131.39	1, 187	1.4642^{20}	1.4773^{20}	−84.8	87	32	0.1 aq; misc alc, chl, eth
t236	Trichloroethylsilane	C$_2$H$_5$SiCl$_3$	163.51	4, 630	1.2373^{20}	1.4256^{20}	−105.6	100.5	22	0.14 aq; s alc, eth
t237	Trichlorofluoromethane	Cl$_3$CF	137.37	Merck: 12, 9770	1.485^{21}	1.384^{20}	−111	23.8		
t238	α,α,2-Trichloro-6-fluorotoluene	ClC$_6$H$_3$(F)CHCl$_2$	213.47	5^3, 701	1.446	1.5506^{20}		228–230	>110	
t239	Trichloroisocyanuric acid		232.41	25, 256			249–251			
t240	Trichloromethanesulfenyl chloride	Cl$_3$CSCl	185.89	3, 135	1.700$^{20}_4$	1.5436^{20}		146–148		s alc, bz, chl, eth
t241	1,1,1-Trichloro-2-methyl-2-propanol	(CH$_3$)$_2$C(OH)CCl$_3$	177.46	1, 382			99 anhyd	167		
t242	Trichloromethylsilane	CH$_3$SiCl$_3$	149.48	4^3, 1896	1.273$^{20}_4$	1.4108^{20}	−90	66	−9	v s bz, eth
t243	1,2,4-Trichloro-5-nitrobenzene	Cl$_3$C$_6$H$_2$NO$_2$	226.45	5, 246	1.790^{20}		49–55	288	>110	
t244	2,4,5-Trichlorophenol	Cl$_3$C$_6$H$_2$OH	197.45	6^2, 180			67–69	253		615 acet; 163 bz; 525 eth; 615 MeOH; i aq
t245	2,4,6-Trichlorophenol	Cl$_3$C$_6$H$_2$OH	197.45	6, 190	1.4901^{75}		69	246	none	525 acet; 113 bz; 354 eth; 525 MeOH; i aq
t246	(2,4,5-Trichlorophenoxy)acetic acid	Cl$_3$C$_6$H$_2$OCH$_2$CO$_2$H	255.49	6^3, 702			154–158			s alc; v sl s aq
t247	1,2,3-Trichloropropane	ClCH$_2$CH(Cl)CH$_2$Cl	147.43	1, 106	1.3889^{20}	1.4854^{20}	−14.7	157	71	misc alc, eth; i aq
t248	2,4,6-Trichloropyrimidine		183.43	23, 90		1.5700^{20}	23–25	>110		
t249	Trichlorosilane	HSiCl$_3$	135.45	Merck: 12, 9776	1.342	1.4000^{20}	−127	31–32	−13	dec aq; s bz, chl
t250	4-(Trichlorosilyl)butyronitrile	Cl$_3$Si(CH$_2$)$_3$CN	202.54	4^4, 4272	1.300	1.4630^{20}		237–238	92	
t251	α,α,α-Trichlorotoluene	C$_6$H$_5$CCl$_3$	195.48	5, 300	1.3723^{20}	1.5580^{20}	−5	219–223	127	s alc, bz, eth
t252	α,2,4-Trichlorotoluene	Cl$_2$C$_6$H$_3$CH$_2$Cl	195.48	5^4, 819	1.407	1.5760^{20}	−2.6	248	>110	v s alc, eth
t253	α,2,6-Trichlorotoluene	Cl$_2$C$_6$H$_3$CH$_2$Cl	195.48	5, 300	1.411	1.5761^{20}	36–39	119^{14mm}	>110	i aq; s alc
t254	α,3,4-Trichlorotoluene	Cl$_2$C$_6$H$_3$CH$_2$Cl	195.48			1.5766^{20}		124^{14mm}	>110	
t255	2,4,6-Trichloro-1,3,5-triazine	Cl$_2$C$_6$H$_3$CH$_2$Cl	184.41	26, 35			146–148	190		

No.	Name	Formula	Mol. wt.	Ref.	Density	n_D	m.p., °C	b.p., °C	Fl. p., °C	Solubility
t256	1,1,1-Trichlorotrifluoroethane	Cl_3CCF_3	187.38		1.579	1.3699^{20}	13–14	46		0.017 aq
t257	1,1,2-Trichlorotrifluoroethane	$Cl_2CFCClF_2$	187.38	1[3], 157	1.5635^{25}	1.3557^{25}	−35	47.7		
t258	Trichlorovinylsilane	$H_2C{=}CHSiCl_3$	161.49	5, 164	1.270	1.4360^{20}	−95	90	10	
t259	Tricyclo[5.2.1.0^{2,6}]decane		136.24				77–79	193	40	
t260	Tricyclo[5.2.1.0^{2,6}]decan-8-one		150.22	7[2], 133	1.063	1.5025^{20}		$132^{30\,\text{mm}}$		
t261	Tridecane	$CH_3(CH_2)_{11}CH_3$	184.37	1, 171	0.7563_4	1.4256^{20}	−5 to −4	235	70	v s alc, eth
t262	Tridecanoic acid	$CH_3(CH_2)_{11}CO_2H$	214.35	2, 364			41–42	$236^{100\,\text{mm}}$	>110	v s alc, eth; i aq
t263	2-Tridecanone	$CH_3(CH_2)_{10}COCH_3$	198.35	1, 715	0.822	1.4350^{20}	29–31	$134^{10\,\text{mm}}$	>110	
t264	7-Tridecanone	$[CH_3(CH_2)_5]_2CO$	198.35	1, 715	0.825		30–32	264	>110	
t265	1-Tridecene	$CH_3(CH_2)_{10}CH{=}CH_2$	182.35	1, 225	0.7658^{20}	1.4340^{20}	−13	232.8	79	s alc; v s eth
t266	Triethanolamine	$(HOCH_2CH_2)_3N$	149.19	4, 285	1.1242^{20}_{4}	1.4853^{20}	20.5	335.4	179	misc aq, alc, acet; 4.5 bz; 1.6 eth; s chl
t267	3,4,5-Triethoxybenzoic acid	$(C_2H_5O)_3C_6H_2CO_2H$	254.29	10, 481			110–112			
t268	Triethoxyborane	$(C_2H_5O)_3B$	145.99	1, 335	0.864	1.3740^{20}		117–118	11	dec aq
t269	Triethoxysilane	$(C_2H_5O)_3SiH$	164.28	1, 334	0.890	1.3770^{20}		134–135	26	
t270	3-(Triethoxysilyl)propionitrile	$(C_2H_5O)_3SiCH_2CH_2CN$	217.34	4[4], 4271	0.979	1.4140^{20}		224	100	
t271	3-(Triethoxysilyl)propyl isocyanate	$(C_2H_5O)_3Si(CH_2)_3NCO$	247.37		0.999	1.4200^{20}		283	77	
t272	Triethoxyvinylsilane	$(C_2H_5O)_3SiCH{=}CH_2$	190.32	4, 643	0.903^{20}_{4}	1.3978^{20}		160–161	34	
t273	Triethylaluminum	$(C_2H_5)_3Al$	114.17	4, 99	0.832^{25}		−50	194	−18	dec aq, air
t274	Triethylamine	$(C_2H_5)_3N$	101.19		0.7275^{20}	1.4010^{20}	−114.7	88.8	−7	5.5 aq; misc alc, eth; s acet, EtOAc
t275	Triethylantimony	$(C_2H_5)_3Sb$	208.94	4, 618	1.324^{16}	1.42	−29	159.5		
t276	Triethylarsine	$(C_2H_5)_3As$	162.11	4, 602	1.150^{20}_{4}			$140^{736\,\text{mm}}$		i aq; misc alc, eth
t277	Triethylborane	$(C_2H_5)_3B$	98.00	4, 641	0.6961^{23}	1.3970^{20}	−2.9	95		i aq; dec by air
t278	Triethyl citrate	$HOC(CO_2C_2H_5)(CH_2CO_2C_2H_5)_2$	276.29	3, 568	1.137	1.4420^{20}		$127^{1\,\text{mm}}$	>110	
t279	Triethylenediamine		112.18	23[3], 484			158–160		62	45 aq; 13 acet; 77 alc; 51 bz
t280	Tri(ethylene glycol)	$(HOCH_2CH_2OCH_2)_2$	150.17	1, 468	1.1274^{15}	1.4550^{20}	−7	285	177	misc aq, alc, bz
t281	Tri(ethylene glycol) dimethacrylate	$[H_2C{=}C(CH_3)CO_2CH_2CH_2OCH_2]_2$	286.33	2[4], 1531	1.092	1.4605^{20}		$172^{5\,\text{mm}}$	>110	
t282	Tri(ethylene glycol) dimethyl ether	$(CH_3OCH_2CH_2OCH_2)_2$	178.23	Merck: 12, 9820	0.990^{20}_{4}	1.4224^{20}	−45	216	111	misc aq, hydrocarbon solvents

(Continued)

TABLE 2.21 Physical Constants of Organic Compounds (*Continued*)

No.	Name	Formula	Formula weight	Beilstein reference	Density, g/mL	Refractive index	Melting point, °C	Boiling point, °C	Flash point, °C	Solubility in 100 parts solvent
t283	Tri(ethylene glycol) divinyl ether	$H_2C{=}CH(OCH_2CH_2)_3OCH{=}CH_2$	202.25	1^3, 2106	0.990	1.4530^{20}		126^{18mm}	>110	
t284	Tri(ethylene glycol) monomethyl ether	$CH_3(OCH_2CH_2)_3OH$	164.20	1^3, 2105	1.026	1.4399^{20}		122^{10mm}	>110	
t285	Triethylenetetramine	$(H_2NCH_2CH_2NHCH_2)_2$	146.24	4, 255	0.982	1.4971	12	266	143	misc alc, chl, eth
t286	Triethylgallium	$(C_2H_5)_3Ga$	156.91	26, 2	1.0576^{30}		−82.3	142.6		dec aq; s alc, eth
t287	1,3,5-Triethylhexahydro-1,3,5-triazine		171.20		0.894	1.4595^{20}		207–208	80	v s alc, eth
t288	Triethylindium	$(C_2H_5)_3In$	202.01		1.260^{20}	1.538^{20}	−32	144	36	s aq(dec), alc, eth
t289	Triethyl orthoacetate	$CH_3C(OC_2H_5)_3$	162.23	2, 129	0.8847^{25}_4	1.3950^{25}		142	30	i aq; misc alc, eth; pyrophoric
t290	Triethyl orthoformate	$HC(OC_2H_5)_3$	148.20	2, 20	0.891^{20}_4	1.3910^{20}	−76	146	60	
t291	Triethyl orthopropionate	$CH_3CH_2C(OC_2H_5)_3$	176.26	2, 240	0.876	1.3995^{20}		155–160		
t292	Triethyl phosphate	$(C_2H_5O)_3P(O)$	182.16	1, 332	1.0695^{20}	1.4058^{20}	−56	215	115	
t293	Triethylphosphine	$(C_2H_5)_3P$	118.16	4, 582	0.800^{15}_4	1.4563^{20}	−88	128–129	−17	
t294	Triethyl phosphite	$(C_2H_5O)_3P$	166.16	1, 330	0.969^{20}_4	1.4130^{20}		156	54	i aq(hyd); misc alc, acet, bz, eth, PE
t295	Triethyl phosphonoacetate	$(CH_3CH_2O)_2P(O)CH_2CO_2C_2H_5$	224.19	4^1, 573	1.130	1.4310^{20}		145^{9mm}	>110	
t296	Triethyl phosphonoformate	$(CH_3CH_2O)_2P(O)CO_2C_2H_5$	212.17	3^2, 103	1.110	1.4320^{20}		135^{12mm}	>110	
t297	Triethylsilane	$(C_2H_5)_3SiH$	116.28	4, 625	0.731^{20}_4	1.412^{20}		107–108	−3	i aq; misc alc, eth
t298	Triethyl thiophosphate	$(C_2H_5O)_3P(S)$	198.22	1, 333	1.082	1.4480^{20}		100^{16mm}	107	
t299	2,2,2-Trifluoroacetamide	CF_3CONH_2	113.04	2^2, 186			70–75	162.5		
t300	Trifluoroacetic acid	CF_3CO_2H	114.02	2^2, 186	1.4890^{20}	1.2850^{20}	−15.3	73		misc aq
t301	Trifluoroacetic anhydride	$[CF_3C(O)]_2O$	210.03	2^2, 186	1.487	<1.300	−65	39–40		
t302	1,1,1-Trifluoroacetone	$CF_3C(O)CH_3$	112.05	1^2, 717	1.252	<1.30		22	−30	
t303	1,3,5-Trifluorobenzene	$C_6H_3F_3$	132.09	6^1, 187	1.277	1.4150^{20}	−5.5	75–76	−7	
t304	α,α,α-Trifluoro-*m*-cresol	$CF_3C_6H_4OH$	162.11	1^3, 1342	1.333	1.4588^{20}	−1.8	178–179	73	
t305	2,2,2-Trifluoroethanol	CF_3CH_2OH	100.04	2^3, 427	1.3842^{24}_4	1.2907^{20}	−43.5	74	29	
t306	2,2,2-Trifluoroethyl trifluoroacetate	$CF_3CH_2O_2CCF_3$	196.05		1.4725^{18}_4	1.2812^{18}	−65.5	55	0	
t307	Trifluoromethane	HCF_3	70.01	1, 59	1.52^{-100}		−160	−84		75 mL aq; 500 mL alc
t308	Trifluoromethanesulfonic acid	CF_3SO_3H	150.07	3^4, 34	1.695^{25}	1.3250^{25}	34	162	none	v s aq; misc eth

676

No.	Name	Formula	Mol. wt.	Beilstein ref.	Density	Melting point	Boiling point	n	Flash point	Solubility
t309	Trifluoromethanesulfonic anhydride	$(CF_3SO_2)_2O$	282.13	3^4, 35	1.677		84	1.3212^{20}	none	dec aq, alc
t310	3-(Trifluoromethyl)aniline	$CF_3C_6H_4NH_2$	161.13	12, 870	1.290	5–6	187	1.4800^{20}	85	
t311	α,α,α-Trifluorotoluene	$C_6H_5CF_3$	146.11	5, 290	1.1886^{20}	−29	102	1.4145^{20}	12	v s alc, eth; i aq
t312	Trihexyl O-acetylcitrate	$CH_3CO_2C[CO_2(CH_2)_5CH_3]\text{-}[CH_2CO_2(CH_2)_5CH_3]_2$	486.65		1.005			1.4470^{20}	>110	
t313	Trihexylamine	$[CH_3(CH_2)_5]_3N$	269.52	4, 188	0.794		163–265	1.4415^{20}	>110	
t314	Trihexyl O-butylcitrate	$C_3H_4CO_2C[CO_2(CH_2)_5CH_3]\text{-}[CH_2CO_2(CH_2)_5CH_3]_2$	514.71		0.993	−55		1.4480^{20}		
t315	Trihexylchlorosilane	$[CH_3(CH_2)_5]_3SiCl$	319.12	4^4, 3915	0.871_4^{20}		155^{5mm}	1.456^{20}		
t316	Trihexylsilane	$[CH_3(CH_2)_5]_3SiH$	284.60		0.799		161	1.448^{20}	>110	
t317	1,2,3-Trihydroxybenzene	$C_6H_3(OH)_3$	126.11	6, 1071	1.45	133	309			59 aq; 77 alc; 62 eth
t318	1,3,5-Trihydroxybenzene	$C_6H_3(OH)_3$	126.11	6, 1092		218–221				1 aq; 10 alc; s eth
t319	3,4,5-Trihydroxybenzoic acid	$(HO)_3C_6H_2CO_2H$	170.12	10, 470		258–265				1.1 aq; 17 alc; 1 eth; 20 acet; i bz, chl, PE
t320	2,3,4-Trihydroxybenzophenone	$(HO)_3C_6H_2COC_6H_5$	230.22	8, 417		140–142				
t321	1,2,6-Trihydroxyhexane	$HO(CH_2)_4CH(OH)CH_2OH$	134.18	1^4, 2784	1.109		178^{5mm}	1.4760^{20}	79	
t322	Triisobutylaluminum	$[(CH_3)_2CHCH_2]_3Al$	198.33	4, 643	0.786	4–6	86^{10mm}	1.4494^{20}	−18	pyrophoric
t323	Triisobutylamine	$[(CH_3)_2CHCH_2]_3N$	185.36	4, 166	0.766		192–193	1.4230^{20}	57	
t324	Triisodecyl phosphite	$[(CH_3)_2CH(CH_2)_7O]_3P$	502.80		0.884	<0	166	1.4600^{20}	235	
t325	Triisopropanolamine	$[CH_3CH(OH)CH_2]_3N$	191.27	4^3, 762	0.9996^{50}_{20}	48–52	305.4	1.3764^{20}	152	
t326	Triisopropoxyborane	$[(CH_3)_2CHO]_3B$	188.08	1, 363	0.815		139–141	1.4880^{20}	10	v s aq
t327	1,3,5-Triisopropylbenzene	$C_6H_3[CH(CH_3)_2]_3$	204.36	5, 458	0.845		232–236		86	
t328	Triisopropyl orthoformate	$CH[OCH(CH_3)_2]_3$	190.29	2^3, 39	0.854		66^{18mm}	1.3970^{20}	42	
t329	Triisopropyl phosphite	$[(CH_3)_2CHO]_3P$	208.24	1, 363	0.914_4^{20}		64^{11mm}	1.4110^{20}	67	i aq(sl hyd)
t330	Triisopropylsilane	$[(CH_3)_2CH]_3SiH$	158.36	4^3, 1851	0.773	73–75	86^{35mm}	1.4344^{20}	37	
t331	3,4,5-Trimethoxybenzaldehyde	$(CH_3O)_3C_6H_2CHO$	196.20	8, 391			165^{10mm}			
t332	1,2,3-Trimethoxybenzene	$C_6H_3(OCH_3)_3$	168.19	6, 1081	1.112	43–45	241		>110	

(Continued)

TABLE 2.21 Physical Constants of Organic Compounds (*Continued*)

No.	Name	Formula	Formula weight	Beilstein reference	Density, g/mL	Refractive index	Melting point, °C	Boiling point, °C	Flash point, °C	Solubility in 100 parts solvent
t333	1,2,4-Trimethoxy-benzene	$C_6H_3(OCH_3)_3$	168.19	6, 1088	1.126	1.5330^{20}		247	>110	
t334	1,3,5-Trimethoxy-benzene	$C_6H_3(OCH_3)_3$	168.19	6, 1101			51–53	255	85	
t335	3,4,5-Trimethoxy-benzoic acid	$(CH_3O)_3C_6H_2CO_2H$	212.20	10, 481			168–171	227^{10mm}		v s alc, eth; s chl
t336	3,4,5-Trimethoxy-benzoyl chloride	$(CH_3O)_3C_6H_2COCl$	230.65	10, 487			81–84	185^{18mm}		
t337	3,4,5-Trimethoxy-benzyl alcohol	$(CH_3O)_3C_6H_2CH_2OH$	198.22	6, 1159	1.233	1.5439^{20}		228^{25mm}	>110	
t338	Trimethoxyborane	$(CH_3O)_3B$	103.91	1, 287	0.920^{23}_4	1.3568^{20}	−34	67–68	−13	hyd aq; misc alc, eth
t339	Trimethoxyboroxine	$[-OB(OCH_3)-]_3$	173.53		1.195	1.3996^{20}	10	130	10	
t340	1,1,2-Trimethoxy-ethane	$CH_3OCH_2CH(OCH_3)_2$	120.15	1^3, 3183	0.932	1.3921^{20}		59^{56mm}	23	
t341	1,1,3-Trimethoxy-propane	$CH_3OCH_2CH_2CH(OCH_3)_2$	134.18	1, 820	0.942	1.4004^{20}		46^{17mm}	40	
t342	1,1,3-Trimethoxy-propylsilane	$CH_3OCH_2CH_2CH_2Si(OCH_3)_3$	164.28		0.932	1.3900^{20}		142	40	
t343	Trimethoxysilane	$(CH_3O)_3SiH$	122.20	1^2, 274	0.960	1.3579^{20}	−115	81	−4	
t344	3-(Trimethoxysilyl)-propylamine	$H_2N(CH_2)_3Si(OCH_3)_3$	179.29		1.027	1.4240^{20}		92^{15mm}	83	
t345	N-[3-(Trimethylsilyl)-propyl]aniline	$C_6H_5NH(CH_2)_3Si(OCH_3)_3$	255.39		1.070	1.5550^{20}		310	>110	
t346	N^1-[3-(Trimethoxysilyl)-propyl]ethylene-diamine	$(CH_3O)_3Si(CH_2)_3NHCH_2CH_2NH_2$	224.36		1.019	1.4450^{20}		146^{15mm}	>110	
t347	3-(Trimethoxysilyl)-propyl methacrylate	$(CH_3O)_3Si(CH_2)_3O_2CC(CH_3)=CH_2$	248.35		1.045^{20}_4	1.4310^{20}		190	92	
t348	[3-(Trimethoxysilyl)-propyl]urea	$(CH_3O)_3Si(CH_2)_3NHCONH_2$	222.32		1.150	1.4600^{20}		217–250	98	
t349	Trimethylacetic acid	$(CH_3)_3CCO_2H$	102.13	2, 319	0.889	1.4090^{20}	33–35	163–164	63	
t350	Trimethylacetic-anhydride	$[(CH_3)_3CCO]_2O$	186.25	2, 320	0.918			193	57	
t351	Trimethylacetyl chloride	$(CH_3)_3CCOCl$	120.58	2, 320	0.979	1.4120^{20}		105–106	8	

No.	Name	Formula	Mol. wt.	Beil. ref.	Density	n_D	mp	bp	Flash	Solubility
t352	Trimethylaluminum	(CH₃)₃Al	72.09	4, 643	0.752^{20}	1.432^{12}	15	125–126	–18	s alk; v sl s alc
t354	Trimethylamine	(CH₃)₃N	59.11	4, 43	0.656	1.3631^{0}	–117	2.9	–7	41 aq; misc alc; s bz, chl, eth
t355	2,4,6-Trimethylaniline	(CH₃)₃C₆H₂NH₂	135.21	12, 1160	0.963	1.5510^{20}		233	96	
t356	1,3,3-Trimethyl-6-aza-bicyclo[3.2.1]octane		153.27		0.902	1.4716^{20}		194	75	
t357	1,2,3-Trimethyl-benzene	C₆H₃(CH₃)₃	120.20	5, 399	0.8944^{20}_4	1.5139^{20}	–25.4	176.1	48	i aq; s alc, eth
t358	1,2,4-Trimethyl-benzene	C₆H₃(CH₃)₃	120.20	5, 400	0.87556^{20}_4	1.5048^{20}	–43.9	169	48	s alc, bz, eth
t359	1,3,5-Trimethyl-benzene	C₆H₃(CH₃)₃	120.20	5, 406	0.8637^{20}_4	1.4994^{20}	–44.7	165	44	misc alc, bz, eth
t360	Trimethyl 1,2,4-benzenetri-carboxylate	C₆H₃(CO₂CH₃)₃	252.22	9¹, 429	1.261	1.5214^{20}	38–40	194^{12mm}	>110	
t361	2,2,3-Trimethylbutane	(CH₃)₂CHC(CH₃)₃	100.20	1², 121	0.6901^{20}_4	1.3890^{20}	–24.9	80.9	–6	s alc, eth
t362	2,3,3-Trimethyl-2-butanol	(CH₃)₃CC(CH₃)₂OH	116.20	1², 447	0.8380^{25}_4	1.4233^{22}	15–17	130.5		misc alc, eth
t363	1,2,4-Trimethylcyclo-hexane	C₆H₉(CH₃)₃	126.24	5, 42	0.786	1.4330^{20}		141–143	18	
t364	3,5,5-Trimethylcyclo-hex-2-ene-1-one		138.2	7, 65	0.918	1.4720^{20}	–8.1	215	80	1.2 aq
t365	2,6,6-Trimethyl-2-cyclohexene-1,4-dione		152.19	7⁴, 2032		1.4910^{20}	26–28	94^{11mm}	96	
t366	Trimethyl-1,6-diiso-cyanatohexane	OCNCH₂CH₂C(CH₃)CH₂-C(CH₃)CH₂CNO	210.28		1.012	1.4620^{20}		149	>110	
t367	2,2,6-Trimethyl-4H-1,3-dioxin-4-one		142.16	19³, 1604	1.088	1.4620^{20}	12–13	67^{2mm}	86	
t368	4,4'-Trimethylenebis-(1-methylpiperidine)		238.42		0.896	1.4820^{20}	13	215^{50mm}	>110	
t369	4,4'-Trimethylene-dipiperidine		210.37				65–58			
t370	3,5,5-Trimethylhexanal	(CH₃)₃CCH₂CH(CH₃)CH₂CHO	142.24	1³, 2894	0.817	1.4215^{20}		$68^{2.4mm}$	46	
t370a	3,5,5-Trimethylhexane	(CH₃)₂CHCH₂CH(CH₃)CH(CH₃)₂	128.26		0.7218^{20}	1.4051^{20}	–128	131		
t371	3,5,5-Trimethyl-1-hexanol	(CH₃)₃CCH₂CH(CH₃)-CH₂CH₂OH	144.25	1³, 1755	0.8236^{20}_4	1.4300^{25}	<–70	193–194	80	s alc, eth
t372	3,5,5-Trimethyl-hexanoyl chloride	(CH₃)₃CCH₂CH(CH₃)-CH₂COCl	176.89	2³, 834	0.930	1.4360^{20}		188–190	140	

(Continued)

TABLE 2.21 Physical Constants of Organic Compounds (*Continued*)

No.	Name	Formula	Formula weight	Beilstein reference	Density, g/mL	Refractive index	Melting point, °C	Boiling point, °C	Flash point, °C	Solubility in 100 parts solvent
t374	Trimethylhydro-quinine	$(CH_3)_3C_6H(OH)_2$	152.19	6, 931			172–174			s aq; v s alc, bz, eth
t375	1,3,3-Trimethyl-2-norbomanol		154.25	6, 70	0.9641_4^{20}		39–45	201	73	s alc, eth
t376	1,3,3-Trimethyl-2-norbomanone		152.24	7, 96	0.948^{18}	1.4635^{18}	5	192–194	52	v s alc, eth
t377	Trimethyl orthoacetate	$CH_3C(OCH_3)_3$	120.15	2^2, 128	0.9428_4^{25}	1.3859^{25}		107–109	16	v s alc, eth
t378	Trimethyl orthoformate	$HC(OCH_3)_3$	106.12	2, 19	0.9676_4^{20}	1.3790^{20}		100.6	15	
t379	2,4,4-Trimethyl-2-oxazoline		113.16		0.887	1.4213^{20}		112–113	12	
t380	2,2,3-Trimethylpentane	$(CH_3)_3CCH(CH_3)CH_2CH_3$	114.23	1^1, 62	0.7160_4^{20}	1.4030^{20}	−112.3	110	<21	s eth; sl s alc
t381	2,2,4-Trimethylpentane	$(CH_3)_3CHCH_2C(CH_3)_3$	114.23	1^2, 127	0.6919_4^{20}	1.3915^{20}	−107.4	99.2	−12	s bz, chl, eth
t382	2,3,4-Trimethylpentane	$(CH_3)_2CH[CH(CH_3)]_2CHCH_3$	114.23	1^3, 500	0.7190_4^{20}	1.4042^{20}	−109.2	113–114	5	s alc, org solv
t383	2,2,4-Trimethyl-1,3-pentanediol	$(CH_3)_2CHCH(OH)C(CH_3)_2CH_2OH$	146.22	1^3, 2225	0.928_5^{55}	1.4513^{15}	52–56	232	113	1.8 aq; 75 alc; 22 bz; 25 acet
t384	2,4,4-Trimethyl-1-pentene	$(CH_3)_3CCH_2C(CH_3)=CH_2$	112.22	1^3, 849	0.7150_4^{20}	1.4112^{20}	−93	101–102	−6	
t385	2,3,5-Trimethylphenol	$(CH_3)_3C_6H_2OH$	136.19	6, 518			92–95	230–231		
t386	2,3,6-Trimethylphenol	$(CH_3)_3C_6H_2OH$	136.19				62–64			
t387	2,4,6-Trimethylphenol	$(CH_3)_3C_6H_2OH$	136.19	6, 518			71–74	220		
t388	2,4,6-Trimethyl-1,3-phenylenediamine	$(CH_3)_3C_6H(NH_2)_2$	152.23	13^1, 190			88–91			
t389	Trimethyl phosphate	$(CH_3O)_3P(O)$	140.08	1, 286	1.197^{20}	1.3967^{20}	−46	197	107	100 aq; s alc
t390	Trimethyl phosphite	$(CH_3O)_3P$	124.08	1, 285	1.046_4^{20}	1.4080^{20}	−78	111–112	27	dec aq; misc alc, acet, bz, PE
t391	Trimethyl phosphono-acetate	$(CH_3O)_2P(O)CH_2CO_2CH_3$	182.11		1.125	1.4370^{20}		$118^{0.85mm}$	>110	
t392	1,2,4-Trimethyl-piperazine		128.22		0.851_{25}^{25}	1.4480^{25}	−50	151^{746mm}		s aq, alc, acet, bz
t393	2,4,6-Trimethylpyridine	$C_5H_2N(CH_3)_3$	121.18	20, 250	0.9166_4^{22}	1.4959^{25}	−46	171	57	3.5 aq; misc eth; s alc, bz, chl
t394	N-(Trimethylsilyl)-	$CH_3CONHSi(CH_3)_3$	131.25				46–49	186	57	

No.	Name	Formula	Mol. wt.	Beilstein	Density	n_D	mp	bp	Flash point	Solubility
t395	Trimethylsilyl acetate	$CH_3CO_2Si(CH_3)_3$	132.24	4^3, 1857	0.882	1.3880^{20}	−32	108	4	
t396	*N*-(Trimethylsilyl)-imidazole		140.26		0.956	1.4751^{20}		94^{14mm}	5	
t397	Trimethylsilyl methacrylate	$H_2C=C(CH_3)CO_2Si(CH_3)_3$	158.28		0.890	1.4150^{20}		51^{20mm}	32	
t398	Trimethylsilyl trifluoromethane sulfonate	$CF_3SO_3Si(CH_3)_3$	222.26		1.228	1.3600^{20}		77^{80mm}	25	
t399	Trimethylsulfonium iodide	$[(CH_3)_3S]I$	204.07					215–220 sublime		
t400	Trimethylsulfoxonium iodide	$[(CH_3)_3S(O)]I$	220.07				169 dec			
t400a	1,7,7-Trimethyltricyclo[2.2.1.O^{2,6}]-heptanes		136.24	5, 164	0.8668^{80}	1.4296^{80}	67.5	152.5		s hot acet; sl s alc
t401	Trimethylvinylsilane	$(CH_3)_3SiCH=CH_2$	100.24		0.649	1.3920^{20}		55	<−34	5.5 alc; 7.1 eth; i aq
t402	2,4,6-Trinitroaniline	$(O_2N)_3C_6H_2NH_2$	228.12	12, 763	1.762^{14}		188–190	explodes		0.035 aq; 1.9 alc; 1.5 eth; 6.2 bz
t403	1,2,4-Trinitrobenzene	$C_6H_3(NO_2)_3$	213.11	5, 271	1.73^{16}		61–62	explodes		
t404	1,3,5-Trinitrobenzene	$C_6H_3(NO_2)_3$	213.11	5, 271	1.688^{20}_{4}		122.5	explodes		1.5 alc; 4 eth; s bz, acet; 0.01 aq
t405	2,4,6-Trinitrotoluene	$(O_2N)_3C_6H_2CH_3$	227.13	5, 347	1.654^{20}_{4}		80.1	explodes		
t406	Trioctylamine	$[CH_3(CH_2)_7]_3N$	353.68	4, 196	0.809	1.4485^{20}		365–367	>110	17.2 aq^{18}; v s alc, bz, eth, EtOAc
t407	1,3,5-Trioxane		90.08	19, 381	1.170^{65}		60.2	115	45	
t408	4,7,10-Trioxa-1,13-tridecanediamine	$O[CH_2CH_2O(CH_2)_3NH_2]_2$	220.31	4^4, 1625	1.005	1.4640^{20}		148^{4mm}	>110	
t409	Tripentaerythritol	$(HOCH_2)_3CCH_2OCH_2\text{–}C(CH_2OH)_2CH_2OCH_2$	372.41				225 dec			
t410	Triphenylamine	$(C_6H_5)_3N$	245.33	12, 181	0.774^{0}_{0}		125–127	347–348		v s bz, eth; sl s alc
t411	Triphenylantimony	$(C_6H_5)_3Sb$	353.07	16, 891	1.4343^{25}		52–54	377		v s bz, eth; s alc
t412	Triphenylarsine	$(C_6H_5)_3As$	306.24	16, 828	1.2225^{48}		60–62	233^{14mm}		v s bz; s abs alc, eth
t413	1,3,5-Triphenylbenzene	$(C_6H_5)_3C_6H_3$	306.41	5, 737	1.205	1.6139^{48}	172–174	460	>110	
t414	Triphenylborane	$(C_6H_5)_3B$	242.13	16^2, 636			145	203^{15mm}		v s hot alc, eth; 49 chl; 7 bz; s PE
t415	Triphenylmethane	$(C_6H_5)_3CH$	244.34	5, 698	1.0134^{99}_{4}		92–94	360		

(Continued)

TABLE 2.21 Physical Constants of Organic Compounds (*Continued*)

No.	Name	Formula	Formula weight	Beilstein reference	Density, g/mL	Refractive index	Melting point, °C	Boiling point, °C	Flash point, °C	Solubility in 100 parts solvent
t416	Triphenylmethanol	$(C_6H_5)_3COH$	260.34	6, 713	1.199^0_4		160–163	360		v s alc, bz, eth; i aq
t417	Triphenylmethyl bromide	$(C_6H_5)_3CBr$	323.24	5, 704			152–154	230^{15mm}		
t418	Triphenylmethyl chloride	$(C_6H_5)_3CCl$	278.78	5, 700			110–112	235^{20mm}		
t419	Triphenyl phosphate	$(C_6H_5O)_3P(O)$	326.29	6, 179			50–52	244^{10mm}	223	misc alc; s bz, acet, chl, eth; i aq
t420	Triphenylphosphine	$(C_6H_5)_3P$	262.29	16, 759	1.0754^{81}_4		79–81	377	181	v s eth; s bz, chl, HOAc; sl alc; i aq
t421	Triphenylphosphine oxide	$(C_6H_5)_3P(O)$	278.29	16, 783			156–158			
t422	Triphenyl phosphite	$(C_6H_5O)_3P$	310.29	6, 177	1.184	1.5903^{20}	22–24	360	218	s alc, bz, chl, eth
t423	Triphenylsilane	$(C_6H_5)_3SiH$	260.41	16^2, 605			42–44	152^{2mm}	76	
t424	Triphenyltin acetate	$CH_3CO_2Sn(C_6H_5)_3$	409.06	16^4, 1606			124–126	$240^{13.5mm}$		s eth; sl s alc, bz
t425	Triphenyltin chloride	$(C_6H_5)_3SnCl$	385.46	16, 914			108 dec			
t426	Triphenyltin hydroxide	$(C_6H_5)_3SnOH$	367.02	16, 914			124–126			
t427	Tripropoxyborane	$(CH_3CH_2CH_2O)_3B$	188.08	1^2, 369	0.8576^{20}_4	1.3948^{20}		175–177	32	v s alc; misc eth
t428	Tripropylaluminum	$(CH_3CH_2CH_2)_3Al$	156.25	4, 643	0.823		−107	84^{2mm}	−18	s aq, alc, eth
t429	Tripropylamine	$(CH_3CH_2CH_2)_3N$	143.27	4, 139	0.753	1.4160^{20}	−93.5	155–158	36	s aq
t430	Tripropylene glycol	$H(OCH_2CH_2CH_2)_3OH$	192.26		1.021	1.442^{25}		273	141	
t431	Tripropylene glycol butyl ether	$HO(CH_2CH_2CH_2O)_3(CH_2)_3CH_3$	248.4		0.932	1.430^{20}		276	135	
t432	Tripropylene glycol monomethyl ether	$HO(CH_2CH_2CH_2O)_3CH_3$	206.29	1^4, 2475	0.967	1.428^{25}	−42	242.4	127	misc aq, alc, eth
t433	Tripropyl orthoformate	$HC(OCH_2CH_2CH_3)_3$	190.28	2, 21	0.8805^{20}_4	1.4072^{20}		108^{40mm}	72	
t434	Tris(2-aminoethyl)-amine	$(H_2NCH_2CH_2)_3N$	146.24	4, 256	0.977	1.4970^{20}		114^{15mm}	>110	
t435	Tris(2-butoxyethyl) phosphate	$(C_4H_9OCH_2CH_2O)_3P(O)$	398.48		1.006	1.4359^{20}		228^{4mm}	110	
t436	Tris(2-chloroethyl) phosphate	$(ClCH_2CH_2O)_3P(O)$	285.49	1^2, 337	1.390	1.4721^{20}		330	232	
t437	Tris(2-chloroethyl) phosphite	$(ClCH_2CH_2O)_3P$	269.49		1.3534^{20}_4	1.4863^{20}		115^{2mm}	190	misc alc, bz, eth
t438	Tris(2-ethylhexyl) phosphate	$[C_4H_9CH(C_2H_5)CH_2O]_3P(O)$	434.65	1^3, 1734	0.924	1.4437^{20}		215^{4mm}	>110	i aq

No.	Name	Formula	MW	Beil./Merck Ref.	Density	n_D	mp (°C)	bp (°C)	fp (°C)	Solubility
t439	Tris(hydroxymethyl)-aminomethane	$(HOCH_2)_3CNH_2$	121.14	4, 303			171–172	220^{10mm}		
t440	1,1,1-Tris(hydroxymethyl)ethane	$CH_3C(CH_2OH)_3$	120.15	1, 520			200–203			
t441	N-[Tris(hydroxymethyl)methyl]-glycine	$(HOCH_2)_3CNHCH_2CO_2H$	179.17	Merck: 12, 9783			187			satd aq^0 is 0.8M
t442	Tris(hydroxymethyl)-nitromethane	$(HOCH_2)_3CNO_2$	151.12	1, 520			214 pure 175 tech			220 aq; v s alc; sl s bz
t443	Tris[2-(2-methoxyethoxy)ethyl] amine	$(CH_3OCH_2CH_2OCH_2CH_2)_3N$	323.43		1.011	1.4486^{20}			>110	
t444	Tris(2-methoxyethoxy)-vinylsilane	$H_2C{=}CHSi(OCH_2CH_2{-}OCH_3)_3$	280.39	4^4, 4257	1.034^{25}_4	1.427^{25}		284–286	>110	
t445	Tris(2-methoxyethyl) borate	$(CH_3OCH_2CH_2O)_3B$	236.08	1^3, 2118	1.010	1.4150^{20}		135^{15mm}	87	
t446	Tris(2-methylallyl)-amine	$[H_2C{=}C(CH_3)CH_2]_3N$	179.31	4^3, 462	0.794	1.4575^{20}		85^{15mm}	53	
t447	Tris(2,2,2-trifluoroethyl) phosphite	$(CF_3CH_2O)_3P$	328.07	1^4, 1371	1.487	1.3245^{20}		131^{743mm}	>110	
t448	Tris [3-(trimethylsilyl)propyl] isocyanurate		615.86		1.170	1.4610^{20}		250	102	
t449	Tris(trimethylsilyl) borate	$[(CH_3)_3SiO]_3B$	278.38	4^3, 1861	0.831	1.3861^{20}		186	42	
t450	1,3,5-Trithiane		138.27	19, 382			216–218	57.8		s bz; sl s alc, eth
t451	Trithiocarbonic acid	$(HS)_2CS$	110.21	3, 221	1.4834^{20}	1.8225^{20}	−26.9			dec aq, alc; sl s eth
t452	Tri-o-tolyl phosphate	$(CH_3C_6H_4O)_3P(O)$	368.37	Merck: 12, 9893	1.1955^{20}	1.5575^{20}	11	410	225	sl s aq, alc; s eth
t453	1,2,4-Trivinylcyclo-hexane	$(H_2C{=}CH)_3C_6H_9$	162.28		0.836	1.4780^{20}		88^{20mm}	68	
t454	L-(−)-Tryptophan		204.23	22, 546			280–285 dec			1.14 aq^{25}; s hot alc, alk; i eth, chl
t455	L-Tyrosine	$(HO)C_6H_4CH_2CH(NH_2)CO_2H$	181.19	14, 605	1.456		342–344			0.045 aq; 0.01 alc; s alk; i eth
u1	Undecanal	$CH_3(CH_2)_9CHO$	170.30	1, 712	0.825	1.4322^{20}	−4	115^{5mm}	96	i aq; s alc, eth
u2	Undecane	$CH_3(CH_2)_9CH_3$	156.31	1, 170	0.7402^{20}_4	1.4173^{20}	−25.6	196	60	i aq; misc alc, eth

(Continued)

TABLE 2.21 Physical Constants of Organic Compounds (*Continued*)

No.	Name	Formula	Formula weight	Beilstein reference	Density, g/mL	Refractive index	Melting point, °C	Boiling point, °C	Flash point, °C	Solubility in 100 parts solvent
u3	Undecanenitrile	$CH_3(CH_2)_9CN$	167.30	2, 358	0.823	1.4330^{20}		253	>110	s alc, chl, eth; i aq
u4	Undecanoic acid	$CH_3(CH_2)_9CO_2H$	186.30	2, 358	0.8907	1.4294^{45}	28.5	228^{160mm}	>110	
u5	Undecanoic γ-lactone		184.28	17, 247	0.949	1.4500^{20}		166^{13mm}	>110	
u6	Undecanoic δ-lactone		184.28	17^3, 4257	0.969	1.4590^{20}		$155^{10.5mm}$	>110	
u7	1-Undecanol	$CH_3(CH_2)_{10}OH$	172.31	1, 427	0.8324	1.4402^{20}	11	242.8	>110	s alc, bz, chl, eth, acet;
u8	2-Undecanol	$CH_3(CH_2)_8CH(OH)CH_3$	172.31	1, 427	0.828	1.4370^{20}	2–3	131^{28mm}	88	i aq
u9	2-Undecanone	$CH_3(CH_2)_8COCH_3$	170.30	1, 173	0.829	1.4300^{20}	11–13	231–232	88 (CC)	i aq; v s alc, eth
u10	3-Undecanone	$CH_3(CH_2)_7COCH_2CH_3$	170.30	1, 173	0.827	1.4291^{20}	12–13	225–229	89	
u11	6-Undecanone	$CH_3(CH_2)_4CO(CH_2)_4CH_3$	170.30	1, 174	0.831	1.4280^{20}	14.6	228	88	
u12	10-Undecenal	$H_2C=CH(CH_2)_8CHO$	168.28	1^3, 3029	0.810	1.4427^{20}		dec >mp	92	
u12a	1-Undecene	$H_2C=CH(CH_2)_8CH_3$	154.30	1, 225	0.7503^{20}	1.4261^{20}	–49	193	71	
u13	10-Undecenoic acid	$H_2C=CH(CH_2)_8CO_2H$	184.28	2, 458	0.907^{24}	1.4493^{20}	24.5	137^{2mm}	148	s alc, chl, eth; i aq
u14	10-Undecen-1-ol	$H_2C=CH(CH_2)_9OH$	170.30	1, 452	0.850^{15}	1.4500^{20}	–2	245	93	
u15	10-Undecenoyl chloride	$H_2C=CH(CH_2)_8COCl$	202.73	2, 459	0.944	1.4540^{20}		122^{10mm}	93	
u16	Urea	$(H_2N)_2CO$	60.06	3, 42	1.335		133–135	dec >mp		100 aq; 20 alc
u17	Uric acid		168.11	26, 513	1.893^{20}		> 300 dec			s alk; i aq, alc, eth
u18	Uridine		244.20	31, 23			166–167			s aq; hot alc, pyr
v1	Valeric anhydride	$[CH_3(CH_2)_3CO]_2O$	186.25	2, 301	0.942	1.4210^{20}	–57	112^{16mm}	101	
v2	γ-Valerolactone		100.12	17, 235	1.057	1.4330^{20}	–31	207–208	81	
v3	δ-Valerolactone		100.12	17, 235	1.079	1.4580^{20}		$60^{0.5mm}$	100	
v4	L-Valine	$(CH_3)_2CHCH(NH)CO_2H$	117.15	4, 427	1.230		>315 subl			8.8 aq; v sl s alc, eth
v5	Vinyl acetate	$H_2C=CHO_2CCH_3$	86.09	2^1, 63	0.9324^{20}	1.3954^{20}	–93	72–73	–8	2 aq; misc alc, eth
v6	Vinyl benzoate	$C_6H_5CO_2CH=CH_2$	148.16	9^1, 65	1.070	1.5290^{20}		96^{20mm}	82	
v7	4-Vinylbenzyl chloride	$H_2C=CHC_6H_4CH_2Cl$	152.62	5^1, 35	1.083	1.5740^{20}		229	104	
v8	Vinylcyclohexene	$C_6H_{11}CH=CH_2$	110.20	5^1, 63	0.805	1.4463^{20}		126–127	20	
v9	4-Vinyl-1-cyclohexene		108.18		0.8034^{20}	1.4640^{20}	–101	127	20	
v10	2-Vinyl-1,3-dioxolane		100.12		1.001	1.4300^{20}		115–116	14	
v11	N-Vinylformamide	$HCONHCH=CH_2$	71.08		1.014	1.4940^{20}	–16	210	102	
v12	1-Vinylimidazole		94.12	23^4, 569	1.039	1.5308^{20}		79^{13mm}	81	
v13	5-Vinyl-2-norbornene		120.20		0.841	1.4802^{20}	–80	141	27	
v14	Vinyl propionate	$CH_3CH_2CO_2CH=CH_2$	100.12	2^3, 532	0.919	1.4030^{20}	–80	94–95	6	
v15	2-Vinylpyridine	$(C_5H_4N)CH=CH_2$	105.14	20, 256	0.975	1.5490^{20}		158–159	46	v s alc, chl, eth
v16	4-Vinylpyridine	$(C_5H_4N)CH=CH_2$	105.14	20^2, 170	0.975	1.5500^{20}		65^{15mm}	51	sl s hot aq, hot alc

No.	Name	Formula	Formula wt.	Beilstein ref.	Density	Refractive index	Melting point	Boiling point	Flash point	Solubility
v17	*N*-Vinyl-2-pyrrolidinone		111.14		1.040	1.5120^{20}		$93^{13\text{mm}}$	93	
v18	Vinyltrimethoxysilane	$H_2C{=}CHSi(OCH_3)_3$	148.24	17, 73	0.968	1.3920^{20}		123	22	s bz, eth; sl s alc, aq
x1	Xanthene		182.22	18[2], 279			101	310–312		s hot alc, eth
x2	Xanthen-9-carboxylic acid		226.23				217 dec			
x3	9-Xanthenone		196.21	17, 354			174–176	$350^{730\text{mm}}$		0.5 alc; v s chl
x4	*m*-Xylene	$C_6H_4(CH_3)_2$	106.17	5, 370	0.8642^{20}	1.4972^{20}	−47.9	139	27	misc alc, eth; 0.02 aq
x5	*o*-Xylene	$C_6H_4(CH_3)_2$	106.17	5, 362	0.8808^{20}	1.5054^{20}	−25.2	144–145	32	misc alc, eth; 0.017 aq
x6	*p*-Xylene	$C_6H_4(CH_3)_2$	106.17	5, 382	0.86114^{20}	1.4958^{20}	13	138	27	v s eth; s alc; 0.02 aq
x7	Xylitol	$HOCH_2(CHOH)_3CH_2OH$	152.15	1, 531	1.52		95–97			64 aq; 1.2 EtOH; 6.0 MeOH
x8	D-(+)-Xylose		150.13	31, 47	1.535^{0}		156–158			117 aq; s hot alc, pyr
x9	*m*-Xylylenediamine	$C_6H_4(CH_2NH_2)_2$	136.20	13, 186	1.032	1.5709^{20}	>110			

TABLE 2.22 Melting Points of Derivatives of Organic Compounds

(a) Derivatives of Alcohols

	3,5-Dinitro-benzoate $\theta_{C,m}/°C$		3,5-Dinitro-benzoate $\theta_{C,m}/°C$
Methanol	109	2-Methylpropan-2-ol	142
Ethanol	94	Pentan-1-ol	46
Propan-1-ol	75	Hexan-1-ol	61
Propan-2-ol	122	Phenylmethanol	113
Butan-1-ol	64	Cyclohexanol	113
2-Methylpropan-1-ol	88	Ethane-1,2-diol (glycol)	169*
Butan-2-ol	76		

(b) Derivatives of Phenols

	3,5-Dinitro benzoate $\theta_{C,m}/°C$	4-Methyl-benzene-sulphonate $\theta_{C,m}/°C$		3,5-Dinitro-benzoate $\theta_{C,m}/°C$	4-Methyl-benzene sulphonate $\theta_{C,m}/°C$
Phenol	146	96	Benzene-1,2-diol	152*	—
2-Methylphenol	138	55	Benzene-1,3-diol	201*	81*
3-Methylphenol	165	51	Benzene-1,4-diol	317*	159*
4-Methylphenol	189	70	2-Nitrophenol	155	83
Naphthalen-1-ol	217	88	3-Nitrophenol	159	113
Naphthalen-2-ol	210	125	4-Nitrophenol	188	97

(c) Derivatives of Aldehydes and Ketones

	2,4-Dinitro-Phenyl-hydrazone $\theta_{C,m}/°C$		2,4-Dinitro-Phenyl-hydrazone $\theta_{C,m}/°C$
Methanal	166	Propanone	126
Ethanal	168	Butanone	116
Propanal	155	Pentan-3-one	156
Butanal	126	Pentan-2-one	144
Benzaldehyde	237	Heptan-4-one	75
2-Hydroxybenzaldehyde	252 dec.	Phenylethanone	250
Ethanedial	327	Diphenylmethanone	239
Trichloroethanal	131	Cyclohexanone	162

(d) Derivatives of Amines

	Ethanoyl derivative $\theta_{C,m}/°C$	Benzoyl derivative $\theta_{C,m}/°C$	4-Methyl-benzene sulphonyl derivative $\theta_{C,m}/°C$
Methylamine	28	80	75
Ethylamine	205*	69	62
Propylamine	47	85	52
Butylamine	229‡	70	65
(Phenylmethyl) amine	60	105	116
Phenylamine	114	163	103
Cyclohexylamine	104	147	87
2-Methylphenylamine	112	143	110
3-Methylphenylamine	66	125	114
4-Methylphenylamine	152	158	118
Dimethylamine	116‡	42	87
Diethylamine	186‡	42	60
Diphenylamine	103	180	142

*Disubstituted derivative.
‡Boiling temperature.

TABLE 2.23 Melting Points of *n*-Paraffins

Number of carbon atoms	Melting point	
	°C	°F
1	−182	−296
2	−183	−297
3	−188	−306
4	−138	−216
5	−130	−202
6	−95	−139
7	−91	−132
8	−57	−71
9	−54	−65
10	−30	−22
11	−26	−15
12	−10	14
13	−5	23
14	6	43
15	10	50
16	18	64
17	22	72
18	28	82
19	32	90
20	36	97
30	66	151
40	82	180
50	92	198
60	99	210

TABLE 2.24 Boiling Point and Density of Alkyl Halides

Name	Chloride		Bromide		Iodide	
	B.p., °C	Density at 20°C	B.p., °C	Density at 20°C	B.p., °C	Density at 20°C
Methyl	−24		5		43	2.279
Ethyl	12.5		38	1.440	72	1.933
n-Propyl	47	.890	71	1.335	102	1.747
n-Butyl	78.5	.884	102	1.276	130	1.617
n-Pentyl	108	.883	130	1.223	157	1.517
n-Hexyl	134	.882	156	1.173	180	1.441
n-Heptyl	160	.880	180		204	1.401
n-Octyl	185	.879	202		225.5	
Isopropyl	36.5	.859	60	1.310	89.5	1.705
Isobutyl	69	.875	91	1.261	120	1.605
sec-Butyl	68	.871	91	1.258	119	1.595
tert-Butyl	51	.840	73	1.222	100*d*	
Cyclohexyl	142.5	1.000	165			
Vinyl(Haloethene)	−14		16		56	
Allyl (3-Halopropene)	45	.938	71	1.398	103	
Crotyl (1-Halo-2-butene)	84				132	

(Continued)

TABLE 2.24 Boiling Point and Density of Alkyl Halides (*Continued*)

Name	Chloride		Bromide		Iodide	
	B.p., °C	Density at 20°C	B.p., °C	Density at 20°C	B.p., °C	Density at 20°C
Methylvinylcarbinyl (3-Halo-1-butene)	64					
Propargyl (3-Halopropyne)	65		90	1.520	115	
Benzyl	179	1.102	201		93[10]	
α-Phenylethyl	92[15]		85[10]			
β-Phenylethyl	92[20]		92[11]		127[19]	
Diphenylmethyl	173[19]		184[20]			
Triphenylmethyl	310		230[15]			
Dihalomethane	40	1.336	99	2.49	180*d*	3.325
Trihalomethane	61	1.489	151	2.89	*subl.*	4.008
Tetrahalomethane	77	1.595	189.5	3.42	*subl.*	4.32
1,1-Dihaloethane	57	1.174	110	2.056	179	2.84
1,2-Dihaloethane	84	1.257	132	2.180	*d*	2.13
Trihaloethylene	87		164	2.708		
Tetrahaloethylene	121				*subl.*	
Benzal halide	205		140[20]			
Benzotrihalide	221	1.38				

TABLE 2.25 Properties of Carboxylic Acids

Name	Formula	M.p., °C	B.p., °C	Solub., g/100 g H$_2$O
Formic	HCOOH	8	100.5	∞
Acetic	CH$_3$COOH	16.6	118	∞
Propionic	CH$_3$CH$_2$COOH	−22	141	∞
Butyric	CH$_3$(CH$_2$)$_2$COOH	−6	164	∞
Valeric	CH$_3$(CH$_2$)$_3$COOH	−34	187	3.7
Caproic	CH$_3$(CH$_2$)$_4$COOH	−3	205	1.0
Caprylic	CH$_3$(CH$_2$)$_6$COOH	16	239	0.7
Capric	CH$_3$(CH$_2$)$_8$COOH	31	269	0.2
Lauric	CH$_3$(CH$_2$)$_{10}$COOH	44	225[100]	i.
Myristic	CH$_3$(CH$_2$)$_{12}$COOH	54	251[100]	i.
Palmitic	CH$_3$(CH$_2$)$_{14}$COOH	63	269[100]	i.
Stearic	CH$_3$(CH$_2$)$_{16}$COOH	70	287[100]	i.
Oleic	*cis*-9-Octadecenoic	16	223[10]	i.
Linoleic	*cis,cis*-9,12-Octadecadienoic	−5	230[16]	i.
Linolenic	*cis,cis,cis*-9,12,15-Octadecatrienoic	−11	232[17]	i.
Cyclohexanecarboxylic	*cyclo*-C$_6$H$_{11}$COOH	31	233	0.20
Phenylacetic	C$_6$H$_5$CH$_2$COOH	77	266	1.66
Benzoic	C$_6$H$_5$COOH	122	250	0.34
o-Toluic	*o*-CH$_3$C$_6$H$_4$COOH	106	359	0.12
m-Toluic	*m*-CH$_3$C$_6$H$_4$COOH	112	263	0.10
p-Toluic	*p*-CH$_3$C$_6$H$_4$COOH	180	275	0.03
o-Chlorobenzoic	*o*-ClC$_6$H$_4$COOH	141		0.22
m-Chlorobenzoic	*m*-ClC$_6$H$_4$COOH	154		0.04
p-Chlorobenzoic	*p*-ClC$_6$H$_4$COOH	242		0.009
o-Bromobenzoic	*o*-BrC$_6$H$_4$COOH	148		0.18
m-Bromobenzoic	*m*-BrC$_6$H$_4$COOH	156		0.04

TABLE 2.25 Properties of Carboxylic Acids (*Continued*)

Name	Formula	M.p., °C	B.p., °C	Solub., g/100 g H₂O
p-Bromobenzoic	*p*-BrC₆H₄COOH	254		0.006
o-Nitrobenzoic	*o*-O₂NC₆H₄COOH	147		0.75
m-Nitrobenzoic	*m*-O₂NC₆H₄COOH	141		0.34
p-Nitrobenzoic	*p*-O₂NC₆H₄COOH	242		0.03
Phthalic	*o*-C₆H₄(COOH)₂	231		0.70
Isophthalic	*m*-C₆H₄(COOH)₂	348		0.01
Terephthalic	*p*-C₆H₄(COOH)₂	300 *subl.*		0.002
Salicylic	*o*-HOC₆H₄COOH	159		0.22
p-Hydroxybenzoic	*p*-HOC₆H₄COOH	213		0.65
Anthranilic	*o*-H₂NC₆H₄COOH	146		0.52
m-Aminobenzoic	*m*-H₂NC₆H₄COOH	179		0.77
p-Aminobenzoic	*p*-H₂NC₆H₄COOH	187		0.3
o-Methoxybenzoic	*o*-CH₃OC₆H₄COOH	101		0.5
m-Methoxybenzoic	*m*-CH₃OC₆H₄COOH	110		
p-Methoxybenzoic (Anisic)	*p*-CH₃OC₆H₄COOH	184		0.04

TABLE 2.26 The Structure, Melting Point, and Boiling Points of Polycyclic Aromatic Hydrocarbons

Structure	IUPAC nomenclature (synonyms)	Molecular weight	Melting point (°C)	Boiling point (°C)^{760}
	Indan Hydrindene 2,3-Dihydroindene	118.18	−51	178
	Indene Indonaphthene	116.16	−2	183
	Naphthalene Tar Camphor White Tar Moth Flakes	128.19	81	218
	2-Methylnaphthalene β-Methylnaphthalene	142.20	35	241
	1-Methylnaphthalene α-Methylnaphthalene	142.20	−22	245
	Biphenyl Diphenyl Phenylbenzene Bibenzene	154.21	71	255
	2-Ethylnaphthalene β-Ethylnaphthalene	156.23	−7	258

(*Continued*)

TABLE 2.26 The Structure, Melting Point, and Boiling Points of Polycyclic Aromatic Hydrocarbons (*Continued*)

Structure	IUPAC nomenclature (synonyms)	Molecular weight	Melting point (°C)	Boiling point (°C)[760]
	1-Ethylnaphthalene	156.23	−14	259
	2,6-Dimethylnaphthalene	156.23	110	262
	2,7-Dimethylnaphthalene	156.23	97	262
	1,7-Dimethylnaphthalene	156.23		263
	1,3-Dimethylnaphthalene	156.23		265
	1,6-Dimethylnaphthalene	156.23		266
	2,3-Dimethylnaphthalene Guaiene	156.23	105	268
	1,4-Dimethylnaphthalene α-Dimethylnaphthalene	156.23	8	268
	4-Methylbiphenyl	168.24	50	268
	1,5-Dimethylnaphthalene	156.23	80	269

TABLE 2.26 The Structure, Melting Point, and Boiling Points of Polycyclic Aromatic Hydrocarbons (*Continued*)

Structure	IUPAC nomenclature (synonyms)	Molecular weight	Melting point (°C)	Boiling point (°C)[760]
	Azulene	128.19	100	270 d
	1,2-Dimethylnaphthalene	156.23	−4	271
	Acenaphthylene	152.21	93	−270 d
	3-Methylbiphenyl	168.24	5	273
	3,5-Dimethylbiphenyl	182.27		275
	Acenaphthene Naphthyleneethylene	154.21	96	279
	1,3,7-Trimethylnaphthalene	170.25	14	280
	2,3,5-Trimethylnaphthalene	170.25	25	285
	2,3,6-Trimethylnaphthalene	170.25	101	286
	Fluorene 2,3-Benzindene Diphenylenemethane	166.23	117	294
	9-Methylfluorene	180.25	47	

TABLE 2.26   The Structure, Melting Point, and Boiling Points of Polycyclic Aromatic Hydrocarbons (*Continued*)

Structure	IUPAC nomenclature (synonyms)	Molecular weight	Melting point (°C)	Boiling point (°C)[760]
	4-Methylfluorene	180.25		
	3-Methylfluorene	180.25	85	316
	2-Methylfluorene	180.25	104	318
	1-Methylfluorene	180.25		−318
	1-Phenylnaphthalene α-Phenylnaphthalene	204.28	−45	334
	Phenanthrene o-Diphenyleneethylene	178.24	101	338
	Anthracene	178.24	216	340
	3-Methylphenanthrene	192.26	65	352
	2-Methylphenanthrene	192.26		355

TABLE 2.26 The Structure, Melting Point, and Boiling Points of Polycyclic Aromatic Hydrocarbons (*Continued*)

Structure	IUPAC nomenclature (synonyms)	Molecular weight	Melting point (°C)	Boiling point (°C)[760]
	9-Methylphenanthrene	192.26	92	355
	2-Methylanthracene	192.26	209	359 sub
	4,5-Methylenephenanthrene 4H-Cyclopenteno[def]phenanthrene 4H-Cyclopenta[def]phenanthrene 4,5-Phenanthrylenemethane	190.24	116	359
	4-Methylphenanthrene	192.26		
	1-Methylphenanthrene	192.26	123	359
	2-Phenylnaphthalene β-Phenylnaphthalene	204.28	104	360
	1-Methylanthracene	192.26	86	363
	3,6-Dimethylphenanthrene	206.29		363
	2,7-Dimethylanthracene	206.29	241	−370

(*Continued*)

TABLE 2.26   The Structure, Melting Point, and Boiling Points of Polycyclic Aromatic Hydrocarbons (*Continued*)

Structure	IUPAC nomenclature (synonyms)	Molecular weight	Melting point (°C)	Boiling point (°C)[760]
	2,6-Dimethylanthracene	206.29	250	~370
	2,3-Dimethylanthracene	206.29	252	
	Fluoranthene Idryl 1,2-Benzacenaphthene Benzo[jk]fluorine Benz[a]acenaphthylene	202.26	111	383
	9,10-Dimethylanthracene	206.29	183	
	Pyrene Benzo[def]phenanthrene	202.26	156	393
	2,7-Dimethylpyrene	230.32		396
	Benzo[b]fluorene 11 H-Benzo[b]fluorene 2,3-Benzofluorene Isonaphthofluorene	216.29	209	402
	Benzo[c]fluorene 7H-Benzo[c]fluorene 3,4-Benzofluorene	216.29		406
	Benzo[a]fluorene 11 H-Benzo[a]fluorene 1,2-Benzofluorene Chrysofluorene	216.29	190	407
	2-Methylpyrene 4-Methylpyren	216.29		410

TABLE 2.26   The Structure, Melting Point, and Boiling Points of Polycyclic Aromatic Hydrocarbons (*Continued*)

Structure	IUPAC nomenclature (synonyms)	Molecular weight	Melting point (°C)	Boiling point (°C)[760]
	1-Methylpyrene 3-Methylpyren	216.29		410
	4-Methylpyrene 1-Methylpyren	216.29		410
	Benzo[ghi]fluoranthene	226.28		432
	Benzo[c]phenanthrene 3,4-Benzophenanthrene	238.30	68	
	Benz[a]anthracene 1,2-Benzanthracene Tetraphene 2,3-Benzophenanthrene Naphthanthracene	228.30	162	435 sub
	Triphenylene 9,10-Benzophenanthrene Isochrysene	228.30	199	439
	Chrysene 1,2-Benzophenanthrene Benzo[a]phenanthrene	228.30	256	441
	6-Methylchrysene	242.32		

(*Continued*)

TABLE 2.26 The Structure, Melting Point, and Boiling Points of Polycyclic Aromatic Hydrocarbons (*Continued*)

Structure	IUPAC nomenclature (synonyms)	Molecular weight	Melting point (°C)	Boiling point (°C)[760]
	1-Methylchrysene	242.32	257	
	Naphthacene Benz[b]anthracene 2,3-Benzanthracene Tetracene	228.30	257	450 sub
	2,2′-Dinaphthyl 2,2′-Binaphthyl β,β'-Binaphthyl β,β'-Dinaphthyl	254.34	188	452[753] sub
	Benzo[b]fluoranthene 2,3-Benzofluoranthene 3,4-Benzofluoranthene Benz[e]acephenanthrylene	252.32	168	481
	Benzo[j]fluoranthene 7,8-Benzofluoranthene 10,11-Benzofluoranthene	252.32	166	~480
	Benzo[k]fluoranthene 8,9-benzofluoranthene 11,12-Benzofluoranthene	252.32	217	481
	Benzo[e]pyrene 4,5-Benzpyrene 1,2-Benzopyrene	252.32	179	493
	Benzo[a]pyrene 1,2-Benzpyrene 3,4-Benzopyrene Benzo[def]chrysene	252.32	177	496

TABLE 2.26   The Structure, Melting Point, and Boiling Points of Polycyclic Aromatic Hydrocarbons (*Continued*)

Structure	IUPAC nomenclature (synonyms)	Molecular weight	Melting point (°C)	Boiling point (°C)[760]
	Perylene peri-Dinaphthalene	252.32	278	
	3-Methylcholanthrene 20-Methylcholanthrene	268.38	180	
	Indeno[1,2,3-cd]pyrene o-Phenylenepyrene	276.34		
	Dibenz[a,c]anthracene 1,2:3,4-Dibenzanthracene Naphtho-2′,3′:9,10-phenanthrene	278.36	205	
	Dibenz[a,h]anthracene 1,2:5,6-Dibenzanthracene	278.36	270	
	Dibenz[a,i]anthracene 1,2:6,7-Dibenzanthracene 1,2-Benzonaphthacene Isopentaphene	278.36	264	
	Dibenz[a,j]anthracene 1,2:7,8-Dibenzanthracene α,α′-Dibenzanthracene Dinaphthanthracene	278.36	198	

(*Continued*)

TABLE 2.26 The Structure, Melting Point, and Boiling Points of Polycyclic Aromatic Hydrocarbons (*Continued*)

Structure	IUPAC nomenclature (synonyms)	Molecular weight	Melting point (°C)	Boiling point (°C)[760]
	Benzo[b]chrysene 1,2:6,7-Dibenzophenanthrene 3,4-Benzotetraphene Naphtho-2′,1′:1,2-anthracene	278.36	294	
	Picene Dibenzo[α,i]phenanthrene 3,4-Benzochrysene 1,2:7,8-Dibenzophenanthrene	278.36	368	519
	Benzo[ghi]perylene 1,12-Benzoperylene	276.34	278	
	Anthanthrene Dibenzo[def, mno]chrysene	276.34		
	Coronene Hexabenzobenzene	300.36	439 cor	525?
	Dibenzo[a,e]pyrene	302.38	234	

*Key: d = decomposes;   sub = sublimes.

TABLE 2.27 Properties of Naturally Occurring Amino Acids

Name	Three-letter code	One-letter code	Side chains $(-R)$ R-CH(NH$_2$)COOH	Mol weight	pK_a	ΔH_{ion} kJ·mol^{-1}	Volume Å^3	ASA$_{mc}$ Å^2	ASA$_{sc}^{npl}$ Å^2	ASA$_{sc}^{pol}$ Å^2
Alanine	Ala	A	-CH$_3$	71.08			88.6	46	67	
Arginine	Arg	R	-(CH$_2$)$_3$-CNH(=NH)NH$_3$	156.20	12	44.9	173.4	45	89	107
Asparagine	Asn	N	-CH$_2$-CONH$_2$	114.11			117.7	45	44	69
Aspartic acid	Asp	D	-CH$_2$-COOH	115.09	4.5	4.6	111.1	45	48	58
Cystein	Cys	C	-CH$_2$-SH	103.14	9.1–9.5	36.0	108.5	36	35	69
Glutamine	Gln	Q	-(CH$_2$)$_2$-CONH$_2$	128.14			143.9	45	53	91
Glutamic acid	Glu	E	-(CH$_2$)$_2$-COOH	129.12	4.6	1.6	138.4	45	61	77
Glycine	Gly	G	-H	57.06			60.1	85		
Histidine	His	H	—CH$_2$— (imidazole ring)	137.15	6.2	43.6	153.2	43	102	49
Isoleucine	Ile	I	-CH(CH$_3$)-C$_2$H$_5$	113.17			166.7	42	140	
Leucine	Leu	L	-CH(CH$_3$)$_2$-CH$_2$	113.17			166.7	43	137	
Lysine	Lys	K	-(CH$_2$)$_4$-NH$_2$	128.18	10.4	53.6	168.6	44	119	48
Methionine	Met	M	-(CH$_2$)$_2$-S-CH$_3$	131.21			162.9	44	117	43
Phenylalanine	Phe	F	—CH$_2$— (phenyl ring)	147.18			189.9	43	175	
Proline	Pro	P	*	97.12			122.7	38	105	
Serine	Ser	S	-CH$_2$-OH	87.08			89.0	42	44	36
Threonine	Thr	T	-CH$_2$-(CH$_3$)-OH	101.11			116.1	44	74	28
Tryptophane	Trp	W	—CH$_2$— (indole ring)	186.21			277.8	42	190	27
Tyrosine	Tyr	Y	—CH$_2$— (phenol ring) —OH	163.18	9.7	25.1	193.6	42	144	43
Valine	Val	V	-CII-(CII$_3$)$_2$	99.14			140	43	117	
			α-amino		6.8–7.9					
			α-carboxyl		3.5–4.3					

[a]Enthalpies of ionization of side chains at 25°C, ΔH_{ion}, are from [20]; van der Waals volume from [21]; ASA$_{mc}$, surface area of the backbone, ASA$_{sc}^{npl}$, nonpolar surface area of the side chains, and ASA$_{sc}^{pol}$, polar surface area of the side chains are taken [17].

TABLE 2.28 Hildebrand Solubility Parameters of Organic Liquids

Solvent	δ (MPa$^{1/2}$)	H-bonding tendency[b]	Solvent	δ (MPa$^{1/2}$)	H-bonding tendency[b]
Acetaldehyde	21.1	m	Ethyl chloride	18.8	m
Acetic acid	20.7	s	Ethylenediamine	25.2	s
Acetone	20.2	m	Ethylene dichloride	20.0	p
Acetonitrile	24.3	p	Ethylene glycol	29.9	s
Acetyl chloride	19.4	m	Ethylene glycol	17.6	m
N-Acetylpiperidine	22.9	s	dimethylether		
Acrylic acid	24.5	s	Ethylene oxide	22.7	m
Allyl acetate	18.8	m	Ethyl formate	19.2	m
Allyl alcohol	24.1	s	Ethyl methacrylate	17.0	m
Ammonia	33.3	s	Formic acid	24.7	s
Benzene	18.8	p	Furan	19.2	m
Bromobenzene	20.2	p	Heptane	15.1	p
1,3-Butadiene	14.5	p	Hexane	14.9	p
Butane	13.9	p	1-Hexene	15.1	p
1,3-Butanediol	23.7	s	Hydrazine	37.0	s
1-Butanol	23.3	s	Hydrogen	6.1	p
2-Butanol	22.1	s	Isobutanol	21.5	s
tert-Butanol	21.7	s	Isobutyl acetate	17.0	m
Butyl acetate	17.4	m	Isobutylene	13.7	p
Butyl amine	17.8	s	Isoprene	15.1	p
Butyl ether	16.0	m	Isopropanol	23.5	s
Butyl lactate	19.2	m	Isopropyl acetate	17.2	m
Carbon disulfide	20.4	p	Methane	11.0	p
Chloroacetonitrile	25.8	p	Methanol	29.6	s
Chlorobenzene	19.4	p	Methyl acetate	19.6	m
Chloroethane	18.8	m	Methyl acrylate	18.2	m
Chloromethane	19.8	m	Methyl butyl ketone	17.0	m
Cyclohexane	16.8	p	Methyl ethyl ketone	19.0	m
Cyclohexanol	23.3	s	Methyl formate	20.9	m
Cyclopentane	17.8	p	Methyl isopropyl ketone	17.4	m
Decalin	18.0	p	Methyl methacrylate	18.0	m
Decane	13.5	p	Nitrobenzene	20.5	p
Diamyl ether	14.9	m	Nitroethane	22.7	p
Dibenzyl ether	19.2	m	Octane	15.6	p
Dibutyl amine	16.6	s	Pentane	14.3	p
Dibutyl fumarate	18.4	m	Propane	13.1	p
Dibutyl phenyl phosphate	17.8	m	1-Propanol	24.3	s
Dibutyl phthalate	19.0	m	2-Propanol	23.5	s
Diethylamine	16.4	s	Pyridine	21.9	s
Diethlene glycol	24.8	s	Quinoline	22.1	s
Diethyl ether	15.1	m	Silicon tetrachloride	15.1	p
Diisopropyl ether	14.1	m	Styrene	19.0	p
Diisopropyl ketone	16.4	m	Succinic anhydride	31.5	s
N,N-Dimethylformamide	24.8	m	Tetra chloromethane	17.6	p
Dimethyl sulfone	29.7	m	Tetrahydrofuran	18.6	m
Dimethylsulfoxide	24.5	m	Toluene	18.2	p
1,4-Dioxane	20.5	m	1,1,2-Trichloroethane	19.6	p
Ethane	12.3	p	Trichloromethane	19.0	p
Ethanol	26.0	s	Water	47.9	s
Ethyl acetate	18.6	m	Xylene	18.0	p
Ethylamine	20.5	s			
Ethylbenzene	18.0	p			

[b]p denotes poor; m, moderate; s, strong.

TABLE 2.29 Hansen Solubility Parameters of Organic Liquids

Solvent	V (cm³/mol)	Solubility parameter (MPa^{1/2})			
		δ_d	δ_p	δ_h	δ_t
Acetic acid	57.1	14.5	8.0	13.5	21.3
Acetone	74.0	15.5	10.4	7.0	20.1
Acetonitrile	52.6	15.3	18.0	6.1	24.6
Acetyl chloride	71.0	15.8	10.6	3.9	19.4
Benzene	29.4	18.4	0.0	2.0	18.6
Benzaldehyde	101.5	19.4	7.4	5.3	21.5
Benzyl chloride	115.0	18.8	7.2	2.7	20.3
Bromoform	87.5	21.5	4.1	6.1	22.7
N-Butane	101.4	14.1	0.0	0.0	14.1
Butyronitrile	27.0	15.3	12.5	5.1	20.5
Carbon tetrachloride	97.1	17.8	0.0	0.6	17.8
Carbon disulfide	60.0	20.5	0.0	0.6	20.5
Chlorobenzene	102.1	19.0	4.3	2.0	19.6
Chloroform	80.7	17.8	3.1	5.7	19.0
Cyclohexanol	106.0	17.4	4.1	13.5	22.5
Cyclohexylamine	115.2	17.4	3.1	6.5	18.8
N-Decane	195.9	15.8	0.0	0.0	15.8
Diacetone alcohol	124.2	15.8	8.2	4.8	20.9
o-Dichlorobenzene	112.8	19.2	6.3	3.3	20.5
Diethyl carbonate	121.0	16.6	3.1	6.1	18.0
Diethyl ketone	106.4	15.8	7.6	4.7	18.2
Dimethyl phthalate	163.0	18.6	4.8	4.9	22.1
Dimethyl sulfoxide	71.3	18.4	16.4	10.2	26.6
Ethanol	58.5	15.8	8.8	19.4	26.6
Ethyl acetate	98.5	15.8	5.3	7.2	18.2
Ethyl bromide	76.9	16.6	8.0	5.1	19.0
Ethyl formate	80.2	15.5	8.4	8.4	19.6
Ethylene carbonate	66.0	19.4	21.7	5.1	29.5
Ethylene dichloride	79.4	19.0	7.4	4.1	20.9
Formic acid	37.8	14.3	11.9	16.6	25.0
Furan	72.5	17.8	1.8	5.3	18.6
Methanol	40.7	15.1	12.3	22.3	29.7
Methyl acetate	79.7	15.5	7.2	7.6	18.8
Methyl chloride	55.4	15.3	6.1	3.9	17.0
Methylene dichloride	63.9	18.2	6.3	6.1	20.3
Nitrobenzene	102.7	20.1	8.6	4.1	22.1
Nitroethane	71.5	16.0	15.5	4.5	22.7
Nitromethane	54.3	15.8	18.8	5.1	25.0
1-Octanol	157.7	17.0	3.3	11.9	20.9
2-Octanol	159.1	16.2	4.9	11.0	20.3
Phenol	87.5	18.0	5.9	14.9	24.1
1-Propanol	75.2	16.0	6.8	17.4	24.6
2-Propanol	76.8	15.8	6.1	16.4	23.5
Quinoline	118.0	19.4	7.0	7.6	22.1
Styrene	115.6	18.6	1.0	4.1	19.0
Tetrahydrofuran	81.7	16.8	5.7	8.0	19.4
Toluene	106.8	18.0	1.4	2.0	18.2
Trimethyl phosphate	99.9	16.8	16.0	10.2	25.4
Water	18.0	15.5	16.0	42.4	47.9

TABLE 2.30   Group Contributions to the Solubility Parameter

Group	F_i (1)	(2)	(3)	Group	F_i (1)	(2)	(3)
—Br	340	258	300	—C≡N	410	355	480
—Cl	250–270	205	230	$\overset{\text{O}}{\overset{\|}{\text{—C—NH}_2}}$	...	...	600
—F	...	41	80				
—H	80–100	...	...	$\overset{\text{O}}{\overset{\|}{\text{NH}_2\text{—C—O—}}}$	...	...	725
—I	425	...	...	—CO—	275	263	335
—NO$_2$	440	...	...	—COO—	310	327	250
—ONO$_2$	440	...	...	—COOH	...	...	319
—O—	70	115	125	—CO$_3$—	...	...	375
—OH	...	226	369	—C≡C—	222	...	...
—PO$_4$	500	...	...	CH≡C—	285	...	...
—S—	225	209	225	$\overset{\text{O O}}{\overset{\|\;\|}{\text{—C—O—C—}}}$	...	567	375
—SH	315	...	...				
$\diagdown\diagup \atop \underset{\diagup\diagdown}{C}$	–93	32	0	—C=C—C=C—	20–30	23	...
—CH=	19	84	40		105–115	21	...
—CF$_2$—	150	115	...				
—CF$_3$	274	156	...	—C$_6$H$_4$	658	705	673
$\underset{\diagdown}{\overset{\diagup}{\text{— CH}}}$	28	86	68	—C$_6$H$_5$	735	683	741
—CH=	111	122	109				
—CH$_2$—	133	131	137				
CH$_2\diagdown$	190	127	...		95–105	–23	...
—CH$_3$	214	148	205	—C$_{10}$H$_7$	1146	...	...

^{a}Adapted from D. W. Van Krevelen, *Properties of Polymers*, 2nd ed. (Elsevier, Amsterdam, 1976), p. 134. The references referred to for the F_i values are (1) P. A. Small, J. Appl. Chem. **3**, 71 (1953); (2) K. L. Hoy, J. Paint Technol. **42**, 76 (1970); (3) D. W. Van Krevelen, *Properties of Polymers*, 2nd ed. (Elsevier, Amsterdam, 1976), p. 134.

2.3   VISCOSITY AND SURFACE TENSION

The *dynamic viscosity*, or coefficient of viscosity, η of a Newtonian fluid is defined as the force per unit area necessary to maintain a unit velocity gradient at right angles to the direction of flow between two parallel planes a unit distance apart. The SI unit is pascal-second or netwon-second per meter squared [N · s · m^{-2}]. The c.g.s. unit of viscosity is the poise [P]; 1 cP≡1 mN · s · m^{-2}.

Kinematic viscosity v is the ratio of the dynamic viscosity to the density of a fluid. The SI unit is meter squared per second [m^2 · s^{-1}]. The c.g.s. units are called stokes [cm^2 · s^{-1}]; poises = stokes × density.

Fluidity ϕ is the reciprocal of the dynamic viscosity.

The primary reference liquid for viscosity measurements is water. The absolute viscosity of water at 20°C is 1.0019 (±0.0003) mN · s · m^{-2} (or centipoise), as determined by Swindells, Coe, and Godfrey, *J. Research Natl. Bur. Standards* **48**:1 (1952). The relative viscosity of water, $\eta/\eta_{20°}$, is 0.8885 at 25°C, 0.7960 at 30°C, and 0.6518 at 40°C. Values at temperatures between 15 and 60°C are best represented by Cragoe's equation:

$$\log\frac{\eta}{\eta_{20°}}=\frac{1.2348(20-t)-0.001\,467(t-20)^2}{t+96}$$

The *Reynolds number* for flow in a tube is defined by $d\bar{v}\rho/\eta$, where d is the diameter of the tube, $\bar{v}$ is the average velocity of the fluid along the tube, ρ is the density of the fluid, and η is its dynamic viscosity. At flow velocities corresponding with values of the Reynolds number of greater than 2000, turbulence is encountered.

The surface tension of a liquid, γ, is the force per unit length on the surface that opposes the expansion of the surface area. In the literature the surface tensions are expressed in dyn · cm^{-1}; 1 dyn · cm^{-1} = 1 mN · m^{-1} in the SI system. For the large majority of compounds the dependence of the surface tension on the temperature can be given as

$$\gamma = a - bt$$

where a and b are constants and t is the temperature in degrees Celsius. The values of a and b given in Tables 2.31 can be used to calculate the values of surface tension for the particular compound within its liquid range. For example, the least-squares constants for acetic anhydride (liquid from −73 to 140°C) are 35.52 and 0.1436, respectively. At 20°C, $\gamma = 35.52 - 0.1436(20) = 32.64$ dyn · cm^{-1}.

TABLE 2.31   Viscosity and Surface Tension of Organic Compounds

For the majority of substances the dependence of the surface tension *g* on the temperature can be given as:

$$\gamma = a - bt$$

where *a* and *b* are constants and *t* is the temperature in degrees Celsius. In the SI system the surface tensions are expressed in $mN \cdot m^{-1}$ (= $dyn \cdot cm^{-1}$).

A compilation of some 2200 liquid compounds has been prepared by J. J. Jasper, *J. Phys. Chem. Reference Data* **1**:841 (1972).

The SI unit of viscosity is pascal-second (Pa · s) or Newton-second per meter squared ($N \cdot s \cdot m^{-2}$). Values tabulated are $mN \cdot s \cdot m^{-2}$ (= centipoise, cP). The temperature in degrees Celsius at which the viscosity of a substance was measured is shown in parentheses after the value.

| Substance | Surface tension, $mN \cdot m^{-1}$ | | Liquid range, °C | Viscosity, $mN \cdot s \cdot m^{-2}$ |
	a	*b*		
Acetaldehyde	23.90	0.1360	−123 to 21	0.2797(0), 0.2557(10), 0.22(20)
Acetaldoxime	34.23	0.1134	12(β) or 46.5(α) to 114.5	
Acetamide	47.66	0.1021	81 to 222	1.63(94), 1.32(105), 1.06(120)
Acetanilide	46.21	0.0912	114 to 304	2.22(120), 1.90(130)
Acetic acid	29.58	0.0994	16.7 to 118	1.056(25), 0.786(50), 0.424(110)
Acetic anhydride	35.52	0.1436	−73 to 139	1.241(0), 0.907(20), 0.699(40)
Acetone	26.26	0.112	−94 to 56	0.395(0), 0.306(25), 0.256(50)
Acetonitrile	29.58	0.1178	−44 to 81.6	0.397(10), 0.329(30), 0.2753(50)
Acetophenone	41.92	0.1154	20 to 202	1.511(30), 1.192(45), 0.634(100)
Acetyl chloride	26.7(15)		−113 to 51	0.368(25), 0.294(50)
Acrylic acid	28.1(30)		14 to 141	
Acrylonitrile	29.58	0.1178	−83.5 to 77.3	
Allyl acetate	28.73	0.1186	up to 104	
Allyl alcohol	27.53	0.0902	−129 to 97	1.218(25), 0.759(50), 0.553(70)
Allylamine	27.49	0.1287	− 88 to 55	
Allyl isothiocyanate	36.76	0.1074	−80 to 152	
2-Aminoethanol	51.11	0.1117	10.3 to 171	
Aniline	44.83	0.1085	−6 to 186	3.847(25), 2.029(50), 1.247(75)
Benzaldehyde	40.72	0.1090	−26 to 179	
Benzamide	47.26	0.0705	129 to 290	
Benzene	28.88(20)	27.56(30)	5.5 to 80	0.649(20), 0.566(30), 0.395(60)
Benzenesulfonyl chloride	45.48	0.1117	14.5 to 251	
Benzenethiol	41.41	0.1202	−14.9 to 169	
Benzonitrile	41.69	0.1159	−12.7 to 191	1.447(15), 1.111(30), 0.883(50)
Benzophenone	46.31	0.1128	48 to 305	
Benzoyl bromide	45.85	0.1397	−24 to 219	
Benzoyl chloride	41.34	0.1084	−1 to 197	
Benzyl alcohol	38.25	0.1381	−15.2 to 205	5.474(25), 2.760(50), 1.618(75)
Benzylamine	42.33	0.1213	10 to 180	1.624(25), 1.080(50), 0.769(75)
Benzyl benzoate	48.07	0.1065	21 to 323	8.454(25)
Benzyl chloride	39.92	0.1227	−43 to 179	

TABLE 2.31   Viscosity and Surface Tension of Organic Compounds (*Continued*)

Substance	Surface tension, mN · m⁻¹		Liquid range, °C	Viscosity, mN · s · m⁻²
	a	b		
Benzyl ethyl ether	32.82(20)	29.97(40)	up to 186	
Biphenyl	41.52	0.0931	69 to 256	
Bis(2-ethoxyethyl) ether	29.74	0.1176	−45 to 188	
Bis(2-hydroxyethyl) ether	46.97	0.0880	−10.4 to 246	
Bis(2-methoxyethyl) ether	32.47	0.1164	−68 to 162	
Bromobenzene	38.14	0.1160	−30.6 to 156	1.196(15), 0.985(30), 0.385(1423)
1-Bromobutane	28.71	0.1126	−112.4 to 101.6	0.633(20), 0.606(25), 0.471(50)
(±)-2-Bromobutane	27.48	0.1107	−112.7 to 91.4	
Bromochloromethane	33.32(20)		−88 to 68	
Bromocyclohexane	36.13	0.1117	up to 165.8	
1-Bromodecane	31.26	0.0856	−30 to 240	
Bromodichloromethane	35.11	0.1294	−55 to 87	
1-Bromododecane	32.58	0.0882	− 11 to bp	
Bromoethane	26.52	0.1159	−119 to 38.2	0.477(10), 0.374(25)
Bromoform	48.14	0.1308	8 to 149	
1-Bromoheptane	30.74	0.0982	−58 to 180	
1-Bromohexadecane	33.37	0.0861	17.8 to 336	
1-Bromohexane	29.81	0.0967	−85 to 158	
Bromomethane	26.52	0.1159	−94 to 3.56	
1-Bromo-3-methylbutane	28.10	0.0996	−112 to 119.7	
1-Bromo-2-methylpropane	26.96	0.1059	−119 to 91.5	
1-Bromonaphthalene	46.44	0.1018	−1.8 to 281	
1-Bromononane	31.36	0.0894	ca. −55 to 201	
1-Bromooctane	31.00	0.0928	−55 to 201	
1-Bromopentane	29.51	0.1049	−88 to 129.6	
p-Bromophenol	48.88	0.1070	64 to 238	
1-Bromopropane	28.30	0.1218	−110.1 to 71	0.539(15), 0.459(30), 0.338(70)
2-Bromopropane	26.21	0.1183	−89 to 59.5	0.536(15), 0.437(30), 0.359(50)
3-Bromopropene	29.45	0.1257	−119 to 70	0.620(0), 0.471(25), 0.373(50)
1-Bromotetradecane	32.93	0.0878	6 to >178	
o-Bromotoluene	36.62	0.0998	−26 to 181	
p-Bromotoluene	36.40	0.0997	28.5 to 184	
1-Bromoundecane	31.94	0.0861	−9 to >138	
Butanal	26.67	0.0925	−99 to 74.8	
Butane	14.87	0.1206	−138.3 to −0.5	
1,3-Butanediol	37.8(25)		<− 50 to 207.5	
2,3-Butanediol	36(25)		25 to 182	
Butanenitrile			−112 to 117.6	0.553(25), 0.418(50), 0.330(75)
Butanesulfonyl chloride	37.33	0.0977		
1-Butanethiol	28.07	0.1142	−116 to 98.5	
Butanoic acid	28.35	0.0920	−6 to 163.5	1.540(20), 0.980(40), 0.323(60)
Butanoic anhydride	28.93(20)	28.44(25)	−66 to 199.5	
1-Butanol	27.18	0.0898	−89.5 to 117.7	5.185(0), 2.948(20), 1.782(40)
(±)-2-Butanol	23.47(20)	22.62(30)	−114.7 to 99.5	3.907(20), 1.332(50), 0.698(75)
2-Butanone	26.77	0.1122	−86.7 to 79.6	0.428(20), 0.349(40), 0.249(75)
1-Butene	15.19	0.1323	−185 to −6.5	
2-Butene	16.11	0.1289	−106 to 0.9	
3-Butenenitrile	31.40	0.1085	−87 to 119	
2-Butoxyethanol	28.18	0.0816	−75 to 168	

(*Continued*)

TABLE 2.31 Viscosity and Surface Tension of Organic Compounds (*Continued*)

Substance	Surface tension, mN · m^{-1}		Liquid range, °C	Viscosity, mN · s · m^{-2}
	a	*b*		
2-(2-Butoxyethoxy)ethanol	30.0(25)		−68.1 to 230.4	
Butyl acetate	27.55	0.1068	−77 to 126	0.734(20), 0.688(25), 0.500(50)
(±)-*sec*-Butyl acetate	23.33(22)	21.24(42)	−99 to 112	0.676(25), 0.493(50), 0.370(75)
tert-Butyl acetate	24.69	0.1102	up to 98	
Butylamine	26.24	0.1122	−50 to 77	0.830(0), 0.574(25), 0.409(50)
sec-Butylamine	23.75	0.1057	−104 to 63	0.770(0), 0.571(25), 0.367(50)
tert-Butylamine	19.44	0.1028	−66 to 44	
Butylbenzene	31.28	0.1025	−88 to 183	1.035(20), 0.683(50), 0.515(75)
sec-Butylbenzene	30.48	0.0979	−82.7 to 173	
tert-Butylbenzene	30.10	0.0985	−58.1 to 168.5	
Butyl butanoate	27.65	0.0965	−91.5 to 166	
Butyl ethyl ether	22.75	0.1049	−124 to 92	
Butyl formate	27.08	0.1026	−91.5 to 106	0.940(0), 0.691(20), 0.472(50)
Butyl methyl ether	22.17	0.1057	−115.5 to 70	
Butyl nitrate	30.35	0.1126	up to 133	
Butyl propanoate	27.37	0.0993	−89 to 146.8	
4-*tert*-Butylpyridine	35.48	0.0951	ca. −44 to 197	
Butyl stearate	33.0(25)	32.7(30)	26 to 343	
Butyl vinyl ether	21.99(20)		−92 to 94.2	
Carbon disulfide	35.29	0.1484	−111.6 to 46.5	0.429(0), 0.363(20), 0.352(25)
Carbon tetrachloride	29.49	0.1224	−23 to 76.7	1.321(0), 0.908(25), 0.656(50)
D-(+)-Carvone	36.54	0.0920	<15 to 230	
Chloroacetic acid	43.27	0.1117	61 to 189	3.15(50), 1.92(75)
o-Chloroaniline	43.41	0.0904	−14 to 208.8	3.316(25), 1.913(50), 1.248(75)
p-Chloroaniline	48.69	0.1099	72.5 to 232	
Chlorobenzene	35.97	0.1191	−45.3 to 131.7	0.799(20), 0.631(40), 0.512(60)
1-Chlorobutane	25.97	0.1117	−123.1 to 78.4	0.556(0), 0.422(25), 0.329(50)
2-Chlorobutane	24.40	0.1118	−131.3 to 68.2	0.439(15)
Chlorocyclohexane	33.90	0.1101	−44 to 142	
1-Chlorododecane	31.56	0.0904	−9 to 116	
1-Chloro-2,3-epoxy propane	39.76	0.1360	−57.2 to 116.1	1.03(25)
Chloroethane	21.18(5)	20.58(10)	−139 to 12.3	0.416(−25), 0.319(0), 0.279(10)
2-Chloroethanol	38.9(20)		−67.5 to 128.6	3.913(15)
Chloroform	29.91	0.1295	− 63.6 to 61.1	0.706(0), 0.596(15), 0.514(30)
1-Chloroheptane	28.94	0.0961	−69 to 161	
1-Chlorohexane	28.32	0.1038		
1-Chloro-3-methylbutane	25.51	0.1076	−104 to 99	
1-Chloro-2-methylpropane	24.40	0.1099	−130.3 to 68.9	0.462(20), 0.373(40)
2-Chloro-2-methylpropane	20.06(15)	18.35(30)	−26 to 50.8	0.543(15)
1-Chloronaphthalene	44.12	0.1035	−2.3 to 259	2.940(25)
o-Chloronitrobenzene	48.10	0.1171	33 to 246	
m-Chloronitrobenzene	49.71	0.1417	44 to 236	
p-Chloronitrobenzene	45.84	0.1046	84 to 242	
1-Chlorooctane	29.64	0.0961	−58 to 182	
1-Chloropentane	27.09	0.1076	−99 to 108	0.580(20)
o-Chlorophenol	42.5	0.1122	9.8 to 175	3.589(25), 1.835(50), 1.131(75)
m-Chlorophenol	43.7	0.1009	33 to 214	11.55(25), 4.725(45), 4.041(50)
p-Chlorophenol	46.0	0.1049	43 to 220	4.99(50)
1-Chloropropane	24.41	0.1246	−122.8 to 47	0.436(0), 0.372(15), 0.318(30)

TABLE 2.31   Viscosity and Surface Tension of Organic Compounds (*Continued*)

Substance	Surface tension, mN · m⁻¹		Liquid range, °C	Viscosity, mN · s · m⁻²
	a	b		
2-Chloropropane	21.37	0.0883	−117 to 36	0.401(0), 0.335(15), 0.299(30)
3-Chloro-1-propene	25.50	0.0946	−134.5 to 45	0.347(15)
o-Chlorotoluene			−35.6 to 159	1.267(25), 0.883(50), 0.662(75)
m-Chlorotoluene			−47.8 to 161.8	0.964(25), 0.710(50), 0.547(75)
p-Chlorotoluene	34.93	0.1082	7.5 to 162.4	0.837(25), 0.621(50), 0.483(75)
Chlorotrimethylsilane	19.51	0.0875	−40 to 57	
o-Cresol	39.43	0.1011	30 to 191	3.035(50), 1.562(75), 0.961(100)
m-Cresol	38.00	0.0924	12 to 202	12.9(25), 4.417(50), 2.093(75)
p-Cresol	38.58	0.0962	34.8 to 202	5.607(45)
Cycloheptanol	35.02	0.0923	2 to 185	
Cyclohexane	27.62	0.1188	6.6 to 80.7	0.980(20), 0.912(25), 0.650(50)
Cyclohexanol	35.33	0.0966	25.4 to 161	57.5(25), 41.07(30), 12.3(50)
Cyclohexanone	37.67	0.1242	−31 to 155.7	2.453(15), 1.803(30), 1.321(50)
Cyclohexene	29.23	0.1223	−103.5 to 83	0.882(0), 0.625(25), 0.467(50)
Cyclohexylamine	34.19	0.1188	−18 to 134	1.079(25), 0.692(50), 0.485(75)
Cyclooctane	32.02	0.1090	14.8 to 151.1	
Cyclopentane	25.53	0.1462	−94 to 50	0.555(0), 0.413(25), 0.321(50)
Cyclopentanol	35.04	0.1011	−19 to 140	0.439(20)
Cyclopentanone	35.55	0.1100	−51 to 130.6	
Cyclopentene	25.94	0.1495	−135.1 to 44.2	
cis-Decahydronaphthalene	32.18(20)	31.01(30)	−43 to 195.8	3.042(25), 1.875(50), 1.271(75)
trans-Decahydronaphthalene	29.89(20)	28.87(30)	−30.4 to 187.3	1.948(25), 1.289(50), 0.917(75)
Decamethylcyclopentasiloxane	19.56	0.0565	−38 to >101	
Decamethyltetrasiloxane	86.20(25)		−68 to 194	1.28(20)
Decane	25.67	0.0920	−29.7 to 174.1	1.277(0), 0.838(25), 0.598(50)
1-Decanol	30.34	0.0732	6.9 to 232	10.9(25), 4.590(50)
1-Decene	25.84	0.0919	−66 to 170.6	0.805(20)
Dibenzylamine	43.27	0.1086	−26 to 300	
Dibenzyl ether	38.2(35)		2 to 298	3.711(25)
p-Dibromobenzene	41.84	0.1007	87.3 to 220	
1,4-Dibromobutane	48.24	0.1190	−20 to 198	
1,2-Dibromoethane	42.85	0.1320	10 to 131.7	1.721(20), 1.286(40), 0.648(100)
1,2-Dibromopropane	36.81	0.1155	55.5 to 142	1.5(25)
Dibromotetrafluoroethane	18.9(20)	18.1(25)	−110.5 to 47	0.72(25)
Dibutylamine	26.50	0.0952	−62 to 159.6	0.918(25), 0.619(50), 0.449(75)
Dibutyl decanedioate			−10 to 345	9.03(25)
Dibutyl ether	24.78	0.0934	−95 to 140	0.637(25), 0.466(50), 0.356(75)
Dibutyl maleate	32.46	0.0865	<−80 to 281	5.62(20), 4.76(25)
Dibutyl *o*-phthalate	33.40(20)		−35 to 340	19.91(20), 11.17(35), 7.85(45)
Dichloroacetic acid	37.8	0.0927	9 to 194	3.23(50), 1.92(75)
o-Dichlorobenzene	35.55(30)		−17 to 180.4	1.324(25), 0.962(50), 0.739(75)
m-Dichlorobenzene	38.30	0.1147	−24.8 to 173.1	1.044(25), 0.783(50), 0.628(75)
p-Dichlorobenzene	34.66	0.0879	53 to 174.1	0.839(55), 0.668(79)
1,4-Dichlorobutane	37.79	0.1174	−38 to 163	
1,1-Dichloroethane	27.03	0.1186	−97 to 57.3	0.505(15), 0.464(25), 0.362(50)
1,2-Dichloroethane	35.43	0.1428	−35.7 to 83.5	1.125(0), 0.779(25), 0.576(50)
1,1-Dichloroethylene			−122.6 to 31.6	0.442(0), 0.358(20)
cis-1,2-Dichloroethylene	28(20)		−80.1 to 60	0.785(−25), 0.575(0), 0.444(25)
trans-1,2-Dichloroethylene	25(20)		−49.8 to 48.7	0.522(−25), 0.398(0), 0.317(25)
2,2′-Dichloroethyl ether	40.57	0.1306	up to 178.5	2.41(20), 2.065(25)

(*Continued*)

TABLE 2.31   Viscosity and Surface Tension of Organic Compounds (*Continued*)

Substance	Surface tension, mN · m⁻¹		Liquid range, °C	Viscosity, mN · s · m⁻²
	a	*b*		
Dichloromethane	30.41	0.1284	−95 to 40	0.533(0), 0.449(15), 0.393(30)
2,4-Dichlorophenol	46.59	0.1221	42 to 210	
1,2-Dichloropropane	31.42	0.1240	−100 to 96	0.865(20), 0.700(25)
1,3-Dichloropropane	36.40	0.1233	−99.5 to 122	
2,2-Dichloropropane	23.60(20)	22.53(30)	−35 to 69	0.769(15), 0.619(30)
α,α-Dichlorotoluene	41.26	0.1035	−16 to 205	
Diethanolamine			28 to 269	368(30), 109.5(50), 28.7(75)
1,1-Diethoxyethane	23.46	0.1030	−100 to 102.2	
1,2-Diethoxyethane			−74 to 121.4	0.65(20)
Dimethoxymethane	23.87	0.1291	up to 88	
Diethylamine	22.71	0.1143	− 50 to 55.5	
N,N-Diethylaniline	36.59	0.1040	−38 to 217	3.838(0), 1.15(50), 0.750(75)
Diethyl carbonate	28.62	0.1100	−43 to 126	0.868(15), 0.748(25)
Diethyl decanedioate	34.68	0.0959		
Diethyl ether	18.92	0.0908	−116 to 34.6	0.283(0), 0.224(25)
Diethyl ethyl phosphonate	30.63	0.0975	up to 198	1.627(15), 0.969(45), 0.743(65)
Di(2-ethylhexyl) *o*-phthalate			−50 to 384	33.67(35), 21.40(45)
Diethyl maleate	34.67	0.1039	−8.8 to 225.3	3.57(20), 3.14(25)
Diethyl 1,3-propanedioate (malonate)	33.91	0.1042	−49.9 to 199.3	2.15(20), 1.94(25)
Diethyl oxalate	34.32	0.1119	−40.6 to 185.4	2.311(15), 1.618(30)
Diethyl *o*-phthalate	38.47	0.0963	−40 to 295	9.18(35), 6.41(45)
Diethyl succinate	33.97	0.1041	−21 to 217.7	
Diethyl sulfate	35.47	0.0976	−25 to 208	
Diethyl sulfide	27.33	0.1106	−104 to 92.1	0.558(0), 0.422(25)
1,2-Dihydroxybenzene	47.6	0.0849	104 to 245.5	
1,3-Dihydroxybenzene	54.8	0.0717	110 to 276	
Diiodomethane	70.21	0.1613	6 to 181	
Diisobutylamine	24.00	0.0912	−77 to 139	
Diisopentyl ether	24.76	0.0871	up to 172.5	1.40(11), 1.012(20)
Diisopropylamine	21.03	0.1077	−61 to 83.5	0.393(25), 0.300(50), 0.237(75)
Diisopropyl ether	19.89	0.1048	−87 to 68	0.379(25)
1,2-Dimethoxybenzene	34.4	0.0642	22.5 to 206	3.281(25), 2.184(40)
1,1-Dimethoxyethane	23.90	0.1159	−113 to 64.5	
1,2-Dimethoxyethane	48.0(25)		− 68 to 85	0.670(− 10), 0.530(10), 0.455(25)
Dimethoxymethane	23.59	0.1199	−104.8 to 42	0.340(15), 0.325(20)
N,N-Dimethylacetamide	32.40(30)	29.50(50)	−20 to 165.5	1.956(25), 1.279(50), 0.896(75)
Dimethylamine	29.50	0.1265	−92 to 6.9	0.300(− 25), 0.232(0)
N,N-Dimethylaniline	38.14	0.1049	2.5 to 194	1.300(25), 0.911(50), 0.675(75)
2,4-Dimethylaniline	39.34	0.0996	−14 to 214	
2,2-Dimethylbutane	18.29	0.0990	−100 to 49.7	0.351(25), 0.330(30)
2,3-Dimethylbutane	19.38	0.1000	−128 to 58	0.361(25), 0.342(30)
2,3-Dimethyl-1-butanol	26.22	0.0992	−14 to 118	
Dimethyl carbonate	31.94	0.1343	0.5 to 91	
1,1-Dimethylcyclopentane	23.78	0.1016	−70 to 87.5	
Dimethyl ether	14.97	0.1478	−141 to −24.9	
N,N-Dimethylformamide	36.76(20)	34.40(40)	−60 to 153	1.176(0), 0.794(25), 0.624(50)
2,4-Dimethylheptane	23.21	0.0929	<− 100 to 133	
2,5-Dimethylheptane	23.21	0.0929	<− 100 to 136	

TABLE 2.31 Viscosity and Surface Tension of Organic Compounds (*Continued*)

Substance	Surface tension, mN · m⁻¹		Liquid range, °C	Viscosity, mN · s · m⁻²
	a	b		
2,6-Dimethylheptane	22.17	0.0887	−103 to 135	
Dimethyl hexanedioate	38.26	0.1138	8 to >112	14(20)
Dimethyl maleate	40.73	0.1220	−19 to 202	3.54(20), 3.21(25)
Dimethyl malonate	39.72	0.1208	−62 to 181	
2,2-Dimethylpentane	19.94	0.0957	−124 to 79	
2,3-Dimethylpentane	21.96	0.0995	up to 90	0.406(20)
2,4-Dimethylpentane	20.09	0.0972	−120 to 80.4	0.361(20)
3,3-Dimethylpentane	21.59	0.0996	−135 to 86	
2,4-Dimethylphenol	34.57	0.0869	24.5 to 211	
2,5-Dimethylphenol	36.72	0.0850	74.5 to 211.5	1.55(80)
3,4-Dimethylphenol	35.75	0.0910	61 to 227	3.00(80)
3,5-Dimethylphenol	34.09	0.0807	64 to 222	2.42(80)
Dimethyl *o*-phthalate			5.5 to 284	14.4(25), 5.309(50), 2.824(75)
2,2-Dimethylpropane	12.05(20)	10.98(30)	−16.6 to 9.5	0.328(0), 0.303(5)
Dimethyl succinate	39.00	0.1191	19 to 196.4	
Dimethyl sulfate	41.26	0.1163	−31.8 to 188	
Dimethyl sulfide	26.07	0.0805	−98 to 37	0.356(0), 0.289(20), 0.265(36)
Dimethyl sulfite	36.48	0.1253	up to 127	0.715(30), 0.436(80)
Dimethyl sulfoxide	43.54(20)	42.41(30)	18.5 to 189	2.47(20), 1.192(55), 0.849(80)
1,4-Dioxane	36.23	0.1391	11.8 to 101.2	1.439(15), 1.087(30), 0.787(50)
Dipentyl ether	26.66	0.0925	− 69 to 190	1.188(15), 0.922(30)
Dipentyl *o*-phthalate	32.56	0.0739		17.03(35), 11.51(45)
Dipentyl sulfide	29.55	0.0876		
Dipentylamine	45.36	0.1017	53 to 302	4.66 (55), 1.04(130)
Diphenyl ether	28.70	0.0780	27 to 258	2.130(50), 1.407(75), 1.023(100)
1,2-Dipropoxyethane	25.03	0.0972		
Dipropoxymethane	25.17	0.0953		
Dipropylamine	24.86	0.1022	−63 to 109	0.517(25), 0.377(50), 0.288(75)
Dipropyl carbonate	28.94	0.1015	up to 168	
Dipropylene glycol butyl ether	28.2(25)		up to >103	4.23(25)
Dipropylene glycol ethyl ether	27.7(25)			3.11(25)
Dipropylene glycol isopropyl ether	25.9(25)		up to 80	386(25)
Dipropylene glycol methyl ether	28.8(25)		−117 to 188	3.1(25)
Dipropyl ether	22.60	0.1047	−126 to 89.6	0.542(0), 0.396(25), 0.304(50)
Dodecane	27.12	0.0884	−10 to 216	2.277(0), 1.378(25), 0.930(50)
1-Dodecanol	31.25	0.0748	24 to 259	
Epichlorohydrin	39.76	0.1360	−26 to 117	1.20(25)
1,2-Epoxybutane	23.9(20)		−150 to 63	0.419(15), 0.358(30)
1,2-Ethanediamine	44.77	0.1398	11 to 117.3	1.54(20), 1.226(30)
1,2-Ethanediol	50.21	0.0890	−12.6 to 197.3	26.09(15), 13.55(30)
Ethanesulfonic acid	45.74	0.0824	− 17 to >123	
Ethanesulfonyl chloride	43.43	0.1177	up to 177	
Ethanethiol	25.06	0.0793	−148 to 35	0.364(0), 0.287(25)
Ethanol	24.05	0.0832	−114 to 78	1.786(0), 1.074(25), 0.694(50)
Ethanolamine	51.11	0.1117	10.5 to 171	21.1(25), 8.560(50), 3.935(75)
Ethoxybenzene (phenetol)	35.17	0.1104	−29.5 to 170	1.364(15), 1.197(25), 0.817(50)
2-Ethoxyethanol	30.59	0.0897	−70 to 135	2.04(20), 1.85(25)
Ethyl acetate	26.29	0.1161	−84 to 77	0.578(0), 0.423(25), 0.325(50)

(*Continued*)

TABLE 2.31 Viscosity and Surface Tension of Organic Compounds (*Continued*)

Substance	Surface tension, mN · m⁻¹		Liquid range, °C	Viscosity, mN · s · m⁻²
	a	b		
Ethyl acetoacetate	34.42	0.1015	−45 to 181	1.419(20), 1.508(25)
Ethylamine	22.63	0.1372	−81 to 16.6	
N-Ethylaniline	39.00	0.1070	−63.5 to 203	2.047(25), 1.231(50), 0.825(75)
Ethylbenzene	31.48	0.1094	−95 to 136	1360.631(25), 0.482(50), 0.380(75)
Ethyl benzoate	37.16	0.1059	−35 to 212	2.407(15), 1.751(30)
Ethyl butanoate	26.55	0.1045	−98 to 121	0.771(15), 0.613(25)
2-Ethylbutanoic acid	26.3(20)		−14 to 194	3.3(20)
2-Ethyl-1-butanol	25.06(15)	24.32(25)	<− 15 to 146	8.021(15), 5.892(25)
Ethyl carbamate			50 to 184	0.916(105), 0.715(120)
Ethyl chloroacetate	34.18	0.1177	−21 to 144	
Ethyl chloroformate	28.90	0.1084	−81 to 93	
Ethyl *trans*-cinnamate	39.99	0.1045	10 to 271	8.7(20)
Ethyl crotonate	29.31	0.1066	up to 138	
Ethyl cyanoacetate	38.80	0.1092	− 22 to 206	3.256(15), 2.148(30)
Ethylcyclohexane	27.78	0.1054	−111 to 132	1.139(0), 0.784(25), 0.579(50)
Ethyl dichloroacetate	34.89	0.1158	up to 155	
Ethyl dodecanoate	30.05	0.0863	−10 to 271	
Ethylene carbonate			36 to 248	1.85(40)
Ethylenediamine	44.77	0.1398	11 to 117	1.540(18)
Ethylene glycol	50.21	0.0890	up to 198	26.09(15), 13.35(30), 6.554(50)
Ethyleneimine	7.9(20)		−78 to 56	0.418(25)
Ethylene oxide	27.66	0.1664	−111 to 10.6	0.3(0)
Ethyl formate	26.47	0.1315	−80 to 54	0.419(15), 0.358(30), 0.300(50)
Ethyl fumarate	33.90	0.1056	68 to >148	
Ethylhexadecanoate	32.86	0.0859	22 to >191	
Ethyl hexanoate	27.73	0.0960	up to 168	
2-Ethyl-1-hexanol	30.0(22)		−70 to 185	6.271(25), 2.631(50), 1.360(75)
Ethyl isobutanoate	25.33	0.1046	−88 to 110	
Ethyl isothiocyanate	38.69	0.1326	−6 to 132	
Ethyl lactate	30.72	0.0983	−26 to 155	2.44(25)
Ethyl 3-methylbutanoate	25.79	0.1006	−99 to 135	
Ethyl methyl ether	18.56	0.1317	−113 to 7.4	
Ethyl methyl sulfide	27.63	0.1286	−106 to 67	0.373(20), 0.354(25)
Ethyl nitrate	30.81	0.1345	−95 to 88	
3-Ethylpentane	22.52	0.1032	−119 to 93.5	
Ethyl pentanoate	27.15	0.0999	−91 to 145	0.847(20)
Ethyl propanoate	26.72	0.1168	−74 to 99	0.564(15), 0.473(30), 0.380(50)
Ethyl propyl ether	21.92	0.1054	−79 to 63	0.401(0), 0.323(20), 0.225(60)
Ethyl salicylate	31.00	0.1091	2 to 234	1.772(45)
Ethyl thiocyanate	37.28	0.1226	up to 145	
o-Ethyltoluene	32.33	0.1060	−81 to 165	
p-Ethyltoluene	30.98	0.1075	−62 to 162	
Ethyl trichloroacetate	32.97	0.1073	up to 168	
Fluorobenzene	29.67	0.1204	−42 to 85	0.620(15), 0.517(30), 0.423(50)
1-Fluorohexane	23.41	0.1001	−103 to 93	
1-Fluoropentane	22.81	0.1315	−120 to 63	
o-Fluorotoluene			−62 to 115	0.680(20), 0.601(30)
m-Fluorotoluene	32.31	0.1257	−87 to 115	0.608(20), 0.534(30)
p-Fluorotoluene	30.44	0.1109	−56 to 117	0.622(20), 0.522(30)

TABLE 2.31 Viscosity and Surface Tension of Organic Compounds (*Continued*)

Substance	Surface tension, mN · m⁻¹		Liquid range, °C	Viscosity, mN · s · m⁻²
	a	*b*		
Formamide	59.13	0.0842	2.6 to 220	4.320(15), 2.296(30), 1.833(50)
Formanilide	44.30	0.0875	47 to 271	1.65(120)
Formic acid	39.87	0.1098	8 to 101	1.966(15), 1.607(25), 1.030(50)
Furan	24.10(20)	23.38(25)	−86 to 31	0.380(20), 0.361(25)
2-Furancarboxaldehyde	46.41	0.1327	−36.5 to 162	2.501(0), 1.587(25), 1.143(50)
2-Furanmethanol	ca. 38(20)		−31 to 171	4.62(25)
Glycerol	63.14(17)	62.5(25)	18 to 290	934(25), 152(50), 39.8(75)
Glycerol tris(acetate)	37.88	0.081		
Glycerol tris(nitrate)	55.74	0.2504	13 to >160	36.0(20), 13.6(40)
Glycerol tris(oleate)	36.03	0.0699	−5 to >233	
Glycerol tris(palmitate)	32.26	0.0672	65 to 320	
Glycerol tris(sterate)	32.73	0.0685		
Heptanal	28.64	0.0920	−43 to 153	0.977(15)
Heptane	22.10	0.0980	−91 to 98	0.523(0), 0.416(20), 0.341(40)
Heptanoic acid	29.88	0.0848	− 8 to 222	3.84(25), 2.282(50), 1.488(75)
1-Heptanol			−34 to 176	8.53(15), 5.810(25), 2.603(50)
2-Heptanol			up to 159	3.955(25), 1.799(50), 0.987(75)
3-Heptanol			−70 to 157	1.957(50), 0.976(75), 0.584(100)
4-Heptanol				4.207(25), 1.695(50), 0.882(75)
2-Heptanone	28.76	0.1056	−35 to 151	0.854(15), 0.686(30), 0.407(50)
4-Heptanone	28.11	0.1060	−32 to 143.7	0.736(20)
1-Heptene	22.28	0.0991	−120 to 93.6	0.441(0), 0.340(25), 0.273(50)
Heptylamine	25.96	0.0783	−23 to 156	1.314(25), 0.865(50), 0.600(75)
Hexadecane	29.18	0.0854	18.2 to 286.8	3.032(25), 1.879(50), 1.260(75)
1,5-Hexadiene	20.93	0.1028	−140.7 to 59.5	0.275(20), 0.244(36)
Hexafluorobenzene	22.6(20)		5.1 to 80.3	2.789(25), 1.730(50), 1.151(75)
Hexamethyldisiloxane	17.01	0.0763	−67 to 101	
Hexamethylphosphoramide	33.8(20)		7 to 232	3.47(20)
Hexane	20.44	0.1022	−95.4 to 68.7	0.405(0), 0.313(20), 0.271(40)
Hexanenitrile	29.64	0.0907	−80 to 163.6	1.041(15), 0.830(30), 0.650(50)
Hexanoic acid	28.05(20)	27.55(25)	−3 to 205	3.525(15), 2.511(30)
1-Hexanol	27.81	0.0801	−44.6 to 157.5	6.203(15), 3.872(30), 2.271(50)
2-Hexanone	28.18	0.1092	−55.5 to 127.6	0.584(25), 0.429(50), 0.329(75)
1-Hexene	20.47	0.1027	−140 to 63.5	0.326(0), 0.252(25), 0.202(50)
Hexyl acetate	28.44	0.0970	−81 to 171	
4-Hydroxy-4-methyl-2-pentanone	31.0(20)		−44 to 168	6.621(0), 2.798(25), 1.829(50)
Iodobenzene	41.52	0.1123	−31 to 188	1.554(25), 1.117(50), 0.854(75)
1-Iodobutane	30.82	0.1031	−103.5 to 131	
2-Iodobutane	30.32	0.1056	−104 to 120	
Iodoethane	31.67	0.1286	−111 to 72.4	0.617(15), 0.540(30), 0.444(50)
1-Iodoheptane	32.18	0.0887	−48 to 204	
1-Iodohexadecane	34.49	0.0880	23 to >207	
1-Iodohexane	31.63	0.0845	up to 180	
Iodomethane	33.42	0.1234	−66.5 to 42.5	0.594(0), 0.500(20), 0.424(40)
1-Iodo-2-methylpropane	30.26	0.1072	−93.5 to 121	0.875(20), 0.697(40)
1-Iodooctane	32.51	0.0915	−46 to 226	
1-Iodopentane	31.41	0.1014	−85 to 155	
1-Iodopropane	31.64	0.1136	−101 to 102.6	0.837(15), 0.670(30), 0.541(50)

(Continued)

TABLE 2.31 Viscosity and Surface Tension of Organic Compounds (*Continued*)

Substance	Surface tension, mN · m^{-1}		Liquid range, °C	Viscosity, mN · s · m^{-2}
	a	*b*		
2-Iodopropane	29.35	0.1107	−90 to 89.5	0.732(15), 0.620(30), 0.506(50)
p-Iodotoluene	39.23	0.0965	up to 211	
α-Ionone	34.10	0.0949	>124	
β-Ionone	35.36	0.0950	>128	
Isobutanenitrile	24.93(20)	23.84(30)	−71.5 to 104	0.551(15), 0.456(30)
Isobutyl acetate	25.59	0.1013	−99 to 116.5	0.676(25), 0.493(50), 0.370(75)
Isobutylamine	24.48	0.1092	−86.6 to 68	0.770(0), 0.571(25), 0.367(50)
Isobutylbenzene	29.39	0.0961	−51.5 to 172.8	
Isobutyl formate	26.14	0.1122	−95.5 to 98.4	0.680(20)
Isobutyl propanoate	30.92	0.1270	−71 to 137	
Isopentyl acetate	26.75	0.0989	−78.5 to 142	0.872(20), 0.790(25)
Isophorone			−8.1 to 215.2	4.201(0), 2.329(25), 1.415(50)
Isopropyl acetate	24.44	0.1072	−73 to 89	0.559(20)
Isopropylamine	19.91	0.0972	−95 to 31.7	0.454(0), 0.325(25)
Isopropylbenzene	30.32	0.1054	−96 to 154	1.075(0), 0.737(25), 0.547(50)
Isopropyl formate	24.56	0.1147		0.512(20)
Lactonitrile	38.31	0.0960	−40 to >103	2.01(30)
D-Limonene	29.50	0.0929	−96.5 to 178	
(±)-Mandelonitrile	45.90	0.0988	−10 to 170	
Methacrylic acid	26.5(25)		16 to 163	1.32(20)
Methacrylonitrile	24.4(20)		−35.8 to 90.3	0.392(20)
Methanesulfonic acid	52.28	0.0893	20 to >167	
Methanethiol	28.09	0.1696	−123 to 6.0	
Methanol	24.00	0.0773	−97.7 to 64.7	0.793(0), 0.676(10), 0.544(25)
o-Methoxybenzaldehyde	45.34	0.1105	37 to 238	
p-Methoxybenzaldehyde	44.69	0.1047	−1 to 248	
Methoxybenzene	38.11	0.1204	−37.5 to 153.8	1.152(15), 1.056(25), 0.747(50)
2-Methoxyethanol	33.30	0.0984	−85.1 to 124	1.71(20), 1.60(25)
2-(2-Methoxyethoxy)ethanol	34.8(25)	29.9(75)	−50 to 194	3.48(25), 1.61(60)
1-Methoxy-2-nitrobenzene	48.62	0.1185	10.5 to 277	
o-Methoxyphenol	41.2	0.0943	28 to 205	
p-Methoxytoluene	36.20	0.1071	up to 174	
N-Methylacetamide	33.67(30)	30.62(50)	30.6 to 206	3.88(30), 2.54(45)
Methyl acetate	27.95	0.1289	−98 to 57	0.477(0), 0.364(25), 0.284(50)
Methyl acetoacetate	34.98	0.0944	27.5 to 171.7	
Methyl acrylate			−76.5 to 80.2	1.398(20)
Methylamine	22.87	0.1488	−93.5 to −6.3	0.319(−25)
N-Methylaniline	39.32	0.0970	−57 to 196	2.042(25), 1.222(50), 0.825(75)
o-Methylaniline				3.823(25), 1.936(50), 1.198(75)
m-Methylaniline				3.306(25), 1.679(50), 1.014(75)
Methyl benzoate	40.10	0.1171	−15 to 199.5	2.298(15), 0.206(20), 1.673(30)
2-Methyl-1,2-butadiene				0.266(0.3), 0.233(20)
2-Methylbutane	17.20	0.1103	up to 30	0.376(−25), 0.277(0), 0.214(25)
Methyl butanoate	27.48	0.1145	−85.8 to 103	0.580(20), 0.459(40), 0.406(50)
3-Methylbutanoic acid	27.28	0.0886	−29.3 to 176.5	2.731(15), 2.411(20)
2-Methyl-1-butanol	21.5(25)		<−70 to 128	5.50(20), 4.453(25), 1.963(50)
2-Methyl-2-butanol	24.18	0.0748	−9.0 to 102.0	5.48(15), 2.81(30)
3-Methyl-1-butanol	25.76	0.0820	−117 to 131	4.81(15), 2.96(30), 1.842(50)
3-Methyl-2-butanol	23.0(25)		up to 112.9	3.51(25)

TABLE 2.31 Viscosity and Surface Tension of Organic Compounds (*Continued*)

Substance	Surface tension, mN · m^{-1}		Liquid range, °C	Viscosity, mN · s · m^{-2}
	a	*b*		
2-Methyl-1-butene	18.81	0.1148	−137.6 to 31	
2-Methyl-2-butene	19.70	0.1271	−133.8 to 38.6	
3-Methyl-1-butene	16.42	0.1031	−168 to 20	
2-Methylbutyl acetate	26.75	0.0989	−99 to 117	0.872(20)
3-Methylbutyronitrile	27.58	0.0827	−101 to 129	
Methyl chloroacetate	37.90	0.1304	−32 to 130	
Methyl cyanoacetate	41.32	0.1074	−22.5 to 201	3.824(50), 3.398(55), 2.687(65)
Methylcyclohexane	26.11	0.1130	−126.6 to 100.9	0.679(25), 0.501(50), 0.390(75)
cis-2-Methylcyclohexanol	32.45	0.0770 (mixed isomers)	7 to 165	18.08(25), 13.60(30)
trans-2-Methylcyclohexanol			−2 to 167.5	37.13(25), 25.14(30)
cis-3-Methylcyclohexanol	29.08	0.0629 (mixed isomers)	−6 to 168	19.7(25), 17.23(30)
trans-3-Methylcyclohexanol	28.80(30)		−0.5 to 167	25.62(16), 15.60(30)
cis-4-Methylcyclohexanol	29.07	0.0690 (mixed isomers)	−9.2 to 173	
2-Methylcyclohexanone	34.06	0.1027	up to 162	
3-Methylcy clohexanone	33.06	0.0925	up to 169	
4-Methylcyclohexanone	32.83	0.0935	up to 171	
Methylcyclopentane	24.63	0.1163	−142.2 to 71.8	0.653(0), 0.478(25), 0.364(50)
Methyl decanoate	30.33	0.0912	−18 to 223	
Methyl dichloroacetate	37.00	0.1219	−52 to 143	
Methyl dodecanoate	31.37	0.0893	4.8 to 262	
N-Methylformamide	37.96(30)	35.02(50)	−4 to 199.5	1.678(25), 1.155(50), 0.824(75)
Methyl formate	28.29	0.1572	−99 to 31.7	0.424(0), 0.360(15), 0.325(25)
Methyl heptanoate	28.95	0.0987	−55.8 to 173.5	
4-Methyl-3-heptanol			−123 to 170	1.085(25), 0.702(50), 0.497(75)
5-Methyl-3-heptanol			−91 to 172	1.178(25), 0.762(50), 0.536(75)
Methyl hexadecanoate (palmitate)	31.50	0.0775	32 to >196	
2-Methylhexane	21.22	0.0966	−118 to 90	0.378(20)
3-Methylhexane	21.73	0.0970	−119 to 92	0.372(20), 0.350(25)
Methyl hexanoate	28.47	0.1045	−71 to 151	
Methyl isobutanoate	25.99	0.1131	−84.7 to 92.5	0.672(0), 0.523(20), 0.419(40)
1-Methyl-4-isopropylbenzene (*p*-cymene)	28.83	0.0877		3.402(20)
Methyl methacrylate	28-29(30)		−48 to 100	0.632(20)
1-Methylnaphthalene	39.96	0.0934	−30.4 to 245	
Methyl octadecanoate	32.20	0.0775	38 to >215	
2-Methyloctane	23.76	0.0940	−80.3 to 143.2	
4-Methyloctane	24.22	0.0940	−113 to 142	
Methyl octanoate	29.93	0.1002	−40 to 192.9	
Methyl oleate	31.3(25)	25.4(100)	−19.9 to >218	4.88(20)
2-Methylpentane	19.37	0.0997	−154 to 60.3	0.372(0), 0.286(25), 0.226(50)
3-Methylpentane	20.26	0.1060	−163 to 63	0.395(0), 0.307(25), 0.292(30)

(*Continued*)

TABLE 2.31   Viscosity and Surface Tension of Organic Compounds (*Continued*)

Substance	Surface tension, mN · m^{-1}		Liquid range, °C	Viscosity, mN · s · m^{-2}
	a	*b*		
4-Methylpentanenitrile	28.89	0.0917	−51.1 to 156.5	0.980(20), 0.843(30)
Methyl pentanoate	27.85	0.1044	up to 128	0.713(20)
2-Methyl-1-pentanol	26.98	0.0819	up to 148	
3-Methyl-1-pentanol	26.92	0.0789	up to 153	
4-Methyl-1-pentanol	25.93	0.0743	up to 152	
2-Methyl-2-pentanol	25.07	0.0861	−103 to 121	
3-Methyl-2-pentanol	27.14	0.0919	up to 134	
4-Methyl-2-pentanol	24.67	0.0821	−90 to 122	4.074(25)
2-Methyl-3-pentanol	26.43	0.0914	up to 126	
3-Methyl-3-pentanol	25.48	0.0888	−23.6 to 123	
4-Methyl-2-pentanone	23.64(20)	19.62(60)	−84 to 116.5	0.585(20), 0.522(30), 0.406(50)
Methyl phenyl sulfide	42.81	0.1238	−15 to 188	
N-Methyl propanamide	31.29(20)	29.12(50)	−43 to >146	6.06(20), 4.58(30), 3.56(40)
2-Methylpropanenitrile			−72 to 108	0.551(15), 0.456(30)
Methyl propanoate	27.58	0.1258	− 88 to 80	0.581(0), 0.431(25), 0.333(50)
2-Methylpropanoic acid	25.55(20)	25.13(25)	−47 to 154	1.857(0), 1.226(25), 0.863(50)
2-Methyl-1-propanol	24.53	0.0795	−108 to 108	4.70(15), 2.876(30)
2-Methyl-2-propanol	20.02(15)	19.10(30)	25.8 to 82.4	1.421(50), 0.678(75)
2-Methylpropene	14.84	0.1319	−140 to −6.9	
1-Methylpropyl acetate	25.72	0.1054		
2-Methyl-1-propylamine	24.48	0.1092	−87 to 68	21.7(25)
2-Methylpropyl formate	26.14	0.1122	−96 to 98	0.680(20)
2-Methylpyridine	36.11	0.1243	−66.7 to 129	0.805(20), 0.710(30)
3-Methylpyridine	37.35	0.1153	−18.3 to 144	
4-Methylpyridine	37.71	0.1141	3.8 to 145	
N-Methyl-2-pyrrolidinone			−24.4 to 202	1.666(25)
Methyl salicylate	42.15	0.1174	−8 to 223	1.102(75), 0.815(100)
Methyl tetradecanoate	31.00	0.0800	18.4 to 323	
2-Methyltetrahydrofuran			<−75 to 78	0.777(−20), 0.601(0), 0.536(10)
Methyl thiocyanate	40.66	0.1305	−5 to 133	64.3(0)
Morpholine	37.63(20)	36.24(30)	−4.9 to 128	2.53(15), 1.79(30), 1.247(50)
Naphthalene			80 to 217.7	0.967(80), 0.780(100)
p-Nitroaniline	60.62	0.0923	147 to 332	
Nitrobenzene	48.62	0.1185	5.8 to 210.8	2.165(15), 1.863(25), 1.262(50)
Nitroethane	35.27	0.1255	−90 to 114	0.940(0), 0.688(25), 0.526(50)
Nitromethane	40.72	0.1678	−28.4 to 101.2	0.692(15), 0.596(30), 0.481(50)
1-Nitro-2-methoxybenzene	48.62	0.1185	95 to 273	
o-Nitrophenol	47.35	0.1174	45 to 216	2.343(45)
1-Nitropropane	32.62	0.1009	−108 to 131.1	0.798(25), 0.589(50), 0.460(75)
2-Nitropropane	32.18	0.1158	−91.3 to 120.3	0.750(25)
o-Nitrotoluene	44.10	0.1174	−10 to 222	2.37(20), 1.63(40)
m-Nitrotoluene	43.54	0.1118	15.5 to 231.9	0.233(20), 1.60(40)
p-Nitrotoluene	42.26	0.0974	52 to 238	1.20(60)
Nonane	24.72	0.0935	−53.5 to 150.8	0.964(0), 0.666(25), 0.488(50)
Nonanoic acid			12.5 to 254.5	7.011(25), 3.712(50), 2.234(75)
1-Nonanol	29.79	0.0789	−5.5 to 215	14.3(20), 9.123(25), 4.032(50)
5-Nonanone	28.72	0.0975	−50 to 187	1.199(25), 0.834(50), 0.619(75)
1-Nonene	24.90	0.0938	−81 to 146	0.620(20), 0.586(25)
Octadecane	29.98	0.0843	28.1 to 316.3	2.487(50), 1.609(75), 1.132(100)

TABLE 2.31   Viscosity and Surface Tension of Organic Compounds (*Continued*)

Substance	Surface tension, mN · m⁻¹		Liquid range, °C	Viscosity, mN · s · m⁻²
	a	b		
Octamethylcyclotetrasiloxane	20.19	0.0811	17 to 176	2.20(20)
Octane	23.52	0.0951	−56.8 to 125.7	0.546(20), 0.433(40), 0.355(60)
Octanenitrile	29.61	0.0802	−45.6 to 205	1.811(15), 1.356(30)
Octanoic acid	29.21(20)	28.7(25)	16.6 to 239	5.020(25), 2.656(50), 1.654(75)
1-Octanol	29.09	0.0795	−15.5 to 195	10.64(15), 6.125(30), 3.232(50)
2-Octanol	27.96	0.0820	−31.6 to 180	
1-Octene	23.68	0.0958	−102 to 121	0.470(20), 0.447(25)
Oleic acid	32.80(20)	27.94(90)	13.4 to 360	38.80(20), 27.64(25)
4-Oxopentanoic acid	41.69	0.0763	33 to 246	
Paraldehyde	28.28	0.1062	12.6 to 124	1.079(25), 0.692(50), 0.485(75)
Parathion	39.2(25)		6 to 375	15.30(25)
Pentachloroethane	37.09	0.1178	−29.9 to 160	2.741(15), 2.070(30), 1.491(50)
Pentadecane	28.78	0.0857	9.9 to 270	2.814(22)
Pentanal	27.96	0.1010	−92 to 103	
Pentane	18.25	0.1121	−129.7 to 36.0	0.351(−25), 0.274(0), 0.224(25)
1,5-Pentanediol	43.2(20)		−18 to 239	128(20)
2,4-Pentanedione	33.28	0.1144	−23.1 to 138	0.6(20)
Pentanenitrile	27.44(20)	26.33(30)	−92 to 141.3	0.779(15), 0.637(30)
Pentanoic acid	28.90	0.0887	−33.7 to 186	2.359(15), 1.774(30), 0.979(70)
1-Pentanol	27.54	0.0874	−79 to 137.5	4.650(15), 3.619(25), 1.820(50)
2-Pentanol	25.96	0.1004	−73 to 119.3	5.130(15), 2.780(30), 1.447(50)
3-Pentanol	24.60(20)	23.76(30)	−69 to 116	7.337(15), 3.306(30), 1.473(50)
2-Pentanone	24.89	0.0655	−76.8 to 102	0.641(0), 0.473(25), 0.362(50)
3-Pentanone	27.36	0.1047	−39.0 to 102	0.592(0), 0.444(25), 0.345(50)
1-Pentene	18.20	0.1099	−165 to 30.1	0.313(−25), 0.241(0), 0.195(25)
cis-2-Pentene	19.71	0.1172	−151 to 37.0	
trans-2-Pentene	18.90	0.0997	−140 to 36.3	
Pentyl acetate	27.66	0.0994	−70.8 to 149.2	0.924(20), 0.862(25)
Pentylamine	24.4(13)		−55 to 104	1.030(0), 0.702(25), 0.493(50)
Phenol	43.54	0.1069	41 to 182	3.437(50), 1.784(75), 1.099(100)
2-Phenylacetamide	46.26	0.0788	157 to bp	
Phenyl acetate			<45 to 196	1.799(45)
Phenylacetonitrile	44.57	0.1155	−23.8 to 233.5	1.93(25)
1-Phenylethanol	42.88	0.1038	20 to 204	
Phenylhydrazine	48.14	0.1292	19.5 to 243	13.0(25), 4.553(50), 1.850(75)
Phenyl isothiocyanate	42.73	0.1086	−30 to 163	
Phenyl salicylate	45.20	0.0976	44 to >173	
(±)-α-Pinene	28.35	0.0944	−64 to 156	1.61(25)
L-β-Pinene	28.26	0.0934	− 61 to 166	1.70(20), 1.41(25)
Piperidine	31.79	0.1153	−11 to 106	1.573(25), 0.958(50), 0.649(75)
1,2-Propanediol (see propylene glycol)				
1,3-Propanediol	47.43	0.0903	−27 to 214	56.0(20), 18.0(40)
Propanenitrile (propionitrile)	29.63	0.1153	−92.8 to 97.2	0.294(25), 0.240(50), 0.202(75)
1-Propanethiol	27.38	0.1272	−113 to 68	0.503(0), 0.385(25)
2-Propanethiol	24.26	0.1174	−131 to 52.6	0.477(0), 0.357(25), 0.280(50)
Propanoic acid	28.68	0.0993	−20.5 to 141.1	1.030(25), 0.749(50), 0.569(75)
Propanoic anhydride	30.30(20)	29.70(25)	−45 to 170	1.144(20), 1.061(25)
1-Propanol	25.26	0.0777	−127 to 97.2	2.522(15), 1.722(30), 1.107(50)

(*Continued*)

TABLE 2.31 Viscosity and Surface Tension of Organic Compounds (*Continued*)

Substance	Surface tension, mN · m^{-1}		Liquid range, °C	Viscosity, mN · s · m^{-2}
	a	*b*		
2-Propanol	22.90	0.0789	−89.5 to 82.4	2.859(15), 1.765(30), 1.028(50)
2-Propen-1-ol (allyl alcohol)	27.53	0.0902	−129 to 98	1.363(20), 0.914(40)
Propionaldehyde (propanal)			−81 to 48	0.357(15), 0.321(25)
Propionamide	39.05	0.0909	79 to 222.2	
Propyl acetate	26.60	0.1120	−93 to 101.6	0.768(0), 0.544(25), 0.406(50)
Propylamine	24.86	0.1243	−83 to 42.2	0.376(25)
Propylbenzene	31.13	0.1075	−99.2 to 159.2	
Propyl benzoate	36.55	0.1069	−51.6 to 98	
Propyl butanoate	27.06	0.1000	−95 to 143	0.831(20)
1,2-Propylene glycol			−60 to 188	40.4(0), 11.3(25), 4.770(50)
Propyleneimine			up to 66	0.491(25)
1,2-Propylene oxide			−112 to 34	0.327(20), 0.28(25)
Propyl formate	26.77	0.1119	−92.9 to 80.9	0.669(0), 0.574(20), 0.417(40)
Propyl isobutanoate	25.83	0.1015	up to 135	0.831(20)
Propyl nitrate	29.67	0.1237	−100 to 110.1	
Propyl pentanoate	27.72	0.0984	−75.9 to 122.5	1.053(20)
Propyl propanoate	26.85	0.1059	−76 to 122.5	0.673(20)
Propyne	14.51	0.1482	−102.8 to −23.2	
2-Propyn-1-ol	38.59	0.1270	−51.8 to 114	1.68(20)
Pyridazine	50.55	0.1036	− 8 to 208	
Pyridine	39.82	0.1306	−41.6 to 115.2	1.361(0), 0.879(25), 0.637(50)
Pyrimidine	32.85	0.1010	22 to 124	
Pyrrole	39.81	0.1100	−23.4 to 130	2.085(0), 1.225(25), 0.828(50)
Pyrrolidine	31.48	0.0900	− 58 to 86.5	1.071(0), 0.704(25), 0.512(50)
2-Pyrrolidone			25 to 251	13.3(25)
Quinoline	45.25	0.1063	−15 to 237	3.337(25), 1.892(50), 1.201(75)
Salicylaldehyde	45.38	0.1242	−7 to 197	2.90(20), 1.71(30), 1.669(45)
Squalane			−38 to 350	6.08(20)
Squalene			−75 to >285	12(25)
Stearic acid			67 to >184	11.6(70)
Styrene	32.0(20)	30.98(30)	−31 to 145	1.050(0), 0.696(25), 0.507(50)
Succinonitrile	53.26	0.1079	54.5 to 266	2.591(60), 2.008(75)
1,1,2,2-Tetrabromoethane	52.37	0.1463	0 to 243.5	13.50(11), 9.797(20)
1,1,2,2-Tetrachlorodifluoro-ethane	26.13	0.1133	26.0 to 92.8	1.21(25), 1.208(30)
1,1,2,2-Tetrachloroethane	38.75	0.1268	−70.2 to 130.5	1.844(15), 1.456(30)
Tetrachloroethylene	32.86(15)	31.27(30)	−22 to 121	1.932(15), 0.798(30), 0.654(53)
Tetradecane	28.30	0.0869	5.5 to 253.6	2.128(25), 1.376(50), 0.953(75)
Tetradecanoic acid	33.90	0.0932	54 to >250	
1-Tetradecanol	32.72	0.0703	39.5 to 289	
Tetraethylene glycol	45(25)		−6 to 328	44.9(25)
Tetraethyl lead	30.50	0.0969	−136 to >85	
Tetraethylsilane	25.22	0.1079	−82 to 154.7	
Tetraethyl silicate	23.63	0.0979	−82.5 to 169	
Tetrahydrofuran	26.5(25)		−108.5 to 65	0.605(0), 0.460(25), 0.359(50)
2,5-Tetrahydrofurandimethanol			<− 50 to 265	225(25)
Tetrahydro-2-furanmethanol	39.96	0.1008	<−80 to 178	6.24(20)
1,2,3,4-Tetrahydronaphthalene	35.55	0.0954	−35.8 to 207.6	2.202(20), 2.003(25)
Tetrahydropyran			−45 to 88	0.826(20), 0.764(25)

TABLE 2.31   Viscosity and Surface Tension of Organic Compounds (*Continued*)

Substance	Surface tension, mN · m^{-1}		Liquid range, °C	Viscosity, mN · s · m^{-2}
	a	*b*		
Tetrahydropyran-2-methanol	34.1(25)		−70 to 187	11.0(20)
Tetrahydrothiophene-1,1-diox-ide (sulfolane)	35.5(30)		27.6 to 287.3	9.87(30), 6.280(50), 3.818(75)
Tetrahydrothiophene oxide				52(30), 19(80)
Thiacyclohexane	36.06(20)	33.74(40)		
Thiacyclopentane	38.44	0.1342		1.042(20), 0.971(25)
2,2′-Thiodiethanol	53.8(20)		−10.2 to 282	65.2(20)
Thiophene	34.00	0.1328	−39.4 to 84	0.871(0), 0.662(20), 0.353(82)
Thymol	33.95	0.0821	49 to 232	
Toluene	30.90	0.1189	−94.9 to 110.6	0.623(15), 0.523(30), 0.424(50)
p-Toluenesulfonyl chloride	42.41	0.0903	67 to >134	
o-Toluidine	42.87	0.1094	−16.5 to 200	5.195(15), 4.39(20)
m-Toluidine	40.33	0.0979	−31 to 203	4.418(15), 2.741(30)
p-Toluidine	39.58	0.0957	43.8 to 200	1.945(45), 1.557(60)
m-Tolunitrile	38.85	0.1013	−23 to 210	
p-Tolunitrile	39.79	0.1100	29.5 to 85	
Tribenzylamine	42.41	0.0953	91–94 to bp	
Tribromomethane	48.14	0.1308	8.1 to 149.6	2.152(15), 1.741(30), 1.367(50)
1,2,3-Tribromopropane	47.99	0.1267	16.5 to 220	
Tributylamine	26.47	0.0831	−70 to 216	1.35(25)
Tributyl borate	26.2(20)	25.8(25)	<−70 to 234	1.776(20), 1.601(25)
Tributyl phosphite	27.57	0.0865	up to >125	1.9(25)
Tributyl phosphate	28.71	0.0666	−79 to 289	11.1(15), 3.39(25)
Trichloroacetaldehyde	27.66	0.1197	−57.5 to 97.8	
Trichloroacetic acid	35.4	0.0895	57.5 to 196.5	
1,1,1-Trichloroethane	28.28	0.1242	−30.4 to 74	0.903(15), 0.725(30), 0.578(50)
1,1,2-Trichloroethane	37.40	0.1351	−37 to 114	0.119(20), 0.110(25)
Trichloroethylene	29.5(20)	28.8(25)	−84.8 to 87	0.703(0), 0.545(25), 0.444(50)
Trichlorofluoromethane	18(25)		−111 to 23.8	0.740(−25), 0.539(0)
2,4,6-Trichlorophenol	43.13	0.0955	69 to 246	
1,2,3-Trichloropropane	37.8(20)	37.05(25)	−14.7 to 157	
Trichlorosilane	20.43	0.1076	−127 to 32	0.332(20), 0.316(25)
α,α,α-Trichlorotoluene			−5 to 223	3.07(10), 2.55(17)
1,1,2-Trichloro-1,2,2-trifluoro-ethane	17.75(20)	16.56(30)	−35 to 47.7	0.711(20), 0.627(30)
Tridecane	27.73	0.0872	−5 to 235	2.909(0), 1.724(25), 1.129(50)
1-Tridecene	28.01	0.0884	−13 to 232.8	
Triethanolamine			20.5 to 335.4	609(25), 114(50), 31.5(75)
Triethylamine	22.70	0.0992	−114.7 to 88.8	0.455(0), 0.347(25), 0.273(50)
Triethylene glycol	47.33	0.0880	−7 to 285	49.0(20), 8.5(60)
Triethyl phosphate	31.81	0.0928	−56 to 215	1.684(40), 1.376(55)
Triethyl phosphite	25.73	0.0878	up to 156	0.72(25)
Trifluoroacetic acid	15.64	0.1844	−15.3 to 73	0.926(20), 0.808(25), 0.571(50)
2,2,2-Trifluoroethanol	20.6(33)		−43.5 to 74	1.996(20)
Trimethylamine	16.24	0.1133	−117 to 2.9	0.321(−33.5)
1,2,3-Trimethylbenzene	30.91	0.1040	−25.4 to 176.1	
1,2,4-Trimethylbenzene	31.76	0.1025	−43.9 to 169	0.894(15), 0.730(30)
1,3,5-Trimethylbenzene	29.79	0.0897	−44.7 to 165	1.154(20)
2,2,3-Trimethylbutane	20.70	0.0973	−24.9 to 80.9	0.579(20)

(*Continued*)

TABLE 2.31   Viscosity and Surface Tension of Organic Compounds (*Continued*)

Substance	Surface tension, mN · m^{-1}		Liquid range, °C	Viscosity, mN · s · m^{-2}
	a	*b*		
cis, cis-1,3,5-Trimethylcyclo-hexane				0.632(20), 0.558(30)
trans-1,3,5-Trimethylcyclo-hexane			−107.4 to 140.5	0.714(20), 0.624(30)
Trimethylene sulfide	36.3(20)	35.0(30)	−73.2 to 95	0.638(20), 0.607(25)
3,5,5-Trimethyl-1-hexanol			<−70 to 194	11.06(25)
2,2,3-Trimethylpentane	22.46	0.0895	−112.3 to 110	0.598(20)
2,2,4-Trimethylpentane	20.55	0.0888	−107.4 to 99.2	0.502(20)
Trimethyl phosphite	27.18(20)	24.88(40)	−78 to 112	0.61(20)
2,4,6-Trimethylpyridine			−46 to 171	1.498(20)
Triphenylamine	46.2	0.0955	125 to 348	
Triphenyl phosphite			22 to 360	6.95(45)
Tripropylamine	24.58	0.0878	−93.5 to 158	
Tripropylene glycol	34(25)		up to 273	56.1(25)
Tripropylene glycol butyl ether	28.8(25)		up to 276	6.58(25)
Tripropylene glycol ethyl ether	28.2(25)			5.17(25)
Tripropylene glycol isopropyl ether	27.4(25)			7.7(25)
Tripropylene glycol methyl ether	30.0(25)		−42 to 242.4	5.96(25)
Tris(*m*-tolyl) phosphite				37.55(15), 9.132(45), 5.075(65)
Tris(*p*-tolyl) phosphite				35.52(15), 8.794(45), 5.017(65)
Tri-*o*-tolyl phosphate	40.9(20)		11 to 410	38.8(35), 16.8(55)
Undecane	26.26	0.0901	−25.6 to 196	1.707(0), 1.098(25), 0.761(50)
Vinyl acetate	23.95(20)	22.54(30)	−93 to 73	0.421(20)
o-Xylene	32.51	0.1101	−25.2 to 145	1.084(0), 0.760(25), 0.561(50)
m-Xylene	31.23	0.1104	−47.9 to 139	0.795(0), 0.581(25), 0.445(50)
p-Xylene	30.69	0.1074	13 to 138	0.603(25), 0.457(50), 0.359(75)

TABLE 2.32 Viscosity of Aqueous Glycerol Solutions

% Weight glycerol	Grams per liter	Relative density 25°/25°C	Viscosity, mN · s · m^{-2}		
			20°C	25°C	30°C
100	1261	1.262 01	1 495	942	622
99	1246	1.259 45	1 194	772	509
98	1231	1.256 85	971	627	423
97	1216	1.254 25	802	521	353
96	1201	1.251 65	659	434	296
95	1186	1.249 10	543.5	365	248
80	966.8	1.209 25	61.8	45.72	34.81
50	563.2	1.127 20	6.032	5.024	4.233
25	265.0	1.061 15	2.089	1.805	1.586
10	102.2	1.023 70	1.307	1.149	1.021

TABLE 2.33 Viscosity of Aqueous Sucrose Solutions

% Weight sucrose	Grams per liter	Relative density 20°/4°C	Viscosity, mN · s · m^{-2}		
			15°C	20°C	25°C
75	1034	1.379 0	4 039	2 328	1 405
70	943.0	1.347 2	746.9	481.6	321.6
65	855.6	1.316 3	211.3	147.2	105.4
60	771.9	1.286 5	79.49	58.49	40.03
50	614.8	1.299 6	19.53	15.43	12.40
40	470.6	1.176 4	7.463	6.617	5.164
30	338.1	1.127 0	3.757	3.187	2.735

2.4   REFRACTION AND REFRACTIVE INDEX

The refractive index n is the ratio of the velocity of light in a particular substance to the velocity of light in vacuum. Values reported refer to the ratio of the velocity in air to that in the substance saturated with air. Usually the yellow sodium doublet lines are used; they have a weighted mean of 589.26 nm and are symbolized by D. When only a single refractive index is available, approximate values over a small temperature range may be calculated using a mean value of 0.00045 per degree for dn/dt, and remembering that n_D decreases with an increase in temperature. If a transition point lies within the temperature range, extrapolation is not reliable.

The *specific refraction* r_D is given by the Lorentz and Lorenz equation,

$$R_D = \frac{n_D^2 - 1}{n_D^2 + 2} \cdot \frac{1}{\rho}$$

where ρ is the density at the same temperature as the refractive index, and is independent of temperature and pressure. The molar refraction is equal to the specific refraction multiplied by the molecular weight. It is a more or less additive property of the groups or elements comprising the compound. A set of atomic refractions is given in Table 1.12; an extensive discussion will be found in Bauer, Fajans, and Lewin, in *Physical Methods of Organic Chemistry*, 3d ed., A. Weissberger (ed.), vol. 1, part II, chap. 28, Wiley-Interscience, New York, 1960.

The empirical Eykman equation

$$\frac{n_D^2 - 1}{n_D + 0.4} \cdot \frac{1}{\rho} = \text{constant}$$

offers a more accurate means for checking the accuracy of experimental densities and refractive indices, and for calculating one from the other, than does the Lorentz and Lorenz equation.

The refractive index of moist air can be calculated from the expression

$$(n-1) \times 10^6 = \frac{103.49}{T} p_1 + \frac{177.4}{T} p_2 + \frac{86.26}{T}\left(1 + \frac{5748}{T}\right) p_3$$

where p_1 is the partial pressure of dry air (in mmHg), p_2 is the partial pressure of carbon dioxide (in mmHg), p_3 is the partial pressure of water vapor (in mmHg), and T is the temperature (in kelvins).

Example: 1-Propynyl acetate has $n_D = 1.4187$ and density = 0.9982 at 20°C; the molecular weight is 98.102. From the Lorentz and Lorenz equation,

$$r_D = \frac{(1.4187)^2 + 1}{(1.4187)^2 + 2} \cdot \frac{1}{0.9982} = 0.2528$$

The molar refraction is

$$Mr_D = (98.102)(0.2528) = 24.80$$

From the atomic and group refractions in Table 5.19, the molar refraction is computed as follows:

6 H	6.600
5 C	12.090
1 C≡C	2.398
1 O(ether)	1.643
1 O(carbonyl)	2.211
	$Mr_D = 24.942$

TABLE 2.34   Atomic and Group Refractions

Group	Mr_D	Group	Mr_D
H	1.100	N (primary aliphatic amine)	2.322
C	2.418	N (*sec*-aliphatic amine)	2.499
Double bond (C=C)	1.733	N (*tert*-aliphatic amine)	2.840
Triple bond (C≡C)	2.398	N (primary aromatic amine)	3.21
Phenyl (C$_6$H$_5$)	25.463	N (*sec*-aromatic amine)	3.59
Naphthyl (C$_{10}$H$_7$)	43.00	N (*tert*-aromatic amine)	4.36
O (carbonyl) (C=O)	2.211	N (primary amide)	2.65
O (hydroxyl) (O—H)	1.525	N (*sec* amide)	2.27
O (ether, ester) (C—O—)	1.643	N (*tert* amide)	2.71
F (one fluoride)	0.95	N (imidine)	3.776
(polyfluorides)	1.1	N (oximido)	3.901
Cl	5.967	N (carbimido)	4.10
Br	8.865	N (hydrazone)	3.46
I	13.900	N (hydroxylamine)	2.48
S (thiocarbonyl) (C=S)	7.97	N (hydrazine)	2.47
S (thiol) (S—H)	7.69	N (aliphatic cyanide) (C≡N)	3.05
S (dithia) (—S—S—)	8.11	N (aromatic cyanide)	3.79
Se (alkyl selenides)	11.17	N (aliphatic oxime)	3.93
3-membered ring	0.71	NO (nitroso)	5.91
4-membered ring	0.48	NO (nitrosoamine)	5.37
		NO$_2$ (alkyl nitrate)	7.59
		(alkyl nitrite)	7.44
		(aliphatic nitro)	6.72
		(aromatic nitro)	7.30
		(nitramine)	7.51

TABLE 2.35   Refractive Indices of Organic Compounds

Substance	Formula	Density, g/mL	Refractive index
Acenaphthene	$C_{12}H_{10}$	1.220	1.6048/98.8°
Acetaldehyde	C_2H_4O	0.788/16°	1.3316
Acetamide	C_2H_5ON	1.159	1.4274/78°
Acetanilide	C_8H_9ON	1.21/4°	
Acetic acid	$C_2H_4O_2$	1.0492	1.3718
Acetic anhydride	$C_4H_6O_3$	1.0850/15°	1.3904
Acetone	C_3H_6O	0.787/25°	1.3620/15°
Acetonitrile	C_2H_3N	0.7828	1.3460
Acetophenone	C_8H_8O	1.0329/15°	1.5342/19°
Acetyl chloride	C_2H_3OCl	1.1051	1.3898
Acetylene	C_2H_2	0.61/–80°	
Adipic acid	$C_6H_{10}O_4$	1.366	
Alloxan + $_4H_2O$	$C_4H_{10}O_8N_2$		
Allyl alcohol	C_3H_6O	0.8573/15°	1.4135
p-Aminobenzoic acid	$C_7H_7O_2N$		
2-Aminopyridine	$C_5H_6N_2$		
n-Amyl alcohol	$C_5H_{12}O$	0.8154	1.414/13°
act-Amyl alcohol	$C_5H_{12}O$	0.816	
sec-Amyl alcohol	$C_5H_{12}O$	0.8103	1.4053
tert-Amyl alcohol	$C_5H_{12}O$	0.809	1.4045
Aniline	C_6H_7N	1.026/15°	1.5863
Aniline hydrochloride	C_6H_8NCl	1.222/4°	
Anisole	C_7H_8O	0.9925/25°	1.5150/22°
Anthracene	$C_{14}H_{10}$	1.243	
Anthraquinone	$C_{14}H_8O$	1.419/4°	
Azobenzene	$C_{12}H_{10}N_2$		
Benzaldehyde	C_7H_6O	1.0504/15°	1.5463/17.6°
Benzene	C_6H_6	0.8790	1.5011
Benzoic acid	$C_7H_6O_2$	1.2656/15°	1.5397/15°
Benzoic anhydride	$C_{14}H_{10}O_3$	1.1989/15°	1.5767/15°
Benzoin	$C_{14}H_{12}O_2$		
Benzonitrile	C_7H_5N	1.0093/15°	1.5289
Benzophenone (*a*)	$C_{13}H_{10}O$	1.085/50°	
Benzoquinone	$C_6H_4O_2$		
Benzoyl chloride	C_7H_5OCl	1.212	1.5537
Benzoyl peroxide	$C_{14}H_{10}O_4$		
Benzyl alcohol	C_7H_8O	1.049/15°	1.5396
Benzyl benzoate	$C_{14}H_{12}O_2$	1.114/18°	1.5681/21°
Benzyl chloride	C_7H_7Cl	1.0983	1.5415/15°
Benzyl cinnamate	$C_{16}H_{14}O_2$		
Borneol (DL)	$C_{10}H_{18}O$	1.01	
a-Bromonaphthalene	$C_{10}H_7Br$	1.4888/16.5°	1.6601/16.5°
Bromobenzene	C_6H_5Br	1.4978/15°	1.5625/15°
Bromoform	$CHBr_3$	2.900/15°	1.6005/15°
n-Butane	C_4H_{10}	0.5788 (at sat. pressure)	
n-Butyl alcohol	$C_4H_{10}O$	0.8098	1.3993
iso-Butyl alcohol	$C_4H_{10}O$	0.8169	1.3968/17.5°
sec-Butyl alcohol	$C_4H_{10}O$	0.808	1.3949/25°
tert-Butyl alcohol	$C_4H_{10}O$	0.7887	1.3878
n-Butyl chloride	C_4H_9Cl	0.9074/0	1.4015
n-Butyric acid	$C_4H_8O_2$	0.9587	1.3991
iso-Butyric acid	$C_4H_8O_2$	0.950	

(Continued)

TABLE 2.35 Refractive Indices of Organic Compounds (*Continued*)

Substance	Formula	Density, g/mL	Refractive index
Camphene (DL)	$C_{10}H_{16}$	0.879	1.4402/80°
Camphor(D)	$C_{10}H_{16}O$	0.992/10°	
Carbitol (Diethyleneglycol-monomethylether)	$C_6H_{14}O_3$	0.9902	
Carbon disulphide	CS_2	1.2927/0°	1.6276
Carbon tetrabromide	CBr_4	2.9109/99.5°	
Carbon tetrachloride	CCl_4	1.6320/0°	1.4607
Cellosolve (Glycolmonoethylether)	$C_4H_{10}O_2$	0.9311	
Chloral hydrate	$C_2H_3O_2Cl_3$	1.9081	
Chloroacetic acid	$C_2H_3O_2Cl$	1.39/75°	1.4297/65°
Chlorobenzene	C_6H_5Cl	1.066	1.5248
Chloroform	$CHCl_3$	1.4985/15°	1.4467
Cholesterol	$C_{27}H_{46}O$	1.067	
Cineol (Eucalyptol)	$C_{10}H_{18}O$	0.9267	1.4584/18°
Cinnamic acid (trans)	$C_9H_8O_2$	1.247	
Cinnamyl alcohol	$C_9H_{10}O$	1.0440	1.5819
Citric acid	$C_6H_8O_7$	1.542/18°	
o-Cresol	C_7H_8O	1.051	1.5372/40°
m-Cresol	C_7H_8O	1.035	1.5406
p-Cresol	C_7H_8O	1.035	1.5316
Cumene	C_9H_{12}	0.8615	1.4909
Cyclohexane	C_6H_{12}	0.7786	1.4262
Cyclohexanol	$C_6H_{12}O$	0.9624	1.4656/22°
Cyclohexanone	$C_6H_{10}O$	0.9478	1.4507
Cyclohexene	C_6H_{10}	0.8108	1.4467
p-Cymene	$C_{10}H_{14}$	0.8766	1.5006
cis-Decalin	$C_{10}H_{18}$	0.8963	1.4811
trans-Decalin	$C_{10}H_{18}$	0.8703/18°	1.4697/18°
Dibenzyl	$C_{14}H_{14}$	0.995	
n-Dibutyl phthalate	$C_{16}H_{22}O_4$	1.0465	
Diethylamine	$C_4H_{11}N$	0.7108/18°	1.3873/18°
Difluorodichloro-methane (Freon 12)	$CC_{12}F_2$		
Difluoromonochloro-methane (Freon 22)	$CHClF_2$		
Dimethylamine	C_2H_7N	0.6804/0°	1.350/17°
Dimethylaniline	$C_8H_{11}N$	0.9557	1.5582
Dioxane	$C_4H_8O_2$	1.0338	1.4224
Diphenyl	$C_{12}H_{10}$	1.180/0°	1.5852/79°
Diphenylamine	$C_{12}H_{11}N$	1.159	
Epichlorhydrin	C_3H_5OCl	1.180	1.4420/11.6°
Ethane	C_2H_6		
Ethanolamine	C_2H_7ON	1.022	1.4539
di-Ethanolamine	$C_4H_{11}O_2N$	1.0966	1.4776
tri-Ethanolamine	$C_6H_{15}O_3N$	1.1242	1.4852
Ether (diethyl)	$C_4H_{10}O$	0.714/20°	1.3538
Ethyl acetate	$C_4H_8O_2$	0.9245	1.3701/25°
Ethyl acetoacetate	$C_6H_{10}O_3$	1.0282	1.4209/16°
Ethyl alcohol	C_2H_6O	0.7893	1.3610/20.5°
Ethylamine	C_2H_7N	0.7057/0°	

TABLE 2.35   Refractive Indices of Organic Compounds (*Continued*)

Substance	Formula	Density, g/mL	Refractive index
Ethylbenzene	C_8H_{10}	0.8669	1.4959
Ethyl benzoate	$C_9H_{10}O_2$	1.0509/15°	1.5068/17.3°
Ethyl bromide	C_2H_5Br	1.4555	1.4239
Ethyl chloride	C_2H_5Cl	0.9214/0°	
Ethylene	C_2H_4		
Ethylenediamine	$C_2H_8N_2$	0.902/15°	1.4540/26.1°
Ethylene dibromide	$C_2H_4Br_2$	2.1785	1.5379
Ethylene dichloride	$C_2H_4Cl_2$	1.2521	1.4443
Ethylene glycol	$C_2H_6O_2$	1.1155	1.4274
Ethylene oxide	C_2H_4O	0.877/7°	1.3597/7°
Ethyl formate	$C_3H_6O_2$	0.9168	1.3598
Ethyl iodide	C_2H_5I	1.9133/30°	1.5168/15°
Ethyl mercaptan	C_2H_6S	0.8315/25°	1.4351
Ethyl nitrate	$C_2H_5O_3N$	1.109	1.3853
Ethyl nitrite	$C_2H_5O_2N$	0.900/15°	
Ethyl oxalate	$C_6H_{10}O_4$	1.0785	1.4101
Ethyl salicylate	$C_9H_{10}O_3$	1.131	1.5226
Ethyl sulphate	$C_4H_{10}O_4S$	1.180/18°	1.4010/18°
Eugenol	$C_{10}H_{12}O_2$	1.0620/25°	1.5439/19°
Fluorescein	$C_{20}H_{12}O_5$		
Fluorobenzene	C_6H_5F	1.0236	1.4677
Formaldehyde	CH_2O	0.815/–20°	
Formamide	CH_3ON	1.1334	1.4472
Formic acid	CH_2O_2	1.220	1.3714
Fructose	$C_6H_{12}O_6$	1.598	
Fumaric acid	$C_4H_4O_4$	1.635	
Furfural	$C_5H_4O_2$	1.1594	1.5261
Furfuryl alcohol	$C_5H_6O_2$	1.1282/23°	1.4852
Furan	C_4H_4O	0.9644/0°	1.4216
Glucose	$C_6H_{12}O_6$	1.544/25°	
Glycerol	$C_3H_8O_3$	1.2604/17.5°	1.4730
Glyceryl trioleate	$C_{57}H_{104}O_6$	0.8992/50°	1.4561/60°
Glyceryl tripalmitate	$C_{51}H_{98}O_6$	0.8752/70°	1.4381/80°
Glyceryl tristearate	$C_{57}H_{110}O_6$	0.8559/90°	1.4385/80°
Glycine	$C_2H_5O_2N$		
Guaiacol	$C_7H_8O_2$	1.1287/21.4°	
n-Heptane	C_7H_{16}	0.6838	1.3877
Hexachlorotethane	C_2Cl_6	2.091	
Hexamine	$C_6H_{12}N_4$		
n-Hexane	C_6H_{14}	0.6594	1.3749
Hippuric acid	$C_9H_9O_3N$	1.371	
Hydroquinone	$C_6H_6O_2$	1.358	
Indene	C_9H_8	0.996	1.5766
Iodoform	CHI_3	4.008	
Isobutane	C_4H_{10}	0.5572 (at sat. press.)	
Isopentane	C_5H_{12}	0.6192	1.3538
isoprene	C_5H_8	0.6806	1.4194
Isooctane	C_8H_{18}	0.6919	1.3915
Isoquinoline	C_9H_7N	1.099	1.6223/25°
Lactic acid	$C_3H_6O_3$	1.2485	1.4414
Lactose + H_2O	$C_{12}H_{24}O_1$	1.525	
Maleic acid	$C_4H_4O_4$	1.5920	

(*Continued*)

TABLE 2.35   Refractive Indices of Organic Compounds (*Continued*)

Substance	Formula	Density, g/mL	Refractive index
Maleic anhydride	$C_4H_2O_3$	0.934	
Malonic acid	$C_3H_4O_4$	1.631/15°	
Maltose + H_2O	$C_{12}H_{24}O_1$	1.540	
Menthol (L)	$C_{10}H_{20}O$	0.903/15°	
Mesitylene	C_9H_{12}	0.8652	1.4994
Metaldehyde	$(C_2H_4O)_n$		
Methane	CH_4		
Methyl acetate	$C_3H_6O_2$	0.9280	1.3593/20°
Methyl alcohol	CH_4O	0.7910	1.3276/25°
Methylamine	CH_5N	0.699/−10.8°	
Methylaniline	C_7H_9N	0.9891	1.5702/21.2°
Methyl anthranilate	$C_8H_9O_2N$	1.1682/18.6°	
Methyl benzoate	$C_8H_8O_2$	1.0937/15°	1.5205/15°
Methyl bromide	CH_3Br	1.732/0°	
Methyl carbonate	$C_3H_6O_3$	1.0694	1.3687
Methyl chloride	CH_3Cl	0.991/−25°	
Methylene bromide	CH_2Br_2	2.8098/15°	
Methylene chloride	CH_2Cl_2	1.3348/15°	1.4237
Methyl ethyl ketone	C_4H_8O	0.8054	1.3814/15°
Methyl formate	$C_2H_4O_2$	0.9867/15°	1.344
Methyl iodide	CH_3I	2.251/30°	1.5293/21°
Methyl methacrylate	$C_5H_8O_2$	0.936	1.413
Methyl sulphate	$C_2H_6O_4S$	1.3348/15°	1.3874
Methyl salicylate	$C_8H_8O_3$	1.1787/25°	1.538/18.1°
Monofluorotrichloromethane (Freon 11)	CCl_3F	1.494/17°	
Morpholine	C_4H_9ON	0.9994	1.4545
Naphthalene	$C_{10}H_8$	1.14	1.5822/100°
α-Naphthol	$C_{10}H_8O$	1.099/99°	1.6206/98.7°
β-Naphthol	$C_{10}H_8O$	1.272	
α-Naphthylamine	$C_{10}H_9N$	1.1196/25°	1.6703/51°
β-Naphthylamine	$C_{10}H_9N$	1.0614/98°	1.6493/98°
Nicotine (L)	$C_{10}H_{14}N_2$	1.0097	1.5280
Nitrobenzene	$C_6H_5O_2N$	1.1732/25°	1.5530
Nitroethane	$C_2H_5O_2N$	1.050	1.3916
Nitromethane	CH_3O_2N	1.137	1.3818
1-Nitropropane	$C_3H_7O_2N$	1.001	1.4015
2-Nitropropane	$C_3H_7O_2N$	0.990	1.3941
n-Octane	C_8H_{18}	0.7025	1.3974
n-Octyl alcohol	$C_8H_{18}O$	0.8270	1.4292
Oleic acid	$C_{18}H_{34}O_2$	0.898	1.4582
Oxalic acid	$C_2H_2O_4$		
Palmitic acid	$C_{16}H_{32}O_2$	0.8527/62°	1.4339/60°
Paraformaldehyde	$(CH_2O)n$		
Paraldehyde	$C_6H_{12}O_3$	0.9943	1.4049
n-Pentane	C_5H_{12}	0.6262	1.3575
Phosgene	$COCl_2$		
Phenanthrene	$C_{14}H_{10}$	1.17	1.6567/129°
Phenol	C_6H_6O	1.073	1.5245/40.6°
Phthalic acid	$C_8H_6O_4$	1.593	
Phthalic anhydride	$C_8H_4O_3$	1.527/4°	
Phthalimide	$C_8H_5O_2N$		

TABLE 2.35   Refractive Indices of Organic Compounds (*Continued*)

Substance	Formula	Density, g/mL	Refractive index
α-Picoline	C_6H_7N	0.9443	1.5010
β-Picoline	C_6H_7N	0.9566	1.5068
γ-Picoline	C_6H_7N	0.9548	1.5058
Picric acid	$C_6H_3O_7N_3$	1.763	
Picryl chloride	$C_6H_2O_6N_3Cl$	1.797	
Pinene (Turpentine)	$C_{10}H_{16}$	0.861	1.4685/15°
Piperidine	$C_5H_{11}N$	0.8606	1.4530
Propane	C_3H_8		
n-Propyl acetate	$C_5H_{10}O_2$	0.887	1.3844
n-Propyl alcohol	C_3H_8O	0.8035	1.3850
iso-Propyl alcohol	C_3H_8O	0.7855	1.3776
Propylene	C_3H_6	0.5139 (at sat. press.)	
Pyridine	C_5H_5N	0.9831	1.5102
Pyrocatechol	$C_6H_6O_2$	1.344	
Pyrogallol	$C_6H_6O_3$		
Quinhydrone	$C_{12}H_{10}O_4$	1.401	
Quinoline	C_9H_7N	1.095	1.6269
Resorcinol	$C_6H_6O_2$	1.285/15°	
Salicylic acid	$C_7H_6O_3$	1.443	
Stearic acid	$C_{18}H_{36}O_2$	0.9408	1.4335/70°
Styrene	C_8H_8	0.9060	1.5469
Succinic acid	$C_4H_6O_4$	1.564/15°	
Succinic anhydride	$C_4H_4O_3$	1.234	
Sucrose	$C_{12}H_{22}O_{11}$	1.588/15°	
Sylvan (2-Methylfuran)	C_5H_6O	0.916	
Tartaric acid (*meso-*)	$C_4H_6O_6$	1.666	
Tartaric acid (racemic) + H_2O	$C_4H_8O_7$	1.697	
Tartaric acid (D)	$C_4H_6O_6$	1.7598	
Tartaric acid (L)	$C_4H_6O_6$	1.7598	
Tetralin	$C_{10}H_{12}$		1.5453/17°
Thiophen	C_4H_4S	1.0644	1.5287
Thiourea	CH_4N_2S	1.405	
Thymol	$C_{10}H_{14}O$	0.969	
Toluene	C_7H_8	0.8670	1.4969
o-Toluidine	C_7H_9N	1.0035	1.5688
m-Toluidine	C_7H_9N	0.987/25°	1.5686
p-Toluidine	C_7H_9N	0.961/50°	1.5532/59.1°
Trichloroethylene	C_2HCl_3	1.4597/15°	1.4782
Tri-o-cresyl phosphate	$C_{21}H_{21}O_4P$		
Tri-p-cresyl phosphate	$C_{21}H_{21}O_4P$		
Triethylamine	$C_6H_{15}N$	0.7495/0°	1.4003
Trimethylamine	C_3H_9N	0.6709/0°	
Trinitrotoluene	$C_7H_5O_6N_3$	1.654	
Triphenylmethane	$C_{19}H_{16}$		
Urea	CH_4ON_2	1.335	
Uric acid	$C_5H_4O_3N_4$	1.893	
n-Valeric acid	$C_5H_{10}O_2$	0.942	1.4086
iso-Valeric acid	$C_5H_{10}O_2$	0.937/15°	1.4018/22.4°
Vanillin	$C_8H_8O_3$		
o-Xylene	C_8H_{10}	0.8802	1.5054
m-Xylene	C_8H_{10}	0.8642	1.4972
p-Xylene	C_8H_{10}	0.8611	1.4958

TABLE 2.36   Solvents Having the Same Refractive Index and the Same Density at 25°C

Solvent 1	Solvent 2	Refractive index		Density, g/mL	
		1	2	1	2
Acetone	Ethanol	1.357	1.359	0.788	0.786
Ethyl formate	Methyl acetate	1.358	1.360	0.916	0.935
Ethanol	Propionitrile	1.359	1.363	0.786	0.777
2,2-Dimethylbutane	2-Methylpentane	1.366	1.369	0.644	0.649
2-Methylpentane	Hexane	1.369	1.372	0.649	0.655
Isopropyl acetate	2-Chloropropane	1.375	1.376	0.868	0.865
3-Butanone	Butyraldehyde	1.377	1.378	0.801	0.799
Butyraldehyde	Butyronitrile	1.378	1.382	0.799	0.786
Dipropyl ether	Butyl ethyl ether	1.379	1.380	0.753	0.746
Propyl acetate	Ethyl propionate	1.382	1.382	0.883	0.888
Propyl acetate	1-Chloropropane	1.382	1.386	0.883	0.890
Butyronitrile	2-Methyl-2-propanol	1.382	1.385	0.786	0.781
Ethyl propionate	1-Chloropropane	1.382	1.386	0.888	0.890
1-Propanol	2-Pentanone	1.383	1.387	0.806	0.804
Isobutyl formate	1-Chloropropane	1.383	1.386	0.881	0.890
1-Chloropropane	Butyl formate	1.386	1.387	0.890	0.888
Butyl formate	Methyl butyrate	1.387	1.391	0.888	0.875
Methyl butyrate	2-Chlorobutane	1.392	1.395	0.875	0.868
Butyl acetate	2-Chlorobutane	1.392	1.395	0.877	0.868
4-Methyl-2-pentanone	Pentanonitrile	1.394	1.395	0.797	0.795
4-Methyl-2-pentanone	1-Butanol	1.394	1.397	0.797	0.812
2-Methyl-1-propanol	Pentanonitrile	1.394	1.395	0.798	0.795
2-Methyl-1-propanol	2-Hexanone	1.394	1.395	0.798	0.810
2-Butanol	2,4-Dimethyl-3-pentanone	1.395	1.399	0.803	0.805
2-Hexanone	1-Butanol	1.395	1.397	0.810	0.812
Pentanonitrile	2,4-Dimethyl-3-pentanone	1.395	1.399	0.795	0.805
2-Chlorobutane	Isobutyl butyrate	1.395	1.399	0.868	0.860
Butyric acid	2-Methoxyethanol	1.396	1.400	0.955	0.960
1-Butanol	3-Methyl-2-pentanone	1.397	1.398	0.812	0.808
1-Chloro-2-methylpropane	Isobutyl butyrate	1.397	1.399	0.872	0.860
1-Chloro-2-methylpropane	Pentyl acetate	1.397	1.400	0.872	0.871
Methyl methacrylate	3-Methyl-2-pentanone	1.398	1.398	0.795	0.808
Triethylamine	2,2,3-Trimethylpentane	1.399	1.401	0.723	0.712
Butylamine	Dodecane	1.399	1.400	0.736	0.746
Isobutyl butyrate	1-Chlorobutane	1.399	1.401	0.860	0.875
1-Nitropropane	Propionic anhydride	1.399	1.400	0.995	1.007
Pentyl acetate	1-Chlorobutane	1.400	1.400	0.871	0.881
Pentyl acetate	Tetrahydrofuran	1.400	1.404	0.871	0.885
Dodecane	Dipropylamine	1.400	1.400	0.746	0.736
1-Chlorobutane	Tetrahydrofuran	1.401	1.404	0.871	0.885
Isopentanoic acid	2-Ethoxyethanol	1.402	1.405	0.923	0.926
Dipropylamine	Cyclopentane	1.403	1.404	0.736	0.740
2-Pentanol	4-Heptanone	1.404	1.405	0.804	0.813
3-Methyl-1-butanol	Hexanonitrile	1.404	1.405	0.805	0.801
3-Methyl-1-butanol	4-Heptanone	1.404	1.405	0.805	0.813
Hexanonitrile	4-Heptanone	1.405	1.405	0.801	0.813
Hexanonitrile	1-Pentanol	1.405	1.408	0.801	0.810
Hexanonitrile	2-Methyl-1-butanol	1.405	1.409	0.801	0.815
4-Heptanone	1-Pentanol	1.405	1.408	0.813	0.810

TABLE 2.36   Solvents Having the Same Refractive Index and the Same Density at 25°C (*Continued*)

Solvent 1	Solvent 2	Refractive index		Density, g/mL	
		1	2	1	2
2-Ethoxyethanol	Pentanoic acid	1.405	1.406	0.926	0.936
2-Heptanone	1-Pentanol	1.406	1.408	0.811	0.810
2-Heptanone	2-Methyl-1-butanol	1.406	1.409	0.811	0.815
2-Heptanone	Dipentyl ether	1.406	1.410	0.811	0.799
2-Pentanol	3-Isopropyl-2-pentanone	1.407	1.409	0.804	0.808
1-Pentanol	Dipentyl ether	1.408	1.410	0.810	0.799
2-Methyl-1-butanol	Dipentyl ether	1.409	1.410	0.815	0.799
Isopentyl isopentanoate	Allyl alcohol	1.410	1.411	0.853	0.847
Dipentyl ether	2-Octanone	1.410	1.414	0.799	0.814
2,4-Dimethyldioxane	3-Chloropentene	1.412	1.413	0.935	0.932
2,4-Dimethyldioxane	Hexanoic acid	1.412	1.415	0.935	0.923
Diethyl malonate	Ethyl cyanoacetate	1.412	1.415	1.051	1.056
3-Chloropentene	Octanoic acid	1.413	1.415	0.932	0.923
2-Octanone	1-Hexanol	1.414	1.416	0.814	0.814
2-Octanone	Octanonitrile	1.414	1.418	0.814	0.810
3-Octanone	3-Methyl-2-heptanone	1.414	1.416	0.830	0.818
3-Methyl-2-heptanone	1-Hexanol	1.415	1.416	0.818	0.814
3-Methyl-2-heptanone	Octanonitrile	1.415	1.418	0.818	0.810
1-Hexanol	Octanonitrile	1.416	1.418	0.814	0.810
Dibutylamine	Allylamine	1.416	1.419	0.756	0.758
Allylamine	Methylcyclohexane	1.419	1.421	0.758	0.765
Butyrolactone	1,3-Propanediol	1.434	1.438	1.051	1.049
Butyrolactone	Diethyl maleate	1.434	1.438	1.051	1.064
2-Chloromethyl-2-propanol	Diethyl maleate	1.436	1.438	1.059	1.064
N-Methylmorpholine	Dibutyl decanedioate	1.436	1.440	0.924	0.932
1,3-Propanediol	Diethyl maleate	1.438	1.438	1.049	1.064
Methyl salicylate	Diethyl sulfide	1.438	1.442	0.836	0.831
Methyl salicylate	1-Butanethiol	1.438	1.442	0.836	0.837
1-Chlorodecane	Mesityl oxide	1.441	1.442	0.862	0.850
Diethylene glycol	Formamide	1.445	1.446	1.128	1.129
Diethylene glycol	Ethylene glycol diglycidyl ether	1.445	1.447	1.128	1.134
Formamide	Ethylene glycol diglycidyl ether	1.446	1.447	1.129	1.134
2-Methylmorpholine	Cyclohexanone	1.446	1.448	0.951	0.943
2-Methylmorpholine	1-Amino-2-propanol	1.446	1.448	0.951	0.961
Dipropylene glycol mono-ethyl ether	Tetrahydrofurfuryl alcohol	1.446	1.450	1.043	1.050
1-Amino-2-methyl-2-pentanol	2-Butylcyclohexanone	1.449	1.453	0.904	0.901
2-Propylcyclohexanone	4-Methylcyclohexanol	1.452	1.454	0.923	0.908
Carbon tetrachloride	4,5-Dichloro-1,3-dioxolane-2-one	1.459	1.461	1.584	1.591
N-Butyldiethanolamine	Cyclohexanol	1.461	1.465	0.965	0.968
D-α-Pinene	*trans*-Decahydro-naphthalene	1.464	1.468	0.855	0.867
Propylbenzene	p-Xylene	1.490	1.493	0.858	0.857
Propylbenzene	Toluene	1.490	1.494	0.858	0.860

(*Continued*)

TABLE 2.36 Solvents Having the Same Refractive Index and the Same Density at 25°C (*Continued*)

Solvent 1	Solvent 2	Refractive index		Density, g/mL	
		1	2	1	2
Phenyl 1-hydroxyphenyl ether	1,3-Dimorpholyl-2-propanol	1.491	1.493	1.081	1.094
Phenetole	Pyridine	1.505	1.507	0.961	0.978
2-Furanmethanol	Thiophene	1.524	1.526	1.057	1.059
m-Cresol	Benzaldehyde	1.542	1.544	1.037	1.041

2.5 VAPOR PRESSURE AND BOILING POINT

The *vapor pressure* is the pressure exerted by a pure component at equilibrium at any temperature when both liquid and vapor phases exist and thus extends from a minimum at the triple point temperature to a maximum at the critical temperature, and the critical pressure is the most important of the basic thermodynamic properties affecting liquids and vapors.

Except at very high total pressures (above about 10 MPa), there is no effect of total pressure on vapor pressure. If such an effect is present, a correction can be applied. The pressure exerted above a solid–vapor mixture may also be called vapor pressure but is normally only available as experimental data for common compounds that sublime.

Numerous mathematical formulas relating the temperature and pressure of the gas phase in equilibrium with the condensed phase have been proposed. The Antoine equation (Eq. 1) gives good correlation with experimental values. Equation 2 is simpler and is often suitable over restricted temperature ranges. In these equations, and the derived differential coefficients for use in the Haggenmacher and Clausius–Clapeyron equations, the p term is the vapor pressure of the compound, the t term is the temperature in degrees Celsius, and the T term is the absolute temperature in kelvins ($t°C + 273.15$).

Eq.	Vapor-pressure equation	dp/dT	$-[d(\ln p)/d(1/T)]$
1	$\log p = A - \dfrac{B}{t+C}$	$\dfrac{2.303pB}{(t+C)^2}$	$\dfrac{2.303BT^2}{(t+C)^2}$
2	$\log p = A - \dfrac{B}{T}$	$\dfrac{2.303pB}{T^2}$	$2.303B$
3	$\log p = A - \dfrac{B}{T} - C \log T$	$p\left(\dfrac{2.303B}{T^2} - \dfrac{C}{T}\right)$	$2.303B - CT$

Equations 1 and 2 are rearranged to calculate the temperature of the normal boiling point:

$$t = \frac{B}{A - \log p} - C$$

$$T = \frac{B}{A - \log P}$$

The constants in the Antoine equation may be estimated by selecting three widely spaced data points and substituting in the following equations in sequence:

$$\left(\frac{y_3 - y_2}{y_2 - y_1}\right)\left(\frac{t_2 - t_1}{t_3 - t_2}\right) = 1 - \left(\frac{t_3 - t_1}{t_3 + C}\right)$$

$$B = \left(\frac{y_3 - y_1}{t_3 - t_1}\right)(t_1 + C)(t_3 + C)$$

$$A = y_2 + \left(\frac{B}{t_2 + C}\right)$$

In these equations, $y_i = \log p_i$.

TABLE 2.37   Vapor Pressures of Various Organic Compounds

Substance	Eq.	Range, °C	A	B	C
Acenaphthene	1	147–187	7.728 19	2 534.234	245.576
	2	147–288	8.033	2834.99	
Acetaldehyde	1	liq	8.005 52	1 600.017	291.809
Acetic acid	1	liq	7.387 82	1 533.313	222.309
Acetic anhydride	1	liq	7.149 48	1 444.718	199.817
Acetone	1	liq	7.117 14	1 210.595	229.664
Acetonitrile	1	liq	7.119 88	1 314.4	230
Acetophenone	2	30–100	9.135 2	2878.8	
Acetyl bromide	1	liq	5.197 02	545.784	150.396
Acetyl chloride	1	liq	6.948 87	1 115.954	223.554
Acetylene	1	−130 to −83	9.140 2	1 232.6	280.9
	1	−82 to −72	7.099 9	711.0	253.4
Acetyl iodide	1	liq	4.181 44	355.452	108.160
Acrylic acid	1	20–70	8.538 67	2305.843	266.547
Acrylonitrile	1	−20 to 140	7.038 55	1 232.53	222.47
Allyl isothiocyanate	1	10–50	5.126 58	791.434	154.019
m-Aminobenzotrifluoride	1	0–96	7.651 86	1 940.6	218.0
		96–300	7.170 30	1 650.21	193.58
p-Aminophenol	1	130–185	−3.357 50	699.157	−331.343
Aniline	1	102–185	7.320 10	1 731.515	206.049
Anthracene	2	100–160	8.91	3 761	
	1	176–380	7.674 01	2 819.63	247.02
9,10-Anthracenedione	2	224–286	12.305	5 747.9	
	2	285–370	8.002	3 341.94	
Benzene	1	−12 to 3	9.106 4	1 885.9	244.2
	1	8–103	6.905 65	1 211.033	220.790
Benzenethiol	1	52–198	6.990 19	1 529.454	203.048
Benzoic acid	2	60–110	9.033	3 333.3	
Benzonitrile	1	liq	6.746 31	1 436.72	181.0
Benzophenone	1	48–202	7.349 66	2 331.4	195.0
	1	200–306	7.162 94	2 051.855	173.074
Benzotrifluoride	1	−20 to 180	7.007 08	1 331.30	220.58
Benzoyl chloride	2	140–200	7.924 5	2 372.1	
Benzyl acetate	1	46–156	8.457 05	2 623.206	259.067
Benzyl alcohol	1	122–205	7.198 17	1 632.593	172.790

(Continued)

TABLE 2.37 Vapor Pressures of Various Organic Compounds (*Continued*)

Substance	Eq.	Range, °C	A	B	C
Biphenyl	1	69–271	7.245 41	1 998.725	202.733
2-(2-Biphenylyloxy)ethanol	1	240–300	8.005 87	2 776.761	206.914
Bromobenzene	1	56–154	6.860 64	1 438.817	205.441
2-Bromobenzyl cyanide	1	85–152	5.044 59	734.821	59.273
1-Bromobutane	1	−78 to 23	5.281 38	685.001	160.880
Bromochloromethane	1	16–68	6.496 06	942.267	192.587
Bromochlorodifluoromethane	1	−95 to 10	6.839 98	935.632	240.330
2-Bromo-2-chloro-1,1,1-trifluoro-ethane	1	−51 to 55	6.945 02	1 127.856	227.341
Bromocyclohexane	1	68–260	6.979 80	1 572.19	217.38
p-Bromodiphenyl ether	1	25–190	7.009 3	1 902.7	153.3
	1	190–400	6.681 43	1 683.84	132.90
Bromoethane	1	28–75	6.988 6	1 121.9	234.7
Bromoethene	1	−88 to 16	6.997 4	1 009.9	251.6
2-Bromoethylbenzene	1	127–217	7.800	2 235.4	238.7
4-Bromoethylbenzene	1	liq	6.982 09	1 632.60	193
2-Bromo-2-methylpropane	1	0–72.8	7.395 9	1 512.7	262.2
1-Bromonaphthalene	1	liq	7.003 50	1 927.05	186.0
o-Bromostyrene	1	liq	6.910 38	1 631.2	195
p-Bromostyrene	1		7.228 38	1 743.67	218.0
4-Bromotoluene	1	85–280	7.007 62	1 612.35	206.36
2-Bromovinylbenzene	1	110–129	0.564 97	82.913	−191.71
4-Bromovinylbenzene	1	119–147	12.504 2	7 349.00	559.02
1,2-Butadiene	1	−69 to −34	7.398 22	1 219.877	259.776
	1	−26 to 30	6.993 81	1 041.117	242.274
1,3-Butadiene	1	−80 to −62	7.035 55	998.106	245.233
	1	−58 to 15	6.849 99	930.546	238.854
n-Butane	1	−77 to 19	6.808 96	935.86	238.73
1-Butanethiol	1	−2 to 123	6.927 54	1 281.018	218.100
2-Butanethiol	1	−13 to 110	6.886 98	1 229.904	222.021
1-Butanol	1	15–131	7.476 80	1 362.39	178.77
2-Butanol	1	25–120	7.474 31	1 314.19	186.55
2-Butanone	1	43–88	7.063 56	1 261.34	221.97
1-Butene	1	−82 to 13	6.792 90	908.80	238.54
2-Butene *cis*	1	−73 to 23	6.884 68	967.32	237.87
trans	1	−76 to 20	6.883 37	967.50	240.84
Butyl acetate	1	60–126	7.127 12	1 430.418	210.745
n-Butylamine trimethylboron	1	0–99	8.465 21	1 980.98	193.60
n-Butylbenzene	1	62–213	6.983 17	1 577.965	201.378
sec-Butylbenzene	1	87–174	6.942 19	1 533.95	204.39
t-Butylbenzene	1	84–170	6.922 55	1 505.987	203.490
n–Butyl borate	1	117–218	7.406 87	1 905.035	186.134
n-Butyl-*t*-butyl ether	1	83–124	6.955 56	1 348.702	206.303
Butyl carbitol	1	50–153	7.741 14	2 056.904	195.655
Butyl cellosolve	1	93–170	6.956 59	1 399.903	172.154
sec-Butylchloroacetate	1	30–172	7.933 38	2 103.30	249.29
n-Butylcyclohexane	1	60–211	6.910 30	1 538.518	200.833
sec-Butylcyclohexane	1	91–180	6.890 96	1 530.70	202.373
t-Butylcyclohexane	1	84–173	6.856 80	1 501.724	206.108
n-Butylcyclopentane	1	41–185	6.899 35	1 457.08	205.99
n-Butyl formate	1	29–112	7.693 6	1 698.7	247.4
sec-Butyl formate	1	30–100	6.493	972.9	176.0
n-Butyl-*α*-hydroxyisobutyrate	1	112–185	8.421 7	2 617.32	287.09

TABLE 2.37 Vapor Pressures of Various Organic Compounds (*Continued*)

Substance	Eq.	Range, °C	A	B	C
1-*n*-Butylnaphthalene	1	25–170	7.434 47	2 227.7	202.2
	1	170–345	7.081 4	1 971.5	180
2-*n*-Butylnaphthalene	1	25–170	7.438 08	2 242.2	202.3
	1	170–345	7.084 8	1 984.3	180
n-Butyl nitrate	1	0–70	8.054 27	1 992.83	254.30
1-Butyl pentafluoropropionate	1	82–116	6.651 00	1 108.02	177.04
2-*sec*-Butylphenol	1	179–240	6.951 93	1 593.74	163.79
2-*t*-Butylphenol	1	135–225	7.217 56	1 822.81	196.23
4-*t*-Butylphenol	1	198–252	7.000 38	1 627.51	155.24
Butyl phenyl ether	1	119–210	7.299 7	1 882.70	215.82
n-Butyl propionate	1	32–93	9.484 89	2 852.58	296.98
n-Butyl trifluoroacetate	1	71–104	8.567 94	2 305.22	301.06
1-Butyl trimethylsilyl ether	1	71–124	7.763 00	1 884.68	261.31
1-Butyne	1	−68 to 27	6.981 98	988.75	233.01
2-Butyne	1	−51 to −34	7.037 91	896.91	199.06
	1	−31 to 47	7.073 38	1 101.71	235.81
n-Butyraldehyde	1	31–74	6.385 44	913.59	185.48
Butyric acid	1	90–163	7.739 9	1 764.7	199.9
Camphor	2	0–180	8.799	2 797.39	
	1	178–232	6.106	1 043.6	116.4
Capric acid	1	153–187	6.255 3	1 106.3	57.96
Caproic acid	1	98–179	6.924 9	1 340.8	126.6
Capronitrile	1	92–164	7.123 1	1 597.2	212.8
Caprylic acid	1	130–206	7.770 64	1 933.05	159.36
Carbazole	1	253–358	7.086 3	2 179.4	163.5
Carbitol	1	40–151	7.640 81	1 801.31	183.97
Chloroacetic acid	1	104–190	7.550 16	1 723.365	179.98
4-Chloroacetophenone	1	122–212	7.084 57	1 693.63	190.95
Chloroacetyl chloride	1	28–107	7.149 77	1 340.79	208.70
N-Chloroaniline	1	61–125	3.037 67	171.35	−14.99
2-Chloroaniline	1	20–108	7.562 65	1 998.6	220.0
	1	108–300	7.192 40	1 762.74	200.0
3-Chloroaniline	1	15–125	7.559 39	2 073.75	215
	1	125–310	7.236 03	1 857.75	196.64
o-Chloroanisole	1	115–186	7.121 36	1 655.80	188.77
Chlorobenzene	1	62–131.7	6.978 08	1 431.05	217.55
o-Chlorobenzotrichloride	1	30–150	7.504 30	2 228.07	220.0
	1	150–350	7.117 94	1 951.37	196.27
1-Chloro-4-bromobenzene	2	23–63	11.629	3 643.30	
1-Chlorobutane	1	−17 to 78.6	6.836 94	1 173.79	218.13
2-Chlorobutane	1	0–40	6.799 23	1 149.12	224.68
1-Chlorodecane	1	86–225.9	6.939 86	1 639.06	177.94
1-Chlorododecane	1	116–246	6.834 08	1 654.82	155.09
Chloroethane	1	−56 to 12.2	6.986 47	1 030.01	238.61
2-Chloroethylbenzene	1		6.981 69	1 556.0	201.0
3-Chloroethylbenzene	1		6.990 82	1 577.3	200
4-Chloroethylbenzene	1		6.983 09	1 577.0	200
Chloroethylene	1	−65 to −13	6.891 17	905.01	239.48
Chloroform	1	−35 to 61	6.493 4	929.44	196.03
1-Chloroheptane	1	34–160	6.916 70	1 453.96	199.83
1-Chlorohexadecane	1	166–327	7.282 03	2 152.61	162.73
1-Chlorohexane	1	15–136	7.051 36	1 461.72	215.57
Chlorohexylisocyanate	1	90–180	7.740 95	2 340.50	241.90

(*Continued*)

TABLE 2.37   Vapor Pressures of Various Organic Compounds (*Continued*)

Substance	Eq.	Range, °C	A	B	C
Chloromethane	1	−75 to −5	7.093 49	948.58	249.34
Chloromethoxytrichlorosilane	1	0–50	7.312 92	1 545.71	226.10
2-Chloro-2-methylpropane	1	22–47	4.896	334.99	114.0
1-Chlorononane	1	69–205	7.046 54	1 655.57	192.26
1-Chlorooctane	1	54–184	7.051 52	1 600.24	200.28
Chloropentafluorobenzene	1	36–140	7.068 83	1 389.19	213.75
p-Chlorophenetole	1	122–212	7.084 57	1 693.63	190.95
2-Chlorophenol	1	80–200	6.877 31	1 471.61	193.17
β-Chloro-*β*-phenylethyl alcohol	1	166–259	6.917 33	1 635.63	145.87
1-Chlorophenylisocyanate	1	50–160	12.265 9	6 532.55	499.59
m-Chlorophenylisocyanate	1	71–158	6.797 29	1 512.43	180.90
Chloroprene	1	20–60	6.161 50	783.45	179.7
1-Chloropropane	1	−25 to 47	6.926 48	1 110.19	227.94
2-Chloropropane	1	0–30	7.771	1 582	288
3-Chloro-1-propene	1	13–44	5.297 16	418.375	128.168
2-Chloropropionitrile	1	0–84	7.329 73	1 732.55	211.79
	1	84–240	7.200 85	1 657.25	205.3
γ-Chloropropyltrichlorosilane	1	87–179	7.156 4	1 679.07	210.38
1-Chlorotetradecane	1	142–296.8	7.200 7	2 018.9	170.6
o-Chlorotoluene	1	0–65	7.367 97	1 735.8	230.0
	1	65–220	6.947 63	1 497.2	209.0
1-Chloro-2,4,6-trinitrobenzene	1	200–270	3.080 9	184.93	−117.9
1-Chloroundecane	1	101–245	6.967 6	1 709.4	172.9
o-Chlorovinylbenzene	1	98–155	6.956 6	1 602.2	204.5
p-Chlorovinylbenzene	1	100–127	9.969 1	4 093.5	392.4
2-Chlorovinyldichloroarsine *cis*	1	68–109	5.487 9	785.09	115.61
trans	1	50–150	6.814 0	1 465.07	178.53
3-Chloro vinyldichloroarsine	1	66–110	2.810 5	97.17	−27.51
o-Cresol	1	120–191	6.911 7	1 435.50	165.16
m-Cresol	1	150–201	7.508 0	1 856.36	199.07
p-Cresol	1	128–202	7.035 08	1 511.08	161.85
Cyanic acid	1	−76 to −6	7.568 59	1 251.86	243.79
Cyclobutane	1	−60 to 12	6.916 31	1 054.54	241.37
Cyclobutanone	1	−24 to 25	6.116 68	933.95	183.19
Cyclobutene	1	−77 to 2	7.305 7	1 166.0	261.06
Cycloheptane	1	68–159	6.853 95	1 331.57	216.35
1,3,5-Cycloheptatriene	1	0–65	6.974 33	1 376.84	220.75
Cyclohexane	1	20–81	6.841 30	1 201.53	222.65
Cyclohexanethiol	1	84–203	6.886 73	1 476.70	209.83
Cyclohexanol	1	94–161	6.255 3	912.87	109.13
Cyclohexene	1		6.886 17	1 229.973	224.10
Cyclohexyl acetate	1	95–172	7.975 86	2 167.99	252.30
Cyclohexylamine	1	61–128	6.689 54	1 229.42	188.80
1-Cyclohexylamino-2-propanol	1	150–238	7.011 56	1 655.02	162.59
Cyclohexylpentafluoropropionate	1	82–155	7.725 5	1 844.73	224.89
Cyclohexyltrifluoroacetate	1	72–147	7.802 35	1 954.66	249.33
Cyclohexyltrimethylsilyl ether	1	91–168	8.090 52	2 276.62	267.94
Cyclooctane	1	97–194	6.861 87	1 437.79	210.02
1,3,5,7-Cyclooctatetraene	1	0–75	7.006 69	1 472.11	215.84
Cyclopentane	1	−40 to 72	6.886 76	1 124.162	231.36
Cyclopentanethiol	1	81–173	6.914 97	1 388.63	212.05
Cyclopentanone	1	0–26	2.902 47	162.90	63.22
Cyclopentene	1		6.920 66	1 121.818	223.45

TABLE 2.37 Vapor Pressures of Various Organic Compounds (*Continued*)

Substance		Eq.	Range, °C	A	B	C
Cyclopentyl-1-thiaethane		1	83–199	6.940 83	1 480.70	208.47
Cyclopropane		1	−90 to −32	6.887 88	856.01	246.50
o-Cymene		1	81–180	7.266 10	1 768.45	224.95
m-Cymene		1	79–176	7.123 74	1 644.95	212.76
p-Cymene		1	107–178	7.050 74	1 608.91	208.72
Decahydronaphthalene	*cis*	1	68–228	6.875 29	1 594.460	203.39
	trans	1	61–219	6.856 81	1 564.683	206.26
Decane		1	58–203	6.943 65	1 495.17	193.86
1-Decanethiol		1	109–271	6.998 1	1 713.6	177.0
1-Decanol		1	25–52	11.560	4 055	273.2
		1	103–230	6.922 44	1 472.01	133.98
1-Decene		1	54–199	6.934 77	1 484.98	195.707
Decylbenzene		1	203–298	7.035 96	1 903.98	160.33
Decylcyclohexane		1	197–298	7.019 37	1 899.33	161.35
Decylcyclopentane		1	182–279	6.999 12	1 822.05	163.05
Deuterodiborane		1	−155 to −94	6.480 83	545.20	244.73
Diacetone alcohol		1	28–115	8.502 42	2 400.56	263.79
1,3-Diacetylbenzene		1	50–145	0.056 24	64.188	−196.97
1,4-Diacetylbenzene		1	116–157	2.803 71	177.25	−46.43
Diacetylene		1	−78 to 0	4.990 79	356.36	143.22
Diallyl sulfide		1	10–40	4.829 30	643.18	142.34
4,4′-Diaminodiphenylmethane		1	198–272	3.172 31	210.49	−137.41
Diamyl ether		1	105–187	7.067 10	1 604.77	196.58
Dibenzyl ketone		2	285–325	8.257	3 244.42	
1,2-Dibromobenzene		1	20–117	7.501 28	2 093.7	230
		1	117–300	7.102 65	1 825.77	207.0
Dibromodichloroethane		1	25–130	5.197 53	763.44	110.81
Dibromodifluoromethane		1	−26 to 23	7.152 22	1 181.612	253.85
1,2-Dibromoethane		1	52–131	6.721 48	1 280.82	201.75
1,2-Dibromoethylene	*cis*	1	26–78	7.038 74	1 349.84	209.26
	trans	1	4–71	4.581 11	393.641	103.56
1,2-Dibromopropane		1	0–50	7.303 98	1 644.4	232.0
		1	50–250	6.891 05	1 419.60	212.0
1,3-Dibromopropane		1	0–71	7.549 84	1 890.56	240.0
		1	71–275	7.198 74	1 678.26	222.0
Di-*n*-butyl ether		1	89–140	6.796 3	1 297.29	191.03
Di-*t*-butyl ether		1	4–109	6.932 9	1 348.53	233.79
Di-*n*-butyl phthalate		1	126–202	6.639 80	1 744.20	113.69
Di-*n*-butyl sebacate		1	128–208	7.587 66	2 364.89	147.54
Di-*n*-butyl sulfide		1	10–40	6.769 3	1 208.80	217.51
1,2-Dichlorobenzene		1	131–181	7.143 78	1 704.49	219.42
1,3-Dichlorobenzene		1	91–173	7.040 1	1 607.05	213.38
1,4-Dichlorobenzene		1	95–174	7.020 8	1 590.9	210.2
Dichlorobenzotrichloride		1	20–167	7.439 54	2 190.0	200
		1	167–340	6.985 24	1 868.91	172.00
Dichlorobenzyl chloride		1	20–138	7.504 57	2 125.9	213.8
		1	138–350	7.147 35	1 881.38	192.93
1,1-Dichloroethane		1	−39 to 18	6.977 0	1 174.02	229.06
1,2-Dichloroethane		1	−31 to 99	7.025 3	1 271.3	222.9
1,1-Dichloroethylene		1	−28 to 32	6.972 2	1 099.4	237.2
1,2-Dichloroethylene	*cis*	1	0–84	7.022 3	1 205.4	230.6
	trans	1	−38 to 85	6.965 1	1 141.9	231.9
2,2′-Dichloroethyl sulfide		1	15–76	8.587 41	2 588.23	246.06

(*Continued*)

TABLE 2.37   Vapor Pressures of Various Organic Compounds (*Continued*)

Substance	Eq.	Range, °C	A	B	C
1,2-Dichloroethyltrichlorosilane	1	102–181	7.826	2 144.9	253.1
Dichloromethane	1	–40 to 40	7.409 2	1 325.9	252.6
2-(2,4-Dichlorophenoxy)-ethanol	1	212–286	7.240 09	2 004.31	157.25
3,4-Dichlorophenylisocyanate	1	60–190	8.679 3	3 312.3	333.9
1,2-Dichloropropane	1	45–96	6.980 7	1 308.1	222.8
3,4-Dichlorotoluene	1	0–105	7.343 94	1 882.5	215.0
	1	105–330	6.979 25	1 655.44	195.0
Diethanolamine	1	194–241	8.138 8	2 327.9	174.4
1,1-Diethoxyethane	1	0–70	6.757 63	1 191.60	203.12
Diethoxymethane	1	0–75	6.908 41	1 229.52	217.01
Diethylaluminum chloride	1	44–125	8.229 70	2 484.53	255.45
Diethylamine	1	31–61	5.801 6	583.30	144.1
N,N-Diethylaniline	1	50–218	7.466 0	1 993.57	218.5
1,2-Diethylbenzene	1	liq	6.987 80	1 576.940	200.51
1,3-Diethylbenzene	1	liq	7.003 60	1 575.310	200.96
1,4-Diethylbenzene	1	liq	6.998 20	1 588.310	201.97
Diethyldichlorosilane	1	48–128	6.862 9	1 346.3	207.7
Diethyl disulfide	1	15–61	7.349 89	1 695.00	227.29
	1	61–230	6.975 07	1 485.970	208.96
Diethylene glycol	1	130–243	7.636 7	1 939.4	162.7
Diethyl ether	1	–61 to 20	6.920 32	1 064.07	228.80
Diethyl ethylphosphate	1	76–134	4.101 6	315.17	15.50
N,N-Diethylformamide	1	30–90	6.395 4	1 203.8	165.6
Diethyl ketone	1		6.857 91	1 216.3	204
3,3-Diethylpentane	1	63–147	6.896 03	1 453.48	215.83
3,5-Diethylphenol	1	114–248	7.651 3	2 228	218.5
Diethylpropylphosphonate	1	87–134	4.558 1	446.50	26.17
Diethyl sulfide	1	0–150	6.928 36	1 257.83	218.66
1,2-*bis*-Difiuoroamino-4-methyl-pentane	1	–20 to 20	8.009 11	1 944.92	245.44
Difiuoromethane	1	–82 to –32	7.138 9	821.7	244.7
1,2-Dihydroxybenzene	1	118–246	7.577	2 054	187
1,3-Dihydroxybenzene	1	151–276	7.889	2 231	169
1,2-Diiodoethylene *cis*	1	29–152	5.522	797.8	106.4
trans	1	77–130	6.093 1	1 197.0	172.3
Diisoamyl sulfide	1	10–80	–1.959 8	390.61	–219.33
p-Diisopropylbenzene	1	120–211	6.993 3	1 663.88	194.41
Diisopropyl ether	1	23–67	6.849 5	1 139.34	218.7
2,4-Diisopropylphenol	1	122–255	6.714	1 506	138
1,2-Dimethoxyethane	1	0–60	6.718 9	1 050.5	209.2
N,N-Dimethylacetamide	1	30–90	9.720 9	3 273.8	334.5
Dimethylamine	1	–72 to 6.9	7.082 12	960.242	221.67
bis-Dimethylaminoborane	1	–25 to 62.5	5.584 52	774.371	170.64
N,N-Dimethylaminodiborane	1	–38 to 14	8.340 1	1 917.35	302.73
bis-Dimethylaminodifluorosilane	1	24–88	5.952	748.7	146.9
N,N-Dimethylaniline	1	71–197	7.367 7	1 857.08	220.36
Dimethyl beryllium	1	100–180	19.089 9	11 535.45	496.64
1,4-Dimethyl-bicyclo(2,2,1)-heptane	1	56–119	6.761 96	1 342.66	213.53
2,3-Dimethyl-bicyclo(2,2,1)-heptane *trans*	1	72–138	6.868 15	1 420.32	212.94
2,3-Dimethyl-1,3-butadiene	1	0–68.5	7.119 7	1 299.69	238.09
2,2-Dimethylbutane	1	–42 to 73	6.754 83	1 081.176	229.34

TABLE 2.37   Vapor Pressures of Various Organic Compounds (*Continued*)

Substance		Eq.	Range, °C	A	B	C
2,3-Dimethylbutane		1	−35 to 81	6.809 83	1 127.187	228.90
2,3-Dimethyl-2-butanethiol		1	56–167	6.839 56	1 354.24	215.96
2,3-Dimethyl-1-butene		1	−36 to 78	6.862 36	1 134.675	229.37
2,3-Dimethyl-2-butene		1	−21 to 97	6.950 58	1 215.428	225.44
3,3-Dimethyl-1-butene		1	−47 to 64	6.677 51	1 010.516	224.91
Dimethyl cadmium		1	−2 to 23	6.490 55	1 126.36	201.07
1,1-Dimethylcyclohexane		1	10–147	6.798 21	1 321.705	217.85
1,2-Dimethylcyclohexane	*cis*	1	18–158	6.837 46	1 367.311	215.84
	trans	1	13–151	6.833 08	1 353.881	219.13
1,3-Dimethylcyclohexane	*cis*	1	11–147	6.838 83	1 338.473	218.07
	trans	1	15–152	6.834 55	1 343.687	215.39
1,4-Dimethylcyclohexane	*cis*	1	15–152	6.832 87	1 345.613	216.15
	trans	1	10–147	6.817 73	1 330.437	218.58
1,1-Dimethylcyclopentane		1	−12 to 113	6.817 24	1 219.474	221.95
1,2-Dimethylcyclopentane	*cis*	1	−3 to 125	6.850 08	1 269.140	220.21
	trans	1	−9 to 117	6.844 22	1 242.748	221.69
1,3-Dimethylcyclopentane	*cis*	1	−10 to 116	6.837 15	1 237.456	222.01
	trans	1	−9 to 117	6.838 17	1 240.023	221.62
Dimethyldichlorosilane		1	28–72	7.062 1	1 280.29	235.65
1,2-Dimethyldisilane		1	−46 to 0	4.024 3	255.4	129.2
Dimethyl ether		1	−71 to −25	6.976 03	889.264	241.96
N,N-Dimethylformamide		1	30–90	6.928 0	1 400.87	196.43
2,2-Dimethylhexane		1		6.837 15	1 273.59	215.07
2,3-Dimethylhexane		1		6.870 04	1 315.50	214.16
2,4-Dimethylhexane		1		6.853 05	1 287.88	214.79
2,5-Dimethylhexane		1		6.859 84	1 287.27	214.41
3,3-Dimethylhexane		1		6.851 21	1 307.88	217.44
3,4-Dimethylhexane		1		6.879 86	1 330.04	214.86
1,1-Dimethylhydrazine		1	−35 to 20	7.408 13	1 305.91	225.53
1,2-Dimethylhydrazine		1	1–25	5.611 9	633.59	143.17
N,N-Dimethylhydroxylamine		1	17–90	7.565 8	1 415.96	201.93
O,N-Dimethylhydroxylamine		1	−45 to 42.2	7.405 4	1 245.58	233.06
Dimethylmalononitrile		1	49–140	7.035 5	1 546.99	202.00
1,3-Dimethylnaphthalene		1	20–148	7.634 7	2 295.4	232.4
		1	148–310	7.269 8	2 076.0	210
1,4-Dimethylnaphthalene		1	20–148	7.634 7	2 345.8	232.6
(same for 1,6- and 1,7-)		1	148–310	7.269 8	2 076.0	210
1,8-Dimethylnaphthalene		1	25–150	7.407 89	2 123.2	201.2
		1	150–320	7.056 4	1 879	180
2,3-Dimethylnaphthalene		1	20–155	7.403 96	2 111.9	201.1
		1	155–315	7.052 7	1 869	180
2,6-Dimethylnaphthalene		1	20–150	7.396 8	2 080.3	200.8
		1	150–310	7.046 0	1 841	180
2,7-Dimethylnaphthalene		1	25–150	7.398 75	2 085.9	200.9
		1	150–310	7.047 8	1 846	180
2,2-Dimethylpentane		1	−19 to 103	6.814 80	1 190.033	223.30
2,3-Dimethylpentane		1	−10 to 115	6.853 82	1 238.017	221.82
2,4-Dimethylpentane		1	−17 to 105	6.826 21	1 192.04	225.32
3,3-Dimethylpentane		1	−14 to 112	6.826 67	1 228.663	225.32
2,4-Dimethyl-3-pentanone		1	48–125	6.968 53	1 382.84	213.06
Dimethyl-o-phthalate		1	82–151	4.522 32	700.31	51.42
2,2-Dimethylpropane		1	−14 to 29	6.604 27	883.42	227.78
2,2-Dimethyl-1-propanol		1	55–115	7.875 3	1 604.7	208.2

(*Continued*)

TABLE 2.37   Vapor Pressures of Various Organic Compounds (*Continued*)

Substance	Eq.	Range, °C	A	B	C
2,5-Dimethylpyrrole	1	100–199	7.203 06	1 509.60	181.76
2,4-Dimethylquinoline	1	185–269	7.025 4	1 830.29	174.44
2,6-Dimethylquinoline	1	188–267	6.931 12	1 748.73	166.37
Dimethyl sulfide	1	−22 to 20	7.150 9	1 195.58	242.68
3,3-Dimethyl-2-thiabutane	1	liq	6.847 09	1 259.648	218.69
2,2-Dimethyl-3-thiapentane	1	liq	6.850 86	1 323.24	212.89
2,4-Dimethyl-3-thiapentane	1	liq	6.871 18	1 327.12	212.55
2,3-Dimethylthiophene	1	50–205	6.924 9	1 430.0	212
2,4-Dimethylthiophene	1	50–205	6.993 9	1 450.7	212.0
2,5-Dimethylthiophene	1	47–200	6.961 1	1 427.7	213.2
3,4-Dimethylthiophene	1	54–205	6.996 1	1 467.1	211.5
1,3-Dinitrobenzene	1	252–292	4.337	229.2	−137
2,4-Dinitrotoluene	1	200–299	5.798	1 118	61.8
2,6-Dinitrotoluene	1	150–260	4.372	380	−43.6
3,5-Dinitrotoluene	1	220–270	1.556	30.59	−302
1,4-Dioxane	1	20–105	7.431 55	1 554.68	240.34
Dipentene	1	21–170	7.111 6	1 613.42	207.8
2,2′-Diphenol	1	171–325	8.193 5	3 067.6	253.1
Diphenyldichlorosilane	1	192–281	6.999 03	1 918.20	161.41
Diphenyl ether	1	204–271	7.011 04	1 799.71	177.74
Diphenylmethane	1	217–282	6.291	1 261	105
Di-*n*-propyl ether	1	26–89	6.947 6	1 256.5	219.0
Disilanyl chloride	1	−46 to 18	7.104 8	1 211.8	245.2
2,3-Dithiabutane	1	6–135	6.977 92	1 346.342	218.86
5,6-Dithiadecane	1	101–263	6.963 8	1 684.1	181.3
3,4-Dithiahexane	1	40–182	6.975 07	1 485.970	208.96
4,5-Dithiaoctane	1	72–226	6.975 29	1 603.793	195.85
Dodecane	1	91–247	6.997 95	1 639.27	181.84
1-Dodecanethiol	1		7.024 4	1 817.8	164.1
Dodecanoic acid	1	106–176	7.860 8	2 159.1	143.2
1-Dodecanol	1	138–214	7.539 86	2 003.29	168.13
1-Dodecene	1	89–244	6.976 07	1 621.11	182.45
Durenol	1	108–249	7.758	2 432	250
Eicosane	1	198–379	7.152 2	2 032.7	132.1
1-Eicosanethiol	1		7.114	2 125	119
1-Eicosene	1	liq	7.135 1	2 043.0	137.9
Ethane	1	−142 to −75	6.829 15	663.72	256.68
Ethanethiol	1	−49 to 56	6.952 06	1 084.531	231.39
Ethanol	1	−2 to 100	8.321 09	1 718.10	237.52
Ethanolamine	1	65–171	7.456 8	1 577.67	173.37
Ethyl acetate	1	15–76	7.101 79	1 244.95	217.88
m-Ethylacetophenone	1	19–143	3.767 2	708.05	182.6
p-Ethylacetophenone	1	21–94	4.274 6	629.34	120.9
Ethylamine	1	−20 to 90	7.054 13	987.31	220.0
N-Ethylaniline	1	50–207	7.422 8	1 903.4	214.3
Ethylbenzene	1	26–164	6.957 19	1 424.255	213.21
2-Ethyl-1-butene	1	−28 to 88	6.997 12	1 218.352	231.30
Ethyl butyl ether	1	38–92	6.944 4	1 256.4	216.9
Ethyl chloroacetate	1	25–146	6.967	1 355.9	188.2
p-Ethylchlorobenzene	1	109–184	6.951 1	1 557.1	198.1
Ethylcyclohexane	1	20–160	6.867 28	1 382.466	214.99
Ethylcyclopentane	1	−0.1 to 129	6.887 09	1 298.599	220.68
Ethylene	1	−153 to −91	6.744 19	594.99	256.16

TABLE 2.37 Vapor Pressures of Various Organic Compounds (*Continued*)

Substance	Eq.	Range, °C	A	B	C
Ethylene glycol	1	50–200	8.090 8	2 088.9	203.5
Ethylene glycol monoethyl ether	1	63–134	7.874 6	1 843.5	234.2
Ethylene glycol monomethyl ether	1	56–124	7.849 8	1 793.9	236.9
Ethylene oxide	1	−49 to 12	7.128 43	1 054.54	237.76
Ethyl formate	1	4–54	7.009 0	1 123.94	218.2
3-Ethylhexane	1		6.890 98	1 327.88	212.60
2-Ethyl-1-hexanol	1	74–184	6.914 7	1 339.7	147.8
2-Ethyl-2-hexenal	1	54–175	6.861 3	1 457.4	190.6
Ethyl iodoacetate	1	29–89	4.073 7	374.64	54.8
Ethyl isothiocyanate	1	10–50	7.106 0	1 567.5	234.2
Ethyl methyl ether	1	5–7.7	5.518	434.5	158
Ethyl methyl ketone	1		6.974 21	1 209.6	216
3-Ethyl-5-methylphenol	1	195–247	7.040 83	1 615.44	152.6
2-Ethyl-4-methyl-1-pentanol	1	70–176	6.582 6	1 134.6	129.2
Ethyl nitrate	1	0–60	7.163 7	1 338.8	224.9
3-Ethylpentane	1	−7 to 119	6.875 64	1 251.827	219.89
2-Ethylphenol	1	86–208	7.800 3	2 140.4	227
3-Ethylphenol	1	97–218	7.468	1 856	187
4-Ethylphenol	1	101–218	8.291	2 423	229
Ethyl phenyl ether	1	117–181	7.021 38	1 508.39	194.49
Ethyl n-propanoate	1	34–98	6.994 9	1 260.6	207.4
Ethyl n-propyl ether	1	20–63	6.985 1	1 188.5	226.4
Ethyl n-propyl ketone	1	75–133	7.000 82	1 365.79	208.01
m-Ethylstyrene	1		7.039 28	1 614.0	198
p-Ethylstyrene	1		6.900 71	1 570.9	198
Ethyl trichloroacetate	1	44–95	7.725 4	1 927.0	233.7
Ethyl trichlorosilane	1	28–96	6.606	1 118	201
Ethyl triethoxysilane	1	64–153	6.886 8	1 377.9	183.0
Ethyl vinyldichlorosilane	1	45–122	6.859	1 331	210.8
Fenchyl alcohol	1	59–200	5.693	797.6	84.6
Fluoranthene	1	197–384	6.373	1 756	118
Fluorene	1	161–300	7.761 8	2 637.1	243.2
Fluorobenzene	1	−18 to 84	7.187 0	1 381.8	235.6
m-Fluorobenzotrifluoride	1	40–137	7.006 59	1 304.35	215.67
bis-(Fluorocarbonyl)-peroxide	1	−47 to −7	9.608 4	2 247.64	319.83
p-Fluorotoluene	1	68–155	6.994 26	1 374.055	217.40
Formaldehyde	1	−109 to −22	7.195 8	970.6	244.1
Formic acid	1	37–101	7.581 8	1 699.2	260.7
Formyl fluoride	1	−95 to −61	5.270	362	175
Furan	1	2–61	6.975 27	1 060.87	227.74
2-Furfuraldehyde	1	56–161	6.575 9	1 198.7	162.8
Glycerol	1	183–260	6.165	1 036	28
Glyceryl-1,3-diacetate	1	100–190	6.407 3	1 092.0	119.3
Guaiacol	1	82–205	6.161	1 051	116
Hemellitenol	1	123–248	6.972	1 563	134
Heptadecane	1	161–337	7.014 3	1 865.1	149.20
1-Heptadecene	1		7.008 67	1 868.9	152.50
Heptane	1	−2 to 124	6.896 77	1 264.90	216.54
1-Heptanethiol	1	58–206	6.952 49	1 525.311	197.70
Heptanoic acid	1	112–150	5.287 4	665.54	42.07
1-Heptanol	1	60–176	6.647 67	1 140.64	126.56
1-Heptene	1	−6 to 118	6.901 87	1 258.345	219.30
Hexadecane	1	149–321	7.028 67	1 830.51	154.45

(*Continued*)

TABLE 2.37   Vapor Pressures of Various Organic Compounds (*Continued*)

Substance	Eq.	Range, °C	A	B	C
1-Hexadecanethiol	1		7.075	1 990	140
1-Hexadecanol	1	50–103	7.281 7	1 909.7	128.1
	1	145–190	6.158 6	1 380.0	91
1-Hexadecene	1		7.040 11	1 840.52	157.57
1,5-Hexadiene	1	0–59	6.574 1	1 013.5	214.8
Hexafluoroacetone	1	−79 to −27	6.650 2	725.90	219.9
Hexafluorobenzene	1	5–114	7.032 95	1 227.98	215.49
Hexafluorodisiloxane	1	−39 to −23	7.471 2	1 169.3	278.1
Hexafluoroethane	1	−93 to −78	6.793 35	657.06	246.2
Hexahydroindane *cis*	1	77–168	6.868 22	1 497.33	207.67
trans	1	71–161	6.861 19	1 475.70	209.66
Hexamethyldisiloxane	1	36–138	6.773 79	1 202.03	208.25
Hexane	1	−25 to 92	6.876 01	1 171.17	224.41
1-Hexanethiol	1	40–181	6.946 64	1 454.004	204.95
1-Hexanol	1	35–157	7.860 45	1 761.26	196.66
2-Hexanol	1	25–142	7.261 0	1 371.7	173.2
3-Hexanol	1	25–138	7.689	1 670.0	211.8
1-Hexene	1	16–64	6.857 70	1 148.62	225.35
3-Hexyne	1	−20 to 24	5.895	863.3	194
Hydroquinone	1	159–286	8.137	2 461	183
3-Hydroxy-3-methyl-2-butanone	1	45–146	7.340 9	1 653.6	227.5
Iodobenzene	1	20–188	7.011 9	1 640.1	208.8
Iodoethane	1	30–60	6.959	1 232	229
Isoamyl acetate	1	41–95	7.436	1 606.6	216
Isobutylbenzene	1	86–174	6.935 56	1 530.05	204.59
Isobutyl borate	1	99–200	7.197	1 745.8	193
Isobutyl cellosolve	1	71–159	7.694 8	1 825.9	219.6
Isobutylcyclohexane	1	85–172	6.867 97	1 493.10	203.16
Isobutyl nitrate	1	0–70	8.164 3	2 022.7	262.4
Isobutyraldehyde	1	13–63	6.735 1	1 053.2	209.1
Isobutyric acid	1	58–152	4.894	382.6	38
Isocaproic acid	1	96–133	6.258	1 038.6	130
Isopropylbenzene	1	39–181	6.936 66	1 460.793	207.78
Isopropyl borate	1	65–139	8.070	2 120	269
o-Isopropylbromobenzene	1	132–210	6.717 8	1 462.7	170.9
Isopropyl caprate	1	90–178	9.959	4 013.9	326.5
Isopropyl caprylate	1	65–146	8.032 2	2 213.6	220.9
Isopropyl cellosolve	1	67–140	7.500 0	1 639.2	213.3
Isopropyl chloroacetate	1	35–153	8.382	2 328	275
Isopropylcyclohexane	1	71–155	6.873 14	1 453.20	209.44
Isopropylcyclopentane	1	47–127	6.887 36	1 380.12	218.05
Isopropyl laurate	1	117–196	8.532 6	2 951.6	240.7
Isopropyl myristate	1	140–193	10.418 0	4 866.48	314.17
Isopropyl nitrate	1	0–70	7.266 6	1 434.4	255.2
Isopropyl palmitate	1	160–197	10.916 4	5 572.0	364.8
o-Isopropylphenol	1	97–215	8.167	2 343	229
p-Isopropylphenol	1	108–228	8.666	2 810	258
Isopropyl phenyl ether	1	72–175	6.517 6	1 238.0	163.0
Isopropyl stearate	1	182–207	0.079 3	10.41	−221
Isopseudocumenol	1	106–233	5.602	768	49
Isoquinoline	1	167–244	6.912 2	1 723.4	184.3
Isovaleric acid	1	86–104	3.946 55	255.41	11.3
Ketene	1	−88 to −49	7.615	1 036	269

TABLE 2.37 Vapor Pressures of Various Organic Compounds (*Continued*)

Substance	Eq.	Range, °C	A	B	C
Laurie acid	1	106–176	7.860 8	2 159.1	143.2
Lepidine	1	199–266	7.271 2	1 946.14	177.64
2,3-Lutidine	1	155–162	7.447 8	1 832.6	240.1
2,4-Lutidine	1	150–160	7.339 0	1 733.4	230.4
2,5-Lutidine	1	85–157	7.081 0	1 539.6	209.6
2,6-Lutidine	1	79–144	7.056 7	1 470.2	208.0
3,4-Lutidine	1	172–180	7.362 0	1 840.1	231.5
3,5-Lutidine	1	163–173	7.333 1	1 783.6	228.7
Mesitol	1	94–221	6.659	1 392	148
Mesityl oxide	1	14–130	6.635 8	1 186.1	186.0
Methacrylonitrile	1		6.980 2	1 274.96	220.7
Methane c	1	−195 to −183	7.193 09	451.64	268.49
liq	1	−181 to −152	6.695 61	405.42	267.78
Methanol	1	−14 to 65	7.897 50	1 474.08	229.13
	1	64–110	7.973 28	1 515.14	232.85
Methoxybenzene	1	110–164	7.052 69	1 489.99	203.57
N-Methylacetamide	1	40–90	2.631 1	121.7	−9.3
Methyl acetate	1	1–56	7.065 2	1 157.63	219.73
Methylal	1	0–35	6.872 2	1 049.2	220.6
Methylamine	1	−83 to −6	7.336 9	1 011.5	233.3
N-Methylaniline	1	50–200	7.081 9	1 631.3	192.4
Methyl benzoate	1	111–199	7.273	1 847	221
Methyl borate	1	31–68	7.646 0	1 491.5	245.5
Methyl boric anhydride	1	0–55	8.004 1	1 726.1	257.9
2-Methyl-1,3-butadiene	1	−52 to −24	7.011 87	1 126.159	238.88
	1	−19 to 55	6.885 64	1 071.578	233.51
3-Methyl-1,2-butadiene	1	−45 to −20	7.151 95	1 194.537	239.47
	1	−20 to 62	6.943 50	1 103.901	230.89
2-Methylbutane	1	−57 to 49	6.833 15	1 040.73	235.45
2-Methyl-1-butanethiol	1	liq	6.913 85	1 347.317	215.07
3-Methyl-1-butanethiol	1	liq	6.914 91	1 342.509	214.45
2-Methyl-2-butanethiol	1	liq	6.828 37	1 254.885	218.76
2-Methyl-1-butanol	1	34–129	7.067 30	1 195.26	156.83
3-Methyl-1-butanol	1	25–153	7.258 21	1 314.36	169.36
2-Methyl-2-butanol	1	25–102	6.519 3	863.4	135.3
3-Methyl-2-butanol	1	25–111	6.942 1	1 090.9	157.2
2-Methyl-1-butene	1	−53 to 52	6.846 37	1 039.69	236.65
3-Methyl-1-butene	1	−63 to 41	6.824 55	1 012.37	236.65
2-Methyl-2-butene	1	−48 to 60	6.966 59	1 124.33	236.63
Methyl butyl ether	1	23–69	6.887 1	1 162.1	219.9
3-Methyl-1-butyne	1	−55 to 47	6.884 80	1 014.81	227.11
2-Methyl-3-butyn-2-ol	1	21–106	6.657 5	976.5	154.1
Methyl *n*-butyrate	1		6.972 11	1 272.73	208.5
Methyl caprate	1	107–188	7.190 0	1 783.8	181.6
Methyl caproate	1	44–105	7.409 3	1 672.74	218.98
Methyl caprylate	1	100–146	6.916 5	1 496.3	176.5
Methyl carbitol	1	112–193	7.424	1 751	192
Methyl cellosolve acetate	1	70–144	7.125 1	1 447.0	196.1
Methyl chloroacetate	1	45–130	7.004 4	1 306.3	187.3
Methylcyclohexane	1	−3 to 127	6.823 00	1 270.763	221.42
Methylcyclopentane	1	−24 to 96	6.862 83	1 186.059	226.04
Methyldichlorosilane	1	1–41	7.027 8	1 167.8	240.7
1-Methyl-2-ethylbenzene	1	48–194	7.003 14	1 535.374	207.30

(*Continued*)

TABLE 2.37   Vapor Pressures of Various Organic Compounds (*Continued*)

Substance		Eq.	Range, °C	A	B	C
1-Methyl-3-ethylbenzene		1	46–190	7.015 82	1 529.184	208.51
1-Methyl-4-ethylbenzene		1	46–191	6.998 02	1 527.113	208.92
1-Methyl-1-ethylcyclopentane		1	43–122	6.859 20	1 347.602	217.21
l-Methyl-2-ethylcyclopentane	*cis*	1	49–129	6.905 88	1 388.412	216.89
2-Methyl-3-ethylpentane		1		6.867 31	1 318.12	215.31
3-Methyl-3-ethylpentane		1		6.867 31	1 347	219.68
3-Methyl-5-ethylphenol		1	111–233	7.958	2 236	208
2-Methyl-5-ethylpyridine		1	52–177	5.050	517	59
N-Methylformamide		1	96–200	7.497 4	1 849.4	201.1
Methyl formate		1	21–32	3.027	3.02	−11.9
2-Methylheptane		1	42–119	6.917 35	1 337.47	213.69
3-Methylheptane		1	43–120	6.899 44	1 331.53	212.41
4-Methylheptane		1		6.900 65	1 327.66	212.57
2-Methylhexane		1	−9 to 115	6.873 18	1 236.026	219.55
3-Methylhexane		1	−8 to 117	6.867 64	1 240.196	219.22
Methylhydrazine		1	2–25	6.576 2	1 007.5	181.4
N-Methylhydroxylamine		1	40–65	7.045 6	1 223.3	172.1
O-Methylhydroxylamine		1	−63 to 48	7.363 9	1 225.3	225.2
Methyl isobutyl ketone		1	22–116	6.672 7	1 168.4	191.9
1-Methyl-2-isopropylbenzene		1	liq	6.940 4	1 548.05	203.15
1-Methyl-3-isopropylbenzene		1	liq	6.940 5	1 539.05	203.93
1-Methyl-4-isopropylbenzene		1	liq	6.923 7	1 537.06	203.05
3-Methylisoquinoline		1	176–225	6.969 2	1 717.3	166.9
Methyl isothiocyanate		1	10–50	2.896 8	103.6	45.4
Methyl laurate		1	158–212	6.767 1	1 589.72	140.5
Methyl linolate		1	166–206	6.111 1	1 660.1	118.8
Methyl methacrylate		1	39–89	8.409 2	2 050.5	274.4
Methyl myristate		1	166–238	7.622 3	2 283.93	184.8
1-Methylnaphthalene		1	108–278	7.035 92	1 826.948	195.00
2-Methylnaphthalene		1	105–274	7.068 50	1 840.268	198.40
Methyl oleate		1	166–205	7.544 1	2 656.9	200.7
Methyl palmitate		1	148–202	9.594 4	4 146.43	297.76
2-Methylpentane		1	−32 to 83	6.839 10	1 135.410	226.57
3-Methylpentane		1	−30 to 87	6.848 87	1 152.368	227.13
2-Methyl-2-pentanethiol		1	56–165	6.858 5	1 343.79	212.8
2-Methyl-1-pentanol		1	25–150	7.520 1	1 564.7	189.2
2-Methyl-4-pentanol		1	25–133	8.467 1	2 174.9	257.8
2-Methyl-1-pentene		1	−30 to 85	6.850 30	1 138.516	224.70
3-Methyl-1-pentene		1	−38 to 77	6.755 23	1 086.316	226.20
4-Methyl-1-pentene		1	−38 to 77	6.835 29	1 121.302	229.68
2-Methyl-2-pentene		1	−26 to 90	6.923 67	1 183.837	225.51
3-Methyl-2-pentene	*cis*	1	−26 to 91	6.910 73	1 186.402	226.70
	trans	1	−23 to 94	6.926 34	1 194.527	224.83
4-Methyl-2-pentene	*cis*	1	−35 to 79	6.841 29	1 120.707	226.59
	trans	1	−33 to 81	6.880 30	1 142.874	227.14
Methyl phenyl ether		1	110–164	7.052 69	1 489.99	203.57
2-Methylpiperidine		1	51–158	6.818 59	1 274.61	205.40
2-Methylpropane		1	−87 to 7	6.910 48	946.35	246.68
2-Methyl-1-propanethiol		1	−10 to 113	6.887 46	1 237.282	220.31
2-Methyl-2-propanethiol		1	1–88	6.787 81	1 115.565	221.31
2-Methyl-1-propanol		1	20–115	7.327 05	1 248.48	172.92
2-Methyl-2-propanol		1	26–83	9.170 6	2 206.4	267.9
2-Methylpropene		1	−82 to 12	6.684 66	866.25	234.64

TABLE 2.37   Vapor Pressures of Various Organic Compounds (*Continued*)

Substance	Eq.	Range, °C	A	B	C
N-Methylpropionamide	1	30–90	−0.9103	119.4	−148.0
Methyl propionate	1	21–79	6.942 4	1 170.2	208.8
2-Methyl-2-propylamine	1	19–75	6.783 2	993.33	210.50
Methyl propyl ether	1	0–39	6.118 6	708.69	179.9
2-Methylpyridine	1	80–168	7.032 4	1 415.73	211.63
3-Methylpyridine	1	74–185	7.050 21	1 481.78	211.25
4-Methylpyridine	1	75–186	7.041 77	1 480.68	210.50
1-Methylpyrrole	1	49–149	7.085 0	1 368.66	212.80
6-Methylquinoline	1	187–266	6.927 2	1 746.08	166.46
7-Methylquinoline	1	238–258	7.597 7	2 229.4	214.9
Methyl salicylate	1	79–220	7.083 3	1 712.8	187.1
Methyl stearate	1	204–240	2.357 0	68.92	−156.5
o-Methylstyrene	1	32–112	7.212 9	1 664.08	214.59
	1	75–255	6.884 61	1 485.41	200.0
m-Methylstyrene	1	10–72	7.275 34	1 695.4	220.0
	1	72–250	6.879 28	1 471.44	200.0
p-Methylstyrene	1	68–170	7.011 2	1 535.1	200.7
α-Methylstyrene	1		6.923 66	1 486.88	202.4
β-Methylstyrene	1		6.923 39	1 499.80	201.0
Methyl sulfoxide	1	20–50	7.763 7	2 048.7	231.6
3-Methyl-2-thiabutane	1	−13 to 109	6.901 96	1 232.170	221.67
2-Methylthiacyclopentane	1	liq	6.944 12	1 409.503	214.41
3-Methylthiacyclopentane	1	67–179	6.949 1	1 431.8	213.6
2-Methyl-3-thiapentane	1	liq	6.891 30	1 293.05	215.04
Methyl-2-thiazole	1	80–128	7.042 1	1 407.05	209.33
2-Methylthiophene	1	9–138	6.938 97	1 326.48	214.31
3-Methylthiophene	1	11–141	6.986 11	1 363.83	216.78
Methyl trichlorosilane	1	13–64	7.088 2	1 289.2	239.9
2-Methyl-5-vinylpyridine	1	69–183	6.156	1 023	129
Morpholine	1	0–44	7.718 13	1 745.8	235.0
	1	44–170	7.160 30	1 447.70	210.0
Naphthalene c	1	86–250	7.010 65	1 733.71	201.86
liq	1	125–218	6.818 1	1 585.86	184.82
1-Naphthol	1	141–282	7.284 21	2 077.56	184.0
2-Naphthol	1	144–288	7.347 14	2 135.00	183.0
Nicotine	1	134–246	6.789	1 650	176
o-Nitroaniline	2	150–260	8.868 4	3 336.50	
m-Nitroaniline	2	170–260	8.818 8	3 440.9	
p-Nitroaniline	2	190–260	9.559 5	4 039.73	
Nitrobenzene	1	134–211	7.115 6	1 746.6	201.8
m-Nitrobenzotrifluoride	1	10–105	7.653 15	2 006.1	220.0
	1	104–280	7.180 25	1 710.60	195.12
Nitromethane	1	56–136	7.281 66	1 446.94	227.60
1-Nitropropane	1	59–131	7.114 6	1 467.45	215.23
o-Nitrotoluene	1	129–222	5.851	946	96
p-Nitrotoluene	1	148–233	6.994 8	1 720.39	184.9
Nonadecane	1	184–366	7.015 3	1 932.8	137.6
1-Nonadecene	1	liq	7.115 1	1 997.4	142.7
Nonafluorocyclopentane	1	17–75	6.945 3	1 051.7	220.1
Nonane	1	39–179	6.938 93	1 431.82	202.01
1-Nonanethiol	1	93–251	6.983 9	1 655.6	183.7
Nonanoic acid	1	137–177	3.235 9	143.97	−75.6
1-Nonanol	1	94–214	7.827 8	1 953.8	181.9

(*Continued*)

TABLE 2.37   Vapor Pressures of Various Organic Compounds (*Continued*)

Substance	Eq.	Range, °C	A	B	C
1-Nonene	1	35–175	6.954 30	1 436.20	205.69
Octadecane	1	172–352	7.002 2	1 894.3	143.30
1-Octadecanethiol	1	liq	7.096	2 061	129
1-Octadecanol	1	120–218	6.461 6	1 599	90
1-Octadecene	1		7.060 65	1 997.4	147.50
Octane	1	19–152	6.918 68	1 351.99	209.15
1-Octanethiol	1	76–229	6.969 09	1 593.0	190.61
1-Octanol	1	0–80	12.070 1	4 506.8	319.9
	1	70–195	6.837 90	1 310.62	136.05
2-Octanol	1	72–180	6.388 8	1 060.4	122.5
3-Octanol	1	76–176	5.221 5	560.3	64.7
4-Octanol	1	71–176	5.739 6	760.5	89.5
1-Octene	1	15–147	6.934 95	1 355.46	213.05
5-Oxyhydrindene	1	120–251	9.213 7	3 665.8	326.4
Pentachloroethane	1	25–162	6.740	1 378	197
Pentadecane	1	136–304	7.023 59	1 789.95	161.38
1-Pentadecene	1		7.022 91	1 788.58	163.347
1,2-Pentadiene	1	−42 to −26	7.259 90	1 250.293	241.96
	1	−21 to 67	6.918 20	1 104.991	228.85
1,3-Pentadiene *cis*	1	−43 to −22	7.193 87	1 223.602	240.62
	1	−18 to 66	6.910 89	1 101.923	229.37
trans	1	−45 to −20	7.102 12	1 185.389	239.41
	1	−18 to 64	6.913 17	1 103.840	231.72
1,4-Pentadiene	1	−57 to −37	7.174 01	1 155.378	244.30
	1	−33 to 47	6.835 43	1 017.995	231.46
2,3-Pentadiene	1	−39 to −18	7.202 53	1 231.768	237.56
	1	−14 to 70	6.962 16	1 126.837	227.84
Pentafluorobenzene	1	49–94	7.036 65	1 254.07	216.02
Pentafluorochloroacetone	1	−40 to 32	6.848 4	925.3	225.4
Pentafluorochlorethane	1	−95 to −39	6.833 34	802.97	242.27
Pentafluorophenol	1	105–155	7.066 0	1 379.15	183.91
2,2,3,3,3-Pentafluoropropanol	1	0–23	6.308 7	830.56	153.8
Pentafluorotoluene	1	39–138	7.084 78	1 392.20	213.67
bis-Pentamethyldisilanoxydisilane	1	169–201	8.556 64	3 051.316	258.85
bis-Pentamethyldisilanyl ether	1	88–183	8.161 44	2 575.250	273.32
Pentane	1	−50 to 58	6.852 96	1 064.84	233.01
Pentanenitrile	1	69–141	7.104 9	1 519.4	218.4
1-Pentanethiol	1	19–153	6.933 11	1 369.479	211.31
Pentanoic acid	1	72–174	5.412	591	60
1-Pentanol	1	37–138	7.177 58	1 314.56	168.11
2-Pentanol	1	25–120	7.275 75	1 271.92	170.37
3-Pentanol	1	21–116	7.414 93	1 354.42	183.41
2-Pentanone	1	56–111	7.021 93	1 313.85	215.01
3-Pentanone	1	56–111	7.025 29	1 310.28	214.19
1-Pentene	1	−55 to 51	6.844 24	1 044.01	233.50
2-Pentene *cis*	1	−49 to 58	6.843 08	1 052.44	228.69
trans	1	−49 to 58	6.899 83	1 080.76	232.57
1-Pentyne	1	−44 to 61	6.967 34	1 092.52	227.18
2-Pentyne	1	−33 to 78	7.046 14	1 189.87	229.60
Perdeuterobenzene	1	10–82	6.892 35	1 198.39	219.43
Perdeuterocyclohexane	1	10–80	6.837 86	1 190.38	222.40
Perfluorobutane	1	−39 to −4	7.035 1	990.27	240.4
Perfluorobutene	1	−28 to 20	9.222	2 401.6	382

TABLE 2.37   Vapor Pressures of Various Organic Compounds (*Continued*)

Substance	Eq.	Range, °C	A	B	C
Perfluorocyclobutane	1	−32 to 0	6.815 29	862.49	225.19
Perfluorocyclohexane	1	19–65	6.04	597	136
Perfluorocyclopentane	1	17–56	7.039 6	1 069.3	234.6
Perfluoroheptane	1	−2 to 106	6.937 72	1 181.14	208.66
Perfluorohexane	1	30–57	6.875 2	1 080.8	213.4
Perfluoromethylcyclohexane	1	33–111	6.824 06	1 133.76	211.22
Perfluorooctane	1	37–105	5.902 5	1 225.93	198.99
Perfluoropentane	1	9–65	7.017 9	1 072.9	230.0
Perfluoropiperidine	1	29–81	6.853 4	1 059.95	217.2
Perfluoropropane	1	−79 to −36	6.919 4	825.8	241.2
Perfluoropropene	1	−41 to 20	7.355	1 012.1	257
Phenanthrene	1	176–379	7.260 82	2 379.04	203.76
Phenol	1	107–182	7.133 0	1 516.79	174.95
β-Phenylethyl acetate	1	149–233	6.834 3	1 555.2	160.8
α-Phenylethyl alcohol	1	82–190	1.508	91	−263
o-Phenylethylphenol	1	169–250	4.506 0	516.8	−32.1
p-Phenylethylphenol	1	174–251	4.304 1	459.3	−52.4
Phenylisocyanate	1	10–80	−0.708 0	106.4	−146.6
4-Phenylphenol	1	177–308	8.657 5	3 022.8	216.1
Phosgene	1	−68 to 68	6.842 97	941.25	230
Phthalic anhydride	2	160–285	8.022	2 868.5	
α-Pinene	1	19–156	6.852 5	1 446.4	208.0
β-Pinene	1	19–166	6.898 4	1 511.7	210.2
Piperidine	1	42–144	6.855 69	1 238.80	205.43
Propadiene	1	−99 to −16	5.713 7	458.06	196.07
Propane	1	−108 to −25	6.803 38	804.00	247.04
1-Propanethiol	1	−25 to 91	6.928 46	1 183.307	224.62
2-Propanethiol	1	−37 to 75	6.877 34	1 113.895	226.16
1-Propanol	1	2–120	7.847 67	1 499.21	204.64
2-Propanol	1	0–101	8.117 78	1 580.92	219.61
2-Propen-l-ol	1	21–97	11.187 0	4 068.5	392.7
Propionic acid	1	56–139.5	6.403	950.2	130.3
Propionic anhydride	1	67–167	5.819 5	810.3	108.7
Propionitrile	1	−84 to 22	5.278 2	665.52	159.10
Propiophenone	1	132–201	7.370	1 894	205
Propyl acetate	1	39–101	7.016 15	1 282.28	208.60
1-Propylamine	1	23–77	6.926 51	1 044.05	210.84
2-Propylamine	1	4–61	6.890 25	985.69	214.07
n-Propylbenzene	1	43–188	6.951 42	1 491.297	207.14
n-Propyl borate	1	85–179	7.399 8	1 741	206
n-Propyl caprate	1	97–186	8.701 22	2 945.99	253.63
n-Propyl caproate	1	43–120	8.667 1	2 556.0	262.9
n-Propyl caprylate	1	70–153	8.516 7	2 599.5	246.2
n-Propyl cellosolve	1	77–149	7.146 4	1 440.6	187.7
n-Propylcyclohexane	1	40–186	6.886 46	1 460.800	207.94
n-Propylcyclopentane	1	21–158	6.903 92	1 384.386	213.16
Propylene	1	−112 to −32	6.778 11	770.85	245.51
1,2-Propylene oxide	1	−35 to 130	7.064 92	1 113.6	232
n-Propyl formate	1	26–82	6.848	1 127	203
n-Propyl laurate	1	124–205	8.068 9	2 692.4	222.5
n-Propyl myristate	1	147–200	9.216 8	3 744.68	272.87
n-Propyl nitrate	1	0–70	6.954 9	1 294.4	206.7
n-Propyl palmitate	1	166–204	14.129 2	9 759.2	539.7

(*Continued*)

TABLE 2.37   Vapor Pressures of Various Organic Compounds (*Continued*)

Substance	Eq.	Range, °C	A	B	C
o-(n-Propyl)phenol	1	104–222	9.215	3 254	292
p-(n-Propyl)phenol	1	0–234	8.329 6	2 661	254
n-Propyl phenyl ether	1	101–190	7.734 3	2 146.2	252.3
Propyne	1	−90 to −6	6.784 85	803.73	229.08
Pseudocumenol	1	107–232	6.915	1 547	152
Pyrene	1	200–395	5.618 4	1 122.0	15.2
Pyridine	1	67–153	7.041 15	1 373.80	214.98
Pyrogallol	1	177–309	6.092	1 031	12
Pyrrole	1	66–166	7.294 70	1 501.56	210.42
Quinaldine	1	178–248	7.179 00	1 857.84	184.50
Quinoline	1	164–238	6.817 59	1 668.73	186.26
Spiropentane	1	3–71	6.917 00	1 090.08	231.10
Styrene	1	32–82	7.140 16	1 574.51	224.09
Terpenyl acetate	1	37–150	6.443 46	1 377.27	143.85
α-Terpineol	1	84–217	8.141 2	2 479.4	253.7
Terpinolene	1	40–179	7.169	1 706	211
Tetrabutyl tin	1	100–300	6.545	1 649	148
1,1,2,2-Tetrachloro-1,2-difluoro-ethane	1	10–91.5	10.995	4 437.1	455.2
1,1,1,2-Tetrachloroethane	1	59–130	6.898 75	1 365.88	209.74
1,1,2,2-Tetrachloroethane	1	25–130	6.631 7	1 228.1	179.9
Tetrachloroethylene	1	37–120	6.976 83	1 386.92	217.53
Tetrachloromethane	1		6.879 26	1 212.021	226.41
Tetradecane	1	122–286	7.013 00	1 740.88	167.72
1-Tetradecanethiol	1		7.048 5	1 909.2	151.9
1-Tetradecanol	1	130–264	6.674 1	1 204.5	54.0
1-Tetradecene	1	119–283	7.030 65	1 754.09	171.52
1,2,3,4-Tetrafluorobenzene	1	6–50	7.084 6	1 339.23	223.49
1,2,3,5-Tetrafluorobenzene	1	6–50	6.986 17	1 245.20	218.35
Tetrafluoroethylene	1	−131 to −65	6.896 59	683.84	245.93
Tetrafluoromethane	1		6.972 31	540.50	260.10
Tetrahydrofuran	1	23–100	6.995 15	1 202.29	226.25
Tetraiodothiophene	1	−65 to 24	5.585 44	871.25	175.59
Tetralin	1	94–206	7.070 55	1 741.30	208.26
1,2,3,4-Tetramethylbenzene	1	80–217	7.059 4	1 690.54	199.48
1,2,3,5-Tetramethylbenzene	1	75–228	7.077 9	1 675.43	201.14
1,2,4,5-Tetramethylbenzene	1	74–227	7.080 0	1 672.43	201.43
2,2,3,3-Tetramethylbutane	1	0–65	6.876 65	1 329.93	226.36
Tetramethyl lead	1	0–60	6.937 7	1 335.3	219.1
2,2,3,3-Tetramethylpentane	1	57–141	6.830 60	1 398.67	213.84
2,2,3,4-Tetramethylpentane	1	52–134	6.834 18	1 375.59	214.94
2,2,4,4-Tetramethylpentane	1	43–123	6.796 20	1 324.59	216.02
Tetramethylsilane	1	−64 to 21	6.822 39	1 033.72	235.62
2-Thiabutane	1	−26 to 90	6.938 49	1 182.562	224.78
Thiacyclobutane	1	−5 to 120	7.016 67	1 321.331	224.51
Thiacyclohexane	1	29–170	6.905 18	1 422.47	211.72
Thiacyclopentane	1	14–148	6.995 40	1 401.939	219.61
Thiacyclopropane	1	−35 to 77	7.037 25	1 194.37	232.42
3-Thiaheptane	1	33–172	6.941 02	1 421.32	205.81
4-Thiaheptane	1	32–170	6.935 77	1 413.44	205.73
2-Thiahexane	1	17–150	6.945 83	1 363.808	212.07
3-Thiahexane	1	14–144	6.933 80	1 341.57	212.51

TABLE 2.37 Vapor Pressures of Various Organic Compounds (*Continued*)

Substance	Eq.	Range, °C	A	B	C
2-Thiapentane	1	−4 to 120	6.955 45	1 284.32	219.66
3-Thiapentane	1	−13 to 109	6.928 36	1 257.833	218.66
2-Thiapropane	1	−47 to 58	6.948 79	1 090.755	230.80
Thiazole	1	63–118	7.142 01	1 425.35	216.26
Thiophene	1	−12 to 108	6.959 26	1 246.02	221.35
Toluene	1	6–137	6.954 64	1 344.800	219.48
o-Toluidine	1	118–200	7.082 03	1 627.72	187.13
m-Toluidine	1	122–203	7.093 67	1 631.43	183.91
p-Toluidine	1		7.260 22	1 758.55	201.0
m-Tolyl pentafluoropropionate	1	98–174	7.427 20	1 707.59	201.70
p-Tolyl pentafluoropropionate	1	99–176	8.078 6	2 223.8	252.1
m-Tolyl trifluoroacetate	1	91–166	7.681 0	1 874.84	223.48
p-Tolyl trifluoroacetate	1	92–169	7.913 8	2 055.41	238.99
Tribromomethane	1	30–101	6.821 8	1 376.7	201.0
1,2,3-Tribromopropane	1	128–205	7.037 2	1 735.32	195.42
Trichloroacetic acid	1	112–198	7.273 0	1 594.3	165.4
Trichloroacetonitrile	1	17–83	7.183 5	1 368.3	232.5
Trichloroacetyl chloride	1	32–119	6.990 75	1 390.47	220.11
1,1,1-Trichloroethane	1	−6 to 17	8.643 4	2 136.6	302.8
1,1,2-Trichloroethane	1	50–114	6.951 85	1 314.41	209.20
Trichloroethylene	1	18–86	6.518 3	1 018.6	192.7
Trichlorofluoromethane	1		6.884 28	1 043.004	236.88
Trichlorosilane	1	2–32	6.773 9	1 009.0	227.2
bis-Trichlorosilylethane	1	91–160	7.835 11	2 241.769	249.84
1,1,1-Trichloro-2,2,2-trifluoro-ethane	1	14–36	4.437 3	204.1	83.9
1,1,2-Trichloro-1,2,2-trifluoro-ethane	1	−25 to 83	6.880 3	1 099.9	227.5
Tridecane	1	107–267	7.007 56	1 690.67	174.22
1-Tridecene	1	105–264	6.981 02	1 672.00	174.95
Triethanolamine	1	252–305	10.067 5	4 542.78	297.76
Triethyl aluminum	1	57–126	11.646 1	4 466.59	322.87
Triethylamine	1	50–95	5.858 8	695.7	144.8
Triethyl borate	1	29–109	7.511 1	1 641.7	236.3
Triethylsilanol	1	24–140	7.793 7	1 756.1	202.4
Trifluoroacetic acid	1	12–72	8.389	1 895	273
Trifluoroacetic anhydride	1	−2 to 39	6.135 8	1 026.1	202.0
Trifluoroacetonitrile	1	−132 to −68	7.127 6	773.82	249.9
1,3,5-Trifluorobenzene	1	6–50	6.919 8	1 197.13	219.12
Trifluorochloroethylene	1	−67 to −11	6.896 16	848.33	293.64
1,1,1-Trifluoroethane	1	−110 to −48	6.903 78	788.20	243.23
2,2,2-Trifluoroethanol	1	−0.5 to 25	6.788 2	978.13	173.06
Trifluoromethane	1	−128 to −82	7.088 6	705.33	249.78
bis-(Trifluoromethyl)-acetoxyphos-phine	1	0–40	7.391 31	1 426.254	220.37
2,2,2-Trifluoro-1-methylbenzene	1	55–139	6.970 45	1 306.35	217.38
bis-(Trifluoromethyl)-chlorophos-phine	1	− 80 to 0	7.661 06	1 386.652	267.14
Trifluoromethylhypofluorite	1	145–189	6.950 6	650.1	−18.4
bis-(Trifluoromethyl)-iodophos-phine	1	0–47	6.901 39	1 180.723	222.95
Triisobutylene	1	56–179	7.002 1	1 613.47	212.5

(*Continued*)

TABLE 2.37 Vapor Pressures of Various Organic Compounds (*Continued*)

Substance	Eq.	Range, °C	A	B	C
Trimethyl aluminum	1	64–127	7.570 29	1 734.72	242.78
Trimethylamine	1	– 80 to 3	6.857 55	955.94	237.52
1,2,3-Trimethylbenzene	1	57–205	7.040 82	1 593.958	207.08
1,2,4-Trimethylbenzene	1	52–198	7.043 83	1 573.257	208.56
1,3,5-Trimethylbenzene	1	49–193	7.074 36	1 569.622	209.58
2,2,3-Trimethylbutane	1	–19 to 106	6.792 30	1 200.563	226.05
Trimethylchlorosilane	1	2–55	7.055 8	1 245.5	240.7
1,1,3-Trimethylcyclohexane	1	55–137	6.839 51	1 394.88	215.73
1,1,2-Trimethylcyclopentane	1	36–115	6.822 38	1 309.81	218.58
1,1,3-Trimethylcyclopentane	1	29–106	6.809 31	1 275.92	219.89
1,2,4-Trimethylcyclopentane					
cis, cis, trans	1	39–118	6.857 38	1 335.69	219.16
cis, trans, cis	1	33–110	6.851 3	1 307.10	219.92
1,3,5-Trimethyl-2-ethylbenzene	1	88–210	6.790 8	1 505.8	174.7
1,4,5-Trimethyl-2-ethylbenzene	1	87–132	3.029 3	116.4	–34.6
2,2,5-Trimethylhexane	1	46–125	6.837 75	1 325.54	210.91
2,4,4-Trimethylhexane	1	51–131	6.856 54	1 371.81	214.40
Trimethylhydrazine	1	–16 to 14	7.106 80	1 189.88	222.06
O,N,N-Trimethylhydroxylamine	1	–79 to 23	6.765 8	979.55	222.2
2,2,3-Trimethylpentane	1		6.825 46	1 294.88	218.42
2,2,4-Trimethylpentane	1	24–100	6.811 89	1 257.84	220.74
2,3,3-Trimethylpentane	1		6.843 53	1 328.05	220.38
2,3,4-Trimethylpentane	1	36–114	6.853 96	1 315.08	217.53
2,4,4-Trimethyl-1-pentene	1	3 to 128	6.834 57	1 273.416	220.62
2,4,4-Trimethyl-2-pentene	1	2–131	6.859 22	1 272.717	214.99
2,3,5-Trimethylphenol	1	186–247	7.080 12	1 685.90	166.14
Trimethylsilanol	1	18–85	8.126 6	1 657.6	219.2
2,4,5-Trimethylstyrene	1	79–216	7.331 5	1 880.7	205.7
2,4,6-Trimethylstyrene	1	90–208	7.089 1	1 702.61	195.93
1,2,4-Trinitrobenzene	1	250–300	3.194	87	–199
1,3,5-Trinitrobenzene	1	202–312	5.534 5	993.6	11.2
2,4,6-Trinitrobenzene	1	249–342	9.621 1	4 987.9	329.9
2,4,6-Trinitrotoluene	1	230–250	7.671 52	2 669.4	205.6
α-Trioxane	1	56–114	7.818 6	1 783.3	247.1
Trivinylarsine	1	22–66	7.894 1	2 115.6	293.9
Trivinyl bismuth	1	20–74	7.237 2	1 667.0	215.1
Trivinylphosphine	1	16–61	7.928 4	2 102.0	301.3
Trivinylstibine	1	20–70	8.322 1	2 446.3	303.8
Undecane	1	75–226	6.972 20	1 569.57	187.70
1-Undecanethiol	1		7.012 2	1 767.4	170.4
1-Undecene	1	72–222	6.966 77	1 563.21	189.87
Urethane	1		7.421 64	1 758.21	205.0
Vinyl acetate	1	22–72	7.210 1	1 296.13	226.66
o-Xylene	1	32–172	6.998 91	1 474.679	213.69
m-Xylene	1	28–166	7.009 08	1 462.266	215.11
p-Xylene	1	27–166	6.990 52	1 453.430	215.31
2,3-Xylenol	1	149–218	7.053 97	1 617.57	170.74
2,4-Xylenol	1	144–212	7.055 39	1 587.46	169.34
2,5-Xylenol	1	144–212	7.051 56	1 592.70	170.74
2,6-Xylenol	1	145–204	7.070 70	1 628.32	187.60
3,4-Xylenol	1	172–229	7.079 19	1 621.45	159.26
3,5-Xylenol	1	155–223	7.130 76	1 639.86	164.16

TABLE 2.38 Boiling Points of Common Organic Compounds at Selected Pressures

Compound Name	Formula	Pressure, mm Hg										Melting point, °C
		1	5	10	20	40	60	100	200	400	760	
		Temperature, °C										
Acenaphthalene	$C_{12}H_{10}$		114.8	131.2	148.7	168.2	181.2	197.5	222.1	250.0	277.5	95
Acetal	$C_6H_{14}O_2$	−23.0	−2.3	+8.0	19.6	31.9	39.8	50.1	66.3	84.0	102.2	
Acetaldehyde	C_2H_4O	−81.5	−65.1	−56.8	−47.8	−37.8	−31.4	−22.6	−10.0	+4.9	20.2	−123.5
Acetamide	C_2H_5NO	65.0	92.0	105.0	120.0	135.8	145.8	158.0	178.3	200.0	222.0	81
Acetanilide	C_8H_9NO	114.0	146.6	162.0	180.0	199.6	211.8	227.2	250.5	277.0	303.8	113.5
Acetic acid	$C_2H_4O_2$	−17.2	+6.3	17.5	29.9	43.0	51.7	63.0	80.0	99.0	118.1	16.7
anhydride	$C_4H_6O_3$	1.7	24.8	36.0	48.3	62.1	70.8	82.2	100.0	119.8	139.6	−73
Acetone	C_3H_6O	−59.4	−40.5	−31.1	−20.8	−9.4	−2.0	+7.7	22.7	39.5	56.5	−94.6
Acetonitrile	C_2H_3N	−47.0	−26.6	−16.3	−5.0	+7.7	15.9	27.0	43.7	62.5	81.8	−41
Acetophenone	C_8H_8O	37.1	64.0	78.0	92.4	109.4	119.9	133.6	154.2	178.0	202.4	20.5
Acetyl chloride	C_2H_3OCl	−50.0	−35.0	−27.6	−19.6	−10.4	−4.5	+3.2	16.1	32.0	50.8	−112.0
Acetylene	C_2H_2	−142.9	−133.0	−128.2	−122.8	−116.7	−112.8	−107.9	−100.3	−92.0	−84.0	−81.5
Acridine	$C_{13}H_9N$	129.4	165.8	184.0	203.5	224.2	238.7	256.0	284.0	314.3	346.0	110.5
Acrolein (2-propenal)	C_3H_4O	−64.5	−46.0	−36.7	−26.3	−15.0	−7.5	+2.5	17.5	34.5	52.5	−87.7
Acrylic acid	$C_3H_4O_2$	+3.5	27.3	39.0	52.0	66.2	75.0	86.1	103.3	122.0	141.0	14
Adipic acid	$C_6H_{10}O_4$	159.5	191.0	205.5	222.0	240.5	251.0	265.0	287.8	312.5	337.5	152
Allene (propadiene)	C_3H_4	−120.6	−108.0	−101.0	−93.4	−85.2	−78.8	−72.5	−61.3	−48.5	−35.0	−136
Allyl alcohol (propen-1-ol-3)	C_3H_6O	−20.0	+0.2	10.5	21.7	33.4	40.3	50.0	64.5	80.2	96.6	−129
chloride (3-chloropropene)	C_3H_5Cl	−70.0	−52.0	−42.9	−32.8	−21.2	−14.1	−4.5	10.4	27.5	44.6	−136.4
isopropyl ether	$C_6H_{12}O$	−43.7	−23.1	−12.9	−1.8	+10.9	18.7	29.0	44.3	61.7	79.5	
isothiocyanate	C_4H_5NS	−2.0	+25.3	38.3	52.1	67.4	76.2	89.5	108.0	129.8	150.7	−80
n-propyl ether	$C_6H_{12}O$	−39.0	−18.2	−7.9	+3.7	16.4	25.0	35.8	52.6	71.4	90.5	
4-Allylveratrole	$C_{11}H_{14}O_2$	85.0	113.9	127.0	142.8	158.3	169.6	183.7	204.0	226.2	248.0	
iso-Amyl acetate	$C_7H_{14}O_2$	0.0	+23.7	35.2	47.8	62.1	71.0	83.2	101.3	121.5	142.0	
n-Amyl alcohol	$C_5H_{12}O$	+13.6	34.7	44.9	55.8	68.0	75.5	85.8	102.0	119.8	137.8	
iso-Amyl alcohol	$C_5H_{12}O$	+10.0	30.9	40.8	51.7	63.4	71.0	80.7	95.8	113.7	130.6	−117.2
sec-Amyl alcohol (2-pentanol)	$C_5H_{12}O$	+1.5	22.1	32.2	42.6	54.1	61.5	70.7	85.7	102.3	119.7	
tert-Amyl alcohol	$C_5H_{12}O$	−12.9	+7.2	17.2	27.9	38.8	46.0	55.3	69.7	85.7	101.7	−11.9
sec-Amylbenzene	$C_{11}H_{16}$	29.0	55.8	69.2	83.8	100.0	110.4	124.1	145.2	168.0	193.0	
iso-Amyl benzoate	$C_{12}H_{16}O_2$	72.0	104.5	121.6	139.7	158.3	171.4	186.8	210.2	235.8	262.0	
bromide (1-bromo-3-methylbutane)	$C_5H_{11}Br$	−20.4	+2.1	13.6	26.1	39.8	48.7	60.4	78.7	99.4	120.4	

(Continued)

TABLE 2.38 Boiling Points of Common Organic Compounds at Selected Pressures (*Continued*)

Name	Formula	\-	\-	\-	\-	Pressure, mm Hg Temperature, °C	\-	\-	\-	\-	\-	Melting point, °C
		1	5	10	20	40	60	100	200	400	760	
n-butyrate	$C_9H_{18}O_2$	21.2	47.1	59.9	74.0	90.0	99.8	113.1	133.2	155.3	178.6	
formate	$C_6H_{12}O_2$	−17.5	+5.4	17.1	30.0	44.0	53.3	65.4	83.2	102.7	123.3	
iodide (1-iodo-3-methylbutane)	$C_5H_{11}I$	−2.5	+21.9	34.1	47.6	62.3	71.9	84.4	103.8	125.8	148.2	
isobutyrate	$C_9H_{18}O_2$	14.8	40.1	52.8	66.6	81.8	91.7	104.4	124.2	146.0	168.8	
Amyl isopropionate	$C_8H_{16}O_2$	+8.5	33.7	46.3	60.0	75.5	85.2	97.6	117.3	138.4	160.2	
iso-Amyl isovalerate	$C_{10}H_{20}O_2$	27.0	54.4	68.6	83.8	100.6	110.3	125.1	146.1	169.5	194.0	
n-Amyl levulinate	$C_{10}H_{18}O_3$	81.3	110.0	124.0	139.7	155.8	165.2	180.5	203.1	227.4	253.2	
iso-Amyl levulinate	$C_{10}H_{18}O_3$	75.6	104.0	118.8	134.4	151.7	162.6	177.0	198.1	222.7	247.9	
nitrate	$C_5H_{11}NO_3$	+5.2	28.8	40.3	53.5	67.6	76.3	88.6	106.7	126.5	147.5	
4-*tert*-Amylphenol	$C_{11}H_{16}O$		109.8	125.5	142.3	160.3	172.6	189.0	213.0	239.5	266.0	93
Anethole	$C_{10}H_{12}O$	62.6	91.6	106.0	121.8	139.3	149.8	164.2	186.1	210.5	235.3	22.5
Angelonitrile	C_5H_7N	−8.0	+15.0	28.0	41.0	55.8	65.2	77.5	96.3	117.7	140.0	
Aniline	C_6H_7N	34.8	57.9	69.4	82.0	96.7	106.0	119.9	140.1	161.9	184.4	−6.2
2-Anilinoethanol	$C_8H_{11}NO$	104.0	134.3	149.6	165.7	183.7	194.0	209.5	230.6	254.5	279.6	
Anisaldehyde	$C_8H_8O_2$	73.2	102.6	117.8	133.5	150.5	161.7	176.7	199.0	223.0	248.0	2.5
o-Anisidine (2-methoxyaniline)	C_7H_9NO	61.0	88.0	101.7	116.1	132.0	142.1	155.2	175.3	197.3	218.5	5.2
Anthracene	$C_{14}H_{10}$	145.0	173.5	187.2	201.9	217.5	231.8	250.0	279.0	310.2	342.0	217.5
Anthraquinone	$C_{14}H_8O_2$	190.0	219.4	234.2	248.3	264.3	273.3	285.0	314.6	346.2	379.9	286
Azelaic acid	$C_9H_{16}O_4$	178.3	210.4	225.5	242.4	260.0	271.8	286.5	309.6	332.8	356.5	106.5
Azelaldehyde	$C_9H_{18}O$	33.3	58.4	71.6	85.0	100.2	110.0	123.0	142.1	163.4	185.0	
Azobenzene	$C_{12}H_{10}N_2$	103.5	135.7	151.5	168.3	187.9	199.8	216.0	240.0	266.1	293.0	68
Benzal chloride (α,α-Dichlorotoluene)	$C_7H_6Cl_2$	35.5	64.0	78.7	94.3	112.1	123.4	138.3	160.7	187.0	214.0	−16.1
Benzaldehyde	C_7H_6O	26.2	50.1	62.0	75.0	90.1	99.6	112.5	131.7	154.1	179.0	−26
Benzanthrone	$C_{17}H_{10}O$	225.0	274.5	297.2	322.5	350.0	368.8	390.0	426.5			174
Benzene	C_6H_6	−36.7	−19.6	−11.5	−2.6	+7.6	15.4	26.1	42.2	60.6	80.1	+5.5
Benzenesulfonylchloride	$C_6H_5ClO_2S$	65.9	96.5	112.0	129.0	147.7	158.2	174.5	198.0	224.0	251.5	14.5
Benzil	$C_{14}H_{10}O_2$	128.4	165.2	183.0	202.8	224.5	238.2	255.8	283.5	314.3	347.0	95
Benzoic acid	$C_7H_6O_2$	96.0	119.5	132.1	146.7	162.6	172.8	186.2	205.8	227.0	249.2	121.7
anhydride	$C_{14}H_{10}O_3$	143.8	180.0	198.0	218.0	239.8	252.7	270.4	299.1	328.8	360.0	42
Benzoin	$C_{14}H_{12}O_2$	135.6	170.2	188.1	207.0	227.9	241.7	258.0	284.4	313.5	343.0	132
Benzonitrile	C_7H_5N	28.2	55.3	69.2	83.4	99.6	109.8	123.5	144.1	166.7	190.6	−12.9
Benzophenone	$C_{13}H_{10}O$	108.2	141.7	157.6	175.8	195.7	208.2	224.4	249.8	276.8	305.4	48.5
Benzotrichloride (α,α,α-Trichlorotoluene)	$C_7H_5Cl_3$	45.8	73.7	87.6	102.7	119.8	130.0	144.3	165.6	189.2	213.5	−21.2

Name	Formula											
Benzotrifluoride (α,α,α-Trifluorotoluene)	C₇H₅F₃	−32.0	−10.3	−0.4	12.2	25.7	34.0	45.3	62.5	82.0	102.2	−29.3
Benzoyl bromide	C₇H₅BrO	47.0	75.4	89.8	105.4	122.6	133.4	147.7	169.2	193.7	218.5	0
chloride	C₇H₅ClO	32.1	59.1	73.0	87.6	103.8	114.7	128.0	149.5	172.8	197.2	−0.5
nitrile	C₈H₅NO	44.5	71.7	85.5	100.2	116.6	127.0	141.0	161.3	185.0	208.0	33.5
Benzyl acetate	C₉H₁₀O₂	45.0	73.4	87.6	102.3	119.6	129.8	144.0	165.5	189.0	213.5	−51.5
alcohol	C₇H₈O	58.0	80.8	92.6	105.8	119.8	129.3	141.7	160.0	183.0	204.7	−15.3
Benzylamine	C₇H₉N	29.0	54.8	67.7	81.8	97.3	107.3	120.0	140.0	161.3	184.5	
Benzyl bromide (α-bromotoluene)	C₇H₇Br	32.2	59.6	73.4	88.3	104.8	115.6	129.8	150.8	175.2	198.5	−4
chloride (α-chlorotoluene)	C₇H₇Cl	22.0	47.8	60.8	75.0	90.7	100.5	114.2	134.0	155.8	179.4	−39
cinnamate	C₁₆H₁₄O₂	173.8	206.3	221.5	239.3	255.8	267.0	281.5	303.8	326.7	350.0	39
Benzyldichlorosilane	C₇H₈Cl₂Si	45.3	70.2	83.2	96.7	111.8	121.3	133.5	152.0	173.0	194.3	
Benzyl ethyl ether	C₉H₁₂O	26.0	52.0	65.0	79.6	95.4	105.5	118.9	139.6	161.5	185.0	
phenyl ether	C₁₃H₁₂O	95.4	127.7	144.0	160.7	180.1	192.6	209.2	233.2	259.8	287.0	
isothiocyanate	C₈H₇NS	79.5	107.8	121.8	137.0	153.0	163.8	177.7	198.0	220.4	243.0	
Biphenyl	C₁₂H₁₀	70.6	101.8	117.0	134.2	152.5	165.2	180.7	204.2	229.4	254.9	69.5
1-Biphenyloxy-2,3-epoxypropane	C₁₅H₁₄O₂	135.5	169.9	187.2	205.8	226.3	239.7	255.0	280.4	309.8	340.0	
d-Bornyl acetate	C₁₂H₂₀O₂	46.9	75.7	90.2	106.0	123.7	135.7	149.8	172.0	197.5	223.0	29
Bornyl n-butyrate	C₁₄H₂₄O₂	74.0	103.4	118.0	133.8	150.7	161.8	176.4	198.0	222.2	247.0	
formate	C₁₁H₁₈O₂	47.0	74.8	89.3	104.0	121.2	131.7	145.8	166.4	190.2	214.0	
isobutyrate	C₁₄H₂₄O₂	70.0	99.8	114.0	130.0	147.2	157.6	172.2	194.2	218.2	243.0	
propionate	C₁₃H₂₂O₂	64.6	93.7	108.0	123.7	140.4	151.2	165.7	187.5	211.2	235.0	
Brassidic acid	C₂₂H₄₂O₂	209.6	241.7	256.0	272.9	290.0	301.5	316.2	336.8	359.6	382.5	61.5
Bromoacetic acid	C₂H₃BrO₂	54.7	81.6	94.1	108.2	124.0	133.8	146.3	165.8	186.7	208.0	49.5
4-Bromoanisole	C₇H₇BrO	48.8	77.8	91.9	107.8	125.0	136.0	150.1	172.7	197.5	223.0	12.5
Bromobenzene	C₆H₅Br	+2.9	27.8	40.0	53.8	68.6	78.1	90.8	110.1	132.3	156.2	−30.7
4-Bromobiphenyl	C₁₂H₉Br	98.0	133.7	150.6	169.8	190.8	204.5	221.8	248.2	277.7	310.0	90.5
1-Bromo-2-butanol	C₄H₉BrO	23.7	45.4	55.8	67.2	79.5	87.0	97.6	112.1	128.3	145.0	
1-Bromo-2-butanone	C₄H₇BrO	+6.2	30.0	41.8	54.2	68.2	77.3	89.2	107.0	126.3	147.0	
cis-1-Bromo-1-butene	C₄H₇Br	−44.0	−23.2	−12.8	−1.4	+11.5	19.8	30.8	47.8	66.8	86.2	−100.3
trans-1-Bromo-1-butene	C₄H₇Br	−38.4	−17.0	−6.4	+5.4	18.4	27.2	38.1	55.7	75.0	94.7	−133.4
2-Bromo-1-butene	C₄H₇Br	−47.3	−27.0	−16.8	−5.3	+7.2	15.4	26.3	42.8	61.9	81.0	−111.2
cis-2-Bromo-2-butene	C₄H₇Br	−39.0	−17.9	−7.2	+4.6	17.7	26.2	37.5	54.5	74.0	93.9	−114.6
trans-2-Bromo-2-butene	C₄H₇Br	−45.0	−24.1	−13.8	−2.4	+10.5	18.7	29.9	46.5	66.0	85.5	
1,4-Bromochlorobenzene	C₆H₄BrCl	32.0	59.5	72.7	87.8	103.8	114.8	128.0	149.5	172.6	196.9	16.6
1-Bromo-1-chloroethane	C₂H₄BrCl	−36.0	−18.0	−9.4	0.0	+10.4	17.0	28.0	44.7	63.4	82.7	

(Continued)

TABLE 2.38 Boiling Points of Common Organic Compounds at Selected Pressures (*Continued*)

Compound Name	Formula	Pressure, mm Hg (Temperature, °C) 1	5	10	20	40	60	100	200	400	760	Melting point, °C
1-Bromo-2-chloroethane	C_2H_4BrCl	−28.8	−7.0	+4.1	16.0	29.7	38.0	49.5	66.8	86.0	106.7	−16.6
2-Bromo-4,6-dichlorophenol	$C_6H_3BrCl_2O$	84.0	115.6	130.8	147.7	165.8	177.6	193.2	216.5	242.0	268.0	68
1-Bromo-4-ethyl benzene	C_8H_9Br	30.4	42.5	74.0	90.2	108.5	121.0	135.5	156.5	182.0	206.0	−45.0
(2-Bromoethyl)-benzene	C_8H_9Br	48.0	76.2	90.5	105.8	123.2	133.8	148.2	169.8	194.0	219.0	
2-Bromoethyl 2-chloroethyl ether	C_4H_8BrClO	36.5	63.2	76.3	90.8	106.6	116.4	129.8	150.0	172.3	195.8	
(2-Bromoethyl)-cyclohexane	$C_8H_{15}Br$	38.7	66.6	80.5	95.8	113.0	123.7	138.0	160.0	186.2	213.0	
1-Bromoethylene	C_2H_3Br	−95.4	−77.8	−68.8	−58.8	−48.1	−41.2	−31.9	−17.2	−1.1	+15.8	−138
Bromoform (tribromomethane)	$CHBr_3$		22.0	34.0	48.0	63.6	73.4	85.9	106.1	127.9	150.5	8.5
1-Bromonaphthalene	$C_{10}H_7Br$	84.2	117.5	133.6	150.2	170.2	183.5	198.8	224.2	252.0	281.1	5.5
2-Bromo-4-phenylphenol	$C_{12}H_9BrO$	100.0	135.4	152.3	171.8	193.8	207.0	224.5	251.0	280.2	311.0	95
3-Bromopyridine	C_5H_4BrN	16.8	42.0	55.2	69.1	84.1	94.1	107.8	127.7	150.0	173.4	
2-Bromotoluene	C_7H_7Br	24.4	49.7	62.3	76.0	91.0	100.0	112.0	133.6	157.3	181.8	−28
3-Bromotoluene	C_7H_7Br	14.8	50.8	64.0	78.1	93.9	104.1	117.8	138.0	160.0	183.7	39.8
4-Bromotoluene	C_7H_7Br	10.3	47.5	61.1	75.2	91.8	102.3	116.4	137.4	160.2	184.5	28.5
3-Bromo-2,4,6-trichlorophenol	$C_6H_2BrCl_3O$	112.4	146.2	163.2	181.8	200.5	213.0	229.3	253.0	278.0	305.8	+9.5
2-Bromo-1,4-xylene	C_8H_9Br	37.5	65.0	78.8	94.0	110.6	121.6	135.7	156.4	181.0	206.7	
1,2-Butadiene (methyl allene)	C_4H_6	−89.5	−72.7	−64.2	−54.9	−44.3	−37.5	−28.3	−14.2	+1.8	18.5	
1,3-Butadiene	C_4H_6	−102.8	−87.6	−79.7	−71.0	−61.3	−55.1	−46.8	−33.9	−19.3	−4.5	−108.9
n-Butane	C_4H_{10}	−101.5	−85.7	−77.8	−68.9	−59.1	−52.8	−44.2	−31.2	−16.3	−0.5	−135
iso-Butane (2-methylpropane)	C_4H_{10}	−109.2	−94.1	−86.4	−77.9	−68.4	−62.4	−54.1	−41.5	−27.1	−11.7	−145
1,3-Butanediol	$C_4H_{10}O_2$	22.2	67.5	85.3	100.0	117.4	127.5	141.2	161.0	183.8	206.5	77
1,2,3-Butanetriol	$C_4H_{10}O_3$	102.0	132.0	146.0	161.0	178.0	188.0	202.5	222.0	243.5	264.0	
1-Butene	C_4H_8	−104.8	−89.4	−81.6	−73.0	−63.4	−57.2	−48.9	−36.2	−21.7	−6.3	−130
cis-2-Butene	C_4H_8	−96.4	−81.1	−73.4	−64.6	−54.7	−48.4	−39.8	−26.8	−12.0	+3.7	−138.9
trans-2-Butene	C_4H_8	−99.4	−84.0	−76.3	−67.5	−57.6	−51.3	−42.7	−29.7	−14.8	+0.9	−105.4
3-Butenenitrile	C_4H_5N	−19.6	+2.9	14.1	26.6	40.0	48.8	60.2	78.0	98.0	119.0	
iso-Butyl acetate	$C_6H_{12}O_2$	−21.2	+1.4	12.8	25.5	39.2	48.0	59.7	77.6	97.5	118.0	−98.9
n-Butyl acrylate	$C_7H_{12}O_2$	−0.5	+23.5	35.5	48.6	63.4	72.6	85.1	104.0	125.2	147.2	−64.6
n-Butyl alcohol	$C_4H_{10}O$	−1.2	+20.0	30.2	41.5	53.4	60.3	70.1	84.3	100.8	117.5	−79.9
iso-Butyl alcohol	$C_4H_{10}O$	−9.0	+11.6	21.7	32.4	44.1	51.7	61.5	75.9	91.4	108.0	−108
sec-Butyl alcohol	$C_4H_{10}O$	−12.2	+7.2	16.9	27.3	38.1	45.2	54.1	67.9	83.9	99.5	−114.7
tert-Butyl alcohol	$C_4H_{10}O$	−20.4	−3.0	+5.5	14.3	24.5	31.0	39.8	52.7	68.0	82.9	25.3

	Formula											
iso-Butyl amine	$C_4H_{11}N$	-50.0	-31.0	-21.0	-10.3	+1.3	8.8	18.8	32.0	50.7	68.6	-85.0
n-Butylbenzene	$C_{10}H_{14}$	22.7	48.8	62.0	76.3	92.4	102.6	116.2	136.9	159.2	183.1	-88.0
iso-Butylbenzene	$C_{10}H_{14}$	14.1	40.5	53.7	67.8	83.3	93.3	107.0	127.2	149.6	172.8	-51.5
sec-Butylbenzene	$C_{10}H_{14}$	18.6	44.2	57.0	70.6	86.2	96.0	109.5	128.8	150.3	173.5	-75.5
tert-Butylbenzene	$C_{10}H_{14}$	13.0	39.0	51.7	65.6	80.8	90.6	103.8	123.7	145.8	168.5	-58
iso-Butyl benzoate	$C_{11}H_{14}O_2$	64.0	93.6	108.6	124.2	141.8	152.0	166.4	188.2	212.8	237.0	
n-Butyl bromide (1-bromobutane)	C_4H_9Br	-33.0	-11.2	-0.3	+11.6	24.8	33.4	44.7	62.0	81.7	101.6	-112.4
iso-Butyl *n*-butyrate	$C_8H_{16}O_2$	+4.6	30.0	42.2	56.1	71.7	81.3	94.0	113.9	135.7	156.9	
carbamate	$C_5H_{11}NO_2$		83.7	96.4	110.1	125.3	134.6	147.2	165.7	186.0	206.5	65
Butyl carbitol (diethylene glycol butyl ether)	$C_8H_{18}O_3$	70.0	95.7	107.8	120.5	135.5	146.0	159.8	181.2	205.0	231.2	
n-Butyl chloride (1-chlorobutane)	C_4H_9Cl	-49.0	-28.9	-18.6	-7.4	+5.0	13.0	24.0	40.0	58.8	77.8	-123.1
iso-Butyl chloride	C_4H_9Cl	-53.8	-34.3	-24.5	-13.8	-1.9	+5.9	16.0	32.0	50.0	68.9	-131.2
sec-Butyl chloride (2-Chlorobutane)	C_4H_9Cl	-60.2	-39.8	-29.2	-17.7	-5.0	+3.4	14.2	31.5	50.0	68.0	-131.3
tert-Butyl chloride	C_4H_9Cl					-19.0	-11.4	-1.0	+14.6	32.6	51.0	-26.5
sec-Butyl chloroacetate	$C_6H_{11}ClO_2$	17.0	41.8	54.6	68.2	83.6	93.0	105.5	124.1	146.0	167.8	
2-*tert*-Butyl-4-cresol	$C_{11}H_{16}O$	70.0	98.0	112.0	127.2	143.9	153.7	167.0	187.8	210.0	232.6	
4-*tert*-Butyl-2-cresol	$C_{11}H_{16}O$	74.3	103.7	118.0	134.0	150.8	161.7	176.2	197.8	221.8	247.0	
iso-Butyl dichloroacetate	$C_6H_{10}Cl_2O_2$	28.6	54.3	67.5	81.4	96.7	106.6	119.8	139.2	160.0	183.0	
2,3-Butylene glycol (2,3-butanediol)	$C_4H_{10}O_2$	44.0	68.4	80.3	93.4	107.8	116.3	127.8	145.6	164.0	182.0	22.5
2-Butyl-2-ethylbutane-1,3-diol	$C_{10}H_{22}O_2$	94.1	122.6	136.8	151.2	167.8	178.0	191.9	212.0	233.5	255.0	
2-*tert*-Butyl-4-ethylphenol	$C_{12}H_{18}O$	76.3	106.2	121.0	137.0	154.0	165.4	179.0	200.3	223.8	247.8	
n-Butyl formate	$C_5H_{10}O_2$	-26.4	-4.7	+6.1	18.0	31.6	39.8	51.0	67.9	86.2	106.0	
iso-Butyl formate	$C_5H_{10}O_2$	-32.7	-11.4	-0.8	+11.0	24.1	32.4	43.4	60.0	79.0	98.2	-95.3
sec-Butyl formate	$C_5H_{10}O_2$	-34.4	-13.3	-3.1	+8.4	21.3	29.6	40.2	56.8	75.2	93.6	
sec-Butyl glycolate	$C_6H_{12}O_3$	28.3	53.6	66.0	79.8	94.2	104.0	116.4	135.5	155.6	177.5	
iso-Butyl iodide (1-iodo-2-methylpropane)	C_4H_9I	-17.0	+5.0	17.0	29.8	42.8	51.8	63.5	81.0	100.3	120.4	-90.7
isobutyrate	$C_8H_{16}O_2$	+4.1	28.0	39.9	52.4	67.2	75.9	88.0	106.3	126.3	147.5	-80.7
isovalerate	$C_9H_{18}O_2$	16.0	41.2	53.8	67.7	82.7	92.4	105.2	124.8	146.4	168.7	
levulinate	$C_9H_{16}O_3$	65.0	92.1	105.9	120.2	136.2	147.0	160.2	181.8	205.5	229.9	
naphthylketone (1-isovaleronaphthone)	$C_{15}H_{16}O$	136.0	167.9	184.0	201.6	219.7	231.5	246.7	269.7	294.0	320.0	
2-*sec*-Butylphenol	$C_{10}H_{14}O$	57.4	86.0	100.8	116.1	133.4	143.9	157.3	179.7	203.8	228.0	
2-*tert*-Butylphenol	$C_{10}H_{14}O$	56.6	84.2	98.1	113.0	129.2	140.0	153.5	173.8	196.3	219.5	
4-*iso*-Butylphenol	$C_{10}H_{14}O$	72.1	100.9	115.5	130.3	147.2	157.0	171.2	192.1	214.7	237.0	
4-*sec*-Butylphenol	$C_{10}H_{14}O$	71.4	100.5	114.8	130.3	147.8	157.9	172.4	194.3	217.6	242.1	

(*Continued*)

TABLE 2.38 Boiling Points of Common Organic Compounds at Selected Pressures (*Continued*)

Name	Formula	_	_	_	_	Pressure, mm Hg — Temperature, °C						Melting point, °C
		1	5	10	20	40	60	100	200	400	760	
4-*tert*-Butylphenol	C₁₀H₁₄O	70.0	99.2	114.0	129.5	146.0	156.0	170.2	191.5	214.0	238.0	99
2-(4-*tert*-Butylphenoxy)ethyl acetate	C₁₄H₂₀O₃	118.0	150.0	165.8	183.3	201.5	212.8	228.0	250.3	277.6	304.4	
4-*tert*-Butylphenyl dichlorophosphate	C₁₀H₁₃Cl₂O₂P	96.0	129.6	146.0	164.0	184.3	197.2	214.3	240.0	268.2	299.0	
tert-Butyl phenyl ketone (pivalophenone)	C₁₁H₁₄O	57.8	85.7	99.0	114.3	130.4	140.8	154.0	175.0	197.7	220.0	
iso-Butyl propionate	C₇H₁₄O₂	−2.3	+20.9	32.3	44.8	58.5	67.6	79.5	97.0	116.4	136.8	−71
4-*tert*-Butyl-2,5-xylenol	C₁₂H₁₈O	88.2	119.8	135.0	151.0	169.8	180.3	195.0	217.5	241.3	265.3	
4-*tert*-Butyl-2,6-xylenol	C₁₂H₁₈O	74.0	103.9	119.0	135.0	152.2	163.6	176.0	196.0	217.8	239.8	
6-*tert*-Butyl-2,4-xylenol	C₁₂H₁₈O	70.3	100.2	115.0	131.0	148.5	158.2	172.0	192.3	214.2	236.5	
6-*tert*-Butyl-3,4-xylenol	C₁₂H₁₈O	83.9	113.6	127.0	143.0	159.7	170.0	184.0	204.5	226.7	249.5	
Butyric acid	C₄H₈O₂	25.5	49.8	61.5	74.0	88.0	96.5	108.0	125.5	144.5	163.5	−74
iso-Butyric acid	C₄H₈O₂	14.7	39.3	51.2	64.0	77.8	86.3	98.0	115.8	134.5	154.5	−47
Butyronitrile	C₄H₇N	−20.0	+2.1	13.4	25.7	38.4	47.3	59.0	76.7	96.8	117.5	
iso-Valerophenone	C₁₁H₁₄O	58.3	87.0	101.4	116.8	133.8	144.6	158.0	180.1	204.2	228.0	50
Camphene	C₁₀H₁₆			47.2	60.4	75.7	85.0	97.9	117.5	138.7	160.5	
Campholenic acid	C₁₀H₁₆O₂	97.6	125.7	139.8	153.9	170.0	180.0	193.7	212.7	234.0	256.0	
d-Camphor	C₁₀H₁₆O	41.5	68.6	82.3	97.5	114.0	124.0	138.0	157.9	182.0	209.2	178.5
Camphylamine	C₁₀H₁₉N	45.3	74.0	83.7	97.6	112.5	122.0	134.6	153.0	173.8	195.0	
Capraldehyde	C₁₀H₂₀O	51.9	78.8	92.0	106.3	122.2	132.0	145.3	164.8	186.3	208.5	
Capric acid	C₁₀H₂₀O₂	125.0	142.0	152.2	165.0	179.9	189.8	200.0	217.1	240.3	268.4	31.5
n-Caproic acid	C₆H₁₂O₂	71.4	89.5	99.5	111.8	125.0	133.3	144.0	160.8	181.0	202.0	−1.5
iso-Caproic acid	C₆H₁₂O₂	66.2	83.0	94.0	107.0	120.4	129.6	141.4	158.3	181.0	207.7	
iso-Caprolactone	C₆H₁₀O₂	38.3	66.4	80.3	95.7	112.3	123.2	137.2	157.8	182.1	207.0	−35
Capronitrile	C₆H₁₁N	9.2	34.6	47.5	61.7	76.9	86.8	99.8	119.7	141.0	163.7	
Capryl alcohol (2-octanol)	C₈H₁₈O	32.8	57.6	70.0	83.3	98.0	107.4	119.8	138.0	157.5	178.5	−38.6
Caprylaldehyde	C₈H₁₆O	73.4	92.0	101.2	110.2	120.0	126.0	133.9	145.4	156.5	168.5	
Caprylic acid (octanoic acid)	C₈H₁₆O₂	92.3	114.1	124.0	136.4	150.6	160.0	172.2	190.3	213.9	237.5	16
Caprylonitrile	C₈H₁₅N	43.0	67.6	80.4	94.6	110.6	121.2	134.8	155.2	179.5	204.5	
Carbazole	C₁₂H₉N						248.2	265.0	292.5	323.0	354.8	244.8
Carbon dioxide	CO₂	−134.3	−124.4	−119.5	−114.4	−108.6	−104.8	−100.2	−93.0	−85.7	−78.2	−57.5
disulfide	CS₂	−73.8	−54.3	−44.7	−34.3	−22.5	−15.3	−5.1	+10.4	28.0	46.5	−110.8
monoxide	CO	−222.0	−217.2	−215.0	−212.8	−210	−208.1	−205.7	−201.3	−196.3	−191.3	−205.0

oxyselenide (carbonyl selenide)	COSe	−117.1	−102.3	−95.0	−86.3	−76.4	−70.2	−61.7	−49.8	−35.6	−21.9	
oxysulfide (carbonyl sulfide)	COS	−132.4	−119.8	−113.3	−106.0	−98.3	−93.0	−85.9	−75.0	−62.7	−49.9	−138.8
tetrabromide	CBr₄					96.3	106.3	119.7	139.7	163.5	189.5	90.1
tetrachloride	CCl₄	−50.0	−30.0	−19.6	−8.2	+4.3	12.3	23.0	38.3	57.8	76.7	−22.6
tetrafluoride	CF₄	−184.6	−174.1	−169.3	−164.3	−158.8	−155.4	−150.7	−143.6	−135.5	−127.7	−183.7
Carvacrol	C₁₀H₁₄O	70.0	98.4	113.2	127.9	145.2	155.3	169.7	191.2	213.8	237.0	+0.5
Carvone	C₁₀H₁₄O	57.4	86.1	100.4	116.1	133.0	143.8	157.3	179.6	203.5	227.5	
Chavibetol	C₁₀H₁₂O₂	83.6	113.3	127.0	143.2	159.8	170.7	185.5	206.8	229.8	254.0	
Chloral (trichloroacetaldehyde)	C₂HCl₃O	−37.8	−16.0	−5.0	+7.2	20.2	29.1	40.2	57.8	77.5	97.7	−57
hydrate (trichloroacetaldehyde hydrate)	C₂H₃Cl₃O₂	−9.8	+10.0	19.5	29.2	39.7	46.2	55.0	68.0	82.1	96.2	51.7
Chloranil	C₆Cl₄O₂	70.7	89.3	97.8	106.4	116.1	122.0	129.5	140.3	151.3	162.6	290
Chloroacetic acid	C₂H₃ClO₂	43.0	68.3	81.0	94.2	109.2	118.3	130.7	149.0	169.0	189.5	61.2
anhydride	C₄H₄Cl₂O₃	67.2	94.1	108.0	122.4	138.2	148.0	159.8	177.8	197.0	217.0	46
2-Chloroaniline	C₆H₆ClN	46.3	72.3	84.8	99.2	115.6	125.7	139.5	160.0	183.7	208.8	0
3-Chloroaniline	C₆H₆ClN	63.5	89.8	102.0	116.7	133.6	144.1	158.0	179.5	203.5	228.5	−10.4
4-Chloroaniline	C₆H₅Cl	59.3	87.9	102.1	117.8	135.0	145.8	159.9	182.3	206.6	230.5	70.5
Chlorobenzene	C₆H₅Cl	−13.0	+10.6	22.2	35.3	49.7	58.3	70.7	89.4	110.0	132.2	−45.2
2-Chlorobenzotrichloride (2-α,α,α-tetrachlorotoluene)	C₇H₄Cl₄	69.0	101.8	117.9	135.8	155.0	167.8	185.0	208.0	233.0	262.1	28.7
2-Chlorobenzotrifluoride (2-chloro-α,α,α-trifluorotoluene)	C₇H₄ClF₃	0.0	24.7	37.1	50.6	65.9	75.4	88.3	108.3	130.0	152.2	−6.0
2-Chlorobiphenyl	C₁₂H₉Cl	89.3	109.8	134.7	151.2	169.9	182.1	197.0	219.6	243.8	267.5	34
4-Chlorobiphenyl	C₁₂H₉Cl	96.4	129.8	146.0	164.0	183.8	196.0	212.5	237.8	264.5	292.9	75.5
α-Chlorocrotonic acid	C₄H₅ClO₂	70.0	95.6	108.0	121.2	135.6	144.4	155.9	173.8	193.2	212.0	
Chlorodifluoromethane	CHClF₂	−122.8	−110.2	−103.7	−96.5	−88.6	−83.4	−76.4	−65.8	−53.6	−40.8	−160
Chlorodimethylphenylsilane	C₈H₁₁ClSi	29.8	56.7	70.0	84.7	101.2	111.5	124.7	145.5	168.6	193.5	
1-Chloro-2-ethoxybenzene	C₈H₉ClO	45.8	72.8	86.5	101.5	117.8	127.8	141.8	162.0	185.0	208.0	
2-(2-Chloroethoxy) ethanol	C₄H₉ClO₂	53.0	78.3	90.7	104.1	118.4	127.5	139.5	157.2	176.5	196.0	
bis-2-Chloroethyl acetacetal	C₆H₁₂Cl₂O₂	56.2	83.7	97.6	112.2	127.8	138.0	150.7	169.8	190.5	212.6	
1-Chloro-2-ethylbenzene	C₈H₉Cl	17.2	43.0	56.1	70.3	86.2	96.4	110.0	130.2	152.2	177.6	−80.2
1-Chloro-3-ethylbenzene	C₈H₉Cl	18.6	45.2	58.1	73.0	89.2	99.6	113.6	133.8	156.7	181.1	−53.3
1-Chloro-4-ethylbenzene	C₈H₉Cl	19.2	46.4	60.0	75.5	91.8	102.0	116.0	137.0	159.8	184.3	−62.6
2-Chloroethyl chloroacetate	C₄H₄ClO₂	46.0	72.1	86.0	100.0	116.0	126.2	140.0	159.8	182.2	205	
2-Chloroethyl 2-chloroisopropyl ether	C₅H₁₀Cl₂O	24.7	50.1	63.0	77.2	92.4	102.2	115.8	135.7	156.5	180.0	
2-Chloroethyl 2-chloropropyl ether	C₅H₁₀Cl₂O	29.8	56.5	70.0	84.8	101.5	111.8	125.6	146.3	169.8	194.1	
2-Chloroethyl α-methylbenzyl ether	C₁₀H₁₃ClO	62.3	91.4	106.0	121.8	139.6	150.0	164.8	186.3	210.8	235.0	
Chloroform (trichloromethane)	CHCl₃	−58.0	−39.1	−29.7	−19.0	−7.1	+0.5	10.4	25.9	42.7	61.3	−63.5

(Continued)

TABLE 2.38 Boiling Points of Common Organic Compounds at Selected Pressures (*Continued*)

Compound		Pressure, mm Hg										Melting point, °C
Name	Formula	1	5	10	20	40	60	100	200	400	760	
		Temperature, °C										
1-Chloronaphthalene	$C_{10}H_7Cl$	80.6	104.8	118.6	134.4	153.2	165.6	180.4	204.2	230.8	259.3	−20
4-Chlorophenethyl alcohol	C_8H_9ClO	84.0	114.3	129.0	145.0	162.0	173.5	188.1	210.0	234.5	259.3	
2-Chlorophenol	C_6H_5ClO	12.1	38.2	51.2	65.9	82.0	92.0	106.0	126.4	149.8	174.5	7
3-Chlorophenol	C_6H_5ClO	44.2	72.0	86.1	101.7	118.0	129.4	143.0	164.8	188.7	214.0	32.5
4-Chlorophenol	C_6H_5ClO	49.8	78.2	92.2	108.1	125.0	136.1	150.0	172.0	196.0	220.0	42
2-Chloro-3-phenylphenol	$C_{12}H_9ClO$	118.0	152.2	169.7	186.7	207.4	219.6	237.0	261.3	289.4	317.5	+6
2-Chloro-6-phenylphenol	$C_{12}H_9ClO$	119.8	153.7	170.7	189.8	208.2	220.0	237.1	261.6	289.5	317.0	
Chloropicrin (trichloronitromethane)	CCl_3NO_2	−25.5	−3.3	+7.8	20.0	33.8	42.3	53.8	71.8	91.8	111.9	−64
1-Chloropropene	C_3H_5Cl	−81.3	−63.4	−54.1	−44.0	−32.7	−25.1	−15.1	+1.3	18.0	37.0	−99.0
2-Chloropyridine	C_5H_4ClN	13.3	38.8	51.7	65.8	81.7	91.6	104.6	125.0	147.7	170.2	
3-Chlorostyrene	C_8H_7Cl	25.3	51.3	65.2	80.0	96.5	107.2	121.2	142.2	165.7	190.0	−15.0
4-Chlorostyrene	C_8H_7Cl	28.0	54.5	67.5	82.0	98.0	108.5	122.0	143.5	166.0	191.0	+0.9
1-Chlorotetradecane	$C_{14}H_{29}Cl$	98.5	131.8	148.2	166.2	187.0	199.8	215.5	240.3	267.5	296.0	
2-Chlorotoluene	C_7H_7Cl	+5.4	30.6	43.2	56.9	72.0	81.8	94.7	115.0	137.1	159.3	
3-Chlorotoluene	C_7H_7Cl	+4.8	30.3	43.2	57.4	73.0	83.2	96.3	116.6	139.7	162.3	
4-Chlorotoluene	C_7H_7Cl	+5.5	31.0	43.8	57.8	73.5	83.3	96.6	117.1	139.8	162.3	+7.3
Chlorotriethylsilane	$C_6H_{15}ClSi$	−4.9	+19.8	32.0	45.5	60.2	69.5	82.3	101.6	123.6	146.3	
1-Chloro-1,2,2-trifluoroethylene	C_2ClF_3	−116.0	−102.5	−95.9	−88.2	−79.7	−74.1	−66.7	−55.0	−41.7	−27.9	−157.5
Chlorotrifluoromethane	$CClF_3$	−149.5	−139.2	−134.1	−128.5	−121.9	−117.3	−111.7	−102.5	−92.7	−81.2	
Chlorotrimethylsilane	C_3H_9ClSi	−62.8	−43.6	−34.0	−23.2	−11.4	−4.0	+6.0	21.9	39.4	57.9	
trans-Cinnamic acid	$C_9H_8O_2$	127.5	157.8	173.0	189.5	207.1	217.8	232.4	253.3	276.7	300.0	133
Cinnamyl alcohol	$C_9H_{10}O$	72.6	102.5	117.8	133.7	151.0	162.0	177.8	199.8	224.6	250.0	33
Cinnamylaldehyde	C_9H_8O	76.1	105.8	120.0	135.7	152.2	163.7	177.7	199.3	222.4	246.0	−7.5
Citraconic anhydride	$C_5H_4O_3$	47.1	74.8	88.9	103.8	120.3	131.3	145.4	165.8	189.8	213.5	
cis-α-Citral	$C_{10}H_{16}O$	61.7	90.0	103.9	119.4	135.9	146.3	160.0	181.8	205.0	228.0	
d-Citronellal	$C_{10}H_{18}O$	44.0	71.4	84.8	99.8	116.1	126.2	140.1	160.0	183.8	206.5	
Citronellic acid	$C_{10}H_{18}O_2$	99.5	127.3	141.4	155.6	171.9	182.1	195.4	214.5	236.6	257.0	
Citronellol	$C_{10}H_{20}O$	66.4	93.6	107.0	121.5	137.2	147.2	159.8	179.8	201.0	221.5	
Citronellyl acetate	$C_{12}H_{22}O_2$	74.7	100.2	113.0	126.0	140.5	149.7	161.0	178.8	197.8	217.0	
Coumarin	$C_9H_6O_2$	106.0	137.8	153.4	170.0	189.0	200.5	216.5	240.0	264.7	291.0	70
o-Cresol (2-cresol; 3-methylphenol)	C_7H_8O	38.2	64.0	76.7	90.5	105.8	115.5	127.4	146.7	168.4	190.8	30.8

754

m-Cresol (3-cresol; 3-methylphenol)	C_7H_8O	52.0	76.0	87.8	101.4	116.0	125.8	138.0	157.3	179.0	202.8	10.9
p-Cresol (4-cresol; 4-methylphenol)	C_7H_8O	53.0	76.5	88.6	102.3	117.7	127.0	140.0	157.3	179.4	201.8	35.5
cis-Crotonic acid	$C_4H_6O_2$	33.5	57.4	69.0	82.0	96.0	104.5	116.3	133.9	152.2	171.9	15.5
trans-Crotonic acid	$C_4H_6O_2$			80.0	93.0	107.8	116.7	128.0	146.0	165.5	185.0	72
cis-Crotononitrile	C_4H_5N	−29.0	−7.1	+4.0	16.4	30.0	38.5	50.1	68.0	88.0	108.0	
trans-Crotononitrile	C_4H_5N	−19.5	+3.5	15.0	27.8	41.8	50.9	62.8	81.1	101.5	122.8	
Cumene	C_9H_{12}	+2.9	26.8	38.3	51.5	66.1	75.4	88.1	107.3	129.2	152.4	−96.0
4-Cumidene	$C_9H_{13}N$	60.0	88.2	102.2	117.8	134.2	145.0	158.0	180.0	203.2	227.0	
Cuminal	$C_{10}H_{12}O$	58.0	87.3	102.0	117.9	135.2	146.0	160.0	182.8	206.7	232.0	
Cuminyl alcohol	$C_{10}H_{14}O$	74.2	103.7	118.0	133.8	150.3	161.7	176.2	197.9	221.7	246.6	
2-Cyano-2-n-butyl acetate	$C_7H_{11}NO_2$	42.0	68.7	82.0	96.2	111.8	121.5	133.8	152.2	173.4	195.2	
Cyanogen	C_2N_2	−95.8	−83.2	−76.8	−70.1	−62.7	−57.9	−51.8	−42.6	−33.0	−21.0	−34.4
bromide	$CBrN$	−35.7	−13.3	−10.0	−1.0	+8.6	14.7	22.6	33.8	46.0	61.5	58
chloride	$CClN$	−76.7	−61.4	−53.8	−46.1	−37.5	−32.1	−24.9	−14.1	−2.3	+13.1	−6.5
iodide	CIN	25.2	47.2	57.7	68.6	80.3	88.0	97.6	111.5	126.1	141.1	
Cyclobutane	C_4H_8	−92.0	−76.0	−67.9	−58.7	−48.4	−41.8	−32.8	−18.9	−3.4	+12.9	−50
Cyclobutene	C_4H_6	−99.1	−83.4	−75.4	−66.6	−56.4	−50.0	−41.2	−27.8	−12.2	+2.4	
Cyclohexane	C_6H_{12}	−45.3	−25.4	−15.9	−5.0	+6.7	14.7	25.5	42.0	60.8	80.7	+6.6
Cyclohexaneethanol	$C_8H_{16}O$	50.4	77.2	90.0	104.0	119.8	129.8	142.7	161.7	183.5	205.4	
Cyclohexanol	$C_6H_{12}O$	21.0	44.0	56.0	68.8	83.0	91.8	103.7	121.7	141.4	161.0	23.9
Cyclohexanone	$C_6H_{10}O$	+1.4	26.4	38.7	52.5	67.8	77.5	90.4	110.3	132.5	155.6	−45.0
2-Cyclohexyl-4,6-dinitrophenol	$C_{12}H_{14}N_2O_5$	132.8	161.8	175.9	191.2	206.7	216.0	229.0	248.7	269.8	291.5	
Cyclopentane	C_5H_{10}	−68.0	−49.6	−40.4	−30.1	−18.6	−11.3	−1.3	+13.8	31.0	49.3	−93.7
Cyclopropane	C_3H_6	−116.8	−104.2	−97.5	−90.3	−82.3	−77.0	−70.0	−59.1	−46.9	−33.5	−126.6
Cymene	$C_{10}H_{14}$	17.3	43.9	57.0	71.1	87.0	97.2	110.8	131.4	153.5	177.2	−68.2
cis-Decalin	$C_{10}H_{18}$	22.5	50.1	64.2	79.8	97.2	108.0	123.2	145.4	169.9	194.6	−43.3
trans-Decalin	$C_{10}H_{18}$	−0.8	+30.6	47.2	65.3	85.7	98.4	114.6	136.2	160.1	186.7	−30.7
Decane	$C_{10}H_{22}$	16.5	42.3	55.7	69.8	85.5	95.5	108.6	128.4	150.6	174.1	−29.7
Decan-2-one	$C_{10}H_{20}O$	44.2	71.9	85.8	100.7	117.1	127.8	142.0	163.2	186.7	211.0	+3.5
1-Decene	$C_{10}H_{20}$	14.7	40.3	53.7	67.8	83.3	93.5	106.5	126.7	149.2	172.0	
Decyl alcohol	$C_{10}H_{22}O$	69.5	97.3	111.3	125.8	142.1	152.0	165.8	186.2	208.8	231.0	+7
Decyltrimethylsilane	$C_{13}H_{30}Si$	67.4	96.4	111.0	126.5	144.0	154.3	169.5	191.0	215.5	240.0	
Dehydroacetic acid	$C_8H_8O_4$	91.7	122.0	137.3	153.0	171.0	181.5	197.5	219.5	244.5	269.0	
Desoxybenzoin	$C_{14}H_{12}O$	123.3	156.2	173.5	192.0	212.0	224.5	241.3	265.2	293.0	321.0	60
Diacetamide	$C_4H_7NO_2$	70.0	95.0	108.0	122.6	138.2	148.0	160.6	180.8	202.0	223.0	78.5
Diacetylene (1,3-butadiyne)	C_4H_2	−82.5	−68.0	−61.2	−53.8	−45.9	−41.0	−34.0	−20.9	−6.1	+9.7	−34.9
Diallyldichlorosilane	$C_6H_{10}Cl_2Si$	+9.5	34.8	47.4	61.3	76.4	86.3	99.7	119.4	142.0	165.3	

(Continued)

TABLE 2.38 Boiling Points of Common Organic Compounds at Selected Pressures (*Continued*)

| Compound | | Pressure, mm Hg | | | | | | | | | | Melting point, °C |
Name	Formula	1	5	10	20	40	60	100	200	400	760	
						Temperature, °C						
Dialyl sulfide	C₆H₁₀S	−9.5	14.4	26.6	39.7	54.2	63.7	75.8	94.8	116.1	138.6	−83
Diisoamyl ether	C₁₀H₂₂O	18.6	44.3	57.0	70.7	86.3	96.0	109.6	129.0	150.3	173.4	
oxalate	C₁₂H₂₂O₄	85.4	116.0	131.4	147.7	165.7	177.0	192.2	215.0	240.0	265.0	
sulfide	C₁₀H₂₂S	43.0	73.0	87.6	102.7	120.0	130.6	145.3	166.4	191.0	216.0	−26
Dibenzylamine	C₁₄H₁₅N	118.3	149.8	165.6	182.2	200.2	212.2	227.3	249.8	274.3	300.0	
Dibenzyl ketone (1,3-diphenyl-2-propanone)	C₁₅H₁₄O	125.5	159.8	177.6	195.7	216.6	229.4	246.6	272.3	301.7	330.5	34.5
1,4-Dibromobenzene	C₆H₄Br₂	61.0	79.3	87.7	103.6	120.8	131.6	146.5	168.5	192.5	218.6	87.5
1,2-Dibromobutane	C₄H₈Br₂	7.5	33.2	46.1	60.0	76.0	86.0	99.8	120.2	143.5	166.3	−64.5
dl-2,3-Dibromobutane	C₄H₈Br₂	+5.0	30.0	41.6	56.4	72.0	82.0	95.3	115.7	138.0	160.5	
meso-2,3-Dibromobutane	C₄H₈Br₂	+1.5	26.6	39.3	53.2	68.0	78.0	91.7	111.8	134.2	157.3	−34.5
1,2-Dibromodecane	C₁₀H₂₀Br₂	95.7	123.6	137.3	151.0	167.4	177.5	190.2	209.6	229.8	250.4	
Di(2-bromoethyl) ether	C₄H₈Br₂O	47.7	75.3	88.5	103.6	119.8	130.0	144.0	165.0	188.0	212.5	
α,β-Dibromomaleie Anhydride	C₄H₂Br₂O₃	50.0	78.0	92.0	106.7	123.5	133.8	147.7	168.0	192.0	215.0	
1,2-Dibromo-2-methylpropane	C₄H₈Br₂	−28.8	−3.0	+10.5	25.7	42.3	53.7	68.8	92.1	119.8	149.0	−70.3
1,3-Dibromo-2-methylpropane	C₄H₈Br₂	14.0	40.0	53.0	67.5	83.5	93.7	107.4	117.8	150.6	174.6	
1,2-Dibromopentane	C₅H₁₀Br₂	19.8	45.4	58.0	72.0	87.4	97.4	110.1	130.2	151.8	175.0	
1,2-Dibromopropane	C₃H₆Br₂	−7.0	+17.3	29.4	42.3	57.2	66.4	78.7	97.8	118.5	141.6	−5.5
1,3-Dibromopropane	C₃H₆Br₂	+9.7	35.4	48.0	62.1	77.8	87.8	101.3	121.7	144.1	167.5	
2,3-Dibromopropene	C₃H₄Br₂	−6.0	+17.9	30.0	43.2	57.8	67.0	79.5	98.0	119.5	141.2	−34.4
2,3-Dibromo-1-propanol	C₃H₆Br₂O	57.0	84.5	98.2	113.5	129.8	140.0	153.0	173.8	196.0	219.0	
Diisobutylamine	C₈H₁₉N	−5.1	+18.4	30.6	43.7	57.8	67.0	79.2	97.6	118.0	139.5	−70
2,6-Ditert-butyl-4-cresol	C₁₅H₂₄O	85.8	116.2	131.0	147.0	164.1	175.2	190.0	212.8	237.6	262.5	
4,6-Ditert-butyl-2-cresol	C₁₅H₂₄O	86.2	117.3	132.4	149.0	167.4	179.0	194.0	217.5	243.4	269.3	
4,6-Ditert-butyl-3-cresol	C₁₅H₂₄O	103.7	135.2	150.0	167.0	185.3	196.1	211.0	233.0	257.1	282.0	
2,6-Ditert-butyl-4-ethylphenol	C₁₆H₂₆O	89.1	121.4	137.0	154.0	172.1	183.9	198.0	220.0	244.0	268.6	
4,6-Ditert-butyl-3-ethylphenol	C₁₆H₂₆O	111.5	142.6	157.4	174.0	192.3	204.4	218.0	241.7	264.6	290.0	
Diisobutyl oxalate	C₁₀H₁₈O₄	63.2	91.2	105.3	120.3	137.5	147.8	161.8	183.5	205.8	229.5	
2,4-Ditert-butylphenol	C₁₄H₂₂O	84.5	115.4	130.0	146.0	164.3	175.8	190.0	212.5	237.0	260.8	
Dibutyl phthalate	C₁₆H₂₂O₄	148.2	182.1	198.2	216.2	235.8	247.8	263.7	287.0	313.5	340.0	−79.7
sulfide	C₈H₁₈S	+21.7	51.8	66.4	80.5	96.0	105.8	118.6	138.0	159.0	182.0	

Diisobutyl *d*-tartrate	$C_{12}H_{22}O_6$	117.8	151.8	169.0	188.0	208.5	221.6	239.5	264.7	294.0	324.0	73.5
Dicarvaryl-mono-(6-chloro-2-xenyl) phosphate	$C_{32}H_{34}ClO_4P$	204.2	234.5	249.3	264.5	280.5	290.7	304.9	323.8	342.0	361.0	
Dicarvacryl-2-tolyl phosphate	$C_{27}H_{33}O_4P$	180.2	209.3	221.8	237.0	251.5	260.3	272.5	290.0	309.8	330.0	
Dichloroacetic acid	$C_2H_2Cl_2O_2$	44.0	69.8	82.6	96.3	111.8	121.5	134.0	152.3	173.7	194.4	9.7
1,2-Dichlorobenzene	$C_6H_4Cl_2$	20.0	46.0	59.1	73.4	89.4	99.5	112.9	133.4	155.8	179.0	−17.6
1,3-Dichlorobenzene	$C_6H_4Cl_2$	12.1	39.0	52.0	66.2	82.0	92.2	105.0	125.9	149.0	173.0	−24.2
1,4-Dichlorobenzene	$C_6H_4Cl_2$			54.8	69.2	84.8	95.2	108.4	128.3	150.2	173.9	53.0
1,2-Dichlorobutane	$C_4H_8Cl_2$	−23.6	−0.3	+11.5	24.5	37.7	47.8	60.2	79.7	100.8	123.5	
2,3-Dichlorobutane	$C_4H_8Cl_2$	−25.2	−3.0	+8.5	21.2	35.0	43.9	56.0	74.0	94.2	116.0	−80.4
1,2-Dichloro-1,2-difluoroethylene	$C_2Cl_2F_2$	−82.0	−65.6	−57.3	−48.3	−38.2	−31.8	−23.0	−10.0	+5.0	20.9	−112
Dichlorodifluoromethane	CCl_2F_2	−118.5	−104.6	−97.8	−90.1	−81.6	−76.1	−68.6	−57.0	−43.9	−29.8	
Dichlorodiphenyl silane	$C_{12}H_{10}Cl_2Si$	109.6	142.4	158.0	176.0	195.5	207.5	223.8	248.0	275.5	304.0	
Dichlorodiisopropyl ether	$C_6H_{12}Cl_2O$	29.6	55.2	68.2	82.2	97.3	106.9	119.7	139.0	159.8	182.7	
Di(2-chloroethoxy) methane	$C_5H_{10}Cl_2O_2$	53.0	80.4	94.0	109.5	125.5	135.8	149.6	170.0	192.0	215.0	
Dichloroethoxymethylsilane	$C_8H_8Cl_2OSi$	−33.8	−12.1	−1.3	+11.3	24.4	32.6	44.1	61.0	80.3	100.6	
1,2-Dichloro-3-ethylbenzene	$C_8H_8Cl_2$	46.0	75.0	90.0	105.9	123.8	135.0	149.8	172.0	197.0	222.1	−40.8
1,2-Dichloro-4-ethylbenzene	$C_8H_8Cl_2$	47.0	77.2	92.3	109.6	127.5	139.0	153.3	176.0	201.7	226.6	−76.4
1,4-Dichloro-2-ethylbenzene	$C_8H_8Cl_2$	38.5	68.0	83.2	99.8	118.0	129.0	144.0	166.2	191.5	216.3	−61.2
cis-1,2-Dichloroethylene	$C_2H_2Cl_2$	−58.4	−39.2	−29.9	−19.4	−7.9	−0.5	+9.5	24.6	41.0	59.0	−80.5
trans-1,2-Dichloro ethylene	$C_2H_2Cl_2$	−65.4	−47.2	−38.0	−28.0	−17.0	−10.0	−0.2	+14.3	30.8	47.8	−50.0
Di(2-chloroethyl) ether	$C_4H_8Cl_2O$	23.5	49.3	62.0	76.0	91.5	101.5	114.5	134.0	155.4	178.5	
Dichlorofluoromethane	$CHCl_2F$	−91.3	−75.5	−67.5	−58.6	−48.8	−42.6	−33.9	−20.9	−6.2	+8.9	−135
1,5-Dichlorohexamethyltrisiloxane	$C_6H_{18}Cl_2O_2Si_3$	26.0	52.0	65.1	79.0	94.8	105.0	118.2	138.3	160.2	184.0	−53.0
Dichloromethylphenylsilane	$C_7H_8Cl_2Si$	35.7	63.5	77.4	92.4	109.5	120.0	134.2	155.5	180.2	205.5	
1,1-Dichloro-2-methylpropane	$C_4H_8Cl_2$	−31.0	−8.4	+2.6	14.6	28.2	37.0	48.2	65.8	85.4	106.0	
1,2-Dichloro-2-methylpropane	$C_4H_8Cl_2$	−25.8	−4.2	+6.7	18.7	32.0	40.2	51.7	68.9	87.8	108.0	
1,3-Dichloro-2-methylpropane	$C_4H_8Cl_2$	−3.0	+20.6	32.0	44.8	58.6	67.5	78.8	96.1	115.4	135.0	
2,4-Dichlorophenol	$C_6H_4Cl_2O$	53.0	80.0	92.8	107.7	123.4	133.5	146.0	165.2	187.5	210.0	45.0
2,6-Dichlorophenol	$C_6H_4Cl_2O$	59.5	87.6	101.0	115.5	131.6	141.8	154.6	175.5	197.7	220.0	
α,α-Dichlorophenylacetonitrile	$C_8H_5Cl_2N$	56.0	84.0	98.1	113.8	130.0	141.0	154.5	176.2	199.5	223.5	
Dichlorophenylarsine	$C_6H_5AsCl_2$	61.8	100.0	116.0	133.1	151.0	163.2	178.9	202.8	228.8	256.5	
1,2-Dichloropropane	$C_3H_6Cl_2$	−38.5	−17.0	−6.1	+6.0	19.4	28.0	39.4	57.0	76.0	96.8	
2,3-Dichlorostyrene	$C_8H_6Cl_2$	61.0	90.1	104.6	120.5	137.8	149.0	163.5	185.7	210.0	235.0	
2,4-Dichlorostyrene	$C_8H_6Cl_2$	53.5	82.2	97.4	111.8	129.2	140.0	153.8	176.0	200.0	225.0	
2,5-Dichlorostyrene	$C_8H_6Cl_2$	55.5	83.9	98.2	114.0	131.0	142.0	155.8	178.0	202.5	227.0	
2,6-Dichlorostyrene	$C_8H_6Cl_2$	47.8	75.7	90.0	105.5	122.4	133.3	147.6	169.0	193.5	217.0	

(Continued)

TABLE 2.38 Boiling Points of Common Organic Compounds at Selected Pressures (*Continued*)

Compound		Pressure, mm Hg										Melting point, °C
Name	Formula	1	5	10	20	40	60	100	200	400	760	
		Temperature, °C										
3,4-Dichlorostyrene	$C_8H_6Cl_2$	57.2	86.0	100.4	116.2	133.7	144.6	158.2	181.5	205.7	230.0	
3,5-Dichlorostyrene	$C_8H_6Cl_2$	53.5	82.2	97.4	111.8	129.2	140.0	153.8	176.0	200.0	225.0	
1,2-Dichlorotetraethylbenzene	$C_{14}H_{20}Cl_2$	105.6	138.7	155.0	172.5	192.2	204.8	220.7	245.6	272.8	302.0	
1,4-Dichlorotetraethylbenzene	$C_{14}H_{20}Cl_2$	91.7	126.1	143.8	162.0	183.2	195.8	212.0	238.5	265.8	296.5	
1,2-Dichloro-1,1,2,2-tetrafluoroethane	$C_2Cl_2F_4$	−95.4	−80.0	−72.3	−63.5	−53.7	−47.5	−39.1	−26.3	−12.0	+3.5	−94
Dichloro-4-tolysilane	$C_7H_8Cl_2Si$	46.2	71.7	84.2	97.8	113.2	122.6	135.5	153.5	175.2	196.3	
3,4-Dichloro-α,α,α-trifluorotoluene	$C_7H_3Cl_2F_3$	11.0	38.3	52.2	67.3	84.0	95.0	109.2	129.0	150.5	172.8	−12.1
Dicyclopentadiene	$C_{10}H_8$		34.1	47.6	62.0	77.9	88.0	101.7	121.8	144.2	166.6	32.9
Diethoxydimethylsilane	$C_6H_{16}O_2Si$	−19.1	+2.4	13.3	25.3	38.0	46.3	57.6	74.2	93.2	113.5	
Diethoxydiphenylsilane	$C_{16}H_{20}O_2Si$	111.5	142.8	157.6	174.3	193.2	205.0	220.0	243.8	259.7	296.0	
Diethyl adipate	$C_{10}H_{18}O_4$	74.0	106.6	123.0	138.3	154.6	165.8	179.0	198.2	219.1	240.0	−21
Diethylamine	$C_4H_{11}N$			−33.0	−22.6	−11.3	−4.0	+6.0	21.0	38.0	55.5	38.9
N-Diethylaniline	$C_{10}H_{15}N$	49.7	78.0	91.9	107.2	123.6	133.8	147.3	168.2	192.4	215.5	−34.4
Diethyl arsanilate	$C_{10}H_{16}AsNO_3$	38.0	62.6	74.8	88.0	102.6	111.8	123.8	141.9	161.0	181.0	
1,2-Diethylbenzene	$C_{10}H_{14}$	22.3	48.7	62.0	76.4	92.5	102.6	116.2	136.7	159.0	183.5	−31.4
1,3-Diethylbenzene	$C_{10}H_{14}$	20.7	46.8	59.9	74.5	90.4	100.7	114.4	134.8	156.9	181.1	−83.9
1,4-Diethylbenzene	$C_{10}H_{14}$	20.7	47.1	60.3	74.7	91.1	101.3	115.3	136.1	159.0	183.8	−43.2
Diethyl carbonate	$C_5H_{10}O_3$	−10.1	+12.3	23.8	36.0	49.5	57.9	69.7	86.5	105.8	125.8	−43
cis-Diethyl citraconate	$C_9H_{14}O_4$	59.8	88.3	103.0	118.2	135.7	146.2	160.0	182.3	206.5	230.3	
Diethyl dioxosuccinate	$C_8H_{10}O_5$	70.0	98.0	112.0	126.8	143.8	153.7	167.7	188.0	210.8	233.5	
Diethylene glycol	$C_4H_{10}O_3$	91.8	120.0	133.8	148.0	164.3	174.0	187.5	207.0	226.5	244.8	
Diethyleneglycol-*bis*-chloroacetate	$C_8H_{12}Cl_2O_5$	148.3	180.0	195.8	212.0	229.0	239.5	252.0	271.5	291.8	313.0	
Diethylene glycol dimethyl ether Di(2-methoxyethyl) ether	$C_6H_{14}O_3$	13.0	37.6	50.0	63.0	77.5	86.8	99.5	118.0	138.5	159.8	
glycol ethyl ether	$C_6H_{14}O_3$	45.3	72.0	85.8	100.3	116.7	126.8	140.3	159.0	180.3	201.9	
Diethyl ether	$C_4H_{10}O$	−74.3	−56.9	−48.1	−38.5	−27.7	−21.8	−11.5	+2.2	17.9	34.6	−116.3
ethylmalonate	$C_9H_{16}O_4$	50.8	77.8	91.6	106.0	122.4	132.4	146.0	166.0	188.7	211.5	
fumarate	$C_8H_{12}O_4$	53.2	81.2	95.3	110.2	126.7	137.7	151.1	172.2	195.8	218.5	+0.6
glutarate	$C_9H_{16}O_4$	65.6	94.7	109.7	125.4	142.8	153.2	167.8	189.5	212.8	237.0	
Diethylhexadecylamine	$C_{20}H_{43}N$	139.8	175.8	194.0	213.5	235.0	248.5	265.5	292.8	324.6	355.0	

Diethyl itaconate	C$_9$H$_{14}$O$_4$	51.3	80.2	95.2	111.0	128.2	139.3	154.3	177.5	203.1	227.9	
ketone (3-pentanone)	C$_5$H$_{10}$O	−12.7	+7.5	17.2	27.9	39.4	46.7	56.2	70.6	86.3	102.7	−42
malate	C$_8$H$_{14}$O$_5$	80.7	110.4	125.3	141.2	157.8	169.0	183.9	205.3	229.5	253.4	
maleate	C$_8$H$_{12}$O$_4$	57.3	85.6	100.0	115.3	131.8	142.4	156.0	177.8	201.7	225.0	
malonate	C$_7$H$_{12}$O$_4$	40.0	67.5	81.3	95.9	113.3	123.0	136.2	155.5	176.8	198.9	−49.8
mesaconate	C$_8$H$_{14}$O$_4$	62.8	91.0	105.3	120.3	137.3	147.9	161.6	183.2	205.8	229.0	
oxalate	C$_6$H$_{10}$O$_4$	47.4	71.8	83.8	96.8	110.6	119.7	130.8	147.9	166.2	185.7	−40.6
phthalate	C$_{12}$H$_{14}$O$_4$	108.8	140.7	156.0	173.6	192.1	204.1	219.5	243.0	267.5	294.0	
sebacate	C$_{14}$H$_{26}$O$_4$	125.3	156.2	172.1	189.8	207.5	218.4	234.4	255.8	280.3	305.5	1.3
2,5-Diethylstyrene	C$_{12}$H$_{16}$	49.7	78.4	92.6	108.5	125.8	136.8	151.0	173.2	198.0	223.0	
Diethyl succinate	C$_8$H$_{14}$O$_4$	54.6	83.0	96.6	111.7	127.8	138.2	151.1	171.7	193.8	216.5	−20.8
isosuccinate	C$_8$H$_{14}$O$_4$	39.8	66.7	80.0	94.7	111.0	121.4	134.8	155.1	177.7	201.3	
sulfate	C$_4$H$_{10}$O$_4$S	47.0	74.0	87.7	102.1	118.0	128.6	142.5	162.5	185.5	209.5	−25.0
sulfide	C$_4$H$_{10}$S	−39.6	−18.6	−8.0	+3.5	16.1	24.2	35.0	51.3	69.7	88.0	−99.5
sulfite	C$_4$H$_{10}$O$_3$S	10.0	34.2	46.4	59.7	74.2	83.8	96.3	115.8	137.0	159.0	
d-Diethyl tartrate	C$_8$H$_{14}$O$_6$	102.0	133.0	148.0	164.2	182.3	194.0	208.5	230.4	254.8	280.0	17
dl-Diethyl tartrate	C$_8$H$_{14}$O$_6$	100.0	131.7	147.2	163.8	181.7	193.2	208.0	230.0	254.3	280.0	
3,5-Diethyltoluene	C$_{11}$H$_{16}$	34.0	61.5	75.3	90.2	107.0	117.7	131.7	152.4	176.5	200.7	
Diethylzinc	C$_4$H$_{10}$Zn	−22.4	0.0	+11.7	24.2	38.0	47.2	59.1	77.0	97.3	118.0	−28
1-Dihydrocarvone	C$_{10}$H$_{16}$O	46.6	75.5	90.0	106.0	123.7	134.7	149.7	171.8	197.0	223.0	
Dihydrocitronellol	C$_{10}$H$_{22}$O	68.0	91.7	103.0	115.0	127.6	136.7	145.9	160.2	176.8	193.5	
1,4-Dihydroxyanthraquinone	C$_{14}$H$_8$O$_4$	196.7	239.8	259.8	282.0	307.4	323.3	344.5	377.8	413.0	450.0	194
Dimethylacetylene (2-butyne)	C$_4$H$_6$	−73.0	−57.9	−50.5	−42.5	−33.9	−27.8	−18.8	−5.0	+10.6	27.2	−32.5
Dimethylamine	C$_2$H$_7$N	−87.7	−72.2	−64.6	−56.0	−46.7	−40.7	−32.6	−20.4	−7.1	+7.4	−96
N,N-Dimethylaniline	C$_8$H$_{11}$N	29.5	56.3	70.0	84.8	101.6	111.9	125.8	146.5	169.2	193.1	+2.5
Dimethyl arsanilate	C$_8$H$_{12}$AsNO$_3$	15.0	39.6	51.8	65.0	79.7	88.6	101.0	119.8	140.3	160.5	
Di(α-methylbenzyl) ether	C$_{16}$H$_{18}$O	96.7	128.3	144.0	160.3	179.6	191.5	206.8	229.7	254.8	281.9	
2,2-Dimethylbutane	C$_6$H$_{14}$	−69.3	−50.7	−41.5	−31.1	−19.5	−12.1	−2.0	+13.4	31.0	49.7	−99.8
2,3-Dimethylbutane	C$_6$H$_{14}$	−63.6	−44.5	−34.9	−24.1	−12.4	−4.9	+5.4	21.1	39.0	58.0	−128.2
Dimethyl citraconate	C$_7$H$_{10}$O$_4$	50.8	78.2	91.8	106.5	122.6	132.7	145.8	165.8	188.0	210.5	
1,1-Dimethylcyclohexane	C$_8$H$_{16}$	−24.4	−1.4	+10.3	23.0	37.3	45.7	57.9	76.2	97.2	119.5	−34
cis-1,2-Dimethylcyclohexane	C$_8$H$_{16}$	−15.9	+7.3	18.4	31.1	45.3	54.4	66.8	85.6	107.0	129.7	−50.0
trans-1,2-Dimethylcyclohexane	C$_8$H$_{16}$	−21.1	+1.7	13.0	25.6	39.7	48.7	61.0	79.6	100.9	123.4	−80.0
trans-1,3-Dimethylcyclohexane	C$_8$H$_{16}$	−19.4	+3.4	14.9	27.4	41.4	50.4	62.5	81.0	102.1	124.4	−92.0
cis-1,3-Dimethylcyclohexane	C$_8$H$_{16}$	−22.7	0.0	+11.2	23.6	37.5	46.4	58.5	76.9	97.8	120.1	−76.2
cis-1,4-Dimethylcyclohexane	C$_8$H$_{16}$	−20.0	+3.2	14.5	27.1	41.1	50.1	62.3	80.8	101.9	124.3	−87.4
trans-1,4-Dimethylcyclohexane	C$_8$H$_{16}$	−24.3	−1.7	+10.1	22.6	36.5	45.4	57.6	76.0	97.0	119.3	−36.9
Dimethyl ether	C$_2$H$_6$O	−115.7	−101.1	−93.3	−85.2	−76.2	−70.4	−62.7	−50.9	−37.8	−23.7	−138.5

(Continued)

TABLE 2.38 Boiling Points of Common Organic Compounds at Selected Pressures (*Continued*)

Name	Formula	1	5	10	20	40	60	100	200	400	760	Melting point, °C
		Temperature, °C										
2,2-Dimethylhexane	C_8H_{18}	−29.7	−7.9	+3.1	15.0	28.2	36.7	48.2	65.7	85.6	106.8	
2,3-Dimethylhexane	C_8H_{18}	−23.0	−1.1	+9.9	22.1	35.6	44.2	56.0	73.8	94.1	115.6	
2,4-Dimethylhexane	C_8H_{18}	−26.9	−5.3	+5.2	17.2	30.5	39.0	50.6	68.1	88.2	109.4	
2,5-Dimethylhexane	C_8H_{18}	−26.7	−5.5	+5.3	17.2	30.4	38.9	50.5	68.0	87.9	109.1	−90.7
3,3-Dimethylhexane	C_8H_{18}	−25.8	−4.4	+6.1	18.2	31.7	40.4	52.5	70.0	90.4	112.0	
3,4-Dimethylhexane	C_8H_{18}	−22.1	+0.2	11.3	23.5	37.1	45.8	57.7	75.6	96.0	117.7	
Dimethyl itaconate	$C_7H_{10}O_4$	69.3	94.0	106.6	119.7	133.7	142.6	153.7	171.0	189.8	208.0	38
1-Dimethyl malate	$C_6H_{10}O_5$	75.4	104.0	118.3	133.8	150.1	160.4	175.1	196.3	219.5	242.6	
Dimethyl maleate	$C_6H_8O_4$	45.7	73.0	86.4	101.3	117.2	127.1	140.4	160.0	182.2	205.0	
malonate	$C_5H_8O_4$	35.0	59.8	72.0	85.0	100.0	109.7	121.9	140.0	159.8	180.7	−62
trans-Dimethyl mesaconate	$C_7H_{10}O_4$	46.8	74.0	87.8	102.1	118.0	127.8	141.5	161.0	183.5	206.0	
2,7-Dimethyloctane	$C_{10}H_{22}$	+6.3	30.5	42.3	55.8	71.2	80.8	93.9	114.0	136.0	159.7	−52.8
Dimethyl oxalate	$C_4H_6O_4$	20.0	44.0	56.0	69.4	83.6	92.8	104.8	123.3	143.3	163.3	
2,2-Dimethylpentane	C_7H_{16}	−49.0	−28.7	−18.7	−7.5	+5.0	13.0	23.9	40.3	59.2	79.2	−123.7
2,3-Dimethylpentane	C_7H_{16}	−42.0	−20.8	−10.3	+1.1	13.9	22.1	33.3	50.1	69.4	89.8	−135
2,4-Dimethylpentane	C_7H_{16}	−48.0	−27.4	−17.1	−5.9	+6.5	14.5	25.4	41.8	60.6	80.5	−119.5
3,3-Dimethylpentane	C_7H_{16}	−45.9	−25.0	−14.4	−2.9	+9.9	18.1	29.3	46.2	65.5	86.1	−135.0
2,3-Dimethylphenol (2,3-xylenol)	$C_8H_{10}O$	56.0	83.8	97.6	112.0	129.2	139.5	152.2	173.0	196.0	218.0	75
2,4-Dimethylphenol (2,4-xylenol)	$C_8H_{10}O$	51.8	78.0	91.3	105.0	121.5	131.0	143.0	161.5	184.2	211.5	25.5
2,5-Dimethylphenol (2,5-xylenol)	$C_8H_{10}O$	51.8	78.0	91.3	105.0	121.5	131.0	143.0	161.5	184.2	211.5	74.5
3,4-Dimethylphenol (3,4-xylenol)	$C_8H_{10}O$	66.2	93.8	107.7	122.0	138.0	148.0	161.0	181.5	203.6	225.2	62.5
3,5-Dimethylphenol (3,5-xylenol)	$C_8H_{10}O$	62.0	89.2	102.4	117.0	133.3	143.5	156.0	176.2	197.8	219.5	68
Dimethylphenylsilane	$C_8H_{12}Si$	+5.3	30.3	42.6	56.2	71.4	81.3	94.2	114.2	136.4	159.3	
Dimethyl phthalate	$C_{10}H_{10}O_4$	100.3	131.8	147.6	164.0	182.8	194.0	210.0	232.7	257.8	283.7	
3,5-Dimethyl-1,2-pyrone	$C_7H_8O_2$	78.6	107.6	122.0	136.4	152.7	163.8	177.5	198.0	221.0	245.0	51.5
4,6-Dimethylresorcinol	$C_8H_{10}O_2$	49.0	76.8	90.7	105.8	122.5	133.2	147.3	167.8	192.0	215.0	
Dimethyl sebacate	$C_{12}H_{22}O_4$	104.0	139.8	156.0	175.8	196.0	208.0	222.6	245.0	269.6	293.5	38
2,4-Dimethylstyrene	$C_{10}H_{12}$	34.2	61.9	75.8	90.8	107.7	118.0	132.3	153.2	177.5	202.0	
2,5-Dimethylstyrene	$C_{10}H_{12}$	29.0	55.9	69.0	84.0	100.2	110.7	124.7	145.6	168.7	193.0	
α,α-Dimethylsuccinic anhydride	$C_6H_8O_3$	61.4	88.1	102.0	116.3	132.6	142.4	155.3	175.8	197.5	219.5	
Dimethyl sulfide	C_2H_6S	−75.6	−58.0	−49.2	−39.4	−28.4	−21.9	−12.0	+2.6	18.7	36.0	−83.2

Compound	Formula											
d-Dimethyl tartrate	$C_6H_{10}O_6$	102.1	133.2	148.2	164.3	182.4	193.8	208.8	230.5	255.0	280.0	61.5
dl-Dimethyl tartrate	$C_6H_{10}O_6$	100.4	131.8	147.5	164.0	182.4	193.8	209.5	232.3	257.4	282.0	89
N,N-Dimethyl-2-toluidine	$C_9H_{13}N$	28.8	54.1	66.2	80.2	95.0	105.2	118.1	138.3	161.5	184.8	-61
N,N-Dimethyl-4-toluidine	$C_9H_{13}N$	50.1	74.3	86.7	100.0	116.3	126.4	140.3	161.6	185.4	209.5	
Di(nitrosomethyl) amine	$C_2H_5N_3O_2$	+3.2	27.8	40.0	53.7	68.2	77.7	90.3	110.0	131.3	153.0	
Diosphenol	$C_{10}H_{16}O_2$	66.7	95.4	109.0	124.0	141.2	151.3	165.6	186.2	209.5	232.0	10
1,4-Dioxane	$C_4H_8O_2$	-35.8	-12.8	-1.2	+12.0	25.2	33.8	45.1	62.3	81.8	101.1	
Dipentene	$C_{10}H_{16}$	14.0	40.4	53.8	68.2	84.3	94.6	108.3	128.2	150.5	174.6	
Diphenylamine	$C_{12}H_{11}N$	108.3	141.7	157.0	175.2	194.3	206.9	222.8	247.5	274.1	302.0	52.9
Diphenyl carbinol (benzhydrol)	$C_{13}H_{12}O$	110.0	145.0	162.0	180.9	200.0	212.0	227.5	250.0	275.6	301.0	68.5
chlorophosphate	$C_{12}H_{10}ClPO_3$	121.5	160.5	182.0	203.8	227.9	244.2	265.0	299.5	337.2	378.0	
disulfide	$C_{12}H_{10}S_2$	131.6	164.0	180.0	197.0	214.8	226.2	241.3	262.6	285.8	310.0	61
1,2-Diphenylethane (dibenzyl)	$C_{14}H_{14}$	86.8	119.8	136.0	153.7	173.7	186.0	202.8	227.8	255.0	284.0	51.5
Diphenyl ether	$C_{12}H_{10}O$	66.1	97.8	114.0	130.8	150.0	162.0	178.8	203.3	230.7	258.5	27
1,1-Diphenylethylene	$C_{14}H_{12}$	87.4	119.6	135.0	151.8	170.8	183.4	198.6	222.8	249.8	277.0	
trans-Diphenylethylene	$C_{14}H_{12}$	113.2	145.8	161.0	179.8	199.0	211.5	227.4	251.7	278.3	306.5	124
1,1-Diphenylhydrazine	$C_{12}H_{12}N_2$	126.0	159.3	176.1	194.0	213.5	225.9	242.5	267.2	294.0	322.2	44
Diphenylmethane	$C_{13}H_{12}$	76.0	107.4	122.8	139.8	157.8	170.2	186.3	210.7	237.5	264.5	26.5
Diphenyl sulfide	$C_{12}H_{10}S$	96.1	129.0	145.0	162.0	182.8	194.8	211.8	236.8	263.9	292.5	
Diphenyl-2-tolyl thiophosphate	$C_{18}H_{17}O_3PS$	159.7	179.6	201.6	215.5	230.6	240.4	252.5	270.3	290.0	310.0	
1,2-Dipropoxyethane	$C_8H_{18}O_2$	-38.8	-10.3	+5.0	22.3	42.3	55.8	74.2	103.8	140.0	180.0	
1,2-Diisopropylbenzene	$C_{12}H_{18}$	40.0	67.8	81.8	96.8	114.0	124.3	138.7	159.8	184.3	209.0	
1,3-Diisopropylbenzene	$C_{12}H_{18}$	34.7	62.3	76.0	91.2	107.9	118.2	132.3	153.7	177.6	202.0	-105
Dipropylene glycol	$C_6H_{14}O_3$	73.8	102.1	116.2	131.3	147.4	156.5	169.9	189.9	210.5	231.8	
Dipropyleneglycol monobutyl ether	$C_{10}H_{22}O_3$	64.7	92.0	106.0	120.4	136.3	146.3	159.8	180.0	203.8	227.0	
isopropyl ether	$C_9H_{20}O_3$	46.0	72.8	86.2	100.8	117.0	126.8	140.3	160.0	183.1	205.6	
Di-*n*-propyl ether	$C_6H_{14}O$	-43.3	-22.3	-11.8	0.0	+13.2	21.6	33.0	50.3	69.5	89.5	-122
Diisopropyl ether	$C_6H_{14}O$	-57.0	-37.4	-27.4	-16.7	-4.5	+3.4	13.7	30.0	48.2	67.5	-60
Di-*n*-propyl ketone (4-heptanone)	$C_7H_{14}O$	23.0	44.4	55.0	66.2	78.1	85.8	96.0	111.2	127.3	143.7	-32.6
Di-*n*-propyl oxalate	$C_8H_{14}O_4$	53.4	80.2	93.9	108.6	124.6	134.8	148.1	168.0	190.3	213.5	
Diisopropyl oxalate	$C_8H_{14}O_4$	43.2	69.0	81.9	95.6	110.5	120.0	132.6	151.2	171.8	193.5	
Di-*n*-propyl succinate	$C_{10}H_{18}O_4$	77.5	107.6	122.2	138.0	154.8	166.0	180.3	202.5	226.5	250.8	
Di-*n*-propyl *d*-tartrate	$C_{10}H_{18}O_6$	115.6	147.7	163.5	180.4	199.7	211.7	227.0	250.1	275.6	303.0	
Diisopropyl *d*-tartrate	$C_{10}H_{18}O_6$	103.7	133.7	148.2	164.0	181.8	192.6	207.3	228.2	251.8	275.0	
Divinyl acetylene (1,5-hexadiene-3-yne)	C_6H_6	-45.1	-24.4	-14.0	-2.8	+10.0	18.1	29.5	46.0	64.4	84.0	
1,3-Divinylbenzene	$C_{10}H_{10}$	32.7	60.0	73.8	88.7	105.5	116.0	130.0	151.4	175.2	199.5	-66.9
Docosanae	$C_{22}H_{46}$	157.8	195.4	213.0	233.5	254.5	268.3	286.0	314.2	343.5	376.0	44.5

(Continued)

TABLE 2.38 Boiling Points of Common Organic Compounds at Selected Pressures (*Continued*)

Name	Formula	Pressure, mm Hg — Temperature, °C										Melting point, °C
		1	5	10	20	40	60	100	200	400	760	
n-Dodecane	C$_{12}$H$_{26}$	47.8	75.8	90.0	104.6	121.7	132.1	146.2	167.2	191.0	216.2	−9.6
1-Dodecene	C$_{12}$H$_{24}$	47.2	74.0	87.8	102.4	118.6	128.5	142.3	162.2	185.5	208.0	−31.5
n-Dodecyl alcohol	C$_{12}$H$_{26}$O	91.0	120.2	134.7	150.0	167.2	177.8	192.0	213.0	235.7	259.0	24
Dodecylamine	C$_{12}$H$_{27}$N	82.8	111.8	127.8	141.6	157.4	168.0	182.1	203.0	225.0	248.0	
Dodecyltrimethylsilane	C$_{15}$H$_{34}$Si	91.2	122.1	137.7	153.8	172.1	184.2	199.5	222.0	248.0	273.0	
Elaidic acid	C$_{18}$H$_{34}$O$_2$	171.3	206.7	223.5	242.3	260.8	273.0	288.0	312.4	337.0	362.0	51.5
Epichlorohydrin	C$_3$H$_5$ClO	−16.5	+5.6	16.6	29.0	42.0	50.6	62.0	79.3	98.0	117.9	−25.6
1,2-Epoxy-2-methylpropane	C$_4$H$_8$O	−69.0	−50.0	−40.3	−29.5	−17.3	−9.7	+1.2	17.5	36.0	55.5	
Erucic acid	C$_{22}$H$_{42}$O$_2$	206.7	239.7	254.5	270.6	289.1	300.2	314.4	336.5	358.8	381.5	33.5
Estragole (p-methoxy allyl benzene)	C$_{10}$H$_{12}$O	52.6	80.0	93.7	108.4	124.6	135.2	148.5	168.7	192.0	215.0	
Ethane	C$_2$H$_6$	−159.5	−148.5	−142.9	−136.7	−129.8	−125.4	−119.3	−110.2	−99.7	−88.6	−183.2
Ethoxydimethylphenylsilane	C$_{10}$H$_{16}$OSi	36.3	63.1	76.2	91.0	107.2	127.5	131.4	151.5	175.0	199.5	
Ethoxytrimethylsilane	C$_5$H$_{14}$OSi	−50.9	−31.0	−20.7	−9.8	+3.7	11.5	22.1	38.1	56.3	75.7	
Ethoxytriphenylsilane	C$_{20}$H$_{20}$OSi	167.0	198.2	213.5	230.0	247.0	258.3	273.5	295.0	319.5	344.0	
Ethyl acetate	C$_4$H$_8$O$_2$	−43.4	−23.5	−13.5	−3.0	+9.1	16.6	27.0	42.0	59.3	77.1	−82.4
acetoacetate	C$_6$H$_{10}$O$_3$	28.5	54.0	67.3	81.1	96.2	106.0	118.5	138.0	158.2	180.8	−45
Ethylacetylene (1-butyne)	C$_4$H$_6$	−92.5	−76.7	−68.7	−59.9	−50.5	−43.4	−34.9	−21.6	−6.9	+8.7	−130
Ethyl acrylate	C$_5$H$_8$O$_2$	−29.5	−8.7	+2.0	13.0	26.0	33.5	44.5	61.5	80.0	99.5	−71.2
α-Ethylacrylic acid	C$_5$H$_8$O$_2$	47.0	70.7	82.0	94.4	108.1	116.7	127.5	144.0	160.7	179.2	
α-Ethylacrylonitrile	C$_5$H$_7$N	−29.0	−6.4	+5.0	17.7	31.8	40.6	53.0	71.6	92.2	114.0	
Ethyl alcohol (ethanol)	C$_2$H$_6$O	−31.3	−12.0	−2.3	+8.0	19.0	26.0	34.9	48.4	63.5	78.4	−112
Ethylamine	C$_2$H$_7$N	−82.3	−66.4	−58.3	−48.6	−39.8	−33.4	−25.1	−12.3	+2.0	16.6	−80.6
4-Ethylaniline	C$_8$H$_{11}$N	52.0	80.0	93.8	109.0	125.7	136.0	149.8	170.6	194.2	217.4	−4
N-Ethylaniline	C$_8$H$_{11}$N	38.5	66.4	80.6	96.0	113.2	123.6	137.3	156.9	180.8	204.0	−63.5
2-Ethylanisole	C$_9$H$_{12}$O	29.7	55.9	69.0	83.1	98.9	109.0	122.3	142.1	164.2	187.1	
3-Ethylanisole	C$_9$H$_{12}$O	33.7	60.3	73.9	88.5	104.8	115.5	129.2	149.7	172.8	196.5	
4-Ethylanisole	C$_9$H$_{12}$O	33.5	60.2	73.9	88.5	104.7	115.4	128.4	149.2	172.3	196.5	
Ethylbenzene	C$_8$H$_{10}$	−9.8	+13.9	25.9	38.6	52.8	61.8	74.1	92.7	113.8	136.2	−94.9
Ethyl benzoate	C$_9$H$_{10}$O$_2$	44.0	72.0	86.0	101.4	118.2	129.0	143.2	164.8	188.4	213.4	−34.6
benzoylacetate	C$_{11}$H$_{12}$O$_3$	107.6	136.4	150.3	166.8	181.8	191.9	205.0	223.8	244.7	265.0	
bromide	C$_2$H$_5$Br	−74.3	−56.4	−47.5	−37.8	−26.7	−19.5	−10.0	+4.5	21.0	38.4	−117.8

Compound	Formula											
α-bromoisobutyrate	$C_6H_{11}BrO_2$	10.6	35.8	48.0	61.8	77.0	86.7	99.8	119.7	141.2	163.6	
n-butyrate	$C_6H_{12}O_2$	−18.4	+4.0	15.3	27.8	41.5	50.1	62.0	79.8	100.0	121.0	−93.3
isobutyrate	$C_6H_{12}O_2$	−24.3	−2.4	+8.4	20.6	33.8	42.3	53.5	71.0	90.0	110.0	−88.2
Ethylcamphoronic anhydride	$C_{11}H_{16}O_5$	113.2	149.8	165.0	181.8	199.8	211.5	226.6	248.5	272.8	298.0	
Ethyl isocaproate	$C_8H_{16}O_2$	11.0	35.8	48.0	61.7	76.3	85.8	98.4	117.8	139.2	160.4	49
carbamate	$C_3H_7NO_2$		65.8	77.8	91.0	105.6	114.8	126.2	144.2	164.0	184.0	
carbanilate	$C_9H_{11}NO_2$	107.8	131.8	143.7	155.5	168.8	177.3	187.9	203.8	220.0	237.0	52.5
Ethylcetylamine	$C_{18}H_{39}N$	133.2	168.2	186.0	205.5	226.5	239.8	256.8	283.3	313.0	342.0	
Ethyl chloride	C_2H_5Cl	−89.8	−73.9	−65.8	−56.8	−47.0	−40.6	−32.0	−18.6	−3.9	+12.3	−139
chloroacetate	$C_4H_7ClO_2$	+1.0	25.4	37.5	50.4	65.2	74.0	86.0	103.8	123.8	144.2	−26
chloroglyoxylate	$C_4H_5ClO_3$	−5.1	+18.0	29.9	42.0	56.0	65.2	76.6	94.5	114.7	135.0	
α-chloropropionate	$C_5H_9ClO_2$	+6.6	30.2	41.9	54.3	68.2	77.3	89.3	107.2	126.2	146.5	
trans-cinnamate	$C_{11}H_{12}O_2$	87.6	108.5	134.0	150.3	169.2	181.2	196.0	219.3	245.0	271.0	12
3-Ethylcumene	$C_{11}H_{16}$	28.3	55.5	68.8	83.6	99.9	110.2	124.3	145.4	168.2	193.0	
4-Ethylcumene	$C_{11}H_{16}$	31.5	58.4	72.0	86.7	103.3	113.8	127.2	148.3	171.8	195.8	
Ethyl cyanoacetate	$C_5H_7NO_2$	67.8	93.5	106.0	119.8	133.8	142.1	152.8	169.8	187.8	206.0	
Ethylcyclohexane	C_8H_{16}	−4.5	+9.2	20.6	33.4	47.6	56.7	69.0	87.8	109.1	131.8	−111.3
Ethylcyclopentane	C_7H_{14}	−32.2	−10.8	−0.1	+11.7	25.0	33.4	45.0	62.4	82.3	103.4	−138.6
Ethyl dichloroacetate	$C_4H_6Cl_2O_2$	9.6	34.0	46.3	59.5	74.0	83.6	96.1	115.2	135.9	156.5	
N,*N*-diethyloxamate	$C_8H_{15}NO_3$	76.0	106.3	121.7	137.7	154.4	166.0	180.3	202.8	226.5	252.0	
N-Ethyldiphenylamine	$C_{14}H_{15}N$	98.3	130.2	146.0	162.8	182.0	193.7	209.8	233.0	258.8	286.0	
Ethylene	C_2H_4	−158.3	−158.3	−153.2	−147.6	−141.3	−137.3	−131.8	−123.4	−113.9	−103.7	−169
Ethylene-*bis*-(chloroacetate)	$C_6H_8Cl_2O_4$	112.0	142.4	158.0	173.5	191.0	201.8	215.0	237.3	259.5	283.5	
Ethylene chlorohydrin (2-chloroethanol)	C_2H_5ClO	−4.0	+19.0	30.3	42.5	56.0	64.1	75.0	91.8	110.0	128.8	−69
diamine (1,2-ethanediamine)	$C_2H_8N_2$	−11.0	+10.5	21.5	33.0	45.8	53.8	62.5	81.0	99.0	117.2	8.5
dibromide (1,2-dibromomethane)	$C_2H_4Br_2$	−27.0	+4.7	18.6	32.7	48.0	57.9	70.4	89.8	110.1	131.5	10
dichloride (1,2-dichloroethane)	$C_2H_4Cl_2$	−44.5	−24.0	−13.6	−2.4	+10.0	18.1	29.4	45.7	64.0	82.4	−35.3
glycol (1,2-ethanediol)	$C_2H_6O_2$	53.0	79.7	92.1	105.8	120.0	129.5	141.8	158.5	178.5	197.3	−15.6
glycol diethyl ether (1,2-diethoxyethane)	$C_6H_{14}O_2$	−33.5	−10.2	+1.6	14.7	29.7	39.0	51.8	71.8	94.1	119.5	
glycol dimethyl ether (1,2-dimethoxyethane)	$C_4H_{10}O_2$	−48.0	−26.2	−15.3	−3.0	+10.7	19.7	31.8	50.0	70.8	93.0	
glycol monomethyl ether (2-methoxyethanol)	$C_3H_8O_2$	−13.5	+10.2	22.0	34.3	47.8	56.4	68.0	85.3	104.3	124.4	
oxide	C_2H_4O	−89.7	−73.8	−65.7	−56.6	−46.9	−40.7	−32.1	−19.5	−4.9	+10.7	−111.3
Ethyl α-ethylacetoacetate	$C_8H_{14}O_3$	40.5	67.3	80.2	94.6	110.3	120.6	133.8	153.2	175.6	198.0	
fluoride	C_2H_5F	−117.0	−103.8	−97.7	−90.0	−81.8	−76.4	−69.3	−58.0	−45.5	−32.0	
formate	$C_3H_6O_2$	−60.5	−42.2	−33.0	−22.7	−11.5	−4.3	−5.4	20.2	37.1	54.3	−79

(Continued)

TABLE 2.38 Boiling Points of Common Organic Compounds at Selected Pressures (*Continued*)

Name	Formula	Pressure, mm Hg										Melting point, °C
		1	5	10	20	40	60	100	200	400	760	
		Temperature, °C										
2-furoate	$C_7H_8O_3$	37.6	63.8	77.1	91.5	107.5	117.5	130.4	150.1	172.5	195.0	34
glycolate	$C_4H_8O_3$	14.3	38.8	50.5	63.9	78.1	87.6	99.8	117.8	138.0	158.2	
3-Ethylhexane	C_8H_{18}	−20.0	+2.1	12.8	25.0	38.5	47.1	58.9	76.7	97.0	118.5	
2-Ethylhexyl acrylate	$C_{11}H_{20}O_2$	50.0	77.7	91.8	106.3	123.7	134.0	147.9	168.2	192.2	216.0	
Ethylidene chloride (1,1-dichloroethane)	$C_2H_4Cl_2$	−60.7	−41.9	−32.3	−21.9	−10.2	−2.9	+7.2	22.4	39.8	57.4	−96.7
fluoride (1,1-difluoroethane)	$C_2H_4F_2$	−112.5	−98.4	−91.7	−84.1	−75.8	−70.4	−63.2	−52.0	−39.5	−26.5	−117
Ethyl iodide	C_2H_5I	−54.4	−34.3	−24.3	−13.1	−0.9	+7.2	18.0	34.1	52.3	72.4	−105
Ethyl l-leucinate	$C_8H_{17}NO_2$	27.8	57.3	72.1	88.0	106.0	117.8	131.8	149.8	167.3	184.0	
Ethyl levulinate	$C_7H_{12}O_3$	47.3	74.0	87.3	101.8	117.7	127.6	141.3	160.2	183.0	206.2	−121
Ethyl mercaptan (ethanethiol)	C_2H_6S	−76.7	−59.1	−50.2	−40.7	−29.8	−22.4	−13.0	+1.5	17.7	35.0	
Ethyl methylcarbamate	$C_4H_9NO_2$	26.5	51.0	63.2	76.1	91.0	100.0	112.0	130.0	149.8	170.0	
Ethyl methyl ether	C_3H_8O	−91.0	−75.6	−67.8	−59.1	−49.4	−43.3	−34.8	−22.0	−7.8	+7.5	
1-Ethylnaphthalene	$C_{12}H_{12}$	70.0	101.4	116.8	133.8	152.0	164.1	180.0	204.6	230.8	258.1	−27
Ethyl α-naphthyl ketone (1-propionaphthone)	$C_{13}H_{12}O$	124.0	155.5	171.0	188.1	206.9	218.2	233.5	255.5	280.2	306.0	
Ethyl 3-nitrobenzoate	$C_9H_9NO_4$	108.1	140.2	155.0	173.6	192.6	205.0	220.3	244.6	270.6	298.0	47
3-Ethylpentane	C_7H_{16}	−37.8	−17.0	−6.8	+4.7	17.5	25.7	36.9	53.8	73.0	93.5	−118.6
4-Ethylphenetole	$C_{10}H_{14}O$	48.5	75.7	89.5	103.8	119.8	129.8	143.5	163.2	185.7	208.0	
2-Ethylphenol	$C_8H_{10}O$	46.2	73.4	87.0	101.5	117.9	127.9	141.8	161.6	184.5	207.5	−45
3-Ethylphenol	$C_8H_{10}O$	60.0	86.8	100.2	114.5	130.0	139.8	152.0	171.8	193.3	214.0	−4
4-Ethylphenol	$C_8H_{10}O$	59.3	86.5	100.2	115.0	131.3	141.7	154.2	175.0	197.4	219.0	46.5
Ethyl phenyl ether (phenetole)	$C_8H_{10}O$	18.1	43.7	56.4	70.3	86.6	95.4	108.4	127.9	149.8	172.0	−30.2
Ethyl propionate	$C_5H_{10}O_2$	−28.0	−7.2	+3.4	14.3	27.2	35.1	45.2	61.7	79.8	99.1	−72.6
Ethyl propyl ether	$C_5H_{12}O$	−64.3	−45.0	−35.0	−24.0	−12.0	−4.0	+6.8	23.3	41.6	61.7	
Ethyl salicylate	$C_9H_{10}O_3$	61.2	90.0	104.2	119.3	136.7	147.6	161.5	183.7	207.0	231.5	1.3
3-Ethylstyrene	$C_{10}H_{12}$	28.3	55.0	68.3	82.8	99.2	109.6	123.2	144.0	167.2	191.5	
4-Ethylstyrene	$C_{10}H_{12}$	26.0	52.7	66.3	80.8	97.3	107.6	121.5	142.0	165.0	189.0	
Ethylisothiocyanate	C_3H_5NS	13.2	+10.6	22.8	36.1	50.8	59.8	71.9	90.0	110.1	131.0	−5.9
2-Ethyltoluene	C_9H_{12}	9.4	34.8	47.6	61.2	76.4	86.0	99.0	119.0	141.4	165.1	
3-Ethyltoluene	C_9H_{12}	7.2	32.3	44.7	58.2	73.3	82.9	95.9	115.5	137.8	161.3	−95.5
4-Ethyltoluene	C_9H_{12}	7.6	32.7	44.9	58.5	73.6	83.2	96.3	116.1	136.4	162.0	

Compound	Formula											
Ethyl trichloroacetate	$C_4H_5Cl_3O_2$	20.7	45.5	57.7	70.6	85.5	94.4	107.4	125.8	146.0	167.0	
Ethyltrimethylsilane	$C_5H_{14}Si$	−60.6	−41.4	−31.8	−21.0	−9.0	−1.2	+9.2	25.0	42.8	62.0	
Ethyltrimethyltin	$C_5H_{14}Sn$	−30.0	−7.6	+3.8	16.1	30.0	38.4	50.0	67.3	87.6	108.8	
Ethyl isovalerate	$C_7H_{14}O_2$	−6.1	+17.0	28.7	41.3	55.2	64.0	75.9	93.8	114.0	134.3	−99.3
2-Ethyl-1,4-xylene	$C_{10}H_{14}$	25.7	52.0	65.6	79.8	96.0	106.2	120.0	140.2	163.1	186.9	
4-Ethyl-1,3-xylene	$C_{10}H_{14}$	26.3	53.0	66.4	80.6	97.2	107.4	121.2	141.8	164.4	188.4	
5-Ethyl-1,3-xylene	$C_{10}H_{14}$	22.1	48.8	62.1	76.5	92.6	103.0	116.5	137.4	159.6	183.7	
Eugenol	$C_{10}H_{12}O_2$	78.4	108.1	123.0	138.7	155.8	167.3	182.2	204.7	228.3	253.5	
iso-Eugenol	$C_{10}H_{12}O_2$	86.3	117.0	132.4	149.5	167.0	178.2	194.0	217.2	242.3	267.5	−10
Eugenyl acetate	$C_{12}H_{14}O_3$	101.6	132.3	148.0	164.2	183.0	194.0	209.7	232.5	257.4	282.0	295
Fencholic acid	$C_{10}H_{16}O_2$	101.7	128.7	142.3	155.8	171.8	181.5	194.0	215.0	237.8	264.1	19
d-Fenchone	$C_{10}H_{16}O$	28.0	54.7	68.3	83.0	99.5	109.8	123.6	144.0	166.8	191.0	5
dl-Fenchyl alcohol	$C_{10}H_{18}O$	45.8	70.3	82.1	95.6	110.8	120.2	132.3	150.0	173.2	201.0	35
Fluorene	$C_{13}H_{10}$		129.3	146.0	164.2	185.2	197.8	214.7	240.3	268.6	295.0	113
Fluorobenzene	C_6H_5F	−43.4	−22.8	−12.4	−1.2	+11.5	19.6	30.4	47.2	65.7	84.7	−42.1
2-Fluorotoluene	C_7H_7F	−24.2	−2.2	+8.9	21.4	34.7	43.7	55.3	73.0	92.8	114.0	−80
3-Fluorotoluene	C_7H_7F	−22.4	−0.3	+11.0	23.4	37.0	45.8	57.5	75.4	95.4	116.0	−110.8
4-Fluorotoluene	C_7H_7F	−21.8	+0.3	11.8	24.0	37.8	46.5	58.1	76.0	96.1	117.0	
Formaldehyde	CH_2O			−88.0	−79.6	−70.6	−65.0	−57.3	−46.0	−33.0	−19.5	−92
Formamide	CH_3NO	70.5	96.3	109.5	122.5	137.5	147.0	157.5	175.5	193.5	210.5	
Formic acid	CH_2O_2	−20.0	−5.0	+2.1	10.3	24.0	32.4	43.8	61.4	80.3	100.6	8.2
trans-Fumaryl chloride	$C_4H_2Cl_2O_2$	+15.0	38.5	51.8	65.0	79.5	89.0	101.0	120.0	140.0	160.0	
Furfural (2-furaldehyde)	$C_5H_4O_2$	18.5	42.6	54.8	67.8	82.1	91.5	103.4	121.8	141.8	161.8	
Furfuryl alcohol	$C_5H_6O_2$	31.8	56.0	68.0	81.0	95.7	104.0	115.9	133.1	151.8	170.0	
Geraniol	$C_{10}H_{18}O$	69.2	96.8	110.0	125.6	141.8	151.5	165.3	185.6	207.8	230.0	
Geranyl acetate	$C_{12}H_{20}O_2$	73.5	102.7	117.9	133.0	150.0	160.3	175.2	196.3	219.8	243.3	
Geranyl n-butyrate	$C_{14}H_{24}O_2$	96.8	125.2	139.0	153.8	170.1	180.2	193.8	214.0	235.0	257.4	
Geranyl isobutyrate	$C_{14}H_{24}O_2$	90.9	119.6	133.0	147.9	164.0	174.0	187.7	207.6	228.5	251.0	
Geranyl formate	$C_{11}H_{18}O_2$	61.8	90.3	104.3	119.8	136.2	147.2	160.7	182.6	205.8	230.0	
Glutaric acid	$C_5H_8O_4$	155.5	183.8	196.0	210.5	226.3	235.5	247.0	265.0	283.5	303.0	97.5
Glutaric anhydride	$C_5H_6O_3$	100.8	133.3	149.5	166.0	185.5	196.2	212.5	236.5	261.0	287.0	
Glutaronitrile	$C_5H_6N_2$	91.3	123.7	140.0	156.5	174.6	189.5	205.5	230.0	257.3	286.2	
Glutaryl chloride	$C_5H_6Cl_2O_2$	56.1	84.0	97.8	112.3	128.3	139.1	151.8	172.4	195.3	217.0	
Glycerol	$C_3H_8O_3$	125.5	153.8	167.2	182.2	198.0	208.0	220.1	240.0	263.0	290.0	17.9
Glycerol dichlorohydrin (1,3-dichloro-2-propanol)	$C_3H_6Cl_2O$	28.0	52.2	64.7	78.0	93.0	102.0	114.8	133.3	153.5	174.3	
Glycol diacetate	$C_6H_{10}O_4$	38.3	64.1	77.1	90.8	106.1	115.8	128.0	147.8	168.3	190.5	−31

(Continued)

TABLE 2.38   Boiling Points of Common Organic Compounds at Selected Pressures (*Continued*)

Compound Name	Formula	Pressure, mm Hg 1	5	10	20	40	60	100	200	400	760	Melting point, °C
		Temperature, °C										
Glycolide (1,4-dioxane-2,6-dione)	C$_4$H$_4$O$_4$		103.0	116.6	132.0	148.6	158.2	173.2	194.0	217.0	240.0	97
Guaiacol (2-methoxyphenol)	C$_7$H$_8$O$_2$	52.4	79.1	92.0	106.0	121.6	131.0	144.1	162.7	184.1	205.0	28.3
Heneicosane	C$_{21}$H$_{44}$	152.6	188.0	205.4	223.2	243.4	255.3	272.0	296.5	323.8	350.5	40.4
Heptacosane	C$_{27}$H$_{56}$	211.7	248.6	266.8	284.6	305.7	318.3	333.5	359.4	385.0	410.6	59.5
Heptadecane	C$_{17}$H$_{36}$	115.0	145.2	160.0	177.7	195.8	207.3	223.0	247.8	274.5	303.0	22.5
Heptaldehyde (enanthaldehyde)	C$_7$H$_{14}$O	12.0	32.7	43.0	54.0	66.3	74.0	84.0	102.0	125.5	155.0	−42
n-Heptane	C$_7$H$_{16}$	−34.0	−12.7	−2.1	+9.5	22.3	30.6	41.8	58.7	78.0	98.4	−90.6
Heptanoic acid (enanthic acid)	C$_7$H$_{14}$O$_2$	78.0	101.3	113.2	125.6	139.5	148.5	160.0	179.5	199.6	221.5	−10
1-Heptanol	C$_7$H$_{16}$O	42.4	64.3	74.7	85.8	99.8	108.0	119.5	136.6	155.6	175.8	34.6
Heptanoyl chloride (enanthyl chloride)	C$_7$H$_{13}$ClO	34.2	54.6	64.6	75.0	86.4	93.5	102.7	116.3	130.7	145.0	
2-Heptene	C$_7$H$_{14}$	−35.8	−14.1	−3.5	+8.3	21.5	30.0	41.3	58.6	78.1	98.5	
Heptylbenzene	C$_{13}$H$_{20}$	64.0	94.6	110.0	126.0	144.0	154.8	170.2	193.3	217.8	244.0	
Heptyl cyanide (enanthonitrile)	C$_7$H$_{13}$N	21.0	47.8	61.6	76.3	92.6	103.0	116.8	137.7	160.0	184.6	
Hexachlorobenzene	C$_6$Cl$_6$	114.4	149.3	166.4	185.7	206.0	219.0	235.5	258.5	283.5	309.4	230
Hexachloroethane	C$_2$Cl$_6$	32.7	49.8	73.5	87.6	102.3	112.0	124.2	143.1	163.8	185.6	186.6
Hexacosane	C$_{26}$H$_{54}$	204.0	240.0	257.4	275.8	295.2	307.8	323.2	348.4	374.6	399.8	56.6
Hexadecane	C$_{16}$H$_{34}$	105.3	135.2	149.8	164.7	181.3	193.2	208.5	231.7	258.3	287.5	18.5
1-Hexadecene	C$_{16}$H$_{32}$	101.6	131.7	146.2	162.0	178.8	190.8	205.3	226.8	250.0	274.0	4
n-Hexadecyl alcohol (cetyl alcohol)	C$_{16}$H$_{34}$O	122.7	158.3	177.8	197.8	219.8	234.3	251.7	280.2	312.7	344.0	49.3
n-Hexadecylamine (cetylamine)	C$_{16}$H$_{35}$N	123.6	157.8	176.0	195.7	215.7	228.8	245.8	272.2	300.4	330.0	
Hexaethylbenzene	C$_{18}$H$_{30}$		134.3	150.3	168.0	187.7	199.7	216.0	241.7	268.5	298.3	130
n-Hexane	C$_6$H$_{14}$	−53.9	−34.5	−25.0	−14.1	−2.3	+5.4	15.8	31.6	49.6	68.7	−95.3
1-Hexanol	C$_6$H$_{14}$O	24.4	47.2	58.2	70.3	83.7	92.0	102.8	119.6	138.0	157.0	−51.6
2-Hexanol	C$_6$H$_{14}$O	14.6	34.8	45.0	55.9	67.9	76.0	87.3	103.7	121.8	139.9	
3-Hexanol	C$_6$H$_{14}$O	+2.5	25.7	36.7	49.0	62.2	70.7	81.8	98.3	117.0	135.5	
1-Hexene	C$_6$H$_{12}$	−57.5	−38.0	−28.1	−17.2	−5.0	+2.8	13.0	29.0	46.8	66.0	−98.5
n-Hexyl levulinate	C$_{11}$H$_{20}$O$_3$	90.0	120.0	134.7	150.2	167.8	179.0	193.6	215.7	241.0	266.8	
n-Hexyl phenyl ketone (enanthophenone)	C$_{13}$H$_{18}$O	100.0	130.3	145.5	161.0	178.9	189.8	204.2	225.0	248.3	271.3	
Hydrocinnamic acid	C$_9$H$_{10}$O$_2$	102.2	133.5	148.7	165.0	183.3	194.0	209.0	230.8	255.0	279.8	48.5
Hydrogen cyanide (hydrocyanic acid)	CHN	−71.0	−55.3	−47.7	−39.7	−30.9	−25.1	−17.8	−5.3	+10.2	25.9	−13.2

Name	Formula											
Hydroquinone	C$_6$H$_6$O$_2$	132.4	153.3	163.5	174.6	192.0	203.0	216.5	238.0	262.5	286.2	170.3
4-Hydroxybenzaldehyde	C$_7$H$_6$O$_2$	121.2	153.2	169.7	186.8	206.0	217.5	233.5	256.8	282.6	310.0	115.5
α-Hydroxyisobutyric acid	C$_4$H$_8$O$_3$	73.5	98.5	110.5	123.8	138.0	146.4	157.7	175.2	193.8	212.0	79
α-Hydroxybutyronitrile	C$_5$H$_9$NO	41.0	65.8	77.8	90.7	104.8	113.9	125.0	142.0	159.8	178.8	
4-Hydroxy-3-methyl-2-butanone	C$_5$H$_{10}$O$_2$	44.6	69.3	81.0	94.0	108.2	117.4	129.0	146.5	165.5	185.0	
4-Hydroxy-4-methyl-2-pentanone	C$_6$H$_{12}$O$_2$	22.0	46.7	58.8	72.0	86.7	96.0	108.2	126.8	147.5	167.9	−47
3-Hydroxypropionitrile	C$_3$H$_5$NO	58.7	87.8	102.0	117.9	134.1	144.7	157.7	178.0	200.0	221.0	
Indene	C$_9$H$_8$	16.4	44.3	58.5	73.9	90.7	100.8	114.7	135.6	157.8	181.6	−2
Iodobenzene	C$_6$H$_5$I	24.1	50.6	64.0	78.3	94.4	105.0	118.3	139.8	163.9	188.6	−28.5
Iodononane	C$_9$H$_{19}$I	70.0	96.2	109.0	123.0	138.1	147.7	159.8	179.0	199.3	219.5	
2-Iodotoluene	C$_7$H$_7$I	37.2	65.9	79.8	95.6	112.4	123.8	138.1	160.0	185.7	211.0	
α-Ionone	C$_{13}$H$_{20}$O	79.5	108.8	123.0	139.0	155.6	166.3	181.2	202.5	225.2	250.0	
Isoprene	C$_5$H$_8$	−79.8	−62.3	−53.3	−43.5	−32.6	−25.4	−16.0	−1.2	+15.4	32.6	−146.7
Lauraldehyde	C$_{12}$H$_{24}$O	77.7	108.4	123.7	140.2	157.8	168.7	184.5	207.8	231.8	257.0	44.5
Lauric acid	C$_{12}$H$_{24}$O$_2$	121.0	150.6	166.0	183.6	201.4	212.7	227.5	249.8	273.8	299.2	48
Levulinaldehyde	C$_5$H$_8$O$_2$	28.1	54.9	68.0	82.7	98.3	108.4	121.8	142.0	164.0	187.0	
Levulinic acid	C$_5$H$_8$O$_3$	102.0	128.1	141.8	154.1	169.5	178.0	190.2	208.3	227.4	245.8	33.5
d-Limonene	C$_{10}$H$_{16}$	14.0	40.4	53.8	68.2	84.3	94.6	108.3	128.5	151.4	175.0	−96.9
Linalyl acetate	C$_{12}$H$_{20}$O$_2$	55.4	82.5	96.0	111.4	127.7	138.1	151.8	173.3	196.2	220.0	
Maleic anhydride	C$_4$H$_2$O$_3$	44.0	63.4	78.7	95.0	111.8	122.0	135.8	155.9	179.5	202.0	58
Menthane	C$_{10}$H$_{20}$	+9.7	35.7	48.3	62.7	78.3	88.6	102.1	122.7	146.0	169.5	
1-Menthol	C$_{10}$H$_{20}$O	56.0	83.2	96.0	110.3	126.1	136.1	149.4	168.3	190.2	212.0	42.5
Menthyl acetate	C$_{12}$H$_{22}$O$_2$	57.4	85.8	100.0	115.4	132.1	143.2	156.7	178.8	202.8	227.0	
benzoate	C$_{17}$H$_{24}$O$_2$	123.2	154.2	170.0	186.3	204.3	215.8	230.4	253.2	277.1	301.0	54.5
formate	C$_{11}$H$_{20}$O$_2$	47.3	75.8	90.0	105.8	123.0	133.8	148.0	169.8	194.2	219.0	
Mesityl oxide	C$_6$H$_{10}$O	−8.7	+14.1	26.0	37.9	51.7	60.4	72.1	90.0	109.8	130.0	−59
Methacrylic acid	C$_4$H$_6$O$_2$	25.5	48.5	60.0	72.7	86.4	95.3	106.6	123.9	142.5	161.0	15
Methacrylonitrile	C$_4$H$_5$N	−44.5	−23.3	−12.5	−0.6	+12.8	21.5	32.8	50.0	70.3	90.3	
Methane	CH$_4$		−205.9	−195.5	−191.8	−187.7	−185.1	−181.4	−175.5	−168.8	−161.5	−182.5
Methanethiol	CH$_4$S	−90.7	−75.3	−67.5	−58.8	−49.2	−43.1	−34.8	−22.1	−7.9	+6.8	−121
Methoxyacetic acid	C$_3$H$_6$O$_3$	52.5	79.3	92.0	106.5	122.0	131.8	144.5	163.5	184.2	204.0	
N-Methylacetanilide	C$_9$H$_{11}$NO		103.8	118.6	135.1	152.2	164.2	179.8	202.3	227.4	253.0	102
Methyl acetate	C$_3$H$_6$O$_2$	−57.2	−38.6	−29.3	−19.1	−7.9	−0.5	+9.4	24.0	40.0	57.8	−98.7
acetylene (propyne)	C$_3$H$_4$	−111.0	−97.5	−90.5	−82.9	−74.3	−68.8	−61.3	−49.8	−37.2	−23.3	−102.7
acrylate	C$_4$H$_6$O$_2$	−43.7	−23.6	−13.5	−2.7	+9.2	17.3	28.0	43.9	61.8	80.2	
alcohol (methanol)	CH$_4$O	−44.0	−25.3	−16.2	−6.0	+5.0	12.1	21.2	34.8	49.9	64.7	−97.8
Methylamine	CH$_5$N	−95.8	−81.3	−73.8	−65.9	−56.9	−51.3	−43.7	−32.4	−19.7	−6.3	−93.5

(Continued)

TABLE 2.38 Boiling Points of Common Organic Compounds at Selected Pressures (*Continued*)

Name	Formula	Pressure, mm Hg Temperature, °C										Melting point, °C
		1	5	10	20	40	60	100	200	400	760	
N-Methylaniline	C_7H_9N	36.0	62.8	76.2	90.5	106.0	115.8	129.8	149.3	172.0	195.5	−57
Methyl anthranilate	$C_8H_9NO_2$	77.6	109.0	124.2	141.5	159.7	172.0	187.8	212.4	238.5	266.5	24
benzoate	$C_8H_8O_2$	39.0	64.4	77.3	91.8	107.8	117.4	130.8	151.4	174.7	199.5	−12.5
2-Methylbenzothiazole	C_8H_7NS	70.0	97.5	111.2	125.5	141.2	150.4	163.9	183.2	204.5	225.5	15.4
α-Methylbenzyl alcohol	$C_8H_{10}O$	49.0	75.2	88.0	102.1	117.8	127.4	140.3	159.0	180.7	204.0	
Methyl bromide	CH_3Br	−96.3	−80.6	−72.8	−64.0	−54.2	−48.0	−39.4	−26.5	−11.9	+3.6	−93
2-Methyl-1-butene	C_5H_{10}	−89.1	−72.8	−64.3	−54.8	−44.1	−37.3	−28.0	−13.8	+2.5	20.2	−135
2-Methyl-2-butene	C_5H_{10}	−75.4	−57.0	−47.9	−37.9	−26.7	−19.4	−9.9	+4.9	21.6	38.5	−133
Methyl isobutyl carbinol (2-methyl-4-pentanol)	$C_6H_{14}O$	−0.3	+22.1	33.3	45.4	58.2	67.0	78.0	94.9	113.5	131.7	
n-butyl ketone (2-hexanone)	$C_6H_{12}O$	+7.7	28.8	38.8	50.0	62.0	69.8	79.8	94.3	111.0	127.5	−56.9
isobutyl ketone (4-methyl-2-pentanone)	$C_6H_{12}O$	−1.4	+19.7	30.0	40.8	52.8	60.4	70.4	85.6	102.0	119.0	−84.7
n-butyrate	$C_5H_{10}O_2$	−26.8	−5.5	+5.0	16.7	29.6	37.4	48.0	64.3	83.1	102.3	
isobutyrate	$C_5H_{10}O_2$	−34.1	−13.0	−2.9	+8.4	21.0	28.9	39.6	55.7	73.6	92.6	−84.7
caprate	$C_{11}H_{22}O_2$	63.7	93.5	108.0	123.0	139.0	148.6	161.5	181.6	202.9	224.0	−18
caproate	$C_7H_{14}O_2$	+5.0	30.0	42.0	55.4	70.0	79.7	91.4	109.8	129.8	150	
caprylate	$C_9H_{18}O_2$	34.2	61.7	74.9	89.0	105.3	115.3	128.0	148.1	170.0	193.0	−40
chloride	CH_3Cl		−99.5	−92.4	−84.8	−76.0	−70.4	−63.0	−51.2	−38.0	−24.0	−97.7
chloroacetate	$C_3H_5ClO_2$	−2.9	19.0	30.0	41.5	54.5	63.0	73.5	90.5	109.5	130.3	−31.9
cinnamate	$C_{10}H_{10}O_2$	77.4	108.1	123.0	140.0	157.9	170.0	185.8	209.6	235.0	263.0	33.4
α-Methylcinnamic acid	$C_{10}H_{10}O_2$	125.7	155.0	169.8	185.2	201.8	212.0	224.8	245.0	266.8	288.0	
Methylcyclohexane	C_7H_{14}	−35.9	−14.0	−3.2	+8.7	22.0	30.5	42.1	59.6	79.6	100.9	−126.4
Methylcyclopentane	C_8H_{12}	−53.7	−33.8	−23.7	−12.8	−0.6	+7.2	17.9	34.0	52.3	71.8	−142.4
Methylcyclopropane	C_4H_8	−96.0	−80.6	−72.8	−64.0	−54.2	−48.0	−39.3	−26.0	−11.3	+4.5	
Methyl *n*-decyl ketone (*n*-dodecan-2-one)	$C_{12}H_{24}O$	77.1	106.0	120.4	136.0	152.4	163.8	177.5	199.0	222.5	246.5	
dichloroacetate	$C_3H_4Cl_2O_2$	3.2	26.7	38.1	50.7	64.7	73.6	85.4	103.2	122.6	143.0	−7.6
N-Methyldiphenylamine	$C_{13}H_{13}N$	103.5	134.0	149.7	165.8	184.0	195.4	210.1	232.8	257.0	282.0	
Methyl *n*-dodecyl ketone (2-tetradecanone)	$C_{14}H_{28}O$	99.3	130.0	145.5	161.3	179.8	191.4	206.0	228.2	253.3	278.0	
Methylene bromide (dibromomethane)	CH_2Br_2	−35.1	−13.2	−2.4	+9.7	23.3	31.6	42.3	58.5	79.0	98.6	−52.8
chloride (dichloromethane)	CH_2Cl_2	−70.0	−52.1	−43.3	−33.4	−22.3	−15.7	−6.3	+8.0	24.1	40.7	−96.7

Compound	Formula										mp, °C
Methyl ethyl ketone (2-butanone)	C_4H_8O	-48.3	-28.0	-17.7	-6.5	14.0	25.0	41.6	60.0	79.6	-85.9
2-Methyl-3-ethylpentane	C_8H_{18}	-24.0	-1.8	+9.5	21.7	43.9	55.7	73.6	94.0	115.6	-114.5
3-Methyl-3-ethylpentane	C_8H_{18}	-23.9	-1.4	+9.9	22.3	45.0	57.1	75.3	96.2	118.3	-90
Methyl fluoride	CH_3F	-147.3	-137.0	-131.6	-125.9	-115.0	-109.0	-99.9	-89.5	-78.2	
formate	$C_2H_4O_2$	-74.2	-57.0	-48.6	-39.2	-21.9	-12.9	+0.8	16.0	32.0	-99.8
α-Methylglutaric anhydride	$C_6H_8O_3$	93.8	125.4	141.8	157.7	189.9	205.0	229.1	255.5	282.5	
Methyl glycolate	$C_3H_6O_3$	+9.6	33.7	45.3	58.1	81.8	93.7	111.8	131.7	151.5	
2-Methylheptadecane	$C_{18}H_{38}$	119.8	152.0	168.7	186.0	216.3	231.5	254.5	279.8	306.5	
2-Methylheptane	C_8H_{18}	-21.0	+1.3	12.3	24.4	46.6	58.3	76.0	96.2	117.6	-109.5
3-Methylheptane	C_8H_{18}	-19.8	+2.6	13.3	25.4	47.6	59.4	77.1	97.4	118.9	-120.8
4-Methylheptane	C_8H_{18}	-20.4	+1.5	12.4	24.5	46.6	58.3	76.1	96.3	117.7	-121.1
2-Methyl-2-heptene	C_8H_{16}	-16.1	+6.7	17.8	30.4	52.8	64.6	82.3	102.2	122.5	
6-Methyl-3-hepten-2-ol	$C_8H_{16}O$	41.6	65.0	76.7	89.3	111.5	122.6	139.5	156.6	175.5	
6-Methyl-5-hepten-2-ol	$C_8H_{16}O$	41.9	66.0	77.8	90.4	112.8	123.8	140.0	156.6	174.3	
2-Methylhexane	C_7H_{16}	-40.4	-19.5	-9.1	+2.3	23.0	34.1	50.8	69.8	90.0	-118.2
3-Methylhexane	C_7H_{16}	-39.0	-18.1	-7.8	+3.6	24.5	35.6	52.4	71.6	91.9	
Methyl iodide	CH_3I	-55.0	-45.8	-35.6	-24.2	-16.9	-7.0	+8.0	25.3	42.4	-64.4
laurate	$C_{13}H_{26}O_2$	87.8	117.9	133.2	149.0	166.0	176.8	190.8			5
levulinate	$C_6H_{10}O_3$	39.8	66.4	79.7	93.7	119.3	133.0	153.4	175.8	197.7	
methacrylate	$C_5H_8O_2$	-30.5	-10.0	+1.0	11.0	34.5	47.0	63.0	82.0	101.0	
myristate	$C_{15}H_{30}O_2$	115.0	145.7	160.8	177.8	195.8	207.5	222.6	245.3	269.8	18.5
α-naphthyl ketone (1-acetonaphthone)	$C_{12}H_{10}O$	115.6	146.3	161.5	178.4	208.6	223.8	246.7	270.5	295.8	
β-naphthyl ketone (2-acetonaphthone)	$C_{12}H_{10}O$	120.2	152.3	168.5	185.7	214.7	229.8	251.6	275.8	301.0	55.5
n-nonyl ketone (undecan-2-one)	$C_{11}H_{22}O$	68.2	95.5	108.9	123.1	148.6	161.0	181.2	202.3	224.0	15
palmitate	$C_{17}H_{34}O_2$	134.3	166.8	184.3	202.0						30
n-pentadecyl ketone (2-heptdecanone)	$C_{17}H_{34}O$	129.6	161.6	178.0	196.4	226.7	242.0	265.8	291.7	319.5	
2-Methylpentane	C_6H_{14}	-60.9	-41.7	-32.1	-21.4	-1.9	+8.1	24.1	41.6	60.3	-154
3-Methylpentane	C_6H_{14}	-59.0	-39.8	-30.1	-19.4	+0.1	+10.5	26.5	44.2	63.3	-118
2-Methyl-1-pentanol	$C_6H_{14}O$	15.4	38.0	49.6	61.6	83.4	94.2	111.3	129.8	147.9	
2-Methyl-2-pentanol	$C_6H_{14}O$	-4.5	+16.8	27.6	38.8	58.8	69.2	85.0	102.6	121.2	-103
Methyl n-pentyl ketone (2-heptanone)	$C_7H_{14}O$	19.3	43.6	55.5	67.7	89.8	100.0	116.1	133.2	150.2	
phenyl ether (anisole)	C_7H_8O	+5.4	30.0	42.2	55.8	80.1	93.0	112.3	133.8	155.5	-37.3
2-Methylpropene	C_4H_8	-105.1	-96.5	-81.9	-73.4	-63.8	-57.7	-36.7	-22.2	-6.9	-140.3
Methyl propionate	$C_4H_8O_2$	-42.0	-21.5	-11.8	-1.0	18.7	29.0	44.2	61.8	79.8	-87.5
4-Methylpropiophenone	$C_{10}H_{12}O$	59.6	89.3	103.8	120.2	149.3	164.2	187.4	212.7	238.5	
2-Methylpropionyl bromide	C_4H_7BrO	13.5	38.4	50.6	64.1	88.8	101.6	120.5	141.7	163.0	
Methyl propyl ether	$C_4H_{10}O$	-72.2	-54.3	-45.4	-35.4	-17.4	-8.1	+6.0	22.5	39.1	

(Continued)

TABLE 2.38 Boiling Points of Common Organic Compounds at Selected Pressures (Continued)

| Compound | | Pressure, mm Hg | | | | | | | | | | Melting point, °C |
Name	Formula	1	5	10	20	40	60	100	200	400	760	
		Temperature, °C										
n-propyl ketone (2-pentanone)	C$_5$H$_{10}$O	−12.0	+8.0	17.9	28.5	39.8	47.3	56.8	71.0	86.8	103.3	−77.8
isopropyl ketone (3-methyl-2-butanone)	C$_5$H$_{10}$O	−19.9	−1.0	+8.3	18.3	29.6	36.2	45.5	59.0	73.8	88.9	−92
2-Methylquinoline	C$_{10}$H$_9$N	75.3	104.0	119.0	134.0	150.8	161.7	176.2	197.8	211.7	246.5	−1
Methyl salicylate	C$_8$H$_8$O$_3$	54.0	81.6	95.3	110.0	126.2	136.7	150.0	172.6	197.5	223.2	−8.3
α-Methyl styrene	C$_9$H$_{10}$	7.4	34.0	47.1	61.8	77.8	88.3	102.2	121.8	143.0	165.4	−23.2
4-Methyl styrene	C$_9$H$_{10}$	16.0	42.0	55.1	69.2	85.0	95.0	108.6	128.7	151.2	175.0	
Methyl n-tetradecyl ketone (2-hexadecanone)	C$_{16}$H$_{32}$O	109.8	151.5	167.3	184.6	203.7	215.0	230.5	254.4	279.8	307.0	
thiocyanate	C$_5$H$_3$NS	−14.0	+9.8	21.6	34.5	49.0	58.1	70.4	89.8	110.8	132.9	−51
isothiocyanate	C$_2$H$_3$NS	−34.7	−8.3	+5.4	20.4	38.2	47.5	59.3	77.5	97.8	119.0	35.5
undecyl ketone (2-tridecanone)	C$_{13}$H$_{26}$O	86.8	117.0	131.8	147.8	165.7	176.6	191.5	214.0	238.3	262.5	28.5
isovalerate	C$_6$H$_{12}$O$_2$	−19.2	+2.9	14.0	26.4	39.8	48.2	59.8	77.3	96.7	116.7	
Monovinylacetylene (butenyne)	C$_4$H$_4$	−93.2	−77.7	−70.0	−61.3	−51.7	−45.3	−37.1	−24.1	−10.1	+5.3	
Myrcene	C$_{10}$H$_{16}$	14.5	40.0	53.2	67.0	82.6	92.6	106.0	126.0	148.3	171.5	
Myristaldehyde	C$_{14}$H$_{28}$O	99.0	132.0	148.3	166.2	186.0	198.3	214.5	240.4	267.9	297.8	23.5
Myristic acid (tetradecanoic acid)	C$_{14}$H$_{28}$O$_2$	142.0	174.1	190.8	207.6	223.5	237.2	250.5	272.3	294.6	318.0	57.5
Napthalene	C$_{10}$H$_8$	52.6	74.2	85.8	101.7	119.3	130.2	145.5	167.7	193.2	217.9	80.2
1-Naphthoic acid	C$_{11}$H$_8$O$_2$	156.0	184.0	196.8	211.2	225.0	234.5	245.8	263.5	281.4	300.0	160.5
2-Naphthoic acid	C$_{11}$H$_8$O$_2$	160.8	189.7	202.8	216.9	231.5	241.3	252.7	270.3	289.5	308.5	184
1-Naphthol	C$_{10}$H$_8$O	94.0	125.5	142.0	158.0	177.8	190.0	206.0	229.6	255.8	282.5	96
2-Naphthol	C$_{10}$H$_8$O		128.6	145.5	161.8	181.7	193.7	209.8	234.0	260.6	288.0	122.5
1-Naphthylamine	C$_{10}$H$_9$N	104.3	137.7	153.8	171.6	191.5	203.8	220.0	244.9	272.2	300.8	50
2-Naphthylamine	C$_{10}$H$_9$N	108.0	141.6	157.6	175.8	195.7	208.1	224.3	249.7	277.4	306.1	111.5
Nicotine	C$_{10}$H$_{14}$N$_2$	61.8	91.8	107.2	123.7	142.1	154.7	169.5	193.8	219.8	247.3	
2-Nitroaniline	C$_6$H$_6$N$_2$O$_2$	104.0	135.7	150.4	167.7	186.0	197.8	213.0	236.3	260.0	284.5	71.5
3-Nitroaniline	C$_6$H$_6$N$_2$O$_2$	119.3	151.5	167.8	185.5	204.2	216.5	232.1	255.3	280.2	305.7	114
4-Nitroaniline	C$_6$H$_6$N$_2$O$_2$	142.4	177.6	194.4	213.2	234.2	245.9	261.8	284.5	310.2	336.0	146.5
2-Nitrobenzaldehyde	C$_7$H$_5$NO$_3$	85.8	117.7	133.4	150.0	168.8	180.7	196.2	220.0	246.8	273.5	40.9
3-Nitrobenzaldehyde	C$_7$H$_5$NO$_3$	96.2	127.4	142.8	159.0	177.7	189.5	204.3	227.4	252.1	278.3	58
Nitrobenzene	C$_6$H$_5$NO$_2$	44.4	71.6	84.9	99.3	115.4	125.8	139.9	161.2	185.8	210.6	+5.7
Nitroethane	C$_2$H$_5$NO$_2$	−21.0	+1.5	12.5	24.8	38.0	46.5	57.8	74.8	94.0	114.0	−90

Nitroglycerin	$C_3H_5N_3O_9$	127	167	188	210	235	251					11
Nitromethane	CH_3NO_2	-29.0	-7.9	+2.8	14.1	27.5	35.5	46.6	63.5	82.0	101.2	-29
2-Nitrophenol	$C_6H_5NO_3$	49.3	76.8	90.4	105.8	122.1	132.6	146.4	167.6	191.0	214.5	45
2-Nitrophenyl acetate	$C_8H_7NO_4$	100.0	128.0	142.0	155.8	172.8	181.7	194.1	213.0	233.5	253.0	
1-Nitropropane	$C_3H_7NO_2$	-9.6	+13.5	25.3	37.9	51.8	60.5	72.3	90.2	110.6	131.6	-108
2-Nitropropane	$C_3H_7NO_2$	-18.8	4.1	15.8	28.2	41.8	50.3	62.0	80.0	99.8	120.3	-93
2-Nitrotoluene	$C_7H_7NO_2$	50.0	79.1	93.8	109.6	126.3	137.6	151.5	173.7	197.7	222.3	-4.1
3-Nitrotoluene	$C_7H_7NO_2$	50.2	81.0	96.0	112.8	130.7	142.5	156.9	180.3	206.8	231.9	15.5
4-Nitrotoluene	$C_7H_7NO_2$	53.7	85.0	100.5	117.7	136.0	147.9	163.0	186.7	212.5	238.3	51.9
4-Nitro-1,3-xylene (4-nitro-m-xylene)	$C_8H_9NO_2$	65.6	95.0	109.8	125.8	143.3	153.8	168.5	191.7	217.5	244.0	+2
Nonacosane	$C_{29}H_{60}$	234.2	260.8	286.4	303.6	323.2	334.8	350.0	373.2	397.2	421.8	63.8
Nonadecane	$C_{19}H_{40}$	133.3	166.3	183.5	200.8	220.0	232.8	248.0	271.8	299.8	330.0	32
n-Nonane	C_9H_{20}	+1.4	25.8	38.0	51.2	66.0	75.5	88.1	107.5	128.2	150.8	-53.7
1-Nonanol	$C_9H_{20}O$	59.5	86.1	99.7	113.8	129.0	139.0	151.3	170.5	192.1	213.5	-5
2-Nonanone	$C_9H_{18}O$	32.1	59.0	72.3	87.2	103.4	113.8	127.4	148.2	171.2	195.0	-19
Octacosane	$C_{28}H_{58}$	226.5	260.3	277.4	295.4	314.2	326.8	341.8	364.8	388.9	412.5	61.6
Octadecane	$C_{18}H_{38}$	119.6	152.1	169.6	187.5	207.4	219.7	236.0	260.6	288.0	317.0	28
n-Octane	C_8H_{18}	-14.0	+8.3	19.2	31.5	45.1	53.8	65.7	83.6	104.0	125.6	-56.8
n-Octanol (1-octanol)	$C_8H_{18}O$	54.0	76.5	88.3	101.0	115.2	123.8	135.2	152.0	173.8	195.2	-15.4
2-Octanone	$C_8H_{18}O$	23.6	48.4	60.9	74.3	89.8	99.8	111.7	130.4	151.0	172.9	-16
n-Octyl acrylate	$C_{11}H_{20}O_2$	58.5	87.7	102.0	117.8	135.6	145.6	159.1	180.2	204.0	227.0	
n-Octyl iodide (1-Iodooctane)	$C_8H_{17}I$	45.8	74.8	90.0	105.9	123.8	135.4	150.0	173.3	199.3	225.5	-45.9
Oleic acid	$C_{18}H_{34}O_2$	176.5	208.5	223.0	240.0	257.2	269.8	286.0	309.8	334.7	360.0	14
Palmitaldehyde	$C_{16}H_{32}O$	121.6	154.6	171.8	190.0	210.0	222.6	239.5	264.1	292.3	321.0	34
Palmitic acid	$C_{16}H_{32}O_2$	153.6	188.1	205.8	223.8	244.4	256.0	271.5	298.7	326.0	353.8	64.0
Palmitonitrile	$C_{16}H_{31}N$	134.3	168.3	185.8	204.2	223.8	236.6	251.5	277.1	304.5	332.0	31
Pelargonic acid	$C_9H_{18}O_2$	108.2	126.0	137.4	149.8	163.7	172.3	184.4	203.1	227.5	253.5	12.5
Pentachlorobenzene	C_6HCl_5	98.6	129.7	144.3	160.0	178.5	190.1	205.5	227.0	251.6	276.0	85.5
Pentachloroethane	C_2HCl_5	+1.0	27.2	39.8	53.9	69.9	80.0	93.5	114.0	137.2	160.5	-22
Pentachloroethylbenzene	$C_6H_5Cl_5$	96.2	130.0	148.0	166.0	186.2	199.0	216.0	241.8	269.3	299.0	
Pentachlorophenol	C_6HCl_5O				192.2	211.2	223.4	239.6	261.8	285.0	309.3	188.5
Pentacosane	$C_{25}H_{52}$	194.2	230.0	248.2	266.1	285.6	298.4	314.0	339.0	365.4	390.3	53.3
Pentadecane	$C_{15}H_{32}$	91.6	121.0	135.4	150.2	167.7	178.4	194.0	216.1	242.8	270.5	10
1,3-Pentadiene	C_5H_8	-71.8	-53.8	-45.0	-34.8	-23.4	-16.5	-6.7	+8.0	24.7	42.1	
1,4-Pentadiene	C_5H_8	-83.5	-66.2	-57.1	-47.7	-37.0	-30.0	-20.6	-6.7	+8.3	26.1	
Pentaethylbenzene	$C_{16}H_{26}$	86.0	120.0	135.8	152.4	171.9	184.2	200.0	224.1	250.2	277.0	
Pentaethylchlorobenzene	$C_{16}H_{25}Cl$	90.0	123.8	140.7	158.1	178.2	191.0	208.0	230.3	257.2	285.0	

(Continued)

TABLE 2.38 Boiling Points of Common Organic Compounds at Selected Pressures (*Continued*)

Compound		Pressure, mm Hg — Temperature, °C										Melting point, °C
Name	Formula	1	5	10	20	40	60	100	200	400	760	
n-Pentane	C_5H_{12}	−76.6	−62.5	−50.1	−40.2	−29.2	−22.2	−12.6	+1.9	18.5	36.1	−129.7
iso-Pentane (2-methylbutane)	C_5H_{12}	−82.9	−65.8	−57.0	−47.3	−36.5	−29.6	−20.2	−5.9	+10.5	27.8	−159.7
neo-Pentane (2,2-dimethylpropane)	C_5H_{12}	−102.0	−85.4	−76.7	−67.2	−56.1	−49.0	−39.1	−23.7	−7.1	+9.5	−16.6
2,3,4-Pentanetriol	$C_5H_{12}O_3$	155.0	159.3	204.5	220.5	239.6	249.8	263.5	284.5	307.0	327.2	
1-Pentene	C_5H_{10}	−80.4	−63.3	−54.5	−46.0	−34.1	−27.1	−17.7	−3.4	+12.8	30.1	
α-Phellandrene	$C_{10}H_{16}$	20.0	45.7	58.0	72.1	87.8	97.6	110.6	130.6	152.0	175.0	
Phenanthrene	$C_{14}H_{10}$	118.2	154.3	173.0	193.7	215.8	229.9	249.0	277.1	308.0	340.2	99.5
Phenethyl alcohol (phenyl cellosolve)	$C_8H_{10}O_2$	58.2	85.9	100.0	114.8	130.5	141.2	154.0	175.0	197.5	219.5	
2-Phenetidine	$C_8H_{11}NO$	67.0	94.7	108.6	123.7	139.9	149.8	163.5	184.0	207.0	228.0	
Phenol	C_6H_6O	40.1	62.5	73.8	86.0	100.1	108.4	121.4	139.0	160.0	181.9	40.6
2-Phenoxyethanol	$C_8H_{10}O_2$	78.0	96.6	121.2	136.0	152.2	163.2	176.5	197.6	221.0	245.3	11.6
2-Phenoxyethyl acetate	$C_{10}H_{12}O_3$	82.6	113.5	128.0	144.5	162.3	174.0	189.2	211.3	235.0	259.7	−6.7
Phenyl acetate	$C_8H_8O_2$	38.2	64.8	78.0	92.3	108.1	118.1	131.6	151.2	173.5	195.9	
Phenylacetic acid	$C_8H_8O_2$	97.0	127.0	141.3	156.0	173.6	184.5	198.2	219.5	243.0	265.5	76.5
Phenylacetonitrile	C_8H_7N	60.0	89.0	103.5	119.4	136.3	147.7	161.8	184.2	208.5	233.5	−23.8
Phenylacetyl chloride	C_8H_7ClO	48.0	75.3	89.0	103.6	119.8	129.8	143.5	163.8	186.0	210.0	
Phenyl benzoate	$C_{13}H_{10}O_2$	106.8	141.5	157.8	177.0	197.6	210.8	237.8	254.0	283.5	314.0	70.5
4-Phenyl-3-buten-2-one	$C_{10}H_{10}O$	81.7	112.2	127.4	143.8	161.3	172.6	187.8	211.0	235.4	261.0	41.5
Phenyl isocyanate	C_7H_5NO	10.6	36.0	48.5	62.5	77.7	87.7	100.6	120.8	142.7	165.6	
Phenyl isocyanide	C_7H_5N	12.0	37.0	49.7	63.4	78.3	88.0	101.0	120.8	142.3	165.0	
Phenylcyclohexane	$C_{12}H_{16}$	67.5	96.5	111.3	126.4	144.0	154.2	169.3	191.3	214.6	240.0	+75
Phenyl dichlorophosphate	$C_6H_5Cl_2O_2P$	66.7	95.9	110.0	125.9	143.4	153.6	168.0	189.8	213.0	239.5	
m-Phenylene diamine (1,3-phenylenediamine)	$C_6H_8N_2$	99.8	131.2	147.0	163.8	182.5	194.0	209.9	233.0	259.0	285.5	62.8
Phenylglyoxal	$C_8H_6O_2$		75.0	87.8	100.7	115.5	124.2	136.2	153.8	173.5	193.5	73
Phenylhydrazine	$C_6H_8N_2$	75.8	101.6	115.8	131.5	148.2	158.7	173.5	195.4	218.2	243.5	19.5
N-Phenyliminodiethanol	$C_{10}H_{15}NO_2$	145.0	170.2	195.8	213.4	233.0	245.3	260.6	284.5	311.3	337.8	
1-Phenyl-1,3-pentanedione	$C_{11}H_{12}O_2$	98.0	128.5	144.0	159.9	178.0	189.8	204.5	226.7	251.2	276.5	
2-Phenylphenol	$C_{12}H_{10}O$	100.0	131.6	146.2	163.3	180.3	192.2	205.9	227.9	251.8	275.0	56.5
4-Phenylphenol	$C_{12}H_{10}O$			176.2	193.8	213.0	225.3	240.9	263.2	285.5	308.0	164.5
3-Phenyl-1-propanol	$C_9H_{12}O$	74.7	102.4	116.0	131.2	147.4	156.8	170.3	191.2	212.8	235.0	

Name	Formula											
Phenyl isothiocyanate	C_7H_5NS	47.2	75.6	89.8	115.5	122.5	133.3	147.7	169.6	194.0	218.5	−21.0
Phorone	$C_9H_{14}O$	42.0	63.3	81.5	95.6	111.3	121.4	134.0	153.5	175.3	197.2	28
iso-Phorone	$C_9H_{14}O$	38.0	66.7	81.2	96.8	114.5	125.6	140.6	163.3	188.7	215.2	
Phosgene (carbonyl chloride)	CCl_2O	−32.9	−77.0	−69.3	−60.3	−50.3	−44.0	−35.6	−22.3	−7.6	+8.3	−104
Phthalic anhydride	$C_8H_4O_3$	96.5	124.3	134.0	151.7	172.0	185.3	202.3	228.0	256.8	284.5	130.8
Phthalide	$C_8H_6O_2$	95.5	127.7	144.0	161.3	181.0	193.5	210.0	234.5	261.8	290.0	73
Phthaloyl chloride	$C_8H_4Cl_2O_2$	86.3	118.3	134.2	151.0	170.0	182.2	197.8	222.0	248.3	275.8	88.5
2-Picoline	C_6H_7N	−11.1	+12.6	24.4	37.4	51.2	59.9	71.4	89.0	108.4	128.8	−70
Pimelic acid	$C_7H_{12}O_4$	163.4	196.2	212.0	229.3	247.0	258.2	272.0	294.5	318.5	342.1	103
α-Pinene	$C_{10}H_{16}$	−1.0	+24.6	37.3	51.4	66.8	76.8	90.1	110.2	132.3	155.0	−55
β-Pinene	$C_{10}H_{16}$	+4.2	30.0	42.3	58.1	71.5	81.2	94.0	114.1	136.1	158.3	
Piperidine	$C_5H_{11}N$		−7.0	+3.9	15.8	29.2	37.7	49.0	66.2	85.7	106.0	−9
Piperonal	$C_8H_6O_3$	87.0	117.4	132.0	148.0	165.7	177.0	191.7	214.3	238.5	263.0	37
Propane	C_3H_8	−128.9	−115.4	−108.5	−100.9	−92.4	−87.0	−79.6	−68.4	−55.6	−42.1	−187.1
Propenylbenzene	C_9H_{10}	17.5	43.8	57.0	71.5	87.7	97.8	111.7	132.0	154.7	179.0	−30.1
Propionamide	C_3H_7NO	65.0	91.0	105.0	119.0	134.8	144.3	156.0	174.2	194.0	213.0	79
Propionic acid	$C_3H_6O_2$	4.6	28.0	39.7	52.0	65.8	74.1	85.8	102.5	122.0	141.1	−22
Propionic anhydride	$C_6H_{10}O_3$	20.6	45.3	57.7	70.4	85.6	94.5	107.2	127.8	146.0	167.0	−45
Propionitrile	C_3H_5N	−35.0	−13.6	−3.0	+8.8	22.0	30.1	41.4	58.2	77.7	97.1	−91.9
Propiophenone	$C_9H_{10}O$	50.0	77.9	92.2	107.6	124.3	135.0	149.3	170.2	194.2	218.0	21
n-Propyl acetate	$C_5H_{10}O_2$	−26.7	−5.4	+5.0	16.0	28.8	37.0	47.8	64.0	82.0	101.8	−92.5
iso-Propyl acetate	$C_5H_{10}O_2$	−38.3	−17.4	−7.2	+4.2	17.0	25.1	35.7	51.7	69.8	89.0	
n-Propyl alcohol (1-propanol)	C_3H_8O	−15.0	+5.0	14.7	25.3	36.4	43.5	52.8	66.8	82.0	97.8	−127
iso-Propyl alcohol (2-propanol)	C_3H_8O	−26.1	−7.0	+2.4	12.7	23.8	30.5	39.5	53.0	67.8	82.5	−85.8
n-Propylamine	C_3H_9N	−64.4	−46.3	−37.2	−27.1	−16.0	−9.0	+0.5	15.0	31.5	48.5	−83
Propylbenzene	C_9H_{12}	6.3	31.3	43.4	56.8	71.6	81.1	94.0	113.5	135.7	159.2	−99.5
Propyl benzoate	$C_{10}H_{12}O_2$	54.6	83.8	98.0	114.3	131.8	143.3	157.4	180.1	205.2	231.0	−51.6
n-Propyl bromide (1-bromopropane)	C_3H_7Br	−53.0	−33.4	−23.3	−12.4	−0.3	+7.5	18.0	34.0	52.0	71.0	−109.9
iso-Propyl bromide (2-bromopropane)	C_3H_7Br	−61.8	−42.5	−32.8	−22.0	−10.1	−2.5	+8.0	23.8	41.5	60.0	−89.0
n-Propyl n-butyrate	$C_7H_{14}O_2$	−1.6	+22.1	34.0	47.0	61.5	70.3	82.6	101.0	121.7	142.7	−95.2
iso-Propyl isobutyrate	$C_7H_{14}O_2$	−6.2	+16.8	28.3	40.6	54.3	63.0	73.9	91.8	112.0	133.9	
iso-Propyl carbamate	$C_4H_9NO_2$	−16.3	+5.8	17.0	29.0	42.4	51.4	62.3	80.2	100.0	120.5	
Propyl carbamate	$C_4H_9NO_2$	52.4	77.6	90.0	103.2	117.7	126.5	138.3	155.8	175.8	195.0	
n-Propyl chloride (1-chloropropane)	C_3H_7Cl	−68.3	−50.0	−41.0	−31.0	−19.5	−12.1	−2.5	+12.2	29.4	46.4	−112.8
iso-Propyl chloride (2-chloropropane)	C_3H_7Cl	−78.8	−61.1	−52.0	−42.0	−31.0	−23.5	−13.7	+1.3	18.1	36.5	−117
iso-Propyl chloroacetate	$C_5H_9ClO_2$	+3.8	28.1	40.2	53.9	68.7	78.0	90.3	108.8	128.0	148.6	
Propyl chloroglyoxylate	$C_5H_7ClO_3$	9.7	32.3	43.5	55.6	68.8	77.2	88.0	104.7	123.0	150.0	

(Continued)

TABLE 2.38 Boiling Points of Common Organic Compounds at Selected Pressures (*Continued*)

| Compound | | Pressure, mm Hg | | | | | | | | | | Melting point, °C |
Name	Formula	1	5	10	20	40	60	100	200	400	760	
						Temperature, °C						
Propylene	C_3H_6	−131.9	−120.7	−112.1	−104.7	−96.5	−91.3	−84.1	−73.3	−60.9	−47.7	−185
Propylene glycol (1,2-Propanediol)	$C_3H_8O_2$	45.5	70.8	83.2	96.4	111.2	119.9	132.0	149.7	168.1	188.2	
Propylene oxide	C_3H_6O	−75.0	−57.8	−49.0	−39.3	−28.4	−21.3	−12.0	+2.1	17.8	34.5	−112.1
n-Propyl formate	$C_4H_8O_2$	−43.0	−22.7	−12.6	−1.7	+10.8	18.8	29.5	45.3	62.6	81.3	−92.9
iso-Propyl formate	$C_4H_8O_2$	−52.0	−32.7	−22.7	−12.1	−0.2	+7.5	17.8	33.6	50.5	68.3	
4,4′-iso-Propylidenebisphenol	$C_{15}H_{16}O_2$	193.0	224.2	240.8	255.5	273.0	282.9	297.0	317.5	339.0	360.5	
n-Propyl iodide (1-iodopropane)	C_3H_7I	−36.0	−13.5	−2.4	+10.0	23.6	32.1	43.8	61.8	81.8	102.5	−98.8
iso-Propyl iodide (2-iodopropane)	C_3H_7I	−43.3	−22.1	−11.7	0.0	+13.2	21.6	32.8	50.0	69.5	89.5	−90
n-Propyl levulinate	$C_8H_{14}O_3$	59.7	86.3	99.9	114.0	130.1	140.6	154.0	175.6	198.0	221.2	
iso-Propyl levulinate	$C_8H_{14}O_3$	48.0	74.5	88.0	102.4	118.1	127.8	141.8	161.6	185.2	208.2	
Propyl mercaptan (1-propanethiol)	C_3H_8S	−56.0	−36.3	−26.3	−15.4	−3.2	+4.6	15.3	31.5	49.2	67.4	−112
2-iso-Propylnaphthalene	$C_{13}H_{14}$	76.0	107.9	123.4	140.3	159.0	171.4	187.6	211.8	238.5	266.0	
iso-Propyl β-naphthyl ketone (2-isobutyronaphthone)	$C_{14}H_{14}O$	133.2	165.4	181.0	197.7	215.6	227.0	242.3	264.0	288.2	313.0	
2-iso-Propylphenol	$C_9H_{12}O$	56.6	83.8	97.0	111.7	127.5	137.7	150.3	170.1	192.6	214.5	15.5
3-iso-Propylphenol	$C_9H_{12}O$	62.0	90.3	104.1	119.8	136.2	146.6	160.2	182.0	205.0	228.0	26
4-iso-Propylphenol	$C_9H_{12}O$	67.0	94.7	108.0	123.4	139.8	149.7	163.3	184.0	206.1	228.2	61
Propyl propionate	$C_6H_{12}O_2$	−14.2	+8.0	19.4	31.6	45.0	53.8	65.2	82.7	102.0	122.4	−76
4-iso-Propylstyrene	$C_{11}H_{14}$	34.7	62.3	76.0	91.2	108.0	118.4	132.8	153.9	178.0	202.5	
Propyl isovalerate	$C_8H_{16}O_2$	+8.0	32.8	45.1	58.0	72.8	82.3	95.0	113.9	135.0	155.9	
Pulegone	$C_{10}H_{16}O$	58.3	82.5	94.0	106.8	121.7	130.2	143.1	162.5	189.8	221.0	
Pyridine	C_5H_5N	−18.9	+2.5	13.2	24.8	38.0	46.8	57.8	75.0	95.6	115.4	−42
Pyrocatechol	$C_6H_6O_2$		104.0	118.3	134.0	150.6	161.7	176.0	197.7	221.5	245.5	105
Pyrocaltechol diacetate (1,2-phenylene diacetate)	$C_{10}H_{10}O_4$	98.0	129.8	145.7	161.8	179.8	191.6	206.5	228.7	253.3	278.0	
Pyrogallol	$C_6H_6O_3$		151.7	167.7	185.3	204.2	216.3	232.0	255.3	281.5	309.0	133
Pyrotartaric anhydride	$C_5H_6O_3$	69.7	99.7	114.2	130.0	147.8	158.6	173.8	196.1	221.0	247.4	
Pyruvic acid	$C_3H_4O_3$	21.4	45.8	57.9	70.8	85.3	94.1	106.5	124.7	144.7	165.0	13.
Quinoline	C_9H_7N	59.7	89.6	103.8	119.8	136.7	148.1	163.2	186.2	212.3	237.7	−15.
iso-Quinoline	C_9H_7N	63.5	92.7	107.8	123.7	141.6	152.0	167.6	190.0	214.5	240.5	24.
Resorcinol	$C_6H_6O_2$	108.4	138.0	152.1	168.0	185.3	195.8	209.8	230.8	253.4	276.5	110.

Name	Formula											
Safrole	C$_{10}$H$_{10}$O$_2$	63.8	93.0	107.6	123.0	140.1	150.3	165.1	186.2	210.0	233.0	11
Salicylaldehyde	C$_7$H$_6$O$_2$	33.0	60.1	73.8	88.7	105.2	115.7	129.4	150.0	173.7	196.5	−7
Salicylic acid	C$_7$H$_6$O$_3$	113.7	136.0	146.2	156.8	172.2	182.0	193.4	210.0	230.5	256.0	159
Sebacic acid	C$_{10}$H$_{18}$O$_4$	183.0	215.7	232.0	250.0	268.2	279.8	294.5	313.2	332.8	352.3	134
Selenophene	C$_4$H$_4$Se	−39.0	−16.0	−4.0	+9.1	24.1	33.8	47.0	66.7	89.8	114.3	
Skatole	C$_9$H$_9$N	95.0	124.2	139.6	154.3	171.9	183.6	197.4	218.8	242.5	266.2	95
Stearaldehyde	C$_{18}$H$_{36}$O	140.0	174.6	192.1	210.6	230.8	244.2	260.0	285.0	313.8	342.5	63.5
Stearic acid	C$_{18}$H$_{36}$O$_2$	173.7	209.0	225.0	243.4	263.3	275.5	291.0	316.5	343.0	370.0	69.3
Stearyl alcohol (1-octadecanol)	C$_{18}$H$_{36}$O	150.3	185.6	202.0	220.0	240.4	252.7	269.4	293.5	320.3	349.5	58.5
Styrene	C$_8$H$_8$	−7.0	+18.0	30.8	44.6	59.8	69.5	82.0	101.3	122.5	145.2	−30.6
Styrene dibromide [(1,2-dibromoethyl)benzene]	C$_8$H$_8$Br$_2$	86.0	115.6	129.8	145.2	161.8	172.2	186.3	207.8	230.0	245.0	
Suberic acid	C$_8$H$_{14}$O$_4$	172.8	205.5	219.5	238.2	254.6	265.4	279.8	300.5	322.8	345.5	142
Succinic anhydride	C$_4$H$_4$O$_3$	92.0	115.0	128.2	145.3	163.0	174.0	189.0	212.0	237.0	261.0	119.6
Succinimide	C$_4$H$_5$NO$_2$	115.0	143.2	157.0	174.0	192.0	203.0	217.4	240.0	263.5	287.5	125.5
Succinyl chloride	C$_4$H$_4$Cl$_2$O$_2$	39.0	65.0	78.0	91.8	107.5	117.2	130.0	149.3	170.0	192.5	17
α-Terpineol	C$_{10}$H$_{18}$O	52.8	80.4	94.3	109.8	126.0	136.3	150.1	171.2	194.3	217.5	35
Terpenoline	C$_{10}$H$_{16}$	32.3	58.0	70.6	84.8	100.0	109.8	122.7	142.0	163.5	185.0	
1,1,1,2-Tetrabromoethane	C$_2$H$_2$Br$_4$	58.0	83.3	95.7	108.5	123.2	132.0	144.0	161.5	181.0	200.0	
1,1,2,2-Tetrabromoethane	C$_2$H$_2$Br$_4$	65.0	95.5	110.0	126.0	144.0	155.1	170.0	192.5	217.5	243.5	
Tetraisobutylene	C$_{16}$H$_{32}$	63.8	93.7	108.5	124.5	142.2	152.6	167.5	190.0	214.6	240.0	
Tetracosane	C$_{24}$H$_{50}$	183.8	219.6	237.6	255.3	276.3	288.4	305.2	330.5	358.0	386.4	51.1
1,2,3,4-Tetrachlorobenzene	C$_6$H$_2$Cl$_4$	68.5	99.6	114.7	131.2	149.2	160.0	175.7	198.0	225.5	254.0	46.5
1,2,3,5-Tetrachlorobenzene	C$_6$H$_2$Cl$_4$	58.2	89.0	104.1	121.6	140.0	152.0	168.0	193.7	220.0	246.0	54.5
1,2,4,5-Tetrachlorobenzene	C$_6$H$_6$Cl$_4$	64.0	95.7	110.1	128.0	146.0	157.7	173.5	196.0	220.5	245.0	139
1,1,2,2-Tetrachloro-1,2-difluoroethane	C$_2$Cl$_4$F$_2$	−37.5	−16.0	−5.0	+6.7	19.8	28.1	33.6	55.0	73.1	92.0	2
1,1,1,2-Tetrachloroethane	C$_2$H$_2$Cl$_4$	−16.3	+7.4	19.3	32.1	46.7	56.0	68.0	87.2	108.2	130.5	−6
1,1,2,2-Tetrachloroethane	C$_2$H$_2$Cl$_4$	−3.8	+20.7	33.0	46.2	60.8	70.0	83.2	102.2	124.0	145.9	
1,2,3,5-Tetrachloro-4-ethylbenzene	C$_8$H$_6$Cl$_4$	77.0	110.0	126.0	143.7	162.1	175.0	191.6	215.3	243.0	270.0	−3
Tetrachloroethylene	C$_2$Cl$_4$	−20.6	+2.4	13.8	26.3	40.1	49.2	61.3	79.8	100.0	120.8	−1
2,3,4,6-Tetrachlorophenol	C$_6$H$_2$Cl$_4$O	100.0	130.3	145.5	161.0	179.1	190.0	205.2	227.2	250.4	275.0	69
3,4,5,6-Tetrachloro-1,2-xylene	C$_8$H$_6$Cl$_4$	94.4	125.0	140.3	156.0	174.2	185.8	200.5	223.0	248.3	273.5	
Tetradecane	C$_{14}$H$_{30}$	76.4	106.0	120.7	135.6	152.7	164.0	178.5	201.8	226.8	252.5	5
Tetradecylamine	C$_{14}$H$_{31}$N	102.6	135.8	152.0	170.0	189.0	200.2	215.7	239.8	264.6	291.2	
Tetradecyltrimethylsilane	C$_{17}$H$_{38}$Si	120.0	150.7	166.2	183.5	201.5	213.3	227.8	250.0	275.0	300.0	
Tetraethoxysilane	C$_8$H$_{20}$O$_4$Si	16.0	40.3	52.6	65.8	81.1	90.7	103.6	123.5	146.2	168.5	
1,2,3,4-Tetraethylbenzene	C$_{14}$H$_{22}$	65.7	96.2	111.6	127.7	145.8	156.7	172.4	196.0	221.4	248.0	11

(Continued)

TABLE 2.38 Boiling Points of Common Organic Compounds at Selected Pressures (*Continued*)

Name	Formula	1	5	10	20	40	60	100	200	400	760	Melting point, °C
		temperature										

Compound — Name	Formula	\multicolumn{10}{c}{Pressure, mm Hg — Temperature, °C}										Melting point, °C
		1	5	10	20	40	60	100	200	400	760	
Tetraethylene glycol	C$_8$H$_{18}$O$_5$	153.9	183.7	197.1	212.3	228.0	237.8	250.0	268.4	288.0	307.8	
Tetraethylene glycol chlorohydrin	C$_8$H$_{17}$ClO$_4$	110.1	141.8	156.1	172.6	190.0	200.5	214.7	236.5	258.2	281.5	
Tetraethyllead	C$_8$H$_{20}$Pb	38.4	63.6	74.8	88.0	102.4	111.7	123.8	142.0	161.8	183.0	−136
Tetraethylsilane	C$_8$H$_{20}$Si	−1.0	+23.9	36.3	50.0	65.3	74.8	88.0	108.0	130.2	153.0	
Tetralin	C$_{10}$H$_{12}$	38.0	65.3	79.0	93.8	110.4	121.3	135.3	157.2	181.8	207.2	−31
1,2,3,4-Tetramethylbenzene	C$_{10}$H$_{14}$	42.6	68.7	81.8	95.8	111.5	121.8	135.7	155.7	180.0	204.4	−6
1,2,3,5-Tetramethylbenzene	C$_{10}$H$_{14}$	40.6	65.8	77.8	91.0	105.8	115.4	128.3	149.9	173.7	197.9	−24
1,2,4,5-Tetramethylbenzene	C$_{10}$H$_{14}$	45.0	65.0	74.6	88.0	104.2	114.8	128.1	149.5	172.1	195.9	79
2,2,3,3-Tetramethylbutane	C$_8$H$_{18}$	−17.4	+3.2	13.5	24.6	36.8	44.5	54.8	70.2	87.4	106.3	−102
Tetramethylene dibromide (1,4-dibromobutane)	C$_4$H$_8$Br$_2$	32.0	58.8	72.4	87.6	104.0	115.1	128.7	149.8	173.8	197.5	−20
Tetramethyllead	C$_4$H$_{12}$Pb	−29.0	−6.8	+4.4	16.6	30.3	39.2	50.8	68.8	89.0	110.0	−27
Tetramethyltin	C$_4$H$_{12}$Sn	−51.3	−31.0	−20.6	−9.3	+3.5	11.7	22.8	39.8	58.5	78.0	
Tetrapropylene glycol monoisopropyl ether	C$_{15}$H$_{32}$O$_{15}$	116.6	147.8	163.0	179.8	197.7	209.0	223.3	245.0	268.3	292.7	−16.5
Thioacetic acid (mercaptoacetic acid)	C$_2$H$_4$O$_2$S	60.0	87.7	101.5	115.8	131.8	142.0	154.0				
Thiodiglycol (2,2′-thiodiethanol)	C$_4$H$_{10}$O$_2$S	42.0	96.0	128.0	165.0	210.0	240.5	285				
Thiophene	C$_4$H$_4$S	−40.7	−20.8	−10.9	0.0	+12.5	20.1	30.5	46.5	64.7	84.4	−38.3
Thiophenol (benzenethiol)	C$_6$H$_6$S	18.6	43.7	56.0	69.7	84.2	93.9	106.6	125.8	146.7	168.0	
α-Thujone	C$_{10}$H$_{16}$O	38.3	65.7	79.3	93.7	110.0	120.2	134.0	154.2	177.8	201.0	
Thymol	C$_{10}$H$_{14}$O	64.3	92.8	107.4	122.6	139.8	149.8	164.1	185.5	209.6	231.8	
Tiglaldehyde	C$_5$H$_8$O	−25.0	−1.6	+10.0	23.2	37.0	45.8	57.7	75.4	95.5	116.8	51
Tiglic acid	C$_5$H$_8$O$_2$	52.0	77.8	90.2	103.8	119.0	127.8	140.5	158.0	179.2	198.5	64
Tiglonitrile	C$_5$H$_7$N	−25.5	−2.4	+9.2	22.1	36.7	46.0	58.2	77.8	99.7	122.0	
Toluene	C$_7$H$_8$	−26.7	−4.4	+6.4	18.4	31.8	40.3	51.9	69.5	89.5	110.6	−95
Toluene-2,4-diamine	C$_7$H$_{10}$N$_2$	106.5	137.2	151.7	167.9	185.7	196.2	211.5	232.8	256.0	280.0	99
2-Toluic nitrile (2-tolunitrile)	C$_8$H$_7$N	36.7	64.0	77.9	93.0	110.0	120.8	135.0	156.0	180.0	205.2	−13
4-Toluic nitrile (4-tolunitrile)	C$_8$H$_7$N	42.5	71.3	85.8	101.7	109.5	130.0	145.2	167.3	193.0	217.6	29
2-Toluidine	C$_7$H$_9$N	44.0	69.3	81.4	95.1	110.0	119.8	133.0	153.0	176.2	199.7	−16
3-Toluidine	C$_7$H$_9$N	41.0	68.0	82.0	96.7	113.5	123.8	136.7	157.6	180.6	203.3	−31
4-Toluidine	C$_7$H$_9$N	42.0	68.2	81.8	95.8	111.5	121.5	133.7	154.0	176.9	200.4	44
2-Tolyl isocyanide	C$_8$H$_7$N	25.2	51.0	64.0	78.2	94.0	104.0	117.7	137.8	159.9	183.5	

4-Tolylhydrazine	C7H10N2	82.4	110.0	123.8	138.6	154.1	165.0	178.0	198.0	219.5	242.0	65.5
Tribromoacetaldehyde	C2HBr3O	18.5	45.0	58.0	72.1	87.8	97.5	110.2	130.0	151.6	174.0	
1,1,2-Tribromobutane	C4H7Br3	45.0	73.5	87.8	103.2	120.2	131.6	146.0	167.8	192.0	216.2	
1,2,2-Tribromobutane	C4H7Br3	41.0	69.0	83.2	98.6	116.0	127.0	141.8	163.5	188.0	213.8	
2,2,3-Tribromobutane	C4H7Br3	38.2	66.0	79.8	94.6	111.8	122.2	136.3	157.8	182.2	206.5	
1,1,2-Tribromoethane	C2H3Br3	32.6	58.0	70.6	84.2	100.0	110.0	123.5	143.5	165.4	188.4	−26
1,2,3-Tribromopropane	C3H5Br3	47.5	75.8	90.0	105.8	122.8	134.0	148.0	170.0	195.0	220.0	16.5
Triisobutylamine	C12H27N	32.3	57.4	69.8	83.0	97.8	107.3	119.7	138.0	157.8	179.0	−22
Triisobutylene	C12H24	18.0	44.0	56.5	70.0	86.7	96.7	110.0	130.2	153.0	179.0	
2,4,6-Tritertbutylphenol	C18H30O	95.2	126.1	142.0	158.0	177.4	188.0	203.0	226.2	250.6	276.3	
Trichloroacetic acid	C2HCl3O2	51.0	76.0	88.2	101.8	116.3	125.9	137.8	155.4	175.2	195.6	57
Trichloroacetic anhydride	C4Cl6O3	56.2	85.3	99.6	114.3	131.2	141.8	155.2	176.2	199.8	223.0	
Trichloroacetyl bromide	C2BrCl3O	−7.4	+16.7	29.3	42.1	57.2	66.7	79.5	98.4	120.2	143.0	
2,4,6-Trichloroaniline	C6H4Cl3N	134.0	157.8	170.0	182.6	195.8	204.5	214.6	229.8	246.4	262.0	78
1,2,3-Trichlorobenzene	C6H3Cl3	40.0	70.0	85.6	101.8	119.8	131.5	146.0	168.2	193.5	218.5	52.5
1,2,4-Trichlorobenzene	C6H3Cl3	38.4	67.3	81.7	97.2	114.8	125.7	140.0	162.0	187.7	213.0	17
1,3,5-Trichlorobenzene	C6H3Cl3		63.8	78.0	93.7	110.8	121.8	136.0	157.7	183.0	208.4	63.5
1,2,3-Trichlorobutane	C4H7Cl3	+0.5	27.2	40.0	55.0	71.5	82.0	96.2	118.0	143.0	169.0	−30.6
1,1,1-Trichloroethane	C2H3Cl3	−52.0	−32.0	−21.9	−10.8	+1.6	9.5	20.0	36.2	54.6	74.1	−36.7
1,1,2-Trichloroethane	C2H3Cl3	−24.0	−2.0	+8.3	21.6	35.2	44.0	55.7	73.3	93.0	113.9	−73
Trichloroethylene	C2HCl3	−43.8	−22.8	−12.4	−1.0	+11.9	20.0	31.4	48.0	67.0	86.7	
Trichlorofluoromethane	CCl3F	−84.3	−67.6	−59.0	−49.7	−39.0	−32.3	−23.0	−9.1	+6.8	23.7	
2,4,5-Trichlorophenol	C6H3Cl3O	72.0	102.1	117.3	134.0	151.5	162.5	178.0	201.5	226.5	251.8	62
2,4,6-Trichlorophenol	C6H3Cl3O	76.5	105.9	120.2	135.8	152.2	163.5	177.8	199.0	222.5	246.0	68
Tri-2-chlorophenylthiophosphate	C18H12Cl3O3 PS	188.2	217.2	231.2	246.7	261.7	271.5	283.8	302.8	322.0	341.3	
1,1,1-Trichloropropane	C3H5Cl3	−28.8	−7.0	+4.2	16.2	29.9	38.3	50.0	67.7	87.5	108.2	−77
1,2,3-Trichloropropane	C3H5Cl3	+9.0	33.7	46.0	59.3	74.0	83.6	96.1	115.6	137.0	158.0	−14
1,1,2-Trichloro-1,2,2-trifluoroethane	C2Cl3F3	−68.0	−49.4	−40.3	−30.0	−18.5	−11.2	−1.7	+13.5	30.2	47.6	−35
Tricosane	C23H48	170.0	206.3	223.0	242.0	261.3	273.8	289.8	313.5	339.8	366.5	47
Tridecane	C13H28	59.4	98.3	104.0	120.2	137.7	148.2	162.5	185.0	209.4	234.0	−6.2
Tridecanoic acid	C13H26O2	137.8	166.3	181.0	195.8	212.4	222.0	236.0	255.2	276.5	299.0	41
Triethoxymethylsilane	C7H18O3Si	−1.5	+22.8	34.6	47.2	61.7	70.4	82.7	101.0	121.8	143.5	
Triethoxyphenylsilane	C12H20O3Si	71.0	98.8	112.6	127.2	143.5	153.2	167.5	188.0	210.5	233.5	
1,2,4-Triethylbenzene	C12H18	46.0	74.2	88.5	104.0	121.7	132.2	146.8	168.3	193.7	218.0	
1,3,4-Triethylbenzene	C12H18	47.9	76.0	90.2	105.8	122.6	133.4	147.7	168.3	193.2	217.5	

(Continued)

TABLE 2.38 Boiling Points of Common Organic Compounds at Selected Pressures (*Continued*)

| Compound | | Pressure, mm Hg | | | | | | | | | | Melting point, °C |
Name	Formula	1	5	10	20	40	60	100	200	400	760	
						Temperature, °C						
Triethylborine	$C_6H_{15}B$			−148.0	−140.6	−131.4	−125.2	−116.0	−101.0	−81.0	−56.2	
Triethyl camphoronate	$C_{15}H_{26}O_6$		150.2	166.0	183.6	201.8	213.5	228.6	250.8	276.0	301.0	135
citrate	$C_{12}H_{20}O_7$	107.0	138.7	144.0	171.1	190.4	202.5	217.8	242.2	267.5	294.0	
Triethyleneglycol	$C_6H_{14}O_4$	114.0	144.0	158.1	174.0	191.3	201.5	214.6	235.2	256.6	278.3	
Triethylheptylsilane	$C_{13}H_{30}Si$	70.0	99.8	114.6	130.3	148.0	158.2	174.0	196.0	221.0	247.0	
Triethyloctylsilane	$C_{14}H_{32}Si$	73.7	104.8	120.6	137.7	155.7	168.0	184.3	208.0	235.0	262.0	
Triethyl orthoformate	$C_7H_{16}O_3$	+5.5	29.2	40.5	53.4	67.5	76.0	88.0	106.0	125.7	146.0	
phosphate	$C_6H_{15}O_4P$	39.6	67.8	82.1	97.8	115.7	126.3	141.6	163.7	187.0	211.0	
Triethylthallium	$C_6H_{15}Tl$	+9.3	37.6	51.7	67.7	85.4	95.7	112.1	136.0	163.5	192.1	−63.0
Trifluorophenylsilane	$C_6H_5F_3Si$	−31.0	−9.7	+0.8	12.3	25.4	33.2	44.2	60.1	78.7	98.3	
Trimethallyl phosphate	$C_{12}H_{21}PO_4$	93.7	131.0	149.8	169.8	192.0	207.0	225.7	255.0	288.5	324.0	
2,3,5-Trimethylacetophenone	$C_{11}H_{14}O$	79.0	108.0	122.3	137.5	154.2	165.7	179.7	201.3	224.3	247.5	
Trimethylamine	C_3H_9N	−97.1	−81.7	−73.8	−65.0	−55.2	−48.8	−40.3	−27.0	−12.5	+2.9	−117
2,4,5-Trimethylaniline	$C_9H_{13}N$	68.4	95.9	109.0	123.7	139.8	149.5	162.0	182.3	203.7	234.5	67
1,2,3-Trimethylbenzene	C_9H_{12}	16.8	42.9	55.9	69.9	85.4	95.3	108.8	129.0	152.0	176.1	−25
1,2,4-Trimethylbenzene	C_9H_{12}	13.6	38.3	50.7	64.5	79.8	89.5	102.8	122.7	145.4	169.2	−44
1,3,5-Trimethylbenzene	C_9H_{12}	9.6	34.7	47.4	61.0	76.1	85.8	98.9	118.6	141.0	164.7	−44
2,2,3-Trimethylbutane	C_7H_{16}			−18.8	−7.5	+5.2	13.3	24.4	41.2	60.4	80.9	−25
Trimethyl citrate	$C_9H_{14}O_7$	106.2	146.2	160.4	177.2	194.2	205.5	219.6	241.3	264.2	287.0	78
Trimethyleneglycol (1,3-propandiol)	$C_3H_8O_2$	59.4	87.2	100.6	115.5	131.0	141.1	153.4	172.8	193.8	214.2	
1,2,4-Trimethyl-5-ethylbenzene	$C_{11}H_{16}$	43.7	71.2	84.6	99.7	106.0	126.3	140.3	160.3	184.5	208.1	
1,3,5-Trimethyl-2-ethylbenzene	$C_{11}H_{16}$	38.8	67.0	80.5	96.0	113.2	123.8	137.9	158.4	183.5	208.0	
2,2,3-Trimethylpentane	C_8H_{18}	−29.0	−7.1	+3.9	16.0	29.5	38.1	49.9	67.8	88.2	109.8	−112
2,2,4-Trimethylpentane	C_8H_{18}	−36.5	−15.0	−4.3	+7.5	20.7	29.1	40.7	58.1	78.0	99.2	−107
2,3,3-Trimethylpentane	C_8H_{18}	−25.8	−3.9	+6.9	19.2	33.0	41.8	53.8	72.0	92.7	114.8	−101
2,3,4-Trimethylpentane	C_8H_{18}	−26.3	−4.1	+7.1	19.3	32.9	41.6	53.4	71.3	91.8	113.5	−109
2,2,4-Trimethyl-3-pentanone	$C_8H_{16}O$	14.7	36.0	46.4	57.6	69.8	77.3	87.6	102.2	118.4	135.0	
Trimethyl phosphate	$C_3H_9O_4P$	26.0	53.7	67.8	83.0	100.0	110.0	124.0	145.0	167.8	192.7	
2,4,5-Trimethylstryene	$C_{11}H_{14}$	48.1	77.0	91.6	107.1	124.2	135.5	149.8	171.8	196.1	221.2	
2,4,6-Trimethylsytrene	$C_{11}H_{14}$	37.5	65.7	79.7	94.8	111.8	122.3	136.8	157.8	182.3	207.0	
Trimethylsuccinic anhydride	$C_7H_{10}O_3$	53.5	82.6	97.4	113.8	131.0	142.2	156.5	179.8	205.5	231.0	

Triphenylmethane	$C_{19}H_{16}$	169.7	188.4	197.0	206.8	215.5	221.2	228.4	239.7	249.8	259.2	93.4
Triphenylphosphate	$C_{18}H_{15}O_4P$	193.5	230.4	249.8	269.7	290.3	305.2	322.5	349.8	379.2	413.5	49.4
Tripropyleneglycol	$C_9H_{20}O_4$	96.0	125.7	140.5	155.8	173.7	184.6	199.0	220.2	244.3	267.2	
Tripropyleneglycol monobutyl ether	$C_{13}H_{38}O_4$	1C1.5	131.6	147.0	161.8	179.8	190.2	204.4	224.4	247.0	269.5	
Tripropyleneglycol monoisopropyl ether	$C_{12}H_{26}O_4$	82.4	112.4	127.3	143.7	161.4	173.2	187.8	209.7	232.8	256.6	
Tritolyl phosphate	$C_{21}H_{21}O_4P$	154.6	184.2	198.0	213.2	229.7	239.8	252.2	271.8	292.7	313.0	
Undecane	$C_{11}H_{24}$	52.7	59.7	73.9	85.6	104.4	115.2	128.1	149.3	171.9	195.8	-25.6
Undecanoic acid	$C_{11}H_{22}O_2$	101.4	133.1	149.0	166.0	185.6	197.2	212.5	237.8	262.8	290.0	29.5
10-Undecenoic acid	$C_{11}H_{20}O_2$	1·4.0	142.8	156.3	172.0	188.7	199.5	213.5	232.8	254.0	275.0	24.5
Undecan-2-ol	$C_{11}H_{24}O$	71.1	99.0	112.8	127.5	143.7	153.7	167.2	187.7	209.8	232.0	
n-Valeric acid	$C_5H_{10}O_2$	42.2	67.7	79.8	93.1	107.8	116.6	128.3	146.0	165.0	184.4	-34.5
iso-Valeric acid	$C_5H_{10}O_2$	34.5	59.6	71.3	84.0	98.0	107.3	118.9	136.2	155.2	175.1	-37.6
γ-Valerolactone	$C_5H_8O_2$	37.5	65.8	79.8	95.2	101.9	122.4	136.5	157.7	182.3	207.5	
Valeronitrile	C_5H_9N	-6.0	+18.1	30.0	43.3	57.8	66.9	78.6	97.7	118.7	140.8	
Vanillin	$C_8H_8O_3$	107.0	138.4	154.0	170.5	188.7	199.8	214.5	237.3	260.0	285.0	81.5
Vinyl acetate	$C_4H_6O_2$	-48.0	-28.0	-18.0	-7.0	+5.3	13.0	23.3	38.4	55.5	72.5	
2-Vinylanisole	$C_9H_{10}O$	41.9	68.0	81.0	94.7	110.0	119.8	132.3	151.0	172.1	194.0	
3-Vinylanisole	$C_9H_{10}O$	43.4	69.9	83.0	97.2	112.5	122.3	135.3	154.0	175.8	197.5	
4-Vinylanisole	$C_9H_{10}O$	45.2	72.0	85.7	100.0	116.0	126.1	139.7	159.0	182.0	204.5	
Vinyl chloride (1-chloroethylene)	C_2H_3Cl	-105.6	-90.8	-83.7	-75.7	-66.8	-61.1	-53.2	-41.3	-28.0	-13.8	-153.7
cyanide (acrylonitrile)	C_3H_3N	-51.0	-30.7	-20.3	-9.0	+3.8	11.8	22.8	38.7	58.3	78.5	-82
fluoride (1-fluoroethylene)	C_2H_3F	-149.3	-138.0	-132.2	-125.4	-118.0	-113.0	-106.2	-95.4	-84.0	-72.2	-160.5
Vinylidene chloride (1,1-dichloroethene)	$C_2H_2Cl_2$	-77.2	-60.0	-51.2	-41.7	-31.1	-24.0	-15.0	-1.0	+14.8	31.7	-122.5
4-Vinylphenetole	$C_{10}H_{12}O$	64.0	91.7	105.6	120.3	136.3	146.4	159.8	180.0	202.8	225.0	
2-Xenyl dichlorophosphate	$C_{12}H_9Cl_2PO$	138.2	171.1	187.0	205.0	223.8	236.0	251.5	275.3	301.5	328.5	
2,4-Xylaldehyde	$C_9H_{10}O$	59.0	85.9	99.0	114.0	129.7	139.8	152.2	172.3	194.1	215.5	75
2-Xylene (2-xylene)	C_8H_{10}	-3.8	+20.2	32.1	45.1	59.5	68.8	81.3	100.2	121.7	144.4	-25.2
3-Xylene (3-xylene)	C_8H_{10}	-6.9	+16.8	28.3	41.1	55.3	64.4	76.8	95.5	116.7	139.1	-47.9
4-Xylene (4-xylene)	C_8H_{10}	-8.1	+15.5	27.3	40.1	54.4	63.5	75.9	94.6	115.9	138.3	+13.3
2,4-Xylidine	$C_8H_{11}N$	52.6	79.8	93.0	107.6	123.8	133.7	146.8	166.4	188.3	211.5	
2,6-Xylidine	$C_8H_{11}N$	44.0	72.6	87.0	102.7	120.2	131.5	146.0	168.0	193.7	217.9	

TABLE 2.39   Organic Solvents Arranged by Boiling Points

Name	BP, °C	Name	BP, °C
Ethylene oxide	10.6	1-Propanol	97.2
Chloroethane	12.3	Heptane	98.4
Furan	31.4	1-Chloro-3-methylbutane	99
Methyl formate	31.5	Ethyl propionate	99.1
Diethyl ether	34.6	2-Butanol	99.6
Propylene oxide	34.5	Formic acid	100.8
Pentane	36.1	Methylcyclohexane	100.9
Bromoethane	38.4	1,4-Dioxane	101.2
Dichloromethane	39.8	Nitromethane	101.2
Dimethoxymethane	42.3	Propyl acetate	101.5
Carbon disulfide	46.3	2-Pentanone	101.7
1-Isopropoxy-2-propanol	47.9	3-Pentanone	102.0
Ethyl formate	54.2	2-Methyl-2-butanol	102.0
Acetone	56.2	1,1-Diethoxyethane	102.7
Methyl acetate	56.3	Butyl formate	106.6
1,1-Dichloroethane	57.3	2-Methyl-1-propanol	107.9
Dichloroethylene	60.6	Toluene	110.6
Chloroform	61.2	sec-Butyl acetate	112.3
Methanol	64.7	1,1,2-Trichloroethane	113.5
Tetrahydrofuran	66.0	Nitroethane	114.1
Diisopropyl ether	68.0	Pyridine	115.2
Hexane	68.7	3-Pentanol	115.6
1-Chloro-2-methylpropane	68.9	4-Methyl-2-pentanone	115.7
1,1,1-Trichloroethane	74.0	1-Chloro-2,3-epoxypro-	116.1
1,3-Dioxolane	74–75	pane	
Carbon tetrachloride	76.7	1-Butanol	117.7
Ethyl acetate	77.1	Acetic acid	117.9
1-Chlorobutane	77.9	Isobutyl acetate	118.0
Ethanol	78.3	2-Pentanol	119.3
2-Butanone	79.6	1-Bromo-3-methylbutane	119.7
2-Methyltetrahydrofuran	80.0	1-Methoxy-2-propanol	120.1
Benzene	80.1	2-Nitropropane	120.3
Cyclohexane	80.7	Tetrachloroethylene	121.1
Propyl formate	80.9	Ethyl butyrate	121.6
Acetonitrile	81.6	3-Hexanone	123
2-Propanol	82.4	2,4-Dimethyl-3-pentanone	124
1,1,-Dimethylethanol	82.4	2-Methoxyethanol	124.6
Cyclohexene	83.0	Octane	125.7
Diisopropylamine	83.5	Butyl acetate	126.1
1,2-Dichloroethane	83.7	Diethyl carbonate	126.8
Thiophene	84.2	2-Hexanone	127.2
Trichloroethylene	87.2	1-Chloro-2-propanol	127.4
Isopropyl acetate	88.2	2-Chloroethanol	128.6
1-Bromo-2-methylpropane	91.5	3-Methyl-1-penten-2-one	129.5
2,5-Dimethylfuran	93–94	1-Nitropropane	131.2
Ethyl chloroformate	94	Chlorobenzene	131.7
Allyl alcohol	96.6	1,2-Dibromoethane	131.7
1,2-Dichloropropane	96.8	4-Methyl-2-pentanol	131.7

TABLE 2.39 Organic Solvents Arranged by Boiling Points (*Continued*)

Name	BP, °C	Name	BP, °C
3-Methyl-1-butanol	132.0	Phenol	181.8
Cyclohexylamine	134.8	2-Ethyl-1-hexanol	184.3
2-Ethoxyethanol	134.8	Aniline	184.4
Ethylbenzene	136.2	Benzyl ethyl ether	185.0
1-Pentanol	138	Diethyl oxalate	185.4
p-Xylene	138.4	1,2-Propanediol	188
m-Xylene	139.1	Bis(2-ethoxyethyl) ether	188.4
Acetic anhydride	140.0	Dimethyl sulfoxide	189.0
2,4-Pentanedione	140.6	1,2-Ethanediol diacetate	190.2
Isopentyl acetate	142	Benzonitrile	191.0
Dibutyl ether	142.4	2,5-Hexanedione	191.4
4-Heptanone	143.7	2-(2-Methoxyethoxy)-	194.1
o-Xylene	144.4	ethanol	
2-Methoxyethyl acetate	144.5	*N*,*N*-Dimethylaniline	194.2
1,1,2,2-Tetrachloroethane	146.3	1-Octanol	195.2
3-Heptanone	147.8	1,2-Ethanediol	197.3
Tribromomethane	149.6	Diethyl malonate	199.3
Nonane	150.8	Methyl benzoate	199.5
2-Heptanone	151	*o*-Toluidine	200.4
Isopropylbenzene	152.4	*p*-Toluidine	200.6
N,*N*-Dimethylformamide	153.0	2-(2-Ethoxyethoxy)-	202
Methoxybenzene	153.8	ethanol	
Ethyl lactate	154.5	Acetophenone	202.1
Cyclohexanone	155.7	1,2-Dibutoxyethane	203.6
Bromobenzene	156.2	1-Phenylethanol	203.9
1,2,3-Trichloropropane	156.9	*m*-Toluidine	203.4
1-Hexanol	157.5	Benzyl alcohol	205.5
Propylbenzene	159.2	Camphor	207
Cyclohexanol	161.1	1,3-Butanediol	207.5
Bis(2-methoxyethyl)ether	160	1,2,3,4-Tetrahydro-	207.6
Isopentyl propionate	160.2	naphthalene	
2-Heptanol	160.4	*γ*-Valerolactone	207–208
Pentachloroethane	160.5	*o*-Chloroaniline	208.8
2-Furaldehyde	161.8	Nitrobenzene	210.8
2,6-Dimethyl-4-heptanone	168.1	Ethyl benzoate	212.4
4-Hydroxy-4-methyl-	169.2	3,5,5-Trimethylcyclo-	215.2
2-pentanone		hex-2-en-1-one	
2-Furanmethanol	170.0	Naphthalene	217.7
Ethoxybenzene	170	2-(2-Ethoxyethoxy)ethyl	218.5
2-Butoxyethanol	170.2	acetate	
Diisopentyl ether	173.4	Acetamide	221.2
Decane	174.2	Methyl salicylate	223.0
1,3-Dichloro-2-propanol	174.3	Diethyl maleate	225.3
Cyclohexyl acetate	174–175	1,4-Butanediol	230
1-Heptanol	175.8	Propyl benzoate	231.2
Furfuryl acetate	175–177	1-Decanol	230.2
1,3,3-Trimethyl-	177.4	Phenylacetonitrile	233.5
2-oxabicyclo-		Quinoline	237
[2.2.2]octane		Tributyl borate	238.5
4-Isopropyl-	177.1	Propylene carbonate	240
1-methylbenzene		2-Phenoxyethanol	240
Isopentyl butyrate	178.6	Bis(2-hydroxyethyl) ether	245
Bis(2-chloroethyl) ether	178.8	Dibutyl oxalate	245.5
2-Octanol	179	Butyl benzoate	250
1,2-Dichlorobenzene	180.4	1,2,3-Propanetriol	258–259
Ethyl acetoacetate	180.8	triacetate	

(*Continued*)

TABLE 2.39   Organic Solvents Arranged by Boiling Points (*Continued*)

Name	BP, °C	Name	BP, °C
1-Chloronaphthalene	259.3	2,2′-(Ethylenedioxy)-bisethanol	285
Isopentyl benzoate	262		
Bis[2-(methoxyethoxy)-ethyl]ether	275.3	Glycerol	290
		Diethyl *o*-phthalate	295
1-Methoxy-2-nitrobenzene	277	Benzyl benzoate	323.5
Isopentyl salicylate	277–278	Dibutyl *o*-phthalate	340.0
1-Bromonaphthalene	281.1	Dibutyl decanedioate	344–345
Dimethyl *o*-phthalate	283.7		

TABLE 2.40   Boiling Points of *n*-Paraffins

Carbon number	Boiling point, °C	Boiling point, °F
5	36	97
6	69	156
7	98	209
8	126	258
9	151	303
10	174	345
11	196	385
12	216	421
13	235	456
14	253	488
15	271	519
16	287	548
17	302	576
18	317	602
19	331	627
20	344	651
21	356	674
22	369	696
23	380	716
24	391	736
25	402	755
26	412	774
27	422	792
28	432	809
29	441	825
30	450	841
31	459	858
32	468	874
33	476	889
34	483	901
35	491	916
36	498	928
37	505	941
38	512	958
39	518	964
40	525	977
41	531	988
42	537	999
43	543	1009
44	548	1018

2.6 FLAMMABILITY PROPERTIES

The *flash point* of a substance is the lowest temperature at which the substance gives off sufficient vapor to form an ignitable mixture with air near its surface or within a vessel. The *fire point* is the temperature at which the flame becomes self-sustained and the burning continues. At the flash point, the flame does not need to be sustained. The fire point is usually a few degrees above the flash point. ASTM test methods include procedures using a closed cup (ASTM D-56, ASTM D-93, and ASTM D-3828), which is preferred, and an open cup (ASTM D-92, ASTM D-I310). When several values are available, the lowest temperature is usually taken in order to assure safe operation of the process.

The *ignition temperature* (or *ignition point*) is the minimum temperature required to initiate self-sustained combustion of a substance (solid, liquid, or gaseous) and independent of external ignition sources or heat.

Flash points, lower and upper flammability limits, and auto-ignition temperatures are the three properties that are used to indicate safe operating limits of temperature when processing organic materials. Prediction methods are somewhat erratic, but, together with comparisons with reliable experimental values for families or similar compounds, they are valuable in setting a conservative value for each of the properties.

The upper and lower flammability limits are the boundary-line mixtures of vapor or gas with air, which, if ignited, will just propagate flame and are given in terms of percent by volume of gas or vapor in the air. Each of these limits also has a temperature at which the flammability limits are reached. The temperature corresponding to the lower-limit partial vapor pressure should equal the flash point. The temperature corresponding to the upper-limit partial vapor pressure is somewhat above the lower limit and is usually considerably below the auto-ignition temperature. Flammability limits are calculated at one atmosphere total pressure and are normally considered synonymous with explosive limits. Limits in oxygen rather than air are sometimes measured and available. Limits are generally reported at 298 K and 1 atmosphere. If the temperature or the pressure is increased, the lower limit will decrease while the upper limit will increase, giving a wider range of compositions over which flame will propagate.

The auto-ignition temperature is the minimum temperature for a substance to initiate self-combustion in air in the absence of a spark or flame. The temperature is no lower than and is generally considerably higher than the temperature corresponding to the upper flammability limit. Large differences can occur in reported values determined by different procedures. The lowest reasonable value should be accepted in order to assure safety. Values are also sometimes given in oxygen rather than in air.

One simple method of estimating auto-ignition temperatures is to compare values for a compound with other members of its homologous series on a plot vs. carbon number as the temperature decreases and carbon number increases.

TABLE 2.41 Boiling Points, Flash Points, and Ignition Temperatures of Organic Compounds

Compound	Boiling point °F (°C)	Flash point, °F (°C)	Ignition point, °F (°C)
Acetal $CH_3CH(OC_2H_5)_2$ (Acetaldehydediethylacetal)	215 (102)	−5 (−21)	446 (230)
Acetaldehyde CH_3CHO (Acetic aldehyde) (Ethanal)	70 (21)	−38 (−39)	347 (175)
Acetaldehydediethylacetal		See Acetal.	
Acetaldel		See Aldol.	
Acetanilide $CH_3CONHC_6H_5$	582 (306)	337 (169) (oc)	985 ± 10 (530)
Acetic Acid, Glacial CH_3COOH	245 (118)	103 (39)	867 (463)
Acetic Acid, Isopropyl Ester		See Isopropyl Acetate.	
Acetic Acid, Methyl Ester		See Methyl Acetate.	
Acetic Acid, n-Propyl Ester		See Propyl Acetate.	
Acetic Aldehyde		See Acetaldehyde	
Acetic Anhydride $(CH_3CO)_2O$ (Ethanoic anhydride)	284 (140)	120 (49)	600 (316)
Acetic Ester		See Ethyl Acetate.	
Acetic Ether		See Ethyl Acetate.	
Acetoacetanilide $CH_3COCH_2CONHC_6H_5$		365 (185)	
o-Acetoacet Anisidide $CH_3COCH_2CONHC_6H_4OCH_3$		325 (168)	
Acetoacetic Acid, Ethyl Ester		See Ethyl acetoacetate.	
Acetoethylamide		See N-Ethylacetamide.	
Acetone CH_3COCH_3 (Dimethyl Ketone) (2-Propanone)	133 (56)	−4 (−20)	869 (465)
Acetone Cyanohydrin $(CH_3)_2C(OH)CN$ (2-Hydroxy2-Methyl Propionitrile)	248 (120) Decomposes	165 (74)	1270 (688)
Acetonitrile CH_3CN (Methyl Cyanide)	179 (82)	42 (6)	975 (524)
Acetonyl Acetone $(CH_2COCH_3)_2$ (2,5-Hexanedione)	378 (192)	174 (79)	920 (499)
Acetophenone $C_6H_5COCH_3$ (Phenyl Methyl Ketone)	396 (202)	170 (77)	1058 (570)
p-Acetotoluidide $CH_3CONHC_6H_4CH_3$	583 (306)	334 (168)	
Acetyl Acetone		See 2,4-Pentanedione.	
Acetyl Chloride CH_3COCl (Ethanoyl Chloride)	124 (51)	40 (4)	734 (390)

TABLE 2.41 Boiling Points, Flash Points, and Ignition Temperatures of Organic Compounds (*Continued*)

Compound	Boiling point °F (°C)	Flash point, °F (°C)	Ignition point, °F (°C)
Acetylene	−118	Gas	581
CH:CH	(−83)		(305)
(Ethine)			
(Ethyne)			
N-Acetyl Ethanolamine	304–308	355	860
CH$_3$C:ONHCH$_2$CH$_2$OH	(151–153)	(179)	(460)
(N-(2-Hydroxyethyl)	@10 mm	(oc)	
acetamide)	Decomposes		
N-Acetyl Morpholine	Decomposes	235	
CH$_3$CONCH$_2$CH$_2$OCH$_2$CH:		(113)	
Acetyl Oxide		See Acetic Anhydride.	
Acetylphenol		See Phenyl Acetate.	
Acrolein	125	−15	428
CH$_2$:CHCHO	(52)	(−26)	(220)
(Acrylic Aldehyde)			Unstable
Acrylic Acid (Glacial)	287	122	820
CH$_2$CHCOOH	(142)	(50)	(438)
Acrylic Aldehyde		See Acrolein.	
Acrylonitrile	171	32	898
CH$_2$:CHCN	(77)	(0)	(481)
(Vinyl Cyanide)			
(Propenenitrile)			
Adipic Acid	509	385	788
HOOC(CH$_2$)$_4$COOH	(265)	(196)	(420)
	@100 mm		
Adipic Ketone		See Cyclopentanone.	
Adiponitrile	563	200	
NC(CH$_2$)$_4$CN	(295)	(93)	
Alcohol		See Ethyl Alcohol, Methyl Alcohol.	
Aldol	174–176	150	482
CH$_3$CH(OH)CH$_2$CHO	(79–80)	(66)	(250)
(3-Hydroxybutanal)	@12 mm		
(β-Hydroxybuteraldehyde)	Decomposes		
	@176		
	(80)		
Allyl Acetate	219	72	705
CH$_3$COCH$_2$CH:CH$_2$	(104)	(22)	(374)
Allyl Alcohol	206	70	713
CH$_2$:CHCH$_2$OH	(97)	(21)	(378)
Allylamine	128	−20	705
CH$_2$:CHCH$_2$NH$_2$	(53)	(−29)	(374)
(2-Propenylamine)			
Allyl Bromide	160	30	563
CH$_2$:CHCH$_2$Br	(71)	(−1)	(295)
(3-Bromopropene)			
Allyl Caproate	367–370	150	
CH$_3$(CH$_2$)$_4$COOCH$_2$CH:Cl	(186–188)	(66)	
(Allyl Hexanoate)			
(2-Propenyl Hexanoate)			

(*Continued*)

TABLE 2.41 Boiling Points, Flash Points, and Ignition Temperatures of Organic Compounds (*Continued*)

Compound	Boiling point °F (°C)	Flash point, °F (°C)	Ignition point, °F (°C)
Allyl Chloride	113	−25	737
CH$_2$:CHCH$_2$Cl	(45)	(−32)	(485)
(3-Chloropropene)			
Allyl Chlorocarbonate		See Allyl Chloroformate.	
Allyl Chloroformate	223–237	88	
CH$_2$:CHCH$_2$OCOCl	(106–114)	(31)	
(Allyl Chlorocarbonate)			
Allylene		See Propyne.	
Allyl Ether	203	20	
(CH$_2$:CHCH$_2$)$_2$O	(95)	(−7)	
(Diallyl Ether)			
Allylidene Diacetate	225	180	
CH$_2$:CHCH(OCOCH$_3$)$_2$	(107)	(82)	
	@50 mm		
Allyl Isothiocyanate		See Mustard Oil.	
Allylpropenyl		See 1,4-Hexadiene.	
Allyl Trichloride		See 1,2,3-Trichloropropane.	
Allyl Vinyl Ether		See Vinyl Allyl Ether.	
Alpha Methyl Pyridine		2-Picoline.	
Aminobenzene		See Aniline.	
2-Aminobiphenyl		See 2-Biphenylamine.	
1-Aminobutane		See Butylamine.	
2-Amino-1-Butanol	352	165	
CH$_3$CH$_2$CHNH$_2$CH$_2$OH	(178)	(74)	
1-Amino-4-Ethoxybenzene		See p-Phenetidine.	
β-Aminoethyl Alcohol		See Ethanolamine.	
Amyl Acetate	300	60	680
CH$_3$COOC$_5$H$_{11}$	(149)	(16)	(360)
(1-Pentanol Acetate)		70	
Comm.		(21)	
sec-Amyl Acetate	249	89	
CH$_3$COOCH(CH$_3$)(CH$_2$)$_2$CH$_3$	(121)	(32)	
(2-Pentanol Acetate)			
Amyl Alcohol	280	91	572
CH$_3$(CH$_2$)$_3$CH$_2$OH	(138)	(33)	(300)
(1-Pentanol)			
sec-Amyl Alcohol	245	94	650
CH$_3$CH$_2$CH$_2$CH(OH)CH$_3$	(118)	(34)	(343)
(Diethyl Carbinol)			
Amylamine	210	30	2.2 22
C$_5$H$_{11}$NH$_2$	(99)	(−1)	
(Pentylamine)			
sec-Amylamine	198	20	
CH$_3$(CH$_2$)$_2$CH(CH$_3$)NH$_2$	(92)	(−7)	
(2-Aminopentane)			
(Methylpropylcarbinylamine)			
p-tert-Amylaniline	498–504	215	
(C$_2$H$_5$)(CH$_2$)$_2$CC$_6$H$_4$NH$_2$	(259–262)	(102)	
Amylbenzene	365	150	
C$_6$H$_5$C$_5$H$_{11}$	(185)	(66)	
(Phenylpentane)		(oc)	

TABLE 2.41 Boiling Points, Flash Points, and Ignition Temperatures of Organic Compounds (*Continued*)

Compound	Boiling point °F (°C)	Flash point, °F (°C)	Ignition point, °F (°C)
Amyl Bromide	128–9	90	
$CH_3CH_2CH_2CH_2CH_2Br$	(53–54)	(32)	
(1-Bromopentane)	@746 mm		
Amyl Butyrate	365	135	
$C_5H_{11}OOCC_3H_7$	(185)	(57)	
Amyl Carbinol		See Hexyl Alcohol.	
Amyl Chloride	223	55	500
$CH_3(CH_2)_3CH_2Cl$	(106)	(13)	(260)
(1-Chloropentane)			
tert-Amyl Chloride	187		653
$CH_3CH_2CCl(CH_3)CH_3$	(86)		(345)
Amyl Chlorides (Mixed)	185–228	38	
$C_5H_{11}Cl$	(85–109)	(3)	
Amylcyclohexane	395		462
$C_5H_{11}C_6H_{11}$	(202)		(239)
Amylene		See 1-Pentene.	
β-Amylene-cis	99	<–4	
$C_2H_5CH:CHCH_3$	(37)	(<–20)	
(2-Pentene-cis)			
β-Amylene-trans	97	<–4	
$C_2H_5CH:CHCH_3$	(36)	(<–20)	
(2-Pentene-trans)			
Amylene Chloride		See 1,5-Dichloropentane.	
Amyl Ether	374	135	338
$C_5H_{11}OC_5H_{11}$	(190)	(57)	(170)
(Diamyl Ether)			
(Pentyloxypentane)			
Amyl Formate	267	79	
$HCOCC_5H_{11}$	(131)	(26)	
Amyl Lactate	237–239	175	
$C_2H_5OCOOCH_2$-	(114–115)	(79)	
$CH(CH_3)C_2II_5$	@36 mm		
Amyl Laurate	554–626	300	
$C_{11}H_{23}COOC_5H_{11}$	(290–330)	(149)	
Amyl Maleate	518–599	270	
$(CHCOOC_5H_{11})_2$	(270–315)	(132)	
Amyl Mercaptan	260	65	
$C_5H_{11}SH$	(127)	(18)	
(1-Pentanethiol)			
Amyl Mercaptans (Mixed)	176–257	65	
$CH_3(CH_2)_4SH$	(80–125)	(18)	
Amyl Naphthalene	550	255	
$C_{10}H_7C_5H_{11}$	(288)	(124)	
Amyl Nitrate	306–315	118	
$CH_3(CH_2)_4NO_2$	(153–157)	(48)	
Amyl Nitrite	220	410	
$CH_3(CH_2)_4NO_2$	(104)	(210)	
Amyl Oleate	392–464	366	
$C_{17}H_{33}COOC_5H_{11}$	(200–240)	(186)	
	@20 mm		
Amyl Oxalate	464–523	245	
$(COOC_5H_{11})_2$	(240–273)	(118)	
(Diamyl Oxalate)			

(*Continued*)

TABLE 2.41 Boiling Points, Flash Points, and Ignition Temperatures of Organic Compounds (*Continued*)

Compound	Boiling point °F (°C)	Flash point, °F (°C)	Ignition point, °F (°C)
o-Amyl Phenol	455–482	219	
$C_5H_{11}C_6H_4OH$	(235–250)	(104)	
p-tert-Amyl Phenol		See Pentaphen.	
p-sec-Amylphenol	482–516	270	
$C_5H_{11}C_6H_4OH$	(250–269)	(132)	
2-(p-tert-Amylphenoxy) Ethanol	567–590	280	
$C_5H_{11}C_6H_4OCH_2CH_2OH$	(297–310)	(138)	
2-(p-tert-Amylphenoxy) Ethyl	464–500	410	
Laurate	(240–260)	(210)	
$C_{11}H_{23}COO(CH_2)_2OC_6H_4C_5H_{11}$	@6 mm		
p-tert-Amylphenyl	507–511	240	
Acetate	(264–266)	(116)	
$CH_3COOC_6H_4C_5H_{11}$			
p-tert-Amylphenyl Butyl	540–550	275	
Ether	(282–288)	(135)	
$C_5H_{11}C_6H_4OC_4H_9$			
Amyl Phenyl Ether	421–444	185	
$CH_3(CH_2)_4OC_6H_5$	(216–229)	(85)	
(Amoxybenzene)			
p-tert-Amylphenyl Methyl	462–469	210	
Ether	(239–243)	(99)	
$C_5H_{11}C_6H_4OCH_3$			
Amyl Phthalate		See Diamyl Phthalate.	
Amyl Propionate	275–347	106	712
$C_2H_5COO(CH_2)_4CH_3$	(135–175)	(41)	(378)
(Pentyl Propionate)			
Amyl Salicylate	512	270	
$HOC_6H_4COOC_5H_{11}$	(267)	(132)	
Amyl Stearate	680	365	
$CH_3(CH_2)_{16}COOC_5H_{11}$	(360)	(185)	
Amyl Sulfides, (Mixed)	338–356	185	
$C_5H_{11}S$	(170–180)	(85)	
Amyl Tolene	400–415	180	
$C_5H_{11}C_6H_4CH_3$	(204–213)	(82)	
Amyl Xylyl Ether	480–500	205	
$C_5H_{11}OC_6H_3(CH_3)_2$	(249–260)	(96)	
Aniline	364	158	1139
$C_6H_5NH_2$	(184)	(70)	(615)
(Aminobenzene)			
(Phenylamine)			
Aniline Hydrochloride	473	380	
$C_6H_5NH_2HCl$	(245)	(193)	
2-Anilinoethanol	547	305	
$C_6H_5NHCH_2CH_2OH$	(286)	(152)	
(β-Anilinoethanol Ethoxyaniline)			
(β-Hydroxyethylaniline)			
β-Anilinoethanol		See 2-Anilinoethanol.	
Ethoxyaniline			
o-Anisaldehyde		See o-Methoxy Benzaldehyde.	
o-Anisidine	435	244	
$H_2NC_6H_4OCH_3$	(224)	(118)	
(2-Methoxyaniline)			

TABLE 2.41   Boiling Points, Flash Points, and Ignition Temperatures of Organic Compounds (*Continued*)

Compound	Boiling point °F (°C)	Flash point, °F (°C)	Ignition point, °F (°C)
Anisole $C_6H_5OCH_3$ (Methoxybenzene) (Methyl Phenyl Ether)	309 (154)	125 (52)	887 (475)
Anol		See Cyclohexanol.	
Anthracene $(C_6H_4CH)_2$	644 (340)	250 (121)	1004 (540)
Anthraquinone $C_6H_4(CO)_2C_6H_4$	716 (380)	365 (185)	
Asphalt (Petroleum Pitch)	>700 (>371)	400+ (204+)	905 (485)
Aziridine		See Ethyleneimine.	
Azobisisobutyronitrile $N:CC(CH_3)_2N:NC(CH_3)_2C:N$	Decomposes	147 (64)	
Benzaldehyde C_6H_5CHO (Benzenecarbonal)	355 (179)	145 (63)	377 (192)
Benzedrine $C_6H_5CH_2CH(CH_3)NH_2$ (1-Phenyl Isopropyl Amine)	392 (200)	<212 (<100)	
Benzene C_6H_6 (Benzol)	176 (80)	12 (−11)	928 (498)
Benzine		See Petroleum Ether.	
Benzocyclobutene	306 (152)	95 (35)	477 (247)
Benzoic Acid C_6H_5COOH	482 (250)	250 (121)	1058 (570)
Benzol		See Benzene.	
p-Benzoquinone $C_6H_4O_2$ (Quinone)	Sublimes	100–200 (38–93)	1040 (560)
Benzotrichloride $C_6H_5CCl_3$ (Toluene, α,α,α-Trichloro) (Phenyl Chloroform)	429 (221)	260 (127)	412 (211)
Benzotrifluoride $C_6H_5CF_3$	216 (102)	54 (12)	
Benzoyl Chloride C_6H_5COCl (Benzene Carbonyl Chloride)	387 (197)	162 (72)	
Benzyl Acetate $CH_3COOCH_2C_6H_5$	417 (214)	195 (90)	860 (460)
Benzyl Alcohol $C_6H_5CH_2OH$ (Phenyl Carbinol)	403 (206)	200 (93)	817 (436)
Benzyl Benzoate $C_6H_5COOCH_2C_6H_5$	614 (323)	298 (148)	896 (480)
Benzyl Butyl Phthalate $C_4H_9COOC_6H_4COOCH_2C_6H_5$ (Butyl Benzyl Phthalate)	698 (370)	390 (199)	

(*Continued*)

TABLE 2.41   Boiling Points, Flash Points, and Ignition Temperatures of Organic Compounds (*Continued*)

Compound	Boiling point °F (°C)	Flash point, °F (°C)	Ignition point, °F (°C)
Benzyl Carbinol		See Phenethyl Alcohol.	
Benzyl Chloride	354	153	1085
C₆H₅CH₂Cl	(179)	(67)	(585)
(α-Chlorotoluene)			
Benzyl Cyanide	452	235	
C₆H₅CH₂CN	(233.5)	(113)	
(Phenyl Acetonitrile)			
(α-Tolunitrile)			
N-Benzyldiethylamine	405–420	170	
C₆H₅CH₂N(C₂H₅)₂	(207–216)	(77)	
Benzyl Ether		See Dibenzyl Ether.	
Benzyl Mercaptan	383	158	
C₆H₅CH₂SH	(195)	(70)	
(α-Toluenethiol)			
Benzyl Sallcilate	406	>212	
OHC₆H₄COOCH₂C₆H₅	(208)	(>100)	
(Salycilic Acid Benzyl Ester)			
Bicyclohexyl	462	165	473
[CH₂(CH₂)₄CH]₂	(239)	(74)	(245)
(Dicyclohexyl)			
Biphenyl	489	235	1004
C₆H₅C₆H₅	(254)	(113)	(540)
(Diphenyl)			
(Phenylbenzene)			
2-Biphenylamine	570	842	
NH₂C₆H₄C₆H₅	(299)	(450)	
(2-Aminobiphenyl)			
Bromobenzene	313	124	1049
C₆H₅Br	(156)	(51)	(565)
(Phenyl Bromide)			
1-Bromo Butane		See Butyl Bromide.	
4-Bromodiphenyl	592	291	
C₆H₅C₆H₄Br	(311)	(144)	
Bromoethane		See Ethyl Bromide.	
Bromomethane		See Methyl Bromide.	
1-Bromopentane		See Amyl Bromide.	
3-Bromopropene		See Allyl Bromide.	
o-Bromotoluene	359	174	
BrC₆H₄CH₃	(182)	(79)	
p-Bromotoluene	363	185	
BrC₆H₄CH₃	(184)	(85)	
1,3-Butadiene	24		788
CH₂:CHCH:CH₂	(−4)	Gas	(420)
Butadiene Monoxide	151	<−58	
CH₂:CHCHOCH₂	(66)	(<−50)	
(Vinylethylene Oxide)			
Butanal		See Butyraldehyde.	
Butanal Oxime		See Butyraldoxime.	
Butane	31	−76	550
CH₃CH₂CH₂CH₃	(−1)	(−60)	(287)
1,3-Butanediamine	289–302	125	
NH₂CH₂CH₂CHNH₂CH₃	(143–150)	(52)	

TABLE 2.41 Boiling Points, Flash Points, and Ignition Temperatures of Organic Compounds (*Continued*)

Compound	Boiling point °F (°C)	Flash point, °F (°C)	Ignition point, °F (°C)
1,2-Butanediol	381	104	
$CH_3CH_2CHOHCH_2OH$	(194)	(40)	
(1,2-Dihydroxybutane)			
(Ethylethylene Glycol)			
1,3-Butanediol		See β-Butylene Glycol.	
1,4-Butanediol	442	250	
$HOCH_2CH_2CH_2CH_2OH$	(228)	(121)	
2,3-Butanediol	363	756	
$CH_3CHOHCHOHCH_3$	(184)	(402)	
2,3-Butanedione	190	80	
$CH_3COCOCH_3$	(88)	(27)	
(Diocetyl)			
1-Butanethiol	208	35	
$CH_3CH_2CH_2CH_2SH$	(98)	(2)	
(Butyl Mercaptan)			
2-Butanethiol	185	−10	
C_4H_9SH	(85)	(−23)	
(sec-Butyl Mercaptan)			
1-Butanol		See Butyl Alcohol.	
2-Butanol		See sec-Butyl Alcohol.	
2-Butanone		See Methyl Ethyl Ketone.	
2-Butenal		See Crotonaldehyde.	
1-Butene	21		725
$CH_3CH_2CH:CH_2$	(−6)		(385)
(α-Butylene)			
2-Butene-cis	38.7		617
$CH_3CH:CHCH_3$	(4)		(325)
2-Butene-trans	−34		615
$CH_3CH:CHCH_3$	(1)		(324)
(β-Butylene)			
Butenediol	286–300	263	
$HOCH_2CH:CHCH_2OH$	(141–149)	(128)	
(2-Butene-1,4-Diol)			
	@20 mm		
2-Butene-1,4-Diol		See Butenediol.	
2-Butene Nitrile		See Crotononitrile.	
Butoxybenzene		See Butyl Phenyl Ether.	
1-Butoxybutane		See Dibutyl Ether.	
2,β-Butoxyethoxyethyl Chloride	392–437	190	
$C_4H_9CH_2CH_2OCH_2CH_2Cl$	(200–225)	(88)	
1-(Butoxyethoxy)2Propanol	445	250	509
	(229)	(121)	(265)
$CH_3CH(OH)CH_2OC_2H_4OC_2H_4C_2H_5$			
β-Butoxyethyl Salicylate	367–378	315	
$OCH_6H_4COOCH_2CH_2OC_4$	(186–192)	(157)	
N-Butyl Acetamide	455–464	240	
$CH_3CONHC_4H_9$	(235–240)	(116)	
N-Butylacetanilide	531–538	286	
$CH_3(CH_2)_3N(C_6H_5)COCH_3$	(277–281)	(141)	
Butyl Acetate	260	72	797
$CH_3COOC_4H_9$	(127)	(22)	(425)
(Butylethanoate)			

(*Continued*)

TABLE 2.41 Boiling Points, Flash Points, and Ignition Temperatures of Organic Compounds (*Continued*)

Compound	Boiling point °F (°C)	Flash point, °F (°C)	Ignition point, °F (°C)
sec-Butyl Acetate	234	88	
$CH_3COOCH(CH_3)C_2H_5$	(112)	(31)	
Butyl Acetoacetate	417	185	
$CH_3COCH_2COO(CH_2)_3CH_3$	(214)	(85)	
Butyl Acetyl Ricinoleate	428	230	725
$C_{17}H_{32}(OCOCH_3)-$	(220)	(110)	(385)
$(COOC_4H_9)$			
Butyl Acrylate	260	84	559
$CH_2:CHCOOC_4H_9$	(127)	(29)	(292)
	Polymerizes		
Butyl Alcohol	243	98	650
$CH_3(CH_2)_2CH_2OH$	(117)	(37)	(343)
(1-Butanol)			
(Propylcarbinol)			
(Propyl Methanol)			
sec-Butyl Alcohol	201	75	761
$CH_3CH_2CHOHCH_3$	(94)	(24)	(405)
(2-Butanol)			
(Methyl Ethyl Carbinol)			
tert-Butyl Alcohol	181	52	892
$(CH_3)_2COHCH_3$	(83)	(11)	(478)
(2-Methyl-2-Propanol)			
(Trimethyl Carbinol)			
Butylamine	172	10	594
$C_4H_9NH_2$	(78)	(−12)	(312)
(1-Amino Butane)			
sec-Butylamine	145	16	
$CH_3CH_2CH(NH_2)CH_3$	(63)	(−9)	
tert-Butylamine	113		716
$(CH_3)_3C:NH_2$	(45)		(380)
Butylamine Oleate		150	
$C_{17}H_{33}COONH_3C_4H_9$		(66)	
tert-Butylaminoethyl	200–221	205	
Methacrylate	(93–105)	(96)	
$(CH_3)_3CNHC_2H_4OOCC(CH_3):CH_2$			
N-Butylaniline	465	225	
$C_6H_5NHC_4H_9$	(241)	(107)	
Butylbenzene	356	160	770
$C_6H_5C_4H_9$	(180)	(71)	(410)
sec-Butylbenzene	344	126	784
$C_6H_5CH(CH_3)C_2H_5$	(173)	(52)	(418)
tert-Butylbenzene	336	140	842
$C_6H_5C(CH_3)_3$	(169)	(60)	(450)
Butyl Benzoate	482	225	
$C_6H_5COOC_4H_9$	(250)	(107)	
2-Butylbiphenyl	−554	>212	806
$C_6H_5C_6H_4C_4H_9$	(−290)	(>100)	(430)
Butyl Bromide	215	65	509
$CH_3(CH_2)_2CH_2Br$	(102)	(18)	(265)
(1-Bromo Butane)			
Butyl Butyrate	305	128	
$CH_3(CH_2)_2COOC_4H_9$	(152)	(53)	

TABLE 2.41   Boiling Points, Flash Points, and Ignition Temperatures of Organic Compounds (*Continued*)

Compound	Boiling point °F (°C)	Flash point, °F (°C)	Ignition point, °F (°C)
Butylcarbamic Acid, Ethyl Ester		See N-Butylurethane.	
tert-Butyl Carbinol	237	98	
$(CH_3)_3CCH_2OH$	(114)	(37)	
(2,2-Dimethyl-1-Propanol)			
Butyl Carbitol		See Diethylene Glycol Monobutyl Ether.	
4-tert-Butyl Catechol	545	266	
$(OH)_2C_6H_3C(CH_3)_3$	(285)	(130)	
Butyl Chloride	170	15	464
C_4H_9Cl	(77)	(−9)	(240)
(1-Chlorobutane)			
sec-Butyl Chloride	155	<32	
$CH_3CHClC_2H_5$	(68)	(<0)	
(2-Chlorobutane)			
tert-Butyl Chloride	124	<32	
$(CH_3)_3CCl$	(51)	(<0)	
(2-Chloro-2-Methyl-Propane)			
4-tert-Butyl-2-	453–484	225	
Chlorophenol	(234–251)	(107)	
$ClC_6H_3(OH)C(CH_3)_3$			
tert-Butyl-m-Cresol	451–469	116	
$C_6H_3(C_4H_9)(CH_3)OH$	(233–243)	(47)	
p-tert-Butyl-o-Cresol	278–280	244	
$(OH)C_6H_3CH_3C(CH_3)_3$	(137–138)	(118)	
Butylcyclohexane	352–356		475
$C_4H_9C_6H_{11}$	(178–180)		(246)
(1-Cyclohexylbutane)			
sec-Butylcyclohexane	351		531
$CH_3CH_2CH(CH_3)C_6H_{11}$	(177)		(277)
(2-Cyclohexylbutane)			
tert-Butylcyclohexane	333–336		648
$(CH_3)_3CC_6H_{11}$	(167–169)		(342)
N-Butylcyclohexylamine	409	200	
$C_6H_{11}NH(C_4H_9)$	(209)	(93)	
Butylcyclopentane	314	480	
$C_4H_9C_5H_9$	(157)	(250)	
Butyl Ether		See Dibutyl Ether.	
Butylethylacetaldehyde		See 2-Ethylhexanal.	
Butyl Ethylene		See 1-Hexene.	
Butyl Ethyl Ether		See Ethyl Butyl Ether.	
Butyl Formate	225	64	612
$HCOOC_4H_9$	(107)	(18)	(322)
(Butyl Methanoate)			
(Formic Acid, Butyl Ester)			
Butyl Glycolate	~356	142	
$CH_2OHCOOC_4H_9$	(~180)	(61)	
tert-Butyl Hydroperoxide		<80	
$(CH_3)_3COOH$		(<27)	
n-Butyl Isocyanate	235	66	
$CH_3(CH_2)_3NCO$	(113)	(19)	
(Butyl Isocyanate)			
Butyl Isovalerate	302	127	
$C_4H_9OOCCH_2CH(CH_3)_2$	(150)	(53)	

(*Continued*)

TABLE 2.41   Boiling Points, Flash Points, and Ignition Temperatures of Organic Compounds (*Continued*)

Compound	Boiling point °F (°C)	Flash point, °F (°C)	Ignition point, °F (°C)
Butyl Lactate	320	160	720
CH$_3$CH(OH)COOC$_4$H$_9$	(160)	(71)	(382)
Butyl Mercaptan		See 1-Butanethiol.	
tert-Butyl Mercaptan		See 2-Methyl-2-Propanethiol.	
Butyl Methacrylate	325	126	
CH$_2$:C(CH$_3$)COO(CH$_2$)$_3$CH$_3$	(163)	(52)	
Butyl Methanoate		See Butyl Formate.	
N-Butyl Monoethanolamine	378	170	
C$_4$H$_9$NHC$_2$H$_4$OH	(192)	(77)	
Butyl Naphthalene		680	
C$_4$H$_9$C$_{10}$H$_7$		(360)	
Butyl Nitrate	277	97	
CH$_3$(CH$_2$)$_3$ONO$_2$	(136)	(36)	
2-Butyloctanol	486	230	
C$_6$H$_{13}$CH(C$_4$H$_9$)CH$_2$OH	(252)	(110)	
Butyl Oleate	440.6–442.4	356	
C$_{17}$H$_{33}$COOC$_4$H$_9$		(180)	
	(227–228)		
	@15 mm		
Butyl Oxalate	472	265	
(COOC$_4$H$_9$)$_2$	(244)	(129)	
(Butyl Ethanedioate)		(oc)	
tert-Butyl Peracetate	Explodes on	<80	
diluted with 25% of benzene	heating.	(<27)	
CH$_3$CO(O$_2$)C(CH$_3$)$_3$			
tert-Butyl Perbenzoate	Explodes on	>190	
C$_6$H$_5$COOOC(CH$_3$)$_3$	heating.	(>88)	
tert-Butyl Peroxypivalate	Explodes on	>155	
diluted with 25% of mineral spirits	heating.	(>68)	
(CH$_3$)$_3$COOCOC(CH$_3$)$_3$			
β-(p-tert-Butyl Phenoxy)	293-313	248	
Ethanol	(145-156)	(120)	
(CH$_3$)$_3$CC$_6$H$_4$OCH$_2$CH$_2$OH			
β-(p-tert-Butylphenoxy)	579–585	324	
Ethyl Acetate	(304–307)	(162)	
(CH$_3$)$_3$CC$_6$H$_6$OCH$_2$CH$_2$OCOCH$_3$			
Butyl Phenyl Ether	410	180	
CH$_3$(CH$_2$)$_3$OC$_6$H$_5$	(210)	(82)	
(Butoxybenzene)			
4-tert-Butyl-2-Phenylphenol	385–388	320	
C$_6$H$_5$C$_6$H$_3$OHC(CH$_3$)$_3$	(196–198)	(160)	
Butyl Propionate	295	90	799
C$_2$H$_5$COOC$_4$H$_9$	(146)	(32)	(426)
Butyl Ricinoleate	790	230	
C$_{18}$H$_{33}$O$_3$C$_4$H$_9$	(421)	(110)	
Butyl Sebacate	653	353	
[(CH$_2$)$_4$COOC$_4$H$_9$]$_2$	(345)	(178)	
Butyl Stearate	650	320	671
C$_{17}$H$_{35}$COOC$_4$H$_9$	(343)	(160)	(355)
tert-Butylstyrene	426	177	
	(219)	(81)	

TABLE 2.41 Boiling Points, Flash Points, and Ignition Temperatures of Organic Compounds (*Continued*)

Compound	Boiling point °F (°C)	Flash point, °F (°C)	Ignition point, °F (°C)
tert-Butyl Tetralin		680	
$C_4H_9C_{10}H_{11}$		(360)	
Butyl Trichlorosilane	300	130	
$CH_3(CH_2)_3SiCl_3$	(149)	(54)	
N-Butylurethane	396–397	197	
$CH_3(CH_2)_3NHCOOC_2H_5$	(202–203)	(92)	
(Butylcarbamic Acid, Ethyl Ester)			
(Ethyl Butylcarbamate)			
Butyl Vinyl Ether		See Vinyl Butyl Ether.	
2-Butyne	81	−4	
$CH_3C:CCH_3$ (Crotonylene)	(27)	(<−20)	
Butyraldehyde	169	−8	425
$CH_3(CH_2)_2CHO$	(76)	(−22)	(218)
(Butanal)			
(Butyric Aldehyde)			
Butyraldol	280	165	
$C_8H_{16}O_2$	(138)	(74)	
	@ 50 mm		
Butyraldoxime	306	136	
C_4H_8NOH	(152)	(58)	
(Butanal Oxime)			
Butyric Acid	327	161	830
$CH_3(CH_2)_2COOH$	(164)	(72)	(443)
Butyric Acid, Ethyl Ester		See Ethyl Butyrate.	
Butyric Aldehyde		See Butyraldehyde.	
Butyric Anhydride	388	180	535
$[CH_3(CH_2)_2CO]_2O$	(196)	(54)	(279)
Butyric Ester		See Ethyl Butyrate.	
Butyrolactone	399	209	
$\overline{CH_2CH_2CH_2COO}$	(204)	(98)	
Butyrone		See 4-Heptanone.	
Butyronitrile	243	76	935
$CH_3CH_2CH_2CN$	(117)	(24)	(501)
Caproic Acid	400	215	716
$(CH_3)(CH_2)_4COOH$	(204)	(102)	(380)
(Hexanoic Acid)			
Carbolic Acid		See Phenol.	
Carbon Bisulfide		See Carbon Disulfide.	
Carbon Disulfide	115	−22	194
CS_2	(46)	(−30)	(90)
(Carbon Bisulfide)			
Cetane		See Hexadecane.	
Chloroacetic Acid	372	259	>932
$CH_2ClCOOH$	(189)	(126)	(>500)
Chloroacetophenone	477	244	
$C_6H_5COCH_2Cl$	(247)	(118)	
(Phenacyl Chloride)			
2-Chloro-4,6-di-tert-Amylphenol	320–354	250	
	(160–179)	(121)	
$(C_5H_{11})_2C_6H_2ClOH$	@ 22 mm		
Chloro-4-tert-Amylphenol	487–509	225	
$C_5H_{11}C_6H_3ClOH$	(253–265)	(107)	

(*Continued*)

TABLE 2.41 Boiling Points, Flash Points, and Ignition Temperatures of Organic Compounds (*Continued*)

Compound	Boiling point °F (°C)	Flash point, °F (°C)	Ignition point, °F (°C)
2-Chloro-4-tert-Amyl-Phenyl Methyl Ether $C_5H_{11}C_6H_3ClOCH_3$	518–529 (270–276)	230 (110)	
p-Chlorobenzaldehyde ClC_6H_4CHO	417 (214)	190 (88)	
Chlorobenzene C_6H_5Cl (Chlorobenzol) (Monochlorobenzene) (Phenyl Chloride)	270 (132)	82 (28)	1099 (593)
Chlorobenzol	See Chlorobenzene.		
o-Chlorobenzotrifluoride $ClC_6H_4CF_3$ (o-Chloro-α,α,α-trifluorotoluene)	306 (152)	138 (59)	
Chlorobutadiene	See 2-Chloro-1,3-Butadiene.		
2-Chloro-1,3-Butadiene $CH_2:CCl:CH:CH_2$ (Chlorobutadiene) (Chloroprene)	138 (59)	–4 (–20)	
1-Chlorobutane	See Butyl Chloride.		
2-Chlorobutene-2 $CH_3CCl:CHCH_3$	143–159 (62–71)	–3 (–19)	
Chlorodinitrobenzene	See Dinitrochlorobenzene.		
Chloroethane	See Ethyl Chloride.		
2-Chloroethanol CH_2ClCH_2OH (2-Chloroethyl Alcohol) (Ethylene Chlorohydrin)	264–266 (129–130)	140 (60)	797 (425)
2-Chloroethyl Acetate $CH_3COOCH_2CH_2Cl$	291 (144)	151 (66)	
2-Chloroethyl Alcohol	See 2-Chloroethanol.		
Chloro-4-Ethylbenzene $C_2H_5C_6H_4Cl$	364 (184)	147 (64)	
Chloroethylene	See Vinyl Chloride.		
2-Chloroethyl Vinyl Ether	See Vinyl 2-Chloroethyl Ether.		
2-Chloroethyl-2-Xenyl Ether $C_6H_5C_6H_4OCH_2CH_2Cl$	613 (323)	320 (160)	
1-Chlorohexane $CH_3(CH_2)_4CH_2Cl$ (Hexyl Chloride)	270 (132)	95 (35)	
Chloroisopropyl Alcohol	See 1-Chloro-2-Propanol.		
Chloromethane	See Methyl Chloride.		
1-Chloro-2-Methyl Propane	See Isobutyl Chloride.		
1-Chloronaphthalene $C_{10}H_7Cl$	505 (263)	250 (121)	>1036 (>558)
2-Chloro-5-Nitrobenzotrifluoride $C_6H_3CF_3(2\text{-}Cl, 5\text{-}NO_2)$ (2-Chloro-α,α,α-Trifluoro-5-Nitrotoluene)	446 (230)	275 (135)	
1-Chloro-1-Nitroethane $C_2H_4NO_2Cl$	344 (173)	133 (56)	

TABLE 2.41 Boiling Points, Flash Points, and Ignition Temperatures of Organic Compounds (*Continued*)

Compound	Boiling point °F (°C)	Flash point, °F (°C)	Ignition point, °F (°C)
1-Chloro-1-Nitropropane	285	144	
$CHNO_2ClC_2H_5$	(141)	(62)	
2-Chloro-2-Nitropropane	273	135	
$CH_3CNO_2ClCH_3$	(134)	(57)	
1-Chloropentane		See Amyl Chloride.	
β-Chlorophenetole	306–311	225	
$C_6H_5OCH_2CH_2Cl$	(152–155)	(107)	
(β-Phenoxyethyl Chloride)			
o-Chlorophenol	347	147	
ClC_6H_4OH	(175)	(64)	
p-Chlorophenol	428	250	
C_6H_4OHCl	(220)	(121)	
2-Chloro-4-Phenylphenol	613	345	
$C_6H_5C_6H_3ClOH$	(323)	(174)	
Chloroprene		See 2-Chloro-1,3-Butadiene.	
1-Chloropropane		See Propyl Chloride.	
2-Chloropropane		See Isopropyl Chloride.	
2-Chloro-1-Propanol	271–273	125	
$CH_3CHClCH_2OH$	(133–134)	(52)	
(β-Chloropropyl Alcohol)			
(Propylene Chlorohydrin			
1-Chloro-2-Propanol	261	125	
$CH_2ClCHOHCH_3$	(127)	(52)	
(Chloroisopropyl Alcohol)			
(sec-Propylene Chlorohydrin)			
1-Chloro-1-Propene		See 1-Chloropropylene.	
3-Chloropropene		See Allyl Chloride.	
α-Chloropropionic Acid	352–374	225	932
$CH_3CHClCOOH$	(178–190)	(107)	(500)
3-Chloropropionitrile	348.8	168	
$ClCH_2CH_2CN$	(176)	(76)	
	Decomposes		
2-Chloropropionyl Chloride	230	88	
	(110)	(31)	
β-Chloropropyl Alcohol		See 2-Chloro-1-Propanol.	
1-Chloropropylene	95–97	<21	
$CH_3CH:CHCl$	(35–36)	(<–6)	
(1-Chloro-1-Propene)			
2-Chloropropylene	73	<–4	
$CH_3CCl:CH_2$	(23)	(<–20)	
(β-Chloropropylene)			
(2-Chloropropene)			
2-Chloropropylene Oxide		See Epichlorohydrin.	
γ-Chloropropylene Oxide		See Epichlorohydrin.	
Chlorotoluene	320	126	
$C_6H_4ClCH_3$	(160)	(52)	
(Tolyl Chloride)			
α-Chlorotoluene		See Benzyl Chloride.	
Chlorotrifluoroethylene		See Trifluorochloroethylene.	
2-Chloro-α,α,α-Trifluoro-5-Nitrotoluene		See 2-Chloro-5-Nitrobenzotrifluoride.	

(*Continued*)

TABLE 2.41   Boiling Points, Flash Points, and Ignition Temperatures of Organic Compounds (*Continued*)

Compound	Boiling point °F (°C)	Flash point, °F (°C)	Ignition point, °F (°C)
o-Chloro-α,α,α-Trifluorotoluene		See o-Chlorobenzotrifluoride.	
Coal Oil		See Fuel Oil No. 1.	
Coal Tar Light Oil		<80 (<27)	
Coal Tar Pitch		405 (207)	
Creosote Oil	382–752 (194–400)	165 (74)	637 (336)
o-Cresol $CH_3C_6H_4OH$ (Cresylic Acid) (o-Hydroxytoluene) (o-Methyl Phenol)	376 (191)	178 (81)	1110 (599)
p-Cresyl Acetate $CH_3C_6H_4OCOCH_3$ (P-Tolyl Acetate)		195 (91)	
Cresyl Diphenyl Phosphate $(C_6H_5O)_2[(CH_3)_2C_6H_4O]\text{-}PO_4$	734 450 (390) (232)		
Cresylic Acid		See o-Cresol.	
Crotonaldehyde $CH_3CH{:}CHCHO$ (2-Butenal) (Crotonic Aldehyde) (Propylene Aldehyde)	216 55 (102) (13)	450 (232)	
Crotonic Acid $CH_3CH{:}CHCOOH$	372 (189)	190 (88)	745 (396)
Crotononitrile $CH_3CH{:}CHCN$ (2-Butenenitrile)	230–240.8 (110–116)	<212 (<100)	
Crotonyl Alcohol $CH_3CH{:}CHCH_2OH$ (2-Buten-1-ol) (Crotyl Alcohol)	250 (121)	81 (27)	660 (349)
1-Crotyl Bromide $CH_3CH{:}CHCH_2Br$ (1-Bromo-2-Butene)			
1-Crotyl Chloride $CH_3CH{:}CHCH_2Cl$ (1-Chloro-2-Butene)			
Cumene $C_6H_5CH(CH_3)_2$ (Cumol) (2-Phenyl Propane) (Isopropyl Benzene)	306 (152)	96 (36)	795 (424)
Cumene Hydroperoxide $C_6H_5C(CH_3)_2OOH$	Explodes on heating.	175 (79)	
Cyanamide NH_2CN	500 (260) Decomposes	286 (141)	
2-Cyanoethyl Acrylate $CH_2CHCOOCH_2CH_2CN$	Polymerizes	255 (124)	
N-(2-Cyanoethyl) Cyclohexylamine $C_6H_{11}NHC_2H_4CN$		255 (124)	

TABLE 2.41 Boiling Points, Flash Points, and Ignition Temperatures of Organic Compounds (*Continued*)

Compound	Boiling point °F (°C)	Flash point, °F (°C)	Ignition point, °F (°C)
Cyclamen Aldehyde $(CH_3)_2CHC_6H_4CH(CH_3)CH_2CHO$ (Methyl Para-Isopropyl Phenyl Propyl Aldehyde)		190 (88)	
Cyclobutane C_4H_8 (Tetramethylene)	55 (13)		
1,5,9-Cyclododecatriene $C_{12}H_{18}$	448 (231)	160 (71)	
Cycloheptane $CH_2(CH_2)_5CH_2$	246 (119)	<70 (<21)	
Cyclohexane C_6H_{12} (Hexahydrobenzene) (Hexamethylene)	179 (82)	−4 (−20)	473 (245)
1,4-Cyclohexane Dimethanol $C_8H_{16}O_2$	525 (274)	332 (167)	600 (316)
Cyclohexanethiol $C_6H_{11}SH$ (Cyclohexylmercaptan)	315–319 (157–159)	110 (43)	
Cyclohexanol $C_6H_{11}OH$ (Anol) (Hexolin) (Hydralin)	322 (161)	154 (68)	572 (300)
Cyclohexanone $C_6H_{10}O$ (Pimelic Ketone)	313 (156)	111 (44)	788 (420)
Cyclohexene $CH_2CH_2CH_2CH_2CH \cdot CH$	181 (83)	<20 (<−7)	471 (244)
3-Cyclohexene-1- Carboxaldehyde		See 1,2,3,6- Tetrahydrobenzaldehyde.	
Cyclohexenone C_6H_8O	313 (156)	93 (34)	
Cyclohexyl Acetate $CH_3CO_2C_6H_{11}$ (Hexolin Acetate)	350 (177)	136 (58)	635 (335)
Cyclohexylamine $C_6H_{11}NH_2$ (Aminocyclohexane) (Hexahydroaniline)	274 (134)	88 (31)	560 (293)
Cyclohexylbenzene $C_6H_5C_6H_{11}$ (Phenylcyclohexone)	459 (237)	210 (99)	
Cyclohexyl Chloride $CH_2(CH_2)_4CHCl$ (Chlorocyclohexane)	288 (142)	90 (32)	
Cyclohexylcyclohexanol $C_6H_{11}C_6H_{10}OH$	304–313 (151–156)	270 (132)	
Cyclohexyl Formate $CH_2(CH_2)_4HCOOCH$	324 (162)	124 (51)	

TABLE 2.41   Boiling Points, Flash Points, and Ignition Temperatures of Organic Compounds (*Continued*)

Compound	Boiling point °F (°C)	Flash point, °F (°C)	Ignition point, °F (°C)
Cyclohexylmethane	See Methylcyclohexane.		
o-Cyclohexylphenol	298	273	
$C_6H_{11}C_6H_4OH$	(148)	(134)	
	@10 mm		
Cyclohexyltrichlorosilane	406	196	
$C_6H_{11}SiCl_3$	(208)	(91)	
1,5-Cyclooctadiene	304	95	
C_8H_{10}	(151)	(35)	
Cyclopentane	121	<20	682
C_5H_{10}	(49)	(<–7)	(361)
Cyclopentene	111	–20	743
CH:CHCH$_2$CH$_2$CH$_2$	(44)	(–29)	(395)
Cyclopentanol	286	124	
CH$_2$(CH$_2$)$_3$CHOH	(141)	(51)	
Cyclopentanone	267	79	
OCCH$_2$CH$_2$CH$_2$CH$_2$	(131)	(26)	
(Adipic Ketone)			
Cyclopropane	–29		928
(CH$_2$)$_3$	(–34)		(498)
(Trimethylene)			
p-Cymene	349	117	817
CH$_3$C$_6$H$_4$CH(CH$_3$)$_2$ Tech.	(176)	(47)	(436)
(4-Isopropyl-1-Methyl		127	833
Benzene)		(53)	(445)
Decahydronaphthalene	382	136	482
$C_{10}H_{18}$	(194)	(58)	(250)
(Decalin)			
Decahydronaphthalene-trans	369	129	491
$C_{10}H_{18}$	(187)	(54)	(255)
Decalin	See Decahydronaphthalene.		
Decane	345	115	410
CH$_3$(CH$_2$)$_8$CH$_3$	(174)	(46)	(210)
Decanol	444.2	180	550
CH$_3$(CH$_2$)$_8$CH$_2$OH	(229)	(82)	(288)
(Decyl Alcohol)			
1-Decene	342	<131	455
CH$_3$(CH$_2$)$_7$CH:CH$_2$	(172)	(<55)	(235)
Decyl Acrylate	316	441	
CH$_3$(CN$_2$)$_9$OCOCH:CH$_2$	(158)	(227)	
	@50 mm		
Decyl Alcohol	See Decanol.		
Decylamine	429	210	
CH$_3$(CH$_2$)$_9$NH$_2$	(221)	(99)	
(1-Aminodecane)			
Decylbenzene	491–536	225	
$C_{10}H_{21}C_6H_5$	(255–280)	(107)	
tert-Decylmercaptan	410–424	190	
$C_{10}H_{21}SH$	(210–218)	(88)	
Decylnaphthalene	635–680	350	
$C_{10}H_{21}C_{10}H_7$	(335–360)	(177)	
Decyl Nitrate	261	235	
CH$_3$(CH$_2$)$_9$ONO$_2$	(127)	(113)	
	@11 mm		

TABLE 2.41 Boiling Points, Flash Points, and Ignition Temperatures of Organic Compounds (*Continued*)

Compound	Boiling point °F (°C)	Flash point, °F (°C)	Ignition point, °F (°C)
Diacetone Alcohol	328	148	1118
$CH_3COCH_2C(CH_3)_2OH$	(164)		
Diacetyl		See 2,3-Butanedione.	
Diallyl Ether		See Allyl Ether.	
Diallyl Phthalate	554	330	
$C_6H_4(CO_2C_3H_5)_2$	(290)	(166)	
1,3-Diaminobutane		See 1,3-Butanediamine.	
1,3-Diamino-2-Propanol	266	270	
$NH_2CH_2CHOHCH_2NH_2$	(130)	(132)	
1,3-Diaminopropane		See 1,3-Propanediamine.	
Diamylamine	356	124	
$(C_5H_{11})_2NII$	(180)	(51)	
Diamylbenzene	491–536	225	
$(C_5H_{11})_2C_6H_4$	(255–280)	(107)	
Diamylbiphenyl	687–759	340	
$C_5H_{11}(C_6H_4)_2C_5H_{11}$	(364–404)	(171)	
(Diaminodiphenyl)			
Di-tert-Amylcyclohexanol	554–572	270	
$(C_5H_{11})_2C_6H_9OH$	(290–300)	(132)	
Diamyidlphenyl		See Diamylbiphenyl.	
Diamylene	302	118	
$C_{10}H_{20}$	(150)	(48)	
Diamyl Ether		See Amyl Ether.	
Diamyl Maleate	505–572	270	
$(CHCOOC_5H_{11})_2$	(263–300)	(132)	
Diamyl Naphthalene	624	315	
$C_{10}H_6(C_5H_{11})_2$	(329)	(159)	
2,4-Diamylphenol	527	260	
$(C_5H_{11})_2C_6H_3OH$	(275)	(127)	
Di-tert-Amylphenoxy Ethanol	615	300	
$C_6H_3(C_5H_{11})_2OC_2H_4OH$	(324)	(149)	
Diamyl Phthalate	475–490	245	
$C_6H_4(COOC_5H_{11})_2$	(246–254)	(118)	
(Amyl Phthalate)	@ 50 mm		
Diamyl Sulfide	338–356	185	
$(C_5H_{11})_2S$	(170–180)	(85)	
o-Dianisldine		403	
$[NH_2(OCH_3)C_6H_3]_2$		(206)	
(o-Dimethoxybenzidine			
Dibenzyl Ether	568	275	
$(C_6H_5CH_2)_2O$	(298)	(135)	
(Benzyl Ether)			
Dibutoxy Ethyl Phthalate	437	407	
$C_6H_4(COOC_2H_4OC_4H_9)_2$	(225)	(208)	
		(oc)	
Dibutoxymethane	330–370	140	
$CH_2(OC_4H_9)_2$	(166–188)	(60)	
Dibutoxy Tetraglycol	635	305	
$(C_4H_9OC_2H_4OC_2H_4)_2O$	(335)	(152)	
(Tetraethylene Glycol Dibutyl			
Ether)			
N,N-Dibutylacetamide	469–482	225	
$CH_3CON(C_4H_9)_2$	(243–250)	(107)	

(Continued)

TABLE 2.41 Boiling Points, Flash Points, and Ignition Temperatures of Organic Compounds (*Continued*)

Compound	Boiling point °F (°C)	Flash point, °F (°C)	Ignition point, °F (°C)
Dibutylamine	322	117	
$(C_4H_9)_2NH$	(161)	(47)	
Di-sec-Butylamine	270–275	75	
$[C_2H_5(CH_3)CH]_2NH$	(132–135)	(24)	
Dibutylaminoethanol	432	200	
$(C_4H_9)_2NC_2H_4OH$	(222)	(93)	
1-Dibutylamino-2-Propanol		See Dibutylisopropanolamine.	
N,N-Dibutylanlline	505–527	230	
$C_6H_5N(CH_2CH_2CH_2CH_3)_2$	(263–275)	(110)	
Di-tert-Butyl-p-Cresol	495–511	261	
$C_6H_2(C_4H_9)_2(CH_3)OH$	(257–266)	(127)	
Dibutyl Ether	286	77	382
$(C_4H_9)_2O$	(141)	(25)	(194)
(1-Butoxybutane)			
(Butyl Ether)			
2,5-Di-tert-Butylhydroquinone		420	790
$[C(CH_3)_3]_2C_6H_2(OH)_2$		(216)	(421)
(DTBHQ)			
Dibutyl Isophthalate		322	
$C_6H_4(CO_2C_4H_9)_2$		(161)	
N,N¹-Di-sec-Butyl-p-		270	625
Phenylenediamine		(132)	(329)
$C_6H_4[-NHCH(CH_3)-$			
$CH_2CH_3]_2$			
Dibutylisopropanolamine	444	205	
$CH_3CHOHCH_2N(C_4H_9)_2$	(229)	(96)	
Dibutyl Maleate	Decomposes	285	
$(-CHCO_2C_4H_9)_2$		(141)	
Dibutyl Oxalate	472	220	
$C_4H_9OOCCOOC_4H_9$	(244)	(104)	
Di-tert-Butyl Peroxide	231	65	
$(CH_3)_3COOC(CH_3)_3$	(111)	(18)	
Dibutyl Phthalate	644	315	757
$C_6H_4(CO_2C_4H_9)_2$	(340)	(157)	(402)
(Dibutyl-o-Phthatate)			
n-Dibutyl Tartrate	650	195	544
$(COOC_4H_9)_2(CHOH)_2$	(343)	(91)	(284)
(Dibutyl-d-2,3-			
Dihydroxybutanedioate)			
N,N-Dibutyltoluene-	392	330	
sulfonamide	(200)	(166)	
$CH_3C_6H_4SO_3N(C_4H_9)_2$	@10 mm		
Dicaproate		See Triethylene Glycol.	
Dicapryl Phthalate	441–453	395	
$C_6H_4[COOCH(CH_3)C_6H_{13}]_2$	(227–234)	(202)	
	@4.5 mm		
Dichloroacetyl Chloride	225–226	151	
$CHCl_2COCl$	(107–108)	(66)	
(Dichloroethanoyl Chloride)			
3,4-Dichloroaniline	522	331	
$NH_2C_6H_3Cl_2$	(272)	(166)	
o-Dichlorobenzene	356	151	1198
$C_6H_4Cl_2$	(180)	(66)	(648)
(o-Dichlorobenzol)			

TABLE 2.41 Boiling Points, Flash Points, and Ignition Temperatures of Organic Compounds (*Continued*)

Compound	Boiling point °F (°C)	Flash point, °F (°C)	Ignition point, °F (°C)
p-Dichlorobenzene	345	150	
$C_6H_4Cl_2$	(174)	(66)	
2,3-Dichlorobutadiene-1,3	212	50	694
$CH_2:C(Cl)C(Cl):CH_2$	(100)	(10)	(368)
1,2-Dichlorobutane		527	
$CH_3CH_2CHClCH_2Cl$		(275)	
1,4-Dichlorobutane	311	126	
$CH_2ClCH_2CH_2CH_2Cl$	(155)	(52)	
2,3-Dichlorobutane	241–253	194	
$CH_3CHClCHClCH_3$	(116–123)	(90)	
1,3-Dichloro-2-Butene	262	80	
$CH_2ClCH:CClCH_3$	(128)	(27)	
3,4-Dichlorobutene-1	316	113	
$CH_2ClCHClCHCH_2$	(158)	(45)	
1,3-Dichlorobutene-2	258	80	
$CH_2ClCH:CClCH_3$	(126)	(27)	
Dichlorodimethylsilane		See Dimethyldichlorosilane.	
1,1-Dichloroethane		See Ethylidene Dichloride.	
1,2-Dichloroethane		See Ethylene Dichloride.	
Dichloroethanoyl Chloride		See Dichloroacetyl Chloride.	
1,1-Dichloroethylene		See Vinylidene Chloride.	
Dichloroisopropyl Ether	369	185	
$ClCH_2CH(CH_3)OCH(CH_3)CH_2Cl$	(187)	(85)	
[Bis (β-Chloroisopropyl) Ether]			
2,2-Dichloro Isopropyl Ether	369	185	
$[ClCH_2CH(CH_3)]_2O$	(187)	(85)	
[Bis(2-Chloro-1-Methylethyl) Ether]			
Dichloromethane		See Methylene Chloride.	
1,1-Dichloro-1-Nitro Ethane	255	168	
$CH_3CCl_2NO_2$	(124)	(76)	
1,1-Dichloro-1-Nitro Propane	289	151	
$C_2H_5CCl_2NO_2$	(143)	(66)	
1,5-Dichloropentane	352–358	>80	
$CH_2Cl(CH_2)_3CH_2Cl$	(178–181)	(>27)	
(Amylene Chloride)			
(Pentamethylene Dichloride)			
2,4-Dichlorophenol	410	237	
$Cl_2C_6H_3OH$	(210)	(114)	
1,2-Dichloropropane		See Propylene Dichloride.	
1,3-Dichloro-2-Propanol	346	165	
$CH_2ClCHOHCH_2Cl$	(174)	(74)	
1,3-Dichloropropene	219	95	
$CHCl:CHCH_2Cl$	(104)	(35)	
2,3-Dichloropropene	201	59	
CH_2CClCH_2Cl	(94)	(15)	
α, β-**Dichlorostyrene**		225	
$C_6H_5CCl:CHCl$		(107)	
Dicyclohexyl		See Bicyclohexyl.	
Dicyclohexylamine	496	>210	
$(C_6H_{11})_2NH$	(258)	(>99)	

(*Continued*)

TABLE 2.41   Boiling Points, Flash Points, and Ignition Temperatures of Organic Compounds (*Continued*)

Compound	Boiling point °F (°C)	Flash point, °F (°C)	Ignition point, °F (°C)
Dicyclopentadiene	342	90	937
$C_{10}H_{12}$	(172)	(32)	(503)
Didecyl Ether		419	
$(C_{10}H_{21})_2O$		(215)	
(Decyl Ether)			
Diesel Fuel Oil		100	
No. 1-D		Min.	
		(38)	
Diesel Fuel Oil		125	
No. 2-D		Min.	
		(52)	
Diesel Fuel Oil		130	
No. 4-D		Min.	
		(54)	
Diethanolomine	514	342	1224
$(HOCH_2CH_2)_2NH$	(268)	(172)	(662)
1,2-Diethoxyethane		See Diethyl Glycol.	
Diethylacetaldehyde		See 2-Ethylbutyraldehyde.	
Diethylacetic Acid		See 2-Ethylbutyric Acid.	
N,N-Diethyl-acetoacetamide	Decomposes	250	
$CH_3COCH_2CON(C_2H_5)_2$		(121)	
Diethyl Acetoacetate	412–424	170	
$CH_3COC(C_2H_5)_2COOC_2H_5$	(211–218)	(77)	
	Decomposes		
Diethylamine	134	−9	594
$(C_2H_5)_2NH$	(57)	(−23)	(312)
2-Diethyl (Amino) Ethanol		See N,N-Diethylethanolamine.	
2-(Diethylamino) Ethyl	Decomposes	195	
Acrylate		(91)	
$CH_2{:}CHCOOCH_2CH_2-$			
$HN(CH_3CH_2)_2$			
3-(Diethylamino)-Propylamine	337	138	
$(C_2H_5)_2NCH_2CH_2CH_2NH_2$	(169)	(59)	
(N,N-Diethyl-1,3-Propanediamine)			
N,N-Diethylaniline	421	185	1166
$C_6H_5N(C_2H_5)_2$	(216)	(85)	(630)
(Phenyldiethylamine)			
o-Diethyl Benzene	362	135	743
$C_6H_4(C_2H_5)_2$	(183)	(57)	(395)
m-Diethyl Benzene	358	133	842
$C_6H_4(C_2H_5)_2$	(181)	(56)	(450)
p-Diethyl Benzene	358	132	806
$C_6H_4(C_2H_5)_2$	(181)	(55)	(430)
N,N-Diethyl-1,3-Butanediamine	354–365	115	
$C_2H_5NHCH_2CH_2CHN(C_2H_5)CH_3$	(179–185)	(46)	
[1,3-Bis(ethylamino) Buiane]			
D1-2-Ethylbutyl Phthalate	662	381	
$C_6H_4[COOCH_2CH(C_2H_5)_2]_2$	350	(194)	
Diethyl Carbamyl Chloride	369–374	325–342	
$(C_2H_5)_2NCOCl$	(187–190)	(163–172)	
Diethyl Carbinol		See sec-Amyl Alcohol.	
Diethyl Carbonate	259	77	
$(C_2H_5)_2CO_3$	(126)	(25)	
(Ethyl Carbonate)			

TABLE 2.41 Boiling Points, Flash Points, and Ignition Temperatures of Organic Compounds (*Continued*)

Compound	Boiling point °F (°C)	Flash point, °F (°C)	Ignition point, °F (°C)
Diethylcyclohexane	344	120	464
$C_{10}H_{20}$	(173)	(49)	(240)
1,3-Diethyl-1,3-Diphenyl Urea	620	302	
$[(C_2H_5)(C_6H_5)N]_2CO$	(327)	(150)	
Diethylene Diamine	299	144	
	(150)	(62)	
Diethylene Dioxide		See p-Dioxane.	
Diethylene Glycol	472	255	435
$O(CH_2CH_2OH)_2$	(244)	(124)	(224)
(2,2-Dihydroxyethyl Ether)			
Diethylene Glycol Methyl Ether	379	205	465
$CH_3OC_2H_4OC_2H_4OH$	(193)	(96)	(240)
(2-(2-Methoxyethoxy) Ethanol)			
Diethylene Glycol Methyl Ether Acetate	410	180	
$CH_3COOC_2H_4OC_2H_4OCH_3$	(210)	(82)	
Diethylene Glycol Monobutyl Ether	448	172	400
$C_4H_9OCH_2CH_2OCH_2CH_2OH$	(231)	(78)	(204)
Diethylene Glycol Monoethyl Ether Acetate	476	240	570
$C_4H_9O(CH_2)_2O(CH_2)_2OOCCH_3$	(247)	(116)	(298.9)
Diethylene Glycol Monoethyl Ether	396	201	400
$CH_2OHCH_2OCH_2\text{-}CH_2OC_2H_5$	(202)	(94)	(204)
Diethylene Glycol Monoethyl Ether Acetate	424	225	680
$C_2H_5O(CH_2)_2O(CH_2)_2OOCCH_3$	(218)	(107)	(360)
Diethylene Glycol Monoisobutyl Ether	422–437	222	452–485
$(CH_3)_2CHCH_2O(CH_2)_2O(CH_2)_2OH$	(217–225)	(106)	(233–252)
Diethylene Glycol Monomethyl Ether	381	205	
$CH_3O(CH_2)O(CH_2)_2OH$	(194)	(96)	
Diethylene Glycol Mono-Methyl Ether Formal	581	310	
$CH_2(CH_3OCH_2CH_2OCH_2CH_2O)_2$	(305)	(154)	
Diethylene Glycol Phthalate		343	
$C_6H_4[COO(CH_2)_2OC_2H_5]_2$		(173)	
Diethylene Oxide		See Tetrahydrofuran.	
Diethylene Triamine	404	208	676
$NH_2CH_2CH_2NHCH_2CH_2NH_2$	(207)	(98)	(358)
N,N-Diethylethanolamine	324	140	608
$(C_2H_5)_2NC_2H_4OH$	(162)	(60)	(320)
(2-(Diethylamino) Ethanol)			
Diethyl Ether		See Ethyl Ether.	
N,N-Diethylethylene-diamine	293	115	
$(C_2H_5)_2NC_2H_4NH_2$	(145)	(46)	
Diethyl Fumarate	442	220	
$C_2H_5OCOCH:CHCOOC_2H_5$	(217)	(104)	
Diethyl Glycol	252	95	401
$(C_2H_5OCH_2)_2$	(122)	(35)	(205)
(1,2-Diethoxyethane)			

(*Continued*)

TABLE 2.41   Boiling Points, Flash Points, and Ignition Temperatures of Organic Compounds (*Continued*)

Compound	Boiling point °F (°C)	Flash point, °F (°C)	Ignition point, °F (°C)
Diethyl Ketone $C_2H_5COC_2H_5$ (3-Pentanone)	217 (103)	55 (13)	842 (450)
N,N-Diethyllauramide $C_{11}H_{23}CON(C_2H_5)_2$	331–351 (166–177) @2 mm	>150 (>66)	
Diethyl Maleate $(-CHCO_2C_2H_3)_2$	438 (226)	250 (121)	662 (350)
Diethyl Malonate $CH_2(COOC_2H_3)_2$ (Ethyl Malonate)	390 (199)	200 (93)	
Diethyl Oxide		See Ethyl Ether.	
3,3-Diethylpentane $CH_3CH_2C(C_2H_5)_2CH_2CH_3$	295 (146)	554 (290)	
Diethyl Phthalate $C_6H_4(COOC_2H_5)_2$	565 (296)	322 (161)	855 (457)
p-Diethyl Phthalate		See Diethyl Terephthalate.	
N,N-Diethylstearamide $C_{17}H_{35}CON(C_2H_5)_2$	246–401 (119–205) @1 mm	375 (191)	
Diethyl Succinate $(CH_2COOCH_2CH_3)_2$	421 (216)	195 (90)	
Diethyl Sulfate $(C_2H_5)_2SO_4$ (Ethyl Sulfate)	Decomposes, giving Ethyl Ether	220 (104)	817 (436)
Diethyl Tartrate $CHOHCOO(C_2H_5)_2$	536 (280)	200 (93)	
Diethyl Terephthalate $C_6H_4(COOC_2H_5)_2$ (p-Diethyl Phthalate)	576 (302)	243 (117)	
3,9-Diethyl-6-tridecanol		See Heptadecanol.	
Diglycol Chlortormate $O:(CH_2CH_2OCOCl)_2$	256–261 (124–127) @5 mm	295 (146)	
Diglycol Chlorohydrin $HOCH_2CH_2OCH_2CH_2Cl$	387 (197)	225 (107)	
Diglycol Diacetate $(CH_3COOCH_2CH_2)_2O$	482 (250)	255 (124)	
Diglycol Dilevulleate $(CH_2CH_2OOC-$ $(CH_2)_2COCH_3)_2:O$		340 (171)	
Diglycol Laurate $C_{16}H_{32}O_4$	559–617 (293–325)	290 (143)	
Dihexyl		See Dodecane.	
Dihexylamine $[CH_3(CH_2)_5]_2NH$	451–469 (233–243)	220 (104)	
Dihexyl Ether		See Hexyl Ether.	
Dihydropyran $CH_2CH_2CH_2:CHCHO$	186 (86)	0 (−18)	
o-Dihydroxybenione $C_6H_4(OH)_2$ (Pyrocalechol)	473 (245)	260 (127)	

TABLE 2.41   Boiling Points, Flash Points, and Ignition Temperatures of Organic Compounds (*Continued*)

Compound	Boiling point °F (°C)	Flash point, °F (°C)	Ignition point, °F (°C)
p-Dihydroxybenione	547	329	959
$C_6H_4(OH)_2$	(286)	(165)	(515)
(Hydroquinone)			
1,2-Dihydroxybenione		See 1,2-Butanediol.	
2,2-Dihydroxyethyl Ether		See Diethylene Glycol.	
2,5-Dihydroxyhexane		See 2,5-Hexanediol.	
Diisobutylamine	273–286	85	
$[(CH_3)_2CHCH_2]_2NH$	(134–141)	(29)	
[Bis(β-Methylpropyl) Amine]			
Diisobutyl Carbinol	353	165	
$[(CH_3)_2CHCH_2]_2CHOH$	(178)	(74)	
(Nonyl Alcohol)			
Diisobutylene		See 2,4,4-Trimethyl-1-Pentene.	
Diisobutylene	214	23	736
$(CH_3)_3CCH_2C(CH_3):CH_2$	(101)	(−5)	(391)
(2,4,4-Trimethyl-H_2-Pentane)			
Diisobutyl Ketone	335	120	745
$[(CH_3)_2CHCH_2]_2CO$	(168)	(49)	(396)
(2,6-Dimethyl-4 Heptanone)			
(Isovalerone)			
Diisobutyl Phthalate	321	365	810
$C_6H_4[COOCH_2OH(CH_3)_2]_2$	(327)	(185)	(432)
Diisodecyl Adipoia	660	225	
$C_{10}H_{21}O_2C(CH_2)_2CO_2\text{-}C_{10}H_{21}$	(349)	(107)	
Diisodecyl Phthalate	182	450	755
$C_6H_4(COOC_{10}H_{21})_2$	(250)	(232)	(402)
Diisooctyl Phthalate	398	450	
$(C_8H_{17}COO)_2C_2H_4$	(370)	(232)	
Diisopropanolamine	480	260	705
$[CH_3CH(OH)\text{-}CH_2]_2NH$	(249)	(127)	(374)
Diisopropyl		See 2,3-Dimethylbutane.	
Diisopropylamine	183	30	600
$[(CH_3)_2CH]_2NH$	(84)	(−1)	(316)
Diisopropyl Benzene	401	170	840
$[(CH_3)_2CH]_2C_6H_4$	(205)	(77)	(449)
N,N-Diisopropyl-ethanolamine	376	175	
$[(CH_3)_2CH]_2NC_2H_4OH$	(191)	(79)	
Diisopropyl Ether		See Isopropyl Ether.	
Diisopropyl Maleate	444	220	
$(CH_3)_2CHOCOCH:$	(229)	(104)	
$CHCOOCH(CH_3)_2$			
Diisopropylmethanol		See 2,4-Dimethyl-3-Pentanol.	
Diisopropyl Peroxydicarbonate	Explodes		
$(CH_3)_2CHOCOOCOOCH(CH_3)_2$	on heating.		
Diketene	261	93	
$CH_2:CCH_2C(O)O$	(127)	(34)	
(Vinylaceto-β-Lactone)			
2,5-Dimethoxyaniline	518	302	735
$NH_2C_6H_3(OCH_3)_2$	(270)	(150)	(391)
2,5-Dimethoxy Chlorobenzene	460–467	243	
$C_8H_9ClO_2$	(238–242)	(117)	
1,2-Dimethoxyethane		See Ethylene Glycol Dimethyl Ether.	

(*Continued*)

TABLE 2.41   Boiling Points, Flash Points, and Ignition Temperatures of Organic Compounds (*Continued*)

Compound	Boiling point °F (°C)	Flash point, °F (°C)	Ignition point, °F (°C)
Dimethoxyethyl Phthalate	644	410	750
$C_6H_4(COOCH_2CH_2OCH_3)_2$	(340)	(210)	(399)
(Bis(2-methoxyethyl) Phthalate)			
Dimethoxymethane		See Methylal.	
Dimethoxy Tetraglycol	528	285	
$CH_3OCH_2(CH_2OCH_2)_3CH_2OCH_3$	(276)	(141)	
(Tetraethylene Glycol Dimethyl Ether)			
Dimethylacetamide	330	158	914
$(CH_3)_2NC{:}OCH_3$	(165)	(70)	(490)
(DMAC)			
Dimethylamine	45	Gos	752
$(CH_3)_2NH$	(7)		(400)
1,2-Dimethylbenzene		See o-Xylene.	
1,3-Dimethylbenzene		See m-Xylene.	
1,4-Dimethylbenzene		See p-Xylene.	
Dimethylbenzylcarbinyl Acetate		205	
$C_6H_5CH_2C(CH_3)_2OOCCH_3$		(96)	
(alpha, alpha-Dimethyl-phenethyl Acelate)			
2,2-Dimethylbutane	122	−54	761
$(CH_3)_3CCH_2CH_3$	(50)	(−48)	(405)
(Neohexane)			
2,3-Dimethylbutane	136	−20	761
$(CH_3)_2CHCH(CH_3)_2$	(58)	(−29)	(405)
(Diisopropyl)			
1,3-Dimethylbutanol		See Methyl Isobutyl Carbinol.	
2,3-Dimethyl-1-Butene	133	<−4	680
$CH_3CH(CH_3)C(CH_3){:}CH_2$	(56)	(<−20)	(360)
2,3-Dimethyl-2-Butene	163	<−4	753
$CH_3C(CH_3){:}C(CH_3)_2$	(73)	(<−20)	(401)
1,3-Dimethylbutyl Acetate	284–297	113	
$CH_3COOCH(CH_3)CH_2CH(CH_3)_2$	(140–147)	(45)	
1,3-Dimethylbutylamine	223–228	55	
$CH_3CHNH_2(CH_2)CH(CH_3)_2$	(106–109)	(13)	
(2-Amino-4-Methylpeniane)			
Dimethyl Carbinol		See Isopropyl Alcohol.	
Dimethyl Carbonate		See Methyl Carbonate.	
Dimethyl Chloracetal	259–270	111	450
$ClCH_2CH(OCH_3)_2$	(126–132)	(44)	(232)
Dimethylcyanamide	320	160	
$(CH_3)_2NCN$	(160)	(71)	
1,2-Dimethylcyclohexane	260		579
$(CH_3)_2C_6H_{10}$	(127)		(304)
1,3-Dimethylcyclohexane	~256	~50	583
$(CH_3)_2C_6H_{10}$	(124)	(10)	(306)
(Hexahydroxylene)			
1,4-Dimethylcyclohexane	248	52	579
$(CH_3)_2C_6H_{10}$	(120)	(11)	(304)
(Hexahydroxylol)			
1,4-Dimethylcyclohexane-cis	255	61	
$C_6H_{10}(CH_3)_2$	(124)	(16)	

TABLE 2.41 Boiling Points, Flash Points, and Ignition Temperatures of Organic Compounds (*Continued*)

Compound	Boiling point °F (°C)	Flash point, °F (°C)	Ignition point, °F (°C)
1,4-Dimethylcyclohexane-trans	246	51	
$C_6H_{10}(CH_3)_2$	(119)	(11)	
Dimethyl Decalin	455	184	455
$C_{10}H_{16}(CH_2)_2$	(235)	(84)	(235)
Dimethyldichlorosilane	158	<70	
$(CH_3)_2SiCl_2$	(70)	(<21)	
(Dichlorodimethylsilane)			
Dimethyldioxane	243	75	
$CH_3CHCH_2OCH_2(CH_3)CHO$	(117)	(24)	
1,3-Dimethyl-1-3-	585–588	289	
Diphenylcyclobutane	(307–309)	(143)	
$(C_6H_5CCH_3)_2(CH_2)_2$			
Dimethylene Oxide		See Ethylene Oxide.	
Dimethyl Ether		See Methyl Ether.	
Dimethyl Ethyl Carbinol		See 2-Methyl-2-Butanol.	
2,4-Dimethyl-3-Ethylpentane	279	734	
$CH_3CH(CH_3)CH(CH_2H_5)$	(137)	(390)	
$CH(CH_3)_2$ (3-Ethyl-2,4-			
Dimethylpentane)			
N,N-Dimethylformamide	307	136	833
$HCON(CH_3)_2$	(153)	(58)	(445)
2,5-Dimethylfuran	200	45	
$OC(CH_3):CHCH:C(CH_3)$	(93)	(7)	
Dimethyl Glycol Phthalate	446	369	
$C_6H_4[COO(CH_2)_2OCH_3]_2$	(230)	(187)	
3,3-Dimethylheptane	279	617	
$CH_3(CH_2)_3C(CH_3)_2CH_2CH_3$	(137)	(325)	
2,6-Dimethyl-4-Heptanone		See Diisobutyl Ketone.	
2,3-Dimethylhexane	237	45	820
$CH_3CH(CH_3)CH(CH_3)C_2H_5CH_3$	(114)	(7)	(438)
2,4-Dimethylhexane	229	50	
$CH_3CH(CH_3)CH(CH_3)C_2H_5CH_3$	(109)	(10)	
Dimethyl Hexynol	302	135	
$C_4H_9CCH_3(OH)C:CH$	(150)	(57)	
(3,5-Dimethyl-1-Hexyn-3-ol)			
1,1-Dimethylhydrazine	145	5	480
$(CH_3)_2NNH_2$	(63)	(−15)	(249)
(Dimethylhydrazine, Unsymmetrical)			
Dimethylisophthalate		280	
$CH_3OOCC_6H_4COOCH_3$		(138)	
N,N-Dimethyliso-	257	95	
propanolamine	(125)	(35)	
$(CH_3)_2NCH_2CH(OH)CH_3$			
Dimethyl Ketone		See Acetone.	
Dimethyl Maleate	393	235	
$(-CHCOOCH_3)_2$	(201)	(113)	
2,6-Dimethylmorpholine	296	112	
$CH(CH_3)CH_2OCH_2CH(CH_3)NH$	(147)	(44)	
2,3-Dimethyloctane	327	<131	437
$CH_3(CH_2)_4CH(CH_3)CH(CH_3)CH_3$	(164)	(<55)	(225)
3,4-Dimethyloctane	324	<131	
$C_3H_7CH(CH_3)CH(CH_3)C_3H_7$	(162)	(<55)	

(*Continued*)

TABLE 2.41   Boiling Points, Flash Points, and Ignition Temperatures of Organic Compounds (*Continued*)

Compound	Boiling point °F (°C)	Flash point, °F (°C)	Ignition point, °F (°C)
2,3-Dimethylpentaldehyde	293	94	
$CH_3CH_2CH(CH_3)CH(CH_3)CHO$	(145)	(34)	
2,3-Dimethylpentane	194	<20	635
$CH_3CH(CH_3)CH(CH_3)CH_2CH_3$	(90)	(<−7)	(335)
2,4-Dimethylpentane	177	10	
$(CH_3)_2CHCH_2CH(CH_3)_2$	(81)	(−12)	
2,4-Dimethyl-3-Pentanol	284	120	
$(CH_3)_2CHCHOHCH(CH_3)_2$	(140)	(49)	
(Diisopropylmethanol)			
Dimethyl Phthalate	540	295	915
$C_6H_4(COOCH_3)_2$	(282)	(146)	(490)
Dimethylpiperazine-cis	329	155	
$C_6H_{14}N_2$	(165)	(68)	
2,2-Dimethylpropane	49		842
$(CH_3)_4C$	(9)		(450)
(Neopentane)			
2,2-Dimethyl-1-Propanol		See tert-Butyl Carbinol.	
2,5-Dimethylpyrazine	311	147	
$CH_3C:CHN:C(CH_3)CH:N$	(155)	(64)	
Dimethyl Sebacate	565	293	
$[-(CH_2)_4COOCH_3]_2$	(296)	(145)	
(Methyl Sebacate)			
Dimethyl Sulfate	370	182	370
$(CH_3)_2SO_4$	(188)	(83)	(188)
(Methyl Sulfate)			
Dimethyl Sulfide	99	<0	403
$(CH_3)_2S$	(37)	(<−18)	(206)
Dimethyl Sulfoxide	372	203	419
$(CH_3)_2SO$	(189)	(95)	(215)
		(oc)	
Dimethyl Terephthalate	543	308	965
$C_6H_4(COOCH_3)_2$	(284)	(153)	(518)
(Dimethyl-1,4-Benzene Dicarboxylate)			
(DMT)			
2,4-Dinitroaniline		435	
$(NO_2)_2C_6H_3NH_2$		(224)	
1,2-Dinitro Benzol	604	302	
$C_6H_4(NO_2)_2$	(318)	(150)	
(o-Dinitrobenzene)			
Dinitrochlorobenzene	599	382	
$C_6H_3Cl(NO_2)_2$	(315)	(194)	
(Chlorodinitrobenzene)			
2,4-Dinitrotoluene	572	404	
$(NO_2)_2C_6H_3CH_3$	(300)	(207)	
Dioctyl Adipate	680	402	710
$[-(CH_2)_2COOCH_2-$	(360)	(206)	(377)
$CH(C_2H_5)C_4-H_9]_2$			
[Bis(2-Ethylhexyl) Adipate]			
[Di(2-Ethylhexyl) Adipate]			
Dioctyl Azelate	709	440	705
$(CH_2)_7[COOCH_2CH(C_2H_5)C_4H_9]_2$	(376)	(227)	(374)
(Bis(2-Ethylhexyl) Azelate)			
(Di(2-Ethylhexyl) Azelate)			

TABLE 2.41 Boiling Points, Flash Points, and Ignition Temperatures of Organic Compounds (*Continued*)

Compound	Boiling point °F (°C)	Flash point, °F (°C)	Ignition point, °F (°C)
Dioctyl Ether	558	>212	401
$(C_8H_{17})_2O$	(292)	(>100)	(205)
(Octyl Ether)			
Dioctyl Phthalate		420	735
$C_6H_4[CO_2CH_2-$		(215)	(390)
$CH(C_2H_5)C_4H_9]_2$			
[Di(2-Ethylhexyl) Phthalate]			
[Bis(2-Ethylhexyl) Phthalate]			
p-Dioxane	214	54	356
$\underline{OCH_2CH_2OCH_2CH_2}$	(101)	(12)	(180)
(Diethylene Dioxide)			
Dioxolane	165	35	
$\underline{OCH_2CH_2OCH_2}$	(74)	(2)	
Dipe ntene	339	113	458
$C_{10}H_{16}$	(170)	(45)	(237)
(Cinene)			
(Limonene)			
Diphenyl		See Biphenyl.	
Diphenylamine	575	307	1173
$(C_6H_5)_2NH$	(302)	(153)	(634)
(Phenylaniline)			
1,1-Diphenylbutane	561	>212	851
$(C_6H_5)_2CHC_3H_7$	(294)	(>100)	(455)
1,3-Diphenyl-2-buten-1-one		See Dypnone.	
Diphenyldichlorosilane	581	288	
$(C_6H_5)_2SiCl_2$	(305)	(142)	
Diphenyldodecyl Phosphite		425	
$(C_6H_5O)_2POC_{10}H_{21}$		(218)	
1,1-Diphenylethane (uns)	546	>212	824
$(C_6H_5)_2CHCH_3$	(286)	(>100)	(440)
1,2-Diphenylethane (sym)	544	264	896
$C_6H_5CH_2CH_2C_6H_5$	(284)	(129)	(480)
Diphenyl Ether		See Diphenyl Oxide.	
Diphenylmethane	508	266	905
$(C_6H_5)_2CH_2$	(264)	(130)	(485)
(Ditane)			
Diphenyl Oxide	496	239	1144
$(C_6H_5)_2O$	(258)	(115)	(618)
(Diphenyl Ether)			
1,1-Diphenylpentane	586	>212	824
$(C_6H_5)_2CHC_4H_9$	(308)	(>100)	(440)
1,1-Diphenylpropane	541	>212	860
$CH_3CH_2CH(C_6H_5)_2$	(283)	(>100)	(460)
Diphenyl Phthalate	761	435	
$C_6H_4(COOC_6H_5)_2$	(405)	(224)	
Dipropylamine	229	63	570
$(C_3H_7)_2NH$	(109)	(17)	(299)
Dipropylene Glycol	449	250	
$(CH_3CHOHCH_2)_2O$	(232)	(121)	
Dipropylene Glycol Methyl Ether	408	186	
$CH_3OC_3H_6OC_3H_6OH$	(209)	(86)	

(*Continued*)

TABLE 2.41 Boiling Points, Flash Points, and Ignition Temperatures of Organic Compounds (*Continued*)

Compound	Boiling point °F (°C)	Flash point, °F (°C)	Ignition point, °F (°C)
Dipropyl Ether		See n-Propyl Ether.	
Dipropyl Ketone		See 4-Heptanone.	
Ditane		See Diphenylmethane.	
Ditridecyl Phthalate	547	470	
$C_6H_4(COOC_{13}H_{27})_2$	@5 mm (286)	(243)	
Divinyl Acetylene	183	<−4	
(⫶CCH:CH$_2$)$_2$	(84)	(<−20)	
(1,5-Hexadien-3-yne)			
Divinylbenzene	392	169	
$C_6H_4(CH:CH_2)_2$	(200)	(76)	
Divinyl Ether	83	<−22	680
(CH$_2$:CH)$_2$O	(28)	(<−30)	(360)
(Ethenylaxyethene)			
(Vinyl Ether)			
Dodecane	421	165	397
CH$_3$(CH$_2$)$_{10}$CH$_3$	(216)	(74)	(203)
(Dihexyl)			
1-Dodecanethiol	289	262	
CH$_3$(CH$_2$)$_{11}$SH	(143)	(128)	
(Dodecyl Mercaptan)	@15 mm		
(Lauryl Mercaptan)			
1-Dodecanol	491	260	527
CH$_3$(CH$_2$)$_{11}$OH	(255)	(127)	(275)
(Louryl Alcohol)			
Dodecyl Bromide		See Lauryl Bromide.	
Dodecylene (α)	406	<212	491
C$_{16}$H$_{21}$CH:CH$_2$	(208)	(<100)	(255)
(1-Dodecane)			
Dodecyl Mercaptan		See 1-Dodecanethiol.	
tert-Dodecyl Mercaptan	428–451	205	
C$_{12}$H$_{25}$SH	(220–233)	(96)	
4-Dodecyloxy-2-Hydroxy-		498	715
Benzophenone		(254)	(379)
C$_{25}$H$_{34}$O$_3$			
Dodecyl Phenol	597–633	325	
C$_{12}$H$_{25}$C$_6$H$_4$OH	(314–334)	(163)	
		(oc)	
Dypnone	475	350	
C$_6$H$_5$COCH:C(CH$_3$)C$_6$H$_5$	(246)	(177)	
(1,3-Diphenyl-2-Buten-1-one)	@50 mm		
Eicosane	651	>212	450
C$_{20}$H$_{42}$	(344)	(>100)	(232)
Epichlorohydrin	239	88	772
CH$_2$CHOCH$_2$Cl	(115)	(31)	(411)
(2-Chloropropylene Oxide)			
(γ-Chloropropylene Oxide)			
1,2-Epoxyethane		See Ethylene Oxide.	
Erythrene		See 1,3-Butadiene.	
Ethanal		See Acetaldehyde.	
Ethane	−128		882
CH$_3$CH$_3$	(−89)		(472)

TABLE 2.41   Boiling Points, Flash Points, and Ignition Temperatures of Organic Compounds (*Continued*)

Compound	Boiling point °F (°C)	Flash point, °F (°C)	Ignition point, °F (°C)
1,2-Ethanediol		See Ethylene Glycol.	
1,2-Ethanediol Diformate	345	200	
HCOOCH$_2$CH$_2$OOCH	(174)	(93)	
(Ethylene Formate)			
(Ethylene Glycol Diformate)			
(Glycol Diformate)			
Ethanethiol		See Ethyl Mercaptan.	
Ethanoic Acid		See Acetic Acid.	
Ethanoic Anhydride		See Acetic Anhydride.	
Ethanol		See Ethyl Alcohol.	
Ethanolamine	342	186	770
NH$_2$CH$_2$CH$_2$OH	(172)	(86)	(410)
(2-Amino Ethanol)			
(β-Aminoethyl Alcohol)			
Ethanoyl Chloride		See Acetyl Chloride.	
Ethene		See Ethylene.	
Ethenyl Ethanoate		See Vinyl Acetate.	
Ethenyloxyethene		See Divinyl Ether.	
Ether		See Ethyl Ether.	
Ethine		See Acetylene.	
Ethoxyacetylene	124	<20	
C$_2$H$_5$OC:CH	(51)	(<−7)	
Ethoxybenzene	342	145	
C$_6$H$_5$OC$_2$H$_5$	(172)	(63)	
(Ethyl Phenyl Ether) (Phenetole)			
2-Ethoxy-3,4-Dihydro-2-Pyran	289	111	
C$_7$H$_{12}$O$_2$	(143)	(44)	
2-Ethoxy Ethanol		See Ethylene Glycol Monoethyl Ether.	
2-Ethoxyethyl Acetate	313	117	716
CH$_3$COOCH$_2$CH$_2$OC$_2$H$_5$	(156)	(47)	(380)
(Ethyl Glycol Acetate)			
3-Ethoxypropanal	275	100	
C$_2$H$_5$OC$_2$H$_4$CHO	(135)	(38)	
(3-Ethoxypropionaldehyde)			
1-Ethoxypropane		See Ethyl Propyl Ether.	
3-Ethoxypropionaldehyde	275	100	
C$_2$H$_5$OCH$_2$CH$_2$CHO	(135)	(38)	
3-Ethoxypropionic Acid	426	225	
C$_2$H$_5$OCH$_2$CH$_2$COOH	(219)	(107)	
Ethoxytriglycol	492	275	
C$_2$H$_5$O(C$_2$H$_4$O)$_3$H	(256)	(135)	
(Triethylene Glycol, Ethyl Ether)			
Ethyl Abietale	662	352	
C$_{19}$H$_{29}$COOC$_2$H$_5$	(350)	(178)	
N-Ethylacetamide	401	230	
CH$_3$CONHC$_2$H$_5$	(205)	(110)	
(Acetoethylamide)			
N-Ethyl Acetanilide	400	126	
CH$_3$CON(C$_2$H$_5$)(C$_6$H$_5$)	(204)	(52)	
Ethyl Acetate	171	24	800
CH$_3$COOC$_2$H$_5$	(77)	(−4)	(426)
(Acetic Ester)			
(Acetic Ether)			
(Ethyl Ethanoate)			

(*Continued*)

TABLE 2.41 Boiling Points, Flash Points, and Ignition Temperatures of Organic Compounds (*Continued*)

Compound	Boiling point °F (°C)	Flash point, °F (°C)	Ignition point, °F (°C)
Ethyl Acetoacetate $C_2H_5CO_2CH_2COCH_3$ (Acetoacetic Acid, Ethyl Ester) (Ethyl 3-Oxobutanoate)	356 (180)	135 (57)	563 (295)
Ethyl Acetyl Glycolate $CH_3COOCH_2COOC_2H_5$ (Ethyl Glycolate Acetate)	−365 (−185)	180 (82)	
Ethyl Acrylate $CH_2{:}CHCOOC_2H_5$	211 (99)	50 (10)	702 (372)
Ethyl Alcohol C_2H_5OH (Grain Alcohol, Ethanol)	173 (78)	55 (13)	685 (363)
Ethylamine $C_2H_5NH_2$ 70% aqueous solution (Aminoethane)	62 (17)	<0 (<−18)	725 (385)
Ethyl Amino Ethanol $C_2H_5NHC_2H_4OH$ [2-(Ethylamino)ethanol]	322 (161)	160 (71)	
Ethylaniline $C_2H_5NH(C_6H_5)$	401 (205)	185 (85)	
Ethylbenzene $C_2H_5C_6H_5$ (Ethylbenzol) (Phenylethane)	277 (136)	70 (21)	810 (432)
Ethyl Benzoate $C_6H_5COOC_2H_5$	414 (212)	190 (88)	914 (490)
Ethylbenzol		See Ethylbenzene.	
Ethyl Bromide C_2H_5Br (Bromoethane)	100 (38)	None	952 (511)
Ethyl Bromoacetate $BrCH_2COOC_2H_5$	318 (159)	118 (48)	
2-Ethylbutanol		See 2-Ethylbutyraldehyde.	
Ethyl Butanoate		See Ethyl Butyrate.	
2-Ethyl-1-Butanol		See 2-Ethylbutyl Alcohol.	
2-Ethyl-1-Butene $(C_2H_5)_2C{:}CH_2$	144 (62)	<−4 (<−20)	599 (315)
3-(2-Ethylbutoxy) Propionic Acid $CH_3CH_2CH(C_2H_5)CH_2{-}OCH_2CH_2COOH$	392 (200) @100 mm	280 (138)	
2-Ethylbutyl Acetate $CH_3COOCH_2CH(C_2H_5)_2$	324 (162)	130 (54)	
2-Ethylbutyl Acrylate $CH_2{:}CHCOOCH_2CH{-}$ $(C_2H_5)C_2H_5$	180 (82) @10 mm	125 (52)	
2-Ethylbutyl Alcohol $(C_2H_5)_2CHCH_2OH$ (2-Ethyl-1-Butanol)	301 (149)	135 (57) (oc)	
Ethylbutylamine $CH_3CH_2CH_2CH_2{-}NHCH_3CH_2$	232 (111)	64 (18)	
Ethyl Butylcarbamate		See N-Butylurethane.	
Ethyl Butyl Carbonate $(C_2H_5)(C_4H_9)CO_3$	275 (135)	122 (50)	

TABLE 2.41   Boiling Points, Flash Points, and Ignition Temperatures of Organic Compounds (*Continued*)

Compound	Boiling point °F (°C)	Flash point, °F (°C)	Ignition point, °F (°C)
Ethyl Butyl Ether	198	40	
$C_2H_5OC_4H_9$	(92)	(4)	
(Butyl Ethyl Ether)			
2-Ethyl Butyl Glycol	386	180	
$(C_2H_5)_2CHCH_2OC_2H_4OH$	(197)	(82)	
[2-(2-Ethylbutoxy)ethanol]			
Ethyl Butyl Ketone	299	115	
$C_2H_5CO(CH_2)_3CH_3$	(148)	(46)	
(3-Heptanone)			
2-Ethyl-2-Butyl-1,3-Propanediol	352	280	
$HOCH_2C(C_2H_5)(C_4H_9)-$	(178)	(138)	
CH_2OH	@50 mm		
2-Ethylbutyraldehyde	242	70	
$(C_2H_5)_2CHCHO$	(117)	(21)	
(Diethyl Acetaldehyde)			
(2-Ethylbutanal)			
Ethyl Butyrate	248	75	865
$CH_3CH_2CH_2COOC_2H_5$	(120)	(24)	(463)
(Butyric Acid, Ethyl Ester)			
(Butyric Ester)			
(Ethyl Butanoate)			
2-Ethylbutyric Acid	380	210	752
$(C_2H_5)_2CHCOOH$	(193)	(99)	(400)
(Diethyl Acetic Acid)			
2-Ethylcaproaldehyde		See 2-Ethylhexanal.	
Ethyl Caproate	333	120	
$C_5H_{11}COOC_2H_5$	(167)	(49)	
(Ethyl Hexoate)			
(Ethyl Hexanoate)			
Ethyl Caprylate	405–408	175	
$CH_3(CH_2)_6COOC_2H_5$	(207–209)	(79)	
(Ethyl Octoate)			
Ethyl Octanoate		See Diethyl Carbonate.	
Ethyl Chloride	54	−58	966
C_2H_5Cl	(12)	(−50)	(519)
(Chloroethane)			
(Hydrochloric Ether)			
(Muriatic Ether)			
Ethyl Chloroacetate	295	147	
$ClCH_2COOC_2H_5$	(146)	(64)	
Ethyl Chlorocarbonate		See Ethyl Chloroformate.	
Ethyl Chloroformate	201	61	932
$ClCOOC_2H_5$	(94)	(16)	(500)
(Ethyl Chlorocarbonate)			
(Ethyl Chloromethanoate)			
Ethyl Chloromethanoate		See Ethyl Chloroformate.	
Ethyl Crotonate	282	36	
$CH_3CH:CHCOOC_2H_5$	(139)	(2)	
Ethyl Cyanoacetate	401–408	230	
$CH_2CNCOOC_2H_5$	(205–209)	(110)	
Ethylcyclobutane	160	<4	410
$C_2H_5C_4H_7$	(71)	(<−16)	(210)

(*Continued*)

TABLE 2.41   Boiling Points, Flash Points, and Ignition Temperatures of Organic Compounds (*Continued*)

Compound	Boiling point °F (°C)	Flash point, °F (°C)	Ignition point, °F (°C)
Ethylcyclohexane	269	95	460
$C_2H_5C_6H_{11}$	(132)	(35)	(238)
N-Ethylcyclohexylamine		86	
$C_6H_{11}NHC_2H_5$		(30)	
Ethylcyclopentane	218	<70	500
$C_2H_5C_5H_9$	(103)	(<21)	(260)
Ethyl Decanoate	469	>212	
$C_9H_{19}COOC_2H_5$	(243)	(>100)	
(Ethyl Caprate)			
N-Ethyldiethanolamine	487	280	
$C_2H_5N(C_2H_4OH)_2$	(253)	(138)	
Ethyl Dimethyl Methane		See Isopentane.	
Ethylene	−155		842
$H_2C{:}CH_2$	(−104)		(450)
(Ethene)			
Ethylene Acetate		See Glycol Diacetate.	
Ethylene Carbonate	351	290	
OCH_2CH_2OCO	(177)	(143)	
	@100 mm		
Ethylene Chlorohydrin		See 2-Chloroethanol.	
Ethylene Cyanohydrin	445	265	
$CH_2(OH)CH_2CN$	(229)	(129)	
(Hydracrylonitrile)	Decomposes		
Ethylenediamine	241	104	725
$H_2NCH_2CH_2NH_2$	(116)	(40)	(385)
Anydrous 76%	239–252	150	
	(115–122)		(66)
Ethylene Dichloride	183	56	775
CH_2ClCH_2Cl	(84)	(13)	(413)
(1,2-Dichloroethone)			
2,2-Ethylenedioxydiethanol		See Triethylene Glycol.	
Ethylene Formate		See 1,2-Ethanediol Diformate.	
Ethylene Glycol	387	232	748
HOC_2H_4OH	(197)	(111)	(398)
(1,2-Ethanediol)			
(Glycol)			
Ethylene Glycol n-Butyl Ether	340	150	
$HOCH_2CH_2OC_4H_9$	(171)	(66)	
Ethylene Glycol Diacetate		See Glycol Diacetate.	
Ethylene Glycol Dibutyl Ether	399	185	
$C_4H_9OC_2H_4OC_4H_9$	(204)	(85)	
Ethylene Glycol Diethyl Ether	251	95	406
$C_2H_5OCH_2CH_2OC_2H_5$	(122)	(35)	
Ethylene Glycol Diformate		See 1,2-Ethanediol Diformate.	
Ethylene Glycol Dimethyl Ether	174	29	395
	(79)	(−2)	(202)
$CH_3O(CH_2)_2OCH_3$	@630 mm		
(1,2-Dimethoxyethane)			
Ethylene Glycol Ethylbutyl Ether	386	180	
	(197)	(85)	
$(C_2H_5)_2CHCH_2OCH_2CH_2OH$			

TABLE 2.41 Boiling Points, Flash Points, and Ignition Temperatures of Organic Compounds (*Continued*)

Compound	Boiling point °F (°C)	Flash point, °F (°C)	Ignition point, °F (°C)
Ethylene Glycol Ethylhexyl Ether $C_4H_9CH(C_2H_5)CH_2OCH_2CH_2OH$	442 (228)	230 (110)	
Ethylene Glycol Isopropyl Ether $(CH_3)_2CHOCH_2CH_2OH$	289 (143)	92 (33)	
Ethylene Glycol Monoacetate $CH_2OHCH_2OOCCH_3$ (Glycol Monoacetate)	357 (181)	215 (102)	
Ethylene Glycol Monoacrylate $CH_2:CHCOOC_2H_4CH$ (2-Hydroxyethylacrylate)	410 (210)	220 (104) (oc)	
Ethylene Glycol Monobenzyl Ether $C_6H_5CH_2OCH_2CH_2OH$	493 (256)	265 (129)	665 (352)
Ethylene Glycol Monobutyl Ether $C_4H_9O(CH_2)_2(OH)$ (2-Butoxyethanol)	340 (171)	143 (62)	460 (238)
Ethylene Glycol Monobutyl Ether Acetate $C_4H_9O(CH_2)_2OOCCH_3$	377 (192)	160 (71)	645 (340)
Ethylene Glycol Monoethyl Ether $HOCH_2CH_2OC_2H_5$ (2-Ethoxyethanol)	275 (135)	110 (43)	455 (235)
Ethylene Glycol Monoethyl Ether Acetate $CH_3COOCH_2CH_2OC_2H_5$ (Cellosolve Acetate)	313 (156)	124 (52)	715 (379)
Ethylene Glycol Monoisobutyl Ether $(CH_3)_2CHCH_2OCH_2CH_2OH$	316–323 (158–162)	136 (58)	540 (282)
Ethylene Glycol Monomethyl Ether $CH_3OCH_2CH_2OH$ (2-Methoxyethanol)	255 (124)	102 (39)	545 (285)
Ethylene Glycol Monomethyl Ether Acetal $CH_3CH(OCH_2CH_2OCH_3)_2$	405 (207)	200 (93)	
Ethylene Glycol Monomethyl Ether Acetate $CH_3O(CH_2)_2OOCCH_3$	293 (145)	120 (49)	740 (392)
Ethylene Glycol Monomethyl Ether Formal $CH_2(OCH_2CH_2OCH_3)_2$	394 (201)	155 (68)	
Ethylene Glycol Phenyl Ether $C_6H_5OC_2H_4OH$ (2-Phenoxyethanol)	473 (245)	260 (127)	

(*Continued*)

TABLE 2.41   Boiling Points, Flash Points, and Ignition Temperatures of Organic Compounds (*Continued*)

Compound	Boiling point °F (°C)	Flash point, °F (°C)	Ignition point, °F (°C)
Ethylene Oxide	51	−20	1058
CH₂OCH₂	(11)		with No Air
(Dimethylene Oxide)			
(1,2-Epoxyethane)			
(Oxirane)			
Ethylenimine	132	12	608
NHCH₂CH₂	(56)	(−11)	(320)
(Aziridine)			
Ethyl Ethanoate		See Ethyl Acetate.	
N-Ethylethanolomine	322	160	
C₂H₅NHC₂H₄OH	(161)	(71)	
Ethyl Ether	95	−49	356
C₂H₅OC₂H₅	(35)	(−45)	(180)
(Diethyl Ether)			
(Diethyl Oxide)			
(Ether)			
(Ethyl Oxide)			
Ethylethylene Glycol		See 1,2-Butanediol.	
Ethyl Fluoride			
C₂H₅F	−36		
(1-Fluoroethane)	(−38)		
Ethyl Formate	130	−4	851
HCO₂C₂H₅	(54)	(−20)	(455)
(Ethyl Methanoate)			
(Formic Acid, Ethyl Ester)			
Ethyl Formate (ortho)	291	86	
(C₂H₅O)₃CH	(144)	(30)	
(Triethyl Orthoformate)			
Ethyl Glycol Acetate		See 2-Ethoxyethyl Acetate.	
2-Ethylhexaldehyde		See 2-Ethylhexanal.	
2-Ethylhexanal	325	112	375
C₄H₉CH(C₂H₅)CHO	(163)	(44)	(190)
(Butylethylacetaldehyde)			
(2-Ethylcaproaldehyde)			
(2-Ethylhexaldehyde)			
2-Ethyl-1,3-Hexanediol	472	260	680
C₃H₇CH(OH)CH(C₂H₅)CH₂OH	(244)	(127)	(360)
2-Ethylhexanoic Acid	440	245	700
C₄H₉CH(C₂H₅)COOH	(227)	(118)	(371)
(2-Ethyl Hexoic Acid)			
2-Ethylhexanol	359	164	448
C₄H₉CH(C₂H₅)CH₂OH	(182)	(73)	(231)
(2-Ethylhexyl Alcohol)			
(Octyl Alcohol)			
2-Ethylhexenyl		See 2-Ethyl-3-Propylacrolein.	
2-Ethylhexoic Acid		See 2-Ethylhexanoic Acid	
2-Ethylhexyl Acetate	390	160	515
CH₃COOCH₂CH(C₂H₅)C₄H₉	(199)	(71)	(268)
(Octyl Acetate)			
2-Ethylhexyl Acrylate	266	180	485
CH:CHCOOCH₂CH—	(130)	(82)	(252)
(C₂H₅)C₄H₉	@50 mm		

TABLE 2.41   Boiling Points, Flash Points, and Ignition Temperatures of Organic Compounds (*Continued*)

Compound	Boiling point °F (°C)	Flash point, °F (°C)	Ignition point, °F (°C)
2-Ethylhexylamine	337	140	
$C_4H_9CH(C_2H_5)CH_2NH_2$	(169)	(60)	
N-2-(Ethylhexyl) Aniline	379	325	
$C_6H_5NHCH_2CH(C_2H_5)C_4H_9$	(193)	(163)	
	@50 mm		
2-Ethylhexyl Chloride	343	140	
$C_4H_9CH(C_2H_5)CH_2Cl$	(173)	(60)	
N-(2-Ethylhexyl)cyclohexylamine	342	265	
$C_6H_{11}NH[CH_2CH—$	(172)	(129)	
$(C_2H_5)C_4H_9]$	@50 mm		
2-Ethylhexyl Ether	517	235	
$[C_4H_9CH(C_2H_5)CH_2]_2O$	(269)	(113)	
1,1-Ethylidene Dichloride	135–138	2	
CH_3CHCl_2	(57–59)	(−17)	
(1,1-Dichloroethane)			
1,2-Ethylidene Dichloride	183	55	824
$ClCH_2CH_2Cl$	(84)	(13)	(440)
Ethyl Isobutyrate	230	<70	
$(CH_3)_2CHCOOC_2H_5$	(110)	(<21)	
2-Ethylisohexanol	343–358	158	600
$(CH_3)_2CHCH_2CH(C_2H_5)CH_2OH$	(173–181)	(70)	(316)
(2-Ethyl Isohexyl Alcohol)			
(2-Ethyl-4-Methyl Pentanol)			
Ethyl Lactate	309	115	752
$CH_3CHOHCOOC_2H_5$	(154)	(46)	(400)
Tech.		131	
		(55)	
Ethyl Malonate		See Diethyl Malonate.	
Ethyl Mercaptan	9	<0	572
C_2H_5SH	(35)	(<−18)	(300)
(Ethanethiol)			
(Ethyl Sulfhydrate)			
Ethyl Methacrylate	239–248	68	
$CH_2:C(CH_3)COOC_2H_5$	(115–120)	(20)	
(Ethyl Methyl Acrylate)			
Ethyl Methanoate		See Ethyl Formate.	
Ethyl Methyl Acrylate		See Ethyl Methacrylate.	
Ethyl Methyl Ether		See Methyl Ethyl Ether.	
7-Ethyl-2-Methyl-4-	507	285	
Hendecanol	(264)	(141)	
$C_4H_9CH(C_2H_5)C_2H_4-$			
$CHOHCH_2CH(CH_3)_2$			
Ethyl Methyl Ketone		See Methyl Ethyl Ketone	
4-Ethylmorpholine	280	90	
$CH_2CH_2OC_2H_4NCH_2CH_3$	(138)	(32)	
1-Ethylnaphthalene	496		896
$C_{10}H_7C_2H_5$	(258)		(480)
Ethyl Nitrate	190	50	
$CH_3CH_2ONO_2$	(88)	(10)	
(Nitric Ether)			
Ethyl Nitrite	63	−31	194
C_2H_5ONO	(17)	(−35)	(90)
(Nitrous Ether)			

TABLE 2.41 Boiling Points, Flash Points, and Ignition Temperatures of Organic Compounds (*Continued*)

Compound	Boiling point °F (°C)	Flash point, °F (°C)	Ignition point, °F (°C)
3-Ethyloctane $C_5H_{11}CH(C_2H_5)C_2H_5$	333 (167)		446 (230)
4-Ethyloctane $C_4H_9CH(C_2H_5)C_3H_7$	328 (164)		445 (229)
Ethyl Oxalate $(COOC_2H_5)_2$ (Oxalic Ether) (Diethyl Oxalate)	367 (186)	168 (76)	
Ethyl Oxide		See Ethyl Ether.	
p-Ethylphenol $HOC_6H_4C_2H_5$	426 (219)	219 (104)	
Ethyl Phenylacetate $C_6H_5CH_2COOC_2H_5$	529 (276)	210 (99)	
Ethyl Phenyl Ether		See Ethoxybenzene.	
Ethyl Phenyl Ketone $C_2H_5COC_6H_5$ (Propiophenone)	425 (218)	210 (99)	
Ethyl Phthalyl Ethyl Glycolate $C_2H_5OCOC_6H_4OCO—$ $CH_2OCOC_2H_5$	608 (320)	365 (185)	
Ethyl Propenyl Ether $CH_3CH{:}CHOCH_2CH_3$	158 (70)	>19 (>−7)	
Ethyl Proplonate $C_2H_5COOC_2H_5$	210 (99)	54 (12)	824 (440)
2-Ethyl-3-Propylacrolein $C_3H_7CH{:}C(C_2H_5)CHO$ (2-Ethylhexenal)	347 (175)	155 (68)	
2-Ethyl-3-Propylacrylic Acid $C_3H_7CH{:}C(C_2H_5)COOH$	450 (232)	330 (166)	
Ethyl Propyl Ether $C_2H_5OC_3H_7$ (1-Ethoxypropane)	147 (64)	<−4 (<−20)	
m-Ethyltoluene $CH_3C_6H_4C_2H_5$ (1-Methyl-3-Ethylbenzene)	322 (161)		896 (480)
o-Ethyltoluene $CH_3C_6H_4C_2H_5$ (1-Methyl-2-Ethylbenzene)	329 (165)		824 (440)
p-Ethyltoluene $CH_3C_6H_4C_2H_5$ (1-Methyl-4-Ethylbenzene)	324 (162)		887 (475)
Ethyl p-Toluene Sulfonamide $C_7H_7SO_2NHC_2H_5$	208 (98) @745 mm	260 (127)	
Ethyl p-Toluene Sulfonate $C_7H_7SO_3C_2H_5$	345 (174)	316 (158)	
Ethyl Vinyl Ether		See Vinyl Ethyl Ether.	
Ethyne		See Acetylene.	
Fluorobenzene C_6H_5F	185 (85)	5 (−15)	
Formal		See Methylal.	
Formalin		See Formaldehyde.	

TABLE 2.41 Boiling Points, Flash Points, and Ignition Temperatures of Organic Compounds (*Continued*)

Compound	Boiling point °F (°C)	Flash point, °F (°C)	Ignition point, °F (°C)
Formaldehyde	−3	Gas	795
HCHO	(−19)	185	(424)
37% Methanol-free	214	(85)	
	(101)		
37%, 15% Methanol		122	
(Formalin)		(50)	
(Methylene Oxide)			
Formamide	410	310	
HCONH$_2$	(210)	(154)	
	Decomposes		
Formic Acid	213	156	1004
HCOOH	(101)	(69)	(539)
90% Solution		122	813
		(50)	(434)
Formic Acid, Butyl Ester		See Butyl Formate.	
Formic Acid, Ethyl Ester		See Ethyl Formate.	
Formic Acid, Methyl Ester		See Methyl Formate.	
Fuel Oil No. 1	304–574	100–162	410
(Kerosene)	(151–301)	(38–72)	(210)
(Range Oil)			
Fuel Oil No. 2		126–204	494
		(52–96)	(257)
Fuel Oil No. 4		142–240	505
		(61–116)	(263)
Fuel Oil No. 5			
Light		156–336	
Heavy		(69–169)	
		160–250	
		(71–121)	
Fuel Oil No. 6		150–270	765
		(66–132)	(407)
2-Furaldehyde		See Furfural.	
Furan	88	<32	
CH:CHCH:CHO	(31)	(<0)	
(Furfuran)			
Furfural	322	140	600
OCH:CHCH:CHCHO	(161)	(60)	(316)
(2-Furaldehyde)			
(Furfuraldehyde)			
(Furol)			
Furfuraldehyde		See Furfural.	
Furfuran		See Furan.	
Furfuryl Acetate	356–367	185	
OCH:CHCH:CCH$_2$OOCCH$_3$	(180–186)	(85)	
Furfuryl Alcohol	340	167	915
OCH:CHCH:CCH$_2$OH	(171)	(75)	(491)
		(oc)	
Furfurylamine	295	99	
C$_4$H$_3$OCH$_2$NH$_2$	(146)	(37)	
Furol		See Furfural.	
Fusel Oil		See Isoamyl Alcohol.	
Gas Oil	500–700	150+	640
	(260–371)	(66+)	(338)

(*Continued*)

TABLE 2.41   Boiling Points, Flash Points, and Ignition Temperatures of Organic Compounds (*Continued*)

Compound	Boiling point °F (°C)	Flash point, °F (°C)	Ignition point, °F (°C)
Gasoline	100–400	−45	
C$_5$H$_{12}$ to C$_9$H$_{20}$	(38–204)	(−43)	
56–60 Octane		−45	536
73 Octane		(−43)	(280)
92 Octane		−36	853
100 Octane		(−38)	(456)
Gasoline			
100–130 (Aviation Grade)		−50	824
		(−46)	(440)
Gasoline			
115–145 (Aviation Grade)		−50	880
		(−46)	(471)
Gasoline (Casinghead)		0	
		(−18)	
Glycerine	340	390	698
HOCH$_2$CHOHCH$_2$OH	(171)	(199)	(370)
(Glycerol)			
α,β-Glycerine Dichiorohydrin	360	200	
CH$_2$ClCHClCH$_2$OH	(182)	(93)	
Glycerol		See Glycerine.	
Glyceryl Triacetate	496	280	812
(C$_3$H$_5$)(OOCCH$_3$)$_3$	(258)	(138)	(433)
(Triacelin)			
Glyceryl Tributyrate	597	356	765
C$_3$H$_5$(OOCC$_3$H$_7$)$_3$	(314)	(180)	(407)
(Tributyrin)			
(Butyrin)			
(Glycerol Tributyrate)			
Glyceryl Trinitrate		See Nitroglycerine.	
Glyceryl Tripropionate	540	332	790
(C$_2$H$_5$COO)$_3$C$_3$H$_5$	(282)	(167)	(421)
(Tripropionin)			
Glycidyl Acrylate	135	141	779
CH$_2$:CHCOOCH$_2$CHCH$_2$O	(57)	(61)	(415)
	@2 mm		
Glycol		See Ethylene Glycol.	
Glycol Diacetate	375	191	900
(CH$_2$OOCCH$_3$)$_2$	(191)	(88)	(482)
(Ethylene Acetate)			
(Ethylene Glycol Diaceate)			
Glycol Dichloride		See Ethylene Dichloride.	
Glycol Diformate		See 1,2-Ethanediol Diformate.	
Glycol Dimercaptoacetate	280	396	
(HSCH$_2$C:OOCH$_2$—)$_2$	(138)	(202)	
(GDMA)	1.2 mm		
Glycol Monoacetate		See Ethylene Glycol Monoacetate.	
Grain Alcohol		See Ethyl Alcohol.	
Hendecane	384	149	
CH$_3$(CH$_2$)$_9$CH$_3$	(196)	(65)	
(Undecane)			
Heptadecanol	588	310	
C$_4$H$_9$CH(C$_2$H$_5$)C$_2$H$_4$–	(309)	(154)	
CH(OH)C$_2$H$_4$CH(C$_2$H$_5$)$_2$			
(3,9-Diethyl-6-Tridecanol)			

TABLE 2.41 Boiling Points, Flash Points, and Ignition Temperatures of Organic Compounds (*Continued*)

Compound	Boiling point °F (°C)	Flash point, °F (°C)	Ignition point, °F (°C)
Heptane	209	25	399
$CH_3(CH_2)_5CH_3$	(98)	(−4)	(204)
2-Heptanol	320	160	
$CH_3(CH_2)_4CH(OH)CH_3$	(160)	(71)	
3-Heptanol	313	140	
$CH_3CH_2CH(OH)C_4H_9$	(156)	(60)	
3-Heptanone		See Ethyl Butyl Ketone.	
4-Heptanone	290	120	
$(C_3H_7)_2CO$	(143)	(49)	
(Butyrone)			
(Dipropyl Ketone)			
1-Heptene		See Heptylene.	
3-Heptene (mixed cis and trans)	203	21	
$C_3H_7CH:CHC_2C_5$	(95)	(−6)	
(3-Heptylene)			
Heptylamine	311	130	
$CH_3(CH_2)_6NH_2$	(155)	(54)	
(1-Aminoheptane)			
Heptylene	201	<32	500
$C_5H_{11}CH:CH_2$	(94)	(<0)	(260)
(1-Heptene)			
Heptylene-2-trans	208	<32	
$C_4H_9CH:CHCH_3$	(98)	(<0)	
(2-Heptene-trans)			
Hexachlorobutadiene			1130
$CCl_2:CClCCl:CCl_2$			(610)
Hexachloro Diphenyl Oxide			1148
$(C_6H_2Cl_3)_2O$			(620)
[Bis(Trichlorophenyl) Ether]			
Hexadecane	549	>212	396
$CH_3(CH_2)_{14}CH_3$	(287)	(>100)	(202)
(Cetane)			
tert-Hexadecanethiol	298–307	(265)	
$C_{16}H_{33}SH$	(148–153)	(129)	
(Hexadecyl-tert-Mercaptan)	@11 mm		
Hexadecylene-1	525	>212	464
$CH_3(CH_2)_{13}CH:CH_2$	(274)	(>100)	(240)
(1-Hexadecene)			
Hexadecyltrichlorosilane	516	295	
$C_{16}H_{33}SiCl_3$	(269)	(146)	
2,4-Hexadienal	339	154	
$CH_3CH:CHCH:CHC(O)H$	(171)	(68)	
1,4-Hexadiene	151	−6	
$CH_3CH:CHCH_2CH:CH_2$	(66)	(−21)	
(Allylpropenyl)			
Hexanal	268	90	
$CH_3(CH_2)_4CHO$	(131)	(32)	
(Caproaldehyde)			
(Hexaldehyde)			
Hexane	156	−7	437
$CH_3(CH_2)_4CH_3$	(69)	(−22)	(225)
(Hexyl Hydride)			

(*Continued*)

TABLE 2.41   Boiling Points, Flash Points, and Ignition Temperatures of Organic Compounds (*Continued*)

Compound	Boiling point °F (°C)	Flash point, °F (°C)	Ignition point, °F (°C)
1,2-Hexanediol		See Hexylene Glycol.	
2,5-Hexanediol	429	230	
$CH_3CH(OH)CH_2$—$CH_2CH(OH)CH_3$	(221)	(110)	
(2,5-Dihydroxyhexane)			
2,5-Hexanedione		See Acetonyl Acetone.	
1,2,6-Hexanetriol	352	375	
$HOCH_2CH(OH)$—$(CH_2)_3CH_2OH$	(178)	(191)	
	@5 mm		
Hexanoic Acid		See Caproic Acid.	
1-Hexanol		See Hexyl Alcohol.	
2-Hexanone		See Methyl Butyl Ketone.	
3-Hexanone	253	95	
$C_2H_5COC_3H_7$	(123)	(35)	
(Ethyl n-Propyl Ketone)			
1-Hexene	146	<20	487
$CH_2{:}CH(CH_2)_3CH_3$	(63)	(<–7)	(253)
(Butyl Ethylene)			
2-Hexene-cis	156	<–4	
$C_3H_7CH{:}CHCH_3$	(69)	(<–20)	
3-Hexenol-cis	313	130	
$CH_3CH_2CH{:}CHCH_2CH_2OH$	(156)	(54)	
(3-Hexen-l-ol)			
(Leaf Alcohol)			
Hexyl Acetate	285	113	
$(CH_3)_2CH(CH_2)_3OOCCH_3$	(141)	(45)	
(Methylamyl Acetate)			
Hexyl Alcohol	311	145	
$CH_3(CH_2)_4CH_2OH$	(155)	(63)	
(Amyl Carbinol)			
(1-Hexanol)			
sec-Hexyl Alcohol	284	136	
$C_4H_9CH(OH)CH_3$	(140)	(58)	
(2-Hexanol)			
Hexylamine	269	85	
$CH_3(CH_2)_5NH_2$	(132)	(29)	
Hexyl Chloride		See 1-Chlorohexane.	
Hexyl Cinnamic Aldehyde	486	>212	
$C_6H_{13}C(CHO){:}CHC_6H_5$	(252)	(>100)	
(Hexyl Cinnamaldehyde)			
Hexylene Glycol	385	215	
$CH_2OHCHOH(CH_2)_3CH_3$	(196)	(102)	
(1,2-Hexanediol)			
Hexyl Ether	440	170	365
$C_6H_{13}OC_6H_{13}$	(227)	(77)	(185)
(Dihexyl Ether)			
Hexyl Methacrylate	388–464	180	
$C_6H_{13}OOCC(CH_3){:}CH_2$	(198–240)	(82)	
Hydracrylonitrile		See Ethylene Cyanohydrin.	
Hydralin		See Cyclohexanol.	
Hydroquinone	547	329	960
$C_6H_4(OH)_2$	(286)	(165)	(516)
(Quinol)			
(Hydroquinol)			

TABLE 2.41 Boiling Points, Flash Points, and Ignition Temperatures of Organic Compounds (*Continued*)

Compound	Boiling point °F (°C)	Flash point, °F (°C)	Ignition point, °F (°C)
Hydroquinone Di-(β-Hydroxyethyl) Ether	365–392	435	875
	@	(224)	(468)
$C_6H_4(-OCH_2CH_2OH)_2$	0.3 mm		
	(185–200)		
Hydroquinone Monomethyl Ether	475	270	790
$CH_3OC_6H_4OH$	(246)	(132)	(421)
(4-Methoxy Phenol)			
(Para-Hydroxyanisole)			
o-Hydroxybenzaldehyde		See Salicylaldehyde.	
3-Hydroxybutanal		See Aldol.	
β-Hydroxybutyraldehyde		See Aldol.	
Hydroxycitronellal	201–205	>212	
$(CH_3)_2C(OH)(CH_2)_3-$	(94–96)	(>100)	
$CH(CH_3)CH_2CHO$	@1 mm		
(Citronellal Hydrate)			
(3,7-Dimethyl-7-Hydroxyoctanal)			
N-(2-Hydroxyethyl)-acetamide		See N-Acetyl Ethanolamine.	
2-Hydroxyethyl Acrylate	410	214	1.8
(HEA)	(210)	(101)	@100°C
β-Hydroxyethylaniline		See 2-Anilinoethanol.	
N-(2-Hydroxyethyl)		249	
Cyclohexylamine		(121)	
$C_6H_{11}NH_2$			
$CH_2OHCH_2NHCH_2CH_2NH_2$			
4-(2-Hydroxyethyl) Morpholine	437	210	
$C_2H_4OC_2H_4NC_2H_4OH$	(225)	(99)	
1-(2-Hydroxyethyl) Piperazine	475	255	
$HOCH_2CH_2-NCH_2CH_2NHCH_2CH_2$	(246)	(124)	
n-(2-Hydroxyethyl) Propylenediamine	465	260	
$CH_3CH(NHC_2H_4OH)CH_2NH_2$	(241)	(127)	
4-Hydroxy-4-Methyl-2-Pentanone		See Diacetone Alcohol.	
2-Hydroxy-2-methylpropionitrile		See Acetone Cyanohydrin.	
Hydroxypropyl Acrylate		See Propylene Glycol Monoacrylate.	
o-Hydroxytoluene		See o-Cresol.	
Ionone Alpha (α-Ionone)	259–262	>212	
$C(CH_3)_2CH_2CH_2CH:C(CH_3)-$	(126–128)	(>100)	
	@12 mm		
$CHCH:CHC(CH_3):O$			
(α-Cyclocitrylideneacetone)			
[4-(2,6,6-Trimethyl-			
2-Cyclohexen-1-yl)-3-Buten-2-one]			
Ionone Beta (β-Ionone)	284	>212	
$C(CH_3)_2CH_2CH_2CH_2-$	(140)	(>100)	
$C(CH_3):CCHCHC(CH_3):O$	@18 mm		
(β-Cyclocitrylidene-acetone)			
[4-(2,6,6-Trimethyl-1-			
Cyclohexen-1-yl)-3-Buten-2-one]			

(*Continued*)

TABLE 2.41   Boiling Points, Flash Points, and Ignition Temperatures of Organic Compounds (*Continued*)

Compound	Boiling point °F (°C)	Flash point, °F (°C)	Ignition point, °F (°C)
Isoamyl Acetate $CH_3COOCH_2CH_2CH(CH_3)_2$ (Banana Oil) (3-Methyl-1-Butanol Acetate) (2-Methyl Butyl Ethanoate)	290 (143)	77 (25)	680 (360)
Isoamyl Alcohol $(CH_3)_2CHCH_2CH_2OH$ (Isobutyl Carbinol) (Fusel Oil) (3-Methyl-1-Butanol)	270 (132)	109 (43)	662 (350)
tert-Isoamyl Alcohol		See 2-Methyl-2-Butanol.	
Isoamyl Butyrate $C_3H_7CO_2(CH_2)_2CH(CH_3)_2$ (Isopentyl Butyrate)	352 (178)	138 (59)	
Isoamyl Chloride $(CH_3)_2CHCH_2CH_2Cl$ (1-Chloro-3-Methylbutane)	212 (100)	<70 (<21)	
Isobornyl Acetate $C_{10}H_{17}OOCCH_3$	428–435 (220–224)	190 (88)	
Isobutane $(CH_3)_3CH$ (2-Methylpropane)	11 (−12)		860 (460)
Isobutyl Acetate $CH_3COOCH_2CH(CH_3)_2$ (β-Methyl Propyl Ethanoate)	244 (118)	64 (18)	790 (421)
Isobutyl Acrylate $(CH_3)_2CHCH_2OOCCH{:}CH_2$	142–145 (61–63) @15 mm	86 (30)	800 (427)
Isobutyl Alcohol $(CH_3)_2CHCH_2OH$ (Isopropyl Carbinol) (2-Methyl-1-Propanol)	225 (107)	82 (28)	780 (415)
Isobutylamine $(CH_3)_2CHCH_2NH_2$	150 (66)	15 (−9)	712 (378)
Isobutylbenzene $(CH_3)_2CHCH_2C_6H_5$	343 (173)	131 (55)	802 (427)
Isobutyl Butyrate $C_3H_7CO_2CH_2(CH_3)_2$	315 (157)	122 (50)	
Isobutyl Carbinol		See Isoamyl Alcohol.	
Isobutyl Chloride $(CH_3)_2CHCH_2Cl$ (1-Chloro-3-Methyl-propane)	156 (69)	<70 (<21)	
Isobutylcyclohexane $(CH_3)_2CHCH_2C_6H_{11}$	336 (169)		525 (274)
Isobutylene		See 2-Methylpropene.	
Isobutyl Formate $HCOOCH_2CH(CH_3)_2$	208 (98)	<70 (<21)	608 (320)
Isobutyl Heptyl Ketone $(CH_3)_2CHCH_2COCH_2-$ $CH(CH_3)CH_2CH(CH_3)_2$ (2,6,8-Trimethyl-4-Non-anone)	412–426 (211–219)	195 (91)	770 (410)
Isobutyl Isobutyrate $(CH_3)_2CHCOOCH_2-CH(CH_3)_2$	291–304 (144–151)	101 (38)	810 (432)

TABLE 2.41 Boiling Points, Flash Points, and Ignition Temperatures of Organic Compounds (*Continued*)

Compound	Boiling point °F (°C)	Flash point, °F (°C)	Ignition point, °F (°C)
Isobutyl Phenylacetate	477	>212	
$(CH_3)_2CHCH_2OOCCH_2C_6H_5$	(247)	(>100)	
Isobutyl Phosphate	302	275	
$PO_4(CH_2CH(CH_3)_2)_3$	(150)	(135)	
(Triisobutyl Phosphate)	@20 mm		
Isobutyl Vinyl Ether		See Vinyl Isobutyl Ether.	
Isobutyraldehyde	142	−1	385
$(CH_3)_2CHCHO$	(61)	(−18)	(196)
(2-Methylpropanal)			
Isobutyric Acid	306	132	900
$(CH_3)_2CHCOOH$	(152)	(56)	(481)
Isobutyric Anhydride	360	139	625
$[(CH_3)_2CHCO]_2O$	(182)	(59)	(329)
Isobutyronitrile	214–216	47	900
$(CH_3)_2CHCN$	(101–102)	(8)	(482)
(2-Methylpropanenitrile)			
(Isopropylcyanide)			
Isodecaldehyde	387	185	
$C_9H_{19}CO$	(197)	(85)	
Isodecane	333		410
$C_7H_{15}CH(CH_3)_2$	(167)		(210)
(2-Methylnonane)			
Isodecanoic Acid	489	300	
$C_9H_{19}COOH$	(254)	(149)	
Isoevgenol	514	>212	
$(CH_3CHCH)C_6H_3OHOCH_3$	(268)	(>100)	
(1-Hydroxy-2 Methoxy-4-Propenylbanzene)			
Isoheptane	194	<0	
$(CH_3)_2CHC_4H_9$	(90)	(−18)	
(2-Methylhexane)			
(Ethylisobutylmethane)			
tert-Isohexyl Alcohol	252	115	
$C_2H_5(CH_3)C(OH)C_2H_5$	(122)	(46)	
(3-Methyl-3-Pentanol)			
Isooctane	210	40	784
$(CH_3)_2CHCH_4C(CH_3)_3$	(99)	(4.5)	(418)
(2,2,4-Trimethylpentane)			
Isooctyl Alcohol	83–91	180	
$C_7H_{15}CH_2OH$	(182–195)	(82)	
(Isooctanol)			
Isooctyl Nitrate	106–109	205	
$C_8H_{17}NO_3$	(41–43)	(96)	
	@1 mm		
Isooctyl Vinyl Ether		See Vinyl Isooctyl Ether.	
Isopentaldehyde	250	48	
$(CH_3)_2CHCH_2CHO$	(121)	(9)	
Isopentane	82	<−60	788
$(CH_3)_2CHCH_2CH_3$	(28)	(<−51)	(420)
(2-Methylbutane)			
(Ethyl Dimethyl Methane)			

(*Continued*)

TABLE 2.41   Boiling Points, Flash Points, and Ignition Temperatures of Organic Compounds (*Continued*)

Compound	Boiling point °F (°C)	Flash point, °F (°C)	Ignition point, °F (°C)
Isopentanoic Acid $(CH_3)_2CHCH_2COOH$ (Isovaleric Acid)	361 (183)		781 (416)
Isophorone $\underline{COCHC(CH_3)CH_2C(CH_3)_2CH_2}$	419 (215)	184 (84)	860 (460)
Isophthaloyl Chloride $C_6H_4(COCl)_2$ (m-Phthalyl Dichloride)	529 (276)	356 (180)	
Isoprene $CH_2{:}C(CH_3)CH{:}CH_2$ (2-Methyl-1,3-Butadiene)	93 (34)	−65 (−54)	743 (395)
Isopropanol		See Isopropyl Alcohol.	
Isopropenyl Acetate $CH_3COOC(CH_3){:}CH_2$ (1-Methylvinyl Acetate)	207 (97)	60 (16)	808 (431)
Isopropenyl Acetylene $CH_2{:}C(CH_3)C{:}CH$	92 (33)	<19 (<−7)	
2-Isopropoxypropane		See Isopropyl Ether.	
3-Isopropoxyproplonitrile $(CH_3)_2CHOCH_2CH_2CN$	149 (65) @10 mm	155 (68)	
Isopropyl Acetate $(CH_3)_2CHOOCCH_3$	194 (90)	35 (2)	860 (460)
Isopropyl Alcohol $(CH_3)_2CHOH$ (Isopropanol) (Dimethyl Carbinol) (2-Propanol) 87.9% iso	181 (83)	53 (12) 57 (14)	750 (399)
Isopropylamine $(CH_3)_2CHNH_2$	89 (32)	−35 (−37)	756 (402)
Isopropylbenzene		See Cumene.	
Isopropyl Benzoate $C_6H_5COOCH(CH_3)_2$	426 (219)	210 (99)	
Isopropyl Bicyclohexyl $C_{15}H_{28}$	530–541 (277–283)	255 (124)	446 (230)
2-Isopropylbiphenyl $C_{15}H_{16}$	518 (270)	285 (141)	815 (435)
Isopropyl Carbinol		See Isobutyl Alcohol.	
Isopropyl Chloride $(CH_3)_2CHCl$ (2-Chloropropane)	95 (35)	−26 (−32)	1100 (593)
Isopropylcyclohexane $(CH_3)_2CHC_6H_{11}$ (Hexahydrocumene) (Normanthane)	310 (154.5)		541 (283)
Isopropylcyclohexylamine $C_6H_{11}NHCHC_2H_6$		93 (34)	
Isopropyl Ether $(CH_3)_2CHOCH(CH_3)_2$ (2-Isopropoxypropane) (Diisopropyl Ether)	156 (69)	−18 (−28)	830 (443)

TABLE 2.41 Boiling Points, Flash Points, and Ignition Temperatures of Organic Compounds (*Continued*)

Compound	Boiling point °F (°C)	Flash point, °F (°C)	Ignition point, °F (°C)
Isopropylethylene		See 3-Methyl-1-Butene.	
Isopropyl Formate	153	22	905
HCOOCH(CH$_3$)$_2$	(67)	(−6)	(485)
(Isopropyl Methanoate)			
4-Isopropylheptane	155		491
C$_3$H$_7$CH(C$_3$H$_7$)C$_3$H$_7$	(68)		(255)
(m-Dihydroxybenzene)			
Isopropyl-2-Hydroxypropanoate		See Isopropyl Lactate.	
Isopropyl Lactate	331–334	130	
CH$_3$CHOHCCOCH(CH$_3$)$_2$	(166–168)	(54)	
(Isopropyl-2-Hydroxypropionate)			
Isopropyl Methanoate		See Isopropyl Formate.	
4-Isopropyl-1-Methyl Benzene		See p-Cymene.	
Isopropyl Vinyl Ether		See Vinyl Isopropyl Ether.	
Isovalerone		See Diisobutyl Ketone.	
Jet Fuel	400–550	110–150	
Jet A and Jet A-1	(204–288)	(43–66)	
Jet Fuel		−10 to +30	
Jet B		(−23 to −1)	
Jet Fuel		−10 to +30	464
JP-4		(−23 to −1)	(240)
Jet Fuel		95–145	475
JP-5		(35–63)	(246)
Jet Fuel	250	100	446
JP-6	(121)	(38)	(230)
Kerosene		See Fuel Oil No. 1.	
Lactonitrile	361	171	
CH$_3$CH(OH)CN	(183)	(77)	
Lanolin		460	833
(Wool Grease)		(238)	(445)
Lard Oil (Commercial or		395	833
Animal)		(202)	(445)
No. 1		440	
		(227)	
Lard Oil (Pure)		500	
		(260)	
No. 2		419	
		(215)	
Mineral		404	
		(207)	
Lauryl Alcohol		See 1-Dodecanol.	
Lauryl Bromide	356	291	
CH$_3$(CH$_2$)$_{10}$CH$_2$Br	(180)	(144)	
(Dodecyl Bromide)	@45 mm		
Lauryl Mercaptan		See 1-Dodecanethiol.	
Linalool	383–390	160	
(CH$_3$)$_2$C:CHCH$_2$CH$_2$C(CH$_3$)—	(195–199)	(71)	
OHCA:CH$_2$			
(3,7-Dimethyl-1,6-Octadiene-3-01)			
Linseed Oil	600+	432	650
	(316+)	(222)	(343)

(*Continued*)

TABLE 2.41   Boiling Points, Flash Points, and Ignition Temperatures of Organic Compounds (*Continued*)

Compound	Boiling point °F (°C)	Flash point, °F (°C)	Ignition point, °F (°C)
Lubricating Oil	680	300–450	500–700
(Paraffin Oil, includes	(360)	(149–232)	(260–371)
Motor Oil)			
Lubricating Oil, Spindle		169	478
(Spindle Oil)		(76)	(248)
Lubricating Oil, Turbine		400	700
(Turbine Oil)		(204)	(371)
Lynalyl Acetate	226–230	185	
$(CH_3)_2C{:}CHCH_2CH_2$—	(108–110)	(85)	
C(—$OOCCH_3$)CH:CH_2			
(Bergamol)			
Maleic Anhydride	396	215	890
$(COCH)_2O$	(202)	(102)	(477)
Marsh Gas		See Methane.	
2-Mercaptoethanol	315	165	
$HSCH_2CH_2OH$	(157)	(74)	
Mesitylene		See 1,3,5-Trimethylbenzene.	
Mesityl Oxide	266	87	652
$(CH_3)_2CCHCOCH_3$	(130)	(31)	(344)
Metaldehyde	subl.	97	
$(C_2H_4O)_4$	233–240	(36)	
	(112–116)		
α-**Methacrolein**		See 2-Methylpropenal.	
Methacrylic Acid	316	171	154
$CH_2{:}C(CH_3)COOH$	(158)	(77)	(68)
Methacrylonitrile	194	34	
C_4H_5N	(90)	(1.1)	
Methallyl Alcohol	237	92	
$CH_2C(CH_3)CH_2OH$	(114)	(33)	
Methallyl Chloride	162	11	
$CH_2C(CH_3)CH_2Cl$	(72)	(−12)	
Methane	−259		999
CH_4	(−162)		(537)
(Marsh Gas)			
Methanol		See Methyl Alcohol.	
Methanethiol		See Methyl Mercaptan.	
o-Methoxybenzaldehyde	275	104	
$CH_3OC_6H_4CHO$	(135)	(40)	
(o-Anisaldehyde)			
Methoxybenzene		See Anisole.	
3-Methoxybutanol	322	165	
$CH_3CH(OCH_3)CH_2CH_2OH$	(161)	(74)	
3-Methoxybutyl Acetate	275–343	170	
$CH_3OCH(CH_3)CH_2CH_2OOCCH_3$	(135–173)	(77)	
(Butoxyl)			
3-Methoxybutyraldehyde	262	140	
$CH_3CH(OCH_3)CH_2CHO$	(128)	(60)	
(Aldol Ether)			
2-Methoxyethanol		See Ethylene Glycol Monomethyl Ether.	

TABLE 2.41   Boiling Points, Flash Points, and Ignition Temperatures of Organic Compounds (*Continued*)

Compound	Boiling point °F (°C)	Flash point, °F (°C)	Ignition point, °F (°C)
2-Methoxyethyl Acrylate	142	180	
$C_2H_3COOC_2H_4OCH_3$	(61)	(82)	
	@17 mm		
Methoxy Ethyl Phthalate	376–412	275	
(Methox)	(191–211)	(135)	
3-Methoxypropionitrile	320	149	
$CH_3OC_2H_4CN$	(160)	(65)	
3-Methoxypropylamine	241	90	
$CH_3OC_3H_6NH_2$	(116)	(32)	
Methoxy Triglycol	480	245	
$CH_3O(C_2H_4O)_3H$	(249)	(118)	
(Triethylene Glycol, Methyl Ether)			
Methoxytriglycol Acetate	266	260	
$CH_3COO(C_2H_4O)_3CH_3$	(130)	(127)	
Methyl Abietate	680–689	356	
$C_{19}H_{29}COOCH_3$	(360–365)	(180)	
(Abalyn)	Decomposes		
Methyl Acetate	140	14	850
CH_3COOCH_3	(60)	(−10)	3.1
(Acetic Acid Methyl Ester)		(454)	16
(Methyl Acetic Ester)			
Methyl Acetic Ester		See Methyl Acetate.	
Methyl Acetoacetate	338	170	536
$CH_3CO_2CH_2COCH_3$	(170)	(77)	(280)
P-Methyl Acetophenone	439	205	
$CH_3C_6H_4COCH_3$	(226)	(96)	
(Methyl-p-Tolyl Ketone)			
(p-Acetotoluene)			
Methylacetylene		See Propyne.	
Methyl Acrylate	176	27	875
$CH_2{:}CHCOOCH_3$	(80)	(−3)	(468)
Methylal	111	−26	459
$CH_3OCH_2OCH_3$	(44)	(−32)	(237)
(Dimethoxymethane)			
(Formal)			
Methyl Alcohol	147	52	867
CH_3OH	(64)	(11)	(464)
(Methanol)			
(Wood Alcohol)			
Methylamine	21	806	4
CH_3NH_2	(−6)	(430)	
2-(Methylamino) Ethanol		See N-Methylethanolamine.	
Methylamyl Acetate		See Hexyl Acetate.	
Methylamyl Alcohol		See Methyl Isobutyl Carbinol.	
Methyl Amyl Ketone	302	102	740
$CH_3CO(CH_2)_4CH_3$	(150)	(39)	(393)
2-Heptanone			
2-Methylaniline		See o-Toluidine.	
4-Methylaniline		See p-Toluidine.	
Methyl Anthranilate	275	>212	
$H_2NC_6H_4CO_2CH_3$	@15 mm	(>100)	
(Methyl-ortho-Amino Benzoate)	(135)		
(Nevoli Oil, Artificial)			

(*Continued*)

TABLE 2.41   Boiling Points, Flash Points, and Ignition Temperatures of Organic Compounds (*Continued*)

Compound	Boiling point °F (°C)	Flash point, °F (°C)	Ignition point, °F (°C)
Methylbenzene	See Toluene.		
Methyl Benzoate	302	181	
C$_6$H$_5$COOCH$_3$	(150)	(83)	
(Niobe Oil)			
α-Methylbenzyl Alcohol	See Phenyl Methyl Carbinol.		
α-Methylbenzylamine	371	175	
C$_6$H$_5$CH(CH$_3$)NH$_2$	(188)	(79)	
α-Methylbenzyl Dimethyl	384	175	
Amine	(196)	(79)	
C$_6$H$_5$CH(CH$_3$)N(CH$_3$)$_2$			
α-Methylbenzyl Ether	548	275	
C$_6$H$_5$CH(CH$_3$)OCH(CH$_3$)C$_6$H$_5$	(287)	(135)	
2-Methylbiphenyl	492	280	936
C$_6$H$_5$C$_6$H$_4$CH$_3$	(255)	(137)	(502)
Methyl Borate	156	<80	
B(OCH$_3$)$_3$	(69)	(<27)	
(Trimethyl Borate)			
Methyl Bromide	38.4	999	
CH$_3$Br	(4)	(537)	
(Bromomethane)			
2-Methyl-1,3-Butadiene	See Isoprene.		
2-Methylbutane	See Isopentane.		
3-Methyl-2-Butanethiol	230	37	
C$_5$H$_{11}$SH	(110)	(3)	
(Sec-Isoamyl Mercaptan)			
2-Methyl-1-Butanol	262	122	725
CH$_3$CH$_2$CH(CH$_3$)CH$_2$OH	(128)	(50)	(385)
2-Methyl-2-Butanol	215	67	819
CH$_3$CH$_2$(CH$_3$)$_2$COH	(102)	(19)	(437)
(tert-Isoamyl Alcohol)			
(Dimethyl Ethyl Carbinol)			
3-Methyl-1-Butanol	See Isoamyl Alcohol.		
3-Methyl-1-Butanol Acetate	See Isoamyl Acetate.		
2-Methyl-1-Butene	88	<20	
CH$_2$:C(CH$_3$)CH$_2$CH$_3$	(31)	(<–7)	
2-Methyl-2-Butene	101	<20	
(CH$_3$)$_2$C:CCHCH$_3$	(38)	(<–7)	
(Trimethylethylene)			
3-Methyl-1-Butene	68	<20	689
(CH$_3$)$_2$CHCH:CH$_2$	(20)	(<–7)	(365)
(Isopropylethylene)			
N-Methylbutylamine	196	55	
CH$_3$CH$_2$CH$_2$CH$_2$NHCH$_3$	(91)	(13)	
2-Methyl Butyl Ethanoate	See Isoamyl Acetate.		
Methyl Butyl Ketone	262	77	795
CH$_3$CO(CH$_2$)$_3$CH$_3$	(128)	(25)	(423)
(2-Hexanone)			
3-Methyl Butynol	218	77	
(CH$_3$)$_2$C(OH)C:CH	(103)	(25)	
2-Methylbutyraldehyde	198–199	49	
CH$_3$CH$_2$CH(CH$_3$)CHO	(92–93)	(9)	
Methyl Butyrate	215	57	
CH$_3$OOCCH$_2$CH$_2$CH$_3$	(102)	(14)	

TABLE 2.41 Boiling Points, Flash Points, and Ignition Temperatures of Organic Compounds (*Continued*)

Compound	Boiling point °F (°C)	Flash point, °F (°C)	Ignition point, °F (°C)
Methyl Carbonate	192	66	
$CO(OCH_3)_2$	(89)	(19)	
(Dimethyl Carbonate)		(oc)	
Methyl Cellosolve Acetate	292	~111	
$CH_3COOC_2H_4OCH_3$	(144)	(~44)	
(2-Methoxyethyl Acetate)			
Methyl Chloride	−11	−50	1170
CH_3Cl	(−24)		(632)
(Chloromethane)			
Methyl Chloroacetate	266	135	
$CH_2ClCOOCH_3$	(130)	(57)	
(Methyl Chloroethanoate)			
Methyl Chloroethanoate		See Methyl Chloroacetate.	
Methyl-p-Cresol		140	
$CH_3C_6H_4OCH_3$		(60)	
(p-Methylanisole)			
Methyl Cyanide		See Acetonitrile.	
Methylcyclohexane	214	25	482
$CH_2(CH_2)_4CHCH_3$	(101)	(−4)	(250)
(Cyclohexylmethane)			
(Hexahydrotoluene)			
2-Methylcyclohexanol	329	149	565
$C_7H_{13}OH$	(165)	(65)	(296)
3-Methylcyclohexonol		158	563
$CH_3C_6H_{10}OH$		(70)	(295)
4-Methylcyclohexanol	343	158	563
$C_7H_{13}OH$	(173)	(70)	(295)
Methylcyclohexanone	325	118	
$C_7H_{12}O$	(163)	(48)	
4-Methylcyclohexene	217	30	
$CH:CHCH_2CH(CH_3)CH_2CH_2$	(103)	(−1)	
Methylcyclohexyl Acetate	351–381	147	
$C_9H_{16}O_2$	(177–194)	(64)	
Methyl Cyclopentadiene	163	120	833
C_6H_8	(73)	(49)	(445)
Methylcyclopentane	161	<20	496
C_6H_{12}	(72)	(<−7)	(258)
2-Methyldecane	374		437
$CH_3(CH_2)_7CH(CH_3)_2$	(190)		(225)
Methyldichlorosilane	106	15	>600
CH_3HSiCl_2	(41)	(−9)	(316)
N-Methyldiethanolamine	464	260	
$CH_3N(C_2H_4OH)_2$	(240)	(127)	
1-Methyl-3,5-Diethyl-benzene	394		851
$(CH_3)C_6H_3(C_2H_5)_2$	(201)		(455)
(3,5-Diethyltoluene)			
Methyl Dihydroabietate	689–698	361	
$C_{19}H_{31}COOCH_3$	(365–370)	(183)	
Methylene Chloride	104		1033
CH_2Cl_2	(40)	None	(556)
(Dichloromethane)			

(*Continued*)

TABLE 2.41   Boiling Points, Flash Points, and Ignition Temperatures of Organic Compounds (*Continued*)

Compound	Boiling point °F (°C)	Flash point, °F (°C)	Ignition point, °F (°C)
Methylenedianiline	748–750	428	
H$_2$NC$_6$H$_4$CH$_2$C$_6$H$_4$NH$_2$	(398–399)		
(MDA)	@78 mm		
(p,p′-DiaminodiPhenylmethane)		(220)	
Methylene Dlisocyanate		185	
CH$_2$(NCO)$_2$		(85)	
Methylene Oxide	See Formaldehyde.		
N-Methylethanolamine			
CH$_3$NHCH$_2$CH$_2$OH	319	165	
(2-(Methylamino) Ethanol)	(159)	(74)	
Methyl Ether	−11	Gas	662
(CH$_3$)$_2$O	(−24)		(350)
(Dimethyl Ether)			
(Methyl Oxide)			
Methyl Ethyl Carbinol	See sec-Butyl Alcohol.		
2-Methyl-2-Ethyl-	244	74	
1,3-Dioxolane	(118)	(23)	
(CH$_3$)(C$_2$H$_5$)COCH$_2$CH$_2$O			
Methyl Ethylene Glycol	See Propylene Glycol.		
Methyl Ethyl Ether	51	−35	374
CH$_3$OC$_2$H$_5$	(11)	(−37)	(190)
(Ethyl Methyl Ether)			
2-Methyl-4-Ethylhexane	273	<70	536
(CH$_3$)$_2$CHCH$_2$CH(C$_2$H$_5$)$_2$	(134)	(<21)	(280)
(4-Ethyl-2-Methylhexane)			
3-Methyl-4-Ethylhexane	284	75	
C$_2$H$_5$CH(CH$_3$)CH(C$_2$H$_5$)$_2$	(140)	(24)	
(3-Ethyl-4-Methylhexane)			
Methyl Ethyl Ketone	176	16	759
C$_2$H$_5$COCH$_3$	(80)	(−9)	(404)
(2-Butanone)			
(Ethyl Methyl Ketone)			
Methyl Ethyl Ketoxime	306–307	156–170	
CH$_3$C(C$_2$H$_5$):HOH	(152–153)	(69–77)	
2-Methyl-3-Ethylpentane	241	<70	860
(CH$_3$)$_2$CHCH(C$_2$H$_5$)$_2$	(116)	(<21)	(460)
(3-Ethyl-2-Methylpentane)			
2-Methyl-5-Ethyl-piperidine	326	126	
NHCH(CH$_3$)CH$_2$CH$_2$CH(C$_2$H$_5$)CH$_2$	(163)	(52)	
2-Methyl-5-Ethylpyridine	353	155	
N:C(CH$_3$)CH:CHC(C$_2$H$_5$):CH	(178)	(68)	
Methyl Formate	90	−2	840
CH$_3$OOCH	(32)	(−19)	(449)
(Formic Acid, Methyl Ether)			
2-Methylfuran	144–147	−22	
C$_4$H$_3$OCH$_3$	(62–64)	(−30)	
(Sylvan)			
Methyl Glycol Acetate		111	
CH$_2$OHCHOHCH$_2$CO$_1$CH$_3$		(44)	
(Propylene Glycol Acetate)			

TABLE 2.41 Boiling Points, Flash Points, and Ignition Temperatures of Organic Compounds (*Continued*)

Compound	Boiling point °F (°C)	Flash point, °F (°C)	Ignition point, °F (°C)
Methyl Heptolocyl Ketone $C_{17}H_{35}COCH_3$	329 (165) @3 mm	255 (124)	
Methylheptenone $(CH_3)_2C{:}CH(CH_2)_2COCH_3$ (6-Methyl-5-Hepten-2-one)	343–345 (173–174)	135 (57)	
Methyl Heptine Carbonate $CH_3(CH_2)_4C{:}CCOOCH_3$ (Methyl 2-Octynoate)		190 (88)	
Methyl Heptyl Ketone $C_7H_{15}COCH_4$ (5-Methyl-2-Octanone)	361–383 (183–195)	140 (60)	680 (360)
2-Methylhexane $(CH_3)_2CH(CH_2)_3CH_3$	194 (90)	<0 (<–18)	536 (280)
3-Methylhexane $CH_3CH_2CH(CH_3)CH_2CH_2CH_3$	198 (92)	25 (–4)	536 (280)
Methyl Hexyl Ketone $CH_3COC_6H_{13}$ (2-Octanone) (Octanone)	344 (173.5)	125 (52)	
Methyl-3-Hydroxybutyrate $CH_3CHOHCH_2COOCH_3$	347 (175)	180 (82)	
Methyl Ionone $C_{14}H_{22}O$ (Irone)	291 (144) @16 mm	>212 (>100)	
Methyl Isoamyl Ketone $CH_3COCH_2CH_2CH(CH_3)_2$	294 (146)	96 (36)	375 (191)
Methyl Isobutyl Carbinol $CH_3CHOHCH_2CHCH_3CH_3$ (1,3-Dimethylbutanol) (4-Methyl-2-Pentanol) (Methylamyl Alcohol)	266–271 (130–133)	106 (41)	
Methylisobutylcarbinol Acetate		See 4-Methyl-2-Pentanol Acetate.	
Methyl Isobutyl Ketone $CH_3COCH_2CH(CH_3)_2$ (Hexone) (4-Methyl-2-Pentanone)	244 (118)	64 (18)	840 (448)
Methyl Isopropenyl Ketone $CH_2COC{:}CH_2(CH_3)$	208 (98)		
Methyl Isocyanate CH_3NCO (Methyl Carbonimide)	102 (39)	19 (–7)	994 (534)
Methyl Iso Eugenol $CH_3CH{:}CHC_6H_3(OCH_3)_2$ (Propenyl Guaiacol)	504–507 (262–264)	>212 (>100)	
Methyl Lactate $CH_3CHOHCOOCH_3$	293 (145) @2.2 mm	121 725 (49) (385)	@2.2 mm 212 (100)
Methyl Mercaptan CH_3SH (Methanethiol)	42.4 (6)		

(*Continued*)

TABLE 2.41   Boiling Points, Flash Points, and Ignition Temperatures of Organic Compounds (*Continued*)

Compound	Boiling point °F (°C)	Flash point, °F (°C)	Ignition point, °F (°C)
β-**Methyl Mercapto-** **propionaldehyde** CH$_3$SC$_2$H$_4$CHO (3-(Methylthio) Propionalde-hyde)	~329 (~165)	142 (61)	491 (255)
Methyl Methacrylate CH$_2$:C(CH$_3$)COOCH$_3$	212 (100)	50 (10)	
Methyl Methanoate		See Methyl Formate.	
4-Methylmorpholine C$_2$H$_4$OC$_2$H$_4$NCH$_3$	239 (115)	75 (24)	
1-Methylnaphthalene C$_{10}$H$_7$CH$_3$	472 (244)		984 (529)
Methyl Nonyl Ketone C$_9$H$_{19}$COCH$_3$	433 (223)	192 (89)	
Methyl Oxide		See Methyl Ether.	
Methyl Pentadecyl Ketone C$_{15}$H$_{31}$COCH$_3$	313 (156) @3 mm	248 (120)	
2-Methyl-1,3-Pentadiene CH$_2$:C(CH$_3$)CH:CHCH$_3$	169 (76)	<–4 (<–20)	
4-Methyl-1,3-Pentadiene CH$_2$:CHCH$_2$:C(CH$_3$)$_2$	168 (76)	–30 (–34)	
Methylpentaldehyde CH$_3$CH$_2$CH$_2$C(CH$_3$)HCHO (Methyl Pentanal)	243 (117)	68 (20)	
Methyl Pentanal		See Methylpentaldehyde.	
2-Methylpentane (CH$_3$)$_2$CH(CH$_2$)$_2$CH$_3$ (Isohexane)	140 (60)	<20 (<–7)	583 (306)
3-Methylpentane CH$_3$CH$_2$CH(CH$_3$)CH$_2$CH$_3$	146 (63)	<20 (<–7)	532 (278)
2-Methyl-1,3-Pentanediol CH$_3$CH$_2$CH(OH)CH(CH$_3$)CH$_2$OH	419 (215)	230 (110)	
2-Methyl-2,4-Pentanediol (CH$_3$)$_2$C(OH)CH$_2$CH(OH)CH$_3$	385 (196)	205 (96)	
2-Methylpentanoic Acid C$_3$H$_7$CH(CH$_3$)COOH	381 (194)	225 (107)	712 (378)
2-Methyl-1-Pentanol CH$_3$(CH$_2$)$_2$CH(CH$_3$)CH$_2$OH	298 (148)	129 (54)	590 (310)
4-Methyl-2-Pentanol		See Methyl Isobutyl Carbinol.	
4-Methyl-2-Pentanol Acetate CH$_3$COOCH(CH$_3$)CH$_2$CH(CH$_3$)$_2$ (Methylisobutylcarbinol Acetate)	295 (146)	110 (43)	660 (349)
4-Methyl-2-Pentanone		See Methyl Isobutyl Ketone.	
2-Methyl-1-Pentene CH$_2$:C(CH$_3$)CH$_2$CH$_2$CH$_3$	143 (62)	<20 (<–7)	572 (300)
4-Methyl-1-Pentene CH$_2$:CHCH$_2$CH(CH$_3$)$_2$	129 (54)	<20 (<–7)	572 (300)

TABLE 2.41 Boiling Points, Flash Points, and Ignition Temperatures of Organic Compounds (*Continued*)

Compound	Boiling point °F (°C)	Flash point, °F (°C)	Ignition point, °F (°C)
2-Methyl-2-Pentene	153	<20	
$(CH_3)_2C{:}CHCH_2CH_3$	(67)	(<−7)	
4-Methyl-2-Pentene	133–137	<20	
$CH_3CH{:}CHCH(CH_3)_2$	(56–58)	(<−7)	
3-Methyl-1-Pentynol	250	101	
$(C_2H_5)(CH_3)C(OH)C{:}CH$	(121)	(38)	
o-Methyl Phenol		See o-Cresol.	
Methyl Phenylacetate	424	195	
$C_6H_5CH_2COOCH_3$	(218)	(91)	
Methylphenyl Carbinol	399	200	
$C_6H_5CH(CH_3)OH$	(204)	(93)	
(α-Methylbenzyl Alcohol)			
(Styralyl Alcohol)			
(sec-Phenethyl Alcohol)			
Methyl Phenyl Carbinyl Acetate		195	
$C_6H_5CH(CH_3)OOCH_3$		(91)	
(α-Methyl-Benzyl Acetate)			
(Styrolyl Acetate)			
(sec-Phenylethyl Acetate)			
(Phenyl Methylcarbinyl Acetate)			
Methyl Phenyl Ether		See Anisole.	
Methyl Phthalyl Ethyl Glycolate	590	380	
$CH_3COOC_6H_4COO{-}$	(310)	(193)	
$CH_2COOC_2H_5$			
1-Methyl Piperazine	280	108	
$CH_3NCH_2CH_2NHCH_2CH_2$	(138)	(42)	
2-Methylpropanal		See Isobutyraldehyde.	
2-Methylpropane		See Isobutane.	
2-Methyl-2-Propanethiol	149–153	<−20	
$(CH_3)_3CSH$	(65–67)	(<−29)	
(tert-Butyl Mercaptan)			
2-Methyl Propanol-1		See Isobutyl Alcohol.	
2-Methyl-2-Propanol		See tert-Butyl Alcohol.	
2-Methylpropenal	154	35	
$CH_2{:}C(CH_3)CHO$	(68)	(2)	
(Methacrolein)			
(α-Methyl Acrolein)			
2-Methylpropene	20		869
$CH_2{:}C(CH_3)CH_3$	(−7)		(465)
(γ-Butylene)			
(Isobutylene)			
Methyl Propionate	176	28	876
$CH_3COOCH_2CH_3$	(80)	(−2)	(469)
Methyl Propyl Acetylene	185	<14	
$CH_3C_2H_4ClCCH_3$	(85)	(<−10)	
(2-Hexyne)			
Methyl Propyl Carbinol	247	105	
$CH_3CHOHC_3H_7$	(119)	(41)	
(2-Pentanol)			

(Continued)

TABLE 2.41 Boiling Points, Flash Points, and Ignition Temperatures of Organic Compounds (*Continued*)

Compound	Boiling point °F (°C)	Flash point, °F (°C)	Ignition point, °F (°C)
Methylpropylcarbinylumine		See sec-Amylamine.	
Methyl n-Propyl Ether	102	<–4	
$CH_3OC_3H_7$	(39)	(<–20)	
Methyl Propyl Ketone	216	45	846
$CH_3COC_3H_7$	(102)	(7)	(452)
(2-Pentanone)			
2-Methylpyrazine		122	
$N:C(CH_3)CH:NCH:CH$		(50)	
2-Methyl Pyridine		See 2-Picoline.	
Methylpyrrole	234	61	
$N(CH_3)CH:CHCH:CH$	(112)	(16)	
Methylpyrrolidine	180	7	
$CH_3NC_4H_5$	(82)	(–14)	
1-Methyl-2-Pyrrolidone	396	204	655
$CH_3NCOCH_2CH_2CH_2$	(202)	(96)	(346)
(N-Methyl-2-Pyrrolidone)			
Methyl Salicylate	432	205	850
$HOC_6H_4COOCH_3$	(222)	(96)	(454)
(Oil of Wintergreen)			
(Gaultheria Oil)			
(Betula Oil)			
(Sweet-Birch Oil)			
Methyl Stearate	421	307	
$C_{17}H_{35}COOCH_3$	(216)	(153)	
α-Methylstyrene	329–331	129	1066
1-Methylethenyl Benzene	(165–166)	(54)	(574)
1-Methyl-1-phenylethene			
Methyl Sulfate		See Dimethyl Sulfate.	
2-Methyltetrahydrofuran	176	12	
$C_4H_7OCH_3$	(80)	(–11)	
Methyl Toluene Sulfonate	315	306	
$CH_3C_6H_4SO_3CH_3$	(157)	(152)	
	@8 mm		
Methyltrichlorosilane	151	15	>760
CH_3SiCl_3	(66)	(–9)	(>404)
(Methyl Silico Chloroform)			
(Trichloromethylsilane)			
Methyl Undecyl Ketone	248	225	
$C_{11}H_{23}COCH_3$	(120)	(107)	
(2-Tridecanone)			
1-Methylvinyl Acetate		See Isopropenyl Acetate.	
Methyl Vinyl Ether		See Vinyl Methyl Ether.	
Methyl Vinyl Ketone	177	20	915
$CH_3COCH:CH_2$	(81)	(–7)	(491)
Mineral Wax		See Wax, Ozocerite.	
Morpholine	262	98	555
$OC_2H_4NHCH_2CH_2$	(128)	(37)	(290)
Mustard Oil	304	115	
$C_3H_5N:C:S$	(151)	(46)	
(Allyl Isothiocyanate)			

TABLE 2.41 Boiling Points, Flash Points, and Ignition Temperatures of Organic Compounds (*Continued*)

Compound	Boiling point °F (°C)	Flash point, °F (°C)	Ignition point, °F (°C)
Naphtha, Coal		107	531
		(42)	(277)
Naphtha, Petroleum		See Petroleum Ether.	
Naphtha V.M. & P., 50° Flash	240–290	50	450
(10)	(116–143)	(10)	(232)
Naphtha V.M. & P., High Flash	280–350	85	450
	(138–177)	(29)	(232)
Naphtha V.M. & P., Regular	212–320	28	450
	(100–160)	(−2)	(232)
Naphthalene	424	174	979
$C_{10}H_8$	(218)	(79)	(526)
β-Naphthol	545	307	
$C_{10}H_7OH$	(285)	(153)	
(β-Hydroxy Naphthalene)			
(2-Naphthol)			
1-Naphthylamine	572	315	
$C_{10}H_7NH_2$	(300)	(157)	
Nechexane		See 2,2-Dimethylbutane.	
Neopentone		See 2,2-Dimethylpropane.	
Neopentyl Glycol	410	265	750
$HOCH_2C(CH_3)_2CH_2OH$	(210)	(129)	(399)
(2,2-Dimethyl 1,3			
Propanediol)			
Nicoline	475		471
$C_{10}H_{14}N_2$	(246)		(244)
Niobe Oil		See Methyl Benzoate.	
Nitric Ether		See Ethyl Nitrate.	
p-Nitroaniline	637	390	
$NO_2C_6H_4NH_2$	(336)	(199)	
Nitrobenzene	412	190	900
$C_6H_5NO_2$	(211)	(88)	(482)
(Nitrobenzol)			
(Oil of Mirbane)			
1,3-Nitrobenzotrifluoride	397	217	
$C_6H_4NO_2CF_3$	(203)	(103)	
α,μ,α-Trifluoronitrotoluene			
Nitrobenzol		See Nitrobenzene.	
Nitrobiphenyl	626	290	
$C_6H_5C_6H_4NO_2$	(330)	(143)	
p-Nitrochlorobenzene	468	261	
$C_6H_4ClNO_2$	(242)	(127)	
(1-Chloro-4-Nitrobenzene)			
Nitrocyclohexane	403	190	
$CH_2(CH_2)_4CHNO_2$	(206)	(88)	
	Decomposes		
Nitroethane	237	82	778
$C_2H_5NO_2$	(114)	(28)	(414)
Nitroglycerine	502	Explodes	518
$C_3H_5(NO_3)_3$	(261)		(270)
(Glyceryl Trinitrate)	Explodes		
Nitromethane	214	95	785
CH_3NO_2	(101)	(35)	(418)
1-Nitronaphthalene	579	327	
$C_{10}H_7NO_2$	(304)	(164)	

(*Continued*)

TABLE 2.41 Boiling Points, Flash Points, and Ignition Temperatures of Organic Compounds (*Continued*)

Compound	Boiling point °F (°C)	Flash point, °F (°C)	Ignition point, °F (°C)
1-Nitropropane	268	96	789
$CH_3CH_2CH_2NO_2$	(131)	(36)	(421)
2-Nitropropane	248	75	802
$CH_3CH(NO_2)CH_3$	(120)	(24)	(428)
(sec-Nitropropane)			
sec-Nitropropane		See 2-Nitropropane.	
m-Nitrotoluene	450	223	
$C_6H_4CH_3NO_2$	(232)	(106)	
o-Nitrotoluene	432	223	
$C_6H_4CH_3NO_2$	(222)	(106)	
p-Nitrotoluene	461	223	
$HO_2C_6H_4CH_3$	(238)	(106)	
2-Nitro-p-toludine		315	
$CH_3C_6H_3(NH_2)NO_2$		(157)	
Nitrous Ether		See Ethyl Nitrite.	
Nonadecane	628	>212	446
$CH_3(CH_2)_{17}CH_3$	(331)	(>100)	(230)
Nonane	303	88	401
C_9H_{20}	(151)	(31)	(205)
Nonane (iso)	290		428
$C_6H_{13}CH(CH_3)_2$	(143)		(220)
(2-Methyloctane)			
Nonane	291		428
$C_5H_{11}CH(CH_3)C_2H_5$	(144)		(220)
(3-Methyloctane)			
Nonane	288		437
$C_4H_9CH(CH_3)C_3H_7$	(142)		(225)
(4-Methyloctane)			
Nonene	270–290	78	
C_9H_{18}	(132–143)	(26)	
(Nonylene)			
Nonyl Acetate	378	155	
$CH_2COOC_9H_{19}$	(192)	(68)	
Nonyl Alcohol		See Diisobutyl Carbinol.	
Nonylbenzene	468–486	210	
$C_9H_{19}C_6H_5$	(242–252)	(99)	
tert-Nonyl Mercaptan	370–385	154	
$C_9H_{19}SH$	(188–196)	(68)	
Nonylnaphthalene	626–653	<200	
$C_9H_{19}C_{10}H_7$	(330–345)	(<93)	
Nonylphenol	559–567	285	
$C_6H_4(C_9H_{19})OH$	(293–297)	(141)	
2,5-Norbornadiene	193	−6	
C_7H_8	(89)	(−21)	
(NBD)			
Octadecane	603	>212	441
$C_{18}H_{38}$	(317)	(>100)	(227)
Octadecylene α	599	>212	482
$CH_3(CH_2)_{15}CH:CH_2$	(315)	(>100)	(250)
(1-Octadecene)			
Octadecyltrichlorosilane	716	193	
$C_{18}H_{37}SiCl_3$	(380)	(89)	
(Trichlorooctadecylsilane)			

TABLE 2.41 Boiling Points, Flash Points, and Ignition Temperatures of Organic Compounds (*Continued*)

Compound	Boiling point °F (°C)	Flash point, °F (°C)	Ignition point, °F (°C)
Octadecyl Vinyl Ether		See Vinyl Octodecyl Ether.	
Octane	258	56	403
$CH_3(CH_2)_6CH_3$	(126)	(13)	(206)
1-Octanethiol	390	156	
$C_8H_{17}SH$	(199)	(69)	
(n-Octyl Mercapian)			
1-Octanol		See Octyl Alcohol.	
2-Octanol	363	190	
$CH_3CHOH(CH_2)_5CH_3$	(184)	(88)	
1-Octene	250	70	446
$CH_2{:}C_7H_{14}$	(121)	(21)	(230)
Octyl Acetate		See 2-Ethylhexyl Acetate.	
Octyl Alcohol	381	178	
$CH_3(CH_2)_6CH_2OH$	(194)	(81)	
(1-Octanol)			
Octylamine	338	140	
$CH_3(CH_2)_6CH_2NH_2$	(170)	(60)	
(1-Aminooctane)			
tert-Octylamine	284	91	
$(CH_3)_3CCH_2C(CH_3)_2NH_2$	(140)	(33)	
(1,1,3,3-Tetramethyl-			
butylamine)			
Octyl Chloride	359	158	
$CH_3(CH_2)_7Cl$	(182)	(70)	
Octylene Glycol	475	230	635
$(CH_3(CH_2)_2CHOH)_2$	(246)	(110)	(335)
tert-Octyl Mercaptan	318–329	115	
$C_8H_{17}SH$	(159–165)	(46)	
		(oc)	
p-Octylphenyl Salicylate		420	780
$C_{21}H_{26}O_3$		(216)	(416)
Oil of Mirbane		See Nitrobenzene.	
Oil of Wintergreen		See Methyl Salicylate.	
Oleic Acid	547	372	685
$C_8H_{17}CH{:}CH(CH_2)_7COOH$	(286)	(189)	(363)
(Red Oil)			
Distilled		364	
		(184)	
Oxalic Ether		See Ethyl Oxalate.	
Oxirane		See Ethylene Oxide.	
Paraffin Oil		444	
(See also Lubricating Oil)		(229)	
Paraformaldehyde		158	572
$HO(CH_2O)_nH$		(70)	(300)
Paraldehyde	255	96	460
$(CH_3CHO)_3$	(124)	(36)	(238)
1,2,3,4,5-Pentamethyl	449	200	800 est
Benzene	(232)	(93)	(427)
$C_6H(CH_3)_5$			
(Pentamethylbenzene)			

(Continued)

TABLE 2.41 Boiling Points, Flash Points, and Ignition Temperatures of Organic Compounds (*Continued*)

Compound	Boiling point °F (°C)	Flash point, °F (°C)	Ignition point, °F (°C)
Pentamethylene Dichloride		See 1,5-Dichloropentane.	
Pentamethylene Glycol		See 1,5-Pentanediol.	
Pentamethylene Oxide	178	–4	
O(CH₂)₄CH₂	(81)	(–20)	
(Tetrahydropyran)			
Pentanal		See Valeraldehyde.	
Pentane	97	<–40	500
CH₃(CH₂)₃CH₃	(36)	(<–40)	(260)
1,5-Pentanediol	468	265	635
HO(CH₂)₅OH	(242)	(129)	(335)
(Pentamethylene Glycol)			
2,4-Pentanedione	284	93	644
CH₃COCH₂COCH₃	(140)	(34)	(340)
(Acetyl Acetone)			
Pentanoic Acid	366	205	752
C₄H₉COOH	(186)	(96)	(400)
(Valeric Acid)			
1-Pentanol		See Amyl Alcohol.	
2-Pentanol		See Methyl Propyl Carbinol.	
3-Pentanol	241	105	815
CH₃CH₂CH(OH)CH₂CH₃	(116)	(41)	(435)
(tert-n-Amyl Alcohol)			
1-Pentanol Acetate		See Amyl Acetate.	
2-Pentanol Acetate		See sec-Amyl Acetate.	
2-Pentanone		See Methyl Propyl Ketone.	
3-Pentanone		See Diethyl Ketone.	
Pentaphen	482	232	
C₅H₁₁C₆H₄OH	(250)	(111)	
(p-tert-Amyl Phenol)			
1-Pentene	86	0	527
CH₃(CH₂)₂CH:CH₂	(30)	(–18)	(275)
(Amylene)			
1-Pentene-cis		See β-Amylene-cis.	
2-Pentene-trans		See β-Amylene-trans.	
Pentylamine		See Amylamine.	
Pentyloxypentane		See Amyl Ether.	
Pentyl Propionate		See Amyl Propionate.	
1-Pentyne	104	<–4	
HC₁CC₃H₇	(40)	(<–20)	
(n-Propyl Acetylene)			
Perchloroethylene	250	None	None
Cl₂C=CCl₂	(121)		
(Tetrachloroethylene)			
Perhydrophenanthrene	187–192		475
C₁₄H₂₄	(86–89)		(246)
(Tetradecahydro Phenanthrene)			
Petroleum, Crude Oil		20–90	
		(–7 to 32)	

TABLE 2.41   Boiling Points, Flash Points, and Ignition Temperatures of Organic Compounds (*Continued*)

Compound	Boiling point °F (°C)	Flash point, °F (°C)	Ignition point, °F (°C)
Petroleum Ether	95–140	<0	550
(Benzine)	(35–60)	(<–18)	(288)
(Petroleum Naphtha)			
Petroleum Pitch		See Asphalt.	
β-**Pheliandrene**	340	120	
$CH_2:CCH:CHCH[CH(CH_3)_2]CH_2CH_2$	(171)	(49)	
(p-Mentha-1(7), 2-Diene)			
Phenanthrene	644	340	
$(C_6H_4CH)_2$	(340)	(171)	
(Phenanthrin)			
Phenethyl Alcohol	430	205	
$C_6H_5CH_2CH_2OH$	(221)	(96)	
(Benzyl Carbinol)			
(Phenylethyl Alcohol)			
o-Phenetidine	442–446	239	
$H_2NC_6H_4OC_2H_5$	(228–230)	(115)	
(2-Ethoxyaniline)			
(o-Amino-Phenetole)			
p-Phenetidine	378–484	241	
$C_2H_5OC_0H_4NH_2$	(192–251)	(116)	
(1-Amino-4-Ethoxy-benzene)			
(p-Aminophenetole)			
Phenetole		See Ethoxybenzene.	
Phenol	358	175	1319
C_6H_5OH	(181)	(79)	(715)
(Carbolic Acid)			
2-Phenoxyethanol		See Ethylene Glycol, Phenyl Ether.	
Phenoxy Ethyl Alcohol	468	250	
$C_6H_5O(CH_2)_2OH$	(242)	(121)	
(2-Phenoxyethanol)			
(Phenyl Cellosolve)			
N-(2-Phenoxyethyl) Aniline	396	338	
$C_6H_5O(CH_2)_3NHC_6H_5$	(202)	(170)	
β-**Phenoxyethyl Chloride**		See *β*-Chlorophenetole.	
Phenylacetaldehyde	383	160	
$C_6H_5CH_2CHO$	(195)	(71)	
(*α*-Toluic Aldehyde)			
Phenyl Acetate	384	176	
$CH_3COOC_6H_5$	(196)	(80)	
(Acetylphenol)			
Phenylocetic Acid	504	>212	
$C_6H_5CH_2COOH$	(262)	(>100)	
(*α*-Toluic Acid)			
Phenylamine		See Aniline.	
N-Phenylaniline		See Diphenylamine.	
Phenylbenzene		See Biphenyl.	
Phenyl Bromide		See Bromobenzene.	
Phenyl Carbinol		See Benzyl Alcohol.	
Phenyl Chloride		See Chlorobenzene.	
Phenyicyclohexane		See Cyclohexylbenzene.	
Phenyl Didecyl Phosphite	425		
$(C_6H_5O)P(OC_{10}H_{21})_2$	(218)		

(*Continued*)

TABLE 2.41 Boiling Points, Flash Points, and Ignition Temperatures of Organic Compounds (*Continued*)

Compound	Boiling point °F (°C)	Flash point, °F (°C)	Ignition point, °F (°C)
N-Phenyldiethanolamine	376	385	730
$C_6H_5N(C_2H_4OH)_2$	(191)	(196)	(387)
Phenyldiethylamine		See N,N-Diethylaniline.	
o-Phenylenediamine	513	313	
$NH_2C_6H_4NH_2$	(267)	(156)	
(1,2-Diaminobenzene)			
Phenylethane		See Ethylbenzene.	
N-Phenylethanolamine	545	305	
$C_6H_5NHC_2H_4OH$	(285)	(152)	
Phenylethyl Acetate (β)	435	230	
$C_6H_5CH_2CH_2OOCCH_3$	(224)	(110)	
Phenylethyl Alcohol		See Phenethyl Alcohol.	
Phenylethylene		See Styrene.	
N-Phenyl-N-Ethyl-	514	270	685
ethanolamine	(268)	(132)	(362)
$C_6H_5N(C_2H_5)C_2H_4OH$	@740 mm	(oc)	
Phenylhydrazine	Decomposes	190	
$C_6H_5NHNH_2$		(88)	
Phenylmethane		See Toluene.	
Phenylmethyl Ethanol Amine	378	280	
$C_6H_5N(CH_3)C_2H_4OH$	(192)	(138)	
(2-(N-Methylaniline)-	@100 mm		
Ethanol)			
Phenyl Methyl Ketone		See Acetophenone.	
4-Phenylmorpholine	518	220	
$C_6H_5NC_2H_4OCH_2CH_2$	(270)	(104)	
		(oc)	
Phenylpentane		See Amylbenzene.	
o-Phenylphenol	547	255	986
$C_6H_5C_6H_4OH$	(286)	(124)	(530)
Phenylpropane		See Propylbenzene.	
2-Phenylpropane		See Cumene.	
Phenylpropyl Alcohol	426	212	
$C_6H_5(CH_2)_3OH$	(219)	(100)	
(Hydrocinnamic Alcohol)			
(3-Phenyl-l-propanol)			
(Phenylethyl Carbinol)			
Phenyl Propyl Aldehyde		205	
$C_6H_5CH_2CH_2CHO$		(96)	
(3-Phenylpropionaldehyde)			
(Hydrocinnamic Aldehyde)			
Phenyl Toluene o	500	>212	923
$C_6H_5C_6H_4CH_3$	(260)	(>100)	(495)
(2-Methylbiphenyl)			
Phorone	388	185	
$(CH_3)_2CCHCOCHC(CH_3)_2$	(198)	(85)	
Phosphine	−126		212
PH_3	(−88)		(100)
Phthalic Acid	552	334	
$C_6H_4(COOH)_2$	(289)	(168)	
Phthalic Anhydride	543	305	1058
$C_6H_4(CO)_2O$	(284)	(152)	(570)

TABLE 2.41 Boiling Points, Flash Points, and Ignition Temperatures of Organic Compounds (*Continued*)

Compound	Boiling point °F (°C)	Flash point, °F (°C)	Ignition point, °F (°C)
m-Phthalyl Dichloride		See Isophthaloyl Chloride.	
2-Picoline	262	102	1000
$CH_3C_5H_4N$	(128)	(39)	(538)
(2-Methylpyridine)		(oc)	
4-Picoline	292	134	
$CH_3C_5H_4N$	(144)	(57)	
(4-Methylpyridine)			
Pinane	336		523
$C_{10}H_{18}$	(151)		(273)
α-Pinene	312	91	491
$C_{10}H_{16}$	(156)	(33)	(255)
Pine Oil	367–439	172	
Steam Distilled	(186–226)	(78)	
		138	
		(59)	
Pine Pitch	490	285	
	(254)	(141)	
Pine Tar	208	130	671
	(98)	(54)	(355)
Pine Tar Oil		144	
(Wood Tar Oil)		(62)	
Piperazine	294	178	
$HNCH_2CH_2NHCH_2CH_2$	(146)	(81)	
		(oc)	
Piperidine	223	61	
$(CH_2)_5NH$	(106)	(16)	
(Hexahydropyridine)			
Polyamyl Naphthalene	667–747	360	
Mixture of Polymers	(353–397)	(182)	
Polyethylene Glycols		360–550	
$OH(C_2H_5O)_nC_2H_4OH$		(182–287)	
Polyoxyethylene Lauryl Ether		>200	
$C_{12}H_{25}O(OCH_2CH_2)_nOH$		(>93)	
Polypropylene Glycols	Decomposes	365	
$OH(C_3H_6O)_nC_3H_4OH$		(185)	
Polyvinyl Alcohol Mixture of		175	
Polymers		(79)	
Potassium Xanthate	392	205	
$KS_2C\text{-}OC_2H_5$	(200)	(96)	
	Decomposes		
Propanal	120	−22	405
CH_3CH_2CHO	(49)	(−30)	(207)
(Propionaldehyde)			
Propane	−44		842
$CH_3CH_2CH_3$	(−42)		(450)
1,3-Propanediamine	276	75	
$NH_2CH_2CH_2CH_2NH_2$	(136)	(24)	
(1,3-Diaminopropane)			
(Trimethylenediamine)			
1,2-Propanediol		See Propylene Glycol.	
1,3-Propanediol		See Trimethylene Glycol.	
1-Propanol		See Propyl Alcohol.	

(*Continued*)

TABLE 2.41   Boiling Points, Flash Points, and Ignition Temperatures of Organic Compounds (*Continued*)

Compound	Boiling point °F (°C)	Flash point, °F (°C)	Ignition point, °F (°C)
2-Propanol		See Isopropyl Alcohol.	
2-Propanone		See Acetone.	
Propanoyl Chloride		See Propionyl Chloride.	
Propargyl Alcohol	239	97	
HC_1CCH_2OH	(115)	(36)	
(2-Propyn-1-ol)			
Propargyl Bromide	192	50	615
HC_1CCH_2Br	(89)	(10)	(324)
(3-Bromopropyne)			
Propene		See Propylene.	
2-Propenylamine		See Allylamine.	
Propenyl Ethyl Ether	158	<20	
$CH_3CH:CHOCH_2CH_3$	(70)	(<–7)	
β-Propiolactone	311	165	
$C_3H_4O_2$	(155)	(74)	
Propionaldehyde		See Propanal.	
Propionic Acid	297	126	870
CH_3CH_2COOH	(147)	(52)	(465)
Propionic Anhydride	336	145	545
$(CH_3CH_2CO)_2O$	(169)	(63)	(285)
Propionic Nitrile	207	36	
CH_3CH_2CN	(97)	(2)	
(Propionitrile)			
Propionic Chloride	176	54	
CH_3CH_2COCl	(80)	(12)	
(Propanoyl Chloride)			
Propyl Acetate	215	55	842
$C_3H_7OOCCH_3$	(102)	(13)	(450)
(Acetic Acid, n-Propyl Ester)			
Propyl Alcohol	207	74	775
$CH_3CH_2CH_2OH$	(97)	(23)	(412)
(1-Propanol)			
Propylamine	120	–35	604
$CH_3(CH_2)_2NH_2$	(49)	(–37)	(318)
Propylbenzene	319	86	842
$C_3H_7C_6H_5$	(159)	(30)	(450)
(Phenylpropane)			
2-Propylbiphenyl	~536	>212	833
$C_6H_5C_6H_4C_3H_7$	(~280)	(>100)	(445)
n-Propyl Bromide	160		914
C_3H_7Br	(71)		(490)
(1-Bromopropane)			
n-Propyl Butyrate	290	99	
$C_3H_7COOC_3H_7$	(143)	(37)	
Propyl Carbinol		See Butyl Alcohol.	
Propyl Chloride	115	<0	968
C_3H_7Cl	(46)	(<–18)	(520)
Propyl Chlorothiolformate	311	145	
C_3H_7SCOCl	(155)	(63)	
Propylcyclohexane	313–315		478
$H_7C_3C_6H_{11}$	(156–157)		(248)
Propylcyclopentane	269		516
$C_3H_7C_5H_9$	(131)		(269)
(1-Cyclopentylpropane)			

TABLE 2.41   Boiling Points, Flash Points, and Ignition Temperatures of Organic Compounds (*Continued*)

Compound	Boiling point °F (°C)	Flash point, °F (°C)	Ignition point, °F (°C)
Propylene	−53	Gas	851
CH$_2$:CHCH$_3$	(−47)		(455)
(Propene)			
Propylene Aldehyde		See Crotonaldehyde.	
Propylene Carbonate	468	275	
OCH$_2$CH$_2$CH$_2$OCO	(242)	(135)	
Propylene Chlorohydrin		See 2-Chloro-1-Propanol.	
sec-Propylene Chlorohydrin		See 1-Chloro-2-Propanol.	
Propylenediamine	246	92	780
CH$_3$CH(NH$_2$)CH$_2$NH$_2$	(119)	(33)	(416)
		(oc)	
Propylene Dichloride	205	60	1035
CH$_3$CHClCH$_2$Cl	(96)	(16)	(557)
(1,2-Dichloropropane)			
Propylene Glycol	370	210	700
CH$_3$CHOHCH$_2$OH	(188)	(99)	(371)
(Methyl Ethylene Glycol)			
(1,2-Propanediol)			
Propylene Glycol Acetate		See Methyl Glycol Acetate.	
Propylene Glycol Isopropyl	283	110	
Ether	(140)	(43)	
Propylene Glycol Methyl Ether	248	90	
CH$_3$OCH$_2$CHOHCH$_3$	(120)	(32)	
(1-Methoxy-2-propanol)			
Propylene Glycol Methyl Ether	295	108	
Acetate	(146)	(42)	
(99% Pure)			
Propylene Glycol Monoacylate	410	207	
CH$_2$:CHCOO(C$_3$H$_6$)OH	(210)	(97)	
(Hydroxypropyl Acrylate)			
Propylene Oxide	94	−35	840
OCH$_2$CHCH$_3$	(35)	(−37)	(449)
n-Propyl Ether	194	70	370
(C$_3$H$_7$)$_2$O	(90)	(21)	(188)
(Dipropyl Ether)			
Propyl Formate	178	27	851
HCOOC$_3$H$_7$	(81)	(−3)	(455)
Propyl Methanol		See Butyl Alcohol.	
Propyl Nitrate	231	68	347
CH$_3$CH$_2$CH$_2$NO$_3$	(111)	(20)	(175)
Propyl Proplonate	245	175	
CH$_3$CH$_2$COOCH$_2$CH$_2$CH$_3$	(118)	(79)	
Propyltrichlorosilane	254	98	
(C$_3$H$_7$)SiCl$_3$	(123.5)	(37)	
Propyne	−10		
CH$_3$C$_1$CH	(−23)		
(Allylene)			
(Methylacetylene)			
Pseudocumene		See 1,2,4-Trimethylbenzene.	
Pyridine	239	68	900
CH < (CHCH)$_2$ >N	(115)	(20)	(482)

(*Continued*)

TABLE 2.41   Boiling Points, Flash Points, and Ignition Temperatures of Organic Compounds (*Continued*)

Compound	Boiling point °F (°C)	Flash point, °F (°C)	Ignition point, °F (°C)
Pyrrole	268	102	
(CHCH)₂NH	(131)	(39)	
(Azole)			
Pyrrolidine	186–189	37	
NHCH₂CH₂CH₂CH₂	(86–87)	(3)	
(Tetrahydropyrrole)			
2-Pyrrolidine	473	265	
NHCOCH₂CH₂CH₂	(245)	(129)	
Quinoline	460		896
C₆H₄N:CHCH:CH	(238)		(480)
Range Oil	See Fuel Oil No. 1.		
Rape Seed Oil		325	836
(Colza Oil)		(163)	(447)
Resorcinol	531	261	1126
C₆H₄(OH)₂	(277)	(127)	(608)
(Dihydroxybenzol)			
Rhodinol	237–239	>212	
CH₂:C(CH₃)(CH₂)₃CH—	(114–115)	(>100)	
(CH₃)(CH₂)₂OH	@12 mm		
Rosin Oil	>680	266	648
	(>360)	(130)	(342)
Salicylaldehyde	384	172	
HOC₆H₄CHO	(196)	(78)	
(o-Hydroxybenzaldehyde)			
Salicylic Acid	Sublimes	315	1004
HOC₆H₄COOH	@169	(157)	(540)
	(76)		
Safrole	451	212	
C₃H₅C₆H₃O₂CH₂	(233)	(100)	
(4-allyl-1,2-Methylenedioxy-			
benzene)			
Santatol	~575	>212	
C₁₅H₂₄O	(~300)	(>100)	
(Arheol)			
Sesame Oil		491	
		(255)	
Soy Bean Oil		540	833
		(282)	(445)
Sperm Oil No. 1		428	586
No. 2		(220)	(308)
		460	
		(238)	
Stearic Acid	726	385	743
CH₃(CH₂)₁₆COOH	(386)	(196)	(395)
Steryl Alcohol	410		842
CH₃(CH₂)₁₇OH	(210)		(450)
(1-Ocladecanol)	@15 mm		
Styrene	295	88	914
C₆H₅CH:CH₂	(146)	(31)	(490)
(Cinnamene)			
(Phenylethylene)			
(Vinyl Benzene)			

TABLE 2.41 Boiling Points, Flash Points, and Ignition Temperatures of Organic Compounds (*Continued*)

Compound	Boiling point °F (°C)	Flash point, °F (°C)	Ignition point, °F (°C)
Styrene Oxide		165	929
$C_6H_5CHOCH_2$		(74)	(498)
Succinonitrile	509–513	270	
$NCCH_2CH_2CN$	(265–267)	(132)	
(Ethylene Dicyanide)			
Sulfolane	545	350	
$CH_2(CH_2)_3SO_2$	(285)	(177)	
(Tetrahydrothiophene-1,1-Dioxide)			
(Tetramethylene Sulfone)			
Tartaric Acid (d, 1)		410	797
$(CHOHCO_2H)_2$		(210)	(425)
		(oc)	
Terephthalle Acid	Sublimes	500	925
$C_6H_4(COOH)_2$	above	(260)	(496)
(para-Phthalic Acid)	572		
(Benzene-para-Dicarboxylic Acid)	(300)		
Terephthaloyl Chloride	498	356	
$C_6H_4(COCl)_2$	(259)	(180)	
(Terephthaloyl Dichloride)			
(p-Phthaloyl Dichloride)			
(1,4-Benzenedicarbonyl Chloride)			
o-Terphenyl	630	325	
$(C_6H_5)_2C_6H_4$	(332)	(163)	
m-Terphenyl	685	375	
$(C_6H_5)_2C_6H_4$	(363)	(191)	
Terpineol	417–435	195	
$C_{10}H_{17}OH$	(214–224)	(91)	
(Terpilenol)			
Terpinyl Acetate	428	200	
$C_{10}H_{17}OOCCH_3$	(220)	(93)	
Tetraamylbenzene	608–662	295	
$(C_5H_{11})_4C_6H_2$	(320–350)	(146)	
1,1,2,2-Tetrabromoethane	275		635
$CHBr_2CHBr_2$	(135)		(335)
(Acetylene Tetrabromide)			
1,2,4,5-Tetrachlorobenzene	472	311	
$C_6H_{12}Cl_4$	(245)	(155)	
Tetradecane	487	212	392
$CH_3(CH_2)_{12}CH_3$	(253)	(100)	(200)
Tetradecanol	507	285	
$C_{14}H_{29}OH$	(264)	(141)	
		(oc)	
1-Tetradecene	493	230	455
$CH_2:CH(CH_2)_{11}CH_3$	(256)	(110)	(235)
tert-Tetradecyl Mercaptan	496–532	250	
$C_{14}H_{29}SH$	(258–278)	(121)	
Tetraethoxypropane	621	190	
$(C_2H_5O)_4C_3H_4$	(327)	(88)	
Tetra (2-Ethylbutyl) Silicate	460	335	
$[C_2H_5CH(C_2H_5)CH_2O]_4Si$	(238)	(168)	
	@50 mm		
Tetraethylene Glycol	Decomposes	360	
$HOCH_2(CH_2OCH_2)_3CH_2OH$		(182)	

(*Continued*)

TABLE 2.41 Boiling Points, Flash Points, and Ignition Temperatures of Organic Compounds (*Continued*)

Compound	Boiling point °F (°C)	Flash point, °F (°C)	Ignition point, °F (°C)
Tetraethylene Glycol, Dimethyl Ether		See Dimethoxy Tetraglycol.	
Tetraethylene Pentamine	631	325	610
$H_2N(C_2H_4NH)_3C_2H_4NH_2$	(333)	(163)	(321)
Tetra (2-Ethylhexyl) Silicate		390	
$[C_4H_9CH(C_2H_5)CH_2O]_4Si$		(199)	
Tetrafluoroethylene	−105		392
$F_2C{:}CF_2$	(−76)		(200)
(TFE)			
(Perfluoroethylene)			
1,2,3,6-Tetrahydrobenzaldehyde	328	135	
$CH_2CH{:}CHCH_2CH_2CHCHO$	(164)	(57)	
(3-Cyclohexene-1-Carboxaldehyde)			
endo-Tetrahydrodicyclopentadiene	379		523
$C_{10}H_{16}$	(193)		(273)
(Tricyclodecane)			
Tetrahydrofuran	151	6	610
$OCH_2CH_2CH_2CH_2$	(66)	(−14)	(321)
(Diethylene Oxide)			
(Tetramethylene Oxide)			
Tetrahydrofurfuryl Alcohol	352	167	540
$C_4H_7OCH_2OH$	(178)	(75)	(282)
	@743 mm		
Tetrahydrofurfuryl Oleale	392–545	390	
$C_4H_7OCH_2OOCC_{17}H_{33}$	(200–285)	(199)	
	@16 mm		
Tetrahydronaphthalene	405	160	725
$C_6H_2(CH_3)_2C_2H_4$	(207)	(71)	(385)
(Tetralin)			
Tetrahydropyran		See Pentamethylene Oxide.	
Tetrahydropyran-2-Methanol	368	200	
$OCH_2CH_2CH_2CH_2CHCH_2OH$	(187)	(93)	
Tetrahydropyrrole		See Pyrrolidine.	
Tetralin		See Tetrahydronaphthalene.	
1,1,3,3-Tetramethoxy-propane	361	170	
$[(CH_3O)_2CH]_2CH_2$	(183)	(77)	
1,2,3,4-Tetramethylbenzene 95%	399–401	166	800
$C_6H_2(CH_3)_4$	(204–205)	(74)	est.
(Prohnitene)			(427)
1,2,3,5-Tetramethylbenzene 85.5%	387–389	160	800
$C_6H_2(CH_3)_4$	(197–198)	(71)	est.
(Isodurene)			(427)
1,2,4,5-Tetramethylbenzene 95%	385	130	
$C_6H_2(CH_3)_4$	(196)	(54)	
(Durene)			
Tetramethylene		See Cyclobutane.	
Tetramethyleneglycol	230	734	
$CH_2OH(CH_2)_2CH_2OH$	(110)	(390)	
Tetramethylene Oxide		See Tetrahydrofuran.	
Tetramethyl Lead, Compounds		100	
$Pb(CH_3)_4$		(38)	

TABLE 2.41 Boiling Points, Flash Points, and Ignition Temperatures of Organic Compounds (*Continued*)

Compound	Boiling point °F (°C)	Flash point, °F (°C)	Ignition point, °F (°C)
2,2,3,3-Tetramethyl Pentane	273	<70	806
$(CH_3)_3CC(CH_3)_2CH_2CH_3$	(134)	(<21)	(430)
2,2,3,4-Tetramethyl-pentane	270	<70	
$(CH_3)_3CCH(CH_3)CH(CH_3)_2$	(132)	(<21)	
	172	<70	
Thialdine	Decomposes	200	
$SCH(CH_3)SCH(CH_3)NHCHCH_3$		(93)	
2,2-Thiodiethanol	540	320	
$(HOCH_2CH_2)_2S$	(282)	(160)	
(Thiodiethylene Glycol)			
Thiodiethylene Glycol		See 2,2-Thiodiethanol.	
Thiodiglycol	541	320	568
$(CH_2CH_2OH)_2S$	(283)	(160)	(298)
(Thiodiethylene Glycol)			
(Beta-bis-Hydroxyethyl Sulfide)			
(Dihydroxyethyl Sulfide)			
Thiophene	184	30	
$SCH:CHCH:CH$	(84)	(−1)	
1,4-Thioxane	300	108	
$O(CH_2CH_2)_2S$	(149)	(42)	
(1,4-Oxathiane)			
Toluene	231	40	896
$C_6H_5CH_3$	(111)	(4)	(480)
(Methylbenzene)			
(Phenylmethane)			
(Toluol)			
Toluene-2,4-Diisocyanate	484	260	
$CH_3C_6H_3(NCO)_2$	(251)	(127)	
p-Toluenesulfonic Acid	295	363	
$C_6H_4(SO_3H)(CH_3)$	(140)	(184)	
	@20 mm		
Toluhydroquinone	545	342	875
$C_6H_3(OH)_2CH_3$	(285)	(172)	(468)
(Methylhydroquinone)			
o-Toluidine	392	185	900
$CH_3C_6H_4NH_2$	(200)	(85)	(482)
(2-Methylaniline)			
p-Toluidine	392	188	900
$CH_3C_6H_4NH_2$	(200)	(87)	(482)
(4-Methylaniline)			
Toluol		See Toluene.	
m-Tolyldiethanolamine	400	740	0.6
$(HOC_2H_4)_2NC_6H_4CH_3$	(204)	(393)	
(MTDEA)			
2,4-Tolylene Diisocyanate		See Toluene-2,4-Diisocyanate.	
o-Tolyl Phosphate		See Tri-o-Cresyl Phosphate.	
o-Tolyl p-Toluene Sulfonate		363	
$C_{14}H_{14}O_3S$		(184)	
Transformer Oil		295	
(Tronsil Oil)		(146)	
Triacetin		See Glyceryl Triacetate.	

TABLE 2.41   Boiling Points, Flash Points, and Ignition Temperatures of Organic Compounds (*Continued*)

Compound	Boiling point °F (°C)	Flash point, °F (°C)	Ignition point, °F (°C)
Triamylamine $(C_5H_{11})_3N$	453 (234)	215 (102)	
Triamylbenzene $(C_5H_{11})_3C_6H_3$	575 (302)	270 (132)	
Tributylamine $(C_4H_9)_3N$	417 (214)	187 (86)	
Tri-n-Butyl Borate $B(OC_4H_9)_3$	446 (230)	200 (93)	
Tributyl Citrate $C_3H_4(OH)(COOC_4H_9)_3$	450 (232)	315 (157)	695 (368)
Tributyl Phosphate $(C_4H_9)_3PO_4$	560 (293)	295 (146)	
Tributylphosphine $(C_4H_9)_3P$	473 (245)		392 (200)
Tributyl Phosphite $(C_4H_9)_3PO_3$	244–250 (118–121) @7 mm	248 (120)	
1,2,4-Trichlorobenzene $C_6H_3Cl_3$	415 (213)	222 (105)	1060 (571)
1,1,1-Trichloroethane CH_3CCl_3 (Methyl Chloroform)	165 (74)		
Trichloroethylene $ClHC:CCl_2$	188 (87)		788 (420)
1,2,3-Trichloropropane $CH_2ClCHClCH_2Cl$ (Allyl Trichloride) (Glyceryl Trichlorohydrin)	313 (156)	160 (71)	
Trichlorosilane $HSiCl_3$	89 (32)	7 (−14)	
Tri-o-Cresyl Phosphate $(CH_3C_6H_4)_3PO_4$ (o-Tolyl Phosphate)	770 (410) Decomposes	437 (225)	725 (385)
Tridecanol $CH_3(CH_2)_{12}OH$	525 (274)	250 (121)	
2-Tridecanone		See Methyl Undecyl Ketone.	
Tridecyl Acrylate $CH_2:CHCOOC_{13}H_{27}$	302 (150) @10 mm	270 (132)	
Tridecyl Alcohol $C_{12}H_{25}CH_2OH$ (Tridecanol)	485–503 (252–262)	180 (82)	
Tridecyl Phosphite $(C_{10}H_{21}O)_3P$	356 (180) @0.1 mm	455 (235)	
Triethanolamine $(CH_2OHCH_2)_3N$ (2,2′,2″-Nitrilotriethanol)	650 (343)	354 (179)	
1,1,3-Triethoxyhexane $CH(OC_2H_5)_2CH_2CH-$ $(OC_2H_5)C_8H_7$	271 (133) @50 mm Decomposes @760 mm	210 (99)	
Triethylamine $(C_2H_5)_3N$	193 (89)	16 (−7)	480 (249)

TABLE 2.41 Boiling Points, Flash Points, and Ignition Temperatures of Organic Compounds (*Continued*)

Compound	Boiling point °F (°C)	Flash point, °F (°C)	Ignition point, °F (°C)
1,2,4-Triethylbenzene	423	181	
$(C_2H_5)_3C_6H_3$	(217)	(83)	
Triethyl Citrate	561	303	
$HOC(CH_2CO_2C_2H_5)CO_2H_2H_5$	(294)	(151)	
Triethylene Glycol	546	350	700
$HOCH_2(CH_2OCH_2)_2CH_2OH$	(286)	(177)	(371)
(Dicaproate)			
Triethylene Glycol Diacetate	572	345	
$CH_3COO(CH_2CH_2O)_3COCH_3$	(300)	(174)	
(TDAC)			
Triethylene Glycol, Dimethyl Ether	421	232	
$CH_3(OCH_2)_3OCH_3$	(216)	(111)	
Triethylene Glycol, Ethyl Ether		See Ethoxytriglycol.	
Triethylene Glycol, Methyl Ether		See Methoxy Triglycol.	
Triethyleneglycol Monobutyl Ether	270	290	
$C_4H_9O(C_2H_4O)_3H$	(132)	(143)	
Triethylenetetramine	532	275	640
$N_2NCH_2(CH_2NHCH_2)_2CH_2NH_2$	(278)	(135)	(338)
Triethyl Phosphate	408–424	240	850
$(C_2H_5)_3PO_4$	(209–218)	(115)	(454)
(Ethyl Phosphate)			
Trifluorochloroethylene	−18		
$CF_2:CFCl$ (R-1113)	(−28)		
(Chlorotrifluoroethylene)			
Triglycol Dichloride	466	250	
$ClCH_2(CH_3OCH_2)_2CH_2Cl$	(241)	(121)	
Trihexyl Phosphite	275–286	320	
$(C_6H_{13})_3PO_3$	(135–141) @2 mm	(160)	
Triisopropanolamine	584	320	608
$[(CH_3)_2COH]_3N$	(307)	(160)	(320)
(1,1′,1″-Nitrolotri-2-propanol)			
Triisopropylbenzene	495	207	
$C_6H_3(CH_3CHCH_3)_3$	(237)	(97)	
Triisopropyl Borate	288	82	
$(C_3H_7O)_3B$	(142)	(28)	
Triiauryl Trithiophosphite		398	
$[CH_3(CH_2)_{11}S]_3P$		(203)	
Trimethylamine	38		374
$(CH_3)_3N$	(3)		(190)
1,2,3-Trimethylbenzene	349	111	878
$C_6H_3(CH_3)_3$	(176)	(44)	(470)
(Hemellitol)			
1,2,4-Trimethylbenzene	329	112	932
$C_6H_3(CH_3)_3$	(165)	(44)	(500)
(Pseudocumene)			
1,3,5-Trimethylbenzene	328	122	1039
$C_6H_3(CH_3)_3$	(164)	(50)	(559)
(Mesitylene)			
Trimethyl Borate		See Methyl Borate.	
2,2,3-Trimethylbutane	178	<32	774
$(CH_3)_3C(CH_3)CHCH_3$	(81)	(<0)	(412)
(Triptane—an isomer of Heptane)			

TABLE 2.41   Boiling Points, Flash Points, and Ignition Temperatures of Organic Compounds (*Continued*)

Compound	Boiling point °F (°C)	Flash point, °F (°C)	Ignition point, °F (°C)
2,3,3-Trimethyl-1-Butene $(CH_3)_3CC(CH_3){:}CH_2$ (Heptylene)	172 (78)	<32 (<0)	707 (375)
Trimethyl Carbinol		See tert-Butyl Alcohol.	
Trimethylchlorosilane $(CH_3)_3SiCl$	135 (57)	−18 (−28)	
1,3,5-Trimethylcyclohexane $(CH_3)_3C_6H_9$ (Hexahydromesitylene)	283 (139)		597 (314)
Trimethylcyclohexanol $CH(OH)CH_2C(CH_3)_2CH_2CH(CH_3)CH_2$	388 (198)	165 (74)	
3,3,5-Trimethyl-1-Cyclohexanol $CH_2CH(CH_3)CH_2C(CH_3)_2CH_2CHOH$	388 (198)	190 (88)	
Trimethylene		See Cyclopropane.	
Trimethylenediamine		See 1,3-Propanediamine.	
Trimethylene Glycol $HO(CH_2)_3OH$ (1,3-Propanediol)	417 (214)		752 (400)
Trimethylethylene		See 2-methyl-2-Butene.	
2,5,5-Trimethylheptane $C_2H_5C(CH_3)_2(CH_2)_2CH(CH_3)_2$	304 (151)	<131 (<55)	527 (275)
2,2,5-Trimethylhexane $(CH_3)_3C(CH_2)_2CH(CH_3)_2$	255 (124)	55 (13) (oc)	
3,5,5-Trimethylhexanol $CH_3C(CH_3)_2CH_2CH(CH_3)CH_2\text{-}$ CH_2OH	381 (194)	200 (93)	
2,4,8-Trimethyl-6-Nonanol $C_4H_9CH(OH)C_7H_{15}$ (2,6,8-Trimethyl-4-nonanol)	491 (255)	199 (93)	
2,6,8-Trimethyl-4-Nonanol $(CH_3)_2CHCH_2CH(OH)CH_2\text{-}$ $CH(CH_3)CH_2CH(CH_3)_2$	438 (226)	200 (93)	
2,6,8-Trimethyl-4-Nonanone $(CH_3)_2CHCH_2CH(CH_3)CH_2\text{-}$ $COCH_2CH(CH_3)_2$	425 (218)	195 (91)	
2,2,4-Trimethylpentane $(CH_3)_3CCH_2CH(CH_3)_2$	211 (99)	10 (−12)	779 (415)
2,3,3-Trimethylpentane $CH_3CH_2C(CH_3)_2CH(CH_3)_2$	239 (115)	<70 (<21)	797 (425)
2,2,4-Trimethyl-1,3-Pentanediol $(CH_3)_2CHCH(OH)C(CH_3)_2\text{-}$ CH_2OH	419–455 (215–235)	235 (113)	655 (346)
2,2,4-Trimethyl pentanediol Diisobutyrate $C_{16}H_{30}O_4$	536 (280)	250 (121)	795 (424)
2,2,4-Trimethyl 1,3-Pentanediol Isobutyrate $(CH_3)_2CHCH(OH)C(CH_3)_2\text{-}$ $CH_2OOCCH(CH_3)_2$	356–360 125 mm (180–182)	248 (120)	740 (393)
2,2,4-Trimethylpentanediol Isobutyrate Benzoate $C_{19}H_{28}O_4$	167 (75) @ 10 mm	325 (163)	

TABLE 2.41   Boiling Points, Flash Points, and Ignition Temperatures of Organic Compounds (*Continued*)

Compound	Boiling point °F (°C)	Flash point, °F (°C)	Ignition point, °F (°C)
2,3,4-Trimethyl-1-Pentene	214	<70	495
$H_2C{:}C(CH_3)CH(CH_3)CH(CH_3)_2$	(101)	(<21)	(257)
2,4,4-Trimethyl-1-Pentene	214	23	736
$CH_2{:}C(CH_3)CH_2C(CH_3)_3$	(101)	(−5)	(391)
(Diisobutylene)			
2,4,4-Trimethyl-2-Pentene	221	35	581
$CH_3CH{:}C(CH_3)C(CH_3)_3$	(105)	(2)	(305)
		(oc)	
3,4,4-Trimethyl-2-Pentene	234	<70	617
$(CH_3)_3CC(CH_3){:}CHCH_3$	(112)	(<21)	(325)
Trimethyl Phosphite	232–234	130	
$(CH_3O)_3P$	(111–112)	(54)	
Trioctyl Phosphite	212	340	
$(C_8H_{17}O)_3P$	(100)	(171)	
[Tris (2-Ethylhexyl)	@0.01 mm		
Phosphite]			
Trioxane	239	113	777
$OCH_2OCH_2OCH_2$	(115)	(45)	(414)
	Sublimes		
Triphenylmethane	678	>212	
$(C_6H_5)_3CH$	(359)	(>100)	
Triphenyl Phosphate	750	428	
$(C_6H_5)_3PO_4$	(399)	(220)	
Triphenylphosphine		See Triphenylphosphorus.	
Triphenyl Phosphite	311–320	425	
$(C_6H_5O)_3PO_3$	(155–160)	(218)	
	@0.1 mm		
Triphenylphosphorus	711	356	
$(C_6H_5)_3P$	(377)	(180)	
(Triphenylphosphine)			
Tripropylamine	313	105	
$(CH_3CH_2CH_2)_3N$	(156)	(41)	
Tripropylene	271–288	75	
C_9H_{18}	(133–142)	(24)	
(Propylene Trimer)			
Tripropylene Glycol	514	285	
$H(OC_3H_6)_3OH$	(268)	(141)	
Tripropylene Glycol	470	250	
Methyl Ether	(243)	(121)	
$HO(C_3H_6O)_2C_3H_6OCH_3$			
Tris (2-Ethylhexyl) Phosphite		See Trioctyl Phosphite.	
Tung Oil		552	855
(China Wood Oil)		(289)	(457)
Turkey Red Oil		476	833
		(247)	(445)
Turpentine	300	95	488
	(149)	(35)	(253)
Undecane		See Hendecane.	
2-Undecanol	437	235	
$C_4H_9CH(C_2H_5)C_2H_4{-}$	(225)	(113)	
$CH(OH)CH_3$			
Valeraldehyde	217	54	432
$CH_3(CH_2)_3CHO$	(103)	(12)	(222)
(Pentanal)			

TABLE 2.41 Boiling Points, Flash Points, and Ignition Temperatures of Organic Compounds (*Continued*)

Compound	Boiling point °F (°C)	Flash point, °F (°C)	Ignition point, °F (°C)
Valeric Acid		See Pentanoic Acid.	
Vinyl Acetate	161	18	756
CH$_2$:CHOOCCH$_3$	(72)	(−8)	(402)
(Ethenyl Ethanoate)			
Vinylaceto-β-Lactone		See Diketene.	
Vinyl Acetylene	41		
CH$_2$:CHC:CH	(5)		
(1-Buten-3-yne)			
Vinyl Allyl Ether	153	<68	
CH$_2$:CHOCH$_2$CH$_2$O(CH$_2$)$_3$CH$_3$	(67)	(<20)	
(Allyl Vinyl Ether)			
Vinylbenzene		See Styrene.	
Vinylbenzylchloride	444	220	
ClCH$_2$H$_6$H$_4$CH:CH$_2$	(229)	(104)	
Vinyl Bromide	60	None	986
	(15.8)		(530)
Vinyl Butyl Ether	202	15	437
CH$_2$:CHOCH$_4$H$_9$	(94)	(−9)	(255)
(Butyl Vinyl Ether)			
Vinyl Butyrate	242	68	
CH$_2$:CHOCOC$_3$H$_7$	(117)	(20)	
Vinyl 2-Chloroethyl Ether	228	80	
CH$_2$:CHOCH$_2$CH$_2$Cl	(109)	(27)	
(2-Chloroethyl Vinyl Ether)			
Vinyl Chloride	7	−108.4	882
CH$_2$CHCl	(−14)	(−78)	(472)
(Chloroethylene)			
Vinyl Crotonate	273	78	
CH$_2$:CHOCOCH:CHCH$_3$	(134)	(26)	
Vinyl Cyanide		See Acrylonitrile.	
4-Vinyl Cyclohexene	266	61	517
C$_8$H$_{12}$	(130)	(16)	(269)
Vinyl Ether		See Divinyl Ether.	
Vinyl Ethyl Alcohol	233	100	
CH$_2$:CH(CH$_2$)$_2$OH	(112)	(38)	
(3-Buten-1-ol)			
Vinyl Ethyl Ether	96	<−50	395
CH$_2$:CHOC$_2$H$_5$	(36)	(<−46)	(202)
(Ethyl Vinyl Ether)			
Vinyl 2-Ethylhexoate	365	165	
CH$_2$:CHOCOCH(C$_2$H$_5$)C$_4$H$_9$	(185)	(74)	
Vinyl 2-Ethylhexyl Ether	352	135	395
C$_{10}$H$_{20}$O	(178)	(57)	(202)
(2-Ethylhexyl Vinyl Ether)			
2-Vinyl-5-Ethylpyridine	248	200	
N:C(CH:CH$_2$)CH:CHC(C$_2$H$_5$):CH	(120)	(93)	
	@50 mm		
Vinyl Fluoride	−97.5		
CH$_2$:CHF	(−72)		
Vinylidene Chloride	89	−19	1058
CH$_2$:CCl$_2$	(32)	(−28)	(570)
(1,1-Dichloroethylene)			
Vinylidene Fluoride	−122.3		
CH$_2$:CF$_2$	(−86)		

TABLE 2.41 Boiling Points, Flash Points, and Ignition Temperatures of Organic Compounds (*Continued*)

Compound	Boiling point °F (°C)	Flash point, °F (°C)	Ignition point, °F (°C)
Vinyl Isobutyl Ether $CH_2:CHOCH_2CH(CH_3)CH_3$ (Isobutyl Vinyl Ether)	182 (83)	15 (−9)	
Vinyl Isooctyl Ether $CH_2:CHO(CH_2)_5CH(CH_3)_2$ (Isooctyl Vinyl Ether)	347 (175)	140 (60)	
Vinyl Isopropyl Ether $CH_2:CHOCH(CH_3)_2$ (Isopropyl Vinyl Ether)	133 (56)	−26 (−32)	522 (272)
Vinyl 2-Methoxyethyl Ether $CH_2:CHOC_2H_4OCH_3$ (1-Methoxy-2-Vinyloxyethane)	228 (109)	64 (18)	
Vinyl Methyl Ether $CH_2:CHOCH_3$ (Methyl Vinyl Ether)	43 (6)		549 (287)
Vinyl Octadecyl Ether $CH_2:CHO(CH_2)_{17}CH_3$ (Octadecyl Vinyl Ether)	297–369 (147–187) @5 mm	350 (177)	
Vinyl Propionate $CH_2:CHOCOC_2H_5$	203 (95)	34 (1)	
1-Vinylpyrrolidone $CH_2:CHNCOCH_2CH_2CH_2$ (Vinyl-2-Pyrrolidone)	205 (96) @14 mm	209 (98)	
Vinyl-2-Pyrrolidone		See 1-Vinylpyrrolidone.	
Vinyl Trichlorosilane $CH_2:CHSiCl_3$	195 (91)	70 (21)	
Wax, Microcrystalline		>400 (>204)	
Wax, Ozocerite (Mineral Wax)		236 (113)	
Wax, Paraffin	>700 (>371)	390 (199)	473 (245)
White Tar		See Naphthalene.	
Wood Alcohol		See Methyl Alcohol.	
Wood Tar Oil		See Pine Tar Oil.	
Wool Grease		See Lanolin.	
m-Xylene $C_6H_4(CH_3)_2$ (1,3-Dimethylbenzene)	282 (139)	81 (27)	982 (527)
o-Xylene $C_6H_4(CH_3)_2$ (1,2-Dimethylbenzene) (o-Xylol)	292 (144)	90 (32)	867 (463)
p-Xylene $C_6H_4(CH_3)_2$ (1,4-Dimethylbenzene)	281 (138)	81 (27)	984 (528)
o-Xylidine $C_6H_3(CH_3)_2NH_2$ (o-Dimethylaniline)	435 (224)	206 (97)	
o-Xylol		See o-Xylene.	

TABLE 2.42   Properties of Combustible Mixtures in Air

The *autoignition temperature* is the minimum temperature required for self-sustained combustion in the absence of an external ignition source. The value depends on specified test conditions. The *flammable (explosive) limits* specify the range of concentration of the vapor in air (in percent by volume) for which a flame can propagate. Below the lower flammable limit, the gas mixture is too lean to burn; above the flammable limit, the mixture is too rich.

Substance	Autoignition temperature, °C	Flammable (explosive) limits, percent by volume of fuel (25°C, 760 mm)	
		Lower	Upper
Acetaldehyde	175	4.0	60
Acetanilide	540		
Acetic acid, glacial	463	4.0	19.9
Acetic anhydride	316	2.7	10.3
Acetone	465	2.5	12.8
Acetonitrile	524	3.0	16.0
Acetophenone	570		
Acetylacetone	340		
Acetylene	305	3.0	65
Acetyl chloride	390		
Acrolein	220	2.8	31.0
Acrylic acid (2-propenoic acid)	438	2.4	8.0
Acrylonitrile	481	3.0	17.0
Adiponitrile	550	2	5
Allyl acetate	374		
Allyl alcohol	378	2.5	18.0
Allylamine	374	2.2	22
Ammonia, anhydrous	651	16	25
Aniline	615	1.3	11
Asphalt	485		
Benzaldehyde	192		
Benzene	498	1.2	7.8
Benzoyl peroxide	80		
Benzyl acetate	460		
Benzyl alcohol	436		
Benzyl benzoate	480		
Benzyl chloride	585	1.1	
Bis(2-aminoethyl)amine	399		
Bis(2-chloroethyl) ether	369	2.7	
Biscyclohexyl	245	0.7	5.1
Bis(2-hydroethyl) ether	229		
Bromobenzene	565		
1-Bromobutane	265	2.6	6.6
Bromoethane	511	6.8	8.0
Bromomethane	537	10	16.0
1-Bromopropane	490		
3-Bromopropene	295	4.4	7.3
1,3-Butadiene	420	2.0	11.5
Butanal (butyraldehyde)	218	1.9	12.5
Butane	287	1.9	8.5
1,3-Butanediol	395		
2,3-Butanediol	402		
Butanenitrile	501	1.65	
Butanoic acid (butyric acid)	443	2.0	10.0
Butanoic anhydride (butyric anhydride)	279	0.9	5.8

TABLE 2.42 Properties of Combustible Mixtures in Air (*Continued*)

Substance	Autoignition temperature, °C	Flammable (explosive) limits, percent by volume of fuel (25°C, 760 mm)	
		Lower	Upper
1-Butanol	343	1.4	11.2
2-Butanol	415	1.7	11
2-Butanone	404	1.4	11.4
trans-2-Butenal (crotonaldehyde)	232	2.1	15.9
1-Butene	384	1.6	9.3
cis-2-Butene	324	1.7	
trans-2-Butene	324	1.8	9.7
1-Butene oxide		1.5	18.3
3-Buten-l-ol		4.7	34
2-Butoxyethanol	238	4	13
2-(2-Butoxyethoxy)ethyl acetate	299		
Butyl acetate	425	1.7	7.6
sec-Butyl acetate		1.7	9.8
Butylamine	312	1.7	9.8
tert-Butylamine	380	1.7	8.9
Butylbenzene	410	0.8	5.8
sec-Butylbenzene	418	0.8	6.9
tert-Butylbenzene	450	0.7	5.7
Butyl formate	322	1.7	8.2
Butyl methyl ketone	423	1	8
Butyl 2-methyl-2-propenoate	294	2	8
Butyl propanoate	427		
Butyl stearate	355		
Butyl vinyl ether	255		
2-Butyne		1.4	
Camphor	466	0.6	3.5
Carbon disulfide	90	1.3	50.0
Carbon monoxide	609	12.5	74.2
Carbonyl sulfide		12	28.5
Chlorobenzene	593	1.3	9.6
1-Chloro-1,3-butadiene		4.0	20.0
1-Chlorobutane	240	1.8	10.1
2-Chloro-2-butene		2.3	9.3
1-Chloro-2,3-epoxypropane	411	4	21
1-Chloro-1,1-difluoroethane		6.2	17.9
1-Chloro-2,4-dinitrobenzene		2.0	22
1-Chloro-2,3-epoxypropane	411	3.8	21
Chloroethane	519	3.8	15.4
2-Chloroethanol	425	4.9	15.9
Chloromethane	632	8.1	17.4
1-Chloro-3-methylbutane		1.5	7.4
1-Chloro-2-methylpropane		2.0	8.8
3-Chloro-2-methyl-1-propene		2.3	9.3
1-Chloronaphthalene	>588		
1-Chloropentane	260	1.6	8.6
1-Chloropropane	520	2.6	11.1
2-Chloropropane	593	2.8	10.7
1-Chloro-1-propene		4.5	16
2-Chloro-1-propene		4.5	16
3-Chloro-1-propene	485	2.9	11.1
Chlorotrifluoroethylene		24	40.3
m-Cresol	558	1.1	

(*Continued*)

TABLE 2.42   Properties of Combustible Mixtures in Air (*Continued*)

Substance	Autoignition temperature, °C	Flammable (explosive) limits, percent by volume of fuel (25°C, 760 mm)	
		Lower	Upper
o-Cresol	599	1.4	
p-Cresol	558	1.1	
Cumene	424	0.9	6.5
Cyanogen		6.6	32
Cyclobutane		1.8	
Cyclohexane	245	1.3	8
Cyclohexanol	300	1	9
Cyclohexanone	420	1.1	9.4
Cyclohexene	244	1.2	
Cyclohexyl acetate	334		
Cyclohexylamine	293	1	9
Cyclopentane	361	1.5	
Cyclopentene	395		
Cyclopropane	500	2.4	10.4
p-Cymene	436	0.7	5.6
trans-Decahydronaphthalene	255	0.7	5.4
Decane	210	0.8	5.4
Decene	235		
Diborane(6)	38–52	0.8	88
Dibutylamine		1.1	6
Dibutyl decanedioate (dibutyl sebacate)	365	0.44	
Dibutyl ether	194	1.5	7.6
Dibutyl *o*-phthalate	402	0.5	
1,2-Dichlorobenzene	648	2.2	9.2
1,1-Dichloroethane	458	5.4	11.4
1,2-Dichloroethane	413	6.2	16
1,1-Dichloroethylene	570	6.5	15.5
cis-1,2-Dichloroethylene	460	3	15
trans-1,2-Dichloroethylene	460	6	13
Dichloromethane	556	13	23
1,2-Dichloropropane	557	3.4	14.5
Diethanolamine [2,2'-iminobis(ethanol)]	662	2	13
1,1-Diethoxyethane (acetal)	230	1.6	10.4
Diethylamine	312	1.8	10.1
Diethylene glycol [bis(2-hydroxyethyl) ether]	224	2	17
Diethylene glycol dibutyl ether	310		
Diethylene glycol monoethyl ether acetate	425		
Diethylene glycol monomethyl ether	240	1.4	22.7
Diethylenetriamine	358	2	6.7
Diethyl ether	180	1.9	36.0
3,3-Diethylpentane	290	0.7	5.7
Diethyl peroxide		2.3	15.9
Diethyl sulfate	436		
1,1-Difluoroethylene		5.5	21.3
1,3-Dihydroxybenzene (resorcinol)	664		
1,4-Dihydroxybenzene	516		
Diisopropylamine	316	1.1	7.1
Diisopropyl ether	443	1.4	7.9
Dimethoxymethane	237	2.2	13.8
N,*N*-Dimethy lacetamide	490	2.0	11.5
Dimethylamine (anhydrous)	400	2.8	14.4
N,*N*-Dimethylaniline	371		

TABLE 2.42 Properties of Combustible Mixtures in Air (*Continued*)

Substance	Autoignition temperature, °C	Flammable (explosive) limits, percent by volume of fuel (25°C, 760 mm)	
		Lower	Upper
2,3-Dimethylaniline		1.0	
2,2-Dimethylbutane	405	1.2	7.0
2,3-Dimethylbutane	405	1.2	7.0
3,3-Dimethyl-2-butanone	423	1	8
cis-1,2-Dimethylcyclohexane	304		
trans-1,2-Dimethylcyclohexane	304		
Dimethyl ether	350	3.4	27.0
N,N-Dimethylformamide	445	2.2	15.2
2,6-Dimethyl-4-heptanol		0.8	6.1
2,6-Dimethyl-4-heptanone	396	0.8	6.2
2,3-Dimethylhexane	438		
1,1-Dimethylhydrazine	249	2	95
2,3-Dimethylpentane	335	1.1	6.7
Dimethyl 1,2-phthalate	490	0.9	
2,2-Dimethylpropane	450	1.4	7.5
Dimethyl sulfate	188		
Dimethyl sulfide	206	2.2	19.7
Dimethyl sulfoxide	215	2.6	42
1,4-Dioxane	180	2.0	22
Dipentene	237		
Dipentyl ether	170		
Diphenylamine	634		
Diphenyl ether	618	0.8	1.5
Dipropylamine	299		
Dipropyl ether	188	1.3	7.0
Divinyl ether	360	1.7	27.0
Dodecane	203	0.6	
1-Dodecanol	275		
1,2-Epoxybutane	439	1.7	19
Ethane	515	3.0	12.5
1,2-Ethanediamine	385	2.5	12.0
1,2-Ethanediol	398	3.2	22
Ethanethiol	299	2.8	18.2
Ethanol	363	3.3	19
Ethanolamine	410	3.0	23.5
2-Ethoxyethanol	235	3	18
2-Ethoxyethyl acetate	379	2	8
1-Ethoxypropane		1.7	9.0
Ethyl acetate	426	2	11.5
Ethyl acetoacetate	295	1.4	9.5
Ethyl acrylate	372	1.4	14
Ethylamine	385	3.5	14.0
Ethylbenzene	432	0.8	6.7
Ethyl benzoate	490		
Ethyl butanoate	463		
2-Ethylbutanoic acid	463		
Ethyl chloroformate	500		
Ethylcyclobutane	210	1.2	7.7
Ethylcyclohexane	238	0.9	6.6
Ethylene	490	2.7	36.0
Ethylene glycol diacetate	482	1.6	8.4

(*Continued*)

TABLE 2.42 Properties of Combustible Mixtures in Air (*Continued*)

Substance	Autoignition temperature, °C	Flammable (explosive) limits, percent by volume of fuel (25°C, 760 mm)	
		Lower	Upper
Ethylene glycol dimethyl ether	202		
Ethylene glycol ethyl ether acetate	379	2	8
Ethylene glycol monobutyl ether	238	4	13
Ethylene glycol methyl ether acetate	392	2	12
Ethylene glycol monoethyl ether	235	3	18
Ethyleneimine	320	3.3	54.8
Ethylene oxide	429	3.0	100
Ethyl formate	455	2.8	16.0
2-Ethylhexanal	197		
2-Ethyl-1,3-hexanediol	360		
2-Ethyl-1-hexanol	231	0.88	9.7
2-Ethylhexyl acetate	268	0.76	8.14
Ethyl lactate	400	1.5	
Ethyl methyl ether		2.0	10.0
3-Ethyl-2-methylpentane	460		
Ethyl nitrate	85 explodes	3.8	
Ethyl nitrite	90 explodes	3.0	50.0
Ethyl propanoate	440	1.9	11
Ethyl vinyl ether	202	1.7	28
Formaldehyde	430	7.0	73.0
Formic acid, 90%	434	18	57
2-Furaldehyde (furfural)	316	2.1	19.3
Furan		2.3	14.3
Furfuryl alcohol	491	1.8	16.3
Gasoline, 50–100 octane	280–456	1.4	7.6
Glycerol	370	3	19
Heptane	204	1.05	6.7
2-Heptanone (methyl pentyl ketone)	393	1.1	7.9
4-Heptanone (diisobutyl ketone)	396	0.8	7.1
1-Heptene	260		
1,1,2,3,4,4-Hexachlorobutadiene	610		
Hexane	225	1.1	7.5
1,6-Hexanedioic acid	420		
Hexanoic acid	380		
2-Hexanone	423	1	8
1-Hexene	253		
Hydrazine	23–270	4.7	100
Hydrogen	400	4.1	74.2
Hydrogen cyanide, 96%	538	5.6	40.0
Hydrogen sulfide	260	4	46
Af-Hydroxy ethyl-1,2-ethanediamine	368		
1-Hydroxy-2-methylbenzene	599	1.4	
1-Hydroxy-3-methylbenzene	559	1.1	
l-Hydroxy-4-methylbenzene (see *p*-cresol)			
4-Hydroxy-4-methyl-2-pentanone	643	1.8	6.9
Isobutanal	196	1.6	10.6
Isobutyl acetate	421	1	10.5
Isobutylamine	378	2	12
Isobutylbenzene	427	0.8	6.0
Isobutyl isobutyrate	432	0.96	7.59
Isopentane	420	1.4	7.6
Isopentyl acetate	360	1.0	7.5

TABLE 2.42 Properties of Combustible Mixtures in Air (*Continued*)

Substance	Autoignition temperature, °C	Flammable (explosive) limits, percent by volume of fuel (25°C, 760 mm)	
		Lower	Upper
Isoprene	220	2	9
Isopropyl acetate	460	1.8	8
Isopropyl alcohol	399	2.5	12.7
Isopropylamine	402	2.3	10.4
Isopropylbenzene (cumene)	424	0.8	6.5
Isopropyl formate	485		
4-Isopropyl-l-methylbenzene	436		
Kerosene	210	0.7	5.0
Maleic anhydride	477	1.4	7.1
Methacrylic acid	68	1.6	8.8
Methacrylonitrile		2	6.8
Methane	650	5.3	15.0
Methanethiol		3.9	21.8
Methanol	464	6.0	36
Methoxybenzene (anisole)	475		
2-Methoxyethanol	285	1.8	14
2-Methoxyethyl acetate	392	1.5	12.3
Methyl acetate	454	3.1	16
Methyl acetoacetate	280		
Methyl acetylacetate	280		
Methyl acrylate	468	2.8	25
Methylamine	430	4.9	20.7
2-Methylbutane		1.4	7.6
2-Methyl-1-butanol	385	1.4	9.0
2-Methyl-2-butanol	437	1.2	9.0
3-Methyl-1-butanol	350	1.2	9.0
3-Methylbutyl acetate	360	1.0	7.5
2-Methyl-2-butene	275	1.6	8.7
3-Methyl-1-butene	365	1.5	9.1
2-Methyl-1-buten-3-one		1.8	9.0
Methyl chloroformate	504		
Methylcyclohexane	250	1.2	6.7
cis-2-Methylcyclohexanol	296		
trans-2-Methylcyclohexanol	296		
cis-4-Methylcyclohexanol	295		
trans-4-Methylcyclohexanol	295		
Methylcyclopentane	258	1.0	8.35
Methyl formate	449	4.5	23
2-Methylhexane	280	1.0	6.0
3-Methylhexane	280		
5-Methyl-2-hexanone	191	1.0	8.2
Methylhydrazine	196	2.5	97.0 ±2
Methyl isobutyl ketone (MIBK)	448	1	8
2-Methyllactonitrile	688		
Methyl methacrylate		1.7	8.2
1-Methyl-4-(1-methylethenyl)-cyclohexene (dipentene)	237		
1-Methylnaphthalene	529		
2-Methylpentane	264	1.0	7.0
3-Methylpentane	278	1.2	7.0
2-Methyl-2,4-pentanediol	306	1	9
2-Methyl-1-pentanol	310	1.1	9.65
4-Methyl-2-pentanol		1.0	5.5

(*Continued*)

TABLE 2.42 Properties of Combustible Mixtures in Air (*Continued*)

Substance	Autoignition temperature, °C	Flammable (explosive) limits, percent by volume of fuel (25°C, 760 mm)	
		Lower	Upper
4-Methyl-2-pentanone	452	2	8.0
4-Methyl-3-penten-2-one	344	1.4	7.2
2-Methylpropanal	223	1.6	10.6
2-Methyl-1-propanamine	378	2	12
2-Methylpropane	460	1.8	8.4
2-Methylpropanenitrile	482		
Methyl propanoate	469	2.5	13
2-Methylpropanoic acid	481	2.0	9.2
2-Methyl-1-propanol	415	1.7	10.6
2-Methyl-2-propanol (*t*-butyl alcohol)	478	2.4	8.0
2-Methyl-1-propene	465	1.8	9.6
2-Methylpropyl acetate	421	1.3	10.5
2-Methylpropyl formate	320	1.7	8
2-Methylpyridine	538		
N-Methyl-2-pyrrolidone	346	1	10
Methyl salicylate	454		
α-Methylstyrene	574	1.9	6.1
Methyl vinyl ether		2.6	39
Morpholine	290	1	11
Naphtha, coal tar	277		
Naphthalene	526	0.9	5.9
Neoprene		4.0	20
Nicotine	244	0.75	4.0
Nitrobenzene	482	1.8	9
2-Nitrobiphenyl	179		
Nitroethane	414	3.4	17
Nitroglycerine	270		
Nitromethane	418	7.3	22
1-Nitropropane	421	2.2	
2-Nitropropane	428	2.6	11
Nonane	205	0.8	2.9
Octadecanoic acid (stearic acid)	395		
cis-9-Octadecenoic acid (oleic acid)	362		
Octane	206	1.0	6.5
1-Octene	230		
Paraldehyde	238	1.3	
Pentaborane(9)		0.42	
Pentanamine		2.2	22
Pentane	260	1.5	7.8
1,5-Pentanediol	335		
Pentanoic acid	400		
1-Pentanol	300	1.2	10.0
2-Pentanol	343		
3-Pentanol	435	1.2	9.0
2-Pentanone (methyl propyl ketone)	452	1.5	8.2
3-Pentanone (diethyl ketone)	450	1.6	
1-Pentene	275	1.5	8.7
Pentyl acetate	360	1.1	7.5
Pentylamine		2.2	22
Petroleum ether (solvent naphtha)	288	1.1	5.9
Phenol	715	1.8	8.6

TABLE 2.42   Properties of Combustible Mixtures in Air (*Continued*)

Substance	Autoignition temperature, °C	Flammable (explosive) limits, percent by volume of fuel (25°C, 760 mm)	
		Lower	Upper
Phosphorus, red	260		
Phosphorus, white	30		
Phosphorus pentasulfide	142		
o-Phthalic anhydride	570	1.7	10.4
Picric acid	300 (explodes)		
α-Pinene	275		
β-Pinene	275		
Piperidine		1	10
1-Propanal	207	2.6	17
1-Propanamine (propylamine)	318	2.0	10.4
Propane	450	2.1	9.5
1,2-Propanediol	371	2.6	12.5
1,3-Propanediol	400		
Propanenitrile	512	3.1	14
1,2,3-Propanetriol (glycerol)	370	3	19
1,2,3-Propanetriol triacetate (triacetin)	433	1.0	
Propanoic acid	465	2.9	12.1
Propanoic anhydride	285	1.3	9.5
1-Propanol	412	2.2	13.7
2-Propanol	399	2.0	12.7
Propene	460	2.4	10.1
Propyl acetate	450	1.7	8
Propylbenzene	450	0.8	6.0
Propyl formate	455		
Propyl nitrate	175	2	100
Propyne		1.7	
Pyridine	482	1.8	12.4
Quinoline	480		
Sodium	115 (dry air)		
Styrene	490	0.9	6.8
Sulfur (di-) dichloride	233		
1,1,2,2-Tetrabromoethane	335		
Tetrabromoethylene	335		
1,1,1,2-Tetrachloroethane		5	12
1,1,2,2-Tetrachloroethane		20	54
Tetrahydrofuran	321	2	11.8
Tetrahydrofurfuryl alcohol	282	1.5	9.7
1,2,3,4-Tetrahydronaphthalene	385	0.8	5.0
2,2,3,3-Tetramethylpentane	430	0.8	4.9
2,2-Thiodiethanol	298		
Titanium, powder	250		
Toluene	480	1.1	7.1
Toluene diisocyanate		0.9	9.5
o-Toluidine (also p-)	482		
Tributylamine		1	5
1,1,1-Trichloroethane	537	7.5	12.5
1,1,2-Trichloroethane	460	6	28
Trichloroethylene	420	8	10.5
(Trichloromethyl)benzene	211		

(*Continued*)

TABLE 2.42 Properties of Combustible Mixtures in Air (*Continued*)

Substance	Autoignition temperature, °C	Flammable (explosive) limits, percent by volume of fuel (25°C, 760 mm)	
		Lower	Upper
Trichloromethylsilane	>404	7.6	>20
1,2,3-Trichloropropane		3.2	12.6
Trichlorosilane	104		
l,l,2-Trichloro-1,2,2-trifluoroethane (Freon 113)	680		
Tri-*o*-cresyl phosphate	385		
Triethanolamine		1	10
Triethylamine	249	1.2	8.0
Triethylene glycol	371	0.9	9.2
Triethyl phosphate	454		
Trimethylamine	190	2.0	11.6
1,2,3-Trimethylbenzene (hemimellitene)	470	0.8	6.6
1,2,4-Trimethylbenzene (pseudocumene)	500	0.9	6.4
1,3,5-Trimethylbenzene	559	1	5
2,2,3-Trimethylbutane	412		
1,1,3-Trimethyl-3-cyclohexen-5-one	462	0.8	3.8
3,5,5-Trimethylcyclohex-2-ene-1-one	460	0.8	3.8
2,2,3-Trimethylpentane	346		
2,2,4-Trimethylpentane	418	1.1	6.0
2,3,3-Trimethylpentane	425		
Trioxane	414	3.6	28.7
Tri-*o*-tolyl phosphate	385		
Turpentine		0.8	
Vinyl acetate	402	2.6	13.4
Vinyl bromide	530	9	15
Vinyl butanoate		1.4	8.8
Vinyl chloride	472	3.6	33.0
4-Vinyl-1-cyclohexene	269		
Vinyl fluoride		2.6	21.7
Vinylidene	573	5.6	16.0
m-Xylene	527	1.1	7.0
o-Xylene	463	0.9	6.7
p-Xylene	528	1.1	7.0

2.7 AZEOTROPIC MIXTURES

An azeotrope is liquid mixture of two or more components that boils at a temperature either higher or lower than the boiling point of any of the individual components. In industrial situation, if the components of a solution are very close in boiling point and cannot be separated by conventional distillation, a substance can be added that forms an azeotrope with one component, modifying its boiling point and making it separable by distillation.

TABLE 2.43   Binary Azeotropic (Constant-Boiling) Mixtures

A. Binary azeotropes containing water

System	BP of azeotrope, °C	Composition, wt%	
		Water	Other component
Inorganic acids			
Hydrogen bromide	126	52.5	47.5
Hydrogen chloride	108.58	79.78	20.22
Hydrogen fluoride	111.35	64.4	35.6
Hydrogen iodide	127	43	57
Hydrogen peroxide	Zeotrope		
Nitric acid	120.7	32.6	67.4
Perchloric acid	203	28.4	71.6
Organic acids			
Formic acid	107.2	22.6	77.4
Acetic acid	Zeotrope		
Propionic acid	99.9	82.3	17.7
Isobutyric acid	99.3	79	21
Butyric acid	99.4	81.6	18.4
Pentanoic acid	99.8	89	11
Isopentanoic acid	99.5	81.6	18.4
Perfluorobutyric acid	97	71	29
Crotonic acid	99.9	97.8	2.2
Alcohols			
Ethanol	78.17	4	96
Allyl alcohol	88.9	27.7	72.3
1-Propanol	71.7	71.7	28.3
2-Propanol	80.3	12.6	87.4
1-Butanol	92.7	42.5	57.5
2-Butanol	87.0	26.8	73.2
2-Methyl-2-propanol	79.9	11.7	88.3
1-Pentanol	95.8	54.4	45.6
2-Pentanol	91.7	36.5	63.5
3-Pentanol	91.7	36.0	64.0
2,2-Dimethyl-2-propanol	87.35	27.5	72.5
1-Hexanol	97.8	67.2	32.8
1-Octanol	99.4	90	10
Cyclopentanol	96.25	58	42
1-Heptanol	98.7	83	17
Phenol	99.52	90.8	9.2
2-Methoxyphenol	99.5	87.5	12.5
1-Phenylphenol	99.95	98.75	1.25
Benzyl alcohol	99.9	91	9
2,3-Dimethyl-2,3-butanediol	Zeotrope		
Furfuryl alcohol	98.5	80	20

(Continued)

TABLE 2.43 Binary Azeotropic (Constant-Boiling) Mixtures (*Continued*)

System	BP of azeotrope, °C	Composition, wt%	
		Water	Other component
Aldehydes			
Propionaldehyde	47.5	2	98
Butyraldehyde	68	6	94
Pentanal	83	19	81
Paraldehyde	90	28.5	71.5
Furaldehyde	97.5	65	35
Amines			
N-Methylbutylamine	82.7	15	85
Furfurylamine	99	74	26
Piperidine	92.8	35	65
Pyridine	93.6	41.3	58.7
2-Methylpyridine	93.5	48	52
3-Methylpyridine	97	60	40
4-Methylpyridine	97.35	62.8	37.2
2,6-Dimethylpyridine	96.02	51.8	48.2
Dibutylamine	97	50.5	49.5
Dihexylamine	99.8	92.8	7.2
Triallylamine	95	38	62
Tributylamine	99.65	79.7	20.3
Aniline	98.6	80.8	19.2
N-Ethylaniline	99.2	83.9	16.1
1-Methyl-2-(2-pyridyl)pyrrolidine	99.85	97.5	2.5
Halogenated hydrocarbons			
Chloroform	56.1	2.8	97.2
Carbon tetrachloride	42.6	2.8	97.2
Trichloroethylene	73.4	17	83
Tetrachloroethylene	88.5	17.2	82.8
1,2-Dichloroethane	72	8.3	91.7
1-Chloropropane	44	2.2	97.8
1,2-Dichloropropane	78	12	88
Chlorobenzene	90.2	28.4	71.6
Esters			
Ethyl formate	52.6	5	95
Isopropyl formate	65.0	3	97
Propyl formate	71.6	2.3	97.7
Isobutyl formate	80.4	7.8	92.2
Butyl formate	83.8	14.5	85.5
Isopentyl formate	90.2	21	79
Pentyl formate	91.6	28.4	71.6
Benzyl formate	99.2	80	20
Ethyl acetate	70.38	8.47	91.53
Allyl acetate	83	14.7	85.3

TABLE 2.43   Binary Azeotropic (Constant-Boiling) Mixtures (*Continued*)

System	BP of azeotrope, °C	Composition, wt%	
		Water	Other component
Esters (*continued*)			
Isopropyl acetate	76.6	10.6	89.4
Propyl acetate	82.4	14	86
Isobutyl acetate	87.4	16.5	83.5
Butyl acetate	90.2	28.7	71.3
Isopentyl acetate	93.8	36.3	63.7
Pentyl acetate	95.2	41	59
Hexyl acetate	97.4	61	39
Phenyl acetate	98.9	75.1	24.9
Benzyl acetate	99.6	87.5	12.5
Methyl propionate	71.4	3.9	96.1
Ethyl propionate	81.2	10	90
Isopropyl propionate	85.2	19.9	80.1
Propyl propionate	88.9	23	77
Isobutyl propionate	92.75	52.2	47.8
Isopentyl propionate	96.55	48.5	51.5
Methyl butyrate	82.7	11.5	88.5
Ethyl butyrate	87.9	21.5	78.5
Propyl butyrate	94.1	36.4	63.6
Isobutyl butyrate	96.3	46	54
Butyl butyrate	97.2	53	47
Isopentyl butyrate	98.05	63.5	36.5
Methyl isobutyrate	77.7	6.8	93.2
Ethyl isobutyrate	85.2	15.2	84.8
Propyl isobutyrate	92.2	30.8	69.2
Isobutyl isobutyrate	95.5	39.4	60.6
Isopentyl isobutyrate	97.4	56.0	44.0
Methyl isopentanoate	87.2	19.2	80.8
Ethyl isopentanoate	92.2	30.2	69.8
Propyl isopentanoate	96.2	45.2	54.8
Isobutyl isopentanoate	97.4	55.8	44.2
Isopentyl isopentanoate	98.8	74.1	25.9
Ethyl pentanoate	94.5	40	60
Ethyl hexanoate	97.2	54	46
Methyl benzoate	99.08	79.2	20.8
Ethyl benzoate	99.4	84.0	16.0
Propyl benzoate	99.7	90.9	9.1
Butyl benzoate	99.9	94	6
Isopentyl benzoate	99.9	95.6	4.4
Ethyl phenylacetate	99.7	91.3	8.7
Methyl cinnamate	99.9	95.5	4.5
Methyl phthalate	99.95	97.5	2.5
Diethyl *o*-phthalate	99.98	98.0	2.0
Ethyl chloroacetate	95.2	45.1	54.9
Butyl chloroacetate	98.12	75.5	24.5
Methyl acrylate	71	7.2	92.8
Isobutyl carbonate	98.6	74	26
Ethyl crotonate	93.5	38	62
Methyl lactate	99	80	20

(*Continued*)

TABLE 2.43   Binary Azeotropic (Constant-Boiling) Mixtures (*Continued*)

System	BP of azeotrope, °C	Composition, wt%	
		Water	Other component
1,2-Ethanediol diacetate	99.7	84.6	15.4
Ethyl nitrate	74.35	22	78
Propyl nitrate	84.8	20	80
Isobutyl nitrate	89.0	25	75
Methyl sulfate	98.6	73	27
Ethers			
Ethyl vinyl ether	34.6	1.5	98.5
Diethyl ether	34.2	1.3	98.7
Ethyl propyl ether	59.5	4	96
Diisopropyl ether	62.2	4.5	95.5
Butyl ethyl ether	76.6	11.9	88.1
Diisobutyl ether	88.6	23	77
Dibutyl ether	92.9	33	67
Diisopentyl ether	97.4	54	46
1,1-Diethoxyethane	82.6	14.5	85.5
Diphenyl ether	99.33	96.75	3.25
Methoxybenzene	95.5	40.5	59.5
Hydrocarbons			
Pentane	34.6	1.4	98.6
Hexane	61.6	5.6	94.4
Heptane	79.2	12.9	87.1
2,2,4-Trimethylpentane	78.8	11.1	88.9
Nonane	94.8	82	18
Undecane	98.85	96.0	4.0
Dodecane	99.45	98	2
Acrolein	52.4	2.6	97.4
Cyclohexene	70.8	8.93	91.07
Cyclohexane	69.5	8.4	91.6
1-Octene	88.0	28.7	71.3
Benzene	69.25	8.83	91.17
Toluene	84.1	13.5	86.5
Ethylbenzene	92.0	33.0	67.0
m-Xylene	92	35.8	64.2
Isopropylbenzene	95	43.8	56.2
Naphthalene	98.8	84	16
Ketones			
Acetone	Zeotrope		
2-Butanone	73.5	11	89
2-Pentanone	83.3	19.5	80.5
Cyclopentanone	94.6	42.4	57.6
4-Methyl-2-pentanone	87.9	24.3	75.7

TABLE 2.43   Binary Azeotropic (Constant-Boiling) Mixtures (*Continued*)

System	BP of azeotrope, °C	Composition, wt%	
		Water	Other component
Ketones (continued)			
2-Heptanone	95	48	52
3-Heptanone	94.6	42.2	57.8
4-Heptanone	94.3	40.5	59.5
4-Hydroxy-4-methyl-2-pentanone	98.8	87.3	12.7
4-Methyl-3-penten-2-one	91.8	34.8	65.2
Nitriles			
Acetonitrile	76.5	16.3	83.7
Isobutyronitrile	82.5	23	177
Butyronitrile	88.7	32.5	67.5
Acrylonitrile	70.6	14.3	85.7
Miscellaneous			
Hydrazine	120	32.3	67.7
Acetamide	Zeotrope		
Nitromethane	83.59	23.6	76.4
Nitroethane	87.22	28.5	71.5
2,5-Dimethylfuran	77.0	11.7	88.3
Trioxane	91.4	30	70
Carbon disulfide	42.6	2.8	97.2

B. Binary azeotropes containing organic acids

System	BP of azeotrope, °C	Composition, wt%	
		Acid	Other component
Formic acid			
2-Methylbutane	27.2	4	96
Pentane	34.2	20	80
Hexane	60.6	28	72
Methylcyclopentane	63.3	29	71
Cyclohexane	70.7	70	30
Methylcyclohexane	80.2	46.5	53.5
Heptane	78.2	56.5	43.5
Octane	90.5	63	37
Benzene	71.05	31	69
Toluene	85.8	50	50
o-Xylene	95.5	74	26
m-Xylene	92.8	71.8	28.2
Styrene	97.8	73	27

(*Continued*)

TABLE 2.43   Binary Azeotropic (Constant-Boiling) Mixtures (*Continued*)

System	BP of azeotrope, °C	Composition, wt%	
		Acid	Other component
Formic acid (continued)			
Iodomethane	42.1	6	94
Chloroform	59.15	15	85
Carbon tetrachloride	66.65	18.5	81.5
Trichloroethylene	74.1	25	75
Tetrachloroethylene	88.2	50	50
Bromoethane	38.2	3	97
1,2-Dibromoethane	94.7	51.5	48.5
1,2-Dichloroethane	77.4	14	86
1-Bromopropane	64.7	27	73
2-Bromopropane	56.0	14	86
1-Chloropropane	45.6	8	92
2-Chloropropane	34.7	1.5	98.5
1-Chloro-2-methylpropane	63.0	19	81
Bromobenzene	98.1	68	32
Chlorobenzene	93.7	59	41
Fluorobenzene	73.0	27	73
o-Chlorotoluene	100.2	83	17
Pyridine	127.43	61.4	38.6
2-Methylpyridine	158.0	25	75
2-Pentanone	105.3	32	68
3-Pentanone	105.4	33	67
Nitromethane	97.07	45.5	54.5
Diethyl sulfide	82.2	35	65
Diisopropyl sulfide	93.5	62	38
Dipropyl sulfide	98.0	83	17
Carbon disulfide	42.55	17	83
Acetic acid			
Hexane	68.3	6.0	94.0
Heptane	91.7	23	67
Octane	105.7	53.7	46.3
Nonane	112.9	69	31
Decane	116.75	79.5	20.5
Undecane	117.9	95	5
Cyclohexane	78.8	9.6	90.4
Methylcyclohexane	96.3	31	69
Benzene	80.05	2.0	98.0
Toluene	100.6	28.1	71.9
o-Xylene	116.6	78	22
m-Xylene	115.35	72.5	27.5
p-Xylene	115.25	72	28
Ethylbenzene	114.65	66	34
Styrene	116.8	85.7	14.3
Isopropylbenzene	116.0	84	16
Triethylamine	163	67	33
Nitromethane	101.2	96	4

TABLE 2.43   Binary Azeotropic (Constant-Boiling) Mixtures (*Continued*)

System	BP of azeotrope, °C	Composition, wt%	
		Acid	Other component
Acetic acid (*continued*)			
Nitroethane	112.4	30	70
Pyridine	138.1	51.1	48.9
2-Methylpyridine	144.1	40.4	59.6
3-Methylpyridine	152.5	30.4	69.6
4-Methylpyridine	154.3	30.3	69.7
2,6-Dimethylpyridine	148.1	22.9	77.1
Carbon tetrachloride	76	98.46	1.54
Trichloroethylene	86.5	96.2	3.8
Tetrachloroethylene	107.4	61.5	38.5
1,2-Dibromoethane	114.4	55	45
2-Iodopropane	88.3	9	91
1-Bromobutane	97.6	18	82
1-Bromo-2-methylpropane	90.2	12	88
Chlorobenzene	114.7	58.5	41.5
Trichloronitromethane	107.65	80.5	19.5
1,4-Dioxane	119.5	77	23
Diisopropyl sulfide	111.5	48	52
Propionic acid			
Heptane	97.8	2	98
Octane	120.9	21.5	78.5
Nonane	134.3	54.0	46.0
Decane	139.8	80.5	19.5
o-Xylene	135.4	43	57
p-Xylene	132.5	34	66
1,3,5-Trimethylbenzene	139.3	77	23
Isopropylbenzene	139.0	65	35
Propylbenzene	139.5	75	25
Camphene	138.0	65	35
α-Pinene	136.4	58.5	41.5
Methoxybenzene	140.8	96	4
Pyridine	148.6	67.2	32.8
2-Methylpyridine	154.5	55.0	45.0
1,2-Dibromoethane	127.8	17.5	82.5
1-Iodo-2-methylpropane	119.5	9	91
Chlorobenzene	128.9	18	82
Dipropyl sulfide	136.5	45	55
Butyric acid			
Undecane	162.4	84.4	15.5
o-Xylene	143.0	10	90
m-Xylene	138.5	6	94
p-Xylene	137.8	5.5	94.5
Ethylbenzene	135.8	4	96

(*Continued*)

TABLE 2.43 Binary Azeotropic (Constant-Boiling) Mixtures (*Continued*)

System	BP of azeotrope, °C	Composition, wt%	
		Acid	Other component
Butyric acid (*continued*)			
Styrene	143.5	15	85
1,2,4-Trimethylbenzene	159.5	45	55
1,3,5-Trimethylbenzene	158.0	38	62
Isopropylbenzene	149.5	20	80
Propylbenzene	154.5	28	72
Butylbenzene	162.5	75	25
Naphthalene	Zeotrope		
Indene	163.7	84	16
Camphene	152.3	2.8	97.2
Methoxybenzene	152.9	12	88
Pyridine	163.2	92.0	8.0
2-Furaldehyde	159.4	42.5	57.5
1,2-Dibromoethane	131.1	3.5	96.5
1-Iodobutane	129.8	2.5	97.5
Chlorobenzene	131.75	2.8	97.2
1,4-Dichlorobenzene	162.0	57	43
o-Bromotoluene	163.0	72	28
m-Bromotoluene	163.6	79.5	20.5
p-Bromotoluene	161.5	75	25
α-Chlorotoluene	160.8	65	35
Ethyl bromoacetate	157.4	84	16
Propyl chloroacetate	160.5	40	60
Isobutyric acid			
2,7-Dimethyloctane	148.6	48	52
o-Xylene	141.0	22	78
m-Xylene	139.9	15	85
p-Xylene	136.4	13	87
Styrene	142.0	27	73
1,2,4-Trimethylbenzene	152.3	63	37
Isopropylbenzene	146.8	35	65
Propylbenzene	149.3	49	51
Camphene	148.1	45	55
D-Limonene	152.5	78	22
Methoxybenzene	149.0	42	58
Ethyl bromoacetate	153.0	40	60
Ethyl 2-oxopropionate	153.0	60	40
1,2-Dibromoethane	130.5	6.5	93.5
1-Iodobutane	128.8	7	93
1-Bromohexane	148.0	35	65
Bromobenzene	148.6	35	65
Chlorobenzene	131.5	8	92
o-Bromo toluene	153.9	85	15
α-Chlorotoluene	153.5	80	20
Diisopentyl ether	154.2	93	7
Ethyl bromoacetate	153.0	40	60

TABLE 2.43   Binary Azeotropic (Constant-Boiling) Mixtures (*Continued*)

C. Binary azeotropes containing alchohols

System	BP of azeotrope, °C	Composition, wt%	
		Alcohol	Other component
Methanol			
Pentane	30.9	7	93
Cyclopentane	38.8	14	86
Cyclohexane	53.9	36.4	63.6
Methylcyclohexane	59.2	54	46
Heptane	59.1	51.5	48.5
Octane	62.8	67.5	32.5
Nonane	64.1	83.4	16.6
Benzene	57.5	39.1	60.9
Fluorobenzene	59.7	32	68
Toluene	63.5	72.5	27.5
Bromomethane	3.55	99.55	0.45
Iodomethane	37.8	95.5	4.5
Bromodichloromethane	63.8	60	40
Chloroform	53.4	87.4	12.6
Carbon tetrachloride	55.7	79.44	20.56
Bromoethane	34.9	5.3	94.7
1,2-Dichloroethane	61.0	32	68
Trichloroethylene	59.3	38	62
1-Bromopropane	54.5	21	79
2-Bromopropane	48.6	15.0	85.0
1-Chloropropane	40.5	9.5	90.5
2-Chloropropane	33.4	6	94
2-Iodopropane	61.0	38	62
1-Chlorobutane	57.0	27	73
Isobutyl formate	64.6	95	5
Methyl acetate	53.5	19	81
Methyl acrylate	62.5	54	46
Methyl nitrate	52.5	73	27
Acetone	55.5	12.1	87.9
1,4-Dioxane	Zeotrope		
Dipropyl ether	63.8	72	28
Methyl *tert*-butyl ether	51.3	14.3	85.7
Diethyl sulfide	61.2	62	38
Carbon disulfide	39.8	71	29
Thiophene	59.7	16.4	83.6
Nitromethane	64.4	9.1	90.9
Ethanol			
Pentane	34.3	5	95
Cyclopentane	44.7	7.5	92.5
Hexane	58.7	21	79
Cyclohexane	64.8	29.2	70.8
Heptane	70.9	49	51

(*Continued*)

TABLE 2.43 Binary Azeotropic (Constant-Boiling) Mixtures (*Continued*)

System	BP of azeotrope, °C	Composition, wt%	
		Alcohol	Other component
Ethanol (*continued*)			
Octane	77.0	78	22
Benzene	67.9	31.7	68.3
Fluorobenzene	70.0	75	25
Toluene	76.7	68	32
Bromodichloromethane	75.5	72	28
Iodomethane	41.2	96.8	3.2
Chloroform	59.3	93	7
Trichloronitromethane	77.5	34	66
Carbon tetrachloride	65.0	84.2	15.8
1,2-Dichloroethane	70.5	37	63
3-Chloro-1-propene	44	5	95
1-Bromopropane	62.8	20.5	79.5
2-Bromopropane	55.6	10.5	89.5
1-Chloropropane	45.0	6	94
2-Chloropropane	35.6	2.8	97.2
1-Iodopropane	75.4	44	56
2-Iodopropane	71.5	27	73
1-Bromobutane	75.0	43	57
1-Chlorobutane	65.7	20.3	79.7
2-Butanone	74.8	40	60
1,1-Diethoxy ethane	78.0	76	24
Dipropyl ether	74.5	44	56
Acetonitrile	72.5	44	56
Acrylonitrile	70.8	41	59
Nitromethane	76.1	29	71
Carbon disulfide	42.6	91	9
Diethyl sulfide	72.6	56	44
1-Propanol			
Hexane	65.7	4	96
Cyclohexane	74.7	18.5	81.5
Methylcyclohexane	87.0	34.7	65.3
Heptane	84.6	34.7	65.3
Octane	93.9	70	30
Benzene	77.1	16.9	83.1
Toluene	92.5	51.2	48.8
o-Xylene	Zeotrope		
m-Xylene	97.1	94	6
p-Xylene	96.9	92.2	7.8
Styrene	97.0	8	92
Propyl formate	80.7	3	97
Butyl formate	95.5	64	36
Propyl acetate	94.7	51	49
Ethyl propionate	93.4	48	52
Methyl butyrate	94.4	49	51
Dipropyl ether	85.7	30	70

TABLE 2.43   Binary Azeotropic (Constant-Boiling) Mixtures (*Continued*)

System	BP of azeotrope, °C	Composition, wt%	
		Alcohol	Other component
1-Propanol (*continued*)			
1,1-Diethoxyethane	92.4	37	63
1,4-Dioxane	95.3	55	45
Chloroform	Zeotrope		
Carbon tetrachloride	73.4	92.1	7.9
Trichloronitromethane	94.1	58.5	41.5
Iodethane	70	93	7
1,2-Dichloroethane	80.7	19	81
Tetrachloroethylene	94.0	52	48
1-Bromopropane	69.7	9	91
1-Chlorobutane	74.8	18	82
Chlorobenzene	96.5	80	20
Fluorobenzene	80.2	18	82
Nitromethane	89.1	48.4	51.6
1-Nitropropane	97.0	8.8	91.2
Carbon disulfide	45.7	94.5	5.5
2-Propanol			
Pentane	35.5	6	94
Hexane	62.7	23	77
Cyclohexane	69.4	32	68
Heptane	76.4	50.5	49.5
Octane	81.6	84	16
Benzene	71.7	33.7	66.3
Fluorobenzene	74.5	30	70
Toluene	80.6	69	31
Chloroform	60.8	4.2	95.8
Trichloronitromethane	81.9	35	65
Carbon tetrachloride	69.0	18	82
1,2 Dichloroethane	74.7	43.5	56.5
Iodoethane	67.1	15	85
3-Bromo-1-propene	66.5	20	80
1-Chloropropane	46.4	2.8	97.2
1-Bromopropane	66.8	20.5	79.5
2-Bromopropane	57.8	12	88
1-Iodopropane	79.8	42	58
2-Iodopropane	76.0	32	68
1-Chlorobutane	70.8	23	77
Ethyl acetate	75.3	25	75
Isopropyl acetate	81.3	60	40
Methyl propionate	76.4	37	63
Acrylonitrile	71.7	56	44
Butylamine	74.7	60	40
2-Butanone	77.5	32	68
1,1-Diethoxyethane	81.3	63	37
Ethyl propyl ether	62.0	10	90
Diisopropyl ether	66.2	14.1	85.9

(*Continued*)

TABLE 2.43   Binary Azeotropic (Constant-Boiling) Mixtures (*Continued*)

System	BP of azeotrope, °C	Composition, wt% Alcohol	Other component
		1-Butanol	
Cyclohexane	79.8	9.5	90.5
Cyclohexene	82.0	5	95
Hexane	68.2	3.2	96.8
Methylcyclohexane	95.3	20	80
Heptane	93.9	18	82
Octane	108.5	45.2	54.8
Nonane	115.9	71.5	28.5
Toluene	105.5	27.8	72.2
o-Xylene	116.8	75	25
m-Xylene	116.5	71.5	28.5
p-Xylene	115.7	68	32
Ethylbenzene	115.9	65.1	34.9
Butyl formate	105.8	23.6	76.4
Isopentyl formate	115.9	69	31
Butyl acetate	117.2	47	53
Isobutyl acetate	114.5	50	50
Ethyl butyrate	115.7	64	36
Ethyl isobutyrate	109.2	17	83
Methyl isopentanoate	113.5	40	60
Ethyl borate	113.0	52	48
Ethyl carbonate	116.5	63	37
Isobutyl nitrate	112.8	45	55
Dibutyl ether	117.8	82.5	17.5
Diisobutyl ether	113.5	48	52
1,1-Diethoxyethane	101.0	13	87
Carbon tetrachloride	76.6	97.6	2.4
Tetrachloroethylene	110.0	68	32
2-Bromo-2-methylpropane	90.2	7	93
2-Iodo-2-methylpropane	110.5	30	70
Chlorobenzene	115.3	56	44
Paraldehyde	115.8	52	48
Hexaldehyde	116.8	77.1	22.9
Ethylenediamine	124.7	35.7	64.3
Pyridine	118.6	69	31
1-Nitropropane	115.3	32.2	67.8
Butyronitrile	113.0	50	50
Diisopropyl sulfide	112.0	45	55
		2-Methyl-2-propanol	
Cyclohexene	80.5	14.2	85.8
Cyclohexane	78.3	14	86
Methylcyclopentane	71.0	5	95
Hexane	68.3	2.5	97.5
Methylcyclohexane	92.6	32	68
Heptane	90.8	27	73
2,5-Dimethylhexane	98.7	42	58
1,3-Dimethylcyclohexane	102.2	56	44
2,2,4-Trimethylpentane	92.0	27	73
Benzene	79.3	7.4	92.6
Chlorobenzene	107.1	63	37
Fluorobenzene	84.0	9	91

TABLE 2.43   Binary Azeotropic (Constant-Boiling) Mixtures (*Continued*)

System	BP of azeotrope, °C	Composition, wt% Alcohol	Composition, wt% Other component
		Alcohol	Other component
2-Methyl-2-propanol (*continued*)			
Toluene	101.2	45	55
Ethylbenzene	107.2	80	20
p-Xylene	107.1	88.6	11.4
Butyl formate	103.0	40	60
Isobutyl formate	97.4	12	88
Propyl acetate	101.0	17	83
Isobutyl acetate	107.6	92	8
Methyl butyrate	101.3	25	75
Ethyl isobutyrate	105.5	52	48
Methyl chloroacetate	107.6	12	88
Dipropyl ether	89.5	10	90
Isobutyl vinyl ether	82.7	6.2	93.8
1,1-Diethoxyethane	98.2	20	80
2-Pentanone	101.8	19	81
3-Pentanone	101.7	20	80
1,2-Dichloroethane	83.5	6.5	93.5
1-Bromobutane	95.0	21	79
1-Chlorobutane	77.7	4	96
2-Bromo-2-methylpropane	88.8	12	88
2-Iodo-2-methylpropane	104.0	36	64
1-Nitropropane	105.3	15.2	84.8
Isobutyl nitrate	105.6	36	64
Diisopropyl sulfide	105.8	73	27
3-Methyl-1-butanol			
Heptane	97.7	7	93
Octane	117.0	30	70
Toluene	109.7	10	90
Ethylbenzene	125.7	49	51
Isopropylbenzene	131.6	94	6
Camphene	130.9	24	76
Bromobenzene	131.7	85	15
o-Fluorotoluene	112.1	14.0	86.0
Butyl acetate	125.9	16.5	83.5
Paraldehyde	123.5	22.0	78.0
Dibutyl ether	129.8	65	35
Cyclohexanol			
o-Xylene	143.0	14	86
m-Xylene	138.9	5	95
Propylbenzene	153.8	40	60
Indene	160.0	75	25
Camphene	151.9	41	59
Cineole	160.6	92	8

(*Continued*)

TABLE 2.43 Binary Azeotropic (Constant-Boiling) Mixtures (*Continued*)

System	BP of azeotrope, °C	Composition, wt%	
		Alcohol	Other component
Allyl alcohol			
Methylcyclohexane	85.0	42	58
Hexane	65.5	4.5	95.5
Cyclohexane	74.0	58	42
2,5-Dimethylhexane	89.3	50	50
Octane	93.4	68	32
Benzene	76.75	17.36	82.64
Toluene	92.4	50	50
Propyl acetate	94.2	53	47
Methyl butyrate	93.8	55	45
1,2-Dichloroethane	79.9	18	82
3-Iodo-l-propene	89.4	28	72
Chlorobenzene	96.2	85	15
Diethyl sulfide	85.1	45	55
Phenol			
2,7-Dimethyloctane	159.5	6	94
Decane	168.0	35	65
Tridecane	180.6	83.1	16.9
Butylbenzene	175.0	46	54
1,2,4-Trimethylbenzene	166.0	25	75
1,3,5-Trimethylbenzene	163.5	21	79
Indene	177.8	47	53
Camphene	156.1	22	78
Benzaldehyde	175.6	51.0	49.0
1-Octanol	195.4	13	87
2-Octanol	184.5	50	50
Dipentyl ether	180.2	78	22
Diisopentyl ether	172.2	15	85
2-Methylpyridine	185.5	75.4	24.6
3-Methylpyridine	188.9	71.2	29.8
4-Methylpyridine	190.0	67.5	32.5
2,4-Dimethylpyridine	193.4	57.0	43.0
2,6-Dimethylpyridine	185.5	72.5	27.5
2,4,6-Trimethylpyridine	195.2	52.3	47.7
Aniline	185.8	41.9	58.1
Ethylene diacetate	195.5	39.2	60.8
Iodobenzene	177.7	53	47
Benzyl alcohol			
Naphthalene	204.1	60	40
D-Limonene	176.4	11	89
1,3,5-Triethylbenzene	203.2	57	43
o-Cresol	Zeotrope		
m-Cresol	207.1	61	39

TABLE 2.43 Binary Azeotropic (Constant-Boiling) Mixtures (*Continued*)

System	BP of azeotrope, °C	Composition, wt%	
		Alcohol	Other component
Benzyl alcohol (*continued*)			
p-Cresol	206.8	62	38
N-Methylaniline	195.8	30	70
N,N-Dimethylaniline	193.9	6.5	93.5
N-Ethylaniline	202.8	50	50
N,N-Diethylaniline	204.2	72	28
Iodobenzene	187.8	12	88
Nitrobenzene	204.0	58	42
o-Bromotoluene	181.3	7	93
Bomeol	205.1	85.8	14.2
2-Ethoxyethanol			
Methylcyclohexane	98.6	15	85
Heptane	96.5	14	86
Octane	116.0	38	62
Toluene	110.2	10.8	89.2
Ethylbenzene	127.8	48	52
p-Xylene	128.6	50	50
Styrene	130.0	55	45
Propylbenzene	134.6	80	20
Isopropylbenzene	133.2	67	33
Camphene	131.0	65	35
Propyl butyrate	133.5	72	28
2-Butoxyethanol			
Dipentene	164.0	53	47
1,3,5-Trimethylbenzene	162.0	32	68
Butylbenzene	169.6	73.4	26.6
Camphene	154.5	30	70
o-Cresol	191.6	15	85
Phenetole	167.1	52	48
Cineole	168.9	58.5	41.5
Benzaldehyde	171.0	91	9
Diisobutyl sulfide	163.8	42	58
1,2-Ethanediol			
Heptane	97.9	3	97
Decane	161.0	23	77
Tridecane	188.0	55	45
Toluene	110.1	2.3	97.7
Styrene	139.5	16.5	83.5
Stilbene	196.8	87	13
m-Xylene	135.1	6.55	93.45
p-Xylene	134.5	6.4	93.6
1,3,5-Trimethylbenzene	156	13	87
Propylbenzene	152	19	81

(*Continued*)

TABLE 2.43 Binary Azeotropic (Constant-Boiling) Mixtures (*Continued*)

System	BP of azeotrope, °C	Composition, wt%	
		Alcohol	Other component
1,2-Ethanediol (continued)			
Isopropylbenzene	147.0	18	82
Naphthalene	183.9	51	49
1-Methylnaphthalene	190.3	60.0	40.0
2-Methylnaphthalene	189.1	57.2	42.8
Anthracene	197	98.3	1.7
Indene	168.4	26	74
Acenaphthene	194.65	74.2	25.8
Fluorene	196.0	82	18
Camphene	152.5	20	80
Camphor	186.2	40	60
Biphenyl	192.3	66.5	33.5
Diphenylmethane	193.3	68.5	31.5
Benzyl alcohol	193.1	56	44
2-Phenylethanol	194.4	69	31
o-Cresol	189.6	27	73
m-Cresol	195.2	60	40
3,4-Dimethylphenol	197.2	89	11
Menthol	188.6	51.5	48.5
Ethyl benzoate	186.1	46.5	53.5
o-Bromotoluene	166.8	25	75
Dibutyl ether	139.5	6.4	93.6
Methoxybenzene	150.5	10.5	89.5
Diphenyl ether	193.1	60	40
Benzyl phenyl ether	195.5	87	13
Acetophenone	185.7	52	48
2,4-Dimethylaniline	188.6	47	53
N,N-Dimethylaniline	175.9	33.5	66.5
m-Toluidine	188.6	42	58
2,4,6-Trimethylpyridine	170.5	9.7	90.3
Quinoline	196.4	79.5	20.5
Tetrachloroethylene	119.1	94	6
1,2-Dibromoethane	129.8	4	96
Chlorobenzene	130.1	94.4	5.6
α-Chlorotoluene	167.0	30	70
Nitrobenzene	185.9	59	41
o-Nitrotoluene	188.5	48.5	51.5
1,2-Ethanediol monoacetate			
Indene	180.0	20	80
1-Octanol	189.5	71	29
Phenol	197.5	65	35
o-Cresol	199.5	51	49
m-Cresol	206.5	31	69
p-Cresol	206.0	33	67
Dipentyl ether	180.8	42	58
Diisopentyl ether	170.2	28	72
m-Bromotoluene	182.0	32	68

TABLE 2.43   Binary Azeotropic (Constant-Boiling) Mixtures (*Continued*)

D. Binary azeotropes containing ketones

System	BP of azeotrope, °C	Composition, wt%	
		Ketone	Other component
Acetone			
Cyclopentane	41.0	36	64
Pentane	32.5	20	80
Cyclohexane	53.0	67.5	32.5
Hexane	49.8	59	41
Heptanc	55.9	89.5	10.5
Diethylamine	51.4	38.2	61.8
Methyl acetate	55.8	48.3	51.7
Diisopropyl ether	54.2	61	39
Chloroform	64.4	78.1	21.9
Carbon tetrachloride	56.1	11.5	88.5
Carbon disulfide	39.3	67	33
Ethylene sulfide	51.5	57	43
2-Butanone			
Cyclohexane	71.8	40	60
Hexane	64.2	28.6	71.4
Heptane	77.0	70	30
2,5-Dimethylhexane	79.0	95	5
Benzene	78.33	44	56
2-Methyl-2-propanol	78.7	69	31
Butylamine	74.0	35	65
Ethyl acetate	77.1	11.8	88.2
Methyl propionate	79.0	60	40
Butyl nitrite	76.7	30	70
1-Chlorobutane	77.0	38	62
Fluorobenzene	79.3	75	25

E. Miscellaneous binary azeotropes

System	BP of azeotrope, °C	Composition, wt %	
		Solvent	Other component
Solvent: acetamide			
Dipentene	169.2	18	82
Biphenyl	213.0	50.5	49.5
Diphenylmethane	215.2	56.5	43.5
1,2-Diphenylethane	218.2	68	32
o-Xylene	142.6	11	89

(*Continued*)

TABLE 2.43   Binary Azeotropic (Constant-Boiling) Mixtures (*Continued*)

System	BP of azeotrope, °C	Composition, wt%	
		Solvent	Other component
Solvent: acetamide (continued)			
m-Xylene	138.4	10	90
p-Xylene	137.8	8	92
Styrene	144	12	88
4-Isopropyl-1-methylbenzene	170.5	19	81
Naphthalene	199.6	27	73
1-Methylnaphthalene	209.8	43.8	56.2
2-Methylnaphthalene	208.3	40	60
Indene	177.2	17.5	82.5
Acenaphthene	217.1	64.2	35.8
Camphene	155.5	12	88
Camphor	199.8	23	77
Benzaldehyde	178.6	6.5	93.5
3,4-Dimethylphenol	221.1	96	4
2-Methoxy-4-(2-propenyl)phenol	220.8	88	12
N-Methylaniline	193.8	14	86
N-Ethylaniline	199.0	18	82
N,N-Diethylaniline	198.1	24	76
Diphenyl ether	214.6	52	48
Safrole	208.8	32	68
Tetrachloroethylene	120.5	97.4	2.6
Solvent: aniline			
Nonane	149.2	13.5	86.5
Decane	167.3	36	64
Undecane	175.3	57.5	42.5
Dodecane	180.4	71.5	28.5
Tridecane	182.9	86.2	13.8
Tetradecane	183.9	95.2	4.8
Butylbenzene	177.8	46	54
1,2,4-Trimethylbenzene	168.6	13.5	86.5
1,3,5-Trimethylbenzene	164.3	12.0	88.0
Indene	179.8	41.5	58.5
1-Octanol	183.9	83	17
o-Cresol	191.3	8	92
Dipentyl ether	177.5	55	45
Diisopentyl ether	169.3	28	72
Hexachloroethane	176.8	66	34
Solvent: pyridine			
Heptane	95.6	25.3	74.7
Octane	109.5	56.1	43.9
Nonane	115.1	89.9	10.1
Toluene	110.1	22.2	77.8
Phenol	183.1	13.1	86.9
Piperidine	106.1	8	92

TABLE 2.43   Binary Azeotropic (Constant-Boiling) Mixtures (*Continued*)

System	BP of azeotrope, °C	Composition, wt%	
		Solvent	Other component
Solvent: thiophene			
Methylcyclopentane	71.5	14	86
Cyclohexane	77.9	41.2	58.8
Hexane	68.5	11.2	88.8
Heptane	83.1	83.2	16.8
2,3-Dimethylpentane	80.9	64	36
2,4-Dimethylpentane	76.6	42.7	57.3
Solvent: benzene			
Methylcyclopentane	71.7	16	84
Cyclohexene	78.9	64.7	35.3
Cyclohexane	77.6	51.9	48.1
Hexane	68.5	4.7	95.3
Heptane	80.1	99.3	0.7
2,2-Dimethylpentane	75.9	46.3	53.7
2,3-Dimethylpentane	79.4	78.8	21.2
2,4-Dimethylpentane	75.2	48.3	51.7
2,2,4-Trimethylpentane	80.1	97.7	2.3
Solvent: bis(2-hydroxyethyl) ether			
Biphenyl	232.7	48	52
Diphenylmethane	236.0	52	48
1,3,5-Trimethylbenzene	210.0	22	78
Naphthalene	212.6	22	78
1-Methylnaphthalene	277.0	45	55
2-Methylnaphthalene	225.5	39	61
Acenaphthene	239.6	62	38
Fluorene	243.0	80	20
Benzyl acetate	214.9	7	93
Bornyl acetate	223.0	18	82
Ethyl fumarate	217.1	10	90
Dimethyl *o*-phthalate	245.4	96.3	3.7
Methyl salicylate	220.6	15	85
2-Hydroxy-1-isopropyl-4-methylbenzene	232.3	13	87
1,2-Dihydroxybenzene	259.5	46	54
Safrole	225.5	33	67
Isosafrole	233.5	46	54
Benzyl phenyl ether	241.5	80	20
Nitrobenzene	210.0	10	90
m-Nitrotoluene	224.2	25	75
o-Nitrophenol	216.0	10.5	89.5
Quinoline	233.6	29	71
p-Dibromobenzene	212.9	13	87

TABLE 2.44 Ternary Azeotropic Mixtures

A. Ternary azeotropes containing water and alcohols

System	BP of azeotrope, °C	Composition, wt%		
		Water	Alcohol	Other component
Methanol				
Chloroform	52.3	1.3	8.2	90.5
2-Methyl-1,3-butadiene	30.2	0.6	5.4	94.0
Methyl chloroacetate	67.9	6.3	81.2	13.5
Ethanol				
Acetonitrile	72.9	1	55	44
Acrylonitrile	69.5	8.7	20.3	71.0
Benzene	64.9	7.4	18.5	74.1
Butylamine	81.8	7.5	42.5	50.0
Butyl methyl ether	62	6.3	8.6	85.1
Carbon disulfide	41.3	1.6	5.0	93.4
Carbon tetrachloride	62	4.5	10.0	85.5
Chloroform	55.3	2.3	3.5	94.2
Crotonaldehyde	78.0	4.8	87.9	7.3
Cyclohexane	62.6	4.8	19.7	75.5
1,2-Dichloroethane	66.7	5	17	78
1,1-Diethoxyethane	77.8	11.4	27.6	61.0
Diethoxymethane	73.2	12.1	18.4	69.5
Ethyl acetate	70.2	9.0	8.4	82.6
Heptane	68.8	6.1	33.0	60.9
Hexane	56.0	3	12	85
Toluene	74.4	12	37	51
Trichloroethylene	67.0	5.5	16.1	78.4
Triethylamine	74.7	9	13	78
1-Propanol				
Benzene	67	7.6	10.1	82.3
Carbon tetrachloride	65.4	5	11	84
Cyclohexane	66.6	8.5	10.0	81.5
1,1-Dipropoxyethane	87.6	27.4	51.6	21.0
Dipropoxymethane	86.4	8.0	44.8	47.2
Dipropyl ether	74.8	11.7	20.2	68.1
3-Pentanone	81.2	20	20	60
Propyl acetate	82.5	17.0	10.0	73.0
Propyl formate	70.8	13	5	82
Tetrachloroethylene	81.2	12.5	20.7	66.8
2-Propanol				
Benzene	66.5	7.5	18.7	73.8
Butylamine	83	12.5	40.5	47.0

TABLE 2.44   Ternary Azeotropic Mixtures (*Continued*)

A. Ternary azeotropes containing water and alcohols

System	BP of azeotrope, °C	Composition, wt%		
		Water	Alcohol	Other component
2-Propanol (*continued*)				
Cyclohexane	64.3	7.5	18.5	74.0
Toluene	76.3	13.1	38.2	48.7
Trichloroethylene	69.4	7	20	73
1-Butanol				
Butyl acetate	89.4	37.3	27.4	35.3
Butyl formate	83.6	21.3	10.0	68.7
Dibutyl ether	90.6	29.9	34.6	35.5
Heptane	78.1	41.4	7.6	51.0
Hexane	61.5	19.2	2.9	77.9
Nonane	90.0	69.9	18.3	11.8
Octane	86.1	60.0	14.6	25.4
2-Butanol				
Carbon tetrachloride	65	4.05	4.95	91.00
Cyclohexane	69.7	8.9	10.8	80.3
Isooctane	76.3	9	19	72
2-Methyl-1-propanol				
Isobutyl acetate	86.8	30.4	23.1	46.5
Isobutyl formate	80.2	17.3	6.7	76.0
Toluene	81.3	17.9	16.4	65.7
2-Methyl-2-propanol				
Benzene	67.3	8.1	21.4	70.5
Carbon tetrachloride	64.7	3.1	11.9	85.0
Cyclohexane	65.0	8	21	71
3-Methyl-1-butanol				
Isopentyl acetate	93.6	44.8	31.2	24.0
Isopentyl formate	89.8	32.4	19.6	48.0
Allyl alcohol				
Benzene	68.2	8.6	9.2	82.2
Carbon tetrachloride	65.2	5	11	84
Cyclohexane	66.2	8	11	81
Hexane	59.7	8.5	5.1	86.4

(*Continued*)

TABLE 2.44 Ternary Azeotropic Mixtures (*Continued*)

			B. Other ternary azeotropes		
System	BP of azeotrope, °C	Composition, wt%	System	BP of azeotrope, °C	Composition, wt%
Water	32.5	0.4	Water	80.7	17.4
Acetone		7.6	Nitromethane		58.3
2-Methyl-1,3-butadiene		92.0	Nonane		24.3
Water	66	8.2	Water	77.4	12.4
Acetonitrile		23.3	Nitromethane		44.3
Benzene		68.5	Octane		43.3
Water	67	6.4	Water	33.1	2.1
Acetonitrile		20.5	Nitromethane		6.5
Trichloroethylene		73.1	Pentane		91.4
Water	68.6	3.5	Water	82.8	20.6
Acetonitrile		9.6	Nitromethane		73.3
Triethylamine		86.9	Undecane		6.1
Water	63.6	5	Water	93.5	40.5
2-Butanone		35	Pyridine		54.5
Cyclohexane		60	Dodecane		5.0
Water	55.0	4	Water	93.1	38.5
Butyraldehyde		21	Pyridine		51.0
Hexane		75	Undecane		10.5
Water	107.6	21.3	Water	92.3	35.5
Formic acid		76.3	Pyridine		45.5
Isopentanoic acid		2.4	Decane		19.0
Water	107.0	15.5	Water	107.6	19.5
Formic acid		66.8	Formic acid		75.9
Isobutyric acid		17.7	Butyric acid		4.6
Water	71.4	7.9	Water	107.2	18.6
Nitromethane		29.7	Formic acid		71.9
Heptane		62.4	Propionic acid		9.5
Water	105	11.0	Pyridine		38.2
Hydrogen bromide		10.4	Decane		30.4
Chlorobenzene		78.6	Acetic acid	129.1	13.5
Water	96.9	20.2	Pyridine		25.2
Hydrogen chloride		5.3	Ethylbenzene		61.3
Chlorobenzene		74.5	Acetic acid	98.5	3.4
Water	107.3	64.8	Pyridine		10.6
Hydrogen chloride		15.8	Heptane		86.0
Phenol		19.4	Acetic acid	128.0	20.7
Water	116.1	54	Pyridine		29.4
Hydrogen fluoride		10	Nonane		49.9
Fluorosilic acid		36	Acetic acid	115.7	10.4
Water	75.1	11.5	Pyridine		20.1
Nitroethane		75.1	Octane		69.5
Heptane		64.0	Water	83.1	21.5
Water	59.5	8.4	Nitromethane		75.3
Nitroethane		9.3	Dodecane		3.2
Hexane		82.3	Acetic acid	129.2	10.2
Water	82.4	19.1	Pyridine		22.5
Nitromethane		68.1	*p*-Xylene		67.3
Decane		12.8	Acetic acid	163.0	75.0
Water	90.5	30.5	2,6-Dimethylpyridine		13.8
Pyridine		37.0	Undecane		11.2
Nonane		32.5	Acetic acid	147.0	12.6
Water	86.7	22.4	2,6-Dimethylpyridine		74.3
Pyridine		25.5	Decane		13.1

TABLE 2.44   Ternary Azeotropic Mixtures (*Continued*)

B. Other ternary azeotropes

System	BP of azeotrope, °C	Composition, wt%	System	BP of azeotrope, °C	Composition, wt%
Octane		52.0	Acetic acid	141.3	19.9
Water	78.6	14.0	2-Methylpyridine		46.8
Pyridine		15.5	Decane		33.3
Heptane		70.5	Acetic acid	135.0	12.8
Acetic acid	134.4	23	2-Methylpyridine		38.4
Pyridine		55	Nonane		48.8
Acetic anhydride		22	Acetic acid	121.3	3.6
Acetic acid	134.1	31.4	2-Methylpyridine		24.8
Octane		71.6	Hexane		34.4
Acetic acid	77.2	7.6	1-Propanol	73.8	15.5
Benzene		34.4	Benzene		30.4
Cyclohexane		58.0	Cyclohexane		54.2
Acetic acid	132	15	2-Propanol	69.1	31.1
2-Methyl-1-butanol		54	Benzene		15.0
Isopentyl acetate		31	Cyclohexane		53.9
Propionic acid	149.3	29.5	1-Butanol	77.4	4
2-Methylpyridine		32.0	Benzene		48
Decane		38.5	Cyclohexane		48
Acetic acid	132.2	17.7	1-Butanol	108.7	11.9
Pyridine		30.5	Pyridine		20.7
o-Xylene		51.8	Toluene		76.4
Methanol	47.4	14.6	Propionic acid	140.1	16.5
Methyl acetate		36.8	2-Methylpyridine		21.5
Hexane		48.6	Nonane		42.0
Ethanol	63.2	10.4	Propionic acid	123.7	4.5
Acetone		24.3	2-Methylpyridine		10.5
Chloroform		65.3	Octane		85.0
Ethanol	70.1	8	Propionic acid	153.4	43.0
Acetonitrile		34	2-Methylpyridine		40.0
Triethylamine		58	Undecane		17.0
Ethanol	64.7	29.6	Propionic acid	147.1	55.5
Benzene		12.8	Pyridine		26.4
Cyclohexane		57.6	Undecane		18.1
Ethanol	57.3	9.5	Methanol	57.5	23
Chloroform		56.1	3-Methylpyridine		16.4
Acetone		30	1,2-Ethanediol	188.6	29.5
Chloroform		47	Phenol		54.8
Methanol	47	14.6	2,4,6-Trimethylpyridine		15.7
Acetone		30.8	Acetone	60.8	3.6
Hexane		59.6	Chloroform		68.8
Methanol	53.7	17.4	Hexane		27.6
Acetone		5.8	Acetone	49.7	51.1
Methyl acetate		76.8	Methyl acetate		5.6
Methanol	50.8	17.8	Hexane		43.3
Methyl acetate		48.6	Chloroform	62.0	79.7
Cyclohexane		33.6	Ethyl formate		5.3
1,2-Ethanediol	185.0	8.7	2-Bromopropane		15.7
Phenol		74.6	1,4-Dioxane	101.8	44.3
2,6-Dimethylpyridine		16.7	2-Methyl-1-propanol		26.7
1,2-Ethanediol	185.1	5.9	Toluene		29.0
Phenol		79.1			
2-Methylpyridine		15.0			
1,2,-Ethanediol	186.4	15.9			
Phenol		67.7			

2.8 FREEZING MIXTURES

A freezing mixture a mixture of substances (such as salt and ice) to obtain a temperature below the freezing point of the solvent (such as water).

TABLE 2.45 Compositions of Aqueous Antifreeze Solutions

		Freezing point of ethyl alcohol–water mixtures*		
Specific gravity 20°/4°C (68°F)	% alcohol by weight	% alcohol by volume	Freezing point	
			°C	°F
0.99363	2.5	3.13	−1.0	30.2
0.98971	4.8	6.00	−2.0	28.4
0.98658	6.8	8.47	−3.0	26.6
0.98006	11.3	14.0	−5.0	23.0
0.97670	13.8	17.0	−6.1	21.0
0.97336	16.4	20.2	−7.5	18.5
0.97194	17.5	21.5	−8.7	16.3
0.97024	18.8	23.1	−9.4	15.1
0.96823	20.3	24.8	−10.6	12.9
0.96578	22.1	27.0	−12.2	10.0
0.96283	24.2	29.5	−14.0	6.8
0.95914	26.7	32.4	−16.0	3.2
0.95400	29.9	36.1	−18.9	−2.0
0.94715	33.8	40.5	−23.6	−10.5
0.93720	39.0	46.3	−28.7	−19.7
0.92193	46.3	53.8	−33.9	−29.0
0.90008	56.1	63.6	−41.0	−41.8
0.86311	71.9	78.2	−51.3	−60.3

		Freezing point of methyl (wood) alcohol-water mixtures*		
Specific gravity 15.6°C (68°F)	% alcohol by weight	% alcohol by volume	Freezing point	
			°C	°F
0.993	3.9	5	−2.2	28
0.986	8.1	10	−5.0	23
0.980	12.2	15	−8.3	17
0.974	16.4	20	−11.7	11
0.968	20.6	25	−15.6	4
0.963	24.9	30	−20.0	−4
0.956	29.2	35	−25.0	−13
0.949	33.6	40	−30.0	−22
0.942	38.0	45	−35.6	−32

*Values are for pure alcohol. Since some commercial antifreezes contain small amounts of water, slightly higher volume concentrations than those given in the table may be required. Antifreezes also contain corrosion inhibitors and other additives to make them function properly as cooling liquids. These affect freezing point slightly and specific gravity to a greater degree.

TABLE 2.45 Compositions of Aqueous Antifreeze Solutions (*Continued*)

Freezing point of Prestone-water mixtures[†]

% Prestone		Specific gravity	Freezing point	
By weight	By volume	15°/15°C (59°F)	°C	°F
10	9.2	1.013	−3.6	25.6
15	13.8	1.019	−5.6	22.0
20	18.3	1.026	−7.9	17.8
25	23.0	1.033	−10.7	12.8
30	28.0	1.040	−14.0	6.8
40	37.8	1.053	−22.3	−8.2
50	47.8	1.067	−33.8	−28.8
60	58.1	1.079	−49.3	−56.7

Freezing point of ethyl alcohol–water mixtures

Specific gravity 15.6°C (60°F)	% alcohol by volume	Freezing point	
		°C	°F
0.990	5	−1.7	29
0.984	10	−3.3	26
0.978	15	−6.1	21
0.972	20	−8.3	17
0.964	25	−11.1	12
0.955	30	−14.4	6
0.945	35	−17.8	0
0.933	40	−18.3	−1
0.922	45	−18.9	−2
0.910	50	−20.0	−4
0.899	55	−21.7	−7
0.887	60	−23.3	−10
0.875	65	−24.4	−12
0.864	70	−26.7	−16
0.852	75	−32.2	−26
0.840	80	−41.7	−43

[†]Eveready Prestone marketed for antifreeze purposes, is 97% ethylene glycol containing fractional percentages of soluble and insoluble ingredients to prevent foaming, creepage and water corrosion in automobile cooling systems.

(*Continued*)

TABLE 2.45   Compositions of Aqueous Antifreeze Solutions (*Continued*)

Freezing point of propylene glycol–water mixtures*

Specific gravity 15.6°C (60°F)	% glycol by volume	Freezing point	
		°C	°F
1.004	5	−1.1	30
1.006	10	−2.2	28
1.012	15	−3.9	25
1.017	20	−6.7	20
1.020	25	−8.9	16
1.024	30	−12.8	9
1.028	35	−16.1	3
1.032	40	−20.6	−5
1.037	45	−26.7	−16
1.040	50	−33.3	−28

Freezing point of glycerol–water mixtures[†]

% Glycerol by weight	Specific gravity 15°/15°C (59°F)	Specific gravity 20°/20°C (68°F)	Freezing point	
			°C	°F
10	1.02415	1.02395	−1.6	29.1
20	1.04935	1.04880	−4.8	23.4
30	1.07560	1.07470	−9.5	14.9
40	1.10255	1.10135	−15.5	4.3
50	1.12985	1.12845	−22.0	−7.4
60	1.15770	1.15605	−33.6	−28.5
70	1.18540	1.18355	−37.8	−36.0
80	1.21290	1.21090	−19.2	−2.3
90	1.23950	1.23755	−1.6	29.1
100	1.26557	1.26362	17.0	62.6

*Values are for pure alcohol. Since some commercial antifreezes contain small amounts of water, slightly higher volume concentrations than those given in the table may be required. Antifreezes also contain corrosion inhibitors and other additives to make them function properly as cooling liquids. These affect freezing point slightly and specific gravity to a greater degree.

[†]The values are those reported by Bosart and Snoddy (*Jour. Ind. Eng. Chem.*, **19**, 506 (1927), and Lane (*Jour. Ind. Eng. Chem.*, **17**, 924 (1925)) but modified by adding 2°F to all temperatures below 0°F.

TABLE 2.45   Compositions of Aqueous Antifreeze Solutions (*Continued*)

Freezing point of magnesium chloride brines

% MgCl$_2$ by weight	Spec. grav. 15.6°C (60°F)	Freezing point		% MgCl$_2$ by weight	Spec. grav. 15.6°C (60°F)	Freezing point	
		°C	°F			°C	°F
5	1.043	−3.11	26.4	18	1.161	−22.1	−7.7
6	1.051	−3.89	25.0	19	1.170	−25.6	−12.2
7	1.060	−4.72	23.5	20	1.180	−27.4	−17.3
8	1.069	−5.67	21.8	21	1.190	−30.6	−23.0
9	1.078	−6.67	20.0	22	1.200	−32.8	−27.0
10	1.086	−7.83	17.9	23	1.210	−28.9	−20.0
11	1.096	−9.05	15.7	24	1.220	−25.6	−14.0
12	1.105	−10.5	13.1	25	1.230	−23.3	−10.0
13	1.114	−12.1	10.3	26	1.241	−21.1	−6.0
14	1.123	−13.7	7.3	27	1.251	−19.4	−3.0
15	1.132	−15.6	4.0	28	1.262	−18.3	−1.0
16	1.142	−17.6	0.4	29	1.273	−17.2	+1.0
17	1.151	−19.7	−3.5	30	1.283	−16.7	2.0

Freezing point of sodium chloride brines

% NaCl by weight	Spec. grav. 15°C (59°F)	Freezing point		% NaCl by weight	Spec. grav. 15°C (59°F)	Freezing point	
		°C	°F			°C	°F
0	1.000	0.00	32.0	15	1.112	−10.88	12.4
1	1.007	−0.58	31.0	16	1.119	−11.90	10.6
2	1.014	−1.13	30.0	17	1.127	−12.93	8.7
3	1.021	−1.72	28.9	18	1.135	−14.03	6.7
4	1.028	−2.35	27.8	19	1.143	−15.21	4.6
5	1.036	−2.97	26.7	20	1.152	−16.46	2.4
6	1.043	−3.63	25.5	21	1.159	−17.78	+0.0
7	1.051	−4.32	24.2	22	1.168	−19.19	−2.5
8	1.059	−5.03	22.9	23	1.176	−20.69	−5.2
9	1.067	−5.77	21.6	23.3 (*E*)	1.179	−21.13	−6.0
10	1.074	−6.54	20.2	24	1.184	−17.0*	+1.4*
11	1.082	−7.34	18.8	25	1.193	−10.4*	13.3*
12	1.089	−8.17	17.3	26	1.201	−2.3*	27.9*
13	1.097	−9.03	15.7	26.3	1.203	0.0*	32.0*
14	1.104	−9.94	14.1				

*Saturation temperatures of sodium chloride dihydrate; at these temperatures NaCl · 2H$_2$O separates leaving the brine of the eutectic composition (*E*).

Propylene glycol, a satisfactory antifreeze with the advantage of being nontoxic, can be combined with glycerol, also an efficient nontoxic antifreeze, to give a mixture that can be tested for freezing point with an ethylene glycol (Prestone) hydrometer. A mixture of 70% propylene glycol and 30% glycerol (% by weight of water-free materials), when diluted, can be tested on the standard instrument used for ethylene glycol solutions.

2.9 BOND LENGTHS AND STRENGTHS

Distances between centers of bonded atoms are called *bond lengths*, or *bond distances*. Bond lengths vary depending on many factors, but in general, they are very consistent. Of course the bond orders affect bond length, but bond lengths of the same order for the same pair of atoms in various molecules are very consistent.

The *bond order* is the number of electron pairs shared between two atoms in the formation of the bond. Bond order for C=C and O=O is 2. The amount of energy required to break a bond is called *bond dissociation energy* or simply *bond energy*. Since bond lengths are consistent, bond energies of similar bonds are also consistent.

Bonds between the same type of atom are *covalent bonds*, and bonds between atoms when their electronegativity differs slightly are also predominant covalent in character. Theoretically, even ionic bonds have some covalent character. Thus, the boundary between ionic and covalent bonds is not a clear line of demarcation.

For covalent bonds, bond energies and bond lengths depend on many factors: electron affinities, sizes of atoms involved in the bond, differences in their electronegativity, and the overall structure of the molecule. There is a general trend in that *the shorter the bond length, the higher the bond energy* but there is no formula to show this relationship, because of the widespread variation in bond character.

TABLE 2.46 Bond Lengths between Carbon and Other Elements

Bond type	Bond length, μm
Carbon–carbon	
Single bond	
Paraffinic: —C—C—	154.1(3)
In presence of —C=C— or of aromatic ring	153(1)
In presence of —C=O bond	151.6(5)
In presence of two carbon–oxygen bonds	149(1)
In presence of two carbon–carbon double bonds	142.6(5)
Aryl-C=O	147(2)
In presence of one carbon–carbon triple bond: —C—C≡C—	146.0(3)
In presence of one carbon–nitrogen triple bond: —C—C≡N	146.6(5)
In compounds with tendency to dipole formation, e.g., C=C—C=O	144(1)
In aromatic compounds	139.5(5)
In presence of carbon–carbon double and triple bounds: —C=C—C≡C—	142.6(5)
In presence of two carbon–carbon triple bounds: —C≡C—C≡C—	137.3(4)
Double bond	
Single: —C=C—	133.7(6)
Conjugated with a carbon–carbon double bond: —C=C—C=C—	133.6(5)
Conjugated with a carbon–oxygen double bond: —C=C—C=O	136(1)
Cumulative: —C=C=C— or —C=C=O	130.9(5)
Triple bond	
Simple: —C≡C—	120.4(2)
Conjugated: —C≡C—C=C—, —C≡C—C=O, or —C≡C—aryl	120.6(4)

Bond type	Bond length, pm			
	Carbon–halogen			
	Fluorine	Chlorine	Bromine	Iodine
Paraffinic: R—X	137.9(5)	176.7(2)	193.8(5)	213.9(1)
Olenfinic: —C=C—X	133.3(5)	171.9(5)	189(1)	209.2(5)
Aromatic: Ar—X	132.8(5)	170(1)	185(1)	205(1)
Acetylenic: —C≡C—X	(127)	163.5(5)	179.5(10)	199(2)

TABLE 2.46 Bond Lengths between Carbon and Other Elements (*Continued*)

Bond type	Bond length, μm
Carbon–carbon	
Paraffinic	
In methane (in CD_4, 109.2)	109.4
In monosubstituted carbon: H—C—Y	109.6(5)
In disubstituted carbon: $H—\overset{\displaystyle X}{\underset{\displaystyle Y}{C}}—$	107.3(5)
In trisubstituted carbon: $H—\overset{\displaystyle X}{\underset{\displaystyle Z}{C}}—Y$	107.0(7)
Olefinic	
Simple: H—C=C—	108.3(5)
Cumulative carbon–carbon double bonds: H—C=C=C—	107(1)
Cumulative carbon–carbon–oxygen double bonds: H—C—C=C=O	108(1)
Aromatic	108.4(5)
Acetylenic (in C_2H_2, 105.9)	105.5(5)
In small rings	108.1(5)
In presence of a carbon triple bond: H—C≡C—	111.5(4)
Carbon–nitrogen	
Single bond	
Paraffinic:	
3-covalent nitrogen: RNH_2, R_2NH, R_3N	147.2(5)
4-covalent nitrogen: RNH_3^+, R_3N-BX_3	147.9(5)
In—C—N=	147.5(10)
In aromatic compounds	143(1)
In conjugated heterocyclic systems (partial double bond)	135.3(5)
In – N –C=O (partial double bond)	132.2(5)
Double bond: —C=N—	132
Triple bond (in CN radical, 117.74): —C≡N	115.7(5)
Carbon–oxygen	
Single bond	
Paraffinic and saturated heterocyclic: —C—O—	142.6(5)
Strained, as in epoxides: $—\underset{\displaystyle O}{C—C}—$	143.5(5)
In aromatic compounds, as Ar—OH	136(1)
Longer bond in carboxylic acids and esters (HCOOH, 131.2)	135.8(5)
In conjugated heterocyclics, as furan	137.1(16)
Double bond	
In CO+	111.5
In CO	112.8
In CO_2^+	117.7
In HCO	119.8(8)
In carbonyls	114.5(10)
In aldehydes and ketones	121.5(5)
In acyl halides: R—CO—X	117.1(4)
Shorter bond in carboxylic acids and esters	123.3(5)
In zwitterion forms	126(1)

(*Continued*)

TABLE 2.46 Bond Lengths between Carbon and Other Elements (*Continued*)

Bond type	Bond length, μm
Carbon–oxygen	
In O=C=	116.0(1)
In isocyanates: RN=C=O	117(1)
In conjugated systems, as in partial triple bond: O=C—C=C	121.5(5)
In 1,4-quinones	115(2)
In metal acetylacetonates	128(2)
In calcite: CaCO$_3$	129(1)
Carbon–selenium	
Single bond	
Paraffinic: —C—Se—	198(2)
In presence of fluorine, as in perfluorocompounds: —CF—Se—	195(2)
Double bond	
In Se=C=, as SeCS and SeCO	170.9(3)
In CSe radical	167
Carbon–silicon	
Alkyl substituent: H$_3$C—Si or H$_2$C—Si	187.0(5)
Aryl substituent: aryl—Si	184.3(5)
Electronegative substituent: R—Si—X	185.4(5)
Carbon–sulfur	
Single bond	
Paraffinic: —C—S—	181.7(5)
In presence of fluorine, as in perfluoro compounds: —CF—S—	183.5(1)
In heterocyclic systems: partial double bonds	171.8(5)
Double bonds	
In S=C; thiophene, S=CR$_2$	171(1)
In sulfoxides and sulfones	180(1)
In presence of second carbon–carbon double bond: S=C—C=C—	155.5(1)
In SC radical [in CS$_2^+$, 155.4(5)]	153.49(2)

Bond type	Bond length, pm	Bond type	Bond length, pm
Other elements and carbon			
C-Al	224(4)	C-Cr	192(4)
C-As	198(1)	C-Fe	184(2)
C-B	156(1)	C-Ge	
C-Be	193	Alkyl	193(3)
C-Bi	230	Aryl	194.5(5)
C-Co	183(2)		
C-Hg	207(1)	C-Sn	
in Hg(CN)2	199(2)	Alkyl	214.3(5)
C-In	216(4)	Electronegative	218(2)
C-Mo	208(4)	substituent	
C-Ni	210.7(5)	C-Te	190.4
C-Pb (alkyl)	230(1)	C-Tl	270.5(5)
C-Pd	227(4)	C-W	206
C-Sb (paraffinic)	220.2(16)		

TABLE 2.47 Bond Dissociation Energies

Bond	ΔHf_{298}, kJ/mol	Bond	ΔHf_{298}, kJ/mol
Carbon		Carbon	
$(CH_3)_2C$—CH_3	335	$C_6H_5CH_2$—$N(CH_3)_2$	255(4)
$(CH_3)_2C$—$C(CH_3)_2$	282.4	CH_3—$(N$=$NCH_3)$	219.7
CH_3—C_6H_5	389	C_2H_5—$(N$=$NC_2H_5)$	209.2
CH_3—$CH_2C_6H_5$	301	$(CH_3)_3C$—N=$NC(CH_3)_3$	182.0
$(CH_3)_3C$—$C(C_6H_5)_3$	63	Aryl—CH_2N=NCH_2—aryl	157
CH_3—allyl	301	CF_3—$(N$=$NCF_3)$	231.0
CH_3—vinyl	121	H_2C=NH	644(21)
CH_3—C≡CH	490	HC≡N	937
CH_2=CH—CH=CH_2	418	CH_3—NO	174.9(38)
HC≡C—C≡CH	628	C_2H_5—NO	175.7(54)
H_2C=CH_2	682	C_3H_7—NO	167.8(75)
HC≡CH	962	$(CH_3)_2CH$—NO	171.5(54)
CH_3—CN	506(21)	n-C_4H_9—NO	215.5(42)
CH_3—CH_2CN	305(8)	C_6H_5—NO	215.5(42)
CH_3—$CH(CH_3)CN$	331(8)	$C_{13}C$—NO	134
CH_3—$C(C_6H_5)CN(CH_3)$	251	F_3C—NO	130
CH_3CH_2—CH_2CN	321.8(71)	C_6F_5—NO	211.3(42)
NC—CN	603(21)	NC—NO	121(13)
$C_6H_5 C_6H_5$	418	CH_3—NO_2	247(13)
CH_3—CF_3	423.4(46)	C_2H_5—NO_2	259
CH_2F—CH_2F	368(8)	C—O	1076.5(4)
CF_3—CF_3	406(13)	CH_3—OCH_3	335
CF_2=CF_2	318(13)	CH_3—OC_6H_5	381
CF_3—CN	501	CH_3—$OCH_2C_6H_5$	280
CH_3—CHO	314	C_2H_5—OC_6H_5	213
CH_3—CO	342.7	$C_6H_5CH_2$—$OCOCH_3$	285
CH_3CO—CF_3	308.8	$C_6H_5CH_2$—$OCOC_6H_5$	289
CH_3CO—$COCH_3$	280(8)	CH_3CO—OCH_3	406
C_6H_5CO—COC_6H_5	277.8	CH_3—$OSOCH_3$	280
Aryl—CH_2COCH_2—aryl	273.6	CH_2=$CHCH_2$—$OSOCH_3$	209
$C_6H_5CH_2$—$COOH$	284.9	$C_6H_5CH_2$—$OSOCH_3$	222
$(C_6H_5CH_2)_2CH$—$COOH$	248.5	C=O	749
C—Cl	397(29)	H_2C=O	732
C—F	536(21)	OC=O	532.2(4)
C—H	337.2(8)	SC=O	628
C—I	209(21)	C≡O	1075
C—N	770(4)	C—P	513(8)
CF_3—NF_2	272(13)	C—S	699(8)
CH_3—NH_2	331(13)	CH_3—SH	305(13)
$C_6H_5CH_2$—NH_2	301(4)	CH_3—SC_6H_5	285(8)
CH_3—NHC_6H_5	285	CH_3—$SCH_2C_6H_5$	247(8)
CH_3—$N(CH_3)C_6H_5$	272	OC—S	310.4
$C_6H_5CH_2$—$NHCH_3$	289(4)	C—Se	582(96)

2.10 DIPOLE MOMENTS AND DIELECTRIC CONSTANTS

The permanent dipole moment of an isolated molecule depends on the magnitude of the charge and on the distance separating the positive and negative charges. It is defined as

$$\mu = \left(\sum_i q_i r_i \right)$$

where the summation extends over all charges (electrons and nuclei) in the molecule. The numerical values of the dipole moment, expressed in the c.g.s. system of units, are in debye units, D, where $1\ D = 10^{-18}$ esu of charge $\times$ centimeters. The conversion factor to SI units is

$$1\ D = 3.335\ 64 \times 10^{-30}\ C \cdot m\ [\text{coulomb-meter}]$$

Tables 2.50 contain a selected group of compounds for which the dipole moment is given. An extensive collection of dipole moments (approximately 7000 entries) is contained in A. L. McClellan, *Tables of Experimental Dipole Moments*, W. H. Freeman, San Francisco, 1963. A critical survey of 500 compounds in the gas phase is given by Nelson, Lide, and Maryott, NSRDS-NBS 10, Washington, D.C., 1967.

If two oppositely charged plates exist in a vacuum, there is a certain force of attraction between them, as stated by Coulomb's law:

$$F = \frac{1}{4\pi\varepsilon_0} \cdot \frac{q_1 q_2}{\varepsilon r^2}$$

where F is the force, in newtons, acting on each of the charges q_1 and q_2, r is the distance between the charges, ε is the dielectric constant of the medium between the plates, and ε_0 is the permittivity of free space. q_1, q_2 are expressed in coulombs and r in meters. If another substance, such as a solvent, is in the space separating these charges (or ions in a solution), their attraction for each other is less. The dielectric constant is a measure of the relative effect a solvent has on the force with which two oppositely charged plates attract each other. The dielectric constant is a unitless number.

Dielectric constants for a selected group of inorganic and organic compounds are included in Tables 2.50 and 1.52. An extensive list has been compiled by Maryott and Smith, *National Bureau Standards Circular 514*, Washington, D.C., 1951.

For gases the values of the dielectric constant can be adjusted to somewhat different conditions of temperature and pressure by means of the equation

$$\frac{(\varepsilon - 1)_{t,p}}{(\varepsilon - 1)_{20^\circ, 1\ \text{atm}}} = \frac{p}{760[1 + 0.003\ 411(t - 20)]}$$

where p is the pressure (in mmHg) and t is the temperature (in °C). The errors associated with this equation probably do not exceed 0.02% for gases between 10 and 30°C and for pressures between 700 and 800 mm. The dielectric constants of selected gases will be found in Table 1.52.

TABLE 2.48 Bond Dipole Moments

	Moment, D*	
Group	Aromatic C—X	Aliphatic C—X
C—CH$_3$	0.37	0.0
C—C$_2$H$_5$	0.37	0.0
C—C(CH$_3$)$_3$	0.5	0.0
C—CH=CH$_2$	<0.4	0.6
C—C≡CH	0.7	0.9
C—F	1.47	1.79

TABLE 2.49 Group Dipole Moments

Group	Moment, D*	
	Aromatic C—X	Aliphatic C—X
C—Cl	1.59	1.87
C—Br	1.57	1.82
C—I	1.40	1.65
C—CH$_2$F	1.77	
C—CF$_3$	2.54	2.32
C—CH$_2$Cl	1.85	1.95
C—CHCl$_2$	2.04	1.94
C—CCl$_3$	2.11	1.57
C—CH$_2$Br	1.86	1.96
C—C≡N	4.05	3.4
C—NC	3.5	3.5
C—CH$_2$CN	1.86	2.0
C—C=O	2.65	2.4
C—CHO	2.96	2.49
C—COOH	1.64	1.63
C—CO—CH$_3$	2.96	2.49
C—CO—OCH$_3$	1.83	1.75
C—CO—OC$_2$H$_5$	1.9	1.8
C—OH	1.6	1.7
C—OCH$_3$	1.28	1.28
C—OCF$_3$	2.36	
C—OCOCH$_3$	1.69	
C—OC$_6$H$_5$	1.16	1.16
C—CH$_2$OH	1.58	1.68
C—NH$_2$	1.53	1.46
C—NHCH$_3$	1.71	
C—N(CH$_3$)$_2$	1.58	0.86
C—NHCOCH$_3$	3.69	
C—N(C$_6$H$_5$)$_2$	(0.3)	−0.3
C—NCO	2.32	2.8
C—N$_3$	1.44	
C—NO	3.09	
C—NO$_2$	4.01	2.70
C—CH$_2$NO$_2$	3.3	3.4
C—SH	1.22	1.55
C—SCH$_3$	1.34	1.40
C—SCF$_3$	2.50	
C—SCN	3.59	3.6
C—NCS	2.9	3.3
C—SC$_6$H$_5$	1.51	1.5
C—SF$_5$	3.4	
C—SOCF$_3$	3.88	
(C—)$_2$SO$_2$	5.05	4.53
(C—)$_2$SO$_2$CH$_3$	4.73	
(C—)$_2$SO$_2$CF$_3$	4.32	
C—SeH	1.08	
C—SeCH$_3$	1.31	1.32
C—Si(CH$_3$)$_3$	0.44	0.4

*To convert debye units D into coulomb-meters, multiply by 3.33564×10^{-30}.

TABLE 2.50   Dielectric Constant (Permittivity) and Dipole Moment of Organic Compounds

The temperature in degrees Celsius at which the dielectric constant and dipole moment were measured is shown in this table in parentheses after the value. In some cases, the dipole moment was determined with the substance dissolved in a solvent, and the solvent used is also shown in parentheses after the temperature.

The dielectric constant (permittivity) tabulated is the relative dielectric constant, which is the ratio of the actual electric displacement to the electric field strength when an external field is applied to the substance, which is the ratio of the actual dielectric constant to the dielectric constant of a vacuum. The table gives the static dielectric constant ϵ, measured in static fields or at relatively low frequencies where no relaxation effects occur.

The dipole moment is given in debye units D. The conversion factor to SI units is I D = 3.33564×10^{-30}C · m.

Alternative names for entries are listed in Table 2.21 at the bottom of each double page.

List of Abbreviations

B, benzene	g, gas
C, CCl$_4$	Hx, hexane
cHex, cyclohexane	lq, liquid
D, 1,4-dioxane	

Substance	Dielectric constant, ϵ	Dipole moment, D
Acetaldehyde	21.8 (10), 21.0 (18)	2.75
Acetaldehyde oxime	4.70 (25)	0.830 (20, lq), 0.90 (25, B)
Acetamide	67.6 (91)	3.76
Acetanilide		3.65 (25, B)
Acetic acid	6.20 (20)	1.70
Acetic anhydride	23.3 (0), 22.45 (20)	2.8
Acetone	21.0 (20), 20.7 (25), 17.6 (56)	2.88
Acetonitrile	36.64 (20), 26.6 (82)	3.924
Acetophenone	17.44 (25), 8.64 (202)	3.02
(±)-*erythro*-2-Acetoxy-2-bromo-butane	7.268 (25)	
(±)-*threo*-2-Acetoxy-2-bromobutane	7.414 (25)	
Acetyl bromide	16.2 (20)	2.43 (20, B)
Acetyl chloride	16.9 (2), 15.8 (22)	2.72
Acetylene	2.484 (−77)	
Acrylonitrile	33.0 (20)	3.87
Allene	2.025 (−4)	
Allylamine		1.2
Allyl alcohol	19.7 (20)	1.61
Allyl isocyanate	15.15 (15)	
Allyl isothiocyanate	17.2 (18)	3.2 (20, B)
Allyl nitrite	9.12 (25)	
2-Aminoethanol	31.94 (20), 37.72 (25)	2.59 (25, D)
2-(2-Aminoethylamino)ethanol	21.81 (20)	
N-(2-Aminoethyl)-1,2-ethane-diamine	12.62 (20)	1.9
Aniline	7.06 (20), 5.93 (70)	1.13
Benzaldehyde	19.7 (0), 17.85 (20)	3.0
Benzaldehyde oxime (mp 30) (mp 128)	3.8 (20)	1.2 (25, B) 1.5 (25, B)
Benzamide		3.42 (25, B)
Benzene	2.292(15), 2.283 (20), 2.274 (25)	0
Benzeneacetonitrile	17.87 (26)	3.5
Benzenesulfonyl chloride	28.90 (50)	4.50 (20, B)
Benzenethiol	4.38 (25), 4.26 (30)	1.13 (25, lq), 1.19 (20, B)
Benzonitrile	25.9 (20), 24.0 (40)	4.18
Benzophenone	14.60(18), 11.4 (50)	3.09 (50, lq), 2.98 (25, B)

TABLE 2.50   Dielectric Constant (Permittivity) and Dipole Moment of Organic Compounds (*Continued*)

Substance	Dielectric constant, ϵ	Dipole moment, D
Benzoyl bromide	21.33 (20), 20.74 (25)	3.40 (20, B)
Benzoyl chloride	29.0 (0), 23 (23)	3.16 (25, B)
Benzoyl fluoride	22.7 (20)	
Benzyl acetate	5.1 (21), 5.34 (930)	1.80 (25, B)
Benzyl alcohol	13.0 (20), 11.92 (30), 9.5 (70)	1.71
Benzylamine	5.5 (1), 5.18 (20)	1.15 (20, lq), 1.38 (25, B)
Benzyl benzoate	5.26 (30)	2.06 (30, B)
Benzyl chloride	7.0 (13), 6.85 (25)	1.83 (20, B)
Benzylethylamine	4.3 (20)	
Benzyl ethyl ether	3.90 (25)	
Benzyl formate	6.34 (30)	
N-Benzylmethylamine	4.4 (19)	
Biphenyl	2.53 (75)	0
Bis(2-aminoethyl)amine	12.62 (20)	
Bis(2-chloroethyl) ether	21.20 (20)	2.6
Bis(3-chloropropyl) ether	10.10 (20)	
Bis(2-ethoxyethyl) ether		1.92 (25, B)
Bis(2-hydroxyethyl) ether	31.69 (20)	2.31 (20, B)
Bis(2-hydroxyethyl)sulfide	28.61 (20)	
Bis(2-hydroxypropyl) ether	20.38 (20)	
Bis(2-methoxyethyl) ether	7.23 (25)	
(±)-Bomyl acetate	4.6 (21)	1.89 (22)
3-Bromoaniline	13.0 (20)	2.67 (20, B)
4-Bromoaniline	7.06 (30)	2.88 (25, B)
2-Bromoanisole	8.96 (30)	
4-Bromoanisole	7.40 (30)	
Bromobenzene	5.45 (20), 5.40 (25)	1.70
1-Bromobutane	7.88 (−10), 7.32 (10), 7.07 (20)	2.08
(±)-2-Bromobutane	8.64 (25)	2.23
2-Bromobutanoic acid	7.2 (20)	
cis-2-Bromo-2-butene	5.38 (20)	
trans-2-Bromo-2-butene	6.76 (20)	
1-Bromo-2-chlorobenzene	6.80 (20)	2.15 (20, B)
1-Bromo-3-chlorobenzene	4.58 (20)	1.52 (22, B)
1-Bromo-4-chlorobenzene		0.1 (25, B)
1-Bromo-2-chloroethane	7.41 (10)	1.09
cis-1-Bromo-2-chloroethene	7.31 (17)	
trans-1-Bromo-2-chloroethene	2.50 (17)	
Bromochlorodifluoromethane	3.92 (−150)	
Bromochloromethane	7.79	1.66 (25, B)
3-Bromo-1-chloro-2-methylpropane	8.90 (30)	
Bromocyclohexane	11 (−65), 8.003(30)	1.08 (25, lq), 2.3 (25, B)
1-Bromodecane	4.75 (1), 4.44 (25)	2.08 (20, lq), 1.90 (25, lq)
Bromodichloromethane		1.31 (25, B)
1-Bromododecane	4.07 (25)	2.01 (25, lq), 1.89 (25, B)
Bromoethane	13.6 (−60), 9.39 (20), 9.01 (25)	2.03 (g), 2.04 (20, lq)
1-Bromo-2-ethoxypentane	6.45 (25)	2.32 (25, B)
2-Bromo-3-ethoxypentane	6.40 (25)	2.07 (25, B)
3-Bromo-2-ethoxypentane	8.24 (25)	2.15 (25, B)
1-Bromo-2-ethylbenzene	5.55 (25)	
1-Bromo-3-ethylbenzene	5.56 (25)	
1-Bromo-4-ethylbenzene	5.42 (25)	

(*Continued*)

TABLE 2.50 Dielectric Constant (Permittivity) and Dipole Moment of Organic Compounds (*Continued*)

Substance	Dielectric constant, ϵ	Dipole moment, D
Bromoethylene	5.63 (5), 4.78 (25)	1.42
1-Bromo-2-fluorobenzene	4.72 (25)	
1-Bromo-3-fluorobenzene	4.85 (25)	
1-Bromo-4-fluorobenzene	2.60 (25)	
Bromoform	4.39 (20)	1.00, 0.92 (25, lq)
1-Bromoheptane	5.33 (25), 4.48 (90)	2.17, 2.02 (20, lq)
2-Bromoheptane	6.46 (22)	2.08 (20, B)
3-Bromoheptane	6.93 (22)	2.06 (20, B)
4-Bromoheptane	6.81 (22)	2.06 (20, B)
1-Bromohexadecane	3.71 (25)	1.98 (20, lq), 1.96 (25, C)
1-Bromohexane	6.30 (1), 5.82 (25)	2.06 (20, lq)
Bromomethane	9.82 (0), 9.71 (3), 1.0068 (100, g)	1.82
(Bromomethyl)benzene	6.658 (20)	
1-Bromo-3-methylbutane	8.04 (−56), 6.33 (18)	1.95 (20, B)
2-Bromo-2-methylbutane	9.21 (25)	
2-Bromo-3-methylbutanoic acid	6.5 (20)	
1-Bromo-2-methylpropane	10.98 (20), 7.2 (25)	1.92 (25, lq), 1.99 (20, B)
2-Bromo-2-methylpropane	10.98 (20)	
1-Bromonaphthalene	5.83 (25), 5.12 (20)	1.29 (25, lq)
3-Bromonitrobenzene	20.2 (55)	
1-Bromononane	5.42 (−20), 4.74 (25)	1.95 (25, lq)
1-Bromooctane	6.35 (−50)	1.99 (20, lq), 1.88 (25, lq)
1-Bromopentadecane	3.9 (20)	
1-Bromopentane	9.9 (−90), 6.32 (25)	2.20
3-Bromopentane	8.37 (25)	
1-Bromopropane	8.09 (20)	2.18
2-Bromopropane	9.46 (20)	2.21
2-Bromopropanoic acid	11.0 (21)	
3-Bromopropene	7.0 (20)	1.9
2-Bromopyridine	23.18 (25)	
1-Bromotetradecane	3.84 (25)	1.92 (20, lq), 1.83 (25, lq)
o-Bromotoluene	4.64 (20), 4.28 (58)	1.45 (20, B)
m-Bromotoluene	5.566 (20), 5.36 (58)	1.77 (20, B)
p-Bromotoluene	5.503 (20), 5.49 (58)	1.95 (20, B)
Bromotrichloromethane	2.40 (20)	
Bromotrifluoromethane	3.73 (−150)	0.65
1-Bromoundecane	4.73 (−9)	
1,3-Butadiene	2.050 (−8)	0.403
Butanal	13.45 (25)	2.72
Butane	1.7697 (22)	0
1,2-Butanediol	22.4 (25)	
1,3-Butanediol	28.8 (25)	
1,4-Butanediol	33 (15), 31.9 (25), 30 (38)	4.07
1,3-Butanediol dinitrate	18.85 (20)	
2,3-Butanediol dinitrate	28.85 (20)	
1,3-Butanedione	4.04 (25)	
Butanenitrile	24.83 (20)	4.07
Butanesulfonyl chloride		3.94 (25, D)
1,2,3,4-Butanetetrol	28.2 (120)	
1-Butanethiol	5.20 (15), 5.07 (25), 4.59 (50)	1.54 (25, lq or B)
2-Butanethiol	5.645 (15)	
Butanoic acid	2.97 (20)	1.65 (30, B)

TABLE 2.50 Dielectric Constant (Permittivity) and Dipole Moment of Organic Compounds (*Continued*)

Substance	Dielectric constant, ϵ	Dipole moment, D
Butanoic anhydride	12.8 (20)	
1-Butanol	17.84 (20), 8.2 (118)	1.66
(±)-2-Butanol	17.26 (20), 16.6 (25)	1.66 (30, B)
2-Butanone	18.56 (20), 15.3 (60)	2.78
2-Butanone oxime	3.4 (20)	
trans-2-Butenal		3.67
1-Butene	2.2195 (−53), 1.0032 (20, g)	0.438
cis-2-Butene	1.960 (23)	0.253
trans-2-Butene		0
3-Butenenitrile	28.1 (20)	4.53
2-Butoxyethanol	9.43 (25)	2.08 (25, B)
Butoxyethyne	6.62 (25)	2.05 (25, lq)
N-Butylacetamide	104.0 (20)	
N-sec-Butylacetamide	100.0 (100)	
Butyl acetate	6.85 (−73), 5.07 (20)	1.86 (22, B)
sec-Butyl acetate	5.135 (20)	1.9
tert-Butyl acetate	5.672 (20)	1.91 (25, B)
tert-Butylacetic acid	2.85 (23)	
Butyl acrylate	5.25 (28)	
Butylamine	4.71 (20)	1.00
sec-Butylamine	4.4 (21)	1.28 (25, B)
tert-Butylamine		1.29 (25, B)
Butylbenzene	2.36 (20)	0
sec-Butylbenzene	2.36 (20)	0
tert-Butylbenzene	2.36 (20)	0.83
Butyl butanoate	4.39 (25)	
Butyl ethyl ether		1.24
Butyl formate	6.10 (30), 2.43 (80)	2.08 (26, lq), 2.03 (25, B)
Butyl isocyanate	12.29 (20)	
Butyl methyl ether		1.25 (25, B)
2-*tert*-Butyl-4-methylphenol		1.31 (20, B)
Butyl nitrate	13.10 (20)	2.99 (20, B)
tert-Butyl nitrite	11.47 (25)	
Butyl oleate	4.00 (25)	
N-Butylpropanamide	100.6 (25)	
Butyl propanoate	4.838 (20)	1.79 (23, B)
4-*tert*-Butylpyridine		2.87 (25, C)
Butylsilane	2.537 (20)	
Butyl stearate	3.11 (30)	1.88 (24, B)
Butyl trichloroacetate	7.480 (20)	
Butyl vinyl ether		1.25 (25, Hx)
4-Butyrolactone	39.0 (20)	4.27
Camphor	11.35 (20)	2.91 (20, B), 3.10 (25, B)
Carbon disulfide	3.0 (−112), 2.64(20)	0
Carbon tetrachloride	2.24 (20), 2.228 (25)	0
Carbon tetrafluoride	1.0006 (25, g)	0
D-(+)-Carvone	11 (22)	2.8 (15, B)
Chloroacetic acid	20 (20), 12.35 (65)	2.31 (30, B)
o-Chloroaniline	13.40 (20)	1.78 (20, B)
m-Chloroaniline	13.3 (20)	2.68 (20, B)
p-Chloroaniline		2.99 (25, B)
Chlorobenzene	5.69 (20), 4.2 (120)	1.69

(*Continued*)

TABLE 2.50 Dielectric Constant (Permittivity) and Dipole Moment of Organic Compounds (*Continued*)

Substance	Dielectric constant, ϵ	Dipole moment, D
2-Chloro-1,3-butadiene	4.914 (20)	
1-Chlorobutane	9.07 (−30), 7.276 (20)	2.05 (g), 2.0 (20, B)
2-Chlorobutane	8.564 (20), 7.09 (30)	2.04 (g), 2.1 (20, B)
Chlorocyclohexane	10.9 (−47), 7.951 (30)	2.2 (25, B)
Chlorodifluoromethane	6.11 (24)	1.42 (g)
2-Chloro-*N*,*N*-dimethylacetamide	39.2 (25)	
1-Chlorododecane	4.2 (20)	2.11 (25, lq), 1.94 (20, B)
1-Chloro-2,3-epoxypropane	25.6 (1), 22.6 (22)	1.8 (25, C)
Chloroethane	1.013 (19, g), 9.45 (20)	2.05
2-Chloroethanol	25.80 (20), 13 (132)	1.78
(2-Chloro)ethylbenzene	4.36 (25)	
(3-Chloro)ethylbenzene	5.18 (25)	
(4-Chloro)ethylbenzene	5.16 (25)	
2-Chlorofluorobenzene	6.10 (25)	
3-Chlorofluorobenzene	4.96 (25)	
4-Chlorofluorobenzene	3.34 (25)	
Chloroform	4.807 (25), 4.31 (50)	1.04
1-Chloroheptane	5.52 (20)	1.86 (22, B)
2-Chloroheptane	6.52 (22)	2.05 (22, B)
3-Chloroheptane	6.70 (22)	2.06 (22, B)
4-Chloroheptane	6.54 (22)	2.06 (22, B)
1-Chlorohexane	6.104 (20)	1.94 (20, B)
6-Chloro-1-hexanol	21.6 (−31)	
1-Chloro-2-isocyanatoethane	29.1 (15)	
Chloromethane	1.0069 (g), 12.6 (−20), 10.0 (22)	1.892
1-Chloro-3-methylbutane	7.63 (−70), 6.05 (20)	1.94 (20, B)
2-Chloro-2-methylbutane	12.31 (−50)	
4-Chloromethyl-1,3-dioxolan-2-one	97.5 (40)	
Chloromethyl methyl ether		1.88 (C)
(Chloromethyl)oxirane	22.6 (20)	1.8
1-Chloro-2-methylpropane	7.87 (−38), 7.027 (20)	2.00
2-Chloro-2-methylpropane	10.95 (0), 9.66 (20)	2.13
1-Chloronaphthalene	5.04 (25)	1.33 (25, lq), 1.52 (25, B)
o-Chloronitrobenzene	37.7 (50), 32 (80)	4.64
m-Chloronitrobenzene	20.9 (50), 18 (80)	3.73
p-Chloronitrobenzene	8.09 (120)	2.83
2-Chloro-2-nitropropane	31.9 (−23)	
4-Chloro-3-nitrotoluene	28.07 (28)	
1-Chlorooctane	5.05 (25)	2.14 (25, lq)
Chloropentafluoroethane		0.52
1-Chloropentane	6.654 (20)	2.16
o-Chlorophenol	7.40 (21), 6.31 (25)	2.19
m-Chlorophenol	6.255 (20)	2.19 (25, B)
p-Chlorophenol	11.18 (41)	2.11
1-Chloropropane	8.59 (20)	2.05
2-Chloropropane	9.82 (20)	2.17
3-Chloro-1,2-propanediol	31.0 (20)	
3-Chloro-1,2-propanediol dinitrate	17.50 (20)	
3-Chloro-1-propanol	36.0 (−58)	
1-Chloro-2-propanol	59.0 (−120)	
1-Chloro-2-propanone	30 (19)	2.22 (g), 2.37 (20, Hx)
2-Chloro-1-propene	8.92 (26)	1.647

TABLE 2.50 Dielectric Constant (Permittivity) and Dipole Moment of Organic Compounds (*Continued*)

Substance	Dielectric constant, ϵ	Dipole moment, D
3-Chloro-1-propene	8.2 (20)	1.94
2-Chloropyridine	27.32 (20)	
4-Chlorothiophenol	3.59 (65)	
o-Chlorotoluene	4.72 (20), 4.2 (55)	1.56
m-Chlorotoluene	5.76 (20), 5.0 (60)	1.77 (20, lq), 1.8 (22, B)
p-Chlorotoluene	6.25 (20), 5.6 (55)	2.21
Chlorotrifluoromethane	1.0013 (29, g), 3.01 (−150)	0.50
2-Chloro-1-trifluoromethyl-5-nitrobenzene	9.8 (30)	
4-Chloro-1-trifluoromethyl-3-nitrobenzene	12.8 (30)	
3-Chloro-1,1,1-trifluoropropane	7.32 (22)	
Chlorotrimethylsilane		2.09 (20, B)
Cineole	4.57 (25)	
Cinnamaldehyde	17 (20), 16.9 (24)	3.74
o-Cresol	6.76 (25)	1.45 (25, B)
m-Cresol	12.44 (25)	1.61 (25, B)
p-Cresol	13.05 (25)	1.54 (20, B)
Crotonic acid		2.13 (30, B)
Cyanoacetic acid	33.4 (4)	
Cyanoacetylene	72.3 (19)	3.724
2-Cyanopyridine	93.77 (30)	
3-Cyanopyridine	20.54 (50)	
4-Cyanopyridine	5.23 (80)	
Cyclobutanone	14.27 (25)	2.89
Cycloheptane	2.078 (30)	
Cycloheptanone	13.16 (25)	
1,3-Cyclohexadiene	2.68 (−89)	0.38 (20, B)
1,4-Cyclohexadiene	2.211 (23)	
Cyclohexane	2.05 (15), 2.02 (25)	0
Cyclohexanecarboxylic acid	2.6 (31)	
1,4-Cyclohexanedione	15.0 (25), 4.40 (78)	1.41
Cyclohexanethiol	5.420 (25)	
Cyclohexanol	16.40 (20), 15.0 (25), 7.24 (100)	1.86 (25, C)
Cyclohexanone	20 (−40), 16.1 (20)	2.87
Cyclohexanone oxime	3.04 (89)	0.83 (25, B)
Cyclohexene	2.6 (−105), 2.218 (20)	0.332
Cyclohexylamine	4.55 (20)	1.22 (20, lq), 1.26 (20, B)
Cyclohexylbenzene		0
Cyclohexylmethanol	9.7 (60), 8.1 (80)	1.68 (20, B)
Cyclohexyl nitrite	9.33 (25)	
o-Cyclohexylphenol	3.97 (55)	
p-Cyclohexylphenol	4.42 (131)	
Cyclooctane	2.116 (22)	0
cis-Cyclooctene	2.306 (23)	
Cyclopentane	1.9687 (20)	0
Cyclopentanecarbonitrile	22.68 (20)	
Cyclopentanol	25 (−20), 18.5 (10)	1.72 (25, C)
Cyclopentanone	16 (−51), 13.58 (25)	3.30
Cyclopentene	2.083 (22)	0.20
p-Cymene	2.243 (20), 2.23 (25)	0
cis-Decahydronaphthalene	2.22 (20)	0

(*Continued*)

TABLE 2.50 Dielectric Constant (Permittivity) and Dipole Moment of Organic Compounds (*Continued*)

Substance	Dielectric constant, ϵ	Dipole moment, D
trans-Decahydronaphthalene	2.18 (20)	0
Decamethylcyclopentasiloxane	2.5 (20)	
Decamethyltetrasiloxane	2.4 (20)	0.79 (25, lq)
Decane	1.991 (20), 1.844 (130)	0
1-Decanol	8.1 (20)	1.71 (20, B), 1.62 (25, B)
1-Decene	2.14 (20)	0
meso-2,3-Diacetoxybutane	6.644 (25)	
Diallyl sulfide	4.9 (20)	1.33 (25, B)
Dibenzofuran	3.0 (100)	0.88 (25, B)
Dibenzylamine	3.6 (20)	0.97 (20, lq), 1.02 (20, B)
Dibenzyl decanedioate	4.6 (25)	
Dibenzyl ether	3.82 (20)	1.39 (21, B)
o-Dibromobenzene	7.86 (20)	2.13 (20, B)
m-Dibromobenzene	4.21 (20)	1.5 (20, B)
p-Dibromobenzene	2.57 (95)	0
1,2-Dibromobutane	4.74 (20)	
1,3-Dibromobutane	9.14 (20)	
1,4-Dibromobutane	8.68 (30)	2.16 (20, lq), 2.06 (20, B)
2,3-Dibromobutane	6.36 (20), 5.75 (25)	2.20
meso-2,3-Dibromobutane	6.245 (25)	
(±)-2,3-Dibromobutane	5.758 (25)	
1,2-Dibromodichloromethane	2.54 (25)	
1,2-Dibromodifluoromethane	2.94 (0)	0.66
1,2-Dibromoethane	4.96 (20), 4.78 (25), 4.09 (131)	1.11
cis-1,2-Dibromoethylene	7.08 (25)	
trans-1,2-Dibromoethylene	2.88 (25)	
Dibromomethane	7.77 (10)	1.43
cis-1,2-Dibromoethylene	7.7 (0), 7.08 (25)	1.35 (B)
trans-1,2-Dibromoethylene	2.9 (0), 2.88 (25)	0
1,2-Dibromoheptane	3.8 (25)	1.78 (25, D)
2,3-Dibromoheptane	5.1 (25)	2.15 (25, B)
3,4-Dibromoheptane	4.7 (25)	2.15 (25, B)
meso-3,4-Dibromohexane	4.67 (25)	
(±)-3,4-Dibromohexane	6.732 (25)	
1,6-Dibromohexane	8.52 (25)	
Dibromomethane	7.77 (10), 6.7 (40)	1.43
1,2-Dibromo-2-methylpropane	4.1 (20)	
1,2-Dibromopentane	4.39 (25)	
(±)-*erythro*-2,3-Dibromopentane	5.43 (25)	
(±)-*threo*-2,3-Dibromopentane	6.507 (25)	
1,4-Dibromopentane	9.05 (20)	
1,5-Dibromopentane	9.14 (30)	
1,2-Dibromopropane	4.60 (10), 4.3 (20)	1.13
1,3-Dibromopropane	9.48 (20)	
Dibromotetrafluoroethane	2.34 (25)	
Dibutylamine	2.78 (20)	1.06 (20, lq), 1.05 (20, B)
Dibutyl decanedioate	4.54 (20)	2.64 (25, B)
Dibutyl ether	3.08 (20)	1.18
Dibutyl maleate		2.70 (25, B)
Dibutyl *o*-phthalate	6.58 (20), 6.436 (30), 5.99 (45)	2.97 (20, lq), 2.85 (30, B)
Dibutyl sulfide	4.29 (25)	1.6
Dichloroacetic acid	8.33 (20), 7.8 (61)	

TABLE 2.50 Dielectric Constant (Permittivity) and Dipole Moment of Organic Compounds (*Continued*)

Substance	Dielectric constant, ϵ	Dipole moment, D
Dichloroacetic anhydride	15.8 (25)	
1,1,-Dichloroacetone	14.6 (20)	
o-Dichlorobenzene	10.12 (20), 9.93 (25), 7.10 (90)	2.50
m-Dichlorobenzene	5.02 (20), 5.04 (25), 4.22 (90)	1.72
p-Dichlorobenzene	2.394 (55)	0
1,2-Dichlorobutane	7.74 (25)	
1,4-Dichlorobutane	9.30 (35)	2.22
Dichlorodifluoromethane	3.50 (−150), 2.13 (29)	0.51
4-Chloro-1,3-dioxalan-2-one	62.0 (40)	
4,5-Dichloro-1,3-dioxalan-2-one	31.8 (40)	
1,1-Dichloroethane	10.10 (20)	2.06
1,2-Dichloroethane	12.7 (−10), 10.42 (20)	1.48
1,1-Dichloroethylene	4.60 (20), 4.60 (25)	1.34
cis-1,2-Dichloroethylene	9.20 (25)	1.90
trans-1,2-Dichloroethylene	2.14 (20)	0
2,2′-Dichloroethyl ether	21.2 (20)	2.61 (20, B)
Dichlorofluoromethane	5.34 (28)	1.29 (g)
1,6-Dichlorohexane	8.60 (35)	
Dichloromethane	9.14 (20), 8.93 (25), 1.0065 (100, g)	1.60
1,3-Dichloroisopropyl nitrate	13.28 (20)	
(Dichloromethyl)benzene	6.9 (20)	2.1
Dichloromethyl isocyanate	7.36 (15)	
1,2-Dichloro-2-methylpropane	7.15 (23)	
2,4-Dichloro-1-nitrobenzene	13.06 (28)	
1,1-Dichloro-1-nitroethane	16.3 (30)	
1,2-Dichloropentane	6.89 (20)	
1,5-Dichloropentane	9.92 (25)	
2,4-Dichlorophenol		1.60 (25, B)
1,2-Dichloropropane	8.37 (20), 8.93 (26), 7.90 (35)	1.87 (25, B)
1,3-Dichloropropane	10.27 (30)	2.08
2,2-Dichloropropane	11.37 (20)	2.62
1,1-Dichloro 2-propanone	14 (20)	
1,2-Dichlorotetrafluoroethane	2.48 (0), 2.26 (25)	0.53
2,4-Dichlorotoluene	5.68 (28)	1.7
2,6-Dichlorotoluene	3.36 (28)	
3,4-Dichlorotoluene	9.39 (28)	3.0
Diethanolamine	25.75 (20)	2.84 (25, B)
1,1-Diethoxy ethane	3.80 (25)	1.08
1,2-Diethoxyethane	3.90 (20)	1.99 (20, B), 1.65 (25, B)
Diethoxymethane	2.527 (20)	
N,*N*-Diethylacetamide	32.1 (20)	
N,*N*-Diethylacetoacetamide	40.8 (25)	
Diethylamine	3.680 (20)	0.92
N,*N*-Diethylaniline	5.5 (19)	1.40 (20, lq), 1.80 (20, B)
Diethyl carbonate	2.82 (24)	1.10
N,*N*-Diethyl-*N*′,*N*′-dimethylurea	17.89 (25)	
Diethyl decanedioate	5.0 (30)	2.38 (20, lq), 2.52 (20, B)
Diethylene glycol	3.182 (20)	2.3
Diethylene glycol diethyl ether	5.70	
Diethyl ether	4.267 (20), 3.97 (40)	1.15
Diethyl ethyl phosphonate	11.00 (15), 9.86 (45)	2.95 (32, lq), 2.91 (20, C)
N,*N*-Diethylform amide	29.6 (20)	

(Continued)

TABLE 2.50   Dielectric Constant (Permittivity) and Dipole Moment of Organic Compounds (*Continued*)

Substance	Dielectric constant, ϵ	Dipole moment, D
Diethyl fumarate	6.56 (23)	2.40 (20, B)
Diethyl glutarate	6.7 (30)	2.46 (30, lq)
Diethyl glycol	31.82 (20)	
Di(2-ethylhexyl) *o*-phthalate	5.3 (20), 4.91 (35), 4.77 (45)	2.8
Diethyl maleate	8.58 (23), 7.56 (25)	2.56 (25, B)
Diethyl methanephosphate	13.405 (40)	
Diethyl 1,3-propanedioate (malonate)	8.03 (25), 7.55 (31)	2.49 (20, lq), 2.54 (25, B)
Diethyl nonanedioate	5.13 (30)	
Diethyl oxalate	8.266 (20)	2.49 (20, D)
Diethyl *o*-phthalate	7.34 (35), 7.13 (45)	2.8 (25, B)
Diethylsilane	2.544 (20)	
Diethyl succinate	6.098 (20)	2.3
Diethyl sulfate	29.2 (20)	4.46 (25, D)
Diethyl sulfide	5.72 (25), 5.24 (50)	1.54
Diethyl sulfite	15.6 (20), 14 (50)	
Diethylzinc	2.55 (20)	0.62 (25, B)
o-Difluorobenzene	13.38 (28)	2.46
m-Difluorobenzene	5.01 (28)	1.51
1,1-Difluoroethane		2.27
Difiuoromethane	53.74 (−121)	1.978
2,3-Dihydropyran	5.136 (35)	
1,2-Dihydroxybenzene	17.57 (115)	2.60 (25, B)
1,3-Dihydroxybenzene	13.55 (120)	2.09 (44, B)
1,4-Dihydroxybenzene		1.4 (44, B)
1,2-Diiodobenzene	5.7 (20), 5.41 (50)	1.70 (20, B)
1,3-Diiodobenzene	4.3 (25), 4.11 (50)	1.22 (20, B)
1,4-Diodobenzene	2.88 (120)	0.19 (20, B)
cis-1,2-Diiodoethylene	4.46 (72)	0.71 (B)
trans-1,2-Diiodoethylene	3.19 (77)	0
Diiodomethane	5.316 (25)	1.08 (25, B)
Diisobutylamine	2.7 (22)	1.10 (25, B)
1,6-Diisocyanatohexane	14.41 (15)	
Diisopentylamine	2.5 (18)	1.48 (30, B)
Diisopentyl ether	2.82 (20)	0.98 (20, lq), 1.23 (25, B)
Diisopropylamine		1.26 (25, B)
Diisopropyl ether	3.88 (25), 3.805 (30)	1.13
1,2-Dimethoxybenzene	4.45 (20), 4.09 (25)	1.32 (25, B)
Dimethoxydimethylsilane	3.663 (25)	
1,2-Dimethoxyethane	7.60 (10), 7.30 (23.5)	1.71 (25, B)
Dimethoxymethane	2.644 (20)	0.74
N,N-Dimethylacetamide	38.85 (21), 37.78 (25)	3.80
2-Dimethylamino-2-methyl-1-propanol	12.36 (25)	
Dimethylamine	6.32 (0), 5.26 (25)	1.01
N,N-Dimethylaniline	4.90 (25), 4.4 (70)	1.68
2,4-Dimethylaniline	4.9 (20)	1.40 (25, B)
2,3-Dimethyl-1,3-butadiene	2.102 (20)	
N,N-Dimethy lbutanamide	29.7 (20)	
2,2-Dimethylbutane	1.869 (20)	0
2,3-Dimethylbutane	1.889 (20)	0
3,3-Dimethyl-2-butanone	12.73 (20)	

TABLE 2.50   Dielectric Constant (Permittivity) and Dipole Moment of Organic Compounds (*Continued*)

Substance	Dielectric constant, ϵ	Dipole moment, D
2,2-Dimethyl-1-butanol	10.5 (20)	
Dimethyl carbonate	3.087 (25)	0.90
cis-1,2-Dimethylcyclohexane	2.06 (25)	0
trans-1,2-Dimethylcyclohexane	2.04 (25)	0
1,1-Dimethylcyclopentane		0
Dimethyl disulfide	9.6 (25)	1.8
Dimethyl ether	6.18 (−15), 5.02 (25), 2.97 (110)	1.30
N,N-Dimethylformamide	38.25 (20), 36.71 (25)	3.82 (25, B)
2,4-Dimethylheptane	1.9 (20)	0
2,5-Dimethylheptane	1.9 (20)	0
2,6-Dimethylheptane	2(20)	0
2,6-Dimethyl-4-heptanone	9.91 (20)	2.66 (25, C)
2,2-Dimethylhexane	1.95 (20)	0
2,5-Dimethylhexane	1.96 (21)	0
3,3-Dimethylhexane	1.96 (20)	0
3,4-Dimethylhexane	1.98 (19)	0
Dimethyl hexanedioate	6.84 (20)	2.28 (20, B)
1,3-Dimethylimidazolidine-2-one	37.60 (25)	
Dimethyl maleate		2.48 (25, C)
Dimethyl malonate	9.82 (20)	2.41 (20, B)
Dimethyl methanephosphate	22.3 (20)	
N,N-Dimethyl methanesulfonamide	80.4 (50)	
1,2-Dimethylnaphthalene	2.61 (25)	0
1,6-Dimethylnaphthalene	2.73 (20)	0
4,4-Dimethyloxazolidine-2-one	39.2 (60)	
N,N-Dimethylpentanamide	26.4 (20)	
2,2-Dimethylpentane	1.915 (20)	0
2,3-Dimethylpentane	1.929 (20)	0
2,4-Dimethylpentane	1.902 (20)	0
3,3-Dimethylpentane	1.942 (20)	0
Dimethyl pentanedioate	7.87 (20)	
2,4-Dimethyl-3-pentanone		2.7
2,3-Dimethylphenol	4.81 (70)	
2,4-Dimethylphenol	5.06 (30)	1.48 (20, B), 1.98 (60, B)
2,5-Dimethylphenol	5.36 (65)	1.43 (20, B), 1.52 (60, B)
2,6-Dimethylphenol	4.90 (40)	1.4
3,4-Dimethy lphenol	9.02 (60)	1.77 (20, B)
3,5-Dimethylphenol	9.06 (50)	1.76 (20, B)
Dimethyl *o*-phthalate	8.66 (20), 8.25 (25), 8.11 (45)	2.8 (25, B)
2,2-Dimethylpropanal	9.051 (20)	2.66
N,N-Dimethylpropanamide	34.6 (20)	
2,2-Dimethylpropanamide	20.13 (25)	
2,2-Dimethylpropane	1.769 (23), 1.678 (98)	0
2,2-Dimethylpropane nitrile	21.1 (20)	3.95
N,N-Dimethylpropanamide	33.1	
2,2-Dimethyl-1-propanol	8.35 (60)	
2,5-Dimethylpyrazine	2.436 (20)	0
2,6-Dimethylpyrazine	2.653 (35)	
2,4-Dimethylpyridine	9.60 (20)	2.3
2,6-Dimethylpyridine	7.33 (20)	1.7
2,6-Dimethylpyridine-1-oxide	46.11 (25)	
2,3-Dimethylquinoxaline	2.3 (25)	0

(*Continued*)

TABLE 2.50 Dielectric Constant (Permittivity) and Dipole Moment of Organic Compounds (*Continued*)

Substance	Dielectric constant, ϵ	Dipole moment, D
Dimethyl succinate	7.19 (20)	2.09 (20, B)
Dimethyl sulfate	55.0 (25)	4.31 (25, D)
Dimethyl sulfide	6.70 (21)	1.554
Dimethyl sulfite	22.5 (23)	2.93 (20, B)
Dimethyl sulfone	47.39 (110)	
Dimethyl sulfoxide	47.24 (20), 41.9 (55)	3.96 (25, B)
cis-2,5-Dimethyltetrahydrofuran	5.03 (23)	
N,N-Dimethylthioformamide	47.5 (25)	
N,N-Dimethyl-*o*-toluidine	3.4 (20)	0.88 (25, B)
N,N-Dimethyl-*p*-toluidine	3.9(20)	1.29 (25, B)
m-Dinitrobenzene	22.9 (92)	
2,2-Dinitropropane	42.4 (52)	
Dinonyl hexanedioate		2.53 (25, B)
Dinonyl *o*-phthalate	4.65 (35), 4.52 (45)	
Dioctyl decanedioate	4.0 (27)	
Dioctyl *o*-phthalate	5.1 (25)	3.06 (25, C)
1,4-Dioxane	2.219 (20), 2.21 (25)	0
1,3-Dioxolane		1.19
1,3-Dioxolan-2-one	89.78 (40)	
Dipentene	2.38 (25)	
Dipentyl ether	2.80 (25)	0.98 (20, lq), 1.24 (25, B)
Dipentyl *o*-phthalate	5.79 (35), 5.62 (45)	2.71 (20, lq)
Dipentyl sulfide	3.83 (25)	1.59 (25, B)
Dipentylamine	3.3 (52)	1.31 (20, C), 1.01 (25, B)
1,2-Diphenylethane	2.4(110)	0 (110, lq), 0.45 (25, B)
Diphenyl ether	3.73 (10), 3.63 (30)	1.3
Diphenylmethane	2.7 (18), 2.57 (26)	0.26 (30, lq), 0.3 (25, B)
Dipropylamine	2.923 (20)	1.01 (20, lq), 1.03 (20, B)
Dipropyl ether	3.38 (24)	1.21
N,N-Dipropylformamaide	23.5 (20)	
Dipropyl sulfone	32.62 (30)	
Dipropyl sulfoxide	30.37 (30)	
Divinyl ether	3.94 (15)	0.78
Dodecamethylcyclohexasiloxane	2.6 (20)	
Dodecamethylpentasiloxane	2.5 (20)	
Dodecane	2.05 (−10), 2.01 (20)	0
1-Dodecanol	5.15 (20), 6.5 (25)	1.52 (20, B)
1-Dodecene	2.15 (20)	0
6-Dodecyne	2.17 (25)	
1,2-Epoxybutane		2.01 (20, B)
Erythritol	28 (128)	
Ethane	1.936 (−178), 1.0015 (0)	0
1,2-Ethanediamine	16.8 (18), 13.82 (20)	1.96
1,2-Ethanediol	41.4 (20), 37.7 (25)	2.28
1,2-Ethanediol diacetate	7.7 (17)	2.34 (30, B)
1,2-Ethanediol dinitrate	28.26 (20)	
1,2-Ethanediol monoacetate	12.95 (30)	
1,2-Ethanedithiol	7.26 (20)	
Ethanesulfonyl chloride		3.89 (25, B)
Ethanethiol	6.9 (15), 6.667 (25)	1.58
Ethanol	25.3 (20), 20.21 (55)	1.69
Ethanolamine	31.94 (20)	

TABLE 2.50 Dielectric Constant (Permittivity) and Dipole Moment of Organic Compounds (*Continued*)

Substance	Dielectric constant, ϵ	Dipole moment, D
Ethoxyacetylene	8.05 (25)	
4-Ethoxyaniline	7.43 (25)	
Ethoxybenzene (phenetol)	4.216(20)	1.45
2-Ethoxyethanol	13.38 (25)	2.24 (30, B)
2-Ethoxyethyl acetate	7.567 (30)	2.25 (30, B)
1-Ethoxy-2-methylbutane	3.96 (20)	
1-Ethoxynaphthalene	3.3 (19)	
1-Ethoxypentane	3.6 (23)	
α-Ethoxytoluene	3.9 (20)	
Ethoxytrimethylsilane	3.013 (25)	
N-Ethylacetamide	135.0 (20)	
Ethyl acetate	6.081 (20), 5.30 (77)	1.78
Ethyl acetoacetate	14.0 (20)	3.22 (18, B, keto form)
		2.04 (−80, CS$_2$, enol form)
Ethyl acrylate	6.05 (30)	2.0
Ethylamine	8.7 (0), 6.94 (10)	1.22
N-Ethylaniline	5.87 (20)	
4-Ethylaniline	4.84 (25)	
Ethylbenzene	2.446 (20)	0.59
Ethyl benzoate	6.20 (20)	2.00
Ethyl 2-bromoacetate	8.75 (30)	
Ethyl α-bromobutanoate	8 (20)	2.40 (25, B)
Ethyl 2-bromo-2-methylpropanoate	8.55 (30)	
Ethyl 2-bromopropanoate	9.4 (20), 8.57 (30)	
N-Ethylbutanamide	107.0 (25)	
Ethyl butanoate	5.18 (28)	1.74 (22, B)
2-Ethylbutanoic acid	2.72 (23)	
2-Ethyl-1-butanol	6.19 (90)	
Ethyl *tert*-butyl ether	7.07 (25)	
Ethyl carbamate	14.2 (50), 14.14 (55)	2.59 (30, D)
Ethyl chloroacetate	11.4 (21)	2.65 (25, B)
Ethyl chlorocarbonate	9.736 (36)	
Ethyl *cis*-3-chlorocrotonate	7.67 (76)	
Ethyl *trans*-3-chlorocrotonate	4.70 (54)	
Ethyl chloroformate	11 (20)	2.56 (35, B)
Ethyl 2-chloropropanoate	11.95 (30)	
Ethyl 3-chloropropanoate	10.19 (30)	
Ethyl *trans*-cinnamate	6.1 (18), 5.83 (20)	1.86 (20, B)
Ethyl crotonate	5.4 (20)	1.95 (24, B)
Ethyl cyanoacetate	31.62 (−10), 26.9 (20)	2.2
Ethylcyclobutane	1.965 (20)	
Ethylcyclohexane	2.054 (20)	0
Ethylcyclopropane	1.933 (20)	
Ethyl dichloroacetate	12 (2), 10 (22)	2.63 (25, B)
Ethyl dodecanoate	3.4 (20), 2.7 (143)	1.3 (20, lq)
Ethylene	1.001 44 (0, g), 1.483 (−3)	0
Ethylene carbonate	89.78 (40), 69.4 (91)	4.87 (25, B)
Ethylenediamine	13.82 (20)	1.98
Ethylene dinitrate	28.3 (20)	3.58 (25, B)
2,2′-(Ethylenedioxy)diethanol	23.69 (20)	5.58 (lq)
Ethylene glycol	41.4 (20), 37.7 (25)	2.28
Ethylene glycol diacetate	7.7 (17)	

(Continued)

TABLE 2.50   Dielectric Constant (Permittivity) and Dipole Moment of Organic Compounds (*Continued*)

Substance	Dielectric constant, ϵ	Dipole moment, D
Ethyleneimine	18.3 (25)	1.90
Ethylene oxide	14 (−1), 12.42 (20)	1.89
Ethylene sulfite	39.6 (25)	
N-Ethylformamide	102.7 (25)	
Ethyl formate	8.57 (15), 7.16 (25)	1.94
Ethyl fumarate	6.5 (23)	
Ethyl furan-2-carboxylate	9.02 (20)	
Ethylhexadecanoate	3.2 (20), 2.71 (104)	1.2 (lq)
3-Ethylhexane	1.96 (20)	0
2-Ethyl-1,2-hexanediol	18.73 (20)	
Ethyl hexanoate	4.45 (20)	1.80 (20, B)
2-Ethyl-1-hexanol	7.58 (25), 4.41 (90)	1.74 (25, B)
2-Ethylhexyl acetate		1.8
Ethyl 2-iodopropanoate	8.6 (20)	
Ethyl isocyanate	19.7 (20)	
Ethyl isopentyl ether	3.96 (20)	
Ethyl isothiocyanate	19.6 (20)	3.67 (20, B)
Ethyl lactate	15.4 (30)	2.4 (20, B)
Ethyl maleate	8.6 (23)	
Ethyl methacrylate	5.68 (30)	
Ethyl 3-methylbutanoate	4.71 (20)	
Ethyl-N-methyl carbamate	21.10 (25)	
Ethyl methyl carbonate	2.985 (20)	
Ethyl methyl ether		1.17
3-Ethyl-2-methylpentane	1.99 (18)	0
Ethyl nitrate	19.7 (20)	2.93 (20, B)
Ethyl 9-octadecanoate	3.2 (25)	1.83 (20, lq)
3-Ethyloxazolidine-2-one	66.8 (25)	
4-Ethyloxazolidine-2-one	42.6 (25)	
Ethyl 4-oxopentanoate	12 (21)	
3-Ethylpentane	1.942 (20)	0
Ethyl pentanoate	4.71 (18)	1.76 (28, B)
3-Ethyl-3-pentanol	3.158 (20)	
Ethyl pentyl ether	3.6 (23)	1.2 (20, B)
Ethyl phenylacetate	5.3 (21)	1.82 (30)
Ethyl phenyl sulfide		4.08 (25, B)
N-Ethyl propanamide	126.8 (25)	
Ethyl propanoate	5.76 (20)	1.75 (22, B)
Ethyl propyl ether		1.16 (25, B)
2-Ethylpyridine	8.33 (20)	
4-Ethylpyridine	10.98 (20)	
Ethyl salicylate	7.99 (30)	2.85 (25, B)
Ethyl stearate	2.98 (40), 2.69 (100)	1.65 (40, lq)
Ethyl thiocyanate	29.3 (21)	3.33 (20, B)
p-Ethyltoluene	2.24 (25)	0
Ethyl trichloroacetate	8.428 (20)	2.56 (25, B)
Ethyltrimethylsilane	2.275 (30)	
Ethyl vinyl ether		1.26 (20, B)
Fluorobenzene	5.465 (20), 5.42 (25), 4.7 (60)	1.60
4-Fluorobenzene sulfonylchloride	12.65 (40)	
2-Fluoroiodobenzene	8.22 (25)	
3-Fluoroiodobenzene	4.62 (25)	

TABLE 2.50 Dielectric Constant (Permittivity) and Dipole Moment of Organic Compounds (*Continued*)

Substance	Dielectric constant, ϵ	Dipole moment, D
4-Fluoroiodobenzene	3.12 (25)	
Fluoromethane	51.0 (−142)	1.858
2-Fluoro-2-methylbutane	5.89 (20)	1.92 (25, B)
1-Fluoropentane	3.93 (20)	1.85 (25, B)
o-Fluorotoluene	4.23 (25), 4.22 (30), 3.9 (60)	1.37
m-Fluorotoluene	5.41 (25), 4.9 (60)	1.82
p-Fluorotoluene	5.88 (25), 5.86 (30), 5.3 (60)	2.00
Formamide	111.0 (20), 103.5 (40)	3.73
Formanilide		3.37 (25, C)
Formic acid	58.5 (15), 57.0 (21), 51.1 (25)	1.41
2-Furaldehyde	42.1 (20), 34.9 (50)	3.63 (25, B)
Furan	2.88 (4)	0.66
2-Furfuryl acetate	5.85 (20)	
Furfuryl alcohol	16.85 (25)	1.92 (25, lq)
Glycerol	46.5 (20), 42.5 (25)	2.68 (25, D)
Glycerol tris(acetate)	7.2 (20)	2.73 (25, B)
Glycerol tris(nitrate)	19.25 (20)	3.38 (25, B)
Glycerol tris(oleate)	3.2 (26)	3.11 (23, B)
Glycerol tris(palmitate)	2.9 (65)	2.80 (23, B)
Glycerol tris(sterate)	2.8 (70)	2.86 (23, B)
1,6-Heptadiene	2.161 (20)	
Heptacosafluorotributylamine	2.15 (20)	
2,2,3,3,4,4,4-Heptafluoro-1-butanol	14.4 (25)	
Heptanal	9.1 (20)	2.26 (40, lq), 2.58 (22, B)
Heptane	1.921 (20), 1.85 (70)	0
1-Heptanethiol	4.194 (20)	
Heptanoic acid	3.04 (15), 2.6 (71)	
1-Heptanol	11.75 (20)	1.73 (20, B)
(±)-2-Heptanol	9.72 (21)	1.73 (20, B)
(±)-3-Heptanol	7.07 (23)	1.73 (20, B)
4-Heptanol	6.18 (23)	1.72 (20, B)
2-Heptanone	11.95 (20), 8.27 (100)	2.61 (22, B)
3-Heptanone	12.7 (20)	2.81 (22, B)
4-Heptanone	12.60 (20), 9.46 (80)	2.74 (20, B)
1-Heptene	2.09 (20)	0
Heptylamine	3.81 (20)	
Hexachloroacetone	3.93 (19)	
Hexachloro-1,3-butadiene	2.55 (20)	
Hexadecamethylcyclooctasiloxane	2.7 (20)	
Hexadecane	2.046 (30)	0
1-Hexadecanol	3.8 (50)	1.67 (25, B)
1,5-Hexadiene	2.125 (26)	
2,4-Hexadiene	2.207 (25)	0.31 (25, B)
cis, cis-2,4-Hexadiene	2.163 (24)	
trans, trans-2,4-Hexadiene	2.123 (24)	
Hexafluoroacetone	2.104 (−71)	
Hexafluorobenzene	2.029 (25)	0
1,1,1,3,3,3-Hexafluoro-2-propanol	16.70 (20)	
Hexamethyldisiloxane	2.2 (20)	0.37 (25, lq)
Hexamethylphosphorotriamide	31.3 (20)	5.5, 4.31 (25, lq)
Hexane	1.904 (15), 1.890 (20)	0
Hexanedinitrile	32.45 (25)	3.8 (25, B)

(*Continued*)

TABLE 2.50   Dielectric Constant (Permittivity) and Dipole Moment of Organic Compounds (*Continued*)

Substance	Dielectric constant, ϵ	Dipole moment, D
Hexanenitrile	17.26 (25)	
1-Hexanethiol	4.436 (20)	
1,2,6-Hexanetriol	31.5 (12)	
Hexanoic acid	2.600 (25)	1.13 (25, lq)
1-Hexanol	13.03 (20), 8.5 (75)	1.55 (20, B)
(±)-2-Hexanol	11.06 (25)	
3-Hexanol	9.66 (25)	
2-Hexanone	14.6 (15), 14.56 (20)	2.68 (22, B)
1-Hexene	2.051 (20)	0
cis-2-Hexene		0
trans-2-HexenQ	1.978 (22)	0
cis-3-Hexene	2.069 (23)	0
trans-3-Hexene	1.954 (20)	0
Hexyl acetate	4.42 (20)	
Hexylamine	4.08 (20)	
1-Hexyne	2.621 (23)	0.83
2-Hydroxyacetophenone	21.33 (25)	
2-Hydroxybutanoic acid	37.7 (23)	
3-Hydroxybutanoic acid	31.5 (23)	
N-(2-Hydroxyethyl)acetamide	96.6 (25)	
4-Hydroxy-4-methyl-2-pentanone	18.2 (25)	3.24 (20, B)
3-Hydroxypropanoic acid	30.0 (23)	
Iodobenzene	4.59 (20)	1.70
1-Iodobutane	6.27 (20), 4.52 (130)	2.10
2-Iodobutane	7.873 (20)	2.12
1-Iodododecane	3.9 (20)	1.87 (20, C)
Iodoethane	10.2 (−50), 7.82 (20)	1.91
1-Iodoheptane	4.92 (22)	1.86 (22, B)
3-Iodoheptane	6.39 (22)	1.95 (22, B)
1-Iodohexadecane	3.5 (20)	
1-Iodohexane	5.37 (20)	1.94 (20, C)
Iodomethane	6.97 (20)	1.62
1-Iodo-3-methylbutane	5.6 (19)	1.85 (20, B)
2-Iodo-2-methylbutane	8.19 (20)	2.20 (20, B)
1-Iodo-2-methylpropane	6.47 (20)	1.89 (20, B)
2-Iodo-2-methylpropane	6.65 (10)	
1-Iodooctane	4.6 (25)	1.80 (25, lq), 1.90 (20, C)
2-Iodooctane	5.8 (20)	2.07 (20, C)
1-Iodopentane	5.78 (20)	1.90 (20, B)
3-Iodopentane	7.432 (20)	
1-Iodopropane	7.07 (20)	2.03
2-Iodopropane	8.19 (25)	2.01 (20, B)
3-Iodopropene	6.1 (19)	
p-Iodotoluene	4.4 (35)	1.72 (22, B)
α-Ionone	11 (18)	
β-Ionone	12 (20)	
Iron pentacarbonyl	2.602 (20)	
Isobutanenitrile	20.4 (24)	3.61 (25, B)
Isobutene	2.1225 (15)	0.503
N-Isobutylacetamide	111.0 (20)	
Isobutyl acetate	5.068 (20)	1.87 (22, B)
Isobutylamine	4.43 (21)	1.27 (25, B)

TABLE 2.50 Dielectric Constant (Permittivity) and Dipole Moment of Organic Compounds (*Continued*)

Substance	Dielectric constant, ϵ	Dipole moment, D
Isobutylbenzene	2.319 (20), 2.298 (30)	0.31 (20, lq)
Isobutyl butanoate	4.1 (20)	1.9
Isobutyl chlorocarbonate	9.1 (20)	
Isobutyl formate	6.41 (20)	1.89 (20, B)
Isobutyl isocyanate	11.64 (20)	
Isobutyl nitrate	2.7 (20)	
Isobutyl pentanoate	3.8 (19)	
Isobutylsilane	2.497 (20)	
Isobutyl trichloroacetate	7.667 (20)	
Isobutyl vinyl ether	3.34 (20)	
Isobutyronitrile	20.4 (24)	3.61 (25, B)
Isopentyl acetate	4.72 (20), 4.63 (30)	1.84 (22, B), 1.76 (30, lq)
Isopentyl butanoate	4.0 (20)	
Isopentyl pentanoate	3.6 (19)	1.8 (28, B)
Isopentyl propanoate	4.2 (20)	
Isopropyl acetate		1.86 (22, B)
Isopropylamine	5.627 (20)	1.19
Isopropylbenzene	2.38 (20)	0.79
Isopropyl carborane	45.0 (20)	
N-Isopropylformamide	65.7 (25)	
1-Isopropyl-4-methylbenzene	2.24 (20)	0
Isopropyl nitrite	13.92 (−13)	
Isoquinoline	11.0 (25)	2.73
Lactic acid	22 (17)	
Lactonitrile	38 (20)	
D-Limonene	2.4 (20), 2.37 (25)	1.57 (25, B)
(±)-Limonene	2.3 (20)	0.63 (25, B)
Maleic anhydride	52.75 (53)	
(±)-Mandelonitrile	17.8 (23)	
D-Mannitol	24.6 (170)	
Menthol		1.55 (20, B)
Methacrylic acid		1.65
Methacrylonitrile		3.69
Methane	1.676 (−182), 1.000 94 (0)	0
Methanesulfonyl chloride	34.0 (20)	
Methanethiol		1.52(g)
Methanol	41.8 (−20), 33.0 (20)	1.70
2-Methoxyaniline	5.230 (30)	
3-Methoxyaniline	8.76 (25)	
4-Methoxyaniline	7.85 (60)	
o-Methoxybenzaldehyde		4.34 (20, B)
p-Methoxybenzaldehyde	22.3 (22), 22.0 (30), 10.4 (248)	3.26 (35, B)
Methoxybenzene	4.30 (21), 3.9 (70)	1.38
2-Methoxyethanol	17.2 (25), 16.0 (30)	2.36
N-(2-Methoxyethyl)acetamide	80.7 (25)	
2-Methoxyethyl acetate	8.25 (20)	2.13 (30, B)
1-Methoxy-2-nitrobenzene	45.75 (20)	4.83
o-Methoxyphenol	11.95 (25)	
m-Methoxyphenol	11.59 (25)	
p-Methoxyphenol	11.05 (60)	
2-Methoxy-4-(2-propenyl)phenol		2.46 (25, B)
o-Methoxytoluene	3.5 (20)	

(*Continued*)

TABLE 2.50   Dielectric Constant (Permittivity) and Dipole Moment of Organic Compounds (*Continued*)

Substance	Dielectric constant, ϵ	Dipole moment, D
m-Methoxytoluene	3.5 (20)	
p-Methoxytoluene	4.0 (20)	
Methoxytrimethylsilane	3.248 (25)	
N-Methylacetamide	178.9 (30), 138.6 (60)	4.39 (20, D)
Methyl acetate	7.07 (15), 7.03 (20), 6.68 (25)	1.72
Methyl acrylate	7.03 (30)	1.77 (25, B)
Methylamine	16.7 (−58), 11.4 (−10), 10.0 (18)	1.31
Methyl 2-aminobenzoate	21.9 (25)	
N-Methylaniline	5.96 (20)	1.67 (25, B)
2-Methylaniline	6.138 (25)	
3-Methylaniline	5.816 (25)	
4-Methylaniline	5.058 (25)	
N-Methylbenzenesulfonamide	67.1 (30)	
Methyl benzoate	6.64 (30)	1.86 (25, B)
2-Methyl-1,2-butadiene	2.1 (25)	0.15
2-Methyl-1,3-butadiene	2.098 (20)	0.25
2-Methylbutane	1.871 (0), 1.845 (20)	0.13
2-Methyl-2-butanethiol	5.083 (20)	
Methyl butanoate	5.6 (20), 5.48 (29)	1.72 (22, B)
3-Methylbutanoic acid	2.64 (20)	0.63 (25)
2-Methyl-1-butanol	15.63 (25)	1.9
2-Methyl-2-butanol	5.78 (25)	1.72 (20, B)
3-Methyl-1-butanol	15.63 (20), 14.7 (25), 5.82 (130)	1.82 (25, B)
3-Methyl-2-butanol	12.1 (25)	
3-Methyl-2-butanone	10.37 (20)	
2-Methyl-1-butene	2.180 (20)	0.52 (20, lq)
2-Methyl-2-butene	1.979 (23)	0.11 (25, lq), 0.34 (25, B)
3-Methyl-1-butene	1.0028 (100, g)	0.320
2-Methyl-1-butene-2-one	10.39 (30)	
2-Methylbutyl acetate	4.63 (30)	1.82 (22)
3-Methylbutyl 3-methylbutanoate	4.39 (15)	
3-Methylbutyronitrile	18 (220)	3.62 (25, C)
Methyl carbamate	18.48 (55)	
Methyl chloroacetate	12.0 (20)	
N-Methyl-2-chloroacetamide	92.3 (50)	
Methyl 4-chlorobutanoate	9.51 (30)	
Methyl crotonate	6.664 (20)	
Methyl cyanoacetate	29.3 (20), 19.23 (50), 17.57 (65)	
Methylcyclohexane	2.024 (20)	0
2-Methylcyclohexanol		1.95 (25, B)
cis-3-Methylcyclohexanol	16.05 (20)	1.91
trans-3-Methylcyclohexanol	8.05 (20)	1.75
4-Methylcyclohexanol		1.9 (25, B)
2-Methylcyclohexanone	16 (−15), 14.0 (20)	2.98 (25, B)
3-Methylcyclohexanone	18 (−80), 12.4 (20)	3.06 (25, B)
4-Methylcyclohexanone	15 (−41), 12.35 (20)	3.07 (25, B)
Methylcyclopentane	1.985 (20)	0
1-Methylcyclopentanol	7.11 (37)	
Methyl decanoate		1.65 (20, Hx)
Methyl dodecanoate		1.70 (20, Hx)
N-Methylformamide	200.1 (15), 189.0 (20), 182.4 (25)	3.83
Methyl formate	9.20 (15), 8.5 (20)	1.77

TABLE 2.50   Dielectric Constant (Permittivity) and Dipole Moment of Organic Compounds (*Continued*)

Substance	Dielectric constant, ϵ	Dipole moment, D
2-Methylfuran	2.76 (20)	0.65
Methyl furan-2-carboxylate	11.01 (20)	
(mono)Methyl glutarate	8.37 (20)	
2-Methylheptane	1.95 (20)	0
2-Methyl-2-heptanol	3.38 (−7), 2.46 (25)	
2-Methyl-3-heptanol	3.37 (20), 3.75 (60)	1.63 (20, B)
2-Methyl-4-heptanol	3.30 (20), 3.65 (60)	
3-Methyl-3-heptanol	3.74 (20), 2.89 (60)	
3-Methyl-4-heptanol	9.1 (−20), 7.4 (20)	
4-Methyl-3 -heptanol	5.25 (20), 4.62 (55)	
4-Methyl-4-heptanol	2.87 (20), 3.27 (60)	
2-Methylhexane	1.922 (20)	0
3-Methylhexane	1.920 (20)	0
Methyl hexanoate	4.615 (20)	1.70 (20, Hx)
2-Methyl-2-hexanol	3.257 (24)	
3-Methyl-2-hexanol	4.990 (24)	
3-Methyl-3-hexanol	3.248 (25)	
5-Methyl-2-hexanone	13.53 (20)	
Methyl isobutanoate		1.98 (20, B)
Methylisocyanate	21.75 (16)	2.8
Methyl methacrylate	6.32 (30)	1.68 (25, B)
N-Methyl methanesulfonamide	104.4 (25)	
Methyl o-methoxybenzene	7.7 (21)	
Methyl p-methoxybenzoate	4.3 (33)	
N-Methyl-2-methylbutanamide	123.0 (34)	
N-Methyl-3-methylbutanamide	114.0 (26)	
Methyl 3-(methylthio)propanoate	8.66 (30)	
1-Methylnaphthalene	2.92 (20)	0
Methyl nitrate	23.9 (20)	
Methyl nitrite	20.77 (−73)	
Methyl o-nitrobenzoate	28 (25)	3.67 (30, B)
2-Methyloctane	1.97 (20)	0
3-Methyloctane		0
4-Methyloctane	1.97 (20)	0
Methyl oleate	3.211 (20)	
2-Methyl-1,3-pentadiene	2.422 (25)	
3-Methyl-1,3-pentadiene	2.426 (25)	
4-Methyl-1,3-pentadiene	2.599 (20)	
N-Methylpentanamide	131.0 (13)	
2-Methylpentane	1.886 (20)	0
3-Methylpentane	1.886 (20)	0
2-Methyl-2,4-pentanediol	23.4 (20)	2.9
4-Methylpentanenitrile	17.5 (22)	3.53 (25, B)
Methyl pentanoate	4.992 (20)	1.62 (22, B)
3-Methyl-1-pentanol	15.2 (25)	
3-Methyl-3-pentanol	4.322 (20)	
4-Methyl-2-pentanone	15.6 (0), 15.1 (20), 11.78 (40)	
4-Methylpentenenitrile	17.5 (22)	3.5
4-Methyl-3-penten-2-one	15.6 (0)	2.8
1-Methyl-1-phenylhydrazine	7.3 (19)	1.84 (15, B)
Methyl phenyl sulfide		1.38 (20, B)
Methyl phenyl sulfone	37.9 (100)	

(*Continued*)

TABLE 2.50   Dielectric Constant (Permittivity) and Dipole Moment of Organic Compounds (*Continued*)

Substance	Dielectric constant, ϵ	Dipole moment, D
2-Methylpropanal		2.6
N-Methylpropanamide	170.0 (20), 151 (40)	3.59
2-Methyl-1-propanamine	4.43 (21)	1.3
2-Methylpropane	1.752 (25)	0.132
2-Methylpropanenitrile	24.42 (20)	4.29
2-Methyl-1-propanethiol	4.961 (25)	
2-Methyl-2-propanethiol	5.475 (20)	1.66
Methyl propanoate	6.200 (20)	1.70 (22, B)
2-Methylpropanoic acid	2.58 (20)	1.08 (25, lq)
2-Methylpropanoic anhydride	13.6 (19)	
2-Methyl-1-propanol	26 (−34), 17.93 (20)	1.64
2-Methyl-2-propanol	12.47 (25), 10.9 (30), 8.49 (50)	1.67 (22, B)
2-Methylpropene		0.50
2-Methyl-2-propenenitrile		3.69
2-Methylpropenoic acid		1.6
2-Methylpropyl acetate	5.07 (20)	1.87 (22, B)
2-Methyl-1-propylamine	4.43 (21)	1.27 (27)
(2-Methylpropyl)benzene	2.32 (20)	0
2-Methylpropyl formate	6.41 (20)	1.88 (22)
2-Methylpyridine	10.18 (20)	1.85
3-Methylpyridine	11.10 (30)	2.41 (25, B)
4-Methylpyridine	12.2 (20)	2.70
2-Methylpyridine-1-oxide	36.4 (50)	
3-Methylpyridine-1-oxide	28.26 (45)	
N-Methylpyrrolidine	32.2 (25)	
N-Methyl-2-pyrrolidinone	32.55 (20), 32.2 (25)	4.09 (30, B)
Methyl salicylate	9.41 (30), 8.80 (41)	2.47 (25, B)
3-Methyl sulfolane	29.4 (25)	
Methyl tetradecanoate		1.62 (25, B)
2-Methyltetrahydrofuran	6.97 (25)	
Methyl tetrahydrothiophene-2-carboxylate	7.30 (20)	
Methyl thiocyanate	4.3 (19)	3.34 (20, B)
2-Methylthiophene		0.674
3-Methylthiophene		0.95
Methyl thiophene-2-carboxylate	8.81 (20)	
Methyl trifluoromethyl sulfone	32.0 (20)	
Morpholine	7.42 (25)	1.55
β-Myrcene	2.3 (25)	
Naphthalene	2.54 (90)	0
1-Naphthonitrile	16 (70)	
2-Naphthonitrile	17 (70)	
o-Nitroaniline	47.3 (80), 34.5 (90)	4.28 (20, B)
m-Nitroaniline	35.6 (125)	
p-Nitroaniline	78.5 (155), 56.3 (160)	6.3 (25, B)
o-Nitroanisole	45.75 (20)	4.83
m-Nitroanisole	25.7 (45)	
p-Nitroanisole	26.95 (65)	
Nitrobenzene	35.6 (20), 34.82 (25), 24.9 (90)	4.22
m-Nitrobenzyl alcohol	22 (20)	
2-Nitrobiphenyl		3.83 (20, B)
Nitroethane	29.11 (15), 28.06 (30), 27.4 (35)	3.23

TABLE 2.50 Dielectric Constant (Permittivity) and Dipole Moment of Organic Compounds (*Continued*)

Substance	Dielectric constant, ϵ	Dipole moment, D
2-Nitro-ethylbenzene	21.9 (0)	
Nitromethane	37.27 (20), 35.87 (30), 35.1 (35)	3.46
1-Nitro-2-methoxybenzene		4.83
o-Nitrophenol	16.50 (50)	3.14 (25, B)
m-Nitrophenol	35.45 (100)	
p-Nitrophenol	42.20 (120)	
1-Nitropropane	24.70 (15), 23.24 (30), 22.7 (35)	3.66
2-Nitropropane	26.74 (15), 25.52 (30)	3.73
N-Nitrosodimethylamine	53 (20)	4.01 (20, B)
o-Nitrotoluene	26.36 (20), 22.0 (58)	3.72 (20, B)
m-Nitrotoluene	24.95 (30), 22 (58)	4.20 (20, B)
p-Nitrotoluene	22.2 (58)	4.47 (25, B)
Nonane	1.972 (20), 1.85 (110)	0
Nonanoic acid	2.48 (22)	0.8
1-Nonanol		1.72 (20, B)
1-Nonene	2.18 (20)	0
(*trans*, *trans*)-9,12-Octadecadienoic acid	2.70 (70), 2.60 (120)	1.40 (18, Hx)
Octamethylcyclotetrasiloxane	2.4 (20)	0.42 (25, lq), 0.67 (25, B)
Octamethyltrisiloxane	2.3 (20)	0.64 (25, lq)
Octane	1.948 (20), 1.83 (110)	0
Octanenitrile	13.90 (20)	
Octanoic acid	2.85 (15), 2.45 (20)	1.15 (25, lq)
1-Octanol	11.3 (10), 10.30 (20)	1.72 (20, B)
2-Octanol	8.13 (20), 6.52 (40)	1.65 (20, B)
2-Octanone	9.51 (20), 7.42 (100)	2.72 (15, B)
1-Octene	2.113 (20)	0
cis-2-Octene	2.06 (25)	0
trans-2-Octene	2.00 (25)	0
Oleic acid	2.34 (20)	1.2
Oxalyl chloride	3.470 (21)	0.93 (20, B)
Palmitic acid	2.3 (70)	
Paraldehyde	13.9 (25)	1.43
Parathion		4.98 (25, B)
Pentachloroethane	3.73 (20), 3.716 (25)	0.92
2,3,4,5,6-Pentachlorotoluene	4.8 (20)	
Pentadecane		0
cis-1,3-Pentadiene	2.32 (25)	0.50 (25, B)
1,4-Pentadiene	2.054 (24)	
Pentanal	10.1 (17), 10.00 (20)	2.59 (20, B)
Pentane	2.011 (−90), 1.837 (20)	0
1,2-Pentanediol	17.31 (24)	
1,4-Pentanediol	26.74 (23)	
1,5-Pentanediol	26.2 (20)	2.45 (20, D)
2,3-Pentanediol	17.37 (24)	
2,4-Pentanediol	24.69 (21)	
2,4-Pentanedione	26.52 (30)	3.03
Pentanenitrile	20.04 (20)	4.12, 3.57 (25, B)
1-Pentanethiol	4.85 (20), 4.55 (25), 4.23 (50)	1.54 (25, lq)
Pentanoic acid	2.66 (21)	1.61 (20, D)
1-Pentanol	16.9 (20), 15.13 (25)	1.71 (20, B)
2-Pentanol	13.71 (25)	1.66 (22, B)

(*Continued*)

TABLE 2.50 Dielectric Constant (Permittivity) and Dipole Moment of Organic Compounds (*Continued*)

Substance	Dielectric constant, ϵ	Dipole moment, D
3-Pentanol	13.35 (25)	1.64 (22, B)
2-Pentanone	15.45 (20), 11.73 (80)	2.72 (22, B)
3-Pentanone	19.4 (−20), 17.00 (20)	2.72 (20, B)
2-Pentanone oxime	3.3 (25)	
1-Pentene	2.011 (20)	0.5
cis-2-Pentene		0
trans-2-Pentene		0
Pentyl acetate	4.79 (20)	1.75
Pentylamine	4.27 (20)	1.55 (30, B)
Pentyl formate	5.7 (19)	1.90
Pentyl nitrate	9.0 (18)	
Pentyl nitrite	7.21 (25)	
tert-Pentyl nitrite	10.88 (25)	
Phenanthrene	2.8 (20)	0
Phenol	12.40 (30), 9.78 (60)	1.224
Phenoxyacetylene	4.76 (25)	1.42 (25, lq)
Phenyl acetate	5.40 (25)	1.54 (22, B)
Phenylacetic acid	3.47 (80)	
Phenylacetonitrile	17.87 (26), 8.5 (234)	3.47 (27, B)
Phenylacetylene	2.98 (20)	0.72 (20, B)
1-Phenylethanol	8.77 (20), 7.6 (90)	1.51 (20, B)
2-Phenylethanol	12.31 (20)	
Phenylhydrazine	7.15 (20)	1.67 (25, B)
Phenyl isocyanate	8.94 (20)	
Phenyl isothiocyanate	10 (20)	
1-Phenylpropene	2.7 (20)	
2-Phenylpropene	2.3 (20)	
3-Phenylpropene	2.6 (20)	
Phenyl salicylate	6.3 (50)	
Phosgene	4.7 (0), 4.3 (22)	
Phthalide	36 (75)	
(±)-α-Pinene	2.64 (25), 2.26 (30)	0.60 (25, B)
L-β-Pinene	2.76 (20)	
Piperidine	4.33 (20)	1.19 (25, B)
Propanal	18.5 (17)	2.52
Propane	1.668 (20)	0.084
1,2-Propanediamine	10.2	
1,3-Propanediamine	9.55	1.96 (25, B)
1,2-Propanediol	32.0 (20), 27.5 (30)	2.27 (25, D)
1,3-Propanediol	35.1 (20)	2.52 (25, D)
1,2-Propanediol dinitrate	26.80 (20)	
1,3-Propanediol dinitrate	18.97 (20)	
1,2-Propanedithiol	7.24 (20)	
1,3-Propanedithiol	8.11 (30)	
Propanenitrile	29.7 (20)	4.05
1-Propanethiol	5.94 (15), 1.55 (25)	1.68
2-Propanethiol	5.95 (25)	1.61
1,2,3-Propanetriol 1-acetate	38.57 (−31), 7.11 (20)	
Propanoic acid	3.30 (10), 3.44 (25)	1.76
Propanoic anhydride	18.30 (20)	
1-Propanol	20.8 (20), 20.33 (25)	1.55
2-Propanol	20.18 (20), 18.3 (25), 16.2 (40)	1.58

TABLE 2.50 Dielectric Constant (Permittivity) and Dipole Moment of Organic Compounds (*Continued*)

Substance	Dielectric constant, ϵ	Dipole moment, D
2-Propenal		3.12
Propene	2.137 (−53), 1.88 (20), 1.44 (90)	0.366
Propenenitrile	33.0 (20)	3.87
2-Propen-l-ol	21.6 (15), 19.7 (20)	1.60
Propionaldehyde (propanal)	18.5 (17)	2.75
Propionamide		3.4 (30, B)
Propyl acetate	5.62 (20)	1.86 (25, B)
N-Propylacetamide	117.8 (25)	
Propylamine	5.31 (20), 5.08 (26)	1.17
Propylbenzene	2.37 (20), 2.351 (30)	0
Propyl benzoate	5.78 (30)	
Propyl butanoate	4.3 (20)	
Propyl carbamate	12.06 (65)	
Propylene carbonate	66.14 (20)	4.9
Propyleneimine		1.77 (*cis*), 1.60 (*trans*)
1,2-Propylene oxide		2.00
Propyl formate	7.72 (19), 6.92 (30)	1.91 (22, B)
Propyl nitrate	14 (18)	3.01 (20, B)
Propyl nitrite	12.35 (−23)	
Propyl pentanoate	4(19)	
N-Propylpropanamide	118.1 (25)	
Propyl propanoate	5.25 (20)	1.79 (22, B)
Propyl trichloroacetate	8.32 (25)	
Propyne	3.218 (−27)	0.784
2-Propyn-l-ol	20.8 (20)	1.13
Pulegone	9.5 (20)	2.00 (25, B)
Pyridazine		4.22
Pyrazine	2.80 (50)	0
Pyridine	13.26 (20), 12.3 (25), 9.4 (116)	2.215
Pyridine-1-oxide	35.94 (70)	
Pyrimidine		2.33
1*H*-Pyrrole	8.00 (20), 8.13 (25)	1.74
Pyrrolidine	8.30 (20)	1.58 (20, B)
2-Pyrrolidone		3.55 (25, B)
Quinoline	9.16 (20), 9.00 (25)	2.29
Safrole	3.1 (21)	
Salicylaldehyde	18.35 (20)	2.86 (20, B)
D-Sorbitol	35.5 (80)	
Squalane	1.911 (100)	0
Squalene		0.68 (25, B)
Stearic acid	2.29 (70), 2.26 (100)	1.76 (25, D)
Styrene	2.47 (20), 2.43 (25), 2.32 (75)	0.13 (25, lq)
Succinonitrile	62.6 (25), 56.5 (57), 54 (68)	3.68 (30, toluene)
α-Terpinene	2.45 (25)	
Terpinolene	2.29 (25)	
1,1,2,2-Tetrabromoethane	8.6 (3), 7.0 (22), 6.72 (30)	1.41
1,1,2,2-Tetrachlorodifluoroethane	2.52 (35)	
1,1,1,2-Tetrachloroethane	9.22 (−66)	
1,1,2,2-Tetrachloroethane	8.50 (20)	1.32
Tetrachloroethylene	2.30 (25), 2.268 (30)	0
1,1,3,4-Tetrachlorohexafluoro- butane	2.86 (20)	

(*Continued*)

TABLE 2.50 Dielectric Constant (Permittivity) and Dipole Moment of Organic Compounds (*Continued*)

Substance	Dielectric constant, ϵ	Dipole moment, D
Tetradecafluorohexane	1.76 (25)	
Tetradecamethylhexasiloxane	2.5 (20)	1.58 (20, lq)
Tetradecane		0
Tetradecanoic acid		0.76 (25, B)
1-Tetradecanol	4.72 (38), 4.40 (48)	1.69 (25, C)
Tetraethylene glycol	20.44 (20)	5.84 (20, lq)
Tetraethyl lead		0.3 (20, B)
Tetraethylsilane	2.09 (20)	0
Tetraethyl silicate	4.1 (20)	1.72 (32, B)
Tetrafluoromethane	1.685 (−147)	
2,2,3,3-Tetrafluoro-1-propanol	21.03 (25)	
Tetrahydrofuran	11.6 (−70), 7.52 (22)	1.75 (25, B)
Tetrahydro-2-furanmethanol	13.61 (23), 13.48 (30)	2.12 (35, lq)
2-Tetrahydrofurfuryl acetate	9.65 (20)	
1,2,3,4-Tetrahydronaphthalene	2.77 (25)	0
1,2,3,4-Tetrahydro-2-naphthol	11.7 (20), 6.7 (90)	
Tetrahydropyran	5.66 (20), 5.61 (25)	1.74
Tetrahydrothiophene		1.9
Tetrahydrothiophene-1,1-dioxide (sulfolane)	43.26 (30)	4.81 (25, B)
Tetrahydrothiophene-*S*-oxide	42.96 (25), 42.5 (30)	
Tetrakis(methylthio)methane	2.818 (70)	
Tetramethoxymethane	2.40 (20)	
Tetramethyl germanium	1.817 (24)	
1,1,3,3-Tetramethylguanidine	11.5 (25)	
Tetramethylsilane	1.921 (20)	0
Tetramethyl silicate	6.0 (20)	
1,1,2,2-Tetramethylurea	23.10 (20)	3.47 (25, B)
Tetranitromethane	2.317 (25)	0
Tetrathiomethylmethane	2.82 (70)	
Thiacyclopentane		1.90 (25, B)
Thioacetic acid	14.30 (25)	
Thiophene	2.74 (20), 2.57 (25)	0.55
Thymol		1.55 (25, B)
Toluene	2.385 (20), 2.364 (30)	0.375
o-Toluidine	6.34 (18), 6.14 (25), 5.71 (58)	1.60 (25, B)
m-Toluidine	5.95 (18), 5.82 (25), 5.45 (58)	1.45 (25, B)
p-Toluidine	5.06 (60)	1.52 (25, B)
m-Tolunitrile		4.21 (22, B)
p-Tolunitrile		4.47 (20, B)
Tribenzylamine		0.65 (20, B)
2,2,2-Tribromoacetaldehyde	7.6 (20)	1.70 (20, C)
Tribromochloromethane	2.60 (60)	
Tribromofluoromethane	3.00 (20)	
Tribromomethane	4.404 (10), 4.39 (20)	0.99
Tribromonitromethane	9.03 (25)	
1,2,3-Tribromopropane	6.45 (20), 6.00 (30)	1.59 (25, B)
Tributylamine	2.34 (20)	0.78 (25, B)
Tributyl borate	2.23 (20)	0.78 (25, C)
Tributyl phosphate	8.34 (20), 7.96 (30)	3.07 (25, B)
Tributyl phosphite		1.92 (20, C)
Trichloroacetaldehyde	7.6 (−40), 6.9 (20), 6.8 (25)	1.96 (25, B)

TABLE 2.50   Dielectric Constant (Permittivity) and Dipole Moment of Organic Compounds (*Continued*)

Substance	Dielectric constant, ϵ	Dipole moment, D
Trichloroacetic acid	4.34 (60)	1.1 (25, B, dimer)
Trichloroacetic anhydride	5.0 (25)	
Trichloroacetonitrile	7.85 (19)	1.93 (19, lq)
4,4,4-Trichlorobutanal	10.0 (18)	
1,2,2-Trichloro-1,1-difluoroethane	4.01 (30)	
1,1,1-Trichloroethane	7.1 (7), 7.24 (20)	1.755
1,1,2-Trichloroethane	7.19 (25)	1.45
Trichloroethylene	3.42 (16), 3.39 (28)	0.77 (30, lq), 0.95 (30, B)
Trichloroethylsilane		2.0
Trichlorofluoromethane	3.00 (25), 2.28 (29)	0.45
(Trichloromethyl)benzene	6.9 (21)	2.0
Trichloromethylsilane		1.87 (25, B)
Trichloronitromethane	7.32 (25)	
2,4,6-Trichlorophenol		1.88 (25, D)
1,2,3-Trichloropropane	7.5 (20)	1.61
Trichlorosilane		0.86
α,α,α-Trichlorotoluene	6.9 (21)	2.17 (20, B)
1,1,2-Trichloro-1,2,2-trifluoroethane	2.41 (25)	
Tridecane	2.02 (20)	0
1-Tridecene	2.14 (20)	0
Triethanolamine	29.36 (25)	3.57 (25, B)
Triethoxymethane	4.779 (20)	
Triethylaluminum	2.9 (20)	
Triethylamine	2.418 (20)	0.66
Triethylborane	1.874 (20)	
Triethylene glycol	23.69 (20)	5.58 (20, lq)
Triethylenetetramine	10.76 (20)	
Triethyl orthovanadate	3.333 (25)	
Triethyl phosphate	13.43 (15), 13.20 (25), 10.93 (65)	3.08 (25, B)
Triethylphosphine oxide	35.5 (50)	
Triethylphosphine sulfide	39.0 (98)	
Triethyl phosphite	5.0	1.82 (25, D)
Trifluoroacetic acid	8.42 (20), 5.76 (50)	2.28
Trifluoroacetic anhydride	2.7 (25)	
1,1,1-Trifluoroethane		2.347
2,2,2-Trifluoroethanol	27.68 (20)	2.03 (25, cHex)
Trifluoromethane	5.2 (26)	1.651
(Trifluoromethyl)benzene	9.22 (25)	2.86
1-Trifluoromethyl-3-nitrobenzene	17.0 (30)	
α,α,α-Trifluorotoluene	9.2 (30), 8.1 (60)	
Trimethoxymethylsilane	4.9 (25)	
Trimethylamine	2.44 (25)	0.612
1,2,3-Trimethylbenzene	2.66 (20), 2.609 (30)	0
1,2,4-Trimethylbenzene	2.38 (20), 2.36 (30)	0
1,3,5-Trimethylbenzene	2.28 (20)	0
Trimethyl borate	2.276 (20)	0.82 (25, C)
2,2,3-Trimethylbutane	1.930 (20)	0
Trimethylchlorosilane	10.21 (0)	
Trimethylene sulfide		1.85
2,2,5-Trimethylhexane		0
2,3,5-Trimethylhexane		0
2,2,3-Trimethylpentane	1.962 (20)	0

(*Continued*)

TABLE 2.50   Dielectric Constant (Permittivity) and Dipole Moment of Organic Compounds (*Continued*)

Substance	Dielectric constant, ϵ	Dipole moment, D
2,2,4-Trimethylpentane	1.940 (20)	0
2,3,3-Trimethylpentane	1.98 (20)	0
2,3,4-Trimethylpentane	1.97 (20)	0
Trimethyl phosphate	20.6 (20)	3.2
Trimethylphosphine sulfide		71.6 (20)
Trimethyl phosphite		1.83 (20, C)
2,4,6-Trimethylpyridine	7.807 (25)	1.95 (25, B)
2,4,6-Trinitrophenol	4.0 (21)	
1,3,5-Trioxane	15.55 (65)	2.08
Triphenyl phosphite	3.67 (45), 3.57 (65)	2.04 (25, B)
Tris(4-ethylphenyl) phosphite	3.74 (15), 3.61 (45)	2.08 (25, B)
Tris(2-methylphenyl) phosphate	6.7 (25)	2.9
Tris(3-methylphenyl) phosphate		3.0
Tris(4-methylphenyl) phosphate		3.2
Tris(*m*-tolyl) phosphite	3.67 (15), 3.53 (45)	1.62 (25, B)
Tris(*p*-tolyl) phosphite	3.88 (15), 3.74 (45)	1.77 (25, B)
Tri-*o*-tolyl phosphate	6.92 (40)	2.84 (40, C)
Undecane	2.00 (20), 1.84 (150)	0
2-Undecanone		2.71 (15, B)
1-Undecene	2.14 (20)	0
Urea		4.59 (25, D)
Vinyl acetate		1.79 (25, B)
Vinyl chloride	6.26 (17)	1.45
Vinyl isocyanate	10.62 (25)	
2-Vinylpyridine	9.126 (20)	
4-Vinylpyridine	10.50 (20)	
o-Xylene	2.562 (20), 2.54 (30)	0.62
m-Xylene	2.359 (20), 2.35 (30)	0.33 (20, lq), 0.37 (20, B)
p-Xylene	2.273 (20), 2.22 (50)	0
Xylitol	40.0 (20)	

2.11   IONIZATION ENERGY

The ionization energy or ionization potential is the energy necessary to remove an electron from the neutral atom. It is a minimum for the alkali metals that have a single electron outside a closed shell. It generally increases across a row on the periodic maximum for the noble gases that have closed shells. For example, sodium requires only 496 kJ/mol or 5.14 eV/atom to ionize it while neon, the noble gas immediately preceding it in the periodic table, requires 2081 kJ/mol or 21.56 eV/atom. The ionization energy is one of the primary energy considerations used in quantifying chemical bonds.

The electron affinity is a measure of the energy change when an electron is added to a neutral atom to form a negative ion. For example, when a neutral chlorine atom in the gaseous form picks up an electron to form a Cl^- ion, it releases energy of 349 kJ/mol or 3.6 eV/atom. It is said to have an electron affinity of −349 kJ/mol and this large number indicates that it forms a stable negative ion. Small numbers indicate that a less stable negative ion is formed. Group VIA and VIIA in the periodic table have the largest electron affinities.

Note: 1 kJ/mol = .010364 eV/atom

TABLE 2.51   Ionization Energy of Molecular and Radical Species

This table gives the first ionization potential in MJ · mol^{-1} and in electron volts. Also listed is the enthalpy of formation of the ion at 25°C (298 K).

Species	Ionization energy		$\Delta_f H$ (ion), kJ · mol^{-1}
	MJ · mol^{-1}	Electron volts	
Acenaphthene	0.741	7.68	896
Acenaphthylene	0.793	8.22(4)	1053
Acetaldehyde	0.98696(7)	10.2290(7)	821
Acetamide	0.931(3)	9.65(3)	693
Acetic acid	1.029(2)	10.66(2)	596
Acetic anhydride	0.965	10.0	398
Acetone	0.9364	9.705	719
Acetonitrile	1.1766(5)	12.194(5)	1252
Acetophenone	0.896(3)	9.29(3)	810
Acetyl chloride	1.047(5)	10.85(5)	804
Acetyl fluoride	1.111(2)	11.51(2)	667
Acetylene	1.1000(2)	11.400(2)	1328
Allene	0.935(1)	9.69(1)	1126
Allyl alcohol	0.933(5)	9.67(5)	808
Allylamine	0.845	8.76	891
3-Amino-I-propanol	0.87	9.0	651
Aniline	0.7449(2)	7.720(2)	832
Anthracene	0.719(3)	7.45(3)	949
Azoxybenzene	0.78	8.1	1123
Azulene	0.715(2)	7.41(2)	1004
Benzaldehyde	0.916(2)	9.49(2)	878
Benzamide	0.912	9.45	811
Benzene	0.89212(2)	9.2459(2)	975
Benzenethiol	0.801(2)	8.30(2)	913
Benzoic acid	0.914	9.47	620
Benzonitrile	0.928	9.62	1146
Benzophenone	0.873(5)	9.05(5)	923
p-Benzoquinone	0.969(2)	10.04(18)	847
Benzoyl chloride	0.920	9.54	816
Benzyl alcohol	0.82	8.5	720
Benzylamine	0.834(5)	8.64(5)	917
Biphenyl	0.767(2)	7.95(2)	950
Bromoacetylene	0.995(2)	10.31(2)	1242
Beomobenzene	0.866(2)	8.98(2)	971
Bromochlorodifluoromethane	1.141	11.83	702
Bromochloromethane	1.039(1)	10.77(1)	1085
Bromodichloromethane	1.02	10.6	973
Bromethane	0.992	10.28	930
Bromethylene	0.946(2)	9.80(2)	1025
Bromomethane	1.0171(3)	10.541(3)	979
1-Bromonaphthalene	0.781	8.09	956
Bromopentafluorobenzene	0.923(2)	9.57(2)	212
1-Bromopropane	0.982(1)	10.18(1)	898
2-Bromopropane	0.972(1)	10.07(1)	874
3-Bromopropene	0.972(1)	10.07(1)	1018
p-Bromotoluene	0.837(1)	8.67(1)	908
Bromotrichloromethane	1.02	10.6	980
Bromotrifluoromethane	1.10	11.4	451
1,2-Butadiene	0.871	9.03	1034

(Continued)

TABLE 2.51 Ionization Energy of Molecular and Radical Species (*Continued*)

Species	Ionization energy		$\Delta_f H$ (ion), kJ · mol^{-1}
	MJ · mol^{-1}	Electron volts	
1,3-Butadiene	0.8750	9.069	985
Butanal	0.949(2)	9.84(2)	742
Butanenitrile	1.08	11.2	1110
2-Butanone	0.918(4)	9.51(4)	677
trans-2-Butenal	0.939(1)	9.73(1)	835
1-Butene	0.924(2)	9.58(2)	924
cis-2-Butene	0.8788(8)	9.108(8)	871
trans-2-Butene	0.8780(8)	9.100(8)	866
l-Buten-3-yne	0.924(2)	9.58(2)	1230
Butyl acetate	0.965	10.0	479
sec-Butyl acetate	0.955	9.90	453
Butyl ethyl ether	0.903	9.36	610
Butylbenzene	0.838(1)	8.69(1)	826
sec-Butylbenzene	0.837(1)	8.68(1)	820
tert-Butylbenzene	0.834(2)	8.64(2)	812
Butylcyclohexane	0.908	9.41	695
Butylcyclopentane	0.960(3)	9.95(3)	793
p-tert-Butylphenol	0.75	7.8	552
p-tert-Butyltoluene	0.799	8.28	745
1-Butyne	0.9821(5)	10.178(5)	1147
2-Butyne	0.9226(5)	9.562(5)	1068
Camphor	0.845(3)	8.76(3)	577
Caprolactam	0.875(2)	9.07(2)	629
Carbazole	0.730(3)	7.57(3)	961
Carbon	1.0865	11.260	1803
Carbon (C_2)	1.188	12.31	2000
Carbon dioxide	1.3289(2)	13.773(2)	935
Carbon monoxide	1.35217	14.0139	1242
Carbon oxyselenide	1.000(1)	10.36(1)	929
Carbon oxysulfide	1.07812(15)	11.1736(15)	936
Carbon sulfide	0.97149(19)	10.0685(20)	1089
Carbon sulfide (CS)	1.093(1)	11.33(1)	1368
Carbonyl fluoride	1.257	13.03	617
Carbonyltrihydroboron (BH_3CO)	1.075(2)	11.14(2)	962
Chloroacetaldehyde	1.011(3)	10.48(3)	815
Chloroacetic acid	0.984	10.2	597
Chloroacetyl chloride	1.06	11.0	815
Chloroacetylene	1.021(2)	10.58(2)	1276
m-Chloroaniline	0.781(10)	8.09(10)	835
o-Chloroaniline	0.820	8.50	883
p-Chloroaniline	0.789	8.18	844
Chlorobenzene	0.874(2)	9.06(2)	929
Chlorodibromomethane	0.1022(1)	10.59(1)	1030
1-Chloro-1,1-difluoroethane	1.156(1)	11.98(1)	626
1-Chloro-2,2-difluoroethylene	0.946(4)	9.80(4)	628
Chlorodifluoromethane	1.18	12.2	693
Chloroethane	1.058(2)	10.97(2)	946
2-Chloroethanol	1.015	10.52	756
Chloroethylene	0.964(2)	9.99(2)	985
Chlorofluoromethane	1.130(1)	11.71(1)	870
Chloromethane	1.083(1)	11.22(1)	1001
Chloromethylene	0.949	9.84	1247

TABLE 2.51   Ionization Energy of Molecular and Radical Species (*Continued*)

Species	Ionization energy MJ · mol⁻¹	Ionization energy Electron volts	$\Delta_f H$ (ion), kJ · mol⁻¹
Chloromethylidine (CCl)	0.86(2)	8.9(2)	1244
1-Chloronaphthalene	0.784	8.13	906
m-Chloronitrobenzene	0.957(10)	9.92(10)	995
p-Chloronitrobenzene	0.961(10)	9.96(10)	999
Chloropentafluorobenzene	0.938(2)	9.72(2)	126
Chloropentafluoroethane	1.22	12.6	99
m-Chlorophenol	0.835	8.65	680
p-Chlorophenol	0.834	8.69	692
1-Chloropropane	1.044(3)	10.82(3)	912
2-Chloropropane	1.040(2)	10.78(2)	895
3-Chloropropene	0.96	9.9	950
m-Chlorotoluene	0.852(2)	8.83(2)	869
o-Chlorotoluene	0.852(2)	8.83(2)	869
p-Chlorotoluene	0.838(2)	8.69(2)	855
Chlorotrifluoroethylene	0.947	9.81(3)	373
Chlorotrifluoromethane	1.195	12.39	485
Chrysene	0.732	7.59(2)	1016
Coronene	0.703	7.29	1026
m-Cresol	0.800	8.29	668
o-Cresol	0.785	8.14	660
p-Cresol	0.784	8.13	659
cis-Crotonic acid	0.973	10.08	625
trans-Crotonic acid	0.96	9.9	604
Cumene	0.842	8.73(1)	847
Cyanamide	1.00	10.4	1137
Cyanate (NCO)	1.135(1)	11.76(1)	1290
Cyanide (CN)	1.360	14.09	1795
Cyanoacetylene	1.123(1)	11.64(1)	1475
Cyanogen	1.290(1)	13.37(1)	1597
Cyanogen chloride	1.191(1)	12.34(1)	1329
Cyanogen fluoride	1.285(1)	13.32(1)	1323
Cyclobutane	0.957(5)	9.92(5)	986
Cyclobutanone	0.9025	9.354	815
Cyclobutene	0.910	9.43	1067
Cycloheptane	0.962	9.97	844
Cyclohexane	0.951(3)	9.86(3)	828
Cyclohexanol	0.941	9.75	651
Cyclohexanone	0.882(1)	9.14(1)	656
Cyclohexene	0.8631(10)	8.945(10)	859
Cyclohexylamine	0.832(23)	8.62(24)	727
Cyclohexylcyclohexane	0.908	9.41	690
Cyclooctane	0.942	9.76	817
Cyclopropane	0.951	9.86	1005
Cyclopropanecarbonitrile	0.989	10.25	1173
Cyclopropanone	0.88(1)	9.1(1)	895
Cyclopropene	0.930	9.67(1)	1209
Cyclopropylamine	0.84	8.7	916
Cyclopropylbenzene	0.806	8.35	956
cis-Decahydronaphthalene	0.893	9.26	724
trans-Decahydronaphthalene	0.892	9.24	710
Decane	0.931	9.65	682
1-Decene	0.909(1)	9.42(1)	786

(*Continued*)

TABLE 2.51 Ionization Energy of Molecular and Radical Species (*Continued*)

Species	Ionization energy		$\Delta_f H$ (ion), kJ · mol^{-1}
	MJ · mol^{-1}	Electron volts	
Diazomethane	0.8683(1)	8.999(1)	1098
1,4-Dibromobutane	0.979	10.15	879
1,2-Dibromoethane	1.001	10.37	963
Dibromofluoromethane	1.069(3)	11.07(3)	687
Dibromomethane	1.013(2)	10.50(2)	1013
1,2-Dibromopropane	0.975	10.1	903
1,3-Dibromopropane	0.990	10.26	919
1,2-Dibromotetrafluoroethane	1.07	11.1	280
Dibutyl ether	0.910	9.43	575
Di-*sec*-butyl ether	0.879	9.11	511
Di-*terf*-butyl ether	0.850	8.81	486
Dibutyl sulfide	0.79	8.2	624
Di-*tert*-butyl sulfide	0.77	8.0	583
Dibutylamine	0.742(3)	7.69(3)	586
Dichloroacetyl chloride	1.06	11.0	819
Dichloroacetylene	0.974	10.09	1183
m-Dichlorobenzene	0.879(1)	9.11(1)	907
o-Dichlorobenzene	0.876(1)	9.08(1)	909
p-Dichlorobenzene	0.856(1)	8.89(1)	882
Dichlorodifluoromethane	1.134(4)	11.75(4)	656
Dichlorodimethylsilane	1.03	10.7	576
1,1-Dichloroethane	1.067	11.06	937
1,2-Dichloroethane	1.065	11.04	931
1,1-Dichloroethylene	0.945(4)	9.79(4)	947
cis-1,2-Dichloroethylene	0.932(1)	9.66(1)	936
trans-1,2-Dichloroethylene	0.931(2)	9.65(2)	935
Dichlorofluoromethane	1.11	11.5	829
Dichloromethane	1.092(1)	11.32(1)	996
Dichloromethylene	1.000	10.36	1163
1,2-Dichloropropane	1.049(5)	10.87(5)	886
1,3-Dichloropropane	1.047(5)	10.85(5)	888
1,2-Dichlorotetrafluoroethane	1.18	12.2	252
Dicyclopropyl ketone	0.88	9.1	1041
1,1-Diethoxyethane	0.944	9.78	490
Diethyl oxalate	0.95	9.8	205
m-Diethylbenzene	0.819(1)	8.49(1)	798
o-Diethylbenzene	0.821	8.51	804
p-Diethylbenzene	0.810	8.40	790
Diethylene glycol dimethyl ether	0.96	9.8	448
m-Difluorobenzene	0.900(1)	9.33(1)	591
o-Difluorobenzene	0.895(1)	9.28(1)	602
p-Difluorobenzene	0.882(1)	9.14(1)	575
1,1-Difluoroethane	1.145(3)	11.87(3)	643
1,1-Difluoroethylene	0.993(1)	10.29(1)	650
cis-1,2-Difluoroethylene	0.987	10.23	690
Difluoromethane	1.226	12.71	774
Difluoromethylene	1.102(1)	11.42(1)	897
2,5-Dihydrothiophene	0.81	8.4	898
Diiodomethane	0.913(2)	9.46(2)	1030
Diisobutyl sulfide	0.807(5)	8.36(5)	627
Diisobutylamine	0.754	7.81	574
Diisopropyl ether	0.888(5)	9.20(5)	569
Diisopropyl sulfide	0.833(5)	8.63(5)	630

TABLE 2.51 Ionization Energy of Molecular and Radical Species (*Continued*)

Species	Ionization energy MJ · mol^{-1}	Ionization energy Electron volts	$\Delta_f H$ (ion), kJ · mol^{-1}
Diisopropylamine	0.746(3)	7.73(3)	602
Diketene	0.93(2)	9.6(2)	736
Dimethoxymethane	0.92	9.5	569
Dimethyl disulfide	0.71	7.4(3)	690
Dimethyl ether	0.9673(23)	10.025(25)	783
Dimethyl oxalate	0.965	10.0	287
o-Dimethyl phthalate	0.930(7)	9.64(7)	277
Dimethyl sulfide	0.838(1)	8.69(1)	801
Dimethyl sulfoxide	0.878	9.01	718
Dimethylamine	0.794(8)	8.23(8)	776
N,N-Dimethylaniline	0.687(2)	7.12(2)	787
2,2-Dimethylbutane	0.971	10.06	787
2,3-Dimethylbutane	0.967	10.02	791
3,3-Dimethyl-2-butanone	0.879(2)	9.11(2)	589
2,3-Dimethyl-1-butene	0.875(1)	9.07(1)	812
2,3-Dimethyl-2-butene	0.798(1)	8.27(1)	729
3,3-Dimethyl-1-butyne	0.946(5)	9.80(5)	1050
1,1-Dimethylcyclohexane	0.909	9.42	728
cis-1,2-Dimethylcyclohexane	<0.944	<9.78	772
cis-1,3-Dimethylcyclohexane	<0.963	<9.98	778
cis-1,4-Dimethylcyclohexane	<0.958	<9.93	782
trans-1,2-Dimethylcyclohexane	0.908	9.41	728
trans-1,3-Dimethylcyclohexane	0.920	9.53	743
trans-1,4-Dimethylcyclohexane	0.922	9.56	738
cis-1,2-Dimethylcyclopentane	0.957(5)	9.92(5)	828
trans-1,2-Dimethylcyclopentane	0.960(5)	9.95(5)	823
N,N-Dimethylformamide	0.881(2)	9.13(2)	689
2,6-Dimethyl-4-heptanone	0.872(3)	9.04(3)	515
1,1-Dimethylhydrazine	0.702(4)	7.28(4)	786
2,4-Dimethyl-3-pentanone	0.864(1)	8.95(1)	552
2,3-Dimethylpyridine	0.854(2)	8.85(2)	922
2,4-Dimethylpyridine	0.854(3)	8.85(3)	918
2,5-Dimethylpyridine	0.849(5)	8.80(5)	916
2,6-Dimethylpyridine	0.847(3)	8.86(3)	913
3,4-Dimethylpyridine	0.883	9.15	953
3,5-Dimethylpyridine	0.893	9.25	965
N,N-Dimethyl-*o*-toluidine	0.714(2)	7.40(2)	814
1,3-Dioxane	0.95	9.8	607
1,4-Dioxane	0.887(1)	9.19(1)	571
1,3-Dioxolane	0.96	9.9	658
Diphenyl ether	0.781(3)	8.09(3)	766
Diphenylacetylene	0.762(2)	7.90(2)	1164
Diphenylamine	0.691(4)	7.16(4)	908
1,2-Diphenylethane	0.84(1)	8.7(1)	983
Diphenylmethane	0.825(3)	8.55(3)	963
Dipropyl ether	0.894(5)	9.27(5)	602
Dipropyl sulfide	0.801(2)	8.30(2)	676
Dipropylamine	0.746(3)	7.73(3)	641
Divinyl ether	0.84	8.7	827
5,7-Dodecadiyne	0.837	8.67	1079
Dodecafluorocyclohexane	1.27	13.2	−1095
Epichlorohydrin	0.98	10.2	875

(*Continued*)

TABLE 2.51 Ionization Energy of Molecular and Radical Species (*Continued*)

Species	Ionization energy		$\Delta_f H$ (ion), kJ · mol^{-1}
	MJ · mol^{-1}	Electron volts	
Ethylene glycol	0.980	10.16	593
Ethylene oxide	1.0195(10)	10.566(10)	967
Ethyleneimine	0.89(1)	9.2(1)	1014
p-Ethylphenol	0.756	7.84	613
Ethynyl (HC≡C)	1.13	11.7	1694
Fluoranthene	0.768(4)	7.95(4)	1057
Fluorene	0.761(3)	7.89(3)	950
Fluoroacetylene	1.086	11.26	1195
Fluorobenzene	0.8877(5)	9.200(5)	772
Fluoroethane	1.12	11.6	856
Fluoroethylene	1.0000(15)	10.363(15)	861
Fluoromethane	1.203(2)	12.47(2)	956
Fluoromethylene	1.012	10.49	1121
Fluoromethylidene (CF)	0.879(1)	9.11(1)	1134
p-Fluoronitrobenzene	0.955	9.90	826
1-Fluoropropane	1.09	11.3	806
2-Fluoropropane	1.069(2)	11.08(2)	776
3-Fluoropropene	0.975	10.11	821
m-Fluorotoluene	0.860(1)	8.91(1)	709
o-Fluorotoluene	0.860(1)	8.91(1)	709
p-Fluorotoluene	0.848(1)	8.79(1)	701
Formaldehyde	1.0492(2)	10.874(2)	940
Formamide	0.980(6)	10.16(6)	796
Formic acid	1.093(1)	11.33(1)	715
Fulminic acid (HCNO)	1.045	10.83	1263
Fulvene	0.807	8.36	1031
Fumaric acid	1.03	10.7	355
Furan	0.8571(3)	8.883(3)	822
Glyoxal	0.975	10.1	763
1-Heptanal	0.931(2)	9.65(2)	668
Heptane	0.957(5)	9.92(5)	770
1-Heptanol	0.949(3)	9.84(3)	614
2-Heptanol	0.936(3)	9.70(3)	580
3-Heptanol	0.934(3)	9.68(3)	578
4-Heptanol	0.927(3)	9.61(3)	572
2-Heptanone	0.897(1)	9.30(1)	596
1-Heptene	0.911	9.44	849
2-Heptene	0.853(2)	8.84(2)	782
3-Heptene	0.861	8.92	790
Hexachlorobenzene	0.866	8.98	822
Hexachloroethane	1.07	11.1	920
1,5-Hexadiene	0.896(5)	9.29(5)	980
Hexafluoroacetone	1.104	11.44	−294
Hexafluorobenzene	0.9558	9.906	10
Hexafluoroethane	1.29	13.4	−50
Hexafluoropropene	1.023(3)	10.60(3)	−103
Hexamethylbenzene	0.757	7.85	670
1-Hexanal	0.933(5)	9.67(5)	686
Hexane	0.977	10.13	810
Hexanoic acid	0.976	10.12	463
1-Hexanol	0.954(3)	9.89(3)	639

TABLE 2.51 Ionization Energy of Molecular and Radical Species (*Continued*)

Species	Ionization energy MJ · mol⁻¹	Ionization energy Electron volts	$\Delta_f H$ (ion), kJ · mol⁻¹
2-Hexanol	0.946(3)	9.80(3)	611
3-Hexanol	0.929(3)	9.63(3)	599
2-Hexanone	0.902(2)	9.35(2)	626
3-Hexanone	0.880(2)	9.12(2)	600
1-Hexene	0.911(4)	9.44(4)	869
cis-2-Hexene	0.865(1)	8.97(1)	818
trans-2-Hexene	0.865(1)	8.97(1)	814
Hexylamine	0.833(5)	8.63(5)	699
1-Hexyne	0.960	9.95(5)	1081
Hydrogen cyanide (HCN)	1.312(1)	13.60(1)	1447
Hydrogen isocyanide (HNC)	1.21(1)	12.5(1)	1407
p-Hydroquinone	0.767(3)	7.95(3)	504
Imidazole	0.850(1)	8.81(1)	997
Indane	0.90	9.3	864
Indene	0.785(1)	8.14(1)	949
Iodobenzene	0.8380	8.685	1003
Iodoethane	0.9018	9.346	893
1-Iodohexane	0.8857	9.179	794
Iodomethane	0.9203	9.538	936
1-Iodopropane	0.8943	9.269	862
2-Iodopropane	0.8853	9.175	844
Isobutylbenzene	0.838(1)	8.68(1)	816
Isocyanic acid	1.120(3)	11.61(3)	1016
Isophthalic acid	0.963(20)	9.98(20)	268
Isopropylcyclohexane	0.900	9.33	704
Isoquinoline	0.8239(3)	8.539(3)	1032
Isoxazole	0.958(5)	9.93(5)	1038
Ketene	0.927(2)	9.61(2)	880
Maleic anhydride	1.04	10.8	645
Mesityl oxide	0.876(3)	9.08(3)	692
Methacrylic acid	0.979	10.15	611
Methane	1.207	12.51	1133
Methanethiol	9.108(5)	9.440(5)	888
Methanol	1.047(1)	10.85(1)	845
Methoxy	0.83	8.6	845
Methoxybenzene (Anisole)	0.792(2)	8.21(2)	724
2-Methoxyethanol	0.93	9.6	562
Methyl	0.949(1)	9.84(1)	1095
Methyl acetate	0.991(2)	10.27(2)	581
Methyl acrylate	0.96	9.9	611
Methyl azide	0.947(2)	9.81(2)	1227
Methyl benzoate	0.899(3)	9.32(3)	611
Methyl chloroacetate	0.99	10.3	575
Methyl 2,2-dimethylpropanoate	0.955(4)	9.90(4)	466
Methyl formate	1.0435(5)	10.815(5)	688
Methyl pentanoate	1.00(2)	10.4(2)	532
Methyl pentyl ether	0.933	9.67	657
Methyl vinyl ether	0.862(2)	8.93(2)	761
Methylacrylonitrile	0.998	10.34	1127
Methylamine	0.865(2)	8.97(2)	843
2-Methylaniline	0.718(2)	7.44(2)	772

(*Continued*)

TABLE 2.51   Ionization Energy of Molecular and Radical Species (*Continued*)

Species	Ionization energy MJ · mol⁻¹	Ionization energy Electron volts	$\Delta_f H$ (ion), kJ · mol⁻¹
3-Methylaniline	0.724(2)	7.50(2)	778
4-Methylaniline	0.698(2)	7.24(2)	753
N-Methylaniline	0.707(2)	7.33(2)	791
Methylcyclohexane	0.930	9.64	775
1-Methylcyclohexanol	0.95(2)	9.8(2)	586
Methylcyclopentane	0.950(3)	9.85(3)	845
Methylcyclopropane	0.913	9.46	936
2-Methyldecane	0.934	9.68	685
Methylene	1.0031(3)	10.396(3)	1386
N-Methylformamide	0.945	9.79	756
2-Methylheptane	0.949	9.84	734
5-Methyl-2-hexanone	0.895(1)	9.28(1)	586
Methylhydrazine	0.740(2)	7.67(2)	835
Methylidyne	1.027(1)	10.64(1)	1622
Methylisocyanate	1.030(2)	10.67(2)	900
1-Methyl-4-isopropylbenzene (*p*-Cymene)	0.800	8.29	771
1-Methylnaphthalene	0.757	7.85	870
2-Methylnaphthalene	0.75	7.8	866
Methyloxirane	0.986(2)	10.22(2)	892
2-Methylpentane	0.976	10.12	802
3-Methylpentane	0.973	10.08	801
2-Methyl-3-pentanone	0.878(1)	9.10(1)	592
3-Methyl-2-pentanone	0.889(1)	9.21(1)	600
4-Methyl-2-pentanone	0.897(1)	9.30(1)	609
2-Methyl-1-pentene	0.876(1)	9.08(1)	817
2-Methyl-2-pentene	0.828	8.58	761
4-Methyl-1-pentene	0.912(1)	9.45(1)	862
4-Methyl-*cis*-2-pentene	0.866(1)	8.98(1)	809
4-Methyl-*trans*-2-pentene	0.865(1)	8.97(1)	804
2-Methylpropanal	0.9364(5)	9.705(5)	721
2-Methylpropanenitrile	1.09	11.3	1115
2-Methylpropenal	0.951	9.86	834
2-Methylpropene (Isobutene)	0.8915(3)	9.239(3)	875
2-Methylpyridine	0.870(3)	9.02(3)	970
3-Methylpyridine	0.872(3)	9.04(3)	979
4-Methylpyridine	0.872(3)	9.04(3)	976
Methylsilane	1.03	10.7	1003
m-Methylstyrene	0.786(2)	8.15(2)	908
o-Methylstyrene	0.888(2)	9.20(2)	908
p-Methylstyrene	0.78(1)	8.1(1)	895
Methyltrichlorosilane	1.096(3)	11.36(3)	548
Naphthalene	0.785(1)	8.14(1)	936
1-Naphthol	0.749(3)	7.76(3)	719
2-Naphthol	0.757(5)	7.85(5)	727
Nickel carbonyl	0.798(4)	8.27(4)	200
m-Nitroaniline	0.802(2)	8.31(2)	865
o-Nitroaniline	0.798(1)	8.27(1)	861
p-Nitroaniline	0.804(1)	8.34(1)	850
Nitrobenzene	0.951(2)	9.86(2)	1019
Nitroethane	1.050(5)	10.88(5)	948

TABLE 2.51 Ionization Energy of Molecular and Radical Species (*Continued*)

Species	Ionization energy		$\Delta_f H$ (ion), kJ · mol^{-1}
	MJ · mol^{-1}	Electron volts	
Nitromethane	1.063(4)	11.02(4)	988
m-Nitrophenol	0.86	9.0	755
o-Nitrophenol	0.88	9.1	782
p-Nitrophenol	0.88	9.1	761
1-Nitropropane	1.043(3)	10.81(3)	919
2-Nitropropane	1.033(5)	10.71(5)	894
m-Nitrotoluene	0.15(2)	9.48(2)	944
o-Nitrotoluene	0.912(4)	9.45(4)	966
p-Nitrotoluene	0.91	9.4	936
Nonane	0.938	9.72	710
2-Nonanone	0.884	9.16	545
5-Nonanone	0.875	9.07	530
Octafluoronaphthalene	0.854	8.85	−368
Octafluoropropane	1.291	13.38	−491
Octafluorotoluene	0.96	9.9	−233
Octane	0.948	9.82	739
1-Octene	0.910(1)	9.43(1)	829
1-Octyne	0.960(2)	9.95(2)	1040
2-Octyne	0.898(1)	9.31(1)	961
3-Octyne	0.890(1)	9.22(1)	952
4-Octyne	0.888(1)	9.20(1)	946
Oxazole	0.93	9.6	910
Oxetane	0.9328(5)	9.668(5)	853
2-Oxetanone	0.936(1)	9.70(1)	653
Oxomethyl (HCO)	0.782(5)	8.10(5)	826
Pentafluorobenzene	0.929	9.63	122
Pentafluorophenol	0.888(2)	9.20(2)	−71
2,3,4,5,6-Pentafluorotoluene	0.91	9.4	64
Pentanchloroethane	1.06	11.0	919
Pentylamine	0.837	8.67	728
Perylene	0.666(1)	6.90(1)	975
Phenanthrene	0.758(2)	7.86(2)	963
Phenetole	0.784(2)	8.13(2)	683
Phenol	0.817	8.47	721
Phenylacetic acid	0.797	8.26	479
m-Phenylenediamine	0.689	7.14	777
o-Phenylenediamine	0.69	7.2	787
p-Phenylenediamine	0.663(5)	6.87(5)	759
Phthalic anhydride	0.96	10.0	593
α-Pinene	0.779	8.07	808
Propanal	0.9603(5)	9.953(5)	773
Propanamide	0.92	9.5	720
Propane	1.057(5)	10.95(5)	952
Propanenitrile	1.142(2)	11.84(2)	1194
1-Propanethiol	0.8872(5)	9.195(5)	819
2-Propanethiol	0.882	9.14	806
Propanoic acid	1.0155(3)	10.525(3)	568
1-Propanol	0.986(3)	10.22(3)	731
2-Propanol	0.976(8)	10.12(8)	704

(*Continued*)

TABLE 2.51 Ionization Energy of Molecular and Radical Species (*Continued*)

Species	Ionization energy		$\Delta_f H$ (ion), kJ · mol^{-1}
	MJ · mol^{-1}	Electron volts	
Propenal	0.975(6)	10.103(6)	900
Propene	0.939(2)	9.73(2)	959
Propenenitrile	1.053(1)	10.91(1)	1237
Propenoic acid	1.023	10.60	701
1-Propylamine	0.847(2)	8.78(2)	777
2-Propylamine	0.841(3)	8.72(3)	758
Propylbenzene	0.841(1)	8.72(1)	849
Propylcyclohexane	0.913	9.46	720
Propylcyclopentane	0.965(4)	10.00(4)	817
Propyleneimine	0.87	9.0	960
Propynal	1.04	10.8	1155
Propyne	1.000(1)	10.36(1)	1186
2-Propyn-1-ol	1.014	10.51	1060
Pyrene	0.715	7.41	933
Pyridazine	0.834	8.64	1112
Pyrimidine	0.891	9.23	1087
Pyrrole	0.7920(5)	8.208(5)	900
2-Pyrrolidone	0.89	9.2	674
Quinoline	0.832(1)	8.62(1)	1041
cis-Stilbene	0.753(2)	7.80(2)	1005
trans-Stilbene	0.743(3)	7.70(3)	977
Styrene	0.813(6)	8.43(6)	961
Succinic anhydride	1.02	10.6	500
Succinonitrile	1.158(24)	12.10(25)	1377
Terephthalic acid	0.951(20)	9.86(20)	232
m-Terphenyl	0.773(1)	8.01(1)	1057
o-Terphenyl	0.77	8.0	1056
p-Terphenyl	0.751(1)	7.78(1)	1035
Tetrabromomethane	0.995(2)	10.31(2)	1079
Tetrachloro-1,2-difluoroethane	1.09	11.3	563
1,1,1,2-Tetrachloroethane	1.07	11.1	920
1,1,2,2-Tetrachloroethane	1.121	11.62	971
Tetrachloroethylene	0.899	9.32	887
Tetrachloromethane	1.107(1)	11.47(1)	1011
Tetraethylsilane	0.86	8.9	595
1,2,3,4-Tetrafluorobenzene	0.920(1)	9.53(1)	284
1,2,3,5-Tetrafluorobenzene	0.920(1)	9.53(1)	263
1,2,4,5-Tetrafluorobenzene	0.902(1)	9.35(1)	254
Tetrafluoroethylene	0.976(2)	10.12(2)	315
Tetrahydrofurane	0.908(2)	9.41(2)	724
1,2,3,4-Tetrahydronaphthalene	0.817	8.47	842
1,2,4,5-Tetramethylbenzene	0.776(1)	8.04(1)	730
2,2,3,3-Tetramethylbutane	0.95	9.8	720
Thiacyclobutane	0.838	8.69	899
Thiophene	0.856(4)	8.87(4)	971
p-Tolualdehyde	0.900(5)	9.33(5)	825
Toluene	0.851(1)	8.82(1)	901
m-Toluic acid	0.910(20)	9.43(20)	579
o-Toluic acid	0.88	9.1	558
p-Toluic acid	0.891(20)	9.23(20)	560

TABLE 2.51 Ionization Energy of Molecular and Radical Species (*Continued*)

Species	Ionization energy MJ · mol^{-1}	Ionization energy Electron volts	$\Delta_f H$ (ion), kJ · mol^{-1}
m-Tolunitrile	0.901	9.34	1085
o-Tolunitrile	0.905	9.38	1085
p-Tolunitrile	0.899	9.32	1083
Tribromomethane	1.011(2)	10.48(2)	1035
Tributylamine	0.71	7.4	492
Trichloroacetyl chloride	1.06	11.0	827
1,2,4-Trichlorobenzene	0.872	9.04	880
1,3,5-Trichlorobenzene	0.899(2)	9.32(2)	899
1,1,1-Trichloroethane	1.06	11.0	917
1,1,2-Trichloroethane	1.06	11.0	911
Trichloroethylene	0.914(1)	9.47(1)	895
Trichlorofluoromethane	1.136(2)	11.77(2)	868
Trichloromethane	1.097(2)	11.37(2)	992
Trichloromethylbenzene	0.926	9.60	914
1,1,2-Trichlorotrifluoroethane	1.157(2)	11.99(2)	429
Triethanolamine	0.76	7.9	206
Triethylamine	0.724	7.50	631
Trifluoroacetic acid	1.106	11.46	75
Trifluoroacetonitrile	1.337	13.86	838
1,1,1-Trifluoro-2-bromo-2-chloroethane	1.06	11.0	362
1,1,1-Trifluoroethane	1.24(1)	12.9(1)	496
Trifluoroethylene	0.978	10.14	489
Trifluoroiodomethane	0.987	10.23	397
Trifluoromethane	1.337	13.86	643
Trifluoromethyl (CF$_3$)	0.86	8.9	399
Trifluoromethylbenzene	0.9345(4)	9.685(4)	335
3,3,3-Trifluoropropene	1.05	10.9	437
Triiodomethane	0.893(2)	9.25(2)	1010
Trimethylamine	0.755462	7.82960	731
1,2,3-Trimethylbenzene	0.812(2)	8.42(2)	803
1,2,4-Trimethylbenzene	0.798(1)	8.27(1)	784
1,3,5-Trimethylbenzene	0.811(1)	8.41(1)	796
Trimethylborate	0.96	10.0	65
Trimethylchlorosilane	0.979	10.15	624
3,5,5-Trimethylcyclohex-2-en-1-one	0.875	9.07	670
2,2,4-Trimethylpentane	0.951	9.86	713
2,2,4-Trimethyl-3-pentanone	0.849(1)	8.80(1)	511
2,4,6-Trimethylpyridine	0.88(1)	8.9(1)	580
Trioxane	0.99	10.3	528
Undecane	0.922	9.56	650
Urea	0.94	9.7	690
Vinyl acetate	0.887	9.19	572
m-Xylene	0.826(1)	8.56(1)	843
o-Xylene	0.826(1)	8.56(1)	844
p-Xylene	0.814(1)	8.44(1)	832
2,3-Xylenol	0.797	8.26	640
2,4-Xylenol	0.77	8.0	609
2,6-Xylenol	0.777(2)	8.05(2)	615
3,4-Xylenol	0.781	8.09	624

2.12 THERMAL CONDUCTIVITY

TABLE 2.52 Thermal Conductivities of Gases as a Function of Temperature

The coefficient k, expressed in $J \cdot sec^{-1} \cdot cm^{-1} \cdot K^{-1}$, is the quantity of heat in joules, transmitted per second through a sample one centimeter in thickness and one square centimeter in area when the temperature difference between the two sides is one degree kelvin (or Celsius). The tabulated values are in microjoules.

Substance	Temperature, °C										
	−40	−20	0	20	40	60	80	100	120	140	160
Acetone		80	95	107	124	140	156	173	190	207	
Acetaldehyde				109	126	142	159	176	195		
Acetonitrile						112	124	137	151	166	
Acetylene			184	205	224	248	269	290			
Air	118^{-75}		242	256	270	284	299	311	324	336	342^{149}
Ammonia	164^{-60}		218	238	259	280	301	321			
Argon			166	176	186	196	206	211		205	266
Benzene						126	146	165	184	241	
Boron trifluoride				186							
Bromine			42	45	50	54	59				
Bromomethane					82	94	104	117			
1-Butanamine			$135^{6.5}$					176^{110}			
Butane			135	154	174	193	213	233			
Carbon dioxide			144	160	176	192	207	215			
Carbon disulfide			67	76	85						
Carbon monoxide			228	245	262	278					
Carbon tetrachloride			59	64	70	75	80	86			109^{184}
Chlorine	64	72	79	85	93	100					
Chlorodifluoromethane		103	110	116	122						
Chloroethane			90	105	120	134	151	167	186	204	
Chloroform					75	84	91	99	107	116	
Chloromethane			84	105	117	130	142	155			
Cyclohexane			77	99	120	141	163				
Cyclopropane											
2-Methyl-2-propanol								225			
Neon	410	433	454	476	497	518	537	556			
Nitric oxide	205	221	238	254	269	285	301	317			
Nitrogen	211	226	241	256	270	282	295	307	320	333	385^{227}

Nitromethane	121	137	152	168	184					139	155
Nitrous oxide				120						190	
Octafluorocyclobutane											
Oxygen	211	228	245	261	278	294	311	328			
Pentane			130					218			
Propane	116	132	151	171	192	215	238	262	330	353	379
2-Propanol				151[31]						250[127]	
Sulfur dioxide			83		163		106				
Sulfur hexafluoride				126			201			275[227]	338[327]
Tetrafluoromethane				235			235				
Thiophene				87			152[110]				
1,1,2-Trichlorotrifluoroethane							133				
Triethylamine				195	216		241		239		
Water	36[−73]	142	159	175	191	207	224	257	239		
Xenon			54			72				89[227]	104[327]
Deuterium	1150	1222	1297	1372	1448	1523					
Deuterium oxide				263						358[220]	
Dibromomethane				74[110]						194[200]	
Dichlorodifluoromethane		81	84	92	100		138				
1,1-Dichloroethane			69	81	93		129	144			
1,2-Dichloroethane					105	117	127	140			
Dichlorofluoromethane		91	94	97	100		161				
Dichloromethane			93	99							
1,2-Dichlorotetrafluoroethane							153			211[227]	
Diethylamine			118	153	179	199	218	243	268		
Diethyl ether			113	135	178	200	222	244	269	351[213]	
1,4-Dioxane				157	167	187	207				
Ethane	137	159	182	204	228	257	288	316			
Ethanol			126	141	155	209	344				
Ethene				230[49]							
Ethyl acetate			115	133	151	170	191	211	234		
Ethylamine			136	153	169	206					
Ethylene	137	158	178	220	241	262	282				
Ethylene oxide			193	256	279						
Ethyl formate			100	142	164	186	206	226			
Ethyl nitrate			121	159	178	197					
Fluorine	212	230	247	264	278	294	309	325			
Helium	1276	1343	1423	1481	1540	1598	1661	1720	1778		
Heptane	100	115	130	174							

(Continued)

TABLE 2.52 Thermal Conductivities of Gases as a Function of Temperature (*Continued*)

	Temperature, °C										
Substance	−40	−20	0	20	40	60	80	100	120	140	160
Hexane			109				178	201	224	247	271
Hydrogen	1494	1607	1724	1828	1925	2025					
Hydrogen bromide	64	70	77	84	90	97	104				
Hydrogen chloride	107	117	128	138	148				191		240[227]
Hydrogen cyanide			110	121	132	143					
Hydrogen sulfide		116	129	143	156	169					
Iodomethane			46	53	60	68	75	82	89		
Krypton		79	85		95			110			
Methane	257	280	307	334	361	387	416	445			
Methanol						174	197	221	241	263	284
Methyl acetate						150[70]		177	195	215	237
2-Methylbutane			122					215			
2-Methylpropane			141	156	176	196	233[93]	271			421

TABLE 2.53   Thermal Conductivity of Various Substances

All values of thermal conductivity, k, are in millijoules $cm^{-1} \cdot s^{-1} \cdot K^{-1}$.

Substance	\multicolumn{7}{c}{Thermal conductivity in $mJ \cdot cm^{-1} \cdot s^{-1} \cdot K^{-1}$}
	−25°C	0°C	20°C	25°C	50°C	75°C	100°C
Acetaldehyde			1.900				
Acetic acid				1.58	1.53	1.49	1.44
Acetic anhydride			2.209				
Acetone	1.987[−80]	1.69	1.61		1.51[40]		
Acetonitrile	2.08	1.98		1.88	1.78	1.68	
Allyl alcohol				1.80[30]			
Aniline			1.77[17]				
Argon	1.259[−189]						
Benzaldehyde				1.51	1.41	1.31	1.21
Benzene				1.411	1.329	1.247	
Bromobenzene			1.113				
Bromoethane			1.029				
1-Bromo-2-methylpropane		1.163[12]					
1-Bromopentane			0.983				
Bromopropane		1.075[12]					
Butanoic acid		1.506[12]					
1-Butanol		1.538		1.54	1.49		
2-Butanone	1.58	1.51		1.45	1.39	1.33	
Butyl acetate			1.368				
2-Butyne	1.37	1.29		1.21			
Carbon disulfide		1.54		1.49			
Carbon tetrachloride	1.100[−20]	1.071	1.029		0.974		
Chlorobenzene	1.36	1.31		1.27	1.22	1.17	1.12
Chloroethane	1.45	1.32		1.19	1.06	0.93	
Chloroform	1.27	1.22		1.17	1.12	1.07	1.02
(Chloromethyl)oxirane	1.42	1.37		1.31	1.25	1.19	1.14
1-Chloro-2-methylpropane		1.163[12]					
1-Chloropentane		1.184[12]					
Chloropropane		1.184[12]					
4-Chlorotoluene			1.297				
m-Cresol			1.498			1.452[80]	
Cyclohexane			1.243	1.23	1.17	1.11	
Cyclohexene	1.42	1.36		1.30	1.24	1.18	
Cyclohexanol				1.34	1.31		
Cyclopentane	1.40	1.33		1.26			
Cyclopentene	1.43	1.36		1.29			
Decane	1.44	1.38		1.32	1.26	1.19	1.13
1-Decanol				1.62	1.56	1.50	1.45
Dibromomethane	1.20	1.14		1.08	1.03	0.97	
Dibutyl phthalate	1.44	1.40		1.36	1.33	1.29	1.25
1,2-Dichloroethane		1.264					
Dichlorofluoromethane	0.134						
Dichloromethane	1.590[−20]	1.564	1.477				
Diethyl ether	1.50	1.40		1.30	1.20	1.10	1.00
Diisopropyl ether			1.096				
2,3-Dimethylbutane				1.038[32]	0.996		
N,N-Dimethylformamide				1.84	1.78	1.71	1.65
Dimethyl phthalate		1.501		1.473	1.443	1.409	1.373

(*Continued*)

TABLE 2.53   Liquid Thermal Conductivity of Various Substances (*Continued*)

Substance	Thermal conductivity in mJ · cm^{-1} · s^{-1} · K^{-1}						
	−25°C	0°C	20°C	25°C	50°C	75°C	100°C
1,4-Dioxane				1.59	1.47	1.35	1.23
Diphenyl ether					1.39	1.35	1.31
Dodecane		1.57		1.52	1.46	1.40	1.35
1-Dodecanol				1.46	1.42	1.39	1.35
Ethanol		1.76		1.69	1.62		
Ethanolamine				2.99	2.86	2.74	2.61
Ethoxybenzene			1.497				
Ethyl acetate	1.62	1.53		1.44	1.35	1.26	
Ethylbenzene				1.30	1.24	1.18	1.12
Ethylene glycol		2.56		2.56	2.56	2.56	2.56
Ethyl formate		1.581[12]					
Furan	1.42	1.34		1.26			
Glycerol				2.92	2.95	2.97	3.00
Heptane	1.378	1.303	1.259	1.228	1.152	1.077	
1-Heptanol		1.66		1.59	1.53	1.47	1.41
Hexadecane				1.40	1.35	1.30	1.25
Hexane	1.37	1.28	1.218	1.20	1.11	1.92	0.93
1-Hexanol	1.59	1.54		1.50	1.45	1.41	1.37
2-Hexanone	1.51	1.45		1.39	1.33	1.27	1.21
1-Hexene	1.37	1.29		1.21	1.13		
Hydrochloric acid, 38%			4.402[32]				
Hydrogen	1.180[−253]						
Iodobenzene	1.063[−20]		1.276			0.937[80]	
Iodoethane				1.109[30]			
1-Iodo-2-methylpropane		0.870[12]					
1-Iodopentane		0.849[12]					
Iodopropane		0.920[12]					
Isopentyl acetate			1.297				
Isopropylbenzene				1.28	1.20	1.12	1.07
Mercury	72.5	77.7		82.5	86.8	90.7	94.3
Methanol	2.14	2.07	2.021	2.00	1.93		
Methoxybenzene	1.70	1.63		1.56	1.50	1.43	1.36
Methyl acetate	1.74	1.64		1.53	1.43	1.33	1.22
Methyl butanoate			1.402				
3-Methylbutanoic acid		1.305					
3-Methyl-1-butanol				1.477[30]			
Methylcyclohexane				1.276[30]			
Methylcyclopentane				1.209	1.151[38]		
N-Methylformamide				2.03	2.01	1.99	1.96
1-Methyl-4-isopropylbenzene	1.32	1.27		1.22	1.17	1.12	1.07
2-Methylpentane				1.084[32]	1.033		
Methyl pentanoate		1.318[12]					
4-Methylpentanoic acid		1.427[12]					
4-Methyl-3-pentene-2-one	1.70	1.63		1.56	1.49	1.42	1.34
2-Methyl-1-propanol		1.423[12]					
2-Methyl-2-propanol				1.159[38]		1.067[77]	
Nitrobenzene			1.510				
Nitromethane				2.151[30]			
Nonane	1.44	1.38		1.31	1.24	1.151[80]	1.11

TABLE 2.53 Liquid Thermal Conductivity of Various Substances (*Continued*)

Substance	Thermal conductivity in mJ · cm^{-1} · s^{-1} · K^{-1}						
	−25°C	0°C	20°C	25°C	50°C	75°C	100°C
1-Nonanol		1.66		1.61	1.55	1.49	1.43
Octadecane					1.46	1.42	1.37
Octane	1.43	1.35		1.28	1.20	1.13	1.06
1-Octanol		1.68	1.657	1.61	1.54	1.47	1.41
Palmitic acid						1.598	
Pentachloroethane			1.251				
Pentane	1.32	1.22	1.138	1.13	1.03	0.95	0.87
Pentanoic acid		1.360[12]					
1-Pentanol		1.57		1.53	1.49	1.45	
1-Pentene	1.31	1.24		1.16			
Pentyl acetate			1.289				
Phenol					1.56	1.53	1.51
Phenylhydrazine				1.724			
1,2-Propanediol		2.02		2.00	1.99	1.98	1.97
Propanoic acid		1.728[12]					
1-Propanol	1.62	1.58		1.54	1.49	1.45	1.41
2-Propanol	1.46	1.41		1.35	1.29	1.24	1.18
1,2-Propylene glycol		2.008					
Propyl formate		1.494[12]					
Pyridine		1.69		1.65	1.61	1.58	
Silicon tetrachloride				0.99	0.96		
Sodium							753.1[300]
Sodium chloride (aq, satd)	5.732						
Stearic acid						1.598	
Styrene	1.48	1.42		1.37	1.31	1.26	1.20
Sulfuric acid, 90%				3.540[32]			
1,1,2,2-Tetrachloroethane		1.138					
Tetrachloroethylene	1.17		1.10	1.04	0.97		
Tetrachloromethane	1.04		0.99	0.93	0.88		
Tetradecane				1.36	1.31	1.26	1.21
1-Tetradecanol					1.67	1.62	1.57
Tetrahydrofuran	1.32	1.26		1.20	1.14		
Thiophene				1.99	1.95	1.91	1.86
Toluene	1.590[−80]	1.386	1.347	1.311	1.236	1.161	
1,1,1-Trichloroethane	1.06		1.01	0.96			
Trichloroethylene	1.35[−60]	1.24		1.160	1.08	1.00	
Trichloromethane	1.27	1.22		1.17	1.12	1.07	
Tridecane				1.37	1.32	1.27	1.22
Triethylamine	1.464[−80]		1.209		1.113[44]		
Trimethylamine	1.43	1.33					
1,3,5-Trimethylbenzene	1.47	1.41		1.36	1.30	1.24	1.18
2,2,4-Trimethylpentane				0.966[38]		0.841[77]	
Undecane				1.40	1.35	1.29	1.23
Water		5.610	5.983	6.071	6.435	6.668	6.791
m-Xylene				1.30	1.24	1.18	1.13
o-Xylene				1.31	1.26	1.20	1.14
p-Xylene				1.30	1.24	1.18	1.12

2.13 ENTHALPIES AND GIBBS ENERGIES OF FORMATION, ENTROPIES, AND HEAT CAPACITIES (CHANGE OF STATE)

The tables in this section contain values of the enthalpy and Gibbs energy of formation, entropy, and heat capacity at 298.15 K (25°C). No values are given in these tables for metal alloys or other solid solutions, for fused salts, or for substances of undefined chemical composition.

The physical state of each substance is indicated in the column headed "State" as crystalline solid (c), liquid (lq), or gaseous (g). Solutions in water are listed as aqueous (aq).

The values of the thermodynamic properties of the pure substances given in these tables are, for the substances in their standard states, defined as follows: For a pure solid or liquid, the standard state is the substance in the condensed phase under a pressure of 1 atm (101, 325 Pa). For a gas, the standard state is the hypothetical ideal gas at unit fugacity, in which state the enthalpy is that of the real gas at the same temperature and at zero pressure.

The values of $\Delta_f H°$ and $\Delta_f G°$ that are given in the tables represent the change in the appropriate thermodynamic quantity when one mole of the substance in its standard state is formed, isothermally at the indicated temperature, from the elements, each in its appropriate standard reference state. The standard reference state at 25°C for each element has been chosen to be the standard state that is thermodynamically stable at 25°C and 1 atm pressure. The standard reference states are indicated in the tables by the fact that the values of $\Delta_f H°$ and $\Delta_f G°$ are exactly zero.

The values of $S°$ represent the virtual or "thermal" entropy of the substance in the standard state at 298.15 K (25°C), omitting contributions from nuclear spins. Isotope mixing effects are also excluded except in the case of the ^{1}H—^{2}H system.

Solutions in water are designated as aqueous, and the concentration of the solution is expressed in terms of the number of moles of solvent associated with 1 mol of the solute. If no concentration is indicated, the solution is assumed to be dilute. The standard state for a solute in aqueous solution is taken as the hypothetical ideal solution of unit molality (indicated as std. state or ss). In this state the partial molal enthalpy and the heat capacity of the solute are the same as in the infinitely dilute real solution.

For some tables the uncertainty of entries is indicated within parentheses immediately following the value; viz., an entry 34.5(4) implies 34.5 ± 0.4 and an entry 34.5(12) implies 34.5 ± 1.2.

References: D. D. Wagman, et al., *The NBS Tables of Chemical Thermodynamic Properties*, in *J. Phys. Chem. Ref. Data*, **11: 2,** 1982; M. W. Chase, et al., *JANAF Thermochemical Tables*, 3rd ed., American Chemical Society and the American Institute of Physics, 1986 (supplements to JANAF appear in *J. Phys. Chem. Ref. Data*); Thermodynamic Research Center, *TRC Thermodynamic Tables*, Texas A&M University, College Station, Texas; I. Barin and O. Knacke, *Thermochemical Properties of Inorganic Substances*, Springer-Verlag, Berlin, 1973; J. B. Pedley, R. D. Naylor, and S. P. Kirby, *Thermochemical Data of Organic Compounds*, 2nd ed., Chapman and Hall, London, 1986; V. Majer and V. Svoboda, *Enthalpies of Vaporization of Organic Compounds*, International Union of Pure and Applied Chemistry, Chemical Data Series No. 32, Blackwell, Oxford, 1985.

2.13.1 Thermodynamic Relations

Enthalpy of Formation. Once standard enthalpies are assigned to the elements, it is possible to determine standard enthalpies for compounds. For the reaction:

$$C(\text{graphite}) + O_2(g) \rightarrow CO_2(g) \qquad \Delta H° = -393.51 \text{ kJ} \tag{6.1}$$

Since the elements are in their standard states, the enthalpy change for the reaction is equal to the standard enthalpy of CO_2 less the standard enthalpies of C and O_2, which are zero in each instance. Thus,

$$\Delta_f H° = -393.51 - 0 - 0 = -393.51 \text{ kJ} \tag{6.2}$$

Tables of enthalpies, such as Tables 2.54 and 1.58, can be used to determine the enthalpy for any reaction at 1 atm and 298.15 K involving the elements and any of the compounds appearing in the tables.

The solution of 1 mole of HCl gas in a large amount of water (infinitely dilute real solution) is represented by:

$$HCl(g) + \text{inf } H_2O \rightarrow H^+(aq) + Cl^-(aq) \tag{6.3}$$

The heat evolved in the reaction is $\Delta H° = -74.84$ kJ. With the value of $\Delta_f H°$ from Table 2.54, one has for the reaction:

$$\Delta_f H° = \Delta_f H°[H^+ (aq)] + \Delta_f H°[Cl^-(aq)] - \Delta_f H°[HCl(g)]$$

for the standard enthalpy of formation of the pair of ions H^+ and Cl^- in aqueous solution (standard state, $m = 1$). To obtain the $\Delta_f H°$ values for individual ions, the enthalpy of formation of $H^+(aq)$ is arbitrarily assigned the value zero at 298.15 K. Thus, from Eq. (6.4):

$$\Delta_f H°[Cl^-(aq)] = -74.84 + (-92.31) = -167.15 \text{ kJ}$$

With similar data from Tables 2.54 and 1.58, the enthalpies of formation of other ions can be determined. Thus, from the $\Delta_f H°[KCl(aq, std.\ state,\ m = 1\ or\ aq,\ ss)]$ of -419.53 kJ and the foregoing value for $\Delta_f H°[Cl^-(aq, ss)]$:

$$\Delta_f H°[K^+(aq,\ ss)] = \Delta_f H°[KCl(aq,\ ss)] - \Delta_f H°[Cl^-(aq, ss)]$$

$$= -419.53 - (-167.15) = -252.38 \text{ kJ}$$

Enthalpy of Vaporization (or Sublimation). When the pressure of the vapor in equilibrium with a liquid reaches 1 atm, the liquid boils and is completely converted to vapor on absorption of the enthalpy of vaporization ΔHv at the normal boiling point T_b. A rough empirical relationship between the normal boiling point and the enthalpy of vaporization (*Trouton's rule*) is:

$$\frac{\Delta Hv}{T_b} = 88 \text{ J} \cdot \text{mol}^{-1} \cdot \text{K}^{-1}$$

It is best applied to nonpolar liquids which form unassociated vapors.

To a first approximation, the enthalpy of sublimation ΔHs at constant temperature is:

$$\Delta Hs = \Delta Hm + \Delta Hv$$

where ΔHm is the enthalpy of melting.

The *Clapeyron* equation expresses the dynamic equilibrium existing between the vapor and the condensed phase of a pure substance:

$$\frac{dP}{dT} = \frac{\Delta Hv}{T\Delta V}$$

where ΔV is the volume increment between the vapor phase and the condensed phase. If the condensed phase is solid, the enthalpy increment is that of sublimation.

Substitution of $V = RT/P$ into the foregoing equation and rearranging gives the *Clausius–Clapeyron* equation,

$$\frac{dP}{p\,dT} = \frac{\Delta Hv}{RT^2}$$

or

$$\Delta Hv = -R\frac{d(\ln P)}{1/T}$$

which may be used for calculating the enthalpy of vaporization of any compound provided its boiling point at any pressure is known. If an Antoine equation is available, differentiation and insertion into the foregoing equation gives:

$$\Delta Hv = \frac{4.5757T^2 B}{(T + C - 273.15)^2}$$

Inclusion of a compressibility factor into the foregoing equation, as suggested by the *Haggenmacher* equation improves the estimate of ΔHv:

$$\Delta Hv = \frac{RT^2}{P}\left(\frac{dP}{dT}\right)\left(1 - \frac{T_c^3 P}{T^3 P_c}\right)^{1/2}$$

where T_c and P_c are critical constants (Table 2.56). Although critical constants may be unknown, the compressibility factor is very nearly constant for all compounds belonging to the same family, and an estimate can be deduced from a related compound whose critical constants are available.

Heat Capacity (or Specific Heat).   The temperature dependence of the heat capacity is complex. If the temperature range is restricted, the heat capacity of any phase may be represented adequately by an expression such as:

$$C_p = a + bT + cT^2$$

in which a, b, and c are empirical constants. These constants may be evaluated by taking three pieces of data: $(T_1, C_{p,1})$, $(T_2, C_{p,2})$, and $(T_3, C_{p,1})$, and substituting in the following expressions:

$$\frac{C_{p,1}}{(T_1 - T_2)(T_1 - T_3)} + \frac{C_{p,2}}{(T_2 - T_1)(T_2 - T_3)} + \frac{C_{p,3}}{(T_3 - T_2)(T_3 - T_1)} = c$$

$$\frac{C_{p,1} - C_{p,2}}{T_1 - T_2} - [(T_1 + T_2)c] = b$$

$$(C_{p,1} - bT_1) - cT_1^2 = a$$

Smoothed data presented at rounded temperatures, such as are available in Tables 2.55 and 1.59, plus the C_p° values at 298 K listed in Table 2.54, are especially suitable for substitution in the foregoing parabolic equations. The use of such a parabolic fit is appropriate for interpolation, but data extrapolated outside the original temperature range should not be sought.

Enthalpy of a System.   The enthalpy increment of a system over the interval of temperature from T_1 to T_2, under the constraint of constant pressure, is given by the expression:

$$H_2 - H_1 = \int_{T_1}^{T_2} C_p \, dT$$

The enthalpy over a temperature range that includes phase transitions, melting, and vaporization, is represented by:

$$H_2 - H_1 = \int_{T_1}^{T_2} C_p(c, \mathrm{II}) \, dT + \Delta Ht + \int_{T_1}^{T_m} C_p(c, \mathrm{I}) \, dT + \Delta Hm$$

$$+ \int_{T_m}^{T_b} C_p(\mathrm{lq}) \, dT + \Delta Hv + \int_{T_b}^{T_2} C_p(g) \, dT$$

Integration of heat capacities, as expressed by Eq. (6.13), leads to:

$$\Delta H = a(T_2 - T_1) + \frac{b(T_2^2 - T_1^2)}{2} + \frac{c(T_2^3 - T_1^3)}{3}$$

Entropy. In the physical change of state,

$$\Delta Sm = \frac{\Delta Hm}{T_m}$$

is the entropy of melting (or fusion),

$$\Delta Sv = \frac{\Delta Hv}{T_b}$$

is the entropy of vaporization, and

$$\Delta Ss = \frac{\Delta Hs}{Ts}$$

is the entropy of sublimation

A general expression for the entropy of a system, involving any phase transitions, is

$$S_2 - S_1 = \int_{T_1}^{T_t} \frac{C_p(c,\mathrm{II})\,dT}{T} + \frac{\Delta Ht}{T} + \int_{T_t}^{T_m} \frac{C_p(c,\mathrm{I})\,dT}{T} + \frac{\Delta Hm}{T}$$

$$+ \int_{T_m}^{T_b} \frac{C_p(\mathrm{lq})\,dT}{T} + \frac{\Delta Hv}{T} + \int_{T_b}^{T_m} \frac{C_p(\mathrm{g})\,dT}{T}$$

If C_p is independent of temperature,

$$\Delta S = C_p(\ln\,T_2 - \ln T_1) = 2.303\,C_p \log\frac{T_2}{T_1}$$

If the heat capacities change with temperature, an empirical equation may be inserted in before integration. Usually the integration is performed graphically from a plot of either C_p/T versus T or C_p versus $\ln T$.

TABLE 2.54 Enthalpies and Gibbs Energies of Formation, Entropies, and Heat Capacities of Organic Compounds

Substance	Physical state	$\Delta_f H°$ kJ · mol^{-1}	$\Delta_f G°$ kJ · mol^{-1}	$S°$ J · deg^{-1} · mol^{-1}	$C_p°$ J · deg^{-1} · mol^{-1}
Acenaphthene	c	70.34		188.9	190.4
Acenaphthylene	c	186.7			166.4
Acetaldehyde	lq	−192.2	−127.6	160.4	89.0
	g	−166.1	−133.0	263.8	55.3
Acetaldoxime	c	−77.9			
	lq	−81.6			
Acetamide	c	−317.0		115.0	91.3
Acetamidoguanidine nitrate	c	−494.0			
1-Acetamido-2-nitroguanidine	c	−193.6			
5-Acetamidotetrazole	c	−5.0			
Acetanilide	c	−210.6			
Acetic acid	lq	−484.4	−390.2	159.9	123.6
	g	−432.2	−374.2	283.5	63.4
ionized; std. state, $m = 1$	aq	−486.34	−369.65	86.7	−6.3
Acetic anhydride	lq	−624.4	−489.14	268.8	168.2[30]

(Continued)

TABLE 2.54 Enthalpies and Gibbs Energies of Formation, Entropies, and Heat Capacities of Organic Compounds (*Continued*)

Substance	Physical state	$\Delta_f H°$ kJ · mol^{-1}	$\Delta_f G°$ kJ · mol^{-1}	$S°$ J · deg^{-1} · mol^{-1}	$C_p°$ J · deg^{-1} · mol^{-1}
Acetone	lq	−248.4	−152.7	198.8	126.3
	g	−217.1	−152.7	295.3	74.5
Acetonitrile	lq	31.4	86.5	149.7	91.5
	g	74.0	91.9	243.4	52.2
Acetophenone	lq	−142.5	−17.0	249.6	204.6
Acetyl bromide	lq	−223.5			
Acetyl chloride	lq	−272.9	−208.2	201.0	117.0
	g	−242.8	−205.8	295.1	67.8
Acetylene	g	227.4	209.0	201.0	44.1
Acetylene-d_2	g	221.5	205.9	208.9	49.3
Acetylenedicarboxylic acid	c	−578.2			
Acetyl fluoride	g	−442.1			
1-Acetylimidazole	c	−574.0			
Acetyl iodide	lq	−163.5			
Acridine	c	179.4			
Adamantane	c	−194.1			
Adenine	c	96.0	299.6	151.1	147.0
(+)-Alanine	c	−561.2	−369.4	132.3	
(−)-Alanine	c	−604.0	−370.5	129.3	
(±)-Alanine	c	−563.6	−372.3	132.3	
β-Alanine	c	−558.0			
(±)-N-Alanylglycine	c	−777.8	−489.9	213.5	
(−)-Alanylglycine	c	−827.0	−533.0	195.2	
Allene	g	190.5			
Alloxan monohydrate	c	−1000.7	−762.3	186.7	
Allylamine	lq	−10.0			
Allyl *tert*-butyl sulfide	lq	−91.0			
Allyl ethyl sulfone	lq	−406.0			
Allyl methyl sulfone	lq	−385.1			
Allyl trichloroacetate	lq	−395.3			
Allyl (*see* Propene)					
Aminetrimethylboron	c	−284.1	−79.3	218.0	
3-Aminoacetophenone	c	−173.3			
4-Aminoacetophenone	c	−182.1			
2-Aminoacridine	c	166.4			
9-Aminoacridine	c	159.2			
2-Aminobenzoic acid	c	−400.9			
3-Aminobenzoic acid	c	−411.6			
4-Aminobenzoic acid	c	−412.9			
2-Aminobiphenyl	c	112.2			
4-Aminobiphenyl	c	81.2			
4-Aminobutanoic acid	c	−581.0			
2-Aminoethanesulfonic acid	c	−785.9	−562.3	154.1	140.7
ionized; std. state, $m = 1$	aq	−719.8	−509.8	200.1	
2-Aminoethanol	lq				195.5
2-Aminohexanoic acid (norleucine)	c	−639.1			
4-Aminohexanoic acid	c	−646.2			
5-Aminohexanoic acid	c	−643.3			
6-Aminohexanoic acid	c	−639.1			
(−)-2-Amino-3-hydroxy-butanoic acid	c	−759.5			

TABLE 2.54 Enthalpies and Gibbs Energies of Formation, Entropies, and Heat Capacities of Organic Compounds (*Continued*)

Substance	Physical state	$\Delta_f H°$ kJ · mol^{-1}	$\Delta_f G°$ kJ · mol^{-1}	$S°$ J · deg^{-1} · mol^{-1}	$C_p°$ J · deg^{-1} · mol^{-1}
2-Amino-2-(hydroxymethyl)-1,1-propanediol	c	717.8			
3-Aminonitroguanidine	c	22.1			
5-Aminopentanoic acid	c	−604.1			
5-Aminotetrazole	c	−207.8			
3-Amino-1,2,4-triazole	c	76.8			
Aniline	lq	31.3	149.2	191.4	191.9
	g	87.5	−7.0	317.9	107.9
Anthracene	c	129.2	286.0	207.6	210.5
9,10-Anthraquinone	c	−207.5			
D-(−)-Arabinose [also (+)-]	c	−1057.9			
(+)-Arginine	c	−623.5	−240.5	250.8	232.0
L-(+)-Ascorbic acid	c	−1164.6			
L-(+)-Asparagine	c	−789.4	−530.6	174.6	
L-(+)-Aspartic acid	c	−973.3	−730.7	170.2	
cis-Azobenzene	c	310.2			
trans-Azobenzene	c	365.2			
Azoisopropane	g	35.8			
Azomethane	g	148.8	239.7	289.9	78.0
Azomethane-d_6	g	119.3	218.3	305.7	90.6
Azopropane	g	51.5			
Azulene	g	289.1	353.4	338.1	128.5
Barbituric acid	c	−637.2			
Benzaldehyde	lq	−87.0	9.4		172.0
Benzamide	c	−202.6			
Benzanilide	c	−93.4			
1,2-Benzanthracene	c	170.9			
2,3-Benzanthracene	c	160.4	359.2	215.5	
1,2-Benzanthracene-9,10-dione	c	−231.9			
Benzene	lq	49.0	124.4	173.4	136.0
	g	82.6	129.7	269.2	82.4
Benzeneboronic acid	c	−720.1			
1,2-Benzenediamine	c	−0.3			
1,3-Benzenediamine	c	−7.8			
1,4-Benzenediamine	c	3.1			
1,3-Benzenedicarboxylic acid	c	803.0			
1,4-Benzenedicarboxylic acid	c	816.1			
1,2,4,5-Benzenetetra-carboxylic acid	c	1571.0			
Benzenethiol (thiophenol)	lq	63.7	134.0	222.8	173.2
	g	111.3	147.6	336.9	104.9
1,2,3-Benzenetricarboxylic acid	c	−1160.0			
1,2,4-Benzenetricarboxylic acid	c	−1179.0			
1,3,5-Benzenetricarboxylic acid	c	−1190.0			
1,2,3-Benzenetriol	c	−551.1			
1,2,4-Benzenetriol	c	−563.8			
1,3,5-Benzenetriol	c	−584.6			
p-Benzidine	c	70.7			
Benzil	c	−153.9			
Benzoic acid	c	−385.2	−245.3	167.6	146.8
Benzoic anhydride	c	−415.4			

(Continued)

TABLE 2.54 Enthalpies and Gibbs Energies of Formation, Entropies, and Heat Capacities of Organic Compounds (*Continued*)

Substance	Physical state	$\Delta_f H°$ kJ · mol^{-1}	$\Delta_f G°$ kJ · mol^{-1}	$S°$ J · deg^{-1} · mol^{-1}	$C_p°$ J · deg^{-1} · mol^{-1}
Benzonitrile	lq	163.2		209.1	165.2
	g	215.8	260.8	321.0	109.1
Benzo [*def*] phenanthrene	c	125.5	269.5	224.8	236.0
Benzophenone	c	−34.5	140.2	245.2	224.8
Benzo[*f*]quinoline	c	150.6			
Benzo [*h*] quinoline	c	149.7			
1,4-Benzoquinone	c	−185.7	−83.6	162.8	129.0
Benzo [*b*] thiophene	c	100.6			
1,2,3-Benzotriazole	c	250.0			
Benzotrifluoride	lq	−636.7			
Benzoyl bromide	lq	−107.3			
Benzoyl chloride	lq	−158.0			
Benzoylformic acid	c	−482.4			
N-Benzoylglycine	c	−609.8	−369.57	239.3	
Benzoyl iodide	lq	−53.5			
3,4-Benzphenanthrene	c	184.9			
Benzylamine	lq	34.2			
Benzyl alcohol	lq	−160.7	−27.5	216.7	218.0
Benzyl bromide	lq	16.0			
Benzyl chloride	lq	−32.6			182.4
N-Benzyldiphenylamine	c	184.7			
Benzyl ethyl sulfide	lq	−4.9			
Benzyl iodide	lq	57.3			
Benzyl methyl ketone	lq	−151.9			
Benzyl methyl sulfide	lq	26.2			
Bicyclo[1.1.0]butane	g	217.1			
Bicyclo[2.2.1]hepta-2,5-dione	lq	213.0			
Bicyclo[2.2.1]heptane	c	−95.1			
Bicyclo[4.1.0]heptane	lq	−36.7			
Bicyclo[2.2.1]heptene	lq	90.0	203.9		130.0
Bicyclo[3.1.0]hexane	g	38.6			
Bicyclohexyl	lq	−273.7			
Bicyclo[2.2.2]octane	c	−146.9			
Bicyclo[4.2.0]octane	g	−26.2			
Bicyclo[5.1.0]octane	g	−16.6			
Bicyclo[2.2.2]oct-2-ene	g	−23.3			
Bicyclopropyl	g	129.3			
Biphenyl	c	99.4	254.2	209.4	198.4
2-Biphenylcarboxylic acid	c	−349.0			
(1,1′-Biphenyl)-4,4′-diamine	c	70.7			
Biphenylene	c	334.0			
Bis(2-chloroethyl) ether	lq				220.9
Bis(dimethylthiocarbonyl) disulfide	c	41.6			
Bis(2-hydroxyethyl) ether	lq	−1621.0		441.0	135.1
	g	−571.1			
Bromoacetone	g	−181.0			
Bromoacetylene	g			253.7	55.7
Bromobenzene	lq	60.9	126.0	219.2	154.3
4-Bromobenzoic acid	c	−378.3			

TABLE 2.54 Enthalpies and Gibbs Energies of Formation, Entropies, and Heat Capacities of Organic Compounds (*Continued*)

Substance	Physical state	$\Delta_f H°$ kJ · mol^{-1}	$\Delta_f G°$ kJ · mol^{-1}	$S°$ J · deg^{-1} · mol^{-1}	$C_p°$ J · deg^{-1} · mol^{-1}
1-Bromobutane	lq	−143.8	−12.9	369.8	109.3
2-Bromobutane	lq	−154.8	−19.25		
	g	−120.3	−25.8	370.3	110.8
Bromochlorodifluoromethane	g	−471.5	−448.4	318.5	74.6
1-Bromo-2-chloroethane	lq				130.1[27]
Bromochlorofluoromethane	g	−295.0	−278.6	304.3	63.2
Bromochloromethane	lq				52.7
	g	−50.2	−39.3	287.6	
1-Bromo-2-chloro-1,1,2-trifluoroethane	g	−644.8			
2-Bromo-2-chloro-1,1,1-trifluoroethane	g	−690.4			
1-Bromodecane	lq	−344.7			
Bromodichlorofluoromethane	g	−269.5	−246.8	330.6	80.0
Bromodichloromethane	g	−58.6	−42.5	316.4	67.4
Bromodifluoromethane	g	−424.9	−447.3	295.1	58.7
Bromoethane	lq	−90.5	−25.8	198.7	100.8
	g	−61.9	−23.9	286.7	64.5
Bromoethylene (vinyl bromide)	lq				107.7[15]
	g	79.2	81.7	275.8	55.4
Bromofluoromethane	g	−252.7	−241.5	276.3	49.2
1-Bromoheptane	lq	−218.4			
1-Bromohexane	lq	−194.2		453.0	203.5
Bromoiodomethane	g	50.2	39.2	307.5	
Bromomethane	lq				78.7[7]
	g	−35.4	−26.3	246.4	42.5
2-Bromo-2-methylpropane	lq	−163.8			151.0
	g	−132.4	−28.2	332.0	116.5
1-Bromooctane	lq	−245.1			
Bromopentafluoroethane	g	−1064.4			
1-Bromopentane	lq	−170.2			132.2
	g	−129.0	−5.7	408.8	
1-Bromopropane	lq	−121.8			86.4
	g	−87.0	−22.5	330.9	
2-Bromopropane	lq	−130.5			132.2
	g	−99.4	−27.2	316.2	89.4
cis-1-Bromopropene	g	40.8			
3-Bromopropene	g	45.2			
N-Bromosuccinimide	c	−335.9			
α-Bromotoluene	lq	23.4			
Bromotrichloromethane	g	−41.1	−12.4	332.8	85.3
Bromotrifluoroethane	g	−694.5			
Bromotrifluoromethane	g	−648.3	−622.6	297.8(5)	69.3
Bromotrimethylsilane	lq	−325.9			
Bromotrinitromethane	g	80.3			
Brucine	c	−496.2			
1,2-Butadiene	g	162.3	199.5	293.0	80.1
1,3-Butadiene	lq	88.5		199.0	123.6
	g	110.0	150.7	278.7	79.5
1,3-Butadiyne	g	472.8	444.0	250.0	73.6
Butanal	lq	−239.2			163.7
	g	−204.9	−114.8	243.7	103.4

TABLE 2.54 Enthalpies and Gibbs Energies of Formation, Entropies, and Heat Capacities of Organic Compounds (*Continued*)

Substance	Physical state	$\Delta_f H°$ kJ · mol^{-1}	$\Delta_f G°$ kJ · mol^{-1}	$S°$ J · deg^{-1} · mol^{-1}	$C_p°$ J · deg^{-1} · mol^{-1}
Butanamide	lq	−346.9			
Butane	lq				104.5$^{-0.5}$
	g	−125.6	−17.2	310.1	97.5
1,2-Butanediamine	lq	−120.2			
(±)-1,2-Butanediol	lq	−523.6			
1,3-Butanediol	lq	−501.0			227.2^{30}
1,4-Butanediol	lq	−503.3		223.4	200.1
2,3-Butanediol	lq	−541.5			213.0
Butanedinitrile	c	139.7			160.5^{62}
	lq				
2,3-Butanedione	lq	−365.8			
1,4-Butanedithiol	lq	−105.7			
Butanenitrile	lq	−5.8			159^{67}
	g	33.6	108.7	325.4	97.0
1-Butanethiol	lq	−124.7	4.1	276.0	171.2
2-Butanethiol	lq	−131.0	−0.17	271.4	
Butanoic acid	lq	−533.8	−377.7	222.2	178.6
Butanoic anhydride	lq				283.7
1-Butanol	lq	−327.3	−162.5	225.8	177.0
	g	−275.0	−150.8	362.8	122.6
(±)-2-Butanol	lq	−342.6	−177.0	214.9	196.9
	g	−292.9	−167.6	359.5	113.3
2-Butanone	lq	−273.3	−151.4	239.1	158.9
	g	−238.5		339.9	101.7
Butanophenone	lq	−188.9			
trans-2-Butenal	lq	−138.7			95.4
cis-Butenedinitrile	c	268.2			
1-Butene	lq	−20.8		227.0	118.0
	g	0.1	71.3	305.6	85.7
cis-2-Butene	lq	−29.8		219.9	127.0
	g	−7.1	65.9	300.8	78.9
trans-2-Butene	g	−11.4	63.0	296.5	87.8
cis-2-Butenenitrile	lq	95.1			
trans-2-Butenenitrile	lq	95.1			
3-Butenenitrile	g	159.7	193.4	298.4	82.1
cis-2-Butenoic acid	lq	−347.0			
trans-2-Butenoic acid	c	−430.5			
cis-2-Butenedioic acid	c	−788.7			
trans-2-Butenedioic acid	c	−811.1			
1-Buten-3-yne	g	304.6	306.0	279.4	73.2
2-Butoxyethanol	lq				281.0
N-Butylacetamide	lq	−380.8			
Butyl acetate	lq	−529.2			227.8
Butylamine	lq	−127.7			179.2
	g	−92.0	49.2	363.3	118.6
sec-Butylamine	lq	−137.5			
	g	−104.6	40.7	351.3	117.2
tert-Butylamine	g	−150.6			192.1
	g	−121.0	28.9	337.9	120.0
Butylbenzene	lq	63.2			243.4
	g	−13.1	144.7	439.5	416.3

TABLE 2.54 Enthalpies and Gibbs Energies of Formation, Entropies, and Heat Capacities of Organic Compounds (*Continued*)

Substance	Physical state	$\Delta_f H°$ kJ · mol^{-1}	$\Delta_f G°$ kJ · mol^{-1}	$S°$ J · deg^{-1} · mol^{-1}	$C_p°$ J · deg^{-1} · mol^{-1}
sec-Butylbenzene	lq	−66.4			
tert-Butylbenzene	lq	−70.7			238.0
sec-Butyl butanoate	lq	−492.6			
Butyl chloroacetate	lq	−538.4			
Butyl 2-chlorobutanoate	lq	−655.2			
Butyl 3-chlorobutanoate	lq	−610.9			
Butyl 4-chlorobutanoate	lq	−618.0			
Butyl 2-chloropropanoate	lq	−572.0			
Butyl 3-chloropropanoate	lq	−558.2			
Butyl crotonate	lq	−467.8			
Butylcyclohexane	lq	−263.1		345.0	271.0
	g	−213.4	56.4	458.5	207.1
Butylcyclopentane	g	−168.3	61.4	456.2	177.5
Butyl dichloroacetate	lq	−550.2			
Butyl ethyl ether	lq				159.0
Butyl ethyl sulfide (3-thiaheptane)	g	−125.2	32.0	453.0	162.0
tert-Butyl ethyl sulfide	lq	−187.3			
Butyl formate	lq				200.2
tert-Butyl hydroperoxide	lq	−293.6			
Butyllithium	lq	−132.2			
Butyl methyl ether	lq	−290.6		295.3	192.7
tert-Butyl methyl ether	lq	−313.6		265.3	187.5
Butyl methyl sulfide (2-thiahexane)	lq	−142.8	17.1	307.5	200.9
tert-Butyl methyl sulfide	lq	−156.9		276.1	199.9
Butyl methyl sulfone	lq	−535.8			
tert-Butyl methyl sulfone	c	−556.0			
cis-Butyl 9-octadecanoate	lq	−816.9			
tert-Butyl peroxide	lq	−380.9			
Butyl trichloroacetate	lq	−545.8			
Butylurea	c	−419.5			
Butyl vinyl ether	lq	−218.8			232.0
1-Butyne	g	165.2	202.1	290.8	81.4
2-Butyne	g	145.7	185.4	283.3	78.0
2-Butynedinitrile	g	529.2			
2-Butynedioic acid	c	−577.4			
3-Butynoic acid	c	−241.8			
γ-Butyrolactone	lq	−420.9			141.4
(+)-Camphor	c	−319.4			271.2
ε-Caprolactam	c	−329.4			
9*H*-Carbazole	c	101.7			
Carbonyl bromide	g	−96.2	−110.9	309.1	61.8
Carbonyl chloride	g	−219.1	−204.9	283.5	57.7
Carbonyl chloride fluoride	g			276.7	52.4
Carbonyl fluoride	g	−639.8			46.8
Chloroacetamide	c	−338.5			
Chloroacetic acid	c	−510.5			
Chloroacetyl chloride	lq	−283.7			
Chloroacetylene	g			242.0	54.3
2-Chlorobenzaldehyde	lq	−118.4			

(*Continued*)

TABLE 2.54 Enthalpies and Gibbs Energies of Formation, Entropies, and Heat Capacities of Organic Compounds (*Continued*)

Substance	Physical state	$\Delta_f H°$ kJ · mol^{-1}	$\Delta_f G°$ kJ · mol^{-1}	$S°$ J · deg^{-1} · mol^{-1}	$C_p°$ J · deg^{-1} · mol^{-1}
3-Chlorobenzaldehyde	lq	−126.0			
4-Chlorobenzaldehyde	c	−146.4			
Chlorobenzene	lq	11.0	89.2	209.2	150.2
2-Chlorobenzoic acid	c	−404.5			
3-Chlorobenzoic acid	c	−423.3			
4-Chlorobenzoic acid	c	−428.9			163.2
Chloro-1,4-benzoquinone	c	−220.6			
1-Chlorobutane	lq	−188.1			175.0
	g	−154.6	−38.8	358.1	107.6
(±)-2-Chlorobutane	lq	−192.8			
	g	−161.2	−53.5	359.6	108.5
2-Chlorobutanoic acid	lq	−575.5			
3-Chlorobutanoic acid	lq	−556.3			
4-Chlorobutanoic acid	lq	−566.3			
Chlorocyclohexane	lq	−207.2			
1-Chloro-1,1-difluoroethane	lq				130.5[21]
	g			307.2	82.5
1-Chloro-2,2-difluoroethylene	g	−315.5	−289.1	303.0	72.1
2-Chloro-1,1-difluoroethylene	g	−331.4	−305.0	302.4	
Chlorodifluoromethane	lq				93.0[41]
	g	−482.6	−450.0	281.0	55.9
2-Chloro-1,4-dihydroxybenzene	c	−382.81			
Chlorodimethylsilane	lq	−79.8			
1-Chloro-2,3-epoxypropane	lq	−148.5			125.1
1-Chloroethane	lq	−136.8	−59.3	190.8	104.3
	g	−112.1	−60.5	275.8	62.6
2-Chloroethanol	lq	−295.4			
1-Chloro-2-ethylbenzene	lq	−54.1			
1-Chloro-4-ethylbenzene	lq	−51.7			
Chloroethylene (vinyl chloride)	lq				89.4
	g	37.3	53.6	263.9	53.7
2-Chloroethyl ethyl ether	g	−301.3			
2-Chloroethyl vinyl ether	g	−170.1			
Chloroethyne	g	213.0	197.0	241.9	54.3
1-Chloro-1-fluoroethane	g	−313.4			
2-Chlorohexane	lq	−246.1			
Chlorofluoromethane	g	−290.8	−265.5	264.3	47.0
Chlorohydroquinone	c	−382.8			
Chloroiodomethane	g	12.6	15.4	296.1	
Chloromethane	lq				75.6[24]
	g	−81.9	−58.5	234.6	40.8
1-Chloro-3-methylbutane	lq	−216.0			175.1
	g	−179.7			
2-Chloro-2-methylbutane	g	−202.2			
2-Chloro-3-methylbutane	g	−185.1			
1-Chloro-2-methylpropane	lq	−191.1			158.6
	g	−159.4	−49.7	355.0	108.5
2-Chloro-2-methylpropane	lq	−211.2			172.8
	g	−182.2	−64.1	322.2	114.2
1-Chloronaphthalene	lq	54.6			212.6
2-Chloronaphthalene	c	55.2			
1-Chlorooctane	lq	−291.3			198.5

TABLE 2.54 Enthalpies and Gibbs Energies of Formation, Entropies, and Heat Capacities of Organic Compounds (*Continued*)

Substance	Physical state	$\Delta_f H°$ kJ · mol⁻¹	$\Delta_f G°$ kJ · mol⁻¹	$S°$ J · deg⁻¹ · mol⁻¹	$C_p°$ J · deg⁻¹ · mol⁻¹
Chloropentafluoroacetone	g	−1121.0			
Chloropentafluoroethane	lq				184.2
	g	−1188.8			
1-Chloropentane	lq	−213.2			
	g	−175.0	−37.4	397.0	130.5
3-Chlorophenol	c	−206.4			
4-Chlorophenol	c	−197.9			
1-Chloropropane	lq	−160.6			132.2
	g	−131.9	−50.7	319.1	84.6
2-Chloropropane	lq	−172.1			
	g	−144.9	−62.5	304.2	87.3
2-Chloro-1,3-propanediol	lq	−517.5			
3-Chloro-1,2-propanediol	lq	−525.3			
2-Chloropropanoic acid	lq	−522.5			131.6
3-Chloropropanoic acid	c	−549.3			
2-Chloro-1-propene	g	−21.0			
3-Chloro-1-propene (allyl chloride)	lq				125.1
	g	−0.63	43.6	306.7	75.4
N-Chlorosuccinimide	c	−358.1			
α-Chlorotoluene	lq	−32.6			
o-Chlorotoluene	lq				166.8
2-Chloro-1,1,1-trifluoro-ethane	g			326.4	154.6
Chlorotrifluoroethylene	g	−505.5	−523.8	322.1	83.9
Chlorotrifluoromethane	g	−707.8	−667.4	285.4	66.9
Chlorotrimethylsilane	lq	−384.1			
Chlorotrinitromethane	lq	− 27.1			
	g	18.4			
Chrysene	c	145.3			
(−)-Cinchonidine	c	29.7			
Cinchonine	c	31.0			
cis-Cinnamic acid	c	−315.0			
trans-Cinnamic acid	c	−338.5			
Cinnamic anhydride	c	−347.7			
Citric acid	c	−1543.9	−1236.4	166.2	
Codeine monohydrate	c	−632.6			
Creatine	c	−537.2			
o-Cresol	c	−204.6		165.4	154.6
	lq				233.6⁴⁰
	g	−128.6	37.1	357.6	130.3
m-Cresol	lq	−194.0		212.6	224.9
	g	−132.3	−40.5	356.8	122.5
p-Cresol	c	−199.3		167.3	150.2
	lq				221.0⁴⁰
	g	−125.4	−30.9	347.6	124.5
Cuban	c	541.3			
Cyanamide	c	58.8			
Cyanide (CN)	g	437.6	407.5	202.6	29.2
Cyanogen	g	306.7	297.2	241.9	56.9
Cyanogen bromide	g	140.5	165.3	248.3	46.9
Cyanogen chloride	g	138.0	131.0	236.2	45.0

(*Continued*)

TABLE 2.54   Enthalpies and Gibbs Energies of Formation, Entropies, and Heat Capacities of Organic Compounds (*Continued*)

Substance	Physical state	$\Delta_f H°$ kJ · mol⁻¹	$\Delta_f G°$ kJ · mol⁻¹	$S°$ J · deg⁻¹ · mol⁻¹	$C_p°$ J · deg⁻¹ · mol⁻¹
Cyanogen fluoride	g	−639.8		224.7	41.8
Cyanogen iodide	c	166.2	185.0	96.2	
	g	205.5	196.6	256.8	48.3
Cyclobutane	g	27.7	110.0	265.4	72.2
Cyclobutanecarbonitrile	lq	103.0			
Cyclobutene	g	156.7	174.7	263.5	67.1
Cyclobutylamine	g	41.2			
Cyclododecane	c	−306.6			
1,3-Cycloheptadiene	g	94.3			
Cycloheptane	lq	−156.6	54.1	242.6	123.1
Cycloheptanone	lq	−299.4			
1,3,5-Cycloheptatriene	lq	142.2	243.1	214.6	162.8
Cycloheptene	g	−9.2			
Cyclohexane	lq	−156.4	26.7	204.4	154.9
	g	−123.4	31.8	298.3	106.3
cis-Cyclohexane-1,2-dicarboxylic acid	c	−961.1			
trans-Cyclohexane-1,2-dicarboxylic acid	c	−970.7			
Cyclohexanethiol	lq	−140.7		255.6	192.6
	g	−96.1			
Cyclohexanol	lq	−348.1	−133.3	199.6	208.2
Cyclohexanone	lq	−271.2		255.6	182.2
	g	−226.1	−90.8	322.2	109.7
Cyclohexene	lq	−38.5	101.6	214.6	148.3
1-Cyclohexenylmethanol	lq	−382.4			
Cyclohexylamine	lq	−147.7			
Cyclohexylbenzene	lq	−76.6			261.3
Cyclohexylcyclohexane	lq	−329.3			
Cyclooctane	lq	−167.7			
Cyclooctanone	lq	−326.0			
1,3,5,7-Cyclooctatetraene	lq	254.5	358.6	220.3	184.0
Cyclooctene	lq	−74.0			
1,3-Cyclopentadiene	g	134.3	179.3	267.8	
Cyclopentane	lq	−105.1	36.4	204.3	128.9
	g	−76.4	38.6	292.9	83.0
cis-1,2-Cyclopentanediol	c	−484.9			
trans-1,2-Cyclopentanediol	c	−489.9			
Cyclopentanethiol	lq	−89.5	46.8	256.9	165.2
Cyclopentanol	lq	−300.1	−127.8	206.3	184.1
Cyclopentanone	lq	−235.7			154.5
Cyclopentene	lq	4.4	108.5	201.3	122.4
	g	34.0	110.8	291.8	75.1
1-Cyclopentenylmethanol	lq	34.3			
Cyclopentylamine	lq	−95.1		241.0	181.2
Cyclopropane	g	53.3	104.4	237.4	55.6
Cyclopropanecarbonitrile	g	182.8			
Cyclopropene	g	277.1	286.3	223.3	
Cyclopropylamine	lq	45.8		187.7	147.1
	g	77.0			
Cyclopropylbenzene	lq	100.3			
(−)-Cysteine	c	−534.1			

TABLE 2.54 Enthalpies and Gibbs Energies of Formation, Entropies, and Heat Capacities of Organic Compounds (*Continued*)

Substance	Physical state	$\Delta_f H°$ kJ · mol^{-1}	$\Delta_f G°$ kJ · mol^{-1}	$S°$ J · deg^{-1} · mol^{-1}	$C_p°$ J · deg^{-1} · mol^{-1}
(−)-Cystine	c	−1032.7			
Cytosine	c	−221.3		132.6	
Decafluorobutane	lq				127.2[20]
cis-Decahydronaphthalene	lq	−219.4	68.9	265.0	232.0
trans-Decahydronaphthalene	lq	−230.6	57.7	265.0	228.5
Decanal	g	−330.9	−66.5	578.6	239.7
Decane	lq	−300.9	17.5	425.5	314.4
Decanedioic acid	c	−1082.8			
1,10-Decanediol	c	−693.5			
1-Decanenitrile	lq	−158.4			
1-Decanethiol	lq	−276.5		476.1	350.4
	g	−211.5	61.4	610.1	255.6
Decanoic acid	c	−713.7			
1-Decanol	lq	−478.1	−132.2	430.5	370.6
1-Decene	lq	−173.8	105.0	425.0	300.8
1-Decyne	g	41.2	252.2	524.5	219.7
Deoxybenzoin	c	−71.0			
Diacetamide	c	−489.0			
Diacetyl peroxide	lq	−535.3			
1,2-Diallyl phthalate	lq	−550.6			
2,2′-Diaminodiethylamine	lq				254[40]
2,6-Diaminopyridine	c	−6.5			
Diazomethane	g	192.5	217.8	242.8	52.5
Dibenz [*de,kl*] anthracene	c	182.8			
1,2-Dibenzoylethane	c	−255.6			
trans-1,2-Dibenzoylethylene	c	−114.7	109.8	319.2	
Dibenzoylmethane	c	−223.5			
Dibenzoyl peroxide	c	−369.6			
Dibenzyl	c	44.1	260.0	269.4	255.2
Dibenzyl sulfide	c	99.0			
Dibenzyl sulfone	c	−282.6			
1,2-Dibromobutane	g	−91.5	−13.1	408.8	127.1
1,3-Dibromobutane	lq	−148.0			
1,4-Dibromobutane	g	−87.8			
2,3-Dibromobutane	g	−102.0			
Dibromochlorofluoromethane	g	−231.8	−223.4	342.8	82.4
Dibromochloromethane	g	−20.9	−18.8	327.7	69.2
1,2-Dibromo-1-chloro-1,2,2-trifluoroethane	lq	−691.7			
	g	−656.6			
1,2-Dibromocycloheptane	lq	−157.6			
1,2-Dibromocyclohexane	lq	−162.8			
1,2-Dibromocyclooctane	lq	−173.3			
Dibromodifluoroethane	g	−36.9		327.7	80.8
Dibromodichloromethane	g	−29.3	−19.5	347.8	87.1
Dibromodifluoromethane	g	−429.7	−419.1	325.3	77.0
1,1-Dibromoethane	lq	−66.2			
1,2-Dibromoethane	lq	−79.2	−20.9	223.3	136.0
	g	−37.5			
cis-1,2-Dibromoethylene	g			313.3	68.8
trans-1,2-Dibromoethylene	g			313.5	70.3
Dibromofluoromethane	g	−223.4	−221.1	316.8	65.1

(*Continued*)

TABLE 2.54 Enthalpies and Gibbs Energies of Formation, Entropies, and Heat Capacities of Organic Compounds (*Continued*)

Substance	Physical state	$\Delta_f H°$ kJ · mol^{-1}	$\Delta_f G°$ kJ · mol^{-1}	$S°$ J · deg^{-1} · mol^{-1}	$C_p°$ J · deg^{-1} · mol^{-1}
Dibromomethane	lq				105.3
	g	−14.8	−16.2	293.2	54.7
1,3-Dibromo-2-methylpropane	g	−137.6			
1,3-Dibromotetrafluoroethane	lq	−817.7			
	g	−789.1			
1,2-Dibromopropane	lq				160.0
	g	−71.5	−17.7	376.1	102.8
1,2-Dibromotetrafluoroethane	lq				180.3
Dibutoxymethane	lq	−549.4			
Dibutylamine	lq	−206.0			292.9
Dibutyl disulfide	g	−160.6	53.9	572.8	231.1
Di-*tert*-butyl disulfide	lq	−255.2			
Dibutyl ether	lq	−377.9			278.2
	g	−332.8	−88.5	500.4	204.0
Di-*sec*-butyl ether	lq	−401.5			
	g	−360.9			
Di-*tert*-butyl ether	lq	−399.6			276.1
	g	−362.0			
Dibutylmercury	lq	−97.9			
Dibutyl peroxide	lq	−380.7			
Dibutyl 1,2-phthalate	c	−842.6			498.0
Dibutyl sulfate	lq	−904.6			
Dibutyl sulfide	lq	−220.7	32.2	405.1	284.3
Di-*tert*-butyl sulfide	lq	−232.4			
Dibutyl sulfite	lq	−693.1			
Dibutyl sulfone	c	−610.2			
Dichloroacetic acid	lq	−496.3			
ionized	aq	−507.1			
Dichloroacetyl chloride	lq	−280.4			
1,2-Dichlorobenzene	lq	−17.5			162.4
	g	30.2	82.7	341.5	113.5
1,3-Dichlorobenzene	lq	−20.7			171
	g	25.7	78.6	343.5	113.8
1,4-Dichlorobenzene	c	−42.3			
	lq			175.4	147.8
	g	22.5	77.2	336.7	113.9
Dichlorodifluoromethane	lq				117.2
	g	−477.4	−439.4	300.8	72.3
1,3-Dichlorobutane	g	−195.0			
1,4-Dichlorobutane	g	−183.4			
Dichlorodimethylsilane	g	−461.1		335.4	101.1
Dichlorodiphenylsilane	lq	−278.2			
1,1-Dichloroethane	lq	−158.4			126.3
	g	−127.7	−73.8	305.1	76.2
1,2-Dichloroethane	lq	−167.4			128.4
	g	−126.4	−73.9	308.4	78.7
1,1-Dichloroethylene	lq	−23.9			111.3
	g	2.8	25.4	289.1	67.0
cis-1,2-Dichloroethylene	g	4.6	24.4	289.5	65.1
trans-1,2-Dichloroethylene	lq	−23.1			116.8
	g	5.0	28.6	289.9	66.7
Dichlorofluoromethane	g	−283.0	−253.0	293.1	61.0

TABLE 2.54 Enthalpies and Gibbs Energies of Formation, Entropies, and Heat Capacities of Organic Compounds (*Continued*)

Substance	Physical state	$\Delta_f H°$ kJ · mol^{-1}	$\Delta_f G°$ kJ · mol^{-1}	$S°$ J · deg^{-1} · mol^{-1}	$C_p°$ J · deg^{-1} · mol^{-1}
1,1-Dichloro-1-fluoroethane	g			320.2	88.7
1,1-Dichlorofluoroethylene	g			313.9	76.5
1,1-Dichlorofluoromethane	lq				112.6
Dichloromethane	lq	−124.2		177.8	101.2
	g	−95.4	−68.9	270.3	51.0
Dichloropentadienyliron	c	141.0			
1,2-Dichloropropane	lq	−198.8			
	g	−162.8	−83.1	354.8	98.2
1,3-Dichloropropane	g	−159.2	−82.6	367.2	99.6
2,2-Dichloropropane	g	−173.2	−84.6	326.0	105.9
1,3-Dichloro-2-propanol	lq	−385.4			
2,3-Dichloro-1-propanol	lq	−381.3			
2,3-Dichloropropene	lq	−73.3			
1,2-Dichlorotetrafluoromethane	lq				164.2
	g	−916.3			
2,2-Dichlorotetrafluoroethane	lq	−960.2			111.7
2,2-Dichloro-1,1,1-trifluoro-ethane	g			352.8	102.5
Dicyanoacetylene	lq	500.4			
Dicyanobenzene	c	275.4			
1,4-Dicyanobutane	lq	85.1			128.7
1,4-Dicyano-2-butyne	c	366.5			
Dicyanodiamide	c	22.6	179.5	129.3	118.8
Dicyclopentadiene	c	116.7			
Diethanolamine	c	−493.8			
	lq				233.5[30]
1,1-Diethoxyethane	lq	−491.4			238.0
1,2-Diethoxyethane	lq	−451.4			259.4
Diethoxymethane	lq	−450.4			
1,3-Diethoxypropane	lq	−482.1			
2,2-Diethoxypropane	lq	−538.5			
Diethylamine	lq	−103.7			169.2
	g	−72.2	72.1	352.2	115.7
Diethylamine hydrochloride	c	−358.6			
Diethylbarbituric acid (veronal)	c	−747.7			
1,2-Diethylbenzene	g	−19.0	141.1	434.3	182.6
1,3-Diethylbenzene	g	−21.8	136.7	439.3	176.9
1,4-Diethylbenzene	g	−22.3	137.9	434.0	176.2
Diethyl carbonate	lq	−681.5			212.4
cis-1,2-Diethylcyclopropane	lq	−79.9			
trans-1,2-Diethylcyclopropane	lq	83.3			
Diethyl disulfide	lq	−120.0	9.5	269.3	171.4
	g	−79.4	22.3	414.5	141.3
Diethylenediamine	c	−13.4	240.2	85.8	
Diethylene glycol	lq	−628.5			244.8
	g	−571.1		441.0	135.1
Diethylene glycol dibutyl ether	lq				452[20]
Diethylene glycol diethyl ether	lq				341.4[15]
Diethylene glycol dimethyl ether	lq				274.1

(*Continued*)

TABLE 2.54 Enthalpies and Gibbs Energies of Formation, Entropies, and Heat Capacities of Organic Compounds (*Continued*)

Substance	Physical state	$\Delta_f H°$ kJ · mol^{-1}	$\Delta_f G°$ kJ · mol^{-1}	$S°$ J · deg^{-1} · mol^{-1}	$C_p°$ J · deg^{-1} · mol^{-1}
Diethylene glycol monoethyl ether	lq				301.0
Diethylene glycol monomethyl ether	lq				271.1
Diethyl ether	lq	−279.5	−116.7	172.4	172.6
	g	−252.1	−122.3	342.7	119.5
Di-2-ethylhexyl phthalate	lq				704.7
Diethyl malonate	lq	−805.5			260.7
Diethylmercury	lq	30.1			182.8
Diethyl oxalate	lq	−805.5			
3,3-Diethylpentane	lq	−275.4			278.2
Diethyl peroxide	lq	−223.3			
Diethyl 1,2-phthalate	lq	−776.6		425.1	366.1
Diethyl selenide	lq	−96.2			
Diethyl sulfate	lq	−813.2			
Diethyl sulfide	lq	−119.4		269.3	171.4
	g	−83.6	17.8	368.0	117.0
Diethyl sulfite	lq	−600.7			
Diethyl sulfone	c	−515.5			
Diethyl sulfoxide	lq	−268.0			
N,N-Diethylurea	c	−372.2			
Diethylzinc	lq	16.7			
1,2-Difluorobenzene	lq	−330.0		222.6	159.0
	g	−293.8	−242.0	321.9	106.5
1,3-Difluorobenzene	lq	−343.9		223.8	159.1
	g	−309.2	−257.0	320.4	106.3
1,4-Difluorobenzene	lq	−342.3			157.5
	g	−306.7	−252.8	315.6	106.9
2,2′-Difluorobiphenyl	c	−295.9			
4,4′-Difluorobiphenyl	c	−296.5			
1,1-Difluoroethane	lq				118.4
	g	−497.0	−443.0	282.4	67.8
1,1-Difluoroethylene	g	−335.0	−321.5	266.2	60.1
Difluoromethane	g	−452.2	−425.4	246.6	42.9
9,10-Dihydroanthracene	c	66.4			
1,2-Dihydronaphthalene	lq	71.5			
1,4-Dihydronaphthalene	lq	84.2			
Dihydro-2H-pyran	lq	−157.4			
5,12-Dihydrotetracene	c	106.4			
2,3-Dihydrothiophene	lq	52.9			
	g	90.7	133.5	303.5	79.8
2,5-Dihydrothiophene	g	86.9	131.6	297.1	83.3
2,5-Dihydrothiophene-1,1-dioxide	c	318.9			
2′,4-Dihydroxyacetophenone	c	−573.6			
1,2-Dihydroxybenzene (pyrocatechol)	c	−354.1	−210.0	150.2	132.2
1,3-Dihydroxybenzene	c	−368.0	−209.2	147.7	131.0
1,4-Dihydroxybenzene (p-hydroquinone)	c	−364.5	−207.0	140.2	136.0
Dihydroxymalonic acid	c	−1216.3			

TABLE 2.54 Enthalpies and Gibbs Energies of Formation, Entropies, and Heat Capacities of Organic Compounds (*Continued*)

Substance	Physical state	$\Delta_f H°$ kJ · mol^{-1}	$\Delta_f G°$ kJ · mol^{-1}	$S°$ J · deg^{-1} · mol^{-1}	$C_p°$ J · deg^{-1} · mol^{-1}
2,4-Dihydroxy-5-methyl-pyrimidine	c	−468.2			
2,4-Dihydroxy-6-methyl-pyrimidine	c	−456.9			
Diiodoacetylene	g			313.1	70.3
1,2-Diiodobenzene	c	172.4			
1,3-Diiodobenzene	c	187.0			
1,4-Diiodobenzene	lq	−30.0			
	c	160.7			
1,2-Diiodoethane	g	75.0	78.5	348.5	82.3
Diiodomethane	lq	66.9	90.4	174.1	134.0
	g	119.5	95.8	309.7	57.7
1,2-Diiodopropane	g	35.6			
1,3-Diiodopropane	lq	−9.0			
Diisobutylamine	lq	−218.5			
Diisopentyl ether	lq				379[100]
Diisopropylamine	lq	−178.5			
Diisopropyl ether	lq	−351.5			216.8
	g	−319.2	−121.9	390.2	158.3
Diisopropylmercury	lq	−13.0			
Diisopropyl sulfide	lq	−181.6		313.0	232.0
	g	−142.1	27.1	415.5	169.2
Diketene	lq	−233.1			
1,2-Dimethoxybenzene	lq	−290.4			
1,1-Dimethoxybutane	lq	−468.1			
2,2-Dimethoxybutane	lq	−485.1			
1,1-Dimethoxy ethane	lq	−420.2			
1,2-Dimethoxy ethane	lq	−376.7			193.3
Dimethoxymethane	lq	−377.8		244.0	161.3
1,1-Dimethoxypentane	lq	−494.6			
2,2-Dimethoxypentane	lq	−509.2			
1,1-Dimethoxypropane	lq	−443.3			
2,2-Dimethoxypropane	lq	−459.0			
1,1-Dimethoxy-2-methyl-propane	lq	−476.2			
N,N-Dimethylacetamide	lq	−278.3			175.6
Dimethylamine	lq	−43.9	70.0	182.3	137.7
	g	−18.5	68.5	273.0	70.7
4-(Dimethylamino)benz-aldehyde	c	−137.6			
Dimethylaminomethanol	lq	−253.6			
N,N-Dimethylaminotri-methylsilane	lq	−279.5			
N,N-Dimethylaniline	lq	47.7			214.6[29]
2,6-Dimethylaniline	lq				238.9
2,3-Dimethylbenzoic acid	c	−450.4			
2,4-Dimethylbenzoic acid	c	−458.5			
2,5-Dimethylbenzoic acid	c	−456.1			
2,6-Dimethylbenzoic acid	c	−440.7			
3,4-Dimethylbenzoic acid	c	−468.8			
3,5-Dimethylbenzoic acid	c	−466.4			
3,3′-Dimethylbiphenyl	lq	20.0			

TABLE 2.54 Enthalpies and Gibbs Energies of Formation, Entropies, and Heat Capacities of Organic Compounds (*Continued*)

Substance	Physical state	$\Delta_f H°$ kJ · mol^{-1}	$\Delta_f G°$ kJ · mol^{-1}	$S°$ J · deg^{-1} · mol^{-1}	$C_p°$ J · deg^{-1} · mol^{-1}
2,2-Dimethylbutane	lq	−213.8		272.5	191.9
	g	−186.1	−9.2	358.2	141.9
2,3-Dimethylbutane	lq	−207.4		287.8	189.7
	g	−178.3	−4.1	365.8	140.5
3,3-Dimethyl-2-butanone	lq	−328.6			
2,3-Dimethyl-1-butene		−62.6	79.0	365.6	143.5
2,3-Dimethyl-2-butene	lq	−101.4		270.2	174.7
	g	−68.2	76.1	364.6	123.6
3,3-Dimethyl-1-butene	g	−60.5	98.2	343.8	126.5
2,3-Dimethyl-2-butenoic acid	c	−455.6			
Dimethylcadmium	lq	63.6	139.3	201.9	132.0
1,1-Dimethylcyclohexane	lq	−218.7	26.5	267.2	209.2
	g	−180.9	35.2	365.0	154.4
cis-1,2-Dimethylcyclohexane	lq	−211.8		274.1	210.2
	g	−172.1	41.2	374.5	165.5
trans-1,2-Dimethylcyclohexane	lq	−218.2		273.2	209.4
	g	−180.0	34.5	370.9	159.0
cis-1,3-Dimethylcyclohexane	lq	−222.9		272.6	209.4
	g	−184.6	29.8	370.5	157.3
trans-1,3-Dimethylcyclohexane	lq	−215.7		276.3	212.8
	g	−176.5	36.3	376.2	157.3
cis-1,4-Dimethylcyclohexane	lq	−215.6		271.1	212.1
	g	−176.6	38.0	370.5	157.3
trans-1,4-Dimethylcyclohexane	lq	−222.4		268.0	210.2
	g	−184.5	31.7	364.8	157.7
1,1-Dimethylcyclopentane	g	−138.2	39.0	359.3	133.3
cis-1,2-Dimethylcyclopentane	lq	−165.3		269.2	
	g	−129.5	45.7	366.1	134.14
trans-1,2-Dimethylcyclopentane	g	−136.6	38.4	366.8	134.5
cis-1,3-Dimethylcyclopentane	g	−135.9	39.2	366.8	134.5
trans-1,3-Dimethylcyclopentane	g	−133.6	41.5	366.8	134.5
1,1-Dimethylcyclopropane	lq	−33.3			
cis-1,2-Dimethylcyclopropane	lq	−26.3			
trans-1,2-Dimethylcyclopropane	lq	−30.7			
cis-2,4-Dimethyl-1,3-dioxane	lq	−465.2			
4,5-Dimethyl-1,3-dioxane	lq	−451.6			
5,5-Dimethyl-1,3-dioxane	lq	−461.3			
4,4′-Dimethyldiphenylamine	c	−11.72			
Dimethyl disulfide	lq	−62.6	7.0	235.4	146.1
Dimethyl ether	g	−184.1	−112.6	266.4	64.4
N,N-Dimethylformamide	lq	−239.3			150.6
Dimethyl fumarate	lq	−729.3			
Dimethylglyoxime	c	−199.7			
2,2-Dimethylheptane	lq	−288.2			
2,6-Dimethyl-4-heptanone	lq	−408.5			297.3
2,2-Dimethylhexane	lq	−261.9	3.0	331.9	
2,3-Dimethylhexane	lq	−252.6	9.1	342.7	
2,4-Dimethylhexane	lq	−257.0	3.7	345.7	
2,5-Dimethylhexane	lq	−260.4	2.5	338.7	249.2
3,3-Dimethylhexane	lq	−257.5	5.2	339.4	246.6
3,4-Dimethylhexane	lq	−251.8	8.5	347.2	

TABLE 2.54 Enthalpies and Gibbs Energies of Formation, Entropies, and Heat Capacities of Organic Compounds (*Continued*)

Substance	Physical state	$\Delta_f H°$ kJ · mol^{-1}	$\Delta_f G°$ kJ · mol^{-1}	$S°$ J · deg^{-1} · mol^{-1}	$C_p°$ J · deg^{-1} · mol^{-1}
Dimethyl hexanedioate	lq	−886.6			
cis-2,2-Dimethyl-3-hexene	lq	−126.4			
trans-2,2-Dimethyl-3-hexene	lq	−144.9			
cis-2,5-Dimethyl-3-hexene	lq	−151.0			
trans-2,5-Dimethyl-3-hexene	lq	−159.2			
5,5-Dimethylhydantoin	c	−533.3			
1,1-Dimethylhydrazine	lq	48.9	206.7	198.0	164.1
1,2-Dimethylhydrazine	lq	52.7	212.6	199.2	171.0
3,5-Dimethylisoxazole	lq	−63.2			
Dimethyl maleate	lq	−703.8			263.2
Dimethylmaleic anhydride	c	−581.6			
Dimethyl malonate	lq	−795.8			
Dimethylmercury	lq	59.8	140.3	209.0	
	g	94.4	146.1	306.0	83.3
6,6-Dimethyl-2-methylene-bicyclo[3.1.1]heptane	lq	−7.7			
Dimethyl oxalate	lq	−756.3			
2,2-Dimethylpentane	lq	−238.3		300.3	221.1
	g	−205.9	0.1	392.9	166.0
2,3-Dimethylpentane	lq	−233.1			218.3
	g	−198.9	0.7	414.0	166.0
2,4-Dimethylpentane	lq	−234.6		303.2	224.2
	g	−201.7	3.1	396.6	166.0
3,3-Dimethylpentane	lq	−234.2			
	g	−201.2	2.6	399.7	166.0
Dimethyl pentanedioate	lq	−205.9			
2,4-Dimethyl-3-pentanone	lq	−352.9		318.0	233.7
	g	−311.5			
2,4-Dimethyl-1-pentene	g	−83.8			
4,4-Dimethyl-1-pentene	g	−81.6			
2,4-Dimethyl-2-pentene	g	−88.7			
cis-4,4-Dimethyl-2-pentene	g	−72.6			
trans-4,4-Dimethyl-2-pentene	g	−88.8			
2,7-Dimethylphenanthrene	c	36.4			
4,5-Dimethylphenanthrene	c	89.0			
9,10-Dimethylphenanthrene	c	47.7			
2,3-Dimethylphenol	c	−241.2			206.9
2,4-Dimethylphenol	lq	−228.7			
2,5-Dimethylphenol	c	−246.6			
2,6-Dimethylphenol	c	−237.4			
3,4-Dimethylphenol	c	−242.3			
3,5-Dimethylphenol	c	−244.4			
Dimethyl 1,2-phthalate	lq	−678			303.1
Dimethyl 1,3-phthalate	c	−730.0			
Dimethyl 1,4-phthalate	c	−732.6			261.1
2,2-Dimethylpropane	lq				163.9[6]
	g	−168.0	−1.5	306.4	121.6
2,2-Dimethylpropanenitrile	lq	−39.8		232.0	179.4
2,2-Dimethyl-1,3-propanediol	c	−551.2			
2,2-Dimethylpropanoic acid	lq	−564.4			
2,2-Dimethylpropanoic anhydride	lq	−779.9			

(*Continued*)

TABLE 2.54 Enthalpies and Gibbs Energies of Formation, Entropies, and Heat Capacities of Organic Compounds (*Continued*)

Substance	Physical state	$\Delta_f H°$ kJ · mol^{-1}	$\Delta_f G°$ kJ · mol^{-1}	$S°$ J · deg^{-1} · mol^{-1}	$C_p°$ J · deg^{-1} · mol^{-1}
2,2-Dimethyl-1-propanol	lq	−399.4			
2,3-Dimethylpyridine	lq	19.4		243.7	189.5
2,4-Dimethylpyridine	lq	16.2		248.5	184.8
2,5-Dimethylpyridine	lq	18.7		248.8	184.7
2,6-Dimethylpyridine	lq	12.7		249.2	185.2
3,4-Dimethylpyridine	lq	18.3		240.7	191.8
3,5-Dimethylpyridine	lq	22.5		241.7	184.5
Dimethyl succinate	lq	−835.1			
2,2-Dimethylsuccinic acid	c	−987.8			
meso-2,3-Dimethylsuccinic acid	c	−977.5			
Dimethyl sulfate	lq	−735.5			
Dimethyl sulfide	lq	−65.4			118.1
	g	−37.5	7.0	285.9	74.1
Dimethyl sulfite	lq	−523.6			
Dimethyl sulfone	c	−450.1	−302.5	142.0	
	lq	−373.1	−272		
	g			310.6	100.0
Dimethyl sulfoxide	lq	−204.2	−99.2	188.3	153.0
1,5-Dimethyltetrazole	c	188.7			
2,2-Dimethylthiacyclopropane	lq	−24.2			
5,5-Dimethyl-4-thia-1-hexene	lq	−90.7			
N,N-Dimethylurea	c	−319.1			
N,N′-Dimethylurea	c	−312.1			
Dimethylzinc	lq	23.4		201.6	129.2
2,3-Dinitroaniline	c	−11.7			
2,4-Dinitroaniline	c	−67.8			
2,5-Dinitroaniline	c	−44.4			
2,6-Dinitroaniline	c	−50.6			
3,4-Dinitroaniline	c	−32.6			
3,5-Dinitroaniline	c	−38.9			
2,4-Dinitroanisole	c	−186.6			
2,6-Dinitroanisole	c	−189.1			
1,2-Dinitrobenzene	c	−1.8	211.5	216.3	
1,3-Dinitrobenzene	c	−27.4	184.6	220.9	
1,4-Dinitrobenzene	c	−38.7			
1,1-Dinitroethane	lq	−148.2			
1,2-Dinitroethane	lq	−165.2			
Dinitromethane	lq	−104.9			
	g	−58.9			
1,5-Dinitronaphthalene	c	30.5			
2,4-Dinitro-1-naphthol	c	−181.4			
2,4-Dinitrophenol	c	−232.6			
2,6-Dinitrophenol	c	−210.0			
1,1-Dinitropropane	lq	−163.2			
1,3-Dinitropropane	lq	−207.1			
2,2-Dinitropropane	lq	−181.2			
2,4-Dinitroresorcinol	c	−415.5			
2,4-Dinitrotoluene	c	−71.6			
2,6-Dinitrotoluene	c	−51.0			
1,3-Dioxane	lq	−379.7			143.9
1,4-Dioxane	lq	−353.9	−188.1	270.2	153.6
	g	−315.8	−180.8	299.8	94.1

TABLE 2.54 Enthalpies and Gibbs Energies of Formation, Entropies, and Heat Capacities of Organic Compounds (*Continued*)

Substance	Physical state	$\Delta_f H°$ kJ · mol^{-1}	$\Delta_f G°$ kJ · mol^{-1}	$S°$ J · deg^{-1} · mol^{-1}	$C_p°$ J · deg^{-1} · mol^{-1}
1,3-Dioxolane	lq	−333.5			118.0
	g	−298.0			
1,3-Dioxolan-2-one	c	−581.6			133.9[50]
1,3-Dioxol-2-one	lq	−459.9			
Dipentene	lq	−50.8			249.4
Dipentyl ether	lq				250
N,N-Diphenylacetamide	c	−43.1			
Diphenylacetylene	c	312.4			225.9
Diphenylamine	c	130.6			
Diphenylboron bromide	lq	−16.1			
cis,cis-1,4-Diphenylbutadiene	c	198.8			
trans,trans-1,4-Diphenyl-butadiene	c	178.8			
Diphenylbutadiyne	c	518.4			
1,4-Diphenylbutane	c	−9.9			
1,4-Diphenyl-1,4-butanedione	c	−256.2	7.8	324.7	
1,4-Diphenyl-2-butene-1,4-dione	c	−114.7	111.5	319.2	
Diphenyl carbonate	c	−401.2	−175.9	278.4	
Diphenyl disulfide	c	−148.5			
Diphenyl disulfone	c	−643.2			
Diphenyleneimine	c	126.8			
1,1-Diphenylethane	lq	48.7	245.1	335.9	
1,2-Diphenylethane	lq	51.5	67.2	270.3	
Diphenylethanedione	c	−154.0			
Diphenyl ether	c	−32.1		233.9	216.6
	lq	−14.9	144.2	291.3	268.6
1,1-Diphenylethylene	lq	172.4			
Diphenylethyne	c	312.4			
6,6-Diphenylfulvene	c	197.4			
1,2-Diphenylhydrazine	c	221.3			
Diphenylmercury	c	279.5			
Diphenylmethane	c	71.7		239.3	
	lq	89.7	276.9		233.1
1,3-Diphenyl-2-propanone	c	−84.0			
Diphenyl sulfide	lq	163.4			
Diphenyl sulfone	c	−225.0			
Diphenyl sulfoxide	c	9.7			
1,3-Diphenylurea	c	−122.6			
Dipropylamine	lq	−156.1			253.0[75]
Dipropyl disulfide	lq	−171.3	19.1	373.6	
Dipropyl ether	lq	−328.8		323.9	221.6
	g	−292.9	−105.6	422.5	158.3
Dipropylmercury	lq	−20.9			
Dipropyl sulfate	lq	−859.0			
Dipropyl sulfide	lq	−171.5			
	g	−125.3	33.2	448.4	161.2
Dipropyl sulfite	lq	−646.8			
Dipropyl sulfone	lq	−548.2			
Dipropyl sulfoxide	lq	−329.4			
2,2′-Dipyridyl ketone	c	−19.7			
1,3-Dithiane	g	−10.0	72.4	333.5	110.4

(*Continued*)

TABLE 2.54 Enthalpies and Gibbs Energies of Formation, Entropies, and Heat Capacities of Organic Compounds (*Continued*)

Substance	Physical state	$\Delta_f H°$ kJ · mol^{-1}	$\Delta_f G°$ kJ · mol^{-1}	$S°$ J · deg^{-1} · mol^{-1}	$C_p°$ J · deg^{-1} · mol^{-1}
1,2-Dithiolane	g	0.0	47.7	313.5	86.5
1,3-Dithiolane	g	10.0	54.7	323.3	84.7
Divinyl ether	lq	−39.8			
	g	−13.6			
Divinyl sulfone	lq	−207.4			
Docosanoic acid	c	−983.0			
cis-13-Docosenic acid	c	−866.0			
trans-13-Docosenic acid	c	−960.7			
Dodecane	lq	−350.9	28.1	490.6	376.0
	g	−289.7	50.0	622.5	280.3
Dodecanedioic acid	c	−1130.0			
Dodecanoic acid	c	−774.6			
	lq	−737.9			404.3
1-Dodecanol	lq	−528.5			438.1
1-Dodecene	lq	−226.2		484.8	360.7
	g	−165.4	137.9	618.3	269.6
1-Dodecyne	g	−0.04	268.6	602.4	265.4
Dulcitol	c	−1346.8			
1,2-Epoxybutane	lq	−168.9		230.9	147.0
Ergosterol	c	−789.9			
Ethane	g	−84.0	−32.0	229.1	52.5
Ethane-d_6	g	−107.4	−47.3	244.5	64.6
1,2-Ethanediamine	lq	−63.0		209.2	172.6
1,2-Ethanediol	lq	−455.3	−323.2	163.2	149.3
	g	−392.2	−304.5	303.8	82.7
Ethanedithioamide	c	−20.8			
Ethanedioyl dichloride	lq	−367.6			
1,2-Ethanedithiol	lq	−54.4			
Ethanethiol	lq	−73.6	−5.5	207.0	117.9
	g	−46.1	−4.8	296.1	72.7
Ethanol	lq	−277.6	−174.8	161.0	112.3
	g	−234.8	−167.9	281.6	65.6
Ethene (*see* Ethylene)					
Ethoxybenzene	lq	−152.6			228.5
2-Ethoxyethyl acetate	lq				376.0
2-Ethoxyethanol	lq				210.8
Ethyl acetate	lq	−479.3	−332.7	257.7	170.7
	g	−443.6	−327.4	362.8	113.6
Ethylamine	lq				130.0
	g	−47.4	36.3	283.8	71.5
Ethyl 4-aminobenzoate	c	−418.0			
N-Ethylaniline	lq	4.0	188.7	239.3	
Ethylbenzene	lq	−12.3			183.2
	g	29.9	130.6	360.5	
Ethyl benzoate	lq				246.0
2-Ethylbenzoic acid	c	−441.3			
3-Ethylbenzoic acid	c	−445.8			
4-Ethylbenzoic acid	c	−460.7			
2-Ethyl-1-butene	g	−56.0	80.0	376.6	133.6
Ethyl *trans*-2-butenoate (ethyl crotonate)	lq	−420.1			228.0
Ethyl carbamate	c	−520.5			
Ethyl 4-chlorobutanoate	lq	−566.5			

TABLE 2.54 Enthalpies and Gibbs Energies of Formation, Entropies, and Heat Capacities of Organic Compounds (*Continued*)

Substance	Physical state	$\Delta_f H°$ kJ $\cdot$ mol^{-1}	$\Delta_f G°$ kJ $\cdot$ mol^{-1}	$S°$ J $\cdot$ deg^{-1} $\cdot$ mol^{-1}	$C_p°$ J $\cdot$ deg^{-1} $\cdot$ mol^{-1}
Ethyl chloroformate	lq	−505.1			
Ethylcyclobutane	g	−27.5			
Ethylcyclohexane	lq	−211.9	29.1	280.9	211.8
	g	−171.7	39.3	382.6	158.8
1-Ethylcyclohexene	lq	−106.7			
Ethylcyclopentane	lq	−163.4	37.3	279.9	185.8
1-Ethylcyclopentene	g	−19.7			
Ethylcyclopropane	lq	−24.8			
Ethyl diethylcarbamate	lq	−592.3			
Ethyl 2,2-dimethylpropanoate	lq	−577.2			
	g	−536.0			
Ethylene	g	52.5	68.4	219.3	42.9
Ethylene-d_4	g	38.2	59.2	230.5	51.9
Ethylene carbonate	c	−581.5			133.9
Ethylenediaminetetra-acetic acid	c	−1759.4			
Ethylenediammonium chloride	c	−513.4			
2,2′-(Ethylenedioxy)bis-ethanol	lq	−804.2			
Ethylene glycol dibutyl ether	lq				350[20]
Ethylene glycol diethyl ether	lq	−451.4			259.4
Ethylene glycol dimethyl ether	lq	−376.6			193.3
Ethyleneimine	lq	91.9			
	g	126.5(9)	178.0	250.6	52.6
Ethylene oxide	lq	−78.0	−11.8	153.9	88.0
	g	−52.6(6)	−13.1	242.4	47.9
Ethyl formate	lq				149.3
2-Ethylhexanal	lq	−342.5			
3-Ethylhexane	lq	−250.4			
	g	−210.7			
2-Ethyl-1-hexanol	lq	−432.8		347.0	317.5
Ethyl hydroperoxide	g	198.9			
Ethylidenecyclohexane	lq	−103.5			
Ethylidenecyclopentane	lq	−56.7			
Ethyl isocyanide	lq	108.4			
Ethyl isopropyl sulfide	lq	−156.1			
Ethyl lactate	lq				254
Ethyllithium	c	−58.6			
Ethylmercury bromide	c	−107.5			
Ethylmercury chloride	c	−141.1			
Ethylmercury iodide	c	−65.7			
1-Ethy 1-2-methylbenzene	g	1.3	131.1	399.2	157.9
2-Ethyl-3-methyl-1-butene	g	−79.5			
Ethyl 2-methylbutanoate	lq	−566.8			
Ethyl 3-methylbutanoate	lq	−570.9			
Ethyl methyl ether	g	−216.4	−117.7	309.2	93.3
3-Ethyl-2-methylpentane	lq	−249.6			
	g	−211.0	21.3	441.1	
3-Ethyl-3-methylpentane	lq	−252.8			
	g	−214.8	19.9	433.0	
3-Ethyl-2-methyl-1-pentene	g	−100.3			

(*Continued*)

TABLE 2.54 Enthalpies and Gibbs Energies of Formation, Entropies, and Heat Capacities of Organic Compounds (*Continued*)

Substance	Physical state	$\Delta_f H°$ kJ · mol^{-1}	$\Delta_f G°$ kJ · mol^{-1}	$S°$ J · deg^{-1} · mol^{-1}	$C_p°$ J · deg^{-1} · mol^{-1}
Ethyl methyl sulfide	lq	−91.6		239.1	144.6
	g	−59.6	11.4	333.1	95.1
Ethyl nitrate	g	−154.1	−36.9	348.3	97.4
Ethyl nitrite	g	−104.2		103.5	99.2
1-Ethyl-2-nitrobenzene	lq	−48.7			
1-Ethyl-4-nitrobenzene	lq	−55.4			
Ethyl 3-oxobutanoate	lq				248.0
3-Ethylpentane	lq	−224.9		314.5	219.6
	g	−189.6	11.0	411.5	166.0
Ethyl pentanoate	lq	−553.0			
2-Ethylphenol	lq		−208.8		
3-Ethylphenol	lq	−214.3			
4-Ethylphenol	c	−224.4			206.9
Ethylphosphonic acid	c	−1051.4			
Ethylphosphonic dichloride	lq	−613.4			
Ethyl propanoate	lq	−502.7			196.1
	g	−463.3	−323.7		
Ethyl propyl ether	g	−272.2		295.0	197.2
Ethyl propyl sulfide	lq	−144.8		309.5	198.4
	g	−104.7	23.6	414.1	139.3
2-Ethylpyridine	lq	7.4			
5-Ethyl thioacetate	lq	−268.2			
2-Ethyltoluene	g	1.3	131.1	399.2	157.9
3-Ethyltoluene	g	−1.8	126.4	404.2	152.2
4-Ethyltoluene	g	−3.2	85.3	398.9	151.5
N-Ethylurea	c	−357.8			
Ethyl β-vinylacrylate	lq	−338.1			
Ethyl vinyl ether	lq	−167.4			
	g	−140.8			
Ethynylbenzene	g	327.3	361.8	321.7	114.9
Ethynylsilane	g			269.4	72.6
Fluoranthene	c	189.9	345.6	230.5	230.2
Fluoroacetamide	c	−496.6			
Fluoroacetic acid	c	−688.3			
Fluoroacetylene	g			269.4	72.6
Fluorobenzene	lq	−150.6		205.9	146.4
	g	−116.0	−69.0	302.6	94.4
2-Fluorobenzoic acid	c	−567.6			
3-Fluorobenzoic acid	c	−582.0			
4-Fluorobenzoic acid	c	−585.7			
Fluoroethane	g	−263.2	−211.0	264.5	58.6
2-Fluoroethanol	lq	−465.7			
Fluoroethylene	g	−138.8			
Fluoromethane	g	−237.8	−213.8	222.8	37.5
1-Fluoropropane	g	−285.9	−200.3	304.2	82.6
2-Fluoropropane	g	−293.5	−204.2	292.1	82.0
Fluorosyltrifluoromethane	g	−766.0	−707.0	322.4	79.4
4-Fluorotoluene	lq	−186.9	−79.8	237.1	171.2
Fluorotribromomethane	g	−190.4	−193.1	345.8	
Fluorotrinitromethane	lq	−220.9			
Formaldehyde	g	−108.6	−102.5	218.8	35.4

TABLE 2.54   Enthalpies and Gibbs Energies of Formation, Entropies, and Heat Capacities of Organic Compounds (*Continued*)

Substance	Physical state	$\Delta_f H°$ kJ · mol^{-1}	$\Delta_f G°$ kJ · mol^{-1}	$S°$ J · deg^{-1} · mol^{-1}	$C_p°$ J · deg^{-1} · mol^{-1}
Formamide	lq	−254.0			107.6
	g	−193.9	−141.0	248.6	45.4
Formanilide	c	−151.5			
Formic acid	lq	−424.7	−361.4	129.0	99.5
	g	−378.7	−351.0	248.7	45.2
Formyl fluoride	g	−376.6	−368.1	246.5(8)	40.0
D-(−)-Fructose	c	−1265.6			
D-(+)-Fucose	c	−1099.1			
Fullerene-C$_{60}$	c	2327.0	2302.0	426.0	520.0
Fumaric acid	c	−811.7	−655.6	168.0	142.0
Fumaronitrile	c	268.2			
Furan	lq	−62.3		177.0	114.8
	g	−34.9	0.88	267.2	65.4
2-Furancarboxaldehyde	lq	−201.6			163.2
2-Furancarboxylic acid	c	−498.4			
2-Furanmethanol	lq	−276.2	−154.2	215.5	204.0
Furfuryl alcohol	lq	−276.2			204.0
Furylacrylic acid	c	−459.0			
Furylethylene	lq	−10.5			
D-(+)-Galactose	c	−1286.3	−918.8	205.4	
D-Gluconic acid	c	−1587.0			
D-(+)-Glucose	c	−1273.3	−910.4	212.1	
D-(−)-Glutamic acid	c	−1009.7	−727.5	191.2	
L-(+)-Glutamic acid	c	−1005.2	−731.3	188.2	
L-Glutamine	c	−826.4			
Glutaric acid	c	−960.0			
Glyceraldehyde	lq	−598.0			
Glycerol	lq	−668.5	−477.0	206.3	218.9
Glyceryl 1-acetate	lq	−909.1			
Glyceryl 1-benzoate	c	−777.3			
Glyceryl 2-benzoate	c	−772.8			
Glyceryl 1,3-diacetate	lq	−1120.7			
Glyceryl 1-dodecanoate	c	−1160.9			
Glyceryl 2-dodecanoate	c	−1152.6			
Glyceryl 1-hexadecanoate	c	−1281.5			
Glyceryl 1-hexanoate	c	−1109.0			
Glyceryl 2-hexanoate	c	−1095.8			
Glyceryl 1-octadecanoate	c	−1324.8			
Glyceryl 1-tetradecanoate	c	−1222.6			
Glyceryl triacetate	lq	−1330.8			
Glyceryl trinitrate	lq	−370.9			
Glyceryl tris(dodecanoate)	c	−2046.0			
Glyceryl tris(tetradecanoate)	c	−2176.0			
Glycine	c	−528.5	−368.6	103.5	99.2
ionized; std. state	aq	−469.8	−315.0	111.0	
$^+$H$_3$NCH$_2$COOH; std. state	aq	−517.9	−384.2	190.2	
Glycylglycine	c	−747.7	−490.6	190.0	
Glyoxal	g	−212.0			
Glyoxime	c	−90.5			
Glyoxylic acid	c	−835.5			
Guanidine	c	−56.0			

(*Continued*)

TABLE 2.54 Enthalpies and Gibbs Energies of Formation, Entropies, and Heat Capacities of Organic Compounds (*Continued*)

Substance	Physical state	$\Delta_f H°$ kJ · mol^{-1}	$\Delta_f G°$ kJ · mol^{-1}	$S°$ J · deg^{-1} · mol^{-1}	$C_p°$ J · deg^{-1} · mol^{-1}
Guanidine carbonate	c	−971.9	−557.4	295.4	258.9
Guanidine nitrate	c	−387.0			
Guanidine sulfate	c	−1205.0			
Guanine	c	−183.9	47.4	160.3	
Guanylurea nitrate	c	−427.2			
L-Gulonic acid-γ-lactone	c	−1219.6			
Heptadecane	g	−393.9	82.1	817.3	394.7
Heptadecanoic acid	c	−924.4			475.7
1-Heptadecene	g	−268.4	179.9	813.1	383.9
Heptanal	lq	−311.5	−100.6	335.4	230.1
	g	−264.0	−86.7	461.7	
Heptane	lq	−224.2			224.9
	g	−187.7	8.0	427.9	166.0
Heptanedioic acid	c	−1009.4			
Heptanenitrile	lq	−82.8			
1-Heptanethiol	g	−150.0	36.2	493.3	186.9
Heptanoic acid	lq	−610.2			265.4
1-Heptanol	lq	−403.3	−142.3	320.1	272.1
	g	−336.4	−120.9	480.3	178.7
2-Heptanone	lq				232.6
1-Heptene	lq	−97.9		327.6	211.8
	g	−62.3	95.8	423.6	155.2
cis-2-Heptene	lq	−105.1			
trans-2-Heptene	lq	−109.5			
cis-3-Heptene	lq	−104.3			
trans-3-Heptene	lq	−109.3			
1-Heptyne	g	103.0	226.7	407.7	151.1
Hexabromoethane	g			441.9	139.3
Hexachlorobenzene	c	−127.6	1.1	260.2	201.3
	g	−35.5	44.2	441.2	173.2
Hexachloroethane	c	−202.8		237.3	198.2
	g	−143.6	−54.9	398.7	136.7
Hexadecafluoroethylcyclo-hexane	lq	−3420.0			
Hexadecafluoroheptane	lq	−3420.8	−3093.0	561.8	419.0
Hexadecane	lq	−456.1			501.6
	g	−374.8	83.7	778.3	371.8
Hexadecanoic acid	c	−891.5	−316.1	452.4	460.7
1-Hexadecanol	c	−686.7	−98.7	451.9	422.0
	lq	−635.4	−96.6	606.7	
1-Hexadecene	lq	−328.7		587.9	488.9
	g	−248.5	171.5	774.1	361.0
1,5-Hexadiene	lq	54.1			
2,4-Hexadienoic acid	c	−390.8			
1,5-Hexadiyne	lq	384.2			
Hexafluoroacetone	g	−1249.3			
Hexafluoroacetylacetone	c	−2286.7			
Hexafluorobenzene	lq	−991.3		280.8	156.6
	g	−955.4	−79.4	383.2	
Hexafluoroethane	g	−1344.2	−1255.8	332.3	106.7
cis-Hexahydroindane	g	−127.2			

TABLE 2.54 Enthalpies and Gibbs Energies of Formation, Entropies, and Heat Capacities of Organic Compounds (*Continued*)

Substance	Physical state	$\Delta_f H°$ kJ · mol^{-1}	$\Delta_f G°$ kJ · mol^{-1}	$S°$ J · deg^{-1} · mol^{-1}	$C_p°$ J · deg^{-1} · mol^{-1}
trans-Hexahydroindane	g	−131.4			
Hexamethylbenzene	c	−162.4	117.4	306.3	245.6
1,1,1,3,3,3-Hexamethyldi-silazane	lq	−518.0			
Hexamethyldisiloxane	lq	−814.6	−541.8	433.8	311.4
	g	−777.7	−534.5	535.0	238.5
Hexamethylenetetramine	c	125.5	434.8	163.4	
Hexamethylphosphoric triamide	lq				321
Hexanal	g	−248.4	−100.1	422.9	148.2
Hexanamide	c	−423.0			
	lq	−397.0			
Hexane	lq	−198.8	−3.8	296.1	195.6
	g	−167.1(8)	−0.25	388.4	143.1
1,6-Hexanedioic acid	lq	−985.4	−207.3		232.2
1,2-Hexandediol	lq	−577.1			
1,6-Hexanediol	c	−569.9			
Hexanedinitrile	lq	85.1			128.7
1-Hexanethiol	g	−129.9	27.8	454.3	164.1
Hexanoic acid	lq	−583.9			225.0
1-Hexanol	lq	−377.5	−152.3	287.4	240.4
	g	−317.6	−135.6	441.4	155.6
2-Hexanol	lq	−392.9			
3-Hexanol	lq	−392.4			286.2
2-Hexanone	lq	−322.0			213.3
3-Hexanone	lq	−320.2		305.3	216.9
1-Hexene	lq	−74.1	83.6	295.1	183.3
	g	−43.5	84.45	384.6	132.3
cis-2-Hexene	lq	−83.9			
	g	−52.3	76.2	386.5	125.7
trans-2-Hexene	lq	−85.5			
	g	−53.9	76.4	380.6	132.4
cis-3-Hexene	lq	−79.0			
	g	−47.6	83.0	379.6	123.6
trans-3-Hexene	lq	−86.1			
Hexyl acetate	lq				282.8
	g	−54.4	77.6	374.8	132.8
1-Hexyne	g	123.6	218.6	368.7	128.2
(−)-Histidine	c	−466.7			
Hydantoin	c	−448.5			
Hydrazine	lq	50.6	149.2	121.2	98.9
Hydrazinecarbothioamide	c	24.7			
Hydrazobenzene	c	221.3			
Hydroxyacetic acid	c	−663.6			
2′-Hydroxyacetophenone	c	−357.7			
3′-Hydroxyacetophenone	c	370.7			
4′-Hydroxyacetophenone	c	−364.4			
2-Hydroxybenzaldehyde	lq	−279.9			
2-Hydroxybenzaldoxime	c	−183.7			
2-Hydroxybenzoic acid	c	−589.9	−421.3	178.2	159.1
3-Hydroxybenzoic acid	c	−584.9	−417.3	177.0	157.3
4-Hydroxybenzoic acid	c	−584.5	−416.5	175.7	155.1

(Continued)

TABLE 2.54   Enthalpies and Gibbs Energies of Formation, Entropies, and Heat Capacities of Organic Compounds (*Continued*)

Substance	Physical state	$\Delta_f H°$ kJ · mol⁻¹	$\Delta_f G°$ kJ · mol⁻¹	$S°$ J · deg⁻¹ · mol⁻¹	$C_p°$ J · deg⁻¹ · mol⁻¹
(±)-2-Hydroxybutanoic acid	lq	−679.1			
2-Hydroxy-2,4,6-cyclohepta-trienone	c	−239.2			
2-Hydroxyisobutanoic acid	c	−744.3			
2-Hydroxy-1-isopropyl-4-methylbenzene	c	−309.6			
3-Hydroxy-4-methoxybenz-aldehyde	c	−453.6			
4-Hydroxy-4-methyl-2-pentanone	lq				221.3
2-Hydroxymethyl-1,3-propane-diol	c	−744.6			
3-Hydroxy-2-naphthalene-carboxylic acid	c	−547.7			
5-Hydroxy-1-pentanal	lq	−479.9			
trans-(−)-4-Hydroxyproline	c	−661.1			
(*S*)-2-Hydroxypropanoic acid	c	−694.0			
2-Hydroxypropanonitrile	lq	−138.9	34.3		
2-Hydroxypyridine	c	−166.3			
3-Hydroxypyridine	c	−132.0			
4-Hydroxypyridine	c	−144.6			
8-Hydroxyquinoline	c	−81.2			
(−)-2-Hydroxysuccinic acid	c	−1103.7	−884.7		
(±)-2-Hydroxysuccinic acid	c	−1105.7			
Hypoxanthene	c	−110.8	76.9	145.6	134.5
Icosane	g	−455.8	117.3	934.1	463.3
Icosanoic acid	c	−1011.9			545.1
Icosene	g	−330.2	205.1	929.9	452.5
Imidazole	c	49.8			
Iminodiacetic acid	c	−932.6			
Indane	lq	11.5	150.8	56.0	190.3
1*H*-Indazole	c	151.9			
Indene	lq	110.6	217.6	215.3	186.9
1*H*-Indole	c	86.7			
Indole-2,3-dione	c	−268.2			
Iodoacetone	g	−130.5			
Iodobenzene	lq	117.1		205.4	158.7
	g	164.9	187.8	334.1	100.8
2-Iodobenzoic acid	c	−302.3			
3-Iodobenzoic acid	c	−316.9			
4-Iodobenzoic acid	c	−316.1			
Iodocyclohexane	lq	−97.2			
Iodoethane	lq	−40.0	14.7	211.7	115.1
	g	−8.1	19.2	306.0	66.9
Iodoethylene	g			285.0	57.9
Iodomethane	g	14.4	15.6	254.1	44.1
2-Iodo-2-methylpropane	lq	−107.5			162.3
	g	−72.0	23.6	342.2	118.3
1-Iodonaphthalene	lq	161.5			
2-Iodonaphthalene	c	144.3			
2-Iodophenol	c	−95.8			
3-Iodophenol	c	−94.5			

TABLE 2.54   Enthalpies and Gibbs Energies of Formation, Entropies, and Heat Capacities of Organic Compounds (*Continued*)

Substance	Physical state	$\Delta_f H°$ kJ · mol^{-1}	$\Delta_f G°$ kJ · mol^{-1}	$S°$ J · deg^{-1} · mol^{-1}	$C_p°$ J · deg^{-1} · mol^{-1}
4-Iodophenol	c	−95.4			
1-Iodopropane	lq	−66.0			126.8
	g	−30.0			
2-Iodopropane	lq	−74.8			91.0
	g	−40.3	20.1	324.5	90.1
3-Iodopropanoic acid	c	−460.0			
3-Iodo-1-propene	g	91.5			
α-Iodotoluene	lq	57.7			
3-Iodotoluene	lq	79.1			
4-Iodotoluene	lq	67.4			
Isobutanenitrile	g	25.4	103.6	313.3	96.4
Isobutylamine	lq	−132.6			183.2
Isobutylbenzene	lq	−69.8			
Isobutyl trichloroacetate	lq	−553.4			
Isocyanomethane	g	163.5	165.7	246.9	52.9
(−)-Isoleucine	c	−637.9	−347.2	208.0	188.3
(±)-Isoleucine	c	−635.3			
Isoxazole	g	78.6			
Isopropenyl acetate	lq	−386.4			
Isopropyl acetate	lq	−518.9			199.4
Isopropylamine	lq	−112.3		218.3	163.8
	g	−83.7	32.2	312.2	97.5
Isopropylbenzene	lq	−41.1	124.3	279.8	210.7
	g	4.0	137.0	388.6	151.7
1-Isopropyl-2-methylbenzene	lq	−73.3			
1-Isopropyl-3-methylbenzene	lq	−78.6			
1-Isopropyl-4-methylbenzene	lq	−78.0	119.1	306.6	
Isopropyl methyl ether	lq	−278.8		253.8	161.9
	g	−252.0	−120.9	332.3	111.1
2-Isopropyl-5-methylphenol	c	−309.7			
Isopropyl methyl sulfide	lq	−105.7		263.1	172.4
	g	−90.5	13.4	359.3	117.2
Isopropyl nitrate	g	−191.0	−40.7	373.2	120.7
2-Isopropylphenol	lq	−233.7			
3-Isopropylphenol	lq	−252.5			
4-Isopropylphenol	lq	−265.9			
Isopropyl thioacetate	lq	−298.2			
Isopropyl trichloroacetate	lq	−536.0			
Isoquinoline	c	144.5			
	lq				196.8
Ketene	g	−47.5	−48.3	247.6	51.8
(+)-Lactic acid	c	−694.1	−522.9	142.3	
(±)-Lactic acid	lq	−674.5	−518.2	192.1	
β-Lactose	c	−2236.7	−1567.0	386.2	
(+)-Leucine	c	−637.3	−347.2	208.0	
(−)-Leucine	c	−637.4	−346.3	211.8	201.0
(+)-Limonene	lq	−54.5			249.0
(±)-Lysine	c	−678.6			
Malic acid	c	−789.4	−625.1	160.8	137.0
Maleic anhydride	c	−469.8			
(R)-Malic acid	c	−1105.7			
(S)-Malic acid	c	−1103.6			

(*Continued*)

TABLE 2.54 Enthalpies and Gibbs Energies of Formation, Entropies, and Heat Capacities of Organic Compounds (*Continued*)

Substance	Physical state	$\Delta_f H°$ kJ · mol^{-1}	$\Delta_f G°$ kJ · mol^{-1}	$S°$ J · deg^{-1} · mol^{-1}	$C_p°$ J · deg^{-1} · mol^{-1}
Malonamide	c	−546.0			
Malonic acid	c	−891.0			
Malonodiamide	c	−546.1			
Malononitrile	c	186.6			
D-(+)-Maltose	c	−2220.9	−1726.3		
(±)-Mandelic acid	c	−579.4			
(+)-Mannitol	c	−1337.1	−942.2	238.5	
D-(+)-Mannose	c	−1263.0			
2-Mercaptopropanoic acid	lq	−468.2	−343.9	228.9	
Methane	g	−74.6	−50.5	186.3	35.7
Methane-d_4	g	−88.2	−59.5	198.9	40.3
Methanethiol	lq	−46.7	−7.7	169.2	90.5
	g	−22.9	−9.9	255.1	50.3
Methanol	lq	−239.1	−166.6	126.8	81.2
	g	−201.0	−162.3	239.9	44.1
(−)-Methionine	c	−577.5	−505.8	231.5	
2-Methoxybenzaldehyde	c	−266.5			
3-Methoxybenzaldehyde	lq	−276.1			
4-Methoxybenzaldehyde	lq	−267.2			
Methoxybenzene	lq	−114.8			199.0
	g	−67.9			
2-Methoxybenzoic acid	c	−538.5			
3-Methoxybenzoic acid	c	−553.5			
4-Methoxybenzoic acid	c	−561.7			
2-Methoxyethanol	lq				171.1
2-Methyoxyethyl acetate	lq				310.0
2-Methoxytetrahydropyran	lq	−442.3			
5-Methoxy tetrazole	c	69.1			
1-Methoxy-2,4,6-trinitro-benzene	c	−157.5			
Methyl (CH$_3$)	g	145.7	147.9	194.2	38.7
Methyl acetate	lq	−445.8			141.9
	g	−413.3		324.4	86.0
Methyl acrylate	lq	−362.2	−243.2	239.5	158.8
	g	−333.0	−237.6		
Methylamine	lq	−47.2	35.7	150.2	102.1
	g	−22.5	32.7	242.9	50.1
N-Methylaniline	lq	32.2			207.1
o-Methylaniline	lq	−6.3			209.6
	g	56.4	167.6	351.0	130.2
m-Methylaniline	lq	−8.1			227.0
	g	54.6	165.4	352.5	125.5
p-Methylaniline	lq	−23.5			
	g	55.3	167.7	347.0	126.2
Methyl benzoate	lq	−343.5			221.3
2-Methylbenzoic acid	c	−416.5			
	lq				174.9
3-Methylbenzoic acid	c	−426.1			
	lq				163.6
4-Methylbenzoic acid	c	−429.2			
	lq				169.0

TABLE 2.54 Enthalpies and Gibbs Energies of Formation, Entropies, and Heat Capacities of Organic Compounds (*Continued*)

Substance	Physical state	$\Delta_f H°$ kJ · mol^{-1}	$\Delta_f G°$ kJ · mol^{-1}	$S°$ J · deg^{-1} · mol^{-1}	$C_p°$ J · deg^{-1} · mol^{-1}
2-Methylbenzoic anhydride	c	−533.5			
4-Methylbenzoic anhydride	c	−520.9			
1-Methylbicyclo[4.1.0]heptane	lq	−59.9			
1-Methylbicyclo[3.1.0]hexane	lq	−33.2			
2-Methylbiphenyl	lq	108.0			
3-Methylbiphenyl	lq	85.4			
4-Methylbiphenyl	c	55.2			
2-Methyl-1,3-butadiene	lq	48.2		229.3	152.6
	g	75.5	145.9	315.6	104.6
3-Methyl-1,2-butadiene	g	129.7	198.6	319.7	105.4
2-Methylbutane	lq	−178.4		260.4	164.8
	g	−154.0	−14.8	343.6	118.8
2-Methyl-2-butanethiol	lq	−162.8		290.1	198.1
	g	−127.1	9.2	386.9	143.5
3-Methyl-1-butanethiol	g	−114.9			
3-Methyl-2-butanethiol	lq	−158.8			
2-Methylbutanoic acid	lq	−554.4			
3-Methylbutanoic acid	lq	−561.6			197.1
2-Methyl-1-butanol	lq	−356.6			220.1
3-Methyl-1-butanol	lq	−356.4			210.0
2-Methyl-2-butanol	lq	−379.5	−175.3	229.3	247.1
(±)-3-Methyl-2-butanol	lq	−366.6			232.2
3-Methyl-2-butanone	lq	−299.5		268.5	179.9
	g	−262.5			
2-Methyl-1-butene	lq	−61.1		254.0	157.2
	g	−35.3	65.6	339.5	110.0
3-Methyl-1-butene	lq	−51.5		253.3	156.1
	g	−27.6	74.8	333.5	118.6
2-Methyl-2-butene	lq	−68.6		251.0	152.8
	g	−41.8	59.7	338.6	105.0
trans-2-Methyl-2-butenedioic acid [also *cis*]	c	−824.4			
cis-2-Methyl-2-butenoic acid	c	−455.6			
trans-2-Methyl-2-butenoic acid	c	−490.8			
3-Methylbutyl acetate	lq				248.5
3-Methyl-1-butyne	g	136.4	205.5	319.0	104.7
Methyl *trans*-2-butenoate	lq	−382.8			
Methylcyclobutane	lq	−44.5			
Methylcyclobutanecarboxylic acid	lq	−395.0			
Methylcyclohexane	lq	−190.1	20.3	247.9	184.9
	g	−154.7	27.3	343.3	135.0
cis-2-Methylcyclohexanol	lq	−390.2			200[17]
trans-2-Methylcyclohexanol	lq	−415.8			200[17]
cis-3-Methylcyclohexanol	lq	−416.1			292[17]
trans-3-Methylcyclohexanol	lq	−394.4			202[17]
cis-4-Methylcyclohexanol	lq	−413.2			202[17]
trans-4-Methylcyclohexanol	lq	−433.3			202[17]
2-Methylcyclohexene	lq	−81.2			
Methylcyclopentane	lq	−138.0	31.5	247.9	158.7
	g	−106.2	35.8	339.9	109.8

(*Continued*)

TABLE 2.54 Enthalpies and Gibbs Energies of Formation, Entropies, and Heat Capacities of Organic Compounds (*Continued*)

Substance	Physical state	$\Delta_f H°$ kJ · mol⁻¹	$\Delta_f G°$ kJ · mol⁻¹	$S°$ J · deg⁻¹ · mol⁻¹	$C_p°$ J · deg⁻¹ · mol⁻¹
1-Methylcyclopentanol	lq	−343.3			
2-Methylcyclopentanone	lq	−265.3			
1-Methylcyclopentene	g	−3.8	102.1	326.4	100.8
3-Methylcyclopentene	g	7.4	115.0	330.5	100.0
4-Methylcyclopentene	g	14.6	121.6	328.9	100.0
1-Methylcyclopropene	lq	1.7			
	g	243.6			
Methylenecyclobutane	g	121.6			
Methylenebutanedioic acid	c	−841.1			
Methylenecyclohexane	lq	−61.3			
Methylenecyclohexene	lq	−12.7			
Methylenecyclopropane	g	200.5			
Methyl decanoate	lq	−640.4			
Methyl 2,2-dimethylpropanoate	lq	−530.0			257.9
2-Methyl-1,3-dioxane	lq	−436.4			
4-Methyl-1,3-dioxane	c	416.1			
N-Methyldiphenylamine	lq	120.5			
4-Methyldiphenylamine	c	49.0			
Methyl dodecanoate	lq	−693.0			
Methylene (CH₂)	g	390.4	372.9	194.9	33.8
Methylenebutanedioic acid	c	−841.1			
Methylenecyclohexane	lq	−61.3			
2-Methylenecyclohexanol	lq	−277.6			
3-Methylenecyclohexene	lq	−12.7			
2-Methylenecyclopentanol	lq	46.9			
Methylenecyclopropane	g	200.5			
Methylenesuccinic acid	c	−841.2			
Methylene sulfate	c	−688.7			
N-Methylformamide	lq				123.8
Methyl formate	lq	−386.1			119.1
	g	−357.4	−297.2	285.3	64.4
Methyl 2-furancarboxylate	lq	−450.0			
2-Methyl-2,5-furandione	lq	−504.5			
α-Methyl-(+)-glucoside	c	−1233.4			
N-Methylglycine	c	−513.3			
Methylglyoxal	g	−27.1			
Methylglyoxime	c	−126.8			
2-Methylheptane	lq	−255.0		356.4	252.0
	g	−215.4	12.8	452.5	
3-Methylheptane	lq	−252.3		362.6	250.2
	g	−212.5	13.7	461.6	
4-Methylheptane	lq	−251.6			251.1
	g	−212.0	16.7	453.3	
Methyl heptanoate	lq	−567.1			285.1
2-Methylhexane	lq	−229.5		323.3	222.9
	g	−194.6	3.2	420.0	166.0
3-Methylhexane	lq	−226.4			214.2
	g	−192.3	4.6	424.1	166.0
Methyl hexanoate	lq	−540.2			
5-Methyl-1-hexene	g	−65.7			
cis-3-Methyl-3-hexene	g	−79.4			

TABLE 2.54 Enthalpies and Gibbs Energies of Formation, Entropies, and Heat Capacities of Organic Compounds (*Continued*)

Substance	Physical state	$\Delta_f H°$ kJ $\cdot$ mol^{-1}	$\Delta_f G°$ kJ $\cdot$ mol^{-1}	$S°$ J $\cdot$ deg$^{-1}\cdot$ mol^{-1}	$C_p°$ J $\cdot$ deg$^{-1}\cdot$ mol^{-1}
trans-3-Methyl-3-hexene	g	−76.8			
Methylhydrazine	lq	54.2	179.9	165.9	134.9
	g	94.7	186.9	278.7	71.1
2-Methyl-1*H*-indole	c	60.7			
3-Methyl-1*H*-indole	c	68.2			
Methyl isocyanate	lq	−92.0			
Methyl isocyanide	g	163.5	165.7	246.8	52.9
1-Methyl-4-isopropylbenzene	lq	−78.0			236.4
Methyl isopropyl sulfide	g	−90.4	13.4	359.3	117.2
Methyl isothiocyanate	c	79.4			
	g	131.0	144.4	252.3	65.5
5-Methylisoxazole	lq	−5.6			
Methylmercury bromide	c	−86.2			
Methylmercury chloride	c	−116.3			
Methylmercury iodide	c	−43.5			
Methyl 2-methylbutanoate	lq	−534.3			
Methyl 3-methylbutanoate	lq	−538.9			
7-Methyl-3-methylene-1,6-octadiene	lq	14.5			
(*R*)-1-Methyl-4-(1-methyl-ethenyl)cyclohexene	lq	−54.5			249[20]
1-Methylnaphthalene	lq	56.3	189.4	254.8	224.4
2-Methylnaphthalene	c	44.9	192.6	220.0	196.0
	g	106.7	216.2	380.0	159.8
Methyl nitrate	lq	−156.3	−43.5	217.2	157.3
	g	−124.4	−39.3	318.5	76.5
Methyl nitrite	g	−66.1	1.0	284.3	63.2
Methyl nitroacetate	lq	−464.0			
2-Methyl-5-nitroaniline	c	−91.3			
4-Methyl-3-nitroaniline	c	−71.7			
1-Methyl-2-nitrobenzene	lq	−9.7			
1-Methyl-3-nitrobenzene	lq	−31.5			
1-Methyl-4-nitrobenzene	c	−48.1			
2-Methyl-2-nitropropane	c	−229.8			
2-Methyl-2-nitro-1,3-propanediol	c	−575.3			
2-Methyl-2-nitro-1-propanol	c	−410.0			
2-Methylnonane	lq	−309.8		420.1	313.3
5-Methylnonane	lq	−307.9		423.8	314.4
Methyl phenylcarbamate	c	−186.7			
Methyl *cis*-9-octadecanoate	lq	−734.5			
Methyl octanoate	lq	−590.3			
2-Methyl-2-oxazoline	g	−130.5			
2-Methylpentane	lq	−204.6		290.6	193.7
	g	−174.8	−5.0	380.5	144.2
3-Methylpentane	lq	−202.4		292.5	190.7
	g	−172.1	2.1	379.8	143.1
2-Methyl-2,4-pentanediol	lq				236.0
Methyl pentanoate	lq	−514.2			229.3
2-Methyl-1-pentanol	lq				248.0
2-Methyl-3-pentanol	lq	−396.4			

TABLE 2.54 Enthalpies and Gibbs Energies of Formation, Entropies, and Heat Capacities of Organic Compounds (*Continued*)

Substance	Physical state	$\Delta_f H°$ kJ · mol^{-1}	$\Delta_f G°$ kJ · mol^{-1}	$S°$ J · deg^{-1} · mol^{-1}	$C_p°$ J · deg^{-1} · mol^{-1}
3-Methyl-2-pentanol	lq				275.9
3-Methyl-3-pentanol	lq				293.4
4-Methyl-2-pentanol	lq	−394.7			273.0
2-Methyl-3-pentanone	lq	−325.9			
4-Methyl-2-pentanone	lq				213.3
2-Methyl-1-pentene	g	−59.4	77.6	382.2	135.6
2-Methyl-2-pentene	g	−66.9	71.2	378.4	126.6
3-Methyl-1-pentene	g	−49.5	86.4	376.8	142.4
cis-3-Methyl-2-pentene	g	−62.3	73.2	378.4	126.6
trans-3-Methyl-2-pentene	g	−63.1	71.3	381.8	126.6
4-Methyl-1-pentene	g	−51.3	90.0	367.7	126.5
cis-4-Methyl-2-pentene	g	−57.5	82.1	373.3	133.6
trans-4-Methyl-2-pentene	g	−61.5	79.6	368.3	141.4
Methyl 2-methylpropenoate	lq				191.2
4-Methyl-3-penten-2-one	lq				212.5
Methyl pentyl sulfide	g	122.9	35.1	450.7	163.7
3-Methyl-1-phenyl-1-butanone	lq	−220.2			
Methyl phenyl sulfide	lq	43.0			
Methyl phenyl sulfone	c	−345.4			
Methylphosphonic acid	c	−1054			
(±)-2-Methylpiperidine	lq	−124.9			
2-Methylpropanal	lq	−247.4			
	g	−215.8			
N-Methylpropan amide	lq				179
2-Methylpropanamine	lq	−132.6			183.2
2-Methylpropane	g	−134.2	−20.9	294.6	130.5^{-12}
2-Methyl-1,2-propanediamine	lq	−133.9			
2-Methyl-1,2-propanediol	lq	−539.7			
2-Methylpropanenitrile	lq	−13.8			
2-Methyl-1-propanethiol	g	−97.3	5.6	362.9	118.3
2-Methyl-2-propanethiol	g	−109.6	0.7	338.0	121.0
2-Methylpropanoic acid	lq				173
2-Methyl-1-propanol	lq	−334.7		214.7	181.2
	g	−283.9	−167.35	359.0	111.3
2-Methyl-2-propanol	lq	−359.2		193.3	219.8
	g	−312.5	−177.7	326.7	113.6
2-Methylpropene	g	−16.9	58.1	293.6	89.1
2-Methylpropenoic acid	lq				161.1
1-Methyl-2-propylbenzene	lq	−72.5			
1-Methyl-3-propylbenzene	lq	−76.2			
1-Methyl-4-propylbenzene	lq	−75.1			
(2-Methylpropyl)benzene	lq	−69.8			240.6
Methyl propyl ether	lq	−266.0		262.9	165.4
	g	−238.2	−109.9	349.5	112.5
Methyl propyl sulfide	g	−82.3	18.4	371.7	117.4
2-Methylpyridine	lq	56.7	166.5	217.9	158.4
	g	99.2	177.1	325.0	100.0
3-Methylpyridine	lq	61.9	214.0	216.3	158.7
	g	106.4	184.3	325.0	99.6
4-Methylpyridine	lq	59.2		209.1	159.0
1-Methyl-1*H*-pyrrole	lq	62.4			

TABLE 2.54 Enthalpies and Gibbs Energies of Formation, Entropies, and Heat Capacities of Organic Compounds (*Continued*)

Substance	Physical state	$\Delta_f H°$ kJ · mol^{-1}	$\Delta_f G°$ kJ · mol^{-1}	$S°$ J · deg^{-1} · mol^{-1}	$C_p°$ J · deg^{-1} · mol^{-1}
2-Methyl-1*H*-pyrrole	lq	23.3			
3-Methyl-1*H*-pyrrole	lq	20.5			
N-Methylpyrrolidone	lq	−262.2			307.8
2-Methylquinoline	c	164.4			
Methyl salicylate	lq	−531.8			249.0
Methylsilane	g			256.5	65.9
α-Methylstyrene	g	113.0	208.5	383.7	145.2
cis-(β)-Methylstyrene	g	121.3	216.9	383.7	145.2
trans-(β)-Methylstyrene	g	117.2	213.7	380.3	146.0
Methylsuccinic acid	c	−958.2			
Methylsuccinic anhydride	lq	−617.6			
Methyl tetradecanoate	lq	−743.9			
2-Methylthiacyclopentane	g	−63.3			
4-Methylthiazole	lq	68.0			
Methylthiirane	g	45.8			
2-Methylthiophene	lq	44.6			149.8
	g	83.5	122.9	320.6	95.4
3-Methylthiophene	lq	43.1			
	g	82.6	121.8	321.3	94.9
Methyl *p*-tolyl sulfone	c	−372.8			
5-Methyluracil	c	−462.8			
Methylurea	c	−332.8			
Morphine monohydrate	c	−711.7			
Morpholine	lq				164.8
Murexide	c	−1212.1			
Naphthalene	c	77.9	201.6	167.4	165.7
	g	150.6	224.1	333.1	131.9
1-Naphthaleneacetic acid	c	−359.2			
2-Naphthaleneacetic acid	c	−371.9			
1-Naphthoic acid	c	333.5			
2-Naphthoic acid	c	−346.1			
1-Naphthol	c	−121.0			166.9
2-Naphthol	lq	−124.2			
1,4-Naphthoquinone	c	−183.4			
1-Naphthyl acetate	c	−288.2			
2-Naphthyl acetate	c	−304.3			
1-Naphthylamine	c	67.8			
2-Naphthylamine	c	59.7			
Nicotine	lq	39.3			
Nitrilotriacetic acid	c	−1311.9	−1307.5		
Nitroacetone	lq	−278.6			
2-Nitroaniline	c	−26.1	178.2	176.2	166.0
3-Nitroaniline	c	−38.3	174.1	176.2	158.8
4-Nitroaniline	c	−42.0	151.0	176.2	167.0
Nitrobenzene	lq	12.5	146.2	224.3	185.8
2-Nitrobenzoic acid	c	−378.5	−196.4	208.4	
3-Nitrobenzoic acid	c	−394.7	−220.5	205.0	
4-Nitrobenzoic acid	c	−392.2	−222.0	210.0	181.2
3-Nitrobiphenyl	c	65.1			
4-Nitrobiphenyl	c	40.5			
1-Nitrobutane	g	−143.9	10.1	394.5	124.9

(*Continued*)

TABLE 2.54 Enthalpies and Gibbs Energies of Formation, Entropies, and Heat Capacities of Organic Compounds (*Continued*)

Substance	Physical state	$\Delta_f H°$ kJ · mol⁻¹	$\Delta_f G°$ kJ · mol⁻¹	$S°$ J · deg⁻¹ · mol⁻¹	$C_p°$ J · deg⁻¹ · mol⁻¹
2-Nitrobutane	g	−163.6	−6.2	383.3	123.5
3-Nitro-2-butanol	lq	−390.0			
N-Nitrodiethylamine	lq	−106.2			
2-Nitrodiphenylamine	c	64.4			
Nitroethane	lq	−143.9			134.4
	g	−102.3	−4.9	315.4	78.2
2-Nitroethanol	lq	−350.7			
2-Nitrofuran	c	−104.1			
5-Nitrofurancarboxylic acid	c	−516.8			
1-Nitroguanidine	c	−92.4			
Nitromethane	lq	−113.1	−14.4	171.8	106.6
	g	−74.3	−6.8	275.0	57.3
(Nitromethyl)benzene	lq	−22.8			
1-Nitronaphthalene	c	42.6			
1-Nitroso-2-naphthol	c	−50.5			
2-Nitroso-1-naphthol	c	−61.8			
4-Nitroso-1-naphthol	c	−107.8			
1-Nitropropane	lq	−167.2			175.3
	g	−123.8			
2-Nitropropane	lq	−180.3			170.3
	g	−139.0			
1-Nitro-2-propanone	c	−294.7			
4-Nitrosodiphenylamine	c	213.0			
β-Nitrostyrene	c	30.5			
4-Nitrotoluene	c	−48.1			172.3
Nonadecane	g	−435.1	108.9	895.2	440.4
1-Nonadecene	g	−309.6	196.7	891.0	429.7
1-Nonanal	g	−310.3	−74.9	539.6	216.8
Nonane	lq	−274.7			284.4
	g	−228.2	24.8	505.7	211.7
1-Nonanethiol	g	−190.8	53.0	571.2	232.7
Nonanoic acid	lq	−659.7			362.4
1-Nonanol	g	−376.3	−110.5	558.6	224.3
2-Nonanone	lq	−397.2			
5-Nonanone	lq	−398.2		401.4	303.6
1-Nonene	g	−103.5	112.7	501.5	201.0
Norleucine	c	−639.1			
Octadecane	c	−567.4		480.2	485.6
	g	−414.6	100.5	856.2	417.6
Octadecanoic acid	c	−947.7			501.5
1,8-Octadecanoic acid	c	−1038.1			
1-Octadecene	g	−289.0	188.3	852.0	406.8
cis-9-Octadecenoic acid	lq	−743.5			577.0⁵⁰
trans-9-Octadecenoic acid	c	−910.9			
1,7-Octadiyne	lq	334.4			
Octafluorocyclobutane	lq				209.8⁻⁶
	g	−1542.6	−1398.8	400.4	156.2
Octafluoropropane	g	−1783.1			
Octafluorotoluene	lq	−1311.1		355.5	262.3
1-Octanal	g	−289.6	−83.3	500.7	194.0
Octanamide	c	−473.2			

TABLE 2.54 Enthalpies and Gibbs Energies of Formation, Entropies, and Heat Capacities of Organic Compounds (*Continued*)

Substance	Physical state	$\Delta_f H°$ kJ · mol^{-1}	$\Delta_f G°$ kJ · mol^{-1}	$S°$ J · deg^{-1} · mol^{-1}	$C_p°$ J · deg^{-1} · mol^{-1}
Octane	lq	−250.1			254.6
	g	−208.6	16.4	466.7	188.9
1-Octanenitrile	lq	−107.3			
1-Octanethiol	g	−44.9	44.6	582.2	209.8
Octanoic acid	lq	−636.0			297.9
1-Octanol	lq	−426.5	−143.1	377.4	305.1
2-Octanol	lq				330.1
2-Octanone	lq	−384.5	−140.3	373.8	273.3
1-Octene	lq	−121.8			241.0
	g	−81.4	104.2	462.5	178.1
cis-2-Octene	lq	−135.7			239.0
trans-2-Octene	lq	−135.7			239.0
1-Octyne	g	82.4	235.4	496.6	174.0
(±)-Ornithine	c	−652.7			
Oxalic acid	c	−821.7	−697.9	109.8	91.0
Oxalic acid dihydrate	c	−1492.0			
Oxaloyl dichloride	lq	−367.6			
Oxaloyl dihydrazide	c	−295.2			
Oxamic acid	c	−661.2			
Oxamide	c	−504.4	−342.7	118.0	
Oxazole	g	−5.5			
2-Oxetanone	lq	−329.9		175.3	122.1
Oxindole	c	−172.4			
2-Oxohexamethyleneimine	c	−329.4	−95.1	168.6	156.8
Oxomethyl (HCO)	g	43.1	28.0	224.7	34.6
2-Oxo-1,5-pentanedioic acid	c	−1026.2			
4-Oxopentanoic acid	c	−697.1			
2-Oxopropanoic acid	lq	−584.5	−463.4	179.5	
8-Oxypurine	c	−64.4			
Papaverine	c	−502.3			
Paraformaldehyde	c	−177.6			
Paraldehyde	lq	−687.0			
Pentachloroethane	lq	−187.6			173.8
	g	−142.0	−70.3	381.5	118.1
Pentachlorofluoroethane	g	−317.2	−234.0	391.8	
Pentachlorophenol	c	−292.4	−144.1	251.9	202.0
Pentacyclo[4.2.0.0^{2,5}.0^{3,8}.0^{4,7}]-octane	c	541.8			
Pentadecane	g	−352.8	75.2	739.4	349.0
Pentadecanoic acid	c	−861.7			443.3
1-Pentadecene	g	−227.2	163.1	735.2	338.2
1-Pentadecyne	g	−61.8	293.9	719.3	33.41
1,2-Pentadiene	g	140.7	210.4	333.5	105.4
cis-1,3-Pentadiene	g	81.5	145.8	324.3	94.6
trans-1,3-Pentadiene	g	76.5	146.73	319.7	103.3
1,4-Pentadiene	g	105.7	170.3	333.5	105.0
2,3-Pentadiene	g	133.1	205.9	324.7	101.3
Pentaerythritol	c	−920.6	−613.8	198.1	190.4
Pentaerythritol tetranitrate	c	−538.6			
Pentafluorobenzoic acid	c	−1239.6			
Pentafluoroethane	g	−1104.6	−1029.3	333.7	95.7

(Continued)

TABLE 2.54 Enthalpies and Gibbs Energies of Formation, Entropies, and Heat Capacities of Organic Compounds (*Continued*)

Substance	Physical state	$\Delta_f H°$ kJ · mol^{-1}	$\Delta_f G°$ kJ · mol^{-1}	$S°$ J · deg^{-1} · mol^{-1}	$C_p°$ J · deg^{-1} · mol^{-1}
Pentafluorophenol	c	−1024.1			
2,3,4,5,6-Pentafluorotoluene	lq	−883.8		306.4	225.8
Pentamethylbenzene	c	−133.6			
	g	−74.5	123.3	443.9	216.5
Pentamethylbenzoic acid	c	−536.1			
Pentanal	g	−228.5	−108.3	383.0	125.4
Pentanamide	c	−379.5			
1-Pentanamine	lq				218.0
Pentane	lq	−173.5	−9.3	262.7	167.2
	g	−146.9	−8.4	349.0	120.2
1,5-Pentanediol	lq	−531.5			321.3
2,4-Pentanedione	lq	−423.8			208.2
	g	−380.6		397.9	120.1
1,5-Pentanedithiol	g	−71.0			
Pentanenitrile	lq	−33.1			180
1-Pentanethiol	lq	−151.3			
Pentanoic acid	lq	−559.4		259.8	210.3
	g	−491.9	−357.2	439.8	
1-Pentanol	lq	−351.6			208.1
	g	−294.7	−146.0	402.5	133.1
2-Pentanol	lq	−365.2			
	g	−311.0			
3-Pentanol	lq	−368.9			239.7
	g	−311.4	−158.2	382.0	
2-Pentanone	lq	−297.3			184.1
	g	−259.0	−137.1	376.2	121.0
3-Pentanone	lq	−296.5		266.0	190.9
1-Pentene	lq	−46.0		262.6	154.0
	g	−21.2	79.1	345.8	109.6
cis-2-Pentene	lq	−53.7		258.6	151.7
	g	−27.6	71.8	346.3	101.8
trans-2-Pentene	lq	−58.2		256.5	157.0
	g	−31.9	69.9	340.4	108.5
cis-2-Pentenenitrile	lq	71.8			
trans-2-Pentenenitrile	lq	74.9			
trans-3-Pentenenitrile	lq	80.9			
2-Pentenoic acid	lq	−446.4			
3-Pentenoic acid	lq	−434.8			
4-Pentenoic acid	lq	−430.6			
cis-3-Penten-1-yne	lq	226.5			
trans-3-Penten-1-yne	lq	228.2			
Pentyl acetate	lq				261.0
1-Pentyne	g	144.4	210.3	329.8	106.7
2-Pentyne	g	128.9	194.2	331.8	98.7
Perfluoropiperidine	lq	−2020.5	−1768.5	393.4	296.8
Perylene	c	182.8			
α-Phellandrene	lq	41.3			
Phenanthrene	c	116.2	268.3	215.1	220.6
9,10-Phenanthrenedione	c	−154.7			
Phenazine	c	237.0			

TABLE 2.54 Enthalpies and Gibbs Energies of Formation, Entropies, and Heat Capacities of Organic Compounds (*Continued*)

Substance	Physical state	$\Delta_f H°$ kJ · mol^{-1}	$\Delta_f G°$ kJ · mol^{-1}	$S°$ J · deg^{-1} · mol^{-1}	$C_p°$ J · deg^{-1} · mol^{-1}
Phenol	c	−165.1	−50.4	144.0	127.4
	lq				199.8[41]
	g	−96.4	−32.9	315.6	103.6
Phenoxyacetic acid	c	−513.8			
Phenyl acetate	lq	−334.9			
Phenylacetic acid	c	−398.7			
Phenylacetylene	g	327.3	363.5	321.7	114.9
(±)-3-Pheny 1-2-Alanine	c	−466.9	−211.7	213.6	203.0
Phenyl benzoate	c	−241.0			
Phenylboron dichloride	lq	−299.4			
1-Phenylcyclohexene	lq	−16.8			
Phenylcyclopropane	lq	100.3			
N-Phenyldiacetamide	c	−362.5			
1,3-Phenylenediamine	c	−7.8		154.5	159.6
Phenyl formate	lq	−268.7			
N-Phenylglycine	c	−402.5			
(±)-2-Phenylglycine	c	−431.8			
Phenylhydrazine	lq	141.0			217.0
Phenyl 2-hydroxybenzoate	c	−436.6			
Phenylmethanethiol	lq	43.5			
Phenylmethyl acetate	lq				148.5
N-Phenyl-2-naphthylamine	c	159.8			
1-Phenyl-1-propanone	lq	−167.2			
1-Phenyl-2-propanone	lq	−151.9			
1-Phenylpyrrole	c	154.3			
2-Phenylpyrrole	c	139.2			
Phenylsuccinic acid	c	−841.0			
S-Phenyl thioacetate	lq	−122.0			
Phenyl vinyl ether	lq	−26.2			
Phosgene	g	−220.9	−206.8	283.8	57.7
Phthalamide	c	−433.1			
1,2-Phthalic acid	c	−782.0	−591.6	207.9	188.3
1,3-Phthalic acid	c	−803.0			
1,4-Phthalic acid	c	−816.1			
Phthalic anhydride	c	−460.1	−331.0	180.0	160.0
Phthalonitrile	c	280.6			
Picric acid	c	−214.4			
α-Pinene	lq	−16.4			
β-Pinene	lq	−7.7			
Piperazine	c	−45.6	240.2	85.8	
2,5-Piperazinedione	c	−446.5			
Piperidine	lq	−86.4		210.0	179.9
2-Piperidone	c	−306.6	−112.1	164.9	(lq 307.8)
L-Proline	c	515.2			
Propadiene	g	190.5	202.4	243.9	59.0
Propanal	lq	−215.3			137.2
	g	−185.6	−130.5	304.5	80.7
Propanamide	c	−338.2			
Propane	lq				98.3[43]
	g	−103.8	−23.4	270.2	73.6
Propanediamide	c	−546.1			
(±)-1,2-Propanediamine	lq	−97.8			

(*Continued*)

TABLE 2.54 Enthalpies and Gibbs Energies of Formation, Entropies, and Heat Capacities of Organic Compounds (*Continued*)

Substance	Physical state	$\Delta_f H°$ kJ · mol^{-1}	$\Delta_f G°$ kJ · mol^{-1}	$S°$ J · deg^{-1} · mol^{-1}	$C_p°$ J · deg^{-1} · mol^{-1}
1,2-Propanediol	lq	−485.7			190.8
1,3-Propanediol	lq	−464.9			
1,2-Propanedione	lq	−309.1			
Propanedinitrile	lq	186.4			
1,2-Propanedithiol	lq	−79.4			
1,3-Propanedithiol	lq	−79.4			
Propanenitrile	lq	15.5	89.2	189.3	119.3
1-Propanethiol	lq	−99.9		242.5	144.6
	g	−67.9	2.2	336.4	94.8
2-Propanethiol	lq	−105.0		233.5	145.3
	g	−76.2	−2.6	324.3	96.0
1,2,3-Propanetriol tris(acetate)	lq	−1330.8		458.3	384.7
Propanoic acid	lq	−510.7	−383.5	191.0	152.8
Propanoic anhydride	lq	−679.1	−475.6		235.0
1-Propanol	lq	−302.6	−170.6	193.6	143.7
	g	−255.1	−161.8	322.7	85.6
2-Propanol	lq	−318.1	−180.3	181.1	155.0
	g	−272.6	−173.4	309.2	89.3
2-Propenal	g	−85.8	−64.6		
Propene	g	20.0	62.8	266.6	64.3
trans-1-Propene-1,2-dicarboxylic acid	c	−824.4			
2-Propenenitrile	lq	147.1			108.8
	g	180.6	195.4	274.1	63.8
cis-1,2,3-Propenetricarboxylic acid	c	−1224.7			
trans-1,2,3-Propenetricarboxylic acid	c	−1233.0			
2-Propenoic acid	lq	−383.8			145.7
	g	−336.5	−286.3	315.2	77.8
2-Propen-1-ol	lq	−171.8			138.9
	g	−124.5	−71.3	307.6	76.0
2-Propenyl acetate	lq	−386.2			184.1
cis-1-Propenylbenzene	g	121.3	216.9	383.7	145.2
trans-1-Propenylbenzene	g	117.2	213.7	380.3	146.0
2-Propenylbenzene	lq	88.0			
Propyl acetate	lq				196.2
Propylamine	lq	−101.5			162.5
	g	−70.2	39.8	325.1	91.2
Propylbenzene	lq	−38.3		287.8	214.7
	g	7.9	137.2	400.7	152.3
Propylcarbamate	c	−552.6			
Propylchloroacetate	lq	−515.6			
Propylchlorocarbonate	g	−492.7			
Propylcyclohexane	lq	−237.4		311.9	242.0
	g	−192.5	47.3	419.5	184.2
Propylcyclopentane	lq	−188.8		310.8	216.8
	g	−147.1	52.6	417.3	154.6
Propylene carbonate	lq	−613.2			218.6
Propylene oxide	lq	−123.0		196.5	120.4
	g	−94.7	−25.8	286.9	72.6

TABLE 2.54 Enthalpies and Gibbs Energies of Formation, Entropies, and Heat Capacities of Organic Compounds (*Continued*)

Substance	Physical state	$\Delta_f H°$ kJ · mol^{-1}	$\Delta_f G°$ kJ · mol^{-1}	$S°$ J · deg^{-1} · mol^{-1}	$C_p°$ J · deg^{-1} · mol^{-1}
Propyl formate	lq	−500.3			171.4
Propyl nitrate	g	−173.9	−27.3	385.4	121.3
S-Propyl thioacetate	lq	−294.1			
Propyl trichloroacetate	lq	−513.0			
Propyl vinyl ether	lq	−190.9			
2-Propynyl-1-amine	lq	205.7			
Propyne	g	184.9	194.4	248.1	60.7
2-Propynoic acid	lq	−193.2			
1*H*-Purine	c	169.4			
Pyrazine	c	139.8			
1*H*-Pyrazole	c	116.0			
	lq	105.4			
Pyrene	c	125.5		224.9	229.7
Pyridazine	lq	224.8			
Pyridine	lq	100.2	181.3	177.9	132.7
	g	140.4	190.2	282.8	78.1
3-Pyridinecarbonitrile	c	193.4			
3-Pyridinecarboxylic acid	c	−344.9			
Pyrimidine	lq	145.9			
1*H*-Pyrrole	lq	63.1		156.4	127.7
Pyrrole-2-carboxaldehyde	c	−106.4			
Pyrrole-2-carboldoxime	c	12.1			
Pyrrolidine	lq	−41.0		204.1	156.6
	g	−3.6	114.7	309.5	81.1
(±)-2-Pyrrolidinecarboxylic acid	c	−524.2			
2-Pyrrolidone	c	−286.2			164.4
Quinhydrone	c	−82.8	−323.0	325.9	277.0
Quinidine	c	−160.3			
Quinine	c	−155.2			
Quinoline	lq	141.2	275.7	217.2	194.9
Raffinose	c	−3184			
L-(+)-Rhamnose	c	−1073.2			
D-(−)-Ribose	c	−1047.2			
Salicylaldehyde	lq	−279.9			222[18]
Salicylaldoxime	c	−183.7			
Salicylic acid	c	−589.5	−418.1	178.2	
Semicarbazide std. state	aq	−166.9	−40.6	297.9	
(−)-Serine	c	−732.7			
(±)-Serine	c	−739.0			
L-(−)-Sorbose	c	−1271.5	−908.4	220.9	
5,5′-Spirobis(1,3-dioxane)	c	−702.1			
Spiro[2.2]pentane	lq	157.5		193.7	134.5
	g	185.2	265.3	282.2	88.1
cis-Stilbene	lq	183.3			
trans-Stilbene	c	136.9	317.6	251.0	
(−)-Strychnine	c	−171.5			
Styrene	lq	103.8	202.4	237.6	182.0
	g	147.9	213.8	345.1	122.1
Succinic acid	c	−940.5	−747.4	167.3	153.1
Succinic acid monoamide	c	−581.2			

(*Continued*)

TABLE 2.54 Enthalpies and Gibbs Energies of Formation, Entropies, and Heat Capacities of Organic Compounds (*Continued*)

Substance	Physical state	$\Delta_f H°$ kJ · mol^{-1}	$\Delta_f G°$ kJ · mol^{-1}	$S°$ J · deg^{-1} · mol^{-1}	$C_p°$ J · deg^{-1} · mol^{-1}
Succinic anhydride	c	−608.6			
Succinimide	c	−459.0			
Succinonitrile	lq	139.7		191.6	145.6
(+)-Sucrose	c	−2226.1	−1544.7	360.2	
(±)-Tartaric acid	c	−1290.8			
(−)-Tartaric acid	c	−1282.4			
meso-Tartaric acid	c	−1279.9			
α-Terpinene	g	−20.5			
1,1,2,2,-Tetrabromoethane	lq				165.7
Tetrabromoethylene	g			387.1	102.7
Tetrabromomethane	c	29.4	47.7	212.5	144.3
	g	83.9	67.0	358.1	91.2
Tetrabutyltin	lq	−304.6			
Tetracene	c	158.8			
Tetrachloro-1,4-benzo-quinone	c	−288.7			
1,1,2,2,-Tetrachloro-1,2-difluoroethane	lq				178.6
	g	−489.9	−407.1	382.8	123.4
1,1,1,2-Tetrachloroethane	lq				153.8
	g	−149.4	−80.3	355.9	102.7
1,1,2,2,-Tetrachloroethane	lq	−195.0	−95.0	246.9	162.3
	g	−149.2	−85.6	362.7	100.8
Tetrachloroethylene	lq	−50.6			143.4
	g	−10.9	3.0	266.9	
Tetrachloromethane	lq	−128.2	−62.6	216.2	130.7
	g	−95.7	−53.6	309.9	83.4
1,1,1,3-Tetrachloropropane	lq	−207.8			
1,2,2,3-Tetrachloropropane	lq	−251.8			
1,1,2,2-Tetracyanocyclo-propane	c	590			
Tetracyanoethylene	c	623.8			
Tetracyanomethane	c	611.6			
Tetradecane	g	−332.1	66.9	700.4	326.1
Tetradecanoic acid	c	−833.5			432.0
1-Tetradecanol	c	−629.6			388.0
1-Tetradecene	g	−206.5	154.8	696.2	315.3
Tetraethylene glycol	lq	−981.6			428.8
Tetraethylgermanium	lq	−210.5			
Tetraethyllead	lq	52.7	336.4	464.6	307.4
Tetraethylsilane	lq				298.1
Tetraethyltin	lq	−95.8			
1,1,1,2-Tetrafluoroethane	g	−895.8	−826.2	316.2	86.3
Tetrafluoroethylene	g	−658.9	−623.7	300.0	80.5
Tetrafluoromethane	g	−933.6	−888.3	261.6	61.0
2,2,3,3-Tetrafluoro-1-propanol	g	−1061.3			
Tetrahydrofuran	lq	−216.2		204.3	124.0
	g	−184.2		302.4	76.3
Tetrahydro-2-furanmethanol	lq	−435.6			181.2
1,2,3,4-Tetrahydronaphthalene	lq	−29.2			217
5,6,7,8-Tetrahydro-1-naphthol	c	−285.3			

TABLE 2.54 Enthalpies and Gibbs Energies of Formation, Entropies, and Heat Capacities of Organic Compounds (*Continued*)

Substance	Physical state	$\Delta_f H°$ kJ · mol^{-1}	$\Delta_f G°$ kJ · mol^{-1}	$S°$ J · deg^{-1} · mol^{-1}	$C_p°$ J · deg^{-1} · mol^{-1}
Tetrahydro-2*H*-pyran	lq	−258.3			156.5
Tetrahydro-2*H*-pyran-2-one	lq	−436.7			
1,2,3,6-Tetrahydropyridine	lq	33.5			
Tetrahydrothiophene	lq	−72.9			
	g	−34.1	−45.8	309.6	92.5
Tetrahydrothiophene-1,1-dioxide	lq				180[20]
Tetraiodoethylene	c	305.0			
Tetraiodomethane	g	474.0	217.1	391.9	95.9
Tetramethylammonium bromide	c	−251.0			
Tetramethylammonium chloride	c	−276.4			
Tetramethylammonium iodide	c	−203.4			
1,2,3,4-Tetramethylbenzene	lq	−90.2	106.7	290.6	
1,2,3,5-Tetramethylbenzene	lq	−96.4	98.7	416.5	240.7
1,2,4,5-Tetramethylbenzene	c	−119.9	101.3	245.6	215.1
2,3,5,6-Tetramethylbenzoic acid	c	−506.1			
2,2,3,3-Tetramethylbutane	c	−269.0		273.7	239.2
	g	−225.6	22.0	389.4	192.5
1,1,2,2-Tetramethylcyclo-propane	lq	−119.7			
Tetramethyllead	lq	97.9	262.8	320.1	
	g	135.9	270.7	420.5	144.0
2,2,3,3-Tetramethylpentane	lq	−278.3			271.5
2,2,3,4-Tetramethylpentane	lq	−277.7			
2,2,4,4-Tetramethylpentane	lq	−280.0			266.3
2,3,3,4-Tetramethylpentane	lq	−277.9			
Tetramethylsilane	lq	−264.0			204.1
	g	−239.1	−100.0	359.1	143.9
Tetramethylsuccinic acid	c	−1012.5			
Tetramethylthiacyclopropane	c	−83.0			
Tetramethyltin	g	−18.8			
Tetranitromethane	lq	38.4			
1,1,1,2-Tetraphenylethane	c	223.0			
1,1,2,2-Tetraphenylethane	c	216.0			
Tetraphenylethylene	c	311.5			
Tetraphenylhydrazine	c	457.9			
Tetraphenylmethane	c	247.1	574.0		
Tetraphenyltin	c	412.1			
Tetrapropylgermanium	g	−229.7			
Tetrapropyltin	lq	−211.3			
1,2,3,4-(1*H*)-Tetrazole	c	237.0			
Theobromine	c	−361.5			
2-Thiaadamantane	c	−143.5			
Thiacyclobutane	g	60.6	107.1	285.0	68.3
Thiacycloheptane	g	−61.3	84.1	361.9	124.6
Thiacyclohexane	lq	−106.3		218.2	163.3
	g	−63.5	53.1	323.0	109.7
Thiacyclopentane	g	−33.8	46.0	309.4	90.9
Thiacyclopropane	g	82.2	96.9	255.3	53.7
Thianthrene	c	−182.5			
Thiirane	g	82.0	96.8	255.2	53.3

(Continued)

TABLE 2.54 Enthalpies and Gibbs Energies of Formation, Entropies, and Heat Capacities of Organic Compounds (*Continued*)

Substance	Physical state	$\Delta_f H°$ kJ · mol^{-1}	$\Delta_f G°$ kJ · mol^{-1}	$S°$ J · deg^{-1} · mol^{-1}	$C_p°$ J · deg^{-1} · mol^{-1}
Thiirene	g	300.0	275.8	255.3	54.7
Thioacetamide	c	−71.7			
Thioacetic acid	lq	−216.9			
	g	−175.1	−154.0	313.2	80.9
1,2-Thiocresol	lq	44.2			
Thiohydantoic acid	c	−554.8			
Thiohydantoin	c	−249.0			
2-Thiolactic acid	lq	−468.4			
Thiophene	lq	80.2	121.2	181.2	123.8
	g	115.0	126.8	278.9	72.9
Thiophenol	lq	64.1	134.0	222.8	173.2
	g	111.6	147.6	336.9	104.9
Thiosemicarbazide	c	25.1			
Thiourea	c	−89.1	21.8	115.9	
	g	22.9			
(−)-Threonine	c	−807.2			
(±)-Threonine	c	−758.8			
Thymine	c	−462.8			150.8
Thymol	c	−309.7			
Toluene	lq	12.4	113.8	221.0	157.0
	g	50.4	122.0	320.7	103.6
1*H*-1,2,4-Triazol-3-amine	c	76.8			
2,4,6-Triamino-1,3,5-triazine	c	−72.4	184.5	149.1	
2-Triazoethanol	lq	94.6			
Tribenzylamine	c	140.6			
Tribromoacetaldehyde	lq	−130.3			
Tribromochloromethane	g	12.6	9.1	357.8	89.4
Tribromofluoromethane	g	−190.0	−193.1	345.9	84.4
Tribromomethane	lq	−28.5	8.0	220.9	130.7
	g	23.8	−5.0	330.9	71.2
Tributoxyborane	lq	−1199.6			
Tributylamine	lq	−281.6			
Tributyl phosphate	lq	−1456			
Tributylphosphine oxide	c	−460			
Trichloroacetaldehyde	lq	−234.5			151.0
2,2,2-Trichloroacetamide	c	−358.2			
Trichloroacetic acid	c	−503.3			
ionized	aq	−517.6			
Trichloroacetonitrile	g			336.6	96.1
Trichloroacetyl chloride	lq	−280.8			
Trichlorobenzoquinone	c	−269.9			
1,1,1-Trichloroethane	lq	−177.4		227.4	144.3
	g	−144.6	−76.2	323.1	93.3
1,1,2-Trichloroethane	lq	−191.5		232.6	150.9
	g	−151.2	−77.5	337.1	89.0
Trichloroethylene	lq	−43.6			124.4
	g	−9.0	19.9	324.8	80.3
Trichlorofluoromethane	lq	−301.3	−236.8	255.4	121.6
	g	−268.3	−249.3	309.7	78.0
Trichloromethane	lq	−134.5	73.7	201.7	114.2
	g	−102.7	−76.0	295.7	65.7

TABLE 2.54 Enthalpies and Gibbs Energies of Formation, Entropies, and Heat Capacities of Organic Compounds (*Continued*)

Substance	Physical state	$\Delta_f H°$ kJ · mol^{-1}	$\Delta_f G°$ kJ · mol^{-1}	$S°$ J · deg^{-1} · mol^{-1}	$C_p°$ J · deg^{-1} · mol^{-1}
1,2,2-Trichloropropane	g	−185.8	−97.8	382.9	112.2
1,2,3-Trichloropropane	lq	−230.6			183.6
	g	−182.9			
1,2,3-Trichloropropene	lq	−101.8			
1,1,2-Trichlorotrifluoroethane	lq	−805.8			170.1
1,1,1-Tricyanoethane	c	351.0			
Tricyanoethylene	c	439.3			
Tridecane	g	−311.5	58.5	661.5	303.2
Tridecanoic acid	c	−806.6			
1-Tridecene	g	−186.0	146.3	657.3	292.4
Triethanolamine	c	−664.2			389.0
Triethoxyborane	lq	−1047.4			
Triethoxymethane	lq	−687.3			
Triethylaluminum	lq	−236.8			
Triethylamine	lq	−127.7			219.9
	g	−92.8	110.3	405.4	160.9
Triethylaminoborane	lq	−198.6			
Triethyl arsenite	lq	−706.7			
Triethylarsine	lq	13.0			
Triethylbismuthine	lq	169.9			
Triethylborane	lq	−194.6	9.4	336.7	241.2
	g	−157.7	16.1	437.8	
Triethylenediamine	c	−14.2	239.7	157.6	
Triethylene glycol	lq	−804.2			
Triethyl phosphate	lq	−1243			
Triethylphosphine	lq	−89.1			
Triethyl phosphite	lq	−861.5			
Triethylstibine	lq	5.0			
Triethylsuccinic acid	c	−1066.5			
Triethyl thiophosphate	lq	−972.8			
Trifluoroacetic acid	lq	−1069.9			
Trifluoroacetonitrile	g	−497.9	−461 9	298.1	77.9
1,1,1-Trifluoroethane	g	−744.6	−678.3	279.9	78.2
1,1,2-Trifluoroethane	g	−730.7			
2,2,2-Trifluoroethanol	lq	−932.4			
Trifluoroethylene	g	−490.4	−469.5	292.6	69.2
Trifluoroiodoethane	g	−644.5			
Trifluoroiodomethane	g	−587.8	−572.0	307.5	70.9
Trifluoromethane	g	−695.4	−658.9	259.6	51.1
(Trifluoromethyl)benzene	g	−599.1	−511.3	372.6	130.4
1,1,1-Trifluoro-2,4-pentane-dione	lq	−1040.2			
3,3,3-Trifluoropropene	g	−614.2			
Trihexylamine	lq	−433.0			
(±)-Trihydroxyglutaric acid	c	−1490			
2,4,6-Trihydroxypryimidine	c	−634.7			
Triiodomethane	g	251.0	178.0	356.2	75.1
Triisopropyl phosphite	lq	−980.3			
Trimethoxyborane	g	−899.1			
Trimethoxyethane	lq	−612.0			
Trimethoxymethane	lq	−570.0			

(Continued)

TABLE 2.54   Enthalpies and Gibbs Energies of Formation, Entropies, and Heat Capacities of Organic Compounds (*Continued*)

Substance	Physical state	$\Delta_f H°$ kJ · mol^{-1}	$\Delta_f G°$ kJ · mol^{-1}	$S°$ J · deg^{-1} · mol^{-1}	$C_p°$ J · deg^{-1} · mol^{-1}
Trimethylacetic acid	lq	−564.4			
Trimethylacetic anhydride	lq	−779.9			
2′,4′,5′-Trimethylacetophenone	lq	−252.3			
2′,4′,6′-Trimethylaceto-phenone	lq	−267.4			
Trimethylaluminum	lq	−136.4	−9.9	209.4	155.6
Trimethylamine	lq	−45.7		208.5	137.9
	g	−23.7	98.9	287.1	91.8
std. state	aq	−76.0	93.0	133.5	
Trimethylamine-aluminum chloride adduct	c	−879.1			
Trimethylamine-borane	c	−142.5	70.7	187.0	
Trimethylammonium ion, std. state	aq	−112.9	37.2	196.7	
Trimethyl arsenite	lq	−590.8			
Trimethylarsine	g	11.7			
1,2,3-Trimethylbenzene	lq	−58.5	107.5	267.8	216.4
1,2,4-Trimethylbenzene	lq	−61.8	102.3	284.2	215.0
1,3,5-Trimethylbenzene	lq	−63.4	103.9	273.6	209.3
2,3,4-Trimethylbenzoic acid	c	−486.6			
2,3,5-Trimethylbenzoic acid	c	−488.7			
2,3,6-Trimethylbenzoic acid	c	−475.7			
2,4,5-Trimethylbenzoic acid	c	−495.7			
2,4,6-Trimethylbenzoic acid	c	−477.9			
3,4,5-Trimethylbenzoic acid	c	−500.9			
2,6,6-Trimethylbicyclo-[3.1.1]-2-heptene	lq	16.4			
Trimethylbismuthine	g	192.9			
Trimethylborane	g	−124.3	−35.9	314.7	88.5
2,2,3-Trimethylbutane	g	−204.5	4.3	383.3	164.6
2,2,3-Trimethylbutane	lq	−236.5		292.2	213.5
2,3,3-Trimethyl-1-butene	lq	−117.7			
Trimethylchlorosilane	lq	−382.8	−246.4	278.2	
	g	−352.8	−243.5	369.1	
cis, cis-1,3,5-Trimethyl-cyclohexane	g	−215.4	33.9	390.4	179.6
1,1,2-Trimethylcyclopropane	lq	−96.2			
Trimethylene oxide (Oxetane)	lq	−110.8			
	g	−80.5	−9.8	273.9	
Trimethylgallium	g	−46.9			
2,3,5-Trimethylhexane	lq	−284.0			
Trimethylindium	g	170.7			
2,2,3-Trimethylpentane	lq	−256.9	9.3	327.6	188.9
	g	−220.0	17.1	425.2	
2,2,4-Trimethylpentane	lq	−259.2	6.9	328.0	239.1
	g	−224.0	13.7	423.2	
2,3,3-Trimethylpentane	lq	−253.5	10.6	334.4	245.6
	g	−216.3	18.9	431.5	
2,3,4-Trimethylpentane	lq	−255.0	10.7	329.3	247.3
2,2,4-Trimethyl-3-pentanone	lq	−381.6			
2,4,4-Trimethyl-1-pentene	lq	−145.9	86.4	306.3	

TABLE 2.54 Enthalpies and Gibbs Energies of Formation, Entropies, and Heat Capacities of Organic Compounds (*Continued*)

Substance	Physical state	$\Delta_f H°$ kJ · mol^{-1}	$\Delta_f G°$ kJ · mol^{-1}	$S°$ J · deg^{-1} · mol^{-1}	$C_p°$ J · deg^{-1} · mol^{-1}
2,4,4-Trimethyl-2-pentene	lq	−142.4	88.0	311.7	
Trimethylphosphine	lq	−122.2			
Trimethylphosphine oxide	c	−477.8			
Trimethyl phosphite	lq	−741.0			
Trimethylsilane	g			331.0	117.9
Trimethylsilanol	lq	−545.0			
Trimethylstibine	g	32.2			
Trimethylsuccinic acid	c	−1000.8			
Trimethylsuccinic anhydride	c	−688.3			
Trimethylthiacyclopropane	lq	−60.5			
Trimethyltin bromide	lq	−185.4			
Trimethyltin chloride	lq	−213.0			
Trimethylurea	c	−330.5			
Trinitroacetonitrile	lq	183.7			
2,4,6-Trinitroanisole	c	−157.3			
1,3,5-Trinitrobenzene	c	−37.2			
1,1,1-Trinitroethane	lq	−96.9			
Trinitroglycerol	lq	−370.9			
Trinitromethane	lq	−32.8			
	g	−0.2			
2,4,6-Trinitrophenetole	c	−204.6			
2,4,6-Trinitrophenol	c	−214.3			
2,4,6-Trinitrophenylhydrazine	c	36.8			
2,4,6-Trinitrotoluene	c	−65.5			
2,4,6-Trinitro-1,3-xylene	c	−102.5			
Trioctylamine	lq	−584.9			
1,3,6-Trioxacyclooctane	lq	−515.9			
1,3,5-Trioxane	c	−522.5		133.0	114.4
Triphenylamine	c	234.7	504.2		
Triphenylarsine	c	310.0			
Triphenylbismuthine	c	469.0			
Triphenylborane	c	48.5			
Triphenylene	c	151.8	329.2	254.7	
1,1,1-Triphenylethane	c	157.2			
1,1,2-Triphenylethane	c	130.2			
Triphenylethylene	c	233.5	514.6		
2,4,6-Triphenylimidazole	c	272			
Triphenylmethane	c	171.2	412.5	312.1	295.0
Triphenylmethanol	c	−3.4	272.8	329.3	
Triphenyl phosphate	c	−757			
Triphenylphosphine	c	232.2			
Triphenylphosphine oxide	c	−60.3			
Triphenylstibine	c	329.3			
Tripropoxyborane	lq	−1127.2			
Tripropylamine	lq	−207.2			
Tripropynylamine	lq	814.2			
Tris(acetylacetonato)-chromium	c	−1533.0			
Tris(diethylamino)phosphine	lq	−289.5			
1,1,1-Tris(hydroxymethyl)-ethane	c	−744.6			

TABLE 2.54 Enthalpies and Gibbs Energies of Formation, Entropies, and Heat Capacities of Organic Compounds (*Continued*)

Substance	Physical state	$\Delta_f H°$ kJ · mol^{-1}	$\Delta_f G°$ kJ · mol^{-1}	$S°$ J · deg^{-1} · mol^{-1}	$C_p°$ J · deg^{-1} · mol^{-1}
Tris(hydroxymethyl)nitro-methane	c	−735.6			
Tris(isopropoxy)borane	lq	−293.3			
Tris(trimethylsilyl)amine	c	−725.1			
(−)-Tryptophane	c	−415.3	−119.4	251.0	238.2
(−)-Tyrosine	c	−685.1	−385.7	214.0	216.4
Undecane	lq	−327.2	22.8	458.1	344.9
Undecanoic acid	c	−735.9			
1-Undecanol	lq	−504.8			
1-Undecene	g	−144.8	129.5	579.4	246.7
10-Undecenoic acid	c	−577			
Uracil	c	−429.4			120.5
Urea	c	−333.1	−196.8	104.6	93.1
	g	−245.8			
Urea nitrate	c	−564.0			
Urea oxalate	c	−1528.4			
5-Ureidohydantoin	c	−718.0	−434.0	195.1	
Uric acid	c	−618.8	−358.8	173.2	166.1
(±)-Valine	c	−628.9	−359.0	178.9	168.8
Valylphenylalanine	c	−767.8			
Vinyl acetate	g	−314.4			
Vinylbenzene	lq	103.8			
Vinylcyclohexane	lq	−88.7			
4-Vinylcyclohexene	lq	26.8			
Vinylcyclopentane	lq	−34.8			
Vinylcyclopropane	lq	122.5			
2-Vinylpyridine	lq	157.1			
Xanthine	c	−379.6	−165.9	161.1	151.3
Xanthone	c	−191.5			
1,2-Xylene	lq	−24.4	110.3	246.5	186.1
	g	19.1	122.1	352.8	133.3
1,3-Xylene	lq	−25.4	107.7	252.2	183.3
	g	17.3	118.9	357.7	127.6
1,4-Xylene	lq	−24.4	110.1	247.4	181.5
	g	18.0	121.1	352.4	126.9
Xylitol	c	−1118.5			
D-(+)-Xylose	c	−1057.8			

TABLE 2.55 Heat of Fusion, Vaporization, Sublimation, and Specific Heat at Various Temperatures of Organic Compounds

Abbreviations Used in the Table

ΔHm, enthalpy of melting (at the melting point) in $kJ \cdot mol^{-1}$

ΔHv, enthalpy of vaporization (at the boiling point) in $kJ \cdot mol^{-1}$

ΔHs, enthalpy of sublimation (or vaporization at 298 K) in $kJ \cdot mol^{-1}$

C_p, specific heat (at temperature specified on the Kelvin scale) for the physical state in existence (or specified: c, lq, g) at that temperature in $J \cdot K^{-1} \cdot mol^{-1}$

ΔHt, enthalpy of transition (at temperature specified, superscript, measured in degrees Celsius) in $kJ \cdot mol^{-1}$

Substance	ΔHm	ΔHv	ΔHs	C_p 400 K	600 K	800 K	1000 K
Acenaphthene	21.54	54.73	86.2				
Acenaphthylene			73.0				
Acetaldehyde	3.24	25.8	25.5	66.3(g)	85.9	101.3	112.5
Acetamide	15.71	56.1	78.7				
Acetanilide		64.7	80.8				
Acetic acid	11.54	23.7	23.4	79.7	106.2	125.5	139.3
Acetic anhydride	10.5	38.2	48.3	129.1	174.1	204.6	226.4
Acetone	5.69	29.1	31.0	92.1	122.8	144.9	162.0
Acetonitrile, $\Delta Ht = 0.22^{-56}$	8.17	29.8	32.9	61.2	76.8	89.0	98.3
Acetophenone		38.8	55.9				
Acetyl bromide			33.1				
Acetyl chloride			30.1	78.9	97.0	110.0	119.7
Acetylene	3.8	17.0	21.3	50.1	58.1	63.5	68.0
Acetylene-d_2				54.8	61.9	67.4	71.8
Acetylenedicarbonitrile			28.8	94.8	106.2	114.1	119.8
Acetyl fluoride			25.1				
Acetyl iodide			38.5				
Acrylic acid	11.16	44.1	54.3	96.0	123.4	142.0	155.3
Acrylonitrile	6.23	32.6	33.5	76.8	96.7	110.6	120.8
Adamantane			59.7				
Adenine			108.8				
α-Alanine			138.1				
Allyl *tert*-butyl sulfide			44.4				
Allyl ethyl sulfone			83.7				
Allyl ethyl sulfoxide			71.6				
Allyl methyl sulfone			79.5				
Allyl trichloroacetate			52.3				
3-Aminoacetophenone	12.1						
4-Aminoacetophenone	15.9						
2-Aminobenzoic acid	20.5		104.9				
3-Aminobenzoic acid	21.8		128.0				
4-Aminobenzoic acid	20.9		116.1				
2-Aminoethanol	20.5	50.9					
Aniline	10.56	42.4	55.8	143.0	192.8	225.1	230.9
Anthracene	28.83	56.5	101.5				
9,10-Anthraquinone		88.5	112.1				
cis-Azobenzene	22.04		92.9				
trans-Azobenzene	22.6	93.8					
Azobutane			49.3				
Azomethane				93.9	123.1	145.7	162.6
Azomethane-d_6				110.7	142.8	165.2	180.6

(Continued)

TABLE 2.55 Heat of Fusion, Vaporization, Sublimation, and Specific Heat at Various Temperatures of Organic Compounds (*Continued*)

Substance	ΔHm	ΔHv	ΔHs	C_p 400 K	600 K	800 K	1000 K
Azoisopropane			36.0				
Azopropane			39.9				
trans-Azoxybenzene	17.93						
Azulene	12.1	55.5	76.8	176.4	248.2	295.4	327.4
Benzaldehyde	9.32	42.5	49.8				
Benzamide	18.49						
1,2-Benzanthracene			123.0				
2,3-Benzanthracene			126				
1,2-Benzanthracene-9,10-dione			82.8				
Benzene	9.95	30.7	33.8	113.5(g)	160.1	190.5	211.4
Benzeneacetic acid	14.49						
1,3-Benzenedicarboxylic acid			106.7				
1,4-Benzenedicarboxylic acid			98.3				
Benzenethiol	11.48	39.9	47.6				
Benzil	23.54						
Benzoic acid	18.06	50.6	91.1	138.4	196.7	234.9	260.7
Benzoic anhydride	17.2		96.4				
Benzonitrile	10.88	45.9	52.5	140.8	187.4	217.9	238.8
Benzo[*def*]phenanthrene	17.1		100.2				
Benzophenone	18.19		94.1				
1,4-Benzoquinone	18.53		62.8				
Benzo[*f*]quinoline			83.1				
Benzo[*h*]quinoline			80.8				
Benzo[*b*]thiophene, $\Delta Ht = 3.0^{-11.6}$	11.8						
Benzotrifluoride			37.6				
Benzoyl bromide			58.6				
Benzoyl chloride			54.8				
Benzoyl iodide			61.9				
4-Benzphenanthrene			106.3				
Benzyl acetate		49.4					
Benzyl alcohol	8.97	50.5	60.3				
Benzylamine			60.2				
Benzyl benzoate		53.6	77.8				
Benzyl bromide			47.3				
Benzyl chloride			51.5				
Benzyl ethyl sulfide			56.9				
Benzyl iodide			47.3				
Benzyl mercaptan			56.6				
Benzyl methyl ketone			49.0				
Benzyl methyl sulfide			53.6				
Bicyclo[1.1.0]butane			23.4				
Bicyclo[2.2.1]hepta-2,5-dione		32.9					
Bicyclo[2.2.1]heptane			40.2				
Bicyclo[4.1.0]heptane			38.0				
Bicyclo[2.2.1]-2-heptene			38.8				
Bicyclo[3.1.0]hexane			32.8				
Bicyclohexyl			58.0				
Bicyclo[2.2.2]octane			48.0				
Bicyclo[4.2.0]octane			42.0				
Bicyclo[5.1.0]octane			43.5				

TABLE 2.55 Heat of Fusion, Vaporization, Sublimation, and Specific Heat at Various Temperatures of Organic Compounds (*Continued*)

Substance	ΔHm	ΔHv	ΔHs	C_p 400 K	600 K	800 K	1000 K
Bicyclo[2.2.2]-2-octene			43.8				
Bicyclopropyl			33.5				
Biphenyl	18.6	45.6	81.8	221.0	307.7	363.7	401.7
Biphenylene			84.3				
Bis(2-butoxyethyl) ether		55.9					
Bis(2-chloroethyl) ether	8.66	45.2					
Bis(2-ethoxyethyl) ether		49.0					
Bis(2-ethoxymethyl) ether		36.2	44.7				
Bis(2-hydroxyethyl) ether		52.3	57.3				
Bis(2-methoxyethyl) ether		43.1					
Bromobenzene	10.62	37.9	44.5	127.4	171.5	199.9	219.2
4-Bromobenzoic acid			87.9				
1-Bromobutane	6.69	32.5	36.7	136.6	180.0	211.2	234.4
(±)-2-Bromobutane	6.89	30.8	34.4	138.1	214.7	238.2	
1-Bromo-2-chloroethane		33.7	38.2				
Bromochloromethane		30.0	32.8				
1-Bromo-3-chloropropane		37.6	44.1				
1-Bromo-2-chloro-1,1,2-trifluoroethane		28.3	30.1				
Bromochloro-2,2,2-trifluoroethane		28.1	29.8				
1-Bromododecane		74.8					
Bromoethane	5.86	27.0	28.0	79.2	102.8	119.6	132.2
Bromoethylene	5.12	23.4	18.2	66.6	83.0	94.1	102.3
1-Bromoheptane			50.6			74.8	
1-Bromohexadecane			94.4				
1-Bromohexane			45.9				
Bromomethane, $\Delta Ht = 0.47^{-99.4}$	5.98	23.9	22.8	50.0	62.7	72.2	79.5
1-Bromo-2-methylpropane		31.3	34.8				
2-Bromo-2-methylpropane	1.97	29.2	31.8	146.1	190.7	220.3	241.6
$\Delta Ht = 5.7^{-64.5}$							
$\Delta Ht = 1.0^{-41.6}$							
1-Bromonaphthalene	15.16	39.3	52.5				
1-Bromooctane			55.8				
1-Bromopentane	11.46	35.0	41.3	165.6	219.0	257.5	286.0
1-Bromopropane	6.53	29.8	32.0	107.5	140.8	164.9	182.8
2-Bromopropane		28.3	30.2	110.2	144.0	167.7	185.2
3-Bromopropene		30.2	32.7				
Bromotrichloromethane	2.54						
Bromotrifluoromethane				79.3	91.3	97.5	100.9
Bromotrimethylsilane			32.6				
1,2-Butadiene	7.0	24.0	23.2	98.4	128.5	150.7	167.4
1,3-Butadiene	7.98	22.5	20.9	101.2	154.1	169.5	
1,3-Butadiyne				84.4	96.8	105.1	111.3
Butanal	11.09	31.5	34.5	126.4	165.7	195.0	216.3
Butanamide	17.6		85.9				
Butane, $\Delta Ht = 2.1^{-165.6}$	4.66	22.4	21.0	123.9	168.6	201.8	226.9
1,2-Butanediamine			46.3				
Butanedinitrile	3.7	48.5	70.0				
1,3-Butanediol		58.5	67.8				
1,4-Butanediol			76.6				
2,3-Butanediol			59.2				

TABLE 2.55   Heat of Fusion, Vaporization, Sublimation, and Specific Heat at Various Temperatures of Organic Compounds (*Continued*)

Substance	ΔHm	ΔHv	ΔHs	C_p 400 K	600 K	800 K	1000 K
2,3-Butanedione			38.7				
1,4-Butanedithiol			55.1				
Butanenitrile	5.02	33.7	39.3	118.8	155.1	181.9	201.8
meso-1,2,3,4-Butanetetrol			135.1				
1,4-Butanedithiol			49.7				
1-Butanethiol	10.46	32.2	36.6	146.2	194.7	233.0	263.4
2-Butanethiol	6.5	30.6	34.0	148.0	194.2	227.2	251.1
1,2,4-Butanetriol		58.6					
Butanoic acid	11.08	41.8	40.5				
Butanoic anhydride		50.0					
1-Butanol	9.28	43.3	52.3	137.2	183.7	218.0	243.8
2-Butanol		40.8	49.7	141.0	187.1	220.4	245.3
2-Butanone	8.44	31.3	34.8	124.7	163.6	192.8	214.8
trans-2-Butenal			34.5				
1-Butene	3.9	22.1	20.2	109.0	147.1	174.9	195.9
cis-2-Butene	7.58	23.3	22.2	101.8	141.4	171.0	193.1
trans-2-Butene	9.8	22.7	21.4	108.9	145.6	184.9	194.9
cis-2-Butenedinitrile			72.0				
cis-2-Butenedioic acid			110.0				
trans-2-Butenedioic acid			136.3				
cis-2-Butene-1,4-diol		66.1					
trans-2-Butene-1,4-diol		69.0					
cis-2-Butenenitrile			38.9				
trans-2-Butenenitrile			40.0				
3-Butenenitrile			40.0				
cis-2-Butenoic acid	12.57						
trans-2-Butenoic acid	12.98						
cis-2-Buten-1-ol		46.4					
1-Buten-3-yne				89.0	111.6	127.2	138.7
2-Butoxyethanol			56.6				
1-*tert*-Butoxy-2-ethoxyethane			50.9				
2-(2-Butoxyethoxy)ethanol		28.0					
2-Butoxyethyl acetate			59.5				
1-*tert*-Butoxy-2-methoxyethane		38.5	47.8				
N-Butylacetamide			76.1				
Butyl acetate		36.3	43.9				
tert-Butyl acetate		33.1	38.0				
Butylamine		31.8	35.7	148.3	197.9	234.4	261.7
sec-Butylamine		29.9	32.8	148.1	199.0	236.1	261.7
tert-Butylamine	0.88	28.3	29.6	152.6	204.5	240.5	266.9
Butylbenzene	11.22	38.9	51.4	229.1	314.6	373.9	416.3
sec-Butylbenzene	9.83	38.0	48.0				
tert-Butylbenzene	8.39	37.6	47.7				
sec-Butyl butanoate			47.3				
Butyl chloroacetate			51.0				
Butyl 2-chlorobutanoate			52.7				
Butyl 3-chlorobutanoate			53.1				
Butyl 4-chlorobutanoate			54.4				
Butyl 2-chloropropanoate			54.4				
Butyl 3-chlorobutanoate			55.4				

TABLE 2.55 Heat of Fusion, Vaporization, Sublimation, and Specific Heat at Various Temperatures of Organic Compounds (*Continued*)

Substance	ΔHm	ΔHv	ΔHs	C_p			
				400 K	600 K	800 K	1000 K
Butyl crotonate			51.9				
sec-Butyl crotonate			49.4				
Butylcyclohexane	14.16	38.5	49.4	276.1	289.5	469.9	525.9
Butylcyclopentane	11.3	36.2	45.9	241.7	336.3	407.3	480.3
N-Butyldiacetimide			64.4				
Butyl dichloroacetate			52.3				
Butylethylamine		34.0	40.2				
Butyl ethyl ether		31.6	36.3				
Butyl ethyl sulfide	12.4	37.0	44.5	202.4	271.8	325.3	367.2
tert Butyl ethyl sulfide	7.1	33.5	39.3				
Butyl formate		36.6	41.1				
tert-Butyl hydroperoxide			47.7				
Butylisopropylamine		34.5	42.1				
Butyllithium			107.1				
Butyl methyl ether		29.6	32.4				
sec-Butyl methyl ether		28.1	30.2				
tert-Butyl methyl ether		27.9	29.8				
Butyl methyl sulfide	12.5	34.5	40.5	174.6	233.0	278.4	314.1
tert-Butyl methyl sulfide	8.4	31.5	35.8				
Butyl methyl sulfone			76.2				
tert-Butyl methyl sulfone			82.4				
Butyl octadecanoate	56.90						
tert-Butyl peroxide			31.8				
Butyl propyl ether		33.7	40.2				
Butyl thiolacetate			48.1				
Butyl trichloroacetate			53.6				
Butyl vinyl ether		31.6	36.2				
1-Butyne	6.0	24.5	23.3	99.9	129.0	150.4	166.7
2-Butyne	9.23	26.5	26.6	94.6	124.2	147.0	164.4
2-Butynedinitrile			28.8				
4-Butyrolactone	9.57	52.2					
Butyrophenone			60.7				
(+)-Camphor	6.84	59.5					
9H-Carbazole	26.9		84.5				
Chloroacetic acid	12.28		75.3				
Chloroacetyl chloride			38.9				
2-Chloroaniline	11.88	44.4	56.8				
2-Chlorobenzaldehyde			53.1				
Chlorobenzene	9.61	35.2	41.0	128.1	172.2	200.4	219.6
2-Chlorobenzoic acid	25.73		79.5				
3-Chlorobenzoic acid			82.0				
4-Chlorobenzoic acid			87.9				
Chloro-1,4-benzoquinone			69.0				
1-Chlorobutane		30.4	33.5	135.1	179.0	210.5	234.0
2-Chlorobutane		29.2	31.5	136.1	180.7	212.7	236.8
Chlorocyclohexane			43.5				
1-Chloro-1,1-difluoroethane	2.69	22.4					
Chlorodifluoromethane	4.12	20.2		65.4	78.9	87.2	92.4
2-Chloro-1,4-dihydroxybenzene			69.0				
Chlorodimethylsilane		26.2					

(*Continued*)

TABLE 2.55   Heat of Fusion, Vaporization, Sublimation, and Specific Heat at Various Temperatures of Organic Compounds (*Continued*)

Substance	ΔHm	ΔHv	ΔHs	C_p			
				400 K	600 K	800 K	1000 K
Chlorodiphenylsilane			69.5				
1-Chloro-2,3-epoxypropane		33.1	40.6				
Chloroethane	4.45	24.7		77.6	101.6	118.8	131.7
2-Chloroethanol		41.4					
1-Chloro-2-ethylbenzene			47.3				
1-Chloro-4-ethylbenzene			48.1				
Chloroethylene	4.75	20.8		65.0	82.1	93.5	101.9
2-Chloroethyl vinyl ether		38.2					
Chloroethyne				60.2	66.8	71.0	74.3
1-Chloroheptane			47.7				
1-Chlorohexane		35.7	42.8				
Chlorohydroquinone			69.0				
Chloromethane	6.43	21.4	18.9	48.2	61.3	71.3	78.9
1-Chloro-2-methylbenzene	8.37	37.5					
1-Chloro-3-methylbenzene	10.46						
1-Chloro-4-methylbenzene		38.7					
1-Chloro-3-methylbutane		32.0	36.2				
1-Chloro-2-methylpropane		29.2	31.7	136.1	180.7	212.7	236.8
2-Chloro-2-methylpropane	2.09	27.6	29.0	142.3	184.9	215.5	238.5
$\Delta Ht = 1.7^{-90.1}$							
$\Delta Ht = 5.8^{-53.6}$							
1-Chloronaphthalene	12.90	52.1	65.3				
2-Chloronaphthalene			82.0				
1-Chloro-3-nitrobenzene	19.37						
1-Chloro-4-nitrobenzene	20.77						
1-Chlorooctane			52.4				
Chloropentafluoroacetone			25.3				
Chloropentafluorobenzene		34.8	41.1				
Chloropentafluoroethane	1.88	19.4					
1-Chloropentane		33.2	38.2	164.2	218.0	256.8	285.6
2-Chloropentane		31.8	36.0				
2-Chlorophenol	12.52						
3-Chlorophenol	14.91		53.1				
4-Chlorophenol	14.07		51.9				
1-Chloropropane	5.54	27.2	28.4	106.1	139.9	164.2	182.4
2-Chloropropane	7.39	26.3	26.9	108.7	143.1	167.1	184.8
3-Chloro-1-propene		29.0	28.2	92.6	111.0	137.8	151.9
Chlorotrifluoroethylene	5.6	20.8					
Chlorotrifluoromethane		15.8		77.5	90.3	96.9	100.5
Chlorotrimethylsilane		27.6	30.1				
Chlorotrinitromethane			45.4				
Chrysene	26.15		124.5				
Coronene	19.2						
1,2-Cresol	13.94	45.2	76.0	166.3	220.8	257.5	287.9
1,3-Cresol	9.41	47.4	61.7	162.1	218.7	256.4	286.6
1,4-Cresol	11.89	47.5	73.9	161.7	218.0	255.7	286.5
Cubane			80.3				
Cyanamide	8.76	68.6					
Cyanogen	8.1	23.3	19.7	61.9(g)	68.2	72.9	76.4
Cyclobutane, $\Delta Ht = 5.8^{-126.8}$	1.1	24.2	23.5	100.0	145.4	177.5	200.7

TABLE 2.55   Heat of Fusion, Vaporization, Sublimation, and Specific Heat at Various Temperatures of Organic Compounds (*Continued*)

Substance	ΔHm	ΔHv	ΔHs	C_p 400 K	600 K	800 K	1000 K
Cyclobutanecarbonitrile		36.9	44.3				
Cyclobutanenitrile			40.0				
Cyclobutene				90.3	126.8	151.7	169.6
Cyclobutylamine			35.6				
Cyclododecane			76.4				
Cycloheptane	1.88	33.2	38.5	175.0	261.2	322.3	365.7
$\Delta Ht = 5.0^{-138.4}$							
$\Delta Ht = 0.3^{-75.0}$							
$\Delta Ht = 0.5^{-60.8}$							
Cycloheptanone			51.9				
1,3,5-Cycloheptatriene	1.2	38.7		155.4	209.5	245.1	270.2
$\Delta Ht = 2.4^{-119.2}$							
Cyclohexane	2.63	30.0	33.0	149.9	225.2	279.3	317.2
$\Delta Ht = 6.7^{-87}$							
Cyclohexanecarbonitrile			51.9				
Cyclohexanethiol		37.1	44.6				
Cyclohexanol	1.76	45.5	62.0	172.1	248.1	302.0	339.5
$\Delta Ht = 8.2^{-9.7}$							
Cyclohexanone		40.3	45.1	150.6	221.3	272.0	305.4
Cyclohexene	3.29	30.5	33.5	144.9	206.9	248.9	278.7
$\Delta Ht = 4.3^{-134.4}$							
1-Cyclohexenecarbonitrile			53.5				
Cyclohexylamine		36.1	43.7				
Cyclohexylbenzene	15.30		59.9				
Cyclohexylcyclohexane		51.9	58.0				
cis,cis-1,5-Cyclooctadiene			43.4				
Cyclooctane	2.41	35.9	43.3	200.1	297.1	365.3	414.3
$\Delta Ht = 6.3^{-106.7}$							
$\Delta Ht = 0.5^{-89.4}$							
Cyclooctanone			54.4				
1,3,5,7-Cyclooctatetraene	11.3	36.4	43.1	160.9	220.8	260.4	288.2
Cyclooctene			47.0				
Cyclopentadiene			28.4				
Cyclopentane	0.61	27.3	28.5	118.7	178.1	220.1	250.4
$\Delta Ht = 4.8^{-150.8}$							
$\Delta Ht = 0.3^{-135.1}$							
Cyclopentanecarbonitrile			43.4				
1-Cyclopentenecarbonitrile			45.0				
Cyclopentanethiol	7.8	35.3	41.4	144.5	203.6	245.2	275.5
Cyclopentanol			57.6				
Cyclopentanone		36.4	42.7				
Cyclopentene	3.36		28.1	104.9	155.6	191.5	217.3
$\Delta Ht = 0.5^{-186.1}$							
Cyclopentylamine	8.31		40.2				
Cyclopropane	5.44	20.1	16.9	76.6	109.4	140.5	148.1
Cyclopropanecarbonitrile		35.6	41.9				
Cyclopropylamine	13.18		31.3				
Cyclopropylbenzene			50.2				
Cyclopropyl methyl ketone		34.1	38.4				
Decafluorobutane		22.9					

(*Continued*)

TABLE 2.55   Heat of Fusion, Vaporization, Sublimation, and Specific Heat at Various Temperatures of Organic Compounds (*Continued*)

Substance	ΔHm	ΔHv	ΔHs	C_p 400 K	600 K	800 K	1000 K
cis-Decahydronaphthalene	9.49	41.0	50.2	237.0	352.0	432.5	489.5
$\Delta Ht = 2.1^{-57.1}$							
trans-Decahydronaphthalene	14.41	40.2	43.5	237.6	352.3	432.6	489.2
Decanal				300.4	400.4	472.8	525.9
Decane	28.78	38.8	51.4	298.1	403.2	480.8	536.4
Decanedioic acid	40.8		160.7				
Decanenitrile			66.8				
1-Decanethiol	31.0	46.4	65.5	320.6	429.4	510.9	573.1
Decanoic acid	28.02		118.8				
1-Decanol	37.7	49.8	81.5	187.2	418.2	495.9	553.3
1-Decene	21.10	38.7	50.4	283.6	381.9	453.0	505.9
$\Delta Ht = 8.0^{-74.8}$							
1-Decyne				274.6	363.8	428.5	476.6
Deoxybenzoin			93.3				
Dibenz[*de,kl*] anthracene			125.5				
Dibenzoyl peroxide	31.4		102.5				
Dibenzyl ether		20.2					
Dibenzyl sulfide			93.3				
Dibenzyl sulfone			125.5				
1,2-Dibromobutane			50.3	153.9	195.4	224.3	244.8
1,4-Dibromobutane			53.1				
2,3-Dibromobutane			37.7				
1,2-Dibromo-1-chloro-1,1,2-trifluoroethane		31.2	35.0				
1,2-Dibromocycloheptane			52.0				
1,2-Dibromocyclohexane			50.5				
1,2-Dibromocyclooctane			54.6				
1,2-Dibromoethane	10.84	34.8	41.7	99.7	122.3	137.8	149.8
1,2-Dibromoheptane			54.4				
Dibromomethane		32.9	37.0	63.0	74.8	82.5	88.0
1,2-Dibromopropane	8.94	35.6	41.7	124.4	157.4	179.5	195.6
1,3-Dibromopropane	13.6		47.5				
1,2-Dibromotetrafluoroethane	7.04	27.0	28.4				
1,2-Dibutoxyethane		47.8	58.8				
Dibutoxymethane			48.1				
Dibutylamine		38.4	49.5				
N,N-Dibutyl-1-butanamine		46.9					
Dibutyl decanedioate		92.9					
Dibutyl disulfide		46.9	64.5	286.1	376.5	442.8	493.1
Di-*tert*-butyl disulfide			54.3				
Dibutyl ether		36.5	45.0	254.3	340.1	403.8	451.3
Di-*sec*-butyl ether		34.1	40.8				
Di-*tert*-butyl ether		32.2	37.6				
Dibutylmercury			63.5				
Di-*tert*-butyl peroxide			31.8				
Dibutyl 1,2-phthalate		79.2	91.6				
Dibutyl sulfate			75.9				
Dibutyl sulfide	19.4	41.3	53.0	259.8	348.6	420.8	475.8
Di-*tert*-butyl sulfide		33.3	43.8				
Dibutyl sulfite			67.8				
Dibutyl sulfone			100.4				

TABLE 2.55 Heat of Fusion, Vaporization, Sublimation, and Specific Heat at Various Temperatures of Organic Compounds (*Continued*)

Substance	ΔHm	ΔHv	ΔHs	C_p 400 K	600 K	800 K	1000 K
Dichloroacetyl chloride			39.3				
1,2-Dichlorobenzene	12.93	39.7	50.2	142.8	184.4	210.4	227.7
1,3-Dichlorobenzene	12.64	38.6	48.6	143.0	184.5	210.4	227.7
1,4-Dichlorobenzene	17.15	38.8	49.0	143.3	184.8	210.7	227.9
2,6-Dichlorobenzoquinone			69.9				
2,2′-Dichlorobiphenyl			96.2				
4,4′-Dichlorobiphenyl			103.8				
1,2-Dichlorobutane		33.9	39.6				
1,4-Dichlorobutane			46.4				
Dichlorodifluoromethane	4.14	20.1		82.4	93.6	99.1	100.0
Dichlorodimethylsilane			34.3				
Dichlorodiphenylsilane			69.5				
1,1-Dichloroethane	8.84	28.9	30.6	91.4	113.7	128.8	139.8
1,2-Dichloroethane	8.83	32.0	35.2	92.1	112.6	127.2	138.1
1,1-Dichloroethylene	6.51	26.1	26.5	78.7	93.9	103.4	110.0
cis-1,2-Dichloroethylene	7.20	30.2	31.0	77.0	93.0	102.9	109.8
trans-1,2-Dichloroethylene	11.98	28.9	29.3	77.7	93.2	102.9	109.8
2,2-Dichloroethyl ether		38.4					
Dichlorofluoromethane		25.2		70.2	82.4	89.6	94.2
1,2-Dichlorohexafluoropropane		26.3	26.9				
1,2-Dichlorohexane			48.2				
Dichloromethane	6.00	28.1	28.8	59.6	72.4	80.8	86.8
1,2-Dichloro-4-methylbenzene	10.68						
1,2-Dichloropentane		36.5	43.9				
1,5-Dichloropentane			50.7				
(±)-1,2-Dichloropropane	6.40	31.8	36.0	119.7	152.6	175.6	192.8
1,3-Dichloropropane		35.2	40.8	120.0	151.5	173.9	190.4
2,2-Dichloropropane		29.3	32.6	127.9	159.2	179.9	194.8
1,3-Dichloro-2-propanol			66.9				
1,2-Dichlorotetrafluoroethane	6.32	23.3					
Dicyanoacetylene			28.8				
Dicyclopentadienyliron			73.6				
Dicyclopropyl ketone			53.7				
Diethanolamine	25.10	65.2					
1,1-Diethoxyethane		36.3	43.2				
1,2-Diethoxyethane		36.3	43.2				
Diethoxymethane		31.3	35.7				
1,3-Diethoxypropane		37.2	45.9				
2,2-Diethoxypropane			31.8				
Diethylamine		29.1	31.3	143.9	197.2	235.0	263.2
1,2-Diethylbenzene	16.8	39.4	52.8	234.4	316.6	374.6	416.3
1,3-Diethylbenzene	11.0	39.4	52.5	230.2	314.6	379.7	415.8
1,4-Diethylbenzene	10.6	39.4	52.5	228.8	313.1	372.5	414.9
Diethyl carbonate		36.2	43.6				
Diethyl disulfide	9.4	37.6	45.2	171.1	218.6	251.8	276.0
Diethylene glycol diethyl ether	13.60	49.0	58.4				
Diethylene glycol dimethyl ether		36.2	44.7				
Diethylene glycol monoethyl ether		47.5					
Diethylene glycol monomethyl ether		46.6					
Diethyl ether	7.27	26.5	27.1	138.1	183.8	218.7	244.8

(*Continued*)

TABLE 2.55   Heat of Fusion, Vaporization, Sublimation, and Specific Heat at Various Temperatures of Organic Compounds (*Continued*)

Substance	ΔHm	ΔHv	ΔHs	C_p 400 K	600 K	800 K	1000 K
Diethyl malonate		54.8					
Diethyl oxalate		42.0	63.5				
Diethyl peroxide			30.5				
3,3-Diethylpentane	10.09	34.6	42.0				
Diethyl 1,2-phthalate			88.3				
Diethyl sulfide	11.90	31.8	35.8	145.0	192.9	229.7	258.5
Diethyl sulfite			48.5				
Diethyl sulfone			86.2				
Diethyl sulfoxide			62.3				
Diethylzinc			40.2				
1,2-Difluorobenzene	11.1	32.2	36.2	137.1	181.3	209.7	229.0
1,3-Difluorobenzene	8.58	31.1	34.6	137.0	180.5	207.8	225.6
1,4-Difluorobenzene		31.8	35.5	137.4	180.1	207.8	225.7
2,2′-Difluorobiphenyl			95.0				
4,4′-Difluorobiphenyl			91.2				
1,1-Difluoroethane		21.6	19.1	83.4	107.5	124.3	136.3
1,1-Difluoroethylene				71.8	89.2	100.2	107.7
Difluoromethane				51.1	65.8	76.2	83.7
9,10-Dihydroanthracene			93.3				
Dihydro-2*H*-pyran			32.2				
5,12-Dihydrotetracene			115.9				
2,3-Dihydrothiophene		33.2	37.7				
2,5-Dihydrothiophene		34.8	40.0				
2,4-Dihydrothiophene-1,1-dioxide			62.8				
1,4-Dihydroxybenzene	27.11		99.2				
1,2-Diiodobenzene			64.9				
1,2-Diiodoethane			65.7	96.0	116.8	131.3	141.6
Diiodomethane	44.80	42.5	51.0	65.9	76.9	83.9	89.1
Diisobutylamine			39.3				
Diisobutyl ether		34.0	40.9				
Diisobutyl sulfide			48.7				
Diisopropylamine		30.4	34.6				
Diisopropyl ether	11.03	29.1	32.1	196.2	262.0	311.3	348.0
Diisopropylmercury			53.6				
Diisopropyl sulfide	10.4	33.8	39.6	211.9	277.1	322.7	356.6
Diketene		36.8	42.9				
1,2-Dimethoxybenzene	16.04	48.2	66.9				
1,1-Dimethoxyethane			30.5				
1,2-Dimethoxyethane	12.60	32.4	36.4				
Dimethoxymethane	8.33		35.1				
2,2-Dimethoxypropane			29.4				
N,*N*-Dimethylacetamide	10.42	43.4	50.2				
Dimethylamine	5.94	26.4	25.0	87.4	118.9	142.0	159.8
Dimethylaminomethanol			50.2				
N,*N*-Dimethylaminotrimethylsilane			31.8				
N,*N*-Dimethylaniline			52.8				
1,4-Dimethylbicyclo[2.2.1]heptane		33.3	38.9				
2,3-Dimethylbicyclo[2.2.1]-2-heptene		34.9	42.2				
2,2-Dimethylbutane	0.58	26.3	27.7	182.8	251.0	298.7	333.5

TABLE 2.55   Heat of Fusion, Vaporization, Sublimation, and Specific Heat at Various Temperatures of Organic Compounds (*Continued*)

Substance	ΔHm	ΔHv	ΔHs	C_p			
				400 K	600 K	800 K	1000 K
$\Delta Ht = 5.4^{-147.3}$							
$\Delta Ht = 0.3^{-132.3}$							
2,3-Dimethylbutane	0.80	27.4	29.1	181.2	247.7	314.6	331.0
$\Delta Ht = 6.5^{-137.1}$							
2,2-Dimethyl-1-butanol		42.6	56.1				
2,3-Dimethyl-1-butanol		47.3					
3,3-Dimethyl-1-butanol		46.4					
2,3-Dimethyl-2-butanol		40.4	51.0				
(±)-3,3-Dimethyl-2-butanol		43.9					
3,3-Dimethyl-2-butanone		33.4	37.9				
2,3-Dimethyl-1-butene		27.4	29.2	178.2	231.8	272.0	302.1
3.3-Dimethyl-1-butene	1.1	25.7	27.1	162.8	223.4	266.1	297.1
$\Delta Ht = 4.3^{-148.3}$							
2,3-Dimethyl-2-butene	5.46	29.6	32.5	156.8	216.7	262.7	297.7
$\Delta Ht = 3.5^{-76.3}$							
Di(3-methylbutyl) ether		35.2					
Dimethylcadmium			38.0				
1,1-Dimethylcyclohexane	2.06	32.5	37.9	212.1	310.0	379.5	427.6
$\Delta Ht = 6.0^{-120.0}$							
cis-1,2-Dimethylcyclohexane	1.64	33.5	39.7	213.8	309.6	377.0	424.3
$\Delta Ht = 8.3^{-100.6}$							
trans-1,2-Dimethylcyclohexane	10.49	33.0	38.4	217.2	312.1	378.7	425.5
cis-1,3-Dimethylcyclohexane	10.82	32.9	38.3	214.2	310.5	378.7	426.8
trans-1,3-Dimethylcyclohexane	9.86	33.4	39.2	213.8	308.8	375.7	423.0
cis-1,4-Dimethylcyclohexane	9.31	33.3	39.0	213.8	308.8	375.7	423.0
trans-1,4-Dimethylcyclohexane	12.33	32.6	37.9	215.9	312.1	378.9	425.7
1,1-Dimethylcyclopentane	1.1	30.3	33.8	182.2	262.6	318.7	359.1
$\Delta Ht = 6.5^{-126.4}$							
cis-1,2-Dimethylcyclopentane	1.7	31.7	35.7	182.7	262.4	317.9	358.0
$\Delta Ht = 6.7^{-131.7}$							
trans-1,2-Dimethylcyclopentane	7.2	30.9	34.6	182.9	262.2	317.3	357.4
cis-1,3-Dimethylcyclopentane	7.4	30.4	34.2	182.9	262.2	317.3	357.4
trans-1,3-Dimethylcyclopentane	7.3	30.8	34.5	182.9	262.2	317.3	357.4
cis-2,4-Dimethyl-1,3-dioxane			39.9				
4,5-Dimethyl-1,3-dioxane			42.5				
5,5-Dimethyl-1,3-dioxane			41.3				
Dimethyl disulfide	9.19	33.8	37.9	110.3	137.4	157.6	172.8
Dimethyl ether	4.94	21.5	18.5	79.6	105.3	125.7	141.4
N,N-Dimethylformamide	16.15	38.4	46.9				
Dimethylglyoxime			97.1				
2,2-Dimethylheptane	8.90						
2,6-Dimethyl-4-heptanone		39.9	50.9				
2,2-Dimethylhexane	6.78	32.1	37.3				
2,3-Dimethylhexane		33.2	38.8				
2,4-Dimethylhexane		32.5	37.8				
2,5-Dimethylhexane	12.95	32.5	37.9				
3,3-Dimethylhexane	6.98	32.3	37.5				
3,4-Dimethylhexane		33.2	39.0				
cis-2,2-Dimethyl-3-hexene			37.2				

(*Continued*)

TABLE 2.55   Heat of Fusion, Vaporization, Sublimation, and Specific Heat at Various Temperatures of Organic Compounds (*Continued*)

Substance	ΔHm	ΔHv	ΔHs	C_p 400 K	600 K	800 K	1000 K
trans-2,2-Dimethyl-3-hexene			37.3				
1,1-Dimethylhydrazine	10.1	32.6	35.0				
1,2-Dimethylhydrazine		35.2	39.3				
3,5-Dimethylisoxazole			45.2				
Dimethyl maleate	14.7		44.3				
Dimethylmercury			34.6				
6,6-Dimethyl-2-methylene-bicyclo[3.1.1] heptane		40.2	46.4				
2,4-Dimethyloctane		36.5	47.1				
Dimethyl oxalate	21.07		47.4				
3,3-Dimethyloxetane		30.9	33.9				
2,2-Dimethylpentane	5.86	29.2	32.4	211.0	285.9	340.7	381.6
2,3-Dimethylpentane		30.5	34.3	211.0	285.9	340.7	381.6
2,4-Dimethylpentane	6.69	29.6	32.9	211.0	285.9	340.7	381.6
3,3-Dimethylpentane	7.07	29.6	33.0	211.0	285.9	340.7	381.6
2,2-Dimethyl-3-pentanone		36.1	42.3				
2,4-Dimethyl-3-pentanone	11.18	34.6	41.5				
2,4-Dimethyl-1-pentene			33.2				
4,4-Dimethyl-1-pentene			29.0				
2,4-Dimethyl-2-pentene			34.4				
cis-4,4-Dimethyl-2-pentene			32.7				
trans-4,4-Dimethyl-2-pentene			32.7				
2,7-Dimethylphenanthrene			106.7				
4,5-Dimethylphenanthrene			104.6				
9,10-Dimethylphenanthrene			119.5				
2,3-Dimethylphenol	21.02		84.0				
2,4-Dimethylphenol		47.1	65.0				
2,5-Dimethylphenol	23.38	46.9	85.0				
2,6-Dimethylphenol	18.90	44.5	75.3				
3,4-Dimethylphenol	18.13	49.7	85.0				
3,5-Dimethylphenol	18.00	49.3	82.0				
Dimethyl 1,2-phthalate	162.7						
2,2-Dimethylpropane	3.10	22.7	21.8	157.1	218.5	254.3	283.7
$\Delta Ht = 2.6^{-133.1}$							
2,2-Dimethylpropanenitrile		32.4	37.3				
2,2-Dimethyl-1-propanol		9.6					
2,3-Dimethylpyridine		39.1	47.7				
2,4-Dimethylpyridine		38.5	47.5				
2,5-Dimethylpyridine			47.8				
2,6-Dimethylpyridine	10.04	37.5	45.4				
3,4-Dimethylpyridine		40.0	50.5				
3,5-Dimethylpyridine		39.5	49.5				
Dimethyl sulfate			48.5				
Dimethyl sulfide	7.99	27.0	27.7	88.4	113.0	132.2	147.2
Dimethyl sulfite			40.2				
Dimethyl sulfone			77.0				
Dimethyl sulfoxide	14.37	43.1	52.9				
2,2-Dimethylthiacyclopropane			35.8				
Dimethylzinc			29.5				
Dinitromethane			46.0				

TABLE 2.55   Heat of Fusion, Vaporization, Sublimation, and Specific Heat at Various Temperatures of Organic Compounds (*Continued*)

Substance	ΔHm	ΔHv	ΔHs	C_p 400 K	600 K	800 K	1000 K
2,4-Dinitrophenol			104.6				
2,6-Dinitrophenol			112.1				
1,1-Dinitropropane			62.5				
1,3-Dioxane		34.4	39.1				
1,4-Dioxane	12.85	34.2	38.6	126.5	181.8	218.2	243.3
$\Delta Ht = 2.4^{-0.3}$							
1,3-Dioxolane	27.48		35.6				
Diphenylamine	17.86		89.1				
Diphenyl carbonate	23.4		90.0				
Diphenyl disulfide			95.0				
Diphenyl disulfone			161.9				
Diphenylenimine			84.5				
1,2-Diphenylethane		51.5	91.4				
1,1-Diphenylethylene			73.2				
Diphenyl ether	17.22	48.2	67.0				
6,6-Diphenylfulvene			104.6				
Diphenylmercury			112.8				
Diphenylmethane	18.2		67.5				
1,3-Diphenyl-2-propanone			89.1				
Diphenyl sulfide			67.8				
Diphenyl sulfone			106.3				
Diphenyl sulfoxide			97.1				
1,2-Dipropoxyethane			50.6				
Dipropylamine		33.5	40.0				
Dipropyl disulfide	13.8	41.9	54.1	186.2	298.3	350.2	390.0
Dipropyl ether	8.83	31.3	35.7	196.2	262.0	311.3	348.0
Dipropylmercury			55.2				
Dipropyl sulfate			66.9				
Dipropyl sulfide	12.1	36.6	44.2	201.7	272.5	328.2	372.6
Dipropyl sulfite			58.6				
Dipropyl sulfone			79.9				
Dipropyl sulfoxide			74.5				
Divinyl ether			26.2				
Divinyl sulfone			56.5				
Dodecane	36.55	44.5	61.5	356.2	481.3	572.2	656.5
Dodecanedioic acid			153.1				
Dodecanenitrile			76.1				
Dodecanoic acid	36.64		132.6				
Dodecanol	31.4	63.5	92.0				
1-Dodecene	17.42	44.0	60.8	341.8	460.0	545.6	608.8
$\Delta Ht = 4.6^{-60.2}$							
1,2-Epoxybutane		30.3					
1,2-Epoxypropane		21.6					
Ergosterol			118.4				
Ethane	2.86	14.7	5.2	65.5	89.3	108.0	122.6
Ethane-d_6				81.7	108.5	127.4	140.5
1,2-Ethanediamine	22.58	38.0	45.0				
1,2-Ethanediol	11.23	50.5	67.8	113.2	136.9	166.9	
1,2-Ethanediol diacetate		45.5	61.4				
1,2-Ethanedithiol		37.9	44.7				

(*Continued*)

TABLE 2.55 Heat of Fusion, Vaporization, Sublimation, and Specific Heat at Various Temperatures of Organic Compounds (*Continued*)

Substance	ΔHm	ΔHv	ΔHs	C_p 400 K	600 K	800 K	1000 K
Ethanethiol	4.98	26.8	27.3	88.2	113.9	133.2	148.0
Ethanol	5.02	38.6	42.3	81.2	107.7	127.2	141.9
Ethanolamine	20.50	49.8					
Ethoxybenzene		40.7	51.0				
2-Ethoxyethanol		39.2	48.2				
2-(2-Ethoxyethoxy)ethanol		47.5					
2-(2-Ethoxyethoxy)ethyl acetate		91.2					
2-Ethoxyethyl acetate			52.7				
1-Ethoxy-2-methoxyethane		34.3	39.8				
N-Ethylacetamide			64.9				
Ethyl acetate	10.48	31.9	35.6	137.4	182.6	213.4	234.5
Ethyl acrylate		34.7					
Ethylamine		28.0	26.6	90.6	119.6	141.8	158.5
N-Ethylaniline			52.3				
Ethylbenzene	9.18	35.6	42.2	170.5	236.1	281.0	312.8
2-Ethylbenzoic acid			100.7				
3-Ethylbenzoic acid			99.1				
4-Ethylbenzoic acid			97.5				
2-Ethyl-1-butanol		43.2	63.2				
Ethyl butanoate		35.5	42.7				
2-Ethylbutanoic acid		51.2					
2-Ethyl-1-butene		28.8	31.1	170.3	228.0	269.5	300.8
Ethyl *trans*-2-butenoate			44.4				
Ethyl chloroacetate		40.4	49.5				
Ethyl 4-chlorobutanoate			52.7				
Ethyl chloroformate			42.3				
Ethyl *trans*-cinnamate		58.6					
Ethyl crotonate			44.3				
Ethyl cyanoacetate		64.4					
Ethylcyclobutane		28.7	31.2				
Ethylcyclohexane	8.33	34.0	40.6	215.9	310.0	377.0	423.8
1-Ethylcyclohexene			43.3				
Ethylcyclopentane	6.9	32.0	36.4	183.6	258.2	314.7	356.3
1-Ethylcyclopentene		38.5					
Ethyl dichloroacetate			50.6				
Ethyl 2,2-dimethylpropanoate		34.5	41.2				
Ethylene	3.35	13.5		53.1	70.7	83.8	93.9
Ethylene-d_4				63.9	82.3	95.6	104.9
Ethylene carbonate	13.19	50.1	73.2				
2,2′-(Ethylenedioxy)bis(ethanol)		71.4	79.1				
Ethylene glycol (*see* 1,2-Ethanediol)							
Ethylene glycol diacetate			61.4				
Ethylene oxide	5.2	25.5	24.8	62.6	86.3	102.9	114.9
Ethylenimine		30.3	34.6	70.4	98.6	117.7	131.6
N-Ethylformamide			58.4				
Ethyl formate	9.20	29.9	32.0				
2-Ethylhexanal			49.0				
2-Ethylhexane		33.6	39.6				
Ethyl hexanoate			51.7				
2-Ethylhexanoic acid		56.0	75.6				

TABLE 2.55   Heat of Fusion, Vaporization, Sublimation, and Specific Heat at Various Temperatures of Organic Compounds (*Continued*)

Substance	ΔHm	ΔHv	ΔHs	C_p 400 K	600 K	800 K	1000 K
2-Ethyl-1-hexanol		45.2					
2-Ethylhexyl acetate		43.5	48.1				
2-Ethyl hydroperoxide			43.1				
Ethylidenecyclohexane			42.0				
Ethylidenecyclopentane		18.1					
Ethyl isocyanide			33.5				
Ethyl isopentanoate	8.7	43.9					
Ethyl isopentyl ether		33.0	39.0				
Ethylisopropylamine		29.9	33.1				
Ethyl isopropyl ether		28.2	30.1				
Ethyl isopropyl sulfide	8.7	32.7	37.8				
Ethyl lactate		46.4	49.4				
Ethyllithium			116.7				
Ethylmercury bromide			76.6				
Ethylmercury chloride			76.1				
Ethylmercury iodide			79.5				
1-Ethyl-2-methylbenzene	10.0	38.9	47.7	202.9	275.3	326.8	363.6
1-Ethyl-3-methylbenzene	7.6	38.5	46.9	198.7	273.6	325.5	363.2
1-Ethyl-4-methylbenzene	13.4	38.4	46.6	197.5	272.0	324.7	362.2
Ethyl 2-methylbutanoate			44.4				
Ethyl 3-methylbutanoate		37.0	43.9				
2-Ethyl-3-methyl-1-butene			34.5				
1-Ethyl-1-methylcyclopentane		33.2	38.9				
Ethyl methyl ether		26.7		109.1	144.7	172.3	193.2
3-Ethyl-2-methylpentane	11.34	32.9	38.5				
3-Ethyl-3-methylpentane	10.84	32.8	38.0				
3-Ethyl-2-methyl-1-pentene			37.5				
Ethyl 2-methylpropanoate		33.7	39.8				
Ethyl methyl sulfide	9.8	29.5	31.9	116.4	152.3	179.6	200.6
Ethyl nitrate	8.5	33.1	36.3	120.2	155.1	178.7	195.4
1-Ethyl-2-nitrobenzene			59.8				
1-Ethyl-4-nitrobenzene			62.8				
3-Ethylpentane	9.55	31.1	35.2	211.0	285.9	340.7	381.6
Ethyl pentanoate		37.0	47.0				
Ethyl pentyl ether		34.4	41.0				
2-Ethylphenol			63.6				
3-Ethylphenol			68.2				
4-Ethylphenol			80.3				
Ethylphosphonic acid			50.6				
Ethylphosphonic dichloride			42.7				
Ethyl propanoate		33.9	39.2				
Ethyl propyl ether		28.9	31.4				
Ethyl propyl sulfide	10.6	34.2	40.0	173.3	232.7	279.0	315.6
Ethyl trichloroacetate			51.0				
S-Ethyl thiolacetate	34.4	40.0					
Ethyl 2-vinylacrylate			48.5				
Ethyl vinyl ether		26.2	26.6				
Fluoranthene	18.87		99.2				
9*H*-Fluorene	19.58						
Fluorobenzene	11.31	31.2	34.6	125.5	171.0	200.1	220.0

TABLE 2.55 Heat of Fusion, Vaporization, Sublimation, and Specific Heat at Various Temperatures of Organic Compounds (*Continued*)

Substance	ΔHm	ΔHv	ΔHs	C_p 400 K	600 K	800 K	1000 K
4-Fluorobenzoic acid			91.2				
Fluoroethane				74.1	98.6	116.4	129.7
Fluoromethane		16.7		44.2	57.9	68.8	77.2
1-Fluorooctane		40.4	49.7				
1-Fluoropropane				102.7	137.3	162.7	181.5
2-Fluoropropane				103.5	138.7	163.8	182.2
2-Fluorotoluene		35.4					
4-Fluorotoluene	9.4	34.1	39.4	152.4	207.9	245.2	271.3
Fluorotrichloromethane		25.0					
Fluorotrinitromethane			34.7				
Formaldehyde		23.3		39.2(g)	48.2	55.9	62.0
Formamide	6.69		60.2				
Formic acid	12.7	22.7	20.1	53.8	67.0	76.8	83.5
Formyl fluoride		21.7		46.4	56.2	63.1	67.9
Fumaric acid			136.0				
Fumaronitrile			72.0				
Furan, $\Delta Ht = 2.1^{-123.2}$	3.80	27.1	27.5	88.7	122.6	164.9	158.5
2-Furancarboxaldehyde	14.35	43.2	50.6				
2-Furancarboxylic acid			108.5				
Furanmethanol	13.13	53.6	64.4				
Glutaric acid	20.9						
Glycerol	18.28	61.0	85.8				
Glyceryl triacetate			85.7				
Glyceryl tributanoate			107.1				
Glyceryl trinitrate	21.87		100.0				
Heptadecane, $\Delta Ht = 11.0^{11.1}$	40.5	52.9	86.0	501.4	676.8	803.7	897.9
Heptadecanoic acid	58.8						
1-Heptadecene	31.4	51.8	85.0	486.9	655.5	777.1	866.9
1-Heptanal	23.6		47.7	213.4	283.3	333.9	371.1
Heptane	14.16	31.8	36.6	211.0	285.9	340.7	381.6
1-Heptanenitrile			51.9				
1-Heptanethiol	25.4	39.8	50.6	233.5	312.1	372.0	418.4
Heptanoic acid			74.0				
1-Heptanol	13.2	48.1	66.8	224.4	300.9	357.0	392.5
2-Heptanol		49.8					
3-Heptanol		42.5					
2-Heptanone		38.3	47.2				
4-Heptanone		36.2					
1-Heptene, $\Delta Ht = 0.3^{-136}$	12.66	31.1	35.5	196.5	264.6	314.1	351.0
trans-2-Heptene	11.72						
Heptylamine			50.0				
Heptyl methyl ether			46.9				
Hexachlorobenzene	23.85		92.6	201.2	233.4	250.9	260.8
Hexachloroethane, $\Delta Ht = 8.0^{71.3}$	9.8	45.9	59.0	151.5	166.6	173.6	177.3
Hexadecafluoroethylcyclohexane			38.5				
Hexadecafluoroheptane			36.4				
Hexadecane	51.8	51.2	81.4	472.3	687.7	757.4	846.0
Hexadecanoic acid	42.04		154.4				
1-Hexadecanol, $\Delta Ht = 16.6^{34}$	34.29		169.5	485.7	652.7	773.6	863.2
1-Hexadecene	30.2	50.4	80.3	457.9	616.4	731.82	815.0

TABLE 2.55 Heat of Fusion, Vaporization, Sublimation, and Specific Heat at Various Temperatures of Organic Compounds (*Continued*)

Substance	ΔHm	ΔHv	ΔHs	C_p			
				400 K	600 K	800 K	1000 K
Hexadienoic acid	13.6						
Hexafluoroacetone		19.8	21.3				
Hexafluoroacetylacetone		27.1	30.6				
Hexafluorobenzene	11.58	31.7	35.7	183.6	219.9	241.1	253.7
Hexafluoroethane, $\Delta Ht = 3.7^{-169.2}$	2.7	16.2		125.6	149.0	160.7	166.8
cis-Hexahydroindane			57.5				
trans-Hexahydroindane			56.1				
Hexamethylbenzene	20.6	48.2	74.7	310.4	406.4	474.9	525.3
$\Delta Ht = 1.1^{-156.7}$							
$\Delta Ht = 1.8^{110.7}$							
1,1,1,3,3,3-Hexamethyldisilazanc			41.4				
Hexamethyldisiloxane			37.2				
Hexamethylphosphoric triamide	14.28						
Hexanal				184.2	243.9	287.4	319.7
Hexanamide	25.1		98.7				
Hexane	13.08	28.9	31.6	181.9	246.8	294.4	330.1
1,6-Hexanedioic acid	34.85		129.3				
1,6-Hexanediol	25.5		83.3				
Hexanenitrile		38.0	47.9				
1-Hexanethiol	18.0(1)	37.2	45.8	204.5	273.1	325.1	366.7
Hexanoic acid	15.40	71.1	72.2				
1-Hexanol	15.40	44.5	61.6	195.3	261.8	310.7	346.9
2-Hexanol		41.0	58.5				
3-Hexanol	44.3	46.0					
2-Hexanone	14.90	36.4	43.1				
3-Hexanone	13.49	35.4	42.5				
1-Hexene	9.35	28.3	30.6	167.5	225.5	267.9	299.3
cis-2-Hexene	8.86	29.1	32.2	161.5	221.8	165.3	297.9
trans-2-Hexene	8.26	28.9	31.6	166.1	223.4	266.1	297.9
cis-3-Hexene	8.25	28.7	31.4	161.1	222.6	265.7	297.9
trans-3-Hexene	11.08	28.9	31.7	168.2	225.5	267.4	298.7
Hexylamine		36.5	45.1				
Hexyl methyl ether		34.9	42.1				
1-Hexyne				158.5	207.5	243.3	270.1
Hydrazine	12.7	45.3					
2-Hydroxybenzaldehyde		38.2					
2-Hydroxybenzoic acid			95.1				
2-Hydroxy-2,4,6-cycloheptatrienone			83.7				
2-Hydroxy-1-isopropyl-4-methylbenzene			91.2				
4-Hydroxy-4-methyl-2-pentanone		28.5	47.7				
3-Hydroxypropanonitrile		56.1					
2-Hydroxypyridine			86.6				
3-Hydroxypyridine			88.3				
4-Hydroxypyridine			103.8				
8-Hydroxyquinoline			108.8				
Icosane	69.88	57.5	100.8	588.5	794.0	942.6	1052.7
Icosanoic acid	72.0		199.6				
1-Icosene	34.3	55.9	99.8	574.0	772.7	916.0	1021.7
Indane		39.6	48.8				
Indene			52.9				

(*Continued*)

TABLE 2.55 Heat of Fusion, Vaporization, Sublimation, and Specific Heat at Various Temperatures of Organic Compounds (*Continued*)

Substance	ΔHm	ΔHv	ΔHs	Cp 400 K	600 K	800 K	1000 K
Indole			69.9				
Iodobenzene	9.76	39.5	47.7	130.1	173.3	201.1	220.1
Iodobenzoic acid			87.9				
1-Iodobutane		34.7	40.6				
2-Iodobutane		33.3	38.5				
Iodocyclohexane			47.3				
Iodoethane		29.4	31.9	80.3	103.1	119.9	132.4
1-Iodohexane			49.8				
Iodomethane		27.3	28.0	51.6	63.9	73.1	80.2
1-Iodo-2-methylpropane		33.5	38.8				
2-Iodo-2-methylpropane	14.5	31.4	35.4	148.8	191.7	221.1	242.3
1-Iodonaphthalene			72.4				
2-Iodonaphthalene			90.8				
1-Iodopentane			45.3				
1-Iodopropane		32.1	36.2	109.9	142.7	166.5	184.2
2-Iodopropane		30.7	34.1	111.2	144.7	168.2	185.5
3-Iodo-1-propene			38.1				
2-Iodotoluene (also 3-, 4-)			54.4				
Isobutanonitrile		32.4	37.2	119.5	156.4	183.0	202.5
Isobutyl acetate		35.9					
Isobutylamine		30.6	33.9				
Isobutylbenzene	12.51	37.8	47.9				
Isobutylcyclohexane			47.6				
Isobutyl dichloroacetate			52.3				
Isobutyl formate		33.6					
Isobutyl isobutanoate		38.2	46.4				
Isobutyl isopropyl ether		31.6	36.6				
Isobutyl methyl ether		28.0	30.1				
Isobutyl propyl ether		28.3	30.3				
Isobutyl trichloroacetate			53.1				
Isobutyl vinyl ether		30.7	34.6				
2-Isopropoxyethanol		40.4	50.1				
Isopropyl acetate		32.9	37.2				
Isopropylamine	7.33	27.8	28.4				
Isopropylbenzene	7.79	37.5	45.1	200.8	277.0	328.9	365.3
Isopropylcyclohexane			44.0				
Isopropylcyclopentane		33.6	39.4				
Isopropylmethylamine		28.7	30.9				
1-Isopropyl-2-methylbenzene	10.0	38.4	50.6				
1-Isopropyl-3-methylbenzene	13.7	38.1	50.0				
1-Isopropyl-4-methylbenzene	9.7	38.2	50.2				
Isopropyl methyl ether		26.1	26.4	138.0	184.8	220.4	247.2
2-Isopropyl-5-methylphenol			91.2				
Isopropyl methyl sulfide	9.4	30.7	34.2	145.1	192.5	229.9	260.6
Isopropyl nitrate		34.9	38.8	150.5	195.9	226.5	247.9
Isopropylpropylamine		32.1	37.2				
Isopropyl propyl sulfide		35.1	41.8				
Isopropyl trichloroacetate			51.9				
Isoquinoline	7.45	49.0	60.3				
Ketene			20.4	59.5	70.7	78.7	86.4

TABLE 2.55 Heat of Fusion, Vaporization, Sublimation, and Specific Heat at Various Temperatures of Organic Compounds (*Continued*)

Substance	ΔHm	ΔHv	ΔHs	C_p 400 K	600 K	800 K	1000 K
(−)-Leucine			150.6				
(+)-Limonene			48.1				
Maleic acid			110.0				
Maleic anhydride			71.5				
Malononitrile			79.1				
D-Mannitol	22.6						
Methacrylonitrile		31.8					
Methane	0.94	8.2		40.5	52.2	62.9	71.8
Methane-d_4				48.6	63.4	74.8	83.0
Methanethiol, $\Delta Ht = 0.22^{-135.6}$	5.91	24.6	23.8	58.7	73.5	85.0	94.1
Methanol, $\Delta Ht = 0.6^{-115.8}$	3.18	35.2	37.4	51.4	67.0	79.7	89.5
4-Methoxybenzaldehyde		56.8	64.5				
Methoxybenzene		39.0	46.9				
2-Methoxybenzoic acid			104.7				
3-Methoxybenzoic acid			107.4				
4-Methoxybenzoic acid			109.8				
3-Methoxy-1-butanol		50.8					
2-Methoxyethanol		37.5	45.2				
2-(2-Methoxyethoxy)ethanol		46.6					
2-Methoxyethyl acetate		43.9	50.3				
2-Methoxy-1-propoxyethane		36.3	43.7				
2-Methoxytetrahydropyran			42.7				
1-Methoxy-2,4,6-trinitrobenzene			133.1				
N-Methylacetamide	9.72	59.4					
Methyl acetate		30.3	32.3				
Methyl acetoacetate		36.0					
Methyl acrylate		33.1	29.2				
Methylamine	6.13	25.6	24.4	60.2	78.9	93.9	105.7
4-Methylaniline	18.22						
Methyl benzoate	9.74	43.2	55.6				
2-Methylbenzoic acid	20.17						
3-Methylbenzoic acid	15.72						
4-Methylbenzoic acid	22.73						
1-Methylbicyclo[4.1.0]heptane			39.2				
1-Methylbicyclo[3.1.0]hexane		31.1	34.8				
2-Methyl-1,3-butadiene	4.79	25.9	26.8	133.1	173.2	200.8	221.3
3-Methyl-1,3-butadiene		27.2	28.0	129.7	168.6	197.5	219.2
2-Methylbutane	5.15	24.7	24.9	152.7	208.7	249.8	280.8
3-Methylbutanenitrile		35.1	41.7				
2-Methylbutanethiol		33.8	39.5				
3-Methyl-1-butanethiol	7.5		39.4				
2-Methyl-2-butanethiol	0.6	31.4	35.7	179.0	236.7	279.4	308.8
$\Delta Ht = 8.0^{-114.0}$							
Methyl butanoate		33.8	39.3				
2-Methylbutanoic acid			46.9				
3-Methylbutanoic acid	7.32	43.2	57.5				
2-Methyl-1-butanol		45.2	55.2				
3-Methyl-1-butanol		44.1	55.6				
2-Methyl-2-butanol, $\Delta Ht = 2.0^{-127.2}$	4.45	39.0	50.1				
3-Methyl-2-butanol		41.8	53.0				

(*Continued*)

TABLE 2.55 Heat of Fusion, Vaporization, Sublimation, and Specific Heat at Various Temperatures of Organic Compounds (*Continued*)

Substance	ΔHm	ΔHv	ΔHs	C_p			
				400 K	600 K	800 K	1000 K
3-Methyl-2-butanone		32.4	36.8				
2-Methyl-1-butene	7.9	25.5	25.9	138.9	187.1	222.4	248.7
3-Methyl-1-butene	5.4	24.1	23.8	147.5	192.1	225.3	250.3
2-Methyl-2-butene	7.6	26.3	27.1	133.6	181.7	217.8	245.0
Methyl 2-butenoate			41.0				
3-Methyl-1-butyne		26.2	25.8	130.1	169.9	198.3	219.2
2-Methylbutyl acetate		37.5					
Methyl chloroacetate		39.2	46.7				
Methyl cyanoacetate		48.2	61.7				
Methyl cyclobutanecarboxylate		37.1	44.7				
Methylcyclohexane	6.75	31.3	35.4	185.6	269.7	329.5	371.5
1-Methylcyclohexanol		79.0	80				
cis-2-Methylcyclohexanol		48.5	63.2				
trans-2-Methylcyclohexanol		53.0	63.2				
cis-3-Methylcyclohexanol			65.3				
trans-3-Methylcyclohexanol			65.3				
cis-4-Methylcyclohexanol			65.7				
trans-4-Methylcyclohexanol			66.1				
1-Methylcyclohexene			37.9				
Methylcyclopentane	6.93	29.1	31.6	151.1	219.4	267.8	303.1
1-Methyl-1-cyclopentene			32.6	136.0	195.8	238.5	269.0
3-Methyl-1-cyclopentene			31.0	136.4	197.1	239.3	269.9
4-Methyl-1-cyclopentene			32.2	136.4	196.7	238.4	269.5
Methyl cyclopropanecarboxylate		35.3	41.3				
2-Methyldecane		40.3	54.3				
4-Methyldecane		40.7	53.8				
Methyl decanoate			66.7				
Methyl dichloroacetate		39.3	47.7				
Methyldichlorosilane			28.0				
Methyl 2,2-dimethylpropanoate		33.4	38.8				
2-Methyl-1,3-dioxane			38.6				
4-Methyl-1,3-dioxane			39.2				
4-Methyl-1,3-dioxolan-2-one	9.62						
Methyl dodecanoate			77.2				
N-Methylethanediamine		37.6	45.2				
1-Methylethyl acetate		32.9	37.3				
1-Methylethyl thiolacetate		35.7	42.3				
N-Methylformamide			56.2				
Methyl formate	7.45	27.9	28.4	81.6	105.4	121.8	133.9
Methyl 2-furancarboxylate			45.2				
Methylglyoxal			38.1				
2-Methylheptane	11.88	33.3	39.7				
3-Methylheptane	11.38	33.7	39.8				
4-Methylheptane	10.84	33.4	39.7				
Methyl heptanoate			51.6				
2-Methylhexane	8.87	30.6	34.9	211.0	285.9	340.7	381.6
3-Methylhexane		30.9	35.1	212.0	285.9	340.7	381.6
Methyl hexanoate		38.6	48.0				
5-Methyl-1-hexene			34.3				
cis-3-Methyl-3-hexene			36.5				

TABLE 2.55 Heat of Fusion, Vaporization, Sublimation, and Specific Heat at Various Temperatures of Organic Compounds (*Continued*)

Substance	ΔHm	ΔHv	ΔHs	C_p 400 K	600 K	800 K	1000 K
trans-3-Methyl-3-hexene			35.9				
Methylhydrazine	10.4	36.1	40.4				
Methyl isobutanoate		32.6	37.3				
Methyl isocyanide			30.8				
1-Methyl-4-isopropylbenzene	9.60	38.2					
3-Methylisoxazole			41.0				
5-Methylisoxazole			41.0				
Methylmercury bromide			67.8				
Methylmercury chloride			64.4				
Methylmercury iodide			65.3				
Methyl methacrylate		36.0	60.7				
Methyl 2-methylbutanoate			41.8				
Methyl-3-methylbutanoate			41.0				
1-Methylnaphthalene	6.94	45.5		212.3	292.0	345.1	381.6
$\Delta Ht = 5.0^{-32.4}$							
2-Methylnaphthalene	11.97	46.0	61.7	211.2	290.0	343.2	381.2
$\Delta Ht = 5.6^{15.4}$							
Methyl nitrate	8.2	31.6	32.1	91.5	115.2	131.7	143.1
Methyl nitrite		20.9	22.6	76.3	97.7	112.8	123.5
1-Methyl-4-nitrobenzene			79.1				
2-Methylnonane		38.2	49.6				
3-Methylnonane		38.3	49.7				
5-Methylnonane		38.1	49.3				
2-Methyloctane	18.00						
Methyl octanoate			56.4				
Methyl oxirane		27.4	27.9				
2-Methylpentane	6.27	27.8	29.9	184.1	211.7	296.2	331.4
3-Methylpentane	5.30	28.1	30.3	181.9	246.9	294.6	330.1
2-Methyl-2,4-pentanediol		57.3					
3-Methylpentanenitrile		35.1	41.6				
Methyl pentanoate		35.4	43.1				
2-Methylpentanoic acid		52.1	57.5				
2-Methyl-1-pentanol		50.2	55.7				
2-Methyl-2-pentanol		39.6	54.8				
2-Methyl-3-pentanol		41.8	54.4				
3-Methyl-1-pentanol		46.3	62.3				
3-Methyl-2-pentanol		43.4	56.9				
4-Methyl-1-pentanol		44.5	60.5				
4-Methyl-2-pentanol		44.2	50.6				
3-Methyl-3-pentanol		41.8					
2-Methyl-3-pentanone		33.8	39.8				
3-Methyl-2-pentanone		34.2	40.5				
4-Methyl-2-pentanone		34.5	40.6				
2-Methyl-1-pentene		28.1	30.5	170.7	227.6	269.5	300.4
3-Methyl-1-pentene		26.9	28.7	177.8	232.6	272.8	302.5
4-Methyl-1-pentene		27.1	28.7	162.8	221.3	264.0	296.2
2-Methyl-2-pentene		29.0	31.6	163.2	222.6	245.2	297.5
cis-3-Methyl-2-pentene		28.8	31.2	163.2	222.6	265.3	297.5
trans-3-Methyl-2-pentene		29.3	31.5	163.2	222.6	265.3	297.5
cis-4-Methyl-2-pentene		27.6	29.5	167.6	226.4	267.8	299.2

(Continued)

TABLE 2.55 Heat of Fusion, Vaporization, Sublimation, and Specific Heat at Various Temperatures of Organic Compounds (*Continued*)

Substance	ΔHm	ΔHv	ΔHs	C_p 400 K	600 K	800 K	1000 K
trans-4-Methyl-2-pentene		28.0	30.0	171.1	229.3	269.9	300.4
4-Methyl-3-penten-2-one		36.1		214.0			
Methyl pentyl ether		32.0	36.9				
Methyl pentyl sulfide		37.4	45.2	203.6	272.2	324.6	366.0
3-Methyl-1-phenyl-1-butanone			59.5				
2-Methyl-1-phenylpropane	12.5	37.8	49.5				
Methyl phenyl sulfide			54.3				
Methyl phenyl sulfone			92.0				
Methylphosphonic acid			48.1				
2-Methylpiperidine			40.5				
2-Methylpropanal			31.5				
2-Methylpropane	4.66	21.3	19.3	124.6	169.5	202.9	227.6
2-Methylpropanenitrile		32.4	37.1				
2-Methyl-1-propanethiol	5.0	31.0	34.6	147.7	193.6	225.0	247.6
2-Methyl-2-propanethiol	2.5	28.5	30.8	151.2	199.2	232.3	256.2
$\Delta Ht = 4.1^{-121.6}$							
$\Delta Ht = 0.7^{-116.2}$							
$\Delta Ht = 1.0^{-73.8}$							
Methyl propanoate		32.2	35.9				
2-Methylpropanoic acid	5.02		35.3				
2-Methyl-1-propanol	6.32	41.8	50.8				
2-Methyl-2-propanol	6.79	39.1	46.7	142.9	189.8	222.9	247.5
$\Delta Ht = 0.8^{13}$							
2-Methylpropene	5.93	22.1	20.6	111.2	147.7	175.1	196.0
Methyl propyl ether		26.8	27.6	138.1	183.8	218.7	244.8
Methyl propyl sulfide	9.9	32.1	36.2	144.9	191.9	227.8	255.8
2-Methylpyridine	9.72	36.2	42.5	133.6	186.4	222.6	243.3
3-Methylpyridine	14.18	37.4	44.4	133.1	186.1	222.3	247.8
4-Methylpyridine	11.57	37.5	44.6				
1-Methyl-1*H*-pyrrole			40.8				
Methyl salicylate		46.7					
α-Methylstyrene				187.4	254.0	300.4	333.9
cis-β-Methylstyrene				187.4	254.0	300.4	333.9
trans-β-Methylstyrene				189.1	256.1	301.3	334.7
Methyl tetradecanoate			37.0				
2-Methylthiacyclopentane		36.4	41.8				
4-Methylthiazole		37.6	43.8				
2-Methylthiophene	9.20	33.9	38.9	123.1	165.6	194.3	214.6
3-Methylthiophene	10.53	34.2	39.4	122.9	164.6	192.3	211.7
Methyl trichloroacetate			48.3				
Methyl tridecanoate			82.7				
Methyl undecanoate			71.4				
5-Methyluracil			134.1				
Morpholine		37.1	44.0				
Naphthalene	18.98	43.2	72.6	180.1(g)	251.5	297.3	329.2
1-Naphthalenecarboxylic acid			110.4				
2-Naphthalenecarboxylic acid			113.6				
1-Naphthol	23.33		91.2				
2-Naphthol	17.51		94.2				
1,4-Naphthoquinone			72.4				

TABLE 2.55 Heat of Fusion, Vaporization, Sublimation, and Specific Heat at Various Temperatures of Organic Compounds (*Continued*)

Substance	ΔHm	ΔHv	ΔHs	C_p 400 K	600 K	800 K	1000 K
1-Naphthylamine			90.0				
2-Naphthylamine			88.3				
2-Nitroaniline	16.11		90.0				
3-Nitroaniline	23.68		96.7				
4-Nitroaniline	21.1		109				
Nitrobenzene	11.59	40.8	55.0				
1-Nitrobutane		38.9	48.6	157.5	210.1	247.0	273.6
2-Nitrobutane		36.8	43.8	157.4	211.1	248.7	276.0
Nitroethane	9.85	38.0	41.6	99.0	131.6	154.0	170.2
Nitromethane	9.70	34.0	38.3	70.3	91.7	106.9	117.9
(Nitromethyl)benzene			53.6				
2-Nitrophenol	17.44						
3-Nitrophenol	19.2						
4-Nitrophenol	18.25						
1-Nitronaphthalene			107.1				
1-Nitropropane		38.5	43.4	128.5	171.0	200.7	222.0
2-Nitropropane		36.8	41.3	129.2	172.3	201.8	222.8
2-Nitroso-1-naphthol			56.5				
4-Nitroso-1-naphthol			87.4				
1-Nitroso-2-naphthol			86.6				
2-Nitrotoluene		16.5	47.2				
3-Nitrotoluene		15.0	49.9				
4-Nitrotoluene	16.81	15.5	50.2				
Nonadecane, $\Delta Ht = 13.8^{22.8}$	45.82	56.0	95.8	559.4	754.9	896.3	1000.8
1-Nonadecene	33.5	54.6	94.9	545.0	733.7	869.7	969.9
1-Nonal			72.3	271.1	361.5	426.4	474.5
Nonane, $\Delta Ht = 6.3^{-56.0}$	15.47	36.9	46.4	269.0	364.1	433.3	484.9
1-Nonanethiol	33.5	44.4		291.6	390.3	464.6	521.5
Nonanoic acid	20.28		82.4				
1-Nonanol		54.4	76.9	282.4	379.1	449.6	501.7
2-Nonanone			56.4				
5-Nonanone	24.93		53.3				
1-Nonene	18.08	36.3	45.5	254.6	342.8	406.8	454.0
cis-Octadecafluorodecahydronaphthalene		35.6	45.2				
trans-Octadecafluorodecahydronaphthalene		35.8	45.4				
Octadecafluoropropylcyclohexane		24.5	43.1				
Octadecafluorooctane		33.4	41.1				
Octadecane	61.39	54.5	152.8	530.4	715.8	850.0	949.4
Octadecanedioic acid	56.6						
Octadecanoic acid	56.59		166.5				
Octadecanol			113.4				
1-Octadecene	32.6	53.3	90.0	516.0	694.5	823.4	918.4
cis-9-Octadecenoic acid		64.7					
Octafluorocyclobutane	2.77	23.2		186.1	225.3	245.4	257.3
Octafluorotoluene	11.58						
Octamethylcyclotetrasiloxane		45.6					
Octanal				242.3	322.2	380.3	422.6
Octanamide			110.5				
Octane	20.65	34.4	41.5	240.0	325.0	387.0	433.5
1,8-Octanedioic acid			143.1				

(*Continued*)

TABLE 2.55   Heat of Fusion, Vaporization, Sublimation, and Specific Heat at Various Temperatures
of Organic Compounds (*Continued*)

Substance	ΔHm	ΔHv	ΔHs	C_p			
				400 K	600 K	800 K	1000 K
Octanenitrile		41.3	56.8				
1-Octanethiol	24.3	42.3		262.6	351.3	418.3	469.9
Octanoic acid	21.36	58.5	81.7				
1-Octanol	42.30	46.9	71.0	253.4	340.0	403.3	450.1
(±)-2-Octanol		44.4					
(±)-3-Octanol		36.5					
4-Octanol		40.5					
2-Octanone	24.42						
1-Octene	15.57	34.1	40.4	225.6	303.7	360.5	402.5
1-Octyne		35.8	42.3	216.5	285.7	336.0	410.9
2-Octyne		37.3	44.5				
3-Octyne		36.9	43.9				
4-Octyne		36.0	42.7				
Oxalic acid			98.0				
Oxaloyl chloride			31.8				
Oxamide			113.0				
Oxetane		28.7	29.9				
2-Oxetanone			47.0				
2-Oxohexamethyleneimine	16.2	54.8	83.3				
4-Oxopentanoic acid	9.22						
1,1′-Oxybis(2-ethoxy)ethane			58.4				
2,2′-Oxybis(ethanol)		52.3	57.3				
Paraldehyde			41.4				
Pentachloroethane	11.34	36.9	45.6	133.7	152.1	162.0	168.1
Pentachlorofluoroethane	1.9						
Pentachlorophenol			67.4				
Pentacyclo-[4.2.0.0²,⁵.0³,⁸.0⁴,⁷]octane			80.3				
Pentadecane, ΔHt = 9.2⁻²·²⁵	34.8	49.5	76.1	443.3	598.6	711.1	794.5
Pentadecanoic acid	50.2		162.7				
1-Pentadecene	28.9	48.7	75.1	428.9	577.3	684.5	763.6
1,2-Pentadiene		27.6	28.7	131.4	170.7	199.6	220.9
cis-1,3-Pentadiene		27.6	28.3	123.4	166.9	196.7	218.4
trans-1,3-Pentadiene		27.0	27.8	130.5	171.1	199.6	220.1
1,4-Pentadiene	6.14	25.2	25.7	131.0	170.2	220.5	
2,3-Pentadiene		28.2	29.5	125.1	164.9	195.0	217.6
Pentaerythritol		92	143.9				
Pentaerythritol tetranitrate			151.9				
Pentafluorobenzene	10.85	32.2	36.3				
Pentafluorobenzoic acid			91.6				
Pentafluoroethane				113.8	137.8	151.1	158.9
Pentafluorophenol	12.85		67.4				
2,3,4,5,6-Pentafluorotoluene	12.99	34.8	41.1				
Pentamethylbenzene	12.3	45.1	60.8	272.0	360.2	423.8	470.0
ΔHt = 2.0²³·⁷							
2,2,4,6,6-Pentamethylheptane			49.0				
Pentanal			38.8	155.2	205.0	241.4	267.8
Pentanamide			89.3				
Pentane	8.42	25.8	26.4	152.8	207.7	248.1	278.5
1,5-Pentanediol		60.7					

TABLE 2.55 Heat of Fusion, Vaporization, Sublimation, and Specific Heat at Various Temperatures of Organic Compounds (*Continued*)

Substance	ΔHm	ΔHv	ΔHs	C_p 400 K	600 K	800 K	1000 K
1,5-Pentanedithiol			59.3				
2,4-Pentanedione		34.3	41.8				
Pentanenitrile	4.73	36.1	43.6				
1-Pentanethiol	17.5	34.9	41.2	175.4	234.0	279.4	315.1
Pentanoic acid	14.16	44.1	62.4				
1-Pentanol	9.83	44.4	57.0	166.3	222.8	264.4	295.4
2-Pentanol		41.4	54.2				
3-Pentanol		43.5	54.0				
2-Pentanone	10.63	33.4	38.4	152.4	202.2	239.0	266.1
3-Pentanonc	11.59	33.5	38.5				
1-Pentene	5.81	25.2	25.5	138.5	186.4	221.5	247.7
cis-2-Pentene	7.12	26.1	26.9	132.1	182.5	218.8	245.9
trans-2-Pentene	8.36	26.1	26.8	136.7	184.2	219.5	246.1
cis-2-Pentenenitrile		36.4	43.2				
trans-2-Pentenenitrile		37.8	44.9				
trans-3-Pentenenitrile		37.1	44.8				
Pentyl acetate		41.0					
Pentylamine		34.0	40.1				
Pentylcyclohexane			53.9				
Pentyl propyl ether		35.0	42.8				
1-Pentyne		27.7	28.4	130.1	169.0	197.1	218.4
2-Pentyne		29.3	30.8	122.2	161.9	192.1	215.1
Perylene	31.75						
α-Phellandrene			50.6				
Phenanthrene	16.46	55.7	75.5				
9,10-Phenanthrenedione			91.6				
Phenazine			99.9				
Phenol	11.29	45.7	57.8	135.8	182.2	211.8	232.2
Phenyl acetate			54.8				
Phenylacetonitrile		52.9					
Phenylacetylene			41.8	150.4	200.9	233.4	255.9
(–)-3-Phenyl-1-alanine			155.2				
α-Phenylbenzeneacetic acid	31.27						
Phenyl benzoate			99.0				
Phenylboron dichloride			33.9				
Phenylcyclopropane			50.2				
N-Phenyldiacetimide			90.0				
Phenyl formate			52.9				
Phenylhydrazine	16.43		61.7				
1-Phenyl-1-propanone			58.5				
1-Phenyl-2-propanone			49.0				
Phenyl salicylate			92.1				
Phenyl vinyl ether			49.9				
Phthalamide			57.3				
1,3-Phthalic acid			106.7				
1,4-Phthalic acid			98.3				
Phthalic anhydride			88.7				
Phthalonitrile			86.9				
Piperidine	14.85	31.7	39.3				
Propadiene		18.6		72.0	92.1	106.4	117.2

(Continued)

TABLE 2.55 Heat of Fusion, Vaporization, Sublimation, and Specific Heat at Various Temperatures of Organic Compounds (*Continued*)

Substance	ΔHm	ΔHv	ΔHs	C_p 400 K	600 K	800 K	1000 K
Propanal		28.3	29.6	96.6	126.4	148.3	164.0
Propanamide	17.6		85.9				
Propane	3.53	19.0	14.8	94.0	128.7	154.8	174.6
1,2-Propanediamine			44.2				
1,3-Propanediamine		40.9	50.2				
Propanedinitrile			79.1				
1,2-Propanediol		54.1	58.0				
1,3-Propanediol		57.9	37.1				
1,2-Propanedione			38.1				
1,2-Propanedithiol			49.7				
Propanenitrile, $\Delta Ht = 1.7^{-96.2}$	5.05	31.8	36.0	88.6	114.7	134.5	149.4
1-Propanethiol, $\Delta Ht = 4.0^{-131.1}$	5.5	29.5	31.9	116.6	153.6	182.4	205.1
2-Propanethiol	5.7	27.9	29.5	118.6	154.9	181.0	200.5
1,2,3-Propanetriol triacetate		57.8	85.7				
1,2,3-Propanetriol trinitrate	21.9						
Propanoic acid	10.66	32.3	32.1				
Propanoic anhydride		41.7	52.6				
1-Propanol	5.20	41.4	47.4	108.2	144.6	171.7	192.2
2-Propanol	5.37	39.9	45.4	112.0	149.6	176.3	195.9
Propanolactone			47.0				
2-Propenal		28.3	31.3				
Propene	3.00	18.4	14.2	80.5	108.0	128.7	144.4
2-Propenenitrile	6.23						
Propenoic acid	11.16						
2-Propen-1-ol		40.0	47.3	95.4	126.0	147.6	163.4
cis-1-Propenylbenzene				187.4	254.0	300.4	333.9
2-Propoxyethanol		41.4	52.1				
Propyl acetate		33.9	39.7				
1-Propylamine	10.97	29.6	31.3	119.3	159.0	188.0	210.1
Propylbenzene	9.27	38.2	46.2	200.1	275.6	327.6	364.7
Propyl benzoate		49.8	51.9				
Propyl carbamate			81.2				
Propyl chloroacetate			48.5				
Propylcyclohexane	10.37	36.1	45.1	247.3	350.6	423.4	474.5
Propylcyclopentane	10.0	34.7	41.1	212.7	297.2	361.0	407.9
Propylene oxide	6.5	27.4	28.3	92.7	125.8	149.3	166.5
Propyl formate		33.6	37.5				
Propyl nitrate		35.9	40.6	149.8	194.5	225.4	247.2
Propyl propanoate		35.5	43.5				
Propyl trichloroacetate			53.1				
Propyl vinyl ether			29.3				
Propyne		22.1		72.5	91.2	105.2	115.9
2-Propyn-1-ol		42.1					
Pyrazine			56.3				
Pyrene	17.11						
Pyridazine			53.5				
Pyridine	8.28	35.1	40.2	106.4	149.5	177.8	197.4
Pyrimidine		49.8	50.0				
1*H*-Pyrrole	7.91	38.8	45.1				
Pyrrolidine, $\Delta Ht = 0.5^{-66}$	8.58	33.0	37.6	114.4	168.7	206.5	233.6

TABLE 2.55 Heat of Fusion, Vaporization, Sublimation, and Specific Heat at Various Temperatures of Organic Compounds (*Continued*)

Substance	ΔHm	ΔHv	ΔHs	C_p 400 K	600 K	800 K	1000 K
Quinoline	10.66	49.7	53.9				
Salicylic acid			95.1				
5,5′-Spirobis(1,3-dioxane)			72.8				
Spiro[2.2]pentane	5.8	26.8	27.5	119.5	167.8	200.5	223.9
cis-Stilbene			69.0				
trans-Stilbene	27.4		99.2				
Styrene	11.0	38.7	43.9	160.3	218.2	256.9	284.2
Succinic acid	32.95		117.5				
Succinic anhydride	20.41						
Succinonitrile	3.92						
p-Terphenyl	35.5						
1,1,2,2-Tetrabromoethane		48.7	70.0				
Tetrabromomethane		45.1	110	97.1	102.6	106.7	105.9
Tetrabutyltin			19.8				
Tetracene			125.5				
Tetrachloro-1,4-benzoquinone			98.7				
1,1,2,2-Tetrachloro-1,2-difluoroethane	3.70	35.0					
1,1,1,2-Tetrachloro-2,2-fluorooctane	3.99						
1,1,1,2-Tetrachloroethane				118.7	139.2	151.6	159.7
1,1,2,2-Tetrachloroethane		37.6	45.7	116.7	137.7	150.0	158.0
Tetrachloroethylene	10.56	34.7	39.7	105.0	116.6	122.6	125.8
Tetrachloromethane	3.28	29.8	32.4	91.7	99.7	103.1	104.8
$\Delta Ht = 4.6^{-47.9}$							
Tetracyanoethylene			81.2				
Tetracyanomethane			61.1				
Tetradecane	45.6	47.6	71.3	414.3	559.5	664.8	743.1
Tetradecanenitrile			85.3				
Tetradecanoic acid	45.38		139.8				
1-Tetradecanol	49.0		102.2				
1-Tetradecene	27.6	46.9	70.2	399.8	538.2	638.2	712.1
Tetraethylene glycol		62.6	98.7				
Tetraethylgermanium			44.8				
Tetraethyllead			56.9				
Tetraethylsilane	13.01						
Tetraethyltin			51.0				
1,1,1,2-Tetrafluoroethane				104.2	128.7	143.1	152.1
Tetrafluoroethylene	7.7	16.8		91.9	106.8	115.5	120.8
Tetrafluoromethane	0.7	12.6		72.4	86.8	94.5	98.8
$\Delta Ht = 1.5^{-196.9}$							
Tetrahydrofuran	8.54	29.8	32.0				
Tetrahydrofuran-2,5-dimethanol		63.6					
Tetrahydrofuran-2-methanol		45.2	51.6				
1,2,3,4-Tetrahydronaphthalene	12.45	43.9	55.2				
Tetrahydropyran		31.2	34.6				
Tetrahydropyran-2-methanol		44.4					
Tetrahydrothiophene		34.7	39.4				
Tetrahydrothiophene-1,1-dioxide	1.43						
Tetraiodomethane				100.4	104.4	105.9	106.7
Tetramethoxysilane		194.6					
1,2,3,4-Tetramethylbenzene	11.2	45.0	57.2	237.7	316.7	374.1	416.2

(*Continued*)

TABLE 2.55 Heat of Fusion, Vaporization, Sublimation, and Specific Heat at Various Temperatures of Organic Compounds (*Continued*)

Substance	ΔHm	ΔHv	ΔHs	C_p 400 K	600 K	800 K	1000 K
1,2,3,5-Tetramethylbenzene	10.7	43.8	53.7	233.3	313.0	371.5	414.3
1,2,4,5-Tetramethylbenzene	21.0	45.5	53.4	232.2	311.2	369.9	413.0
2,2,3,3-Tetramethylbutane	7.54	31.4	42.9				
$\Delta Ht = 2.0^{-120.7}$							
Tetramethylene sulfone	1.4	61.5					
Tetramethyllead			38.1				
2,2,3,3-Tetramethylpentane	2.33						
2,2,3,4-Tetramethylpentane	0.50						
2,2,4,4-Tetramethylpentane	9.75	32.5	38.5				
2,3,3,4-Tetramethylpentane	9.00						
Tetramethylsilane	6.88						
Tetramethyltin			33.1				
1,1,3,3-Tetramethylurea	14.10	45.6					
Tetranitromethane		40.7	49.9				
Tetraphenylmethane			150.6				
Tetraphenyltin			66.3				
Tetrapropylgermanium			61.5				
Tetrapropyltin			66.9				
1,2,3,4-(1H)-Tetrazole			97.5				
Thiacyclobutane		32.3	36.0				
Thiacycloheptane			47.3	175.7	272.0	330.5	368.2
Thiacyclohexane	2.5	36.0	42.6	149.4	219.1	267.8	302.7
$\Delta Ht = 1.1^{-71.8}$							
$\Delta Ht = 7.8^{-33.1}$							
Thiacyclopentane	7.4	34.7	39.5	121.1	167.5	199.4	222.3
Thiacyclopropane		29.2	30.3	69.2	92.0	107.2	118.0
Thioacetamide			83.3				
Thioacetic acid			37.2	93.1	111.8	127.2	136.5
1,2-Thiocresol			51.5				
2,2′-Thiodiethanol		66.8					
Thiophene, $\Delta Ht = 0.6^{-101.6}$	5.09	31.5	34.7	96.3	129.5	150.7	165.4
Thiophenol	11.5	39.9	47.6	137.1	184.6	215.9	237.6
Thymol	17.27						
Toluene	6.85	33.2	38.0	140.1	197.5	236.9	264.9
o-Toluidine		44.6	56.7				
m-Toluidine	3.89	44.9	57.3				
p-Toluidine	18.22	44.3					
Triacetamide			60.4				
2,4,6-Triamino-1,3,5-triazine			124.3				
Tribromomethane		39.7	46.1	78.7	88.0	93.3	96.7
Tributoxyborane		56.1	52.3				
Tributyl phosphate		61.4	72.0				
Trichloroacetic acid	5.88						
Trichloroacetonitrile		34.1					
Trichloroacetyl chloride			41.0				
1,3,5-Trichlorobenzene	18.2						
Trichlorobenzoquinone			88.7				
1,1,1-Trichloroethane	2.73	29.9	32.5	107.6	128.4	141.1	149.8
$\Delta Ht = 7.5^{-49.0}$							
1,1,2-Trichloroethane	11.54	34.8	40.2	104.7	126.1	139.2	148.2

TABLE 2.55 Heat of Fusion, Vaporization, Sublimation, and Specific Heat at Various Temperatures of Organic Compounds (*Continued*)

Substance	ΔHm	ΔHv	ΔHs	C_p 400 K	600 K	800 K	1000 K
Trichloroethylene		31.4	34.5	91.2	104.9	112.7	117.8
Trichloromethane	8.8	29.2	31.3	74.3	85.3	91.5	95.5
Trichloromethylsilane	8.94						
1,2,3-Trichloropropane	8.9	37.1		31.7	38.9	43.8	47.3
1,1,1-Trichlorotrifluoroethane		26.9	28.1				
1,1,2-Trichlorotrifluoroethane	2.47	27.0	28.4				
1,1,1-Trichloro-3,3,3-trifluoropropane		32.2	36.8				
Tricyanoethylene			81.2				
Tridecane, $\Delta Ht = 7.7^{-18.2}$	28.50	45.7	66.4	385.2	520.4	618.5	691.2
Tridecanenitrile			85.3				
Tridecanoic acid	43.1		146.4				
1-Tridecene	22.83	45.0	65.3	370.8	499.1	592.0	660.2
Triethanolamine	27.2	67.5					
Triethoxyborane			43.9				
Triethoxymethane			46.0				
Triethylaluminum			73.2				
Triethylamine		31.0	34.8	203.8	276.6	328.7	367.4
Triethylaminoborane			60.7				
Triethylarsine			43.1				
Triethyl arsenite			50.6				
Triethylbismuthine			46.0				
Triethylborane			36.8				
Triethylenediamine	6.1		61.9				
$\quad \Delta Ht = 9.6^{79.8}$							
Triethylene glycol		71.4	79.1				
Triethylphosphine			39.8				
Triethyl phosphate			57.3				
Triethyl phosphite			41.8				
Triethylstibine			43.5				
Trifluoroacetic acid		33.3	38.5				
$\quad \Delta H(\text{dimer dissoc}) = 58.8^{100}$							
Trifluoroacetonitrile	5.0						
1,1,1-Trifluoro-2-bromo-2-chloroethane		28.1	29.6				
1,1,1-Trifluoroethane	6.19	19.2		95.2	118.7	133.8	144.1
2,2,2-Trifluoroethanol		40.0					
Trifluoroethylene				81.1	97.5	107.5	113.9
Trifluoromethane	4.1	16.7		61.1	76.0	85.1	91.0
(Trifluoromethyl)benzene	13.46	32.6	37.6	169.8	226.8	262.6	286.4
Triiodomethane	16.3		69.9	82.0	90.0	94.7	97.8
Triisopropylborane			41.8				
Triisopropyl phosphite			46.0				
Trimethoxyborane			34.7				
1,1,1-Trimethoxyethane			39.2				
Trimethoxymethane			38.1				
2′,4′,5′-Trimethylacetophenone			63.2				
2′,4′,6′-Trimethylacetophenone			62.3				
Trimethylaluminum			63.2				
Trimethylamine	6.55	22.9	21.7	117.5	160.4	190.9	213.3
Trimethyl arsenite			42.3				
Trimethylarsine			28.9				

(Continued)

TABLE 2.55   Heat of Fusion, Vaporization, Sublimation, and Specific Heat at Various Temperatures of Organic Compounds (*Continued*)

Substance	ΔHm	ΔHv	ΔHs	C_p 400 K	600 K	800 K	1000 K
1,2,3-Trimethylbenzene	8.37	40.0	49.1	196.2	267.8	320.9	359.4
$\Delta Ht = 0.7^{-54.5}$							
$\Delta Ht = 1.3^{-42.9}$							
1,2,4-Trimethylbenzene		39.3	47.9	196.5	269.0	321.9	360.2
1,3,5-Trimethylbenzene	9.51	39.0	47.5	194.2	268.1	321.5	360.1
2,6,6-Trimethylbicyclo[3.1.1]-2-heptene			44.8				
Trimethylbismuthine			34.7				
Trimethylborane			20.2				
2,2,3-Trimethylbutane	2.20	28.9	32.0	212.7	291.3	346.1	386.3
$\Delta Ht = 2.5^{-151.8}$							
2,3,3-Trimethyl-1-butene			32.2				
cis,cis-1,3,5-Trimethylcyclohexane				242.9	351.2	427.6	482.0
Trimethylene oxide		28.7	29.9				
Trimethylene sulfide	8.3	32.3	36.0	91.6	127.4	152.3	170.2
$\Delta Ht = 0.7^{-96.5}$							
Trimethylgallium			38.1				
2,2,5-Trimethylhexane	6.2	33.7	40.2				
2,3,5-Trimethylhexane	10.00	34.4	41.4				
Trimethylindium			48.5				
2,4,7-Trimethyloctane		38.2	49.9				
2,2,3-Trimethylpentane	8.62	31.9	36.9				
2,2,4-Trimethylpentane	9.04	30.8	35.1				
2,3,3-Trimethylpentane	0.86	32.1	37.3				
$\Delta Ht = 7.7^{-109.0}$							
2,3,4-Trimethylpentane	9.27	32.4	37.7				
2,2,4-Trimethyl-1,3-pentanediol	8.6	55.7					
2,2,4-Trimethyl-3-pentanone		35.6	43.3				
2,4,4-Trimethyl-1-pentene		31.4	35.8				
2,4,4-Trimethyl-2-pentene		32.6	37.5				
Trimethylphosphine			28.0				
Trimethylphosphine oxide			50.2				
Trimethyl phosphate			36.8				
2,3,6-Trimethylpyridine		40.0	50.6				
2,4,6-Trimethylpyridine	9.53	39.9	50.3				
Trimethylsilanol			45.6				
Trimethylstibine			31.4				
Trimethylsuccinic anhydride			74.1				
Trimethylthiacyclopropane			39.3				
Trimethyltin bromide			47.3				
2,4,6-Trinitroanisole			133.1				
1,3,5-Trinitrobenzene	16.7		99.6				
Trinitromethane		32.6	46.7				
2,4,6-Trinitrophenetole			120.5				
2,4,6-Trinitrotoluene			104.7				
1,3,6-Trioxacycloactane			48.8				
1,3,5-Trioxane	15.11		56.6				
Triphenylarsine			99.3				
Triphenylbismuthine			110.9				
Triphenylborane			81.6				
Triphenylene			118.0				

TABLE 2.55 Heat of Fusion, Vaporization, Sublimation, and Specific Heat at Various Temperatures of Organic Compounds (*Continued*)

Substance	ΔHm	ΔHv	ΔHs	C_p 400 K	600 K	800 K	1000 K
Triphenylmethane			100.0				
Triphenylphosphine			96				
Triphenylstibine			106.3				
Tripropoxyborane			49.4				
Tris(diethylamino)phosphine			60.7				
Tris(trimethylsilyl)amine			54.4				
Tropolone			83.7				
Undecane	22.32	41.5	56.4	327.1	442.7	525.9	588.3
$\Delta Ht = 6.9^{-36.6}$							
Undecanenitrile			71.1				
Undecanoic acid	25.9		121.3				
1-Undecene, $\Delta Ht = 9.2^{-55.8}$	16.99	40.9	55.4	312.7	421.1	499.3	557.3
Uracil			126.5				
Urea	15.1	87.9					
(−)-Valine			162.8				
Vinyl acetate		34.4	34.8				
Vinyl benzene			39.6				
Vinylcyclohexane			39.7				
4-Vinyl-1-cyclohexene		33.5	38.3				
1,2-Xylene	13.61	36.2	43.4	171.7	234.2	278.8	311.1
1,3-Xylene	11.55	35.7	42.7	167.5	232.2	277.9	310.6
1,4-Xylene	16.81	35.7	42.4	166.1	230.8	276.7	309.7

2.14 CRITICAL PROPERTIES

Critical temperature (T_c), critical pressure (P_c), and critical volume (V_c) represent three widely used pure component constants. These critical constants are very important properties in chemical engineering field because almost all other thermo chemical properties are predictable from boiling point and critical constants with using corresponding state theory. Therefore, precise prediction of critical constants is very necessary.

2.14.1 Critical Temperature

The critical temperature of a compound is the temperature above which a liquid phase cannot be formed no matter what the pressure on the system. The critical temperature is important in determining the phase boundaries of any compound and is a required input parameter for most phase equilibrium thermal property or volumetric property calculations using analytic equations of state or the theorem of corresponding states. Critical temperatures are predicted by various empirical methods according to the type of compound or mixture being considered.

2.14.2 Critical Pressure

The critical pressure of a compound is the vapor pressure of that compound at the critical temperature. Below the critical temperature, any compound above its vapor pressure will be a liquid.

2.14.3 Critical Volume

The critical volume of a compound is the volume occupied by a specified mass of a compound at its critical temperature and critical pressure.

2.14.4   Critical Compressibility Factor

The critical compressibility factor of a compound is used as a characterization parameter in corresponding states methods to predict volumetric and thermal properties. The factor varies from approximately 0.23 for water to 0.26–0.28 for most hydrocarbons to above 0.30 for light gases.

TABLE 2.56   Critical Properties

Substance	T_c, °C	P_c, atm	P_c, MPa	V_c, cm$^3 \cdot$ mol^{-1}	ρ_c, g $\cdot$ cm^{-3}
Acetaldehyde	193	55	5.57	154	0.286
Acetic acid	319.56	57.1	5.786	171.3	0.351
Acetic anhydride	333	39.5	4.0	290	0.352
Acetone	235.0	46.4	4.700	209	0.278
Acetonitrile	272.4	47.7	4.85	173	0.237
Acetophenone	436.4	38	3.85	386	0.311
Acetyl chloride	235	58	5.88	204	0.325
Acetylene	35.2	60.6	6.14	113	0.231
Acrylic acid	342	56	5.67	210	0.343
Acrylonitrile	263	45	4.56	210	0.253
Allene	120	54.0	5.47	162	0.247
Allyl alcohol	272.0	56.4	5.71	203	0.286
2-Aminoethanol	341	44	4.46	196	0.312
Aniline	426	49.5	4.89	287	0.324
Anthracene	610	28.6	2.90	554	0.333
Benzaldehyde	422	45.9	4.65	324	0.327
Benzene	288.90	48.31	4.895	255	0.306
Benzoic acid	479	41.55	4.21	341	0.358
Benzonitrile	426.3	41.55	4.21	339	0.304
Benzyl alcohol	422	42.4	4.3	334	0.324
Biphenyl	516	38.0	3.85	502	0.307
Bromobenzene	397	44.6	4.52	324	0.485
Bromochlorodifluoromethane	158.8	41.98	4.254	246	0.672
Bromoethane	230.8	61.5	6.23	215	0.507
Bromomethane	173.4	85	8.61	156	0.609
Bromopentafluorobenzene	397	44.6	4.52		
1-Bromopropane	−1.8				0.462
2-Bromopropane	−14.2				0.462
Bromotrifluoromethane	67.1	39.2	3.97	200	0.76
1,2-Butadiene	170.6	44.4	4.50	219	0.247
1,3-Butadiene	152	42.7	4.33	221	0.245
Butanal	264.1	42.6	4.32	258	0.279
Butane	151.97	37.34	3.784	255	0.228
Butanenitrile	312.3	38.3	3.88	285	0.242
Butanoic acid	351	39.8	4.03	290	0.304
1-Butanol	289.9	43.56	4.414	275	0.270
2-Butanol	263.1	41.47	4.202	269	0.276
2-Butanone	263.63	41.52	4.207	267	0.270
1-Butene	146.5	39.7	4.02	240	0.234
cis-2-Butene	147.5	40.5	4.10	238	0.240
trans-2-Butene	147.5	40.5	4.10	238	0.236
3-Butenenitrile	312.3	38.3	3.88	265	0.253
1-Buten-3-yne	182	49	4.96	202	0.258
Butyl acetate	306.7	31	3.14	400	0.290
1-Butylamine	258.8	41.9	4.25	277	0.264
sec-Butylamine	241.2	41.4	4.20	278	0.263
tert-Butylamine	210.8	37.9	3.84	292	0.250

TABLE 2.56 Critical Properties (*Continued*)

Substance	T_c, °C	P_c, atm	P_c, MPa	V_c, cm$^3 \cdot$ mol^{-1}	ρ_c, g $\cdot$ cm^{-3}
Butylbenzene	387.4	28.5	2.89	497	0.270
sec-Butylbenzene	391	29.1	2.94	510	0.263
tert-Butylbenzene	387	29.3	2.97	490	0.273
Butyl benzoate	450	26	2.63	561	0.318
Butyl butanoate	338				0.292
Butylcyclohexane	394	31.1	3.15	534	0.63
sec-Butylcyclohexane	396	26.4	2.67		
tert-Butylcyclohexane	385.9	26.3	2.66		
Butylcyclopentane	357.9				
Butyl ethyl ether	257.9	30	3.04	390	0.262
2-Butylhexadecafluoro- tetrahydrofuran	227.1	15.86	1.607	588	0.707
Butylisopropylamine	290.5				
tert-Butyl methyl sulfide	296.7				
1-Butyne	190.6	46.5	4.71	220	0.246
2-Butyne	215.5	50.2	5.09	221	0.246
4-Butyrolactone	436				
Carbon tetrachloride	283.3	45.0	4.56	276	0.558
Carbon tetrafluoride	−45.7	36.9	3.74	140	0.629
Chlorobenzene	359.3	44.6	4.52	308	0.365
1-Chlorobutane	268.9	36.4	3.69	312	0.297
2-Chlorobutane	247.5	39	3.95	305	0.303
1-Chloro-1,1-difluoroethane	137.1	40.7	4.12	231	0.435
2-Chloro-1,1-difluoroethylene	127.5	44.0	4.46	197	0.499
Chlorodifluoromethane	96.1	49.1	4.98	165	0.525
1-Chloro-2,3-epoxypropane	351				
Chloroethane	187.3	52.0	5.27	199	0.324
Chloroform	263.3	54.0	5.47	239	0.504
1-Chlorohexane	321.5				
Chloromethane	143.1	65.9	6.679	139	0.353
2-Chloro-2-methylpropane	234	39	3.95	295	0.314
Chloropentafluoroacetone	137.6	28.4	2.88		
Chloropentafluorobenzene	297.9	31.8	3.22		
Chloropentafluoroethane	80.1	31.9	3.229	252	0.613
1-Chloropentane	295.4				
1-Chloropropane	230	45.2	4.58	254	0.309
2-Chloropropane	212	46.6	4.72	230	0.341
3-Chloropropene	241	47	4.76	234	0.336
Chlorotrifluoromethane	29	38.98	3.946	180	0.579
Chlorotrifluorosilane	35.4	34.2	3.47		
Chlorotrimethylsilane	224.7	31.6	3.20		
1,2-Cresol	424.5	49.4	5.01	282	0.384
1,3-Cresol	432.7	45.0	4.56	309	0.346
1,4-Cresol	431.5	50.8	5.15	277	0.391
Cyanogen	126.7	62.2	6.30	145	0.360
Cyclobutane	186.8	49.2	4.99	210	0.267
Cycloheptane	316	36.7	3.72	390	0.252
Cyclohexane	280.4	40.2	4.07	308	0.273
trans-Cyclohexanedimethanol	451	34.85	3.531		
Cyclohexanethiol	390.9				
Cyclohexanol	376.9	42.0	4.26	327	0.306
Cyclohexanone	379.9	39.5	4.0	312	0.315
Cyclohexene	287.33	42.9	4.35	292	0.281

(*Continued*)

TABLE 2.56 Critical Properties (*Continued*)

Substance	T_c, °C	P_c, atm	P_c, MPa	V_c, cm³ · mol⁻¹	ρ_c, g · cm⁻³
Cyclohexylamine	341.5				
Cyclopentane	238.6	44.49	4.508	260	0.27
Cyclopentanethiol	360.4				
Cyclopentanone	353	53	5.37	268	0.314
Cyclopentene	232.9				
1-Cyclopentylheptane	406	19.2	1.94	649	0.260
1-Cyclopentylpentadecane	506.9	10.1	1.02	1096	0.256
Cyclopropane	124.7	54.2	5.49	170	0.248
p-Cymene	379	2.80	2.84	492	0.273
Decafluorobutane	113.3	22.93	2.323	378	0.629
cis-Decahydronaphthalene	429.2	31.6	3.20	480	0.288
trans-Decahydronaphthalene	414.0	31	3.14	480	0.288
Decane	344.6	20.8	2.11	624	0.228
Decanenitrile	348.8	32.1	3.25		
1-Decanol	413.9	22	2.23	600	0.264
1-Decene	343.3	21.89	2.218	585	0.240
Dibutyl sulfide	380				
Decylcyclohexane	477	13.4	1.36		
Decylcyclopentane	450	15.0	1.52		
Diallyl sulfide	380				
1,2-Dibromo-2-chlorotrifluoro-ethane	287.6				
Dibromodifluoromethane	198.3	40.8	4.13	249	0.843
1,2-Dibromoethane	309.9	71.1	7.2	242	0.776
Dibromomethane	310	71	7.19		
1,2-Dibromotetrafluoroethane	214.7	33.49	3.393	329	0.790
Dibutylamine	334.4	30.7	3.11	517	0.250
Dibutyl ether	311.0	29.7	3.01	500	0.260
Dibutyl sulfide	377	24.7	2.50	537	0.272
1,2-Dichlorobenzene	424.2	40.5	4.10	360	0.408
1,3-Dichlorobenzene	411	38	3.85	359	0.408
1,4-Dichlorobenzene	412	39	3.95	372	0.395
Dichlorodifluoromethane	111.80	40.82	4.136	217	0.558
1,1-Dichloroethane	250	50.0	5.07	236	0.419
Dichlorodifluorosilane	95.8	34.5	3.50		
1,2-Dichloroethane	288	53	5.4	225	0.440
1,1-Dichloroethylene	222	51.3	5.20	218	0.445
cis-1,2-Dichloroethylene	271.1			224	0.433
trans-1,2-Dichloroethylene	234.4	54.4	5.51	224	0.433
Dichlorofluoromethane	178.43	51.1	5.18	196	0.522
1,2-Dichlorohexafluoropropane	172.9				
Dichloromethane	237	60.2	6.10	193	0.440
1,2-Dichloropropane	304	44	4.49	226	0.500
Dichlorosilane	176	46.1	4.67		
1,1-Dichlorotetrafluoroethane	145.5	32.6	3.30	294	0.582
1,2-Dichlorotetrafluoroethane	145.63	32.1	3.252	297	0.582
Dideuterium oxide (D₂O)	371.0	215.7	21.86		0.363
Diethanolamine	442.0	32.3	3.27	349	0.301
1,1-Diethoxyethane (Acetal)	254				
Diethylamine	226.84	37.3	3.758	301	0.243
1,4-Diethylbenzene	384.8	27.7	2.81	480	0.280
Diethyl disulfide	368.9				
Diethylene glycol	408	46	4.66	316	0.336
Diethyl ether	193.59	35.9	3.638	280	0.265

TABLE 2.56 Critical Properties (*Continued*)

Substance	T_c, °C	P_c, atm	P_c, MPa	V_c, cm$^3 \cdot$ mol^{-1}	ρ_c, g $\cdot$ cm^{-3}
3,3-Diethyl-2-methylpentane	366.8	25.0	2.53	501	0.284
3,3-Diethylpentane	337	26.4	2.67		
Diethyl sulfide	284	39.1	3.96	318	0.284
Difluoroamine (HNF$_2$)	130	93	9.42		
1,2-Difluorobenzene	284.2			300	0.381
cis-Difluorodiazine	−1	70	7.09		
trans-Difluorodiazine	−13	55	5.57		
1,1-Difluoroethane	113.6	44.4	4.50	181	0.365
1,1-Difluoroethylene	29.8	44.0	4.46	154	0.417
Dihexyl ether	384	18	1.82	720	0.259
Diisopropyl sulfide	391				
Diisopropyl ether	227.17	27.9	2.832	386	0.265
1,2-Dimethoxyethane	263	38.2	3.87	271	0.333
Dimethoxymethane	242.1	44.2	4.48		
N,N-Dimethylacetamide	364	38.7	3.92		
Dimethylamine	164.07	52.7	5.340	187	0.241
N,N-Dimethylaniline	414	35.8	3.63		
2,2-Dimethylbutane	215.7	30.49	3.090	359	0.240
2,3-Dimethylbutane	499.9	30.90	3.131	358	0.241
3,3-Dimethyl-2-butanone	289.8				
2,3-Dimethyl-1-butene	228	32.0	3.24	343	0.245
3,3-Dimethyl-1-butene	217	32.1	3.25	340	0.248
2,3-Dimethyl-2-butene	250.9	33.2	3.36	351	0.240
1,1-Dimethylcyclohexane	318	29.3	2.97	416	0.378
cis-1,2-Dimethylcyclohexane	333.0	29.0	2.94	460	0.244
trans-1,2-Dimethylcyclohexane	323.0	29.3	2.97	460	0.244
cis-1,3-Dimethylcyclohexane	317.9	29.3	2.97	450	0.249
trans-1,3-Dimethylcyclohexane	325	29.3	2.97	460	0.244
cis-1,4-Dimethylcyclohexane	325.0	29.0	2.94	460	0.244
trans-1,4-Dimethylcyclohexane	317.0	29.0	2.94	459	0.249
1,1-Dimethylcyclopentane	274	34.0	3.44	360	0.273
cis-1,2-Dimethylcyclopentane	291.7	34.0	3.44	368	0.267
trans-1,2-Dimethylcyclopentane	277.2	34.0	3.44	362	0.271
cis-1,3-Dimethylcyclopentane	318.9				
Dimethyl disulfide	59.5				
Dimethyl ether	126.9	53.0	5.37	190	0.242
N,N-Dimethylformamide	376.5	51.5	5.22	262	0.279
2,2-Dimethylheptane	303.7	23.19	2.350	519	0.247
2,2-Dimethylhexane	276.8	25.0	2.529	478	0.239
2,3-Dimethylhexane	290.4	25.94	2.628	468	0.244
2,4-Dimethylhexane	280.5	25.22	2.556	472	0.242
2,5-Dimethylhexane	277.0	24.54	2.487	482	0.237
3,3-Dimethylhexane	289.0	26.19	2.654	443	0.258
3,4-Dimethylhexane	295.8	26.57	2.692	466	0.245
1,1-Dimethylhydrazine	250	53.6	5.43	230	0.261
2,4-Dimethyl-3-iso- pentane	341.3	23.1	2.34	521	0.273
2,3-Dimethyloctane	340.1	21.6	2.19	567	0.251
2,4-Dimethyloctane	326.3	21.1	2.14	566	0.251

(*Continued*)

TABLE 2.56 Critical Properties (*Continued*)

Substance	T_c, °C	P_c, atm	P_c, MPa	V_c, cm$^3 \cdot$ mol^{-1}	ρ_c, g $\cdot$ cm^{-3}
2,5-Dimethyloctane	330	21.2	2.15	569	0.250
2,6-Dimethyloctane	330	21.1	2.15	576	0.247
2,7-Dimethyloctane	329.8	20.7	2.10	590	0.241
3,3-Dimethyloctane	339	21.9	2.22	557	0.255
3,4-Dimethyloctane	341	22.1	2.24	551	0.258
3,5-Dimethyloctane	333.2	21.6	2.19	555	0.256
3,6-Dimethyloctane	335.2	21.6	2.19	562	0.253
4,5-Dimethyloctane	333.8	21.8	2.21	548	0.260
4,5-Dimethyloctane	339.1	22.1	2.24	546	0.261
Dimethyl oxalate	355	39.2	3.97		
2,2-Dimethylpentane	247.4	27.4	2.773	416	0.241
2,3-Dimethylpentane	264.3	28.70	2.908	393	0.255
2,4-Dimethylpentane	246.7	27.01	2.737	418	0.240
3,3-Dimethylpentane	263.3	29.07	2.946	414	0.242
2,3-Dimethylphenol	449.7	48	4.86	470	0.26
2,4-Dimethylphenol	434.5	43	4.36	509	0.24
2,5-Dimethylphenol	433.8	48	4.86	470	0.26
2,6-Dimethylphenol	427.9	42	4.26	509	0.24
3,4-Dimethylphenol	456.7	49	4.96	552	0.27
3,5-Dimethylphenol	442.5	36	3.65	611	0.25
2,2-Dimethylpropane	160.7	31.55	3.197	307	0.238
2,2-Dimethyl-1-propanol	276	39	3.95	319	
2,3-Dimethylpyridine	382.3				
2,4-Dimethylpyridine	373.9				
2,5-Dimethylpyridine	371				
2,6-Dimethylpyridine	350.7			316	0.339
3,4-Dimethylpyridine	410.7				
3,5-Dimethylpyridine	394.1				
Dimethyl sulfide	229.9	54.6	5.53	201	0.309
N,N-Dimethyl-1,2-toluidine	395	30.8	3.12		
1,4-Dioxane	314	51.5	5.21	238	0.370
Diphenyl ether	493.7	31	3.14		
Diphenylmethane	494	29.4	2.98		
Dipropylamine	282.7	35.8	3.63	407	0.249
Dipropyl ether	257.5	29.91	3.028		
Docosafluorodecane	269	14.3	1.45		
Dodecafluorocyclohexane	184.1	24	2.43		
Dodecafluorocyclohexene	188.7				
Dodecafluoro-1-hexene	181.3				
Dodecafluoropentane	149	20.1	2.03		
Dodecane	385	18.0	1.82	754	0.226
1-Dodecanol	405.9	19	1.92	718	0.260
1-Dodecene	384.5	18.3	1.85		
Dodecylbenzene	501	15.6	1.58	1000	0.246
Dodecylcyclopentane	477	12.8	1.30		
Ethane	32.3	48.2	4.90	148	0.203
1,2-Ethanediamine	319.8	62.1	6.29	206	0.292
1,2-Ethanediol	445	76	7.7	186	0.334
Ethanethiol	225.5	54.2	5.49	207	0.300
Ethanol	240.9	60.57	6.137	167	0.276
Ethoxybenzene	374.0	33.8	3.42		
Ethyl acetate	250.2	38.31	3.882	286	0.308

TABLE 2.56   Critical Properties (*Continued*)

Substance	T_c, °C	P_c, atm	P_c, MPa	V_c, cm$^3 \cdot$ mol^{-1}	ρ_c, g $\cdot$ cm^{-3}
Ethyl acetoacetate	400				
Ethyl acrylate	279	37.0	3.75	320	0.313
Ethylamine	183	55.5	5.62	182	0.248
Ethylbenzene	344.00	35.61	3.609	374	0.284
Ethyl benzoate	424	32	3.24	451	0.111
Ethylbutanoate	293	30.2	3.06	421	0.28
2-Ethyl-1-butanol	145.7				
Ethyl crotonate	326				
Ethylcyclohexane	336	29.9	3.03	450	0.249
Ethylcyclopentane	296.4	33.5	3.39	375	0.262
3-Ethyl-2,2-dimethylhexane	338.6	22.8	2.31	526	0.271
4-Ethyl-2,2-dimethylhexane	321.5	21.9	2.22	539	0.264
3-Ethyl-2,3-dimethylhexane	353.7	23.9	2.42	516	0.276
4-Ethyl-2,3-dimethylhexane	344.2	23.1	2.34	524	0.271
3-Ethyl-2,4-dimethylhexane	343.0	23.1	2.34	522	0.273
4-Ethyl-2,4-dimethylhexane	347.8	24.4	2.47	524	0.271
3-Ethyl-2,5-dimethylhexane	330.4	22.1	2.24	537	0.265
3-Ethyl-3,4-dimethylhexane	351.4	23.9	2.42	511	0.278
Ethylene	9.3	49.7	5.036	129	0.218
Ethylene glycol dimethyl ether	263	38.2	3.87	271	0.333
Ethylene glycol ethyl ether acetate	334.2	31.25	3.166	443	0.298
Ethylene glycol monobutyl ether	360.8			424	0.279
Ethylene oxide	196	71.0	7.275	140	0.314
Ethyl formate	235.4	46.8	4.74	229	0.323
3-Ethylhexane	292.4	25.74	2.608	455	0.251
2-Ethyl-1-hexanol	367.5	27.2	2.76	494	0.264
Ethyl isopentanoate	315				
Ethyl isopropyl ether	217.2				
2-Ethyl-1-methylbenzene	378	30.0	3.04	460	0.26
3-Ethyl-1-methylbenzene	364	28.0	2.84	490	0.24
4-Ethyl-1-methylbenzene	367	29.0	2.94	470	0.26
Ethyl 3-methylbutanoate	314.9				
1-Ethyl-1-methylcyclopentane	319	29.5	2.99		
Ethyl methyl ether	164.8	43.4	4.40	221	0.272
3-Ethyl-2-methylheptane	337.8	22.0	2.23	544	0.262
4-Ethyl-2-methylheptane	328.7	21.6	2.19	545	0.261
5-Ethyl-2-methylheptane	333.6	21.6	2.19	555	0.256
3-Ethyl-3-methylheptane	347.0	22.8	2.31	532	0.267
4-Ethyl-3-methylheptane	341.2	22.5	2.28	530	0.269
5-Ethyl-3-methylheptane	333.5	22.0	2.23	541	0.263
3-Ethyl-4-methylheptane	342.4	22.5	2.28	533	0.267
4-Ethyl-4-methylheptane	342.4	22.8	2.31	525	0.271
Ethyl methyl ketone	262.4	41.0	4.154	267	0.270
3-Ethyl-2-methylpentane	294.0	26.65	2.700	443	0.258
3-Ethyl-3-methylpentane	303.5	27.71	2.808	455	0.351
Ethyl 2-methylpropanoate	280	30	3.04	410	0.28
Ethyl methyl sulfide	260	42	4.26		
2-Ethylnaphthalene	502	31.0	3.14	521	0.300
Ethyl nonanoate	401				

(*Continued*)

TABLE 2.56   Critical Properties (*Continued*)

Substance	T_c, °C	P_c, atm	P_c, MPa	V_c, cm$^3 \cdot$ mol^{-1}	ρ_c, g $\cdot$ cm^{-3}
3-Ethyloctane	340	21.6	2.19	561	0.241
4-Ethyloctane	337	21.5	2.18	552	0.258
Ethyl octanoate	386				
3-Ethylpentane	267.6	28.53	2.891	416	0.241
1,2-Ethylphenol	429.9				
1,3-Ethylphenol	443.3				
1,4-Ethylphenol	443.3				
Ethyl propanoate	272.9	33.18	3.362	345	0.296
Ethyl propyl ether	227.1	32.1	3.25	244	0.361
m-Ethyltoluene	364.0	28.1	2.837	490	0.245
o-Ethyltoluene	378.0	30.1	3.04	460	0.261
p-Ethyltoluene	367	29.0	2.94	479	0.256
3-Ethyl-2,2,3-trimethyl- pentane	372.9	25.4	2.57	503	0.283
3-Ethyl-2,2,4-trimethyl- pentane	342.2	23.4	2.37	518	0.275
3-Ethyl-2,3,4-trimethyl- pentane	369.2	25.1	2.54	506	0.281
Ethyl vinyl ether	202	40.17	4.07	260	0.277
Fluorobenzene	286.94	44.91	4.551	357	0.269
Fluoroethane	102.2	49.6	5.03	169	0.284
Fluoromethane	44.7	58.0	5.88	124	0.274
4-Fluorotoluene	316.4				
Formaldehyde	135	65	6.6	105	0.286
Formic acid	315				
2-Furaldehyde	397	58.1	5.89		
Furan	217.1	54.3	5.50	218	0.312
Glycerol	453	66	6.69	255	0.361
Heptadecane	460	13.0	1.32	1006	0.140
1-Heptadecanol	736	14.0	1.42	960	0.267
Heptane	267.1	27.0	2.74	428	0.232
1-Heptanol	359.5	30.18	3.058	435	0.267
2-Heptanol	335.2	29.81	3.021	432	0.269
3-Heptanol	332.3				
2-Heptanone	338.4	33.91	3.436	421	0.271
1-Heptene	264.2	28.83	2.921	402	0.246
Heptylcyclopentane	406	19.2	1.945		
Hexadecafluoroheptane	201.7	16.0	1.62	664	0.584
Hexadecane	444	14	1.42	930	0.243
1-Hexadecene	444	13.2	1.34	933	0.241
Hexadecylcyclopentane	518	9.6	0.97		
1,5-Hexadiene	234	34	3.44	328	0.250
Hexafluoroacetone	84.1	29.0	2.94	329	0.505
Hexafluorobenzene	243.6	32.30	3.273	335	0.505
Hexafluoroethane	19.7			224	0.617
Hexamethylbenzene	494			600	0.271
Hexane	234.5	29.85	3.025	368	0.233
Hexanenitrile	360.7	32.57	3.30		
Hexanoic acid	389	31.6	3.20		
1-Hexanol	337.2	33.72	3.417	381	0.268
2-Hexanol	312.8	32.67	3.310		

TABLE 2.56   Critical Properties (*Continued*)

Substance	T_c, °C	P_c, atm	P_c, MPa	V_c, cm³·mol⁻¹	ρ_c, g·cm⁻³
3-Hexanol	309.3	33.2	3.36		
2-Hexanone	313.9	32.8	3.32		
3-Hexanone	309.7	32.76	3.320		
1-Hexene	231.0	31.64	3.206	348	0.242
cis-2-Hexene	245	32.4	3.28	351	0.240
trans-2-Hexene	243	32.3	3.27	351	0.240
cis-3-Hexene	244	32.4	3.28	350	0.240
trans-3-Hexene	246.8	32.1	3.25	350	0.240
Hexylcyclopentane	387.0	21.1	2.14		
Icosafluorononane	251	15.4	1.56		
Icosane	494	10.3	1.04	1190	0.237
1-Icosanol	497	12.0	1.22		
Indane	411.8	39.0	3.95	381	0.310
Iodine	546	115	11.7	155	0.164
Iodobenzene	448	44.6	4.52	351	0.581
Iodoethane	281.0				
Iodomethane	255	65	6.59	190	0.75
1-Iodopropane	323				
Isobutyl acetate	288	31.2	3.16	414	0.281
Isobutylamine	246	40.2	4.07	284	0.258
Isobutylbenzene	377	30.1	3.05	480	0.280
Isobutyl bromide	294.1				
Isobutyl butanoate	338				
Isobutylcyclohexane	386	30.8	3.12		
Isobutyl formate	278	38.3	3.88	350	0.29
Isobutyl isobutanoate	329				
Isobutyl 3-methylbutanoate	348				
Isobutyl propanoate	319				
Isopentyl acetate	326				
Isopentyl butanoate	346				
Isopentyl propanoate	338				
Isopropyl acetate	258				
Isopropylamine	198.7	44.8	4.54	221	0.267
Isopropylbenzene	357.9	31.67	3.209	429	0.281
Isopropylcycloheptane	334.5				
Isopropylcyclohexane	367	28	2.84		
Isopropylcyclopentane	328	29.6	3.00		
4-Isopropylheptane	334.5	22.0	2.23	537	0.265
Isopropylmethylamine	217.6				
2-Isopropyl-1-methylbenzene	397	28.6	2.90		
3-Isopropyl-1-methylbenzene	393	29.0	2.94		
4-Isopropyl-1-methylbenzene	380	27.9	2.83		
3-Isopropyl-2-methylhexane	359.3	22.6	2.29	529	0.269
Isopropyl methyl sulfide	276.4				
Isoquinoline	530	50.3	5.10	374	0.345
Isoxazole	278.9				
Ketene	380	64	6.5	145	0.290
Methane	-82.60	45.44	4.604	99.0	0.162
Methanethiol	196.8	71.4	7.23	145	0.332
Methanol	239.4	79.78	8.084	118	0.272
Methoxybenzene	372.5	41.9	4.25		0.321
Methyl acetamide	417				
Methyl acetate	233.40	46.9	4.75	228	0.325

(*Continued*)

TABLE 2.56 Critical Properties (*Continued*)

Substance	T_c, °C	P_c, atm	P_c, MPa	V_c, cm$^3 \cdot$ mol^{-1}	ρ_c, g $\cdot$ cm^{-3}
Methyl acrylate	263	42	4.26	265	0.325
Methylamine	157.6	75.14	7.614	140	0.222
N-Methylaniline	428	51.3	5.20	373	0.287
Methyl benzoate	438	36	3.65	396	0.344
2-Methyl-1,3-butadiene	211	38.0	3.85	276	0.247
3-Methyl-1,3-butadiene	223	40.6	4.11	267	0.255
2-Methylbutane	187.3	33.4	3.38	306	0.236
2-Methyl-1-butanethiol	318.8				
2-Methyl-2-butanethiol	297.0				
Methyl butanoate	281.3	34.3	3.475	340	0.300
3-Methylbutanoic acid	356	33.6	3.40		
2-Methyl-1-butanol	302.3	38.9	3.94	322	0.274
3-Methyl-1-butanol	304.1	38.8	3.93	329	0.268
2-Methyl-2-butanol	270.6	36.6	3.71	319	0.276
3-Methyl-2-butanol	283.0	38.2	3.87		
3-Methyl-2-butanone	280.3	38.0	3.85	310	0.278
2-Methyl-1-butene	196.9	34.0	3.445	294	0.239
3-Methyl-1-butene	191.6	34.7	3.52	300	0.234
2-Methyl-2-butene	207.9	34.0	3.445	318	0.221
Methylcyclohexane	299.1	34.26	3.471	368	0.267
Methylcyclopentane	259.58	37.35	3.784	319	0.264
Methyl dodecanoate	439			758	0.283
N-Methylethylamine	223.5	36.6	3.71	243	0.243
Methyl formate	214.1	59.20	5.998	172	0.349
2-Methylfuran	254	46.6	4.72	247	0.333
2-Methylheptane	286.6	24.52	2.484	488	0.234
3-Methylheptane	290.6	25.13	2.546	464	0.246
4-Methylheptane	288.7	25.09	2.542	476	0.240
2-Methylhexane	257.3	26.98	2.734	421	0.238
3-Methylhexane	262.2	27.77	2.814	404	0.248
Methylhydrazine	294	79.3	8.035	271	0.170
Methyl 2-hydroxybenzoate	436				
Methyl isobutanoate	267.7	33.9	3.43	339	0.301
Methyl isocyanate	218	55	5.57		
1-Methylnaphthalene	499	35.5	3.60	445	0.320
2-Methylnaphthalene	488	34.6	3.51	462	0.308
2-Methyloctane	313.9	22.80	2.310		
2-Methylpentane	224.6	29.91	3.031	367	0.235
3-Methylpentane	231.4	30.85	3.126	367	0.235
2-Methyl-2,4-pentanediol	405	33.9	3.43		
Methyl pentanoate	294				
2-Methyl-2-pentanol	286.4				
2-Methyl-3-pentanol	302.9	34.1	3.46		
3-Methyl-3-pentanol	302.5	34.7	3.52		
4-Methyl-1-pentanol	330.4				
4-Methyl-2-pentanol	301.3	42.4	4.30	380	0.269
3-Methyl-2-pentanone	298.8				
4-Methyl-2-pentanone	298	32.3	3.27	371	0.270
2-Methyl-2-pentene	245	32.4	3.28	351	0.240
cis-3-Methyl-2-pentene	245	32.4	3.28	351	0.240
trans-3-Methyl-2-pentene	248	32.3	3.27	350	0.240
cis-4-Methyl-2-pentene	217	30	3.04	360	0.234
trans-4-Methyl-2-pentene	220	30	3.04	360	0.234

TABLE 2.56   Critical Properties (*Continued*)

Substance	T_c, °C	P_c, atm	P_c, MPa	V_c, cm$^3 \cdot$ mol^{-1}	ρ_c, g $\cdot$ cm^{-3}
2-Methylpropanal	240	41	4.15	274	0.263
2-Methyl-1-propanamine	246	40.2	4.07	278	0.263
N-Methylpropanamide	412				
2-Methylpropane	134.70	35.83	3.630	263	0.221
2-Methyl-1-propanethiol	286.4				
2-Methyl-2-propanethiol	257.0				
Methyl propanoate	257.5	39.5	4.00	282	0.312
2-Methylpropanoic acid	332	36.5	3.7	292	0.302
2-Methyl-1-propanol	274.6	42.39	4.295	273	0.272
2-Methyl-2-propanol	233.1	39.20	3.972	275	0.270
2-Methylpropene	144.73	39.48	4.000	239	0.235
2-Methylpropyl acetate	288	31.2	3.16	414	0.281
Methyl propyl ether	203.2				
Methyl propyl sulfide	301.0				
2-Methylpyridine	347.9	45.4	4.60	292	0.319
3-Methylpyridine	371.9	44.2	4.48	288	0.323
4-Methylpyridine	373	46.4	4.70	292	0.319
1-Methyl-2-pyrrolidinone	448.7			311	0.319
1-Methylstyrene	381	33.6	3.40	397	0.298
2-Methyltetrahydrofuran	264	37.1	3.76	267	0.322
2-Methylthiophene	333.1	47.9	4.85	275	0.356
3-Methylthiophene	337.7	48.9	4.95	275	0.356
Methyl vinyl ether	163	47	4.76	205	0.283
Morpholine	345	54	54.7	253	0.344
Naphthalene	475.3	39.98	4.051	407	0.31
Nitrobenzene	459				
Nitroethane	284	37	3.75		
Nitromethane	315	57.9	5.87	173	0.352
1-Nitropropane	402.0				
2-Nitropropane	344.8				
Nonadecane	483	11.0	1.12	1130	0.238
Nonane	321.5	22.6	2.29	555	0.231
Nonanoic acid	438	23.7	2.40		
1-Nonanol	404			546	0.264
1-Nonene	319	23.1	2.34	580	0.218
Nonylbenzene	468	18.7	1.89	790	0.259
Nonylcyclopentane	437.4	16.3	1.65		
Octadecafluorooctane	229	16.4	1.66		
Octadecane	472.3	12.73	1.29	1070	0.238
1-Octadecanol	474	14	1.42		
1-Octadecene	466	11.2	1.13		
Octafluorocyclobutane	115.31	27.48	2.784	325	0.616
Octafluoronaphthalene	399.9				
Octafluoropropane	72.7	26.5	2.69	299	0.628
Octamethylcyclotetrasiloxane	313	13.2	1.33	970	0.306
Octane	295.6	24.6	2.49	492	0.232
Octanenitrile	401.3	28.1	2.85		
Octanoic acid	422	26.1	2.64		
1-Octanol	379.4	27.41	2.777	490	0.266
2-Octanol	356.5	27.18	2.754	494	0.278
1-Octene	293.6	26.40	2.675	464	0.242
cis-2-Octene	307	27.3	2.77		
Octylcyclopentane	421	17.7	1.79		

(*Continued*)

TABLE 2.56 Critical Properties (*Continued*)

Substance	T_c, °C	P_c, atm	P_c, MPa	V_c, cm³ · mol⁻¹	ρ_c, g · cm⁻³
Pentachloroethane	373.0				
Pentadecane	433.9	15	1.52	880	0.241
1-Pentadecene	431	14.4	1.46		
Pentadecylcyclopentane	507	10.1	1.02		
1,2-Pentadiene	230	40.2	4.07	276	0.248
cis-1,3-Pentadiene	223	39.4	3.99	275	0.248
1,4-Pentadiene	205	37.4	3.79	276	0.248
Pentafluorobenzene	258.9	34.7	3.52		
2,3,4,5,6-Pentafluorotoluene	275.5				
2,2,3,3,4-Pentamethyl-pentane	370.7	25.5	2.58	508	0.280
2,2,3,4,4-Pentamethyl-pentane	354.2	23.7	2.40	521	0.273
Nonadecane	483	11.0	1.12	1130	0.238
Nonane	321.5	22.6	2.29	555	0.231
Nonanoic acid	438	23.7	2.40		
1-Nonanol	404			546	0.264
1-Nonene	319	23.1	2.34	580	0.218
Nonylbenzene	468	18.7	1.89	790	0.259
Nonylcyclopentane	437.4	16.3	1.65		
Octadecafluorooctane	229	16.4	1.66		
Octadecane	472.3	12.73	1.29	1070	0.238
1-Octadecanol	474	14	1.42		
1-Octadecene	466	11.2	1.13		
Octafluorocyclobutane	115.31	27.48	2.784	325	0.616
Octafluoronaphthalene	399.9				
Octafluoropropane	72.7	26.5	2.69	299	0.628
Octamethylcyclotetrasiloxane	313	13.2	1.33	970	0.306
Octane	295.6	24.6	2.49	492	0.232
Octanenitrile	401.3	28.1	2.85		
Octanoic acid	422	26.1	2.64		
1-Octanol	379.4	27.41	2.777	490	0.266
2-Octanol	356.5	27.18	2.754	494	0.278
1-Octene	293.6	26.40	2.675	464	0.242
cis-2-Octene	307	27.3	2.77		
Octylcyclopentane	421	17.7	1.79		
Osmium tetroxide	132	170	17.2		
Oxygen	−118.56	49.77	5.043	73.4	0.436
Oxygen difluoride	−58.0	48.9	4.95	97.7	0.553
Ozone	−12.10	53.8	5.45	88.9	0.540
Pentachloroethane	373.0				
Pentadecane	433.9	15	1.52	880	0.241
1-Pentadecene	431	14.4	1.46		
Pentadecylcyclopentane	507	10.1	1.02		
1,2-Pentadiene	230	40.2	4.07	276	0.248
cis-1,3-Pentadiene	223	39.4	3.99	275	0.248
1,4-Pentadiene	205	37.4	3.79	276	0.248
Pentafluorobenzene	258.9	34.7	3.52		
2,3,4,5,6-Pentafluorotoluene	275.5				
2,2,3,3,4-Pentamethyl-pentane	370.7	25.5	2.58	508	0.280
2,2,3,4,4-Pentamethyl-pentane	354.2	23.7	2.40	521	0.273

TABLE 2.56 Critical Properties (*Continued*)

Substance	T_c, °C	P_c, atm	P_c, MPa	V_c, cm$^3 \cdot$ mol^{-1}	ρ_c, g $\cdot$ cm^{-3}
3-Pentanol	286.5				
2-Pentanone	287.93	36.46	3.694	301	0.286
3-Pentanone	288.31	36.9	3.729	336	0.256
1-Pentene	191.63	34.81	3.527	293	0.239
cis-2-Pentene	202	36.4	3.69		
trans-2-Pentene	198	34.7	3.52	304	0.231
Pentyl acetate	332				
Pentylbenzene	406.8	25.7	2.60	550	0.269
Pentyl formate	303				
1-Pentyne	220.3	40	4.05	278	0.245
Perchloryl fluoride	95.3	53.0	5.37	161	0.637
Phenanthrene	596			554	0.322
Phenol	421.1	60.5	6.13	229	0.41
1-Phenylhexadecane	535	12.7	1.29	1200	0.252
1-Phenylpentadecane	526.9	13.3	1.35	1140	0.253
1-Phenyltetradecane	519	14.0	1.42	1110	0.247
Phthalic anhydride	537	47	4.76	368	0.402
Piperidine	321.0	48.8	4.94	288	0.296
Propadiene	120	54.0	5.47	162	0.247
Propanal	231.3	52.0	5.27	204	0.285
Propane	96.68	41.92	4.248	200	0.217
1,2-Propanediol	352	60	6.08	237	0.321
1,3-Propanediol	385	59	5.98	241	0.316
Propanenitrile	288.2	42.0	4.26	230	0.240
1-Propanethiol	262.5				
2-Propanethiol	244.2				
Propanoic acid	331	44.7	4.53	222	0.32
1-Propanol	263.7	51.01	5.169	218.5	0.275
2-Propanol	235.2	47.02	4.764	220	0.273
2-Propenal	233	51	5.17	197	0.285
Propene	91.9	45.6	4.62	181	0.233
2-Propen-1-ol	272.0			208	0.279
Propyl acetate	276.6	33.2	3.36	345	0.296
Propylamine	223.9	46.6	4.72	233	0.254
Propylbenzene	365.20	31.58	3.200	440	0.273
Propyl butanoate	327				
Propylcyclopentane	358.7	29.6	3.00	425	0.264
Propylcyclohexane	336.7	27.7	2.81		
Propylene oxide	209.1	48.6	4.92	186	0.312
Propyl formate	264.9	40.1	4.06	285	0.309
Propyl 2-methylpropanoate	316				
Propyl 3-methylpropanoate	336				
Propyl propanoate	305				
Propyne	129.3	55.5	5.62	164	0.245
Pyridine	346.9	55.96	5.67	243	0.325
Pyrrole	366.6	62.6	6.34	200	0.335
Pyrrolidine	295.1	55.2	5.59	238	0.300
Quinoline	509	48.0	4.86	437	0.300
Spiro[2.2]pentane	233.3				
Styrene	363.8	36.3	3.68	347	0.300

(*Continued*)

TABLE 2.56   Critical Properties (*Continued*)

Substance	T_c, °C	P_c, atm	P_c, MPa	V_c, cm$^3 \cdot$ mol^{-1}	ρ_c, g $\cdot$ cm^{-3}
1,2-Terphenyl	617.9	38.5	3.90	755	0.305
1,3-Terphenyl	651.7	34.6	3.51	768	0.300
1,4-Terphenyl	652.9	32.8	3.32	762	0.302
1,1,2,2-Tetrachlorodifluoro-ethane	278	34	3.44	371	0.549
1,1,2,2-Tetrachloroethane	388.00				
Tetrachloroethylene	347.1	44.3	4.49	290	0.572
Tetrachloromethane	283.5	44.57	4.516	276	0.557
Tetradecafluoro-1-heptene	205.1				
Tetradecafluorohexane	174.5	18.8	1.90		
Tetradecafluoromethylcyclohexane	213.7	23	2.33		
Tetradecane	420.9	16	1.62	830	0.239
1-Tetradecene	416	15.4	1.56		
Tetradecylcyclopentane	499	11.1	1.12		
Tetraethylsilane	330.6	25.68	2.602		
Tetrafluoroethylene	33.4	38.9	3.91	175	0.58
Tetrafluorohydrazine	33.3	37	3.75		
Tetrafluoromethane	−45.5	36.9	3.74	140	0.629
Tetrahydrofuran	267.0	51.22	5.19	224	0.322
1,2,3,4-Tetrahydronaphthalene	447	36.0	3.65	408	0.324
Tetrahydropyran	299.1	47.1	4.77	263	0.328
Tetrahydrothiophene	358.9				
1,2,4,5-Tetramethylbenzene	402	29	2.94	480	0.280
2,2,3,3-Tetramethylbutane	294.7	28.3	2.87	461	0.248
2,2,3,3-Tetramethylhexane	350.0	24.8	2.51	573	0.248
2,2,3,4-Tetramethylhexane	347.3	23.4	2.37	525	0.271
2,2,3,5-Tetramethylhexane	328.2	22.4	2.27	540	0.263
2,2,4,4-Tetramethylhexane	337.1	22.2	2.25	535	0.266
2,2,4,5-Tetramethylhexane	325.4	21.9	2.22	544	0.262
2,2,5,5-Tetramethylhexane	308.4	21.6	2.19	573	0.248
2,3,3,4-Tetramethylhexane	360.0	24.5	2.48	514	0.277
2,3,3,5-Tetramethylhexane	337.0	22.9	2.32	531	0.268
2,3,4,4-Tetramethylhexane	353.5	23.9	2.42	518	0.275
2,3,4,5-Tetramethylhexane	340.1	23.1	2.34	530	0.269
3,3,4,4-T etramethylhexane	373.6	25.4	2.57	506	0.281
2,2,3,3-Tetramethylpentane	334.6	27.05	2.741		
2,2,3,4-T etramethylpentane	319.6	25.68	2.602		
2,2,4,4-Tetramethylpentane	301.6	24.52	2.485		
2,3,3,4-Tetramethylpentane	334.6	26.80	2.716		
Tetramethylsilane	175.49	27.84	2.821	362	0.244
Thiacyclopentane	358.8				
2-Thiapropane	230.0	54.6	5.53	201	0.309
Thiophene	306.3	56.16	5.69	219	0.385
Thiophenol	416.4				
Thymol	425				
Toluene	318.60	40.54	4.108	316	0.292
1,2-Toluidine	434	43.1	4.37	343	0.312
1,3-Toluidine	434	42.2	4.28	343	0.312
1,4-Toluidine	433	45.2	4.58		
Toluonitrile	450				
Tributoxyborane	472	19.6	1.99	863	0.267

TABLE 2.56 Critical Properties (*Continued*)

Substance	T_c, °C	P_c, atm	P_c, MPa	V_c, cm³ · mol⁻¹	ρ_c, g · cm⁻³
Tributylamine	365.3	18	1.82		
1,1,1-Trichloroethane	272	42.4	4.30		
1,1,2-Trichloroethane	329	41	4.15	294	0.454
Trichloroethylene	271.1	49.5	5.02	256	0.513
Trichlorofluoromethane	198.1	43.5	4.41	248	0.554
Trichlorofluorosilane	165.4	35.3	3.57		
Trichloromethane	263.3	54.0	5.47	239	0.500
Trichloromethylsilane	244	32.4	3.28	348	0.430
1,2,3-Trichloropropane	378	39	3.95	348	0.424
1,2,2-Trichlorotrifluoroethane	214.2	33.7	3.42	325	0.576
Tridecane	402	16.6	1.68	780	0.236
1-Tridecene	401	16.8	1.70		
Tridecylcyclopentane	488	11.9	1.21		
Triethanolamine	514.3	24.2	2.45		
Triethylamine	262.5	29.92	3.032	389	0.26
Trifluoroacetic acid	218.2	32.15	3.258	204	0.559
Trifluoroamine oxide (NOF₃)	29.5			169	0.593
1,1,1-Trifluoroethane	73.2	37.1	3.76	194	0.434
Trifluoromethane	25.8	47.7	4.83	133	0.525
(Trifluoromethyl)benzene	286.8				
Trimethylamine	159.64	40.34	4.087	254	0.233
1,2,3-Trimethylbenzene	391.4	34.09	3.454	430	0.280
1,2,4-Trimethylbenzene	376.0	31.90	3.232	430	0.280
1,3,5-Trimethylbenzene	364.2	30.86	3.127	433	0.278
2,2,3-Trimethylbutane	258.1	29.15	2.954	398	0.252
2,2,3-Trimethyl-1-butene	260	28.6	2.90	400	0.245
1,1,2-Trimethylcyclopentane	306.4	29.0	2.94		
1,1,3-Trimethylcyclopentane	296.4	27.9	2.83		
cis, *trans*, *cis*-1,2,4-Trimethyl-cyclopentane	298	27.7	2.81		
cis, *cis*, *trans*-1, 2,4-Trimethyl-cyclopentane	306	28.4	2.88		
2,2,3-Trimethylheptane	338.6	22.4	2.27	546	0.261
2,2,4-Trimethylheptane	321.4	21.4	2.17	552	0.258
2,2,5-Trimethylheptane	325.0	21.4	2.17	559	0.256
2,2,6-Trimethylheptane	320.3	21.0	2.13	573	0.248
2,3,3-Trimethylheptane	344.4	22.9	2.32	538	0.265
2,3,4-Trimethylheptane	340.6	22.6	2.29	538	0.265
2,3,5-Trimethylheptane	339.7	22.1	2.24	547	0.260
2,3,6-Trimethylheptane	331.0	21.6	2.19	560	0.254
2,4,4-Trimethylheptane	327.2	21.9	2.22	541	0.263
2,4,5-Trimethylheptane	333.8	22.1	2.24	544	0.262
2,4,6-Trimethylheptane	317.2	21.2	2.15	560	0.254
2,5,5-Trimethylheptane	329.8	21.9	2.22	550	0.259
3,3,4-Trimethylheptane	349.4	23.4	2.37	526	0.271
3,3,5-Trimethylheptane	336.5	22.9	2.32	579	0.246
3,4,4-Trimethylheptane	347.8	23.4	2.37	524	0.271
3,4,5-Trimethylheptane	339.7	22.1	2.24	547	0.261
2,2,3-Trimethylhexane	315	24.6	2.49		
2,2,4-Trimethylhexane	300.6	23.4	2.37		

(*Continued*)

TABLE 2.56   Critical Properties (*Continued*)

Substance	T_c, °C	P_c, atm	P_c, MPa	V_c, cm$^3 \cdot$ mol^{-1}	ρ_c, g $\cdot$ cm^{-3}
2,2,5-Trimethylhexane	295	23.0	2.33	519	0.247
2,4,7-Trimethyloctane	335.7				
2,2,3-Trimethylpentane	290.4	26.94	2.730	436	0.262
2,2,4-Trimethylpentane	270.9	25.34	2.568	468	0.244
2,3,3-Trimethylpentane	300.5	27.83	2.820	455	0.251
2,3,4-Trimethylpentane	293.4	26.94	2.730	461	0.248
2,2,4-Trimethyl-1,3-pentanediol	398	25.6	2.59	364.6	0.4010
2,3,6-Trimethylpyridine	381.4				
2,4,6-Trimethylpyridine	379.9				
2,4,6-Trimethyl-1,3,5-trioxane	290				
1*H*-Undecafluoropentane	170.8				
Undecane	365.7	19.4	1.97	657	0.238
1-Undecene	364	19.7	2.00		0.240
Vinyl acetate	228.4	22.4	2.27	265	0.325
Vinyl chloride	156.6	55.3	5.60	169	0.370
Vinyl fluoride	54.7	51.7	5.24	114	0.320
Vinyl formate	202	57	5.78	210	0.343
1,2-Xylene	357.2	36.83	3.732	370	0.288
1,3-Xylene	343.9	34.95	3.541	375	0.282
1,4-Xylene	343.1	34.65	3.511	379	0.280

TABLE 2.57   Lydersen's Critical Property Increments

	Δ_T	Δ_p	Δ_v
Nonring increments			
—CH₃	0.020	0.227	55
—CH₂	0.020	0.227	55
—CH	0.012	0.210	51
—C—	0.00	0.210	41
=CH₂	0.018	0.198	45
=CH	0.018	0.198	45
=CH—	0.0	0.198	36
=C=	0.0	0.198	36
≡CH	0.005	0.153	(36)
≡C—	0.005	0.153	(36)
Ring increments			
—CH₂—	0.013	0.184	44.5
—CH	0.012	0.192	46
—C—	(−0.007)	(0.154)	(31)
=CH	0.011	0.154	37
=CH—	0.011	0.154	36
=C=	0.011	0.154	36
Halogen increments			
—F	0.018	0.224	18
—Cl	0.017	0.320	49
—Br	0.010	(0.50)	(70)
—I	0.012	(0.83)	(95)
Oxygen increments			
—OH (alcohols)	0.082	0.06	(18)
—OH (phenols)	0.031	(−0.02)	(3)
—O— (nonring)	0.021	0.16	20
—O— (ring)	(0.014)	(0.12)	(8)
—C=O (nonring)	0.040	0.29	60
—C=O (ring)	(0.033)	(0.2)	(50)
HC=O (aldehyde)	0.048	0.33	73
—COOH (acid)	0.085	(0.4)	80
—COO— (ester)	0.047	0.47	80
=O (except for combinations above)	(0.02)	(0.12)	(11)
Nitrogen increments			
—NH₂	0.031	0.095	28
—NH (nonring)	0.031	0.135	(37)

(*Continued*)

TABLE 2.57 Lydersen's Critical Property Increments (*Continued*)

	Δ_T	Δ_p	Δ_v
Nitrogen increments (*continued*)			
—NH (ring)	(0.024)	(0.09)	(27)
—NH— (nonring)	0.014	0.17	(42)
—N— (ring)	(0.007)	(0.13)	(32)
—CN	(0.060)	(0.36)	(80)
—NO$_2$	(0.055)	(0.42)	(78)
Sulfur increments			
—SH	0.015	0.27	55
—S— (nonring)	0.015	0.27	55
—S— (ring)	(0.008)	(0.24)	(45)
=S	(0.003)	(0.24)	(47)
Miscellaneous			
—Si—	0.03	(0.54)	
—B—	(0.03)		
Nonring:			

†There are no increments for hydrogen. All bonds shown as free are connected with atoms other than hydrogen. Values in parentheses are based upon too few experimental values to be reliable. From vapor-pressure measurements and a calculational technique similar to Fishtine [6], it has been suggested that the C—H ring increment common to two condensed saturated rings be given the value of $\Delta_T = 0.064$.

TABLE 2.58 Vetere Group Contribution to Estimate Critical Volume

Group	ΔV_i	Group	ΔV_i
Nonring:		—C=O (nonring)	1.765
In linear chain:			
CH$_3$, CH$_2$, CH, C	3.360	—C=O (ring)	1.500
In side chain			
CH$_3$, CH$_2$, CH, C	2.888	—HC=O (aldehyde)	2.333
=CH$_2$, =CH, =C—	2.940	—COOH	1.652
=C=	2.908	—COO—	1.607
≡CH, ≡C—	2.648		
Ring:		—NH$_2$	2.184
CH$_2$, CH, C	2.813	—NH (nonring)	2.333
=CH, =C—	2.538	—NH (ring)	1.736
F	0.770	—N— (nonring)	1.793
Cl	1.237		
Br	0.899	—N— (ring)	1.883
I	0.702	—CN	2.784
		—NO$_2$	1.559
—OH (alcohols)	0.704		
—OH (phenols)	1.553		
—O—(nonring)	1.075	—SH	1.537
—O—(ring)	0.790	—S— (nonring)	0.591
—O—(epoxy)	−0.252	—S— (ring)	0.911

TABLE 2.59 Van der Waals' Constants for Gases

The van der Waals' equation of state for a real gas is:

$$\left(P + \frac{n^2 a}{V^2}\right)(V - nb) = nRT \qquad \text{for } n \text{ moles}$$

where P is the pressure. V the volume (in liters per mole = 0.001 m^3 per mole in the SI system), T the temperature (in kelvins), n the amount of substance (in moles), and R the gas constant. To use the values of a and b in the table, P must be expressed in the same units as in the gas constant. Thus, the pressure of a standard atmosphere may be expressed in the SI system as follows:

$$1 \text{ atm} = 101{,}325 \text{ N} \cdot \text{m}^{-2} = 101{,}325 \text{ Pa} = 1.01325 \text{ bar}$$

The appropriate value for the gas constant is:

$$0.083\,144\,1 \text{ L} \cdot \text{bar} \cdot \text{K}^{-1} \cdot \text{mol}^{-1} \quad \text{or} \quad 0.082\,056 \text{ L} \cdot \text{atm} \cdot \text{K}^{-1} \cdot \text{mol}^{-1}$$

The van der Waals' constants are related to the critical temperature and pressure, T_c and P_c, in Table 2.56 by:

$$a = \frac{27 \, R^2 T_c^2}{64 \, P_c} \quad \text{and} \quad b = \frac{RT_c}{8 \, P_c}$$

Substance	a, L$^2 \cdot$ bar $\cdot$ mol^{-2}	b, L $\cdot$ mol^{-1}
Acetaldehyde	11.37	0.08695
Acetic acid	17.71	0.1065
Acetic anhydride	26.8	0.157
Acetone	16.02	0.1124
Acetonitrile	17.89	0.1169
Acetyl chloride	12.80	0.08979
Acetylene	4.516	0.05218
Acrylic acid	19.45	0.1127
Acrylonitrile	18.37	0.1222
Allene	8.235	0.07467
Allyl alcohol	15.17	0.1036
Aluminum trichloride	42.63	0.2450
2-Aminoethanol	7.616	0.0431
Ammonia	4.225	0.03713
Ammonium chloride	2.380	0.00734
Aniline	29.14	0.1486
Antimony tribromide	42.08	0.1658
Argon	1.355	0.03201
Arsenic trichloride	17.23	0.1039
Arsine	6.327	0.06048
Benzaldehyde	30.30	0.1553
Benzene	18.82	0.1193
Benzonitrile	33.89	0.1727
Benzyl alcohol	34.7	0.173
Biphenyl	47.16	0.2130
Bismuth trichloride	33.89	0.1025
Boron trichloride	15.60	0.1222
Boron trifluoride	3.98	0.05443
Bromine (Br$_2$)	9.75	0.0591
Bromobenzene	28.96	0.1541
Bromochlorodifluoromethane	12.79	0.1055
Bromoethane	11.89	0.08406
Bromomethane	6.753	0.05390
Bromotrifluoromethane	8.502	0.0891

(Continued)

TABLE 2.59   Van der Waalls' Constants for Gases (*Continued*)

Substance	a, L$^2 \cdot$ bar $\cdot$ mol^{-2}	b, L $\cdot$ mol^{-1}
1,2-Butadiene	12.76	0.1025
1,3-Butadiene	12.17	0.1020
Butanal	19.48	0.1292
Butane	13.93	0.1168
Butanenitrile	25.76	0.1568
Butanoic acid	28.18	0.1609
1-Butanol	20.90	0.1323
2-Butanol	20.94	0.1326
2-Butanone	19.97	0.1326
1-Butene	12.76	0.1084
cis-2-Butene	12.58	0.1066
trans-2-Butene	12.58	0.1066
3-Butenenitrile	25.76	0.1568
Butyl acetate	31.22	0.1919
1-Butylamine	19.41	0.1301
sec-Butylamine	18.37	0.1273
tert-Butylamine	17.78	0.1310
Butylbenzene	44.071	0.2378
sec-Butylbenzene	43.74	0.2347
tert-Butylbenzene	42.77	0.2310
Butyl benzoate	57.97	0.2857
Butylcyclohexane	41.19	0.2201
sec-Butylcyclohexane	48.89	0.2604
tert-Butylcyclohexane	48.34	0.2614
Butyl ethyl ether	27.05	0.1815
2-Butylhexadecafluorotetrahydrofuran	45.41	0.3235
1-Butyne	13.31	0.1023
2-Butyne	13.68	0.0998
Carbon dioxide	3.658	0.04284
Carbon disulfide	11.25	0.07262
Carbon monoxide	1.472	0.03948
Carbon oxysulfide (COS)	6.975	0.06628
Carbon tetrachloride	20.01	0.1281
Carbon tetrafluoride	4.029	0.06319
Carbonyl chloride	10.65	0.08340
Carbonyl sulfide	3.933	0.05817
Chlorine	6.343	0.05422
Chlorine pentafluoride	9.581	0.08214
Chlorobenzene	25.80	0.1454
1-Chlorobutane	23.22	0.1527
2-Chlorobutane	20.01	0.1370
1-Chloro-1,1-difluoroethane	11.91	0.1035
2-Chloro-1,1-difluoroethylene	10.49	0.09335
Chloroethane	11.7	0.090
Chloroform	15.34	0.1019
Chloromethane	7.566	0.06477
2-Chloro-2-methylpropane	18.98	0.1334
Chloropentafluoroacetone	17.08	0.1482
Chloropentafluorobenzene	29.53	0.1843
Chloropentafluoroethane	11.27	0.1137
1-Chloropropane	16.11	0.1141
2-Chloropropane	14.53	0.1068
Chlorotrifluoromethane	6.873	0.08110

TABLE 2.59   Van der Waalls' Constants for Gases (*Continued*)

Substance	a, $L^2 \cdot bar \cdot mol^{-2}$	b, $L \cdot mol^{-1}$
Chlorotrifluorosilane	7.994	0.09240
Chlorotrimethylsilane	22.58	0.1617
m-Cresol	31.86	0.1609
o-Cresol	28.33	0.1447
p-Cresol	28.11	0.1422
Cyanogen	7.803	0.06952
Cyclobutane	12.39	0.0960
Cycloheptane	27.20	0.1645
Cyclohexane	21.95	0.1413
Cyclohexanol	28.93	0.1586
Cyclohexanone	31.1	0.170
Cyclohexene	75.04	0.1339
Cyclopentane	16.94	0.1180
Cyclopentanone	75.84	0.1211
Cyclopentene	15.61	0.1097
Cyclopropane	8.293	0.07420
p-Cymene	43.65	0.2386
Decane	52.88	0.3051
Decanenitrile	34.71	0.1988
1-Decanol	57.45	0.2971
1-Decene	49.96	0.2888
Deuterium (normal)	0.2583	0.02397
Deuterium oxide	5.584	0.03090
Diborane (B_2H_6)	6.048	0.07437
Dibromodifluoromethane	15.69	0.1186
1,2-Dibromoethane	13.98	0.08664
1,2-Dibromotetrafluoroethane	20.45	0.1494
Dibutylamine	34.61	0.2030
Dibutyl ether	33.06	0.2017
Dibutyl sulfide	49.3	0.2702
1,2-Dichlorobenzene	34.59	0.1767
1,3-Dichlorobenzene	35.44	0.1846
1,4-Dichlorobenzene	34.64	0.1802
Dichlorodifluoromethane	10.45	0.09672
Dichlorodifluorosilane	11.34	0.1095
1,1-Dichloroethane	15.73	0.1072
1,2-Dichloroethane	17.0	0.108
1,1-Dichloroethylene	13.74	0.09893
trans-1,2-Dichloroethylene	13.63	0.09573
Dichlorofluoromethane	11.48	0.09060
Dichloromethane	12.44	0.08689
1,2-Dichloropropane	21.62	0.1335
Dichlorosilane	12.59	0.09992
1,1-Dichlorotetrafluoroethane	15.49	0.1318
1,2-Dichlorotetrafluoroethane	15.72	0.1338
Dideuterium oxide	5.535	0.03062
Diethanolamine	45.61	0.2273
Diethylamine	19.40	0.1383
1,4-Diethylbenzene	45.03	0.2439
Diethylene glycol	29.02	0.1519
Diethyl ether	17.46	0.1333
3,3-Diethylhexane	47.69	0.2707
3,4-Diethylhexane	47.93	0.2760

(*Continued*)

TABLE 2.59   Van der Waalls' Constants for Gases (*Continued*)

Substance	a, L$^2 \cdot$ bar $\cdot$ mol^{-2}	b, L $\cdot$ mol^{-1}
3,3-Diethyl-2-methylpentane	47.20	0.2629
3,3-Diethylpentane	40.64	0.2374
Diethyl sulfide	22.85	0.1462
Difluoroamine	5.028	0.04446
cis-Difluorodiazine	3.043	0.03987
trans-Difluorodiazine	3.539	0.04851
1,1-Difluoroethane	9.691	0.08931
1,1-Difluoroethylene	6.000	0.07058
Difluoromethane	6.184	0.06268
Dihexyl ether	69.17	0.3752
Dihydrogen disulfide	16.15	0.1006
Diisopropyl ether	25.26	0.1836
Dimethoxyethane	21.65	0.1439
Dimethoxymethane	17.28	0.1195
N,N-Dimethoxyacetamide	30.19	0.1689
Dimethylamine	10.44	0.08510
N,N-Dimethylaniline	37.92	0.1967
2,2-Dimethylbutane	22.55	0.1644
2,3-Dimethylbutane	23.29	0.1660
2,3-Dimethyl-1-butene	22.59	0.2566
3,3-Dimethyl-1-butene	21.55	0.1567
2,3-Dimethyl-2-butene	23.83	0.1621
1,1-Dimethylcyclohexane	34.30	0.2068
cis-1,2-Dimethylcyclohexane	36.44	0.2143
trans-1,2-Dimethylcyclohexane	34.89	0.2086
cis-1,3-Dimethylcyclohexane	34.30	0.2068
trans-1,3-Dimethylcyclohexane	35.11	0.2093
cis-1,4-Dimethylcyclohexane	35.47	0.2114
trans-1,4-Dimethylcyclohexane	34.54	0.2086
1,1-Dimethylcyclopentane	25.37	0.1653
cis-1,2-Dimethylcyclopentane	27.04	0.1706
trans-1,2-Dimethylcyclopentane	25.67	0.1663
Dimethyl ether	8.690	0.07742
N,N-Dimethylformamide	23.57	0.1293
2,2-Dimethylheptane	41.29	0.2551
2,2-Dimethylhexane	34.87	0.2260
2,3-Dimethylhexane	35.24	0.2228
2,4-Dimethylhexane	34.97	0.2251
2,5-Dimethylhexane	35.49	0.2299
3,3-Dimethylhexane	34.72	0.2201
3,4-Dimethylhexane	35.06	0.2196
1,1-Dimethylhydrazine	14.69	0.1001
2,4-Dimethyl-3-isopentane	47.05	0.2729
Dimethyl oxalate	28.97	0.1644
2,2-Dimethylpentane	28.49	0.1951
2,3-Dimethylpentane	28.96	0.1921
2,4-Dimethylpentane	28.79	0.1974
3,3-Dimethylpentane	28.48	0.1892
2,3-Dimethylphenol	31.35	0.1545
2,4-Dimethylphenol	33.49	0.1687
2,5-Dimethylphenol	29.99	0.1512
2,6-Dimethylphenol	33.64	0.1710
3,4-Dimethylphenol	31.32	0.1529

TABLE 2.59 Van der Waalls' Constants for Gases (*Continued*)

Substance	a, $L^2 \cdot bar \cdot mol^{-2}$	b, $L \cdot mol^{-1}$
3,5-Dimethylphenol	40.92	0.2037
2,2-Dimethylpropane	17.17	0.1410
2,3-Dimethylpropane	23.13	0.1669
2,2-Dimethyl-1-propanol	22.25	0.1444
Dimethyl sulfide	13.34	0.09453
N,N-Dimethyl-1,2-toluidine	41.71	0.2225
1,4-Dioxane	19.29	0.1171
Diphenyl ether	54.61	0.2538
Diphenylmethane	60.46	0.2798
Dipropylamine	24.82	0.1591
Dipropyl ether	27.12	0.1821
Dodecafluorocyclohexane	25.09	0.1955
Dodecafluoropentane	25.58	0.2161
Dodecane	69.14	0.3741
1-Dodecanol	72.69	0.3598
1-Dodecene	68.17	0.3694
Ethane	5.570	0.06499
1,2-Ethanediamine	16.30	0.09796
Ethanethiol	13.23	0.09447
Ethanol	12.56	0.08710
Ethoxybenzene	35.70	0.1996
Ethyl acetate	20.57	0.1401
Ethyl acrylate	23.70	0.1530
Ethylamine	10.79	0.08433
Ethylbenzene	30.86	0.1782
Ethyl benzoate	43.73	0.2236
Ethyl butanoate	30.53	0.1922
Ethylcyclohexane	35.70	0.2089
Ethylcyclopentane	27.90	0.1746
3-Ethyl-2,2-dimethylhexane	47.24	0.2752
4-Ethyl-2,2-dimethylhexane	46.45	0.2784
3-Ethyl-2,3-dimethylhexane	47.35	0.2692
4-Ethyl-2,3-dimethylhexane	47.49	0.2742
3-Ethyl-2,4-dimethylhexane	47.31	0.2736
4-Ethyl-2,4-dimethylhexane	45.52	0.2613
3-Ethyl-2,5-dimethylhexane	47.42	0.2800
3-Ethyl-3,4-dimethylhexane	47.00	0.2682
Ethylene	4.612	0.05821
Ethylene glycol dimethyl ether	21.65	0.1439
Ethylene glycol ethyl ether acetate	33.97	0.05594
Ethylene oxide	8.922	0.06779
Ethyl formate	15.91	0.1115
3-Ethylhexane	35.76	0.2253
Ethyl mercaptan	11.24	0.08098
2-Ethyl-1-methylbenzene	40.66	0.2226
3-Ethyl-1-methylbenzene	41.67	0.2331
4-Ethyl-1-methylbenzene	40.63	0.2262
1-Ethyl-1-methylcyclopentane	34.18	0.2058
Ethyl methyl ether	12.70	0.1034
3-Ethyl-2-methylheptane	48.81	0.2847
Ethyl methyl ketone	20.13	0.1340
3-Ethyl-2-methylpentane	34.74	0.2183
3-Ethyl-2-methylpentane	34.53	0.2134

(*Continued*)

TABLE 2.59 Van der Waalls' Constants for Gases (*Continued*)

Substance	a, $L^2 \cdot bar \cdot mol^{-2}$	b, $L \cdot mol^{-1}$
Ethyl 2-methylpropanoate	29.05	0.1872
Ethyl methyl sulfide	19.45	0.1300
3-Ethylpentane	29.49	0.1944
Ethyl phenyl ether	35.16	0.1963
Ethyl propanoate	25.86	0.1688
Ethyl propyl ether	22.45	0.1600
m-Ethyltoluene	41.73	0.2334
o-Ethyltoluene	40.67	0.2226
p-Ethyltoluene	40.63	0.2262
Ethyl vinyl ether	16.17	0.1213
Fluorine	1.171	0.02896
Fluorobenzene	20.10	0.1279
Fluoroethane	8.170	0.07758
Fluoroethylene	5.984	0.06504
Fluoromethane	5.009	0.05617
Formaldehyde	7.356	0.06425
Furan	12.74	0.0926
2-Furaldehyde (furfural)	22.23	0.1182
Germanium tetrachloride	23.12	0.1489
Germanium tetrahydride	5.743	0.06555
Glycerol	22.98	0.07037
Hafnium tetrachloride	26.01	0.1282
Helium (equilibrium)	0.0346	0.02356
Heptane	30.89	0.2038
1-Heptanol	37.22	0.2097
2-Heptanol	35.72	0.2093
2-Heptanone	31.78	0.1850
1-Heptene	28.82	0.09400
Hexadecafluoroheptane	40.58	0.3046
1,5-Hexadiene	21.79	0.1532
Hexafluoroacetone	12.66	0.1264
Hexafluorobenzene	26.63	0.1641
Hexane	24.97	0.1753
Hexanenitrile	35.50	0.1996
Hexanoic acid	39.94	0.2150
1-Hexanol	31.35	0.1829
2-Hexanol	30.25	0.1840
3-Hexanol	29.44	0.1803
2-Hexanone	30.27	0.1837
3-Hexanone	29.84	0.1824
1-Hexene	23.12	0.1634
cis-2-Hexene	23.86	0.1641
trans-2-Hexene	23.75	0.1640
cis-3-Hexene	23.77	0.1638
trans-3-Hexene	24.25	0.1663
Hexylcyclopentane	59.38	0.3206
Hydrazine	8.46	0.0462
Hydrogen (normal)	0.2484	0.02651
Hydrogen bromide	4.500	0.04415
Hydrogen chloride	3.700	0.04061
Hydrogen cyanide	11.29	0.08806
Hydrogen deuteride	0.2527	0.02516
Hydrogen fluoride	9.565	0.0739

TABLE 2.59 Van der Waalls' Constants for Gases (*Continued*)

Substance	a, $L^2 \cdot bar \cdot mol^{-2}$	b, $L \cdot mol^{-1}$
Hydrogen iodide	6.309	0.05303
Hydrogen selenide	5.523	0.0479
Hydrogen sulfide	4.544	0.04339
Indane	34.63	0.1802
Iodobenzene	33.54	0.1658
Iodomethane	12.34	0.08327
Isobutyl acetate	29.05	0.1845
Isobutylamine	19.30	0.1325
Isobutylbenzene	40.40	0.2215
Isobutylcyclohexane	40.39	0.2195
Isobutyl formate	22.82	0.1476
Isopropylamine	14.30	0.1080
Isopropylbenzene	36.20	0.2044
Isopropylcyclohexane	42.06	0.2342
Isopropylcyclopentane	35.11	0.2082
4-Isopropylheptane	48.28	0.2832
2-Isopropyl-1-methylbenzene	45.14	0.2401
3-Isopropyl-1-methylbenzene	44.00	0.2354
4-Isopropyl-1-methylbenzene	43.94	0.2398
3-Isopropyl-2-methylhexane	50.93	0.2870
Ketene	19.1	0.1044
Krypton	2.325	0.0396
Mercury	5.193	0.01057
Methane	2.300	0.04301
Methanethiol	8.911	0.06756
Methanol	9.472	0.06584
Methoxybenzoate	28.60	0.1579
Methyl acetate	15.75	0.1108
Methyl acrylate	19.67	0.1308
Methylamine	7.106	0.05879
2-Methyl-1,3-butadiene	17.74	0.1307
3-Methyl-1,3-butadiene	17.46	0.1245
2-Methylbutane	18.29	0.1415
Methyl butanoate	25.83	0.1661
3-Methylbutanoic acid	33.94	0.1923
2-Methyl-1-butanol	24.51	0.1518
3-Methyl-1-butanol	24.72	0.1526
2-Methyl-2-butanol	23.24	0.1523
3-Methyl-2-butanol	23.30	0.1493
3-Methyl-2-butanone	23.20	0.1494
2-Methyl-1-butene	16.9	0.129
3-Methyl-1-butene	18.08	0.1405
2-Methyl-2-butene	17.26	0.1279
Methylcyclohexane	27.51	0.1713
Methylcyclopentane	21.87	0.1463
N-Methylethylamine	19.39	0.1391
Methyl formate	11.54	0.08406
2-Methylfuran	14.67	0.1160
2-Methylheptane	36.78	0.2342
3-Methylheptane	36.40	0.2301
4-Methylheptane	36.21	0.2297
2-Methylhexane	30.01	0.2016
3-Methylhexane	29.70	0.1977

(*Continued*)

TABLE 2.59   Van der Waalls' Constants for Gases (*Continued*)

Substance	a, L$^2 \cdot$ bar $\cdot$ mol^{-2}	b, L $\cdot$ mol^{-1}
Methylhydrazine	11.67	0.07334
Methyl isobutanoate	24.87	0.1639
Methyl isocyanate	12.6	0.09161
1-Methyl-2-isopropylbenzene	42.7	0.234
1-Methyl-4-isopropylbenzene	45.27	0.2478
Methyl 2-methylpropanoate	24.50	0.1637
2-Methyloctane	43.50	0.2641
2-Methylpentane	23.83	0.1707
3-Methylpentane	23.75	0.1677
2-Methyl-2,4-pentanediol	39.05	0.2054
Methyl pentanoate	29.39	0.1847
2-Methyl-3-pentanol	27.96	0.1730
3-Methyl-3-pentanol	27.45	0.1699
4-Methyl-2-pentanol	22.38	0.1388
4-Methyl-2-pentanone	29.08	0.1815
2-Methyl-2-pentene	23.86	0.1641
cis-3-Methyl-2-pentene	23.86	0.1641
trans-3-Methyl-2-pentene	24.60	0.1656
cis-4-Methyl-2-pentene	23.03	0.1675
trans-4-Methyl-2-pentene	23.32	0.1685
2-Methylpropanal	18.49	0.1285
2-Methyl-1-propanamine	19.30	0.1325
2-Methylpropane (isobutane)	13.36	0.1168
Methyl propanoate	20.51	0.1377
2-Methylpropanoic acid	28.9	0.170
2-Methyl-1-propanol	20.35	0.1324
2-Methyl-2-propanol	18.81	0.1324
2-Methylpropene	12.73	0.1086
2-Methylpropyl acetate	29.05	0.1845
2-Methylpropyl formate	22.54	0.1476
2-Methylpyridine	24.45	0.1403
3-Methylpyridine	27.08	0.1496
4-Methylpyridine	25.89	0.1428
1-Methylstyrene	36.69	0.1999
2-Methyltetrahydrofuran	22.37	0.1484
2-Methylthiophene	22.10	0.1299
3-Methylthiophene	21.98	0.1282
Methyl vinyl ether	11.65	0.09520
Morpholine	20.36	0.1174
Naphthalene	40.32	0.1920
Neon	0.208	0.01709
Niobium pentafluoride	25.22	0.1220
Nitric oxide (NO)	1.46	0.0289
Nitroethane	24.13	0.1544
Nitrogen-14	15.18	0.1288
Nitrogen chloride difluoride	6.447	0.06089
Nitrogen dioxide (NO_2)	5.36	0.0443
Nitrogen trifluoride	3.58	0.05364
Nitrous oxide (N_2O)	3.852	0.04435
Nitromethane	17.18	0.1041
Nitrosyl chloride	6.191	0.05014
Nonane	45.11	0.2702
1-Nonanol	50.00	0.2634

TABLE 2.59   Van der Waalls' Constants for Gases (*Continued*)

Substance	a, L$^2 \cdot$ bar $\cdot$ mol^{-2}	b, L $\cdot$ mol^{-1}
1-Nonene	43.68	0.2629
Octadecafluorooctane	44.27	0.3143
Octafluorocyclobutane	15.81	0.1450
Octafluoropropane	12.96	0.1338
Octamethylcyclotetrasiloxane	75.30	0.4579
Octane	37.86	0.2370
1-Octanol	44.71	0.2371
2-Octanol	41.98	0.2376
1-Octene	35.01	0.2227
cis-2-Octene	35.42	0.2176
Osmium tetraoxide	2.79	0.2447
Oxygen	1.382	0.03186
Oxygen difluoride	2.726	0.04516
Ozone	3.570	0.04977
Pentadecane	95.91	0.4834
1-Pentadecene	99.00	0.5011
1,2-Pentadiene	18.13	0.1284
cis-1,3-Pentadiene	17.98	0.1292
1,4-Pentadiene	17.58	0.1311
Pentafluorobenzene	23.45	0.1571
2,2,3,3,4-Pentamethylpentane	46.85	0.2593
2,2,3,4,4-Pentamethylpentane	47.82	0.2716
Pentanal	25.21	0.1622
Pentane	19.13	0.1449
Pentanenitrile	34.16	0.1772
Pentanoic acid	33.68	0.1867
1-Pentanol	25.81	0.1572
2-Pentanol	24.89	0.1585
2-Pentanone	24.85	0.1578
3-Pentanone	24.65	0.1565
1-Pentene	17.86	0.1370
cis-2-Pentene	17.83	0.1338
trans-2-Pentene	18.30	0.1391
Pentylbenzene	51.85	0.2718
Pentyl formate	27.97	0.1730
1-Pentyne	17.53	0.1266
Perchloryl fluoride (ClO_3F)	7.371	0.07130
Phenol	22.93	0.1177
Phosgene	10.65	0.08340
Phosphine	4.693	0.05155
Phosphonium chloride	4.111	0.04545
Phosphorus	53.6	0.157
Phosphorus chloride difluoride	8.47	0.0833
Phosphorus dichloride fluoride	12.50	0.0962
Phosphorus trifluoride	4.954	0.06510
Phosphoryl chloride difluoride	11.90	0.1001
Phosphoryl trifluoride	8.26	0.0849
Piperidine	20.84	0.1250
Propadiene	8.23	0.0747
Propanal	14.08	0.0995
Propane	9.385	0.09044
1,2-Propanediol	18.74	0.1068
1,3-Propanediol	21.11	0.1143

(*Continued*)

TABLE 2.59 Van der Waalls' Constants for Gases (*Continued*)

Substance	a, L$^2 \cdot$ bar $\cdot$ mol^{-2}	b, L $\cdot$ mol^{-1}
Propanenitrile	21.57	0.1369
Propanoic acid	23.49	0.1386
1-Propanol	16.26	0.1080
2-Propanol	15.82	0.1109
2-Propenal	14.44	0.1017
Propene	8.411	0.08211
Propyl acetate	26.23	0.1700
Propylamine	15.26	0.1095
Propylbenzene	37.14	0.2073
Propylcyclopentane	38.80	0.2189
Propylcyclohexane	38.59	0.2255
Propylene oxide	13.78	0.1019
Propyl formate	20.79	0.1377
Propyne	8.40	0.0744
Pyridine	19.77	0.1136
Pyrrole	18.82	0.1049
Pyrrolidine	16.84	0.1056
Quinoline	36.70	0.1672
Radon	6.601	0.06239
Selenium	33.4	0.0675
Silicon chloride trifluoride	7.95	0.0921
Silicon tetrachloride	20.96	0.1470
Silicon tetrafluoride	5.259	0.072361
Silicon tetrahydride (silane)	4.30	0.0579
Styrene	32.15	0.1799
Sulfur (S)	24.3	0.0660
Sulfur dioxide	6.714	0.05636
Sulfur hexafluoride (SF$_6$)	7.857	0.08786
Sulfur trioxide	8.57	0.0622
1,1,2,2-Tetrachlorodifluoroethane	25.74	0.1665
Tetrachloroethylene	24.98	0.1435
Tetrachloromethane	20.01	0.1281
Tetradecafluorohexane	30.75	0.2448
Tetradecafluoromethylcyclohexane	29.66	0.2171
1-Tetradecanol	89.91	0.4289
Tetraethylsilane	40.85	0.2411
Tetrafluoroethylene	6.954	0.08085
Tetrafluorohydrazine (N$_2$F$_4$)	7.426	0.08564
Tetrafluoromethane	4.040	0.06325
Tetrahydrofuran	16.39	0.1082
Tetrahydropyran	20.02	0.1247
1,2,4,5-Tetramethylbenzene	45.8	0.2422
2,2,3,3-Tetramethylbutane	32.76	0.2056
2,2,3,3-Tetramethylhexane	45.11	0.2580
2,2,3,4-Tetramethylhexane	47.36	0.2721
2,2,3,5-Tetramethylhexane	46.45	0.2753
2,2,4,4-Tetramethylhexane	48.26	0.2819
2,2,4,5-Tetramethylhexane	47.05	0.2802
2,2,5,5-Tetramethylhexane	45.03	0.2760
2,3,3,4-Tetramethylhexane	47.13	0.2653
2,3,3,5-Tetramethylhexane	46.79	0.2733
2,3,4,4-Tetramethylhexane	47.32	0.2691
2,3,4,5-Tetramethylhexane	46.86	0.2723

TABLE 2.59 Van der Walls' Constants for Gases (*Continued*)

Substance	a, $L^2 \cdot bar \cdot mol^{-2}$	b, $L \cdot mol^{-1}$
3,3,4,4-Tetramethylhexane	47.46	0.2615
2,2,3,3-Tetramethylpentane	39.29	0.2304
2,2,3,4-Tetramethylpentane	39.37	0.2367
2,2,4,4-Tetramethylpentane	38.76	0.2403
2,3,3,4-Tetramethylpentane	39.65	0.2325
Tetramethylsilane	20.81	0.1653
Thiophene	17.21	0.1058
Tin(IV) chloride	27.25	0.1641
Titanium(IV) chloride	25.47	0.1423
Toluene	24.89	0.1499
1,2-Toluidine	33.36	0.1681
1,3-Toluidine	34.06	0.1717
1,4-Toluidine	31.74	0.1602
Tributoxyborane	81.34	0.3891
Tributylamine	65.31	0.3645
1,1,1-Trichloroethane	20.14	0.1317
1,1,2-Trichloroethane	25.47	0.1508
Trichloroethylene	17.21	0.1127
Trichlorofluoromethane	14.68	0.1111
Trichlorofluorosilane	15.67	0.1277
Trichloromethane	15.34	0.1019
Trichloromethylsilane	23.77	0.1638
1,2,3-Trichloropropane	31.29	0.1713
1,1,2-Trichlorotrifluoroethane	20.25	0.1481
1,2,2-Trichlorotrifluoroethane	20.25	0.1481
Tridecane	79.09	0.4176
1-Tridecanol	81.20	0.3942
1-Tridecene	77.93	0.4121
Tridecylcyclopentane	139.6	0.6536
Triethanolamine	32.14	0.3340
Triethylamine	27.59	0.1836
Trifluoroacetic acid	21.61	0.1567
1,1,1-Trifluoroethane	9.302	0.09572
Trifluoromethane	5.378	0.06403
Trimethylamine	13.37	0.1101
1,2,3-Trimethylbenzene	37.28	0.1999
1,2,4-Trimethylbenzene	38.03	0.2088
1,3,5-Trimethylbenzene	37.87	0.2118
2,2,3-Trimethylbutane	27.86	0.1869
2,2,3-Trimethyl-1-butene	28.57	0.1910
1,1,2-Trimethylcyclopentane	33.31	0.2048
1,1,3-Trimethylcyclopentane	33.42	0.2091
2,2,3-Trimethylheptane	48.07	0.2801
2,2,4-Trimethylheptane	47.49	0.2847
2,3,4-Trimethylheptane	47.96	0.2785
3,3,4-Trimethylheptane	47.68	0.2730
2,2,3-Trimethylhexane	40.5	0.2452
2,2,4-Trimethylhexane	40.50	0.2516
2,2,5-Trimethylhexane	40.38	0.2533
2,2,3-Trimethylpentane	33.92	0.2145
2,2,4-Trimethylpentane	33.61	0.2202
2,3,3-Trimethylpentane	34.03	0.2114
2,3,4-Trimethylpentane	34.28	0.2157

(*Continued*)

TABLE 2.59   Van der Waalls' Constants for Gases (*Continued*)

Substance	a, L$^2 \cdot$ bar $\cdot$ mol^{-2}	b, L $\cdot$ mol^{-1}
2,2,4-Trimethyl-1,3-pentanediol	19.96	0.2692
Tungsten(VI) fluoride (WF$_6$)	13.25	0.1063
Undecane	60.88	0.3396
1-Undecene	59.17	0.3310
Uranium(VI) fluoride (UF$_6$)	16.01	0.1128
Vinyl acetate	32.31	0.2296
Vinyl chloride	9.62	0.07975
Vinyl fluoride	5.98	0.06502
Vinyl formate	11.38	0.08541
Xenon	4.192	0.05156
Xenon difluoride	12.46	0.7037
Xenon tetrafluoride	15.52	0.09035
m-Xylene	31.41	0.1814
o-Xylene	31.06	0.1756
p-Xylene	31.54	0.1824
Water	5.537	0.03052
Zirconium(IV) chloride	30.59	0.1401

2.15   EQUILIBRIUM CONSTANTS

The equilibrium constant, K, relates to a chemical reaction at equilibrium. It can be calculated if the equilibrium concentration of each reactant and product in a reaction at equilibrium is known.

There are several types of equilibrium constants. *Each is constant at a constant temperature.*

TABLE 2.60 pK, Values of Organic Materials in Water at 25°C
Ionic strength μ is zero unless otherwise indicated. Protonated cations are designated by (+ 1), (+ 2), etc., after the pK_a value; neutral species by (0), if not obvious; and negatively charged acids by (−1), (−2), etc.

Substance	pK_1	pK_2	pK_3	pK_4
Abietic acid	7.62			
Acetamide	−0.37(+1)			
Acetamidine	1.60(+1)			
N-(2-Acetamido)-2-aminoethane-sulfonic acid (20°C)	6.88			
2-Acetamidobenzoic acid	3.63			
3-Acetamidobenzoic acid	4.07			
4-Acetamidobenzoic acid	4.28			
2-(Acetamido)butanoic acid	3.716			
N-(2-Acetamido)iminodiacetic acid (20°C)	6.62			
3-Acetamidopyridine	4.37(+1)			
Acetanilide	0.4(+1)	13.39(0)$^{40°C}$		
Acetic acid	4.756			
Acetic acid-*d* (in D$_2$O)	5.32			

TABLE 2.60   *pK*, Values of Organic Materials in Water at 25°C (*Continued*)

Substance	pK_1	pK_2	pK_3	pK_4
Acetoacetic acid (18°C)	3.58			
Acetohydrazine	3.24(+1)			
Acetone oxime	12.2			
2-Acetoxybenzoic acid (acetylsali- cyclic acid)	3.48			
3-Acetoxybenzoic acid	4.00			
4-Acetoxybenzoic acid	4.38			
Acetylacetic acid (18°C)	3.58			
N-Acetyl-α-alanine	3.715			
N-Acetyl-β-alanine	4.455			
2-Acetylaminobutanoic acid	3.72			
3-Acetylaminopropionic acid	4.445			
2-Acetylbenzoic acid	4.13			
3-Acetylbenzoic acid	3.83			
4-Acetylbenzoic acid	3.70			
2-Acetylcyclohexanone	14.1			
N-Acetylcysteine (30°C)	9.52			
Acetylenedicarboxylic acid	1.75	4.40		
N-Acetylglycine	3.670			
N-Acetylguanidine	8.23(+1)			
N-α-Acetyl-L-histidine	7.08			
Acetylhydroxamic acid (20°C)	9.40			
N-Acetyl-2-mercaptoethylamine	9.92(SH)			
4-Acetyl-β-mercaptoisoleucine (30°C)	10.30			
2-Acetyl-1-naphthol (30°C)	13.40			
N-Acetylpenicillamine (30°C)	9.90			
2-Acetylphenol	9.19			
4-Acetylphenol	8.05			
2-Acetylpyridine	2.643(+1)			
3-Acetylpyridine	3.256(+1)			
4-Acetylpyridine	3.505(+1)			
Aconitine	8.11(+1)			
Acridine	5.60(+1)			
Acrylic acid	4.26			
Adenine	4.17(+1)	9.75(0)		
Adeninedeoxyriboside-5'-phos- phoric acid	—	4.4	6.4	
Adenine-*N*-oxide	2.69(+1)	8.49(0)		
Adenosine	3.5(+1)	12.34(0)		
Adenosine-5'-diphosphoric acid	—	4.2(−1)	7.20(−2)	
Adenosine-2'-phosphoric acid	3.81(+1)	6.17(0)		
Adenosine-3'-phosphoric acid	3.65(0)	5.88(−1)		
Adenosine-5'-phosphoric acid	3.74(0)	6.05(−1)	13.06(−2)	
Adenosine-5'-triphosphoric acid	—	4.00(−1)	6.48(−2)	
Adipamic acid (adipic acid monoamide)	4.629			
Adipic acid	4.418	5.412		
α-Alanine	2.34(+1)	9.69(0)		
β-Alanine	3.55(+1)	10.238(0)		
α-Alanine, methyl ester ($\mu = 0.10$)	7.743(+1)			

(*Continued*)

TABLE 2.60 *pK*, Values of Organic Materials in Water at 25°C (*Continued*)

Substance	pK_1	pK_2	pK_3	pK_4
β-Alanine, methyl ester ($\mu = 0.10$)	9.170(+1)			
N-D-Alanyl-α-D-alanine ($\mu = 0.1$)	3.32(+1)	8.13(0)		
N-L-Alanyl-α-L-alanine ($\mu = 0.1$)	3.32(+1)	8.13(0)		
N-L-Alanyl-α-D-alanine	3.12(+1)	8.30(0)		
N-α-Alanylglycine	3.11(+1)	8.11(0)		
Alanylglycylglycine	3.190(+1)	8.15(0)		
β-Alanylhistidine	2.64	6.86	9.40	
Albumin (bovine serum ($\mu = 0.15$)	10–10.3			
2-Aldoxime pyridine	3.42(+1)	10.22(0)		
Alizarin Black SN	5.79	12.8		
Alizarin-3-sulfonic acid	5.54	11.01		
Allantoin	8.96			
Allothreonine	2.108(+1)	9.096(0)		
Alloxanic acid	6.64			
Allylacetic acid	4.68			
Allylamine	9.69(+1)			
5-Allylbarbituric acid	4.78(+1)			
5-Allyl-5-(-methylbutyl)barbituric acid	8.08			
2-Allylphenol	10.28			
1-Allylpiperidine	9.65(+1)			
2-Allylpropionic acid	4.72			
3 -Amidotetrazoline	3.95(+1)			
2-Aminoacetamide	7.95(+1)			
Aminoacetonitrile	5.34(+1)			
9-Aminoacridine (20°C)	9.95(+1)			
4-Aminoantipyrine	4.94(+1)			
2-Aminobenzenesulfonic acid	2.459(0)			
3-Aminobenzenesulfonic acid	3.738(0)			
4-Aminobenzenesulfonic acid	3.227(0)			
2-Aminobenzoic acid	2.09(+1)	4.79(0)		
3-Aminobenzoic acid	3.07(+1)	4.79(0)		
4-Aminobenzoic acid	2.41(+1)	4.85(0)		
2-Aminobenzoic acid, methyl ester	2.36(+1)			
3-Aminobenzoic acid, methyl ester	3.58(+1)			
4-Aminobenzoic acid, methyl ester	2.45(+1)			
3-Aminobenzonitrile	2.75(+1)			
4-Aminobenzonitrile	1.74(+1)			
4-Aminobenzophenone	2.15(+1)			
2-Aminobenzothiazole (20°C)	4.48(+1)			
2-Aminobenzoylhydrazide	1.85	3.47	12.80	
2-Aminobiphenyl	3.78(+1)			
3-Aminobiphenyl	4.18(+1)			
4-Aminobiphenyl	4.27(+1)			
4-Amino-3-bromomethylpyridine	7.47(+1)			
4-Amino-3-bromopyridine (20°C)	7.04(+1)			
2-Aminobutanoic acid	2.286(+1)	9.830(0)		
3-Aminobutanoic acid	—	10.14(0)		
4-Aminobutanoic acid	4.031(+1)	10.556(0)		
2-Aminobutanoic acid, methyl ester ($\mu = 0.1$)	7.640(+1)			

TABLE 2.60   *pK*, Values of Organic Materials in Water at 25°C (*Continued*)

Substance	pK_1	pK_2	pK_3	pK_4
4-Aminobutanoic acid, methyl ester ($\mu = 0.1$)	9.838(+1)			
D-(+)-2-Amino-1-butanol	9.52(+1)			
3-Amino-*N*-butyl-3-methyl-2-butanone oxime	9.09(+1)			
4-Aminobutylphosphonic acid	2.55	7.55	10.9	
2-Amino-*N*-carbamoylbutanoic acid	3.886(+1)			
4-Amino-*N*-carbamoylbutanoic acid	4.683(+1)			
2-Amino-*N*-carbamoyl-2-methyl-propanoic acid	4.463			
1-Amino-1-cycloheptanecarboxylic acid	2.59(+1)	10.46(0)		
1-Amino-1-cyclohexanecarboxylic acid	2.65(+1)	10.03(0)		
2-Amino-1-cyclohexanecarboxylic acid	3.56(+1)	10.21(0)		
1-Aminocyclopentane	10.65(+1)			
1-Aminocyclopropane	9.10(+1)			
10-Aminodecylphosphonic acid	—	8.0	11.25	
10-Aminodecylsulfonic acid	2.65(+1)			
1-Amino-2-di(aminomethyl)butane	3.58(+3)	8.59(+2)	9.66(+1)	
2-Amino-*N,N*-dihydroxyethyl-2-hydroxyl-1,3-propanediol	6.484(+1)			
2-Amino-*N,N* dimethylbenzoic acid	1.63(+1)	8.42(0)		
4-Amino-2,5-dimethylphenol	5.28(+1)	10.40(0)		
4-Amino-3,5-dimethylpyridine (20°C)	9.54(+1)			
12-Aminododecanoic acid	4.648(+1)			
2-Aminoethane-1-phosphoric acid	5.838	10.64		
1-Aminoethanesulfonic acid	−0.33	9.06		
2-Aminoethanesulfonic acid	1.5	9.061		
2-Aminoethanethiol (cysteamine) ($\mu = 0.01$)	8.23(+1)			
2-Aminoethanol (ethanolamine)	9.50(+1)			
2-[2-(2-Aminoethyl)amino-ethyl]pyridine	3.50	6.59	9.51	
2-Amino-2-ethyl-1-butanol	9.82(+1)			
3-(2-Aminoethyl)indole	—	10.2		
3-Amino-*N*-ethyl-3-methyl-2-buta-none oxime	9.23(+1)			
N-(2-Aminoethyl)morpholine	4.06(+2)	9.15(+1)		
p-(2-Aminoethyl)phenol	9.3	10.9		
2-Aminoethylphosphonic acid	2.45(+1)	7.0(0)	10.8(−1)	
N-(2-Aminoethyl)piperidine (30°C)	6.38	9.89		
2-(2-Aminoethyl)pyridine ($\mu = 0.5$)	4.24(+2)	9.78(+1)		
4-Amino-3-ethylpyridine (20°C)	9.51(+1)			
N-(2-Aminoethyl)pyrrolidine (30°C)	6.56(+2)	9.74(+1)		

(*Continued*)

TABLE 2.60   *pK*, Values of Organic Materials in Water at 25°C (*Continued*)

Substance	pK_1	pK_2	pK_3	pK_4
2-Aminofluorine	10.34(+1)			
2-Amino-D-β-glucose ($\mu = 0.05$)	2.20(+1)	9.08(0)		
2-Amino-*N*-glycylbutanoic acid	3.155(+1)	8.331(0)		
7-Aminoheptanoic acid	4.502			
2-Aminohexanoic acid	2.335(+1)	9.834(0)		
6-Aminohexanoic acid	4.373(+1)	10.804(0)		
C-Amino-*C*-hydrazinocarbonyl-methane	2.38(+2)	7.69(+1)		
2-Amino-3-hydroxybenzoic acid	2.5(+1)	5.192(0)	10.118(OH)	
L-2-Amino-3-hydroxybutanoic acid (threonine)	2.088(+1)	9.100(0)		
DL-2-Amino-4-hydroxybutanoic acid ($\mu = 0.1$)	2.265(+1)	9.257(0)		
DL-4-Amino-3-hydroxybutanoic acid ($\mu = 0.1$)	3.834(+1)	9.487(0)		
2-Amino-2′-hydroxydiethyl sulfide	9.27(+1)			
4-Amino-2-hydroxypyrimidine (cytosine)	4.58(+1)	12.15(0)		
3-Amino-*N*-isopropyl-3-methyl-2-butanone oxime	9.09(+1)			
4-Amino-3-isopropylpyridine (20°C)	9.54(+1)			
1-Aminoisoquinoline (20°C, $\mu = 0.01$)	7.62(+1)			
3-Aminoisoquinoline (20°C, $\mu = 0.005$)	5.05(+1)			
4-Aminoisoxazolidine-3-one	7.4(+1)			
Aminomalonic acid	3.32(+1)	9.83(0)		
DL-2-Amino-4-mercaptobutanoic acid	2.22(+1)	8.87(0)	10.86(SH)	
2-Amino-3-mercapto-3-Methylbutanoic acid	1.8(+1)	7.9(0)	10.5(SH)	
2-Amino-6-methoxybenzothiazole	4.50(+1)			
3-Amino-4-methylbenzenesulfonic acid	3.633			
4-Amino-3-methylbenzenesulfonic acid	3.125			
2-Amino-4-methylbenzothiazole	4.7(+1)			
1-Amino-3-methylbutane	10.64(+1)			
3-Amino-3-methyl-2-butanone oxime	9.09(+1)			
3-Amino-*N*-methyl-3-methyl-2-butanone oxime	9.23(+1)			
2-Amino-3-methylpentanoic acid	2.320(+1)	9.758(0)		
3-Aminomethyl-6-methylpyridine (30°C)	8.70(+1)			
Aminomethylphosphonic acid	2.35	5.9	10.8	
2-Amino-2-methyl-1,3-propanediol	8.801			
2-Amino-2-methyl-1-propanol	9.694(+1)			
2-Amino-2-methylpropanoic acid	2.357(+1)	10.205(0)		
(2-Aminomethyl(pyridine ($\mu = 0.5$)	2.31(+2)	8.79(+1)		

TABLE 2.60   *pK*, Values of Organic Materials in Water at 25°C (*Continued*)

Substance	pK_1	pK_2	pK_3	pK_4
2-Amino-3-methylpyridine	7.24(+1)			
4-Amino-3-methylpyridine	9.43(+1)			
2-Amino-4-methylpyridine	7.48(+1)			
2-Amino-5-methylpyridine	7.22(+1)			
2-Amino-6-methylpyridine	7.41(+1)			
2-Amino-4-methylpyrimidine (20°C)	4.11(+1)			
Aminomethylsulfonic acid	5.57(+1)			
N-Aminomorpholine	4.19(+1)			
4-Amino-1-naphthalenesulfonic acid	2.81			
1-Amino-2-naphthalenesulfonic acid	1.71			
1-Amino-3-naphthalenesulfonic acid	3.20			
1-Amino-5-naphthalenesulfonic acid	3.69			
1-Amino-6-naphthalenesulfonic acid	3.80			
1-Amino-7-naphthalenesulfonic acid	3.66			
1-Amino-8-naphthalenesulfonic acid	5.03			
2-Amino-1-naphthalenesulfonic acid	2.35			
2-Amino-4-naphthalenesulfonic acid	3.79			
2-Amino-6-naphthalenesulfonic acid	3.79	8.94		
2-Amino-8-naphthalenesulfonic acid	3.89			
3-Amino-1-naphthoic acid	2.61	4.39		
4-Amino-2-naphthoic acid	2.89	4.46		
8-Amino-2-naphthol	4.20(+1)			
DL-2-Aminopentanoic acid (DL-norvaline)	2.318(+1)	9.808		
3-Aminopentanoic acid	4.02(+1)	10.399(0)		
4-Aminopentanoic acid	3.97(+1)	10.46(0)		
5-Aminopentanoic acid	4.20(+1)	9.758(0)		
5-Aminopentanoic acid, ethyl ester	10.151			
2-Aminophenol	9.28	9.72		
3-Aminophenol	9.83	9.87		
4-Aminophenol	8.50	10.30		
4-Aminophenylacetic acid (20°C)	3.60	5.26		
2-Aminophenylarsonic acid	ca 2	3.77	8.66	
3-Aminophenylarsonic acid	ca 2	4.02	8.92	
4-Aminophenylarsonic acid	ca 2	4.02	8.62	
3-Aminophenylboric acid	4.46	8.81		
4-Aminophenylboric acid	3.71	9.17		
4-Aminophenyl (4-chlorophenyl) sulfone	1.38			
2-Aminophenylphosphonic acid	—	4.10	7.29	
3-Aminophenylphosphonic acid	—	—	7.16	

(*Continued*)

TABLE 2.60   *pK*, Values of Organic Materials in Water at 25°C (*Continued*)

Substance	pK_1	pK_2	pK_3	pK_4
4-Aminophenylphosphonic acid	—	—	7.53	
1-Amino-1,2,3-propanetricarbox-ylic acid ($\mu = 2.2$)	2.10(+1)	3.60(0)	4.60(−1)	9.82(−2)
3-Aminopropanoic acid	3.551(+1)	10.235(0)		
1-Amino-1-propanol	9.96(+1)			
DL-2-Amino-1-propanol	9.469(+1)			
3-Amino-1-propanol	9.96(+1)			
3-Aminopropene	9.691(+1)			
3-Amino-*N*-propyl-3-methyl-2-butanone oxime	9.09(+1)			
2-Aminopropylsulfonic acid	—	9.15		
2-Aminopyridine	6.71(+1)			
3-Aminopyridine	6.03(+1)			
4-Aminopyridine	9.114(+1)			
2-Aminopyridine-1-oxide	2.58(+1)			
3-Aminopyridine-1-oxide	1.47(+1)			
4-Aminopyridine-1-oxide	3.54(+1)			
8-Aminoquinaldine	4.86(+1)			
2-Aminoquinoline (20°C, $\mu = 0.01$)	7.34(+1)			
3-Aminoquinoline (20°C, $\mu = 0.01$)	4.95(+1)			
4-Aminoquinoline (20°C, $\mu = 0.01$)	9.17(+1)			
5-Aminoquinoline (20°C, $\mu = 0.01$)	5.46(+1)			
6-Aminoquinoline (20°C, $\mu = 0.01$)	5.63(+1)			
8-Aminoquinoline (20°C, $\mu = 0.01$)	3.99(+1)			
4-Aminosalicyclic acid	1.991(+1)	3.917(0)	13.74	
5-Aminosalicyclic acid	2.74(+1)	5.84(0)		
2-Amino-3-sulfopropanoic acid	1.89(+1)	8.70(0)		
4-Amino-2,3,5,6-tetramethylpyri-dine (20°C)	10.58(+1)			
5-Amino-1,2,3,4-tetrazole (20°C)	1.76	6.07		
2-Aminothiazole (20°C)	5.36(+1)			
1-Amino-3-thiobutane (30°C)	9.18(+1)			
5-Amino-3-thio-1-pentanol (30°C)	9.12(+1)			
2-Aminothiophenol	<2(+1)	7.90(0)		
2-Amino-4,4,4-trifluorobutanoic acid		8.171(0)		
3-Amino-4,4,4-trifluorobutanoic acid		5.831(0)		
3-Amino-2,4,6-trinitrotoluene		9.5(+1)		
Angiotensin II	10.37			
Anhydroplatynecine	9.40			
Aniline	4.60(+1)			
2-Anilinoethylsulfonic acid	3.80(+1)			
3-Anilinoethylsulfonic acid	4.85(+1)			
Anthracene-1-carboxylie acid	3.68			
Anthracene-2-carboxylic acid	4.18			
Anthracene-9-carboxylic acid	3.65			

TABLE 2.60 *pK*, Values of Organic Materials in Water at 25°C (*Continued*)

Substance	pK_1	pK_2	pK_3	pK_4
Anthraquinone-1-carboxylic acid (20°C)	3.37			
Anthraquinone-2-carboxylic acid (20°C)	3.42			
9,10-Anthraquinone monoxime	9.78			
9,10-Anthraquinone-1-sulfonic acid	0.27			
9,10-Anthraquinone-2-sulfonic acid	0.38			
Antipyrine	1.45(+1)			
Apomorphine (15°C)		8.92		
D-(−)-Arabinose	12.34			
L-(+)-Arginine	2.17	9.04(+1)	12.47(−1)	
Arsenazo III [pK$_5$ 10.5(−4); pK$_6$ 12.0(−5)]		1.2	2.7	7.9(−3)
Arsenoacetic acid		4.67	7.68	
Arsenoacrylic acid		4.23	8.60	
Arsenobutanoic acid		4.92	7.64	
2-Arsenocrotonic acid		4.61	8.75	
3-Arsenocrotonic acid		4.03	8.81	
Arsenopentanoic acid		4.89	7.75	
L-(+)-Ascorbic acid (vitamin C)	4.17	11.57		
L-(+)-Asparagine	2.01(0)	8.80(+1)		
L-Asparaginylglycine		4.53	9.07	
D-Aspartic acid	1.89(0)	3.65	9.60	
Aspartic diamide ($\mu = 0.2$)	7.00			
Aspartylaspartic acid		3.40	4.70	8.26
α-Aspartylhistidine (38°C, $\mu = 0.1$)		3.02	6.82	7.98
β-Aspartylhistidine (38°C, $\mu = 0.1$)		2.95	6.93	8.72
N-Aspartyl-p-tyrosine ($\mu - 0.01$)		3.57	8.92	10.23(QH)
Aspidospermine	7.65			
Atropine (17°C)	4.35(+1)			
1-Azacycloheptane	11.11(+1)			
1-Azacyclooctane	11.1(+1)			
Azetidine	11.29(+1)			
Aziridine	8.04(+1)			
Barbituric acid		8.372(0)		
m-Benzbetaine	3.217(+1)			
p-Benzbetaine	3.245(+1)			
Benzenearsonic acid (22°C)		8.48(−1)		
Benzene-1-arsonic acid-4-carboxylic acid		4.22 (COOH)	5.59	
Benzeneboronic acid	13.7			
Benzene-1-carboxylic acid-2-phosphoric acid		3.78	9.17	
Benzene-1-carboxylic acid-3-phosphoric acid		4.03	7.03	
Benzene-1-carboxylic acid-4-phosphoric acid	1.50	3.95	6.89	
Benzenediazine	11.08(+1)			
1,3-Benzenedicarboxylic acid (isophthalic acid)	3.62(0)	4.60(−1)		

(*Continued*)

TABLE 2.60 *pK*, Values of Organic Materials in Water at 25°C (*Continued*)

Substance	pK_1	pK_2	pK_3	pK_4
1,4-Benzenedicarboxylic acid (tere-phthalic acid)	3.54(0)	4.46(−1)		
1,3-Benzenedicarboxylic acid mononitrile	3.60(0)			
1,4-Benzenedicarboxylic acid mononitrile	3.55(0)			
Benzenehexacarboxylic acid (pK_5 6.32; pK_6 7.49)	0.68	2.21	3.52	5.09
Benzenepentacarboxylic acid (pK_5 6.46)	1.80	2.73	3.96	5.25
Benzenesulfinic acid	1.50			
Benzenesulfonic acid	2.554			
1,2,3,4-Benzenetetracarboxylic acid	2.05	3.25	4.73	6.21
1,2,3,5-Benzenetetracarboxylic acid	2.38	3.51	4.44	5.81
1,2,4,5-Benzenetetracarboxylic acid	1.92	2.87	4.49	5.63
1,2,3-Benzenetricarboxylic acid	2.88	4.75	7.13	
1,2,4-Benzenetricarboxylic acid	2.52	3.84	5.20	
1,3,5-Benzenetricarboxylic acid	2.12	4.10	5.18	
Benzil-α-dioxime	12.0			
Benzilic acid	3.09			
Benzimidazole	5.53(+1)	12.3(0)		
Benzohydroxamic acid (20°C)	8.89(0)			
Benzoic acid	4.204			
5,6-Benzoquinoline (20°C)	5.00(+1)			
7,8-Benzoquinoline (20°C)	4.15(+1)			
1,4-Benzoquinone monoxime	6.20			
Benzosulfonic acid	0.70			
1,2,3-Benzotriazole	8.38(+1)			
1-Benzoylacetone	8.23			
Benzoylamine	9.34(+1)			
2-Benzoylbenzoic acid	3.54			
Benzoylglutamic acid	3.49	4.99		
N-Benzoylglycine (hippuric acid)	3.65			
Benzoylhydrazine	3.03(+2)	12.45(+1)		
Benzoylpyruvic acid	6.40	12.10		
3-Benzoyl-1,1,1-trifluoroacetone	6.35			
Benzylamine	9.35(+1)			
Benzylamine-4-carboxylic acid	3.59	9.64		
2-Benzyl-2-phenylsuccinic acid (20°C)	3.69	6.47		
2-Benzylpyridine	5.13(+1)			
4-Benzylpyridine-1-oxide	−1.018(+1)			
1-Benzylpyrrolidine	9.51(+1)			
2-Benzylpyrrolidine	10.31(+1)			
Benzylsuccinic acid (20°C)	4.11	5.65		
3-(Benzylthio)propanoic acid	4.463			
Berberine (18°C)	11.73(+1)			
Betaine	1.832(+1)			
Biguanide	2.96(+2)	11.51(+1)		
2,2′-Biimidazolyl ($\mu = 0.3$)	5.01(+1)			
2-Biphenylcarboxylic acid	3.46			
(1,1′-Biphenyl)-4,4′-diamine	3.63(+2)	4.70(+1)		
Bis(2-aminoethyl) ether (30°C)	8.62(+2)	9.59(+1)		

TABLE 2.60   *pK*, Values of Organic Materials in Water at 25°C (*Continued*)

Substance	pK_1	pK_2	pK_3	pK_4
N,N'-Bis(2-aminoethyl)-ethylenediamine (20°C)	3.32(+4)	6.67(+3)	9.20(+2)	9.92(+1)
N,N-Bis(2-hydroxyethyl)-2-aminoethane sulfonic acid (BES) (20°C)	7.15			
N,N-Bis(2-hydroxyethyl)glycine (bicine) (20°C)	8.35			
Bis(2-hydroxyethyl)iminotris (hydroxymethyl)methane (bis-tris)	6.46(+1)			
1,3-Bis [tris(hydroxymethyl)methylamino]propane (20°C)	6.80(+1)			
Bromoacetic acid	2.902			
2-Bromoaniline	2.53(+1)			
3-Bromoaniline	3.53(+1)			
4-Bromoaniline	3.88(+1)			
2-Bromobenzoic acid	2.85			
3-Bromobenzoic acid	3.810			
4-Bromobenzoic acid	3.99			
2-Bromobutanoic acid (35°C)	2.939			
erythro-2-Bromo-3-chlorosuccinic acid (19°C, $\mu = 0.1$)	1.4	2.6		
threo-2-Bromo-chlorosuccinic acid (19°C, $\mu = 0.1$)	1.5	2.8		
trans-2-Bromocinnamic acid	4.41			
3-Bromo-4-(dimethylamino)pyridine (20°C)	6.52(+1)			
2-Bromo-4,6-dinitroaniline	−6.94(+1)			
3-Bromo-2-hydroxymethylbenzoic acid (20°C)	3.28			
6-Bromo-2-hydroxymethylbenzoic acid (20°C)	2.25			
7-Bromo-8-hydroxyquinoline-5-sulfonic acid	2.51	6.70		
3-Bromomandelic acid	3.13			
3-Bromo-4-methylaminopyridine (20°C)	7.49(+1)			
(2-Bromomethyl)butanoic acid	3.92			
Bromomethylphosphonic acid	1.14	6.52		
2-Bromo-6-nitrobenzoic acid	1.37			
2-Bromophenol	8.452			
3-Bromophenol	9.031			
4-Bromophenol	9.34			
2-(2′-Bromophenoxy)acetic acid	3.12			
2-(3′-Bromophenoxy)acetic acid	3.09			
2-(4′-Bromophenoxy)acetic acid	3.13			
2-Bromo-2-phenylacetic acid	2.21			
2-(Bromophenyl)acetic acid	4.054			
4-(Bromophenyl)acetic acid	4.188			
4-Bromophenylarsonic acid	3.25	8.19		
4-Bromophenylphosphinic acid (17°C)	2.1			
2-Bromophenylphosphonic acid	1.64	7.00		

(*Continued*)

TABLE 2.60 pK, Values of Organic Materials in Water at 25°C (*Continued*)

Substance	pK_1	pK_2	pK_3	pK_4
3-Bromophenylphosphonic acid	1.45	6.69		
4-Bromophenylphosphonic acid	1.60	6.83		
3-Bromophenylselenic acid	4.43			
4-Bromophenylselenic acid	4.50			
2-Bromopropanoic acid	2.971			
3-Bromopropanoic acid	3.992			
Bromopropynoic acid	1.855			
2-Bromopyridine	0.71(+1)			
3-Bromopyridine	2.85(+1)			
4-Bromopyridine	3.71(+1)			
3-Bromoquinoline	2.69(+1)			
Bromosuccinic acid	2.55	4.41		
2-Bromo-*p*-tolylphosphonic acid	1.81	7.15		
Brucine (15°C)	2.50(+2)	8.16(+1)		
2-Butanamine (*sec*-butylamine)	10.56(+1)			
1,2-Butanediamine	6.399(+2)	9.388(+1)		
1,4-Butanediamine	9.35(+2)	10.82(+1)		
2,3-Butanediamine	6.91(+2)	10.00(+1)		
1,2,3,4-Butanetetracarboxylic acid	3.43	4.58	5.85	7.16
cis-2-Butenoic acid (isocrotonic acid)	4.44			
trans-2-Butenoic acid (*trans*-crotonic acid) (35°C)	4.676			
3-Butenoic acid (vinylacetic acid)	4.68			
3-Butoxybenzoic acid (20°C)	4.25			
Butylamine	10.64(+1)			
tert-Butylamine	10.685(+1)			
4-*tert*-Butylaniline	3.78(+1)			
N-*tert*-Butylaniline	7.10(+1)			
Butylarsonic acid (18°C)	4.23	8.91		
2-*tert*-Butylbenzoic acid	3.57			
3-*tert*-Butylbenzoic acid	4.199			
4-*tert*-Butylbenzoic acid	4.389			
N-Butylethylenediamine	7.53(+2)	10.30(+1)		
N-Butylglycine	2.35(+1)	10.25(0)		
tert-Butylhydroperoxide	12.80			
1-(*tert*-Butyl)-2-hydroxybenzene	10.62			
1-(*tert*-Butyl)-3-hydroxybenzene	10.119			
1-(*tert*-Butyl)-4-hydroxybenzene	10.23			
Butylmethylamine	10.90(+1)			
2-Butyl-1-methyl-2-pyrroline	11.84(+1)			
4-*tert*-Butylphenylactic acid	4.417			
Butylphosphinic acid	3.41			
tert-Butylphosphinic acid	4.24			
tert-Butylphosphonic acid	2.79	8.88		
1-Butylpiperidine ($\mu = 0.02$)	10.43(+1)			
2-*tert*-Butylpyridine	5.76(+1)			
3-*tert*-Butylpyridine	5.82(+1)			
4-*tert*-Butylpyridine	5.99(+1)			
2-*tert*-Butylthiazole ($\mu = 0.1$)	3.00(+1)			
4-*tert*-Butylthiazole ($\mu = 0.1$)	3.04(+1)			
2-Butyn-1,4-dioic acid	1.75	4.40		
2-Butynoic acid (tetrolic acid)	2.620			

TABLE 2.60 *pK*, Values of Organic Materials in Water at 25°C (*Continued*)

Substance	pK_1	pK_2	pK_3	pK_4
Butyric acid	4.817			
4-Butyrobetaine (20°C)	3.94(+1)			
Caffeine (+0°C)	10.4			
Calcein ($pK_5 > 12$)	<4	5.4	9.0	10.5
Calmagite	8.14	12.35		
D-Camphoric acid	4.57	5.10		
Canaline	2.40	3.70	9.20	
Canavanine	2.50(+2)	6.60(+1)	9.25(0)	
N-Carbamoylacetic acid	3.64			
N-Carbamoyl-α-D-alanine	3.89(+1)			
N-Carbamoyl-β-alanine	4.99(+1)			
DL-*N*-Carbamoylalanine	3.892(+1)			
N-Carbamoylglycine	3.876			
2-Carbamoylpyridine (20°C)	2.10(+1)			
3-Carbamoylpyridine	3.328(+1)			
4-Carbamoylpyridine (20°C)	3.61(+1)			
β-Carboxymethylaminopropanoic acid	3.61(+1)	9.46(0)		
Chloroacetic acid	2.867			
N-(2'-Chloroacetyl)glycine	3.38(0)			
cis-3-Chloroacrylic acid (18°C, $\mu = 0.1$)	3.32			
trans-3-chloroacrylic acid (18°C, $\mu = 0.1$)	3.65			
2-Chloroaniline	2.64(+1)			
3-Chloroaniline	3.52(+1)			
4-Chloroaniline	3.99(+1)			
2-Chlorobenzoic acid	2.877			
3-Chlorobenzoic acid	3.83			
4-Chlorobenzoic acid	3.986			
2-Chlorobutanoic acid	2.86			
3-Chlorobutanoic acid	4.05			
4-Chlorobutanoic acid	4.50			
2-Chloro-3-butenoic acid	2.54			
3-Chlorobutylarsonic acid (18°C)	3.95	8.85		
trans-2'-Chlorocinnamic acid	4.234			
trans-3'-Chlorocinnamic acid	4.294			
trans-4'-Chlorocinnamic acid	4.413			
2-Chlorocrotonic acid	3.14			
3-Chlorocrotonic acid	3.84			
Chlorodifluoroacetic acid	0.46			
1-Chloro-1,2-dihydroxybenzene	8.522			
1-Chloro-2,6-dimethyl-4-hydroxy-benzene	9.549			
4-Chloro-2,6-dinitrophenol	2.97			
2-Chloroethylarsonic acid	3.68	8.37		
3-Chlorohexyl-1-arsonic acid (18°C)	3.51	8.31		
2-Chloro-3-hydroxybutanoic acid	2.59			
3-Chloro-2-(hydroxy-methyl)benzoic acid (20°C)	3.27			

(*Continued*)

TABLE 2.60 *pK*, Values of Organic Materials in Water at 25°C (*Continued*)

Substance	pK_1	pK_2	pK_3	pK_4
6-Chloro-2-(hydroxy-methyl)benzoic acid (20°C)	2.26			
7-Chloro-8-hydroxyquinoline-5-sulfonic acid	2.92	6.80		
2-Chloroisocrotonic acid	2.80			
3-Chloroisocrotonic acid	4.02			
3-Chlorolactic acid	3.12			
3-Chloromandelic acid	3.237			
3-Chloro-4-methoxyphenyl-phosphonic acid	2.25	6.7		
3-Chloro-4-methylaniline	4.05(+1)			
4-Chloro-N-methylaniline	3.9(+1)			
4-Chloro-3-methylphenol	9.549			
Chloromethylphosphonic acid	1.40	6.30		
2-Chloro-2-methylpropanoic acid	2.975			
2-Chloro-6-nitroaniline	−2.41(+1)			
4-Chloro-2-nitroaniline	−1.10(+1)			
2-Chloro-3-nitrobenzoic acid	2.02			
2-Chloro-4-nitrobenzoic acid	1.96			
2-Chloro-5-nitrobenzoic acid	2.17			
2-Chloro-6-nitrobenzoic acid	1.342			
4-Chloro-2-nitrophenol	6.48			
2-Chlorophenol	8.55			
3-Chlorophenol	9.10			
4-Chlorophenol	9.43			
(4-Chloro-3-nitrophenoxy)acetic acid	2.959			
2-Chloro-4-nitrophenylphosphonic acid	1.12	6.14		
3-Chloropentyl-1-arsonic acid (18°C)	3.71	8.77		
2-Chlorophenoxyacetic acid	3.05			
3-Chlorophenoxyacetic acid	3.07			
4-Chlorophenoxyacetic acid	3.10			
4-Chlorophenoxy-2-methylacetic acid	3.26			
2-Chlorophenylacetic acid	4.066			
3-Chlorophenylacetic acid	4.140			
4-Chlorophenylacetic acid	4.190			
2-Chlorophenylalanine	2.23(+1)	8.94(0)		
3-Chlorophenylalanine	2.17(+1)	8.91(0)		
DL-4-Chlorophenylalanine	2.08(+1)	8.96(0)		
4-Chlorophenylarsonic acid	3.33	8.25		
2-Chlorophenylphosphonic acid	1.63	6.98		
3-Chlorophenylphosphonic acid	1.55	6.65		
4-Chlorophenylphosphonic acid	1.66	6.75		
3-(2′-Chlorophenyl)propanoic acid	4.577			
3-(3′-Chlorophenyl)propanoic acid	4.585			
3-(4′-Chlorophenyl)propanoic acid	4.607			
3-Chlorophenylselenic acid	4.47			
4-Chlorophenylselenic acid	4.48			
4-Chloro-1,2-phthalic acid	1.60			

TABLE 2.60   *pK*, Values of Organic Materials in Water at 25°C (*Continued*)

Substance	pK_1	pK_2	pK_3	pK_4
2-Chloropropanoic acid	2.84			
3-Chloropropanoic acid	3.992			
2-Chloropropylarsonic acid (18°C)	3.76	8.39		
3-Chloropropylarsonic acid (18°C)	3.63	8.53		
Chloropropynoic acid	1.854			
2-Chloropyridine	0.49(+1)			
3-Chloropyridine	2.84(+1)			
4-Chloropyridine	3.83(+1)			
7-Chlorotetracycline	3.30(+1)	7.44	9.27	
4-Chloro-2-(2′-thiazolylazo)phenol	7.09			
4-Chlorothiophenol	5.9			
N-Chloro-*p*-toluenesulfonamide	4.54(+1)			
3-Chloro-*o*-toluidine	2.49(+1)			
4-Chloro-*o*-toluidine	3.385(+1)			
5-Chloro-*o*-toluidine	3.85(+1)			
6-Chloro-*o*-toludine	3.62(+1)			
Chrome Azurol S	2.45	4.86	11.47	
Chrome Dark Blue	7.56	9.3	12.4	
Cinchonine	5.85(+2)	9.92(+1)		
cis-Cinnamic acid	3.879			
trans-Cinnamic acid	4.438			
Citraconic acid	2.29(0)	6.15(−1)		
Citric acid	3.128	4.761	6.396	
L-(+)-Citrulline	2.43(+1)	9.41(0)		
Cocaine	8.41(+1)			
Codeine	7.95(+1)			
Colchicine	1.65(+1)			
Coniine ($\mu = 0.5$)	11.24(+1)			
Creatine (40°C)	3.28(+1)			
Creatinine	3.57(+1)			
o-Cresol	10.26			
m-Cresol	10.00			
p-Cresol	10.26			
Cumene hydroperoxide	12.60			
Cupreine	7.63(+1)			
Cyanamide	10.27			
Cyanoacetic acid	2.460			
Cyanoacetohydrazide	2.34(+2)	11.17(+1)		
2-Cyanobenzoic acid	3.14			
3-Cyanobenzoic acid	3.60			
4-Cyanobenzoic acid	3.55			
4-Cyanobutanoic acid	4.44			
trans-1-Cyanocyclohexane-2-carboxylic acid	3.865			
4-Cyano-2,6-dimethylphenol	8.27			
4-Cyano-3,5-dimethylphenol	8.21			
2-Cyanoethylamine	7.7(+1)			
N-(2-Cyano)ethylnorcodeine	5.68(+1)			
Cyanomethylamine	5.34(+1)			
2-Cyano-2-methyl-2-phenylacetic acid	2.290			
1-Cyanomethylpiperidine	4.55(+1)			
2-Cyano-2-methylpropanoic acid	2.422			

(*Continued*)

TABLE 2.60 *pK*, Values of Organic Materials in Water at 25°C (*Continued*)

Substance	pK_1	pK_2	pK_3	pK_4
3-Cyanophenol	8.61			
o-Cyanophenoxyacetic acid	2.98			
m-Cyanophenoxyacetic acid	3.03			
p-Cyanophenoxyacetic acid	2.93			
2-Cyanopropanoic acid	2.37			
3-Cyanopropanoic acid	3.99			
2-Cyanopyridine	−0.26(+1)			
3-Cyanopyridine	1.45(+1)			
4-Cyanopyridine	1.90(+1)			
Cyanuric acid	6.78			
Cyclobutanecarboxylic acid	4.785			
1,1-Cyclobutanedicarboxylic acid	3.13	5.88		
cis-1,2-Cyclobutanedicarboxylic acid	3.90	5.89		
trans-1,2-Cyclobutanedicarboxylic acid	3.79	5.61		
cis-1,3-Cyclobutanedicarboxylic acid	4.04	5.31		
trans-1,3-Cyclobutanedicarboxylic acid	3.81	5.28		
Cyclohexanecarboxylic acid	4.90			
1,1-Cyclohexanediacetic acid	3.49	6.96		
cis-1,2-Cyclohexanediacetic acid (20°C)	4.42	5.45		
trans-1,2-Cyclohexanediacetic acid (20°C)	4.38	5.42		
cis-1,2-Cyclohexanediamine	6.43(+2)	9.93(+1)		
trans-1,2-Cyclohexanediamine	6.34(+2)	9.74(+1)		
1,1-Cyclohexanedicarboxylic acid	3.45	4.11		
cis-1,2-Cyclohexanedicarboxylic acid (20°C)	4.34	6.76		
trans-1,2-Cyclohexanedicarboxylic acid (20°C)	4.18	5.93		
cis-1,3-Cyclohexanedicarboxylic acid (16°C)	4.10	5.46		
trans-1,3-Cyclohexanedicarboxylic acid (19°C)	4.31	5.73		
trans-1,4-Cyclohexanedicarboxylic acid (16°C)	4.18	5.42		
1,3-Cyclohexanedione	5.26			
cis,cis-1,3,5-Cyclohexanetriamine	6.9(+3)	8.7(+2)	10.4(+1)	
Cyclohexanonimine	9.15			
cis-4-Cyclohexene-1,2-dicarboxylic acid (20°C)	3.89	6.79		
trans-4-Cyclohexene-1,2-dicarboxylic acid (20°C)	3.95	5.81		
Cyclohexylacetic acid	4.51			
Cyclohexylamine	10.64(+1)			
2-(Cyclohexylamino)ethanesulfonic acid (CHES) (20°C)	9.55			
3-Cyclohexylamino-1-propanesulfonic acid (CAPS) (20°C)	10.40			
4-Cyclohexylbutanoic acid	4.95			

TABLE 2.60 *pK*, Values of Organic Materials in Water at 25°C (*Continued*)

Substance	pK_1	pK_2	pK_3	pK_4
Cyclohexylcyanoacetic acid	2.367			
1,2-Cyclohexylenedinitriloacetic acid ($\mu = 0.1$)	2.4	3.5	6.16	12.35
3-Cyclohexylpropanoic acid	4.91			
2-Cyclohexylpyrrolidine	10.76(+1)			
2-Cyclohexyl-2-pyrroline	7.91(+1)			
Cyclohexylthioacetic acid	3.488			
Cyclopentanecarboxylic acid	4.905			
cis-Cyclopentane-1-carboxylic acid-2-acetic acid	4.40	5.79		
trans-Cyclopentane-1-carboxylic acid-2-acetic acid	4.39	5.67		
Cyclopentane-1,2-diamine-N',N',N'-tetraacetic acid ($\mu = 0.1$)	—	—	—	10.20
Cyclopentane-1,1-dicarboxylic acid	3.23	4.08		
cis-Cyclopentane-1,2-dicarboxylic acid	4.43	6.67		
trans-Cyclopentane-1,2-dicarboxylic acid	3.96	5.85		
cis-Cyclopentane-1,3-dicarboxylic acid	4.26	5.51		
trans-Cyclopentane-1,3-dicarboxylic acid	4.32	5.42		
Cyclopentylamine	10.65(+1)			
1,1-Cyclopentyldiacetic acid	3.80	6.77		
cis-Cyclopentyl-1,2-diacetic acid	4.42	5.42		
trans-Cyclopentyl-1,2-diacetic acid	4.43	5.43		
Cyclopropanecarboxylic acid	4.827			
Cyclopropane-1,1-dicarboxylic acid	1.82	5.43		
cis-Cyclopropane-1,2-dicarboxylic acid	3.33	6.47		
trans-Cyclopropane-1,2-dicarboxylic acid	3.65	5.13		
Cyclopropylamine	9.10(+1)			
5-Cyclopropyl-1,2,3,4-tetrazole	4.90(+1)			
L-Cysteic acid (3-sulfo-L-alanine)	1.89(+1)	8.7(0)		
L-(+)-Cysteine	1.96	8.18	10.29(SH)	
L-(+)-Cysteine, ethyl ester	6.69 (NH$_3^+$)	9.17(SH)		
L-(+)-Cysteine, methyl ester	6.56 (NH$_3^+$)	8.99(SH)		
L-Cysteinyl-L-asparagine	2.97	7.09	8.47	
L-Cystine (35°C)	1.6(+2)	2.1(+1)	8.02(0)	8.71(−1)
Cystinylglycylglycine (35°C)	3.12	3.21	6.01	6.87
Cytidine	4.08(+1)	12.24(0)		
Cytidine-2′-phosphoric acid	0.8(+1)	4.36(0)	6.17(−1)	
Cytidine-3′-phosphoric acid	0.80(+1)	4.31(0)	6.04(−1)	13.2(sugar)
Cytidine-5′-phosphoric acid	—	4.39(0)	6.62(−1)	
Cytosine	4.58(+1)	12.15(0)		
Decanedioic acid (sebacic acid)	4.59	5.59		
Dehydroascorbic acid (20°C)	3.21	7.92	10.3	
2′-Deoxy adenosine ($\mu = 0.1$)	3.8(+1)			

(Continued)

TABLE 2.60   *pK*, Values of Organic Materials in Water at 25°C (*Continued*)

Substance	pK_1	pK_2	pK_3	pK_4
Deoxycholic acid	6.58			
2-Deoxyglucose	12.52			
2-Deoxyguanosine ($\mu = 0.1$)	2.5(+1)			
5-Desoxypyridoxal ($\mu = 0$)	4.17(+1)	8.14(OH)		
1,1-Diacetic acid semicarbazide (30°C, $\mu = 0.1$)	2.96	4.04		
Diacetylacetone	7.42			
Diallylamine ($\mu = 0.02$)	9.29(+1)			
5,5-Diallybarbituric acid	7.78(0)			
1,3-Diamino-2-aminomethylpro-pane	6.44(+3)	8.56(+2)	10.38(+1)	
3,5-Diaminobenzoic acid	5.30			
1,3-Diamino-N,N'-bis-(2-amino-ethyl)propane ($\mu = 0.5$)	6.01(+4)	7.26(+3)	9.49(+2)	10.23(+1)
2,4-Diaminobutanoic acid (20°C)	1.85(+2)	8.24(+1)	10.40(0)	
2,2′-Diaminodiethyl sulfide (30°C)	8.84(+2)	9.64(+1)		
1,8-Diamino-3,6-dithiooctane (30°C)	8.43(+2)	9.31(+1)		
2,7-Diaminooctanedioic acid (20°C, $\mu = 0.1$)	1.84(+2)	2.64(+1)	9.23(0)	9.89(−1)
1,8-Diamino-3,6-octanedione (30°C)	8.60(+2)	9.57(+1)		
1,8-Diamino-3-oxa-6-thiooctane	8.54(+2)	9.46(+1)		
2,3-Diaminopropanoic acid ($\mu = 0.1$)	1.33(+2)	6.674(+1)	9.623(0)	
2,3-Diaminopropanoic acid, methyl ester ($\mu = 0.1$)	4.412(+1)	8.250(0)		
1,3-Diamino-2-propanol (20°C)	7.93(+2)	9.69(+1)		
2,5-Diaminopyridine (20°C)	2.13(+2)	6.48(+1)		
1,4-Diazabicyclo[2.2.2]octane	2.90(+2)	8.60(+1)		
Dibenzylamine	8.52(+1)			
Dibenzylsuccinic acid (20°C)	3.96	6.66		
Dibromoacetic acid	1.39			
3,5-Dibromoaniline	2.35(+1)			
3,5-Dibromophenol	8.056			
2,2-Dibromopropanoic acid	1.48			
2,3-Dibromopropanoic acid	2.33			
rac-2,3-Dibromosuccinic acid (20°C)	1.43	2.24		
meso-2,3-Dibromosuccinic acid (20°C)	1.51	2.71		
3,5-Dibromo-*p*-L-tyrosine	2.17(+1)	6.45(0)	7.60(−1)	
Dibutylamine	11.25(+1)			
Di-*sec*-butylamine	10.91(+1)			
2,6-Di-*tert*-butylpyridine	3.58(+1)			
rac-2,3-Di-*tert*-butylsuccinic acid ($\mu = 0.1$)	3.58	10.2		
1,12-Dicarboxydodecaborane	9.07	10.23		
Dichloroacetic acid	1.26			
Dichloroacetylacetic acid	2.11			
3,5-Dichloroaniline	2.37(+1)			
1,3-Dichloro-2,5-dihydroxybenzene ($\mu = 0.65$)	7.30	9.99		

TABLE 2.60   *pK*, Values of Organic Materials in Water at 25°C (*Continued*)

Substance	pK_1	pK_2	pK_3	pK_4
2,5-Dichloro-3,6-dihydroxy-*p*-benzoquinone	1.09	2.42		
Dichloromethylphosphonic acid	1.14	5.61		
2,4-Dichloro-6-nitroaniline	−3.00(+1)			
2,5-Dichloro-4-nitroaniline	−1.74(+1)			
2,6-Dichloro-4-nitroaniline	−3.31(+1)			
2,3-Dichlorophenol	7.44			
2,4-Dichlorophenol	7.85			
2,6-Dichlorophenol	6.78			
3,4-Dichlorophenol	8.630			
3,5-Dichlorophenol	8.179			
2,4-Dichlorophenoxyacetic acid (2,4-D)	2.64			
4,6-Dichlorophenoxy-2-methylacetic acid	3.13			
3,6-Dichlorophthalic acid	1.46			
2,2-Dichloropropanoic acid	2.06			
2,3-Dichloropropanoic acid	2.85			
rac-2,3-Dichlorosuccinic (20°C)	1.43	2.81		
meso-2,3-Dichlorosuccinic acid	1.49	2.97		
3,5-Dichloro-*p*-tyrosine	2.12	6.47	7.62	
2-Dicyanoethylamine	5.14(+1)			
2,2-Dicyanopropanoic acid	−2.8			
Dicyclohexylamine	11.25(+1)			
Dicyclopentylamine	10.93(+1)			
Didodecylamine	10.99(+1)			
Diethanolamine	8.88(+1)			
Di(ethoxyethyl)amine	8.47(+1)			
3,5-Diethoxyphenol	9.370			
3-(Diethoxyphosphinyl)benzoic acid	3.65			
4-(Diethoxyphosphinyl)benzoic acid	3.60			
3-(Diethoxyphosphinyl)phenol	8.66			
4-(Diethoxyphosphinyl)phenol	8.28			
Diethylamine	10.8(+1)			
2-(Diethylamino)ethyl-4-aminobenzoate	8.85(+1)			
α-(Diethylamino)toluene	9.44(+1)			
N,*N*-Diethylaniline	6.56(+1)			
5,5-Diethylbarbituric acid (veronal)	8.020(0)			
N,*N*-Diethylbenzylamine	9.48(+1)			
Diethylbiguanide (30°C)	2.53(+1)	11.68(0)		
Diethylenetriamine	4.42(+3)	9.21(+2)	10.02(+1)	
Diethylenetriaminepentaacetic acid (pK_5, 10.58)	1.80(0)	2.55(−1)	4.33(−2)	8.60(−3)
N,*N*-Diethylethylenediamine	7.70(+2)	10.46(+1)		
2,2-Diethylglutaric acid	3.62	7.12		
N,*N*-Diethylglycine	2.04(+1)	10.47(0)		
Diethylglycolic acid (18°C)	3.804			
Diethylmalonic acid	2.151	7.417		
Diethylmethylamine	10.43(+1)			
rac-2,3-Diethylsuccinic acid	3.63	6.46		

(Continued)

TABLE 2.60   *pK*, Values of Organic Materials in Water at 25°C (*Continued*)

Substance	pK_1	pK_2	pK_3	pK_4
meso-2,3-Diethylsuccinic acid	3.54	6.59		
N,N-Diethyl-*o*-toluidine	7.18(+1)			
Difluoroacetic acid	1.33			
3,3-Difluoroacrylic acid	3.17			
Diglycolic acid	2.96			
Diguanidine	12.8			
Dihexylamine	11.0(+1)			
Dihydroarecaidine	9.70			
Dihydroarecaidine, methyl ester	8.39			
Dihydrocodeine	8.75(+1)			
Dihydroergonovine	7.38(+1)			
α-Dihydrolysergic acid	3.57	8.45		
γ-Dihydrolysergic acid	3.60	8.71		
α-Dihydrolysergol	8.30			
β-Dihydrolysergol	8.23			
Dihydromorphine	9.35			
3,4-Dihydroxy alanine	2.32(+1)	8.68(0)	9.87(−1)	
1,2-Dihydroxyanthraquinone-3-sulfonic acid (alizarin-3-sulfonic acid)	—	5.54(−1)	11.01(−2)	
3,4-Dihydroxybenzaldehyde	7.55			
1,2-Dihydroxybenzene (pyrocatechol) ($\mu = 0.1$)	9.356(0)	12.98(−1)		
1,3-Dihydroxybenzene (resorcinol)	9.44(0)	12.32(−1)		
1,4-Dihydroxybenzene (hydroquinone)	9.91(0)	12.04(−1)		
4,5-Dihydroxybenzene-1,3-disulfonic acid	—	—	7.66(−2)	12.6(−3)
2,3-Dihydroxybenzoic acid (30°C)	2.98	10.14		
2,4-Dihydroxybenzoic acid (β-resorcyclic acid)	3.29	8.98		
2,5-Dihydroxybenzoic acid	2.97	10.50		
2,6-Dihydroxybenzoic acid	1.30			
3,4-Dihydroxybenzoic acid	4.48	8.67	11.74	
3,5-Dihydroxybenzoic acid	4.04			
2,5-Dihydroxy-*p*-benzoquinone	2.71	5.18		
3,4-Dihydroxy-3-cyclobutene-1,2-dione	0.541	3.480		
2,3-Dihydroxy-2-cyclopenten-1-one (20°C)	4.72			
1,4-Dihydroxy-2,6-dinitrobenzene	4.42	9.14		
Di(2,2′-hydroxy ethyl)amine	8.8(+1)			
N,N-Di(2-hydroxyethyl)glycine	8.333			
Dihydroxymaleic acid	1.10			
Dihydroxymalic acid	1.92			
1,3-Dihydroxy-2-methylbenzene ($\mu = 0.65$)	10.05	11.64		
2,2-Di(hydroxymethyl)-3-hydroxypropanoic acid	4.460			
2,4-Dihydroxy-5-methylpyrimidine	9.90			
2,4-Dihydroxy-6-methylpyrimidine	9.52			
1,4-Dihydroxynaphthalene (26°C, $\mu = 0.65$)	9.37	10.93		
1,2-Dihydroxy-3-nitrobenzene	6.68			

TABLE 2.60 *pK*, Values of Organic Materials in Water at 25°C (*Continued*)

Substance	pK_1	pK_2	pK_3	pK_4
1,2-Dihydroxy-4-nitrobenzene ($\mu = 0.1$)	6.701			
2,4-Dihydroxy-1-phenylazobenzene ($\mu = 0.1$)	11.98			
2,4-Dihydroxyoxazolidine	6.11(+1)			
2,4-Dihydroxypteridine	<1.3	7.92		
2,6-Dihydroxypurine	7.53(0)	11.84(−1)		
2,4-Dihydroxypyridine (20°C)	1.37(+1)	6.45(0)	13(−1)	
Dihydroxytartaric acid	1.95	4.00		
1,4-Dihydroxy-2,3,5,6-tetramethyl-benzene ($\mu = 0.65$)	11.25	12.70		
3,5-Diiodoaniline	2.37(+1)			
2,5-Diiodohistamine	2.31(+2)	8.20(+1)	10.11(0)	
2,5-Diiodohistidine ($\mu = 0.1$)	2.72	8.18	9.76	
3,5-Diiodophenol	8.103			
3,5-Diiodotyrosine	2.117(+1)	6.479(0)	7.821(−1)	
Diisopropylmalonic acid	2.124	8.848		
Dilactic acid	2.955			
threo-1,4-Dimercapto-2,3-butanediol	8.9			
meso-2,3-Dimercaptosuccinic acid	2.71	3.48	8.89(SH)	10.79(SH)
3,5-Dimethoxy aniline	3.86(+1)			
2,6-Dimethoxybenzoic acid	3.44			
1,10-Dimethoxy-3,8-dimethyl-4,7-phenanthroline	7.21			
Di(2-methoxyethyl)amine	9.51(+1)			
3,5-Dimethoxyphenol	9.345			
(3,4-Dimethoxy)phenylacetic acid	4.333			
Dimethylamine	10.77(+1)			
4-Dimethylaminobenzaldehyde	1.647(+1)			
N,N-Dimethylaminocyclohexane	10.72(+1)			
4-Dimethylamino-2,3-dimethyl-1-phenyl-3-pyrazolin-5-one	4.18(+1)			
4-Dimethylamino-3,5-dimethylpyridine (20°C)	8.15(+1)			
2-(Dimethylamino)ethanol	9.26(+1)			
2-[2-(Dimethyl-amino)ethyl]pyridine	3.46(+2)	8.75(+1)		
3-(Dimethylaminoethyl)pyridine	4.30(+2)	8.86(+1)		
4-(Dimethylaminoethyl)pyridine	4.66(+2)	8.70(+1)		
4-(Dimethylamino)-3-ethylpyridine (20°C)	8.66(+1)			
4-(Dimethylamino)-3-isopropylpyridine (20°C)	8.27(+1)			
2-(Dimethylaminomethyl)pyridine	2.58(+2)	8.12(+1)		
3-(Dimethylaminomethyl)pyridine	3.17(+2)	8.00(+1)		
4-(Dimethylaminomethyl)pyridine	3.39(+2)	7.66(+1)		
4-(Dimethylamino)-3-methylpyridine (20°C)	8.68(+1)			
4-(Dimethylamino-phenyl)phosphonic acid	2.0(+1)	4.2	7.35	
3-(Dimethylamino)propanoic acid	9.85(+1)			
4-(Dimethylamino)pyridine (20°C)	6.09(+1)			

(*Continued*)

TABLE 2.60   *pK*, Values of Organic Materials in Water at 25°C (*Continued*)

Substance	pK_1	pK_2	pK_3	pK_4
N,N-Dimethylaniline	5.15(+1)			
2,3-Dimethylaniline	4.70(+1)			
2,4-Dimethylaniline	4.89(+1)			
2,5-Dimethylaniline	4.53(+1)			
2,6-Dimethylaniline	3.95(+1)			
3,4-Dimethylaniline	5.17(+1)			
3,5-Dimethylaniline	4.765(+1)			
N,N-Dimethylaniline-4-phosphonic acid (17°C)	2.0(+1)	4.2	7.39	
Dimethylarsinic acid (cacodylic acid)	1.67	6.273		
1,3-Dimethylbarbituric acid	4.68(+1)			
2,3-Dimethylbenzoic acid	3.771			
2,4-Dimethylbenzoic acid	4.217			
2,5-Dimethylbenzoic acid	3.990			
2,6-Dimethylbenzoic acid	3.362			
3,4-Dimethylbenzoic acid	4.41			
3,5-Dimethylbenzoic acid	4.302			
N,N-Dimethylbenzylamine	9.02(+1)			
Dimethylbiguanide	2.77(+1)	11.52		
2,2-Dimethylbutanoic acid (18°C)	5.03			
Dimethylchlorotetracycline ($\mu = 0.01$)	3.30(+1)			
2,6-Dimethyl-4-cyanophenol	8.27			
3,5-Dimethyl-4-cyanophenol	8.21			
5,5-Dimethyl-1,3-cyclohexanedione	5.15			
cis-3,3-Dimethyl-1,2-cyclopropanedicarboxylic acid	2.34	8.31		
trans-3,3-Dimethyl-1,2-cyclopropanedicarboxylic acid	3.92	5.32		
3,5-Dimethyl-4-(dimethylamino)-pyridine (20°C)	8.12(+1)			
2,2-Dimethyl-1,3-dioxane-4,6-dione	5.1			
1,1-Dimethylethanethiol ($\mu = 0.1$)	11.22			
N,N-Dimethylethylenediamine-*N,N*-diacetic acid	6.63	9.53		
N,N'-Dimethylethylenediamine-*N,N'*-diacetic acid	7.40	10.16		
N,N-Dimethylethylenediamine-*N,N'*-diacetic acid	5.99	9.97		
N,N-Dimethylglycine	2.146(+1)	9.940(0)		
Dimethylglycolic acid (18°C)	4.04			
N,N-Dimethylglycylglycine	3.11(+1)	8.09(0)		
Dimethylglyoxime	10.60			
5,5-Dimethyl-2,4-hexanedione	10.01			
5,5-Dimethylhydantoin	9.19			
2,4-Dimethyl-8-hydroxyquinoline	6.20(+1)	10.60(0)		
3,4-Dimethyl-8-hydroxyquinoline	5.80(+1)	10.05(0)		
2,4-Dimethyl-8-hydroxyquinoline-7-sulfonic acid	3.20 (NH^+)	10.14(OH)		
Dimethylhydroxytetracycline	7.5	9.4		
2,4-Dimethylimidazole	8.38(+1)			

TABLE 2.60 *pK*, Values of Organic Materials in Water at 25°C (*Continued*)

Substance	pK_1	pK_2	pK_3	pK_4
Dimethylmalic acid	3.17	6.06		
2,2-Dimethylmalonic acid	3.17	6.06		
3,5-Dimethyl-4-(methylamino) pyridine (20°C)	9.96(+1)			
2,3-Dimethylnaphthalene-1-carboxylic acid	3.33			
2,6-Dimethyl-4-nitrophenol	7.190			
3,5-Dimethyl-4-nitrophenol	8.245			
α,α-Dimethyloxaloacetic acid	1.77	4.62		
3,3-Dimethylpentanedioic acid	3.70	6.34		
2,2-Dimethylpentanoic acid	4.969			
4,4-Dimethylpentanoic acid (18°C)	4.79			
2,3-Dimethylphenol	10.50			
2,4-Dimethylphenol	10.58			
2,5-Dimethylphenol	10.22			
2,6-Dimethylphenol	10.59			
3,4-Dimethylphenol	10.32			
3,5-Dimethylphenol	10.15			
2,6-Dimethylphenoxyacetic acid	3.356			
Dimethylphenylsilylacetic acid	5.27			
N,N'-Dimethylpiperazine	4.630(+2)	8.539(+1)		
1,2-Dimethylpiperidine	10.22			
cis-2,6-Dimethylpiperidine	11.07(+1)			
2,2-Dimethylpropanoic acid (pivalic acid)	5.031			
2,2'-Dimethylpropylphosphonic acid	2.84	8.65		
2,4-Dimethylpyridine (2,4-lutidine)	6.74(+1)			
2,5-Dimethylpyridine (2,5-lutidine)	6.43(+1)			
2,6-Dimethylpyridine (2,6-lutidine)	6.71(+1)			
3,4-Dimethylpyridine (3,4-lutidine)	6.47(+1)			
3,5-Dimethylpyridine (3,5-lutidine)	6.09(+1)			
2,4-Dimethylpyridine-1-oxide	1.627(+1)			
2,5-Dimethylpyridine-1-oxide	1.208(+1)			
2,6-Dimethylpyridine-1-oxide	1.366(+1)			
3,4-Dimethylpyridine-1-oxide	1.493(+1)			
3,5-Dimethylpyridine-1-oxide	1.181(+1)			
2,3-Dimethylquinoline	4.94(+1)			
2,6-Dimethylquinoline	5.46(+1)			
meso-2,2-Dimethylsuccinic acid	3.77	5.936		
rac-2,2-Dimethylsuccinic acid	3.93	6.20		
D-2,3-Dimethylsuccinic acid	3.82	5.93		
meso-2,3-Dimethylsuccinic acid	3.67	5.30		
rac-2,3-Dimethylsuccinic acid	3.94	6.20		
2,4-Dimethylthiazole ($\mu = 0.1$)	3.98			
2,5-Dimethylthiazole ($\mu = 0.1$)	3.91			
4,5-Dimethylthiazole ($\mu = 0.1$)	3.73			
N,N-Dimethyl-*o*-toluidine	5.86(+1)			
N,N-Dimethyl-*p*-toluidine	7.24(+1)			
2,4-Dinitroaniline	−4.25(+1)			
2,6-Dinitroaniline	−5.23(+1)			
3,5-Dinitroaniline	0.229(+1)			
2,3-Dinitrobenzoic acid	1.85			

(*Continued*)

TABLE 2.60   *pK*, Values of Organic Materials in Water at 25°C (*Continued*)

Substance	pK_1	pK_2	pK_3	pK_4
2,4-Dinitrobenzoic acid	1.43			
2,5-Dinitrobenzoic acid	1.62			
2,6-Dinitrobenzoic acid	1.14			
3,4-Dinitrobenzoic acid	2.82			
3,5-Dinitrobenzoic acid	2.85			
1,1-Dinitrobutane (20°C)	5.90			
1,1-Dinitrodecane	3.60			
1,1-Dinitroethane (20°C)	5.21			
Dinitromethane (20°C)	3.60			
1,1-Dinitropentane	5.337			
2,4-Dinitrophenol	4.08			
2,5-Dinitrophenol	5.216			
2,6-Dinitrophenol	3.713			
3,4-Dinitrophenol	5.424			
3,5-Dinitrophenol	6.732			
2,4-Dinitrophenylacetic acid	3.50			
1,1-Dinitropropane (20°C)	5.5			
2,6-Dioxo-1,2,3,6-tetrahydro-4-pyrimidinecarboxylic acid (orotic acid)	1.8(+1)	9.55(0)		
Diphenylacetic acid	3.939			
Diphenylamine	0.9(+1)			
2,2-Diphenylglutaric acid (20°C)	3.91	5.38		
1,3-Diphenylguanidine	10.12			
2,2-Diphenylheptanedioic acid (20°C)	4.28	5.39		
2,2-Diphenylhexanedioic acid (20°C)	4.17	5.40		
3,3-Diphenylhexanedioic acid	4.22	5.19		
Diphenylhydroxyacetic acid (35°C)	3.05			
Diphenylketimine	6.82			
2,2-Diphenylnonanedioic acid (20°C)	4.33	5.38		
meso-2,2-Diphenylsuccinic acid	3.48			
rac-2,2-Diphenylsuccinic acid	3.58			
2,2-Diphenylsuccinic acid, 1-methyl ester (20°C)	4.47			
2,2-Diphenylsuccinic acid, 4-methyl ester (20°C)	3.900			
Diphenylthiocarbazone	4.50	15		
Dipropylamine	10.91(+1)			
Dipropylenetriamine	7.72(+3)	9.56(+2)	10.65(+1)	
2,2-Dipropylglutaric acid	3.688	7.31		
Dipropylmalonic acid	2.04	7.51		
2,2′-Dipyridyl	−0.52(+2)	4.352(+1)		
2,3′-Dipyridyl (20°C)	1.52(+2)	4.42(+1)		
2,4′-Dipyridyl (20°C)	1.19(+2)	4.77(+1)		
3,3′-Dipyridyl (20°C, $\mu = 0.2$)	3.0(+2)	4.60(+1)		
3,4′-Dipyridyl (20°C, $\mu = 0.2$)	3.0(+2)	4.85(+1)		
4,4′-Dipyridyl	3.17(+2)	4.82(+1)		
Dithiodiacetic acid (18°C)	3.075	4.201		
1,4-Dithioerythritol	9.5			

TABLE 2.60 *pK*, Values of Organic Materials in Water at 25°C (*Continued*)

Substance	pK_1	pK_2	pK_3	pK_4
Dithiooxamide (rubeanic acid)	10.89			
Dulcitol	13.46			
Ecgonine	10.91			
Emetine	7.36(+1)	8.23(0)		
Epinephrine enantiomorph	9.39(+1)			
Epinephrine, pseudo	9.53(+1)			
Ergometrinine	7.32(+1)			
Ergonovine	6.73(+1)			
Eriochrome Black T	6.3	11.55		
1,2-Ethanediamine	6.85(+2)	9.92(+1)		
Ethane-1,2-diamino-*N,N'*-dimethyl-*N,N'*-diacetic acid (20°C)	6.047(0)	10.068(−1)		
1,2-Ethanedithiol	8.96	10.54		
Ethanethiol ($\mu = 0.015$)	10.61			
Ethoxyacetic acid (18°C)	3.65			
2-Ethoxyaniline (*o*-phenetidine)	4.47(+1)			
3-Ethoxyaniline	4.17(+1)			
4-Ethoxyaniline	5.25(+1)			
2-Ethoxybenzoic acid (20°C)	4.21			
3-Ethoxybenzoic acid (20°C)	4.17			
4-Ethoxybenzoic acid (20°C)	4.80			
Ethoxycarbonylethylamine	9.13(+1)			
2-Ethoxyethanethiol	9.38			
2-Ethoxyethylamine	6.26(+1)			
2-Ethoxyphenol	10.109			
3-Ethoxyphenol	9.655			
(4-Ethoxyphenyl)phosphonic acid	2.06	7.28		
4-Ethoxypyridine	6.67(+1)			
Ethyl acetoacetate	10.68			
3-Ethylacrylic acid	4.695			
N-Ethylalanine	2.22(+1)	10.22(0)		
Ethylamine	10.63(+1)			
(3-Ethylamino)phenylphosphonic acid	1.1(+1)	4.90(0)	7.24(−1)	
N-Ethylaniline	5.11(+1)			
2-Ethylaniline	4.42(+1)			
3-Ethylaniline	4.70(+1)			
4-Ethylaniline	5.00(+1)			
Ethylarsonic acid (18°C)	3.89	8.35		
Ethylbarbituric acid	3.69(+1)			
2-Ethylbenzimidazole ($\mu = 0.16$)	6.27(+1)			
2-Ethylbenzoic acid	3.79			
4-Ethylbenzoic acid	4.35			
Ethylbiguanide	2.09(+1)	11.47(0)		
2-Ethylbutanoic acid (20°C)	4.710			
S-Ethyl-L-cysteine ($\mu = 0.1$)	2.03(+1)	8.60(0)		
Ethylenebiguanide (30°C)	1.74	2.88	11.34	11.76
Ethylenebis(thioacetic acid) (18°C)	3.382(0)	4.352(−1)		
Ethylenediamine-*N,N'*-diacetic acid	6.42	9.46		
Ethylenediamine-*N,N*-dimethyl-*N',N'*-diacetic acid	6.047	10.068		

(*Continued*)

TABLE 2.60 *pK*, Values of Organic Materials in Water at 25°C (*Continued*)

Substance	pK_1	pK_2	pK_3	pK_4
Ethylenediamine-*N*,*N*-dipropanoic acid (30°C)	6.87	9.60		
Ethylenediamine-*N*,*N*,*N'*,*N'*-tetra-acetic acid ($\mu = 0.1$)	1.99	2.67	6.16	10.26
Ethylenediamine-*N*,*N*,*N'*,*N'*-tetra-propanoic acid (30°C)	3.00	3.43	6.77	9.60
Ethylene glycol	14.22			
Ethyleneimine	8.04(+1)			
cis-Ethylene oxide dicarboxylic acid	1.93	3.92		
trans-Ethylene oxide dicarboxylic acid	1.93	3.25		
N-Ethylethylenediamine	7.63(+2)	10.56(+1)		
N-Ethylglycine ($\mu = 0.1$)	2.34(+1)	10.23(0)		
3-Ethylglutaric acid	4.28	5.33		
Ethylhydroperoxide	11.80			
Ethylhydrogen malonate	3.55			
3-Ethyl-2-hydroxypyridine	5.00(+1)			
Ethylmalonic acid	2.90(0)	5.55(−1)		
N-Ethyl mercaptoacetamide	8.14(SH)			
Ethyl 2-mercaptoacetate	7.95(SH)			
Ethyl 3-mercaptopropanoate	9.48(SH)			
3-Ethyl-4-(methylamino)pyridine (20°C)	9.90(+1)			
5-Ethyl-5-(1-methylbutyl)barbituric acid	8.11(0)			
Ethyl methyl ketoxime	12.45			
Ethylmethylmalonic acid	2.86(0)	6.41(−1)		
1-Ethyl-2-methylpiperidine	10.66(+1)			
3-Ethyl-6-methylpyridine (20°C)	6.51(+1)			
3-Ethyl-4-methylpyridine-1-oxide	−1.534(+1)			
5-Ethyl-2-methylpyridine-1-oxide	−1.288(+1)			
1-Ethyl-2-methyl-2-pyrroline	11.84(+1)			
Ethylmorphine (15°C)	8.08			
Ethyl nitroacetate	5.85			
3-Ethylpentane-2,4-dione	11.34			
2-Ethylpentanoic acid (18°C)	4.71			
5-Ethyl-5-pentylbarbituric acid	7.960			
2-Ethylphenol	10.2			
3-Ethylphenol	10.07			
4-Ethylphenol	10.0			
4-Ethylphenylacetic acid	4.373			
5-Ethyl-5-phenylbarbituric acid	7.445			
Ethylphosphinic acid	3.29			
Ethylphosphonic acid	2.43	8.05		
1-Ethylpiperidine ($\mu = 0.01$)	10.45(+1)			
2,2-Ethylpropylglutaric acid	3.511			
Ethylpropylmalonic acid	3.14	7.43		
2-Ethylpyridine	5.89(+1)			
3-Ethylpyridine (20°C)	5.80(+1)			
4-Ethylpyridine	5.87(+1)			
Ethyl 3-pyridinecarboxylate	3.35(+1)			

TABLE 2.60   *pK*, Values of Organic Materials in Water at 25°C (*Continued*)

Substance	pK_1	pK_2	pK_3	pK_4
Ethyl 4-pyridinecarboxylate	3.45(+1)			
2-Ethylpyridine-1-oxide	−1.19(+1)			
3-Ethylpyridine-1-oxide	−0.965(+1)			
Ethylpyrrolidine	10.43(+1)			
2-Ethyl-2-pyrroline	7.87(+1)			
Ethylsuccinic acid	4.08(0)			
S-Ethylthioacetic acid	5.06			
N-Ethyl-o-toluidine	4.92(+1)			
N-Ethylveratramine	7.40(+1)			
β-Eucaine	9.35(+1)			
Fluoroacetic acid	2.586			
2-Fluoroacrylic acid	2.55			
2-Fluoroaniline	3.20(+1)			
3-Fluoroaniline	3.58(+1)			
4-Fluoroaniline	4.65(+1)			
2-Fluorobenzoic acid	3.27			
3-Fluorobenzoic acid	3.865			
4-Fluorobenzoic acid	4.14			
Fluoromandelic acid	4.244			
2-Fluorophenol	8.73			
3-Fluorophenol	9.29			
4-Fluorophenol	9.89			
2-Fluorophenoxyacetic acid	3.08			
3-Fluorophenoxyacetic acid	3.08			
4-Fluorophenoxyacetic acid	3.13			
4-Fluorophenylacetic acid	4.25			
2′-Fluorophenylalanine	2.14(+1)	9.01(0)		
3′-Fluorophenylalanine	2.10(+1)	8.98(0)		
4-Fluorophenylalanine	2.13(+1)	9.05(0)		
2-Fluorophenylphosphonic acid	1.64	6.80		
3-Fluorophenylselenic acid	4.34			
4-Fluorophenylselenic acid	4.50			
2-Fluoropyridine	−0.44(+1)			
3-Fluoropyridine	2.97(+1)			
5-Fluorouracil	8.00(0)	ca 13(−1)		
Folic acid (pteroylglutamic acid)	8.26			
Formic acid	3.751			
N-Formylglycine	3.43			
2-Formyl-3-hydroxypyridine (20°C)	3.40(+1)	6.95(OH)		
4-Formyl-3-hydroxypyridine	4.05(+1)	6.77(OH)		
2-Formyl-3-methoxypyridine (20°C)	3.89(+1)	12.95		
Formyl-3-methoxypyridine (20°C)	4.45(+1)	11.7		
D-(−)-Fructose	12.03			
Fumaric acid	3.10	4.60		
2-Furancarboxylic acid (2-furoic acid)	3.164			
D-(+)-Galactose	12.35			
Galactose-1-phosphoric acid	1.00	6.17		
Glucoascorbic acid	4.26	11.58		
D-Gluconic acid	3.86			

(Continued)

TABLE 2.60   *pK*, Values of Organic Materials in Water at 25°C (*Continued*)

Substance	pK_1	pK_2	pK_3	pK_4
α-D-Glucose-1-phosphate	1.11(0)	6.504(−1)		
trans-Glutaconic acid	3.77	5.08		
D-(−)-Glutamic acid	2.162(+1)	4.272(0)	9.358(−1)	
L-Glutamic acid	2.19(+1)	4.25(0)	9.67(−1)	
Glutamic acid, 1-ethyl ester	3.85(+1)	7.84(0)		
Glutamic acid, 5-ethyl ester	2.15(+1)	9.19(0)		
L-Glutamine ($\mu = 0.2$)	2.17(+1)	9.13(0)		
Glutaric acid	3.77	6.08		
Glutaric acid monoamide	4.600(0)			
Glutarimide	11.43			
Glutathione	2.12(+1)	3.53(0)	8.66	9.12
DL-Glyceric acid	3.64			
Glycerol	14.15			
Glyceryl-1-phosphoric acid	—	6.656(−1)		
Glyceryl-2-phosphoric acid	1.335(0)	6.650(−1)		
Glycine	2.341(+1)	9.60(0)		
Glycine amide	8.03(+1)			
Glycine, ethyl ester	7.66(+1)			
Glycine hydroxamic acid	7.10	9.10		
Glycine, methyl ester	7.59(+1)			
Glycine-*O*-phenylphosphorylserine	2.96	8.07		
Glycolic acid	3.831			
N-Glycl-α-alanine	3.15(+1)	8.33(0)		
Glycylalanylalanine	3.38(+1)	8.10(0)		
N-Glycylasparagine	2.942			
Glycyclaspartic acid	2.81(+1)	4.45(0)	8.60(−1)	
Glycyl-DL-glutamine (18°C)	2.88(+1)	8.33(0)		
N-Glycylglycine	3.126(+1)	8.252(0)		
Glycylglycylcysteine (35°C)	2.71	2.71	7.94	7.94
Glycylglycylglycine	3.225(+1)	8.090(0)		
Glycyl-L-histidine ($\mu = 0.16$)	6.79	8.20		
Glycylisoleucine	8.00			
N-Glycyl-L-leucine	3.180(+1)	8.327(0)		
Glycyl-*O*-phosphorylserine	2.90	6.02	8.43	
L-Glycylproline ($\mu = 0.1$)	2.81(+1)	8.65(0)		
N-Glycylsarcosine ($\mu = 0.1$)	2.98(+1)	8.55(0)		
N-Glycylserine	2.98(+1)	8.38(0)		
Glycylserylglycine	3.32	7.99		
Glycyltyrosine	2.93	8.45	10.49	
Glycylvaline	3.15	8.18		
Glyoxaline	7.03(+1)			
Glyoxylic acid	3.30(0)			
Guanidineacetic acid	2.82(+1)			
Guanine	3.3(+1)	9.2	12.3	
Guanine deoxyriboside-3′-phos-phoric acid	—	2.9	6.4	9.7
Guanosine	1.9(+1)	9.25(0)	12.33(OH)	
Guanosine-5′-diphosphoric acid ($\mu = 0.1$; pK_5 9.6)	—	—	2.9	6.3
Guanosine-3′-phosphoric acid	0.7	2.3	5.92	9.38
Guanosine-5′-phosphoric acid ($\mu = 0.1$)	—	2.4	6.1	9.4

TABLE 2.60 *pK*, Values of Organic Materials in Water at 25°C (*Continued*)

Substance	pK_1	pK_2	pK_3	pK_4
Guanosine-5′-triphosphoric acid [$\mu = 0.1$; pK$_5$ 7.10(-3); pK$_6$ 9.3(-4)]	—	—	—	3.0(-2)
Guanylurea	1.80	8.20		
Harmine (20°C)	7.61(+1)			
Heptafluorobutanoic acid	0.17			
4,4,5,5,6,6,6-Heptafluorohexanoic acid	4.18			
4,4,5,5,6,6,6-Heptafluoro-2-hexen-oic acid	3.23			
Heptanedioic acid (pimelic acid)	4.484	5.424		
2,4-Heptanedione	8.43(keto); 9.15(enol)			
Heptanoic acid	4.893			
Heroin	7.6(+1)			
2,4-Hexadienoic acid (sorbic acid)	4.77			
1,1,1,3,3,3-Hexafluoro-2,2-pro-panediol	8.801			
1,1,1,3,3,3-Hexafluoro-2-propanol	9.42			
Hexahydroazepine	11.07			
Hexamethyldisilazine	7.55			
1,2,3,8,9,10-Hexamethyl-4,7-phen-anthroline (20°C)	7.26			
1,6-Hexanediamine	9.830(+2)	10.930(+1)		
1,6-Hexanedioic acid	4.418	5.412		
2,4-Hexanedione	8.49 (enol); 9.32 (keto)			
2,2′,4,4′,6,6′-Hexanitrodipheny-lamine	5.42(+1)			
Hexanoic acid (20°C)	4.849			
trans-2-Hexenoic acid	4.74			
trans-3-Hexenoic acid	4.72			
3-Hexen-4-oic acid	4.58			
4-Hexen-5-oic acid	4.74			
Hexylamine	10.64(+1)			
Hexylarsonic acid	4.16	9.19		
Hexylphosphonic acid	2.6	7.9		
DL-Histidine	1.82(+2)	6.00(+1)	9.16(0)	
Histidine amide ($\mu = 0.2$)	5.78(+2)	7.64(+1)		
Histidine, methyl ester ($\mu = 0.1$)	5.01(+2)	7.23(+1)		
Histidylglycine	2.40(+2)	5.80(+1)	7.82(0)	
Histidylhistidine($\mu = 0.16$)	5.40(+2)	6.80(+1)	7.95(0)	
DL-Homatropine	9.7(+1)			
DL-Homocysteine	2.222(+1)	8.87	10.86	
Homocysteine ($\mu = 0.1$)	1.593(+2)	2.523(+1)	8.676(0)	9.413(-1)
Hydantoin	9.12			
Hydrastine	6.23(+1)			
Hydrazine-*N*,*N*-diacetic acid	<0.1	2.8	3.8	
Hydrazine-*N*′-*N*′-diacetic acid	2.40	3.12	7.32	
4-Hydrazinocarbonylpyridine (20°C)	1.82	3.52	10.79	
N-Hydroxyacetamide	9.40			

(*Continued*)

TABLE 2.60   *pK*, Values of Organic Materials in Water at 25°C (*Continued*)

Substance	pK_1	pK_2	pK_3	pK_4
2′-Hydroxyacetophenone	9.90			
3′-Hydroxyacetophenone	9.19			
4′-Hydroxyacetophenone	8.05			
1-Hydroxyacridine (15°C)	5.72			
2-Hydroxyacridine (15°C)	5.62			
3-Hydroxyacridine (15°C)	5.30			
α-Hydroxyasparagine	2.28(+1)	7.20(0)		
β-Hydroxyasparagine	2.09(+1)	8.29(0)		
Hydroxyaspartic acid	1.91(+1)	3.51(0)	9.11(−1)	
2-Hydroxybenzaldehyde (salicyl-aldehyde)	8.34			
3-Hydroxybenzaldehyde	9.00			
4-Hydroxybenzaldehyde	7.620			
2-Hydroxybenzaldehyde oxime	1.37(+1)	9.18	12.11	
2-Hydroxybenzamide	8.36			
2-Hydroxybenzenemethanol (2-hydroxybenzyl alcohol)	9.92			
3-Hydroxybenzenemethanol	9.83			
4-Hydroxybenzenemethanol	9.82			
4-Hydroxybenzenesulfonic acid	—	9.055(−1)		
2-Hydroxybenzohydroxamic acid	5.19			
2-Hydroxybenzoic acid (salicyclic acid)	2.98	12.38		
3-Hydroxybenzoic acid	4.076	9.85		
4-Hydroxybenzoic acid	4.582	9.23		
4-Hydroxybenzonitrile	7.95			
2-Hydroxy-5-bromobenzoic acid	2.61			
2-Hydroxybutanoic acid (30°C)	3.65			
L-3-Hydroxybutanoic acid (30°C)	4.41			
4-Hydroxybutanoic acid (30°C)	4.71			
2-Hydroxy-5-chlorobenzoic acid	2.63			
trans-2′-Hydroxycinnamic acid	4.614			
trans-3′-Hydroxycinnamic acid	4.40			
10-Hydroxycodeine	7.12			
cis-2-Hydroxycyclohexane-1-carboxylic acid	4.796			
trans-2-Hydroxycyclohexane-1-carboxylic acid	4.682			
cis-3-Hydroxycyclohexane-1-carboxylic acid	4.602			
trans-3-Hydroxycyclohexane-1-carboxylic acid	4.815			
cis-4-Hydroxycyclohexane-1-carboxylic acid	4.836			
trans-4-Hydroxycyclohexane-1-carboxylic acid	4.687			
1-Hydroxy-2,4-dihydroxymethyl-benzene	9.79			
N-(Hydroxy ethyl)biguanide	2.8(+2)	11.53(+1)		
N-(2-Hydroxy-ethyl)ethylenediamine	7.21(+2)	10.12(+1)		
N′-(2-Hydroxyethyl)ethylenediam-ine-*N*,*N*,*N*′-triacetic acid	2.39	5.37	9.93	

TABLE 2.60 *pK*, Values of Organic Materials in Water at 25°C (*Continued*)

Substance	pK_1	pK_2	pK_3	pK_4
N-(2-Hydroxyethyl)iminodiacetic acid ($\mu = 0.1$)	2.2	8.65		
N-(2-Hydroxyethyl)piperazine-N'-ethansulfonic acid (20°C)	7.55			
4'-(2-Hydroxy ethyl)-1'-piperazine-propanesulfonic acid (20°C)	8.00			
2-Hydroxyethyltrimethylamine	8.94(+1)			
L-β-Hydroxyglutamic acid	2.09	4.18	9.20	
1-Hydroxy-4-hydroxymethylbenzene	9.84			
5-Hydroxy-2-(hydroxymethyl)-4H-pyran-4-one	7.90	8.03		
3-Hydroxy-2-hydroxymethylpyridine (20°C, $\mu = 0.2$)	5.00(+1)	9.07(OH)		
3-Hydroxy-4-hydroxymethylpyridine (20°C, $\mu = 0.2$)	5.00(+1)	8.95(OH)		
8-Hydroxy-7-iodoquinoline-5-sulfonic acid	2.51(0)	7.417(−1)		
Hydroxylysine (38°C, $\mu = 0.1$)	2.13(+2)	8.62(+1)	9.67(0)	
2-Hydroxy-3-methoxybenzaldehyde	7.912			
3-Hydroxy-4-methoxybenzaldehyde (isovanillin)	8.889			
4-Hydroxy-3-methoxybenzaldehyde (vanillin)	7.396			
4-Hydroxy-3-methoxybenzoic acid	4.355			
1-Hydroxy-2-methoxybenzylamine	8.70(+1)	10.52(0)		
2-Hydroxy-1-methoxybenzylamine	8.89(+1)	10.52(0)		
3-Hydroxy-2-methoxybenzylamine	8.94(+1)	10.42(0)		
2-Hydroxymethyl-2-benzeneacetic acid	4.12			
(2-Hydroxy-5-methylbenzene)-methanol	10.15			
2-Hydroxy-3-methylbenzoic acid	2.99			
2-Hydroxy-4-methylbenzoic acid	3.17			
2-Hydroxy-5-methylbenzoic acid	4.08			
2-Hydroxy-6-methylbenzoic acid	3.32			
2-Hydroxy-2-methylbutanoic acid (18°C)	3.991			
3-Hydroxy-2-methylbutanoic acid (18°C)	4.648			
4-Hydroxy-4-methylpentanoic acid (18°C)	4.873			
1-Hydroxymethylphenol	9.95			
Hydroxymethylphosphoric acid	1.91	7.15		
2-Hydroxy-2-methylpropanoic acid ($\mu = 0.1$)	3.717			
2-Hydroxy-4-methylpyridine	4.529(+1)			
8-Hydroxy-2-methylquinoline	5.55(+1)	10.31(0)		
8-Hydroxy-4-methylquinoline	5.56(+1)	10.00(0)		
8-Hydroxy-2-methylquinoline-5-sulfonic acid	4.80(0)	9.30(−1)		

(*Continued*)

TABLE 2.60 *pK*, Values of Organic Materials in Water at 25°C (*Continued*)

Substance	pK_1	pK_2	pK_3	pK_4
8-Hydroxy-4-methylquinoline-7-sulfonic acid	4.78(0)	10.01(−1)		
8-Hydroxy-6-methylquinoline-5-sulfonic acid	4.20(0)	8.7(−1)		
2-Hydroxy-1-naphthoic acid (20°C)	3.29	9.68		
2-Hydroxy-2-nitrobenzoic acid	2.23			
2-Hydroxy-3-nitrobenzoic acid	1.87			
2-Hydroxy-5-nitrobenzoic acid	2.12			
2-Hydroxy-6-nitrobenzoic acid	2.24			
2-Hydroxy-4-nitrophenylphosphonic acid	1.22	5.39		
8-Hydroxy-7-nitroquinoline-5-sulfonic acid	1.94(0)	5.750(−1)		
3-Hydroxy-4-nitrotoluene (μ = 0.1)	7.41			
4-Hydroxypentanoic acid (18°C)	4.686			
4-Hydroxy-3-pentenoic acid	4.30			
3-Hydroxyphenazine (15°C)	2.67			
4-Hydroxyphenylarsonic acid	3.89	8.37 (phenol)	10.05	
3-Hydroxyphenylboric acid	8.55	10.84		
2-Hydroxy-2-phenylpropanoic acid	3.532			
2-(2-Hydroxyphenyl)pyridine (20°C)	4.19(+1)	10.64		
trans-4-Hydroxyproline	1.818(+1)	9.662(0)		
Hydroxypropanedioic acid (tartronic acid)	2.37	4.74		
2-Hydroxypropanoic acid	3.858			
1-Hydroxy-2-propylbenzene	10.50			
4-Hydroxypteridine	1.3(+1)	7.89(0)		
2-Hydroxypyridine	1.25(+1)	11.62(0)		
3-Hydroxypyridine	4.80(+1)	8.72(0)		
4-Hydroxypyridine	3.23(+1)	11.09(0)		
2-Hydroxypyridine-*N*-oxide	−0.62(+1)	5.97(0)		
2-Hydroxypyrimidine	2.24(+1)	9.17(0)		
4-Hydroxypyrimidine	1.85(+1)	8.59(0)		
8-Hydroxyquinazoline	3.41(+1)	8.65(0)		
2-Hydroxyquinoline (20°C)	−0.31(+1)	11.74		
3-Hydroxyquinoline (20°C)	4.30(+1)	8.06(0)		
4-Hydroxyquinoline (20°C)	2.27(+1)	11.25(0)		
5-Hydroxyquinoline (20°C)	5.20(+1)	8.54(0)		
6-Hydroxyquinoline (20°C)	5.17(+1)	8.88(0)		
7-Hydroxyquinoline (20°C)	5.48(+1)	8.85(0)		
8-Hydroxyquinoline (20°C)	4.91(+1)	9.81(0)		
8-Hydroxyquinoline-5-sulfonic acid	4.092(+1)	8.776(0)		
DL-Hydroxysuccinic acid (malic acid)	3.458	5.097		
L-Hydroxysuccinic acid	3.40	5.05		
Hydroxytetracycline	3.27(+1)	7.32(0)	9.11(−1)	
5-Hydroxy-1,2,3,4-tetrazole	3.32			
4-Hydroxy-3-(2′-thiazolylazo)toluene	8.36			

TABLE 2.60 pK, Values of Organic Materials in Water at 25°C (*Continued*)

Substance	pK_1	pK_2	pK_3	pK_4
2-Hydroxytoluene	10.33			
3-Hydroxytoluene	10.10			
4-Hydroxytoluene	10.276			
4-Hydroxy-α,α,α-trifluorotoluene	8.675			
1-Hydroxy-2,4,6-trihydroxymethyl-benzene	9.56			
Hydroxyuracil	8.64			
Hydroxyvaline	2.55(+1)	9.77(0)		
Hyoscyamine	9.68(+1)			
Hypoxanthene	1.79(+1)	8.91(0)	12.07(−1)	
Hypoxanthine	5.3			
Imidazole	6.993(+1)	10.58(0)		
Imidazolidinetrione (parabanic acid)	6.10			
4-(4-Imidazolyl)butanoic acid ($\mu = 0.1$)	4.26(+1)	7.26(0)		
2-(4-Imidazolyl)ethylamine	5.784(+2)	9.756(+1)		
3-(4-Imidazolyl)propanoic acid ($\mu = 0.16$)	3.96(+1)	7.57(0)		
3,3′-Iminobispropanoic acid	4.11(0)	9.61(−1)		
3,3′-Iminobispropylamine (30°C)	8.02(+2)	9.70(+1)	10.70(0)	
2,2′-Iminodiacetic acid (diglycine) (30°C, $\mu = 0.1$)	2.54(0)	9.12(−1)		
4-Indanol	10.32			
Indole-3-acetic acid	4.75			
Inosine	ca 1.5(+1)	8.96(0)	12.36	
Inosine-5′-phosphoric acid	1.54(0)	6.66(−1)		
Inosine-5′-triphosphoric acid [pK_5 7.68(−4)]	—	—	2.2(−2)	6.92(−3)
Iodoacetic acid	3.175			
2-Iodoaniline	2.54(+1)			
3-Iodoaniline	3.58(+1)			
4-Iodoanilinc	3.82(+1)			
2-Iodobenzoic acid	2.86			
3-Iodobenzoic acid	3.86			
4-Iodobenzoic acid	4.00			
5-Iodohistamine	4.06(+1) (imidazole)	9.20(+1) (NH_3^+)	11.88(0) (imino)	
7-Iodo-8-hydroxy quinoline-5-sulfonic acid	2.514	7.417		
Iodomandelic acid	3.264			
Iodomethylphosphoric acid	1.30	6.72		
2-Iodophenol	8.464			
3-Iodophenol	8.879			
4-Iodophenol	9.200			
2-Iodophenoxyacetic acid	3.17			
3-Iodophenoxyacetic acid	3.13			
4-Iodophenoxyacetic acid	3.16			
2-Iodophenylacetic acid	4.038			
3-Iodophenylacetic acid	4.159			
4-Iodophenylacetic acid	4.178			

(*Continued*)

TABLE 2.60   *pK*, Values of Organic Materials in Water at 25°C (*Continued*)

Substance	pK_1	pK_2	pK_3	pK_4
2-Iodophenylphosphoric acid	1.74	7.06		
2-Iodopropanoic acid	3.11			
3-Iodopropanoic acid	4.08			
2-Iodopyridine	1.82(+1)			
3-Iodopyridine	3.25(+1)			
4-Iodopyridine (20°C)	4.02(+1)			
Isoasparagine	2.97(+1)	8.02(0)		
Isobutylacetic acid (18°C)	4.79			
Isobutylamine	10.41(+1)			
Isochlorotetracycline	3.1(+1)	6.7(0)	8.3(−1)	
Isocreatine	2.84(+1)			
Isogluatamine	3.81(+1)	7.88(0)		
Isohistamine ($\mu = 0.1$)	6.036(+2)	9.274(+1)		
L-Isoleucine	2.35(+1)	9.68(0)		
Isolysergic acid	3.33(0)	8.46(NH)		
Isopilocarpine (15°C)	7.18(+1)			
2-(Isopropoxy)benzoic acid (20°C)	4.24			
3-(Isopropoxy)benzoic acid (20°C)	4.15			
4-(Isopropoxy)benzoic acid (20°C)	4.68	L		
Isopropylamine	10.64(+1)			
N-Isopropylaniline	5.50(+1)			
5-Isopropylbarbituric acid	4.907(+1)			
2-Isopropylbenzene acid	3.64			
4-Isopropylbenzene acid	4.36			
N-Isopropylglycine ($\mu = 0.1$)	2.36(+1)	10.06(0)		
Isopropylmalonic acid	2.94	5.88		
Isopropylmalonic acid mononitrile	2.401			
3-Isopropyl-4-(methylam-ino)pyridine (20°C)	9.96(+1)			
3-Isopropylpentanedioic acid	4.30	5.51		
4-Isopropylphenylacetic acid	4.391			
Isopropylphosphinic acid	3.56			
Isopropylphosphonic acid	2.66	8.44		
2-Isopropylpyridine	5.83(+1)			
3-Isopropylpyridine (20°C)	5.72(+1)			
4-Isopropylpyridine	6.02(+1)			
DL-Isoproterenol	8.64(+1)			
Isoquinoline	5.40(+1)			
Isoretronecanol	10.83			
L-Isoserine ($\mu = 0.16$)	2.72(+1)	9.25(0)		
Isothiocyanatoacetic acid	6.62			
L-(+)-Lactic acid	3.858			
L-Leucine	2.33(+1)	9.60(0)		
Leucine amide	7.80(+1)			
Leucine, ethyl ester ($\mu = 0.1$)	7.57(+1)			
L-Leucyl-L-asparagine	3.00(+1)	8.12(0)		
L-Leucyl-L-glutamine	2.99(+1)	8.11(0)		
DL-Leucylglycine	3.25(+1)	8.28(0)		
Leucylisoserine (20°C)	3.188(+1)	8.207(0)		
D-Leucyl-L-tyrosine	3.12(+1)	8.38(0)	10.35(−1)	
L-Leucyl-L-tyrosine	3.46(+1)	7.84(0)	10.09(−1)	
Lysergic acid	3.44(+1)	7.68(0)		

TABLE 2.60   *pK*, Values of Organic Materials in Water at 25°C (*Continued*)

Substance	pK_1	pK_2	pK_3	pK_4
L-(+)-Lysine	2.18(+2)	8.94(+1)	10.53(0)	
Lysine, methyl ester ($\mu = 0.1$)	6.965(+1)	10.251(0)		
L-Lysyl-L-alanine	3.22(+1)	7.62(0)	10.70(−1)	
L-Lysyl-D-alanine	3.00(+1)	7.74(0)	10.63(−1)	
Lysylglutamic acid	2.93(+2)	4.47(+1)	7.75(0)	10.50(+1)
L-Lysyl-L-lysine ($\mu = 0.1$)	3.01(+2)	7.53(+1)	10.05(0)	10.01(−1)
L-Lysyl-D-lysine ($\mu = 0.1$)	2.85(+2)	7.53(+1)	9.92(0)	10.89(−1)
L-Lysyl-L-lysyl-L-lysine ($\mu = 0.1$)	3.08(+2)	7.34(+1)	9.80(0)	10.54(−1)
L-Lysyl-D-lysyl-L-lysine ($\mu = 0.1$)	2.91(+2)	7.29(+1)	9.79(0)	10.54(−1)
L-Lysyl-D-lysyl-lysine ($\mu = 0.1$)	2.94(+2)	7.15(+1)	9.60(0)	10.38(−1)
α-D-Lyxose	12.11			
Maleic acid	1.910	6.33		
Malonamic acid	3.641(0)			
Malonic acid	2.826	5.696		
Malonitrile (cyanoacetic acid)	2.460			
Mandelic acid	3.411			
D-(+)-Mannose	12.08			
Mercaptoacetic acid (thioglycolic acid)	3.60(0)	10.56(SH)		
2-Mercaptobenzoic acid (20°C)	4.05(0)			
2-Mercaptobutanoic acid	3.53(0)			
Mercaptodiacetic acid	3.32	4.29		
2-Mercaptoethanesulfonic acid (20°C)		9.5(−1)		
2-Mercaptoethanol	9.88			
2-Mercaptoethylamine	8.27(+1)	10.53(0)		
2-Mercaptohistidine	1.84(+1)	8.47(0)	11.4(SH)	
Mercapto-*S*-phenylacetic acid ($\mu = 0.1$)	3.9			
2-Mercaptopropane ($\mu = 0.1$)	10.86			
3-Mercapto-1,2-propanediol ($\mu = 0.5$)	9.43			
2-Mercaptopropanoic acid	4.32(0)	10.20(SH)		
3-Mercaptopropanoic acid		10.84(SH)		
2-Mercaptopyridine (20°C)	−1.07(+1)	10.00(0)		
3-Mercaptopyridine (20°C)	2.26(+1)	7.03(0)		
4-Mercaptopyridine (20°C)	1.43(+1)	8.86(0)		
2-Mercaptoquinoline (20°C)	−1.44(+1)	10.21(0)		
3-Mercaptoquinoline (20°C)	2.33(+1)	6.13(0)		
4-Mercaptoquinoline (20°C)	0.77(+1)	8.83(0)		
Mercaptosuccinic acid	3.30(0)	4.94(−1)	10.94(SH)	
Mesitylenic acid	4.32			
Mesoxaldialdehyde	3.60			
Methacrylic acid	4.66			
Methanethiol	10.70			
DL-Methionine	2.28(+1)	9.21(0)		
2-(*N*-Methoxyacetamido)pyridine	2.01(+1)			
3-(*N*-Methoxy acetamido)pyridine	3.52(+1)			
4-(*N*-Methoxyacetamido)pyridine	4.62(+1)			
Methoxyacetic acid	3.570			
3-Methoxy-D-α-alanine	2.037(+1)	9.176(0)		

(Continued)

TABLE 2.60   *pK*, Values of Organic Materials in Water at 25°C (*Continued*)

Substance	pK_1	pK_2	pK_3	pK_4
2-Methoxyaniline	4.53(+1)			
3-Methoxy aniline	4.20(+1)			
4-Methoxyaniline	5.36(+1)			
2-Methoxybenzoic acid	4.09			
3-Methoxybenzoic acid	4.08			
4-Methoxybenzoic acid	4.49			
N,N-Methoxybenzylamine	9.68(+1)			
2-Methoxycarbonylaniline	2.23(+1)			
3-Methoxycarbonylaniline	3.64(+1)			
4-Methoxycarbonylaniline	2.38(+1)			
Methoxycarbonylmethylamine	7.66(+1)			
2-Methoxycarbonylpyridine	2.21(+1)			
3-Methoxycarbonylpyridine	3.13(+1)			
4-Methoxycarbonylpyridine	3.26(+1)			
trans-2-Methoxycinnamic acid	4.462			
trans-3-Methoxycinnamic acid	4.376			
trans-4-Methoxycinnamic acid	4.539			
2-Methoxyethylamine	9.45(+1)			
2-Methoxy-4-nitrophenylphos- phonic acid	1.53	6.96		
2-Methoxyphenol	9.99			
3-Methoxyphenol	9.652			
4-Methoxyphenol	10.20			
(2′-Methoxy)phenoxyacetic acid	3.231			
(3′-Methoxy)phenoxyacetic acid	3.141			
(4′-Methoxy)phenoxyacetic acid	3.213			
4′-Methoxypheny lacetic acid	4.358			
(4-Methoxyphenyl)phosphinic acid (17°C)	2.35			
(2-Methoxyphenyl)phosphonic acid	2.16	7.77		
(4-Methoxyphenyl)phosphonic acid (17°C)	2.4	7.15		
3-(2′-Methoxyphenyl)propanoic acid	4.804			
3-(3′-Methoxyphenyl)propanoic acid	4.654			
3-(4′-Methoxyphenyl)propanoic acid	4.689			
3-Methoxyphenylselenic acid	4.65			
4-Methoxyphenylselenic acid	5.05			
2-Methoxy-4-(2-propenyl)phenol	10.0			
2-Methoxypyridine	3.06(+1)			
3-Methoxypyridine	4.91(+1)			
4-Methoxypyridine	6.47(+1)			
4-Methoxy-2-(2′-thiazoy- lazo)phenol	7.83			
2-Methylacrylic acid (18°C)	4.66			
N-Methylalanine	2.22(+1)	10.19(0)		
O-Methylallothreonine ($\mu = 0.1$)	1.92(+1)	8.90(0)		
Methylamine	10.62(+1)			
2-(N-Methylamino)benzoic acid	1.93(+1)	5.34(0)		
3-(N-Methylamino)benzoic acid	—	5.10(0)		
4-(N-Methylamino)benzoic acid	—	5.05		

TABLE 2.60 *pK*, Values of Organic Materials in Water at 25°C (*Continued*)

Substance	pK_1	pK_2	pK_3	pK_4
Methylaminodiacetic acid (20°C)	2.146	10.088		
2-(Methylamino)ethanol	9.88(+1)			
2-(2-Methylaminoethyl)pyridine (30°C)	3.58(+2)	9.65(+1)		
2-(Methylaminomethyl)6-methyl-pyridine ($\mu = 0.5$)	3.03(+2)	9.15(+1)		
2-(Methylaminomethyl)pyridine (30°C)	2.92(+2)	8.82(+1)		
4-Methylamino-3-methylpyridine (20°C)	9.83(+1)			
(3-Methylamino)phenylphosphonic acid	1.1(+1)	4.72(+1)	7.30(−1)	
(4-Methylamino)phenylphosphonic acid	—	—	7.85(−1)	
3-(Methylamino)pyridine (30°C)	8.70(+1)			
4-(Methylamino)pyridine (20°C)	9.65(+1)			
4-(Methylamino)-2,3,5,6-tetra-methylpyridine (20°C)	10.06(+1)			
N-Methylaniline	4.85(+1)			
Methylarsonic acid (18°C)	3.41	8.18		
1-Methylbarbituric acid	4.35(+1)			
5-Methylbarbituric acid	3.386(+1)			
2-(*N*-Methylbenzamido)pyridine	1.44(+1)			
3-(*N*-Methylbenzamido)pyridine	3.66(+1)			
4-(*N*-Methylbenzamido)pyridine	4.68(+1)			
2-Methylbenzimidazole ($\mu = 0.16$)	6.29(+1)			
2-Methylbenzoic acid (*o*-toluic acid)	3.90			
3-Methylbenzoic acid	4.269			
4-Methylbenzoic acid	4.362			
N-Methyl-1-benzoylecgonine	8.65			
Methylbiguanidine	3.00(+2)	11.44(+1)		
2-Methyl-2-butanethiol	11.35			
2-Methylbutanoic acid	4,761			
3-Methylbutanoic acid (20°C)	4.767			
(*E*)-2-Methyl-2-butendioic acid (mesaconic acid)	3.09	4.75		
3-Methyl-2-butenoic acid	5.12			
(*E*)-2-Methyl-2-butenoic acid (tiglic acid)	4.96			
(*Z*)-2-Methyl-2-butenoic acid (angelic acid)	4.30			
4-Methylcarboxylphenol	8.47			
(*E*)-2-Methylcinnamic acid	4.500			
(*E*)-3-Methylcinnamic acid	4.442			
(*E*)-4-Methylcinnamic acid	4.564			
1-Methylcyclohexane-1-carboxylic acid	5.13			
cis-2-Methylcyclohexane-1-carboxylic acid	5.03			
trans-2-Methylcy clohexane-1-carboxylic acid	5.73			
cis-3-Methylcyclohexane-1-carboxylic acid	4.88			

(*Continued*)

TABLE 2.60   *pK*, Values of Organic Materials in Water at 25°C (*Continued*)

Substance	pK_1	pK_2	pK_3	pK_4
trans-3-Methylcyclohexane-1-carboxylic acid	5.02			
cis-4-Methylcyclohexane-1-carboxylic acid	5.04			
trans-4-Methylcyclohexane-1-carboxylic acid	4.89			
2-Methylcyclohexyl-1,1-diacetic acid	3.53	6.89		
3-Methylcyclohexyl-1,1-diacetic acid	3.49	6.08		
4-Methylcyclohexyl-1,1,1-diacetic acid	3.49	6.10		
3-Methylcyclopentyl-1,1-diacetic acid	3.79	6.74		
S-Methyl-L-cysteine	8.97			
N-Methylcytidine	3.88			
5-Methylcytidine	4.21			
N-Methyl-2′-deoxycytidine	3.97			
5-Methyl-2′-deoxycytidine	4.33			
2-Methyl-3,5-dinitrobenzoic acid	2.97			
5-Methyldipropylenetriamine (30°C)	6.32(+3)	9.19(+2)	10.33(+1)	
2,2′-Methylenebis(4-chlorophenol)	7.6	11.5		
2,2′-Methylenebis(4,6-dichlorophenol)	5.6	10.56		
Methylenebis(thioacetic acid (18°C)	3.310	4.345		
3,3′-(Methylenedithio)dialanine	2.200(+1)	8.16(0)		
Methylenesuccinic acid	3.85	5.45		
N-Methylethylamine	4.23(+1)			
N-Methylethylenediamine	6.86(+1)	10.15(+1)		
α-Methylglucoside	13.71			
3-Methylglutaric acid	4.24	5.41		
N-Methylglycine (sarcosine)	2.12(+1)	10.20(0)		
5-Methyl-2,4-heptanedione	8.52(enol); 9.10(keto)			
5-Methyl-2,4-hexanedione	8.66(enol); 9.31(keto)			
5-Methyl-4-hexenoic acid	4.80			
3-Methylhistamine	5.80(+1)	9.90(0)		
1-Methylhistidine	1.69	6.48	8.85	
2-Methylhistidine (18°C)	1.7	7.2	9.5	
2-Methyl-8-hydroxyquinoline ($\mu = 0.005$)	4.58(+1)	11.71(0)		
4-Methyl-8-hydroxyquinoline	4.67(+1)	11.62(0)		
1-Methylimidazole	7.06(+1)			
4-Methylimidazole	7.55(+1)			
N-Methyliminodiacetic acid	2.15	10.09		
S-Methylisothiourea	9.83(+1)			
O-Methylisourea	9.72(+1)			
Methylmalonic acid	3.07	5.87		
2-(*N*-Methylmethanesulfonamido)pyridine	1.73(+1)			

TABLE 2.60 *pK*, Values of Organic Materials in Water at 25°C (*Continued*)

Substance	pK_1	pK_2	pK_3	pK_4
3-(*N*-Methylmethanesulfonam-ido)pyridine	3.94(+1)			
4-(*N*-Methylmethanesulfonam-ido)pyridine	5.14(+1)			
2-Methyl-6-methylaminopyridine (20°C)	3.17(+1)	8.84(0)		
3-Methyl-4-methylaminopyridine (20°C)	—	9.84(0)		
4-Methyl-2,2′-(4-methylpyri-dyl)pyridine	5.32(+1)			
N-Methylmorpholine	7.13(+1)			
2-Methyl-1-naphthoic acid	3.11			
N-Methyl-1-naphthylamine	3.70(+1)			
2-Methyl-4-nitrobenzoic acid	1.86			
2-Methyl-6-nitrobenzoic acid	1.87			
1-Methyl-2-nitroterephthalic acid	3.11			
4-Methyl-2-nitroterephthalic acid	1.82			
3-Methylpentanedioic acid	4.25	5.41		
3-Methylpentane-2,4-dione	10.87			
2-Methylpentanoic acid	4.782			
3-Methylpentanoic acid	4.766			
4-Methylpentanoic acid	4.845			
cis-3-Methyl-2-pentenoic acid	5.15			
trans-3-Methyl-2-pentenoic acid	5.13			
4-Methyl-2-pentenoic acid	4.70			
4-Methyl-3-pentenoic acid	4.60			
6-Methyl-1,10-phenanthroline	5.11(+1)			
(2-Methylphenoxy)acetic acid	3.227			
(3-Methylphenoxy)acetic acid	3.203			
(4-Methylphenoxy)acetic acid	3.215			
(2-Methylphenyl)acetic acid (18°C)	4.35			
(4-Methylphenyl)acetic acid	4.370			
5-Methyl-5-phenylbarbituric acid	8.011(0)			
3-(2-Methylphenyl)propanoic acid	4.66			
3-(3-Methylphenyl)propanoic acid	4.677			
3-(4-Methylphenyl)propanoic acid	4.684			
1-Methyl-2-phenylpyrrolidine	8.80			
5-Methyl-1-phenyl-1,2,3-triazole-4-carboxylic acid	3.73			
Methylphosphinic acid	3.08			
Methylphosphonic acid	2.38	7.74		
3-Methyl-*o*-phthalic acid	3.18			
4-Methyl-*o*-phthalic acid	3.89			
N-Methylpiperazine ($\mu = 0.1$)	4.94(+2)	9.09(+1)		
2-Methylpiperazine	5.62(+2)	9.60(+1)		
N-Methylpiperidine	10.19(+1)			
2-Methylpiperidine	10.95(+1)			
3-Methylpiperidine	11.07(+1)			
4-Methylpiperidine ($\mu = 0.5$)	11.23(+1)			
2-Methyl-1,2-propanediamine	6.178(+2)	9.420(+1)		
2-Methyl-2-propanethiol	11.2			
2-Methylpropanoic acid	4.853			

(*Continued*)

TABLE 2.60 pK, Values of Organic Materials in Water at 25°C (*Continued*)

Substance	pK_1	pK_2	pK_3	pK_4
2-Methyl-2-propylamine	10.682(+1)			
2-Methyl-2-propylglutaric acid	3.626			
2-Methylpyridine	5.96(+1)			
3-Methylpyridine	5.68(+1)			
4-Methylpyridine	6.00(+1)			
Methyl 4-pyridinecarboxylate	3.26(+1)			
6-Methylpyridine-2-carboxylic acid	5.83			
2-Methylpyridine-1-oxide	1.029(+1)			
3-Methylpyridine-1-oxide	10.921(+1)			
4-Methylpyridine-1-oxide	1.258(+1)			
O-Methylpyridoxal ($\mu = 0.16$)	4.74			
Methyl-2-pyridyl ketoxime	9.97			
1-Methyl-2-(3-pyridyl)pyrrolidine	3.41	7.94		
1-Methylpyrrolidine	10.46(+1)			
1-Methyl-3-pyrroline	9.88(+1)			
5-Methylquinoline	4.62(+1)			
Methylsuccinic acid	4.13	5.64		
Methylsulfonylacetic acid	2.36			
3-Methylsulfonylaniline	2.68(+1)			
4-Methylsulfonylaniline	1.48(+1)			
3-Methylsulfonylbenzoic acid	3.52			
4-Methylsulfonylbenzoic acid	3.64			
4-Methylsulfonyl-3,5-dimethyl-phenol	8.13			
3-Methylsulfonylphenol	9.33			
4-Methylsulfonylphenol	7.83			
1-Methyl-1,2,3,4-tetrahydro-3-pyri-dinecarboxylic acid (arecaidine; isoguvacine)	9.07			
5-Methyl-1,2,3,4-tetrazole	3.32			
2-Methylthiazole ($\mu = 0.1$)	3.40(+1)			
4-Methylthiazole ($\mu = 0.1$)	3.16(+1)			
5-Methylthiazole ($\mu = 0.1$)	3.03(+1)			
Methylthioacetic acid	3.72			
4-Methylthioaniline	4.40(+1)			
2-Methylthioethylamine (30°C)	9.18(+1)			
Methylthioglycolic acid	7.68			
3-(S-Methylthio)phenol	9.53			
4-(S-Methylthio)phenol	9.53			
2-Methylthiopyridine (20°C)	3.59(+1)			
3-Methylthiopyridine (20°C)	4.42(+1)			
4-Methylthiopyridine (20°C)	5.94(+1)			
5-Methylthio-1,2,3,4-tetrazole	4.00(+1)			
O-Methylthreonine	2.02(+1)	9.00(0)		
O-Methyltyrosine	2.21(+1)	9.35(0)		
1-Methylxanthine	7.70	12.0		
3-Methylxanthine	8.10	11.3		
7-Methylxanthine	8.33	ca 13		
9-Methylxanthine	6.25			
Morphine (20°C)	7.87(+1)	9.85(0)		
Morpholine	8.492(+1)			
2-(N-Morpholino)ethanesulfonic acid (MES) (20°C)	6.15			

TABLE 2.60 *pK*, Values of Organic Materials in Water at 25°C (*Continued*)

Substance	pK_1	pK_2	pK_3	pK_4
3-(*N*-Morpholino)-2-hydroxypro-panesulfonic acid (37°C)	6.75			
3-(*N*-Morpholino)propanesulfonic acid (20°C)	7.20			
Murexide	0.0	9.20	10.50	
Myosmine	5.26			
1-Naphthalenecarboxylic acid (1-naphthoic acid)	3.695			
2-Naphthalenecarboxylic acid	4.161			
1-Naphthol (20°C)	9.30			
2-Naphthol (20°C)	9.57			
Naphthoquinone monoxime	8.01			
1-Naphthylacetic acid	4.236			
2-Naphthylacetic acid	4.256			
1-Naphthylamine	3.92(+1)			
2-Naphthylamine	4.11(+1)			
1-Naphthylarsonic acid	3.66	8.66		
1-Naphthysulfonic acid	0.57			
Narceine (15°C)	3.5(+1)	9.3		
Narcotine	6.18(+1)			
Nicotine	3.15(+1)	7.87(0)		
Nicotyrine	4.76(+1)			
Nitrilotriacetic acid (NTA) (20°C)	1.65	2.94	10.33	
Nitroacetic acid	1.68			
2-Nitroaniline	−0.28(+1)			
3-Nitroaniline	2.46(+1)			
4-Nitroaniline	1.01(+1)			
2-Nitrobenzene-1,4-dicarboxylic acid	1.73			
3-Nitrobenzene-1,2-dicarboxylic acid	1.88			
4-Nitrobenzene-1,2-dicarboxylic acid	2.11			
2-Nitrobenzoic acid	2.18			
3-Nitrobenzoic acid	3.46			
4-Nitrobenzoic acid	3.441			
trans-2-Nitrocinnamic acid	4.15			
trans-3-Nitrocinnamic acid	4.12			
trans-4-Nitrocinnamic acid	4.05			
Nitroethane	8.57			
2-Nitrohydroquinone	7.63	10.06		
N-Nitroiminodiacetic acid	2.21	3.33		
3-Nitromesitol	8.984			
Nitromethane	10.12			
1-Nitro-6,7-phenanthroline($\mu = 0.2$)	3.23(+1)			
5-Nitro-1,10-phenanthroline	3.232(+1)			
6-Nitro-1,10-phenanthroline	3.23(+1)			
2-Nitrophenol	7.222			
3-Nitrophenol	8.360			
4-Nitrophenol	7.150			
(2-Nitrophenoxy)acetic acid	2.896			

(*Continued*)

TABLE 2.60 *pK*, Values of Organic Materials in Water at 25°C (*Continued*)

Substance	pK_1	pK_2	pK_3	pK_4
(3-Nitrophenoxy)acetic acid	2.951			
(4-Nitrophenoxy)acetic acid	2.893			
2-Nitrophenylacetic acid	4.00			
3-Nitrophenylacetic acid	3.97			
4-Nitrophenylacetic acid	3.85			
2-Nitrophenylarsonic acid	3.37	8.54		
3-Nitrophenylarsonic acid	3.41	7.80		
4-Nitrophenylarsonic acid	2.90	7.80		
7-(4-Nitrophenylazo)-8-hydroxy-5-quinolinesulfonic acid	3.14(0)	7.495(−1)		
3-Nitrophenylphosphonic acid	1.30	6.27		
4-Nitrophenylphosphonic acid	1.24	6.23		
3-(2′-Nitrophenyl)propanoic acid	4.504			
3-(4′-Nitrophenyl)propanoic acid	4.473			
3-Nitrophenylselenic acid	4.07			
4-Nitrophenylselenic acid	4.00			
1-Nitropropane	8.98			
2-Nitropropane	7.675			
2-Nitropropanoic acid	3.79			
2-Nitropyridine ($\mu = 0.02$)	−2.06(+1)			
3-Nitropyridine ($\mu = 0.02$)	0.79(+1)			
4-Nitropyridine ($\mu = 0.02$)	1.23(+1)			
N-Nitrosoiminodiacetic acid	2.28	3.38		
4-Nitrosophenol	6.48			
Nitrourea	4.15(+1)			
1,9-Nonanedioic acid (azelaic acid)	4.53	5.40		
Nonanoic acid (pelargonic acid)	4.95			
DL-Norleucine	2.335(+1)	9.834(0)		
Novocaine	8.85(+1)			
2,2,3,3,4,4,5,5-Octafluoropentanoic acid	2.65			
1,8-Octanedioic acid (suberic acid)	4.512	5.404		
Octanoic acid (caprylic acid)	4.895			
Octopine-DD	1.35	2.30	8.68	11.25
Octopine-LD	1.40	2.30	8.72	11.34
Octylamine	10.65(+1)			
L-(+)-Ornithine	1.94(+2)	8.65(+1)	10.76(0)	
Oxalic acid	1.271	4.272		
3,6-Oxaoctanedioic acid ($\mu = 1.0$)	3.055	3.676		
Oxoacetic acid	3.46			
2-Oxabutanedioic acid (oxaloacetic acid)	2.56	4.37		
2-Oxobutanoic acid	2.50			
5-Oxohexanoic acid (5-ketohexanoic acid) (18°C)	4.662			
3-Oxo-1,5-pentanedioic acid	3.10			
4-Oxopentanoic acid (levulinic acid)	4.59			
2-Oxopropanoic acid (pyruvic acid)	2.49			
Oxytetracycline	3.10(+1)	7.26	9.11	
Papaverine	5.90(+1)			

TABLE 2.60   *pK*, Values of Organic Materials in Water at 25°C (*Continued*)

Substance	pK_1	pK_2	pK_3	pK_4
Pentamethylenebis(thioacetic acid) (18°C)	3.485	4.413		
3,3-Pentamethylenepentanedioic acid	3.49	6.96		
1,5-Pentanediamine	10.05(+2)	10.916(+1)		
2,4-Pentanedione	8.24(enol); 8.95(keto)			
1-Pentanoic acid (valeric acid)	4.842			
2-Pentenoic acid	4.70			
3-Pentenoic acid	4.52			
4-Pentenoic acid	4.677			
Pentylarsonic acid	4.14	9.07		
N-Pentylveratramine	7.28(+1)			
Perhydrodiphenic acid (20°C)	4.96	6.68		
Perlolidine (18°C)	4.01	11.39		
Peroxyacetic acid	8.20			
1,7-Phenanthroline	4.30(+1)			
1,10-Phenanthroline	4.857(+1)			
6,7-Phenanthroline	4.857(+1)			
Phenazine	1.2(+1)			
Phenethylthioacetic acid	3.795			
Phenol	9.99			
Phenol-3-phosphoric acid	1.78	7.03	10.2	
Phenol-4-phosphoric acid	1.99	7.25	9.9	
Phenolphthalein	9.4			
3-Phenolsulfonic acid	—	9.05(−1)		
Phenosulsulfonephthalein	7.9			
Phenoxyactic acid	3.171			
2-Phenoxybenzoic acid	3.53			
3-Phenoxybenzoic acid	3.95			
4-Phenoxybenzoic acid	4.52			
5-Phenoxy-1,2,3,4-tetrazole	3.49(+1)			
Phenylacetic acid	4.312			
L-3-Phenyl-α-alanine	1.83(+1)	9.12(0)		
3-Phenyl-α-alanine, methyl ester	7.05(+1)			
Phenylalanylarginine ($\mu = 0.01$)	2.66(+1)	7.57(0)	12.40(−1)	
Phenylalanylglycine ($\mu = 0.01$)	3.10(+1)	7.71(0)		
7-Phenylazo-8-hydroxy-5-quinolinesulfonic acid	3.41(0)	7.850(−1)		
5-Phenylbarbituric acid	2.544(+1)			
2-Phenyl-2-benzylsuccinic acid	3.69	6.47		
1-Phenylbiguanide	2.13(+2)	10.76(+1)		
4-Phenylbutanoic acid	4.757			
Phenylbutazone	4.5(+1)			
2-Phenylenediamine	<2(+2)	4.47(+1)		
3-Phenylenediamine	2.65(+2)	4.88(+1)		
4-Phenylenediamine	3.29(+2)	6.08(+1)		
2-Phenylethylamine	9.83(+1)			
β-Phenylethylboronic acid	10.0			
DL-α-Phenylglycine	1.83(+1)	4.39(0)		
Phenylguanidine	10.77(+1)			

(*Continued*)

TABLE 2.60   *pK*, Values of Organic Materials in Water at 25°C (*Continued*)

Substance	pK_1	pK_2	pK_3	pK_4
Phenylhydrazine	5.20(+1)			
2-Phenyl-3-hydroxypropanoic acid	3.53			
3-Phenyl-3-hydroxypropanoic acid	4.40			
Phenyliminodiacetic acid (20°C)	2.40	4.98		
Phenylmalonic acid	2.58	5.03		
Phenylmethanethiol	10.70			
2-Phenyl-2-phenethylsuccinic acid (20°C)	3.74	6.52		
2-Phenylphenol	9.55			
3-Phenylphenol	9.63			
4-Phenylphenol	9.55			
Phenylphosphinic acid (17°C)	2.1			
Phenylphosphonic acid	1.83	7.07		
O-Phenylphosphorylserine	2.13(+1)	8.79		
O-Phenylphosphorylserylglycine	3.18(+1)	6.95(0)		
O-Phenylphosphoryl-L-seryl-L-leucine	3.16(+1)	7.12(0)		
N-Phenylpiperazine ($\mu = 0.1$)	8.71(+1)			
2-Phenylpropanoic acid	4.38			
3-Phenylpropanoic acid (35°C)	4.664			
3-Phenyl-1-propylamine	10.39(+1)			
Phenylpropynoic acid (35°C)	2.269			
Phenylselenic acid	4.79			
Phenylselenoacetic acid ($\mu = 0.1$)	3.75			
β-Phenylserine ($\mu = 0.16$)	8.79(0)			
Phenylsuccinic acid (20°C)	3.78	5.55		
Phenylsulfenylacetic acid	2.66			
Phenylsulfonylacetic acid	2.44			
5-Phenyl-1,2,3,4-tetrazole	4.38(+1)			
1-Phenyl-1,2,3-triazole-4-carboxylic acid	2.88			
1-Phenyl-1,2,3-triazole-4,5-dicarboxylic acid	2.13	4.93		
Phosphoramidic acid	3.08	8.63		
O-Phosphorylethanolamine	5.838(+1)	10.638(0)		
O-Phosphorylserylglycine	3.13	5.41	8.01	
O-Phosphoryl-L-seryl-L-leucine	3.11	5.47	8.26	
Phosphoserine	2.08	5.65	9.74	
Phthalamide	3.79(0)			
Phthalazine	3.47(+1)			
o-Phthalic acid	2.950	5.408		
Phthalimide	9.90(0)			
Physostigmine	1.76(+1)	7.88(0)		
Picric acid (2,4,6-trinitrophenol) (18°C)	0.419			
Pilocarpine	1.3(+1)	6.85(0)		
Piperazine	5.333(+2)	9.781(+1)		
1,4-Piperazinebis(ethanesulfonic acid) (20°C)	6.80			
Piperazine-2-carboxylic acid	1.5	5.41	9.53	
Piperdine	11.123(+1)			
2-Piperidinecarboxylic acid	2.12(+1)	10.75(0)		
3-Piperidinecarboxylic acid	3.35(+1)	10.64(0)		

TABLE 2.60   *pK*, Values of Organic Materials in Water at 25°C (*Continued*)

Substance	pK_1	pK_2	pK_3	pK_4
4-Piperidinecarboxylic acid	3.73(+1)	10.72(0)		
1-(2-Piperidinyl)-2-propanone (15°C)	9.45			
Piperine (15°C)	1.98(+1)			
Proline	1.99(+1)	10.96(0)		
1,2-Propanediamine	6.607(+2)	9.702(+1)		
1,3-Propanediamine	8.49(+2)	10.47(+1)		
1-Propanethiol	10.86			
1,2,3-Propanetriamine	3.72(+3)	7.95(+2)	9.59(+1)	
1,2,3-Propanetricarboxylic acid	3.67	4.87	6.38	
Propanoic acid	4.874			
Propenoic acid	4.247			
N-Propionylglycine	3.718(0)			
2-Propoxybenzoic acid (20°C)	4.24			
3-Propoxybenzoic acid (20°C)	4.20			
4-Propoxybenzoic acid (20°C)	4.78			
N-Propylalanine	2.21(+1)	10.19(0)		
Propylamine	10.568(+1)			
Propylarsonic acid (18°C)	4.21	9.09		
Propylenimine	8.18(+1)			
N-Propylglycine ($\mu = 0.1$)	2.38(+1)	10.03(0)		
L-Propylglycine	3.19(+1)	8.97(0)		
Propylmalonic acid	2.97	5.84		
Propylphosphinic acid	3.46			
Propylphosphonic acid	2.49	8.18		
2-Propylpyridine	6.30(+1)			
N-Propylveratramine	7.20(+1)			
2-Propynoic acid	1.887			
Pseudoecgonine	9.70			
Pseudoisocyanine ($\mu = 0.2$)	4.59(+2)			
Pseudotropine	9.86(+1)			
Pteroylglutamic acid	8.26			
Purine	2.52(+1)	8.92(0)		
Pyrazine	0.6(+1)			
Pyrazinecarboxamide	0.5(+1)			
Pyrazole	2.61(+1)			
Pyridazine	2.33(+1)			
Pyridine	5.17(+1)			
Pyridine-d_5	5.83(+1)			
2-Pyridinealdoxime	3.56(+1)	10.17(0)		
3-Pyridinealdoxime	4.07(+1)	10.39(0)		
4-Pyridinealdoxime	4.73(+1)	10.03(0)		
2-Pyridinecarbaldehyde	3.84(+1)			
3-Pyridinecarbaldehyde	3.80(+1)			
4-Pyridinecarbaldehyde	4.74(+1)			
3-Pyridinecarbamide (nicotin-amide)	3.33(+1)			
3-Pyridinecarbonitrile	1.35(+1)			
Pyridine-2-carboxylic acid (picol-inic acid)	1.01(+1)	5.29(0)		
Pyridine-3-carboxylic acid (nico-tinic acid)	2.07(+1)	4.75(0)		

(*Continued*)

TABLE 2.60 pK, Values of Organic Materials in Water at 25°C (*Continued*)

Substance	pK_1	pK_2	pK_3	pK_4
Pyridine-4-carboxylic acid (isoni-cotinic acid)	1.84(+1)	4.86(0)		
Pyridine-2,3-dicarboxylic acid	2.36(+1)	7.08(0)		
Pyridine-2,4-dicarboxylic acid	2.23(+1)	7.02(0)		
Pyridine-2,6-dicarboxylic acid	2.16(+1)	6.92(0)		
Pyridine-1-oxide	0.688(+1)			
Pyridoxal	4.20(+1)	8.66(ring OH)		
Pyridoxal-5-phosphate ($\mu = 0.15$)	<2.5	4.14	6.20	8.69
Pyridoxamine ($\mu = 0.1$)	3.37(+2)	8.01(+1)	10.13(ring OH)	
Pyridoxamine-5-phosphate ($\mu = 0.15$; pK_5 10.92)	2.5	3.69	5.76	8.61
Pyridoxine (vitamin B$_6$) (18°C)	5.00(+1)	8.96(ring OH)		
3-(2′-Pyridyl)alanine	1.37(+2)	4.02(+1)	9.22(0)	
3-(3′-Pyridyl)alanine	1.77(+2)	4.64(+1)	9.10(0)	
2-(2′-Pyridyl)benzimidazole ($\mu = 0.16$)	5.58(+1)			
2-(2′-Pyridyl)imidazole ($\mu = 0.005$)	8.98(+1)			
4-(2′-Pyridyl)imidazole ($\mu = 0.1$)	5.49(+1)			
Pyrimidine	1.30(+1)			
2,4(1H,3H)-Pyrimidinedione (ura-cil)	0.6(+1)	9.46(0)		
2,4,5,6(1H,3H)-Pyrimidinetetrone-5-oxime	4.57(0)			
Pyrocatecholsulfonephthaleine	7.82	9.76	11.73	
Pyroxilidine	11.11(+1)			
Pyrrole-1-carboxylic acid	4.45			
Pyrrole-2-carboxylic acid	4.45			
Pyrrole-3-carboxylic acid	4.453			
Pyrrolidine	11.305(+1)			
Pyrrolidine-2-carboxylic acid (pro-line)	1.952(+1)	10.640(0)		
2-[2-(N-Pyrrolidinyl)ethyl]pyridine	3.60(+2)	9.39(+1)		
3-[2-(N-Pyrrolidinyl)ethyl]pyridine	4.28(+2)	9.28(+1)		
4-[2-(N-Pyrrolidinyl)ethyl]pyridine	4.65(+2)	9.27(+1)		
2-(1-Pyrrolidinylmethyl)pyridine	2.54(+1)	8.56(+1)		
3-(1-Pyrrolidinylmethyl)pyridine	3.14(+2)	8.36(+1)		
4-(1-Pyrrolidinylmethyl)pyridine	3.38(+2)	8.16(+1)		
3-Pyrroline	−0.27(+1)			
Quinidine	4.0(+1)	8.54(0)		
Quinine	4.11(+1)	8.52(0)		
Quinoline	4.80(+1)			
Quinoxaline	0.72(+1)			
D-Raffinose	12.74			
Riboflavin (vitamin B$_2$) ($\mu = 0.01$)	ca −0.2	9.69		
α-D-Ribofuranose	12.11			
D-Ribose-5′-phosphonic acid	—	6.70(−1)	13.05(−2)	

TABLE 2.60   *pK*, Values of Organic Materials in Water at 25°C (*Continued*)

Substance	pK_1	pK_2	pK_3	pK_4
D-Saccharic acid	5.00(0)			
Saccharin (*o*-benzoic sulfimide)	2.32			
Sarcosine	2.12(+1)	10.20(0)		
Sarcosine amide	8.35(+1)			
Sarcosine dimethylamide	8.86(+1)			
Sarcosine methylamide	8.28(+1)			
Sarcosylglycine ($\mu = 0.16$)	3.15(+1)	8.56(0)		
Sarcosylleucine	3.15(+1)	8.67(0)		
Sarcosylsarcosine	2.92(+1)	9.15(0)		
Sarcosylserine	3.17(+1)	8.63(0)		
3-Selenosemicarbazide ($\mu = 0.1$)	0.8(+1)			
Semicarbazide ($\mu = 0.1$)	3.53(+1)			
L-Serine	2.21(+1)	9.15(0)	13.6	
Serine, methyl ester ($\mu = 0.1$)	7.03(+1)			
Serylglycine ($\mu = 0.15$)	2.10(+1)	7.33(0)		
L-Seryl-L-leucine	3.08(+1)	7.45(0)		
Solanine	7.34(+1)			
D-Sorbitol (17.5°C)	13.60			
L-(−)-Sorbose (18°C)	11.55			
Sparteine	4.49(+1)	11.76(0)		
Spinaceamine ($\mu = 0.1$)	4.895(+2)	8.90(+1)		
Spinacine	1.649(+2)	4.936(+1)	8.663(0)	
L-Strychnine (15°C)	2.50	8.20		
Succinamic acid (succinic acid monoamide)	4.39(0)			
Succinic acid	4.207	5.635		
DL-Succinimide	9.623			
β-(4′-Sulfaminophenyl)alanine	1.99(+1)	8.64(0)	10.26(−1)	
3-Sulfamylbenzoic acid	3.54			
4-Sulfamylbenzoic acid	3.47			
4-Sulfamylphenylphosphoric acid	1.42	6.38	10.0	
Sulfanilamide	10.43(+1)			
Sulfoacetic acid	—	4.0		
3-Sulfobenzoic acid	—	3.78		
4-Sulfobenzoic acid	—	3.72		
3-Sulfophenol	0.39	9.07		
4-Sulfophenol	0.58	8.70		
2-Sulfopropanoic acid	1.99			
5-Sulfosalicyclic acid	2.49	12.00		
Sylvie acid	7.62			
D-Tartaric acid	3.036	4.366		
meso-Tartaric acid	3.22	4.81		
Tetracycline ($\mu = 0.005$)	3.30(+1)	7.68	9.69	
Tetradehydroyohimbine	10.59(+1)			
Tetraethylenepentamine [$\mu = 0.1$; pK_5 9.67(+1)]	2.98(+5)	4.72(+4)	8.08(+3)	9.10(+2)
1,4,5,6-Tetrahydro-1,2-dimethyl-pyridine	11.38(+1)			
1,4,5,6-Tetrahydro-2-methylpyri-dine	9.53(+1)			
cis-Tetrahydronaphthalene-2,3-dicarboxylic acid (20°C)	3.98	6.47		

(*Continued*)

TABLE 2.60   *pK*, Values of Organic Materials in Water at 25°C (*Continued*)

Substance	pK_1	pK_2	pK_3	pK_4
trans-Tetrahydronaphthalene-2,3-dicarboxylic acid (20°C)	4.00	5.70		
5,6,7,8-Tetrahydro-1-naphthol	10.28			
5,6,7,8-Tetrahydro-2-naphthol	10.48			
Tetrahydroserpentine	10.55(+1)			
2,3,5,6-Tetramethylbenzoic acid	3.415			
Tetramethylenebis(thioacetic acid) (18°C)	3.463	4.423		
Tetramethylenediamine	9.22(+2)	10.75(+1)		
N,N,N′,N′-Tetramethylethylenediamine	2.20(+2)	6.35(+1)		
2,3,5,6-Tetramethyl-4-methylaminopyridine	0.07(+1)			
2,2,6,6-Tetramethylpiperidine ($\mu = 0.5$)	1.24(+1)			
2,3,5,6-Tetramethylpyridine (20°C)	7.90(+1)			
Tetramethylsuccinic acid	3.50	7.28		
1,2,3,4-Tetrazole	4.90			
Thebaine	7.95(+1)			
2-Thenoyltrifluoroacetone	5.70(0)			
Theobromine	0.68(+1)	7.89		
Theophylline	<1(+1)	8.80		
Thiazoline	2.53(+1)			
Thioacetic acid	3.33			
o-Thiocresol	6.64			
m-Thiocresol	6.58			
p-Thiocresol	6.52			
Thiocyanatoacetic acid	2.58			
2,2′-Thiodiacetic acid	3.32	4.29		
4,4′-Thiodibutanoic acid (18°C)	4.351	5.275		
3,3′-Thiodipropanoic acid (18°C)	4.085	5.075		
3-Thio-*S*-methylcarbazide ($\mu = 0.1$)	7.563(+1)			
1-Thionylcarboxylic acid	3.53			
2-Thionylcarboxylic acid	4.10			
2-Thiophenecarboxylic acid (30°C)	3.529			
3-Thiophenecarboxylic acid (3-thenoic acid)	4.10			
Thiophenol	6.50			
3-Thiosemicarbazide ($\mu = 0.1$)	1.5(+1)			
3-Thiosemicarbazide-1,1-diacetic acid (30°C)	2.94	4.07		
Thiourea	2.03(+1)			
Thorin	3.7	8.3	11.8	
Thymidine	9.79	12.85		
p-Toluenesulfinic acid	1.7			
Toluhydroquinone	10.03	11.62		
o-Toluidine	4.45(+1)			
m-Toluidine	4.71(+1)			
p-Toluidine	5.08(+1)			
o-Tolylacetic acid (18°C)	4.36			
p-Tolylacetic acid (18°C)	4.36			
o-Tolylarsonic acid	3.82	8.85		

TABLE 2.60 *pK*, Values of Organic Materials in Water at 25°C (*Continued*)

Substance	pK_1	pK_2	pK_3	pK_4
m-Tolylarsonic acid	3.82	8.60		
p-Tolylarsonic acid	3.70	8.68		
o-Tolylphosphonic acid	2.10	7.68		
m-Tolylphosphonic acid	1.88	7.44		
p-Tolylphosphonic acid	1.84	7.33		
3-Tolylselenic acid	4.80			
4-Tolylselenic acid	4.88			
Triacetylmethane	5.81			
Triallylamine	8.31(+1)			
1,3,5-Triazine-2,4,6-triol	7.20	11.10		
1*H*-1,2,3-Triazole	—	9.26		
1*H*-1,2,4-Triazole	2.386(+1)	9.972		
1,2,3-Triazole-4-carboxylic acid	3.22	8.73		
1,2,3-Triazole-4,5-dicarboxylic acid	1.86	5.90	9.30	
1,2,4-Triazolidine-3,5-dione (urazole)	5.80			
Tribromoacetic acid	−0.147			
2,4,6-Tribromobenzoic acid	1.41			
Trichloroacetic acid	0.52			
Trichloroacrylic acid	1.15			
3,3,3-Trichlorolactic acid	2.34			
Trichloromethylphosphonic acid	1.63	4.81		
2,4,5-Trichlorophenol	7.37			
3,4,5-Trichlorophenol	7.839			
Tricine (20°C)	8.15			
Triethanolamine	7.76(+1)			
Triethylamine	10.72(+1)			
Triethylenediamine	4.18(+2)	8.19(+1)		
Triethylenetetramine (20°C)	3.32(+4)	6.67(+3)	9.20(+2)	9.92(+1)
Triethylsuccinic acid	2.74			
Trifluoroacetic acid	0.50			
Trifluoroacrylic acid	1.79			
4,4,4-Trifluoro-2-aminobutanoic acid	1.600(+1)	8.169(0)		
4,4,4-Trifluoro-3-aminobutanoic acid	2.756(+1)	5.822(0)		
4,4,4-Trifluorobutanoic acid	4.16			
α,α,α-Trifluoro-*m*-cresol	8.950			
4,4,4-Trifluorocrotonic acid	3.15			
5,5,5-Trifluoroleucine	2.045(+1)	8.942(0)		
3-(Trifluoromethyl)aniline	3.5(+1)			
4-(Trifluoromethyl)aniline	2.6(+1)			
3-Trifluoromethylphenol	8.950			
5-Trifluoromethyl-1,2,3,4-tetrazole	1.70			
6,6,6-Trifluoronorleucine	2.164(+1)	9.463(0)		
5,5,5-Trifluoronorvaline	2.042(+1)	8.916(0)		
5,5,5-Trifluoropentanoic acid	4.50			
3,3,3-Trifluoropropanoic acid	3.06			
4,4,4-Trifluorothreonine	1.554(+1)	7.822(0)		
4,4,4-Trifluorovaline	1.537(+1)	8.098(0)		

(*Continued*)

TABLE 2.60   *pK*, Values of Organic Materials in Water at 25°C (*Continued*)

Substance	pK_1	pK_2	pK_3	pK_4
1,2,3-Trihydroxybenzene (pyrogal-lol)	9.03(0)	11.63(−1)		
1,3,5-Trihydroxybenzene (phloro-glucinol)	8.45(0)	8.88(−1)		
2,4,6-Trihydroxybenzoic acid	1.68(0)			
3,4,5-Trihydroxybenzoic acid	4.19(0)	8.85(−1)		
3,4,5-Trihydroxycyclohex-1-ene-1-carboxylic acid [D-(−)-shikimic acid]	4.15			
2,4,6-Tri(hydroxymethyl)phenol	9.56			
Triisobutylamine	10.42(+1)			
Trimethylamine	9.80(+1)			
3-(Trimethylamino)phenol	8.06			
4-(Trimethylamino)phenol	8.35			
2,4,6-Trimethylaniline	4.38(+1)			
2,4,6-Trimethylbenzoic acid	3.448			
Trimethylenebis(thioacetic acid) (18°C)	3.435	5.383		
2,3,4-Trimethylphenol	10.59			
2,4,5-Trimethylphenol	10.57			
2,4,6-Trimethylphenol	10.88			
3,4,5-Trimethylphenol	10.25			
2,3,6-Trimethylpyridine ($\mu = 0.5$)	7.60(+1)			
2,4,6-Trimethylpyridine	7.43(+1)			
2,4,6-Trimethylpyridine-1-oxide	1.990(+1)			
3-(Trimethylsilyl)benzoic acid	4.089			
4-(Trimethylsilyl)benzoic acid	4.192			
2,4,5-Trimethylthiazole ($\mu = 0.1$)	4.55			
2,4,6-Trinitroaniline (picramide)	−10.23(+1)			
2,4,6-Trinitrobenzene acid	0.654			
2,2,2-Trinitroethanol	2.36			
Trinitromethane (20°C)	0.17			
Triphenylacetic acid	3.96			
Tripropylamine	10.66(+1)			
Tris(2-hydroxyethyl)amine	7.762(+1)			
Tri(hydroxymethyl)aminomethane (TRIS)	8.08(+1)			
2-[Tris(hydroxymethyl)methyl amino]-1-ethanesulfonic acid (TES)	7.50			
3-[Tris(hydroxymethyl)methyl amino]-1-propanesulfonic acid (TAPS) (20°C)	8.4			
N-[Tris(hydroxymethyl)methyl]-glycine (tricine)	2.023(+1)	8.135		
Tris(trimethylsilyl)amine	4.70(+1)			
Trithiocarbonic acid (20°C)	2.64			
Tropacocaine (15°C)	9.88(+1)			
3-Tropanol (tropine)	10.33(+1)			
Trypsin ($\mu = 0.1$)	6.25			
L-Tryptophan	2.38(+1)	9.39(0)		
DL-Tyrosine	2.18(+1)	9.11(0)	10.6(OH)	

TABLE 2.60   *pK*, Values of Organic Materials in Water at 25°C (*Continued*)

Substance	pK_1	pK_2	pK_3	pK_4
Tyrosine amide	7.48	9.89		
Tyrosine, ethyl ester	7.33	9.80		
Tyrosylarginine ($\mu = 0.01$)	2.65(+1)	7.39(0)	9.36(−1)	11.62(−2)
Tyrosyltyrosine	3.52(+1)	7.68(0)	9.80(−1)	10.26(−2)
α-Ureidobutanoic acid	3.886(0)			
γ-Ureidobutanoic acid	4.683(0)			
β-Ureidopropanoic acid	4.487(0)			
Uric acid	5.40	5.53		
Uridine	9.30			
Uridine-5'-diphosphoric acid	7.16			
Uridine-5'-phosphoric acid (5'-uridylic acid)	6.63			
Uridine-5'-triphosphoric acid	7.58			
DL-Valine	2.32(+1)	9.61(0)		
L-Valine	2.296(+1)	9.79(0)		
Valine amide ($\mu = 0.2$)	8.00			
L-Valine, methyl ester	7.49(+1)			
L-Valylglycine	3.23(+1)	8.00(0)		
Vetramine	7.49(+1)			
Veratrine	8.85(+1)			
Vinylmethylamine	9.69(+1)			
2-Vinylpyridine	4.98(+1)			
4-Vinylpyridine	5.62(+1)			
Vitamin B$_{12}$	7.64(+1)			
Xanthine (40°C)	0.68(+1)			
Xanthosine	<2.5(+1)	5.67(0)	12.00(−1)	
Xylenol Orange [pK$_5$ 10.46(−4); pK$_6$ 12.28(−5)]	—	2.58(−1)	3.23(−2)	6.37(−3)
D-(+)-Xylose	12.15(0)			
Zincon	—	4	7.85	15

TABLE 2.61 Selected Equilibrium Constants in Aqueous Solution at Various Temperatures

Abbreviations Used in the Table

(+1), protonated cation
(0), neutral molecule
(−1), singly ionized anion
(−2), doubly ionized anion
pK_{auto}, negative logarithm (base 10) of autoprotolysis constant
pK_{sp}, negative logarithm (base 10) of solubility product

Substance	*Temperature, °C* 0	5	10	15	20	25	30	35	40	50
Acetic acid (0)	4.780	4.770	4.762	4.758	4.757	4.756	4.757	4.762	4.769	4.787
DL-N-Acetylalanine (+1)		3.699	3.699	3.703	3.708	3.715	3.725	3.733	3.745	3.774
β-Acetylaminopropionic (+1)		4.479	4.465	4.465	4.449	4.445	4.444	4.443	4.445	4.457
N-Acetylglycine (+1)		3.682	3.676	3.673	3.667	3.670	3.673	3.678	3.685	3.706
α-Alanine										
(+1)	2.42		2.39		2.35	2.34	2.33	2.33	2.33	2.33
(0)	10.59		10.29		10.01	9.87	9.74	9.62	9.49	9.26
2-Aminobenzenesulfonic acid (0), pK_2	2.633	2.591	2.556	2.521	2.448	2.459	2.431	2.404	2.380	2.338
3-Aminobenzenesulfonic acid (0), pK_2	4.075	4.002	3.932	3.865	3.799	3.738	3.679	3.622	3.567	3.464
4-Aminobenzenesulfonic acid (0), pK_2	3.521	3.457	3.398	3.338	3.283	3.227	3.176	3.126	3.079	2.989
3-Aminobenzoic acid (0)					4.90	4.79	4.75		4.68	4.60
4-Aminobenzoic acid (0)					4.95	4.85	4.90		4.95	5.10
2-Aminobutyric acid										
(+1)			2.334			2.286		2.289$^{37.5°C.}$		2.297
(0)			10.530			9.380		9.518$^{37.5°C}$		9.234
4-Aminobutyric acid										
(+1)			4.057	4.046	4.038	4.031	4.027	4.025	4.027	4.032
(0)			11.026	10.867	10.706	10.556	10.409	10.269	10.114	9.874
2-Aminoethylsulfonic acid (0)			9.452	9.316	9.186	9.061	8.940	8.824	8.712	9.499
2-Amino-3-methylpentanoic acid										
(+1)	2.365$^{1°C}$		2.338$^{12.5°C}$			2.320		2.317$^{37.5°C}$		2.332
(0)	10.460$^{1°C}$		10.100$^{12.5°C}$			9.758		9.439$^{37.5°C}$		9.157

2-Amino-2-methyl-1,3-propanediol	9.612	9.433	9.266	9.104	8.951	8.801	8.659	8.519	8.385	8.132
2-Amino-2-methylpropionic acid										
(+1)	$2.419^{1°C}$		$2.380^{12.5°C}$			2.357		$2.351^{37.5°C}$		2.356
(0)	$10.960^{1°C}$		$10.580^{12.5°C}$			10.205		$9.872^{37.5°C}$		9.561
2-Aminopentanoic acid										
(+1)	$2.376^{1°C}$		2.347			2.318			2.309	2.313
(0)	$10.508^{1°C}$		$10.154^{12.5°C}$			9.808		$9.490^{37.5°C}$		9.198
3-Aminopropionic acid										
(+1)	3.656	3.627	3.583			3.551		3.524	3.517	
(0)	11.000	10.830	10.526			10.235		9.963	9.842	
4-Aminopyridine (+1)	9.873	9.704	9.549	9.398	9.252	9.114	8.978	8.846	8.717	8.477
Ammonium ion (+1)	10.081	9.904	9.731	9.564	9.400	9.245	9.093	8.947	8.805	8.539
Arginine										
(+1)	1.914	1.885	1.870	1.849	1.837	1.823	1.814	1.801	1.800	1.787
(0)	9.718	9.563	9.407	9.270	9.123	8.994	8.859	8.739	8.614	8.385
Barbituric acid										
(+1)				3.969	3.980	4.02	4.00	4.008	4.017	4.032
(0)				8.493	8.435	8.372	8.302	8.227	8.147	7.974
Benzoic acid (0)		4.231	4.220	4.215	4.206	4.204	4.203	4.207	4.219	4.223
Boric acid (0)	9.508	9.439	9.380	9.327	9.280	9.236	9.197	9.161	9.132	9.080
Bromoacetic acid (0)				2.875	2.887	2.902	2.918	2.936		
3-Bromobenzoic acid (0)				3.818	3.813	3.810	3.808	3.810	3.813	
4-Bromobenzoic acid (0)				4.011	4.005	3.99	4.001	4.001	4.003	
Bromopropynoic acid (0)		1.786		1.814	1.839	1.855	1.879	1.900	1.919	
3-tert-Butylbenzoic acid (0)				4.266	4.231	4.199	4.170	4.143	4.119	
4-tert-Butylbenzoic acid (0)				4.463	4.425	4.389	4.354	4.320	4.287	
2-Butynoic acid (0)		2.618		2.626	2.611	2.620	2.618	2.621	2.631	
Butyric acid (0)	4.806	4.804	4.803	4.805	4.810	4.817	4.827	4.840	4.854	4.885
DL-N-Carbamoylalanine (+1)		3.898	3.894	3.891	3.890	3.892	3.896	3.902	3.908	3.931
N-Carbamoylglycine (+1)		3.911	3.900	3.889	3.879	3.876	3.874	3.873	3.875	3.888
Carbon dioxide + water										
(0)	6.577	6.517	6.465	6.429	6.382	6.352	6.327	6.309	6.296	6.285
(−1)	10.627	10.558	10.499	10.431	10.377	10.329	10.290	10.250	10.220	10.172
Chloroacetic acid (0)				2.845	2.856	2.867	2.883	2.900		
3-Chlorobenzoic acid (0)				3.838	3.831	3.83	3.825	3.826	3.829	

(Continued)

TABLE 2.61 Selected Equilibrium Constants in Aqueous Solution at Various Temperatures (*Continued*)

Substance	0	5	10	15	20	25	30	35	40	50
4-Chlorobenzoic acid (0)				4.000	3.991	3.986	3.981	3.980	3.981	
Chloropropynoic acid (0)			1.766	1.796	1.820	1.845	1.864	1.879	1.893	
Citric acid										
(0)	3.220	3.200	3.176	3.160	3.142	3.128	3.116	3.109	3.099	3.095
(−1)	4.837	4.813	4.797		4.769	4.761	4.755	4.751	4.750	4.757
(−2)	6.393	6.386	6.383	6.384	6.388	6.396	6.406	6.423	6.439	6.484
Cyanoacetic acid (0)		2.445	2.447	2.452	2.460	2.460	2.482	2.496	2.511	
2-Cyano-2-methylpropionic acid (0)		2.342	2.360	2.379	2.400	2.422	2.446	2.471	2.498	
5,5-Diethylbarbituric acid (0)	8.40	8.30	8.22	8.169	8.094	8.020	7.948	7.877	7.808	7.673
Diethylmalonic acid										
(0)			2.129	2.136	2.144	2.151	2.160	2.172	2.187	
(−1)			7.400	7.401	7.408	7.417	7.428	7.441	7.457	
2,3-Dimethylbenzoic acid (0)				3.663	3.687	3.771	3.726	3.762	3.788	
2,4-Dimethylbenzoic acid (0)				4.154	4.187	4.217	4.244	4.268	4.290	
2,5-Dimethylbenzoic acid (0)				3.911	3.954	3.990	4.020	4.045	4.065	
2,6-Dimethylbenzoic acid (0)				3.234	3.304	3.362	3.409	3.445	3.472	
3,5-Dimethylbenzoic acid (0)				4.292	4.299	4.302	4.304	4.306	4.306	
N,N′-Dimethylethyleneamine-N,N′-diacetic acid										
(0)	6.294		6.169		6.047		5.926		5.803	
(−1)	10.446		10.268		10.068		9.882		9.684	
N,N-Dimethylglycine (0)		10.34		10.14		9.94		9.76		
3,5-Dinitrobenzoic acid (0)			2.60		2.73		2.85		2.96	3.07
2-Ethylbutyric acid (0)	4.623		4.664		4.710	4.751	4.758		4.812	4.869
5-Ethyl-5-phenylbarbituric acid (0)				7.592	7.517	7.445	7.377	7.311	7.248	7.130
Fluoroacetic acid (0)				2.555	2.571	2.586	2.604	2.624		
Formic acid (0)	3.786	3.772	3.762	3.757	3.753	3.751	3.752	3.758	3.766	3.782
2-Furancarboxylic acid (0)						3.164	3.200	3.216	3.239	
Glucose-1-phosphate (0)		6.506	6.500	6.499	6.500	6.504	6.510	6.519	6.531	6.561
Glycerol-1-phosphoric acid (−1)		6.642	6.641	6.643	6.648	6.656	6.666	6.679	6.695	6.733
Glycerol-2-phosphoric acid (0)		1.223	1.245	1.271	1.301	1.335	1.372	1.413	1.457	1.554

Temperature, °C

Species										
(−1)	6.712	6.679	6.666	6.657	6.650	6.646	6.646	6.650	6.657	
Glycine (+1)	2.32	2.327	2.33	2.34	2.351	2.36	2.380	2.397		
Glycine (0)	9.19	9.412	9.53	9.65	9.780	9.91	10.044	10.193	10.34	
Glycolic acid (0)	3.849		3.833$^{37.5°C}$		3.831			3.844$^{12.5°C}$		3.875
Glycylasparagine (+1)	2.959	2.947	2.944	2.942	2.942	2.943	2.952	2.958	2.968	
N-Glycylglycine (+1)	3.159				3.126					3.201
N-Glycylglycine (0)	7.668		7.948$^{37.5°C}$		8.252			8.594$^{12.5°C}$		
Hexanoic acid (0)	4.920	4.890		4.865		4.849		4.839		4.840
Hydrogen cyanide (0)		8.88	8.99	9.11	9.21	9.36	9.49	9.63		
Hydrogen peroxide (0)	11.21		11.45	11.55	11.65	11.75	11.86			12.23
Hydrogen sulfide (0)	6.69	6.79	6.82	6.90	6.97	7.05	7.13	7.24	7.33	
Hydrogen sulfide (−1)			12.6	12.75	12.90		13.2		13.5	
4-Hydroxybenzoic acid (0)		4.578	4.576	4.577	4.582	4.586	4.596			
Hydroxylamine (0)			5.730		5.948	6.063	6.186			
2-Hydroxy-1-naphthoic acid (0)	3.26	3.19		3.24		3.29				
2-Hydroxy-1-naphthoic acid (−1)	9.58	9.61		9.65		9.68				
4-Hydroxyproline (+1)	1.796		1.798$^{37.5°C}$		1.818			1.850$^{12.5°C}$		1.900$^{1°C}$
4-Hydroxyproline (0)	9.138		9.394$^{37.5°C}$		9.662			9.958$^{12.5°C}$		10.274$^{1°C}$
2-Hydroxypropionic acid (0)	3.895	3.873	3.867	3.861	3.858	3.857	3.861	3.868	3.873	3.880
DL-2-Hydroxysuccinic acid (0)	3.445	3.444	3.446	3.452	3.458	3.472	3.482	3.494	3.520	3.537
DL-2-Hydroxysuccinic acid (−1)	5.149	5.117	5.104	5.099	5.097	5.096	5.096	5.098	5.108	5.119
Hypobromous acid (0)		8.37$^{45°C}$	8.47		8.60		8.83			
Hypochlorous acid (0)	7.05		7.46	7.50	7.54	7.58	7.63	7.69	7.75	7.82
Imidazole (+1)	6.497	6.685	6.784	6.887	6.993	7.103	7.216	7.334	7.467	7.581
Iodoacetic acid (0)			3.213	3.193	3.175	3.158	3.143			
DL-Isoleucine (+1)	2.332		2.317$^{37.5°C}$		2.318			2.338$^{12.5°C}$		2.365
DL-Isoleucine (0)	9.157		9.439$^{37.5°C}$		9.758			10.100$^{12.5°C}$		10.460
Isopropylmalonic acid, mononitrile (0)		2.481	2.452	2.427	2.401	2.365	2.343	2.320	2.259	
Lactic acid (0)	3.895	3.873	3.867	3.861	3.858	3.857	3.862	3.868	3.873	3.880
Lead sulfate, pK_{sp}	7.63		7.73		7.80		7.87			8.01
DL-Leucine (+1)	2.333		2.327$^{37.5°C}$		2.328			2.348$^{12.5°C}$		2.383$^{1°C}$
DL-Leucine (0)	9.142		9.434$^{37.5°C}$		9.744			10.095$^{12.5°C}$		10.458$^{1°C}$

(Continued)

TABLE 2.61 Selected Equilibrium Constants in Aqueous Solution at Various Temperatures (*Continued*)

Substance	Temperature, °C									
	0	5	10	15	20	25	30	35	40	50
Malonic acid (−1)	5.670	5.665	5.667	5.673	5.683	5.696	5.710	5.730	5.753	5.803
Mannose (0)			12.45			12.08			11.81	
Mercury(I) chloride, pK_{sp}		17.12	18.65	18.48	18.27	17.88		16.79		
Methanol (solvent), pK_{auto}				16.84		16.71	16.65	16.53		
Methylamine (+1)	11.496		11.130		10.787	10.62	10.466		10.161	9.876
Methylaminodiacetic acid										
(0)	2.138		2.142		2.146		2.150		2.154	
(−1)	10.474		10.287		10.088		9.920		9.763	
3-Methylbenzoic acid (0)				4.303	4.285	4.269	4.256	4.244	4.235	
4-Methylbenzoic acid (0)				4.390	4.376	4.362	4.349	4.336	4.322	
3-Methylbutyric acid (0)	4.726		4.742		4.767		4.794		4.831	4.871
4-Methylpentanoic acid (0)	4.827		4.827		4.837		4.853		4.879	4.908
5-Methyl-5-phenylbarbituric acid										
(0)				8.104	8.057	8.011	7.966	7.922	7.879	7.797
2-Methylpropionic acid (0)	4.825		4.827		4.840	4.853	4.886		4.918	4.955
2-Methyl-2-propylamine (+1)		11.439	11.240	11.048	10.862	10.682	10.511	10.341		
Nitric acid (0)	−1.65					−1.38				−1.20
Nitrilotriacetic acid										
(0)	1.69		1.65	1.65	1.65		1.66		1.67	
(−1)	2.95		2.95	2.94	2.94		2.96		2.98	
(−2)	10.59		10.45	10.33	10.33		10.23			
4-Nitrobenzoic acid (0)				3.448	3.444	3.441	3.441	3.442	3.445	
Nitrous acid (0)				3.244	3.177	3.138		3.100		
DL-Norleucine										
(+1)	2.394		$2.356^{12.5°C}$			2.335		$2.324^{37.5°C}$		2.328
(0)	10.564		$10.190^{12.5°C}$			9.834		$9.513^{37.5°C}$		9.224
Oxalic acid (−1)	4.210	4.216	4.227	4.240	4.254	4.272	4.295	4.318	4.349	4.409
2,4-Pentanedione (0)	9.07					8.95			8.90	
Pentanoic acid (0)	4.823		4.763		4.835	4.842	4.851		4.861	4.906
Phenylalanine (0)			9.75			9.31			8.96	
Phosphoric acid (0)	2.056	2.073	2.088	2.107	2.127	2.148	2.171	2.196	2.224	2.277
(−1)	7.313	7.282	7.254	7.231	7.213	7.198	7.189	7.185	7.181	7.183

Compound										
o-Phthalic acid (0)	2.925	2.927	2.931	2.937	2.943	2.950	2.958	2.967	2.978	3.001
(−1)	5.432	5.418	5.410	5.405	5.405	5.408	5.416	5.427	5.442	5.485
Piperidine (+1)	11.963	11.786	11.613	11.443	11.280	11.123	10.974	10.818	10.670	10.384
Proline (+1)	2.011		$1.964^{12.5°C}$		1.952			$1.950^{37.5°C}$		1.958
(0)	11.296		$10.972^{12.5°C}$		10.640			$10.342^{37.5°C}$		10.064
Propenoic acid (0)				4.267	4.250	4.247	4.249	4.267	4.301	
N-Propionylglycine (+1)		3.728	3.723	3.718	3.716	3.718	3.721	3.725	3.731	3.750
Propynoic acid (0)			1.791	1.829	1.867	1.887	1.940	1.932	1.963	
Pyrrolidine (+1)	12.17	11.98	11.81	11.63	11.43	11.30	11.15	10.99	10.84	11.56
Serine (+1)	$2.296^{1°C}$		$2.232^{12.5°C}$			2.186		$2.154^{37.5°C}$		2.132
(0)	$9.880^{1°C}$		$9.542^{12.5°C}$			9.208		$8.904^{37.5°C}$		8.628
Silver bromide, pK_{sp}	13.33			12.83	12.57	12.30	12.07	11.83	11.61	11.19
Silver chloride, pK_{sp}	10.595			10.152		9.749		9.381	9.21	8.88
Succinic acid (0)	4.285	4.263	4.245	4.232	4.218	4.207	4.198	4.191	4.188	4.186
(−1)	5.674	5.660	5.649	5.642	5.639	5.635	5.641	5.647	5.654	5.680
Sulfuric acid (−1)	1.778	$1.812^{4.3°C}$	1.74	1.894	1.894	1.987	2.05	2.095	2.17	2.246
Sulfurous acid (0)	1.63					1.89		1.98		2.12
D-Tartaric acid (0)	3.118	3.095	3.075	3.057	3.044	3.036	3.025	3.019	3.018	3.021
(−1)	4.426	4.407	4.391	4.381	4.372	4.366	4.365	4.367	4.372	4.391
2,3,5,6-Tetramethylbenzoic acid (0)				3.310	3.367	3.415	3.453	3.483	3.505	
Threonine (+1)	$2.200^{1°C}$		$2.132^{12.5°C}$			2.088		$2.070^{37.5°C}$		2.055
(0)	$9.748^{1°C}$		$9.420^{12.5°C}$			9.100		$8.812^{37.5°C}$		8.548
o-Toluidine (0)				4.58	4.495	4.45	4.345	4.28	4.20	
1,2,4-Triazole (+1)				2.451	2.418	2.386	2.327			
(0)				10.205	10.083	9.972	9.768			
3,4,5-Trihydroxybenzoic acid (0)				4.19	4.19	4.19	4.30		4.38	4.53
Tris(2-hydroxyethyl)amine (+1)	8.290	8.173	8.067	7.963	7.861	7.762	7.666	7.570	7.477	7.299
2,4,6-Trimethylbenzoic (0)				3.325	3.391	3.448	3.498	3.541	3.577	
3-Trimethylsilylbenzoic acid (0)				4.142	4.116	4.089	4.060	4.029	3.996	
4-Trimethylsilylbenzoic acid (0)				4.270	4.230	4.192	4.155	4.119	4.084	
β-Ureidopropionic acid (0)	4.514		4.505	4.497	4.490	4.487	4.486	4.486	4.488	4.500
DL-Valine (+1)	2.320		$2.297^{12.5°C}$			2.296		$2.292^{37.5°C}$		2.310
(0)	10.413		$10.064^{12.5°C}$			9.719		$9.405^{37.5°C}$		9.124

TABLE 2.62 pK, Values for Proton-Transfer Reactions in Non-Aqueous Solvents

Acid	Methanol	Ethanol	Other solvents
Acetic acid	9.52	10.32	11.4[a], 9.75[d]
p-Aminobenzoic acid	10.25		
Ammonium ion	10.7		6.40[b]
Anilinium ion	6.0	5.70	
Benzoic acid		10.72	10.0[a]
Bromocresol purple	11.3	11.5	
Bromocresol green	9.8	10.65	
Bromophenol blue	8.9	9.5	
Bromothymol blue	12.4	13.2	
Di-n-butylammonium ion			10.3[a]
o-Chloroanilinium ion	3.4		
Cyanoacetic acid		7.49	
2,5-Dichloroanilinium ion			9.48[b]
Dimethylaminoazobenzene		5.2	6.32[b]
N,N'-Dimethylanilinium ion		4.37	
Formic acid		9.15	
Hydrobromic acid			5.5[c]
Hydrochloric acid			8.55[b], 8.9[c]
Methyl orange	3.8	3.4	
Methyl red (acid range)	4.1	3.55	
(alkaline range)	9.2	10.45	
Methyl yellow	3.4	3.55	
Neutral red	8.2	8.2	
o-Nitrobenzoic acid	7.6		
m-Nitrobenzoic acid	8.3		
p-Nitrobenzoic acid	8.4		
Perchloric acid			4.87[b]
Phenol	14.0		
Phenol red	12.8	13.4	
Phthalic acid, pK_2	11.65		11.5[d], 6.10[d] (pK_1)
Picric acid	3.8	3.8	8.9[c]
Pyridinium ion			6.1[b]
Salicylic acid	8.7	7.9	
Stearic acid	10.0		
Succinic acid, pK_2	11.4		
Sulfuric acid, pK_1			7.24[b,c]
Tartaric acid, pK_2	9.9		
Thymol blue (alkaline range)	14.0	15.2	
(acid range)	4.7	5.35	
Thymolbenzein (acid range)	3.5		
(alkaline range)	13.1		
p-Toluenesulfonic acid			8.44[b]
p-Toluidinium ion		6.24	
Tribenzylammonium ion			5.40[b]
Tropeoline 00	2.2		
Urea (protonated cation)			6.96[b]
Veronal	12.6		

[a] Dimethylsulfoxide. [b] Glacial acetic acid. [c] Acetonitrile. [d] Acetone + 10% water.

2.16 INDICATORS

An acid–base indicator is a conjugate acid–base pair of which the acid form and the base form are of different colors. These indicators are used to show the relative acidity or alkalinity of the test material.

Acid–base indicators are dyes that are themselves weak acids and bases. The conjugate acid–base forms of the dye are of different colors. An indicator does not change color from pure acid to pure alkaline at specific hydrogen ion concentration, but, rather, color change occurs over a range of hydrogen ion concentrations. This range is termed the *color change interval* and is expressed as a pH range. The chemical structures of the dyes are often complex but can be represented chemically by the symbol HIn. The acid–base indicator reaction is represented as:

$$HIn + H_2O \rightarrow H_3O^+ + In \qquad (1)$$

TABLE 2.63 Acid–Base Indicators

Indicator	pH range		Color	
	Minimum	Maximum	Acid	Alkaline
Brilliant cresyl blue	0.0	1.0	Red-orange	Blue
Methyl violet	0.0	1.6	Yellow	Blue
Crystal violet	0.0	1.8	Yellow	Blue
Ethyl violet	0.0	2.4	Yellow	Blue
Methyl violet 6B	0.1	1.5	Yellow	Blue
Cresyl red	0.2	1.8	Red	Yellow
2-(p-Dimethylaminophenylazo) pyridine	0.2	1.8	Yellow	Blue
Malachite green	0.2	1.8	Yellow	Blue-green
Methyl green	0.2	1.8	Yellow	Blue
Cresol red (o-Cresolsulfonephthalein)	1.0	2.0	Red	Yellow
Quinaldine red	1.0	2.2	Colorless	Red
p-Methyl red	1.0	3.0	Red	Yellow
Metanil yellow	1.2	2.3	Red	Yellow
Pentamethoxy red	1.2	2.3	Red-violet	Colorless
Metanil yellow	1.2	2.4	Red	Yellow
p-Phenylazodiphenylamine	1.2	2.6	Red	Yellow
Thymol blue (Thymolsulfonephthalein)	1.2	2.8	Red	Yellow
m-Cresol purple	1.2	2.8	Red	Yellow
p-Xylenol blue	1.2	2.8	Red	Yellow
Benzopurpurin 4B	1.2	3.8	Violet	Red
Tropeolin OO	1.3	3.2	Red	Yellow
Orange IV	1.4	2.8	Red	Yellow
4-o-Tolylazo-o-toluidine	1.4	2.8	Orange	Yellow
Methyl violet 6B	1.5	3.2	Blue	Violet
Phloxine B	2.1	4.1	Colorless	Pink
Erythrosine, disodium salt	2.2	3.6	Orange	Red
Benzopupurine 4B	2.2	4.2	Violet	Red
N,N-dimethyl-p-(m-tolylazo) aniline	2.6	4.8	Red	Yellow
2,4-Dinitrophenol	2.8	4.0	Colorless	Yellow
N,N-Dimethyl-p-phenylazoaniline	2.8	4.4	Red	Yellow
Methyl yellow	2.9	4.0	Red	Yellow
Bromophenol blue	3.0	4.6	Yellow	Blue-violet
Tetrabromophenol blue	3.0	4.6	Yellow	Blue
Direct purple	3.0	4.6	Blue-purple	Red

(Continued)

TABLE 2.63 Acid–Base Indicators (*Continued*)

Indicator	pH range		Color	
	Minimum	Maximum	Acid	Alkaline
Congo red	3.1	4.9	Blue	Red
Methyl orange	3.1	4.4	Red	Yellow
Bromochlorophenol blue	3.2	4.8	Yellow	Blue
Ethyl orange	3.4	4.8	Red	Yellow
p-Ethoxychrysoidine	3.5	5.5	Red	Yellow
Alizarin sodium sulfonate	3.7	5.2	Yellow	Violet
α-Naphthyl red	3.7	5.7	Red	Yellow
Bromocresol green	3.8	5.4	Yellow	Blue
Resazurin	3.8	6.4	Orange	Violet
Bromophenol green	4.0	5.6	Yellow	Blue
2,5-Dinitrophenol	4.0	5.8	Colorless	Yellow
Methyl red	4.2	6.2	Red	Yellow
2-(p-Dimethylaminophenylazo) pyridine	4.4	5.6	Red	Yellow
Lacmoid	4.4	6.2	Red	Blue
Azolitmin	4.5	8.3	Red	Blue
Litmus	4.5	8.3	Red	Blue
Alizarin red S	4.6	6.0	Yellow	Red
Chlorophenol red	4.8	6.4	Yellow	Red
Cochineal	4.8	6.2	Red	Violet
Propyl red	4.8	6.6	Red	Yellow
Hematoxylin	5.0	6.0	Red	Blue
Bromocresol purple	5.2	6.8	Yellow	Violet
Bromophenol red	5.2	7.0	Yellow	Red
Chlorophenol red	5.4	6.8	Yellow	Red
p-Nitrophenol	5.6	6.6	Colorless	Yellow
Alizarin	5.6	7.2	Yellow	Red
Bromothymol blue	6.0	7.6	Yellow	Blue
Indo-oxine	6.0	8.0	Red	Blue
Bromophenol blue	6.2	7.6	Yellow	Blue
m-Dinitrobenzoylene urea	6.4	8.0	Colorless	Yellow
Phenol red (Phenolsulfonephthalein)	6.4	8.0	Yellow	Red
Rosolic acid	6.4	8.0	Yellow	Red
Brilliant yellow	6.6	7.9	Yellow	Orange
Quinoline blue	6.6	8.6	Colorless	Blue
Neutral red	6.8	8.0	Red	Orange
Phenol red	6.8	8.4	Yellow	Yellow
m-Nitrophenol	6.8	8.6	Colorless	Yellow
Cresol red (o-Cresolsulfonephthalein)	7.0	8.8	Yellow	Red
α-Naphtholphthalein	7.3	8.8	Yellow	Blue
Curcumin	7.4	8.6	Yellow	Red
m-Cresol purple (m-Cresolsulfonephthalein)	7.4	9.0	Yellow	Violet
Tropeolin OOO	7.6	8.9	Yellow	Rose-red
2,6-Divanillydenecyclohexanone	7.8	9.4	Yellow	Red
Thymol blue (Thymolsulfonephthalein)	8.0	9.6	Yellow	Purple
p-Xylenol blue	8.0	9.6	Yellow	Blue
Turmeric	8.0	10.0	Yellow	Orange
Phenolphthalein	8.0	10.0	Colorless	Red
o-Cresolphthalein	8.2	9.8	Colorless	Red
p-Naphtholphthalein	8.2	10.0	Colorless	Pink
Ethyl bis(2,4-dimethylphenyl acetate)	8.4	9.6	Colorless	Blue

TABLE 2.63   Acid–Base Indicators (*Continued*)

Indicator	pH range		Color	
	Minimum	Maximum	Acid	Alkaline
Ethyl bis(2,4-dinitrophenyl acetate)	8.4	9.6	Colorless	Blue
α-Naphtholbenzein	8.5	9.8	Yellow	Green
Thymolphthalein	9.4	10.6	Colorless	Blue
Nile blue A	10.0	11.0	Blue	Purple
Alizarin yellow CG	10.0	12.0	Yellow	Lilac
Alizarin yellow R	10.2	12.0	Yellow	Orange red
Salicyl yellow	10.0	12.0	Yellow	Orange-brown
Diazo violet	10.1	12.0	Yellow	Violet
Nile blue	10.1	11.1	Blue	Red
Curcumin	10.2	11.8	Yellow	Red
Malachite green hydrochloride	10.2	12.5	Green-blue	Colorless
Methyl blue	10.6	13.4	Blue	Pale violet
Brilliant cresyl blue	10.8	12.0	Blue	Yellow
Alizarin	11.0	12.4	Red	Purple
Nitramine	11.0	13.0	Colorless	Orange brown
Poirier's blue	11.0	13.0	Blue	Violet-pink
Tropeolin O	11.0	13.0	Yellow	Orange
Indigo carmine	11.4	13.0	Blue	Yellow
Sodium indigosulfonate	11.4	13.0	Blue	Yellow
Orange G	11.5	14.0	Yellow	Pink
2,4,6-Trinitrotoluene	11.7	12.8	Colorless	Orange
1,3,5-Trinitrobenzene	12.0	14.0	Colorless	Orange
2,4,6-Trinitrobenzoic acid	12.0	13.4	Blue	Violet-pink
Clayton yellow	12.2	13.2	Yellow	Amber

TABLE 2.64   Mixed Indicators

Mixed indicators give sharp color changes and are especially useful in titrating to a given titration exponent (*pI*). The information given in this table is from the two-volume work *Volumetric Analysis* by Kolthoff and Stenger, published by Interscience Publishers, Inc., New York, 1942 and 1947, and reproduced with their permission.

Composition of indicator solution	*pI*	Color		Notes
		Acid	Alkaline	
1 part 0.1% methyl yellow in alc. 1 part 0.1% methylene blue in alc. *	3.25	Blue-violet	Green	Still green at pH 3.4, blue-violet at 3.2†
1 part 0.14% xylene cyanol FF in alc. 1 part 0.1% methyl orange in aq. *	3.8	Violet	Green	Color is gray at pH 3.8
1 part 0.1% methyl orange in aq. 1 part 0.25% indigo carmine in aq. *	4.1	Violet	Green	Good indicator, especially in artificial light
1 part 0.1% methyl orange in aq. 1 part 0.1% aniline blue in aq.	4.3	Violet	Green	Yellow at pH 3.5, greenish yellow at 4.0, weakly green at 4.3
1 part 0.1% bromcresol green sodium salt in aq. 1 part 0.02% methyl orange in aq.	4.3	Orange	Blue-green	
3 parts 0.1% bromcresol green in alc. 1 part 0.2% methyl red in alc.	5.1	Wine-red	Green	Very sharp color change†
1 part 0.2% methyl red in alc. 1 part 0.1% methylene blue in alc. *	5.4	Red-violet	Green	Color is red-violet at pH 5.2, a dirty blue at 5.4, and a dirty green at 5.6
1 part 0.1% chlorphenol red sodium salt in aq. 1 part 0.1% aniline blue in water	5.8	Green	Violet	Pale violet at pH 5.8
1 part 0.1% bromcresol green sodium salt in aq. 1 part 0.1% chlorphenol red sodium salt in aq.	6.1	Yellow-green	Blue-violet	Blue-green at pH 5.4, blue at 5.8, blue with a touch of violet at 6.0, blue-violet at 6.2
1 part 0.1% bromcresol purple sodium salt in aq. 1 part 0.1% bromthymol blue sodium salt in aq.	6.7	Yellow	Violet-blue	Yellow-violet at pH 6.2, violet at 6.6, blue-violet at 6.8
2 parts 0.1% bromthymol blue sodium salt in aq. 1 part 0.1% azolitmin in aq.	6.9	Violet	Blue	

Composition		pH			
1 part 0.1% neutral red in alc. 1 part 0.1% methylene blue in alc.	*	7.0	Violet-blue	Green	Violet blue at pH 7.0†
1 part 0.1% neutral red in alc. 1 part 0.1% bromthymol blue in alc.		7.2	Rose	Green	Dirty green at pH 7.4, pale rose at 7.2, clear rose at 7.0
2 parts 0.1% cyanine in 50% alc. 1 part 0.1% phenol red in 50% alc.		7.3	Yellow	Violet	Orange at pH 7.2, beautiful violet at 7.4, color fades on standing
1 part 0.1% bromthymol blue sodium salt in aq. 1 part 0.1% phenol red sodium salt in aq.		7.5	Yellow	Violet	Dirty green at pH 7.2, pale violet at 7.4, strong violet at 7.6†
1 part 0.1% cresol red sodium salt in aq. 3 parts 0.1% thymol blue sodium salt in aq.		8.3	Yellow	Violet	Rose at pH 8.2, distinctly violet at 8.4†
2 parts 0.1% α-naphtholphthalein in alc. 1 part 0.1% cresol red in alc.		8.3	Pale rose	Violet	Pale violet at pH 8.2, strong violet at 8.4
1 part 0.1% α-naphtholphthalein in alc. 3 parts 0.1% phenolphthalein in alc.		8.9	Pale rose	Violet	Pale green at pH 8.6, violet at 9.0
1 part 0.1% phenolphthalein in alc. 2 parts 0.1% methyl green in alc.	*	8.9	Green	Violet	Pale blue at pH 8.8, violet at 9.0
1 part 0.1% thymol blue in 50% alc. 3 parts 0.1% phenolphthalein in 50% alc.		9.0	Yellow	Violet	From yellow thru green to violet†
1 part 0.1% phenolphthalein in alc. 1 part 0.1% thymolphthalein in alc.		9.9	Colorless	Violet	Rose at pH 9.6, violet at 10; sharp color change
1 part 0.1% phenolphthalein in alc. 2 parts 0.2% Nile blue in alc.		10.0	Blue	Red	Violet at pH 10†
2 parts 0.1% thymolphthalein in alc. 1 part 0.1% alizarin yellow in alc.		10.2	Yellow	Violet	Sharp color change
2 parts 0.2% Nile blue in aq. 1 part 0.1% alizarin yellow in alc.		10.8	Green	Red-brown	

*Store in a dark bottle.　†Excellent indicator.

TABLE 2.65 Fluorescent Indicators

Name	pH range	Color change acid to base	Indicator solution
Benzoflavine	−0.3–1.7	Yellow to green	1
3,6-Dihydroxyphthalimide	0–2.4	Blue to green	1
	6.0–8.0	Green to yellow/green	
Eosin (tetrabromofluorescein)	0–3.0	Non-fl to green	4, 1%
4-Ethoxyacridone	1.2–3.2	Green to blue	1
3,6-Tetramethyldiaminoxanthone	1.2–3.4	Green to blue	1
Esculin	1.5–2.0	Weak blue to strong blue	
Anthranilic acid	1.5–3.0	Non-fl to light blue	2 (50% ethanol)
	4.5–6.0	Light blue to dark blue	
	12.5–14	Dark blue to non-fl	
3-Amino-1-naphthoic acid	1.5–3.0	Non-fl to green	2 (as sulfate
	4.0–6.0	Green to blue	in 50% ethanol)
	11.6–13.0	Blue to non-fl	
1-Naphthylamino-6-sulfonamide	1.9–3.9	Non-fl to green	3
(also the 1-, 7-)	9.6–13.0	Green to non-fl	
2-Naphthylamino-6-sulfonamide	1.9–3.9	Non-fl to dark blue	3
(also the 2-, 8-)	9.6–13.0	Dark blue to non-fl	
1-Naphthylamino-5-sulfonamide	2.0–4.0	Non-fl to yellow/orange	3
	9.5–13.0	Yellow/orange to non-fl	
1-Naphthoic acid	2.5–3.5	Non-fl to blue	4
Salicylic acid	2.5–4.0	Non-fl to dark blue	4 (0.5%)
Phloxin BA extra (tetrachlorotetrabromofluorescein)	2.5–4.0	Non-fl to dark blue	2
Erythrosin B (tetraiodofluorescein)	2.5–4.0	Non-fl to light green	4 (0.2%)
2-Naphthylamine	2.8–4.4	Non-fl to violet	1
Magdala red	3.0–4.0	Non-fl to purple	
p-Aminophenylbenzenesulfonamide	3.0–4.0	Non-fl to light blue	3
2-Hydroxy-3-naphthoic acid	3.0–6.8	Blue to green	4 (0.1%)
Chromotropic acid	3.1–4.4	Non-fl to light blue	4 (5%)
1-Naphthionic acid	3–4	Non-fl to blue	4
	10–12	Blue to yellow-green	
1-Naphthylamine	3.4–4.8	Non-fl to blue	1
5-Aminosalicylic acid	3.1–4.4	Non-fl to light green	1 (0.2% fresh)
Quinine	3.0–5.0	Blue to weak violet	1 (0.1%)
	9.5–10.0	Weak violet to non-fl	
o-Methoxybenzaldehyde	3.1–4.4	Non-fl to green	4 (0.2%)
o-Phenylenediamine	3.1–4.4	Green to non-fl	5
p-Phenylenediamine	3.1–4.4	Non-fl to orange/yellow	5
Morin (2′,4′,3,5,7-pentahydroxyflavone)	3.1–4.4	Non-fl to green	6 (0.2%)
	8–9.8	Green to yellow/green	
Thioflavine S	3.1–4.4	Dark blue to light blue	6 (0.2%)
Fluorescein	4.0–4.5	Pink/green to green	4 (1%)
Dichlorofluorescein	4.0–6.6	Blue green to green	1
β-Methylesculetin	4.0–6.2	Non-fl to blue	1
	9.0–10.0	Blue to light green	
Quininic acid	4.0–5.0	Yellow to blue	6 (satd)
β-Naphthoquinoline	4.4–6.3	Blue to non-fl	3
Resorufin (7-oxyphenoxazone)	4.4–6.4	Yellow to orange	

TABLE 2.65   Fluorescent Indicators (*Continued*)

Name	pH range	Color change acid to base	Indicator solution
Acridine	5.2–6.6	Green to violet	2
3,6-Dihydroxyxanthone	5.4–7.6	Non-fl to blue/violet	1
5,7-Dihydroxy-4-methylcoumarin	5.5–5.8	Light blue to dark blue	
3,6-Dihydroxyphthalic acid dinitrile	5.8–8.2	Blue to green	1
1,4-Dihydroxybenzenedisulfonic acid	6–7	Non-fl to light blue	4 (0.1%)
Luminol	6–7	Non-fl to blue	
2-Naphthol-6-sulfonic acid	5–7 to 8–9	Non-fl to blue	4
Quinoline	6.2–7.2	Blue to non-fl	6 (satd)
1-Naphthol-5-sulfonic acid	6.5–7.5	Non-fl to green	6 (satd)
Umbelliferone	6.5–8.0	Non-fl to blue	
Magnesium-8-hydroxyquinolinate	6.5–7.5	Non-fl to yellow	6 (0.1% in 0.01 M HCl)
Orcinaurine	6.5–8.0	Non-fl to green	6 (0.03%)
Diazo brilliant yellow	6.5–7.5	Non-fl to blue	
Coumaric acid	7.2–9.0	Non-fl to green	1
β-Methylumbelliferone	>7.0	Non-fl to blue	2 (0.3%)
Harmine	7.2–8.9	Blue to yellow	
2-Naphthol-6,8-disulfonic acid	7.5–9.1	Blue to light blue	4
Salicylaldehyde semicarbazone	7.6–8.0	Yellow to blue	2
1-Naphthol-2-sulfonic acid	8.0–9.0	Dark blue to light blue	4
Salicylaldehyde acetylhydrazone	8.3	Non-fl to green/blue	2
Salicylaldehyde thiosemicarbazone	8.4	Non-fl to blue/green	2
1-Naphthol-4-sulfonic acid	8.2	Dark blue to light blue	4
Naphthol AS	8.2–10.3	Non-fl to yellow/green	4
2-Naphthol	8.5–9.5	Non-fl to blue	2
Acridine orange	8.4–10.4	Non-fl to yellow/green	1
Orcinsulfonephthalein	8.6–10.0	Non-fl to yellow	
2-Naphthol-3,6-disulfonic acid	9.0–9.5	Dark blue to light blue	4
Ethoxyphenylnaphthostilbazonium chloride	9–11	Green to non-fl	1
o-Hydroxyphenylbenzothiazole	9.3	Non-fl to blue green	2
o-Hydroxyphenylbenzoxazole	9.3	Non-fl to blue/violet	2
o-Hydroxyphenylbenzimidazole	9.9	Non-fl to blue/violet	2
Coumarin	9.5–10.5	Non-fl to light green	
6,7-Dimethoxyisoquinoline-1-carboxylic acid	9.5–11.0	Yellow to blue	0.1% in glycerine/ ethanol/water in 2:2:18 ratio
1-Naphthylamino-4-sulfonamide	9.5–13.0	Dark blue to white/blue	3

Indicator solutions: 1, 1% solution in ethanol; 2, 0.1% solution in ethanol; 3, 0.05% solution in 90% ethanol; 4, sodium or potassium salt in distilled water; 5, 0.2% solution in 70% ethanol; 6, distilled water.

TABLE 2.66 Selected List of Oxidation–Reduction Indicators

Name	Reduction potential (30°C) in Volts at		Suitable pH range	Color change upon oxidation
	pH = 0	pH = 7		
Bis(5-bromo-1,10-phenanthroline) ruthenium(II) dinitrate	1.41*			Red to faint blue
Tris(5-nitro-1,10-phenanthroline) iron(II) sulfate	1.25*			Red to faint blue
Iron(II)-2,2′:2″-tripyridine sulfate	1.25*			Pink to faint blue
Tris(4,7-diphenyl-1,10-phenanthroline) iron(II) disulfate	1.13 (4.6 M H_2SO_4)* 0.87 (1.0 M H_2SO_4)*			Red to faint blue
o,m'-Diphenylaminedicarboxylic acid	1.12			Colorless to blue-violet
Setopaline	1.06 (*trans*)†			Yellow to orange
p-Nitrodiphenylamine	1.06			Colorless to violet
Tris(1,10-phenanthroline)-iron(II) sulfate	1.06 (1.00 M H_2SO_4)* 1.00 (3.0 M H_2SO_4)* 0.89 (6.0 M H_2SO_4)*			Red to faint blue
Setoglaucine O	1.01 (*trans*)†			Yellow-green to yellow-red
Xylene cyanole FF	1.00 (*trans*)†			Yellow-green to pink
Erioglaucine A	1.00 (*trans*)†			Green-yellow to bluish red
Eriogreen	0.99 (*trans*)†			Green-yellow to orange
Tris(2,2′-bipyridine)-iron(II) hydrochloride	0.97*			Red to faint blue
2-Carboxydiphenylamine [N-phenyl-anthranilic acid]	0.94			Colorless to pink
Benzidine dihydrochloride	0.92			Colorless to blue
o-Toluidine	0.87			Colorless to blue
Bis(1,10-phenanthroline)-osmium(II) perchlorate	0.859 (0.1 M H_2SO_4)			Green to pink
Diphenylamine-4-sulfonate (Na salt)	0.85			Colorless to violet
3,3′-Dimethoxybenzidine dihydrochloride [o-dianisidine]	0.85			Colorless to red
Ferrocyphen	0.81			Yellow to violet
4′-Ethoxy-2,4-diaminoazobenzene	0.76			Red to pale yellow
N,N-Diphenylbenzidine	0.76			Colorless to violet

Indicator				Color change
Diphenylamine	0.76			Colorless to violet
N,N-Dimethyl-p-phenylenediamine	0.76			Colorless to red
Variamine blue B hydrochloride	0.712‡	0.310	1.5–6.3	Colorless to blue
N-Phenyl-1,2,4-benzenetriamine	0.70			Colorless to red
Bindschedler's green	0.680‡	0.224	2–9.5	Colorless to blue
2,6-Dichloroindophenol (Na salt)	0.668‡	0.217	6.3–11.4	Colorless to blue
2,6-Dibromophenolindophenol	0.668‡	0.216	7.0–12.3	Colorless to blue
Brilliant cresyl blue [3-amino-9-dimethyl-amino-10-methylphenoxyazine chloride]	0.583	0.047	0–11	Red to faint blue
Iron(II)-tetrapyridine chloride	0.59			Colorless to violet
Thionine [Lauth's violet]	0.563‡	0.064	1–13	Colorless to blue
Starch (soluble potato, I_3 present)	0.54			Colorless to violet-blue
Gallocyanine (25°C)	0.532‡	0.021		Colorless to blue
Methylene blue	0.406‡	0.011	1–13	Colorless to blue
Nile blue A [aminonaphthodiethylamino-phenoxazine sulfate]	0.365‡	−0.119	1.4–12.3	Colorless to blue
Indigo-5,5',7,7'-tetrasulfonic acid (Na salt)	0.332‡	−0.046	<9	Colorless to blue
Indigo-5,5',7-trisulfonic acid (Na salt)	0.291‡	−0.081	<9	Colorless to blue
Indigo-5,5'-disulfonic acid (Na salt)	0.280‡	−0.125	<9	Colorless to violet-blue
Phenosafranine	0.262‡	−0.252	1–11	Colorless to blue
Indigo-5-monosulfonic acid (Na salt)	0.24‡	−0.157	<9	Colorless to violet-blue
Safranine T		−0.289	1–12	Colorless to violet-blue
Bis(dimethylglyoximato)-iron(II) chloride	0.155		6–10	Red to colorless
Induline scarlet	0.047‡	−0.299	3–8.6	Colorless to red
Neutral red		−0.323	2–11	Colorless to red-violet

*Transition point is at higher potential than the tabulated formal potential because the molar absorptivity of the reduced form is very much greater than that of the oxidized form.

†Trans = first noticeable color transition; often 60 mV less than $E°$.

‡Values of $E°$ are obtained by extrapolation from measurements in weakly acid or weakly alkaline systems.

TABLE 2.67 Indicators for Approximate pH Determination

No. 1. Dissolve 60 mg methyl yellow, 40 mg methyl red, 80 mg bromthymol blue, 100 mg thymol blue and 20 mg phenolphthalein in 100 ml of ethanol and add enough 0.1N NaOH to produce a yellow color.

No. 2. Dissolve 18.5 mg methyl red, 60 mg bromthymol blue and 64 mg phenolphthalein in 100 ml of 50% ethanol and add enough 0.1N NaOH to produce a green color.

| pH | Color | | pH | Color | |
	No. 1	No. 2		No. 1	No. 2
1	Cherry-red	Red	7	Yellowish-green	Greenish-yellow
2	Rose	Red	8	Green	Green
3	Red-orange	Red	9	Bluish-green	Greenish-blue
4	Orange-red	Deeper red	10	Blue	Violet
5	Orange	Orange-red	11	—	Reddish-violet
6	Yellow	Orange-yellow			

TABLE 2.68 Oxidation–Reduction Indicators

| Common name | Reference | Transition potential, volts (N hydrogen electrode = 0.000) | Color | |
			Reduced form	Oxidized form
p-Ethoxychrysoidine	1	0.76	Red	Yellow
Diphenylamine	2	0.776	Colorless	Purple
Diphenylbenzidine	3	0.776	Colorless	Purple
Diphenylamine-sulfonic acid or barium salt	4	0.84	Colorless	Purple
Naphthidine	5	—	Colorless	Red
Dimethylferroin	6	0.97	Red	Yellowish-green
Eriogreen B	7	0.99	Yellow	Orange
Erioglaucin A	7	1.0	Yellowish-green	Red
Xylene cyanole FF	11	1.0		
2,2′-Dipyridyl ferrous ion	6	1.03	Red	Colorless
N-Phenylanthranilic acid	8	1.08	Colorless	Pink
Methylferroin	6	1.08	Red	Pale-blue
Ferroin (o-phenanthrolineferrous ion)	9	1.12	Red	Pale-blue
Chloroferroin	6	1.17	Red	Pale-blue
Nitroferroin	6	1.31	Red	Pale greenish-blue
α-Naphtolflavone	10	—	Pale straw	Brownish-orange

2.17   *ELECTRODE POTENTIALS*

The potential of a polarographic or voltammetric *indicator electrode* at the point, on the rising part of a polarographic or voltammetric wave, where the difference between the total current and the *residual current* is equal to one-half of the *limiting current*. The quarter-wave potential, the three-quarter-wave potential, etc., may be similarly defined.

TABLE 2.69   Half-Wave Potentials (vs. Saturated Calomel Electrode) of Organic Compounds at 25°C
The solvent system in this table are listed below:

A, acetonitrile and a perchlorate salt such as $LiClO_4$ or a tetraalkyl ammonium salt
B, acetic acid and an alkali acetate, often plus a tetraalkyl ammonium iodide
C, 0.05 to 0.175M tetraalkyl ammonium halide and 75% 1,4-dioxane
D, buffer plus 50% ethanol (EtOH)

Abbreviations Used in the Table

Bu, butyl	Me, methyl
Et, ethyl	MeOH, methanol
EtOH, ethanol	PrOh, propanol
M, molar	

Compound	Solvent system	$E_{1/2}$
	Unsaturated aliphatic hydrocarbons	
Acrylonitrile	C but 30% EtOH	−1.94
Allene	C	−2.29
1,3-Butadiene	A	−2.03
	C	−2.59
1,3-Butadiyne	C	−1.89
1-Buten-2-yne	C	−2.40
1,4-Cyclohexadiene	A	1.6
Cyclohexene	A	−1.89
1,3,5,7-Cyclooctatetraene	B	−1.42
	C	−1.51
Diethyl fumarate	B, pH 4.0	−0.84
Diethyl maleate	B, pH 4.0	−0.95
2,3-Dimethyl-1,3-butadiene	A	−1.83
Dimethylfulvene	C	−1.89
Diphenylacetylene	C	−2.20
1,1-Diphenylethylene	B	−1.52
	C	−2.19
Ethyl methacrylate	0.1 N LiCl + 25% EtOH	−1.9
2-Methyl-1,3-butadiene	A	−1.84
2-Methyl-1-butene	A	−1.97
1-Piperidino-4-cyano-4-phenyl-1,3-butadiene	$LiClO_4$ in dimethylformamide	−0.16
trans-Stilbene	B	−1.51
Tetrakis(dimethylamino)ethylene	A	−0.75

(Continued)

TABLE 2.69 Half-Wave Potentials (vs. Saturated Calomel Electrode) of Organic Compounds at 25°C (*Continued*)

Compound	Solvent system	$E_{1/2}$
	Aromatic hydrocarbons	
Acenaphthene	A	−0.95
	B	−1.36
	C	−2.58
Anthracene	A	−0.84
	B	−1.20
	C	−1.94
Azulene	A	−0.71
	C	−1.66, −2.26, −2.56
	Aromatic hydrocarbons (*continued*)	
1,2-Benzanthracene	C	−2.03, −2.54
2,3-Benzanthracene	A	−0.54, −1.20
Benzene	A	−2.08
1,2-Benzo[*a*]pyrene	A	−0.76
Biphenyl	A	−1.48
	B	−1.91
	C	−2.70
Chrysene	A	−1.22
1,2,5,6-Dibenzanthracene	A	−1.00, −1.26
1,2-Dihydronaphthalene	C	−2.57
9,10-Dimethylanthracene	A	−0.65
2,3-Dimethylnaphthalene	A	−1.08, −1.34
9,10-Diphenylanthracene	A	−0.92
Fluorene	A	−1.25
	B	−1.65
	C	−2.65
Hexamethylbenzene	A	−1.16
	B	−1.52
Indan	A	−1.59, −2.02
Indene	A	−1.23
	C	−2.81
1-Methylnaphthalene	A	−1.24
	B	−1.53
	C	−2.46
2-Methylnaphthalene	A	−1.22
	B	−1.55
	C	−2.46
Naphthalene	A	−1.34
	B	−1.72
Pentamethylbenzene	A	−1.28
	B	−1.62
Phenanthrene	A	−1.23
	B	−1.68
	C	−2.46, −2.71
Phenylacetylene	C	−2.37
Pyrene	A	−1.06, −1.24
trans-Stilbene	B	−1.51
	C	−2.26

TABLE 2.69 Half-Wave Potentials (vs. Saturated Calomel Electrode) of Organic Compounds
at 25°C (*Continued*)

Compound	Solvent system	$E_{1/2}$
	Aromatic hydrocarbons (*continued*)	
Styrene	C	−2.35
1,2,3,5-Tetramethylbenzene	A	−1.50, −1.99
1,2,4,5-Tetramethylbenzene	A	−1.29
Tetraphenylethylene	C	−2.05
1,4,5,8-Tetraphenylnaphthalene	A	−1.39
Toluene	A	−1.98
1,2,3-Trimethylbenzene	A	−1.58
1,2,4-Trimethylbenzene	A	−1.41
1,3,5-Trimethylbenzene	A	−1.50
	B	−1.90
Triphenylene	A	−1.46, −1.55
Triphenylmethane	C	−1.01, −1.68, −1.96
o-Xylene	A	−1.58, −2.04
m-Xylene	A	−1.58
p-Xylene	A	−1.56
	Aldehydes	
Acetaldehyde	B, pH 6.8–13	−1.89
Benzaldehyde	McIlvaine buffer, pH 2.2	−0.96, −1.32
Bromoacetaldehyde	pH 8.5	−0.40
	pH 9.8	−1.58, −1.82
Chloroacetaldehyde	Ammonia buffer, pH 8.4	−1.06, −1.66
Cinnamaldehyde	Buffer + EtOH, pH 6.0	−0.9, −1.5, −1.7
Crotonaldehyde	B, pH 1.3–2.0	−0.92
	Ammonia buffer, pH 8.0	−1.30
Dichloroacetaldehyde	Ammonia buffer, pH 8.4	−1.03, −1.67
3,7-Dimethyl-2,6-octadienal	0.1 M Et$_4$NI	−1.56, −2.22
Formaldehyde	0.05 M KOH + 0.1 M KCl, pH 12.7	−1.59
2-Furaldehyde	pH 1–8	−0.86 to 0.07 pH
	pH 10	−1.43
Glucose	Phosphate buffer, pH 7	−1.55
Glyceraldehyde	Britton–Robinson buffer, pH 5.0	−1.47
	Britton–Robinson buffer, pH 8.0	−1.55
Glycolaldehyde	0.1 M KOH, pH 13	−1.70
Glyoxal	B, pH 3.4	−1.41
4-Hydroxybenzaldehyde	Britton–Robinson buffer, pH 1.8	−1.16
	Britton–Robinson buffer, pH 6.8	−1.45
4-Hydroxy-2-methoxybenzaldehyde	McIlvaine buffer, pH 2.2	−1.05
	McIlvaine buffer, pH 5.0	−1.16, −1.36
	McIlvaine buffer, pH 8.0	−1.47
o-Methoxybenzaldehyde	Britton–Robinson buffer, pH 1.8	−1.02
	Britton–Robinson buffer, pH 6.8	−1.49
p-Methoxybenzaldehyde	Britton–Robinson buffer, pH 1.8	−1.17
	Britton–Robinson buffer, pH 6.8	−1.48
Methyl glyoxal	A, pH 4.5	−0.83
m-Nitrobenzaldehyde	Buffer + 10% EtOH, pH 2.0	−0.28, −1.20

(*Continued*)

TABLE 2.69   Half-Wave Potentials (vs. Saturated Calomel Electrode) of Organic Compounds at 25°C (*Continued*)

Compound	Solvent system	$E_{1/2}$
Aldehydes (*continued*)		
Phthalaldehyde	Buffer, pH 3.1	−0.64, −1.07
	Buffer, pH 7.3	−0.89, −1.29
2-Propenal (acrolein)	pH 4.5	−1.36
	pH 9.0	−1.1
Propionaldehyde	0.1 M LiOH, pH 13	−1.93
Pyrrole-2-carbaldehyde	0.1 M HCl + 50% EtOH	−1.25
Salicylaldehyde	McIlvaine buffer, pH 2.2	−0.99, −1.23
	McIlvaine buffer, pH 5.0	−1.20, −1.30
	McIlvaine buffer, pH 8.0	−1.32
Trichloroacetaldehyde	Ammonia buffer, pH 8.4	−1.35, −1.66
	0.1 M KCl + 50% EtOH	−1.55
Ketones		
Acetone	B, pH 9.3	−1.52
	C	−2.46
Acetophenone	D + McIlvaine buffer, pH 4.9	−1.33
	D + McIlvaine buffer, pH 7.2	−1.58
	D + McIlvaine buffer, pH 1.3	−1.08
7H-Benz[*de*]anthracen-7-one	0.1 N H$_2$SO$_4$ + 75% MeOH	−0.96
Benzil	D + McIlvaine buffer, pH 1.3	−0.27
	D + McIlvaine buffer, pH 4.9	−0.50
Benzoin	D + McIlvaine buffer, pH 1.3	−0.90
	D + McIlvaine buffer, pH 8.6	−1.49
Benzophenone	D + McIlvaine buffer, pH 1.3	−0.94
	D + McIlvaine buffer, pH 8.6	−1.36
Benzoylacetone	Buffer, pH 2.6	−1.60
	Buffer, pH 5.3 and pH 7.6	−1.68
	Buffer, pH 9.7	−1.72
Bromoacetone	0.1 M LiCl	−0.29
2,3-Butanedione	0.1 M HCl	−0.84
3-Buten-2-one	0.1 M KCl	−1.42
Butyrophenone	0.1 M NH$_4$Cl + 50% EtOH	−1.55
D-Carvone	0.1 M Et$_4$NI + 80% EtOH	−1.71
Chloroacetone	0.1 M LiCl	−1.18
Coumarin	McIlvaine buffer, pH 2.0	−0.95
	McIlvaine buffer, pH 5.0	−1.11, −1.44
Cyclohexanone	C	−2.45
cis-Dibenzoylethylene	D, pH 1	−0.30
	D, pH 11	−0.62, −1.65
trans-Dibenzoylethylene	D, pH 1	−0.12
	D, pH 11	−0.57, −1.52
Dibenzoylmethane	D, pH 1.3	−0.59
	D, pH 11.3	−1.30, −1.62
9,10-Dihydro-9-oxoanthracerie	D, pH 2.0	−0.93
1,5-Diphenyl-1,5-pentanedione	A	−2.10
1,5-Diphenylthiocarbazone	D, pH 7.0	−0.6
Flavanone	Acetate buffer + Me$_4$NOH + 50% 2-PrOH, pH 6.1	−1.30
	Acetate buffer + Me$_4$NOH + 50% 2-PrOH, pH 9.6	−1.51

TABLE 2.69   Half-Wave Potentials (vs. Saturated Calomel Electrode) of Organic Compounds at 25°C (*Continued*)

Compound	Solvent system	$E_{1/2}$
	Ketones (*continued*)	
Fluorescein	Acetate buffer, pH 2.0	−0.50
	Phthalate buffer, pH 5.0	−0.65
	Borate buffer, pH 10.1	−1.18, −1.44
Fructose	0.02 M LiCl	−1.76
Girard derivatives of aliphatic ketones	pH 8.2	−1.52
o-Hydroxyacetophenone	D, pH 5	−1.36
p-Hydroxyacetophenone	D, pH 5	−1.46
1,2,3-Indantrione (ninhydrin)	Britton–Robinson buffer, pH 2.5	−0.67, −0.83
	Britton–Robinson buffer, pH 4.5	−0.73, −1.01
	Britton–Robinson buffer, pH 6.8	−0.10, −0.90, −1.20
	Britton–Robinson buffer, pH 9.2	−1.35
α-Ionone	C	−1.59, −2.08
Isatin	Phosphate buffer + citrate buffer, pH 2.9	−0.3, −0.5
	Phosphate buffer + citrate buffer, pH 4.3	−0.3, −0.5, −0.8
	Phosphate buffer + citrate buffer, pH 5.4	−0.8
4-Methyl-3,5-heptadien-2-one	A	−0.64
4-Methyl-2,6-heptanedione	A	−1.28
4-Methyl-3-penten-2-one	D + McIlvaine buffer, pH 1.3	−1.01
	D + McIlvaine buffer, pH 11.3	−1.60
4-Phenyl-3-buten-2-one	D, pH 1.3	−0.72
	D, pH 8.6	−1.27
Phthalide	0.1 M Bu$_4$NI + 50% dioxane	−0.20
Phthalimide	pH 4.2	−1.1, −1.5
	pH 9.7	−1.2, −1.4
Pulegone	C	−1.74
Quinalizarin	Phosphate buffer + 1% EtOH, pH 8.0	−0.56
Testosterone	D + Britton–Robinson buffer, pH 2.6	−1.20
	D + Britton–Robinson buffer, pH 5.8	−1.40
	D + Britton–Robinson buffer, pH 8.8	−1.53, −1.79
	Quinones	
Anthraquinone	Acetate buffer + 40% dioxane, pH 5.6	−0.51
	Phosphate buffer + 40% dioxane, pH 7.9	−0.71
o-Benzoquinone	Britton–Robinson buffer, pH 7.0	+0.20
	Britton–Robinson buffer, pH 9.0	+0.08
2,3-Dimethylnaphthoquinone	D, pH 5.4	−0.22
1,2-Naphthoquinone	Phosphate buffer, pH 5.0	−0.03
	Phosphate buffer, pH 7.0	−0.13
1,4-Naphthoquinone	Britton–Robinson buffer, pH 7.0	−0.07
	Britton–Robinson buffer, pH 9.0	−0.19

(*Continued*)

TABLE 2.69 Half-Wave Potentials (vs. Saturated Calomel Electrode) of Organic Compounds at 25°C (*Continued*)

Compound	Solvent system	$E_{1/2}$
	Acids	
Acetic acid	A	−2.3
Acrylic acid	pH 5.6	−0.85
Adenosine-5′-phosphoric acid	$HClO_4 + KClO_4$, pH 2.2	−1.13
4-Aminobenzenesulfonic acid	0.05 M Me$_4$NI	−1.58
3-Aminobenzoic acid	pH 5.6	−0.67
Anthranilic acid	pH 5.6	−0.67
Ascorbic acid	Britton–Robinson buffer, pH 3.4	+0.17
	Britton–Robinson buffer, pH 7.0	−0.06
Barbituric acid	Borate buffer, pH 9.3	−0.04
Benzoic acid	A	−2.1
Benzoylformic acid	Britton–Robinson buffer, pH 2.2	−0.48
	Britton–Robinson buffer, pH 5.5	−0.85, −1.26
	Britton–Robinson buffer, pH 7.2	−0.98, −1.25
	Britton–Robinson buffer, pH 9.2	−1.25
Bromoacetic acid	pH 1.1	−0.54
2-Bromopropionic acid	pH 2.0	−0.39
Crotonic acid	C	−1.94
Dibromoacetic acid	pH 1.1	−0.03, −0.59
Dichloroacetic acid	pH 8.2	−1.57
5,5-Diethylbarbituric acid	Borate buffer, pH 9.3	0.00
Flavanol	D, pH 5.6	−1.25
	D, pH 7.7	−1.40
Folic acid	Britton–Robinson buffer, pH 4.6	−0.73
Formic acid	0.1 M KCl	−1.66
Fumaric acid	HCl + KCl, pH 2.6	−0.83
	Acetate buffer, pH 4.0	−0.93
	Acetate buffer, pH 5.9	−1.20
2,4-Hexadienedioic acid	Acetate buffer, pH 4.5	−0.97
Iodoacetic acid	pH 1	−0.16
Maleic acid	Britton–Robinson buffer, pH 2.0	−0.70
	Britton–Robinson buffer, pH 4.0	−0.97
	Britton–Robinson buffer, pH 6.0	−1.11, −1.30
	Britton–Robinson buffer, pH 10.0	−1.51
Mercaptoacetic acid	B, pH 6.8	−0.38
Methacrylic acid	D + 0.1 M LiCl	−1.69
Nitrobenzoic acids	Buffer + 10% EtOH, pH 2.0	−0.2, −0.7
Oxalic acid	B, pH 5.4–6.1	−1.80
2-Oxo-1,5-pentanedioic acid	HCl + KCl, pH 1.8	−0.59
	Ammonia buffer, pH 8.2	−1.30
2-Oxopropionic acid	Britton–Robinson buffer, pH 5.6	−1.17
	Britton–Robinson buffer, pH 6.8	−1.22, −1.53
	Britton–Robinson buffer, pH 9.7	−1.51
Phenolphthalein	Phthalate buffer, pH 2.5	−0.67
	Phthalate buffer, pH 4.7	−0.80
	D, pH 9.6	−0.98, −1.35
Picric acid	pH 4.2	−0.34
	pH 11.7	−0.36, −0.56, −0.96

TABLE 2.69 Half-Wave Potentials (vs. Saturated Calomel Electrode) of Organic Compounds at 25°C (*Continued*)

Compound	Solvent system	$E_{1/2}$
| Acids (*continued*)			
1,2,3-Propenetricarboxylic acid	pH 7.0	−2.1
Trichloroacetic acid	Ammonia buffer, pH 8.2	−0.84, −1.57
	Phosphate buffer, pH 10.4	−0.9, −1.6
3,4,5-Trihydroxybenzoic acid	Phosphate buffer, pH 2.9	+0.50
	Phosphate buffer, pH 8.8	+0.1
p-Aminophenol	Britton–Robinson buffer, pH 6.3	+0.14
	Britton–Robinson buffer, pH 8.6	−0.04
	Britton–Robinson buffer, pH 12.0	−0.16
o-Chlorophenol	pH 5.6	−0.63
m-Chlorophenol	pH 5.6	−0.73
p-Chlorophenol	pH 5.6	−0.65
o-Cresol	pH 5.6	−0.56
m-Cresol	pH 5.6	−0.61
p-Cresol	pH 5.6	−0.54
1,2-Dihydroxybenzene	pH 5.6	−0.35
1,3-Dihydroxybenzene	pH 5.6	−0.61
1,4-Dihydroxybenzene	pH 5.6	−0.23
o-Methoxyphenol	pH 5.6	−0.46
m-Methoxyphenol	pH 5.6	−0.62
p-Methoxyphenol	pH 5.6	−0.41
1-Naphthol	A	−0.74
2-Naphthol	A	−0.82
1,2,3-Trihydroxybenzene	Britton–Robinson buffer, pH 3.1	+0.35
	Britton–Robinson buffer, pH 6.5	+0.10
	Britton–Robinson buffer, pH 9.5	−0.10
| Halogen compounds			
Bromobenzene	A	−1.98
	C	−2.32
1-Bromobutane	C	−2.27
Bromoethane	C	−2.08
Bromomethane	C	−1.63
1-Bromonaphthalene (also 2-bromonaphthalene)	A	−1.55, −1.60
3-Bromo-1-propene	C	−1.29
p-Bromotoluene	A	−1.72
Carbon tetrachloride	C	−0.78, −1.71
Chlorobenzene	A	−2.07
Chloroform	C	−1.63
Chloromethane	C	−2.23
3-Chloro-1-propene	C	−1.91
α-Chlorotoluene	C	−1.81
p-Chlorotoluene	A	−1.76
N-Chloro-*p*-toluenesulfonamide	0.5 *M* K_2SO_4	−0.13
9,10-Dibromoanthracene	A	−1.15, −1.47
p-Dibromobenzene	C	−2.10
1,2-Dibromobutane	D + 1% Na_2SO_3	−1.45

(Continued)

TABLE 2.69 Half-Wave Potentials (vs. Saturated Calomel Electrode) of Organic Compounds at 25°C (*Continued*)

Compound	Solvent system	$E_{1/2}$
Halogen compounds (*continued*)		
Dibromoethane	C	−1.48
meso-2,3-Dibromosuccinic acid	Acetate buffer, pH 4.0	−0.23, −0.89
Dichlorobenzenes	C	−2.5
Dichloromethane	C	−1.60
Diiodomethane	C	−1.12, −1.53
Hexabromobenzene	C	−0.8, −1.5
Hexachlorobenzene	C	−1.4, −1.7
Iodobenzene	A	−1.72
Iodoethane	C	−1.67
Iodomethane	A	−2.12
	C	−1.63
Tetrabromomethane	C	−0.3, −0.75, −1.49
Tetraidomethane	C	−0.45, −1.05, −1.46
Tribromomethane	C	−0.64, −1.47
α,α,α-Trichlorotoluene	C	−0.68, −1.65, −2.00
Nitro and nitroso compounds		
1,2-Dinitrobenzene	Phthalate buffer, pH 2.5	−0.12, −0.32, −1.26
	Borate buffer, pH 9.2	−0.38, −0.74
1,3-Dinitrobenzene	Phthalate buffer, pH 2.5	−0.17, −0.29
	Borate buffer, pH 9.2	−0.46, −0.68
1,4-Dinitrobenzene	Phthalate buffer, pH 2.5	−0.12, −0.33
	Borate buffer, pH 9.2	−0.35, −0.80
Methyl nitrobenzoates	Buffer + 10% EtOH, pH 2.0	−0.20 to −0.25
		−0.68 to −0.74
p-Nitroacetophenone	Britton–Robinson buffer, pH 2.2	−0.16, −0.61, −1.09
	Britton–Robinson buffer, pH 10.0	−0.51, −1.40, −1.73
o-Nitroaniline	0.03 *M* LiCl + 0.02 *M* benzoic acid in EtOH	−0.88
m-Nitroaniline	Britton–Robinson buffer, pH 4.3	−0.3, −0.8
	Britton–Robinson buffer, pH 7.2	−0.5
	Britton–Robinson buffer, pH 9.2	−0.7
p-Nitroaniline	pH 2.0	−0.36
	Acetate buffer, pH 4.6	−0.5
o-Nitroanisole	Buffer + 10% EtOH, pH 2.0	−0.29, −0.58
p-Nitroanisole	Buffer + 10% EtOH, pH 2.0	−0.35, −0.64
1-Nitroanthraquinone	Britton–Robinson buffer, pH 7.0	−0.16
Nitrobenzene	HCl + KCl + 8% EtOH, pH 0.5	−0.16, −0.76
	Phthalate buffer, pH 2.5	−0.30
	Borate buffer, pH 9.2	−0.70
Nitrocresols	Britton–Robinson buffer, pH 2.2	−0.2 to −0.3
	Britton–Robinson buffer, pH 4.5	−0.4 to −0.5
	Britton–Robinson buffer, pH 8.0	−0.6
Nitroethane	Britton–Robinson buffer + 30% MeOH, pH 1.8	−0.7
	Britton–Robinson buffer + 30% MeOH, pH 4.6	−0.8

TABLE 2.69   Half-Wave Potentials (vs. Saturated Calomel Electrode) of Organic Compounds at 25°C (*Continued*)

Compound	Solvent system	$E_{1/2}$
Nitro and nitroso compounds (*continued*)		
2-Nitrohydroquinone	Phosphate buffer+citrate buffer, pH 2.1	−0.2
	Phosphate buffer+citrate buffer, pH 5.2	−0.4
	Phosphate buffer+citrate buffer, pH 8.0	−0.5
Nitromethane	Britton–Robinson buffer+30% MeOH, pH 1.8	−0.8
	Britton–Robinson buffer+30% MeOH, pH 4.6	−0.85
o-Nitrophenol	Britton–Robinson buffer +10% EtOH, pH 2.0	−0.23
	Britton–Robinson buffer +10% EtOH, pH 4.0	−0.4
	Britton–Robinson buffer +10% EtOH, pH 8.0	−0.65
	Britton–Robinson buffer +10% EtOH, pH 10.0	−0.80
m-Nitrophenol	Britton–Robinson buffer+10% EtOH, pH 2.0	−0.37
	Britton–Robinson buffer +10% EtOH, pH 4.0	−0.40
	Britton–Robinson buffer +10% EtOH, pH 8.0	−0.64
	Britton–Robinson buffer +10% EtOH, pH 10.0	−0.76
p-Nitrophenol	Britton–Robinson buffer +10% EtOH, pH 2.0	−0.35
	Britton–Robinson buffer +10% EtOH, pH 4.0	−0.50
	Britton–Robinson buffer +10% EtOH, pH 8.0	−0.82
1-Nitropropane	Britton–Robinson buffer+30% MeOH, pH 1.8	−0.73
	Britton–Robinson buffer+30% MeOH, pH 8.6	−0.88
	Britton–Robinson buffer+30% MeOH, pH 8.0	−0.95
2-Nitropropane	McIlvaine buffer, pH 2.1	−0.53
	McIlvaine buffer, pH 5.1	−0.81
Nitrosobenzene	McIlvaine buffer, pH 6.0	−0.03
	McIlvaine buffer, pH 8.0	−0.14
1-Nitroso-2-naphthol	D+buffer, pH 4.0	+0.02
	D+buffer, pH 7.0	−0.20
	D+buffer, pH 9.0	−0.31
N-Nitrosophenylhydroxylamine	pH 2.0	−0.84
o-Nitrotoluene	Phthalate buffer, pH 2.5	−0.35, −0.66
	Phthalate buffer, pH 7.4	−0.60, −1.06

(Continued)

TABLE 2.69 Half-Wave Potentials (vs. Saturated Calomel Electrode) of Organic Compounds at 25°C (*Continued*)

Compound	Solvent system	$E_{1/2}$
Nitro and nitroso compounds (continued)		
m-Nitrotoluene (also *p*-nitrotoluene)	Phthalate buffer, pH 2.5	−0.30, −0.53
	Phthalate buffer, pH 7.4	−0.58, −1.06
Tetranitromethane	pH 12.0	−0.41
1,3,5-Trinitrobenzene	Phthalate buffer, pH 4.1	−0.20, −0.29, −0.34
	Borate buffer, pH 9.2	−0.34, −0.48, −0.65
Heterocyclic compounds containing nitrogen		
Acridine	D, pH 8.3	−0.80, −1.45
Cinchonine	B, pH 3	−0.90
2-Furanmethanol	Britton–Robinson buffer, pH 2.0	−0.96
	Britton–Robinson buffer, pH 5.8	−1.38, −1.70
2-Hydroxyphenazine	Britton–Robinson buffer, pH 4.0	−0.24
8-Hydroxyquinoline	B, pH 5.0	−1.12
	Phosphate buffer, pH 8.0	−1.18, −1.71
3-Methylpyridine	D+0.1 M LiCl	−1.76
4-Methylpyridine	D+0.1 M LiCl	−1.87
Phenazine	Phosphate buffer+citrate buffer, pH 7.0	−0.36
Pyridine	Phosphate buffer+citrate buffer, pH 7.0	−1.75
Pyridine-2-carboxylic acid	B, pH 4.1	−1.10
	B, pH 9.3	−1.48, −1.94
Pyridine-3-carboxylic acid	0.1 M HCl	−1.08
Pyridine-4-carboxylic acid	Britton–Robinson buffer, pH 6.1	−1.14
	pH 9.0	−1.39, −1.68
Pyrimidine	Citrate buffer, pH 3.6	−0.92, −1.24
	Ammonia buffer, pH 9.2	−1.54
Quinoline-8-carboxylic acid	pH 9	−1.11
Quinoxaline	Phosphate buffer+citrate buffer, pH 7.0	−0.66, −1.52
Azo, hydrazine, hydroxylamine, and oxime compounds		
Azobenzene	D, pH 4.0	−0.20
	D, pH 7.0	−0.50
Azoxybenzene	Buffer+20% EtOH, pH 6.3	−0.30
Benzoin 1-oxime	Buffer, pH 2.0	−0.88
	Buffer, pH 5.6	−1.08
	Buffer, pH 8.2	−1.67
Benzoylhydrazine	0.13 M NaOH, pH 13.0	−0.30
Dimethylglyoxime	Ammonia buffer, pH 9.6	−1.63
Hydrazine	Britton–Robinson buffer, pH 9.3	−0.09
Hydroxylamine	Britton–Robinson buffer, pH 4.6	−1.42
	Britton–Robinson buffer, pH 9.2	−1.65

TABLE 2.69 Half-Wave Potentials (vs. Saturated Calomel Electrode) of Organic Compounds at 25°C (*Continued*)

Compound	Solvent system	$E_{1/2}$
Azo, hydrazine, hydroxylamine, and oxime compounds (*continued*)		
Oxamide	Acetate buffer	−1.55
Phenylhydrazine	McIlvaine buffer, pH 2	+0.19
	0.13 M NaOH, pH 13.0	−0.36
Phenylhydroxylamine	McIlvaine buffer + 10% EtOH, pH 2	−0.68
	McIlvaine buffer + 10 EtOH, pH 4–10	−0.33 0.061 pH
Salicylaldoxime	Phosphate buffer, pH 5.4	−1.02
Thiosemicarbazide	Borate buffer, pH 9.3	−0.26
Thiourea	0.1 M sulfuric acid	+0.02
Indicators and dyestuffs		
Brilliant Green	HCl + KCl, pH 2.0	−0.2, −0.5
Indigo carmine	pH 2.5	−0.24
Indigo disulfonate	pH 7.0	−0.37
Malachite Green G	HCl + KCl, pH 2.0	−0.2, −0.5
Metanil yellow	Phosphate buffer + 1% EtOH, pH 7.0	−0.51
Methylene blue	Britton–Robinson buffer, pH 4.9	−0.15
	Britton–Robinson buffer, pH 9.2	−0.30
Methylene green	Phosphate buffer + 1% EtOH, pH 7.0	−0.12
Methyl orange	Phosphate buffer + 1% EtOH, pH 7.0	−0.51
Morin	D, pH 7.6	−1.7
Neutral red	Britton–Robinson buffer, pH 2.0	−0.21
	Britton–Robinson buffer, pH 7.0	−0.57
Peroxide		
Ethyl peroxide	0.02 M HCl	−0.2

2.18 ELECTRICAL CONDUCTIVITY

TABLE 2.70   Electrical Conductivity of Various Pure Liquids

Liquid	Temp. °C	mhos/cm or ohm^{-1} · cm^{-1}	Liquid	Temp. °C	mhos/cm or ohm^{-1} · cm^{-1}
Acetaldehyde	15	1.7×10^{-6}	Epichlorohydrin	25	3.4×10^{-8}
Acetamide	100	$<4.3 \times 10^{-5}$	Ethyl acetate	25	$<1 \times 10^{-9}$
Acetic acid	0	5×10^{-9}	Ethyl acetoacetate	25	4×10^{-8}
	25	1.12×10^{-8}	Ethyl alcohol	25	1.35×10^{-9}
Acetic anhydride	0	1×10^{-6}	Ethylamine	0	4×10^{-7}
	25	4.8×10^{-7}	Ethyl benzoate	25	$<1 \times 10^{-9}$
Acetone	18	2×10^{-8}	Ethyl bromide	25	$<2 \times 10^{-8}$
	25	6×10^{-8}	Ethylene bromide	19	$<2 \times 10^{-10}$
Acetonitrile	20	7×10^{-6}	Ethylene chloride	25	3×10^{-8}
Acetophenone	25	6×10^{-9}	Ethyl ether	25	$<4 \times 10^{-13}$
Acetyl bromide	25	2.4×10^{-6}	Ethylidene chloride	25	$<1.7 \times 10^{-8}$
Acetyl chloride	25	4×10^{-7}	Ethyl iodide	25	$<2 \times 10^{-8}$
Alizarin	233	1.45×10^{-6} (?)	Ethyl isothiocyanate	25	1.26×10^{-7}
Allyl alcohol	25	7×10^{-6}	Ethyl nitrate	25	5.3×10^{-7}
Ammonia	−79	1.3×10^{-7}	Ethyl thiocyanate	25	1.2×10^{-6}
Aniline	25	2.4×10^{-8}	Eugenol	25	$<1.7 \times 10^{-8}$
Anthracene	230	3×10^{-10}			
Arsenic tribromide	35	1.5×10^{-6}	Formamide	25	4×10^{-6}
Arsenic trichloride	25	1.2×10^{-6}	Formic acid	18	5.6×10^{-5}
				25	6.4×10^{-5}
Benzaldehyde	25	1.5×10^{-7}	Furfural	25	1.5×10^{-6}
Benzene	. . .	7.6×10^{-8}			
Benzoic acid	125	3×10^{-9}	Gallium	30	36,800
Benzonitrile	25	5×10^{-8}	Glycerol	25	6.4×10^{-8}
Benzyl alcohol	25	1.8×10^{-6}	Glycol	25	3×10^{-7}
Benzylamine	25	$<1.7 \times 10^{-8}$	Guaiacol	25	2.8×10^{-7}
Benzyl benzoate	25	$<1 \times 10^{-9}$			
Bromine	17.2	1.3×10^{-13}	Heptane	. . .	$<1 \times 10^{-13}$
Bromobenzene	25	$<2 \times 10^{-11}$	Hexane	18	$<1 \times 10^{-18}$
Bromoform	25	$<2 \times 10^{-8}$	Hydrogen bromide	−80	8×10^{-9}
iso-Butyl alcohol	25	8×10^{-8}	Hydrogen chloride	−96	1×10^{-8}
			Hydrogen cyanide	0	3.3×10^{-6}
Capronitrile	25	3.7×10^{-6}	Hydrogen iodide	B.P.	2×10^{-7}
Carbon disulfide	1	7.8×10^{-18}	Hydrogen sulfide	B.P.	1×10^{-11}
Carbon tetrachloride	18	4×10^{-18}			
Chlorine	−70	$<1 \times 10^{-16}$	Iodine	110	1.3×10^{-10}
Chloroacetic acid	60	1.4×10^{-6}			
m-Chloroaniline	25	5×10^{-8}	Kerosene	25	$<1.7 \times 10^{-8}$
Chloroform	25	$<2 \times 10^{-8}$			
Chlorohydrin	25	5×10^{-7}	Mercury	0	10,629.6
m-Cresol	25	$<1.7 \times 10^{-8}$	Methyl acetate	25	3.4×10^{-6}
Cyanogen	. . .	$<7 \times 10^{-9}$	Methyl alcohol	18	4.4×10^{-7}
Cymene	25	$<2 \times 10^{-8}$	Methyl ethyl ketone	25	1×10^{-7}
			Methyl iodide	25	$<2 \times 10^{-8}$
Dichloroacetic acid	25	7×10^{-8}	Methyl nitrate	25	4.5×10^{-6}
Dichlorohydrin	25	1.2×10^{-5}	Methyl thiocyanate	25	1.5×10^{-6}
Diethylamine	−33.5	2.2×10^{-9}			
Diethyl carbonate	25	1.7×10^{-8}	Naphthalene	82	4×10^{-10}
Diethyl oxalate	25	7.6×10^{-7}	Nitrobenzene	0	5×10^{-9}
Diethyl sulfate	25	2.6×10^{-7}	Nitromethane	18	6×10^{-7}
Dimethyl sulfate	0	1.6×10^{-7}	o- or m-Nitrotoluene	25	$<2 \times 10^{-7}$
			Nonane	25	$<1.7 \times 10^{-8}$

TABLE 2.70 Electrical Conductivity of Various Pure Liquids (*Continued*)

Liquid	Temp. °C	mhos/cm or ohm^{-1} · cm^{-1}	Liquid	Temp. °C	mhos/cm or ohm^{-1} · cm^{-1}
Oleic acid	15	$<2 \times 10^{-10}$	Salicylaldehyde	25	1.6×10^{-7}
			Stearic acid	80	$<4 \times 10^{-13}$
Pentane	19.5	$<2 \times 10^{-10}$	Sulfonyl chloride,	25	2×10^{-6}
Petroleum	$\cdots$	3×10^{-13}	$SOCl_2$		
Phenetole	25	$<1.7 \times 10^{-8}$	Sulfur	115	1×10^{-12}
Phenol	25	$<1.7 \times 10^{-8}$		130	5×10^{-12}
Phenyl isothiocyanate	25	1.4×10^{-6}		440	1.2×10^{-7}
Phosgene	25	7×10^{-9}	Sulfur dioxide	35	1.5×10^{-8}
Phosphorus	25	4×10^{-7}	Sulfuric acid	25	1×10^{-2}
Phosphorus oxychloride	25	2.2×10^{-6}	Sulfuryl chloride,	25	3×10^{-8}
Pinene	23	$<2 \times 10^{-10}$	SO_2Cl_2		
Piperidine	25	$<2 \times 10^{-7}$			
Propionaldehyde	25	8.5×10^{-7}	Toluene	$\cdots$	$<1 \times 10^{-14}$
Propionic acid	25	$<1 \times 10^{-9}$	o-Toluidine	25	$<2 \times 10^{-6}$
Propionitrile	25	$<1 \times 10^{-7}$	p-Toluidine	100	6.2×10^{-8}
n-Propyl alcohol	18	5×10^{-8}	Trichloroacetic acid	25	3×10^{-9}
	25	2×10^{-8}	Trimethylamine	−33.5	2.2×10^{-10}
iso-Propyl alcohol	25	3.5×10^{-6}	Turpentine	$\cdots$	2×10^{-13}
n-Propyl bromide	25	$<2 \times 10^{-8}$	iso-Valeric acid	80	$<4 \times 10^{-13}$
Pyridine	18	5.3×10^{-8}	Water	18	4×10^{-8}
Quinoline	25	2.2×10^{-8}	Xylene	$\cdots$	$<1 \times 10^{-5}$

TABLE 2.71 Limiting Equivalent Ionic Conductances in Aqueous Solutions

Ion	Temperature, °C		
	0	18	25
Fluoroacetate$^-$			44.4
Fluorobenzoate$^-$			33
Formate$^-$		47	54.6
Fumarate(2−)			61.8
Glutarate(2−)			52.6
Hydrogenoxalate (1−)			40.2
Iodoacetate$^-$			40.6
Lactate(1−)			38.8
Malate(2−)			58.8
Malonate(1−)			63.5
3-Methylbutanoate$^-$			32.7
Methylsulfonate$^-$			48.8
Naphthylacetate$^-$			28.4
1,8-Octanedioate(2−)			36
Octylsulfonate$^-$			29
Oxalate(2−)			74.11
Phenylacetate$^-$			30.6
m-Phthalate(2−)			54.7
o-Phthalate(2−)			52.3
Picrate$^-$			30.37
Propanoate$^-$			35.8
Propylsulfonate$^-$			37.1
Salicylate$^-$			36
Succinate(2−)			58.8
Tartrate(2−)		55	59.6
Trichloroacetate$^-$			36.6
Trimethylacetate$^-$			31.9

TABLE 2.72 Properties of Organic Semi-Conductors

Substance	Formula	Resis-tivity, ohm-cm	Band gap	
			Conduc-tivity, eV	Photo conduct, eV
POLYACENES				
Anthracene		300	0.83	—
Tetracene		10	0.85	3.6
Pyrene		300	1.01	3.2
Perylene		10	0.98	—
Chrysene		100	1.10	3.2
Coronene		0.2	1.15	—
Pyranthrene		10^7	0.54	0.85

TABLE 2.72   Properties of Organic Semi-Conductors (*Continued*)

Substance	Formula	Resistivity, ohm-cm	Band gap	
			Conductivity, eV	Photo conduct, eV
POLYACENES WITH QUINONOID ATTACHEMENTS				
Violanthrone		1000	0.39	0.84
Pyranthrone		10^6	0.54	1.14
AZO-AROMATIC COMPOUNDS				
Indanthrone black		300	0.28	—
1,9,4,10-Anthradipyrimidine		1000	1.61	—

(*Continued*)

TABLE 2.72 Properties of Organic Semi-Conductors (*Continued*)

Substance	Formula	Resis-tivity, ohm-cm	Band gap	
			Conduc-tivity, eV	Photo conduct, eV
PHTHALOCYANINES		10^4	1.2	1.56
FREE RADICALS α,α-Diphenyl β-pieryl hydrazyl		10^6	0.74	–

2.19 LINEAR FREE ENERGY RELATIONSHIPS

Many equilibrium and rate processes can be systematized when the influence of each substituent on the reactivity of substrates is assigned a characteristic constant σ and the reaction parameter ρ is known or can be calculated. The Hammett equation

$$\log \frac{K}{K^\circ} = \sigma\rho$$

describes the behavior of many *meta-* and *para*-substituted aromatic species. In this equation K° is the acid dissociation constant of the reference in aqueous solution at 25°C and K is the corresponding constant for the substituted acid. Separate sigma values are defined by this reaction for *meta* and *para* substituents and provide a measure of the total electronic influence (polar, inductive, and resonance effects) in the absence of conjugation effects. Sigma constants are not valid of substituents *ortho* to the reaction center because of anomalous (mainly steric) effects. The inductive effect is transmitted about equally to the *meta* and *para* positions. Consequently, σ_m is an approximate measure of the size of the inductive effect of a given substituent and $\sigma_p - \sigma_m$ is an approximate measure of a substituent's resonance effect. Values of Hammett sigma constants are listed in Table 2.73.

Taft sigma values σ^* perform a similar function with respect to aliphatic and alicyclic systems. Values of σ^* are listed in Table 2.73.

The reaction parameter ρ depends upon the reaction series but not upon the substituents employed. Values of the reaction parameter for some aromatic and aliphatic system are given in Tables 2.74 and 2.75.

Since substituent effects in aliphatic systems and in *meta* positions in aromatic systems are essentially inductive in character, σ^* and σ_m values are often related by the expression.

$\sigma_m = 0.217 \ \sigma^* - 0.106$. Substituent effects fall off with increasing distance from the reaction center; generally a factor of 0.36 corresponds to the interposition of a —CH_2— group, which enables σ^* values to be estimated for R—CH_2— groups not otherwise available.

Two modified sigma constants have been formulated for situations in which the substituent enters into resonance with the reaction center in an electron-demanding transition state (σ^+) or for an electron-rich transition state (σ^-). σ^- constants give better correlations in reactions involving phenols, anilines, and pyridines and in nucleophilic substitutions. Values of some modified sigma constants are given in Table 2.76.

TABLE 2.73   Hammett and Taft Substituent Constants

Substituent	Hammett constants		Taft constant
	σ_m	σ_p	σ^*
—AsO$_3$H$^-$	−0.09	−0.02	0.06
—B(OH)$_2$	0.01	0.45	
—Br	0.39	0.23	2.84
—CH$_2$Br			1.00
m-BrC$_6$H$_4$—		0.09	
p-BrC$_6$H$_4$—		0.08	
—CH$_3$	−0.07	−0.17	0.0
—CH$_2$CH$_3$	−0.07	−0.15	−0.10
—CH$_2$CH$_2$CH$_3$	−0.05	−0.15	−0.12
—CH(CH$_3$)$_2$	−0.07	−0.15	−0.19
—CH$_2$CH$_2$CH$_2$CH$_3$	−0.07	−0.16	−0.13
—CH$_2$CH(CH$_3$)$_2$	−0.07	−0.12	−0.13
—CH(CH$_3$)CH$_2$CH$_3$		−0.12	−0.19
—C(CH$_3$)$_3$	−0.10	−0.20	−0.30
—CH$_2$CH$_2$CH$_2$CH$_2$CH$_3$			−0.25
—CH$_2$CH$_2$CH(CH$_3$)$_2$			−0.17
—CH$_2$C(CH$_3$)$_3$		−0.23	−0.12
—CH$_2$CH$_2$CH$_2$CH$_2$CH$_2$CH$_2$CH$_3$			−0.37
Cyclopropyl—	−0.07	−0.21	
Cyclohexyl—			−0.15
—3,4-(CH$_2$)$_2$ (fused)		−0.26	
—3,4 (CH$_2$)$_3$— (fused ring)		−0.48	
—3,4-(CH)$_4$— (fused ring)	0.06	0.04	
—CH=CH$_2$	0.02		0.56
—CH=C(CH$_3$)$_2$			0.19
—CH=CHCH$_3$, *trans*			0.36
—CH$_2$—CH=CH$_2$			0.0
—CH=CHC$_6$H$_5$	0.14	−0.05	0.41
—C≡CH	0.21	0.23	2.18
—C≡CC$_6$H$_5$	0.14	0.16	1.35
—CH$_2$—C≡CH			0.81
—C$_6$H$_5$	0.06	−0.01	0.60
p-CH$_3$C$_6$H$_4$—		−0.5	
Naphthyl— (both 1- and 2-)			0.75
—CH$_2$C$_6$H$_5$		0.46	0.22
—CH$_2$CH$_2$—C$_6$H$_5$			−0.06
—CH(CH$_3$)C$_6$H$_5$			0.37
—CH(C$_6$H$_5$)$_2$			0.41
—CH$_2$—C$_{10}$H$_7$			0.44
2-Furoyl—			0.25
3-Indolyl—			−0.06
2-Thienyl—			1.31

(Continued)

TABLE 2.73 Hammett and Taft Substituent Constants (*Continued*)

Substituent	Hammett constants		Taft constant
	σ_m	σ_p	σ^*
2-Thienylmethylene—			0.31
—CHO	0.36	0.22	
—COCH$_3$	0.38	0.50	1.65
—COCH$_2$CH$_2$		0.48	
—COCH(CH$_3$)$_2$		0.47	
—COC(CH$_3$)$_3$		0.32	
—COCF$_3$	0.65		3.7
—COC$_6$H$_5$	0.34	0.46	2.2
—CONH$_2$	0.28	0.36	1.68
—CONHC$_6$H$_5$			1.56
—CH$_2$COCH$_3$			0.60
—CH$_2$CONH$_2$			0.31
—CH$_2$CH$_2$CONH$_2$			0.19
—CH$_2$CH$_2$CH$_2$CONH$_2$			0.12
—CH$_2$CONHC$_6$H$_5$			0.0
—COO$^-$	−0.1	0.0	−1.06
—COOH	0.36	0.43	2.08
—CO—OCH$_3$	0.32	0.39	2.00
—CO—OCH$_2$CH$_3$	0.37	0.45	2.12
—CH$_2$CO—OCH$_3$			1.06
—CH$_2$CO—OCH$_2$CH$_3$			0.82
—CH$_2$COO			−0.06
—CH$_2$CH$_2$COOH	−0.03	−0.07	
—Cl	0.37	0.23	2.96
—CCl$_3$	0.47		2.65
—CHCl$_2$			1.94
—CH$_2$Cl	0.12	0.18	1.05
—CH$_2$CH$_2$Cl			0.38
—CH$_2$CCl$_3$			0.75
—CH$_2$CH$_2$CCl$_3$			0.25
—CH=CCl$_2$			1.00
—CH$_2$CH=CCl$_2$			0.19
p-ClC$_6$H$_4$—		0.08	
—F	0.34	0.06	3.21
—CF$_3$	0.43	0.54	2.61
—CHF$_2$			2.05
—CH$_2$F			1.10
—CH$_2$CF$_3$			0.90
—CH$_2$CF$_2$CF$_2$CF$_3$			0.87
—C$_6$F$_5$	−0.12	−0.03	
—Ge(CH$_3$)$_3$		0.0	
—Ge(CH$_2$CH$_3$)$_3$		0.0	
—H	0.00	0.00	0.49
—I	0.35	0.28	2.46
—CH$_2$I			0.85
—IO$_2$	0.70	0.76	
—N$_2^+$	1.76	1.91	
—N$_3$ (azide)	0.33	0.08	2.62
—NH$_2$	−0.16	−0.66	0.62
—NH$_3^+$	1.13	1.70	3.76
—CH$_2$—NH$_2$			0.50
—CH$_2$—NH$_3^+$			2.24
—NH—CH$_3$	−0.30	−0.84	

TABLE 2.73   Hammett and Taft Substituent Constants (*Continued*)

Substituent	Hammett constants		Taft constant
	σ_m	σ_p	σ^*
—NH—C_2H_5	0.24	−0.61	
—NH—C_4H_9	0.34	−0.51	
—NH$(CH_3)_2^+$			4.36
—NH$_2$—CH$_3^+$	0.96		3.74
—NH$_2$—$C_2H_5^+$	0.96		3.74
—N$(CH_3)_3^+$	0.88	0.82	4.55
—N$(CH_3)_2$	−0.2	−0.83	0.32
—CH$_2$—N$(CH_3)_3^+$			1.90
—N$(CF_3)_2$	0.45	0.53	
p-H$_2$N—C_6H_5—		−0.30	
—NH—CO—CH$_3$	0.21	0.00	1.40
—NH—CO—C_2H_5			1.56
—NH—CO—C_6H_5	0.22	0.08	1.68
—NH—CHO	0.25		1.62
—NH—CO—NH$_2$	0.18		1.31
—NH—OH	−0.04	−0.34	
—NH—CO—OC$_2H_5$	0.33		1.99
—CH$_2$—NH—CO—CH$_3$			0.43
—NH—SO$_2$—C_6H_5			1.99
—NH—NH$_2$	−0.02	−0.55	
—CN	0.56	0.66	3.30
—CH$_2$—CN	0.17	0.01	1.30
—NO		0.12	
—NO$_2$	0.71	0.78	4.0
—CH$_2$—NO$_2$			1.40
—CH$_2$—CH$_2$—NO$_2$			0.50
—CH=CHNO$_2$	0.33	0.26	
m-O$_2$N—C_6H_4		0.18	
p-O$_2$N—C_6H_4		0.24	
$(NO_2)_3C_6H_2$— (picryl)	0.43	0.41	
—N(CO—CH$_3$)(CO—C_6H_5)			1.37
—N(CO—CH$_3$)(naphthyl)			1.65
—O$^-$	−0.71	−0.52	
—OH	0.12	−0.37	1.34
—O—CH$_3$	0.12	−0.27	1.81
—O—C_2H_5	0.10	−0.24	1.68
—O—C_3H_7	0.00	−0.25	1.68
—O—CH$(CH_3)_2$	0.05	−0.45	1.62
—O—C_4H_9	−0.05	−0.32	1.68
—O—cyclopentyl			1.62
—O—cyclohexyl	0.29		1.81
—O—CH$_2$—cyclohexyl	0.18		1.31
—O—C_6H_5	0.25	−0.32	2.43
—O—CH$_2$—C_6H_5		−0.42	
—OCF$_3$	0.40	0.35	
3,4-O—CH$_2$—O—		−0.27	
3,4-O—(CH$_2$—)$_2$O—		−0.12	
—O—CO—CH$_3$	0.39	0.31	
—ONO$_2$			3.86
—O—N=C$(CH_3)_2$			1.81
—ONH$_3^+$			2.92
—CH$_2$—O$^-$			0.27

(*Continued*)

TABLE 2.73 Hammett and Taft Substituent Constants (*Continued*)

Substituent	Hammett constants		Taft constant
	σ_m	σ_p	σ^*
—CH₂—OH	0.08	0.08	0.31
—CH₂—O—CH₃			0.52
—CH(OH)—CH₃			0.12
—CH(OH)—C₆H₅			0.50
p-HO—C₆H₄—		−0.24	
p-CH₃O—C₆H₄—		−0.10	
—CH₂—CH(OH)—CH₃			−0.06
—CH₂—C(OH)(CH₃)₂			−0.25
—P(CH₃)₂	0.1	0.05	
—P(CH₃)₃⁺	0.8	0.9	
—P(CF₃)₂	0.6	0.7	
—PO₃H⁻	0.2	0.26	
—PO(OC₂H₅)₂	0.55	0.60	
—SH	0.25	0.15	1.68
—SCH₃	0.15	0.00	1.56
—S(CH₃)₂⁺	1.0	0.9	
—SCH₂CH₃	0.23	0.03	1.56
—SCH₂CH₂CH₃			1.49
—SCH₂CH₂CH₂CH₃			1.44
—S—cyclohexyl			1.93
—SC₆H₅	0.30		1.87
—SC(C₆H₅)₃			0.69
—SCH₂C₆H₅			1.56
—SCH₂CH₂C₆H₅			1.44
—CH₂SH	0.03		0.62
—CH₂SCH₂C₆H₅			0.37
—SCF₃	0.40	0.50	
—SCN	0.63	0.52	3.43
—S—CO—CH₃	0.39	0.44	
—S—CONH₂	0.34		2.07
—SO—CH₃	0.52	0.49	
—SO—C₆H₅			3.24
—CH₂—SO—CH₃			1.33
—SO₂—CH₃	0.60	0.68	3.68
—SO₂—CH₂CH₃			3.74
—SO₂—CH₂CH₂CH₃			3.68
—SO₂—C₆H₅	0.67		3.55
—SO₂—CF₃	0.79	0.93	
—SO₂—NH₂	0.46	0.57	
—CH₂—SO₂—CH₃			1.38
—SO₃⁻	0.05	0.09	0.81
—SO₃H		0.50	
—SeCH₃	0.1	0.0	
—Se—cyclohexyl			2.37
—SeCN	0.67	0.66	3.61
—Si(CH₃)₃	−0.04	−0.07	−0.81
—Si(CH₂CH₃)₃		0.0	
—Si(CH₃)₂C₆H₅			−0.87
—Si(CH₃)₂—O—Si(CH₃)₃			−0.81
—CH₂Si(CH₃)₃	−0.16	−0.22	−0.25
—CH₂CH₂Si(CH₃)₃			−0.25
—Sn(CH₃)₃		0.0	
—Sn(CH₂CH₃)₃		0.0	

TABLE 2.74 pK_a° and Rho Values for Hammett Equation

Acid	pK_a°	ρ
Arenearsonic acids		
pK_1	3.54	1.05
pK_2	8.49	0.87
Areneboronic acids (in aqueous 25% ethanol)	9.70	2.15
Arenephosphonic acids		
pK_1	1.84	0.76
pK_2	6.97	0.95
α-Aryladoximes	10.70	0.86
Benzeneseleninic acids	4.78	1.03
Benzenesulfonamides (20°C)	10.00	1.06
Benzenesulfonanilides (20°C)		
X—C$_6$H$_4$—SO$_2$—NH—C$_6$H$_5$	8.31	1.16
C$_6$H$_5$—SO$_2$—NH—C$_6$H$_4$—X	8.31	1.74
Benzoic acids	4.21	1.00
Cinnamic acids	4.45	0.47
Phenols	9.92	2.23
Phenylacetic acids	4.30	0.49
Phenylpropiolic acids (in aqueous 35% dioxane)	3.24	0.81
Phenylpropionic acids	4.45	0.21
Phenyltrifluoromethylcarbinols	11.90	1.01
Pyridine-1-oxides	0.94	2.09
2-Pyridones	11.65	4.28
4-Pyridones	11.12	4.28
Pyrroles	17.00	4.28
5-Substituted pyrrole-2carboxylic acids	2.82	1.40
Thiobenzoic acids	2.61	1.0
Thiophenols	6.50	2.2
Trifluoroacetophenone hydrates	10.00	1.11
5-Substituted topolones	6.42	3.10
Protonated cations of		
Acetophenones	−6.0	2.6
Anilines	4.60	2.90
C-Aryl-N-dibutylamidines (in aqueous 50% ethanol)	11.14	1.41
N,N-Dimethylanilines	5.07	3.46
Isoquinolines	5.32	5.90
1-Naphthylamines	3.85	2.81
2-Naphthylamines	4.29	2.81
Pyridines	5.18	5.90
Quinolines	4.88	5.90

TABLE 2.75 pK_a° and Rho Values for Taft Equation

Acid	pK_a°	ρ
RCOOH	4.66	1.62
RCH$_2$COOH	4.76	0.67
RC≡C—COOH	2.39	1.89
H$_2$C=C(R)—COOH	4.39	0.64
(CH$_3$)$_2$C=C(R)—COOH	4.65	0.47
cis-C$_6$H$_5$—CH=C(R)—COOH	3.77	0.63
trans-C$_6$H$_5$—CH=C(R)—COOH	4.61	0.47
R—CO—CH$_2$—COOH	4.12	0.43
HON=C(R)—COOH	4.84	0.34
RCH$_2$OH	15.9	1.42
RCH(OH)$_2$	14.4	1.42
R$_1$CO—NHR$_2$	22.0	3.1*
CH$_3$CO—C(R)=C(OH)CH$_3$	9.25	1.78
CH$_3$CO—CH(R)—CO—OC$_2$H$_5$	12.59	3.44
R—CO—NHOH	9.48	0.98
R$_1$R$_2$C=NOH (R$_1$, R$_2$ not acyl groups)	12.35	1.18
(R)(CH$_3$CO)C=NOH	9.00	0.94
RC(NO$_2$)$_2$H	5.24	3.60
RSH	10.22	3.50
RCH$_2$SH	10.54	1.47
R—CO—SH	3.52	1.62
Protonated cations of		
RNH$_2$	10.15	3.14
R$_1$R$_2$NH	10.59	3.23
R$_1$R$_2$R$_3$N	9.61	3.30
R$_1$R$_2$PH	3.59	2.61
R$_1$R$_2$R$_3$P	7.85	2.67

σ for R$_1$CO and R$_2$.

TABLE 2.76 Special Hammett Sigma Constants

Substituent	σ_m^+	σ_p^+	σ_p^-
—CH₃	−0.07	−0.31	−0.17
—C(CH₃)₃	−0.06	−0.26	
—C₆H₅	0.11	−0.18	
—CF₃	0.52	0.61	0.74
—F	0.35	−0.07	0.02
—Cl	0.40	0.11	0.23
—Br	0.41	0.15	0.26
—I	0.36	0.14	
—CN	0.56	0.66	0.88
—CHO			1.13
—CONH₂			0.63
—COCH₃			0.85
—COOH	0.32	0.42	0.73
—CO—OCH₃	0.37	0.49	0.66
—CO—OCH₂CH₃	0.37	0.48	0.68
—N₂⁺			3.2
—NH₂	0.16	−1.3	−0.66
—N(CH₃)₂		−1.7	
—N(CH₃)₃⁺	0.36	0.41	
—NH—CO—CH₃		−0.60	
—NO₂	0.67	0.79	1.25
—OH		−0.92	
—O⁻			−0.81
—OCH₃	0.05	−0.78	−0.27
—SF₅			0.70
—SCF₃			0.57
—SO₂CH₃			1.05
—SO₂CF₃			1.36

2.20 POLYMERS

Polymers are mixtures of macromolecules with similar structures and molecular weights that exhibit some average characteristic properties. In some polymers long segments of linear polymer chains are oriented in a regular manner with respect to one another. Such polymers have many of the physical characteristics of crystals and are said to be *crystalline*. Polymers that have polar functional groups show a considerable tendency to be crystalline. Orientation is aided by alignment of dipoles on different chains. Van der Waals' interactions between long hydrocarbon chains may provide sufficient total attractive energy to account for a high degree of regularity within the polymers.

Irregularities such as branch points, comonomer units, and cross-links lead to *amorphous* polymers. They do not have true melting points but instead have glass transition temperatures at which the rigid and glasslike material becomes a viscous liquid as the temperature is raised.

Elastomers. Elastomers is a generic name for polymers that exhibit rubberlike elasticity. Elastomers are soft yet sufficiently elastic that they can be stretched several hundred percent under tension. When the stretching force is removed, they retract rapidly and recover their original dimensions.

Polymers that soften or melt and then solidify and regain their original properties on cooling are called *thermoplastic*. A thermoplastic polymer is usually a single strand of linear polymer with few if any cross-links.

Thermosetting Polymers.   Polymers that soften or melt on warming and then become infusible solids are called *thermosetting*. The term implies that thermal decomposition has not taken place. Thermosetting plastics contain a cross-linked polymer network that extends through the finished article, making it stable to heat and insoluble in organic solvents. Many molded plastics are shaped while molten and are then heated further to become rigid solids of desired shapes.

Synthetic Rubbers.   Synthetic rubbers are polymers with rubberlike characteristics that are prepared from dienes or olefins. Rubbers with special properties can also be prepared from other polymers, such as polyacrylates, fluorinated hydrocarbons, and polyurethanes.

Structural Differences.   Polymers exhibit structural differences. A *linear* polymer consists of long segments of single strands that are oriented in a regular manner with respect to one another. *Branched* polymers have substituents attached to the repeating units that extend the polymer laterally. When these units participate in chain propagation and link together chains, a *cross-linked* polymer is formed. A *ladder* polymer results when repeating units have a tetravalent structure such that a polymer consists of two backbone chains regularly cross-linked at short intervals.

Generally polymers involve bonding of the most substituted carbon of one monomeric unit to the least substituted carbon atom of the adjacent unit in a *head-to-tail* arrangement. Substituents appear on alternate carbon atoms. *Tacticity* refers to the configuration of substituents relative to the backbone axis. In an *isotactic* arrangement, substituents are on the same plane of the backbone axis; that is, the configuration at each chiral center is identical.

$$\begin{array}{c} \text{Y} \quad \text{Y} \quad \text{Y} \quad \text{Y} \\ | \quad\; | \quad\; | \quad\; | \\ -\text{C}-\text{C}-\text{C}-\text{C}- \end{array}$$

In a *syndiotactic* arrangement, the substituents are in an ordered alternating sequence, appearing alternately on one side and then on the other side of the chain, thus

$$\begin{array}{c} \text{Y} \qquad\quad \text{Y} \\ | \qquad\quad\; | \\ -\text{C}-\;\text{C}-\;\text{C}-\;\text{C}- \\ \qquad | \qquad\quad\; | \\ \qquad \text{Y} \qquad\quad \text{Y} \end{array}$$

In an *atactic* arrangement, substituents are in an unordered sequence along the polymer chains.

Copolymerization. Copolymerization occurs when a mixture of two or more monomer types polymerizes so that each kind of monomer enters the polymer chain. The fundamental structure resulting from copolymerization depends on the nature of the monomers and the relative rates of monomer reactions with the growing polymer chain. A tendency toward alternation of monomer units is common.

$$-\text{X}-\text{Y}-\text{X}-\text{Y}-\text{X}-\text{Y}-$$

Random copolymerization is rather unusual. Sometimes a monomer which does not easily form a homopolymer will readily add to a reactive group at the end of a growing polymer chain. In turn, that monomer tends to make the other monomer much more reactive.

In *graft copolymers* the chain backbone is composed of one kind of monomer and the branches are made up of another kind of monomer.

$$\begin{array}{c} -\text{X}-\text{X}-\text{X}-\text{X}-\text{X}-\text{X}- \\ | \qquad\qquad\;\; | \\ \text{Y} \qquad\qquad\; \text{Y} \\ | \qquad\qquad\;\; | \\ \text{Y} \qquad\qquad\; \text{Y} \end{array}$$

The structure of a *block copolymer* consists of a homopolymer attached to chains of another homopolymer.

$$—XXXX—YYY—XXXX—YYY—$$

Configurations around any double bond give rise to cis and trans stereoisomerism.

2.20.1 Additives

Antioxidants.   Antioxidants markedly retard the rate of autoxidation throughout the useful life of the polymer. Chain-terminating antioxidants have a reactive —NH or —OH functional group and include compounds such as secondary aryl amines or hindered phenols. They function by transfer of hydrogen to free radicals, principally to peroxy radicals. Butylated hydroxytoluene is a widely used example.

Peroxide-decomposing antioxidants destroy hydroperoxides, the sources of free radicals in polymers. Phosphites and thioesters such as tris(nonylphenyl) phosphite, distearyl pentaerythritol diphosphite, and dialkyl thiodipropionates are examples of peroxide-decomposing antioxidants.

Antistatic Agents.   External antistatic agents are usually quaternary ammonium salts of fatty acids and ethoxylated glycerol esters of fatty acids that are applied to the plastic surface. Internal antistatic agents are compounded into plastics during processing. Carbon blacks provide a conductive path through the bulk of the plastic. Other types of internal agents must bloom to the surface after compounding in order to be active. These latter materials are ethoxylated fatty amines and ethoxylated glycerol esters of fatty acids, which often must be individually selected to match chemically each plastic type.

Antistatic agents require ambient moisture to function. Consequently their effectiveness is dependent on the relative humidity. They provide a broad range of protection at 50% relative humidity. Much below 20% relative humidity, only materials which provide a conductive path through the bulk of the plastic to ground (such as carbon black) will reduce electrostatic charging.

Chain-Transfer Agents.   Chain-transfer agents are used to regulate the molecular weight of polymers. These agents react with the developing polymer and interrupt the growth of a particular chain. The products, however, are free radicals that are capable of adding to monomers and initiating the formation of new chains. The overall effect is to reduce the average molecular weight of the polymer without reducing the rate of polymerization. Branching may occur as a result of chain transfer between a growing but rather short chain with another and longer polymer chain. Branching may also occur if the radical end of a growing chain abstracts a hydrogen from a carbon atom four or five carbons removed from the end. Thiols are commonly used as chain-transfer agents.

Coupling Agents.   Coupling agents are molecular bridges between the interface of an inorganic surface (or filler) and an organic polymer matrix. Titanium-derived coupling agents interact with the free protons at the inorganic interface to form organic monomolecular layers on the inorganic surface. The titanate-coupling-agent molecule has six functions:

$$
\overset{\displaystyle 1}{(RO)_m} — Ti — \overset{\displaystyle 2\ \ \ 3\ \ \ 4\ \ \ 5\ 6}{(O — Y — R^2 — Z)_n}
$$

where

Type	m	n
Monoalkoxy	1	3
Coordinate	4	2
Chelate	1	2

Function 1 is the attachment of the hydrolyzable portion of the molecule to the surface of the inorganic (or proton-bearing) species.

Function 2 is the ability of the titanate molecule to transesterify.

Function 3 affects performance as determined by the chemistry of alkylate, carboxyl, sulfonyl, phenolic, phosphate, pyrophosphate, and phosphite groups.

Function 4 provides van der Waals' entanglement via long carbon chains.

Function 5 provides thermoset reactivity via functional groups such as methacrylates and amines.

Function 6 permits the presence of two or three pendent organic groups. This allows all functionality to be controlled to the first-, second-, or third-degree levels.

Silane coupling agents are represented by the formula

$$Z\text{---}R\text{---}SiY_3$$

where Y represents a hydrolyzable group (typically alkoxy); Z is a functional organic group, such as amino, methacryloxy, epoxy; and R typically is a small aliphatic linkage that serves to attach the functional organic group to silicon in a stable fashion. Bonding to surface hydroxy groups of inorganic compounds is accomplished by the $\text{---}SiY_3$ portion, either by direct bonding of this group or more commonly via its hydrolysis product $\text{---}Si(OH)_3$. Subsequent reaction of the functional organic group with the organic matrix completes the coupling reaction and establishes a covalent chemical bond from the organic phase through the silane coupling agent to the inorganic phase.

Flame Retardants.   Flame retardants are thought to function via several mechanisms, dependent upon the class of flame retardant used. Halogenated flame retardants are thought to function principally in the vapor phase either as a diluent and heat sink or as a free-radical trap that stops or slows flame propagation. Phosphorus compounds are thought to function in the solid phase by forming a glaze or coating over the substrate that prevents the heat and mass transfer necessary for sustained combustion. With some additives, as the temperature is increased, the flame retardant acts as a solvent for the polymer, causing it to melt at lower temperatures and flow away from the ignition source.

Mineral hydrates, such as alumina trihydrate and magnesium sulfate heptahydrate, are used in highly filled thermoset resins.

Foaming Agents (Chemical Blowing Agents).   Foaming agents are added to polymers during processing to form minute gas cells throughout the product. Physical foaming agents include liquids and gases. Compressed nitrogen is often used in injection molding. Common liquid foaming agents are short-chain aliphatic hydrocarbons in the C_5 to C_7 range and their chlorinated or fluorinated analogs.

The chemical foaming agent used varies with the temperature employed during processing. At relatively low temperatures (15–200°C), the foaming agent is often 4,4′-oxybis-(benzenesulfonylhydrazide) or *p*-toluenesulfonylhydrazide. In the midrange (160–232°C), either sodium hydrogen carbonate or 1,1′ azobisformamide is used. For the high range (200–285°C), there are *p*-toluenesulfonyl semicarbazide, 5-phenyltetrazole and analogs, and trihydrazinotriazine.

Inhibitors. Inhibitors slow or stop polymerization by reacting with the initiator or the growing polymer chain. The free radical formed from an inhibitor must be sufficiently unreactive that it does not function as a chain-transfer agent and begin another growing chain. Benzoquinone is a typical free-radical chain inhibitor. The resonance-stabilized free radical usually dimerizes or disproportionates to produce inert products and end the chain process.

Lubricants. Materials such as fatty acids are added to reduce the surface tension and improve the handling qualities of plastic films.

Plasticizers. Plasticizers are relatively nonvolatile liquids which are blended with polymers to alter their properties by intrusion between polymer chains. Diisooctyl phthalate is a common plasticizer. A plasticizer must be compatible with the polymer to avoid bleeding out over long periods of time. Products containing plasticizers tend to be more flexible and workable.

Ultraviolet Stabilizers. 2-Hydroxybenzophenones represent the largest and most versatile class of ultraviolet stabilizers that are used to protect materials from the degradative effects of ultraviolet radiation. They function by absorbing ultraviolet radiation and by quenching electronically excited states.

Hindered amines, such as 4-(2,2,6,6-tetramethylpiperidinyl) decanedioate, serve as radical scavengers and will protect thin films under conditions in which ultraviolet absorbers are ineffective. Metal salts of nickel, such as dibutyldithiocarbamate, are used in polyolefins to quench singlet oxygen or electronically excited states of other species in the polymer. Zinc salts function as peroxide decomposers.

Vulcanization and Curing. Originally, vulcanization implied heating natural rubber with sulfur, but the term is now also employed for curing polymers. When sulfur is employed, sulfide and disulfide cross-links form between polymer chains. This provides sufficient rigidity to prevent *plastic flow*. Plastic flow is a process in which coiled polymers slip past each other under an external deforming force; when the force is released, the polymer chains do not completely return to their original positions.

Organic peroxides are used extensively for the curing of unsaturated polyester resins and the polymerization of monomers having vinyl unsaturation. The —O—O— bond is split into free radicals which can initiate polymerization or cross-linking of various monomers or polymers.

2.20.2 Plastics

Homopolymer. Acetal homopolymers are prepared from formaldehyde and consist of high-molecular-weight linear polymers of formaldehyde.

$$H-\overset{\overset{\displaystyle H}{|}}{C}=O \rightarrow \left[-\overset{\overset{\displaystyle H}{|}}{\underset{\underset{\displaystyle H}{|}}{C}}-O- \right]_n$$

The good mechanical properties of this homopolymer result from the ability of the oxymethylene chains to pack together into a highly ordered crystalline configuration as the polymers change from the molten to the solid state.

Key properties include high melt point, strength and rigidity, good frictional properties, and resistance to fatigue. Higher molecular weight increases toughness but reduces melt flow.

Copolymer. Acetal copolymers are prepared by copolymerization of 1,3,5-trioxane with small amounts of a comonomer. Carbon–carbon bonds are distributed randomly in the polymer chain. These carbon–carbon bonds help to stabilize the polymer against thermal, oxidative, and acidic attack.

Acrylics

Poly(methyl Methacrylate).   The monomer used for poly(methyl methacrylate), 2-hydroxy-2-methylpropanenitrile, is prepared by the following reaction:

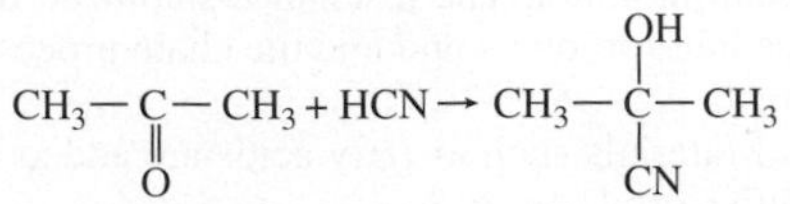

$$CH_3-\underset{\underset{O}{\|}}{C}-CH_3 + HCN \longrightarrow CH_3-\underset{\underset{CN}{|}}{\overset{\overset{OH}{|}}{C}}-CH_3$$

2-Hydroxy-2-methylpropanenitrile is then reacted with methanol (or other alcohol) to yield methacrylate ester. Free-radical polymerization is initiated by peroxide or azo catalysts and produce poly(methyl methacrylate) resins having the following formula:

$$\left[-CH_2-\underset{\underset{COOCH_3}{|}}{\overset{\overset{CH_3}{|}}{C}}- \right]_n$$

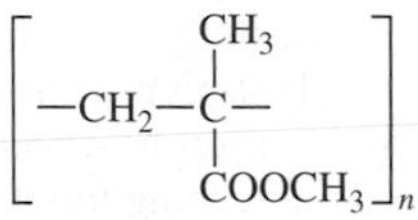

Key properties are improved resistance to heat, light, and weathering. This polymer is unaffected by most detergents, cleaning agents, and solutions of inorganic acids, alkalies, and aliphatic hydrocarbons. Poly(methyl methacrylate) has light transmittance of 92% with a haze of 1 to 3% and its clarity is equal to glass.

Poly(methyl Acrylate).   The monomer used for preparing poly(methyl acrylate) is produced by the oxidation of propylene. The resin is made by free-radical polymerization initiated by peroxide or azo catalysts and has the following formula:

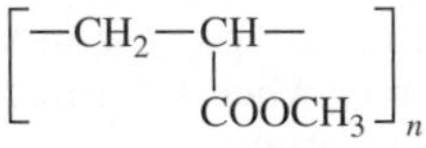

$$\left[-CH_2-\underset{\underset{COOCH_3}{|}}{CH}- \right]_n$$

Resins vary from soft, elastic, film-forming materials to hard plastics.

Poly(acrylic Acid) and Poly(methacrylic Acid).   Glacial acrylic acid and glacial methacrylic acid can be polymerized to produce water-soluble polymers having the following structures:

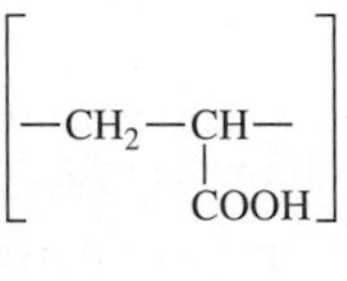

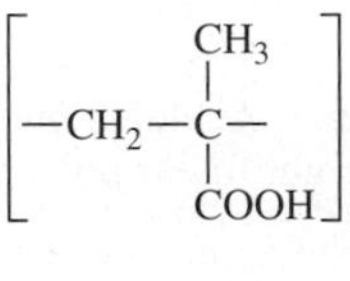

$$\left[-CH_2-\underset{\underset{COOH}{|}}{CH}- \right]_n \qquad \left[-CH_2-\underset{\underset{COOH}{|}}{\overset{\overset{CH_3}{|}}{C}}- \right]_n$$

These monomers provide a means for introducing carboxyl groups into copolymers. In copolymers these acids can improve adhesion properties, improve freeze–thaw and mechanical stability of polymer dispersions, provide stability in alkalies (including ammonia), increase resistance to attack by oils, and provide reactive centers for cross-linking by divalent metal ions, diamines, or epoxides.

Functional Group Methacrylate Monomers.   Hydroxyethyl methacrylate and dimethylaminoethyl methacrylate produce polymers having the following formulas:

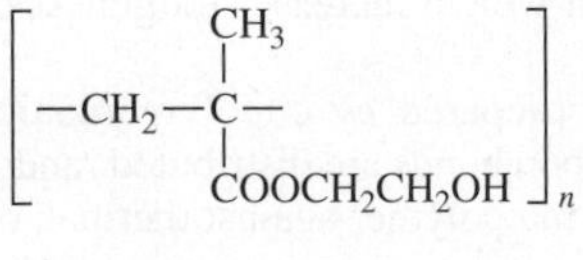

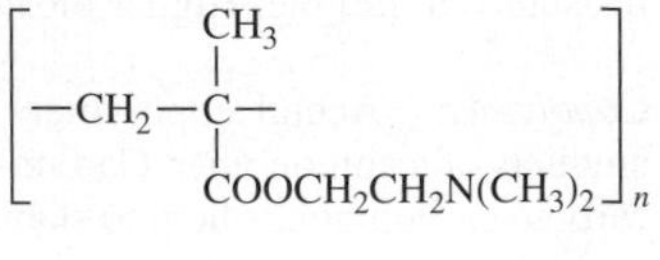

$$\left[-CH_2-\underset{\underset{COOCH_2CH_2OH}{|}}{\overset{\overset{CH_3}{|}}{C}}- \right]_n \qquad \left[-CH_2-\underset{\underset{COOCH_2CH_2N(CH_3)_2}{|}}{\overset{\overset{CH_3}{|}}{C}}- \right]_n$$

The use of hydroxyethyl (also hydroxypropyl) methacrylate as a monomer permits the introduction of reactive hydroxyl groups into the copolymers. This offers the possibility for subsequent cross-linking with an HO-reactive difunctional agent (diisocyanate, diepoxide, or melamine–formaldehyde resin). Hydroxyl groups promote adhesion to polar substrates.

Use of dimethylaminoethyl (also *tert*-butylaminoethyl) methacrylate as a monomer permits the introduction of pendent amino groups which can serve as sites for secondary cross-linking, provide a way to make the copolymer acid-soluble, and provide anchoring sites for dyes and pigments.

Poly(acrylonitrile).   Poly(acrylonitrile) polymers have the following formula:

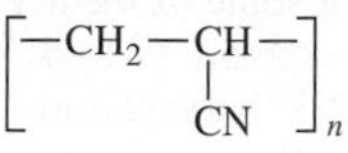

$$\left[-CH_2-\underset{\underset{CN}{|}}{CH}-\right]_n$$

Alkyds.   Alkyds are formulated from polyester resins, cross-linking monomers, and fillers of mineral or glass. The unsaturated polyester resins used for thermosetting alkyds are the reaction products of poly-functional organic alcohols (glycols) and dibasic organic acids.

Key properties of alkyds are dimensional stability, colorability, and arc track resistance. Chemical resistance is generally poor.

Alloys.   Polymer alloys are physical mixtures of structurally different homopolymers or copolymers. The mixture is held together by secondary intermolecular forces such as dipole interaction, hydrogen bonding, or van der Waals' forces.

Homogeneous alloys have a single glass transition temperature which is determined by the ratio of the components. The physical properties of these alloys are averages based on the composition of the alloy.

Heterogeneous alloys can be formed when graft or block copolymers are combined with a compatible polymer. Alloys of incompatible polymers can be formed if an interfacial agent can be found.

Allyls
Diallyl Phthalate (and Diallyl 1,3-Phthalate).   These allyl polymers are prepared from

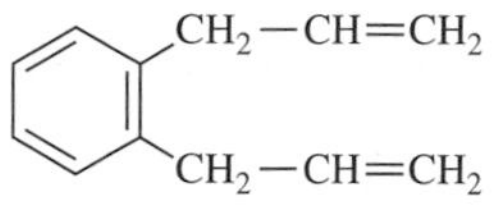

These resulting polymers are solid, linear, internally cyclized, thermoplastic structures containing unreacted allylic groups spaced at regular intervals along the polymer chain.

Molding compounds with mineral, glass, or synthetic fiber filling exhibit good electrical properties under high humidity and high temperature conditions, stable low-loss factors, high surface and volume resistivity, and high arc and track resistance.

Cellulosics
Cellulose Triacetate.   Cellulose triacetate is prepared according to the following reaction:

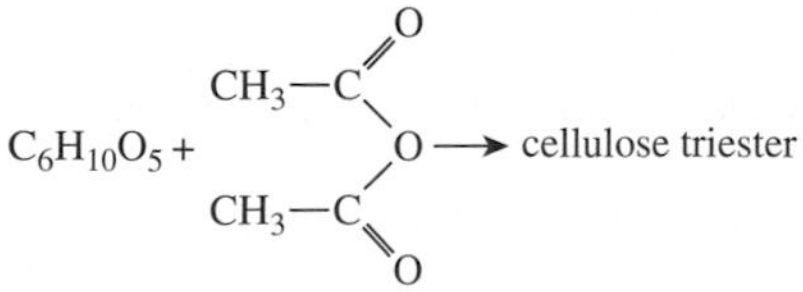

$$C_6H_{10}O_5 + \quad \begin{array}{c} CH_3-C\diagup^{O} \\ \diagdown O \\ CH_3-C\diagup \\ \diagdown_{O} \end{array} \longrightarrow \text{cellulose triester}$$

Because cellulose triacetate has a high softening temperature, it must be processed in solution. A mixture of dichloromethane and methanol is a common solvent.

Cellulose triacetate sheeting and film have good gauge uniformity and good optical clarity. Cellulose triacetate products have good dimensional stability and resistance to water and have good folding endurance and burst strength. It is highly resistant to solvents such as acetone. Cellulose triacetate products have good heat resistance and a high dielectric constant.

Cellulose Acetate, Propionate, and Butyrate.   Cellulose acetate is prepared by hydrolyzing the triester to remove some of the acetyl groups; the plastic-grade resin contains 38 to 40% acetyl. The propionate and butyrate esters are made by substituting propionic acid and its anhydride (or butyric acid and its anhydride) for some of the acetic acid and acetic anhydride. Plastic grades of cellulose-acetate–propionate resin contain 39 to 47% propionyl and 2 to 9% acetyl; cellulose-acetate-butyr-ate resins contain 26 to 39% butyryl and 12 to 15% acetyl.

These cellulose esters form tough, strong, stiff, hard plastics with almost unlimited color possibilities. Articles made from these plastics have a high gloss and are suitable for use in contact with food.

Cellulose Nitrate.   Cellulose nitrate is prepared according to the following reaction:

$$C_6H_{10}O_5 + HNO_3 \longrightarrow [-C_6H_7O_2(OH)(ONO_2)_2-]_n$$

The nitrogen content for plastics is usually about 11%, for lacquers and cement base it is 12%, and for explosives it is 13%. The standard plasticizer added is camphor.

Key properties of cellulose nitrate are good dimensional stability, low water absorption, and toughness. Its disadvantages are its flammability and lack of stability to heat and sunlight.

Ethyl Cellulose.   Ethyl cellulose is prepared by reacting cellulose with caustic to form caustic cellulose, which is then reacted with chloroethane to form ethyl cellulose. Plastic-grade material contains 44 to 48% ethoxyl.

Although not as resistant as cellulose esters to acids, it is much more resistant to bases. An outstanding feature is its toughness at low temperatures.

Rayon.   Viscose rayon is obtained by reacting the hydroxy groups of cellulose with carbon disulfide in the presence of alkali to give xanthates. When this solution is poured (spun) into an acid medium, the reaction is reserved and the cellulose is regenerated (coagulated).

Epoxy.   Epoxy resin is prepared by the following condensation reaction:

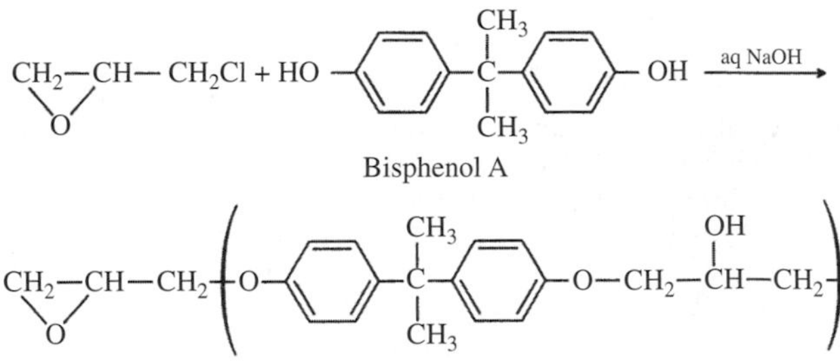

The condensation leaves epoxy end groups that are then reacted in a separate step with nucleophilic compounds (alcohols, acids, or amines). For use as an adhesive, the epoxy resin and the curing resin (usually an aliphatic polyamine) are packaged separately and mixed together immediately before use.

Epoxy novolac resins are produced by glycidation of the low-molecular-weight reaction products of phenol (or cresol) with formaldehyde. Highly cross-linked systems are formed that have superior performance at elevated temperatures.

Fluorocarbon

Poly(tetrafluoroethylene).   Poly(tetrafluoroethylene) is prepared from tetrafluoroethylene and consists of repeating units in a predominantly linear chain:

$$F_2C{=\!=}CF_2 \rightarrow [-CF_2-CF_2-]_n$$

Tetrafluoroethylene polymer has the lowest coefficient of friction of any solid. It has remarkable chemical resistance and a very low brittleness temperature ($-100°C$). Its dielectric constant and loss factor are low and stable across a broad temperature and frequency range. Its impact strength is high.

Fluorinated Ethylene–Propylene Resin.   Polymer molecules of fluorinated ethylene–propylene consist of predominantly linear chains with this structure:

$$\left[-CF_2-CF_2-CF_2-\underset{\underset{\displaystyle CF_3}{|}}{CF}- \right]_n$$

Key properties are its flexibility, translucency, and resistance to all known chemicals except molten alkali metals, elemental fluorine and fluorine precursors at elevated temperatures, and concentrated perchloric acid. It withstands temperatures from -270 to $250°C$ and may be sterilized repeatedly by all known chemical and thermal methods.

Perfluoroalkoxy Resin.   Perfluoroalkoxy resin has the following formula:

$$\left[-CF_2-CF_2-\underset{\underset{\underset{\displaystyle R}{|}}{\overset{\overset{\displaystyle |}{O}}{CF}}}{}-CF_2-CF_2- \right]_n \qquad \text{where R is} - C_nF_{2n+1}$$

It resembles polytetrafluoroethylene and fluorinated ethylene propylene in its chemical resistance, electrical properties, and coefficient of friction. Its strength, hardness, and wear resistance are about equal to the former plastic and superior to that of the latter at temperatures above $150°C$.

Poly(vinylidene Fluoride).   Poly(vinylidene fluoride) consists of linear chains in which the predominant repeating unit is

$$[-CH_2-CF_2-]_n$$

It has good weathering resistance and does not support combustion. It is resistant to most chemicals and solvents and has greater strength, wear resistance, and creep resistance than the preceding three fluorocarbon resins.

Poly(1-Chloro-1,2,2-Trifluoroethylene).   Poly(1-chloro-1,2,2-trifluoroethylene consists of linear chains in which the predominant repeating unit is

$$\left[-CF_2-\underset{\underset{\displaystyle Cl}{|}}{CF}- \right]_n$$

It possesses outstanding barrier properties to gases, especially water vapor. It is surpassed only by the fully fluorinated polymers in chemical resistance. A few solvents dissolve it at temperatures above $100°C$, and it is swollen by a number of solvents, especially chlorinated solvents. It is harder and stronger than perfluorinated polymers, and its impact strength is lower.

Ethylene–Chlorotrifluoroethylene Copolymer.   Ethylene–chlorotrifluoroethylene copolymer consists of linear chains in which the predominant 1:1 alternating copolymer is

$$\left[-CH_2-CH_2-CF_2-\underset{\underset{Cl}{|}}{CF}-\right]_n$$

This copolymer has useful properties from cryogenic temperatures to 180°C. Its dielectric constant is low and stable over a broad temperature and frequency range.

Ethylene–Tetrafluoroethylene Copolymer.   Ethylene–tetrafluoroethylene copolymer consists of linear chains in which the repeating unit is

$$[-CH_2-CH_2-CF_2-CF_2-]_n$$

Its properties resemble those of ethylene–chlorotrifluoroethylene copolymer.

Poly(vinyl Fluoride).   Poly(vinyl fluoride) consists of linear chains in which the repeating unit is

$$[-CH_2-CHF-]_n$$

It is used only as a film, and it has good resistance to abrasion and resists staining. It also has outstanding weathering resistance and maintains useful properties from −100 to 150°C.

Nitrile Resins.   The principal monomer of nitrile resins is acrylonitrile, which constitutes about 70% by weight of the polymer and provides the polymer with good gas barrier and chemical resistance properties. The remainder of the polymer is 20 to 30% methylacrylate (or styrene), with 0 to 10% butadiene to serve as an impact-modifying termonomer.

Melamine Formaldehyde.   The monomer used for preparing melamine formaldehyde is formed as follows:

Hexamethylolmelamine

Hexamethylolmelamine can further condense in the presence of an acid catalyst; ether linkages can also form (see "Urea Formaldehyde" directly before Sec. 2.20.3). A wide variety of resins can be obtained by careful selection of pH, reaction temperature, reactant ratio, amino monomer, and extent of condensation. Liquid coating resins are prepared by reacting methanol or butanol with the initial methylolated products. These can be used to produce hard, solvent-resistant coatings by heating with a variety of hydroxy, carboxyl, and amide functional polymers to produce a cross-linked film.

Phenolics
Phenol–Formaldehyde Resin.   Phenol–formaldehyde resin is prepared as follows:

$$C_6H_5OH + H_2C{=}O \rightarrow [-C_6H_2(OH)CH_2-]_n$$

One-Stage Resins.   The ratio of formaldehyde to phenol is high enough to allow the thermosetting process to take place without the addition of other sources of cross-links.

Two-Stage Resins. The ratio of formaldehyde to phenol is low enough to prevent the thermosetting reaction from occurring during manufacture of the resin. At this point the resin is termed *novolac* resin. Subsequently, hexamethylenetetramine is incorporated into the material to act as a source of chemical cross-links during the molding operation (and conversion to the thermoset or cured state).

Polyamides

Nylon 6, 11, and 12. This class of polymers is polymerized by addition reactions of ring compounds that contain both acid and amine groups on the monomer.

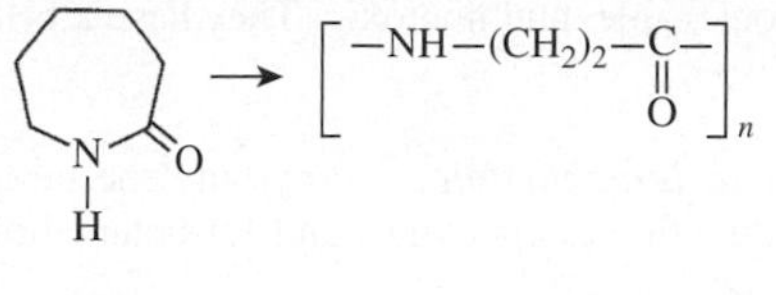

$$\left[-NH-(CH_2)_2-\underset{\underset{O}{\|}}{C}- \right]_n$$

Nylon 6 is polymerized from 2-oxohexamethyleneimine (6 carbons); nylon 11 and 12 are made this way from 11- and 12-carbon rings, respectively.

Nylon 6/6, 6/9, and 6/12. As illustrated below, nylon 6/6 is polymerized from 1,6-hexanedioic acid (six carbons) and 1,6-hexanediamine (six carbons).

$$HOOC-(CH_2)_4-COOH \ + \ H_2N-CH_2-(CH_2)_4-CH_2-NH_2 \ \rightarrow$$

1,6-Hexanedioic acid   1,6-Hexanediamine

$$\left[-NH-(CH_2)_6-NH-\underset{\underset{O}{\|}}{C}-(CH_2)_4-\underset{\underset{O}{\|}}{C}- \right]_n$$

Poly(hexamethylene 1,6-hexanediamide)

Other nylons are made this way from direct combinations of monomers to produce types 6/9, 6/10, and 6/12.

Nylon 6 and 6/6 possess the maximum stiffness, strength, and heat resistance of all the types of nylon. Type 6/6 has a higher melt temperature, whereas type 6 has a higher impact resistance and better processibility. At a sacrifice in stiffness and heat resistance, the higher analogs of nylon are useful primarily for improved chemical resistance in certain environments (acids, bases, and zinc chloride solutions) and for lower moisture absorption.

Aromatic nylons, [—NH—C_6H_4—CO—]$_n$ (also called aramids), have specialty uses because of their improved clarity.

Poly(amide-imide). Poly(amide-imide) is the condensation polymer of 1,2,4-benzenetricarboxylic anhydride and various aromatic diamines and has the general structure:

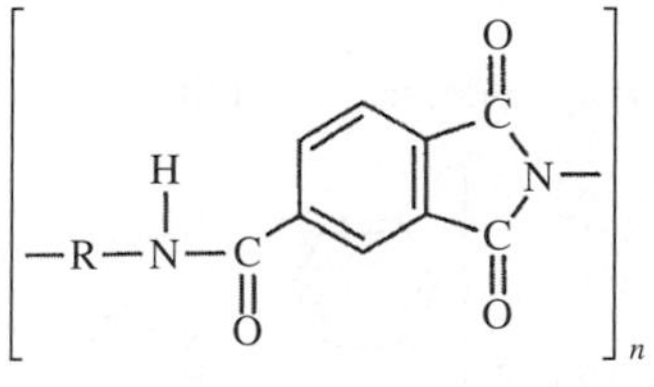

It is characterized by high strength and good impact resistance, and retains its physical properties at temperatures up to 260°C. Its radiation (gamma) resistance is good.

Polycarbonate.   Polycarbonate is a polyester in which dihydric (or polyhydric) phenols are joined through carbonate linkages. The general-purpose type of polycarbonate is based on 2,2-bis(4′-hydroxybenzene)propane (bisphenol A) and has the general structure:

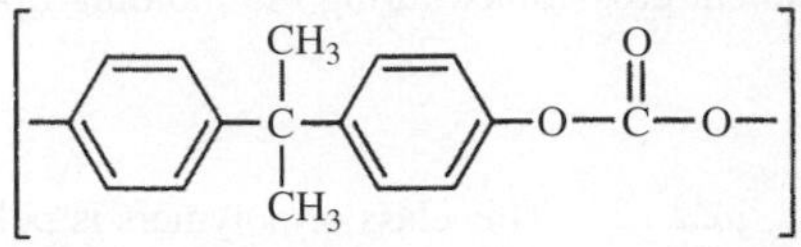

Polycarbonates are the toughest of all thermoplastics. They are window-clear, amazingly strong and rigid, autoclavable, and nontoxic. They have a brittleness temperature of $-135°C$.

Polyester

Poly(butylene Terephthalate).   Poly(butylene terephthalate) is prepared in a condensation reaction between dimethyl terephthalate and 1,4-butanediol and its repeating unit has the general structure

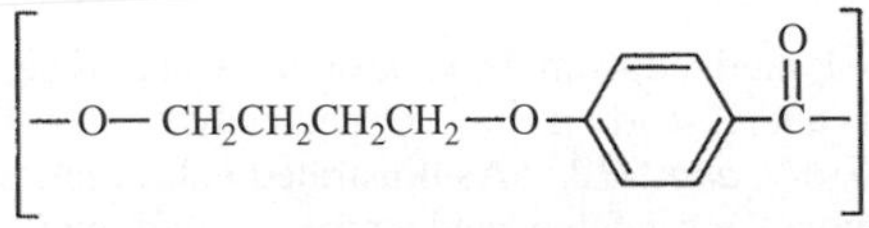

This thermoplastic shows good tensile strength, toughness, low water absorption, and good frictional properties, plus good chemical resistance and electrical properties.

Poly(ethylene Terephthalate).   Poly(ethylene terephthalate) is prepared by the reaction of either terephthalic acid or dimethyl terephthalate with ethylene glycol, and its repeating unit has the general structure.

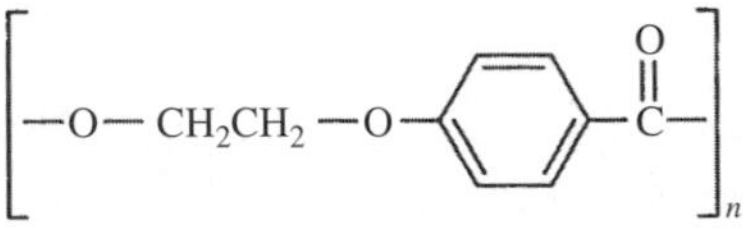

The resin has the ability to be oriented by a drawing process and crystallized to yield a high-strength product.

Unsaturated Polyesters.   Unsaturated polyesters are produced by reaction between two types of dibasic acids, one of which is unsaturated, and an alcohol to produce an ester. Double bonds in the body of the unsaturated dibasic acid are obtained by using maleic anhydride or fumaric acid.

PCTA Copolyester.   Poly(1,4-cyclohexanedimethylene terephthalic acid) (PCTA) copolyester is a polymer of cyclohexanedimethanol and terephthalic acid, with another acid substituted for a portion of the terephthalic acid otherwise required. It has the following formula:

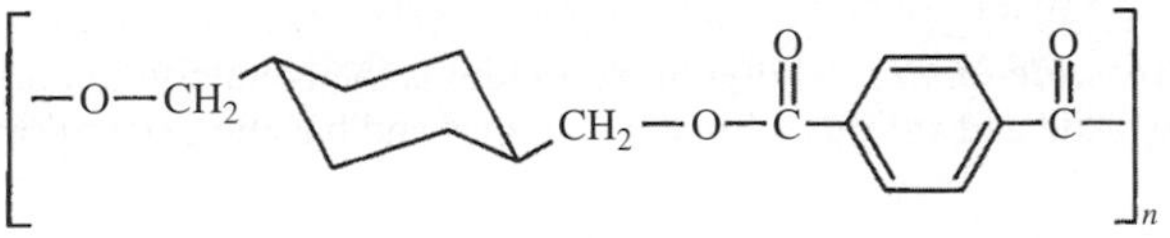

Polyimides.   Polyimides have the following formula:

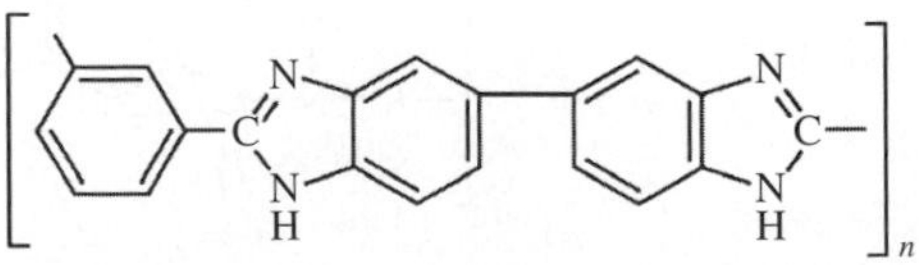

They are used as high-temperature structural adhesives since they become rubbery rather than melt at about 300°C.

Poly(methylpentene). Poly(methylpentene) is obtained by a Ziegler-type catalytic polymerization of 4-methyl-1-pentene.

Its key properties are its excellent transparency, rigidity, and chemical resistance, plus its resistance to impact and to high temperatures. It withstands repeated autoclaving, even at 150°C.

Polyolefins

Polyethylene. Polymerization of ethylene results in an essentially straight-chain high-molecular-weight hydrocarbon.

$$CH_2{=}CH_2 \rightarrow [-CH_2{-}CH_2{-}]_n$$

Branching occurs to some extent and can be controlled. Minimum branching results in a "high-density" polyethylene because of its closely packed molecular chains. More branching gives a less compact solid known as "low-density" polyethylene.

A key property is its chemical inertness. Strong oxidizing agents eventually cause some oxidation, and some solvents cause softening or swelling, but there is no known solvent for polyethylene at room temperature. The brittleness temperature is −100°C for both types. Polyethylene has good low-temperature toughness, low water absorption, and good flexibility at subzero temperatures.

Polypropylene. The polymerization of propylene results in a polymer with the following structure:

$$CH_2{=}CH{-}CH_3 \rightarrow \left[\begin{array}{c} -CH_2{-}CH{-} \\ | \\ CH_3 \end{array} \right]_n$$

The desired form in homopolymers is the isotactic arrangement (at least 93% is required to give the desired properties). Copolymers have a random arrangement. In block copolymers a secondary reactor is used where active polymer chains can further polymerize to produce segments that use ethylene monomer.

Polypropylene is translucent and autoclavable and has no known solvent at room temperature. It is slightly more susceptible to strong oxidizing agents than polyethylene.

Polybutylene. Polybutylene is composed of linear chains having an isotactic arrangement of ethyl side groups along the chain backbone.

$$CH_2{=}CH{-}CH_2{-}CH_3 \rightarrow \left[\begin{array}{c} -CH_2{-}CH{-} \\ | \\ CH_2 \\ | \\ CH_3 \end{array} \right]_n$$

It has a helical conformation in the stable crystalline form.

Polybutylene exhibits high tear, impact, and puncture resistance. It also has low creep, excellent chemical resistance, and abrasion resistance with coilability.

Ionomer. Ionomer is the generic name for polymers based on sodium or zinc salts of ethylene-methacrylic acid copolymers in which interchain ionic bonding, occurring randomly between the long-chain polymer molecules, produces solid-state properties.

The abrasion resistance of ionomers is outstanding, and ionomer films exhibit optical clarity. In composite structures ionomers serve as a heat-seal layer.

Poly(phenylene Sulfide). Poly(phenylene sulfide) has the following formula:

The recurring *para*-substituted benzene rings and sulfur atoms form a symmetrical rigid backbone.

The high degree of crystallization and the thermal stability of the bond between the benzene ring and sulfur are the two properties responsible for the polymer's high melting point, thermal stability, inherent flame retardance, and good chemical resistance. There are no known solvents of poly(phenylene sulfide) that can function below 205°C.

Polyurethane

Foams. Polyurethane foams are prepared by the polymerization of polyols with isocyanates.

Commonly used isocyanates are toluene diisocyanate, methylene diphenyl isocyanate, and polymeric isocyanates. Polyols used are macroglycols based on either polyester or polyether. The former [poly(ethylene phthalate) or poly(ethylene 1,6-hexanedioate)] have hydroxyl groups that are free to react with the isocyanate. Most flexible foam is made form 80/20 toluene diisocyanate (which refers to the ratio of 2,4-toluene diisocyanate to 2,6-toluene diisocyanate). High-resilience foam contains about 80% 80/20 toluene diisocyanate and 20% poly(methylene diphenyl isocyanate), while semi-flexible foam is almost always 100% poly(methylene diphenyl isocyanate). Much of the latter reacts by trimerization to form isocyanurate rings.

Flexible foams are used in mattresses, cushions, and safety applications. Rigid and semiflexible foams are used in structural applications and to encapsulate sensitive components to protect them against shock, vibration, and moisture. Foam coatings are tough, hard, flexible, and chemically resistant.

Elastomeric Fiber. Elastomeric fibers are prepared by the polymerization of polymeric polyols with diisocyanates.

Polymeric polyols   Diisocyanate   essentially linear polymers

The structure of elastomeric fibers is similar to that illustrated for polyurethane foams.

Silicones. Silicones are formed in the following multistage reaction:

$$R_2SiCl_2 + 2H_2O \rightarrow R_2Si(OH)_2 + 2HCl$$
$$\downarrow$$
$$[-Si(R)_2-O-]_n$$

The silanols formed above are unstable and under dehydration. On polycondensation, they give polysiloxanes (or silicones) which are characterized by their three-dimensional branched-chain structure. Various organic groups introduced within the polysiloxane chain impart certain characteristics and properties to these resins.

Methyl groups impart water repellency, surface hardness, and noncombustibility.

Phenyl groups impart resistance to temperature variations, flexibility under heat, resistance to abrasion, and compatibility with organic products.

Vinyl groups strengthen the rigidity of the molecular structure by creating easier cross-linkage of molecules.

Methoxy and alkoxy groups facilitate cross-linking at low temperatures.

Oils and gums are nonhighly branched- or straight-chain polymers whose viscosity increases with the degree of polycondensation.

Styrenics

Polystyrene.   Polystyrene has the following formula:

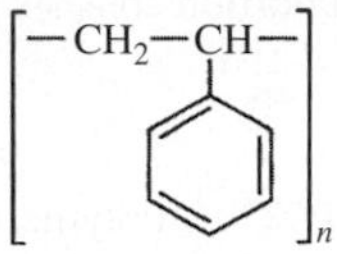

Polystyrene is rigid with excellent dimensional stability, has good chemical resistance to aqueous solutions, and is an extremely clear material.

Impact polystyrene contains polybutadiene added to reduce brittleness. The polybutadiene is usually dispersed as a discrete phase in a continuous polystyrene matrix. Polystyrene can be grafted onto rubber particles, which assures good adhesion between the phases.

Acrylonitrile-Butadiene-Styrene (ABS) Copolymers.   This basic three-monomer system can be tailored to yield resins with a variety of properties. Acrylonitrile contributes heat resistance, high strength, and chemical resistance. Butadiene contributes impact strength, toughness, and retention of low-temperature properties. Styrene contributes gloss, processibility, and rigidity. ABS polymers are composed of discrete polybutadiene particles grafted with the styrene-acrylonitrile copolymer; these are dispersed in the continuous matrix of the copolymer.

Styrene-Acrylonitrile (SAN) Copolymers.   SAN resins are random, amorphous copolymers whose properties vary with molecular weight and copolymer composition. An increase in molecular weight or in acrylonitrile content generally enhances the physical properties of the copolymer but at some loss in case of processing and with a slight increase in polymer color.

SAN resins are rigid, hard, transparent thermoplastics which process easily and have good dimensional stability—a combination of properties unique in transparent polymers.

Sulfones.   Below are the formulas for three polysulfones.

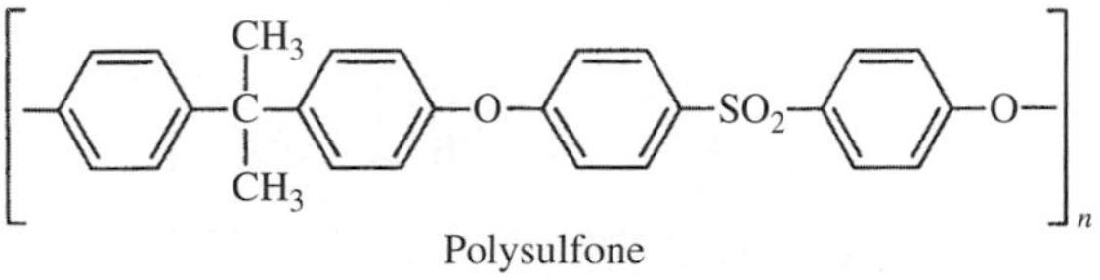

Polysulfone

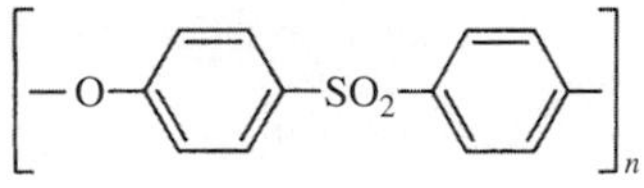

Poly(ester sulfone)

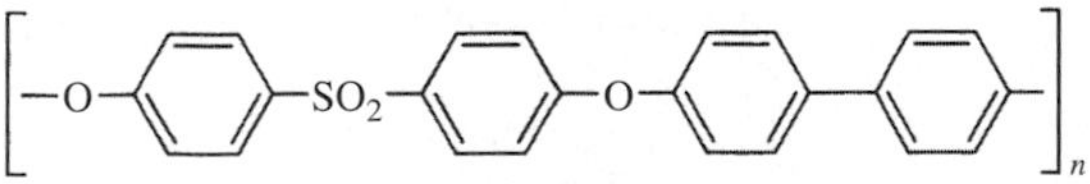

Poly(phenyl sulfone)

The isopropylidene linkage imparts chemical resistance, the ether linkage imparts temperature resistance, and the sulfone linkage imparts impact strength. The brittleness temperature of polysulfones is $-100°C$. Polysulfones are clear, strong, nontoxic, and virtually unbreakable. They do not hydrolyze during autoclaving and are resistant to acids, bases, aqueous solutions, aliphatic hydrocarbons, and alcohols.

Thermoplastic Elastomers

Polyolefins.   In these thermoplastic elastomers the hard component is a crystalline polyolefin, such as polyethylene or polypropylene, and the soft portion is composed of ethylene-propylene rubber. Attractive forces between the rubber and resin phases serve as labile cross-links. Some contain a chemically cross-linked rubber phase that imparts a higher degree of elasticity.

Styrene-Butadiene-Styrene Block Copolymers.   Styrene blocks associate into domains that form hard regions. The midblock, which is normally butadiene, ethylene–butene, or isoprene blocks, forms the soft domains. Polystyrene domains serve as cross-links.

Polyurethanes.   The hard portion of polyurethane consists of a chain extender and polyisocyanate. The soft component is composed of polyol segments.

Polyesters.   The hard portion consists of copolyester, and the soft portion is composed of polyol segments.

Vinyl

Poly(vinyl Chloride) (PVC).   Polymerization of vinyl chloride results in the formation of a polymer with the following formula:

$$CH_2 = CHCl \rightarrow \left[-CH_2 - \underset{\underset{Cl}{|}}{CH} - \right]_n$$

When blended with phthalate ester plasticizers, PVC becomes soft and pliable.

Its key properties are good resistance to oils and a very low permeability to most gases.

Poly(vinyl Acetate).   Poly(vinyl acetate) has the following formula:

$$\left[-CH_2 - \underset{\underset{O-CO-CH_3}{|}}{CH} - \right]_n$$

Poly(vinyl acetate) is used in latex water paints because of its weathering, quick-drying, recoatability, and self-priming properties. It is also used in hot-melt and solution adhesives.

Poly(vinyl Alcohol).   Poly(vinyl alcohol) has the following formula:

$$\left[-CH_2 - \underset{\underset{OH}{|}}{CH} - \right]_n$$

It is used in adhesives, paper coating and sizing, and textile warp size and finishing applications.

Poly(vinyl Butyral).   Poly(vinyl butyral) is prepared according to the following reaction:

$$\left[-CH_2 - \underset{\underset{OH}{|}}{CH} - \right]_n + CH_3CH_2CH_2CHO \rightarrow \left[\begin{array}{c} -CH_2 - CH - CH_2 - CH - \\ O - CH - O \\ | \\ CH_2 - CH_2 - CH_3 \end{array} \right]_n$$

Its key characteristics are its excellent optical and adhesive properties. It is used as the interlayer film for safety glass.

Poly(vinylidene Chloride).   Poly(vinylidene chloride) is prepared according to the following reaction:

$$CH_2 = CCl_2 + CH_2 = CHCl \rightarrow [-CH_2 - CCl_2 - CH_2 - CHCl -]_n$$
$$\text{Random copolymer}$$

Urea Formaldehyde.
The reaction of urea with formaldehyde yields the following products, which are used as monomers in the preparation of urea formaldehyde resin.

$$H_2N - CO - NH_2 + H_2CO \rightarrow H_2N - CO - NH - CH_2OH$$
$$+ HOCH_2 - NH - CO - NH - CH_2OH$$

The reaction conditions can be varied so that only one of those monomers is formed. 1-Hydroxy-methylurea and 1,3-bis(hydroxymethyl)urea condense in the presence of an acid catalyst to produce urea formaldehyde resins. A wide variety of resins can be obtained by careful selection of the pH, reaction temperature, reactant ratio, amino monomer, and degree of polymerization. If the reaction is carried far enough, an infusible polymer network is produced.

Liquid coating resins are prepared by reacting methanol or butanol with the initial hydroxy-methylureas. Ether exchange reactions between the amino resin and the reactive sites on the polymer produce a cross-linked film.

2.20.3 Rubber

Natural rubber (also called India rubber or *caoutchouc*) consists of polymers of isoprene [2-methyl-1,3-butadiene, $CH_2{=}C(CH_3)CH{=}CH_2$] with minor impurities of other organic compounds plus water. Forms of poly-isoprene that are used as natural rubbers are classified as elastomers. Currently, rubber is harvested mainly in the form of the latex from certain trees. The latex is a sticky, milky colloid drawn off by making incisions into the bark and collecting the fluid in vessels (tapping). The latex then is refined into rubber ready for commercial processing. Natural rubber has a large stretch ratio and high resilience, and is extremely waterproof.

Synthetic rubber is made by the polymerization of a variety of petroleum-based precursors. The most prevalent synthetic rubbers are styrene-butadiene rubbers (SBRs) derived from the copolymerization of styrene ($C_6H_5CH{=}CH_2$) and 1,3-butadiene ($CH_2{=}CHCH{=}CH_2$). Other synthetic rubbers are prepared from isoprene (2-methyl-1,3-butadiene), chloroprene (2-chloro-1,3-butadiene), and isobutylene (methylpropene) with a small percentage of isoprene for cross-linking. These and other monomers can be mixed in various proportions to be copolymerized to produce products with a range of physical, mechanical, and chemical properties. The monomers can be produced pure and the addition of impurities or additives can be controlled by design to give optimal properties. Polymerization of pure monomers can be better controlled to give a desired proportion of *cis* and *trans* double bonds.

Gutta Percha. Gutta percha is a natural polymer of isoprene (3-methyl-1,3-butadiene) in which the configuration around each double bond is *trans*. It is hard and horny and has the following formula:

$$\left[\begin{array}{c} CH_3 \\ | \\ C \\ CH_2{\diagup}{\diagdown}CH{\diagdown}CH_2 \end{array}\right]_n$$

Natural Rubber. Natural rubber is a polymer of isoprene in which the configuration around each double bond is *cis* (or *Z*):

$$\left[\begin{array}{c} H_3C \\ {\diagdown} \\ C{=}CH \\ {-}CH_2{\diagup}{\diagdown}CH_2{-} \end{array}\right]_n$$

Its principal advantages are high resilience and good abrasion resistance.

Chlorosulfonated Polyethylene. Chlorosulfonated polyethylene is prepared as follows:

$$[{-}CH_2{-}CH_2{-}]_n + HSO_3Cl \rightarrow \left[\begin{array}{c} {-}CH_2{-}CH{-} \\ | \\ SO_3H \end{array}\right]_n + HCl$$

Cross-linking, which can occur as a result of side reactions, causes an appreciable gel content in the final product.

The polymer can be vulcanized to give a rubber with very good chemical (solvent) resistance, excellent resistance to aging and weathering, and good color retention in sunlight.

Epichlorohydrin. Epichlorohydrin is a product of covulcanization of epichlorohydrin (epoxy) polymers with rubbers, especially *cis*-polybutadiene.

Its advantages include impermeability to air, excellent adhesion to metal, and good resistance to oils, weathering, and low temperature.

Nitrile Rubber (NBR, GRN, Buna N). Nitrile rubber can be prepared as follows:

$$CH_2= CH-CH=CH_2 + CH_2 =CH-CN \rightarrow$$

$$\text{2 parts} \qquad\qquad \text{1 part}$$

$$\left[-CH_2-CH=CH-CH_2-CH_2-\underset{\underset{CN}{|}}{CH}-CH_2-CH=CH-CH_2-\right]_n$$

Nitrile rubber is also known as nitrile-butadiene rubber (NBR), government rubber nitrile (GRN), and Buna N.

It possesses resistance to oils up to 120°C and excellent abrasion resistance and adhesion to metal.

Polyacrylate. Polyacrylate has the following formula:

$$\left[-CH_2-\underset{\underset{CN}{|}}{CH}-\right]_n$$

It possesses oil and heat resistance to 175°C and excellent resistance to ozone.

cis-Polybutadiene Rubber (BR). *cis*-Polybutadiene is prepared by polymerization of butadiene by mostly, 1,4-addition.

$$CH_2=CH-CH=CH_2 \rightarrow [-CH_2-CH=CH-CH_2-]_n$$

The polybutadiene produced is in the Z (or *cis*) configuration.

cis-Polybutadiene has good abrasion resistance, is useful at low temperature, and has excellent adhesion to metal.

Polychloroprene (Neoprene). Polychloroprene is prepared as follows:

$$CH_2=CH-\underset{\underset{Cl}{|}}{C}=CH_2 \rightarrow [-CH_2-CH=C(Cl)-CH_2-]_n$$

It has very good weathering characteristics, is resistant to ozone and to oil, and is heat-resistant to 100°C.

Ethylene-Propylene-Diene Rubber (EPDM). Ethylene-propylene-diene rubber is polymerized from 60 parts ethylene, 40 parts propylene, and a small amount of nonconjugated diene. The nonconjugated diene permits sulfur vulcanization of the polymer instead of using peroxide.

It is a very lightweight rubber and has very good weathering and electrical properties, excellent adhesion, and excellent ozone resistance.

Polyisobutylene (Butyl Rubber).   Polyisobutylene is prepared as follows:

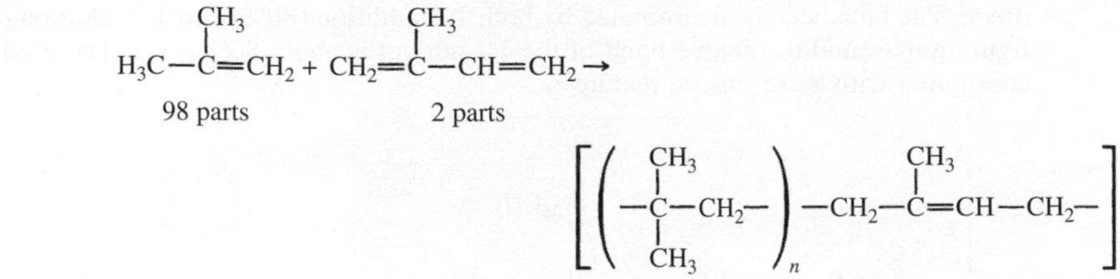

$$\underset{\text{98 parts}}{H_3C-\overset{\overset{\displaystyle CH_3}{|}}{C}=CH_2} + \underset{\text{2 parts}}{CH_2=\overset{\overset{\displaystyle CH_3}{|}}{C}-CH=CH_2} \rightarrow$$

$$\left[\left(-\overset{\overset{\displaystyle CH_3}{|}}{\underset{\underset{\displaystyle CH_3}{|}}{C}}-CH_2- \right)_n -CH_2-\overset{\overset{\displaystyle CH_3}{|}}{C}=CH-CH_2- \right]$$

It possesses excellent ozone resistance, very good weathering and electrical properties, and good heat resistance.

(Z)-Polyisoprene (Synthetic Natural Rubber).   Polymerization of isoprene by 1,4-addition produces polyisoprene that has a *cis* (or *Z*) configuration.

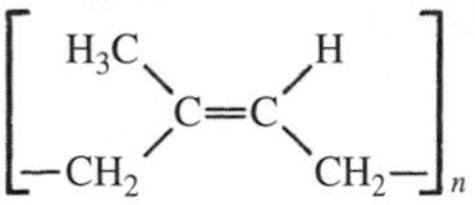

$$\left[\overset{H_3C}{\underset{-CH_2}{\diagdown}}C=C\overset{H}{\underset{CH_2-}{\diagup}} \right]_n$$

Polysulfide Rubbers.   Polysulfide rubbers are prepared as follows:

$$Cl-R-Cl + Na-S-S-S-S-Na \rightarrow HS[-R-S-S-S-S-]_nR-SH$$

where R can be

$$-CH_2CH_2-, \quad -CH_2CH_2-O-CH_2CH_2-,$$

or

$$-CH_2CH_2-O-CH_2-O-CH_2CH_2-.$$

Polysulfide rubbers posses excellent resistance to weathering and oils and have very good electrical properties.

Poly(vinyl Chloride) (PVC).   Poly(vinyl chloride) has the following structures:

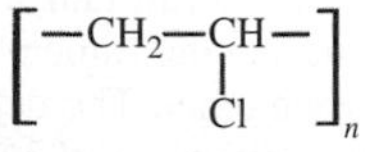

$$\left[-CH_2-\overset{\displaystyle CH}{\underset{\underset{\displaystyle Cl}{|}}{}}- \right]_n$$

PVC polymer plus special plasticizers are used to produce flexible tubing which has good chemical resistance.

Silicone Rubbers.   Silicone rubbers are prepared as follows:

$$Cl-\overset{\overset{\displaystyle CH_3}{|}}{\underset{\underset{\displaystyle CH_3}{|}}{Si}}-Cl \xrightarrow{H_2O} HO-\overset{\overset{\displaystyle CH_3}{|}}{\underset{\underset{\displaystyle CH_3}{|}}{Si}}-OH \xrightarrow{\text{polymerize}} \left[-\overset{\overset{\displaystyle CH_3}{|}}{\underset{\underset{\displaystyle CH_3}{|}}{Si}}-O- \right]_n$$

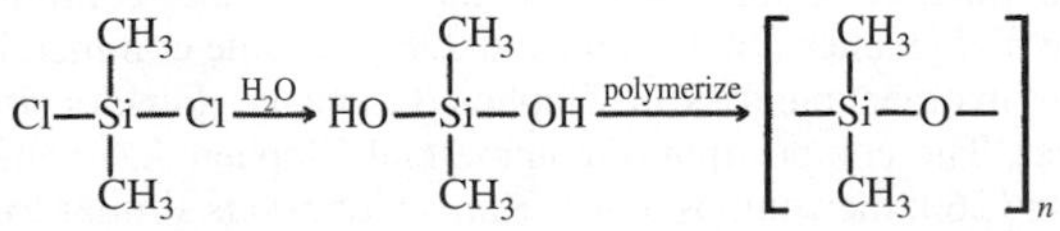

Other groups may replace the methyl groups.

Silicone rubbers have excellent ozone and weathering resistance, good electrical properties, and good adhesion to metal.

Styrene-Butadiene Rubber (GRS, SBR, Buna S).   Styrene-butadiene rubber is prepared from the free-radical copolymerization of one part by weight of styrene and three parts by weight of 1,3-butadiene. The butadiene is incorporated by both 1,4-addition (80%) and 1,2-addition (20%). The configuration around the double bond of the 1,4-adduct is about 80% *trans*. The product is a random copolymer with these general features:

| *trans*-1,4-Adduct | 1,2-Adduct | *trans*-1,4-Adduct | Styrene | *cis*-1,4-Adduct |

Styrene-butadiene rubber (SBR) is also known as government rubber styrene (GRS) and Buna S.

Urethane.   See Table 2.81.

2.21 POLYCHLOROBIPHENYLS

Polychlorobiphenyls (PCBs) are a class of chemical compounds in which as few as 2 and as many as 10 chlorine atoms (the maximum—5 per ring) are attached to the biphenyl molecule. Monochlorinated biphenyls (i.e., one chlorine atom attached to the biphenyl molecule) are often included when describing polychlorobiphenyl mixtures. The general chemical structure of chlorinated biphenyls is

The benzene rings can rotate around the inter-ring bond—the two extreme configurations are planar (the two benzene rings in the same plane) and the nonplanar in which the benzene rings are at a ninety-degree angle to each other. The degree of planarity is largely determined by the number of substitutions in the ortho positions (the 2,6 and the 2′6′ positions). The replacement of hydrogen atoms in the ortho positions with the relatively bulky chlorine atoms (in place of the hydrogen atoms) forces the benzene rings to rotate out of the planar configuration. The benzene rings of non-ortho substituted polychlorobiphenyls, as well as mono-ortho substituted polychlorobiphenyls, may assume a planar configuration whereas the benzene rings that contain bulky substituents on the 2,6 and 2′6′ positions cannot assume a planar or coplanar configuration and are nonplanar.

The trade names (Tables 2.98 and 2.99) of some commercial polychlorobiphenyl mixtures manufactured in other countries are Clophen (Germany), Fenclor (Italy), Kanechlor (Japan), and Phenoclor (France). The composition of commercial Clophen A-60 and Phenoclor DP-6 is similar to that of Aroclor 1260; the composition of Kanechlor 500 is similar to that of Aroclor 1254. Fenclor contains 100% decachlorobiphenyl.

TABLE 2.77 Names and Structures of Polymers

Common name	Acronym, alternate name	Class	Structure of repeat unit
Amylose		Polysaccharide	
Cellulose	Rayon Cellophane Regenerated cellulose	Polysaccharide	
Cellulose acetate	CA	Cellulose ester	$R = -\overset{\overset{\displaystyle O}{\|\|}}{C} - CH_3$
Cellulose nitrate	CN	Cellulose ester	$R = -NO_2$
Hydroxypropylcellulose	HPC	Cellulose ester	$R = -(CH_2)_3 - OH$
Ladder polymer	Double-strand polymer		
Phenol-formaldehyde	Bakelite	Phenolic polymer	

(*Continued*)

TABLE 2.77   Names and Structures of Polymers (*Continued*)

Common name	Acronym, alternate name	Class	Structure of repeat unit
Polyacetal		Polyacetal	
Polyacetylene		Polyalkyne	$\left[\!\!\!-CH=CH-\!\!\!\right]_n$
Polyacrylamide		Vinyl polymer	
Poly(acrylic acid)		Vinyl polymer	
Polyacrylonitrile	PAN	Vinyl polymer	
Poly(L-alanine)		Polypeptide	
Polyamide	Nylon	Polyamide	
Polyaniline		Polyamine	
Polybenzimidazole	PBI	Polyhetero-aromatic	
Polybenzobisoxazole	PBO	Polyhetero-aromatic	
Polybenzobisthiazole	PBT	Polyhetero-aromatic	

TABLE 2.77   Names and Structures of Polymers (*Continued*)

Common name	Acronym, alternate name	Class	Structure of repeat unit
Poly(γ-benzyl-L-glutamate)	PBLG	Polypeptide	
1,2-Polybutadiene	PBD	Diene polymer	
cis-1,4-Polybutadiene	PBD	Diene polymer	
trans-1,4-Polybutadiene	PBD	Diene polymer	
Poly(butene-1)	PB-1	Poly(α-olefin)	
Polybutylene-terephthalate	PBT	Polyester	
Poly(ε-caprolactam)	Nylon-6	Polyamide	
Poly(ε-caprolactone)		Polyester	
Polycarbonate	PC	Polyester	
*cis, trans-*1,4-Polychloroprene	Neoprene	Diene polymer	
Polychlorotrifluoroethylene	PCTFE	Vinyl polymer	

(*Continued*)

TABLE 2.77 Names and Structures of Polymers (*Continued*)

Common name	Acronym, alternate name	Class	Structure of repeat unit
Polydiethylsiloxane	PDES	Polysiloxane	$\left[\begin{array}{c} CH_2CH_3 \\ Si-O \\ CH_2CH_3 \end{array}\right]_n$
Polydimethylsiloxane	PDMS	Polysiloxane	$\left[\begin{array}{c} CH_3 \\ Si-O \\ CH_3 \end{array}\right]_n$
Polydiphenylsiloxane	PDPS	Polysiloxane	$\left[\,Si-O\,\right]_n$ (diphenyl substituents)
Polyester		Polyester	$\left[O-R-O-\overset{O}{\overset{\|}{C}}-R'-\overset{O}{\overset{\|}{C}} \right]_n$
Polyetheretherketone	PEEK	Polyketone	$\left[O-\!\!\bigcirc\!\!-\overset{O}{\overset{\|}{C}}-\!\!\bigcirc\!\! \right]_n$
Polyethylene	PE	Polyolefin	$\left[CH_2-CH_2 \right]_n$
Poly(ethylene imine)		Polyamine	$\left[CH_2-CH_2-NH \right]_n$
Poly(ethylene oxide) [Poly(ethylene glycol)]	PEO (PEG)	Polyether	$\left[CH_2-CH_2-O \right]_n$
Polyethylene-terephthalate	PET	Polyester	$\left[(CH_2)_2-O-\overset{O}{\overset{\|}{C}}-\!\!\bigcirc\!\!-\overset{O}{\overset{\|}{C}} \right]_n$
Polyglycine		Polypeptide	$\left[NH-CH_2-\overset{O}{\overset{\|}{C}} \right]_n$
Poly(hexamethylene adipamide)	Nylon-66	Polyamide	$\left[NH-(CH_2)_6-NH-\overset{O}{\overset{\|}{C}}-(CH_2)_4-\overset{O}{\overset{\|}{C}} \right]_n$
Polyhydroxybutyrate	PHB	Polyester	$\left[O-\overset{CH_3}{\overset{\|}{CH}}-CH_2-\overset{O}{\overset{\|}{C}} \right]_n$

TABLE 2.77 Names and Structures of Polymers (*Continued*)

Common name	Acronym, alternate name	Class	Structure of repeat unit
Polyimide	PI	Polyimide	
Poly(imino-1,3-phenylene iminoisophthaloyl) (Nomex)		Polyaramide	
Poly(imino-1,4-phenylene iminoterephthaloyl) (Kevlar)		Polyaramide	
Polyisobutylene	Butyl rubber	Vinylidene polymer	
Polyisocyanate	PIC	Polyamide	
Polyisocyanide		Polyisocyanide	
cis-1,4-Polyisoprene	*cis* PIP, Natural rubber	Diene polymer	
tran-1,4-Polyisoprene	*trans*-PIP, Gutta percha	Diene polymer	
Polylactam		Polyamide	
Polylactone		Polyester	
Poly(*p*-methyl styrene)		Vinyl polymer	

(*Continued*)

TABLE 2.77   Names and Structures of Polymers (*Continued*)

Common name	Acronym, alternate name	Class	Structure of repeat unit
Poly(methyl acrylate)	PMA	Vinyl polymer	
Poly(methyl methacrylate)	PMMA	Vinylidene polymer	
Poly(α-methyl styrene)		Vinylidene polymer	
Poly(methylene oxide)	PMO	Polyether	
Polymethylphenyl-siloxane	PMPS	Polysiloxane	
Polynitrile		Polyimine	
Polynucleotide		Polynucleotide	
Poly(n-pentene-2)		Poly(α-olefin)	
Poly(n-pentene-1)		Poly(α-olefin)	
Polypeptides [Poly(α-amino acid)]		Polypeptide	
Poly(p-phenylene oxide)	PPO	Polyether	

TABLE 2.77 Names and Structures of Polymers (*Continued*)

Common name	Acronym, alternate name	Class	Structure of repeat unit
Poly(*p*-phenylene sulfide)	PPS	Polysulfide	
Poly(*p*-phenylene vinylene)		Polyaromatic	
Poly(*p*-phenylene)	PP	Polyaromatic	
Polyphosphate		Inorganic polymer	
Polyphosphazene		Inorganic polymer	
Polyphosphonate		Inorganic polymer	
Polypropylene	PP	Poly(α-olefin)	
Poly(propylene oxide)	PPO	Polyether	
Poly(pyromellitimide-1,4-diphenyl ether) (Kapton)		Polyimide	
Polypyrrole		Polyhetero-cyclic	
Polysilane		Inorganic polymer	

(*Continued*)

TABLE 2.77 Names and Structures of Polymers (*Continued*)

Common name	Acronym, alternate name	Class	Structure of repeat unit
Polyailazane		Inorganic polymer	$\left[\begin{array}{c} R \\ \mid \\ -Si-N- \\ \mid\quad\mid \\ R'\quad R'' \end{array}\right]_n$
Polysiloxane	Silicones	Inorganic polymer	$\left[\begin{array}{c} R \\ \mid \\ -Si-O- \\ \mid \\ R' \end{array}\right]_n$
Polystyrene	PS Styrofoam	Vinyl polymer	$\left[\begin{array}{c} -CH-CH_2- \\ \mid \\ C_6H_5 \end{array}\right]_n$
Polysulfide	Thiokol	Polysulfide	$\left[\!-R-S_m-\!\right]_n$
Polysulfur		Polysulfur	$\left[\!-S-\!\right]_{8n}$
Polytetrafluoroethylene (Teflon)	PTFE	Poly(α-olefin)	$\left[\begin{array}{c} F\quad F \\ \mid\quad\mid \\ -C-C- \\ \mid\quad\mid \\ F\quad F \end{array}\right]_n$
Poly(tetramethylene oxide)	PTMO	Polyether	$\left[\!-CH_2-CH_2-CH_2-CH_2-O-\!\right]_n$
Polythiophene		Polyhetero-cyclic	$\left[\!-C_4H_2S-\!\right]_n$
Polyurea		Polyurea	$\left[\!-NH-R-NH-\overset{\displaystyle O}{\overset{\displaystyle \|}{C}}-NH-R'-NH-\overset{\displaystyle O}{\overset{\displaystyle \|}{C}}-\!\right]_n$
Polyurethane	Adiprene	Polyurethane	$\left[\!-O-R-O-\overset{\displaystyle O}{\overset{\displaystyle \|}{C}}-NH-R'-NH-\overset{\displaystyle O}{\overset{\displaystyle \|}{C}}-\!\right]_n$
Poly(L-valine)		Polypeptide	$\left[\begin{array}{c} \quad\quad\overset{\displaystyle O}{\overset{\displaystyle \|}{}} \\ -NH-CH-C- \\ \mid \\ CH(CH_3)_2 \end{array}\right]_n$
Poly(vinyl acetate)	PVAc	Vinyl polymer	$\left[\begin{array}{c} -CH-CH_2- \\ \mid \\ O-\underset{\displaystyle \underset{\displaystyle O}{\|}}{C}-CH_3 \end{array}\right]_n$

TABLE 2.77 Names and Structures of Polymers (*Continued*)

Common name	Acronym, alternate name	Class	Structure of repeat unit
Poly(vinyl alcohol)	PVA	Vinyl polymer	$\left[\text{CH}-\text{CH}_2\right]_n$ with OH
Poly(vinyl chloride)	PVC	Vinyl polymer	$\left[\text{CH}-\text{CH}_2\right]_n$ with Cl
Poly(vinyl fluoride)	PVF	Vinyl polymer	$\left[\text{CH}-\text{CH}_2\right]_n$ with F
Poly(2-vinyl pyridine)	PVP	Vinyl polymer	$\left[\text{CH}-\text{CH}_2\right]_n$ with pyridine (N) ring
Poly(N-vinyl pyrrolidone)		Vinyl polymer	$\left[\text{CH}-\text{CH}_2\right]_n$ with N, pyrrolidone ring (=O)
Poly(vinylidene chloride)	PVDC Saran	Vinylidene polymer	$\left[\begin{array}{c}\text{Cl}\\\text{C}-\text{CH}_2\\\text{Cl}\end{array}\right]_n$
Poly(vinylidiene fluoride)	PVDF	Vinylidiene polymer	$\left[\begin{array}{c}\text{F}\\\text{C}-\text{CH}_2\\\text{F}\end{array}\right]_n$
Vinyl polymer		Vinyl polymer	$\left[\begin{array}{cc}\text{R}&\text{R}''\\\text{C}-\text{C}\\\text{R}'&\text{R}'''\end{array}\right]_n$

TABLE 2.78   Plastics

Acetals	Fluorocarbons (*continued*)
Acrylics	Poly(vinylidene fluoride) (PVDF)
Poly(methyl methacrylate) (PMMA)	Ethylene-chlorotrifluoroethylene copolymer
Poly(acrylonitrile)	Ethylene-tetrafluoroethylene copolymer
Alkyds	Poly(vinyl fluoride) (PVF)
Alloys	Melamine formaldehyde
Acrylic-poly(vinyl chloride) alloy	Melamine phenolic
Acrylonitrile-butadiene-styrene-poly(vinyl chloride)	Nitrile resins
alloy (ABS-PVC)	Phenolics
Acrylonitrile-butadiene-styrene-polycarbonate alloy	Polyamides
(ABS-PC)	Nylon 6
Allyls	Nylon 6/6
Allyl-diglycol-carbonate polymer	Nylon 6/9
Diallyl phthalate (DAP) polymer	Nylon 6/12
Cellulosics	Nylon 11
Cellulose acetate resin	Nylon 12
Cellulose-acetate-propionate resin	Aromatic nylons
Cellulose-acetate-butyrate resin	Poly(amide-imide)
Cellulose nitrate resin	Poly(aryl ether)
Ethyl cellulose resin	Polycarbonate (PC)
Rayon	Polyesters
Chlorinated polyether	Poly(butylenes terephthalate) (PBT) [also called
Epoxy	polytetramethylene terephthalate (PTMT)]
Fluorocarbons	Poly(ethylene terephthalate) (PET)
Poly(tetrafluoroethylene) (PTFE)	Unsaturated polyesters (SMC, BMC)
Poly(chlorotrifluoroethylene) (PCTFE)	Butadiene-maleic acid copolymer (BMC)
Perfluoroalkoxy (PFA) resin	Styrene-maleic acid copolymer (SMC)
Fluorinated ethylene-propylene (FEP) resin	Polyimide
Poly(methylpentene)	Sulfones (*continued*)
Polyolefins (PO)	Poly(ether sulfone)
Low-density polyethylene (LDPE)	Poly(phenyl sulfone)
High-density polyethylene (HDPE)	Thermoplastic elastomers
Ultrahigh-molecular-weight polyethylene (UHMWPE)	Polyolefin
Polypropylene (PP)	Polyester
Polybutylene (PB)	Block copolymers
Polyallomers	Styrene-butadiene block copolymer
Poly(phenylene oxide)	Styrene-isoprene block copolymer
Poly(phenylene sulfide) (PPS)	Styrene-ethylene block copolymer
Polyurethanes	Styrene-butylene block copolymer
Silicones	Urea formaldehyde
Styrenics	Vinyls
Polystyrene (PS)	Poly(vinyl chloride) (PVC)
Acrylonitrile-butadiene-styrene (ABS) copolymer	Poly(vinyl acetate) (PVAC)
Styrene-acrylonitrile (SAN) copolymer	Poly(vinylidene chloride)
Styrene-butadiene copolymer	Poly(vinyl butyrate) (PVB)
Sulfones	Poly(vinyl formal)
Polysulfone (PSF)	Poly(vinyl alcohol) (PVAL)

TABLE 2.79 Properties of Commercial Plastics

Properties	Acetal				
	Homopolymer	Copolymer	20% glass-reinforced homopolymer	25% glass-reinforced copolymer	21% poly(tetrafluoroethylene)-filled homopolymer
Physical					
Melting temperature, °C					
Crystalline	175	175	181	175	181
Amorphous					
Specific gravity	1.42	1.41	1.56	1.61	1.54
Water absorption (24 h), %	0.25–0.40	0.22	0.25	0.29	0.20
Dielectric strength, kV · mm⁻¹	19.7	19.7	19.3	22.8	15.7
Electrical					
Volume (dc) resistivity, ohm-cm	10^{15}	10^{15}	5×10^{14}		3×10^{16}
Dielectric constant (60 Hz)	3.7	3.7	3.9		3.1
Dielectric constant (10⁶ Hz)	3.7	3.7	3.9		3.1
Dissipation (power) factor (60 Hz)	0.005	0.005	0.005		0.005
Dissipation factor (10⁶ Hz)					
Mechanical					
Compressive modulus, 10³ lb · in⁻²	670	450			
Compressive strength, rupture or 1% yield, 10³ lb · in⁻²	5.29	16 (10% yield)	18 (10% yield)	17 (10% yield)	13 (10% yield)
Elongation at break, %	25–75	40–75	7	3	15–22
Flexural modulus at 23°C, 10³ lb · in²	380–430	375	730	1100	340–350
Flexural strength, rupture or yield, 10³ lb · in⁻²	14	13	15	28	
Hardness, Rockwell (or Shore)	M94	M78	M90	M79	M78
Impact strength (Izod) at 23°C, J · m⁻¹	69–123	53–80	43	96	37–64
Tensile modulus, 10³ lb · in⁻²	520	410	1000	1250	
Tensile strength at break, 10³ lb · in⁻²	10	10			7.6
Tensile yield strength, 10³ lb · in⁻²	9.5–12	8.5	8.5	18.5	6.9–7.6
Thermal					
Burning rate, mm · min⁻¹	27.9				
Coefficient of linear thermal expansion, 10⁻⁶ °C	100	85	36–81		75
Deflection temperature under flex ural load (264 lb · in⁻²), °C	124	110	157	163	100
Maximum recommended service temperature, °C	84				
Specific heat, cal · g⁻¹	0.35				
Thermal conductivity, W · m⁻¹ · K⁻¹	0.23	0.23			

(Continued)

TABLE 2.79 Properties of Commercial Plastics (*Continued*)

Properties	Acrylic				Alkyd, molded	Alloy	
	Poly(methyl methacrylate)	Cast sheet	Impact-modified	Heat-resistant		Acrylic poly(vinyl chloride) alloy	Acrylonitrile-butadiene-styrene-poly(vinyl chloride) alloy
Physical							
Melting temperature, °C							
Crystalline							
Amorphous	90–105	90–105	80–100	100–125		105	
Specific gravity	1.17–1.20	1.18–1.20	1.11–1.18	1.16–1.19	2.22–2.24		
Water absorption (24 h), %	0.1–0.4	0.2–0.4	0.2–0.8	0.2–0.3		0.06	
Dielectric strength, kV · mm^{-1}	15.7–19.9	17.7–21.7	15.0–19.9	15.7–19.9		>15.7	19.7
Electrical							
Volume (dc) resistivity, ohm-cm	>10^{14}	>10^{14}					
Dielectric constant (60 Hz)	3.3–4.5	3.5–4.5			3.8–5.0		
Dielectric constant (10^6 Hz)		3.0–3.5			3.6–4.7		
Dissipation (power) factor (60 Hz)		0.04–0.06			0.012–0.026		
Dissipation factor (10^6 Hz)		0.02–0.03			0.01–0.016		
Mechanical							
Compressive modulus, 10^4 lb · in^{-2}	370–460	390–475	240–370	350–460		330–400	
Compressive strength, rupture or 1% yield, 10^3 lb · in^{-2}	12–18	11–19	4–14	17	16–20	8.4	
Elongation at break, %	2–10	2–7	20–70	3–5		100	
Flexural modulus at 23°C, 10^3 lb · in^{-2}	420–460	390–475	200–380	460–500		330–400	340
Flexural strength, rupture or yield, 10^3 lb · in^{-2}	13–19	12–17	7–13	12–16		10.7	9.6
Hardness, Rockwell (or Shore)	M85–M105	M80–M100	R105–R120	M95–M105	E76	R99–R105	R100
Impact strength (Izod) at 23°C, J · m^{-1}	16–27	16–21	43–133	16–21	27–240	800	560

Tensile modulus, 10^3 lb · in^{-2}	330	330–335		350–460	200–400	350–450	380–450
Tensile strength at break, 10^3 lb · in^{-2}	5.8	6.5	4.5–6.5	10	5–9	8–11	7–11
Tensile yield strength, 10^3 lb · in^{-2}			10–13				
Thermal							
Burning rate, mm · min^{-1}			Self-extinguishing			0.5–2.2	
Coefficient of linear thermal expansion, 10^{-6} °C	46		40–55	50–60	50–80	50–90	50–90
Deflection temperature under flexural load (264 lb · in^{-2}), °C		71	177–204	88–104	74–95	71–102	74–99
Maximum recommended service temperature, °C			220			60–71	
Specific heat, cal · g^{-1}						0.35	0.36
Thermal conductivity, W · m^1, K^{-1}				0.19	0.17–0.21	0.17–0.25	0.17–0.25

(Continued)

TABLE 2.79 Properties of Commercial Plastics (*Continued*)

	Alloy	Allyl			Cellulosic		
			Diallyl phthalate molding		Cellulose acetate		
Properties	Polycarbonate acrylonitrile-butadiene-styrene alloy	Allyl-diglycol-carbonate polymer	Glass-filled	Mineral-filled	Sheet	Molding	Cellulose-acetate-butyrate resin, Sheet
Physical							
Melting temperature, °C							
Crystalline							
Amorphous	150	Thermoset	Thermoset	Thermoset	230	230	140
Specific gravity	1.12–1.20	1.3–1.4	1.7–2.0	1.65–1.85	1.27–1.34	1.29–1.34	1.15–1.22
Water absorption (24 h), %	0.21–0.24	0.2	0.12–0.35	0.2–0.5	2–7	1.7–6.5	0.9–2.2
Dielectric strength, kV · mm^{-1}	17.7	15.0	15.7–17.7	15.7–17.7	11–24	9–24	9–18
Electrical							
Volume (dc) resistivity, ohm-cm					10^{10}–10^{13}	10^{10}–10^{13}	10^{10}–10^{12}
Dielectric constant (60 Hz)					3.4–7.4	3.5–7.5	3.7–4.3
Dielectric constant (10^6 Hz)					3.2–7.0	3.2–7.0	3.3–3.8
Dissipation (power) factor (60 Hz)					0.01–0.06	0.01–0.06	0.01–0.04
Dissipation factor (10^6 Hz)					0.01–0.06	0.01–0.10	0.01–0.04
Mechanical							
Compressive modulus, 10^3 lb · in^{-2}		300					
Compressive strength, rupture or 1% yield, 10^3 lb · in^{-2}	11	21–23	25–35	20–32	22–33	25–36	50–100
Elongation at break, %	10–15		3–5	3–5	17–40	6–40	
Flexural modulus at 23°C, 10^3 lb · in^{-2}	300–400	250–330	1200–1500	1000–1400			740–1300

	13.0–13.7 R117	6–13 M95–M100	9–20 E80–E87	8.5–11 E61	6–10 R85–R120	2–16 R100–R123	4–9 R50–R95
Flexural strength, rupture or yield, 10^3 lb·in^{-2}	13.0–13.7	6–13	9–20	8.5–11	6–10	2–16	4–9
Hardness, Rockwell (or Shore)	R117	M95–M100	E80–E87	E61	R85–R120	R100–R123	R50–R95
Impact strength (Izod) at 23°C, J·m^{-1}	560	11–21	21–800	16–43	107–454	53–214	133–288
Tensile modulus, 10^3 lb·in^{-2}	370–380	300	1400–2200	1200–2200			200–250
Tensile strength at break, 10^3 lb·in^{-2}	7.0–7.3	5–6	6–11	5–8	4.5–8.0	1.9–9.0	2.6–6.9
Tensile yield strength, 10^3 lb·in^{-2}	8.5				2.2–7.4	4.1–7.6	
Thermal							
Burning rate, mm·min^{-1}						1.3–3.8	1.3–3.8
Coefficient of linear thermal expansion, 10^{-6}°C	63–67	5.4–9.6	0.68–2.4	2.8	100–150	80–180	110–170
Deflection temperature under flexural load (264 lb·in^{-2}), °C	104–116	60–88	165–288+	160–288	44–91	51–98	49–58
Maximum recommended service temperature, °C							
Specific heat, cal·g^{-1}					0.3–0.4	0.3–0.42	0.3–0.4
Thermal conductivity, W·m^{-1}·K^{-1}	0.25–0.38	0.20–0.21	0.21–0.63	0.30–1.04	0.17–0.34	0.17–0.34	0.17–0.34

(Continued)

TABLE 2.79 Properties of Commercial Plastics (*Continued*)

Properties	Cellulosic Cellulose-acetate butyrate resin, molding	Cellulose-acetate propionate resin, molding	Ethyl cellulose	Cellulose nitrate	Chlorinated polyether	Epoxy / Bisphenol Glass-fiber-reinforced	Mineral-filled
Physical							
Melting temperature, °C							
Crystalline							
Amorphous	140	190	135		125	Thermoset	Thermoset
Specific gravity	1.15–1.22	1.17–1.24	1.09–1.17	1.35–1.40	1.4	1.6–2.0	1.6–2.1
Water absorption (24 h), %	0.9–2.2	1.2–2.8	0.8–1.8			0.04–0.20	0.03–0.20
Dielectric strength, kV · mm^{-1}	9–13	12–17.7	13.8–19.7			9.8–15.7	9.8–15.7
Electrical							
Volume (dc) resistivity, ohm-cm	10^{10}–10^{12}			10^{10}			
Dielectric constant (60 Hz)	3.5–6.4			7.0–7.5			
Dielectric constant (10^6 Hz)	3.2–6.2		3.01	6.6			
Dissipation (power) factor (60 Hz)	0.01–0.04						
Dissipation factor (10^6 Hz)	0.01–0.04						
Mechanical							
Compressive modulus, 10^3 lb · in^{-2}						3000	
Compressive strength, rupture or 1% yield, 10^3 lb · in^{-2}	2.1–7.5	2.4–7.0		2.1–8.0	600–800	18,000–40,000	18,000–40,000
Elongation at break, %	40–88	29–100	5–40	40–45		4	
Flexural modulus at 23°C, 10^3 lb · in^{-2}	90–300	120–350				2–4.5	

Flexural strength, rupture or yield, 10^3 lb · in^{-2}	1.8–9.3	2.9–11.4	4–12	9–11	5	8–30	6–18
Hardness, Rockwell (or Shore)	R31–R116	R10–R122	R50–R115	R95–R115	R100	M100–M112	M100–M112
Impact strength (Izod) at 23°C, J · m^{-1}	53–582	27 to no break	21		21	16–533	16–22
Tensile modulus, 10^3 lb · in^{-2}	50–200	60–215		267–374 190–220		3	
Tensile strength at break, 10^3 lb · in^{-2}	2.6–6.9	2.0–7.8	2–8	7–8	1.5–1.8	5–20	4–10
Tensile yield strength, 10^3 lb · in^{-2}							
Thermal							
Burning rate, mm · min^{-1}	1.3–3.8				Self-extinguishing		
Coefficient of linear thermal expansion, 10^{-6} °C	110–170	110–170	100–200	80–120	6.6	11–50	20–60
Deflection temperature under flexural load (264 lb · in^{-2}), °C	44–94	44–109	45–88	60–71	185	107–260	107–260
Maximum recommended service temperature, °C					255		
Specific heat, cal · g^{-1}	0.3–0.4			0.31–0.41			
Thermal conductivity, W · m^{-1} · K^{-1}	0.17–0.30	0.17–0.30	0.16–0.30	0.23		0.17–0.42	0.17–1.48

(Continued)

TABLE 2.79 Properties of Commercial Plastics (*Continued*)

Properties	Epoxy			Fluorocarbon			
	Casting resin		Novolac resin	Poly(tetrafluoroethylene)		Poly(chloro-trifluoro-ethylene)	Perfluoroalkoxy
	Unfilled	Flexible	Mineral-filled	Granular	Glass-fiber-reinforced		
Physical							
Melting temperature, °C							
Crystalline				327	327	220	310
Amorphous	Thermoset	Thermoset	Thermoset				
Specific gravity	1.11–1.40	1.05–1.35	1.7–2.1	2.14–2.20	2.2–2.3	2.1–2.2	2.12–2.17
Water absorption (24 h), %	0.08–0.15	0.27–0.50	0.05–0.2	0.01		0.03	
Dielectric strength, kV · mm^{-1}	11.8–19.7	9.3–15.8	11.8–13.8	18.9	12.6	19.7–23	19.7
Electrical							
Volume (dc) resistivity, ohm-cm	10^{12}–10^{17}			10^{18}		10^{18}	
Dielectric constant (60 Hz)	3.5–5.0			2.1		2.3–2.7	
Dielectric constant (10^6 Hz)	3.5–5.0			2.1		2.3–2.5	
Dissipation (power) factor (60 Hz)				0.0002		0.001	
Dissipation factor (10^6 Hz)				0.0002		0.005	
Mechanical							
Compressive modulus, 10^3 lb · in^{-2}				60			
Compressive strength, rupture or 1% yield, 10^3 lb · in^{-2}	15–25	1–14	30	1.7		4.6–7.4	
Elongation at break, %	3–6	20–70	2–4	200–400	200–300	80–250	300
Flexural modulus at 23°C, 10^3 lb · in^{-2}			2000	80	235	120	

Property							
Flexural strength, rupture or yield, 10^{-3} lb · in^{-2}	13–21	1–13	16–20		2	7.4–9.3	
Hardness. Rockwell (or Shore)	M80–M110			(D50–D55)	(D60–D70)	R75–R95	(D64)
Impact strength (Izod) at 23°C, J · m^{-1}	10.7–53	187–267	21	160	144	133–160	No break
Tensile modulus, 10^3 lb · in^{-2}	350	1–350		58–80		150–300	
Tensile strength at break, 10^3 lb · in^{-2}	4–13	2–10	6–12	2–5	2–2.7	4.5–6	4–4.3
Tensile yield strength, 10^3 lb · in^{-2}			30				
Thermal							
Burning rate, mm · min^{-1}				Self-extinguishing	Self-extinguishing	Self-extinguishing	
Coefficient of linear thermal expansion, 10^{-6} °C	45–65	20–100	22–30	100	77–100	70	
Deflection temperature under flexural load (264 lb · in^{-2}), °C	46–288	23–121	149–260	121 (66 lb · in^{-2})		126 (66 lb · in^{-2})	74 (66 lb · in^{-2})
Maximum recommended service temperature, °C				260		200	
Specific heat, cal · g^{-1}				0.25		0.22	
Thermal conductivity, W · m^{-1} · K^{-1}	0.17–0.21			0.25	0.34–0.40	0.19–0.22	0.25

(Continued)

TABLE 2.79 Properties of Commercial Plastics (*Continued*)

Properties	Fluorinated ethylene-propylene resin	Poly(vinylidene fluoride)	Fluorocarbon — Ethylene-tetrafluoroethylene copolymer — Unfilled	Fluorocarbon — Ethylene-tetrafluoroethylene copolymer — Glass-fiber-reinforced	Ethylene-chlorotrifluoroethylene copolymer	Melamine formaldehyde — Cellulose-filled	Melamine formaldehyde — Glass-fiber-reinforced
Physical							
Melting temperature, °C							
Crystalline	275	156	270	270	245	Thermoset	Thermoset
Amorphous							
Specific gravity	2.14–2.17	1.75–1.78	1.7	1.8	1.68	1.47–1.52	1.5–2.0
Water absorption (24 h), %	<0.01	0.04–0.06	0.03	0.02	0.01	0.1–0.8	0.09–1.3
Dielectric strength, kV · mm^{-1}	20–24	10	16	17	19	11–16	5–15
Electrical							
Volume (dc) resistivity, ohm-cm							
Dielectric constant (60 Hz)	2.1	8–9	2.6		2.6		
Dielectric constant (10^6 Hz)	2.1	8–9	2.6		2.6		
Dissipation (power) factor (60 Hz)		High					
Dissipation factor (10^6 Hz)		High					
Mechanical							
Compressive modulus, 10^3 lb · in^{-2}		120	120	1200	240		
Compressive strength, rupture or 1% yield, 10^3 lb · in^{-2}	2.2	8.7–10	7.1	10		33–45	20–35
Elongation at break, %	250–330	25–500	100–400	8	200–300	0.6–1.0	0.6
Flexural modulus at 23°C, 10^3 lb · in^{-2}	80–95	200	200	950	240	1100	
Flexural strength, rupture or yield, 10^3 lb · in^{-2}		8.6–11	5.5	10.7	7	9–16	14–23

Property							
Hardness, Rockwell (or Shore)	(D60–D65)	(D80)	R50 (D75)	R74	R95	M115–M125	M115
Impact strength (Izod) at 23°C, J · m⁻¹	No break	192–214	No break	480	No break	11–21	32–961
Tensile modulus, 10^3 lb · in⁻²	50	120	120	1200	240	1.1–1.4	1.6–2.4
Tensile strength at break, 10^3 lb · in⁻²	2.7–3.1	5.5–7.4	6.5	12	7	5–13	5–10.5
Tensile yield strength, 10^3 lb · in⁻²							
Thermal							
Burning rate, mm · min⁻¹	Not combustible	Not combustible	Not combustible	Not combustible	Not combustible	Self-extinguishing	Self-extinguishing
Coefficient of linear thermal expansion, 10^{-6} °C	83–105	85	59	10–32	80	40–45	15–28
Deflection temperature under flexural load (264 lb · in⁻²), °C	70 (66 lb · in⁻²)	80–90	71	210	77	177–199	190–204
Maximum recommended service temperature, °C	205	150				210	
Specific heat, cal · g⁻¹	0.28						
Thermal conductivity, W · m⁻¹ · K⁻¹	0.25	0.19–0.24	0.24		0.16	0.27–0.41	0.41–0.49

(Continued)

TABLE 2.79 Properties of Commercial Plastics (*Continued*)

Properties	Melamine phenolic, woodflour- and cellulose-filled	Nitrile	Phenolic				
			Unfilled	Woodflour-filled	Glass-fiber-reinforced	Cellulose-filled	Mineral-filled
Physical							
Melting temperature, °C							
Crystalline		95					
Amorphous	Thermoset		Thermoset	Thermoset	Thermoset	Thermoset	Thermoset
Specific gravity	1.5–1.7	1.15	1.24–1.32	1.37–1.46	1.69–2.0	1.38–1.42	1.42–1.84
Water absorption (24 h), %	0.3–0.65	0.28	0.1–0.36	0.3–1.2	0.03–1.2	0.5–0.9	0.1–0.3
Dielectric strength, kV · mm^{-1}	8.7–12.8	8.7–9.5	9.8–15.8	10.2–15.8	5.5–15.8	11.8–15	7.9–13.8
Electrical							
Volume (dc) resistivity, ohm-cm		1.9×10^{15}	1×10^{12} to 7×10^{12}				
Dielectric constant (60 Hz)			6.5–7.5				
Dielectric constant (10^6 Hz)			4.0–5.5				
Dissipation (power) factor (60 Hz)			0.10–0.15				
Dissipation factor (10^6 Hz)			0.04–0.05				
Mechanical							
Compressive modulus, 10^3 lb · in^{-2}							
Compressive strength, rupture or 1% yield, 10^3 lb · in^{-2}	26–30	12	18–32	25–31	26–70	22–31	22.5–34.6
Elongation at break, %	0.4–0.8	3–4	1.5–2.0	0.4–0.8	0.2	1–2	0.1–0.5
Flexural modulus at 23°C, 10^3 lb · in^{-2}	1000–2000	500–590	700–1500	1000–1200	2000–33,000	900–1300	1000–2000
Flexural strength, rupture or yield, 10^3 lb · in^{-2}	8–10	14	11–17	7–14	15–60	5.5–11	11–14

	E95–E100	M72–M76	M93–M120	M100–M115	E54–E101	M95–115	E88
Hardness, Rockwell (or Shore)	E95–E100	M72–M76	M93–M120	M100–M115	E54–E101	M95–115	E88
Impact strength (Izod) at 23°C, $J \cdot m^{-1}$	11–21	80–256	13–21	11–32	27–960	21–59	14–19
Tensile modulus, $10^3 \, lb \cdot in^{-2}$	800–1700	510–580	700–1500	800–1700	1900–3300		2400
Tensile strength at break, $10^3 \, lb \cdot in^{-2}$	6–8	9	6–9	5–9	7–18	3.5–6.5	6–9.7
Tensile yield strength, $10^3 \, lb \cdot in^{-2}$			12–15				
Thermal							
Burning rate, $mm \cdot min^{-1}$			Self-extinguishing				
Coefficient of linear thermal expansion, $10^{-6}°C$	10–40	66	68	30–45	8–21	20–31	19–26
Deflection temperature under flexural load (264 $lb \cdot in^{-2}$), °C	140–154	73	74–80	149–188	177–316	149–177	320–246
Maximum recommended service temperature, °C							
Specific heat, $cal \cdot g^{-1}$							
Thermal conductivity, $W \cdot m^{-1} \cdot K^{-1}$	0.17–0.30	0.26	0.15	0.17–0.34	0.34–0.59	0.25–0.38	0.42–0.57

(Continued)

1181

TABLE 2.79 Properties of Commercial Plastics (*Continued*)

Properties	Polyamide						
	Nylon 6			Nylon 6/6			Nylon 6/6-nylon 6 copolymer
	Molding and extrusion	30–35% glass-fiber-reinforced	High-impact copolymer	Molding	33% glass-fiber-reinforced	Molybdenum disulfide-filled	
Physical							
Melting temperature, °C							
Crystalline	216	216	216	265	265	265	240
Amorphous							
Specific gravity	1.12–1.14	1.35–1.42	1.08–1.17	1.13–1.15	1.38	1.15–1.17	1.08–1.14
Water absorption (24 h), %	2.9	1.2	1.3–1.5	1.0–1.3	1.0	0.8–1.1	1.5–2.0
Dielectric strength, kV · mm⁻¹	15.8	15.8	22	24		14	15.8
Electrical							
Volume (dc) resistivity, ohm-cm	10^{12}			10^{12}–10^{15}			10^{10}
Dielectric constant (60 Hz)	9.8			4.0			16
Dielectric constant (10^6 Hz)	3.7			3.6			4
Dissipation (power) factor (60 Hz)	0.14			0.01–0.02			0.4
Dissipation factor (10^6 Hz)	0.12			0.02–0.03			0.1
Mechanical							
Compressive modulus, 10^3 lb · in⁻²	250						
Compressive strength, rupture or 1% yield, 10^3 lb · in⁻²	13–16	19		15 (yield)	24.9	12.5	
Elongation at break, %	30–100	3–6	150–270	60	3	15	40
Flexural modulus at 23°C, 10^3 lb · in⁻²	390	1500	110–320	420	1300	450	150–410
Flexural strength, rupture or yield, 10^3 lb · in⁻²	14	33	5–12	17	41	17	

	R119	M101	R81–R110	R120	M100	R119	R119
Hardness, Rockwell (or Shore)							
Impact strength (Izod) at 23°C, J·m⁻¹	32–53	160	96 to no break	43–53	117	240	37
Tensile modulus, 10^3 lb·in⁻²	380	1450				550	150–410
Tensile strength at break, 10^3 lb·in⁻²	11.8	25	7.5–11	12	28	13.7	7.4–12.4
Tensile yield strength, 10^3 lb·in⁻²	8			8			
Thermal							
Burning rate, mm·min⁻¹	Self-extinguishing	Self-extinguishing	Self-extinguishing	Self-extinguishing	Self-extinguishing	Self-extinguishing	Self-extinguishing
Coefficient of linear thermal expansion, 10^{-6}°C	80–90	20–30	30–40	80	15–20	54	
Deflection temperature under flexural load (264 lb·in⁻²), °C	68–85	210	45–54	75	249	127	77
Maximum recommended service temperature, °C	107			135			
Specific heat, cal·g⁻¹	0.4			0.4			
Thermal conductivity, W·m⁻¹·K⁻¹	0.24	0.24		0.24	0.22		

(Continued)

1183

TABLE 2.79 Properties of Commercial Plastics (*Continued*)

Properties	Polyamide						Poly(amide-imide), unfilled
	Nylon 6/9 molding and extrusion	Nylon 6/12 Molding	Nylon 6/12 30–35% glass-fiber-reinforced	Nylon 11, molding and extrusion	Nylon 12, molding and extrusion	Aromatic nylon (aramid), molded and unfilled	
Physical							
Melting temperature, °C							
Crystalline	205	217	217	194	179	275	275
Amorphous							
Specific gravity	1.08–1.10	1.06–1.08	1.31–1.38	1.03–1.05	1.01–1.02	1.30	1.40
Water absorption (24 h), %	0.5	0.4	0.2	0.3	0.25	0.6	0.28
Dielectric strength, kV · mm^{-1}	24	16	21	17	18	31	24
Electrical							
Volume (dc) resistivity, ohm-cm		10^{15}			10^{14}		
Dielectric constant (60 Hz)		4.0			3.8		
Dielectric constant (10^6 Hz)		3.5			3.0		
Dissipation (power) factor (60 Hz)		0.02			0.07		
Dissipation factor (10^6 Hz)		0.02			0.04		
Mechanical							
Compressive modulus, 10^3 lb · in^{-2}	1125			180		290	413
Compressive strength, rupture or 1% yield, 10^3 lb · in^{-2}		2.4			7.5	30	40
Elongation at break, %		150	4	300	300	5	12–18
Flexural modulus at 23°C, 10^3 lb · in^{-2}	290	290	1120	150	165	640	664
Flexural strength, rupture or yield, 10^3 lb · in^{-2}					1.5	25.8	30

	R111	R114	E40–E50	R108	R106–R109	E90	E78
Hardness, Rockwell (or Shore)	59	53	139	96	107–300	75	133
Impact strength (Izod) at 23°C, J · m⁻¹	275	290	1200	185	180		730
Tensile modulus, 10³ lb · in⁻²	8.5	8.8	24	8	8–9	17.5	26.9
Tensile strength at break, 10³ lb · in⁻²		8.8					
Tensile yield strength, 10³ lb · in⁻²							
Thermal							
Burning rate, mm · min⁻¹				Self-extinguishing			
Coefficient of linear thermal expansion, 10^{-6}°C	57–60	90		55–100	67–100	40	36
Deflection temperature under flexural load (264 lb · in⁻²), °C	82	82	93–218	54	54	260	274
Maximum recommended service temperature, °C				100–120			260
Specific heat, cal · g⁻¹		0.4		0.58			
Thermal conductivity, W · m⁻¹ · K⁻¹	0.22	0.22		0.34	0.22	0.22	0.25

(Continued)

TABLE 2.79 Properties of Commercial Plastics (*Continued*)

Properties	Poly(aryl ether), unfilled	Polycarbonate		Thermoplastic polyester			
				Poly(butylene terephthalate)		Poly(ethylene terephthalate)	
		Low viscosity	30% glass-fiber reinforced	Unfilled	30% glass-fiber-reinforced	Unfilled	30% glass-fiber-reinforced
Physical							
Melting temperature, °C							
Crystalline	160	140	150	232–267	232–267	245	245
Amorphous							
Specific gravity	1.14	1.2	1.4	1.31–1.38	1.52	1.34–1.39	1.27
Water absorption (24 h), %	0.25	0.15	0.14	0.08–0.09	0.06–0.08	0.1–0.2	0.05
Dielectric strength, kV · mm^{-1}	17	15	19	16–22	18–22		22
Electrical							
Volume (dc) resistivity, ohm-cm		2×10^{16}	$>10^{16}$		10^{16}	10^{16}	
Dielectric constant (60 Hz)		3.17	3.35				
Dielectric constant (10^6 Hz)		2.96	3.31			3.25	
Dissipation (power) factor (60 Hz)		0.0009	0.011				
Dissipation factor (10^6 Hz)		0.010	0.007				
Mechanical							
Compressive modulus, 10^3 lb · in^{-2}		350	1300				
Compressive strength, rupture or 1% yield, 10^3 lb · in^{-2}	80	12.5	18	8.6–14.5	18–23.5	11–15	25
Elongation at break, %		110	3–5	50–300	2–4	50–300	3
Flexural modulus at 23°C, 10^3 lb · in^{-2}	300	340	1100	330–400	1100–1200	35–450	1440
Flexural strength, rupture or yield, 10^3 lb · in^{-2}	11	13.5	23	12–16.7	26–29	14–18	33.5

	R117	M70	M92	M68–M78	M90	M94–M101	M100
Hardness, Rockwell (or Shore)	R117	M70	M92	M68–M78	M90	M94–M101	M100
Impact strength (Izod) at 23°C, $J \cdot m^{-1}$	427	14	107	43–53	69–85	13–32	101
Tensile modulus, $10^3 \, lb \cdot in^{-2}$	320	345	1250	280	1300	400–600	1440
Tensile strength at break, $10^3 \, lb \cdot in^{-2}$	7.5	9.5	19	8.2	17–19	8.5–10.5	23
Tensile yield strength, $10^3 \, lb \cdot in^{-2}$		9.0					
Thermal							
Burning rate, $mm \cdot min^{-1}$		Self-extinguishing	Self-extinguishing				
Coefficient of linear thermal expansion, 10^{-6}°C	65	68	22	60–95	25	65	29
Deflection temperature under flexural load (264 $lb \cdot in^{-2}$), °C	149	138–145	146	50–85	220	38–41	224
Maximum recommended service temperature, °C		143					
Specific heat, $cal \cdot g^{-1}$		0.3				0.27	
Thermal conductivity, $W \cdot m^{-1} \cdot K^{-1}$	0.30	0.20	0.22	0.18–0.30	0.30	0.15	

(Continued)

TABLE 2.79 Properties of Commercial Plastics (*Continued*)

| Properties | Thermoplastic polyester | | Thermosetting and alkyd polyester | | | | Polyimide, unfilled |
| | Aromatic polyester | | Unsaturated polyester | | Alkyd molding compounds | | |
	Extrusion-transparent	Injection molding	Styrene-maleic acid copolymer, low-shrink	Butadiene-maleic acid copolymer	Putty, mineral-filled	Glass-fiber-reinforced	
Physical							
Melting temperature, °C							
Crystalline							
Amorphous	81		Thermoset	Thermoset	Thermoset	Thermoset	310–365
Specific gravity		1.39					1.36–1.43
Water absorption (24 h), %		0.01					0.24
Dielectric strength, kV · mm^{-1}		14					22
Electrical							
Volume (dc) resistivity, ohm-cm							>10^{16}
Dielectric constant (60 Hz)							3–4
Dielectric constant (10^6 Hz)							
Dissipation (power) factor (60 Hz)							
Dissipation factor (10^6 Hz)							
Mechanical							
Compressive modulus, 10^3 lb · in^{-2}					2000–3000		
Compressive strength, rupture or 1% yield, 10^3 lb · in^{-2}	225	10	15–30	14–30	12–38	15–36	30–40
Elongation at break, %		7–10	3–5				8–10
Flexural modulus at 23°C, 10^3 lb · in^{-2}	290	700	1000–2500		2000	2000	450–500
Flexural strength, rupture or yield, 10^3 lb · in^{-2}	10.6	12	9–35	16–24	6–17	8.5–26	19–28.8
Hardness, Rockwell (or Shore)	R105		40–70 (Barcol)	50–60 (Barcol)	E98	E95	E52–E99

Property							
Impact strength (Izod) at 23°C, $J \cdot m^{-1}$	101		133–800	214–694	16–27	27–854	80
Tensile modulus, $10^3 \, lb \cdot in^{-2}$		300	1000–2500	1500–2500	500–3000		300
Tensile strength at break, $10^3 \, lb \cdot in^{-2}$	6	11	4.5–20	5–10	3–9	4–9.5	10.5–17.1
Tensile yield strength, $10^3 \, lb \cdot in^{-2}$	7						12.5
Thermal							
Burning rate, $mm \cdot min^{-1}$							
Coefficient of linear thermal expansion, $10^{-6}°C$		29	6–30		20–50	15–33	45–56
Deflection temperature under flexural load ($264 \, lb \cdot in^{-2}$), °C	63	282	190–260	160–177	177–260	204–260	277–360
Maximum recommended service temperature, °C							
Specific heat, $cal \cdot g^{-1}$							0.27
Thermal conductivity, $W \cdot m^{-1} \cdot K^{-1}$		0.29	0.76–0.93	0.76–0.93	0.51–0.89	0.6–0.89	0.10–0.11

(Continued)

TABLE 2.79 Properties of Commercial Plastics (*Continued*)

Properties	Poly(methyl pentene), unfilled	Polyolefin — Polyethylene					Ethylene-vinyl acetate copolymer
		Low-density	Medium-density	High-density	Ultra high-molecular-weight	Glass-fiber-reinforced, high-density	
Physical							
Melting temperature, °C							
Crystalline	230–240	95–130	120–140	120–140	125–135	120–140	65–90
Amorphous							
Specific gravity	0.84	0.910–0.925	0.926–0.94	0.941–0.965	0.94	1.28	0.92–0.95
Water absorption (24 h), %	0.01	<0.01	<0.01	<0.01	<0.01	0.02	0.05–0.13
Dielectric strength, kV · mm^{-1}		18–39	18–39	18–39	28	20	24–30
Electrical							
Volume (dc) resistivity, ohm-cm		>10^{15}	>10^{15}	<10^{15}			
Dielectric constant (60 Hz)		2.3	2.3	2.3			
Dielectric constant (10^6 Hz)		2.3	2.3	2.3			
Dissipation (power) factor (60 Hz)		<0.0005	<0.0005	<0.0005			
Dissipation factor (10^6 Hz)		<0.0005	<0.0005	<0.0005			
Mechanical							
Compressive modulus, 10^3 lb · in^{-2}	114–171						
Compressive strength, rupture or 1% yield, 10^3 lb · in^{-2}	5–6.6			2.7–3.6		7	
Elongation at break, %	10–50	90–800	50–600	20–130	450–525	1.5	550–900
Flexural modulus at 23°C, 10^3 lb · in^{-2}	110–260	8–60	60–115	100–260	130–140	800	
Flexural strength, rupture or yield, 10^3 lb · in^{-2}	4–6.5					11	1–20

	L67–L74	(D40–D51)	(D50–D60)	R30–R50	R50	R75	
Hardness, Rockwell (or Shore) at 23°C	L67–L74	(D40–D51)	(D50–D60)	R30–R50	R50	R75	
Impact strength (Izod) at 23°C, J · m^{-1}	16–64	No break	27–854	27–1068	No break	59	No break
Tensile modulus, 10^3 lb · in^{-2}	160–280	14–38	25–55	60–180		9	20–120
Tensile strength at break, 10^3 lb · in^{-2}	3.5–4	0.6–2.3	1.2–3.5	3.1–5.5	5.6		1.4–2.8
Tensile yield, strength, 10^3 lb · in^{-2}		0.8–1.2	1.0–2.2	3–4	3.1–4.0		
Thermal							
Burning rate, mm · min^{-1}		1.0	1.0	1.0			
Coefficient of linear thermal expansion, 10^{-6} °C	117	100–200	140–160	110–130	130	48	160–200
Deflection temperature under flexural load (264 lb · in^{-2}), °C	41	32–41	41–49	43–54	43–49	121	34
Maximum recommended service temperature, °C	175	70	93	200			
Specific heat, cal · g^{-1}		0.55	0.55	0.46–0.55			
Thermal conductivity, W · m^{-1} · K^{-1}	0.17	0.34	0.34–0.42	0.46–0.51		0.46	

(Continued)

TABLE 2.79 Properties of Commercial Plastics (*Continued*)

Properties	Polybutylene extrusion	Polyolefin — Polypropylene Homopolymer	Copolymer	Impact copolymer	Polyallomer	Poly(phenylene sulfide) Injection molding	40% glass-fiber-reinforced
Physical							
Melting temperature, °C							
Crystalline	126	168	160–168		120–135	290	290
Amorphous							
Specific gravity	0.91–0.925	0.90–0.91	0.89–0.905	0.90	0.90	1.3	1.6
Water absorption (24 h),%	0.01–0.02	0.01–0.03	0.03	<0.03	<0.01	<0.02	0.05
Dielectric strength, kV · mm^{-1}	18	24	24	24	31	15	18
Electrical							
Volume (dc) resistivity, ohm-cm		10^{17}	10^{17}	10^{17}			
Dielectric constant (60 Hz)		2.2–2.6	2.3	2.3			
Dielectric constant (10^6 Hz)		2.2–2.6	2.3				
Dissipation (power) factor (60 Hz)		<0.0005	0.0001–0.0005				
Dissipation factor (10^6 Hz)		0.0005–0.002	0.0001–0.0002	0.0003			
Mechanical							
Compressive modulus, 10^3 lb · in^{-2}	31	150–300					
Compressive strength, rupture or 1% yield, 10^3 lb · in^{-2}		5.5–8.0	3.5–8.0			16	21
Elongation at break, %	300–380	100–600	200–700	8–20	400–500	1–2	1
Flexural modulus at 23°C, 10^3 lb · in^{-2}	45–50	170–250	130–200	130–190	70–110	550	1700
Flexural strength, rupture or yield, 10^3 lb · in^{-2}	2–2.3	6–8	5–7			14	29

	R80–R102	R50–R96	R40–R90	R50–R85	R123	R123
Hardness, Rockwell (or Shore)	R80–R102	R50–R96	R40–R90	R50–R85	R123	R123
Impact strength (Izod) at 23°C, J·m⁻¹	No break	53–1068	80–900	91–203	<27	75
Tensile modulus, 10³ lb·in⁻²	30–40	100–170			480	1100
Tensile strength at break, 10³ lb·in⁻²	3.8–4.4	4–5.5		3–3.8	9.5	19.5
Tensile yield strength, 10³ lb·in⁻²	1.7–2.5	3.5–4.3	2.5–3.1	3–3.4		
Thermal						
Burning rate, mm·min⁻¹						
Coefficient of linear thermal expansion, 10⁻⁶°C	128–150	68–95	60–90	83–100	49	22
Deflection temperature under flexural load (264 lb·in⁻²), °C	54–60	45–57	90–105 (66 lb·in⁻²)	51–56	135	249
Maximum recommended service temperature, °C	160	240	140–160			
Specific heat, cal·g⁻¹	0.44–0.46	0.45–0.50	0.45–0.50			
Thermal conductivity, W·m⁻¹K⁻¹	0.22	0.15–0.17	0.12–0.17	0.09–0.17	0.29	0.29

(Continued)

TABLE 2.79 Properties of Commercial Plastics (*Continued*)

Properties	Polyurethane Casting resin — Liquid	Polyurethane Casting resin — Unsaturated	Thermoplastic elastomer	Cast resin, flexible	Silicone Mineral- and/or glass-filled	Epoxy molding and encapsulating compound	Styrenic Polystyrene — Crystal
Physical							
Melting temperature, °C							
Crystalline			120–160				
Amorphous	Thermoset	Thermoset		Thermoset	Thermoset	Thermoset	85–105
Specific gravity	1.1–1.5	1.05	1.05–1.25	0.99–1.5	1.8–1.94	1.84	1.04–1.05
Water absorption (24 h), %	0.02–1.5	0.1–0.2	0.7–0.9		8–15	10	0.03–0.10
Dielectric strength, kV · mm^{-1}	12–20		13–25	22			24
Electrical							
Volume (dc) resistivity, ohm-cm	10^{11}–10^{15}		10^{11}–10^{13}	10^{14}–10^{15}			$>10^{16}$
Dielectric constant (60 Hz)	4.0–7.5		5.4–7.6	2.7–4.2			
Dielectric constant (10^6 Hz)							2.5
Dissipation (power) factor (60 Hz)							
Dissipation factor (10^6 Hz)							
Mechanical							
Compressive modulus, 10^3 lb · in^{-2}	10–100		4–9				
Compressive strength, rupture or 1% yield, 10^3 lb · in^{-2}	20	3–6	20		10–16	28	11.5–16
Elongation at break, %	100–1000		100–1100	100–700			1–2
Flexural modulus at 23°C, 10^3 lb · in^{-2}	10–100	610	10–350		1000–2500		380–450

Flexural strength, rupture or yield, 10^3 lb · in^{-2}	0.7–4.5	19	0.7–9		9–14	17	8–14
Hardness, Rockwell (or Shore)		21	(A65–D80)	(A15–A65)	M80–M90	16	M60–M75
Impact strength (Izod) at 23°C, J · m^{-1}	1334 to flexible		No break 10–350				13–21
Tensile modulus, 10^3 lb · in^{-2}	10–100				13–427		350–485
Tensile strength at break, 10^3 lb · in^{-2}	0.175–10	10–11	1.5–8.4	0.35–1.0	4–6.5	6–8	5.3–7.9
Tensile yield strength, 10^3 lb · in^{-2}							
Thermal							
Burning rate, mm · min^{-1}					0–78		
Coefficient of linear thermal expansion, 10^{-6} °C	100–200		100–200	300–800	20–50	30	70–80
Deflection temperature under flexural load (264 lb · in^{-2}), °C	Varies over wide range		Varies over wide range		260	74–100	
Maximum recommended service temperature, °C		87–93			371		93
Specific heat, cal · g^{-1}	0.43		0.43				0.3
Thermal conductivity, W · m^{-1} · K^{-1}	0.21		0.07–0.31	0.15–0.31	0.30	0.68	0.09–0.13

(Continued)

TABLE 2.79 Properties of Commercial Plastics (*Continued*)

	Polystyrene	Styrenic — Acrylonitrile-butadiene-styrene copolymer					
				Molding			
Properties	Heat-resistant	Extrusion	Heat-resistant	High-impact	Flame-retarded	Platable	20% glass-reinforced
Physical							
Melting temperature, °C							
Crystalline							
Amorphous	110–125	88–120	110–125	100–110	110–125	100–110	
Specific gravity	1.05–1.09	1.02–1.06	1.05–1.08	1.01–1.04	1.16–1.21	1.06–1.07	1.22
Water absorption (24 h), %	0.03–0.12	0.20–0.45	0.20–0.45	0.20–0.45	0.2–0.6		
Dielectric strength, kV · mm^{-1}	20	14–20	14–20	14–20	14–20	16–22	18
Electrical							
Volume (dc) resistivity, ohm-cm							
Dielectric constant (60 Hz)				2.4–5.0			
Dielectric constant (10^6 Hz)				2.4–3.8			
Dissipation (power) factor (60 Hz)				0.003–0.008			
Dissipation factor (10^6 Hz)				0.007–0.015			
Mechanical							
Compressive modulus, 10^3 lb · in^{-2}		150–390	190–440	140–300	130–310		
Compressive strength, rupture or 1% yield, 10^3 lb · in^{-2}	11.5–16	5.2–10	7.2–10	4.5–8	6.5–7.5		14
Elongation at break, %	2–60	20–100	3–20	5–70	5–25		
Flexural modulus at 23°C, 10^3 lb · in^{-2}	340–470	130–420	300–400	250–350	300–400	340–390	710

	8.9–14 L80–L108	4–14 R75–R115	10–13 R100–R115	8–11 R85–R105	9–14 R100–R120	10.5–11.5 R103–R109	15.5 M85
Flexural strength, rupture or yield, 10^3 lb · in^{-2}	8.9–14	4–14	10–13	8–11	9–14	10.5–11.5	15.5
Hardness, Rockwell (or Shore)	L80–L108	R75–R115	R100–R115	R85–R105	R100–R120	R103–R109	M85
Impact strength (Izod) at 23°C, J · m^{-1}	21–181	133–640	107–347	347–400	160–640	267–283	64
Tensile modulus, 10^3 lb · in^{-2}	320–460	130–380	300–350	230–330	320–400	330–380	740
Tensile strength at break, 10^3 lb · in^{-2}	5–7.8	2.5–8.0	6–7.5	4.8–6.3	5–8	6–6.4	11
Tensile yield strength, 10^3 lb · in^{-2}			5.5–7	4–5.5	4–6		
Thermal							
Burning rate, mm · min^{-1}		1.3		1.3			
Coefficient of linear thermal expansion, 10^{-6}°C	60–70	60–130	60–93	95–110	65–95	47–53	21
Deflection temperature under flexural load (264 lb · in^{-2}), °C	93–120	77–104 annealed	104–116 annealed	96–102 annealed	90–107 annealed	96–102 annealed	99
Maximum recommended service temperature, °C				110			
Specific heat, cal g^{-1}				0.3–0.4			
Thermal conductivity, W · m^{-1} · K^{-1}			0.19–0.34				

(Continued)

TABLE 2.79 Properties of Commercial Plastics (*Continued*)

Properties	Styrenic			Sulfone			
	Styrene-acrylonitrile copolymer		Styrene-butadiene copolymer, high-impact	Polysulfone		Poly(ether sulfone)	Poly(phenyl sulfone)
	Unfilled	20% glass-fiber-reinforced		Unfilled	20% glass-fiber-reinforced		
Physical							
Melting temperature, °C							
Crystalline							
Amorphous	115–125	115–125	90–110	200	200	230	220
Specific gravity	1.07–1.08	1.22	1.03–1.06	1.24	1.46	1.37	1.29
Water absorption (24 h), %	0.2–0.3	0.15–0.20	0.05–0.10	0.22	0.23	0.43	1.1–1.3 (saturated)
Dielectric strength, kV · mm^{-1}	16–20	20	18	17	17	17	16
Electrical							
Volume (dc) resistivity, ohm-cm				10^{15}			
Dielectric constant (60 Hz)				3.14	3.7		
Dielectric constant (10^6 Hz)				3.26	3.7		
Dissipation (power) factor (60 Hz)				0.004	0.002		
Dissipation factor (10^6 Hz)				0.008	0.009		
Mechanical							
Compressive modulus, 10^3 lb · in^{-2}	530			370			
Compressive strength, rupture of 1% yield, 10^3 lb · in^{-2}	14–17	19	4–9	13.9	22		
Elongation at break, %	1–4	1–2	13–50	50–100	2	30–80	60
Flexural modulus at 23°C, 10^3 lb · in^{-2}	550	100–1100	280–450	390	1000	375	330
Flexural strength, rupture or yield, 10^3 lb · in^{-2}	14–17	20	5.3–9.4	15.4	23	18.7	12.4

	M80–M90	R122	M10–M68	M69, R120	M123	M88	
Hardness, Rockwell (or Shore)	M80–M90	R122	M10–M68	M69, R120	M123	M88	
Impact strength (Izod) at 23°C, J · m⁻¹	19–27	53	32–192	64	59	85	640
Tensile modulus, 10³ lb · in⁻²	400–560	1150–1200	280–465	360	1200	350	310
Tensile strength at break, 10³ lb · in⁻²	9–12	15.8–18	3.2–4.9		17		
Tensile yield strength, 10³ lb · in⁻²			2.9–4.9	10.2		12.2	10.4
Thermal							
Burning rate, mm · min⁻¹							
Coefficient of linear thermal expansion, 10⁻⁶°C	36–38	38–40	70–101	52–56	25	55	31
Deflection temperature under flexural load (264 lb · in⁻²), °C	88–104	99	74–93	174	182	203	204
Maximum recommended service temperature, °C				149			
Specific heat, cal · g⁻¹							
Thermal conductivity, W · m⁻¹ · K⁻¹	0.12	0.26–0.28	0.12–0.21	0.12	0.38	0.14–0.19	

(Continued)

TABLE 2.79 Properties of Commercial Plastics (*Continued*)

Properties	Thermoplastic elastomers				Urea formaldehyde, alpha-cellulose filled	Poly(vinyl chloride) and poly(vinyl acetate) (Vinyl)	
	Polyolefin	Polyester	Block copolymers of styrene and butadiene or styrene and isoprene	Block copolymers of styrene and ethylene or styrene and butylene		Rigid	Flexible and unfilled
Physical							
Melting temperature, °C							
Crystalline		168–206			Thermoset		
Amorphous						75–105	75–105
Specific gravity	0.88–0.90	1.17–1.25	0.9–1.2	0.9–1.2	1.47–1.52	1.30–1.58	1.16–1.35
Water absorption (24 h), %	0.01		0.19–0.39		0.4–0.8	0.04–0.4	0.15–0.75
Dielectric strength, kV · mm^{-1}	24–26		16–21		12–16	14–20	12–16
Electrical							
Volume (dc) resistivity, ohm-cm					0.5–5.0	10^{12}–10^{15}	10^{11}–10^{14}
Dielectric constant (60 Hz)					7.7–9.5	3.2–4.0	5.0–9.0
Dielectric constant (10^6 Hz)					6.7–8.0	3.0–4.0	3.0–4.0
Dissipation (power) factor (60 Hz)					0.036–0.043	0.01–0.02	0.03–0.05
Dissipation factor (10^6 Hz)					0.025–0.035	0.006–0.02	0.06–0.1
Mechanical							
Compressive modulus, 10^3 lb · in^{-2}			3.6–120				
Compressive strength, rupture or 1% yield, 10^3 lb · in^{-2}					25–45	8–13	0.9–1.7
Elongation at break, %	150–300	350–450	500–1350	600–800	<1	40–80	200–450
Flexural modulus at 23°C, 10^3 lb · in^{-2}	1.5–2.0	7–75	4–150	4–100	1300–1600	300–500	

Flexural strength, rupture or yield, 10^3 lb·in^{-2}					10–18	10–16	
Hardness, Rockwell (or Shore)	(A65–A92)	(D40–D72)	(A40–A90)	(A50–A90)	M110–M120	(D65–D95)	(A50–A100)
Impact strength (Izod) at 23°C, J·m^{-1}	No break	208 to no break	No break	No break	13–21	21–1068	Varies over wide range
Tensile modulus, 10^3 lb·in^{-2}		1.1–2.5	0.8–50		1000–1500	350–600	
Tensile strength at break, 10^3 lb·in^{-2}	0.65–2.0	3.7–5.7	0.6–3.0	1–3	5.5–13	6–75	1.5–3.5
Tensile yield strength, 10^3 lb·in^{-2}							
Thermal							
Burning rate, mm·min^{-1}					Self-extinguishing	Self-extinguishing	Slow to self-extinguishing
Coefficient of linear thermal expansion, 10^{-6}°C	130–170		130–137		22–36	50–100	70–250
Deflection temperature under flexural load (264 lb·in^{-2}), °C			<0–49		127–143	60–77	
Maximum recommended service temperature, °C					77	70–74	80–105
Specific heat, cal·g^{-1}					0.6	0.2–0.28	0.36–0.5
Thermal conductivity, W·m^{-1}·K^{-1}	0.19–0.21		0.15		0.30–0.42	0.15–0.21	0.13–0.17

(Continued)

TABLE 2.79 Properties of Commercial Plastics (*Continued*)

Properties	Vinyl					
	Poly(vinyl chloride) and poly(vinyl acetate) — Flexible and filled	Poly(vinyl chloride), 15% glass-fiber-reinforced	Poly(vinylidene chloride)	Poly(vinyl formal)	Chlorinated poly(vinyl chloride)	Poly(vinyl butyral), flexible
Physical						
Melting temperature, °C						
Crystalline						
Amorphous	75–105	75–105	210	105	110	49
Specific gravity	1.3–1.7	1.54	1.65–1.72	1.2–1.4	1.49–1.56	1.05
Water absorption (24 h),%	0.5–1.0	0.01	0.1	0.5–3.0	0.02–0.15	1.0–2.0
Dielectric strength, kV · mm^{-1}	9.8–12	24–31	16–24	19		14
Electrical						
Volume (dc) resistivity, ohm-cm			10^{14}–10^{16}			
Dielectric constant (60 Hz)			4.5–6.0			
Dielectric constant (10^6 Hz)						
Dissipation (power) factor (60 Hz)						
Dissipation factor (10^6 Hz)						
Mechanical						
Compressive modulus, 10^3 lb · in^{-2}	1.0–1.8	9			335–600	
Compressive strength, rupture or 1% yield, 10^3 lb · in^{-2}		2–3	2–2.7		9–22	
Elongation at break, %	200–400		50–250	5–20	4–65	150–450
Flexural modulus at 23°C, 10^3 lb · in^{-2}		750			380–450	

Flexural strength, rupture or yield, 10^3 lb · in^{-2}		13.5	4.2–6.2	17–18	14.5–17	
Hardness, Rockwell (or Shore)	(A50–A100) Varies over wide range	R118	M50–M65	M85	R117–R122	A10–A100 Varies over wide range
Impact strength (Izod) at 23°C, J · m^{-1}		53	16–53	43–75	53–299	
Tensile modulus, 10^3 lb · in^{-2}		870	50–80	350–600	360–475	
Tensile strength at break, 10^3 lb · in^{-2}	1–3.5	9.5	3–5	10–12	7.5–9	0.5–3.0
Tensile yield strength, 10^3 lb · in^{-2}						
Thermal						
Burning rate, mm · min^{-1}			Self-extinguishing			Slow
Coefficient of linear thermal expansion, 10^{-6}°C			190	64	68–78	
Deflection temperature under flexural load (264 lb · in^{-2}), °C	68		54–71	71–77	94–112	
Maximum recommended service temperature, °C			100			
Specific heat, cal · g^{-1}			0.32			
Thermal conductivity, W · m^{-1} · K^{-1}	0.13–0.17		0.13	0.16	0.14	

TABLE 2.80   Glass Transition Temperatures and Melting
Temperatures of Selected Polymers

	T_g °C	T_m °C
Nylon 6,6	57	265
Polycarbonate	150	265
Polyester	73	265
Polyethylene		
High density	−90	137
Low density	−110	115
Polymethylmethacrylate	105	
Polypropylene	−14	176
Polystyrene	100	239
Polytetrafluoroethylene (Teflon)	−90	327
Polyvinyl chloride	87	212
Rubber	−73	

TABLE 2.81 Properties of Natural and Synthetic Rubbers

Rubber	Specific gravity	Durometer hardness (or Shore)	Ultimate elongation % (23°C)	Tensile strength, lb · in^{-2} (23°C)	Service temperature, °C Minimum	Service temperature, °C Maximum
Gutta percha (hard rubber)	1.2–1.95	(65–95)	3–8	4000–10,000	−56	104
Natural rubber (NR)	0.93	20–100	750–850	3000–4500	−54	82
Chlorosulfonated polyethylene	1.10	50–95	100–500	500–3000	−54	121
Epichlorohydrin	1.27	60–90	100–400	1000–2500	−46	121
Fluoroelastomers	1.4–1.95	60–90	100–350	2000–3000	−40	232
Isobutene-isoprene rubber (IIR) [also known as government rubber I(GR-I)]	0.91	(40–70)	750–950	2300–3000		121
Nitrile rubber (butadiene-acrylonitrile rubber) (also known as Buna N and NBR)	1.00	30–100	100–600	500–4000	−54	121
Polyacrylate	1.10	40–100	100–400	1000–2200	−18	149
Polybutadiene rubber (BR)	0.93	30–100	100–700	2500–3000	−62	79–100
Polychloroprene (neoprene)	1.23	20–90	800–1000	2000–3500	−54	121
Poly(ethylene-propylene-diene) (EPDM)	0.85	30–100	100–300	1000–3000	−40	149
Polyisobutylene (butyl rubber)	0.92	30–100	100–700	1000–3000	−54	100
Polyisoprene	0.94	20–100	100–750	2000–3000	−54	79–82
Polysulfide (Thiokol ST)	1.34	20–80	100–400	700–1250	−54	82–100
Poly(vinyl chloride) (Koroseal)	1.32	(80–90)		2400–3000		71
Silicone, high-temperature				700–800		316
Silicone	0.98	20–95	50–800	500–1500	−84	232
Styrene-butadiene rubber (SBR) (also known as Buna S)	0.94	40–100	400–600	1600–3700	−60	107
Urethane	0.85	62–95	100–700	1000–8000	−54	100

TABLE 2.82   Density of Polymers Listed by Trade Name

Common or trade name	$\rho(\mathrm{g/cm^3})$
Acetate Rayon	1.32
Acrylic	1.16
Acrylonitrile-styrene copolymer	1.075–1.10
Acrylonitrile-styrene-butadiene copolymer (ABS)	1.04–1.07
Aniline-formaldehyde	1.22–1.25
Benzylcellulose	1.22
Bisphenol-*A* polycarbonate (BPAPC)	1.20
Butyl rubber	0.92
Cellulose I	1.582–1.630
Cellulose II	1.583–1.62
Cellulose III	1.61
Cellulose IV	1.61
Cellulose acetate	1.28–1.32
Cellulose acetate-butyrate	1.14–1.22
Cellulose formate fiber	1.45
Cellulose nitrate	1.35–1.40
Cellulose propionate	1.18–1.24
Cellulose triacetate	1.28–1.33
Cellulose tributyrate	1.16
Chlorinated polyether	1.40
Cotton	1.50–1.54
Cotton, acetylated	1.43
Ethylcellulose	1.09–1.17
Ethylene-propylene copolymer (EPM)	0.86
Glass	3.54
Glass and asbestos	2.5
Kevlar	1.44
Lignocellulose	1.45
Maleic anhydride-styrene copolymer	1.286
Melamine-formaldehyde	1.16
Methyl polyvinyl ketone	1.12
Methylcellulose	1.362
Nomex	1.38
Nylon 6	1.12–1.24
Nylon 66	1.13–1.15, 1.22–1.25
Nylon-610	1.156
Nylon-12	1.02–1.034
Rubber, butyl	0.92
Rubber (unvulcanized)	0.91
Rubber (hard) (Ebonite)	1.11–1.17
Rubber, chlorinated (Neoprene) (CR), unvulcanized	1.23
Rubber, chlorinated (Neoprene) (CR), vulcanized	1.32–1.42
Rubber, fluorinated silicone	1.0
Rubber, silicone	0.80
Rubber, silicone (vulcanized)	1.3–2.3
Rubber, styrene-butadiene (SBR), (unvulcanized)	0.93–0.94
Rubber, styrene-butadiene (SBR), (vulcanized)	0.961
Silk	1.25–1.35
Toluene-sulfonamide-formaldehyde	1.21–1.35
Urea-formaldehyde	1.16
Urea-thiourea-formaldehyde	1.477
Viscose Rayon	1.5
Wool	1.28–1.33

TABLE 2.83 Density of Polymers Listed by Chemical Name

Chemical name	ρ (g/cm^3)
Poly-	
acetaldehyde	1.07
acrolein	1.322
acrylic acid	1.22
acrylonitrile (PAN)	1.01–1.17,
	1.20
acrylonitrile-vinyl acetate	1.14
amide-6 (PA-6)	1.12–1.24
amide-66 (PA-66)	1.13–1.15,
	1.22–1.25
amide-610 (PA-610)	1.156
amide-12 (PA-12)	1.02–1.034
aryl ether ether ketone (PEEK)	1.20
arylate	1.21
bisphenol carbonate (BPAPC)	1.20
butadiene-1,2, isotactic	0.96
butadiene-1,2, syndiotactic	0.96
butadiene-1,4-*cis*	1.01
butadiene-1,4-*trans*	0.93–0.97,
	1.01
1-butene	0.85
butene	0.91–0.92
butyl acrylate	1.08
sec.-butyl acrylate	1.05
butylene	0.60
tert.-butyl methacrylate	1.03
-*n*-butyl methacrylate	1.055
sec.-butyl methacrylate	1.04
tert.-butylstyrene	0.957
caprolactam, nylon	0.985
carbonate (PC)	1.14–1.2
chlorobutadiene	1.25
chloroprene (Neoprene rubber)	1.23
(CR), unvulcanized	
chloroprene (Neoprene rubber)	1.32–1.42
(CR), vulcanized	
chlorotrifluoroethylene	2.03
dichlorostyrene	1.38
2,2-dimethylpropyl acrylate	1.04
dimethylsiloxane	0.970
dodecyl methacrylate	0.93
1-ethylpropyl acrylate	1.04
etheretherketone (PEEK)	1.27
ethyl acrylate	1.095, 1.12
ethyl methacrylate	1.11, 1.12
ethylbutadiene	0.891
ethylene	0.870,
	0.910–0.965
ethylene (amorphous)	0.85
ethylene (crystalline)	0.99
ethylene (high density: HDPE)	0.941–0.965
ethylene (linear low density: LLDPE)	0.918–0.935
ethylene (low density: LDPE)	0.910–0.925
ethylene (medium density: MDPE)	0.926–0.940
ethylene glycol	1.0951
ethylene glycol fumarate	1.385
ethylene glycol isophthalate, cryst.	1.358

(Continued)

TABLE 2.83 Density of Polymers Listed by Chemical Name (*Continued*)

Chemical name	$\rho\,(g/cm^3)$
ethylene glycol phthalate	1.352
ethylene glycol waxes	1.15–1.20
ethylene isophthalate	1.34
ethylene phthalate	1.34
ethylene terephthalate (PETP)	1.33–1.42
formaldehyde	1.425
−*n*-hexyl methacrylate	1.01
imide	1.43
isobutene	0.917
isobutyl methacrylate	1.02–1.04
isobutylene	0.87–0.93
isoprene (1,4−)	0.900–0.913
−*N*-isopropylacrylamide	1.070–1.118
isopropyl acrylate	1.08
isopropyl methacrylate	1.04
methacrylonitrile	1.10
methyl acrylate	1.07–1.223
methyl methacrylate (PMMA)	1.16–1.20
4-methyl-1-pentene	0.84
myrcene	0.895
oxymethylene (POM)	1.41–1.435
phenylene oxide	1.00–1.06
polysulfide (Thiokol A)	1.60
polysulfide (Thiokol B)	1.65
propyl methacrylate	1.06–1.08
propylene (PP)	0.85–0.92
propylene, amorphous	0.87
propylene, head-to-head	0.878
propylene, isotactic	0.90–0.92
propylene, isotactic (crystalline)	0.92–0.939
propylene, syndiotactic (crystalline)	0.93
propylene oxide	1.00
styrene (PS)	1.04–1.09
styrene, crystalline	1.08–1.111
styrene-butadiene thermoplastic elastomer	0.93–1.10
sulfone	1.24
tetrafluoroethylene (PTFE)	2.28–2.344
trifluorochloroethylene	2.11–2.13
vinyl acetate (PVAC)	1.08–1.25
vinyl alcohol (PVA)	1.21–1.31
vinyl butyral	1.07–1.20
vinyl chloride	1.37–1.44
vinyl chloride-co-methyl acrylate	1.34
vinyl chloride, flexible	1.25–1.35
vinyl chloride, rigid	1.35–1.55
vinyl chloride acrylonitrile (60/40)	1.28
vinylethylene	0.889
vinyl formal	1.2–1.4
vinyl pyrrolidone (PVP)	1.25
vinyl-vinylidene chloride	1.70
vinylcarbazole	1.20
vinylidene chloride (PVDC)	1.65–1.875
vinylidene fluoride (PVDF)	1.75–1.78
vinylisobutyl ether	0.91–0.92
−*m*-xylene adipamide	1.22

TABLE 2.84 Density of Polymers at Various Temperatures

Temperature (°C)	0	20	40	60	80	100	120	140	160	180	200	220	240	260	280	300	320	340	360	380
Natural rubber, unvulcanized	0.9283	0.9162																		
Natural rubber, cured	0.9211	0.9093																		
Polyamide, Nylon 6													1.176	1.165	1.154	1.143				
Polyamide, Nylon 6,6													0.963		1.100	1.086	1.071			
Poly(butene-1), isotactic								0.797	0.786	0.776	0.765	0.755	0.745							
Poly(n-butyl methacrylate)	g1.063[a]	1.057	1.043	1.030	1.017	1.005	0.993													
			1.045	1.032	1.018	1.004	0.990	0.975	0.961	0.947	0.933									
Poly(e-caprolactone)						1.037	1.023	1.010												
Polycarbonate, (with Bisphenol A)			g1.192	g1.186	g1.180	g1.174	g1.167	g1.161	1.150	1.136	1.123	1.109	1.095	1.081	1.067	1.053	1.039	1.025		
Poly(cyclohexyl methacrylate)		g1.101	g1.095	g1.090	g1.084		1.066	1.054	1.041	1.028	1.015									
Poly(2,6-dimethylphenylene ether)		g1.061	g1.057	g1.052	g1.048	g1.043	1.039	g1.035	g1.030	g1.026		1.012	0.997	0.983	0.968	0.953	0.939			
Poly(dimethyl siloxane)	0.9742	0.9566	0.9393	0.9222	0.9053	0.8887	0.8722	0.8560	0.8400	0.8242										
		0.9566	0.9389																	
Polyetheretherketone																	1.113	1.098	1.084	
Polyethylene, branched						0.785	0.774	0.763	0.752											
							0.801	0.790	0.780	0.769	0.759	0.749								
Polyethylene, linear						0.7847	0.7735	0.7624	0.7514											
						0.789	0.778	0.766	0.753											
Poly(ethylene terephthalate)															1.172	1.156	1.140	1.125		
Poly(ethyl methacrylate)	g1.131	g1.125	g1.119	g1.113	1.103															
Polyisobutylene	0.9297	0.9195	0.9093	0.8992	0.8891	0.8791	0.8691	0.8592												
Poly(methyl methacrylate)			g1.181	g1.177	g1.171	g1.166	1.153	1.139	1.126	1.112	1.097	1.082	1.067	1.052						
			g1.184	g1.179	g1.174	g1.168	1.148	1.136	1.123											
							1.153	1.141	1.129	1.117	1.106	1.094								
Poly(methyl methacrylate), isotactic	g1.175	g1.170	g1.165	g1.160	g1.155	g1.150	1.140	1.128												
		g1.220	1.204	1.189	1.174	1.160	1.146	1.132	1.119											
Poly(o-methyl styrene)			g1.016	1.011	g1.006			0.9881	0.9777	0.9674	0.9571									
Polyoxyethylene				1.063	1.048	1.033	1.018	1.004	0.990	0.976										
Polyoxymethylene									1.167	1.151										
Polypropylene, atactic				0.827	0.816	0.802														

(Continued)

TABLE 2.84 Density of Polymers at Various Temperatures (*Continued*)

Temperature (°C)	0	20	40	60	80	100	120	140	160	180	200	220	240	260	280	300	320	340	360	380
Polypropylene, isotactic										0.764	0.754	0.744	0.734	0.724	0.714	0.705				
										0.763	0.753	0.743								
Polystyrene				1.0260	1.0142	1.0025	0.9909	0.9795	0.9681											
			g1.044	g1.040	g1.035															
						1.0125	1.0021	0.9919	0.9818	0.9717										
Polysulfone (with Bisphenol A)			g1.040	g1.034	1.026	1.016	1.005	0.994	0.984	0.973	0.961									
Polytetrafluoroethylene			g1.232	g1.226	g1.221	g1.216	g1.211	g1.206	g1.201	g1.195	1.183	1.170	1.157	1.144	1.130	1.117	1.104	1.091	1.078	
																		1.548	1.504	
Polytetrahydrofuran				0.944	0.931	0.919	0.907	0.895												
Poly(vinyl acetate)		g1.196	g1.189	1.1783	1.1615	1.1449	1.1285													
Poly(vinyl chloride)				1.352	1.338	1.322														
Poly(vinyl methyl ether)			1.0580	1.0436	1.0294	1.0152	1.0011	0.9871												

$^a g$ = glass.

TABLE 2.85 Surface Tension (Liquid Phase) of Polymers

Polymer	MW	γ_{LV} at 20°C (mN/m)	$-d\gamma/dT$ [mN/(mK)]
Poly(oxyhexafluoropropylene)	∞	18.4 (25°C)	0.059 ($M_n \sim 7000$)
Poly[heptadecafluorodecyl)methylsiloxane]	$M_n \sim 19600$	18.5 (25°C)	⋯
Poly(dimethylsiloxane)	∞	21.3 (20°C)	0.048 (10^6 cS)
Poly[methyl(trifluoropropyl)siloxane]	∞	24.4 (25°C)	⋯
Poly(tetrafluoroethylene)	∞	25.6	0.053 ($M_n = 1038$)
Poly(oxyisobutylene)	$M \sim 30000$	27.5	0.066
Poly(vinyl octanoate)	⋯	28.7	0.061
Polypropylene, atactic	Melt index ~1000	29.4	0.056
Paraffin wax	⋯	30.0 (20°C)	~0.06
Poly(1,2-butadiene)	$M_n \sim 1000$	30.4 (25°C)	⋯
Poly(t-butyl methacrylate)	$M_v \sim 6000$	30.5	0.059
Poly(oxypropylene)	$M_n \sim 4100$	30.7 (25°C)	0.073
Poly(i-butyl methacrylate)	$M_v \sim 35000$	30.9	0.060
Poly(chlorotrifluoroethylene)	$M_n \sim 1280$	30.9	0.067
Poly(vinyl hexadecanoate)	⋯	30.9	0.066
Poly(n-butyl methacrylate)	$M_v \sim 37000$	31.2	0.059
Poly(oxytetramethylene)	$M_n \sim 32000$	31.8	0.060
Poly(methoxyethylene)	$M_n \sim 46500$	31.8	0.075
Poly(n-butyl acrylate)	$M \sim 32000$	33.7	0.070
Polyethylene, branched	$M_n \sim 7000$	34.3	0.060
Poly(isobutylene)	∞	35.6 (24°C)	0.064 ($M_n \sim 2700$)
Polyethylene, linear	$M_w \sim 67000$	35.7	0.057
Poly(oxydecamethylene)	⋯	36.1	0.068
Poly(vinyl acetate)	$M_w \sim 120000$	36.5	0.066
Poly(2-methylstyrene)	$M_n \sim 3000$	38.7	0.058
Poly(oxydodecamethyleneoxyisophthaloyl)	⋯	40.0	0.070
Polystyrene	$M_v \sim 44000$	40.7	0.072
Poly(methyl acrylate)	$M_n \sim 25000$	41.0	0.070
Poly(methyl methacrylate)	$M_v \sim 3000$	41.1	0.076
Poly(epichlorohydrin)	$M_n \sim 1500$	43.2 (25°C)	⋯
Polychloroprene	$M_v \sim 30000$	43.6	0.086
Poly(oxyethyleneoxyterephthaloyl)	$M_n \sim 16000$	44.5	0.064
Poly(oxyethylene)	∞	45.0 (24°C)	0.076 ($M_n \sim 6000$)
Poly(hexamethylene adipamide)	$M_n \sim 17000$	46.4	0.064
Poly(oxyisophthaloyloxypropylene)	⋯	49.3	0.083

TABLE 2.86 Interfacial Tension (Liquid Phase) of Polymers

Polymer pair	γ_{12} at 20°C (mN/m)	$-d\gamma/dT$ [mN/(mK)]
Polychloroprene/polystyrene	0.5 (140°C)	···
Polychloroprene/poly(n-butyl methacrylate)	1.6 (140°C)	···
Poly(methyl methacrylate)/poly (t-butyl methacrylate)	3.0	0.005
Poly(methyl methacrylate)/polystyrene	3.2	0.013
Poly(dimethylsiloxane)/polypropylene	3.2	0.002
Poly(methyl methacrylate)/poly(n-butyl methacrylate)	3.4	0.012
Poly(dimethylsiloxane)/poly(t-butyl methacrylate)	3.6	0.003
Polybutadiene/poly(dimethylsiloxane)	4.0	0.009
Poly(methyl acrylate)/poly(n-butyl acrylate)	4.0	0.008
Poly(dimethylsiloxane)/poly(isobutylene)	4.0	0.016
Poly(n-butyl methacrylate)/poly(vinyl acetate)	4.2	0.011
Poly(dimethylsiloxane)/poly(n-butyl methacrylate)	4.2	0.004
Polystyrene/poly(vinyl acetate)	4.2	0.004
Polyethylene/polystyrene	4.4 (200°C)	···
Poly(oxyethylene)/poly(oxtetramethylene)	4.5	0.005
Polychloroprene/polyethylene, branched	4.6	0.008
Polyethylene, linear/poly(n-butyl acrylate)	5.0	0.014
Polyethylene, branched/poly(oxytetramethylene)	5.0	0.007
Poly(dimethylsiloxane)/polyethylene, branched	5.3	0.002
Poly(oxytetramethylene)/poly(vinyl acetate)	5.5	0.008
Polyethylene, branched/poly(i-butyl methacrylate)	5.5	0.010
Polyethylene, branched/poly(oxydodecamethyleneoxyisophthaloyl)	5.9	0.011
Polyethylene, branched/poly(t-butyl methacrylate)	5.9	0.016
Poly(dimenthylsiloxane)/polystyrene	6.1	~0
Poly(dimethylsiloxane)/poly(oxytetramethylene)	6.4	0.001
Poly(dimethylsiloxane)/polychloroprene	7.1	0.005
Polyethylene, linear/poly(n-butyl methacrylate)	7.1	0.015
Polyethylene, linear/polystyrene	8.3	0.020
Poly(dimentylsiloxane)/poly(vinyl acetate)	8.4	0.008
Poly(isobutylene)/poly(vinyl acetate)	9.9	0.020
Polyethylene, linear/poly(methyl acrylate)	10.6	0.018
Polyethylene/poly(caprolactam)	10.7 (250°C)	···
Poly(dimethylsiloxane)/poly(oxyethylene)	10.9	0.008
Polyethylene, branched/poly(oxyethylene)	11.6	0.016
Polyethylene, linear/poly(methyl methacrylate)	11.9	0.018
Polyethylene, linear/poly(vinyl acetate)	14.5	0.027
Polyethylene, linear/poly(hexamethylene adipamide)	14.9	0.018
Polyethylene, branched/poly(oxyisophthaloyloxpropylene)	15.4	0.030

TABLE 2.87 Thermal Expansion Coefficients of Polymers

Temperature (°C)	0	20	40	60	80	100	120	140	160	180	200	220	240	260	280	300	320	340	360	380
Natural Rubber, unvulcanized	6.6	6.6																		
Natural Rubber, cured	6.5	6.4																		
		6.7																		
Polyamide, Nylon 6													4.7	4.7	4.7					
Polyamide, Nylon 6,6														6.6	6.6					
															6.8					
Poly(butene-1), isotactic								6.7	6.7	6.7	6.7	6.7	6.7							
Poly(n-butyl methacrylate)	g3.8[a]	6.4	6.2	6.5	6.8	7.0	7.2	7.3	7.4	7.4	7.4									
Poly(ε-caprolactone)							6.1	6.3												
							6.4													
Polycarbonate, (with Bisphenol A)			g2.6	g2.6	g2.6	g2.6	g2.6	g2.6	5.8	5.9	6.1	6.2	6.3	6.4	6.6	6.7	6.8	6.9		
Poly(cyclohexyl methacrylate)		g2.4	g2.5	g2.5	g2.5		5.9	6.0	6.2	6.3	6.4									
Poly(2,6-dimethylphenylene ether)			g2.1	g2.1	g2.1	g2.1	g2.1	g2.1	g2.1	g2.1	g2.1	7.1	7.3	7.4	7.6	7.7	7.8			
Poly(dimethyl siloxane)		9.06	9.11	9.17	9.23	9.29	9.35	9.41	9.47	9.53	9.59									
		9.0	9.4	9.2																
Polyetheretherketone																	6.7		6.7	6.7
Polyethylene, branched							6.7				6.7	6.7								
								7.5	7.2	6.9										
Polyethylene, linear								7.14	7.18	7.24	7.32									
									7.6	7.9										
										7.0	7.0									
Poly(ethylene terephthalate)															6.8	6.8	6.8	6.8		
Poly(ethyl methacrylate)	g2.7	g2.7	g2.7	g2.7																
Polyisobutylene	5.51	5.54	5.58	5.61	5.65	5.68	5.72	5.75	6.1	6.4	6.7	7.0	7.2	7.5						
Poly(methyl methacrylate)	g2.1	g2.1	g2.1	g2.1	g2.1	g2.1	5.2	5.2	5.2	5.2	5.2	5.2								
		g1.8	g1.8	g2.1	g2.4		5.4	5.7	6.0	6.0										
			g2.2	g2.5	g2.9	g2.7	5.5	5.8		6.4										

(Continued)

TABLE 2.87 Thermal Expansion Coefficients of Polymers (*Continued*)

Temperature (°C)	0	20	40	60	80	100	120	140	160	180	200	220	240	260	280	300	320	340	360	380
Poly(methyl methacrylate), isotactic		g2.2		6.4	6.3	6.2	6.1	5.9	5.8	5.7										
Poly(o-methyl styrene)			g2.6	g2.6	g2.6			5.3	5.3	5.3										
Polyoxyethylene						7.1	7.1	7.1	7.1	7.1	7.1	7.1								
Polyoxymethylene										7.9	6.8	6.8								
Polypropylene, atactic					6.1	7.7	9.3													
Polypropylene, isotactic										6.6	6.6	6.6								
										6.7	6.7	6.7	6.7	6.7	6.7	6.7				
										6.7	6.7									
Polystyrene		g2.0				5.78	5.79	5.81	5.82	5.84	5.85	5.87								
			g2.3	g2.5	5.0	5.2	5.1	5.1	5.1	5.1	5.1									
			g2.9	g2.9			5.3	5.4	5.6	5.7	5.8	5.9	6.0	6.2	6.3	6.4	6.5	6.1	6.1	
Polysulfone, (with Bisphenol A)			g2.1	g2.1	g2.1	g2.1	g2.1	g2.1	g2.2	g2.2	5.5	5.6	5.7	5.8	5.8	5.9	6.0	6.1	6.1	
Polytetrafluoroethylene																		14.4	14.7	
Polytetrahydrofuran					6.7	6.7	6.7	6.7	6.7											
Poly(vinyl acetate)		g2.8	7.13	7.17	7.20	7.23														
Poly(vinyl chloride)	g2.8					4.7	5.5	6.2												
Poly(vinyl methyl ether)			6.87	6.92	6.96	7.01	7.06													

[a] g = glass

TABLE 2.88 Heat Capacities of Polymers

Polymer	Abbre-viations	Molecular[a] weight g/mol	T_g (K)	Temp. (K)	C_p^b kJ/kg·K	J/mol·K	ΔC_p^c J/mol·K
				1. Main-chain carbon polymers			
				Poly(acrylics)			
Poly(*iso*-butyl acrylate)	PiBA	128.17	249	220	1.2156	155.80	36.60
				240	1.3365	171.30	
				300	1.8108	232.09	
				500	2.3388	299.77	
Poly(*n*-butyl acrylate)	PnBA	128.17	218	80	0.5598	71.75	45.40
				180	1.0632	136.27	
				300	1.8201	233.28	
				440	2.1803	279.45	
Poly(ethyl acrylate)	PEA	100.12	249	90	0.5792	57.99	45.60
				200	1.0301	103.13	
				300	1.7867	178.88	
				500	2.2189	222.16	
Poly(methyl acrylate)	PMA	86.09	279	100	0.6154	52.98	42.30
				200	0.9816	84.51	
				300	1.765	151.99	
				500	2.143	184.49	
				Poly(dienes)			
1,4-Poly(butadiene)	PBD	54.09					
cis-			171	50	0.3694	19.98	29.10
				150	0.8967	48.50	
				300	1.960	106.00	
				350	2.214	114.90	
trans-			180	50	0.3465	18.74	28.20
				150	0.9057	48.99	
				300	NA	NA	
				500	2.616	141.50	
Poly(1-butene)	PB	56.11	249	100	0.6733	37.78	23.06
				200	1.2190	68.40	
				300	2.086	117.02	
				600	3.071	172.31	
Poly(1-butenylene)	PBUT	55.10					
cis-			171	30	0.2140	11.79	28.91
				130	0.7775	42.838	
				300	1.924	106.03	
				450	2.409	132.73	
trans-			190	30	0.1761	9.704	26.48
				130	0.7898	43.516	
				300	1.924	106.03	
				450	2.409	132.73	
				Poly(alkenes)			
Poly(ethylene)	PE	14.03	252	100	0.674	9.45 (c)	10.1
				200	1.110	15.57	
				300	1.555	21.81 (s)	
					2.202	30.89 (m)	
				600	3.127	43.87	
Poly(1-hexene)	PHE	84.16	223	100	0.7020	59.08 (a)	25.1
				200	1.3319	112.09	
				250	1.903	160.18 (a)	
				290	2.079	174.98 (a)	

(Continued)

TABLE 2.88 Heat Capacities of Polymers (*Continued*)

Polymer	Abbre-viations	Molecular[a] weight g/mol	T_g (K)	Temp. (K)	C_p[b] kJ/kg·K	C_p[b] J/mol·K	ΔC_p[c] J/mol·K
Poly(isobutene)	PiB	56.11	200	50	0.2440	13.69 (a)	22.29
				150	0.8660	48.59	
				300	1.962	110.09 (a)	
				380	2.311	129.66	
Poly(2-methylbutadiene) *cis-*	PMBD	68.12	200	50	0.3573	24.34	30.87 (a)
				150	0.9025	61.48	
				300	1.911	130.20	
				360	2.216	144.80	
Poly(4-methyl-1-pentene)	P4MPE	84.16	303	80	0.5610	47.21	33.7 (a)
				180	1.090	91.75	
				250	1.4449	121.60	
				300	1.728	145.40	
Poly(1-pentene)	PPE	70.14	233	200	1.253	87.90	27.03 (a)
				220	1.338	93.82	
				300	2.058	144.34	
				470	2.770	194.32	
Poly(propylene)	PP	42.08	260	100	0.6238	26.25 (c)	17.37
				200	1.132	47.63 (c)	
				300	1.622	68.24 (s)	
					2.099	88.34 (m)	
				600	3.178	133.73 (a)	

Poly(methacrylics)

Polymer	Abbre-viations	Molecular[a] weight g/mol	T_g (K)	Temp. (K)	C_p[b] kJ/kg·K	C_p[b] J/mol·K	ΔC_p[c] J/mol·K
Poly(n-butyl methacrylate)	PnBMA	142.20	293	80	0.5472	77.81	29.70
				200	1.1557	164.34	
				300	1.8524	263.41	
				450	2.3673	336.63	
Poly(i-butyl methacrylate)	PiBMA	142.20	326	230	1.2229	173.90	39.00
				300	1.5710	223.40	
				350	2.0190	287.10	
				400	2.1127	300.43	
Poly(ethyl methacrylate)	PEMA	114.15	338	80	0.5155	58.84	31.70
				300	1.4666	167.42	
				350	1.9489	222.47	
				380	2.0462	233.57	
Poly(hexyl methacrylate)	PHMA	170.25	268	270	1.8264	310.77	—
				300	1.9091	324.83	
				420	2.2396	381.06	
Poly(methacrylic acid)	PMAA	86.09	—	100	0.5248	45.18	—
				200	0.9456	81.41	
				300	1.307	112.50	
Poly(methacrylamide)	PMAM	85.11	—	100	0.5904	50.25	—
				200	1.032	87.81	
				300	1.395	118.70	
Poly(methyl methacrylate)	PMMA	100.12	378	100	0.5742	57.49	33.5
				300	1.3755	137.72	
				400	2.0766	207.91	
				550	2.4323	243.52	

TABLE 2.88   Heat Capacities of Polymers (*Continued*)

Polymer	Abbre-viations	Molecular[a] weight g/mol	T_g (K)	Temp. (K)	C_p^b kJ/kg·K	J/mol·K	ΔC_p^c J/mol·K
				Poly(styrenes)			
Poly(styrene)	PS	104.15	373	100	0.4548	47.37 (g)	30.7 (a)
				300	1.2230	127.38	
					1.2730	132.58	
				400	1.9322	201.24	
				600	2.4417	254.30	
—,α-methyl	PαMS	118.18	441	100	0.4712	55.69	25.3
				300	1.2752	150.70 (g)	
				460	2.1868	258.44	
				490	2.3331	275.72	
—, p-bromo-	PBS	183.05	410	300	0.79650	145.800	31.9
				350	0.92349	169.045	
				420	1.2651	231.582	
				550	1.4641	267.995	
—, p-chloro-	PCS	138.60	406	300	1.0229	141.780	31.1
				350	1.19848	166.110	
				410	1.6331	226.345	
				550	1.9134	265.195	
—, p-fluoro-	PFS	122.14	384	130	0.47611	58.152	33.3
				200	0.62048	75.786	
				300	0.93079	113.687	
				380	1.2672	154.773	
—, p-iodo-	PIS	230.05	424	300	0.67607	155.53	37.9
				400	0.89102	204.980	
				430	1.1145	256.41	
				550	1.2570	289.17	
—, p-methyl-	PMS	118.18	380	300	1.2743	150.600	34.6
				350	1.4917	176.290	
				390	1.9449	229.846	
				500	2.2766	269.05	
				Poly(vinyl halides) and Poly(vinyl nitriles)			
Poly(acrylonitrile)	PAN	53.06	378	100	0.5695	30.22	—
				200	0.9286	49.27	
				300	1.297	68.83	
				370	1.624	86.16	
Poly(chlorotrifluoroethylene)	PC3FE	116.47	325	80	0.2787	32.46	—
				200	0.6257	72.87	
				300	0.85945	100.10	
				320	0.90667	105.60	
Poly(tetrafluoroethylene)	PTFE	50.01	240	100	0.3873	19.37	7.82
				200	0.6893	34.47	
				300	0.9016	45.09 (s)	
					1.028	51.42 (m)	
				700	1.454	72.69	
Poly(trifluoroethylene)	P3FE	82.02	304	100	0.4049	33.21	21.00
				200	0.7128	58.46	
				300	1.078	88.40	
Poly(vinyl chloride)	PVC	62.50	354	100	0.4291	26.82 (g)	19.37 (a)
				300	0.9496	59.35 (g)	
				360	1.457	91.08	
				380	1.569	98.05	

(*Continued*)

TABLE 2.88   Heat Capacities of Polymers (*Continued*)

Polymer	Abbre-viations	Molecular[a] weight g/mol	T_g (K)	Temp. (K)	C_p^b kJ/kg·K	C_p^b J/mol·K	ΔC_p^c J/mol·K
Poly(vinylidene chloride)	PVC2	96.95	255	100	0.3745	36.31	70.26
				200	0.5932	57.51	
				250	0.7115	68.98	
				300	NA	NA	
Poly(vinylidene fluoride)	PVF2	64.03	233	100	0.4435	28.40	22.80
				150	0.6185	39.60	
				230	0.8918	57.10	
				250	0.7856	50.30	
				300	NA	NA	
Poly(vinyl fluoride)	PVF	46.04	314	100	0.5204	23.96	17.80(a)
				200	0.8692	40.02	
				300	1.301	59.91	
				310	1.353	62.29	

Others

Polymer	Abbre-viations	Molecular[a] weight g/mol	T_g (K)	Temp. (K)	C_p^b kJ/kg·K	C_p^b J/mol·K	ΔC_p^c J/mol·K
Poly(*p*-phenylene)	PPP	76.10	—	80	0.3708	28.22 (sc)	—
				150	0.58135	44.241 (sc)	
				250	0.92926	70.717 (sc)	
				300	1.117	85.040 (sc)	
Poly(vinyl acetate)	PVAc	86.09	304	80	0.3230	27.81	53.7
				300	1.183	101.86	
				320	1.8409	158.48	
				370	1.898	163.37	
Poly(vinyl alcohol)	PVA	44.05	358	60	0.2674	11.78	—
				150	0.7187	31.66	
				250	1.185	52.21	
				300	1.546	68.11	
Poly(vinyl benzoate)	PVBZ	148.16	347	190	0.71808	106.39	69.5
				300	1.1025	163.35	
				400	1.8390	272.47	
				500	2.0333	301.25	
Poly(*p*-xylylene)	PPX	104.15	286	220	0.91445	95.241 (sc)	37.6 (a)
				250	1.0576	110.149 (sc)	
				300	1.3022	135.622 (sc)	
				410	1.8686	194.619 (sc)	

2. Main-chain heteroatom polymers
Poly(amides)

Polymer	Abbre-viations	Molecular[a] weight g/mol	T_g (K)	Temp. (K)	C_p^b kJ/kg·K	C_p^b J/mol·K	ΔC_p^c J/mol·K
Poly(iminoadipoy-liminododecamethylene)	Nylon 612	310.48	319	230	1.2296	381.78	214.8 (a)
				300	1.5926	494.48	
				400	2.4842	771.30	
				600	3.1596	980.986	
Poly(imioadipoy-liminohexamethylene)	Nylon 66	226.32	323	230	1.1139	252.10	145.0 (a)
				300	1.4638	331.30	
				400	2.3794	538.50	
				600	2.793	632.1	
Poly(iminohexamethylene-iminoazelaoyl)	Nylon 69	268.40	331	230	1.1980	321.53	—
				300	1.5204	408.080	
				400	2.3840	639.874	
				600	3.0720	824.534	
Poly(iminohexamethylene-iminosebacoyl)	Nylon 610	282.43	323	230	1.2069	340.870	—
				300	1.5644	441.820	
				400	2.3975	677.125	
				600	3.1041	876.685	

TABLE 2.88 Heat Capacities of Polymers (*Continued*)

Polymer	Abbre-viations	Molecular[a] weight g/mol	T_g (K)	Temp. (K)	C_p^b kJ/kg·K	C_p^b J/mol·K	ΔC_p^c J/mol·K
Poly(imino-(1-oxohexamethylene))	Nylon 6	113.16	313	70	0.4400	49.78	93.6 (a)
				300	1.5023	170.00	
				400	2.5186	285.00	
				600	2.7881	315.50	
Poly(imino-1-oxododecamethylene)	Nylon 12	197.32	314	230	1.2874	254.020	—
				300	1.6952	334.49	
				400	2.4709	487.565	
				600	3.2786	646.945	
Poly(imino-1-oxoundecamethylene)	Nylon 11	183.30	316	230	1.2996	238.21	—
				300	1.7507	320.91	
				400	2.4567	450.314	
				600	3.2449	594.794	
Poly(methacrylamide)	PMAM	85.11	—	100	0.5904	50.25	—
				200	1.032	87.81	
				250	1.214	103.30	
				300	1.395	118.70	
Poly(amino acids)							
Poly(L-alanine)	PALA	71.08	—	230	1.102	78.33	—
				300	1.315	93.47	
				350	1.498	106.5	
				390	1.622	115.3	
Poly(L-asparagine)	PASN	114.10	—	230	0.958	109.3	—
				300	1.218	139.0	
				350	1.397	159.4	
				390	1.537	175.4	
Polyglycine	PGLY	57.05	—	230	0.929	53.00	—
				300	1.170	66.75	
				350	1.356	77.36	
				390	1.516	86.49	
Poly(L-methionine)	PMET	131.19	—	220	0.936	122.8	—
				300	1.347	176.7	
				350	1.595	209.3	
				390	1.768	232.0	
Poly(L-phenylalanine)	PPHE	147.18	—	220	0.830	122.1	—
				300	1.153	169.7	
				350	1.382	203.4	
				390	1.548	227.8	
Poly(L-serine)	PSER	87.08	—	220	0.959	83.50	—
				300	1.297	112.9	
				350	1.541	134.2	
				390	1.747	152.1	
Poly(L-valine)	PVAL	99.13	—	230	1.213	120.2	—
				300	1.455	144.2	
				350	1.647	163.3	
				390	1.802	178.6	
Poly(esters)							
Poly(butylene adipate)	PBAD	200.24	199	80	0.54302	108.734	140.046
				150	0.87449	175.107	
				300	1.9706	394.595	
				450	2.2147	443.470	

(*Continued*)

TABLE 2.88 Heat Capacities of Polymers (*Continued*)

Polymer	Abbre-viations	Molecular[a] weight g/mol	T_g (K)	Temp. (K)	C_p^b kJ/kg·K	C_p^b J/mol·K	ΔC_p^c J/mol·K
Poly(butylene terephthalate)	PBT	220.23	248	150	0.61075	134.505 (sc)	106.77
			320	200	0.82262	181.166	77.812
				300	1.6134	355.311	
				400	1.8187	400.532	
				570	2.1678	477.407	
Poly(ethylene terephthalate)	PET	192.16	342	100	0.4393	84.42	77.8 (a)
				300	1.172	225.2	
				400	1.8203	349.80	
				600	2.1136	406.15	
Poly(tridecanolactone)	PTDL	212.34	237	185	0.95	202	—
				260	1.45	308	
				300	1.79	380	
				395	2.15	457	
Poly(trimethylene adipate)	PTMA	186.21	—	300	NA	NA	—
				310	1.8710	348.401	
				330	1.9137	356.341	
				360	1.9776	368.252	
Poly(trimethylene succinate)	PTMS	158.15	—	300	NA	NA	—
				310	1.8401	291.014	
				330	1.8721	296.074	
				360	1.9201	303.664	
Poly(γ-butyrolactone)	PBL	86.09	214	100	0.6012	51.760	57.4
				210	1.024	88.170	
				300	1.810	155.858 (m)	
				350	1.870	161.031 (m)	
Poly(ε-caprolactone)	PCL	114.15	209	100	0.62322	71.140	59.5
				200	1.0243	116.923	
				300	1.4229	162.42	
					1.8138	207.04 (s)	
				350	1.9415	221.62 (m)	
Poly(glycolide)	PGL	58.04	318	100	0.5250	30.470	44.4
				300	1.127	65.42	
				400	1.999	116.039 (m)	
				550	2.098	121.75 (m)	
Poly(β-propiolactone)	PPL	72.07	249	100	0.5568	40.130	50.4
				240	1.044	75.220	
				300	1.878	135.354 (m)	
				400	2.081	149.994 (m)	
Poly(ethylene oxalate)	PEOL	116.07	306	100	0.49910	57.930	56.23
				300	1.1175	129.705	
				320	1.6395	190.295 (m)	
				360	1.7012	197.456 (m)	
Poly(ethylene sebacate)	PES	228.29	245	120	0.66292	151.338 (s)	154.059
				200	0.95269	217.490 (sc)	
				300	1.9245	439.34 (m)	
				410	2.1923	500.500 (m)	
Poly(oxides)							
Poly(oxy-2,6-dimethyl-1,4-phenylene)	PPO	120.15	482	80	0.4418	53.08	31.9 (a)
				300	1.2459	149.70	
				500	2.1232	255.10	
				570	2.2555	271.00	

TABLE 2.88 Heat Capacities of Polymers (*Continued*)

Polymer	Abbre-viations	Molecular[a] weight g/mol	T_g (K)	Temp. (K)	C_p^b kJ/kg·K	C_p^b J/mol·K	ΔC_p^c J/mol·K
Poly(oxyethylene)	POE	44.05	206	100	0.6114	26.93 (s)	38.96
				200	0.9507	41.88 (s)	
				300	1.257	55.36 (s)	
					1.995	87.89 (m)	
				450	2.223	97.91	
Polyoxymethylene	POM	30.03	190	100	0.5554	16.68 (s)	27.47
				150	0.7266	21.82 (s)	
				300	1.283	38.52 (s)	
					1.920	57.67 (m)	
				600	2.292	68.83	
Poly(oxy-1,4-phenylene)	POPh	92.10	358	300	1.185	109.10 (s)	21.4 (a)
				350	1.367	125.90 (s)	
				400	1.694	156.00 (m)	
				600	2.003	184.50 (m)	
Poly(oxypropylene)	POPP	58.08	198	80	0.537	31.21 (s)	32.15
				180	1.014	58.89 (s)	
				300	1.915	111.23 (m)	
				370	2.105	122.27 (m)	
Poly(oxytetramethylene)	PO4M	72.11	189	80	0.5465	39.41 (s)	46.49
				180	1.033	74.52 (s)	
				300	1.985	143.15 (m)	
				340	2.081	150.04 (m)	
Poly(oxytrimethylene)	PO3M	58.08	195	80	0.5095	29.59 (s)	50.73
				180	0.9464	54.97 (s)	
				300	1.373	79.73 (s)	
					2.055	119.34 (m)	
				330	2.107	122.37	
Others							
Poly(diethyl siloxane)	PDES	102.21	135	50	0.38820	39.678 (sc)	30.189
				100	0.73995	75.630 (sc)	
				300	1.6184	165.417 (m)	
				360	1.7525	179.125 (m)	
Poly(dimethyl itaconate)	PDMI	158.16	377	110	0.59700	94.419 (a)	54.23
				300	1.3183	208.507 (a)	
				400	1.9282	304.968 (m)	
				450	2.0009	316.463 (m)	
Poly(dimethyl siloxane)	PDMS	74.15	146	50	0.3672	27.23	27.7 (a)
				100	0.7131	52.88	
				300	1.591	118.0	
				340	1.657	122.9	
Poly(4-hydroxybenzoic acid)	PHBA	120.11	434	170	0.58914	70.762	34
				300	1.0207	122.60	
				400	1.3662	164.091	
				434	1.4686	176.399	
Poly(4,4'-isopropylidene diphenylenecarbonate)	PC	254.27	418	100	0.43143	109.70 (s)	48.5
				300	1.207	306.8 (s)	
				450	1.9570	497.60 (m)	
				560	2.207	561.3 (m)	

(*Continued*)

TABLE 2.88   Heat Capacities of Polymers (*Continued*)

Polymer	Abbre-viations	Molecular[a] weight g/mol	T_g (K)	Temp. (K)	C_p^b kJ/kg·K	J/mol·K	ΔC_p^c J/mol·K
Poly(oxy-1,4-phenylene-oxy-1,4-phenylene-carbonyl-1,4-phenylene)	PEEK	288.30	419	300	NA	NA	78.1
				419	1.789	515.8	
				500	1.928	555.9	
				750	2.358	679.8	
Poly(oxy-1,4-phenylene-sulphonyl-1,4-phenylene-oxy-1,4-phenylene-(1-methylidene)-1,4-phenylene)	PBISP	442.54	458.5	200	0.75870	335.754	102.482
				300	1.1161	493.934	
				500	1.9436	860.132	
				540	2.0251	896.19	
Poly(1,4-phenylene sulphonyl)	PAS	140.16	492.6	150	0.597	83.7	—
				300	1.009	141.4	
				500	1.571	220.2	
				620	1.642	230.1	
Poly(1-propene sulphone)	P1PS	106.14		10	0.01580	1.677	—
				30	1.165	123.7	
Trigonal selenium	SEt	78.96	303.4	100	0.2304	18.19 (s)	13.29
				300	0.318	25.11	
				400	0.3338	26.36 (s)	
					0.4777	37.72 (m)	
				600	0.4343	34.29	

[a]This is the molecular weight of the repeat unit of the polymer.
[b]Except the data for PTDL and P1PS, C_p data reported in the unit of kJ/kg·K were converted from the C_p data which were directly cited from the literature, using the molecular weight of the repeat unit.
[c]Specific heat increment at T_g.

TABLE 2.89   Thermal Conductivity of Polymers

Polymer	Temperature (K)	k (W/m K)
Polyamides		
Polylauryllactam (nylon-12)		0.25
		0.19
Polycaprolactam (nylon-6)		
Moldings	293	0.24
Crystalline	303	0.43
Amorphous	303	0.36
Melt	523	0.210
Poly(hexamethylene adipamide) (nylon-6,6)		
Moldings	293	0.24
Crystalline	303	0.43
Amorphous	303	0.36
Melt	523	0.15
Poly(hexamethylene dodecanediamide) (nylon-6, 12)		0.22
Poly(hexamethylene sebacamide) (nylon-6, 10)		0.22
Polyundecanolactam (nylon-11)		0.23
Polycarbonates, polyesters, polyethers, and polyketones		
Polyacetal		0.23
		0.3
Polyaryletherketone	293	0.30
Poly(butylene terephthalate) (PBT)	293	0.29
		0.16
Polycarbonate (Biphenol A)	293	0.20
Temperature dependence	300–573	
	150–400	
Poly(dially carbonate)		0.21
Poly(2,6-dimethyl-1,4-phenylene ether)		0.12
Polyester		
Cast, rigid		0.17
Chlorinated		0.33
Polyetheresteramide	303	0.24–0.34
	353	0.20–0.26
Polyetheretherketone (PEEK)		0.25
Poly(ethylene terephthalate) (PET)	293	0.15
Temperature dependence	200–350	
Poly(oxymethylene)	293	0.292
	293	0.44
Temperature dependence	100–400	
Poly(phenylene oxide)		
Molding grade		0.23
Epoxides		
Epoxy resin		
Casting grade	293	0.19
Temperature dependence	300–500	0.19–0.34
Halogenated olefin polymers		
Polychlorotrifluoroethylene	293	0.29
	311–460	0.146–0.248
Poly(ethylene-tetrafluoroethylene) copolymer		0.238
Polytetrafluoroethylene	293	0.25
	298	0.25
	345	0.34
Low-temperature dependence	5–20.8	

(Continued)

TABLE 2.89   Thermal Conductivity of Polymers (*Continued*)

Polymer	Temperature (K)	k (W/m K)
Poly(tetrafluoroethylene-hexafluoropropylene) copolymer (Teflon EEP)		0.202
Poly(vinyl chloride)		
Rigid	293	0.21
Flexible	293	0.17
Chlorinated	293	0.14
Temperature dependence	103	0.129
	273	0.158
	373	0.165
Poly(vinylidene chloride)	293	0.13
Poly(vinylidene fluoride)	293	0.13
	298–433	0.17–0.19
Hydrocarbon polymers		
Polybutene		0.22
Polybutadiene		
Extrusion grade	293	0.22
Poly(butadiene-styrene) copolymer (SBR)		
23.5% styrene content		
Pure gum vulcanizate		0.190–0.250
Carbon black vulcanizate		0.300
Polychloroprene (neoprene)		
Unvulcanized	293	0.19
Pure gum vulcanizate		0.192
Carbon black vulcanizate		0.210
Poly(1,3-cyclopentylenevinylene) [poly(2-norbornene)]		0.29
Polyethylene		
Low density		0.33
Medium density		0.42
High density		0.52
Temperature dependence	20–573	
Molecular-weight dependence		
Poly(ethylene-propylene) copolymer		0.355
Polyisobutylene		0.13
Polyisoprene (natural rubber)		
Unvulcanized		0.13
Pure gum vulcanizate		0.15
Carbon black vulcanizate		0.28
Poly(4-methyl-1-pentene)		0.167
Polypropylene	293	0.12
		0.2
Temperature dependence		
Polystyrene	273	0.105
	373	0.128
	473	0.13
	573	0.14
	673	0.160
Poly(*p*-xylylene) (PPX)		12
Polyimides		
Polyetherimide		0.07
Polyimide		
Thermoplastic	293	0.11
Thermoset		0.23–0.50
Temperature dependence	300–500	

TABLE 2.89 Thermal Conductivity of Polymers (*Continued*)

Polymer	Temperature (K)	k (W/m K)
Phenolic resins		
Poly(phenol-formaldehyde) resin		
Casting grade		0.15
Molding grade		0.25
Poly(phenol-furfural) resin		
Molding grade	293	0.25
Polysaccharides		
Cellulose		
Cotton		0.071
Rayon		0.054–0.07
Sulfite pulp, wet		0.8
Sulfite pulp, dry		0.067
Laminated Kraft paper		0.13
Alkali cellulose		0.046–0.067
Different papers	303–333	0.029–0.17
Cellulose acetate	293	0.20
Cellulose acetate butyrate	293	0.33
Cellulose nitrate		0.23
Cellulose propionate		0.20
Ethylcellulose		0.21
Polysiloxanes		
Poly(dimethylsiloxane)	230	0.25
	290	0.22
	340	0.20
	410	0.17
Poly(methylphenylsiloxane)		
9.5% phenyl, $d = 1110$ kg/m^3	273	0.158
	323	0.150
	373	0.144
48% phenyl, $d = 1070$ kg/m^3	273	0.143
	323	0.136
	373	0.127
62% phenyl, $d = 1110$ kg/m^3	273	0.141
	323	0.137
	373	0.132
Polysulfide and polysulfones		
Polyarylsulfone		0.18
Polyethersulfone		0.18
Poly(phenylene sulfide)	293	0.29
	240–310	0.288
Poly(phenylene sulfone)		0.18
Udel polysulfone		0.26
Polyurethanes		
Polyurethane		
Casting resin	293	0.21
Elastomer	293	0.31
Vinyl Polymers		
Polyacrylonitrile	293	0.26
Poly(acrylonitrile-butadiene) copolymer (NBR)		
35% acrylonitrile	333	0.251
	413	0.184

(*Continued*)

TABLE 2.89   Thermal Conductivity of Polymers (*Continued*)

Polymer	Temperature (K)	k (W/m K)
Poly(acrylonitrile-butadiene-styrene) copolymer (ABS)		
Injection molding grade		0.33
Poly(acrylonitrile-styrene) copolymer	293	0.18
Poly(*i*-butyl methacrylate)		
At 0.82 atm		0.13
Poly(*n*-butyl methacrylate)		
At 0.82 atm		0.45
Poly(butyl methacrylate-triethylene glycol		
dimethacrylate) copolymer		0.15
Poly(chloroethylene-vinyl acetate) copolymer	293	0.134
	325	0.146
	375	0.218
Poly(dially phthalate)		0.21
Poly(ethyl acrylate)	310.9	0.213
	422.1	0.230
	533.2	0.213
Poly(ethyl methacrylate)		
At 0.82 atm	273	0.175
Poly(ethylene vinyl acetate)		0.34
Poly(methyl methacrylate)	293	0.21
Poly(methyl methacrylate-acrylonitrile) copolymer		0.18
Poly(methyl methacrylate-styrene) copolymer		0.21–0.21
Poly(vinyl acetate)		0.159
Poly(vinyl acetate-vinyl chloride) copolymer		0.167
Poly(vinyl alcohol)		0.2
Poly(*N*-vinyl carbozole)	293	0.126
	443	0.168
Poly(vinyl fluoride)	243	0.14
	333	0.17
Poly(vinyl formal) Molding grade	293	0.27

TABLE 2.90   Thermal Conductivity of Foamed Polymers

Name	k (W/m K)
Poly(acrylonitrile-butadiene) copolymer	
$\quad d = 160–400$ kg/m^3	0.036–0.043
Cellulose acetate	
$\quad d = 96–128$ kg/m^3	0.045–0.46
Polychloroprene (Neoprene)	
$\quad d = 112$ kg/m^3	0.040
$\quad d = 192$ kg/m^3	0.065
Poly(dimethylsiloxane)	
$\quad$Sheet, $d = 160$ kg/m^3	0.086
Epoxy	
$\quad d = 32–48$ kg/m^3	0.016–0.022
$\quad d = 80–128$ kg/m^3	0.035–0.040
Polythylene	
$\quad$Extruded plank	
$\quad d = 35$ kg/m^3	0.053
$\quad d = 64$ kg/m^3	0.058
$\quad d = 96$ kg/m^3	0.058
$\quad d = 144$ kg/m^3	0.058
$\quad$Sheet, extruded, $d = 43$ kg/m^3	0.040–0.049
$\quad$Sheet, crosslinked, $d = 26–38$ kg/m^3	0.036–0.040
Polyisocyanurate	
$\quad d = 24–56$ kg/m^3	0.012–0.02
Polyisoprene (natural rubber)	
$\quad d = 56$ kg/m^3	0.036
$\quad d = 320$ kg/m^3	0.043
Phenolic resin	
$\quad d = 32–64$ kg/m^3	0.029–0.032
$\quad d = 112–160$ kg/m^3	0.035–0.040
Polypropylene	
$\quad d = 64–96$ kg/m^3	0.039
Polystyrene	
$\quad d = 16$ kg/m^3	0.040
$\quad d = 32$ kg/m^3	0.036
$\quad d = 64$ kg/m^3	0.033
$\quad d = 96$ kg/m^3	0.036
$\quad d = 160$ kg/m^3	0.039
Poly(styrene-butadiene) copolymer (SBR)	
$\quad d = 72$ kg/m^3	0.030
Poly(urea-formaldehyde) resin	
$\quad d = 13–19$ kg/m^3	0.026–0.030
Polyurethane	
$\quad$Air blown, $d = 20–70$ kg/m^3	
$\quad$At 0°C	0.033
$\quad$At 20°C	0.036
$\quad$At 70°C	0.040
$\quad$CO$_2$ blown, $d = 64$ kg/m^3, at 20°C	0.016
$\quad$20% closed cells, at 20°C	0.033
$\quad$90% closed cells, at 20°C	0.016
$\quad$500 μm cell size, at 20°C	0.024
$\quad$100 μm cell size, at 20°C	0.016
Poly(vinyl chloride)	
$\quad d = 56$ kg/m^3	0.035
$\quad d = 112$ kg/m^3	0.040

TABLE 2.91   Thermal Conductivity of Polymers with Fillers

Name	k (W/m K)	Name	k (W/m K)
Polyacetal		Polyisoprene (natural rubber)	
5–20% polytetrafluoroethylene (PTFE)	0.20	33% carbon black	0.28
Poly(acrylonitrile-butadiene-styrene) copolymer (ABS)		Poly(melamine-formaldehyde) resin	
20% glass fiber	0.20	Asbestor	0.544–0.73
Polyaryletherketone		Cellulose fiber	0.27–0.42
40% glass fiber	0.44	Glass fiber	0.42–0.48
Poly(butylene terephthalate) (PBT)		Macerated fabric	0.443
30% glass fiber	0.29	Wood flour/cellulose	0.17–0.48
	0.21	Poly(melamine-phenolic) resin	
40–45% glass fiber	0.42	Cellulose fiber	0.17–0.29
Polycarbonate		Wood flour	0.17–0.29
10% glass fiber	0.22	Nylon-6 (polycaprolactam)	
30% glass fiber	0.32	30–35% glass fiber	0.24–0.28
Polychloroprene (Neoprene)		Nylon-6,6 [poly(hexamethylene adipamide)]	
33% carbon black	0.210	30–33% glass fiber	0.21–0.49
Poly(dially phthalate)		40% glass fiber and mineral	0.46
Glass fiber	0.21–0.62	30% graphite or polyacrylonitrile (PAN) carbon fiber	1.0
Epoxy resin		Nylon-6,12 [poly(hexamethylenedodecanediamide)]	
50% aluminum	1.7–3.4	30–35% glass fiber	0.427
25% Al_2O_3	0.35–0.52	Poly(phenylene oxide)	
50% Al_2O_3	0.52–0.69	30% glass fiber	0.16
75% Al_2O_3	1.4–1.7	Poly(phenylene sulfide)	
30% mica	0.24	40% glass fiber	0.288
50% mica	0.39	30% carbon fiber	0.28–0.75
Silica	0.42–0.84	Polypropylene	
Polyetheretherketone (PEEK)		40% talc	0.32
30% glass fiber	0.21	40% $CaCO_3$	0.29
30% carbon fiber	0.21	40% glass fiber	0.37
Polyethylene		Polystyrene	
30% glass fiber	0.36–0.46	20% glass fiber	0.25
Poly(ethylene terephthalate) (PET)		Poly(styrene-acrylonitrile) copolymer	
30% glass fiber	0.29	20% glass fiber	0.28
45% glass fiber	0.31	Poly(styrene-butadiene) copolymer (SBR)	
30% graphite fiber	0.71	33% carbon black	0.300
40% polyacrylonitrile (PAN) carbon fiber	0.72	Polytetrafluoroethylene	
Polyimide		25% glass fiber	0.33–0.41
Thermoplastic, 15% graphite	0.87	Poly(urea-formaldehyde) resin	
Thermoplastic, 40% graphite	1.73	33% α-cellulose	0.423
Thermoset, 50% glass fiber	0.41		

TABLE 2.92 Resistance of Selected Polymers and Rubber to Various Chemicals at 20°C

The information in this table is intended to be used only as a general guide. The chemical resistance classifications are E = excellent (30 days of exposure causes no damage), G = good (some damage after 30 days), F = fair (exposure may cause crazing, softening, swelling, or loss of strength), N = not recommended (immediate damage may occur).

Polymers

	Acids, dilute or weak	Acids, strong and concentrated	Alcohols, aliphatic	Aldehydes	Alkalies, concentrated	Esters	Ethers	Glycols	Hydrocarbons, aliphatic	Hydrocarbons, aromatic	Hydrocarbons, halogenated	Ketones	Oxidizing agents, strong
Acetals	F	N	F	N	N	N	N	G	N	N	N	N	N
Acrylics: poly (methyl methacrylate)	G	N	E	—	N	N	E	E	G	N	N	N	N
Allyls: diallyl phthalate	G	—	—	N	N	—	—	—	E	G	G	N	—
Cellulosics: cellulose-acetate-butyrate and cellulose-acetate-propionate polymers	F	N	N	N	N	N	N	G	F	N	N	N	—
Fluorocarbons	E	E	E	E	E	E	E	E	E	E	E	E	E
Polyamides	N	N	G	E	E	G	—	G	G	F	F	G	N
Polycarbonates	G	N	N	F	N	N	N	G	N	G	G	N	N
Polyesters	G	G	N	—	N	N	F	G	G	F	F	N	N
Poly(methyl pentene)	E	E	E	G	E	G	N	E	F	G	N	F	F
Low-density polyethylene	E	E	E	G	E	G	N	E	F	F	F	G	F
High-density polyethylene	E	E	E	E	E	G	N	E	G	G	N	G	F
Polybutadiene	G	F	E	—	—	—	—	—	—	—	—	E	—
Polypropylene and polyallomer	E	E	E	E	E	G	N	E	G	E	E	G	F
Polystyrene	N	N	E	E	N	N	N	E	N	N	N	N	N
Styrene-acrylonitrile copolymers	—	—	N	—	N	—	—	F	N	—	—	—	—
Styrene-acrylonitrile-butadiene copolymers	—	N	G	G	G	N	—	—	F	N	N	N	G
Sulfones: polysulfone	G	N	F	F	E	N	F	G	F	N	N	N	G
Vinyls: poly(vinyl chloride)	E	G	E	G	G	N	F	F	G	N	N	N	G

Rubbers

	Acids, dilute or weak	Acids, strong and concentrated	Alcohols, aliphatic	Aldehydes	Alkalies, concentrated	Esters	Ethers	Glycols	Hydrocarbons, aliphatic	Hydrocarbons, aromatic	Hydrocarbons, halogenated	Ketones	Oxidizing agents, strong
Natural rubber	—	—	E	—	—	N	N	E	N	N	N	N	—
Nitrile rubber	—	—	E	—	—	N	G	E	E	N	N	N	—
Polychloroprene	—	—	E	—	—	N	F	E	F	N	N	N	—
Polyisobutylene	—	—	E	—	—	F	F	E	E	F	F	N	—
Polysulfide rubbers: Thiokol	—	—	E	—	—	E	E	E	E	E	E	N	—
Styrene-butadiene rubber	—	—	E	—	—	N	N	E	N	N	N	N	—

TABLE 2.93 Gas Permeability Constants ($10^{10}\,P$) at 25°C for Polymers and Rubber

The gas permeability constant P is

$$P = \frac{\text{amount of permeant}}{(\text{area}) \times (\text{time}) \times (\text{driving forced across the film})}$$

The gas permeability constant is the amount of gas expressed in cubic centimeters passed in 1 s through a 1-cm² area of film when the pressure across a film thickness of 1 cm is 1 cm Hg and the temperature is 25°C. All tabulated values are multiplied by 10^{10} and are in units of seconds^{-1} (centimeters of Hg)$^{-1}$. Other temperatures are indicated by exponents and are expressed in degrees Celsius.

Polymer or rubber	Gas						
	He	N_2	H_2	O_2	CO_2	H_2O	Other
Cellulose (cellophane)	0.005^{20}	0.0032	0.0065	0.0021	0.0047	1900	0.006^{45} (H_2S); 0.001 7 (SO_2)
Cellulose acetate	13.6^{20}	0.28^{30}	3.5^{20}	0.78^{30}	22.7^{30}	5500	3.5^{30} (H_2S); 17^{0} (ethylene oxide); 6.8^{60} (bromomethane)
Cellulose nitrate	6.9	0.12	2.0^{20}	1.95	2.12	6290	57.1 (NH_3); 1.76 (SO_2)
Ethyl cellulose	400^{30}	8.4^{30}	87^{20}	26.5^{30}	41.0^{30}	$12{,}000^{20}$	705 (NH_3); 204 (SO_2); 420^{0} (ethylene oxide)
Gutta percha		2.17	14.4	6.16	35.4	510	15.7 (CO); 30.1 (CH_4); 1.68 (C_3H_8); 98.9 (C_2H_2);
Natural rubber		9.43	52.0	23.3	15.3	2290	550 ($CH_3C\!\equiv\!CH$); 3.59 (SF_6)
Nylon 6	0.53^{20}	0.0095^{30}		0.038^{30}	0.10^{30}	177	0.33^{30} (H_2S); 1.2^{20} (NH_3); 0.84^{60} (CH_3Br)
Nylon 11	1.95^{30}		1.78^{30}		1.00^{40}		0.344^{30} (Ne); 0.189^{40} (Ar); 13.6^{50} (propyne)
Poly(acrylonitrile)				0.0002	0.0008	300	
Acrylonitrile-styrene copolymer (66:34)				0.048	0.21	2000	
Poly(1,3-butadiene)	32.6	6.42	41.9	19.0	138.0	5070	19.2 (Ne); 41.0 (Ar)
Poly(cis-1,4-butadiene)		19.2	15.9				
Butadiene-acrylonitrile copolymer (80:20)	12.2	1.06		3.85	30.8		24.8 (C_2H_2); 7.7 (propyne)

Butadiene-styrene copolymer (80:20)	13.4	1.71		4.0	25.8		5.01 (Ne); 4.49 (Ar)
Butadiene-styrene copolymer (92:8)	22.9	5.11		2.88	12.6		9.70 (Ne); 12.7 (Ar)
Polychloroprene		1.2	13.6				3.79 (Ar); 3.27 (CH_4)
Polyethylene, low-density	4.9	0.969	12.0[30]	0.403	0.36	90	2.88 (CH_4); 6.81 (C_2H_6); 9.43 (C_3H_8); 1.48 (CO); 49[0] (ethylene oxide); 14.4 (propene); 42.2 (propyne); 0.170 (SF_6); 472[60] (CH_3Br)
Polyethylene, high-density	1.14	0.143	3.0[20]			12.0	0.388 (CH_4); 0.590 (C_2H_6); 0.537 (C_3H_8); 0.008 3 (SF_6); 1.69 (Ar); 4.01 (propene)
Poly(ethylene terephthalate)							
Crystalline	1.32	0.006 5	3.70[20]	0.035	0.17	130	0.003 2 (CH_4); 0.08[60] (CH_3Br)
Amorphous	3.28	0.013		0.059	0.30		0.009 (CH_4)
Poly(ethyl methacrylate)	6.82	0.220		1.15	5.00	3200	2.98 (Ne); 0.565 (Ar); 0.370 (Kr); 3.83 (H_2S); 0.000 001 65 (SF_6)
Isobutene-isoprene copolymer (98:2)	8.38	0.324	7.20	1.30	5.16	110[38]	13.6[50] (C_3H_8)
Isoprene-acrylonitrile copolymer (76:24)	7.77	0.181	7.41	0.852	4.32		
Isoprene-methacrylonitrile copolymer (76:24)		0.596	13.6	2.34	14.1		
Methacrylonitrile-styrene-butadiene copolymer (88:7:5)	101	7.83		0.004 8	0.014	600	
Poly(methylpentene)	38[20]	0.44[30]	136	32.0	92.6		0.33[20] (H_2S); 9.2[20] (NH_3)
Polypropylene			41[20]	2.3[30]	9.2[30]	51	191[0] (Ne); 550[0] (Ar); 1020[0] (Kr); 2550[0] (Xe); 19,000[0] (butane)
Silicone rubber, 10% filler	233[0]	227[0]	464[0]	489[0]	3240	43,000[35]	
Polystyrene	18.7	0.788	23.3	2.63	10.5	1200	15.7 (NO_2); 37.5 (N_2O_4)
Poly(tetrafluoroethylene)	6.8[20]	1.4	9.8	4.2	11.7		1.2[0] (ethylene oxide); 4.6[60] (CH_3Br)
Poly(trifluoroethylene)		0.003	0.94[20]	0.025[40]	0.048[40]	0.29	
Poly(vinyl acetate)	12.6[30]		89[30]	0.50[30]			2.64[30] (Ne); 0.19[30] (Ar); 0.078[30] (Kr); 0.050[30] (CH_4)
Poly(vinyl alcohol)	0.001[30]	<0.001[14]	0.009	0.008 9	0.001[23]		0.007 (H_2S); 0.002[0] (ethylene oxide)
Poly(vinyl chloride)	2.05	0.011 8	1.70	0.045 3	0.157	275	3.92 (Ne); 0.011 5 (Ar); 0.028 6 (CH_4)
Poly(vinylidene chloride)	0.31[34]	0.000 94[30]	0.000 94[30]	0.005 3[30]	0.03[30]	0.5	0.03[30] (H_2S); 0.008[60] (CH_3Br)

TABLE 2.94 Vapor Permeability Constants ($10^{10} P$) at 35°C for Polymers

| Polymer | Vapor | | | | |
	Benzene	Hexane	Carbon tetrachloride	Ethanol	Ethyl acetate
Cellulose	1.4	0.912	0.836	85.8	13.4
Cellulose acetate	512	2.80	3.74	2980	3595
Poly(acrylonitrile)	2.61	1.59	1.47	0	0
Polyethylene, low-density	5300	2910	3810	55.9	513
Polystyrene	10,600		6820	0	soluble
Poly(vinyl alcohol)	3.58	2.34	1.61	32.7	2.53

TABLE 2.95 Hildebrand Solubility Parameters of Polymers

Polymer	δ (MPa$^{1/2}$)	T (°C)	Method
Cellulose	32.02		
Cellulose diacetate	23.22		Calc.
Cellulose nitrate (11.83% N)	21.44		Calc.
Epoxy resin	22.3		
Natural rubber	16.2		
	17.09		
Poly(4-acetoxystyrene)	22.7	25	Visc.
Poly(acrylic acid)			
—, butyl ester	18.0	35	
	18.52		Swelling
—, methyl ester	20.77		Swelling
	20.7		Swelling
Poly(acrylonitrilc)	26.09	25	Calc.
Poly(butadiene)	16.2	75	IPGC
	17.15		Calc.
Poly(butadiene-co-acrylonitrile)			
BUNA N (72/55)	18.93	25	Calc.
(61/39)	20.5	75	IPGC
Poly(butadiene-co-styrene)			
BUNA S (85/15)	17.41		Calc.
	17.39		Obs.
Poly(butadiene-co-vinylpyridine)			
(75/25)	19.13		
Poly(chloroprene)	18.42	25	
	19.19		Calc.
	17.6		Swelling
Poly(dimethyl siloxane)	14.9	30	Calc.
Poly(ethylene)	16.6		Calc.
Poly(ethylene)	16.4		Calc.
	16.2		Obs.
Poly(ethylene-co-vinyl-acetate)	18.6	25	IPGC
	17.0	75	IPGC
Poly(*tetra*-fluoroethylene)	12.7		Calc.
Poly(heptamethylene p,p'-bibenzoate)	19.50	25	Visc.
Poly(4-hydroxystyrene)	23.9	25	Visc.
Poly(isobutene)	16.06	35	Av.
	16.47		Swelling
	16.06	25	
Poly(isobutene-co-isoprene) butyl rubber	16.47		
Poly(isoprene)			
1,4-*cis*	15.18	25	Calc.
	16.68	25	
	16.57	35	
	20.46	35	Swelling
	16.6		Swelling
	16.68	25	Calc.
Poly(methacrylic acid)			
—, isobutyl ester	14.7	140	IPGC
—, ethyl ester	18.31		Swelling
—, methyl ester	18.58	25	
Poly(methacrylonitrile)	21.9		Calc.
Poly(methylene)	14.3	20	Extrap.
Poly(α-methyl styrene)	18.75	30	Visc.

(*Continued*)

TABLE 2.95 Hildebrand Solubility Parameters of Polymers (*Continued*)

Polymer	$\delta\,(\text{MPa}^{1/2})$	$T\,(^{\circ}\text{C})$	Method
Poly(σ-methylstyrene-co-acrylonitrile)	16.4	180	IPGC
Poly(oxyethylene)	20.2	25	IPGC
Poly(propylene)	18.8	25	
Poly(styrene)	18.72	35	
Poly(styrene-co-*n*-butyl-methacrylate)	15.1	140	IPGC
Poly(thioethylene)	19.19		Swelling
Poly(vinyl acetate)	19.62	25	Calc.
Poly(vinyl alcohol)	25.78		
Poly(vinyl chloride)	19.28		Calc.
	19.8		Obs.
Poly(vinyl chloride), chlorinated	19.0	25	Visc.
Poly(vinyl propionate)	18.01	35	

TABLE 2.96 Hansen Solubility Parameters of Polymers

Polymer (trade name, supplier)	Solubility parameter (MPa$^{1/2}$)			
	δ_d	δ_p	δ_h	δ_t
Acrylonitrile-butadiene elastomer (Hycar 1052, BF Goodrich)	18.6	8.8	4.2	21.0
Alcohol soluble resin (Pentalyn 255, Hercules)	17.5	9.3	14.3	24.4
Alcohol soluble resin (Pentalyn 830, Hercules)	20.5	5.8	10.9	23.5
Alkyd, long oil (66% oil length, Plexal P65, Polyplex)	20.42	3.44	4.56	21.20
Alkyd, short oil (Coconut oil 34% phthalic anhydride; Plexal C34)	18.50	9.21	4.91	21.24
Blocked isocyanate (Phenol, Suprasec F5100, ICI)	20.19	13.16	13.07	27.42
Cellulose acetate (Cellidore A, Bayer)	18.60	12.73	11.01	25.08
Cellulose nitrate (1/2 s; H-23, Hagedon)	15.41	14.73	8.84	23.08
Epoxy (Epikote 1001, Shell)	20.36	12.03	11.48	26.29
Ester gum (Ester gum BL, Hercules)	19.64	4.73	7.77	21.65
Furfuryl alcohol resin (Durez 14383, Hooker Chemical)	21.16	13.56	12.81	28.21
Hexamethoxymethyl melamine (Cymel 300 American Cyanimid)	20.36	8.53	10.64	24.51
Isoprene elastomer (Cariflex IR 305, Shell)	16.57	1.41	−0.82	16.65
Methacrylonitrile/methacrylic acid copolymer	17.39	14.32	12.28	25.78
Nylon 66	18.62	5.11	12.28	22.87
Nylon 66 (Zytel, DuPont)	18.62	0.00	14.12	23.37
Petroleum hydrocarbon resin (Piceopale 110, Penn. Ind. Chem.)	17.55	11.19	3.60	17.96

TABLE 2.96 Hansen Solubility Parameters of Polymers (*Continued*)

Polymer (trade name, supplier)	Solubility parameter (MPa$^{1/2}$)			
	δ_d	δ_p	δ_h	δ_t
Phenolic resin				
(Resole, Phenodur 373 U Chemische Werke Albert)	19.74	11.62	14.59	27.15
Phenolic resin, pure				
(Super Beckacite 1001, Reichhold)	23.26	6.55	8.35	25.57
Poly(4-acetoxy,α-acetoxy styrene)	17.80	10.23	7.37	21.89
Poly(4-acetoxystyrene)	17.80	9.00	8.39	21.69
Poly(acrylonitrile)	18.21	16.16	6.75	25.27
Polyamid, thermoplastic				
(Versamid 930, General Mills)	17.43	−1.92	14.89	23.02
Poly(*p*-benzamide)	18.0	11.9	7.9	23.0
cis-Poly(butadiene) clastomer				
(Bunahuls CB10, Chemische Werke Huels)	17.53	2.25	3.42	18.00
Poly(isobutylene)				
(Lutonal IC/123, BASF)	14.53	2.52	4.66	15.47
Poly(ethyl methacrylate)				
(Lucite 2042, DuPont)	17.60	9.66	3.97	20.46
Poly(ethylene terephthalate)	19.44	3.48	8.59	21.54
Poly(4-hydroxystyrene)	17.60	10.03	13.71	24.55
Poly(methacrylic acid)	17.39	12.48	15.96	26.80
Poly(methacrylonitrile)	18.00	15.96	7.98	25.37
Poly(methyl methacrylate)				
Poly(sulfone), Bisphenol A				
(Polystyrene LG, BASF)	21.28	5.75	4.30	22.47
Poly(sulfone), Bisphenol A				
(Udel)	19.03	0.00	6.96	20.26
Poly(vinyl acetate)				
(Mowilith 50, Hoechst)	20.93	11.27	9.66	25.66
Poly(vinyl butyral)				
(Butvar B76, Shawinigan)	18.60	4.36	13.03	23.12
Poly(vinyl chloride)				
(Vipla KR $K = 50$, Montecatini)	18.23	7.53	8.35	21.42
Poly(vinyl chloride)	18.72	10.03	3.07	21.46
Poly(vinyl chloride)	18.82	10.03	3.07	21.54
Saturated polyester				
(Desmophen 850, Bayer)	21.54	14.94	12.28	28.95
Styrene-butadiene (SBR) raw elastomer				
(Polysar 5630, Polymer Corp.)	17.55	3.36	2.70	18.07
Terpene resin				
(Piccolyte S-1000, Penn. Ind. Chem.)	16.47	0.37	2.84	16.72
Urea-formaldehyde resin				
(Plastopal H, BASF)	20.81	8.29	12.71	25.74
Vinylidene cyanide/4-acetoxy,α-acetoxy styrene copolymer	21.48	11.25	7.16	21.89
Vinylidene cyanide/4-chloro-styrene copolymer	16.98	12.07	8.18	22.38
(Rohm and Haas)	18.64	10.52	7.51	22.69
Poly(styrene)				

TABLE 2.97 Refractive Indices of Polymers

Polymer name	Refractive index (20°C, 68°F)	Polymer name	Refractive index (20°C, 68°F)
Acetal homopolymer	1.48	Polyethylene (medium density)	1.52
Acrylics	1.49–1.52	Polyethylene (high density)	1.54
Ally diglycol carbonate	1.50	Polyethylene dimethacrylate	1.51
Cellulose acetate	1.46–1.50	Poly(ethylene terephthalate)	1.57–1.58
Cellulose acetate butyrate	1.46–1.49	Poly(methyl-α-chloroacrylate)	1.52
Cellulose ester	1.47–1.50	Poly(methyl methacrylate)	1.49
Cellulose nitrate	1.49–1.51	Polypropylene	1.49
Cellulose propionate	1.46–1.49	Poly(propyl methacrylate)	1.48
Chlorotrifluoroethylene (CTFE)	1.42	Polystyrene	1.57–1.60
Diallyl isophthalate	1.57	Polysulfone	1.63
Epoxies	1.55–1.65	Poly(tetrafluoroethylene) (PTFE)	1.35
Ethyl cellulose	1.47	Poly(trifluorochloroethylene)	1.43
Fluorinated ethylene-propylene	1.34	Poly(trifluoroethylene)	1.35–1.37
Methylpentene polymer	1.485	Poly(vinyl alcohol)	1.49–1.53
Nylon	1.52–1.53	Poly(vinyl acetal)	1.48
Phenol formaldehyde	1.50–1.70	Poly(vinyl acetate)	1.46–1.47
Phenoxy polymer	1.60	Poly(vinyl butyral)	1.49
Polyacetal	1.48	Poly(vinyl chloride)	1.52–1.55
Polyallomer	1.49	Poly(vinyl cyclohexene dioxide)	1.53
Polyallyl methacrylate	1.52	Poly(vinyl formal)	1.60
Polyamide nylon 6/6	1.53	Poly(vinyl naphthalene)	1.68
Polyamide nylon 11	1.52	Poly(vinylidene chloride)	1.60–1.63
Polybutylene	1.50	Poly(vinylidene fluoride)	1.42
Polycarbornate	1.57–1.59	Silicone polymer	1.43
Poly(cyclohexyl methacrylate)	1.51	Styrene acrylonitrile copolymer	1.56–1.57
Poly(diallyl phthalate)	1.57	Styrene butadiene thermoplastic	1.52–1.55
Polyester	1.53–1.58	Styrene methacrylate copolymer	1.53
Poly(ester-styrene)	1.54–1.57	Urea formaldehyde	1.54–1.58
Polyethylene (low density)	1.51	Urethane	1.50–1.60

TABLE 2.98 Chemical Identity of Selected Polychlorinated Biphenyl Derivatives (Aroclor Compounds)

Characteristic	Aroclor 1016	Aroclor 1221	Aroclor 1232	Aroclor 1242	Aroclor 1248
Synonym(s)	PCB-1016; Polychlorinated biphenyl mixture with 41.5% chlorine	PCB-1221; Polychlorinated biphenyl mixture with 21% chlorine	PCB-1232; Polychlorinated biphenyl mixture with 32% chlorine	PCB-1242; Polychlorinated biphenyl mixture with 41.5% chlorine	PCB-1248; Polychlorinated biphenyl mixture with 48% chlorine
Registered trade name(s)	Aroclor	Aroclor	Aroclor	Aroclor	Aroclor

Characteristic	Aroclor 1254	Aroclor 1260	Aroclor 1262	Aroclor 1268
Synonym(s)	PCB-1254; Polychlorinated biphenyl mixture with 54% chlorine	PCB-1260; Polychlorinated biphenyl mixture with 60% chlorine	PCB-1262; Polychlorinated biphenyl mixture with 61.5–62.5% chlorine	PCB-1268; Polychlorinated biphenyl mixture with 68% chlorine
Registered trade name(s)	Aroclor	Aroclor	Aroclor	Aroclor

TABLE 2.99   Physical and Chemical Properties of Aroclor Derivatives

Property	Aroclor 1016	Aroclor 1221	Aroclor 1232	Aroclor 1242
Molecular weight	257.9	200.7	232.2	266.5
Color	Clear	Clear	Clear	Clear
Physical state	Oil	Oil	Oil	Oil
Melting point, °C	No data	1[d]	No data	No data
Boiling point, °C	325–356	275–320	290–325	325–366
Density, g/cm^3 at 25°C	1.37	1.18	1.26	1.38
Odor	No data	No data	No data	Mild hydrocarbon
Odor threshold:				
Water	No data	No data	No data	No data
Air	No data	No data	No data	No data
Solubility:				
Water, mg/L	0.42 (25°C)	0.59 (24°C)	0.45 (25°C)	0.24, 0.34 (25°C) 0.10 (24°C)
Organic solvent(s)	Very soluble	Very soluble	Very soluble	Very soluble
Partition coefficients:				
Log K_{ow}	5.6	4.7	5.1	5.6
Log K_{oc}	No data	No data	No data	No data
Vapor pressure, mm Hg at 25°C	4×10^{-4}	6.7×10^{-3}	4.06×10^{-3}	4.06×10^{-4}
Henry's law constant, atm-m^3/mol at 25°C	2.9×10^{-4}	3.5×10^{-3}	No data	5.2×10^{-4}
Autoignition temperature	No data	No data	No data	No data
Flashpoint °C (Cleveland open cup)	170	141–150	152–154	176–180
Flammability limits, °C	None to boiling point	176	328	None to boiling point
Conversion factors, Air (25°C)	1 mg/m^3 = 0.095 ppm	1 mg/m^3 = 0.12 ppm	1 mg/m^3 = 0.105 ppm	1 mg/m^3 = 0.092 ppm
Explosive limits	No data	No data	No data	No data

Property	Aroclor 1254	Aroclor 1260	Aroclor 1262	Aroclor 1268
Molecular weight	328	357.7	389	453
Color	Light yellow	Light yellow	No data	Clear
Physical state	Viscous liquid	Sticky resin	No data	Viscous liquid
Melting point	No data	No data	No data	No data
Boiling point, °C	365–390	385–420	390–425	435–450
Density, g/cm^3 at 25°C	1.54	1.62	1.64	1.81
Odor	Mild hydrocarbon	No data	No data	No data
Odor threshold:				
Water	No data	No data	No data	No data
Air	No data	No data	No data	No data
Solubility				
Water, mg/L	10.012; 0.057 (24°C)	0.0027; 0.08 (24°C)	0.052 (24°C)	0.300 (24°C)
Organic solvent(s)	Very soluble	Very soluble	No data	Soluble
Partition coefficients:				
Log K_{ow}	6.5	6.8	No data	No data
Log K_{oc}	No data	No data	No data	No data
Vapor pressure, mm Hg at 25°C	7.71×10^{-5}	4.05×10^{-5}	No data	No data
Henry's law constant, atm-m^3/mol at 25°C	2.0×10^{-3}	4.6×10^{-3}	No data	No data
Autoignition temperature	No data	No data	No data	No data
Flashpoint °C (Cleveland open cup)	No data	No data	195°C	195°C
Flammability limits, °C	None to boiling point	None to boiling point	None to boiling point	None to boiling point
Conversion factors, Air (25°C)	1 mg/m^3 = 0.075 ppm	1 mg/m^3 = 0.065 ppm	1 mg/m^3 = 0.061 ppm	1 mg/m^3 = 0.052 ppm
Explosive limits	No data	No data	No data	No data

SECTION 3

NATURALLY OCCURRING CHEMICALS AND CHEMICAL SOURCES

3.1 COAL

Coal is a fossil fuel formed in swamp ecosystems where plant remains were saved by water and mud from oxidization and biodegradation. Coal is a combustible organic sedimentary rock (composed primarily of carbon, hydrogen, and oxygen) formed from ancient vegetation and consolidated between other rock strata to form coal seams. The harder forms, such as anthracite coal, can be regarded as organic metamorphic rocks because of a higher degree of maturation.

Coal occurs in different forms or *types*. Variations in the nature of the source material and local or regional variations in the coalification processes cause the vegetal matter to evolve differently. Thus, various classification systems exist to define the different types of coal; the most commonly used system in North American is the system developed by ASTM International (ASTM D388). Thus, as the geological processes increased their effect over time, the coal precursors were transformed into the following:

1. Lignite—also referred to as *brown coal*—is the lowest rank of coal and used almost exclusively as fuel for steam-electric power generation. Jet is a compact form of lignite that is sometimes polished and has been used as an ornamental stone since the Iron Age.

2. Sub-bituminous coal—whose properties range from those of lignite to those of bituminous coal—is used primarily as fuel for steam-electric power generation.

3. Bituminous coal—a dense coal, usually black, sometimes dark brown, often with well-defined bands of bright and dull material—is used primarily as fuel in steam-electric power generation, with substantial quantities also used for heat and power applications in manufacturing and to make coke.

4. Anthracite—the highest rank; a harder, glossy, black coal—is used primarily for residential and commercial space heating.

The *rank* of a coal indicates the progressive changes in carbon, volatile matter, and probably ash and sulfur that take place as coalification progresses from the lower-rank lignite through the higher ranks of sub-bituminous, high-volatile bituminous, low-volatile bituminous, and anthracite. The rank of a coal should not be confused with its grade. A high rank (e.g., anthracite) represents coal from a deposit that has undergone the greatest degree of metamorphosis and contains very little mineral matter, ash, and moisture. On the other hand, any rank of coal when cleaned of impurities through coal preparation will be of a higher grade. Other designations, such as coking coal and steam coal, have been applied to coals, but they tend to differ from country to country.

TABLE 3.1   Differentiation of Coal Rank, Coal Type, and Coal Grade

Rank
• Indicative of the degree of metamorphism (or coalification) to which the original mass of plant debris (peat) has been subjected during its burial history.   • Dependent on the maximum temperature to which the *proto-coal* has been exposed and the time it has been held at that temperature.   • Also reflects the depth of burial and the geothermal gradient prevailing at the time of coalification in the basin concerned.

Type
• Indicative of the nature of the plant debris (*proto-coal*) from which the coal was derived, including the mixture of plant components (wood, leaves, and algae) involved and the degree of degradation before burial.   • The individual plant components occurring in coal, and in some cases fragments or other materials derived from them, are referred to as *macerals*.   • The kind and distribution of the various macerals are the starting point for most coal petrology studies.

Grade
• Indicative of the extent to which the accumulation of plant debris has been kept free of contamination by inorganic material (mineral matter), before burial (i.e., during peat accumulation), after burial, and during coalification.   • A high-grade coal is coal, regardless of its rank or type, with a low overall content of mineral matter.

TABLE 3.2   General Description of Different Coal Types and Peat

Rank (excluding peat)	Properties
Peat	A mass of recently accumulated to partially carbonized plant debris and is not classed as coal; a carbon content of less than 60% on a dry ash-free basis.
Lignite	Lignite is the lowest rank of coal and sometimes contains recognizable plant structures; a heating value of less than 8300 Btu/lb on a mineral matter-free basis. It has a carbon content between 60 and 70% on a dry ash-free basis; known as *brown coal* in Europe, Australia, and the United Kingdom.
Sub-bituminous	Carbon content between 71 and 77% (dry ash-free basis) and a heating value between 8300 and 13,000 Btu/lb on a mineral matter free basis; subdivided (on the basis of heating value) into sub-bituminous A, sub-bituminous B, and sub-bituminous C ranks.
Bituminous	Carbon content between 77 and 87% on a dry ash-free basis and a heating value that is much higher than lignite or sub-bituminous coal; on the basis of volatile matter content, bituminous coals are subdivided into low-volatile bituminous, medium-volatile bituminous, and high-volatile bituminous; often referred to as *soft coal* (layman's term) with little to do with the hardness of the coal.
Anthracite	Highest rank of coal and has a carbon content of over 87% on a dry ash-free basis; often subdivided into semianthracite, anthracite, and meta-anthracite on the basis of carbon content; often referred to as *hard coal* (layperson's term) with little to do with the hardness of the coal.

TABLE 3.3   Typical Properties of Coal

Sulfur content in coal
- Anthracite: 0.6–0.77% w/w
- Bituminous coal: 0.7–3.0% w/w
- Lignite: 0.4% w/w

Moisture content
- Anthracite: 2.8–16.3% w/w
- Bituminous coal: 2.2–15.9% w/w
- Lignite: 39% w/w

Fixed carbon
- Anthracite: 80.5–85.7% w/w
- Bituminous coal: 43.9–78.2% w/w
- Lignite: 31.4% w/w

Bulk density
- Anthracite: 50–58 (lb/ft^3), 800–929 (kg/m^3)
- Bituminous coal: 42–57 (lb/ft^3), 673–913 (kg/m^3)
- Lignite: 40–54 (lb/ft^3), 641–865 (kg/m^3)

Mineral matter content (as mineral ash)
- Anthracite: 9.7–20.2% w/w
- Bituminous coal: 3.3–11.7% w/w
- Lignite: 3.2% w/w

TABLE 3.4   Typical Analysis for Various Coal Types

Sulfur content
- Anthracite coal: 0.6–0.77% w/w
- Bituminous coal: 0.7–3.0% w/w
- Lignite coal: 0.4% w/w

Moisture content
- Anthracite coal: 2.8–16.3% w/w
- Bituminous coal: 2.2–15.9% w/w
- Lignite coal: 39% w/w

Fixed carbon
- Anthracite coal: 80.5–85.7% w/w
- Bituminous coal: 43.9–78.2% w/w
- Lignite coal: 31.4% w/w

Bulk density
- Anthracite coal: 50–58 (lb/ft^3), 800–929 (kg/m^3)
- Bituminous coal: 42–57 (lb/ft^3), 673–913 (kg/m^3)
- Lignite coal: 40–54 (lb/ft^3), 641–865 (kg/m^3)

Ash production
- Anthracite coal: 9.7–20.2% w/w
- Bituminous coal: 3.3–11.7% w/w
- Lignite coal: 3.2% w/w

TABLE 3.5   Classification According to Rank (ASTM System*)

Class/groups	Fixed carbon limits (dry mineral-matter–free basis), %		Volatile matter limits (dry mineral-matter–free basis), %		Gross calorific value limits (moist,[†] mineral-matter–free basis), MJ/kg		Agglomerating character
	Equal or greater than	Less than	Greater than	Equal or less than	Equal or greater than	Less than	
Anthracitic:							
Meta-anthracite	98			2			
Anthracite	92	98	2	8			Agglomerating
Semianthracite[‡]	86	92	8	14			
Bituminous:							
Low-volatile bituminous coal	78	86	14	22			
Medium-volatile bituminous	69	78	22	31			Commonly
High-volatile A bituminous coal		69	31		32.6[§]		agglomerating[¶]
High-volatile B bituminous coal					32.6[§]	32.6	
High-volatile C bituminous coal					26.7	30.2	
					24.4	26.7	Agglomerating
Subbituminous:							
Subbituminous A coal					24.4	26.7	
Subbituminous B coal					22.1	24.4	
Subbituminous C coal					19.3	22.1	Non agglomerating
Lignitic:							
Lignite A					14.7	19.3	
Lignite B						14.7	

*This classification does not apply to certain coals.

[†]Moist refers to coal containing its natural inherent moisture but not including visible water on the surface of the coal.

[‡]If agglomerating classify in low-volatile group of bituminous class.

[§]Coals having 69% or more fixed carbon on the dry, mineral-matter-free basis shall be classified according to fixed carbon, regardless of gross calorific value.

[¶]It is recognized that there may be nonagglomerating varieties in these groups of the bituminous class, and that there are notable exceptions in the high-volatile C bituminous group.

TABLE 3.6   Typical ASTM Test Methods Used for the Analysis of Coal

ASTM D189. Test Method for Conradson Carbon Residue of Petroleum Products. Annual Book of Standards. ASTM International, West Conshohocken, Pennsylvania.

ASTM D197. Method of Sampling and Fineness Test of Pulverized Coal. Annual Book of Standards. ASTM International, West Conshohocken, Pennsylvania.

ASTM D346. Practice for Collection and Preparation of Coke Samples for Laboratory Analysis. Annual Book of Standards. ASTM International, West Conshohocken, Pennsylvania.

ASTM D388. Classification of Coals by Rank. Annual Book of Standards. ASTM International, West Conshohocken, Pennsylvania.

ASTM D1412. Test Method for Equilibrium Moisture of Coal at 96 to 97 Percent Relative Humidity and 30°C. Annual Book of Standards. ASTM International, American Society for Testing and Materials, West Conshohocken, Pennsylvania.

TABLE 3.6 Typical ASTM Test Methods Used for the Analysis of Coal (*Continued*)

ASTM D1857. Test Method for Fusibility of Coal and Coke Ash. Annual Book of Standards. ASTM International, West Conshohocken, Pennsylvania.

ASTM D2013. Method for Preparing Coal Samples for Analysis. Annual Book of Standards. ASTM International, West Conshohocken, Pennsylvania.

ASTM D2015. Test Method for Gross Calorific Value of Coal and Coke by the Adiabatic Bomb Calorimeter. Annual Book of Standards. ASTM International, West Conshohocken, Pennsylvania.

ASTM D2233. Method for Collection of a Gross Sample of Coal. Annual Book of Standards. ASTM International, West Conshohocken, Pennsylvania.

ASTM D2361. Test Method for Chlorine in Coal. Annual Book of ASTM Standards. ASTM International, West Conshohocken, Pennsylvania.

ASTM D2492. Test Method for Forms of Sulfur in Coal. Annual Book of Standards. ASTM International, West Conshohocken, Pennsylvania.

ASTM D2795. Method for Analysis of Coal and Coke Ash. Annual Book of Standards. ASTM International, West Conshohocken, Pennsylvania.

ASTM D3172. Practice for Proximate Analysis of Coal and Coke. Annual Book of Standards. ASTM International, Conshohocken, Pennsylvania.

ASTM D3173. Test Method for Moisture in the Analysis Sample of Coal and Coke. Annual Book of Standards. ASTM International, West Conshohocken, Pennsylvania.

ASTM D3173. Test Method for Ash in the Analysis Sample of Coal and Coke from Coal. Annual Book of Standards. ASTM International, West Conshohocken, Pennsylvania.

ASTM D3175. Test Method for Volatile Matter in the Analysis Sample of Coal and Coke. Annual Book of Standards. ASTM International, West Conshohocken, Pennsylvania.

ASTM D3176. Practice for Ultimate Analysis of Coal and Coke. Annual Book of Standards. ASTM International, West Conshohocken, Pennsylvania.

ASTM D3177. Test Method for Total Sulfur in the analysis Sample of Coal and Coke. Annual Book of Standards. ASTM International, West Conshohocken, Pennsylvania.

ASTM D3178. Test Method for Carbon and Hydrogen in the Analysis Sample of Coal and Coke. Annual Book of Standards. ASTM International, West Conshohocken, Pennsylvania.

ASTM D3179. Test Method for Nitrogen in the Analysis Sample of Coal and Coke. Annual Book of ASTM Standards. ASTM International, West Conshohocken, Pennsylvania.

ASTM D3180. Practice for Calculating Coal and Coke Analyses from As-Determined to Different Bases. Annual Book of Standards. ASTM International, West Conshohocken, Pennsylvania.

ASTM D3286. Test Method for Gross Calorific Value by the Isoperibol Bomb Calorimeter. Annual Book of Standards. ASTM International, West Conshohocken, Pennsylvania.

ASTM D3302. Test Method for Total Moisture in Coal. Annual Book of Standards. ASTM International, West Conshohocken, Pennsylvania.

ASTM D3682. Test Method for Major and Minor Elements in Coal and Coke Ash by the Atomic Absorption Method. Annual Book of Standards. ASTM International, West Conshohocken, Pennsylvania.

ASTM D3683. Test Method for Trace Elements in Coal and Coke Ash by the Atomic Absorption Method. Annual Book of Standards. ASTM International, West Conshohocken, Pennsylvania.

ASTM D3683. Test Method for Mercury in Coal by the Oxygen Bomb Combustion/Atomic Absorption Method. Annual Book of Standards. West Conshohocken, Pennsylvania.

TABLE 3.7 Minerals Commonly Associated with Coal

Group	Species	Formula
Shale	Muscovite	$(K, Na, H_2O, Ca)_2(Al, Mg, Fe, Ti)_4$
	Hydromuscovite	$(Al, Si)_8O_{20}(OH, F)_4$ (general formula)
	Illite	$(HO)_4K_2(Si_6 \cdot Al_2)Al_4O_{20}$
	Montmorillonite	$Na_2(Al\ Mg)Si_4O_{10}(OH)_2$
Kaolin	Kaolinite	$Al_2(Si_2O_5)(OH)_4$
	Livesite	$Al_2(Si_2O_5)(OH)_4$
	Metahalloysite	$Al_2(Si_2O_5)(OH)_4$
Sulfide	Pyrite	FeS_2
	Marcasite	FeS_2
Carbonate	Ankerite	$CaCO_3 \cdot (Mg, Fe, Mn)CO_3$
	Calcite	$CaCO_3$
	Dolomite	$CaCO_3 \cdot MgCO_3$
	Siderite	$FeCO_3$
Chloride	Sylvite	KCl
	Halite	$NaCl$
Accessory minerals	Quartz	SiO_2
	Feldspar	$(K, Na)_2O \cdot Al_2O_3 \cdot 6\ SiO_2$
	Garnet	$3\ CaO \cdot Al_2O_3 \cdot 3\ SiO_2$
	Hornblende	$CaO \cdot 3\ FeO \cdot 4\ SiO_2$
	Gypsum	$CaSO_4 \cdot 2\ H_2O$
	Apatite	$9\ CaO \cdot 3\ P_2O_5 \cdot CaF_2$
	Zircon	$ZrSiO_4$
	Epidote	$4\ CaO \cdot 3\ Al_2O_3 \cdot 6\ SiO_2 \cdot H_2O$
	Biotite	$K_2O \cdot MgO \cdot Al_2O_3 \cdot 3\ SiO_2 \cdot H_2O$
	Augite	$CaO \cdot MgO \cdot 2\ SiO_2$
	Prochlorite	$2\ FeO \cdot 2\ MgO \cdot Al_2O_3 \cdot 2\ SiO_2 \cdot 2\ H_2O$
	Diaspore	$Al_2O_3 \cdot H_2O$
	Lepidocrocite	$Fe_2O_3 \cdot H_2O$
	Magnetite	Fe_3O_4
	Kyanite	$Al_2O_3 \cdot SiO_2$
	Staurolite	$2\ FeO \cdot 5\ Al_2O_3 \cdot 4\ SiO_2 \cdot H_2O$
	Topaz	$2\ AlFO \cdot SiO_2$
	Tourmaline	$3\ Al_2O_3 \cdot 4\ BO(OH) \cdot 8\ SiO_2 \cdot 9\ H_2O$
	Hematite	Fe_2O_3
	Penninite	$5\ MgO \cdot Al_2O_3 \cdot 3\ SiO_2 \cdot 2\ H_2O$
	Sphalerite	ZnS
	Chlorite	$10(Mg, Fe)O \cdot 2\ Al_2O_3 \cdot 6\ SiO_2 \cdot 8\ H_2O$
	Barite	$BaSO_4$
	Pyrophillite	$Al_2O_3 \cdot 4\ SiO_2 \cdot H_2O$

TABLE 3.8 Proximate Analysis and Ultimate Analysis of Solid Fuels (Dry, Ash-Free)

	Fuel type				
	Wood	Peat	Lignite	Bituminous coal	Refuse-derived fuel
Proximate analysis (wt%)					
Volatile matter	81	65	55	40	85
Fixed carbon	19	35	45	60	15
Ultimate analysis (wt%)					
Hydrogen	6	6	5	5	7
Carbon	50	55	68	78	52
Sulfur	0.1	0.4	1	2	0.3
Nitrogen	0.1	1	1	2	0.6
Oxygen	44	38	25	13	40
Higher heating value (Btu lib)	8700	9500	10,000	14,000	9700

TABLE 3.9 Chemical Products from Coal

Feedstock	Primary products	Secondary products	Tertiary products
Coal	Gas	Fuel gas	
		Sulfur	
	Light oil	Naphtha	Cyclopentadiene
			Indene
		Benzol	Benzene
		Toluols	Toluene
		Xylols	o-Xylene
			m-Xylene
			p-Xylene
	Coal tar	Pitch	
		Tar	Fluorene
			Diphenylene oxide
			Acenaphthene
			Methyl-naphthalene
		Creosote	
		Tar acids	Phenol
			o-Cresol
			m-Cresol
			p-Cresol
			3,5-Xylenol
		Tar bases	Pyridine
			Alpha-picoline
			Beta-picoline
			Gamma-picoline
			2,6-Lutidine
			Quinoline
		Naphthalene	
		Crude anthracene	Anthracene
			Phenanthrene
			Carbazole
	Coke		Carbon

TABLE 3.10 Chemicals from Synthesis Gas

Feedstock	Reaction type	Product
Synthesis gas (CO + H$_2$)	Oxo reaction	Oxo products
	Shift reaction	Hydrogen
	Shift reaction	Methyl alcohol
	Shift reaction	Ammonia
	Shift reaction/methanation	Substitute natural gas
	Organic synthesis	Hydroquinone
	Homologation	Ethyl alcohol
	Carbonylation	Acetic acid
	Fischer-Tropsch	Ethylene
	Fischer-Tropsch	Paraffins
	Glycol synthesis	Ethylene glycol

3.1.1 Bibliography

ASTM D388. 2015. Standard Classification of Coals by Rank. *Annual Book of Standards*. ASTM International, West Conshohocken, Pennsylvania.

BP. 2015. Statistical Review of World Energy 2015. British Petroleum Company, London, England. June.

Speight, J. G. 2013a. *The Chemistry and Technology of Coal*, 3rd ed. CRC–Taylor and Francis Group, Boca Raton, Florida.

Speight, J. G. 2013b. *The Chemistry and Technology of Petroleum*, 5th ed. CRC–Taylor and Francis Group, Boca Raton, Florida.

3.2 CRUDE OIL AND HEAVY OIL

Crude oil, and the equivalent term *petroleum oil*, covers a wide assortment of materials consisting of mixtures of hydrocarbons and other compounds containing gaseous, liquid, and solid hydrocarbon compounds that occur in sedimentary rock deposits throughout the world and also contains small quantities of nitrogen-, oxygen-, and sulfur-containing compounds as well as trace amounts of metallic constituents. Metal-containing constituents, notably the compounds that contain vanadium and nickel, usually occur in the more viscous crude oils in amounts up to several thousand parts per million and can have serious consequences during processing of these feedstocks. Petroleum is a mixture of widely varying constituents and proportions, which cause variation in the physical properties and the color (from colorless to black).

3.2.1 Heavy Oil

Heavy oil is a *type* of petroleum that is different from conventional petroleum insofar as they are much more difficult to recover from the subsurface reservoir. Heavy oil, particularly heavy oil formed by biodegradation of organic deposits, is found in shallow reservoirs, formed by unconsolidated sands. This characteristic, which brings about difficulties during well drilling and completion operations, may become a production advantage due to higher permeability. In simple terms, heavy oil is a type of crude oil, which is very viscous and does not flow easily. The common characteristic properties (relative to conventional crude oil) are high specific gravity, low hydrogen-to-carbon ratios, high carbon residues, and high contents of asphaltenes, heavy metal, sulfur, and nitrogen.

The arbitrary definition of heavy oil is usually based on American Petroleum Institute (API) gravity or viscosity, although there have been attempts to rationalize the definition based upon viscosity, API gravity, and density.

Extra heavy oil is a nondescript term (related to viscosity) of little scientific meaning which is usually applied to tar sands bitumen, which is generally incapable of free flow under reservoir conditions. The general difference is that extra heavy oil may have properties similar to tar sands bitumen, but unlike tar sands bitumen, it has some degree of mobility in the reservoir or deposit. Extra heavy oils can flow at reservoir temperature and can be produced economically, without additional viscosity-reduction techniques, through variants of conventional processes such as long horizontal wells, or multilaterals. This is the case, for instance, in the Orinoco (Venezuela) or in offshore reservoirs of the coast of Brazil, but, once outside of the influence of the high reservoir temperature, these oils are too viscous at surface to be transported through conventional pipelines and require heated pipelines for transportation. Alternatively, the oil must be partially upgraded or fully upgraded or diluted with a light hydrocarbon (such as aromatic naphtha) to create a mix that is suitable for transportation (Speight, 2014a).

TABLE 3.11   Generic Boiling Fractions of Crude Oil

Fraction	Boiling range*	
	°C	°F
Light naphtha	−1 to 150	30–300
Gasoline	−1 to 180	30–355
Heavy naphtha	150–205	300–400
Kerosene	205–260	400–500
Light gas oil	260–315	400–600
Heavy gas oil	315–425	600–800
Lubricating oil	>400	>750
Vacuum gas oil	425–600	800–1100
Residuum	>510	>950

*For convenience, boiling ranges are converted to the nearest 5°.

TABLE 3.12   Production and Uses of Naphtha

Process	Primary product	Secondary process	Secondary product
Atmospheric distillation	Light naphtha	Cracking	Petrochemical Feedstocks
	Heavy naphtha	Catalytic cracking	Light naphtha
	Gas oil	Catalytic cracking	Light naphtha
	Gas oil	Hydrocracking	Light naphtha
Vacuum distillation	Gas oil	Catalytic cracking	Light naphtha
		Hydrocracking	Light naphtha
	Residuum	Coking	Light naphtha
		Hydrocracking	Light naphtha

TABLE 3.13 Component Streams for Gasoline

Stream	Producing process	Boiling range °C	Boiling range °F
Paraffinic:			
Butane	Distillation Conversion	0	32
Iso-pentane	Distillation Conversion Isomerization	27	81
Alkylate	Alkylation	40–150	105–300
Isomerate	Isomerization	40–70	105–160
Naphtha	Distillation	30–100	85–212
Hydrocrackate	Hydrocracking	40–200	105–390
Olefinic:			
Catalytic naphtha	Catalytic cracking	40–200	105–390
Cracked naphtha	Steam cracking	40–200	105–390
Polymer	Polymerization	60–200	140–390
Aromatic:			
Catalytic reformate	Catalytic reforming	40–200	105–390

TABLE 3.14 API Gravity and Sulfur Content of Selected Crude Oils

Country	Crude oil	API	Sulfur % w/w
Abu Dhabi (U.A.E.)	Abu Al Bu Khoosh	31.6	2.00
Abu Dhabi (U.A.E.)	Abu Mubarras	38.1	0.93
Abu Dhabi (U.A.E.)	El Bunduq	38.5	1.12
Abu Dhabi (U.A.E.)	Murban	40.5	0.78
Abu Dhabi (U.A.E.)	Umm Shaif	37.4	1.51
Abu Dhabi (U.A.E.)	Zakum (Lower)	40.6	1.05
Abu Dhabi (U.A.E.)	Zakum (Upper)	33.1	2.00
Algeria	Zarzaitine	43.0	0.07
Angola	Cabinda	31.7	0.17
Angola	Palanca	40.1	0.11
Angola	Takula	32.4	0.09
Australia	Airlie	44.7	0.01
Australia	Barrow Island	37.3	0.05
Australia	Challis	39.5	0.070
Australia	Cooper Basin	45.2	0.02
Australia	Gippsland	47.0	0.09
Australia	Griffin	55.0	0.03
Australia	Harriet	37.9	0.05
Australia	Jabiru	42.3	0.05
Australia	Jackson	43.8	0.03
Australia	Saladin	48.2	0.02
Australia	Skua	41.9	0.06
Brazil	Garoupa	30.0	0.68
Brazil	Sergipano Platforma	38.4	0.19
Brazil	Sergipano Terra	24.1	0.41

TABLE 3.14 API Gravity and Sulfur Content of Selected Crude Oils (*Continued*)

Country	Crude oil	API	Sulfur % w/w
Brunei	Champion Export	23.9	0.12
Brunei	Seria	40.5	0.06
Cameroon	Kole Marine	32.6	0.33
Cameroon	Lokele	20.7	0.46
Canada (Alberta)	Bow River Heavy	26.7	2.10
Canada (Alberta)	Pembina	38.8	0.20
Canada (Alberta)	Rainbow	40.7	0.50
Canada (Alberta)	Rangeland South	39.5	0.75
Canada (Alberta)	Wainwright-Kinsella	23.1	2.58
China	Daqing (Taching)	32.6	0.09
China	Nanhai Light	40.6	0.06
China	Shengli	23.2	1.00
China	Weizhou	39.7	0.08
Colombia	Cano Limon	29.3	0.51
Congo (Brazzaville)	Emeraude	23.6	0.600
Dubai (U.A.E.)	Fateh	31.1	2.000
Dubai (U.A.E.)	Margham Light	50.3	0.040
Ecuador	Oriente	29.2	0.880
Egypt	Belayim	27.5	2.200
Egypt	Gulf of Suez	31.9	1.520
Egypt	Ras Gharib	21.5	3.640
Gabon	Gamba	31.4	0.090
Gabon	Rabi-Kounga	33.5	0.070
Ghana	Salt Pond	37.4	0.097
India	Bombay High	39.2	0.150
Indonesia	Anoa	45.2	0.040
Indonesia	Ardjuna	35.2	0.105
Indonesia	Attaka	43.3	0.040
Indonesia	Badak	49.5	0.032
Indonesia	Bekapai	41.2	0.080
Indonesia	Belida	45.1	0.020
Indonesia	Bima	21.1	0.250
Indonesia	Cinta	33.4	0.080
Indonesia	Duri (Sumatran Heavy)	21.3	0.180
Indonesia	Ikan Pari	48.0	0.020
Indonesia	Kakap	51.5	0.050
Indonesia	Katapa	50.8	0.060
Indonesia	Lalang (Malacca Straits)	39.7	0.050
Indonesia	Minas (Sumatran Light)	34.5	0.081
Indonesia	Udang	38.0	0.050
Iran	Aboozar (Ardeshir)	26.9	2.480
Iran	Bahrgansar/Nowruz	27.1	2.450
Iran	Dorrood (Darius)	33.6	2.350
Iran	Foroozan (Fereidoon)	31.3	2.500
Iran	Iranian Heavy	30.9	1.730
Iran	Iranian Light	33.8	1.350
Iran	Rostam	35.9	1.550
Iran	Salmon (Sassan)	33.9	1.910
Iraq	Basrah Heavy	24.7	3.500
Iraq	Basrah Light	33.7	1.950
Iraq	Basrah Medium	31.1	2.580
Iraq	North Rumaila	33.7	1.980

(*Continued*)

TABLE 3.14 API Gravity and Sulfur Content of Selected Crude Oils (*Continued*)

Country	Crude oil	API	Sulfur % w/w
Ivory Coast	Espoir	32.3	0.340
Kazakhstan	Kumkol	42.5	0.07
Kuwait	Kuwait Export	31.4	2.52
Libya	Amna	36.0	0.15
Libya	Brega	40.4	0.21
Libya	Bu Attifel	43.3	0.04
Libya	Buri	26.2	1.76
Libya	Es Sider	37.0	0.45
Libya	Sarir	38.4	0.16
Libya	Sirtica	41.3	0.45
Libya	Zueitina	41.3	0.28
Malaysia	Bintulu	28.1	0.08
Malaysia	Dulang	39.0	0.12
Malaysia	Labuan	32.2	0.07
Malaysia	Miri Light	32.6	0.04
Malaysia	Tembungo	37.4	0.04
Mexico	Isthmus	33.3	1.49
Mexico	Maya	22.2	3.30
Mexico	Olmeca	39.8	0.80
Neutral Zone	Burgan	23.3	3.37
Neutral Zone	Eocene	18.6	4.55
Neutral Zone	Hout	32.8	1.91
Neutral Zone	Khafji	28.5	2.85
Neutral Zone	Ratawi	23.5	4.07
Nigeria	Antan	32.1	0.32
Nigeria	Bonny Light	33.9	0.14
Nigeria	Bonny Medium	25.2	0.23
Nigeria	Brass River	42.8	0.06
Nigeria	Escravos	36.4	0.12
Nigeria	Forcados	29.6	0.18
Nigeria	Pennington	36.6	0.07
Nigeria	Qua Iboe	35.8	0.12
North Sea (Denmark)	Danish North Sea	34.5	0.260
North Sea (Norway)	Ekofisk	39.2	0.169
North Sea (Norway)	Emerald	22.0	0.750
North Sea (Norway)	Oseberg	33.7	0.310
North Sea (U.K.)	Alba	20.0	1.330
North Sea (U.K.)	Duncan	38.5	0.180
North Sea (U.K.)	Forties Blend	40.5	0.350
North Sea (U.K.)	Innes	45.7	0.130
North Sea (U.K.)	Kittiwake	37.0	0.65
North Yemen	Alif	40.3	0.10
Oman	Oman Export	34.7	0.94
Papua New Guinea	Kubutu	44.0	0.04
Peru	Loreto Peruvian	33.1	0.23
Qatar	Dukhan (Qatar Land)	40.9	1.27
Qatar	Qatar Marine	36.0	1.42
Ras Al Khaiman (U.A.E.)	Ras Al Khaiman	44.3	0.15
Russia	Siberian Light	37.8	0.42
Saudi Arabia	Arab Extra Light (Berri)	37.2	1.15
Saudi Arabia	Arab Heavy (Safaniya)	27.4	2.80

TABLE 3.14 API Gravity and Sulfur Content of Selected Crude Oils (*Continued*)

Country	Crude oil	API	Sulfur % w/w
Saudi Arabia	Arab Light	33.4	1.77
Saudi Arabia	Arab Medium (Zuluf)	28.8	2.49
Sharjah (U.A.E.)	Mubarek	37.0	0.62
Sumatra	Duri	20.3	0.21
Syria	Souedie	24.9	3.82
Timor Sea (Indonesia)	Hydra	37.5	0.08
Trinidad Tobago	Galeota Mix	32.8	0.27
Tunisia	Ashtart	30.0	0.99
U.S.A. (Alaska)	Alaskan North Slope	27.5	1.11
U.S.A. (Alaska)	Cook Inlet	35.0	0.10
U.S.A. (Alaska)	Drift River	35.3	0.09
U.S.A. (Alaska)	Nikiski Terminal	34.6	0.10
U.S.A. (California)	Hondo Sandstone	35.2	0.21
U.S.A. (California)	Huntington Beach	20.7	1.38
U.S.A. (Florida)	Sunniland	24.9	3.25
U.S.A. (Louisiana)	Grand Isle	33.2	0.35
U.S.A. (Louisiana)	Lake Arthur	41.9	0.06
U.S.A. (Louisiana)	Louisiana Light Sweet	36.1	0.45
U.S.A. (Louisiana)	Ostrica	32.0	0.30
U.S.A. (Louisiana)	South Louisiana	32.8	0.28
U.S.A. (Michigan)	Lakehead Sweet	47.0	0.31
U.S.A. (New Mexico)	New Mexico Intermediate	37.6	0.17
U.S.A. (New Mexico)	New Mexico Light	43.3	0.07
U.S.A. (Oklahoma)	Basin-Cushing Composite	34.0	1.95
U.S.A. (Texas)	Coastal B-2	32.2	0.22
U.S.A. (Texas)	East Texas	37.0	0.21
U.S.A. (Texas)	Sea Breeze	37.9	0.10
U.S.A. (Texas)	West Texas Intermediate	40.8	0.34
U.S.A. (Texas)	West Texas Semi-Sweet	39.0	0.27
U.S.A. (Texas)	West Texas Sour	34.1	1.640
U.S.A. (Wyoming)	Tom Brown	38.2	0.100
U.S.A. (Wyoming)	Wyoming Sweet	37.2	0.330
Venezeula	Lago Medio	32.2	1.010
Venezeula	Leona	24.4	1.510
Venezeula	Mesa	29.8	1.010
Venezuela	Ceuta Export	27.8	1.370
Venezuela	Guanipa	30.3	0.850
Venezuela	La Rosa Medium	25.3	1.730
Venezuela	Lago Treco	26.7	1.500
Venezuela	Oficina	33.3	0.780
Venezuela	Temblador	21.0	0.830
Venezuela	Tia Juana	25.8	1.630
Venezuela	Tia Juana Light	31.8	1.160
Venezuela	Tia Juana Medium 24	24.8	1.610
Venezuela	Tia Juana Medium 26	26.9	1.540
Viet Nam	Bach Ho (White Tiger)	38.6	0.030
Viet Nam	Dai Hung (Big Bear)	36.9	0.080
Yemen	Masila	30.5	0.670
Zaire	Zaire	31.7	0.130

TABLE 3.15 API Gravity and Sulfur Content of Selected Heavy Oils and Bitumen

Country	Crude oil	API	Sulfur % w/w
Brazil	Albacor Leste	18.9	0.66
Canada (Alberta)	Athabasca	8.0	4.8
Canada (Alberta)	Cold Lake	13.2	4.11
Canada (Alberta)	Lloydminster	16.0	2.60
Canada (Alberta)	Wabasca	19.6	3.90
Chad	Bolobo	16.8	0.14
Chad	Kome	18.5	0.20
China	Bozhong	16.7	0.30
China	Qinhuangdao	16.00	0.26
China	Zhao Dong	18.4	0.25
Colombia	Castilla	13.3	0.22
Colombia	Chichimene	19.8	1.12
Congo	Yombo	17.7	0.33
Ecuador	Ecuador Heavy	18.2	2.23
Ecuador	Napo	19.2	1.98
Guatemala	Xan-Coban	18.7	6.00
Indonesia	Kulin (South)	19.8	0.30
Iran	Soroosh (Cyrus)	18.1	3.30
Kuwait	Eocene	18.4	4.00
U.S.A. (California)	Arroyo Grande/Edna	14.9	2.03
U.S.A. (California)	Belridge (South)	13.7	1.00
U.S.A. (California)	Beta Offshore	15.9	3.60
U.S.A. (California)	Beta Offshore	16.9	3.30
U.S.A. (California)	Hondo Monterey	19.4	4.70
U.S.A. (California)	Hondo Monterey	17.2	4.70
U.S.A. (California)	Huntington Beach	19.4	2.00
U.S.A. (California)	Huntington Beach	14.4	0.90
U.S.A. (California)	Kern River	13.3	1.10
U.S.A. (California)	Kern River	14.4	1.02
U.S.A. (California)	Lost Hills	18.4	1.00
U.S.A. (California)	Midway Sunset	12.6	1.60
U.S.A. (California)	Midway Sunset	11.0	1.55
U.S.A. (California)	Monterey	12.2	2.30
U.S.A. (California)	Mount Poso	16.0	0.70
U.S.A. (California)	Newport Beach	15.1	2.00
U.S.A. (California)	Point Arguello	19.5	3.50
U.S.A. (California)	Point Pedernales	15.9	5.10
U.S.A. (California)	San Ardo	12.2	2.30
U.S.A. (California)	San Joaquin Valley	15.7	1.20
U.S.A. (California)	Santa Maria	13.7	5.20
U.S.A. (California)	Sockeye	19.6	5.30
U.S.A. (California)	Sockeye	15.9	5.40
U.S.A. (California)	Torrance	18.2	1.80
U.S.A. (California)	Wilmington	18.6	1.59
U.S.A. (California)	Wilmington	16.9	1.70
U.S.A. (Mississippi)	Baxterville	16.3	3.02
Venezuela	Bachaquero	16.3	2.35
Venezuela	Bachaquero	12.2	2.80
Venezuela	Bachaquero	14.4	2.52
Venezuela	Bachaquero Heavy	10.7	2.78
Venezuela	Boscan	10.1	5.50

TABLE 3.15 API Gravity and Sulfur Content of Selected Heavy Oils and Bitumen (*Continued*)

Country	Crude oil	API	Sulfur % w/w
Venezuela	Hamaca	8.4	3.82
Venezuela	Jobo	9.2	4.10
Venezuela	Laguna	10.9	2.66
Venezuela	Lagunillas Heavy	17.0	2.19
Venezuela	Merey	18.0	2.28
Venezuela	Morichal	12.2	2.78
Venezuela	Pilon	14.1	1.91
Venezuela	Tia Juana Pesado	12.1	2.70
Venezuela	Tia Juana Heavy	18.2	2.24
Venezuela	Tia Juana Heavy	11.6	2.68
Venezuela	Tremblador	19.0	0.80
Venezuela	Zuata	15.7	2.69

For reference, Athabasca tar sands bitumen has API = 8° and sulfur content = 4.8–5% w/w.

TABLE 3.16 Hydrocarbon and Heteroatom Types in Crude Oil

Class	Compound types
Saturated hydrocarbons	n-Paraffins iso-Paraffins and other branched paraffins Cycloparaffins (naphthenes) Condensed cycloparaffins (including steranes, hopanes) Alkyl side chains on ring systems
Unsaturated hydrocarbons	Olefins not indigenous to petroleum; present in products of thermal reactions
Aromatic hydrocarbons	Benzene systems Condensed aromatic systems Condensed aromatic-cycloalkyl systems Alkyl side chains on ring systems
Saturated heteroatomic systems	Alkyl sulfides Cycloalkyl sulfides Alkyl side chains on ring systems
Aromatic heteroatomic systems	Furans (single- and multiring systems) Thiophenes (single- and multiring systems) Pyrroles (single- and multiring systems) Pyridines (single- and multiring systems) Mixed heteroatomic systems Amphoteric (acid-base) systems Alkyl side chains on ring systems

TABLE 3.17 Recommended Inspection Data Required for Petroleum and Heavy Feedstocks (Heavy Oil, Extra Heavy Oil, and Tar Sands Bitumen)

Petroleum	Heavy feedstocks
Density, specific gravity	Density, specific gravity
API gravity	API gravity
Carbon, wt%	Carbon, wt%
Hydrogen, wt%	Hydrogen, wt%
Nitrogen, wt%	Nitrogen, wt%
Sulfur, wt%	Sulfur, wt%
	Nickel, ppm
	Vanadium, ppm
	Iron, ppm
Pour point	Pour point
Wax content	
Wax appearance temperature	
Viscosity (various temperatures)	Viscosity (various temperatures)
Carbon residue of residuum	Carbon residue*
	Ash, wt%
Distillation profile	Fractional composition
All fractions plus vacuum residue	Asphaltenes, wt%
	Resins, wt%
	Aromatics, wt%
	Saturates, wt%

*Conradson carbon residue

TABLE 3.18 General Properties of Bulk Products from Petroleum

Product	Lower carbon limit	Upper carbon limit	Lower boiling point °C	Upper boiling point °C	Lower boiling point °F	Upper boiling point °F
Refinery gas	C1	C4	−161	−1	−259	31
Liquefied petroleum gas	C3	C4	−42	−1	−44	31
Naphtha	C5	C17	36	302	97	575
Gasoline	C4	C12	−1	216	31	421
Kerosene/diesel fuel	C8	C18	126	258	302	575
Aviation turbine fuel	C8	C16	126	287	302	548
Fuel oil	C12	>C20	216	421	>343	>649
Lubricating oil	>C20		>343		>649	
Wax	C17	>C20	302	>343	575	>649
Asphalt	>C20		>343		>649	
Coke	>C50*		>1000*		>1832*	

*Carbon number and boiling point difficult to assess; inserted for illustrative purposes only.

TABLE 3.19 Density of Petroleum and Petroleum Products

Petroleum (light crude oil)	0.8505
Natural gas liquids	0.6502
Additives	0.7893
Refinery gas	0.6975
Ethane	0.3113
LPG	0.5625
Naphtha	0.718
Motor gasoline	0.7449
Aviation gasoline	0.7172
Gasoline type jet fuel	0.7699
Kerosene type jet fuel	0.797
Kerosene	0.8036
Gas/diesel oil	0.8397
Fuel oil (including Bunker C oil)	0.9471
White spirit	0.7699
Lubricating oil	0.891
Asphalt	1.0132
Paraffin wax	0.8654
Petroleum coke	0.9654

TABLE 3.20 General Properties of Liquid Products from Petroleum

	Molecular weight	Specific gravity	Boiling point °F	Ignition temperature °F	Flash point °F	Flammability limits in air % v/v
Benzene	78.1	0.879	176.2	1040	12	1.35–6.65
Diesel fuel	170–198	0.875			100–130	
Fuel oil no. 1		0.875	304–574	410	100–162	0.7–5.0
Fuel oil no. 2		0.920		494	126–204	
Fuel oil no. 4	198.0	0.959		505	142–240	
Fuel oil no. 5		0.960			156–336	
Fuel oil no. 6		0.960			150	
Gasoline	113.0	0.720	100–400	536	−45	1.4–7.6
n-Hexane	86.2	0.659	155.7	437	−7	1.25–7.0
n-Heptane	100.2	0.668	419.0	419	25	1.00–6.00
Kerosene	154.0	0.800	304–574	410	100–162	0.7–5.0
Neohexane	86.2	0.649	121.5	797	−54	1.19–7.58
Neopentane	72.1		49.1	841	Gas	1.38–7.11
n-Octane	113.2	0.707	258.3	428	56	0.95–3.2
iso-Octane	113.2	0.702	243.9	837	10	0.79–5.94
n-Pentane	72.1	0.626	97.0	500	−40	1.40–7.80
iso-Pentane	72.1	0.621	82.2	788	−60	1.31–9.16
n-Pentene	70.1	0.641	86.0	569	−	1.65–7.70
Toluene	92.1	0.867	321.1	992	40	1.27–6.75
Xylene	106.2	0.861	281.1	867	63	1.00–6.00

TABLE 3.21 Chemical and Physical Requirements for JP-4, JP-5, and JP-8

Issuing Agency:	USAF	USAF	USAF
Grade: Designation:	JP-4 (NATO F-40)	JP-5 (NATO F-44)	JP-8 (NATO F-34/F-35)
Fuel type:	Wide-cut, gasoline type	Kerosene type	Kerosene type
Composition maximums:			
Acidity, total (mg KOH/g)	0.015	0.015	0.015
Aromatics (% v/v)	25.0	25.0	25.0
Sulfur, mercaptan (wt%)	0.002	0.002	0.002
Sulfur, total (wt%)	0.40	0.40	0.30
Volatility:			
Flash point (°C) min.		60	38
Density range (kg/L, 15°C)	0.751–0.802	0.788–0.845	0.755–0.840
V. P. at 37.8°C, kPa	14–21		
Fluidity:			
Freezing point °C, max.	−58	−46	−47
Viscosity @ −20°C, max.		8.5 centistokes	8.0 centistokes
Contaminant maximums:			
Existent gum (mg/100 mL)	7.0	7.0	7.0
Particulate matter (mg/L)	1.0	1.0	1.0

Note: JP-8+100 is identical to JP-8 except for the addition of an additive package (antioxidant, dispersant/detergent, metal deactivator, and solvent) injected at 256 ppm, which increases its thermal stability from 325 to 425°C.

TABLE 3.22 Heat Content of Fuel Oil

Grade	Heating Value (Btu/gal)	Comments
Fuel oil no. 1	132,900–137,000	Small space heaters
Fuel oil no. 2	137,000–141,800	Residential heating
Fuel oil no. 4	143,100–148,100	Industrial burners
Fuel oil no. 5 (Light)	146,800–150,000	Preheating generally required
Fuel oil no. 5 (Heavy)	149,400–152,000	Heating required
Fuel oil no. 6	151,300–155,900	Bunker C

TABLE 3.23   Properties of Some Alternate Liquid Fuels

Fuel	Cetane number	Research octane number	Motor octane number	Density (lb/gal)	LHV (Btu/gal)	LHV (Btu/lb)	DEE (gal)	DEE (lb)
100% Ethanol	8	109	90	6.6	75,600	11,500	0.59	0.66
85% Ethanol (E85)		105	89	6.5	83,600	12,855	0.64	0.72
10% Ethanol/gasoline		96.5	86	6.1	111,000	18,000	0.86	1.39
10% Ethanol/diesel	45			7	123,000	17,500	0.95	1.01
100% Methanol	5	109	89	6.7	56,200	8,400	0.44	0.48
100% Soy methyl-ester	49			7.3	120,200	16,500	0.93	0.95
100% Biodiesel (B100)	54				117,000	15,800		
20% Biodiesel, 80% petrodiesel (B20)	46					18,100		
#2 Diesel	44			6.7–7.4	126,000–130,000	18,000–19,000	1	1
#1 Diesel	44			7.6	125,800	16,600	0.98	0.95
Gasoline		90–100	80–90	6	115,400	19,200	0.89	1.1
CNG—Compressed Natural Gas	<0	>127	122			20,400		1.17
LNG—Liquefied Natural Gas	<0	>127	122	3.5	78,000	22,300	0.6	1.28
Liquefied petroleum gas		109	96	3.2	83,600	19,900	0.65	1.14
DME—Dimethyl Ether	55–60			5.6	74,800	13,600	0.58	0.78

TABLE 3.24   Properties of Conventional and Alternative Fuels

Property	Gasoline	No. 2 diesel	Methanol	Ethanol
Chemical formula	C_4 to C_{12}	C_3 to C_{25}	CH_3OH	C_2H_5OH
Physical state	Liquid	Liquid	Liquid	Liquid
Molecular weight	100–105	≈200	32.04	46.07
Composition (wt%)				
Carbon	85–88	84–87	37.5	52.2
Hydrogen	12–15	33–16	12.6	13.1
Oxygen	0	0	49.9	34.7
Main fuel source(s)	Crude oil	Crude oil	Natural gas, coal, or woody biomass	Corn, grains, or agricultural waste
Specific gravity (60°F/60°F)	0.72–0.78	0.81–0.89	0.796	0.796
Density (lb/gal @ 60°F)	6.0–6.5	6.7–7.4	6.63	6.61
Boiling temperature (°F)	80–437	370–650	149	172
Freezing point (°F)	−40	−40–30	−143.5	−173.2
Autoignition temperature (°F)	495	≈600	867	793
Reid vapor pressure (psi)	8–15	0.2	4.6	2.3

(Continued)

TABLE 3.24   Properties of Conventional and Alternative Fuels (*Continued*)

Property	Propane	Compressed natural gas (CNG)	Hydrogen
Chemical formula	C_3H_8	CH_4	H_2
Physical state	Compressed gas	Compressed gas	Compressed gas or liquid
Molecular weight	44.1	16.04	2.02
Composition (wt%)			
Carbon	82	75	0
Hydrogen	18	25	100
Oxygen	n/a	n/a	0
Main fuel source	Underground reserves	Underground reserves	Natural gas, methanol, and other energy sources
Specific gravity (60°F/60°F)	0.508	0.424	0.07
Density (lb/gal @ 60°F)	3.22	1.07	n/a
Boiling temperature (°F)	−44	−259	−423
Freezing point (°F)	−305.8	−296	−435
Autoignition temperature (°F)	850–950	1004	1050–1080
Reid vapor pressure (psi)	208	2400	n/a

Note: n/a = not applicable.

TABLE 3.25   Production of Feedstocks for the Manufacture of Petrochemicals

Starting feedstock	Process	Product
Petroleum	Distillation	Light ends
		Methane
		Ethane
		Propane
		Butane
	Catalytic cracking	Ethylene
		Propylene
		Butylenes
		Higher olefins
	Coking	Ethylene
		Propylene
		Butylenes
		Higher olefins
Natural gas	Refining	Methane
		Ethane
		Propane
		Butane

3.2.2 Bibliography

ASTM D613. 2015. Standard Test Method for Cetane Number of Diesel Fuel Oil. *Annual Book of Standards*, ASTM International, West Conshohocken, Pennsylvania.

ASTM D975. 2015. Standard Specification for Diesel Fuel Oils. *Annual Book of Standards*, ASTM International, West Conshohocken, Pennsylvania.

ASTM D1835. 2015. Standard Specification for Liquefied Petroleum (LP) Gases. *Annual Book of Standards*, ASTM International, West Conshohocken, Pennsylvania.

ASTM D2699. 2015. Standard Test Method for Research Octane Number of Spark-Ignition Engine Fuel. *Annual Book of Standards*, ASTM International, West Conshohocken, Pennsylvania.

ASTM D2700. 2015. Standard Test Method for Motor Octane Number of Spark-Ignition Engine Fuel. *Annual Book of Standards*, ASTM International, West Conshohocken, Pennsylvania.

ASTM D4175. 2015. Standard Terminology Relating to Petroleum, Petroleum Products, and Lubricants. *Annual Book of Standards*, ASTM International, West Conshohocken, Pennsylvania.

Speight, J. G. 2011. *The Refinery of the Future*. Gulf Professional Publishing, Elsevier, Oxford, United Kingdom.

Speight, J. G. (ed.). 2011. *Biofuels Handbook*. Royal Society of Chemistry, London, United Kingdom.

Speight, J. G. 2012a. *Crude Oil Assay Database*. Knovel, Elsevier, New York. Online version available at: http://www.knovel.com/web/portal/browse/display?_EXT_KNOVEL_DISPLAY_bookid=5485&VerticalID=0

Speight, J. G. 2012b. *Shale Oil Production Processes*. Gulf Professional Publishing, Elsevier, Oxford, United Kingdom.

Speight, J. G. 2013a. *Heavy Oil Production Processes*. Gulf Professional Publishing, Elsevier, Oxford, United Kingdom.

Speight, J. G. 2013b. *Oil Sand Production Processes*. Gulf Professional Publishing, Elsevier, Oxford, United Kingdom.

Speight, J. G. 2013c. *Heavy and Extra Heavy Oil Upgrading Technologies*. Gulf Professional Publishing, Elsevier, Oxford, United Kingdom.

Speight, J. G. 2014a. *The Chemistry and Technology of Petroleum*, 5th ed. CRC–Taylor and Francis Group, Boca Raton, Florida.

Speight, J. G. 2014b. *High Acid Crudes*. Gulf Professional Publishing, Elsevier, Oxford, United Kingdom.

Speight, J. G. 2014c. *Oil and Gas Corrosion Prevention*. Gulf Professional Publishing, Elsevier, Oxford, United Kingdom.

Speight, J. G. 2015a. *Handbook of Hydraulic Fracturing*. John Wiley & Sons, Hoboken, New Jersey.

Speight, J. G. 2015b. *Handbook of Petroleum Product Analysis*, 2nd ed. John Wiley & Sons, Hoboken, New Jersey.

Speight, J. G. 2105c. *Asphalt Materials Science and Technology*. Butterworth-Heinemann, Elsevier, Oxford, United Kingdom.

Speight, J. G. 2016. *Introduction to Enhanced Recovery Methods for Heavy Oil and Tar Sands*, 2nd ed. Gulf Professional Publishing, Elsevier, Oxford, United Kingdom.

3.3 NATURAL GAS AND NATURAL GAS HYDRATES

Natural gas (also called *marsh gas* and *swamp gas* in older texts and more recently *landfill gas*) is a naturally occurring gaseous fossil fuel that is found in oil fields and natural gas fields, and in coal beds. On the other hand, gaseous fuels is a collective name for any one of a number of fuels that, standard conditions of temperature and pressure (STP), exist in the gaseous state. Many fuel gases are composed of hydrocarbons (such as methane or propane), hydrogen, carbon monoxide, or mixtures thereof. Such gases are sources of light in the form of heat or light that can be readily transported and distributed (typically through a pipeline system) from the point of origin directly to the place of consumption.

TABLE 3.26   Range of Composition of Natural Gas

Methane	CH_4	70–90%
Ethane	C_2H_6	
Propane	C_3H_8	0–20%
Butane	C_4H_{10}	
Pentane and higher hydrocarbons	C_5H_{12}	0–10%
Carbon dioxide	CO_2	0–8%
Oxygen	O_2	0–0.2%
Nitrogen	N_2	0–5%
Hydrogen sulfide, carbonyl sulfide	H_2S, COS	0–5%
Rare gases: Argon, helium, neon, xenon	A, He, Ne, Xe	Trace

TABLE 3.27   General Properties of the C_1—C_8 n-Hydrocarbons

	Molecular weight	Specific gravity	Vapor density, air = 1	Boiling point, °C	Ignition temperature, °C	Flash point, °C
Methane	16	0.553	0.56	−160	537	−221
Ethane	30	0.572	1.04	−89	515	−135
Propane	44	0.504	1.50	−42	468	−104
Butane	58	0.601	2.11	−1	405	−60
Pentane	72	0.626	2.48	36	260	−40
Hexane	86	0.659	3.00	69	225	−23
Benzene	78	0.879	2.80	80	560	−11
Heptane	100	0.668	3.50	98	215	−4
Octane	114	0.707	3.90	126	220	13

TABLE 3.28   Standard Test Methods for Natural Gas*

Test number	Title
ASTM D1070	Standard Test Methods for Relative Density of Gaseous Fuels
ASTM D1071	Standard Test Methods for Volumetric Measurement of Gaseous Fuel Samples
ASTM D1072	Standard Test Method for Total Sulfur in Fuel Gases
ASTM D1142	Standard Test Method for Water Vapor Content of Gaseous Fuels by Measurement of Dew-Point Temperature
ASTM D1826	Standard Test Method for Calorific (Heating) Value of Gases in Natural Gas Range by Continuous Recording Calorimeter
ASTM D1945	Standard Test Method for Analysis of Natural Gas by Gas Chromatography
ASTM D1946	Standard Practice for Analysis of Reformed Gas by Gas Chromatography
ASTM D1988	Standard Test Method for Mercaptans in Natural Gas Using Length-of-Stain Detector Tubes
ASTM D3588	Standard Practice for Calculating Heat Value, Compressibility Factor, and Relative Density of Gaseous Fuels
ASTM D3956	Standard Specification for Methane Thermophysical Property Tables
ASTM D3984	Standard Specification for Ethane Thermophysical Property Tables
ASTM D4084	Standard Test Method for Analysis of Hydrogen Sulfide in Gaseous Fuels (Lead Acetate Reaction Rate Method)
ASTM D4150	Standard Terminology Relating to Gaseous Fuels
ASTM D4362	Standard Specification for Propane Thermophysical Property Tables
ASTM D4468	Standard Test Method for Total Sulfur in Gaseous Fuels by Hydrogenolysis and Rateometric Colorimetry
ASTM D4650	Standard Specification for Normal Butane Thermophysical Property Tables
ASTM D4651	Standard Specification for Isobutane Thermophysical Property Tables
ASTM D4784	Standard for LNG Density Calculation Models
ASTM D4810	Standard Test Method for Hydrogen Sulfide in Natural Gas Using Length-of-Stain Detector Tubes
ASTM D4888	Standard Test Method for Water Vapor in Natural Gas Using Length-of-Stain Detector Tubes
ASTM D4891	Standard Test Method for Heating Value of Gases in Natural Gas Range by Stoichiometric Combustion
ASTM D4984	Standard Test Method for Carbon Dioxide in Natural Gas Using Length-of-Stain Detector Tubes
ASTM D5287	Standard Practice for Automatic Sampling of Gaseous Fuels
ASTM D5454	Standard Test Method for Water Vapor Content of Gaseous Fuels Using Electronic Moisture Analyzers
ASTM D5503	Standard Practice for Natural Gas Sample-Handling and Conditioning Systems for Pipeline Instrumentation
ASTM D5504	Standard Test Method for Determination of Sulfur Compounds in Natural Gas and Gaseous Fuels by Gas Chromatography and Chemiluminescence
ASTM D5954	Standard Test Method for Mercury Sampling and Measurement in Natural Gas by Atomic Absorption Spectroscopy
ASTM D6228	Standard Test Method for Determination of Sulfur Compounds in Natural Gas and Gaseous Fuels by Gas Chromatography and Flame Photometric Detection
ASTM D6273	Standard Test Methods for Natural Gas Odor Intensity
ASTM D6350	Standard Test Method for Mercury Sampling and Analysis in Natural Gas by Atomic Fluorescence Spectroscopy

*Annual Book of Standards. ASTM International, West Conshohocken, Pennsylvania.

TABLE 3.29 Olamines Used for Gas Processing

Olamine	Formula	Derived name	Molecular weight	Specific gravity	Melting point, °C	Boiling point, °C	Flash point, °C	Relative capacity, %
Ethanolamine (monoethanolamine)	$HOC_2H_4NH_2$	MEA	61.08	1.01	10	170	85	100
Diethanolamine	$(HOC_2H_4)_2NH$	DEA	105.14	1.097	27	217	169	58
Triethanolamine	$(HOC_2H_4)_3N$	TEA	148.19	1.124	18	335, d	185	41
Diglycolamine (hydroxyethanolamine)	$H(OC_2H_4)_2NH_2$	DGA	105.14	1.057	−11	223	127	58
Diisopropanolamne	$(HOC_3H_6)_2NH$	DIPA	133.19	0.99	42	248	127	46
Methyldiethanolamine	$(HOC_2H_4)_2NCH_3$	MDEA	119.17	1.03	−21	247	127	51

Note: d = with decomposition.

TABLE 3.30 Composition of Gaseous Fuels

Fuel	Composition, %									
	Carbon dioxide	Carbon monoxide	Methane	Butane	Ethane	Propane	Hydrogen	Hydrogen sulfide	Oxygen	Nitrogen
Carbon monoxide		100								
Coal gas	3.8	28.4	0.2				17.0			50.6
Coke oven gas	2.0	5.5	32				51.9		0.3	4.8
Digester gas	30		64				0.7	0.8		2.0
Hydrogen							100			
Landfill gas	47	0.1	47				0.1	0.01	0.8	3.7
Natural gas	0–0.8	0–0.45	82–93		0–15.8		0–1.8	0–0.18	0–0.35	0.5–8.4
Propane				0.5–0.8	2.0–2.2	73–97				

TABLE 3.31 Calorific Value of Gaseous Fuels

Gaseous fuels (dry)	Gross calorific value
Coal gas coke oven (benzene-free)	540 Btu ft^3, 20 MJ m^{-3}
Coal gas continuous vertical retort	480 Btu ft^3, 18 MJ m^{-3}
Coal gas low temperature	910 Btu ft^3, 34 MJ m^{-3}
Butane	3170 Btu ft^3, 118 MJ m^{-3}
Propane	2520 Btu ft^3, 94 MJ m^{-3}
Natural gas (North Sea)	1045 Btu ft^3, 39 MJ m^{-3}
Producer gas (from coal)	160 Btu ft^3, 6 MJ m^{-3}
Producer gas (from coke)	135 Btu ft^3, 5 MJ m^{-3}
Carbureted water gas	240 Btu ft^3, 9 MJ m^{-3}
Blue water gas blue	26 Btu ft^3, 1 MJ m^{-3}

3.3.1   Bibliography

BP. 2015. *Statistical Review of World Energy*. 2005. British Petroleum. London, United Kingdom. June.

Luque, R., and Speight, J. G. (eds.). 2015. *Gasification for Synthetic Fuel Production: Fundamentals, Processes, and Applications*. Woodhead Publishing, Elsevier, Cambridge, United Kingdom.

Mokhatab, S., Poe, W. A., and Speight, J. G. 2006. *Handbook of Natural Gas Transmission and Processing*. Elsevier, Amsterdam, Netherlands.

Speight, J. G. 2005. *Environmental Analysis and Technology for the Refining Industry*. John Wiley & Sons, Hoboken, New Jersey.

Speight, J. G. 2007. *Natural Gas: A Basic Handbook*. GPC Books, Gulf Publishing Company, Houston, Texas.

Speight, J. G. (ed.). 2011. *The Biofuels Handbook*. Royal Society of Chemistry, London, United Kingdom.

Speight, J. G. 2013. *The Chemistry and Technology of Coal*, 3rd ed. CRC–Taylor and Francis Group, Boca Raton, Florida.

Speight, J. G. 2014. *The Chemistry and Technology of Petroleum*, 5th ed. CRC–Taylor and Francis Group, Boca Raton, Florida.

3.4   TAR SANDS AND TAR SANDS BITUMEN

Tar sands (also called *oil sands* in Canada) is the term used to describe a sandstone reservoir that is impregnated with bitumen, a naturally occurring material that is solid or near solid and is substantially immobile under reservoir conditions. The more geologically correct term is *bituminous sand*. Tar sands deposits are also a combination of clay, sand, water, and bitumen, a heavy black viscous oil. Tar sands can be mined and processed to extract the oil-rich bitumen, which is then refined into oil. The bitumen in tar sands deposits cannot be pumped from the ground in its natural state; instead, tar sands deposits are mined, usually using strip mining or open pit techniques, or the oil is extracted by underground heating with additional upgrading.

Tar sands are actually a mixture of sand, water, and bitumen, but many of the tar sands deposits in countries other than Canada lack the water layer that is believed to facilitate the hot water recovery process. The heavy bituminous material has a high viscosity under reservoir conditions and cannot be retrieved through a well by conventional production techniques. Tar sands are defined (FE-76-4) in the United States as:

> The several rock types that contain an extremely viscous hydrocarbon which is not recoverable in its natural state by conventional oil well production methods including currently used enhanced recovery techniques. The hydrocarbon-bearing rocks are variously known as bitumen-rocks oil, impregnated rocks, tar sands, and rock asphalt.

Tar sands deposits near the surface can be recovered by open pit mining techniques. New methods introduced in the 1990s considerably improved the efficiency of tar sands mining, thus reducing the cost. These systems use large hydraulic and electrically powered shovels to dig up tar sands and load them into enormous trucks that can carry up to 320 tons of tar sands per load. After mining, the tar sands are transported to an extraction plant, where a hot water process separates the bitumen from sand, water, and minerals. The separation takes place in separation cells. Hot water is added to the sand, and the resulting slurry is piped to the extraction plant where it is agitated. The combination of hot water and agitation releases bitumen from the oil sands, and causes tiny air bubbles to attach to the bitumen droplets that float to the top of the separation vessel, where the bitumen can be skimmed off. Further processing removes residual water and solids. The bitumen is then transported and eventually upgraded into synthetic crude oil. Approximately 2 tons of tar sands are required to produce one barrel (42 U.S. gallons, 35.5 Imperial gallons) of oil. Approximately 75% v/v of the bitumen can be recovered from sand. After oil extraction, the spent sand and other materials are then returned to the mine, which is eventually reclaimed. In situ production methods are used on bitumen deposits buried too deep to be economically recovered by mining. These techniques include steam injection and solvent injection of which the steam injection methods have been favored.

TABLE 3.32   Distillation Data (Cumulative % w/w Distilled) for Tar Sands Bitumen and Crude Oil

Cut point °C	°F	Cumulative % by weight distilled		
		Athabasca	PR Spring	Leduc (Canada)
200	390	3	1	35
225	435	5	2	40
250	480	7	3	45
275	525	9	4	51
300	570	14	5	
325	615	26	7	
350	660	18	8	
375	705	22	10	
400	750	26	13	
425	795	29	16	
450	840	33	20	
475	885	37	23	
500	930	40	25	
525	975	43	29	
538	1000	45	35	
538+	1000+	55	65	

TABLE 3.33   Properties of Selected Atmospheric (>650°F) and Vacuum (>1050°F) Residua

Feedstock	Gravity API	Sulfur wt%	Nitrogen wt%	Nickel ppm	Vanadium ppm	Asphaltene (heptane) wt%	Carbon residue (conradson) wt%
Arabian Light, >650°F	17.7	3.0	0.2	10.0	26.0	1.8	7.5
Arabian Light, >1050°F	8.5	3.4	0.5	24.0	66.0	4.3	14.2
Arabian Heavy, > 650°F	11.9	3.4	0.3	27.0	103.0	8.0	14.0
Arabian Heavy, >1050°F	7.3	5.1	0.3	40.0	174.0	10.0	19.0
Alaska, North Slope, >650°F	15.2	1.6	0.4	18.0	30.0	2.0	8.5
Alaska, North Slope, >1050°F	8.2	2.2	0.6	47.0	82.0	4.0	18.0
Lloydminster (Canada), >650°F	10.3	4.1	0.3	65.0	141.0	14.0	12.1
Lloydminster (Canada), >1050°F	8.5	3.4	0.6	115.0	252.0	18.0	21.4
Kuwait, >650°F	13.9	3.4	0.3	14.0	50.0	2.4	12.2
Kuwait, >1050°F	5.5	5.5	0.4	32.0	102.0	7.1	23.1
Tia Juana, >650°F	17.3	1.8	0.3	25.0	185.0		9.3
Tia Juana, >1050°F	7.1	2.6	0.6	64.0	450.0		21.6
Taching, >650°F	27.3	0.2	0.2	5.0	1.0	3.4	3.8
Taching, >1050°F	21.5	0.3	0.4	9.0	2.0	7.6	7.9
Maya, >650°F	10.5	3.4	0.5	70.0	370.0	16.0	15.0

TABLE 3.34 Properties of Synthetic Crude Oil from Athabasca Bitumen

Property	Bitumen	Synthetic crude oil
Gravity, °API	8	32
Sulfur, % by w/w	4.8	0.2
Nitrogen, % w/w	0.4	0.1
Viscosity cP @ 100°F	500,000	10
Distillation profile, % by weight		

°C	°F		
0	30	0	5
30	85	0	30
220	430	1	60
345	650	17	90
550	1020	45	100

TABLE 3.35 Comparison of Brent Crude Oil and Syncrude Sweet Blend

	Brent	Sweet blend
API	38.6	31.8
Sulfur, % w/w	0.3	0.1
C_4^-	2.9	3.4
C_5–350°F	28.6	15.0
350–650°F	29.6	44.0
650–1050°F	29.8	37.6
1050°F+	9.1	0.0

3.4.1 Bibliography

Speight, J. G. 1990. In: *Fuel Science and Technology Handbook*, J. G. Speight (ed.). Marcel Dekker, New York. Part II. Chapters 12–16.

Speight, J. G. 2000. *The Desulfurization of Heavy Oils and Residua*, 2nd ed. Marcel Dekker, New York.

Speight, J. G. 2005. Natural Bitumen (Tar Sands) and Heavy Oil. In: *Coal, Oil Shale, Natural Bitumen, Heavy Oil and Peat, from Encyclopedia of Life Support Systems (EOLSS)*, Developed under the Auspices of UNESCO, EOLSS Publishers, Oxford, UK, [http://www.eolss.net].

Speight, J. G. 2008. *Synthetic Fuels Handbook: Properties, Processes, and Performance*. McGraw-Hill, New York.

Speight, J. G. 2014. *Chemistry and Technology of Petroleum*, 5th ed. CRC–Taylor and Francis Group, Boca Raton, Florida.

U.S. Congress. 1976. Public Law FEA-76-4. Washington, D.C.

3.5 OIL SHALE

Oil shale is an inorganic, nonporous sedimentary marlstone rock containing various amounts of solid organic material (known as *kerogen*) that yields hydrocarbons, along with nonhydrocarbons, and a variety of solid products, when subjected to pyrolysis (a treatment that consists of heating the rock at high temperature). Thus, by definition, *kerogen* is naturally occurring insoluble organic matter found in shale deposits. However, *shale oil* is the synthetic fuel produced by the thermal decomposition of *kerogen* at high temperature (>500°C, >930°F). Shale oil is referred to as *synthetic crude oil* after hydrotreating.

TABLE 3.36 Composition (% w/w) of the Organic Matter in the Mahogany Zone and New Albany Shale

Component	Green River Mahogany Zone	New Albany
Carbon	80.5	82.0
Hydrogen	10.3	7.4
Nitrogen	2.4	2.3
Sulfur	1.0	2.0
Oxygen	5.8	6.3
Total	100.0	100.0
H/C atomic ratio	1.54	1.08

TABLE 3.37 Porosity and Permeability of Raw and Treated Oil Shale

Fischer assay	Porosity Raw	Porosity Heated to 815°C	Permeability Heated to 815°C
1.0*	9.0†	11.9	0.36§ 0.56
6.5	5.5	12.5	0.21 0.65
13.5	0.5	16.4	3.53 8.02

*Fischer Assay in gal/ton.
†Numbers in percentages of the initial bulk volume. Porosity was taken as an isotropic property, i.e., property that is independent of measurement direction.
§Units in milliDarcy.

TABLE 3.38 Properties of Shale Oil from Various Sources

Location	Specific gravity (API)	Elemental analysis, % w/w C	H	O	N	S	Analysis of <350°C distillate, % w/w Saturates	Olefins	Aromatics
Colorado, USA	0.943 (18.6)	84.90	11.50	0.80	2.19	0.61	27	44	29
Kukersite, Estonia	1.010 (8.5)	82.85	9.20	6.79	0.30	0.86	22	25	53
Stuart, Australia		82.70	12.40	3.34	0.91	0.65			
Rundle, Australia	0.636 (0.91)	79.50	11.50	7.60	0.99	0.41	48	2	50
Irati, Brazil	0.919 (22.5)	84.30	12.00	1.96	1.06	0.68	23	41	36
Maoming, China	0.903 (25.2)	84.82	11.40	2.20	1.10	0.48	55	20	25
Fushun, China	0.912 (23.2)	85.39	12.09	0.71	1.27	0.54	38	37	25

TABLE 3.39 Major Compound Types in Shale Oil

Saturates	Heteroatom systems
Paraffin derivatives	Benzothiophene derivatives
Cycloparaffin derivatives	Dibenzothiophene derivatives
Olefins	Phenol derivatives
Aromatics	Carbazole derivatives
Benzene derivatives	Pyridine derivatives
Indan derivatives	Quinoline derivatives
Tetralin derivatives	Nitrile derivatives
Naphthalene derivatives	Ketone derivatives
Biphenyl derivatives	Pyrrole derivatives
Phenanthrene derivatives	
Chrysene derivatives	

TABLE 3.40 Challenges for Oil Shale Processing

Particulates	Plugging on processing
	Product quality
Arsenic content	Toxicity
	Catalyst poison
High pour point	Oil not pipeline quality
Nitrogen content	Catalyst poison
	Contributes to instability
	Toxicity
Diolefins	Contributes to instability
	Plugging on processing

3.5.1 Bibliography

Baldwin, R. M. 2002. *Oil Shale: A Brief Technical Overview*. Colorado School of Mines, Golden, Colorado. July.

Bartis, J. T., LaTourrette, T., and Dixon, L. 2005. *Oil Shale Development in the United States: Prospects and Policy Issues*. Prepared for the National Energy Technology of the United States Department of Energy. Rand Corporation, Santa Monica, California.

Scouten, C. 1990. In: *Fuel Science and Technology Handbook*. J. G. Speight (ed.). Marcel Dekker, New York.

Speight, J. G. 2008. *Synthetic Fuels Handbook: Properties, Processes, and Performance*. McGraw-Hill, New York.

Speight, J. G. 2014. *The Chemistry and Technology of Petroleum*, 5th ed. CRC–Taylor and Francis Group, Boca Raton, Florida.

3.6 BIOMASS

Biomass is the collective name for *renewable materials* that includes (1) energy crops grown specifically for use as fuel, such as wood or various grasses, (2) agricultural residues and by-products, such as straw, sugarcane fiber, rice hulls animal waste, and (3) residues from forestry, construction, and other wood-processing industries.

The production of fuels and chemicals from renewable plant-based feedstocks utilizing state-of-the-art conversion technologies presents an opportunity to maintain competitive advantage and contribute to the attainment of national environmental targets. Bioprocessing routes have a number of compelling advantages over conventional petrochemicals production; however, it is only in the last decade that rapid progress in biotechnology has facilitated the commercialization of a number of plant-based chemical processes.

TABLE 3.41 Composition of Biogas from Different Sources

Component	Agricultural biogas	Sewage gas	Landfill gas
Methane	55–75%	55–65%	40–45%
Carbon dioxide	25–45%	30–40%	35–50%
Nitrogen	0–10%	0–10%	0–20%
Hydrogen sulfide	0–1.5%	≤200 ppm	Approx. 200 ppm
Water	Saturated	Saturated	Saturated
Halogens	Trace amount	Up to 4 ppm	Dependent on the landfill
Higher hydrocarbons	Trace amount	Trace amount	Up to 200 ppm

TABLE 3.42 Elemental Composition (with Ash Content) of Biofuels (% w/w, Dry Basis)

Fuel	C	H	O	N	S	Ash
Birch wood	48.8	6.0	44.2	0.5	0.01	20.0
Pine wood	49.3	6.0	44.2	0.5	0.01	20.1
Bark	47.2	5.6	46.9	0.3	0.07	20.9
Wheat straw	49.6	6.2	43.6	0.6	n/a	18.6
Miscanthus	49.5	6.2	43.7	0.6	n/a	18.5
Sugar cane	49.5	6.2	43.8	0.5	n/a	18.5
Reed grass	49.4	6.3	42.7	1.6	n/a	18.8
Peat	53.1	5.5	38.1	1.3	0.2	20.5
Coal	80.4	5.0	6.7	1.3	0.53	30.4

Note: n/a = not applicable.

TABLE 3.43 Heating Value of Selected Fuels

Fuel	Btu/lb
Natural gas	23,000
Gasoline	20,000
Crude oil	18,000
Heavy oil	16,000
Coal (anthracite)	14,000
Coal (bituminous)	11,000
Wood (farmed trees, dry)	8400
Coal (lignite)	8000
Biomass (herbaceous, dry)	7400
Biomass (corn stover, dry)	7000
Wood (forest residue, dry)	6600
Bagasse (sugar cane)	6500
Wood	6000

TABLE 3.44   Typical Plants Used as a Source of Energy

Biomass	Plant species	Predominant form of energy use
Wood	Various types (bushes, shrubs, trees) depending upon the location	Firewood (ca. 50% w/w of harvest)
Starch	Cereals, millets, root and tuber crops, e.g., potato	Bioethanol manufacture
Sugar	Sugarcane, sugar beet	Bioethanol manufacture
Hydrocarbons	Various types depending upon the location	Biodiesel manufacture
Wastes	Crop residues, animal/human refuse, sewage	Biogas manufacture

TABLE 3.45   Chemical Composition of Different Biomass Types (% w/w, Dry Basis)

Type	Cellulose	Hemicellulose	Lignin	Others	Ash
Soft wood	41	24	28	2	0.4
Hard wood	39	35	20	3	0.3
Pine bark	34	16	34	14	2
Straw (wheat)	40	28	17	11	7
Rice husks	30	25	12	18	16
Peat	10	32	44	11	6

TABLE 3.46   Selected Properties of Common Bio-Feedstocks and Biofuels (cf Coal and Crude Oil Distillate)

		Chemical characteristics			
		Ash % w/w	Sulfur % w/w	Potassium % w/w	Ash melting temperature °C
Bio-feedstocks	Corn stover	5.6			
	Sweet sorghum	5.5			
	Sugarcane bagasse	0.2–5.5	0.10–0.15	0.73–0.97	
	Sugarcane leaves	7.7			
	Hardwood	0.45	0.009	0.04	900
	Softwood	0.3	0.01		
	Hybrid poplar	0.5–1.5	0.03	0.3	1350
	Bamboo	0.8–2.5	0.03–0.05	0.15–0.50	
	Switchgrass	4.5–5.8	0.12		1016
	Miscanthus	1.5–4.5	0.1	0.37–1.12	1090
	Giant cane	5–6	0.07		
Liquid biofuels	Bioethanol		<0.01		n/a
	Biodiesel	<0.02	<0.05	<0.0001	n/a
Fossil fuels	Coal—low rank; lignite/sub-bituminous	5–20	1.0–3.0	0.02–0.3	~1300
	Coal—high rank; bituminous/anthracite	1–10	0.5–1.5	0.06–0.15	~1300
	Crude oil distillate	0.5–1.5	0.2–1.2		n/a

Note: n/a = not applicable.

TABLE 3.47 Physical Properties of Fats and Oils

Fat or oil	Solidification point, °C	Specific gravity (15°C/15°C)	Refractive index	Acid value	Saponification value	Iodine value
Animal origin						
Butterfat	20–23	$0.91^{40°C}_{15°C}$	$1.45^{40°C}$	0.5–35	210–230	26–38
Chicken fat	21–27	0.924		1.2	193–205	66–72
Cod–liver oil	–3	0.92–0.93	$1.481^{25°C}$	5.6	171–189	137–166
Deer fat		0.96–0.97		0.8–5.3	195–200	26–36
Dolphin	–3 to +5	0.91–0.93		2–12	203 (body); 290 (jaw)	127 (body); 33 (jaw)
Goat butter	22–24	$0.91–0.94^{38°C}_{38°C}$		0.6	233–236	25–37
Goose fat		0.92–0.93		1.8–44	191–193	58–67
Herring oil		0.92–0.94	$1.4610^{60°C}$	0–2.4	170–194	102–149
Horse fat	20–45	0.92–0.93			195–200	75–86
Human fat	15	0.903	1.460		193–200	57–73
Lard oil	–2 to +4	0.913–0.915	1.462	0.1–2.5	193–198	63–79
Lard oil, fatty tissue	27–30	0.93–0.94	1.462	0.5–0.8	195–203	47–67
Menhaden oil	–5	0.92–0.93	$1.465^{60°C}$	3–12	189–193	148–185
Neat's–foot oil	–2 to +10	0.91–0.92	$1.464^{25°C}$	0.1–0.6	193–199	58–75
Porpoise, body oil	–16	0.926		1.2	203	127
Rabbit fat	17–23	0.93–0.94	$1.466^{60°C}$	1.4–7.2	199–203	70–100
Sardine oil	20–22	0.92–0.93		4–25	188–196	130–152
Seal	3	0.915–0.926		1.9–40	188–196	130–152
Shark		0.916–0.919			157–164	115–139
Sperm oil	15.5	0.878–0.884		13	120–137	80–84
Tallow, beef	31–38	0.895	$1.457^{40°C}$	0.25	196–200	35–42
Tallow, mutton	32–41	0.937–0.953		2–14	195–196	48–61
Whale oil	–2 to 0	0.917–0.924	$1.460^{60°C}$	1.9	160–202	90–146

Plant origin

Plant origin						
Acorn	−10	0.916	$1.443^{60°C}$	0.5–3.5	199	100
Almond	−20 to −15	0.914–0.921			183–208	93–103
Babassu oil	22–26	$0.893^{60°C}$			247	16
Beechnut oil	−17	0.922			191–196	97–111
Castor oil	−18 to −17	0.960–0.967	1.477	0.1–0.8	175–183	84
Chaulmoogra oil, USP	<−25	$0.950^{25°C}$			196–213	98–110
Chinese vegetable tallow	24–34	0.918–0.922	$1.457^{40°C}$	2.4	179–206	23–41
Cocoa butter	21.5–23	0.964–0.974	$1.449^{40°C}$	1.1–1.9	193–195	33–42
Coconut oil	14–22	0.926		2.5–10	153–262	6–10
Corn (maize) oil	−20 to −10	0.921–0.928	$1.473^{40°C}$	1.4–2.0	187–193	111–128
Cottonseed oil	−13 to +12	$0.918g^{25°C}$	$1.474^{40°C}$	0.6–0.9	194–196	103–111
Hazelnut oil	−18 to −17	0.917			191–197	87
Hemp-seed oil	−28 to −15	0.928–0.934		0.45	190–195	145–162
Linseed oil	−27 to −19	0.930–0.938	$1.478^{25°C}$	1–3.5	188–195	175–202
Mustard, black, oil	16	0.918–0.921	$1.475^{40°C}$	5.7–7.3	173–175	99–110
Neem oil	−3	0.917	$1.462^{40°C}$		195	71
Niger-seed oil		0.925	$1.471^{40°C}$		190	129
Oiticica oil		$0.974^{25°C}$				140–180
Olive oil	−6	0.914–0.918	$1.468^{40°C}$	0.3–1.0	185–196	79–88
Palm oil	35–42	0.915	$1.458^{40°C}$	10	200–205	49–59
Palm kernel oil	24	0.918–0.925	$1.457^{40°C}$	0.3–0.6	220–231	26–32
Peanut oil	3	0.917–0.926	$1.469^{40°C}$	0.8	186–194	88–98
Perilla oil		0.930–0.937	$1.481^{25°C}$		188–194	185–206
Pistachio–nut oil	−10 to −5	0.913–0.919			191	83–87
Poppy–seed oil	−18 to −16	0.924–0.926	$1.469^{40°C}$	2.5	193–195	128–141
Pumpkin–seed oil	−15	0.923–0.925			188–193	121–130
Rapeseed oil	−10	0.913–0.917	$1.471^{40°C}$	0.36–1.0	168–179	94–105
Safflower oil	−18 to −13	0.925–0.928	$1.462^{60°C}$	0.6	188–203	122–141
Sesame oil	−6 to −4	$0.919^{25°C}$	$1.465^{40°C}$	9.8	188–193	103–117
Soybean oil	−16 to −10	0.924–0.927	$1.473^{40°C}$	0.3–1.8	189–194	122–134
Sunflower–seed oil	−17	0.924–0.926	$1.469^{40°C}$	11.2	188–193	129–136
Tung oil	−2.5	0.94–0.95	$1.517^{25°C}$	2	190–197	163–171
White–mustard–seed oil				5.4	171–174	94–98
Wheat-germ oil	−16 to −8	0.912–0.916				125

TABLE 3.48 Physical Properties of Waxes

Wax	Melting point, °C	Specific gravity (15°C/15°C)	Refractive index	Acid value	Saponification value	Iodine value
Bamboo leaf	79–80	0.961$^{25°C}$		14–15	43–44	7.8
Bayberry (myrtle)	47–49	0.99		3–4	205–212	4–9.5
Beeswax, ordinary	62–66	0.95–0.97	1.44–1.48$^{40°C}$	17–21	88–100	8–11
Beeswax, East Indian	61–67	0.95–0.97	1.44$^{40°C}$	5–10.5	87–117	4–10.5
Beeswax, white, USP	61–69	0.95–0.98	1.45–1.47$^{65°C}$	17–24	90–96	7–11
Candelilla	73–77	0.98–0.99	1.45–1.46$^{85°C}$	19–24	55–64	14–20
Cape berry	40–45	1.01	1.45$^{45°C}$	2.5–4.0	211–215	0.5–2.5
Caranda	80–85	0.99–1.00		5.0–9.5	64–79	8–9
Carnauba, No. 1, yellow	86–88	0.99–1.00		1.5–2.5	75–86	
Carnauba, No. 3, crude	86–90	0.99–1.01		3.0–8.5	75–89	
Carnauba, No. 3, refined	86–89	0.96–0.97	1.47$^{40°C}$	3.0–5.0	76–85	7–13.5
Castor oil, hydrogenated	83–88	0.98–0.99$^{20°C}$		1.0–5.0	177–181	2.5–8.5
Chinese insect	80–85	0.95–0.97	1.46$^{40°C}$	2–9	78–93	1.0–2.5
Cotton	68–71	0.96		32	71	25
Cranberry	207–218	0.97–0.98		42–59	131–134	44–53
Esparto	75–79	0.985–0.995		22–27	58–73	7–15
Flax	61–70	0.91–0.99		17–48	37–102	22–29
Japan	49–56	0.97–1.00		4–15	210–235	4–15
Jojoba	11–12	0.86–0.90$^{25°C}$	1.465$^{25°C}$	0.2–0.6	92–95	82–88
Microcrystalline, amber	64–91	0.91–0.94	1.42–1.45$^{80°C}$	0	0	0
Microcrystalline, white	71–89	0.93–0.94	1.441$^{80°C}$	0	0	0
Montan, crude	76–86	1.01–1.02$^{25°C}$		22–31	59–92	14–18
Montan, refined	77–84	1.02–1.04		23–45	72–115	10–14
Ouricury	86–89	0.99–1.01		12–19	88–96	6.9–7.8
Ozokerite	56–82	0.90–1.00		0	0	4–8
Palm	74–86	0.99–1.05		5–11	64–104	9–17
Paraffin, American	49–63	0.896–0.925	1.44–1.48$^{80°C}$	0	0	0
Shellac	79–82	0.97–0.98		12–24	64–83	6–9
Sisal hemp	74–81	1.007–1.010		16–19	56–58	28–29
Spermaceti	41–49	0.905–0.960		0.5–3.0	121–135	2.5–8.5
Sugarcane, refined	76–82	0.96–0.98	1.51$^{25°C}$	8–23	55–70	13–29
Wool	38–40	0.97	1.48$^{40°C}$	6–22	82–130	15–47

TABLE 3.49 Physical and Chemical Properties of Ethanol, Methanol, and Gasoline

Property	Methanol CH_3OH	Ethanol C_2H_5OH	Gasoline C_4–C_{12}
Molecular weight (g/mol)	32	46	~114
Specific gravity	0.789 (298 K)	0.788 (298 K)	0.739 (288.5 K)
Vapor density rel. to air	1.10	1.59	3.0 to 4.0
Liquid density (g·cm^{-3} at 298 K)	0.79	0.79	0.74
Boiling point (K)	338	351	300 to 518
Melting point (K)	175	129	
Vapor pressure @311 K (psia)	3.6	2.5	8–10
Heat of evaporation (Btu/lb)	472	410	135
Heating value (kBtu·gal^{-1})			
Lower	58	74	111
Upper	65	85	122
Tank design pressure (psig)	15	15	15
Viscosity (cp)	0.54	1.20	0.56
Flash point (K)	284	287	228
Flammability/explosion limits			
(%) Lower (LFL)	6.7	3.3	1.3
(%) Upper (UFL)	36	19	7.6
Auto ignition temperature (K)	733	636	523–733
Solubility in H_2O (%)	Miscible (100%)	Miscible (100%)	Negligible (~0.01)
Azeotrope with H_2O	None	95% EtOH	Immiscible
Peak flame temperature (K)	2143	2193	2303
Minimum ignition energy in air (mJ)	0.14		0.23

TABLE 3.50 Specifications of Diesel and Biodiesel Fuels

Property	Diesel	Biodiesel
Standard	ASTM D975	ASTM D6751
Composition	(C_{10}–C_2) Hydrocarbons	FAME* (C_{12}–C_{22})
Specific gravity (g/mL)	0.85	0.88
Flash point (K)	333–353	373–443
Cloud point (K)	258–278	270–285
Pour point (K)	238–258	258–289
Water, % v/v	0.05	0.05
Carbon, % w/w	87	77
Hydrogen, % w/w	13	12
Oxygen, % w/w	0	11
Sulfur, % w/w	0.05	0.05
Cetane number	40–55	48–60

*FAME: Fatty acid methyl esters.

TABLE 3.51 Typical Properties of Bio-Oil from Wood Pyrolysis and No. 2 Diesel Fuel

	Bio-oil	No 2. diesel fuel
Moisture content	15–30	n.a.
pH	2.5	1
Specific gravity	1.20	0.847
Elemental analysis C (wt%)	55–58	86
H	5.5–7.0	11.1
O	35–40	0
N	0–0.2	1
S	n.d.	0.8
HHV (MJ/kg) as produces	16–19	44.7
Viscosity	40–100 cp (315 K, 25% water)	<2.39 (325 K)

TABLE 3.52 Properties of Vegetable Oil Biodiesel and Diesel Fuel

Vegetable oil	Cetane number	Flash point	Cloud point, °C	Density	Pour point, °C
Peanut	64	176	5	0.883	—
Soybean	45	178	1	0.885	–7
Sunflower	63	127	4	0.875	—
Palm	62	164	13	0.880	—
Babassu	49	183	1	0.860	—
Tallow	—	96	12		9
Diesel	50	76	—	0.855	–16

TABLE 3.53 Properties of Fischer-Tropsch Diesel Fuel and No. 2 Diesel Fuel

Property	FT diesel fuel	No. 2 diesel fuel
Density, g/cm^3	0.78	0.83
Aromatics, %	0–0.1	8–16
Cetane number	76–80	40–44
Sulfur content, ppm	0–0.1	25–125

TABLE 3.54 Moisture, Ash, Heat Content, and Chemical Composition of Selected Biomass Fuels

Fuel type	Clean wood	Verge grass	Organic domestic waste	Demolition wood	Sludge
Moisture content, % w/w wet fuel	50	60	54	20	20
Ash content, % w/w dry fuel	1.3	8.4	18.9	0.9	37.5
LHV (as received), MJ/kg	7.7	5.4	6.4	13.9	8.8
HHV (as received), MJ/kg	9.6	7.4	8.3	15.4	9.9
Composition, % w/w (maf*)					
C	49.10	48.70	51.90	48.40	52.50
H	6.00	6.40	6.70	5.20	7.20
O	44.30	42.50	38.70	45.20	30.30
N	0.48	1.90	2.20	0.15	6.99
S	0.01	0.14	0.50	0.03	2.74
Cl	0.10	0.39	0.3	0.08	0.19

*Moisture ash free.

3.6.1 Bibliography

Speight, J. G. 2008. *Synthetic Fuels Handbook: Properties, Processes, and Performance*. McGraw-Hill, New York.

Speight, J. G. 2011a. *The Refinery of the Future*. Gulf Professional Publishing, Elsevier, Oxford, United Kingdom.

Speight, J. G. (ed.). 2011b. *Biofuels Handbook*. Royal Society of Chemistry, London, United Kingdom.

3.7 MINERALS

Minerals are substances formed naturally in the earth. They have a definite chemical composition and structure. There are over 3000 minerals known. Some are rare and precious such as gold and diamond, while others are more ordinary, such as quartz.

Minerals are chemical compounds, sometimes specified by crystalline structure as well as by composition, which are found in rocks (or pulverized rocks, known as sand). On the other hand, rocks consist of one or more minerals and fall into three main types depending on their origin and maturation history: (1) igneous rocks, which are rocks that have solidified directly from a molten state, such as volcanic lava, (2) sedimentary rocks, which are rocks that have been re-manufactured from previously existing rocks, usually from the products of chemical weathering or mechanical erosion, without melting, and (3) metamorphic rocks, which are rocks that have resulted from processing, by heat and pressure (but not melting), of previously existing sedimentary or igneous rocks.

TABLE 3.55 Elements in the Crust of the Earth

Element name	Symbol	% w/w of the earth's crust
Oxygen	O	47
Silicon	Si	28
Aluminum	Al	8
Iron	Fe	5
Calcium	Ca	3.5
Sodium	Na	3
Potassium	K	2.5
Magnesium	Mg	2
All other elements		1

TABLE 3.56 Common Elements and Uses

Aluminum	The most abundant metal element in earth's crust. Aluminum originates as an oxide called alumina. Bauxite ore is the main source of aluminum and must be imported from Jamaica, Guinea, Brazil, Guyana, etc. Used in transportation (automobiles), packaging, building/construction, electrical, machinery, and other uses.
Antimony	A native element and the metal is extracted from stibnite ore and other minerals. Used as a hardening alloy for lead, especially storage batteries and cable sheaths; also used in bearing metal, type metal, solder, collapsible tubes and foil, sheet and pipes, and semiconductor technology. Antimony is used as a flame retardant in fireworks, and the antimony salts are used in the rubber, chemical, and textile industries as well as medicine and glassmaking.
Barium	A heavy metal contained in barite. Used as a heavy additive in oil well drilling; in the paper and rubber industries; as a filler or extender in cloth, ink, and plastics products; in radiography ("barium milkshake"); as a deoxidizer for copper; a sparkplug in alloys; and in making expensive white pigments.
Bauxite	Rock composed of hydrated aluminum oxides. In the United States, it is primarily converted to alumina (Al_2O_3).
Beryllium	Used in the nuclear industry to make light and very strong alloys used in the aircraft industry. Beryllium salts are used in fluorescent lamps, in X-ray tubes, and as a deoxidizer in bronze metallurgy. Beryl is the gemstone emerald and aquamarine. It is used in computers, telecommunication products, aerospace and defense applications, appliances and automotive, and consumer electronics. Also used in medical equipment.
Chromite	The United States consumes about 6% of world chromite ore production in various forms of imported materials, such as chromite ore, chromite chemicals, chromium ferroalloys, chromium metal, and stainless steel. Used as an alloy and in stainless and heat resisting steel products. Used in chemical and metallurgical industries (chrome fixtures, etc.). Super alloys require chromium. It is produced in South Africa, Kazakhstan, and India.
Clay	Used in floor and wall tile as an absorbent, in sanitation, mud drilling, foundry sand bond, iron pelletizing, brick, lightweight aggregate, and cement. It is produced in 40 states. Ball clay is used in floor and wall tile. Bentonite is used for drilling mud, pet waste absorbent, iron ore pelletizing, and foundry sand bond. Kaolin is used for paper coating and filling, refractory products, fiberglass, paint, rubber, and catalyst manufacture. Common clay is used in brick, light aggregate, and cement.
Cobalt	Used primarily in super alloys for aircraft gas turbine engines, in cemented carbides for cutting tools and wear-resistant applications, chemicals (paint dryers, catalysts, magnetic coatings), and permanent magnets. The United States has cobalt resources in Minnesota, Alaska, California, Idaho, Missouri, Montana, and Oregon. Cobalt production comes principally from Congo, China, Canada, Russia, Australia, and Zambia.
Copper	Used in building construction, electric, and electronic products (cables and wires, switches, plumbing, and heating); transportation equipment; roofing; chemical and pharmaceutical machinery; and alloys (brass, bronze, and beryllium alloyed with copper are particularly vibration resistant); alloy castings; electroplated protective coatings and undercoats for nickel, chromium, zinc, etc. More recently copper is being used in medical equipment due to its anti-microbial properties. The United States has mines in Arizona, Utah, New Mexico, Nevada, and Montana. Leading producers are Chile, Peru, China, United States, and Australia.
Feldspar	A rock-forming mineral; industrially important in glass and ceramic industries; patter and enamelware; soaps; bond for abrasive wheels; cements; insulating compositions; fertilizer; tarred roofing materials; and as a sizing, or filler, in textiles and paper. In pottery and glass, feldspar functions as a flux. End-uses for feldspar in the United States include glass (70%) and pottery and other uses (30%).
Fluorite (fluorspar)	Used in production of hydrofluoric acid, which is used in the pottery, ceramics, optical, electroplating, and plastics industries; in the metallurgical treatment of bauxite; as a flux in open hearth steel furnaces and in metal smelting; in carbon electrodes; emery wheels; electric arc welders; toothpaste; and paint pigment. It is a key ingredient in the processing of aluminum and uranium.
Gallium	Gallium is used in integrated circuits, light-emitting diodes (LEDs), photodetectors, and solar cells. It has a new use in chemotherapy for some types of cancer. Integrated circuits are used in defense applications, high-performance computers, and telecommunications. Optoelectronic devices were used in areas such as aerospace, consumer goods, industrial equipment, medical equipment, and telecommunications. Leading sources are Germany, UK, China, and Canada.

TABLE 3.56 Common Elements and Uses (*Continued*)

Gold	Used in jewelry and arts; dentistry and medicine; in medallions and coins; in ingots as a store of value; for scientific and electronic instruments; as an electrolyte in the electroplating industry. Mined in Alaska and several western states. Leading producers are China, Australia, United States, Russia, and Canada.
Gypsum	Processed and used as prefabricated wallboard or an industrial or building plaster; used in cement manufacturing; agriculture, and other uses.
Halite (sodium chloride—salt)	Used in human and animal diet, food seasoning, and food preservation; used to prepare sodium hydroxide, soda ash, caustic soda, hydrochloric acid, chlorine, metallic sodium; used in ceramic glazes; metallurgy, curing of hides; mineral waters; soap manufacturing; home water softeners; highway de-icing; photography; in scientific equipment for optical parts. Single crystals used for spectroscopy, ultraviolet, and infrared transmission.
Indium	Indium tin oxide is used for electrical conductivity purposes in flat panel devices—most commonly in liquid crystal displays (LCDs). It is also used in solders, alloys, compounds, electrical components, semiconductors, and research. Indium ore is not recovered from ores in the United States, China is the leading producer. It is also produced in Canada, Japan, and Belgium.
Iron ore	Used to manufacture steels of various types. Powdered iron: used in metallurgy products; magnets; high-frequency cores; auto parts; catalyst. Radioactive iron (Fe^{59}): in medicine; tracer element in biochemical and metallurgical research. Iron blue: in paints, printing inks, plastics, cosmetics, paper dyeing. Black iron oxide: as pigment, in polishing compounds, metallurgy, medicine, and magnetic inks. Most United States production is from Michigan and Minnesota. China, Australia, Brazil, and Russia are the major producers.
Lead	Used in lead-acid batteries, gasoline additives (now being eliminated) and tanks, and solders, seals or bearing; used in electrical and electronic applications; TV tubes and glass, construction, communications and protective coatings; in ballast or weights; ceramics or crystal glass; X-ray and gamma radiation shielding; soundproofing material in construction industry; and ammunition. Industrial type batteries are used as a source of uninterruptible power equipment for computer and telecommunications networks and mobile power. United States mines lead mainly in Missouri, but also in Alaska and Idaho.
Lithium	Compounds are used in ceramics and glass, batteries, lubricating greases, air treatment, in primary aluminum production, in the manufacture of lubricants and greases, rocket propellants, vitamin A synthesis, silver solder, batteries, and medicine. Lithium ion batteries have become a substitute for nickel-cadmium batteries in hand held/portable electronic devices. There is one brine operation in Nevada. Australia, Chile, and China are major producers.
Manganese	Ore is essential to iron and steel production. Also used in the making of manganese ferroalloys. Construction, machinery, and transportation end uses account for most United States consumption of manganese. Manganese ore has not been produced in the United States since 1970. Major producers are South Africa, Australia, China, Gabon, and Brazil.
Mica	Micas commonly occur as flakes, scales, or shreds. Ground mica is used in paints, as joint cement, as a dusting agent, in oil well-drilling muds; and in plastics, roofing, rubber and welding rods. Sheet mica is fabricated into parts for electronic and electronic equipment. China and Russia are leading producers.
Molybdenum	Used in alloy steels to make automotive parts, construction equipment, gas transmission pipes; stainless steels; tool steels; cast irons; super alloys; and chemicals and lubricants. As a pure metal, molybdenum is used because of its high melting temperatures (2610°C, 4730°F) as filament supports in light bulbs, metalworking dies and furnace parts. Major producers are China, the United States, Chile, and Peru.
Nickel	Vital as an alloy to stainless steel; plays key role in the chemical and aerospace industries. End uses were transportation, fabricated metal products, electrical equipment, petroleum and chemical industries, household appliances and industrial machinery. Major producers are the Philippines, Indonesia, Russia, Australia, and Canada.
Perlite	Expanded perlite is used in building construction products like roof insulation boards; as fillers, for horticulture aggregate and filter aids. It is produced in New Mexico and other western states and is processed in over 20 states. Leading producers are the United States, Greece, and Turkey.

(*Continued*)

TABLE 3.56 Common Elements and Uses (*Continued*)

Platinum group metals (PGMs)	Includes platinum, palladium, rhodium, iridium, osmium, and ruthenium. Commonly occur together in nature and are among the scarcest of the metallic elements. Platinum is used principally in catalysts for the control of automobile and industrial plant emissions; in jewelry; in catalysts to produce acids, organic chemicals, and pharmaceuticals. PGMs are used in bushings for making glass fibers used in fiber-reinforced plastic and other advanced materials, in electrical contacts, in capacitors, in conductive and resistive films used in electronic circuits, in dental alloys used for making crowns and bridges. South Africa, Russia, the United States, and Canada are major producers.
Phosphate rock	Used to produce phosphoric acid for ammoniated phosphate fertilizers, feed additives for livestock, elemental phosphorus, and a variety of phosphate chemicals for industrial and home consumers. United States production occurs in Florida, North Carolina, Idaho, and Utah.
Potash	A carbonate of potassium; used as a fertilizer, in medicine, in the chemical industry and to produce decorative color effects on brass, bronze, and nickel. The leading producers are Canada, Russia, and Belarus.
Pyrite	Used in the manufacture of sulfur, sulfuric acid, and sulfur dioxide; pellets of pressed pyrite dust are used to recover iron, gold, copper, cobalt, nickel; used to make inexpensive jewelry.
Quartz (silica)	As a crystal, quartz is used as a semiprecious gem stone. Crystalline varieties include amethyst, citrine, rose quartz, smoky quartz, etc. Cryptocrystalline forms include agate, jasper, onyx, etc. Because of its piezoelectric properties, quartz is used for pressure gauges, oscillators, resonators, and wave stabilizes; because of its ability to rotate the plane of polarization of light and its transparency in ultraviolet rays, it is used in heat-ray lamps, prism, and spectrographic lenses. Also used in manufacturing glass, paints, abrasives, refractory materials, and precision instruments.
Rare earth elements (lanthanum, cerium, praseodymium, neodymium, promethium, samarium, europium, gadolinium, terbium, dysprosium, holmium, erbium, thulium, ytterbium, and lutetium)	Used mainly in petroleum fluid cracking catalysts, metallurgical additives and alloys, glass polishing and ceramics, permanent magnets, and phosphors. It is estimated that 40 pounds of rare earths are used in a hybrid car for rechargeable battery, permanent magnet motor and the regenerative braking system. The United States now has one rare earth (Bastnasite) mine in California. More than 85% of global production is in China.
Silica	Aluminum and aluminum alloy producers and the chemical industry are major users of silicon metal. Silica is also used in manufacture of computer chips, glass, and refractory materials; ceramics; abrasives; water filtration; component of hydraulic cements; filler in cosmetics, pharmaceutical, paper, insecticides; anti-caking agent in foods; flatting agent in paints; thermal insulator; and photovoltaic cells. China is the leading producer.
Silver	Used in coins and medals, electrical and electronic devices, industrial applications, jewelry, silverware, and photography. The physical properties of silver include ductility, electronics conductivity, malleability, and reflectivity. Used in lining vats and other equipment for chemical reaction vessels, water distillation, etc.; a catalyst in manufacture of ethylene; mirrors; silver plating; table cutlery; dental, medical and scientific equipment; bearing metal; magnet windings; brazing alloys, solder. Also used in catalytic converters, cell phone covers, electronics, circuit boards, bandages for wound care and batteries. Silver is produced in the United States at over 30 base and precious metal mines primarily in Alaska and Nevada. The leading global producers include Mexico, China, Peru, Chile, Australia, Bolivia, and the United States.
Sodium carbonate (soda ash or trona)	Used in glass container manufacture; in fiberglass and specialty glass; also used in production of flat glass; in liquid detergents; in medicine; as a food additive; photography; cleaning and boiler compounds; pH control of water. Most United States production comes from Wyoming.
Sulfur	Used in the manufacture of sulfuric acid, fertilizers, petroleum refining; and metal mining. Elemental sulfur and byproduct sulfuric acid were produced in over 100 operations in 26 state and the Virgin Islands. The United States, Canada, China, and Germany are major producers.

TABLE 3.56 Common Elements and Uses (*Continued*)

Tantalum	A refractory metal with unique electrical, chemical, and physical properties used to produce electronic components, tantalum capacitors (in auto electronics, pagers, personal computers, and portable telephones) ; for high-purity tantalum metals in products ranging from weapon systems to superconductors; high-speed tools; catalyst; sutures and body implants; electronic circuitry; thin-film components. Used in optical glass and electroplating devices. Leading producers are Mozambique, Brazil, and Congo.
Titanium	Titanium mineral concentrates are used primarily by titanium dioxide pigment producers. A small amount is used in welding rod coatings and for manufacturing carbides, chemicals, and metals. It is produced in Florida and Virginia. Leading producing countries are South Africa, Australia, Canada, and China.
	Titanium and titanium dioxide are used in aerospace applications (in jet engines, airframes and space and missile applications). It is also used in armor, chemical processing, marine, medical, power generation, sporting goods, and other non-aerospace applications. Titanium sponge metal was produced in three operations in Nevada and Utah. The leading global producers are China, Japan, Russia, and Kazakhstan.
Tungsten	More than half of the tungsten consumed in the United States was used in cemented carbide parts for cutting and wear-resistant materials, primarily in the construction, metalworking, mining, and oil- and gas-drilling industries. The remaining tungsten was consumed to make tungsten heavy alloys for applications requiring high density; electrodes, filaments, wires, and other components for electrical, electronic, heating, lighting, and welding applications; steels, super alloys, and wear-resistant alloys; and chemicals for various applications. China is by far the leading producer. Russia, Canada, Austria, and Bolivia also produce tungsten.
Uranium	Nearly 20% of electricity in the United States is produced using uranium in nuclear generation. It is also used for nuclear medicine, atomic dating, powering nuclear submarines, and other uses in the U.S. defense system.
Vanadium	Metallurgical use, primarily as an alloying agent for iron and steel, accounted for about 93% of the domestic vanadium consumption. Of the other uses for vanadium, the major nonmetallurgical use was in catalysts for the production of maleic anhydride and sulfuric acid. China, South Africa, and Russia are largest producers.
Zeolites	Used in animal feed, cat litter, cement, aquaculture (fish hatcheries for removing ammonia from the water); water softener and purification; in catalysts; odor control; and for removing radioactive ions from nuclear plant effluent.
Zinc	Of the total zinc consumed in the United States, about 55% is used in galvanizing, 21% in zinc-based alloys, 16% in brass and bronze, and 8% in other uses. Zinc compounds and dust were used principally by the agriculture, chemical, paint, and rubber industries. Major co-products of zinc mining and smelting, in order of decreasing tonnage, were lead, sulfuric acid, cadmium, silver, gold, and germanium. Zinc is used as protective coating on steel, as die casting, as an alloying metal with copper to make brass and as chemical compounds in rubber and paints; used as sheet zinc and for galvanizing iron, electroplating, metal spraying, automotive parts, electrical fuses, anodes, dry cell batteries, nutrition, chemicals, roof gutter, engravers' plates, cable wrappings, organ pipes, and pennies. Zinc oxide used in medicine, paints, in vulcanizing rubber, sun block. Zinc dust used for primers, paints, precipitation of noble metals; removal of impurities from solution in zinc electro-winning. United States production is in 3 states and 13 mines. Leading producers are China, Australia, Peru, and the United States.

Sources: The United States Geological Survey; Facts about Minerals (National Mining Association); Mineral Information Institute; the Energy Information Administration.

TABLE 3.57   Common Minerals and Uses

Bauxite	Aluminum, foil, airplane parts
Borax	Antiseptic soaps, welding flux or cleaner (found in dry lake beds)
Calcite	Medicine, toothpaste, building; materials (hard water deposit, ancient sea beds)
Copper	Tubing, electrical wires, sculptures
Diamond	Cutting tools/blades/saws
Feldspar	Ceramics and porcelain, colors in granites (not black)
Galena	Source of lead
Graphite	Pencils, lubricant in machinery
Gypsum	Wall board, plaster of paris
Halite	Salt
Hematite	Source of iron
Jade	Jewelry, figurines
Limonite/taconite	Source of iron (around Cedar City)
Muscovite (mica)	White, gray material in electrical insulators
Quartz (massive type), quartz crystal	Glass manufacturing, radios, computers and electronic equipment
Silver	Jewelry, photography, electrical equipment
Sulfur	Fungicides, kills bacteria, vulcanizes rubber, in coal and fuels, fertilizer
Talc	Baby powder, soapstone, gymnastics to grasp bars

3.7.1   Bibliography

Pellant, C. 2002. *Rocks and Minerals*. DK Books, New York.

Speight, J. G. 2015. *Asphalt Materials Science and Technology*. Butterworth-Heinemann, Elsevier, Oxford, United Kingdom.

INDEX